PETERSON'S™
THOMSON LEARNING

DECISION •••••
Graduate School
GUIDES •••••

GRADUATE PROGRAMS IN
Biology 2002

**A compact,
easy-to-use guide
to graduate
and professional
programs in
the U.S.**

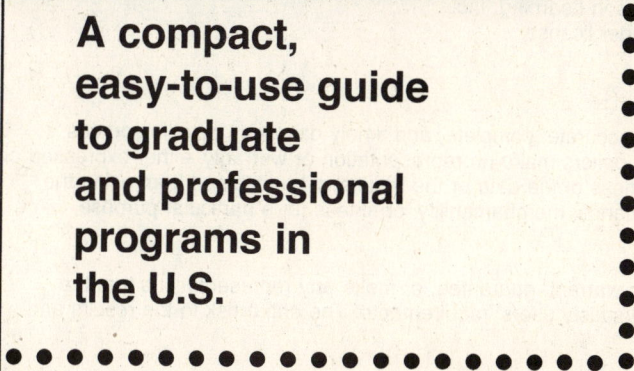

PETERSON'S™
THOMSON LEARNING

Australia • Canada • Mexico • Singapore • Spain • United Kingdom • United States

PETERSON'S

™

THOMSON LEARNING

About Peterson's

Founded in 1966, Peterson's, a division of Thomson Learning, is the nation's largest and most respected provider of lifelong learning online resources, software, reference guides, and books. The Education Supersite^{SM} at petersons. com—the Web's most heavily traveled education resource—has searchable databases and interactive tools for contacting U.S.-accredited institutions and programs. CollegeQuest® (CollegeQuest.com) offers a complete solution for every step of the college decision-making process. GradAdvantage^{TM} (GradAdvantage.org), developed with Educational Testing Service, is the only electronic admissions service capable of sending official graduate test score reports with a candidate's online application. Peterson's serves more than 55 million education consumers annually.

Thomson Learning is among the world's leading providers of lifelong learning, serving the needs of individuals, learning institutions, and corporations with products and services for both traditional classrooms and for online learning. For more information about the products and services offered by Thomson Learning, please visit www.thomsonlearning.com. Headquartered in Stamford, Connecticut, with offices worldwide, Thomson Learning is part of The Thomson Corporation (www.thomson.com), a leading e-information and solutions company in the business, professional, and education marketplaces. The Corporation's common shares are listed on the Toronto and London stock exchanges.

Contents

Contents

How to Use This Book

The graduate and professional programs in this *Decision Guide* are offered by colleges, universities, and professional schools and specialized institutions in the United States and U.S. territories. They are accredited by U.S. accrediting bodies recognized by the Department of Education or the Council on Higher Education Accreditation.

This volume is divided into the subject fields in which graduate degrees are offered. The eighteen major sections of the book are Biological and Biomedical Sciences; Anatomy; Biochemistry; Biophysics; Botany and Plant Sciences; Cell, Molecular, and Structural Biology; Ecology, Environmental Biology, and Evolutionary Biology; Entomology; Genetics, Developmental Biology, and Reproductive Biology; Marine Biology; Microbiological Sciences; Neuroscience; Nutrition; Parasitology; Pathology; Pharmacology and Toxicology; Physiology; and Zoology. Many of these major sections are subdivided into narrower subject areas.

How Information Is Organized

Graduate program information in this *Decision Guide* is presented in profile form. The format of the profiles is constant, making it easy to compare one institution with another and one program with another. Any item that does not apply to or was not provided by a graduate unit is omitted from its listing. The following outline describes the profile information.

Identifying Information. In the conventional university-college-department organizational structure, the parent institution's name is followed by the name of the administrative unit or units under which the degree program is offered and then the specific unit that offers the degree program. (For example, University of Notre Dame, College of Arts and Letters, Division of Humanities, Department of Art, Art History, and Design, Concentration in Design.) The last unit listed is the one to which all information in the profile pertains. The institution's city, state, and postal code follow.

Awards. Each postbaccalaureate degree awarded is listed; fields of study offered by the unit may also be listed. Frequently, fields of study are divided into subspecializations, and those appear following the degrees awarded. Students enrolled in the graduate program would be able to specialize in any of the fields mentioned.

Part-Time and Evening/Weekend Programs. When information regarding the availability of part-time or evening/weekend study appears in the profile, it means that students are able to earn a degree exclusively through such study.

Postbaccalaureate Distance Learning Degrees. A postbaccalaureate distance learning degree program signifies that course requirements can be fulfilled off the main campus. If these programs require minimal on-campus study or no on-campus study, it may be indicated here.

Faculty. Figures on the number of faculty members actively involved with graduate students through teaching or research are separated into full- and part-time as well as men and women whenever the information has been supplied.

Students. Figures for the number of students enrolled in graduate and professional programs pertain to the semester of highest enrollment from the 1999–2000 academic year. These figures are divided into full- and part-time and men and women whenever the data have been supplied. Information on the number of students who are members of a minority group or are international students appears here. The average age of the students is followed by the number of applicants and the percentage accepted for fall 1999.

This section also includes the number of degrees awarded in the 1999 calendar year and information on the percentages of students who have gone on to continue full-time study, entered university research or teaching, or chosen other work related to their field. Many doctoral programs offer a terminal master's degree if students leave the program after completing only part of the requirements for a doctoral degree; that is indicated here. All degrees are classified into one of four types: master's, doctoral, first-professional, and other advanced degrees. A unit may award one or several degrees at a given level; however, the data are only collected by type and may therefore represent several different degree programs.

Degree Requirements. The information in this section is also broken down by type of degree, and all information for a degree level pertains to all degrees of that type unless otherwise specified. Degree requirements are collected in a simplified form to provide some very basic information on the nature of the program and on foreign language, computer, and thesis or dissertation requirements. Many units also provide a short list of additional requirements, such as fieldwork or internships. Information on the average amount of time required to earn the degree for full-time and part-time students is also included. No information is listed on the number of courses or credits required for completion or whether a minimum or maximum number of years or semesters is needed. For complete information on graduation requirements, contact the graduate school or program directly.

Entrance Requirements. Entrance requirements are divided into the levels of master's, doctoral, first-professional, and other advanced degrees. Within each level, information may be provided in two basic categories, entrance exams and other requirements. The entrance exams use the standard acronyms used by the testing agencies, unless they are not well known. Additional information on each of the common tests is provided in the "Taking the Entrance Exams" article in this volume. More information on the scale and other aspects of the test may be obtained directly from the testing agency. Other entrance requirements are quite varied, but they often contain an undergraduate or graduate grade point average (GPA). Unless otherwise stated, the GPA is calculated on a 4.0 scale and is listed as a minimum required for admission.

Application. The standard application **deadline,** any nonrefundable application **fee,** and whether electronic applications are accepted may be listed here. Note that the deadline should be used for reference only; these dates are subject to change, and students interested in applying should contact the graduate unit directly about application procedures and deadlines.

Expenses. The cost of study for the 1999–2000 academic year is given in two basic categories, tuition and fees. It is not possible to represent the complete tuition and fees schedule for each graduate unit, so a simplified version of the cost of study in that unit is provided. In general, the costs of both full- and part-time study are listed if the unit offers both and lists separate costs. For public institutions, the tuition and fees are listed for both state residents and nonresidents. Cost of study may be quite complex at a graduate institution. There are often sliding scales for part-time study, a different cost for first-year students, and other variables that make it impossible to completely cover the cost of study for each graduate program. To provide the most usable information, figures are given for full-time study for a full year where available and for part-time study in terms of a per-unit rate (per credit, per semester hour, etc.). Expenses are usually subject to change; for exact costs at any given time, contact your chosen schools and programs directly.

Financial Aid. This section contains data on the number of awards that are administered by the institution and were given to graduate students during the 1999–2000 academic year. The first figure given represents the total number of students enrolled in that unit who received financial aid. If the unit has provided information on graduate appointments, these are broken down into three major categories: *fellowships* give money to graduate students to cover the cost of study and living expenses and are not based on a work obligation or research commitment, *research assistantships* provide stipends to graduate students for assistance in a formal research project with a faculty member, and *teaching assistantships* provide stipends to graduate students for teaching or for assisting faculty members in teaching undergraduate classes.

In addition to graduate appointments, the availability of several other financial aid sources is covered in this section. *Career-related internships* or *fieldwork* offer money to students who are participating in a formal off-campus research project or practicum. *Federal Work-Study* is made available to students who demonstrate need and meet the federal guidelines; this form of aid normally includes 10 or more hours of work per week in an office of the institution. *Tuition waivers* are routinely part of a graduate appointment, but units sometimes waive part or all of a student's tuition even if a graduate appointment is not available. *Institutionally sponsored loans* are low-interest loans available to graduate students to cover both educational and living expenses. The availability of grants, scholarships, traineeships, unspecified assistantships, and financial aid to part-time students is also indicated here.

Some programs list the financial aid application deadline and the forms that need to be completed for students to be eligible for financial aid. There are two forms: FAFSA, the Free Application for Federal Student Aid, which is required for federal aid; and the CSS Financial Aid PROFILE.

Faculty Research. Each unit has the opportunity to list several keyword phrases describing the current research involving faculty members and graduate students. Space limitations prevent the unit from listing complete information on all research programs. The total expenditure for funded research from the previous academic year may also be included.

Unit Head and Application Contact. The head of the graduate program for each unit is listed with the academic title and telephone and fax numbers and e-mail addresses, if available. In addition to the unit head, many graduate programs list separate contacts for application and admission information. If no unit head or application contact is given, you should contact the overall institution for information.

For Further Information

Many programs offer more in-depth, narrative style information that can be located at www.petersons.com/graduate. There is a notation to this effect at the end of those program profiles.

Data Collection

The information published in this book was collected through *Peterson's Annual Survey of Graduate and Professional Institutions.* Each spring and summer, this survey is sent to more than 1,700 institutions offering postbaccalaureate degree programs, including accredited institutions in the United States and U.S. territories. Deans and other administrators provide information on specific programs as well as overall institutional information. Peterson's editorial staff then goes over each returned survey carefully and verifies or revises responses after further research and discussion with administrators at the institutions. Extensive files on past responses are kept from year to year.

While every effort is made to ensure the accuracy and completeness of the data, information is sometimes unavailable or changes occur after publication deadlines. The omission of any particular item from a directory or profile signifies either that the item is not applicable to the institution or program or that information was not available. If no usable information was submitted by an institution, its name, address, and program name are still included in order to indicate the existence of graduate work.

An Overview
of Graduate Degree Programs

Some Major Differences in Degrees

Traditionally, graduate education has been either academic or professional in orientation. Academic graduate education emphasizes performing and evaluating research. Those going for the more professionally oriented graduate degrees learn the skills and knowledge necessary to practice a profession. A graduate student getting a degree in psychology, for instance, has a choice between two distinctly different paths toward a degree. On the academic side, the student may emphasize experimental psychology and conduct significant research on the relationship between aerobic exercise and stress. On the professional side, a graduate student in clinical psychology will learn the skills to provide psychotherapy to patients.

The Academically Oriented Advanced Degree

Graduate education in an academic field such as history, English literature, or biochemistry involves acquiring, evaluating, and communicating knowledge in a narrow aspect of a broad subject. The first postbaccalaureate academic degree is the Master of Art (M.A.) or Master of Science (M.S.). To earn such a degree, the student takes courses and conducts research, whether in the library, in the laboratory, or in the field. For some master's degrees, the student must also write a thesis. An academic master's degree may qualify you to go on to doctoral study or it may simply improve your chances of employment. Many employers consider a master's degree an indicator of good critical thinking, communication, and research skills. A master's degree takes a year or two to earn on a full-time basis.

The highest academic degree, the Doctor of Philosophy (Ph.D.), requires course work beyond the master's level as well as original, specialized research culminating in a dissertation. Because there is so much work involved, earning a Ph.D. can take from four to ten years. People who earn academic Ph.D.'s usually hope to conduct research and/or teach at the university level. Denise Kaiser, who earned a Ph.D. in medieval history, has a broad knowledge of this time period from courses she has taken. From researching and writing her dissertation, she also has specialized knowledge of a small aspect of the subject: what thirteenth-century sermons can tell us about the education of the preacher and the concerns of his audience.

The Professionally Oriented Advanced Degree

In contrast to academic graduate education, professional education emphasizes the practical application of knowledge and skills. For some professions, a master's degree may be preferred or even required for employment. Social workers need a Master in Social Work (M.S.W.). Librarians need a Master of Library Science (M.L.S.). In other professions, one of the prerequisites is a doctoral degree. To practice medicine, you need a Doctor of Medicine degree (M.D.); to become an optometrist, a Doctor of Optometry degree (O.D.). But in many fields, professional master's degrees and doctoral degrees are optional. People who pursue them do so to advance their careers. Degrees in business administration, journalism, fine arts, and environmental science are not required for employment but may help people find work and prosper in these fields.

Professional and Academic Paths Begin to Cross

While there are clear differences between academic and professional degrees, you should be aware that the distinction between academic and professional graduate education is beginning to blur. Some academic programs have begun to integrate aspects of professional education in order to make their Ph.D. students more marketable. Some now require students to do internships to gain practical experience in related fields. At the University of Texas in Dallas, doctoral students in chemistry take courses for three semesters, then intern in a chemicals firm for three semesters, and finally return to campus to write a thesis.

Tailor Your Own Degree

Many institutions now offer combined-degree programs in which students can study both professional and academic subjects. Boston College offers a combined-degree program in business administration and Russian and East European studies (M.B.A./M.A.), while Arizona State offers a combined degree in anthropology and justice studies (M.A./M.S.). At many institutions, you can design your own combined-degree program tailored to your academic and professional interests.

Certificate Programs

A rapidly growing type of graduate education is the certificate program. Certificate programs are usually aimed at working professionals who seek to upgrade job skills or meet the requirements of a professional credentialing body or a master's program. To serve these students, certificate programs are offered part-time and they are short, typically 18 hours for a postbaccalaureate certificate and 24 hours for a post-master's certificate. Almost half of all certificates are granted in the field of education. Most of these fulfill state requirements for elementary or secondary teaching. Other popular fields are health sciences, particularly in areas of interest to nurse practitioners, and social sciences, mainly in psychology and counseling. Finally, some certificate programs are in the arts and sciences, including area and ethnic studies, and they are often interdisciplinary. One of the most popular of these is the women's studies certificate.

Research and Teaching Institutions

Degrees through the doctoral level are generally offered at universities, both public and private. Some universities are research universities and others are teaching universities. What's the difference? Funding, for one thing. Large research universities receive money—often millions of dollars a year from the federal government and private sources—to support the research efforts of their faculties. In these institutions, graduate education focuses on preparing you for a career in academia and in field or laboratory research. At teaching universities, the emphasis is on preparing you for a career in teaching at the college level. Many universities combine the features of both research and teaching institutions.

Many universities also offer professional degrees from their professional schools. In addition to a school of arts and sciences, a university may also have schools of medicine, law, journalism, education, engineering, and social work, to name a few. These usually offer master's and doctoral degrees.

Innovative Alternatives

To accommodate the wide variety of students now seeking graduate education, some institutions have developed innovative options to traditional courses of study. Although many programs, especially the traditional academic ones, still require full-time enrollment and your presence on campus, many others do not. Some programs offer part-time enrollment, allowing you to work full-time and take longer to complete the requirements for a degree. Others have established satellite locations to make getting to classes easier if you live far from the main campus. And some have established distance learning programs, for which it doesn't really matter where you are. In these programs, most instruction and communication is done through telecommunications, with periodic face-to-face meetings. In fact, there is even a new type of institution—the virtual university—that operates primarily through telecommunications. For more information about these alternatives, visit the distance learning channel at www.petersons.com/dlearn.

What Is a
Graduate Degree Worth?

The true value of a graduate degree is known only to the person who possesses it. It may be priceless if it has opened the door to a subject or work you love. Graduate degrees have economic benefits as well. The U.S. Census Bureau has found that higher degrees lead to higher income. On average, individuals with master's degrees earn 24 percent more than those with bachelor's degrees, Ph.D. holders earn 35 percent more than master's recipients, and those with professional degrees earn 15 percent more than Ph.D. recipients. Moreover, graduate degree recipients today are finding rewarding positions not just in the academic sector but throughout the economy, including private industry and government.

Will Graduate Work Advance Your Career?

Many people pursue graduate degrees as a career-enhancing move, says Kevin Boyer, former Executive Director, National Association of Graduate-Professional Students (NAGPS). Graduate degrees are increasingly important in the business world and in the nonprofit and public sectors. Graduate degrees have always been a key to success in academic employment. However, a graduate degree will not automatically enhance your career opportunities unless it is the right degree. Take the time to research the degree requirements of the career you are considering. Talk with people already in that field. Question human resource managers or career placement professionals, counsels Boyer.

Looking at Your Degree from the Financial Side

Earning power is one way to measure the value of a degree. For many people, the desire to make more money is a prime motivation for acquiring a graduate degree. But be aware that another factor may be that people who go to grad school are simply more ambitious and driven than people who do not. These personality traits influence the success of their careers as much as—and maybe more than—the degree itself.

Another fact that contributes to the value of a graduate degree is its relative scarcity. If you look at the U.S. population as a whole, very few people have postbaccalaureate degrees. Roughly 1 percent of Americans over age 25 have an academic doctoral degree. Only about 5 percent hold a master's degree.

Will Your Degree Be Worth the Cost?

Boyer observes that whether you go to school full- or part-time, you may incur student loan or credit card debt. Will your graduate degree "net" you back enough to cover your debt? Graduate student loan debt is rising as more students choose student loans to pay for tuition, books, fees, and room and board.

Remember that you will lose wages if you quit work to go back to school full-time, so plan your budget accordingly. Graduate and professional programs are great mind-expanding experiences. Get your graduate degree, but keep your eyes on your wallet so that the costs don't become prohibitive.

Student Employment

Boyer raises some cautions about being a graduate student employee. Serving as a teaching assistant (TA) or research assistant (RA) can be one of the most fulfilling parts of your graduate education. Most campuses pay a stipend and waive tuition for graduate teaching and research assistants. Being a TA or RA can provide real financial and educational benefits. However, there are trade-offs. For example, work requirements may vary by department. Some questions you might want to ask: How many hours will you have to work per week and what kind of benefits are offered? Some universities offer excellent benefits to graduate student employees, while many other universities offer none. Will your stipend be eroded because you have to pay your own health insurance and parking-permit fees? Often there is a contractual relationship between you and your department. Does this contract offer you any protection? You may enter a program with no previous teaching experience. Does the university or department offer TA training and resources? Is there a collective bargaining unit for graduate student employees? Being a TA or an RA is usually a great experience. Take the position, but remember to keep the lines of communication open with your faculty mentors so that you can address problems should they arise.

NOTE: This article was based, in part, on input from Kevin Boyer, former Executive Director of the National Association of Graduate-Professional Students (NAGPS). NAGPS is a nonprofit organization designed to provide a mechanism for exchange of information among graduate/professional students, foster the development of graduate/professional student organizations, and improve the quality of graduate/professional education and student life in general. NAGPS offers many resources to students at member campuses and to student members. Student memberships are available for $22.50 per year and include a subscription to the NAGPS news publication, a student discount card, and access to NAGPS' endorsed health insurance and dental plan, auto insurance discounts, and many other benefits. To learn more about NAGPS, contact NAGPS, 209 Pennsylvania Avenue, SE; Washington, DC 20003-1107; telephone: 888-88-NAGPS; fax: 202-454-5298; e-mail: office@nagps.org; World Wide Web: http://www.nagps.org/NAGPS/.

Choosing a
Biology Program

Prevailing wisdom says that to do a really thorough job of researching a graduate department of biology you should take a close look at the research being done there and scope out the faculty members. However, graduate biology students might also suggest following the smell of brownies. "Those people who can do microbiology and other molecular work often turn out to be good cooks," jokes graduate student Jennifer Cole of the microbiology program in the biology department at Indiana University. She adds that biologists are used to looking at a protocol and following it, and this is the same method that works with recipes.

She's not the only one advocating that great biology and great brownies often appear together. Graduate students in biology love to heat up the oven while spicing their conversations with talk about the latest in science. Nichole Broderick, working toward her master's in the entomology and plant pathology departments at the University of Wisconsin, says she's a member of an eating group that cooks for each other, enjoying the food as well as the talk.

Culinary skills aside, there are nonedible factors in choosing a graduate department in biology. Faculty members, their research projects, and their labs must be considered in making a good decision. But before you get to the inner workings of labs, you've got some fundamental choices to make.

Master's or Ph.D.?

One of the first decisions you will have to make is whether you're going for a terminal master's or Ph.D. degree. Traditionally, if your aim is academic, you'll be on the Ph.D. track from the beginning. If it's industry, then the master's is all you'll need. But each institution has its own rules.

At Johns Hopkins University, Keith Byrd, who is studying for his Ph.D. in molecular biology, reports that if applicants state that they want only a master's they will not be accepted. However, by the end of the second year, students can leave the program and get a master's after taking comprehensive exams. At Stanford University, Marwah Helmy is getting her master's in biology in a one-year program and may go on to a Ph.D. in cancer biology, but she will have to reapply for a doctorate if she decides to stay at Stanford. At the University of Wisconsin, Nichole Broderick is also undecided about her career path in biology, but she is using the master's to get familiar with the research in entomology and plant pathology before committing to a Ph.D.

On the other hand, Scott Berggren at Colorado State University's Department of Biology started out on the Ph.D. track and dropped to a master's to get experience in several labs where he's studying systematic botany. "I decided to get my master's first and get a doctorate under another professor. That way I'll have more lab experience and work under 2 different advisers. The advantage is that I'll have 2 people supporting my career," he explains.

Graduate students at Florida State University's Department of Biological Science can switch from the master's to the Ph.D. track. Some students are highly motivated and know how far they want to go in the discipline. Others aren't sure about a Ph.D. "We encourage them to make that decision in their first year. The student who does go on for a Ph.D. has to have a definite desire to do independent research. Not all students who start in grad school have that motivation," says department chairman Thomas Roberts, alluding to a competitive job market for Ph.D.'s in biology and the commitment that must be made to getting a postdoctoral degree.

Students must also consider whether they want a biology program that is attached to a medical school or one that is part of the arts and sciences. As Roberts observes, not only do the environments differ, but those on the medical side tend to be specialized. By contrast, a biology department in the arts and sciences offers more breadth in the curriculum, which can range from ecology to molecular biology.

Research Interests Under the Microscope

• •
■ **GRAD TIP: Before choosing a grad school, look for areas of research that interest you.**
• •

"It's just as important to find a person whose work you're interested in as going to a school with a good name," says Nichole Broderick. "You've got to be happy in that environment. You'll be in it for four to six years if you're going for your Ph.D." The grad school experience can be grueling. If you base your decision on a school name only, you could be miserable. Current grad students emphasize, above everything else, the need to match your research interests and find a lab director with whom you will work well. Each time you change labs, you're adding more time to finish your degree.

Your Project

Berggren recommends that incoming graduate students have a few project ideas in mind. It's not difficult to do. The Web is a rich source of detailed information about professors and the labs they support. Once broad research areas are known, applicants should contact individual professors at each school they're considering. Not only will they discover the individual research each professor is involved in, but also they will find out which labs are active and, even more important, which are funded.

You don't need to know exactly what type of research path you want to follow. Coming in with a broad idea about the research you think you'd like can be to your advantage. Focused on developmental genetics at the California Institute of Technology's Biology Department, Ph.D. candidate Martha Kirouac has seen many incoming grad students. Some know exactly what they want to pursue. Others come with an open mind and thus are able explore options before settling on one area. It's better to look before you leap because of the specialization that exists now in biology. "It's frustrating to be so specialized, but it's where things are."

The Size Factor

If you are not sure of your niche, you need to consider the size of a biology department. "Caltech is a small department so it's good in a couple of key areas but has holes in others," acknowledges Kirouac. For that reason, applicants who want to be able to explore various avenues in biology should look for departments with a broad range of faculty members who will offer more choices.

Hand-Holding or Sink or Swim?

• •
■ **GRAD TIP: Some biology departments follow a rigid curriculum and lab track. Others are more flexible in what grad students can choose.**
• •

With no predetermined curriculum at Caltech, Kirouac is free to address the questions she wants to research. Students are responsible for learning the material they need to pass oral and written exams. But this kind of flexibility is not for everyone. "As they evaluate a graduate program, students need to ascertain how much hand-holding they want. With freedom comes foundering," Kirouac warns. Not knowing if she was headed toward teaching or industry, Cole at Indiana University deliberately sought a program without a specific track to follow. At Indiana, biology grads start out with the same core set of classes but then branch out on their own.

Choosing the Right Lab

The choice of labs also can be a critical factor when choosing a department. Some programs ask incoming students to pick a lab before they arrive for the first semester, which can be a big mistake for those who are not absolutely sure what area they want to pursue. Cole points out that the lab environment is not just dependent on the research being done there. The professor heading the lab, the way it's set up, and those in the lab can greatly influence the quality of your experience. Labs can center on one or two techniques or involve dozens of approaches. Many programs send their first-year Ph.D. students through lab rotations in which they might work on four to six small projects, which is ideal for those who haven't chosen a research direction. "There's more pressure for master's students to make the choice of lab early," says Roberts. "If you spend a year looking for a lab, that will only add to the length of time to get your degree."

The Case for Getting Along

Having watched people come and go through the labs he's in at Johns Hopkins, Byrd suggests that getting along with the lab adviser can make or break your grad experience. If the research is fascinating, a personality clash can be overlooked. But if not, it can amount to a big deal. Adds Broderick, "You need to be able to work with a professor you can approach and be in a supportive lab environment where you're happy with your lab mates. It's a stressful time of life. In a graduate lab you want something to show for your work, especially if it will take you to the next level."

Lab Environment

The lab environment can be as critical a factor as the research being done in it. Cole wasn't expecting to find a convivial atmosphere. She anticipated she'd be working at a bench alone. Instead, she found the interaction between lab mates was a big part of the experience. That's why Kirouac stresses that students in the process of picking a lab should note such things as whether the lab director allows music to be played. Academically, labs can be the same, but the environment can differ drastically.

Adds Broderick, "One key to whether you will have a successful graduate career is if you're happy in a lab situation. The professor can set the whole tone of a lab and can establish the whole lab dynamic." Her point is that if you don't get along in that environment, you've limited your future options. It's the lab director who will write recommendations in a job search or connect you to resources. Kirouac tips incoming students that some advisers expect their students to be in the lab whenever they are, which can be all the time. Others don't care as long as the work is done. From her experience, Broderick describes the differences in two labs she has. One is strictly 9 to 5. Come in, do your work, and leave. In the other, people work crazy hours, but they also bring in the proverbial brownies, go to lunch together, and gather for the ubiquitous Friday beers.

How to Get Honest Answers

Experienced grad students are your best source for the kind of inside information you will need to make decisions about a lab. "Press them to be honest," Berggren suggests. "Ask how they get along with the lab adviser socially and in an academic setting. What is expected of the lab participants? Do they like being in the lab, and would they make the same decisions again?" Kirouac felt so strongly about this aspect of choosing a graduate department that her search included finding at least 3 people at each institution with whom she would feel comfortable working. "Get an idea of what goes on—how people interact and what projects are available. You might get into a lab where you dislike the people and the research doesn't interest you," she notes.

. . . .
■ **GRAD TIP: Ask current and former students how the lab professor handles failed experiments.**
. . . .

A well-placed question can cut right to the core. Broderick finds that asking just about the lab and professor can produce bland answers. Ask current lab students what happens when they have to go to the professor and report that an experiment failed, or results turned out poorly, or they messed up something because they were new to the protocol. That should produce some straight talk.

The Bottom Line of Funding

To grad students in other fields, funding can be a significant factor in their choice of school. For biology students, this is generally not much of an issue, says Roberts. Many graduate departments in biology are well funded and, depending on student progress, often grant tuition waivers and stipends to Ph.D. students. Teaching and research assistantships (TA and RA) often accompany the funding, but Roberts warns students that the more teaching they do, the slower their own progress toward the degree will be.

At Stanford, terminal master's students get no funding at all, and there are no TA positions available for that level. Ph.D. students are a different category. Berggren reports that in science, schools usually try to give their Ph.D. candidates some sort of support so they can devote most of their time to research. As can be expected, they only accept the number of students who can be funded. At Caltech, Kirouac says that, to her knowledge, most natural science and chemistry fields are funded with special grants that come to the university. She is also required to be a TA. Cole is currently being funded through a professor with whom she works, although she has also been funded by TA positions.

Though funding is widely available, biology students might have to dig for it by finding professors whose labs have grants or through teaching internships. In addition to an RA position, Broderick got a stipend that was covered through a grant from the Department of Natural Resources to study gypsy moths. It provides her with a monthly income. "It's more common for students to be funded through grants written by professors," she says.

Teaching Comes with the Territory

As to the responsibilities of TA positions, they vary from school to school. At Indiana University, Cole teaches entry-level undergraduate classes and helps out in labs and discussion groups. At Johns Hopkins, second-year graduate students have to teach one class per semester. When assessing the amount and kind of funding available from different programs, make sure to look at the fine print. How long will you get funding? Will you have to be a TA during your entire academic career? Stipends can vary from $8000 to $20,000 a year. Choosing a program based on the dollar amount can land you in a city with a high cost of living, thereby nullifying the greater figure. Housing is another factor to consider, says Helmy, noting how expensive it is around Stanford.

Some Finer Points to Consider

When these larger aspects of choosing a graduate department—faculty members, research, labs, and funding—are put on the table, you should also compare the facilities. Berggren visited several campuses not only to talk to professors but also to see what kind of equipment was available for students. "The more equipment they have, the more easily you can answer questions," he states. "You won't have to send experiments off somewhere else."

Location and quality of life should also figure prominently in the mix of items to consider. The research in that northern school might be spectacular, but will you be as thrilled at the four feet of snow you'll have to deal with on a regular basis? As a grad student with a few years of experience behind her, Kirouac often talks to prospective students at Caltech. They usually want to know about the quality of life. She advises applicants to get a really well rounded idea of student life.

Getting the Information

Potential grad students today have a number of resources to get information about programs before applying or making a final decision. Obviously, guide books and Web resources will give you immense amounts of data about professors and their research interests, as well as a vehicle to contact them. Students themselves are an abundant source for straight talk about a program. However, get plenty of opinions. You might be catching someone on a bad day.

Kirouac advises talking to other students who are interviewing for the department to get their impressions as well as comparing notes on the questions they're asking. And, of course, on any visit, you'll be interviewing faculty members. You should go armed with lots of questions for them from all the research you've done on each school. Kirouac tried to visit all ten schools she was interested in while still in college. Even though she was gone Friday through Monday for ten weeks, she emphasizes the importance of visits. "Sit down on campus and just watch. Find out where grad students live. Jot down your impressions before they start to blur," she says.

Taking the
Entrance Exams

The prospect of a graduate admissions test is enough to make some students put aside their graduate school plans indefinitely. You may be anxious about taking the Graduate Record Examinations (GRE) or one of the professional exams but usually there is no way of avoiding it. Most programs require one of the major standardized exams, and they may also require a subject area test, writing assessment, or test of English language proficiency if you are not a native speaker of English. So unless you've selected a program that does not require an examination, you are going to have to take at least one—and do well.

How Graduate Programs Use the Test Results

It is helpful to understand how an admissions committee might use your score. The role played by a graduate admissions test is similar to the one played by the SAT or ACT at the undergraduate level. It provides a benchmark. Essentially, it is one of the few objective bits of information in your application that can be used to gauge where you fall in the range of applicants. Some programs, especially the top professional programs that receive many more applicants than they can admit, may use the score as a means of reducing the applicant pool: if your score is below their cutoff, they will not even look at the rest of your application. But most programs are much more flexible in the way they evaluate scores. If your score is low, you still may be considered for admission, especially if your grade point average is high or your application is otherwise strong.

But, let's face it, a low or average score will not help your case. When Heather Helms-Erikson applied to master's degree programs in marriage and family therapy, she took the GRE with no preparation, during finals week. "I did well enough to get in but not well enough to get funding," she says. A few years later, when applying to Ph.D. programs in human development and family studies, Helms-Erikson was determined to get the best funding package she could. To that end, she studied 2 to 5 hours a week for four months preparing for the GRE. Needless to say, her score was considerably higher and she was admitted to several programs with funding.

One story does not prove that a good score will open all doors. But you should regard the test as an opportunity to improve your application. And that means you must take the test in plenty of time to meet application deadlines. That way, if you take the test early and are disappointed with the results, you will have time to retake it.

As exemplified by Helms-Erikson, you can squeak by. But why not put some time into preparing for the exams by refreshing your memory and getting the practice with test taking. Preparation is especially important for applicants who have been out of school for years. You may need to do a quick recap of high school mathematics, for example, to do well on the mathematics portion of the test. And you may have forgotten what test taking is like. Study and practice will help you overcome any weaknesses you may have.

There are three types of Graduate Record Examinations: the General Test, which is usually referred to as the GRE; the Subject Tests; and the Writing Assessment. Each of these tests has a different purpose, and you may need to take more than one of them. If so, try not to schedule two tests on the same day. The experience may be more arduous than you anticipate.

The General Test (GRE)

According to the Educational Testing Service (ETS), the GRE "measures verbal, quantitative, and analytical reasoning skills that have been developed over a long period of time and are not necessarily related to any field of study."

Like the SAT, the GRE is a test designed to assess whether you have the aptitude for higher-level study. Even though the GRE may not have subject-area relevance, it can indicate that you are capable of doing the difficult reading, synthesizing, and writing demanded of most graduate students.

The GRE is a computer-adaptive test (CAT). It is divided into three separately timed parts, and all the questions are multiple-choice. The three sections are a 30-minute verbal section consisting of thirty questions, a 45-minute quantitative section with twenty-eight questions, and a 60-minute analytical section of thirty-five questions. The parts may be presented in any order. In addition, an unidentified verbal, quantitative, or analytical section that doesn't count in your score may be included. You don't have any way to tell which of the duplicated sections is the "real" one, so you should complete both carefully. Finally, another section, on which ETS is still doing research, may also appear. This section will be identified as such and will also not count in your score. ETS tells test takers to plan to spend about 4½ hours at the testing site.

Verbal Section

The thirty questions in the verbal section of the GRE test your ability to recognize relationships between words and concepts, analyze sentences, and analyze and evaluate written material. In other words, they test your vocabulary and your reading and thinking skills. The words and reading material on which you are tested in this section come from a wide range of subjects, ranging from daily life to the sciences and humanities. There are four main types of questions.

- In sentence-completion questions, sentences are presented with missing word(s). You are asked to select the words that best complete the sentences. Answering correctly involves figuring out the meanings of the missing words from their context in the sentence.
- Analogy questions present a pair of words or phrases that are related to one another. Your task is to figure out the relationship between the two words or phrases. Then you must select the pair of words or phrases whose relationship is most similar to that of the given pair.
- In antonym questions, you are given a word and asked to select the word that is most opposite in meaning.
- Reading comprehension questions test your ability to understand a reading passage and synthesize information on the basis of what you've read.

Quantitative Section

The quantitative questions test your knowledge of arithmetic and high school algebra and geometry, as well as data analysis. They do not cover trigonometry or calculus. You will be tested on your ability to reason quantitatively and solve quantitative problems.

- Quantitative comparison questions require that you determine which of two quantities is the larger, if possible. If such a determination is not possible, then you must so indicate.
- Data analysis questions provide you with a graph or a table on which to base your solution to a problem.
- Problem-solving questions test a variety of mathematical concepts. They may be word problems or symbolic problems.

Analytical Section

According to ETS, the analytical section of the GRE "tests your ability to understand structured sets of relationships, deduce new information from sets of relationships, analyze and evaluate arguments, identify central issues and hypotheses, draw sound inferences, and identify plausible causal relationships." In other words, can you reason analytically and logically? The subject matter in the analytical section is drawn from all fields of study as well as everyday life. There are two main types of questions in this section.

- Analytical reasoning questions appear in groups, and they are all based on the same set of conditions or rules. A situation is described and you are told how many people or things you will be manipulating. Then you are asked to manipulate the items according to the conditions. For example, you may be given information about a group of people and then asked to rank them in order of age.
- Logical reasoning questions consist of arguments that you must analyze and evaluate. Each argument has assumptions, facts, and conclusions, and you must answer questions that test your ability to assess these.

Computer-Adaptive Tests (CATs)

The GRE is now given only in computer format in most locations around the world and is somewhat different from the old paper-and-pencil test. At the start of each section, you are given questions of moderate difficulty. The computer uses your responses to each question and its knowledge of the test's structure to decide which question to give you next. If your responses continue to be correct, how does the computer reward you? It typically gives you a harder question. On the other hand, if you answer incorrectly, the next question will typically be easier. In short, the computer uses a cumulative assessment of your performance along with information about the test's design to decide which question you get next.

. . . .

■ **GRAD TIP: On the CAT, you cannot skip a question. The computer needs your answer to a question before it can give you the next one.**

. . . .

You have no choice. You must answer in order to move to the next question. In addition, this format means you cannot go back to a previous question to change your answer. The computer has already taken your answer and used it to give you subsequent questions. No backtracking is possible once you've entered and confirmed your answer.

On computer-adaptive tests, each person's test is different. Even if two people start with the same item set in the basic test section, once they differ on an answer, the subsequent portion of the test will branch differently.

According to ETS, even though people take different tests, their scores are comparable. This is because the characteristics of the questions answered correctly and incorrectly, including their difficulty levels, are taken into account in the calculation of the score. In addition, ETS has conducted research that indicates that the computer-based test scores are also comparable to the old paper-and-pencil test scores.

One benefit of the computer-based format is that when you finish the test you can cancel the results—before seeing them—if you feel you've done poorly. If you do decide to keep the test, then you can see your unofficial scores right away. In addition, official score reporting is relatively fast—ten to fifteen days.

A drawback of the format, in addition to the fact that you cannot skip around, is that some of the readings, graphs, and questions are too large to appear on the screen in their entirety. You have to scroll up and down to see the whole item. Likewise, referring to a passage or graph while answering a question means that you must scroll. In addition, you can't underline sentences in a passage or make marks in the margin as you could on the paper test. To make up for this, ETS provides scratch paper that you can use to make notes and do calculations.

To help test takers accustom themselves to the computerized format, ETS provides a tutorial that you complete before starting the actual test. The tutorial familiarizes you with the use of a mouse; the conventions of pointing, clicking, and scrolling; and the format of the test. If you are familiar with computers, the tutorial will take you less than half an hour. If you are not, you are permitted to spend more time on it. According to ETS, the system is easy to use, even for a person with no previous computer experience. However, if you are not accustomed to computers, you would be far better off practicing your basic skills before you get to the testing site. If it's any consolation, knowledge of the keyboard is not required—everything is accomplished by pointing and clicking.

More information

Test takers can take a GRE and GMAT computer adaptive test at http://www.petersons.com. Test takers can buy *Peterson's GRE CAT Success* and *Peterson's GMAT CAT Success* with a CD that enables them to launch Peterson's CAT site, register for the tests, and take them without fees. Arco's *Master the GMAT CAT* and *Master the GRE CAT* offer more comprehensive test preparation and also include a CD to launch Peterson's CAT site. If the books are bought without the CD, users will have to pay to take the CAT on petersons com. Users who do not buy the books but access Peterson's site will also be able to take the GRE and GMAT CAT after paying a fee.

GRE-ETS, P.O. Box 6000, Princeton, New Jersey 08541-6000. Telephone: 609-771-7670. Web site: http://www.gre.org

Peterson's offers *GRE Success*, a complete guide to the GRE, and also Arco's *30 Days to the GMAT CAT* and *30 Days to the GRE CAT.* Visit your local bookstore for these titles or contact Peterson's at 800-225-0261 or http://www.petersons.com/ for Peterson's online store.

Subject Tests

The Subject Tests test your content knowledge in a particular subject. There are currently eight Subject Tests, and they are given in paper-and-pencil format only. The subjects are biochemistry and cell and molecular biology, biology, chemistry, computer science, literature in English, mathematics, physics, and psychology. The Subject Tests assume a level of knowledge consistent with majoring in a subject or at least having an extensive background in it. ETS suggests allowing about 3½ hours at the testing site when taking a Subject Test.

Unlike the General Test, which is given many times all year round, the Subject Tests are given only three times a year. Keep in mind that because the tests are paper-based, it takes four to

six weeks for your scores to be mailed to your designated institutions.

. . . .

■ **GRAD TIP: Because the tests are given infrequently and score reporting is slow, be sure you plan ahead carefully so your test results will arrive before your deadlines.**

. . . .

The Writing Assessment

Introduced in 1999, the Writing Assessment is a performance-based assessment of critical reasoning skills and analytical writing. It can be taken in computer or paper formats and consists of two parts:

- In the 45-minute task, called "Present Your Perspective on an Issue," you must address an issue from any point of view and provide examples and reasons to explain and support your perspective. You are given a choice of two topics.
- In the 30-minute task, "Analyze an Argument," you must critique an argument by saying whether it is well reasoned. There is no choice of topics in this section.

Scoring of the Writing Assessment is done according to a 6-point scale by college and university faculty with experience in teaching writing or writing-intensive courses. Each essay is scored independently by two readers. If the two scores are not identical or adjacent, a third reader is used. The reported score is the average of your two essay scores.

Other Tests

Miller Analogies Test

The Miller Analogies Test (MAT), which is administered by The Psychological Corporation, is accepted by over 2,300 graduate school programs. It is a test of mental ability given entirely in the form of analogies. The MAT tests your store of general information on a variety of subjects through the different types of analogies you must complete. For example, the analogies may tap your knowledge of fine arts, literature, mathematics, natural science, and social science.

On the MAT, you have 50 minutes to solve 100 problems. The test is given on an as-needed basis at more than 600 test centers in the United States.

For more information

The Psychological Corporation, 555 Academic Court, San Antonio, Texas 78204. Telephone: 210-299-1061 or 800-211-8378 (7 a.m. to 7 p.m., Monday through Friday, Central time).

Tests of English Language Proficiency

If your native language is not English, you may be required to take the Test of English as a Foreign Language (TOEFL) or Test of Spoken English (TSE) in order to determine your proficiency in English. Both tests are administered by ETS.

The TOEFL is given in computer-based form throughout most of the world. Like the computer-based GRE, the TOEFL does not require previous computer experience. You are given the opportunity to practice on the computer before the test begins. The TOEFL has four sections—listening, reading, structure, and writing—and it lasts about 4 hours.

The TSE evaluates your ability to speak English. During the test, which takes about 20 minutes, you answer questions that are presented in written and recorded form. Your responses are recorded; there is no writing required on this test. The TSE is not given in as many locations as the TOEFL, so you may have to travel a considerable distance to take it.

For more information

TOEFL, P.O. Box 6151, Princeton, New Jersey 08541-6151. Telephone: 609-771-7100. E-mail: toefl@ets.org Web site: http://www.toefl.org

Preparing for the Tests

At the very least, preparation will mean that you are familiar with the test instructions and the types of questions you will be asked. At the most, your preparation will lead to improved scores and reduced anxiety. Your time will be well invested.

If your computer skills need improvement, adequate preparation will mean you can focus on the questions rather than struggle with the mouse when you take the computer-based tests. For the Subject Tests, you will actually need to study content. There are many ways you can prepare for the tests, but whichever method you choose, start early.

Jim Lipuma, who is earning a Ph.D. in environmental science at New Jersey Institute of Technology, favored practice tests. "My only advice for the tests is to read and practice with old tests. . . . Though I did not use a review course or study, I did know exactly what to expect by reviewing sample exams. . . . Cramming will not work for more than a few points. Others I know who also have done well have pretested using old tests to hone skills."

. . . .
■ **GRAD TIP: You can check the Web sites of the various tests to download or request practice tests, or you can buy practice-test books at a bookstore.**
. . . .

Other students used workbooks that give information and test-taking strategies, as well as practice items. Bob Connelly, who earned an Ed.D. in educational administration from Seton Hall University in New Jersey, used a workbook to prepare for the MAT. "I'm glad I did," says Connelly. "If I had gone in cold, I would not have recognized the patterns of the analogies, and the test would have been more stressful." There are many workbooks, some with CDs, that will help you prepare for a graduate admissions test.

Many students don't trust themselves to stick with a self-study program using practice tests, workbooks, or software. If this sounds like you, you may prefer the structure and discipline of a professional review course. Although the courses are much more expensive than the do-it-yourself approach, they may be worth it if they make you study.

If you are still in college, your professors may be able to help you prepare. You can ask your professors if they would be willing to help you and other students prepare for a Subject Test.

Reducing Test Anxiety

The best way to reduce test anxiety is to be thoroughly prepared. If you are well acquainted with the format, directions, and types of questions you will encounter, you will not need to waste precious time puzzling over these aspects of the exam. In addition to thorough preparation, here are some suggestions to reduce the stress of taking the exam.

- Get a good night's rest and don't tank up on caffeinated beverages. They will only make you feel more stressed.
- Make sure you've got all the things you will need, including your admission ticket, proper identification, and pencils and erasers if you are taking a paper-based test.
- Dress in layers so you will be prepared for a range of room temperatures.
- Get to the testing site at least a half hour early. Make sure you know the way and leave yourself plenty of time to get there.
- Pace yourself during the exam. It is to your benefit to answer each question and complete each section.

- Keep things in perspective. The exam is just one part of a much larger application process.

Bad News/Good News

If the "worst" happens and you do not do well the first time, don't despair. Some programs will admit you conditionally despite a poor score, but they expect you to retake the test and improve your performance. One applicant from Italy missed a question on the TSE, lowering her score to 220 (out of 300). She was admitted on a conditional basis, and when she retook the test, she scored 300.

When applying to the public communications program at the University of Alaska in Fairbanks, Jenn Wagaman had a similar experience with the GRE. "I took the GRE the first time when I was living in my hometown of New Orleans. I did miserably," says Wagaman. "I was lucky enough to be admitted to a graduate program at a small school that knew my capabilities and was willing to let me take another stab at my scores." The second time she took the GRE, Wagaman's scores increased significantly.

Applying for Your Graduate Degree

Preparing a thorough, focused, and well-written application is one of the most important tasks you will ever undertake. In addition to gaining you admission to a graduate program that can help you achieve your goals, a good application may win you enough monetary support to finance your degree. With these benefits in mind, work on your applications as if they are the most important work you can possibly be doing—because they are.

If you have not already done so, request an application and information packet from each program to which you plan to apply. When you look over these materials, you will see that there is a lot of work involved in applying to graduate school. It may take you a year or more to assemble and submit all the necessary information, especially if you're an international student or you've been out of school for a few years. Because the process is complicated and time consuming, start well ahead of time.

Timetable

In general, it's advisable to start the application process at least a year and a half before you plan to enroll. Allow yourself even more time if you are applying for national fellowships or if you are applying to a health-care program through your college's evaluation committee. In these cases, you may need to start two years before matriculation in order to meet all the deadlines for test scores, letters of recommendation, and so on.

Application deadlines for fall admission may range from August, one full year prior to your planned enrollment, to late spring or summer for programs with rolling admissions. However, most programs require that you submit your application between January and March of the year in which you wish to start. Be careful to check application deadlines. Different programs at a university may have different deadlines.

- **GRAD TIP: Deadlines are really important if you want to get funding.**

If you are applying for financial aid, leave yourself extra time to assemble all the financial information you'll need to support your request for assistance. Applicants for aid usually have to send in the entire application by an earlier date. Be certain that you understand which deadline applies to you. After all, what's the point of being admitted if you cannot afford to attend?

- **GRAD TIP: Applying early indicates your interest.**

An early application demonstrates strong interest and motivation on your part, especially when a program uses rolling admissions. Even more important, however, is that applying early means that the department or program will evaluate your application when it still has a full budget of funding to award. When you apply late, you may not be awarded full or even partial funding because the department has already used up its resources. Says Suzette Vandeburg, Assistant Vice Provost for Graduate Studies at the State University of New York at Binghamton, "You may be highly qualified but lose out if you miss a deadline."

This does not mean that you will necessarily miss out on funding if you just meet the program's deadline, but given the competition for financial aid, why gamble? While you could get lucky, you may be in for some weeks of nail-biting until a program makes all its awards. "When I applied to graduate school I noticed that many positions were offered early, especially in departments that were very small and very competitive," says Cindy Liutkus, a Ph.D. candidate in geology at Rutgers University in New Jersey. "Because I hadn't applied as early as most people, I had to wait until two of the departments obtained rejections from their early offers before I knew whether I would receive a teaching assistantship or a research assistantship." Don't rely on luck for something so important. Apply early.

Who Has the Power?

University graduate admission offices usually act as clearinghouses for applications, but in some cases they have the authority to reject an applicant or to waive a university requirement for an exceptional candidate. For example, they can turn down an applicant whose qualifications are clearly below university standards (someone with an extremely low GPA for all four college years). They can also bar an application from further consideration if it is incomplete.

Once your application is accepted, the members of the admission committee are the people on whom your future depends. They are the small group of department or program faculty members and administrators who review and evaluate each applicant and decide not only who gets in but who gets funding. Admission committees usually have at least four sources of information on which to base their decisions: your transcripts, your test scores, your personal essay, and your letters of recommendation. The importance of each of these sources will vary from admission committee to admission committee and, indeed, will vary among the members of a committee. Their decision-making processes will vary as well. Let's take a look at how two actual admission committees work to give you an idea of what goes on behind those closed doors.

A Peek Behind Closed Doors in the Social Sciences

In this program, the admission committee receives about 120 applications per year. A staff member extracts certain data, including undergraduate

school, degree, and area of concentration; grade point average; GRE scores; and field of interest. This information is placed on a cover sheet and attached to the application. This cover sheet is the first thing the admission committee members will see when they pick up an application. A few weeks after the application deadline, the committee meets for a day-long marathon of reading and assessing applications. The applications are divided into two groups—master's degree candidates and doctoral degree candidates—and they are handled separately. Each application is passed around for each committee member to read. While the applicant's essay and letters of recommendation are fresh in each person's mind, the committee makes a decision on the candidate.

Usually there is agreement on accepting or rejecting an applicant, but occasionally members of the committee have a difference of opinion on a particular candidate. In that case, a decision on the candidate may be deferred until the candidate can be interviewed. Or the candidate may be accepted on a conditional basis. At the same time the accept/reject decision is being made, a tentative decision on department funding is also made. After all the applications have been evaluated, the committee goes through the applications in the acceptance pile again, adjusting the funding decisions that they made in the first round.

■ **GRAD TIP: Essays that indicate applicants know what the strengths of a department are and how that department matches their goals carry a lot of clout.**

In this committee, a great deal of weight is placed on the personal essay. Members of the committee are looking for evidence that a candidate is focused and committed. "We look to see whether the applicant knows why he or she is applying to our graduate program in a specific way, not just as a next step in life while they're figuring out what to do," comments one member of the committee. "When a student knows what our strengths are and how their interests fit into our program, we are impressed."

A Peek Behind Closed Doors in the "Hard" Sciences

In this department, the admission committee consists of 5 faculty members, one from each major division of the department. Each faculty member reviews the applications of the students interested in his or her area of specialization. In addition, the chair of the committee reviews all the applications. Periodically, the committee meets to discuss and make decisions on the applicants. Since the department has the resources to fund all first-year students, an acceptance automatically means the student will have financial support.

■ **GRAD TIP: Letters of recommendation tell admission committees about your research experience and relevant summer internships.**

In this committee, grade point average and letters of recommendation are weighted heavily. A minimum GPA of 3.0 is required, although extenuating circumstances are considered if the GPA is uneven; a typical example is a low freshman-year GPA, which the committee may decide to overlook. The letters of recommendation are important because the committee learns about the student's undergraduate research experience and relevant summer internships from them. Of the four main elements of the application, the essay is the least important to this committee. As long as it is coherent and gives an indication of the student's interest in research, the members pay little attention to it.

■ **GRAD TIP: Conduct research on faculty member interests. In some programs, if no faculty member shares your focus, you won't be accepted.**

This particular committee does not try to make a match between every single applicant and a faculty member with similar research interests. "Students often change their minds once they get here," comments a member of the committee.

This is unlike other programs, where the admission committees may turn down an excellent applicant solely because there is no faculty member to work with the student.

Putting the Pieces Together

From our description of the various admission committees, you can see that you cannot always tell which parts of your application will be considered the most important. For that reason, work hard to make each element of your application the best it can possibly be. For each program to which you apply, you will have to submit a number of items to make your application complete. For most programs, these include:

- Application forms
- Undergraduate and other transcripts
- Graduate admission test scores
- Letters of recommendation
- Personal essay(s)
- Application fees

Application Forms

■ **GRAD TIP: You will be competing against people whose applications are complete, legible, and error-free.**

Do not omit information, and double-check for spelling errors. If possible, type the application or fill it out online at the program's Web site. If neither of these options is available, then print your entries neatly.

■ **GRAD TIP: READ THE INSTRUCTIONS!**

"Take your time filling out all the necessary information, no matter how tedious it may be," advises Tammy Hammershoy, who is earning a master's degree in English at Western Connecticut State University. "Read everything very carefully and follow all instructions. If you really want to get into the program of your choice, be patient and careful when filling out application forms and other materials."

Transcripts

- **GRAD TIP: Allow enough time—two or three months—for your transcripts to be processed.**

To request official transcripts, contact the registrars of your undergraduate college and other institutions you have attended. It will save time if you call to find out what the fee for each transcript is and what information they need to pull your file and send the transcript to the proper recipient. Then you can enclose a check for that amount with your written request. You will need to submit official transcripts from each college and university you have attended, even if you have taken just one course from that institution.

- **GRAD TIP: Look for weaknesses in your application that may need explaining.**

For example, a low GPA one semester, a very poor grade in a course, or even a below-average overall GPA may hurt your chances of acceptance unless you have good reasons for them. You can explain any shortfalls in your transcripts in your personal essay, cover letter, or addendum to the application.

- **GRAD TIP: Your undergraduate grades count, no matter how long ago you earned them.**

If you have been out of school for years and have been successful in your professional and postgraduate endeavors, do not assume that a poor undergraduate GPA will not count against you because it's "ancient history." For example, one 58-year-old prospective graduate student who had an A- average in his previous master's program but a C average as an undergraduate found that the A- did not cancel out the C. He had to take a

semester of master's-level courses and achieve a B average before he was admitted to the new master's program as a matriculating student.

Test Scores

Like your GPA, your admission test scores are numbers that pop right out of your application and tell the admission committee something about you before they have even begun reading your file. Your scores give the admission committee a way to compare your performance to that of every other applicant, even though you attended very different colleges with very different instructional and grading standards.

- **GRAD TIP: Reading, writing, and analytical thinking skills count heavily in all fields.**

Although your GRE scores may not directly relate to the field in which you are planning to work, the scores do predict how well you can cope with the types of tasks graduate students face all the time—reading, writing, and analytical thinking. "Over the years we have found that students with poor verbal scores do not have the ability to read and write at the graduate level," says Gail Ashley, Professor of Geological Sciences at Rutgers University in New Jersey. Still, it is rare for an admission committee to reject an applicant solely on the basis of poor test scores. In fact, the committee may scrutinize an application with low scores even more thoroughly to see if other qualifications compensate for poor test performance.

- **GRAD TIP: Plan on taking the graduate admission test about a year before you plan to enroll—earlier if you are taking the MCAT.**

Taking the test early will give you plenty of time for score reports to be submitted and plenty of time to retake the test if your first set of scores is lower than you had hoped. When you register for a graduate admission test, you can request that the testing service send your official scores to the institutions you designate on the registration form. If you decide later to apply to additional programs and need more score reports, you can request them in writing.

Letters of Recommendation

. . . .

■ **GRAD TIP: Good letters of recommendation can tremendously increase your chances of admission and funding. Lukewarm letters can harm your application.**

. . . .

You will have to provide letters of recommendation for each program to which you apply. These letters are important because, like the personal essay, they give the members of the admission committee a more personal view of you than is possible from your grades and test scores. So it's important to approach the task of choosing and preparing your letter-writers in a thoughtful and timely fashion.

. . . .

■ **GRAD TIP: Start asking recommenders at least six months before your application deadline.**

. . . .

"Contact the people who will be writing letters of recommendation well in advance of application deadlines," suggests Felecia Bartow, an M.S.W. candidate at Washington University in St. Louis. "Many professionals and academics are extremely busy, and the more time that you can give them to work on your recommendation, the more it will reflect who you are." Starting early will also give you an opportunity to follow up with your recommenders well before the application deadlines.

Choosing People to Write Recommendations

Most of your recommendations should be from faculty members because they are in the best position to judge you as a potential graduate student, and members of the admission committee will consider them peers and will be more inclined to trust their judgment of you. Having professors

write your letters is absolutely essential if you are applying to academic programs.

If you cannot make up the full complement of letters from faculty members or if you are applying to professional programs, ask employers or people who know you in a professional capacity to write references for you.

. . . .

■ **GRAD TIP: It won't do you much good to have a glowing letter of recommendation from your manager at the insurance company if you are applying to a program in history or social work.**

. . . .

When you are trying to decide whom to ask for recommendations, keep these criteria in mind. The people you ask should:
- have a high opinion of you
- know you well, preferably in more than one context
- be familiar with your field
- be familiar with the programs to which you are applying
- have taught a large number of students (or have managed a large number of employees) so they have a good basis upon which to compare you (favorably!) to your peers
- be known by the admission committee as someone whose opinion can be trusted
- have good writing skills
- be reliable enough to write and mail the letter on time

A tall order? Yes. It's likely that no one person you choose will meet all these criteria, but try to find people who come close to this ideal.

"The most important thing to remember is that you want the writers of these letters to be very familiar with you and your work," advises Cindy Liutkus. "As I was choosing professors to ask for letters, many people gave me advice as to who would write the best letter. Some suggested that the chair of the department carries the most weight, even if he or she doesn't know you very well. Others said to ask the dean of the school. But once again, since he didn't know me very well,

I was skeptical as to the quality of the letter. Instead, I chose a professor from each of my major disciplines, namely my thesis adviser and my favorite undergraduate geology professor. I needed a third and had a lot of trouble deciding whom to ask. I eventually chose the woman in the geology department whom I respected the most. Although I had only one class with her, I felt she would give the most honest and straightforward account of my undergraduate accomplishments, my personality and work habits, and goals for the future."

Approaching Your Letter Writers

Once you've decided whom you plan to ask for references, be diplomatic. Don't simply show up in their offices, ask them to write a letter, and give them the letter of recommendation forms. Plan your approach so that you leave the potential recommender, as well as yourself, a graceful "out" in case the recommender reacts less than enthusiastically.

- -
■ **GRAD TIP: A confidential letter usually has more validity in the eyes of the admission committee.**
- -

On your first approach, remind the person about who you are (if necessary) and then ask whether they think they can write you a good letter of recommendation. This gives the person a chance to say no. If the person says yes but hesitates or seems to be less than enthusiastic, you can thank them for agreeing to help you. Later, you can write them a note saying that you won't need a letter of recommendation after all. On the other hand, if the person seems genuinely pleased to help you, you can then make an appointment to give that person the letter of recommendation forms and the other information he or she will need.

The letter of recommendation forms in your application packets contain a waiver. If you sign the waiver, you give up your right to see the letter of recommendation. Before you decide whether to sign it, discuss the waiver with each person who is writing you a reference. Some people will write you a reference only if you agree to sign the waiver and they can be sure the letter is confidential. This does not necessarily mean they intend to write a negative letter; instead, it means that they think a confidential letter will carry more weight with the admission committee. From the committee's point of view, an "open" letter may be less than candid because the letter writer knew you were going to read it. So, in general, it's better for you to waive your right to see a letter. If this makes you anxious with regard to a particular recommender, don't choose that person to write a letter.

- -
■ **GRAD TIP: Provide letter writers with information about yourself.**
- -

Once a faculty member or employer has agreed to write a letter of recommendation for you, he or she wants to write something positive on your behalf. No matter how great you are, this won't be possible if the letter writer cannot remember you and your accomplishments very well. "Help faculty members write a more effective letter by reminding them of what you've done," advises Teresa Shaw, Associate Dean for Arts and Humanities at Claremont Graduate University in California. "Letters that are not specific are ineffective."

Bring a short resume that highlights your academic, professional, and personal accomplishments when you meet with your letter writers. List the course or courses you took with them, the grades you got, and any significant work you did, such as a research paper or lab project. "Many of the people I asked to write recommendation letters found it helpful if I wrote down a list of my accomplishments and my plans," recalls Jenn Wagaman, a master's candidate in public communications at the University of Alaska at Fairbanks. "Even though these people knew me, they wrote better letters because they had the exact information right in front of them."

What should you do if the letter writer asks you to draft the letter? Accept gracefully. Then pretend you are the writer and craft a letter extolling your virtues and accomplishments in detail. Remember, if the letter writer does not like what you've written, he or she is free to change it in the final draft.

■ **GRAD TIP: Do everything you can to make it easy for the letter writer, including providing stamps and preaddressed envelopes.**

You can help your letter writers by filling in as much of the information as you can on the letter of recommendation forms. Also be sure to provide stamped, addressed envelopes for the letters if they are to be mailed directly to the programs or to you for inclusion in your application. Be sure your letter writers understand what their deadlines are. In other words, do everything you can to expedite the process, especially since you may be approaching your professors at the beginning of the fall semester, when they are busiest. Last, send thank you notes to professors and employers who have come through for you with letters of recommendation. Remember that you are hoping someday to be their colleague in academia or a profession. Cementing good relationships now can only help you in the future.

If you are unsure of your plans for graduate school, you can ask your professors to write you letters of recommendation now, when you are still fresh in their minds. Have the letters placed in your file in the College Placement Office, and ask that your file be kept active. Although there may be a fee for this service, it's worth it. When you do apply to graduate school a few years down the road, you will already have several letters of recommendation that you can use.

If You've Been Out of School for Years

What should you do if you have lost touch with your professors? If you established a file of letters of recommendation at the placement office when you were an undergraduate, you will now reap the benefit of your foresight. But if you did not, there are several things you can do to overcome the problems associated with the passage of time.

First, if a professor is still teaching at your alma mater, you can make contact, reminding the person of who you are, and describing what you've done since graduation and what your plans for graduate school are. Include a resume. Tell the professor what you remember most about the

courses you took with him or her. Most professors keep their course records for at least a few years and can look up your grades. If you are still near your undergraduate institution, you can make your approach in person. "I arranged to meet one of my college professors for coffee to talk about what I had been doing in the five years since she had had me as a student," says Felecia Bartow. "It gave me a chance to bring her up to date on my experience and it gave her a lot more information with which to write her recommendation."

Another strategy if you've been out of school for a while is to obtain letters of recommendation from faculty members teaching in the programs to which you plan to apply. In order to obtain such a letter, you may have to take a course in the program before you enroll so that the faculty member gets to know you. Members of an admission committee will hesitate to reject a candidate who has been strongly recommended by one of their colleagues.

Finally, if you are having trouble recruiting professors to recommend you, call the programs to which you are applying and ask what their policy is for applicants in your situation. They may waive the letters of recommendation, allow you to substitute letters from employers, or ask you to take relevant courses at a nearby institution in order to obtain faculty letters. Remember, if you are applying to an academic rather than a professional program, letters from employers will not carry as much weight with the admission committee as letters from faculty members. In fact, many academics are not at all impressed by work experience because they feel it does not predict how successful you will be as a graduate student.

Fees

Each application must be accompanied by a fee. If you cannot afford the fee, you can ask the admission office and your undergraduate financial aid office for a fee waiver.

The application process may cost hundreds of dollars, even more if you are applying to many schools. If you are applying to half a dozen schools, you can see that the costs will mount quickly. In addition to the program application fees, you must pay transcript fees, test fees, score report fees, photocopying, mailing costs, and travel costs if you are interviewing or auditioning. "Put aside some money because the process will cost more than you expect, especially if you are interviewing," suggests Jennifer Cheavens, a Ph.D. candidate in clinical psychology at the University of Kansas at Lawrence.

Submitting Your Application

Submit your completed applications well before they are due. Be sure to keep a copy of everything. You can either mail the application to the admission offices, or you can file portions of it online through the program's Web site. Remember, however, that some elements of the application, such as the fee and official transcripts, will still need to be mailed. Note also that most schools that accept online applications simply print them and process them as if they had come in by regular mail.

■ **GRAD TIP: Submit all your materials at once. This simplifies the task of compiling and tracking your application at the admission office.**

If that's impossible, as it is for many students, keep track of missing items and forward them as soon as possible. Remember that if items are missing, your application is likely to just sit in the admission office. According to Suzette Vandeburg, at the State University of New York at Binghamton, incomplete applications are held for a year and then they are tossed.

Graduate School Interviews

Interviews are usually required by medical schools and sometimes required by business schools and other programs. But, in most cases, an interview is not necessary. However, if you think you do well in interviews, you can call each program and ask for an interview. A good interview may be an opportunity to sway the admission committee in your favor. Human nature being what it is, an excellent half-hour interview may loom larger in

the minds of admission staff and faculty than four years of average grades.

■ **GRAD TIP: Graduate program interviewers are interested more in how you think than in what you think.**

Most interviewers are interested in the way you approach problems, think, and articulate your ideas, so they will concentrate on questions that will reveal these aspects of your character. They may ask you controversial questions or give you hypothetical problems to solve. Or they may ask about your professional goals, motivation for graduate study, and areas of interest.

When you prepare for an interview, it will be helpful if you have already written your personal essay, because the thought processes involved in preparing the essay will help you articulate many of the issues that are likely to come up in an interview. It is also helpful to do your homework on the program, so if the opportunity arises for you to ask questions, you can do so intelligently. Last, be sure you are dressed properly. That means dressing as if you are going to a professional job interview.

Following Up

■ **GRAD TIP: Proactively check on the status of your applications. Don't assume everything is okay.**

Give the admission office a couple of weeks to process your application and then call to find out whether everything was received. Usually the missing items are transcripts or letters of recommendation. "Don't assume anything; follow up," warns Rose Ann Trantham, Assistant Director of Graduate Admission and Records for the University of Tennessee at Knoxville. Suzette Vandeburg of the State University of New York at Binghamton agrees. She advises applicants to be proactive about their applications. "Check in periodically," says Vandeburg. "E-mail is a great way to check on your application."

Cindy Liutkus remembers how anxious she was about her applications. "The application process is definitely nerve-wracking," Liutkus says. "I was always worried that something wouldn't make it on time. I eventually sent stamped postcards along in every application and asked that the department secretary check the package and send the card along if everything was okay." Not content with that, Liutkus made doubly sure by following up with e-mail as well.

Paying for Your Graduate Degree

Pursuing a master's, a Ph.D., or a professional degree requires a substantial investment of time and money. How are you going to pay for graduate school and support yourself at the same time? Lots of people who don't have the "work full-time, go to school part-time" option get stuck at this point when they are applying to graduate programs. Even though you may have a good chance of gaining admittance to the program of your choice, you could conclude that you can't afford to attend. But that pessimistic conclusion may be unwarranted. Admittedly, finding money to help you pay for your graduate education can be difficult, in part because there are so many types and sources of funds and because information about them is scattered. Yet financial help for graduate students is available.

Financial Aid Overview

Merit-Based Versus Need-Based Aid

Financial aid for undergraduates is usually based on a calculation of need, but aid for graduate students is generally based on academic excellence, especially in the sciences, humanities, and arts. And excellence, for an incoming student, is judged on the basis of your application package.

. . . .
- **GRAD TIP: Devote the time and effort to make all aspects of your application as good as they can be. A lot of money may depend on it.**
. . . .

There is need-based aid for graduate students, but it usually comes in the form of federal student loans, which of course must be repaid, or Federal Work-Study programs. A university bases its assessment of your need on the cost of attendance—the amount a graduate student spends on tuition, fees, books and supplies, transportation, living expenses, personal expenses, child care, credit card and other debt payments, summer costs, and miscellaneous expenses—minus the amount you (and your spouse, if you are married) can be expected to contribute. The resulting figure is your need. Any given school may or may not be able or choose to give you enough aid to cover your need.

Many schools include a sample cost of attendance in their catalog or application packet. When you apply for financial aid, you can use the sample as a basis for developing your own budget and cost of attendance. Note that most schools use a nine- or ten-month academic year as the basis of their cost of attendance. When you figure your own budget, you must account for your expenses during the summer months as well.

Internal Versus External Funding

Excluding loans, there are two basic sources of financial assistance for graduate students. The first is internal funding, which comes from the university, college, and department or program. This internal funding may take the form of fellowships, scholarships, grants, assistantships, work-study programs, and tuition waivers. If you receive any of this type of funding, you must use it at the school that is awarding it.

The second source of financial assistance is external funding, which comes from private foundations, corporations, and other organizations. External funding usually comes in the form of fellowships, scholarships, and grants. Each award has a purpose, usually to further research in a particular area or to promote the educational opportunities of a particular group, such as women and minorities, often in a specific field. Some of the best known of the national fellowships come from the National Institutes of Health, the National Science Foundation, the Ford Foundation, and the Woodrow Wilson Fellowship Foundation, which awards Mellon Fellowships in the Humanities. You must apply to the awarding organization for each fellowship individually; your program application does not cover them. If you are awarded a fellowship, you may use it at whichever school you attend. If you receive a large external fellowship, graduate programs may find you a much more attractive candidate, since the university or department need not use its own resources to fund you.

- **GRAD TIP: The lion's share of your effort should be directed to making your program application outstanding, since that application is more likely to yield funding than applications to national fellowship programs.**

Most graduate students who receive nonloan financial assistance receive it from their own departments and universities, not from outside sources. However, this fact should not discourage you from applying for external sources of funding if you think you qualify.

Funding Priorities: Ph.D. or Master's Degree. Ph.D. candidates are not expected to be able to finance themselves for the six to ten years it may take to earn their degrees, so most programs give financial aid priority to doctoral candidates over master's candidates. If there is any money left after the doctoral awards have been made, then master's degree students may be given financial help. If this happens, then second-year master's students, who have already proved themselves, are more likely to be given financial assistance than incoming master's students. But, in general, since their degrees take less time to earn, master's students are expected to pay for their graduate education themselves or borrow money if their own resources are insufficient.

Academic Versus Professional Degree

Most nonloan funding goes to doctoral candidates in academic fields. Students pursuing professional degrees, such as business, law, and medicine, do not generally receive merit-based funding such as fellowships. In addition, their services as teaching assistants are not usually needed, because courses at the professional schools are taught by professional faculty. Instead, graduate students in most professional programs are expected to borrow if

Sample Costs of Attendance for a Single Student with No Dependents for Nine-Month Academic Year at Private and Public Universities

	Tuition	Fees	Books and supplies	Other expenses*	TOTAL
Private University	$20,250	$ 130	$648	$11,086	**$32,114**
Public University, state resident	$ 3,446	$ 1,790	$720	$ 9,700	**$15,656**
Public University, out-of-state resident	$ 9,850	$ 1,790	$720	$10,150	**$22,510**

*This figure can vary considerably depending on your personal circumstances. Note that it does not include expenses for the summer months.

they need help paying for their education and living expenses.

One rationale for this is that professional students can expect to make large salaries after they receive their degrees. Therefore, accumulating debt is not as risky for them as it is for academic students, whose employment prospects are less certain and generally less remunerative.

Sciences and Engineering Versus Humanities and Arts

Students in the sciences and engineering are more likely to be generously funded than students in the humanities and arts. Professors in the sciences and engineering often have large, ongoing grants from the federal government or private organizations in order to conduct their research. Part of the grant money is often allocated to hiring graduate students as research assistants. At large universities, students in the sciences may also receive teaching assistantships to help teach introductory science courses. In addition, there are more sources of external fellowships for science and engineering students, both from the federal government and private corporations.

Full-Time Versus Part-Time Enrollment

Full-time students are more likely than part-time students to get financial assistance. Full-time students are seen as more committed to their education, and they are not expected to work at an outside job to support themselves. Thus a department or program usually funds the full-time students first. If there is money left, then part-time students may be given help.

. . . .
■ **GRAD TIP: If you are planning to borrow money, be sure you are taking enough courses to qualify for the loan program you have in mind.**
. . . .

Part-time students must be careful to understand the ramifications of their status on their eligibility for financial aid. The definition of part-time varies from university to university, and some forms of assistance, such as student loans, may require at least half-time enrollment.

Of course, there are financial advantages to attending part-time: you spread your costs out over a longer period, making them easier to pay. In addition, if you are working full-time while going to school part-time, your employer may reimburse part or all of your tuition.

Public Versus Private University

A top private university may have more resources with which to support its graduate students than a public university. If you think you have the academic credentials to be admitted to a program at one of the Ivy League schools or other top private institutions, then you should apply. If you are admitted, your chances of receiving adequate funding are good. On the other hand, the private universities usually have fewer undergraduates than the large public universities, so they have fewer teaching assistantships to award. In addition, if, for some reason, part or all of your funding is discontinued, you will be faced with the prospect of coming up with $25,000 to $30,000 per year.

Large public universities cost less than the private universities, especially for in-state residents. They have more teaching assistantships to offer, because of the large number of undergraduates, but fewer fellowships than the private universities. If you think you'll be financing all or most of your graduate education, attending a public university is the best way to reduce your costs—dramatically. Almost 70 percent of all graduate students attend public universities. They are getting a bargain.

In-State Versus Out-of-State Residency

If you are planning to attend a public university, your costs will be much lower if you are a state resident. For example, at the University of Michigan, full-time tuition and fees for out-of-state graduate students are $21,700 a year; for in-state residents they are $10,800.

With so much money at stake, it is definitely worth your while to find out how you can establish residency in the state in which you are planning to get your graduate degree. You may simply have to reside in the state for a year—your first year of graduate school—in order to be considered a legal resident. But residence while a student may not count, and you may have to move to the state a year before you plan to enroll. The legal residency requirements of each state vary, so be sure you have the right information.

Types of Financial Aid

Now that you have an overview of some of the factors involved in financial aid at the graduate level, let's examine the various types of aid that are available. They are fellowships and scholarships, assistantships, federal work-study and other work programs, loans, and tuition reimbursement.

Fellowships and Scholarships

Fellowships and scholarships are cash awards given by a department, university, or outside organization. They are usually awarded on the basis of merit, but some are awarded on the basis of need or are reserved for minority or women applicants. In addition, there are fellowships that are awarded simply because you have the particular qualifications that the philanthropist wanted to reward: for example, you are an Eagle Scout studying labor relations. (Needless to say, getting one of these is a long shot!) The words fellowship and scholarship are used somewhat interchangeably; there is no real difference between them, except that scholarships are usually awarded to undergraduates and fellowships to graduate students.

First, any amount of money that you don't have to borrow is a plus, and small grants can add up. And second, having a history of receiving small fellowships will make your applications more attractive when you apply for the large fellowships in your later years of graduate school.

■ **GRAD TIP: Many of the grants and fellowships that entering graduate students are eligible for have small cash awards, but you should apply for them anyway.**

Fellowships are excellent because in return for the award you are not expected to do anything

but keep your grades up and make progress toward your degree. If the fellowship is substantial, it can free you to study and do research. If the fellowship is small, it may still add enough to your total aid package to enable you to attend school without borrowing. Fellowships may range from a low one-time award of $250 to a generous amount that covers tuition, fees, and living expenses and is renewable for several years.

"At Penn State . . . I had to ask about fellowships," says Heather Helms-Erikson, who is earning a Ph.D. in human development and family studies. "I was very assertive—in fact, I hope I wasn't a pain in the neck. I made it very clear that I wasn't going unless I was fully funded," she recalls. "I was offered a research assistantship and tuition waiver for the entire time. . . . They nominated me for a university fellowship, and I got an additional $8,000 over two years. Since then, I've applied for every source of funding I can find."

■ **GRAD TIP: If you want to pursue university fellowships, you must take the initiative and ask the department, graduate school, college, and financial aid offices.**

Fellowships may be awarded by the department to which you are applying, the graduate school, the university, or an outside organization, such as the federal government or a private foundation. Your program application takes care of departmental fellowships. But you have to apply for external fellowships separately.

Assistantships

If you are offered an assistantship, you will be expected to work for the university in exchange for a stipend or salary, which is taxable. You may also receive a partial or full tuition waiver along with the assistantship.

■ **GRAD TIP: Most financial aid from large public universities is granted in the form of assistantships.**

The value of assistantships varies widely from one university to another and from one field to another. In some cases, an assistantship and a tuition waiver provide enough for you to go to school and pay your living expenses if you are single. "I had teaching assistantships and sometimes a research assistantship, plus a tuition waiver. I never had to pay anything," recalls a woman who earned a Ph.D. in Italian from a large, private Midwestern university. "What I received was enough to live on my own, although not always very comfortably." However, in other cases, an assistantship provides only partial funding and you will have to make up the balance of your school and living costs from other sources. "Assistantships are a source of professional development as well as funding," says Martha J. Johnson, Assistant Dean of the Graduate School at Virginia Tech. Assistantships may draw you into the department's academic life because you are usually assigned to work with faculty members.

There are three major types of assistantships: teaching, research, and administrative.

Teaching Assistantships. Large universities need many teaching assistants (TAs), particularly in departments, such as English and psychology, in which many undergraduates take courses. Teaching assistantships are awarded by the department to which you are applying. As a TA, you usually help a professor by conducting small discussion classes, grading papers and exams, counseling students, and supervising laboratory groups. Some TAs teach a section of an introductory course or are permitted to design and teach an upper-level course on their own. At many universities or departments, you will be given an orientation course to prepare you for teaching introductory classes, but at some institutions all you will get is on-the-job training. TAs usually work 15 to 20 hours a week and, as a consequence, may take a lighter course load.

Although working as a TA may slow down progress toward your degree, in most cases students feel that the experience they are gaining more than compensates for the extra time it takes, especially if their ultimate goal is to teach at the university level. "I find that my teaching assistantship is extremely rewarding," says Cindy Liutkus, a Ph.D. candidate in geology at Rutgers University in New Jersey. For teaching one course per semester, Liutkus receives a full tuition waiver and a stipend that covers fees, an off-campus apartment, and other living costs. Liutkus continues, "Not only does teaching provide me with the opportunity to strengthen my skills in the field of geology, but I enjoy interacting with the students. I hope that my enthusiasm for the subject is translated to them and that they will continue on in the field."

Liutkus recommends a teaching assistantship for any student who has an interest in teaching as a career. "Teaching . . . requires a number of skills: information preparation and organization, public speaking, discipline, and time management. Graduate study will teach you all of these things as well, but teaching your own class sharpens your skills and makes you appreciate the dedication that your professors have for their work."

Some people consider teaching assistantships less attractive than fellowships because you must earn your money rather than getting it "free." However, fellowship recipients lack the close contact with departmental faculty and students that teaching assistants enjoy. This close contact helps TAs keep abreast of events and changes in their departments and makes it easier to know what's really going on. A teaching assistantship may also be a welcome change from the lonely life of doing solo research for many years. A department's TAs, who often share office space and take courses together, may find they enjoy the resulting camaraderie and competitive spirit.

. . . .
■ **GRAD TIP: Most research assistantships are offered to students in the hard sciences and social sciences.**
. . . .

Research Assistantships. A research assistant (RA) helps a faculty member with his or her research. Generally, research assistantships are awarded by a department and are paid from grant money obtained by a professor from the federal government or private organizations. Some research assistantships are funded by university endowments or state money, and some are funded through grants obtained by the graduate student.

An RA in the sciences works under the direction of a faculty member, assisting with laboratory research or field work. The professor who has the grant(s) gets to select the students he or she wants as RAs, generally choosing promising candidates with similar research interests.

There are also research assistantships in the humanities and arts, although they are fewer, tend to be of shorter duration, and have less monetary value. A humanities RA might perform research in libraries, assemble bibliographies, or check citations for a professor. In many cases, they may be doing less rewarding clerical work, such as data entry or photocopying. Such research assistantships are rarely offered to incoming students but are given to students who have proven they have the ability or experience to do the job.

A research assistantship can be very rewarding or very frustrating. The benefit of receiving a research assistantship is that you are often able to work on research that is related to your own degree, especially if you are working with your adviser or mentor. Another bonus is that if you have done a lot of the research for a project, the faculty member may reward you with coauthorship of a publication—one of your first professional credentials. So if you've been matched up with a faculty member in your area of interest or have the opportunity to work with other graduate students on a research team, the experience can be professionally rewarding. But if you are working for a faculty member whose interests do not match yours, a research assistantship helps pay your way but does not further your own educational or professional development.

Administrative Assistantships. Some schools offer assistantships in the university's administrative offices. You work 10 to 20 hours a week, and, ideally, the work you do is related to your field of interest. For example, if you are a computer sciences student, you might do computer-related work for the university, or if you are a library and information sciences student, you might work in one of the university libraries.

Since administrative assistantships are sometimes outside your department and graduate school, they are often not awarded on the basis of your application, as are teaching and research assistantships. Instead, you have to look for them. You can find information about administrative assistantships in the school catalog or by contacting the university departments in which you'd like to work.

If You Are Offered an Assistantship

If you are offered an assistantship by a department or program, be sure to ask what the likelihood is that it will continue in subsequent academic years. Some programs routinely offer their assistantships to incoming students in order to get them to enroll; a year or two later, when the student is committed to the degree, they take away the assistantship and offer it to a new incoming student.

You should also determine whether a full or partial tuition waiver comes with the assistantship. A tuition waiver is worth thousands of dollars and can make the difference between having enough to cover all your costs and scrambling to make up the shortfall.

Federal Work-Study and Other Work Programs

Federal Work-Study Program. The Federal Work-Study Program provides students who demonstrate financial need with jobs in public and private nonprofit organizations. The government pays up to 75 percent of your wages, and your employer pays the balance. The value of a work-study job depends on your need, the other elements in your financial aid package, and the amount of money the school has to offer. Not all universities have work-study funds, and some limit the use of funds to undergraduates.

If you receive work-study funds, you may be able to work in a job related to your field. Check with the financial aid office to find out what jobs are available, whether you can use the funds in a job you find elsewhere, and what bureaucratic requirements you will have to satisfy.

Internships and Cooperative Education Programs. Internships with organizations outside the university can provide money as well as practical experience in your field. As an intern, you are usually paid by the outside organization, but you may or may not get credit for the work you do. Although they have been popular for years in professional programs, such as law and business, internships have recently been growing in popularity in academic programs as well.

In cooperative education programs, you usually alternate periods of full-time work in your field with periods of full-time study. You are paid for the work you do, but you may or may not get academic credit for it as well. Internship and cooperative education programs may be administered in your department or by a separate university office, so you must ask to find out.

Loans

Unfortunately, at some point most graduate students do have to take out loans to finance their education. Only wealthy students and Ph.D. students in certain fields are likely to be fully funded for the duration of their studies. Other students must borrow, whether they do it just one year to make up the amount not covered by other types of aid, or whether they borrow each year to finance most of their graduate education. Still, even if you expect to be making a huge salary after you receive your degree, it pays to minimize your indebtedness if you can. Financial aid counselors recommend that your total student debt payment should not exceed 8 to 15 percent of your projected monthly income after you receive your degree.

Debt is manageable only when considered in terms of five things:

1. Your future income
2. The amount of time it takes to repay the loan
3. The interest rate you are being charged
4. Your personal lifestyle and expenses after graduation
5. Unexpected circumstances that change your income or your ability to repay what you owe

The approximate monthly installments for repaying borrowed principal at 5, 8, 9, 10, and 12 percent are indicated above.

Estimated Loan Repayment Schedule
Monthly Payments for Every $1,000 Borrowed

Rate	5 years	10 years	15 years	20 years	25 years
5%	$18.87	$10.61	$ 7.91	$ 6.60	$ 5.85
8%	20.28	12.13	9.56	8.36	7.72
9%	20.76	12.67	10.14	9.00	8.39
10%	21.74	13.77	10.75	9.65	9.09
12%	22.24	14.35	12.00	11.01	10.53

Use this table to estimate your monthly payments on a loan for any of the five repayment periods (5, 10, 15, 20, and 25 years). The amounts listed are the monthly payments for a $1,000 loan for each of the interest rates. To estimate your monthly payment, choose the closest interest rate and multiply the amount of the payment listed by the total amount of your loan and then divide by 1,000. For example, for a total loan of $15,000 at 9 percent to be paid back over ten years, multiply $12.67 times 15,000 (190,050) divided by 1,000. This yields $190.05 per month.

If you're wondering just how much of a loan payment you can afford monthly without running into payment problems, consult the following chart.

How Much Can You Afford to Repay?

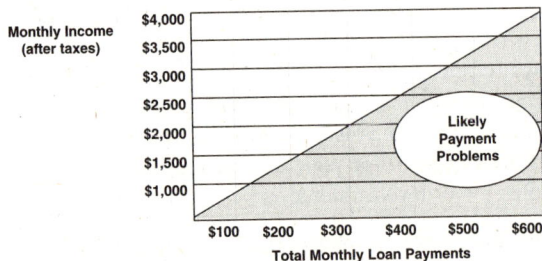

The graph above shows the monthly cash-flow outlook based on your total monthly loan payments in comparison with your monthly income earned after taxes. Ideally, to eliminate likely payment problems, your monthly loan payment should be less than 15 percent of your monthly income.

Ways to Borrow Less

Ask Your Family for Help. Although the federal government considers you "independent," your parents and family may still be willing and able to help pay for your graduate education. If your family is not open to just giving you money, they may be open to making a low-interest (or deferred-interest) loan. Family loans usually have more attractive interest rates and repayment terms than commercial loans. They may also have tax consequences, so you may want to check with a tax adviser.

Push to Graduate Early. It's possible to reduce your total indebtedness by completing your program ahead of schedule. You can either take more courses per semester or during the summer. Keep in mind, though, that this option reduces the time you have available to work.

Work More, Attend Less. Another alternative is to enroll part-time, leaving more time to work. Remember, though, to qualify for aid, you must be enrolled at least half-time, which is usually considered 6 credits per term. And if you're enrolled less than half-time, you'll have to start repaying your loans once the grace period has expired.

Both graduate students and administrators have very strong opinions about loans. "I advise students not to take out loans if they can possibly help it," says Martha Johnson of Virginia Tech. "Consider the graduate student lifestyle as a short-term situation. If you are single, you can live in a dorm, use the university food service, sell your car, and concentrate on finishing as quickly as possible. With a simplified lifestyle, a single person can live on an assistantship." Denise Kaiser, who earned a Ph.D. in history from Columbia University, recalls how little money she had during her grad school years. "Boy, did I eat a lot of chicken during those years! . . . But the scrimping made it possible to afford my own place on a fellowship. . . . I didn't borrow any money while I was at Columbia."

Jean Godby, a Ph.D. candidate in linguistics at Ohio State University, thinks that loans are a last resort for good reason. "All of my schooling was paid for by fellowships and assistantships. I was very leery of going into debt," comments Godby. "I noticed that my friends who went into debt didn't like starting their work life with a financial burden." Another student, Tom Fuchs, who is earning a Psy.D. at the California School of Professional Psychology, had this comment: "Student loans are very appealing and often easy to get, but they accumulate very easily and are probably the major source of anxiety for students as they get closer to the end of the program." Fuchs adds, "I am glad I decided to continue to work, even though it means I will need more time to finish the program."

On the other hand, some students do not share this aversion to indebtedness. "I have tons of loans," says one student with a master's in education who is currently pursuing an Ed.D. "I have a student loan debt that is more than the value of my house at this point." Nestor Montilla, who is earning a master's degree in public administration from John Jay College of Criminal Justice in New York, also thinks loans are worth it. "If you have to take loans to get your education, do it," Montilla advises. "A federal loan is an investment in your education. You will pay later when you are a productive citizen."

Others advise taking loans if necessary to get started, and then looking for other sources of funding once you have been in the program a short time. "I began paying for grad school with student loans, and I worked part-time to support myself," says Jenn Wagaman, a master's candidate in public communications at the University of Alaska at Fairbanks. "The second year I got an assistantship." Kimani Toussaint, a Ph.D. candidate in electrical engineering at Boston University, adds: "Even if you have to take out loans during the first semester or year of graduate school, during that time you have the opportunity to approach many professors in your department to find out if they have any assistantships available or pending."

The question of taking out loans is more than just a calculation of present need and how much future indebtedness you can afford based on salary projections. It is also a question of your feelings about borrowing and your attitudes toward debt. In addition, your credit history will affect whether or not you will be able to borrow.

If you *do* decide to borrow, there are two basic sources of student loans, the federal government and private loan programs. If you are a homeowner, you might find it advantageous to use a home equity loan to help pay your educational costs. Whatever you do, do not use your credit cards to borrow money for school. The interest rates and finance charges will be astronomically high, and unless you can pay the balance in full, the charges will accrue rapidly.

Before tackling any loan application paperwork, look at the eligibility criteria. In addition, the terms of these loans will differ, so compare them, and make sure you understand what you are agreeing to before you sign on the dotted line. Keep in mind that most student loans have a guarantee fee (which insures the lender against your default) and an origination fee (which covers the administrative costs of the loan), both of which are a percentage of the amount you are borrowing. So, for example, if you are borrowing $5,000 with a guarantee fee of 1 percent and an origination fee of 3 percent, you will actually receive only $4,800.

Federal Student Loans

There are two basic types of loans offered to graduate students by the federal government: the Stafford Loan (and the similar Direct Student Loan) and the Perkins Loan. Up-to-date information about federal loan programs can be found at the Department of Education's Web site, http://www.ed.gov or by calling 1-800-4FEDAID.

Stafford Loan Program. The federal government sponsors the Stafford Loan Program, which provides low-interest loans to graduate students through banks, credit unions, savings and loan institutions, and the universities themselves (through the Department of Education's Direct Lending Program). There are two types of Stafford Loans: **Subsidized Stafford Loans** and **Unsubsidized Stafford Loans.** To get a Subsidized Stafford Loan, you must demonstrate financial need. With a subsidized loan, the government pays the interest that is accruing while you are enrolled at least half-time in a graduate program.

If you cannot demonstrate financial need according to government criteria, you may still borrow, but your Stafford Loan will be unsubsidized. This means you are responsible for paying the interest on the loan while you are still in school.

In both types of Stafford Loans, repayment of the principal as well as future interest begins six months after you are last enrolled on at least a half-time basis. You may borrow up to $18,500 per year up to a maximum of $138,500, which includes any undergraduate loans you may still have. The interest rate varies annually and is set each July. Right now it is capped at 8.25 percent.

Perkins Loan Program. Another source of federal funds is the Perkins Loan Program. The Perkins Loan is available to students who demonstrate exceptional financial need, and it is administered by the university itself. In some cases, universities reserve Perkins Loans for undergraduates. If you are eligible for a Perkins Loan, you may borrow up to $5,000 per year, up to a maximum of $30,000, including undergraduate borrowing. Currently, the interest rate is 5 percent, and no interest accrues while you are enrolled in school at least half-time. You must start repaying the loan nine months after you are last enrolled on a half-time basis.

Consolidating Your Federal Loans. When you leave school (with a degree, we hope!), you can consolidate all your outstanding federal loans into one loan. Having one loan to repay will minimize the chances of administrative error and allow you to write one check per month rather than several.

Private Student Loans

In addition to the federal loan programs, there are many private loan programs that can help graduate students. Most private loan programs disburse funds based on your creditworthiness rather than your financial need. Some loan programs target all types of graduate students; others are designed specifically for business, law, or medical students. In addition, you can use other types of private loans not specifically designed for education to help finance your graduate degree.

There are many private loan programs designed to help graduate students in all fields. The loans are generally unsubsidized, and your creditworthiness, as well as the limits of the program, will determine the amount you can borrow.

CitiAssist Loans. Offered by Citibank, these no-fee loans help graduate students fill the gap between the financial aid they receive and the money they need for school.

EXCEL Loan. This program, sponsored by Nellie Mae, is designed for students who are not ready to borrow on their own and wish to borrow with a creditworthy cosigner.

GradAchiever Loan. This is a credit-based loan for borrowers who are enrolled at least half-time, sponsored by Key Education Resources. Borrowers make no payments while in school and for a nine-month grace period. Interest is added to the loan principal only once, at the beginning of repayment. Applicants can apply online.

Graduate Access Loan. Sponsored by the Access Group, this is for graduate students enrolled at least half-time.

Signature Student Loan. A loan program for students who are enrolled at least half-time, this is sponsored by Sallie Mae.

Home Equity Loans

For students who own their own homes, a home equity loan or line of credit can be an attractive financing alternative to private loan programs. Some of these loans are offered at low rates and allow you to defer payment of the principal for years. In addition, if you use the loan to pay for educational expenses, the interest on the loan is tax deductible.

Tuition Reimbursement

If you are working full-time and attending school part-time, you may be reimbursed for part or all of your tuition by your employer. Andrea Edwards Myers, who is earning a Master of Arts in public communications on a part-time basis from the College of St. Rose in Albany, New York, received tuition reimbursement through her company. "They offer a capped total of $3,000 a year for

tuition. . . . One thing to be aware of, however, is that the federal government taxes graduate tuition reimbursement money, so you will receive a reduced amount in your check. You must make up the balance when paying the college," she cautions.

. . . .
■ **GRAD TIP: Check with your employer before you enroll; some employers reimburse tuition only for job-related courses.**
. . . .

For Women, Minority Students, and Veterans

There are many sources of financial assistance that target qualified women, minority, or veteran graduate students. For women, much of this aid is available for graduate study in fields in which women have been traditionally underrepresented (the physical sciences and engineering). For example, in the physical sciences, fewer than one third of graduate students are women. In engineering, fewer than one fifth are women. To help achieve a gender balance in these fields, many fellowship programs are offered only to qualified women. For instance, the Zonta International Foundation offers fellowships for study in aerospace-related sciences or engineering, and the National Science Foundation offers special grants to women studying in the sciences, engineering, or mathematics.

In addition, there are funds earmarked for women in other fields of study. For example, the American Association of University Women awards several grants and fellowships each year to women pursuing graduate study in any field, and the Women's Research and Education Institute offers fellowships for study in fields related to public policy.

Qualified minority students are in great demand in graduate schools in all fields. Historically, minority students have been underrepresented at the graduate level, and there are now many programs, including some from the federal government, that seek to increase the number of minority students by offering financial assistance. The Indian Fellowship Program, Minority Access to Research Careers, and the National Science Foundation Minority Graduate Fellowships are just a few examples of federal programs that target minority students. In addition, many private and corporate sponsors have developed programs to help minority students finance their graduate education. These include the American Fund for Dental Health, the American Geological Institute, the American Planning Association, and the Ford Foundation.

Veterans who have contributed to one of the Veteran Educational Benefits Programs are entitled to use their benefits for graduate education. You do not have to show financial need to participate; they are a benefit of your service in the armed forces. The amount of assistance you will receive depends on length of service, the number of dependents you have, and the number of courses you are taking.

For International Students

Unfortunately, financial assistance for international students is limited. About 75 percent of international students do not receive any aid from U.S. sources. If you are an international student, you do not qualify to receive federal loans or work-study assistance, although if you can find a willing and creditworthy U.S. citizen to cosign a loan, you may be able to borrow from the private loan programs described above. If you do get financial help, it is likely to come from the department or program in which you are enrolled, although there are some government programs that underwrite graduate students to promote cultural exchange. Finding help to finance your graduate education is a challenge, but it's not impossible. You can start at home by contacting your country's U.S. educational advising center, which can help you identify institutions that fund international students.

Finding More Information

There is no central clearinghouse for information about financial aid for graduate study. You are going to have to check a number of different sources to get the full picture.

The University

Even at the university, there is more than one source of information about financial aid. Each university has a different administrative structure, so you will have to figure out the likely offices you will need to contact. These may include:

- The program or department to which you are applying. If you cannot find this information in the printed materials you've been sent, then call the program and ask.
- The financial aid office is generally the best source of information about federal and private loan programs as well as work-study assistance. They may also be able to steer you to other sources of information.
- The next place to check is the administrative office of the college in which your program is located. For example, you may be applying for a Ph.D. in English literature. The English Department is likely to be under the jurisdiction of the College of Arts and Sciences. That office may administer fellowships and grants to the students of the college. Call them to find out.
- The office of the graduate school is another administrative office that may have funding to award. If they do, the fellowships or grants are likely to be awarded on a university-wide, competitive basis.

It's important to check with all these offices to see what's available. "A program's general tendency is to broadcast news of outside fellowships, but keep close to the vest about its own funds," says Jonathan Roberts, Manager of Enrollment Services at Pepperdine University's Graduate School of Education and Psychology in California. "Students need to be proactive and call the program as well as the financial aid office to find out their chances of receiving aid."

The Government

A good source of information on federal aid for graduate students is the federal government itself. Most need-based aid is administered by the

Department of Education. You can contact them through their Web site, by telephone, or by mail. Remember, however, that not all universities participate in each federal program, so if a particular program interests you, you will have to contact the university financial aid office to make sure it's available.

■ **GRAD TIP: Many agencies of the federal government offer fellowships to graduate students in related fields. Contact the agencies that are relevant to your field of study.**

It also pays to check whether your state offers support to graduate students. Some states, like California, New York, Michigan, Oklahoma, and Texas, have large aid programs for their residents. Other states may have little or nothing to offer. Contact your state scholarship office directly to find out what's available and whether you are eligible to apply.

The Internet

The Internet is an excellent source of information about all types of financial aid. One of the best places to start your Internet search for financial aid is the Financial Aid Information Page at http://www.finaid.org. This site has a great deal of information about different types of financial aid and provides links to other relevant sites as well. It provides a good overview of the financial aid situation. In addition, the site offers several calculators that enable you to estimate many useful figures, including projected costs of attendance, loan payments, and the amount you will be expected to contribute to your education and living expenses if you are applying for need-based aid.

There are also a number of searchable databases of national scholarships and fellowships on the Internet. The best known of these is FastWeb at http://www.fastweb.com. It takes about half an hour to answer the FastWeb questionnaire about your educational background, field of study, and personal characteristics. When you are done, FastWeb searches its database to match your data with eligibility requirements of several hundred

thousand fellowships and scholarships. You are then given a list of possible fellowships and scholarships to pursue on your own. There is no cost for this service.

There are a few things you should beware of when using Internet search services. First, a searchable database is only as good as its index, so you may find yourself getting some odd matches. In addition, most searchable databases of scholarships and fellowships are designed primarily for undergraduates, so the number of potential matches for a graduate student is far fewer than the several hundred thousand sources of aid that a database may contain. Finally, some of these Internet search services charge a fee. Given the amount of free information that's available, both on the Internet and in libraries, it's not necessary to pay for this type of research.

Print Directories

Although the searchable databases on the Internet are easy to use, it's still a good idea to check print directories of national fellowships, grants, and scholarships. These directories have indexes that make locating potential sources of funds easy. Fellowships and grants are indexed by field of study as well as by type of student. So, for example, you can search for all funding related to the study of Latin America or landscape architecture. Or you can search for funding that is targeted to Hispanic students, students with disabilities, or entering students. It's a good idea just to browse, too, in case something catches your eye.

There are quite a few directories that you can consult. *The Annual Register of Grant Support: A Directory of Funding Sources*, published by the National Register Publishing Company, is a comprehensive guide to awards from the government, foundations, and business and professional organizations. *Peterson's Grants for Graduate and Postdoctoral Study* is a directory of 1,400 fellowships and scholarships that covers all fields of study. There are also directories that specialize in fellowships for particular fields of study and for particular types of students.

Applying for Financial Aid

Depending on your personal situation and the requirements of the graduate school, you may have to submit just one or a number of applications for financial aid. The simplest situation is that of a student applying only for merit-based departmental or program funding. However, if you are applying for need-based aid, university fellowships, national fellowships, or private loan programs, you will have several application forms to deal with. Even if you have only one application to deal with, start the process early.

Timetable

"I cannot overemphasize the importance of applying early," says Emerelle McNair, Director of Scholarships and Financial Aid at Southern Polytechnic State University in Georgia. "Most awards are made in spring for the following academic year." Be sure you've picked the correct deadlines from your program application information packet.

If you are looking for sources of funding outside the program and university, such as national fellowships, then it is even more important to start early—a full year or more before you plan to enroll. "You have to fill out your program applications concurrently with your fellowship applications," advises Martha J. Johnson, Assistant Dean of the Graduate School at Virginia Tech. "Everything is due around the same time."

Remember, it can easily take months to fill out applications and assemble all the supporting data for a financial aid request. You may need to submit income tax forms, untaxed income verification, asset verification, and documents that support any special circumstances you are claiming. Give yourself plenty of time to submit the initial application. Later, if you are asked to provide additional information or supporting documents by the financial aid office, do so as quickly as possible.

The Program Application

For many graduate schools, the program application is the main financial aid application as well. As was mentioned before, much of the funding for incoming graduate students is determined by the admission committee's assessment of the merit of program applications.

■ **GRAD TIP: A strong program application, submitted on time, will improve your chances of getting funding from your department.**

"I was offered a scholarship to attend my program based on my previous experience as well as my personal essays," says Felecia Bartow, an M.S.W. candidate at Washington University in St. Louis. "Put a lot of time and effort into your personal essays as they are often used to award scholarship money." Other students suspected that their GRE scores helped them get aid. You'll probably never know on what basis the funding decisions were made.

Financial Aid Application Forms

In addition to the program application, there may be a separate financial aid application. This will often be the case if you are applying for need-based aid. If you do not see such a form in the program application packet, call the graduate school to find out whether you need to obtain it from another office.

Some schools require you to submit a standardized form, the College Scholarship Service's Financial Aid PROFILE. This form is similar to the FAFSA, described below, but it is used to award university aid.

FAFSA

You may remember the Free Application for Federal Student Aid (FAFSA) from your undergraduate days. FAFSA is also used by graduate students who are applying for need-based federal aid. The FAFSA is issued annually by the Department of Education right after January 1 (see http://www.fafsa.ed.gov). It requires financial data from the previous year so that you can be considered for aid in the school year starting the following fall.

. . . .

■ **GRAD TIP:** It's much easier to fill out
the FAFSA if you have already done
your federal income tax forms for the
year.

. . . .

For purposes of need-based federal financial
aid, all graduate students are considered financially
independent of their parents. Because the FAFSA
is designed for undergraduate students who are
dependent on their parents, you may find you are
having difficulty interpreting some of the questions
or that the questions do not cover all your
circumstances. If there is information about your
financial situation that is not elicited by the
FAFSA but that you feel is germane to your
application, then explain the circumstances in a
separate letter to the financial aid office.

Suppose, for example, that you have been
working full-time for a few years but you are planning to quit your job and attend graduate school
full-time. You would complete the FAFSA using
the previous year's full-time income figures, but
this would not be an accurate reflection of your
financial situation during the following school year
because your income will drop precipitously. In
this case, you would notify the financial aid office
so that they can make a professional judgment as
to whether your aid should be revised upward.

After you submit the FAFSA, you will
receive an acknowledgment that includes a sum-
mary of the data you have sent them. Check to
make sure the information is accurate and that the
schools to which you have chosen to have the data
sent are correctly listed. If there are errors, make
corrections right away. Your acknowledgment will
also show your Expected Family Contribution, the
amount you and your spouse can be expected to
contribute. This information is used by each
school to calculate your need (cost of attendance
minus Expected Family Contribution) and to
award need-based aid.

You can do a rough calculation of your
Estimated Family Contribution. All you need is
your previous year's tax return and a program's
cost of attendance figures. Use one of the EFC
calculators on the Internet, such as the one at
http://www.finaid.org.

Fellowship Applications

If you are applying for university or national fellowships, you will have to submit separate applications for each one. Follow instructions carefully,
making sure you meet all deadlines. Fellowship
applications can be as elaborate as program
applications, including letters of recommendation
and essays, so allow yourself a lot of time to
complete them.

Follow Up

You must follow up with your financial aid applications, just as you do with your program application. If you do not receive an acknowledgment
that your FAFSA was received within a couple of
weeks, check on its status. In addition, call the
university offices with which you are dealing to
make sure everything is proceeding smoothly.
"Politely check on your application every so
often—making as many friends as you can in the
process," recommends Neill Kipp, a Ph.D.
candidate in computer science at Virginia Tech.
"As with any large organization, things fall in the
cracks. Having friends in the financial aid office,
your own academic department, and the graduate
school helps immensely."

It *Is* Possible

You can see that it is possible to find the financial
aid that will help you pay for graduate school. You
will have to be persistent in your search for funds.
You may have to spend months working on
financial aid research and applications. You may
have to borrow money. And once you enter a
graduate program you may have to simplify your
lifestyle in order to cut your expenses.

But if you really want to go to graduate
school, you can find the financial help that will
make it possible. Be realistic about your needs,
leave yourself enough time to complete all the
paperwork, and do your homework. Now is a good
time to look back on all the reasons you want to
attend graduate school—to remind yourself why
it's worth it.

Students Tell
What It's Like

When choosing among graduate programs in biology, first-person advice can be very useful. In fact, many people who have been through graduate school suggest that one way to really find out about a department is to talk to currently enrolled students. Though graduate students in biology share many issues, a student studying how cells communicate with each other and a student focusing on the bacteria inside the gut of the gypsy moth, for instance, probably approach the rigors of graduate school in different ways. Programs differ and geographic locations and types of institutions each contribute their own factors to make the graduate experience unique. To get a personal perspective that *Decision Guide* readers can learn from, graduate students in various departments, stretching across the broad spectrum of biology, were interviewed.

These graduate students are either currently enrolled in a variety of representative departments or they have recently graduated. All were asked the same questions. The intent was not to discuss the curriculums but rather to get at the heart of what it's like to be a graduate student in biology. Outside of the academic details, what had each student looked for in a program? What is the work load like? How was the transition to graduate school? Though you may not find your exact academic focus represented by the students who were interviewed, their experiences will give you an idea of what to look for and what to expect.

Scott Berggren
Colorado State University
College of Natural Sciences
Department of Biology
Focus: Systematic botany

Nichole Broderick
University of Wisconsin–Madison
Department of Entomology and Department of Plant Pathology
Focus: Gut bacteria of insects

Keith Byrd
Johns Hopkins University
Department of Biology
Focus: Organization of DNA in a nucleus

Jennifer Cole
Indiana University
Biology Department
Focus: Microbiology

Marwah Helmy
Stanford University
Department of Biology
Focus: Molecular biology

Martha Kirouac
California Institute of Technology
Department of Biology
Focus: Developmental genetics

Adjusting to Graduate School

As soon as the euphoria of getting accepted—hopefully to the graduate program you wanted—wears off, you're facing a whole new set of hurdles. Adjusting to the rigors of graduate school can be a rough experience or a smooth ride. You could be going to a program with a well-thought-out structure in place to ease you into the immediate pool of work. Or you might be tossed into the water to swim to the end of the first semester.

Much of the transition experience depends on where you're coming from. Keith Byrd found that his prior 8- to 9-hour days in the workplace adequately prepared him for the Johns Hopkins molecular biology program. He welcomed the independence of graduate school and the fact that he didn't have a boss peering over his shoulder.

But there are as many arguments against taking time out as there are for getting work experience first. Martha Kirouac, now at Caltech for her Ph.D. in developmental genetics, cites the difference in lifestyle between the comfort zone of being employed and that of a grad student living on a stipend.

"There are benefits both ways," states Thomas Roberts, Chairman of the Department of Biological Science at Florida State University. In fact, he says it's typical for graduate biology students to have worked in a lab as technicians before seeking an advanced degree. While in the lab, they become intrigued with research and want the authority and independence a graduate degree will give them. Nicole Broderick came directly from college to the University of Wisconsin–Madison's Departments of Entomology and Plant Pathology for her joint master's. She didn't feel the need for a break because she knew exactly what she wanted to accomplish. Why waste the time, she reasoned.

But even if you have a few years of experience between college and a higher degree, you will be inundated with course work and reading. Some programs offer breathing space and provide advisers to help you get started. Some send their students on lab rotations to assist others in their research projects. Either way, you can have some control over how swamped you'll feel at first.

. . . .
■ **GRAD TIP: Don't overload yourself with courses the first semester.**
. . . .

No matter how high your enthusiasm, it's unwise to pile on courses and reading until you're acclimated to graduate school. There's a reason why most students feel such a wrenching start to their grad school experience. The bar has been raised. You're supposed to rise to the top and

become the expert over time. "Go into the situation like a blank slate," suggests Scott Berggren, who is getting his master's at Colorado State University's Department of Biology. He came to grad school thinking it would be an extension of college. Instead, he found it was completely different and that he had to be much more willing to be pushed in ways he was not at all used to.

Before the realities of work become overwhelming, it's helpful to look at the benefits of grad school. Keep your eyes on why you're there. For Kirouac at Caltech, the delight of the academic freedom to delve into questions that intrigue her keeps her going. "I have friends who have traditional jobs and have to get up at 6 a.m., but I can get up at noon and work until 6 a.m. I can personally make the call about what courses I need to take and what my most immediate concerns should be," she says.

Berggren is also happy with the amount of independent thought and direction of research he has. When he worked as a technician, he didn't have those choices. For Marwah Helmy at Stanford University, it's the breadth of classes available to her. Though she must adhere to a few broad requirements, she describes the selection of classes as a workout for her brain since there are no restrictions on taking courses outside the biology discipline. "People come here from around the world and talk about their research and they are available to students," she says. Within the roster of Stanford's top-tier faculty members, Helmy says she gets to hear about research that's on the cutting edge of biology.

"I love the fact that I can spend as much time doing research as I want," states Broderick. With fewer classes to take, she's able to be in the lab and take full advantage of the environment that she describes as supportive and intellectually stimulating.

But with the highs come the lows. "I wish I'd known about the amount of failure that can happen—that things don't always work out," reflects Berggren.

A Whole New Way of Thinking and Working

Berggren, like other incoming students, finds it's tough to prepare for the new way of learning they all encounter. "There's so much done in the field of biology," says Kirouac. "In order to get your Ph.D., you have to come up with new research. And in order to get to something new, you have to get through a lot. Just the publication standards alone have risen."

According to Helmy, experimental and scientific theory isn't stressed that much in undergrad classes. When students coming from other institutions see the amount and quality of the serious research done at Stanford, they are forced into a whole new way of thinking about biology. No more sitting in class and throwing back to the professor what you've learned. Grad students are expected to analyze new material, take it apart, and put it together in unique ways.

Berggren was surprised by the pace at which work is accomplished. Before getting into grad school, he joked that he was putting off the real world by going back to school. He realizes now that grad school is very much the real world, with all its stresses and frenetic whirl of research. Professors take for granted that their students will dive head first into a project and spend inordinate amounts of time in the lab. "You're expected to live and breathe your research project," reports Berggren.

Working in a lab as an undergrad, Broderick thought she knew what lay ahead. Her biggest adjustment to grad school was being in control of research and the assumption that she would have more to contribute to lab meetings.

From his position as chairman, Roberts notes that bad time-management is a factor that often knocks students out of grad school. They may be initiating research projects, teaching for the first time, and taking graduate-level courses that entail a lot more information and mastery of complex concepts. Berggren paints those first months in grad school as his "clueless time." But he also had a good adviser who guided him into the literature he needed to be reading and the seminars he needed to take.

■ **GRAD TIP: If you don't have a strong background, get some experience before graduate school. Take some graduate-level classes in college, and do some preliminary reading on your own to acclimate to graduate thinking.**

The work load in graduate school can vary from department to department, but it will be rigorous. "I'm taking one class at the graduate level, and we have reading assignments due every week and three to four scientific papers to read, plus answering questions at the edge of comprehension," states Berggren. Problem sets and lab work take time, as do seminars and lab meetings. Students getting their funding through teaching assistantships have schedules that are even more demanding.

"What Am I Doing Here?"

■ **GRAD TIP: Appearances are deceiving. What looks like only a few courses can translate into hours of outside work for each one.**

What makes the reality of the actual work load hard to comprehend is that from the outside, it doesn't look like much. Most times, new graduate students only take two to three classes. What is hidden is the sheer amount of work that must be done outside the class. Students will have 2 hours (at minimum) of reading to do for every hour of class. Some professors will specify what they want you to read; others will expect you to ferret out sources on your own. Helmy adds that the onus is on the student to learn.

However, you will be paddling the same boat with other graduate students in your class, and they have historically come up with some ingenious ways to keep up. Broderick reports from the University of Wisconsin that she has a journal club in one of her labs that helps her and others to keep up with the background reading they all have to do. Lectures and conferences are another source of information. "Any chance you get, go to meetings," advises Cole. "Hearing others speak is a big part of science. In our department, we have a seminar series with people from other universities."

Adding to the pressure are the preliminary or qualifier exams for those students on Ph.D. tracks. Master's students generally have a written exam or must defend a thesis at the end of their second year. But Kirouac admits she almost enjoyed the Caltech prelim. "It's more like a rite of passage for yourself," she says. "You feel you've earned your place." She quips that Ph.D. students spend their first three years getting into the program and the next three years trying to get out.

But if the shock of the work load doesn't get to you the first semester, the "I don't belong here" syndrome will probably hit somewhere in that time frame. From Caltech, Kirouac says the traditional questions making the rounds are, "What am I doing here? Am I doing the right thing? Is all this work worth it? Am I living up to the standards expected of me?"

Take your pick of the pressures you're likely to put on yourself. Berggren remembers being in classes and thinking he just wasn't going to pick up the information quickly enough. Fortunately, his adviser said she went through the same process. She sat in seminars and discussion groups and felt stupid. Though there's reassurance in knowing others go through the same experience, it can be a very humbling time, Berggren contends. Keith Byrd at Johns Hopkins emphasizes keeping a positive attitude and reminding yourself how much you like what you're doing. When the "I'm not good enough" feelings hit, he suggests talking to some post-docs and advisers. Then go to seminars and read journals to become familiar with the lingo.

. . . .

■ GRAD TIP: Establish relationships with other students and look for social outlets. Grad students can feel very isolated.

. . . .

Feeling constantly behind is the norm, reports Broderick, because graduate students tend to look ahead. The nature of lab work is to jump ahead to the next experiment, or write up a proposal for master's research, or wonder how you can correct bad experiment results.

. . . .

■ GRAD TIP: Don't be shy about tapping into the professors as resources.

. . . .

To stop those wheels from spinning out of control, Berggren and others urge new grad students to find a balance between work and fun. And take a look around at what an exciting environment of amazingly smart people you're in, offers Kirouac. As you will soon find out, relationships between graduate students play a big role in both academic success and keeping your sanity. And, as Berggren found out, the parties, get-togethers, picnics, and meals that seem to sprout up in biology labs are an excellent way to keep in the loop of department happenings. Collaboration is essential. You will need to rely on others to tell you you're going to get through the semester. You will want to bounce ideas off each other. "Graduate school is about learning and growing. It's not about an end product," states Kirouac.

Despite the collegial side of graduate departments, the graduate experience can be very isolating. Students often have families and work different hours. Classes are scheduled at odd times. There's a lot of pressure to work by yourself. "And," says Kirouac, "science is always in your head. Be really careful not to isolate yourself."

Broderick says that the weekly lab meetings in which students share recent experiments and toss around ideas are a great way to connect. She points to the trend in collaborative efforts going on in science now. Thirty years ago, papers were commonly authored by 1 or 2 people. Now articles have 5 or more authors. "There's no way you can know all there is to know without collaboration," adds Byrd. "Success is not being a hermit." Kirouac mentions the way students in her lab often swap articles they've found that they know will interest someone else working on that particular research.

Collaboration with professors is also taken to another intellectual level in graduate school. Students are being primed to be colleagues not subordinates. Professors generally don't lead classes as much as guide the interaction that takes place between students. Kirouac was surprised when she discovered that professors considered her more of a peer and began to respect her ideas.

The Job Market

It might be hard to imagine there's a life outside the ivory tower, much less one that you'll be involved with. Fortunately for graduate students in biology, what awaits after graduation are some solid career paths. According to Thomas Roberts, Chairman of Florida State University's Department of Biological Science, the triumvirate of Ph.D. students going into academia, industry, and government is still in operation. Master's students also have flexibility in their choices. Among them, Roberts cites the pharmaceutical industry's need for armies of people to do research in genome-sequencing projects.

Accreditation
and Accrediting Agencies

Colleges and universities in the United States, and their individual academic and professional programs, are accredited by nongovernmental agencies concerned with monitoring the quality of education in this country. Agencies with both regional and national jurisdictions grant accreditation to institutions as a whole, while specialized bodies acting on a nationwide basis—often national professional associations—grant accreditation to departments and programs in specific fields.

Institutional and specialized accrediting agencies share the same basic concerns: the purpose an academic unit—whether university or program—has set for itself and how well it fulfills that purpose, the adequacy of its financial and other resources, the quality of its academic offerings, and the level of services it provides. Agencies that grant institutional accreditation take a broader view, of course, and examine university-wide or college-wide services that a specialized agency may not concern itself with.

Both types of agencies follow the same general procedures when considering an application for accreditation. The academic unit prepares a self-evaluation, focusing on the concerns mentioned above and usually including an assessment of both its strengths and weaknesses; a team of representatives of the accrediting body reviews this evaluation, visits the campus, and makes its own report; and finally, the accrediting body makes a decision on the application. Often, even when accreditation is granted, the agency makes a recommendation regarding how the institution or program can improve. All institutions and programs are also reviewed every few years to determine whether they continue to meet established standards; if they do not, they may lose their accreditation.

Accrediting agencies themselves are reviewed and evaluated periodically by the U.S. Department of Education and the Council for Higher Education Accreditation (CHEA). Agencies recognized adhere to certain standards and practices, and their authority in matters of accreditation is widely accepted in the educational community.

This does not mean, however, that accreditation is a simple matter, either for schools wishing to become accredited or for students deciding where to apply. Indeed, in certain fields the very meaning and methods of accreditation are the subject of a good deal of debate. For their part, those applying to graduate school should be aware of the safeguards provided by regional accreditation, especially in terms of degree acceptance and institutional longevity. Beyond this, applicants should understand the role that specialized accreditation plays in their field, as this varies considerably from one discipline to another. In certain professional fields, it is necessary to have graduated from a program that is accredited in order to be eligible for a license to practice, and in some fields the federal government also makes this a hiring requirement. In other disciplines, however, accreditation is not as essential, and there can be excellent programs that are not accredited. In fact, some programs choose not to seek accreditation, although most do.

Institutions and programs that present themselves for accreditation are sometimes granted the status of candidate for accreditation, or what is known as "preaccreditation." This may happen, for example, when an academic unit is too new to have met all the requirements for accreditation. Such status signifies initial recognition and indicates that the school or program in question is working to fulfill all requirements; it does not, however, guarantee that accreditation will be granted.

Readers are advised to contact agencies directly for answers to their questions about accreditation. The names and addresses of all agencies recognized by the U.S. Department of Education and the Council for Higher Education Accreditation are listed below.

Institutional Accrediting Agencies—Regional

MIDDLE STATES ASSOCIATION OF COLLEGES AND SCHOOLS

Accredits institutions in Delaware, District of Columbia, Maryland, New Jersey, New York, Pennsylvania, Puerto Rico, and the Virgin Islands.

Jean Avnet Morse, Executive Director
Commission on Higher Education
3624 Market Street
Philadelphia, Pennsylvania 19104-2680
Telephone: 215-662-5606
Fax: 215-662-5501
E-mail: jamorse@msache.org
World Wide Web: http://www.msache.org

NEW ENGLAND ASSOCIATION OF SCHOOLS AND COLLEGES

Accredits institutions in Connecticut, Maine, Massachusetts, New Hampshire, Rhode Island, and Vermont.

Charles M. Cook, Director
Commission on Institutions of Higher Education
209 Burlington Road
Bedford, Massachusetts 01730-1433
Telephone: 781-271-0022
Fax: 781-271-0950
E-mail: ccook@neasc.org
World Wide Web: http://www.neasc.org

NORTH CENTRAL ASSOCIATION OF COLLEGES AND SCHOOLS

Accredits institutions in Arizona, Arkansas, Colorado, Illinois, Indiana, Iowa, Kansas, Michigan, Minnesota, Missouri, Nebraska, New Mexico, North Dakota, Ohio, Oklahoma, South Dakota, West Virginia, Wisconsin, and Wyoming.

Steve Crow, Executive Director
Commission on Institutions of Higher Education
30 North LaSalle, Suite 2400
Chicago, Illinois 60602-2504
Telephone: 312-263-0456
Fax: 312-263-7462
E-mail: crow@ncacihe.org
World Wide Web: http://www.ncacihe.org

NORTHWEST ASSOCIATION OF SCHOOLS AND COLLEGES

Accredits institutions in Alaska, Idaho, Montana, Nevada, Oregon, Utah, and Washington.

Sandra E. Elman, Executive Director
Commission on Colleges
11130 Northeast 33rd Place, Suite 120
Bellevue, Washington 98004
Telephone: 425-827-2005
Fax: 425-827-3395
E-mail: pjarnold@cocnasc.org
World Wide Web: http://www.cocnasc.org

SOUTHERN ASSOCIATION OF COLLEGES AND SCHOOLS

Accredits institutions in Alabama, Florida, Georgia, Kentucky, Louisiana, Mississippi, North Carolina, South Carolina, Tennessee, Texas, and Virginia.

James T. Rogers, Executive Director
Commission on Colleges
1866 Southern Lane
Decatur, Georgia 30033-4097
Telephone: 404-679-4500
Fax: 404-679-4558
E-mail: jrogers@sacscoc.org
World Wide Web: http://www.sacscoc.org

WESTERN ASSOCIATION OF SCHOOLS AND COLLEGES

Accredits institutions in California, Guam, and Hawaii.

Ralph A. Wolff, Executive Director
Accrediting Commission for Senior Colleges and Universities
985 Atlantic Avenue, Suite 100
Alameda, California 94501
Telephone: 510-748-9001
E-mail: wascsr@wascsenior.org
World Wide Web: http://www.wascweb.org

Institutional Accrediting Agencies—Other

ACCREDITING COUNCIL FOR INDEPENDENT COLLEGES AND SCHOOLS

Dr. Steven A. Eggland, Executive Director
750 First Street, NE, Suite 980
Washington, D.C. 20002-4241
Telephone: 202-336-6780
Fax: 202-842-2593
E-mail: steve@acics.org
World Wide Web: http://www.acics.org

DISTANCE EDUCATION AND TRAINING COUNCIL

Michael P. Lambert, Executive Secretary
1601 Eighteenth Street, NW
Washington, D.C. 20009-2529
Telephone: 202-234-5100
Fax: 202-332-1386
E-mail: mike@detc.org
World Wide Web: http://www.detc.org

Graduate Programs in
Biology

Biological and Biomedical Sciences

BIOLOGICAL AND BIOMEDICAL SCIENCES—GENERAL

■ ADELPHI UNIVERSITY

Graduate School of Arts and Sciences, Department of Biology, Garden City, NY 11530

AWARDS MS. Part-time and evening/weekend programs available.

Students: Average age 28. In 1999, 8 degrees awarded.

Degree requirements: For master's, thesis or alternative required, foreign language not required. *Application deadline:* Applications are processed on a rolling basis. *Application fee:* $50.

Expenses: Tuition: Full-time $16,600; part-time $500 per credit. Required fees: $150 per semester. Part-time tuition and fees vary according to course load and program.

Financial aid: In 1999–00, 14 students received aid; teaching assistantships available. Financial aid application deadline: 2/15; financial aid applicants required to submit FAFSA.

Faculty research: Cell and molecular biology, genetics, physiology, ecology and evolution, microbiology.

Dr. James Dooley, Chairperson, 516-877-4200.

■ ALABAMA AGRICULTURAL AND MECHANICAL UNIVERSITY

School of Graduate Studies, School of Arts and Sciences, Department of Natural and Physical Sciences, Area in Biology, Normal, AL 35762

AWARDS MS. Part-time and evening/weekend programs available.

Faculty: 5 full-time (1 woman).

Students: In 1999, 7 degrees awarded (100% found work related to degree).

Degree requirements: For master's, computer language, thesis, comprehensive exam required, foreign language not required.

Entrance requirements: For master's, GRE General Test. *Application deadline:* For fall admission, 5/1. *Application fee:* $15 ($20 for international students).

Expenses: Tuition, state resident: full-time $1,932. Tuition, nonresident: full-time $3,864. Tuition and fees vary according to course load.

Financial aid: Fellowships, research assistantships, Federal Work-Study available. Financial aid application deadline: 4/1.

Faculty research: Radiation and chemical mutagenesis, human cytogenetics, microbial biotechnology, microbial metabolism, environmental toxicology. *Total annual research expenditures:* $40,000.

Dr. Charles McMillan, Chair, 256-851-4927.

■ ALABAMA STATE UNIVERSITY

School of Graduate Studies, College of Arts and Sciences, Department of Biology, Montgomery, AL 36101-0271

AWARDS Biology (MS); biology education (Ed S). Part-time programs available.

Faculty: 6 full-time (0 women).

Students: 14 full-time (8 women), 31 part-time (25 women); includes 35 minority (34 African Americans, 1 Hispanic American). In 1999, 3 master's awarded.

Degree requirements: For master's, thesis, comprehensive exam required; for Ed S, thesis required.

Entrance requirements: For master's, GRE General Test, GRE Subject Test, Graduate Writing Competency Test; for Ed S, Graduate Writing Competency Test, GRE, MAT. *Application deadline:* For fall admission, 7/15; for spring admission, 12/15. Applications are processed on a rolling basis. *Application fee:* $10.

Expenses: Tuition, state resident: full-time $2,880; part-time $120 per credit. Tuition, nonresident: full-time $5,760; part-time $240 per credit.

Financial aid: In 1999–00, 4 research assistantships with tuition reimbursements (averaging $12,000 per year) were awarded.

Faculty research: *Salmonella pseudomonas,* cancer cells. *Total annual research expenditures:* $125,000.

Dr. Shiva P. Singh, Acting Chair, 334-229-4467, *Fax:* 334-229-1007.

Application contact: Dr. Annette Marie Allen, Dean of Graduate Studies, 334-229-4275, *Fax:* 334-229-4928, *E-mail:* aallen@asunet.alasu.edu.

■ ALBANY MEDICAL COLLEGE

Graduate Programs in the Biological Sciences, Albany, NY 12208-3479

AWARDS MS, PhD. Part-time programs available.

Faculty: 54 full-time (13 women), 35 part-time/adjunct (6 women).

Students: 125 full-time (69 women); includes 16 minority (4 African Americans, 10 Asian Americans or Pacific Islanders, 2 Hispanic Americans), 24 international. Average age 26. *160 applicants, 36% accepted.* In 1999, 29 master's, 3 doctorates awarded. Terminal master's awarded for partial completion of doctoral program.

Degree requirements: For master's, thesis required, foreign language not required; for doctorate, dissertation, oral qualifying exam required, foreign language not required. *Average time to degree:* Master's–2.5 years full-time; doctorate–5 years full-time.

Entrance requirements: For master's and doctorate, GRE General Test, TOEFL. *Application deadline:* Applications are processed on a rolling basis.

Expenses: Tuition: Full-time $13,367; part-time $446 per credit hour.

Financial aid: In 1999–00, 10 fellowships with full tuition reimbursements (averaging $15,000 per year), 62 research assistantships with full tuition reimbursements (averaging $15,000 per year) were awarded; teaching assistantships, career-related internships or fieldwork, Federal Work-Study, grants, institutionally sponsored loans, scholarships, and tuition waivers (full) also available. Aid available to part-time students.

Dr. Henry S. Pohl, Senior Associate Dean for Education Programs, 518-262-5253, *Fax:* 518-262-5183.

Application contact: Jean M. Connelly, Admissions Coordinator, 518-262-5253, *Fax:* 518-262-5183, *E-mail:* graduate-studies@mail.amc.edu.

Find an in-depth description at www.petersons.com/graduate.

■ ALCORN STATE UNIVERSITY

School of Graduate Studies, School of Arts and Sciences, Department of Biology, Alcorn State, MS 39096-7500

AWARDS MS.

Students: 9 full-time (4 women), 6 part-time (1 woman); all minorities (all African Americans). Average age 26. *Application deadline:* For fall admission, 7/1 (priority date); for spring admission, 12/1. Applications are processed on a rolling basis. *Application fee:* $10.

Expenses: Tuition, state resident: full-time $1,072; part-time $120 per semester hour. Required fees: $24 per semester.

Dr. Alice Russell Powell, Chairperson, 601-877-6236.

■ AMERICAN UNIVERSITY

College of Arts and Sciences, Department of Biology, Program in Biology, Washington, DC 20016-8001

AWARDS MA, MS. Part-time programs available.

Faculty: 8 full-time (3 women).
Students: 4 full-time (3 women), 5 part-time (4 women). *12 applicants, 83% accepted.* In 1999, 6 degrees awarded.
Degree requirements: For master's, comprehensive written exam, tool of research exam required.
Entrance requirements: For master's, GRE General Test, GRE Subject Test, TOEFL, minimum GPA of 3.0. *Application deadline:* For fall admission, 2/1; for spring admission, 10/1. *Application fee:* $50.
Expenses: Tuition: Part-time $721 per credit hour. Required fees: $90 per semester. Tuition and fees vary according to program.
Financial aid: Fellowships, research assistantships, teaching assistantships, career-related internships or fieldwork, Federal Work-Study, and institutionally sponsored loans available. Financial aid application deadline: 2/1.
Application contact: Dr. Richard Fox, Chair, 202-885-2176, *Fax:* 202-885-2182.
Find an in-depth description at www.petersons.com/graduate.

■ ANDREWS UNIVERSITY

School of Graduate Studies, College of Arts and Sciences, Department of Biology, Berrien Springs, MI 49104

AWARDS MAT, MS.

Faculty: 7 full-time (0 women).
Students: 9 full-time (5 women), 8 part-time (4 women); includes 3 minority (2 African Americans, 1 Asian American or Pacific Islander), 9 international. In 1999, 3 degrees awarded.
Degree requirements: For master's, thesis, comprehensive exam required.
Entrance requirements: For master's, GRE Subject Test. *Application deadline:* Applications are processed on a rolling basis. *Application fee:* $40.
Expenses: Tuition: Full-time $11,040; part-time $300 per credit. Required fees: $80 per quarter. Tuition and fees vary according to degree level, campus/location and program.
Financial aid: Fellowships, research assistantships, teaching assistantships, career-related internships or fieldwork, Federal Work-Study, and institutionally sponsored loans available. Financial aid application deadline: 3/15.
Dr. John F. Stout, Chairman, 616-471-3243.

■ ANGELO STATE UNIVERSITY

Graduate School, College of Sciences, Department of Biology, San Angelo, TX 76909

AWARDS MS. Part-time and evening/weekend programs available.

Faculty: 16 full-time (3 women).
Students: 3 full-time (1 woman), 12 part-time (6 women); includes 4 minority (1 Asian American or Pacific Islander, 3 Hispanic Americans), 1 international. Average age 31. *8 applicants, 75% accepted.* In 2000, 2 degrees awarded.
Degree requirements: For master's, thesis, comprehensive exam required, foreign language not required.
Entrance requirements: For master's, GRE General Test, minimum GPA of 3.0. *Application deadline:* For fall admission, 8/7 (priority date); for spring admission, 1/2. Applications are processed on a rolling basis. *Application fee:* $25 ($50 for international students).
Expenses: Tuition, state resident: part-time $38 per semester hour. Tuition, nonresident: part-time $249 per semester hour. Required fees: $40 per semester hour. $71 per semester. Tuition and fees vary according to degree level.
Financial aid: In 2000–01, 12 fellowships, 3 teaching assistantships were awarded; career-related internships or fieldwork, Federal Work-Study, tuition waivers (partial), and unspecified assistantships also available. Aid available to part-time students. Financial aid application deadline: 8/1.
Dr. Bonnie B. Amos, Head, 915-942-2287.

■ ANNA MARIA COLLEGE

Graduate Division, Programs in Biological Sciences, Paxton, MA 01612

AWARDS Biological studies (MA). Part-time and evening/weekend programs available.

Faculty: 20.
Students: Average age 33. In 1999, 2 degrees awarded.
Degree requirements: For master's, oral comprehensive exam required, foreign language and thesis not required.
Entrance requirements: For master's, bachelor's degree in natural sciences, minimum GPA of 2.7. *Application deadline:* Applications are processed on a rolling basis. *Application fee:* $30.
Expenses: Tuition: Part-time $775 per course.
Financial aid: Career-related internships or fieldwork available. Aid available to part-time students. Financial aid applicants required to submit FAFSA.
Faculty research: Molecular biology, microbiology, enzymes, cancer research, cell membrane pharmacology.

Dr. Lorraine Popowicz, Director, 508-849-3382, *Fax:* 508-849-3339, *E-mail:* 1popowicz@annamaria.edu.
Application contact: Beth Ann Carey, Assistant Director of Admissions for Graduate Programs and the Department of Professional Studies, 508-849-3361, *Fax:* 508-849-3362, *E-mail:* bcarey@annamaria.edu.

■ APPALACHIAN STATE UNIVERSITY

Cratis D. Williams Graduate School, College of Arts and Sciences, Department of Biology, Boone, NC 28608

AWARDS MA, MS. Part-time programs available.

Faculty: 25 full-time (6 women).
Students: 24 full-time (7 women), 10 part-time (4 women), 3 international. *22 applicants, 68% accepted.* In 1999, 7 degrees awarded.
Degree requirements: For master's, thesis, comprehensive exam required. *Average time to degree:* Master's–2 years full-time.
Entrance requirements: For master's, GRE General Test, GRE Subject Test. *Application deadline:* For fall admission, 7/1 (priority date); for spring admission, 11/1. *Application fee:* $35.
Expenses: Tuition, state resident: full-time $1,909. Tuition, nonresident: full-time $9,179. Tuition and fees vary according to course load and degree level.
Financial aid: In 1999–00, fellowships (averaging $2,000 per year), research assistantships (averaging $6,500 per year), 27 teaching assistantships (averaging $6,500 per year) were awarded; career-related internships or fieldwork, scholarships, and unspecified assistantships also available. Aid available to part-time students. Financial aid application deadline: 7/1; financial aid applicants required to submit FAFSA.
Faculty research: Aquatic and terrestrial ecology, animal and plant physiology, behavior and systematics, immunology and cell biology, molecular biology and microbiology.
Vicki Martin, Chairman, 828-262-3025.
Application contact: Dr. Howard Neufeld, Graduate Director, 828-262-2683, *E-mail:* neufeldhs@appstate.edu.

■ ARIZONA STATE UNIVERSITY

Graduate College, College of Liberal Arts and Sciences, Department of Biology, Tempe, AZ 85287

AWARDS Behavior (MS, PhD); biology (MNS); biology education (MS, PhD); cell and developmental biology (MS, PhD);

Arizona State University (continued)
computational, statistical, and mathematical biology (MS, PhD); conservation (MS, PhD); ecology (MS, PhD); evolution (MS, PhD); genetics (MS, PhD); history and philosophy of biology (MS, PhD); molecular and cellular biology (MS, PhD); neuroscience (MS, PhD); physiology (MS, PhD).

Faculty: 41 full-time (9 women), 20 part-time/adjunct (5 women).
Students: 95 full-time (38 women), 10 part-time (2 women); includes 3 African Americans, 2 Asian Americans or Pacific Islanders, 1 Native American, 8 international. Average age 30. *150 applicants, 15% accepted.* In 1999, 6 master's awarded (17% entered university research/teaching, 83% found other work related to degree); 11 doctorates awarded (100% entered university research/teaching). Terminal master's awarded for partial completion of doctoral program.
Degree requirements: For master's, thesis required, foreign language not required; for doctorate, dissertation, oral exam required, foreign language not required.
Entrance requirements: For master's and doctorate, GRE General Test, GRE Subject Test. *Application deadline:* For fall admission, 12/15. *Application fee:* $45. Electronic applications accepted.
Expenses: Tuition, state resident: part-time $115 per credit hour. Tuition, nonresident: part-time $389 per credit hour. Required fees: $18 per semester. Tuition and fees vary according to program.
Financial aid: In 1999–00, 12 fellowships with full tuition reimbursements, 30 research assistantships with partial tuition reimbursements, 43 teaching assistantships with partial tuition reimbursements were awarded; career-related internships or fieldwork, Federal Work-Study, grants, institutionally sponsored loans, scholarships, and tuition waivers (partial) also available. Financial aid application deadline: 12/15.
Faculty research: Behavioral genetics, comparative endocrinology, invertebrate neurophysiology.
Dr. James P. Collins, Chair, 480-965-3571, *Fax:* 480-965-2519.
Application contact: Dr. Michael C. Moore, Director, 480-965-0386, *Fax:* 480-965-2519.

Find an in-depth description at www.petersons.com/graduate.

■ ARKANSAS STATE UNIVERSITY

Graduate School, College of Arts and Sciences, Department of Biological Sciences, Jonesboro, State University, AR 72467

AWARDS Biology (MS); biology education (MSE, SCCT); environmental sciences (PhD). Part-time programs available.

Faculty: 17 full-time (3 women), 1 part-time/adjunct (0 women).
Students: 7 full-time (1 woman), 29 part-time (15 women), 2 international. Average age 29. In 1999, 9 master's awarded.
Degree requirements: For master's, thesis (for some programs), comprehensive exam required; for doctorate, dissertation, comprehensive exam required; for SCCT, comprehensive exam required, thesis not required.
Entrance requirements: For master's, GRE General Test, appropriate bachelor's degree; for doctorate, GRE General Test, interview, master's degree; for SCCT, GRE General Test or MAT, interview, master's degree. *Application deadline:* For fall admission, 7/1 (priority date); for spring admission, 11/15 (priority date). Applications are processed on a rolling basis. *Application fee:* $15 ($25 for international students).
Expenses: Tuition, state resident: full-time $2,976; part-time $124 per hour. Tuition, nonresident: full-time $7,488; part-time $312 per hour. Required fees: $506; $19 per hour. $25 per semester.
Financial aid: Teaching assistantships available. Aid available to part-time students. Financial aid application deadline: 7/1; financial aid applicants required to submit FAFSA.
Dr. Roger Buchanan, Chair, 870-972-3082, *Fax:* 870-972-2638, *E-mail:* rbuck@navajo.astate.edu.

■ AUBURN UNIVERSITY

College of Veterinary Medicine and Graduate School, Graduate Program in Veterinary Medicine, Interdepartmental Program in Biomedical Sciences, Auburn, Auburn University, AL 36849-0002

AWARDS PhD.

Faculty: 84 full-time (20 women).
Students: *6 applicants, 0% accepted.* In 1999, 6 degrees awarded.
Degree requirements: For doctorate, dissertation required.
Entrance requirements: For doctorate, GRE General Test, GRE Subject Test. *Application deadline:* For fall admission, 7/17; for spring admission, 11/24. Applications are processed on a rolling basis.

Application fee: $25 ($50 for international students). Electronic applications accepted.
Expenses: Tuition, state resident: full-time $2,895; part-time $80 per credit hour. Tuition, nonresident: full-time $8,685; part-time $240 per credit hour.
Financial aid: Federal Work-Study available. Aid available to part-time students. Financial aid application deadline: 3/15.
Faculty research: Animal biotechnology, mechanisms of disease, cell biology, diagnosis and therapy of disease.
Application contact: Dr. John F. Pritchett, Dean of the Graduate School, 334-844-4700.

■ AUBURN UNIVERSITY

Graduate School, College of Sciences and Mathematics, Department of Biological Sciences, Auburn, Auburn University, AL 36849-0002

AWARDS Botany (MS, PhD); microbiology (MS, PhD); zoology (MS, PhD).

Faculty: 31 full-time (6 women).
Students: 38 full-time (23 women), 35 part-time (18 women); includes 6 minority (5 African Americans, 1 Hispanic American), 11 international. *66 applicants, 48% accepted.* In 1999, 12 master's, 1 doctorate awarded.
Entrance requirements: For master's and doctorate, GRE General Test, TOEFL. *Application deadline:* For fall admission, 7/7; for spring admission, 11/24. Electronic applications accepted.
Expenses: Tuition, state resident: full-time $2,895; part-time $80 per credit hour. Tuition, nonresident: full-time $8,685; part-time $240 per credit hour.
Financial aid: Research assistantships, teaching assistantships available.
Dr. Alfred E. Brown, Interim Chair, 334-844-4830, *Fax:* 334-844-1645.

Find an in-depth description at www.petersons.com/graduate.

■ AUSTIN PEAY STATE UNIVERSITY

Graduate School, College of Arts and Sciences, Department of Biology, Clarksville, TN 37044-0001

AWARDS MS. Part-time programs available.

Faculty: 1 (woman) full-time, 2 part-time/adjunct (0 women).
Students: 3 full-time (1 woman), 13 part-time (7 women); includes 4 minority (3 African Americans, 1 Asian American or Pacific Islander), 1 international. In 1999, 4 degrees awarded.
Degree requirements: For master's, thesis optional, foreign language not required.
Entrance requirements: For master's, GRE General Test, minimum GPA of 2.5.

Application deadline: For fall admission, 7/31 (priority date); for spring admission, 12/4. Applications are processed on a rolling basis. *Application fee:* $25.

Expenses: Tuition, state resident: full-time $3,276; part-time $137 per credit hour. Tuition, nonresident: full-time $8,392; part-time $361 per credit hour. Tuition and fees vary according to course load.

Financial aid: In 1999–00, research assistantships (averaging $6,450 per year); career-related internships or fieldwork, Federal Work-Study, institutionally sponsored loans, scholarships, and unspecified assistantships also available. Aid available to part-time students. Financial aid application deadline: 4/1.

Faculty research: Nonpaint source pollution, amphibian biomonitoring, aquatic toxicology, biological indicators of water quality, taxonomy, survey and range of threatened fauna and flora, distribution/ecology of amphibians and reptiles, recombining DNA probes, scanning electron microscopy of human chromosomes, endocrinology.

Dr. David Snyder, Chair, 931-221-7781, *E-mail:* snyderd@apsu.edu.

Application contact: Dr. Cindy Taylor, Associate Professor, 931-221-7784, *E-mail:* taylorc@apsu.edu.

■ BALL STATE UNIVERSITY

Graduate School, College of Sciences and Humanities, Department of Biology, Muncie, IN 47306-1099

AWARDS Biology (MA, MAE, MS); biology education (Ed D).

Faculty: 21.

Students: 9 full-time (3 women), 24 part-time (9 women); includes 4 minority (3 African Americans, 1 Asian American or Pacific Islander), 4 international. Average age 27. *16 applicants, 100% accepted.* In 1999, 25 master's, 1 doctorate awarded.

Degree requirements: For master's, foreign language not required; for doctorate, dissertation required, foreign language not required.

Entrance requirements: For doctorate, GRE General Test, minimum graduate GPA of 3.2. *Application fee:* $25 ($35 for international students).

Expenses: Tuition, state resident: full-time $3,024. Tuition, nonresident: full-time $7,482. Tuition and fees vary according to course load.

Financial aid: Teaching assistantships with full tuition reimbursements, career-related internships or fieldwork available. Financial aid application deadline: 3/1.

Faculty research: Aquatics and fisheries, tumors, water and air pollution, developmental biology and genetics.

Dr. Carl E. Warnes, Chairman, 765-285-8820, *E-mail:* cwarnes@bsu.edu.

■ BARRY UNIVERSITY

School of Natural and Health Sciences, Program in Biology and Biomedical Sciences, Miami Shores, FL 33161-6695

AWARDS Biology (MS); biomedical sciences (MS). Part-time and evening/weekend programs available.

Faculty: 8 full-time (3 women), 12 part-time/adjunct (5 women).

Students: 104 full-time (55 women), 56 part-time (37 women); includes 101 minority (51 African Americans, 18 Asian Americans or Pacific Islanders, 31 Hispanic Americans, 1 Native American), 6 international. Average age 26. In 1999, 73 degrees awarded.

Degree requirements: For master's, thesis (for some programs), comprehensive exam required, foreign language not required.

Entrance requirements: For master's, GRE General Test or Florida Teacher's Certification Exam (biology); GRE General Test, MCAT, or DAT (biomedical sciences). *Application deadline:* For fall admission, 8/1 (priority date); for spring admission, 12/1 (priority date). Applications are processed on a rolling basis. *Application fee:* $30.

Expenses: Tuition: Full-time $11,040; part-time $460 per credit. Tuition and fees vary according to degree level and program.

Financial aid: Application deadline: 5/1;

Faculty research: Genetics, immunology, anthropology.

Dr. Ralph Laudan, Associate Dean, 305-899-3229, *Fax:* 305-899-3225, *E-mail:* laudan@mail.barry.edu.

Application contact: Dr. Jocelyn Goulet, Director, Health Services Admissions Operation, 305-899-3541, *Fax:* 305-899-3232, *E-mail:* jgoulet@mail.barry.edu.

■ BAYLOR COLLEGE OF MEDICINE

Graduate School of Biomedical Sciences, Houston, TX 77030-3498

AWARDS PhD, MD/PhD.

Faculty: 276 full-time (58 women).

Students: 388 full-time (186 women); includes 50 minority (12 African Americans, 23 Asian Americans or Pacific Islanders, 15 Hispanic Americans), 139 international. Average age 27. *869 applicants, 20% accepted.* In 1999, 49 doctorates awarded.

Degree requirements: For doctorate, dissertation, public defense, qualifying exam required, foreign language not required. *Average time to degree:* Doctorate–5.64 years full-time.

Entrance requirements: For doctorate, GRE General Test; GRE Subject Test (strongly recommended), TOEFL, minimum GPA of 3.0. *Application deadline:* For fall admission, 2/1 (priority date). Applications are processed on a rolling basis. *Application fee:* $30. Electronic applications accepted.

Expenses: Tuition: Full-time $8,200. Required fees: $175. Full-time tuition and fees vary according to student level.

Financial aid: In 1999–00, 388 students received aid, including 215 fellowships (averaging $16,000 per year), 171 research assistantships (averaging $16,000 per year), 2 teaching assistantships; career-related internships or fieldwork, Federal Work-Study, institutionally sponsored loans, and tuition waivers (full and partial) also available. Financial aid applicants required to submit FAFSA.

Faculty research: Cell and molecular biology of cardiac muscle, structural biophysics, gene expression and regulation, human genomes, viruses.

Dr. William R. Brinkley, Dean of Graduate Sciences, 713-798-5263, *Fax:* 713-798-6325, *E-mail:* brinkley@bcm.tmc.edu.

Application contact: Donna Otwell, Administrative Associate, 713-798-4029, *Fax:* 713-798-6325, *E-mail:* dotwell@bcm.tmc.edu.

Find an in-depth description at www.petersons.com/graduate.

■ BAYLOR UNIVERSITY

Graduate School, College of Arts and Sciences, Department of Biology, Waco, TX 76798

AWARDS Biology (MA, MS, PhD); environmental biology (MS); limnology (MSL). Part-time programs available.

Faculty: 13 full-time (3 women).

Students: 10 full-time (3 women), 4 part-time (3 women); includes 2 minority (both Asian Americans or Pacific Islanders), 2 international. In 1999, 7 degrees awarded.

Degree requirements: For master's, thesis required (for some programs); for doctorate, dissertation required.

Entrance requirements: For master's and doctorate, GRE General Test. *Application deadline:* For fall admission, 1/31 (priority date). Applications are processed on a rolling basis. *Application fee:* $25.

Expenses: Tuition: Part-time $329 per semester hour. Tuition and fees vary according to program.

Financial aid: Teaching assistantships, career-related internships or fieldwork, Federal Work-Study, institutionally sponsored loans, and tuition waivers (full and partial) available. Aid available to part-time students. Financial aid application deadline: 2/28.

Baylor University (continued)
Faculty research: Terrestrial ecology, aquatic ecology, genetics.
Dr. Richard E. Duhrkopf, Director of Graduate Studies, 254-710-2911, *Fax:* 254-710-2969, *E-mail:* rick_duhrkopf@baylor.edu.
Application contact: Sandy Tighe, Administrative Assistant, 254-710-2911, *Fax:* 254-710-2969, *E-mail:* sandy_tighe@baylor.edu.

■ **BAYLOR UNIVERSITY**
Graduate School, Institute of Biomedical Studies, Waco, TX 76798
AWARDS MS, PhD.

Students: 22 full-time (12 women), 3 part-time; includes 4 minority (all Asian Americans or Pacific Islanders), 5 international. In 1999, 3 master's, 3 doctorates awarded.
Entrance requirements: For master's and doctorate, GRE General Test. *Application deadline:* Applications are processed on a rolling basis. *Application fee:* $25.
Expenses: Tuition: Part-time $329 per semester hour. Tuition and fees vary according to program.
Financial aid: Research assistantships, teaching assistantships available.
Dr. Darden Powers, Director, 254-710-2514, *Fax:* 254-710-3878, *E-mail:* darden_powers@baylor.edu.
Application contact: Suzanne Keener, Administrative Assistant, 254-710-3588, *Fax:* 254-710-3870, *E-mail:* graduate_school@baylor.edu.

■ **BEMIDJI STATE UNIVERSITY**
Graduate Studies, College of Social and Natural Sciences, Department of Biology, Bemidji, MN 56601-2699
AWARDS MA. Part-time programs available.

Faculty: 7 part-time/adjunct (2 women).
Students: 3 full-time (1 woman), 1 part-time. Average age 34. In 1999, 2 degrees awarded.
Degree requirements: For master's, thesis or alternative, departmental qualifying exam required.
Application deadline: For fall admission, 5/1. *Application fee:* $20.
Expenses: Tuition, state resident: part-time $140 per credit. Tuition, nonresident: part-time $222 per credit. Required fees: $43 per credit. Tuition and fees vary according to course load, campus/location, program and reciprocity agreements.
Financial aid: In 1999–00, teaching assistantships with partial tuition reimbursements (averaging $5,500 per year); career-related internships or fieldwork and Federal Work-Study also available. Aid available to part-time

students. Financial aid application deadline: 5/1.
Kerry L. Openshaw, Chair, 218-755-2799, *Fax:* 218-755-4107, *E-mail:* klopenshaw@vax1.bemidji.msus.edu.

■ **BENNINGTON COLLEGE**
Graduate Programs, Premedical Program, Bennington, VT 05201-9993
AWARDS Certificate.
Expenses: Tuition: Part-time $19,500 per degree program.

■ **BLOOMSBURG UNIVERSITY OF PENNSYLVANIA**
School of Graduate Studies, College of Arts and Sciences, Department of Biological and Allied Health Sciences, Program in Biology, Bloomsburg, PA 17815-1905
AWARDS MS.

Faculty: 18 full-time (7 women), 1 part-time/adjunct (0 women).
Students: 2 full-time (0 women), 11 part-time (5 women). Average age 29. *5 applicants, 100% accepted.* In 1999, 4 degrees awarded.
Degree requirements: For master's, thesis or alternative required, foreign language not required.
Entrance requirements: For master's, GRE General Test, minimum QPA of 2.5. *Application deadline:* Applications are processed on a rolling basis. *Application fee:* $30.
Expenses: Tuition, state resident: full-time $3,780; part-time $210 per credit. Tuition, nonresident: full-time $6,610; part-time $367 per credit. Required fees: $21 per credit. Part-time tuition and fees vary according to course load.
Dr. Margaret Till, Coordinator, 570-389-4780, *Fax:* 570-389-3028, *E-mail:* till@bloomu.edu.

■ **BOISE STATE UNIVERSITY**
Graduate College, College of Arts and Sciences, Department of Biology, Boise, ID 83725-0399
AWARDS Biology (MA, MS); raptor biology (MS). Part-time programs available.

Faculty: 16 full-time (2 women), 35 part-time/adjunct (8 women).
Students: 23 full-time (12 women), 15 part-time (7 women); includes 1 minority (Asian American or Pacific Islander). Average age 32. *18 applicants, 50% accepted.* In 1999, 2 degrees awarded.
Degree requirements: For master's, thesis required.
Entrance requirements: For master's, GRE General Test, minimum GPA of 3.0. *Application deadline:* For fall admission,

7/21 (priority date); for spring admission, 11/22 (priority date). Applications are processed on a rolling basis. *Application fee:* $20 ($30 for international students). Electronic applications accepted.
Expenses: Tuition, state resident: part-time $145 per credit. Tuition, nonresident: full-time $5,880; part-time $145 per credit. Required fees: $3,217. Tuition and fees vary according to course load.
Financial aid: In 1999–00, 24 students received aid, including 20 research assistantships with full tuition reimbursements available; career-related internships or fieldwork, Federal Work-Study, institutionally sponsored loans, and unspecified assistantships also available. Aid available to part-time students. Financial aid application deadline: 3/1.
Faculty research: Soil and stream microbial ecology, avian ecology.
Dr. James Munger, Chairperson, 208-426-3262, *Fax:* 208-426-3117.
Application contact: Dr. Alfred Dufty, Coordinator, 208-426-3263, *Fax:* 208-426-3117.

■ **BOSTON COLLEGE**
Graduate School of Arts and Sciences, Department of Biology, Chestnut Hill, MA 02467-3800
AWARDS Biochemistry (MS, PhD); biology (MS, PhD).

Faculty: 22 full-time (4 women).
Students: 11 full-time (8 women), 30 part-time (13 women); includes 4 minority (3 Asian Americans or Pacific Islanders, 1 Hispanic American), 7 international. *79 applicants, 16% accepted.* In 1999, 3 master's, 3 doctorates awarded. Terminal master's awarded for partial completion of doctoral program.
Degree requirements: For master's and doctorate, thesis/dissertation required, foreign language not required.
Entrance requirements: For master's and doctorate, GRE General Test, GRE Subject Test. *Application deadline:* For fall admission, 2/1. *Application fee:* $40.
Expenses: Tuition: Part-time $656 per credit. Tuition and fees vary according to program.
Financial aid: Fellowships with full tuition reimbursements, research assistantships with full tuition reimbursements, teaching assistantships with full tuition reimbursements, Federal Work-Study and scholarships available. Aid available to part-time students. Financial aid application deadline: 3/15; financial aid applicants required to submit FAFSA.
Faculty research: DNA replication in mammalian cells, control of the cell cycle, immunology, plant genetics.
Dr. William Petri, Chairperson, 617-552-3540, *E-mail:* william.petri@bc.edu.

Application contact: Dr. Clare O'Connor, Graduate Director, Admissions, 617-552-3540, *E-mail:* clare.oconnor@bc.edu.

Find an in-depth description at www.petersons.com/graduate.

■ BOSTON UNIVERSITY

Graduate School of Arts and Sciences, Department of Biology, Boston, MA 02215

AWARDS Botany (MA, PhD); cell and molecular biology (MA, PhD); cell biology (MA, PhD); ecology (PhD); ecology, behavior, and evolution (MA, PhD); ecology/physiology, endocrinology and reproduction (MA); marine biology (MA, PhD); molecular biology, cell biology and biochemistry (MA, PhD); neurobiology, neuroendocrinology and reproduction (MA, PhD); physiology, endocrinology, and neurobiology (MA, PhD); zoology (MA, PhD). Part-time programs available.

Faculty: 41 full-time (8 women).
Students: 131 full-time (74 women), 11 part-time (7 women); includes 10 minority (7 Asian Americans or Pacific Islanders, 3 Hispanic Americans), 33 international. Average age 27. *238 applicants, 39% accepted.* In 1999, 61 master's, 45 doctorates awarded. Terminal master's awarded for partial completion of doctoral program.
Degree requirements: For master's, one foreign language, thesis not required; for doctorate, one foreign language, dissertation, qualifying exam required. *Average time to degree:* Master's–1 year full-time, 3 years part-time; doctorate–5.75 years full-time.
Entrance requirements: For master's and doctorate, GRE General Test, GRE Subject Test, TOEFL. *Application deadline:* For fall admission, 1/1 (priority date); for spring admission, 11/1. *Application fee:* $50.
Expenses: Tuition: Full-time $23,770; part-time $743 per credit. Required fees: $220. Tuition and fees vary according to class time, course level, campus/location and program.
Financial aid: In 1999–00, 82 students received aid, including 1 fellowship with full tuition reimbursement available (averaging $12,000 per year), 28 research assistantships with full tuition reimbursements available (averaging $11,500 per year), 43 teaching assistantships with full tuition reimbursements available (averaging $11,500 per year); Federal Work-Study, grants, institutionally sponsored loans, scholarships, and traineeships also available. Financial aid application deadline: 1/15; financial aid applicants required to submit FAFSA.
Faculty research: Marine science, endocrinology, behavior. *Total annual research expenditures:* $5 million.

Geoffrey M. Cooper, Chairman, 617-353-2432, *Fax:* 617-353-6340, *E-mail:* gmcooper@bu.edu.
Application contact: Yolanta Kovalko, Senior Staff Assistant, 617-353-2432, *Fax:* 617-353-6340, *E-mail:* yolanta@bu.edu.

Find an in-depth description at www.petersons.com/graduate.

■ BOSTON UNIVERSITY

School of Medicine, Division of Graduate Medical Sciences, Boston, MA 02215

AWARDS MA, PhD, MBA/MA, MD/MA, MD/PhD, MPH/MA. Part-time programs available.

Faculty: 80 full-time (20 women), 134 part-time/adjunct (19 women).
Students: 320 full-time (144 women), 23 part-time (8 women); includes 69 minority (6 African Americans, 55 Asian Americans or Pacific Islanders, 7 Hispanic Americans, 1 Native American), 57 international. Average age 27. In 1999, 5 master's, 1 doctorate awarded. Terminal master's awarded for partial completion of doctoral program.
Degree requirements: For master's, qualifying exam required; for doctorate, dissertation, qualifying exam required, foreign language not required.
Entrance requirements: For master's and doctorate, TOEFL. *Application deadline:* For fall admission, 1/15 (priority date); for spring admission, 10/15 (priority date). *Application fee:* $50. Electronic applications accepted.
Expenses: Tuition: Full-time $24,700; part-time $772 per credit. Required fees: $220.
Financial aid: In 1999–00, 38 fellowships with tuition reimbursements, 121 research assistantships with tuition reimbursements, 6 teaching assistantships with tuition reimbursements were awarded; Federal Work-Study, scholarships, and traineeships also available.
Dr. Carl Franzblau, Chairman, 617-638-5120, *Fax:* 617-638-4842.

■ BOWLING GREEN STATE UNIVERSITY

Graduate College, College of Arts and Sciences, Department of Biological Sciences, Bowling Green, OH 43403

AWARDS Applied biology (Specialist); biological sciences (MAT, MS, PhD). Part-time programs available.

Degree requirements: For master's, thesis or alternative required, foreign language not required; for doctorate, dissertation required, foreign language not required; for Specialist, foreign language and thesis not required.

Entrance requirements: For master's and doctorate, GRE General Test, GRE Subject Test, TOEFL. Electronic applications accepted.
Expenses: Tuition, state resident: full-time $6,362. Tuition, nonresident: full-time $11,910. Tuition and fees vary according to course load.
Faculty research: Aquatic ecology, helminth energetics, endocrinology and neurophysiology, nitrogen fixation, photosynthesis.

Find an in-depth description at www.petersons.com/graduate.

■ BRADLEY UNIVERSITY

Graduate School, College of Liberal Arts and Sciences, Department of Biology, Peoria, IL 61625-0002

AWARDS MS. Part-time programs available.

Degree requirements: For master's, thesis, comprehensive exam required.
Entrance requirements: For master's, TOEFL.

■ BRANDEIS UNIVERSITY

Graduate School of Arts and Sciences, Postbaccalaureate Premedical Program, Waltham, MA 02454-9110

AWARDS Certificate.

Students: 6 full-time (2 women), 1 part-time. Average age 29. *35 applicants, 63% accepted.*
Degree requirements: For Certificate, thesis not required.
Entrance requirements: For degree, GRE, SAT. *Application deadline:* For fall admission, 5/1 (priority date). Applications are processed on a rolling basis. *Application fee:* $60. Electronic applications accepted.
Expenses: Tuition: Full-time $25,392; part-time $3,174 per course. Required fees: $509. Tuition and fees vary according to class time, degree level, program and student level.
Financial aid: Application deadline: 4/15. Tish Schilling, Health Professions Advisor, 781-736-3458, *Fax:* 781-736-3469, *E-mail:* tschilling@brandeis.edu.

■ BRANDEIS UNIVERSITY

Graduate School of Arts and Sciences, Programs in Biological Sciences, Waltham, MA 02454-9110

AWARDS Biochemistry (MS, PhD); biophysics and structural biology (PhD); molecular and cell biology (MS, PhD), including cell biology (PhD), developmental biology (PhD), genetics (PhD), microbiology (PhD), molecular and cell biology (MS), molecular biology (PhD), neurobiology (PhD); neuroscience (MS, PhD). Part-time programs available.

Brandeis University (continued)

Faculty: 56 full-time (15 women).

Students: 156 full-time (67 women). *473 applicants, 16% accepted.* In 1999, 21 master's, 20 doctorates awarded.

Degree requirements: For doctorate, dissertation required.

Entrance requirements: For doctorate, GRE General Test. *Application deadline:* Applications are processed on a rolling basis. *Application fee:* $60. Electronic applications accepted.

Expenses: Tuition: Full-time $25,392; part-time $3,174 per course. Required fees: $509. Tuition and fees vary according to class time, degree level, program and student level.

Financial aid: Fellowships, research assistantships, teaching assistantships, career-related internships or fieldwork, scholarships, and tuition waivers (full and partial) available. Aid available to part-time students. Financial aid application deadline: 4/15; financial aid applicants required to submit CSS PROFILE or FAFSA.

Application contact: Margaret Haley, Assistant Dean, Graduate Admissions, 781-736-3410, *Fax:* 781-736-3412, *E-mail:* haley@brandeis.edu.

Find an in-depth description at www.petersons.com/graduate.

■ BRIGHAM YOUNG UNIVERSITY

Graduate Studies, College of Biological and Agricultural Sciences, Provo, UT 84602-1001

AWARDS MS, PhD, MS/PhD. Part-time programs available.

Faculty: 92 full-time (9 women), 7 part-time/adjunct (3 women).

Students: 154 full-time (57 women), 10 part-time (4 women); includes 9 minority (5 Asian Americans or Pacific Islanders, 4 Hispanic Americans), 10 international. Average age 27. *100 applicants, 62% accepted.* In 1999, 33 master's, 8 doctorates awarded. Terminal master's awarded for partial completion of doctoral program.

Degree requirements: For doctorate, dissertation required.

Entrance requirements: For master's and doctorate, GRE General Test. *Application deadline:* Applications are processed on a rolling basis. *Application fee:* $30. Electronic applications accepted.

Expenses: Tuition: Full-time $3,330; part-time $185 per credit hour. Tuition and fees vary according to program and student's religious affiliation.

Financial aid: In 1999–00, 104 students received aid; fellowships, research assistantships, teaching assistantships, career-related internships or fieldwork,

institutionally sponsored loans, scholarships, tuition waivers (partial), and tuition awards available. Aid available to part-time students.

Dr. R. Kent Crookston, Dean, 801-378-2007, *Fax:* 801-378-7499.

■ BROOKLYN COLLEGE OF THE CITY UNIVERSITY OF NEW YORK

Division of Graduate Studies, Department of Biology, Brooklyn, NY 11210-2889

AWARDS Applied biology (MA); biology (MA, PhD).

Degree requirements: For master's, one foreign language (computer language can substitute), thesis, comprehensive exam required.

Entrance requirements: For master's, GRE General Test, GRE Subject Test, TOEFL, minimum GPA of 3.0.

Expenses: Tuition, state resident: full-time $4,350; part-time $185 per credit. Tuition, nonresident: full-time $7,600; part-time $320 per credit.

Faculty research: Evolutionary biology, molecular biology of development, cell biology, comparative endocrinology, ecology.

■ BROWN UNIVERSITY

Graduate School, Division of Biology and Medicine, Providence, RI 02912

AWARDS M Med Sc, MA, MS, Sc M, PhD, MD/PhD. Part-time programs available.

Faculty: 165 full-time (56 women), 2 part-time/adjunct (1 woman).

Students: 136 full-time (72 women), 5 part-time (3 women); includes 13 minority (4 African Americans, 8 Asian Americans or Pacific Islanders, 1 Hispanic American), 32 international. Average age 24. *334 applicants, 9% accepted.* In 1999, 12 master's, 14 doctorates awarded. Terminal master's awarded for partial completion of doctoral program.

Degree requirements: For master's, foreign language not required; for doctorate, dissertation required.

Entrance requirements: For master's and doctorate, GRE General Test. *Application deadline:* For fall admission, 1/2 (priority date). Applications are processed on a rolling basis. *Application fee:* $60. Electronic applications accepted.

Financial aid: Fellowships with full tuition reimbursements, research assistantships with full tuition reimbursements, teaching assistantships with full tuition reimbursements, institutionally sponsored loans, traineeships, and tuition waivers (full) available. Financial aid application deadline: 1/2.

Dr. Donald Marsh, Dean, 401-863-3330.

Application contact: Dr. Peter Shank, Associate Dean, 401-863-3281, *Fax:* 401-863-7411, *E-mail:* peter_shank@brown.edu.

■ BRYN MAWR COLLEGE

Graduate School of Arts and Sciences, Department of Biology, Bryn Mawr, PA 19010-2899

AWARDS Biochemistry (MA, PhD); biology (MA, PhD); neural and behavioral science (PhD).

Students: 2 full-time (both women), 1 (woman) part-time, 2 international. *18 applicants, 11% accepted.* In 1999, 1 doctorate awarded.

Degree requirements: For master's, thesis required; for doctorate, dissertation required.

Entrance requirements: For master's and doctorate, GRE General Test, GRE Subject Test. *Application deadline:* For fall admission, 6/30; for spring admission, 12/7. Applications are processed on a rolling basis. *Application fee:* $40.

Expenses: Tuition: Full-time $20,790; part-time $3,530 per course.

Financial aid: Fellowships, research assistantships, teaching assistantships, Federal Work-Study and institutionally sponsored loans available. Aid available to part-time students. Financial aid application deadline: 1/2.

Margaret Hollyday, Chair, 610-526-5097.

Application contact: Graduate School of Arts and Sciences, 610-526-5072.

■ BUCKNELL UNIVERSITY

Graduate Studies, College of Arts and Sciences, Department of Biology, Lewisburg, PA 17837

AWARDS MA, MS.

Faculty: 19 full-time (4 women).

Students: 5 full-time (2 women), 3 part-time (all women).

Degree requirements: For master's, thesis required.

Entrance requirements: For master's, GRE General Test, GRE Subject Test, TOEFL, minimum GPA of 2.8. *Application deadline:* For fall admission, 6/1 (priority date); for spring admission, 12/1 (priority date). Applications are processed on a rolling basis. *Application fee:* $25.

Expenses: Tuition: Part-time $2,600 per course. Tuition and fees vary according to course load.

Financial aid: Unspecified assistantships available. Financial aid application deadline: 3/1.

Dr. Wayne McDiffett, Head, 570-577-1124.

■ CALIFORNIA INSTITUTE OF TECHNOLOGY

Division of Biology, Pasadena, CA 91125-0001

AWARDS Biochemistry (PhD); cell biology and biophysics (PhD); developmental biology (PhD); genetics (PhD); immunology (PhD); molecular biology (PhD); neurobiology (PhD).

Faculty: 40 full-time (7 women).
Students: 79 full-time (38 women); includes 8 minority (1 African American, 6 Hispanic Americans, 1 Native American), 21 international. *436 applicants, 12% accepted.* In 1999, 16 degrees awarded (100% entered university research/teaching).
Degree requirements: For doctorate, dissertation, qualifying exam required, foreign language not required.
Entrance requirements: For doctorate, GRE General Test. *Application deadline:* For fall admission, 1/1. *Application fee:* $0.
Expenses: Tuition: Full-time $19,260. Required fees: $24. One-time fee: $100 full-time.
Financial aid: In 1999–00, fellowships with full tuition reimbursements (averaging $14,688 per year), research assistantships with full tuition reimbursements (averaging $14,688 per year), teaching assistantships with full tuition reimbursements (averaging $3,729 per year) were awarded; institutionally sponsored loans also available. Financial aid application deadline: 1/1.
Faculty research: Molecular genetics of differentiation and development, structure of biological macromolecules, molecular and integrative neurobiology.
Dr. Melvin Simon, Chairman, 626-395-4951, *Fax:* 626-683-3343.
Application contact: Elizabeth Ayala, Graduate Option Coordinator, 626-395-4497, *Fax:* 626-449-0756, *E-mail:* biograd@cco.caltech.edu.

Find an in-depth description at www.petersons.com/graduate.

■ CALIFORNIA POLYTECHNIC STATE UNIVERSITY, SAN LUIS OBISPO

College of Science and Mathematics, Department of Biological Sciences, San Luis Obispo, CA 93407

AWARDS MS.

Faculty: 32 full-time (4 women), 31 part-time/adjunct (19 women).
Students: 20 full-time (15 women), 16 part-time (7 women). *26 applicants, 58% accepted.* In 1999, 8 degrees awarded (25% entered university research/teaching, 50% found other work related to degree, 25% continued full-time study).

Degree requirements: For master's, thesis optional, foreign language not required. *Average time to degree:* Master's–3.2 years full-time.
Entrance requirements: For master's, GRE General Test, minimum GPA of 3.0 in last 90 quarter units. *Application deadline:* For fall admission, 7/1; for winter admission, 11/1; for spring admission, 3/1. *Application fee:* $55.
Expenses: Tuition, nonresident: part-time $164 per unit. Required fees: $526 per quarter.
Financial aid: In 1999–00, 25 teaching assistantships (averaging $5,400 per year) were awarded; career-related internships or fieldwork and Federal Work-Study also available. Aid available to part-time students. Financial aid application deadline: 3/2; financial aid applicants required to submit FAFSA.
Faculty research: Ancient fossil DNA, restoration ecology microbe biodiversity indices, biological inventories.
Dr. V. L. Holland, Chair, 805-756-2788, *E-mail:* vholland@calpoly.edu.
Application contact: Dennis F. Frey, Graduate Coordinator, 805-756-2802, *Fax:* 805-756-1419, *E-mail:* dfrey@calpoly.edu.

■ CALIFORNIA STATE POLYTECHNIC UNIVERSITY, POMONA

Academic Affairs, College of Science, Program in Biological Sciences, Pomona, CA 91768-2557

AWARDS MS. Part-time programs available.

Students: 32 full-time (19 women), 36 part-time (20 women). Average age 30. *37 applicants, 46% accepted.* In 1999, 12 degrees awarded.
Degree requirements: For master's, thesis required.
Entrance requirements: For master's, GRE General Test. *Application deadline:* Applications are processed on a rolling basis. *Application fee:* $55.
Expenses: Tuition, nonresident: part-time $164 per unit. Required fees: $306 per quarter.
Financial aid: Career-related internships or fieldwork, Federal Work-Study, and institutionally sponsored loans available. Aid available to part-time students. Financial aid application deadline: 3/2; financial aid applicants required to submit FAFSA.
Dr. David J. Moriarty, Coordinator, 909-869-4055, *E-mail:* djmoriarty@csupomona.edu.

■ CALIFORNIA STATE UNIVERSITY, CHICO

Graduate School, College of Natural Sciences, Department of Biological Sciences, Chico, CA 95929-0722

AWARDS Biological sciences (MS); botany (MS); nutritional science (MS), including nutrition education.

Degree requirements: For master's, thesis required, foreign language not required.
Entrance requirements: For master's, GRE General Test.
Expenses: Tuition, nonresident: part-time $246 per credit. Required fees: $2,108; $1,442 per year.

■ CALIFORNIA STATE UNIVERSITY, DOMINGUEZ HILLS

College of Arts and Sciences, Department of Biology, Carson, CA 90747-0001

AWARDS Biology (MA); human cytogenic technology (Certificate). Part-time and evening/weekend programs available.

Faculty: 9 full-time, 9 part-time/adjunct.
Students: 11 full-time (9 women), 11 part-time (8 women); includes 16 minority (5 African Americans, 3 Asian Americans or Pacific Islanders, 8 Hispanic Americans), 1 international. Average age 30. *9 applicants, 67% accepted.* In 1999, 8 master's awarded.
Degree requirements: For master's, thesis required (for some programs).
Entrance requirements: For master's, GRE General Test, GRE Subject Test, minimum GPA of 2.5. *Application deadline:* For fall admission, 6/1. *Application fee:* $55.
Expenses: Tuition, nonresident: part-time $246 per unit. Required fees: $1,904; $1,230 per year.
Dr. John Roberts, Chair, 310-243-3381.

■ CALIFORNIA STATE UNIVERSITY, FRESNO

Division of Graduate Studies, College of Science and Mathematics, Department of Biology, Fresno, CA 93740

AWARDS MA. Part-time and evening/weekend programs available.

Faculty: 21 full-time (6 women).
Students: 10 full-time (6 women), 23 part-time (12 women); includes 8 minority (1 African American, 5 Asian Americans or Pacific Islanders, 2 Hispanic Americans), 2 international. Average age 31. *18 applicants, 89% accepted.* In 1999, 4 degrees awarded.
Degree requirements: For master's, thesis required, foreign language not required. *Average time to degree:* Master's–3.5 years full-time.

California State University, Fresno (continued)

Entrance requirements: For master's, GRE General Test, GRE Subject Test, TOEFL. *Application deadline:* For fall admission, 8/1 (priority date); for spring admission, 12/1. Applications are processed on a rolling basis. *Application fee:* $55. Electronic applications accepted.
Expenses: Tuition, nonresident: part-time $246 per unit. Required fees: $1,906; $620 per semester.
Financial aid: In 1999–00, 2 fellowships, 24 teaching assistantships were awarded; career-related internships or fieldwork, Federal Work-Study, scholarships, and unspecified assistantships also available. Financial aid application deadline: 3/1; financial aid applicants required to submit FAFSA.
Faculty research: Genome neuroscience, ecology conflict resolution, biomechanics, cell death.
Dr. Fred Schreiber, Chair, 559-278-5466, *Fax:* 559-278-3936, *E-mail:* fred_schreiber@csufresno.edu.
Application contact: Dr. Brian Tsukimura, Graduate Coordinator, 559-278-4074, *Fax:* 559-278-3936, *E-mail:* brian_tsukimua@csufresno.edu.

■ **CALIFORNIA STATE UNIVERSITY, FULLERTON**

Graduate Studies, College of Natural Science and Mathematics, Department of Biological Science, Fullerton, CA 92834-9480

AWARDS Biological science (MA); botany (MA); microbiology (MA). Part-time programs available.

Faculty: 23 full-time (7 women), 41 part-time/adjunct.
Students: 4 full-time (all women), 60 part-time (35 women); includes 20 minority (1 African American, 8 Asian Americans or Pacific Islanders, 10 Hispanic Americans, 1 Native American), 3 international. Average age 29. *51 applicants, 45% accepted.* In 1999, 13 degrees awarded.
Degree requirements: For master's, thesis required, foreign language not required.
Entrance requirements: For master's, DAT, GRE General Test and GRE Subject Test, or MCAT, minimum GPA of 3.0 in biology. *Application fee:* $55.
Expenses: Tuition, nonresident: part-time $264 per unit. Required fees: $1,887; $629 per year.
Financial aid: Teaching assistantships, career-related internships or fieldwork, Federal Work-Study, grants, and institutionally sponsored loans available. Aid available to part-time students. Financial aid application deadline: 3/1.

Faculty research: Glycosidase release and the block to polyspermy in ascidian eggs. Dr. Eugene Jones, Chair, 714-278-3614.
Application contact: Dr. Michael Horn, Adviser, 714-278-3707.

■ **CALIFORNIA STATE UNIVERSITY, HAYWARD**

Graduate Programs, School of Science, Department of Biological Sciences, Hayward, CA 94542-3000

AWARDS Biological sciences (MS); marine sciences (MS), including marine sciences. Part-time programs available.

Students: 25 full-time (17 women), 21 part-time (14 women); includes 23 minority (3 African Americans, 13 Asian Americans or Pacific Islanders, 7 Hispanic Americans). *22 applicants, 68% accepted.* In 1999, 6 degrees awarded.
Degree requirements: For master's, thesis required, foreign language not required.
Entrance requirements: For master's, GRE Subject Test, minimum GPA of 3.0 in field, 2.75 overall. *Application deadline:* For fall admission, 6/15; for winter admission, 10/29; for spring admission, 1/7. *Application fee:* $55.
Expenses: Tuition, nonresident: part-time $164 per unit. Required fees: $587 per quarter.
Financial aid: Career-related internships or fieldwork, Federal Work-Study, and institutionally sponsored loans available. Aid available to part-time students. Financial aid application deadline: 3/1.
Dr. Stephen Benson, Chair, 510-885-3471.
Application contact: Jennifer Rice, Graduate Program Coordinator, 510-885-3286, *Fax:* 510-885-4795, *E-mail:* gradprograms@csuhayward.edu.

■ **CALIFORNIA STATE UNIVERSITY, LONG BEACH**

Graduate Studies, College of Natural Sciences, Department of Biological Sciences, Long Beach, CA 90840

AWARDS Biological sciences (MS); microbiology (MPH, MS), including medical technology (MPH), microbiology (MS), nurse epidemiology (MPH). Part-time programs available.

Faculty: 18 full-time (2 women).
Students: 17 full-time (12 women), 52 part-time (31 women); includes 36 minority (1 African American, 5 Asian Americans or Pacific Islanders, 30 Hispanic Americans), 5 international. Average age 31. *57 applicants, 40% accepted.* In 1999, 20 degrees awarded.
Degree requirements: For master's, foreign language not required.

Entrance requirements: For master's, GRE Subject Test, minimum GPA of 3.0. *Application deadline:* For fall admission, 8/1; for spring admission, 12/1. Applications are processed on a rolling basis. *Application fee:* $55. Electronic applications accepted.
Expenses: Tuition, nonresident: part-time $246 per credit. Required fees: $569 per semester. Tuition and fees vary according to course load.
Financial aid: Teaching assistantships, Federal Work-Study, grants, institutionally sponsored loans, traineeships, and unspecified assistantships available. Financial aid application deadline: 3/2.
Dr. Laura Kingsford, Acting Chair, 562-985-4807, *Fax:* 562-985-8878, *E-mail:* lkingsfo@csulb.edu.
Application contact: Dr. Terry Shuster, Coordinator, 562-985-4820, *Fax:* 562-985-8878, *E-mail:* tshuster@csulb.edu.

■ **CALIFORNIA STATE UNIVERSITY, LOS ANGELES**

Graduate Studies, School of Natural and Social Sciences, Department of Biology and Microbiology, Los Angeles, CA 90032-8530

AWARDS Biology (MS). Part-time and evening/weekend programs available.

Faculty: 19 full-time, 24 part-time/adjunct.
Students: 22 full-time (10 women), 54 part-time (25 women); includes 36 minority (4 African Americans, 14 Asian Americans or Pacific Islanders, 18 Hispanic Americans), 12 international. In 1999, 6 degrees awarded.
Degree requirements: For master's, comprehensive exam or thesis required.
Entrance requirements: For master's, TOEFL. *Application deadline:* For fall admission, 6/30; for spring admission, 2/1. Applications are processed on a rolling basis. *Application fee:* $55.
Expenses: Tuition, nonresident: full-time $7,703; part-time $164 per unit. Required fees: $1,799; $387 per quarter.
Financial aid: In 1999–00, 37 students received aid. Federal Work-Study available. Aid available to part-time students. Financial aid application deadline: 3/1.
Faculty research: Ecology, environmental biology, cell and molecular biology, physiology, medical microbiology.
Dr. Alan Muchlinski, Chair, 323-343-2050.

■ **CALIFORNIA STATE UNIVERSITY, NORTHRIDGE**

Graduate Studies, College of Science and Mathematics, Department of Biology, Northridge, CA 91330

AWARDS Biology (MS); genetic counseling (MS).

Faculty: 65 full-time, 19 part-time/adjunct.
Students: 32 full-time (26 women), 34 part-time (20 women); includes 19 minority (1 African American, 7 Asian Americans or Pacific Islanders, 11 Hispanic Americans), 7 international. Average age 31. 72 *applicants, 74% accepted.* In 1999, 17 degrees awarded.
Degree requirements: For master's, comprehensive exam or thesis required.
Entrance requirements: For master's, GRE Subject Test, TOEFL. *Application fee:* $55.
Expenses: Tuition, nonresident: part-time $246 per unit. International tuition: $7,874 full-time. Required fees: $1,970. Tuition and fees vary according to course load.
Financial aid: Research assistantships, teaching assistantships, Federal Work-Study, institutionally sponsored loans, tuition waivers (partial), and unspecified assistantships available. Aid available to part-time students. Financial aid applicants required to submit FAFSA.
Faculty research: Cell adhesion, cancer research, fishery research.
Dr. Jim Dole, Chair, 818-677-3356.
Application contact: Dr. Andrew Starrett, Graduate Coordinator, 818-677-3353.

■ CALIFORNIA STATE UNIVERSITY, SACRAMENTO

Graduate Studies, School of Natural Sciences and Mathematics, Department of Biological Sciences, Sacramento, CA 95819-6048

AWARDS Biological sciences (MA, MS); immunohematology (MS); marine science (MS). Part-time programs available.

Students: 38 full-time, 39 part-time.
Degree requirements: For master's, thesis, writing proficiency exam required, foreign language not required.
Entrance requirements: For master's, TOEFL, bachelor's degree in biology or equivalent; minimum GPA of 3.0 in biology, 2.75 overall during previous 2 years. *Application deadline:* For fall admission, 4/15; for spring admission, 11/1. *Application fee:* $55.
Expenses: Tuition, nonresident: full-time $5,904; part-time $246 per unit. Required fees: $1,945; $1,315 per year.
Financial aid: Research assistantships, teaching assistantships, career-related internships or fieldwork and Federal Work-Study available. Aid available to part-time students. Financial aid application deadline: 3/1.
Dr. Laurel Heffernan, Chair, 916-278-6535.
Application contact: Dr. Michael Baad, Coordinator, 916-278-6494.

■ CALIFORNIA STATE UNIVERSITY, SAN BERNARDINO

Graduate Studies, School of Natural Sciences, Department of Biology, San Bernardino, CA 92407-2397

AWARDS MS. Part-time programs available.

Degree requirements: For master's, thesis or alternative required, foreign language not required.
Entrance requirements: For master's, minimum GPA of 3.0.
Faculty research: Ecology, molecular biology, physiology, cell biology, neurobiology.

■ CALIFORNIA UNIVERSITY OF PENNSYLVANIA

School of Graduate Studies, School of Science and Technology, Department of Biological and Environmental Sciences, California, PA 15419-1394

AWARDS Biology (M Ed, MS). Part-time and evening/weekend programs available.

Faculty: 5 part-time/adjunct (0 women).
Students: 20 full-time (9 women), 8 part-time (2 women), 1 international. *22 applicants, 55% accepted.* In 1999, 12 degrees awarded.
Degree requirements: For master's, thesis, comprehensive exam required, foreign language not required.
Entrance requirements: For master's, GRE General Test, TOEFL, minimum GPA of 2.5, teaching certificate. *Application deadline:* Applications are processed on a rolling basis. *Application fee:* $25.
Expenses: Tuition, state resident: full-time $3,780; part-time $210 per credit. Tuition, nonresident: full-time $6,610; part-time $367 per credit. Required fees: $1,012. Full-time tuition and fees vary according to campus/location and program. Part-time tuition and fees vary according to course load and campus/location.
Financial aid: Tuition waivers (full) and unspecified assistantships available.
Dr. Barry Hunter, Coordinator, 724-938-4200, *E-mail:* hunter@cup.edu.

■ CARNEGIE MELLON UNIVERSITY

Mellon College of Science, Department of Biological Sciences, Pittsburgh, PA 15213-3891

AWARDS Biochemistry (PhD); biophysics (PhD); cell biology (PhD); developmental biology (PhD); genetics (PhD); molecular biology (PhD).

Faculty: 38 full-time (17 women), 1 (woman) part-time/adjunct.
Students: 35 full-time (24 women); includes 2 minority (1 African American, 1 Asian American or Pacific Islander), 18

international. Average age 27. In 1999, 4 degrees awarded.
Degree requirements: For doctorate, dissertation required, foreign language not required.
Entrance requirements: For doctorate, GRE General Test, GRE Subject Test, interview. *Application deadline:* For fall admission, 2/1 (priority date). Applications are processed on a rolling basis. *Application fee:* $0.
Expenses: Tuition: Full-time $22,100; part-time $307 per unit. Required fees: $200. Tuition and fees vary according to program.
Financial aid: Fellowships, research assistantships, teaching assistantships, traineeships available.
Faculty research: Genetic structure, function, and regulation; protein structure and function; biological membranes; biological spectroscopy. *Total annual research expenditures:* $4.4 million.
Dr. William E. Brown, Head, 412-268-3416, *Fax:* 412-268-7129, *E-mail:* wb02@andrew.cmu.edu.
Application contact: Stacey L. Young, Assistant Head, 412-268-7372, *Fax:* 412-268-7129, *E-mail:* sf38+@andrew.cmu.edu.
Find an in-depth description at www.petersons.com/graduate.

■ CASE WESTERN RESERVE UNIVERSITY

School of Graduate Studies, Department of Biology, Cleveland, OH 44106

AWARDS MS, PhD. Part-time programs available. Terminal master's awarded for partial completion of doctoral program.

Degree requirements: For master's, thesis or alternative required; for doctorate, dissertation required.
Entrance requirements: For master's and doctorate, GRE General Test and GRE Subject Test or MCAT, TOEFL.
Faculty research: Cellular, developmental, and molecular biology; genetics; genetic engineering; biotechnology; ecology.
Find an in-depth description at www.petersons.com/graduate.

■ CASE WESTERN RESERVE UNIVERSITY

School of Medicine, Biomedical Sciences Training Program, Cleveland, OH 44106

AWARDS PhD.

Faculty: 198 full-time (32 women).
Students: 33 full-time (19 women); includes 2 minority (1 Asian American or Pacific Islander, 1 Hispanic American), 8 international. Average age 24. *500 applicants, 20% accepted.*

Case Western Reserve University (continued)

Degree requirements: For doctorate, dissertation required.

Entrance requirements: For doctorate, GRE General Test, TOEFL. *Application deadline:* For fall admission, 2/1 (priority date). Applications are processed on a rolling basis. *Application fee:* $25.

Financial aid: In 1999–00, 33 students received aid, including 33 fellowships with full tuition reimbursements available (averaging $16,000 per year).

Faculty research: Biochemistry, molecular biology, immunology, genetics, neurosciences. *Total annual research expenditures:* $113 million.

Dr. Martin D. Snider, Chairman, 216-368-5572, *Fax:* 216-368-0795, *E-mail:* mds5@po.cwru.edu.

Application contact: Debbie Noureddine, Coordinator, 216-368-3347, *Fax:* 216-368-0795, *E-mail:* drn2@po.cwru.edu.

Find an in-depth description at www.petersons.com/graduate.

■ CASE WESTERN RESERVE UNIVERSITY

School of Medicine and School of Graduate Studies, Graduate Programs in Medicine, Cleveland, OH 44106

AWARDS MA, MPH, MS, PhD, MD/MS, MD/PhD. Part-time programs available.

Students: 338 full-time (173 women), 171 part-time (101 women); includes 45 minority (10 African Americans, 23 Asian Americans or Pacific Islanders, 8 Hispanic Americans, 4 Native Americans), 141 international. Average age 28. *928 applicants, 31% accepted.* In 1999, 71 master's, 47 doctorates awarded.

Degree requirements: For doctorate, dissertation required.

Entrance requirements: For master's and doctorate, GRE General Test, TOEFL. *Application deadline:* Applications are processed on a rolling basis. *Application fee:* $25.

Financial aid: Fellowships, research assistantships, teaching assistantships, career-related internships or fieldwork, Federal Work-Study, institutionally sponsored loans, and tuition waivers (full and partial) available. Aid available to part-time students.

Dr. Joyce E. Jentoft, Dean, Graduate Studies, 216-368-4400, *Fax:* 216-368-4250.

■ THE CATHOLIC UNIVERSITY OF AMERICA

School of Arts and Sciences, Department of Biology, Washington, DC 20064

AWARDS Cell and microbial biology (MS, PhD), including cell biology, microbiology; clinical laboratory science (MS, PhD). Part-time programs available.

Faculty: 9 full-time (2 women), 1 (woman) part-time/adjunct.

Students: 7 full-time (5 women), 17 part-time (7 women); includes 6 minority (3 African Americans, 3 Hispanic Americans), 6 international. *41 applicants, 59% accepted.* In 1999, 4 master's awarded. Terminal master's awarded for partial completion of doctoral program.

Degree requirements: For master's, thesis or alternative, comprehensive exam required, foreign language not required; for doctorate, dissertation, comprehensive exam required, foreign language not required.

Entrance requirements: For master's, GRE General Test, GRE Subject Test, TOEFL; for doctorate, GRE General Test, GRE Subject Test. *Application deadline:* For fall admission, 8/1 (priority date); for spring admission, 12/1. Applications are processed on a rolling basis. *Application fee:* $55. Electronic applications accepted.

Expenses: Tuition: Full-time $18,200; part-time $700 per credit hour. Required fees: $378 per semester. Part-time tuition and fees vary according to campus/location and program.

Financial aid: Fellowships, research assistantships, teaching assistantships, career-related internships or fieldwork, institutionally sponsored loans, and tuition waivers (full and partial) available. Aid available to part-time students. Financial aid application deadline: 2/1.

Faculty research: Cell differentiation, regulation of cell growth, drug resistance, gene cloning and sequencing, developmental biology and neurobiology.

Dr. John Golin, Chair, 202-319-5279, *Fax:* 202-319-5721.

■ CENTRAL CONNECTICUT STATE UNIVERSITY

School of Graduate Studies, School of Arts and Sciences, Department of Biological Sciences, New Britain, CT 06050-4010

AWARDS Biological sciences (MA), including anesthesia, health science, professional education. Part-time and evening/weekend programs available.

Faculty: 16 full-time (5 women), 12 part-time/adjunct (6 women).

Students: 93 full-time (52 women), 25 part-time (16 women); includes 9 minority (1 African American, 5 Asian Americans or Pacific Islanders, 1 Hispanic American, 2 Native Americans). Average age 31. *38 applicants, 68% accepted.* In 1999, 37 degrees awarded.

Degree requirements: For master's, thesis or alternative, comprehensive exam required, foreign language not required.

Entrance requirements: For master's, TOEFL, minimum GPA of 2.7. *Application deadline:* For fall admission, 6/1 (priority date); for spring admission, 12/1. Applications are processed on a rolling basis. *Application fee:* $40.

Expenses: Tuition, state resident: full-time $2,568; part-time $175 per credit. Tuition, nonresident: full-time $7,156. Required fees: $1,672. One-time fee: $45 full-time. Tuition and fees vary according to course level.

Financial aid: In 1999–00, 2 research assistantships (averaging $4,800 per year), teaching assistantships (averaging $4,800 per year) were awarded; Federal Work-Study also available. Financial aid application deadline: 3/15; financial aid applicants required to submit FAFSA.

Faculty research: Environmental science, anesthesia, health sciences, zoology, animal behavior, botany, physiology.

Dr. Leeds M. Carluccio, Chair, 860-832-2645.

■ CENTRAL MICHIGAN UNIVERSITY

College of Graduate Studies, College of Science and Technology, Department of Biology, Mount Pleasant, MI 48859

AWARDS Biology (MS); conservation biology (MS).

Faculty: 39 full-time (6 women).

Students: 20 full-time (10 women), 38 part-time (17 women). Average age 27. In 1999, 18 degrees awarded.

Degree requirements: For master's, thesis or alternative required, foreign language not required.

Entrance requirements: For master's, bachelor's degree in biology, minimum GPA of 3.0. *Application deadline:* Applications are processed on a rolling basis. *Application fee:* $30.

Expenses: Tuition, state resident: part-time $144 per credit hour. Tuition, nonresident: part-time $285 per credit hour. Required fees: $240 per semester. Tuition and fees vary according to degree level and program.

Financial aid: In 1999–00, 2 fellowships with tuition reimbursements, 11 research assistantships with tuition reimbursements, 26 teaching assistantships with tuition

reimbursements were awarded; career-related internships or fieldwork and Federal Work-Study also available. Financial aid application deadline: 3/7.
Faculty research: Vertebrates, morphology and taxonomy of aquatic plants, molecular biology and genetics, microbials and invertebrate ecology.
Dr. John Scheide, Chairperson, 517-774-3227, *Fax:* 517-774-3462, *E-mail:* john.iver.scheide@cmich.edu.

■ CENTRAL MISSOURI STATE UNIVERSITY

School of Graduate Studies, College of Arts and Sciences, Department of Biology, Warrensburg, MO 64093
AWARDS MS. Part-time programs available.
Faculty: 14 full-time (1 woman), 2 part-time/adjunct (1 woman).
Students: 4 full-time (0 women), 4 part-time (3 women); includes 2 minority (1 African American, 1 Hispanic American). Average age 29. In 1999, 1 degree awarded.
Degree requirements: For master's, thesis or alternative, oral exam required.
Entrance requirements: For master's, GRE Subject Test, 30 hours in biology, minimum undergraduate GPA of 2.5. *Application deadline:* Applications are processed on a rolling basis. *Application fee:* $25 ($50 for international students).
Expenses: Tuition, state resident: full-time $3,576; part-time $149 per credit hour. Tuition, nonresident: full-time $7,152; part-time $298 per credit hour. Tuition and fees vary according to course load and campus/location.
Financial aid: Federal Work-Study, grants, scholarships, unspecified assistantships, and laboratory assistantships available. Aid available to part-time students. Financial aid application deadline: 3/1; financial aid applicants required to submit FAFSA.
Faculty research: Molecular time keeping mechanism, plant/animal systematics, insects systematics, plant ecology, systematics of fungi. *Total annual research expenditures:* $15,000.
Dr. Steven Mills, Chair, 660-543-8827, *Fax:* 660-543-8006, *E-mail:* smills@cmsu1.cmsu.edu.

■ CENTRAL WASHINGTON UNIVERSITY

Graduate Studies and Research, College of the Sciences, Department of Biology, Ellensburg, WA 98926
AWARDS MS. Part-time programs available.
Faculty: 17 full-time (4 women).
Students: 7 full-time (4 women), 2 part-time (both women). *6 applicants, 67% accepted.* In 1999, 4 degrees awarded.

Degree requirements: For master's, thesis or alternative required, foreign language not required.
Entrance requirements: For master's, GRE General Test, minimum GPA of 3.0. *Application deadline:* For fall admission, 4/1 (priority date); for winter admission, 10/1; for spring admission, 1/1. Applications are processed on a rolling basis. *Application fee:* $35.
Expenses: Tuition, state resident: full-time $4,389; part-time $146 per credit. Tuition, nonresident: full-time $13,365; part-time $446 per credit. Tuition and fees vary according to course load.
Financial aid: In 1999–00, 7 teaching assistantships with partial tuition reimbursements (averaging $6,664 per year) were awarded; research assistantships, Federal Work-Study also available. Financial aid application deadline: 3/1; financial aid applicants required to submit FAFSA.
Dr. David Hosford, Chair, 509-963-2731.
Application contact: Barbara Sisko, Office Assistant, Graduate Studies and Research, 509-963-3103, *Fax:* 509-963-1799, *E-mail:* masters@cwu.edu.

■ CHICAGO STATE UNIVERSITY

Graduate Studies, College of Arts and Sciences, Department of Biological Sciences, Chicago, IL 60628
AWARDS MS. Part-time and evening/weekend programs available.
Faculty: 10 full-time (1 woman), 1 part-time/adjunct (0 women).
Students: 27 (18 women); includes 23 minority (all African Americans) 3 international. *15 applicants, 67% accepted.* In 1999, 5 degrees awarded (40% entered university research/teaching, 40% found other work related to degree).
Degree requirements: For master's, thesis required, foreign language not required. *Average time to degree:* Master's–3.5 years part-time.
Entrance requirements: For master's, minimum GPA of 2.75, 15 credit hours in biological sciences. *Application deadline:* For fall admission, 7/1; for spring admission, 11/10. Applications are processed on a rolling basis. *Application fee:* $20.
Expenses: Tuition, state resident: full-time $1,212; part-time $101 per credit hour. Tuition, nonresident: full-time $3,636; part-time $303 per credit hour. Required fees: $147 per term. Tuition and fees vary according to campus/location and program.
Financial aid: In 1999–00, 3 fellowships with full tuition reimbursements (averaging $5,300 per year), 3 research assistantships with full tuition reimbursements (averaging $9,000 per year) were awarded.

Faculty research: Molecular genetics of gene complexes, mammalian immune cell function, genetics of agriculturally important microbes, environmental toxicology, neuromuscular physiology. *Total annual research expenditures:* $350,000.
Dr. Marian Wilson-Comer, Chairperson, 773-995-2183, *Fax:* 773-995-3759, *E-mail:* bacomer@csu.edu.
Application contact: Graduate Studies Office, 773-995-2404.

■ CITY COLLEGE OF THE CITY UNIVERSITY OF NEW YORK

Graduate School, College of Liberal Arts and Science, Division of Science, Department of Biology, New York, NY 10031-9198
AWARDS MA, PhD. Part-time programs available.
Students: 7 full-time (4 women), 38 part-time (24 women); includes 19 minority (8 African Americans, 6 Asian Americans or Pacific Islanders, 4 Hispanic Americans, 1 Native American), 14 international. *32 applicants, 81% accepted.* In 1999, 14 degrees awarded. Terminal master's awarded for partial completion of doctoral program.
Degree requirements: For master's, thesis or alternative required, foreign language not required; for doctorate, one foreign language (computer language can substitute), dissertation, teaching experience required.
Entrance requirements: For master's, GRE General Test (recommended), TOEFL; for doctorate, GRE General Test, TOEFL. *Application deadline:* For fall admission, 5/1; for spring admission, 12/1. *Application fee:* $40.
Expenses: Tuition, state resident: full-time $4,350; part-time $185 per credit. Tuition, nonresident: full-time $7,600; part-time $320 per credit. Required fees: $20 per semester.
Financial aid: Fellowships, research assistantships, teaching assistantships, career-related internships or fieldwork and grants available.
Faculty research: Animal behavior, ecology, genetics, neurobiology, molecular biology.
Dr. John Lee, Chair, 212-650-6800.
Application contact: Ralph Zuzulo, Graduate Adviser, 212-650-6800.

■ CLARION UNIVERSITY OF PENNSYLVANIA

College of Graduate Studies, College of Arts and Sciences, Department of Biology, Clarion, PA 16214
AWARDS MS.
Faculty: 14 full-time (3 women).

Clarion University of Pennsylvania (continued)

Students: 10 full-time (7 women), 2 part-time. *17 applicants, 88% accepted.* In 1999, 3 degrees awarded.

Degree requirements: For master's, thesis, comprehensive exam required, foreign language not required.

Entrance requirements: For master's, GRE General Test, minimum QPA of 2.75. *Application deadline:* For fall admission, 8/1 (priority date); for spring admission, 12/1 (priority date). Applications are processed on a rolling basis. *Application fee:* $30.

Expenses: Tuition, state resident: full-time $3,780; part-time $210 per credit. Tuition, nonresident: full-time $6,610; part-time $367 per credit. Required fees: $982; $77 per credit. Part-time tuition and fees vary according to course load.

Financial aid: In 1999–00, research assistantships with full tuition reimbursements (averaging $4,002 per year) Aid available to part-time students. Financial aid application deadline: 5/1.

Dr. Steve Harris, Chairman, 814-393-2273, *Fax:* 814-393-2731.

Application contact: Dr. Charles Williams, Graduate Coordinator, 814-393-1936, *Fax:* 814-393-2731, *E-mail:* cwilliams@clarion.edu.

■ CLARK ATLANTA UNIVERSITY

School of Arts and Sciences, Department of Biology, Atlanta, GA 30314

AWARDS MS, PhD. Part-time programs available. Terminal master's awarded for partial completion of doctoral program.

Degree requirements: For master's, one foreign language (computer language can substitute), thesis required; for doctorate, 2 foreign languages (computer language can substitute for one), dissertation required.

Entrance requirements: For master's, GRE General Test, minimum GPA of 2.5; for doctorate, GRE General Test, minimum graduate GPA of 3.0.

Expenses: Tuition: Full-time $10,250.

Faculty research: Regulation of amino-DNA, cellular regulations.

■ CLARK UNIVERSITY

Graduate School, Department of Biology, Worcester, MA 01610-1477

AWARDS MA, PhD.

Students: 16 full-time (7 women), 8 international. *24 applicants, 17% accepted.* In 1999, 2 master's, 3 doctorates awarded.

Degree requirements: For master's and doctorate, thesis/dissertation required, foreign language not required.

Entrance requirements: For master's, GRE General Test, TOEFL; for doctorate, GRE, TOEFL. *Application deadline:* For fall admission, 3/1 (priority date). Applications are processed on a rolling basis. *Application fee:* $40.

Expenses: Tuition: Full-time $22,400; part-time $2,800 per course.

Financial aid: Fellowships, research assistantships, teaching assistantships, scholarships and tuition waivers (full and partial) available.

Faculty research: Human genetic diseases, lichenology, cytoskeletal proteins, taste physiology.

Dr. Tom Leonard, Chairman, 508-793-7173.

Application contact: Rene Baril, Department Secretary, 528-793-7173.

Find an in-depth description at www.petersons.com/graduate.

■ CLEMSON UNIVERSITY

Graduate School, College of Agriculture, Forestry and Life Sciences, School of Animal, Biomedical and Biological Sciences, Department of Biological Sciences, Clemson, SC 29634

AWARDS Biochemistry (MS, PhD); botany (MS); genetics (MS, PhD); zoology (MS, PhD).

Students: 64 full-time (40 women), 21 part-time (11 women); includes 4 minority (1 African American, 2 Asian Americans or Pacific Islanders, 1 Hispanic American), 24 international. *75 applicants, 24% accepted.* In 1999, 8 master's, 7 doctorates awarded.

Degree requirements: For doctorate, dissertation required.

Entrance requirements: For master's and doctorate, GRE General Test, TOEFL. *Application deadline:* Applications are processed on a rolling basis. *Application fee:* $40.

Expenses: Tuition, state resident: full-time $3,480; part-time $174 per credit hour. Tuition, nonresident: full-time $9,256; part-time $388 per credit hour. Required fees: $5 per term. Full-time tuition and fees vary according to course level, course load and campus/location.

Financial aid: Fellowships, research assistantships, teaching assistantships available. Financial aid application deadline: 3/15; financial aid applicants required to submit FAFSA. *Total annual research expenditures:* $1.3 million.

Dr. James K. Zimmerman, Chair, 864-656-3600, *Fax:* 864-656-0435, *E-mail:* jkzmm@clemson.edu.

Application contact: Dr. Thomas McInnis, Coordinator of Graduate Studies, 864-656-3587, *Fax:* 864-656-0435, *E-mail:* botany@clemson.edu.

■ CLEVELAND STATE UNIVERSITY

College of Graduate Studies, College of Arts and Sciences, Department of Biological, Geological, and Environmental Sciences, Cleveland, OH 44115

AWARDS MS, PhD. Part-time programs available.

Faculty: 18 full-time (3 women), 20 part-time/adjunct (4 women).

Students: 53 full-time (29 women), 13 part-time (4 women). Average age 32. *30 applicants, 30% accepted.* In 1999, 5 master's, 5 doctorates awarded. Terminal master's awarded for partial completion of doctoral program.

Degree requirements: For master's, thesis required (for some programs), foreign language not required; for doctorate, dissertation required, foreign language not required.

Entrance requirements: For master's and doctorate, GRE General Test, GRE Subject Test (Biology or Biochemistry). *Application deadline:* For fall admission, 9/1 (priority date). Applications are processed on a rolling basis. *Application fee:* $25.

Expenses: Tuition, state resident: part-time $215 per credit hour. Tuition, nonresident: part-time $425 per credit hour. Tuition and fees vary according to program.

Financial aid: In 1999–00, 12 research assistantships with tuition reimbursements (averaging $12,000 per year), 15 teaching assistantships with tuition reimbursements (averaging $12,000 per year) were awarded; institutionally sponsored loans and unspecified assistantships also available.

Faculty research: Physiology, biochemistry/neurochemistry, immunology, taxonomic botany, molecular parasitology, molecular biology, environmental science. *Total annual research expenditures:* $150,000.

Dr. Michael Gates, Chairperson, 216-687-3917, *Fax:* 216-687-6972, *E-mail:* m.gates@csuohio.edu.

Application contact: Dr. Jeffrey Dean, Graduate Program Director, 216-687-2440, *E-mail:* gpd.bges@csuchio.edu.

■ COLD SPRING HARBOR LABORATORY, WATSON SCHOOL OF BIOLOGICAL SCIENCES

Graduate Program, Cold Spring Harbor, NY 11724

AWARDS Biological sciences (PhD).

Faculty: 45 full-time (6 women).

Students: 6 full-time (3 women), 3 international. Average age 23. *130 applicants, 5% accepted.*
Degree requirements: For doctorate, dissertation, lab rotations, teaching experience, qualifying exam, research and postdoctoral proposals required.
Entrance requirements: For doctorate, GRE General Test, GRE Subject Test, TOEFL. *Application deadline:* For winter admission, 1/1. *Application fee:* $50.
Expenses: All accepted students are fully funded.
Faculty research: Genetics, molecular, cellular and structural biology, neurobiology, cancer, plant biology.
Dr. Winship Herr, Dean, 516-367-8401, *Fax:* 516-367-6919, *E-mail:* herr@cshl.org.
Application contact: Janet Duffy, Admissions and Academic Records Administrator, 516-367-6890, *Fax:* 516-367-6919, *E-mail:* duffy@cshl.org.
Find an in-depth description at www.petersons.com/graduate.

■ THE COLLEGE OF WILLIAM AND MARY

Faculty of Arts and Sciences, Department of Biology, Williamsburg, VA 23187-8795
AWARDS MA.

Faculty: 23 full-time (9 women), 1 (woman) part-time/adjunct.
Students: 20 full-time (13 women), 3 part-time (1 woman), Average age 25. *28 applicants, 64% accepted.* In 1999, 11 degrees awarded.
Degree requirements: For master's, thesis (for some programs), comprehensive exam required, foreign language not required.
Entrance requirements: For master's, GRE Subject Test, GRE General Test, minimum GPA of 3.0. *Application deadline:* For fall admission, 3/1 (priority date). Applications are processed on a rolling basis. *Application fee:* $30.
Expenses: Tuition, state resident: full-time $2,974; part-time $165 per hour. Tuition, nonresident: full-time $13,820; part-time $510 per hour. Required fees: $2,308. Tuition and fees vary according to program.
Financial aid: Teaching assistantships with full tuition reimbursements, Federal Work-Study, institutionally sponsored loans, and unspecified assistantships available. Financial aid application deadline: 3/1; financial aid applicants required to submit FAFSA.
Faculty research: Cellular and molecular biology, genetics, ecology, organismic biology, physiology. *Total annual research expenditures:* $685,715.
Dr. L. L. Wiseman, Chair, 757-221-5433.

Application contact: Dr. Stanton F. Hoegerman, Graduate Director, 757-221-5433, *E-mail:* sfhoeg@facstaff.wm.edu.

■ COLORADO STATE UNIVERSITY

Graduate School, College of Natural Sciences, Department of Biology, Fort Collins, CO 80523-0015
AWARDS Botany (MS, PhD); zoology (MS, PhD). Part-time programs available.

Faculty: 25 full-time (7 women).
Students: 27 full-time (13 women), 14 part-time (7 women). Average age 29. *86 applicants, 19% accepted.* In 1999, 8 master's, 6 doctorates awarded.
Degree requirements: For doctorate, dissertation required.
Entrance requirements: For master's and doctorate, GRE General Test, GRE Subject Test, TOEFL, minimum GPA of 3.0. *Application deadline:* For fall admission, 2/1 (priority date); for spring admission, 11/1. Applications are processed on a rolling basis. *Application fee:* $30. Electronic applications accepted.
Expenses: Tuition, state resident: full-time $2,694; part-time $150 per credit. Tuition, nonresident: full-time $10,460; part-time $581 per credit. Required fees: $32 per semester. Tuition and fees vary according to program.
Financial aid: In 1999–00, 5 research assistantships, 23 teaching assistantships were awarded; fellowships, career-related internships or fieldwork, Federal Work-Study, institutionally sponsored loans, and traineeships also available.
Faculty research: Aquatic and terrestrial ecology, cytology/cell biology, genetics, plant/animal physiology, animal behavior, developmental biology. *Total annual research expenditures:* $3 million.
Joan Herbers, Chair, 970-491-7011.
Application contact: Tina Sund, Graduate Coordinator, 970-491-1923, *Fax:* 970-491-0649, *E-mail:* tsund@lamar.colostate.edu.

■ COLUMBIA UNIVERSITY

College of Physicians and Surgeons and Graduate School of Arts and Sciences, Graduate School of Arts and Sciences at the College of Physicians and Surgeons, New York, NY 10032
AWARDS M Phil, MA, PhD, MD/PhD. Only candidates for the PhD are admitted. Terminal master's awarded for partial completion of doctoral program.

Degree requirements: For master's, foreign language and thesis not required; for doctorate, dissertation required, foreign language not required.

Entrance requirements: For master's and doctorate, GRE General Test, TOEFL.
Expenses: Tuition: Full-time $25,072.
Faculty research: Molecular and cellular biology, neurobiology, biochemistry, genetics and developmental biology, immunology.
Find an in-depth description at www.petersons.com/graduate.

■ COLUMBIA UNIVERSITY

Graduate School of Arts and Sciences, Division of Natural Sciences, Department of Biological Sciences, New York, NY 10027
AWARDS M Phil, MA, PhD, MD/PhD.

Degree requirements: For master's, teaching experience, written exam required, foreign language and thesis not required; for doctorate, dissertation required, foreign language not required.
Entrance requirements: For master's and doctorate, GRE General Test, GRE Subject Test, TOEFL.
Expenses: Tuition: Full-time $25,072. Full-time tuition and fees vary according to course load and program.
Find an in-depth description at www.petersons.com/graduate.

■ CREIGHTON UNIVERSITY

School of Medicine and Graduate School, Graduate Programs in Medicine, Department of Biomedical Sciences, Omaha, NE 68178-0001
AWARDS MS, PhD, MD/PhD.

Faculty: 26 full-time (1 woman), 26 part-time/adjunct (5 women).
Students: 52 full-time. Average age 26. *52 applicants, 48% accepted.* In 1999, 7 master's, 4 doctorates awarded. Terminal master's awarded for partial completion of doctoral program.
Degree requirements: For master's and doctorate, thesis/dissertation required. *Average time to degree:* Master's–2 years full-time; doctorate–3 years full-time.
Entrance requirements: For master's and doctorate, GRE General Test. *Application deadline:* For fall admission, 4/1. *Application fee:* $30.
Expenses: Tuition: Full-time $8,940; part-time $447 per credit hour. Required fees: $598; $50 per semester.
Financial aid: Fellowships with tuition reimbursements, research assistantships with tuition reimbursements, teaching assistantships with tuition reimbursements, institutionally sponsored loans and tuition waivers (full) available.
Faculty research: Molecular biology and gene transfection. *Total annual research expenditures:* $1.7 million.

Creighton University (continued)
Dr. Richard Murphy, Chairman, 402-280-2918.
Application contact: Dr. Sandor Lovas, Graduate Coordinator, 402-280-5753, *Fax:* 402-280-2690, *E-mail:* slovas@ creighton.edu.
Find an in-depth description at www.petersons.com/graduate.

■ DARTMOUTH COLLEGE

School of Arts and Sciences, Department of Biological Sciences, Hanover, NH 03755

AWARDS Biology (PhD).

Faculty: 18 full-time (5 women), 2 part-time/adjunct (0 women).
Students: 41 full-time (17 women); includes 1 minority (African American), 9 international. Average age 23. *32 applicants, 16% accepted.* In 1999, 5 doctorates awarded (60% entered university research/teaching, 40% found other work related to degree).
Degree requirements: For doctorate, dissertation, teaching experience required. *Average time to degree:* Doctorate–5 years full-time.
Entrance requirements: For doctorate, GRE General Test, GRE Subject Test, TOEFL. *Application deadline:* For fall admission, 1/15 (priority date). Electronic applications accepted.
Expenses: Tuition: Full-time $24,624. Required fees: $916. One-time fee: $15 full-time. Full-time tuition and fees vary according to program.
Financial aid: In 1999–00, 39 students received aid, including 34 fellowships with full tuition reimbursements available (averaging $16,500 per year), 5 research assistantships with full tuition reimbursements available (averaging $16,500 per year); grants, institutionally sponsored loans, scholarships, traineeships, tuition waivers (full), and unspecified assistantships also available. Financial aid applicants required to submit FAFSA.
Faculty research: Population, community and ecosystem ecology; development (Coelegans, Drosophila and Arabidopsis); circadian rhythms and plant nutrition (Arabidopsis); cell motility and the cytoskeleton (squid and sea urchins).
Dr. Mark McPeek, Chair, 603-646-2378.
Application contact: Amy F. Layne, Administrative Assistant, 603-646-3847, *Fax:* 603-646-1347, *E-mail:* amy.f.layne@ dartmouth.edu.
Find an in-depth description at www.petersons.com/graduate.

■ DELAWARE STATE UNIVERSITY

Graduate Programs, Department of Biology, Dover, DE 19901-2277

AWARDS Biology (MS); biology education (MS). Part-time and evening/weekend programs available.

Degree requirements: For master's, thesis required (for some programs), foreign language not required.
Entrance requirements: For master's, GRE, minimum GPA of 3.0 in major, 2.75 overall.
Faculty research: Cell biology, immunology, microbiology, genetics, ecology.

■ DELTA STATE UNIVERSITY

Graduate Programs, College of Arts and Sciences, Department of Biological Sciences, Cleveland, MS 38733-0001

AWARDS MSNS. Part-time programs available.

Faculty: 7 full-time (1 woman), 1 part-time/adjunct (0 women).
Students: Average age 26.
Degree requirements: For master's, research project or thesis required. *Average time to degree:* Master's–2 years full-time, 4 years part-time.
Entrance requirements: For master's, GRE General Test. *Application deadline:* For fall admission, 8/1 (priority date); for spring admission, 12/1 (priority date). Applications are processed on a rolling basis. *Application fee:* $0.
Expenses: Tuition, state resident: full-time $2,596; part-time $121 per hour. Tuition, nonresident: full-time $5,546; part-time $285 per hour.
Financial aid: In 1999–00, 5 students received aid; research assistantships, career-related internships or fieldwork, Federal Work-Study, and institutionally sponsored loans available. Aid available to part-time students. Financial aid application deadline: 6/1.
Dr. Grady E. Williams, Chair, 601-846-4240.
Application contact: Dr. John G. Thornell, Dean of Graduate Studies and Continuing Education, 662-846-4310, *Fax:* 662-846-4313, *E-mail:* thornell@ dsu.deltast.edu.

■ DEPAUL UNIVERSITY

College of Liberal Arts and Sciences, Department of Biological Sciences, Chicago, IL 60604-2287

AWARDS MS.

Faculty: 9 full-time (3 women), 4 part-time/adjunct (1 woman).
Students: 18 full-time (11 women), 3 part-time (1 woman); includes 3 minority (1

African American, 2 Asian Americans or Pacific Islanders). Average age 25. *35 applicants, 40% accepted.* In 1999, 4 degrees awarded.
Degree requirements: For master's, thesis (for some programs), oral exam required, foreign language not required.
Entrance requirements: For master's, GRE, minimum GPA of 2.7. *Application deadline:* Applications are processed on a rolling basis. *Application fee:* $25.
Expenses: Tuition: Part-time $332 per credit hour. Required fees: $10 per term. Part-time tuition and fees vary according to program.
Financial aid: Teaching assistantships with full tuition reimbursements, Federal Work-Study, grants, institutionally sponsored loans, and tuition waivers (full and partial) available. Aid available to part-time students. Financial aid application deadline: 4/1.
Faculty research: Blood oxygen transport in vertebrates, cell motility, detoxification in plant cells, molecular biology of fungi, B-lymphocyte development, vertebrate embryology.
Dr. Leigh Maginniss, Chair, 773-325-7595, *Fax:* 773-325-7596.
Application contact: Director of Graduate Admissions, 773-325-7315.
Find an in-depth description at www.petersons.com/graduate.

■ DREXEL UNIVERSITY

Graduate School, College of Arts and Sciences, Department of Bioscience and Biotechnology, Philadelphia, PA 19104-2875

AWARDS Biological science (MS, PhD); nutrition and food sciences (MS, PhD), including food science (MS), nutrition science. Part-time programs available.

Faculty: 13 full-time (7 women), 1 part-time/adjunct (0 women).
Students: 3 full-time (all women), 55 part-time (33 women); includes 4 minority (1 African American, 2 Asian Americans or Pacific Islanders, 1 Hispanic American), 16 international. Average age 28. *163 applicants, 37% accepted.* In 1999, 11 master's, 4 doctorates awarded. Terminal master's awarded for partial completion of doctoral program.
Degree requirements: For master's, thesis required (for some programs), foreign language not required; for doctorate, dissertation required, foreign language not required.
Entrance requirements: For master's and doctorate, GRE General Test, TOEFL. *Application deadline:* For fall admission, 8/21. Applications are processed on a rolling basis. *Application fee:* $35. Electronic applications accepted.

Expenses: Tuition: Part-time $585 per credit.

Financial aid: Research assistantships, teaching assistantships, Federal Work-Study and unspecified assistantships available. Financial aid application deadline: 2/1.

Faculty research: Genetic engineering, physiological ecology.

Dr. Shortie McKinney, Head, 215-895-2418, *Fax:* 215-895-2624.

Application contact: Director of Graduate Admissions, 215-895-6700, *Fax:* 215-895-5939, *E-mail:* enroll@drexel.edu.

Find an in-depth description at www.petersons.com/graduate.

■ DREXEL UNIVERSITY

Graduate School, School of Biomedical Engineering, Science and Health Systems, Philadelphia, PA 19104-2875

AWARDS Biomedical engineering (MS, PhD); biomedical science (MS, PhD); biostatistics (MS); clinical/rehabilitation engineering (MS).

Faculty: 7 full-time (1 woman), 3 part-time/adjunct (0 women).

Students: 16 full-time (5 women), 52 part-time (17 women); includes 11 minority (7 African Americans, 4 Asian Americans or Pacific Islanders), 27 international. Average age 29. *168 applicants, 59% accepted.* In 1999, 14 master's, 7 doctorates awarded.

Degree requirements: For master's, thesis required (for some programs); for doctorate, dissertation, 1 year of residency, qualifying exam required.

Entrance requirements: For master's, TOEFL, minimum GPA of 3.0; for doctorate, TOEFL, minimum GPA of 3.0, MS. *Application deadline:* For fall admission, 8/21. Applications are processed on a rolling basis. *Application fee:* $35. Electronic applications accepted.

Expenses: Tuition: Full-time $15,795; part-time $585 per credit. Required fees: $375; $67 per term. Tuition and fees vary according to program.

Financial aid: Research assistantships, teaching assistantships, career-related internships or fieldwork, Federal Work-Study, institutionally sponsored loans, tuition waivers (full and partial), and unspecified assistantships available. Financial aid application deadline: 2/1.

Faculty research: Cardiovascular dynamics, diagnostic and therapeutic ultrasound. Dr. Banu Onaral, Director, 215-895-2215.

Application contact: 215-895-6700, *Fax:* 215-895-5939, *E-mail:* enroll@drexel.edu.

Find an in-depth description at www.petersons.com/graduate.

■ DUKE UNIVERSITY

Graduate School, Department of Biology, Durham, NC 27708

AWARDS PhD.

Faculty: 57 full-time, 19 part-time/adjunct.

Students: 110 full-time, 1 part-time; includes 2 minority (both Asian Americans or Pacific Islanders), 25 international. *145 applicants, 24% accepted.* In 1999, 9 doctorates awarded.

Degree requirements: For doctorate, dissertation required.

Entrance requirements: For doctorate, GRE General Test, GRE Subject Test. *Application deadline:* For fall admission, 12/31. *Application fee:* $75.

Expenses: Tuition: Full-time $21,406; part-time $760 per unit. Required fees: $3,136; $3,136 per year. One-time fee: $30. Tuition and fees vary according to program.

Financial aid: Fellowships, research assistantships, teaching assistantships, Federal Work-Study available. Financial aid application deadline: 12/31. Bill Morris, Contact, 919-684-5257, *E-mail:* wfmorris@duke.edu.

Find an in-depth description at www.petersons.com/graduate.

■ DUQUESNE UNIVERSITY

Bayer School of Natural and Environmental Sciences, Department of Biological Sciences, Pittsburgh, PA 15282-0001

AWARDS MS, MS/MS. Part-time and evening/weekend programs available.

Faculty: 13 full-time (2 women), 9 part-time/adjunct (7 women).

Students: 33 full-time (17 women), 4 part-time (2 women). Average age 26. *30 applicants, 73% accepted.* In 1999, 12 degrees awarded (25% entered university research/teaching, 25% found other work related to degree, 50% continued full-time study).

Degree requirements: For master's, thesis (for some programs), final exam required, foreign language not required. *Average time to degree:* Master's–2.3 years full-time, 4 years part-time.

Entrance requirements: For master's, GRE General Test, TOEFL, BS in biological sciences or related field. *Application deadline:* For fall admission, 3/1 (priority date); for spring admission, 10/1 (priority date). Applications are processed on a rolling basis. *Application fee:* $40.

Expenses: Tuition: Part-time $511 per credit. Required fees: $46 per credit. $50 per year. One-time fee: $125 part-time. Tuition and fees vary according to program.

Financial aid: In 1999–00, 1 fellowship with full tuition reimbursement (averaging $13,500 per year), 1 research assistantship with full tuition reimbursement (averaging $13,500 per year), 20 teaching assistantships with full tuition reimbursements (averaging $13,500 per year) were awarded; scholarships, tuition waivers (partial), and unspecified assistantships also available. Financial aid application deadline: 5/15; financial aid applicants required to submit FAFSA.

Faculty research: Reproductive biology, genetics/epigenetics, microbial physiology, immunology, environmental microbiology. *Total annual research expenditures:* $336,673. Dr. Richard P. Elinson, Chair, 412-396-5640, *Fax:* 412-396-5907.

Application contact: Mary Ann Quinn, Assistant to the Dean Graduate Affairs, 412-396-6339, *Fax:* 412-396-4881, *E-mail:* gradinfo@duq.edu.

Find an in-depth description at www.petersons.com/graduate.

■ EAST CAROLINA UNIVERSITY

Graduate School, College of Arts and Sciences, Department of Biology, Greenville, NC 27858-4353

AWARDS Biology (MS); molecular biology/biotechnology (MS). Part-time programs available.

Faculty: 19 full-time (5 women).

Students: 29 full-time (14 women), 50 part-time (22 women); includes 9 minority (5 African Americans, 2 Asian Americans or Pacific Islanders, 2 Native Americans), 1 international. Average age 28. *52 applicants, 65% accepted.* In 1999, 17 degrees awarded.

Degree requirements: For master's, one foreign language (computer language can substitute), thesis, comprehensive exams required.

Entrance requirements: For master's, GRE General Test, GRE Subject Test, TOEFL. *Application deadline:* For fall admission, 6/1 (priority date); for spring admission, 10/15. Applications are processed on a rolling basis. *Application fee:* $40.

Expenses: Tuition, state resident: full-time $1,012. Tuition, nonresident: full-time $8,578. Required fees: $1,006. Full-time tuition and fees vary according to degree level. Part-time tuition and fees vary according to course load.

Financial aid: Fellowships with partial tuition reimbursements, research assistantships with partial tuition reimbursements, teaching assistantships with partial tuition reimbursements, career-related internships

East Carolina University (continued)
or fieldwork, Federal Work-Study, scholarships, and unspecified assistantships available. Aid available to part-time students. Financial aid application deadline: 6/1.
Faculty research: Biochemistry, microbiology, cell biology.
Dr. Gerhard W. Kalmus, Director of Graduate Studies, 252-328-6722, *Fax:* 252-328-4178, *E-mail:* kalmusg@mail.ecu.edu.
Application contact: Dr. Paul D. Tschetter, Senior Associate Dean, 252-328-6012, *Fax:* 252-328-6071, *E-mail:* grad@mail.ecu.edu.

Find an in-depth description at www.petersons.com/graduate.

■ EAST CAROLINA UNIVERSITY

School of Medicine, Department of Pathology and Laboratory Medicine, Greenville, NC 27858-4353

AWARDS Interdisciplinary biological sciences (PhD).

Faculty: 10 full-time (1 woman).
Students: 2 full-time (0 women), 1 part-time. Average age 30. *5 applicants, 60% accepted.*
Degree requirements: For doctorate, dissertation required, foreign language not required.
Entrance requirements: For doctorate, GRE General Test, GRE Subject Test, TOEFL, bachelor's degree in biological chemistry or physical science. *Application deadline:* For fall admission, 6/1 (priority date). Applications are processed on a rolling basis. *Application fee:* $40.
Expenses: Tuition, state resident: full-time $1,012. Tuition, nonresident: full-time $8,578. Required fees: $1,006. Full-time tuition and fees vary according to degree level. Part-time tuition and fees vary according to course load.
Financial aid: Fellowships available. Financial aid application deadline: 6/1.
Faculty research: Immunochemistry and allergens, immunological disorders, cell biology of tumors, membrane antigens, microphage biology.
Dr. Peter Kragel, Chairperson, 252-816-2801, *Fax:* 252-816-3616, *E-mail:* pkragel@brody.med.ecu.edu.
Application contact: Dr. Donald Hoffman, Director of Graduate Studies, 252-816-2816, *Fax:* 252-816-3616, *E-mail:* mdhoffman@eastnet.ecu.edu.

■ EAST CAROLINA UNIVERSITY

School of Medicine and Graduate School, Graduate Programs in Medicine, Greenville, NC 27858-4353
AWARDS PhD.

Degree requirements: For doctorate, dissertation required.

Entrance requirements: For doctorate, GRE General Test, GRE Subject Test, TOEFL. *Application deadline:* Applications are processed on a rolling basis. *Application fee:* $40.
Expenses: Tuition, state resident: full-time $1,012. Tuition, nonresident: full-time $8,578. Required fees: $1,006. Full-time tuition and fees vary according to degree level. Part-time tuition and fees vary according to course load.
Financial aid: Fellowships available. Financial aid application deadline: 6/1.
Dr. Sam Pennington, Associate Dean for Research and Graduate Studies, 252-816-2827, *Fax:* 252-816-3260, *E-mail:* snpennington@brody.med.ecu.edu.
Application contact: Dr. Paul D. Tschetter, Senior Associate Dean of the Graduate School, 252-328-6012, *Fax:* 252-328-6071, *E-mail:* grad@mail.ecu.edu.

Find an in-depth description at www.petersons.com/graduate.

■ EASTERN ILLINOIS UNIVERSITY

Graduate School, College of Sciences, Program in Biological Sciences, Charleston, IL 61920-3099

AWARDS Biological sciences (MS); botany (MS); environmental biology (MS); zoology (MS).

Degree requirements: For master's, exam required, foreign language and thesis not required.

■ EASTERN KENTUCKY UNIVERSITY

The Graduate School, College of Natural and Mathematical Sciences, Department of Biological Sciences, Richmond, KY 40475-3102

AWARDS Biological sciences (MS); ecology (MS). Part-time programs available.

Faculty: 17 full-time (3 women).
Students: 32; includes 1 minority (African American), 1 international. In 1999, 10 degrees awarded.
Degree requirements: For master's, thesis required.
Entrance requirements: For master's, GRE General Test, minimum GPA of 2.5. *Application deadline:* For fall admission, 8/1; for spring admission, 12/1. Applications are processed on a rolling basis. *Application fee:* $0.
Expenses: Tuition, state resident: full-time $2,390; part-time $145 per credit hour. Tuition, nonresident: full-time $6,430; part-time $391 per credit hour.
Financial aid: Research assistantships, teaching assistantships, career-related internships or fieldwork and Federal Work-Study available. Aid available to

part-time students. Financial aid applicants required to submit FAFSA.
Faculty research: Systematics, ecology, and biodiversity; animal behavior; protein structure and molecular genetics; biomonitoring and aquatic toxicology; pathogenesis of microbes and parasites. *Total annual research expenditures:* $45,000.
Dr. Ross Clark, Chair, 606-622-1531, *Fax:* 606-622-1020, *E-mail:* bioclark@acs.eku.edu.

■ EASTERN MICHIGAN UNIVERSITY

Graduate School, College of Arts and Sciences, Department of Biology, Ypsilanti, MI 48197

AWARDS MS. Evening/weekend programs available.

Faculty: 18 full-time (3 women).
Students: 7 full-time, 37 part-time; includes 6 minority (4 Asian Americans or Pacific Islanders, 2 Hispanic Americans), 5 international. *36 applicants, 53% accepted.* In 1999, 15 degrees awarded.
Degree requirements: For master's, thesis required (for some programs), foreign language not required.
Entrance requirements: For master's, GRE General Test, GRE Subject Test, TOEFL. *Application deadline:* For fall admission, 5/15; for spring admission, 3/15. Applications are processed on a rolling basis. *Application fee:* $30.
Expenses: Tuition, state resident: part-time $157 per credit. Tuition, nonresident: part-time $350 per credit. Required fees: $17 per credit. $40 per semester. Tuition and fees vary according to course level, degree level and reciprocity agreements.
Financial aid: Fellowships, research assistantships with full tuition reimbursements, teaching assistantships with full tuition reimbursements, Federal Work-Study and scholarships available. Aid available to part-time students. Financial aid application deadline: 3/15; financial aid applicants required to submit FAFSA.
Dr. Robert Neely, Interim Head, 734-487-4242.

■ EASTERN NEW MEXICO UNIVERSITY

Graduate School, College of Liberal Arts and Sciences, Department of Biology, Portales, NM 88130

AWARDS MS. Part-time programs available.

Faculty: 8 full-time (2 women).
Students: 2 full-time (both women), 14 part-time (7 women); includes 2 minority (1 African American, 1 Hispanic American). Average age 31. *12 applicants, 83% accepted.* In 1999, 6 degrees awarded.

Degree requirements: For master's, one foreign language required, (computer language can substitute), thesis optional.
Entrance requirements: For master's, minimum GPA of 2.5. *Application deadline:* For fall admission, 8/20 (priority date). Applications are processed on a rolling basis. *Application fee:* $10. Electronic applications accepted.
Expenses: Tuition, state resident: full-time $2,040; part-time $85 per credit hour. Tuition, nonresident: full-time $6,918; part-time $288 per credit hour.
Financial aid: In 1999–00, 2 fellowships (averaging $7,200 per year), 9 teaching assistantships (averaging $7,000 per year) were awarded; research assistantships, Federal Work-Study also available. Aid available to part-time students. Financial aid application deadline: 3/1.
Dr. Gary Pfaffenberger, Graduate Coordinator, 505-562-2495, *E-mail:* gary.pfaffenberger@enmu.edu.

■ EASTERN VIRGINIA MEDICAL SCHOOL

Doctoral Program in Biomedical Sciences, Norfolk, VA 23501-1980
AWARDS PhD, MD/PhD.

Degree requirements: For doctorate, dissertation required, foreign language not required.
Entrance requirements: For doctorate, GRE General Test, TOEFL.
Faculty research: Reproductive physiology, cancer, molecular biology and immunology, cytogenetics and development, systems biology.

■ EASTERN VIRGINIA MEDICAL SCHOOL

Master's Program in Biomedical Sciences, Norfolk, VA 23501-1980
AWARDS MS.

■ EASTERN WASHINGTON UNIVERSITY

Graduate School, College of Science, Mathematics and Technology, Department of Biology, Cheney, WA 99004-2431
AWARDS MS.

Faculty: 23 full-time (9 women).
Students: 14 full-time (9 women), 9 part-time (4 women); includes 2 minority (1 Hispanic American, 1 Native American). *10 applicants, 30% accepted.* In 1999, 6 degrees awarded.
Degree requirements: For master's, thesis, comprehensive oral exam required, foreign language not required.
Entrance requirements: For master's, GRE General Test, minimum GPA of 3.0.

Application deadline: For fall admission, 4/1 (priority date); for spring admission, 1/15. Applications are processed on a rolling basis. *Application fee:* $35.
Expenses: Tuition, state resident: full-time $4,326. Tuition, nonresident: full-time $13,161.
Financial aid: Research assistantships, teaching assistantships, career-related internships or fieldwork, Federal Work-Study, and institutionally sponsored loans available. Financial aid application deadline: 2/1.
Faculty research: Ecology of Eastern Washington Scablands, Columbia River fisheries, biotechnology applied to vaccines, role of mycorrhiza in plant nutrition, exercise and estrous cycles.
Dr. Prakash Bhuta, Chair, 509-359-2339.
Application contact: Dr. Ross Black, Graduate Adviser, 509-359-2339.

■ EAST STROUDSBURG UNIVERSITY OF PENNSYLVANIA

Graduate School, School of Arts and Sciences, Department of Biology, East Stroudsburg, PA 18301-2999
AWARDS Biology (M Ed, MS). Part-time and evening/weekend programs available.

Degree requirements: For master's, thesis or alternative, comprehensive exam required, foreign language not required.
Entrance requirements: For master's, undergraduate major in life sciences, previous course work in organic chemistry.
Expenses: Tuition, state resident: full-time $3,780; part-time $210 per credit. Tuition, nonresident: full-time $6,610; part-time $367 per credit. Required fees: $724; $40 per credit.

■ EAST TENNESSEE STATE UNIVERSITY

James H. Quillen College of Medicine and School of Graduate Studies, Biomedical Science Graduate Program, Johnson City, TN 37614
AWARDS Anatomy and cell biology (MS, PhD); biochemistry and molecular biology (MS, PhD); biophysics (MS, PhD); microbiology (MS, PhD); pharmacology (MS, PhD); physiology (MS, PhD). Part-time programs available.

Faculty: 40 full-time (9 women).
Students: 23 full-time (10 women), 2 part-time (1 woman); includes 1 minority (Asian American or Pacific Islander), 4 international. Average age 30. *83 applicants, 20% accepted.* In 1999, 5 master's, 4 doctorates awarded. Terminal master's awarded for partial completion of doctoral program.
Degree requirements: For master's, one foreign language (computer language can

substitute), thesis, comprehensive qualifying exam required; for doctorate, 2 foreign languages (computer language can substitute for one), dissertation required.
Entrance requirements: For master's, GRE General Test, minimum GPA of 3.0, bachelor's degree in biological or related science; for doctorate, GRE General Test, GRE Subject Test. *Application deadline:* For fall admission, 3/15 (priority date); for spring admission, 3/1. *Application fee:* $25 ($35 for international students).
Expenses: Tuition, state resident: full-time $10,342. Tuition, nonresident: full-time $21,080. Required fees: $532.
Financial aid: In 1999–00, 16 research assistantships, 5 teaching assistantships were awarded; fellowships, career-related internships or fieldwork, Federal Work-Study, grants, institutionally sponsored loans, and tuition waivers (full) also available.
Dr. Mitchell Robinson, Assistant Dean for Graduate Studies, 423-439-4658, *E-mail:* robinson@etsu.edu.

■ EAST TENNESSEE STATE UNIVERSITY

School of Graduate Studies, College of Arts and Sciences and Biomedical Science Graduate Program, Department of Biological Sciences, Johnson City, TN 37614
AWARDS MS.

Faculty: 13 full-time (2 women).
Students: 27 full-time (15 women), 9 part-time (5 women); includes 4 minority (1 African American, 1 Asian American or Pacific Islander, 1 Hispanic American, 1 Native American), 2 international. Average age 26. *25 applicants, 72% accepted.* In 1999, 8 degrees awarded.
Degree requirements: For master's, thesis or alternative, oral and written comprehensive exams required, foreign language not required.
Entrance requirements: For master's, GRE General Test or GRE Subject Test, TOEFL, minimum GPA of 3.0. *Application deadline:* For fall admission, 7/15 (priority date); for spring admission, 11/1. Applications are processed on a rolling basis. *Application fee:* $25 ($35 for international students).
Expenses: Tuition, state resident: full-time $2,404; part-time $123 per semester hour. Tuition, nonresident: full-time $2,558; part-time $224 per semester hour. International tuition: $7,400 full-time. Required fees: $172 per hour.
Financial aid: In 1999–00, 5 teaching assistantships were awarded; research assistantships, career-related internships or fieldwork and institutionally sponsored loans also available.

East Tennessee State University (continued)
Dr. Dan Johnson, Chair, 423-439-4329, *Fax:* 423-439-5958, *E-mail:* johnsodm@etsu.edu.

■ EDINBORO UNIVERSITY OF PENNSYLVANIA

Graduate Studies, School of Science, Management and Technology, Department of Biology and Health Sciences, Edinboro, PA 16444

AWARDS Biology (MS). Part-time and evening/weekend programs available.

Faculty: 12 full-time (4 women).
Students: 8 full-time (4 women), 7 part-time (4 women). Average age 26. In 1999, 2 degrees awarded.
Degree requirements: For master's, thesis or alternative, competency exam required, foreign language not required. *Average time to degree:* Master's–2 years full-time, 4 years part-time.
Entrance requirements: For master's, GRE or MAT, minimum QPA of 2.5. *Application deadline:* Applications are processed on a rolling basis. *Application fee:* $25.
Expenses: Tuition, state resident: full-time $3,780; part-time $210 per credit. Tuition, nonresident: full-time $6,610; part-time $367 per credit. Required fees: $945; $53 per credit. Part-time tuition and fees vary according to course load.
Financial aid: In 1999–00, 6 students received aid. Career-related internships or fieldwork, Federal Work-Study, institutionally sponsored loans, scholarships, and unspecified assistantships available. Aid available to part-time students. Financial aid application deadline: 5/1; financial aid applicants required to submit FAFSA.
Faculty research: Microbiology, molecular biology, zoology, botany, ecology.
Dr. Craig Steele, Head, 814-732-2353, *Fax:* 814-732-2792, *E-mail:* cstelle@edinboro.edu.
Application contact: Dr. Terry L. Smith, Acting Dean of Graduate Studies, 814-732-2856, *Fax:* 814-732-2611, *E-mail:* tlsmith@edinboro.edu.

■ EMORY UNIVERSITY

Graduate School of Arts and Sciences, Division of Biological and Biomedical Sciences, Atlanta, GA 30322-1100

AWARDS PhD.

Faculty: 260 full-time (44 women).
Students: 282 full-time (162 women). *754 applicants, 17% accepted.* In 1999, 49 doctorates awarded.
Degree requirements: For doctorate, dissertation required, foreign language not required.

Entrance requirements: For doctorate, GRE General Test, TOEFL, minimum GPA of 3.0 in science course work. *Application deadline:* For fall admission, 1/20 (priority date). *Application fee:* $45.
Expenses: Tuition: Full-time $22,770. Tuition and fees vary according to program.
Financial aid: In 1999–00, 175 fellowships with full tuition reimbursements (averaging $18,000 per year) were awarded; scholarships also available.
Faculty research: Biochemistry and genetics, immunology and microbiology, neuroscience and pharmacology, nutrition, population biology and ecology.
Dr. Bryan D. Noe, Director, 404-727-2545, *Fax:* 404-727-3322.
Application contact: 404-727-2547, *Fax:* 404-727-3322, *E-mail:* gdbbs@emory.edu.

Find an in-depth description at www.petersons.com/graduate.

■ EMPORIA STATE UNIVERSITY

School of Graduate Studies, College of Liberal Arts and Sciences, Division of Biological Sciences, Emporia, KS 66801-5087

AWARDS Botany (MS); environmental biology (MS); general biology (MS); microbial and cellular biology (MS); zoology (MS). Part-time programs available.

Faculty: 15 full-time (3 women), 4 part-time/adjunct (0 women).
Students: 20 full-time (10 women), 8 part-time (3 women), 2 international. *8 applicants, 75% accepted.* In 1999, 12 degrees awarded.
Degree requirements: For master's, comprehensive exam or thesis required.
Entrance requirements: For master's, TOEFL, written exam. *Application deadline:* For fall admission, 8/15 (priority date). Applications are processed on a rolling basis. *Application fee:* $30 ($75 for international students). Electronic applications accepted.
Expenses: Tuition, state resident: full-time $2,410; part-time $108 per credit hour. Tuition, nonresident: full-time $6,212; part-time $266 per credit hour.
Financial aid: In 1999–00, 2 fellowships (averaging $1,396 per year), 1 research assistantship (averaging $5,390 per year), 11 teaching assistantships with full tuition reimbursements (averaging $5,047 per year) were awarded; career-related internships or fieldwork, Federal Work-Study, and institutionally sponsored loans also available. Financial aid application deadline: 3/15; financial aid applicants required to submit FAFSA.
Faculty research: Fisheries, range, and wildlife management; aquatic, plant, grassland, vertebrate, and invertebrate

ecology; mammalian and plant systematics, taxonomy, and evolution; immunology, virology, and molecular biology.
Dr. Marshall Sundberg, Chair, 316-341-5311, *Fax:* 316-341-5607, *E-mail:* sundberm@emporia.edu.

■ FAIRLEIGH DICKINSON UNIVERSITY, FLORHAM-MADISON CAMPUS

Maxwell Becton College of Arts and Sciences, Department of Biological and Allied Health Services, Madison, NJ 07940-1099

AWARDS Biology (MS).

Students: 2 full-time (both women), 11 part-time (6 women).
Degree requirements: For master's, one foreign language required, (computer language can substitute), thesis optional.
Entrance requirements: For master's, GRE General Test. *Application deadline:* Applications are processed on a rolling basis. *Application fee:* $35.
Expenses: Tuition: Full-time $9,396; part-time $522 per credit. Required fees: $69 per semester.
Financial aid: Fellowships, research assistantships, teaching assistantships available.
Dr. R. Gordon Perry, Chairperson, 973-443-8746.

■ FAIRLEIGH DICKINSON UNIVERSITY, TEANECK–HACKENSACK CAMPUS

University College: Arts, Sciences, and Professional Studies, School of Natural Sciences, Program in Biology, Teaneck, NJ 07666-1914

AWARDS MS.

Degree requirements: For master's, thesis or alternative required, foreign language not required.
Entrance requirements: For master's, GRE General Test.
Faculty research: Aquatic ecology, toxicology, desalination, lightwave technology, neuroendocrinology.

■ FAYETTEVILLE STATE UNIVERSITY

Graduate School, Department of Natural Sciences, Fayetteville, NC 28301-4298

AWARDS Biology (MS). Part-time and evening/weekend programs available.

Students: 1 (woman) full-time, 14 part-time (11 women); includes 10 minority (7 African Americans, 3 Hispanic Americans). Average age 27. *5 applicants, 100% accepted.* In 1999, 1 degree awarded.

Degree requirements: For master's, computer language, thesis, comprehensive exams, internship required, foreign language not required.
Entrance requirements: For master's, GRE General Test. *Application deadline:* For fall admission, 8/1; for spring admission, 12/15. Applications are processed on a rolling basis. *Application fee:* $25.
Expenses: Tuition, area resident: Full-time $982. Tuition, nonresident: full-time $8,252. Required fees: $580.
Dr. P. V. Murthy, Chairperson, 910-486-1691.

■ FINCH UNIVERSITY OF HEALTH SCIENCES/THE CHICAGO MEDICAL SCHOOL

School of Graduate and Postdoctoral Studies, North Chicago, IL 60064-3095

AWARDS MS, PhD, MD/MS, MD/PhD. Part-time programs available.

Faculty: 59 full-time, 30 part-time/adjunct.
Students: 263 full-time (128 women), 3 part-time (2 women); includes 83 minority (5 African Americans, 67 Asian Americans or Pacific Islanders, 11 Hispanic Americans), 18 international. Average age 24. *701 applicants, 34% accepted.* In 1999, 84 master's, 23 doctorates awarded.
Degree requirements: For doctorate, dissertation required. *Average time to degree:* Master's–2 years full-time, 2.5 years part-time; doctorate–4 years full-time.
Entrance requirements: For master's and doctorate, GRE General Test, TOEFL, TWE. *Application deadline:* For fall admission, 6/1 (priority date). Applications are processed on a rolling basis. *Application fee:* $25.
Expenses: Tuition: Full-time $14,054; part-time $391 per credit hour. Tuition and fees vary according to program.
Financial aid: In 1999–00, fellowships (averaging $15,500 per year); research assistantships, teaching assistantships, career-related internships or fieldwork, grants, and tuition waivers (full and partial) also available. Financial aid application deadline: 6/9; financial aid applicants required to submit FAFSA.
Faculty research: Extracellular matrix, nutrition and mood, neuropsychopharmacology, membrane transport, brain metabolism.
Dr. Velayudhan Nair, Dean, 847-578-3250, *Fax:* 847-578-3332.
Application contact: Dana Frederick, Admissions Officer, 847-578-3209.
Find an in-depth description at www.petersons.com/graduate.

■ FISK UNIVERSITY

Graduate Programs, Department of Biology, Nashville, TN 37208-3051

AWARDS MA.

Faculty: 4 full-time (2 women).
Students: 4 full-time (2 women); all minorities (all African Americans). Average age 27. In 1999, 5 degrees awarded (20% entered university research/teaching, 80% continued full-time study).
Degree requirements: For master's, thesis, comprehensive exam required, foreign language not required. *Average time to degree:* Master's–2 years full-time. *Application deadline:* For fall admission, 6/15 (priority date). Applications are processed on a rolling basis. *Application fee:* $25.
Expenses: Tuition: Full-time $8,480; part-time $471 per semester hour. Required fees: $540; $270 per semester.
Financial aid: In 1999–00, 3 research assistantships with partial tuition reimbursements (averaging $7,000 per year) were awarded; fellowships.
Faculty research: Embryology, plant morphology.
Dr. Mary McKelvey-Welch, Chairperson, 615-329-8796.
Application contact: Mark Adkins, Director of Admissions and Financial Aid, 615-329-8665, *Fax:* 615-329-8774.

■ FLORIDA AGRICULTURAL AND MECHANICAL UNIVERSITY

Division of Graduate Studies, Research, and Continuing Education, College of Arts and Sciences, Department of Biology, Tallahassee, FL 32307-3200

AWARDS MS.

Students: 10 (4 women); all minorities (all African Americans). In 1999, 2 degrees awarded.
Degree requirements: For master's, thesis required.
Entrance requirements: For master's, GRE General Test, minimum GPA of 3.0. *Application deadline:* For fall admission, 5/13. *Application fee:* $20.
Expenses: Tuition, state resident: full-time $2,644; part-time $147 per credit hour. Tuition, nonresident: full-time $9,137; part-time $508 per credit hour. Required fees: $52 per semester. Tuition and fees vary according to course load.
Dr. James Adams, Chairperson, 850-599-3907, *Fax:* 850-561-2996.

■ FLORIDA ATLANTIC UNIVERSITY

Charles E. Schmidt College of Science, Department of Biological Sciences, Boca Raton, FL 33431-0991

AWARDS MBS, MS, MST. Part-time programs available.

Faculty: 17 full-time (2 women).
Students: 31 full-time (20 women), 21 part-time (14 women); includes 5 minority (1 African American, 1 Asian American or Pacific Islander, 3 Hispanic Americans), 7 international. Average age 29. *41 applicants, 32% accepted.* In 1999, 14 degrees awarded.
Degree requirements: For master's, thesis required (for some programs), foreign language not required.
Entrance requirements: For master's, GRE General Test, minimum GPA of 3.0. *Application deadline:* For fall admission, 6/1. *Application fee:* $20.
Expenses: Tuition, state resident: full-time $2,663; part-time $148 per credit hour. Tuition, nonresident: full-time $9,156; part-time $509 per credit hour.
Financial aid: In 1999–00, 3 research assistantships, 30 teaching assistantships with tuition reimbursements were awarded; fellowships, career-related internships or fieldwork and Federal Work-Study also available.
Faculty research: Salt regulation in birds, ecology of non-native fish, tropical plant ecology, shrimp systematics, molecular biology and biotechnology. *Total annual research expenditures:* $410,000.
Dr. Peter Lutz, Chair, 561-297-3320, *Fax:* 561-297-2749, *E-mail:* lutz@fau.edu.
Application contact: Dr. Randy Brooks, Graduate Coordinator, 561-297-3320, *Fax:* 561-297-2749, *E-mail:* wbrooks@fau.edu.

■ FLORIDA INSTITUTE OF TECHNOLOGY

Graduate School, College of Science and Liberal Arts, Department of Biological Sciences, Melbourne, FL 32901-6975

AWARDS Biology (PhD); biotechnology (MS); cell and molecular biology (MS); ecology (MS); marine biology (MS). Part-time programs available.

Faculty: 16 full-time (1 woman), 5 part-time/adjunct (0 women).
Students: 16 full-time (10 women), 48 part-time (22 women); includes 6 minority (2 African Americans, 3 Asian Americans or Pacific Islanders, 1 Hispanic American), 12 international. Average age 30. *152 applicants, 19% accepted.* In 1999, 8 master's, 2 doctorates awarded.
Degree requirements: For master's, thesis required, foreign language not required; for doctorate, comprehensive and

Florida Institute of Technology (continued) departmental qualifying exams, oral defense of dissertation required.
Entrance requirements: For master's, GRE General Test, minimum GPA of 3.0, resume; for doctorate, GRE General Test, GRE Subject Test, minimum GPA of 3.2. *Application deadline:* Applications are processed on a rolling basis. *Application fee:* $50. Electronic applications accepted.
Expenses: Tuition: Part-time $575 per credit hour. Required fees: $50. Tuition and fees vary according to campus/location and program.
Financial aid: In 1999–00, 30 students received aid, including 6 research assistantships with full and partial tuition reimbursements available (averaging $5,344 per year), 21 teaching assistantships with full and partial tuition reimbursements available (averaging $3,762 per year); career-related internships or fieldwork and tuition remissions also available. Financial aid application deadline: 3/1; financial aid applicants required to submit FAFSA.
Faculty research: Reactions and components in initiation of protein synthesis in eukaryotic cells, fixation of radioactive carbon, changes in DNA molecule and differential expression of genetic information during aging, endangered or threatened avian and mammalian species, hydroacoustics and feeding preference of the West Indian manatee. *Total annual research expenditures:* $608,325.
Dr. Gary N. Wells, Head, 321-674-8034, *Fax:* 321-674-7238, *E-mail:* gwells@fit.edu.
Application contact: Carolyn P. Farrior, Associate Dean of Graduate Admissions, 321-674-7118, *Fax:* 321-674-9468, *E-mail:* cfarrior@fit.edu.

■ FLORIDA INTERNATIONAL UNIVERSITY

College of Arts and Sciences, Department of Biological Sciences, Miami, FL 33199
AWARDS MS, PhD. Part-time programs available.
Faculty: 30 full-time (6 women).
Students: 64 full-time (29 women), 31 part-time (16 women); includes 26 minority (7 African Americans, 3 Asian Americans or Pacific Islanders, 16 Hispanic Americans), 11 international. Average age 33. *103 applicants, 19% accepted.* In 1999, 104 master's, 19 doctorates awarded.
Degree requirements: For master's, thesis, 1 foreign language, computer language, or 2 semesters of graduate statistics required; for doctorate, dissertation, 1 foreign language, computer language, or 2 semesters of graduate

statistics, teaching experience required. *Average time to degree:* Master's–2.9 years full-time; doctorate–4.4 years full-time.
Entrance requirements: For master's and doctorate, GRE General Test, TSE, minimum GPA of 3.0. *Application deadline:* For fall admission, 2/11 (priority date); for spring admission, 8/25 (priority date). Applications are processed on a rolling basis. *Application fee:* $20.
Expenses: Tuition, state resident: full-time $3,479; part-time $145 per credit hour. Tuition, nonresident: full-time $12,137; part-time $506 per credit hour. Required fees: $158; $158 per year.
Financial aid: In 1999–00, 3 fellowships (averaging $15,000 per year), 10 research assistantships with partial tuition reimbursements (averaging $15,000 per year), 44 teaching assistantships with partial tuition reimbursements (averaging $15,000 per year) were awarded; Federal Work-Study, institutionally sponsored loans, and tuition waivers (partial) also available. Aid available to part-time students. Financial aid application deadline: 4/1; financial aid applicants required to submit FAFSA.
Faculty research: Aquaculture, aquatic plant biology, anatomy, vertebrate anatomy, cellular biology, molecular biology, microbiology, bioluminescence. *Total annual research expenditures:* $1.8 million.
Dr. David N. Kuhn, Graduate Program Director, 305-348-4130, *Fax:* 305-348-1986, *E-mail:* kuhnd@fiu.edu.
Application contact: Maribel Herrera, Graduate Program Secretary, 305-348-4130, *E-mail:* herreram@fiu.edu.

■ FLORIDA STATE UNIVERSITY

Graduate Studies, College of Arts and Sciences, Department of Biological Science, Tallahassee, FL 32306
AWARDS Cell biology (MS, PhD); developmental biology (MS, PhD); ecology (MS, PhD); evolutionary biology (MS, PhD); genetics (MS, PhD); immunology (MS, PhD); marine biology (MS, PhD); microbiology (MS, PhD); molecular biology (MS, PhD); neuroscience (PhD); physiology (MS, PhD); plant sciences (MS, PhD); radiation biology (MS, PhD).
Faculty: 47 full-time (6 women).
Students: 82 full-time (40 women); includes 10 minority (1 African American, 5 Asian Americans or Pacific Islanders, 4 Hispanic Americans), 13 international. *123 applicants, 31% accepted.* In 1999, 11 master's awarded (27% entered university research/teaching, 55% found other work related to degree, 18% continued full-time study).
Degree requirements: For master's and doctorate, thesis/dissertation, teaching experience required. *Average time to degree:*

Master's–3 years full-time; doctorate–5 years full-time.
Entrance requirements: For master's, GRE General Test, TOEFL; for doctorate, GRE General Test, GRE Subject Test, TOEFL. *Application deadline:* For fall admission, 1/15; for spring admission, 10/15. *Application fee:* $20. Electronic applications accepted.
Expenses: Tuition, state resident: full-time $3,504; part-time $146 per credit hour. Tuition, nonresident: full-time $12,162; part-time $507 per credit hour. Tuition and fees vary according to program.
Financial aid: In 1999–00, 3 fellowships with full tuition reimbursements (averaging $13,740 per year), 22 research assistantships with full tuition reimbursements (averaging $13,740 per year), 49 teaching assistantships with full tuition reimbursements (averaging $13,740 per year) were awarded; traineeships also available. Financial aid application deadline: 1/15; financial aid applicants required to submit FAFSA.
Dr. Thomas C. S. Keller, Associate Professor and Associate Chairman, 850-644-3023, *Fax:* 850-644-9829.
Application contact: Judy Bowers, Coordinator, Graduate Affairs, 850-644-3023, *Fax:* 850-644-9829, *E-mail:* bowers@bio.fsu.edu.

Find an in-depth description at www.petersons.com/graduate.

■ FORDHAM UNIVERSITY

Graduate School of Arts and Sciences, Department of Biological Sciences, New York, NY 10458
AWARDS Biological sciences (MS, PhD), including cell and molecular biology, ecology. Part-time and evening/weekend programs available.
Faculty: 17 full-time (1 woman).
Students: 29 full-time (17 women), 10 part-time (6 women); includes 2 minority (both Asian Americans or Pacific Islanders), 6 international. *57 applicants, 35% accepted.* In 1999, 11 master's, 2 doctorates awarded. Terminal master's awarded for partial completion of doctoral program.
Degree requirements: For master's, comprehensive exam required, thesis optional; for doctorate, 2 foreign languages (computer language can substitute for one), dissertation, comprehensive exam required.
Entrance requirements: For master's and doctorate, GRE General Test, GRE Subject Test (recommended). *Application deadline:* For fall admission, 1/16 (priority date); for spring admission, 12/1. *Application fee:* $60. Electronic applications accepted.

Expenses: Tuition: Full-time $14,400; part-time $600 per credit. Required fees: $125 per semester. Tuition and fees vary according to program.
Financial aid: In 1999–00, 29 students received aid, including 4 fellowships (averaging $15,000 per year), 3 research assistantships (averaging $15,000 per year), teaching assistantships (averaging $15,000 per year); institutionally sponsored loans, tuition waivers (full and partial), and unspecified assistantships also available. Aid available to part-time students. Financial aid application deadline: 1/15. *Total annual research expenditures:* $365,196. Dr. Berish Rubin, Chair, 718-817-3641, *Fax:* 718-817-3645, *E-mail:* rubin@fordham.edu.
Application contact: Dr. Craig W. Pilant, Assistant Dean, 718-817-4420, *Fax:* 718-817-3566, *E-mail:* pilant@fordham.edu.
Find an in-depth description at www.petersons.com/graduate.

■ FORT HAYS STATE UNIVERSITY

Graduate School, College of Health and Life Sciences, Department of Biological Sciences and Allied Health, Program in Biology, Hays, KS 67601-4099

AWARDS MS. Part-time programs available.

Faculty: 14 full-time (3 women).
Students: 12 full-time (5 women), 7 part-time. *15 applicants, 80% accepted.* In 1999, 6 degrees awarded.
Degree requirements: For master's, foreign language and thesis not required. *Application deadline:* For fall admission, 7/1 (priority date). Applications are processed on a rolling basis. *Application fee:* $25 ($35 for international students).
Expenses: Tuition, state resident: part-time $95 per credit hour. Tuition, nonresident: part-time $254 per credit hour. Full-time tuition and fees vary according to course level and course load.
Financial aid: Research assistantships, teaching assistantships, tuition waivers (full) available.
Dr. Robert Nicholson, Chairman, Department of Biological Sciences and Allied Health, 785-628-4214.

■ FROSTBURG STATE UNIVERSITY

Graduate School, College of Liberal Arts and Sciences, Department of Biology, Frostburg, MD 21532-1099

AWARDS Applied ecology and conservation biology (MS); fisheries and wildlife management (MS). Part-time and evening/weekend programs available.
Faculty: 12 full-time (4 women).

Students: 22 full-time (4 women), 25 part-time (8 women); includes 1 minority (Hispanic American), 3 international. Average age 29. In 1999, 5 degrees awarded.
Degree requirements: For master's, thesis required, foreign language not required.
Entrance requirements: For master's, GRE General Test, resume. *Application deadline:* For fall admission, 7/15 (priority date). Applications are processed on a rolling basis. *Application fee:* $30.
Expenses: Tuition, state resident: full-time $3,132; part-time $174 per credit hour. Tuition, nonresident: full-time $3,636; part-time $202 per credit hour. Required fees: $31 per credit hour. $8 per semester.
Financial aid: In 1999–00, 17 research assistantships with full tuition reimbursements (averaging $5,000 per year) were awarded; career-related internships or fieldwork and Federal Work-Study also available. Financial aid application deadline: 4/1; financial aid applicants required to submit FAFSA.
Faculty research: Molecular and morphological evolution, ecology and behavior of birds, binecology of forest nematodes and associated insects, conservation genetics of amphibians and fishes, biology of endangered species.
Dr. David Morton, Chair, 301-687-4166.
Application contact: Robert E. Smith, Assistant Dean for Graduate Services, 301-687-7053, *Fax:* 301-687-4597, *E-mail:* rsmith@frostburg.edu.

■ GEORGE MASON UNIVERSITY

College of Arts and Sciences, Department of Biology, Master's Program in Biology, Fairfax, VA 22030-4444

AWARDS Bioinformatics (MS); ecology, systematics and evolution (MS); environmental science and public policy (MS); interpretive biology (MS); molecular, microbial, and cellular biology (MS); organismal biology (MS). Part-time programs available.
Faculty: 30 full-time (11 women), 32 part-time/adjunct (20 women).
Students: 5 full-time (3 women), 55 part-time (37 women); includes 7 minority (1 African American, 4 Asian Americans or Pacific Islanders, 2 Hispanic Americans), 8 international. Average age 34. *36 applicants, 44% accepted.* In 1999, 18 degrees awarded.
Degree requirements: For master's, thesis or alternative required, foreign language not required.
Entrance requirements: For master's, GRE General Test, GRE Subject Test, bachelor's degree in biology or equivalent. *Application deadline:* For fall admission, 5/1; for spring admission, 11/1. *Application fee:* $30. Electronic applications accepted.

Expenses: Tuition, state resident: full-time $4,416; part-time $184 per credit hour. Tuition, nonresident: full-time $12,516; part-time $522 per credit hour. Tuition and fees vary according to program.
Financial aid: Available to part-time students. Application deadline: 3/1.
Dr. George E. Andrykovitch, Director, 703-993-1027, *Fax:* 703-993-1046.

■ GEORGETOWN UNIVERSITY

Graduate School of Arts and Sciences, Department of Biology, Washington, DC 20057

AWARDS MS, PhD.

Degree requirements: For master's, thesis, comprehensive exam required, foreign language not required; for doctorate, dissertation, comprehensive exam required.
Entrance requirements: For master's and doctorate, GRE General Test, GRE Subject Test (biology), TOEFL.
Find an in-depth description at www.petersons.com/graduate.

■ GEORGETOWN UNIVERSITY

Graduate School of Arts and Sciences, Programs in Biomedical Sciences, Washington, DC 20057

AWARDS MS, PhD, MD/PhD, MS/PhD.

Entrance requirements: For doctorate, GRE General Test, TOEFL.
Find an in-depth description at www.petersons.com/graduate.

■ THE GEORGE WASHINGTON UNIVERSITY

Columbian School of Arts and Sciences, Department of Biological Sciences, Washington, DC 20052

AWARDS Biology (MS, PhD). Part-time and evening/weekend programs available.

Faculty: 17 full-time (4 women).
Students: 16 full-time (10 women), 23 part-time (15 women); includes 5 minority (4 Asian Americans or Pacific Islanders, 1 Hispanic American), 12 international. Average age 29. *54 applicants, 24% accepted.* In 1999, 4 master's, 2 doctorates awarded. Terminal master's awarded for partial completion of doctoral program.
Degree requirements: For master's, comprehensive exam required, thesis not required; for doctorate, dissertation, general exam required.
Entrance requirements: For master's, GRE General Test, appropriate bachelor's degree, interview, minimum GPA of 3.0; for doctorate, GRE General Test, interview, minimum GPA of 3.0. *Application deadline:* For fall admission, 1/2. Applications are processed on a rolling

The George Washington University (continued)

basis. *Application fee:* $55. Electronic applications accepted.

Expenses: Tuition: Full-time $16,836; part-time $702 per credit hour. Required fees: $828; $35 per credit hour. Tuition and fees vary according to campus/location and program.

Financial aid: In 1999–00, 15 fellowships with full tuition reimbursements (averaging $16,000 per year), 10 teaching assistantships with full tuition reimbursements (averaging $12,000 per year) were awarded; Federal Work-Study also available. Financial aid application deadline: 1/2.

Faculty research: Systematics, evolution, ecology, developmental biology, cell/molecular biology.

Dr. Robert P. Donaldson, Chair, 202-994-6090.

Application contact: Dr. John R. Burns, Professor, 202-994-7149, *Fax:* 202-994-6100, *E-mail:* jrburns@gwu.edu.

Find an in-depth description at www.petersons.com/graduate.

■ THE GEORGE WASHINGTON UNIVERSITY

Columbian School of Arts and Sciences, Institute for Biomedical Sciences, Washington, DC 20052

AWARDS Biochemistry (PhD); genetics (MS, PhD); immunology (PhD); molecular and cellular oncology (PhD); neuroscience (PhD); pharmacology (PhD). Part-time and evening/weekend programs available.

Students: 24 full-time (13 women), 77 part-time (51 women); includes 20 minority (2 African Americans, 13 Asian Americans or Pacific Islanders, 5 Hispanic Americans), 16 international. Average age 26. *194 applicants, 34% accepted.* In 1999, 10 master's, 12 doctorates awarded.

Degree requirements: For doctorate, dissertation required.

Entrance requirements: For doctorate, GRE General Test, minimum GPA of 3.0. *Application fee:* $55.

Expenses: Tuition: Full-time $16,836; part-time $702 per credit hour. Required fees: $828; $35 per credit hour. Tuition and fees vary according to campus/location and program.

Financial aid: Fellowships, Federal Work-Study and institutionally sponsored loans available.

Application contact: 202-994-2179, *Fax:* 202-994-0967.

Find an in-depth description at www.petersons.com/graduate.

■ GEORGIA COLLEGE AND STATE UNIVERSITY

Graduate School, College of Arts and Sciences, Department of Biology, Milledgeville, GA 31061

AWARDS MS. Part-time programs available.

Students: 9 full-time (8 women), 6 part-time (2 women); includes 3 minority (2 African Americans, 1 Asian American or Pacific Islander). Average age 26. In 1999, 7 degrees awarded.

Degree requirements: For master's, thesis optional.

Entrance requirements: For master's, GRE. *Application deadline:* For fall admission, 7/15 (priority date). Applications are processed on a rolling basis. *Application fee:* $10.

Expenses: Tuition, state resident: full-time $2,080; part-time $91 per hour. Tuition, nonresident: full-time $6,510; part-time $272 per hour. Required fees: $408; $97 per hour. Tuition and fees vary according to course load.

Financial aid: In 1999–00, 10 research assistantships were awarded; career-related internships or fieldwork, Federal Work-Study, and unspecified assistantships also available. Aid available to part-time students. Financial aid application deadline: 3/1; financial aid applicants required to submit FAFSA.

Faculty research: Vertebrate collecting and monitoring, paleontologic expedition.

Dr. William Wall, Chairman, 912-445-0811.

Application contact: Dr. Harold Reed, Coordinator, *E-mail:* hreed@mail.gcsu.edu.

■ GEORGIA INSTITUTE OF TECHNOLOGY

Graduate Studies and Research, College of Sciences, School of Biology, Atlanta, GA 30332-0001

AWARDS MS, MS Biol, PhD. Part-time programs available.

Faculty: 15 full-time (2 women), 2 part-time/adjunct (1 woman).

Students: 33 full-time (14 women), 9 part-time (3 women); includes 7 minority (1 African American, 5 Asian Americans or Pacific Islanders, 1 Native American), 10 international. Average age 23. *69 applicants, 22% accepted.* In 1999, 7 master's, 6 doctorates awarded (100% entered university research/teaching). Terminal master's awarded for partial completion of doctoral program.

Degree requirements: For master's, thesis required, foreign language not required; for doctorate, dissertation, qualifying exams required, foreign language not required. *Average time to*

degree: Master's–2.5 years full-time; doctorate–5.5 years full-time.

Entrance requirements: For master's, GRE General Test, TOEFL, minimum GPA of 2.9; for doctorate, GRE General Test, TOEFL, minimum GPA of 3.0. *Application deadline:* For fall admission, 8/1 (priority date). Applications are processed on a rolling basis. *Application fee:* $50.

Financial aid: In 1999–00, 1 fellowship, 15 research assistantships, 15 teaching assistantships were awarded; career-related internships or fieldwork, Federal Work-Study, institutionally sponsored loans, and tuition waivers (partial) also available. Aid available to part-time students. Financial aid application deadline: 2/15.

Faculty research: Biotechnology, molecular microbiology, molecular genetics, environmental science. *Total annual research expenditures:* $500,000.

Dr. Roger M. Wartell, Chair, 404-894-3735.

Find an in-depth description at www.petersons.com/graduate.

■ GEORGIAN COURT COLLEGE

Graduate School, Program in Biology, Lakewood, NJ 08701-2697

AWARDS MS. Part-time and evening/weekend programs available.

Students: Average age 31. In 1999, 2 degrees awarded.

Degree requirements: For master's, thesis optional, foreign language not required.

Entrance requirements: For master's, GRE General Test, GRE Subject Test. *Application deadline:* For fall admission, 8/25; for spring admission, 1/15. Applications are processed on a rolling basis. *Application fee:* $40.

Expenses: Tuition: Full-time $6,750; part-time $375 per credit. Required fees: $66; $66 per year. Tuition and fees vary according to course load.

Dr. Michael Gross, Chairperson, Biology Department, 732-364-2200 Ext. 345, *Fax:* 732-905-8571, *E-mail:* gross@georgian.edu.

Application contact: Renee Loew, Director of Graduate Admissions and Records, 732-367-1717, *Fax:* 732-364-4516, *E-mail:* admissions-grad@georgian.edu.

■ GEORGIA SOUTHERN UNIVERSITY

Jack N. Averitt College of Graduate Studies, Allen E. Paulson College of Science and Technology, Department of Biology, Statesboro, GA 30460

AWARDS MS. Part-time programs available.

Faculty: 25 full-time (6 women), 1 (woman) part-time/adjunct.

Students: 5 full-time (3 women), 36 part-time (18 women); includes 1 minority (Asian American or Pacific Islander), 3 international. Average age 28. *32 applicants, 63% accepted.* In 1999, 18 degrees awarded.
Degree requirements: For master's, thesis, terminal exam required. *Average time to degree:* Master's–1.63 years full-time, 2.48 years part-time.
Entrance requirements: For master's, GRE General Test, minimum GPA of 2.75, BS in biology. *Application deadline:* For fall admission, 7/1 (priority date); for spring admission, 11/15. Applications are processed on a rolling basis. *Application fee:* $0. Electronic applications accepted.
Expenses: Tuition, state resident: full-time $1,820; part-time $91 per semester hour. Tuition, nonresident: full-time $7,260; part-time $363 per semester hour. Required fees: $312 per semester. Tuition and fees vary according to course load and campus/location.
Financial aid: In 1999–00, 20 students received aid, including 6 research assistantships with partial tuition reimbursements available (averaging $4,900 per year), 14 teaching assistantships with partial tuition reimbursements available (averaging $4,900 per year); career-related internships or fieldwork, Federal Work-Study, and unspecified assistantships also available. Aid available to part-time students. Financial aid application deadline: 4/15; financial aid applicants required to submit FAFSA.
Faculty research: Cellular and molecular biology, ecology and evolution, medical-veterinary entomology, conservation biology. *Total annual research expenditures:* $500,000.
Dr. John E. Averett, Chair, 912-681-5487, *Fax:* 912-681-0845, *E-mail:* averett@gasou.edu.
Application contact: Dr. John R. Diebolt, Associate Graduate Dean, 912-681-5384, *Fax:* 912-681-0740, *E-mail:* gradschool@gasou.edu.

■ GEORGIA STATE UNIVERSITY

College of Arts and Sciences, Department of Biology, Atlanta, GA 30303-3083

AWARDS Applied and environmental microbiology (MS, PhD); cell biology and physiology (MS, PhD); molecular genetics and biochemistry (MS, PhD); neurobiology (MS, PhD). Part-time and evening/weekend programs available.

Faculty: 34 full-time (12 women).
Students: 124 full-time (76 women), 46 part-time (28 women); includes 53 minority (40 African Americans, 9 Asian Americans or Pacific Islanders, 3 Hispanic Americans, 1 Native American), 46 international. Average age 30. *155*

applicants, 40% accepted. In 1999, 26 master's, 8 doctorates awarded.
Degree requirements: For master's, one foreign language (computer language can substitute), thesis or alternative, exam required; for doctorate, dissertation required. *Average time to degree:* Master's–2.5 years full-time, 4 years part-time; doctorate–5 years full-time, 8 years part-time.
Entrance requirements: For master's and doctorate, GRE General Test, TOEFL, minimum GPA of 3.0. *Application deadline:* For fall admission, 7/18; for spring admission, 2/13. Applications are processed on a rolling basis. *Application fee:* $25.
Expenses: Tuition, state resident: full-time $2,896; part-time $121 per credit hour. Tuition, nonresident: full-time $11,584; part-time $483 per credit hour. Required fees: $228. Full-time tuition and fees vary according to course load and program.
Financial aid: Fellowships, research assistantships, teaching assistantships, career-related internships or fieldwork, Federal Work-Study, institutionally sponsored loans, and tuition waivers (partial) available. Aid available to part-time students. Financial aid application deadline: 2/6.
Faculty research: Physiological biochemistry, gene expression, molecular virology, microbial ecology, integration in neural systems. *Total annual research expenditures:* $3 million.
Dr. P. C. Tai, Chair, 404-651-3409, *Fax:* 404-651-2509, *E-mail:* biopct@panther.gsu.edu.
Application contact: Latesha Morrison, Graduate Administrative Coordinator, 404-651-2759, *Fax:* 404-651-2509, *E-mail:* biolxm@langate.gsu.edu.

Find an in-depth description at www.petersons.com/graduate.

■ GOUCHER COLLEGE

Premedical Studies Concentration, Baltimore, MD 21204-2794

AWARDS Certificate.

Faculty: 9 full-time (2 women).
Students: 26 full-time (17 women), 1 part-time; includes 4 minority (2 African Americans, 2 Asian Americans or Pacific Islanders). Average age 25. *159 applicants, 26% accepted.*
Degree requirements: For Certificate, foreign language and thesis not required. *Application deadline:* Applications are processed on a rolling basis. *Application fee:* $50.
Expenses: Tuition: Full-time $17,400.
Financial aid: In 1999–00, 18 students received aid. Application deadline: 3/1.

Liza Thompson, Director, 800-414-3437, *Fax:* 410-337-6085, *E-mail:* lthompso@goucher.edu.

■ GRADUATE SCHOOL AND UNIVERSITY CENTER OF THE CITY UNIVERSITY OF NEW YORK

Graduate Studies, Program in Biology, New York, NY 10016-4039

AWARDS PhD.

Degree requirements: For doctorate, dissertation, teaching experience required.
Entrance requirements: For doctorate, GRE General Test.
Expenses: Tuition, state resident: full-time $4,350; part-time $245 per credit hour. Tuition, nonresident: full-time $7,600; part-time $425 per credit hour.

Find an in-depth description at www.petersons.com/graduate.

■ GRADUATE SCHOOL AND UNIVERSITY CENTER OF THE CITY UNIVERSITY OF NEW YORK

Graduate Studies, Program in Biomedical Science, New York, NY 10016-4039

AWARDS PhD, MD/PhD.

Degree requirements: For doctorate, dissertation required.
Entrance requirements: For doctorate, GRE General Test.
Expenses: Tuition, state resident: full-time $4,350; part-time $245 per credit hour. Tuition, nonresident: full-time $7,600; part-time $425 per credit hour.

■ HAMPTON UNIVERSITY

Graduate College, Department of Biological Sciences, Hampton, VA 23668

AWARDS MA, MS. Part-time and evening/weekend programs available.

Degree requirements: For master's, thesis optional, foreign language not required.
Entrance requirements: For master's, GRE General Test.
Expenses: Tuition: Full-time $9,490; part-time $230 per semester hour. Required fees: $60; $35 per semester. Tuition and fees vary according to course load.
Faculty research: Marine ecology, microbial and chemical pollution, pesticide problems.

■ HARVARD UNIVERSITY

Extension School, Cambridge, MA 02138-3722

AWARDS Applied sciences (CAS); English for graduate and professional studies (DGP);

Harvard University (continued)
information technology (ALM); liberal arts (ALM); museum studies (CMS); premedical studies (Diploma); public health (CPH); publication and communication (CPC); special studies in administration and management (CSS). Part-time and evening/weekend programs available.

Faculty: 450 part-time/adjunct.
Students: Average age 35. In 1999, 92 master's, 292 Diploma's awarded.
Degree requirements: For master's, thesis required, foreign language not required; for other advanced degree, computer language required, foreign language and thesis not required.
Entrance requirements: For master's and other advanced degree, TOEFL, TWE. *Application deadline:* Applications are processed on a rolling basis. *Application fee:* $75.
Expenses: Tuition: Part-time $1,145 per semester. Required fees: $35 per semester. Part-time tuition and fees vary according to program.
Financial aid: In 1999–00, 194 students received aid. Scholarships available. Aid available to part-time students. Financial aid application deadline: 8/16; financial aid applicants required to submit FAFSA. Michael Shinagel, Dean.
Application contact: Program Director, 617-495-4024, *Fax:* 617-495-9176.

■ HARVARD UNIVERSITY
Graduate School of Arts and Sciences, Department of Organismic and Evolutionary Biology, Cambridge, MA 02138

AWARDS Biology (PhD).

Students: 57 full-time (22 women). *98 applicants, 13% accepted.* In 1999, 11 doctorates awarded.
Degree requirements: For doctorate, public presentation of thesis research, exam required.
Entrance requirements: For doctorate, GRE General Test, GRE Subject Test (recommended), TOEFL, 7 courses in biology, chemistry, physics, mathematics, computer science, or geology. *Application deadline:* For fall admission, 12/15. *Application fee:* $60.
Expenses: Tuition: Full-time $22,054. Required fees: $711. Tuition and fees vary according to program.
Financial aid: Fellowships, research assistantships, teaching assistantships, career-related internships or fieldwork, Federal Work-Study, and institutionally sponsored loans available. Financial aid application deadline: 12/30.
Josephine Ferraro, Officer, 617-495-5396.
Application contact: Departmental Office, 617-495-2305.

■ HARVARD UNIVERSITY
Graduate School of Arts and Sciences, Program in Biological and Biomedical Sciences, Boston, MA 02115

AWARDS Biological chemistry and molecular pharmacology (PhD); cell biology (PhD); genetics (PhD); microbiology and molecular genetics (PhD); pathology (PhD), including experimental pathology.

Students: 433 full-time (210 women). In 1999, 83 doctorates awarded.
Degree requirements: For doctorate, dissertation required, foreign language not required.
Entrance requirements: For doctorate, GRE General Test, GRE Subject Test, TOEFL. *Application fee:* $60.
Expenses: Tuition: Full-time $22,054. Required fees: $711. Tuition and fees vary according to program.
Financial aid: Fellowships, research assistantships, teaching assistantships, institutionally sponsored loans and tuition waivers (full) available. Financial aid application deadline: 1/1.
Dr. Tom Roberts, Chair, 617-432-0884.
Application contact: Leah Simons, Manager of Student Affairs, 617-432-0162.
Find an in-depth description at www.petersons.com/graduate.

■ HARVARD UNIVERSITY
Graduate School of Arts and Sciences, Program in Biological Sciences in Public Health (BPH), Boston, MA 02115

AWARDS PhD.

Degree requirements: For doctorate, dissertation, qualifying exam required, foreign language not required.
Entrance requirements: For doctorate, GRE General Test, GRE Subject Test, TOEFL. *Application deadline:* For fall admission, 12/15. *Application fee:* $60.
Expenses: Tuition: Full-time $22,054. Required fees: $711. Tuition and fees vary according to program.
Financial aid: Fellowships, research assistantships, teaching assistantships, institutionally sponsored loans and tuition waivers (full) available. Financial aid application deadline: 1/1.
Faculty research: Nutrition biochemistry, molecular and cellular toxicology, cardiovascular disease, cancer biology, tropical public health, environmental health physiology.
Josephine Ferraro, Officer, 617-495-5396.
Application contact: Leah Simons, Manager of Student Affairs, 617-432-0162.
Find an in-depth description at www.petersons.com/graduate.

■ HARVARD UNIVERSITY
Medical School and Graduate School of Arts and Sciences, Division of Medical Sciences, Boston, MA 02115

AWARDS PhD, MD/PhD.

Degree requirements: For doctorate, dissertation, qualifying exam required, foreign language not required.
Entrance requirements: For doctorate, GRE General Test, GRE Subject Test, TOEFL.
Expenses: Tuition: Full-time $27,000. Required fees: $380. One-time fee: $25 full-time. Full-time tuition and fees vary according to student level.

Find an in-depth description at www.petersons.com/graduate.

■ HARVARD UNIVERSITY
School of Public Health, Division of Biological Sciences, Boston, MA 02115-6096

AWARDS PhD.

Faculty: 3 part-time/adjunct (1 woman).
Students: 37 full-time (21 women); includes 18 minority (4 African Americans, 10 Asian Americans or Pacific Islanders, 4 Hispanic Americans), 4 international. Average age 31. *69 applicants, 22% accepted.* In 1999, 2 degrees awarded.
Degree requirements: For doctorate, dissertation, qualifying exam required, foreign language not required.
Entrance requirements: For doctorate, GRE, TOEFL. *Application deadline:* For fall admission, 12/15. *Application fee:* $60.
Expenses: Tuition: Full-time $22,950; part-time $574 per credit.
Financial aid: Fellowships with full tuition reimbursements, teaching assistantships, grants, scholarships, traineeships, tuition waivers (partial), and unspecified assistantships available.
Faculty research: Cancer biology, epidemiology, environmental science and physiology, toxicology, tropical public health.
Dr. Dyann F. Wirth, Director, 617-432-4269, *Fax:* 617-432-4766, *E-mail:* dfwirth@hsph.harvard.edu.
Application contact: Ruth Kenworthy, 617-432-4470, *Fax:* 617-432-0433, *E-mail:* kenworthy@cvlab.harvard.edu.

■ HOFSTRA UNIVERSITY
College of Liberal Arts and Sciences, Division of Natural Sciences, Mathematics, Engineering, and Computer Science, Department of Biology, Hempstead, NY 11549

AWARDS Biology (MA); human cytogenetics (MS). Part-time and evening/weekend programs available.

Degree requirements: For master's, thesis, internship (MS) required.
Entrance requirements: For master's, bachelor's degree in biology or equivalent.
Expenses: Tuition: Full-time $11,400. Required fees: $670. Tuition and fees vary according to course load and program.
Faculty research: Evolution, genetics, aquaculture, molecular and cellular biology, marine sciences.
Find an in-depth description at www.petersons.com/graduate.

■ HOOD COLLEGE

Graduate School, Program in Biomedical Science, Frederick, MD 21701-8575

AWARDS MS. Part-time and evening/weekend programs available.

Students: 5 full-time (4 women), 93 part-time (67 women); includes 14 minority (5 African Americans, 6 Asian Americans or Pacific Islanders, 2 Hispanic Americans, 1 Native American), 2 international. Average age 33. In 1999, 12 degrees awarded.
Degree requirements: For master's, thesis or alternative required, foreign language not required.
Entrance requirements: For master's, bachelor's degree in biology; minimum GPA of 2.5; undergraduate course work in cell biology, chemistry, organic chemistry, and genetics. *Application deadline:* Applications are processed on a rolling basis. *Application fee:* $30.
Expenses: Tuition: Full-time $5,310; part-time $295 per credit hour. Required fees: $5 per credit hour.
Financial aid: Career-related internships or fieldwork, institutionally sponsored loans, and tuition waivers (partial) available. Aid available to part-time students. Financial aid applicants required to submit FAFSA.
Faculty research: Cell cycle regulation, developmental biology, DNA repair and recombination, oncogene expression, vaccine development.
Dr. Ricky Hirschhorn, Program Director, 301-696-3649, *Fax:* 301-696-3597, *E-mail:* hirschhorn@hood.edu.
Application contact: 301-696-3600, *Fax:* 301-696-3597, *E-mail:* hoodgrad@hood.edu.

■ HOWARD UNIVERSITY

Graduate School of Arts and Sciences, Department of Biology, Washington, DC 20059-0002

AWARDS MS, PhD. Part-time programs available.

Faculty: 19.
Students: 60; includes 51 minority (50 African Americans, 1 Asian American or

Pacific Islander), 7 international. Average age 33. *71 applicants, 44% accepted.* In 1999, 14 master's awarded (20% found work related to degree, 70% continued full-time study); 2 doctorates awarded (50% entered university research/teaching, 50% found other work related to degree).
Degree requirements: For master's and doctorate, thesis/dissertation, qualifying exams required, foreign language not required. *Average time to degree:* Master's–2 years full-time, 5 years part-time; doctorate–4 years full-time, 6 years part-time.
Entrance requirements: For master's and doctorate, GRE General Test, TOEFL, minimum GPA of 3.0. *Application deadline:* For fall admission, 2/1 (priority date); for spring admission, 11/1. *Application fee:* $45. Electronic applications accepted.
Expenses: Tuition: Full-time $10,500; part-time $583 per credit hour. Required fees: $405; $203 per semester.
Financial aid: In 1999–00, 36 students received aid, including 6 fellowships, 4 research assistantships, 14 teaching assistantships; career-related internships or fieldwork, institutionally sponsored loans, and teaching associateships also available. Financial aid application deadline: 4/1.
Faculty research: Physiology, molecular biology, cell biology, microbiology, environmental biology.
Dr. Arthur L. Williams, Chairman, 202-806-6933, *Fax:* 202-806-4564, *E-mail:* alwilliams@fac.howard.edu.
Application contact: Lila Stroud, Administrative Secretary, 202-806-6933, *Fax:* 202-806-4564.

■ HUMBOLDT STATE UNIVERSITY

Graduate Studies, College of Natural Resources and Sciences, Department of Biological Sciences, Arcata, CA 95521-8299

AWARDS MA.

Faculty: 23 full-time (3 women), 4 part-time/adjunct (2 women).
Students: 26 full-time (14 women), 16 part-time (7 women); includes 3 minority (2 Asian Americans or Pacific Islanders, 1 Hispanic American), 1 international. Average age 30. *39 applicants, 46% accepted.* In 1999, 8 degrees awarded.
Degree requirements: For master's, project or thesis required.
Entrance requirements: For master's, GRE General Test, TOEFL, appropriate bachelor's degree, minimum GPA of 2.5. *Application deadline:* Applications are processed on a rolling basis. *Application fee:* $55.

Expenses: Tuition, nonresident: full-time $5,904; part-time $246 per unit. Required fees: $1,936; $1,306 per year.
Financial aid: Application deadline: 3/1;
Faculty research: Plant ecology, DNA sequencing, invertebrates.
Dr. Tim Lawlor, Coordinator, 707-826-3245, *E-mail:* tel1@axe.humboldt.edu.

■ HUNTER COLLEGE OF THE CITY UNIVERSITY OF NEW YORK

Graduate School, Division of Sciences and Mathematics, Department of Biological Sciences, New York, NY 10021-5085

AWARDS MA, PhD. Part-time programs available.

Degree requirements: For master's, one foreign language (computer language can substitute), comprehensive exam or thesis required.
Entrance requirements: For master's, GRE General Test, TOEFL.
Expenses: Tuition, state resident: full-time $4,350; part-time $185 per credit. Tuition, nonresident: full-time $7,600; part-time $320 per credit. Required fees: $8 per term.
Faculty research: Analysis of prokaryotic and eukaryotic DNA, protein structure, mammalian DNA replication, oncogene expression, neuroscience.

■ ICR GRADUATE SCHOOL

Graduate Programs, Santee, CA 92071

AWARDS Astro/geophysics (MS); biology (MS); geology (MS); science education (MS). Part-time programs available.

Faculty: 6 full-time (0 women), 9 part-time/adjunct (1 woman).
Students: 18 full-time (8 women), 14 part-time (3 women). Average age 30. In 1999, 8 degrees awarded (38% entered university research/teaching, 62% found other work related to degree).
Degree requirements: For master's, thesis required (for some programs), foreign language not required. *Average time to degree:* Master's–3 years full-time, 5 years part-time.
Application deadline: Applications are processed on a rolling basis. *Application fee:* $30.
Expenses: Tuition: Full-time $1,350; part-time $150 per semester hour. Required fees: $200 per term. Tuition and fees vary according to course load.
Faculty research: Age of the earth, limits of variation, catastrophe, optimum methods for teaching. *Total annual research expenditures:* $20,000.
Kenneth B. Cumming, Dean, 619-448-0900, *Fax:* 619-448-3469.

ICR Graduate School (continued)
Application contact: Dr. Jack Kriege, Registrar, 619-448-0900, *Fax:* 619-448-3469.

■ IDAHO STATE UNIVERSITY

Office of Graduate Studies, College of Arts and Sciences, Department of Biological Sciences, Pocatello, ID 83209

AWARDS Biology (MS, DA, PhD); microbiology (MS); natural science (MNS).

Faculty: 37 full-time (10 women), 6 part-time/adjunct (2 women).

Students: 61 full-time (19 women), 10 part-time (5 women), 4 international. Average age 30. In 1999, 7 master's, 4 doctorates awarded.

Degree requirements: For master's, one foreign language (computer language can substitute), thesis required; for doctorate, 2 foreign languages (computer language can substitute for one), dissertation required.

Entrance requirements: For master's, GRE General Test; for doctorate, GRE General Test, GRE Subject Test. *Application deadline:* For fall admission, 7/1; for spring admission, 12/1. Applications are processed on a rolling basis. *Application fee:* $30.

Expenses: Tuition, nonresident: full-time $6,240; part-time $90 per credit. Required fees: $3,384; $147 per credit.

Financial aid: In 1999–00, 8 fellowships, 20 research assistantships, 19 teaching assistantships were awarded; Federal Work-Study and institutionally sponsored loans also available.

Faculty research: Ecology and evolutionary biology, plant and animal physiology, plant and animal developmental biology, immunology, molecular biology. *Total annual research expenditures:* $800,000.
Dr. Rod R. Seeley, Chairman, 208-282-3765, *Fax:* 208-282-4570.

Find an in-depth description at www.petersons.com/graduate.

■ ILLINOIS INSTITUTE OF TECHNOLOGY

Graduate College, Armour College of Engineering and Sciences, Department of Biological, Chemical and Physical Sciences, Biology Division, Chicago, IL 60616-3793

AWARDS Biochemistry (MS); biology (PhD); biotechnology (MS); cell biology (MS); microbiology (MS). Part-time and evening/weekend programs available.

Faculty: 8 full-time (0 women), 2 part-time/adjunct (0 women).

Students: 38 full-time (15 women), 80 part-time (52 women); includes 34 minority (27 African Americans, 5 Asian Americans or Pacific Islanders, 2 Hispanic Americans), 39 international. *155 applicants, 36% accepted.* In 1999, 10 master's, 2 doctorates awarded. Terminal master's awarded for partial completion of doctoral program.

Degree requirements: For master's, thesis (for some programs), comprehensive exam required, foreign language not required; for doctorate, dissertation, comprehensive exam required, foreign language not required.

Entrance requirements: For master's and doctorate, GRE General Test, TOEFL, minimum undergraduate GPA of 3.0. *Application deadline:* For fall admission, 7/1; for spring admission, 11/1. Applications are processed on a rolling basis. *Application fee:* $30. Electronic applications accepted.

Expenses: Tuition: Part-time $590 per credit hour. Required fees: $100. Tuition and fees vary according to course load and program.

Financial aid: In 1999–00, 7 fellowships, 1 research assistantship, 5 teaching assistantships were awarded; Federal Work-Study, institutionally sponsored loans, scholarships, and unspecified assistantships also available. Aid available to part-time students. Financial aid application deadline: 3/1; financial aid applicants required to submit FAFSA.

Faculty research: Genetics, molecular biology.
Dr. Benjamin Stark, Associate Chair, 312-567-3980, *Fax:* 312-567-3494, *E-mail:* starkb@iit.edu.

Application contact: Dr. S. Mohammad Shahidehpour, Dean of Graduate College, 312-567-3024, *Fax:* 312-567-7517, *E-mail:* gradstu@alpha1.ais.iit.edu.

■ ILLINOIS STATE UNIVERSITY

Graduate School, College of Arts and Sciences, Department of Biological Sciences, Normal, IL 61790-2200

AWARDS Biological sciences (MS); biology (PhD); botany (PhD); ecology (PhD); genetics (PhD); microbiology (PhD); physiology (PhD); zoology (PhD). Part-time programs available.

Faculty: 24 full-time (4 women).

Students: 57 full-time (25 women), 27 part-time (10 women); includes 6 minority (1 African American, 3 Asian Americans or Pacific Islanders, 2 Hispanic Americans), 13 international. *58 applicants, 59% accepted.* In 1999, 12 master's, 3 doctorates awarded.

Degree requirements: For master's, thesis or alternative required; for doctorate, variable foreign language requirement (computer language can substitute for one), dissertation, 2 terms of residency required.

Entrance requirements: For master's, GRE General Test, minimum GPA of 2.6 in last 60 hours; for doctorate, GRE General Test. *Application deadline:* Applications are processed on a rolling basis. *Application fee:* $0.

Expenses: Tuition, state resident: full-time $2,526; part-time $105 per credit hour. Tuition, nonresident: full-time $7,578; part-time $316 per credit hour. Required fees: $1,082; $38 per credit hour. Tuition and fees vary according to course load and program.

Financial aid: In 1999–00, 8 research assistantships, 63 teaching assistantships were awarded; Federal Work-Study, tuition waivers (full), and unspecified assistantships also available. Financial aid application deadline: 4/1.

Faculty research: Phenotypic plasticity in reproduction: molecular mechanisms, physiological control, adaptive significance; molecular stress physiology of *Listeria monocytogenes*; enzymology of eggshell formation in *Schistosoma mansoni*, compensatory adaptation in the rat midbrain after neurodegeneration, analysis of staphylococcal virulence germs. *Total annual research expenditures:* $586,648.
Dr. Hou Cheung, Chairperson, 309-438-3669.

Application contact: Derek A. McCracken, Graduate Adviser, 309-438-3664.

Find an in-depth description at www.petersons.com/graduate.

■ INDIANA STATE UNIVERSITY

School of Graduate Studies, College of Arts and Sciences, Department of Life Sciences, Terre Haute, IN 47809-1401

AWARDS Clinical laboratory sciences (MS); ecology (MA, MS, PhD); microbiology (MA, MS, PhD); physiology (MA, MS, PhD).

Degree requirements: For doctorate, computer language, dissertation required.

Entrance requirements: For master's and doctorate, GRE General Test. Electronic applications accepted.

Expenses: Tuition, state resident: full-time $3,552; part-time $148 per hour. Tuition, nonresident: full-time $8,088; part-time $337 per hour.

Find an in-depth description at www.petersons.com/graduate.

■ INDIANA UNIVERSITY BLOOMINGTON

Graduate School, College of Arts and Sciences, Department of Biology, Bloomington, IN 47405

AWARDS Biology teaching (MAT); evolution, ecology, and behavior (MA, PhD), including ecology, evolutionary biology, zoology; genetics (PhD); microbiology (MA, PhD);

molecular, cellular, and developmental biology (PhD); plant sciences, molecular and organismal biology (MA, PhD). PhD offered through the University Graduate School. Part-time programs available.

Faculty: 35 full-time (5 women), 10 part-time/adjunct (2 women).

Students: 200 full-time (89 women), 2 part-time (1 woman). In 1999, 8 master's awarded (90% entered university research/teaching); 20 doctorates awarded (99% entered university research/teaching, 1% found other work related to degree). Terminal master's awarded for partial completion of doctoral program.

Degree requirements: For master's and doctorate, thesis/dissertation, oral defense required.

Entrance requirements: For master's and doctorate, GRE General Test, TOEFL. *Application deadline:* For fall admission, 1/5 (priority date); for spring admission, 9/1 (priority date). Applications are processed on a rolling basis. *Application fee:* $45. Electronic applications accepted.

Expenses: Tuition, state resident: full-time $3,853; part-time $161 per credit hour. Tuition, nonresident: full-time $11,226; part-time $468 per credit hour. Required fees: $360 per year. Tuition and fees vary according to course load and program.

Financial aid: In 1999–00, 130 students received aid, including 15 fellowships with tuition reimbursements available (averaging $16,500 per year), 25 research assistantships with tuition reimbursements available (averaging $15,000 per year), 70 teaching assistantships with tuition reimbursements available (averaging $15,000 per year); grants, scholarships, and tuition waivers (full) also available.

Dr. Jeffrey D. Palmer, Chair, 812-855-6283.

Application contact: Gretchen Clearwater, Advisor for Graduate Affairs, 812-855-1861, *Fax:* 812-855-6705, *E-mail:* biograd@bio.indiana.edu.

Find an in-depth description at www.petersons.com/graduate.

■ INDIANA UNIVERSITY BLOOMINGTON

Medical Sciences Program, Bloomington, IN 47405

AWARDS Anatomy and cell biology (MA, PhD); pharmacology (MS, PhD); physiology (MA, PhD).

Faculty: 12 full-time (3 women).

Students: 10 full-time (6 women), 6 part-time (2 women); includes 1 minority (Asian American or Pacific Islander), 3 international. In 1999, 1 master's, 1 doctorate awarded.

Entrance requirements: For master's, GRE, TOEFL, minimum GPA of 3.0; for

doctorate, GRE, TOEFL. *Application deadline:* For fall admission, 1/15. *Application fee:* $45.

Expenses: Tuition, state resident: full-time $3,853; part-time $161 per credit hour. Tuition, nonresident: full-time $11,226; part-time $468 per credit hour. Required fees: $360 per year. Tuition and fees vary according to course load and program. Dr. Talmage Bosin, Assistant Dean, 812-855-8118, *E-mail:* bosin@indiana.edu.

Application contact: Kimberly Bunch, Director of Graduate Admissions, 812-855-1119, *E-mail:* kbunch@indiana.edu.

■ INDIANA UNIVERSITY OF PENNSYLVANIA

Graduate School and Research, College of Natural Sciences and Mathematics, Department of Biology, Indiana, PA 15705-1087

AWARDS MS.

Students: 16 full-time (9 women), 2 part-time (both women), 2 international. Average age 26. *26 applicants, 58% accepted.* In 1999, 6 degrees awarded.

Degree requirements: For master's, thesis optional, foreign language not required.

Entrance requirements: For master's, TOEFL. *Application deadline:* For fall admission, 7/1 (priority date); for spring admission, 11/1. Applications are processed on a rolling basis. *Application fee:* $30.

Expenses: Tuition, state resident: full-time $3,780; part-time $210 per credit hour. Tuition, nonresident: full-time $6,610; part-time $367 per credit hour. Required fees: $705; $138 per semester.

Financial aid: Research assistantships, Federal Work-Study available. Aid available to part-time students. Financial aid application deadline: 3/15.

Dr. W. Barkley Butler, Chairperson, 724-357-2352, *E-mail:* bbutler@grove.iup.edu.

Application contact: Dr. Robert Hinrichsen, Graduate Coordinator, 724-357-2612, *E-mail:* bhinrich@grove.iup.edu.

Find an in-depth description at www.petersons.com/graduate.

■ INDIANA UNIVERSITY–PURDUE UNIVERSITY FORT WAYNE

School of Arts and Sciences, Department of Biological Sciences, Fort Wayne, IN 46805-1499

AWARDS Biology (MS). Part-time and evening/weekend programs available.

Faculty: 10 full-time (2 women), 1 (woman) part-time/adjunct.

Students: 11 full-time (4 women), 9 part-time (7 women). Average age 27. *13 applicants, 85% accepted.* In 1999, 6 degrees awarded.

Degree requirements: For master's, thesis optional, foreign language not required. *Average time to degree:* Master's–2 years full-time, 4 years part-time.

Entrance requirements: For master's, GRE General Test, minimum GPA of 2.8, major or minor in biology. *Application deadline:* For fall admission, 4/15 (priority date); for spring admission, 12/1. Applications are processed on a rolling basis. *Application fee:* $30.

Expenses: Tuition, state resident: full-time $2,471; part-time $137 per credit hour. Tuition, nonresident: full-time $5,528; part-time $307 per credit hour. Required fees: $207; $1,650 per credit hour.

Financial aid: In 1999–00, 7 students received aid, including research assistantships with partial tuition reimbursements available (averaging $7,000 per year), teaching assistantships with partial tuition reimbursements available (averaging $7,000 per year); Federal Work-Study and grants also available. Aid available to part-time students. Financial aid application deadline: 3/1; financial aid applicants required to submit FAFSA.

Faculty research: Behavior, aquatic and physiological ecology, molecular biology and immunology, toxicology, conservation. *Total annual research expenditures:* $620,000. Dr. Frank V. Paladino, Chairperson, 219-481-6305, *Fax:* 219-481-6087, *E-mail:* paladino@ipfw.edu.

■ INDIANA UNIVERSITY–PURDUE UNIVERSITY INDIANAPOLIS

School of Medicine, Graduate Programs in Medicine, Indianapolis, IN 46202-5114

AWARDS MS, PhD, MD/MS, MD/PhD.

Students: 111 full-time (62 women), 69 part-time (30 women); includes 10 minority (3 African Americans, 6 Asian Americans or Pacific Islanders, 1 Hispanic American), 57 international. Average age 26. In 1999, 39 master's, 25 doctorates awarded. Terminal master's awarded for partial completion of doctoral program.

Degree requirements: For doctorate, dissertation required. *Average time to degree:* Master's–2.5 years full-time; doctorate–6 years full-time.

Entrance requirements: For master's and doctorate, GRE General Test. *Application fee:* $35 ($55 for international students).

Expenses: Tuition, state resident: full-time $13,245; part-time $158 per credit hour. Tuition, nonresident: full-time $30,330; part-time $455 per credit hour. Required fees: $121 per year. Tuition and fees vary according to course load and degree level.

Financial aid: In 1999–00, 148 students received aid, including 30 fellowships with full and partial tuition reimbursements

Indiana University–Purdue University Indianapolis (continued)

available, 90 research assistantships with full and partial tuition reimbursements available, 7 teaching assistantships with full tuition reimbursements available; career-related internships or fieldwork, Federal Work-Study, grants, institutionally sponsored loans, scholarships, traineeships, tuition waivers (full and partial), unspecified assistantships and stipends also available. Aid available to part-time students. **Application contact:** Dr. William Bosron, Assistant Dean for Graduate Studies, 317-274-3441, *E-mail:* wbosron@iupui.edu.

■ INDIANA UNIVERSITY–PURDUE UNIVERSITY INDIANAPOLIS

School of Science, Department of Biology, Indianapolis, IN 46202-2896

AWARDS MS, PhD. Part-time and evening/weekend programs available.

Students: 54 full-time (20 women), 28 part-time (16 women); includes 11 minority (5 African Americans, 3 Asian Americans or Pacific Islanders, 3 Hispanic Americans), 3 international. Average age 25. In 1999, 49 degrees awarded. Terminal master's awarded for partial completion of doctoral program.
Degree requirements: For master's, thesis required (for some programs), foreign language not required; for doctorate, dissertation required, foreign language not required. *Average time to degree:* Master's–1.5 years full-time, 3 years part-time.
Entrance requirements: For master's and doctorate, GRE General Test. *Application deadline:* For fall admission, 6/1. *Application fee:* $35 ($55 for international students).
Expenses: Tuition, state resident: part-time $158 per credit hour. Tuition, nonresident: part-time $455 per credit hour. Required fees: $121 per year. Tuition and fees vary according to course load, degree level and program.
Financial aid: In 1999–00, 6 fellowships with partial tuition reimbursements (averaging $15,000 per year), 4 research assistantships with partial tuition reimbursements (averaging $11,000 per year), 14 teaching assistantships with partial tuition reimbursements (averaging $11,000 per year) were awarded; career-related internships or fieldwork also available. Financial aid application deadline: 4/1.
Faculty research: Cell and model membranes, cell and molecular biology, immunology, oncology, developmental biology.
Dr. N. Douglas Lees, Chair, 317-274-0588, *Fax:* 317-274-2846.

■ INSTITUTE OF PAPER SCIENCE AND TECHNOLOGY

Graduate Programs, Program in Biology, Atlanta, GA 30318-5794

AWARDS MS, PhD. Part-time programs available. Terminal master's awarded for partial completion of doctoral program.

Degree requirements: For master's, industrial experience, research project required, foreign language and thesis not required; for doctorate, dissertation required, foreign language not required.
Entrance requirements: For master's and doctorate, GRE, minimum GPA of 3.0.
Expenses: Tuition: Full-time $15,000; part-time $625 per credit hour.

■ JACKSON STATE UNIVERSITY

Graduate School, School of Science and Technology, Department of Biology, Jackson, MS 39217

AWARDS Biology education (MST); environmental science (MS, PhD). Part-time and evening/weekend programs available.

Degree requirements: For master's, comprehensive exam, thesis (alternative accepted for MST) required; for doctorate, dissertation, comprehensive exam required.
Entrance requirements: For master's, GRE General Test, TOEFL; for doctorate, MAT.
Expenses: Tuition, state resident: full-time $2,688. Tuition, nonresident: full-time $2,994. Part-time tuition and fees vary according to course load.
Faculty research: Comparative studies on the carbohydrate composition of marine macroalgae, host-parasite relationship between the spruce budworm and entomepathogen fungus.

■ JACKSONVILLE STATE UNIVERSITY

College of Graduate Studies and Continuing Education, College of Arts and Sciences, Department of Biology, Jacksonville, AL 36265-1602

AWARDS MS.

Faculty: 13 full-time (2 women).
Students: 9 full-time (8 women), 21 part-time (10 women); includes 3 minority (2 African Americans, 1 Native American), 3 international. In 1999, 13 degrees awarded.
Degree requirements: For master's, thesis optional.
Entrance requirements: For master's, GRE General Test or MAT. *Application deadline:* Applications are processed on a rolling basis. *Application fee:* $20.
Expenses: Tuition, area resident: Part-time $122 per credit hour.
Financial aid: Available to part-time students. Application deadline: 4/1.

Application contact: 256-782-5329.

■ JAMES MADISON UNIVERSITY

Graduate School, College of Science and Mathematics, Department of Biology, Harrisonburg, VA 22807

AWARDS MS. Part-time programs available.

Faculty: 5 full-time (0 women).
Students: 8 full-time (5 women), 2 part-time (1 woman); includes 2 minority (1 Asian American or Pacific Islander, 1 Native American), 1 international. Average age 29. In 1999, 1 degree awarded.
Degree requirements: For master's, thesis required, foreign language not required.
Entrance requirements: For master's, GRE General Test, GRE Subject Test. *Application deadline:* For fall admission, 7/1 (priority date). Applications are processed on a rolling basis. *Application fee:* $50.
Expenses: Tuition, state resident: full-time $3,240; part-time $135 per credit hour. Tuition, nonresident: full-time $9,960; part-time $415 per credit hour.
Financial aid: In 1999–00, 2 research assistantships with full tuition reimbursements (averaging $6,010 per year) were awarded; teaching assistantships, Federal Work-Study and unspecified assistantships also available. Financial aid application deadline: 2/15; financial aid applicants required to submit FAFSA.
Faculty research: Evolutionary ecology, gene regulation, microbial ecology, plant development, biomechanics.
Find an in-depth description at www.petersons.com/graduate.

■ JOAN AND SANFORD I. WEILL MEDICAL COLLEGE AND GRADUATE SCHOOL OF MEDICAL SCIENCES OF CORNELL UNIVERSITY

Graduate School of Medical Sciences, New York, NY 10021

AWARDS MS, PhD, MD/PhD.

Faculty: 214 full-time (50 women), 2 part-time/adjunct (0 women).
Students: 216 full-time (111 women); includes 51 minority (7 African Americans, 37 Asian Americans or Pacific Islanders, 7 Hispanic Americans), 62 international. Average age 27. *456 applicants, 19% accepted.* In 2000, 8 master's, 25 doctorates awarded.
Degree requirements: For doctorate, dissertation, final exam required.
Entrance requirements: For doctorate, GRE General Test, GRE Subject Test, MCAT (MD/PhD). *Application deadline:* For fall admission, 1/15. *Application fee:* $50.

Expenses: All students in good standing receive an annual stipend of $22,880.
Financial aid: Fellowships, grants, tuition waivers (full), and stipends available. Dr. David P. Hajjar, Dean, 212-746-6565.
Application contact: Liliana Montano, Graduate Field Assistant, 607-254-4340.

Find an in-depth description at www.petersons.com/graduate.

■ JOAN AND SANFORD I. WEILL MEDICAL COLLEGE AND GRADUATE SCHOOL OF MEDICAL SCIENCES OF CORNELL UNIVERSITY

Medical College, MD/PhD Program, New York, NY 10021-4896

AWARDS MD/PhD. Offered through the Tri-Institutional Program with Rockefeller University and Sloan-Kettering Institute.

Faculty: 234.
Students: 86 full-time (26 women); includes 30 minority (4 African Americans, 20 Asian Americans or Pacific Islanders, 6 Hispanic Americans), 13 international. *290 applicants, 2% accepted.*
Application deadline: For fall admission, 10/15. *Application fee:* $0.
Expenses: All students are fully funded.
Financial aid: In 2000–01, 86 students received aid.
Faculty research: Neuroscience, pharmacology, immunology, structural biology, genetics.
Dr. Olaf S. Andersen, Director, 212-746-6023, *E-mail:* mdphd@mail.med.cornell.edu.
Application contact: Ruth Gotian, Coordinator, 212-746-6023, *Fax:* 212-746-8678, *E-mail:* mdphd@mail.med.cornell.edu.

■ JOHN CARROLL UNIVERSITY

Graduate School, Department of Biology, University Heights, OH 44118-4581

AWARDS MA, MS. Part-time programs available.

Faculty: 10 full-time (3 women).
Students: 16 full-time (9 women), 12 part-time (8 women); includes 5 minority (1 African American, 4 Asian Americans or Pacific Islanders). Average age 25. *8 applicants, 75% accepted.* In 1999, 6 degrees awarded (50% entered university research/teaching, 50% found other work related to degree).
Degree requirements: For master's, essay or thesis required. *Average time to degree:* Master's–2 years full-time, 3 years part-time.
Entrance requirements: For master's, undergraduate major in biology, 1 semester

of biochemistry. *Application deadline:* For fall admission, 8/15 (priority date); for spring admission, 1/3. Applications are processed on a rolling basis. *Application fee:* $25 ($35 for international students).
Expenses: Tuition: Part-time $498 per credit hour. Part-time tuition and fees vary according to program.
Financial aid: In 1999–00, 13 students received aid, including 13 teaching assistantships with full tuition reimbursements available Financial aid application deadline: 3/1; financial aid applicants required to submit FAFSA.
Faculty research: Algal ecology, wetlands rehabilitation, systematics, molecular genetics, neurophysiology. *Total annual research expenditures:* $300,000.
Dr. Miles M. Coburn, Chairperson, 216-397-4253, *Fax:* 216-397-4482, *E-mail:* coburn@jcu.edu.
Application contact: Dr. Jeffrey Johansen, Graduate Coordinator, 216-397-3077, *Fax:* 216-397-4482, *E-mail:* johansen@jcu.edu.

■ JOHNS HOPKINS UNIVERSITY

School of Medicine, Graduate Programs in Medicine, Baltimore, MD 21218-2699

AWARDS MA, MS, PhD, MD/PhD.

Faculty: 201 full-time (57 women), 32 part-time/adjunct (9 women).
Students: 469 full-time (215 women); includes 85 minority (10 African Americans, 66 Asian Americans or Pacific Islanders, 8 Hispanic Americans, 1 Native American), 151 international. *768 applicants, 12% accepted.* In 1999, 12 master's, 48 doctorates awarded. Terminal master's awarded for partial completion of doctoral program.
Degree requirements: For master's, thesis required, foreign language not required; for doctorate, dissertation required.
Application deadline: Applications are processed on a rolling basis. *Application fee:* $50.
Expenses: Tuition: Full-time $23,660.
Financial aid: Fellowships, research assistantships, teaching assistantships, career-related internships or fieldwork, Federal Work-Study, institutionally sponsored loans, and tuition waivers (full) available. Financial aid applicants required to submit FAFSA.
Dr. James Hildreth, Associate Dean for Graduate Student Affairs, 410-614-3385.

■ JOHNS HOPKINS UNIVERSITY

Zanvyl Krieger School of Arts and Sciences, Department of Biology, Baltimore, MD 21218-2699

AWARDS Biochemistry (PhD); cell biology (PhD); developmental biology (PhD); genetic biology (PhD); molecular biology (PhD).

Faculty: 23 full-time (4 women).
Students: 85 full-time (50 women); includes 23 minority (4 African Americans, 14 Asian Americans or Pacific Islanders, 5 Hispanic Americans), 18 international. Average age 24. *202 applicants, 27% accepted.* In 1999, 17 doctorates awarded.
Degree requirements: For doctorate, dissertation required. *Average time to degree:* Doctorate–6 years full-time.
Entrance requirements: For doctorate, GRE General Test, GRE Subject Test. *Application deadline:* For fall admission, 12/15 (priority date). Applications are processed on a rolling basis. *Application fee:* $55.
Expenses: Tuition: Full-time $24,930. Tuition and fees vary according to program.
Financial aid: In 1999–00, 73 students received aid, including 12 fellowships, 61 research assistantships, 12 teaching assistantships; Federal Work-Study and institutionally sponsored loans also available. Financial aid application deadline: 12/15; financial aid applicants required to submit FAFSA.
Faculty research: Protein and nucleic acid biochemistry and biophysical chemistry, molecular biology and development. *Total annual research expenditures:* $9.3 million.
Dr. Victor G. Corces, Chair, 410-516-4693, *Fax:* 410-516-5213, *E-mail:* corces_v@jhuvms.hct.jhu.edu.
Application contact: Joan Miller, Graduate Admissions Coordinator, 410-516-5502, *Fax:* 410-516-5213, *E-mail:* joan@jhu.edu.

Find an in-depth description at www.petersons.com/graduate.

■ KANSAS STATE UNIVERSITY

Graduate School, College of Arts and Sciences, Division of Biology, Manhattan, KS 66506

AWARDS Cell biology (MS, PhD); developmental biology and physiology (MS, PhD); microbiology and immunology (MS, PhD); molecular biology and genetics (MS, PhD); systematics and ecology (MS, PhD); virology and oncology (MS, PhD). Terminal master's awarded for partial completion of doctoral program.

Degree requirements: For master's and doctorate, thesis/dissertation required, foreign language not required.

Kansas State University (continued)

Entrance requirements: For master's and doctorate, GRE General Test. Electronic applications accepted.

Expenses: Tuition, state resident: part-time $103 per credit hour. Tuition, nonresident: part-time $338 per credit hour. Required fees: $17 per credit hour. One-time fee: $64 part-time.

Faculty research: Immune cell function, prairie ecology.

■ **KECK GRADUATE INSTITUTE OF APPLIED LIFE SCIENCES**

Program in Biosciences, Claremont, CA 91711

AWARDS MBS.

Faculty: 6 full-time (1 woman), 3 part-time/adjunct (0 women).

Entrance requirements: For master's, GRE General Test. *Application deadline:* For fall admission, 2/15 (priority date). *Application fee:* $60. Electronic applications accepted.

Financial aid: Fellowships with full and partial tuition reimbursements, grants, institutionally sponsored loans, and scholarships available.

Faculty research: Computational biology, drug discovery and development, molecular and cellular biology, biomedical engineering.

Henry E. Riggs, President, 909-607-7855, *Fax:* 909-607-8086.

Application contact: John Friesman, Director of Admissions and Student Services, 909-607-8590, *Fax:* 909-607-8598, *E-mail:* admissions@kgi.edu.

Find an in-depth description at www.petersons.com/graduate.

■ **KENT STATE UNIVERSITY**

College of Arts and Sciences, Department of Biological Sciences, Kent, OH 44242-0001

AWARDS Botany (MA, MS, PhD); ecology (MS, PhD); physiology (MS, PhD); zoology (MA, PhD).

Faculty: 28 full-time.

Students: 31 full-time (17 women), 10 part-time (9 women); includes 1 minority (African American), 12 international. *30 applicants, 77% accepted.* In 1999, 6 master's, 5 doctorates awarded.

Degree requirements: For master's and doctorate, thesis/dissertation required, foreign language not required.

Entrance requirements: For master's, GRE General Test, minimum GPA of 2.75; for doctorate, GRE General Test, minimum GPA of 3.0. *Application deadline:* For fall admission, 7/12. Applications are processed on a rolling basis. *Application fee:* $30.

Expenses: Tuition, state resident: full-time $5,334; part-time $243 per hour. Tuition, nonresident: full-time $10,238; part-time $466 per hour.

Financial aid: Fellowships, research assistantships, teaching assistantships, Federal Work-Study, institutionally sponsored loans, and tuition waivers (full) available. Financial aid application deadline: 2/1.

Dr. Brent C. Bruot, Chairman, 330-672-3613, *Fax:* 330-672-3713.

Application contact: Dr. John R. D. Stalvey, Coordinator of Graduate Studies, 330-672-2819.

■ **KENT STATE UNIVERSITY**

School of Biomedical Sciences, Kent, OH 44242-0001

AWARDS MS, PhD. Offered in cooperation with Northeastern Ohio Universities College of Medicine. Terminal master's awarded for partial completion of doctoral program.

Degree requirements: For master's and doctorate, thesis/dissertation required.

Entrance requirements: For master's and doctorate, GRE General Test.

Expenses: Tuition, state resident: full-time $5,334; part-time $243 per hour. Tuition, nonresident: full-time $10,238; part-time $466 per hour.

Find an in-depth description at www.petersons.com/graduate.

■ **LAMAR UNIVERSITY**

College of Graduate Studies, College of Arts and Sciences, Department of Biology, Beaumont, TX 77710

AWARDS MS. Part-time and evening/weekend programs available.

Faculty: 7 full-time (2 women).

Students: 3 full-time (all women), 4 part-time (3 women); includes 1 minority (African American). Average age 29.

Degree requirements: For master's, thesis required, foreign language not required.

Entrance requirements: For master's, GRE General Test, TOEFL, minimum GPA of 2.5 in last 60 hours of undergraduate course work. *Application deadline:* For fall admission, 8/1; for spring admission, 12/1. Applications are processed on a rolling basis. *Application fee:* $0.

Expenses: Tuition, area resident: Part-time $62 per hour. Tuition, state resident: full-time $1,488; part-time $62 per hour. Tuition, nonresident: full-time $6,672; part-time $278 per hour. Required fees: $536. Tuition and fees vary according to program.

Financial aid: In 1999–00, 1 research assistantship (averaging $5,200 per year), 3 teaching assistantships (averaging $5,200

per year) were awarded. Financial aid application deadline: 4/1.

Faculty research: Physiology, ichthyology, microbiology, vertebrate evolution, behavior.

Dr. Michael E. Warren, Chair, 409-880-8262, *Fax:* 409-880-1827.

Application contact: Dr. R. C. Harrel, Graduate Adviser, 409-880-8255, *Fax:* 409-880-1827.

■ **LEHIGH UNIVERSITY**

College of Arts and Sciences, Department of Biological Sciences, Bethlehem, PA 18015-3094

AWARDS Behavioral and evolutionary bioscience (PhD); behavioral neuroscience (PhD); biochemistry (PhD); biology (PhD); molecular biology (PhD). Part-time programs available. Postbaccalaureate distance learning degree programs offered (no on-campus study).

Students: 31 full-time (22 women), 77 part-time (47 women); includes 11 minority (3 African Americans, 5 Asian Americans or Pacific Islanders, 3 Hispanic Americans), 8 international. *142 applicants, 22% accepted.* In 1999, 5 doctorates awarded.

Degree requirements: For doctorate, dissertation, comprehensive exam required, foreign language not required. *Average time to degree:* Doctorate–6.3 years full-time.

Entrance requirements: For doctorate, GRE General Test, GRE Subject Test, TOEFL. *Application deadline:* For fall admission, 7/15; for spring admission, 12/1. Applications are processed on a rolling basis. *Application fee:* $40. Electronic applications accepted.

Expenses: Tuition: Part-time $860 per credit. Required fees: $6 per term. Tuition and fees vary according to program.

Financial aid: In 1999–00, 30 students received aid, including 4 fellowships, 6 research assistantships, 15 teaching assistantships; career-related internships or fieldwork, institutionally sponsored loans, and tuition waivers (full and partial) also available. Financial aid application deadline: 1/15.

Faculty research: Gene expression, cell biology, virology, bacteriology, developmental biology.

Dr. Neal G. Simon, Chairperson, 610-758-3680, *Fax:* 610-758-4004.

Application contact: Dr. Jennifer J. Swann, Graduate Coordinator, 610-758-5884, *Fax:* 610-758-4004, *E-mail:* jms5@lehigh.edu.

■ LEHMAN COLLEGE OF THE CITY UNIVERSITY OF NEW YORK

Division of Natural and Social Sciences, Department of Biological Sciences, Bronx, NY 10468-1589

AWARDS Biology (MA); plant sciences (PhD). Part-time and evening/weekend programs available.

Faculty: 7 full-time (1 woman), 1 part-time/adjunct (0 women).
Students: 2 full-time (0 women), 8 part-time (4 women). Average age 30. Terminal master's awarded for partial completion of doctoral program.
Degree requirements: For master's, thesis not required; for doctorate, dissertation required.
Entrance requirements: For doctorate, GRE General Test. *Application deadline:* For fall admission, 4/1; for spring admission, 11/1. Applications are processed on a rolling basis. *Application fee:* $40.
Expenses: Tuition, state resident: full-time $4,350; part-time $185 per credit. Tuition, nonresident: full-time $7,600; part-time $320 per credit.
Financial aid: Fellowships, research assistantships, teaching assistantships, Federal Work-Study and tuition waivers (full and partial) available. Aid available to part-time students. Financial aid application deadline: 5/15; financial aid applicants required to submit FAFSA.
Faculty research: Taste receptors, ultrastructure of C&I1 bacteria, hydroxyproline rich proteins and morphogenesis, foraging in fish, water relations and abscission in plants.
Thomas Jensen, Chairperson, 718-960-8235.
Application contact: Jack Valdovinos, Adviser, 718-960-8235.

■ LOMA LINDA UNIVERSITY

Graduate School, Department of Biology, Loma Linda, CA 92350

AWARDS MS, PhD. Part-time programs available. Terminal master's awarded for partial completion of doctoral program.

Degree requirements: For master's and doctorate, thesis/dissertation required.
Entrance requirements: For master's and doctorate, GRE General Test. *Application deadline:* For fall admission, 8/31. *Application fee:* $40.
Expenses: Tuition: Part-time $395 per unit.
Financial aid: Fellowships, Federal Work-Study and tuition waivers (full and partial) available. Aid available to part-time students.
Dr. David Cowles, Coordinator, 909-824-4530.

■ LOMA LINDA UNIVERSITY

Graduate School, Graduate Programs in Medicine, Loma Linda, CA 92350

AWARDS MS, PhD, MD/MS, MD/PhD. Part-time programs available.

Degree requirements: For doctorate, dissertation required.
Entrance requirements: For master's and doctorate, GRE General Test. *Application deadline:* Applications are processed on a rolling basis. *Application fee:* $40.
Expenses: Tuition: Part-time $395 per unit.
Financial aid: Tuition waivers (full and partial) available. Aid available to part-time students.
Dr. Daniel Giang, Associate Dean, 909-824-4466.

■ LONG ISLAND UNIVERSITY, BROOKLYN CAMPUS

Richard L. Conolly College of Liberal Arts and Sciences, Department of Biology, Brooklyn, NY 11201-8423

AWARDS MS. Part-time and evening/weekend programs available.

Degree requirements: For master's, thesis or alternative required, foreign language not required.
Electronic applications accepted.
Expenses: Tuition: Part-time $505 per credit. Full-time tuition and fees vary according to course load, degree level and program.

■ LONG ISLAND UNIVERSITY, C.W. POST CAMPUS

College of Liberal Arts and Sciences, Department of Biology, Brookville, NY 11548-1300

AWARDS MS. Part-time and evening/weekend programs available.

Faculty: 8 full-time (1 woman), 8 part-time/adjunct (1 woman).
Students: 3 full-time (2 women), 9 part-time (5 women); includes 4 minority (1 African American, 3 Asian Americans or Pacific Islanders), 3 international. *17 applicants, 71% accepted.* In 1999, 2 degrees awarded.
Degree requirements: For master's, thesis optional, foreign language not required.
Entrance requirements: For master's, GRE General Test, minimum GPA of 2.75 in major. *Application deadline:* Applications are processed on a rolling basis. *Application fee:* $30. Electronic applications accepted.
Expenses: Tuition: Part-time $405 per credit. Required fees: $310; $65 per year. Tuition and fees vary according to course load and program.

Financial aid: In 1999–00, 2 teaching assistantships were awarded; Federal Work-Study also available. Aid available to part-time students. Financial aid application deadline: 5/15; financial aid applicants required to submit FAFSA.
Faculty research: Recombinant DNA, molecular genetics, membrane receptors, biochemistry, animal ecology, plant systematics and ecology, animal behavior.
Michael Shodell, Chairman, 516-299-2481, *Fax:* 516-299-2768, *E-mail:* shodell@aurora.liunet.edu.
Application contact: Orland J. Blanchard, Graduate Adviser, 516-299-3041, *E-mail:* oblancha@titan.liunet.edu.

■ LONG ISLAND UNIVERSITY, C.W. POST CAMPUS

School of Health Professions, Department of Biomedical Sciences, Program in Medical Biology, Brookville, NY 11548-1300

AWARDS Hematology (MS); immunology (MS); medical chemistry (MS); microbiology (MS). Part-time and evening/weekend programs available.

Faculty: 3 full-time (1 woman), 12 part-time/adjunct (6 women).
Students: 4 full-time (2 women), 41 part-time (21 women); includes 25 minority (10 African Americans, 10 Asian Americans or Pacific Islanders, 5 Hispanic Americans). *37 applicants, 89% accepted.* In 1999, 11 degrees awarded.
Degree requirements: For master's, computer language, thesis required, foreign language not required. *Average time to degree:* Master's–2 years full-time, 4 years part-time.
Entrance requirements: For master's, TOEFL, minimum GPA of 2.75 in major. *Application deadline:* For fall admission, 9/1 (priority date); for spring admission, 1/20 (priority date). Applications are processed on a rolling basis. *Application fee:* $30. Electronic applications accepted.
Expenses: Tuition: Part-time $405 per credit. Required fees: $310; $65 per year. Tuition and fees vary according to course load and program.
Financial aid: In 1999–00, 20 students received aid; fellowships with partial tuition reimbursements available, teaching assistantships with partial tuition reimbursements available, career-related internships or fieldwork, Federal Work-Study, and institutionally sponsored loans available. Aid available to part-time students. Financial aid application deadline: 5/15; financial aid applicants required to submit FAFSA.
Faculty research: Hematopoiesis, growth factors in cancer, interleukins in allergy,

Long Island University, C.W. Post Campus (continued)
PCR techniques. *Total annual research expenditures:* $15,000.
Application contact: Robin Steadman, Graduate Adviser, 516-299-2337, *Fax:* 516-299-2527, *E-mail:* rsteadman@phoenix.liu.edu.

■ LOUISIANA STATE UNIVERSITY AND AGRICULTURAL AND MECHANICAL COLLEGE

Graduate School, College of Basic Sciences, Department of Biological Sciences, Baton Rouge, LA 70803

AWARDS Biochemistry (MS, PhD); microbiology (MS, PhD); plant biology (MS, PhD); zoology (MS, PhD). Part-time programs available.

Faculty: 79 full-time (7 women), 2 part-time/adjunct (0 women).
Students: 91 full-time (44 women), 22 part-time (10 women); includes 9 minority (3 African Americans, 1 Asian American or Pacific Islander, 5 Hispanic Americans), 29 international. Average age 28. *98 applicants, 28% accepted.* In 1999, 13 master's, 7 doctorates awarded. Terminal master's awarded for partial completion of doctoral program.
Degree requirements: For master's, foreign language not required; for doctorate, dissertation required.
Entrance requirements: For master's and doctorate, GRE General Test, minimum GPA of 3.0. *Application deadline:* Applications are processed on a rolling basis. *Application fee:* $25.
Expenses: Tuition, state resident: full-time $2,881. Tuition, nonresident: full-time $7,081. Part-time tuition and fees vary according to course load and program.
Financial aid: In 1999–00, 12 fellowships, 20 research assistantships with partial tuition reimbursements, 62 teaching assistantships with partial tuition reimbursements were awarded; Federal Work-Study, institutionally sponsored loans, and unspecified assistantships also available. Aid available to part-time students. *Total annual research expenditures:* $2.2 million.
Dr. Harold Silverman, Chairman, 225-388-2601, *Fax:* 225-388-2597, *E-mail:* cxsiv@unix1.sncc.lsu.edu.
Find an in-depth description at www.petersons.com/graduate.

■ LOUISIANA STATE UNIVERSITY HEALTH SCIENCES CENTER

School of Graduate Studies in New Orleans, New Orleans, LA 70112-2223
AWARDS MPH, MS, PhD, MD/PhD. Part-time and evening/weekend programs available.

Faculty: 166 full-time (31 women), 20 part-time/adjunct (4 women).
Students: 92 full-time (39 women), 6 part-time (1 woman); includes 26 minority (14 African Americans, 8 Asian Americans or Pacific Islanders, 4 Hispanic Americans), 16 international. Average age 26. *197 applicants, 24% accepted.* In 1999, 7 master's awarded (100% entered university research/teaching); 7 doctorates awarded. Terminal master's awarded for partial completion of doctoral program.
Degree requirements: For master's and doctorate, thesis/dissertation required, foreign language not required. *Average time to degree:* Master's–2.5 years full-time, 7 years part-time; doctorate–4.5 years full-time, 7 years part-time.
Entrance requirements: For master's and doctorate, GRE General Test, TOEFL. *Application deadline:* Applications are processed on a rolling basis. *Application fee:* $30.
Expenses: Tuition, state resident: full-time $2,878; part-time $126 per hour. Tuition, nonresident: full-time $6,003; part-time $265 per hour. Required fees: $2,272. Tuition and fees vary according to course load, degree level and program.
Financial aid: In 1999–00, 2 fellowships with full tuition reimbursements, 28 research assistantships with full tuition reimbursements, 27 teaching assistantships with full tuition reimbursements were awarded; Federal Work-Study, scholarships, and tuition waivers (full) also available.
Dr. Joseph M. Moerschbaecher, Head, 504-568-4740, *Fax:* 504-568-2361.
Application contact: Nancy W. Rhodes, Director, Student Affairs, 504-568-2211, *Fax:* 504-568-5588, *E-mail:* nrhode@lsumc.edu.

■ LOUISIANA STATE UNIVERSITY HEALTH SCIENCES CENTER

School of Graduate Studies in Shreveport, Shreveport, LA 71130-3932
AWARDS MS, PhD, MD/PhD. Terminal master's awarded for partial completion of doctoral program.

Degree requirements: For master's and doctorate, thesis/dissertation required, foreign language not required.
Entrance requirements: For master's and doctorate, GRE General Test, TOEFL.
Expenses: Tuition, state resident: full-time $2,878; part-time $126 per hour. Tuition, nonresident: full-time $6,003; part-time $265 per hour. Required fees: $2,272. Tuition and fees vary according to course load, degree level and program.
Find an in-depth description at www.petersons.com/graduate.

■ LOUISIANA TECH UNIVERSITY

Graduate School, College of Applied and Natural Sciences, Department of Biological Sciences, Ruston, LA 71272

AWARDS MS. Part-time programs available.

Degree requirements: For master's, computer language, thesis or alternative required, foreign language not required.
Entrance requirements: For master's, GRE General Test, GRE Subject Test.
Faculty research: Genetics, animal biology, plant biology, physiology biocontrol.

■ LOYOLA UNIVERSITY CHICAGO

Graduate School, Department of Biology, Chicago, IL 60611-2196

AWARDS MS.

Faculty: 20 full-time (4 women), 6 part-time/adjunct (3 women).
Students: 26 full-time (15 women), 6 part-time (4 women); includes 5 minority (4 Asian Americans or Pacific Islanders, 1 Hispanic American), 4 international. Average age 25. *30 applicants, 57% accepted.* In 1999, 6 degrees awarded (50% found work related to degree, 50% continued full-time study).
Degree requirements: For master's, thesis required, foreign language not required. *Average time to degree:* Master's–3 years full-time.
Entrance requirements: For master's, GRE General Test, GRE Subject Test. *Application deadline:* For fall admission, 7/1; for spring admission, 12/1. Applications are processed on a rolling basis. *Application fee:* $35. Electronic applications accepted.
Expenses: Tuition: Part-time $500 per credit hour. Required fees: $42 per term.
Financial aid: In 1999–00, 15 students received aid, including 15 fellowships with tuition reimbursements available; Federal Work-Study and institutionally sponsored loans also available. Financial aid application deadline: 2/1; financial aid applicants required to submit FAFSA.
Faculty research: Molecular biology and genetics, biochemistry, cell biology and physiology, population biology and aquatic ecology. *Total annual research expenditures:* $1.1 million.
Dr. Jeffrey Doering, Chairman, 773-508-3620, *Fax:* 773-508-3646, *E-mail:* jdoerin@luc.edu.
Application contact: Dr. Diane E. Suter, Graduate Program Director, 773-508-3285, *Fax:* 773-508-3646, *E-mail:* dsuter@luc.edu.
Find an in-depth description at www.petersons.com/graduate.

■ MARQUETTE UNIVERSITY

Graduate School, College of Arts and Sciences, Department of Biology, Milwaukee, WI 53201-1881

AWARDS Cell biology (MS, PhD); developmental biology (MS, PhD); ecology (MS, PhD); endocrinology (MS, PhD); evolutionary biology (MS, PhD); genetics (MS, PhD); microbiology (MS, PhD); molecular biology (MS, PhD); muscle and exercise physiology (MS, PhD); neurobiology (MS, PhD); reproductive physiology (MS, PhD).

Faculty: 16 full-time (4 women), 2 part-time/adjunct (0 women).
Students: 34 full-time (20 women), 3 part-time; includes 3 minority (all Asian Americans or Pacific Islanders), 2 international. Average age 31. *42 applicants, 29% accepted.* In 1999, 1 master's, 4 doctorates awarded. Terminal master's awarded for partial completion of doctoral program.
Degree requirements: For master's, thesis, 1 year of teaching experience or equivalent, comprehensive exam required, foreign language not required; for doctorate, dissertation, 1 year of teaching experience or equivalent, qualifying exam required, foreign language not required.
Entrance requirements: For master's and doctorate, GRE General Test, GRE Subject Test, TOEFL. *Application fee:* $40.
Expenses: Tuition: Part-time $510 per credit hour. Tuition and fees vary according to program.
Financial aid: In 1999–00, 4 fellowships, 22 teaching assistantships were awarded; research assistantships, Federal Work-Study, institutionally sponsored loans, scholarships, and tuition waivers (full and partial) also available. Aid available to part-time students. Financial aid application deadline: 2/15.
Faculty research: Microbial and invertebrate ecology, evolution of gene function, DNA methylation, DNA arrangement. *Total annual research expenditures:* $1.5 million.
Dr. Brian Unsworth, Chairman, 414-288-7355, *Fax:* 414-288-7357.
Application contact: Barbara DeNoyer, Graduate Studies Coordinator, 414-288-7355, *Fax:* 414-288-7357.

Find an in-depth description at www.petersons.com/graduate.

■ MARSHALL UNIVERSITY

Graduate College, College of Science, Department of Biological Science, Huntington, WV 25755

AWARDS MA, MS.

Faculty: 18 full-time (4 women), 2 part-time/adjunct (0 women).

Students: 39 full-time (21 women), 17 part-time (10 women); includes 1 minority (Asian American or Pacific Islander), 2 international. In 1999, 17 degrees awarded.
Degree requirements: For master's, thesis (MS) required.
Entrance requirements: For master's, GRE General Test, GRE Subject Test.
Expenses: Tuition, state resident: part-time $112 per credit. Tuition, nonresident: part-time $372 per credit. Required fees: $25 per credit. Tuition and fees vary according to campus/location, program and reciprocity agreements.
Financial aid: Career-related internships or fieldwork available.
Dr. Marcia Harrison, Chairperson, 304-696-4867, *E-mail:* harrison@marshall.edu.
Application contact: Ken O'Neal, Assistant Vice President, Adult Student Services, 304-746-2500 Ext. 1907, *Fax:* 304-746-1902, *E-mail:* oneal@marshall.edu.

■ MARSHALL UNIVERSITY

School of Medicine and Graduate College, Program in Biomedical Sciences, Huntington, WV 25755

AWARDS MS, PhD. Part-time programs available.

Faculty: 28 full-time (6 women), 3 part-time/adjunct (0 women).
Students: 41 full-time (13 women), 2 part-time (1 woman); includes 2 minority (1 African American, 1 Native American), 4 international. Average age 26. *50 applicants, 74% accepted.* In 1999, 1 master's awarded (100% continued full-time study); 3 doctorates awarded (100% found work related to degree). Terminal master's awarded for partial completion of doctoral program.
Degree requirements: For master's, thesis optional, foreign language not required; for doctorate, dissertation required, foreign language not required.
Entrance requirements: For master's, GRE General Test, or MCAT, 1 year of course work in biology, physics, chemistry, and organic chemistry; for doctorate, GRE General Test, MCAT, 1 year of course work in biology, physics, chemistry, and organic chemistry. *Application deadline:* For fall admission, 3/15 (priority date); for spring admission, 8/15 (priority date). Applications are processed on a rolling basis.
Expenses: Tuition, state resident: part-time $112 per credit. Tuition, nonresident: part-time $372 per credit. Required fees: $25 per credit. Tuition and fees vary according to reciprocity agreements.
Financial aid: Research assistantships, career-related internships or fieldwork, Federal Work-Study, and institutionally sponsored loans available. Aid available to

part-time students. Financial aid application deadline: 5/1; financial aid applicants required to submit FAFSA.
Faculty research: Neurosciences, cardiopulmonary science, molecular biology, toxicology, endocrinology.
Dr. Louis H. Aulick, Associate Dean for Research and Graduate Education, 304-696-7330, *Fax:* 304-696-7171, *E-mail:* aulick@marshall.edu.
Application contact: Marlene P. Gruetter, Senior Administrative Assistant, 304-696-7326, *Fax:* 304-696-7171, *E-mail:* gruettem@marshall.edu.

Find an in-depth description at www.petersons.com/graduate.

■ MASSACHUSETTS INSTITUTE OF TECHNOLOGY

School of Science, Department of Biology, Cambridge, MA 02139-4307

AWARDS Biochemistry (PhD); biological oceanography (PhD, Sc D); biophysics (PhD); cellular and developmental biology (PhD); genetics (PhD); immunology (PhD); microbiology (PhD); neurobiology (PhD).

Faculty: 63 full-time (12 women).
Students: 204 full-time (96 women); includes 40 minority (2 African Americans, 33 Asian Americans or Pacific Islanders, 5 Hispanic Americans), 16 international. Average age 22. *473 applicants, 22% accepted.* In 1999, 30 doctorates awarded (90% entered university research/teaching, 10% found other work related to degree).
Degree requirements: For doctorate, dissertation, general exam required, foreign language not required.
Entrance requirements: For doctorate, GRE General Test. *Application deadline:* For fall admission, 12/31. *Application fee:* $55.
Expenses: Tuition: Full-time $25,000. Full-time tuition and fees vary according to degree level, program and student level.
Financial aid: In 1999–00, 195 students received aid, including 111 fellowships with full tuition reimbursements available (averaging $19,300 per year), 84 research assistantships with full tuition reimbursements available (averaging $19,300 per year); teaching assistantships, Federal Work-Study and institutionally sponsored loans also available. Financial aid application deadline: 12/31.
Faculty research: DNA recombination, replication, and repair; transcription and gene regulation; signal transduction; cell cycle; neuronal cell fate. *Total annual research expenditures:* $92.1 million.
Dr. Robert T. Sauer, Head, 617-253-4701, *Fax:* 617-253-9810, *E-mail:* mitbio@mit.edu.

Massachusetts Institute of Technology (continued)

Application contact: Dr. Janice D. Chang, Educational Administrator, 617-253-3717, *Fax:* 617-258-9329, *E-mail:* gradbio@mit.edu.

Find an in-depth description at www.petersons.com/graduate.

■ MASSACHUSETTS INSTITUTE OF TECHNOLOGY

Whitaker College of Health Sciences and Technology, Division of Health Sciences and Technology, Program in Medical Sciences, Cambridge, MA 02139-4307

AWARDS MD, MD/MS, MD/PhD.

Students: 191 full-time (47 women); includes 95 minority (6 African Americans, 83 Asian Americans or Pacific Islanders, 6 Hispanic Americans), 20 international. *625 applicants, 7% accepted.* In 1999, 31 degrees awarded.
Application deadline: For fall admission, 11/15. Applications are processed on a rolling basis. *Application fee:* $75.
Expenses: Tuition: Full-time $25,000. Full-time tuition and fees vary according to degree level, program and student level.
Financial aid: Application deadline: 1/15. Dr. Joseph Bonventre, Director, 617-432-1738.
Application contact: Dr. Daniel C. Shannon, Director of MD Admissions, 617-726-5576.

■ MAYO GRADUATE SCHOOL

Graduate Programs in Biomedical Sciences, Rochester, MN 55905

AWARDS PhD, MD/PhD.

Faculty: 206 full-time.
Students: 147 full-time (72 women). Average age 26. In 1999, 19 degrees awarded.
Degree requirements: For doctorate, oral defense of dissertation, qualifying oral and written exam required.
Entrance requirements: For doctorate, GRE, TOEFL, 2 years of chemistry; 1 year of biology, calculus, and physics. *Application deadline:* For fall admission, 12/31 (priority date). Applications are processed on a rolling basis. *Application fee:* $0.
Expenses: Tuition: Full-time $17,900.
Financial aid: Fellowships available. Paul J. Leibson, MD PhD, Dean, 507-284-3163, *Fax:* 507-284-0999.
Application contact: Sherry Kallies, Information Contact, 507-266-0122, *Fax:* 507-284-0999, *E-mail:* phd.training@mayo.edu.

Find an in-depth description at www.petersons.com/graduate.

■ MCNEESE STATE UNIVERSITY

Graduate School, College of Science, Department of Biological and Environmental Sciences, Lake Charles, LA 70609

AWARDS Biology (MS); environmental sciences (MS). Evening/weekend programs available.

Faculty: 16 full-time (2 women).
Students: 9 full-time (7 women), 6 part-time (1 woman). In 1999, 5 degrees awarded.
Degree requirements: For master's, thesis or alternative required, foreign language not required.
Entrance requirements: For master's, GRE General Test. *Application deadline:* For fall admission, 7/15 (priority date). Applications are processed on a rolling basis. *Application fee:* $10 ($25 for international students).
Expenses: Tuition, state resident: full-time $2,118. Tuition, nonresident: full-time $5,870. Tuition and fees vary according to course load.
Financial aid: Application deadline: 5/1. Dr. Mark Wygoda, Head, 318-475-5674.

■ MCP HAHNEMANN UNIVERSITY

School of Medicine, Biomedical Graduate Programs, Philadelphia, PA 19102-1192

AWARDS MBS, MLAS, MMS, MS, PhD, MD/MS, MD/PhD. Part-time programs available. Terminal master's awarded for partial completion of doctoral program.

Degree requirements: For master's, comprehensive exam required; for doctorate, dissertation, qualifying exam required.
Entrance requirements: For master's and doctorate, GRE General Test, TOEFL.

Find an in-depth description at www.petersons.com/graduate.

■ MEDICAL COLLEGE OF GEORGIA

School of Graduate Studies, Augusta, GA 30912

AWARDS MHE, MN, MPT, MS, MSMI, MSN, PhD. Part-time programs available. Postbaccalaureate distance learning degree programs offered (no on-campus study).

Faculty: 187 full-time (52 women).
Students: 319 full-time (189 women), 86 part-time (63 women); includes 42 minority (21 African Americans, 16 Asian Americans or Pacific Islanders, 5 Hispanic Americans), 33 international. *320 applicants, 61% accepted.* In 1999, 101 master's, 14 doctorates awarded. Terminal master's awarded for partial completion of doctoral program.

Degree requirements: For master's, foreign language not required; for doctorate, dissertation required, foreign language not required.
Entrance requirements: For master's and doctorate, TOEFL. *Application deadline:* For fall admission, 6/30 (priority date). Applications are processed on a rolling basis. *Application fee:* $25.
Expenses: Tuition, state resident: full-time $2,896; part-time $121 per hour. Tuition, nonresident: full-time $11,584; part-time $483 per hour. Required fees: $286; $143 per semester. Tuition and fees vary according to program.
Financial aid: In 1999–00, 156 students received aid, including 60 research assistantships with partial tuition reimbursements available (averaging $15,500 per year); fellowships, teaching assistantships, career-related internships or fieldwork, Federal Work-Study, grants, institutionally sponsored loans, traineeships, and unspecified assistantships also available. Aid available to part-time students. Financial aid application deadline: 3/31; financial aid applicants required to submit FAFSA.
Faculty research: Genetics, sickle cell anemia, hormone actions, kinins, AIDS. Dr. Matthew J. Kluger, Dean, 706-721-3278, *Fax:* 706-721-6829, *E-mail:* mkluger@mail.mcg.edu.
Application contact: Elizabeth Griffin, Director of Academic Admissions, 706-721-2725, *Fax:* 706-721-7279, *E-mail:* gradadm@mail.mcg.edu.

Find an in-depth description at www.petersons.com/graduate.

■ MEDICAL COLLEGE OF OHIO

Graduate School, Toledo, OH 43614-5805

AWARDS MOT, MPH, MS, MSBS, MSN, PhD, Certificate, Post Master's Certificate, MD/MPH, MD/MS, MD/PhD. Part-time and evening/weekend programs available.

Faculty: 154 full-time (35 women), 24 part-time/adjunct (6 women).
Students: 208 full-time (131 women), 158 part-time (105 women); includes 36 minority (9 African Americans, 23 Asian Americans or Pacific Islanders, 4 Hispanic Americans), 49 international. Average age 32. *418 applicants, 41% accepted.* In 1999, 80 master's, 18 doctorates, 6 other advanced degrees awarded. Terminal master's awarded for partial completion of doctoral program.
Degree requirements: For master's, thesis required, foreign language not required; for doctorate, dissertation, qualifying exam required, foreign language not required; for degree, foreign language not required. *Average time to degree:*

Master's–2 years full-time, 3 years part-time; doctorate–5 years full-time; other advanced degree–1 year part-time.

Entrance requirements: For master's and doctorate, GRE General Test, minimum undergraduate GPA of 3.0. *Application deadline:* Applications are processed on a rolling basis. *Application fee:* $30.

Expenses: Tuition, state resident: part-time $193 per hour. Tuition, nonresident: part-time $445 per hour. Tuition and fees vary according to degree level.

Financial aid: Research assistantships, Federal Work-Study, institutionally sponsored loans, and scholarships available. Financial aid applicants required to submit FAFSA.

Faculty research: Cardiovascular research, neuroscience, cancer, immunology, molecular genetics. *Total annual research expenditures:* $11.9 million.

Dr. Keith K. Schlender, Dean, 419-383-4112, *Fax:* 419-383-6140, *E-mail:* kschlender@mco.edu.

Application contact: Joann Braatz, Clerk, 419-383-4117, *Fax:* 419-383-6140, *E-mail:* mcogradschool@mco.edu.

Find an in-depth description at www.petersons.com/graduate.

■ MEDICAL COLLEGE OF WISCONSIN

Graduate School of Biomedical Sciences, Milwaukee, WI 53226-0509

AWARDS MA, MS, PhD, MD/MA, MD/MS, MD/PhD. Part-time and evening/weekend programs available. Terminal master's awarded for partial completion of doctoral program.

Degree requirements: For master's, foreign language not required; for doctorate, dissertation required, foreign language not required.

Entrance requirements: For master's, TOEFL; for doctorate, GRE General Test, TOEFL (average 600).

Expenses: Tuition, state resident: full-time $9,318. Tuition, nonresident: full-time $9,318. Required fees: $115.

Find an in-depth description at www.petersons.com/graduate.

■ MEDICAL UNIVERSITY OF SOUTH CAROLINA

College of Graduate Studies, Charleston, SC 29425-0002

AWARDS MS, PhD, DMD/PhD, MD/PhD.

Students: 297 full-time (141 women); includes 24 minority (13 African Americans, 10 Asian Americans or Pacific Islanders, 1 Hispanic American), 32 international. Average age 30. *219 applicants, 58% accepted.* In 1999, 6

master's, 30 doctorates awarded. Terminal master's awarded for partial completion of doctoral program.

Degree requirements: For master's, thesis, research seminar required; for doctorate, dissertation, teaching and research seminar, oral and written exams required. *Average time to degree:* Master's–3 years full-time; doctorate–5 years full-time.

Entrance requirements: For master's and doctorate, GRE General Test, TOEFL, interview. *Application deadline:* Applications are processed on a rolling basis. *Application fee:* $55. Electronic applications accepted.

Expenses: Tuition, state resident: full-time $3,470; part-time $160 per semester hour. Tuition, nonresident: full-time $4,426; part-time $213 per semester hour. Required fees: $408 per semester. One-time fee: $160. Tuition and fees vary according to program.

Financial aid: In 1999–00, 81 fellowships (averaging $16,000 per year), 27 research assistantships, 7 teaching assistantships were awarded; Federal Work-Study and tuition waivers (partial) also available. Aid available to part-time students. Financial aid application deadline: 4/1; financial aid applicants required to submit FAFSA.

Dr. Barry E. Ledford, Interim Dean, 843-792-3391, *Fax:* 843-792-6590.

Application contact: Judy Yost, Director of Admissions, 843-792-8710, *Fax:* 843-792-3764.

Find an in-depth description at www.petersons.com/graduate.

■ MEHARRY MEDICAL COLLEGE

School of Graduate Studies, Division of Biomedical Sciences, Nashville, TN 37208-9989

AWARDS PhD.

Students: 13 full-time (8 women); all minorities (12 African Americans, 1 Asian American or Pacific Islander). Average age 30. *4 applicants, 100% accepted.* In 1999, 3 doctorates awarded.

Degree requirements: For doctorate, dissertation, oral and written comprehensive exams required, foreign language not required. *Average time to degree:* Doctorate–6 years full-time.

Entrance requirements: For doctorate, GRE General Test, GRE Subject Test. *Application deadline:* For fall admission, 6/1. Applications are processed on a rolling basis. *Application fee:* $45.

Expenses: Tuition: Full-time $8,732. Required fees: $2,133.

Financial aid: Fellowships, research assistantships available. Financial aid application deadline: 4/15; financial aid applicants required to submit FAFSA.

Faculty research: Molecular mechanisms of biological systems and their relationship

to human diseases, regulatory biological and cellular structure and function, genetic regulation of growth and cellular metabolisms.

Dr. Shirley Russell, Chair, 615-327-6193, *E-mail:* srussell@mmc.edu.

■ MIAMI UNIVERSITY

Graduate School, College of Arts and Sciences, Department of Botany, Oxford, OH 45056

AWARDS Biological sciences (MAT); botany (MA, MS, PhD). Part-time programs available.

Faculty: 12 full-time (3 women).

Students: 12 full-time (7 women), 25 part-time (10 women), 7 international. *27 applicants, 96% accepted.* In 1999, 20 master's, 3 doctorates awarded.

Degree requirements: For master's, thesis (for some programs), final exam required; for doctorate, dissertation, comprehensive and final exams required.

Entrance requirements: For master's, GRE General Test, GRE Subject Test, minimum undergraduate GPA of 3.0 during previous 2 years or 2.75 overall; for doctorate, GRE General Test, GRE Subject Test, minimum GPA of 2.75 (undergraduate) or 3.0 (graduate). *Application deadline:* For fall admission, 3/1. *Application fee:* $35.

Expenses: Tuition, state resident: part-time $260 per hour. Tuition, nonresident: full-time $3,125; part-time $538 per hour. International tuition: $6,452 full-time. Required fees: $18 per semester. Tuition and fees vary according to campus/location.

Financial aid: In 1999–00, 22 fellowships, 3 research assistantships, 11 teaching assistantships were awarded; Federal Work-Study and tuition waivers (full) also available. Financial aid application deadline: 3/1.

Dr. David Francko, Chair, 513-529-4200.

Application contact: Dr. James Hickey, Director of Graduate Studies, 513-529-4200, *Fax:* 513-529-4243, *E-mail:* botany@muohio.edu.

Find an in-depth description at www.petersons.com/graduate.

■ MICHIGAN STATE UNIVERSITY

College of Human Medicine and Graduate School, Graduate Programs in Human Medicine, East Lansing, MI 48824

AWARDS Biochemistry (MS, PhD), including biochemistry, biochemistry-environmental toxicology (PhD); epidemiology (MS); human pathology (MS, PhD); microbiology (MS, PhD); pharmacology/toxicology (MS, PhD); physiology (MS, PhD); surgery (MS). Part-time programs available.

Michigan State University (continued)
Students: 58 (32 women); includes 10 minority (2 African Americans, 6 Asian Americans or Pacific Islanders, 1 Hispanic American, 1 Native American) 16 international. *133 applicants, 28% accepted.* In 1999, 4 master's, 4 doctorates awarded.
Entrance requirements: For master's and doctorate, GRE General Test, minimum GPA of 3.0. *Application deadline:* Applications are processed on a rolling basis. *Application fee:* $30 ($40 for international students). Electronic applications accepted.
Expenses: Tuition, state resident: full-time $10,868; part-time $229 per credit. Tuition, nonresident: full-time $23,168; part-time $464 per credit.
Financial aid: Fellowships with tuition reimbursements, research assistantships with tuition reimbursements, teaching assistantships with tuition reimbursements, institutionally sponsored loans available. Aid available to part-time students. Financial aid applicants required to submit FAFSA.
Application contact: Dr. Lynda Farquhar, Assistant to the Dean for Research, 517-353-8858, *Fax:* 517-432-0021, *E-mail:* farquhal@pilot.msu.edu.

■ MICHIGAN STATE UNIVERSITY

College of Human Medicine, Human Medicine—Dual Degree Medical Scientist Training Program, East Lansing, MI 48824

AWARDS MD/PhD.

Students: 4 (1 woman); includes 2 minority (1 African American, 1 Hispanic American).
Application fee: $30 ($40 for international students).
Expenses: Tuition, state resident: full-time $7,544. Tuition, nonresident: full-time $15,664.
Financial aid: Fellowships, research assistantships, teaching assistantships, Federal Work-Study and institutionally sponsored loans available. Financial aid application deadline: 4/1; financial aid applicants required to submit FAFSA. Dr. Loran Bieber, Associate Dean for Graduate Studies and Research, 517-353-8858, *E-mail:* mdadmissions@msu.edu.
Application contact: Dr. Lynda Farquhar, Director, 517-353-8858, *Fax:* 517-432-0148, *E-mail:* farquhal@pilot.msu.edu.

■ MICHIGAN STATE UNIVERSITY

College of Osteopathic Medicine and Graduate School, Graduate Programs in Osteopathic Medicine, East Lansing, MI 48824

AWARDS Anatomy (MS, PhD); biochemistry (MS, PhD); microbiology (PhD); pathology (MS, PhD); pharmacology/toxicology (MS,

PhD), including pharmacology; physiology (MS, PhD), including environmental toxicology (PhD), neuroscience (PhD). Part-time programs available.

Students: 17 full-time (9 women), 1 part-time; includes 5 minority (3 Asian Americans or Pacific Islanders, 1 Hispanic American, 1 Native American), 5 international. Average age 26. *33 applicants, 9% accepted.* In 1999, 2 doctorates awarded.
Degree requirements: For doctorate, dissertation required.
Entrance requirements: For master's and doctorate, GRE. *Application deadline:* For fall admission, 3/1 (priority date). Applications are processed on a rolling basis. *Application fee:* $30 ($40 for international students).
Expenses: Tuition, area resident: Full-time $15,879. Tuition, state resident: full-time $33,797.
Financial aid: Fellowships, research assistantships, teaching assistantships, career-related internships or fieldwork, Federal Work-Study, and institutionally sponsored loans available. Financial aid application deadline: 4/2. *Total annual research expenditures:* $5 million.
Application contact: Dr. Veronica M. Maher, Associate Dean for Graduate Studies, 517-353-7785, *Fax:* 517-353-9004, *E-mail:* maher@com.msu.edu.

■ MICHIGAN STATE UNIVERSITY

College of Osteopathic Medicine, Osteopathic Medicine—Dual Degree Medical Scientist Training Program, East Lansing, MI 48824

AWARDS DO/PhD.

Students: 9 full-time (4 women); includes 2 minority (both Asian Americans or Pacific Islanders), 2 international. Average age 26.
Expenses: Tuition, area resident: Full-time $15,879. Tuition, state resident: full-time $33,797.
Financial aid: Fellowships, research assistantships available. Financial aid application deadline: 12/1.
Dr. Veronica H. Maher, Associate Dean for Graduate Studies, 517-353-7785, *Fax:* 517-353-9004.

■ MICHIGAN STATE UNIVERSITY

Graduate School, College of Natural Science, Biological Sciences Interdepartmental Programs, East Lansing, MI 48824

AWARDS MAT, MS.

Students: 7 applicants, 0% accepted.
Entrance requirements: For master's, GRE General Test. *Application deadline:* Applications are processed on a rolling

basis. *Application fee:* $30 ($40 for international students).
Expenses: Tuition, state resident: part-time $229 per credit. Tuition, nonresident: part-time $464 per credit. Required fees: $241 per semester. Tuition and fees vary according to course load, degree level and program.
Financial aid: Teaching assistantships available. Financial aid applicants required to submit FAFSA.
Dr. Richard W. Hill, Director, 517-355-4640.

■ MICHIGAN TECHNOLOGICAL UNIVERSITY

Graduate School, College of Sciences and Arts, Department of Biological Sciences, Houghton, MI 49931-1295

AWARDS MS, PhD. Part-time programs available.

Faculty: 16 full-time (5 women).
Students: 31 full-time (12 women), 4 part-time (2 women); includes 1 minority (Asian American or Pacific Islander), 12 international. Average age 31. *47 applicants, 57% accepted.* In 1999, 1 master's, 1 doctorate awarded.
Degree requirements: For master's and doctorate, thesis/dissertation required, foreign language not required. *Average time to degree:* Master's–3.7 years full-time; doctorate–4 years full-time.
Entrance requirements: For master's and doctorate, GRE General Test, TOEFL. *Application deadline:* For fall admission, 3/15 (priority date). Applications are processed on a rolling basis. *Application fee:* $30 ($35 for international students).
Expenses: Tuition, state resident: full-time $4,377. Tuition, nonresident: full-time $9,108. Required fees: $126. Tuition and fees vary according to course load.
Financial aid: In 1999–00, 3 fellowships with full tuition reimbursements (averaging $13,500 per year), 11 research assistantships with full tuition reimbursements (averaging $11,900 per year), 14 teaching assistantships with full tuition reimbursements (averaging $8,950 per year) were awarded; career-related internships or fieldwork, Federal Work-Study, grants, institutionally sponsored loans, traineeships, unspecified assistantships, and Co-op also available. Aid available to part-time students. Financial aid application deadline: 2/1; financial aid applicants required to submit FAFSA.
Faculty research: Aquatic ecology, biological control, predator-prey interactions, plant enzymes, plant-microbe interactions, membrane biochemistry. *Total annual research expenditures:* $1 million.

Dr. John H. Adler, Chair, 906-487-2025, *Fax:* 906-487-3167, *E-mail:* jhadler@mtu.edu.
Application contact: Dr. Donald F. Lueking, Director of Graduate Studies, 906-487-2027, *Fax:* 906-487-3167, *E-mail:* drluekin@mtu.edu.

■ MIDDLE TENNESSEE STATE UNIVERSITY

College of Graduate Studies, College of Basic and Applied Sciences, Department of Biology, Murfreesboro, TN 37132

AWARDS MS, MST. Part-time programs available.

Faculty: 28 full-time (6 women), 1 (woman) part-time/adjunct.
Students: 15 full-time (10 women), 34 part-time (22 women); includes 4 minority (2 African Americans, 2 Asian Americans or Pacific Islanders). Average age 29. *28 applicants, 64% accepted.* In 1999, 12 degrees awarded.
Degree requirements: For master's, one foreign language, thesis, comprehensive exams required.
Entrance requirements: For master's, GRE or MAT. *Application deadline:* For fall admission, 8/1 (priority date). Applications are processed on a rolling basis. *Application fee:* $25. Electronic applications accepted.
Expenses: Tuition, state resident: full-time $1,356; part-time $137 per semester hour. Tuition, nonresident: full-time $3,914; part-time $361 per semester hour.
Financial aid: Teaching assistantships available. Aid available to part-time students. Financial aid application deadline: 5/1; financial aid applicants required to submit FAFSA. *Total annual research expenditures:* $18,387.
Dr. George Murphy, Chair, 615-898-2847, *E-mail:* gmurphy@mtsu.edu.

■ MIDWESTERN STATE UNIVERSITY

Graduate Studies, College of Science and Mathematics, Program in Science, Wichita Falls, TX 76308

AWARDS Biology (MS).

Faculty: 9 full-time (1 woman).
Students: 16 full-time (10 women), 10 part-time (5 women); includes 7 minority (2 African Americans, 3 Asian Americans or Pacific Islanders, 2 Hispanic Americans). Average age 35. *10 applicants, 100% accepted.* In 1999, 6 degrees awarded.
Degree requirements: For master's, thesis required, foreign language not required.
Entrance requirements: For master's, GRE General Test, MAT, TOEFL. *Application deadline:* For fall admission, 8/7

(priority date); for spring admission, 12/15. *Application fee:* $0 ($50 for international students).
Expenses: Tuition, state resident: full-time $1,542; part-time $46 per hour. Tuition, nonresident: full-time $5,376; part-time $304 per hour. Tuition and fees vary according to course load.
Financial aid: In 1999–00, 7 research assistantships, 5 teaching assistantships were awarded. Aid available to part-time students.
Faculty research: Ecology and systematics of spiders and mammals, plant physiology and molecular biology, Drosophila genetics.
Dr. Norman V. Horner, Dean, 940-397-4253.
Application contact: Dr. John V. Grimes, Graduate Coordinator, 940-397-4251.

■ MILLERSVILLE UNIVERSITY OF PENNSYLVANIA

Graduate School, School of Science and Mathematics, Department of Biology, Millersville, PA 17551-0302

AWARDS MS. Part-time and evening/weekend programs available.

Faculty: 19 full-time (5 women), 4 part-time/adjunct (3 women).
Students: 1 (woman) full-time, 3 part-time (2 women); includes 1 minority (Asian American or Pacific Islander). Average age 26. *7 applicants, 43% accepted.*
Degree requirements: For master's, departmental exam required, thesis optional, foreign language not required.
Entrance requirements: For master's, GRE General Test, GRE Subject Test, minimum undergraduate GPA of 2.75. *Application deadline:* For fall admission, 5/1 (priority date). Applications are processed on a rolling basis. *Application fee:* $25.
Expenses: Tuition, state resident: full-time $3,780; part-time $210 per credit. Tuition, nonresident: full-time $6,610; part-time $367 per credit. Required fees: $977; $41 per credit.
Financial aid: Research assistantships with full tuition reimbursements, Federal Work-Study, institutionally sponsored loans, and unspecified assistantships available. Aid available to part-time students. Financial aid application deadline: 3/15; financial aid applicants required to submit FAFSA.
Faculty research: Reproductive physiology, gene expression, motor neuron analysis, plant anatomy and systemics, ecosystems anaylsis. *Total annual research expenditures:* $150,000.
Dr. Daniel H. Yocom, Coordinator, 717-872-3338, *Fax:* 717-872-3985, *E-mail:* daniel.yocom@millersville.edu.
Application contact: 717-872-3030, *Fax:* 717-871-2022.

■ MILLS COLLEGE

Graduate Studies, Program in Pre-Med, Oakland, CA 94613-1000

AWARDS Certificate. Part-time programs available.

Faculty: 8 full-time (2 women), 9 part-time/adjunct (8 women).
Students: 60 full-time (46 women); includes 16 minority (2 African Americans, 11 Asian Americans or Pacific Islanders, 2 Hispanic Americans, 1 Native American). Average age 26. *89 applicants, 80% accepted.* In 1999, 29 degrees awarded (100% continued full-time study).
Degree requirements: For Certificate, thesis not required. *Average time to degree:* 2 years full-time, 3 years part-time.
Entrance requirements: For degree, GRE General Test, TOEFL, bachelor's degree in a non-science area. *Application deadline:* For fall admission, 2/1. Applications are processed on a rolling basis. *Application fee:* $50. Electronic applications accepted.
Expenses: Tuition: Full-time $11,130; part-time $2,690 per credit. One-time fee: $977. Tuition and fees vary according to course load and program.
Financial aid: In 1999–00, fellowships with partial tuition reimbursements (averaging $500 per year), 18 teaching assistantships with partial tuition reimbursements (averaging $5,565 per year) were awarded; institutionally sponsored loans and scholarships also available. Aid available to part-time students. Financial aid application deadline: 2/1; financial aid applicants required to submit FAFSA.
Faculty research: Bacterial and viral genetics, microbiology, lipid biochemistry, inorganic nitrogen chemistry.
Dr. Chuck Lutz, Director, 510-430-2202, *Fax:* 510-430-3314, *E-mail:* grad-studies@mills.edu.
Application contact: Ron Clement, Assistant Director of Graduate Studies, 510-430-2355, *Fax:* 510-430-2159, *E-mail:* rclement@mills.edu.

■ MINNESOTA STATE UNIVERSITY, MANKATO

College of Graduate Studies, College of Science, Engineering and Technology, Department of Biological Sciences, Mankato, MN 56001

AWARDS Biology (MS); environmental science (MS), including ecology, economic and political systems, human ecosystems, physical science, technology. Part-time programs available.

Faculty: 22 full-time (3 women).
Students: 27 full-time (16 women), 7 part-time (3 women); includes 3 minority (2

Minnesota State University, Mankato
(continued)

Asian Americans or Pacific Islanders, 1 Hispanic American), 3 international. Average age 31. In 1999, 11 degrees awarded.

Degree requirements: For master's, thesis or alternative, comprehensive exam required.

Entrance requirements: For master's, minimum GPA of 3.0 during previous 2 years. *Application deadline:* For fall admission, 7/9 (priority date); for spring admission, 11/27. Applications are processed on a rolling basis. *Application fee:* $20.

Expenses: Tuition, state resident: part-time $152 per credit hour. Tuition, nonresident: part-time $228 per credit hour.

Financial aid: Research assistantships with partial tuition reimbursements, teaching assistantships with partial tuition reimbursements, career-related internships or fieldwork, Federal Work-Study, and institutionally sponsored loans available. Aid available to part-time students. Financial aid application deadline: 3/15; financial aid applicants required to submit FAFSA.

Faculty research: Limnology, enzyme analysis, membrane engineering, converters.

Dr. Gregg Marg, Chairperson, 507-389-2786.

Application contact: Joni Roberts, Admissions Coordinator, 507-389-2321, *Fax:* 507-389-5974, *E-mail:* grad@mankato.msus.edu.

Find an in-depth description at www.petersons.com/graduate.

■ **MISSISSIPPI COLLEGE**

Graduate School, College of Arts and Sciences, Program in Combined Sciences, Major in Biology, Clinton, MS 39058

AWARDS MCS.

Faculty: 5 full-time (1 woman), 4 part-time/adjunct.

Students: 3 full-time (1 woman), 20 part-time (11 women); includes 4 minority (3 African Americans, 1 Asian American or Pacific Islander). In 1999, 7 degrees awarded.

Degree requirements: For master's, comprehensive exam required.

Entrance requirements: For master's, GRE General Test, minimum GPA of 2.5. *Application deadline:* For fall admission, 8/15 (priority date). Applications are processed on a rolling basis. *Application fee:* $25 ($75 for international students).

Expenses: Tuition: Full-time $5,274; part-time $293 per hour. Required fees: $250. Tuition and fees vary according to course load.

Financial aid: Career-related internships or fieldwork available. Aid available to part-time students. Financial aid application deadline: 4/1.

Dr. Ted Snazelle, Head, 601-925-3339.

■ **MISSISSIPPI STATE UNIVERSITY**

College of Arts and Sciences, Department of Biological Sciences, Mississippi State, MS 39762

AWARDS MS, PhD.

Faculty: 22 full-time (4 women).

Students: 23 full-time (10 women), 9 part-time (4 women); includes 4 minority (all African Americans), 1 international. Average age 29. *14 applicants, 7% accepted.* In 1999, 5 master's, 3 doctorates awarded. Terminal master's awarded for partial completion of doctoral program.

Degree requirements: For master's and doctorate, thesis/dissertation, comprehensive oral or written exam required.

Entrance requirements: For master's and doctorate, GRE General Test, TOEFL. *Application deadline:* For fall admission, 7/1; for spring admission, 11/1. Applications are processed on a rolling basis. *Application fee:* $25 for international students.

Expenses: Tuition, state resident: full-time $3,017; part-time $168 per credit. Tuition, nonresident: full-time $6,119; part-time $340 per credit. Part-time tuition and fees vary according to course load and program.

Financial aid: In 1999–00, 5 fellowships, 21 teaching assistantships were awarded; Federal Work-Study, grants, institutionally sponsored loans, and unspecified assistantships also available. Financial aid applicants required to submit FAFSA.

Faculty research: Botany, zoology, microbiology, ecology. *Total annual research expenditures:* $337,000.

Dr. Donald N. Downer, Head, 662-325-3483, *Fax:* 662-325-7939, *E-mail:* gloria@biology.msstate.edu.

Application contact: Jerry B. Inmon, Director of Admissions, 662-325-2224, *Fax:* 662-325-7360, *E-mail:* admit@admissions.msstate.edu.

■ **MONTANA STATE UNIVERSITY–BOZEMAN**

College of Graduate Studies, College of Letters and Science, Department of Biology, Bozeman, MT 59717

AWARDS Biological sciences (MS, PhD); fish and wildlife biology (PhD); fish and wildlife management (MS). Part-time programs available.

Students: 8 full-time (3 women), 50 part-time (14 women); includes 1 minority

(Native American). Average age 31. *11 applicants, 82% accepted.* In 1999, 14 master's, 4 doctorates awarded.

Degree requirements: For master's and doctorate, thesis/dissertation or alternative required, foreign language not required.

Entrance requirements: For master's and doctorate, GRE General Test, TOEFL, minimum GPA of 3.0. *Application deadline:* For fall admission, 6/1; for spring admission, 11/1. Applications are processed on a rolling basis. *Application fee:* $50. Electronic applications accepted.

Expenses: Tuition, state resident: full-time $2,674. Tuition, nonresident: full-time $6,986. International tuition: $7,136 full-time. Tuition and fees vary according to course load and program.

Financial aid: In 1999–00, 42 students received aid, including 10 research assistantships with partial tuition reimbursements available, 8 teaching assistantships with full tuition reimbursements available (averaging $1,600 per year); career-related internships or fieldwork, Federal Work-Study, and scholarships also available. Financial aid application deadline: 3/1; financial aid applicants required to submit FAFSA.

Faculty research: Applied ecology, cell biology/modeling, neurobiology and human anatomy, paleontology, wildlife management. *Total annual research expenditures:* $2.9 million.

Dr. Ernest Vyse, Interim Head, 406-994-4548, *Fax:* 406-994-3190, *E-mail:* biology@montana.edu.

■ **MONTCLAIR STATE UNIVERSITY**

Office of Graduate Studies, College of Science and Mathematics, Department of Biology, Upper Montclair, NJ 07043-1624

AWARDS MS. Part-time and evening/weekend programs available.

Degree requirements: For master's, thesis or alternative required, foreign language not required.

Entrance requirements: For master's, GRE General Test.

■ **MOREHEAD STATE UNIVERSITY**

Graduate Programs, College of Science and Technology, Department of Biological and Environmental Sciences, Morehead, KY 40351

AWARDS Biology (MS). Part-time programs available.

Faculty: 9 full-time (0 women).

Students: 13 full-time (7 women), 2 part-time (1 woman); includes 1 minority

(African American), 2 international. Average age 25. *2 applicants, 100% accepted.* In 1999, 2 degrees awarded.

Degree requirements: For master's, oral and written final exams required, thesis optional, foreign language not required.
Entrance requirements: For master's, GRE General Test, minimum GPA of 3.0 in biology, 2.5 overall; undergraduate major/minor in biology, environmental science, or equivalent. *Application deadline:* For fall admission, 8/1 (priority date); for spring admission, 12/1 (priority date). Applications are processed on a rolling basis. *Application fee:* $0.
Expenses: Tuition, state resident: full-time $2,640; part-time $147 per hour. Tuition, nonresident: full-time $7,080; part-time $394 per hour. Full-time tuition and fees vary according to course level and course load.
Financial aid: In 1999–00, 5 research assistantships (averaging $5,000 per year), 5 teaching assistantships (averaging $5,000 per year) were awarded; Federal Work-Study also available. Financial aid application deadline: 4/1; financial aid applicants required to submit FAFSA.
Faculty research: Atherosclerosis, RNA evolution, cancer biology, water quality/ecology, immunoparasitology.
Dr. Joe E. Winstead, Chair, 606-783-2944, *E-mail:* j.winstead@morehead-st.edu.
Application contact: Betty R. Cowsert, Graduate Admissions Officer, 606-783-2039, *Fax:* 606-783-5061, *E-mail:* b.cowsert@morehead-st.edu.

■ MOREHOUSE SCHOOL OF MEDICINE

Program in Biomedical Sciences, Atlanta, GA 30310-1495
AWARDS PhD, MD/PhD.

Faculty: 59 full-time (24 women), 3 part-time/adjunct (2 women).
Students: 13 full-time (11 women); includes 10 African Americans, 2 international. Average age 30. *12 applicants, 33% accepted.* In 1999, 1 degree awarded (100% found work related to degree).
Degree requirements: For doctorate, computer language, dissertation required, foreign language not required. *Average time to degree:* Doctorate–6 years full-time.
Entrance requirements: For doctorate, GRE General Test. *Application deadline:* For fall admission, 2/1. *Application fee:* $25.
Expenses: Tuition: Full-time $19,200. Required fees: $2,574. Tuition and fees vary according to degree level, program and student level.
Financial aid: In 1999–00, 11 students received aid, including 8 fellowships with full and partial tuition reimbursements available (averaging $12,833 per year), 4

research assistantships with full and partial tuition reimbursements available (averaging $12,840 per year); grants, institutionally sponsored loans, and tuition waivers (full) also available. Financial aid application deadline: 5/1; financial aid applicants required to submit FAFSA. *Total annual research expenditures:* $5.8 million.
Dr. Douglas Paulsen, Director, 404-752-1559.
Application contact: Karen A. Lewis, Assistant Director of Admissions, 404-752-1650, *Fax:* 404-752-1512, *E-mail:* karen@msm.edu.

Find an in-depth description at www.petersons.com/graduate.

■ MORGAN STATE UNIVERSITY

School of Graduate Studies, School of Computer, Mathematical, and Natural Sciences, Interdisciplinary Program in Science, Baltimore, MD 21251
AWARDS MS.

Students: 17 (12 women); includes 16 minority (all African Americans) 1 international. In 1999, 2 degrees awarded.
Degree requirements: For master's, thesis required.
Entrance requirements: For master's, minimum GPA of 2.5. *Application deadline:* For fall admission, 7/1; for spring admission, 11/1. Applications are processed on a rolling basis. *Application fee:* $0.
Expenses: Tuition, state resident: part-time $160 per credit hour. Tuition, nonresident: part-time $286 per credit hour. Required fees: $174 per semester.
Dr. T. Joan Robinson, Dean, School of Computer, Mathematical, and Natural Sciences, 443-885-4515, *E-mail:* jrobinson@moac.morgan.edu.

■ MOUNT SINAI SCHOOL OF MEDICINE OF NEW YORK UNIVERSITY

Graduate School of Biological Sciences, New York, NY 10029-6504
AWARDS PhD, MD/PhD.

Faculty: 167 full-time.
Students: 192 full-time (82 women); includes 51 minority (9 African Americans, 38 Asian Americans or Pacific Islanders, 4 Hispanic Americans), 62 international. Average age 30. *581 applicants, 7% accepted.* In 1999, 31 degrees awarded.
Degree requirements: For doctorate, dissertation required, foreign language not required.
Entrance requirements: For doctorate, GRE General Test, GRE Subject Test, MCAT, TOEFL. *Application deadline:* For fall admission, 4/15. *Application fee:* $60.
Expenses: Tuition: Full-time $21,750; part-time $725 per credit. Required fees:

$750; $25 per credit. Full-time tuition and fees vary according to student level.
Financial aid: In 1999–00, 36 fellowships with full tuition reimbursements (averaging $17,800 per year) were awarded; grants also available.
Faculty research: Molecular genetics of cancer, molecular virology, neurochemistry, image analysis in pathology cytogenetics.
Dr. Terry Ann Krulwich, Dean, 212-241-6546, *Fax:* 212-241-0651, *E-mail:* terry.krulwich@mssm.edu.
Application contact: C. Gita Bosch, Administrative Manager and Assistant Dean, 212-241-6546, *Fax:* 212-241-0651, *E-mail:* grads@mssm.edu.

Find an in-depth description at www.petersons.com/graduate.

■ MURRAY STATE UNIVERSITY

College of Science, Department of Biological Sciences, Murray, KY 42071-0009
AWARDS MAT, MS, PhD. Part-time programs available.

Students: 11 full-time (5 women), 11 part-time (6 women), 5 international. *10 applicants, 90% accepted.* In 1999, 5 degrees awarded.
Degree requirements: For master's, thesis required (for some programs), foreign language not required.
Entrance requirements: For master's, GRE General Test, TOEFL. *Application deadline:* Applications are processed on a rolling basis. *Application fee:* $20.
Expenses: Tuition, state resident: full-time $2,600; part-time $130 per hour. Tuition, nonresident: full-time $7,040; part-time $374 per hour. Required fees: $90 per semester. Part-time tuition and fees vary according to course load and program.
Financial aid: Research assistantships, teaching assistantships, Federal Work-Study available. Financial aid application deadline: 4/1.
Dr. Thomas Timmons, Chairman, 270-762-6754, *Fax:* 270-762-2788, *E-mail:* tom.timmons@murraystate.edu.
Application contact: Dr. Timothy Johnston, Graduate Coordinator, 270-762-6367, *Fax:* 270-762-2788, *E-mail:* tim.johnston@murraystate.edu.

■ NEW JERSEY INSTITUTE OF TECHNOLOGY

Office of Graduate Studies, Department of Life Sciences, Newark, NJ 07102-1982
AWARDS Biology (MS, PhD); computational biology (MS).

Degree requirements: For master's, foreign language not required.

New Jersey Institute of Technology (continued)

Entrance requirements: For master's, GRE General Test. *Application deadline:* For fall admission, 6/5 (priority date); for spring admission, 10/15. Applications are processed on a rolling basis. *Application fee:* $35 ($50 for international students). Electronic applications accepted.
Expenses: Tuition, state resident: full-time $5,508; part-time $206 per credit. Tuition, nonresident: full-time $9,852; part-time $424 per credit. Required fees: $972.
Financial aid: Application deadline: 3/15.
Faculty research: Technological, computational, and mathematical aspects of biology and bioengineering.
Dr. Michael Recce, Director, 973-596-6597, *E-mail:* lifesciences@njit.edu.
Application contact: Kathy Kelly, Director of Admissions, 973-596-3300, *Fax:* 973-596-3461, *E-mail:* admissions@njit.edu.

■ **NEW MEXICO HIGHLANDS UNIVERSITY**

Graduate Studies, College of Arts and Sciences, Department of Life Sciences, Las Vegas, NM 87701

AWARDS Biology (MS); environmental science and management (MS). Part-time programs available.

Faculty: 9 full-time (2 women).
Students: 11 full-time (5 women), 7 part-time (1 woman); includes 5 minority (all Hispanic Americans), 3 international. Average age 36. *13 applicants, 85% accepted.* In 1999, 2 degrees awarded.
Degree requirements: For master's, thesis or alternative required, foreign language not required.
Entrance requirements: For master's, minimum undergraduate GPA of 3.0. *Application deadline:* For fall admission, 8/1 (priority date). Applications are processed on a rolling basis. *Application fee:* $15.
Expenses: Tuition, state resident: full-time $1,988; part-time $83 per credit hour. Tuition, nonresident: full-time $8,034; part-time $83 per credit hour. Tuition and fees vary according to course load.
Financial aid: In 1999–00, 9 research assistantships with full and partial tuition reimbursements (averaging $6,500 per year) were awarded; Federal Work-Study also available. Financial aid application deadline: 3/1.
Dr. Maureen Romine, Chair, 505-454-3264, *Fax:* 505-454-3063, *E-mail:* romine_m@nmhu.edu.
Application contact: Dr. Glen W. Davidson, Provost, 505-454-3311, *Fax:* 505-454-3558, *E-mail:* glendavidson@nmhu.edu.

■ **NEW MEXICO INSTITUTE OF MINING AND TECHNOLOGY**

Graduate Studies, Department of Biology, Socorro, NM 87801

AWARDS MS, PhD.

Faculty: 5 full-time (2 women).
Students: 9 full-time (6 women), 1 part-time; includes 2 minority (1 Hispanic American, 1 Native American). Average age 30. *12 applicants, 50% accepted.* In 1999, 1 degree awarded.
Degree requirements: For master's, thesis required, foreign language not required. *Average time to degree:* Master's–3 years full-time.
Entrance requirements: For master's, GRE General Test, TOEFL. *Application deadline:* For fall admission, 3/1 (priority date); for spring admission, 6/1 (priority date). Applications are processed on a rolling basis. *Application fee:* $16. Electronic applications accepted.
Expenses: Tuition, state resident: full-time $1,693; part-time $94 per credit hour. Tuition, nonresident: full-time $6,978; part-time $388 per credit hour. Required fees: $727; $79 per credit hour. Tuition and fees vary according to course load.
Financial aid: In 1999–00, 1 research assistantship (averaging $9,670 per year), 5 teaching assistantships (averaging $9,670 per year) were awarded; fellowships, Federal Work-Study and institutionally sponsored loans also available. Financial aid application deadline: 3/1; financial aid applicants required to submit CSS PROFILE or FAFSA.
Dr. J. A. Smoake, Chairman, 505-835-5612, *Fax:* 505-835-5668, *E-mail:* smoake@nmt.edu.
Application contact: Dr. David B. Johnson, Dean of Graduate Studies, 505-835-5513, *Fax:* 505-835-5476, *E-mail:* graduate@nmt.edu.

■ **NEW MEXICO STATE UNIVERSITY**

Graduate School, College of Arts and Sciences, Department of Biology, Las Cruces, NM 88003-8001

AWARDS MS, PhD. Part-time programs available.

Faculty: 27.
Students: 43 full-time (17 women), 10 part-time (5 women); includes 8 minority (2 African Americans, 6 Hispanic Americans), 4 international. *27 applicants, 33% accepted.* In 1999, 9 master's, 1 doctorate awarded.
Degree requirements: For master's, thesis required (for some programs), foreign language not required; for doctorate, dissertation required, foreign language not required.

Application deadline: For fall admission, 2/15 (priority date); for spring admission, 11/1. Applications are processed on a rolling basis. *Application fee:* $15 ($35 for international students). Electronic applications accepted.
Expenses: Tuition, state resident: full-time $2,682; part-time $112 per credit. Tuition, nonresident: full-time $8,376; part-time $349 per credit. Tuition and fees vary according to course load.
Financial aid: Fellowships, teaching assistantships, Federal Work-Study available. Aid available to part-time students. Financial aid application deadline: 3/1.
Faculty research: Microbiology, cell and organismal physiology, ecology and ethology, evolution, genetics, developmental biology.
Dr. Laura Huenneke, Head, 505-646-3611, *Fax:* 505-646-5665, *E-mail:* lhuenek@nmsu.edu.
Application contact: Dr. Naida Zucker, Associate Head, 505-646-3611, *Fax:* 505-646-5665, *E-mail:* nzucker@nmsu.edu.

■ **NEW YORK MEDICAL COLLEGE**

Graduate School of Basic Medical Sciences, Valhalla, NY 10595-1691

AWARDS MS, PhD, MD/PhD. Part-time and evening/weekend programs available.

Faculty: 83 full-time (19 women).
Students: 82 full-time (42 women), 53 part-time (32 women); includes 26 minority (5 African Americans, 17 Asian Americans or Pacific Islanders, 4 Hispanic Americans), 31 international. Average age 30. *184 applicants, 34% accepted.* In 1999, 31 master's, 12 doctorates awarded. Terminal master's awarded for partial completion of doctoral program.
Degree requirements: For master's and doctorate, computer language, thesis/dissertation required, foreign language not required.
Entrance requirements: For master's, GRE General Test, TOEFL; for doctorate, GRE General Test, GRE Subject Test, TOEFL. *Application deadline:* For fall admission, 7/1 (priority date); for spring admission, 12/1 (priority date). Applications are processed on a rolling basis. *Application fee:* $35 ($60 for international students).
Expenses: Tuition: Part-time $430 per credit. Required fees: $15 per semester. One-time fee: $100.
Financial aid: In 1999–00, 49 research assistantships with full tuition reimbursements were awarded; career-related internships or fieldwork, Federal Work-Study, grants, institutionally sponsored loans, and tuition waivers (full) also available. Aid

available to part-time students. Financial aid applicants required to submit FAFSA. Dr. Francis L. Belloni, Dean, 914-594-4110, *Fax:* 914-594-4944, *E-mail:* francis_belloni@nymc.edu.

Application contact: Nina Pelella-Doyle, Admission Coordinator, 914-594-4110, *Fax:* 914-594-4944.

Find an in-depth description at www.petersons.com/graduate.

■ NEW YORK UNIVERSITY

Graduate School of Arts and Science, Department of Basic Medical Sciences, New York, NY 10012-1019

AWARDS Biochemistry (MS, PhD); cell biology (MS, PhD); microbiology (MS, PhD); parasitology (PhD); pathology (MS, PhD); pharmacology (PhD); physiology (MS, PhD). Part-time programs available.

Faculty: 23 full-time (2 women), 3 part-time/adjunct (1 woman).
Students: 182 full-time (73 women), 4 part-time (1 woman); includes 52 minority (11 African Americans, 34 Asian Americans or Pacific Islanders, 7 Hispanic Americans), 48 international. Average age 26. *671 applicants, 11% accepted.* In 1999, 15 master's, 32 doctorates awarded. Terminal master's awarded for partial completion of doctoral program.
Degree requirements: For master's, thesis or alternative, written comprehensive exam required, foreign language not required; for doctorate, one foreign language, dissertation, oral and written comprehensive exams required.
Entrance requirements: For master's and doctorate, GRE General Test, GRE Subject Test, TOEFL. *Application deadline:* For fall admission, 2/1 (priority date). *Application fee:* $60.
Expenses: Tuition: Full-time $17,880; part-time $745 per credit. Required fees: $1,140; $35 per credit. Tuition and fees vary according to course load and program.
Financial aid: Fellowships with tuition reimbursements, research assistantships with tuition reimbursements, teaching assistantships with tuition reimbursements, career-related internships or fieldwork, Federal Work-Study, institutionally sponsored loans, and tuition waivers (full and partial) available. Financial aid application deadline: 2/1; financial aid applicants required to submit FAFSA. Dr. Joel D. Oppenheim, Director, 212-263-5648, *Fax:* 212-263-7600, *E-mail:* sackler-info@nyumed.med.nyu.edu.

■ NEW YORK UNIVERSITY

Graduate School of Arts and Science, Department of Biology, New York, NY 10012-1019

AWARDS Applied recombinant DNA technology (MS); biochemistry (PhD); biomedical journalism (MA); cell biology (PhD); computers in biological research (MS); environmental biology (PhD); general biology (MS); neural sciences and physiology (PhD); oral biology (MS); population and evolutionary biology (PhD). Part-time programs available.

Faculty: 22 full-time (5 women), 8 part-time/adjunct.
Students: 99 full-time (48 women), 61 part-time (31 women); includes 30 minority (4 African Americans, 22 Asian Americans or Pacific Islanders, 4 Hispanic Americans), 52 international. Average age 24. *371 applicants, 41% accepted.* In 1999, 54 master's, 4 doctorates awarded. Terminal master's awarded for partial completion of doctoral program.
Degree requirements: For master's, thesis or alternative, qualifying paper required, foreign language not required; for doctorate, dissertation, oral and written comprehensive exams required, foreign language not required.
Entrance requirements: For master's, GRE General Test, TOEFL; for doctorate, GRE General Test, GRE Subject Test, TOEFL. *Application deadline:* For fall admission, 1/4 (priority date). *Application fee:* $60.
Expenses: Tuition: Full-time $17,880; part-time $745 per credit. Required fees: $1,140; $35 per credit. Tuition and fees vary according to course load and program.
Financial aid: Fellowships with tuition reimbursements, research assistantships with tuition reimbursements, teaching assistantships with tuition reimbursements, career-related internships or fieldwork, Federal Work-Study, institutionally sponsored loans, and tuition waivers (full and partial) available. Financial aid application deadline: 1/4; financial aid applicants required to submit FAFSA.
Faculty research: Development and genetics, neurobiology, plant sciences, molecular and cell biology.
Philip Furmanski, Chairman, 212-998-8200.

Application contact: Gloria Coruzzi, Director of Graduate Studies, 212-998-8200, *Fax:* 212-995-4015, *E-mail:* biology@nyu.edu.

Find an in-depth description at www.petersons.com/graduate.

■ NEW YORK UNIVERSITY

Graduate School of Arts and Science, Nelson Institute of Environmental Medicine, New York, NY 10012-1019

AWARDS Environmental health sciences (MS, PhD), including environmental health (MS), ergonomics and biomechanics (PhD), occupational biomechanics (PhD). Part-time programs available.

Faculty: 26 full-time (7 women).
Students: 16 full-time (6 women), 11 part-time (1 woman); includes 4 minority (1 African American, 2 Asian Americans or Pacific Islanders, 1 Hispanic American), 8 international. Average age 33. *50 applicants, 22% accepted.* In 1999, 4 master's, 4 doctorates awarded. Terminal master's awarded for partial completion of doctoral program.
Degree requirements: For master's, thesis or alternative required, foreign language not required; for doctorate, one foreign language, dissertation, oral and written exams required.
Entrance requirements: For master's and doctorate, GRE General Test, GRE Subject Test, TOEFL, minimum GPA of 3.0; bachelor's degree in biological, physical, or engineering science. *Application deadline:* For fall admission, 2/1. *Application fee:* $60.
Expenses: Tuition: Full-time $17,880; part-time $745 per credit. Required fees: $1,140; $35 per credit. Tuition and fees vary according to course load and program.
Financial aid: Fellowships with tuition reimbursements, teaching assistantships with tuition reimbursements, career-related internships or fieldwork, Federal Work-Study, institutionally sponsored loans, and tuition waivers (partial) available. Financial aid application deadline: 2/1; financial aid applicants required to submit FAFSA.
Faculty research: Biomathematics, inhalation toxicology, radiation epidemiology, aerosol science, epidemiology of AIDS.
Dr. Max Costa, Director, 914-885-5281.
Application contact: Richard Schlesinger, Director of Graduate Studies, 914-885-5281, *E-mail:* ehs@charlotte.med.nyu.edu.

Find an in-depth description at www.petersons.com/graduate.

■ NEW YORK UNIVERSITY

School of Medicine and Graduate School of Arts and Science, Medical Scientist Training Program, New York, NY 10012-1019

AWARDS Biochemistry (MD/PhD); cell biology (MD/PhD); environmental health sciences (MD/PhD); microbiology (MD/PhD); parasitology (MD/PhD); pathology (MD/PhD); pharmacology (MD/PhD). Students must be

New York University (continued)
accepted by both the School of Medicine and the Graduate School of Arts and Science.

Faculty: 150 full-time (35 women).
Students: 80 full-time (18 women); includes 37 minority (3 African Americans, 32 Asian Americans or Pacific Islanders, 2 Hispanic Americans), 1 international. Average age 25. *195 applicants, 18% accepted.*
Degree requirements: One foreign language.
Application deadline: For fall admission, 11/15. Applications are processed on a rolling basis. *Application fee:* $60.
Expenses: Students receive full tuition support and an annual stipend of $17,500.
Financial aid: Application deadline: 5/1.
Faculty research: Genetics, tumor biology, cardiovascular biology, neuroscience, host defense mechanisms.
Dr. James Salzer, Director, 212-263-0758.
Application contact: Arlene Kohler, Administrative Officer, 212-263-5649.

■ **NEW YORK UNIVERSITY**

School of Medicine and Graduate School of Arts and Science, Sackler Institute of Graduate Biomedical Sciences, New York, NY 10012-1019

AWARDS PhD, MD/PhD.

Faculty: 150 full-time (35 women).
Students: 178 full-time (90 women); includes 53 minority (9 African Americans, 37 Asian Americans or Pacific Islanders, 7 Hispanic Americans), 41 international. Average age 25. *662 applicants, 11% accepted.* In 1999, 42 degrees awarded (100% entered university research/teaching).
Degree requirements: For doctorate, one foreign language, dissertation, qualifying exam required. *Average time to degree:* Doctorate–5.5 years full-time.
Entrance requirements: For doctorate, GRE General Test, GRE Subject Test, TOEFL. *Application deadline:* For fall admission, 2/1 (priority date). Applications are processed on a rolling basis. *Application fee:* $60.
Expenses: Tuition: Full-time $17,880; part-time $745 per credit. Required fees: $1,140; $35 per credit. Tuition and fees vary according to course load and program.
Financial aid: In 1999–00, 122 research assistantships (averaging $22,000 per year), 56 teaching assistantships (averaging $22,000 per year) were awarded; fellowships, tuition waivers (full) also available. *Total annual research expenditures:* $59 million.
Dr. Joel D. Oppenheim, Associate Dean for Graduate Studies, 212-263-8001, *Fax:* 212-263-7600.

Application contact: Debra E. Stalk, Administrative Assistant, 212-263-5648, *Fax:* 212-263-7600, *E-mail:* stalkd01@popmail.med.nyu.edu.

Find an in-depth description at www.petersons.com/graduate.

■ **NORTH CAROLINA AGRICULTURAL AND TECHNICAL STATE UNIVERSITY**

Graduate School, College of Arts and Sciences, Department of Biology, Greensboro, NC 27411

AWARDS MS. Part-time and evening/weekend programs available.

Degree requirements: For master's, thesis (for some programs), comprehensive exam, qualifying exam required, foreign language not required.
Entrance requirements: For master's, GRE General Test, minimum GPA of 2.6.
Expenses: Tuition, state resident: full-time $982; part-time $368 per semester. Tuition, nonresident: full-time $8,252; part-time $3,095 per semester. Required fees: $464 per semester.
Faculty research: Physical ecology, cytochemistry, botany, parasitology, microbiology.

■ **NORTH CAROLINA CENTRAL UNIVERSITY**

Division of Academic Affairs, College of Arts and Sciences, Department of Biology, Durham, NC 27707-3129
AWARDS MS.

Faculty: 16 full-time (6 women), 6 part-time/adjunct (1 woman).
Students: 15 full-time (10 women), 22 part-time (14 women); includes 36 minority (all African Americans). Average age 28. *18 applicants, 67% accepted.* In 1999, 5 degrees awarded.
Degree requirements: For master's, one foreign language (computer language can substitute), thesis, comprehensive exam required.
Entrance requirements: For master's, minimum GPA of 3.0 in major, 2.5 overall. *Application deadline:* For fall admission, 8/1. *Application fee:* $30.
Expenses: Tuition, state resident: full-time $982. Tuition, nonresident: full-time $8,252. Required fees: $873. Full-time tuition and fees vary according to program.
Financial aid: Federal Work-Study and institutionally sponsored loans available. Aid available to part-time students. Financial aid application deadline: 5/1.
Dr. Sandra L. White, Chairperson, 919-560-6407, *Fax:* 919-530-7773, *E-mail:* swhite@wpo.ncc.edu.

Application contact: Dr. Bernice D. Johnson, Dean, College of Arts and Sciences, 919-560-6368, *Fax:* 919-560-5361, *E-mail:* bjohnson@wpo.ncc.edu.

■ **NORTH CAROLINA STATE UNIVERSITY**

Graduate School, College of Agriculture and Life Sciences, Raleigh, NC 27695

AWARDS M Ag, M Econ, M Ed, M Soc, M Tox, MA, MAEE, MAWB, MBAE, MLS, MS, MSA, Ed D, PhD. Part-time programs available.

Faculty: 448 full-time (55 women), 236 part-time/adjunct (19 women).
Students: 516 full-time (247 women), 256 part-time (124 women); includes 62 minority (36 African Americans, 11 Asian Americans or Pacific Islanders, 12 Hispanic Americans, 3 Native Americans), 138 international. Average age 31. *705 applicants, 35% accepted.* In 1999, 131 master's, 84 doctorates awarded.
Application fee: $45.
Expenses: Tuition, state resident: full-time $1,578. Tuition, nonresident: full-time $10,744. Required fees: $892. Full-time tuition and fees vary according to program.
Financial aid: In 1999–00, 49 fellowships (averaging $5,876 per year), 388 research assistantships (averaging $5,039 per year), 62 teaching assistantships (averaging $5,330 per year) were awarded; career-related internships or fieldwork, Federal Work-Study, institutionally sponsored loans, traineeships, and tuition waivers (partial) also available. Aid available to part-time students. *Total annual research expenditures:* $171.4 million.
Dr. James L. Oblinger, Interim Dean, 919-515-2668, *Fax:* 919-515-6980, *E-mail:* james_oblinger@ncsu.edu.
Application contact: Bee Smith, Administrative Assistant, 919-515-2668, *Fax:* 919-515-6980, *E-mail:* bee_smith@ncsu.edu.

Find an in-depth description at www.petersons.com/graduate.

■ **NORTHEASTERN ILLINOIS UNIVERSITY**

Graduate College, College of Arts and Sciences, Department of Biology, Program in Biology, Chicago, IL 60625-4699

AWARDS MS. Part-time and evening/weekend programs available.

Degree requirements: For master's, thesis optional, foreign language not required.
Entrance requirements: For master's, minimum GPA of 2.75.

Expenses: Tuition, state resident: full-time $2,626; part-time $109 per credit. Tuition, nonresident: full-time $7,234; part-time $301 per credit.

Faculty research: Paleoecology and freshwater biology; protein biosynthesis and targeting; microbial growth and physiology; molecular biology of antibody production; reptilian neurobiology.

■ NORTHEASTERN UNIVERSITY

Bouvé College of Health Sciences Graduate School, Programs in Biomedical Sciences, Boston, MA 02115-5096

AWARDS Biomedical sciences (MS); medical laboratory science (PhD); medicinal chemistry (PhD); pharmaceutics (PhD); pharmacology (PhD); toxicology (MS, PhD).

Faculty: 14 full-time (1 woman), 14 part-time/adjunct (5 women).
Students: 47 full-time (30 women), 6 part-time (4 women). Average age 31. *162 applicants, 47% accepted.* In 1999, 15 master's, 6 doctorates awarded. Terminal master's awarded for partial completion of doctoral program.
Degree requirements: For master's, comprehensive exam required, thesis optional, foreign language not required; for doctorate, dissertation, qualifying exam required, foreign language not required.
Entrance requirements: For master's and doctorate, GRE General Test, TOEFL. *Application deadline:* For fall admission, 3/15. *Application fee:* $50.
Expenses: Tuition: Full-time $16,560; part-time $460 per quarter hour. Required fees: $150; $25 per year. Tuition and fees vary according to course load and program.
Financial aid: In 1999–00, 40 students received aid, including 10 research assistantships with full tuition reimbursements available, 12 teaching assistantships with full tuition reimbursements available (averaging $12,650 per year); tuition waivers (partial) also available. Financial aid applicants required to submit FAFSA.
Faculty research: Neuropharmacology, cardiovascular pharmacology, steroid chemistry, anti-infectives, behavioral pharmacology.
Dr. Roger W. Giese, Director, 617-373-3227, *Fax:* 617-266-6756, *E-mail:* rgiese@lynx.neu.edu.
Application contact: Bill Purnell, Director of Graduate Admissions, 617-373-2708, *Fax:* 617-373-4701, *E-mail:* w.purnell@nunet.neu.edu.

Find an in-depth description at www.petersons.com/graduate.

■ NORTHEASTERN UNIVERSITY

College of Arts and Sciences, Department of Biology, Boston, MA 02115-5096

AWARDS MS, PhD. Part-time programs available.

Faculty: 23 full-time (6 women), 4 part-time/adjunct (2 women).
Students: 63 full-time (31 women), 21 part-time (8 women). Average age 29. *150 applicants, 35% accepted.* In 1999, 7 master's, 5 doctorates awarded. Terminal master's awarded for partial completion of doctoral program.
Degree requirements: For master's, thesis required, foreign language not required; for doctorate, dissertation, qualifying exam required, foreign language not required.
Entrance requirements: For master's and doctorate, GRE General Test, TOEFL. *Application deadline:* For fall admission, 2/1 (priority date). *Application fee:* $50.
Expenses: Tuition: Full-time $16,560; part-time $460 per quarter hour. Required fees: $150; $25 per year. Tuition and fees vary according to course load and program.
Financial aid: In 1999–00, research assistantships with tuition reimbursements (averaging $12,150 per year); fellowships with tuition reimbursements, teaching assistantships with tuition reimbursements, career-related internships or fieldwork, Federal Work-Study, and tuition waivers (full and partial) also available. Financial aid application deadline: 2/1; financial aid applicants required to submit FAFSA.
Faculty research: Biochemistry, cell and systems physiology, ecology, marine sciences, molecular biology. *Total annual research expenditures:* $2 million.
Dr. Edward Jarroll, Chairman, 617-373-2260, *Fax:* 617-373-3724.
Application contact: Florence Lewis, Graduate Secretary, 617-373-2262, *Fax:* 617-373-3724, *E-mail:* f.lewis@nunet.neu.edu.

Find an in-depth description at www.petersons.com/graduate.

■ NORTHERN ARIZONA UNIVERSITY

Graduate College, College of Arts and Sciences, Department of Biological Sciences, Flagstaff, AZ 86011

AWARDS Biology (MS, PhD); biology education (MAT).

Faculty: 42 full-time (11 women), 24 part-time/adjunct (7 women).
Students: 66 full-time (35 women), 27 part-time (10 women); includes 12 minority (1 African American, 8 Hispanic Americans, 3 Native Americans), 3 international. Average age 30. *74 applicants, 50% accepted.* In 1999, 19 master's, 6 doctorates awarded.
Degree requirements: For master's, foreign language not required; for doctorate, dissertation required.
Entrance requirements: For master's, GRE General Test, GRE Subject Test; for doctorate, GRE General Test. *Application deadline:* For fall admission, 2/15. *Application fee:* $45. Electronic applications accepted.
Expenses: Tuition, state resident: full-time $2,261; part-time $125 per credit hour. Tuition, nonresident: full-time $8,377; part-time $356 per credit hour.
Financial aid: In 1999–00, 37 research assistantships, 38 teaching assistantships were awarded; fellowships, Federal Work-Study, institutionally sponsored loans, and tuition waivers (full) also available. Aid available to part-time students. Financial aid application deadline: 2/15.
Faculty research: Genetic levels of trophic levels, plant hybrid zones, insect biodiversity, natural history and cognition of wild jays. *Total annual research expenditures:* $2.2 million.
Dr. Lee Drickamer, Chairman, 520-523-2381, *E-mail:* lee.drickamer@nau.edu.
Application contact: Dr. Ronald Markle, Graduate Coordinator, 520-523-7531, *E-mail:* ronald.markle@nau.edu.

■ NORTHERN ILLINOIS UNIVERSITY

Graduate School, College of Liberal Arts and Sciences, Department of Biological Sciences, De Kalb, IL 60115-2854

AWARDS MS, PhD. Part-time programs available.

Faculty: 33 full-time (6 women), 6 part-time/adjunct (0 women).
Students: 49 full-time (24 women), 25 part-time (11 women); includes 9 minority (1 African American, 6 Asian Americans or Pacific Islanders, 1 Hispanic American, 1 Native American), 3 international. Average age 28. *62 applicants, 47% accepted.* In 1999, 10 master's, 2 doctorates awarded. Terminal master's awarded for partial completion of doctoral program.
Degree requirements: For master's, comprehensive exam required, thesis optional, foreign language not required; for doctorate, variable foreign language requirement, dissertation, candidacy exam, dissertation defense required.
Entrance requirements: For master's, GRE General Test or MCAT, GRE Subject Test, TOEFL, bachelor's degree in related field, minimum GPA of 2.75; for doctorate, GRE General Test or MCAT, TOEFL, bachelor's or master's degree in

Northern Illinois University (continued)
related field; minimum GPA of 2.75 (undergraduate), 3.2 (graduate). *Application deadline:* For fall admission, 6/1; for spring admission, 11/1. Applications are processed on a rolling basis. *Application fee:* $30.
Expenses: Tuition, state resident: part-time $169 per credit hour. Tuition, nonresident: part-time $295 per credit hour. Tuition and fees vary according to campus/location and program.
Financial aid: In 1999–00, 12 research assistantships with full tuition reimbursements, 35 teaching assistantships with full tuition reimbursements were awarded; fellowships with full tuition reimbursements, career-related internships or fieldwork, Federal Work-Study, tuition waivers (full), and unspecified assistantships also available. Aid available to part-time students. Dr. Michael Parrish, Chair, 815-753-1753, *Fax:* 815-753-0461.
Application contact: Dr. Carl von Ende, Director of Graduate Studies, 815-753-7826.

Find an in-depth description at www.petersons.com/graduate.

■ NORTHERN MICHIGAN UNIVERSITY

College of Graduate Studies, College of Arts and Sciences, Department of Biology, Marquette, MI 49855-5301

AWARDS MS. Part-time programs available.

Degree requirements: For master's, thesis or alternative required, foreign language not required.
Entrance requirements: For master's, GRE, minimum GPA of 2.75.
Expenses: Tuition, state resident: full-time $3,348; part-time $140 per credit. Tuition, nonresident: full-time $5,400; part-time $225 per credit. Required fees: $31 per credit. Tuition and fees vary according to course level, course load and campus/location.
Faculty research: Molecular genetics of sex-linked genes, biology of protozoan parasites, wildlife ecology, organochlorines in the environment, insect development.

■ NORTHWESTERN UNIVERSITY

The Graduate School, Division of Interdepartmental Programs, Combined MD/MPH Program in Public Health, Chicago, IL 60611

AWARDS MD/MPH. Application must be made to both The Graduate School and the Medical School.

Faculty: 13 full-time (7 women), 28 part-time/adjunct (7 women).
Students: 38 full-time (22 women); includes 14 minority (1 African American, 12 Asian Americans or Pacific Islanders, 1

Hispanic American). Average age 24. *18 applicants, 78% accepted.*
Application deadline: For fall admission, 8/1 (priority date). Applications are processed on a rolling basis. *Application fee:* $50 ($55 for international students).
Expenses: Tuition: Full-time $23,301. Full-time tuition and fees vary according to program.
Financial aid: Career-related internships or fieldwork and institutionally sponsored loans available. Financial aid application deadline: 1/15; financial aid applicants required to submit FAFSA.
Faculty research: Cardiovascular epidemiology, cancer epidemiology, nutritional interventions for the prevention of cardiovascular disease and cancer, women's health, outcomes research. *Total annual research expenditures:* $13.1 million.
Rowland W. Chang, Director, 312-503-2952, *E-mail:* rwchang@northwestern.edu.
Application contact: Maureen Moran, Associate Director, 312-503-0500, *Fax:* 312-908-9588, *E-mail:* rmph_prog@ northwestern.edu.

■ NORTHWESTERN UNIVERSITY

The Graduate School, Division of Interdepartmental Programs, Combined MD/PhD Medical Scientist Training Program, Chicago, IL 60611

AWARDS MD/PhD. Application must be made to both The Graduate School and the Medical School.

Students: 48 full-time (22 women); includes 25 minority (1 African American, 16 Asian Americans or Pacific Islanders, 7 Hispanic Americans, 1 Native American). Average age 25. *6 applicants, 100% accepted.*
Application deadline: For fall admission, 10/15. Applications are processed on a rolling basis. *Application fee:* $50 ($55 for international students).
Expenses: Tuition: Full-time $23,301. Full-time tuition and fees vary according to program.
Financial aid: In 1999–00, 8 fellowships with full tuition reimbursements (averaging $17,000 per year) were awarded.
Faculty research: Cardiovascular epidemiology, cancer epidemiology, nutritional interventions for the prevention of cardiovascular disease and cancer, women's health, outcomes research. David M. Engman, Director, 312-503-1288, *E-mail:* d-engman@ northwestern.edu.
Application contact: Sharon McBride, Program Assistant, 312-503-5232, *Fax:* 312-908-5253, *E-mail:* mstp@ northwestern.edu.

■ NORTHWESTERN UNIVERSITY

The Graduate School, Division of Interdepartmental Programs and Medical School, Integrated Graduate Programs in the Life Sciences, Chicago, IL 60611

AWARDS Cancer biology (PhD); cell biology (PhD); developmental biology (PhD); evolutionary biology (PhD); immunology and microbial pathogenesis (PhD); molecular biology and genetics (PhD); neurobiology (PhD); pharmacology and toxicology (PhD); structural biology and biochemistry (PhD).

Degree requirements: For doctorate, dissertation, written and oral qualifying exams required, foreign language not required.
Entrance requirements: For doctorate, GRE General Test, TOEFL.
Expenses: Tuition: Full-time $23,301. Full-time tuition and fees vary according to program.

Find an in-depth description at www.petersons.com/graduate.

■ NORTHWESTERN UNIVERSITY

The Graduate School, Judd A. and Marjorie Weinberg College of Arts and Sciences, Interdepartmental Biological Sciences Program (IBiS), Evanston, IL 60208

AWARDS Biochemistry, molecular biology, and cell biology (PhD), including cell and molecular biology, molecular biophysics; biotechnology (PhD); cell and molecular biology (PhD); genetics and developmental biology (PhD); integrative biology (PhD); reproductive biology (PhD); structural biology (PhD). Participants in the Interdepartmental Biological Sciences Program include the Departments of Biochemistry, Molecular Biology, and Cell Biology; Chemistry; Neurobiology and Physiology; Chemical Engineering; Civil Engineering; and Evanston Hospital.

Faculty: 59 full-time (11 women).
Students: 79 full-time (46 women); includes 5 minority (2 African Americans, 3 Hispanic Americans), 14 international. *236 applicants, 19% accepted.* In 1999, 1 degree awarded.
Degree requirements: For doctorate, dissertation, 2 quarters of teaching experience required, foreign language not required.
Average time to degree: Doctorate–5.5 years full-time.
Entrance requirements: For doctorate, GRE General Test, TOEFL, TSE.
Application deadline: For fall admission, 1/15. Applications are processed on a rolling basis. *Application fee:* $50 ($55 for international students).
Expenses: Tuition: Full-time $23,301. Full-time tuition and fees vary according to program.

Financial aid: In 1999–00, 15 fellowships with full tuition reimbursements (averaging $12,078 per year), 64 research assistantships with partial tuition reimbursements (averaging $17,000 per year), 15 teaching assistantships with full tuition reimbursements (averaging $16,620 per year) were awarded; Federal Work-Study, institutionally sponsored loans, and traineeships also available. Financial aid application deadline: 12/31; financial aid applicants required to submit FAFSA.

Faculty research: Developmental genetics, gene regulation, DNA-protein interactions, biological clocks, bioremediation, ion channels, neurobiology.

Richard Gaber, Director, 800-545-1761, *Fax:* 847-467-1380, *E-mail:* ibis@northwestern.edu.

Application contact: Latonia Trimuel, Program Assistant, 800-546-1761, *E-mail:* ibis@northwestern.edu.

Find an in-depth description at www.petersons.com/graduate.

■ NORTHWEST MISSOURI STATE UNIVERSITY

Graduate School, College of Arts and Sciences, Department of Biology, Maryville, MO 64468-6001

AWARDS MS. Part-time programs available.

Faculty: 7 full-time (2 women).

Students: *1 applicant, 100% accepted.*

Degree requirements: For master's, thesis, comprehensive exam required, foreign language not required.

Entrance requirements: For master's, GRE General Test, TOEFL, minimum GPA of 3.0 in last 60 hours or 2.75 overall, writing sample. *Application deadline:* Applications are processed on a rolling basis. *Application fee:* $0 ($50 for international students).

Expenses: Tuition, state resident: full-time $2,282; part-time $127 per credit. Tuition, nonresident: full-time $3,893; part-time $216 per credit. Tuition and fees vary according to course level and course load.

Financial aid: Research assistantships, teaching assistantships, tutorial assistantships available. Financial aid application deadline: 3/1.

Dr. Edward Farquhar, Chairperson, 660-562-1209.

Application contact: Dr. Frances Shipley, Dean of Graduate School, 660-562-1145, *E-mail:* gradsch@mail.nwmissouri.edu.

■ NOVA SOUTHEASTERN UNIVERSITY

Health Professions Division, College of Medical Sciences, Fort Lauderdale, FL 33314-7721

AWARDS Biomedical sciences (MBS).

Faculty: 22 full-time (6 women), 1 (woman) part-time/adjunct.

Students: 31 full-time (9 women); includes 9 minority (1 African American, 5 Asian Americans or Pacific Islanders, 3 Hispanic Americans), 5 international. Average age 28. *53 applicants, 57% accepted.* In 1999, 1 degree awarded. *Average time to degree:* Master's–2 years full-time.

Entrance requirements: For master's, minimum GPA of 2.5 required. *Application deadline:* For spring admission, 5/1. Applications are processed on a rolling basis. *Application fee:* $50.

Expenses: Tuition: Full-time $20,500. Required fees: $100.

Faculty research: Neurophysiology, mucosal immunology, allergies involving the lungs, cardiovascular physiology parasitology. *Total annual research expenditures:* $25,000.

Dr. Harold E. Laubach, Dean, 954-262-1303, *Fax:* 954-262-1802, *E-mail:* harold@nova.edu.

Application contact: Tammy Gibson, Admissions Coordinator, 954-262-1112, *E-mail:* tgibson@nova.edu.

■ OAKLAND UNIVERSITY

Graduate Studies, College of Arts and Sciences, Department of Biological Sciences, Rochester, MI 48309-4401

AWARDS Biological sciences (MS); cellular biology of aging (MS).

Faculty: 19 full-time (4 women), 1 (woman) part-time/adjunct.

Students: 8 full-time (all women), 4 part-time (3 women); includes 1 minority (Asian American or Pacific Islander), 3 international. Average age 29. In 1999, 2 degrees awarded.

Degree requirements: For master's, thesis required, foreign language not required.

Entrance requirements: For master's, GRE Subject Test, minimum GPA of 3.0 for unconditional admission. *Application deadline:* For fall admission, 7/15; for spring admission, 3/15. *Application fee:* $30.

Expenses: Tuition, state resident: full-time $5,294; part-time $221 per credit hour. Tuition, nonresident: full-time $11,720; part-time $488 per credit hour. Required fees: $214 per semester. Tuition and fees vary according to campus/location and program.

Financial aid: Federal Work-Study, institutionally sponsored loans, and tuition waivers (full) available. Financial aid application deadline: 3/1; financial aid applicants required to submit FAFSA.

Dr. Virinder K. Moudgil, Chair, 248-370-3553.

Application contact: Dr. George Gamboa, Coordinator, 248-370-3550.

■ OCCIDENTAL COLLEGE

Graduate Studies, Department of Biology, Los Angeles, CA 90041-3392

AWARDS MA. Part-time programs available.

Faculty: 9 full-time (4 women), 2 part-time/adjunct (both women).

Students: 1 full-time (0 women). Average age 23. In 1999, 1 degree awarded.

Degree requirements: For master's, thesis, final exam required. *Average time to degree:* Master's–2 years full-time.

Entrance requirements: For master's, GRE General Test, GRE Subject Test, TOEFL, minimum GPA of 3.0. *Application deadline:* For fall admission, 3/1; for spring admission, 10/1. Applications are processed on a rolling basis. *Application fee:* $40.

Expenses: Tuition: Full-time $22,698; part-time $925 per unit. Required fees: $498; $249 per semester. Tuition and fees vary according to program.

Financial aid: Fellowships, Federal Work-Study, institutionally sponsored loans, and scholarships available. Aid available to part-time students. Financial aid application deadline: 3/1; financial aid applicants required to submit FAFSA.

Chair, 323-259-2697.

Application contact: Susan Molik, Administrative Assistant, Graduate Office, 323-259-2921, *E-mail:* molik@oxy.edu.

■ THE OHIO STATE UNIVERSITY

College of Medicine and Public Health and Graduate School, Graduate Programs in the Basic Medical Sciences, Columbus, OH 43210

AWARDS MHA, MPH, MS, PhD, JD/MHA, MD/MHA, MD/MPH, MD/MS, MD/PhD, MHA/MBA, MHA/MPA, MHA/MS. Part-time and evening/weekend programs available.

Faculty: 292 full-time, 74 part-time/adjunct.

Students: 220 full-time (135 women), 71 part-time (53 women); includes 36 minority (11 African Americans, 19 Asian Americans or Pacific Islanders, 2 Hispanic Americans, 4 Native Americans), 48 international. *542 applicants, 29% accepted.* In 1999, 83 master's, 13 doctorates awarded. Terminal master's awarded for partial completion of doctoral program.

Degree requirements: For master's, foreign language not required; for doctorate, dissertation required, foreign language not required.

Application deadline: Applications are processed on a rolling basis. Electronic applications accepted.

Expenses: Tuition, state resident: full-time $5,400. Tuition, nonresident: full-time $14,535. Part-time tuition and fees vary according to course load.

Financial aid: Fellowships with full tuition reimbursements, research assistantships

The Ohio State University (continued)
with full tuition reimbursements, teaching assistantships with full tuition reimbursements, Federal Work-Study, grants, institutionally sponsored loans, scholarships, traineeships, and unspecified assistantships available. Aid available to part-time students. Financial aid applicants required to submit FAFSA.
Dr. Susan L. Huntington, Vice Provost and Dean of the Graduate School, 614-292-6031, *Fax:* 614-292-3656.
Application contact: Amy Edgar, Data Analyst, 614-292-6031, *Fax:* 614-292-3656, *E-mail:* edgar.1@osu.edu.

Find an in-depth description at www.petersons.com/graduate.

■ **THE OHIO STATE UNIVERSITY**

Graduate School, College of Biological Sciences, Columbus, OH 43210

AWARDS MS, PhD. Part-time programs available.

Faculty: 254 full-time, 114 part-time/adjunct.
Students: 422 full-time (190 women), 24 part-time (11 women); includes 32 minority (9 African Americans, 17 Asian Americans or Pacific Islanders, 5 Hispanic Americans, 1 Native American), 177 international. *727 applicants, 26% accepted.* In 1999, 40 master's, 39 doctorates awarded.
Degree requirements: For doctorate, dissertation required.
Entrance requirements: For master's and doctorate, GRE General Test. *Application deadline:* For fall admission, 8/15. Applications are processed on a rolling basis. *Application fee:* $30 ($40 for international students).
Expenses: Tuition, state resident: full-time $5,400. Tuition, nonresident: full-time $14,535. Part-time tuition and fees vary according to course load and program.
Financial aid: Fellowships, research assistantships, teaching assistantships, career-related internships or fieldwork, Federal Work-Study, and institutionally sponsored loans available. Aid available to part-time students.
Alan G. Goodridge, Dean, 614-292-8772, *Fax:* 614-292-1538, *E-mail:* goodridge.4@osu.edu.

Find an in-depth description at www.petersons.com/graduate.

■ **OHIO UNIVERSITY**

Graduate Studies, College of Arts and Sciences, Department of Biological Sciences, Athens, OH 45701-2979

AWARDS Biological sciences (MS, PhD); microbiology (MS, PhD); zoology (MS, PhD).

Faculty: 24 full-time (7 women), 11 part-time/adjunct (6 women).
Students: 53 full-time (17 women), 5 part-time (2 women); includes 1 minority (African American), 18 international. Average age 24. *85 applicants, 21% accepted.* In 1999, 3 master's awarded (100% continued full-time study); 5 doctorates awarded (100% found work related to degree).
Degree requirements: For master's, thesis, 1 quarter of teaching experience required, foreign language not required; for doctorate, dissertation, 2 quarters of teaching experience required.
Entrance requirements: For master's and doctorate, GRE General Test. *Application deadline:* For fall admission, 1/15. *Application fee:* $30.
Expenses: Tuition, state resident: full-time $5,754; part-time $238 per credit hour. Tuition, nonresident: full-time $11,055; part-time $457 per credit hour. Tuition and fees vary according to course load, degree level and campus/location.
Financial aid: In 1999–00, 1 fellowship with full tuition reimbursement, 1 research assistantship with full tuition reimbursement (averaging $15,000 per year), 33 teaching assistantships with full tuition reimbursements (averaging $13,500 per year) were awarded; Federal Work-Study, institutionally sponsored loans, and tuition waivers (full) also available. Financial aid application deadline: 1/15.
Faculty research: Ecology and evolutionary biology, exercise physiology and muscle biology, endocrinology and metabolic shysiology, neurobiology. *Total annual research expenditures:* $2.8 million.
Dr. Anne Loucks, Chair, 740-593-2290, *Fax:* 740-593-0300, *E-mail:* loucks@ohiou.edu.
Application contact: Dr. William R. Holmes, Graduate Chair, 740-593-2334, *Fax:* 740-593-0300, *E-mail:* holmes@ohiou.edu.

Find an in-depth description at www.petersons.com/graduate.

■ **OKLAHOMA STATE UNIVERSITY COLLEGE OF OSTEOPATHIC MEDICINE**

Program in Biomedical Sciences, Tulsa, OK 74107-1898

AWARDS PhD, DO/PhD.

Faculty: 23 full-time (3 women), 7 part-time/adjunct (2 women).
Students: 5 full-time (3 women), 10 part-time (9 women). Average age 27. 7 *applicants, 43% accepted.*
Degree requirements: For doctorate, dissertation required, foreign language not required. *Average time to degree:* Doctorate–4 years full-time, 7 years part-time.

Entrance requirements: For doctorate, GRE General Test, MCAT, TOEFL. *Application deadline:* Applications are processed on a rolling basis. *Application fee:* $25. Electronic applications accepted.
Expenses: Tuition, state resident: full-time $9,552. Tuition, nonresident: full-time $24,244. Full-time tuition and fees vary according to student level.
Financial aid: In 1999–00, 3 research assistantships with partial tuition reimbursements (averaging $15,000 per year) were awarded; grants, scholarships, and tuition waivers (partial) also available.
Dr. Thomas Wesley Allen, Vice President for Health Affairs and Dean of the College of Osteopathic Medicine, 918-582-1972 Ext. 8201, *Fax:* 918-561-8413.
Application contact: Dr. Gary H. Watson, Associate Dean for Research and Sponsored Programs, 918-561-8241, *Fax:* 918-699-8629, *E-mail:* wgary@osu-com.okstate.edu.

■ **OLD DOMINION UNIVERSITY**

College of Sciences, Department of Biological Sciences, Norfolk, VA 23529

AWARDS Biology (MS); biomedical sciences (PhD); ecological sciences (PhD). Part-time programs available.

Faculty: 27 full-time (5 women).
Students: 77 full-time (38 women), 51 part-time (31 women); includes 23 minority (16 African Americans, 4 Asian Americans or Pacific Islanders, 1 Hispanic American, 2 Native Americans), 22 international. Average age 31. *55 applicants, 58% accepted.* In 1999, 14 master's awarded (100% continued full-time study); 14 doctorates awarded. Terminal master's awarded for partial completion of doctoral program.
Degree requirements: For master's, comprehensive exam required, thesis optional, foreign language not required; for doctorate, dissertation, comprehensive exam required.
Entrance requirements: For master's, GRE General Test, GRE Subject Test, MCAT, TOEFL, minimum GPA of 3.0 in major, 2.7 overall; for doctorate, GRE General Test, TOEFL, minimum GPA of 3.0. *Application fee:* $30. Electronic applications accepted.
Expenses: Tuition, state resident: full-time $4,440; part-time $185 per credit. Tuition, nonresident: full-time $11,784; part-time $477 per credit. Required fees: $1,612. Tuition and fees vary according to program.
Financial aid: In 1999–00, 2 fellowships (averaging $6,575 per year), 34 research assistantships with tuition reimbursements (averaging $8,841 per year), 15 teaching assistantships with tuition reimbursements (averaging $9,750 per year) were awarded;

career-related internships or fieldwork, grants, and tuition waivers (partial) also available. Aid available to part-time students. Financial aid application deadline: 2/15; financial aid applicants required to submit FAFSA.
Faculty research: Wetland ecology, systematics and ecology of vertebrates, marine biology, biotechnology. *Total annual research expenditures:* $1.7 million.
Dr. Mark Butler, Chair, 757-683-3595, *Fax:* 757-683-5283, *E-mail:* biolgpd@ odu.edu.

■ OREGON HEALTH SCIENCES UNIVERSITY

School of Medicine, Graduate Programs in Medicine, Portland, OR 97201-3098

AWARDS MS, PhD, Certificate, MD/PhD. Part-time programs available.

Faculty: 395 full-time (180 women), 120 part-time/adjunct (40 women).
Students: 268 full-time (117 women), 65 part-time (30 women); includes 76 minority (4 African Americans, 61 Asian Americans or Pacific Islanders, 7 Hispanic Americans, 4 Native Americans), 15 international. Average age 26. *558 applicants, 6% accepted.* Terminal master's awarded for partial completion of doctoral program.
Degree requirements: For master's, thesis, capstone experience required, foreign language not required; for doctorate, dissertation, qualifying exam required, foreign language not required.
Entrance requirements: For master's and doctorate, GRE General Test. *Application deadline:* Applications are processed on a rolling basis.
Expenses: Tuition, state resident: full-time $3,132; part-time $174 per credit hour. Tuition, nonresident: full-time $5,256; part-time $292 per credit hour. Required fees: $8.5 per credit hour. $146 per term. Part-time tuition and fees vary according to course load.
Financial aid: In 1999–00, 198 research assistantships with full tuition reimbursements (averaging $16,000 per year) were awarded; fellowships, teaching assistantships, career-related internships or fieldwork, Federal Work-Study, institutionally sponsored loans, scholarships, and tuition waivers (full) also available. Aid available to part-time students. Financial aid application deadline: 3/1; financial aid applicants required to submit FAFSA.
Dr. Richard Maurer, Associate Dean for Graduate Studies, 503-494-7566.

Application contact: Wendy R. Doggett, Admissions Coordinator, 503-494-6222, *Fax:* 503-494-3400, *E-mail:* grdstudy@ ohsu.edu.

Find an in-depth description at www.petersons.com/graduate.

■ OREGON HEALTH SCIENCES UNIVERSITY

School of Medicine, Integrative Biomedical Sciences Program, Portland, OR 97201-3098

AWARDS PhD, MD/PhD.

Faculty: 48 full-time (12 women).
Students: 20 full-time (12 women); includes 1 minority (Asian American or Pacific Islander), 3 international. Average age 29. *49 applicants, 14% accepted.* In 1999, 1 degree awarded (100% found work related to degree).
Degree requirements: For doctorate, dissertation required, foreign language not required. *Average time to degree:* Doctorate–5 years full-time.
Entrance requirements: For doctorate, GRE General Test, TOEFL. *Application deadline:* For fall admission, 1/15. Applications are processed on a rolling basis. *Application fee:* $60.
Expenses: Tuition, state resident: full-time $14,688. Tuition, nonresident: full-time $30,960. Required fees: $1,949. Full-time tuition and fees vary according to program and student level. Part-time tuition and fees vary according to degree level and program.
Financial aid: In 1999–00, 13 research assistantships with full tuition reimbursements (averaging $16,000 per year) were awarded; Federal Work-Study, institutionally sponsored loans, and scholarships also available. Financial aid application deadline: 3/1; financial aid applicants required to submit FAFSA.
Faculty research: Physiology, giophysics, phanmacology, electrophysiology, molecular/cell biology.
Dr. David C. Dawson, Chairman, 503-494, *Fax:* 503-494-4352, *E-mail:* dawschda@ ohsu.edu.
Application contact: Andrea P. Ilg, Graduate Coordinator, 503-494-8982, *Fax:* 503-494-4352, *E-mail:* ibms@ohsu.edu.

Find an in-depth description at www.petersons.com/graduate.

■ THE PENNSYLVANIA STATE UNIVERSITY MILTON S. HERSHEY MEDICAL CENTER

Graduate School, Hershey, PA 17033-2360

AWARDS MS, PhD, MD/MBA, MD/PhD, PhD/MBA. Part-time programs available.

Students: 141 full-time (72 women), 6 part-time (3 women). Average age 27. *Application deadline:* For fall admission, 7/26. *Application fee:* $50.
Expenses: Tuition, state resident: full-time $6,886; part-time $291 per credit. Tuition, nonresident: full-time $14,118; part-time $588 per credit. Required fees: $43 per semester. Part-time tuition and fees vary according to course load.
Financial aid: Fellowships, research assistantships, teaching assistantships available. Aid available to part-time students. Financial aid applicants required to submit FAFSA.
Dr. D. Eugene Rannels, Assistant Dean, 717-531-8892.

Find an in-depth description at www.petersons.com/graduate.

■ THE PENNSYLVANIA STATE UNIVERSITY UNIVERSITY PARK CAMPUS

Graduate School, Eberly College of Science, Department of Biology, State College, University Park, PA 16802-1503

AWARDS Biology (MS, PhD); molecular evolutionary biology (MS, PhD).

Students: 28 full-time (12 women), 10 part-time (4 women). In 1999, 6 master's, 2 doctorates awarded.
Entrance requirements: For master's and doctorate, GRE General Test. *Application fee:* $50.
Expenses: Tuition, state resident: full-time $6,886; part-time $291 per credit. Tuition, nonresident: full-time $14,118; part-time $588 per credit. Required fees: $46 per semester. Part-time tuition and fees vary according to course load and program.
Financial aid: Fellowships, research assistantships, teaching assistantships available.
Dr. Charles Fisher, Head, 814-865-7034.
Application contact: Dr. Stephen Schaeffer, Chair, 814-863-7034.

Find an in-depth description at www.petersons.com/graduate.

■ THE PENNSYLVANIA STATE UNIVERSITY UNIVERSITY PARK CAMPUS

Graduate School, Intercollege Graduate Programs, Intercollege Graduate Program in Integrative Biosciences, State College, University Park, PA 16802-1503

AWARDS Integrative biosciences (MS, PhD), including biomolecular transport dynamics, cell and developmental biology, cellular and molecular mechanisms of toxicity, chemical biology, ecological and molecular plant

The Pennsylvania State University University Park Campus (continued) physiology, immunobiology, molecular medicine, neuroscience, nutrition science.

Students: 39 full-time (24 women), 1 part-time.

Entrance requirements: For master's and doctorate, GRE General Test. *Application fee:* $50.

Expenses: Tuition, state resident: full-time $6,886; part-time $291 per credit. Tuition, nonresident: full-time $14,118; part-time $588 per credit. Required fees: $46 per semester. Part-time tuition and fees vary according to course load and program.

Financial aid: Fellowships available. Dr. C. R. Matthews, Co-Director, 814-863-3650.

Application contact: Admissions Committee, 814-865-3155, *Fax:* 814-863-1357, *E-mail:* lscgradadm@mail.biotec.psu.edu.

Find an in-depth description at www.petersons.com/graduate.

■ PHILADELPHIA COLLEGE OF OSTEOPATHIC MEDICINE

Graduate and Professional Programs, Program in Biomedical Sciences, Philadelphia, PA 19131-1694

AWARDS MS.

Faculty: 15 full-time (7 women).
Students: 63 full-time (30 women); includes 22 minority (10 African Americans, 11 Asian Americans or Pacific Islanders, 1 Hispanic American). Average age 23. *118 applicants, 48% accepted.* In 1999, 10 degrees awarded.
Degree requirements: For master's, thesis required.
Entrance requirements: For master's, GRE or MCAT, minimum GPA of 3.0, previous course work in biology, chemistry, English, physics. *Application deadline:* For fall admission, 7/15. Applications are processed on a rolling basis. *Application fee:* $50.
Expenses: Tuition: Full-time $12,650. One-time fee: $250 full-time.
Faculty research: Developmental biology, cytokine and inflammation, neurobiology of aging, pain mechanisms, cell death. *Total annual research expenditures:* $50,000.
Dr. Richard Kriebel, Assistant Dean, 215-871-6527.
Application contact: Carol A. Fox, Associate Dean for Admissions and Enrollment Management, 215-871-6700, *Fax:* 215-871-6719, *E-mail:* admissions@pcom.edu.

■ PITTSBURG STATE UNIVERSITY

Graduate School, College of Arts and Sciences, Department of Biology, Pittsburg, KS 66762

AWARDS MS.

Students: 8 full-time (2 women), 10 part-time (6 women). In 1999, 10 degrees awarded.
Degree requirements: For master's, thesis or alternative required, foreign language not required.
Application fee: $0 ($40 for international students).
Expenses: Tuition, state resident: full-time $2,466; part-time $105 per credit hour. Tuition, nonresident: full-time $6,268; part-time $264 per credit hour.
Financial aid: Research assistantships, teaching assistantships, career-related internships or fieldwork and Federal Work-Study available.
Dr. James Triplett, Chairperson, 316-235-4730.
Application contact: Marvene Darraugh, Administrative Officer, 316-235-4220, *Fax:* 316-235-4219, *E-mail:* mdarraug@pittstate.edu.

■ PONCE SCHOOL OF MEDICINE

Program in Biomedical Sciences, Ponce, PR 00732-7004

AWARDS PhD.

Students: 15 full-time (9 women), 2 part-time (both women); includes 16 minority (all Hispanic Americans). Average age 27. *9 applicants, 44% accepted.* In 1999, 1 degree awarded.
Degree requirements: For doctorate, one foreign language, computer language, dissertation required. *Average time to degree:* Doctorate–8 years full-time.
Entrance requirements: For doctorate, GRE General Test. *Application deadline:* For fall admission, 5/1. Applications are processed on a rolling basis. *Application fee:* $100.
Expenses: Tuition: Full-time $4,500; part-time $225 per credit. Required fees: $1,001 per semester. One-time fee: $250.
Dr. Carmen Mercado, Assistant Dean of Admissions, 787-840-2575 Ext. 251.

■ PORTLAND STATE UNIVERSITY

Graduate Studies, College of Liberal Arts and Sciences, Department of Biology, Portland, OR 97207-0751

AWARDS MA, MS, PhD.

Faculty: 12 full-time (2 women), 7 part-time/adjunct (3 women).
Students: 22 full-time (14 women), 15 part-time (8 women); includes 3 minority

(2 Hispanic Americans, 1 Native American), 1 international. Average age 29. *26 applicants, 62% accepted.* In 1999, 10 degrees awarded.
Degree requirements: For master's, one foreign language, computer language, thesis required; for doctorate, dissertation required.
Entrance requirements: For master's, GRE General Test, GRE Subject Test, TOEFL, minimum GPA of 3.0 in upper-division course work or 2.75 overall; for doctorate, GRE General Test, GRE Subject Test, minimum GPA of 3.5 in science course work. *Application deadline:* For fall admission, 4/1 (priority date); for spring admission, 11/1. Applications are processed on a rolling basis. *Application fee:* $50.
Expenses: Tuition, state resident: full-time $5,514; part-time $204 per credit. Tuition, nonresident: full-time $9,987; part-time $370 per credit. Required fees: $260 per term. Full-time tuition and fees vary according to program. Part-time tuition and fees vary according to course load.
Financial aid: In 1999–00, 3 research assistantships with full tuition reimbursements (averaging $6,600 per year), 15 teaching assistantships with full tuition reimbursements (averaging $7,400 per year) were awarded; Federal Work-Study and institutionally sponsored loans also available. Aid available to part-time students. Financial aid application deadline: 3/1; financial aid applicants required to submit FAFSA.
Faculty research: Genetic diversity and natural population, vertebrate temperature regulation, water balance and sensory physiology, trace elements and aquatic ecology, molecular genetics.
Dr. Stanley Hillman, Co-Chair, 505-725-3851.
Application contact: Dr. Deborah Duffield, Coordinator, 503-725-3851, *Fax:* 503-725-3888, *E-mail:* duffield@pdx.edu.

■ PRAIRIE VIEW A&M UNIVERSITY

Graduate School, College of Arts and Sciences, Department of Biology, Prairie View, TX 77446-0188

AWARDS MS.

Faculty: 1 full-time (0 women).
Students: 4 full-time (3 women), 3 part-time (all women); includes 6 minority (all African Americans). Average age 27.
Degree requirements: For master's, thesis required, foreign language not required. *Average time to degree:* Master's–2.5 years full-time, 4 years part-time.
Entrance requirements: For master's, GRE General Test, BS in biology or equivalent. *Application deadline:* For fall

admission, 7/1 (priority date); for spring admission, 11/1. Applications are processed on a rolling basis. *Application fee:* $25.
Expenses: Tuition, state resident: full-time $756; part-time $40 per credit hour. Tuition, nonresident: full-time $4,572; part-time $254 per credit hour. Required fees: $1,108.
Financial aid: Career-related internships or fieldwork, Federal Work-Study, and institutionally sponsored loans available. Financial aid application deadline: 8/1. Dr. George E. Brown, Head, 409-857-3912, *Fax:* 409-857-3928.

■ PRINCETON UNIVERSITY

Graduate School, Department of Ecology and Evolutionary Biology, Princeton, NJ 08544-1019

AWARDS Biology (PhD); neuroscience (PhD).

Degree requirements: For doctorate, dissertation required, foreign language not required.
Entrance requirements: For doctorate, GRE General Test, GRE Subject Test.
Expenses: Tuition: Full-time $25,050.

Find an in-depth description at www.petersons.com/graduate.

■ PURDUE UNIVERSITY

Graduate School, School of Science, Department of Biological Sciences, West Lafayette, IN 47907

AWARDS Biochemistry (PhD); biophysics (PhD); cell and developmental biology (PhD); ecology, evolutionary and population biology (MS, PhD), including ecology, evolutionary biology, population biology; genetics (MS, PhD); microbiology (MS, PhD); molecular biology (PhD); neurobiology (MS, PhD); plant physiology (PhD).

Faculty: 44 full-time (7 women), 1 (woman) part-time/adjunct.
Students: 80 full-time (40 women), 20 part-time (9 women); includes 17 minority (7 African Americans, 2 Asian Americans or Pacific Islanders, 8 Hispanic Americans), 40 international. *248 applicants, 24% accepted.* In 1999, 6 master's, 19 doctorates awarded. Terminal master's awarded for partial completion of doctoral program.
Degree requirements: For master's, thesis required (for some programs), foreign language not required; for doctorate, dissertation, seminars, teaching experience required. *Average time to degree:* Master's–3.11 years full-time; doctorate–5.4 years full-time.
Entrance requirements: For master's and doctorate, GRE General Test, TOEFL, TSE. *Application deadline:* For fall admission, 2/15. *Application fee:* $30. Electronic applications accepted.

Expenses: Tuition, state resident: full-time $4,530; part-time $130 per credit hour. Tuition, nonresident: full-time $15,310; part-time $404 per credit hour. Tuition and fees vary according to campus/location and program.
Financial aid: In 1999–00, 15 fellowships, 60 research assistantships, 53 teaching assistantships were awarded. Aid available to part-time students. Financial aid application deadline: 2/15; financial aid applicants required to submit FAFSA. Dr. A. E. Konopka, Acting Head, 765-494-4407.
Application contact: Nancy Konopka, Graduate Studies Office Manager, 765-494-8142, *Fax:* 765-494-0876, *E-mail:* njk@bilbo.bio.purdue.edu.

Find an in-depth description at www.petersons.com/graduate.

■ PURDUE UNIVERSITY CALUMET

Graduate School, School of Engineering, Mathematics, and Science, Department of Biological Sciences, Hammond, IN 46323-2094

AWARDS Biology (MS); biology teaching (MS).

Degree requirements: For master's, foreign language and thesis not required.
Entrance requirements: For master's, GRE, TOEFL. Electronic applications accepted.
Faculty research: Cell biology, protein chemistry, molecular biology, DNA fingerprinting, gene cloning, cancer biology, neurobiology.

■ QUEENS COLLEGE OF THE CITY UNIVERSITY OF NEW YORK

Division of Graduate Studies, Mathematics and Natural Sciences Division, Department of Biology, Flushing, NY 11367-1597

AWARDS MA. Part-time and evening/weekend programs available.

Faculty: 17 full-time (5 women).
Students: 1 full-time (0 women), 15 part-time (10 women); includes 7 minority (2 African Americans, 4 Asian Americans or Pacific Islanders, 1 Hispanic American), 1 international. *36 applicants, 81% accepted.* In 1999, 4 degrees awarded.
Degree requirements: For master's, thesis or alternative, comprehensive exam, qualifying exam required, foreign language not required.
Entrance requirements: For master's, TOEFL, minimum GPA of 3.0. *Application deadline:* For fall admission, 4/1; for spring admission, 11/1. Applications are processed on a rolling basis. *Application fee:* $40.

Expenses: Tuition, state resident: full-time $4,350; part-time $185 per credit. Tuition, nonresident: full-time $7,600; part-time $320 per credit. Required fees: $114; $57 per semester. Tuition and fees vary according to course load and program.
Financial aid: Career-related internships or fieldwork, Federal Work-Study, institutionally sponsored loans, tuition waivers (partial), and unspecified assistantships available. Aid available to part-time students. Financial aid application deadline: 4/1; financial aid applicants required to submit FAFSA.
Faculty research: Cell biology, evolutionary biology, environmental biology, microbiology.
Dr. Harold Magazine, Chairperson, 718-997-3400.
Application contact: Dr. Jeanne Szalay, Graduate Adviser, 718-997-3400, *E-mail:* jeanne_szalay@qc.edu.

■ QUINNIPIAC UNIVERSITY

School of Health Sciences, Programs in Medical Laboratory Sciences, Hamden, CT 06518-1940

AWARDS Biomedical sciences (MHS); laboratory management (MHS); microbiology (MHS). Part-time and evening/weekend programs available.

Faculty: 6 full-time (1 woman), 2 part-time/adjunct (1 woman).
Students: 3 full-time (all women), 27 part-time (22 women); includes 2 minority (both Asian Americans or Pacific Islanders). Average age 28. *10 applicants, 70% accepted.* In 1999, 10 degrees awarded (100% found work related to degree).
Degree requirements: For master's, comprehensive exam required, thesis optional. *Average time to degree:* Master's–2 years full-time, 4 years part-time.
Entrance requirements: For master's, minimum GPA of 2.5. *Application deadline:* For fall admission, 8/1 (priority date); for spring admission, 12/15 (priority date). Applications are processed on a rolling basis. *Application fee:* $45. Electronic applications accepted.
Expenses: Tuition: Part-time $410 per credit hour. Required fees: $20 per term. Tuition and fees vary according to program.
Financial aid: Available to part-time students. Applicants required to submit FAFSA.
Faculty research: Microbial physiology, fermentation technology.
Dr. Kenneth Kaloustian, Director, 203-582-8676, *Fax:* 203-582-3443, *E-mail:* ken.kaloustian@quinnipiac.edu.

Quinnipiac University (continued)
Application contact: Scott Farber, Director of Graduate Admissions, 800-462-1944, *Fax:* 203-582-3443, *E-mail:* graduate@quinnipiac.edu.

Find an in-depth description at www.petersons.com/graduate.

■ RENSSELAER POLYTECHNIC INSTITUTE

Graduate School, School of Science, Department of Biology, Troy, NY 12180-3590

AWARDS Biochemistry (MS, PhD); biophysics (MS, PhD); cell biology (MS, PhD); developmental biology (MS, PhD); microbiology (MS, PhD); molecular biology (MS, PhD); plant science (MS, PhD). Part-time programs available.

Faculty: 14 full-time (5 women).
Students: 16 full-time (8 women), 2 part-time (1 woman); includes 1 minority (Asian American or Pacific Islander), 4 international. *37 applicants, 41% accepted.* In 1999, 6 master's, 3 doctorates awarded. Terminal master's awarded for partial completion of doctoral program.
Degree requirements: For master's and doctorate, thesis/dissertation required, foreign language not required.
Entrance requirements: For master's and doctorate, GRE General Test, TOEFL. *Application deadline:* For fall admission, 2/1 (priority date). Applications are processed on a rolling basis. *Application fee:* $35.
Expenses: Tuition: Part-time $665 per credit hour. Required fees: $980.
Financial aid: In 1999–00, 8 research assistantships with partial tuition reimbursements (averaging $15,000 per year), 11 teaching assistantships with full tuition reimbursements (averaging $15,000 per year) were awarded; fellowships, career-related internships or fieldwork and institutionally sponsored loans also available. Financial aid application deadline: 2/1.
Faculty research: Applied environmental biology, genetics, environmental science, fresh water ecology, microbial ecology.
Dr. John Salerno, Chair, 518-276-2699, *Fax:* 518-276-2344.
Application contact: Dr. Jackie L. Collier, Assistant Professor, 518-276-6446, *Fax:* 518-276-2344, *E-mail:* collij3@rpi.edu.

Find an in-depth description at www.petersons.com/graduate.

■ RHODE ISLAND COLLEGE

School of Graduate Studies, Faculty of Arts and Sciences, Department of Biology, Providence, RI 02908-1924
AWARDS MA, MAT.

Faculty: 13 full-time (1 woman), 1 (woman) part-time/adjunct.
Students: 1 full-time (0 women), 6 part-time (3 women). In 1999, 4 degrees awarded.
Degree requirements: For master's, thesis (MA) required.
Entrance requirements: For master's, GRE General Test and GRE Subject Test or MAT. *Application deadline:* For fall admission, 4/1. Applications are processed on a rolling basis. *Application fee:* $25.
Expenses: Tuition, state resident: part-time $162 per credit. Tuition, nonresident: part-time $328 per credit. Required fees: $18 per credit. One-time fee: $40. Tuition and fees vary according to program and reciprocity agreements.
Financial aid: Career-related internships or fieldwork available. Financial aid application deadline: 4/1.
Dr. Kenneth Kinsey, Chair, 401-456-8010, *E-mail:* kkinsey@ric.edu.

■ THE ROCKEFELLER UNIVERSITY

Program in Biomedical Sciences, New York, NY 10021-6399

AWARDS PhD, MD/PhD. MD/PhD offered through the Tri-Institutional Program with Cornell University Medical College and Sloan-Kettering Institute.

Faculty: 221 full-time (59 women), 177 part-time/adjunct (40 women).
Students: 137 full-time (50 women); includes 36 minority (5 African Americans, 19 Asian Americans or Pacific Islanders, 12 Hispanic Americans), 57 international. Average age 25. *610 applicants, 10% accepted.* In 1999, 24 degrees awarded.
Degree requirements: For doctorate, dissertation required. *Average time to degree:* Doctorate–5.5 years full-time.
Application deadline: For fall admission, 1/1. *Application fee:* $60.
Financial aid: In 1999–00, 137 fellowships with full tuition reimbursements (averaging $20,500 per year) were awarded; grants, institutionally sponsored loans, and traineeships also available.
Dr. Frederick Cross, Dean of Graduate Studies, 212-327-8086, *Fax:* 212-327-8505, *E-mail:* phd@rockvax.rockefeller.edu.
Application contact: Kristen Cullen, Admissions and Records Administrator, 212-327-8088, *Fax:* 212-327-8505, *E-mail:* cullenk@rockvax.rockefeller.edu.

Find an in-depth description at www.petersons.com/graduate.

■ RUTGERS, THE STATE UNIVERSITY OF NEW JERSEY, CAMDEN

Graduate School, Program in Biology, Camden, NJ 08102-1401

AWARDS MS, MST. Part-time and evening/weekend programs available.

Faculty: 13 full-time (2 women), 1 part-time/adjunct (0 women).
Students: 8 full-time (4 women), 47 part-time (23 women); includes 10 minority (3 African Americans, 6 Asian Americans or Pacific Islanders, 1 Native American). *41 applicants, 85% accepted.* In 1999, 10 degrees awarded.
Degree requirements: For master's, foreign language and thesis not required.
Entrance requirements: For master's, GRE General Test, GRE Subject Test (recommended). *Application deadline:* For fall admission, 7/1 (priority date); for spring admission, 12/1. Applications are processed on a rolling basis. *Application fee:* $50. Electronic applications accepted.
Expenses: Tuition, state resident: full-time $6,776; part-time $279 per credit. Tuition, nonresident: full-time $9,936; part-time $412 per credit. Required fees: $151 per semester. Part-time tuition and fees vary according to course load and program.
Financial aid: In 1999–00, 10 students received aid, including 1 fellowship (averaging $2,000 per year), 1 research assistantship with full tuition reimbursement available (averaging $16,500 per year), 7 teaching assistantships with full tuition reimbursements available (averaging $16,500 per year); unspecified assistantships also available. Financial aid application deadline: 3/15; financial aid applicants required to submit FAFSA.
Faculty research: Neurobiology, biochemistry, ecology, developmental biology, biological signalling mechanisms. *Total annual research expenditures:* $117,000.
Dr. Joseph V. Martin, Director, 856-225-6335, *Fax:* 856-225-6312, *E-mail:* gradbio@crab.rutgers.edu.

■ RUTGERS, THE STATE UNIVERSITY OF NEW JERSEY, NEWARK

Graduate School, Department of Biological Sciences, Newark, NJ 07102

AWARDS Biology (MS, PhD). Part-time and evening/weekend programs available.

Faculty: 18 full-time (4 women), 9 part-time/adjunct (1 woman).
Students: 34 full-time (11 women), 70 part-time (39 women); includes 37 minority (8 African Americans, 24 Asian Americans or Pacific Islanders, 5 Hispanic Americans). *168 applicants, 55% accepted.* In

1999, 16 master's, 3 doctorates awarded. Terminal master's awarded for partial completion of doctoral program.

Degree requirements: For master's, comprehensive exam required, thesis optional; for doctorate, dissertation, qualifying exam required, foreign language not required. *Average time to degree:* Master's–1.5 years full-time, 3 years part-time; doctorate–5 years full-time, 6 years part-time.

Entrance requirements: For master's, GRE General Test, minimum undergraduate B average; for doctorate, GRE General Test, GRE Subject Test, minimum B average. *Application deadline:* 2/15 (priority date); for spring admission, 12/1. Applications are processed on a rolling basis. *Application fee:* $50. Electronic applications accepted.

Expenses: Tuition, state resident: full-time $6,776; part-time $279 per credit hour. Tuition, nonresident: full-time $9,936; part-time $412 per credit hour. Required fees: $201 per semester. Tuition and fees vary according to course load and program.

Financial aid: In 1999–00, 36 students received aid, including 2 fellowships with full tuition reimbursements available (averaging $12,000 per year), 4 research assistantships, 22 teaching assistantships with full tuition reimbursements available (averaging $13,350 per year); Federal Work-Study and unspecified assistantships also available. Aid available to part-time students. Financial aid application deadline: 3/1.

Faculty research: Cell-cytoskeletal elements, development and regeneration in the nervous system, cellular trafficking, environmental stressors and their impact on development, opportunistic parasitic infections in AIDS.

Dr. Gene Miller Jonakait, Program Director, 973-353-1355, *Fax:* 973-353-5518, *E-mail:* jonakait@andromeda.rutgers.edu.

Application contact: Amy Trimarco, Department Administrator, 973-353-1235, *Fax:* 973-353-5518, *E-mail:* trimarco@andromeda.rutgers.edu.

■ ST. CLOUD STATE UNIVERSITY

School of Graduate Studies, College of Science and Engineering, Department of Biological Sciences, St. Cloud, MN 56301-4498

AWARDS MA, MS.

Faculty: 23 full-time (8 women), 2 part-time/adjunct (0 women).
Students: 13 full-time (5 women), 1 part-time. *9 applicants, 89% accepted.* In 1999, 9 degrees awarded.
Degree requirements: For master's, thesis or alternative required, foreign language not required.

Entrance requirements: For master's, GRE General Test, minimum GPA of 2.75. *Application fee:* $20.
Expenses: Tuition, state resident: part-time $149 per semester hour. Tuition, nonresident: part-time $225 per semester hour. Required fees: $17 per semester hour.
Financial aid: Federal Work-Study and unspecified assistantships available. Financial aid application deadline: 3/1.

Dr. David DeGroote, Chairperson, 320-255-2036, *Fax:* 320-255-4166.
Application contact: Ann Anderson, Graduate Studies Office, 320-255-2113, *Fax:* 320-654-5371, *E-mail:* aeanderson@stcloudstate.edu.

■ SAINT FRANCIS COLLEGE

Medical Science Program, Loretto, PA 15940-0600

AWARDS MMS. Part-time and evening/weekend programs available. Postbaccalaureate distance learning degree programs offered (no on-campus study).

Faculty: 1 full-time (0 women), 10 part-time/adjunct (1 woman).
Students: Average age 32. *24 applicants, 100% accepted.* In 1999, 10 degrees awarded (100% found work related to degree).
Degree requirements: For master's, thesis or alternative, clinical residency project required, foreign language not required. *Average time to degree:* Master's–2 years part-time.

Entrance requirements: For master's, National Commission on Certification of Physician Assistants certification. *Application deadline:* For fall admission, 9/1 (priority date); for spring admission, 1/1. Applications are processed on a rolling basis. *Application fee:* $50.
Expenses: Tuition: Part-time $483 per credit. Part-time tuition and fees vary according to course level and degree level.
Financial aid: In 1999–00, 1 teaching assistantship with full tuition reimbursement was awarded; career-related internships or fieldwork also available.
Faculty research: Health care policy, physician assistant practice roles, health promotion/disease prevention, public health epidemiology.

Dr. William Duryea, Director, 814-472-3132, *Fax:* 814-472-3137, *E-mail:* bduryea@sfcpa.edu.

■ ST. JOHN'S UNIVERSITY

College of Liberal Arts and Sciences, Department of Biological Sciences, Jamaica, NY 11439

AWARDS MS, PhD. Part-time and evening/weekend programs available.

Faculty: 16 full-time (3 women), 8 part-time/adjunct (4 women).
Students: 8 full-time (4 women), 44 part-time (22 women); includes 17 minority (5 African Americans, 7 Asian Americans or Pacific Islanders, 5 Hispanic Americans), 12 international. Average age 29. *53 applicants, 60% accepted.* In 1999, 8 master's, 6 doctorates awarded.
Degree requirements: For master's, comprehensive exam, internship required, thesis optional, foreign language not required; for doctorate, 2 foreign languages, computer language, dissertation, comprehensive exam, internship required.

Entrance requirements: For master's, GRE General Test, GRE Subject Test, minimum GPA of 3.0; for doctorate, GRE General Test, GRE Subject Test, minimum GPA of 3.0 (undergraduate), 3.5 (graduate). *Application deadline:* Applications are processed on a rolling basis. *Application fee:* $40.
Expenses: Tuition: Full-time $13,200; part-time $550 per credit. Required fees: $150; $75 per term. Tuition and fees vary according to degree level, program and student level.
Financial aid: In 1999–00, 15 fellowships were awarded; research assistantships, scholarships also available. Aid available to part-time students. Financial aid application deadline: 3/1; financial aid applicants required to submit FAFSA.
Faculty research: Regulation of gene transcription, molecular control of development in yeast, physiology of aging, cellular signal transduction.

Dr. Irvin Hirschfield, Chair, 718-990-1679, *E-mail:* hirshfii@stjohns.edu.
Application contact: Patricia G. Armstrong, Director, Office of Admission, 718-990-2000, *Fax:* 718-990-2096, *E-mail:* armstrop@stjohns.edu.

Find an in-depth description at www.petersons.com/graduate.

■ SAINT JOSEPH COLLEGE

Graduate Division, Field of Natural Sciences, Department of Biology, West Hartford, CT 06117-2700

AWARDS MS, Certificate. Part-time and evening/weekend programs available.

Faculty: 2 full-time (both women).
Students: 1 full-time (0 women), 7 part-time (all women). Average age 33. In 1999, 1 degree awarded.
Degree requirements: For master's, thesis or alternative required, foreign language not required.
Entrance requirements: For master's, GRE General Test or MAT. *Application deadline:* For fall admission, 7/15 (priority date); for spring admission, 12/1 (priority

Saint Joseph College (continued)
date). Applications are processed on a rolling basis. *Application fee:* $25.
Expenses: Tuition: Part-time $420 per credit hour. Required fees: $25 per course.
Financial aid: In 1999–00, 1 research assistantship with full tuition reimbursement was awarded. Financial aid application deadline: 7/15; financial aid applicants required to submit FAFSA.

■ SAINT JOSEPH'S UNIVERSITY

College of Arts and Sciences, Program in Biology, Philadelphia, PA 19131-1395

AWARDS MA, MS.

Expenses: Tuition: Part-time $470 per credit. One-time fee: $100 part-time. Part-time tuition and fees vary according to program.

■ SAINT LOUIS UNIVERSITY

Graduate School, College of Arts and Sciences, Department of Biology, St. Louis, MO 63103-2097

AWARDS MS, MS(R), PhD.

Faculty: 25 full-time (8 women), 8 part-time/adjunct (1 woman).
Students: 20 full-time (9 women), 19 part-time (10 women); includes 3 minority (1 African American, 2 Asian Americans or Pacific Islanders), 8 international. Average age 28. *31 applicants, 58% accepted.* In 1999, 5 master's, 4 doctorates awarded.
Degree requirements: For master's, comprehensive oral exam required; for doctorate, one foreign language (computer language can substitute), dissertation, preliminary exams required.
Entrance requirements: For master's and doctorate, GRE General Test. *Application deadline:* For fall admission, 6/1; for spring admission, 11/1. Applications are processed on a rolling basis. *Application fee:* $40.
Expenses: Tuition: Full-time $20,520; part-time $570 per credit hour. Required fees: $38 per term. Tuition and fees vary according to program.
Financial aid: In 1999–00, 34 students received aid, including 9 research assistantships, 12 teaching assistantships; fellowships. Financial aid application deadline: 4/1; financial aid applicants required to submit FAFSA.
Faculty research: Molecular systematics, pathogen-host interactions, developmental regulation.
Dr. Robert I. Bolla, Chair, 314-977-3910, *Fax:* 314-977-3658, *E-mail:* bollari@sluvca.slu.edu.
Application contact: Dr. Marcia Buresch, Assistant Dean of the Graduate School, 314-977-2240, *Fax:* 314-977-3943, *E-mail:* bureschm@slu.edu.

■ SAINT LOUIS UNIVERSITY

School of Medicine and Graduate School, Graduate Programs in Biomedical Sciences, St. Louis, MO 63103-2097

AWARDS MS(R), PhD, MD/JD, MD/PhD.

Faculty: 126 full-time (43 women), 43 part-time/adjunct (10 women).
Students: 28 full-time (19 women), 51 part-time (19 women); includes 9 minority (4 African Americans, 3 Asian Americans or Pacific Islanders, 2 Hispanic Americans), 13 international. Average age 28. *35 applicants, 100% accepted.* In 1999, 2 master's, 21 doctorates awarded. Terminal master's awarded for partial completion of doctoral program.
Degree requirements: For master's, thesis, comprehensive exam required; for doctorate, dissertation required.
Entrance requirements: For master's and doctorate, GRE General Test. *Application deadline:* Applications are processed on a rolling basis. Electronic applications accepted.
Expenses: Tuition: Part-time $507 per credit hour. Required fees: $38 per term.
Financial aid: In 1999–00, 73 students received aid, including 34 fellowships, 9 research assistantships, 3 teaching assistantships; career-related internships or fieldwork, Federal Work-Study, institutionally sponsored loans, and tuition waivers (full and partial) also available. Aid available to part-time students. Financial aid application deadline: 8/1; financial aid applicants required to submit FAFSA.
Dr. Willis K. Samson, Director, 314-577-8633.

■ SALEM-TEIKYO UNIVERSITY

Graduate School, Department of Bioscience, Salem, WV 26426-0500

AWARDS Biotechnology/molecular biology (MS).

Degree requirements: For master's, thesis required, foreign language not required.
Entrance requirements: For master's, GRE, minimum undergraduate GPA of 3.0. Electronic applications accepted.
Expenses: Tuition: Full-time $10,000; part-time $165 per credit hour. Required fees: $55; $55 per year.
Faculty research: Genetic engineering of seed storage proteins, virus replication and infection, gene therapy, programmed cell death, cell protocols for creation of gene transfer.

■ SAM HOUSTON STATE UNIVERSITY

College of Arts and Sciences, Department of Biological Sciences, Huntsville, TX 77341

AWARDS M Ed, MA, MS. Part-time programs available.

Students: 3 full-time (2 women), 11 part-time (7 women); includes 1 minority (Hispanic American). Average age 29. In 1999, 5 degrees awarded.
Degree requirements: For master's, thesis required (for some programs). *Average time to degree:* Master's–3 years full-time.
Entrance requirements: For master's, GRE General Test, TOEFL. *Application fee:* $20.
Expenses: Tuition, state resident: full-time $684; part-time $38 per credit hour. Tuition, nonresident: full-time $4,572; part-time $254 per credit hour. Required fees: $906; $906 per year.
Financial aid: Research assistantships, teaching assistantships available.
Faculty research: Genetics/cell biology, ecology, microbiology, mammalogy, plant systematics.
Dr. Andrew Dewees, Chair, 409-294-1538.

■ SAN DIEGO STATE UNIVERSITY

Graduate and Research Affairs, College of Sciences, Department of Biological Sciences, San Diego, CA 92182

AWARDS Biology (MA, MS), including ecology (MS), molecular biology (MS), physiology (MS), systematics/evolution (MS); cell and molecular biology (PhD); ecology (PhD); microbiology (MS).

Students: 37 full-time (23 women), 86 part-time (46 women); includes 13 minority (1 African American, 10 Asian Americans or Pacific Islanders, 2 Hispanic Americans), 4 international. Average age 26. *135 applicants, 33% accepted.* In 1999, 20 master's, 4 doctorates awarded. Terminal master's awarded for partial completion of doctoral program.
Degree requirements: For master's, thesis required; for doctorate, dissertation required.
Entrance requirements: For master's, GRE General Test, GRE Subject Test, TOEFL. *Application deadline:* For fall admission, 7/1 (priority date); for spring admission, 12/1. Applications are processed on a rolling basis. *Application fee:* $55.
Expenses: Tuition, nonresident: part-time $246 per unit. Required fees: $1,932; $633 per semester. Tuition and fees vary according to course load.

Financial aid: Fellowships, research assistantships, teaching assistantships, career-related internships or fieldwork and unspecified assistantships available. *Total annual research expenditures:* $11.5 million. Sanford Bernstein, Chair, 619-594-5629, *Fax:* 619-594-5676, *E-mail:* sanford.bernstein@sdsu.edu.
Application contact: Ken Johnson, Graduate Coordinator, 619-594-6919, *Fax:* 619-594-5676, *E-mail:* kjohnson@ sunstroke.sdsu.edu.

■ SAN FRANCISCO STATE UNIVERSITY

Graduate Division, College of Science and Engineering, Department of Biology, San Francisco, CA 94132-1722

AWARDS Cell and molecular biology (MA); conservation biology (MA); ecology and systematic biology (MA); marine biology (MA); microbiology (MA); physiology and behavioral biology (MA).

Entrance requirements: For master's, minimum GPA of 2.5 in last 60 units.
Expenses: Tuition, nonresident: full-time $5,904; part-time $246 per unit. Required fees: $1,904; $637 per semester. Tuition and fees vary according to course load.

■ SAN JOSE STATE UNIVERSITY

Graduate Studies, College of Science, Department of Biological Sciences, San Jose, CA 95192-0001

AWARDS MA, MS. Part-time programs available.

Degree requirements: For master's, foreign language not required.
Entrance requirements: For master's, GRE.
Expenses: Tuition, nonresident: part-time $246 per unit. Required fees: $1,939; $1,309 per year.
Faculty research: Systemic physiology, molecular genetics, SEM studies, toxicology, large mammal ecology.

■ SETON HALL UNIVERSITY

College of Arts and Sciences, Department of Biology, South Orange, NJ 07079-2697

AWARDS Biology (MS); microbiology (MS). Part-time and evening/weekend programs available.

Faculty: 11 full-time (5 women).
Students: 16 full-time (7 women), 61 part-time (36 women); includes 10 minority (1 African American, 6 Asian Americans or Pacific Islanders, 3 Hispanic Americans), 3 international. In 1999, 10 degrees awarded (90% found work related to degree, 10% continued full-time study).

Degree requirements: For master's, research paper or thesis, seminar required.
Entrance requirements: For master's, minimum GPA of 3.0. *Application deadline:* For fall admission, 7/15; for spring admission, 12/15. Applications are processed on a rolling basis. *Application fee:* $50.
Expenses: Tuition: Full-time $10,404; part-time $578 per credit. Required fees: $185 per year. Tuition and fees vary according to course load, campus/location, program and student's religious affiliation.
Financial aid: Teaching assistantships, career-related internships or fieldwork available.
Faculty research: Neurobiology, genetics, immunology, molecular biology, cellular physiology.
Dr. Suli Chang, Chairperson, 973-761-9044, *Fax:* 973-761-9596.
Application contact: Dr. Eliot Krause, Graduate Adviser, 973-761-9532, *E-mail:* krauseel@lanmail.shu.edu.

■ SHIPPENSBURG UNIVERSITY OF PENNSYLVANIA

School of Graduate Studies and Research, College of Arts and Sciences, Department of Biology, Shippensburg, PA 17257-2299

AWARDS M Ed, MS. Part-time and evening/weekend programs available.

Faculty: 7 full-time (1 woman).
Students: 8 full-time (2 women), 8 part-time (4 women), 1 international. Average age 31. *13 applicants, 46% accepted.* In 1999, 14 degrees awarded.
Degree requirements: For master's, thesis required (for some programs), foreign language not required.
Entrance requirements: For master's, TOEFL, GRE or minimum GPA of 2.75, 33 credit hours in biology, 16 credit hours in chemistry. *Application deadline:* Applications are processed on a rolling basis. *Application fee:* $30. Electronic applications accepted.
Expenses: Tuition, state resident: full-time $3,780; part-time $210 per credit hour. Tuition, nonresident: full-time $6,610; part-time $367 per credit hour. Required fees: $692. Part-time tuition and fees vary according to course load and degree level.
Financial aid: Research assistantships with full tuition reimbursements, career-related internships or fieldwork, Federal Work-Study, and unspecified assistantships available. Aid available to part-time students. Financial aid application deadline: 3/1; financial aid applicants required to submit FAFSA.
Dr. Michael McNichols, Chairperson, 717-477-1401, *Fax:* 717-477-4064, *E-mail:* mjmcni@ship.edu.

Application contact: Renee Payne, Assistant Dean of Graduate Studies, 717-477-1213, *Fax:* 717-477-4038, *E-mail:* rmpayn@ship.edu.

■ SMITH COLLEGE

Graduate Studies, Department of Biological Sciences, Northampton, MA 01063

AWARDS MA, MAT, PhD. Part-time programs available.

Faculty: 12 full-time (5 women), 3 part-time/adjunct (2 women).
Students: 2 full-time (both women), 2 part-time (both women). Average age 22. *4 applicants, 75% accepted.* In 1999, 2 master's awarded.
Degree requirements: For master's, thesis required (for some programs); for doctorate, dissertation required. *Average time to degree:* Master's–2 years full-time, 4 years part-time.
Entrance requirements: For master's, GRE General Test, GRE Subject Test, MAT; for doctorate, GRE General Test, GRE Subject Test. *Application deadline:* For fall admission, 4/15; for spring admission, 12/1. *Application fee:* $50.
Expenses: Tuition: Full-time $23,400.
Financial aid: In 1999–00, 4 teaching assistantships with full tuition reimbursements (averaging $9,270 per year) were awarded; institutionally sponsored loans and scholarships also available. Aid available to part-time students. Financial aid application deadline: 1/15; financial aid applicants required to submit CSS PROFILE or FAFSA.
Robert Merritt, Chair, 413-585-3819, *E-mail:* rmerritt@smith.edu.
Application contact: Steven Williams, Graduate Student Adviser, 413-585-3826, *E-mail:* swilliams@smith.edu.

Find an in-depth description at www.petersons.com/graduate.

■ SONOMA STATE UNIVERSITY

School of Natural Sciences, Department of Biology, Rohnert Park, CA 94928-3609

AWARDS Environmental biology (MA); general biology (MA). Part-time programs available.

Faculty: 13 full-time (3 women), 16 part-time/adjunct (8 women).
Students: 12 full-time (8 women), 16 part-time (13 women); includes 1 minority (Hispanic American). Average age 28. *18 applicants, 39% accepted.* In 1999, 4 degrees awarded.
Degree requirements: For master's, thesis or alternative, oral exam required, foreign language not required.
Entrance requirements: For master's, GRE General Test, GRE Subject Test,

Sonoma State University (continued)
minimum GPA of 3.0. *Application deadline:*
For fall admission, 11/30. Applications are
processed on a rolling basis. *Application fee:*
$55.

Expenses: Tuition, nonresident: part-time
$246 per unit. Required fees: $2,064; $715
per semester. Tuition and fees vary according to course load.

Financial aid: In 1999–00, 11 students
received aid, including 2 research assistantships, 9 teaching assistantships; career-
related internships or fieldwork and
Federal Work-Study also available.
Financial aid application deadline: 3/2.

Faculty research: Molecular biology,
genetics, riparian and wetland ecology,
plant ecology, feeding mechanisms of
invertebrates, ichthyology, microbiology.
Total annual research expenditures: $32,000.
Dr. Philip Northen, Chairperson, 707-
664-2189, *E-mail:* philip.northen@
sonoma.edu.

Application contact: John Hopkirk,
Graduate Adviser, 707-664-2180.

■ SOUTH DAKOTA STATE UNIVERSITY

**Graduate School, College of
Agriculture and Biological Sciences,
Department of Biology/Microbiology,
Brookings, SD 57007**

AWARDS Biology (MS); microbiology (MS).

Degree requirements: For master's,
thesis, oral exam required, foreign
language not required.

Entrance requirements: For master's,
GRE, TOEFL.

Faculty research: Plant tissue culture,
molecular biology studies of metabolic
regulation in plants, mechanisms of mammalian gene expression, aquatic-wetland
ecosystem ecology, stress-induced
immunosuppression on parasite-induced
pathology.

■ SOUTH DAKOTA STATE UNIVERSITY

**Graduate School, College of
Agriculture and Biological Sciences,
Program in Biological Sciences,
Brookings, SD 57007**

AWARDS PhD.

Degree requirements: For doctorate, dissertation, preliminary oral and written
exams required.

Entrance requirements: For doctorate,
TOEFL.

■ SOUTHEASTERN LOUISIANA UNIVERSITY

**College of Arts and Sciences,
Department of Biological Sciences,
Hammond, LA 70402**

AWARDS MS. Part-time programs available.

Faculty: 12.

Students: 7 full-time (6 women), 28 part-
time (16 women); includes 4 minority (all
African Americans). Average age 26. In
1999, 6 degrees awarded.

Degree requirements: For master's,
thesis, oral and written exams required.

Entrance requirements: For master's,
GRE General Test, minimum GPA of 3.0.
Application deadline: For fall admission,
7/15 (priority date); for spring admission,
12/15 (priority date). Applications are
processed on a rolling basis. *Application fee:*
$10 ($25 for international students).
Electronic applications accepted.

Expenses: Tuition, state resident: full-time
$2,100. Tuition, nonresident: full-time
$6,096. Tuition and fees vary according to
course load.

Financial aid: In 1999–00, 18 research
assistantships with full tuition reimbursements (averaging $2,200 per year), 6
teaching assistantships with full tuition
reimbursements (averaging $2,200 per
year) were awarded; fellowships, career-
related internships or fieldwork, Federal
Work-Study, and unspecified assistantships
also available. Aid available to part-time
students. Financial aid application
deadline: 5/1; financial aid applicants
required to submit FAFSA.

Faculty research: Wetlands restoration
and creation, ecology, environmental sciences, evolution, conservation.
Dr. Nick Norton, Head, 504-549-3740,
Fax: 504-549-3851.

Application contact: Stephen C. Soutullo,
Registrar and Director of Enrollment
Services, 504-549-2066, *Fax:* 504-549-
5632, *E-mail:* ssoutullo@selu.edu.

■ SOUTHEAST MISSOURI STATE UNIVERSITY

**Graduate School, Department of
Biology, Cape Girardeau, MO 63701-
4799**

AWARDS MNS. Part-time programs available.

Degree requirements: For master's,
thesis or alternative required, foreign
language not required.

■ SOUTHERN CONNECTICUT STATE UNIVERSITY

**School of Graduate Studies, School of
Arts and Sciences, Department of
Biology, New Haven, CT 06515-1355**

AWARDS Biology (MS); biology for nurse
anesthetists (MS). Part-time and evening/
weekend programs available.

Faculty: 8 full-time (1 woman).

Students: 17 full-time (11 women), 52
part-time (30 women); includes 7 minority
(4 African Americans, 3 Asian Americans
or Pacific Islanders), 1 international. *124
applicants, 31% accepted.* In 1999, 18
degrees awarded.

Degree requirements: For master's,
thesis optional, foreign language not
required.

Entrance requirements: For master's,
previous course work in biology, chemistry,
and mathematics; interview. *Application
deadline:* For fall admission, 7/15 (priority
date). Applications are processed on a rolling basis. *Application fee:* $40.

Expenses: Tuition, state resident: part-
time $198 per credit. Tuition, nonresident:
part-time $214 per credit. Required fees:
$5 per credit. $45 per semester. Part-time
tuition and fees vary according to
program.

Financial aid: Application deadline: 4/15.
Dr. Vernon Nelson, Chairperson, 203-392-
6211, *Fax:* 203-392-5364, *E-mail:* nelson@
southernct.edu.

Application contact: Dr. Dwight Smith,
Graduate Coordinator, 203-392-6220, *Fax:*
203-392-5364, *E-mail:* smith_d@
southernct.edu.

■ SOUTHERN ILLINOIS UNIVERSITY CARBONDALE

**Graduate School, College of Science,
Biological Sciences Program,
Carbondale, IL 62901-6806**

AWARDS MS.

Students: 3 full-time (1 woman). Average
age 25. *1 applicant, 100% accepted.* In 1999,
2 degrees awarded.

Degree requirements: For master's,
thesis or alternative required, foreign
language not required. *Average time to
degree:* Master's–2.5 years full-time.

Entrance requirements: For master's,
GRE General Test, TOEFL, minimum
GPA of 2.7. *Application deadline:* Applica-
tions are processed on a rolling basis.
Application fee: $0.

Expenses: Tuition, state resident: full-time
$2,902. Tuition, nonresident: full-time
$5,810. Tuition and fees vary according to
course load.

Financial aid: In 1999–00, 3 students
received aid; fellowships with full tuition
reimbursements available, research

assistantships with full tuition reimbursements available, teaching assistantships with full tuition reimbursements available, Federal Work-Study, institutionally sponsored loans, and tuition waivers (full) available. Aid available to part-time students.

Faculty research: Molecular mechanisms of mutagenesis, reproductive endocrinology, avian energetics and nutrition, developmental plant physiology.
Philip Robertson, Director, 618-536-2032.

■ SOUTHERN ILLINOIS UNIVERSITY CARBONDALE

School of Medicine and Graduate School, Graduate Program in Medicine, Carbondale, IL 62901-6806
AWARDS Pharmacology (MS, PhD); physiology (MS, PhD).

Faculty: 31 full-time (5 women).
Students: 24 full-time (10 women), 6 part-time (1 woman); includes 3 minority (1 African American, 1 Asian American or Pacific Islander, 1 Native American), 9 international. *16 applicants, 50% accepted.* In 1999, 3 master's, 6 doctorates awarded. Terminal master's awarded for partial completion of doctoral program.
Degree requirements: For master's, thesis required, foreign language not required; for doctorate, dissertation required.
Entrance requirements: For master's, TOEFL, minimum GPA of 3.0; for doctorate, TOEFL, minimum GPA of 3.25. *Application fee:* $0.
Expenses: Tuition, state resident: full-time $2,604. Tuition, nonresident: full-time $5,208. Required fees: $380 per semester.
Financial aid: In 1999–00, 27 students received aid, including 12 fellowships with full tuition reimbursements available, 1 research assistantship with full tuition reimbursement available, 10 teaching assistantships with full tuition reimbursements available; institutionally sponsored loans and tuition waivers (full) also available.
Faculty research: Cardiovascular physiology, neurophysiology of hearing.
Application contact: Graduate Program Committee, 618-536-5513.

■ SOUTHERN ILLINOIS UNIVERSITY EDWARDSVILLE

Graduate Studies and Research, College of Arts and Sciences, Department of Biological Sciences, Edwardsville, IL 62026-0001
AWARDS MA, MS.

Students: 12 full-time (7 women), 23 part-time (16 women); includes 3 minority (1 Asian American or Pacific Islander, 1 Hispanic American, 1 Native American), 3 international. *24 applicants, 45% accepted.* In 1999, 13 degrees awarded.
Degree requirements: For master's, variable foreign language requirement, thesis or alternative, final exam, thesis required.
Entrance requirements: For master's, GRE General Test, GRE Subject Test, TOEFL. *Application deadline:* For fall admission, 7/24. *Application fee:* $25.
Expenses: Tuition, state resident: full-time $1,814; part-time $100 per credit hour. Tuition, nonresident: full-time $3,631; part-time $201 per credit hour. Required fees: $477 per term. Tuition and fees vary according to course load and program.
Financial aid: In 1999–00, 1 fellowship with full tuition reimbursement, 7 research assistantships with full tuition reimbursements, 14 teaching assistantships with full tuition reimbursements were awarded; Federal Work-Study, institutionally sponsored loans, and unspecified assistantships also available. Aid available to part-time students. Financial aid application deadline: 3/1.
Richard Brugam, Chair, 618-650-2377, *E-mail:* rbrugam@siue.edu.

■ SOUTHERN METHODIST UNIVERSITY

Dedman College, Department of Biological Sciences, Dallas, TX 75275
AWARDS MA, MS, PhD. Terminal master's awarded for partial completion of doctoral program.

Degree requirements: For master's, thesis (MS), oral exam required; for doctorate, dissertation, qualifying exam required, dissertation, qualifying exam required.
Entrance requirements: For master's and doctorate, GRE General Test, minimum GPA of 3.0.
Expenses: Tuition: Part-time $686 per credit hour. Required fees: $88 per credit hour. Part-time tuition and fees vary according to course load and program.
Faculty research: Aging, gene expression, microbial metabolism, molecular parasitology, membrane protein structure and function.

Find an in-depth description at www.petersons.com/graduate.

■ SOUTHERN UNIVERSITY AND AGRICULTURAL AND MECHANICAL COLLEGE

Graduate School, College of Sciences, Department of Biology, Baton Rouge, LA 70813
AWARDS MS.

Faculty: 17 full-time (6 women).

Students: 7 full-time (3 women), 6 part-time (4 women); includes 12 minority (all African Americans). Average age 30. *2 applicants, 100% accepted.* In 1999, 5 degrees awarded (10% entered university research/teaching, 90% found other work related to degree).
Degree requirements: For master's, thesis required, foreign language not required. *Average time to degree:* Master's–3 years full-time, 4 years part-time.
Entrance requirements: For master's, GRE General Test, TOEFL. *Application deadline:* For fall admission, 6/1 (priority date); for spring admission, 11/1. Applications are processed on a rolling basis. *Application fee:* $5.
Expenses: Tuition, state resident: full-time $2,304. Tuition, nonresident: full-time $7,470. Tuition and fees vary according to course load, campus/location and program.
Financial aid: In 1999–00, 4 teaching assistantships (averaging $7,000 per year) were awarded; research assistantships Financial aid application deadline: 4/15; financial aid applicants required to submit FAFSA.
Faculty research: Toxicology, neuroendocrinology, mycotoxin, virology.
Dr. Dorothy P. Thompson, Chair, 225-771-5210, *Fax:* 225-771-5386.

■ SOUTHWEST MISSOURI STATE UNIVERSITY

Graduate College, College of Natural and Applied Sciences, Department of Biology, Springfield, MO 65804-0094
AWARDS Biology (MS); biology education (MS).

Faculty: 15 full-time (3 women).
Students: 17 full-time (10 women), 13 part-time (5 women); includes 1 minority (Native American). In 1999, 17 degrees awarded.
Degree requirements: For master's, thesis or alternative, oral and written comprehensive exams required, foreign language not required. *Average time to degree:* Master's–2.5 years full-time, 5 years part-time.
Entrance requirements: For master's, GRE General Test, 24 hours of biology, minimum undergraduate GPA of 3.0 in biology, 2.75 overall. *Application deadline:* For fall admission, 8/2 (priority date); for spring admission, 12/28 (priority date). Applications are processed on a rolling basis. *Application fee:* $25. Electronic applications accepted.
Expenses: Tuition, state resident: full-time $2,070; part-time $115 per credit. Tuition, nonresident: full-time $4,140; part-time $230 per credit. Required fees: $91 per credit. Tuition and fees vary according to course level, course load and program.

Southwest Missouri State University
(continued)
Financial aid: In 1999–00, research assistantships with full tuition reimbursements (averaging $6,150 per year), 17 teaching assistantships with full tuition reimbursements (averaging $6,150 per year) were awarded; Federal Work-Study, grants, scholarships, and unspecified assistantships also available. Financial aid application deadline: 3/31.
Faculty research: Field biology, organismal biology, microbiology.
Dr. Steven Jensen, Head, 417-836-5126, *Fax:* 417-836-6934, *E-mail:* biology@mail.smsu.edu.
Application contact: Dr. Thomas Tomasi, Graduate Adviser, 417-836-5126, *Fax:* 417-836-6934, *E-mail:* tet962f@mail.smsu.edu.

■ SOUTHWEST TEXAS STATE UNIVERSITY

Graduate School, College of Science, Department of Biology, San Marcos, TX 78666

AWARDS Aquatic biology (MS); biology (M Ed, MA, MS). Part-time programs available.

Faculty: 19 full-time (2 women), 1 part-time/adjunct (0 women).
Students: 42 full-time (20 women), 63 part-time (27 women); includes 15 minority (1 African American, 1 Asian American or Pacific Islander, 11 Hispanic Americans, 2 Native Americans). Average age 29. In 1999, 29 degrees awarded.
Degree requirements: For master's, thesis (for some programs), 3 seminars, comprehensive exam required, foreign language not required.
Entrance requirements: For master's, GRE General Test, TOEFL, previous course work in biology, minimum GPA of 2.75 in last 60 hours. *Application deadline:* For fall admission, 6/15 (priority date); for spring admission, 10/15 (priority date). Applications are processed on a rolling basis. *Application fee:* $25 ($75 for international students).
Expenses: Tuition, state resident: full-time $720; part-time $40 per semester hour. Tuition, nonresident: full-time $4,608; part-time $256 per semester hour. Required fees: $1,470; $122.
Financial aid: Teaching assistantships, career-related internships or fieldwork, Federal Work-Study, institutionally sponsored loans, and laboratory instructorships available. Aid available to part-time students. Financial aid application deadline: 4/1; financial aid applicants required to submit FAFSA.
Faculty research: Genetics, wildlife management, anatomy, microbiology.

Dr. Francis L. Rose, Chair, 512-245-2178, *Fax:* 512-245-8713, *E-mail:* fr02@swt.edu.
Application contact: Dr. J. Michael Willoughby, Dean of the Graduate School, 512-245-2581, *Fax:* 512-245-8365, *E-mail:* jw02swt.edu.

■ STANFORD UNIVERSITY

School of Humanities and Sciences, Department of Biological Sciences, Stanford, CA 94305-9991

AWARDS MAT, MS, PhD.

Faculty: 42 full-time (10 women).
Students: 124 full-time (72 women), 32 part-time (12 women); includes 33 minority (3 African Americans, 24 Asian Americans or Pacific Islanders, 6 Hispanic Americans), 17 international. Average age 26. *373 applicants, 16% accepted.* In 1999, 36 master's, 16 doctorates awarded.
Degree requirements: For doctorate, dissertation, oral exam required, foreign language not required.
Entrance requirements: For master's, GRE General Test, TOEFL; for doctorate, GRE General Test, GRE Subject Test, TOEFL. *Application deadline:* For fall admission, 12/15. *Application fee:* $65 ($80 for international students). Electronic applications accepted.
Expenses: Tuition: Full-time $24,441. Required fees: $171. Full-time tuition and fees vary according to program. Part-time tuition and fees vary according to course load.
Financial aid: Fellowships, research assistantships, teaching assistantships, Federal Work-Study and institutionally sponsored loans available.
H. Craig Heller, Chair, 650-725-4818, *Fax:* 650-725-5807.
Application contact: Student Services Office, 650-723-5413.

Find an in-depth description at www.petersons.com/graduate.

■ STANFORD UNIVERSITY

School of Medicine, Graduate Programs in Medicine, Stanford, CA 94305-9991

AWARDS MS, PhD.

Faculty: 84 full-time (20 women).
Students: 236 full-time (102 women), 120 part-time (48 women); includes 104 minority (11 African Americans, 75 Asian Americans or Pacific Islanders, 16 Hispanic Americans, 2 Native Americans), 37 international. Average age 27. *847 applicants, 15% accepted.* In 1999, 12 master's, 52 doctorates awarded. Terminal master's awarded for partial completion of doctoral program.
Degree requirements: For master's and doctorate, thesis/dissertation required.

Entrance requirements: For doctorate, TOEFL. *Application fee:* $65 ($80 for international students). Electronic applications accepted.
Expenses: Tuition: Full-time $23,058. Required fees: $152. Part-time tuition and fees vary according to course load.
Financial aid: Research assistantships, teaching assistantships available.
Application contact: Admissions Office, 650-723-2460.

■ STATE UNIVERSITY OF NEW YORK AT ALBANY

College of Arts and Sciences, Department of Biological Sciences, Albany, NY 12222-0001

AWARDS Biodiversity, conservation, and policy (MS); ecology, evolution, and behavior (MS, PhD); molecular, cellular, developmental, and neural biology (MS, PhD). Evening/weekend programs available.

Students: 38 full-time (17 women), 33 part-time (16 women); includes 5 minority (3 African Americans, 2 Asian Americans or Pacific Islanders), 16 international. Average age 30. *140 applicants, 29% accepted.* In 1999, 2 master's, 3 doctorates awarded.
Degree requirements: For master's, one foreign language required; for doctorate, dissertation required.
Entrance requirements: For master's and doctorate, GRE General Test. *Application deadline:* For fall admission, 8/1; for spring admission, 11/1. *Application fee:* $50.
Expenses: Tuition, state resident: full-time $5,100; part-time $214 per credit. Tuition, nonresident: full-time $8,416; part-time $352 per credit. Required fees: $31 per credit.
Financial aid: Fellowships, research assistantships, teaching assistantships, unspecified assistantships and minority assistantships available. Financial aid application deadline: 5/1.
Faculty research: Interferon, neural development, RNA self-splicing, behavioral ecology, DNA repair enzymes.
Dr. David Shub, Chair, 518-442-4300.

Find an in-depth description at www.petersons.com/graduate.

■ STATE UNIVERSITY OF NEW YORK AT ALBANY

School of Public Health, Department of Biomedical Sciences, Albany, NY 12222-0001

AWARDS Biochemistry, molecular biology, and genetics (MS, PhD); cell and molecular structure (MS, PhD); immunobiology and immunochemistry (MS, PhD); molecular pathogenesis (MS, PhD); neuroscience (MS, PhD).

Students: 10 full-time (4 women), 38 part-time (17 women); includes 4 minority (3 Asian Americans or Pacific Islanders, 1 Native American), 17 international. Average age 30. *62 applicants, 37% accepted.* In 1999, 2 master's, 3 doctorates awarded.
Degree requirements: For master's and doctorate, thesis/dissertation required.
Entrance requirements: For master's and doctorate, GRE General Test, GRE Subject Test. *Application deadline:* For fall admission, 1/15 (priority date); for spring admission, 11/1 (priority date). *Application fee:* $50.
Expenses: Tuition, state resident: full-time $5,100; part-time $214 per credit. Tuition, nonresident: full-time $8,416; part-time $352 per credit. Required fees: $31 per credit.
Financial aid: Fellowships, research assistantships available. Financial aid application deadline: 2/1.
Dr. Harry Taber, Chair, 518-474-2662.
Find an in-depth description at www.petersons.com/graduate.

■ STATE UNIVERSITY OF NEW YORK AT BINGHAMTON

Graduate School, School of Arts and Sciences, Department of Biological Sciences, Binghamton, NY 13902-6000
AWARDS MA, PhD.
Faculty: 23 full-time (6 women), 11 part-time/adjunct (7 women).
Students: 41 full-time (21 women), 14 part-time (4 women); includes 5 minority (1 African American, 2 Asian Americans or Pacific Islanders, 1 Hispanic American, 1 Native American), 4 international. Average age 29. *59 applicants, 54% accepted.* In 1999, 5 master's, 2 doctorates awarded. Terminal master's awarded for partial completion of doctoral program.
Degree requirements: For master's, thesis, oral exam, seminar presentation required, foreign language not required; for doctorate, dissertation, comprehensive exams required, foreign language not required.
Entrance requirements: For master's and doctorate, GRE General Test, GRE Subject Test, TOEFL. *Application deadline:* For fall admission, 4/15 (priority date); for spring admission, 11/1. Applications are processed on a rolling basis. *Application fee:* $50. Electronic applications accepted.
Expenses: Tuition, state resident: full-time $5,100; part-time $213 per credit. Tuition, nonresident: full-time $8,416; part-time $351 per credit. Required fees: $77 per credit. Part-time tuition and fees vary according to course load.
Financial aid: In 1999–00, 35 students received aid, including 1 fellowship with full tuition reimbursement available

(averaging $9,691 per year), 3 research assistantships with full tuition reimbursements available (averaging $15,437 per year), 28 teaching assistantships with full tuition reimbursements available (averaging $9,216 per year); career-related internships or fieldwork, Federal Work-Study, institutionally sponsored loans, and unspecified assistantships also available. Aid available to part-time students. Financial aid application deadline: 2/15.
Dr. Robert Van Buskirk, Chairperson, 607-777-6746.

■ STATE UNIVERSITY OF NEW YORK AT BUFFALO

Graduate School, College of Arts and Sciences, Department of Biological Sciences, Buffalo, NY 14260
AWARDS MS, PhD. Part-time programs available.
Faculty: 26 full-time (2 women), 4 part-time/adjunct (2 women).
Students: 33 full-time (18 women), 22 part-time (8 women); includes 2 minority (1 Asian American or Pacific Islander, 1 Hispanic American), 27 international. Average age 24. *291 applicants, 11% accepted.* In 1999, 6 master's, 8 doctorates awarded (100% entered university research/teaching). Terminal master's awarded for partial completion of doctoral program.
Degree requirements: For master's, research rotation, seminar, written exam required, foreign language and thesis not required; for doctorate, dissertation, candidacy exam, oral and written exams, research rotation, seminar required, foreign language not required.
Entrance requirements: For master's and doctorate, GRE General Test, TOEFL, 2 semesters of calculus. *Application deadline:* For fall admission, 2/1 (priority date); for spring admission, 11/1. Applications are processed on a rolling basis. *Application fee:* $35.
Expenses: Tuition, state resident: full-time $5,100; part-time $213 per credit hour. Tuition, nonresident: full-time $8,416; part-time $351 per credit hour. Required fees: $935; $75 per semester. Tuition and fees vary according to course load and program.
Financial aid: In 1999–00, 4 fellowships with tuition reimbursements (averaging $4,000 per year), 32 research assistantships with tuition reimbursements (averaging $16,000 per year), 25 teaching assistantships with tuition reimbursements (averaging $15,000 per year) were awarded; Federal Work-Study, institutionally sponsored loans, and tuition waivers (full) also available. Financial aid application

deadline: 2/15; financial aid applicants required to submit FAFSA.
Faculty research: Nucleic acid biochemistry, hormones and receptors, osmotic regulation, marine and terrestrial systems, ion channels, transcriptional control. *Total annual research expenditures:* $2.1 million.
Dr. Mary Bisson, Chairman, 716-645-2363, *Fax:* 716-645-2975, *E-mail:* bisson@ acsu.buffalo.edu.
Application contact: Dr. Gerald Koudelka, Director of Graduate Studies, 716-645-3489, *Fax:* 716-645-2975, *E-mail:* koudelka@acsu.buffalo.edu.

Find an in-depth description at www.petersons.com/graduate.

■ STATE UNIVERSITY OF NEW YORK AT BUFFALO

Graduate School, Graduate Programs in Biomedical Sciences at Roswell Park Cancer Institute, Buffalo, NY 14260
AWARDS MS, PhD. Part-time programs available.
Faculty: 79 full-time (13 women), 6 part-time/adjunct (2 women).
Students: 116 full-time (48 women), 68 part-time (28 women); includes 27 minority (5 African Americans, 20 Asian Americans or Pacific Islanders, 2 Hispanic Americans), 45 international. Average age 24. *201 applicants, 49% accepted.* In 1999, 24 master's, 8 doctorates awarded. Terminal master's awarded for partial completion of doctoral program.
Degree requirements: For master's and doctorate, thesis/dissertation required.
Average time to degree: Master's–1.5 years full-time, 3 years part-time; doctorate–5.5 years full-time.
Entrance requirements: For master's, GRE General Test, TOEFL, TSE, TWE; for doctorate, GRE General Test, GRE Subject Test, TOEFL, TSE, TWE.
Application deadline: Applications are processed on a rolling basis. *Application fee:* $35. Electronic applications accepted.
Expenses: Tuition, state resident: full-time $5,100; part-time $213 per credit hour. Tuition, nonresident: full-time $8,416; part-time $351 per credit hour. Required fees: $935; $75 per semester. Tuition and fees vary according to course load and program.
Financial aid: In 1999–00, 23 fellowships with full tuition reimbursements (averaging $15,000 per year), 100 research assistantships with full tuition reimbursements (averaging $15,000 per year) were awarded; teaching assistantships, Federal Work-Study, institutionally sponsored loans, and unspecified assistantships also

State University of New York at Buffalo (continued)

available. Financial aid applicants required to submit FAFSA.

Faculty research: Basic and biomedical cancer research, cell and molecular biology, biophysics, biochemistry, pharmacology, chemistry, immunology and physiology. *Total annual research expenditures:* $26.4 million.
Dr. Arthur M. Michalek, Dean, 716-845-2339, *Fax:* 716-845-8178, *E-mail:* rpgradapp@sc3103.med.buffalo.edu.
Application contact: Craig R. Johnson, Director of Graduate Studies, 716-845-2339, *Fax:* 716-845-8178, *E-mail:* rpgradapp@sc3103.med.buffalo.edu.

Find an in-depth description at www.petersons.com/graduate.

■ STATE UNIVERSITY OF NEW YORK AT BUFFALO

Graduate School, School of Medicine and Biomedical Sciences, Graduate Programs in Medicine and Biomedical Sciences, Buffalo, NY 14260
AWARDS MA, MS, PhD, MD/PhD. Degrees awarded through participating departments.

Faculty: 137 full-time (38 women), 12 part-time/adjunct (3 women).
Students: 85 full-time (40 women), 77 part-time (48 women); includes 13 minority (3 African Americans, 5 Asian Americans or Pacific Islanders, 5 Hispanic Americans), 56 international. *500 applicants, 8% accepted.* In 1999, 24 master's, 20 doctorates awarded. Terminal master's awarded for partial completion of doctoral program.
Degree requirements: For master's, thesis required (for some programs); for doctorate, dissertation required.
Entrance requirements: For doctorate, GRE General Test, TOEFL. *Application deadline:* Applications are processed on a rolling basis. *Application fee:* $35. Electronic applications accepted.
Expenses: Tuition, state resident: full-time $5,100. Tuition, nonresident: full-time $8,416. Required fees: $935.
Financial aid: In 1999–00, 64 students received aid, including 3 fellowships with full tuition reimbursements available (averaging $16,000 per year), 30 research assistantships with full tuition reimbursements available (averaging $15,000 per year), 31 teaching assistantships with full tuition reimbursements available; career-related internships or fieldwork, Federal Work-Study, grants, institutionally sponsored loans, scholarships, traineeships, tuition waivers (full and partial), and unspecified assistantships also available. Financial aid applicants required to submit FAFSA.

Faculty research: Neuroscience, molecular and cell biology, microbial pathogenesis, cardiopulmonary physiology, social and preventive medicine. *Total annual research expenditures:* $41.7 million.
Dr. Bruce A. Holm, Associate Dean for Research and Graduate Studies, 716-829-3393, *Fax:* 716-829-2437, *E-mail:* bholm@ubmedc.buffalo.edu.
Application contact: Elizabeth Hayden, Staff Associate, 716-829-3898, *Fax:* 716-829-2487, *E-mail:* ehayden@buffalo.edu.

Find an in-depth description at www.petersons.com/graduate.

■ STATE UNIVERSITY OF NEW YORK AT BUFFALO

Interdisciplinary Graduate Program in Biomedical Sciences, Buffalo, NY 14214
AWARDS PhD. PhD awarded through participating departments.

Students: 27 full-time (14 women), 1 (woman) part-time; includes 1 minority (Hispanic American), 13 international. Average age 23. *443 applicants, 11% accepted.*
Degree requirements: For doctorate, dissertation required.
Entrance requirements: For doctorate, GRE General Test, TOEFL. *Application deadline:* For fall admission, 2/1 (priority date). Applications are processed on a rolling basis. *Application fee:* $35. Electronic applications accepted.
Expenses: Tuition, state resident: full-time $5,100; part-time $213 per credit hour. Tuition, nonresident: full-time $8,416; part-time $351 per credit hour. Required fees: $935; $75 per semester. Tuition and fees vary according to course load and program.
Financial aid: In 1999–00, 21 students received aid, including 3 fellowships with full tuition reimbursements available (averaging $16,000 per year), 21 teaching assistantships with full tuition reimbursements available (averaging $14,000 per year); Federal Work-Study, scholarships, and unspecified assistantships also available. Financial aid application deadline: 2/1; financial aid applicants required to submit FAFSA.
Faculty research: Molecular and cell biology, pharmacology and toxicology, neurosciences, microbiology, pathogenesis and disease. *Total annual research expenditures:* $71.8 million.
Dr. William Ruyechan, Director, 716-829-3398, *Fax:* 716-829-2437, *E-mail:* smbs-gradprog@buffalo.edu.

Application contact: Elizabeth Hayden, Staff Associate, 716-829-3898, *Fax:* 716-829-2487, *E-mail:* chayden@buffalo.edu.

Find an in-depth description at www.petersons.com/graduate.

■ STATE UNIVERSITY OF NEW YORK AT NEW PALTZ

Graduate School, Faculty of Liberal Arts and Sciences, Department of Biology, New Paltz, NY 12561
AWARDS MA, MAT, MS Ed.

Students: 2 full-time (1 woman), 7 part-time (3 women); includes 1 minority (Native American). In 1999, 6 degrees awarded.
Degree requirements: For master's, thesis (for some programs), comprehensive exam required.
Entrance requirements: For master's, GRE General Test, GRE Subject Test, minimum GPA of 3.0. *Application deadline:* For fall admission, 3/15 (priority date). Applications are processed on a rolling basis. *Application fee:* $50.
Expenses: Tuition, state resident: full-time $5,100; part-time $213 per credit. Tuition, nonresident: full-time $8,416; part-time $351 per credit. Required fees: $1,025; $513 per semester.
Financial aid: Research assistantships, teaching assistantships, Federal Work-Study and institutionally sponsored loans available.
Dr. Hon Hing Ho, Chairman, 914-257-3770.

■ STATE UNIVERSITY OF NEW YORK AT STONY BROOK

Graduate School, College of Arts and Sciences, Program in Biological and Biomedical Sciences, Stony Brook, NY 11794
AWARDS PhD.

Expenses: Tuition, state resident: full-time $5,100; part-time $213 per credit hour. Tuition, nonresident: full-time $8,416; part-time $351 per credit hour. Required fees: $492. Tuition and fees vary according to program.

Find an in-depth description at www.petersons.com/graduate.

■ STATE UNIVERSITY OF NEW YORK AT STONY BROOK

Health Sciences Center, School of Medicine and Graduate School, Graduate Programs in Medicine, Stony Brook, NY 11794
AWARDS PhD, MD/PhD.

Students: 61 full-time (32 women), 52 part-time (19 women); includes 25 minority (5 African Americans, 14 Asian Americans or Pacific Islanders, 6 Hispanic Americans), 31 international. *203 applicants, 20% accepted.* In 1999, 17 doctorates awarded.
Degree requirements: For doctorate, dissertation, exam required.
Entrance requirements: For doctorate, GRE General Test, TOEFL. *Application deadline:* For fall admission, 1/15. *Application fee:* $50. Electronic applications accepted.
Expenses: Tuition, state resident: full-time $5,100. Tuition, nonresident: full-time $8,416. Required fees: $492.
Financial aid: In 1999–00, 41 fellowships, 84 research assistantships, 6 teaching assistantships were awarded; career-related internships or fieldwork and Federal Work-Study also available. Financial aid application deadline: 3/15. *Total annual research expenditures:* $18.1 million.
Application contact: Dr. William Jungers, Chairman, Committee on Admissions, 631-444-2113, *Fax:* 631-444-6032, *E-mail:* admissions@dean.som.sunysb.edu.

■ **STATE UNIVERSITY OF NEW YORK AT STONY BROOK**

Health Sciences Center, School of Medicine, Medical Scientist Training Program, Stony Brook, NY 11794
AWARDS MD/PhD.

Application deadline: For fall admission, 1/15.
Expenses: Tuition, state resident: full-time $10,840. Tuition, nonresident: full-time $21,940. Tuition and fees vary according to program.
Financial aid: Tuition waivers (full) available.
Application contact: Dr. William Jungers, Chairman, Committee on Admissions, 631-444-2113, *Fax:* 631-444-6032, *E-mail:* admissions@dean.som.sunysb.edu.
Find an in-depth description at www.petersons.com/graduate.

■ **STATE UNIVERSITY OF NEW YORK COLLEGE AT BROCKPORT**

School of Letters and Sciences, Department of Biological Sciences, Brockport, NY 14420-2997
AWARDS MS. Part-time programs available.

Faculty: 12 full-time (3 women), 1 (woman) part-time/adjunct.
Students: 4 full-time (2 women), 7 part-time (3 women). Average age 31. *5 applicants, 100% accepted.* In 1999, 4 degrees awarded.

Degree requirements: For master's, thesis or alternative, comprehensive exam required, foreign language not required.
Application deadline: Applications are processed on a rolling basis. *Application fee:* $50.
Expenses: Tuition, state resident: full-time $5,100; part-time $213 per credit. Tuition, nonresident: full-time $8,416; part-time $351 per credit. Required fees: $464; $25 per credit.
Financial aid: In 1999–00, 8 teaching assistantships were awarded; career-related internships or fieldwork and Federal Work-Study also available. Aid available to part-time students. Financial aid application deadline: 4/1; financial aid applicants required to submit FAFSA.
Faculty research: Avian ecology, limnology, ichthyology, neurobiology, molecular and cell biology, terrestrial ecology.
Dr. Larry Kline, Chairperson, 716-395-2193, *Fax:* 716-395-2741, *E-mail:* lkline@acspr1.acs.brockport.edu.
Application contact: Dr. Christopher Norment, Graduate Program Director, 716-395-5748, *Fax:* 716-395-2741, *E-mail:* cnorment@brockport.edu.

■ **STATE UNIVERSITY OF NEW YORK COLLEGE AT BUFFALO**

Graduate Studies and Research, Faculty of Natural and Social Sciences, Department of Biology, Buffalo, NY 14222-1095
AWARDS Biology (MA); secondary education (MS Ed), including biology. Evening/weekend programs available.

Degree requirements: For master's, thesis (MA) required.
Entrance requirements: For master's, minimum GPA of 2.5 in last 60 hours.
Expenses: Tuition, state resident: full-time $5,100; part-time $213 per credit hour. Tuition, nonresident: full-time $8,416; part-time $351 per credit hour. Required fees: $195; $8.6 per credit hour. Tuition and fees vary according to course load.

■ **STATE UNIVERSITY OF NEW YORK COLLEGE AT CORTLAND**

Graduate Studies, Division of Arts and Sciences, Department of Biology, Cortland, NY 13045
AWARDS MAT, MS Ed.
Students: 16 full-time (5 women), 16 part-time (8 women). In 1999, 8 degrees awarded.
Degree requirements: For master's, comprehensive exam required.
Entrance requirements: *Application deadline:* Applications are processed on a rolling basis. *Application fee:* $50.

Expenses: Tuition, state resident: part-time $213 per credit. Tuition, nonresident: full-time $5,100; part-time $351 per credit. International tuition: $8,416 full-time. Required fees: $352.
Financial aid: Applicants required to submit CSS PROFILE or FAFSA.
Dr. Elliott B. Mason, Chair, 607-753-2715, *E-mail:* masone@cortland.edu.
Application contact: Mark Yacavone, Assistant Director of Admissions, 607-753-4711, *Fax:* 607-753-5998, *E-mail:* marky@em.cortland.edu.

■ **STATE UNIVERSITY OF NEW YORK COLLEGE AT FREDONIA**

Graduate Studies, Department of Biology, Fredonia, NY 14063
AWARDS MS, MS Ed. Part-time and evening/weekend programs available.

Faculty: 5 full-time (0 women).
Students: 4 full-time (2 women), 9 part-time (7 women); includes 1 minority (Asian American or Pacific Islander). *7 applicants, 86% accepted.* In 1999, 6 degrees awarded.
Degree requirements: For master's, thesis required, foreign language not required.
Application deadline: For fall admission, 7/5. *Application fee:* $50.
Expenses: Tuition, state resident: full-time $5,100; part-time $213 per credit hour. Tuition, nonresident: full-time $8,416; part-time $351 per credit hour. Required fees: $775; $32 per credit hour.
Financial aid: In 1999–00, 6 teaching assistantships with partial tuition reimbursements (averaging $5,500 per year) were awarded; research assistantships, tuition waivers (full and partial) also available. Aid available to part-time students. Financial aid application deadline: 3/15.
Faculty research: Limnology.
Dr. Robert Byrne, Chairman, 716-673-3282.

■ **STATE UNIVERSITY OF NEW YORK COLLEGE AT ONEONTA**

Graduate Studies, Department of Biology, Oneonta, NY 13820-4015
AWARDS MA. Part-time and evening/weekend programs available.

Students: In 1999, 2 degrees awarded.
Entrance requirements: For master's, GRE General Test, GRE Subject Test.
Application deadline: For fall admission, 4/15. *Application fee:* $50.
Expenses: Tuition, state resident: full-time $5,100; part-time $213 per semester hour.

State University of New York College at Oneonta (continued)
Tuition, nonresident: full-time $8,416; part-time $351 per semester hour. Required fees: $582; $154 per semester. Part-time tuition and fees vary according to course load.
Dr. William Pietraface, Chair, 607-436-3703.

■ STATE UNIVERSITY OF NEW YORK HEALTH SCIENCE CENTER AT BROOKLYN

MD/PhD Program, Brooklyn, NY 11203-2098

AWARDS MD/PhD.

Expenses: Tuition, state resident: full-time $5,100; part-time $213 per credit. Tuition, nonresident: full-time $8,416; part-time $351 per credit. Required fees: $200. Full-time tuition and fees vary according to program and student level.

■ STATE UNIVERSITY OF NEW YORK HEALTH SCIENCE CENTER AT BROOKLYN

School of Graduate Studies, Brooklyn, NY 11203-2098

AWARDS PhD, MD/PhD.

Degree requirements: For doctorate, one foreign language, dissertation required.
Entrance requirements: For doctorate, GRE.
Expenses: Tuition, state resident: full-time $5,100; part-time $213 per credit. Tuition, nonresident: full-time $8,416; part-time $351 per credit. Required fees: $200. Full-time tuition and fees vary according to program and student level.
Faculty research: Cellular and molecular neurobiology, role of oncogenes in early cardiogenesis, mechanism of gene regulation, cardiovascular physiology, yeast molecular genetics.

Find an in-depth description at www.petersons.com/graduate.

■ STATE UNIVERSITY OF NEW YORK UPSTATE MEDICAL UNIVERSITY

College of Graduate Studies, Syracuse, NY 13210-2334

AWARDS MS, PhD, MD/PhD.

Students: 90 full-time (38 women), 6 part-time (2 women); includes 19 minority (1 African American, 16 Asian Americans or Pacific Islanders, 2 Hispanic Americans), 19 international. *114 applicants, 18% accepted.* In 1999, 2 master's awarded (100% found work related to degree); 8 doctorates awarded (12% entered

university research/teaching, 25% found other work related to degree, 63% continued full-time study). Terminal master's awarded for partial completion of doctoral program.
Degree requirements: For master's, thesis required, foreign language not required; for doctorate, dissertation, comprehensive exam required, foreign language not required. *Average time to degree:* Master's–3 years full-time; doctorate–5 years full-time.
Entrance requirements: For master's and doctorate, GRE General Test, GRE Subject Test, TSE. *Application deadline:* For fall admission, 4/1 (priority date). Applications are processed on a rolling basis. *Application fee:* $40.
Expenses: Tuition, state resident: full-time $5,100; part-time $213 per credit. Tuition, nonresident: full-time $8,416; part-time $351 per credit. Required fees: $410; $25 per credit. Part-time tuition and fees vary according to course load and program.
Financial aid: Fellowships, research assistantships, teaching assistantships, Federal Work-Study and institutionally sponsored loans available. Aid available to part-time students.
Dr. Maxwell M. Mozell, Dean, 315-464-4538, *Fax:* 315-464-4544.
Application contact: Office of Graduate Studies, 315-464-4538, *Fax:* 315-464-4544, *E-mail:* gradstud@vax.cs.hscsyr.edu.

Find an in-depth description at www.petersons.com/graduate.

■ STATE UNIVERSITY OF WEST GEORGIA

Graduate School, College of Arts and Sciences, Department of Biology, Carrollton, GA 30118

AWARDS MS. Part-time programs available.

Faculty: 7 full-time (1 woman).
Students: 11 full-time (7 women), 9 part-time (5 women); includes 4 minority (3 African Americans, 1 Hispanic American). Average age 28. In 1999, 7 degrees awarded.
Degree requirements: For master's, one foreign language (computer language can substitute), thesis, comprehensive exam required.
Entrance requirements: For master's, GRE General Test, GRE Subject Test, minimum GPA of 2.5, undergraduate degree in biology. *Application deadline:* For fall admission, 8/1. *Application fee:* $20.
Expenses: Tuition, state resident: full-time $2,252; part-time $94 per credit hour. Tuition, nonresident: full-time $6,756; part-time $282 per credit hour. Part-time tuition and fees vary according to course level.

Financial aid: Research assistantships, career-related internships or fieldwork and unspecified assistantships available. Aid available to part-time students. Financial aid applicants required to submit FAFSA.
Faculty research: Reaction of: SO&I2 with NAC, pyrethroid insecticides. *Total annual research expenditures:* $489,396.
Gregory J. Stewart, Chairman, 770-836-6547.
Application contact: Dr. Jack O. Jenkins, Dean, Graduate School, 770-836-6419, *Fax:* 770-836-2301, *E-mail:* jjenkins@westga.edu.

■ STEPHEN F. AUSTIN STATE UNIVERSITY

Graduate School, College of Sciences and Mathematics, Department of Biology, Nacogdoches, TX 75962

AWARDS MS.

Faculty: 16 full-time (0 women).
Students: 23 full-time (10 women), 9 part-time (3 women); includes 4 minority (1 African American, 1 Asian American or Pacific Islander, 2 Hispanic Americans). *18 applicants, 100% accepted.* In 1999, 22 degrees awarded.
Degree requirements: For master's, comprehensive exam required, thesis optional, foreign language not required.
Entrance requirements: For master's, GRE General Test, TOEFL, minimum GPA of 2.8 in last 60 hours, 2.5 overall. *Application deadline:* For fall admission, 8/1 (priority date); for spring admission, 12/15. Applications are processed on a rolling basis. *Application fee:* $0 ($50 for international students).
Expenses: Tuition, area resident: Part-time $38 per hour. Tuition, state resident: full-time $912; part-time $38 per hour. Tuition, nonresident: full-time $6,096; part-time $254 per hour. International tuition: $6,096 full-time. Required fees: $1,154; $48 per hour. Tuition and fees vary according to course level.
Financial aid: In 1999–00, teaching assistantships (averaging $6,375 per year). Financial aid application deadline: 3/1.
Dr. Don A. Hay, Chair, 936-468-3601.

■ SUL ROSS STATE UNIVERSITY

School of Arts and Sciences, Department of Biology, Alpine, TX 79832

AWARDS MS. Part-time programs available.

Degree requirements: For master's, thesis optional, foreign language not required.
Entrance requirements: For master's, GRE General Test, minimum GPA of 2.5 in last 60 hours of undergraduate work.

Faculty research: Plant-animal interaction, Chihuahuan desert biology, insect biological control, plant and animal systematics, wildlife biology.

■ SYRACUSE UNIVERSITY

Graduate School, College of Arts and Sciences, Department of Biology, Syracuse, NY 13244-0003

AWARDS Biology (MS, PhD); biophysics (PhD).

Faculty: 26.
Students: 31 full-time (10 women), 5 part-time (4 women); includes 1 minority (Asian American or Pacific Islander), 13 international. Average age 29. *131 applicants, 14% accepted.* In 1999, 2 master's, 2 doctorates awarded. Terminal master's awarded for partial completion of doctoral program.
Degree requirements: For master's and doctorate, thesis/dissertation required, foreign language not required.
Entrance requirements: For master's and doctorate, GRE General Test, GRE Subject Test. *Application deadline:* Applications are processed on a rolling basis. *Application fee:* $40.
Expenses: Tuition: Full-time $13,992; part-time $583 per credit hour.
Financial aid: Fellowships, research assistantships, teaching assistantships, Federal Work-Study and tuition waivers (partial) available. Financial aid application deadline: 3/1.
Dr. Richard Levy, Chairperson, 315-443-3984.

Find an in-depth description at www.petersons.com/graduate.

■ TARLETON STATE UNIVERSITY

College of Graduate Studies, College of Arts and Sciences, Department of Biological Sciences, Tarleton Station, TX 76402

AWARDS MS. Part-time and evening/weekend programs available.

Students: 9 full-time (all women), 5 part-time; includes 1 minority (Hispanic American). *4 applicants, 75% accepted.* In 1999, 4 degrees awarded.
Degree requirements: For master's, thesis (for some programs), comprehensive exam required.
Entrance requirements: For master's, GRE General Test, minimum GPA of 2.9 during last 60 hours. *Application deadline:* For fall admission, 8/5 (priority date); for spring admission, 12/1. Applications are processed on a rolling basis. *Application fee:* $25 ($100 for international students).
Expenses: Tuition, state resident: part-time $72 per hour. Tuition, nonresident: part-time $278 per hour. Required fees: $269 per course.

Financial aid: In 1999–00, 1 research assistantship (averaging $12,000 per year), 8 teaching assistantships (averaging $12,000 per year) were awarded; career-related internships or fieldwork and Federal Work-Study also available. Aid available to part-time students. Financial aid application deadline: 5/1; financial aid applicants required to submit FAFSA.
Dr. John Calahan, Head, 254-968-9159.

■ TEMPLE UNIVERSITY

Graduate School, College of Science and Technology, Department of Biology, Philadelphia, PA 19122-6096

AWARDS MA, PhD.

Faculty: 19 full-time (4 women).
Students: 29 full-time (16 women); includes 9 minority (1 African American, 6 Asian Americans or Pacific Islanders, 2 Hispanic Americans), 5 international. *29 applicants, 55% accepted.* In 1999, 55 master's, 17 doctorates awarded. Terminal master's awarded for partial completion of doctoral program.
Degree requirements: For master's and doctorate, thesis/dissertation required, foreign language not required.
Entrance requirements: For master's and doctorate, GRE General Test, GRE Subject Test, minimum GPA of 3.0 during previous 2 years, 2.8 overall. *Application deadline:* For fall admission, 4/1 (priority date); for spring admission, 11/1. Applications are processed on a rolling basis. *Application fee:* $40.
Expenses: Tuition, state resident: full-time $6,030; part-time $335 per credit. Tuition, nonresident: full-time $8,298; part-time $461 per credit. Required fees: $230. One-time fee: $10. Tuition and fees vary according to program.
Financial aid: In 1999–00, 2 fellowships, 1 research assistantship (averaging $12,400 per year), 26 teaching assistantships (averaging $12,400 per year) were awarded; Federal Work-Study and tuition waivers (full) also available.
Faculty research: Membrane proteins, genetics, molecular biology, neuroscience, aquatic biology.
Dr. Joel Sheffield, Chair, 215-204-8854, *Fax:* 215-204-6646, *E-mail:* jbs@ sqibio.chem.temple.edu.
Application contact: Dr. Edward R. Gruber, Graduate Admissions Chair, 215-204-1920, *Fax:* 215-204-6646, *E-mail:* v5381e@vm.temple.edu.

Find an in-depth description at www.petersons.com/graduate.

■ TEMPLE UNIVERSITY

Health Sciences Center, School of Medicine and Graduate School, Graduate Programs in Medicine, Philadelphia, PA 19122-6096

AWARDS MS, PhD, MD/PhD.

Faculty: 77 full-time (13 women).
Students: 143 full-time (61 women), 2 part-time; includes 35 minority (7 African Americans, 28 Asian Americans or Pacific Islanders), 5 international. *180 applicants, 33% accepted.* In 1999, 1 master's, 17 doctorates awarded. Terminal master's awarded for partial completion of doctoral program.
Degree requirements: For master's, thesis required; for doctorate, dissertation, research seminars required.
Entrance requirements: For master's and doctorate, GRE General Test. *Application fee:* $40. Electronic applications accepted.
Expenses: Tuition, state resident: full-time $6,030. Tuition, nonresident: full-time $8,298. Required fees: $230. One-time fee: $10 full-time.
Financial aid: Fellowships, research assistantships, career-related internships or fieldwork, Federal Work-Study, grants, institutionally sponsored loans, and tuition waivers (full and partial) available. Aid available to part-time students. Financial aid applicants required to submit FAFSA.
Faculty research: Molecular biology and biochemistry; cardiovascular, renal, and neurophysiological pharmacology; reproductive and developmental biology; immunology and microbiology; cancer research. *Total annual research expenditures:* $8.3 million.
Dr. Laurie G. Paavola, Assistant Dean for Graduate Studies, 215-707-3252, *Fax:* 215-707-2940.
Application contact: Shirley Burton, Coordinator, Graduate Student Services, 215-707-7006, *Fax:* 215-707-2940, *E-mail:* sburto@astro.ocis.temple.edu.

■ TENNESSEE STATE UNIVERSITY

Graduate School, College of Arts and Sciences, Department of Biological Sciences, Nashville, TN 37209-1561

AWARDS MS, PhD.

Faculty: 10 full-time (3 women).
Students: 27 full-time (20 women), 15 part-time (9 women); includes 41 minority (36 African Americans, 5 Asian Americans or Pacific Islanders). Average age 27. *21 applicants, 90% accepted.* In 1999, 7 degrees awarded.
Degree requirements: For master's, thesis required; for doctorate, dissertation required.

Tennessee State University (continued)
Entrance requirements: For master's, GRE General Test, GRE Subject Test, minimum GPA of 2.5. *Application deadline:* Applications are processed on a rolling basis. *Application fee:* $15.
Expenses: Tuition, state resident: full-time $3,134; part-time $191 per credit hour. Tuition, nonresident: full-time $8,250; part-time $415 per credit hour.
Financial aid: In 1999–00, 7 research assistantships (averaging $20,734 per year), 7 teaching assistantships (averaging $20,370 per year) were awarded; unspecified assistantships also available. Aid available to part-time students. Financial aid application deadline: 5/1.
Faculty research: Soybean tissue culture, neurochemistry, microbial genetics.
Dr. Terrence Johnson, Head, 615-963-5748.

Find an in-depth description at www.petersons.com/graduate.

■ **TENNESSEE TECHNOLOGICAL UNIVERSITY**

Graduate School, College of Arts and Sciences, Department of Biology, Cookeville, TN 38505

AWARDS Environmental biology (MS); fish, game, and wildlife management (MS). Part-time programs available.

Faculty: 22 full-time (2 women).
Students: 23 full-time (8 women), 7 part-time (3 women); includes 3 minority (2 Asian Americans or Pacific Islanders, 1 Hispanic American). Average age 25. *16 applicants, 13% accepted.* In 1999, 15 degrees awarded.
Degree requirements: For master's, thesis required, foreign language not required.
Entrance requirements: For master's, GRE General Test, TOEFL. *Application deadline:* For fall admission, 3/1 (priority date); for spring admission, 8/1. *Application fee:* $25 ($30 for international students).
Expenses: Tuition, state resident: full-time $3,082; part-time $154 per hour. Tuition, nonresident: full-time $7,908; part-time $365 per hour. Required fees: $1,541; $154 per hour. Tuition and fees vary according to course load.
Financial aid: In 1999–00, 24 students received aid, including 17 research assistantships (averaging $7,000 per year), 7 teaching assistantships (averaging $5,320 per year). Financial aid application deadline: 4/1.
Faculty research: Aquatics, environmental studies.
Dr. Daniel Combs, Interim Chairperson, 931-372-3134, *Fax:* 931-372-6257, *E-mail:* dcombs@tntech.edu.

Application contact: Dr. Rebecca F. Quattlebaum, Dean of the Graduate School, 931-372-3233, *Fax:* 931-372-3497, *E-mail:* rquattlebaum@tntech.edu.

■ **TEXAS A&M UNIVERSITY**

College of Science, Department of Biology, College Station, TX 77843

AWARDS Biology (MS, PhD); botany (MS, PhD); microbiology (MS, PhD); zoology (MS, PhD).

Faculty: 39 full-time (6 women), 4 part-time/adjunct (2 women).
Students: 85 full-time (34 women), 22 part-time (16 women); includes 7 minority (4 Asian Americans or Pacific Islanders, 3 Hispanic Americans), 38 international. Average age 28. *87 applicants, 26% accepted.* In 1999, 6 master's, 17 doctorates awarded.
Degree requirements: For master's, foreign language not required; for doctorate, dissertation required, foreign language not required.
Entrance requirements: For master's and doctorate, GRE General Test, TOEFL. *Application fee:* $50 ($75 for international students).
Expenses: Tuition, state resident: part-time $76 per semester hour. Tuition, nonresident: part-time $292 per semester hour. Required fees: $11 per semester hour. Tuition and fees vary according to program.
Financial aid: Fellowships, research assistantships, teaching assistantships available. Financial aid application deadline: 4/1; financial aid applicants required to submit FAFSA.
Dr. Terry L. Thomas, Head, 979-845-0184, *Fax:* 979-845-2891.
Application contact: Graduate Adviser, 979-845-7755.

Find an in-depth description at www.petersons.com/graduate.

■ **TEXAS A&M UNIVERSITY–COMMERCE**

Graduate School, College of Arts and Sciences, Department of Biological and Earth Sciences, Commerce, TX 75429-3011

AWARDS M Ed, MS.

Faculty: 8 full-time (0 women), 2 part-time/adjunct (0 women).
Students: 10 full-time, 19 part-time; includes 6 minority (3 African Americans, 1 Asian American or Pacific Islander, 2 Hispanic Americans). Average age 36. *10 applicants, 100% accepted.* In 1999, 5 master's awarded.
Degree requirements: For master's, thesis (for some programs), comprehensive exam required.

Entrance requirements: For master's, GRE General Test. *Application deadline:* For fall admission, 6/1 (priority date); for spring admission, 11/1 (priority date). Applications are processed on a rolling basis. *Application fee:* $0 ($25 for international students). Electronic applications accepted.
Expenses: Tuition, state resident: full-time $2,558; part-time $365 per semester. Tuition, nonresident: full-time $7,740; part-time $1,007 per semester. Tuition and fees vary according to course load.
Financial aid: In 1999–00, research assistantships (averaging $7,875 per year), teaching assistantships (averaging $7,875 per year) were awarded; Federal Work-Study, institutionally sponsored loans, and scholarships also available. Financial aid application deadline: 5/1; financial aid applicants required to submit FAFSA.
Faculty research: Microbiology, botany, environmental science, birds. *Total annual research expenditures:* $3,000.
Dr. Don R. Lee, Interim Head, 903-886-5378, *Fax:* 903-886-5997, *E-mail:* don_lee@tamu-commerce.edu.
Application contact: Janet Swart, Graduate Admissions Adviser, 903-886-5167, *Fax:* 903-886-5165, *E-mail:* jan_swart@tamu-commerce.edu.

■ **TEXAS A&M UNIVERSITY–CORPUS CHRISTI**

Graduate Programs, College of Science and Technology, Program in Sciences, Corpus Christi, TX 78412-5503

AWARDS Biology (MS); environmental sciences (MS); mariculture (MS). Part-time and evening/weekend programs available.

Students: 38 full-time (14 women), 64 part-time (34 women); includes 25 minority (2 African Americans, 3 Asian Americans or Pacific Islanders, 20 Hispanic Americans), 1 international. Average age 31. In 1999, 28 degrees awarded.
Degree requirements: For master's, thesis required (for some programs), foreign language not required.
Entrance requirements: For master's, GRE General Test. *Application deadline:* For fall admission, 7/15 (priority date); for spring admission, 11/15. Applications are processed on a rolling basis. *Application fee:* $10 ($30 for international students). Electronic applications accepted.
Expenses: Tuition, state resident: full-time $1,134; part-time $70 per credit hour. Tuition, nonresident: full-time $5,022; part-time $285 per credit hour.
Financial aid: Research assistantships, teaching assistantships, career-related internships or fieldwork, Federal Work-Study, and institutionally sponsored loans

available. Aid available to part-time students. Financial aid application deadline: 3/15; financial aid applicants required to submit FAFSA.

Dr. Claudia Johnston, Assistant Dean, 361-825-2712, *E-mail:* claudia.johnston@mail.tamacc.edu.

Application contact: Mary Margaret Dechant, Director of Admissions, 361-825-2624, *Fax:* 361-825-5887, *E-mail:* margaret.dechant@mail.tamucc.edu.

■ TEXAS A&M UNIVERSITY–KINGSVILLE

College of Graduate Studies, College of Arts and Sciences, Department of Biology, Kingsville, TX 78363

AWARDS MS. Part-time programs available.

Faculty: 7 full-time (0 women).
Students: 6 full-time (4 women), 15 part-time (8 women); includes 12 minority (all Hispanic Americans), 2 international. Average age 28. In 1999, 4 degrees awarded.
Degree requirements: For master's, thesis or alternative, comprehensive exam required, foreign language not required.
Entrance requirements: For master's, GRE General Test, TOEFL, minimum GPA of 3.0. *Application deadline:* For fall admission, 6/1; for spring admission, 11/15. Applications are processed on a rolling basis. *Application fee:* $15 ($25 for international students).
Expenses: Tuition, state resident: full-time $2,062; part-time $102 per hour. Tuition, nonresident: full-time $7,246; part-time $316 per hour. Tuition and fees vary according to course load.
Financial aid: Research assistantships, teaching assistantships, Federal Work-Study and institutionally sponsored loans available. Financial aid application deadline: 5/15.
Faculty research: Venom physiology, monoclonal research with venom, shore bird ecology, metabolism of foreign amino acids.
Dr. James Pierce, Chair, 361-593-3803.

■ TEXAS A&M UNIVERSITY SYSTEM HEALTH SCIENCE CENTER

Baylor College of Dentistry, Graduate Division, Department of Biomedical Sciences, College Station, TX 77840-7896

AWARDS MS, PhD. Part-time programs available.

Faculty: 19 full-time (1 woman), 14 part-time/adjunct (1 woman).
Students: 2 full-time (1 woman), 9 part-time (5 women); includes 1 minority (Hispanic American), 5 international. Average age 31. *6 applicants, 67% accepted.* In

1999, 1 master's, 1 doctorate awarded. Terminal master's awarded for partial completion of doctoral program.
Degree requirements: For master's and doctorate, thesis/dissertation required, foreign language not required. *Average time to degree:* Doctorate–3 years full-time.
Entrance requirements: For master's, GRE General Test, TOEFL; for doctorate, GRE General Test, TOEFL, DDS or DMD. *Application deadline:* For fall admission, 5/1 (priority date); for spring admission, 12/1. Applications are processed on a rolling basis. *Application fee:* $35.
Expenses: Tuition, state resident: part-time $51 per quarter hour. Tuition, nonresident: part-time $169 per quarter hour. Required fees: $31 per quarter hour. $30 per quarter. One-time fee: $15 part-time.
Financial aid: In 1999–00, 2 students received aid, including research assistantships (averaging $12,000 per year), teaching assistantships (averaging $12,000 per year); fellowships, institutionally sponsored loans also available. Aid available to part-time students. Financial aid application deadline: 2/23; financial aid applicants required to submit FAFSA.
Faculty research: Craniofacial biology, aging, neuroscience, physiology, molecular/cellular biology.
Dr. David S. Carlson, Chair, 214-828-8270, *Fax:* 214-828-8951, *E-mail:* dscarlson@tambcd.edu.

Application contact: Dr. Paul C. Dechow, Chair, Graduate Program Committee and Program Director, 214-828-8277, *Fax:* 214-828-8951, *E-mail:* pcdechow@tambcd.edu.

■ TEXAS A&M UNIVERSITY SYSTEM HEALTH SCIENCE CENTER

College of Medicine, Graduate School of Biomedical Sciences, College Station, TX 77840-7896

AWARDS Human anatomy and medical neurobiology (PhD); medical biochemistry and genetics (PhD); medical microbiology and immunology (PhD), including immunology, microbiology, molecular biology, virology; medical pharmacology and toxicology (PhD); medical physiology (PhD); neuroscience (PhD); pathology and laboratory medicine (PhD), including molecular pathology.

Faculty: 128 full-time (17 women), 38 part-time/adjunct (9 women).
Students: 65 full-time (36 women), 1 part-time; includes 14 minority (1 African American, 11 Asian Americans or Pacific Islanders, 2 Hispanic Americans). Average age 28. *12 applicants, 33% accepted.* In 1999, 4 degrees awarded (6% entered university research/teaching).

Degree requirements: For doctorate, dissertation required, foreign language not required.
Entrance requirements: For doctorate, GRE General Test, minimum GPA of 3.0. *Application deadline:* For fall admission, 2/1 (priority date). Applications are processed on a rolling basis. *Application fee:* $50 ($75 for international students).
Expenses: Tuition, area resident: Full-time $1,368. Tuition, state resident: part-time $76 per credit. Tuition, nonresident: full-time $5,256; part-time $292 per credit. International tuition: $5,256 full-time. Required fees: $678; $38 per credit. Full-time tuition and fees vary according to course load and student level.
Financial aid: In 1999–00, 26 research assistantships (averaging $17,600 per year) were awarded; fellowships, teaching assistantships, institutionally sponsored loans also available. Financial aid applicants required to submit FAFSA.
Dr. Fuller W. Bazer, Interim Vice President and Dean, 713-677-7716, *Fax:* 713-677-7725, *E-mail:* fbazer@cvm.tamu.edu.

Find an in-depth description at www.petersons.com/graduate.

■ TEXAS A&M UNIVERSITY SYSTEM HEALTH SCIENCE CENTER

Institute of Biosciences and Technology, Houston, TX 77030-3303

AWARDS Medical sciences (PhD).

Faculty: 9 full-time (0 women), 7 part-time/adjunct (0 women).
Students: 6 full-time (4 women); includes 1 minority (Hispanic American), 4 international. Average age 30. *5 applicants, 100% accepted.* In 1999, 4 degrees awarded (75% entered university research/teaching, 25% continued full-time study).
Degree requirements: For doctorate, dissertation, required course proficiency required, foreign language not required.
Entrance requirements: For doctorate, GRE General Test. *Application deadline:* For fall admission, 3/1 (priority date); for winter admission, 6/1 (priority date); for spring admission, 8/1 (priority date). Applications are processed on a rolling basis. *Application fee:* $50 ($75 for international students).
Expenses: Tuition, state resident: part-time $76 per credit hour.
Financial aid: In 1999–00, research assistantships (averaging $18,500 per year); fellowships, unspecified assistantships also available.
Faculty research: Cancer biology, DNA structure, extracellular matrix biology, development, human genetic disease. *Total annual research expenditures:* $4.5 million.

Texas A&M University System Health Science Center (continued)

Dr. Fuller W. Bazer, Director, 713-677-7716, *Fax:* 713-677-7725, *E-mail:* fbazer@cvm.tamu.edu.

Application contact: Dr. Richard R. Sinden, Graduate Program Director, 713-677-7716, *Fax:* 713-677-7725, *E-mail:* gradprog@ibt.tamu.edu.

Find an in-depth description at www.petersons.com/graduate.

■ TEXAS CHRISTIAN UNIVERSITY

Add Ran College of Arts and Sciences, Department of Biology, Fort Worth, TX 76129-0002

AWARDS Biology (MA, MS); environmental sciences (MS). Part-time and evening/weekend programs available.

Students: 5 full-time (1 woman), 16 part-time (10 women); includes 2 minority (both Asian Americans or Pacific Islanders), 2 international. In 1999, 8 degrees awarded.

Entrance requirements: For master's, GRE General Test, GRE Subject Test, TOEFL. *Application deadline:* For fall admission, 3/1; for spring admission, 12/1. Applications are processed on a rolling basis. *Application fee:* $0.

Expenses: Tuition: Full-time $6,570; part-time $365 per credit hour. Required fees: $50 per credit hour.

Financial aid: Unspecified assistantships available. Financial aid application deadline: 3/1.

Dr. Wayne Barcellona, Chairperson, 817-257-7165, *E-mail:* w.barcellona@tcu.edu.

■ TEXAS SOUTHERN UNIVERSITY

Graduate School, College of Arts and Sciences, Department of Biology, Houston, TX 77004-4584

AWARDS MS.

Faculty: 10 full-time (5 women), 7 part-time/adjunct (3 women).

Students: 19 full-time (15 women); all minorities (18 African Americans, 1 Asian American or Pacific Islander). In 1999, 9 degrees awarded (100% found work related to degree).

Degree requirements: For master's, one foreign language (computer language can substitute), thesis, comprehensive exam required. *Average time to degree:* Master's–2 years full-time.

Entrance requirements: For master's, GRE General Test, TOEFL, minimum GPA of 2.5. *Application deadline:* For fall admission, 7/15 (priority date). Applications are processed on a rolling basis. *Application fee:* $35 ($75 for international students).

Expenses: Tuition, area resident: Part-time $296 per credit hour. Tuition, nonresident: part-time $449 per credit hour.

Financial aid: In 1999–00, 7 fellowships, 12 teaching assistantships were awarded; career-related internships or fieldwork, Federal Work-Study, and institutionally sponsored loans also available. Financial aid application deadline: 5/1.

Faculty research: Microbiology, cell and molecular biology, biochemistry, biochemical virology, biophysics.

Dr. Sunday Fadulu, Chairman, 713-313-7005.

■ TEXAS TECH UNIVERSITY

Graduate School, College of Arts and Sciences, Department of Biological Sciences, Lubbock, TX 79409

AWARDS Biology (MS, PhD); environmental toxicology (MS); microbiology (MS); zoology (MS, PhD). Part-time programs available.

Faculty: 36 full-time (4 women), 1 part-time/adjunct (0 women).

Students: 86 full-time (41 women), 26 part-time (10 women); includes 5 minority (1 Asian American or Pacific Islander, 4 Hispanic Americans), 28 international. Average age 30. *54 applicants, 46% accepted.* In 2000, 26 master's, 11 doctorates awarded.

Degree requirements: For master's, thesis required (for some programs), foreign language not required; for doctorate, dissertation required, foreign language not required.

Entrance requirements: For master's and doctorate, GRE General Test. *Application deadline:* For fall admission, 4/15 (priority date); for spring admission, 11/1 (priority date). Applications are processed on a rolling basis. *Application fee:* $25 ($50 for international students). Electronic applications accepted.

Expenses: Tuition, state resident: full-time $2,376; part-time $99 per credit hour. Tuition, nonresident: full-time $7,560; part-time $315 per credit hour. Required fees: $464 per semester. Part-time tuition and fees vary according to course load, program and reciprocity agreements.

Financial aid: In 2000–01, 60 students received aid, including 40 research assistantships (averaging $10,405 per year), 53 teaching assistantships (averaging $11,067 per year); fellowships, career-related internships or fieldwork, Federal Work-Study, and institutionally sponsored loans also available. Aid available to part-time students. Financial aid application deadline: 5/15; financial aid applicants required to submit FAFSA.

Faculty research: Development of strains of transgenic plants, ecological studies of Arctic tundra and Puerto Rican rain forests, genome organization and evolution. *Total annual research expenditures:* $2.1 million.

Dr. Carleton Phillips, Chairman, 806-742-2715, *Fax:* 806-742-2963.

Application contact: Graduate Adviser, 806-742-2715, *Fax:* 806-742-2963.

■ TEXAS TECH UNIVERSITY HEALTH SCIENCES CENTER

Graduate School of Biomedical Sciences, Lubbock, TX 79430

AWARDS MS, PhD, MD/PhD.

Faculty: 84 full-time (21 women), 4 part-time/adjunct (1 woman).

Students: 47 full-time (22 women); includes 5 minority (3 Asian Americans or Pacific Islanders, 2 Hispanic Americans), 21 international. Average age 27. *101 applicants, 47% accepted.* In 1999, 4 doctorates awarded. Terminal master's awarded for partial completion of doctoral program.

Degree requirements: For master's and doctorate, thesis/dissertation required, foreign language not required. *Average time to degree:* Doctorate–5 years full-time.

Entrance requirements: For master's and doctorate, GRE General Test, TOEFL, minimum GPA of 3.0. *Application deadline:* For fall admission, 4/15 (priority date); for spring admission, 9/15 (priority date). Applications are processed on a rolling basis. *Application fee:* $30 ($55 for international students). Electronic applications accepted.

Expenses: Tuition, state resident: part-time $38 per credit hour. Tuition, nonresident: part-time $254 per credit hour. Part-time tuition and fees vary according to program.

Financial aid: In 1999–00, 6 fellowships (averaging $14,500 per year), 37 research assistantships (averaging $14,500 per year) were awarded; institutionally sponsored loans and scholarships also available. Financial aid applicants required to submit FAFSA.

Faculty research: Genetics of neurological disorders, hemodynamics to prevent DVT, toxin A synthesis, DA neurons, peroxidases.

Dr. Barbara C. Pence, Associate Dean, 806-743-2556, *Fax:* 806-743-2656, *E-mail:* acagsbs@ttuhsc.edu.

Application contact: Pamela Johnson, Director of Graduate Programs, 806-743-2556, *Fax:* 806-743-2656, *E-mail:* acapj@ttuhsc.edu.

■ TEXAS WOMAN'S UNIVERSITY

Graduate School, College of Arts and Sciences, Department of Biology, Denton, TX 76204

AWARDS Biology (MS); biology teaching (MS); molecular biology (PhD). Part-time programs available.

Faculty: 11 full-time (4 women), 2 part-time/adjunct (1 woman).
Students: 27 full-time (23 women), 35 part-time (26 women); includes 20 minority (7 African Americans, 8 Asian Americans or Pacific Islanders, 5 Hispanic Americans), 6 international. Average age 32. *33 applicants, 70% accepted.* In 1999, 7 master's, 1 doctorate awarded (100% continued full-time study). Terminal master's awarded for partial completion of doctoral program.
Degree requirements: For master's, thesis required (for some programs), foreign language not required; for doctorate, variable foreign language requirement (computer language can substitute for one), dissertation, residency required. *Average time to degree:* Master's–2.5 years full-time, 4 years part-time; doctorate–5 years full-time, 8 years part-time.
Entrance requirements: For master's and doctorate, GRE General Test, minimum GPA of 3.0. *Application deadline:* For fall admission, 4/1 (priority date); for spring admission, 8/1. Applications are processed on a rolling basis. *Application fee:* $30.
Expenses: Tuition, state resident: full-time $2,045; part-time $83 per semester hour. Tuition, nonresident: full-time $5,933; part-time $279 per semester hour. Required fees: $500 per semester. Tuition and fees vary according to course load.
Financial aid: In 1999–00, 8 research assistantships with partial tuition reimbursements (averaging $10,000 per year), 18 teaching assistantships with partial tuition reimbursements (averaging $9,000 per year) were awarded; career-related internships or fieldwork, Federal Work-Study, institutionally sponsored loans, and tuition waivers (partial) also available. Aid available to part-time students. Financial aid application deadline: 4/1. *Total annual research expenditures:* $2 million.
Dr. Fritz E. Schwalm, Chair, 940-898-2352, *Fax:* 940-898-2382, *E-mail:* d_schwalm@venus.twu.edu.

Find an in-depth description at www.petersons.com/graduate.

■ THOMAS JEFFERSON UNIVERSITY

College of Graduate Studies, Philadelphia, PA 19107

AWARDS MS, PhD, Certificate, MD/PhD. Part-time and evening/weekend programs available.

Faculty: 204.
Students: 245 full-time (150 women), 213 part-time (152 women); includes 70 minority (30 African Americans, 31 Asian Americans or Pacific Islanders, 8 Hispanic Americans, 1 Native American), 24 international. Average age 29. *660 applicants, 36% accepted.* In 1999, 146 master's, 27 doctorates awarded. Terminal master's awarded for partial completion of doctoral program.
Degree requirements: For master's, foreign language not required; for doctorate, dissertation required, foreign language not required. *Average time to degree:* Master's–1.5 years full-time, 3 years part-time; doctorate–5 years full-time.
Entrance requirements: For master's, minimum GPA of 3.0; for doctorate, TOEFL, minimum GPA of 3.2. *Application deadline:* Applications are processed on a rolling basis. *Application fee:* $40.
Expenses: Tuition: Full-time $12,670. Tuition and fees vary according to degree level and program.
Financial aid: In 1999–00, 187 students received aid, including 118 fellowships with full tuition reimbursements available; research assistantships, Federal Work-Study, institutionally sponsored loans, traineeships, and training grants also available. Aid available to part-time students. Financial aid application deadline: 5/1; financial aid applicants required to submit FAFSA.
Faculty research: Developmental biology, immunology, genetics, oncology, molecular biology.
Dr. Jussi J. Saukkonen, Dean, 215-503-8986, *Fax:* 215-503-6690.
Application contact: Jessie F. Pervall, Director of Admissions, 215-503-4400, *Fax:* 215-503-3433, *E-mail:* cgs-info@mail.tju.edu.

Find an in-depth description at www.petersons.com/graduate.

■ TOURO COLLEGE

Barry Z. Levine School of Health Sciences, Biomedical Sciences Program, New York, NY 10010

AWARDS MS, MD/MS. Professional course work completed at the Technion Faculty of Medicine, Israel.

Degree requirements: For master's, comprehensive exam or project required.

Entrance requirements: For master's, GRE Subject Test or MCAT.

■ TOWSON UNIVERSITY

Graduate School, Program in Biology, Towson, MD 21252-0001

AWARDS MS. Part-time and evening/weekend programs available.

Faculty: 18 full-time (3 women).
Students: 15 full-time, 20 part-time. In 1999, 9 degrees awarded.
Degree requirements: For master's, exam required, thesis optional, foreign language not required.
Entrance requirements: For master's, GRE General Test, GRE Subject Test. *Application deadline:* For fall admission, 3/1 (priority date); for spring admission, 10/1. Applications are processed on a rolling basis. *Application fee:* $40.
Expenses: Tuition, state resident: full-time $3,510; part-time $195 per credit. Tuition, nonresident: full-time $6,948; part-time $386 per credit. Required fees: $40 per credit.
Financial aid: Teaching assistantships, Federal Work-Study and unspecified assistantships available. Financial aid application deadline: 4/1; financial aid applicants required to submit FAFSA.
Faculty research: Zoosporic fungi, herpetology, fishes, ornithology, genetic analysis.
Larry E. Wimmers, Co-Director, 410-830-2766, *Fax:* 410-830-3434, *E-mail:* lwimmer@towson.edu.
Application contact: Phil Adams, Assistant Director of Graduate School, 410-830-2501, *Fax:* 410-830-4675, *E-mail:* petgrad@towson.edu.

■ TRUMAN STATE UNIVERSITY

Graduate School, Division of Science, Program in Biology, Kirksville, MO 63501-4221

AWARDS MS.

Faculty: 27 full-time (6 women).
Students: 4 full-time (0 women), 2 international. Average age 23. *22 applicants, 27% accepted.* In 1999, 4 degrees awarded.
Degree requirements: For master's, thesis, comprehensive exam required, foreign language not required.
Entrance requirements: For master's, GRE General Test, minimum GPA of 3.0. *Application deadline:* For fall admission, 6/15 (priority date); for spring admission, 11/1. Applications are processed on a rolling basis. *Application fee:* $0 ($25 for international students).
Expenses: Tuition, state resident: full-time $2,844; part-time $158 per credit. Tuition, nonresident: full-time $5,094; part-time $283 per credit. Required fees: $9 per

Truman State University (continued)
semester. Tuition and fees vary according to course load.

Financial aid: Research assistantships, career-related internships or fieldwork and Federal Work-Study available. Financial aid application deadline: 5/1; financial aid applicants required to submit FAFSA. Dr. Cynthia Cooper, Director.

Application contact: Peggy Orchard, Graduate Office Secretary, 660-785-4109, *Fax:* 660-785-7460.

■ TUFTS UNIVERSITY

Division of Graduate and Continuing Studies and Research, Graduate School of Arts and Sciences, Department of Biology, Medford, MA 02155

AWARDS MS, PhD. Part-time programs available.

Faculty: 19 full-time.
Students: 32 (20 women). *113 applicants, 19% accepted.* In 1999, 4 master's, 1 doctorate awarded. Terminal master's awarded for partial completion of doctoral program.
Degree requirements: For master's, computer language, thesis required (for some programs), foreign language not required; for doctorate, computer language, dissertation required, foreign language not required.
Entrance requirements: For master's and doctorate, GRE General Test, TOEFL. *Application deadline:* For fall admission, 2/15. Applications are processed on a rolling basis. *Application fee:* $50. Electronic applications accepted.
Expenses: Tuition: Full-time $24,804; part-time $2,480 per course. Required fees: $485; $40 per year. Full-time tuition and fees vary according to program. Part-time tuition and fees vary according to course load.
Financial aid: Research assistantships with full and partial tuition reimbursements, teaching assistantships with full and partial tuition reimbursements, Federal Work-Study, scholarships, and tuition waivers (partial) available. Financial aid application deadline: 2/15; financial aid applicants required to submit FAFSA. Dr. Harry Bernheim, Chair, 617-627-3195.
Application contact: Dr. Barry Trimmer, Information Contact, 617-627-3195, *E-mail:* btrimmer@emerald.tufts.edu.

Find an in-depth description at www.petersons.com/graduate.

■ TUFTS UNIVERSITY

Division of Graduate and Continuing Studies and Research, Professional and Continuing Studies, Premedical Studies Program, Medford, MA 02155
AWARDS Certificate.

Students: 38 full-time (19 women). Average age 26. *65 applicants, 57% accepted.* In 1999, 12 degrees awarded. *Average time to degree:* 2 years part-time.
Application deadline: For fall admission, 6/1; for spring admission, 11/1. *Application fee:* $40.
Expenses: Tuition: Part-time $1,990 per course.
Financial aid: Available to part-time students. Application deadline: 5/1;

■ TUFTS UNIVERSITY

Sackler School of Graduate Biomedical Sciences, Medford, MA 02155

AWARDS PhD, DVM/PhD, MD/PhD.

Faculty: 122 full-time (32 women).
Students: 201 full-time (112 women); includes 23 minority (1 African American, 17 Asian Americans or Pacific Islanders, 5 Hispanic Americans), 45 international. *901 applicants, 10% accepted.* In 1999, 22 doctorates awarded.
Degree requirements: For doctorate, dissertation required, foreign language not required. *Average time to degree:* Doctorate–6 years full-time.
Entrance requirements: For doctorate, GRE General Test, TOEFL. *Application deadline:* For fall admission, 1/15 (priority date). Applications are processed on a rolling basis. *Application fee:* $45.
Expenses: Tuition: Full-time $19,325.
Financial aid: In 1999–00, research assistantships with full tuition reimbursements (averaging $18,805 per year); fellowships with full tuition reimbursements, career-related internships or fieldwork, scholarships, and tuition waivers (full and partial) also available. Financial aid applicants required to submit FAFSA.
Faculty research: Cell biology, molecular biology, biochemistry, genetics, immunology.
Dr. Louis Lasagna, Dean, 617-636-6767.
Application contact: Staff Assistant, 617-636-6767, *Fax:* 617-636-0375, *E-mail:* sackler-school@tufts.edu.

Find an in-depth description at www.petersons.com/graduate.

■ TULANE UNIVERSITY

Graduate School, Department of Ecology, Evolution, and Organismal Biology, New Orleans, LA 70118-5669
AWARDS Biology (MS, PhD). Part-time programs available.

Students: 21 full-time (8 women), 1 part-time; includes 2 minority (both African Americans), 6 international. *20 applicants, 35% accepted.* In 1999, 3 master's, 1 doctorate awarded. Terminal master's awarded for partial completion of doctoral program.
Degree requirements: For master's, thesis or alternative required, foreign language not required; for doctorate, dissertation required, foreign language not required.
Entrance requirements: For master's, GRE General Test, TSE, minimum B average in undergraduate course work; for doctorate, GRE General Test, TSE. *Application deadline:* For fall admission, 2/1. *Application fee:* $45.
Expenses: Tuition: Full-time $23,500. Tuition and fees vary according to program.
Financial aid: Fellowships, research assistantships, teaching assistantships, career-related internships or fieldwork available. Financial aid application deadline: 2/1.
Faculty research: Ichthyology, plant systematics, crustacean endocrinology, ecotoxicology, ornithology.
Dr. David Heins, Chair, 504-865-5191, *Fax:* 504-862-8706.

■ TULANE UNIVERSITY

School of Medicine and Graduate School, Graduate Programs in Medicine, New Orleans, LA 70118-5669
AWARDS MS, MSPH, PhD, MD/PhD.

Students: 138 full-time (66 women), 8 part-time (6 women). *381 applicants, 16% accepted.* In 1999, 4 master's, 25 doctorates awarded.
Degree requirements: For doctorate, dissertation required.
Entrance requirements: For master's, GRE General Test, TOEFL, or TSE, minimum B average in undergraduate course work; for doctorate, GRE General Test, TOEFL, or TSE. *Application deadline:* For fall admission, 2/1. *Application fee:* $45.
Expenses: Tuition: Full-time $23,030.
Financial aid: Fellowships, research assistantships, teaching assistantships, career-related internships or fieldwork, Federal Work-Study, and institutionally sponsored loans available. Financial aid application deadline: 2/1.
Application contact: Gayle A. Sayas, Administrative Assistant, 504-588-5187.

■ TUSKEGEE UNIVERSITY

Graduate Programs, College of Agricultural, Environmental and Natural Sciences, Department of Biology, Tuskegee, AL 36088
AWARDS MS.

Faculty: 12 full-time (3 women).
Students: 10 full-time (6 women), 2 part-time (both women); includes 11 minority (all African Americans), 1 international. Average age 24. In 1999, 2 degrees awarded.
Degree requirements: For master's, computer language, thesis required, foreign language not required.
Entrance requirements: For master's, GRE General Test, GRE Subject Test. *Application deadline:* For fall admission, 7/15. Applications are processed on a rolling basis. *Application fee:* $25 ($35 for international students).
Expenses: Tuition: Full-time $9,500. Tuition and fees vary according to course load and degree level.
Financial aid: Fellowships, teaching assistantships, Federal Work-Study and institutionally sponsored loans available. Aid available to part-time students. Financial aid application deadline: 4/15.
Dr. Roberta Troy, Head, 334-727-8829.

■ UNIFORMED SERVICES UNIVERSITY OF THE HEALTH SCIENCES

School of Medicine, Division of Basic Medical Sciences, Bethesda, MD 20814-4799

AWARDS Anatomy and cell biology (PhD), including cell biology, developmental biology, and neurobiology; biochemistry (PhD), including emerging infectious diseases; emerging infectious diseases (PhD); medical and clinical psychology (PhD), including clinical psychology, medical psychology; medical history (MMH); microbiology and immunology (PhD); molecular and cell biology (PhD); neuroscience (PhD); pathology (PhD), including molecular pathobiology; pharmacology (PhD); physiology (PhD); preventive medicine/biometrics (MPH, MSPH, MTMH, Dr PH, PhD), including public health (MPH, MSPH, Dr PH), tropical medicine and hygiene (MTMH), zoology (PhD).

Faculty: 142 full-time (40 women), 335 part-time/adjunct (73 women).
Students: 119 full-time (62 women), 15 part-time (2 women); includes 20 minority (10 African Americans, 6 Asian Americans or Pacific Islanders, 3 Hispanic Americans, 1 Native American). Average age 26. *183 applicants, 28% accepted.* In 1999, 37 master's, 9 doctorates awarded. Terminal master's awarded for partial completion of doctoral program.
Degree requirements: For master's, comprehensive exam required; for doctorate, dissertation, qualifying exam required. *Average time to degree:* Master's–1 year full-time.
Entrance requirements: For master's and doctorate, GRE General Test, U.S.

citizenship. *Application deadline:* For fall admission, 1/15 (priority date). Applications are processed on a rolling basis. *Application fee:* $0.
Financial aid: In 1999–00, fellowships with full tuition reimbursements (averaging $15,000 per year), research assistantships with full tuition reimbursements (averaging $15,000 per year) were awarded; career-related internships or fieldwork and tuition waivers (full) also available.
Dr. Michael N. Sheridan, Associate Dean, 800-772-1747, *Fax:* 301-295-6772, *E-mail:* msheridan@usuhs.mil.
Application contact: Janet M. Anastasi, Graduate Program Coordinator, 301-295-9474, *Fax:* 301-295-6772, *E-mail:* janastasi@usuhs.mil.

Find an in-depth description at www.petersons.com/graduate.

■ THE UNIVERSITY OF AKRON

Graduate School, Buchtel College of Arts and Sciences, Department of Biology, Akron, OH 44325-0001

AWARDS MS. Part-time programs available.

Degree requirements: For master's, oral defense of thesis, oral exam, seminars required.
Entrance requirements: For master's, TOEFL, minimum GPA of 2.75.
Expenses: Tuition, state resident: part-time $189 per credit. Tuition, nonresident: part-time $353 per credit. Required fees: $7.3 per credit.
Faculty research: Genetics, immunology, molecular biology, physiology, virology.

■ THE UNIVERSITY OF ALABAMA

Graduate School, College of Arts and Sciences, Department of Biological Sciences, Tuscaloosa, AL 35487
AWARDS MS, PhD.

Faculty: 28 full-time (6 women).
Students: 63 full-time (28 women). Average age 24. *46 applicants, 67% accepted.* In 1999, 10 master's, 3 doctorates awarded. Terminal master's awarded for partial completion of doctoral program.
Degree requirements: For master's, written exam required; for doctorate, dissertation, oral and written final exams required. *Average time to degree:* Master's–3 years full-time; doctorate–5 years full-time.
Entrance requirements: For master's and doctorate, GRE General Test, minimum GPA of 3.0. *Application deadline:* For fall admission, 7/6 (priority date). Applications are processed on a rolling basis. *Application fee:* $25.
Expenses: Tuition, state resident: full-time $2,872. Tuition, nonresident: full-time $7,722. Part-time tuition and fees vary according to course load and program.

Financial aid: In 1999–00, 53 students received aid, including 6 fellowships (averaging $11,000 per year), 32 teaching assistantships (averaging $11,000 per year); Federal Work-Study and institutionally sponsored loans also available. Aid available to part-time students. Financial aid application deadline: 8/14; financial aid applicants required to submit FAFSA.
Faculty research: Developmental genetics, limnology, taxonomy, microbiology, teratology. *Total annual research expenditures:* $1.2 million.
Dr. Martha J. Powell, Chair, 205-348-5960, *Fax:* 205-348-1786, *E-mail:* mpowell@biology.as.ua.edu.
Application contact: Dr. Keller F. Suberkropp, Graduate Director, 205-348-1795, *Fax:* 205-348-1403, *E-mail:* ksuberkp@biology.as.ua.edu.

■ THE UNIVERSITY OF ALABAMA AT BIRMINGHAM

Graduate School and School of Medicine and School of Dentistry, Graduate Programs in Joint Health Sciences, Birmingham, AL 35294

AWARDS Basic medical sciences (MSBMS); biochemistry and molecular genetics (PhD), including biochemistry; biophysical sciences (PhD); cell biology (PhD), including anatomy; medical genetics (PhD); microbiology (PhD); neurobiology (PhD); pathology (PhD); pharmacology (PhD); physiology and biophysics (PhD).

Students: 352 full-time (164 women), 12 part-time (1 woman); includes 35 minority (23 African Americans, 10 Asian Americans or Pacific Islanders, 1 Hispanic American, 1 Native American), 116 international. Average age 33. *498 applicants, 31% accepted.* In 1999, 1 master's, 49 doctorates awarded.
Entrance requirements: For master's, GRE; for doctorate, GRE, interview. *Application deadline:* Applications are processed on a rolling basis. *Application fee:* $35 ($60 for international students). Electronic applications accepted.
Expenses: Tuition, state resident: part-time $104 per semester hour. Tuition, nonresident: part-time $208 per semester hour. Required fees: $17 per semester hour. $57 per quarter. Tuition and fees vary according to program.
Financial aid: Fellowships, career-related internships or fieldwork available.
Dr. William B. Deal, Dean, School of Medicine, 205-934-1111, *Fax:* 205-934-0333, *E-mail:* wdeal@uab.edu.

■ THE UNIVERSITY OF ALABAMA AT BIRMINGHAM

Graduate School, School of Natural Sciences and Mathematics, Department of Biology, Birmingham, AL 35294

AWARDS Comparative and cellular biology (PhD); comparative and cellular physiology (MS); marine science (MS, PhD); microbial ecology and physiology (MS, PhD); reproduction and development (MS, PhD).

Students: 34 full-time (19 women), 6 international. *61 applicants, 38% accepted.* In 1999, 1 master's, 3 doctorates awarded. Terminal master's awarded for partial completion of doctoral program.
Degree requirements: For master's and doctorate, thesis/dissertation required.
Entrance requirements: For master's and doctorate, GRE General Test, TOEFL, previous course work in biology, calculus, organic chemistry, physics. *Application deadline:* Applications are processed on a rolling basis. *Application fee:* $35 ($60 for international students). Electronic applications accepted.
Expenses: Tuition, state resident: part-time $104 per semester hour. Tuition, nonresident: part-time $208 per semester hour. Required fees: $17 per semester hour. $57 per quarter. Tuition and fees vary according to program.
Financial aid: In 1999–00, 22 students received aid, including 3 fellowships with full tuition reimbursements available (averaging $14,000 per year), 19 teaching assistantships with full tuition reimbursements available (averaging $14,000 per year); research assistantships, career-related internships or fieldwork, Federal Work-Study, institutionally sponsored loans, and tuition waivers (full) also available. Aid available to part-time students.
Faculty research: Invertebrate physiology, marine biology, environmental biology.
Dr. Daniel D. Jones, Chairman, 205-934-4290, *Fax:* 205-975-6097, *E-mail:* ddjones@uab.edu.

Find an in-depth description at www.petersons.com/graduate.

■ THE UNIVERSITY OF ALABAMA IN HUNTSVILLE

School of Graduate Studies, College of Science, Department of Biological Sciences, Huntsville, AL 35899

AWARDS MS. Part-time and evening/weekend programs available.

Faculty: 10 full-time (2 women).
Students: 19 full-time (12 women), 17 part-time (13 women); includes 7 minority (4 African Americans, 3 Asian Americans

or Pacific Islanders), 2 international. Average age 30. *23 applicants, 91% accepted.* In 1999, 5 degrees awarded.
Degree requirements: For master's, oral and written exams required, thesis optional, foreign language not required.
Entrance requirements: For master's, GRE General Test, previous course work in biochemistry and organic chemistry, minimum GPA of 3.0. *Application deadline:* For fall admission, 7/24 (priority date); for spring admission, 11/15 (priority date). Applications are processed on a rolling basis. *Application fee:* $35.
Expenses: Tuition, area resident: Full-time $3,880. Tuition, state resident: part-time $144 per hour. Tuition, nonresident: full-time $7,956; part-time $296 per hour. Tuition and fees vary according to course load.
Financial aid: In 1999–00, 15 students received aid, including 13 teaching assistantships with full and partial tuition reimbursements available (averaging $6,691 per year); fellowships with full and partial tuition reimbursements available, research assistantships with full and partial tuition reimbursements available, career-related internships or fieldwork, Federal Work-Study, grants, institutionally sponsored loans, scholarships, and tuition waivers (full and partial) also available. Aid available to part-time students. Financial aid application deadline: 4/1; financial aid applicants required to submit FAFSA.
Faculty research: Cellular and developmental biology, reproductive physiology, immunology, biology of crustacean. *Total annual research expenditures:* $246,567.
Dr. P. Samuel Campbell, Chair, 256-890-6260, *Fax:* 256-890-6305, *E-mail:* campbellp@email.uah.edu.

■ UNIVERSITY OF ALASKA ANCHORAGE

College of Arts and Sciences, Department of Biological Sciences, Anchorage, AK 99508-8060

AWARDS MS. Part-time programs available.

Degree requirements: For master's, thesis or alternative required, foreign language not required.
Entrance requirements: For master's, GRE General Test, GRE Subject Test.
Expenses: Tuition, state resident: full-time $3,006; part-time $167 per credit. Tuition, nonresident: full-time $5,868; part-time $326 per credit. Required fees: $280; $5 per credit. $60 per semester. Tuition and fees vary according to campus/location.
Faculty research: Taxonomy and vegetative analysis in Alaskan ecosystems, fish environment and seafood, biochemistry, arctic ecology, vertebrate ecology.

■ UNIVERSITY OF ALASKA FAIRBANKS

Graduate School, College of Science, Engineering and Mathematics, Department of Biology and Wildlife, Program in Biological Sciences, Fairbanks, AK 99775

AWARDS Biology (MAT, MS, PhD); botany (MS, PhD); zoology (MS, PhD). Part-time programs available.

Faculty: 24 full-time (2 women), 2 part-time/adjunct (0 women).
Students: 39 full-time (19 women), 11 part-time (6 women); includes 1 minority (Native American), 6 international. Average age 31. *32 applicants, 38% accepted.* In 1999, 5 master's, 5 doctorates awarded.
Degree requirements: For master's, thesis, comprehensive exam required, foreign language not required; for doctorate, one foreign language (computer language can substitute), dissertation, comprehensive exam required.
Entrance requirements: For master's and doctorate, GRE General Test, GRE Subject Test, TOEFL. *Application deadline:* For fall admission, 8/1. Applications are processed on a rolling basis. *Application fee:* $35.
Expenses: Tuition, state resident: full-time $3,006; part-time $167 per credit. Tuition, nonresident: full-time $5,868; part-time $326 per credit. Required fees: $370; $10 per credit. $140 per semester.
Financial aid: Research assistantships, teaching assistantships, career-related internships or fieldwork available. Financial aid application deadline: 6/1.
Faculty research: Plant insect interactions, wildlife ecology, adaptations to winter/cold, cell/molecular biology, ecology.
Dr. Ed Murphy, Acting Dean, College of Science, Engineering and Mathematics, 907-474-7941.

■ THE UNIVERSITY OF ARIZONA

College of Medicine, Graduate Programs in Medicine, Tucson, AZ 85721

AWARDS MPH, MS, PhD, MD/PhD. Part-time programs available. Terminal master's awarded for partial completion of doctoral program.

Degree requirements: For doctorate, dissertation required.
Entrance requirements: For master's and doctorate, GRE General Test.
Expenses: Tuition, nonresident: full-time $4,814; part-time $274 per unit. Required fees: $1,094; $115 per unit.

■ UNIVERSITY OF ARKANSAS

Graduate School, J. William Fulbright College of Arts and Sciences, Department of Biological Sciences, Fayetteville, AR 72701-1201

AWARDS Biology (MA, MS, PhD).

Faculty: 22 full-time (5 women).
Students: 54 full-time (20 women), 5 part-time (3 women); includes 3 minority (1 Asian American or Pacific Islander, 2 Hispanic Americans), 6 international. *44 applicants, 43% accepted.* In 1999, 10 master's, 3 doctorates awarded.
Degree requirements: For master's, foreign language not required; for doctorate, dissertation required.
Entrance requirements: For master's and doctorate, GRE Subject Test. *Application fee:* $40 ($50 for international students).
Expenses: Tuition, state resident: full-time $3,186; part-time $177 per credit. Tuition, nonresident: full-time $7,560; part-time $420 per credit. Required fees: $756; $21 per credit. One-time fee: $22 part-time. Tuition and fees vary according to course load and program.
Financial aid: In 1999–00, 5 research assistantships, 32 teaching assistantships were awarded; career-related internships or fieldwork and Federal Work-Study also available. Aid available to part-time students. Financial aid application deadline: 4/1; financial aid applicants required to submit FAFSA.
Dr. Donald Roufa, Chair, 501-575-3251.
Application contact: Dr. Mack Ivey, Graduate Coordinator, 501-575-3251, *E-mail:* mivey@comp.uark.edu.

■ UNIVERSITY OF ARKANSAS FOR MEDICAL SCIENCES

College of Medicine and Graduate School, Graduate Programs in Medicine, Little Rock, AR 72205-7199

AWARDS MS, PhD, MD/PhD.

Students: 77 full-time (28 women), 28 part-time (13 women). In 1999, 6 master's, 9 doctorates awarded.
Degree requirements: For master's, foreign language not required; for doctorate, dissertation required, foreign language not required.
Entrance requirements: For master's and doctorate, GRE General Test. *Application fee:* $0.
Expenses: Tuition: Full-time $8,928.
Financial aid: In 1999–00, 95 research assistantships were awarded; fellowships, teaching assistantships, unspecified assistantships also available. Aid available to part-time students.
Dr. Barry D. Lindley, Associate Dean, Graduate School, 501-686-5454.

Application contact: Paul Carter, Assistant to the Vice Chancellor for Academic Affairs, 501-686-5454.
Find an in-depth description at www.petersons.com/graduate.

■ UNIVERSITY OF CALIFORNIA, BERKELEY

Graduate Division, College of Letters and Science, Department of Integrative Biology, Berkeley, CA 94720-1500

AWARDS Endocrinology (PhD); integrative biology (PhD).

Degree requirements: For doctorate, dissertation, oral qualifying exam required.
Entrance requirements: For doctorate, GRE General Test, GRE Subject Test (biology).
Expenses: Tuition, nonresident: full-time $9,804. Required fees: $4,268. Tuition and fees vary according to program.
Faculty research: Morphology, physiology, development of plants and animals, behavior, ecology.

■ UNIVERSITY OF CALIFORNIA, DAVIS

Graduate Studies, Programs in the Biological Sciences, Davis, CA 95616

AWARDS Animal behavior (MS, PhD); biochemistry and molecular biology (MS, PhD); biophysics (MS, PhD); cell and developmental biology (PhD); comparative pathology (MS, PhD); ecology (MS, PhD); entomology (MS, PhD); epidemiology (MS, PhD); exercise science (MS); genetics (MS, PhD); horticulture, agronomy, and vegetable crops (MS); immunology (MS, PhD); microbiology (MS, PhD); neuroscience (PhD); nutrition (MS, PhD); pharmacology/toxicology (MS, PhD); physiology (MS, PhD); plant biology (MS, PhD); plant pathology (MS, PhD); population biology (PhD). Part-time programs available.

Students: 970 full-time (543 women), 14 part-time (9 women); includes 167 minority (10 African Americans, 102 Asian Americans or Pacific Islanders, 46 Hispanic Americans, 9 Native Americans), 150 international.
Degree requirements: For doctorate, dissertation required.
Entrance requirements: For doctorate, GRE General Test. *Application fee:* $40. Electronic applications accepted.
Expenses: Tuition, nonresident: full-time $9,804. Tuition and fees vary according to program and student level.
Financial aid: Fellowships with full and partial tuition reimbursements, research assistantships with full and partial tuition reimbursements, teaching assistantships

with full and partial tuition reimbursements, career-related internships or fieldwork, Federal Work-Study, grants, institutionally sponsored loans, scholarships, and tuition waivers (full and partial) available. Aid available to part-time students. Financial aid application deadline: 1/15; financial aid applicants required to submit FAFSA.
Application contact: Rosemarie H. Kraft, Associate Dean, 530-752-0655, *Fax:* 530-752-6222.

■ UNIVERSITY OF CALIFORNIA, IRVINE

College of Medicine and Office of Research and Graduate Studies, Graduate Programs in Medicine, Irvine, CA 92697

AWARDS MS, PhD, MD/PhD.

Students: 39 full-time (23 women), 2 part-time (1 woman). *438 applicants, 23% accepted.* In 1999, 2 master's, 5 doctorates awarded.
Entrance requirements: For master's, GRE; for doctorate, GRE General Test, GRE Subject Test. *Application deadline:* Applications are processed on a rolling basis. *Application fee:* $40. Electronic applications accepted.
Expenses: Tuition, nonresident: full-time $10,322; part-time $1,720 per quarter. Required fees: $5,354; $1,300 per quarter. Tuition and fees vary according to program.
Financial aid: Fellowships, research assistantships, teaching assistantships, career-related internships or fieldwork, institutionally sponsored loans, and tuition waivers (full and partial) available. Financial aid application deadline: 3/2; financial aid applicants required to submit FAFSA.
Dr. Thomas Cesario, Dean, College of Medicine, 949-824-5926.

■ UNIVERSITY OF CALIFORNIA, IRVINE

Office of Research and Graduate Studies, School of Biological Sciences, Irvine, CA 92697

AWARDS MS, PhD, MD/PhD.

Students: 154 full-time (81 women), 19 part-time (7 women). Average age 27. *410 applicants, 28% accepted.* In 1999, 4 master's, 3 doctorates awarded. Terminal master's awarded for partial completion of doctoral program.
Degree requirements: For master's, one foreign language required; for doctorate, dissertation required.
Entrance requirements: For master's, GRE General Test, GRE Subject Test, minimum GPA of 3.0; for doctorate, GRE

University of California, Irvine (continued)
General Test, GRE Subject Test. *Application deadline:* Applications are processed on a rolling basis. *Application fee:* $40. Electronic applications accepted.
Expenses: Tuition, nonresident: full-time $10,244; part-time $1,720 per quarter. Required fees: $5,252; $1,300 per quarter. Tuition and fees vary according to course load and program.
Financial aid: Fellowships with full tuition reimbursements, research assistantships with full tuition reimbursements, teaching assistantships with full tuition reimbursements, career-related internships or fieldwork, grants, institutionally sponsored loans, scholarships, and tuition waivers (full and partial) available. Financial aid application deadline: 3/2; financial aid applicants required to submit FAFSA.
Faculty research: Molecular biology and biochemistry, developmental and cell biology, physiology and biophysics, neurosciences, ecology and evolutionary biology.
Dr. Susan V. Bryant, Dean, 949-824-5316.
Application contact: Kimberly McKinney, Administrator, 949-824-8145, *Fax:* 949-824-7407, *E-mail:* gp-mbgb@uci.edu.

■ **UNIVERSITY OF CALIFORNIA, LOS ANGELES**

Graduate Division, College of Letters and Science, Department of Organic Biology, Ecology and Evolution, Los Angeles, CA 90095
AWARDS Biology (MA, PhD); plant molecular biology (PhD).
Students: 78 full-time (29 women); includes 11 minority (1 African American, 7 Asian Americans or Pacific Islanders, 2 Hispanic Americans, 1 Native American), 8 international. *60 applicants, 37% accepted.*
Degree requirements: For master's, comprehensive exam or thesis required; for doctorate, dissertation, oral and written qualifying exams required, foreign language not required.
Entrance requirements: For master's, GRE General Test, GRE Subject Test (biology), minimum GPA of 3.0; for doctorate, GRE General Test, GRE Subject Test (biology), minimum undergraduate GPA of 3.0. *Application deadline:* For fall admission, 1/1. *Application fee:* $40. Electronic applications accepted.
Expenses: Tuition, nonresident: full-time $9,804. Required fees: $4,405. Full-time tuition and fees vary according to program and student level.
Financial aid: In 1999–00, 50 fellowships, 55 research assistantships were awarded; teaching assistantships, Federal Work-Study, institutionally sponsored loans, scholarships, tuition waivers (full and

partial), and federal fellowships also available. Financial aid application deadline: 3/1.
Faculty research: Molecular, cell, and developmental biology; interactive biology; organisms and populations.
Dr. Park S. Nobel, Chair, 310-825-1959, *E-mail:* jocelyny@lifesci.ucla.edu.
Application contact: Departmental Office, 310-825-1959, *Fax:* 310-206-5280, *E-mail:* jocelyny@lifesci.ucla.edu.

■ **UNIVERSITY OF CALIFORNIA, LOS ANGELES**

School of Medicine and Graduate Division, Graduate Programs in Medicine, Los Angeles, CA 90095
AWARDS MA, MS, PhD, MD/PhD.
Students: 351 full-time (154 women); includes 114 minority (4 African Americans, 90 Asian Americans or Pacific Islanders, 16 Hispanic Americans, 4 Native Americans), 38 international. *327 applicants, 21% accepted.* Terminal master's awarded for partial completion of doctoral program.
Degree requirements: For master's, foreign language not required; for doctorate, dissertation, qualifying exams required, foreign language not required.
Entrance requirements: For master's, GRE General Test. *Application fee:* $40.
Expenses: Tuition, nonresident: full-time $9,804. Required fees: $4,405.
Financial aid: In 1999–00, 267 fellowships, 217 research assistantships, 94 teaching assistantships were awarded; career-related internships or fieldwork, Federal Work-Study, institutionally sponsored loans, scholarships, and tuition waivers (full and partial) also available. Financial aid application deadline: 3/1.
Application contact: School of Medicine Admissions Office, 310-825-6081.

■ **UNIVERSITY OF CALIFORNIA, RIVERSIDE**

Graduate Division, College of Natural and Agricultural Sciences, Department of Biology, Riverside, CA 92521-0102
AWARDS MS, PhD.
Faculty: 24 full-time (7 women).
Students: 45 full-time (22 women); includes 3 minority (all Hispanic Americans), 4 international. Average age 29. In 1999, 3 master's, 5 doctorates awarded. Terminal master's awarded for partial completion of doctoral program.
Degree requirements: For master's, oral defense of thesis required; for doctorate, dissertation, 3 quarters of teaching experience, qualifying exams required, foreign language not required. *Average time to*

degree: Master's–3 years full-time; doctorate–7.3 years full-time.
Entrance requirements: For master's and doctorate, GRE General Test, GRE Subject Test, TOEFL, minimum GPA of 3.2. *Application deadline:* For fall admission, 5/1; for winter admission, 9/1; for spring admission, 12/1. Applications are processed on a rolling basis. *Application fee:* $40. Electronic applications accepted.
Expenses: Tuition, nonresident: full-time $9,804. Required fees: $4,758. Full-time tuition and fees vary according to program.
Financial aid: Fellowships, research assistantships, teaching assistantships, career-related internships or fieldwork, Federal Work-Study, institutionally sponsored loans, and tuition waivers (full and partial) available. Financial aid application deadline: 1/5; financial aid applicants required to submit FAFSA.
Faculty research: Molecular genetics, neurophysiology, evolutionary biology, physiology and organismal biology, cell biology, signal transduction, membrane biophysics.
Dr. Mark Chappell, Chair, 909-787-5901, *Fax:* 909-787-4286, *E-mail:* chappell@ucrac1.ucr.edu.
Application contact: Helene Serewis, Graduate Student Affairs Officer, 800-735-0717, *Fax:* 909-787-5517, *E-mail:* biopgrad@pep.ycr.edu.

■ **UNIVERSITY OF CALIFORNIA, RIVERSIDE**

Graduate Division, College of Natural and Agricultural Sciences, Program in Biomedical Sciences, Riverside, CA 92521-0102
AWARDS PhD.
Faculty: 25 full-time (7 women).
Students: 10 full-time (7 women); includes 2 minority (1 Asian American or Pacific Islander, 1 Hispanic American), 6 international. In 1999, 2 degrees awarded.
Degree requirements: For doctorate, dissertation, qualifying exams required, foreign language not required. *Average time to degree:* Doctorate–5 years full-time.
Entrance requirements: For doctorate, GRE General Test, TOEFL, minimum GPA of 3.2.
Expenses: Tuition, nonresident: full-time $9,804. Required fees: $4,758. Full-time tuition and fees vary according to program.
Financial aid: Fellowships, research assistantships, teaching assistantships available. Financial aid application deadline: 2/1; financial aid applicants required to submit FAFSA.
Faculty research: Regulation of cell proliferation; signal transduction in

endocrine, nervous, reproductive, and immune tissues; human genetic disorders; microbiology of human pathogens.
Dr. Michael B. Stemerman, Divisional Dean and Director, 909-787-5705, *Fax:* 909-787-5504, *E-mail:* mstema@ ucrac1.ucr.edu.
Application contact: Mary Jane Ragus, Graduate Program Assistant, 909-787-5707, *Fax:* 909-787-5504, *E-mail:* bmspasst@ucrac1.ucr.edu.

Find an in-depth description at www.petersons.com/graduate.

■ UNIVERSITY OF CALIFORNIA, SAN DIEGO

Graduate Studies and Research, Department of Biology, La Jolla, CA 92093

AWARDS Biochemistry (PhD); cell and developmental biology (PhD); computational neurobiology (PhD); ecology, behavior, and evolution (PhD); genetics and molecular biology (PhD); immunology, virology, and cancer biology (PhD); molecular and cellular biology (PhD); neurobiology (PhD); plant molecular biology (PhD); signal transduction (PhD). Offered in association with the Salk Institute.

Faculty: 101.
Students: 222 (103 women). *410 applicants, 35% accepted.* In 1999, 30 doctorates awarded.
Degree requirements: For doctorate, dissertation required.
Entrance requirements: For doctorate, GRE General Test, pre-application beginning in September. *Application deadline:* For fall admission, 1/8. *Application fee:* $40.
Expenses: Tuition, nonresident: full-time $14,691. Required fees: $4,697. Full-time tuition and fees vary according to program.
Dr. Suresh Subramami, Chair.
Application contact: Biology Graduate Admissions Committee, 858-534-3835.

Find an in-depth description at www.petersons.com/graduate.

■ UNIVERSITY OF CALIFORNIA, SAN DIEGO

School of Medicine and Graduate Studies and Research, Graduate Studies in Biomedical Sciences, La Jolla, CA 92093-0685

AWARDS Cell and molecular biology (PhD); molecular pathology (PhD); neuroscience (PhD); pharmacology (PhD); physiology (PhD); regulatory biology (PhD).

Faculty: 106.
Students: 219. *241 applicants, 23% accepted.* In 1999, 16 doctorates awarded.

Degree requirements: For doctorate, dissertation, qualifying exam required, foreign language not required.
Entrance requirements: For doctorate, GRE General Test, TOEFL. *Application deadline:* For fall admission, 1/5. *Application fee:* $40.
Expenses: Program pays tuition, fees, health insurance, and stipend for all students in good standing.
Financial aid: Fellowships, research assistantships, career-related internships or fieldwork, tuition waivers (full), and stipends available.
Faculty research: Molecular and cellular biology, molecular and cellular pharmacology, cell and organ physiology.
Kim Barrett, Chair, 858-543-3726.
Application contact: Gina Butcher, Graduate Program Representative, 858-534-3982.

Find an in-depth description at www.petersons.com/graduate.

■ UNIVERSITY OF CALIFORNIA, SAN FRANCISCO

Graduate Division, Biomedical Sciences Graduate Group, San Francisco, CA 94143

AWARDS Anatomy (PhD); endocrinology (PhD); experimental pathology (PhD); physiology (PhD).

Students: In 1999, 5 degrees awarded.
Degree requirements: For doctorate, dissertation required.
Entrance requirements: For doctorate, GRE General Test. *Application fee:* $40.
Financial aid: Fellowships, research assistantships, teaching assistantships available. Financial aid application deadline: 1/10.
Dr. Donald Ganem, Director, 415-476-2826.
Application contact: Pamela Humphrey, Program Administrator, 415-476-8467.

■ UNIVERSITY OF CALIFORNIA, SANTA CRUZ

Graduate Division, Division of Natural Sciences, Department of Biology, Santa Cruz, CA 95064

AWARDS Molecular, cellular, and developmental biology (PhD), including biology.

Faculty: 38 full-time.
Students: 90 full-time (47 women); includes 18 minority (1 African American, 13 Asian Americans or Pacific Islanders, 4 Hispanic Americans), 5 international. *196 applicants, 20% accepted.* In 1999, 11 doctorates awarded.
Degree requirements: For doctorate, one foreign language (computer language can

substitute), dissertation, oral and written qualifying exams required.
Entrance requirements: For doctorate, GRE General Test, GRE Subject Test. *Application deadline:* For fall admission, 1/1. *Application fee:* $40.
Expenses: Tuition, state resident: full-time $4,925. Tuition, nonresident: full-time $14,919.
Financial aid: Fellowships, research assistantships, teaching assistantships, career-related internships or fieldwork, Federal Work-Study, and institutionally sponsored loans available. Financial aid application deadline: 1/1.
Faculty research: Neurophysiology and psychophysiology; plant sciences; population, environmental, and evolutionary biology.
Barry Bowman, Chairperson, 831-459-2385.
Application contact: Graduate Admissions, 831-459-2301.

■ UNIVERSITY OF CENTRAL ARKANSAS

Graduate School, College of Natural Sciences and Math, Department of Biological Science, Conway, AR 72035-0001

AWARDS MS. Part-time programs available.

Faculty: 19 full-time (2 women), 1 part-time/adjunct (0 women).
Students: 10 full-time (3 women), 7 part-time (5 women); includes 4 minority (all African Americans). Average age 25. *10 applicants, 90% accepted.* In 1999, 5 degrees awarded.
Degree requirements: For master's, comprehensive exam required, thesis optional, foreign language not required. *Average time to degree:* Master's–2 years full-time, 4 years part-time.
Entrance requirements: For master's, GRE General Test, minimum GPA of 2.7. *Application deadline:* For fall admission, 3/1 (priority date); for spring admission, 10/1 (priority date). Applications are processed on a rolling basis. *Application fee:* $25 ($40 for international students).
Expenses: Tuition, state resident: part-time $144 per credit hour. Tuition, nonresident: part-time $297 per credit hour. Required fees: $17 per hour. $15 per term. Tuition and fees vary according to program.
Financial aid: In 1999–00, 10 students received aid, including 4 research assistantships with partial tuition reimbursements available (averaging $8,000 per year), 4 teaching assistantships with partial tuition reimbursements available (averaging $8,000 per year); unspecified assistantships also available. Financial aid application deadline: 2/15.

University of Central Arkansas (continued)
Dr. Paul Hamilton, Chairperson, 501-450-3146, *Fax:* 501-450-5914, *E-mail:* paulh@cc1.mail.edu.

Application contact: Nancy Gage, Co-Admissions Secretary, 501-450-3124, *Fax:* 501-450-5066, *E-mail:* nancyg@ecom.uca.edu.

■ **UNIVERSITY OF CENTRAL FLORIDA**

College of Arts and Sciences, Program in Biological Sciences, Orlando, FL 32816

AWARDS Biological sciences (MS); conservation biology (Certificate). Part-time and evening/weekend programs available.

Faculty: 15 full-time, 2 part-time/adjunct.
Students: 17 full-time (7 women), 36 part-time (22 women); includes 3 minority (2 Asian Americans or Pacific Islanders, 1 Hispanic American), 4 international. Average age 30. *38 applicants, 55% accepted.* In 1999, 7 degrees awarded.
Degree requirements: For master's, thesis or alternative, comprehensive exam, biology field exam required, foreign language not required.
Entrance requirements: For master's, GRE General Test, TOEFL, minimum GPA of 3.0 in last 60 hours. *Application deadline:* For fall admission, 3/1 (priority date); for spring admission, 10/15. *Application fee:* $20.
Expenses: Tuition, state resident: full-time $2,054; part-time $137 per credit. Tuition, nonresident: full-time $7,207; part-time $480 per credit. Required fees: $47 per term.
Financial aid: In 1999–00, 17 fellowships with partial tuition reimbursements (averaging $2,676 per year), 50 research assistantships with partial tuition reimbursements (averaging $2,316 per year), 48 teaching assistantships with partial tuition reimbursements (averaging $4,028 per year) were awarded; career-related internships or fieldwork, Federal Work-Study, institutionally sponsored loans, tuition waivers (partial), and unspecified assistantships also available. Financial aid application deadline: 3/1; financial aid applicants required to submit FAFSA.
Dr. D. H. Vickers, Chair, 407-823-2141, *Fax:* 407-823-5769, *E-mail:* dvickers@pegasus.cc.ucf.edu.

Application contact: Dr. David Kuhn, Coordinator, 407-823-2141, *Fax:* 407-823-5769, *E-mail:* dkuhn@pegasus.cc.ucf.edu.

■ **UNIVERSITY OF CENTRAL OKLAHOMA**

Graduate College, College of Mathematics and Science, Department of Biology, Edmond, OK 73034-5209

AWARDS MS. Part-time programs available.

Faculty: 17 full-time (4 women).
Students: 2 full-time (both women), 13 part-time (8 women); includes 2 minority (both Asian Americans or Pacific Islanders), 2 international. Average age 34. *5 applicants, 80% accepted.* In 1999, 2 degrees awarded.
Degree requirements: For master's, thesis required, foreign language not required.
Entrance requirements: For master's, GRE General Test, GRE Subject Test (biology). *Application deadline:* Applications are processed on a rolling basis. *Application fee:* $15.
Expenses: Tuition, state resident: part-time $66 per hour. Tuition, nonresident: part-time $84 per hour. Full-time tuition and fees vary according to course level and course load.
Financial aid: Federal Work-Study and unspecified assistantships available. Financial aid application deadline: 3/31; financial aid applicants required to submit FAFSA.
Faculty research: Environmental (*legionella*), aquatic biology (ecological), mammalogy field studies, microbiology, genetics.
Dr. Peggy Guthrie, Chairperson, 404-974-5017, *Fax:* 405-974-3824, *E-mail:* pguthrie@ucok.edu.

■ **UNIVERSITY OF CHICAGO**

Division of the Biological Sciences, Chicago, IL 60637-1513

AWARDS PhD, MD/PhD.

Faculty: 458 full-time (99 women), 18 part-time/adjunct (8 women).
Students: 325 full-time (124 women). Average age 25. *1009 applicants, 13% accepted.* In 1999, 45 doctorates awarded.
Degree requirements: For doctorate, dissertation required. *Average time to degree:* Doctorate–5.5 years full-time.
Entrance requirements: For doctorate, GRE General Test, TOEFL. *Application deadline:* For fall admission, 1/5 (priority date). Applications are processed on a rolling basis. *Application fee:* $55.
Expenses: Tuition: Full-time $24,804; part-time $3,422 per course. Required fees: $390. Tuition and fees vary according to program.
Financial aid: In 1999–00, 324 students received aid, including fellowships with full tuition reimbursements available (averaging $17,178 per year); research assistantships with full tuition reimbursements available, grants, institutionally sponsored loans, and traineeships also available.
Dr. Glenn D. Steele, Dean, 773-702-9000.

Application contact: Parag M. Shah, Administrator, Graduate Affairs, 773-702-5853, *Fax:* 773-834-1618, *E-mail:* parag@prufrock.bsd.uchicago.edu.

■ **UNIVERSITY OF CINCINNATI**

Division of Research and Advanced Studies, College of Medicine, Graduate Programs in Medicine, Cincinnati, OH 45221-0091

AWARDS MS, D Sc, PhD.

Students: In 1999, 36 master's, 29 doctorates awarded. Terminal master's awarded for partial completion of doctoral program.
Degree requirements: For master's, thesis required, foreign language not required; for doctorate, dissertation, qualifying exam required.
Entrance requirements: For master's and doctorate, GRE General Test. *Application deadline:* For fall admission, 2/1 (priority date). Applications are processed on a rolling basis. *Application fee:* $30.
Expenses: Tuition, state resident: full-time $5,139; part-time $196 per credit hour. Tuition, nonresident: full-time $10,326; part-time $369 per credit hour. Required fees: $561; $187 per quarter.
Financial aid: Career-related internships or fieldwork, Federal Work-Study, tuition waivers (full), and unspecified assistantships available. Financial aid application deadline: 5/1.

Application contact: Bridgette Harrison, Director, Graduate Affairs, 513-558-5625, *E-mail:* bridgette.harrison@uc.edu.

■ **UNIVERSITY OF CINCINNATI**

Division of Research and Advanced Studies, College of Medicine, Physician Scientist Training Program, Cincinnati, OH 45267

AWARDS MD/PhD.

Students: *41 applicants, 17% accepted. Application deadline:* For fall admission, 11/15. *Application fee:* $30.
Expenses: Tuition, state resident: full-time $5,880; part-time $196 per credit hour. Tuition, nonresident: full-time $11,067; part-time $369 per credit hour. Required fees: $741; $247 per quarter. Tuition and fees vary according to program.
Financial aid: Unspecified assistantships available. Financial aid application deadline: 5/1.

Dr. Leslie Myatt, Director, 513-558-6587, *Fax:* 513-558-2850, *E-mail:* leslie.myatt@uc.edu.

Find an in-depth description at www.petersons.com/graduate.

■ UNIVERSITY OF CINCINNATI

Division of Research and Advanced Studies, McMicken College of Arts and Sciences, Department of Biological Sciences, Cincinnati, OH 45221-0091

AWARDS MS, PhD.

Faculty: 30 full-time.
Students: 54 full-time (17 women), 12 part-time (6 women), 11 international. *75 applicants, 17% accepted.* In 1999, 31 master's, 4 doctorates awarded.
Degree requirements: For master's, thesis or alternative required, foreign language not required; for doctorate, dissertation required. *Average time to degree:* Master's–3.1 years full-time; doctorate–6.2 years full-time.
Entrance requirements: For master's and doctorate, GRE General Test, GRE Subject Test, BS in biology, chemistry, or equivalent. *Application deadline:* For fall admission, 2/5. *Application fee:* $30. Electronic applications accepted.
Expenses: Tuition, state resident: full-time $5,880; part-time $196 per credit hour. Tuition, nonresident: full-time $11,067; part-time $369 per credit hour. Required fees: $741; $247 per quarter. Tuition and fees vary according to program.
Financial aid: Fellowships, tuition waivers (full) and unspecified assistantships available. Aid available to part-time students. Financial aid application deadline: 2/5. *Total annual research expenditures:* $1.2 million.
Guy Cameron, Head, 513-556-9700, *Fax:* 513-556-5299, *E-mail:* guy.cameron@uc.edu.
Application contact: Carol Gundrum, Graduate Program Secretary, 513-556-2497, *Fax:* 513-556-5299, *E-mail:* biosecr2@email.uc.edu.

Find an in-depth description at www.petersons.com/graduate.

■ UNIVERSITY OF COLORADO AT DENVER

Graduate School, College of Liberal Arts and Sciences, Program in Biology, Denver, CO 80217-3364

AWARDS MA. Part-time programs available.

Faculty: 12 full-time (6 women).
Students: 6 full-time (5 women), 16 part-time (10 women); includes 3 minority (1 Asian American or Pacific Islander, 2 Hispanic Americans), 3 international. Average age 26. *15 applicants, 67% accepted.* In 1999, 2 degrees awarded.

Degree requirements: For master's, thesis or alternative required.
Entrance requirements: For master's, GRE General Test. *Application deadline:* For fall admission, 4/15; for spring admission, 10/15. Applications are processed on a rolling basis. *Application fee:* $50 ($60 for international students). Electronic applications accepted.
Expenses: Tuition, state resident: part-time $185 per credit hour. Tuition, nonresident: part-time $735 per credit hour. Required fees: $3 per credit hour. $130 per year. One-time fee: $25 part-time. Tuition and fees vary according to program.
Financial aid: Research assistantships, teaching assistantships, Federal Work-Study available. Financial aid application deadline: 3/1; financial aid applicants required to submit FAFSA. *Total annual research expenditures:* $210,085.
Leo Bruederie, Chair, 303-556-2658, *Fax:* 303-556-4352.
Application contact: Peggy Burress, Administrative Assistant, 303-556-8440, *Fax:* 303-556-4352.

■ UNIVERSITY OF COLORADO HEALTH SCIENCES CENTER

Graduate School, Programs in Biological and Medical Sciences, Denver, CO 80262

AWARDS MS, PhD, MD/PhD. Terminal master's awarded for partial completion of doctoral program.

Degree requirements: For master's, foreign language not required; for doctorate, dissertation required, foreign language not required.
Entrance requirements: For master's and doctorate, GRE.
Expenses: Tuition, state resident: full-time $1,512; part-time $56 per hour. Tuition, nonresident: full-time $7,209; part-time $267 per hour. Full-time tuition and fees vary according to course load and program.

■ UNIVERSITY OF CONNECTICUT

Graduate School, College of Liberal Arts and Sciences, Biological Sciences Group, Storrs, CT 06269

AWARDS Ecology and evolutionary biology (MS, PhD), including botany, ecology, entomology, systematics, zoology; molecular and cell biology (MS, PhD), including biochemistry, biophysics, biotechnology (MS), cell and developmental biology, genetics, microbiology, plant molecular and cell biology; physiology and neurobiology (MS, PhD), including neurobiology, physiology.

Degree requirements: For doctorate, dissertation required.

Entrance requirements: For master's and doctorate, GRE General Test, GRE Subject Test, TOEFL.
Expenses: Tuition, state resident: full-time $5,118. Tuition, nonresident: full-time $13,298. Required fees: $1,022.

■ UNIVERSITY OF CONNECTICUT

Graduate School, Field of Biomedical Science, Storrs, CT 06269

AWARDS PhD.

Degree requirements: For doctorate, dissertation required.
Entrance requirements: For doctorate, GRE General Test, GRE Subject Test, TOEFL.
Expenses: Tuition, state resident: full-time $5,118. Tuition, nonresident: full-time $13,298. Required fees: $1,022.

■ UNIVERSITY OF CONNECTICUT HEALTH CENTER

Graduate School and School of Medicine, Combined Degree Program in Biomedical Sciences, Farmington, CT 06030

AWARDS MD/PhD.

Students: 24 full-time (10 women); includes 4 minority (all Asian Americans or Pacific Islanders), 1 international. *Application deadline:* For fall admission, 2/1 (priority date); for spring admission, 10/1. Applications are processed on a rolling basis.
Financial aid: In 1999–00, research assistantships (averaging $17,000 per year) Dr. Dominic Cinti, Director, 860-679-4571.
Application contact: Marizta Barta, Information Contact, 860-679-4306, *Fax:* 860-679-1282, *E-mail:* barta@adp.uchc.edu.

■ UNIVERSITY OF CONNECTICUT HEALTH CENTER

Graduate School, Programs in Biomedical Sciences, Farmington, CT 06030

AWARDS PhD, DMD/PhD, MD/PhD, PhD/Certificate. Part-time and evening/weekend programs available.

Faculty: 100.
Students: 130 full-time (65 women); includes 1 African American, 7 Asian Americans or Pacific Islanders, 1 Hispanic American, 46 international. Average age 27. In 1999, 10 doctorates awarded.
Degree requirements: For doctorate, one foreign language (computer language can substitute), dissertation required. *Average time to degree:* Doctorate–5 years full-time.
Entrance requirements: For doctorate, GRE General Test, TOEFL. *Application*

University of Connecticut Health Center (continued)

deadline: For fall admission, 2/1 (priority date); for spring admission, 10/1. Applications are processed on a rolling basis. *Application fee:* $40 ($45 for international students).

Expenses: Tuition, state resident: full-time $5,272; part-time $293 per credit. Tuition, nonresident: full-time $13,696; part-time $761 per credit. Required fees: $320; $198 per semester. One-time fee: $50 full-time. Full-time tuition and fees vary according to course load, program and reciprocity agreements.

Financial aid: In 1999–00, research assistantships (averaging $17,000 per year); fellowships, teaching assistantships, Federal Work-Study also available.

Application contact: Marizta Barta, Information Contact, 860-679-4306, *Fax:* 860-679-1282, *E-mail:* barta@ adp.uchc.edu.

Find an in-depth description at www.petersons.com/graduate.

■ **UNIVERSITY OF DAYTON**

Graduate School, College of Arts and Sciences, Department of Biology, Dayton, OH 45469-1300

AWARDS MS, PhD.

Faculty: 13 full-time (4 women).
Students: 15 full-time (8 women); includes 1 minority (Asian American or Pacific Islander). Average age 24. *22 applicants, 18% accepted.* In 1999, 1 doctorate awarded (100% entered university research/ teaching). Terminal master's awarded for partial completion of doctoral program.
Degree requirements: For master's and doctorate, thesis/dissertation required, foreign language not required. *Average time to degree:* Master's–2 years full-time; doctorate–5 years full-time.
Entrance requirements: For master's and doctorate, GRE General Test, minimum undergraduate GPA of 3.0. *Application deadline:* For fall admission, 3/15 (priority date). Applications are processed on a rolling basis. *Application fee:* $30.
Expenses: Tuition: Part-time $438 per semester hour. Required fees: $25 per term. Tuition and fees vary according to program.
Financial aid: In 1999–00, 14 students received aid, including 14 teaching assistantships with tuition reimbursements available (averaging $10,000 per year); research assistantships, institutionally sponsored loans also available. Financial aid application deadline: 3/15.
Faculty research: Plant and animal physiology; cell, molecular, and developmental biology; animal behavior and ecology; community ecology and

environmental biology; genetics and microbiology. *Total annual research expenditures:* $500,000.
Dr. John J. Rowe, Chairperson, 937-229-2521, *Fax:* 937-229-2021.
Application contact: Dr. Robert Kearns, Graduate Director, 937-229-2521, *Fax:* 937-229-2021, *E-mail:* kearns@ neelix.udayton.edu.

Find an in-depth description at www.petersons.com/graduate.

■ **UNIVERSITY OF DELAWARE**

College of Arts and Science, Department of Biological Sciences, Newark, DE 19716

AWARDS Biotechnology (MS, PhD); cell and extracellular matrix biology (MS, PhD); cell and systems physiology (MS, PhD); ecology and evolution (MS, PhD); microbiology (MS, PhD); molecular biology and genetics (MS, PhD); plant biology (MS, PhD).

Faculty: 37 full-time (10 women).
Students: 22 full-time (11 women), 1 part-time; includes 2 minority (both African Americans), 9 international. Average age 25. *37 applicants, 27% accepted.* In 2000, 9 doctorates awarded.
Degree requirements: For master's and doctorate, thesis/dissertation required, foreign language not required. *Average time to degree:* Master's–2.5 years full-time; doctorate–6 years full-time.
Entrance requirements: For master's and doctorate, GRE General Test, GRE Subject Test (advanced biology). *Application deadline:* For fall admission, 6/15. Applications are processed on a rolling basis. *Application fee:* $50. Electronic applications accepted.
Expenses: Tuition, state resident: full-time $4,380; part-time $243 per credit. Tuition, nonresident: full-time $12,750; part-time $708 per credit. Required fees: $15 per term. Tuition and fees vary according to program.
Financial aid: In 2000–01, 18 students received aid, including 2 fellowships with full tuition reimbursements available (averaging $18,000 per year), 4 research assistantships with full tuition reimbursements available (averaging $18,000 per year), 11 teaching assistantships with full tuition reimbursements available (averaging $18,000 per year); tuition waivers (partial) also available. Financial aid application deadline: 6/15.
Faculty research: Cell interactions, molecular mechanisms, microorganisms, embryo implantation. *Total annual research expenditures:* $1.8 million.
Dr. Daniel D. Carson, Chair, 302-831-6977, *Fax:* 302-831-2281, *E-mail:* dcarson@udel.edu.

Application contact: Norman Karin, Graduate Coordinator, 302-831-1841, *Fax:* 302-831-2281, *E-mail:* ccoletta@udel.edu.

Find an in-depth description at www.petersons.com/graduate.

■ **UNIVERSITY OF DENVER**

Graduate Studies, Faculty of Natural Sciences, Mathematics and Engineering, Department of Biological Sciences, Denver, CO 80208

AWARDS MS, PhD. Part-time programs available.

Faculty: 13.
Students: 18 (10 women); includes 3 minority (all Hispanic Americans) 2 international. *31 applicants, 61% accepted.* In 1999, 4 master's, 2 doctorates awarded. Terminal master's awarded for partial completion of doctoral program.
Degree requirements: For master's, thesis required, foreign language not required; for doctorate, one foreign language (computer language can substitute), dissertation required.
Entrance requirements: For master's and doctorate, GRE General Test, GRE Subject Test, TOEFL. *Application deadline:* For fall admission, 3/1. Applications are processed on a rolling basis. *Application fee:* $40 ($45 for international students).
Expenses: Tuition: Full-time $18,936; part-time $526 per credit hour. Required fees: $159; $4 per credit hour. Part-time tuition and fees vary according to course load and program.
Financial aid: In 1999–00, 1 fellowship, 13 teaching assistantships with full and partial tuition reimbursements (averaging $12,006 per year) were awarded; research assistantships with full and partial tuition reimbursements, Federal Work-Study and institutionally sponsored loans also available. Aid available to part-time students. Financial aid application deadline: 3/1; financial aid applicants required to submit FAFSA.
Faculty research: Molecular biology, cell biology, neurobiology, ecology, molecular evolution. *Total annual research expenditures:* $634,961.
Dr. Robert Dores, Chairperson, 303-871-3661.
Application contact: Dr. James Fogleman, Graduate Adviser, 303-871-3661.

■ **UNIVERSITY OF DETROIT MERCY**

College of Engineering and Science, Department of Biology, Detroit, MI 48219-0900

AWARDS MS.

Degree requirements: For master's, thesis or alternative required, foreign language not required.
Faculty research: Genetics, histology, ecology.

■ UNIVERSITY OF FLORIDA

College of Medicine and Graduate School, Interdisciplinary Program in Biomedical Sciences, Gainesville, FL 32611

AWARDS MS, PhD, MD/PhD. Terminal master's awarded for partial completion of doctoral program.

Degree requirements: For master's and doctorate, thesis/dissertation required, foreign language not required.
Entrance requirements: For master's and doctorate, GRE General Test, TOEFL, minimum GPA of 3.0. Electronic applications accepted.
Expenses: Tuition, state resident: part-time $144 per credit hour. Tuition, nonresident: part-time $505 per credit hour. Tuition and fees vary according to course level, course load and program.

Find an in-depth description at www.petersons.com/graduate.

■ UNIVERSITY OF GUAM

Graduate School and Research, College of Arts and Sciences, Program in Biology, Mangilao, GU 96923

AWARDS Tropical marine biology (MS).

Degree requirements: For master's, thesis, oral comprehensive exam required, foreign language not required.
Entrance requirements: For master's, GRE General Test, GRE Subject Test, TOEFL.
Expenses: Tuition, state resident: part-time $99 per credit hour. Tuition, nonresident: part-time $246 per credit hour. Required fees: $170 per semester. Tuition and fees vary according to course load.
Faculty research: Maintenance and ecology of coral reefs.

■ UNIVERSITY OF HARTFORD

College of Arts and Sciences, Program in Biology, West Hartford, CT 06117-1599

AWARDS MS. Part-time and evening/weekend programs available.

Faculty: 5 full-time (1 woman), 1 (woman) part-time/adjunct.
Students: 5 full-time (3 women), 2 part-time, 1 international. Average age 25. *14 applicants, 57% accepted.* In 1999, 2 degrees awarded.

Degree requirements: For master's, comprehensive and oral exams required, thesis optional, foreign language not required.
Entrance requirements: For master's, GRE or MCAT, TOEFL. *Application deadline:* Applications are processed on a rolling basis. *Application fee:* $40 ($55 for international students). Electronic applications accepted.
Expenses: Tuition: Full-time $6,570; part-time $365 per hour. Required fees: $50 per term. Full-time tuition and fees vary according to degree level, program and student level.
Financial aid: Research assistantships, teaching assistantships, Federal Work-Study and tuition waivers (partial) available. Aid available to part-time students. Financial aid application deadline: 6/1; financial aid applicants required to submit FAFSA.
Faculty research: Neurobiology of aging, central actions of neural steroids, neuroendocrine control of reproduction, retinopatheis in sharks, plasticity in the central nervous system. *Total annual research expenditures:* $15,900.
Dr. William Coleman, Chairman, 860-768-4533, *Fax:* 860-768-5002, *E-mail:* wcoleman@mail.hartford.edu.
Application contact: Nancy Clubb-Lazzerini, Coordinator of Graduate Applications, 860-768-4373, *Fax:* 860-768-5160, *E-mail:* gettoknow@mail.hartford.edu.

■ UNIVERSITY OF HAWAII AT MANOA

John A. Burns School of Medicine and Graduate Division, Graduate Programs in Biomedical Sciences, Honolulu, HI 96822

AWARDS MS, PhD. Part-time programs available.

Faculty: 52 full-time (0 women), 1 (woman) part-time/adjunct.
Students: 25 full-time (10 women), 11 part-time (10 women). Average age 31. *54 applicants, 54% accepted.* In 1999, 14 master's, 2 doctorates awarded.
Degree requirements: For doctorate, dissertation required.
Application fee: $25 ($50 for international students).
Expenses: Tuition, state resident: part-time $168 per credit. Tuition, nonresident: part-time $415 per credit. Required fees: $51 per semester. Part-time tuition and fees vary according to course load.
Financial aid: In 1999–00, 13 research assistantships (averaging $15,868 per year), 1 teaching assistantship (averaging $12,786 per year) were awarded; fellowships, career-related internships or fieldwork,

Federal Work-Study, institutionally sponsored loans, and tuition waivers (full and partial) also available. Aid available to part-time students.
Dr. Edwin L. Cadmen, Dean, John A. Burns School of Medicine, 808-956-8287, *Fax:* 808-956-5506.

Find an in-depth description at www.petersons.com/graduate.

■ UNIVERSITY OF HOUSTON

College of Natural Sciences and Mathematics, Department of Biology and Biochemistry, Houston, TX 77004

AWARDS Biochemistry (MS, PhD); biology (MS, PhD).

Faculty: 28 full-time (4 women), 6 part-time/adjunct (1 woman).
Students: 86 full-time (38 women), 13 part-time (10 women); includes 14 minority (1 African American, 10 Asian Americans or Pacific Islanders, 3 Hispanic Americans), 38 international. Average age 29. *188 applicants, 12% accepted.* In 1999, 10 master's, 5 doctorates awarded. Terminal master's awarded for partial completion of doctoral program.
Degree requirements: For master's, thesis required (for some programs), foreign language not required; for doctorate, dissertation, oral and written comprehensive exam required, foreign language not required.
Entrance requirements: For master's and doctorate, GRE General Test, TOEFL, TSE. *Application deadline:* For fall admission, 4/1 (priority date); for spring admission, 11/1. Applications are processed on a rolling basis. *Application fee:* $0 ($75 for international students).
Expenses: Tuition, state resident: full-time $1,296; part-time $72 per credit. Tuition, nonresident: full-time $4,932; part-time $274 per credit. Required fees: $1,162. Tuition and fees vary according to program.
Financial aid: In 1999–00, 83 students received aid, including 2 fellowships, 40 research assistantships, 43 teaching assistantships; Federal Work-Study and institutionally sponsored loans also available. Financial aid application deadline: 4/1.
Faculty research: Evolutionary biology, neuroscience, infectious diseases, circadian rhythm, ion channels. *Total annual research expenditures:* $5.5 million.
Dr. Arnold Eskin, Chairman, 713-743-8386.
Application contact: Marcie Newton, Graduate Adviser and Office Coordinator, 713-743-2633, *Fax:* 713-743-2899, *E-mail:* mnewton@dna.bchs.uh.edu.

Find an in-depth description at www.petersons.com/graduate.

UNIVERSITY OF HOUSTON–CLEAR LAKE

School of Natural and Applied Sciences, Program in Biological Sciences, Houston, TX 77058-1098

AWARDS MS.

Faculty: 6 full-time (3 women), 3 part-time/adjunct (1 woman).
Students: 12 full-time (9 women), 21 part-time (15 women); includes 10 minority (4 African Americans, 2 Asian Americans or Pacific Islanders, 4 Hispanic Americans), 4 international. Average age 32.
Degree requirements: For master's, foreign language not required.
Entrance requirements: For master's, GRE General Test. *Application deadline:* Applications are processed on a rolling basis. *Application fee:* $30 ($70 for international students).
Expenses: Tuition, state resident: full-time $1,368. Tuition, nonresident: full-time $4,572. Tuition and fees vary according to course load.
Financial aid: Research assistantships, teaching assistantships, career-related internships or fieldwork and Federal Work-Study available. Aid available to part-time students. Financial aid application deadline: 5/1.
Dr. Cynthia Howard, Chair, 281-283-3770, *Fax:* 281-283-3707.
Application contact: Dr. Robert Ferebee, Associate Dean, 281-283-3700, *Fax:* 281-283-3707, *E-mail:* ferebee@uhcl4.cl.uh.edu.

UNIVERSITY OF IDAHO

College of Graduate Studies, College of Letters and Science, Department of Biological Sciences, Moscow, ID 83844-4140

AWARDS Biological sciences (M Nat Sci); botany (MS, PhD); zoology (MS, PhD).

Faculty: 11 full-time (3 women), 3 part-time/adjunct (2 women).
Students: 21 full-time (13 women), 10 part-time (4 women); includes 3 minority (all Asian Americans or Pacific Islanders), 7 international. *24 applicants, 25% accepted.* In 1999, 5 master's, 1 doctorate awarded.
Degree requirements: For master's, foreign language not required; for doctorate, dissertation required.
Entrance requirements: For master's, GRE, minimum GPA of 2.8; for doctorate, GRE, minimum undergraduate GPA of 2.8, 3.0 graduate. *Application deadline:* For fall admission, 8/1; for spring admission, 12/15. *Application fee:* $35 ($45 for international students).
Expenses: Tuition, nonresident: full-time $6,000; part-time $239 per credit hour.

Required fees: $2,888; $144 per credit hour. Tuition and fees vary according to program.
Financial aid: In 1999–00, 4 research assistantships (averaging $12,113 per year), 13 teaching assistantships (averaging $11,670 per year) were awarded. Financial aid application deadline: 2/15.
Dr. Rolf L. Ingermann, Interim Chair, 208-885-7764.

UNIVERSITY OF ILLINOIS AT CHICAGO

College of Medicine and Graduate College, Graduate Programs in Medicine, Chicago, IL 60607-7128

AWARDS Anatomy and cell biology (MS, PhD); biochemistry and molecular biology (MS, PhD); genetics (PhD), including molecular genetics; health professions education (MHPE); microbiology and immunology (PhD); pathology (MS, PhD); pharmacology (PhD), including pharmacology; physiology and biophysics (MS, PhD); surgery (MS). Part-time programs available.

Students: 169 full-time (98 women), 40 part-time (22 women); includes 29 minority (4 African Americans, 21 Asian Americans or Pacific Islanders, 3 Hispanic Americans, 1 Native American), 100 international. Average age 29. *648 applicants, 15% accepted.* In 1999, 22 master's, 22 doctorates awarded. Terminal master's awarded for partial completion of doctoral program.
Degree requirements: For master's and doctorate, thesis/dissertation required, foreign language not required.
Entrance requirements: For master's and doctorate, GRE General Test. *Application deadline:* For fall admission, 6/1; for spring admission, 11/1. *Application fee:* $40 ($50 for international students).
Expenses: Tuition, state resident: full-time $3,750. Tuition, nonresident: full-time $10,588. Tuition and fees vary according to course load.
Financial aid: In 1999–00, 119 students received aid; fellowships, research assistantships, teaching assistantships, career-related internships or fieldwork, Federal Work-Study, institutionally sponsored loans, scholarships, traineeships, and tuition waivers (full) available. Financial aid application deadline: 3/1; financial aid applicants required to submit FAFSA.
Gerald S. Moss, Dean, College of Medicine, 312-996-3500.

Find an in-depth description at www.petersons.com/graduate.

UNIVERSITY OF ILLINOIS AT CHICAGO

Graduate College, College of Liberal Arts and Sciences, Department of Biological Sciences, Chicago, IL 60607-7128

AWARDS Cell and developmental biology (PhD); ecology and evolution (MS, DA, PhD); genetics and development (PhD); molecular biology (MS, PhD); neurobiology (MS, PhD); plant biology (MS, DA, PhD).

Faculty: 40 full-time (5 women).
Students: 100 full-time (47 women), 14 part-time (10 women); includes 13 minority (11 Asian Americans or Pacific Islanders, 2 Hispanic Americans), 42 international. Average age 29. *99 applicants, 36% accepted.* In 1999, 3 master's, 9 doctorates awarded.
Degree requirements: For master's, thesis required, foreign language not required; for doctorate, dissertation, preliminary exam required, foreign language not required.
Entrance requirements: For master's and doctorate, GRE General Test, GRE Subject Test, TOEFL, previous course work in physics, calculus, and organic chemistry; minimum GPA of 3.75 on a 5.0 scale. *Application deadline:* For fall admission, 6/1. Applications are processed on a rolling basis. *Application fee:* $40 ($50 for international students). Electronic applications accepted.
Expenses: Tuition, state resident: full-time $3,750; part-time $1,250 per semester. Tuition, nonresident: full-time $10,588; part-time $3,530 per semester. Required fees: $507 per semester. Tuition and fees vary according to course load and program.
Financial aid: In 1999–00, 87 students received aid; fellowships, research assistantships, teaching assistantships, career-related internships or fieldwork, Federal Work-Study, traineeships, and tuition waivers (full) available. Financial aid application deadline: 3/1; financial aid applicants required to submit FAFSA.
Dr. Lon Kaufman, Head, 312-996-2213.
Application contact: Dr. Leo Miller, Director of Graduate Studies, 312-996-2220.

Find an in-depth description at www.petersons.com/graduate.

UNIVERSITY OF ILLINOIS AT SPRINGFIELD

Graduate Programs, College of Liberal Arts and Sciences, Program in Biology, Springfield, IL 62794-9243

AWARDS MA. Part-time and evening/weekend programs available.

Faculty: 7 full-time (2 women).

Students: 14 full-time (4 women), 26 part-time (15 women); includes 3 minority (2 African Americans, 1 Asian American or Pacific Islander), 1 international. Average age 30. *29 applicants, 72% accepted.* In 1999, 4 degrees awarded.
Degree requirements: For master's, thesis or alternative required, foreign language not required.
Entrance requirements: For master's, GRE General Test, GRE Subject Test, BA in biology, minimum undergraduate GPA of 3.0. *Application deadline:* Applications are processed on a rolling basis. *Application fee:* $0.
Expenses: Tuition, state resident: part-time $105 per credit hour. Tuition, nonresident: part-time $314 per credit hour.
Financial aid: In 1999–00, 20 students received aid, including 4 research assistantships with full and partial tuition reimbursements available (averaging $6,300 per year); career-related internships or fieldwork, Federal Work-Study, grants, tuition waivers (partial), and unspecified assistantships also available. Aid available to part-time students. Financial aid application deadline: 6/1; financial aid applicants required to submit FAFSA. Anne Larson, Convener, 217-206-7337.

■ UNIVERSITY OF ILLINOIS AT URBANA–CHAMPAIGN

Graduate College, College of Liberal Arts and Sciences, School of Life Sciences, Urbana, IL 61801
AWARDS MS, PhD, MD/PhD.

Faculty: 102 full-time (16 women), 36 part-time/adjunct (5 women).
Students: 309 full-time (123 women); includes 38 minority (7 African Americans, 21 Asian Americans or Pacific Islanders, 9 Hispanic Americans, 1 Native American), 88 international. *361 applicants, 12% accepted.* In 1999, 37 master's, 20 doctorates awarded.
Degree requirements: For doctorate, dissertation required.
Entrance requirements: For master's, minimum GPA of 4.0 on a 5.0 scale. *Application deadline:* Applications are processed on a rolling basis. *Application fee:* $40 ($50 for international students).
Expenses: Tuition, state resident: full-time $4,616. Tuition, nonresident: full-time $11,768. Full-time tuition and fees vary according to course load.
Financial aid: Fellowships, research assistantships, teaching assistantships, career-related internships or fieldwork, Federal Work-Study, institutionally sponsored loans, and tuition waivers (full and partial) available. Financial aid application deadline: 2/15.

Ed Brown, Associate Director for Academic Affairs, 217-333-4944, *Fax:* 217-244-1224, *E-mail:* e-brown1@uiuc.edu.
Application contact: Carol Hall, Graduate Advising and Records, 217-333-8208, *Fax:* 217-244-1224, *E-mail:* e-hall@uiuc.edu.

■ UNIVERSITY OF INDIANAPOLIS

Graduate School, College of Arts and Sciences, Department of Biology, Indianapolis, IN 46227-3697
AWARDS MS. Part-time and evening/weekend programs available.

Degree requirements: For master's, foreign language and thesis not required.
Entrance requirements: For master's, GRE Subject Test.

■ THE UNIVERSITY OF IOWA

College of Medicine and Graduate College, Biosciences Program, Iowa City, IA 52242-1316
AWARDS PhD.

Degree requirements: For doctorate, dissertation required.
Entrance requirements: For doctorate, GRE General Test, TOEFL, minimum GPA of 3.0. *Application deadline:* For fall admission, 2/1 (priority date). Applications are processed on a rolling basis. Electronic applications accepted.
Dr. Andrew F. Russo, Professor, 319-335-7872, *Fax:* 319-335-7330, *E-mail:* andrew-russo@uiowa.edu.
Application contact: Jodi M. Graff, Program Associate, 319-335-8305, *Fax:* 319-335-7656, *E-mail:* jodi-hamel@uiowa.edu.
Find an in-depth description at www.petersons.com/graduate.

■ THE UNIVERSITY OF IOWA

College of Medicine and Graduate College, Graduate Programs in Medicine, Iowa City, IA 52242-1316
AWARDS MA, MHA, MPAS, MPH, MPT, MS, PhD, JD/MHA, MBA/MHA, MD/PhD, MHA/MA, MHA/MS, MPH/MHA, MS/MA, MS/MS. Part-time programs available.

Students: In 1999, 115 master's, 57 doctorates awarded.
Degree requirements: For doctorate, dissertation required.
Application fee: $30 ($50 for international students). Electronic applications accepted.
Expenses: Tuition, state resident: full-time $3,308. Tuition, nonresident: full-time $10,662. Tuition and fees vary according to course load and program.
Financial aid: Fellowships, research assistantships, teaching assistantships, career-related internships or fieldwork,

Federal Work-Study, institutionally sponsored loans, and tuition waivers (full and partial) available. Aid available to part-time students. Financial aid applicants required to submit FAFSA.

■ THE UNIVERSITY OF IOWA

College of Medicine and Graduate College, Medical Scientist Training Program, Iowa City, IA 52242-1316
AWARDS MD/PhD.

Faculty: 40 full-time (10 women), 76 part-time/adjunct (26 women).
Students: 42 full-time (14 women); includes 13 minority (2 African Americans, 9 Asian Americans or Pacific Islanders, 1 Hispanic American, 1 Native American). Average age 25. *96 applicants, 26% accepted.* *Application deadline:* For fall admission, 11/1 (priority date). Applications are processed on a rolling basis. *Application fee:* $50. Electronic applications accepted.
Expenses: Tuition, state resident: full-time $9,840. Tuition, nonresident: full-time $26,356. Tuition and fees vary according to course load and program.
Financial aid: In 1999–00, 42 students received aid, including 14 fellowships (averaging $16,277 per year), 28 research assistantships with full tuition reimbursements available (averaging $16,277 per year); traineeships also available.
Dr. Pamela Geyer, Director, 319-335-6844, *Fax:* 319-335-6887, *E-mail:* gary-koretzky@uiowa.edu.

Find an in-depth description at www.petersons.com/graduate.

■ THE UNIVERSITY OF IOWA

Graduate College, College of Liberal Arts, Department of Biological Sciences, Iowa City, IA 52242-1316
AWARDS MS, PhD.

Faculty: 31 full-time, 3 part-time/adjunct.
Students: 33 full-time (16 women), 21 part-time (11 women); includes 2 minority (1 African American, 1 Asian American or Pacific Islander), 27 international. *151 applicants, 17% accepted.* In 1999, 6 master's, 2 doctorates awarded.
Degree requirements: For master's, exam required, thesis optional; for doctorate, dissertation, comprehensive exam required.
Entrance requirements: For master's and doctorate, GRE General Test, minimum GPA of 3.0. *Application deadline:* For fall admission, 2/1 (priority date); for spring admission, 12/1 (priority date). Applications are processed on a rolling basis.
Application fee: $30 ($50 for international students). Electronic applications accepted.
Expenses: Tuition, state resident: full-time $3,308; part-time $184 per semester hour. Tuition, nonresident: full-time $10,662;

The University of Iowa (continued)
part-time $184 per semester hour.
Required fees: $93 per semester. Tuition and fees vary according to course load and program.
Financial aid: In 1999–00, 2 fellowships, 26 research assistantships, 25 teaching assistantships were awarded. Financial aid applicants required to submit FAFSA. George D. Cain, Interim Chair, 319-335-1058, *Fax:* 319-335-1069.
Application contact: 319-335-1058, *Fax:* 319-335-1069, *E-mail:* biology-admissions@uiowa.edu.

Find an in-depth description at www.petersons.com/graduate.

■ UNIVERSITY OF KANSAS

Graduate School, College of Liberal Arts and Sciences, Division of Biological Sciences, Lawrence, KS 66045

AWARDS MA, MS, PhD. Part-time programs available.

Faculty: 60.
Students: 32 full-time (17 women), 38 part-time (13 women); includes 3 minority (1 African American, 2 Asian Americans or Pacific Islanders), 13 international. *122 applicants, 17% accepted.* In 1999, 11 master's, 10 doctorates awarded.
Degree requirements: For master's, foreign language not required; for doctorate, dissertation required.
Entrance requirements: For master's and doctorate, GRE General Test, GRE Subject Test, TOEFL. *Application fee:* $25.
Expenses: Tuition, state resident: full-time $2,482; part-time $103 per credit hour. Tuition, nonresident: full-time $8,104; part-time $338 per credit hour. Required fees: $428; $31 per credit hour. Tuition and fees vary according to program.
Financial aid: In 1999–00, research assistantships (averaging $11,588 per year), teaching assistantships (averaging $11,588 per year) were awarded; fellowships, career-related internships or fieldwork, Federal Work-Study, and institutionally sponsored loans also available. Aid available to part-time students. Financial aid application deadline: 3/1; financial aid applicants required to submit FAFSA. *Total annual research expenditures:* $4.1 million. James Orr, Chair, 785-864-4301, *Fax:* 785-864-5321, *E-mail:* jorr@alive.bio.ukans.edu.

■ UNIVERSITY OF KANSAS

Graduate Studies Medical Center, Graduate Programs in Biomedical and Basic Sciences, Lawrence, KS 66045

AWARDS MA, MPH, MS, PhD, MD/MPH, MD/MS, MD/PhD. Part-time and evening/weekend programs available.

Faculty: 91 full-time (15 women), 4 part-time/adjunct (0 women).
Students: 36 full-time (21 women), 95 part-time (59 women); includes 11 minority (4 African Americans, 6 Asian Americans or Pacific Islanders, 1 Native American), 27 international. Average age 32. *161 applicants, 15% accepted.* In 1999, 21 master's, 8 doctorates awarded. Terminal master's awarded for partial completion of doctoral program.
Degree requirements: For master's, thesis required; for doctorate, dissertation, comprehensive oral exam required.
Entrance requirements: For master's and doctorate, GRE, TOEFL, TSE. *Application deadline:* Applications are processed on a rolling basis. *Application fee:* $0. Electronic applications accepted.
Expenses: Tuition, state resident: full-time $2,482; part-time $103 per credit hour. Tuition, nonresident: full-time $8,104; part-time $338 per credit hour. Required fees: $428; $31 per credit hour. Tuition and fees vary according to program.
Financial aid: In 1999–00, 13 students received aid; fellowships, research assistantships, teaching assistantships, Federal Work-Study, institutionally sponsored loans, traineeships, and unspecified assistantships available. Aid available to part-time students. Financial aid application deadline: 3/31; financial aid applicants required to submit FAFSA.
Faculty research: Cardiovascular biology, neurosciences, signal transduction and cancer biology, molecular biology and genetics, and developmental biology. *Total annual research expenditures:* $19.3 million. Dr. Michael P. Sarras, Director, 913-588-2039, *Fax:* 913-588-2711, *E-mail:* igpbs@kumc.edu.

Application contact: David Brown, Coordinator, 913-588-2719, *Fax:* 913-588-2711, *E-mail:* dbrown2@kumc.edu.

■ UNIVERSITY OF KENTUCKY

Graduate School and College of Medicine, Graduate Programs in Medicine, Lexington, KY 40506-0032

AWARDS MS, MSHP, MSPH, MSRMP, PhD, MD/PhD. MSHP and MSRMP offered in cooperation with the Program in Radiation Sciences. Terminal master's awarded for partial completion of doctoral program.

Degree requirements: For master's, comprehensive exam required; for doctorate, dissertation, comprehensive exam required.
Entrance requirements: For master's, GRE General Test, minimum undergraduate GPA of 2.5; for doctorate, GRE General Test, minimum graduate GPA of 3.0.
Expenses: Tuition, state resident: full-time $3,596; part-time $188 per credit hour.

Tuition, nonresident: full-time $10,116; part-time $550 per credit hour.

■ UNIVERSITY OF KENTUCKY

Graduate School, Graduate School Programs from the College of Arts and Sciences, Program in Biological Sciences, Lexington, KY 40506-0032

AWARDS MS, PhD.

Degree requirements: For master's, comprehensive exam required, thesis optional, foreign language not required; for doctorate, dissertation, comprehensive exam required, foreign language not required.
Entrance requirements: For master's, GRE General Test, minimum undergraduate GPA of 2.5; for doctorate, GRE General Test, minimum graduate GPA of 3.0.
Expenses: Tuition, state resident: full-time $3,596; part-time $188 per credit hour. Tuition, nonresident: full-time $10,116; part-time $550 per credit hour.
Faculty research: General biology, microbiology, *Drosophila* molecular genetics, molecular virology, multiple loci inheritance.

■ UNIVERSITY OF LOUISIANA AT LAFAYETTE

Graduate School, College of Sciences, Department of Biology, Lafayette, LA 70504

AWARDS Biology (MS); environmental and evolutionary biology (PhD).

Faculty: 40 full-time (7 women).
Students: 59 full-time (25 women), 8 part-time (4 women); includes 2 minority (1 African American, 1 Asian American or Pacific Islander), 17 international. *62 applicants, 52% accepted.* In 1999, 7 master's, 6 doctorates awarded. Terminal master's awarded for partial completion of doctoral program.
Degree requirements: For master's, thesis required, foreign language not required; for doctorate, 2 foreign languages (computer language can substitute for one), dissertation required.
Entrance requirements: For master's, GRE General Test, minimum GPA of 2.75; for doctorate, GRE General Test, GRE Subject Test, minimum GPA of 3.0. *Application deadline:* For fall admission, 5/15. *Application fee:* $20 ($30 for international students).
Expenses: Tuition, state resident: full-time $2,021; part-time $287 per credit. Tuition, nonresident: full-time $7,253; part-time $287 per credit. Part-time tuition and fees vary according to course load.
Financial aid: In 1999–00, 14 fellowships with full tuition reimbursements (averaging

$14,572 per year), 7 research assistantships with full tuition reimbursements (averaging $4,565 per year), 24 teaching assistantships with full tuition reimbursements (averaging $5,377 per year) were awarded; Federal Work-Study and institutionally sponsored loans also available. Financial aid application deadline: 5/1.

Faculty research: Structure and ultrastructure, system biology, ecology, processes, environmental physiology. Dr. Darryl L. Felder, Head, 337-482-6748. **Application contact:** Dr. Karl Hasenstein, Graduate Coordinator, 337-482-6750.

Find an in-depth description at www.petersons.com/graduate.

■ UNIVERSITY OF LOUISIANA AT MONROE

Graduate Studies and Research, College of Pure and Applied Sciences, Department of Biology, Monroe, LA 71209-0001

AWARDS MS.

Faculty: 12 full-time (3 women).
Students: 20 full-time (11 women), 8 part-time (3 women); includes 4 minority (all African Americans), 1 international. Average age 26. In 1999, 5 degrees awarded.
Degree requirements: For master's, thesis required, foreign language not required.
Entrance requirements: For master's, GRE General Test, minimum GPA of 2.8 overall or 3.0 during last 21 hours of biology. *Application deadline:* For fall admission, 7/1 (priority date); for spring admission, 11/1. Applications are processed on a rolling basis. *Application fee:* $15 ($25 for international students).
Expenses: Tuition, state resident: full-time $1,650. Tuition, nonresident: full-time $7,608. Required fees: $380.
Financial aid: Teaching assistantships, Federal Work-Study and unspecified assistantships available. Financial aid application deadline: 7/1.
Faculty research: Fish systematics and zoogeography, taxonomy and distribution of Louisiana plants, aquatic biology, secondary succession, microbial ecology. Dr. Kim Marie Tolson, Head, 318-342-1790.

■ UNIVERSITY OF LOUISVILLE

Graduate School, College of Arts and Sciences, Department of Biology, Program in Biology, Louisville, KY 40292-0001

AWARDS MS.

Degree requirements: For master's, thesis required.
Entrance requirements: For master's, GRE General Test.

Expenses: Tuition, state resident: full-time $3,260; part-time $182 per hour. Tuition, nonresident: full-time $9,780; part-time $544 per hour. Required fees: $143; $28 per hour. Tuition and fees vary according to program.

Find an in-depth description at www.petersons.com/graduate.

■ UNIVERSITY OF LOUISVILLE

School of Medicine and Graduate School, Integrated Programs in Biomedical Sciences, Louisville, KY 40292-0001

AWARDS MS, PhD, MD/MS, MD/PhD.

Degree requirements: For doctorate, dissertation required.
Entrance requirements: For master's and doctorate, GRE General Test.
Expenses: Tuition, state resident: full-time $3,260; part-time $182 per hour. Tuition, nonresident: full-time $9,780; part-time $544 per hour. Required fees: $143; $28 per hour.

Find an in-depth description at www.petersons.com/graduate.

■ UNIVERSITY OF MAINE

Graduate School, College of Natural Sciences, Forestry, and Agriculture, Department of Biological Sciences, Program in Biological Sciences, Orono, ME 04469

AWARDS PhD.

Degree requirements: For doctorate, dissertation required.
Entrance requirements: For doctorate, GRE General Test, TOEFL.
Expenses: Tuition, state resident: full-time $3,564. Tuition, nonresident: full-time $10,116. Required fees: $378. Tuition and fees vary according to course load.

■ UNIVERSITY OF MARYLAND

Graduate School, Graduate Programs in Medicine, Baltimore, MD 21201-1627

AWARDS MA, MS, PhD, MD/MS, MD/PhD. Part-time and evening/weekend programs available.

Degree requirements: For doctorate, dissertation required.
Entrance requirements: For master's and doctorate, GRE General Test, TOEFL, minimum GPA of 3.0.
Expenses: Tuition, state resident: part-time $261 per credit hour. Tuition, nonresident: part-time $468 per credit hour. Tuition and fees vary according to program.

■ UNIVERSITY OF MARYLAND, BALTIMORE COUNTY

Graduate School, Department of Biological Sciences, Baltimore, MD 21250-5398

AWARDS Applied molecular biology (MS); biological sciences (MS, PhD); molecular and cell biology (PhD); neurosciences and cognitive sciences (MS, PhD). Part-time programs available.

Faculty: 35 full-time (9 women), 3 part-time/adjunct (1 woman).
Students: 65 full-time (34 women), 15 part-time (7 women); includes 9 minority (2 African Americans, 6 Asian Americans or Pacific Islanders, 1 Hispanic American), 22 international. *135 applicants, 42% accepted.* In 1999, 16 master's, 5 doctorates awarded.
Degree requirements: For master's, foreign language not required; for doctorate, dissertation required, foreign language not required.
Entrance requirements: For master's and doctorate, GRE General Test, TOEFL, minimum GPA of 3.0. *Application deadline:* Applications are processed on a rolling basis. *Application fee:* $45.
Expenses: Tuition, state resident: part-time $268 per credit hour. Tuition, nonresident: part-time $470 per credit hour. Required fees: $38 per credit hour. $557 per semester.
Financial aid: In 1999–00, 2 fellowships with tuition reimbursements (averaging $18,122 per year), 17 research assistantships with tuition reimbursements (averaging $13,000 per year), 25 teaching assistantships with tuition reimbursements (averaging $12,193 per year) were awarded; career-related internships or fieldwork also available.
Faculty research: Molecular genetics, neurobiology, metabolism. Dr. Lasse Lindahl, Chairman, 410-455-2261.
Application contact: Richard E. Wolf, Director, Graduate Program, 410-455-3669, *Fax:* 410-455-3875, *E-mail:* biograd@umbc.edu.

Find an in-depth description at www.petersons.com/graduate.

■ UNIVERSITY OF MARYLAND, COLLEGE PARK

Graduate Studies and Research, College of Life Sciences, Department of Biology, College Park, MD 20742

AWARDS Biology (MS, PhD); sustainable development and conservation biology (MS). Part-time and evening/weekend programs available.

Faculty: 58 full-time (23 women), 9 part-time/adjunct (5 women).

University of Maryland, College Park
(continued)

Students: 68 full-time (36 women), 28 part-time (19 women); includes 11 minority (1 African American, 5 Asian Americans or Pacific Islanders, 4 Hispanic Americans, 1 Native American), 16 international. *111 applicants, 19% accepted.* In 1999, 8 master's, 7 doctorates awarded.

Degree requirements: For doctorate, dissertation required.

Entrance requirements: For master's, GRE General Test, minimum GPA of 3.0; for doctorate, GRE General Test, GRE Subject Test, TOEFL, minimum GPA of 3.0. *Application deadline:* For spring admission, 12/1. Applications are processed on a rolling basis. *Application fee:* $50 ($70 for international students). Electronic applications accepted.

Expenses: Tuition, state resident: part-time $272 per credit hour. Tuition, nonresident: part-time $415 per credit hour. Required fees: $632; $379 per year.

Financial aid: In 1999–00, 9 fellowships with full tuition reimbursements (averaging $9,627 per year), 13 research assistantships with tuition reimbursements (averaging $12,348 per year), 65 teaching assistantships with tuition reimbursements (averaging $11,381 per year) were awarded; Federal Work-Study, grants, and scholarships also available. Aid available to part-time students. Financial aid application deadline: 2/1; financial aid applicants required to submit FAFSA.

Faculty research: Physiology, cell biology, evolution, ecology, ethology, histotechnology. *Total annual research expenditures:* $2.6 million.

Dr. William Jeffery, Chairman, 301-405-6884, *Fax:* 301-314-9358.

Application contact: Trudy Lindsey, Director, Graduate Admissions and Records, 301-405-4198, *Fax:* 301-314-9305, *E-mail:* grschool@deans.umd.edu.

■ UNIVERSITY OF MASSACHUSETTS AMHERST

Graduate School, College of Natural Sciences and Mathematics, Department of Biology, Amherst, MA 01003

AWARDS MA, MS, PhD. Part-time programs available.

Faculty: 39 full-time (8 women).

Students: 2 full-time (1 woman), 7 part-time (2 women), 1 international. Average age 34. *2 applicants, 0% accepted.* In 1999, 4 doctorates awarded. Terminal master's awarded for partial completion of doctoral program.

Degree requirements: For master's, thesis or alternative required; for doctorate, dissertation required.

Entrance requirements: For master's and doctorate, GRE General Test, GRE Subject Test. *Application deadline:* For fall admission, 1/15; for spring admission, 10/1. Applications are processed on a rolling basis. *Application fee:* $40.

Expenses: Tuition, state resident: full-time $2,640; part-time $165 per credit. Tuition, nonresident: full-time $9,756; part-time $407 per credit. Required fees: $1,221 per term. One-time fee: $110. Full-time tuition and fees vary according to course load, campus/location and reciprocity agreements.

Financial aid: In 1999–00, 1 fellowship with full tuition reimbursement (averaging $1,000 per year), research assistantships with full tuition reimbursements (averaging $11,055 per year), teaching assistantships with full tuition reimbursements (averaging $11,020 per year) were awarded; career-related internships or fieldwork, Federal Work-Study, grants, scholarships, traineeships, and unspecified assistantships also available. Aid available to part-time students. Financial aid application deadline: 1/15.

Dr. Christopher Woodcock, Chair, 413-545-2602, *Fax:* 413-545-3243, *E-mail:* chris@bio.umass.edu.

■ UNIVERSITY OF MASSACHUSETTS BOSTON

Office of Graduate Studies and Research, College of Arts and Sciences, Faculty of Sciences, Program in Biology, Boston, MA 02125-3393

AWARDS MS. Part-time and evening/weekend programs available.

Students: 10 full-time (4 women), 19 part-time (15 women); includes 3 minority (1 African American, 2 Asian Americans or Pacific Islanders), 3 international. *40 applicants, 20% accepted.* In 1999, 7 degrees awarded.

Degree requirements: For master's, thesis, oral exams required, foreign language not required.

Entrance requirements: For master's, GRE General Test, GRE Subject Test, minimum GPA of 2.75. *Application deadline:* For fall admission, 3/1 (priority date); for spring admission, 11/1. *Application fee:* $25 ($40 for international students).

Expenses: Tuition, state resident: full-time $2,590; part-time $108 per credit. Tuition, nonresident: full-time $4,758; part-time $407 per credit. Required fees: $150; $159 per term.

Financial aid: In 1999–00, 5 research assistantships with full tuition reimbursements (averaging $13,000 per year), 26 teaching assistantships with full tuition reimbursements (averaging $10,000 per

year) were awarded; career-related internships or fieldwork, Federal Work-Study, and unspecified assistantships also available. Aid available to part-time students. Financial aid application deadline: 3/1; financial aid applicants required to submit FAFSA.

Faculty research: Microbial ecology, population and conservation genetics energetics of insect locomotion, science education, evolution and ecology of marine invertebrates.

Dr. Michael Shiaris, Director, 617-287-6600.

Application contact: Lisa Lavely, Director of Graduate Admissions and Records, 617-287-6400, *Fax:* 617-287-6236, *E-mail:* bos.gadm@dpc.umassp.edu.

■ UNIVERSITY OF MASSACHUSETTS BOSTON

Office of Graduate Studies and Research, College of Arts and Sciences, Faculty of Sciences, Program in Biotechnology and Biomedical Science, Boston, MA 02125-3393

AWARDS MS. Part-time and evening/weekend programs available.

Students: 5 full-time, 2 part-time; includes 1 minority (African American), 2 international. *16 applicants, 25% accepted.* In 1999, 3 degrees awarded.

Degree requirements: For master's, comprehensive exams, oral exams required, thesis optional, foreign language not required.

Entrance requirements: For master's, GRE General Test, GRE Subject Test, minimum GPA of 2.75, 3.0 in science and math. *Application deadline:* For fall admission, 3/1 (priority date); for spring admission, 11/1. *Application fee:* $25 ($40 for international students).

Expenses: Tuition, state resident: full-time $2,590; part-time $108 per credit. Tuition, nonresident: full-time $4,758; part-time $407 per credit. Required fees: $150; $159 per term.

Financial aid: Research assistantships with full tuition reimbursements, teaching assistantships with full tuition reimbursements, career-related internships or fieldwork, Federal Work-Study, and unspecified assistantships available. Aid available to part-time students. Financial aid application deadline: 3/1; financial aid applicants required to submit FAFSA.

Faculty research: Evolutionary and molecular immunology, molecular genetics, tissue culture, computerized laboratory technology.

Dr. Michael Shiaris, Director, 617-287-6600.

Application contact: Lisa Lavely, Director of Graduate Admissions and Records, 617-287-6400, *Fax:* 617-287-6236, *E-mail:* bos.gadm@dpc.umassp.edu.

■ UNIVERSITY OF MASSACHUSETTS DARTMOUTH

Graduate School, College of Arts and Sciences, Department of Biology, North Dartmouth, MA 02747-2300

AWARDS Biology (MS); marine biology (MS). Part-time programs available.

Faculty: 18 full-time (5 women).
Students: 5 full-time (3 women), 17 part-time (10 women); includes 4 minority (all Asian Americans or Pacific Islanders). Average age 26. *29 applicants, 34% accepted.* In 1999, 10 degrees awarded.
Degree requirements: For master's, thesis or alternative required, foreign language not required.
Entrance requirements: For master's, GRE General Test, GRE Subject Test, TOEFL. *Application deadline:* For fall admission, 5/7; for spring admission, 11/15 (priority date). *Application fee:* $40 for international students.
Expenses: Tuition, area resident: Full-time $2,071; part-time $86 per credit. Tuition, state resident: full-time $2,071; part-time $86 per credit. Tuition, nonresident: full-time $7,845; part-time $327 per credit. Required fees: $127 per credit. Full-time tuition and fees vary according to program and reciprocity agreements. Part-time tuition and fees vary according to course load and reciprocity agreements.
Financial aid: In 1999–00, 2 research assistantships with full tuition reimbursements (averaging $5,615 per year), 13 teaching assistantships with full tuition reimbursements (averaging $7,158 per year) were awarded; Federal Work-Study also available. Aid available to part-time students. Financial aid application deadline: 3/1; financial aid applicants required to submit FAFSA.
Faculty research: Bacterial regulatory, gene, phytoplankton analysis, estimating age from teeth, microbial community structure, cranberry production and pesticicles. *Total annual research expenditures:* $944,000.
Dr. Nancy O'Connor, Director, 508-999-8217, *Fax:* 508-999-8196, *E-mail:* noconnor@unmassd.edu.
Application contact: Carol A. Novo, Graduate Admissions Office, 508-999-8026, *Fax:* 508-999-8183, *E-mail:* graduate@umassd.edu.

■ UNIVERSITY OF MASSACHUSETTS LOWELL

Graduate School, College of Fine Arts/Humanities/Social Sciences, Department of Biological Sciences, Lowell, MA 01854-2881

AWARDS Biochemistry (PhD); biological sciences (MS); biotechnology (MS). Part-time programs available.

Faculty: 12 full-time (3 women).
Students: 18 full-time (12 women), 46 part-time (26 women); includes 6 minority (3 African Americans, 1 Asian American or Pacific Islander, 2 Hispanic Americans), 9 international. Average age 33. *35 applicants, 71% accepted.* In 1999, 22 degrees awarded.
Degree requirements: For master's, thesis required, foreign language not required; for doctorate, computer language, dissertation required. *Average time to degree:* Master's–2 years full-time, 3 years part-time.
Entrance requirements: For master's and doctorate, GRE General Test. *Application deadline:* For fall admission, 4/1 (priority date); for spring admission, 10/1. Applications are processed on a rolling basis. *Application fee:* $20 ($35 for international students). Electronic applications accepted.
Expenses: Tuition, state resident: full-time $1,610; part-time $89 per credit. Tuition, nonresident: full-time $5,610; part-time $312 per credit. Required fees: $2,100; $120 per credit. Tuition and fees vary according to reciprocity agreements.
Financial aid: In 1999–00, 10 teaching assistantships with tuition reimbursements were awarded; research assistantships with tuition reimbursements, career-related internships or fieldwork, Federal Work-Study, scholarships, and traineeships also available. Financial aid application deadline: 4/1.
Dr. Robert Lynch, Chair, 978-934-2891, *E-mail:* robert_lynch@woods.uml.edu.
Application contact: Dr. Ilze Skare, Coordinator, 978-934-2885, *E-mail:* ilze_skare@woods.uml.edu.

■ UNIVERSITY OF MASSACHUSETTS WORCESTER

Graduate School of Biomedical Sciences, Worcester, MA 01655-0115
AWARDS PhD, DVM/PhD, MD/PhD.

Faculty: 186 full-time (32 women).
Students: 184 full-time (87 women); includes 15 minority (1 African American, 9 Asian Americans or Pacific Islanders, 4 Hispanic Americans, 1 Native American), 53 international. Average age 28. *290 applicants, 14% accepted.* In 1999, 11 doctorates awarded.
Degree requirements: For doctorate, dissertation required, foreign language not

required. *Average time to degree:* Doctorate–6.3 years full-time.
Entrance requirements: For doctorate, GRE General Test, 1 year of calculus, physics, organic chemistry and biology. *Application deadline:* For fall admission, 1/15 (priority date). Applications are processed on a rolling basis. *Application fee:* $25 ($50 for international students).
Expenses: Tuition, state resident: full-time $2,640. Tuition, nonresident: full-time $9,756. Required fees: $825. Full-time tuition and fees vary according to program.
Financial aid: In 1999–00, 60 fellowships with full tuition reimbursements (averaging $17,500 per year), 124 research assistantships with full tuition reimbursements (averaging $17,500 per year) were awarded; institutionally sponsored loans, tuition waivers (full), and unspecified assistantships also available.
Faculty research: Cancer, diabetes, insulin action, genetics, neurosciences, immunology. *Total annual research expenditures:* $91 million.
Dr. Thomas B. Miller, Dean, 508-856-4135, *E-mail:* gsbs@umassmed.edu.
Find an in-depth description at www.petersons.com/graduate.

■ UNIVERSITY OF MEDICINE AND DENTISTRY OF NEW JERSEY

Graduate School of Biomedical Sciences, Graduate Programs in Biomedical Sciences, Piscataway, NJ 08854-5635

AWARDS Biochemistry and molecular biology (MS, PhD); biomedical engineering (MS, PhD); cell and developmental biology (MS, PhD), including cell biology, developmental biology, immunology (MS); molecular genetics and microbiology (MS, PhD); neuroscience and cell biology (MS, PhD); pharmacology (PhD), including cellular and molecular pharmacology; toxicology (MS, PhD), including environmental toxicology, industrial-occupational toxicology, nutritional toxicology, pharmaceutical toxicology.

Students: 603 full-time (286 women), 5 part-time (2 women); includes 312 minority (22 African Americans, 248 Asian Americans or Pacific Islanders, 42 Hispanic Americans). In 1999, 6 master's, 19 doctorates awarded. Terminal master's awarded for partial completion of doctoral program.
Degree requirements: For master's, thesis, qualifying exam required; for doctorate, dissertation, qualifying exam required, foreign language not required.
Entrance requirements: For master's and doctorate, GRE General Test, TOEFL. *Application deadline:* For spring admission,

University of Medicine and Dentistry of New Jersey (continued)

10/1. Applications are processed on a rolling basis. *Application fee:* $40.

Expenses: Tuition, state resident: part-time $270 per credit hour. Tuition, nonresident: part-time $407 per credit hour. Part-time tuition and fees vary according to campus/location and program.

Financial aid: Fellowships, research assistantships, teaching assistantships, career-related internships or fieldwork, Federal Work-Study, institutionally sponsored loans, traineeships, and tuition waivers (full and partial) available. Financial aid application deadline: 5/1. Dr. Michael J. Leibowitz, Associate Dean, Graduate School, 732-235-5016, *Fax:* 732-235-4720, *E-mail:* gsbspisc@umdnj.edu.

Find an in-depth description at www.petersons.com/graduate.

■ UNIVERSITY OF MEDICINE AND DENTISTRY OF NEW JERSEY

Graduate School of Biomedical Sciences, Graduate Programs in Biomedical Sciences, Stratford, NJ 08084-5634

AWARDS MS, PhD, DO/PhD.

Students: 11 full-time (5 women), 7 part-time (5 women); includes 10 minority (1 African American, 9 Asian Americans or Pacific Islanders).

Degree requirements: For master's, thesis required; for doctorate, dissertation, qualifying exam required, foreign language not required.

Entrance requirements: For master's and doctorate, GRE General Test, TOEFL. *Application deadline:* For fall admission, 2/1; for spring admission, 10/1. Applications are processed on a rolling basis. *Application fee:* $40.

Expenses: Tuition, state resident: part-time $270 per credit hour. Tuition, nonresident: part-time $407 per credit hour. Part-time tuition and fees vary according to campus/location and program.

Financial aid: Fellowships, Federal Work-Study available. Financial aid application deadline: 5/1.

■ UNIVERSITY OF MEDICINE AND DENTISTRY OF NEW JERSEY

Graduate School of Biomedical Sciences, Graduate Programs in Biomedical Sciences, Newark, NJ 07107

AWARDS Anatomy, cell biology, and injury sciences (MS, PhD); biochemistry and molecular biology (MS, PhD); molecular genetics and microbiology (MS, PhD);

neurosciences (MS, PhD); oral biology (MS); pathology and laboratory medicine (MS, PhD); pharmacology and physiology (MS, PhD).

Students: 142 full-time (74 women), 24 part-time (15 women); includes 78 minority (13 African Americans, 52 Asian Americans or Pacific Islanders, 13 Hispanic Americans). In 1999, 10 master's, 19 doctorates awarded. Terminal master's awarded for partial completion of doctoral program.

Degree requirements: For master's, thesis required; for doctorate, dissertation, qualifying exam required, foreign language not required.

Entrance requirements: For master's and doctorate, GRE General Test, TOEFL. *Application deadline:* For fall admission, 2/1; for spring admission, 10/1. Applications are processed on a rolling basis. *Application fee:* $40.

Expenses: Tuition, state resident: part-time $270 per credit hour. Tuition, nonresident: part-time $407 per credit hour. Part-time tuition and fees vary according to campus/location and program.

Financial aid: Fellowships, research assistantships, teaching assistantships, career-related internships or fieldwork, Federal Work-Study, institutionally sponsored loans, and tuition waivers (full and partial) available. Financial aid application deadline: 5/1.

Application contact: Dr. Henry E. Brezenoff, Dean, Graduate School of Biomedical Sciences, 973-972-5333, *Fax:* 973-972-7148, *E-mail:* hbrezeno@umdnj.edu.

Find an in-depth description at www.petersons.com/graduate.

■ THE UNIVERSITY OF MEMPHIS

Graduate School, College of Arts and Sciences, Department of Biology, Memphis, TN 38152

AWARDS MS, PhD.

Faculty: 10 full-time (0 women).

Students: 52 full-time (26 women), 35 part-time (10 women); includes 18 minority (11 African Americans, 3 Asian Americans or Pacific Islanders, 3 Hispanic Americans, 1 Native American), 10 international. Average age 29. *73 applicants, 48% accepted.* In 1999, 17 master's, 1 doctorate awarded.

Degree requirements: For master's, thesis or alternative, oral and written comprehensive exams required, foreign language not required; for doctorate, dissertation, comprehensive exam required, foreign language not required, dissertation, comprehensive exam required, foreign language not required.

Entrance requirements: For master's, GRE General Test, GRE Subject Test, minimum GPA of 2.5; for doctorate, GRE General Test, GRE Subject Test, master's degree. *Application deadline:* For fall admission, 8/1; for spring admission, 12/1. Applications are processed on a rolling basis. *Application fee:* $25 ($50 for international students).

Expenses: Tuition, state resident: full-time $3,410; part-time $178 per credit hour. Tuition, nonresident: full-time $8,670; part-time $408 per credit hour. Tuition and fees vary according to program.

Financial aid: In 1999–00, 22 students received aid, including 1 fellowship, 4 research assistantships, 17 teaching assistantships with full tuition reimbursements available (averaging $9,000 per year). Financial aid application deadline: 2/1.

Faculty research: Ecotoxicology, river and stream ecology, wetland ecology, increasing catfish production, coyote studies, small mammal behavior. *Total annual research expenditures:* $346,000.

Dr. Jerry O. Wolff, Chairman, 901-678-2581, *Fax:* 901-678-4746, *E-mail:* jwolff@memphis.edu.

Application contact: Chris Powless, Information Contact, 901-678-4757, *Fax:* 901-678-4746, *E-mail:* apowless@memphis.edu.

Find an in-depth description at www.petersons.com/graduate.

■ UNIVERSITY OF MIAMI

Graduate School, College of Arts and Sciences, Department of Biology, Coral Gables, FL 33124

AWARDS Biology (MS, PhD); genetics and evolution (MS, PhD); tropical biology, ecology, and behavior (MS, PhD).

Faculty: 24 full-time (3 women), 5 part-time/adjunct (1 woman).

Students: 42 full-time (23 women); includes 7 minority (1 African American, 3 Asian Americans or Pacific Islanders, 3 Hispanic Americans), 2 international. Average age 26. *56 applicants, 13% accepted.* In 1999, 3 doctorates awarded (34% entered university research/teaching, 66% found other work related to degree). Terminal master's awarded for partial completion of doctoral program.

Degree requirements: For master's, oral defense required, thesis optional, foreign language not required; for doctorate, dissertation, oral and written qualifying exam required, foreign language not required. *Average time to degree:* Doctorate–5.5 years full-time.

Entrance requirements: For master's and doctorate, GRE General Test, GRE Subject Test, TOEFL. *Application deadline:* For fall admission, 3/15. Applications are

processed on a rolling basis. *Application fee:* $50. Electronic applications accepted.
Expenses: Tuition: Full-time $15,336; part-time $852 per credit. Required fees: $174. Tuition and fees vary according to program.
Financial aid: In 1999–00, 37 students received aid, including 6 fellowships with tuition reimbursements available (averaging $15,000 per year), 10 research assistantships with tuition reimbursements available (averaging $13,000 per year), 21 teaching assistantships with tuition reimbursements available (averaging $13,202 per year); career-related internships or fieldwork and institutionally sponsored loans also available. Financial aid application deadline: 3/15.
Faculty research: Population biology, behavioral ecology, plant-animal and plant-environment interactions and genetic co-evolution, biogeography, conservation biology.
Dr. Julian C. Lee, Director, 305-284-6420, *Fax:* 305-284-3039, *E-mail:* jlee@fig.cox.miami.edu.
Application contact: 305-284-3973, *Fax:* 305-284-3039, *E-mail:* gaac@fig.cox.miami.edu.
Find an in-depth description at www.petersons.com/graduate.

■ **UNIVERSITY OF MIAMI**

Graduate School, College of Arts and Sciences, Program in Tropical Biology, Ecology, and Behavior, Coral Gables, FL 33124
AWARDS MS, PhD.
Faculty: 21 full-time (4 women), 2 part-time/adjunct (1 woman).
Students: 52 full-time (20 women); includes 9 minority (2 African Americans, 3 Asian Americans or Pacific Islanders, 4 Hispanic Americans), 14 international. Average age 27. *80 applicants, 14% accepted.* Terminal master's awarded for partial completion of doctoral program.
Degree requirements: For master's, thesis optional, foreign language not required; for doctorate, dissertation, oral and written qualifying exam required, foreign language not required. *Average time to degree:* Master's–3 years full-time; doctorate–6 years full-time.
Entrance requirements: For master's and doctorate, GRE General Test, GRE Subject Test, TOEFL. *Application deadline:* For fall admission, 2/1. Applications are processed on a rolling basis. *Application fee:* $35.
Expenses: Tuition: Full-time $15,336; part-time $852 per credit. Required fees: $174. Tuition and fees vary according to program.

Financial aid: In 1999–00, 8 fellowships, 12 research assistantships, 23 teaching assistantships were awarded; career-related internships or fieldwork and institutionally sponsored loans also available. Financial aid application deadline: 3/15.
Faculty research: Behavioral ecology, plant-animal and plant-environmental interactions and coevolution, biogeography, conservation biology, genetics.
Jean Crawford, Graduate Coordinator, 305-284-3973, *Fax:* 305-284-3039, *E-mail:* jean@fig.cox.miami.edu.

■ **UNIVERSITY OF MIAMI**

School of Medicine and Graduate School, Graduate Programs in Medicine, Coral Gables, FL 33124
AWARDS MPH, MS, MSPT, PhD, JD/MPH, MD/MPH, MD/PhD, MPA/MPH. Part-time and evening/weekend programs available.
Faculty: 208.
Students: 378. Terminal master's awarded for partial completion of doctoral program.
Degree requirements: For master's, foreign language not required; for doctorate, dissertation required.
Entrance requirements: For master's and doctorate, GRE General Test, TOEFL. *Application deadline:* Applications are processed on a rolling basis. *Application fee:* $35.
Expenses: Tuition, area resident: Part-time $899 per credit.
Financial aid: Fellowships, research assistantships, teaching assistantships, career-related internships or fieldwork, Federal Work-Study, institutionally sponsored loans, and tuition waivers (full and partial) available. Aid available to part-time students.
Dr. Richard J. Bookman, Associate Dean for Graduate Studies, 305-243-6406, *Fax:* 305-243-3593, *E-mail:* biomedgrad@miami.edu.
Application contact: Maria T. Zayas, Administrative Assistant, 305-243-6406, *Fax:* 305-243-3593, *E-mail:* mzayas@miami.edu.

■ **UNIVERSITY OF MICHIGAN**

Horace H. Rackham School of Graduate Studies, College of Literature, Science, and the Arts, Department of Biology, Ann Arbor, MI 48109
AWARDS Biology (MS, PhD); plant biology (MS).
Faculty: 72 full-time (14 women).
Students: 148 full-time (74 women); includes 9 minority (2 African Americans, 5 Asian Americans or Pacific Islanders, 2 Hispanic Americans), 33 international. Average age 29. *231 applicants, 29%*

accepted. In 1999, 19 master's, 10 doctorates awarded.
Degree requirements: For master's, thesis not required; for doctorate, oral defense of dissertation, preliminary exam required.
Entrance requirements: For master's and doctorate, GRE General Test, TOEFL. *Application deadline:* Applications are processed on a rolling basis. *Application fee:* $55.
Expenses: Tuition, state resident: full-time $10,316. Tuition, nonresident: full-time $20,922. Required fees: $185. Part-time tuition and fees vary according to course load and program.
Financial aid: In 1999–00, 30 fellowships with full tuition reimbursements (averaging $10,750 per year), 33 research assistantships with full tuition reimbursements (averaging $11,810 per year), 129 teaching assistantships with full tuition reimbursements (averaging $11,810 per year) were awarded; career-related internships or fieldwork, grants, traineeships, and unspecified assistantships also available.
Faculty research: Evolution, ecology and organismal biology; development and genetics; neurobiology and animal physiology; molecular plant biology; microbiology and gene action. *Total annual research expenditures:* $5.1 million.
Dr. Julian Adams, Chair, 734-764-7427, *Fax:* 734-747-0884, *E-mail:* chair@biology.lsa.umich.edu.
Application contact: Lisa Herring, Graduate Coordinator, 734-764-1443, *Fax:* 734-764-0884, *E-mail:* gradcord@biology.lsa.umich.edu.
Find an in-depth description at www.petersons.com/graduate.

■ **UNIVERSITY OF MICHIGAN**

Medical School and Horace H. Rackham School of Graduate Studies, Medical Scientist Training Program, Ann Arbor, MI 48109
AWARDS MD/PhD.
Students: 63 full-time (21 women); includes 22 minority (4 African Americans, 17 Asian Americans or Pacific Islanders, 1 Hispanic American).
Application deadline: For fall admission, 11/15. Applications are processed on a rolling basis. *Application fee:* $55. Electronic applications accepted.
Expenses: Tuition, state resident: full-time $18,020. Tuition, nonresident: full-time $27,776.
Financial aid: Fellowships with full tuition reimbursements, research assistantships with full tuition reimbursements, teaching assistantships with full tuition reimbursements, institutionally sponsored loans and traineeships available.

University of Michigan (continued)
Ronald J. Koenig, Director, 734-764-6176, *Fax:* 734-764-8180, *E-mail:* rkoenig@umich.edu.
Application contact: Carol Kruise, Program Secretary, 734-764-6176, *Fax:* 734-764-8180, *E-mail:* ckruise@umich.edu.

■ UNIVERSITY OF MICHIGAN

Medical School and Horace H. Rackham School of Graduate Studies, Program in Biomedical Sciences (PIBS), Ann Arbor, MI 48109

AWARDS MS, PhD, MD/PhD, Pharm D/PhD.

Students: 66 full-time (38 women); includes 11 minority (4 African Americans, 3 Asian Americans or Pacific Islanders, 4 Hispanic Americans), 5 international. Average age 23. *508 applicants, 31% accepted.*
Degree requirements: For doctorate, oral defense of dissertation, preliminary exam required.
Entrance requirements: For master's, GRE General Test; for doctorate, GRE General Test, research experience. *Application deadline:* For fall admission, 1/5. *Application fee:* $55. Electronic applications accepted.
Expenses: Tuition, state resident: full-time $10,316. Tuition, nonresident: full-time $20,922.
Financial aid: In 1999–00, 66 fellowships with full tuition reimbursements (averaging $17,000 per year) were awarded; research assistantships, teaching assistantships, grants, institutionally sponsored loans, scholarships, traineeships, tuition waivers (full), and unspecified assistantships also available.
Faculty research: Genetics, cellular and molecular biology, microbial pathogenesis, cancer biology, neuroscience.
Dr. David R. Engelke, PIBS Director/Professor of Biological Chemistry, 734-615-7005, *Fax:* 734-647-7022, *E-mail:* engelke@umich.edu.
Application contact: Janine Leah Capsouras, Student Services Representative, 734-615-1660, *Fax:* 734-647-7022, *E-mail:* jleahcap@umich.edu.
Find an in-depth description at www.petersons.com/graduate.

■ UNIVERSITY OF MINNESOTA, DULUTH

Graduate School, College of Science and Engineering, Program in Biology, Duluth, MN 55812-2496

AWARDS MS. Part-time programs available.

Faculty: 9 full-time (2 women), 21 part-time/adjunct (6 women).
Students: 20 full-time (10 women), 4 part-time (3 women); includes 4 minority (all

Native Americans), 2 international. Average age 30. *27 applicants, 52% accepted.* In 1999, 6 degrees awarded (100% found work related to degree).
Degree requirements: For master's, thesis (for some programs), seminar required, foreign language not required. *Average time to degree:* Master's–3 years full-time.
Entrance requirements: For master's, GRE General Test, minimum GPA of 3.0; previous course work in chemistry, mathematics, biology, and physical sciences. *Application deadline:* For fall admission, 7/15; for spring admission, 11/15. Applications are processed on a rolling basis. *Application fee:* $50 ($55 for international students).
Expenses: Tuition, state resident: full-time $5,040; part-time $420 per credit. Tuition, nonresident: full-time $9,900; part-time $825 per credit. Required fees: $509. Tuition and fees vary according to course load and program.
Financial aid: In 1999–00, 18 students received aid, including 5 research assistantships with full tuition reimbursements available (averaging $11,000 per year), 13 teaching assistantships with full tuition reimbursements available (averaging $10,210 per year); fellowships with full tuition reimbursements available, career-related internships or fieldwork, Federal Work-Study, institutionally sponsored loans, traineeships, and tuition waivers (full and partial) also available. Aid available to part-time students. Financial aid application deadline: 3/15; financial aid applicants required to submit FAFSA.
Faculty research: Cell biology, developmental biology, forest ecology, freshwater ecology, landscape ecology. *Total annual research expenditures:* $2.5 million.
Application contact: Graduate School Office, 218-726-7523.

■ UNIVERSITY OF MINNESOTA, TWIN CITIES CAMPUS

Graduate School, College of Biological Sciences, Biological Science Program, Minneapolis, MN 55455-0213

AWARDS MBS. Part-time and evening/weekend programs available.

Faculty: 19 full-time (6 women).
Students: *21 applicants, 90% accepted.* In 1999, 2 degrees awarded (2% found work related to degree). *Average time to degree:* Master's–1.5 years full-time.
Entrance requirements: For master's, 2 years of work experience. *Application deadline:* For fall admission, 6/15 (priority date); for spring admission, 10/15 (priority date). Applications are processed on a rolling basis. *Application fee:* $50 ($55 for

international students). Electronic applications accepted.
Expenses: Tuition, state resident: part-time $631 per credit.
Application contact: Carol Jane Gross, Program Administrator, 612-625-3133, *Fax:* 612-624-2785, *E-mail:* cgross@cbs.umn.edu.

■ UNIVERSITY OF MINNESOTA, TWIN CITIES CAMPUS

Medical School and Graduate School, Graduate Programs in Medicine, Minneapolis, MN 55455-0213

AWARDS MA, MS, PhD. Part-time and evening/weekend programs available.

Expenses: Tuition, state resident: full-time $11,984; part-time $1,498 per semester. Tuition, nonresident: full-time $22,264; part-time $2,783 per semester. Full-time tuition and fees vary according to program and student level. Part-time tuition and fees vary according to course load and program.

■ UNIVERSITY OF MISSISSIPPI

Graduate School, College of Liberal Arts, Department of Biology, Oxford, University, MS 38677

AWARDS MS, PhD.

Faculty: 18 full-time (4 women).
Students: 27 full-time (12 women), 3 part-time (1 woman); includes 4 minority (2 African Americans, 1 Asian American or Pacific Islander, 1 Native American), 6 international. In 1999, 7 master's, 7 doctorates awarded.
Degree requirements: For master's and doctorate, thesis/dissertation required, foreign language not required.
Entrance requirements: For master's and doctorate, GRE General Test, GRE Subject Test, TOEFL, minimum GPA of 3.0. *Application deadline:* For fall admission, 8/1. Applications are processed on a rolling basis. *Application fee:* $0 ($25 for international students).
Expenses: Tuition, state resident: full-time $3,053; part-time $170 per credit hour. Tuition, nonresident: full-time $6,155; part-time $342 per credit hour. Tuition and fees vary according to program.
Financial aid: Research assistantships, teaching assistantships available. Financial aid application deadline: 3/1.
Faculty research: Freshwater biology, including ecology and evolutionary biology; environmental and applied biology.
Dr. Gary L. Miller, Chairman, 662-915-7203, *Fax:* 662-915-5144, *E-mail:* bymiller@olemiss.edu.

■ UNIVERSITY OF MISSISSIPPI MEDICAL CENTER

Graduate Programs in Biomedical Sciences, Jackson, MS 39216-4505

AWARDS MS, PhD, MD/PhD.

Faculty: 89 full-time (13 women), 16 part-time/adjunct (1 woman).
Students: 84 full-time (39 women), 35 part-time (27 women); includes 9 minority (all African Americans), 53 international. Average age 28. *135 applicants, 22% accepted.* In 1999, 5 master's awarded (20% entered university research/teaching, 60% found other work related to degree, 20% continued full-time study); 10 doctorates awarded (40% entered university research/ teaching, 60% continued full-time study). Terminal master's awarded for partial completion of doctoral program.
Degree requirements: For master's, thesis required, foreign language not required; for doctorate, dissertation, first authored publication required, foreign language not required. *Average time to degree:* Master's–4 years full-time; doctorate–5 years full-time.
Entrance requirements: For master's and doctorate, GRE General Test, minimum GPA of 3.0. *Application deadline:* For fall admission, 3/1. Applications are processed on a rolling basis. *Application fee:* $10.
Expenses: Tuition, state resident: full-time $2,378; part-time $132 per hour. Tuition, nonresident: full-time $4,697; part-time $261 per hour. Tuition and fees vary according to program.
Financial aid: In 1999–00, 71 students received aid, including 71 research assistantships (averaging $16,234 per year); Federal Work-Study and grants also available.
Faculty research: Immunology; protein chemistry and biosynthesis; cardiovascular, renal, and endocrine physiology; rehabilitation therapy on immune system/ hypothalamic/adrenal axis interaction. *Total annual research expenditures:* $3.7 million.
Dr. I. K. Ho, Interim Associate Vice Chancellor, 601-984-1600, *Fax:* 601-984-1637.
Application contact: Dr. Billy M. Bishop, Director, Student Services and Records, 601-984-1080, *Fax:* 601-984-1079, *E-mail:* bbishop@registrar.umsmed.edu.

Find an in-depth description at www.petersons.com/graduate.

■ UNIVERSITY OF MISSOURI– COLUMBIA

Graduate School, College of Arts and Sciences, Division of Biological Sciences, Columbia, MO 65211

AWARDS Biological sciences (MA, PhD); genetics (MA, PhD). Terminal master's awarded for partial completion of doctoral program.

Degree requirements: For master's, thesis required; for doctorate, dissertation, comprehensive exam required.
Entrance requirements: For master's and doctorate, GRE General Test, minimum GPA of 3.0.
Expenses: Tuition, state resident: full-time $3,020; part-time $168 per hour. Tuition, nonresident: full-time $6,066; part-time $505 per hour. Required fees: $445; $18 per hour. Tuition and fees vary according to course load and program.

Find an in-depth description at www.petersons.com/graduate.

■ UNIVERSITY OF MISSOURI– COLUMBIA

School of Medicine and Graduate School, Graduate Programs in Medicine, Columbia, MO 65211

AWARDS MA, MS, MSPH, PhD, MBA/MSPH, MD/MS, MD/PhD, MPA/MSPH. Part-time programs available.

Degree requirements: For doctorate, dissertation required.
Entrance requirements: For master's and doctorate, GRE General Test, minimum GPA of 3.0.
Expenses: Tuition, state resident: full-time $3,020; part-time $168 per hour. Tuition, nonresident: full-time $6,066; part-time $505 per hour. Required fees: $445; $18 per hour.

■ UNIVERSITY OF MISSOURI– KANSAS CITY

School of Biological Sciences, Kansas City, MO 64110-2499

AWARDS Biology (MA); cell biology and biophysics (PhD); cellular and molecular biology (MS, PhD); molecular biology and biochemistry (PhD). Part-time and evening/ weekend programs available.

Faculty: 33 full-time (9 women), 6 part-time/adjunct (1 woman).
Students: 12 full-time (5 women), 23 part-time (8 women); includes 1 minority (African American). Average age 32. In 1999, 11 master's, 3 doctorates awarded. Terminal master's awarded for partial completion of doctoral program.

Degree requirements: For master's, foreign language not required; for doctorate, dissertation required.
Entrance requirements: For master's, GRE, minimum GPA of 3.0; for doctorate, GRE General Test, TOEFL. *Application fee:* $25.
Expenses: Tuition, state resident: part-time $173 per hour. Tuition, nonresident: part-time $348 per hour. Required fees: $22 per hour. $15 per term. Part-time tuition and fees vary according to course load and program.
Financial aid: Research assistantships, teaching assistantships, Federal Work-Study, institutionally sponsored loans, and tuition waivers (full and partial) available. Aid available to part-time students.
Faculty research: Structural biology and molecular genetics, protein gene regulation, structure-function studies, molecular, cell and neurobiology. *Total annual research expenditures:* $4 million.
Dr. Marino Martinez-Carrion, Dean, 816-235-1388.
Application contact: Dorothy Stringer, Graduate Programs Office, 816-235-2352, *Fax:* 816-235-5158, *E-mail:* stringerd@ umkc.edu.

■ UNIVERSITY OF MISSOURI–ST. LOUIS

Graduate School, College of Arts and Sciences, Department of Biology, St. Louis, MO 63121-4499

AWARDS Biology (MS, PhD), including animal behavior (MS), biochemistry, biotechnology (MS), conservation biology (MS), development (MS), ecology (MS), environmental studies (PhD), evolution (MS), genetics (MS), molecular biology and biotechnology (PhD), molecular/cellular biology (MS), physiology (MS), plant systematics, population biology (MS), tropical biology (MS); biotechnology (Certificate); tropical biology and conservation (Certificate). Part-time programs available.

Faculty: 46.
Students: 21 full-time (11 women), 75 part-time (44 women); includes 13 minority (2 African Americans, 2 Asian Americans or Pacific Islanders, 8 Hispanic Americans, 1 Native American), 23 international. In 1999, 14 master's, 4 doctorates awarded.
Degree requirements: For master's, thesis or alternative required, foreign language not required; for doctorate, one foreign language, dissertation, 1 semester of teaching experience required.
Entrance requirements: For doctorate, GRE General Test. *Application deadline:* For fall admission, 7/1 (priority date); for spring admission, 11/1 (priority date). Applications are processed on a rolling basis. *Application fee:* $25 ($40 for

University of Missouri–St. Louis (continued)

international students). Electronic applications accepted.

Expenses: Tuition, state resident: full-time $4,932; part-time $173 per credit hour. Tuition, nonresident: full-time $13,279; part-time $521 per credit hour. Required fees: $775; $33 per credit hour. Tuition and fees vary according to degree level and program.

Financial aid: In 1999–00, 8 research assistantships with partial tuition reimbursements (averaging $10,635 per year), 14 teaching assistantships with partial tuition reimbursements (averaging $11,488 per year) were awarded; career-related internships or fieldwork and Federal Work-Study also available. Aid available to part-time students. Financial aid application deadline: 2/1. *Total annual research expenditures:* $908,828.

Application contact: Graduate Admissions, 314-516-5458, *Fax:* 314-516-6759, *E-mail:* gradadm@umsl.edu.

■ **THE UNIVERSITY OF MONTANA–MISSOULA**

Graduate School, Division of Biological Sciences, Missoula, MT 59812-0002

AWARDS MS, PhD.

Faculty: 41 full-time (4 women).

Students: 41 full-time (13 women), 28 part-time (14 women); includes 7 minority (3 Asian Americans or Pacific Islanders, 4 Native Americans). *118 applicants, 19% accepted.* In 1999, 9 master's, 5 doctorates awarded.

Degree requirements: For doctorate, dissertation required.

Entrance requirements: For master's and doctorate, GRE General Test. *Application deadline:* For fall admission, 2/1 (priority date). Applications are processed on a rolling basis. *Application fee:* $45.

Expenses: Tuition, state resident: full-time $2,484; part-time $151 per credit. Tuition, nonresident: full-time $8,000; part-time $305 per credit. Required fees: $1,600. Full-time tuition and fees vary according to degree level and program.

Financial aid: In 1999–00, 30 research assistantships with tuition reimbursements (averaging $9,400 per year), 25 teaching assistantships with full tuition reimbursements (averaging $9,400 per year) were awarded; Federal Work-Study, institutionally sponsored loans, and tuition waivers (full and partial) also available. Financial aid application deadline: 3/1.

Faculty research: Biochemistry/microbiology, organismal biology, ecology. *Total annual research expenditures:* $6 million.

Dr. Don Christian, Associate Dean, 406-243-5122.

Application contact: Janean Clark, Graduate Programs Secretary, 406-243-5222, *Fax:* 406-243-4184, *E-mail:* jmclark@selway.umt.edu.

■ **UNIVERSITY OF NEBRASKA AT KEARNEY**

College of Graduate Study, College of Natural and Social Sciences, Department of Biology, Kearney, NE 68849-0001

AWARDS MS. Part-time and evening/weekend programs available.

Faculty: 9 full-time (2 women).

Students: 6 full-time (3 women), 4 part-time (1 woman), 2 international. *6 applicants, 50% accepted.* In 1999, 2 degrees awarded.

Degree requirements: For master's, thesis optional, foreign language not required.

Entrance requirements: For master's, GRE General Test. *Application deadline:* For fall admission, 8/1 (priority date); for spring admission, 12/15 (priority date). Applications are processed on a rolling basis. *Application fee:* $35.

Expenses: Tuition, state resident: full-time $1,575; part-time $88 per credit. Tuition, nonresident: full-time $2,983; part-time $166 per credit. Required fees: $477; $11 per credit. $55 per semester. Tuition and fees vary according to course load and reciprocity agreements.

Financial aid: In 1999–00, research assistantships with full tuition reimbursements (averaging $4,870 per year), 4 teaching assistantships with full tuition reimbursements (averaging $4,870 per year) were awarded; career-related internships or fieldwork and scholarships also available. Aid available to part-time students. Financial aid application deadline: 3/1; financial aid applicants required to submit FAFSA.

Faculty research: Pollution injury, molecular biology-viral gene expression, praire range condition modeling, evolution of symbiotic nitrogen fixation.

Dr. Charles Bicak, Chair, 308-865-8548.

■ **UNIVERSITY OF NEBRASKA AT OMAHA**

Graduate Studies and Research, College of Arts and Sciences, Department of Biology, Omaha, NE 68182

AWARDS MA, MS. Part-time programs available.

Faculty: 17 full-time (2 women).

Students: 8 full-time (6 women), 28 part-time (13 women); includes 4 minority (1

African American, 1 Asian American or Pacific Islander, 1 Hispanic American, 1 Native American), 1 international. Average age 34. In 1999, 14 degrees awarded.

Degree requirements: For master's, thesis, comprehensive exam required, foreign language not required.

Entrance requirements: For master's, GRE General Test, minimum GPA of 3.0. *Application deadline:* For fall admission, 3/1 (priority date); for spring admission, 10/15. Applications are processed on a rolling basis. *Application fee:* $35.

Expenses: Tuition, state resident: part-time $100 per credit hour. Tuition, nonresident: part-time $239 per credit hour. Required fees: $5 per credit hour. $91 per semester. Tuition and fees vary according to course load.

Financial aid: In 1999–00, 27 students received aid; fellowships, teaching assistantships, institutionally sponsored loans and tuition waivers (full) available. Aid available to part-time students. Financial aid application deadline: 3/1; financial aid applicants required to submit FAFSA.

Dr. William de Graw, Chairperson, 402-554-2641.

■ **UNIVERSITY OF NEBRASKA– LINCOLN**

Graduate College, College of Agricultural Sciences and Natural Resources, Department of Veterinary and Biomedical Sciences, Lincoln, NE 68588

AWARDS MS, PhD.

Faculty: 24 full-time (3 women).

Students: 3 full-time (1 woman), 8 part-time (6 women), 4 international. Average age 31. *10 applicants, 30% accepted.* In 1999, 4 degrees awarded.

Degree requirements: For master's, thesis optional, foreign language not required; for doctorate, dissertation, comprehensive exams required.

Entrance requirements: For master's, GRE General Test, TOEFL; for doctorate, GRE General Test, MCAT, or VAT; TOEFL. *Application deadline:* For fall admission, 3/1 (priority date). Applications are processed on a rolling basis. *Application fee:* $35. Electronic applications accepted.

Expenses: Tuition, state resident: part-time $116 per credit hour. Tuition, nonresident: part-time $285 per credit hour. Required fees: $119 per semester. Tuition and fees vary according to course load and program.

Financial aid: In 1999–00, 5 fellowships, 25 research assistantships were awarded;

teaching assistantships, Federal Work-Study also available. Aid available to part-time students. Financial aid application deadline: 2/15.

Faculty research: Virology, immunobiology, molecular biology, mycotoxins, ocular degeneration. Dr. John A. Schmitz, Head, 402-472-2952, *Fax:* 402-472-9690.

Find an in-depth description at www.petersons.com/graduate.

■ UNIVERSITY OF NEBRASKA–LINCOLN

Graduate College, School of Biological Sciences, Lincoln, NE 68588

AWARDS MA, MS, PhD.

Faculty: 31 full-time (7 women), 2 part-time/adjunct (0 women).
Students: 50 full-time (22 women), 13 part-time (10 women); includes 2 Hispanic Americans, 22 international. Average age 29. *92 applicants, 21% accepted.* In 1999, 12 master's, 6 doctorates awarded.
Degree requirements: For master's, thesis optional, foreign language not required; for doctorate, dissertation, comprehensive exams required.
Entrance requirements: For master's and doctorate, GRE General Test, GRE Subject Test, TOEFL. *Application deadline:* For fall admission, 1/1; for spring admission, 10/1. *Application fee:* $35. Electronic applications accepted.
Expenses: Tuition, state resident: part-time $116 per credit hour. Tuition, nonresident: part-time $285 per credit hour. Required fees: $119 per semester. Tuition and fees vary according to course load and program.
Financial aid: In 1999–00, 1 fellowship, 10 research assistantships, 31 teaching assistantships were awarded; Federal Work-Study also available. Aid available to part-time students. Financial aid application deadline: 1/1.
Faculty research: Behavior, botany, and zoology; ecology and evolutionary biology; genetics; cellular and molecular biology; microbiology.
Dr. T. Jack Morris, Director, 402-472-2720.

Find an in-depth description at www.petersons.com/graduate.

■ UNIVERSITY OF NEBRASKA MEDICAL CENTER

Graduate College, Biomedical Research Training Program, Omaha, NE 68198

Expenses: Tuition, state resident: part-time $116 per semester hour. Tuition,

nonresident: part-time $270 per semester hour. Tuition and fees vary according to program.

Find an in-depth description at www.petersons.com/graduate.

■ UNIVERSITY OF NEBRASKA MEDICAL CENTER

Graduate College, Medical Sciences Interdepartmental Area, Omaha, NE 68198

AWARDS MS, PhD.

Faculty: 190.
Students: 19 full-time (11 women), 29 part-time (13 women). In 1999, 4 master's, 9 doctorates awarded (100% entered university research/teaching). Terminal master's awarded for partial completion of doctoral program.
Degree requirements: For master's, thesis required, foreign language not required; for doctorate, dissertation required.
Entrance requirements: For master's and doctorate, GRE General Test, TOEFL. *Application fee:* $35.
Expenses: Tuition, state resident: part-time $116 per semester hour. Tuition, nonresident: part-time $270 per semester hour. Tuition and fees vary according to program.
Financial aid: Fellowships, research assistantships with full tuition reimbursements, teaching assistantships, institutionally sponsored loans available. Aid available to part-time students. Financial aid application deadline: 3/1.
Faculty research: Molecular genetics, oral biology, veterinary pathology, newborn medicine, immunology.
Dr. M. Patricia Leuschen, Graduate Committee Chair, 402-559-6750, *Fax:* 402-359-7341, *E-mail:* pleusche@unmc.edu.
Application contact: Jo Wagner, Associate Director of Admissions, 402-559-6468.

■ UNIVERSITY OF NEVADA, LAS VEGAS

Graduate College, College of Science, Department of Biological Sciences, Las Vegas, NV 89154-9900

AWARDS Biological sciences (MS); environmental biology (PhD). Part-time programs available.

Faculty: 31 full-time (8 women).
Students: *15 applicants, 53% accepted.* In 1999, 6 master's awarded.
Degree requirements: For master's, thesis, oral exam required, foreign language not required; for doctorate, one foreign language, computer language, dissertation required.
Entrance requirements: For master's, GRE General Test, GRE Subject Test,

minimum GPA of 3.0 during previous 2 years, 2.75 overall; for doctorate, GRE General Test, GRE Subject Test, minimum GPA of 3.5. *Application deadline:* For fall admission, 6/15. *Application fee:* $40 ($95 for international students).
Expenses: Tuition, state resident: part-time $97 per credit. Tuition, nonresident: full-time $6,347; part-time $198 per credit. Required fees: $62; $31 per semester.
Financial aid: In 1999–00, 3 research assistantships with full tuition reimbursements (averaging $10,434 per year), 10 teaching assistantships with partial tuition reimbursements (averaging $8,500 per year) were awarded. Financial aid application deadline: 3/1.
Dr. Dawn Neuman, Chair, 702-895-3399.
Application contact: Graduate College Admissions Evaluator, 702-895-3320.

■ UNIVERSITY OF NEVADA, RENO

Graduate School, College of Arts and Science, Department of Biology, Reno, NV 89557

AWARDS MS.

Faculty: 17 full-time (5 women).
Students: 18 full-time (12 women), 5 part-time (3 women); includes 3 minority (2 Asian Americans or Pacific Islanders, 1 Hispanic American). Average age 27. *22 applicants, 32% accepted.* In 1999, 3 degrees awarded.
Degree requirements: For master's, thesis optional, foreign language not required.
Entrance requirements: For master's, GRE General Test, TOEFL, minimum GPA of 2.75. *Application deadline:* For fall admission, 3/1 (priority date); for spring admission, 11/1. Applications are processed on a rolling basis. *Application fee:* $40.
Expenses: Tuition, area resident: Part-time $3,173 per semester. Tuition, nonresident: full-time $6,347. Required fees: $101 per credit. $101 per credit.
Financial aid: In 1999–00, 18 research assistantships, 18 teaching assistantships were awarded; Federal Work-Study and institutionally sponsored loans also available. Financial aid application deadline: 3/1.
Faculty research: Gene expression, stress protein genes, secretory proteins, conservation biology, behavioral ecology.
Dr. Lee Weber, Chair, 775-784-4484.
Application contact: Dr. Ardythe McCracken, Graduate Director, 775-784-6188, *E-mail:* mccracke@cmo.unr.edu.

Find an in-depth description at www.petersons.com/graduate.

■ UNIVERSITY OF NEVADA, RENO

School of Medicine and Graduate School, Graduate Programs in Medicine, Reno, NV 89557

AWARDS MS, PhD.

Faculty: 40.

Students: 54 full-time (43 women), 2 part-time (1 woman); includes 6 minority (4 Asian Americans or Pacific Islanders, 2 Hispanic Americans), 3 international. *128 applicants, 23% accepted.* In 1999, 18 master's, 4 doctorates awarded. Terminal master's awarded for partial completion of doctoral program.

Degree requirements: For master's, foreign language not required; for doctorate, dissertation required.

Entrance requirements: For master's, GRE General Test, minimum GPA of 2.75; for doctorate, GRE General Test, minimum GPA of 3.0. *Application deadline:* For fall admission, 3/1. Applications are processed on a rolling basis. *Application fee:* $40.

Expenses: Tuition, state resident: full-time $7,782. Tuition, nonresident: full-time $22,808. Required fees: $1,918.

Financial aid: Fellowships, research assistantships, teaching assistantships, Federal Work-Study available. Aid available to part-time students. Financial aid application deadline: 3/1.

Kenneth Hunter, Vice President of Research and Dean of Graduate School, 775-784-6869, *Fax:* 775-784-6064, *E-mail:* gradschool@unr.edu.

■ UNIVERSITY OF NEW HAMPSHIRE

Graduate School, College of Life Sciences and Agriculture, Graduate Programs in the Biological Sciences and Natural Resources, Durham, NH 03824

AWARDS Animal and nutritional sciences (MS, PhD); biochemistry and molecular biology (MS, PhD); genetics (MS, PhD); microbiology (MS, PhD); natural resources (MS, PhD), including environmental conservation (MS), forestry (MS), natural resources (PhD), soil science (MS), water resources management (MS), wildlife (MS); plant biology (MS, PhD); zoology (MS, PhD). Part-time programs available.

Faculty: 169 full-time.

Students: 121 full-time (71 women), 111 part-time (62 women); includes 5 minority (3 Asian Americans or Pacific Islanders, 2 Hispanic Americans), 34 international. Average age 30. *196 applicants, 45% accepted.* In 1999, 24 master's, 9 doctorates awarded. Terminal master's awarded for partial completion of doctoral program.

Degree requirements: For master's, foreign language not required; for doctorate, dissertation required.

Entrance requirements: For master's, GRE General Test. *Application deadline:* Applications are processed on a rolling basis. *Application fee:* $50.

Expenses: Tuition, area resident: Full-time $5,750; part-time $319 per credit. Tuition, state resident: full-time $8,625; part-time $478. Tuition, nonresident: full-time $14,640; part-time $598 per credit. Required fees: $224 per semester. Tuition and fees vary according to course load, degree level and program.

Financial aid: In 1999–00, 10 fellowships, 51 research assistantships, 76 teaching assistantships were awarded; career-related internships or fieldwork, Federal Work-Study, scholarships, and tuition waivers (full and partial) also available. Aid available to part-time students. Financial aid application deadline: 2/15.

Dr. William Mautz, Dean, College of Life Sciences and Agriculture, 603-862-1450.

■ UNIVERSITY OF NEW MEXICO

Graduate School, College of Arts and Sciences, Department of Biology, Albuquerque, NM 87131-2039

AWARDS Biology (MS, PhD), including air land ecology, behavioral ecology, botany, cellular and molecular biology, community ecology, comparative immunology, comparative physiology, conservation biology, ecology, ecosystem ecology, evolutionary biology, evolutionary genetics, microbiology, molecular genetics, parasitology, physiological ecology, physiology, population biology, vertebrate and invertebrate zoology. Part-time programs available.

Faculty: 35 full-time (5 women), 18 part-time/adjunct (11 women).

Students: 71 full-time (37 women), 28 part-time (11 women); includes 8 minority (2 Asian Americans or Pacific Islanders, 5 Hispanic Americans, 1 Native American), 11 international. Average age 33. *93 applicants, 30% accepted.* In 1999, 11 master's, 12 doctorates awarded. Terminal master's awarded for partial completion of doctoral program.

Degree requirements: For master's, one foreign language (computer language can substitute), thesis required (for some programs); for doctorate, 2 foreign languages (computer language can substitute for one), dissertation required.

Entrance requirements: For master's and doctorate, GRE General Test, GRE Subject Test, minimum GPA of 3.2. *Application deadline:* For fall admission, 1/15. *Application fee:* $25.

Expenses: Tuition, state resident: full-time $2,514; part-time $105 per credit hour.

Tuition, nonresident: full-time $10,304; part-time $417 per credit hour. International tuition: $10,304 full-time. Required fees: $516; $22 per credit hour. Tuition and fees vary according to program.

Financial aid: In 1999–00, 58 students received aid, including 24 fellowships (averaging $1,645 per year), 26 research assistantships with tuition reimbursements available (averaging $8,921 per year), 40 teaching assistantships with tuition reimbursements available (averaging $11,066 per year); career-related internships or fieldwork, Federal Work-Study, institutionally sponsored loans, and tuition waivers (full and partial) also available. Aid available to part-time students. Financial aid applicants required to submit FAFSA.

Faculty research: Developmental biology, immunobiology. *Total annual research expenditures:* $4.5 million.

Dr. Kathryn Vogel, Chair, 505-277-3411, *Fax:* 505-277-0304, *E-mail:* kgvogel@unm.edu.

Application contact: Vivian Kent, Information Contact, 505-277-1712, *Fax:* 505-277-0304, *E-mail:* vkent@unm.edu.

■ UNIVERSITY OF NEW MEXICO

Graduate School, Biomedical Sciences Graduate Program, Albuquerque, NM 87131-2039

AWARDS MS, PhD. Part-time programs available.

Faculty: 54 full-time (15 women), 11 part-time/adjunct (2 women).

Students: 57 full-time (38 women), 17 part-time (9 women); includes 16 minority (1 African American, 3 Asian Americans or Pacific Islanders, 12 Hispanic Americans), 7 international. Average age 30. *55 applicants, 60% accepted.* In 1999, 6 master's, 6 doctorates awarded. Terminal master's awarded for partial completion of doctoral program.

Degree requirements: For master's, thesis required, foreign language not required; for doctorate, dissertation, comprehensive exam, qualifying exam required, foreign language not required.

Entrance requirements: For master's and doctorate, GRE General Test, TOEFL, minimum undergraduate GPA of 3.0. *Application deadline:* For fall admission, 6/1. *Application fee:* $25.

Expenses: Tuition, state resident: full-time $2,514; part-time $105 per credit hour. Tuition, nonresident: full-time $10,304; part-time $417 per credit hour. International tuition: $10,304 full-time. Required fees: $516; $22 per credit hour. Tuition and fees vary according to program.

Financial aid: In 1999–00, 4 fellowships (averaging $3,475 per year), 49 research assistantships with tuition reimbursements (averaging $11,063 per year) were awarded; teaching assistantships with tuition reimbursements, career-related internships or fieldwork, Federal Work-Study, and institutionally sponsored loans also available. Financial aid application deadline: 2/1; financial aid applicants required to submit FAFSA.

Faculty research: Signal transduction, molecular epidemiology, extracellular matrix, toxicology, neurosciences.
William R. Galey, Director, 505-272-1887, *Fax:* 505-272-8738, *E-mail:* bgaley@salud.unm.edu.

Application contact: Kathy Hayden, Administrative Assistant, 505-272-1887, *Fax:* 505-272-8738, *E-mail:* khayden@salud.unm.edu.

■ UNIVERSITY OF NEW ORLEANS

Graduate School, College of Sciences, Department of Biological Sciences, New Orleans, LA 70148

AWARDS Biological sciences (MS); conservation biology (PhD).

Faculty: 28 full-time (10 women), 3 part-time/adjunct (all women).
Students: 14 full-time (7 women), 6 part-time (2 women); includes 3 minority (1 African American, 2 Hispanic Americans), 2 international. Average age 29. *33 applicants, 21% accepted.* In 1999, 7 degrees awarded.
Degree requirements: For master's, one foreign language (computer language can substitute), thesis required.
Entrance requirements: For master's, GRE General Test. *Application deadline:* For fall admission, 7/1 (priority date). Applications are processed on a rolling basis. *Application fee:* $20.
Expenses: Tuition, state resident: full-time $2,362. Tuition, nonresident: full-time $7,888. Part-time tuition and fees vary according to course load.
Faculty research: Biochemistry, genetics, vertebrate and invertebrate systematics and ecology, cell and mammalian physiology, morphology.
Dr. Sam Rogers, Chairman, 504-280-6307, *Fax:* 504-280-6121, *E-mail:* jsrogers@uno.edu.

Application contact: Dr. Jerry Howard, Graduate Coordinator, 504-280-7059, *Fax:* 504-280-6121, *E-mail:* jjhoward@uno.edu.

Find an in-depth description at www.petersons.com/graduate.

■ THE UNIVERSITY OF NORTH CAROLINA AT CHAPEL HILL

Graduate School, College of Arts and Sciences, Department of Biology, Chapel Hill, NC 27599

AWARDS Botany (MA, MS, PhD); cell biology, development, and physiology (MA, MS, PhD); ecology and behavior (MA, MS, PhD); genetics and molecular biology (MA, MS, PhD); morphology, systematics, and evolution (MA, MS, PhD).

Degree requirements: For master's, thesis (for some programs), comprehensive exams required; for doctorate, dissertation, comprehensive exams required.
Entrance requirements: For master's and doctorate, GRE General Test, GRE Subject Test. Electronic applications accepted.
Expenses: Tuition, state resident: full-time $1,578. Tuition, nonresident: full-time $10,744. Required fees: $827. One-time fee: $15 full-time. Tuition and fees vary according to program.
Faculty research: Gene expression, biomechanics, yeast genetics, plant ecology, plant molecular biology.

Find an in-depth description at www.petersons.com/graduate.

■ THE UNIVERSITY OF NORTH CAROLINA AT CHAPEL HILL

School of Medicine and Graduate School, Graduate Programs in Medicine, Chapel Hill, NC 27599

AWARDS Allied health sciences (MPT, MS, PhD), including human movement science (PhD), occupational science (MS), physical therapy (MPT), rehabilitation psychology and counseling (MS), speech and hearing sciences (MS); biochemistry and biophysics (MS, PhD); biomedical engineering (MS, PhD); cell and molecular physiology (PhD); cell biology and anatomy (PhD); genetics and molecular biology (MS, PhD); microbiology and immunology (MS, PhD), including immunology, microbiology; neurobiology (PhD); pathology and laboratory medicine (PhD), including experimental pathology; pharmacology (PhD). Postbaccalaureate distance learning degree programs offered.

Faculty: 210 full-time (69 women), 64 part-time/adjunct (16 women).
Students: 538 (377 women). In 1999, 160 master's, 43 doctorates awarded. Terminal master's awarded for partial completion of doctoral program.
Degree requirements: For master's, comprehensive exam required; for doctorate, dissertation required.
Entrance requirements: For master's and doctorate, GRE General Test. *Application deadline:* Applications are processed on a

rolling basis. *Application fee:* $55. Electronic applications accepted.
Expenses: Tuition, state resident: full-time $1,966. Tuition, nonresident: full-time $11,026. Required fees: $8,940. One-time fee: $15 full-time. Part-time tuition and fees vary according to course load.
Financial aid: In 1999–00, 86 fellowships with full and partial tuition reimbursements, 187 research assistantships with full tuition reimbursements, 29 teaching assistantships with full tuition reimbursements were awarded; career-related internships or fieldwork, Federal Work-Study, institutionally sponsored loans, traineeships, tuition waivers (full and partial), and unspecified assistantships also available. Aid available to part-time students. Financial aid applicants required to submit FAFSA.
Dr. Jeffrey L. Houpt, Dean, 919-966-4161, *Fax:* 919-966-6354.

Find an in-depth description at www.petersons.com/graduate.

■ THE UNIVERSITY OF NORTH CAROLINA AT CHARLOTTE

Graduate School, College of Arts and Sciences, Department of Biology, Charlotte, NC 28223-0001

AWARDS MA, MS, PhD. Part-time and evening/weekend programs available.

Faculty: 26 full-time (5 women), 5 part-time/adjunct (2 women).
Students: 15 full-time (9 women), 34 part-time (17 women); includes 3 minority (all Asian Americans or Pacific Islanders), 9 international. Average age 27. *36 applicants, 47% accepted.* In 1999, 7 degrees awarded.
Degree requirements: For master's and doctorate, thesis/dissertation required.
Entrance requirements: For master's, GRE General Test, minimum GPA of 3.0 in undergraduate major, 2.75 overall. *Application deadline:* For fall admission, 7/15; for spring admission, 11/15. Applications are processed on a rolling basis. *Application fee:* $35. Electronic applications accepted.
Expenses: Tuition, state resident: full-time $982; part-time $246 per year. Tuition, nonresident: full-time $8,252; part-time $2,064 per year. Required fees: $958; $252 per year. Part-time tuition and fees vary according to course load.
Financial aid: In 1999–00, 2 fellowships (averaging $4,000 per year), 6 research assistantships, 23 teaching assistantships were awarded; career-related internships or fieldwork, Federal Work-Study, and unspecified assistantships also available. Financial aid application deadline: 4/1.
Faculty research: Microbiology, immnology, cell biology, ecology, physiology.

The University of North Carolina at Charlotte (continued)
Dr. Mark G. Clemens, Chair, 704-547-2318, *Fax:* 704-547-3128.

Application contact: Kathy Barringer, Assistant Director of Graduate Admissions, 704-547-3366, *Fax:* 704-547-3279, *E-mail:* gradadm@email.uncc.edu.

■ THE UNIVERSITY OF NORTH CAROLINA AT GREENSBORO

Graduate School, College of Arts and Sciences, Department of Biology, Greensboro, NC 27412-5001

AWARDS M Ed, MS.

Faculty: 18 full-time (5 women), 3 part-time/adjunct (2 women).
Students: 10 full-time (3 women), 20 part-time (13 women); includes 6 minority (4 African Americans, 1 Asian American or Pacific Islander, 1 Hispanic American), 2 international. *21 applicants, 86% accepted.* In 1999, 6 degrees awarded.
Degree requirements: For master's, thesis required, foreign language not required. *Average time to degree:* Master's–2 years full-time, 3 years part-time.
Entrance requirements: For master's, GRE General Test, GRE Subject Test, TOEFL. *Application deadline:* For fall admission, 3/1 (priority date); for spring admission, 11/1. Applications are processed on a rolling basis. *Application fee:* $35.
Expenses: Tuition, state resident: full-time $2,200; part-time $182 per semester. Tuition, nonresident: full-time $10,600; part-time $1,238 per semester. Tuition and fees vary according to course load and program.
Financial aid: In 1999–00, 3 research assistantships with full tuition reimbursements (averaging $7,333 per year), 27 teaching assistantships with full tuition reimbursements (averaging $8,259 per year) were awarded; career-related internships or fieldwork, grants, scholarships, traineeships, and unspecified assistantships also available.
Faculty research: Environmental biology, biochemistry, animal ecology, vertebrate reproduction.
Dr. Anne Hershey, Head, 336-334-5391, *Fax:* 336-334-5839, *E-mail:* anne_hershey@uncg.edu.
Application contact: Dr. James Lynch, Director of Graduate Recruitment and Information Services, 336-334-4881, *Fax:* 336-334-4424, *E-mail:* jmlynch@office.uncg.edu.

■ THE UNIVERSITY OF NORTH CAROLINA AT WILMINGTON

College of Arts and Sciences, Department of Biological Sciences, Wilmington, NC 28403-3201

AWARDS Biology (MS); marine biology (MS). Part-time programs available.

Faculty: 19 full-time (2 women).
Students: 8 full-time (all women), 56 part-time (30 women); includes 3 minority (all Asian Americans or Pacific Islanders). Average age 29. *100 applicants, 23% accepted.* In 1999, 25 degrees awarded.
Degree requirements: For master's, thesis, oral and written comprehensive exams required.
Entrance requirements: For master's, GRE General Test, GRE Subject Test, minimum B average in undergraduate major. *Application deadline:* For fall admission, 3/15. Applications are processed on a rolling basis. *Application fee:* $45.
Expenses: Tuition, state resident: full-time $982. Tuition, nonresident: full-time $2,252. Required fees: $1,106. Part-time tuition and fees vary according to course load.
Financial aid: In 1999–00, 5 research assistantships, 31 teaching assistantships were awarded; career-related internships or fieldwork and Federal Work-Study also available. Aid available to part-time students. Financial aid application deadline: 3/15.
Faculty research: Marine processes, estuaries studies, biotechnology, underwater research, acid rain.
Dr. L. Scott Quackenbush, Chairman, 910-962-3470.
Application contact: Dr. Neil F. Hadley, Dean, Graduate School, 910-962-4117, *Fax:* 910-962-3787, *E-mail:* hadleyn@uncwil.edu.

■ UNIVERSITY OF NORTH DAKOTA

Graduate School, College of Arts and Sciences, Department of Biology, Grand Forks, ND 58202

AWARDS Botany (MS, PhD); ecology (MS, PhD); entomology (MS, PhD); environmental biology (MS, PhD); fisheries/wildlife (MS, PhD); genetics (MS, PhD); zoology (MS, PhD).

Faculty: 18 full-time (3 women).
Students: 21 full-time (8 women). *13 applicants, 62% accepted.* In 1999, 3 master's awarded. Terminal master's awarded for partial completion of doctoral program.
Degree requirements: For master's, thesis, final exam required; for doctorate, dissertation, comprehensive exam, final exam required.

Entrance requirements: For master's, GRE General Test, GRE Subject Test, TOEFL, minimum GPA of 3.0; for doctorate, GRE General Test, GRE Subject Test, TOEFL, minimum GPA of 3.5. *Application deadline:* For fall admission, 3/1 (priority date). Applications are processed on a rolling basis. *Application fee:* $25.
Expenses: Tuition, state resident: full-time $3,166; part-time $158 per credit. Tuition, nonresident: full-time $7,658; part-time $345 per credit. International tuition: $7,658 full-time. Required fees: $46 per credit. Tuition and fees vary according to program and reciprocity agreements.
Financial aid: In 1999–00, 6 research assistantships with full tuition reimbursements (averaging $11,250 per year), 15 teaching assistantships with full tuition reimbursements (averaging $11,250 per year) were awarded; fellowships, Federal Work-Study, institutionally sponsored loans, scholarships, and tuition waivers (full and partial) also available. Aid available to part-time students. Financial aid application deadline: 3/15; financial aid applicants required to submit FAFSA.
Faculty research: Population biology, wildlife ecology, RNA processing, hormonal control of behavior.
Dr. Jeff Lang, Director, 701-777-2621, *Fax:* 701-777-2623, *E-mail:* jlang@badlands.nodak.edu.

■ UNIVERSITY OF NORTH DAKOTA

School of Medicine and Graduate School, Graduate Programs in Medicine, Grand Forks, ND 58202

AWARDS MPT, MS, PhD, MD/PhD. Postbaccalaureate distance learning degree programs offered (minimal on-campus study).

Faculty: 44 full-time (10 women).
Students: 93 full-time (58 women), 16 part-time (12 women). *92 applicants, 68% accepted.* In 1999, 61 master's, 9 doctorates awarded.
Degree requirements: For doctorate, dissertation, comprehensive final exam, final exam required.
Entrance requirements: For master's, TOEFL, minimum GPA of 3.0; for doctorate, TOEFL, minimum GPA of 3.5. *Application deadline:* For fall admission, 3/1 (priority date). Applications are processed on a rolling basis. *Application fee:* $20.
Expenses: Tuition, state resident: full-time $2,690; part-time $112 per credit. Tuition, nonresident: full-time $7,182; part-time $299 per credit. Required fees: $46 per semester.
Financial aid: In 1999–00, 43 students received aid, including 17 research

assistantships with full tuition reimbursements available (averaging $10,586 per year), 21 teaching assistantships with full tuition reimbursements available (averaging $10,586 per year); fellowships, Federal Work-Study, institutionally sponsored loans, scholarships, and tuition waivers (full and partial) also available. Aid available to part-time students. Financial aid application deadline: 3/15; financial aid applicants required to submit FAFSA. *Total annual research expenditures:* $4.7 million.
Dr. H. David Wilson, Dean, 701-777-2514, *Fax:* 701-777-3527, *E-mail:* hdwilson@mail.med.und.nodak.edu.

■ **UNIVERSITY OF NORTHERN COLORADO**

Graduate School, College of Arts and Sciences, Department of Biological Sciences, Greeley, CO 80639

AWARDS Biological education (PhD); biological sciences (MA).

Faculty: 13 full-time (4 women).
Students: 28 full-time (17 women), 1 part-time; includes 4 minority (1 African American, 2 Asian Americans or Pacific Islanders, 1 Hispanic American). Average age 30. *14 applicants, 71% accepted.* In 1999, 9 master's, 1 doctorate awarded.
Degree requirements: For master's, thesis or alternative, comprehensive exams required; for doctorate, dissertation, comprehensive exams required.
Entrance requirements: For doctorate, GRE General Test. *Application deadline:* Applications are processed on a rolling basis. *Application fee:* $35.
Expenses: Tuition, state resident: full-time $2,382; part-time $132 per credit hour. Tuition, nonresident: full-time $8,997; part-time $500 per credit hour. Required fees: $686; $38 per credit hour.
Financial aid: In 1999–00, 28 students received aid, including 6 fellowships (averaging $383 per year), 5 research assistantships (averaging $10,452 per year), 23 teaching assistantships (averaging $8,585 per year); unspecified assistantships also available. Financial aid application deadline: 3/1.
Dr. Curt Peterson, Chairperson, 970-351-2921.

■ **UNIVERSITY OF NORTHERN IOWA**

Graduate College, College of Natural Sciences, Department of Biology, Cedar Falls, IA 50614

AWARDS MA, MS. Part-time programs available.

Faculty: 16 full-time (2 women).
Students: 9 full-time (7 women), 2 part-time; includes 1 minority (African

American). Average age 34. *11 applicants, 73% accepted.* In 1999, 3 degrees awarded.
Degree requirements: For master's, thesis or alternative required, foreign language not required.
Application deadline: For fall admission, 8/1 (priority date). Applications are processed on a rolling basis. *Application fee:* $20 ($50 for international students).
Expenses: Tuition, state resident: full-time $3,308; part-time $184 per hour. Tuition, nonresident: full-time $8,156; part-time $454 per hour. Required fees: $202; $101 per semester. Tuition and fees vary according to course load.
Financial aid: Scholarships available. Financial aid application deadline: 3/1.
Dr. Barbara A. Hetrick, Head, 319-273-2456, *Fax:* 319-.273-7125, *E-mail:* barbara.hetrick@uni.edu.

■ **UNIVERSITY OF NORTH TEXAS**

Robert B. Toulouse School of Graduate Studies, College of Arts and Sciences, Department of Biological Sciences, Denton, TX 76203

AWARDS Biochemistry (MS, PhD); biology (MA, MS, PhD); environmental science (MS, PhD); molecular biology (MA, MS, PhD). Terminal master's awarded for partial completion of doctoral program.

Degree requirements: For master's, oral defense of thesis required; for doctorate, dissertation, oral and written comprehensive exams required.
Entrance requirements: For master's and doctorate, GRE General Test, minimum GPA of 3.0.
Expenses: Tuition, state resident: full-time $2,865; part-time $600 per semester. Tuition, nonresident: full-time $8,049; part-time $1,896 per semester. Required fees: $26 per hour.
Faculty research: Network neuroscience, toxicology with earthworm model, aquatic toxicology, biodegradation of toxic chemicals, physiology with roundworm parasite.

■ **UNIVERSITY OF NORTH TEXAS HEALTH SCIENCE CENTER AT FORT WORTH**

Graduate School of Biomedical Sciences, Fort Worth, TX 76107-2699

AWARDS Anatomy and cell biology (MS, PhD); biochemistry and molecular biology (MS, PhD); biotechnology (MS); integrative physiology (MS, PhD); microbiology and immunology (MS, PhD); pharmacology (MS, PhD).

Faculty: 65 full-time (9 women), 11 part-time/adjunct (1 woman).

Students: 59 full-time (27 women), 44 part-time (20 women); includes 30 minority (10 African Americans, 9 Asian Americans or Pacific Islanders, 11 Hispanic Americans), 23 international. *70 applicants, 70% accepted.* In 1999, 5 master's awarded (40% found work related to degree, 60% continued full-time study); 5 doctorates awarded (80% entered university research/teaching, 20% found other work related to degree).
Degree requirements: For doctorate, computer language, dissertation required. *Average time to degree:* Master's–2.5 years full-time, 4 years part-time; doctorate–5 years full-time.
Entrance requirements: For master's and doctorate, GRE General Test, TOEFL. *Application deadline:* For fall admission, 5/1; for spring admission, 11/1. Applications are processed on a rolling basis. *Application fee:* $25 ($50 for international students).
Expenses: Tuition, state resident: full-time $1,188; part-time $66 per credit. Tuition, nonresident: full-time $5,058; part-time $281 per credit. Required fees: $366; $183 per semester.
Financial aid: In 1999–00, 11 fellowships, 70 research assistantships (averaging $16,500 per year) were awarded; teaching assistantships, career-related internships or fieldwork, Federal Work-Study, grants, institutionally sponsored loans, and traineeships also available. Aid available to part-time students. Financial aid application deadline: 4/1; financial aid applicants required to submit FAFSA.
Faculty research: Alzheimer's disease, diabetes, eye diseases, cancer, cardiovascular physiology.
Dr. Thomas Yorio, Dean, 817-735-2560, *Fax:* 817-735-0243, *E-mail:* yoriot@hsc.unt.edu.
Application contact: Jan Sharp, Administrative Assistant, 817-735-0258, *Fax:* 817-735-0243, *E-mail:* gsbs@hsc.unt.edu.

Find an in-depth description at www.petersons.com/graduate.

■ **UNIVERSITY OF NOTRE DAME**

Graduate School, College of Science, Department of Biological Sciences, Notre Dame, IN 46556

AWARDS Aquatic ecology, evolution and environmental biology (MS, PhD); cellular and molecular biology (MS, PhD); developmental biology (MS, PhD); genetics (MS, PhD); physiology (MS, PhD); vector biology and parasitology (MS, PhD).

Faculty: 33 full-time (5 women), 5 part-time/adjunct (1 woman).
Students: 78 full-time (37 women); includes 4 minority (1 African American, 2 Asian Americans or Pacific Islanders, 1

University of Notre Dame (continued)
Hispanic American), 24 international. *224 applicants, 17% accepted.* In 1999, 1 master's, 10 doctorates awarded (70% entered university research/teaching, 20% found other work related to degree). Terminal master's awarded for partial completion of doctoral program.

Degree requirements: For master's and doctorate, thesis/dissertation required, foreign language not required. *Average time to degree:* Master's–3 years full-time; doctorate–6 years full-time.

Entrance requirements: For master's and doctorate, GRE General Test, GRE Subject Test, TOEFL. *Application deadline:* For fall admission, 2/1 (priority date); for spring admission, 11/1. Applications are processed on a rolling basis. *Application fee:* $50.

Expenses: Tuition: Full-time $21,930; part-time $1,218 per credit. Required fees: $95. Tuition and fees vary according to program.

Financial aid: In 1999–00, 78 students received aid, including 26 fellowships with full tuition reimbursements available (averaging $18,000 per year), 14 research assistantships with full tuition reimbursements available (averaging $13,000 per year), 38 teaching assistantships with full tuition reimbursements available (averaging $13,000 per year); traineeships and tuition waivers (full) also available. Financial aid application deadline: 2/1. *Total annual research expenditures:* $5.5 million.

Dr. Frederick W. Goetz, Director of Graduate Studies, 219-631-6552, *Fax:* 219-631-7413, *E-mail:* biology.biosadm.1@nd.edu.

Application contact: Dr. Terrence J. Akai, Director of Graduate Admissions, 219-631-7706, *Fax:* 219-631-4183, *E-mail:* gradad@nd.edu.

Find an in-depth description at www.petersons.com/graduate.

■ UNIVERSITY OF OKLAHOMA HEALTH SCIENCES CENTER

College of Medicine and Graduate College, Graduate Programs in Medicine, Oklahoma City, OK 73190

AWARDS Biochemistry and molecular biology (MS, PhD), including biochemistry, molecular biology; cell biology (MS, PhD); medical sciences (MS); microbiology and immunology (MS, PhD), including immunology, microbiology; molecular medicine (PhD); neuroscience (MS, PhD); pathology (PhD); physiology (MS, PhD); psychiatry and behavioral sciences (MS, PhD), including biological psychology; radiological sciences (MS, PhD), including medical radiation physics. Part-time programs available.

Faculty: 132 full-time (25 women), 46 part-time/adjunct (14 women).

Students: 44 full-time (24 women), 56 part-time (28 women); includes 13 minority (1 African American, 8 Asian Americans or Pacific Islanders, 1 Hispanic American, 3 Native Americans), 20 international. Average age 27. *186 applicants, 25% accepted.* In 1999, 15 master's, 15 doctorates awarded. Terminal master's awarded for partial completion of doctoral program.

Degree requirements: For master's, foreign language not required; for doctorate, dissertation required.

Entrance requirements: For master's and doctorate, GRE General Test, TOEFL. *Application fee:* $25 ($50 for international students).

Expenses: Tuition, state resident: part-time $90 per semester hour. Tuition, nonresident: part-time $264 per semester hour. Tuition and fees vary according to program.

Financial aid: Fellowships, research assistantships, teaching assistantships, career-related internships or fieldwork, Federal Work-Study, institutionally sponsored loans, and tuition waivers (full and partial) available. Aid available to part-time students.

Faculty research: Behavior and drugs, structure and function of endothelium, genetics and behavior, gene structure and function, action of antibiotics.

Dr. O. Ray Kling, Dean, 405-271-2085, *Fax:* 405-271-1155, *E-mail:* ray-kling@uokhsc.edu.

Find an in-depth description at www.petersons.com/graduate.

■ UNIVERSITY OF OREGON

Graduate School, College of Arts and Sciences, Department of Biology, Eugene, OR 97403

AWARDS Ecology and evolution (MA, MS, PhD); marine biology (MA, MS, PhD); molecular, cellular and genetic biology (PhD); neuroscience and development (PhD).

Faculty: 42 full-time (14 women), 5 part-time/adjunct (2 women).

Students: 69 full-time (33 women), 8 part-time (5 women); includes 6 minority (2 Asian Americans or Pacific Islanders, 3 Hispanic Americans, 1 Native American), 5 international. *18 applicants, 83% accepted.* In 1999, 11 master's, 7 doctorates awarded. Terminal master's awarded for partial completion of doctoral program.

Degree requirements: For master's, thesis required (for some programs); for doctorate, dissertation required, foreign language not required.

Entrance requirements: For master's and doctorate, GRE General Test, TOEFL,

minimum GPA of 3.2. *Application deadline:* For fall admission, 1/10. *Application fee:* $50.

Expenses: Tuition, state resident: full-time $6,750. Tuition, nonresident: full-time $11,409. Part-time tuition and fees vary according to course load.

Financial aid: In 1999–00, 36 teaching assistantships were awarded; research assistantships, Federal Work-Study, grants, and institutionally sponsored loans also available. Financial aid application deadline: 2/1.

Faculty research: Developmental neurobiology; evolution, population biology, and quantitative genetics; regulation of gene expression; biochemistry of marine organisms.

Janis C. Weeks, Head, 541-346-4502, *Fax:* 541-346-6056.

Application contact: Donna Overall, Graduate Program Coordinator, 541-346-4503, *Fax:* 541-346-6056, *E-mail:* doverall@oregon.uoregon.edu.

Find an in-depth description at www.petersons.com/graduate.

■ UNIVERSITY OF PENNSYLVANIA

School of Arts and Sciences, Graduate Group in Biology, Philadelphia, PA 19104

AWARDS Cell, molecular, and developmental biology (PhD); ecology and population biology (PhD); neurobiology/physiology and behavior (PhD); plant science (PhD).

Students: 62 full-time (33 women), 1 (woman) part-time, 26 international. Average age 25. *113 applicants, 28% accepted.* In 1999, 1 degree awarded.

Degree requirements: For doctorate, dissertation required, foreign language not required.

Entrance requirements: For doctorate, GRE General Test, GRE Subject Test, TOEFL. *Application fee:* $65.

Expenses: Tuition: Full-time $23,670. Required fees: $1,546. Full-time tuition and fees vary according to degree level and program.

Financial aid: Fellowships, research assistantships, teaching assistantships, tuition waivers (full and partial) available. Financial aid application deadline: 1/2.

Dr. Fevzi Daldal, Graduate Group Chair, 215-898-6786, *Fax:* 215-898-8780, *E-mail:* fdaldal@mail.sas.upenn.edu.

Application contact: Allan Aiken, Graduate Group Coordinator, 215-898-6786, *Fax:* 215-898-8780, *E-mail:* aaiken@sas.upenn.edu.

Find an in-depth description at www.petersons.com/graduate.

■ UNIVERSITY OF PENNSYLVANIA

School of Medicine, Biomedical Graduate Studies, Philadelphia, PA 19104

AWARDS MS, PhD, DMD/PhD, MD/PhD, VMD/PhD. Part-time programs available.

Faculty: 584 full-time (105 women), 6 part-time/adjunct (2 women).
Students: 502 full-time (222 women), 3 part-time (1 woman); includes 130 minority (20 African Americans, 92 Asian Americans or Pacific Islanders, 17 Hispanic Americans, 1 Native American), 53 international. *853 applicants, 19% accepted.* In 1999, 8 master's, 61 doctorates awarded. Terminal master's awarded for partial completion of doctoral program.
Degree requirements: For master's, thesis, comprehensive exam required, foreign language not required; for doctorate, dissertation required, foreign language not required.
Entrance requirements: For master's and doctorate, GRE General Test, TOEFL. *Application deadline:* For fall admission, 1/2 (priority date). Applications are processed on a rolling basis. *Application fee:* $65. Electronic applications accepted.
Expenses: Tuition: Full-time $17,256; part-time $2,991 per course. Required fees: $2,588; $363 per course. $726 per term.
Financial aid: In 1999–00, 419 students received aid, including 278 fellowships (averaging $38,517 per year), 203 research assistantships (averaging $38,517 per year); teaching assistantships, grants, institutionally sponsored loans, and unspecified assistantships also available. Financial aid application deadline: 1/2.
Dr. Michael E. Selzer, Director, 215-898-1030.
Application contact: Evelyn P. Olivieri, Admissions Coordinator, 215-898-1030, *E-mail:* oliviere@mail.med.upenn.edu.

■ UNIVERSITY OF PITTSBURGH

Faculty of Arts and Sciences, Department of Biological Sciences, Pittsburgh, PA 15260

AWARDS Ecology and evolution (MS, PhD); molecular biophysics (PhD); molecular, cellular, and developmental biology (PhD). Part-time programs available.

Faculty: 43 full-time (15 women), 9 part-time/adjunct (7 women).
Students: 50 full-time (26 women); includes 5 minority (3 African Americans, 2 Hispanic Americans), 17 international. *68 applicants, 15% accepted.* In 1999, 5 master's, 6 doctorates awarded.
Degree requirements: For master's and doctorate, thesis/dissertation required,

foreign language not required. *Average time to degree:* Master's–3 years full-time; doctorate–6 years full-time.
Entrance requirements: For master's and doctorate, GRE General Test, GRE Subject Test, TOEFL. *Application deadline:* For fall admission, 2/1 (priority date). Applications are processed on a rolling basis. *Application fee:* $40.
Expenses: Tuition, state resident: full-time $8,338; part-time $342 per credit. Tuition, nonresident: full-time $17,168; part-time $707 per credit. Required fees: $480; $90 per semester. Tuition and fees vary according to program.
Financial aid: In 1999–00, 22 fellowships with tuition reimbursements (averaging $1,525 per year), 45 research assistantships with tuition reimbursements (averaging $1,325 per year), 41 teaching assistantships with tuition reimbursements (averaging $1,325 per year) were awarded; Federal Work-Study also available.
Faculty research: Biochemistry, molecular biology/genetics, structural biophysics. *Total annual research expenditures:* $5.1 million.
Dr. James M. Pipas, Chairman, 412-624-4350, *Fax:* 412-624-4759, *E-mail:* pipas+@pitt.edu.
Application contact: Cathleen M. Barr, Graduate Administrator, 412-624-4268, *Fax:* 412-624-4759, *E-mail:* cbarr+@pitt.edu.

■ UNIVERSITY OF PITTSBURGH

School of Medicine, Graduate Programs in Medicine, Interdisciplinary Biomedical Sciences Program, Pittsburgh, PA 15260

AWARDS PhD.

Students: 58 full-time (26 women); includes 9 minority (1 African American, 5 Asian Americans or Pacific Islanders, 3 Hispanic Americans), 18 international. *325 applicants, 22% accepted.*
Degree requirements: For doctorate, dissertation required, foreign language not required.
Entrance requirements: For doctorate, GRE General Test, GRE Subject Test, TOEFL, minimum QPA of 3.0. *Application deadline:* For fall admission, 1/15 (priority date). Applications are processed on a rolling basis. *Application fee:* $30 ($40 for international students).
Expenses: Tuition, state resident: full-time $9,778; part-time $403 per credit. Tuition, nonresident: full-time $20,146; part-time $830 per credit. Required fees: $480; $90 per semester.
Financial aid: Research assistantships with full tuition reimbursements, teaching assistantships with full tuition reimbursements, Federal Work-Study, institutionally

sponsored loans, scholarships, traineeships, and unspecified assistantships available.
Faculty research: Biochemistry and molecular genetics, cell biology and molecular physiology, cellular and molecular pathology, immunology, molecular pharmacology. *Total annual research expenditures:* $76.3 million.
Application contact: Graduate Studies Administrator, 412-648-8957, *Fax:* 412-648-1236, *E-mail:* biomed_phd@fsl.dean_med.pitt.edu.

Find an in-depth description at www.petersons.com/graduate.

■ UNIVERSITY OF PUERTO RICO, MAYAGÜEZ CAMPUS

Graduate Studies, College of Arts and Sciences, Department of Biology, Mayagüez, PR 00681-9000

AWARDS MS. Part-time programs available.

Degree requirements: For master's, one foreign language, thesis, comprehensive exam required.
Faculty research: Herpetology, entomology, microbiology, immunology, botany.

■ UNIVERSITY OF PUERTO RICO, MEDICAL SCIENCES CAMPUS

School of Medicine, Division of Graduate Studies, San Juan, PR 00936-5067

AWARDS MS, PhD.

Faculty: 60 full-time (18 women).
Students: 65 full-time (40 women); all minorities (all Hispanic Americans). Average age 23. Terminal master's awarded for partial completion of doctoral program.
Degree requirements: For master's and doctorate, one foreign language, thesis/dissertation required.
Entrance requirements: For master's and doctorate, GRE General Test, GRE Subject Test, interview. *Application deadline:* For fall admission, 2/15. *Application fee:* $15.
Expenses: Tuition, state resident: full-time $5,500. Tuition, nonresident: full-time $8,400. Required fees: $600. Tuition and fees vary according to class time, course load, degree level and program.
Financial aid: Fellowships, research assistantships, teaching assistantships, career-related internships or fieldwork, Federal Work-Study, institutionally sponsored loans, and tuition waivers (full and partial) available. Aid available to part-time students. Financial aid application deadline: 4/30.
Dr. Guillermo Vázquez, Director, 787-758-2525 Ext. 1309, *Fax:* 787-758-4808.

Find an in-depth description at www.petersons.com/graduate.

■ UNIVERSITY OF PUERTO RICO, RÍO PIEDRAS

Faculty of Natural Sciences, Department of Biology, San Juan, PR 00931

AWARDS MS, PhD. Part-time and evening/weekend programs available.

Faculty: 26 full-time (10 women).
Students: 130 full-time (78 women); includes 129 minority (3 Asian Americans or Pacific Islanders, 126 Hispanic Americans), 1 international. *41 applicants, 46% accepted.* In 1999, 5 master's, 5 doctorates awarded.
Degree requirements: For master's and doctorate, one foreign language, thesis/dissertation, comprehensive exam required. *Average time to degree:* Master's–6 years full-time; doctorate–8 years full-time.
Entrance requirements: For master's, GRE Subject Test, interview, minimum GPA of 3.0; for doctorate, GRE Subject Test, interview, master's degree, minimum GPA of 3.0. *Application deadline:* For fall admission, 2/1. *Application fee:* $17.
Expenses: Tuition, state resident: full-time $1,200; part-time $75 per credit. Tuition, nonresident: full-time $3,500; part-time $219 per credit. Required fees: $70; $70 per year. Tuition and fees vary according to course load.
Financial aid: Fellowships, research assistantships, teaching assistantships, Federal Work-Study, institutionally sponsored loans, and tuition waivers (partial) available. Financial aid application deadline: 5/31.
Faculty research: Environmental, poblational and systematic biology.
Dr. José E. Garcia-Arraás, Coordinator, 787-764-0000 Ext. 3551, *E-mail:* jegarcia@upracd.upr.clu.edu.

Find an in-depth description at www.petersons.com/graduate.

■ UNIVERSITY OF RICHMOND

Graduate School, Department of Biology, Richmond, University of Richmond, VA 23173

AWARDS MS, JD/MS.

Degree requirements: For master's, thesis required, foreign language not required.
Entrance requirements: For master's, GRE General Test, GRE Subject Test, undergraduate major in biology or related area.
Expenses: Tuition: Full-time $19,440; part-time $335 per hour. Part-time tuition and fees vary according to course load.
Faculty research: DNA repair to gene regulation in microorganisms, phylogenetic relationships among microhylid subfamilies.

■ UNIVERSITY OF ROCHESTER

The College, Arts and Sciences, Department of Biology, Rochester, NY 14627-0250

AWARDS Cellular and molecular biology (MS, PhD); ecology and evolutionary biology (MS, PhD); genetics and developmental biology (MS, PhD).

Faculty: 17.
Students: 42 full-time (15 women), 1 part-time; includes 1 minority (Hispanic American), 18 international. *227 applicants, 14% accepted.* In 1999, 9 master's, 3 doctorates awarded. Terminal master's awarded for partial completion of doctoral program.
Degree requirements: For master's, thesis not required; for doctorate, dissertation, qualifying exam required, foreign language not required.
Entrance requirements: For master's and doctorate, GRE General Test, GRE Subject Test, TOEFL. *Application deadline:* For fall admission, 2/1 (priority date). *Application fee:* $25.
Expenses: Tuition: Part-time $697 per credit hour. Tuition and fees vary according to program.
Financial aid: Fellowships, research assistantships, teaching assistantships, tuition waivers (full and partial) available. Financial aid application deadline: 2/1. Robert Angerer, Chair, 716-275-3835.
Application contact: Cindy Landry, Graduate Program Secretary, 716-275-7991.

Find an in-depth description at www.petersons.com/graduate.

■ UNIVERSITY OF ROCHESTER

School of Medicine and Dentistry, Graduate Programs in Medicine and Dentistry, Rochester, NY 14627-0250

AWARDS MA, MPH, MS, PhD, Certificate, MBA/MPH, MBA/MS, MD/MPH, MD/MS, MD/PhD, MPH/MS, MPH/PhD. Part-time programs available.

Faculty: 141.
Students: 352 full-time (159 women), 88 part-time (43 women); includes 53 minority (16 African Americans, 21 Asian Americans or Pacific Islanders, 15 Hispanic Americans, 1 Native American), 111 international. *937 applicants, 22% accepted.* In 1999, 52 master's, 28 doctorates awarded.
Degree requirements: For doctorate, dissertation, qualifying exam required.
Entrance requirements: For master's and doctorate, GRE General Test. *Application deadline:* For fall admission, 2/1 (priority date). *Application fee:* $25. Electronic applications accepted.

Expenses: Tuition: Part-time $697 per credit hour. Tuition and fees vary according to program.
Financial aid: Fellowships, research assistantships, teaching assistantships, scholarships and tuition waivers (full and partial) available. Financial aid application deadline: 2/1.
Dr. Shey-Shing Sheu, Associate Dean, 716-275-2933.

Find an in-depth description at www.petersons.com/graduate.

■ UNIVERSITY OF SAN FRANCISCO

College of Arts and Sciences, Department of Biology, San Francisco, CA 94117-1080

AWARDS MS.

Faculty: 5 full-time (2 women), 2 part-time/adjunct (0 women).
Students: 3 full-time (all women), 2 part-time (both women); includes 1 minority (Asian American or Pacific Islander), 2 international. Average age 30. *14 applicants, 57% accepted.* In 1999, 1 degree awarded.
Degree requirements: For master's, thesis required, foreign language not required.
Entrance requirements: For master's, GRE General Test, GRE Subject Test, BS or equivalent in biology. *Application deadline:* For fall admission, 4/15; for spring admission, 10/15. *Application fee:* $55 ($65 for international students).
Expenses: Tuition: Full-time $12,618; part-time $701 per unit. Tuition and fees vary according to course load, degree level, campus/location and program.
Financial aid: In 1999–00, 4 students received aid; teaching assistantships, career-related internships or fieldwork, Federal Work-Study, institutionally sponsored loans, and tuition waivers available. Financial aid application deadline: 3/2.
Dr. Deneb Karentz, Chair, 415-422-6755, *E-mail:* karentzd@usfca.edu.

■ UNIVERSITY OF SOUTH ALABAMA

College of Medicine and Graduate School, Program in Basic Medical Sciences, Mobile, AL 36688-0002

AWARDS Biochemistry and molecular biology (PhD); cancer biology (PhD); cellular biology and neuroscience (PhD); microbiology and immunology (PhD); pharmacology (PhD); physiology (PhD).

Faculty: 48 full-time (6 women), 1 (woman) part-time/adjunct.
Students: 40 full-time (16 women), 6 part-time (3 women); includes 3 minority (1 Asian American or Pacific Islander, 2

Hispanic Americans), 12 international. In 1999, 9 degrees awarded.

Degree requirements: For doctorate, dissertation required, foreign language not required.

Application deadline: For fall admission, 4/1. Applications are processed on a rolling basis. *Application fee:* $25.

Expenses: Tuition, state resident: part-time $116 per semester hour. Tuition, nonresident: part-time $230 per semester hour. Required fees: $121 per semester.

Financial aid: Fellowships, research assistantships, institutionally sponsored loans available. Financial aid application deadline: 4/1.

Faculty research: Microcirculation, molecular biology, cell biology, growth control.

Lanette Flagge, Coordinator, 334-460-6153.

Find an in-depth description at www.petersons.com/graduate.

■ UNIVERSITY OF SOUTH ALABAMA

Graduate School, College of Arts and Sciences, Department of Biological Sciences, Mobile, AL 36688-0002

AWARDS MS. Part-time programs available.

Faculty: 15 full-time (3 women), 1 (woman) part-time/adjunct.

Students: 9 full-time (7 women), 12 part-time (10 women); includes 6 minority (4 African Americans, 1 Asian American or Pacific Islander, 1 Hispanic American), 1 international. *16 applicants, 63% accepted.* In 1999, 7 degrees awarded.

Degree requirements: For master's, one foreign language (computer language can substitute), comprehensive exams required, thesis optional.

Entrance requirements: For master's, GRE Subject Test, minimum GPA of 3.0. *Application deadline:* For fall admission, 9/1 (priority date). Applications are processed on a rolling basis. *Application fee:* $25.

Expenses: Tuition, state resident: part-time $116 per semester hour. Tuition, nonresident: part-time $230 per semester hour. Required fees: $121 per semester. Part-time tuition and fees vary according to course load and program.

Financial aid: In 1999–00, 9 research assistantships, 5 teaching assistantships were awarded. Aid available to part-time students. Financial aid application deadline: 4/1.

Faculty research: Aquatic and marine biology, molecular biochemistry, plant and animal taxonomy.

Dr. John Freeman, Chair, 334-460-6331.

■ UNIVERSITY OF SOUTH CAROLINA

Graduate School, College of Science and Mathematics, Department of Biological Sciences, Columbia, SC 29208

AWARDS Biology (MS, PhD); biology education (IMA, MAT); ecology, evolution and organismal biology (MS, PhD); molecular, cellular, and developmental biology (MS, PhD). IMA and MAT offered in cooperation with the College of Education.

Faculty: 42 full-time (6 women).

Students: 65 full-time (35 women), 8 part-time (3 women); includes 7 minority (2 African Americans, 3 Asian Americans or Pacific Islanders, 2 Hispanic Americans), 13 international. Average age 29. *415 applicants, 10% accepted.* In 1999, 12 master's awarded (100% continued full-time study); 4 doctorates awarded (100% continued full-time study). Terminal master's awarded for partial completion of doctoral program.

Degree requirements: For master's, thesis required (for some programs); for doctorate, dissertation required. *Average time to degree:* Master's–2.5 years full-time; doctorate–5.5 years full-time.

Entrance requirements: For master's and doctorate, GRE General Test, minimum GPA of 3.0 in science. *Application deadline:* For fall admission, 2/15 (priority date). Applications are processed on a rolling basis. *Application fee:* $35. Electronic applications accepted.

Expenses: Tuition, state resident: full-time $4,014; part-time $202 per credit hour. Tuition, nonresident: full-time $8,528; part-time $428 per credit hour. Required fees: $100; $4 per credit hour. Tuition and fees vary according to program.

Financial aid: In 1999–00, 1 fellowship with partial tuition reimbursement (averaging $15,000 per year), 24 research assistantships with partial tuition reimbursements (averaging $13,500 per year), 34 teaching assistantships with partial tuition reimbursements (averaging $13,333 per year) were awarded. Financial aid application deadline: 2/15.

Faculty research: Marine ecology, population and evolutionary biology, molecular biology and genetics, development. *Total annual research expenditures:* $5.9 million.

Dr. Franklin Berger, Chair, 803-777-4141, *Fax:* 803-777-4002, *E-mail:* berger@biol.sc.edu.

Application contact: Dr. Franklyn F. Bolander, Director of Graduate Studies, 803-777-2755, *Fax:* 803-777-4002, *E-mail:* bolander@sc.edu.

■ UNIVERSITY OF SOUTH CAROLINA

School of Medicine and Graduate School, Graduate Programs in Medicine, Columbia, SC 29208

AWARDS Biomedical science (MBS, PhD), including biomedical science, nurse anesthesia (MBS); genetic counseling (MS); rehabilitation counseling (MRC).

Faculty: 47 full-time (16 women), 20 part-time/adjunct (7 women).

Students: 108 full-time (66 women), 22 part-time (10 women); includes 19 minority (10 African Americans, 8 Asian Americans or Pacific Islanders, 1 Hispanic American), 6 international. Average age 29. *288 applicants, 21% accepted.* In 1999, 39 master's awarded (3% entered university research/teaching, 92% found other work related to degree, 3% continued full-time study); 2 doctorates awarded (100% entered university research/teaching). Terminal master's awarded for partial completion of doctoral program.

Degree requirements: For master's, foreign language not required; for doctorate, dissertation required, foreign language not required. *Average time to degree:* Master's–2.2 years full-time, 3 years part-time; doctorate–3.8 years full-time.

Entrance requirements: For doctorate, GRE General Test. *Application deadline:* Applications are processed on a rolling basis. *Application fee:* $35. Electronic applications accepted.

Expenses: Tuition, state resident: full-time $4,014; part-time $234 per credit hour. Tuition, nonresident: full-time $8,529; part-time $483 per credit hour. Required fees: $100; $4 per credit hour. Tuition and fees vary according to program.

Financial aid: Fellowships, research assistantships, teaching assistantships, career-related internships or fieldwork, Federal Work-Study, institutionally sponsored loans, and unspecified assistantships available.

Faculty research: Cardiovascular diseases, oncology, neuroscience, psychiatric rehabilitation, genetics. *Total annual research expenditures:* $6.7 million.

Dr. Larry R. Faulkner, Dean, School of Medicine, 803-733-3200.

■ UNIVERSITY OF SOUTH CAROLINA

School of Medicine and Graduate School, Graduate Programs in Medicine, Graduate Program in Biomedical Science, Doctoral Program in Biomedical Science, Columbia, SC 29208

AWARDS PhD.

University of South Carolina (continued)
Faculty: 38 full-time (11 women), 13 part-time/adjunct (2 women).
Students: 32 full-time (15 women), 1 part-time; includes 4 minority (1 African American, 3 Asian Americans or Pacific Islanders), 5 international. Average age 27. *34 applicants, 38% accepted.* In 1999, 2 degrees awarded (100% entered university research/teaching).
Degree requirements: For doctorate, dissertation required, foreign language not required. *Average time to degree:* Doctorate–3.8 years full-time.
Entrance requirements: For doctorate, GRE General Test. *Application deadline:* For fall admission, 4/15 (priority date); for spring admission, 11/15 (priority date). Applications are processed on a rolling basis. *Application fee:* $35. Electronic applications accepted.
Expenses: Tuition, state resident: full-time $4,014; part-time $234 per credit hour. Tuition, nonresident: full-time $8,529; part-time $483 per credit hour. Required fees: $100; $4 per credit hour. Tuition and fees vary according to program.
Financial aid: In 1999–00, 2 fellowships (averaging $13,500 per year), 30 research assistantships with partial tuition reimbursements (averaging $13,500 per year) were awarded. Financial aid application deadline: 4/15; financial aid applicants required to submit FAFSA.
Faculty research: Cancer, neuroscience, cardiovascular, developmental, reproductive, vision, immunology. *Total annual research expenditures:* $6.7 million.
Dr. James Buggy, Assistant Dean for Graduate Studies, 803-733-3100, *Fax:* 803-733-3168, *E-mail:* buggy@med.sc.edu.
Application contact: Jennifer Long, Program Coordinator, 803-733-3100, *Fax:* 803-733-3168, *E-mail:* biomed@ med.sc.edu.

Find an in-depth description at www.petersons.com/graduate.

■ UNIVERSITY OF SOUTH CAROLINA

School of Medicine and Graduate School, Graduate Programs in Medicine, Graduate Program in Biomedical Science, Master's Program in Biomedical Science, Columbia, SC 29208

AWARDS MBS.

Faculty: 38 full-time (11 women), 13 part-time/adjunct (2 women).
Students: 5 full-time (1 woman), 5 part-time (3 women); includes 5 minority (2 African Americans, 3 Asian Americans or Pacific Islanders). Average age 26. *13 applicants, 54% accepted.* In 1999, 4 degrees awarded (25% entered university research/

teaching, 25% found other work related to degree, 25% continued full-time study). *Average time to degree:* Master's–1.8 years full-time.
Entrance requirements: For master's, GRE General Test. *Application deadline:* For fall admission, 4/15 (priority date); for spring admission, 11/15 (priority date). Applications are processed on a rolling basis. *Application fee:* $35. Electronic applications accepted.
Expenses: Tuition, state resident: full-time $4,014; part-time $234 per credit hour. Tuition, nonresident: full-time $8,529; part-time $483 per credit hour. Required fees: $100; $4 per credit hour. Tuition and fees vary according to program.
Financial aid: In 1999–00, 9 students received aid, including 2 research assistantships with partial tuition reimbursements available (averaging $10,000 per year). Financial aid application deadline: 4/15; financial aid applicants required to submit FAFSA.
Faculty research: Cardiovascular diseases, oncology, reproductive biology, vision, neuroscience. *Total annual research expenditures:* $6.7 million.
Dr. James Buggy, Assistant Dean for Graduate Studies, 803-733-3100, *Fax:* 803-733-3168, *E-mail:* buggy@med.sc.edu.
Application contact: Jennifer Long, Program Coordinator, 803-733-3100, *Fax:* 803-733-3168, *E-mail:* biomed@ med.sc.edu.

Find an in-depth description at www.petersons.com/graduate.

■ UNIVERSITY OF SOUTH DAKOTA

Graduate School, College of Arts and Sciences, Department of Biology, Vermillion, SD 57069-2390

AWARDS MA, MNS, MS, PhD.

Faculty: 10 full-time (0 women), 3 part-time/adjunct (0 women).
Students: 16 full-time (8 women), 3 part-time (1 woman), 1 international. *28 applicants, 46% accepted.* In 1999, 8 master's, 1 doctorate awarded.
Degree requirements: For master's and doctorate, thesis/dissertation required, foreign language not required.
Entrance requirements: For master's, GRE Subject Test; for doctorate, GRE General Test, GRE Subject Test. *Application fee:* $15.
Expenses: Tuition, state resident: full-time $2,126; part-time $89 per credit. Tuition, nonresident: full-time $6,270; part-time $261 per credit. Required fees: $1,194; $50 per credit. Full-time tuition and fees vary according to degree level, program and reciprocity agreements.

Financial aid: Research assistantships, teaching assistantships, Federal Work-Study available. Aid available to part-time students.
Dr. Karen Olmstead, Chair, 605-677-5211.
Application contact: Dr. Paula Mabee, Graduate Advisor, *E-mail:* pmabee@ usd.edu.

■ UNIVERSITY OF SOUTH DAKOTA

School of Medicine and Graduate School, Biomedical Sciences Graduate Program, Vermillion, SD 57069-2390

AWARDS Cardiovascular research (MA, PhD); cellular and molecular biology (MA, PhD); molecular microbiology and immunology (MA, PhD); neuroscience (MA, PhD); physiology and pharmacology (MA, PhD). Part-time programs available.

Faculty: 34 full-time (9 women), 4 part-time/adjunct (0 women).
Students: 29 full-time (9 women); includes 1 minority (Asian American or Pacific Islander), 9 international. Average age 33. *197 applicants, 4% accepted.* In 1999, 3 master's awarded (100% found work related to degree); 3 doctorates awarded (100% continued full-time study). Terminal master's awarded for partial completion of doctoral program.
Degree requirements: For master's, thesis required, foreign language not required; for doctorate, dissertation required. *Average time to degree:* Master's–2.3 years full-time; doctorate–5 years full-time.
Entrance requirements: For master's and doctorate, GRE General Test, TOEFL, minimum GPA of 3.0. *Application deadline:* For fall admission, 3/15 (priority date). Applications are processed on a rolling basis. *Application fee:* $15.
Expenses: Tuition, state resident: full-time $2,126; part-time $89 per credit. Tuition, nonresident: full-time $6,270; part-time $261 per credit. Required fees: $1,194; $50 per credit. Tuition and fees vary according to course load and reciprocity agreements.
Financial aid: In 1999–00, 28 research assistantships with full tuition reimbursements (averaging $13,500 per year) were awarded; fellowships, teaching assistantships, Federal Work-Study and tuition waivers (partial) also available.
Faculty research: Molecular biology, microbiology, neuroscience, cellular biology, physiology.
Dr. Steven B. Waller, Head, 605-677-5157, *Fax:* 605-677-6381, *E-mail:* swaller@ usd.edu.
Application contact: Luana Johnson, Senior Secretary, 605-677-5254, *Fax:* 605-677-6381, *E-mail:* lsjohnso@usd.edu.

■ UNIVERSITY OF SOUTHERN CALIFORNIA

Graduate School, College of Letters, Arts and Sciences, Department of Biological Sciences, Los Angeles, CA 90089

AWARDS Marine biology and biological oceanography (MS, PhD); molecular biology (MS, PhD); neurobiology (PhD).

Students: 101 full-time (37 women), 1 part-time; includes 14 minority (2 African Americans, 10 Asian Americans or Pacific Islanders, 2 Hispanic Americans), 47 international. Average age 30. *168 applicants, 22% accepted.* In 1999, 3 master's, 17 doctorates awarded.
Degree requirements: For doctorate, dissertation required.
Entrance requirements: For master's and doctorate, GRE General Test. *Application deadline:* For fall admission, 2/1 (priority date). *Application fee:* $55.
Expenses: Tuition: Full-time $17,952; part-time $748 per unit. Required fees: $406; $203 per unit. Tuition and fees vary according to program.
Financial aid: In 1999–00, 13 fellowships, 35 research assistantships, 33 teaching assistantships were awarded; Federal Work-Study, institutionally sponsored loans, and scholarships also available. Aid available to part-time students. Financial aid application deadline: 2/15; financial aid applicants required to submit FAFSA.
Dr. Donal Manahan, Chair, 213-740-1109.

■ UNIVERSITY OF SOUTHERN CALIFORNIA

Keck School of Medicine and Graduate School, Graduate Programs in Medicine, Los Angeles, CA 90089
AWARDS MPH, MS, PhD, MD/PhD.

Faculty: 187 full-time (51 women), 12 part-time/adjunct (4 women).
Students: 319 full-time (176 women), 1 part-time; includes 83 minority (4 African Americans, 67 Asian Americans or Pacific Islanders, 12 Hispanic Americans), 144 international. Average age 28. *584 applicants, 18% accepted.* In 1999, 31 master's, 42 doctorates awarded. Terminal master's awarded for partial completion of doctoral program.
Degree requirements: For master's, foreign language not required; for doctorate, dissertation required.
Entrance requirements: For master's, GRE General Test, TOEFL, minimum GPA of 3.0; for doctorate, GRE General Test, TOEFL. *Application fee:* $55. Electronic applications accepted.
Expenses: Tuition: Full-time $22,198; part-time $748 per unit. Required fees: $406.

Financial aid: In 1999–00, 193 students received aid, including 17 fellowships with full tuition reimbursements available, 149 research assistantships with full tuition reimbursements available (averaging $17,580 per year), 29 teaching assistantships with full tuition reimbursements available (averaging $17,580 per year); career-related internships or fieldwork, Federal Work-Study, grants, institutionally sponsored loans, traineeships, and tuition waivers (full and partial) also available. Aid available to part-time students.
Dr. Richard N. Lolley, Associate Dean for Scientific Affairs, 323-442-1607, *Fax:* 323-442-1610, *E-mail:* lolley@hsc.usc.edu.

Find an in-depth description at www.petersons.com/graduate.

■ UNIVERSITY OF SOUTHERN MISSISSIPPI

Graduate School, College of Science and Technology, Department of Biological Sciences, Hattiesburg, MS 39406

AWARDS Environmental biology (MS, PhD); marine biology (MS, PhD); microbiology (MS, PhD); molecular biology (MS, PhD).

Degree requirements: For master's, thesis required, foreign language not required; for doctorate, 2 foreign languages (computer language can substitute for one), dissertation required.
Entrance requirements: For master's, GRE General Test, TOEFL, minimum GPA of 3.0; for doctorate, GRE General Test, TOEFL, minimum GPA of 3.5.
Expenses: Tuition, state resident: full-time $2,250; part-time $137 per semester hour. Tuition, nonresident: full-time $3,102; part-time $172 per semester hour. Required fees: $602.

■ UNIVERSITY OF SOUTH FLORIDA

College of Medicine and Graduate School, Graduate Programs in Medical Sciences, Tampa, FL 33620-9951

AWARDS PhD.

Degree requirements: For doctorate, dissertation required.
Entrance requirements: For doctorate, GRE General Test, minimum GPA of 3.0.
Expenses: Tuition, state resident: part-time $148 per credit hour. Tuition, nonresident: part-time $509 per credit hour.

Find an in-depth description at www.petersons.com/graduate.

■ UNIVERSITY OF SOUTH FLORIDA

Graduate School, College of Arts and Sciences, Department of Biology, Tampa, FL 33620-9951

AWARDS Biology (PhD); botany (MS); ecology (PhD); marine biology (MS, PhD); microbiology (MS); physiology (PhD); zoology (MS). Part-time programs available.

Degree requirements: For master's, foreign language not required; for doctorate, 2 foreign languages (computer language can substitute for one), dissertation required.
Entrance requirements: For master's, GRE General Test, minimum GPA of 3.0 in last 60 hours; for doctorate, GRE General Test, GRE Subject Test in biology. Electronic applications accepted.
Expenses: Tuition, state resident: part-time $148 per credit hour. Tuition, nonresident: part-time $509 per credit hour.

■ THE UNIVERSITY OF TENNESSEE

Graduate School, College of Arts and Sciences, Program in Life Sciences, Knoxville, TN 37996

AWARDS Genome science and technology (MS, PhD); plant physiology and genetics (MS, PhD).

Faculty: 4 full-time (1 woman).
Students: 12 full-time (7 women), 5 part-time (2 women); includes 1 minority (Hispanic American), 9 international. *4 applicants, 0% accepted.* In 1999, 2 master's, 1 doctorate awarded.
Degree requirements: For doctorate, dissertation required.
Entrance requirements: For master's and doctorate, GRE General Test, TOEFL, minimum GPA of 2.7. *Application deadline:* For fall admission, 2/1 (priority date). Applications are processed on a rolling basis. *Application fee:* $35. Electronic applications accepted.
Expenses: Tuition, state resident: full-time $3,806; part-time $184 per credit hour. Tuition, nonresident: full-time $9,874; part-time $522 per credit hour. Tuition and fees vary according to program.
Financial aid: Fellowships, unspecified assistantships available. Financial aid application deadline: 2/1; financial aid applicants required to submit FAFSA.
Dr. Jeffrey M. Becker, Chairperson, 865-974-6841, *Fax:* 865-974-4057, *E-mail:* russellg@utk.edu.

■ THE UNIVERSITY OF TENNESSEE

Graduate School, Intercollegiate Programs, Program in Comparative and Experimental Medicine, Knoxville, TN 37996

AWARDS MS, PhD.

Students: 17 full-time (10 women), 10 part-time (6 women), 14 international. *18 applicants, 33% accepted.* In 1999, 3 master's, 8 doctorates awarded.
Degree requirements: For master's and doctorate, thesis/dissertation required, foreign language not required.
Entrance requirements: For master's and doctorate, GRE General Test, TOEFL, minimum GPA of 2.7. *Application deadline:* For fall admission, 2/1 (priority date). Applications are processed on a rolling basis. *Application fee:* $35. Electronic applications accepted.
Expenses: Tuition, state resident: full-time $3,806; part-time $184 per credit hour. Tuition, nonresident: full-time $9,874; part-time $522 per credit hour. Tuition and fees vary according to program.
Financial aid: Application deadline: 2/1.
Dr. L. N. D. Potgieter, Director, 865-974-5576, *Fax:* 865-974-5640, *E-mail:* potgieter@utk.edu.

■ THE UNIVERSITY OF TENNESSEE HEALTH SCIENCE CENTER

College of Graduate Health Sciences, Memphis, TN 38163-0002

AWARDS MS, PhD, MD/PhD, Pharm D/PhD. Part-time programs available. Terminal master's awarded for partial completion of doctoral program.

Degree requirements: For master's, thesis, oral and written comprehensive exams required, foreign language not required; for doctorate, dissertation, oral and written preliminary and comprehensive exams required, foreign language not required.
Entrance requirements: For master's and doctorate, GRE General Test, TOEFL, minimum GPA of 3.0.

■ THE UNIVERSITY OF TENNESSEE–OAK RIDGE NATIONAL LABORATORY GRADUATE SCHOOL OF GENOME SCIENCE AND TECHNOLOGY

Graduate Program, Oak Ridge, TN 37830-8026

AWARDS MS, PhD.

Faculty: 1 full-time (0 women), 49 part-time/adjunct (10 women).
Students: 9 full-time (6 women), 3 international. Average age 32. *10 applicants, 20% accepted.* In 1999, 1 degree awarded (100% entered university research/teaching).
Degree requirements: For master's and doctorate, thesis/dissertation required, foreign language not required. *Average time to degree:* Doctorate–6 years full-time.
Entrance requirements: For master's and doctorate, GRE General Test. *Application deadline:* For fall admission, 4/30 (priority date); for spring admission, 11/1. Applications are processed on a rolling basis. *Application fee:* $35.
Expenses: Tuition, state resident: full-time $3,306; part-time $184 per credit hour. Tuition, nonresident: full-time $9,374; part-time $522 per credit hour. Required fees: $22 per credit hour. Tuition and fees vary according to course load.
Financial aid: Fellowships, research assistantships, institutionally sponsored loans and tuition waivers (full) available. Financial aid application deadline: 3/31.
Faculty research: Mamalian genomics, structural biology, proteomics, computational biology and bioinformatics, bioanalytical technologies.
Dr. Jeffrey M. Becker, Director, 865-574-1227, *Fax:* 865-576-4149, *E-mail:* jbecker@utk.edu.
Application contact: Kay Gardner, Program/Resource Specialist, 865-574-1227, *Fax:* 865-576-4149, *E-mail:* gardnerk@utkux.utcc.utk.edu.

Find an in-depth description at www.petersons.com/graduate.

■ THE UNIVERSITY OF TEXAS AT ARLINGTON

Graduate School, College of Science, Department of Biology, Arlington, TX 76019

AWARDS MS, PhD. Part-time programs available.

Faculty: 22 full-time (3 women).
Students: 40 full-time (23 women), 22 part-time (12 women); includes 8 minority (1 African American, 3 Asian Americans or Pacific Islanders, 3 Hispanic Americans, 1 Native American), 3 international. *37 applicants, 51% accepted.* In 1999, 7 master's, 5 doctorates awarded.
Degree requirements: For master's, thesis optional, foreign language not required; for doctorate, one foreign language (computer language can substitute), dissertation required.
Entrance requirements: For master's and doctorate, GRE General Test, minimum GPA of 3.0. *Application deadline:* For fall admission, 6/16. Applications are processed

on a rolling basis. *Application fee:* $25 ($50 for international students).
Expenses: Tuition, state resident: full-time $2,052. Tuition, nonresident: full-time $6,138. Tuition and fees vary according to course load.
Financial aid: Research assistantships, teaching assistantships, Federal Work-Study available. Financial aid application deadline: 6/1; financial aid applicants required to submit FAFSA.
Dr. John D. Bacon, Chair, 817-272-2871, *Fax:* 817-272-2855, *E-mail:* jdbacon@exchange.uta.edu.
Application contact: Dr. Daniel R. Formanowicz, Graduate Adviser, 817-272-2871, *Fax:* 817-272-2855, *E-mail:* formanow@uta.edu.

■ THE UNIVERSITY OF TEXAS AT AUSTIN

Graduate School, College of Natural Sciences, School of Biological Sciences, Austin, TX 78712-1111

AWARDS MA, PhD.

Entrance requirements: For master's and doctorate, GRE General Test. *Application deadline:* Applications are processed on a rolling basis. *Application fee:* $50 ($75 for international students). Electronic applications accepted.
Expenses: Tuition, state resident: part-time $114 per semester hour. Tuition, nonresident: part-time $330 per semester hour. Tuition and fees vary according to program.
Financial aid: Fellowships, research assistantships, teaching assistantships available. Financial aid application deadline: 2/1.
Application contact: Dr. James Mauseth, Graduate Adviser, 512-471-3189.

Find an in-depth description at www.petersons.com/graduate.

■ THE UNIVERSITY OF TEXAS AT BROWNSVILLE

Graduate Studies and Sponsored Programs, College of Science, Mathematics and Technology, Brownsville, TX 78520-4991

AWARDS Biological sciences (MSIS). Part-time and evening/weekend programs available.

Faculty: 20.
Students: 13 (9 women); includes 8 minority (1 African American, 7 Hispanic Americans).
Degree requirements: For master's, thesis optional, foreign language not required.
Entrance requirements: For master's, GRE General Test, TOEFL. *Application deadline:* For fall admission, 8/1 (priority

date); for spring admission, 12/15 (priority date). Applications are processed on a rolling basis. *Application fee:* $15.

Expenses: Tuition, state resident: full-time $1,080; part-time $36 per hour. Tuition, nonresident: full-time $7,830; part-time $261 per hour. Tuition and fees vary according to course load and degree level.

Financial aid: Federal Work-Study, scholarships, and tuition waivers (partial) available. Aid available to part-time students. Financial aid application deadline: 4/3; financial aid applicants required to submit FAFSA.

Faculty research: Fish, insects, barrier islands, algae, curlits.

Dr. José G. Martin, Dean, 956-574-6700, *Fax:* 956-574-8988.

Application contact: David McNeely, Chair, Department of Biology, 956-544-8289, *Fax:* 956-983-7115, *E-mail:* mcneely@utb1.utb.edu.

■ THE UNIVERSITY OF TEXAS AT EL PASO

Graduate School, College of Science, Department of Biological Sciences, El Paso, TX 79968-0001

AWARDS MS, PhD. Part-time and evening/weekend programs available.

Students: 47; includes 22 minority (2 Asian Americans or Pacific Islanders, 20 Hispanic Americans), 8 international. Average age 34. In 1999, 13 master's awarded.

Degree requirements: For master's, thesis required, foreign language not required.

Entrance requirements: For master's, GRE General Test, TOEFL, minimum GPA of 3.0; for doctorate, GRE General Test, TOEFL. *Application deadline:* For fall admission, 7/1 (priority date); for spring admission, 11/1 (priority date). Applications are processed on a rolling basis. *Application fee:* $15 ($65 for international students). Electronic applications accepted.

Expenses: Tuition, state resident: full-time $2,217; part-time $96 per credit hour. Tuition, nonresident: full-time $5,961; part-time $304 per credit hour. Required fees: $245 per semester. One-time fee: $10. Tuition and fees vary according to course level, course load, program and reciprocity agreements.

Financial aid: In 1999–00, research assistantships with partial tuition reimbursements (averaging $22,500 per year), teaching assistantships with partial tuition reimbursements (averaging $18,000 per year) were awarded; fellowships with partial tuition reimbursements, Federal Work-Study, institutionally sponsored loans, and tuition waivers (partial) also available. Financial aid application

deadline: 3/15; financial aid applicants required to submit FAFSA.

Dr. Eppie D. Rael, Chairperson, 915-747-5844, *Fax:* 915-747-5808, *E-mail:* erael@miners.utep.edu.

Application contact: Dr. Charles H. Ambler, Associate Vice President for Graduate Studies, 915-747-5491, *Fax:* 915-747-5788, *E-mail:* cambler@miners.utep.edu.

Find an in-depth description at www.petersons.com/graduate.

■ THE UNIVERSITY OF TEXAS AT SAN ANTONIO

College of Sciences and Engineering, Division of Life Sciences, San Antonio, TX 78249-0617

AWARDS Biology (PhD), including neurobiology; biology and biotechnology (MS), including biology, biotechnology. Part-time programs available. Terminal master's awarded for partial completion of doctoral program.

Degree requirements: For master's, comprehensive exam required, thesis optional; for doctorate, dissertation, comprehensive exam required.

Entrance requirements: For master's, GRE General Test, minimum GPA of 3.0; for doctorate, GRE General Test, TOEFL, minimum GPA of 3.0.

Expenses: Tuition, state resident: full-time $2,640; part-time $110 per credit hour. Tuition, nonresident: full-time $7,824; part-time $326 per credit hour. Tuition and fees vary according to course load.

Faculty research: Hippocampus, learning and memory, developmental neurobiology, neuroplasticity.

Find an in-depth description at www.petersons.com/graduate.

■ THE UNIVERSITY OF TEXAS AT TYLER

Graduate Studies, College of Sciences and Mathematics, Department of Biology, Tyler, TX 75799-0001

AWARDS Biology (MS); interdisciplinary studies (MS).

Faculty: 6 full-time (0 women).

Students: 4 full-time, 5 part-time; includes 2 minority (1 African American, 1 Asian American or Pacific Islander). In 1999, 2 degrees awarded.

Degree requirements: For master's, oral qualifying exam, thesis defense required.

Entrance requirements: For master's, GRE General Test. *Application deadline:* Applications are processed on a rolling basis. *Application fee:* $0.

Expenses: Tuition, state resident: part-time $245 per credit hour. Tuition, nonresident: part-time $379 per credit hour.

Financial aid: In 1999–00, 4 teaching assistantships (averaging $6,000 per year) were awarded. Financial aid application deadline: 7/1.

Faculty research: Phenotypic plasticity and heritability of life history traits, bioremediation, metabolism of hydrocarbons by bacteria, serum growth factors substitution.

Dr. Don Killebrew, Chair, 903-566-7252, *E-mail:* dkille@mail.uttyl.edu.

Application contact: Carol A. Hodge, Office of Graduate Studies, 903-566-7142, *Fax:* 903-566-7068, *E-mail:* chodge@mailuttly.edu.

■ THE UNIVERSITY OF TEXAS HEALTH SCIENCE CENTER AT SAN ANTONIO

Graduate School of Biomedical Sciences, San Antonio, TX 78229-3900

AWARDS Pharm D, MS, MSN, PhD, Certificate. Part-time and evening/weekend programs available.

Entrance requirements: For Pharm D, master's, and doctorate, GRE General Test.

Expenses: Tuition, state resident: part-time $38 per credit hour. Tuition, nonresident: part-time $249 per credit hour.

■ THE UNIVERSITY OF TEXAS–HOUSTON HEALTH SCIENCE CENTER

Graduate School of Biomedical Sciences, Houston, TX 77225-0036

AWARDS MS, PhD, MD/PhD.

Faculty: 436 full-time (98 women).

Students: 426 full-time (222 women); includes 86 minority (10 African Americans, 47 Asian Americans or Pacific Islanders, 27 Hispanic Americans, 2 Native Americans), 131 international. Average age 27. *356 applicants, 49% accepted.* In 1999, 36 master's, 61 doctorates awarded. Terminal master's awarded for partial completion of doctoral program.

Degree requirements: For master's and doctorate, thesis/dissertation required, foreign language not required.

Entrance requirements: For master's and doctorate, GRE General Test, TOEFL, TWE. *Application deadline:* Applications are processed on a rolling basis. *Application fee:* $10. Electronic applications accepted.

Financial aid: In 1999–00, 14 fellowships, 335 research assistantships, 12 teaching assistantships were awarded; grants and

The University of Texas–Houston Health Science Center (continued)
institutionally sponsored loans also available. Financial aid application deadline: 1/15.
Dr. Paul E. Darlington, Associate Dean, 713-500-9855, *Fax:* 713-500-9877, *E-mail:* pdarling@gsbs.gs.uth.tmc.edu.
Application contact: Anne Baronitis, Director of Admissions, 713-500-9860, *Fax:* 713-500-9877, *E-mail:* abaron@gsbs.gs.uth.tmc.edu.

Find an in-depth description at www.petersons.com/graduate.

■ **THE UNIVERSITY OF TEXAS MEDICAL BRANCH AT GALVESTON**

Graduate School of Biomedical Sciences, Galveston, TX 77555

AWARDS MA, MMS, MS, PhD, JD/PhD, MD/MA, MD/PhD. Part-time programs available.

Students: 235 full-time (116 women), 51 part-time (33 women); includes 48 minority (8 African Americans, 15 Asian Americans or Pacific Islanders, 21 Hispanic Americans, 4 Native Americans), 74 international. Average age 32. *288 applicants, 43% accepted.* In 1999, 13 master's, 52 doctorates awarded.
Degree requirements: For master's, foreign language not required; for doctorate, dissertation required, foreign language not required.
Entrance requirements: For doctorate, GRE General Test. *Application deadline:* Applications are processed on a rolling basis. *Application fee:* $25 ($50 for international students). Electronic applications accepted.
Expenses: Tuition, state resident: full-time $684; part-time $38 per credit hour. Tuition, nonresident: full-time $4,572; part-time $254 per credit hour. Required fees: $29; $7.5 per credit hour. One-time fee: $55. Tuition and fees vary according to program.
Financial aid: In 1999–00, 81 students received aid; fellowships, research assistantships, teaching assistantships, career-related internships or fieldwork, Federal Work-Study, institutionally sponsored loans, traineeships, and unspecified assistantships available. Aid available to part-time students. Financial aid applicants required to submit FAFSA.
Faculty research: Tumor virology, gene expression and recombinant DNA, neuropharmacology, cellular physiology, aging.
Dr. Cary W. Cooper, Dean, 409-772-2665, *Fax:* 409-747-0772, *E-mail:* ccooper@utmb.edu.

Application contact: Robert C. Bennett, Associate Dean for Administration and Student Affairs, 409-772-2665, *Fax:* 409-747-0772, *E-mail:* rbennett@utmb.edu.

■ **THE UNIVERSITY OF TEXAS OF THE PERMIAN BASIN**

Graduate School, College of Arts and Sciences, Department of Sciences and Mathematics, Program in Biology, Odessa, TX 79762-0001

AWARDS MS.

Degree requirements: For master's, thesis or alternative required, foreign language not required.
Entrance requirements: For master's, GRE General Test.

■ **THE UNIVERSITY OF TEXAS– PAN AMERICAN**

College of Science and Engineering, Department of Biology, Edinburg, TX 78539-2999

AWARDS MS. Part-time and evening/weekend programs available.

Faculty: 10 full-time.
Students: 15 full-time (3 women), 12 part-time (4 women); all minorities (all Hispanic Americans). Average age 23. *6 applicants, 100% accepted.*
Degree requirements: For master's, comprehensive exam required, foreign language and thesis not required.
Entrance requirements: For master's, GRE General Test, minimum GPA of 2.75 in biology. *Application fee:* $0.
Expenses: Tuition, state resident: full-time $1,392; part-time $98 per hour. Tuition, nonresident: full-time $6,576; part-time $314 per hour. Required fees: $956. Tuition and fees vary according to course load and degree level.
Financial aid: Teaching assistantships, Federal Work-Study, institutionally sponsored loans, and tuition waivers (partial) available. Aid available to part-time students. Financial aid application deadline: 6/1.
Faculty research: Flora and fauna of South Padre Island, plant taxonomy of Rio Grande Valley.
Dr. Mohammed Farooqui, Chair, 956-381-3537.

■ **THE UNIVERSITY OF TEXAS SOUTHWESTERN MEDICAL CENTER AT DALLAS**

Southwestern Graduate School of Biomedical Sciences, Division of Cell and Molecular Biology, Medical Scientist Training Program, Dallas, TX 75390

AWARDS PhD, MD/PhD.

Application deadline: For fall admission, 1/5. *Application fee:* $0.
Expenses: Tuition, state resident: full-time $912. Tuition, nonresident: full-time $6,096. Required fees: $216. Full-time tuition and fees vary according to course load and program.
Financial aid: Application deadline: 3/15.
Dr. Rodney E. Ulane, Associate Dean, 214-648-6764.
Application contact: Nancy McKinney, Education Coordinator, 214-648-8099, *Fax:* 214-648-2978, *E-mail:* dcmbinfo@utsouthwestern.edu.

Find an in-depth description at www.petersons.com/graduate.

■ **THE UNIVERSITY OF TEXAS SOUTHWESTERN MEDICAL CENTER AT DALLAS**

Southwestern Graduate School of Biomedical Sciences, Division of Cell and Molecular Biology, Program in Integrative Biology, Dallas, TX 75390

AWARDS PhD.

Faculty: 29 full-time (3 women).
Students: 16 full-time (7 women); includes 2 minority (1 Asian American or Pacific Islander, 1 Hispanic American), 4 international. Average age 25.
Degree requirements: For doctorate, dissertation required, foreign language not required.
Entrance requirements: For doctorate, GRE General Test, minimum GPA of 3.0. *Application deadline:* For fall admission, 1/5 (priority date). *Application fee:* $0. Electronic applications accepted.
Expenses: Tuition, state resident: full-time $912. Tuition, nonresident: full-time $6,096. Required fees: $216. Full-time tuition and fees vary according to course load and program.
Financial aid: In 1999–00, 10 students received aid; fellowships, research assistantships, institutionally sponsored loans and traineeships available. Financial aid application deadline: 3/15; financial aid applicants required to submit FAFSA.
Faculty research: Muscle physiology, ion transport in secretory cells, nuclear hormone receptors, contractile protein phosphorylation, cardiovascular homeostasis.
Dr. James T. Stull, Chair, 214-648-6849, *Fax:* 214-648-2974, *E-mail:* jstull@mednet.swmed.edu.
Application contact: Nancy McKinney, Education Coordinator, 214-648-8099, *Fax:* 214-648-2978, *E-mail:* dcmbinfo@utsouthwestern.edu.

Find an in-depth description at www.petersons.com/graduate.

■ UNIVERSITY OF THE INCARNATE WORD

School of Graduate Studies and Research, School of Mathematics, Sciences, and Engineering, Program in Biology, San Antonio, TX 78209-6397

AWARDS MA, MS. Part-time and evening/weekend programs available.

Students: 7 full-time (2 women), 8 part-time (5 women); includes 10 minority (1 African American, 8 Hispanic Americans, 1 Native American). Average age 30. *12 applicants, 92% accepted.* In 1999, 11 degrees awarded.
Degree requirements: For master's, thesis optional, foreign language not required.
Entrance requirements: For master's, GRE General Test, TOEFL, minimum GPA of 3.0. *Application deadline:* For fall admission, 8/15 (priority date); for spring admission, 12/31. Applications are processed on a rolling basis. *Application fee:* $20.
Expenses: Tuition: Part-time $395 per hour. Required fees: $15 per hour. One-time fee: $130 part-time. Tuition and fees vary according to degree level.
Financial aid: Research assistantships, teaching assistantships, career-related internships or fieldwork, Federal Work-Study, and institutionally sponsored loans available. Aid available to part-time students. Financial aid application deadline: 9/12.
Faculty research: Mammalogy, zoogeography, cell biology, physical chemistry, molecular genetics, parasitology. *Total annual research expenditures:* $185,000.
Dr. John Sullivan, Coordinator, 210-283-5033, *Fax:* 210-829-3154, *E-mail:* sullivan@universe.uiwtx.edu.
Application contact: Andrea Cyterski, Director of Admissions, 210-829-6005, *Fax:* 210-829-3921, *E-mail:* cyterski@universe.uiwtx.edu.

■ UNIVERSITY OF THE PACIFIC

Graduate School, Department of Biological Sciences, Stockton, CA 95211-0197

AWARDS MS.

Faculty: 5 full-time (0 women).
Students: 2 full-time (0 women), 13 part-time (7 women), 1 international. In 1999, 7 degrees awarded.
Degree requirements: For master's, thesis required.
Entrance requirements: For master's, GRE General Test, GRE Subject Test. *Application deadline:* For fall admission, 3/1 (priority date); for spring admission, 10/15.

Applications are processed on a rolling basis. *Application fee:* $50.
Expenses: Tuition: Full-time $19,570; part-time $612 per unit. Required fees: $260. Tuition and fees vary according to program.
Financial aid: In 1999–00, 12 teaching assistantships were awarded; institutionally sponsored loans also available. Aid available to part-time students. Financial aid application deadline: 3/1.
Dr. Paul Richmond, Chairman, 209-946-2181.

■ UNIVERSITY OF TOLEDO

Graduate School, College of Arts and Sciences, Department of Biology, Toledo, OH 43606-3398

AWARDS MES, MS, PhD. Part-time programs available.

Faculty: 15 full-time (3 women).
Students: 52 (28 women); includes 3 minority (2 African Americans, 1 Asian American or Pacific Islander) 22 international. Average age 28. *66 applicants, 30% accepted.* In 1999, 9 master's, 1 doctorate awarded.
Degree requirements: For master's, thesis or alternative required, foreign language not required; for doctorate, 2 foreign languages (computer language can substitute for one), dissertation required.
Entrance requirements: For master's and doctorate, GRE General Test, GRE Subject Test. *Application deadline:* For fall admission, 8/1 (priority date). *Application fee:* $30. Electronic applications accepted.
Expenses: Tuition, state resident: full-time $2,741; part-time $228 per credit hour. Tuition, nonresident: full-time $5,926; part-time $494 per credit hour. Required fees: $402; $34 per credit hour.
Financial aid: In 1999–00, 9 research assistantships, 29 teaching assistantships were awarded; fellowships, Federal Work-Study and tuition waivers (full) also available. Aid available to part-time students. Financial aid application deadline: 4/1.
Faculty research: Biochemical parasitology, physiological ecology, animal physiology.
Dr. Patricia Komuniecki, Chair, 419-530-2066.

■ UNIVERSITY OF TULSA

Graduate School, College of Engineering and Applied Sciences, Department of Biological Sciences, Tulsa, OK 74104-3189

AWARDS MS, PhD, JD/MS. Part-time programs available.

Faculty: 14 full-time (4 women).
Students: 7 full-time (5 women); includes 1 minority (Hispanic American), 2

international. Average age 28. *18 applicants, 28% accepted.*
Degree requirements: For master's, thesis, oral exams required, foreign language not required; for doctorate, dissertation required.
Entrance requirements: For master's and doctorate, GRE General Test, GRE Subject Test, TOEFL. *Application deadline:* Applications are processed on a rolling basis. *Application fee:* $30. Electronic applications accepted.
Expenses: Tuition: Full-time $9,000; part-time $500 per credit. Required fees: $3 per hour. One-time fee: $200 full-time. Tuition and fees vary according to course load.
Financial aid: In 1999–00, 3 fellowships with full tuition reimbursements (averaging $3,000 per year), teaching assistantships with full tuition reimbursements (averaging $7,341 per year) were awarded; research assistantships, tuition waivers (partial) also available. Aid available to part-time students. Financial aid application deadline: 2/1; financial aid applicants required to submit FAFSA.
Faculty research: Molecular biology, botany, neurological biology.
Dr. Glen E. Collier, Chairperson, 918-631-2204.
Application contact: Dr. Charles R. Brown, Adviser, 918-631-3943, *Fax:* 918-631-2762.

■ UNIVERSITY OF UTAH

Graduate School, College of Science, Department of Biology, Salt Lake City, UT 84112-1107

AWARDS Biology (M Phil); ecology and evolutionary biology (MS, PhD); genetics (MS, PhD); molecular biology (PhD). Part-time programs available. Terminal master's awarded for partial completion of doctoral program.

Degree requirements: For master's and doctorate, thesis/dissertation required, foreign language not required.
Entrance requirements: For master's and doctorate, GRE General Test, GRE Subject Test, TOEFL, minimum GPA of 3.0.
Expenses: Tuition, state resident: full-time $1,663. Tuition, nonresident: full-time $5,201. Tuition and fees vary according to course load and program.
Faculty research: Behavioral ecology, cellular neurobiology, DNA replication, ecological genetics, herpetology.

Find an in-depth description at www.petersons.com/graduate.

■ UNIVERSITY OF UTAH

School of Medicine and Graduate School, Graduate Programs in Medicine, Salt Lake City, UT 84112-1107

AWARDS M Phil, M Stat, MPH, MS, MSPH, PhD. Part-time programs available.

Faculty: 146 full-time (39 women), 77 part-time/adjunct (10 women).
Students: 235 full-time (67 women), 48 part-time (19 women); includes 30 minority (4 African Americans, 19 Asian Americans or Pacific Islanders, 7 Hispanic Americans), 32 international. Average age 28. In 1999, 59 master's, 13 doctorates awarded. Terminal master's awarded for partial completion of doctoral program.
Degree requirements: For master's, foreign language not required; for doctorate, dissertation required, foreign language not required.
Electronic applications accepted.
Expenses: Tuition, state resident: full-time $2,105. Tuition, nonresident: full-time $6,312.
Financial aid: In 1999–00, 9 fellowships with full tuition reimbursements, 105 research assistantships with full tuition reimbursements, 4 teaching assistantships were awarded; career-related internships or fieldwork, Federal Work-Study, grants, institutionally sponsored loans, scholarships, traineeships, and tuition waivers (full and partial) also available. Aid available to part-time students.
Dr. T. Samuel Shomaker, Interim Dean, School of Medicine, 801-581-6436.

■ UNIVERSITY OF VERMONT

College of Medicine and Graduate College, Graduate Programs in Medicine, Burlington, VT 05405

AWARDS Anatomy and neurobiology (PhD); biochemistry (MS, PhD); microbiology and molecular genetics (MS, PhD); molecular physiology and biophysics (MS, PhD); pathology (MS); pharmacology (MS, PhD).

Degree requirements: For master's and doctorate, thesis/dissertation required.
Entrance requirements: For master's and doctorate, GRE General Test, TOEFL.
Expenses: Tuition, state resident: full-time $7,464; part-time $311 per credit. Tuition, nonresident: full-time $18,672; part-time $778 per credit. Full-time tuition and fees vary according to degree level and program.

■ UNIVERSITY OF VERMONT

Graduate College, College of Arts and Sciences, Department of Biology, Burlington, VT 05405

AWARDS Biology (MS, PhD); biology education (MAT, MST).

Degree requirements: For master's, thesis required, foreign language not required; for doctorate, dissertation required.
Entrance requirements: For master's and doctorate, GRE General Test, TOEFL.
Expenses: Tuition, state resident: full-time $7,464; part-time $311 per credit. Tuition, nonresident: full-time $18,672; part-time $778 per credit. Full-time tuition and fees vary according to degree level and program.

Find an in-depth description at www.petersons.com/graduate.

■ UNIVERSITY OF VIRGINIA

College and Graduate School of Arts and Sciences, Department of Biology, Charlottesville, VA 22903

AWARDS Biology (MA, MS, PhD); biology education (MAT).

Faculty: 37 full-time (9 women), 2 part-time/adjunct (0 women).
Students: 53 full-time (34 women), 5 part-time (2 women); includes 2 minority (both African Americans), 19 international. Average age 28. *69 applicants, 42% accepted.* In 1999, 5 master's, 5 doctorates awarded.
Degree requirements: For master's, thesis required, foreign language not required; for doctorate, dissertation required.
Entrance requirements: For master's and doctorate, GRE General Test, GRE Subject Test. *Application fee:* $40. Electronic applications accepted.
Expenses: Tuition, state resident: full-time $3,832. Tuition, nonresident: full-time $15,519. Required fees: $1,084. Tuition and fees vary according to course load and program.
Financial aid: Application deadline: 2/1. Raymond R. Keller, Chairman, 804-924-7118.
Application contact: Duane J. Osheim, Associate Dean, 804-924-7184, *E-mail:* microbiology@virginia.edu.

Find an in-depth description at www.petersons.com/graduate.

■ UNIVERSITY OF VIRGINIA

College and Graduate School of Arts and Sciences, Program in Biological and Physical Sciences, Charlottesville, VA 22903

AWARDS MS.

Students: 1 full-time (0 women). Average age 24. In 1999, 5 degrees awarded. *Application fee:* $40. Electronic applications accepted.
Expenses: Tuition, state resident: full-time $3,832. Tuition, nonresident: full-time

$15,519. Required fees: $1,084. Tuition and fees vary according to course load and program.
Financial aid: Applicants required to submit FAFSA.
Application contact: Duane J. Osheim, Associate Dean, 804-924-7184, *E-mail:* microbiology@virginia.edu.

■ UNIVERSITY OF WASHINGTON

School of Medicine and Graduate School, Graduate Programs in Medicine, Seattle, WA 98195

AWARDS MOT, MPT, MS, MSE, PhD. Part-time programs available.

Faculty: 676 full-time (107 women), 146 part-time/adjunct (24 women).
Students: 509 full-time (221 women), 20 part-time (15 women). *3031 applicants, 8% accepted.* In 1999, 15 master's, 48 doctorates awarded.
Degree requirements: For doctorate, dissertation required.
Entrance requirements: For doctorate, GRE. *Application fee:* $35. Electronic applications accepted.
Expenses: Tuition, state resident: full-time $9,210; part-time $236 per credit. Tuition, nonresident: full-time $23,256; part-time $596 per credit.
Financial aid: Fellowships with full tuition reimbursements, research assistantships with full tuition reimbursements, teaching assistantships with full tuition reimbursements, career-related internships or fieldwork, Federal Work-Study, grants, institutionally sponsored loans, traineeships, tuition waivers (full and partial), and stipends available. Aid available to part-time students. Financial aid applicants required to submit FAFSA.
Application contact: Patricia T. Fero, Admissions Officer, 206-543-7212, *E-mail:* askuwsom@u.washington.edu.

■ UNIVERSITY OF WEST FLORIDA

College of Arts and Sciences: Sciences, Department of Biology, Program in General Biology, Pensacola, FL 32514-5750

AWARDS Biology (MS); biology education (MST).

Students: 8 full-time (6 women), 19 part-time (15 women); includes 3 minority (2 Asian Americans or Pacific Islanders, 1 Hispanic American), 1 international. Average age 30. In 1999, 6 degrees awarded.
Degree requirements: For master's, thesis required, foreign language not required.
Entrance requirements: For master's, GRE General Test. *Application deadline:*

For fall admission, 7/1; for spring admission, 11/1. Applications are processed on a rolling basis. *Application fee:* $20.
Expenses: Tuition, state resident: full-time $3,582; part-time $149 per credit hour. Tuition, nonresident: full-time $12,240; part-time $510 per credit hour.
Dr. J. Riehm, Chairperson, Department of Biology, 850-474-2748.

■ UNIVERSITY OF WISCONSIN–EAU CLAIRE

College of Arts and Sciences, Program in Biology, Eau Claire, WI 54702-4004
AWARDS MS.

Faculty: 19 full-time (5 women), 1 (woman) part-time/adjunct.
Students: 3 full-time (2 women), 5 part-time (3 women). Average age 31. *2 applicants, 100% accepted.* In 1999, 1 degree awarded.
Degree requirements: For master's, thesis, oral and written comprehensive exams required, foreign language not required.
Entrance requirements: For master's, bachelor's degree in biology or related field, minimum GPA of 3.0. *Application deadline:* For fall admission, 7/1; for spring admission, 12/1. Applications are processed on a rolling basis. *Application fee:* $45.
Expenses: Tuition, state resident: full-time $3,904; part-time $217 per credit. Tuition, nonresident: full-time $12,262; part-time $682 per credit. Tuition and fees vary according to program and reciprocity agreements.
Financial aid: In 1999–00, 2 teaching assistantships (averaging $7,780 per year) were awarded; Federal Work-Study also available. Financial aid application deadline: 4/15; financial aid applicants required to submit FAFSA.
Michael Weil, Chair, 715-836-4166, *Fax:* 715-836-5089, *E-mail:* weilmr@uwec.edu.

■ UNIVERSITY OF WISCONSIN–LA CROSSE

Graduate Studies, College of Science and Allied Health, Department of Biology, La Crosse, WI 54601-3742
AWARDS Biology (MS); clinical microbiology (MS); nurse anesthetist (MS). Part-time programs available.

Degree requirements: For master's, thesis (for some programs), oral comprehensive exam required, foreign language not required.
Entrance requirements: For master's, GRE General Test, minimum GPA of 3.0 during previous 2 years or 2.85 overall.
Expenses: Tuition, state resident: full-time $3,737; part-time $208 per credit. Tuition,

nonresident: full-time $11,380; part-time $633 per credit. Tuition and fees vary according to course load, program and reciprocity agreements.
Faculty research: Ecology, river studies, aquatic toxicology, aquatic microbiology, molecular biology, physiology.

■ UNIVERSITY OF WISCONSIN–MADISON

Graduate School, College of Agricultural and Life Sciences, Department of Animal Health and Biomedical Sciences, Program in Comparative Biosciences, Madison, WI 53706-1380
AWARDS Anatomy (MS, PhD); biochemistry (MS, PhD); cellular and molecular biology (MS, PhD); environmental toxicology (MS, PhD); neurosciences (MS, PhD); pharmacology (MS, PhD); physiology (MS, PhD).

Degree requirements: For doctorate, dissertation required.
Expenses: Tuition, state resident: full-time $5,406; part-time $339 per credit. Tuition, nonresident: full-time $17,110; part-time $1,071 per credit. Full-time tuition and fees vary according to program and reciprocity agreements. Part-time tuition and fees vary according to course load and program.

■ UNIVERSITY OF WISCONSIN–MADISON

Medical School and Graduate School, Graduate Programs in Medicine, Madison, WI 53706-1380
AWARDS Biomolecular chemistry (MS, PhD); cancer biology (PhD); genetics and medical genetics (MS, PhD), including genetics (PhD), medical genetics (MS); medical physics (MS, PhD), including health physics (MS), medical physics; microbiology (PhD); molecular and cellular pharmacology (PhD); oncology (PhD); pathology and laboratory medicine (PhD); physiology (PhD), including neurophysiology, physiology; population health (MS, PhD). Part-time programs available.
Postbaccalaureate distance learning degree programs offered (minimal on-campus study). Terminal master's awarded for partial completion of doctoral program.

Degree requirements: For master's, foreign language not required.
Application fee: $45. Electronic applications accepted.
Expenses: Tuition, state resident: full-time $5,406; part-time $339 per credit. Tuition, nonresident: full-time $17,110; part-time $1,071 per credit.
Financial aid: Fellowships with full tuition reimbursements, research assistantships with full tuition reimbursements, teaching

assistantships with full tuition reimbursements, grants, scholarships, traineeships, and tuition waivers (full) available.
Dr. Paul M. DeLuca, Associate Dean of Research and Graduate Studies, 608-265-0524, *Fax:* 608-265-0522, *E-mail:* pmdeluca@facstaff.wisc.edu.

■ UNIVERSITY OF WISCONSIN–MADISON

Medical School, MD/PhD Medical Scientist Training Program, Madison, WI 53706
AWARDS MD/PhD. PhD awarded through 23 different departments in the life sciences.

Faculty: 15 full-time (3 women).
Students: 31 full-time (15 women); includes 1 minority (Native American). *43 applicants, 23% accepted.*
Application deadline: For fall admission, 12/7 (priority date). Applications are processed on a rolling basis. *Application fee:* $45.
Expenses: Tuition, state resident: full-time $17,006. Tuition, nonresident: full-time $25,716. Full-time tuition and fees vary according to program.
Dr. Deane Mosher, Director, 608-262-1576, *Fax:* 608-263-4969, *E-mail:* dfmosher@facstaff.wisc.edu.
Application contact: Paul Cook, Program Administrator, 608-262-6321, *Fax:* 608-262-5253, *E-mail:* pscook@facstaff.wisc.edu.

■ UNIVERSITY OF WISCONSIN–MILWAUKEE

Graduate School, College of Letters and Sciences, Department of Biological Sciences, Milwaukee, WI 53201-0413
AWARDS MS, PhD.

Faculty: 30 full-time (10 women).
Students: 25 full-time (11 women), 38 part-time (15 women); includes 6 minority (3 Asian Americans or Pacific Islanders, 3 Hispanic Americans), 4 international. *33 applicants, 36% accepted.* In 1999, 5 master's, 5 doctorates awarded.
Degree requirements: For master's, thesis required; for doctorate, dissertation, 1 foreign language or data analysis proficiency required.
Entrance requirements: For master's and doctorate, GRE General Test. *Application deadline:* For fall admission, 3/1 (priority date). Applications are processed on a rolling basis. *Application fee:* $45 ($75 for international students).
Expenses: Tuition, state resident: full-time $5,363; part-time $134 per credit. Tuition, nonresident: full-time $16,537; part-time $493 per credit. Required fees: $168 per credit. $214 per credit. Full-time tuition

University of Wisconsin–Milwaukee (continued)

and fees vary according to program and reciprocity agreements. Part-time tuition and fees vary according to course load and program.

Financial aid: In 1999–00, 7 fellowships, 7 research assistantships, 27 teaching assistantships were awarded; career-related internships or fieldwork and unspecified assistantships also available. Aid available to part-time students. Financial aid application deadline: 4/15.

James Coggins, Chair, 414-229-4214.

■ UNIVERSITY OF WISCONSIN–OSHKOSH

Graduate School, College of Letters and Science, Department of Biology and Microbiology, Oshkosh, WI 54901

AWARDS Biology (MS), including botany, microbiology, zoology.

Degree requirements: For master's, thesis required, foreign language not required.

Entrance requirements: For master's, GRE General Test, minimum GPA of 3.0, BS in biology.

Expenses: Tuition, state resident: full-time $3,917; part-time $219 per credit. Tuition, nonresident: full-time $12,375; part-time $684 per credit. Part-time tuition and fees vary according to course load and program.

■ UTAH STATE UNIVERSITY

School of Graduate Studies, College of Science, Department of Biology, Logan, UT 84322

AWARDS Biology (MS, PhD); ecology (MS, PhD). Part-time programs available.

Faculty: 39 full-time (7 women).
Students: 39 full-time (15 women), 25 part-time (11 women); includes 2 minority (both Asian Americans or Pacific Islanders), 8 international. Average age 26. *47 applicants, 23% accepted.* In 1999, 9 master's, 3 doctorates awarded.

Degree requirements: For master's and doctorate, thesis/dissertation required, foreign language not required.

Entrance requirements: For master's and doctorate, GRE General Test, TOEFL, minimum GPA of 3.0. *Application deadline:* For fall admission, 6/15 (priority date); for spring admission, 10/15. Applications are processed on a rolling basis. *Application fee:* $40.

Expenses: Tuition, state resident: full-time $1,553. Tuition, nonresident: full-time $5,436. International tuition: $5,526 full-time. Required fees: $447. Tuition and fees vary according to course load and program.

Financial aid: In 1999–00, 3 fellowships with partial tuition reimbursements, 27 research assistantships with partial tuition reimbursements (averaging $10,250 per year), 44 teaching assistantships with partial tuition reimbursements (averaging $10,250 per year) were awarded; career-related internships or fieldwork, Federal Work-Study, and institutionally sponsored loans also available. Aid available to part-time students. Financial aid application deadline: 3/1.

Faculty research: Plant, insect, microbial, and animal biology.

Dr. Edmund D. Brodie, Head, 435-797-2483.

Application contact: Nancy Kay Harrison, Coordinator of Graduate Studies, 435-797-1770, *Fax:* 435-797-1575, *E-mail:* nancykay@biology.usu.edu.

■ VANDERBILT UNIVERSITY

Graduate School, Department of Biological Sciences, Nashville, TN 37240-1001

AWARDS MS, PhD, MD/PhD.

Degree requirements: For doctorate, dissertation required.
Application deadline: For fall admission, 1/15. *Application fee:* $40.

Expenses: Tuition: Full-time $17,244; part-time $958 per hour. Required fees: $242; $121 per semester. Tuition and fees vary according to program.

Financial aid: In 1999–00, fellowships (averaging $17,270 per year), teaching assistantships (averaging $17,270 per year) were awarded.

Dr. James V. Staros, Chair, 615-322-2008, *Fax:* 615-343-6707, *E-mail:* starosjv@ctrvax.vanderbilt.edu.

Application contact: Todd R. Graham, Director of Graduate Studies, 615-343-1835, *Fax:* 615-343-6707, *E-mail:* grahamtr@ctrvax.vanderbilt.edu.

Find an in-depth description at www.petersons.com/graduate.

■ VANDERBILT UNIVERSITY

Graduate School, Department of Biology, Nashville, TN 37240-1001

AWARDS MAT, MS, PhD.

Faculty: 14 full-time (1 woman).
Students: 14 full-time (4 women), 5 international. Average age 28. *23 applicants, 30% accepted.* In 1999, 1 doctorate awarded.

Degree requirements: For master's, thesis required, foreign language not required; for doctorate, dissertation, preliminary, qualifying, and final exams required, foreign language not required.

Entrance requirements: For master's and doctorate, GRE General Test, GRE

Subject Test (recommended). *Application deadline:* For fall admission, 1/15. *Application fee:* $40.

Expenses: Tuition: Full-time $17,244; part-time $958 per hour. Required fees: $242; $121 per semester. Tuition and fees vary according to program.

Financial aid: In 1999–00, 6 teaching assistantships with full tuition reimbursements (averaging $17,270 per year) were awarded; research assistantships, career-related internships or fieldwork, Federal Work-Study, and institutionally sponsored loans also available. Financial aid application deadline: 1/15.

Faculty research: Plant and animal physiology, ecology, evolution, neurobiology and biological clocks, parasitology.

Terry L. Page, Chair, 615-322-2961, *Fax:* 615-343-0336, *E-mail:* terry.l.page@vanderbilt.edu.

Application contact: N. Olof Pellmyr, Director of Graduate Studies, 615-322-2961, *Fax:* 615-343-0336, *E-mail:* olle.pellmyr@vanderbilt.edu.

■ VANDERBILT UNIVERSITY

Graduate School and School of Medicine, Interdisciplinary Graduate Program in the Biomedical Sciences, Nashville, TN 37240-1001

AWARDS PhD. First-year students in biomedical sciences enter program. Degrees awarded through participating departments of biochemistry, cell biology, cellular and molecular pathology, microbiology and immunology, molecular biology, molecular physiology and biophysics, and pharmacology.

Faculty: 162 full-time (24 women).
Students: 66 full-time (36 women), 2 part-time (both women); includes 5 minority (2 African Americans, 1 Asian American or Pacific Islander, 2 Hispanic Americans), 16 international. Average age 24. *269 applicants, 40% accepted.*

Degree requirements: For doctorate, dissertation, final and qualifying exams required.

Entrance requirements: For doctorate, GRE General Test. *Application deadline:* For fall admission, 1/15. *Application fee:* $40.

Expenses: Tuition: Full-time $17,244; part-time $958 per hour. Required fees: $242; $121 per semester. Tuition and fees vary according to program.

Financial aid: In 1999–00, 56 fellowships with full tuition reimbursements (averaging $17,000 per year) were awarded; Federal Work-Study, institutionally sponsored loans, and tuition waivers (partial) also available. Financial aid application deadline: 1/15.

Faculty research: Genetics; immunology; neurobiology; cell and developmental biology; signal transduction; endocrine, cell, and gene regulation.
James G. Patton, Director, 615-343-4611, *Fax:* 615-343-0749, *E-mail:* james.g.patton@vanderbilt.edu.
Application contact: Ellen Carter, Program Administrator, 615-343-4611, *Fax:* 615-343-0749, *E-mail:* ellen.carter@mcmail.vanderbilt.edu.
Find an in-depth description at www.petersons.com/graduate.

■ VANDERBILT UNIVERSITY

School of Medicine and Graduate School, Medical Scientist Training Program, Nashville, TN 37240-1001
AWARDS MD/PhD.

Expenses: Tuition: Part-time $958 per hour. Required fees: $121 per semester.
Financial aid: Traineeships available.
Dr. David Robertson, Director.
Application contact: Ellen Carter, Program Administrator, 615-343-4611, *Fax:* 615-343-0749, *E-mail:* ellen.carter@mcmail.vanderbilt.edu.
Find an in-depth description at www.petersons.com/graduate.

■ VASSAR COLLEGE

Graduate Programs, Department of Biology, Poughkeepsie, NY 12604
AWARDS MS.

Degree requirements: For master's, thesis required, foreign language not required.
Entrance requirements: For master's, GRE General Test, bachelor's degree in biology. *Application deadline:* For fall admission, 1/1. *Application fee:* $60.
Expenses: Tuition: Full-time $23,700; part-time $2,790 per unit. Required fees: $330.
David Jemiolo, Chairman, 914-437-7411, *E-mail:* jemiolo@vassar.edu.

■ VILLANOVA UNIVERSITY

Graduate School of Liberal Arts and Sciences, Department of Biology, Villanova, PA 19085-1699
AWARDS MA, MS. Part-time and evening/weekend programs available.

Students: 22 full-time (13 women), 8 part-time (5 women); includes 2 minority (both Asian Americans or Pacific Islanders), 4 international. Average age 26. *33 applicants, 67% accepted.* In 1999, 11 degrees awarded.
Degree requirements: For master's, thesis (for some programs), comprehensive exam required, foreign language not required.

Entrance requirements: For master's, GRE General Test, GRE Subject Test, minimum GPA of 3.0. *Application deadline:* For fall admission, 8/1 (priority date); for spring admission, 12/1. *Application fee:* $40.
Expenses: Tuition: Full-time $20,470.
Financial aid: Research assistantships with tuition reimbursements, Federal Work-Study and scholarships available. Aid available to part-time students. Financial aid application deadline: 4/1; financial aid applicants required to submit FAFSA.
Dr. Wilber W. Baker, Chair, 610-519-4830.
Find an in-depth description at www.petersons.com/graduate.

■ VIRGINIA COMMONWEALTH UNIVERSITY

School of Graduate Studies, College of Humanities and Sciences, Department of Biology, Richmond, VA 23284-9005
AWARDS MS.

Students: 10 full-time (7 women), 37 part-time (20 women); includes 4 minority (2 African Americans, 2 Asian Americans or Pacific Islanders). In 1999, 12 degrees awarded.
Degree requirements: For master's, thesis required, foreign language not required.
Entrance requirements: For master's, GRE General Test, GRE Subject Test, BS in biology or related field. *Application deadline:* For fall admission, 7/1 (priority date); for spring admission, 11/15. Applications are processed on a rolling basis. *Application fee:* $30.
Expenses: Tuition, state resident: full-time $4,031; part-time $224 per credit hour. Tuition, nonresident: full-time $11,946; part-time $664 per credit hour. Required fees: $1,081; $40 per credit hour. Tuition and fees vary according to campus/location and program.
Financial aid: Fellowships, research assistantships, teaching assistantships, Federal Work-Study, institutionally sponsored loans, and tuition waivers (full and partial) available. Aid available to part-time students.
Faculty research: Molecular and cellular biology; terrasteral and acquatic biology; systematics and physiology and developmental biology.
Dr. Leonard A. Smock, Chair, 804-828-1562, *Fax:* 804-828-0503, *E-mail:* lasmock@vcu.edu.
Application contact: Dr. Donald Young, Graduate Program Director, 804-828-1562, *Fax:* 804-828-0503, *E-mail:* dryoung@vcu.edu.

■ VIRGINIA COMMONWEALTH UNIVERSITY

School of Graduate Studies and School of Medicine, School of Medicine Graduate Programs, Richmond, VA 23284-9005

AWARDS MPH, MS, PhD, CBHS, MD/MPH, MD/MS, MD/PhD. Part-time programs available.

Students: 259 full-time, 111 part-time; includes 20 African Americans, 84 Asian Americans or Pacific Islanders, 10 Hispanic Americans. Average age 26. *760 applicants, 38% accepted.* Terminal master's awarded for partial completion of doctoral program.
Degree requirements: For master's, foreign language not required; for doctorate, dissertation, comprehensive oral and written exams required, foreign language not required.
Entrance requirements: For doctorate, GRE General Test. *Application fee:* $30.
Expenses: Tuition, state resident: full-time $4,031; part-time $224 per credit hour. Tuition, nonresident: full-time $11,946; part-time $664 per credit hour. Required fees: $1,081; $40 per credit hour. Tuition and fees vary according to campus/location and program.
Financial aid: Fellowships, research assistantships, teaching assistantships, career-related internships or fieldwork, Federal Work-Study, institutionally sponsored loans, and tuition waivers (full) available.
Dr. Hermes A. Kontos, Vice President for Health Sciences and Dean, School of Medicine, 804-828-9771, *Fax:* 804-828-8002, *E-mail:* hakontos@vcu.edu.
Application contact: Dr. Jan F. Chlebowski, Associate Dean for Graduate Education, 804-828-1023, *Fax:* 804-828-1473, *E-mail:* jfchlebo@vcu.edu.
Find an in-depth description at www.petersons.com/graduate.

■ VIRGINIA POLYTECHNIC INSTITUTE AND STATE UNIVERSITY

Graduate School, College of Arts and Sciences, Department of Biology, Blacksburg, VA 24061

AWARDS Botany (MS, PhD); ecology (MS, PhD); genetics (PhD); microbiology (MS, PhD); zoology (MS, PhD).

Faculty: 44 full-time (3 women).
Students: 67 full-time (30 women), 12 part-time (8 women); includes 5 minority (2 African Americans, 2 Asian Americans or Pacific Islanders, 1 Hispanic American), 14 international. Average age 25. *98*

Virginia Polytechnic Institute and State University (continued)
applicants, 26% accepted. In 1999, 14 master's, 10 doctorates awarded.

Degree requirements: For master's and doctorate, thesis/dissertation required, foreign language not required.

Entrance requirements: For master's, GRE General Test, TOEFL; for doctorate, GRE General Test, GRE Subject Test, TOEFL. *Application fee:* $25.

Expenses: Tuition, state resident: full-time $4,122; part-time $229 per credit hour. Tuition, nonresident: full-time $6,930; part-time $385 per credit hour. Required fees: $828; $107 per semester. Part-time tuition and fees vary according to course load.

Financial aid: In 1999–00, 19 research assistantships, 39 teaching assistantships were awarded; fellowships, unspecified assistantships also available.

Faculty research: Evolution, molecular biology, systematics.

Dr. Joe R. Cowles, Chairman, 540-231-8928, *E-mail:* cowlesjr@vt.edu.

Find an in-depth description at www.petersons.com/graduate.

■ **VIRGINIA STATE UNIVERSITY**

School of Graduate Studies, Research, and Outreach, School of Agriculture, Science and Technology, Department of Life Sciences, Petersburg, VA 23806-0001

AWARDS Biology (MS).

Faculty: 4 full-time (1 woman).
Students: In 1999, 2 degrees awarded.
Degree requirements: For master's, thesis required.
Entrance requirements: For master's, GRE General Test. *Application deadline:* For fall admission, 8/15. Applications are processed on a rolling basis. *Application fee:* $25.
Expenses: Tuition, state resident: full-time $2,306; part-time $106 per credit hour. Tuition, nonresident: full-time $7,824; part-time $346 per credit hour. Required fees: $29 per credit hour.
Financial aid: Application deadline: 5/1.
Faculty research: Schwann cell cultures, selection of apios as an alternative crop, systematic botany, flowers of three species of wild ginger.
Dr. Regina Knight-Mason, Chair, 804-524-5025, *E-mail:* rknight@vsu.edu.
Application contact: Dr. Wayne F. Virag, Dean, Graduate Studies, Research, and Outreach, 804-524-5985, *Fax:* 804-524-5104, *E-mail:* wvirag@vsu.edu.

■ **WAGNER COLLEGE**

Division of Graduate Studies, Department of Biological Sciences, Staten Island, NY 10301-4495

AWARDS Microbiology (MS). Part-time and evening/weekend programs available.

Faculty: 7 full-time (4 women), 2 part-time/adjunct (1 woman).
Students: 9 full-time (7 women), 6 part-time (4 women); includes 1 minority (Hispanic American). *10 applicants, 30% accepted.* In 1999, 7 degrees awarded.
Degree requirements: For master's, comprehensive exam or thesis required.
Entrance requirements: For master's, minimum GPA of 2.5, proficiency in statistics, undergraduate major in science. *Application deadline:* For fall admission, 8/1 (priority date); for spring admission, 12/10. Applications are processed on a rolling basis. *Application fee:* $50 ($65 for international students).
Expenses: Tuition: Part-time $580 per credit.
Financial aid: In 1999–00, 1 fellowship, 4 teaching assistantships with full tuition reimbursements (averaging $2,400 per year) were awarded; tuition waivers (partial) and alumni fellowships also available.
Dr. Donald Stearns, Head, 718-390-3197.
Application contact: 718-390-3411.

■ **WAKE FOREST UNIVERSITY**

Graduate School, Department of Biology, Winston-Salem, NC 27109

AWARDS MS, PhD. Part-time programs available.

Faculty: 20 full-time (3 women), 3 part-time/adjunct (1 woman).
Students: 35 full-time (18 women); includes 1 minority (Hispanic American), 3 international. Average age 26. *33 applicants, 52% accepted.* In 1999, 8 master's awarded (100% continued full-time study).
Degree requirements: For master's, one foreign language (computer language can substitute), thesis required; for doctorate, 2 foreign languages (computer language can substitute for one), dissertation required.
Entrance requirements: For master's and doctorate, GRE General Test, GRE Subject Test. *Application deadline:* For fall admission, 2/1. *Application fee:* $25.
Expenses: Tuition: Full-time $18,300. Full-time tuition and fees vary according to program.
Financial aid: Research assistantships, teaching assistantships, scholarships available. Aid available to part-time students. Financial aid application deadline: 2/15.
Faculty research: Cell biology, ecology, parasitology, immunology.
Dr. Ray Kuhn, Director, 336-758-5322.

■ **WAKE FOREST UNIVERSITY**

School of Medicine and Graduate School, Graduate Programs in Medicine, Winston-Salem, NC 27109

AWARDS MS, PhD, MD/MS, MD/PhD.

Degree requirements: For master's and doctorate, thesis/dissertation required.
Entrance requirements: For master's and doctorate, GRE General Test, GRE Subject Test. Electronic applications accepted.
Expenses: Tuition: Full-time $18,300.
Faculty research: Atherosclerosis, cardiovascular physiology, pharmacology, neuroanatomy, endocrinology.

■ **WALLA WALLA COLLEGE**

Graduate School, Department of Biological Science, College Place, WA 99324-1198

AWARDS MS. Part-time programs available.

Faculty: 5 full-time (2 women), 1 part-time/adjunct (0 women).
Students: 5 full-time (1 woman). Average age 24. *5 applicants, 60% accepted.* In 1999, 3 degrees awarded (33% found work related to degree, 66% continued full-time study).
Degree requirements: For master's, thesis, marine station experience required, foreign language not required.
Entrance requirements: For master's, GRE General Test and GRE Subject Test or Undergraduate Assessment Program, minimum GPA of 2.75. *Application deadline:* For fall admission, 4/1. Applications are processed on a rolling basis. *Application fee:* $40. Electronic applications accepted.
Expenses: Tuition: Full-time $14,235; part-time $365 per credit. Tuition and fees vary according to program.
Financial aid: In 1999–00, teaching assistantships with full tuition reimbursements (averaging $3,906 per year); Federal Work-Study also available. Financial aid application deadline: 4/1; financial aid applicants required to submit FAFSA.
Faculty research: Marine biology, physiology, animal behavior, plant development.
Dr. Scott Ligman, Chair, 509-527-2602, *E-mail:* ligmsc@wwc.edu.
Application contact: Dr. Joe G. Galusha, Dean of Graduate Studies, 509-527-2421, *Fax:* 509-527-2253, *E-mail:* galujo@wwc.edu.

■ **WASHINGTON STATE UNIVERSITY**

Graduate School, College of Sciences, Program in Biology, Pullman, WA 99164

AWARDS MS.

Faculty: 2 full-time (1 woman), 11 part-time/adjunct (1 woman).
Students: Average age 25. In 1999, 3 degrees awarded (100% continued full-time study).
Degree requirements: For master's, oral exam required, foreign language and thesis not required. *Average time to degree:* Master's–2 years full-time.
Entrance requirements: For master's, GRE General Test, GRE Subject Test (biology), minimum GPA of 3.0. *Application deadline:* For fall admission, 3/1 (priority date). Applications are processed on a rolling basis. *Application fee:* $35.
Expenses: Tuition, state resident: full-time $5,654. Tuition, nonresident: full-time $13,850. International tuition: $13,850 full-time. Tuition and fees vary according to program.
Financial aid: Research assistantships, teaching assistantships, Federal Work-Study, institutionally sponsored loans, and tuition waivers (partial) available. Financial aid application deadline: 4/1; financial aid applicants required to submit FAFSA. *Total annual research expenditures:* $212,888.
Dr. John Paznokas, Chair, 509-335-8649.

■ WASHINGTON STATE UNIVERSITY

Graduate School, College of Sciences, School of Biological Sciences, Pullman, WA 99164

AWARDS Biology (MS); botany (MS, PhD); zoology (MS, PhD).

Students: 41 full-time (16 women), 4 part-time (all women); includes 3 minority (all Asian Americans or Pacific Islanders), 1 international. In 1999, 12 master's, 3 doctorates awarded.
Degree requirements: For master's, oral exam required; for doctorate, dissertation, oral exam required, foreign language not required.
Entrance requirements: For master's and doctorate, GRE General Test, minimum GPA of 3.0. *Application fee:* $35.
Expenses: Tuition, state resident: full-time $5,654. Tuition, nonresident: full-time $13,850. International tuition: $13,850 full-time. Tuition and fees vary according to program.
Financial aid: In 1999–00, 6 research assistantships with full and partial tuition reimbursements, 39 teaching assistantships with full and partial tuition reimbursements were awarded.
Dr. Gary Thorgaard, Director, 509-335-3553, *Fax:* 509-335-3184, *E-mail:* zoology@wsu.edu.

■ WASHINGTON UNIVERSITY IN ST. LOUIS

Graduate School of Arts and Sciences, Division of Biology and Biomedical Sciences, St. Louis, MO 63130-4899

AWARDS Biochemistry (PhD); bioorganic chemistry (PhD); developmental biology (PhD); evolutionary and population biology (PhD), including ecology, environmental biology, evolutionary biology, genetics; immunology (PhD); molecular biophysics (PhD); molecular cell biology (PhD); molecular genetics (PhD); molecular microbiology and microbial pathogenesis (PhD); neurosciences (PhD); plant biology (PhD).

Faculty: 283 full-time (29 women).
Students: 443 full-time (195 women); includes 76 minority (6 African Americans, 56 Asian Americans or Pacific Islanders, 12 Hispanic Americans, 2 Native Americans), 87 international. *1025 applicants, 21% accepted.* In 1999, 53 doctorates awarded.
Degree requirements: For doctorate, dissertation required, foreign language not required.
Entrance requirements: For doctorate, GRE General Test, GRE Subject Test. *Application deadline:* For fall admission, 1/1 (priority date). Applications are processed on a rolling basis. *Application fee:* $0.
Expenses: Tuition: Full-time $23,400; part-time $975 per credit. Tuition and fees vary according to program.
Financial aid: Fellowships, research assistantships, tuition waivers (full) available. Financial aid application deadline: 1/1.
Application contact: Rosemary Garagneni, Director of Admissions, 800-852-9074, *E-mail:* admissions@dbbs.wustl.edu.

Find an in-depth description at www.petersons.com/graduate.

■ WAYNE STATE UNIVERSITY

Graduate School, College of Science, Department of Biological Sciences, Detroit, MI 48202

AWARDS Biological sciences (MS, PhD); molecular biotechnology (MS). Terminal master's awarded for partial completion of doctoral program.

Degree requirements: For master's, thesis required (for some programs), foreign language not required; for doctorate, dissertation required, foreign language not required.
Entrance requirements: For master's, GRE General Test, minimum GPA of 3.0; for doctorate, GRE General Test, GRE Subject Test, minimum GPA of 3.2.

Faculty research: Cell biology, molecular genetics, development, microbiology, ecology.

Find an in-depth description at www.petersons.com/graduate.

■ WAYNE STATE UNIVERSITY

School of Medicine and Graduate School, Graduate Programs in Medicine, Detroit, MI 48202

AWARDS MS, PhD, Certificate, MD/PhD. Part-time and evening/weekend programs available.

Degree requirements: For doctorate, dissertation required.
Entrance requirements: For master's and doctorate, GRE.

■ WESLEYAN UNIVERSITY

Graduate Programs, Department of Biology, Middletown, CT 06459-0260

AWARDS Cell biology (PhD); comparative physiology (PhD); developmental biology (PhD); genetics (PhD); neurophysiology (PhD); population biology (PhD).

Faculty: 12 full-time (3 women).
Students: 24 full-time (12 women); includes 1 minority (African American), 11 international. Average age 28. *125 applicants, 4% accepted.* In 1999, 2 doctorates awarded.
Degree requirements: For doctorate, one foreign language (computer language can substitute), dissertation required.
Entrance requirements: For doctorate, GRE Subject Test. *Application deadline:* For fall admission, 1/15. Applications are processed on a rolling basis. *Application fee:* $0.
Expenses: Tuition: Full-time $24,876. Required fees: $650. Tuition and fees vary according to program.
Financial aid: Research assistantships, teaching assistantships, stipends available.
Faculty research: Microbial population genetics, genetic basis of evolutionary adaptation, genetic regulation of differentiation and pattern formation in *drosophila*.
Dr. Fred Cohan, Chairman, 860-685-3489.
Application contact: Marina J. Melendez, Director of Graduate Student Services, 860-685-2390, *Fax:* 860-685-2439, *E-mail:* mmelendez@wesleyan.edu.

Find an in-depth description at www.petersons.com/graduate.

■ WEST CHESTER UNIVERSITY OF PENNSYLVANIA

Graduate Studies, College of Arts and Sciences, Department of Biology, West Chester, PA 19383

AWARDS MS. Part-time and evening/weekend programs available.

West Chester University of Pennsylvania (continued)
Faculty: 8.
Students: 6 full-time (4 women), 12 part-time (7 women), 1 international. *12 applicants, 83% accepted.* In 1999, 3 degrees awarded.
Degree requirements: For master's, thesis, comprehensive exam required, foreign language not required.
Entrance requirements: For master's, GRE General Test, GRE Subject Test. *Application deadline:* For fall admission, 4/15 (priority date); for spring admission, 10/15. Applications are processed on a rolling basis. *Application fee:* $25.
Expenses: Tuition, state resident: full-time $3,780; part-time $210 per credit. Tuition, nonresident: full-time $6,610; part-time $367 per credit. Required fees: $660; $39 per credit. Tuition and fees vary according to course load.
Financial aid: In 1999–00, 2 research assistantships with full tuition reimbursements (averaging $5,000 per year) were awarded; unspecified assistantships also available. Aid available to part-time students. Financial aid application deadline: 2/15; financial aid applicants required to submit FAFSA.
Faculty research: Cell physiology of insect ovarian follicles, field inventory of reptiles and amphibians.
Dr. Martha Potvin, Chair, 610-436-2538.
Application contact: Dr. Leslie Slusher, Graduate Coordinator, 610-436-2751, *E-mail:* lslusher@wcupa.edu.

■ **WESTERN CAROLINA UNIVERSITY**

Graduate School, College of Arts and Sciences, Department of Biology, Cullowhee, NC 28723

AWARDS MA Ed, MAT, MS. Part-time and evening/weekend programs available.

Faculty: 19.
Students: 21 full-time (9 women), 5 part-time (2 women), 2 international. *28 applicants, 54% accepted.* In 1999, 8 degrees awarded.
Degree requirements: For master's, thesis, comprehensive exam required, foreign language not required.
Entrance requirements: For master's, GRE General Test. *Application deadline:* For fall admission, 5/1 (priority date); for spring admission, 10/1 (priority date). Applications are processed on a rolling basis. *Application fee:* $35.
Expenses: Tuition, area resident: Part-time $147 per hour. Tuition, state resident: full-time $962; part-time $147 per hour. Tuition, nonresident: full-time $8,232; part-time $1,056 per hour. Required fees: $975.

Financial aid: In 1999–00, 24 students received aid, including 4 research assistantships with full and partial tuition reimbursements available (averaging $6,156 per year), 20 teaching assistantships with full and partial tuition reimbursements available (averaging $4,800 per year); fellowships, Federal Work-Study, grants, and institutionally sponsored loans also available. Financial aid application deadline: 3/15; financial aid applicants required to submit FAFSA.
Henry Mainwaring, Head, 828-227-7244.
Application contact: Kathleen Owen, Assistant to the Dean, 828-227-7398, *Fax:* 828-227-7480, *E-mail:* kowen@wcu.edu.

■ **WESTERN CONNECTICUT STATE UNIVERSITY**

Division of Graduate Studies, School of Arts and Sciences, Department of Biological and Environmental Sciences, Danbury, CT 06810-6885

AWARDS MA. Part-time and evening/weekend programs available.

Faculty: 8 full-time (1 woman).
Students: In 1999, 5 degrees awarded.
Degree requirements: For master's, comprehensive exam or thesis required.
Entrance requirements: For master's, minimum GPA of 2.5. *Application deadline:* For fall admission, 8/1 (priority date). Applications are processed on a rolling basis. *Application fee:* $40.
Expenses: Tuition, state resident: full-time $2,568; part-time $178 per credit. Tuition, nonresident: full-time $7,156; part-time $178 per credit. Required fees: $240; $30 per semester.
Financial aid: Fellowships, career-related internships or fieldwork and Federal Work-Study available. Aid available to part-time students. Financial aid application deadline: 5/1; financial aid applicants required to submit FAFSA.
Dr. Howard Russock, Chair, 203-837-8798.
Application contact: Chris Shankle, Associate Director of Graduate Admissions, 203-837-8244, *Fax:* 203-837-8338, *E-mail:* shanklec@wcsu.edu.

■ **WESTERN ILLINOIS UNIVERSITY**

School of Graduate Studies, College of Arts and Sciences, Department of Biological Sciences, Macomb, IL 61455-1390

AWARDS MS. Part-time programs available.

Faculty: 20 full-time (5 women), 3 part-time/adjunct (0 women).
Students: 37 full-time (19 women), 19 part-time (14 women); includes 3 minority (2 African Americans, 1 Hispanic

American), 5 international. Average age 27. *31 applicants, 65% accepted.* In 1999, 12 degrees awarded.
Degree requirements: For master's, thesis or alternative required, foreign language not required.
Entrance requirements: For master's, minimum GPA of 2.7. *Application deadline:* Applications are processed on a rolling basis. *Application fee:* $0 ($25 for international students).
Expenses: Tuition, state resident: full-time $2,376; part-time $99 per semester hour. Tuition, nonresident: full-time $4,752; part-time $198 per semester hour. Required fees: $29 per semester hour. Tuition and fees vary according to student level.
Financial aid: In 1999–00, 18 students received aid, including 18 research assistantships with full tuition reimbursements available (averaging $4,880 per year). Financial aid applicants required to submit FAFSA.
Faculty research: White-sided dolphins, study of macroinvertebrate species, pollinator abundance, plankton in lakes, environmental education.
Dr. Larrance M. O'Flaherty, Chairperson, 309-298-2408.
Application contact: Barbara Baily, Director of Graduate Studies, 309-298-1806, *Fax:* 309-298-2345, *E-mail:* grad_office@ccmail.wiu.edu.

Find an in-depth description at www.petersons.com/graduate.

■ **WESTERN KENTUCKY UNIVERSITY**

Graduate Studies, Ogden College of Science, Technology, and Health, Department of Biology, Bowling Green, KY 42101-3576

AWARDS Biology (MA Ed, MS).

Students: 11 full-time (4 women), 11 part-time (6 women), 7 international. Average age 30. *21 applicants, 67% accepted.* In 1999, 7 degrees awarded.
Degree requirements: For master's, research tool, comprehensive exam required, thesis optional, foreign language not required. *Average time to degree:* Master's–2 years full-time.
Entrance requirements: For master's, GRE General Test, minimum GPA of 2.75. *Application deadline:* For fall admission, 8/1 (priority date); for spring admission, 12/1. Applications are processed on a rolling basis. *Application fee:* $30.
Expenses: Tuition, state resident: full-time $2,590; part-time $140 per hour. Tuition, nonresident: full-time $6,430; part-time $387 per hour. Required fees: $370. Part-time tuition and fees vary according to course load.

Financial aid: In 1999–00, 10 students received aid, including 8 teaching assistantships with partial tuition reimbursements available (averaging $4,500 per year); Federal Work-Study, institutionally sponsored loans, and service awards also available. Aid available to part-time students. Financial aid application deadline: 4/1; financial aid applicants required to submit FAFSA.
Faculty research: Biotechnology, biodiversity, ecology, immunoparasitology, genetics. *Total annual research expenditures:* $250,000.
Blaine Ferrell, Acting Head, 270-745-3696, *Fax:* 270-745-6856, *E-mail:* blaine.ferrell@wku.edu.

■ WESTERN MICHIGAN UNIVERSITY

Graduate College, College of Arts and Sciences, Department of Biological Sciences, Kalamazoo, MI 49008-5202

AWARDS MS, PhD.

Students: 31 full-time (16 women), 17 part-time (8 women); includes 6 minority (2 African Americans, 2 Asian Americans or Pacific Islanders, 2 Hispanic Americans), 9 international. *30 applicants, 30% accepted.* In 1999, 3 degrees awarded.
Degree requirements: For master's, oral exam required, foreign language and thesis not required; for doctorate, computer language, dissertation, oral exam required, foreign language not required.
Entrance requirements: For master's and doctorate, GRE General Test. *Application deadline:* For fall admission, 2/15 (priority date). Applications are processed on a rolling basis. *Application fee:* $25.
Expenses: Tuition, state resident: full-time $3,831; part-time $160 per credit hour. Tuition, nonresident: full-time $9,221; part-time $384 per credit hour. Required fees: $602; $602 per year. Full-time tuition and fees vary according to course load, degree level and program.
Financial aid: Fellowships, research assistantships, teaching assistantships, institutionally sponsored loans available. Financial aid application deadline: 2/15; financial aid applicants required to submit FAFSA.
Dr. Leonard Beuving, Chair, 616-387-5600.
Application contact: Paula J. Boodt, Coordinator, Graduate Admissions and Recruitment, 616-387-2000, *Fax:* 616-387-2355, *E-mail:* paula.boodt@wmich.edu.

■ WESTERN WASHINGTON UNIVERSITY

Graduate School, College of Arts and Sciences, Department of Biology, Bellingham, WA 98225-5996

AWARDS MS. Part-time programs available.
Faculty: 18.
Students: 13 full-time (6 women), 2 part-time (1 woman); includes 2 minority (1 African American, 1 Asian American or Pacific Islander). *20 applicants, 75% accepted.* In 1999, 7 degrees awarded.
Degree requirements: For master's, thesis required.
Entrance requirements: For master's, GRE General Test, GRE Subject Test (biology), TOEFL, minimum GPA of 3.0 in last 60 semester hours or last 90 quarter hours. *Application deadline:* For fall admission, 2/1 (priority date); for winter admission, 10/1; for spring admission, 2/1. Applications are processed on a rolling basis. *Application fee:* $35.
Expenses: Tuition, state resident: full-time $3,247; part-time $146 per credit hour. Tuition, nonresident: full-time $13,364; part-time $445 per credit hour. Required fees: $254; $85 per quarter.
Financial aid: In 1999–00, 2 research assistantships with partial tuition reimbursements (averaging $8,325 per year), 11 teaching assistantships with partial tuition reimbursements (averaging $8,325 per year) were awarded; Federal Work-Study, institutionally sponsored loans, scholarships, and tuition waivers (partial) also available. Aid available to part-time students. Financial aid application deadline: 2/15; financial aid applicants required to submit FAFSA.
Dr. Roger Anderson, Chair, 360-650-3992.
Application contact: Dr. David Morgan, Adviser, 360-650-6575.

■ WEST TEXAS A&M UNIVERSITY

College of Agriculture, Nursing, and Natural Sciences, Department of Life, Earth, and Environmental Sciences, Program in Biology, Canyon, TX 79016-0001

AWARDS MS. Part-time programs available.

Degree requirements: For master's, comprehensive exam required, thesis optional, foreign language not required.
Entrance requirements: For master's, GRE General Test. Electronic applications accepted.
Expenses: Tuition, state resident: full-time $1,152; part-time $48 per credit. Tuition, nonresident: full-time $6,336; part-time $264 per credit. Required fees: $1,063; $531 per semester.

■ WEST VIRGINIA UNIVERSITY

Eberly College of Arts and Sciences, Department of Biology, Morgantown, WV 26506

AWARDS Animal behavior (MS); cellular and molecular biology (MS, PhD); environmental plant biology (MS, PhD); plant systematics (MS); population genetics (MS).

Faculty: 18 full-time (4 women), 2 part-time/adjunct (1 woman).
Students: 19 full-time (10 women), 3 part-time (2 women); includes 2 minority (1 Asian American or Pacific Islander, 1 Hispanic American), 4 international. Average age 27. *50 applicants, 10% accepted.* In 1999, 4 master's, 1 doctorate awarded. Terminal master's awarded for partial completion of doctoral program.
Degree requirements: For master's, thesis, final exam required, foreign language not required; for doctorate, dissertation, preliminary and final exams required, foreign language not required. *Average time to degree:* Master's–2.5 years full-time; doctorate–5 years full-time.
Entrance requirements: For master's, GRE General Test, GRE Subject Test, TOEFL, minimum GPA of 3.0; for doctorate, GRE General Test, TOEFL, minimum GPA of 3.0. *Application deadline:* For fall admission, 4/1; for spring admission, 10/1. Applications are processed on a rolling basis. *Application fee:* $45.
Expenses: Tuition, state resident: full-time $2,910; part-time $154 per credit hour. Tuition, nonresident: full-time $8,368; part-time $457 per credit hour.
Financial aid: In 1999–00, 3 research assistantships, 13 teaching assistantships were awarded; Federal Work-Study and institutionally sponsored loans also available. Financial aid application deadline: 4/1; financial aid applicants required to submit FAFSA.
Faculty research: Environmental biology, genetic engineering, developmental biology, global change, biodiversity.
Dr. Keith Garbutt, Chair, 304-293-5394.
Application contact: Dr. James B. McGraw, Director of Graduate Studies, 304-293-5201, *Fax:* 304-293-6363.

Find an in-depth description at www.petersons.com/graduate.

■ WEST VIRGINIA UNIVERSITY

School of Medicine, Graduate Programs in Health Sciences, Morgantown, WV 26506

AWARDS MOT, MPH, MPT, MS, PhD, MD/PhD. Part-time and evening/weekend programs available. Postbaccalaureate distance learning degree programs offered (minimal on-campus study).

West Virginia University (continued)
Students: 128 full-time (65 women), 143 part-time (106 women); includes 23 minority (5 African Americans, 14 Asian Americans or Pacific Islanders, 2 Hispanic Americans, 2 Native Americans), 26 international. In 1999, 116 master's, 12 doctorates awarded.
Application fee: $45.
Expenses: Tuition, state resident: full-time $3,564. Tuition, nonresident: full-time $10,230.
Financial aid: In 1999–00, 33 research assistantships, 48 teaching assistantships were awarded; fellowships, career-related internships or fieldwork, Federal Work-Study, institutionally sponsored loans, tuition waivers (full and partial), and graduate administrative assistantships also available. Financial aid applicants required to submit FAFSA.
Dr. George A. Hedge, Associate Dean/ Graduate Coordinator, 304-293-7206, *Fax:* 304-293-7038.
Application contact: Claire Noel, Graduate Adviser, 304-293-7116, *Fax:* 304-293-7038, *E-mail:* cnoel@wvu.edu.
Find an in-depth description at www.petersons.com/graduate.

■ WICHITA STATE UNIVERSITY

Graduate School, Fairmount College of Liberal Arts and Sciences, Department of Biological Sciences, Wichita, KS 67260

AWARDS MS. Part-time programs available.
Faculty: 12 full-time (1 woman), 1 (woman) part-time/adjunct.
Students: 6 full-time (2 women), 17 part-time (10 women); includes 2 minority (1 African American, 1 Asian American or Pacific Islander), 5 international. Average age 30. *16 applicants, 50% accepted.* In 1999, 7 degrees awarded.
Degree requirements: For master's, variable foreign language requirement, comprehensive exam required, thesis optional.
Entrance requirements: For master's, GRE Subject Test, TOEFL. *Application deadline:* For fall admission, 7/1 (priority date); for spring admission, 1/1. Applications are processed on a rolling basis. *Application fee:* $25 ($40 for international students). Electronic applications accepted.
Expenses: Tuition, state resident: full-time $1,769; part-time $98 per credit. Tuition, nonresident: full-time $5,906; part-time $328 per credit. Required fees: $338; $19 per credit. One-time fee: $17. Tuition and fees vary according to course load.
Financial aid: In 1999–00, 4 research assistantships (averaging $8,000 per year), 10 teaching assistantships with full tuition reimbursements (averaging $7,000 per

year) were awarded; Federal Work-Study, institutionally sponsored loans, and unspecified assistantships also available. Financial aid application deadline: 4/1; financial aid applicants required to submit FAFSA.
Faculty research: Molecular biology, environmental science, reproductive endocrinology, cancer, plant ecology.
Dr. J. David McDonald, Chairperson, 316-978-3111, *Fax:* 316-978-3772.
Application contact: Dr. William Hendry, Graduate Coordinator, 316-978-3111, *Fax:* 316-978-3772, *E-mail:* whendry@ twsuvm.uc.twsu.edu.
Find an in-depth description at www.petersons.com/graduate.

■ WILLIAM PATERSON UNIVERSITY OF NEW JERSEY

College of Science and Health, Department of Biology, General Biology Program, Wayne, NJ 07470-8420

AWARDS General biology (MA); limnology and terrestrial ecology (MA); molecular biology (MA); physiology (MA). Part-time and evening/weekend programs available.
Students: 1 (woman) full-time, 6 part-time (3 women); includes 1 minority (Hispanic American). Average age 25. *9 applicants, 44% accepted.* In 1999, 1 degree awarded.
Degree requirements: For master's, comprehensive exam, independent study or thesis required.
Entrance requirements: For master's, GRE General Test, minimum GPA of 2.75. *Application deadline:* For fall admission, 4/1; for spring admission, 10/15. Applications are processed on a rolling basis. *Application fee:* $35. Electronic applications accepted.
Expenses: Tuition, state resident: part-time $244 per credit. Tuition, nonresident: part-time $350 per credit.
Financial aid: In 1999–00, 2 teaching assistantships (averaging $9,000 per year) were awarded; career-related internships or fieldwork and unspecified assistantships also available. Financial aid application deadline: 4/1; financial aid applicants required to submit FAFSA.
Application contact: Office of Graduate Studies, 973-720-2237, *Fax:* 973-720-2035.
Find an in-depth description at www.petersons.com/graduate.

■ WINTHROP UNIVERSITY

College of Arts and Sciences, Department of Biology, Rock Hill, SC 29733

AWARDS MS. Part-time programs available.
Faculty: 11 full-time (3 women).

Students: 9 full-time (5 women), 2 part-time (both women). Average age 27. In 1999, 6 degrees awarded.
Degree requirements: For master's, thesis optional.
Entrance requirements: For master's, GRE General Test, minimum GPA of 3.0. *Application deadline:* For fall admission, 7/15 (priority date); for spring admission, 12/1. Applications are processed on a rolling basis. *Application fee:* $35.
Expenses: Tuition, state resident: full-time $4,020; part-time $168 per semester hour. Tuition, nonresident: full-time $7,240; part-time $302 per semester hour.
Financial aid: Federal Work-Study, scholarships, and unspecified assistantships available. Aid available to part-time students. Financial aid application deadline: 2/1; financial aid applicants required to submit FAFSA.
Dr. Ralph A. Gustafuson, Chairman, 803-323-2111, *Fax:* 803-323-2246, *E-mail:* gustafusonr@winthrop.edu.
Application contact: Sharon Johnson, Director of Graduate Studies, 803-323-2204, *Fax:* 803-323-2292, *E-mail:* johnsons@winthrop.edu.

■ WORCESTER POLYTECHNIC INSTITUTE

Graduate Studies, Department of Biology and Biotechnology, Worcester, MA 01609-2280

AWARDS Biology (MS); biomedical sciences (PhD); biotechnology (MS, PhD).
Faculty: 12 full-time (4 women), 1 part-time/adjunct (0 women).
Students: 27 full-time (15 women), 2 part-time (both women); includes 1 minority (Asian American or Pacific Islander), 4 international. *51 applicants, 31% accepted.* In 1999, 17 master's, 3 doctorates awarded.
Degree requirements: For master's, thesis required, foreign language not required; for doctorate, dissertation, qualifying exam required, foreign language not required.
Entrance requirements: For master's and doctorate, GRE General Test, TOEFL. *Application deadline:* For fall admission, 2/1 (priority date); for spring admission, 10/15 (priority date). Applications are processed on a rolling basis. *Application fee:* $50. Electronic applications accepted.
Expenses: Tuition: Full-time $13,220; part-time $703 per credit. Required fees: $50.
Financial aid: In 1999–00, 17 students received aid, including 2 fellowships with full tuition reimbursements available (averaging $12,750 per year), 2 research assistantships with full tuition reimbursements available (averaging $15,950 per year), 12 teaching assistantships with full

tuition reimbursements available (averaging $12,330 per year); career-related internships or fieldwork, grants, institutionally sponsored loans, scholarships, and tuition waivers (full and partial) also available. Financial aid application deadline: 2/15; financial aid applicants required to submit FAFSA.
Faculty research: Genetic engineering, microbial genetics, immunology, DNA technology, fermentation genetics, pharmaceutical production. *Total annual research expenditures:* $287,251.
Dr. Ronald D. Cheetham, Head, 508-831-5582, *Fax:* 508-831-5936, *E-mail:* cheetham@wpi.edu.
Application contact: Dr. Dainiel Gibson, Graduate Coordinator, 508-831-5543, *Fax:* 508-831-5936, *E-mail:* digibson@wip.edu.

Find an in-depth description at www.petersons.com/graduate.

■ WRIGHT STATE UNIVERSITY

School of Graduate Studies, College of Science and Mathematics, Department of Biological Sciences, Dayton, OH 45435

AWARDS Biological sciences (MS); environmental sciences (MS).

Students: 36 full-time (14 women), 19 part-time (12 women); includes 10 minority (5 African Americans, 2 Asian Americans or Pacific Islanders, 1 Hispanic American, 2 Native Americans), 7 international. Average age 27. *24 applicants, 92% accepted.* In 1999, 20 degrees awarded.
Degree requirements: For master's, thesis optional, foreign language not required.
Entrance requirements: For master's, TOEFL. *Application fee:* $25.
Expenses: Tuition, state resident: full-time $5,568; part-time $175 per quarter hour. Tuition, nonresident: full-time $9,696; part-time $302 per quarter hour. Full-time tuition and fees vary according to course load, campus/location and program.
Financial aid: Fellowships, research assistantships, teaching assistantships, career-related internships or fieldwork, institutionally sponsored loans, and unspecified assistantships available. Aid available to part-time students. Financial aid applicants required to submit FAFSA.
Dr. Michele Wheatly, Chair, 937-775-2655, *Fax:* 937-775-3320, *E-mail:* michele.whently@wright.edu.
Application contact: Dr. David L. Goldstein, Director, 937-775-3430, *Fax:* 937-775-3320, *E-mail:* david.goldstein@wright.edu.

■ WRIGHT STATE UNIVERSITY

School of Graduate Studies, College of Science and Mathematics and School of Medicine, Program in Biomedical Sciences, Dayton, OH 45435

AWARDS PhD.

Students: 46 full-time (26 women), 3 part-time (1 woman); includes 4 minority (2 African Americans, 2 Asian Americans or Pacific Islanders), 10 international. Average age 30. *58 applicants, 28% accepted.* In 1999, 6 degrees awarded.
Degree requirements: For doctorate, dissertation required.
Entrance requirements: For doctorate, TOEFL. *Application deadline:* For fall admission, 7/15. *Application fee:* $25.
Expenses: Tuition, state resident: full-time $5,568; part-time $175 per quarter hour. Tuition, nonresident: full-time $9,696; part-time $302 per quarter hour. Full-time tuition and fees vary according to course load, campus/location and program.
Financial aid: Fellowships, research assistantships, Federal Work-Study, institutionally sponsored loans, tuition waivers (full), and unspecified assistantships available. Aid available to part-time students. Financial aid applicants required to submit FAFSA.
Dr. Robert E. W. Fyffe, Director, 937-775-2504, *Fax:* 937-775-3485, *E-mail:* robert.fyffe@wright.edu.

Find an in-depth description at www.petersons.com/graduate.

■ YALE UNIVERSITY

School of Medicine and Graduate School of Arts and Sciences, Combined Program in Biological and Biomedical Sciences (BBS), New Haven, CT 06520

AWARDS PhD, MD/PhD.

Degree requirements: For doctorate, dissertation required.
Entrance requirements: For doctorate, GRE General Test, TOEFL. *Application deadline:* For fall admission, 1/2. *Application fee:* $65. Electronic applications accepted.
Expenses: All students receive full tuition of $22,330 and an annual stipend of $17,600 .
Financial aid: Fellowships, research assistantships, teaching assistantships available.
Dr. Ira Mellman, Director, 203-785-4302, *Fax:* 203-785-7226, *E-mail:* bbs@yale.edu.
Application contact: Dr. John Alvaro, Administrative Director, 203-785-3735, *Fax:* 203-785-3734, *E-mail:* bbs@yale.edu.

Find an in-depth description at www.petersons.com/graduate.

■ YESHIVA UNIVERSITY

Albert Einstein College of Medicine, Medical Scientist Training Program, New York, NY 10033-3201

AWARDS MD/PhD. Students must apply through the Albert Einstein College of Medicine.

Students: 108 full-time (36 women); includes 31 minority (9 African Americans, 18 Asian Americans or Pacific Islanders, 4 Hispanic Americans), 8 international. Average age 25. *16 applicants, 56% accepted. Application deadline:* For fall admission, 11/15. *Application fee:* $75.
Expenses: Tuition: Part-time $525 per credit. Tuition and fees vary according to degree level and program.
Financial aid: In 1999–00, 108 fellowships were awarded.
Dr. Betty Diamond, Director, 718-430-2128, *Fax:* 718-430-8655.
Application contact: Sheila Cleeton, Assistant Director, 718-430-2128, *Fax:* 718-430-8655, *E-mail:* phd@aecom.yu.edu.

Find an in-depth description at www.petersons.com/graduate.

■ YESHIVA UNIVERSITY

Albert Einstein College of Medicine, Sue Golding Graduate Division of Medical Sciences, Bronx, NY 10461

AWARDS PhD, MD/PhD.

Faculty: 172 full-time, 17 part-time/adjunct.
Students: 229 full-time (105 women); includes 45 minority (13 African Americans, 24 Asian Americans or Pacific Islanders, 8 Hispanic Americans), 101 international. Average age 25. *213 applicants, 20% accepted.* In 1999, 37 degrees awarded.
Degree requirements: For doctorate, dissertation required, foreign language not required.
Entrance requirements: For doctorate, GRE General Test, TOEFL. *Application deadline:* For fall admission, 1/15 (priority date). *Application fee:* $0.
Expenses: Tuition: Part-time $525 per credit. Tuition and fees vary according to degree level and program.
Financial aid: In 1999–00, 229 fellowships were awarded.
Dr. Anne M. Etgen, Director, 718-430-2345, *Fax:* 718-430-8655.
Application contact: Sheila Cleeton, Assistant Director, 718-430-2345, *Fax:* 718-430-8655, *E-mail:* phd@aecom.yu.edu.

Find an in-depth description at www.petersons.com/graduate.

■ YOUNGSTOWN STATE UNIVERSITY

Graduate School, College of Arts and Sciences, Department of Biological Sciences, Youngstown, OH 44555-0001

AWARDS MS. Part-time programs available.

Faculty: 14 full-time (3 women), 1 part-time/adjunct (0 women).

Students: 22 full-time (12 women), 10 part-time (5 women); includes 1 minority (Hispanic American), 1 international. *17 applicants, 100% accepted.* In 1999, 9 degrees awarded.

Degree requirements: For master's, thesis, oral review, written comprehensive exam required, foreign language not required.

Entrance requirements: For master's, GRE General Test, TOEFL, minimum GPA of 2.7. *Application deadline:* For fall admission, 7/15 (priority date); for spring admission, 12/15 (priority date). Applications are processed on a rolling basis. *Application fee:* $30 ($75 for international students).

Expenses: Tuition, state resident: part-time $109 per credit hour. Tuition, nonresident: part-time $235 per credit hour. Required fees: $21 per credit hour. $41 per quarter. Tuition and fees vary according to program.

Financial aid: In 1999–00, 23 students received aid, including 7 research assistantships with full tuition reimbursements available (averaging $7,500 per year), 14 teaching assistantships with full tuition reimbursements available (averaging $7,500 per year); Federal Work-Study, institutionally sponsored loans, and scholarships also available. Aid available to part-time students. Financial aid application deadline: 3/1.

Faculty research: Cell biology, neurophysiology, molecular biology, neurobiology, gene regulation.
Dr. Paul Peterson, Chair, 330-742-3601.

Application contact: Dr. Peter J. Kasvinsky, Dean of Graduate Studies, 330-742-3091, *Fax:* 330-742-1580, *E-mail:* amgrad03@ysub.ysu.edu.

Anatomy

ANATOMY

■ AUBURN UNIVERSITY

College of Veterinary Medicine and Graduate School, Graduate Program in Veterinary Medicine, Department of Anatomy, Physiology and Pharmacology, Auburn, Auburn University, AL 36849-0002

AWARDS Anatomy and histology (MS); physiology and pharmacology (MS). Part-time programs available.

Students: *9 applicants, 44% accepted.*

Degree requirements: For master's, thesis required, foreign language not required.

Entrance requirements: For master's, GRE General Test. *Application deadline:* For fall admission, 7/1; for spring admission, 12/1. Applications are processed on a rolling basis. *Application fee:* $25 ($50 for international students).

Expenses: Tuition, state resident: full-time $2,895; part-time $80 per credit hour. Tuition, nonresident: full-time $8,685; part-time $240 per credit hour.

Financial aid: Research assistantships, teaching assistantships available. Aid available to part-time students. Financial aid application deadline: 3/15.

Faculty research: Chemosensory systems, embryo transfer, osteoarthritis, neurosenses, audiology, cardiovascular physiology, molecular endocrinology. *Total annual research expenditures:* $400,000.
Dr. Philip Posner, Head, 334-844-4427.

Application contact: Dr. John F. Pritchett, Dean of the Graduate School, 334-844-4700.

■ BOSTON UNIVERSITY

Sargent College of Health and Rehabilitation Sciences, Department of Health Sciences, Boston, MA 02215

AWARDS Applied anatomy and physiology (MS, D Sc); nutrition (MS). Part-time programs available.

Faculty: 8 full-time (4 women), 2 part-time/adjunct (1 woman).

Students: 49 full-time (41 women), 7 part-time (5 women); includes 4 minority (3 Asian Americans or Pacific Islanders, 1 Hispanic American), 6 international. Average age 26. *86 applicants, 66% accepted.* In 1999, 8 master's, 3 doctorates awarded.

Degree requirements: For master's, thesis or alternative required, foreign language not required; for doctorate, computer language, dissertation required, foreign language not required.

Entrance requirements: For master's, GRE General Test, minimum GPA of 3.0; for doctorate, GRE General Test, master's degree. *Application deadline:* For fall admission, 4/1 (priority date); for spring admission, 10/1. Applications are processed on a rolling basis. *Application fee:* $60.

Expenses: Tuition: Full-time $23,770; part-time $743 per credit. Required fees: $220. Tuition and fees vary according to class time, course level, campus/location and program.

Financial aid: In 1999–00, 20 fellowships, 7 research assistantships, 6 teaching assistantships were awarded; career-related internships or fieldwork, Federal Work-Study, institutionally sponsored loans, and scholarships also available. Aid available to part-time students. Financial aid application deadline: 4/15.

Faculty research: Muscle metabolism, body acid-base balance, human performance, physical conditioning, diabetes.
Dr. Gary Skrinar, Chairman, 617-353-2717.

Application contact: Judy Skeffington, Senior Admissions Coordinator, 617-353-2713, *Fax:* 617-353-7500, *E-mail:* jaskeff@bu.edu.

■ BOSTON UNIVERSITY

School of Medicine, Division of Graduate Medical Sciences, Department of Anatomy and Neurobiology, Boston, MA 02118

AWARDS MA, PhD, MD/PhD. Part-time programs available.

Faculty: 13 full-time (3 women), 6 part-time/adjunct (0 women).

Students: 20 full-time (10 women); includes 7 minority (1 African American, 4 Asian Americans or Pacific Islanders, 2 Hispanic Americans), 1 international. Average age 27. Terminal master's awarded for partial completion of doctoral program.

Degree requirements: For master's and doctorate, thesis/dissertation, qualifying exam required, foreign language not required.

Entrance requirements: For master's and doctorate, GRE General Test, GRE

Subject Test, TOEFL. *Application deadline:* For fall admission, 1/15 (priority date); for spring admission, 10/15 (priority date). *Application fee:* $50. Electronic applications accepted.

Expenses: Tuition: Full-time $24,700; part-time $772 per credit. Required fees: $220.

Financial aid: Fellowships with tuition reimbursements, research assistantships with tuition reimbursements, Federal Work-Study, scholarships, and traineeships available.

Faculty research: Neuroanatomy, development of the nervous system, aging, respiratory system, reproductive system. Mark Moss, Chairman, 617-638-4200, *Fax:* 617-638-4216 Ext. -, *E-mail:* mmos@cajal-1.bu.edu.

■ CASE WESTERN RESERVE UNIVERSITY

School of Medicine and School of Graduate Studies, Graduate Programs in Medicine, Department of Anatomy, Cleveland, OH 44106

AWARDS Applied anatomy (MS); biological anthropology (MS, PhD); cellular biology (MS, PhD); developmental biology (PhD); molecular biology (PhD). Part-time programs available.

Faculty: 14 full-time (5 women), 14 part-time/adjunct (4 women).
Students: 17 full-time (9 women), 12 part-time (1 woman); includes 11 minority (1 African American, 9 Asian Americans or Pacific Islanders, 1 Hispanic American), 2 international. Average age 26. *16 applicants, 50% accepted.* In 1999, 5 master's awarded (50% found work related to degree, 50% continued full-time study); 1 doctorate awarded (100% entered university research/teaching).
Degree requirements: For master's, thesis required (for some programs); for doctorate, dissertation required. *Average time to degree:* Master's–5 years full-time; doctorate–1 year full-time.
Entrance requirements: For master's, GRE General Test, TOEFL; for doctorate, GRE General Test, GRE Subject Test, TOEFL. *Application deadline:* For fall admission, 5/1 (priority date); for spring admission, 8/1 (priority date). Applications are processed on a rolling basis. *Application fee:* $25.
Financial aid: In 1999–00, 7 research assistantships with full tuition reimbursements (averaging $16,000 per year) were awarded; fellowships, grants also available.
Faculty research: Hypoxia, cell injury, biochemical aberration occurrences in ischemic tissue, human functional morphology, evolutionary morphology. *Total annual research expenditures:* $562,716.

Joseph C. LaManna, Acting Chairman, 216-368-1100, *Fax:* 216-368-8669, *E-mail:* jcl4@po.cwru.edu.
Application contact: Laila Boesinger, Administrator, 216-368-3430, *Fax:* 216-368-8669, *E-mail:* lvb2@po.cwru.edu.

■ COLORADO STATE UNIVERSITY

College of Veterinary Medicine and Biomedical Sciences and Graduate School, Graduate Programs in Veterinary Medicine and Biomedical Sciences, Department of Anatomy and Neurobiology, Fort Collins, CO 80523-0015

AWARDS MS, PhD.

Faculty: 19 full-time (7 women).
Students: 42 full-time (25 women), 5 part-time (3 women); includes 6 minority (2 Asian Americans or Pacific Islanders, 3 Hispanic Americans, 1 Native American), 1 international. Average age 28. *47 applicants, 68% accepted.* In 1999, 33 master's, 4 doctorates awarded.
Degree requirements: For master's, thesis required (for some programs), foreign language not required; for doctorate, dissertation required, foreign language not required.
Entrance requirements: For master's and doctorate, GRE General Test, GRE Subject Test, TOEFL. *Application deadline:* For fall admission, 6/1 (priority date). Applications are processed on a rolling basis. *Application fee:* $30. Electronic applications accepted.
Expenses: Tuition, state resident: full-time $2,694; part-time $150 per credit. Tuition, nonresident: full-time $10,460; part-time $581 per credit. Required fees: $32 per semester.
Financial aid: In 1999–00, 6 research assistantships, 3 teaching assistantships were awarded; fellowships, traineeships also available.
Faculty research: Structural biology, integrative neuroscience, developmental neurobiology. *Total annual research expenditures:* $4 million.
F. Edward Dudek, Chair, 970-491-5847.
Application contact: Dr. Robert Handa, Graduate Coordinator, 970-491-7130, *Fax:* 970-491-7907, *E-mail:* dadams@cvmbs.colostate.edu.

■ COLUMBIA UNIVERSITY

College of Physicians and Surgeons and Graduate School of Arts and Sciences, Graduate School of Arts and Sciences at the College of Physicians and Surgeons, Department of Anatomy and Cell Biology, New York, NY 10032

AWARDS Anatomy (M Phil, MA, PhD); anatomy and cell biology (PhD). Only candidates for the PhD are admitted. Terminal master's awarded for partial completion of doctoral program.

Degree requirements: For master's, foreign language and thesis not required; for doctorate, dissertation, oral exam required, foreign language not required.
Entrance requirements: For master's and doctorate, GRE General Test, TOEFL.
Expenses: Tuition: Full-time $25,072.
Faculty research: Protein sorting, membrane biophysics, muscle energetics, neuroendocrinology, developmental biology, cytoskeleton, transcription factors.

Find an in-depth description at www.petersons.com/graduate.

■ CORNELL UNIVERSITY

Graduate School, Graduate Fields of Veterinary Medicine, Field of Veterinary Medicine, Ithaca, NY 14853-0001

AWARDS Anatomy (MS, PhD); cancer biology (MS, PhD); clinical sciences (MS, PhD); infectious diseases (MS, PhD); pathology (MS, PhD); pharmacology (MS, PhD); veterinary physiology (MS, PhD); virology (MS, PhD).

Faculty: 83 full-time.
Students: 25 full-time (13 women); includes 2 minority (1 African American, 1 Asian American or Pacific Islander), 12 international. *28 applicants, 18% accepted.* In 1999, 1 master's, 12 doctorates awarded.
Degree requirements: For master's and doctorate, thesis/dissertation required, foreign language not required.
Entrance requirements: For master's and doctorate, GRE General Test, TOEFL. *Application deadline:* For fall admission, 1/15; for spring admission, 10/1. Applications are processed on a rolling basis. *Application fee:* $65. Electronic applications accepted.
Expenses: Tuition: Full-time $12,400.
Financial aid: In 1999–00, 24 students received aid, including 7 fellowships with full tuition reimbursements available, 17 research assistantships with full tuition reimbursements available; teaching assistantships with full tuition reimbursements available, institutionally sponsored loans, scholarships, tuition waivers (full and partial), and unspecified assistantships

Cornell University (continued)
also available. Financial aid applicants required to submit FAFSA.

Faculty research: Receptors and signal transduction, viral and bacterial infectious diseases, tumor metastasis, clinical sciences/nutritional disease, development/neurologic disorders.

Application contact: Graduate Field Assistant, 607-253-3276, *E-mail:* vetgradpgms@cornell.edu.

■ DUKE UNIVERSITY

Graduate School, Department of Biological Anthropology and Anatomy, Durham, NC 27708-0586

AWARDS Cellular and molecular biology (PhD); gross anatomy and physical anthropology (PhD), including comparative morphology of human and non-human primates, primate social behavior, vertebrate paleontology; neuroanatomy (PhD).

Faculty: 14 full-time, 1 part-time/adjunct.
Students: 20 full-time (12 women); includes 3 minority (2 African Americans, 1 Hispanic American), 1 international. *62 applicants, 6% accepted.* In 1999, 2 doctorates awarded.
Degree requirements: For doctorate, dissertation required.
Entrance requirements: For doctorate, GRE General Test. *Application deadline:* For fall admission, 12/31. *Application fee:* $75.
Expenses: Tuition: Full-time $21,406; part-time $760 per unit. Required fees: $3,136; $3,136 per year. One-time fee: $30. Tuition and fees vary according to program.
Financial aid: Fellowships, teaching assistantships, Federal Work-Study available. Financial aid application deadline: 12/31.
Kathleen Smith, Director of Graduate Studies, 919-684-4124, *Fax:* 919-684-8034, *E-mail:* rachel_hougom@baa.mc.duke.edu.

■ EAST CAROLINA UNIVERSITY

School of Medicine, Department of Anatomy and Cell Biology, Greenville, NC 27858-4353

AWARDS PhD.

Faculty: 13 full-time (2 women).
Students: 2 full-time (1 woman), 4 part-time (2 women); includes 1 minority (Asian American or Pacific Islander), 1 international. Average age 32. *2 applicants, 50% accepted.* In 1999, 1 degree awarded.
Degree requirements: For doctorate, one foreign language (computer language can substitute), dissertation required.
Entrance requirements: For doctorate, GRE General Test, GRE Subject Test, TOEFL. *Application deadline:* For fall

admission, 6/1 (priority date). Applications are processed on a rolling basis. *Application fee:* $40.

Expenses: Tuition, state resident: full-time $1,012. Tuition, nonresident: full-time $8,578. Required fees: $1,006. Full-time tuition and fees vary according to degree level. Part-time tuition and fees vary according to course load.
Financial aid: Fellowships available. Financial aid application deadline: 6/1.
Faculty research: Diabetes mellitus, cellular biology of pancreas, autonomic nervous system, neural cytoarchitecture, CNS neuropeptides.
Dr. Jack Brinn, Chairman, 252-816-2851, *Fax:* 252-816-2850, *E-mail:* brinnj@mail.ecu.edu.
Application contact: Dr. David Terrian, Senior Director of Graduate Studies, 252-816-3284, *Fax:* 252-816-2850, *E-mail:* grad@mail.ecu.edu.

■ EAST TENNESSEE STATE UNIVERSITY

James H. Quillen College of Medicine and School of Graduate Studies, Biomedical Science Graduate Program, Johnson City, TN 37614

AWARDS Anatomy and cell biology (MS, PhD); biochemistry and molecular biology (MS, PhD); biophysics (MS, PhD); microbiology (MS, PhD); pharmacology (MS, PhD); physiology (MS, PhD). Part-time programs available.

Faculty: 40 full-time (9 women).
Students: 23 full-time (10 women), 2 part-time (1 woman); includes 1 minority (Asian American or Pacific Islander), 4 international. Average age 30. *83 applicants, 20% accepted.* In 1999, 5 master's, 4 doctorates awarded. Terminal master's awarded for partial completion of doctoral program.
Degree requirements: For master's, one foreign language (computer language can substitute), thesis, comprehensive qualifying exam required; for doctorate, 2 foreign languages (computer language can substitute for one), dissertation required.
Entrance requirements: For master's, GRE General Test, minimum GPA of 3.0, bachelor's degree in biological or related science; for doctorate, GRE General Test, GRE Subject Test. *Application deadline:* For fall admission, 3/15 (priority date); for spring admission, 3/1. *Application fee:* $25 ($35 for international students).
Expenses: Tuition, state resident: full-time $10,342. Tuition, nonresident: full-time $21,080. Required fees: $532.
Financial aid: In 1999–00, 16 research assistantships, 5 teaching assistantships were awarded; fellowships, career-related

internships or fieldwork, Federal Work-Study, grants, institutionally sponsored loans, and tuition waivers (full) also available.
Dr. Mitchell Robinson, Assistant Dean for Graduate Studies, 423-439-4658, *E-mail:* robinson@etsu.edu.

■ FINCH UNIVERSITY OF HEALTH SCIENCES/THE CHICAGO MEDICAL SCHOOL

School of Graduate and Postdoctoral Studies, Department of Cell Biology and Anatomy, Program in Anatomy, North Chicago, IL 60064-3095

AWARDS MS, PhD, MD/MS, MD/PhD. Part-time programs available.

Faculty: 4 full-time.
Students: 6 full-time (3 women), 2 part-time (1 woman); includes 1 minority (Asian American or Pacific Islander). Terminal master's awarded for partial completion of doctoral program.
Degree requirements: For master's, computer language, thesis, qualifying exam required, foreign language not required; for doctorate, computer language, dissertation, comprehensive exam, original research project required, foreign language not required.
Entrance requirements: For master's and doctorate, GRE General Test, TOEFL, TWE, minimum GPA of 3.0. *Application deadline:* For fall admission, 6/1 (priority date). Applications are processed on a rolling basis. *Application fee:* $25.
Expenses: Tuition: Full-time $14,054; part-time $391 per credit hour. Tuition and fees vary according to program.
Financial aid: In 1999–00, fellowships (averaging $15,500 per year); tuition waivers (full and partial) also available. Financial aid application deadline: 6/9; financial aid applicants required to submit FAFSA.
Faculty research: Molecular neuroscience, cell biology.
Application contact: Dr. Monica M. Oblinger, Graduate Student Coordinator, 847-578-3440.

Find an in-depth description at www.petersons.com/graduate.

■ HOWARD UNIVERSITY

Graduate School of Arts and Sciences, Department of Anatomy, Washington, DC 20059-0002

AWARDS MS, PhD. Part-time programs available.

Faculty: 18.
Students: 6; includes 5 minority (4 African Americans, 1 Asian American or Pacific Islander), 1 international. Average age 30. *10 applicants, 10% accepted.* In 1999, 1

master's awarded (100% continued full-time study).

Degree requirements: For master's and doctorate, one foreign language, thesis/dissertation, comprehensive exam, teaching experience required. *Average time to degree:* Master's–2 years full-time; doctorate–4 years full-time.

Entrance requirements: For master's and doctorate, GRE General Test, minimum GPA of 2.5. *Application deadline:* For fall admission, 6/1 (priority date); for spring admission, 11/1. Applications are processed on a rolling basis. *Application fee:* $45.

Expenses: Tuition: Full-time $10,500; part-time $583 per credit hour. Required fees: $405; $203 per semester.

Financial aid: In 1999–00, 3 students received aid, including 3 teaching assistantships with partial tuition reimbursements available; fellowships, research assistantships, career-related internships or fieldwork, grants, and institutionally sponsored loans also available. Financial aid application deadline: 6/1.

Faculty research: Neural control of function, mammalian evolution and paleontology, cellular differentiation, cellular communication, ovarian cancer.

Dr. James H. Baker, Chairman, 202-806-6555, *Fax:* 202-267-7055.

Application contact: K. Baldwin, Co-Chair of the Graduate Committee, 202-806-5275, *Fax:* 202-267-7055, *E-mail:* kbaldwin@fac.howard.edu.

■ **INDIANA UNIVERSITY BLOOMINGTON**

Medical Sciences Program, Bloomington, IN 47405

AWARDS Anatomy and cell biology (MA, PhD); pharmacology (MS, PhD); physiology (MA, PhD).

Faculty: 12 full-time (3 women).
Students: 10 full-time (6 women), 6 part-time (2 women); includes 1 minority (Asian American or Pacific Islander), 3 international. In 1999, 1 master's, 1 doctorate awarded.

Entrance requirements: For master's, GRE, TOEFL, minimum GPA of 3.0; for doctorate, GRE, TOEFL. *Application deadline:* For fall admission, 1/15. *Application fee:* $45.

Expenses: Tuition, state resident: full-time $3,853; part-time $161 per credit hour. Tuition, nonresident: full-time $11,226; part-time $468 per credit hour. Required fees: $360 per year. Tuition and fees vary according to course load and program.
Dr. Talmage Bosin, Assistant Dean, 812-855-8118, *E-mail:* bosin@indiana.edu.

Application contact: Kimberly Bunch, Director of Graduate Admissions, 812-855-1119, *E-mail:* kbunch@indiana.edu.

■ **INDIANA UNIVERSITY–PURDUE UNIVERSITY INDIANAPOLIS**

School of Medicine, Graduate Programs in Medicine, Department of Anatomy, Indianapolis, IN 46202-2896

AWARDS MS, PhD, MD/MS, MD/PhD.

Students: 4 full-time (3 women), 7 part-time (3 women), 6 international. Average age 25.

Degree requirements: For master's, thesis or alternative required, foreign language not required; for doctorate, dissertation required, foreign language not required.

Entrance requirements: For master's and doctorate, GRE General Test. *Application deadline:* For fall admission, 1/15 (priority date). *Application fee:* $35 ($55 for international students).

Expenses: Tuition, state resident: full-time $13,245; part-time $158 per credit hour. Tuition, nonresident: full-time $30,330; part-time $455 per credit hour. Required fees: $121 per year. Tuition and fees vary according to course load and degree level.

Financial aid: In 1999–00, 1 fellowship was awarded; research assistantships, Federal Work-Study, institutionally sponsored loans, tuition waivers (partial), and stipends also available. Financial aid application deadline: 2/15.

Faculty research: Acoustic reflex control, osteoarthritis, and bone disease; diabetes; kidney diseases; cellular and molecular neurobiology; nervous disorders.
Dr. David B. Burr, Chairman, 317-274-7494, *Fax:* 317-278-2040, *E-mail:* dburr@indyvax.iupui.edu.

Application contact: Dr. James Williams, Graduate Adviser, 317-274-3423, *Fax:* 317-278-2040, *E-mail:* williams@anatomy.iupui.edu.

■ **IOWA STATE UNIVERSITY OF SCIENCE AND TECHNOLOGY**

College of Veterinary Medicine and Graduate College, Graduate Programs in Veterinary Medicine, Department of Biomedical Sciences, Ames, IA 50011

AWARDS Veterinary anatomy (MS, PhD); veterinary physiology (MS, PhD).

Faculty: 11 full-time, 1 part-time/adjunct.
Students: 14 full-time (7 women), 4 part-time (3 women); includes 1 minority (Asian American or Pacific Islander), 12 international. *12 applicants, 33% accepted.* In 1999, 1 master's, 4 doctorates awarded.
Degree requirements: For master's, thesis or alternative required; for doctorate, dissertation required.
Entrance requirements: For master's and doctorate, GRE General Test, TOEFL. *Application deadline:* For fall admission, 6/1 (priority date); for spring admission, 11/1

(priority date). *Application fee:* $20 ($50 for international students). Electronic applications accepted.

Expenses: Tuition, state resident: full-time $3,308. Tuition, nonresident: full-time $9,744. Tuition and fees vary according to course load and program.

Financial aid: In 1999–00, 7 research assistantships with partial tuition reimbursements (averaging $10,355 per year), 6 teaching assistantships with partial tuition reimbursements (averaging $9,450 per year) were awarded; scholarships also available.
Dr. Richard J. Martin, Chair, 515-294-2440, *Fax:* 515-294-2315, *E-mail:* biomedsci@iastate.edu.

■ **JOHNS HOPKINS UNIVERSITY**

School of Medicine, Graduate Programs in Medicine, Department of Cell Biology and Anatomy, Baltimore, MD 21218-2699

AWARDS PhD.

Faculty: 5 full-time (1 woman).
Students: 10 full-time (7 women), 4 international. Average age 25. In 1999, 2 degrees awarded.

Degree requirements: For doctorate, one foreign language, dissertation, comprehensive and oral exams required.

Entrance requirements: For doctorate, GRE. *Application deadline:* For fall admission, 1/15. *Application fee:* $50.

Expenses: Tuition: Full-time $23,660.

Financial aid: In 1999–00, 7 students received aid, including 7 teaching assistantships; career-related internships or fieldwork and institutionally sponsored loans also available.

Faculty research: Vertebrate evolution, functional anatomy, primate evolution, vertebrate paleobiology.
Dr. Joan T. Richtsmeier, Director, 410-955-7892, *Fax:* 410-955-4129, *E-mail:* jtr@welchlink.welch.jhu.edu.

Application contact: Catherine L. Will, Coordinator, Graduate Student Affairs, 410-614-3385, *E-mail:* grad_study@som.adm.jhu.edu.

■ **KANSAS STATE UNIVERSITY**

College of Veterinary Medicine and Graduate School, Graduate Programs in Veterinary Medicine, Department of Anatomy and Physiology, Manhattan, KS 66506

AWARDS Anatomy (MS); physiology (MS, PhD).

Degree requirements: For master's, thesis required, foreign language not required; for doctorate, dissertation required.

Expenses: Tuition, state resident: part-time $103 per credit hour. Tuition,

Kansas State University (continued)
nonresident: part-time $338 per credit
hour. Required fees: $17 per credit hour.
One-time fee: $64 part-time.

■ LOMA LINDA UNIVERSITY

**Graduate School, Graduate Programs
in Medicine, Department of Anatomy,
Loma Linda, CA 92350**

AWARDS MS, PhD. Part-time programs available. Terminal master's awarded for partial
completion of doctoral program.

Degree requirements: For master's,
thesis required, foreign language not
required; for doctorate, 2 foreign
languages (computer language can
substitute for one), dissertation required.
Entrance requirements: For master's and
doctorate, GRE General Test. *Application
deadline:* Applications are processed on a
rolling basis. *Application fee:* $40.
Expenses: Tuition: Part-time $395 per
unit.
Financial aid: Tuition waivers (full and
partial) available. Aid available to part-time
students.
Faculty research: Neuroendocrine system,
histochemistry and image analysis, effect of
age and diabetes on PNS, electron
microscopy, histology.
Dr. Paul McMillan, Coordinator, 909-824-
4301.

■ LOUISIANA STATE UNIVERSITY HEALTH SCIENCES CENTER

**School of Graduate Studies in New
Orleans, Department of Cell Biology
and Anatomy, New Orleans, LA
70112-2223**

AWARDS Cell biology and anatomy (MS,
PhD), including cell biology, developmental
biology, neurobiology and anatomy.

Faculty: 18 full-time (3 women), 1 part-
time/adjunct (0 women).
Students: 9 full-time (2 women); includes
1 minority (African American), 4
international. Average age 26. *6 applicants,
50% accepted.* In 1999, 3 doctorates
awarded (34% entered university research/
teaching, 33% found other work related to
degree, 33% continued full-time study).
Degree requirements: For master's and
doctorate, thesis/dissertation required,
foreign language not required. *Average
time to degree:* Doctorate–4.5 years full-
time.
Entrance requirements: For master's and
doctorate, GRE General Test, GRE
Subject Test, TOEFL, minimum
undergraduate GPA of 3.0. *Application
deadline:* For fall admission, 3/1 (priority
date); for spring admission, 10/15. Applications are processed on a rolling basis.
Application fee: $30.

Expenses: Tuition, state resident: full-time
$2,878; part-time $126 per hour. Tuition,
nonresident: full-time $6,003; part-time
$265 per hour. Required fees: $2,272.
Tuition and fees vary according to course
load, degree level and program.
Financial aid: In 1999–00, 1 fellowship
with full tuition reimbursement (averaging
$15,000 per year), teaching assistantships
with full tuition reimbursements (averaging
$14,000 per year) were awarded; research
assistantships, career-related internships or
fieldwork, Federal Work-Study, grants,
institutionally sponsored loans, tuition
waivers (full), and unspecified assistantships
also available. Aid available to part-time
students. Financial aid application
deadline: 4/1.
Faculty research: Visual system organization, neural development, plasticity of
sensory systems, information processing
through the nervous system, visuomotor
integration. *Total annual research
expenditures:* $772,550.
Dr. R. Ranney Mize, Head, 504-599-1458,
Fax: 504-568-4392, *E-mail:* rmize@
lsumc.edu.
Application contact: Dr. Mark C.
Alliegro, Director of Graduate Studies,
504-568-7618, *Fax:* 504-568-4392, *E-mail:*
mallie@lsumc.edu.

**Find an in-depth description at
www.petersons.com/graduate.**

■ LOUISIANA STATE UNIVERSITY HEALTH SCIENCES CENTER

**School of Graduate Studies in
Shreveport, Department of Cellular
Biology and Anatomy, Shreveport, LA
71130-3932**

AWARDS MS, PhD, MD/PhD. MD/PhD open
only to residents of Louisiana. Terminal
master's awarded for partial completion of
doctoral program.

Degree requirements: For master's and
doctorate, thesis/dissertation required,
foreign language not required.
Entrance requirements: For master's and
doctorate, GRE General Test, TOEFL.
Expenses: Tuition, state resident: full-time
$2,878; part-time $126 per hour. Tuition,
nonresident: full-time $6,003; part-time
$265 per hour. Required fees: $2,272.
Tuition and fees vary according to course
load, degree level and program.
Faculty research: Alcohol and immunity,
neuroscience, olfactory physiology,
extracellular matrix, cancer cell biology
and gene therapy.

■ LOYOLA UNIVERSITY CHICAGO

**Graduate School, Department of Cell
Biology, Neurobiology and Anatomy,
Maywood, IL 60153**

AWARDS MS, PhD, MD/PhD. Part-time
programs available.

Faculty: 13 full-time, 13 part-time/
adjunct.
Students: 26 full-time (12 women), 3 part-
time (all women); includes 4 minority (all
Asian Americans or Pacific Islanders), 1
international. Average age 26. *42 applicants,
14% accepted.* In 1999, 1 master's awarded
(100% continued full-time study); 3
doctorates awarded (100% entered
university research/teaching).
Degree requirements: For master's,
thesis or alternative, comprehensive exams
required, foreign language not required;
for doctorate, dissertation, comprehensive
exams required, foreign language not
required. *Average time to degree:* Master's–2
years full-time; doctorate–5 years full-time.
Entrance requirements: For master's and
doctorate, GRE General Test, GRE
Subject Test (biology), minimum GPA of
3.0. *Application deadline:* For fall admission,
5/1 (priority date). Applications are
processed on a rolling basis. *Application fee:*
$35.
Expenses: Tuition: Part-time $500 per
credit hour. Required fees: $42 per term.
Financial aid: In 1999–00, 8 fellowships
with full tuition reimbursements (averaging
$17,500 per year), 8 research assistantships
with full tuition reimbursements (averaging
$17,500 per year) were awarded; Federal
Work-Study and institutionally sponsored
loans also available. Aid available to part-
time students. Financial aid application
deadline: 5/1; financial aid applicants
required to submit FAFSA.
Faculty research: Brain steroids,
immunology, neuroregeneration, cytokines.
Total annual research expenditures: $1.4 mil-
lion.
Dr. John Clancy, Chair, 708-216-3352.
Application contact: Thackery S. Gray,
Graduate Program Director, 708-216-
3352.

**Find an in-depth description at
www.petersons.com/graduate.**

■ MCP HAHNEMANN UNIVERSITY

**School of Medicine, Biomedical
Graduate Programs, Department of
Anatomy and Neurobiology,
Philadelphia, PA 19102-1192**

AWARDS MS, PhD, MD/PhD. Terminal
master's awarded for partial completion of
doctoral program.

Degree requirements: For master's,
thesis, comprehensive exam required,

foreign language not required; for doctorate, one foreign language (computer language can substitute), dissertation, qualifying exam required.
Entrance requirements: For master's, GRE General Test, TOEFL, minimum GPA of 2.75; for doctorate, GRE General Test, TOEFL, minimum GPA of 3.0.
Faculty research: Cell biology, anatomy of brain tumors, membrane excitability.

■ MEDICAL COLLEGE OF GEORGIA

School of Graduate Studies, Department of Cellular Biology and Anatomy, Augusta, GA 30912-1500
AWARDS MS, PhD.

Faculty: 18 full-time (2 women).
Students: 11 full-time (6 women); includes 1 minority (African American), 3 international. *10 applicants, 20% accepted.* In 1999, 5 doctorates awarded. Terminal master's awarded for partial completion of doctoral program.
Degree requirements: For master's and doctorate, thesis/dissertation required, foreign language not required.
Entrance requirements: For master's and doctorate, GRE General Test, TOEFL. *Application deadline:* For fall admission, 6/30 (priority date). Applications are processed on a rolling basis. *Application fee:* $25.
Expenses: Tuition, state resident: full-time $2,896; part-time $121 per hour. Tuition, nonresident: full-time $11,584; part-time $483 per hour. Required fees: $286; $143 per semester. Tuition and fees vary according to program.
Financial aid: In 1999–00, 11 research assistantships with partial tuition reimbursements (averaging $15,500 per year) were awarded; teaching assistantships, grants and institutionally sponsored loans also available. Aid available to part-time students. Financial aid application deadline: 3/31; financial aid applicants required to submit FAFSA.
Faculty research: Stereocilia, biomedicine, neuroanatomy, gross anatomy, histology.
Dr. Dale E. Bockman, Chair, 706-721-3731, *E-mail:* dbockman@mail.mcg.edu.
Application contact: Dr. Dale Sickles, Director, 706-721-3913, *E-mail:* dsickles@mail.mcg.edu.

■ MEDICAL COLLEGE OF OHIO

Graduate School, Department of Anatomy and Neurobiology, Toledo, OH 43614-5805
AWARDS MS. Part-time programs available.
Faculty: 12 full-time (2 women), 1 part-time/adjunct (0 women).

Students: 2 full-time (1 woman), 1 (woman) part-time. Average age 33. *9 applicants, 56% accepted.* In 1999, 1 degree awarded.
Degree requirements: For master's, thesis, qualifying exam required, foreign language not required. *Average time to degree:* Master's–4 years part-time.
Entrance requirements: For master's, GRE General Test, minimum undergraduate GPA of 3.0. *Application fee:* $30.
Expenses: Tuition, state resident: part-time $193 per hour. Tuition, nonresident: part-time $445 per hour. Tuition and fees vary according to degree level.
Financial aid: Fellowships, Federal Work-Study and institutionally sponsored loans available. Financial aid applicants required to submit FAFSA.
Faculty research: Organization, development and path of the cardiovascular and nervous systems. *Total annual research expenditures:* $1.5 million.
Dr. Robert W. Rhoades, Chairman, 419-383-4117, *Fax:* 419-383-6140, *E-mail:* mcogradschool@mco.edu.
Application contact: Dr. Richard D. Mooney, Coordinator, 419-383-4117, *Fax:* 419-383-6140, *E-mail:* mcogradschool@mco.edu.

■ MEDICAL UNIVERSITY OF SOUTH CAROLINA

College of Graduate Studies, Program in Molecular and Cellular Biology and Pathobiology, Charleston, SC 29425-0002
AWARDS Cell biology and anatomy (PhD); marine biomedicine (PhD).

Faculty: 72 part-time/adjunct (8 women).
Students: 84 full-time (36 women). Average age 29. *56 applicants, 64% accepted.* In 1999, 9 degrees awarded.
Degree requirements: For doctorate, dissertation, teaching and research seminar, oral and written exams required.
Entrance requirements: For doctorate, GRE General Test, TOEFL, interview, minimum GPA of 3.2. *Application deadline:* Applications are processed on a rolling basis. *Application fee:* $55. Electronic applications accepted.
Expenses: Tuition, state resident: full-time $3,470; part-time $160 per semester hour. Tuition, nonresident: full-time $4,426; part-time $213 per semester hour. Required fees: $408 per semester. One-time fee: $160. Tuition and fees vary according to program.
Financial aid: In 1999–00, 27 fellowships (averaging $16,000 per year) were awarded; research assistantships, teaching assistantships, Federal Work-Study and tuition waivers (partial) also available.

Financial aid application deadline: 4/1; financial aid applicants required to submit FAFSA.
Faculty research: Structural biology, marine biology, neurobiology.
Dr. Barry E. Ledford, Interim Dean, 843-792-3391, *Fax:* 843-792-6590.
Application contact: Julie Johnston, Director of Admissions, 843-792-8710, *Fax:* 843-792-3764.

■ MICHIGAN STATE UNIVERSITY

College of Osteopathic Medicine and Graduate School, Graduate Programs in Osteopathic Medicine, East Lansing, MI 48824
AWARDS Anatomy (MS, PhD); biochemistry (MS, PhD); microbiology (PhD); pathology (MS, PhD); pharmacology/toxicology (MS, PhD), including pharmacology; physiology (MS, PhD), including environmental toxicology (PhD), neuroscience (PhD). Part-time programs available.

Students: 17 full-time (9 women), 1 part-time; includes 5 minority (3 Asian Americans or Pacific Islanders, 1 Hispanic American, 1 Native American), 5 international. Average age 26. *33 applicants, 9% accepted.* In 1999, 2 doctorates awarded.
Degree requirements: For doctorate, dissertation required.
Entrance requirements: For master's and doctorate, GRE. *Application deadline:* For fall admission, 3/1 (priority date). Applications are processed on a rolling basis. *Application fee:* $30 ($40 for international students).
Expenses: Tuition, area resident: Full-time $15,879. Tuition, state resident: full-time $33,797.
Financial aid: Fellowships, research assistantships, teaching assistantships, career-related internships or fieldwork, Federal Work-Study, and institutionally sponsored loans available. Financial aid application deadline: 4/2. *Total annual research expenditures:* $5 million.
Application contact: Dr. Veronica M. Maher, Associate Dean for Graduate Studies, 517-353-7785, *Fax:* 517-353-9004, *E-mail:* maher@com.msu.edu.

■ MICHIGAN STATE UNIVERSITY

College of Veterinary Medicine and Graduate School, Graduate Programs in Veterinary Medicine, East Lansing, MI 48824
AWARDS Anatomy (MS, PhD); large animal clinical sciences (MS, PhD); microbiology (MS, PhD); pathology (MS, PhD); pharmacology/toxicology (MS, PhD), including pharmacology; small animal clinical sciences (MS).

Michigan State University (continued)
Students: 52 full-time (28 women); includes 6 minority (4 African Americans, 2 Asian Americans or Pacific Islanders), 18 international. Average age 32. *45 applicants, 20% accepted.* In 1999, 3 master's, 3 doctorates awarded.
Degree requirements: For master's, thesis or alternative required, foreign language not required; for doctorate, dissertation required, foreign language not required.
Application deadline: Applications are processed on a rolling basis. *Application fee:* $30 ($40 for international students). Electronic applications accepted.
Expenses: Tuition, state resident: full-time $9,766. Tuition, nonresident: full-time $20,082. Tuition and fees vary according to program.
Financial aid: Fellowships, research assistantships available.
Faculty research: Molecular genetics, food safety/toxicology, comparative orthopedics, airway disease, population medicine.
Dr. John C. Baker, Associate Dean for Research and Graduate Studies, 517-432-2388, *Fax:* 517-432-1037, *E-mail:* baker@cvm.msu.edu.
Application contact: Victoria Hoelzer-Maddox, Administrative Assistant, 517-353-3118, *Fax:* 517-432-1037, *E-mail:* hoelzer-maddox@cvm.msu.edu.

■ NEW YORK MEDICAL COLLEGE

Graduate School of Basic Medical Sciences, Department of Cell Biology and Anatomy, Valhalla, NY 10595-1691
AWARDS Cell biology and neuroscience (MS, PhD). Part-time and evening/weekend programs available.

Faculty: 18 full-time (6 women).
Students: 8 full-time (5 women), 1 (woman) part-time; includes 1 minority (Asian American or Pacific Islander), 3 international. Average age 30. *8 applicants, 38% accepted.* In 1999, 2 master's, 2 doctorates awarded. Terminal master's awarded for partial completion of doctoral program.
Degree requirements: For master's and doctorate, computer language, thesis/dissertation required, foreign language not required.
Entrance requirements: For master's, GRE General Test, TOEFL; for doctorate, GRE General Test, GRE Subject Test, TOEFL. *Application deadline:* For fall admission, 7/1 (priority date); for spring admission, 12/1 (priority date). Applications are processed on a rolling basis. *Application fee:* $35 ($60 for international students).

Expenses: Tuition: Part-time $430 per credit. Required fees: $15 per semester. One-time fee: $100.
Financial aid: In 1999–00, 7 research assistantships with full tuition reimbursements were awarded; career-related internships or fieldwork, Federal Work-Study, grants, institutionally sponsored loans, and tuition waivers (full) also available. Aid available to part-time students. Financial aid applicants required to submit FAFSA.
Faculty research: Mechanisms of growth control in skeletal muscle, cartilage differentiation, cytoskeletal functions, signal transduction pathways, neuronal development and plasticity.
Dr. Anna B. Drakontides, Director, 914-594-4036.

Find an in-depth description at www.petersons.com/graduate.

■ THE OHIO STATE UNIVERSITY

College of Medicine and Public Health and Graduate School, Graduate Programs in the Basic Medical Sciences, Department of Anatomy and Medical Education, Columbus, OH 43210
AWARDS Anatomy (MS, PhD).

Faculty: 11 full-time (2 women), 2 part-time/adjunct (0 women).
Students: 5 full-time (3 women), 1 (woman) part-time. In 1999, 2 doctorates awarded (100% entered university research/teaching). Terminal master's awarded for partial completion of doctoral program.
Degree requirements: For master's, foreign language and thesis not required; for doctorate, dissertation required, foreign language not required. *Average time to degree:* Doctorate–4 years full-time.
Entrance requirements: For master's and doctorate, GRE General Test, GRE Subject Test. *Application deadline:* For fall admission, 5/31. Applications are processed on a rolling basis. *Application fee:* $30 ($40 for international students). Electronic applications accepted.
Expenses: Tuition, state resident: full-time $5,400. Tuition, nonresident: full-time $14,535. Part-time tuition and fees vary according to course load.
Financial aid: In 1999–00, 1 student received aid, including research assistantships with tuition reimbursements available (averaging $12,600 per year), teaching assistantships with tuition reimbursements available (averaging $12,600 per year); fellowships, Federal Work-Study also available. Financial aid application deadline: 1/10.
Faculty research: Neuroscience, molecular biology. *Total annual research expenditures:* $1.4 million.

Dr. Robert M. DePhilip, Interim Chair, 614-292-4769, *Fax:* 614-292-7659.
Application contact: Dr. Kenneth H. Jones, Graduate Studies Committee Chair, 614-292-9665, *Fax:* 614-292-7659, *E-mail:* jones.4@osu.edu.

■ THE OHIO STATE UNIVERSITY

College of Veterinary Medicine and Graduate School, Graduate Programs in Veterinary Medicine, Department of Veterinary Biosciences, Columbus, OH 43210
AWARDS Anatomy and cellular biology (MS, PhD); pathobiology (MS, PhD); pharmacology (MS, PhD); toxicology (MS, PhD); veterinary physiology (MS, PhD).

Faculty: 28 full-time (8 women).
Students: 49 full-time (20 women); includes 2 minority (1 African American, 1 Asian American or Pacific Islander), 18 international.
Degree requirements: For master's and doctorate, thesis/dissertation, final exam required.
Entrance requirements: For master's, GRE General Test; for doctorate, GRE General Test, master's degree. *Application fee:* $25.
Expenses: Tuition, state resident: full-time $5,757. Tuition, nonresident: full-time $14,892.
Financial aid: Fellowships, research assistantships, teaching assistantships available.
Faculty research: Microvasculature, muscle biology, neonatal lung and bone development.
Charles C. Capen, Interim Chair, 614-292-4489.
Application contact: Graduate Admission Committee, 614-292-4489.

■ PALMER COLLEGE OF CHIROPRACTIC

Institute of Graduate Studies, Davenport, IA 52803-5287
AWARDS Anatomy (MS).

Faculty: 12.
Students: 12 full-time (4 women), 2 part-time (1 woman); includes 2 minority (1 African American, 1 Asian American or Pacific Islander), 1 international. 7 *applicants, 71% accepted.*
Degree requirements: For master's, thesis, comprehensive exam required, foreign language not required. *Average time to degree:* Master's–2 years full-time.
Entrance requirements: For master's, GRE General Test, TOEFL, TSE, minimum GPA of 2.5. *Application deadline:* For fall admission, 9/1; for spring admission, 5/28. Applications are processed on a rolling basis. *Application fee:* $25.

Expenses: Tuition: Full-time $4,950; part-time $275 per credit hour. Required fees: $40.

Financial aid: In 1999–00, 5 students received aid, including teaching assistant-ships with full and partial tuition reimbursements available (averaging $6,269 per year); research assistantships, Federal Work-Study, institutionally sponsored loans, and stipends also available. Aid available to part-time students. Financial aid application deadline: 4/1; financial aid applicants required to submit FAFSA.

Dr. Iftikhar H. Bhatti, Dean of Graduate and Undergraduate Studies, 319-884-5868, *Fax:* 319-884-5226.

Application contact: Marilyn Olson, Graduate Studies Coordinator, 319-884-5867, *Fax:* 319-884-5226.

Find an in-depth description at www.petersons.com/graduate.

■ **THE PENNSYLVANIA STATE UNIVERSITY MILTON S. HERSHEY MEDICAL CENTER**

Graduate School, Department of Neuroscience and Anatomy, Program in Anatomy, Hershey, PA 17033-2360
AWARDS MS, PhD, MD/PhD.

Students: 10 full-time (0 women). Average age 27. In 1999, 2 master's awarded.
Degree requirements: For doctorate, dissertation required.
Entrance requirements: For master's and doctorate, GRE General Test or MCAT. *Application deadline:* For fall admission, 7/26. *Application fee:* $50.
Expenses: Tuition, state resident: full-time $6,886; part-time $291 per credit. Tuition, nonresident: full-time $14,118; part-time $588 per credit. Required fees: $43 per semester. Part-time tuition and fees vary according to course load.
Dr. Robert J. Milner, Head, Department of Neuroscience and Anatomy, 717-531-8650.

■ **PURDUE UNIVERSITY**

School of Veterinary Medicine and Graduate School, Graduate Programs in Veterinary Medicine, Department of Basic Medical Sciences, West Lafayette, IN 47907
AWARDS Anatomy (MS, PhD); pharmacology (MS, PhD); physiology (MS, PhD). Part-time programs available.

Faculty: 19 full-time (3 women).
Students: 26 full-time (13 women). Average age 27. 26 applicants, 27% accepted. In 1999, 4 master's, 5 doctorates awarded. Terminal master's awarded for partial completion of doctoral program.

Degree requirements: For master's and doctorate, thesis/dissertation required, foreign language not required. *Average time to degree:* Master's–3 years full-time; doctorate–3 years full-time.
Entrance requirements: For master's and doctorate, GRE General Test, TOEFL. *Application deadline:* For fall admission, 7/1 (priority date); for spring admission, 12/1 (priority date). *Application fee:* $30. Electronic applications accepted.
Expenses: Tuition, state resident: full-time $3,732. Tuition, nonresident: full-time $8,732.
Financial aid: In 1999–00, 6 fellowships, 14 research assistantships, 3 teaching assistantships were awarded. Financial aid application deadline: 3/1; financial aid applicants required to submit FAFSA.
Faculty research: Development and regeneration, tissue injury and shock, biomedical engineering, ovarian function, bone and cartilage biology, cell and molecular biology. *Total annual research expenditures:* $764,843.
Dr. Gordon L. Coppoc, Head, 765-494-8632, *Fax:* 765-494-0781, *E-mail:* coppoc@vet.purdue.edu.
Application contact: Dr. Ronald Hullinger, Chairman, Graduate Committee, 765-494-8580, *Fax:* 765-494-0781, *E-mail:* bmsgrad@vet.purdue.edu.

■ **RUSH UNIVERSITY**

Graduate College, Division of Anatomical Sciences, Chicago, IL 60612-3832
AWARDS MS, PhD, MD/PhD. Terminal master's awarded for partial completion of doctoral program.

Degree requirements: For master's, thesis required; for doctorate, dissertation, preliminary exam required, foreign language not required.
Entrance requirements: For master's, GRE General Test, TOEFL, minimum GPA of 3.0, bachelor's degree in biology or chemistry preferred, interview; for doctorate, GRE General Test, TOEFL, minimum GPA of 3.0, interview.
Expenses: Tuition: Full-time $13,020; part-time $390 per credit. Tuition and fees vary according to program.
Faculty research: Osteoarthritis, nerve injury and regeneration, hematology-RBC and megakaryocyte, eye-retina and lens, bone biology.

■ **SAINT LOUIS UNIVERSITY**

School of Medicine and Graduate School, Graduate Programs in Biomedical Sciences, Department of Anatomy and Neurobiology, St. Louis, MO 63103-2097
AWARDS Anatomy (MS(R), PhD); neurobiology (PhD).

Faculty: 11 full-time (1 woman).
Students: Average age 29. 13 applicants, 100% accepted. In 1999, 2 degrees awarded.
Degree requirements: For master's, thesis, comprehensive oral exam required, foreign language not required; for doctorate, dissertation, departmental qualifying exams required.
Entrance requirements: For master's and doctorate, GRE General Test, GRE Subject Test. *Application deadline:* For fall admission, 4/15 (priority date). Applications are processed on a rolling basis. *Application fee:* $40.
Expenses: Tuition: Part-time $507 per credit hour. Required fees: $38 per term.
Financial aid: In 1999–00, 11 students received aid, including 2 fellowships, 2 teaching assistantships; research assistantships, Federal Work-Study and institutionally sponsored loans also available. Aid available to part-time students. Financial aid application deadline: 8/1; financial aid applicants required to submit FAFSA.
Faculty research: Systems neurobiology.
Dr. Paul Young, Chairman, 314-577-8274, *Fax:* 314-268-5127.
Application contact: Kris Sherman, Director, 314-577-8275, *Fax:* 314-268-5127, *E-mail:* shermankb@slu.edu.

■ **STATE UNIVERSITY OF NEW YORK AT BUFFALO**

Graduate School, School of Medicine and Biomedical Sciences, Graduate Programs in Medicine and Biomedical Sciences, Department of Anatomy and Cell Biology, Buffalo, NY 14214
AWARDS MA, PhD.

Faculty: 10 full-time (2 women), 2 part-time/adjunct (1 woman).
Students: 3 full-time (1 woman), 2 part-time, 3 international. Average age 28. 7 applicants, 14% accepted. In 1999, 1 master's, 1 doctorate awarded.
Degree requirements: For master's, thesis, exam required; for doctorate, one foreign language (computer language can substitute), dissertation, exam required.
Entrance requirements: For master's, GRE General Test, TOEFL, previous course work in biology, chemistry, or physics; for doctorate, GRE General Test, TOEFL, previous course work in biology, chemistry, and physics. *Application deadline:* For fall admission, 2/1 (priority date).

State University of New York at Buffalo (continued)

Applications are processed on a rolling basis. *Application fee:* $35. Electronic applications accepted.

Expenses: Tuition, state resident: full-time $5,100. Tuition, nonresident: full-time $8,416. Required fees: $935.

Financial aid: In 1999–00, 4 students received aid, including 3 teaching assistantships with full tuition reimbursements available (averaging $14,700 per year); fellowships with full tuition reimbursements available, research assistantships with full tuition reimbursements available, Federal Work-Study and institutionally sponsored loans also available. Financial aid application deadline: 2/28; financial aid applicants required to submit FAFSA.

Faculty research: Developmental biology, neuroscience, digital imaging, cell biology. *Total annual research expenditures:* $350,000. Dr. Frank Mendel, Interim Chairperson, 716-829-2912, *Fax:* 716-829-2915, *E-mail:* fcmendel@buffalo.edu.

Application contact: Debbie Tomasulo, Keyboard Specialist, 716-829-2912, *Fax:* 716-829-2915.

■ **STATE UNIVERSITY OF NEW YORK AT STONY BROOK**

Health Sciences Center, School of Medicine and Graduate School, Graduate Programs in Medicine, Department of Anatomical Sciences, Stony Brook, NY 11794

AWARDS PhD.

Faculty: 9 full-time (3 women).
Students: 3 full-time (1 woman), 5 part-time (1 woman); includes 1 minority (Asian American or Pacific Islander). *11 applicants, 9% accepted.* In 1999, 1 degree awarded.

Degree requirements: For doctorate, dissertation, comprehensive exam required, foreign language not required.

Entrance requirements: For doctorate, GRE General Test, GRE Subject Test, TOEFL, BA in life sciences, minimum GPA of 3.0. *Application deadline:* For fall admission, 1/15. *Application fee:* $50.

Expenses: Tuition, state resident: full-time $5,100. Tuition, nonresident: full-time $8,416. Required fees: $492.

Financial aid: In 1999–00, 4 research assistantships, 1 teaching assistantship were awarded; fellowships, Federal Work-Study also available. Financial aid application deadline: 3/15.

Faculty research: Biological membranes, biomechanics of locomotion, systematics and evolutionary history of primates. *Total annual research expenditures:* $447,620.

Dr. Jack Stern, Chair, 631-444-2350, *Fax:* 631-444-3947, *E-mail:* jstern@ mail.som.sunnysb.edu.

Application contact: Dr. Catherine Forster, Graduate Program Director, 631-444-8203, *Fax:* 631-444-3947, *E-mail:* cforster@mail.som.sunnysb.edu.

■ **STATE UNIVERSITY OF NEW YORK HEALTH SCIENCE CENTER AT BROOKLYN**

School of Graduate Studies, Department of Anatomy and Cell Biology, Brooklyn, NY 11203-2098

AWARDS PhD, MD/PhD.

Degree requirements: For doctorate, one foreign language, dissertation required.
Entrance requirements: For doctorate, GRE.

Expenses: Tuition, state resident: full-time $5,100; part-time $213 per credit. Tuition, nonresident: full-time $8,416; part-time $351 per credit. Required fees: $200. Full-time tuition and fees vary according to program and student level.

Faculty research: Role of oncogenes in early cardiogenesis, transepithelial migration of human neutrophils, biochemical pathways of platelet activation, mechanism of action of interferon-T, nerve growth factor's role in inner ear development.

■ **STATE UNIVERSITY OF NEW YORK UPSTATE MEDICAL UNIVERSITY**

College of Graduate Studies, Department of Anatomy and Cell Biology, Syracuse, NY 13210-2334

AWARDS MS, PhD, MD/PhD.

Faculty: 33.
Students: 12 full-time (6 women); includes 3 minority (1 African American, 2 Asian Americans or Pacific Islanders), 2 international. *22 applicants, 41% accepted.* In 1999, 2 doctorates awarded (50% found work related to degree, 50% continued full-time study). Terminal master's awarded for partial completion of doctoral program.

Degree requirements: For master's, thesis required, foreign language not required; for doctorate, dissertation, comprehensive exam required, foreign language not required.

Entrance requirements: For master's and doctorate, GRE General Test, GRE Subject Test, TSE. *Application deadline:* For fall admission, 4/1 (priority date). Applications are processed on a rolling basis. *Application fee:* $40.

Expenses: Tuition, state resident: full-time $5,100; part-time $213 per credit. Tuition, nonresident: full-time $8,416; part-time

$351 per credit. Required fees: $410; $25 per credit. Part-time tuition and fees vary according to course load and program.

Financial aid: Fellowships, research assistantships, Federal Work-Study and institutionally sponsored loans available. Aid available to part-time students. Dr. James Schwob, Interim Chair, 315-464-5120.

Application contact: Dr. Ira Ames, Professor, 315-464-5120.

■ **TEMPLE UNIVERSITY**

Health Sciences Center, School of Medicine and Graduate School, Graduate Programs in Medicine, Department of Anatomy and Cell Biology, Philadelphia, PA 19140

AWARDS PhD.

Faculty: 18 full-time (6 women).
Students: 4 full-time (2 women); all minorities (1 African American, 3 Asian Americans or Pacific Islanders). *8 applicants, 50% accepted.*

Degree requirements: For doctorate, dissertation, research seminars required, foreign language not required.

Entrance requirements: For doctorate, GRE General Test, GRE Subject Test, minimum GPA of 3.0. *Application deadline:* For fall admission, 9/1 (priority date); for spring admission, 2/1. *Application fee:* $40. Electronic applications accepted.

Expenses: Tuition, state resident: full-time $6,030. Tuition, nonresident: full-time $8,298. Required fees: $230. One-time fee: $10 full-time.

Financial aid: Fellowships, Federal Work-Study available.

Faculty research: Neurobiology, reproductive biology, cardiovascular system, musculoskeletal biology, developmental biology. *Total annual research expenditures:* $632,373. Dr. Steve Popoff, Chair, 215-707-3161, *E-mail:* spopoff@vm.temple.edu.

Application contact: Dr. Judith Litvin, Admissions Chair, 215-707-2070, *Fax:* 215-707-2966, *E-mail:* jl1@ astro.ocis.temple.edu.

■ **TEXAS A&M UNIVERSITY**

College of Veterinary Medicine and Office of Graduate Studies, Graduate Programs in Veterinary Medicine, Department of Veterinary Anatomy and Public Health, College Station, TX 77843

AWARDS Anatomy (MS, PhD); epidemiology (MS); genetics (PhD); toxicology (PhD); veterinary public health (MS).

Faculty: 23 full-time (6 women), 13 part-time/adjunct (6 women).
Students: 36 full-time (21 women), 2 part-time (both women); includes 13 minority

(2 African Americans, 9 Asian Americans or Pacific Islanders, 2 Hispanic Americans). Average age 27. *19 applicants, 47% accepted.* In 1999, 4 master's, 10 doctorates awarded. Terminal master's awarded for partial completion of doctoral program.
Degree requirements: For master's and doctorate, thesis/dissertation required, foreign language not required.
Entrance requirements: For master's and doctorate, GRE General Test, TOEFL. *Application deadline:* For fall admission, 7/15 (priority date); for spring admission, 10/1. Applications are processed on a rolling basis. *Application fee:* $50 ($75 for international students).
Expenses: Tuition, state resident: part-time $76 per semester hour. Tuition, nonresident: part-time $292 per semester hour. Required fees: $11 per semester hour.
Financial aid: In 1999–00, 2 fellowships (averaging $12,000 per year), 18 research assistantships (averaging $13,500 per year), 5 teaching assistantships (averaging $14,000 per year) were awarded; Federal Work-Study, institutionally sponsored loans, and clinical associateships also available. Financial aid application deadline: 7/15; financial aid applicants required to submit FAFSA.
Faculty research: Metal toxicology, reproductive biology, genetics of neural development, developmental biology, environmental toxicology. *Total annual research expenditures:* $3.4 million.
Dr. Evelyn Tiffany-Castiglioni, Head, 979-845-2828, *Fax:* 979-847-8981, *E-mail:* ecastiglioni@cvm.tamu.edu.

■ **TEXAS A&M UNIVERSITY SYSTEM HEALTH SCIENCE CENTER**
College of Medicine, Graduate School of Biomedical Sciences, Department of Human Anatomy and Medical Neurobiology, College Station, TX 77840-7896
AWARDS PhD.
Faculty: 11 full-time (2 women).
Students: 5 full-time (2 women); includes 1 minority (Hispanic American). Average age 26. *30 applicants, 7% accepted.* In 1999, 2 degrees awarded (100% found work related to degree).
Degree requirements: For doctorate, dissertation required, foreign language not required. *Average time to degree:* Doctorate–4.5 years full-time.
Entrance requirements: For doctorate, GRE General Test. *Application deadline:* For fall admission, 2/1 (priority date). Applications are processed on a rolling basis. *Application fee:* $50 ($75 for

international students). Electronic applications accepted.
Expenses: Tuition, area resident: Full-time $1,368. Tuition, state resident: part-time $76 per credit. Tuition, nonresident: full-time $5,256; part-time $292 per credit. International tuition: $5,256 full-time. Required fees: $678; $38 per credit. Full-time tuition and fees vary according to course load and student level.
Financial aid: In 1999–00, 4 research assistantships (averaging $17,400 per year) were awarded; fellowships, teaching assistantships Financial aid application deadline: 3/1; financial aid applicants required to submit FAFSA.
Faculty research: Fetal alcohol syndrome, circadian rhythms, neuroendocrinology, bone and joint diseases, molecular neurobiology. *Total annual research expenditures:* $850,000.
Dr. James R. West, Head, 979-845-4915, *Fax:* 979-845-0790, *E-mail:* jrwest@tamu.edu.
Application contact: Dr. Farida Sohrabji, Assistant Professor, 979-845-4072, *Fax:* 979-845-0790, *E-mail:* sohrabji@medicine.tamu.edu.

Find an in-depth description at www.petersons.com/graduate.

■ **TEXAS TECH UNIVERSITY HEALTH SCIENCES CENTER**
Graduate School of Biomedical Sciences, Department of Cell Biology and Biochemistry, Program in Anatomy/Cell Biology, Lubbock, TX 79430
AWARDS MS, PhD, MD/PhD.
Faculty: 12 full-time (4 women).
Students: 8 full-time (6 women), 4 international. Average age 27. *17 applicants, 12% accepted.* In 1999, 1 doctorate awarded (100% entered university research/teaching). Terminal master's awarded for partial completion of doctoral program.
Degree requirements: For master's and doctorate, thesis/dissertation required, foreign language not required. *Average time to degree:* Doctorate–6 years full-time.
Entrance requirements: For master's and doctorate, GRE General Test, TOEFL, minimum GPA of 3.0. *Application deadline:* For fall admission, 4/15 (priority date). *Application fee:* $30 ($55 for international students).
Expenses: Tuition, state resident: part-time $38 per credit hour. Tuition, nonresident: part-time $254 per credit hour. Part-time tuition and fees vary according to program.
Financial aid: In 1999–00, 2 fellowships with full tuition reimbursements (averaging

$14,688 per year), 6 research assistantships (averaging $14,500 per year) were awarded.
Faculty research: Biochemical endocrinology, neurobiology, molecular biology, reproductive biology, biology of developing systems.
Dr. James Hutson, Graduate Director, 806-743-2700, *Fax:* 806-743-2990, *E-mail:* james.hutson@ttmc.ttuhsc.edu.

Find an in-depth description at www.petersons.com/graduate.

■ **THE UNIVERSITY OF ALABAMA AT BIRMINGHAM**
Graduate School and School of Medicine and School of Dentistry, Graduate Programs in Joint Health Sciences, Department of Cell Biology, Birmingham, AL 35294
AWARDS Anatomy (PhD). The department participates in the Cellular and Molecular Biology Graduate Program.
Students: 49 full-time (19 women), 1 part-time; includes 2 minority (1 Asian American or Pacific Islander, 1 Hispanic American), 28 international. *6 applicants, 100% accepted.* In 1999, 10 degrees awarded.
Degree requirements: For doctorate, dissertation, qualifying exam required.
Entrance requirements: For doctorate, GRE General Test, interview. *Application deadline:* Applications are processed on a rolling basis. Electronic applications accepted.
Expenses: Tuition, state resident: part-time $104 per semester hour. Tuition, nonresident: part-time $208 per semester hour. Required fees: $17 per semester hour. $57 per quarter. Tuition and fees vary according to program.
Financial aid: In 1999–00, 4 fellowships were awarded.
Faculty research: Neuroscience, immunology.
Dr. Richard B. Marchase, Chair, 205-934-9672, *Fax:* 205-934-0950, *E-mail:* marchase@uab.edu.
Application contact: Injformation Contact, 205-975-7145, *Fax:* 205-975-6748.

Find an in-depth description at www.petersons.com/graduate.

■ **THE UNIVERSITY OF ARIZONA**
College of Medicine, Graduate Programs in Medicine, Department of Cell Biology and Anatomy, Tucson, AZ 85721
AWARDS PhD.
Degree requirements: For doctorate, dissertation required, foreign language not required.

The University of Arizona (continued)
Entrance requirements: For doctorate, GRE General Test, GRE Subject Test (optional).
Expenses: Tuition, nonresident: full-time $4,814; part-time $274 per unit. Required fees: $1,094; $115 per unit.
Faculty research: Heart development, neural development, cellular toxicology and microcirculation; membrane traffic and cytoskeleton; cell-surface receptors.
Find an in-depth description at www.petersons.com/graduate.

■ UNIVERSITY OF ARKANSAS FOR MEDICAL SCIENCES

College of Medicine and Graduate School, Graduate Programs in Medicine, Department of Anatomy, Little Rock, AR 72205-7199
AWARDS MS, PhD, MD/PhD.

Faculty: 18 full-time (3 women), 1 part-time/adjunct (0 women).
Students: 3 full-time (0 women), 2 part-time; includes 2 minority (both Asian Americans or Pacific Islanders), 1 international.
Degree requirements: For master's and doctorate, thesis/dissertation required, foreign language not required.
Entrance requirements: For master's, GRE General Test; for doctorate, GRE General Test, GRE Subject Test. *Application fee:* $0.
Expenses: Tuition: Full-time $8,928.
Financial aid: In 1999–00, 6 research assistantships were awarded. Aid available to part-time students.
Dr. Gwen Childs, Chair, 501-686-5180.
Application contact: Dr. M. D. Cave, Graduate Coordinator, 501-686-5180.
Find an in-depth description at www.petersons.com/graduate.

■ UNIVERSITY OF CALIFORNIA, IRVINE

College of Medicine and Office of Research and Graduate Studies, Graduate Programs in Medicine and School of Biological Sciences, Department of Anatomy and Neurobiology, Irvine, CA 92697
AWARDS Biological sciences (MS, PhD).

Faculty: 14 full-time (3 women), 8 part-time/adjunct (4 women).
Students: 9 full-time (2 women); includes 1 minority (Asian American or Pacific Islander), 1 international. *292 applicants, 25% accepted.*
Degree requirements: For doctorate, dissertation required.
Entrance requirements: For master's and doctorate, GRE General Test, GRE

Subject Test. *Application deadline:* For fall admission, 1/15 (priority date). Applications are processed on a rolling basis. *Application fee:* $40. Electronic applications accepted.
Expenses: Tuition, nonresident: full-time $10,322; part-time $1,720 per quarter. Required fees: $5,354; $1,300 per quarter. Tuition and fees vary according to program.
Financial aid: Fellowships, research assistantships, teaching assistantships, institutionally sponsored loans and tuition waivers (full and partial) available. Financial aid application deadline: 3/2; financial aid applicants required to submit FAFSA.
Faculty research: Neurotransmitter immunocytochemistry, intracellular physiology, molecular neurobiology, forebrain organization and development, structure and function of sensory and motor systems.
Dr. Richard T. Robertson, Professor and Chair, 949-824-6553, *Fax:* 949-824-1105, *E-mail:* rtrobert@uci.edu.
Application contact: Kimberly McKinney, Biological Sciences Contact, 949-824-8145, *Fax:* 949-824-7407, *E-mail:* kamckinn@uci.edu.

■ UNIVERSITY OF CALIFORNIA, LOS ANGELES

School of Medicine and Graduate Division, Graduate Programs in Medicine, Department of Neurobiology, Los Angeles, CA 90095
AWARDS Anatomy and cell biology (PhD).

Students: 16 full-time (8 women); includes 7 minority (1 African American, 2 Asian Americans or Pacific Islanders, 3 Hispanic Americans, 1 Native American), 1 international. *8 applicants, 25% accepted.*
Degree requirements: For doctorate, dissertation, oral and written qualifying exams required, foreign language not required.
Entrance requirements: For doctorate, GRE General Test, GRE Subject Test, bachelor's degree in physical or biological science. *Application fee:* $40.
Expenses: Tuition, nonresident: full-time $9,804. Required fees: $4,405.
Financial aid: In 1999–00, 6 fellowships, 8 research assistantships, 1 teaching assistantship were awarded; Federal Work-Study, institutionally sponsored loans, scholarships, and tuition waivers (full and partial) also available. Financial aid application deadline: 3/1.
Faculty research: Neuroendocrinology, neurophysiology.
Dr. Jack Feldman, Chair, 310-825-9558.

Application contact: UCLA Access Coordinator, 800-284-8252, *Fax:* 310-206-5280, *E-mail:* uclaaccess@ibes.medsch.ucla.edu.

■ UNIVERSITY OF CALIFORNIA, SAN FRANCISCO

Graduate Division, Biomedical Sciences Graduate Group, Program in Anatomy, San Francisco, CA 94143
AWARDS PhD.

Degree requirements: For doctorate, dissertation required.
Entrance requirements: For doctorate, GRE General Test. *Application deadline:* For fall admission, 2/1. *Application fee:* $40.
Financial aid: Fellowships, research assistantships, teaching assistantships available. Financial aid application deadline: 1/10.
Faculty research: Cell biology, neurohistology, mammalian development.
Application contact: Pamela Humphrey, Program Administrator, 415-476-8467.

■ UNIVERSITY OF CHICAGO

Division of the Biological Sciences, Darwinian Sciences: Ecological, Integrative and Evolutionary Biology, Department of Organismal Biology and Anatomy, Chicago, IL 60637-1513
AWARDS Functional and evolutionary biology (PhD); organismal biology and anatomy (PhD).

Faculty: 12 full-time (1 woman), 5 part-time/adjunct (3 women).
Students: 18 full-time (5 women); includes 2 minority (1 Hispanic American, 1 Native American), 4 international. Average age 23. *20 applicants, 20% accepted.* In 1999, 3 doctorates awarded (100% entered university research/teaching).
Degree requirements: For doctorate, dissertation required, foreign language not required. *Average time to degree:* Doctorate–6 years full-time.
Entrance requirements: For doctorate, GRE General Test, TOEFL. *Application deadline:* For fall admission, 1/5 (priority date). *Application fee:* $55.
Expenses: Tuition: Full-time $24,804; part-time $3,422 per course. Required fees: $390. Tuition and fees vary according to program.
Financial aid: In 1999–00, 18 students received aid, including 18 fellowships with tuition reimbursements available (averaging $17,050 per year). Financial aid application deadline: 6/1.
Faculty research: Ecological physiology, evolution of fossil reptiles, vertebrate paleontology.
Dr. Neil Shubin, Chairman, 773-702-9228, *Fax:* 773-702-0037.

Application contact: Carolyn Johnson, Graduate Administrative Director, 773-702-9474, *Fax:* 773-702-4699, *E-mail:* cs-johnson@uchicago.edu.

■ UNIVERSITY OF CINCINNATI

Division of Research and Advanced Studies, College of Medicine, Graduate Programs in Medicine, Department of Cell Biology, Neurobiology and Anatomy, Cincinnati, OH 45221-0091

AWARDS Anatomy (PhD); cell biology (PhD); neurobiology (PhD).

Faculty: 14 full-time.
Students: 31 full-time (17 women), 2 part-time (both women); includes 3 minority (1 African American, 2 Asian Americans or Pacific Islanders), 12 international. 77 *applicants, 18% accepted.* In 1999, 2 degrees awarded.
Degree requirements: For doctorate, dissertation, qualifying exam required. *Average time to degree:* Doctorate–5.6 years full-time.
Entrance requirements: For doctorate, GRE General Test, TOEFL. *Application deadline:* For fall admission, 2/1 (priority date). Applications are processed on a rolling basis. *Application fee:* $30.
Expenses: Tuition, state resident: full-time $5,139; part-time $196 per credit hour. Tuition, nonresident: full-time $10,326; part-time $369 per credit hour. Required fees: $561; $187 per quarter.
Financial aid: Tuition waivers (full) and unspecified assistantships available. Financial aid application deadline: 5/1.
Faculty research: Cell structure, molecular genetics. *Total annual research expenditures:* $5.3 million.
Dr. Peter Stambrook, Head, 513-558-5685, *Fax:* 513-556-4454, *E-mail:* peter.stambrook@uc.edu.
Application contact: Robert Brackenbury, Graduate Program Director, 513-558-6080, *Fax:* 513-556-4454, *E-mail:* robert.brackenbury@uc.edu.

■ UNIVERSITY OF FLORIDA

College of Medicine and Graduate School, Interdisciplinary Program in Biomedical Sciences, Department of Anatomy and Cell Biology, Gainesville, FL 32610

AWARDS PhD.

Degree requirements: For doctorate, dissertation required, foreign language not required.
Entrance requirements: For doctorate, GRE General Test, TOEFL, minimum GPA of 3.0. Electronic applications accepted.
Expenses: Tuition, state resident: part-time $144 per credit hour. Tuition,

nonresident: part-time $505 per credit hour. Tuition and fees vary according to course level, course load and program.
Faculty research: Structure and function of intracellular organelles, cell adhesion, differentiation.

■ UNIVERSITY OF GEORGIA

College of Veterinary Medicine and Graduate School, Graduate Programs in Veterinary Medicine, Department of Veterinary Anatomy and Radiology, Athens, GA 30602

AWARDS Veterinary anatomy (MS).

Degree requirements: For master's, thesis required, foreign language not required.
Entrance requirements: For master's, GRE General Test. Electronic applications accepted.
Expenses: Tuition, state resident: full-time $7,516; part-time $431 per credit hour. Tuition, nonresident: full-time $12,204; part-time $793 per credit hour.

■ UNIVERSITY OF ILLINOIS AT CHICAGO

College of Medicine and Graduate College, Graduate Programs in Medicine, Department of Anatomy and Cell Biology, Chicago, IL 60607-7128

AWARDS MS, PhD, MD/PhD.

Faculty: 13 full-time (5 women).
Students: 10 full-time (5 women); includes 1 minority (Asian American or Pacific Islander), 7 international. Average age 29. *26 applicants, 15% accepted.* In 1999, 1 doctorate awarded.
Degree requirements: For master's and doctorate, thesis/dissertation required, foreign language not required.
Entrance requirements: For master's and doctorate, GRE General Test, TOEFL. *Application deadline:* For fall admission, 6/1; for spring admission, 11/1. *Application fee:* $40 ($50 for international students).
Expenses: Tuition, state resident: full-time $3,750. Tuition, nonresident: full-time $10,588. Tuition and fees vary according to course load.
Financial aid: In 1999–00, 7 students received aid; fellowships, research assistantships, teaching assistantships, Federal Work-Study, scholarships, traineeships, and tuition waivers (full) available. Financial aid application deadline: 3/1; financial aid applicants required to submit FAFSA.
Faculty research: Neuroanatomy, functional morphology, cytoskeleton, synapses, neural transplants.
Rochelle Cohen, Acting Head.

Application contact: Conwell Anderson, Director of Graduate Studies, 312-996-3360, *E-mail:* conwell@uic.edu.

Find an in-depth description at www.petersons.com/graduate.

■ THE UNIVERSITY OF IOWA

College of Medicine and Graduate College, Graduate Programs in Medicine, Department of Anatomy and Cell Biology, Iowa City, IA 52242-1316

AWARDS PhD.

Faculty: 20 full-time (7 women).
Students: 12 full-time (5 women); includes 1 minority (Hispanic American), 8 international. Average age 28. *23 applicants, 26% accepted.* In 1999, 2 degrees awarded (100% entered university research/teaching).
Degree requirements: For doctorate, dissertation required, foreign language not required. *Average time to degree:* Doctorate–5 years full-time.
Entrance requirements: For doctorate, GRE General Test, TOEFL, minimum GPA of 3.0. *Application deadline:* For fall admission, 2/1 (priority date). Applications are processed on a rolling basis. *Application fee:* $30 ($50 for international students). Electronic applications accepted.
Expenses: Tuition, state resident: full-time $3,308. Tuition, nonresident: full-time $10,662. Tuition and fees vary according to course load and program.
Financial aid: In 1999–00, 11 students received aid, including 11 teaching assistantships with full tuition reimbursements available (averaging $16,277 per year); fellowships with full tuition reimbursements available, research assistantships with full tuition reimbursements available, Federal Work-Study, grants, institutionally sponsored loans, and scholarships also available. Financial aid application deadline: 3/1.
Faculty research: Biology of differentiation and transformation, developmental and vascular cell biology, neurobiology. *Total annual research expenditures:* $4.1 million.
Dr. Mary J. C. Hendrix, Head, 319-335-7755, *Fax:* 319-335-7198, *E-mail:* mary-hendrix@uiowa.edu.
Application contact: Tracy M. Middleton, Program Assistant, 319-335-7744, *Fax:* 319-335-7198, *E-mail:* tracy-middleton@uiowa.edu.

Find an in-depth description at www.petersons.com/graduate.

■ UNIVERSITY OF KANSAS

Graduate Studies Medical Center, Graduate Programs in Biomedical and Basic Sciences, Department of Anatomy and Cell Biology, Lawrence, KS 66045

AWARDS MA, PhD, MD/PhD. Part-time programs available.

Faculty: 14 full-time (3 women).
Students: 1 full-time (0 women), 9 part-time (4 women), 2 international. Average age 31. *0 applicants, 0% accepted.* In 1999, 2 doctorates awarded. Terminal master's awarded for partial completion of doctoral program.
Degree requirements: For master's, comprehensive oral exam, oral defense of thesis required; for doctorate, dissertation, comprehensive oral exam required.
Entrance requirements: For master's and doctorate, GRE General Test, GRE Subject Test, TOEFL, TSE. *Application deadline:* For fall admission, 1/31 (priority date). Applications are processed on a rolling basis. *Application fee:* $0. Electronic applications accepted.
Expenses: Tuition, state resident: full-time $2,482; part-time $103 per credit hour. Tuition, nonresident: full-time $8,104; part-time $338 per credit hour. Required fees: $428; $31 per credit hour. Tuition and fees vary according to program.
Financial aid: Fellowships, research assistantships, teaching assistantships, Federal Work-Study, institutionally sponsored loans, and scholarships available. Aid available to part-time students. Financial aid application deadline: 3/31; financial aid applicants required to submit FAFSA.
Faculty research: Neuropeptides, exocrine cell biology, immunobiology of pregnancy, platelet activating facor, signal transduction across nuclear membrane. *Total annual research expenditures:* $2.1 million.
Dr. Dale R. Abrahamson, Chairman, 913-588-7000, *Fax:* 913-588-2710, *E-mail:* dabrahamson@kume.edu.
Application contact: Dr. Robert C. De Lisle, Graduate Adviser, 913-588-2742, *Fax:* 913-588-2710, *E-mail:* rdelisle@kumc.edu.

■ UNIVERSITY OF KENTUCKY

Graduate School and College of Medicine, Graduate Programs in Medicine, Program in Anatomy and Neurobiology, Lexington, KY 40506-0032

AWARDS PhD, MD/PhD.

Degree requirements: For doctorate, dissertation, comprehensive exam required, foreign language not required.

Entrance requirements: For doctorate, GRE General Test, minimum undergraduate GPA of 3.0.
Expenses: Tuition, state resident: full-time $3,596; part-time $188 per credit hour. Tuition, nonresident: full-time $10,116; part-time $550 per credit hour.
Faculty research: Neuroendocrinology, developmental neurobiology, neurotrophic substances, neural plasticity and trauma, neurobiology of aging.

■ UNIVERSITY OF LOUISVILLE

School of Medicine and Graduate School, Integrated Programs in Biomedical Sciences, Department of Anatomical Sciences and Neurobiology, Louisville, KY 40292-0001

AWARDS MS, PhD.

Degree requirements: For master's and doctorate, thesis/dissertation required.
Entrance requirements: For master's and doctorate, GRE General Test, TOEFL. Electronic applications accepted.
Expenses: Tuition, state resident: full-time $3,260; part-time $182 per hour. Tuition, nonresident: full-time $9,780; part-time $544 per hour. Required fees: $143; $28 per hour.

■ UNIVERSITY OF MARYLAND

Graduate School, Graduate Programs in Medicine, Department of Anatomy and Neurobiology, Baltimore, MD 21201-1627

AWARDS MS, PhD, MD/PhD. Part-time and evening/weekend programs available.

Degree requirements: For master's, thesis optional; for doctorate, one foreign language (computer language can substitute), dissertation required.
Entrance requirements: For master's, GRE General Test, TOEFL, minimum GPA of 3.0; for doctorate, GRE General Test, GRE Subject Test (recommended), TOEFL, minimum GPA of 3.0.
Expenses: Tuition, state resident: part-time $261 per credit hour. Tuition, nonresident: part-time $468 per credit hour. Tuition and fees vary according to program.
Faculty research: Neural networks, chemical sensory pathways, electrophysiology, developmental neurobiology.

■ UNIVERSITY OF MEDICINE AND DENTISTRY OF NEW JERSEY

Graduate School of Biomedical Sciences, Graduate Programs in Biomedical Sciences, Department of Anatomy, Cell Biology, and Injury Sciences, Newark, NJ 07107

AWARDS MS, PhD. Terminal master's awarded for partial completion of doctoral program.

Degree requirements: For master's, thesis required, foreign language not required; for doctorate, dissertation, qualifying exam required, foreign language not required.
Entrance requirements: For master's and doctorate, GRE General Test, TOEFL. *Application deadline:* For fall admission, 2/1; for spring admission, 10/1. *Application fee:* $40.
Expenses: Tuition, state resident: part-time $270 per credit hour. Tuition, nonresident: part-time $407 per credit hour. Part-time tuition and fees vary according to campus/location and program.
Financial aid: Fellowships, research assistantships, Federal Work-Study, institutionally sponsored loans, and tuition waivers (full and partial) available. Financial aid application deadline: 5/1. Dr. John Siegel, Chairperson, 973-972-4414.
Application contact: Dr. Henry E. Brezenoff, Dean, Graduate School of Biomedical Sciences, 973-972-5333, *Fax:* 973-972-7148, *E-mail:* hbrezeno@umdnj.edu.

Find an in-depth description at www.petersons.com/graduate.

■ UNIVERSITY OF MINNESOTA, DULUTH

School of Medicine, Department of Anatomy and Cell Biology, Duluth, MN 55812-2496

AWARDS MS, PhD. Part-time programs available.

Faculty: 4 full-time (1 woman).
Students: 1 full-time (0 women), 1 international. Average age 25. *10 applicants, 10% accepted.* In 1999, 1 master's awarded (100% found work related to degree). Terminal master's awarded for partial completion of doctoral program.
Degree requirements: For master's and doctorate, thesis/dissertation required, foreign language not required. *Average time to degree:* Master's–2 years full-time.
Entrance requirements: For master's and doctorate, GRE, biology/chemistry background. *Application deadline:* For fall admission, 7/15. *Application fee:* $30.

Expenses: Tuition, state resident: full-time $5,040; part-time $420 per credit. Tuition, nonresident: full-time $9,900; part-time $825 per credit. Required fees: $509. Tuition and fees vary according to course load and program.

Financial aid: Research assistantships, teaching assistantships, Federal Work-Study available. Financial aid application deadline: 4/1.

Faculty research: Cytoskeleton, neurobiology, cancer biology, cardiovascular biology, developmental biology. *Total annual research expenditures:* $45,000.

Dr. Arlen R. Severson, Head, 218-726-7903, *Fax:* 218-726-6235, *E-mail:* aseverso@d.umn.edu.

Application contact: Dr. David J. Schimpf, Director, Graduate Studies in Biology, 218-726-7265, *Fax:* 218-726-8142, *E-mail:* dschimpf@d.umn.edu.

■ UNIVERSITY OF MISSISSIPPI MEDICAL CENTER

Graduate Programs in Biomedical Sciences, Department of Anatomy, Jackson, MS 39216-4505

AWARDS MS, PhD, MD/PhD.

Faculty: 15 full-time (3 women), 6 part-time/adjunct (1 woman).

Students: 11 full-time (7 women), 8 international. Average age 27. *15 applicants, 20% accepted.* In 1999, 3 doctorates awarded (100% continued full-time study). Terminal master's awarded for partial completion of doctoral program.

Degree requirements: For master's, thesis required, foreign language not required; for doctorate, dissertation, first authored publication required, foreign language not required. *Average time to degree:* Master's–3 years full-time; doctorate–5 years full-time.

Entrance requirements: For master's and doctorate, GRE General Test, TOEFL, minimum GPA of 3.0. *Application deadline:* For fall admission, 3/15 (priority date). *Application fee:* $10.

Expenses: Tuition, state resident: full-time $2,378; part-time $132 per hour. Tuition, nonresident: full-time $4,697; part-time $261 per hour. Tuition and fees vary according to program.

Financial aid: In 1999–00, 11 students received aid, including 11 research assistantships (averaging $16,234 per year); Federal Work-Study also available. Financial aid application deadline: 3/15.

Faculty research: Systems neuroscience with emphasis on motor and sensory, cell biology with emphasis on cell-matrix interactions, development of cardiovascular system, biology of glial cells. *Total annual research expenditures:* $600,000.

Dr. James C. Lynch, Coordinator, 601-984-1657, *Fax:* 601-984-1655, *E-mail:* jcl@anat.umsmed.edu.

■ UNIVERSITY OF NEBRASKA MEDICAL CENTER

Graduate College, Department of Cell Biology and Anatomy, Omaha, NE 68198

AWARDS MS, PhD. Part-time programs available.

Faculty: 16 full-time (1 woman), 2 part-time/adjunct (1 woman).

Students: 5 full-time (4 women), 1 (woman) part-time. Average age 28. *4 applicants, 0% accepted.* In 1999, 4 doctorates awarded. Terminal master's awarded for partial completion of doctoral program.

Degree requirements: For master's, thesis required, foreign language not required; for doctorate, dissertation required.

Entrance requirements: For master's and doctorate, GRE General Test. *Application deadline:* For fall admission, 3/1 (priority date). Applications are processed on a rolling basis. *Application fee:* $35.

Expenses: Tuition, state resident: part-time $116 per semester hour. Tuition, nonresident: part-time $270 per semester hour. Tuition and fees vary according to program.

Financial aid: In 1999–00, 1 fellowship, 1 research assistantship, 4 teaching assistantships were awarded; institutionally sponsored loans also available. Aid available to part-time students. Financial aid application deadline: 3/1.

Faculty research: Hematology, immunology, developmental biology, cardiovascular biology, neuroscience.

Dr. J. B. Turpen, Graduate Committee Chair, 402-559-4388, *Fax:* 402-559-7328, *E-mail:* jturpen@unmc.edu.

Application contact: Jo Wagner, Associate Director of Admissions, 402-559-6468.

Find an in-depth description at www.petersons.com/graduate.

■ THE UNIVERSITY OF NORTH CAROLINA AT CHAPEL HILL

School of Medicine and Graduate School, Graduate Programs in Medicine, Department of Cell Biology and Anatomy, Chapel Hill, NC 27599

AWARDS PhD.

Faculty: 27 full-time (7 women).

Students: 25 full-time (14 women); includes 6 minority (all Asian Americans or Pacific Islanders). Average age 24. *36 applicants, 19% accepted.* In 1999, 5 degrees awarded (100% entered university research/teaching).

Degree requirements: For doctorate, dissertation, comprehensive exams required. *Average time to degree:* Doctorate–5 years full-time.

Entrance requirements: For doctorate, GRE General Test, GRE Subject Test. *Application deadline:* For fall admission, 1/1 (priority date). Applications are processed on a rolling basis. *Application fee:* $55. Electronic applications accepted.

Expenses: Tuition, state resident: full-time $1,966. Tuition, nonresident: full-time $11,026. Required fees: $8,940. One-time fee: $15 full-time. Part-time tuition and fees vary according to course load.

Financial aid: In 1999–00, 25 students received aid, including 13 research assistantships with full tuition reimbursements available (averaging $17,000 per year), 12 teaching assistantships with full tuition reimbursements available (averaging $17,000 per year); fellowships, tuition waivers (full) and unspecified assistantships also available. Financial aid application deadline: 2/1; financial aid applicants required to submit FAFSA.

Faculty research: Cell adhesion, motility and cytoskeleton; molecular analysis of signal transduction; development biology and toxicology; reproductive biology; cell and molecular imaging. *Total annual research expenditures:* $4.5 million.

Dr. Michael G. O'Rand, Interim Chair, 919-966-3026.

Application contact: Dr. Ellen Weiss, Director of Graduate Studies, 919-966-7683, *E-mail:* erweiss@med.unc.edu.

Find an in-depth description at www.petersons.com/graduate.

■ UNIVERSITY OF NORTH DAKOTA

School of Medicine and Graduate School, Graduate Programs in Medicine, Department of Anatomy, Grand Forks, ND 58202

AWARDS MS, PhD.

Faculty: 7 full-time (0 women).

Students: 12 full-time (9 women). 7 *applicants, 100% accepted.* In 1999, 2 master's, 1 doctorate awarded.

Degree requirements: For master's, thesis, final exam required, foreign language not required; for doctorate, dissertation, comprehensive exam, final exam required, foreign language not required.

Entrance requirements: For master's, GRE General Test, TOEFL, minimum GPA of 3.0; for doctorate, GRE General Test, TOEFL, minimum GPA of 3.5. *Application deadline:* For fall admission, 3/1

University of North Dakota (continued) (priority date). Applications are processed on a rolling basis. *Application fee:* $25.

Expenses: Tuition, state resident: full-time $2,690; part-time $112 per credit. Tuition, nonresident: full-time $7,182; part-time $299 per credit. Required fees: $46 per semester.

Financial aid: In 1999–00, 12 students received aid, including 4 research assistantships with full tuition reimbursements available (averaging $10,586 per year), 8 teaching assistantships with full tuition reimbursements available (averaging $10,586 per year); fellowships, Federal Work-Study, institutionally sponsored loans, scholarships, and tuition waivers (full and partial) also available. Aid available to part-time students. Financial aid applicants required to submit FAFSA.

Faculty research: Coronary vessel, vasculogenesis, acellular glomerular and retinal microvessel membranes, ependymal cells, cardiac muscle.
Dr. John McCormack, Director, 701-777-2101, *Fax:* 701-777-3527, *E-mail:* johnm@ medicine.nodak.edu.

■ UNIVERSITY OF NORTH TEXAS HEALTH SCIENCE CENTER AT FORT WORTH

Graduate School of Biomedical Sciences, Fort Worth, TX 76107-2699

AWARDS Anatomy and cell biology (MS, PhD); biochemistry and molecular biology (MS, PhD); biotechnology (MS); integrative physiology (MS, PhD); microbiology and immunology (MS, PhD); pharmacology (MS, PhD).

Faculty: 65 full-time (9 women), 11 part-time/adjunct (1 woman).

Students: 59 full-time (27 women), 44 part-time (20 women); includes 30 minority (10 African Americans, 9 Asian Americans or Pacific Islanders, 11 Hispanic Americans), 23 international. *70 applicants, 70% accepted.* In 1999, 5 master's awarded (40% found work related to degree, 60% continued full-time study); 5 doctorates awarded (80% entered university research/teaching, 20% found other work related to degree).

Degree requirements: For doctorate, computer language, dissertation required. *Average time to degree:* Master's–2.5 years full-time, 4 years part-time; doctorate–5 years full-time.

Entrance requirements: For master's and doctorate, GRE General Test, TOEFL. *Application deadline:* For fall admission, 5/1; for spring admission, 11/1. Applications are processed on a rolling basis. *Application fee:* $25 ($50 for international students).

Expenses: Tuition, state resident: full-time $1,188; part-time $66 per credit. Tuition,

nonresident: full-time $5,058; part-time $281 per credit. Required fees: $366; $183 per semester.

Financial aid: In 1999–00, 11 fellowships, 70 research assistantships (averaging $16,500 per year) were awarded; teaching assistantships, career-related internships or fieldwork, Federal Work-Study, grants, institutionally sponsored loans, and traineeships also available. Aid available to part-time students. Financial aid application deadline: 4/1; financial aid applicants required to submit FAFSA.

Faculty research: Alzheimer's disease, diabetes, eye diseases, cancer, cardiovascular physiology.
Dr. Thomas Yorio, Dean, 817-735-2560, *Fax:* 817-735-0243, *E-mail:* yoriot@ hsc.unt.edu.

Application contact: Jan Sharp, Administrative Assistant, 817-735-0258, *Fax:* 817-735-0243, *E-mail:* gsbs@ hsc.unt.edu.

Find an in-depth description at www.petersons.com/graduate.

■ UNIVERSITY OF PUERTO RICO, MEDICAL SCIENCES CAMPUS

School of Medicine, Division of Graduate Studies, Department of Anatomy, San Juan, PR 00936-5067

AWARDS MS, PhD.

Degree requirements: For master's and doctorate, one foreign language, thesis/dissertation required.

Entrance requirements: For master's and doctorate, GRE General Test, GRE Subject Test, interview.

Expenses: Tuition, state resident: full-time $5,500. Tuition, nonresident: full-time $8,400. Required fees: $600. Tuition and fees vary according to class time, course load, degree level and program.

Faculty research: Neurobiology, primatology, evolution, visual system, muscle structure.

■ UNIVERSITY OF ROCHESTER

School of Medicine and Dentistry, Graduate Programs in Medicine and Dentistry, Department of Neurobiology and Anatomy, Program in Anatomy, Rochester, NY 14627-0250

AWARDS MS, PhD.

Students: 9 full-time (3 women); includes 1 minority (Asian American or Pacific Islander). In 1999, 2 master's, 1 doctorate awarded. Terminal master's awarded for partial completion of doctoral program.

Degree requirements: For doctorate, one foreign language, dissertation, qualifying exam required.

Entrance requirements: For master's and doctorate, GRE General Test. *Application deadline:* For fall admission, 2/1. *Application fee:* $25.

Expenses: Tuition: Part-time $697 per credit hour. Tuition and fees vary according to program.

Financial aid: Fellowships, research assistantships, teaching assistantships, tuition waivers (full and partial) available. Financial aid application deadline: 2/1.
Dr. John Olschowka, Director, 716-275-8238.

Application contact: Jennifer Dwyer, Graduate Program Secretary, 716-275-5788.

■ UNIVERSITY OF SOUTHERN CALIFORNIA

Keck School of Medicine and Graduate School, Graduate Programs in Medicine, Department of Cell and Neurobiology, Los Angeles, CA 90089

AWARDS Anatomy and cell biology (MS, PhD), including anatomy (PhD), cell biology (PhD); cell and neurobiology (MS, PhD); pharmacology and nutrition (MS, PhD); preventive nutrition (MS).

Faculty: 21 full-time (3 women), 8 part-time/adjunct (2 women).

Students: 19 full-time (9 women); includes 10 minority (all Asian Americans or Pacific Islanders). Average age 23. *54 applicants, 22% accepted.* In 1999, 5 master's awarded (40% found work related to degree, 60% continued full-time study). Terminal master's awarded for partial completion of doctoral program.

Degree requirements: For master's, thesis or alternative required, foreign language not required; for doctorate, dissertation required. *Average time to degree:* Master's–2 years full-time; doctorate–5 years full-time.

Entrance requirements: For master's, GRE General Test, TOEFL, minimum GPA of 3.0; for doctorate, GRE General Test, TOEFL. *Application fee:* $55. Electronic applications accepted.

Expenses: Tuition: Full-time $22,198; part-time $748 per unit. Required fees: $406.

Financial aid: In 1999–00, 13 students received aid; fellowships, research assistantships, teaching assistantships, Federal Work-Study, institutionally sponsored loans, and tuition waivers (partial) available. Aid available to part-time students.

Faculty research: Neurobiology and development, circaulian rhythm, gene therapy in vision, lacrimal glands, neuroendocrinology, signal transduction mechanisms.

Dr. Cheryl Craft, Chair, 323-442-1881, *Fax:* 323-442-2709, *E-mail:* ccraft@hsc.usc.edu.
Application contact: Darlene Marie Campbell, Administrative Assistant, 323-442-1881, *Fax:* 323-442-0466, *E-mail:* dmc@hsc.usc.edu.

■ UNIVERSITY OF SOUTH FLORIDA

College of Medicine and Graduate School, Graduate Programs in Medical Sciences, Department of Anatomy, Tampa, FL 33620-9951

AWARDS PhD.

Degree requirements: For doctorate, computer language, dissertation required, foreign language not required.
Entrance requirements: For doctorate, GRE General Test, minimum GPA of 3.0.
Expenses: Tuition, state resident: part-time $148 per credit hour. Tuition, nonresident: part-time $509 per credit hour.
Faculty research: Central nervous system, muscle aging, neuroendocrinology, cardiovascular anatomy.

Find an in-depth description at www.petersons.com/graduate.

■ THE UNIVERSITY OF TENNESSEE

Graduate School, College of Agricultural Sciences and Natural Resources, Department of Animal Science, Knoxville, TN 37996

AWARDS Animal anatomy (PhD); breeding (MS, PhD); management (MS, PhD); nutrition (MS, PhD); physiology (MS, PhD). Part-time programs available.

Faculty: 19 full-time (2 women).
Students: 24 full-time (15 women), 10 part-time (4 women); includes 3 minority (1 African American, 1 Asian American or Pacific Islander, 1 Hispanic American), 7 international. *27 applicants, 56% accepted.* In 1999, 5 master's, 4 doctorates awarded.
Degree requirements: For master's and doctorate, thesis/dissertation required, foreign language not required.
Entrance requirements: For master's and doctorate, GRE General Test, TOEFL, minimum GPA of 2.7. *Application deadline:* For fall admission, 2/1 (priority date). Applications are processed on a rolling basis. *Application fee:* $35. Electronic applications accepted.
Expenses: Tuition, state resident: full-time $3,806; part-time $184 per credit hour. Tuition, nonresident: full-time $9,874; part-time $522 per credit hour. Tuition and fees vary according to program.
Financial aid: In 1999–00, 19 research assistantships, 3 teaching assistantships

were awarded; fellowships, career-related internships or fieldwork, Federal Work-Study, institutionally sponsored loans, and unspecified assistantships also available. Financial aid application deadline: 2/1; financial aid applicants required to submit FAFSA.
Dr. Kelly Robbins, Head, 865-974-7286, *Fax:* 865-974-7297, *E-mail:* krobbins@utk.edu.
Application contact: Dr. James Godkin, Graduate Representative, *E-mail:* jgodkin@utk.edu.

■ THE UNIVERSITY OF TENNESSEE HEALTH SCIENCE CENTER

College of Graduate Health Sciences, Department of Anatomy and Neurobiology, Memphis, TN 38163-0002

AWARDS PhD.

Degree requirements: For doctorate, dissertation, oral and written preliminary and comprehensive exams required, foreign language not required.
Entrance requirements: For doctorate, GRE General Test, minimum GPA of 3.0.

Find an in-depth description at www.petersons.com/graduate.

■ UNIVERSITY OF UTAH

School of Medicine and Graduate School, Graduate Programs in Medicine, Department of Neurology and Anatomy, Salt Lake City, UT 84112-1107

AWARDS M Phil, MS, PhD. Part-time programs available.

Faculty: 10 full-time (3 women), 3 part-time/adjunct (1 woman).
Students: 8 full-time (4 women), 1 (woman) part-time. Average age 26. *2 applicants, 100% accepted.* Terminal master's awarded for partial completion of doctoral program.
Degree requirements: For master's and doctorate, one foreign language, computer language, thesis/dissertation required.
Entrance requirements: For master's and doctorate, GRE. *Application deadline:* For fall admission, 2/15. Applications are processed on a rolling basis. *Application fee:* $30 ($50 for international students).
Expenses: Tuition, state resident: full-time $2,105. Tuition, nonresident: full-time $6,312.
Financial aid: In 1999–00, 9 students received aid, including 9 research assistantships with tuition reimbursements available (averaging $17,000 per year); fellowships, teaching assistantships Financial aid application deadline: 2/15.

Faculty research: Neuroscience, neuroanatomy, developmental neurobiology, neurogenetics. *Total annual research expenditures:* $1.5 million.
Tom Parks, Chair, 801-581-5494, *Fax:* 801-585-9736, *E-mail:* tom.parks@hsc.utah.edu.
Application contact: Kathleen A. Kjaglien, Administrative Officer, 801-581-5494, *Fax:* 801-585-9736, *E-mail:* kathleen.kjaglien@usc.utah.edu.

■ UNIVERSITY OF VERMONT

College of Medicine and Graduate College, Graduate Programs in Medicine, Department of Anatomy and Neurobiology, Burlington, VT 05405

AWARDS PhD, MD/PhD.

Degree requirements: For doctorate, dissertation required.
Entrance requirements: For doctorate, GRE General Test, TOEFL.
Expenses: Tuition, state resident: full-time $7,464; part-time $311 per credit. Tuition, nonresident: full-time $18,672; part-time $778 per credit. Full-time tuition and fees vary according to degree level and program.
Faculty research: Autonomic neurobiology, developmental neurobiology, neurotransmitter expression and release, plasticity and regeneration.

Find an in-depth description at www.petersons.com/graduate.

■ UNIVERSITY OF WISCONSIN–MADISON

Graduate School, College of Agricultural and Life Sciences, Department of Animal Health and Biomedical Sciences, Program in Comparative Biosciences, Madison, WI 53706-1380

AWARDS Anatomy (MS, PhD); biochemistry (MS, PhD); cellular and molecular biology (MS, PhD); environmental toxicology (MS, PhD); neurosciences (MS, PhD); pharmacology (MS, PhD); physiology (MS, PhD).

Degree requirements: For doctorate, dissertation required.
Expenses: Tuition, state resident: full-time $5,406; part-time $339 per credit. Tuition, nonresident: full-time $17,110; part-time $1,071 per credit. Full-time tuition and fees vary according to program and reciprocity agreements. Part-time tuition and fees vary according to course load and program.

■ VIRGINIA COMMONWEALTH UNIVERSITY

School of Graduate Studies, School of Allied Health Professions, Department of Physical Therapy and Department of Anatomy, Program in Anatomy and Physical Therapy, Richmond, VA 23284-9005

AWARDS PhD.

Degree requirements: For doctorate, dissertation required, foreign language not required.
Entrance requirements: For doctorate, GRE General Test. *Application deadline:* For fall admission, 5/1. *Application fee:* $30.
Expenses: Tuition, state resident: full-time $4,031; part-time $224 per credit hour. Tuition, nonresident: full-time $11,946; part-time $664 per credit hour. Required fees: $1,081; $40 per credit hour. Tuition and fees vary according to campus/location and program.
Application contact: Dr. Sheryl Finucane, Assistant Professor, 804-828-0234.

■ VIRGINIA COMMONWEALTH UNIVERSITY

School of Graduate Studies and School of Medicine, School of Medicine Graduate Programs, Department of Anatomy, Richmond, VA 23284-9005

AWARDS Anatomy (MS, PhD, CBHS), including neuroscience (MS, PhD); anatomy and physical therapy (PhD).

Students: 10 full-time (4 women), 35 part-time (11 women); includes 12 minority (1 African American, 10 Asian Americans or Pacific Islanders, 1 Hispanic American). In 1999, 8 master's, 4 doctorates, 7 other advanced degrees awarded.
Degree requirements: For master's, thesis required, foreign language not required; for doctorate, dissertation, comprehensive oral and written exams required, foreign language not required.
Entrance requirements: For master's, DAT, GRE General Test, or MCAT; for doctorate, GRE General Test. *Application deadline:* For fall admission, 5/15. *Application fee:* $30.
Expenses: Tuition, state resident: full-time $4,031; part-time $224 per credit hour. Tuition, nonresident: full-time $11,946; part-time $664 per credit hour. Required fees: $1,081; $40 per credit hour. Tuition and fees vary according to campus/location and program.
Financial aid: Fellowships available.
Dr. John T. Povlishock, Chair, 804-828-9535, *Fax:* 804-828-9477.
Application contact: Dr. George R. Leichnetz, Graduate Program Director,

804-828-9512, *Fax:* 804-828-9477, *E-mail:* grleichn@vcu.edu.

■ WAKE FOREST UNIVERSITY

School of Medicine and Graduate School, Graduate Programs in Medicine, Department of Neurobiology and Anatomy, Winston-Salem, NC 27109

AWARDS PhD.

Degree requirements: For doctorate, one foreign language (computer language can substitute), dissertation required.
Entrance requirements: For doctorate, GRE General Test, GRE Subject Test. Electronic applications accepted.
Expenses: Tuition: Full-time $18,300.
Faculty research: Sensory neurobiology, reproductive endocrinology, regulatory processes in cell biology.

Find an in-depth description at www.petersons.com/graduate.

■ WAYNE STATE UNIVERSITY

School of Medicine and Graduate School, Graduate Programs in Medicine, Department of Anatomy and Cell Biology, Detroit, MI 48202

AWARDS MS, PhD, MD/PhD. Terminal master's awarded for partial completion of doctoral program.

Degree requirements: For master's, thesis required (for some programs), foreign language not required; for doctorate, dissertation required, foreign language not required.
Entrance requirements: For master's and doctorate, GRE General Test, minimum GPA of 3.0.
Faculty research: Cytoskeletal proteins, neuronal plasticity, neural connections, glial cells, receptor interaction.

Find an in-depth description at www.petersons.com/graduate.

■ WEST VIRGINIA UNIVERSITY

School of Medicine, Graduate Programs in Health Sciences, Department of Anatomy, Morgantown, WV 26506

AWARDS Developmental anatomy (MS); gross anatomy (MS, PhD); microscopic anatomy (MS, PhD); molecular and deevlo9pmental anatomy (PhD); neuroanatomy (MS, PhD).

Students: 7 full-time (2 women), 3 international. Average age 27. *10 applicants, 20% accepted.* Terminal master's awarded for partial completion of doctoral program.
Degree requirements: For master's, thesis required, foreign language not

required; for doctorate, dissertation, written and oral comprehensive exams required, foreign language not required.
Entrance requirements: For master's, GRE General Test, TOEFL, minimum GPA of 3.0; for doctorate, GRE General Test, TOEFL, TSE. *Application deadline:* Applications are processed on a rolling basis. *Application fee:* $45.
Expenses: Tuition, state resident: full-time $3,564. Tuition, nonresident: full-time $10,230.
Financial aid: In 1999–00, 1 research assistantship, 5 teaching assistantships were awarded; Federal Work-Study and institutionally sponsored loans also available. Financial aid application deadline: 2/1; financial aid applicants required to submit FAFSA.
Faculty research: Microcirculation, molecular biology, neurochemistry, neuropeptides, nutrition, astrocytes, x-ray microanalysis, brain development, auditory systems. *Total annual research expenditures:* $819,851.
Dr. Richard C. Wiggins, Chair, 304-293-2211.
Application contact: Dr. Morton H. Friedman, Admissions Officer, 304-293-2212.

■ WRIGHT STATE UNIVERSITY

School of Graduate Studies, College of Science and Mathematics, Department of Anatomy, Dayton, OH 45435

AWARDS MS.

Students: 18 full-time (5 women), 2 part-time; includes 5 minority (all African Americans). Average age 26. *12 applicants, 67% accepted.* In 1999, 8 degrees awarded.
Degree requirements: For master's, thesis optional, foreign language not required.
Entrance requirements: For master's, TOEFL. *Application fee:* $25.
Expenses: Tuition, state resident: full-time $5,568; part-time $175 per quarter hour. Tuition, nonresident: full-time $9,696; part-time $302 per quarter hour. Full-time tuition and fees vary according to course load, campus/location and program.
Financial aid: Fellowships, research assistantships, teaching assistantships available. Aid available to part-time students. Financial aid applicants required to submit FAFSA.
Faculty research: Reproductive cell biology, neurobiology of pain, neurohistochemistry.
Dr. Jane N. Scott, Chair, 937-775-3067, *Fax:* 937-775-3391, *E-mail:* jane.scott@wright.edu.
Application contact: Dr. Larry J. Ream, Director, 937-775-3188, *Fax:* 937-775-3391, *E-mail:* larry.ream@wright.edu.

■ YESHIVA UNIVERSITY

Albert Einstein College of Medicine, Sue Golding Graduate Division of Medical Sciences, Department of Anatomy and Structural Biology, Bronx, NY 10461

AWARDS Anatomy (PhD); cell and developmental biology (PhD).

Faculty: 13 full-time.
Students: 14 full-time (8 women); includes 3 minority (1 African American, 2 Asian Americans or Pacific Islanders), 8 international. In 1999, 1 degree awarded.

Degree requirements: For doctorate, dissertation required, foreign language not required.
Entrance requirements: For doctorate, GRE General Test, TOEFL. *Application deadline:* For fall admission, 1/15. *Application fee:* $0. Electronic applications accepted.
Expenses: Tuition: Part-time $525 per credit. Tuition and fees vary according to degree level and program.
Financial aid: In 1999–00, 14 fellowships were awarded.

Faculty research: Cell motility, cell membranes and membrane-cytoskeletal interactions as applied to processing of pancreatic hormones, mechanisms of secretion.
Dr. Peter Satir, Chairperson, 718-430-2836.
Application contact: Sheila Cleeton, Assistant Director, 718-430-2128, *Fax:* 718-430-8655, *E-mail:* phd@aecom.yu.edu.

Biochemistry

BIOCHEMISTRY
••••••••••••••••••••••••••••••••••

■ ARIZONA STATE UNIVERSITY

Graduate College, College of Liberal Arts and Sciences, Department of Chemistry and Biochemistry, Tempe, AZ 85287

AWARDS MNS, MS, PhD.

Faculty: 43 full-time (5 women).
Students: 85 full-time (31 women), 9 part-time (2 women); includes 6 minority (2 African Americans, 3 Asian Americans or Pacific Islanders, 1 Hispanic American), 28 international. Average age 27. *82 applicants, 76% accepted.* In 1999, 16 master's, 10 doctorates awarded.
Degree requirements: For master's, thesis required, foreign language not required; for doctorate, one foreign language, dissertation required.
Entrance requirements: For master's and doctorate, GRE, TOEFL, TSE. *Application deadline:* For fall admission, 2/1 (priority date). *Application fee:* $45.
Expenses: Tuition, state resident: part-time $115 per credit hour. Tuition, nonresident: part-time $389 per credit hour. Required fees: $18 per semester. Tuition and fees vary according to program.
Financial aid: Research assistantships, teaching assistantships, tuition waivers (full) available.
Faculty research: Meteorite chemistry, structure of biopolymers, electron microprobe analysis of air pollutants, x-ray crystallography.
Dr. J. Devens Gust, Chair, 480-965-3461.

Application contact: Dr. Ana Moore, Director of Graduate Studies, 480-965-4664, *Fax:* 480-965-2747, *E-mail:* chmgrad@asu.edu.
Find an in-depth description at www.petersons.com/graduate.

■ BAYLOR COLLEGE OF MEDICINE

Graduate School of Biomedical Sciences, Department of Biochemistry, Houston, TX 77030-3498
AWARDS PhD, MD/PhD.

Faculty: 32 full-time (6 women).
Students: 40 full-time (21 women); includes 2 minority (both Asian Americans or Pacific Islanders), 27 international. Average age 27. *105 applicants, 16% accepted.* In 1999, 4 doctorates awarded.
Degree requirements: For doctorate, dissertation, public defense, qualifying exam required, foreign language not required. *Average time to degree:* Doctorate–7.02 years full-time.
Entrance requirements: For doctorate, GRE General Test (average 80th percentile), GRE Subject Test (strongly recommended), TOEFL, minimum GPA of 3.0. *Application deadline:* For fall admission, 2/1 (priority date). *Application fee:* $30. Electronic applications accepted.
Expenses: Tuition: Full-time $8,200. Required fees: $175. Full-time tuition and fees vary according to student level.
Financial aid: In 1999–00, 40 students received aid, including 20 fellowships (averaging $20,000 per year), 20 research assistantships (averaging $20,000 per year); Federal Work-Study, institutionally sponsored loans, and tuition waivers (full) also available. Financial aid applicants required to submit FAFSA.

Faculty research: Mechanisms of enzyme action, nucleic acid enzymology, and mutagenesis; biochemistry of connective tissue, proteins, and polysaccharides; chemical metabolism of lipids and lipoproteins.
Dr. Adam Kuspa, Director, 713-798-4527.
Application contact: Carol Rodriguez, Graduate Program Administrator, 713-798-4527, *Fax:* 713-796-9438, *E-mail:* carolr@bcm.tmc.edu.
Find an in-depth description at www.petersons.com/graduate.

■ BOSTON COLLEGE

Graduate School of Arts and Sciences, Department of Biology, Program in Biochemistry, Chestnut Hill, MA 02467-3800

AWARDS MS, PhD. Terminal master's awarded for partial completion of doctoral program.

Degree requirements: For master's and doctorate, thesis/dissertation required, foreign language not required.
Entrance requirements: For master's and doctorate, GRE General Test, GRE Subject Test. *Application deadline:* For fall admission, 2/1. *Application fee:* $40.
Expenses: Tuition: Part-time $656 per credit. Tuition and fees vary according to program.
Financial aid: Fellowships, research assistantships, teaching assistantships, Federal Work-Study and scholarships available. Aid available to part-time students. Financial aid application deadline: 3/15; financial aid applicants required to submit FAFSA.
Application contact: Dr. Thomas Chiles, Graduate Director, Admissions, 617-552-3540, *E-mail:* thomas.chiles@bc.edu.

■ BOSTON COLLEGE

Graduate School of Arts and Sciences, Department of Chemistry, Program in Biochemistry, Chestnut Hill, MA 02467-3800

AWARDS MS, PhD.

Degree requirements: For master's, thesis, comprehensive exam required; for doctorate, 2 foreign languages (computer language can substitute for one), dissertation, comprehensive exam required.
Entrance requirements: For master's and doctorate, GRE General Test, GRE Subject Test. *Application deadline:* For fall admission, 2/1. *Application fee:* $40.
Expenses: Tuition: Part-time $656 per credit. Tuition and fees vary according to program.
Financial aid: Fellowships, research assistantships, teaching assistantships, Federal Work-Study and tuition waivers (partial) available. Aid available to part-time students. Financial aid application deadline: 3/15.
Application contact: Dr. Mary Roberts, Graduate Program Director, 617-552-3616, *E-mail:* mary.roberts@bc.edu.

■ BOSTON UNIVERSITY

Graduate School of Arts and Sciences, Department of Chemistry, Boston, MA 02215

AWARDS Biochemistry (MA, PhD); chemical physics (MA, PhD); inorganic chemistry (MA, PhD); organic chemistry (MA, PhD); photochemistry (MA, PhD); physical chemistry (MA, PhD); theoretical chemistry (MA, PhD).

Faculty: 21 full-time (1 woman).
Students: 92 full-time (33 women); includes 4 minority (1 African American, 2 Asian Americans or Pacific Islanders, 1 Hispanic American), 62 international. Average age 27. *199 applicants, 37% accepted.* In 1999, 6 master's, 14 doctorates awarded. Terminal master's awarded for partial completion of doctoral program.
Degree requirements: For master's, one foreign language, thesis not required; for doctorate, one foreign language, dissertation, exams required. *Average time to degree:* Master's–3 years full-time; doctorate–5 years full-time.
Entrance requirements: For master's and doctorate, GRE General Test, GRE Subject Test (recommended), TOEFL. *Application deadline:* For fall admission, 7/1. Applications are processed on a rolling basis. *Application fee:* $50.
Expenses: Tuition: Full-time $23,770; part-time $743 per credit. Required fees: $220. Tuition and fees vary according to class time, course level, campus/location and program.

Financial aid: In 1999–00, 67 students received aid, including 2 fellowships with tuition reimbursements available (averaging $12,500 per year), 18 research assistantships with tuition reimbursements available (averaging $12,500 per year), 43 teaching assistantships with tuition reimbursements available (averaging $12,500 per year); Federal Work-Study, scholarships, and tuition waivers (full) also available. Aid available to part-time students. Financial aid application deadline: 1/15; financial aid applicants required to submit FAFSA.
Thomas D. Tullius, Chairman, 617-353-4277, *Fax:* 617-353-6466, *E-mail:* tullius@chem.bu.edu.
Application contact: Kevin Burgoyne, Academic Administrator, 617-353-2503, *Fax:* 617-353-6466, *E-mail:* burgoyne@chem.bu.edu.

■ BOSTON UNIVERSITY

Graduate School of Arts and Sciences, Molecular Biology, Cell Biology, and Biochemistry Program (MCBB), Boston, MA 02215

AWARDS MA, PhD. Part-time programs available.

Faculty: 45 full-time (10 women).
Students: 31 full-time (16 women), 2 part-time (1 woman); includes 7 minority (5 Asian Americans or Pacific Islanders, 2 Hispanic Americans), 10 international. Average age 26. *87 applicants, 30% accepted.* Terminal master's awarded for partial completion of doctoral program.
Degree requirements: For master's, one foreign language, thesis required (for some programs); for doctorate, one foreign language, dissertation, qualifying exam required. *Average time to degree:* Master's–2 years full-time.
Entrance requirements: For master's and doctorate, GRE General Test, GRE Subject Test, TOEFL. *Application deadline:* For fall admission, 1/1 (priority date). *Application fee:* $50.
Expenses: Tuition: Full-time $23,770; part-time $743 per credit. Required fees: $220. Tuition and fees vary according to class time, course level, campus/location and program.
Financial aid: In 1999–00, 16 students received aid, including 9 research assistantships with full tuition reimbursements available (averaging $11,500 per year), 7 teaching assistantships with full tuition reimbursements available (averaging $11,500 per year); fellowships, Federal Work-Study, scholarships, and traineeships also available. Financial aid application deadline: 1/15; financial aid applicants required to submit FAFSA.
Faculty research: Signal transduction, gene expression, protein and nucleic acid

biochemistry, genomics, modular physiology and development. *Total annual research expenditures:* $4.5 million.
Gary R. Jacobson, Director, 617-353-2432, *Fax:* 617-353-6340, *E-mail:* jacobson@bio.bu.edu.
Application contact: Michelle Brodkowitz, Academic Administrator, 617-353-2432, *Fax:* 617-353-6340, *E-mail:* mcbb@bio.bu.edu.

Find an in-depth description at www.petersons.com/graduate.

■ BOSTON UNIVERSITY

School of Medicine, Division of Graduate Medical Sciences, Department of Biochemistry, Boston, MA 02118

AWARDS MA, PhD, MD/PhD. Part-time programs available.

Faculty: 22 full-time (6 women), 29 part-time/adjunct (2 women).
Students: 48 full-time (21 women); includes 7 minority (all Asian Americans or Pacific Islanders), 14 international. Average age 27. In 1999, 11 master's, 7 doctorates awarded. Terminal master's awarded for partial completion of doctoral program.
Degree requirements: For master's, thesis or alternative, qualifying exam required, foreign language not required; for doctorate, dissertation, qualifying exam required, foreign language not required.
Entrance requirements: For master's and doctorate, GRE General Test, GRE Subject Test, TOEFL. *Application deadline:* For fall admission, 1/15 (priority date); for spring admission, 10/15 (priority date). *Application fee:* $50. Electronic applications accepted.
Expenses: Tuition: Full-time $24,700; part-time $772 per credit. Required fees: $220.
Financial aid: In 1999–00, 13 fellowships, 32 research assistantships were awarded; Federal Work-Study, scholarships, and traineeships also available.
Faculty research: Extracellular matrix, gene expression, receptors, growth control.
Application contact: Dr. Barbara Schreiber, 617-638-5094, *Fax:* 617-638-5339, *E-mail:* schreibe@biochem.bumc.bu.edu.

Find an in-depth description at www.petersons.com/graduate.

■ BRANDEIS UNIVERSITY

Graduate School of Arts and Sciences, Programs in Biological Sciences, Program in Biochemistry, Waltham, MA 02454-9110

AWARDS MS, PhD. Part-time programs available.

Faculty: 11 full-time (5 women).

Students: 28 full-time (12 women); includes 1 minority (Hispanic American), 13 international. Average age 24. *35 applicants, 11% accepted.* In 1999, 6 degrees awarded (100% entered university research/teaching).

Degree requirements: For doctorate, dissertation, area exams required.

Entrance requirements: For doctorate, GRE General Test. *Application deadline:* For fall admission, 1/15 (priority date). Applications are processed on a rolling basis. *Application fee:* $60. Electronic applications accepted.

Expenses: Tuition: Full-time $25,392; part-time $3,174 per course. Required fees: $509. Tuition and fees vary according to class time, degree level, program and student level.

Financial aid: Fellowships, research assistantships, teaching assistantships, career-related internships or fieldwork, scholarships, and tuition waivers (full and partial) available. Financial aid application deadline: 4/15; financial aid applicants required to submit CSS PROFILE or FAFSA.

Faculty research: Enzyme mechanisms, genetics, molecular developmental biology, structural biology, neurobiology.

Dr. Daniel Oprian, Chair, 781-736-2322.

Application contact: Dr. Chris Miller, Chair, 781-736-2340, *Fax:* 781-736-2365, *E-mail:* dr.miller@brandeis.edu.

■ **BRIGHAM YOUNG UNIVERSITY**

Graduate Studies, College of Physical and Mathematical Sciences, Department of Chemistry and Biochemistry, Provo, UT 84602-1001

AWARDS Analytical chemistry (MS, PhD); biochemistry (MS, PhD); inorganic chemistry (MS, PhD); organic chemistry (MS, PhD); physical chemistry (MS, PhD).

Faculty: 28 full-time (1 woman), 5 part-time/adjunct (0 women).

Students: 69 full-time (21 women); includes 2 minority (both Asian Americans or Pacific Islanders), 37 international. Average age 30. *231 applicants, 17% accepted.* In 1999, 7 master's, 9 doctorates awarded.

Degree requirements: For master's, thesis required, foreign language not required; for doctorate, dissertation, degree qualifying exam required, foreign language not required. *Average time to degree:* Master's–3 years full-time; doctorate–4.9 years full-time.

Entrance requirements: For master's and doctorate, chemistry entrance exam, minimum GPA of 3.0 in last 60 hours. *Application deadline:* For fall admission, 2/1 (priority date); for winter admission, 8/15. Applications are processed on a rolling basis. *Application fee:* $30.

Expenses: Tuition: Full-time $3,330; part-time $185 per credit hour. Tuition and fees vary according to program and student's religious affiliation.

Financial aid: In 1999–00, 69 students received aid, including 13 fellowships (averaging $15,468 per year), 25 research assistantships (averaging $15,468 per year), 31 teaching assistantships (averaging $15,468 per year); institutionally sponsored loans, scholarships, and tuition waivers (full) also available. Financial aid application deadline: 2/1.

Faculty research: Separation science, molecular recognition, organic synthesis and biomedical application, biochemistry and molecular biology, molecular spectroscopy. *Total annual research expenditures:* $3.9 million.

Dr. Francis R. Nordmeyer, Chair, 801-378-3667, *Fax:* 801-378-5474, *E-mail:* fran_nordmeyer@byu.edu.

Application contact: N. Kent Dalley, Coordinator, Graduate Admissions, 801-378-3434, *Fax:* 801-378-5474, *E-mail:* chemgrad@byu.edu.

Find an in-depth description at www.petersons.com/graduate.

■ **BROWN UNIVERSITY**

Graduate School, Department of Chemistry, Providence, RI 02912

AWARDS Biochemistry (PhD); chemistry (Sc M, PhD).

Degree requirements: For master's, thesis required, foreign language not required; for doctorate, dissertation, cumulative exam required.

■ **BROWN UNIVERSITY**

Graduate School, Division of Biology and Medicine, Program in Molecular Biology, Cell Biology, and Biochemistry, Providence, RI 02912

AWARDS Biochemistry (M Med Sc, Sc M, PhD), including biochemistry (Sc M, PhD), biology (Sc M, PhD), medical science (M Med Sc, PhD); biology (MA); cell biology (M Med Sc, Sc M, PhD), including biochemistry (Sc M, PhD), biology (Sc M, PhD), medical science (M Med Sc, PhD); developmental biology (M Med Sc, Sc M, PhD), including biochemistry (Sc M, PhD), biology (Sc M, PhD), medical science (M Med Sc, PhD); immunology (M Med Sc, Sc M, PhD), including biochemistry (Sc M, PhD), biology (Sc M, PhD), medical science (M Med Sc, PhD); molecular microbiology (M Med Sc, Sc M, PhD), including biochemistry (Sc M, PhD), biology (Sc M, PhD), medical science (M Med Sc, PhD). Part-time programs available.

Faculty: 50 full-time (14 women).

Students: 61 full-time (34 women); includes 4 minority (1 African American, 3 Asian Americans or Pacific Islanders), 21 international. Average age 25. *106 applicants, 28% accepted.* In 1999, 1 master's, 3 doctorates awarded. Terminal master's awarded for partial completion of doctoral program.

Degree requirements: For master's, thesis required (for some programs), foreign language not required; for doctorate, one foreign language, dissertation, preliminary exam required. *Average time to degree:* Doctorate–5 years full-time.

Entrance requirements: For master's and doctorate, GRE General Test, GRE Subject Test. *Application deadline:* For fall admission, 1/2 (priority date). Applications are processed on a rolling basis. *Application fee:* $60. Electronic applications accepted.

Financial aid: In 1999–00, 58 students received aid, including 11 fellowships (averaging $18,916 per year), 9 research assistantships (averaging $18,916 per year), 13 teaching assistantships (averaging $12,690 per year); institutionally sponsored loans and traineeships also available. Financial aid application deadline: 1/2.

Faculty research: Molecular genetics, gene regulation.

Dr. Gary Wessel, Director, 401-863-1051, *E-mail:* chet@brown.edu.

Application contact: Mary C. Esser, Graduate Program Coordinator, 401-863-1661, *Fax:* 401-863-1348, *E-mail:* mary_esser@brown.edu.

Find an in-depth description at www.petersons.com/graduate.

■ **BRYN MAWR COLLEGE**

Graduate School of Arts and Sciences, Department of Biology and Department of Chemistry, Program in Biochemistry, Bryn Mawr, PA 19010-2899

AWARDS MA, PhD.

Faculty: 4.

Degree requirements: For master's, thesis required; for doctorate, dissertation required.

Entrance requirements: For master's and doctorate, GRE General Test, GRE Subject Test. *Application deadline:* For fall admission, 6/30; for spring admission, 12/7. *Application fee:* $40.

Expenses: Tuition: Full-time $20,790; part-time $3,530 per course.

Financial aid: Application deadline: 1/2.

Application contact: Graduate School of Arts and Sciences, 610-526-5075.

■ BRYN MAWR COLLEGE

Graduate School of Arts and Sciences, Department of Chemistry, Bryn Mawr, PA 19010-2899

AWARDS Biochemistry (MA, PhD); chemistry (MA, PhD).

Students: 6 full-time (4 women), 3 part-time (1 woman), 4 international. *17 applicants, 65% accepted.* In 1999, 4 master's awarded.

Degree requirements: For master's, thesis required; for doctorate, dissertation required.

Entrance requirements: For master's and doctorate, GRE General Test, GRE Subject Test. *Application deadline:* For fall admission, 6/30; for spring admission, 12/7. *Application fee:* $40.

Expenses: Tuition: Full-time $20,790; part-time $3,530 per course.

Financial aid: In 1999–00, 5 teaching assistantships were awarded; fellowships, research assistantships, Federal Work-Study and institutionally sponsored loans also available. Aid available to part-time students. Financial aid application deadline: 1/2.

Dr. Sharon Burgmayer, Chairman, 610-526-5104.

Application contact: Graduate School of Arts and Sciences, 610-526-5072.

■ CALIFORNIA INSTITUTE OF TECHNOLOGY

Division of Biology and Division of Chemistry and Chemical Engineering, Biochemistry Graduate Option, Pasadena, CA 91125-0001

AWARDS PhD.

Faculty: 35 full-time (7 women).

Students: 26 full-time (7 women); includes 12 minority (1 African American, 10 Asian Americans or Pacific Islanders, 1 Hispanic American), 5 international. Average age 28. *131 applicants, 40% accepted.* In 1999, 1 degree awarded (100% entered university research/teaching).

Degree requirements: For doctorate, dissertation, qualifying exam required, foreign language not required.

Entrance requirements: For doctorate, GRE General Test. *Application deadline:* For fall admission, 1/1. *Application fee:* $0.

Expenses: Tuition: Full-time $19,260. Required fees: $24. One-time fee: $100 full-time.

Financial aid: In 1999–00, 11 fellowships with full tuition reimbursements (averaging $14,688 per year), 7 research assistantships with full tuition reimbursements (averaging $14,688 per year), 18 teaching assistantships with full tuition reimbursements (averaging $3,729 per year) were awarded. Financial aid application deadline: 1/1.

Application contact: Alison Ross, Graduate Option Coordinator, 626-395-6446, *Fax:* 626-564-9571, *E-mail:* biochemop@ cco.caltech.edu.

■ CALIFORNIA INSTITUTE OF TECHNOLOGY

Division of Chemistry and Chemical Engineering, Pasadena, CA 91125-0001

AWARDS Biochemistry (PhD); chemical engineering (MS, PhD); chemistry (PhD).

Faculty: 35 full-time (5 women).

Students: 249 full-time (64 women); includes 32 minority (3 African Americans, 22 Asian Americans or Pacific Islanders, 7 Hispanic Americans), 75 international. Average age 24. *732 applicants, 19% accepted.* In 1999, 24 master's awarded (17% entered university research/teaching, 46% found other work related to degree, 37% continued full-time study); 44 doctorates awarded (55% entered university research/teaching, 43% found other work related to degree, 2% continued full-time study). Terminal master's awarded for partial completion of doctoral program.

Degree requirements: For master's and doctorate, thesis/dissertation required, foreign language not required. *Average time to degree:* Master's–1.75 years full-time; doctorate–5.7 years full-time. *Application deadline:* For fall admission, 1/15. *Application fee:* $0.

Expenses: Tuition: Full-time $19,260. Required fees: $24. One-time fee: $100 full-time.

Financial aid: In 1999–00, 77 fellowships, 73 research assistantships, 65 teaching assistantships were awarded; Federal Work-Study and institutionally sponsored loans also available. Financial aid application deadline: 1/15.

Faculty research: Molecular structure and interactions in chemical and biological systems, theoretical studies of fundamental chemical processes, chemical reaction engineering, biochemical engineering, catalysis.

Peter B. Dervan, Chairman, 626-395-3646, *Fax:* 626-568-8824.

■ CALIFORNIA STATE UNIVERSITY, FULLERTON

Graduate Studies, College of Natural Science and Mathematics, Department of Chemistry and Biochemistry, Fullerton, CA 92834-9480

AWARDS Analytical chemistry (MS); biochemistry (MS); geochemistry (MS); inorganic chemistry (MS); organic chemistry (MS); physical chemistry (MS). Part-time programs available.

Faculty: 21 full-time (5 women), 11 part-time/adjunct.

Students: 4 full-time (3 women), 34 part-time (13 women); includes 15 minority (2 African Americans, 9 Asian Americans or Pacific Islanders, 3 Hispanic Americans, 1 Native American), 8 international. Average age 28. *33 applicants, 64% accepted.* In 1999, 4 degrees awarded.

Degree requirements: For master's, thesis, departmental qualifying exam required, foreign language not required.

Entrance requirements: For master's, minimum GPA of 2.5 in last 60 units, major in chemistry or related field. *Application fee:* $55.

Expenses: Tuition, nonresident: part-time $264 per unit. Required fees: $1,887; $629 per year.

Financial aid: Teaching assistantships, career-related internships or fieldwork, Federal Work-Study, grants, and institutionally sponsored loans available. Aid available to part-time students. Financial aid application deadline: 3/1.

Dr. John Olmsted, Chair, 714-278-3621.

Application contact: Dr. Gregory Williams, Adviser, 714-278-2170.

■ CALIFORNIA STATE UNIVERSITY, HAYWARD

Graduate Programs, School of Science, Department of Chemistry, Option in Biochemistry, Hayward, CA 94542-3000

AWARDS MS.

Degree requirements: For master's, comprehensive exam or thesis required.

Entrance requirements: For master's, minimum GPA of 2.5 in field during previous 2 years. *Application fee:* $55.

Expenses: Tuition, nonresident: part-time $164 per unit. Required fees: $587 per quarter.

Financial aid: Federal Work-Study and institutionally sponsored loans available. Aid available to part-time students. Financial aid application deadline: 3/1.

Dr. Charles Perrino, Coordinator, 510-885-3449.

Application contact: Jennifer Rice, Graduate Program Coordinator, 510-885-3286, *Fax:* 510-885-4795, *E-mail:* gradprograms@csuhayward.edu.

■ CALIFORNIA STATE UNIVERSITY, LONG BEACH

Graduate Studies, College of Natural Sciences, Department of Chemistry and Biochemistry, Program in Biochemistry, Long Beach, CA 90840

AWARDS MS. Part-time programs available.

Students: 7 full-time (6 women), 7 part-time; includes 6 minority (1 African

American, 5 Asian Americans or Pacific Islanders), 2 international. Average age 30. *22 applicants, 55% accepted.* In 1999, 2 degrees awarded.

Degree requirements: For master's, thesis, departmental qualifying exam required, foreign language not required. *Application deadline:* For fall admission, 8/1; for spring admission, 12/1. Applications are processed on a rolling basis. *Application fee:* $55. Electronic applications accepted.

Expenses: Tuition, nonresident: part-time $246 per credit. Required fees: $569 per semester. Tuition and fees vary according to course load.

Financial aid: Federal Work-Study, grants, and institutionally sponsored loans available. Financial aid application deadline: 3/2.

Dr. Jeffrey Cohlberg, Graduate Coordinator, 562-985-4944, *Fax:* 562-985-2315, *E-mail:* cohlberg@csulb.edu.

■ CALIFORNIA STATE UNIVERSITY, LOS ANGELES

Graduate Studies, School of Natural and Social Sciences, Department of Chemistry and Biochemistry, Option in Biochemistry, Los Angeles, CA 90032-8530

AWARDS MS. Part-time and evening/weekend programs available.

Students: In 1999, 2 degrees awarded.
Degree requirements: For master's, one foreign language (computer language can substitute), comprehensive exam or thesis required.
Entrance requirements: For master's, TOEFL. *Application deadline:* For fall admission, 6/30; for spring admission, 2/1. Applications are processed on a rolling basis. *Application fee:* $55.
Expenses: Tuition, nonresident: full-time $7,703; part-time $164 per unit. Required fees: $1,799; $387 per quarter.
Financial aid: In 1999–00, 1 student received aid. Federal Work-Study available. Aid available to part-time students. Financial aid application deadline: 3/1.
Faculty research: Biosynthesis of NAD in bacteria, kinetic and regulatory properties of enzymes, regulation of lipoprotein by dietary cholesterol.
Dr. Scott Grover, Acting Chair, Department of Chemistry and Biochemistry, 323-343-2300.

■ CARNEGIE MELLON UNIVERSITY

Mellon College of Science, Department of Biological Sciences, Pittsburgh, PA 15213-3891

AWARDS Biochemistry (PhD); biophysics (PhD); cell biology (PhD); developmental biology (PhD); genetics (PhD); molecular biology (PhD).

Faculty: 38 full-time (17 women), 1 (woman) part-time/adjunct.
Students: 35 full-time (24 women); includes 2 minority (1 African American, 1 Asian American or Pacific Islander), 18 international. Average age 27. In 1999, 4 degrees awarded.
Degree requirements: For doctorate, dissertation required, foreign language not required.
Entrance requirements: For doctorate, GRE General Test, GRE Subject Test, interview. *Application deadline:* For fall admission, 2/1 (priority date). Applications are processed on a rolling basis. *Application fee:* $0.
Expenses: Tuition: Full-time $22,100; part-time $307 per unit. Required fees: $200. Tuition and fees vary according to program.
Financial aid: Fellowships, research assistantships, teaching assistantships, traineeships available.
Faculty research: Genetic structure, function, and regulation; protein structure and function; biological membranes; biological spectroscopy. *Total annual research expenditures:* $4.4 million.
Dr. William E. Brown, Head, 412-268-3416, *Fax:* 412-268-7129, *E-mail:* wb02@andrew.cmu.edu.
Application contact: Stacey L. Young, Assistant Head, 412-268-7372, *Fax:* 412-268-7129, *E-mail:* sf38+@andrew.cmu.edu.

Find an in-depth description at www.petersons.com/graduate.

■ CASE WESTERN RESERVE UNIVERSITY

School of Medicine and School of Graduate Studies, Graduate Programs in Medicine, Department of Biochemistry, Cleveland, OH 44106

AWARDS Biochemical research (MS); biochemistry (MS, PhD). Part-time programs available.

Faculty: 17 full-time (2 women), 21 part-time/adjunct (5 women).
Students: 38 full-time (18 women); includes 3 minority (1 African American, 2 Hispanic Americans), 24 international. Average age 24. *800 applicants, 11% accepted.* In 1999, 3 master's, 3 doctorates awarded. Terminal master's awarded for partial completion of doctoral program.

Degree requirements: For master's, thesis required (for some programs), foreign language not required; for doctorate, dissertation required, foreign language not required. *Average time to degree:* Master's–3 years full-time; doctorate–5 years full-time.
Entrance requirements: For master's and doctorate, GRE General Test, GRE Subject Test, TOEFL. *Application deadline:* For fall admission, 3/1 (priority date). Applications are processed on a rolling basis. *Application fee:* $25.
Financial aid: In 1999–00, 38 fellowships with full tuition reimbursements (averaging $18,000 per year) were awarded; research assistantships with full tuition reimbursements, tuition waivers (full) also available.
Faculty research: Regulation of metabolism, regulation of gene expression and protein synthesis, cell biology, molecular biology, structural biology. *Total annual research expenditures:* $2.9 million.
Dr. Michael A. Weiss, Chairman, 216-368-5991, *Fax:* 216-368-3419, *E-mail:* maw21@po.cwru.edu.
Application contact: Dr. Pieter L. deHaseth, Chairman, Graduate Admissions Committee, 216-368-3684, *Fax:* 216-368-3419, *E-mail:* pld2@po.cwru.edu.

Find an in-depth description at www.petersons.com/graduate.

■ CASE WESTERN RESERVE UNIVERSITY

School of Medicine and School of Graduate Studies, Graduate Programs in Medicine, Department of Nutrition, Cleveland, OH 44106

AWARDS Dietetics (MS); nutrition (MS, PhD), including nutrition and biochemistry (PhD); public health nutrition (MS). Part-time programs available. Terminal master's awarded for partial completion of doctoral program.

Degree requirements: For master's, thesis required (for some programs), foreign language not required; for doctorate, dissertation required, foreign language not required.
Entrance requirements: For master's and doctorate, GRE General Test, GRE Subject Test, TOEFL.
Faculty research: Fatty acid metabolism, application of gene therapy to nutritional problems, dietary intake methodology, nutrition and physical fitness, metabolism during infancy and pregnancy.

Find an in-depth description at www.petersons.com/graduate.

■ CITY COLLEGE OF THE CITY UNIVERSITY OF NEW YORK

Graduate School, College of Liberal Arts and Science, Division of Science, Department of Chemistry, Program in Biochemistry, New York, NY 10031-9198

AWARDS MA, PhD.

Students: 9 (4 women). *15 applicants, 53% accepted.* In 1999, 5 degrees awarded. Terminal master's awarded for partial completion of doctoral program.
Degree requirements: For master's, foreign language and thesis not required; for doctorate, dissertation required.
Entrance requirements: For master's, TOEFL; for doctorate, GRE, TOEFL. *Application deadline:* For fall admission, 5/1; for spring admission, 12/1. *Application fee:* $40.
Expenses: Tuition, state resident: full-time $4,350; part-time $185 per credit. Tuition, nonresident: full-time $7,600; part-time $320 per credit. Required fees: $20 per semester.
Financial aid: Application deadline: 6/1.
Faculty research: Fatty acid metabolism, lectins, gene structure.
Hurst Schulz, Chairman, 212-650-8323.

■ CLARK UNIVERSITY

Graduate School, Concentration in Biochemistry/Molecular Biology, Worcester, MA 01610-1477

AWARDS MA, PhD.

Students: 16 full-time (7 women), 8 international. *24 applicants, 17% accepted.* In 1999, 2 master's, 3 doctorates awarded.
Degree requirements: For doctorate, dissertation required.
Entrance requirements: For master's and doctorate, GRE General Test, TOEFL. *Application deadline:* For fall admission, 2/15 (priority date). Applications are processed on a rolling basis. *Application fee:* $40.
Expenses: Tuition: Full-time $22,400; part-time $2,800 per course.
Faculty research: Molecular genetics, neurochemistry, protein chemistry.
Dr. David Thurlow, Head, 508-793-7173.
Application contact: Rene Baril, Department Secretary, 528-793-7173.

■ CLEMSON UNIVERSITY

Graduate School, College of Agriculture, Forestry and Life Sciences, School of Animal, Biomedical and Biological Sciences, Department of Biological Sciences, Program in Biochemistry, Clemson, SC 29634

AWARDS MS, PhD.

Students: 7 full-time (5 women), 1 (woman) part-time; includes 1 minority (Asian American or Pacific Islander), 4 international. *26 applicants, 12% accepted.* In 1999, 1 master's awarded.
Degree requirements: For master's, thesis required; for doctorate, dissertation, comprehensive exam required.
Entrance requirements: For master's and doctorate, GRE General Test, TOEFL. *Application deadline:* For fall admission, 6/1 (priority date). Applications are processed on a rolling basis. *Application fee:* $40.
Expenses: Tuition, state resident: full-time $3,480; part-time $174 per credit hour. Tuition, nonresident: full-time $9,256; part-time $388 per credit hour. Required fees: $5 per term. Full-time tuition and fees vary according to course level, course load and campus/location.
Financial aid: Fellowships, research assistantships, teaching assistantships available. Financial aid application deadline: 3/15; financial aid applicants required to submit FAFSA.
Faculty research: Biomembranes, protein structure, heme and tetrapyrrole biosynthesis in plants, molecular biology of plants, APYA and stress response. *Total annual research expenditures:* $670,000.

■ COLORADO STATE UNIVERSITY

Graduate School, College of Natural Sciences, Department of Biochemistry and Molecular Biology, Fort Collins, CO 80523-0015

AWARDS MS, PhD.

Faculty: 20 full-time (6 women), 2 part-time/adjunct (0 women).
Students: 18 full-time (9 women), 12 part-time (7 women); includes 1 minority (Asian American or Pacific Islander), 8 international. Average age 29. *181 applicants, 5% accepted.* In 1999, 14 master's, 4 doctorates awarded (100% entered university research/teaching). Terminal master's awarded for partial completion of doctoral program.
Degree requirements: For master's, thesis or alternative required, foreign language not required; for doctorate, dissertation required, foreign language not required.
Entrance requirements: For master's, GRE General Test, TOEFL, minimum GPA of 3.0; for doctorate, GRE General Test, TOEFL, minimum GPA of 3.2. *Application deadline:* For fall admission, 1/15 (priority date). Applications are processed on a rolling basis. *Application fee:* $30. Electronic applications accepted.
Expenses: Tuition, state resident: full-time $2,694; part-time $150 per credit. Tuition, nonresident: full-time $10,460; part-time

$581 per credit. Required fees: $32 per semester. Tuition and fees vary according to program.
Financial aid: In 1999–00, 1 fellowship, 21 research assistantships, 9 teaching assistantships were awarded; Federal Work-Study, institutionally sponsored loans, traineeships, and tuition waivers (partial) also available. Financial aid application deadline: 1/15; financial aid applicants required to submit FAFSA.
Faculty research: Cellular biology, molecular gene expression, structure and function of macromolecules, neurobiology, transcriptional control mechanisms.
Norman P. Curthoys, Chair, 970-491-5566, *Fax:* 970-491-0494, *E-mail:* ncurth@lamar.colostate.edu.
Application contact: Diane Keith, Graduate Recruitment Committee, 970-491-6841, *Fax:* 970-491-0494, *E-mail:* dkeith@vines.colostate.edu.

Find an in-depth description at www.petersons.com/graduate.

■ COLUMBIA UNIVERSITY

College of Physicians and Surgeons and Graduate School of Arts and Sciences, Graduate School of Arts and Sciences at the College of Physicians and Surgeons, Department of Biochemistry and Molecular Biophysics, New York, NY 10032

AWARDS Biochemistry and molecular biophysics (M Phil, PhD); biophysics (PhD). Only candidates for the PhD are admitted.

Degree requirements: For master's, foreign language and thesis not required; for doctorate, one foreign language, dissertation required.
Entrance requirements: For master's and doctorate, GRE General Test, TOEFL.
Expenses: Tuition: Full-time $25,072.

■ CORNELL UNIVERSITY

Graduate School, Graduate Fields of Agriculture and Life Sciences, Field of Biochemistry, Molecular and Cell Biology, Ithaca, NY 14853-0001

AWARDS Biochemistry (PhD); biophysics (PhD); cell biology (PhD); molecular and cell biology (PhD); molecular biology (PhD).

Faculty: 43 full-time.
Students: 75 full-time (37 women); includes 15 minority (1 African American, 6 Asian Americans or Pacific Islanders, 8 Hispanic Americans), 19 international. *288 applicants, 21% accepted.* In 1999, 9 doctorates awarded.
Degree requirements: For doctorate, dissertation, 2 semesters of teaching experience required.
Entrance requirements: For doctorate, GRE General Test, GRE Subject Test

(biology, chemistry, physics, biochemistry or, cell and molecular biology), TOEFL. *Application deadline:* For fall admission, 1/5. *Application fee:* $65. Electronic applications accepted.
Expenses: Tuition: Full-time $12,100.
Financial aid: In 1999–00, 74 students received aid, including 35 fellowships with full tuition reimbursements available, 32 research assistantships with full tuition reimbursements available, 7 teaching assistantships with full tuition reimbursements available; institutionally sponsored loans, scholarships, tuition waivers (full and partial), and unspecified assistantships also available. Financial aid applicants required to submit FAFSA.
Faculty research: Biophysics, structural biology.
Application contact: Graduate Field Assistant, 607-255-2317, *E-mail:* bmcb@cornell.edu.

Find an in-depth description at www.petersons.com/graduate.

■ **DARTMOUTH COLLEGE**

School of Arts and Sciences, Program in Biochemistry, Hanover, NH 03755
AWARDS PhD, MD/PhD.

Faculty: 31 full-time (4 women).
Students: 75 full-time (37 women); includes 10 minority (8 Asian Americans or Pacific Islanders, 2 Hispanic Americans), 17 international. Average age 28. *392 applicants, 14% accepted.* In 1999, 6 degrees awarded (17% entered university research/teaching, 83% continued full-time study).
Degree requirements: For doctorate, dissertation required, foreign language not required. *Average time to degree:* Doctorate–5 years full-time.
Entrance requirements: For doctorate, GRE General Test, GRE Subject Test, TOEFL. *Application deadline:* For fall admission, 1/5 (priority date). *Application fee:* $25. Electronic applications accepted.
Expenses: Tuition: Full-time $24,624. Required fees: $916. One-time fee: $15 full-time. Full-time tuition and fees vary according to program.
Financial aid: In 1999–00, 63 students received aid, including 25 fellowships with full tuition reimbursements available (averaging $16,750 per year), 30 research assistantships with full tuition reimbursements available (averaging $16,750 per year); Federal Work-Study, grants, scholarships, traineeships, and unspecified assistantships also available. Financial aid application deadline: 4/15.
Faculty research: Gene transcription, hormone regulation, cellular immunology, neurobiology, lipid metabolism. *Total annual research expenditures:* $2.8 million.

Dr. William T. Wickner, Chair, 603-650-1701, *Fax:* 603-650-1128.
Application contact: T. Y. Chang, Chair, Graduate Committee, 603-650-1622.

Find an in-depth description at www.petersons.com/graduate.

■ **DUKE UNIVERSITY**

Graduate School, Department of Biochemistry, Durham, NC 27708-0586
AWARDS Crystallography of macromolecules (PhD); enzyme mechanisms (PhD); lipid biochemistry (PhD); membrane structure and function (PhD); molecular genetics (PhD); neurochemistry (PhD); nucleic acid structure and function (PhD); protein structure and function (PhD).

Faculty: 31 full-time, 10 part-time/adjunct.
Students: 51 full-time (17 women); includes 3 minority (2 Asian Americans or Pacific Islanders, 1 Hispanic American), 10 international. *106 applicants, 19% accepted.* In 1999, 6 doctorates awarded.
Degree requirements: For doctorate, dissertation required, foreign language not required.
Entrance requirements: For doctorate, GRE General Test, GRE Subject Test (recommended). *Application deadline:* For fall admission, 12/31. *Application fee:* $75.
Expenses: Tuition: Full-time $21,406; part-time $760 per unit. Required fees: $3,136; $3,136 per year. One-time fee: $30. Tuition and fees vary according to program.
Financial aid: Fellowships, research assistantships, teaching assistantships, Federal Work-Study available. Financial aid application deadline: 12/31.
Terry Oas, Director of Graduate Studies, 919-681-8770, *Fax:* 919-684-8885, *E-mail:* pwilkison@biochem.duke.edu.

Find an in-depth description at www.petersons.com/graduate.

■ **DUKE UNIVERSITY**

Graduate School, Program in Biological Chemistry, Durham, NC 27708-0586
AWARDS Certificate. Students must be enrolled in a participating PhD program.

Faculty: 16 full-time.
Students: 2 full-time (1 woman); includes 1 minority (African American). *13 applicants, 54% accepted.*
Entrance requirements: For degree, GRE General Test, GRE Subject Test. *Application deadline:* For fall admission, 12/31. *Application fee:* $75.
Expenses: Tuition: Full-time $21,406; part-time $760 per unit. Required fees:

$3,136; $3,136 per year. One-time fee: $30. Tuition and fees vary according to program.
Financial aid: Application deadline: 12/31. Eric J. Toone, Director of Graduate Studies, 919-681-8825, *Fax:* 919-684-8346, *E-mail:* agw@biochem.duke.edu.

■ **DUQUESNE UNIVERSITY**

Bayer School of Natural and Environmental Sciences, Department of Chemistry and Biochemistry, Pittsburgh, PA 15282-0001
AWARDS Biochemistry (MS, PhD); chemistry (MS, PhD). Part-time programs available.

Faculty: 15 full-time (0 women), 6 part-time/adjunct (1 woman).
Students: 44 full-time (13 women), 1 part-time. Average age 27. *51 applicants, 37% accepted.* In 1999, 1 master's awarded (100% entered university research/teaching); 3 doctorates awarded (33% entered university research/teaching, 67% found other work related to degree). Terminal master's awarded for partial completion of doctoral program.
Degree requirements: For master's, thesis required (for some programs), foreign language not required; for doctorate, dissertation required, foreign language not required. *Average time to degree:* Master's–3.2 years full-time, 6 years part-time.
Entrance requirements: For master's and doctorate, GRE General Test, TOEFL, TSE (for international students seeking assistantships). *Application deadline:* For fall admission, 2/15 (priority date); for spring admission, 10/1. Applications are processed on a rolling basis. *Application fee:* $40.
Expenses: Tuition: Part-time $511 per credit. Required fees: $46 per credit. $50 per year. One-time fee: $125 part-time. Tuition and fees vary according to program.
Financial aid: In 1999–00, 1 fellowship with full tuition reimbursement (averaging $17,000 per year), 5 research assistantships with full tuition reimbursements (averaging $15,500 per year), 25 teaching assistantships with full tuition reimbursements (averaging $15,500 per year) were awarded; scholarships and tuition waivers (partial) also available. Financial aid application deadline: 5/15; financial aid applicants required to submit FAFSA.
Faculty research: Computational physical chemistry, bioinorganic chemistry, analytical chemistry, biophysics, synthetic organic chemistry. *Total annual research expenditures:* $1.4 million.
Dr. Thomas L. Isenhour, Chair, 412-396-6341, *Fax:* 412-396-5683, *E-mail:* isenhour@duq.edu.
Application contact: Mary Ann Quinn, Assistant to the Dean Graduate Affairs,

Duquesne University (continued)
412-396-6339, *Fax:* 412-396-4881, *E-mail:* gradinfo@duq.edu.

Find an in-depth description at www.petersons.com/graduate.

■ EAST CAROLINA UNIVERSITY

School of Medicine, Department of Biochemistry, Greenville, NC 27858-4353

AWARDS PhD.

Faculty: 9 full-time (0 women).
Students: 2 full-time (0 women), 6 part-time (2 women), 1 international. Average age 29. *3 applicants, 67% accepted.* In 1999, 2 degrees awarded.
Degree requirements: For doctorate, dissertation required, foreign language not required.
Entrance requirements: For doctorate, GRE General Test, GRE Subject Test, TOEFL. *Application deadline:* For fall admission, 6/1 (priority date). Applications are processed on a rolling basis. *Application fee:* $40.
Expenses: Tuition, state resident: full-time $1,012. Tuition, nonresident: full-time $8,578. Required fees: $1,006. Full-time tuition and fees vary according to degree level. Part-time tuition and fees vary according to course load.
Financial aid: Fellowships available. Financial aid application deadline: 6/1.
Faculty research: Metabolic regulation, metalloproteins, cellular differentiation and gene expression, prostaglandin biosynthesis, flavins and folate biochemistry.
Dr. Joseph Cory, Chairman, 252-816-2675, *Fax:* 252-816-3383, *E-mail:* cory@brody.med.ecu.edu.
Application contact: Dr. George Kasperek, Director of Graduate Studies, 252-816-2681, *Fax:* 252-816-3383, *E-mail:* kasperek@brody.med.ecu.edu.

■ EAST TENNESSEE STATE UNIVERSITY

James H. Quillen College of Medicine and School of Graduate Studies, Biomedical Science Graduate Program, Department of Biochemistry and Molecular Biology, Johnson City, TN 37614

AWARDS MS, PhD.

Faculty: 8 full-time (2 women), 6 part-time/adjunct (2 women).
Students: 9 full-time (3 women); includes 2 minority (both Asian Americans or Pacific Islanders). In 1999, 1 doctorate awarded.
Degree requirements: For master's, one foreign language (computer language can

substitute), thesis, comprehensive qualifying exam required; for doctorate, 2 foreign languages (computer language can substitute for one), dissertation required.
Entrance requirements: For master's, GRE General Test, minimum GPA of 3.0, bachelor's degree in biological or related science; for doctorate, GRE General Test, GRE Subject Test. *Application deadline:* For fall admission, 3/15. Applications are processed on a rolling basis. *Application fee:* $25 ($30 for international students).
Expenses: Tuition, state resident: full-time $10,342. Tuition, nonresident: full-time $21,080. Required fees: $532.
Financial aid: In 1999–00, 9 students received aid; research assistantships, teaching assistantships, institutionally sponsored loans and tuition waivers (full) available.
Faculty research: Molecular approach to the biochemistry of proteins, nucleic acids, and lipids.
Application contact: Dr. Mitchell Robinson, Director of Graduate Studies, 423-439-7157, *E-mail:* robinson@etsu.edu.

■ EMORY UNIVERSITY

Graduate School of Arts and Sciences, Division of Biological and Biomedical Sciences, Program in Biochemistry, Cell and Developmental Biology, Atlanta, GA 30322-1100

AWARDS PhD.

Faculty: 66 full-time (12 women).
Students: 52 full-time (33 women); includes 4 African Americans, 1 Hispanic American, 13 international. In 1999, 12 degrees awarded.
Degree requirements: For doctorate, dissertation required, foreign language not required.
Entrance requirements: For doctorate, GRE General Test, TOEFL, minimum GPA of 3.0 in science course work. *Application deadline:* For fall admission, 1/20 (priority date). *Application fee:* $45.
Expenses: Tuition: Full-time $22,770. Tuition and fees vary according to program.
Financial aid: In 1999–00, fellowships with full tuition reimbursements (averaging $18,000 per year).
Faculty research: Signal transduction, molecular biology, enzymes and cofactors, receptor and ion channel function, membrane biology.
Steven L'Hernault, Director, 404-727-3924, *Fax:* 404-727-3949, *E-mail:* jfridov@emory.edu.
Application contact: 404-727-2547, *Fax:* 404-727-3322, *E-mail:* gdbbs@gsas.emory.edu.

Find an in-depth description at www.petersons.com/graduate.

■ FINCH UNIVERSITY OF HEALTH SCIENCES/THE CHICAGO MEDICAL SCHOOL

School of Graduate and Postdoctoral Studies, Department of Biochemistry and Molecular Biology, Program in Biochemistry, North Chicago, IL 60064-3095

AWARDS MS, PhD, MD/MS, MD/PhD. Part-time programs available.

Faculty: 7 full-time.
Students: 13 full-time (6 women), 1 (woman) part-time; includes 3 minority (2 Asian Americans or Pacific Islanders, 1 Hispanic American), 4 international. In 1999, 3 master's, 2 doctorates awarded. Terminal master's awarded for partial completion of doctoral program.
Degree requirements: For master's, computer language, thesis, comprehensive exam required, foreign language not required; for doctorate, computer language, dissertation, comprehensive exam required.
Entrance requirements: For master's and doctorate, GRE General Test, TOEFL, TWE, minimum GPA of 3.0. *Application deadline:* For fall admission, 6/1 (priority date). Applications are processed on a rolling basis. *Application fee:* $25.
Expenses: Tuition: Full-time $14,054; part-time $391 per credit hour. Tuition and fees vary according to program.
Financial aid: In 1999–00, fellowships (averaging $15,500 per year); research assistantships, tuition waivers (full and partial) also available. Financial aid application deadline: 6/9; financial aid applicants required to submit FAFSA.
Application contact: Dana Frederick, Admissions Officer, 847-578-3209.

Find an in-depth description at www.petersons.com/graduate.

■ FLORIDA ATLANTIC UNIVERSITY

Charles E. Schmidt College of Science, Department of Chemistry and Biochemistry, Boca Raton, FL 33431-0991

AWARDS MS, MST. Part-time programs available.

Faculty: 13 full-time (2 women).
Students: 31 full-time (16 women), 9 part-time (1 woman); includes 8 minority (3 African Americans, 2 Asian Americans or Pacific Islanders, 3 Hispanic Americans), 12 international. Average age 31. *19 applicants, 74% accepted.* In 1999, 5 degrees awarded.
Degree requirements: For master's, thesis required, foreign language not required.

Entrance requirements: For master's, GRE General Test, minimum GPA of 3.0. *Application deadline:* For fall admission, 6/1. *Application fee:* $20.

Expenses: Tuition, state resident: full-time $2,663; part-time $148 per credit hour. Tuition, nonresident: full-time $9,156; part-time $509 per credit hour.

Financial aid: In 1999–00, 24 teaching assistantships were awarded; fellowships, research assistantships, Federal Work-Study also available.

Faculty research: Polymer synthesis and characterization, spectroscopy, geochemistry, environmental chemistry, biomedical chemistry. *Total annual research expenditures:* $50,000.
Dr. Donald Baird, Chair, 561-297-3390, *E-mail:* baird@fau.edu.

Application contact: Dr. Earl Baker, Professor, 561-297-3308, *Fax:* 561-297-2759, *E-mail:* baker@acc.fau.edu.

■ FLORIDA STATE UNIVERSITY

Graduate Studies, College of Arts and Sciences, Department of Chemistry, Specialization in Biochemistry, Tallahassee, FL 32306

AWARDS MS, PhD. Part-time programs available.

Faculty: 10 full-time (4 women).
Students: 18 full-time (7 women), 6 international. Average age 25. In 1999, 1 master's, 2 doctorates awarded. Terminal master's awarded for partial completion of doctoral program.

Degree requirements: For master's and doctorate, thesis/dissertation, cumulative and diagnostic exams required, foreign language not required. *Average time to degree:* Master's–3 years full-time; doctorate–4.6 years full-time.

Entrance requirements: For master's and doctorate, GRE General Test, minimum B average in undergraduate course work. *Application deadline:* For fall admission, 4/15. *Application fee:* $20.

Expenses: Tuition, state resident: full-time $3,504; part-time $146 per credit hour. Tuition, nonresident: full-time $12,162; part-time $507 per credit hour. Tuition and fees vary according to program.

Financial aid: In 1999–00, 5 research assistantships with tuition reimbursements (averaging $16,500 per year), 2 teaching assistantships with tuition reimbursements (averaging $16,500 per year) were awarded; fellowships, career-related internships or fieldwork, Federal Work-Study, and institutionally sponsored loans also available. Financial aid application deadline: 2/15; financial aid applicants required to submit FAFSA.

Faculty research: Metalloenzymes, gene regulation, DNA structure, NMR of synthetic membranes, secondary metabolites.

Application contact: Dr. Joseph Schlenoff, Chair, Graduate Admissions Committee, 850-644-3398, *Fax:* 850-644-8281, *E-mail:* gradinfo@chem.fsu.edu.

■ GEORGETOWN UNIVERSITY

Graduate School of Arts and Sciences, Department of Chemistry, Washington, DC 20057

AWARDS Analytical chemistry (MS, PhD); biochemistry (MS, PhD); chemical physics (MS, PhD); inorganic chemistry (MS, PhD); organic chemistry (MS, PhD); physical chemistry (MS, PhD); theoretical chemistry (MS, PhD). Terminal master's awarded for partial completion of doctoral program.

Degree requirements: For master's, thesis (for some programs), qualifying exam required, foreign language not required; for doctorate, dissertation, comprehensive exam required.
Entrance requirements: For master's and doctorate, GRE General Test, TOEFL.

Find an in-depth description at www.petersons.com/graduate.

■ GEORGETOWN UNIVERSITY

Graduate School of Arts and Sciences, Programs in Biomedical Sciences, Department of Biochemistry and Molecular Biology, Washington, DC 20057

AWARDS PhD, MD/PhD.

Degree requirements: For doctorate, dissertation, comprehensive exam required.
Entrance requirements: For doctorate, GRE General Test, TOEFL.

Find an in-depth description at www.petersons.com/graduate.

■ THE GEORGE WASHINGTON UNIVERSITY

Columbian School of Arts and Sciences, Department of Biochemistry and Molecular Biology, Washington, DC 20037

AWARDS MS, PhD, MD/PhD.

Faculty: 11 full-time (4 women).
Students: 3 full-time (2 women), 11 part-time (3 women); includes 4 minority (3 Asian Americans or Pacific Islanders, 1 Hispanic American), 4 international. Average age 29. *7 applicants, 100% accepted.* In 1999, 5 master's, 1 doctorate awarded.
Degree requirements: For master's, thesis (for some programs), comprehensive exam required; for doctorate, dissertation, general exam required.
Entrance requirements: For master's, GRE General Test, interview, minimum

GPA of 3.0; for doctorate, GRE General Test, minimum GPA of 3.0. *Application fee:* $55.

Expenses: Tuition: Full-time $16,836; part-time $702 per credit hour. Required fees: $828; $35 per credit hour. Tuition and fees vary according to campus/location and program.

Financial aid: In 1999–00, 6 students received aid, including 6 fellowships. Financial aid application deadline: 2/1. Dr. Allan L. Goldstein, Chair, 202-994-3517.

Application contact: Dr. Glenn Walker, Director of Graduate Studies, 202-994-2919.

Find an in-depth description at www.petersons.com/graduate.

■ THE GEORGE WASHINGTON UNIVERSITY

Columbian School of Arts and Sciences, Institute for Biomedical Sciences, Program in Biochemistry, Washington, DC 20052

AWARDS PhD.

Students: 3 full-time (2 women), 11 part-time (3 women); includes 4 minority (3 Asian Americans or Pacific Islanders, 1 Hispanic American), 4 international. Average age 28. *7 applicants, 100% accepted.* In 1999, 1 doctorate awarded.
Degree requirements: For doctorate, dissertation required.
Entrance requirements: For doctorate, GRE General Test, minimum GPA of 3.0. *Application fee:* $55.
Expenses: Tuition: Full-time $16,836; part-time $702 per credit hour. Required fees: $828; $35 per credit hour. Tuition and fees vary according to campus/location and program.
Financial aid: Fellowships available.
Application contact: 202-994-2179, *Fax:* 202-994-0967.

■ GEORGIA INSTITUTE OF TECHNOLOGY

Graduate Studies and Research, College of Sciences, School of Chemistry and Biochemistry, Atlanta, GA 30332-0001

AWARDS MS, MS Chem, PhD.

Faculty: 24 full-time (0 women), 6 part-time/adjunct (2 women).
Students: 109 full-time (47 women), 8 part-time (6 women); includes 26 minority (16 African Americans, 4 Asian Americans or Pacific Islanders, 5 Hispanic Americans, 1 Native American), 28 international. Average age 24. *69 applicants, 49% accepted.* In 1999, 22 master's awarded (100% found work related to degree); 6 doctorates

Georgia Institute of Technology (continued) awarded. Terminal master's awarded for partial completion of doctoral program.

Degree requirements: For master's, thesis required (for some programs), foreign language not required; for doctorate, dissertation required, foreign language not required.

Entrance requirements: For master's and doctorate, GRE General Test, GRE Subject Test, TOEFL, minimum GPA of 2.7. *Application deadline:* For fall admission, 8/1. Applications are processed on a rolling basis. *Application fee:* $50. Electronic applications accepted.

Financial aid: In 1999–00, 3 fellowships, 72 research assistantships, 45 teaching assistantships were awarded; career-related internships or fieldwork, Federal Work-Study, and institutionally sponsored loans also available. Aid available to part-time students. Financial aid application deadline: 2/15.

Faculty research: Inorganic, organic, physical, and analytical chemistry. *Total annual research expenditures:* $4.5 million.
Dr. Laren Tolbert, Chair, 404-894-4002.
Application contact: Dr. Larry Bottomley, Graduate Coordinator, 404-894-4014.

Find an in-depth description at www.petersons.com/graduate.

■ **GEORGIA STATE UNIVERSITY**

College of Arts and Sciences, Department of Biology, Program in Molecular Genetics and Biochemistry, Atlanta, GA 30303-3083

AWARDS MS, PhD.

Degree requirements: For master's, one foreign language (computer language can substitute), thesis or alternative, exam required; for doctorate, dissertation required.

Entrance requirements: For master's and doctorate, GRE General Test, TOEFL, minimum GPA of 3.0. *Application deadline:* For fall admission, 7/18; for spring admission, 2/13. Applications are processed on a rolling basis. *Application fee:* $25.

Expenses: Tuition, state resident: full-time $2,896; part-time $121 per credit hour. Tuition, nonresident: full-time $11,584; part-time $483 per credit hour. Required fees: $228. Full-time tuition and fees vary according to course load and program.

Financial aid: Application deadline: 2/6.
Application contact: Latesha Morrison, Graduate Administrative Coordinator, 404-651-2759, *Fax:* 404-651-2509, *E-mail:* biolxm@langate.gsu.edu.

Find an in-depth description at www.petersons.com/graduate.

■ **GRADUATE SCHOOL AND UNIVERSITY CENTER OF THE CITY UNIVERSITY OF NEW YORK**

Graduate Studies, Program in Biochemistry, New York, NY 10016-4039

AWARDS PhD.

Degree requirements: For doctorate, dissertation, field experience required, foreign language not required.

Entrance requirements: For doctorate, GRE General Test.

Expenses: Tuition, state resident: full-time $4,350; part-time $245 per credit hour. Tuition, nonresident: full-time $7,600; part-time $425 per credit hour.

■ **HARVARD UNIVERSITY**

Graduate School of Arts and Sciences, Department of Chemistry and Chemical Biology, Cambridge, MA 02138

AWARDS Biochemical chemistry (AM, PhD); inorganic chemistry (AM, PhD); organic chemistry (AM, PhD); physical chemistry (AM, PhD).

Students: 188 full-time (36 women). *346 applicants, 24% accepted.* In 1999, 20 master's, 33 doctorates awarded.

Degree requirements: For doctorate, dissertation, cumulative exams required, foreign language not required.

Entrance requirements: For master's, GRE General Test, TOEFL; for doctorate, GRE General Test, GRE Subject Test, TOEFL. *Application deadline:* For fall admission, 12/31. *Application fee:* $60.

Expenses: Tuition: Full-time $22,054. Required fees: $711. Tuition and fees vary according to program.

Financial aid: Fellowships, research assistantships, teaching assistantships, career-related internships or fieldwork, Federal Work-Study, and institutionally sponsored loans available. Financial aid application deadline: 12/30.
Josephine Ferraro, Officer, 617-495-5396.
Application contact: Graduate Admissions Office, 617-496-3208.

Find an in-depth description at www.petersons.com/graduate.

■ **HARVARD UNIVERSITY**

Graduate School of Arts and Sciences, Program in Biological and Biomedical Sciences, Department of Biological Chemistry and Molecular Pharmacology, Boston, MA 02115

AWARDS PhD. Applications through the Program in Biological and Biomedical Sciences (BBS).

Degree requirements: For doctorate, dissertation, qualifying exam required, foreign language not required.

Entrance requirements: For doctorate, GRE General Test, GRE Subject Test, TOEFL. *Application deadline:* For fall admission, 12/15. *Application fee:* $60.

Expenses: Tuition: Full-time $22,054. Required fees: $711. Tuition and fees vary according to program.

Financial aid: Fellowships, research assistantships, teaching assistantships, institutionally sponsored loans and tuition waivers (full) available. Financial aid application deadline: 1/1.

Faculty research: Cellular and molecular mechanisms of drug action with emphasis on basic approaches to chemotherapy, neuropharmacology, membrane biology, endocrinology, and toxicology; molecular mechanisms of receptor and drug enzyme interactions; genetics and molecular biology of DNA replication and transcription; molecular aspects of membrane protein function.
Dr. Kevin Struhl, Acting Chairman, 617-432-2104.
Application contact: Leah Simons, Manager of Student Affairs, 617-432-0162.

Find an in-depth description at www.petersons.com/graduate.

■ **HOWARD UNIVERSITY**

College of Medicine, Department of Biochemistry and Molecular Biology, Washington, DC 20059-0002

AWARDS Biochemistry and molecular biology (PhD); biotechnology (MS). Part-time programs available.

Faculty: 16.
Students: 10; includes 8 minority (6 African Americans, 2 Asian Americans or Pacific Islanders), 2 international. *15 applicants, 33% accepted.* In 1999, 1 doctorate awarded.

Degree requirements: For master's, externship required, foreign language and thesis not required; for doctorate, dissertation, oral and written comprehensive exams required, foreign language not required. *Average time to degree:* Doctorate–4 years full-time.

Entrance requirements: For master's and doctorate, GRE General Test, minimum GPA of 3.0. *Application deadline:* For fall admission, 4/1; for spring admission, 11/1. Applications are processed on a rolling basis. *Application fee:* $45.

Financial aid: In 1999–00, research assistantships with full tuition reimbursements (averaging $10,000 per year), teaching assistantships with full tuition reimbursements (averaging $10,000 per year) were awarded.

Faculty research: Cellular and molecular biology of olfaction, gene regulation and expression, enzymology, NMR spectroscopy of molecular structure, hormone regulation/metabolism. *Total annual research expenditures:* $3.4 million. Dr. Matthew George, Interim Chair, 202-806-6289, *Fax:* 202-806-5784, *E-mail:* mgeorge@howard.edu.
Application contact: Dr. Cynthia K. Abrams, Director of Graduate Studies, 202-806-6289, *Fax:* 202-806-5784, *E-mail:* cabrams@fac.howard.edu.

Find an in-depth description at www.petersons.com/graduate.

■ HOWARD UNIVERSITY

Graduate School of Arts and Sciences, Department of Chemistry, Washington, DC 20059-0002

AWARDS Analytical chemistry (MS, PhD); biochemistry (MS, PhD); inorganic chemistry (MS, PhD); organic chemistry (MS, PhD); physical chemistry (MS, PhD); polymer chemistry (MS, PhD); theoretical chemistry (MS, PhD). Part-time programs available.

Faculty: 25.
Students: 36; includes 30 minority (all African Americans), 6 international. Average age 28. *17 applicants, 6% accepted.* In 1999, 1 master's, 8 doctorates awarded.
Degree requirements: For master's, one foreign language, computer language, thesis, comprehensive exam, teaching experience required; for doctorate, 2 foreign languages, computer language, dissertation, comprehensive exam, teaching experience required. *Average time to degree:* Master's–2 years full-time, 4 years part-time; doctorate–4 years full-time, 7 years part-time.
Entrance requirements: For master's, GRE General Test, minimum GPA of 2.7; for doctorate, GRE General Test, minimum GPA of 3.0. *Application deadline:* For fall admission, 4/1 (priority date); for spring admission, 11/1. Applications are processed on a rolling basis. *Application fee:* $45.
Expenses: Tuition: Full-time $10,500; part-time $583 per credit hour. Required fees: $405; $203 per semester.
Financial aid: In 1999–00, 4 fellowships, 20 research assistantships, 16 teaching assistantships were awarded; grants and institutionally sponsored loans also available. Financial aid application deadline: 4/1.
Faculty research: Stratospheric aerosols, liquid crystals, polymer coatings, terrestrial and extraterrestrial atmospheres, amidogen reaction. *Total annual research expenditures:* $1.7 million.
Dr. Jesse M. Nicholson, Chairman, 202-806-6900.

Application contact: Dr. Marlene M. Sherrill, Director, Student Relations and Enrollment Management, 202-806-7469, *Fax:* 202-462-4053.

■ HUNTER COLLEGE OF THE CITY UNIVERSITY OF NEW YORK

Graduate School, Division of Sciences and Mathematics, Department of Chemistry, Program in Biochemistry, New York, NY 10021-5085

AWARDS MA. Part-time programs available.

Degree requirements: For master's, comprehensive exam or thesis required.
Entrance requirements: For master's, GRE General Test, TOEFL.
Expenses: Tuition, state resident: full-time $4,350; part-time $185 per credit. Tuition, nonresident: full-time $7,600; part-time $320 per credit. Required fees: $8 per term.
Faculty research: Protein/nucleic acid interactions, physical properties of iron-sulfur proteins, neurotransmitter receptors in *Drosophila*, requirements of DNA synthesis, oncogenes.

■ ILLINOIS INSTITUTE OF TECHNOLOGY

Graduate College, Armour College of Engineering and Sciences, Department of Biological, Chemical and Physical Sciences, Biology Division, Chicago, IL 60616-3793

AWARDS Biochemistry (MS); biology (PhD); biotechnology (MS); cell biology (MS); microbiology (MS). Part-time and evening/weekend programs available.

Faculty: 8 full-time (0 women), 2 part-time/adjunct (0 women).
Students: 38 full-time (15 women), 80 part-time (52 women); includes 34 minority (27 African Americans, 5 Asian Americans or Pacific Islanders, 2 Hispanic Americans), 39 international. *155 applicants, 36% accepted.* In 1999, 10 master's, 2 doctorates awarded. Terminal master's awarded for partial completion of doctoral program.
Degree requirements: For master's, thesis (for some programs), comprehensive exam required, foreign language not required; for doctorate, dissertation, comprehensive exam required, foreign language not required.
Entrance requirements: For master's and doctorate, GRE General Test, TOEFL, minimum undergraduate GPA of 3.0. *Application deadline:* For fall admission, 7/1; for spring admission, 11/1. Applications are processed on a rolling basis. *Application fee:* $30. Electronic applications accepted.

Expenses: Tuition: Part-time $590 per credit hour. Required fees: $100. Tuition and fees vary according to course load and program.
Financial aid: In 1999–00, 7 fellowships, 1 research assistantship, 5 teaching assistantships were awarded; Federal Work-Study, institutionally sponsored loans, scholarships, and unspecified assistantships also available. Aid available to part-time students. Financial aid application deadline: 3/1; financial aid applicants required to submit FAFSA.
Faculty research: Genetics, molecular biology.
Dr. Benjamin Stark, Associate Chair, 312-567-3980, *Fax:* 312-567-3494, *E-mail:* starkb@iit.edu.
Application contact: Dr. S. Mohammad Shahidehpour, Dean of Graduate College, 312-567-3024, *Fax:* 312-567-7517, *E-mail:* gradstu@alpha1.ais.iit.edu.

■ INDIANA UNIVERSITY BLOOMINGTON

Graduate School, College of Arts and Sciences, Department of Chemistry, Bloomington, IN 47405

AWARDS Analytical chemistry (PhD); biological chemistry (PhD); chemistry (MAT, MS); inorganic chemistry (PhD); physical chemistry (PhD). PhD offered through the University Graduate School.

Faculty: 25 full-time (0 women).
Students: 106 full-time (37 women), 61 part-time (7 women); includes 15 minority (1 African American, 13 Asian Americans or Pacific Islanders, 1 Hispanic American), 42 international. *309 applicants, 41% accepted.* In 1999, 9 master's, 17 doctorates awarded. Terminal master's awarded for partial completion of doctoral program.
Degree requirements: For master's and doctorate, thesis/dissertation required, foreign language not required. *Average time to degree:* Master's–2.2 years full-time; doctorate–5.8 years full-time.
Entrance requirements: For master's and doctorate, GRE General Test, GRE Subject Test, TOEFL. *Application deadline:* For fall admission, 1/15 (priority date); for spring admission, 9/1 (priority date). Applications are processed on a rolling basis. *Application fee:* $45.
Expenses: Tuition, state resident: full-time $3,853; part-time $161 per credit hour. Tuition, nonresident: full-time $11,226; part-time $468 per credit hour. Required fees: $360 per year. Tuition and fees vary according to course load and program.
Financial aid: In 1999–00, 23 fellowships with full tuition reimbursements (averaging $15,091 per year), 57 research assistantships with full tuition reimbursements (averaging $14,844 per year), 78 teaching

Indiana University Bloomington (continued)

assistantships with full tuition reimbursements (averaging $15,588 per year) were awarded; Federal Work-Study and institutionally sponsored loans also available.

Faculty research: Synthesis of complex natural products, organic reaction mechanisms, organic electrochemistry, transitive-metal chemistry, solid-state and surface chemistry. *Total annual research expenditures:* $7.7 million.

Dr. Gary M. Hieftje, Chairperson, 812-855-6239, *Fax:* 812-855-8300, *E-mail:* cemchair@indiana.edu.

Application contact: Dr. Jack K. Crandall, Chairperson of Admissions, 812-855-2068, *Fax:* 812-855-8300, *E-mail:* chemgrad@indiana.edu.

■ INDIANA UNIVERSITY BLOOMINGTON

Graduate School, College of Arts and Sciences, Interdepartmental Program in Biochemistry and Molecular Biology, Bloomington, IN 47405

AWARDS MS, PhD. PhD offered through the University Graduate School.

Students: 1 full-time (0 women), 1 (woman) part-time; includes 1 minority (Asian American or Pacific Islander). Terminal master's awarded for partial completion of doctoral program.

Degree requirements: For master's and doctorate, thesis/dissertation required, foreign language not required. *Average time to degree:* Doctorate–6.7 years full-time.

Entrance requirements: For doctorate, GRE General Test, GRE Subject Test (biochemistry or chemistry), TOEFL, BA or BE in biochemistry or chemistry. *Application deadline:* For fall admission, 1/15 (priority date); for spring admission, 9/1 (priority date). *Application fee:* $45.

Expenses: Tuition, state resident: full-time $3,853; part-time $161 per credit hour. Tuition, nonresident: full-time $11,226; part-time $468 per credit hour. Required fees: $360 per year. Tuition and fees vary according to course load and program.

Financial aid: In 1999–00, fellowships with full tuition reimbursements (averaging $15,750 per year), research assistantships with full tuition reimbursements (averaging $15,696 per year), teaching assistantships with full tuition reimbursements (averaging $15,720 per year) were awarded.

Faculty research: Biological membranes, enzymology, bioanalytical chemistry, photosynthesis.

Dr. John P. Richardson, Director, 812-855-1520, *E-mail:* richardj@indiana.edu.

Application contact: Dr. Jack K. Crandall, Chairperson of Admissions, 812-855-2068, *Fax:* 812-855-8300, *E-mail:* chemgrad@indiana.edu.

■ INDIANA UNIVERSITY–PURDUE UNIVERSITY INDIANAPOLIS

School of Medicine, Graduate Programs in Medicine, Department of Biochemistry and Molecular Biology, Indianapolis, IN 46202-2896

AWARDS MS, PhD, MD/MS, MD/PhD.

Students: 19 full-time (9 women), 19 part-time (8 women); includes 2 minority (1 African American, 1 Asian American or Pacific Islander), 25 international. Average age 28. *173 applicants, 16% accepted.* In 1999, 3 master's, 7 doctorates awarded. Terminal master's awarded for partial completion of doctoral program.

Degree requirements: For master's and doctorate, thesis/dissertation required, foreign language not required. *Average time to degree:* Master's–3.3 years full-time; doctorate–7 years full-time.

Entrance requirements: For master's and doctorate, GRE General Test, GRE Subject Test (recommended), previous course work in organic chemistry. *Application deadline:* For fall admission, 1/15 (priority date). Applications are processed on a rolling basis. *Application fee:* $35 ($55 for international students).

Expenses: Tuition, state resident: full-time $13,245; part-time $158 per credit hour. Tuition, nonresident: full-time $30,330; part-time $455 per credit hour. Required fees: $121 per year. Tuition and fees vary according to course load and degree level.

Financial aid: In 1999–00, 15 fellowships with tuition reimbursements (averaging $18,000 per year), 28 research assistantships with tuition reimbursements (averaging $18,000 per year) were awarded; teaching assistantships, Federal Work-Study, grants, institutionally sponsored loans, and tuition waivers (partial) also available. Aid available to part-time students. Financial aid application deadline: 2/1.

Faculty research: Metabolic regulation, enzymology, peptide and protein chemistry, cell biology, signal transduction, cancer, diabetes, structural biology.

Dr. Robert A. Harris, Chairman, 317-274-7151, *E-mail:* rharris@iupui.edu.

Application contact: Dr. David W. Allmann, Chairperson, Admissions Committee, 317-274-4096, *Fax:* 317-274-4686, *E-mail:* dallman@iupui.edu.

Find an in-depth description at www.petersons.com/graduate.

■ IOWA STATE UNIVERSITY OF SCIENCE AND TECHNOLOGY

Graduate College, College of Agriculture and College of Liberal Arts and Sciences, Department of Biochemistry, Biophysics, and Molecular Biology, Ames, IA 50011

AWARDS Biochemistry (MS, PhD); biophysics (MS, PhD); genetics (MS, PhD); molecular, cellular, and developmental biology (MS, PhD); toxicology (MS, PhD).

Faculty: 19 full-time, 1 part-time/adjunct.

Students: 57 full-time (24 women), 6 part-time (1 woman); includes 1 minority (African American), 20 international. *22 applicants, 59% accepted.* In 2000, 4 master's, 4 doctorates awarded.

Degree requirements: For master's and doctorate, thesis/dissertation required.

Entrance requirements: For master's and doctorate, GRE General Test, TOEFL. *Application deadline:* For fall admission, 6/15 (priority date); for spring admission, 11/15 (priority date). *Application fee:* $20 ($50 for international students). Electronic applications accepted.

Expenses: Tuition, state resident: full-time $3,308. Tuition, nonresident: full-time $9,744. Part-time tuition and fees vary according to course load, campus/location and program.

Financial aid: In 2000–01, 44 research assistantships with partial tuition reimbursements (averaging $12,314 per year), 2 teaching assistantships with partial tuition reimbursements (averaging $12,375 per year) were awarded; scholarships also available.

Dr. Marit Nilsen-Hamilton, Chair, 515-294-2231, *E-mail:* biochem@iastate.edu.

Find an in-depth description at www.petersons.com/graduate.

■ JOAN AND SANFORD I. WEILL MEDICAL COLLEGE AND GRADUATE SCHOOL OF MEDICAL SCIENCES OF CORNELL UNIVERSITY

Graduate School of Medical Sciences, Department of Biochemistry and Structural Biology, New York, NY 10021

AWARDS PhD, MD/PhD.

Faculty: 19 full-time (3 women).

Students: 14 full-time (9 women); includes 2 minority (1 African American, 1 Asian American or Pacific Islander), 5 international. *41 applicants, 15% accepted.* In 2000, 1 degree awarded.

Degree requirements: For doctorate, dissertation, final exam required.

Entrance requirements: For doctorate, GRE General Test, GRE Subject Test,

MCAT (MD/PhD). *Application deadline:* For fall admission, 1/15. *Application fee:* $50.

Expenses: All students in good standing receive an annual stipend of $22,880.

Financial aid: Fellowships, stipends available.

Dr. Timothy McGraw, Director, 212-746-4982.

■ **JOHNS HOPKINS UNIVERSITY**

School of Hygiene and Public Health, Department of Biochemistry and Molecular Biology, Baltimore, MD 21205

AWARDS Biochemistry (PhD); biophysics (PhD), including basic mechanisms of carcinogenesis, interferon induction and herpes simplex viral transformation of mammalian cells, nucleic acid structure, function, and interaction, structure and function of mammalian genetic apparatus, structure and interaction of biopolymers and cells; reproductive biology (MHS, Sc M, PhD).

Faculty: 25 full-time, 3 part-time/adjunct. **Students:** 6 full-time (3 women), 42 part-time (20 women); includes 7 minority (1 African American, 6 Asian Americans or Pacific Islanders), 15 international. *24 applicants, 50% accepted.* In 1999, 7 master's, 2 doctorates awarded.

Degree requirements: For master's, thesis required, foreign language not required; for doctorate, dissertation, 1 year full-time residency, oral and written exams required, foreign language not required.

Entrance requirements: For master's and doctorate, GRE General Test, TOEFL. *Application deadline:* For fall admission, 2/1 (priority date). Applications are processed on a rolling basis. *Application fee:* $60. Electronic applications accepted.

Expenses: Tuition: Full-time $23,660; part-time $493 per unit. Full-time tuition and fees vary according to degree level, campus/location and program.

Financial aid: Federal Work-Study, institutionally sponsored loans, scholarships, and stipends available. Aid available to part-time students. Financial aid application deadline: 4/15.

Faculty research: DNA replication, recombination, repair, mutation, and structure; glycoprotein synthesis, enzyme catalysis, carcinogenesis, and protein structure. *Total annual research expenditures:* $2.9 million.

Dr. Roger McMacken, Chairman, 410-955-3671, *E-mail:* rmcmacke@jhsph.edu.

Application contact: Gerry Graziano, Admissions Coordinator, 410-955-3671, *Fax:* 410-955-2926, *E-mail:* ggrazian@jhsph.edu.

Find an in-depth description at www.petersons.com/graduate.

■ **JOHNS HOPKINS UNIVERSITY**

School of Hygiene and Public Health, Program in Public Health, Baltimore, MD 21218-2699

AWARDS Biochemistry (MPH); biostatistics (MPH); environmental health sciences (MPH); epidemiology (MPH); health policy and management (MPH); international health (MPH); mental hygiene (MPH); molecular microbiology and immunology (MPH); population and family health sciences (MPH). Part-time and evening/weekend programs available. Postbaccalaureate distance learning degree programs offered.

Students: *694 applicants, 73% accepted.* In 1999, 196 degrees awarded.

Degree requirements: For master's, foreign language and thesis not required.

Entrance requirements: For master's, GRE General Test, TOEFL, 2 years of work related experience. *Application deadline:* For fall admission, 12/1 (priority date). Applications are processed on a rolling basis. *Application fee:* $60. Electronic applications accepted.

Expenses: Tuition: Full-time $23,660; part-time $493 per unit. Full-time tuition and fees vary according to degree level, campus/location and program.

Financial aid: Federal Work-Study, institutionally sponsored loans, and scholarships available. Aid available to part-time students. Financial aid application deadline: 4/15.

Dr. Miriam Alexander, Director, 410-955-1291, *Fax:* 410-955-4749.

Application contact: Lenora Davis, Administrator, 410-955-1291, *Fax:* 410-955-4749, *E-mail:* lrdavis@jhsph.edu.

■ **JOHNS HOPKINS UNIVERSITY**

School of Medicine, Graduate Programs in Medicine, Department of Biological Chemistry, Baltimore, MD 21205

AWARDS PhD.

Degree requirements: For doctorate, dissertation required, foreign language not required.

Entrance requirements: For doctorate, GRE General Test, GRE Subject Test. *Application deadline:* For fall admission, 10/1. *Application fee:* $50.

Expenses: Tuition: Full-time $23,660.

Financial aid: Application deadline: 1/1.

Dr. Denise Montell, Associate Professor, 410-955-1199, *Fax:* 410-955-5759.

Application contact: Jan Railey, Admissions Coordinator, 410-614-2976, *Fax:* 410-614-8375.

Find an in-depth description at www.petersons.com/graduate.

■ **JOHNS HOPKINS UNIVERSITY**

School of Medicine, Graduate Programs in Medicine, Program in Biochemistry, Cellular and Molecular Biology, Baltimore, MD 21205

AWARDS PhD.

Faculty: 83 full-time (23 women). **Students:** 173 full-time (83 women); includes 29 minority (2 African Americans, 24 Asian Americans or Pacific Islanders, 2 Hispanic Americans, 1 Native American), 46 international. Average age 25. *296 applicants, 23% accepted.* In 1999, 17 doctorates awarded.

Degree requirements: For doctorate, dissertation, comprehensive oral exams required, foreign language not required.

Entrance requirements: For doctorate, GRE General Test, GRE Subject Test. *Application deadline:* For fall admission, 1/15. Applications are processed on a rolling basis. *Application fee:* $50.

Expenses: Tuition: Full-time $23,660.

Financial aid: In 1999–00, 49 fellowships with full tuition reimbursements (averaging $16,760 per year), 124 research assistantships with full tuition reimbursements (averaging $16,760 per year) were awarded. Financial aid application deadline: 1/15.

Faculty research: DNA topology, protein folding, enzyme mechanisms and glycoproteins, bioenergetics, gene transcription.

Dr. Craig Montell, Director, 410-955-3225, *Fax:* 410-614-8842.

Application contact: Dr. Daniel Raben, Admissions Chairman, 410-955-3506, *Fax:* 410-614-8842, *E-mail:* pantol@jhmi.edu.

Find an in-depth description at www.petersons.com/graduate.

■ **JOHNS HOPKINS UNIVERSITY**

Zanvyl Krieger School of Arts and Sciences, Department of Biology, Baltimore, MD 21218-2699

AWARDS Biochemistry (PhD); cell biology (PhD); developmental biology (PhD); genetic biology (PhD); molecular biology (PhD).

Faculty: 23 full-time (4 women). **Students:** 85 full-time (50 women); includes 23 minority (4 African Americans, 14 Asian Americans or Pacific Islanders, 5 Hispanic Americans), 18 international. Average age 24. *202 applicants, 27% accepted.* In 1999, 17 doctorates awarded.

Degree requirements: For doctorate, dissertation required. *Average time to degree:* Doctorate–6 years full-time.

Entrance requirements: For doctorate, GRE General Test, GRE Subject Test. *Application deadline:* For fall admission, 12/15 (priority date). Applications are processed on a rolling basis. *Application fee:* $55.

Johns Hopkins University (continued)
Expenses: Tuition: Full-time $24,930. Tuition and fees vary according to program.

Financial aid: In 1999–00, 73 students received aid, including 12 fellowships, 61 research assistantships, 12 teaching assistantships; Federal Work-Study and institutionally sponsored loans also available. Financial aid application deadline: 12/15; financial aid applicants required to submit FAFSA.

Faculty research: Protein and nucleic acid biochemistry and biophysical chemistry, molecular biology and development. *Total annual research expenditures:* $9.3 million.
Dr. Victor G. Corces, Chair, 410-516-4693, *Fax:* 410-516-5213, *E-mail:* corces_v@jhuvms.hct.jhu.edu.

Application contact: Joan Miller, Graduate Admissions Coordinator, 410-516-5502, *Fax:* 410-516-5213, *E-mail:* joan@jhu.edu.
Find an in-depth description at www.petersons.com/graduate.

■ KANSAS STATE UNIVERSITY

Graduate School, College of Arts and Sciences, Department of Biochemistry, Manhattan, KS 66506
AWARDS MS, PhD.

Degree requirements: For master's and doctorate, thesis/dissertation required, foreign language not required.
Entrance requirements: For master's and doctorate, GRE General Test.
Expenses: Tuition, state resident: part-time $103 per credit hour. Tuition, nonresident: part-time $338 per credit hour. Required fees: $17 per credit hour. One-time fee: $64 part-time.
Faculty research: Protein structure, gene expression, enzyme mechanism, protease inhibitors.

■ KENT STATE UNIVERSITY

College of Arts and Sciences, Department of Chemistry, Kent, OH 44242-0001
AWARDS Analytical chemistry (MS, PhD); biochemistry (PhD); chemistry (MA, MS, PhD); inorganic chemistry (MS, PhD); organic chemistry (MS, PhD); physical chemistry (MS, PhD).

Faculty: 23 full-time.
Students: 33 full-time (20 women), 5 part-time (1 woman); includes 1 minority (African American), 19 international. *33 applicants, 91% accepted.* In 1999, 3 master's, 1 doctorate awarded.
Degree requirements: For master's and doctorate, thesis/dissertation required, foreign language not required.

Entrance requirements: For master's, minimum GPA of 2.75; for doctorate, minimum GPA of 3.0. *Application deadline:* For fall admission, 7/12; for spring admission, 11/29. Applications are processed on a rolling basis. *Application fee:* $30.
Expenses: Tuition, state resident: full-time $5,334; part-time $243 per hour. Tuition, nonresident: full-time $10,238; part-time $466 per hour.
Financial aid: Fellowships, research assistantships, teaching assistantships, Federal Work-Study, institutionally sponsored loans, and tuition waivers (full) available. Financial aid application deadline: 2/1.
Dr. Rathindra N. Bose, Chairman, 330-672-2032, *Fax:* 330-672-3816.
Find an in-depth description at www.petersons.com/graduate.

■ LEHIGH UNIVERSITY

College of Arts and Sciences, Department of Biological Sciences, Bethlehem, PA 18015-3094
AWARDS Behavioral and evolutionary bioscience (PhD); behavioral neuroscience (PhD); biochemistry (PhD); biology (PhD); molecular biology (PhD). Part-time programs available. Postbaccalaureate distance learning degree programs offered (no on-campus study).

Students: 31 full-time (22 women), 77 part-time (47 women); includes 11 minority (3 African Americans, 5 Asian Americans or Pacific Islanders, 3 Hispanic Americans), 8 international. *142 applicants, 22% accepted.* In 1999, 5 doctorates awarded.
Degree requirements: For doctorate, dissertation, comprehensive exam required, foreign language not required. *Average time to degree:* Doctorate–6.3 years full-time.
Entrance requirements: For doctorate, GRE General Test, GRE Subject Test, TOEFL. *Application deadline:* For fall admission, 7/15; for spring admission, 12/1. Applications are processed on a rolling basis. *Application fee:* $40. Electronic applications accepted.
Expenses: Tuition: Part-time $860 per credit. Required fees: $6 per term. Tuition and fees vary according to program.
Financial aid: In 1999–00, 30 students received aid, including 4 fellowships, 6 research assistantships, 15 teaching assistantships; career-related internships or fieldwork, institutionally sponsored loans, and tuition waivers (full and partial) also available. Financial aid application deadline: 1/15.
Faculty research: Gene expression, cell biology, virology, bacteriology, developmental biology.

Dr. Neal G. Simon, Chairperson, 610-758-3680, *Fax:* 610-758-4004.
Application contact: Dr. Jennifer J. Swann, Graduate Coordinator, 610-758-5884, *Fax:* 610-758-4004, *E-mail:* jms5@lehigh.edu.

■ LEHIGH UNIVERSITY

College of Arts and Sciences, Department of Chemistry, Bethlehem, PA 18015-3094
AWARDS Biochemistry and analytical chemistry (MS, PhD); chemistry (DA); clinical chemistry (MS); inorganic chemistry (MS, PhD); organic chemistry (MS, PhD); pharmaceutical chemistry (PhD); physical chemistry (MS, PhD). Part-time programs available. Postbaccalaureate distance learning degree programs offered (no on-campus study).

Students: 30 full-time (14 women), 92 part-time (46 women); includes 10 minority (7 Asian Americans or Pacific Islanders, 3 Hispanic Americans), 10 international. *42 applicants, 93% accepted.* In 1999, 18 master's, 3 doctorates awarded. Terminal master's awarded for partial completion of doctoral program.
Degree requirements: For master's, foreign language and thesis not required; for doctorate, dissertation required. *Average time to degree:* Master's–2.5 years full-time, 4 years part-time; doctorate–6 years full-time.
Entrance requirements: For master's and doctorate, GRE General Test, TSE. *Application deadline:* For fall admission, 7/15; for spring admission, 12/1. Applications are processed on a rolling basis. *Application fee:* $40.
Expenses: Tuition: Part-time $860 per credit. Required fees: $6 per term. Tuition and fees vary according to program.
Financial aid: In 1999–00, 26 students received aid, including 4 fellowships, 3 research assistantships, 16 teaching assistantships; career-related internships or fieldwork and institutionally sponsored loans also available. Financial aid application deadline: 1/15.
Dr. Keith J. Schray, Chairman, 610-758-3474, *Fax:* 610-758-6536, *E-mail:* kjs0@lehigh.edu.
Application contact: Dr. James E. Roberts, Graduate Coordinator, 610-758-4847, *Fax:* 610-758-6536, *E-mail:* jer1@lehigh.edu.

■ LOMA LINDA UNIVERSITY

Graduate School, Graduate Programs in Medicine, Department of Biochemistry, Loma Linda, CA 92350
AWARDS MS, PhD. Part-time programs available.

Degree requirements: For master's, thesis or alternative required; for doctorate, dissertation required.

Entrance requirements: For master's and doctorate, GRE General Test. *Application deadline:* Applications are processed on a rolling basis. *Application fee:* $40.

Expenses: Tuition: Part-time $395 per unit.

Financial aid: Tuition waivers (full and partial) available. Aid available to part-time students.

Faculty research: Physical chemistry of macromolecules, biochemistry of endocrine system, biochemical mechanism of bone volume regulation.

Dr. Charles Slattery, Coordinator, 909-824-4527.

■ LOUISIANA STATE UNIVERSITY AND AGRICULTURAL AND MECHANICAL COLLEGE

Graduate School, College of Basic Sciences, Department of Biological Sciences, Baton Rouge, LA 70803

AWARDS Biochemistry (MS, PhD); microbiology (MS, PhD); plant biology (MS, PhD); zoology (MS, PhD). Part-time programs available.

Faculty: 79 full-time (7 women), 2 part-time/adjunct (0 women).

Students: 91 full-time (44 women), 22 part-time (10 women); includes 9 minority (3 African Americans, 1 Asian American or Pacific Islander, 5 Hispanic Americans), 29 international. Average age 28. *98 applicants, 28% accepted.* In 1999, 13 master's, 7 doctorates awarded. Terminal master's awarded for partial completion of doctoral program.

Degree requirements: For master's, foreign language not required; for doctorate, dissertation required.

Entrance requirements: For master's and doctorate, GRE General Test, minimum GPA of 3.0. *Application deadline:* Applications are processed on a rolling basis. *Application fee:* $25.

Expenses: Tuition, state resident: full-time $2,881. Tuition, nonresident: full-time $7,081. Part-time tuition and fees vary according to course load and program.

Financial aid: In 1999–00, 12 fellowships, 20 research assistantships with partial tuition reimbursements, 62 teaching assistantships with partial tuition reimbursements were awarded; Federal Work-Study, institutionally sponsored loans, and unspecified assistantships also available. Aid available to part-time students. *Total annual research expenditures:* $2.2 million.

Dr. Harold Silverman, Chairman, 225-388-2601, *Fax:* 225-388-2597, *E-mail:* cxsiv@unix1.sncc.lsu.edu.

Find an in-depth description at www.petersons.com/graduate.

■ LOUISIANA STATE UNIVERSITY HEALTH SCIENCES CENTER

School of Graduate Studies in New Orleans, Department of Biochemistry and Molecular Biology, New Orleans, LA 70112-2223

AWARDS MS, PhD, MD/PhD.

Faculty: 12 full-time (2 women), 8 part-time/adjunct (2 women).

Students: 13 full-time (5 women); includes 5 minority (2 African Americans, 2 Asian Americans or Pacific Islanders, 1 Hispanic American), 2 international. Average age 29. *69 applicants, 10% accepted.* Terminal master's awarded for partial completion of doctoral program.

Degree requirements: For master's and doctorate, thesis/dissertation required, foreign language not required.

Entrance requirements: For master's and doctorate, GRE General Test, TOEFL, previous course work in physics, chemistry, and calculus. *Application deadline:* For fall admission, 6/1 (priority date); for spring admission, 11/1. Applications are processed on a rolling basis. *Application fee:* $30.

Expenses: Tuition, state resident: full-time $2,878; part-time $126 per hour. Tuition, nonresident: full-time $6,003; part-time $265 per hour. Required fees: $2,272. Tuition and fees vary according to course load, degree level and program.

Financial aid: In 1999–00, 1 fellowship with full tuition reimbursement, research assistantships with full tuition reimbursements (averaging $16,000 per year), 10 teaching assistantships with full tuition reimbursements (averaging $14,500 per year) were awarded; scholarships and tuition waivers (full) also available. Financial aid application deadline: 4/1.

Faculty research: Signal transduction; enzymology; gene structure, regulation, and cloning; cancer biology. *Total annual research expenditures:* $2.2 million.

Dr. Robert Roskoski, Head, 504-619-8568, *Fax:* 504-619-8775, *E-mail:* biocrr@lsumc.edu.

Application contact: Mildred Williams, Administrative Assistant, 504-619-8568, *Fax:* 504-619-8775, *E-mail:* biocmcw@lsumc.edu.

Find an in-depth description at www.petersons.com/graduate.

■ LOUISIANA STATE UNIVERSITY HEALTH SCIENCES CENTER

School of Graduate Studies in Shreveport, Department of Biochemistry and Molecular Biology, Shreveport, LA 71130-3932

AWARDS MS, PhD.

Degree requirements: For master's and doctorate, thesis/dissertation required, foreign language not required.

Entrance requirements: For master's and doctorate, GRE General Test, TOEFL.

Expenses: Tuition, state resident: full-time $2,878; part-time $126 per hour. Tuition, nonresident: full-time $6,003; part-time $265 per hour. Required fees: $2,272. Tuition and fees vary according to course load, degree level and program.

Faculty research: Metabolite transport, regulation of translation and transcription, procaryotic molecular genetics, cell matrix biochemistry, yeast molecular genetics, oncogenes.

■ LOYOLA UNIVERSITY CHICAGO

Graduate School, Department of Molecular and Cellular Biochemistry, Chicago, IL 60611-2196

AWARDS Biochemistry (MS, PhD); molecular biology (PhD); neurochemistry (PhD).

Faculty: 12 full-time (2 women), 1 (woman) part-time/adjunct.

Students: 16 full-time (9 women); includes 3 minority (1 African American, 2 Asian Americans or Pacific Islanders), 3 international. Average age 29. *40 applicants, 33% accepted.* In 1999, 1 master's awarded (100% found work related to degree); 2 doctorates awarded (100% entered university research/teaching).

Degree requirements: For master's, oral and written reports required, foreign language and thesis not required; for doctorate, dissertation, oral and written comprehensive exams required, foreign language not required. *Average time to degree:* Master's–2 years full-time; doctorate–4.5 years full-time.

Entrance requirements: For master's and doctorate, GRE General Test. *Application deadline:* For fall admission, 2/15 (priority date). Applications are processed on a rolling basis. *Application fee:* $35. Electronic applications accepted.

Expenses: Tuition: Part-time $500 per credit hour. Required fees: $42 per term.

Financial aid: In 1999–00, 11 students received aid, including 10 fellowships with full tuition reimbursements available, 4 research assistantships with full tuition reimbursements available; Federal Work-Study, institutionally sponsored loans, and

Loyola University Chicago (continued) scholarships also available. Financial aid application deadline: 3/15.

Faculty research: Molecular oncology; molecular neurochemical mechanisms of brain development and alcohol addiction; biochemistry of RNA and protein synthesis and intracellular protein degradation; developmentally regulated genes; cell membranes, neurotransmitters, and cell-cell interactions.

Dr. Richard M Schultz, Chairman, 708-216-3360, *Fax:* 708-216-8523.

Application contact: Dr. Michael A. Collins, Admissions Committee, 708-216-3361, *Fax:* 708-216-8523, *E-mail:* mcollin@luc.edu.

Find an in-depth description at www.petersons.com/graduate.

■ MASSACHUSETTS INSTITUTE OF TECHNOLOGY

School of Science, Department of Biology, Program in Biochemistry, Cambridge, MA 02139-4307

AWARDS PhD.

Degree requirements: For doctorate, dissertation, general exam required, foreign language not required.

Entrance requirements: For doctorate, GRE General Test. *Application deadline:* For fall admission, 1/1. *Application fee:* $55.

Expenses: Tuition: Full-time $25,000. Full-time tuition and fees vary according to degree level, program and student level.

Financial aid: Fellowships, research assistantships available. Financial aid application deadline: 1/1.

Application contact: Dr. Janice D. Chang, Educational Administrator, 617-253-3717, *Fax:* 617-258-9329, *E-mail:* gradbio@mit.edu.

■ MASSACHUSETTS INSTITUTE OF TECHNOLOGY

School of Science, Department of Chemistry, Cambridge, MA 02139-4307

AWARDS Biochemistry (PhD); biological chemistry (PhD, Sc D); inorganic chemistry (PhD, Sc D); organic chemistry (PhD, Sc D); physical chemistry (PhD, Sc D).

Faculty: 30 full-time (4 women).

Students: 200 full-time (74 women); includes 24 minority (3 African Americans, 6 Asian Americans or Pacific Islanders, 14 Hispanic Americans, 1 Native American), 66 international. Average age 26. *469 applicants, 36% accepted.* In 1999, 41 doctorates awarded.

Degree requirements: For doctorate, dissertation, comprehensive oral exam, oral presentation, written exams required, foreign language not required. *Average time to degree:* Doctorate–5 years full-time.

Application deadline: For fall admission, 1/15. *Application fee:* $55.

Expenses: Tuition: Full-time $25,000. Full-time tuition and fees vary according to degree level, program and student level.

Financial aid: In 1999–00, 30 fellowships with full tuition reimbursements (averaging $15,930 per year), 108 research assistantships with full tuition reimbursements (averaging $17,640 per year), 53 teaching assistantships with full tuition reimbursements (averaging $18,990 per year) were awarded; career-related internships or fieldwork and traineeships also available. Financial aid application deadline: 1/14.

Faculty research: Synthetic organic chemistry, enzymatic reaction mechanisms, inorganic and organometallic spectroscopy, high resolution NMR spectroscopy. *Total annual research expenditures:* $11.9 million.

Dr. Stephen J. Lippard, Chairman, 617-253-1845, *Fax:* 617-258-7500.

Application contact: Susan Brighton, Graduate Administrator, 617-253-1845, *Fax:* 617-258-0241, *E-mail:* brighton@mit.edu.

■ MAYO GRADUATE SCHOOL

Graduate Programs in Biomedical Sciences, Program in Biochemistry and Molecular Biology, Rochester, MN 55905

AWARDS Biochemistry (PhD); molecular biology (PhD).

Faculty: 70 full-time (10 women).

Students: 33 full-time (19 women); includes 3 minority (1 Asian American or Pacific Islander, 2 Hispanic Americans), 7 international. In 1999, 6 degrees awarded.

Degree requirements: For doctorate, oral defense of dissertation, qualifying oral and written exam required.

Entrance requirements: For doctorate, GRE, TOEFL, 2 years of chemistry; 1 year of biology, calculus, and physics. *Application deadline:* For fall admission, 12/31 (priority date). Applications are processed on a rolling basis. *Application fee:* $0.

Expenses: Tuition: Full-time $17,900.

Financial aid: In 1999–00, 28 students received aid, including 28 fellowships with full tuition reimbursements available; tuition waivers (full) also available.

Faculty research: Gene structure and function, membranes and receptors/cytoskeleton, oncogenes and growth factors, protein structure and function, steroid hormonal action.

Dr. Edward B. Leof, Education Coordinator, 507-284-5717, *E-mail:* leof.edward@mayo.edu.

Application contact: Sherry Kallies, Information Contact, 507-266-0122, *Fax:* 507-284-0999, *E-mail:* phd.training@mayo.edu.

Find an in-depth description at www.petersons.com/graduate.

■ MCP HAHNEMANN UNIVERSITY

School of Medicine, Biomedical Graduate Programs, Department of Biochemistry, Program in Biochemistry, Philadelphia, PA 19102-1192

AWARDS MS, PhD, MD/PhD. Part-time programs available. Terminal master's awarded for partial completion of doctoral program.

Degree requirements: For master's, thesis, comprehensive exam required, foreign language not required; for doctorate, dissertation, qualifying exam required, foreign language not required.

Entrance requirements: For master's, GRE General Test, TOEFL, minimum GPA of 2.75; for doctorate, GRE General Test, TOEFL, minimum GPA of 3.0.

Find an in-depth description at www.petersons.com/graduate.

■ MEDICAL COLLEGE OF GEORGIA

School of Graduate Studies, Department of Biochemistry and Molecular Biology, Augusta, GA 30912-1500

AWARDS MS, PhD.

Faculty: 14 full-time (1 woman).

Students: 9 full-time (2 women), 1 (woman) part-time, 5 international. *9 applicants, 33% accepted.* In 1999, 6 degrees awarded. Terminal master's awarded for partial completion of doctoral program.

Degree requirements: For master's and doctorate, thesis/dissertation required, foreign language not required.

Entrance requirements: For master's and doctorate, GRE General Test, TOEFL. *Application deadline:* For fall admission, 6/30 (priority date). Applications are processed on a rolling basis. *Application fee:* $25.

Expenses: Tuition, state resident: full-time $2,896; part-time $121 per hour. Tuition, nonresident: full-time $11,584; part-time $483 per hour. Required fees: $286; $143 per semester. Tuition and fees vary according to program.

Financial aid: In 1999–00, 9 research assistantships with partial tuition reimbursements (averaging $15,500 per year) were awarded; grants and institutionally sponsored loans also available. Aid available to part-time students. Financial

aid application deadline: 3/31; financial aid applicants required to submit FAFSA.
Faculty research: Cancer biology, chemical biology, chemical kinetics, physical methods of structure determination, physical biochemistry.
Dr. Frederick H. Leibach, Chair, 706-721-7661, *Fax:* 706-721-9947, *E-mail:* fleibach@mail.mcg.edu.
Application contact: Dr. Eugene Howard, Director, 706-721-7647, *Fax:* 706-721-9947, *E-mail:* ehoward@mail.mcg.edu.

■ MEDICAL COLLEGE OF OHIO

Graduate School, Department of Biochemistry and Molecular Biology, Toledo, OH 43614-5805

AWARDS Medical sciences (MS), including biochemistry, genetics, molecular biology. Part-time programs available.

Faculty: 8 full-time (2 women).
Students: 4 full-time (2 women), 8 part-time (1 woman), 9 international. Average age 30. In 1999, 3 degrees awarded (100% found work related to degree).
Degree requirements: For master's, thesis, qualifying exam required, foreign language not required. *Average time to degree:* Master's–3 years part-time.
Entrance requirements: For master's, GRE General Test, TOEFL, minimum undergraduate GPA of 3.0. *Application fee:* $30.
Expenses: Tuition, state resident: part-time $193 per hour. Tuition, nonresident: part-time $445 per hour. Tuition and fees vary according to degree level.
Financial aid: Fellowships, Federal Work-Study and institutionally sponsored loans available. Financial aid applicants required to submit FAFSA.
Faculty research: Gene regulation, protein structure, receptors, protein phosphorylation, peptides. *Total annual research expenditures:* $81,820.
James P. Trempe, Interim Chairman, 419-383-4117, *Fax:* 419-383-6140, *E-mail:* mcogradschool@mco.edu.
Application contact: Joann Braatz, Clerk, 419-383-4117, *Fax:* 419-383-6140, *E-mail:* mcogradschool@mco.edu.

■ MEDICAL COLLEGE OF WISCONSIN

Graduate School of Biomedical Sciences, Department of Biochemistry, Milwaukee, WI 53226-0509

AWARDS MS, PhD. Terminal master's awarded for partial completion of doctoral program.

Degree requirements: For master's and doctorate, thesis/dissertation required, foreign language not required.

Entrance requirements: For master's and doctorate, GRE General Test, GRE Subject Test, TOEFL.
Expenses: Tuition, state resident: full-time $9,318. Tuition, nonresident: full-time $9,318. Required fees: $115.
Faculty research: Enzymology, macromolecular structure and synthesis, nucleic acids, molecular and cell biology.

■ MEDICAL UNIVERSITY OF SOUTH CAROLINA

College of Graduate Studies, Department of Biochemistry and Molecular Biology, Charleston, SC 29425-0002

AWARDS MS, PhD, MD/PhD.

Faculty: 14 part-time/adjunct (4 women).
Students: 17 full-time (3 women); includes 3 minority (2 African Americans, 1 Hispanic American), 7 international. Average age 31. *15 applicants, 67% accepted.* In 1999, 1 degree awarded. Terminal master's awarded for partial completion of doctoral program.
Degree requirements: For master's, thesis, research seminar required, foreign language not required; for doctorate, dissertation, teaching and research seminar, oral and written exams required, foreign language not required.
Entrance requirements: For master's and doctorate, GRE General Test, TOEFL, interview. *Application deadline:* Applications are processed on a rolling basis. *Application fee:* $55. Electronic applications accepted.
Expenses: Tuition, state resident: full-time $3,470; part-time $160 per semester hour. Tuition, nonresident: full-time $4,426; part-time $213 per semester hour. Required fees: $408 per semester. One-time fee: $160. Tuition and fees vary according to program.
Financial aid: In 1999–00, 5 students received aid, including 2 fellowships (averaging $16,000 per year), 3 research assistantships; teaching assistantships, Federal Work-Study and tuition waivers (partial) also available. Financial aid application deadline: 4/1; financial aid applicants required to submit FAFSA.
Faculty research: Protein chemistry, polypeptide hormones, gene expression, development, antineoplastic agents. *Total annual research expenditures:* $2.5 million.
Dr. Y. A. Hannun, Chairman, 843-792-2331.
Application contact: Julie Johnston, Director of Admissions, 843-792-8710, *Fax:* 843-792-3764.

■ MEHARRY MEDICAL COLLEGE

School of Graduate Studies, Department of Biochemistry, Nashville, TN 37208-9989

AWARDS PhD.

Faculty: 7 full-time (1 woman), 1 part-time/adjunct (0 women).
Students: 4 full-time (2 women); all minorities (all African Americans). Average age 26. *9 applicants, 67% accepted.* In 1999, 2 degrees awarded.
Degree requirements: For doctorate, dissertation, oral and written comprehensive exams required, foreign language not required. *Average time to degree:* Doctorate–6 years full-time.
Entrance requirements: For doctorate, GRE. *Application deadline:* For fall admission, 6/1. Applications are processed on a rolling basis. *Application fee:* $45.
Expenses: Tuition: Full-time $8,732. Required fees: $2,133.
Financial aid: Fellowships available. Financial aid application deadline: 4/15.
Faculty research: Regulation of metabolism, enzymology, signal transduction, physical biochemistry.
Dr. Samuel Adunyah, Chair, 615-327-6345, *Fax:* 615-327-6440, *E-mail:* sadymya@mmc.edu.

■ MIAMI UNIVERSITY

Graduate School, College of Arts and Sciences, Department of Chemistry and Biochemistry, Oxford, OH 45056

AWARDS Analytical chemistry (MS, PhD); biochemistry (MS, PhD); chemical education (MS, PhD); chemistry (MS, PhD); inorganic chemistry (MS, PhD); organic chemistry (MS, PhD); physical chemistry (MS, PhD). Part-time programs available.

Faculty: 24 full-time (2 women).
Students: 42 full-time (17 women), 3 part-time (1 woman); includes 6 minority (2 African Americans, 4 Asian Americans or Pacific Islanders), 19 international. *119 applicants, 98% accepted.* In 1999, 4 master's, 5 doctorates awarded.
Degree requirements: For master's, thesis, final exam required; for doctorate, dissertation, comprehensive and final exams required.
Entrance requirements: For master's, minimum undergraduate GPA of 3.0 during previous 2 years or 2.75 overall; for doctorate, minimum GPA of 2.75 (undergraduate), 3.0 (graduate). *Application deadline:* For fall admission, 3/1. *Application fee:* $35.
Expenses: Tuition, state resident: part-time $260 per hour. Tuition, nonresident: full-time $3,125; part-time $538 per hour. International tuition: $6,452 full-time.

Miami University (continued)
Required fees: $18 per semester. Tuition and fees vary according to campus/location.

Financial aid: In 1999–00, 26 fellowships, 2 research assistantships, 18 teaching assistantships were awarded; Federal Work-Study and tuition waivers (full) also available. Financial aid application deadline: 3/1.

Dr. Michael Novak, Chair, 513-529-2813.

Application contact: Dr. Chris Makaroff, Director of Graduate Studies, 513-529-2813, *Fax:* 513-529-5715, *E-mail:* chemistry@muohio.edu.

Find an in-depth description at www.petersons.com/graduate.

■ MICHIGAN STATE UNIVERSITY

College of Human Medicine and Graduate School, Graduate Programs in Human Medicine, East Lansing, MI 48824

AWARDS Biochemistry (MS, PhD), including biochemistry, biochemistry-environmental toxicology (PhD); epidemiology (MS); human pathology (MS, PhD); microbiology (MS, PhD); pharmacology/toxicology (MS, PhD); physiology (MS, PhD); surgery (MS). Part-time programs available.

Students: 58 (32 women); includes 10 minority (2 African Americans, 6 Asian Americans or Pacific Islanders, 1 Hispanic American, 1 Native American) 16 international. *133 applicants, 28% accepted.* In 1999, 4 master's, 4 doctorates awarded.

Entrance requirements: For master's and doctorate, GRE General Test, minimum GPA of 3.0. *Application deadline:* Applications are processed on a rolling basis. *Application fee:* $30 ($40 for international students). Electronic applications accepted.

Expenses: Tuition, state resident: full-time $10,868; part-time $229 per credit. Tuition, nonresident: full-time $23,168; part-time $464 per credit.

Financial aid: Fellowships with tuition reimbursements, research assistantships with tuition reimbursements, teaching assistantships with tuition reimbursements, institutionally sponsored loans available. Aid available to part-time students. Financial aid applicants required to submit FAFSA.

Application contact: Dr. Lynda Farquhar, Assistant to the Dean for Research, 517-353-8858, *Fax:* 517-432-0021, *E-mail:* farquhal@pilot.msu.edu.

■ MICHIGAN STATE UNIVERSITY

College of Osteopathic Medicine and Graduate School, Graduate Programs in Osteopathic Medicine, East Lansing, MI 48824

AWARDS Anatomy (MS, PhD); biochemistry (MS, PhD); microbiology (PhD); pathology (MS, PhD); pharmacology/toxicology (MS, PhD), including pharmacology; physiology (MS, PhD), including environmental toxicology (PhD), neuroscience (PhD). Part-time programs available.

Students: 17 full-time (9 women), 1 part-time; includes 5 minority (3 Asian Americans or Pacific Islanders, 1 Hispanic American, 1 Native American), 5 international. Average age 26. *33 applicants, 9% accepted.* In 1999, 2 doctorates awarded.

Degree requirements: For doctorate, dissertation required.

Entrance requirements: For master's and doctorate, GRE. *Application deadline:* For fall admission, 3/1 (priority date). Applications are processed on a rolling basis. *Application fee:* $30 ($40 for international students).

Expenses: Tuition, area resident: Full-time $15,879. Tuition, state resident: full-time $33,797.

Financial aid: Fellowships, research assistantships, teaching assistantships, career-related internships or fieldwork, Federal Work-Study, and institutionally sponsored loans available. Financial aid application deadline: 4/2. *Total annual research expenditures:* $5 million.

Application contact: Dr. Veronica M. Maher, Associate Dean for Graduate Studies, 517-353-7785, *Fax:* 517-353-9004, *E-mail:* maher@com.msu.edu.

■ MICHIGAN STATE UNIVERSITY

Graduate School, College of Natural Science and Graduate Programs in Human Medicine and Graduate Programs in Osteopathic Medicine, Department of Biochemistry, East Lansing, MI 48824

AWARDS Biochemistry (MS, PhD); biochemistry-environmental toxicology (PhD).

Faculty: 27.

Students: 31 full-time (9 women), 9 part-time (5 women); includes 5 minority (2 African Americans, 2 Asian Americans or Pacific Islanders, 1 Hispanic American), 20 international. Average age 28. *80 applicants, 14% accepted.* In 1999, 2 master's, 11 doctorates awarded.

Entrance requirements: For master's, GRE, TOEFL; for doctorate, GRE, TOEFL, MS. *Application deadline:* Applications are processed on a rolling basis. *Application fee:* $30 ($40 for international students).

Expenses: Tuition, state resident: part-time $229 per credit. Tuition, nonresident: part-time $464 per credit. Required fees: $241 per semester. Tuition and fees vary according to course load, degree level and program.

Financial aid: In 1999–00, 40 research assistantships with tuition reimbursements (averaging $12,690 per year) were awarded. Financial aid applicants required to submit FAFSA.

Faculty research: Membrane protein structure, gene activity and regulation, human disease processes, tumor suppressor activity. *Total annual research expenditures:* $6.4 million.

Dr. William L. Smith, Chairperson, 517-355-1600, *Fax:* 517-353-9334.

Find an in-depth description at www.petersons.com/graduate.

■ MISSISSIPPI STATE UNIVERSITY

College of Agriculture and Life Sciences, Department of Biochemistry and Molecular Biology, Mississippi State, MS 39762

AWARDS Biochemistry (MS); molecular biology (PhD).

Faculty: 8 full-time (1 woman), 3 part-time/adjunct (all women).

Students: 10 full-time (5 women), 3 part-time (1 woman); includes 1 minority (African American), 8 international. Average age 27. *42 applicants, 17% accepted.* In 1999, 1 master's, 2 doctorates awarded. Terminal master's awarded for partial completion of doctoral program.

Degree requirements: For master's, thesis, comprehensive oral or written exam required, foreign language not required; for doctorate, dissertation, comprehensive oral and written exam required, foreign language not required.

Entrance requirements: For master's, GRE General Test, TOEFL, minimum GPA of 2.75; for doctorate, GRE, TOEFL. *Application deadline:* For fall admission, 7/1; for spring admission, 11/1. Applications are processed on a rolling basis. *Application fee:* $25 for international students.

Expenses: Tuition, state resident: full-time $3,017; part-time $168 per credit. Tuition, nonresident: full-time $6,119; part-time $340 per credit. Part-time tuition and fees vary according to course load and program.

Financial aid: Federal Work-Study, institutionally sponsored loans, and unspecified assistantships available. Financial aid applicants required to submit FAFSA.

Faculty research: Fish nutrition, plant and animal molecular biology, plant biochemistry, enzymology, lipid metabolism. *Total annual research expenditures:* $50,000.

Dr. John A. Boyle, Head, 662-325-2640, *Fax:* 662-325-8664, *E-mail:* jab@ra.msstate.edu.

Application contact: Jerry B. Inmon, Director of Admissions, 662-325-2224, *Fax:* 662-325-7360, *E-mail:* admit@admissions.msstate.edu.

■ MONTANA STATE UNIVERSITY–BOZEMAN

College of Graduate Studies, College of Letters and Science, Department of Chemistry, Bozeman, MT 59717

AWARDS Biochemistry (MS, PhD); chemistry (MS, PhD). Part-time programs available.

Students: 43 full-time (12 women), 26 part-time (5 women); includes 3 minority (1 Asian American or Pacific Islander, 1 Hispanic American, 1 Native American). Average age 28. *22 applicants, 100% accepted.* In 1999, 4 master's, 6 doctorates awarded.

Degree requirements: For master's and doctorate, thesis/dissertation required, foreign language not required.

Entrance requirements: For master's and doctorate, GRE General Test, TOEFL, minimum GPA of 3.0. *Application deadline:* For fall admission, 6/1; for spring admission, 11/1. Applications are processed on a rolling basis. *Application fee:* $50. Electronic applications accepted.

Expenses: Tuition, state resident: full-time $2,674. Tuition, nonresident: full-time $6,986. International tuition: $7,136 full-time. Tuition and fees vary according to course load and program.

Financial aid: In 1999–00, 69 students received aid, including research assistantships with tuition reimbursements available (averaging $16,000 per year), teaching assistantships with tuition reimbursements available (averaging $16,000 per year); career-related internships or fieldwork, Federal Work-Study, and scholarships also available. Financial aid application deadline: 3/1; financial aid applicants required to submit FAFSA.

Faculty research: Protein structure and function, organic synthesis, optical materials, biochemistry, metallo-proteins. *Total annual research expenditures:* $3.1 million.

Dr. Paul Grieco, Interim Head, 406-994-4801, *Fax:* 406-994-5407, *E-mail:* pgrieco@montana.edu.

■ MOUNT SINAI SCHOOL OF MEDICINE OF NEW YORK UNIVERSITY

Graduate School of Biological Sciences, Molecular, Cellular, Biochemical and Developmental Sciences (MCBDS) Training Area, New York, NY 10029-6504

AWARDS PhD, MD/PhD.

Students: 48 full-time (18 women).

Degree requirements: For doctorate, dissertation required, foreign language not required.

Entrance requirements: For doctorate, GRE General Test, GRE Subject Test, MCAT, TOEFL. *Application deadline:* For fall admission, 4/15. *Application fee:* $60.

Expenses: Tuition: Full-time $21,750; part-time $725 per credit. Required fees: $750; $25 per credit. Full-time tuition and fees vary according to student level.

Financial aid: Fellowships with full tuition reimbursements, grants available.

Dr. Gillian Small, Program Director, *E-mail:* small@msvax.mssm.edu.

Application contact: C. Gita Bosch, Administrative Manager and Assistant Dean, 212-241-6546, *Fax:* 212-241-0651, *E-mail:* grads@mssm.edu.

■ NEW MEXICO INSTITUTE OF MINING AND TECHNOLOGY

Graduate Studies, Department of Chemistry, Socorro, NM 87801

AWARDS Biochemistry (MS); chemistry (MS); environmental chemistry (PhD); explosives technology and atmospheric chemistry (PhD).

Faculty: 7 full-time (1 woman).

Students: 17 full-time (4 women), 1 part-time; includes 2 minority (both Hispanic Americans), 7 international. Average age 30. *61 applicants, 34% accepted.* In 1999, 5 master's, 1 doctorate awarded.

Degree requirements: For master's and doctorate, thesis/dissertation required, foreign language not required. *Average time to degree:* Master's–4 years full-time; doctorate–7 years full-time.

Entrance requirements: For master's, GRE General Test, TOEFL; for doctorate, GRE General Test, GRE Subject Test, TOEFL. *Application deadline:* For fall admission, 3/1 (priority date); for spring admission, 6/1. Applications are processed on a rolling basis. *Application fee:* $16.

Expenses: Tuition, state resident: full-time $1,693; part-time $94 per credit hour. Tuition, nonresident: full-time $6,978; part-time $388 per credit hour. Required fees: $727; $79 per credit hour. Tuition and fees vary according to course load.

Financial aid: In 1999–00, 5 research assistantships (averaging $9,670 per year), 10 teaching assistantships (averaging $9,670 per year) were awarded; fellowships, Federal Work-Study and institutionally sponsored loans also available. Financial aid application deadline: 3/1; financial aid applicants required to submit CSS PROFILE or FAFSA.

Faculty research: Organic, analytical, environmental, and explosives chemistry.

Dr. Lawrence Werbelow, Chairman, 505-835-5263, *Fax:* 505-835-5364, *E-mail:* werbelow@jupiter.nmt.edu.

Application contact: Dr. David B. Johnson, Dean of Graduate Studies, 505-835-5513, *Fax:* 505-835-5476, *E-mail:* graduate@nmt.edu.

■ NEW MEXICO STATE UNIVERSITY

Graduate School, College of Arts and Sciences, Department of Chemistry and Biochemistry, Las Cruces, NM 88003-8001

AWARDS MS, PhD. Part-time programs available.

Faculty: 22.

Students: 45 full-time (8 women), 3 part-time (2 women); includes 4 minority (3 Hispanic Americans, 1 Native American), 25 international. Average age 30. *43 applicants, 42% accepted.* In 1999, 4 master's, 6 doctorates awarded.

Degree requirements: For master's and doctorate, thesis/dissertation required, foreign language not required.

Entrance requirements: For master's and doctorate, GRE, TOEFL, BS in chemistry or biochemistry, minimum GPA of 3.0. *Application deadline:* For fall admission, 7/1 (priority date); for spring admission, 11/1. Applications are processed on a rolling basis. *Application fee:* $15 ($35 for international students).

Expenses: Tuition, state resident: full-time $2,682; part-time $112 per credit. Tuition, nonresident: full-time $8,376; part-time $349 per credit. Tuition and fees vary according to course load.

Financial aid: Fellowships, research assistantships, teaching assistantships, career-related internships or fieldwork and Federal Work-Study available. Aid available to part-time students. Financial aid application deadline: 3/1.

Faculty research: Clays, surfaces, and water structure; electroanalytical and environmental chemistry; organometallic synthesis and organobiomimetics; molecular genetics and enzymology of stress; spectroscopy and reaction kinetics.

Dr. Wolfgang Mueller, Head, 505-646-5877, *Fax:* 505-646-2649, *E-mail:* wmueller@nmsu.edu.

Application contact: Dr. Jeff Arterburn, Assistant Professor, Chemistry, 505-646-2738, *Fax:* 505-646-2649.

■ NEW YORK MEDICAL COLLEGE

Graduate School of Basic Medical Sciences, Program in Biochemistry and Molecular Biology, Valhalla, NY 10595-1691

AWARDS MS, PhD, MD/PhD. Part-time and evening/weekend programs available.

Faculty: 9 full-time (4 women).
Students: 17 full-time (10 women), 2 part-time (both women); includes 3 minority (1 African American, 2 Asian Americans or Pacific Islanders), 7 international. Average age 30. *22 applicants, 36% accepted.* In 1999, 1 doctorate awarded. Terminal master's awarded for partial completion of doctoral program.
Degree requirements: For master's and doctorate, computer language, thesis/dissertation required, foreign language not required.
Entrance requirements: For master's, GRE General Test, TOEFL; for doctorate, GRE General Test, GRE Subject Test, TOEFL. *Application deadline:* For fall admission, 7/1 (priority date); for spring admission, 12/1 (priority date). Applications are processed on a rolling basis. *Application fee:* $35 ($60 for international students).
Expenses: Tuition: Part-time $430 per credit. Required fees: $15 per semester. One-time fee: $100.
Financial aid: In 1999–00, 14 research assistantships with full tuition reimbursements were awarded; career-related internships or fieldwork, Federal Work-Study, grants, institutionally sponsored loans, and tuition waivers (full) also available. Aid available to part-time students. Financial aid applicants required to submit FAFSA.
Faculty research: Mechanisms of control of blood coagulation, molecular neurobiology, molecular probes for infectious disease, protein-DNA interactions, molecular biology and biochemistry of double-stranded RNA-dependent enzymes.
Dr. Joseph Wu, Director, 914-594-4062.

■ NEW YORK UNIVERSITY

Graduate School of Arts and Science, Department of Basic Medical Sciences, New York, NY 10012-1019

AWARDS Biochemistry (MS, PhD); cell biology (MS, PhD); microbiology (MS, PhD); parasitology (PhD); pathology (MS, PhD); pharmacology (PhD); physiology (MS, PhD). Part-time programs available.

Faculty: 23 full-time (2 women), 3 part-time/adjunct (1 woman).
Students: 182 full-time (73 women), 4 part-time (1 woman); includes 52 minority (11 African Americans, 34 Asian Americans or Pacific Islanders, 7 Hispanic Americans), 48 international. Average age 26. *671 applicants, 11% accepted.* In 1999, 15 master's, 32 doctorates awarded. Terminal master's awarded for partial completion of doctoral program.
Degree requirements: For master's, thesis or alternative, written comprehensive exam required, foreign language not required; for doctorate, one foreign language, dissertation, oral and written comprehensive exams required.
Entrance requirements: For master's and doctorate, GRE General Test, GRE Subject Test, TOEFL. *Application deadline:* For fall admission, 2/1 (priority date). *Application fee:* $60.
Expenses: Tuition: Full-time $17,880; part-time $745 per credit. Required fees: $1,140; $35 per credit. Tuition and fees vary according to course load and program.
Financial aid: Fellowships with tuition reimbursements, research assistantships with tuition reimbursements, teaching assistantships with tuition reimbursements, career-related internships or fieldwork, Federal Work-Study, institutionally sponsored loans, and tuition waivers (full and partial) available. Financial aid application deadline: 2/1; financial aid applicants required to submit FAFSA.
Dr. Joel D. Oppenheim, Director, 212-263-5648, *Fax:* 212-263-7600, *E-mail:* sackler-info@nyumed.med.nyu.edu.

■ NEW YORK UNIVERSITY

Graduate School of Arts and Science, Department of Biology, New York, NY 10012-1019

AWARDS Applied recombinant DNA technology (MS); biochemistry (PhD); biomedical journalism (MA); cell biology (PhD); computers in biological research (MS); environmental biology (PhD); general biology (MS); neural sciences and physiology (PhD); oral biology (MS); population and evolutionary biology (PhD). Part-time programs available.

Faculty: 22 full-time (5 women), 8 part-time/adjunct.
Students: 99 full-time (48 women), 61 part-time (31 women); includes 30 minority (4 African Americans, 22 Asian Americans or Pacific Islanders, 4 Hispanic Americans), 52 international. Average age 24. *371 applicants, 41% accepted.* In 1999, 54 master's, 4 doctorates awarded. Terminal master's awarded for partial completion of doctoral program.
Degree requirements: For master's, thesis or alternative, qualifying paper required, foreign language not required; for doctorate, dissertation, oral and written comprehensive exams required, foreign language not required.

Entrance requirements: For master's, GRE General Test, TOEFL; for doctorate, GRE General Test, GRE Subject Test, TOEFL. *Application deadline:* For fall admission, 1/4 (priority date). *Application fee:* $60.
Expenses: Tuition: Full-time $17,880; part-time $745 per credit. Required fees: $1,140; $35 per credit. Tuition and fees vary according to course load and program.
Financial aid: Fellowships with tuition reimbursements, research assistantships with tuition reimbursements, teaching assistantships with tuition reimbursements, career-related internships or fieldwork, Federal Work-Study, institutionally sponsored loans, and tuition waivers (full and partial) available. Financial aid application deadline: 1/4; financial aid applicants required to submit FAFSA.
Faculty research: Development and genetics, neurobiology, plant sciences, molecular and cell biology.
Philip Furmanski, Chairman, 212-998-8200.
Application contact: Gloria Coruzzi, Director of Graduate Studies, 212-998-8200, *Fax:* 212-995-4015, *E-mail:* biology@nyu.edu.

Find an in-depth description at www.petersons.com/graduate.

■ NEW YORK UNIVERSITY

School of Medicine and Graduate School of Arts and Science, Medical Scientist Training Program, New York, NY 10012-1019

AWARDS Biochemistry (MD/PhD); cell biology (MD/PhD); environmental health sciences (MD/PhD); microbiology (MD/PhD); parasitology (MD/PhD); pathology (MD/PhD); pharmacology (MD/PhD). Students must be accepted by both the School of Medicine and the Graduate School of Arts and Science.

Faculty: 150 full-time (35 women).
Students: 80 full-time (18 women); includes 37 minority (3 African Americans, 32 Asian Americans or Pacific Islanders, 2 Hispanic Americans), 1 international. Average age 25. *195 applicants, 18% accepted.*
Degree requirements: One foreign language.
Application deadline: For fall admission, 11/15. Applications are processed on a rolling basis. *Application fee:* $60.
Expenses: Students receive full tuition support and an annual stipend of $17,500.
Financial aid: Application deadline: 5/1.
Faculty research: Genetics, tumor biology, cardiovascular biology, neuroscience, host defense mechanisms.
Dr. James Salzer, Director, 212-263-0758.
Application contact: Arlene Kohler, Administrative Officer, 212-263-5649.

■ NEW YORK UNIVERSITY

School of Medicine and Graduate School of Arts and Science, Sackler Institute of Graduate Biomedical Sciences, Department of Biochemistry, New York, NY 10012-1019

AWARDS PhD, MD/PhD.

Faculty: 16 full-time (4 women).
Students: 14 full-time (5 women); includes 1 minority (Asian American or Pacific Islander), 7 international. Average age 25. In 1999, 3 degrees awarded (100% entered university research/teaching).
Degree requirements: For doctorate, one foreign language, dissertation, qualifying exam required. *Average time to degree:* Doctorate–5.5 years full-time.
Entrance requirements: For doctorate, GRE General Test, GRE Subject Test, TOEFL. *Application deadline:* For fall admission, 2/1 (priority date). Applications are processed on a rolling basis. *Application fee:* $60.
Expenses: Tuition: Full-time $17,880; part-time $745 per credit. Required fees: $1,140; $35 per credit. Tuition and fees vary according to course load and program.
Financial aid: Fellowships, research assistantships, teaching assistantships available. Financial aid application deadline: 1/15.
Faculty research: Signal transduction, DNA replication, differentiation, cell cycle regulation, protein folding. *Total annual research expenditures:* $2.1 million.
Dr. G. Nigel Godson, Chairman, 212-263-5622.
Application contact: Dr. Jim Boroweic, Graduate Adviser, 212-263-8453, *E-mail:* borowj01@mcrcr6.med.nyu.edu.

■ NORTH CAROLINA STATE UNIVERSITY

Graduate School, College of Agriculture and Life Sciences, Department of Biochemistry, Raleigh, NC 27695

AWARDS MS, PhD.

Faculty: 13 full-time (3 women), 8 part-time/adjunct (1 woman).
Students: 26 full-time (17 women), 2 part-time; includes 3 minority (1 African American, 1 Asian American or Pacific Islander, 1 Hispanic American), 8 international. Average age 33. *46 applicants, 26% accepted.* In 1999, 6 doctorates awarded. Terminal master's awarded for partial completion of doctoral program.
Degree requirements: For master's and doctorate, thesis/dissertation required.
Entrance requirements: For master's and doctorate, GRE General Test. *Application*

deadline: For fall admission, 1/15 (priority date). *Application fee:* $45.
Expenses: Tuition, state resident: full-time $1,578. Tuition, nonresident: full-time $10,744. Required fees: $892. Full-time tuition and fees vary according to program.
Financial aid: In 1999–00, 24 research assistantships (averaging $5,666 per year) were awarded; fellowships, teaching assistantships, career-related internships or fieldwork, Federal Work-Study, and institutionally sponsored loans also available. Financial aid application deadline: 1/15.
Faculty research: Regulation of gene expression, structure and function of proteins and nucleic acids, molecular biology, high-field NMR, bioinorganic chemistry. *Total annual research expenditures:* $4.1 million.
Dr. Dennis T. Brown, Director of Graduate Programs, 919-515-5802, *Fax:* 919-515-2047, *E-mail:* dbrown@bchserver.bch.ncsu.edu.

Find an in-depth description at www.petersons.com/graduate.

■ NORTH DAKOTA STATE UNIVERSITY

Graduate Studies and Research, College of Science and Mathematics, Department of Biochemistry and Molecular Biology, Fargo, ND 58105

AWARDS MS, PhD. Part-time programs available.

Faculty: 6 full-time (0 women), 4 part-time/adjunct (0 women).
Students: 9 full-time (4 women), 3 international. Average age 26. *10 applicants, 0% accepted.* Terminal master's awarded for partial completion of doctoral program.
Degree requirements: For master's and doctorate, thesis/dissertation required, foreign language not required. *Average time to degree:* Master's–2.5 years full-time; doctorate–5 years full-time.
Entrance requirements: For master's and doctorate, TOEFL. *Application deadline:* For fall admission, 4/15 (priority date). Applications are processed on a rolling basis. *Application fee:* $25.
Expenses: Tuition, state resident: full-time $3,096; part-time $112 per credit hour. Tuition, nonresident: full-time $7,588; part-time $299 per credit hour. Tuition and fees vary according to course load, campus/location and reciprocity agreements.
Financial aid: Research assistantships with full tuition reimbursements, teaching assistantships with full tuition reimbursements, career-related internships or fieldwork, Federal Work-Study, and

institutionally sponsored loans available. Financial aid application deadline: 4/15.
Faculty research: Biotechnology, molecular biology, metabolism and enzymology, pesticide metabolism, protein structures.
Dr. S. Derek Killilea, Chair, 701-231-7946, *Fax:* 701-231-8324, *E-mail:* skillile@badlands.nodak.edu.

■ NORTHERN ILLINOIS UNIVERSITY

Graduate School, College of Liberal Arts and Sciences, Department of Chemistry and Biochemistry, De Kalb, IL 60115-2854

AWARDS MS, PhD. Part-time programs available.

Faculty: 17 full-time (1 woman).
Students: 39 full-time (20 women), 6 part-time (4 women); includes 3 minority (2 African Americans, 1 Hispanic American), 6 international. Average age 29. *35 applicants, 46% accepted.* In 1999, 4 master's, 4 doctorates awarded. Terminal master's awarded for partial completion of doctoral program.
Degree requirements: For master's, comprehensive exam, research seminar required, thesis optional, foreign language not required; for doctorate, one foreign language (computer language can substitute), dissertation, candidacy exam, dissertation defense, research seminar required.
Entrance requirements: For master's, GRE General Test, TOEFL, bachelor's degree in mathematics or science, minimum GPA of 2.75; for doctorate, GRE General Test, TOEFL, bachelor's degree in mathematics or science, minimum GPA of 2.75 (undergraduate), 3.2 (graduate). *Application deadline:* For fall admission, 6/1; for spring admission, 11/1. Applications are processed on a rolling basis. *Application fee:* $30.
Expenses: Tuition, state resident: part-time $169 per credit hour. Tuition, nonresident: part-time $295 per credit hour. Tuition and fees vary according to campus/location and program.
Financial aid: In 1999–00, 3 research assistantships with full tuition reimbursements, 32 teaching assistantships with full tuition reimbursements were awarded; fellowships with full tuition reimbursements, career-related internships or fieldwork, Federal Work-Study, tuition waivers (full), and unspecified assistantships also available. Aid available to part-time students.
Dr. James Erman, Chair, 815-753-1181, *Fax:* 815-753-4802.

Northern Illinois University (continued)
Application contact: Dr. Jon Carnahan, Director, Graduate Studies, 815-753-6879.

Find an in-depth description at www.petersons.com/graduate.

■ NORTHWESTERN UNIVERSITY

The Graduate School, Division of Interdepartmental Programs and Medical School, Integrated Graduate Programs in the Life Sciences, Chicago, IL 60611

AWARDS Cancer biology (PhD); cell biology (PhD); developmental biology (PhD); evolutionary biology (PhD); immunology and microbial pathogenesis (PhD); molecular biology and genetics (PhD); neurobiology (PhD); pharmacology and toxicology (PhD); structural biology and biochemistry (PhD).

Degree requirements: For doctorate, dissertation, written and oral qualifying exams required, foreign language not required.
Entrance requirements: For doctorate, GRE General Test, TOEFL.
Expenses: Tuition: Full-time $23,301. Full-time tuition and fees vary according to program.

Find an in-depth description at www.petersons.com/graduate.

■ NORTHWESTERN UNIVERSITY

The Graduate School, Judd A. and Marjorie Weinberg College of Arts and Sciences, Interdepartmental Biological Sciences Program (IBiS), Department of Biochemistry, Molecular Biology, and Cell Biology, Evanston, IL 60208

AWARDS Cell and molecular biology (PhD); molecular biophysics (PhD). Department participates in the Interdepartmental Biological Sciences Program (IBiS).

Faculty: 59 full-time (11 women).
Students: 79 full-time (46 women); includes 5 minority (2 African Americans, 3 Hispanic Americans), 14 international. *Application fee:* $50 ($55 for international students).
Expenses: Tuition: Full-time $23,301. Full-time tuition and fees vary according to program.
Financial aid: In 1999–00, 15 fellowships (averaging $12,078 per year), 64 research assistantships (averaging $17,000 per year), 16 teaching assistantships (averaging $16,620 per year) were awarded.
Application contact: Latonia Trimuel, Program Assistant, 800-546-1761, *E-mail:* ibis@northwestern.edu.

Find an in-depth description at www.petersons.com/graduate.

■ THE OHIO STATE UNIVERSITY

Graduate School, College of Biological Sciences, Biochemistry Program, Columbus, OH 43210

AWARDS MS, PhD.

Faculty: 76 full-time.
Students: 78 full-time (37 women), 2 part-time; includes 4 minority (1 African American, 3 Asian Americans or Pacific Islanders), 29 international. *253 applicants, 15% accepted.* In 1999, 1 master's, 2 doctorates awarded. Terminal master's awarded for partial completion of doctoral program.
Degree requirements: For master's and doctorate, thesis/dissertation required, foreign language not required.
Entrance requirements: For master's and doctorate, GRE General Test. *Application deadline:* For fall admission, 8/15. Applications are processed on a rolling basis. *Application fee:* $30 ($40 for international students).
Expenses: Tuition, state resident: full-time $5,400. Tuition, nonresident: full-time $14,535. Part-time tuition and fees vary according to course load and program.
Financial aid: Fellowships, research assistantships, teaching assistantships, Federal Work-Study and institutionally sponsored loans available.
Faculty research: Biotechnology, carbohydrates, carcinogenesis, enzymology, intermediary metabolism.
Dr. C. Russell Hille, Director, 614-292-1463, *Fax:* 614-292-6511, *E-mail:* hille.l@osu.edu.
Application contact: Charlene K. Maxwell, Administrative Assistant, 614-292-1463, *Fax:* 614-292-6511, *E-mail:* maxwell.8@osu.edu.

Find an in-depth description at www.petersons.com/graduate.

■ THE OHIO STATE UNIVERSITY

Graduate School, College of Biological Sciences, Department of Biochemistry, Columbus, OH 43210

AWARDS MS.

Faculty: 15 full-time, 4 part-time/adjunct.
Students: 24 full-time (9 women), 2 part-time; includes 3 minority (all Asian Americans or Pacific Islanders), 18 international. *11 applicants, 73% accepted.* In 1999, 10 master's awarded.
Degree requirements: For master's, thesis optional, foreign language not required.
Entrance requirements: For master's, GRE General Test. *Application deadline:* For fall admission, 8/15. Applications are processed on a rolling basis. *Application fee:* $30 ($40 for international students).

Expenses: Tuition, state resident: full-time $5,400. Tuition, nonresident: full-time $14,535. Part-time tuition and fees vary according to course load and program.
Financial aid: Fellowships, research assistantships, teaching assistantships, career-related internships or fieldwork, Federal Work-Study, and institutionally sponsored loans available. Aid available to part-time students.
Dr. George A. Marzluf, Chairman, 614-292-6771, *Fax:* 614-292-6773, *E-mail:* marzluf.1@osu.edu.

■ OHIO UNIVERSITY

Graduate Studies, College of Arts and Sciences, Department of Chemistry and Biochemistry, Athens, OH 45701-2979

AWARDS MS, PhD.

Faculty: 21 full-time (5 women), 3 part-time/adjunct (1 woman).
Students: 53 full-time (17 women), 7 part-time (1 woman); includes 3 minority (2 African Americans, 1 Asian American or Pacific Islander), 27 international. *39 applicants, 54% accepted.* In 1999, 6 master's, 6 doctorates awarded.
Degree requirements: For master's and doctorate, thesis/dissertation, exam required, foreign language not required. *Average time to degree:* Master's–2.2 years full-time; doctorate–4.5 years full-time. *Application deadline:* For fall admission, 2/15 (priority date). Applications are processed on a rolling basis. *Application fee:* $30.
Expenses: Tuition, state resident: full-time $5,754; part-time $238 per credit hour. Tuition, nonresident: full-time $11,055; part-time $457 per credit hour. Tuition and fees vary according to course load, degree level and campus/location.
Financial aid: In 1999–00, 28 teaching assistantships with full tuition reimbursements (averaging $15,000 per year) were awarded; fellowships, research assistantships with full tuition reimbursements, Federal Work-Study, institutionally sponsored loans, and tuition waivers (full and partial) also available. Financial aid application deadline: 3/15.
Dr. Kenneth Brown, Chair, 740-593-9831, *Fax:* 740-593-0148, *E-mail:* brownk3@oak.cats.ohiou.edu.
Application contact: Dr. Howard Dewald, Graduate Chair, 740-593-1755, *Fax:* 740-593-0143, *E-mail:* hdewald1@ohiou.edu.

■ OKLAHOMA STATE UNIVERSITY

Graduate College, College of Agricultural Sciences and Natural Resources, Department of Biochemistry and Molecular Biology, Stillwater, OK 74078

AWARDS MS, PhD.

Faculty: 5 full-time (1 woman).
Students: 27 full-time (11 women), 20 part-time (8 women); includes 4 minority (3 African Americans, 1 Hispanic American), 31 international. Average age 28. In 1999, 4 master's, 4 doctorates awarded.
Degree requirements: For master's, thesis required, foreign language not required; for doctorate, dissertation required.
Entrance requirements: For master's and doctorate, TOEFL. *Application deadline:* For fall admission, 6/1 (priority date). *Application fee:* $25.
Expenses: Tuition, state resident: part-time $86 per credit hour. Tuition, nonresident: part-time $275 per credit hour. Required fees: $17 per credit hour. $14 per semester. One-time fee: $20 full-time. Tuition and fees vary according to course load.
Financial aid: In 1999–00, 36 students received aid, including 5 research assistantships (averaging $12,810 per year), 33 teaching assistantships (averaging $13,300 per year); career-related internships or fieldwork, Federal Work-Study, and tuition waivers (partial) also available. Aid available to part-time students. Financial aid application deadline: 3/1.
Dr. James Blair, Head, 405-744-6189.
Find an in-depth description at www.petersons.com/graduate.

■ OLD DOMINION UNIVERSITY

College of Sciences, Department of Chemistry and Biochemistry, Norfolk, VA 23529

AWARDS Analytical chemistry (MS); biochemistry (MS); biomedical sciences (PhD); clinical chemistry (MS); environmental chemistry (MS); organic chemistry (MS); physical chemistry (MS). Part-time and evening/weekend programs available.

Faculty: 11 full-time (6 women).
Students: 7 full-time (2 women), 9 part-time (4 women); includes 1 minority (Hispanic American), 1 international. Average age 28. *36 applicants, 39% accepted.* In 1999, 7 master's, 3 doctorates awarded. Terminal master's awarded for partial completion of doctoral program.
Degree requirements: For master's and doctorate, thesis/dissertation,

comprehensive exam required, foreign language not required.
Entrance requirements: For master's, GRE General Test, TOEFL, minimum GPA of 3.0 in major, 2.5 overall; for doctorate, GRE General Test, TOEFL. *Application deadline:* For fall admission, 7/1; for spring admission, 11/1. Applications are processed on a rolling basis. *Application fee:* $30.
Expenses: Tuition, state resident: full-time $4,440; part-time $185 per credit. Tuition, nonresident: full-time $11,784; part-time $477 per credit. Required fees: $1,612. Tuition and fees vary according to program.
Financial aid: In 1999–00, 2 students received aid, including 8 research assistantships with tuition reimbursements available (averaging $11,612 per year), 9 teaching assistantships with tuition reimbursements available (averaging $8,973 per year); fellowships, career-related internships or fieldwork and grants also available. Financial aid application deadline: 2/15; financial aid applicants required to submit FAFSA.
Faculty research: Electroanalytical chemistry, clinical applications of trace element analysis, environmental-atmospheric chemistry, cancer biochemistry, drug chemistry. *Total annual research expenditures:* $854,968.
Dr. John R. Donet, Director, 757-683-4078, *E-mail:* chemgpd@odu.edu.

■ OREGON GRADUATE INSTITUTE OF SCIENCE AND TECHNOLOGY

Graduate Studies, Department of Biochemistry and Molecular Biology, Beaverton, OR 97006-8921

AWARDS MS, PhD. Part-time programs available. Terminal master's awarded for partial completion of doctoral program.

Degree requirements: For master's, thesis optional, foreign language not required; for doctorate, comprehensive exam, oral defense of dissertation required.
Entrance requirements: For master's, TOEFL; for doctorate, GRE General Test, GRE Subject Test, TOEFL. Electronic applications accepted.
Expenses: Tuition: Full-time $17,860; part-time $425 per credit. Full-time tuition and fees vary according to degree level, program and reciprocity agreements.
Faculty research: Biotechnology, membrane transport, metallobiochemistry.

Find an in-depth description at www.petersons.com/graduate.

■ OREGON HEALTH SCIENCES UNIVERSITY

School of Medicine, Graduate Programs in Medicine, Department of Biochemistry and Molecular Biology, Portland, OR 97201-3098

AWARDS PhD, MD/PhD. Part-time programs available.

Faculty: 9 full-time (2 women), 18 part-time/adjunct (8 women).
Students: 18 full-time (8 women); includes 2 minority (both Asian Americans or Pacific Islanders), 6 international. Average age 29. In 1999, 2 degrees awarded (100% entered university research/teaching).
Degree requirements: For doctorate, dissertation required, foreign language not required. *Average time to degree:* Doctorate–5 years full-time.
Entrance requirements: For doctorate, GRE General Test, MCAT (MD/PhD). *Application deadline:* For fall admission, 1/15. Applications are processed on a rolling basis. *Application fee:* $0.
Expenses: Tuition, state resident: full-time $3,132; part-time $174 per credit hour. Tuition, nonresident: full-time $5,256; part-time $292 per credit hour. Required fees: $8.5 per credit hour. $146 per term. Part-time tuition and fees vary according to course load.
Financial aid: In 1999–00, 6 fellowships with tuition reimbursements (averaging $16,500 per year), 12 research assistantships with tuition reimbursements (averaging $16,500 per year) were awarded; Federal Work-Study, institutionally sponsored loans, and scholarships also available. Financial aid application deadline: 3/1; financial aid applicants required to submit FAFSA.
Faculty research: Protein structure and function, enzymology, metabolism, membranes transport.
Dr. Jack H. Kaplan, Chairperson, 503-494-1655.
Application contact: Dr. Richard G. Brennan, Director of Graduate Admissions, 503-494-7781, *E-mail:* brennanr@ohsu.edu.

Find an in-depth description at www.petersons.com/graduate.

■ OREGON STATE UNIVERSITY

Graduate School, College of Science, Department of Biochemistry and Biophysics, Corvallis, OR 97331-6503

AWARDS MA, MAIS, MS, PhD.

Faculty: 16 full-time (2 women), 13 part-time/adjunct (1 woman).
Students: 28 full-time (12 women), 1 (woman) part-time; includes 3 minority (2 Asian Americans or Pacific Islanders, 1 Hispanic American), 12 international.

Oregon State University (continued)
Average age 30. In 1999, 3 master's, 4 doctorates awarded.

Degree requirements: For master's, thesis optional, foreign language not required; for doctorate, dissertation, exams required, foreign language not required.

Entrance requirements: For master's, GRE General Test, TOEFL, minimum GPA of 3.0; for doctorate, GRE Subject Test, TOEFL, minimum GPA of 3.0. *Application deadline:* For fall admission, 4/15 (priority date). Applications are processed on a rolling basis. *Application fee:* $50.

Expenses: Tuition, state resident: full-time $6,489. Tuition, nonresident: full-time $11,061. Tuition and fees vary according to program.

Financial aid: Research assistantships, teaching assistantships, institutionally sponsored loans available. Aid available to part-time students. Financial aid application deadline: 2/1.

Faculty research: DNA and deoxyribonucleotide metabolism, cell growth control, receptors and membranes, protein structure and function.
Dr. Christopher K. Mathews, Chairman, 541-737-1865, *Fax:* 541-737-0481, *E-mail:* mathewsc@ucs.orst.edu.

Application contact: Dr. W. Curtis Johnson, Chairman, Graduate Committee, 541-737-4143, *Fax:* 541-737-0481, *E-mail:* johnsowc@ucs.orst.edu.

■ THE PENNSYLVANIA STATE UNIVERSITY MILTON S. HERSHEY MEDICAL CENTER

Graduate School, Department of Biochemistry and Molecular Biology, Hershey, PA 17033-2360

AWARDS MS, PhD, MD/PhD.

Students: 22 full-time (9 women), 1 (woman) part-time. Average age 27. In 1999, 4 degrees awarded.

Degree requirements: For master's, foreign language and thesis not required; for doctorate, dissertation required.

Entrance requirements: For master's and doctorate, GRE General Test, TOEFL. *Application deadline:* For fall admission, 7/26. *Application fee:* $50.

Expenses: Tuition, state resident: full-time $6,886; part-time $291 per credit. Tuition, nonresident: full-time $14,118; part-time $588 per credit. Required fees: $43 per semester. Part-time tuition and fees vary according to course load.
Dr. Judith S. Bond, Assistant Dean, 717-531-8585.

■ THE PENNSYLVANIA STATE UNIVERSITY UNIVERSITY PARK CAMPUS

Graduate School, Eberly College of Science, Department of Biochemistry and Molecular Biology, Program in Biochemistry, Microbiology, and Molecular Biology, State College, University Park, PA 16802-1503

AWARDS MS, PhD.

Students: 83 full-time (45 women), 12 part-time (6 women). In 1999, 4 master's, 12 doctorates awarded.

Degree requirements: For master's, thesis required; for doctorate, dissertation, comprehensive exam required.

Entrance requirements: For master's and doctorate, GRE General Test, GRE Subject Test. *Application fee:* $50.

Expenses: Tuition, state resident: full-time $6,886; part-time $291 per credit. Tuition, nonresident: full-time $14,118; part-time $588 per credit. Required fees: $46 per semester. Part-time tuition and fees vary according to course load and program.

Financial aid: Fellowships, unspecified assistantships available.
Ronald D. Porter, Director, 814-865-2538.

Find an in-depth description at www.petersons.com/graduate.

■ THE PENNSYLVANIA STATE UNIVERSITY UNIVERSITY PARK CAMPUS

Graduate School, Intercollege Graduate Programs, Intercollege Graduate Program in Integrative Biosciences, State College, University Park, PA 16802-1503

AWARDS Integrative biosciences (MS, PhD), including biomolecular transport dynamics, cell and developmental biology, cellular and molecular mechanisms of toxicity, chemical biology, ecological and molecular plant physiology, immunobiology, molecular medicine, neuroscience, nutrition science.

Students: 39 full-time (24 women), 1 part-time.

Entrance requirements: For master's and doctorate, GRE General Test. *Application fee:* $50.

Expenses: Tuition, state resident: full-time $6,886; part-time $291 per credit. Tuition, nonresident: full-time $14,118; part-time $588 per credit. Required fees: $46 per semester. Part-time tuition and fees vary according to course load and program.

Financial aid: Fellowships available.
Dr. C. R. Matthews, Co-Director, 814-863-3650.

Application contact: Admissions Committee, 814-865-3155, *Fax:* 814-863-1357, *E-mail:* lscgradadm@mail.biotec.psu.edu.

Find an in-depth description at www.petersons.com/graduate.

■ PURDUE UNIVERSITY

Graduate School, Interdisciplinary Biochemistry and Molecular Biology Program, West Lafayette, IN 47907

AWARDS PhD.

Faculty: 60 full-time (11 women).

Students: 58 full-time (16 women); includes 8 minority (1 African American, 3 Asian Americans or Pacific Islanders, 4 Hispanic Americans), 19 international. Average age 25. *90 applicants, 30% accepted.*

Degree requirements: For doctorate, dissertation, preliminary and qualifying exams required, foreign language not required. *Average time to degree:* Doctorate–6.25 years full-time.

Entrance requirements: For doctorate, GRE General Test, TOEFL, TWE. *Application deadline:* For fall admission, 2/15 (priority date). *Application fee:* $30. Electronic applications accepted.

Expenses: Tuition, state resident: full-time $4,530; part-time $130 per credit hour. Tuition, nonresident: full-time $15,310; part-time $404 per credit hour. Tuition and fees vary according to campus/location and program.

Financial aid: In 1999–00, 8 fellowships with partial tuition reimbursements (averaging $16,800 per year), 50 research assistantships with partial tuition reimbursements (averaging $16,800 per year), 11 teaching assistantships with partial tuition reimbursements (averaging $16,800 per year) were awarded; grants also available. Financial aid application deadline: 4/15.

Faculty research: Structure of macromolecules, regulatory mechanisms in metabolism, gene expression and DNA/RNA processing, drug design, cell development and differentiation.
Dr. S. S. Broyles, Chair of the Executive Committee, 765-494-0745, *Fax:* 765-496-1475.

Application contact: Elizabeth A. Chandler, Coordinator, 765-494-1634, *Fax:* 765-496-1475.

Find an in-depth description at www.petersons.com/graduate.

■ PURDUE UNIVERSITY

Graduate School, School of Agriculture, Department of Biochemistry, West Lafayette, IN 47907

AWARDS MS, PhD.

Faculty: 16 full-time (3 women).

Students: 21 full-time (10 women), 3 part-time; includes 7 minority (2 African Americans, 4 Asian Americans or Pacific Islanders, 1 Hispanic American), 11 international. Average age 26. *48 applicants, 40% accepted.* In 1999, 3 master's, 8 doctorates awarded. Terminal master's awarded for partial completion of doctoral program.
Degree requirements: For master's, thesis required; for doctorate, dissertation, preliminary and qualifying exams required. *Average time to degree:* Doctorate–6 years full-time.
Entrance requirements: For master's, TOEFL; for doctorate, GRE General Test, TOEFL. *Application deadline:* For fall admission, 1/15 (priority date). Applications are processed on a rolling basis. *Application fee:* $30. Electronic applications accepted.
Expenses: Tuition, state resident: full-time $4,530; part-time $130 per credit hour. Tuition, nonresident: full-time $15,310; part-time $404 per credit hour. Tuition and fees vary according to campus/location and program.
Financial aid: In 1999–00, 10 fellowships (averaging $16,200 per year), 19 research assistantships (averaging $16,200 per year), 4 teaching assistantships (averaging $16,200 per year) were awarded. Aid available to part-time students. Financial aid application deadline: 4/15; financial aid applicants required to submit FAFSA.
Faculty research: Molecular biology and post-translational modifications of neuropeptides, membrane transport proteins.
Dr. M. A. Hermodson, Head, 765-494-1637, *Fax:* 765-494-7987, *E-mail:* hermodson@biochem.purdue.edu.
Application contact: Jonathan LeBowitz, Chair, Graduate Admissions Committee, 765-494-0112, *Fax:* 765-494-7987, *E-mail:* lebowitz@biochem.purdue.edu.
Find an in-depth description at www.petersons.com/graduate.

■ **PURDUE UNIVERSITY**

School of Pharmacy and Pharmacal Sciences and Graduate School, Graduate Programs in Pharmacy and Pharmacal Sciences, Department of Medicinal Chemistry and Molecular Pharmacology, West Lafayette, IN 47907

AWARDS Analytical medicinal chemistry (PhD); computational and biophysical medicinal chemistry (PhD); medicinal and bioorganic chemistry (PhD); medicinal biochemistry and molecular biology (PhD); molecular pharmacology and toxicology (PhD); natural products and pharmacognosy (PhD); nuclear pharmacy (MS);

radiopharmaceutical chemistry and nuclear pharmacy (PhD).
Faculty: 24 full-time (2 women).
Students: 48 full-time (26 women), 3 part-time (1 woman); includes 4 minority (1 African American, 1 Asian American or Pacific Islander, 2 Hispanic Americans), 13 international. Average age 29. *139 applicants, 19% accepted.* In 1999, 3 master's, 11 doctorates awarded. Terminal master's awarded for partial completion of doctoral program.
Degree requirements: For master's and doctorate, thesis/dissertation required, foreign language not required.
Entrance requirements: For master's, GRE General Test, TOEFL, minimum B average; BS in biology, chemistry, or pharmacy; for doctorate, GRE General Test, TOEFL, minimum B average; BS in biology, chemistry, or pharmacology. *Application deadline:* Applications are processed on a rolling basis. *Application fee:* $30. Electronic applications accepted.
Expenses: Tuition, state resident: full-time $4,530; part-time $130 per credit hour. Tuition, nonresident: full-time $15,310; part-time $404 per credit hour. Tuition and fees vary according to campus/location and program.
Financial aid: Fellowships, research assistantships, teaching assistantships, traineeships available. Aid available to part-time students. Financial aid applicants required to submit FAFSA.
Faculty research: Drug design and development, cancer research, drug synthesis and analysis, chemical pharmacology, environmental toxicology.
Dr. R. F. Borch, Graduate Head, 765-494-1403.
Application contact: Dr. D. E. Bergstrom, Graduate Committee, 765-494-6275, *E-mail:* bergstrom@pharmacy.purdue.edu.
Find an in-depth description at www.petersons.com/graduate.

■ **QUEENS COLLEGE OF THE CITY UNIVERSITY OF NEW YORK**

Division of Graduate Studies, Mathematics and Natural Sciences Division, Department of Chemistry, Flushing, NY 11367-1597

AWARDS Biochemistry (MA); chemistry (MA). Part-time and evening/weekend programs available.

Faculty: 16 full-time (2 women).
Students: *8 applicants, 50% accepted.* In 1999, 1 degree awarded.
Degree requirements: For master's, comprehensive exam required, foreign language and thesis not required.
Entrance requirements: For master's, GRE, TOEFL, previous course work in

calculus and physics, minimum GPA of 3.0. *Application deadline:* For fall admission, 4/1; for spring admission, 11/1. Applications are processed on a rolling basis. *Application fee:* $40.
Expenses: Tuition, state resident: full-time $4,350; part-time $185 per credit. Tuition, nonresident: full-time $7,600; part-time $320 per credit. Required fees: $114; $57 per semester. Tuition and fees vary according to course load and program.
Financial aid: Career-related internships or fieldwork, Federal Work-Study, institutionally sponsored loans, tuition waivers (partial), and adjunct lectureships available. Aid available to part-time students. Financial aid application deadline: 4/1; financial aid applicants required to submit FAFSA.
Dr. A. David Baker, Chairperson, 718-997-4100, *E-mail:* a.davidbaker@qc.edu.
Application contact: Graduate Adviser, 718-997-4100.

■ **RENSSELAER POLYTECHNIC INSTITUTE**

Graduate School, School of Science, Department of Biochemistry and Biophysics, Troy, NY 12180-3590

AWARDS Biochemistry (MS); biophysics (MS).

Students: 3 full-time (1 woman); includes 2 minority (1 African American, 1 Asian American or Pacific Islander). *9 applicants, 44% accepted.* In 1999, 3 degrees awarded.
Entrance requirements: For master's, GRE General Test, TOEFL. *Application deadline:* For fall admission, 2/1 (priority date). Applications are processed on a rolling basis. *Application fee:* $35.
Expenses: Tuition: Part-time $665 per credit hour. Required fees: $980.
Financial aid: Application deadline: 2/1.
Faculty research: Biopolymers, photosynthesis, cellular bioengineering.
Dr. Jane Koretz, Head, 518-276-6492, *E-mail:* koretj@rpi.edu.

■ **RENSSELAER POLYTECHNIC INSTITUTE**

Graduate School, School of Science, Department of Biology, Troy, NY 12180-3590

AWARDS Biochemistry (MS, PhD); biophysics (MS, PhD); cell biology (MS, PhD); developmental biology (MS, PhD); microbiology (MS, PhD); molecular biology (MS, PhD); plant science (MS, PhD). Part-time programs available.

Faculty: 14 full-time (5 women).
Students: 16 full-time (8 women), 2 part-time (1 woman); includes 1 minority (Asian American or Pacific Islander), 4 international. *37 applicants, 41% accepted.* In

Rensselaer Polytechnic Institute (continued)

1999, 6 master's, 3 doctorates awarded. Terminal master's awarded for partial completion of doctoral program.
Degree requirements: For master's and doctorate, thesis/dissertation required, foreign language not required.
Entrance requirements: For master's and doctorate, GRE General Test, TOEFL. *Application deadline:* For fall admission, 2/1 (priority date). Applications are processed on a rolling basis. *Application fee:* $35.
Expenses: Tuition: Part-time $665 per credit hour. Required fees: $980.
Financial aid: In 1999–00, 8 research assistantships with partial tuition reimbursements (averaging $15,000 per year), 11 teaching assistantships with full tuition reimbursements (averaging $15,000 per year) were awarded; fellowships, career-related internships or fieldwork and institutionally sponsored loans also available. Financial aid application deadline: 2/1.
Faculty research: Applied environmental biology, genetics, environmental science, fresh water ecology, microbial ecology.
Dr. John Salerno, Chair, 518-276-2699, *Fax:* 518-276-2344.
Application contact: Dr. Jackie L. Collier, Assistant Professor, 518-276-6446, *Fax:* 518-276-2344, *E-mail:* collij3@rpi.edu.
Find an in-depth description at www.petersons.com/graduate.

■ RENSSELAER POLYTECHNIC INSTITUTE

Graduate School, School of Science, Department of Chemistry, Troy, NY 12180-3590
AWARDS Analytical chemistry (MS, PhD); biochemistry (MS, PhD); inorganic chemistry (MS, PhD); organic chemistry (MS, PhD); physical chemistry (MS, PhD); polymer chemistry (MS, PhD). Part-time and evening/weekend programs available.

Faculty: 19 full-time (4 women).
Students: 67 full-time (27 women), 4 part-time (1 woman); includes 2 minority (1 Asian American or Pacific Islander, 1 Hispanic American), 44 international. *118 applicants, 40% accepted.* In 1999, 4 master's, 7 doctorates awarded.
Degree requirements: For master's, thesis required (for some programs), foreign language not required; for doctorate, dissertation required, foreign language not required.
Entrance requirements: For master's and doctorate, GRE General Test, TOEFL. *Application deadline:* For fall admission, 2/1 (priority date). Applications are processed on a rolling basis. *Application fee:* $35.

Expenses: Tuition: Part-time $665 per credit hour. Required fees: $980.
Financial aid: In 1999–00, 18 research assistantships with full tuition reimbursements (averaging $13,000 per year), 45 teaching assistantships with full tuition reimbursements (averaging $13,000 per year) were awarded; fellowships with full tuition reimbursements, institutionally sponsored loans and tuition waivers (full and partial) also available. Financial aid application deadline: 2/1.
Faculty research: Materials chemistry, biochemistry, polymersynthesis, organic synthesis, interfacial chemistry. *Total annual research expenditures:* $6.5 million.
Dr. Thomas Apple, Chair, 518-276-6344, *Fax:* 518-276-4887.
Application contact: Dr. Wilfredo Colon, Chair, Graduate Committee, 518-276-6340, *Fax:* 518-276-6434, *E-mail:* colonw@rpi.edu.
Find an in-depth description at www.petersons.com/graduate.

■ RICE UNIVERSITY

Graduate Programs, Wiess School of Natural Sciences, Department of Biochemistry and Cell Biology, Houston, TX 77251-1892
AWARDS MA, PhD.

Faculty: 21 full-time (5 women), 1 part-time/adjunct (0 women).
Students: 54 full-time (29 women); includes 3 minority (all Hispanic Americans), 15 international. Average age 22. *295 applicants, 5% accepted.* In 1999, 2 master's awarded (50% found work related to degree, 50% continued full-time study); 12 doctorates awarded (90% found work related to degree, 10% continued full-time study).
Degree requirements: For master's and doctorate, thesis/dissertation required, foreign language not required. *Average time to degree:* Master's–3 years full-time; doctorate–5 years full-time.
Entrance requirements: For master's and doctorate, GRE. *Application deadline:* For fall admission, 2/1 (priority date). Applications are processed on a rolling basis. *Application fee:* $0. Electronic applications accepted.
Expenses: Tuition: Full-time $16,100. Required fees: $300.
Financial aid: In 1999–00, 21 fellowships, 32 research assistantships were awarded; traineeships and tuition waivers (full) also available.
Faculty research: Steroid metabolism, protein structure NMR, biophysics, cell growth and movement. *Total annual research expenditures:* $5 million.
Dr. F. B. Rudolph, Chair, 713-348-4015, *Fax:* 713-348-5154, *E-mail:* fbr@rice.edu.

Application contact: Dolores M. Schwartz, Office Manager, 713-348-4230, *Fax:* 713-348-5154, *E-mail:* dolores@rice.edu.
Find an in-depth description at www.petersons.com/graduate.

■ RUSH UNIVERSITY

Graduate College, Division of Biochemistry, Chicago, IL 60612-3832
AWARDS PhD, MD/PhD.

Faculty: 14 full-time (6 women), 17 part-time/adjunct (3 women).
Students: 22 full-time (9 women); includes 1 minority (Asian American or Pacific Islander), 6 international. Average age 29. *80 applicants, 5% accepted.* In 1999, 2 degrees awarded (80% entered university research/teaching).
Degree requirements: For doctorate, dissertation, preliminary exam required, foreign language not required. *Average time to degree:* Doctorate–6 years full-time.
Entrance requirements: For doctorate, GRE General Test, TOEFL. *Application deadline:* For fall admission, 4/15 (priority date). Applications are processed on a rolling basis. *Application fee:* $25.
Expenses: Tuition: Full-time $13,020; part-time $390 per credit. Tuition and fees vary according to program.
Financial aid: In 1999–00, research assistantships with tuition reimbursements (averaging $14,500 per year); Federal Work-Study and institutionally sponsored loans also available. Aid available to part-time students. Financial aid application deadline: 4/15.
Faculty research: Biochemistry of extracellular matrix, connective tissue biosynthesis and degradation, molecular biology of connective tissue components. *Total annual research expenditures:* $1.5 million.
Dr. A. Bezkorovainy, Director, 312-942-5429, *Fax:* 312-942-3053.
Application contact: Thyra Jackson, Coordinator of Admissions, 312-942-6247, *Fax:* 312-942-2100, *E-mail:* tjackson@rushu.rush.edu.
Find an in-depth description at www.petersons.com/graduate.

■ RUTGERS, THE STATE UNIVERSITY OF NEW JERSEY, NEWARK

Graduate School, Department of Chemistry, Newark, NJ 07102
AWARDS Analytical chemistry (MS, PhD); biochemistry (MS, PhD); inorganic chemistry (MS, PhD); organic chemistry (MS, PhD); physical chemistry (MS, PhD). Part-time and evening/weekend programs available.

Faculty: 22 full-time (3 women).
Students: 31 full-time (13 women), 34 part-time (15 women); includes 37 minority (1 African American, 33 Asian Americans or Pacific Islanders, 3 Hispanic Americans). *109 applicants, 32% accepted.* In 1999, 8 master's, 7 doctorates awarded. Terminal master's awarded for partial completion of doctoral program.
Degree requirements: For master's, cumulative exams required, thesis optional, foreign language not required; for doctorate, dissertation, exams, research proposal required, foreign language not required.
Entrance requirements: For master's and doctorate, GRE General Test, TOEFL, minimum undergraduate B average. *Application deadline:* For fall admission, 8/1 (priority date); for spring admission, 12/15. Applications are processed on a rolling basis. *Application fee:* $50. Electronic applications accepted.
Expenses: Tuition, state resident: full-time $6,776; part-time $279 per credit hour. Tuition, nonresident: full-time $9,936; part-time $412 per credit hour. Required fees: $201 per semester. Tuition and fees vary according to course load and program.
Financial aid: In 1999–00, 35 students received aid, including 7 fellowships with full tuition reimbursements available (averaging $12,000 per year), 3 research assistantships, 18 teaching assistantships with full tuition reimbursements available (averaging $13,350 per year); Federal Work-Study and institutionally sponsored loans also available. Financial aid application deadline: 3/1.
Faculty research: Medicinal chemistry, natural products, isotope effects, biophysics and biorganic approaches to enzyme mechanisms, organic and organometallic synthesis.
Prof. Piotr Piotrowiak, Coordinator, 973-353-5318, *Fax:* 973-353-1264, *E-mail:* piotr@andromeda.rutgers.edu.

Find an in-depth description at www.petersons.com/graduate.

■ RUTGERS, THE STATE UNIVERSITY OF NEW JERSEY, NEW BRUNSWICK

Graduate School, Program in Biochemistry, New Brunswick, NJ 08901-1281

AWARDS Biochemistry (MS, PhD); molecular biology (MS, PhD).

Faculty: 100 full-time (26 women).
Students: 47 full-time (14 women), 13 part-time (8 women); includes 5 minority (all Asian Americans or Pacific Islanders), 34 international. Average age 25. *101 applicants, 19% accepted.* In 1999, 2 master's, 7 doctorates awarded. Terminal

master's awarded for partial completion of doctoral program.
Degree requirements: For master's, thesis, qualifying exam required, foreign language not required; for doctorate, dissertation, written qualifying exam required, foreign language not required.
Entrance requirements: For master's, GRE General Test, GRE Subject Test; for doctorate, GRE General Test, GRE Subject Test, TOEFL, minimum GPA of 3.0. *Application deadline:* For fall admission, 2/15. Applications are processed on a rolling basis. *Application fee:* $50.
Expenses: Tuition, state resident: full-time $6,776; part-time $279 per credit. Tuition, nonresident: full-time $9,936; part-time $412 per credit. Required fees: $20 per credit. $89 per semester. Tuition and fees vary according to course load, campus/location and program.
Financial aid: In 1999–00, 21 fellowships with full tuition reimbursements (averaging $15,000 per year), 24 research assistantships with full tuition reimbursements (averaging $15,000 per year), 3 teaching assistantships with full tuition reimbursements (averaging $14,000 per year) were awarded. Financial aid application deadline: 3/1; financial aid applicants required to submit FAFSA.
Faculty research: Chemistry of biological systems, mechanisms of enzyme action.
Dr. Alice Y. Liu, Director, 732-445-2730, *Fax:* 732-445-6370, *E-mail:* liu@biology.rutgers.edu.
Application contact: Carolyn Ambrose, Administrative Assistant, 732-445-3430, *Fax:* 732-445-6370, *E-mail:* ambrose@biology.rutgers.edu.

■ RUTGERS, THE STATE UNIVERSITY OF NEW JERSEY, NEW BRUNSWICK

Graduate School, Program in Chemistry, New Brunswick, NJ 08901-1281

AWARDS Analytical chemistry (MS, PhD); biological chemistry (PhD); chemistry education (MST); inorganic chemistry (MS, PhD); organic chemistry (MS, PhD); physical chemistry (MS, PhD). Part-time and evening/weekend programs available.

Faculty: 44 full-time (9 women), 4 part-time/adjunct (2 women).
Students: 97 full-time (41 women), 56 part-time (17 women); includes 23 minority (2 African Americans, 19 Asian Americans or Pacific Islanders, 2 Hispanic Americans), 77 international. Average age 25. *160 applicants, 33% accepted.* In 1999, 14 master's, 21 doctorates awarded (48% entered university research/teaching, 52% found other work related to degree).

Terminal master's awarded for partial completion of doctoral program.
Degree requirements: For master's, thesis or alternative, exam required, foreign language not required; for doctorate, dissertation, cumulative exams required, foreign language not required. *Average time to degree:* Master's–3 years full-time, 5 years part-time; doctorate–5 years full-time, 8 years part-time.
Entrance requirements: For master's and doctorate, GRE General Test, GRE Subject Test, TOEFL. *Application deadline:* For fall admission, 4/15 (priority date); for spring admission, 11/1. Applications are processed on a rolling basis. *Application fee:* $50.
Expenses: Tuition, state resident: full-time $6,776; part-time $279 per credit. Tuition, nonresident: full-time $9,936; part-time $412 per credit. Required fees: $20 per credit. $89 per semester. Tuition and fees vary according to course load, campus/location and program.
Financial aid: In 1999–00, 110 students received aid, including 37 fellowships, 26 research assistantships with full tuition reimbursements available, 47 teaching assistantships with full tuition reimbursements available; Federal Work-Study also available. Financial aid application deadline: 3/1; financial aid applicants required to submit FAFSA.
Faculty research: Biophysical organic/bioorganic, inorganic/bioinorganic, theoretical, and solid-state/surface chemistry. *Total annual research expenditures:* $6.5 million.
Dr. Roger Jones, Director, 732-445-2259, *Fax:* 732-445-5312, *E-mail:* jones@rutchem.rutgers.edu.

Find an in-depth description at www.petersons.com/graduate.

■ SAINT JOSEPH COLLEGE

Graduate Division, Field of Natural Sciences, Department of Chemistry, West Hartford, CT 06117-2700

AWARDS Chemistry (Certificate); chemistry and biological chemistry (MS). Part-time and evening/weekend programs available.

Faculty: 5 full-time (1 woman).
Students: Average age 33. In 1999, 4 degrees awarded.
Degree requirements: For master's, thesis or alternative required, foreign language not required.
Entrance requirements: For master's, GRE or MAT. *Application deadline:* For fall admission, 7/15 (priority date); for spring admission, 12/1 (priority date). Applications are processed on a rolling basis. *Application fee:* $25.
Expenses: Tuition: Part-time $420 per credit hour. Required fees: $25 per course.

Saint Joseph College (continued)
Financial aid: In 1999–00, 1 research assistantship with full tuition reimbursement was awarded. Financial aid application deadline: 7/15; financial aid applicants required to submit FAFSA.
Harold T. McKone, Chair, 860-231-5241, *Fax:* 860-233-5695.

■ **SAINT LOUIS UNIVERSITY**

School of Medicine and Graduate School, Graduate Programs in Biomedical Sciences, Department of Biochemistry and Molecular Biology, St. Louis, MO 63103-2097
AWARDS PhD.

Faculty: 26 full-time (9 women), 1 part-time/adjunct (0 women).
Students: 2 full-time (both women), 8 part-time, 1 international. Average age 26. In 1999, 6 degrees awarded.
Degree requirements: For doctorate, dissertation, departmental qualifying exams required, foreign language not required.
Entrance requirements: For doctorate, GRE General Test. *Application deadline:* Applications are processed on a rolling basis. *Application fee:* $40.
Expenses: Tuition: Part-time $507 per credit hour. Required fees: $38 per term.
Financial aid: In 1999–00, 9 students received aid, including 8 fellowships, 2 research assistantships; teaching assistantships. Aid available to part-time students. Financial aid application deadline: 8/1; financial aid applicants required to submit FAFSA.
Faculty research: Biochemical genetics, lipid metabolism, cell biology, structure and function of proteins, gene expression.
Dr. William S. Sly, Chairman, 314-577-8131.

Find an in-depth description at www.petersons.com/graduate.

■ **SAN JOSE STATE UNIVERSITY**

Graduate Studies, College of Science, Department of Chemistry, San Jose, CA 95192-0001
AWARDS Analytical chemistry (MS); biochemistry (MS); chemistry (MA); inorganic chemistry (MS); organic chemistry (MS); physical chemistry (MS); polymer chemistry (MS); radiochemistry (MS). Part-time and evening/weekend programs available.

Degree requirements: For master's, thesis or alternative required, foreign language not required.
Entrance requirements: For master's, GRE, minimum B average.
Expenses: Tuition, nonresident: part-time $246 per unit. Required fees: $1,939; $1,309 per year.

Faculty research: Intercalated compounds, organic/biochemical reaction mechanisms, complexing agents in biochemistry, DNA repair, metabolic inhibitors.

■ **THE SCRIPPS RESEARCH INSTITUTE**

Office of Graduate Studies, Macromolecular and Cellular Structure and Chemistry Graduate Program, La Jolla, CA 92037
AWARDS PhD.

Faculty: 84 full-time (12 women).
Students: 73 full-time (29 women). *121 applicants, 26% accepted.* In 1999, 10 degrees awarded.
Degree requirements: For doctorate, dissertation required, foreign language not required. *Average time to degree:* Doctorate–5 years full-time.
Entrance requirements: For doctorate, GRE General Test (average 90th percentile), GRE Subject Test (average 72nd percentile), TOEFL. *Application deadline:* For fall admission, 1/1. *Application fee:* $0.
Expenses: All students are fully funded.
Financial aid: Institutionally sponsored loans and stipends available.
Faculty research: Biocatalysis and enzyme engineering, molecular structure and function, neurosciences, immunology, plant biology.
Application contact: Marylyn Rinaldi, Graduate Program Administrator, 858-784-8469, *Fax:* 858-784-2802, *E-mail:* mrinaldi@scripps.edu.

■ **SETON HALL UNIVERSITY**

College of Arts and Sciences, Department of Chemistry and Biochemistry, South Orange, NJ 07079-2694
AWARDS Analytical chemistry (MS, PhD); biochemistry (MS, PhD); chemistry (MS); inorganic chemistry (MS, PhD); organic chemistry (MS, PhD); physical chemistry (MS, PhD). Part-time and evening/weekend programs available.

Faculty: 14 full-time (2 women).
Students: 25 full-time (12 women), 123 part-time (35 women); includes 41 minority (11 African Americans, 26 Asian Americans or Pacific Islanders, 4 Hispanic Americans), 8 international. Average age 32. *50 applicants, 50% accepted.* In 1999, 5 master's, 6 doctorates awarded (100% found work related to degree). Terminal master's awarded for partial completion of doctoral program.
Degree requirements: For master's, formal seminar required, thesis optional; for doctorate, dissertation, comprehensive

exams, annual seminars required. *Average time to degree:* Master's–2 years full-time, 4 years part-time; doctorate–5 years full-time, 7 years part-time.
Entrance requirements: For master's, TOEFL, undergraduate major in chemistry or related field with minimum 30 credits in chemistry, including 2 semesters of physical chemistry; for doctorate, oral matriculation exam based on proposed doctoral research, minimum GPA of 3.0 in course distribution requirements, formal seminar. *Application deadline:* For fall admission, 7/1; for spring admission, 11/1. Applications are processed on a rolling basis. *Application fee:* $50.
Expenses: Tuition: Full-time $10,404; part-time $578 per credit. Required fees: $185 per year. Tuition and fees vary according to course load, campus/location, program and student's religious affiliation.
Financial aid: In 1999–00, 1 research assistantship, 19 teaching assistantships were awarded.
Faculty research: DNA metal reactions; chromatography; bioinorganic, biophysical, organometallic, polymer chemistry; heterogeneous catalyst.
Dr. Richard D. Sheardy, Chair, 973-761-9416, *Fax:* 973-761-9772.

■ **SOUTH DAKOTA STATE UNIVERSITY**

Graduate School, College of Arts and Science and College of Agriculture and Biological Sciences, Department of Chemistry, Brookings, SD 57007
AWARDS Analytical chemistry (MS, PhD); biochemistry (MS, PhD); chemistry (MS, PhD); inorganic chemistry (MS, PhD); organic chemistry (MS, PhD); physical chemistry (MS, PhD).

Degree requirements: For master's, thesis, oral exam required, foreign language not required; for doctorate, dissertation, preliminary oral and written exams, research tool required.
Entrance requirements: For master's, TOEFL, bachelor's degree in chemistry or equivalent; for doctorate, TOEFL.
Faculty research: Environmental chemistry, computational chemistry, organic synthesis and photochemistry, novel material development and characterization.

■ **SOUTHERN ILLINOIS UNIVERSITY CARBONDALE**

Graduate School, College of Science, Department of Chemistry and Biochemistry, Carbondale, IL 62901-6806
AWARDS MS, PhD. Part-time programs available.

Faculty: 18 full-time (1 woman), 2 part-time/adjunct (0 women).
Students: 25 full-time (12 women), 4 part-time (2 women); includes 2 minority (both African Americans), 16 international. Average age 25. *34 applicants, 38% accepted.* In 1999, 3 master's, 2 doctorates awarded. Terminal master's awarded for partial completion of doctoral program.
Degree requirements: For master's, one foreign language, thesis required; for doctorate, variable foreign language requirement, dissertation required. *Average time to degree:* Master's–5 years full-time; doctorate–3 years full-time.
Entrance requirements: For master's, TOEFL, minimum GPA of 2.7; for doctorate, GRE General Test, TOEFL, minimum GPA of 3.25. *Application deadline:* Applications are processed on a rolling basis. *Application fee:* $0.
Expenses: Tuition, state resident: full-time $2,902. Tuition, nonresident: full-time $5,810. Tuition and fees vary according to course load.
Financial aid: Fellowships with full tuition reimbursements, research assistantships with full tuition reimbursements, teaching assistantships with full tuition reimbursements, Federal Work-Study, institutionally sponsored loans, and tuition waivers (full) available. Aid available to part-time students.
Faculty research: Materials, separations, computational chemistry, synthetics.
David Rosten, Chair, 618-453-6475, *Fax:* 618-453-6408.
Application contact: Steve Scheiner, Chair, Graduate Admissions Committee, 618-453-6476, *Fax:* 618-453-6408, *E-mail:* scheiner@chem.siu.edu.

■ SOUTHERN ILLINOIS UNIVERSITY CARBONDALE

Graduate School, College of Science, Program in Molecular Biology, Microbiology, and Biochemistry, Carbondale, IL 62901-6806

AWARDS MS, PhD.

Faculty: 16 full-time (2 women).
Students: 51 full-time (24 women), 11 part-time (8 women); includes 4 minority (2 African Americans, 1 Asian American or Pacific Islander, 1 Hispanic American), 24 international. Average age 25. *34 applicants, 59% accepted.* In 1999, 7 master's, 5 doctorates awarded.
Degree requirements: For master's and doctorate, thesis/dissertation required, foreign language not required.
Entrance requirements: For master's, GRE, TOEFL, minimum GPA of 2.7; for doctorate, GRE, TOEFL, minimum GPA

of 3.25. *Application deadline:* Applications are processed on a rolling basis. *Application fee:* $20.
Expenses: Tuition, state resident: full-time $2,902. Tuition, nonresident: full-time $5,810. Tuition and fees vary according to course load.
Financial aid: In 1999–00, 40 students received aid, including 3 fellowships with full tuition reimbursements available, 24 research assistantships with full tuition reimbursements available, 12 teaching assistantships with full tuition reimbursements available; Federal Work-Study and institutionally sponsored loans also available. Aid available to part-time students. Financial aid application deadline: 3/1.
Faculty research: Prokaryotic gene regulation and expression; eukaryotic gene regulation; microbial, phylogenetic, and metabolic diversity; immune responses to tumors, pathogens, and autoantigens; protein folding and structure.
Dr. John Martinko, Director, 618-536-2349, *Fax:* 618-453-8036, *E-mail:* martinko.mbmb@science.siu.edu.
Application contact: Donna Mueller, Office Systems Specialist II, 618-536-2349, *Fax:* 618-453-8036, *E-mail:* mueller@micro.siu.edu.

■ SOUTHERN UNIVERSITY AND AGRICULTURAL AND MECHANICAL COLLEGE

Graduate School, College of Sciences, Department of Chemistry, Baton Rouge, LA 70813

AWARDS Analytical chemistry (MS); biochemistry (MS); environmental sciences (MS); inorganic chemistry (MS); organic chemistry (MS); physical chemistry (MS).

Faculty: 9 full-time (2 women), 3 part-time/adjunct (2 women).
Students: 74 full-time (40 women), 102 part-time (60 women); includes 135 minority (130 African Americans, 3 Asian Americans or Pacific Islanders, 2 Hispanic Americans), 32 international. Average age 23. *61 applicants, 72% accepted.* In 1999, 61 degrees awarded (100% entered university research/teaching).
Degree requirements: For master's, thesis required, foreign language not required. *Average time to degree:* Master's–2 years full-time, 2 years part-time.
Entrance requirements: For master's, GMAT or GRE General Test, TOEFL. *Application deadline:* For fall admission, 6/1 (priority date); for spring admission, 11/1. Applications are processed on a rolling basis. *Application fee:* $5.
Expenses: Tuition, state resident: full-time $2,304. Tuition, nonresident: full-time $7,470. Tuition and fees vary according to course load, campus/location and program.

Financial aid: In 1999–00, 31 research assistantships (averaging $7,000 per year), 10 teaching assistantships (averaging $7,000 per year) were awarded; scholarships also available. Financial aid application deadline: 4/15.
Faculty research: Synthesis of macrocyclic ligands, latex accelerators, anticancer drugs, biosensors, absorption isotheums, isolation of specific enzymes from plants. *Total annual research expenditures:* $400,000.
Dr. Robert Harvey Miller, Chairman, 225-771-3990, *Fax:* 225-771-3992.

■ SOUTHWEST TEXAS STATE UNIVERSITY

Graduate School, College of Science, Department of Chemistry and Biochemistry, Program in Biochemistry, San Marcos, TX 78666

AWARDS MS.

Degree requirements: For master's, thesis required, foreign language not required.
Entrance requirements: For master's, GRE General Test, bachelor's degree in chemistry, minimum GPA of 2.75 in last 60 hours. *Application deadline:* For fall admission, 6/15 (priority date); for spring admission, 10/15 (priority date). Applications are processed on a rolling basis. *Application fee:* $25 ($75 for international students).
Expenses: Tuition, state resident: full-time $720; part-time $40 per semester hour. Tuition, nonresident: full-time $4,608; part-time $256 per semester hour. Required fees: $1,470; $122.
Dr. Carl Joseph Carrano, Chair, 512-245-3117, *Fax:* 512-245-2374, *E-mail:* cc05@swt.edu.
Application contact: Dr. James Irvin, Graduate Adviser, 512-245-3113, *E-mail:* ji02@swt.edu.

■ STANFORD UNIVERSITY

School of Medicine, Graduate Programs in Medicine, Department of Biochemistry, Stanford, CA 94305-9991

AWARDS PhD.

Faculty: 11 full-time (1 woman).
Students: 23 full-time (8 women), 7 part-time (5 women); includes 6 minority (all Asian Americans or Pacific Islanders), 3 international. Average age 25. *141 applicants, 7% accepted.* In 1999, 7 doctorates awarded.
Degree requirements: For doctorate, dissertation required.
Entrance requirements: For doctorate, GRE General Test, GRE Subject Test (biology or chemistry), TOEFL. *Application deadline:* For fall admission, 12/15.

Stanford University (continued)
Application fee: $65 ($80 for international students). Electronic applications accepted.
Expenses: Tuition: Full-time $23,058. Required fees: $152. Part-time tuition and fees vary according to course load.
Financial aid: Research assistantships available. Financial aid application deadline: 1/1.
Faculty research: DNA replication, recombination, and gene regulation; methods of isolating, analyzing, and altering genes and genomes; protein structure, protein folding, and protein processing; protein targeting and transport in the cell; intercellular signaling.
Suzanne Pfeffer, Chair, 650-723-6161, *Fax:* 650-723-6783, *E-mail:* pfeffer@ cmgm.stanford.edu.
Application contact: Graduate Admissions Secretary, 650-723-6161.
Find an in-depth description at www.petersons.com/graduate.

■ STATE UNIVERSITY OF NEW YORK AT ALBANY

School of Public Health, Department of Biomedical Sciences, Program in Biochemistry, Molecular Biology, and Genetics, Albany, NY 12222-0001

AWARDS MS, PhD.

Degree requirements: For master's and doctorate, thesis/dissertation required.
Entrance requirements: For master's and doctorate, GRE General Test, GRE Subject Test. *Application deadline:* For fall admission, 1/15 (priority date); for spring admission, 11/1 (priority date). *Application fee:* $50.
Expenses: Tuition, state resident: full-time $5,100; part-time $214 per credit. Tuition, nonresident: full-time $8,416; part-time $352 per credit. Required fees: $31 per credit.
Financial aid: Application deadline: 2/1. Dr. Harry Taber, Chair, Department of Biomedical Sciences, 518-474-2662.

■ STATE UNIVERSITY OF NEW YORK AT BUFFALO

Graduate School, Graduate Programs in Biomedical Sciences at Roswell Park Cancer Institute, Department of Biochemistry at Roswell Park, Buffalo, NY 14263

AWARDS PhD.

Degree requirements: For doctorate, dissertation required.
Entrance requirements: For doctorate, GRE General Test, GRE Subject Test (chemistry, biology, or biochemistry, cell and molecular biology).
Expenses: Tuition, state resident: full-time $5,100; part-time $213 per credit hour.

Tuition, nonresident: full-time $8,416; part-time $351 per credit hour. Required fees: $935; $75 per semester. Tuition and fees vary according to course load and program.
Financial aid: Fellowships, research assistantships, teaching assistantships available.
Application contact: Director of Graduate Studies, 716-829-2727.
Find an in-depth description at www.petersons.com/graduate.

■ STATE UNIVERSITY OF NEW YORK AT BUFFALO

Graduate School, School of Medicine and Biomedical Sciences, Graduate Programs in Medicine and Biomedical Sciences, Department of Biochemistry, Buffalo, NY 14214

AWARDS MA, PhD.

Faculty: 8 full-time (2 women), 1 part-time/adjunct (0 women).
Students: 8 full-time (0 women), 10 part-time (3 women); includes 1 minority (Asian American or Pacific Islander), 10 international. Average age 29. *3 applicants, 33% accepted.* In 1999, 6 degrees awarded. Terminal master's awarded for partial completion of doctoral program.
Degree requirements: For master's, thesis not required; for doctorate, dissertation required.
Entrance requirements: For master's and doctorate, GRE General Test, GRE Subject Test (biology, chemistry, or biochemistry and molecular biology), TOEFL. *Application deadline:* For fall admission, 2/1 (priority date). Applications are processed on a rolling basis. *Application fee:* $35.
Expenses: Tuition, state resident: full-time $5,100. Tuition, nonresident: full-time $8,416. Required fees: $935.
Financial aid: In 1999–00, 1 fellowship with full tuition reimbursement (averaging $16,000 per year), 8 research assistantships with full tuition reimbursements (averaging $14,770 per year), 3 teaching assistantships with full tuition reimbursements (averaging $15,360 per year) were awarded; grants, institutionally sponsored loans, and unspecified assistantships also available. Financial aid application deadline: 2/1; financial aid applicants required to submit FAFSA.
Faculty research: Gene expression, proteins and metalloenzymes, biochemical endocrinology. *Total annual research expenditures:* $1.4 million.
Dr. Murray J. Ettinger, Interim Chair, 716-829-3390, *Fax:* 716-829-2725, *E-mail:* ettinger@acsu.buffalo.edu.

Application contact: Dr. Daniel Kosman, Director of Graduate Studies, 716-829-2842, *Fax:* 716-829-2725.

■ STATE UNIVERSITY OF NEW YORK AT STONY BROOK

Graduate School, College of Arts and Sciences, Department of Biochemistry and Cell Biology, Program in Biochemistry and Structural Biology, Stony Brook, NY 11794

AWARDS PhD.

Expenses: Tuition, state resident: full-time $5,100; part-time $213 per credit hour. Tuition, nonresident: full-time $8,416; part-time $351 per credit hour. Required fees: $492. Tuition and fees vary according to program.
Find an in-depth description at www.petersons.com/graduate.

■ STATE UNIVERSITY OF NEW YORK AT STONY BROOK

Graduate School, College of Arts and Sciences, Molecular and Cellular Biology Program, Specialization in Biochemistry and Molecular Biology, Stony Brook, NY 11794

AWARDS PhD.

Faculty: 98.
Students: 27 full-time (11 women), 23 part-time (10 women); includes 5 minority (4 Asian Americans or Pacific Islanders, 1 Hispanic American), 22 international.
Degree requirements: For doctorate, dissertation, comprehensive exam, teaching experience required.
Entrance requirements: For doctorate, GRE General Test, GRE Subject Test, TOEFL. *Application deadline:* For fall admission, 1/15. *Application fee:* $50.
Expenses: Tuition, state resident: full-time $5,100; part-time $213 per credit hour. Tuition, nonresident: full-time $8,416; part-time $351 per credit hour. Required fees: $492. Tuition and fees vary according to program.
Financial aid: In 2000–01, 10 fellowships, 47 research assistantships, 27 teaching assistantships were awarded; Federal Work-Study also available.
Application contact: Dr. James Quigley, Director, Graduate Program, 631-632-8533, *Fax:* 631-632-9730, *E-mail:* jquigley@path.som.sunysb.edu.
Find an in-depth description at www.petersons.com/graduate.

■ STATE UNIVERSITY OF NEW YORK HEALTH SCIENCE CENTER AT BROOKLYN

School of Graduate Studies, Department of Biochemistry, Brooklyn, NY 11203-2098

AWARDS PhD, MD/PhD.

Degree requirements: For doctorate, one foreign language, dissertation required.
Entrance requirements: For doctorate, GRE.

Expenses: Tuition, state resident: full-time $5,100; part-time $213 per credit. Tuition, nonresident: full-time $8,416; part-time $351 per credit. Required fees: $200. Full-time tuition and fees vary according to program and student level.
Faculty research: Molecular virology, molecular endocrinology, protein chemistry.

■ STATE UNIVERSITY OF NEW YORK UPSTATE MEDICAL UNIVERSITY

College of Graduate Studies, Department of Biochemistry and Molecular Biology, Syracuse, NY 13210-2334

AWARDS MS, PhD, MD/PhD.

Faculty: 10 full-time.
Students: 13 full-time (9 women), 1 part-time; includes 3 minority (all Asian Americans or Pacific Islanders), 6 international. *48 applicants, 10% accepted.* In 1999, 1 master's awarded (100% found work related to degree). Terminal master's awarded for partial completion of doctoral program.
Degree requirements: For master's, thesis required, foreign language not required; for doctorate, dissertation, comprehensive exam required, foreign language not required.
Entrance requirements: For master's and doctorate, GRE General Test, GRE Subject Test, TSE. *Application deadline:* For fall admission, 4/1 (priority date). Applications are processed on a rolling basis. *Application fee:* $40.
Expenses: Tuition, state resident: full-time $5,100; part-time $213 per credit. Tuition, nonresident: full-time $8,416; part-time $351 per credit. Required fees: $410; $25 per credit. Part-time tuition and fees vary according to course load and program.
Financial aid: Fellowships, research assistantships, Federal Work-Study and institutionally sponsored loans available. Aid available to part-time students. Financial aid application deadline: 4/15.
Faculty research: Enzymology, membrane structure and functions, developmental biochemistry.

Dr. Richard L. Cross, Chairperson, 315-464-5127.
Application contact: Dr. David Gilbert, Associate Professor, 315-464-8723.

■ STEVENS INSTITUTE OF TECHNOLOGY

Graduate School, School of Applied Sciences and Liberal Arts, Department of Chemistry and Chemical Biology, Hoboken, NJ 07030

AWARDS Chemistry (MS, PhD, Certificate), including analytical chemistry, chemical biology, chemical physiology (Certificate), instrumental analysis (Certificate), organic chemistry (MS, PhD), physical chemistry (MS, PhD), polymer chemistry. Part-time and evening/weekend programs available. Terminal master's awarded for partial completion of doctoral program.

Degree requirements: For master's, thesis or alternative required, foreign language not required; for doctorate, one foreign language, dissertation required; for Certificate, computer language, project or thesis required.
Entrance requirements: For master's and doctorate, TOEFL. Electronic applications accepted.

■ TEMPLE UNIVERSITY

Health Sciences Center, School of Medicine and Graduate School, Graduate Programs in Medicine, Department of Biochemistry, Philadelphia, PA 19140

AWARDS MS, PhD.

Faculty: 13 full-time (2 women).
Students: 24 full-time (8 women); includes 7 minority (1 African American, 6 Asian Americans or Pacific Islanders), 2 international. *10 applicants, 60% accepted.* In 1999, 2 doctorates awarded.
Degree requirements: For master's, thesis, research seminar required, foreign language not required; for doctorate, dissertation, research seminars required, foreign language not required.
Entrance requirements: For master's and doctorate, GRE General Test, GRE Subject Test, minimum GPA of 3.0. *Application deadline:* For fall admission, 7/11 (priority date); for spring admission, 11/1. *Application fee:* $40. Electronic applications accepted.
Expenses: Tuition, state resident: full-time $6,030. Tuition, nonresident: full-time $8,298. Required fees: $230. One-time fee: $10 full-time.
Financial aid: Fellowships, research assistantships, Federal Work-Study and institutionally sponsored loans available. Financial aid application deadline: 3/1.

Faculty research: Metabolism, enzymology, molecular biology, membranology, biophysics.
Dr. Parkson Chong, Chair, 215-707-4307, *Fax:* 215-707-7536, *E-mail:* reddy@ sgi1.fels.temple.edu.
Application contact: Dr. Charles Grubmeyer, Admissions Chair, 215-707-3263, *Fax:* 215-707-7536, *E-mail:* ctg@ ariel.fels.temple.edu.

Find an in-depth description at www.petersons.com/graduate.

■ TEXAS A&M UNIVERSITY

College of Agriculture and Life Sciences, Department of Biochemistry and Biophysics, College Station, TX 77843

AWARDS Biochemistry (MS, PhD); biophysics (MS).

Faculty: 29 full-time (5 women).
Students: 84 full-time (31 women), 5 part-time (1 woman); includes 4 minority (1 African American, 2 Asian Americans or Pacific Islanders, 1 Hispanic American), 26 international. Average age 27. In 1999, 6 master's, 13 doctorates awarded.
Entrance requirements: For master's and doctorate, GRE General Test, TOEFL. *Application deadline:* For fall admission, 3/1 (priority date). *Application fee:* $50 ($75 for international students).
Expenses: Tuition, state resident: part-time $76 per semester hour. Tuition, nonresident: part-time $292 per semester hour. Required fees: $11 per semester hour. Tuition and fees vary according to program.
Financial aid: In 1999–00, 2 fellowships (averaging $15,000 per year), 67 research assistantships with tuition reimbursements (averaging $17,650 per year), 16 teaching assistantships (averaging $17,650 per year) were awarded. Financial aid application deadline: 4/1; financial aid applicants required to submit FAFSA.
Faculty research: Enzymology, gene expression, protein structure, plant biochemistry.
Dr. James R. Wild, Head, 979-845-1011, *Fax:* 979-845-9274.
Application contact: Susan Whiting, Staff Assistant, 979-845-1779, *Fax:* 979-845-9274, *E-mail:* susanw@bioch.tamu.edu.

Find an in-depth description at www.petersons.com/graduate.

■ TEXAS A&M UNIVERSITY SYSTEM HEALTH SCIENCE CENTER

College of Medicine, Graduate School of Biomedical Sciences, Department of Medical Biochemistry and Genetics, College Station, TX 77840-7896

AWARDS PhD.

Faculty: 13 full-time (1 woman).
Students: 24 full-time (9 women), 4 international. In 1999, 1 degree awarded.
Degree requirements: For doctorate, dissertation required, foreign language not required.
Entrance requirements: For doctorate, GRE General Test. *Application deadline:* For fall admission, 2/1 (priority date). Applications are processed on a rolling basis. *Application fee:* $50 ($75 for international students).
Expenses: Tuition, area resident: Full-time $1,368. Tuition, state resident: part-time $76 per credit. Tuition, nonresident: full-time $5,256; part-time $292 per credit. International tuition: $5,256 full-time. Required fees: $678; $38 per credit. Full-time tuition and fees vary according to course load and student level.
Financial aid: Fellowships, research assistantships available. Financial aid applicants required to submit FAFSA.
Faculty research: Immunology, cell and membrane biology, protein biochemistry, molecular genetics, parasitology, vertebrate embryogenesis and microbiology.
Dr. Hagan Bayley, Head, 979-845-2726, *Fax:* 979-847-9481.
Application contact: Janis Chmiel, Graduate Secretary, 800-298-2260, *Fax:* 979-847-9481, *E-mail:* jchmiel@tamu.edu.

Find an in-depth description at www.petersons.com/graduate.

■ TEXAS TECH UNIVERSITY HEALTH SCIENCES CENTER

Graduate School of Biomedical Sciences, Department of Cell Biology and Biochemistry, Program in Medical Biochemistry, Lubbock, TX 79430

AWARDS MS, PhD, MD/PhD.

Faculty: 18 full-time (4 women), 8 part-time/adjunct (4 women).
Students: 9 full-time (4 women); includes 1 minority (Asian American or Pacific Islander), 4 international. Average age 27. *26 applicants, 8% accepted.* In 1999, 2 doctorates awarded (50% entered university research/teaching, 50% continued full-time study). Terminal master's awarded for partial completion of doctoral program.
Degree requirements: For master's and doctorate, thesis/dissertation, preliminary,

comprehensive, and final exams required, foreign language not required. *Average time to degree:* Doctorate–6 years full-time.
Entrance requirements: For master's and doctorate, GRE General Test, TOEFL, minimum GPA of 3.0. *Application deadline:* For fall admission, 4/15 (priority date). *Application fee:* $30 ($55 for international students).
Expenses: Tuition, state resident: part-time $38 per credit hour. Tuition, nonresident: part-time $254 per credit hour. Part-time tuition and fees vary according to program.
Financial aid: In 1999–00, 1 fellowship with full tuition reimbursement (averaging $14,688 per year), 8 research assistantships (averaging $14,500 per year) were awarded.
Faculty research: Reproductive endocrinology, immunology, developmental biochemistry.
Dr. Charles Faust, Graduate Director, 806-743-2500, *Fax:* 806-743-2990, *E-mail:* charles.faust@ttmc.ttuhsc.edu.

Find an in-depth description at www.petersons.com/graduate.

■ THOMAS JEFFERSON UNIVERSITY

College of Graduate Studies, Program in Biochemistry and Molecular Biology, Philadelphia, PA 19107

AWARDS PhD.

Faculty: 18 full-time (5 women), 17 part-time/adjunct (2 women).
Students: 21 full-time (14 women); includes 4 minority (2 African Americans, 1 Asian American or Pacific Islander, 1 Hispanic American). Average age 24. *82 applicants, 18% accepted.* In 1999, 9 degrees awarded.
Degree requirements: For doctorate, dissertation, preliminary exam required, foreign language not required. *Average time to degree:* Doctorate–4.5 years full-time.
Entrance requirements: For doctorate, GRE General Test, TOEFL, minimum GPA of 3.2. *Application deadline:* For fall admission, 3/1 (priority date). Applications are processed on a rolling basis. *Application fee:* $40.
Expenses: Tuition: Full-time $12,670. Tuition and fees vary according to degree level and program.
Financial aid: Fellowships with full tuition reimbursements, research assistantships, Federal Work-Study, institutionally sponsored loans, traineeships, and training grants available. Aid available to part-time students. Financial aid application deadline: 5/1; financial aid applicants required to submit FAFSA.

Faculty research: Human mitochondrial genetics, molecular biology of protein-RNA interaction, mammalian mitochondrial biogenesis and function, glucocorticoid-induced programmed cell death, molecular cytogenetics of aneuploidy syndromes, cri-du-chat syndrome. *Total annual research expenditures:* $1.9 million.
Dr. Gerald Litwack, Chairman, Graduate Committee, 215-503-4634, *Fax:* 215-503-5393, *E-mail:* gerry.litwack@mail.tju.edu.
Application contact: Jessie F. Pervall, Director of Admissions, 215-503-4400, *Fax:* 215-503-3433, *E-mail:* cgs-info@mail.tju.edu.

Find an in-depth description at www.petersons.com/graduate.

■ THOMAS JEFFERSON UNIVERSITY

College of Graduate Studies, Program in Biomedical Chemistry, Philadelphia, PA 19107

AWARDS MS. Part-time and evening/weekend programs available.

Faculty: 9 full-time (1 woman).
Students: 4 full-time (1 woman), 32 part-time (20 women); includes 12 minority (6 African Americans, 6 Asian Americans or Pacific Islanders), 3 international. Average age 26. *30 applicants, 77% accepted.* In 1999, 10 degrees awarded.
Degree requirements: For master's, thesis required, foreign language not required.
Entrance requirements: For master's, GRE General Test, minimum GPA of 3.0. *Application deadline:* Applications are processed on a rolling basis. *Application fee:* $40.
Expenses: Tuition: Full-time $17,625; part-time $610 per credit.
Financial aid: In 1999–00, 13 students received aid. Federal Work-Study and institutionally sponsored loans available. Aid available to part-time students. Financial aid application deadline: 5/1; financial aid applicants required to submit FAFSA.
Faculty research: Genetics, protein chemistry, clinical enzymology, chemical carcinogenesis, autoimmunity.
Dr. Georganne K. Buescher, Associate Dean, 215-503-5799, *Fax:* 215-503-3433, *E-mail:* georganne.buescher@mail.tju.edu.
Application contact: Jessie F. Pervall, Director of Admissions, 215-503-4400, *Fax:* 215-503-3433, *E-mail:* cgs-info@mail.tju.edu.

■ TUFTS UNIVERSITY

Sackler School of Graduate Biomedical Sciences, Department of Biochemistry, Medford, MA 02155

AWARDS PhD.

Faculty: 19 full-time (4 women), 2 part-time/adjunct (1 woman).
Students: 21 full-time (13 women); includes 6 minority (all Asian Americans or Pacific Islanders), 4 international. Average age 28. *103 applicants, 4% accepted.* In 1999, 3 degrees awarded (67% found work related to degree, 33% continued full-time study).
Degree requirements: For doctorate, dissertation required, foreign language not required. *Average time to degree:* Doctorate–6 years full-time.
Entrance requirements: For doctorate, GRE General Test, GRE Subject Test, TOEFL. *Application deadline:* For fall admission, 1/15 (priority date). Applications are processed on a rolling basis. *Application fee:* $45.
Expenses: Tuition: Full-time $19,325.
Financial aid: In 1999–00, 21 students received aid, including 21 research assistantships with full tuition reimbursements available (averaging $18,805 per year). Financial aid application deadline: 2/1.
Faculty research: Enzymes and mechanisms, signal transduction, NMR spectroscopy, DNA biosynthesis, membrane function. *Total annual research expenditures:* $3.3 million.
Dr. Brian S. Schaffhausen, Program Director, 617-636-6876, *Fax:* 617-636-2409, *E-mail:* bschaffh_pol@opal.tufts.edu.
Application contact: Staff Assistant, 617-636-6767, *Fax:* 617-636-0375, *E-mail:* sackler-school@tufts.edu.

■ TULANE UNIVERSITY

School of Medicine and Graduate School, Graduate Programs in Medicine, Department of Biochemistry, New Orleans, LA 70118-5669

AWARDS MS, PhD, MD/PhD. MS and PhD offered through the Graduate School.

Students: 11 full-time (6 women), 2 part-time (both women); includes 3 minority (all African Americans), 3 international. *34 applicants, 18% accepted.* In 1999, 3 doctorates awarded.
Degree requirements: For master's, thesis required, foreign language not required; for doctorate, 2 foreign languages (computer language can substitute for one), dissertation required.
Entrance requirements: For master's, GRE General Test, GRE Subject Test, TOEFL or TSE, minimum B average in undergraduate course work; for doctorate, GRE General Test, GRE Subject Test,

TOEFL or TSE. *Application deadline:* For fall admission, 2/1. *Application fee:* $45.
Expenses: Tuition: Full-time $23,030.
Financial aid: Fellowships, research assistantships available. Financial aid application deadline: 2/1.
Dr. Jim Karam, Chairman, 504-588-5291.

Find an in-depth description at www.petersons.com/graduate.

■ UNIFORMED SERVICES UNIVERSITY OF THE HEALTH SCIENCES

School of Medicine, Division of Basic Medical Sciences, Department of Biochemistry, Bethesda, MD 20814-4799

AWARDS PhD.

Faculty: 13 full-time (4 women), 2 part-time/adjunct (1 woman).
Students: In 1999, 1 degree awarded (100% found work related to degree).
Degree requirements: For doctorate, one foreign language (computer language can substitute), dissertation, qualifying exam required.
Entrance requirements: For doctorate, GRE General Test, minimum GPA of 3.0, U.S. citizenship. *Application deadline:* For fall admission, 1/15 (priority date). Applications are processed on a rolling basis. *Application fee:* $0.
Financial aid: Fellowships, tuition waivers (full) available.
Faculty research: Biochemistry of hemoglobin, nutritional biochemistry, basic mechanisms of cell replication, immunological mechanisms and molecular structure.
Dr. Isaiahu Shechter, Chair, 301-295-3550.
Application contact: Janet M. Anastasi, Graduate Program Coordinator, 301-295-9474, *Fax:* 301-295-6772, *E-mail:* janastasi@usuhs.mil.

■ THE UNIVERSITY OF AKRON

Graduate School, Buchtel College of Arts and Sciences, Department of Chemistry, Akron, OH 44325-0001

AWARDS Analytical chemistry (MS, PhD); biochemistry (MS, PhD); chemistry (MS, PhD); inorganic chemistry (MS, PhD); organic chemistry (MS, PhD); physical chemistry (MS, PhD). Part-time and evening/weekend programs available. Terminal master's awarded for partial completion of doctoral program.

Degree requirements: For master's, one foreign language (computer language can substitute), thesis, seminar presentation required; for doctorate, 2 foreign languages (computer language can substitute for one), dissertation, cumulative exams required.

Entrance requirements: For master's, TOEFL, minimum GPA of 2.75; for doctorate, TOEFL.
Expenses: Tuition, state resident: part-time $189 per credit. Tuition, nonresident: part-time $353 per credit. Required fees: $7.3 per credit.
Faculty research: NMR studies, catalyzing organic reactions, free radical chemistry, natural product synthesis, laser spectroscopy.

■ THE UNIVERSITY OF ALABAMA AT BIRMINGHAM

Graduate School and School of Medicine and School of Dentistry, Graduate Programs in Joint Health Sciences, Department of Biochemistry and Molecular Genetics, Birmingham, AL 35294

AWARDS Biochemistry (PhD). The department participates in the Cellular and Molecular Biology Graduate Program.

Students: 61 full-time (23 women); includes 4 minority (2 African Americans, 2 Asian Americans or Pacific Islanders), 20 international. *19 applicants, 58% accepted.* In 1999, 10 degrees awarded.
Degree requirements: For doctorate, dissertation required, foreign language not required.
Entrance requirements: For doctorate, GRE General Test, interview. *Application deadline:* Applications are processed on a rolling basis. Electronic applications accepted.
Expenses: Tuition, state resident: part-time $104 per semester hour. Tuition, nonresident: part-time $208 per semester hour. Required fees: $17 per semester hour. $57 per quarter. Tuition and fees vary according to program.
Financial aid: In 1999–00, 8 fellowships were awarded.
Dr. Jeffrey Engler, Interim Chair, 205-934-4753, *E-mail:* engler@uab.edu.
Application contact: Information Contact, 205-934-6034, *Fax:* 205-975-2547.

Find an in-depth description at www.petersons.com/graduate.

■ UNIVERSITY OF ALASKA FAIRBANKS

Graduate School, College of Science, Engineering and Mathematics, Department of Chemistry, Program in Biochemistry, Fairbanks, AK 99775

AWARDS MS, PhD.

Students: 8 full-time (5 women), 3 part-time (2 women); includes 1 minority (Asian American or Pacific Islander), 4 international. Average age 31. *8 applicants,*

University of Alaska Fairbanks (continued)

50% accepted. In 1999, 1 master's, 1 doctorate awarded.

Degree requirements: For master's, thesis, comprehensive exam required, foreign language not required; for doctorate, one foreign language (computer language can substitute), dissertation, comprehensive exam required.
Entrance requirements: For master's and doctorate, GRE General Test, GRE Subject Test (biology or chemistry), TOEFL. *Application deadline:* For fall admission, 3/1. *Application fee:* $35.
Expenses: Tuition, state resident: full-time $3,006; part-time $167 per credit. Tuition, nonresident: full-time $5,868; part-time $326 per credit. Required fees: $370; $10 per credit. $140 per semester.
Financial aid: Research assistantships, teaching assistantships available. Financial aid application deadline: 3/1.

■ THE UNIVERSITY OF ARIZONA

College of Medicine, Graduate Programs in Medicine and Graduate College, Department of Biochemistry, Tucson, AZ 85721

AWARDS MS, PhD. Terminal master's awarded for partial completion of doctoral program.

Degree requirements: For master's, thesis optional, foreign language not required; for doctorate, dissertation required, foreign language not required.
Entrance requirements: For master's and doctorate, GRE General Test.
Expenses: Tuition, nonresident: full-time $4,814; part-time $274 per unit. Required fees: $1,094; $115 per unit.
Faculty research: Membrane biochemistry, lipid biochemistry, neurobiochemistry, protein synthesis and degradation, bioenergetics.

Find an in-depth description at www.petersons.com/graduate.

■ UNIVERSITY OF ARKANSAS FOR MEDICAL SCIENCES

College of Medicine and Graduate School, Graduate Programs in Medicine, Department of Biochemistry and Molecular Biology, Little Rock, AR 72205-7199

AWARDS MS, PhD, MD/PhD.

Faculty: 25 full-time (5 women), 7 part-time/adjunct (3 women).
Students: 28 full-time (11 women), 5 part-time (3 women); includes 6 minority (all Asian Americans or Pacific Islanders), 5 international. In 1999, 2 master's, 5 doctorates awarded.

Degree requirements: For master's, thesis, comprehensive exam required, foreign language not required; for doctorate, dissertation, qualifying exam required, foreign language not required.
Entrance requirements: For master's, GRE General Test, TOEFL, bachelor's degree in biology, chemistry, or related field; for doctorate, GRE General Test. *Application fee:* $0.
Expenses: Tuition: Full-time $8,928.
Financial aid: In 1999–00, 27 research assistantships were awarded; unspecified assistantships also available. Aid available to part-time students.
Faculty research: Gene regulation, growth factors, oncogenes, metabolic diseases, hormone regulation.
Dr. Alan D. Elbein, Chairman, 501-686-5185.
Application contact: Dr. Richard Drake, Graduate Coordinator, 501-686-5185.

■ UNIVERSITY OF CALIFORNIA, BERKELEY

Graduate Division, Group in Comparative Biochemistry, Berkeley, CA 94720-1500

AWARDS PhD.

Degree requirements: For doctorate, dissertation, qualifying exam required, foreign language not required.
Entrance requirements: For doctorate, GRE General Test, GRE Subject Test (biology, chemistry, or biochemistry), TOEFL, minimum GPA of 3.0.
Expenses: Tuition, nonresident: full-time $9,804. Required fees: $4,268. Tuition and fees vary according to program.

■ UNIVERSITY OF CALIFORNIA, DAVIS

Graduate Studies, Programs in the Biological Sciences, Program in Biochemistry and Molecular Biology, Davis, CA 95616

AWARDS MS, PhD.

Faculty: 80 full-time (14 women).
Students: 59 full-time (28 women); includes 20 minority (2 African Americans, 16 Asian Americans or Pacific Islanders, 2 Hispanic Americans), 2 international. Average age 28. *105 applicants, 38% accepted.* In 1999, 8 doctorates awarded.
Degree requirements: For master's and doctorate, thesis/dissertation required.
Entrance requirements: For master's and doctorate, GRE General Test, GRE Subject Test. *Application deadline:* For fall admission, 4/1. *Application fee:* $40. Electronic applications accepted.
Expenses: Tuition, nonresident: full-time $9,804. Tuition and fees vary according to program and student level.

Financial aid: In 1999–00, 55 students received aid, including 13 fellowships with full and partial tuition reimbursements available, 28 research assistantships with full and partial tuition reimbursements available, 9 teaching assistantships with partial tuition reimbursements available; Federal Work-Study, institutionally sponsored loans, and scholarships also available. Financial aid application deadline: 1/15; financial aid applicants required to submit FAFSA.
Faculty research: Gene expression, protein structure, molecular virology, protein synthesis, enzymology, membrane transport and structural biology.
J. Clark Lagarias, Chair, 530-752-1865, *E-mail:* jclagarias@ucdavis.edu.
Application contact: Dawne Shell, Graduate Program Assistant, 530-752-9091, *Fax:* 530-752-8822, *E-mail:* drshell@ucdavis.edu.

Find an in-depth description at www.petersons.com/graduate.

■ UNIVERSITY OF CALIFORNIA, IRVINE

College of Medicine and Office of Research and Graduate Studies, Graduate Programs in Medicine and School of Biological Sciences, Department of Biological Chemistry, Irvine, CA 92697

AWARDS Biological sciences (PhD). Students apply through the Graduate Program in Molecular Biology, Genetics, and Biochemistry.

Faculty: 10 full-time (2 women).
Students: 10 full-time (6 women); includes 3 minority (all Asian Americans or Pacific Islanders), 2 international. *292 applicants, 25% accepted.*
Degree requirements: For doctorate, dissertation required.
Entrance requirements: For doctorate, GRE General Test, GRE Subject Test. *Application deadline:* For fall admission, 2/1 (priority date). *Application fee:* $40. Electronic applications accepted.
Expenses: Tuition, nonresident: full-time $10,322; part-time $1,720 per quarter. Required fees: $5,354; $1,300 per quarter. Tuition and fees vary according to program.
Financial aid: Fellowships, research assistantships, teaching assistantships, institutionally sponsored loans and tuition waivers (full and partial) available. Financial aid application deadline: 3/2; financial aid applicants required to submit FAFSA.
Faculty research: RNA splicing, mammalian chromosomal organization,

membrane-hormone interactions, regulation of protein synthesis, molecular genetics of metabolic processes.

Suzanne Sandmeyer, Chair, 949-824-7571, *Fax:* 949-824-7407, *E-mail:* gp-mbgb@uci.edu.

Application contact: Administrator, 949-824-8145, *Fax:* 949-824-7407, *E-mail:* gp-mbgb@uci.edu.

Find an in-depth description at www.petersons.com/graduate.

■ **UNIVERSITY OF CALIFORNIA, IRVINE**

Office of Research and Graduate Studies, School of Biological Sciences, Department of Molecular Biology and Biochemistry, Irvine, CA 92697

AWARDS Biological sciences (PhD).

Faculty: 22 full-time (4 women).
Students: 12 full-time (6 women), 2 part-time. *252 applicants, 34% accepted.* In 1999, 2 doctorates awarded.
Degree requirements: For doctorate, dissertation required.
Entrance requirements: For doctorate, GRE General Test, GRE Subject Test. *Application deadline:* For fall admission, 2/1 (priority date). Applications are processed on a rolling basis. *Application fee:* $40. Electronic applications accepted.
Expenses: Tuition, nonresident: full-time $10,244; part-time $1,720 per quarter. Required fees: $5,252; $1,300 per quarter. Tuition and fees vary according to course load and program.
Financial aid: Fellowships, research assistantships, teaching assistantships, institutionally sponsored loans and tuition waivers (full and partial) available. Financial aid application deadline: 3/2; financial aid applicants required to submit FAFSA.
Faculty research: Structure and synthesis of nucleic acids and proteins, regulation, virology, biochemical genetics, gene organization.

Jerry Manning, Chair, 949-824-5578, *Fax:* 949-824-8551, *E-mail:* gp-mbgb@uci.edu.
Application contact: Administrator, 949-824-8145, *Fax:* 949-824-7407, *E-mail:* gp-mbgb@uci.edu.

Find an in-depth description at www.petersons.com/graduate.

■ **UNIVERSITY OF CALIFORNIA, IRVINE**

Office of Research and Graduate Studies, School of Biological Sciences and College of Medicine, Graduate Program in Molecular Biology, Genetics, and Biochemistry, Irvine, CA 92697-1450

AWARDS Biological sciences (PhD).

Faculty: 102 full-time (20 women).
Students: 30 full-time (14 women); includes 7 minority (4 Asian Americans or Pacific Islanders, 3 Hispanic Americans), 2 international. Average age 25. *287 applicants, 37% accepted.*
Degree requirements: For doctorate, dissertation, teaching assignment required. *Average time to degree:* Doctorate–5 years full-time.
Entrance requirements: For doctorate, GRE General Test, GRE Subject Test (biochemistry, cell and molecular biology; biology; or chemistry), TOEFL, TSE, minimum GPA of 3.0, research experience. *Application deadline:* For fall admission, 1/1. *Application fee:* $40. Electronic applications accepted.
Expenses: Tuition, nonresident: full-time $9,384. Required fees: $5,178. Full-time tuition and fees vary according to program.
Financial aid: In 1999–00, 30 students received aid, including 30 fellowships with full tuition reimbursements available (averaging $33,995 per year); grants, institutionally sponsored loans, scholarships, and tuition waivers (full) also available. Financial aid application deadline: 3/2; financial aid applicants required to submit FAFSA.
Faculty research: Cellular biochemistry; gene structure and expression; protein structure, function, and design; molecular genetics; pathogenesis and inherited disease; molecular virology and immunology.

Dr. Rozanne Sandri-Goldin, Director, 949-824-7570, *Fax:* 949-824-7407, *E-mail:* gp-mbgb@uci.edu.
Application contact: Kimberly McKinney, Administrator, 949-824-8145, *Fax:* 949-824-7407, *E-mail:* gp-mbgb@uci.edu.

Find an in-depth description at www.petersons.com/graduate.

■ **UNIVERSITY OF CALIFORNIA, LOS ANGELES**

Graduate Division, College of Letters and Science, Department of Chemistry and Biochemistry, Program in Biochemistry and Molecular Biology, Los Angeles, CA 90095

AWARDS MS, PhD. MS admission to program only under exceptional circumstances.

Students: 94 full-time (29 women); includes 28 minority (2 African Americans, 22 Asian Americans or Pacific Islanders, 3 Hispanic Americans, 1 Native American), 19 international. *111 applicants, 29% accepted.*
Entrance requirements: For master's, GRE General Test, GRE Subject Test, minimum GPA of 3.0; for doctorate, GRE General Test, GRE Subject Test, minimum undergraduate GPA of 3.0. *Application deadline:* For fall admission, 1/15. *Application fee:* $40. Electronic applications accepted.
Expenses: Tuition, nonresident: full-time $9,804. Required fees: $4,405. Full-time tuition and fees vary according to program and student level.
Financial aid: In 1999–00, 67 fellowships, 82 research assistantships, 32 teaching assistantships were awarded; scholarships also available.
Application contact: Departmental Office, 310-825-3150, *E-mail:* grad@chem.ucla.edu.

■ **UNIVERSITY OF CALIFORNIA, LOS ANGELES**

School of Medicine and Graduate Division, Graduate Programs in Medicine, Department of Biological Chemistry, Los Angeles, CA 90095
AWARDS MS, PhD.

Students: 61 full-time (30 women); includes 25 minority (22 Asian Americans or Pacific Islanders, 2 Hispanic Americans, 1 Native American), 11 international. *6 applicants, 100% accepted.*
Degree requirements: For master's, comprehensive exam or thesis required; for doctorate, dissertation, oral and written qualifying exams required, foreign language not required.
Entrance requirements: For master's and doctorate, GRE General Test. *Application fee:* $40.
Expenses: Tuition, nonresident: full-time $9,804. Required fees: $4,405.
Financial aid: In 1999–00, 53 students received aid, including 53 fellowships, 50 research assistantships, 13 teaching assistantships; Federal Work-Study, institutionally sponsored loans, scholarships, and tuition waivers (full and partial)

University of California, Los Angeles (continued)
also available. Financial aid application deadline: 3/1.
Dr. Elizabeth Neufeld, Chair, 310-825-2762.
Application contact: UCLA Access Coordinator, 800-284-8252, *Fax:* 310-206-5280, *E-mail:* uclaaccess@ibes.medsch.ucla.edu.

Find an in-depth description at www.petersons.com/graduate.

■ UNIVERSITY OF CALIFORNIA, RIVERSIDE

Graduate Division, College of Natural and Agricultural Sciences, Program in Biochemistry and Molecular Biology, Riverside, CA 92521-0102

AWARDS MS, PhD. Part-time programs available.

Faculty: 39 full-time (11 women).
Students: 41 full-time (15 women), 2 part-time (both women); includes 10 minority (9 Asian Americans or Pacific Islanders, 1 Hispanic American), 10 international. Average age 26. In 1999, 17 master's, 4 doctorates awarded. Terminal master's awarded for partial completion of doctoral program.
Degree requirements: For master's, comprehensive exams or thesis required; for doctorate, dissertation, 2 quarters of teaching experience, qualifying exams required, foreign language not required. *Average time to degree:* Master's–1 year full-time; doctorate–6 years full-time.
Entrance requirements: For master's and doctorate, GRE General Test, TOEFL, minimum GPA of 3.2. *Application deadline:* For fall admission, 5/1; for winter admission, 9/1; for spring admission, 12/1. Applications are processed on a rolling basis. *Application fee:* $40.
Expenses: Tuition, nonresident: full-time $9,804. Required fees: $4,758. Full-time tuition and fees vary according to program.
Financial aid: In 1999–00, 5 fellowships with full tuition reimbursements (averaging $16,000 per year) were awarded; research assistantships, career-related internships or fieldwork, Federal Work-Study, institutionally sponsored loans, and tuition waivers (full and partial) also available. Financial aid application deadline: 2/1; financial aid applicants required to submit FAFSA.
Faculty research: Structural biology and molecular biophysics, signal transduction, plant biochemistry and molecular biology, gene expression and metabolic regulation, molecular toxicology and pathogenesis, molecular endocrinology. *Total annual research expenditures:* $5.6 million.

Dr. Stephen Spindler, Chair, 909-787-4227, *Fax:* 909-787-3590.
Application contact: Janet Fast, Graduate Program Assistant, 909-787-5093, *Fax:* 909-787-3590, *E-mail:* janet.fast@ucr.edu.

Find an in-depth description at www.petersons.com/graduate.

■ UNIVERSITY OF CALIFORNIA, SAN DIEGO

Graduate Studies and Research, Department of Biology, Program in Biochemistry, La Jolla, CA 92093-0348

AWARDS PhD. Offered in association with the Salk Institute.

Degree requirements: For doctorate, dissertation required.
Entrance requirements: For doctorate, GRE General Test, pre-application beginning in September. *Application deadline:* For fall admission, 1/7. *Application fee:* $40.
Expenses: Tuition, nonresident: full-time $14,691. Required fees: $4,697. Full-time tuition and fees vary according to program.
Financial aid: Tuition waivers (full) and stipends available.
Application contact: 858-534-3835.

■ UNIVERSITY OF CALIFORNIA, SAN DIEGO

Graduate Studies and Research, Department of Chemistry and Biochemistry, La Jolla, CA 92093

AWARDS Chemistry (PhD).

Faculty: 52.
Students: 169 (61 women). *410 applicants, 37% accepted.* In 1999, 25 doctorates awarded.
Degree requirements: For doctorate, dissertation required.
Entrance requirements: For doctorate, GRE General Test, GRE Subject Test. *Application fee:* $40.
Expenses: Tuition, nonresident: full-time $14,691. Required fees: $4,697. Full-time tuition and fees vary according to program.
Mark Thiemens, Chair.
Application contact: Applications Coordinator, 858-534-6871.

Find an in-depth description at www.petersons.com/graduate.

■ UNIVERSITY OF CALIFORNIA, SAN FRANCISCO

Graduate Division and School of Medicine, Department of Biochemistry and Biophysics, Program in Biochemistry and Molecular Biology, San Francisco, CA 94143

AWARDS PhD, MD/PhD.

Students: In 1999, 22 degrees awarded.
Degree requirements: For doctorate, dissertation required, foreign language not required.
Entrance requirements: For doctorate, GRE General Test, GRE Subject Test, TOEFL. *Application deadline:* For fall admission, 1/5. *Application fee:* $40.
Expenses: All students guaranteed an annual stipend of $17,600.
Financial aid: Fellowships available. Financial aid application deadline: 2/1.
Faculty research: Structural biology, genetics, cell biology, cell physiology, metabolism.
Keith R. Yamamoto, Director, 415-476-3941.
Application contact: Ray Herrman, Graduate Admissions Assistant, 415-476-3941, *E-mail:* admissions@biochem.ucsf.edu.

Find an in-depth description at www.petersons.com/graduate.

■ UNIVERSITY OF CALIFORNIA, SAN FRANCISCO

School of Pharmacy and Graduate Division, Chemistry and Chemical Biology Graduate Program, San Francisco, CA 94143

AWARDS PhD, Pharm D/PhD.

Faculty: 36 full-time (5 women), 2 part-time/adjunct (0 women).
Students: 45 full-time (20 women); includes 13 minority (2 African Americans, 11 Asian Americans or Pacific Islanders), 7 international. Average age 27. *71 applicants, 20% accepted.* In 1999, 7 degrees awarded.
Degree requirements: For doctorate, dissertation required, foreign language not required. *Average time to degree:* Doctorate–5 years full-time.
Entrance requirements: For doctorate, GRE General Test, TOEFL, minimum GPA of 3.0. *Application deadline:* For fall admission, 1/15. Applications are processed on a rolling basis. *Application fee:* $40. Electronic applications accepted.
Expenses: Tuition, nonresident: full-time $98,042. Required fees: $7,757. Full-time tuition and fees vary according to program and student level.
Financial aid: In 1999–00, 7 fellowships, 7 research assistantships, 7 teaching assistantships were awarded; career-related internships or fieldwork, grants, institutionally sponsored loans, scholarships, and tuition waivers (full) also available. Financial aid application deadline: 1/10.
Faculty research: Biochemistry; macromolecular structure; cellular and molecular pharmacology; physical chemistry and computational biology; synthetic chemistry.

Charles S. Craik, Director, 415-476-1913, *Fax:* 415-502-4690.

Application contact: Christine Olson, Graduate Program Coordinator, 415-476-1914, *Fax:* 415-502-4690, *E-mail:* ccb@picasso.ucsf.edu.

Find an in-depth description at www.petersons.com/graduate.

■ **UNIVERSITY OF CALIFORNIA, SANTA BARBARA**

Graduate Division, College of Letters and Sciences, Division of Mathematics, Life, and Physical Sciences, Interdepartmental Program in Biochemistry and Molecular Biology, Santa Barbara, CA 93106

AWARDS PhD.

Faculty: 21 part-time/adjunct (3 women). **Students:** 15 full-time (7 women). Average age 27. *55 applicants, 27% accepted.* In 1999, 1 doctorate awarded.
Degree requirements: For doctorate, dissertation required, foreign language not required. *Average time to degree:* Doctorate–6 years full-time.
Entrance requirements: For doctorate, GRE General Test, TOEFL. *Application deadline:* For fall admission, 12/15 (priority date). *Application fee:* $40. Electronic applications accepted.
Expenses: Tuition, state resident: full-time $14,637. Tuition, nonresident: full-time $24,441.
Financial aid: In 1999–00, 15 students received aid, including 5 fellowships with full tuition reimbursements available (averaging $17,500 per year); research assistantships, teaching assistantships with partial tuition reimbursements available, career-related internships or fieldwork, Federal Work-Study, institutionally sponsored loans, and tuition waivers (full and partial) also available. Financial aid application deadline: 12/15; financial aid applicants required to submit FAFSA.
Faculty research: Genetics and biochemistry of bacterial gene expression; structure-function relationships in proteins and nucleic acids, protein chemistry; biochemistsry and biophysics of marine adhesion.
James Cooper, Director, 805-893-8028, *Fax:* 805-893-7558.
Application contact: *E-mail:* bmbprog@lifesci.ucsb.edu.

Find an in-depth description at www.petersons.com/graduate.

■ **UNIVERSITY OF CHICAGO**

Division of the Biological Sciences, Molecular Biosciences: Biochemistry, Genetics, Cell and Developmental Biology, Department of Biochemistry and Molecular Biology, Chicago, IL 60637-1513

AWARDS PhD, MD/PhD.

Faculty: 32 full-time (3 women). **Students:** 32 full-time (10 women); includes 12 minority (1 African American, 8 Asian Americans or Pacific Islanders, 3 Hispanic Americans). Average age 26. In 1999, 5 doctorates awarded (60% entered university research/teaching, 40% found other work related to degree).
Degree requirements: For doctorate, one foreign language, dissertation, qualifying exam required. *Average time to degree:* Doctorate–6 years full-time.
Entrance requirements: For doctorate, GRE General Test, GRE Subject Test, TOEFL. *Application deadline:* For fall admission, 1/5 (priority date). *Application fee:* $55.
Expenses: Tuition: Full-time $24,804; part-time $3,422 per course. Required fees: $390. Tuition and fees vary according to program.
Financial aid: In 1999–00, 32 students received aid; fellowships, research assistantships, institutionally sponsored loans and traineeships available. Financial aid application deadline: 6/1.
Faculty research: Molecular biology, gene expression, and DNA-protein interactions; membrane biochemistry, molecular endocrinology, and transmembrane signaling; enzyme mechanisms, physical biochemistry. and structural biology. *Total annual research expenditures:* $5 million.
Dr. Anthony A. Kossiakoff, Chairman, 773-702-9297, *Fax:* 773-702-0439, *E-mail:* koss@cummings.uchicago.edu.
Application contact: Anna P. Wright, Graduate Student Administrator, 773-702-0571, *Fax:* 773-702-0439, *E-mail:* anna@cummings.uchicago.edu.

■ **UNIVERSITY OF CINCINNATI**

Division of Research and Advanced Studies, College of Medicine, Graduate Programs in Medicine, Department of Molecular Genetics, Biochemistry, Microbiology and Immunology, Cincinnati, OH 45267

AWARDS PhD.

Faculty: 22 full-time.
Students: 42 full-time (21 women), 3 part-time (2 women); includes 2 minority (both Asian Americans or Pacific Islanders), 3 international. *109 applicants, 17% accepted.* In 1999, 8 doctorates awarded.
Degree requirements: For doctorate, dissertation, qualifying exam required, foreign

language not required. *Average time to degree:* Doctorate–7.3 years full-time.
Entrance requirements: For doctorate, GRE General Test, GRE Subject Test. *Application deadline:* For fall admission, 2/1 (priority date). Applications are processed on a rolling basis. *Application fee:* $30.
Expenses: Tuition, state resident: full-time $5,139; part-time $196 per credit hour. Tuition, nonresident: full-time $10,326; part-time $369 per credit hour. Required fees: $561; $187 per quarter.
Financial aid: Tuition waivers (full) and unspecified assistantships available. Financial aid application deadline: 5/1. *Total annual research expenditures:* $6.8 million.
Dr. Jerry Lingrel, Head, 513-558-5324, *Fax:* 513-558-1190, *E-mail:* jerry.lingrel@uc.edu.
Application contact: Iain Cartwright, Graduate Program Director, 513-558-5532, *Fax:* 513-558-8474, *E-mail:* iain.cartwright@uc.edu.

Find an in-depth description at www.petersons.com/graduate.

■ **UNIVERSITY OF CINCINNATI**

Division of Research and Advanced Studies, McMicken College of Arts and Sciences, Department of Chemistry, Cincinnati, OH 45221-0091

AWARDS Analytical chemistry (MS, PhD); biochemistry (MS, PhD); inorganic chemistry (MS, PhD); organic chemistry (MS, PhD); physical chemistry (MS, PhD); polymer chemistry (MS, PhD). Part-time and evening/weekend programs available.

Faculty: 25 full-time.
Students: 87 full-time (26 women), 26 part-time (3 women); includes 12 minority (1 African American, 11 Asian Americans or Pacific Islanders), 34 international. *91 applicants, 25% accepted.* In 1999, 13 master's, 15 doctorates awarded. Terminal master's awarded for partial completion of doctoral program.
Degree requirements: For master's, thesis optional, foreign language not required; for doctorate, dissertation required, foreign language not required. *Average time to degree:* Master's–3.4 years full-time; doctorate–6.3 years full-time.
Entrance requirements: For master's and doctorate, GRE General Test, GRE Subject Test. *Application deadline:* For fall admission, 2/1. *Application fee:* $30.
Expenses: Tuition, state resident: full-time $5,880; part-time $196 per credit hour. Tuition, nonresident: full-time $11,067; part-time $369 per credit hour. Required fees: $741; $247 per quarter. Tuition and fees vary according to program.

University of Cincinnati (continued)
Financial aid: Fellowships, tuition waivers (full) and unspecified assistantships available. Aid available to part-time students. Financial aid application deadline: 5/1.
Faculty research: Biomedical chemistry, laser chemistry, surface science, chemical sensors, synthesis. *Total annual research expenditures:* $669,373.
Dr. Marshall Wilson, Head, 513-556-9200, *Fax:* 513-556-9239, *E-mail:* marshall.wilson@uc.edu.
Application contact: Thomas Ridgway, Graduate Program Director, 513-556-9200, *Fax:* 513-556-9239, *E-mail:* thomas.ridgway@uc.edu.

■ UNIVERSITY OF COLORADO AT BOULDER

Graduate School, College of Arts and Sciences, Department of Chemistry and Biochemistry, Boulder, CO 80309
AWARDS Biochemistry (PhD); chemical physics (PhD); chemistry (MS, PhD).

Faculty: 37 full-time (8 women).
Students: 104 full-time (41 women), 85 part-time (42 women); includes 17 minority (10 Asian Americans or Pacific Islanders, 4 Hispanic Americans, 3 Native Americans), 11 international. Average age 25. 72 *applicants,* 49% *accepted.* In 1999, 8 master's, 30 doctorates awarded.
Degree requirements: For master's, thesis or alternative, comprehensive exam or thesis required, foreign language not required; for doctorate, dissertation, oral comprehensive, cumulative exam required, foreign language not required.
Entrance requirements: For master's and doctorate, GRE General Test, GRE Subject Test (chemistry or biochemistry), minimum GPA of 3.0. *Application deadline:* For fall admission, 3/1 (priority date). Applications are processed on a rolling basis. *Application fee:* $40 ($60 for international students).
Expenses: Tuition, state resident: part-time $181 per credit hour. Tuition, nonresident: part-time $542 per credit hour. Required fees: $99 per term. Tuition and fees vary according to course load and program.
Financial aid: In 1999–00, 22 fellowships with full tuition reimbursements (averaging $1,657 per year), 110 research assistantships with full tuition reimbursements (averaging $10,958 per year), 46 teaching assistantships with full tuition reimbursements (averaging $11,171 per year) were awarded; institutionally sponsored loans, traineeships, and tuition waivers (full) also available. Aid available to part-time students. Financial aid application deadline: 2/28.

Faculty research: Environmental chemistry, biochemistry, inorganic, organic, and physical chemistry, atmospheric chemistry. *Total annual research expenditures:* $8.2 million.
Carl A. Koval, Chair, 303-492-6533, *Fax:* 303-492-5894.
Application contact: Hilary Oppermann, Graduate Program Assistant, 303-492-8978, *Fax:* 303-492-5894, *E-mail:* hilary.oppermann@colorado.edu.

■ UNIVERSITY OF COLORADO HEALTH SCIENCES CENTER

Graduate School, Programs in Biological and Medical Sciences, Program in Biochemistry, Denver, CO 80262
AWARDS PhD.

Degree requirements: For doctorate, dissertation required, foreign language not required.
Entrance requirements: For doctorate, GRE General Test, TOEFL, minimum GPA of 2.75.
Expenses: Tuition, state resident: full-time $1,512; part-time $56 per hour. Tuition, nonresident: full-time $7,209; part-time $267 per hour. Full-time tuition and fees vary according to course load and program.
Find an in-depth description at www.petersons.com/graduate.

■ UNIVERSITY OF CONNECTICUT

Graduate School, College of Liberal Arts and Sciences, Biological Sciences Group, Storrs, CT 06269
AWARDS Ecology and evolutionary biology (MS, PhD), including botany, ecology, entomology, systematics, zoology; molecular and cell biology (MS, PhD), including biochemistry, biophysics, biotechnology (MS), cell and developmental biology, genetics, microbiology, plant molecular and cell biology; physiology and neurobiology (MS, PhD), including neurobiology, physiology.

Degree requirements: For doctorate, dissertation required.
Entrance requirements: For master's and doctorate, GRE General Test, GRE Subject Test, TOEFL.
Expenses: Tuition, state resident: full-time $5,118. Tuition, nonresident: full-time $13,298. Required fees: $1,022.

■ UNIVERSITY OF CONNECTICUT

Graduate School, College of Liberal Arts and Sciences, Biological Sciences Group, Department of Molecular and Cell Biology, Field of Biochemistry, Storrs, CT 06269
AWARDS MS, PhD.

Degree requirements: For doctorate, dissertation required.
Entrance requirements: For master's and doctorate, GRE General Test, GRE Subject Test, TOEFL.
Expenses: Tuition, state resident: full-time $5,118. Tuition, nonresident: full-time $13,298. Required fees: $1,022.
Find an in-depth description at www.petersons.com/graduate.

■ UNIVERSITY OF CONNECTICUT HEALTH CENTER

Graduate School, Programs in Biomedical Sciences, Program in Genetics, Molecular Biology, and Biochemistry, Farmington, CT 06030
AWARDS PhD, DMD/PhD, MD/PhD.

Faculty: 29.
Students: 31 full-time (18 women); includes 2 minority (1 African American, 1 Asian American or Pacific Islander), 10 international. In 1999, 3 degrees awarded.
Degree requirements: For doctorate, one foreign language (computer language can substitute), dissertation required.
Entrance requirements: For doctorate, GRE General Test, GRE Subject Test, TOEFL. *Application deadline:* For fall admission, 2/1 (priority date); for spring admission, 10/1. Applications are processed on a rolling basis. *Application fee:* $40 ($45 for international students).
Expenses: Tuition, state resident: full-time $5,272; part-time $293 per credit. Tuition, nonresident: full-time $13,696; part-time $761 per credit. Required fees: $320; $198 per semester. One-time fee: $50 full-time. Full-time tuition and fees vary according to course load, program and reciprocity agreements.
Financial aid: In 1999–00, research assistantships (averaging $17,000 per year); fellowships also available.
Dr. Steve Pfeiffer, Director, 860-679-3395.
Application contact: Marizta Barta, Information Contact, 860-679-4306, *Fax:* 860-679-1282, *E-mail:* barta@adp.uchc.edu.

Find an in-depth description at www.petersons.com/graduate.

■ UNIVERSITY OF DELAWARE

College of Arts and Science, Department of Chemistry and Biochemistry, Newark, DE 19716
AWARDS Biochemistry (MA, MS, PhD); chemistry (MA, MS, PhD). Part-time programs available.

Faculty: 30 full-time (5 women), 1 part-time/adjunct (0 women).
Students: 107 full-time (29 women), 4 part-time (1 woman); includes 9 minority (4 African Americans, 3 Asian Americans

or Pacific Islanders, 2 Hispanic Americans), 27 international. Average age 25. *73 applicants, 70% accepted.* In 1999, 8 master's, 13 doctorates awarded. Terminal master's awarded for partial completion of doctoral program.

Degree requirements: For master's, one foreign language, thesis required (for some programs); for doctorate, one foreign language, dissertation, comprehensive exams required. *Average time to degree:* Master's–3.3 years full-time; doctorate–4.7 years full-time.

Entrance requirements: For master's and doctorate, GRE General Test, GRE Subject Test, TOEFL, TSE. *Application deadline:* For fall admission, 3/31 (priority date). Applications are processed on a rolling basis. *Application fee:* $45. Electronic applications accepted.

Expenses: Tuition, state resident: full-time $4,380; part-time $243 per credit. Tuition, nonresident: full-time $12,750; part-time $708 per credit. Required fees: $15 per term. Tuition and fees vary according to program.

Financial aid: In 1999–00, 89 students received aid, including 7 fellowships with full tuition reimbursements available (averaging $17,800 per year), 44 research assistantships with full tuition reimbursements available (averaging $17,800 per year), 33 teaching assistantships with full tuition reimbursements available (averaging $17,800 per year). Financial aid application deadline: 3/31.

Faculty research: Protein studies; mechanism of enzymes; synthesis, electronic structure, and bonding of inorganic and organometallic compounds; spectroscopy studies. *Total annual research expenditures:* $3.9 million.

Dr. Steven D. Brown, Chairman, 302-831-1247, *Fax:* 302-831-6335, *E-mail:* sdb@udel.edu.

Application contact: Dr. Charles Riordan, Graduate Coordinator, 302-831-1247, *Fax:* 302-831-6335, *E-mail:* riordan@udel.edu.

■ UNIVERSITY OF DETROIT MERCY

College of Engineering and Science, Department of Chemistry and Biochemistry, Detroit, MI 48219-0900

AWARDS Economic aspects of chemistry (MSEC); macromolecular chemistry (MS, PhD). Evening/weekend programs available.

Degree requirements: For master's and doctorate, thesis/dissertation required, foreign language not required.

Entrance requirements: For master's, GRE General Test, minimum GPA of 3.0; for doctorate, GRE Subject Test, minimum GPA of 3.0.

Faculty research: Polymer and physical chemistry, industrial aspects of chemistry.

■ UNIVERSITY OF FLORIDA

College of Medicine and Graduate School, Interdisciplinary Program in Biomedical Sciences, Concentration in Biochemistry and Molecular Biology, Gainesville, FL 32611

AWARDS PhD.

Degree requirements: For doctorate, dissertation required, foreign language not required.

Entrance requirements: For doctorate, GRE General Test, TOEFL (average 550), minimum GPA of 3.0. Electronic applications accepted.

Expenses: Tuition, state resident: part-time $144 per credit hour. Tuition, nonresident: part-time $505 per credit hour. Tuition and fees vary according to course level, course load and program.

Faculty research: Gene expression, metabolic regulation, structural biology, enzyme mechanism, membrane transporters.

Find an in-depth description at www.petersons.com/graduate.

■ UNIVERSITY OF FLORIDA

College of Medicine and Graduate School, Interdisciplinary Program in Biomedical Sciences, Department of Biochemistry and Molecular Biology, Gainesville, FL 32610

AWARDS PhD.

Degree requirements: For doctorate, dissertation required, foreign language not required.

Entrance requirements: For doctorate, GRE General Test, TOEFL, minimum GPA of 3.0. Electronic applications accepted.

Expenses: Tuition, state resident: part-time $144 per credit hour. Tuition, nonresident: part-time $505 per credit hour. Tuition and fees vary according to course level, course load and program.

Faculty research: Gene expression, metabolic regulation, structural biology, enzyme mechanism, membrane transporters.

Find an in-depth description at www.petersons.com/graduate.

■ UNIVERSITY OF GEORGIA

Graduate School, College of Arts and Sciences, Department of Biochemistry and Molecular Biology, Athens, GA 30602

AWARDS MS, PhD.

Degree requirements: For master's, one foreign language, thesis required; for

doctorate, one foreign language (computer language can substitute), dissertation required.

Entrance requirements: For master's and doctorate, GRE General Test. Electronic applications accepted.

Expenses: Tuition, state resident: full-time $7,516; part-time $431 per credit hour. Tuition, nonresident: full-time $12,204; part-time $793 per credit hour. Tuition and fees vary according to program.

■ UNIVERSITY OF HAWAII AT MANOA

John A. Burns School of Medicine and Graduate Division, Graduate Programs in Biomedical Sciences, Department of Biochemistry and Biophysics, Honolulu, HI 96822

AWARDS Biochemistry (MS, PhD); biophysics (MS, PhD).

Faculty: 13 full-time (1 woman), 1 part-time/adjunct (0 women).

Students: 5 full-time (1 woman), 1 (woman) part-time, 5 international. Average age 27. *0 applicants, 0% accepted.* In 1999, 1 master's awarded (100% continued full-time study); 2 doctorates awarded. Terminal master's awarded for partial completion of doctoral program.

Degree requirements: For master's, thesis required (for some programs), foreign language not required; for doctorate, dissertation required, foreign language not required. *Average time to degree:* Master's–6 years full-time.

Entrance requirements: For master's and doctorate, GRE General Test, GRE Subject Test. *Application deadline:* For fall admission, 2/1; for spring admission, 9/1. *Application fee:* $25 ($50 for international students).

Expenses: Tuition, state resident: part-time $168 per credit. Tuition, nonresident: part-time $415 per credit. Required fees: $51 per semester. Part-time tuition and fees vary according to course load.

Financial aid: In 1999–00, research assistantships (averaging $16,511 per year); fellowships, teaching assistantships, Federal Work-Study and tuition waivers (full) also available. Financial aid application deadline: 2/1.

Faculty research: Protein and nucleic acid structure, function, and metabolism; endocrine metabolism; molecular basis of cancer; clinical biochemistry. *Total annual research expenditures:* $199,817.

Dr. N. V. Bhagavan, Chairperson, 808-956-8490, *Fax:* 808-956-9498, *E-mail:* bhagavan@jabsom.biomed.hawaii.edu.

■ UNIVERSITY OF HOUSTON

College of Natural Sciences and Mathematics, Department of Biology and Biochemistry, Houston, TX 77004

AWARDS Biochemistry (MS, PhD); biology (MS, PhD).

Faculty: 28 full-time (4 women), 6 part-time/adjunct (1 woman).
Students: 86 full-time (38 women), 13 part-time (10 women); includes 14 minority (1 African American, 10 Asian Americans or Pacific Islanders, 3 Hispanic Americans), 38 international. Average age 29. *188 applicants, 12% accepted.* In 1999, 10 master's, 5 doctorates awarded. Terminal master's awarded for partial completion of doctoral program.
Degree requirements: For master's, thesis required (for some programs), foreign language not required; for doctorate, dissertation, oral and written comprehensive exam required, foreign language not required.
Entrance requirements: For master's and doctorate, GRE General Test, TOEFL, TSE. *Application deadline:* For fall admission, 4/1 (priority date); for spring admission, 11/1. Applications are processed on a rolling basis. *Application fee:* $0 ($75 for international students).
Expenses: Tuition, state resident: full-time $1,296; part-time $72 per credit. Tuition, nonresident: full-time $4,932; part-time $274 per credit. Required fees: $1,162. Tuition and fees vary according to program.
Financial aid: In 1999–00, 83 students received aid, including 2 fellowships, 40 research assistantships, 43 teaching assistantships; Federal Work-Study and institutionally sponsored loans also available. Financial aid application deadline: 4/1.
Faculty research: Evolutionary biology, neuroscience, infectious diseases, circadian rhythm, ion channels. *Total annual research expenditures:* $5.5 million.
Dr. Arnold Eskin, Chairman, 713-743-8386.
Application contact: Marcie Newton, Graduate Adviser and Office Coordinator, 713-743-2633, *Fax:* 713-743-2899, *E-mail:* mnewton@dna.bchs.uh.edu.

Find an in-depth description at www.petersons.com/graduate.

■ UNIVERSITY OF IDAHO

College of Graduate Studies, College of Agriculture, Department of Microbiology, Molecular Biology and Biochemistry, Program in Biochemistry, Moscow, ID 83844-4140

AWARDS MS, PhD.

Degree requirements: For master's, thesis required, foreign language not required; for doctorate, dissertation required.
Entrance requirements: For master's, minimum GPA of 2.8; for doctorate, minimum undergraduate GPA of 2.8, 3.0 graduate. *Application deadline:* For fall admission, 8/1; for spring admission, 12/15. *Application fee:* $35 ($45 for international students).
Expenses: Tuition, nonresident: full-time $6,000; part-time $239 per credit hour. Required fees: $2,888; $144 per credit hour. Tuition and fees vary according to program.
Financial aid: Application deadline: 2/15.
Faculty research: Enzymology and proteins, nucleic acids, lipid and membrane biochemistry.
Dr. Greg Bohach, Head, Department of Microbiology, Molecular Biology and Biochemistry, 208-885-6666.

■ UNIVERSITY OF IDAHO

College of Graduate Studies, College of Agriculture, Department of Microbiology, Molecular Biology and Biochemistry, Program in Microbiology, Molecular Biology and Biochemistry, Moscow, ID 83844-4140

AWARDS MS, PhD.

Students: 28 full-time (7 women), 5 part-time (1 woman); includes 1 minority (Asian American or Pacific Islander), 14 international.
Degree requirements: For master's, thesis required, foreign language not required; for doctorate, dissertation required.
Entrance requirements: For master's, minimum GPA of 2.8; for doctorate, minimum undergraduate GPA of 2.8, 3.0 graduate. *Application deadline:* For fall admission, 8/1; for spring admission, 12/15. *Application fee:* $35 ($45 for international students).
Expenses: Tuition, nonresident: full-time $6,000; part-time $239 per credit hour. Required fees: $2,888; $144 per credit hour. Tuition and fees vary according to program.
Financial aid: Application deadline: 2/15.
Dr. Greg Bohach, Head, Department of Microbiology, Molecular Biology and Biochemistry, 208-885-6666.

■ UNIVERSITY OF ILLINOIS AT CHICAGO

College of Medicine and Graduate College, Graduate Programs in Medicine, Department of Biochemistry and Molecular Biology, Chicago, IL 60607-7128

AWARDS MS, PhD.

Faculty: 17 full-time (5 women).
Students: 19 full-time (13 women), 1 (woman) part-time; includes 4 minority (3 Asian Americans or Pacific Islanders, 1 Hispanic American), 10 international. Average age 27. *107 applicants, 13% accepted.* In 1999, 4 master's, 2 doctorates awarded. Terminal master's awarded for partial completion of doctoral program.
Degree requirements: For master's and doctorate, thesis/dissertation required, foreign language not required.
Entrance requirements: For master's and doctorate, GRE General Test. *Application deadline:* For fall admission, 4/1. *Application fee:* $40 ($50 for international students).
Expenses: Tuition, state resident: full-time $3,750. Tuition, nonresident: full-time $10,588. Tuition and fees vary according to course load.
Financial aid: In 1999–00, 18 students received aid; fellowships, research assistantships, teaching assistantships, career-related internships or fieldwork, Federal Work-Study, institutionally sponsored loans, and traineeships available. Financial aid application deadline: 3/1; financial aid applicants required to submit FAFSA.
Faculty research: Nature of cellular components, control of metabolic processes, regulation of gene expression.
Dr. Donald Chambers, Head, 312-996-7670.
Application contact: Dr. James Vary, Contact, 312-996-8444.

Find an in-depth description at www.petersons.com/graduate.

■ UNIVERSITY OF ILLINOIS AT URBANA–CHAMPAIGN

Graduate College, College of Liberal Arts and Sciences, School of Chemical Sciences, Department of Biochemistry, Urbana, IL 61801

AWARDS MS, PhD.

Faculty: 13 full-time (0 women), 8 part-time/adjunct (2 women).
Students: 78 full-time (33 women); includes 16 minority (1 African American, 12 Asian Americans or Pacific Islanders, 3 Hispanic Americans), 19 international. *75 applicants, 20% accepted.* In 1999, 3 master's, 12 doctorates awarded.
Degree requirements: For master's, foreign language and thesis not required; for doctorate, dissertation required.
Entrance requirements: For master's, GRE General Test, GRE Subject Test, minimum GPA of 3.0; for doctorate, GRE General Test, GRE Subject Test. *Application deadline:* Applications are processed on a rolling basis. *Application fee:* $40 ($50 for international students).

Expenses: Tuition, state resident: full-time $4,616. Tuition, nonresident: full-time $11,768. Full-time tuition and fees vary according to course load.

Financial aid: In 1999–00, 61 research assistantships, 22 teaching assistantships were awarded; fellowships, traineeships and tuition waivers (full and partial) also available. Financial aid application deadline: 2/15.

Dr. John A. Gerlt, Head, 217-244-0205, *Fax:* 217-244-5858, *E-mail:* j-gerlt@uiuc.edu.

Application contact: Louise R. Cox, Director of Graduate Studies, 217-333-7149, *Fax:* 217-244-5858, *E-mail:* l-cox1@uiuc.edu.

Find an in-depth description at www.petersons.com/graduate.

■ **THE UNIVERSITY OF IOWA**

College of Medicine and Graduate College, Graduate Programs in Medicine, Department of Biochemistry, Iowa City, IA 52242-1316

AWARDS MS, PhD, MD/PhD.

Faculty: 20 full-time (5 women), 4 part-time/adjunct (2 women).

Students: 43 full-time (19 women); includes 4 minority (3 Asian Americans or Pacific Islanders, 1 Hispanic American), 20 international. Average age 22. *207 applicants, 5% accepted.* In 1999, 2 master's, 11 doctorates awarded (100% found work related to degree). Terminal master's awarded for partial completion of doctoral program.

Degree requirements: For master's, thesis required, foreign language not required; for doctorate, dissertation, research project required, foreign language not required. *Average time to degree:* Doctorate–6 years full-time.

Entrance requirements: For master's, GRE General Test, TOEFL. *Application deadline:* For fall admission, 4/15 (priority date). *Application fee:* $30 ($50 for international students). Electronic applications accepted.

Expenses: Tuition, state resident: full-time $3,308. Tuition, nonresident: full-time $10,662. Tuition and fees vary according to course load and program.

Financial aid: In 1999–00, 40 research assistantships with full tuition reimbursements (averaging $17,297 per year), 1 teaching assistantship with full tuition reimbursement (averaging $17,297 per year) were awarded; grants, institutionally sponsored loans, scholarships, and tuition waivers also available.

Faculty research: Regulation of gene expression, protein structure, membrane structure/function, DNA structure and replication. *Total annual research expenditures:* $7 million.

Dr. John E. Donelson, Head, 319-335-7934, *Fax:* 319-335-9570, *E-mail:* john-donelson@uiowa.edu.

Application contact: Admissions Committee, 319-335-7933, *Fax:* 319-335-9570, *E-mail:* cay-wieland@uiowa.edu.

Find an in-depth description at www.petersons.com/graduate.

■ **UNIVERSITY OF KANSAS**

Graduate School, College of Liberal Arts and Sciences, Division of Biological Sciences, Department of Molecular Biosciences, Lawrence, KS 66045

AWARDS Biochemistry and biophysics (MA, MS, PhD); microbiology (MA, MS, PhD); molecular, cellular, and developmental biology (MA, MS, PhD).

Students: 13 full-time (7 women), 8 part-time (2 women); includes 1 minority (Asian American or Pacific Islander), 5 international. *57 applicants, 21% accepted.* In 2000, 11 master's, 10 doctorates awarded.

Degree requirements: For master's, thesis required, foreign language not required; for doctorate, dissertation required.

Entrance requirements: For master's and doctorate, GRE General Test, GRE Subject Test, TOEFL. *Application deadline:* For fall admission, 1/15 (priority date). *Application fee:* $25.

Expenses: Tuition, state resident: full-time $2,482; part-time $103 per credit hour. Tuition, nonresident: full-time $8,104; part-time $338 per credit hour. Required fees: $428; $31 per credit hour. Tuition and fees vary according to program.

Financial aid: In 2000–01, research assistantships (averaging $11,588 per year), teaching assistantships (averaging $11,588 per year) were awarded; fellowships. Financial aid application deadline: 3/1.

Paul Kelly, Chair, 785-864-4311, *Fax:* 785-864-5294.

Application contact: Information Contact, 785-864-4311, *Fax:* 785-864-5294.

Find an in-depth description at www.petersons.com/graduate.

■ **UNIVERSITY OF KANSAS**

Graduate Studies Medical Center, Graduate Programs in Biomedical and Basic Sciences, Department of Biochemistry and Molecular Biology, Lawrence, KS 66045

AWARDS MS, PhD, MD/PhD. Part-time programs available.

Faculty: 12 full-time (0 women).

Students: 1 full-time (0 women), 10 part-time (5 women), 5 international. Average age 28. *0 applicants, 0% accepted.* Terminal master's awarded for partial completion of doctoral program.

Degree requirements: For master's, oral defense of thesis required; for doctorate, one foreign language (computer language can substitute), dissertation, comprehensive oral and written exam required.

Entrance requirements: For master's and doctorate, GRE, TOEFL, TSE, minimum GPA of 3.0. *Application deadline:* For fall admission, 4/15 (priority date). Applications are processed on a rolling basis. *Application fee:* $0. Electronic applications accepted.

Expenses: Tuition, state resident: full-time $2,482; part-time $103 per credit hour. Tuition, nonresident: full-time $8,104; part-time $338 per credit hour. Required fees: $428; $31 per credit hour. Tuition and fees vary according to program.

Financial aid: Fellowships, research assistantships, teaching assistantships, Federal Work-Study, institutionally sponsored loans, scholarships, and unspecified assistantships available. Aid available to part-time students. Financial aid application deadline: 3/31; financial aid applicants required to submit FAFSA.

Faculty research: Regulation of gene expression, molecular genetics of kidney disease, molecular chaperones, glycoproteins, cell cycle control. *Total annual research expenditures:* $3 million.

Dr. Billy G. Hudson, Chairman, 913-588-7008, *Fax:* 913-588-7440, *E-mail:* bhudson@kumc.edu.

Application contact: Dr. Glen Andrews, Director of Graduate Studies, 913-588-6935, *Fax:* 913-588-7440, *E-mail:* gandrews@kumc.edu.

■ **UNIVERSITY OF KENTUCKY**

Graduate School and College of Medicine, Graduate Programs in Medicine, Program in Biochemistry, Lexington, KY 40506-0032

AWARDS PhD, MD/PhD.

Degree requirements: For doctorate, dissertation, comprehensive exam required, foreign language not required.

Entrance requirements: For doctorate, GRE General Test, minimum graduate GPA of 3.0.

Expenses: Tuition, state resident: full-time $3,596; part-time $188 per credit hour. Tuition, nonresident: full-time $10,116; part-time $550 per credit hour.

Faculty research: Gene structure, receptor signaling pathways, lipid signaling pathways, cell cycle, structural biology.

■ UNIVERSITY OF LOUISVILLE

School of Medicine and Graduate School, Integrated Programs in Biomedical Sciences, Department of Biochemistry and Molecular Biology, Louisville, KY 40292-0001

AWARDS MS, PhD, MD/MS, MD/PhD.

Degree requirements: For master's, thesis required; for doctorate, dissertation, comprehensive exams required.
Entrance requirements: For master's and doctorate, GRE General Test, TOEFL, minimum GPA of 3.0. Electronic applications accepted.
Expenses: Tuition, state resident: full-time $3,260; part-time $182 per hour. Tuition, nonresident: full-time $9,780; part-time $544 per hour. Required fees: $143; $28 per hour.
Faculty research: Protein structure and function, genetic and metabolic regulation, molecular endocrinology.

■ UNIVERSITY OF MAINE

Graduate School, College of Natural Sciences, Forestry, and Agriculture, Department of Biochemistry, Molecular Biology, and Microbiology, Orono, ME 04469

AWARDS Biochemistry (MPS, MS); biochemistry and molecular biology (PhD); microbiology (MPS, MS, PhD).

Degree requirements: For doctorate, dissertation required.
Entrance requirements: For master's and doctorate, GRE General Test, TOEFL.
Expenses: Tuition, state resident: full-time $3,564. Tuition, nonresident: full-time $10,116. Required fees: $378. Tuition and fees vary according to course load.

■ UNIVERSITY OF MARYLAND

Graduate School, Graduate Programs in Medicine, Department of Biochemistry and Molecular Biology, Baltimore, MD 21201-1627

AWARDS Biochemistry (PhD).

Degree requirements: For doctorate, dissertation required, foreign language not required.
Entrance requirements: For doctorate, GRE General Test, GRE Subject Test, TOEFL.
Expenses: Tuition, state resident: part-time $261 per credit hour. Tuition, nonresident: part-time $468 per credit hour. Tuition and fees vary according to program.
Faculty research: Membrane transport, hormonal regulation, protein structure, molecular virology.

■ UNIVERSITY OF MARYLAND, BALTIMORE COUNTY

Graduate School, Department of Chemistry and Biochemistry, Program in Biochemistry, Baltimore, MD 21250-5398

AWARDS Biochemistry (PhD); neuroscience (PhD).

Faculty: 6 full-time, 1 part-time/adjunct. **Students:** 6 full-time (4 women), 1 part-time; includes 1 minority (Asian American or Pacific Islander), 2 international. *24 applicants, 25% accepted.* In 1999, 1 degree awarded.
Degree requirements: For doctorate, dissertation, comprehensive exams required, foreign language not required.
Entrance requirements: For doctorate, GRE General Test, GRE Subject Test, TOEFL, minimum GPA of 3.0. *Application deadline:* For fall admission, 4/15. Applications are processed on a rolling basis. *Application fee:* $45.
Expenses: Tuition, state resident: part-time $268 per credit hour. Tuition, nonresident: part-time $470 per credit hour. Required fees: $38 per credit hour. $557 per semester.
Financial aid: In 1999–00, research assistantships with full tuition reimbursements (averaging $17,000 per year), teaching assistantships with tuition reimbursements (averaging $17,000 per year) were awarded; fellowships.
Faculty research: Biochemical genetics, chemistry of proteins.
Dr. Michael Summers, Graduate Coordinator, 410-455-2527.

Find an in-depth description at www.petersons.com/graduate.

■ UNIVERSITY OF MARYLAND, COLLEGE PARK

Graduate Studies and Research, College of Life Sciences, Department of Chemistry and Biochemistry, Biochemistry Program, College Park, MD 20742

AWARDS MS, PhD. Part-time and evening/weekend programs available.

Students: 29 full-time (13 women), 3 part-time (2 women); includes 7 minority (2 African Americans, 5 Asian Americans or Pacific Islanders), 9 international. *48 applicants, 31% accepted.* In 1999, 3 master's, 7 doctorates awarded. Terminal master's awarded for partial completion of doctoral program.
Degree requirements: For master's, thesis optional, foreign language not required; for doctorate, dissertation, 2 seminar presentations, oral exam required.
Entrance requirements: For master's, GRE General Test, TOEFL, GRE Subject Test (recommended), minimum GPA of 3.1; for doctorate, GRE General Test, minimum GPA of 3.1. *Application deadline:* For fall admission, 7/1; for spring admission, 12/1. Applications are processed on a rolling basis. *Application fee:* $50 ($70 for international students). Electronic applications accepted.
Expenses: Tuition, state resident: part-time $272 per credit hour. Tuition, nonresident: part-time $415 per credit hour. Required fees: $632; $379 per year.
Financial aid: Fellowships, research assistantships, teaching assistantships with partial tuition reimbursements, Federal Work-Study available. Aid available to part-time students. Financial aid applicants required to submit FAFSA.
Faculty research: Analytical biochemistry, immunochemistry, drug metabolism, fermentation.
Application contact: Trudy Lindsey, Director, Graduate Admissions and Records, 301-405-4198, *Fax:* 301-314-9305, *E-mail:* grschool@deans.umd.edu.

■ UNIVERSITY OF MASSACHUSETTS AMHERST

Graduate School, College of Natural Sciences and Mathematics, Department of Biochemistry and Molecular Biology, Amherst, MA 01003

AWARDS Biochemistry (MS, PhD). PhD offered through the Molecular and Cellular Biology Graduate Program. Part-time programs available.

Faculty: 17 full-time (7 women). **Students:** 9 full-time (5 women), 5 part-time (1 woman); includes 2 minority (both Asian Americans or Pacific Islanders), 1 international. Average age 23. *50 applicants, 22% accepted.* In 2000, 1 master's awarded. Terminal master's awarded for partial completion of doctoral program.
Degree requirements: For master's, thesis or alternative required, foreign language not required; for doctorate, one foreign language, dissertation required. *Application deadline:* For fall admission, 1/15 (priority date). Applications are processed on a rolling basis. *Application fee:* $40.
Expenses: Tuition, state resident: full-time $2,640; part-time $165 per credit. Tuition, nonresident: full-time $9,756; part-time $407 per credit. Required fees: $1,221 per term. One-time fee: $110. Full-time tuition and fees vary according to course load, campus/location and reciprocity agreements.
Financial aid: In 2000–01, research assistantships with full tuition reimbursements (averaging $11,055 per year), 9 teaching assistantships with full tuition reimbursements (averaging $6,122 per

year) were awarded; fellowships with full tuition reimbursements, career-related internships or fieldwork, Federal Work-Study, grants, scholarships, traineeships, and unspecified assistantships also available. Aid available to part-time students. Financial aid application deadline: 1/15. Dr. Lila M. Gierasch, Director, 413-545-2318, *Fax:* 413-545-4490, *E-mail:* gierasch@chem.umass.edu.

Find an in-depth description at www.petersons.com/graduate.

■ UNIVERSITY OF MASSACHUSETTS AMHERST

Graduate School, Interdisciplinary Programs, Molecular and Cellular Biology Graduate Program, Amherst, MA 01003

AWARDS Biological chemistry (PhD); cell and developmental biology (PhD). Part-time programs available.

Students: 29 full-time (19 women), 33 part-time (21 women); includes 4 minority (1 African American, 3 Asian Americans or Pacific Islanders), 28 international. Average age 28. *288 applicants, 15% accepted.* In 2000, 8 degrees awarded.
Degree requirements: For doctorate, one foreign language, dissertation required.
Entrance requirements: For doctorate, GRE General Test. *Application deadline:* For fall admission, 1/15 (priority date). Applications are processed on a rolling basis. *Application fee:* $40.
Expenses: Tuition, state resident: full-time $2,640; part-time $165 per credit. Tuition, nonresident: full-time $9,756; part-time $407 per credit. Required fees: $1,221 per term. One-time fee: $110. Full-time tuition and fees vary according to course load, campus/location and reciprocity agreements.
Financial aid: In 2000–01, 17 research assistantships with full tuition reimbursements (averaging $3,796 per year), 3 teaching assistantships with full tuition reimbursements (averaging $5,510 per year) were awarded; fellowships with full tuition reimbursements, career-related internships or fieldwork, Federal Work-Study, grants, scholarships, traineeships, and unspecified assistantships also available. Aid available to part-time students. Financial aid application deadline: 1/15. Dr. Rodney K. Murphey, Acting Head, 413-545-3246, *Fax:* 413-545-1812, *E-mail:* rmurphey@bio.umass.edu.

Find an in-depth description at www.petersons.com/graduate.

■ UNIVERSITY OF MASSACHUSETTS LOWELL

Graduate School, College of Fine Arts/Humanities/Social Sciences, Department of Biological Sciences, Lowell, MA 01854-2881

AWARDS Biochemistry (PhD); biological sciences (MS); biotechnology (MS). Part-time programs available.

Faculty: 12 full-time (3 women).
Students: 18 full-time (12 women), 46 part-time (26 women); includes 6 minority (3 African Americans, 1 Asian American or Pacific Islander, 2 Hispanic Americans), 9 international. Average age 33. *35 applicants, 71% accepted.* In 1999, 22 degrees awarded.
Degree requirements: For master's, thesis required, foreign language not required; for doctorate, computer language, dissertation required. *Average time to degree:* Master's–2 years full-time, 3 years part-time.
Entrance requirements: For master's and doctorate, GRE General Test. *Application deadline:* For fall admission, 4/1 (priority date); for spring admission, 10/1. Applications are processed on a rolling basis. *Application fee:* $20 ($35 for international students). Electronic applications accepted.
Expenses: Tuition, state resident: full-time $1,610; part-time $89 per credit. Tuition, nonresident: full-time $5,610; part-time $312 per credit. Required fees: $2,100; $120 per credit. Tuition and fees vary according to reciprocity agreements.
Financial aid: In 1999–00, 10 teaching assistantships with tuition reimbursements were awarded; research assistantships with tuition reimbursements, career-related internships or fieldwork, Federal Work-Study, scholarships, and traineeships also available. Financial aid application deadline: 4/1.
Dr. Robert Lynch, Chair, 978-934-2891, *E-mail:* robert_lynch@woods.uml.edu.
Application contact: Dr. Ilze Skare, Coordinator, 978-934-2885, *E-mail:* ilze_skare@woods.uml.edu.

■ UNIVERSITY OF MASSACHUSETTS LOWELL

Graduate School, College of Fine Arts/Humanities/Social Sciences, Department of Chemistry, Lowell, MA 01854-2881

AWARDS Biochemistry (PhD); chemistry (MS, PhD); environmental studies (PhD); polymer sciences (MS, PhD).

Faculty: 24 full-time (2 women).
Students: 36 full-time (12 women), 29 part-time (19 women). Average age 33. *90 applicants, 42% accepted.* In 1999, 11 master's, 8 doctorates awarded. Terminal master's awarded for partial completion of doctoral program.
Degree requirements: For master's, thesis required, foreign language not required; for doctorate, 2 foreign languages, computer language, dissertation required.
Entrance requirements: For master's and doctorate, GRE General Test. *Application deadline:* For fall admission, 4/1 (priority date); for spring admission, 10/1. Applications are processed on a rolling basis. *Application fee:* $20 ($35 for international students). Electronic applications accepted.
Expenses: Tuition, state resident: full-time $1,610; part-time $89 per credit. Tuition, nonresident: full-time $5,610; part-time $312 per credit. Required fees: $2,100; $120 per credit. Tuition and fees vary according to reciprocity agreements.
Financial aid: In 1999–00, 13 fellowships, 13 research assistantships with tuition reimbursements, 30 teaching assistantships with tuition reimbursements were awarded; career-related internships or fieldwork, grants, and traineeships also available. Financial aid application deadline: 4/1.
Dr. Edwin Johngen, Chair, 978-934-3663.
Application contact: Dr. Melissa McDonald, Coordinator, 978-934-3683, *E-mail:* melissa_mcdonald@woods.uml.edu.

■ UNIVERSITY OF MASSACHUSETTS WORCESTER

Graduate School of Biomedical Sciences, Department of Biochemistry and Molecular Biology, Worcester, MA 01655-0115

AWARDS PhD.

Faculty: 29 full-time (1 woman).
Students: 39 full-time (16 women); includes 1 minority (Asian American or Pacific Islander), 14 international. In 1999, 3 degrees awarded. *Average time to degree:* Doctorate–5.9 years full-time.
Entrance requirements: For doctorate, GRE General Test, 1 year of calculus, physics, organic chemistry and biology. *Application deadline:* For fall admission, 1/15 (priority date). Applications are processed on a rolling basis. *Application fee:* $25 ($50 for international students).
Expenses: Tuition, state resident: full-time $2,640. Tuition, nonresident: full-time $9,756. Required fees: $825. Full-time tuition and fees vary according to program.
Financial aid: In 1999–00, research assistantships with full tuition reimbursements (averaging $17,500 per year); unspecified assistantships also available.
Faculty research: Metabolic regulation, membrane structure, nutrition, hormone action, cloning.

University of Massachusetts Worcester (continued)

Dr. Anthony Carruthers, Acting Chair, 508-856-2254, *Fax:* 508-856-6231, *E-mail:* anthony.carruthers@umassmed.edu.
Application contact: Dr. William Royer, Graduate Director, 508-856-6912.

■ UNIVERSITY OF MASSACHUSETTS WORCESTER

Graduate School of Biomedical Sciences, Department of Biochemistry and Molecular Pharmacology, Worcester, MA 01655-0115

AWARDS Biochemistry and molecular biology (PhD); pharmacology and molecular toxicology (PhD).

Faculty: 25 full-time (2 women).
Students: 12 full-time (6 women); includes 2 minority (both Asian Americans or Pacific Islanders), 3 international.
Degree requirements: For doctorate, dissertation required.
Entrance requirements: For doctorate, GRE General Test, GRE Subject Test, 1 year of calculus, physics, organic chemistry and biology. *Application deadline:* For fall admission, 1/15 (priority date). Applications are processed on a rolling basis. *Application fee:* $25 ($50 for international students).
Expenses: Tuition, state resident: full-time $2,640. Tuition, nonresident: full-time $9,756. Required fees: $825. Full-time tuition and fees vary according to program.
Financial aid: In 1999–00, research assistantships with full tuition reimbursements (averaging $17,500 per year); unspecified assistantships also available.
Faculty research: Neuropharmacology, control of DNA metabolism, clinical pharmacology and toxicology.
Dr. C. Robert Matthews, Chair, 508-856-2251, *Fax:* 508-856-2151.
Application contact: Dr. Alonzo Ross, Graduate Director, 508-856-8016, *Fax:* 508-856-2151, *E-mail:* alonzo.ross@umassmed.edu.

Find an in-depth description at www.petersons.com/graduate.

■ UNIVERSITY OF MEDICINE AND DENTISTRY OF NEW JERSEY

Graduate School of Biomedical Sciences, Graduate Programs in Biomedical Sciences, Department of Biochemistry and Molecular Biology, Newark, NJ 07107

AWARDS MS, PhD.

Degree requirements: For master's, thesis required; for doctorate, dissertation, qualifying exam required, foreign language not required.

Entrance requirements: For master's and doctorate, GRE General Test, TOEFL. *Application deadline:* For fall admission, 2/1; for spring admission, 10/1. *Application fee:* $40.
Expenses: Tuition, state resident: part-time $270 per credit hour. Tuition, nonresident: part-time $407 per credit hour. Part-time tuition and fees vary according to campus/location and program.
Financial aid: Fellowships, research assistantships, Federal Work-Study, institutionally sponsored loans, and tuition waivers (full and partial) available. Financial aid application deadline: 5/1. Dr. Michael B. Mathews, Chairperson, 973-972-4411.
Application contact: Dr. Henry E. Brezenoff, Dean, Graduate School of Biomedical Sciences, 973-972-5333, *Fax:* 973-972-7148, *E-mail:* hbrezeno@umdnj.edu.

Find an in-depth description at www.petersons.com/graduate.

■ UNIVERSITY OF MEDICINE AND DENTISTRY OF NEW JERSEY

Graduate School of Biomedical Sciences, Graduate Programs in Biomedical Sciences, Program in Biochemistry and Molecular Biology, Piscataway, NJ 08854-5635

AWARDS MS, PhD. Terminal master's awarded for partial completion of doctoral program.

Degree requirements: For master's, thesis, qualifying exam required; for doctorate, dissertation, qualifying exam required, foreign language not required.
Entrance requirements: For master's and doctorate, GRE General Test, TOEFL. *Application deadline:* For fall admission, 2/1; for spring admission, 10/1. *Application fee:* $40.
Expenses: Tuition, state resident: part-time $270 per credit hour. Tuition, nonresident: part-time $407 per credit hour. Part-time tuition and fees vary according to campus/location and program.
Financial aid: Fellowships, research assistantships, teaching assistantships available. Financial aid application deadline: 5/1.
Faculty research: Signal transduction, regulation of RNA, polymerase II transcribed genes, developmental gene expression.
Dr. Masayori Inouye, Director, 732-235-4115.
Application contact: Dr. Michael J. Leibowitz, Associate Dean, Graduate School, 732-235-5016, *Fax:* 732-235-4720, *E-mail:* gsbspisc@umdnj.edu.

■ UNIVERSITY OF MIAMI

School of Medicine and Graduate School, Graduate Programs in Medicine, Department of Biochemistry and Molecular Biology, Coral Gables, FL 33124

AWARDS PhD, MD/PhD. Part-time programs available.

Faculty: 24 full-time (5 women).
Students: 31 full-time (15 women); includes 3 minority (all Hispanic Americans), 22 international. Average age 28. *70 applicants, 11% accepted.* In 1999, 2 doctorates awarded.
Degree requirements: For doctorate, dissertation, comprehensive and proposition exams required, foreign language not required.
Entrance requirements: For doctorate, GRE General Test, TOEFL. *Application deadline:* For fall admission, 4/15. Applications are processed on a rolling basis. *Application fee:* $35.
Expenses: Tuition, area resident: Part-time $899 per credit.
Financial aid: In 1999–00, 29 students received aid; fellowships, research assistantships, Federal Work-Study and scholarships available.
Faculty research: Macromolecule metabolism, molecular genetics, protein folding and 3-D structure, regulation of gene expression and enzyme function, signal transduction and developmental biology. *Total annual research expenditures:* $2.9 million.
Dr. Murray P. Deutscher, Chairman, 305-243-3150, *Fax:* 305-243-3955, *E-mail:* mdeutsch@mednet.med.miami.edu.
Application contact: Dr. Rudolf Werner, Director, 305-243-6998, *Fax:* 305-547-3955, *E-mail:* rwerner@mednet.med.miami.edu.

Find an in-depth description at www.petersons.com/graduate.

■ UNIVERSITY OF MICHIGAN

Horace H. Rackham School of Graduate Studies, College of Literature, Science, and the Arts, Department of Chemistry, Ann Arbor, MI 48109

AWARDS Analytical chemistry (PhD); chemical biology (PhD); inorganic chemistry (PhD); organic chemistry (PhD); physical chemistry (PhD).

Faculty: 48 full-time (5 women), 1 part-time/adjunct (0 women).
Students: 180 full-time (56 women); includes 7 minority (3 African Americans, 3 Asian Americans or Pacific Islanders, 1 Hispanic American), 41 international. Average age 26. *379 applicants, 44% accepted.* In 1999, 27 degrees awarded.

Degree requirements: For doctorate, oral defense of dissertation, preliminary exam, organic cumulative proficiency exams required, foreign language not required. *Average time to degree:* Doctorate–5 years full-time.

Entrance requirements: For doctorate, GRE General Test, GRE Subject Test (recommended), statement of prior research. *Application deadline:* For fall admission, 2/15 (priority date). Applications are processed on a rolling basis. *Application fee:* $55.

Expenses: Tuition, state resident: full-time $10,316. Tuition, nonresident: full-time $20,922. Required fees: $185. Part-time tuition and fees vary according to course load and program.

Financial aid: In 1999–00, 10 fellowships with full tuition reimbursements (averaging $17,750 per year), 64 research assistantships with full tuition reimbursements (averaging $16,000 per year), 110 teaching assistantships with full tuition reimbursements (averaging $16,000 per year) were awarded. Financial aid applicants required to submit FAFSA.

Faculty research: Biological catalysis, protein engineering, chemical sensors, de novo metalloprotein design, supramolecular architecture, organic zeolites. *Total annual research expenditures:* $8 million.

Dr. Joseph P. Marino, Chair, 734-763-9681, *Fax:* 734-647-4847, *E-mail:* jpmarino@umich.edu.

Application contact: Holly Bender, Assistant Director Graduate Studies, 734-764-7278, *Fax:* 734-647-4864, *E-mail:* chemadmissions@umich.edu.

■ UNIVERSITY OF MICHIGAN

Medical School and Horace H. Rackham School of Graduate Studies, Program in Biomedical Sciences (PIBS), Department of Biological Chemistry, Ann Arbor, MI 48109

AWARDS PhD.

Degree requirements: For doctorate, oral defense of dissertation, preliminary exam required.

Entrance requirements: For doctorate, GRE General Test, GRE Subject Test, minimum GPA of 3.0. Electronic applications accepted.

Expenses: Tuition, state resident: full-time $10,316. Tuition, nonresident: full-time $20,922.

Faculty research: Nucleic acids, gene regulation, mechanistic enzymology, protein biochemistry, signal transduction.

Find an in-depth description at www.petersons.com/graduate.

■ UNIVERSITY OF MINNESOTA, DULUTH

School of Medicine, Department of Biochemistry and Molecular Biology, Duluth, MN 55812-2496

AWARDS MS, PhD.

Faculty: 6 full-time (1 woman), 3 part-time/adjunct (2 women).

Students: 4 full-time (2 women), 1 (woman) part-time; includes 2 minority (1 Asian American or Pacific Islander, 1 Native American). Average age 24. *22 applicants, 9% accepted.* In 1999, 1 master's awarded (100% found work related to degree). Terminal master's awarded for partial completion of doctoral program.

Degree requirements: For master's and doctorate, thesis/dissertation required, foreign language not required. *Average time to degree:* Master's–2 years full-time.

Entrance requirements: For master's and doctorate, GRE General Test, TOEFL. *Application deadline:* For fall admission, 2/15 (priority date). *Application fee:* $50 ($55 for international students).

Expenses: Tuition, state resident: full-time $5,040; part-time $420 per credit. Tuition, nonresident: full-time $9,900; part-time $825 per credit. Required fees: $509. Tuition and fees vary according to course load and program.

Financial aid: In 1999–00, 5 students received aid, including 4 research assistantships with tuition reimbursements available (averaging $12,487 per year), 1 teaching assistantship with tuition reimbursement available (averaging $13,947 per year); fellowships with tuition reimbursements available, institutionally sponsored loans also available. Financial aid application deadline: 4/1.

Faculty research: Fish liver metabolism, yeast molecular genetics. *Total annual research expenditures:* $750,000.

Dr. Lester R. Drewes, Head, 218-726-7925, *Fax:* 218-726-8014, *E-mail:* ldrewes@d.umn.edu.

■ UNIVERSITY OF MINNESOTA, TWIN CITIES CAMPUS

Medical School and Graduate School, Graduate Programs in Medicine, Department of Biochemistry, Molecular Biology and Biophysics, Minneapolis, MN 55455-0213

AWARDS PhD.

Faculty: 67 full-time (13 women).

Students: 77 full-time (39 women); includes 1 African American, 1 Hispanic American, 5 international. *189 applicants, 22% accepted.* In 1999, 16 degrees awarded (94% entered university research/teaching, 6% found other work related to degree).

Degree requirements: For doctorate, dissertation required, foreign language not required. *Average time to degree:* Doctorate–5 years full-time.

Entrance requirements: For doctorate, GRE General Test. *Application deadline:* For fall admission, 1/15 (priority date). *Application fee:* $50 ($55 for international students).

Expenses: Program provides tuition for all students.

Financial aid: In 1999–00, 12 fellowships with full tuition reimbursements (averaging $16,500 per year), research assistantships with full tuition reimbursements (averaging $16,500 per year) were awarded.

Faculty research: Physical biochemistry, enzymology, physiological chemistry. *Total annual research expenditures:* $10.9 million.

Dr. Douglas Ohlendorf, Director of Admissions, 612-625-6100.

Application contact: Mary F. Petrie-Terry, Student Support Assistant, 612-625-5179, *Fax:* 612-625-2163, *E-mail:* petri008@umn.edu.

Find an in-depth description at www.petersons.com/graduate.

■ UNIVERSITY OF MISSISSIPPI MEDICAL CENTER

Graduate Programs in Biomedical Sciences, Department of Biochemistry, Jackson, MS 39216-4505

AWARDS PhD, MD/PhD.

Faculty: 13 full-time (1 woman).

Students: 13 full-time (5 women), 10 international. Average age 28. *36 applicants, 8% accepted.* In 1999, 1 doctorate awarded (100% entered university research/teaching).

Degree requirements: For doctorate, dissertation, first authored publication required, foreign language not required. *Average time to degree:* Doctorate–5 years full-time.

Entrance requirements: For doctorate, GRE General Test, TOEFL, minimum GPA of 3.0. *Application deadline:* For fall admission, 3/1. *Application fee:* $10.

Expenses: Tuition, state resident: full-time $2,378; part-time $132 per hour. Tuition, nonresident: full-time $4,697; part-time $261 per hour. Tuition and fees vary according to program.

Financial aid: In 1999–00, 13 students received aid, including 13 research assistantships (averaging $16,234 per year); Federal Work-Study also available. Financial aid application deadline: 4/1.

Faculty research: Protein chemistry and biosynthesis, developmental molecular biochemistry, enzymology, molecular biology. *Total annual research expenditures:* $700,000.

University of Mississippi Medical Center (continued)
Dr. Mark O. J. Olson, Chairman, 601-984-1500.
Application contact: Dr. David T. Brown, Director of Graduate Studies, 601-984-1500, *Fax:* 601-984-1501, *E-mail:* dbrown@biochem.umsmed.edu.

Find an in-depth description at www.petersons.com/graduate.

■ UNIVERSITY OF MISSOURI–COLUMBIA

School of Medicine and Graduate School, Graduate Programs in Medicine and College of Agriculture, Department of Biochemistry, Columbia, MO 65211

AWARDS MS, PhD, MD/MS, MD/PhD. Terminal master's awarded for partial completion of doctoral program.

Degree requirements: For master's and doctorate, thesis/dissertation required, foreign language not required.
Entrance requirements: For master's and doctorate, GRE General Test, minimum GPA of 3.0.
Expenses: Tuition, state resident: full-time $3,020; part-time $168 per hour. Tuition, nonresident: full-time $6,066; part-time $505 per hour. Required fees: $445; $18 per hour.
Faculty research: Enzymology, plant biochemistry, molecular biochemistry.

Find an in-depth description at www.petersons.com/graduate.

■ UNIVERSITY OF MISSOURI–KANSAS CITY

School of Biological Sciences, Program in Molecular Biology and Biochemistry, Kansas City, MO 64110-2499

AWARDS PhD. PhD offered through the School of Graduate Studies.

Students: In 2000, 1 degree awarded.
Degree requirements: For doctorate, dissertation required.
Entrance requirements: For doctorate, GRE General Test, TOEFL, bachelor's degree in chemistry, biology, or a related discipline; minimum GPA of 3.0. *Application deadline:* For fall admission, 3/1. *Application fee:* $25.
Expenses: Tuition, state resident: part-time $173 per hour. Tuition, nonresident: part-time $348 per hour. Required fees: $22 per hour. $15 per term. Part-time tuition and fees vary according to course load and program.
Financial aid: Fellowships with tuition reimbursements, research assistantships,

teaching assistantships, grants, scholarships, tuition waivers (full and partial), and unspecified assistantships available.
Application contact: Graduate Adviser, 816-235-2352, *Fax:* 816-235-5158.

Find an in-depth description at www.petersons.com/graduate.

■ UNIVERSITY OF MISSOURI–ST. LOUIS

Graduate School, College of Arts and Sciences, Department of Biology, St. Louis, MO 63121-4499

AWARDS Biology (MS, PhD), including animal behavior (MS), biochemistry, biotechnology (MS), conservation biology (MS), development (MS), ecology (MS), environmental studies (PhD), evolution (MS), genetics (MS), molecular biology and biotechnology (PhD), molecular/cellular biology (MS), physiology (MS), plant systematics, population biology (MS), tropical biology (MS); biotechnology (Certificate); tropical biology and conservation (Certificate). Part-time programs available.

Faculty: 46.
Students: 21 full-time (11 women), 75 part-time (44 women); includes 13 minority (2 African Americans, 2 Asian Americans or Pacific Islanders, 8 Hispanic Americans, 1 Native American), 23 international. In 1999, 14 master's, 4 doctorates awarded.
Degree requirements: For master's, thesis or alternative required, foreign language not required; for doctorate, one foreign language, dissertation, 1 semester of teaching experience required.
Entrance requirements: For doctorate, GRE General Test. *Application deadline:* For fall admission, 7/1 (priority date); for spring admission, 11/1 (priority date). Applications are processed on a rolling basis. *Application fee:* $25 ($40 for international students). Electronic applications accepted.
Expenses: Tuition, state resident: full-time $4,932; part-time $173 per credit hour. Tuition, nonresident: full-time $13,279; part-time $521 per credit hour. Required fees: $775; $33 per credit hour. Tuition and fees vary according to degree level and program.
Financial aid: In 1999–00, 8 research assistantships with partial tuition reimbursements (averaging $10,635 per year), 14 teaching assistantships with partial tuition reimbursements (averaging $11,488 per year) were awarded; career-related internships or fieldwork and Federal Work-Study also available. Aid available to part-time students. Financial aid application deadline: 2/1. *Total annual research expenditures:* $908,828.

Application contact: Graduate Admissions, 314-516-5458, *Fax:* 314-516-6759, *E-mail:* gradadm@umsl.edu.

■ UNIVERSITY OF MISSOURI–ST. LOUIS

Graduate School, College of Arts and Sciences, Department of Chemistry, St. Louis, MO 63121-4499

AWARDS Chemistry (MS, PhD), including biochemistry, inorganic chemistry, organic chemistry, physical chemistry. Part-time and evening/weekend programs available.

Faculty: 18.
Students: 29 full-time (12 women), 22 part-time (8 women); includes 7 minority (1 African American, 5 Asian Americans or Pacific Islanders, 1 Hispanic American), 18 international. In 1999, 15 master's, 9 doctorates awarded. Terminal master's awarded for partial completion of doctoral program.
Degree requirements: For master's, thesis optional, foreign language not required; for doctorate, dissertation required, foreign language not required.
Entrance requirements: For doctorate, GRE General Test, GRE Subject Test. *Application deadline:* For fall admission, 7/1 (priority date); for spring admission, 12/7 (priority date). Applications are processed on a rolling basis. *Application fee:* $25 ($40 for international students). Electronic applications accepted.
Expenses: Tuition, state resident: full-time $4,932; part-time $173 per credit hour. Tuition, nonresident: full-time $13,279; part-time $521 per credit hour. Required fees: $775; $33 per credit hour. Tuition and fees vary according to degree level and program.
Financial aid: In 1999–00, 1 fellowship with partial tuition reimbursement (averaging $12,000 per year), 11 research assistantships with partial tuition reimbursements (averaging $12,400 per year), 17 teaching assistantships with partial tuition reimbursements (averaging $12,400 per year) were awarded.
Faculty research: Metallaborane chemistry, serum transferrin chemistry, natural products chemistry, organic synthesis. *Total annual research expenditures:* $1.2 million.
Dr. R. E. K. Winter, Director of Graduate Studies, 314-516-5337, *Fax:* 314-516-5342.
Application contact: Graduate Admissions, 314-516-5458, *Fax:* 314-516-6759, *E-mail:* gradadm@umsl.edu.

■ THE UNIVERSITY OF MONTANA–MISSOULA

Graduate School, Division of Biological Sciences, Program in Biochemistry and Microbiology, Missoula, MT 59812-0002

AWARDS MS, PhD.

Students: 17 full-time (5 women), 12 part-time (5 women); includes 3 minority (2 Asian Americans or Pacific Islanders, 1 Native American). *41 applicants, 27% accepted.* In 1999, 1 master's, 2 doctorates awarded. Terminal master's awarded for partial completion of doctoral program.
Degree requirements: For master's, thesis required (for some programs); for doctorate, dissertation required.
Entrance requirements: For master's and doctorate, GRE General Test. *Application deadline:* For fall admission, 2/1. *Application fee:* $45.
Expenses: Tuition, state resident: full-time $2,484; part-time $151 per credit. Tuition, nonresident: full-time $8,000; part-time $305 per credit. Required fees: $1,600. Full-time tuition and fees vary according to degree level and program.
Financial aid: In 1999–00, research assistantships with tuition reimbursements (averaging $9,400 per year), teaching assistantships with full tuition reimbursements (averaging $9,400 per year) were awarded; Federal Work-Study and tuition waivers (full and partial) also available. Financial aid application deadline: 3/1.
Faculty research: Ribosome structure, medical microbiology/pathogenesis, microbial ecology/environmental microbiology.
Application contact: Janean Clark, Graduate Programs Secretary, 406-243-5222, *Fax:* 406-243-4184, *E-mail:* jmclark@selway.umt.edu.

■ UNIVERSITY OF NEBRASKA–LINCOLN

Graduate College, Center for Biological Chemistry, Lincoln, NE 68588

AWARDS MS, PhD.

Faculty: 10 full-time (2 women), 1 part-time/adjunct (0 women).
Students: 29 full-time (12 women); includes 1 minority (Hispanic American), 23 international. Average age 27. *90 applicants, 12% accepted.* In 1999, 4 master's awarded.
Degree requirements: For master's, thesis optional; for doctorate, dissertation, comprehensive exams required.
Entrance requirements: For master's and doctorate, GRE General Test, GRE Subject Test, TOEFL. *Application deadline:*

For fall admission, 2/1. *Application fee:* $35. Electronic applications accepted.
Expenses: Tuition, state resident: part-time $116 per credit hour. Tuition, nonresident: part-time $285 per credit hour. Required fees: $119 per semester. Tuition and fees vary according to course load and program.
Financial aid: In 1999–00, 1 fellowship, 20 research assistantships, 4 teaching assistantships were awarded; Federal Work-Study also available. Aid available to part-time students. Financial aid application deadline: 2/15.
Faculty research: Photosynthesis, enzymology, nitrogen fixation, molecular genetics, molecular virology.
Dr. Robert Klucas, Head, 402-472-3212, *Fax:* 402-472-7842, *E-mail:* rklucas1@unl.edu.

Find an in-depth description at www.petersons.com/graduate.

■ UNIVERSITY OF NEBRASKA MEDICAL CENTER

Graduate College, Department of Biochemistry and Molecular Biology, Omaha, NE 68198

AWARDS MS, PhD.

Faculty: 13 full-time, 23 part-time/adjunct.
Students: 21 full-time (11 women), 3 part-time (2 women); includes 2 minority (1 Asian American or Pacific Islander, 1 Native American), 11 international. Average age 30. *25 applicants, 32% accepted.* In 1999, 1 master's, 4 doctorates awarded (100% entered university research/teaching).
Degree requirements: For master's and doctorate, thesis/dissertation required, foreign language not required. *Average time to degree:* Master's–1.5 years part-time.
Entrance requirements: For master's and doctorate, GRE General Test. *Application fee:* $35.
Expenses: Tuition, state resident: part-time $116 per semester hour. Tuition, nonresident: part-time $270 per semester hour. Tuition and fees vary according to program.
Financial aid: In 1999–00, 24 students received aid, including 4 fellowships with tuition reimbursements available (averaging $15,000 per year), 18 research assistantships with tuition reimbursements available (averaging $15,000 per year); institutionally sponsored loans also available. Aid available to part-time students. Financial aid application deadline: 3/1.
Faculty research: Recombinant DNA, cancer biology, diabetes and drug metabolism, biochemical endocrinology.

Dr. Pi-Wan Cheng, Chairman, Graduate Committee, 402-559-4419, *Fax:* 402-559-6650, *E-mail:* broberts@unmc.edu.
Application contact: Jo Wagner, Associate Director of Admissions, 402-559-6468.

■ UNIVERSITY OF NEVADA, RENO

Graduate School, Interdisciplinary Program in Biochemistry, Reno, NV 89557

AWARDS MS, PhD. Offered through the College of Arts and Science, the M. C. Fleischmann College of Agriculture, and the School of Medicine.

Faculty: 3 full-time (0 women).
Students: 22 full-time (11 women), 2 part-time (both women); includes 2 minority (both Asian Americans or Pacific Islanders), 7 international. Average age 30. *12 applicants, 25% accepted.* In 1999, 1 master's, 1 doctorate awarded. Terminal master's awarded for partial completion of doctoral program.
Degree requirements: For master's and doctorate, thesis/dissertation required, foreign language not required.
Entrance requirements: For master's, GRE General Test, TOEFL, minimum GPA of 2.75; for doctorate, GRE General Test, TOEFL, minimum GPA of 3.0. *Application deadline:* For fall admission, 3/1. *Application fee:* $40.
Expenses: Tuition, area resident: Part-time $3,173 per semester. Tuition, nonresident: full-time $6,347. Required fees: $101 per credit. $101 per credit.
Financial aid: In 1999–00, 20 research assistantships were awarded; teaching assistantships, Federal Work-Study and institutionally sponsored loans also available. Financial aid application deadline: 2/15.
Faculty research: Cancer research, insect biochemistry, plant biochemistry, enzymology.
Dr. Jeffrey R. Seemann, Chair, 775-784-6031.
Application contact: Dr. John H. Frederick, Director of Graduate Studies, 775-784-6031, *E-mail:* jhf@chem.unr.edu.

Find an in-depth description at www.petersons.com/graduate.

■ UNIVERSITY OF NEW HAMPSHIRE

Graduate School, College of Life Sciences and Agriculture, Graduate Programs in the Biological Sciences and Natural Resources, Department of Biochemistry and Molecular Biology, Durham, NH 03824

AWARDS MS, PhD. Part-time programs available.

University of New Hampshire (continued)
Faculty: 12 full-time.
Students: 13 full-time (8 women), 11 part-time (6 women), 8 international. Average age 28. *37 applicants, 30% accepted.* In 1999, 1 master's, 1 doctorate awarded. Terminal master's awarded for partial completion of doctoral program.
Degree requirements: For master's, thesis required, foreign language not required; for doctorate, dissertation required.
Entrance requirements: For master's and doctorate, GRE General Test. *Application deadline:* For fall admission, 4/1 (priority date). Applications are processed on a rolling basis. *Application fee:* $50.
Expenses: Tuition, area resident: Full-time $5,750; part-time $319 per credit. Tuition, state resident: full-time $8,625; part-time $478. Tuition, nonresident: full-time $14,640; part-time $598 per credit. Required fees: $224 per semester. Tuition and fees vary according to course load, degree level and program.
Financial aid: In 1999–00, 2 fellowships, 8 research assistantships, 8 teaching assistantships were awarded; career-related internships or fieldwork, Federal Work-Study, scholarships, and tuition waivers (full and partial) also available. Aid available to part-time students. Financial aid application deadline: 2/15.
Faculty research: Developmental biochemistry, biochemistry of natural products, physical biochemistry, biochemical genetics, structure and metabolism of macromolecules.
Dr. Clyde Denis, Chairperson, 603-862-2427, *E-mail:* cdenis@cisunix.unh.edu.
Application contact: Dr. Stacia Sower, Graduate Coordinator, 603-862-2103, *E-mail:* sasower@cisunix.unh.edu.

■ **THE UNIVERSITY OF NORTH CAROLINA AT CHAPEL HILL**
School of Medicine and Graduate School, Graduate Programs in Medicine, Department of Biochemistry and Biophysics, Chapel Hill, NC 27599
AWARDS MS, PhD.

Faculty: 27 full-time (6 women).
Students: 41 full-time (12 women); includes 1 minority (Asian American or Pacific Islander), 17 international. *86 applicants, 33% accepted.* In 1999, 2 master's, 7 doctorates awarded.
Degree requirements: For master's, thesis, comprehensive exam required, foreign language not required; for doctorate, dissertation, comprehensive exams required, foreign language not required. *Average time to degree:* Master's–4.5 years full-time; doctorate–6 years full-time.

Entrance requirements: For master's and doctorate, GRE General Test, GRE Subject Test (recommended), minimum GPA of 3.0. *Application deadline:* For fall admission, 1/1. *Application fee:* $55. Electronic applications accepted.
Expenses: Tuition, state resident: full-time $1,966. Tuition, nonresident: full-time $11,026. Required fees: $8,940. One-time fee: $15 full-time. Part-time tuition and fees vary according to course load.
Financial aid: In 1999–00, research assistantships with full tuition reimbursements (averaging $16,000 per year), 5 teaching assistantships with full tuition reimbursements (averaging $16,000 per year) were awarded; fellowships also available.
Dr. David Lee, Chair, 919-962-8326, *Fax:* 919-966-2852, *E-mail:* dclee@med.unc.edu.
Application contact: Diane M. Harris, Assistant to the Director of Graduate Studies, 919-966-4683, *E-mail:* diane_harris@med.unc.edu.

Find an in-depth description at www.petersons.com/graduate.

■ **UNIVERSITY OF NORTH DAKOTA**
School of Medicine and Graduate School, Graduate Programs in Medicine, Department of Biochemistry, Grand Forks, ND 58202
AWARDS MS, PhD.

Faculty: 9 full-time (1 woman).
Students: 11 full-time (4 women). *2 applicants, 100% accepted.*
Degree requirements: For master's, thesis, final exam required, foreign language not required; for doctorate, computer language, dissertation, comprehensive exam, final exam required.
Entrance requirements: For master's, GRE General Test, GRE Subject Test, TOEFL, minimum GPA of 3.0; for doctorate, GRE General Test, GRE Subject Test, TOEFL, minimum GPA of 3.5. *Application deadline:* For fall admission, 3/1 (priority date). Applications are processed on a rolling basis. *Application fee:* $25.
Expenses: Tuition, state resident: full-time $2,690; part-time $112 per credit. Tuition, nonresident: full-time $7,182; part-time $299 per credit. Required fees: $46 per semester.
Financial aid: In 1999–00, 11 students received aid, including 6 research assistantships with full tuition reimbursements available (averaging $10,586 per year), 5 teaching assistantships with full tuition reimbursements available (averaging $10,586 per year); fellowships, Federal Work-Study, institutionally sponsored loans, scholarships, and tuition waivers

(full and partial) also available. Aid available to part-time students. Financial aid application deadline: 3/15; financial aid applicants required to submit FAFSA.
Faculty research: Glucose-6-phosphatase, guanine nucleotides, carbohydrate and lipid metabolism, cytoskeletal proteins, chromatin structure.
Dr. John Shabb, Director, 701-777-3937, *Fax:* 701-777-3527, *E-mail:* jshabb@mail.med.und.nodak.edu.

■ **UNIVERSITY OF NORTH TEXAS**
Robert B. Toulouse School of Graduate Studies, College of Arts and Sciences, Department of Biological Sciences, Division of Biochemistry, Denton, TX 76203
AWARDS MS, PhD. Terminal master's awarded for partial completion of doctoral program.

Degree requirements: For master's, comprehensive exam, oral defense of thesis required; for doctorate, one foreign language, computer language, dissertation, oral and written comprehensive exams required.
Entrance requirements: For master's, GRE General Test, placement exams in 3 areas, minimum GPA of 3.0; for doctorate, GRE General Test, placement exams in 4 areas, minimum GPA of 3.0.
Expenses: Tuition, state resident: full-time $2,865; part-time $600 per semester. Tuition, nonresident: full-time $8,049; part-time $1,896 per semester. Required fees: $26 per hour.
Faculty research: Protein and nucleic acid structure, aging and nutritional biochemical changes, biochemical parasitology, plasma lipoprotein metabolism.

■ **UNIVERSITY OF NORTH TEXAS HEALTH SCIENCE CENTER AT FORT WORTH**
Graduate School of Biomedical Sciences, Fort Worth, TX 76107-2699
AWARDS Anatomy and cell biology (MS, PhD); biochemistry and molecular biology (MS, PhD); biotechnology (MS); integrative physiology (MS, PhD); microbiology and immunology (MS, PhD); pharmacology (MS, PhD).

Faculty: 65 full-time (9 women), 11 part-time/adjunct (1 woman).
Students: 59 full-time (27 women), 44 part-time (20 women); includes 30 minority (10 African Americans, 9 Asian Americans or Pacific Islanders, 11 Hispanic Americans), 23 international. *70 applicants, 70% accepted.* In 1999, 5 master's awarded (40% found work related to degree, 60% continued full-time study); 5 doctorates awarded (80% entered

university research/teaching, 20% found other work related to degree).

Degree requirements: For doctorate, computer language, dissertation required. *Average time to degree:* Master's–2.5 years full-time, 4 years part-time; doctorate–5 years full-time.

Entrance requirements: For master's and doctorate, GRE General Test, TOEFL. *Application deadline:* For fall admission, 5/1; for spring admission, 11/1. Applications are processed on a rolling basis. *Application fee:* $25 ($50 for international students).

Expenses: Tuition, state resident: full-time $1,188; part-time $66 per credit. Tuition, nonresident: full-time $5,058; part-time $281 per credit. Required fees: $366; $183 per semester.

Financial aid: In 1999–00, 11 fellowships, 70 research assistantships (averaging $16,500 per year) were awarded; teaching assistantships, career-related internships or fieldwork, Federal Work-Study, grants, institutionally sponsored loans, and traineeships also available. Aid available to part-time students. Financial aid application deadline: 4/1; financial aid applicants required to submit FAFSA.

Faculty research: Alzheimer's disease, diabetes, eye diseases, cancer, cardiovascular physiology.
Dr. Thomas Yorio, Dean, 817-735-2560, *Fax:* 817-735-0243, *E-mail:* yoriot@ hsc.unt.edu.
Application contact: Jan Sharp, Administrative Assistant, 817-735-0258, *Fax:* 817-735-0243, *E-mail:* gsbs@ hsc.unt.edu.

Find an in-depth description at www.petersons.com/graduate.

■ **UNIVERSITY OF NOTRE DAME**

Graduate School, College of Science, Department of Chemistry and Biochemistry, Notre Dame, IN 46556

AWARDS Biochemistry (MS, PhD); inorganic chemistry (MS, PhD); organic chemistry (MS, PhD); physical chemistry (MS, PhD).

Faculty: 32 full-time (3 women).
Students: 102 full-time (39 women); includes 3 minority (2 African Americans, 1 Asian American or Pacific Islander), 52 international. *180 applicants, 36% accepted.* In 1999, 10 master's, 15 doctorates awarded. Terminal master's awarded for partial completion of doctoral program.
Degree requirements: For master's, thesis, comprehensive exam required, foreign language not required; for doctorate, dissertation, qualifying exam required, foreign language not required. *Average time to degree:* Master's–3.3 years full-time; doctorate–6 years full-time.
Entrance requirements: For master's, GRE General Test, GRE General Subject

Test (strongly recommended), TOEFL; for doctorate, GRE General Test, GRE Subject Test (strongly recommended), TOEFL. *Application deadline:* For fall admission, 2/1 (priority date). Applications are processed on a rolling basis. *Application fee:* $50.
Expenses: Tuition: Full-time $21,930; part-time $1,218 per credit. Required fees: $95. Tuition and fees vary according to program.
Financial aid: In 1999–00, 102 students received aid, including 28 fellowships with full tuition reimbursements available (averaging $18,000 per year), 32 research assistantships with full tuition reimbursements available (averaging $13,500 per year), 42 teaching assistantships with full tuition reimbursements available (averaging $13,500 per year); tuition waivers (full) also available. Financial aid application deadline: 2/1.
Faculty research: Protein, carbohydrate and lipid metabolism, structure and function; synthesis, structure, and reactivity of organometallic and cluster complexes; natural products synthesis; surface chemistry; theoretical chemistry. *Total annual research expenditures:* $8.3 million.
Dr. A. Graham Lappin, Chairman, 219-631-7058, *Fax:* 219-631-6652, *E-mail:* lappin.1@nd.edu.
Application contact: Dr. Terrence J. Akai, Director of Graduate Admissions, 219-631-7706, *Fax:* 219-631-4183, *E-mail:* gradad@nd.edu.

■ **UNIVERSITY OF OKLAHOMA**

Graduate College, College of Arts and Sciences, Department of Chemistry and Biochemistry, Norman, OK 73019-0390

AWARDS MS, PhD. Part-time programs available.

Faculty: 23 full-time (2 women).
Students: 69 full-time (26 women), 5 part-time; includes 5 minority (2 African Americans, 1 Asian American or Pacific Islander, 2 Native Americans), 31 international. *54 applicants, 50% accepted.* In 1999, 15 master's, 6 doctorates awarded. Terminal master's awarded for partial completion of doctoral program.
Degree requirements: For master's, thesis optional; for doctorate, dissertation required.
Entrance requirements: For master's, GRE, TOEFL, BS in chemistry; for doctorate, GRE, TOEFL. *Application deadline:* For fall admission, 4/1 (priority date); for spring admission, 9/1 (priority date). Applications are processed on a rolling basis. *Application fee:* $25.
Expenses: Tuition, state resident: full-time $2,064; part-time $86 per credit hour. Tuition, nonresident: full-time $6,588;

part-time $275 per credit hour. Required fees: $468; $12 per credit hour. $94 per semester. Tuition and fees vary according to course level, course load and program.
Financial aid: In 1999–00, 4 fellowships, 22 research assistantships, 48 teaching assistantships were awarded; career-related internships or fieldwork, Federal Work-Study, grants, scholarships, traineeships, and tuition waivers (partial) also available. Aid available to part-time students. Financial aid application deadline: 4/1.
Faculty research: 7600000.
Dr. Glenn Dryhurst, Chair, 405-325-4811.
Application contact: Ariene Crawford, Graduate Recruiting Secretary, 405-325-2946, *Fax:* 405-325-6111, *E-mail:* admission@chemdept.chem.ou.edu.

Find an in-depth description at www.petersons.com/graduate.

■ **UNIVERSITY OF OKLAHOMA HEALTH SCIENCES CENTER**

College of Medicine and Graduate College, Graduate Programs in Medicine, Department of Biochemistry and Molecular Biology, Oklahoma City, OK 73190

AWARDS Biochemistry (MS, PhD); molecular biology (MS, PhD). Part-time programs available.

Faculty: 22 full-time (4 women), 8 part-time/adjunct (3 women).
Students: 7 full-time (4 women), 23 part-time (10 women); includes 4 minority (2 Asian Americans or Pacific Islanders, 2 Native Americans), 9 international. Average age 25. *53 applicants, 21% accepted.* In 1999, 5 master's, 4 doctorates awarded. Terminal master's awarded for partial completion of doctoral program.
Degree requirements: For master's and doctorate, thesis/dissertation required, foreign language not required.
Entrance requirements: For master's and doctorate, GRE General Test, GRE Subject Test, TOEFL. *Application deadline:* For fall admission, 3/31. *Application fee:* $25 ($50 for international students).
Expenses: Tuition, state resident: part-time $90 per semester hour. Tuition, nonresident: part-time $264 per semester hour. Tuition and fees vary according to program.
Financial aid: Research assistantships, institutionally sponsored loans available. Aid available to part-time students. Financial aid application deadline: 3/31; financial aid applicants required to submit FAFSA.
Faculty research: Gene expression, regulation of transcription, enzyme evolution, melanogenesis, signal transduction.
Dr. Paul H. Weigel, Chairman, 405-271-2227.

University of Oklahoma Health Sciences Center (continued)

Application contact: Dr. A. C. Cox, Committee Chairperson, Biochemistry Admissions, 405-271-2227, *Fax:* 405-271-3092, *E-mail:* biochemistry@uokhsc.edu.

Find an in-depth description at www.petersons.com/graduate.

■ UNIVERSITY OF OREGON

Graduate School, College of Arts and Sciences, Department of Chemistry, Eugene, OR 97403

AWARDS Biochemistry (MA, MS, PhD); chemistry (MA, MS, PhD).

Faculty: 34 full-time (7 women), 7 part-time/adjunct (3 women).
Students: 70 full-time (24 women), 4 part-time (1 woman); includes 8 minority (1 African American, 5 Asian Americans or Pacific Islanders, 2 Hispanic Americans), 7 international. *18 applicants, 89% accepted.* In 1999, 8 master's awarded (100% found work related to degree); 17 doctorates awarded. Terminal master's awarded for partial completion of doctoral program.
Degree requirements: For master's, foreign language and thesis not required; for doctorate, dissertation required, foreign language not required.
Entrance requirements: For master's and doctorate, GRE General Test, TOEFL, minimum GPA of 3.0. *Application deadline:* For fall admission, 1/10. Applications are processed on a rolling basis. *Application fee:* $50.
Expenses: Tuition, state resident: full-time $6,750. Tuition, nonresident: full-time $11,409. Part-time tuition and fees vary according to course load.
Financial aid: In 1999–00, 54 teaching assistantships were awarded; Federal Work-Study and institutionally sponsored loans also available. Financial aid application deadline: 4/15.
Faculty research: Organic chemistry, organometallic chemistry, inorganic chemistry, physical chemistry, materials science, biochemistry, chemical physics, molecular or cell biology.
Frederick W. Dahlquist, Head, 541-346-4601.
Application contact: Lynde Ritzow, Graduate Admissions Coordinator, 541-346-4789.

■ UNIVERSITY OF PENNSYLVANIA

School of Medicine, Biomedical Graduate Studies, Graduate Group in Biochemistry and Molecular Biophysics, Philadelphia, PA 19104
AWARDS PhD, MD/PhD.
Faculty: 66 full-time (10 women).

Students: 75 full-time (30 women); includes 27 minority (2 African Americans, 25 Asian Americans or Pacific Islanders), 7 international. *123 applicants, 23% accepted.* In 1999, 15 doctorates awarded.
Degree requirements: For doctorate, dissertation required, foreign language not required.
Entrance requirements: For doctorate, GRE General Test, TOEFL. *Application deadline:* For fall admission, 1/2 (priority date). Applications are processed on a rolling basis. *Application fee:* $65. Electronic applications accepted.
Expenses: Tuition: Full-time $17,256; part-time $2,991 per course. Required fees: $2,588; $363 per course. $726 per term.
Financial aid: In 1999–00, 45 students received aid, including 28 fellowships, 46 research assistantships; grants and institutionally sponsored loans also available. Financial aid application deadline: 1/2.
Faculty research: Biochemistry of cell differentiation, tissue culture, intermediary metabolism, structure of proteins and nucleic acids, biochemical genetics.
Dr. Kim Sharp, Chairperson, 215-898-3506.
Application contact: Ruth Keris, Graduate Group Administrator, 215-898-4639, *Fax:* 215-573-2085, *E-mail:* keris@mail.med.upenn.edu.

Find an in-depth description at www.petersons.com/graduate.

■ UNIVERSITY OF PITTSBURGH

School of Medicine, Graduate Programs in Medicine, Program in Biochemistry and Molecular Genetics, Pittsburgh, PA 15260
AWARDS MS, PhD.

Faculty: 29 full-time (9 women), 1 part-time/adjunct (0 women).
Students: 16 full-time (5 women); includes 1 minority (Asian American or Pacific Islander), 1 international. *325 applicants, 22% accepted.* In 1999, 3 master's, 4 doctorates awarded.
Degree requirements: For doctorate, dissertation required, foreign language not required. *Average time to degree:* Doctorate–5.3 years full-time.
Entrance requirements: For doctorate, GRE General Test, GRE Subject Test, TOEFL, minimum QPA of 3.0. *Application deadline:* For fall admission, 1/15 (priority date). Applications are processed on a rolling basis. *Application fee:* $30 ($40 for international students).
Expenses: Tuition, state resident: full-time $9,778; part-time $403 per credit. Tuition, nonresident: full-time $20,146; part-time

$830 per credit. Required fees: $480; $90 per semester.
Financial aid: Research assistantships with full tuition reimbursements, teaching assistantships with full tuition reimbursements, Federal Work-Study, institutionally sponsored loans, traineeships, and unspecified assistantships available.
Faculty research: Cell cycle control and DNA replication, gene expression, human gene therapy, molecular genetics of inherited diseases, protein structure and function. *Total annual research expenditures:* $76.3 million.
Application contact: Graduate Studies Administrator, 412-648-8957, *Fax:* 412-648-1236, *E-mail:* biomed_phd@fs1.dean-med.pitt.edu.

Find an in-depth description at www.petersons.com/graduate.

■ UNIVERSITY OF PUERTO RICO, MEDICAL SCIENCES CAMPUS

School of Medicine, Division of Graduate Studies, Department of Biochemistry, San Juan, PR 00936-5067
AWARDS MS, PhD.

Faculty: 9 full-time (3 women), 3 part-time/adjunct (1 woman).
Students: 19 full-time (13 women); all minorities (all Hispanic Americans). Average age 23. *4 applicants, 50% accepted.* In 1999, 2 doctorates awarded (50% entered university research/teaching, 50% found other work related to degree).
Degree requirements: For master's, one foreign language, thesis required; for doctorate, one foreign language, dissertation, comprehensive exam required. *Average time to degree:* Doctorate–5 years full-time.
Entrance requirements: For master's, GRE General Test, GRE Subject Test, interview, minimum GPA of 3.0; for doctorate, GRE General Test, GRE Subject Test, interview, minimum GPA of 3.0 required. *Application deadline:* For fall admission, 9/15; for spring admission, 2/15. *Application fee:* $15.
Expenses: Tuition, state resident: full-time $5,500. Tuition, nonresident: full-time $8,400. Required fees: $600. Tuition and fees vary according to class time, course load, degree level and program.
Financial aid: In 1999–00, 14 students received aid, including 1 fellowship (averaging $12,000 per year), 13 research assistantships (averaging $9,350 per year); Federal Work-Study, grants, and institutionally sponsored loans also available. Financial aid application deadline: 4/30.
Faculty research: Aging, plasma proteins, vitamins and minerals, protein structure/

function, glycosilation of proteins, tumorigenesis. *Total annual research expenditures:* $747,913.

Dr. José R. Rodriguez-Medina, Director, 787-758-2525 Ext. 1633, *Fax:* 787-274-8724, *E-mail:* jo_rodriguez@ rcmaca.upr.clu.edu.

Application contact: Dr. Barulio D. Jimenez, Coordinator, 787-758-2525 Ext. 1235, *Fax:* 787-274-8724, *E-mail:* b_jimenez@rcmaca.upr.clu.edu.

■ UNIVERSITY OF RHODE ISLAND

Graduate School, College of Arts and Sciences, Department of Biochemistry, Microbiology, and Molecular Genetics, Kingston, RI 02881

AWARDS Biochemistry (MS, PhD); microbiology (MS, PhD), including biodegradation (MS), cellular development (MS), electron microscopy and ultrastructure (MS), genetics and molecular biology (MS), immunology (MS), marine and freshwater ecosystems (MS), microbial pathogenesis (MS), microbial physiology (MS), protozoology (MS), virology (MS), water-pollution microbiology (MS).

Degree requirements: For master's and doctorate, thesis/dissertation required.
Entrance requirements: For master's and doctorate, GRE General Test, TOEFL.
Expenses: Tuition, state resident: full-time $3,540; part-time $197 per credit. Tuition, nonresident: full-time $10,116; part-time $197 per credit. Required fees: $1,352; $37 per credit. $65 per term.
Financial aid: Fellowships, research assistantships, teaching assistantships available.
Dr. David Laux, Chairperson, 401-874-2201.

■ UNIVERSITY OF ROCHESTER

School of Medicine and Dentistry, Graduate Programs in Medicine and Dentistry, Department of Biochemistry and Biophysics, Program in Biochemistry, Rochester, NY 14627-0250

AWARDS MS, PhD.

Students: 27 full-time (18 women), 1 part-time; includes 2 minority (1 Asian American or Pacific Islander, 1 Hispanic American), 12 international. In 1999, 10 doctorates awarded. Terminal master's awarded for partial completion of doctoral program.
Degree requirements: For master's, foreign language not required; for doctorate, dissertation, qualifying exam required, foreign language not required.

Entrance requirements: For master's and doctorate, GRE General Test. *Application deadline:* For fall admission, 2/1. *Application fee:* $25.
Expenses: Tuition: Part-time $697 per credit hour. Tuition and fees vary according to program.
Financial aid: Fellowships, research assistantships, teaching assistantships, tuition waivers (full and partial) available. Financial aid application deadline: 2/1.
Application contact: Rose Burgholzer, Graduate Program Secretary, 716-275-3417.

Find an in-depth description at www.petersons.com/graduate.

■ THE UNIVERSITY OF SCRANTON

Graduate School, Department of Chemistry, Program in Biochemistry, Scranton, PA 18510

AWARDS MA, MS. Part-time and evening/weekend programs available.

Faculty: 10 full-time (3 women), 2 part-time/adjunct (1 woman).
Students: 4 full-time (3 women), 13 part-time (6 women); includes 1 minority (African American), 2 international. Average age 25. *14 applicants, 100% accepted.* In 1999, 6 degrees awarded.
Degree requirements: For master's, thesis (for some programs), capstone experience required, foreign language not required.
Entrance requirements: For master's, TOEFL, minimum GPA of 2.75. *Application deadline:* Applications are processed on a rolling basis. *Application fee:* $35.
Expenses: Tuition: Part-time $490 per credit. Required fees: $25 per semester. Tuition and fees vary according to program.
Financial aid: Teaching assistantships, career-related internships or fieldwork, Federal Work-Study, and teaching fellowships available. Aid available to part-time students. Financial aid application deadline: 3/1.
Dr. Christopher A. Baumann, Director, 570-941-6389, *Fax:* 570-941-7510, *E-mail:* cab@tiger.uofs.edu.

■ UNIVERSITY OF SOUTH ALABAMA

College of Medicine and Graduate School, Program in Basic Medical Sciences, Specialization in Biochemistry and Molecular Biology, Mobile, AL 36688-0002

AWARDS PhD.

Faculty: 8 full-time (1 woman).
Students: In 1999, 5 degrees awarded.

Degree requirements: For doctorate, dissertation required, foreign language not required.
Entrance requirements: For doctorate, GRE General Test or MCAT. *Application deadline:* For fall admission, 4/1. Applications are processed on a rolling basis. *Application fee:* $25.
Expenses: Tuition, state resident: part-time $116 per semester hour. Tuition, nonresident: part-time $230 per semester hour. Required fees: $121 per semester.
Financial aid: Fellowships, institutionally sponsored loans available. Financial aid application deadline: 4/1.
Faculty research: Biochemistry of aging, mechanisms of metabolic regulation and oxygen metabolism.
Dr. Nathan N. Aronson, Chair, 334-460-6402.
Application contact: Lanette Flagge, Coordinator, 334-460-6153.

■ UNIVERSITY OF SOUTH CAROLINA

Graduate School, College of Science and Mathematics, Department of Chemistry and Biochemistry, Columbia, SC 29208

AWARDS IMA, MAT, MS, PhD. IMA and MAT offered in cooperation with the College of Education. Part-time programs available.

Faculty: 28 full-time (5 women).
Students: 117 full-time (51 women), 9 part-time (4 women); includes 14 minority (7 African Americans, 5 Asian Americans or Pacific Islanders, 2 Hispanic Americans), 31 international. Average age 27. *376 applicants, 18% accepted.* In 1999, 9 master's, 16 doctorates awarded. Terminal master's awarded for partial completion of doctoral program.
Degree requirements: For master's and doctorate, thesis/dissertation required, foreign language not required.
Entrance requirements: For master's and doctorate, GRE General Test. *Application deadline:* For fall admission, 4/15. Applications are processed on a rolling basis. *Application fee:* $35. Electronic applications accepted.
Expenses: Tuition, state resident: full-time $4,014; part-time $202 per credit hour. Tuition, nonresident: full-time $8,528; part-time $428 per credit hour. Required fees: $100; $4 per credit hour. Tuition and fees vary according to program.
Financial aid: In 1999–00, 13 fellowships, 54 teaching assistantships were awarded; Federal Work-Study, institutionally sponsored loans, and tuition waivers (full) also available. Financial aid application deadline: 4/15.
Faculty research: Spectroscopy, crystallography, organic and organometallic

University of South Carolina (continued) synthesis, analytical chemistry, materials, optical sensing.

Dr. R. Bruce Dunlap, Chair, 803-777-5264, *Fax:* 803-777-9521.

Application contact: Dr. John Dawson, Chairman, Graduate Admissions, 803-777-7234.

■ UNIVERSITY OF SOUTHERN CALIFORNIA

Keck School of Medicine and Graduate School, Graduate Programs in Medicine, Department of Biochemistry and Molecular Biology, Los Angeles, CA 90089

AWARDS MS, PhD.

Faculty: 39 full-time (7 women).
Students: 84 full-time (39 women); includes 23 minority (1 African American, 20 Asian Americans or Pacific Islanders, 2 Hispanic Americans), 44 international. Average age 23. *205 applicants, 13% accepted.* In 1999, 8 master's awarded (75% entered university research/teaching, 25% found other work related to degree); 8 doctorates awarded (100% entered university research/teaching). Terminal master's awarded for partial completion of doctoral program.
Degree requirements: For master's and doctorate, thesis/dissertation required, foreign language not required.
Entrance requirements: For master's and doctorate, GRE General Test, GRE Subject Test, TOEFL, minimum GPA of 3.0. *Application deadline:* For fall admission, 1/15 (priority date). *Application fee:* $55. Electronic applications accepted.
Expenses: Tuition: Full-time $22,198; part-time $748 per unit. Required fees: $406.
Financial aid: In 1999–00, 57 students received aid, including 5 fellowships with tuition reimbursements available, 51 research assistantships with tuition reimbursements available (averaging $1,465 per year), 1 teaching assistantship with tuition reimbursement available; career-related internships or fieldwork, grants, institutionally sponsored loans, and tuition waivers (full) also available. Aid available to part-time students. Financial aid application deadline: 4/15.
Faculty research: Molecular genetics, gene expression, membrane biochemistry, metabolic regulation, cancer biology.
Dr. Laurence H. Kedes, Chairman, 323-442-1144, *Fax:* 323-442-2764, *E-mail:* kedes@zygote.hsc.usc.edu.
Application contact: Anne L. Vazquez, Admissions Counselor, 323-442-1145, *Fax:* 323-442-1224, *E-mail:* annvazqu@hsc.usc.edu.

■ UNIVERSITY OF SOUTHERN MISSISSIPPI

Graduate School, College of Science and Technology, Department of Chemistry and Biochemistry, Hattiesburg, MS 39406

AWARDS Analytical chemistry (MS, PhD); biochemistry (MS, PhD); inorganic chemistry (MS, PhD); organic chemistry (MS, PhD); physical chemistry (MS, PhD).

Degree requirements: For master's, thesis required, foreign language not required; for doctorate, 2 foreign languages (computer language can substitute for one), dissertation required.
Entrance requirements: For master's, GRE General Test, TOEFL, minimum GPA of 2.75; for doctorate, GRE General Test, TOEFL, minimum GPA of 3.5.
Expenses: Tuition, state resident: full-time $2,250; part-time $137 per semester hour. Tuition, nonresident: full-time $3,102; part-time $172 per semester hour. Required fees: $602.
Faculty research: Plant biochemistry, photo chemistry, polymer chemistry, x-ray analysis, enzyme chemistry.

■ UNIVERSITY OF SOUTH FLORIDA

College of Medicine and Graduate School, Graduate Programs in Medical Sciences, Department of Biochemistry and Molecular Biology, Tampa, FL 33620-9951

AWARDS PhD.

Degree requirements: For doctorate, dissertation required, foreign language not required.
Entrance requirements: For doctorate, GRE General Test, minimum GPA of 3.0.
Expenses: Tuition, state resident: part-time $148 per credit hour. Tuition, nonresident: part-time $509 per credit hour.
Faculty research: Eukaryotic systems.
Find an in-depth description at www.petersons.com/graduate.

■ UNIVERSITY OF SOUTH FLORIDA

Graduate School, College of Arts and Sciences, Department of Chemistry, Tampa, FL 33620-9951

AWARDS Analytical chemistry (MS, PhD); biochemistry (MS, PhD); inorganic chemistry (MS, PhD); organic chemistry (MS, PhD); physical chemistry (MS, PhD). Part-time programs available. Terminal master's awarded for partial completion of doctoral program.

Degree requirements: For master's, thesis required; for doctorate, 2 foreign languages (computer language can substitute for one), dissertation, colloquium required.
Entrance requirements: For master's and doctorate, GRE General Test, minimum GPA of 3.0 in last 30 hours of chemistry course work. Electronic applications accepted.
Expenses: Tuition, state resident: part-time $148 per credit hour. Tuition, nonresident: part-time $509 per credit hour.
Faculty research: Synthesis, bioorganic chemistry, bioinorganic chemistry, environmental chemistry, NMR.

■ THE UNIVERSITY OF TENNESSEE

Graduate School, College of Arts and Sciences, Department of Biochemistry and Cellular and Molecular Biology, Knoxville, TN 37996

AWARDS MS, PhD.

Faculty: 23 full-time (4 women).
Students: 34 full-time (16 women); includes 1 minority (Asian American or Pacific Islander), 8 international. *35 applicants, 26% accepted.* In 1999, 1 doctorate awarded. Terminal master's awarded for partial completion of doctoral program.
Degree requirements: For master's and doctorate, thesis/dissertation required, foreign language not required.
Entrance requirements: For master's and doctorate, GRE General Test, TOEFL, minimum GPA of 2.7. *Application deadline:* For fall admission, 2/1 (priority date). Applications are processed on a rolling basis. *Application fee:* $35. Electronic applications accepted.
Expenses: Tuition, state resident: full-time $3,806; part-time $184 per credit hour. Tuition, nonresident: full-time $9,874; part-time $522 per credit hour. Tuition and fees vary according to program.
Financial aid: In 1999–00, 3 fellowships, 25 research assistantships, 32 teaching assistantships were awarded; Federal Work-Study, institutionally sponsored loans, and unspecified assistantships also available. Financial aid application deadline: 2/1; financial aid applicants required to submit FAFSA.
Dr. John Koontz, Head, 865-974-5148, *Fax:* 865-974-6306, *E-mail:* jkoontz@utk.edu.

■ THE UNIVERSITY OF TENNESSEE HEALTH SCIENCE CENTER

College of Graduate Health Sciences, Department of Biochemistry, Memphis, TN 38163-0002

AWARDS MS, PhD. Terminal master's awarded for partial completion of doctoral program.

Degree requirements: For master's, thesis, oral and written comprehensive exams required, foreign language not required; for doctorate, dissertation, oral and written preliminary and comprehensive exams required, foreign language not required.
Entrance requirements: For master's and doctorate, GRE General Test, TOEFL, minimum GPA of 3.0.

Find an in-depth description at www.petersons.com/graduate.

■ THE UNIVERSITY OF TEXAS AT AUSTIN

Graduate School, College of Natural Sciences, Department of Chemistry and Biochemistry, Program in Biochemistry, Austin, TX 78712-1111

AWARDS MA, PhD.

Students: 43 (18 women); includes 5 minority (3 Asian Americans or Pacific Islanders, 2 Hispanic Americans) 7 international. *36 applicants, 25% accepted.* In 1999, 4 master's, 4 doctorates awarded.
Entrance requirements: For master's and doctorate, GRE General Test. *Application fee:* $50 ($75 for international students).
Expenses: Tuition, state resident: part-time $114 per semester hour. Tuition, nonresident: part-time $330 per semester hour. Tuition and fees vary according to program.
Financial aid: Fellowships, research assistantships, teaching assistantships, scholarships available. Financial aid application deadline: 2/1.
Application contact: Dr. G. Barrie Kitto, Graduate Adviser, 512-471-3279.

■ THE UNIVERSITY OF TEXAS AT AUSTIN

Graduate School, College of Natural Sciences, Institute for Cellular and Molecular Biology, Program in Biochemistry, Austin, TX 78712-1111

AWARDS PhD.

Expenses: Tuition, state resident: part-time $114 per semester hour. Tuition, nonresident: part-time $330 per semester hour. Tuition and fees vary according to program.

Marvin L. Hackert, Chair, 512-471-3949.
Find an in-depth description at www.petersons.com/graduate.

■ THE UNIVERSITY OF TEXAS HEALTH SCIENCE CENTER AT SAN ANTONIO

Graduate School of Biomedical Sciences, Department of Biochemistry, San Antonio, TX 78229-3900

AWARDS MS, PhD. Terminal master's awarded for partial completion of doctoral program.

Degree requirements: For master's and doctorate, thesis/dissertation required.
Entrance requirements: For master's and doctorate, GRE General Test.
Expenses: Tuition, state resident: part-time $38 per credit hour. Tuition, nonresident: part-time $249 per credit hour.
Faculty research: Protein structure and function, lipid biochemistry, metabolic regulation, immunology, membrane assembly.

■ THE UNIVERSITY OF TEXAS–HOUSTON HEALTH SCIENCE CENTER

Graduate School of Biomedical Sciences, Program in Biochemistry and Molecular Biology, Houston, TX 77225-0036

AWARDS MS, PhD, MD/PhD.

Faculty: 17 full-time (3 women).
Students: 8 full-time (3 women), 6 international. Average age 27. *38 applicants, 61% accepted.* In 1999, 1 master's, 6 doctorates awarded. Terminal master's awarded for partial completion of doctoral program.
Degree requirements: For master's and doctorate, thesis/dissertation required, foreign language not required.
Entrance requirements: For master's and doctorate, GRE General Test, TOEFL, TWE. *Application deadline:* For fall admission, 1/15 (priority date); for spring admission, 11/1. Applications are processed on a rolling basis. *Application fee:* $10. Electronic applications accepted.
Financial aid: Fellowships, research assistantships, teaching assistantships, institutionally sponsored loans available. Financial aid application deadline: 1/15.
Faculty research: Protein structure and function, regulation of gene expression, cellular signal transduction, regulation of cell growth, membrane protein assembly and function.

Dr. John Putkey, Director, 713-500-6061, *Fax:* 713-500-0652, *E-mail:* john.putkey@ uth.tmc.edu.
Application contact: Anne Baronitis, Director of Admissions, 713-500-9860, *Fax:* 713-500-9877, *E-mail:* abaron@ gsbs.gs.uth.tmc.edu.

Find an in-depth description at www.petersons.com/graduate.

■ THE UNIVERSITY OF TEXAS MEDICAL BRANCH AT GALVESTON

Graduate School of Biomedical Sciences, Program in Human Biological Chemistry and Genetics, Galveston, TX 77555

AWARDS Biochemistry (MS, PhD); cell biology (MS, PhD); genetics (MS, PhD); molecular biology (PhD); structural biology (PhD).

Faculty: 61 full-time (4 women).
Students: 39 full-time (12 women); includes 2 minority (both Hispanic Americans), 12 international. Average age 28. *41 applicants, 66% accepted.* In 1999, 1 master's awarded (100% found work related to degree); 4 doctorates awarded (100% found work related to degree).
Degree requirements: For master's and doctorate, thesis/dissertation required, foreign language not required.
Entrance requirements: For master's and doctorate, GRE General Test. *Application deadline:* For fall admission, 2/1 (priority date). Applications are processed on a rolling basis. *Application fee:* $25 ($50 for international students). Electronic applications accepted.
Expenses: Tuition, state resident: full-time $684; part-time $38 per credit hour. Tuition, nonresident: full-time $4,572; part-time $254 per credit hour. Required fees: $29; $7.5 per credit hour. One-time fee: $55. Tuition and fees vary according to program.
Financial aid: Fellowships, research assistantships, Federal Work-Study and institutionally sponsored loans available. Financial aid application deadline: 2/1; financial aid applicants required to submit FAFSA.
Faculty research: Cell surfaces and transmembrane signaling, structural biology, growth factors, DNA repair and aging.
Dr. James C. Lee, Director, 409-772-2769, *Fax:* 409-747-0552, *E-mail:* gradprog.hbcg@utmb.edu.

The University of Texas Medical Branch at Galveston (continued)
Application contact: Becky L. Hansen, Senior Administrative Secretary, 409-772-2769, *Fax:* 409-747-0552, *E-mail:* bhansen@utmb.edu.

Find an in-depth description at www.petersons.com/graduate.

■ THE UNIVERSITY OF TEXAS SOUTHWESTERN MEDICAL CENTER AT DALLAS

Southwestern Graduate School of Biomedical Sciences, Division of Cell and Molecular Biology, Program in Biological Chemistry, Dallas, TX 75390
AWARDS PhD.

Faculty: 46 full-time (4 women).
Students: 17 full-time (5 women); includes 3 minority (2 Asian Americans or Pacific Islanders, 1 Hispanic American), 4 international. Average age 25. In 1999, 4 doctorates awarded.
Degree requirements: For doctorate, dissertation required, foreign language not required.
Entrance requirements: For doctorate, GRE General Test, minimum GPA of 3.0. *Application deadline:* For fall admission, 1/5 (priority date). *Application fee:* $0. Electronic applications accepted.
Expenses: Tuition, state resident: full-time $912. Tuition, nonresident: full-time $6,096. Required fees: $216. Full-time tuition and fees vary according to course load and program.
Financial aid: Fellowships, research assistantships, institutionally sponsored loans available. Financial aid application deadline: 3/15; financial aid applicants required to submit FAFSA.
Faculty research: Regulation of gene expression, protein trafficking, molecular neurobiology, protein structure and function, metabolic regulation.
Dr. Margaret A. Phillips, Chair, 214-648-3637, *Fax:* 214-648-9961, *E-mail:* philli01@utsw.swmed.edu.
Application contact: Nancy McKinney, Education Coordinator, 214-648-8099, *Fax:* 214-648-2978, *E-mail:* dcmbinfo@utsouthwestern.edu.

Find an in-depth description at www.petersons.com/graduate.

■ UNIVERSITY OF THE PACIFIC

Graduate School, Department of Chemistry, Stockton, CA 95211-0197
AWARDS Biochemistry (MS, PhD); chemistry (MS, PhD).

Faculty: 7 full-time (1 woman).
Students: 1 full-time (0 women), 12 part-time (3 women), 14 international. In 1999, 4 master's awarded. Terminal master's awarded for partial completion of doctoral program.
Degree requirements: For master's, thesis required, foreign language not required; for doctorate, one foreign language (computer language can substitute), dissertation required.
Entrance requirements: For master's and doctorate, GRE General Test, GRE Subject Test. *Application deadline:* For fall admission, 3/1 (priority date); for spring admission, 10/15. Applications are processed on a rolling basis. *Application fee:* $50.
Expenses: Tuition: Full-time $19,570; part-time $612 per unit. Required fees: $260. Tuition and fees vary according to program.
Financial aid: Teaching assistantships, institutionally sponsored loans available. Aid available to part-time students. Financial aid application deadline: 3/1.
Dr. Pat Jones, Chairman, 209-946-2241, *E-mail:* pjones@uop.edu.

■ UNIVERSITY OF TOLEDO

Graduate School, College of Arts and Sciences, Department of Chemistry, Toledo, OH 43606-3398
AWARDS Analytical chemistry (MES, MS, PhD); biological chemistry (MES, MS, PhD); inorganic chemistry (MES, MS, PhD); organic chemistry (MES, MS, PhD); physical chemistry (MES, MS, PhD). Part-time programs available.

Faculty: 16 full-time (3 women).
Students: 34 (20 women); includes 11 minority (2 African Americans, 3 Asian Americans or Pacific Islanders, 6 Hispanic Americans) 18 international. Average age 28. *36 applicants, 33% accepted.* In 1999, 8 master's, 5 doctorates awarded.
Degree requirements: For master's and doctorate, thesis/dissertation required, foreign language not required.
Entrance requirements: For master's and doctorate, GRE General Test, GRE Subject Test, TOEFL. *Application deadline:* For fall admission, 8/1 (priority date). Applications are processed on a rolling basis. *Application fee:* $30. Electronic applications accepted.
Expenses: Tuition, state resident: full-time $2,741; part-time $228 per credit hour. Tuition, nonresident: full-time $5,926; part-time $494 per credit hour. Required fees: $402; $34 per credit hour.
Financial aid: In 1999–00, 9 research assistantships were awarded; fellowships, teaching assistantships, Federal Work-Study and institutionally sponsored loans also available. Aid available to part-time students. Financial aid application deadline: 4/1; financial aid applicants required to submit FAFSA.

Faculty research: Enzymology, materials chemistry, crystallography, theoretical chemistry.
Dr. Alan Pinkerton, Interim Chair, 419-530-2109, *Fax:* 419-530-4033.

■ UNIVERSITY OF TOLEDO

Graduate School, College of Pharmacy, Graduate Programs in Pharmacy, Program in Medicinal and Biological Chemistry, Toledo, OH 43606-3398
AWARDS MS, PhD.

Faculty: 8 full-time (2 women).
Students: 18 full-time (6 women), 14 international. Average age 30. *59 applicants, 5% accepted.* In 1999, 3 master's, 1 doctorate awarded. Terminal master's awarded for partial completion of doctoral program.
Degree requirements: For master's and doctorate, thesis/dissertation required, foreign language not required. *Average time to degree:* Master's–2 years full-time; doctorate–5 years full-time.
Entrance requirements: For master's and doctorate, GRE General Test, TOEFL. *Application deadline:* For fall admission, 2/1 (priority date). *Application fee:* $30. Electronic applications accepted.
Expenses: Tuition, state resident: full-time $2,741; part-time $228 per credit hour. Tuition, nonresident: full-time $5,926; part-time $494 per credit hour. Required fees: $402; $34 per credit hour.
Financial aid: Research assistantships, teaching assistantships, tuition waivers (full) available.
Faculty research: Neuroscience, molecular modeling, immunotoxicology, organic synthesis, peptide biochemistry.
Dr. Richard Hudson, Chairman, 419-530-1979, *Fax:* 419-530-7946.
Application contact: Dr. Steven Peseckis, Coordinator, 419-530-1944, *Fax:* 419-530-7946, *E-mail:* speseck@ofto2.utoledo.edu.

■ UNIVERSITY OF UTAH

Graduate School, College of Science and Graduate Programs in Medicine, Program in Biological Chemistry, Salt Lake City, UT 84112-1107
AWARDS PhD.

Degree requirements: For doctorate, dissertation required.
Entrance requirements: For doctorate, GRE General Test, GRE Subject Test, TOEFL.
Expenses: Tuition, state resident: full-time $1,663. Tuition, nonresident: full-time $5,201. Tuition and fees vary according to course load and program.

Find an in-depth description at www.petersons.com/graduate.

■ UNIVERSITY OF UTAH

School of Medicine and Graduate School, Graduate Programs in Medicine, Department of Biochemistry, Salt Lake City, UT 84112-1107

AWARDS MS, PhD.

Faculty: 9 full-time (2 women), 13 part-time/adjunct (1 woman).
Students: 27 full-time (10 women); includes 3 minority (1 African American, 2 Asian Americans or Pacific Islanders), 9 international. Average age 30. *601 applicants, 16% accepted.* In 1999, 1 master's, 6 doctorates awarded (17% found work related to degree, 83% continued full-time study). Terminal master's awarded for partial completion of doctoral program.
Degree requirements: For master's and doctorate, thesis/dissertation required, foreign language not required. *Average time to degree:* Master's–4 years full-time; doctorate–7.3 years full-time.
Entrance requirements: For doctorate, GRE Subject Test, minimum GPA of 3.0. *Application deadline:* For fall admission, 1/15 (priority date). Applications are processed on a rolling basis. *Application fee:* $0. Electronic applications accepted.
Expenses: Tuition, state resident: full-time $2,105. Tuition, nonresident: full-time $6,312.
Financial aid: In 1999–00, 27 students received aid, including 1 fellowship with full tuition reimbursement available, 26 research assistantships with full tuition reimbursements available; tuition waivers (full) also available.
Faculty research: Protein structure and function, nucleic acid structure and function, nucleic acid enzymology, RNA modification, protein turnover. *Total annual research expenditures:* $1.5 million.
Dr. Dana Carroll, Chair, 801-581-5977, *Fax:* 801-581-7959, *E-mail:* tami.brunson@path.med.utah.edu.
Application contact: 801-581-5207, *Fax:* 801-585-2465, *E-mail:* tami.brunson@path.med.utah.edu.

■ UNIVERSITY OF VERMONT

College of Medicine and Graduate College, Graduate Programs in Medicine, Department of Biochemistry, Burlington, VT 05405

AWARDS MS, PhD, MD/MS, MD/PhD.

Degree requirements: For master's and doctorate, thesis/dissertation required, foreign language not required.
Entrance requirements: For master's and doctorate, GRE General Test, GRE Subject Test, TOEFL.
Expenses: Tuition, state resident: full-time $7,464; part-time $311 per credit. Tuition, nonresident: full-time $18,672; part-time

$778 per credit. Full-time tuition and fees vary according to degree level and program.
Faculty research: Endocrinology, protein chemistry, cell-surface signaling.

Find an in-depth description at www.petersons.com/graduate.

■ UNIVERSITY OF VERMONT

Graduate College, College of Agriculture and Life Sciences, Department of Agricultural Biochemistry, Burlington, VT 05405

AWARDS MS, PhD.

Degree requirements: For doctorate, dissertation required.
Entrance requirements: For master's and doctorate, GRE General Test, TOEFL.
Expenses: Tuition, state resident: full-time $7,464; part-time $311 per credit. Tuition, nonresident: full-time $18,672; part-time $778 per credit. Full-time tuition and fees vary according to degree level and program.

■ UNIVERSITY OF VIRGINIA

College and Graduate School of Arts and Sciences, Department of Biochemistry and Molecular Genetics, Charlottesville, VA 22903

AWARDS Biochemistry (PhD); molecular genetics (PhD).

Faculty: 21 full-time (4 women), 1 part-time/adjunct (0 women).
Students: 24 full-time (11 women); includes 3 minority (2 African Americans, 1 Asian American or Pacific Islander), 9 international. Average age 26. *43 applicants, 21% accepted.* In 1999, 1 degree awarded.
Degree requirements: For doctorate, dissertation required, foreign language not required.
Entrance requirements: For doctorate, GRE General Test, GRE Subject Test. *Application fee:* $40. Electronic applications accepted.
Expenses: Tuition, state resident: full-time $3,832. Tuition, nonresident: full-time $15,519. Required fees: $1,084. Tuition and fees vary according to course load and program.
Financial aid: Application deadline: 2/1.
Joyce L. Hamlin, Chairman, 804-924-5139.
Application contact: Duane J. Osheim, Associate Dean, 804-924-7184, *E-mail:* microbiology@virginia.edu.

Find an in-depth description at www.petersons.com/graduate.

■ UNIVERSITY OF WASHINGTON

School of Medicine and Graduate School, Graduate Programs in Medicine, Department of Biochemistry, Seattle, WA 98195

AWARDS PhD.

Faculty: 22 full-time (3 women), 10 part-time/adjunct (3 women).
Students: 50 full-time (21 women); includes 3 minority (2 Asian Americans or Pacific Islanders, 1 Native American), 13 international. Average age 23. *95 applicants, 23% accepted.* In 1999, 4 degrees awarded (50% entered university research/teaching, 50% found other work related to degree).
Degree requirements: For doctorate, dissertation required, foreign language not required. *Average time to degree:* Doctorate–6 years full-time.
Entrance requirements: For doctorate, GRE General Test, GRE Subject Test (biology, chemistry or biochemistry, cell and molecular biology), TOEFL, minimum GPA of 3.0. *Application deadline:* For fall admission, 12/15. *Application fee:* $50. Electronic applications accepted.
Expenses: Tuition, state resident: full-time $9,210; part-time $236 per credit. Tuition, nonresident: full-time $23,256; part-time $596 per credit.
Financial aid: In 1999–00, 22 fellowships (averaging $17,304 per year), 28 research assistantships (averaging $17,304 per year) were awarded; teaching assistantships, institutionally sponsored loans, traineeships, and tuition waivers (full and partial) also available.
Faculty research: Blood coagulation, structure and function of enzymes, fertilization events, interaction of plants with bacteria, protein structure.
Dr. Richard A. Palmiter, Acting Chairman, 206-543-1768, *Fax:* 206-685-1792.
Application contact: Lynett Kimmel, Program Coordinator, 206-543-0485, *Fax:* 206-685-1792, *E-mail:* biochem@u.washington.edu.

Find an in-depth description at www.petersons.com/graduate.

■ UNIVERSITY OF WISCONSIN–MADISON

Graduate School, College of Agricultural and Life Sciences, Department of Animal Health and Biomedical Sciences, Program in Comparative Biosciences, Madison, WI 53706-1380

AWARDS Anatomy (MS, PhD); biochemistry (MS, PhD); cellular and molecular biology (MS, PhD); environmental toxicology (MS, PhD); neurosciences (MS, PhD); pharmacology (MS, PhD); physiology (MS, PhD).

University of Wisconsin–Madison (continued)

Degree requirements: For doctorate, dissertation required.

Expenses: Tuition, state resident: full-time $5,406; part-time $339 per credit. Tuition, nonresident: full-time $17,110; part-time $1,071 per credit. Full-time tuition and fees vary according to program and reciprocity agreements. Part-time tuition and fees vary according to course load and program.

■ **UNIVERSITY OF WISCONSIN– MADISON**

Graduate School, College of Agricultural and Life Sciences, Department of Biochemistry, Madison, WI 53706-1380

AWARDS MS, PhD.

Faculty: 33 full-time (7 women).
Students: 100 full-time (40 women); includes 2 minority (1 Asian American or Pacific Islander, 1 Hispanic American), 27 international. Average age 25. *227 applicants, 22% accepted.* In 1999, 2 master's, 22 doctorates awarded. Terminal master's awarded for partial completion of doctoral program.
Degree requirements: For master's and doctorate, thesis/dissertation required. *Average time to degree:* Master's–3 years full-time; doctorate–5.5 years full-time.
Entrance requirements: For master's and doctorate, GRE General Test, GRE Subject Test. *Application deadline:* For fall admission, 1/1 (priority date). Applications are processed on a rolling basis. *Application fee:* $45. Electronic applications accepted.
Expenses: Tuition, state resident: full-time $5,406; part-time $339 per credit. Tuition, nonresident: full-time $17,110; part-time $1,071 per credit. Full-time tuition and fees vary according to program and reciprocity agreements. Part-time tuition and fees vary according to course load and program.
Financial aid: In 1999–00, 15 fellowships with tuition reimbursements (averaging $16,000 per year), 85 research assistantships with tuition reimbursements (averaging $16,000 per year) were awarded; traineeships and tuition waivers (full) also available. Financial aid application deadline: 1/1.
Faculty research: Molecular structure of vitamins and hormones, enzymology, NMR spectroscopy, protein structure, molecular genetics. *Total annual research expenditures:* $13 million.
Dr. Hector F. DeLuca, Chair, 608-262-9832, *Fax:* 608-262-3453, *E-mail:* deluca@ biochem.wisc.edu.
Application contact: Colleen A. Clary, Program Assistant, 608-262-3899, *Fax:*

608-265-9661, *E-mail:* cclary@ biochem.wisc.edu.

Find an in-depth description at www.petersons.com/graduate.

■ **UNIVERSITY OF WISCONSIN– MADISON**

Medical School and Graduate School, Graduate Programs in Medicine, Department of Biomolecular Chemistry, Madison, WI 53706-1380

AWARDS MS, PhD.

Faculty: 10 full-time (3 women).
Students: 44 full-time (27 women); includes 6 minority (2 Asian Americans or Pacific Islanders, 4 Hispanic Americans), 7 international. Terminal master's awarded for partial completion of doctoral program.
Degree requirements: For master's and doctorate, thesis/dissertation required, foreign language not required. *Average time to degree:* Master's–2 years full-time; doctorate–5.7 years full-time.
Entrance requirements: For doctorate, GRE. *Application deadline:* For fall admission, 1/1 (priority date). *Application fee:* $45. Electronic applications accepted.
Expenses: Tuition, state resident: full-time $5,406; part-time $339 per credit. Tuition, nonresident: full-time $17,110; part-time $1,071 per credit.
Financial aid: In 1999–00, 16 fellowships with full tuition reimbursements (averaging $16,300 per year), 23 research assistantships with full tuition reimbursements (averaging $16,300 per year) were awarded; traineeships and tuition waivers (full) also available.
Faculty research: Membrane biochemistry, protein folding and translocation, gene expression, signal transduction, cell growth and differentiation.
Dr. Elizabeth A. Craig, Chair, 608-262-1347, *Fax:* 608-262-5253, *E-mail:* ecraig@ facstaff.wisc.edu.
Application contact: Mary Smith, Student Services Coordinator, 608-262-1347, *Fax:* 608-262-5253, *E-mail:* msmith@ facstaff.wisc.edu.

Find an in-depth description at www.petersons.com/graduate.

■ **UTAH STATE UNIVERSITY**

School of Graduate Studies, College of Science, Department of Chemistry and Biochemistry, Logan, UT 84322

AWARDS Biochemistry (MS, PhD); chemistry (MS, PhD). Part-time programs available.

Faculty: 19 full-time (2 women), 3 part-time/adjunct (0 women).
Students: 40 full-time (10 women), 8 part-time (3 women); includes 2 minority (both Asian Americans or Pacific Islanders), 17

international. Average age 28. *36 applicants, 53% accepted.* In 1999, 3 master's, 9 doctorates awarded. Terminal master's awarded for partial completion of doctoral program.
Degree requirements: For master's and doctorate, thesis/dissertation, oral and written exams required, foreign language not required.
Entrance requirements: For master's and doctorate, GRE General Test, TOEFL, minimum GPA of 3.0. *Application deadline:* For fall admission, 4/15 (priority date); for spring admission, 10/15. Applications are processed on a rolling basis. *Application fee:* $40.
Expenses: Tuition, state resident: full-time $1,553. Tuition, nonresident: full-time $5,436. International tuition: $5,526 full-time. Required fees: $447. Tuition and fees vary according to course load and program.
Financial aid: In 1999–00, 31 research assistantships with partial tuition reimbursements, 22 teaching assistantships with partial tuition reimbursements were awarded; fellowships, Federal Work-Study, institutionally sponsored loans, and tuition waivers (partial) also available. Aid available to part-time students. Financial aid application deadline: 4/15.
Faculty research: Analytical, inorganic, organic, and physical chemistry.
Dr. Vernon Parker, Head, 435-797-1619, *Fax:* 435-797-3390.
Application contact: Dr. Alvan C. Hengge, Admissions Chair, 435-797-3442, *Fax:* 435-797-3390.

Find an in-depth description at www.petersons.com/graduate.

■ **VANDERBILT UNIVERSITY**

Graduate School and School of Medicine, Department of Biochemistry, Nashville, TN 37240-1001

AWARDS MS, PhD, MD/PhD.

Faculty: 20 full-time (2 women), 8 part-time/adjunct (0 women).
Students: 27 full-time (13 women); includes 1 minority (African American). Average age 27. In 1999, 1 master's, 9 doctorates awarded.
Degree requirements: For master's, thesis required, foreign language not required; for doctorate, dissertation, preliminary, qualifying, and final exams required, foreign language not required.
Entrance requirements: For master's, GRE General Test; for doctorate, GRE General Test, GRE Subject Test (recommended). *Application deadline:* For fall admission, 1/15. *Application fee:* $40.
Expenses: Tuition: Full-time $17,244; part-time $958 per hour. Required fees:

$242; $121 per semester. Tuition and fees vary according to program.

Financial aid: In 1999–00, fellowships with full tuition reimbursements (averaging $17,000 per year), research assistantships (averaging $17,000 per year) were awarded; institutionally sponsored loans, traineeships, and tuition waivers (partial) also available. Financial aid application deadline: 1/15.

Faculty research: Protein chemistry, carcinogenesis, metabolism, toxicology, receptors and signaling, DNA recognition and transcription. *Total annual research expenditures:* $5.6 million.

Michael R. Waterman, Chair, 615-322-3315, *Fax:* 615-322-4349, *E-mail:* michael.waterman@vanderbilt.edu.

Application contact: Ronald M. Wisdom, Director of Graduate Studies, 615-343-3315, *Fax:* 615-322-4349, *E-mail:* ronald.m.wisdom@vanderbilt.edu.

Find an in-depth description at www.petersons.com/graduate.

■ VIRGINIA COMMONWEALTH UNIVERSITY

School of Graduate Studies and School of Medicine, School of Medicine Graduate Programs, Department of Biochemistry and Molecular Biophysics, Richmond, VA 23284-9005

AWARDS Biochemistry (PhD); biochemistry and molecular biophysics (MS, CBHS); molecular biology and genetics (PhD); neurosciences (PhD).

Students: 5 full-time (3 women), 21 part-time (10 women); includes 12 minority (9 Asian Americans or Pacific Islanders, 1 Hispanic American, 2 Native Americans). In 1999, 1 master's, 10 doctorates awarded.

Degree requirements: For master's, thesis required, foreign language not required; for doctorate, dissertation, comprehensive oral and written exams required, foreign language not required.

Entrance requirements: For master's and doctorate, GRE General Test. *Application deadline:* For fall admission, 5/1. *Application fee:* $30.

Expenses: Tuition, state resident: full-time $4,031; part-time $224 per credit hour. Tuition, nonresident: full-time $11,946; part-time $664 per credit hour. Required fees: $1,081; $40 per credit hour. Tuition and fees vary according to campus/location and program.

Financial aid: Fellowships, research assistantships available.

Faculty research: Molecular biology, peptide/protein chemistry, neurochemistry, enzyme mechanisms, macromolecular structure determination. *Total annual research expenditures:* $3.5 million.

Chair, 804-828-9762, *Fax:* 804-828-1473.

Application contact: Dr. Zendra E. Zehner, Program Director, 804-828-8753, *Fax:* 804-828-1473, *E-mail:* zezehner@vcu.edu.

■ VIRGINIA POLYTECHNIC INSTITUTE AND STATE UNIVERSITY

Graduate School, College of Agriculture and Life Sciences, Department of Biochemistry, Blacksburg, VA 24061

AWARDS Biochemistry (MS, PhD); cell and molecular biology (PhD).

Students: 17 full-time (8 women), 4 part-time (all women), 9 international. *40 applicants, 35% accepted.* In 1999, 2 master's, 2 doctorates awarded.

Degree requirements: For doctorate, dissertation required, foreign language not required.

Entrance requirements: For master's, TOEFL; for doctorate, GRE General Test, TOEFL. *Application deadline:* For fall admission, 12/1 (priority date). Applications are processed on a rolling basis. *Application fee:* $25.

Expenses: Tuition, state resident: full-time $4,122; part-time $229 per credit hour. Tuition, nonresident: full-time $6,930; part-time $385 per credit hour. Required fees: $828; $107 per semester. Part-time tuition and fees vary according to course load.

Financial aid: In 1999–00, research assistantships with full tuition reimbursements (averaging $15,815 per year), 8 teaching assistantships with full tuition reimbursements (averaging $15,815 per year) were awarded; fellowships. Financial aid application deadline: 4/1.

Dr. John Hess, Head, 540-231-6315, *E-mail:* jlhess@vt.edu.

Find an in-depth description at www.petersons.com/graduate.

■ WAKE FOREST UNIVERSITY

School of Medicine and Graduate School, Graduate Programs in Medicine, Department of Biochemistry, Winston-Salem, NC 27109

AWARDS PhD.

Degree requirements: For doctorate, dissertation required.

Entrance requirements: For doctorate, GRE General Test, GRE Subject Test. Electronic applications accepted.

Expenses: Tuition: Full-time $18,300.

Faculty research: Biomembranes, cancer, biophysics.

Find an in-depth description at www.petersons.com/graduate.

■ WAKE FOREST UNIVERSITY

School of Medicine and Graduate School, Graduate Programs in Medicine, Interdisciplinary Studies in Bioorganic and Macromolecular Structure, Winston-Salem, NC 27109

AWARDS PhD.

Degree requirements: For doctorate, dissertation required.

Entrance requirements: For doctorate, GRE General Test, GRE Subject Test.

Expenses: Tuition: Full-time $18,300.

■ WASHINGTON STATE UNIVERSITY

Graduate School, College of Sciences, School of Molecular Biosciences, Department of Biochemistry and Biophysics, Pullman, WA 99164

AWARDS MS, PhD.

Faculty: 23.

Students: 43 full-time (14 women), 3 part-time (2 women); includes 3 minority (all Asian Americans or Pacific Islanders), 9 international. In 1999, 4 master's, 6 doctorates awarded. Terminal master's awarded for partial completion of doctoral program.

Degree requirements: For master's, oral exam required, thesis optional, foreign language not required; for doctorate, dissertation, oral exam, written exam required, foreign language not required. *Average time to degree:* Master's–2 years full-time; doctorate–4 years full-time.

Entrance requirements: For master's and doctorate, GRE General Test, minimum GPA of 3.0. *Application deadline:* For fall admission, 3/1 (priority date). Applications are processed on a rolling basis. *Application fee:* $35.

Expenses: Tuition, state resident: full-time $5,654. Tuition, nonresident: full-time $13,850. International tuition: $13,850 full-time. Tuition and fees vary according to program.

Financial aid: In 1999–00, 33 research assistantships with full and partial tuition reimbursements, 9 teaching assistantships with full and partial tuition reimbursements were awarded; fellowships, career-related internships or fieldwork, Federal Work-Study, institutionally sponsored loans, traineeships, and tuition waivers (partial) also available. Financial aid application deadline: 4/1; financial aid applicants required to submit FAFSA.

Faculty research: Gene regulation, signal transduction, protein export, reproductive

Washington State University (continued) biology, DNA repair. *Total annual research expenditures:* $3.4 million.
Dr. Gerald Hazelbauer, Chairman, 509-335-1276.
Application contact: Dr. Kirk McMichael, Graduate Coordinator, 509-335-8866, *Fax:* 509-335-8867.

Find an in-depth description at www.petersons.com/graduate.

■ **WASHINGTON UNIVERSITY IN ST. LOUIS**

Graduate School of Arts and Sciences, Division of Biology and Biomedical Sciences, Program in Biochemistry, St. Louis, MO 63130-4899

AWARDS PhD.

Degree requirements: For doctorate, dissertation required, foreign language not required.
Entrance requirements: For doctorate, GRE General Test, GRE Subject Test. *Application deadline:* For fall admission, 1/1 (priority date). Applications are processed on a rolling basis. *Application fee:* $0.
Expenses: Tuition: Full-time $23,400; part-time $975 per credit. Tuition and fees vary according to program.
Financial aid: Application deadline: 1/1.
Dr. David Cistola, Coordinator, 314-362-4393.
Application contact: Rosemary Garagneni, Director of Admissions, 800-852-9074, *E-mail:* admissions@dbbs.wustl.edu.

■ **WAYNE STATE UNIVERSITY**

School of Medicine and Graduate School, Graduate Programs in Medicine, Department of Biochemistry and Molecular Biology, Detroit, MI 48202

AWARDS Biochemistry (MS, PhD). Terminal master's awarded for partial completion of doctoral program.

Degree requirements: For master's, thesis required, foreign language not required; for doctorate, one foreign language, dissertation required.
Entrance requirements: For master's and doctorate, GRE General Test, GRE Subject Test.
Faculty research: Protein structure, molecular biology, molecular genetics, enzymology, x-ray crystallography.

■ **WESLEYAN UNIVERSITY**

Graduate Programs, Department of Chemistry, Middletown, CT 06459-0260
AWARDS Biochemistry (MA, PhD); chemical physics (MA, PhD); inorganic chemistry (MA,

PhD); organic chemistry (MA, PhD); physical chemistry (MA, PhD); theoretical chemistry (MA, PhD).

Faculty: 13 full-time (2 women), 2 part-time/adjunct (1 woman).
Students: 29 full-time (11 women); includes 2 minority (1 African American, 1 Native American), 16 international. Average age 28. *217 applicants, 4% accepted.* In 1999, 3 master's, 5 doctorates awarded. Terminal master's awarded for partial completion of doctoral program.
Degree requirements: For master's and doctorate, one foreign language (computer language can substitute), thesis/dissertation required.
Entrance requirements: For master's, GRE General Test, GRE Subject Test; for doctorate, GRE Subject Test. *Application deadline:* For fall admission, 3/1. Applications are processed on a rolling basis. *Application fee:* $0.
Expenses: Tuition: Full-time $24,876. Required fees: $650. Tuition and fees vary according to program.
Financial aid: Fellowships, research assistantships, teaching assistantships, institutionally sponsored loans available. Stewart Novick, Graduate Adviser, 860-685-2649.
Application contact: Marina J. Melendez, Director of Graduate Student Services, 860-685-2390, *Fax:* 860-685-2439, *E-mail:* mmelendez@wesleyan.edu.

Find an in-depth description at www.petersons.com/graduate.

■ **WESLEYAN UNIVERSITY**

Graduate Programs, Department of Molecular Biology and Biochemistry, Middletown, CT 06459-0260

AWARDS Biochemistry (PhD); molecular biology (PhD).

Faculty: 9 full-time (2 women).
Students: 22 full-time (11 women); includes 2 minority (both Hispanic Americans), 16 international. Average age 28. *89 applicants, 3% accepted.* In 1999, 1 doctorate awarded.
Degree requirements: For doctorate, computer language, dissertation required.
Entrance requirements: For doctorate, GRE General Test, GRE Subject Test. *Application deadline:* For fall admission, 3/15. Applications are processed on a rolling basis. *Application fee:* $0.
Expenses: Tuition: Full-time $24,876. Required fees: $650. Tuition and fees vary according to program.
Financial aid: Research assistantships, teaching assistantships, institutionally sponsored loans available.
Faculty research: Genome organization, regulation of gene expression, molecular

biology of development, physical biochemistry.
Dr. Irina Russu, Chair, 860-685-3556.
Application contact: Marina J. Melendez, Director of Graduate Student Services, 860-685-2390, *Fax:* 860-685-2439, *E-mail:* mmelendez@wesleyan.edu.

Find an in-depth description at www.petersons.com/graduate.

■ **WEST VIRGINIA UNIVERSITY**

School of Medicine, Graduate Programs in Health Sciences, Department of Biochemistry, Morgantown, WV 26506

AWARDS Developmental biology (MS); energy transduction (MS); enzymes and serum proteins (PhD); enzymology (MS); gene expression (MS, PhD); hormonal regulation/metabolism (MS, PhD); membrane biogenesis (MS, PhD); molecular virology (MS); nucleic acids (MS, PhD); nutritional oncology (PhD); protein chemistry (MS); secretory mechanisms (PhD).

Students: 14 full-time (7 women), 2 part-time, 7 international. Average age 27. *30 applicants, 10% accepted.* In 1999, 2 master's, 2 doctorates awarded (100% entered university research/teaching). Terminal master's awarded for partial completion of doctoral program.
Degree requirements: For master's, thesis required, foreign language not required; for doctorate, dissertation, comprehensive exam required, foreign language not required. *Average time to degree:* Master's–2 years full-time; doctorate–5 years full-time.
Entrance requirements: For master's, GRE General Test, TOEFL, minimum GPA of 3.0; for doctorate, GRE General Test, TOEFL. *Application deadline:* For fall admission, 6/1. Applications are processed on a rolling basis. *Application fee:* $45.
Expenses: Tuition, state resident: full-time $3,564. Tuition, nonresident: full-time $10,230.
Financial aid: In 1999–00, 16 students received aid, including 10 research assistantships, 5 teaching assistantships; Federal Work-Study, institutionally sponsored loans, and tuition waivers (full and partial) also available. Financial aid application deadline: 2/1; financial aid applicants required to submit FAFSA.
Faculty research: Import, assembly, and bioenergetics of mitochondrial proteins; regulation of gene expression; DNA metabolism; structural and mechanistic characterization of metal-containing enzymes. *Total annual research expenditures:* $620,062.
Dr. Diana S. Beattie, Chair, 304-293-7758, *Fax:* 304-293-6846.

Application contact: Dr. Lisa Salati, Graduate Program Coordinator, 304-293-7759, *Fax:* 304-293-6846, *E-mail:* lsalati@hsc.wvu.edu.

■ WORCESTER POLYTECHNIC INSTITUTE

Graduate Studies, Department of Chemistry and Biochemistry, Worcester, MA 01609-2280

AWARDS Biochemistry (MS, PhD); chemistry (MS, PhD).

Faculty: 11 full-time (1 woman), 1 part-time/adjunct (0 women).
Students: 21 full-time (6 women), 1 part-time; includes 3 minority (2 Asian Americans or Pacific Islanders, 1 Hispanic American), 12 international. *64 applicants, 23% accepted.* In 1999, 8 master's, 2 doctorates awarded.
Degree requirements: For master's and doctorate, thesis/dissertation required, foreign language not required. *Average time to degree:* Master's–2 years full-time; doctorate–5 years full-time.
Entrance requirements: For master's and doctorate, GRE General Test, TOEFL. *Application deadline:* For fall admission, 2/1 (priority date); for spring admission, 10/15 (priority date). Applications are processed on a rolling basis. *Application fee:* $50. Electronic applications accepted.
Expenses: Tuition: Full-time $13,220; part-time $703 per credit. Required fees: $50.
Financial aid: In 1999–00, 17 students received aid, including 1 fellowship with full tuition reimbursement available (averaging $11,250 per year), 1 research assistantship with full tuition reimbursement available (averaging $13,200 per year), 13 teaching assistantships with full tuition reimbursements available (averaging $12,330 per year); career-related internships or fieldwork, grants, institutionally sponsored loans, and scholarships also available. Financial aid application deadline: 2/15; financial aid applicants required to submit FAFSA.
Faculty research: Biomembrane studies, computational photochemistry, laser flash, photolysis, organic photochemistry, alkaloid synthesis. *Total annual research expenditures:* $543,618.
Dr. James P. Dittami, Head, 508-831-5149, *Fax:* 508-831-5933, *E-mail:* jdittami@wpi.edu.
Application contact: Dr. W. Grant McGimpsey, Graduate Coordinator, 508-831-5486, *Fax:* 508-831-5933, *E-mail:* wgm@wpi.edu.

■ WRIGHT STATE UNIVERSITY

School of Graduate Studies, College of Science and Mathematics, Department of Biochemistry and Molecular Biology, Dayton, OH 45435

AWARDS MS.

Students: 7 full-time (1 woman), 1 (woman) part-time; includes 1 minority (African American), 1 international. *10 applicants, 40% accepted.* In 1999, 3 degrees awarded.
Degree requirements: For master's, thesis required, foreign language not required.
Entrance requirements: For master's, TOEFL. *Application deadline:* For fall admission, 8/20. *Application fee:* $25.
Expenses: Tuition, state resident: full-time $5,568; part-time $175 per quarter hour. Tuition, nonresident: full-time $9,696; part-time $302 per quarter hour. Full-time tuition and fees vary according to course load, campus/location and program.
Financial aid: Fellowships, research assistantships, teaching assistantships, unspecified assistantships available. Aid available to part-time students. Financial aid application deadline: 6/15; financial aid applicants required to submit FAFSA.
Faculty research: Regulation of gene expression, macromolecular structural function, NMR imaging, visual biochemistry.
Dr. Daniel T. Organisciak, Chair, 937-775-3041, *Fax:* 937-775-3730, *E-mail:* dto@wright.edu.
Application contact: Dr. John J. Turchi, Director, 937-775-2853, *Fax:* 937-775-3730, *E-mail:* john.turchi@wright.edu.

Find an in-depth description at www.petersons.com/graduate.

■ YALE UNIVERSITY

Graduate School of Arts and Sciences, Department of Molecular Biophysics and Biochemistry, New Haven, CT 06520

AWARDS MS, PhD.

Faculty: 95 full-time (24 women), 2 part-time/adjunct (1 woman).
Students: 71 full-time (22 women); includes 9 minority (2 African Americans, 6 Asian Americans or Pacific Islanders, 1 Hispanic American), 16 international. *130 applicants, 41% accepted.* In 1999, 14 degrees awarded.
Degree requirements: For doctorate, dissertation required. *Average time to degree:* Doctorate–6 years full-time.
Entrance requirements: For doctorate, GRE General Test, GRE Subject Test. *Application deadline:* For fall admission, 1/4. *Application fee:* $65.

Expenses: Tuition: Full-time $22,300. Full-time tuition and fees vary according to program.
Financial aid: Fellowships, research assistantships, teaching assistantships, Federal Work-Study and institutionally sponsored loans available. Aid available to part-time students.
Application contact: Admissions Information, 203-432-2770.

Find an in-depth description at www.petersons.com/graduate.

■ YALE UNIVERSITY

School of Medicine and Graduate School of Arts and Sciences, Combined Program in Biological and Biomedical Sciences (BBS), Molecular Biophysics and Biochemistry Track, New Haven, CT 06520

AWARDS PhD, MD/PhD.

Degree requirements: For doctorate, dissertation required.
Entrance requirements: For doctorate, GRE General Test, TOEFL. *Application deadline:* For fall admission, 1/2. *Application fee:* $65. Electronic applications accepted.
Expenses: All students receive full tuition of $22,330 and an annual stipend of $17,600 .
Financial aid: Fellowships, research assistantships available.
Dr. Nigel Grindley, Director, 203-432-8891, *Fax:* 203-432-9782, *E-mail:* mbb.grad@yale.edu.
Application contact: Nessie Stewart, Graduate Registrar, 203-432-5662, *Fax:* 203-432-6178, *E-mail:* mbb.grad@yale.edu.

Find an in-depth description at www.petersons.com/graduate.

■ YESHIVA UNIVERSITY

Albert Einstein College of Medicine, Sue Golding Graduate Division of Medical Sciences, Department of Biochemistry, Bronx, NY 10461-1602

AWARDS PhD, MD/PhD.

Faculty: 21 full-time.
Students: 27 full-time (6 women); includes 5 minority (2 African Americans, 1 Asian American or Pacific Islander, 2 Hispanic Americans), 7 international. In 1999, 3 degrees awarded.
Degree requirements: For doctorate, dissertation required, foreign language not required.
Entrance requirements: For doctorate, GRE General Test, TOEFL. *Application deadline:* For fall admission, 1/15. *Application fee:* $0.
Expenses: Tuition: Part-time $525 per credit. Tuition and fees vary according to degree level and program.

Yeshiva University (continued)
Financial aid: In 1999–00, 27 fellowships were awarded.
Faculty research: Biochemical mechanisms, enzymology, protein chemistry, bio-organic chemistry, molecular genetics.

Dr. V. Schramm, Chairperson, 718-430-2814.
Application contact: Sheila Cleeton, Assistant Director, 718-430-2128, *Fax:* 718-430-8655, *E-mail:* phd@aecom.yu.edu.

Find an in-depth description at www.petersons.com/graduate.

Biophysics

BIOPHYSICS

■ BAYLOR COLLEGE OF MEDICINE

Graduate School of Biomedical Sciences, Department of Molecular Physiology and Biophysics, Houston, TX 77030-3498

AWARDS PhD, MD/PhD.

Faculty: 21 full-time (2 women).
Students: 13 full-time (8 women); includes 1 minority (Hispanic American), 8 international. Average age 27. *26 applicants, 27% accepted.*
Degree requirements: For doctorate, dissertation, public defense, qualifying exam required, foreign language not required.
Entrance requirements: For doctorate, GRE General Test (average 80th percentile), GRE Subject Test (strongly recommended), TOEFL, minimum GPA of 3.0. *Application deadline:* For fall admission, 2/1 (priority date). *Application fee:* $30. Electronic applications accepted.
Expenses: Tuition: Full-time $8,200. Required fees: $175. Full-time tuition and fees vary according to student level.
Financial aid: In 1999–00, 13 students received aid, including 8 fellowships (averaging $16,000 per year), 5 research assistantships (averaging $16,000 per year); Federal Work-Study, institutionally sponsored loans, and tuition waivers (full) also available. Financial aid applicants required to submit FAFSA.
Faculty research: Membrane ion channels, ion transport, recombinant DNA.
Dr. Brian Knoll, Director, 713-798-5107, *Fax:* 713-798-3475.
Application contact: Becky Pullen, Graduate Program Administrator, 713-798-5107, *Fax:* 713-798-3475, *E-mail:* rpullen@bcm.tmc.edu.

Find an in-depth description at www.petersons.com/graduate.

■ BAYLOR COLLEGE OF MEDICINE

Graduate School of Biomedical Sciences, Program in Structural and Computational Biology and Molecular Biophysics, Houston, TX 77030-3498

AWARDS PhD, MD/PhD.

Faculty: 44 full-time (4 women).
Students: 15 full-time (1 woman); includes 2 minority (1 African American, 1 Asian American or Pacific Islander), 7 international. Average age 27. *55 applicants, 20% accepted.*
Degree requirements: For doctorate, dissertation, public defense, qualifying exam required, foreign language not required.
Entrance requirements: For doctorate, GRE General Test (average 80th percentile), GRE Subject Test (strongly recommended), TOEFL, minimum GPA of 3.0. *Application deadline:* For fall admission, 2/1. *Application fee:* $30. Electronic applications accepted.
Expenses: Tuition: Full-time $8,200. Required fees: $175. Full-time tuition and fees vary according to student level.
Financial aid: In 1999–00, 15 students received aid, including 13 fellowships (averaging $16,000 per year), 2 research assistantships (averaging $16,000 per year); Federal Work-Study, institutionally sponsored loans, and tuition waivers (full) also available. Financial aid applicants required to submit FAFSA.
Faculty research: X-ray and electron crystallography, light and electron microscopy, computer image reconstruction, molecular spectroscopy.
Dr. Wah Chiu, Director, 713-798-6985.
Application contact: Wanda Waguespack, Graduate Program Administrator, 713-798-5197, *Fax:* 713-798-6325, *E-mail:* wandaw@bcm.tmc.edu.

Find an in-depth description at www.petersons.com/graduate.

■ BOSTON UNIVERSITY

Graduate School of Arts and Sciences, Program in Cellular Biophysics, Boston, MA 02215

AWARDS PhD.

Students: 7 full-time (4 women), 4 international. Average age 24. *6 applicants, 0% accepted.* In 1999, 1 degree awarded.
Degree requirements: For doctorate, one foreign language, dissertation, comprehensive exam required.
Entrance requirements: For doctorate, GRE General Test, GRE Subject Test, TOEFL. *Application deadline:* For fall admission, 2/1; for spring admission, 11/1. *Application fee:* $50.
Expenses: Tuition: Full-time $23,770; part-time $743 per credit. Required fees: $220. Tuition and fees vary according to class time, course level, campus/location and program.
Financial aid: Career-related internships or fieldwork available. Aid available to part-time students. Financial aid application deadline: 1/15; financial aid applicants required to submit FAFSA.
Faculty research: Membrane biophysics, biomolecular structure, ionic transport, visual transduction, muscular contraction.
Dr. M. Carter Cornwall, Director, 617-638-4256, *Fax:* 617-638-7228.

■ BOSTON UNIVERSITY

School of Medicine, Division of Graduate Medical Sciences, Department of Physiology and Biophysics, Boston, MA 02118

AWARDS MA, PhD, MD/PhD.

Faculty: 28.
Students: 27 full-time. Average age 28. In 2000, 1 master's, 1 doctorate awarded.
Degree requirements: For master's and doctorate, thesis/dissertation, qualifying exam required, foreign language not required.
Entrance requirements: For master's and doctorate, GRE General Test, GRE Subject Test in related fields (strongly recommended), TOEFL. *Application deadline:* For fall admission, 1/15 (priority

date); for spring admission, 10/15 (priority date). *Application fee:* $50. Electronic applications accepted.

Expenses: Tuition: Full-time $24,700; part-time $772 per credit. Required fees: $220.

Financial aid: Fellowships with tuition reimbursements, research assistantships with tuition reimbursements, scholarships and traineeships available.

Faculty research: X-ray scattering, NMR spectroscopy, protein crystallography, structural electron microscopy, molecular modeling.

Dr. Donald M. Small, Chairman.

Application contact: Dr. Christopher Akey, Chair of Admissions and Student Affairs Committee, 617-638-4051, *Fax:* 617-638-4041, *E-mail:* cakey@bu.edu.

Find an in-depth description at www.petersons.com/graduate.

■ **BRANDEIS UNIVERSITY**

Graduate School of Arts and Sciences, Programs in Biological Sciences, Program in Biophysics and Structural Biology, Waltham, MA 02454-9110

AWARDS PhD.

Faculty: 16 full-time (4 women).

Students: 25 full-time (10 women); includes 8 minority (all Asian Americans or Pacific Islanders), 9 international. Average age 24. *15 applicants, 7% accepted.* In 1999, 1 degree awarded (100% entered university research/teaching).

Degree requirements: For doctorate, one foreign language (computer language can substitute), dissertation required. *Average time to degree:* Doctorate–6 years full-time.

Entrance requirements: For doctorate, GRE General Test. *Application deadline:* For fall admission, 1/15 (priority date). Applications are processed on a rolling basis. *Application fee:* $60. Electronic applications accepted.

Expenses: Tuition: Full-time $25,392; part-time $3,174 per course. Required fees: $509. Tuition and fees vary according to class time, degree level, program and student level.

Financial aid: In 1999–00, 3 fellowships with tuition reimbursements (averaging $17,000 per year), 20 research assistantships with tuition reimbursements (averaging $17,000 per year), 1 teaching assistantship with tuition reimbursement (averaging $2,000 per year) were awarded; scholarships and tuition waivers (full and partial) also available. Aid available to part-time students. Financial aid application deadline: 4/15; financial aid applicants required to submit CSS PROFILE or FAFSA.

Faculty research: Biophysical chemistry, structural biology, protein crystallography, neuroscience, photobiology.

Dr. Jeff Gelles, Chair, 781-736-2377, *Fax:* 781-736-3107.

Application contact: Marcia Cabral, Information Officer, 781-736-3100, *Fax:* 781-736-3107, *E-mail:* cabral@brandeis.edu.

■ **CALIFORNIA INSTITUTE OF TECHNOLOGY**

Division of Biology, Program in Cell Biology and Biophysics, Pasadena, CA 91125-0001

AWARDS PhD.

Degree requirements: For doctorate, dissertation, qualifying exam required, foreign language not required.

Entrance requirements: For doctorate, GRE General Test. *Application deadline:* For fall admission, 1/1. *Application fee:* $0.

Expenses: Tuition: Full-time $19,260. Required fees: $24. One-time fee: $100 full-time.

Financial aid: Application deadline: 1/1.

Application contact: Elizabeth Ayala, Graduate Option Coordinator, 626-395-4497, *Fax:* 626-449-0756, *E-mail:* biograd@cco.caltech.edu.

■ **CARNEGIE MELLON UNIVERSITY**

Mellon College of Science, Department of Biological Sciences, Pittsburgh, PA 15213-3891

AWARDS Biochemistry (PhD); biophysics (PhD); cell biology (PhD); developmental biology (PhD); genetics (PhD); molecular biology (PhD).

Faculty: 38 full-time (17 women), 1 (woman) part-time/adjunct.

Students: 35 full-time (24 women); includes 2 minority (1 African American, 1 Asian American or Pacific Islander), 18 international. Average age 27. In 1999, 4 degrees awarded.

Degree requirements: For doctorate, dissertation required, foreign language not required.

Entrance requirements: For doctorate, GRE General Test, GRE Subject Test, interview. *Application deadline:* For fall admission, 2/1 (priority date). Applications are processed on a rolling basis. *Application fee:* $0.

Expenses: Tuition: Full-time $22,100; part-time $307 per unit. Required fees: $200. Tuition and fees vary according to program.

Financial aid: Fellowships, research assistantships, teaching assistantships, traineeships available.

Faculty research: Genetic structure, function, and regulation; protein structure and function; biological membranes; biological spectroscopy. *Total annual research expenditures:* $4.4 million.

Dr. William E. Brown, Head, 412-268-3416, *Fax:* 412-268-7129, *E-mail:* wb02@andrew.cmu.edu.

Application contact: Stacey L. Young, Assistant Head, 412-268-7372, *Fax:* 412-268-7129, *E-mail:* sf38+@andrew.cmu.edu.

Find an in-depth description at www.petersons.com/graduate.

■ **CASE WESTERN RESERVE UNIVERSITY**

School of Medicine and School of Graduate Studies, Graduate Programs in Medicine, Department of Physiology and Biophysics, Cleveland, OH 44106

AWARDS Biophysics and bioengineering (PhD); cell physiology (PhD); physiology and biophysics (PhD); systems physiology (PhD).

Degree requirements: For doctorate, dissertation required, foreign language not required.

Entrance requirements: For doctorate, GRE General Test, TOEFL. Electronic applications accepted.

Faculty research: Cardiovascular physiology, calcium metabolism, epithelial cell biology.

Find an in-depth description at www.petersons.com/graduate.

■ **CLEMSON UNIVERSITY**

Graduate School, College of Engineering and Science, Department of Physics and Astronomy, Clemson, SC 29634

AWARDS Physics (MS, PhD), including astronomy and astrophysics, atmospheric physics, biophysics. Part-time programs available.

Students: 36 full-time (8 women), 6 part-time; includes 1 minority (African American), 14 international. Average age 25. *49 applicants, 76% accepted.* In 1999, 6 master's, 1 doctorate awarded. Terminal master's awarded for partial completion of doctoral program.

Degree requirements: For master's, thesis or alternative required, foreign language not required; for doctorate, dissertation required, foreign language not required.

Entrance requirements: For master's and doctorate, GRE General Test, TOEFL. *Application deadline:* For fall admission, 2/15 (priority date). Applications are processed on a rolling basis. *Application fee:* $40.

Expenses: Tuition, state resident: full-time $3,480; part-time $174 per credit hour.

Clemson University (continued)
Tuition, nonresident: full-time $9,256; part-time $388 per credit hour. Required fees: $5 per term. Full-time tuition and fees vary according to course level, course load and campus/location.
Financial aid: Fellowships, research assistantships, teaching assistantships available. Financial aid application deadline: 6/1; financial aid applicants required to submit FAFSA.
Faculty research: Radiation physics, solid-state physics, nuclear physics, radar and lidar studies of atmosphere.
Dr. Peter J. McNulty, Chair, 864-656-3419, *Fax:* 864-656-0805, *E-mail:* mpeter@clemson.edu.
Application contact: Dr. Miguel F. Larsen, Graduate Coordinator, 864-656-5309, *Fax:* 864-656-0805, *E-mail:* mlarson@clemson.edu.
Find an in-depth description at www.petersons.com/graduate.

■ COLUMBIA UNIVERSITY

College of Physicians and Surgeons and Graduate School of Arts and Sciences, Graduate School of Arts and Sciences at the College of Physicians and Surgeons, Department of Biochemistry and Molecular Biophysics, New York, NY 10032

AWARDS Biochemistry and molecular biophysics (M Phil, PhD); biophysics (PhD). Only candidates for the PhD are admitted.

Degree requirements: For master's, foreign language and thesis not required; for doctorate, one foreign language, dissertation required.
Entrance requirements: For master's and doctorate, GRE General Test, TOEFL.
Expenses: Tuition: Full-time $25,072.

■ COLUMBIA UNIVERSITY

College of Physicians and Surgeons and Graduate School of Arts and Sciences, Graduate School of Arts and Sciences at the College of Physicians and Surgeons, Department of Physiology and Cellular Biophysics, New York, NY 10032

AWARDS M Phil, MA, PhD, MD/PhD. Only candidates for the PhD are admitted. Terminal master's awarded for partial completion of doctoral program.

Degree requirements: For master's, foreign language and thesis not required; for doctorate, dissertation required, foreign language not required.
Entrance requirements: For master's and doctorate, GRE General Test, TOEFL.
Expenses: Tuition: Full-time $25,072.

Faculty research: Membrane physiology, cellular biology, cardiovascular physiology, neurophysiology.

■ COLUMBIA UNIVERSITY

College of Physicians and Surgeons and Graduate School of Arts and Sciences, Graduate School of Arts and Sciences at the College of Physicians and Surgeons, Integrated Program in Cellular, Molecular and Biophysical Studies, New York, NY 10032

AWARDS M Phil, MA, PhD, MD/PhD. Only candidates for the PhD are admitted. Terminal master's awarded for partial completion of doctoral program.

Degree requirements: For master's, foreign language and thesis not required; for doctorate, dissertation required, foreign language not required.
Entrance requirements: For master's, GRE General Test, TOEFL; for doctorate, GRE General Test, GRE Subject Test, TOEFL.
Expenses: Tuition: Full-time $25,072.
Faculty research: Transcription, macromolecular sorting, gene expression during development, cellular interaction.

■ CORNELL UNIVERSITY

Graduate School, Graduate Field of Biophysics, Ithaca, NY 14853-0001
AWARDS PhD.

Faculty: 18 full-time.
Students: 8 full-time (3 women); includes 1 minority (African American), 4 international. *41 applicants, 10% accepted.* In 1999, 1 doctorate awarded.
Degree requirements: For doctorate, dissertation required, foreign language not required.
Entrance requirements: For doctorate, GRE General Test, GRE Subject Test (physics or chemistry), TOEFL. *Application deadline:* For fall admission, 1/15. *Application fee:* $65. Electronic applications accepted.
Expenses: Tuition: Full-time $23,760. Required fees: $48. Full-time tuition and fees vary according to program.
Financial aid: In 1999–00, 7 students received aid, including 4 fellowships with full tuition reimbursements available, 2 research assistantships with full tuition reimbursements available, 1 teaching assistantship with full tuition reimbursement available; institutionally sponsored loans, scholarships, tuition waivers (full and partial), and unspecified assistantships also available. Financial aid applicants required to submit FAFSA.

Faculty research: Protein structure and function, biomolecular and cellular function, membrane biophysics, signal transduction, computational biology.
Application contact: Graduate Field Assistant, 610-255-0962, *E-mail:* biophysics@cornell.edu.

■ CORNELL UNIVERSITY

Graduate School, Graduate Fields of Agriculture and Life Sciences, Field of Biochemistry, Molecular and Cell Biology, Ithaca, NY 14853-0001

AWARDS Biochemistry (PhD); biophysics (PhD); cell biology (PhD); molecular and cell biology (PhD); molecular biology (PhD).

Faculty: 43 full-time.
Students: 75 full-time (37 women); includes 15 minority (1 African American, 6 Asian Americans or Pacific Islanders, 8 Hispanic Americans), 19 international. *288 applicants, 21% accepted.* In 1999, 9 doctorates awarded.
Degree requirements: For doctorate, dissertation, 2 semesters of teaching experience required.
Entrance requirements: For doctorate, GRE General Test, GRE Subject Test (biology, chemistry, physics, biochemistry or, cell and molecular biology), TOEFL. *Application deadline:* For fall admission, 1/5. *Application fee:* $65. Electronic applications accepted.
Expenses: Tuition: Full-time $12,100.
Financial aid: In 1999–00, 74 students received aid, including 35 fellowships with full tuition reimbursements available, 32 research assistantships with full tuition reimbursements available, 7 teaching assistantships with full tuition reimbursements available; institutionally sponsored loans, scholarships, tuition waivers (full and partial), and unspecified assistantships also available. Financial aid applicants required to submit FAFSA.
Faculty research: Biophysics, structural biology.
Application contact: Graduate Field Assistant, 607-255-2317, *E-mail:* bmcb@cornell.edu.

Find an in-depth description at www.petersons.com/graduate.

■ DUKE UNIVERSITY

Graduate School, Program in Molecular Biophysics, Durham, NC 27708-0586

AWARDS Certificate. Students must be enrolled in a participating PhD program.

Faculty: 22 full-time.
Students: 4 full-time (0 women). *17 applicants, 47% accepted.*
Entrance requirements: For degree, GRE General Test, GRE Subject Test.

Application deadline: For fall admission, 12/31. *Application fee:* $75.
Expenses: Tuition: Full-time $21,406; part-time $760 per unit. Required fees: $3,136; $3,136 per year. One-time fee: $30. Tuition and fees vary according to program.
Financial aid: Application deadline: 12/31. Harold Erickson, Director of Graduate Studies, 919-681-8825, *Fax:* 919-684-8346, *E-mail:* agw@biochem.duke.edu.

■ EAST CAROLINA UNIVERSITY

Graduate School, College of Arts and Sciences, Department of Physics, Greenville, NC 27858-4353

AWARDS Applied and biomedical physics (MS), including applied and biomedical physics; medical physics (MS). Part-time programs available.

Faculty: 12 full-time (0 women).
Students: 15 full-time (4 women); includes 2 minority (both African Americans), 7 international. Average age 28. *18 applicants, 50% accepted.* In 1999, 3 master's awarded.
Degree requirements: For master's, one foreign language (computer language can substitute), comprehensive exam required.
Entrance requirements: For master's, GRE General Test, TOEFL. *Application deadline:* Applications are processed on a rolling basis. *Application fee:* $40.
Expenses: Tuition, state resident: full-time $1,012. Tuition, nonresident: full-time $8,578. Required fees: $1,006. Full-time tuition and fees vary according to degree level. Part-time tuition and fees vary according to course load.
Financial aid: Research assistantships with partial tuition reimbursements, teaching assistantships with partial tuition reimbursements, Federal Work-Study available. Aid available to part-time students. Financial aid application deadline: 6/1.
Dr. Larry Toburen, Director of Graduate Studies, 252-328-6739, *Fax:* 252-328-6314, *E-mail:* toburenl@mail.ecu.edu.
Application contact: Dr. Paul D. Tschetter, Senior Associate Dean, 252-328-6012, *Fax:* 252-328-6071, *E-mail:* grad@mail.ecu.edu.

■ EAST TENNESSEE STATE UNIVERSITY

James H. Quillen College of Medicine and School of Graduate Studies, Biomedical Science Graduate Program, Johnson City, TN 37614

AWARDS Anatomy and cell biology (MS, PhD); biochemistry and molecular biology (MS, PhD); biophysics (MS, PhD); microbiology (MS, PhD); pharmacology (MS, PhD); physiology (MS, PhD). Part-time programs available.

Faculty: 40 full-time (9 women).
Students: 23 full-time (10 women), 2 part-time (1 woman); includes 1 minority (Asian American or Pacific Islander), 4 international. Average age 30. *83 applicants, 20% accepted.* In 1999, 5 master's, 4 doctorates awarded. Terminal master's awarded for partial completion of doctoral program.
Degree requirements: For master's, one foreign language (computer language can substitute), thesis, comprehensive qualifying exam required; for doctorate, 2 foreign languages (computer language can substitute for one), dissertation required.
Entrance requirements: For master's, GRE General Test, minimum GPA of 3.0, bachelor's degree in biological or related science; for doctorate, GRE General Test, GRE Subject Test. *Application deadline:* For fall admission, 3/15 (priority date); for spring admission, 3/1. *Application fee:* $25 ($35 for international students).
Expenses: Tuition, state resident: full-time $10,342. Tuition, nonresident: full-time $21,080. Required fees: $532.
Financial aid: In 1999–00, 16 research assistantships, 5 teaching assistantships were awarded; fellowships, career-related internships or fieldwork, Federal Work-Study, grants, institutionally sponsored loans, and tuition waivers (full) also available.
Dr. Mitchell Robinson, Assistant Dean for Graduate Studies, 423-439-4658, *E-mail:* robinson@etsu.edu.

■ EMORY UNIVERSITY

Graduate School of Arts and Sciences, Department of Physics, Atlanta, GA 30322-1100

AWARDS Physics (PhD), including biophysics, radiological physics, solid-state physics.

Faculty: 17 full-time (1 woman), 3 part-time/adjunct (0 women).
Students: 10 full-time (3 women), 5 international. Average age 24. *24 applicants, 33% accepted.* In 1999, 2 doctorates awarded.
Degree requirements: For doctorate, dissertation, comprehensive exams required, foreign language not required.
Entrance requirements: For doctorate, GRE General Test, TOEFL, minimum GPA of 3.0. *Application deadline:* For fall admission, 1/20 (priority date). *Application fee:* $45.
Expenses: Tuition: Full-time $22,770. Tuition and fees vary according to program.
Financial aid: Fellowships, teaching assistantships, institutionally sponsored loans, scholarships, and tuition waivers (partial) available. Financial aid application deadline: 1/20; financial aid applicants required to submit FAFSA.

Faculty research: Theory of semiconductors and superlattices, experimental laser optics and submillimeter spectroscopy theory, neural networks and stereoscopic vision, experimental studies of the structure and function of metalloproteins.
Dr. Vincent Huynh, Director of Graduate Studies, 404-727-4295, *Fax:* 404-727-8073, *E-mail:* phsbhh@physics.emory.edu.
Application contact: Brenda J. Wingo, Coordinator of Academic Services, 404-727-8037, *Fax:* 404-727-8073, *E-mail:* phsbw@physics.emory.edu.

Find an in-depth description at www.petersons.com/graduate.

■ FLORIDA STATE UNIVERSITY

Graduate Studies, College of Arts and Sciences, Program in Molecular Biophysics, Tallahassee, FL 32306

AWARDS PhD.

Faculty: 37 full-time (6 women).
Students: 23 full-time (10 women); includes 1 minority (Hispanic American), 12 international. Average age 30. *120 applicants, 3% accepted.* In 1999, 6 degrees awarded (67% entered university research/teaching, 33% found other work related to degree).
Degree requirements: For doctorate, dissertation, comprehensive exam required, foreign language not required. *Average time to degree:* Doctorate–5.2 years full-time.
Entrance requirements: For doctorate, GRE General Test, minimum GPA of 3.0. *Application deadline:* For fall admission, 2/1. Applications are processed on a rolling basis. *Application fee:* $20. Electronic applications accepted.
Expenses: Tuition, state resident: full-time $3,504; part-time $146 per credit hour. Tuition, nonresident: full-time $12,162; part-time $507 per credit hour. Tuition and fees vary according to program.
Financial aid: In 1999–00, 20 students received aid, including 20 research assistantships with full tuition reimbursements available (averaging $14,320 per year). Financial aid applicants required to submit FAFSA.
Faculty research: Protein and nucleic acid structure and function, membrane protein structure, computational biophysics, 3-D image reconstruction. *Total annual research expenditures:* $3.4 million.
Dr. Laura R. Keller, Director, MOB Graduate Program, 850-644-5780, *Fax:* 850-561-1406, *E-mail:* lkeller@bio.fsu.edu.
Application contact: Dale E. Leonard, Academic Coordinator, 850-644-1012, *Fax:* 850-561-1406, *E-mail:* mob@sb.fsu.edu.

Find an in-depth description at www.petersons.com/graduate.

■ GEORGETOWN UNIVERSITY

Graduate School of Arts and Sciences, Programs in Biomedical Sciences, Department of Physiology and Biophysics, Washington, DC 20057

AWARDS MS, PhD, MD/PhD.

Degree requirements: For master's, foreign language not required; for doctorate, dissertation required, foreign language not required.

Entrance requirements: For master's, GRE General Test, MCAT, TOEFL; for doctorate, GRE General Test, TOEFL.

■ HARVARD UNIVERSITY

Graduate School of Arts and Sciences, Committee on Biophysics, Cambridge, MA 02138

AWARDS PhD.

Students: 9 full-time (0 women). *60 applicants, 25% accepted.* In 1999, 8 degrees awarded.

Degree requirements: For doctorate, dissertation, exam, qualifying paper required, foreign language not required.

Entrance requirements: For doctorate, GRE General Test, GRE Subject Test (recommended), TOEFL. *Application deadline:* For fall admission, 12/15. *Application fee:* $60.

Expenses: Tuition: Full-time $22,054. Required fees: $711. Tuition and fees vary according to program.

Financial aid: Fellowships, research assistantships, teaching assistantships, career-related internships or fieldwork, Federal Work-Study, and institutionally sponsored loans available. Financial aid application deadline: 12/30.

Faculty research: Structural molecular biology, cell and membrane biophysics, molecular genetics, physical biochemistry, mathematical biophysics.
Josephine Ferraro, Officer, 617-495-5396.
Application contact: Office of Admissions and Financial Aid, 617-495-5315.

Find an in-depth description at www.petersons.com/graduate.

■ HOWARD UNIVERSITY

Graduate School of Arts and Sciences, Department of Physiology and Biophysics, Program in Biophysics, Washington, DC 20059-0002

AWARDS PhD. Part-time programs available.

Degree requirements: For doctorate, dissertation, comprehensive exam required, foreign language not required.

Entrance requirements: For doctorate, GRE General Test, minimum B average in field.

Expenses: Tuition: Full-time $10,500; part-time $583 per credit hour. Required fees: $405; $203 per semester.

■ INDIANA UNIVERSITY–PURDUE UNIVERSITY INDIANAPOLIS

School of Medicine, Graduate Programs in Medicine, Department of Physiology and Biophysics, Indianapolis, IN 46202-2896

AWARDS MS, PhD, MD/MS, MD/PhD.

Students: 39 full-time (17 women), 5 part-time (3 women). Average age 24. *22 applicants, 18% accepted.* In 1999, 2 doctorates awarded (100% entered university research/teaching). Terminal master's awarded for partial completion of doctoral program.

Degree requirements: For master's, foreign language and thesis not required; for doctorate, dissertation required, foreign language not required. *Average time to degree:* Master's–2 years full-time; doctorate–5.5 years full-time.

Entrance requirements: For master's, GRE General Test, GRE Subject Test, previous course work in biology, calculus, physical chemistry, and physics; for doctorate, GRE General Test, GRE Subject Test. *Application deadline:* For fall admission, 4/15 (priority date). Applications are processed on a rolling basis. *Application fee:* $35 ($55 for international students).

Expenses: Tuition, state resident: full-time $13,245; part-time $158 per credit hour. Tuition, nonresident: full-time $30,330; part-time $455 per credit hour. Required fees: $121 per year. Tuition and fees vary according to course load and degree level.

Financial aid: In 1999–00, 14 students received aid, including 4 fellowships with partial tuition reimbursements available (averaging $16,000 per year), 7 research assistantships with partial tuition reimbursements available (averaging $16,000 per year); Federal Work-Study and institutionally sponsored loans also available.

Faculty research: Cardiovascular physiology, cell growth and development, respiratory biology, cell signaling mechanisms, cytoskeleton function.
Dr. Rodney A. Rhoades, Chair, 317-274-7772, *Fax:* 317-274-3318.
Application contact: Dr. Fred M. Pavalko, Graduate Director, 317-274-3140, *Fax:* 317-274-3318, *E-mail:* fpavalko@iupui.edu.

Find an in-depth description at www.petersons.com/graduate.

■ INDIANA UNIVERSITY–PURDUE UNIVERSITY INDIANAPOLIS

School of Medicine, Graduate Programs in Medicine, Program in Medical Biophysics, Indianapolis, IN 46202-2896

AWARDS MS, PhD, MD/MS, MD/PhD.

Students: 3 full-time (1 woman), 1 part-time; includes 1 minority (Asian American or Pacific Islander), 1 international. Average age 24. *5 applicants, 40% accepted.* Terminal master's awarded for partial completion of doctoral program.

Degree requirements: For master's and doctorate, thesis/dissertation required, foreign language not required. *Average time to degree:* Master's–2 years full-time; doctorate–5.5 years full-time.

Entrance requirements: For master's and doctorate, GRE General Test, GRE Subject Test. *Application deadline:* For fall admission, 1/15 (priority date). Applications are processed on a rolling basis. *Application fee:* $35 ($55 for international students).

Expenses: Tuition, state resident: full-time $13,245; part-time $158 per credit hour. Tuition, nonresident: full-time $30,330; part-time $455 per credit hour. Required fees: $121 per year. Tuition and fees vary according to course load and degree level.

Financial aid: In 1999–00, 3 students received aid, including 1 fellowship with full tuition reimbursement available (averaging $16,000 per year), 1 research assistantship with full tuition reimbursement available (averaging $16,000 per year), 1 teaching assistantship with full tuition reimbursement available (averaging $12,000 per year); Federal Work-Study, institutionally sponsored loans, and tuition waivers (partial) also available. Financial aid application deadline: 4/15.

Faculty research: Membrane biophysics, protein structure and function, biological magnetic resonance, smooth muscle biophysics, photobiology. *Total annual research expenditures:* $2 million.
Chair, 317-274-3772.
Application contact: Dr. Richard Haak, Graduate Adviser, 317-274-7626, *Fax:* 317-274-4090, *E-mail:* rhaak@iupui.edu.

■ IOWA STATE UNIVERSITY OF SCIENCE AND TECHNOLOGY

Graduate College, College of Agriculture and College of Liberal Arts and Sciences, Department of Biochemistry, Biophysics, and Molecular Biology, Ames, IA 50011

AWARDS Biochemistry (MS, PhD); biophysics (MS, PhD); genetics (MS, PhD); molecular, cellular, and developmental biology (MS, PhD); toxicology (MS, PhD).

Faculty: 19 full-time, 1 part-time/adjunct. **Students:** 57 full-time (24 women), 6 part-time (1 woman); includes 1 minority (African American), 20 international. *22 applicants, 59% accepted.* In 2000, 4 master's, 4 doctorates awarded. **Degree requirements:** For master's and doctorate, thesis/dissertation required. **Entrance requirements:** For master's and doctorate, GRE General Test, TOEFL. *Application deadline:* For fall admission, 6/15 (priority date); for spring admission, 11/15 (priority date). *Application fee:* $20 ($50 for international students). Electronic applications accepted. **Expenses:** Tuition, state resident: full-time $3,308. Tuition, nonresident: full-time $9,744. Part-time tuition and fees vary according to course load, campus/location and program. **Financial aid:** In 2000–01, 44 research assistantships with partial tuition reimbursements (averaging $12,314 per year), 2 teaching assistantships with partial tuition reimbursements (averaging $12,375 per year) were awarded; scholarships also available. Dr. Marit Nilsen-Hamilton, Chair, 515-294-2231, *E-mail:* biochem@iastate.edu.

Find an in-depth description at www.petersons.com/graduate.

■ **JOAN AND SANFORD I. WEILL MEDICAL COLLEGE AND GRADUATE SCHOOL OF MEDICAL SCIENCES OF CORNELL UNIVERSITY**

Graduate School of Medical Sciences, Program in Physiology, Biophysics, and Molecular Medicine, New York, NY 10021

AWARDS PhD, MD/PhD.

Faculty: 21 full-time (4 women). **Students:** 24 full-time (9 women); includes 3 minority (2 Asian Americans or Pacific Islanders, 1 Hispanic American), 10 international. *63 applicants, 10% accepted.* In 2000, 2 degrees awarded. **Degree requirements:** For doctorate, dissertation, final exam required. **Entrance requirements:** For doctorate, GRE General Test, GRE Subject Test, MCAT (MD/PhD), introductory courses in biology, inorganic and organic chemistry, physics, and mathematics. *Application deadline:* For fall admission, 1/15. *Application fee:* $50. **Expenses:** All students in good standing receive an annual stipend of $22,880. **Financial aid:** Fellowships, stipends available. Lawrence Palmer, Director, 212-746-6350.

■ **JOHNS HOPKINS UNIVERSITY**

Program in Molecular Biophysics, Baltimore, MD 21218-2699

AWARDS PhD.

Faculty: 36 full-time (4 women). **Students:** 30 full-time (12 women); includes 11 minority (2 African Americans, 6 Asian Americans or Pacific Islanders, 3 Hispanic Americans), 4 international. Average age 26. *25 applicants, 68% accepted.* In 1999, 2 degrees awarded (100% continued full-time study). **Degree requirements:** For doctorate, computer language, dissertation required, foreign language not required. *Average time to degree:* Doctorate–5.5 years full-time. **Entrance requirements:** For doctorate, GRE General Test, GRE Subject Test. *Application deadline:* For fall admission, 1/15 (priority date). *Application fee:* $55. **Expenses:** Tuition: Full-time $24,930. Tuition and fees vary according to program. **Financial aid:** In 1999–00, 30 students received aid, including 16 fellowships with full tuition reimbursements available (averaging $17,150 per year), 14 research assistantships with full tuition reimbursements available (averaging $17,150 per year); institutionally sponsored loans and tuition waivers (full) also available. Financial aid application deadline: 1/15. **Faculty research:** NMR and EPR spectroscopy, computer modeling, x-ray crystallography, calorimetry. Dr. David E. Draper, Director, 410-516-7448, *Fax:* 410-516-0199, *E-mail:* ipmb@jhu.edu. **Application contact:** Ranice Crosby, Coordinator, Graduate Admissions, 410-516-5197, *Fax:* 410-516-5199, *E-mail:* tcjenkin@jhu.edu.

Find an in-depth description at www.petersons.com/graduate.

■ **JOHNS HOPKINS UNIVERSITY**

School of Hygiene and Public Health, Department of Biochemistry and Molecular Biology, Baltimore, MD 21205

AWARDS Biochemistry (PhD); biophysics (PhD), including basic mechanisms of carcinogenesis, interferon induction and herpes simplex viral transformation of mammalian cells, nucleic acid structure, function, and interaction, structure and function of mammalian genetic apparatus, structure and interaction of biopolymers and cells; reproductive biology (MHS, Sc M, PhD).

Faculty: 25 full-time, 3 part-time/adjunct. **Students:** 6 full-time (3 women), 42 part-time (20 women); includes 7 minority (1 African American, 6 Asian Americans or

Pacific Islanders), 15 international. *24 applicants, 50% accepted.* In 1999, 7 master's, 2 doctorates awarded. **Degree requirements:** For master's, thesis required, foreign language not required; for doctorate, dissertation, 1 year full-time residency, oral and written exams required, foreign language not required. **Entrance requirements:** For master's and doctorate, GRE General Test, TOEFL. *Application deadline:* For fall admission, 2/1 (priority date). Applications are processed on a rolling basis. *Application fee:* $60. Electronic applications accepted. **Expenses:** Tuition: Full-time $23,660; part-time $493 per unit. Full-time tuition and fees vary according to degree level, campus/location and program. **Financial aid:** Federal Work-Study, institutionally sponsored loans, scholarships, and stipends available. Aid available to part-time students. Financial aid application deadline: 4/15. **Faculty research:** DNA replication, recombination, repair, mutation, and structure; glycoprotein synthesis, enzyme catalysis, carcinogenesis, and protein structure. *Total annual research expenditures:* $2.9 million. Dr. Roger McMacken, Chairman, 410-955-3671, *E-mail:* rmcmacke@jhsph.edu. **Application contact:** Gerry Graziano, Admissions Coordinator, 410-955-3671, *Fax:* 410-955-2926, *E-mail:* ggrazian@jhsph.edu.

Find an in-depth description at www.petersons.com/graduate.

■ **JOHNS HOPKINS UNIVERSITY**

School of Medicine, Graduate Programs in Medicine, Department of Biophysics, Baltimore, MD 21218-2699

AWARDS MS, PhD.

Faculty: 16. **Students:** 10. **Entrance requirements:** For master's, GRE. *Application fee:* $55. **Expenses:** Tuition: Full-time $23,660. **Financial aid:** Grants available. Dr. Jeremy Borg, Head.

■ **JOHNS HOPKINS UNIVERSITY**

Zanvyl Krieger School of Arts and Sciences, Thomas C. Jenkins Department of Biophysics, Baltimore, MD 21218-2699

AWARDS MA, PhD.

Faculty: 6 full-time (1 woman). **Students:** 25 full-time (6 women); includes 6 minority (2 African Americans, 3 Asian Americans or Pacific Islanders, 1 Hispanic American), 3 international. Average age 24. *43 applicants, 53% accepted.* In 1999, 1 master's, 4 doctorates awarded. Terminal

Johns Hopkins University (continued)
master's awarded for partial completion of doctoral program.

Degree requirements: For doctorate, dissertation required, foreign language not required. *Average time to degree:* Master's–4.2 years full-time; doctorate–5.5 years full-time.

Entrance requirements: For master's and doctorate, GRE General Test. *Application deadline:* For fall admission, 1/15 (priority date). Applications are processed on a rolling basis. *Application fee:* $55.

Expenses: Tuition: Full-time $24,930. Tuition and fees vary according to program.

Financial aid: In 1999–00, 6 fellowships with full tuition reimbursements (averaging $17,500 per year), 16 research assistantships with full tuition reimbursements (averaging $16,000 per year), 1 teaching assistantship were awarded; Federal Work-Study and institutionally sponsored loans also available. Financial aid application deadline: 3/14; financial aid applicants required to submit FAFSA.

Faculty research: Macromolecular interactions, x-ray crystallography, cell contractility and motility, cellular neurophysiology, mucosal protectionary reproductive health. *Total annual research expenditures:* $1.6 million.

Dr. Eaton E. Lattman, Chair, 410-516-0151, *Fax:* 410-516-4118, *E-mail:* lattman@jhu.edu.

Application contact: Ranice Crosby, Graduate Admissions Coordinator, 410-516-5197, *Fax:* 410-516-4118, *E-mail:* tcjenkin@jhu.edu.

Find an in-depth description at www.petersons.com/graduate.

■ MASSACHUSETTS INSTITUTE OF TECHNOLOGY

School of Science, Department of Biology, Program in Biophysics, Cambridge, MA 02139-4307

AWARDS PhD.

Degree requirements: For doctorate, dissertation, general exam required, foreign language not required.

Entrance requirements: For doctorate, GRE General Test. *Application deadline:* For fall admission, 1/1. *Application fee:* $55.

Expenses: Tuition: Full-time $25,000. Full-time tuition and fees vary according to degree level, program and student level.

Financial aid: Application deadline: 1/1.

Application contact: Dr. Janice D. Chang, Educational Administrator, 617-253-3717, *Fax:* 617-258-9329, *E-mail:* gradbio@mit.edu.

■ MEDICAL COLLEGE OF WISCONSIN

Graduate School of Biomedical Sciences, Program in Biophysics, Milwaukee, WI 53226-0509

AWARDS PhD, MD/PhD.

Faculty: 34 full-time (2 women), 1 part-time/adjunct (0 women).

Students: 17 full-time (4 women), 2 part-time; includes 8 minority (all Asian Americans or Pacific Islanders). Average age 28. *19 applicants, 26% accepted.* In 1999, 3 doctorates awarded (100% found work related to degree).

Degree requirements: For doctorate, dissertation, oral exam required, foreign language not required. *Average time to degree:* Doctorate–6 years full-time.

Entrance requirements: For doctorate, GRE General Test, TOEFL. *Application deadline:* For fall admission, 2/15 (priority date). Applications are processed on a rolling basis. *Application fee:* $40.

Expenses: Tuition, state resident: full-time $9,318. Tuition, nonresident: full-time $9,318. Required fees: $115.

Financial aid: In 1999–00, 19 students received aid, including 5 fellowships (averaging $15,000 per year), 14 research assistantships (averaging $15,000 per year).

Faculty research: X-ray crystallography, electron spin resonance and membrane structure, protein and membrane dynamics, magnetic resonance imaging.

Dr. B. Kalyanaraman, Interim Chairman, 414-456-4000, *Fax:* 414-456-6512, *E-mail:* balarama@mcw.edu.

Application contact: Dr. Robert W. Cox, Director of Recruitment, 414-456-4038, *Fax:* 414-456-6512, *E-mail:* rwcox@mcw.edu.

Find an in-depth description at www.petersons.com/graduate.

■ MOUNT SINAI SCHOOL OF MEDICINE OF NEW YORK UNIVERSITY

Graduate School of Biological Sciences, Biophysics, Structural Biology and Biomathematics (BSBB) Training Area, New York, NY 10029-6504

AWARDS PhD, MD/PhD.

Students: 12 full-time (3 women).

Degree requirements: For doctorate, dissertation required, foreign language not required.

Entrance requirements: For doctorate, GRE General Test, GRE Subject Test, MCAT, TOEFL. *Application deadline:* For fall admission, 4/15. *Application fee:* $60.

Expenses: Tuition: Full-time $21,750; part-time $725 per credit. Required fees: $750; $25 per credit. Full-time tuition and fees vary according to student level.

Financial aid: Fellowships with full tuition reimbursements, grants available.

Dr. Rami Osman, Program Director, *E-mail:* osman@msvax.mssm.edu.

Application contact: C. Gita Bosch, Administrative Manager and Assistant Dean, 212-241-6546, *Fax:* 212-241-0651, *E-mail:* grads@mssm.edu.

■ MOUNT SINAI SCHOOL OF MEDICINE OF NEW YORK UNIVERSITY

Graduate School of Biological Sciences, Department of Physiology and Biophysics, New York, NY 10029-6504

AWARDS PhD, MD/PhD.

Faculty: 19 full-time.

Students: 20 full-time (9 women); includes 1 minority (Hispanic American), 9 international. *25 applicants, 12% accepted.* In 1999, 1 degree awarded.

Degree requirements: For doctorate, dissertation required, foreign language not required.

Entrance requirements: For doctorate, GRE General Test, GRE Subject Test, TOEFL. *Application deadline:* For fall admission, 4/15. *Application fee:* $35.

Expenses: Tuition: Full-time $21,750; part-time $725 per credit. Required fees: $750; $25 per credit. Full-time tuition and fees vary according to student level.

Financial aid: Fellowships, grants available.

Dr. Harel Weinstein, Chairman, 212-241-7018.

Application contact: C. Gita Bosch, Administrative Manager and Assistant Dean, 212-241-6546, *Fax:* 212-241-0651, *E-mail:* grads@mssm.edu.

Find an in-depth description at www.petersons.com/graduate.

■ NORTHWESTERN UNIVERSITY

The Graduate School, Judd A. and Marjorie Weinberg College of Arts and Sciences, Interdepartmental Biological Sciences Program (IBiS), Department of Biochemistry, Molecular Biology, and Cell Biology, Evanston, IL 60208

AWARDS Cell and molecular biology (PhD); molecular biophysics (PhD). Department participates in the Interdepartmental Biological Sciences Program (IBiS).

Faculty: 59 full-time (11 women).

Students: 79 full-time (46 women); includes 5 minority (2 African Americans, 3 Hispanic Americans), 14 international. *Application fee:* $50 ($55 for international students).

Expenses: Tuition: Full-time $23,301. Full-time tuition and fees vary according to program.

Financial aid: In 1999–00, 15 fellowships (averaging $12,078 per year), 64 research assistantships (averaging $17,000 per year), 16 teaching assistantships (averaging $16,620 per year) were awarded.

Application contact: Latonia Trimuel, Program Assistant, 800-546-1761, *E-mail:* ibis@northwestern.edu.

Find an in-depth description at www.petersons.com/graduate.

■ THE OHIO STATE UNIVERSITY

Graduate School, College of Biological Sciences, Program in Biophysics, Columbus, OH 43210

AWARDS MS, PhD.

Faculty: 39 full-time.

Students: 36 full-time (12 women), 1 part-time; includes 1 minority (Asian American or Pacific Islander), 25 international. *81 applicants, 33% accepted.* In 1999, 4 master's, 4 doctorates awarded.

Degree requirements: For master's, thesis optional, foreign language not required; for doctorate, dissertation required, foreign language not required.

Entrance requirements: For master's and doctorate, GRE General Test. *Application deadline:* For fall admission, 8/15. Applications are processed on a rolling basis. *Application fee:* $30 ($40 for international students).

Expenses: Tuition, state resident: full-time $5,400. Tuition, nonresident: full-time $14,535. Part-time tuition and fees vary according to course load and program.

Financial aid: Fellowships, research assistantships, teaching assistantships, Federal Work-Study and institutionally sponsored loans available. Aid available to part-time students.

Dr. Elizabeth L. Gross, Director, 614-292-9480, *Fax:* 614-292-1538, *E-mail:* gross.3@osu.edu.

Find an in-depth description at www.petersons.com/graduate.

■ OREGON STATE UNIVERSITY

Graduate School, College of Science, Department of Biochemistry and Biophysics, Corvallis, OR 97331-6503

AWARDS MA, MAIS, MS, PhD.

Faculty: 16 full-time (2 women), 13 part-time/adjunct (1 woman).

Students: 28 full-time (12 women), 1 (woman) part-time; includes 3 minority (2 Asian Americans or Pacific Islanders, 1 Hispanic American), 12 international. Average age 30. In 1999, 3 master's, 4 doctorates awarded.

Degree requirements: For master's, thesis optional, foreign language not required; for doctorate, dissertation, exams required, foreign language not required.

Entrance requirements: For master's, GRE General Test, TOEFL, minimum GPA of 3.0; for doctorate, GRE Subject Test, TOEFL, minimum GPA of 3.0. *Application deadline:* For fall admission, 4/15 (priority date). Applications are processed on a rolling basis. *Application fee:* $50.

Expenses: Tuition, state resident: full-time $6,489. Tuition, nonresident: full-time $11,061. Tuition and fees vary according to program.

Financial aid: Research assistantships, teaching assistantships, institutionally sponsored loans available. Aid available to part-time students. Financial aid application deadline: 2/1.

Faculty research: DNA and deoxyribonucleotide metabolism, cell growth control, receptors and membranes, protein structure and function.

Dr. Christopher K. Mathews, Chairman, 541-737-1865, *Fax:* 541-737-0481, *E-mail:* mathewsc@ucs.orst.edu.

Application contact: Dr. W. Curtis Johnson, Chairman, Graduate Committee, 541-737-4143, *Fax:* 541-737-0481, *E-mail:* johnsowc@ucs.orst.edu.

■ PRINCETON UNIVERSITY

Graduate School, Graduate Program in Molecular Biophysics, Princeton, NJ 08544-1019

AWARDS PhD. Program administered by the Departments of Molecular Biology, Chemistry, and Physics.

Degree requirements: For doctorate, dissertation required.

Entrance requirements: For doctorate, GRE General Test, GRE Subject Test (biology, chemistry, or physics).

Expenses: Tuition: Full-time $25,050.

Find an in-depth description at www.petersons.com/graduate.

■ PURDUE UNIVERSITY

Graduate School, School of Science, Department of Biological Sciences, Program in Biophysics, West Lafayette, IN 47907

AWARDS PhD.

Faculty: 13 full-time (3 women).

Degree requirements: For doctorate, dissertation, seminars, teaching experience required, foreign language not required.

Entrance requirements: For doctorate, GRE General Test, TOEFL, TSE. *Application deadline:* For fall admission, 2/15. *Application fee:* $30. Electronic applications accepted.

Expenses: Tuition, state resident: full-time $4,530; part-time $130 per credit hour. Tuition, nonresident: full-time $15,310; part-time $404 per credit hour. Tuition and fees vary according to campus/location and program.

Financial aid: Fellowships, research assistantships, teaching assistantships available. Aid available to part-time students. Financial aid application deadline: 2/15; financial aid applicants required to submit FAFSA.

Faculty research: Protein atomic structure, molecular virology, structural biology of proteins and viruses, molecular biophysics and computer analysis of molecular structure.

Application contact: Nancy Konopka, Graduate Studies Office Manager, 765-494-8142, *Fax:* 765-494-0876, *E-mail:* njk@bilbo.bio.purdue.edu.

■ RENSSELAER POLYTECHNIC INSTITUTE

Graduate School, School of Science, Department of Biochemistry and Biophysics, Troy, NY 12180-3590

AWARDS Biochemistry (MS); biophysics (MS).

Students: 3 full-time (1 woman); includes 2 minority (1 African American, 1 Asian American or Pacific Islander). *9 applicants, 44% accepted.* In 1999, 3 degrees awarded.

Entrance requirements: For master's, GRE General Test, TOEFL. *Application deadline:* For fall admission, 2/1 (priority date). Applications are processed on a rolling basis. *Application fee:* $35.

Expenses: Tuition: Part-time $665 per credit hour. Required fees: $980.

Financial aid: Application deadline: 2/1.

Faculty research: Biopolymers, photosynthesis, cellular bioengineering.

Dr. Jane Koretz, Head, 518-276-6492, *E-mail:* koretj@rpi.edu.

■ RENSSELAER POLYTECHNIC INSTITUTE

Graduate School, School of Science, Department of Biology, Troy, NY 12180-3590

AWARDS Biochemistry (MS, PhD); biophysics (MS, PhD); cell biology (MS, PhD); developmental biology (MS, PhD); microbiology (MS, PhD); molecular biology (MS, PhD); plant science (MS, PhD). Part-time programs available.

Faculty: 14 full-time (5 women).

Students: 16 full-time (8 women), 2 part-time (1 woman); includes 1 minority (Asian American or Pacific Islander), 4 international. *37 applicants, 41% accepted.* In 1999, 6 master's, 3 doctorates awarded.

Rensselaer Polytechnic Institute (continued)
Terminal master's awarded for partial completion of doctoral program.
Degree requirements: For master's and doctorate, thesis/dissertation required, foreign language not required.
Entrance requirements: For master's and doctorate, GRE General Test, TOEFL. *Application deadline:* For fall admission, 2/1 (priority date). Applications are processed on a rolling basis. *Application fee:* $35.
Expenses: Tuition: Part-time $665 per credit hour. Required fees: $980.
Financial aid: In 1999–00, 8 research assistantships with partial tuition reimbursements (averaging $15,000 per year), 11 teaching assistantships with full tuition reimbursements (averaging $15,000 per year) were awarded; fellowships, career-related internships or fieldwork and institutionally sponsored loans also available. Financial aid application deadline: 2/1.
Faculty research: Applied environmental biology, genetics, environmental science, fresh water ecology, microbial ecology.
Dr. John Salerno, Chair, 518-276-2699, *Fax:* 518-276-2344.
Application contact: Dr. Jackie L. Collier, Assistant Professor, 518-276-6446, *Fax:* 518-276-2344, *E-mail:* collij3@rpi.edu.
Find an in-depth description at www.petersons.com/graduate.

■ STANFORD UNIVERSITY

School of Medicine, Graduate Programs in Medicine, Program in Biophysics, Stanford, CA 94305-9991
AWARDS PhD.

Students: 21 full-time (8 women), 11 part-time (4 women); includes 7 minority (1 African American, 6 Asian Americans or Pacific Islanders), 3 international. Average age 26. *69 applicants, 12% accepted.* In 1999, 5 degrees awarded.
Degree requirements: For doctorate, dissertation, oral exam required, foreign language not required.
Entrance requirements: For doctorate, GRE General Test, GRE Subject Test, TOEFL. *Application deadline:* For fall admission, 12/15. *Application fee:* $65 ($80 for international students). Electronic applications accepted.
Expenses: Tuition: Full-time $23,058. Required fees: $152. Part-time tuition and fees vary according to course load.
Financial aid: Fellowships, research assistantships, career-related internships or fieldwork, Federal Work-Study, and institutionally sponsored loans available.
William Weis, Chairman, 650-723-7576, *Fax:* 650-723-8464, *E-mail:* bill.weis@stanford.edu.

Application contact: Graduate Admissions Coordinator, 650-725-0261.

■ STATE UNIVERSITY OF NEW YORK AT BUFFALO

Graduate School, Graduate Programs in Biomedical Sciences at Roswell Park Cancer Institute, Department of Molecular and Cellular Biophysics at Roswell Park Cancer Institute, Buffalo, NY 14214
AWARDS MS, PhD.

Faculty: 20 full-time (2 women), 3 part-time/adjunct (0 women).
Students: 12 full-time (4 women), 11 part-time (3 women); includes 3 minority (all Asian Americans or Pacific Islanders), 8 international. Average age 25. *20 applicants, 40% accepted.* In 1999, 2 master's, 3 doctorates awarded. Terminal master's awarded for partial completion of doctoral program.
Degree requirements: For master's, thesis, exam, project required; for doctorate, dissertation required.
Entrance requirements: For master's, GRE General Test, TOEFL, TSE, TWE; for doctorate, GRE General Test, GRE Subject Test, TOEFL, TSE, TWE. *Application deadline:* For fall admission, 2/1 (priority date). Applications are processed on a rolling basis. *Application fee:* $35. Electronic applications accepted.
Expenses: Tuition, state resident: full-time $5,100; part-time $213 per credit hour. Tuition, nonresident: full-time $8,416; part-time $351 per credit hour. Required fees: $935; $75 per semester. Tuition and fees vary according to course load and program.
Financial aid: In 1999–00, 4 fellowships with full tuition reimbursements (averaging $15,000 per year), 21 research assistantships with full tuition reimbursements (averaging $15,000 per year) were awarded; Federal Work-Study and institutionally sponsored loans also available. Financial aid application deadline: 6/1; financial aid applicants required to submit FAFSA.
Faculty research: MRI research, structural and function of biomolecules, photodynamic therapy, DNA damage and repair, heat-shock proteins and vaccine research. *Total annual research expenditures:* $2.5 million.
Dr. John Subjeck, Chairman, 716-845-3147, *Fax:* 716-845-8389, *E-mail:* subjeck@sc3101.med.buffalo.edu.
Application contact: Craig R. Johnson, Director of Graduate Studies, 716-845-2339, *Fax:* 716-845-8178, *E-mail:* rpgradapp@sc3103.med.buffalo.edu.
Find an in-depth description at www.petersons.com/graduate.

■ STATE UNIVERSITY OF NEW YORK AT BUFFALO

Graduate School, School of Medicine and Biomedical Sciences, Graduate Programs in Medicine and Biomedical Sciences, Department of Physiology and Biophysics, Buffalo, NY 14260
AWARDS Biophysical sciences (MS, PhD); physiology (MA, PhD).

Faculty: 26 full-time (3 women), 2 part-time/adjunct (both women).
Students: 6 full-time (2 women), 17 part-time (9 women), 8 international. Average age 26. *7 applicants, 57% accepted.* In 1999, 5 master's, 1 doctorate awarded. Terminal master's awarded for partial completion of doctoral program.
Degree requirements: For master's, thesis (for some programs), oral exam, project required, foreign language not required; for doctorate, dissertation, oral and written qualifying exam or 2 research proposals required, foreign language not required.
Entrance requirements: For master's and doctorate, GRE General Test, TOEFL. *Application deadline:* For fall admission, 2/1 (priority date). *Application fee:* $35.
Expenses: Tuition, state resident: full-time $5,100. Tuition, nonresident: full-time $8,416. Required fees: $935.
Financial aid: In 1999–00, 17 research assistantships with tuition reimbursements (averaging $16,000 per year), 1 teaching assistantship with tuition reimbursement were awarded; fellowships, Federal Work-Study, institutionally sponsored loans, and unspecified assistantships also available. Financial aid application deadline: 2/1; financial aid applicants required to submit FAFSA.
Faculty research: Neurosciences, ion channels, cardiac physiology, renal/epithelial transport, cardiopulmonary exercise. *Total annual research expenditures:* $4.9 million.
Dr. Harold C. Strauss, Chair, 716-829-2738, *Fax:* 716-829-2344, *E-mail:* hstrauss@buffalo.edu.
Application contact: Dr. Malcolm M. Slaughter, Director of Graduate Studies, 716-829-3240, *Fax:* 716-829-2344, *E-mail:* pgy-bph@acsu.buffalo.edu.

■ STATE UNIVERSITY OF NEW YORK AT STONY BROOK

Health Sciences Center, School of Medicine and Graduate School, Graduate Programs in Medicine, Department of Molecular Physiology and Biophysics, Stony Brook, NY 11794
AWARDS Physiology and biophysics (PhD).
Faculty: 38.

Students: 12 full-time (1 woman), 18 part-time (8 women); includes 7 minority (2 African Americans, 4 Asian Americans or Pacific Islanders, 1 Hispanic American), 11 international. Average age 28. *33 applicants, 21% accepted.* In 1999, 2 degrees awarded.
Degree requirements: For doctorate, computer language, dissertation, comprehensive exams required, foreign language not required.
Entrance requirements: For doctorate, GRE General Test, GRE Subject Test, TOEFL, BS in related field, minimum GPA of 3.0. *Application deadline:* For fall admission, 1/15. *Application fee:* $50.
Expenses: Tuition, state resident: full-time $5,100. Tuition, nonresident: full-time $8,416. Required fees: $492.
Financial aid: In 1999–00, 3 fellowships, 12 research assistantships were awarded; teaching assistantships, Federal Work-Study also available. Financial aid application deadline: 3/15.
Faculty research: Cellular electrophysiology, membrane permeation and transport, metabolic endocrinology. *Total annual research expenditures:* $3.7 million.
Dr. Peter Brink, Chair, 631-444-2287, *Fax:* 631-444-3432.
Application contact: Dr. Leon C. Moore, Graduate Adviser, 631-444-2287, *Fax:* 631-444-3432, *E-mail:* moore@ pofvax.pnb.sunysb.edu.
Find an in-depth description at www.petersons.com/graduate.

■ **STATE UNIVERSITY OF NEW YORK HEALTH SCIENCE CENTER AT BROOKLYN**

School of Graduate Studies, Program in Biophysics and Physiology, Brooklyn, NY 11203-2098

AWARDS PhD, MD/PhD.

Degree requirements: For doctorate, one foreign language, dissertation required.
Entrance requirements: For doctorate, GRE.
Expenses: Tuition, state resident: full-time $5,100; part-time $213 per credit. Tuition, nonresident: full-time $8,416; part-time $351 per credit. Required fees: $200. Full-time tuition and fees vary according to program and student level.
Faculty research: Cardiovascular physiology, neurophysiology, developmental physiology, membrane transport, molecular basis of muscle contraction.

■ **SYRACUSE UNIVERSITY**

Graduate School, College of Arts and Sciences, Department of Biology and Department of Physics, Program in Biophysics, Syracuse, NY 13244-0003

AWARDS PhD.

Faculty: 72.
Students: 1 full-time (0 women). Average age 25. *3 applicants, 0% accepted.* In 1999, 2 degrees awarded.
Degree requirements: For doctorate, dissertation, exam required, foreign language not required.
Entrance requirements: For doctorate, GRE General Test, GRE Subject Test, TOEFL. *Application deadline:* Applications are processed on a rolling basis. *Application fee:* $40.
Expenses: Tuition: Full-time $13,992; part-time $583 per credit hour.
Financial aid: Fellowships, research assistantships, teaching assistantships, Federal Work-Study and tuition waivers (partial) available. Financial aid application deadline: 3/1.
James Dabrowiak, Director, 315-443-4601.

■ **SYRACUSE UNIVERSITY**

Graduate School, College of Arts and Sciences, Department of Physics, Syracuse, NY 13244-0003

AWARDS Biophysics (PhD); physics (MS, PhD).

Faculty: 23.
Students: 32 full-time (5 women), 7 part-time, 24 international. Average age 28. *106 applicants, 20% accepted.* In 1999, 4 doctorates awarded. Terminal master's awarded for partial completion of doctoral program.
Degree requirements: For master's, thesis or alternative required, foreign language not required; for doctorate, dissertation required, foreign language not required.
Entrance requirements: For master's and doctorate, GRE General Test, GRE Subject Test, TOEFL. *Application deadline:* Applications are processed on a rolling basis. *Application fee:* $40.
Expenses: Tuition: Full-time $13,992; part-time $583 per credit hour.
Financial aid: Fellowships, research assistantships, teaching assistantships, Federal Work-Study and tuition waivers (partial) available. Financial aid application deadline: 3/1.
Eric Schiff, Chair, 315-443-3901.
Application contact: Edward Lipson, Graduate Program Director, 315-443-5958.

■ **TEXAS A&M UNIVERSITY**

College of Agriculture and Life Sciences, Department of Biochemistry and Biophysics, College Station, TX 77843

AWARDS Biochemistry (MS, PhD); biophysics (MS).

Faculty: 29 full-time (5 women).
Students: 84 full-time (31 women), 5 part-time (1 woman); includes 4 minority (1 African American, 2 Asian Americans or Pacific Islanders, 1 Hispanic American), 26 international. Average age 27. In 1999, 6 master's, 13 doctorates awarded.
Entrance requirements: For master's and doctorate, GRE General Test, TOEFL. *Application deadline:* For fall admission, 3/1 (priority date). *Application fee:* $50 ($75 for international students).
Expenses: Tuition, state resident: part-time $76 per semester hour. Tuition, nonresident: part-time $292 per semester hour. Required fees: $11 per semester hour. Tuition and fees vary according to program.
Financial aid: In 1999–00, 2 fellowships (averaging $15,000 per year), 67 research assistantships with tuition reimbursements (averaging $17,650 per year), 16 teaching assistantships (averaging $17,650 per year) were awarded. Financial aid application deadline: 4/1; financial aid applicants required to submit FAFSA.
Faculty research: Enzymology, gene expression, protein structure, plant biochemistry.
Dr. James R. Wild, Head, 979-845-1011, *Fax:* 979-845-9274.
Application contact: Susan Whiting, Staff Assistant, 979-845-1779, *Fax:* 979-845-9274, *E-mail:* susanw@bioch.tamu.edu.
Find an in-depth description at www.petersons.com/graduate.

■ **THE UNIVERSITY OF ALABAMA AT BIRMINGHAM**

Graduate School and School of Medicine and School of Dentistry, Graduate Programs in Joint Health Sciences, Department of Physiology and Biophysics, Birmingham, AL 35294

AWARDS PhD.

Students: 26 full-time (13 women); includes 3 minority (all African Americans), 9 international. *59 applicants, 25% accepted.* In 1999, 6 degrees awarded.
Entrance requirements: For doctorate, GRE General Test, TOEFL, interview, minimum GPA of 3.0. *Application deadline:* For fall admission, 3/1. Applications are processed on a rolling basis. *Application fee:* $35 ($60 for international students). Electronic applications accepted.

The University of Alabama at Birmingham (continued)

Expenses: All doctoral students receive a full fellowship, stipend, and tuition.
Financial aid: Fellowships available.
Faculty research: Standard physiology (neurological, endocrine, cardiovascular, respiratory, and renal), cell and membrane biology.
Dr. Dale J. Benos, Interim Chair, 205-934-6220, *Fax:* 205-934-2377.
Application contact: Director of Graduate Studies, 205-934-3969, *Fax:* 205-975-9028.
Find an in-depth description at www.petersons.com/graduate.

■ THE UNIVERSITY OF ALABAMA AT BIRMINGHAM

Graduate School and School of Medicine and School of Dentistry, Graduate Programs in Joint Health Sciences, Program in Biophysical Sciences, Birmingham, AL 35294
AWARDS PhD.

Students: *1 applicant, 100% accepted.* In 1999, 3 degrees awarded.
Entrance requirements: For doctorate, GRE, interview. *Application deadline:* Applications are processed on a rolling basis. *Application fee:* $35 ($60 for international students). Electronic applications accepted.
Expenses: Tuition, state resident: part-time $104 per semester hour. Tuition, nonresident: part-time $208 per semester hour. Required fees: $17 per semester hour. $57 per quarter. Tuition and fees vary according to program.
Dr. Herbert C. Cheung, Director, 205-934-2485, *Fax:* 205-975-4621, *E-mail:* hccheung@uab.edu.

■ UNIVERSITY OF ARKANSAS FOR MEDICAL SCIENCES

College of Medicine and Graduate School, Graduate Programs in Medicine, Department of Physiology and Biophysics, Little Rock, AR 72205-7199
AWARDS MS, PhD, MD/PhD.

Faculty: 24 full-time (3 women).
Students: 23 full-time (9 women), 6 part-time (2 women); includes 7 minority (4 African Americans, 3 Asian Americans or Pacific Islanders), 4 international. In 1999, 2 master's awarded.
Degree requirements: For master's and doctorate, thesis/dissertation required, foreign language not required.
Entrance requirements: For master's and doctorate, GRE General Test. *Application fee:* $0.
Expenses: Tuition: Full-time $8,928.

Financial aid: In 1999–00, 16 research assistantships were awarded. Aid available to part-time students.
Dr. Michael L. Jennings, Chairman, 501-686-5123.
Application contact: Dr. Parimal Chowdhury, Graduate Coordinator, 501-686-5123.

■ UNIVERSITY OF CALIFORNIA, BERKELEY

Graduate Division, Group in Biophysics, Berkeley, CA 94720-1500
AWARDS MA, PhD.

Degree requirements: For doctorate, dissertation, qualifying exam required.
Entrance requirements: For master's and doctorate, GRE General Test, GRE Subject Test, minimum GPA of 3.0.
Expenses: Tuition, nonresident: full-time $9,804. Required fees: $4,268. Tuition and fees vary according to program.
Find an in-depth description at www.petersons.com/graduate.

■ UNIVERSITY OF CALIFORNIA, DAVIS

Graduate Studies, Programs in the Biological Sciences, Program in Biophysics, Davis, CA 95616
AWARDS MS, PhD.

Faculty: 20.
Students: 8 full-time (3 women); includes 2 minority (both Hispanic Americans), 2 international. Average age 26. *7 applicants, 57% accepted.*
Degree requirements: For master's and doctorate, thesis/dissertation required.
Entrance requirements: For master's and doctorate, GRE General Test, GRE Subject Test. *Application deadline:* For fall admission, 3/15. *Application fee:* $40. Electronic applications accepted.
Expenses: Tuition, nonresident: full-time $9,804. Tuition and fees vary according to program and student level.
Financial aid: In 1999–00, 7 students received aid, including 6 fellowships with full and partial tuition reimbursements available, 2 research assistantships with full and partial tuition reimbursements available; Federal Work-Study, grants, institutionally sponsored loans, scholarships, and tuition waivers (full and partial) also available. Financial aid application deadline: 1/15; financial aid applicants required to submit FAFSA.
Faculty research: Molecular structure, protein structure/function relationships, spectroscopy.
Thomas Jue, Chair, 530-752-4569, *Fax:* 530-752-8822, *E-mail:* tjue@ucdavis.edu.
Application contact: Tori Hollowell, Administrative Assistant, 530-752-9092,

Fax: 530-752-8822, *E-mail:* trhollowell@ucdavis.edu.
Find an in-depth description at www.petersons.com/graduate.

■ UNIVERSITY OF CALIFORNIA, IRVINE

College of Medicine and Office of Research and Graduate Studies, Graduate Programs in Medicine and School of Biological Sciences, Department of Physiology and Biophysics, Irvine, CA 92697

AWARDS Biological sciences (PhD). Students apply through the Graduate Program in Molecular Biology, Genetics, and Biochemistry.
Faculty: 13 full-time (2 women), 4 part-time/adjunct (1 woman).
Students: 10 full-time (3 women); includes 1 minority (Hispanic American), 1 international. *292 applicants, 25% accepted.* In 1999, 1 doctorate awarded.
Degree requirements: For doctorate, dissertation required.
Entrance requirements: For doctorate, GRE General Test, GRE Subject Test. *Application deadline:* For fall admission, 2/1 (priority date). *Application fee:* $40. Electronic applications accepted.
Expenses: Tuition, nonresident: full-time $10,322; part-time $1,720 per quarter. Required fees: $5,354; $1,300 per quarter. Tuition and fees vary according to program.
Financial aid: Fellowships, research assistantships, institutionally sponsored loans and tuition waivers (full and partial) available. Financial aid application deadline: 3/2; financial aid applicants required to submit FAFSA.
Faculty research: Membrane physiology, exercise physiology, regulation of hormone biosynthesis and action, endocrinology, ion channels and signal transduction.
Dr. Janos K. Lanyi, PhD, Chair, 949-824-7788, *Fax:* 949-824-8540, *E-mail:* jklanyi@uci.edu.
Application contact: Kimberly McKinney, Administrator, 949-824-8145, *Fax:* 949-824-7407, *E-mail:* gp-mbgb@uci.edu.

■ UNIVERSITY OF CALIFORNIA, SAN DIEGO

Graduate Studies and Research, Department of Physics, La Jolla, CA 92093
AWARDS Biophysics (MS, PhD); physics (MS, PhD); physics/materials physics (MS).

Faculty: 47.
Students: 111 (13 women). *199 applicants, 33% accepted.* In 1999, 13 master's, 28 doctorates awarded.

Degree requirements: For doctorate, dissertation required.

Entrance requirements: For master's and doctorate, GRE General Test, GRE Subject Test, TOEFL. *Application deadline:* For fall admission, 1/15. *Application fee:* $40.

Expenses: Tuition, nonresident: full-time $14,691. Required fees: $4,697. Full-time tuition and fees vary according to program.
Thomas M. O'Neil, Chair, 858-534-4176, *E-mail:* toneil@ucsd.edu.

Application contact: Debra Bomar, Graduate Coordinator, 858-534-3293, *E-mail:* dbomar@physics.ucsd.edu.

Find an in-depth description at www.petersons.com/graduate.

■ UNIVERSITY OF CALIFORNIA, SAN FRANCISCO

Graduate Division, Graduate Group in Biophysics, San Francisco, CA 94143
AWARDS PhD.

Students: In 1999, 5 degrees awarded.
Degree requirements: For doctorate, dissertation required, foreign language not required.
Entrance requirements: For doctorate, GRE General Test, GRE Subject Test. *Application deadline:* For fall admission, 1/15. *Application fee:* $40.
Financial aid: Tuition waivers (full) and stipends available. Financial aid application deadline: 1/10.
Dr. David Agard, Director, 415-476-2521.
Application contact: Julie Ransom, Program Administrator, 415-476-6671, *E-mail:* ransom@cgl.ucsf.edu.

Find an in-depth description at www.petersons.com/graduate.

■ UNIVERSITY OF CINCINNATI

Division of Research and Advanced Studies, College of Medicine, Graduate Programs in Medicine, Department of Molecular and Cellular Physiology, Cincinnati, OH 45267
AWARDS Biophysics (MS, PhD); physiology (MS, PhD).

Faculty: 13 full-time.
Students: 19 full-time (6 women), 1 (woman) part-time; includes 4 minority (2 African Americans, 2 Asian Americans or Pacific Islanders), 3 international. *64 applicants, 16% accepted.* In 1999, 1 master's, 1 doctorate awarded.
Degree requirements: For master's, thesis required, foreign language not required; for doctorate, dissertation, qualifying exam required, foreign language not required. *Average time to degree:* Master's–2.8 years full-time; doctorate–4.8 years full-time.

Entrance requirements: For master's and doctorate, GRE General Test, GRE Subject Test. *Application deadline:* For fall admission, 2/1 (priority date). Applications are processed on a rolling basis. *Application fee:* $30.
Expenses: Tuition, state resident: full-time $5,139; part-time $196 per credit hour. Tuition, nonresident: full-time $10,326; part-time $369 per credit hour. Required fees: $561; $187 per quarter.
Financial aid: Unspecified assistantships available. Financial aid application deadline: 5/1.
Faculty research: Neurobiology, electrophysiology, muscle physiology, cardiovascular physiology, endocrinology. *Total annual research expenditures:* $5.1 million.
Dr. David E. Millhorn, Head, 513-558-5636, *Fax:* 513-558-5738, *E-mail:* david.millhorn@uc.edu.
Application contact: John Dedman, Graduate Program Director, 513-558-4145, *Fax:* 513-558-5738, *E-mail:* john.dedman@uc.edu.

■ UNIVERSITY OF CINCINNATI

Division of Research and Advanced Studies, College of Medicine, Graduate Programs in Medicine, Department of Molecular, Cellular, and Biochemical Pharmacology, Cincinnati, OH 45267
AWARDS Cell biophysics (PhD); pharmacology (PhD).

Faculty: 7 full-time.
Students: 13 full-time (5 women), 1 (woman) part-time; includes 1 minority (African American), 2 international. *27 applicants, 15% accepted.* In 1999, 2 degrees awarded.
Degree requirements: For doctorate, dissertation, qualifying exam required, foreign language not required. *Average time to degree:* Doctorate–6.3 years full-time.
Entrance requirements: For doctorate, GRE General Test, GRE Subject Test. *Application deadline:* For fall admission, 2/1 (priority date). Applications are processed on a rolling basis. *Application fee:* $30.
Expenses: Tuition, state resident: full-time $5,139; part-time $196 per credit hour. Tuition, nonresident: full-time $10,326; part-time $369 per credit hour. Required fees: $561; $187 per quarter.
Financial aid: Tuition waivers (full) and unspecified assistantships available. Financial aid application deadline: 5/1.
Faculty research: Lipoprotein research, enzyme regulation, electrophysiology, gene actuation. *Total annual research expenditures:* $3.1 million.
Dr. John Maggio, Chair, 513-558-4723, *Fax:* 513-558-1190, *E-mail:* john.maggio@uc.edu.

Application contact: Dr. Robert Rapoport, Director of Graduate Studies, 513-558-2376, *Fax:* 513-558-1169, *E-mail:* robert.rapoport@uc.edu.

Find an in-depth description at www.petersons.com/graduate.

■ UNIVERSITY OF COLORADO HEALTH SCIENCES CENTER

Graduate School, Programs in Biological and Medical Sciences, Programs in Biophysics and Genetics, Denver, CO 80262

AWARDS Biophysics and genetics (PhD); genetic counseling (MS). Terminal master's awarded for partial completion of doctoral program.

Degree requirements: For master's, thesis optional, foreign language not required; for doctorate, dissertation required, foreign language not required.
Entrance requirements: For master's, GRE General Test, minimum GPA of 2.75; for doctorate, GRE General Test, TOEFL, minimum GPA of 2.75.
Expenses: Tuition, state resident: full-time $1,512; part-time $56 per hour. Tuition, nonresident: full-time $7,209; part-time $267 per hour. Full-time tuition and fees vary according to course load and program.

■ UNIVERSITY OF CONNECTICUT

Graduate School, College of Liberal Arts and Sciences, Biological Sciences Group, Storrs, CT 06269

AWARDS Ecology and evolutionary biology (MS, PhD), including botany, ecology, entomology, systematics, zoology; molecular and cell biology (MS, PhD), including biochemistry, biophysics, biotechnology (MS), cell and developmental biology, genetics, microbiology, plant molecular and cell biology; physiology and neurobiology (MS, PhD), including neurobiology, physiology.

Degree requirements: For doctorate, dissertation required.
Entrance requirements: For master's and doctorate, GRE General Test, GRE Subject Test, TOEFL.
Expenses: Tuition, state resident: full-time $5,118. Tuition, nonresident: full-time $13,298. Required fees: $1,022.

■ UNIVERSITY OF CONNECTICUT

Graduate School, College of Liberal Arts and Sciences, Biological Sciences Group, Department of Molecular and Cell Biology, Field of Biophysics, Storrs, CT 06269
AWARDS MS, PhD.

Degree requirements: For doctorate, dissertation required.

University of Connecticut (continued)

Entrance requirements: For master's and doctorate, GRE General Test, GRE Subject Test, TOEFL.

Expenses: Tuition, state resident: full-time $5,118. Tuition, nonresident: full-time $13,298. Required fees: $1,022.

Find an in-depth description at www.petersons.com/graduate.

■ **UNIVERSITY OF HAWAII AT MANOA**

John A. Burns School of Medicine and Graduate Division, Graduate Programs in Biomedical Sciences, Department of Biochemistry and Biophysics, Honolulu, HI 96822

AWARDS Biochemistry (MS, PhD); biophysics (MS, PhD).

Faculty: 13 full-time (1 woman), 1 part-time/adjunct (0 women).
Students: 5 full-time (1 woman), 1 (woman) part-time, 5 international. Average age 27. *0 applicants, 0% accepted.* In 1999, 1 master's awarded (100% continued full-time study); 2 doctorates awarded. Terminal master's awarded for partial completion of doctoral program.
Degree requirements: For master's, thesis required (for some programs), foreign language not required; for doctorate, dissertation required, foreign language not required. *Average time to degree:* Master's–6 years full-time.
Entrance requirements: For master's and doctorate, GRE General Test, GRE Subject Test. *Application deadline:* For fall admission, 2/1; for spring admission, 9/1. *Application fee:* $25 ($50 for international students).
Expenses: Tuition, state resident: part-time $168 per credit. Tuition, nonresident: part-time $415 per credit. Required fees: $51 per semester. Part-time tuition and fees vary according to course load.
Financial aid: In 1999–00, research assistantships (averaging $16,511 per year); fellowships, teaching assistantships, Federal Work-Study and tuition waivers (full) also available. Financial aid application deadline: 2/1.
Faculty research: Protein and nucleic acid structure, function, and metabolism; endocrine metabolism; molecular basis of cancer; clinical biochemistry. *Total annual research expenditures:* $199,817.
Dr. N. V. Bhagavan, Chairperson, 808-956-8490, *Fax:* 808-956-9498, *E-mail:* bhagavan@jabsom.biomed.hawaii.edu.

■ **UNIVERSITY OF ILLINOIS AT CHICAGO**

College of Medicine and Graduate College, Graduate Programs in Medicine, Department of Physiology and Biophysics, Chicago, IL 60607-7128

AWARDS MS, PhD.

Faculty: 19 full-time (6 women).
Students: 23 full-time (12 women), 1 (woman) part-time; includes 4 minority (1 African American, 3 Asian Americans or Pacific Islanders), 4 international. Average age 29. *55 applicants, 22% accepted.* In 1999, 5 doctorates awarded. Terminal master's awarded for partial completion of doctoral program.
Degree requirements: For master's and doctorate, thesis/dissertation required, foreign language not required.
Entrance requirements: For master's and doctorate, GRE General Test, TOEFL. *Application deadline:* For fall admission, 6/1; for spring admission, 11/1. *Application fee:* $40 ($50 for international students).
Expenses: Tuition, state resident: full-time $3,750. Tuition, nonresident: full-time $10,588. Tuition and fees vary according to course load.
Financial aid: In 1999–00, 10 students received aid; fellowships, research assistantships, teaching assistantships, Federal Work-Study and traineeships available. Financial aid application deadline: 3/1; financial aid applicants required to submit FAFSA.
Faculty research: Neuroscience, endocrinology and reproduction, cell physiology, exercise physiology, NMR.
R. John Solaro, Head, 312-996-7620.

Find an in-depth description at www.petersons.com/graduate.

■ **UNIVERSITY OF ILLINOIS AT URBANA–CHAMPAIGN**

Graduate College, Program in Biophysics and Computational Biology, Urbana, IL 61801

AWARDS PhD.

Students: 51 full-time (15 women); includes 8 minority (7 Asian Americans or Pacific Islanders, 1 Hispanic American), 27 international. *112 applicants, 12% accepted.* In 1999, 8 doctorates awarded.
Degree requirements: For doctorate, dissertation required, foreign language not required.
Entrance requirements: For doctorate, TOEFL, minimum GPA of 4.0 on a 5.0 scale. *Application deadline:* Applications are processed on a rolling basis. *Application fee:* $40 ($50 for international students).
Expenses: Tuition, state resident: full-time $4,040. Tuition, nonresident: full-time $11,192. Full-time tuition and fees vary according to program.
Financial aid: Fellowships, research assistantships, teaching assistantships, grants and traineeships available. Financial aid application deadline: 1/15.
Dr. Erik Jakobsson, Chairman, 217-333-1630, *Fax:* 217-244-1224, *E-mail:* jake@ncsa.uiuc.edu.
Application contact: Cynthia Dodds, Director of Graduate Studies, 217-333-2446, *Fax:* 217-244-1224, *E-mail:* dodds@uiuc.edu.

Find an in-depth description at www.petersons.com/graduate.

■ **THE UNIVERSITY OF IOWA**

College of Medicine and Graduate College, Graduate Programs in Medicine, Department of Physiology and Biophysics, Iowa City, IA 52242-1316

AWARDS Physiology and biophysics (PhD); physiology and biophysiology (MS).

Faculty: 21 full-time (2 women), 3 part-time/adjunct (1 woman).
Students: 16 full-time (4 women); includes 2 minority (1 African American, 1 Asian American or Pacific Islander), 5 international. Average age 25. *61 applicants, 7% accepted.* In 1999, 4 doctorates awarded (100% entered university research/teaching). Terminal master's awarded for partial completion of doctoral program.
Degree requirements: For master's and doctorate, thesis/dissertation, comprehensive exam, teaching experience required, foreign language not required. *Average time to degree:* Doctorate–5 years full-time.
Entrance requirements: For master's and doctorate, GRE General Test, minimum GPA of 3.0. *Application deadline:* For fall admission, 4/1. Applications are processed on a rolling basis. *Application fee:* $20 ($30 for international students).
Expenses: Tuition, state resident: full-time $3,308. Tuition, nonresident: full-time $10,662. Tuition and fees vary according to course load and program.
Financial aid: In 1999–00, 7 fellowships with full tuition reimbursements (averaging $16,787 per year), 9 research assistantships with full tuition reimbursements (averaging $16,787 per year) were awarded; traineeships also available. Financial aid application deadline: 4/1.
Faculty research: Cellular and molecular endocrinology, membrane structure and function, cardiac cell electrophysiology, regulation of gene expression, neurophysiology.
Robert E. Fellows, Head, 319-335-7802, *Fax:* 319-335-7330, *E-mail:* robert-fellows@uiowa.edu.

Application contact: Dr. Toshinori Hoshi, Chairman of Graduate Admissions, 319-335-7845, *Fax:* 319-335-7330, *E-mail:* toshinori-hoshi@uiowa.edu.

Find an in-depth description at www.petersons.com/graduate.

■ UNIVERSITY OF KANSAS

Graduate School, College of Liberal Arts and Sciences, Division of Biological Sciences, Department of Molecular Biosciences, Lawrence, KS 66045

AWARDS Biochemistry and biophysics (MA, MS, PhD); microbiology (MA, MS, PhD); molecular, cellular, and developmental biology (MA, MS, PhD).

Students: 13 full-time (7 women), 8 part-time (2 women); includes 1 minority (Asian American or Pacific Islander), 5 international. *57 applicants, 21% accepted.* In 2000, 11 master's, 10 doctorates awarded.

Degree requirements: For master's, thesis required, foreign language not required; for doctorate, dissertation required.

Entrance requirements: For master's and doctorate, GRE General Test, GRE Subject Test, TOEFL. *Application deadline:* For fall admission, 1/15 (priority date). *Application fee:* $25.

Expenses: Tuition, state resident: full-time $2,482; part-time $103 per credit hour. Tuition, nonresident: full-time $8,104; part-time $338 per credit hour. Required fees: $428; $31 per credit hour. Tuition and fees vary according to program.

Financial aid: In 2000–01, research assistantships (averaging $11,588 per year), teaching assistantships (averaging $11,588 per year) were awarded; fellowships. Financial aid application deadline: 3/1. Paul Kelly, Chair, 785-864-4311, *Fax:* 785-864-5294.

Application contact: Information Contact, 785-864-4311, *Fax:* 785-864-5294.

Find an in-depth description at www.petersons.com/graduate.

■ UNIVERSITY OF LOUISVILLE

School of Medicine and Graduate School, Integrated Programs in Biomedical Sciences, Department of Physiology and Biophysics, Louisville, KY 40292-0001

AWARDS MS, PhD.

Degree requirements: For master's and doctorate, thesis/dissertation required, foreign language not required.

Entrance requirements: For master's, GRE General Test, 36 hours of graduate course work; for doctorate, GRE General

Test, minimum GPA of 3.0. Electronic applications accepted.

Expenses: Tuition, state resident: full-time $3,260; part-time $182 per hour. Tuition, nonresident: full-time $9,780; part-time $544 per hour. Required fees: $143; $28 per hour.

Faculty research: Control of small blood vessels, neuroendocrine physiology, renal blood flow, regulation of heart and lungs, neurophysiological control of breathing and circulation.

■ UNIVERSITY OF MIAMI

School of Medicine and Graduate School, Graduate Programs in Medicine, Department of Physiology and Biophysics, Coral Gables, FL 33124

AWARDS PhD, MD/PhD.

Faculty: 16 full-time (2 women).

Students: 13 full-time (9 women); includes 5 minority (4 Asian Americans or Pacific Islanders, 1 Hispanic American). Average age 25. In 1999, 1 doctorate awarded (100% entered university research/teaching).

Degree requirements: For doctorate, dissertation, qualifying exam required, foreign language not required. *Average time to degree:* Doctorate–5 years full-time.

Entrance requirements: For doctorate, GRE General Test, TOEFL, minimum GPA of 3.0 in sciences. *Application deadline:* Applications are processed on a rolling basis. *Application fee:* $0.

Expenses: Tuition, area resident: Part-time $899 per credit.

Financial aid: Fellowships, research assistantships, career-related internships or fieldwork and institutionally sponsored loans available.

Faculty research: Cell and membrane physiology, cell-to-cell communication, molecular neurobiology, neuroimmunology, neural development. Dr. Karl Magleby, Chairman, 305-243-6821, *Fax:* 305-243-6898, *E-mail:* kmagleby@mednet.med.miami.edu.

Application contact: Dr. Nirupa Chaudhari, Adviser, 305-243-3187, *Fax:* 305-243-5931, *E-mail:* nchaudha@newssun.med.miami.edu.

Find an in-depth description at www.petersons.com/graduate.

■ UNIVERSITY OF MICHIGAN

Horace H. Rackham School of Graduate Studies, Interdepartmental Program in Biophysics, Ann Arbor, MI 48109

AWARDS PhD.

Faculty: 37 full-time (7 women).

Students: 15 full-time (2 women); includes 2 minority (1 African American, 1 Asian

American or Pacific Islander), 2 international. *33 applicants, 27% accepted.* In 1999, 3 degrees awarded.

Degree requirements: For doctorate, oral defense of dissertation, preliminary exam required. *Average time to degree:* Doctorate–5.7 years full-time.

Entrance requirements: For doctorate, GRE General Test, GRE Subject Test. *Application deadline:* For fall admission, 1/10. *Application fee:* $55. Electronic applications accepted.

Expenses: Tuition, state resident: full-time $10,316. Tuition, nonresident: full-time $20,922. Required fees: $185. Part-time tuition and fees vary according to course load and program.

Financial aid: In 1999–00, 7 fellowships with full tuition reimbursements (averaging $17,688 per year), 4 research assistantships with full tuition reimbursements (averaging $17,688 per year), 4 teaching assistantships with full tuition reimbursements (averaging $17,688 per year) were awarded; grants also available. Financial aid application deadline: 3/15.

Faculty research: Structural biology, computational biophysics, physical chemistry, cellular biophysics. Dr. Rowena G. Matthews, Chair, 734-764-5257, *Fax:* 734-764-3323.

Application contact: Ruby A. Hogue, Student Services Associate I, 734-763-6722, *Fax:* 734-764-3323, *E-mail:* rhogue@biop.umich.edu.

Find an in-depth description at www.petersons.com/graduate.

■ UNIVERSITY OF MINNESOTA, TWIN CITIES CAMPUS

Graduate School, Program in Biophysical Sciences and Medical Physics, Minneapolis, MN 55455-0213

AWARDS MS, PhD. Part-time programs available.

Degree requirements: For master's, research paper, oral exam required, thesis optional, foreign language not required; for doctorate, dissertation, oral/written preliminary exam, oral final exam required, foreign language not required.

Expenses: Tuition, state resident: full-time $5,040; part-time $420 per credit. Tuition, nonresident: full-time $9,900; part-time $825 per credit. Full-time tuition and fees vary according to course load, program and reciprocity agreements.

Faculty research: Theoretical biophysics, radiological physics, cellular and molecular biophysics.

Find an in-depth description at www.petersons.com/graduate.

■ UNIVERSITY OF MINNESOTA, TWIN CITIES CAMPUS

Medical School and Graduate School, Graduate Programs in Medicine, Department of Biochemistry, Molecular Biology and Biophysics, Minneapolis, MN 55455-0213

AWARDS PhD.

Faculty: 67 full-time (13 women). **Students:** 77 full-time (39 women); includes 1 African American, 1 Hispanic American, 5 international. *189 applicants, 22% accepted.* In 1999, 16 degrees awarded (94% entered university research/teaching, 6% found other work related to degree). **Degree requirements:** For doctorate, dissertation required, foreign language not required. *Average time to degree:* Doctorate–5 years full-time. **Entrance requirements:** For doctorate, GRE General Test. *Application deadline:* For fall admission, 1/15 (priority date). *Application fee:* $50 ($55 for international students). **Expenses:** Program provides tuition for all students. **Financial aid:** In 1999–00, 12 fellowships with full tuition reimbursements (averaging $16,500 per year), research assistantships with full tuition reimbursements (averaging $16,500 per year) were awarded. **Faculty research:** Physical biochemistry, enzymology, physiological chemistry. *Total annual research expenditures:* $10.9 million. Dr. Douglas Ohlendorf, Director of Admissions, 612-625-6100. **Application contact:** Mary F. Petrie-Terry, Student Support Assistant, 612-625-5179, *Fax:* 612-625-2163, *E-mail:* petri008@umn.edu.

Find an in-depth description at www.petersons.com/graduate.

■ UNIVERSITY OF MISSISSIPPI MEDICAL CENTER

Graduate Programs in Biomedical Sciences, Department of Physiology and Biophysics, Jackson, MS 39216-4505

AWARDS MS, PhD, MD/PhD.

Faculty: 16 full-time (1 woman). **Students:** 8 full-time (6 women). Average age 25. *6 applicants, 50% accepted.* **Degree requirements:** For master's, thesis required, foreign language not required; for doctorate, dissertation, first authored publication required, foreign language not required. *Average time to degree:* Master's–4 years full-time; doctorate–5 years full-time. **Entrance requirements:** For master's and doctorate, GRE General Test, minimum GPA of 3.0. *Application deadline:* For fall

admission, 8/1. Applications are processed on a rolling basis. *Application fee:* $10. **Expenses:** Tuition, state resident: full-time $2,378; part-time $132 per hour. Tuition, nonresident: full-time $4,697; part-time $261 per hour. Tuition and fees vary according to program. **Financial aid:** In 1999–00, 5 students received aid, including 5 research assistantships (averaging $16,234 per year); Federal Work-Study also available. Financial aid application deadline: 4/1. **Faculty research:** Cardiovascular, renal, endocrine, and cellular neurophysiology; molecular physiology. *Total annual research expenditures:* $1.4 million. Dr. John E. Hall, Chairman, 601-984-1801, *Fax:* 601-984-1817. **Application contact:** Dr. Michael W. Brands, Director of Graduate Programs, 601-984-1820, *Fax:* 601-984-1817, *E-mail:* mbrands@physiology.umsmed.edu.

■ UNIVERSITY OF MISSOURI– KANSAS CITY

School of Biological Sciences, Program in Cell Biology and Biophysics, Kansas City, MO 64110-2499

AWARDS PhD. PhD offered through School of Graduate Studies.

Degree requirements: For doctorate, dissertation required. **Entrance requirements:** For doctorate, GRE General Test, TOEFL, minimum GPA of 3.0, bachelor's degree in chemistry, biology or related field. *Application deadline:* For fall admission, 3/1 (priority date). Applications are processed on a rolling basis. *Application fee:* $25. Electronic applications accepted. **Expenses:** Tuition, state resident: part-time $173 per hour. Tuition, nonresident: part-time $348 per hour. Required fees: $22 per hour. $15 per term. Part-time tuition and fees vary according to course load and program. **Financial aid:** In 1999–00, research assistantships with full tuition reimbursements (averaging $16,000 per year), teaching assistantships with full and partial tuition reimbursements (averaging $16,000 per year) were awarded; grants, scholarships, and unspecified assistantships also available. **Application contact:** Graduate Adviser, 816-235-2580.

Find an in-depth description at www.petersons.com/graduate.

■ THE UNIVERSITY OF NORTH CAROLINA AT CHAPEL HILL

School of Medicine and Graduate School, Graduate Programs in Medicine, Department of Biochemistry and Biophysics, Chapel Hill, NC 27599

AWARDS MS, PhD.

Faculty: 27 full-time (6 women). **Students:** 41 full-time (12 women); includes 1 minority (Asian American or Pacific Islander), 17 international. *86 applicants, 33% accepted.* In 1999, 2 master's, 7 doctorates awarded. **Degree requirements:** For master's, thesis, comprehensive exam required, foreign language not required; for doctorate, dissertation, comprehensive exams required, foreign language not required. *Average time to degree:* Master's–4.5 years full-time; doctorate–6 years full-time. **Entrance requirements:** For master's and doctorate, GRE General Test, GRE Subject Test (recommended), minimum GPA of 3.0. *Application deadline:* For fall admission, 1/1. *Application fee:* $55. Electronic applications accepted. **Expenses:** Tuition, state resident: full-time $1,966. Tuition, nonresident: full-time $11,026. Required fees: $8,940. One-time fee: $15 full-time. Part-time tuition and fees vary according to course load. **Financial aid:** In 1999–00, research assistantships with full tuition reimbursements (averaging $16,000 per year), 5 teaching assistantships with full tuition reimbursements (averaging $16,000 per year) were awarded; fellowships also available. Dr. David Lee, Chair, 919-962-8326, *Fax:* 919-966-2852, *E-mail:* dclee@med.unc.edu. **Application contact:** Diane M. Harris, Assistant to the Director of Graduate Studies, 919-966-4683, *E-mail:* diane_harris@med.unc.edu.

Find an in-depth description at www.petersons.com/graduate.

■ UNIVERSITY OF PENNSYLVANIA

School of Medicine, Biomedical Graduate Studies, Graduate Group in Biochemistry and Molecular Biophysics, Philadelphia, PA 19104

AWARDS PhD, MD/PhD.

Faculty: 66 full-time (10 women). **Students:** 75 full-time (30 women); includes 27 minority (2 African Americans, 25 Asian Americans or Pacific Islanders), 7 international. *123 applicants, 23% accepted.* In 1999, 15 doctorates awarded. **Degree requirements:** For doctorate, dissertation required, foreign language not required.

Entrance requirements: For doctorate, GRE General Test, TOEFL. *Application deadline:* For fall admission, 1/2 (priority date). Applications are processed on a rolling basis. *Application fee:* $65. Electronic applications accepted.

Expenses: Tuition: Full-time $17,256; part-time $2,991 per course. Required fees: $2,588; $363 per course. $726 per term.

Financial aid: In 1999–00, 45 students received aid, including 28 fellowships, 46 research assistantships; grants and institutionally sponsored loans also available. Financial aid application deadline: 1/2.

Faculty research: Biochemistry of cell differentiation, tissue culture, intermediary metabolism, structure of proteins and nucleic acids, biochemical genetics. Dr. Kim Sharp, Chairperson, 215-898-3506.

Application contact: Ruth Keris, Graduate Group Administrator, 215-898-4639, *Fax:* 215-573-2085, *E-mail:* keris@ mail.med.upenn.edu.

Find an in-depth description at www.petersons.com/graduate.

■ UNIVERSITY OF PITTSBURGH

Faculty of Arts and Sciences, Department of Biological Sciences, Program in Molecular Biophysics, Pittsburgh, PA 15260

AWARDS PhD.

Degree requirements: For doctorate, dissertation required, foreign language not required.

Entrance requirements: For doctorate, GRE General Test, GRE Subject Test, TOEFL. *Application fee:* $40.

Expenses: Tuition, state resident: full-time $8,338; part-time $342 per credit. Tuition, nonresident: full-time $17,168; part-time $707 per credit. Required fees: $480; $90 per semester. Tuition and fees vary according to program.

Application contact: Cathleen M. Barr, Graduate Administrator, 412-624-4268, *Fax:* 412-624-4759, *E-mail:* cbarr+@ pitt.edu.

Find an in-depth description at www.petersons.com/graduate.

■ UNIVERSITY OF ROCHESTER

School of Medicine and Dentistry, Graduate Programs in Medicine and Dentistry, Department of Biochemistry and Biophysics, Program in Biophysics, Rochester, NY 14627-0250

AWARDS MS, PhD.

Students: 25 full-time (5 women), 16 international. Terminal master's awarded for partial completion of doctoral program.

Degree requirements: For master's, foreign language not required; for doctorate, dissertation, qualifying exam required, foreign language not required.

Entrance requirements: For master's and doctorate, GRE General Test. *Application deadline:* For fall admission, 2/1. *Application fee:* $25.

Expenses: Tuition: Part-time $697 per credit hour. Tuition and fees vary according to program.

Financial aid: Fellowships, research assistantships, teaching assistantships, tuition waivers (full and partial) available. Financial aid application deadline: 2/1.

Application contact: Rose Burgholzer, Graduate Program Secretary, 716-275-3417.

Find an in-depth description at www.petersons.com/graduate.

■ UNIVERSITY OF SOUTHERN CALIFORNIA

Keck School of Medicine and Graduate School, Graduate Programs in Medicine, Department of Physiology and Biophysics, Los Angeles, CA 90089

AWARDS MS, PhD, MD/PhD.

Faculty: 12 full-time (4 women).

Students: 22 full-time (13 women); includes 5 minority (all Asian Americans or Pacific Islanders), 10 international. Average age 25. *26 applicants, 23% accepted.* In 1999, 4 degrees awarded (100% entered university research/teaching). Terminal master's awarded for partial completion of doctoral program.

Degree requirements: For master's, foreign language not required; for doctorate, dissertation required, foreign language not required. *Average time to degree:* Doctorate–5 years full-time.

Entrance requirements: For master's and doctorate, GRE General Test, minimum GPA of 3.0. *Application deadline:* For fall admission, 2/1 (priority date). Applications are processed on a rolling basis. *Application fee:* $55. Electronic applications accepted.

Expenses: Tuition: Full-time $22,198; part-time $748 per unit. Required fees: $406.

Financial aid: In 1999–00, 3 fellowships with partial tuition reimbursements (averaging $18,598 per year), 14 research assistantships with full tuition reimbursements (averaging $18,598 per year), 2 teaching assistantships with full tuition reimbursements (averaging $18,598 per year) were awarded; grants, traineeships, and tuition waivers (full) also available. Financial aid application deadline: 2/1.

Faculty research: Endocrinology, metabolism, cell transport, molecular biology, mathematical modelling. *Total annual research expenditures:* $2.1 million. Dr. Alicia McDonough, Director, Graduate Studies, 323-442-1238, *Fax:* 323-442-2283, *E-mail:* mcdonoug@hsc.usc.edu.

Application contact: Elena Camarena, Graduate Coordinator, 323-442-1039, *Fax:* 323-442-2283, *E-mail:* physiol@ hsc.usc.edu.

Find an in-depth description at www.petersons.com/graduate.

■ UNIVERSITY OF SOUTH FLORIDA

College of Medicine and Graduate School, Graduate Programs in Medical Sciences, Department of Physiology and Biophysics, Tampa, FL 33620-9951

AWARDS PhD.

Degree requirements: For doctorate, 2 foreign languages (computer language can substitute for one), dissertation required.

Entrance requirements: For doctorate, GRE General Test, minimum GPA of 3.0.

Expenses: Tuition, state resident: part-time $148 per credit hour. Tuition, nonresident: part-time $509 per credit hour.

Faculty research: Cardiovascular, neurorespiratory, and endocrine physiology; Alzheimer's disease; cell membrane biophysics.

Find an in-depth description at www.petersons.com/graduate.

■ THE UNIVERSITY OF TENNESSEE HEALTH SCIENCE CENTER

College of Graduate Health Sciences, Department of Physiology, Memphis, TN 38163-0002

AWARDS MS, PhD.

Degree requirements: For master's, thesis, oral and written comprehensive exams required, foreign language not required; for doctorate, dissertation, oral and written preliminary and comprehensive exams required, foreign language not required.

Entrance requirements: For master's, GRE General Test, TOEFL, minimum GPA of 3.0; for doctorate, GRE General Test, GRE Subject Test, TOEFL, minimum GPA of 3.0.

Find an in-depth description at www.petersons.com/graduate.

■ THE UNIVERSITY OF TEXAS–HOUSTON HEALTH SCIENCE CENTER

Graduate School of Biomedical Sciences, Program in Medical Physics, Houston, TX 77225-0036

AWARDS MS, PhD, MD/PhD.

Faculty: 27 full-time (3 women).
Students: 9 full-time; includes 1 minority (Asian American or Pacific Islander), 1 international. Average age 27. *19 applicants, 42% accepted.* In 1999, 2 master's awarded.
Degree requirements: For master's and doctorate, thesis/dissertation required, foreign language not required.
Entrance requirements: For master's and doctorate, GRE General Test, TOEFL, TWE. *Application deadline:* For fall admission, 1/15 (priority date); for spring admission, 11/1. Applications are processed on a rolling basis. *Application fee:* $10. Electronic applications accepted.
Financial aid: Fellowships, research assistantships, institutionally sponsored loans available. Financial aid application deadline: 1/15.
Faculty research: Three-dimensional dose optimization, photon and electron conformal radiotherapy, electron beam dose calculations, stereotactic radiosurgery, brachytherapy dosimetry, quality assurance, digital imaging and PACS, magnetic resonance imaging, image guided therapies, internal dosimetry, PET instrumentation development.
Dr. Kenneth R. Hogstrom, Director, 713-792-3216, *Fax:* 713-794-5272, *E-mail:* khogstro@mdanderson.org.
Application contact: Anne Baronitis, Director of Admissions, 713-500-9860, *Fax:* 713-500-9877, *E-mail:* abaron@ gsbs.gs.uth.tmc.edu.
Find an in-depth description at www.petersons.com/graduate.

■ THE UNIVERSITY OF TEXAS MEDICAL BRANCH AT GALVESTON

Graduate School of Biomedical Sciences, Program in Cellular Physiology and Molecular Biophysics, Galveston, TX 77555

AWARDS MS, PhD.

Faculty: 32 full-time (4 women).
Students: 14 full-time (5 women), 1 part-time; includes 1 minority (Asian American or Pacific Islander), 11 international. Average age 30. *15 applicants, 47% accepted.* In 1999, 1 master's, 3 doctorates awarded.
Degree requirements: For master's, thesis or alternative required, foreign language not required; for doctorate, dissertation required, foreign language not required.
Entrance requirements: For master's and doctorate, GRE General Test. *Application deadline:* For fall admission, 3/2 (priority date). Applications are processed on a rolling basis. *Application fee:* $25 ($50 for international students). Electronic applications accepted.
Expenses: Tuition, state resident: full-time $684; part-time $38 per credit hour. Tuition, nonresident: full-time $4,572; part-time $254 per credit hour. Required fees: $29; $7.5 per credit hour. One-time fee: $55. Tuition and fees vary according to program.
Financial aid: Fellowships, research assistantships, Federal Work-Study and institutionally sponsored loans available. Financial aid applicants required to submit FAFSA.
Faculty research: Molecular biology of surface proteins, membrane transportation, channel biophysics, cell signaling, structural biology.
Dr. Simon A. Lewis, Director, 409-772-3397, *Fax:* 409-772-3381, *E-mail:* slewis@ utmb.edu.
Application contact: Lori Ann Stewart, Secretary, 409-772-5445, *Fax:* 409-772-3381, *E-mail:* lastewart@utmb.edu.

■ THE UNIVERSITY OF TEXAS SOUTHWESTERN MEDICAL CENTER AT DALLAS

Southwestern Graduate School of Biomedical Sciences, Division of Cell and Molecular Biology, Program in Molecular Biophysics, Dallas, TX 75390

AWARDS PhD.

Faculty: 28 full-time (5 women).
Students: 25 full-time (10 women); includes 1 minority (Asian American or Pacific Islander), 6 international. Average age 28. In 1999, 3 doctorates awarded (100% entered university research/ teaching).
Degree requirements: For doctorate, dissertation required, foreign language not required.
Entrance requirements: For doctorate, GRE General Test, minimum GPA of 3.0. *Application deadline:* For fall admission, 1/5 (priority date). *Application fee:* $0. Electronic applications accepted.
Expenses: Tuition, state resident: full-time $912. Tuition, nonresident: full-time $6,096. Required fees: $216. Full-time tuition and fees vary according to course load and program.
Financial aid: Fellowships, research assistantships, institutionally sponsored loans and traineeships available. Financial aid application deadline: 3/15; financial aid applicants required to submit FAFSA.
Faculty research: Optical spectroscopy, x-ray crystallography, protein chemistry, ion channels, contractile and cytoskeletal proteins.
Dr. Stephen R. Sprang, Chair, 214-648-5008, *Fax:* 214-648-6336, *E-mail:* sprang@ howie.swmed.edu.
Application contact: Nancy McKinney, Education Coordinator, 214-648-8099, *Fax:* 214-648-2978, *E-mail:* dcmbinfo@ utsouthwestern.edu.
Find an in-depth description at www.petersons.com/graduate.

■ UNIVERSITY OF VERMONT

College of Medicine and Graduate College, Graduate Programs in Medicine, Department of Molecular Physiology and Biophysics, Burlington, VT 05405

AWARDS MS, PhD, MD/MS, MD/PhD.

Degree requirements: For master's and doctorate, thesis/dissertation required.
Entrance requirements: For master's and doctorate, GRE General Test, TOEFL.
Expenses: Tuition, state resident: full-time $7,464; part-time $311 per credit. Tuition, nonresident: full-time $18,672; part-time $778 per credit. Full-time tuition and fees vary according to degree level and program.
Find an in-depth description at www.petersons.com/graduate.

■ UNIVERSITY OF VIRGINIA

College and Graduate School of Arts and Sciences, Department of Molecular Physiology and Biological Physics, Charlottesville, VA 22903

AWARDS PhD.

Faculty: 22 full-time (5 women).
Students: 20 full-time (9 women); includes 1 minority (Asian American or Pacific Islander), 12 international. Average age 28. *6 applicants, 50% accepted.* In 1999, 3 degrees awarded.
Degree requirements: For doctorate, dissertation required.
Entrance requirements: For doctorate, GRE General Test, GRE Subject Test. *Application deadline:* For fall admission, 7/15; for spring admission, 12/1. Applications are processed on a rolling basis. *Application fee:* $40. Electronic applications accepted.
Expenses: Tuition, state resident: full-time $3,832. Tuition, nonresident: full-time $15,519. Required fees: $1,084. Tuition and fees vary according to course load and program.
Financial aid: Application deadline: 2/1.

Dr. Andrew P. Somlyo, Chairman, 804-924-5108.

Find an in-depth description at www.petersons.com/graduate.

■ **UNIVERSITY OF VIRGINIA**

College and Graduate School of Arts and Sciences, Interdisciplinary Program in Biophysics, Charlottesville, VA 22903

AWARDS PhD.

Faculty: 37 full-time (3 women).
Students: 21 full-time (7 women); includes 1 minority (African American), 12 international. *29 applicants, 17% accepted.* In 1999, 3 degrees awarded (100% entered university research/teaching).
Degree requirements: For doctorate, dissertation, research proposal and defense required, foreign language not required. *Average time to degree:* Doctorate–5.3 years full-time.
Entrance requirements: For doctorate, GRE General Test, GRE Subject Test (recommended), TOEFL. *Application deadline:* For fall admission, 2/15 (priority date); for spring admission, 10/15. Applications are processed on a rolling basis. *Application fee:* $40.
Expenses: Tuition, state resident: full-time $3,832. Tuition, nonresident: full-time $15,519. Required fees: $1,084. Tuition and fees vary according to course load and program.
Financial aid: In 1999–00, 20 students received aid, including fellowships with full tuition reimbursements available (averaging $18,750 per year), research assistantships with full tuition reimbursements available (averaging $18,750 per year), teaching assistantships with full tuition reimbursements available (averaging $18,750 per year); tuition waivers (full) also available. Financial aid application deadline: 2/15.
Faculty research: Membrane biophysics, macromolecular structure, thermodynamics, radiological physics and imaging, neurobiology.
Dr. Lukas Tamm, Director, 804-924-2181, *Fax:* 804-924-0140, *E-mail:* medgpo@virginia.edu.
Application contact: 804-924-2181, *Fax:* 804-924-0140, *E-mail:* medgpo@virginia.edu.

Find an in-depth description at www.petersons.com/graduate.

■ **UNIVERSITY OF WASHINGTON**

School of Medicine and Graduate School, Graduate Programs in Medicine, Department of Physiology and Biophysics, Seattle, WA 98195

AWARDS PhD.

Faculty: 37 full-time (6 women), 3 part-time/adjunct (1 woman).
Students: 21 full-time (11 women); includes 3 minority (1 African American, 1 Asian American or Pacific Islander, 1 Hispanic American), 2 international. Average age 28. *39 applicants, 18% accepted.* In 1999, 2 doctorates awarded.
Degree requirements: For doctorate, dissertation required, foreign language not required. *Average time to degree:* Doctorate–5.5 years full-time.
Entrance requirements: For doctorate, GRE General Test, TOEFL. *Application deadline:* For fall admission, 2/1.
Expenses: Tuition, state resident: full-time $9,210; part-time $236 per credit. Tuition, nonresident: full-time $23,256; part-time $596 per credit.
Financial aid: In 1999–00, 20 students received aid, including fellowships with tuition reimbursements available (averaging $17,304 per year), research assistantships with tuition reimbursements available (averaging $17,304 per year); Federal Work-Study, institutionally sponsored loans, and tuition waivers (full) also available. Financial aid applicants required to submit FAFSA.
Faculty research: Membrane and cell biophysics, neuroendocrinology, cardiovascular and respiratory physiology, systems neurophysiology and behavior, molecular physiology. *Total annual research expenditures:* $4.7 million.
Dr. Albert Berger, Acting Chair, 206-543-0954, *Fax:* 206-685-0619, *E-mail:* berger@u.washington.edu.
Application contact: Alyssa Kern, Graduate Coordinator, 206-685-0519, *Fax:* 206-685-0619, *E-mail:* pbio@u.washington.edu.

■ **UNIVERSITY OF WISCONSIN–MADISON**

Graduate School, Program in Biophysics, Madison, WI 53706-1380

AWARDS PhD.

Degree requirements: For doctorate, dissertation required, foreign language not required.
Entrance requirements: For doctorate, GRE General Test, minimum GPA of 3.0.
Expenses: Tuition, state resident: full-time $5,406; part-time $339 per credit. Tuition, nonresident: full-time $17,110; part-time $1,071 per credit. Full-time tuition and fees vary according to program and reciprocity agreements. Part-time tuition and fees vary according to course load and program.
Faculty research: NMR spectroscopy, high-speed automated DNA sequencing, digital imaging microscopy, x-ray crystallography.

■ **VANDERBILT UNIVERSITY**

Graduate School and School of Medicine, Department of Molecular Physiology and Biophysics, Nashville, TN 37240-1001

AWARDS PhD, MD/PhD.

Faculty: 20 full-time (2 women).
Students: 23 full-time (11 women), 1 part-time; includes 3 minority (1 Asian American or Pacific Islander, 2 Hispanic Americans), 4 international. Average age 29. In 1999, 11 degrees awarded.
Degree requirements: For doctorate, dissertation, preliminary, qualifying, and final exams required, foreign language not required.
Entrance requirements: For doctorate, GRE General Test, GRE Subject Test (recommended). *Application deadline:* For fall admission, 1/15. *Application fee:* $40.
Expenses: Tuition: Full-time $17,244; part-time $958 per hour. Required fees: $242; $121 per semester. Tuition and fees vary according to program.
Financial aid: In 1999–00, fellowships with full tuition reimbursements (averaging $17,000 per year), research assistantships with full tuition reimbursements (averaging $17,000 per year) were awarded; Federal Work-Study, institutionally sponsored loans, traineeships, and tuition waivers (partial) also available. Financial aid application deadline: 1/15.
Faculty research: Molecular endocrinology, membrane transport biophysics, metabolic regulation, neurobiology.
Alan D. Cherrington, Chair, 615-322-7000, *Fax:* 615-322-7236, *E-mail:* alan.cherrington@mcmail.vanderbilt.edu.
Application contact: Roger J. Colbran, Director of Graduate Studies, 615-322-7000, *Fax:* 615-322-7236, *E-mail:* roger.j.colbran@vanderbilt.edu.

Find an in-depth description at www.petersons.com/graduate.

■ **VIRGINIA COMMONWEALTH UNIVERSITY**

School of Graduate Studies and School of Medicine, School of Medicine Graduate Programs, Department of Biochemistry and Molecular Biophysics, Richmond, VA 23284-9005

AWARDS Biochemistry (PhD); biochemistry and molecular biophysics (MS, CBHS); molecular biology and genetics (PhD); neurosciences (PhD).

Students: 5 full-time (3 women), 21 part-time (10 women); includes 12 minority (9 Asian Americans or Pacific Islanders, 1 Hispanic American, 2 Native Americans). In 1999, 1 master's, 10 doctorates awarded.

Virginia Commonwealth University (continued)

Degree requirements: For master's, thesis required, foreign language not required; for doctorate, dissertation, comprehensive oral and written exams required, foreign language not required.

Entrance requirements: For master's and doctorate, GRE General Test. *Application deadline:* For fall admission, 5/1. *Application fee:* $30.

Expenses: Tuition, state resident: full-time $4,031; part-time $224 per credit hour. Tuition, nonresident: full-time $11,946; part-time $664 per credit hour. Required fees: $1,081; $40 per credit hour. Tuition and fees vary according to campus/location and program.

Financial aid: Fellowships, research assistantships available.

Faculty research: Molecular biology, peptide/protein chemistry, neurochemistry, enzyme mechanisms, macromolecular structure determination. *Total annual research expenditures:* $3.5 million. Chair, 804-828-9762, *Fax:* 804-828-1473.

Application contact: Dr. Zendra E. Zehner, Program Director, 804-828-8753, *Fax:* 804-828-1473, *E-mail:* zezehner@ vcu.edu.

■ WASHINGTON STATE UNIVERSITY

Graduate School, College of Sciences, School of Molecular Biosciences, Department of Biochemistry and Biophysics, Pullman, WA 99164

AWARDS MS, PhD.

Faculty: 23.

Students: 43 full-time (14 women), 3 part-time (2 women); includes 3 minority (all Asian Americans or Pacific Islanders), 9 international. In 1999, 4 master's, 6 doctorates awarded. Terminal master's awarded for partial completion of doctoral program.

Degree requirements: For master's, oral exam required, thesis optional, foreign language not required; for doctorate, dissertation, oral exam, written exam required, foreign language not required. *Average time to degree:* Master's–2 years full-time; doctorate–4 years full-time.

Entrance requirements: For master's and doctorate, GRE General Test, minimum GPA of 3.0. *Application deadline:* For fall admission, 3/1 (priority date). Applications are processed on a rolling basis. *Application fee:* $35.

Expenses: Tuition, state resident: full-time $5,654. Tuition, nonresident: full-time $13,850. International tuition: $13,850 full-time. Tuition and fees vary according to program.

Financial aid: In 1999–00, 33 research assistantships with full and partial tuition reimbursements, 9 teaching assistantships with full and partial tuition reimbursements were awarded; fellowships, career-related internships or fieldwork, Federal Work-Study, institutionally sponsored loans, traineeships, and tuition waivers (partial) also available. Financial aid application deadline: 4/1; financial aid applicants required to submit FAFSA.

Faculty research: Gene regulation, signal transduction, protein export, reproductive biology, DNA repair. *Total annual research expenditures:* $3.4 million.

Dr. Gerald Hazelbauer, Chairman, 509-335-1276.

Application contact: Dr. Kirk McMichael, Graduate Coordinator, 509-335-8866, *Fax:* 509-335-8867.

Find an in-depth description at www.petersons.com/graduate.

■ WASHINGTON UNIVERSITY IN ST. LOUIS

Graduate School of Arts and Sciences, Division of Biology and Biomedical Sciences, Program in Molecular Biophysics, St. Louis, MO 63130-4899

AWARDS PhD.

Degree requirements: For doctorate, dissertation required, foreign language not required.

Entrance requirements: For doctorate, GRE General Test, GRE Subject Test. *Application deadline:* For fall admission, 1/1 (priority date). Applications are processed on a rolling basis. *Application fee:* $0.

Expenses: Tuition: Full-time $23,400; part-time $975 per credit. Tuition and fees vary according to program.

Financial aid: Application deadline: 1/1. Dr. Kathleen Hall, Coordinator.

Application contact: Rosemary Garagneni, Director of Admissions, 800-852-9074, *E-mail:* admissions@ dbbs.wustl.edu.

■ WRIGHT STATE UNIVERSITY

School of Graduate Studies, College of Science and Mathematics, Department of Physiology and Biophysics, Dayton, OH 45435

AWARDS MS.

Students: 3 full-time (2 women), 1 part-time; includes 1 minority (Asian American or Pacific Islander), 2 international. Average age 29. *5 applicants, 100% accepted.* In 1999, 3 degrees awarded.

Degree requirements: For master's, thesis required, foreign language not required.

Entrance requirements: For master's, TOEFL. *Application deadline:* For fall admission, 7/15. *Application fee:* $25.

Expenses: Tuition, state resident: full-time $5,568; part-time $175 per quarter hour. Tuition, nonresident: full-time $9,696; part-time $302 per quarter hour. Full-time tuition and fees vary according to course load, campus/location and program.

Financial aid: Fellowships, research assistantships, teaching assistantships, unspecified assistantships available. Aid available to part-time students. Financial aid applicants required to submit FAFSA.

Faculty research: Membrane transport, ion channels, pH regulation, lipid dynamics, neurophysiology.

Dr. Peter K. Lauf, Chair, 937-775-2360, *Fax:* 937-775-3769, *E-mail:* peter.lauf@ wright.edu.

Application contact: Dr. Noel S. Nussbaum, Director, 937-775-2081, *Fax:* 937-775-3769, *E-mail:* noel.nussbaum@ wright.edu.

■ YALE UNIVERSITY

Graduate School of Arts and Sciences, Department of Molecular Biophysics and Biochemistry, New Haven, CT 06520

AWARDS MS, PhD.

Faculty: 95 full-time (24 women), 2 part-time/adjunct (1 woman).

Students: 71 full-time (22 women); includes 9 minority (2 African Americans, 6 Asian Americans or Pacific Islanders, 1 Hispanic American), 16 international. *130 applicants, 41% accepted.* In 1999, 14 degrees awarded.

Degree requirements: For doctorate, dissertation required. *Average time to degree:* Doctorate–6 years full-time.

Entrance requirements: For doctorate, GRE General Test, GRE Subject Test. *Application deadline:* For fall admission, 1/4. *Application fee:* $65.

Expenses: Tuition: Full-time $22,300. Full-time tuition and fees vary according to program.

Financial aid: Fellowships, research assistantships, teaching assistantships, Federal Work-Study and institutionally sponsored loans available. Aid available to part-time students.

Application contact: Admissions Information, 203-432-2770.

Find an in-depth description at www.petersons.com/graduate.

■ YALE UNIVERSITY

School of Medicine and Graduate School of Arts and Sciences, Combined Program in Biological and Biomedical Sciences (BBS), Molecular Biophysics and Biochemistry Track, New Haven, CT 06520

AWARDS PhD, MD/PhD.

Degree requirements: For doctorate, dissertation required.
Entrance requirements: For doctorate, GRE General Test, TOEFL. *Application deadline:* For fall admission, 1/2. *Application fee:* $65. Electronic applications accepted.
Expenses: All students receive full tuition of $22,330 and an annual stipend of $17,600 .
Financial aid: Fellowships, research assistantships available.
Dr. Nigel Grindley, Director, 203-432-8891, *Fax:* 203-432-9782, *E-mail:* mbb.grad@yale.edu.
Application contact: Nessie Stewart, Graduate Registrar, 203-432-5662, *Fax:* 203-432-6178, *E-mail:* mbb.grad@yale.edu.

Find an in-depth description at www.petersons.com/graduate.

■ YESHIVA UNIVERSITY

Albert Einstein College of Medicine, Sue Golding Graduate Division of Medical Sciences, Department of Physiology and Biophysics, Bronx, NY 10461

AWARDS PhD, MD/PhD.

Faculty: 30 full-time.
Students: 8 full-time (3 women); includes 1 minority (Asian American or Pacific Islander), 4 international. In 1999, 2 degrees awarded.
Degree requirements: For doctorate, dissertation required, foreign language not required.
Entrance requirements: For doctorate, GRE General Test, TOEFL. *Application deadline:* For fall admission, 1/15. *Application fee:* $0.
Expenses: Tuition: Part-time $525 per credit. Tuition and fees vary according to degree level and program.
Financial aid: In 1999–00, 8 fellowships were awarded.
Faculty research: Biophysical and biochemical basis of body function at the subcellular, cellular, organ, and whole-body level.
Dr. Denis M. Rousseau, Chairperson, 718-430-4264.
Application contact: Sheila Cleeton, Assistant Director, 718-430-2128, *Fax:* 718-430-8655, *E-mail:* phd@aecom.yu.edu.

Find an in-depth description at www.petersons.com/graduate.

RADIATION BIOLOGY

■ AUBURN UNIVERSITY

College of Veterinary Medicine and Graduate School, Graduate Program in Veterinary Medicine, Department of Radiology, Auburn, Auburn University, AL 36849-0002

AWARDS MS. Part-time programs available.

Faculty: 5 full-time (2 women).
Degree requirements: For master's, thesis required, foreign language not required.
Entrance requirements: For master's, GRE General Test. *Application deadline:* For fall admission, 9/1; for spring admission, 12/1. Applications are processed on a rolling basis. *Application fee:* $25 ($50 for international students).
Expenses: Tuition, state resident: full-time $2,895; part-time $80 per credit hour. Tuition, nonresident: full-time $8,685; part-time $240 per credit hour.
Financial aid: Application deadline: 3/15.
Dr. Jan E. Bartels, Head, 334-844-5045.
Application contact: Dr. John F. Pritchett, Dean of the Graduate School, 334-844-4700.

■ COLORADO STATE UNIVERSITY

College of Veterinary Medicine and Biomedical Sciences and Graduate School, Graduate Programs in Veterinary Medicine and Biomedical Sciences, Department of Radiological Health Sciences, Fort Collins, CO 80523-0015

AWARDS Cellular and molecular biology (MS, PhD); health physics (MS, PhD); mammalian radiobiology (MS, PhD); nuclear-waste management (MS); radiobiology (MS); radioecology (MS, PhD); radiology (MS, PhD); veterinary radiology (MS).

Faculty: 21 full-time (3 women).
Students: 20 full-time (4 women), 4 part-time (1 woman), 5 international. Average age 32. *19 applicants, 53% accepted.* In 1999, 3 master's, 4 doctorates awarded (100% found work related to degree).
Degree requirements: For master's, thesis required, foreign language not required; for doctorate, dissertation required.
Entrance requirements: For master's and doctorate, GRE General Test, TOEFL. *Application deadline:* For fall admission, 2/1 (priority date). Applications are processed on a rolling basis. *Application fee:* $30. Electronic applications accepted.
Expenses: Tuition, state resident: full-time $2,694; part-time $150 per credit. Tuition,

nonresident: full-time $10,460; part-time $581 per credit. Required fees: $32 per semester.
Financial aid: In 1999–00, 1 fellowship, 5 research assistantships were awarded; teaching assistantships, career-related internships or fieldwork, Federal Work-Study, and traineeships also available.
Faculty research: Radiation therapy; cell, molecular, and tissue mechanisms in radiation biology; diagnostic radiology.
Dr. F. W. Whicker, Interim Chairman, 970-491-0555, *Fax:* 970-491-0623, *E-mail:* wiedeman@cvmbs.colostate.edu.

■ FLORIDA STATE UNIVERSITY

Graduate Studies, College of Arts and Sciences, Department of Biological Science, Program in Radiation Biology, Tallahassee, FL 32306

AWARDS MS, PhD.

Faculty: 1 full-time (0 women).
Degree requirements: For master's and doctorate, thesis/dissertation, teaching experience required.
Entrance requirements: For master's, GRE General Test, TOEFL; for doctorate, GRE General Test, GRE Subject Test, TOEFL. *Application deadline:* For fall admission, 1/15; for spring admission, 10/15. *Application fee:* $20.
Expenses: Tuition, state resident: full-time $3,504; part-time $146 per credit hour. Tuition, nonresident: full-time $12,162; part-time $507 per credit hour. Tuition and fees vary according to program.
Financial aid: In 1999–00, fellowships with full tuition reimbursements (averaging $13,740 per year), research assistantships with full tuition reimbursements (averaging $13,740 per year), teaching assistantships with full tuition reimbursements (averaging $13,740 per year) were awarded. Financial aid application deadline: 1/15; financial aid applicants required to submit FAFSA.
Faculty research: Cancer.
Dr. Thomas C. S. Keller, Associate Professor and Associate Chairman, 850-644-3023, *Fax:* 850-644-9829.
Application contact: Judy Bowers, Coordinator, Graduate Affairs, 850-644-3023, *Fax:* 850-644-9829, *E-mail:* bowers@bio.fsu.edu.

■ GEORGETOWN UNIVERSITY

Graduate School of Arts and Sciences, Programs in Biomedical Sciences, Department of Health Physics, Washington, DC 20057

AWARDS Health physics (MS); radiobiology (MS).

Degree requirements: For master's, thesis required.
Entrance requirements: For master's, TOEFL.

■ UNIVERSITY OF CALIFORNIA, IRVINE

College of Medicine and Office of Research and Graduate Studies, Graduate Programs in Medicine, Department of Radiological Sciences, Irvine, CA 92697

AWARDS MS, PhD.

Faculty: 5 full-time (0 women).
Students: 1 full-time (0 women); minority (Asian American or Pacific Islander). In 1999, 1 doctorate awarded.
Entrance requirements: For master's, GRE; for doctorate, GRE General Test, GRE Subject Test. *Application deadline:* For fall admission, 1/15 (priority date). Applications are processed on a rolling basis. *Application fee:* $40. Electronic applications accepted.
Expenses: Tuition, nonresident: full-time $10,322; part-time $1,720 per quarter. Required fees: $5,354; $1,300 per quarter. Tuition and fees vary according to program.
Financial aid: Fellowships, research assistantships, institutionally sponsored loans and tuition waivers (full and partial) available. Financial aid application deadline: 3/2; financial aid applicants required to submit FAFSA.
Faculty research: Medical imaging, medical physics, bioengineering, radiobiology, radiological engineering.
Anton N. Hasso, Chair, 714-456-6595, *Fax:* 714-456-7864.
Application contact: Dr. Sabee Molloi, Departmental Office of Student Affairs, 949-824-5904, *Fax:* 949-824-2837, *E-mail:* symolloi@uci.edu.

■ THE UNIVERSITY OF IOWA

College of Medicine and Graduate College, Graduate Programs in Medicine, Department of Free Radical and Radiation Biology, Iowa City, IA 52242-1316

AWARDS MS, PhD. Part-time programs available.

Faculty: 6 full-time (0 women).
Students: 19 full-time (13 women), 2 part-time, 16 international. Average age 31. *11 applicants, 55% accepted.* In 1999, 3 master's awarded (33% entered university research/teaching, 66% continued full-time study); 2 doctorates awarded (100% entered university research/teaching). Terminal master's awarded for partial completion of doctoral program.
Degree requirements: For master's, thesis required (for some programs),

foreign language not required; for doctorate, dissertation required, foreign language not required. *Average time to degree:* Master's–2.3 years full-time, 3 years part-time; doctorate–4.5 years full-time.
Application deadline: For fall admission, 5/31 (priority date); for spring admission, 10/31. Applications are processed on a rolling basis. *Application fee:* $30 ($50 for international students).
Expenses: Tuition, state resident: full-time $3,308. Tuition, nonresident: full-time $10,662. Tuition and fees vary according to course load and program.
Financial aid: In 1999–00, 20 students received aid, including 3 fellowships with partial tuition reimbursements available (averaging $14,688 per year), 16 research assistantships (averaging $16,800 per year); teaching assistantships
Faculty research: Radiation injury and cellular repair, cell proliferation kinetics, free radical biology, tumor control, PET imaging, EPR.
Larry W. Oberley, Head, 319-335-8015, *Fax:* 319-335-8039, *E-mail:* larry-oberley@uiowa.edu.

■ UNIVERSITY OF OKLAHOMA HEALTH SCIENCES CENTER

College of Medicine and Graduate College, Graduate Programs in Medicine, Department of Radiological Technology, Oklahoma City, OK 73190

AWARDS Medical radiation physics (MS, PhD), including diagnostic radiology, nuclear medicine, radiation therapy, ultrasound. Part-time programs available.

Faculty: 12 full-time (0 women), 2 part-time/adjunct (1 woman).
Students: 1 full-time (0 women), 1 part-time; includes 1 minority (Asian American or Pacific Islander). Average age 33. *13 applicants, 54% accepted.* Terminal master's awarded for partial completion of doctoral program.
Degree requirements: For master's and doctorate, thesis/dissertation required, foreign language not required.
Entrance requirements: For master's and doctorate, GRE General Test, TOEFL. *Application deadline:* For fall admission, 7/1 (priority date); for spring admission, 12/1. Applications are processed on a rolling basis. *Application fee:* $25 ($50 for international students).
Expenses: Tuition, state resident: part-time $90 per semester hour. Tuition, nonresident: part-time $264 per semester hour. Tuition and fees vary according to program.

Financial aid: Fellowships, research assistantships, career-related internships or fieldwork and institutionally sponsored loans available. Aid available to part-time students. Financial aid application deadline: 7/1.
Faculty research: Monte Carlo applications in radiation therapy, observer performed studies in diagnostic radiology, error analysis in gated cardiac nuclear medicine studies, nuclear medicine absorbed fraction determinations.
Dr. Bob Eaton, Head, 405-271-5132.
Application contact: Dr. Robert Y. L. Chu, Graduate Liaison, 405-271-5641.

■ THE UNIVERSITY OF TEXAS SOUTHWESTERN MEDICAL CENTER AT DALLAS

Southwestern Graduate School of Biomedical Sciences, Radiological Sciences Program, Dallas, TX 75390

AWARDS MS, PhD.

Faculty: 20 full-time (1 woman), 1 part-time/adjunct (0 women).
Students: 6 full-time (1 woman); includes 1 minority (Hispanic American). Average age 33. *8 applicants, 13% accepted.* In 1999, 1 doctorate awarded (100% entered university research/teaching). Terminal master's awarded for partial completion of doctoral program.
Degree requirements: For master's and doctorate, thesis/dissertation required. *Average time to degree:* Doctorate–6 years full-time.
Entrance requirements: For master's and doctorate, GRE General Test, TOEFL. *Application deadline:* For fall admission, 5/15 (priority date). *Application fee:* $0.
Expenses: Tuition, state resident: full-time $912. Tuition, nonresident: full-time $6,096. Required fees: $216. Full-time tuition and fees vary according to course load and program.
Financial aid: In 1999–00, 2 research assistantships were awarded; grants and institutionally sponsored loans also available. Financial aid application deadline: 3/15; financial aid applicants required to submit FAFSA.
Faculty research: Medical physics, nuclear medicine, noninvasive NMR methods, ultrasound, pathophysiological *in vivo* investigation.
Dr. Peter P. Antich, Chair, 214-648-2856, *Fax:* 214-648-2991, *E-mail:* pantic@mednet.swmed.edu.
Application contact: Kay Emerson, 214-648-2503, *Fax:* 214-648-2991, *E-mail:* kaywana.emerson@email.swmed.edu.

Find an in-depth description at www.petersons.com/graduate.

Botany and Plant Sciences

BOTANY AND PLANT SCIENCES
....

■ ALABAMA AGRICULTURAL AND MECHANICAL UNIVERSITY

School of Graduate Studies, School of Agricultural and Environmental Sciences, Department of Plant and Soil Sciences, Normal, AL 35762

AWARDS Animal sciences (MS); environmental science (MS); plant and soil science (PhD). Evening/weekend programs available.

Faculty: 10 full-time (1 woman).
Students: 6 full-time (3 women), 24 part-time (3 women); includes 19 minority (18 African Americans, 1 Asian American or Pacific Islander), 11 international. In 1999, 2 master's awarded (100% found work related to degree); 1 doctorate awarded (100% entered university research/ teaching). Terminal master's awarded for partial completion of doctoral program.
Degree requirements: For master's, thesis required, foreign language not required; for doctorate, one foreign language, computer language, dissertation required. *Average time to degree:* Master's–3 years full-time; doctorate–7 years full-time.
Entrance requirements: For master's, GRE General Test, BS in agriculture; for doctorate, GRE General Test, MS. *Application deadline:* For fall admission, 5/1. Applications are processed on a rolling basis. *Application fee:* $15 ($20 for international students).
Expenses: Tuition, state resident: full-time $1,932. Tuition, nonresident: full-time $3,864. Tuition and fees vary according to course load.
Financial aid: In 1999–00, 1 fellowship with tuition reimbursement (averaging $18,000 per year), 9 research assistantships with tuition reimbursements (averaging $9,000 per year) were awarded; career-related internships or fieldwork and Federal Work-Study also available. Financial aid application deadline: 4/1.
Faculty research: Plant breeding, cytogenetics, crop production, soil chemistry and fertility, remote sensing. *Total annual research expenditures:* $113,000.
Dr. Govind Sharma, Chair, 256-851-5462.

■ ARIZONA STATE UNIVERSITY

Graduate College, College of Liberal Arts and Sciences, Department of Plant Biology, Tempe, AZ 85287
AWARDS MNS, MS, PhD.

Faculty: 21 full-time (5 women), 2 part-time/adjunct (1 woman).
Students: 47 full-time (26 women), 17 part-time (11 women); includes 5 minority (2 African Americans, 3 Hispanic Americans), 7 international. Average age 31. *30 applicants, 43% accepted.* In 1999, 9 master's, 5 doctorates awarded.
Degree requirements: For master's, thesis required, foreign language not required; for doctorate, one foreign language, dissertation, oral and written exams required.
Entrance requirements: For master's and doctorate, GRE. *Application fee:* $35.
Expenses: Tuition, state resident: part-time $115 per credit hour. Tuition, nonresident: part-time $389 per credit hour. Required fees: $18 per semester. Tuition and fees vary according to program.
Faculty research: Effects of air pollution on plants, ecological changes by man, floristic studies of Arizona and the Southwest.
Dr. J. Kenneth Hoober, Chair, 480-965-3414, *E-mail:* botany@asuvm.inre.asu.edu.
Application contact: Graduate Secretary, 480-965-7730, *E-mail:* bchafe@asu.edu.

■ AUBURN UNIVERSITY

Graduate School, College of Sciences and Mathematics, Department of Biological Sciences, Auburn, Auburn University, AL 36849-0002
AWARDS Botany (MS, PhD); microbiology (MS, PhD); zoology (MS, PhD).

Faculty: 31 full-time (6 women).
Students: 38 full-time (23 women), 35 part-time (18 women); includes 6 minority (5 African Americans, 1 Hispanic American), 11 international. *66 applicants, 48% accepted.* In 1999, 12 master's, 1 doctorate awarded.
Entrance requirements: For master's and doctorate, GRE General Test, TOEFL. *Application deadline:* For fall admission, 7/7; for spring admission, 11/24. Electronic applications accepted.
Expenses: Tuition, state resident: full-time $2,895; part-time $80 per credit hour. Tuition, nonresident: full-time $8,685; part-time $240 per credit hour.
Financial aid: Research assistantships, teaching assistantships available.

Dr. Alfred E. Brown, Interim Chair, 334-844-4830, *Fax:* 334-844-1645.
Find an in-depth description at www.petersons.com/graduate.

■ BOSTON UNIVERSITY

Graduate School of Arts and Sciences, Department of Biology, Boston, MA 02215

AWARDS Botany (MA, PhD); cell and molecular biology (MA, PhD); cell biology (MA, PhD); ecology (PhD); ecology, behavior, and evolution (MA, PhD); ecology/physiology, endocrinology and reproduction (MA); marine biology (MA, PhD); molecular biology, cell biology and biochemistry (MA, PhD); neurobiology, neuroendocrinology and reproduction (MA, PhD); physiology, endocrinology, and neurobiology (MA, PhD); zoology (MA, PhD). Part-time programs available.

Faculty: 41 full-time (8 women).
Students: 131 full-time (74 women), 11 part-time (7 women); includes 10 minority (7 Asian Americans or Pacific Islanders, 3 Hispanic Americans), 33 international. Average age 27. *238 applicants, 39% accepted.* In 1999, 61 master's, 45 doctorates awarded. Terminal master's awarded for partial completion of doctoral program.
Degree requirements: For master's, one foreign language, thesis not required; for doctorate, one foreign language, dissertation, qualifying exam required. *Average time to degree:* Master's–1 year full-time, 3 years part-time; doctorate–5.75 years full-time.
Entrance requirements: For master's and doctorate, GRE General Test, GRE Subject Test, TOEFL. *Application deadline:* For fall admission, 1/1 (priority date); for spring admission, 11/1. *Application fee:* $50.
Expenses: Tuition: Full-time $23,770; part-time $743 per credit. Required fees: $220. Tuition and fees vary according to class time, course level, campus/location and program.
Financial aid: In 1999–00, 82 students received aid, including 1 fellowship with full tuition reimbursement available (averaging $12,000 per year), 28 research assistantships with full tuition reimbursements available (averaging $11,500 per year), 43 teaching assistantships with full tuition reimbursements available (averaging $11,500 per year); Federal Work-Study, grants, institutionally sponsored loans, scholarships, and traineeships also available. Financial aid application

Boston University (continued)
deadline: 1/15; financial aid applicants required to submit FAFSA.
Faculty research: Marine science, endocrinology, behavior. *Total annual research expenditures:* $5 million.
Geoffrey M. Cooper, Chairman, 617-353-2432, *Fax:* 617-353-6340, *E-mail:* gmcooper@bu.edu.
Application contact: Yolanta Kovalko, Senior Staff Assistant, 617-353-2432, *Fax:* 617-353-6340, *E-mail:* yolanta@bu.edu.
Find an in-depth description at www.petersons.com/graduate.

■ BRIGHAM YOUNG UNIVERSITY

Graduate Studies, College of Biological and Agricultural Sciences, Department of Botany and Range Science, Provo, UT 84602-1001

AWARDS Biological science education (MS); botany (MS, PhD); range science (MS); wildlife and range resources (MS, PhD).

Faculty: 17 full-time (1 woman).
Students: 32 full-time (12 women); includes 1 minority (Asian American or Pacific Islander), 1 international. Average age 28. *21 applicants, 67% accepted.* In 1999, 4 master's, 1 doctorate awarded.
Degree requirements: For master's and doctorate, thesis/dissertation required.
Entrance requirements: For master's, GRE General Test, minimum GPA of 3.2 during previous 2 years; for doctorate, GRE General Test. *Application deadline:* For fall admission, 2/15; for winter admission, 9/15. Applications are processed on a rolling basis. *Application fee:* $30. Electronic applications accepted.
Expenses: Tuition: Full-time $3,330; part-time $185 per credit hour. Tuition and fees vary according to program and student's religious affiliation.
Financial aid: In 1999–00, 20 students received aid; fellowships with partial tuition reimbursements available, research assistantships with partial tuition reimbursements available, teaching assistantships with partial tuition reimbursements available, career-related internships or fieldwork, institutionally sponsored loans, and tuition waivers (partial) available. Financial aid application deadline: 2/15.
Faculty research: Plant classification (seed plants, lichens, and diatoms); ethnobotany; plant physiology; ecology (wildlife and plant); genetics.
Dr. Bruce A. Roundy, Chair, 801-378-2582, *Fax:* 801-378-7499.
Application contact: Rex G. Cates, Graduate Coordinator, 801-378-4281, *Fax:* 801-378-7499.

■ CALIFORNIA STATE UNIVERSITY, CHICO

Graduate School, College of Natural Sciences, Department of Biological Sciences, Program in Botany, Chico, CA 95929-0722

AWARDS MS.

Degree requirements: For master's, thesis, seminar presentation required, foreign language not required.
Entrance requirements: For master's, GRE General Test.
Expenses: Tuition, nonresident: part-time $246 per credit. Required fees: $2,108; $1,442 per year.

■ CALIFORNIA STATE UNIVERSITY, FRESNO

Division of Graduate Studies, College of Agricultural Sciences and Technology, Department of Plant Science, Fresno, CA 93740

AWARDS MS. Part-time programs available.

Faculty: 9 full-time (1 woman).
Students: 5 full-time (1 woman), 11 part-time (4 women); includes 3 minority (all Hispanic Americans), 2 international. Average age 31. *5 applicants, 60% accepted.* In 1999, 3 degrees awarded.
Degree requirements: For master's, thesis required, foreign language not required. *Average time to degree:* Master's–3.5 years full-time.
Entrance requirements: For master's, GRE General Test, TOEFL, minimum GPA of 2.50. *Application deadline:* For fall admission, 6/1 (priority date); for spring admission, 11/1. Applications are processed on a rolling basis. *Application fee:* $55. Electronic applications accepted.
Expenses: Tuition, nonresident: part-time $246 per unit. Required fees: $1,906; $620 per semester.
Financial aid: In 1999–00, 4 students received aid; fellowships, career-related internships or fieldwork, Federal Work-Study, and scholarships available. Financial aid application deadline: 3/1; financial aid applicants required to submit FAFSA.
Faculty research: Crop patterns, small watershed management, postharvest techniques, larval control, electronic monitoring of feedlot cattle.
Dr. Maylon Hile, Chair, 559-278-2861, *Fax:* 559-278-7413.
Application contact: Dr. Arthur Olney, Graduate Program Coordinator, 559-278-2861, *Fax:* 559-278-7413, *E-mail:* arthur_olney@csufresno.edu.

■ CALIFORNIA STATE UNIVERSITY, FULLERTON

Graduate Studies, College of Natural Science and Mathematics, Department of Biological Science, Fullerton, CA 92834-9480

AWARDS Biological science (MA); botany (MA); microbiology (MA). Part-time programs available.

Faculty: 23 full-time (7 women), 41 part-time/adjunct.
Students: 4 full-time (all women), 60 part-time (35 women); includes 20 minority (1 African American, 8 Asian Americans or Pacific Islanders, 10 Hispanic Americans, 1 Native American), 3 international. Average age 29. *51 applicants, 45% accepted.* In 1999, 13 degrees awarded.
Degree requirements: For master's, thesis required, foreign language not required.
Entrance requirements: For master's, DAT, GRE General Test and GRE Subject Test, or MCAT, minimum GPA of 3.0 in biology. *Application fee:* $55.
Expenses: Tuition, nonresident: part-time $264 per unit. Required fees: $1,887; $629 per year.
Financial aid: Teaching assistantships, career-related internships or fieldwork, Federal Work-Study, grants, and institutionally sponsored loans available. Aid available to part-time students. Financial aid application deadline: 3/1.
Faculty research: Glycosidase release and the block to polyspermy in ascidian eggs.
Dr. Eugene Jones, Chair, 714-278-3614.
Application contact: Dr. Michael Horn, Adviser, 714-278-3707.

■ CLAREMONT GRADUATE UNIVERSITY

Graduate Programs, Program in Botany, Claremont, CA 91711-6160

AWARDS Systematics and evolution of higher plants (MA, PhD). Part-time programs available.

Faculty: 4 full-time (1 woman).
Students: 2 full-time (1 woman), 7 part-time (4 women), 3 international. Average age 31. In 1999, 1 doctorate awarded. Terminal master's awarded for partial completion of doctoral program.
Degree requirements: For master's, thesis required; for doctorate, dissertation, 2 research tools required.
Entrance requirements: For master's and doctorate, GRE General Test. *Application deadline:* For fall admission, 2/15 (priority date). Applications are processed on a rolling basis. *Application fee:* $40. Electronic applications accepted.
Expenses: Tuition: Full-time $20,950; part-time $913 per unit. Required fees:

$65 per semester. Tuition and fees vary according to program.

Financial aid: Fellowships, research assistantships, Federal Work-Study, institutionally sponsored loans, and tuition waivers (full) available. Aid available to part-time students. Financial aid application deadline: 2/15; financial aid applicants required to submit FAFSA.

Roy L. Taylor, Chair, 909-625-8767 Ext. 220, *Fax:* 909-626-3489.

Application contact: Ann Joslin, Secretary, 909-625-8767 Ext. 251, *Fax:* 909-621-8390, *E-mail:* botany@cgu.edu.

■ CLEMSON UNIVERSITY

Graduate School, College of Agriculture, Forestry and Life Sciences, School of Animal, Biomedical and Biological Sciences, Department of Biological Sciences, Program in Botany, Clemson, SC 29634

AWARDS MS.

Students: 7 full-time (5 women), 3 part-time (2 women). *5 applicants, 60% accepted.* In 1999, 2 degrees awarded.

Degree requirements: For master's, thesis required, foreign language not required.

Entrance requirements: For master's, GRE General Test, TOEFL, bachelor's degree in biological science or chemistry. *Application deadline:* For fall admission, 6/1. *Application fee:* $40.

Expenses: Tuition, state resident: full-time $3,480; part-time $174 per credit hour. Tuition, nonresident: full-time $9,256; part-time $388 per credit hour. Required fees: $5 per term. Full-time tuition and fees vary according to course level, course load and campus/location.

Financial aid: Teaching assistantships available. Financial aid application deadline: 3/15; financial aid applicants required to submit FAFSA.

Faculty research: Systematics, aquatic botany, plant ecology, plant-fungus interactions, plant developmental genetics.

Dr. Thomas McInnis, Coordinator of Graduate Studies, 864-656-3587, *Fax:* 864-656-0435, *E-mail:* botany@clemson.edu.

■ COLORADO STATE UNIVERSITY

Graduate School, College of Agricultural Sciences, Department of Bioagricultural Sciences and Pest Management, Plant Pathology and Weed Science, Fort Collins, CO 80523-0015

AWARDS Plant pathology (MS, PhD); weed science (MS, PhD).

Faculty: 10 full-time (1 woman), 2 part-time/adjunct (0 women).

Students: 12 full-time (4 women), 9 part-time (5 women), 3 international. Average age 31. *24 applicants, 29% accepted.* In 1999, 5 master's, 3 doctorates awarded.

Degree requirements: For master's, thesis required (for some programs), foreign language not required; for doctorate, dissertation required, foreign language not required.

Entrance requirements: For master's and doctorate, GRE General Test, TOEFL, minimum GPA of 3.0. *Application deadline:* For fall admission, 2/1 (priority date). Applications are processed on a rolling basis. *Application fee:* $30. Electronic applications accepted.

Expenses: Tuition, state resident: full-time $2,694; part-time $150 per credit. Tuition, nonresident: full-time $10,460; part-time $581 per credit. Required fees: $32 per semester. Tuition and fees vary according to program.

Financial aid: In 1999–00, 10 research assistantships, 2 teaching assistantships were awarded; fellowships, traineeships also available.

Faculty research: Biological control of plant pathogens and weeds, integrated pest management, weed ecology/biology, seed physiology/technology, molecular biology of plant stress. *Total annual research expenditures:* $2 million.

Thomas O. Holtzer, Head, 970-491-5261, *Fax:* 970-491-3862, *E-mail:* tholtzer@lamar.agsci.colostate.edu.

■ COLORADO STATE UNIVERSITY

Graduate School, College of Natural Sciences, Department of Biology, Program in Botany, Fort Collins, CO 80523-0015

AWARDS MS, PhD.

Faculty: 10 full-time (4 women).

Students: 10 full-time (3 women), 5 part-time (2 women); includes 1 minority (Hispanic American), 2 international. Average age 28. *29 applicants, 3% accepted.* In 1999, 2 master's, 1 doctorate awarded.

Degree requirements: For master's, thesis or alternative required, foreign language not required; for doctorate, dissertation required, foreign language not required.

Entrance requirements: For master's and doctorate, GRE General Test, GRE Subject Test, TOEFL, minimum GPA of 3.0. *Application deadline:* For fall admission, 2/1 (priority date); for spring admission, 11/1. Applications are processed on a rolling basis. *Application fee:* $30. Electronic applications accepted.

Expenses: Tuition, state resident: full-time $2,694; part-time $150 per credit. Tuition, nonresident: full-time $10,460; part-time $581 per credit. Required fees: $32 per semester. Tuition and fees vary according to program.

Financial aid: In 1999–00, 2 research assistantships, 9 teaching assistantships were awarded; fellowships, career-related internships or fieldwork, Federal Work-Study, and traineeships also available.

Faculty research: Tissue culture, disturbed land ecology, *mycorrhizae synaptonemal* complex, algae, molecular genetics.

Application contact: Tina Sund, Graduate Coordinator, 970-491-1923, *Fax:* 970-491-0649, *E-mail:* tsund@lamar.colostate.edu.

■ CONNECTICUT COLLEGE

Graduate School, Department of Botany, New London, CT 06320-4196

AWARDS MA, MAT. Part-time programs available.

Degree requirements: For master's, thesis required, foreign language not required.

Entrance requirements: For master's, GRE or MAT.

Faculty research: Tidal marsh ecology, upland vegetation dynamics, plant development, halophyte physiology.

■ CORNELL UNIVERSITY

Graduate School, Graduate Fields of Agriculture and Life Sciences, Field of Plant Biology, Ithaca, NY 14853-0001

AWARDS Paleobotany (PhD); plant cell biology (PhD); plant ecology (PhD); plant molecular biology (PhD); plant morphology, anatomy and biomechanics (PhD); plant physiology (PhD); systematic botany (PhD).

Faculty: 38 full-time.

Students: 51 full-time (26 women); includes 10 minority (1 African American, 5 Asian Americans or Pacific Islanders, 4 Hispanic Americans), 16 international. *94 applicants, 24% accepted.* In 1999, 11 doctorates awarded.

Degree requirements: For doctorate, dissertation required, foreign language not required.

Entrance requirements: For doctorate, GRE General Test, GRE Subject Test (biology) (recommended), TOEFL. *Application deadline:* For fall admission, 1/15. *Application fee:* $65. Electronic applications accepted.

Expenses: Tuition: Full-time $12,100.

Financial aid: In 1999–00, 47 students received aid, including 12 fellowships with full tuition reimbursements available, 19 research assistantships with full tuition reimbursements available, 16 teaching

Cornell University (continued)
assistantships with full tuition reimbursements available; institutionally sponsored loans, scholarships, tuition waivers (full and partial), and unspecified assistantships also available. Financial aid applicants required to submit FAFSA.

Faculty research: Plant cell biology/cytology, plant molecular biology plant morphology/anatomy/biomechanics, plant physiology, systematic botany, paleobotany, plant ecology, ethnobotany, plant biochemistry, photosynthesis.

Application contact: Graduate Field Assistant, 607-255-8507, *E-mail:* plbio@cornell.edu.

■ **CORNELL UNIVERSITY**

Graduate School, Graduate Fields of Agriculture and Life Sciences, Field of Plant Breeding, Ithaca, NY 14853-0001

AWARDS Plant breeding (MPS, MS, PhD); plant genetics (MPS, MS, PhD).

Faculty: 23 full-time.
Students: 36 full-time (14 women); includes 1 minority (Asian American or Pacific Islander), 20 international. *31 applicants, 26% accepted.* In 1999, 2 master's, 9 doctorates awarded. Terminal master's awarded for partial completion of doctoral program.
Degree requirements: For master's, thesis (MS), project paper (MPS) required; for doctorate, dissertation required, foreign language not required.
Entrance requirements: For master's and doctorate, GRE General Test, TOEFL. *Application deadline:* For fall admission, 1/15. *Application fee:* $65. Electronic applications accepted.
Expenses: Tuition: Full-time $12,100.
Financial aid: In 1999–00, 26 students received aid, including 8 fellowships with full tuition reimbursements available, 16 research assistantships with full tuition reimbursements available, 2 teaching assistantships with full tuition reimbursements available; institutionally sponsored loans, scholarships, tuition waivers (full and partial), and unspecified assistantships also available. Financial aid applicants required to submit FAFSA.
Faculty research: Disease, insect, and stress resistance; molecular biology and biotechnology; crop genetics and breeding; international crop improvement; germ plasm conservation and improvement.
Application contact: Graduate Field Assistant, 607-255-3101, *E-mail:* clm15@cornell.edu.

■ **CORNELL UNIVERSITY**

Graduate School, Graduate Fields of Agriculture and Life Sciences, Field of Plant Protection, Ithaca, NY 14853-0001

AWARDS MPS.

Faculty: 25 full-time.
Students: 1 (woman) full-time. *1 applicant, 0% accepted.* In 1999, 1 degree awarded.
Degree requirements: For master's, internship, final exam required.
Entrance requirements: For master's, GRE General Test, TOEFL. *Application deadline:* For fall admission, 4/1. *Application fee:* $65. Electronic applications accepted.
Expenses: Tuition: Full-time $12,100.
Financial aid: In 1999–00, 1 student received aid, including 1 research assistantship with full tuition reimbursement available; fellowships with full tuition reimbursements available, teaching assistantships with full tuition reimbursements available, institutionally sponsored loans, scholarships, tuition waivers (full and partial), and unspecified assistantships also available. Financial aid applicants required to submit FAFSA.
Faculty research: Fruit and vegetable crop insects and diseases, plant disease epidemiology, systems modeling, biological control, plant protection economics.
Application contact: Graduate Field Assistant, 607-255-3259, *E-mail:* plprotection@cornell.edu.

■ **CORNELL UNIVERSITY**

Graduate School, Graduate Fields of Agriculture and Life Sciences, Field of Pomology, Ithaca, NY 14853-0001

AWARDS MPS, MS, PhD.

Faculty: 13 full-time.
Students: 10 full-time (5 women), 5 international. *6 applicants, 50% accepted.* In 1999, 4 doctorates awarded.
Degree requirements: For master's, thesis (MS), project paper (MPS) required; for doctorate, dissertation required, foreign language not required.
Entrance requirements: For master's and doctorate, GRE General Test, TOEFL, interview (recommended). *Application deadline:* For fall admission, 1/15; for spring admission, 8/15. *Application fee:* $65. Electronic applications accepted.
Expenses: Tuition: Full-time $12,100.
Financial aid: In 1999–00, 8 students received aid, including 6 research assistantships with full tuition reimbursements available, 2 teaching assistantships with full tuition reimbursements available; fellowships with full tuition reimbursements available, institutionally sponsored loans, scholarships, tuition waivers (full and partial), and unspecified assistantships also

available. Financial aid applicants required to submit FAFSA.
Faculty research: Fruit and vegetable crop insects and diseases, plant disease epidemiology, systems modeling, biological control, plant protection economics.
Application contact: Graduate Field Assistant, 607-255-4568, *Fax:* 607-255-0599, *E-mail:* cah8@cornell.edu.

■ **CORNELL UNIVERSITY**

Graduate School, Graduate Fields of Agriculture and Life Sciences, Field of Vegetable Crops, Ithaca, NY 14853-0001

AWARDS MPS, MS, PhD.

Faculty: 17 full-time.
Students: 11 full-time (3 women), 6 international. *10 applicants, 30% accepted.* In 1999, 1 master's, 2 doctorates awarded. Terminal master's awarded for partial completion of doctoral program.
Degree requirements: For master's, thesis (MS), project paper (MPS) required; for doctorate, dissertation, teaching experience required, foreign language not required.
Entrance requirements: For master's, TOEFL; for doctorate, GRE General Test, TOEFL. *Application deadline:* Applications are processed on a rolling basis. *Application fee:* $65. Electronic applications accepted.
Expenses: Tuition: Full-time $12,100.
Financial aid: In 1999–00, 9 students received aid, including 2 fellowships with full tuition reimbursements available, 5 research assistantships with full tuition reimbursements available, 2 teaching assistantships with full tuition reimbursements available; institutionally sponsored loans and tuition waivers (full and partial) also available. Financial aid applicants required to submit FAFSA.
Faculty research: Vegetable nutrition and physiology, post-harvest physiology and storage, new technologies including genetic, sustainable vegetable production, weed management and IPM.
Application contact: Graduate Field Assistant, 607-255-4568, *E-mail:* cah8@cornell.edu.

■ **EASTERN ILLINOIS UNIVERSITY**

Graduate School, College of Sciences, Program in Biological Sciences, Charleston, IL 61920-3099

AWARDS Biological sciences (MS); botany (MS); environmental biology (MS); zoology (MS).

Degree requirements: For master's, exam required, foreign language and thesis not required.

■ EMPORIA STATE UNIVERSITY

School of Graduate Studies, College of Liberal Arts and Sciences, Division of Biological Sciences, Emporia, KS 66801-5087

AWARDS Botany (MS); environmental biology (MS); general biology (MS); microbial and cellular biology (MS); zoology (MS). Part-time programs available.

Faculty: 15 full-time (3 women), 4 part-time/adjunct (0 women).
Students: 20 full-time (10 women), 8 part-time (3 women), 2 international. *8 applicants, 75% accepted.* In 1999, 12 degrees awarded.
Degree requirements: For master's, comprehensive exam or thesis required.
Entrance requirements: For master's, TOEFL, written exam. *Application deadline:* For fall admission, 8/15 (priority date). Applications are processed on a rolling basis. *Application fee:* $30 ($75 for international students). Electronic applications accepted.
Expenses: Tuition, state resident: full-time $2,410; part-time $108 per credit hour. Tuition, nonresident: full-time $6,212; part-time $266 per credit hour.
Financial aid: In 1999–00, 2 fellowships (averaging $1,396 per year), 1 research assistantship (averaging $5,390 per year), 11 teaching assistantships with full tuition reimbursements (averaging $5,047 per year) were awarded; career-related internships or fieldwork, Federal Work-Study, and institutionally sponsored loans also available. Financial aid application deadline: 3/15; financial aid applicants required to submit FAFSA.
Faculty research: Fisheries, range, and wildlife management; aquatic, plant, grassland, vertebrate, and invertebrate ecology; mammalian and plant systematics, taxonomy, and evolution; immunology, virology, and molecular biology.
Dr. Marshall Sundberg, Chair, 316-341-5311, *Fax:* 316-341-5607, *E-mail:* sundberm@emporia.edu.

■ FLORIDA STATE UNIVERSITY

Graduate Studies, College of Arts and Sciences, Department of Biological Science, Program in Plant Sciences, Tallahassee, FL 32306

AWARDS MS, PhD.

Faculty: 7 full-time (2 women).
Students: 12 full-time (6 women); includes 2 minority (1 African American, 1 Asian American or Pacific Islander), 6 international.
Degree requirements: For master's and doctorate, thesis/dissertation, teaching experience required.

Entrance requirements: For master's, GRE General Test, TOEFL; for doctorate, GRE General Test, GRE Subject Test, TOEFL. *Application deadline:* For fall admission, 1/15; for spring admission, 10/15. *Application fee:* $20.
Expenses: Tuition, state resident: full-time $3,504; part-time $146 per credit hour. Tuition, nonresident: full-time $12,162; part-time $507 per credit hour. Tuition and fees vary according to program.
Financial aid: In 1999–00, fellowships with full tuition reimbursements (averaging $13,740 per year), research assistantships with full tuition reimbursements (averaging $13,740 per year), teaching assistantships with full tuition reimbursements (averaging $13,740 per year) were awarded. Financial aid application deadline: 1/15; financial aid applicants required to submit FAFSA.
Faculty research: Photosynthetic mechanisms, cell development, physiology, ecology, cytogenetics, metabolism.
Dr. Thomas C. S. Keller, Associate Professor and Associate Chairman, 850-644-3023, *Fax:* 850-644-9829.
Application contact: Judy Bowers, Coordinator, Graduate Affairs, 850-644-3023, *Fax:* 850-644-9829, *E-mail:* bowers@bio.fsu.edu.

■ ILLINOIS STATE UNIVERSITY

Graduate School, College of Arts and Sciences, Department of Biological Sciences, Normal, IL 61790-2200

AWARDS Biological sciences (MS); biology (PhD); botany (PhD); ecology (PhD); genetics (PhD); microbiology (PhD); physiology (PhD); zoology (PhD). Part-time programs available.

Faculty: 24 full-time (4 women).
Students: 57 full-time (25 women), 27 part-time (10 women); includes 6 minority (1 African American, 3 Asian Americans or Pacific Islanders, 2 Hispanic Americans), 13 international. *58 applicants, 59% accepted.* In 1999, 12 master's, 3 doctorates awarded.
Degree requirements: For master's, thesis or alternative required; for doctorate, variable foreign language requirement (computer language can substitute for one), dissertation, 2 terms of residency required.
Entrance requirements: For master's, GRE General Test, minimum GPA of 2.6 in last 60 hours; for doctorate, GRE General Test. *Application deadline:* Applications are processed on a rolling basis. *Application fee:* $0.
Expenses: Tuition, state resident: full-time $2,526; part-time $105 per credit hour. Tuition, nonresident: full-time $7,578; part-time $316 per credit hour. Required fees: $1,082; $38 per credit hour. Tuition and fees vary according to course load and program.

Financial aid: In 1999–00, 8 research assistantships, 63 teaching assistantships were awarded; Federal Work-Study, tuition waivers (full), and unspecified assistantships also available. Financial aid application deadline: 4/1.
Faculty research: Phenotypic plasticity in reproduction: molecular mechanisms, physiological control, adaptive significance; molecular stress physiology of *Listeria monocytogenes*; enzymology of eggshell formation in *Schistosoma mansoni*, compensatory adaptation in the rat midbrain after neurodegeneration, analysis of staphylococcal virulence germs. *Total annual research expenditures:* $586,648.
Dr. Hou Cheung, Chairperson, 309-438-3669.
Application contact: Derek A. McCracken, Graduate Adviser, 309-438-3664.

Find an in-depth description at www.petersons.com/graduate.

■ INDIANA UNIVERSITY BLOOMINGTON

Graduate School, College of Arts and Sciences, Department of Biology, Programs in Plant Sciences, Molecular and Organismal Biology, Bloomington, IN 47405

AWARDS MA, PhD. PhD offered through the University Graduate School. Part-time programs available.

Faculty: 18 full-time.
Students: 14 full-time (6 women); includes 2 minority (1 Asian American or Pacific Islander, 1 Hispanic American), 1 international. Terminal master's awarded for partial completion of doctoral program.
Degree requirements: For master's, thesis or alternative required; for doctorate, dissertation required.
Entrance requirements: For master's and doctorate, GRE General Test, TOEFL. *Application deadline:* For fall admission, 1/5 (priority date); for spring admission, 9/1 (priority date). Applications are processed on a rolling basis. *Application fee:* $45. Electronic applications accepted.
Expenses: Tuition, state resident: full-time $3,853; part-time $161 per credit hour. Tuition, nonresident: full-time $11,226; part-time $468 per credit hour. Required fees: $360 per year. Tuition and fees vary according to course load and program.
Financial aid: Fellowships with tuition reimbursements, research assistantships with tuition reimbursements, teaching assistantships with tuition reimbursements available. Financial aid application deadline: 1/15.
Faculty research: Molecular biology, physiology, systematics, genetics and evolutionary biology.

Indiana University Bloomington (continued)
Dr. Loren Rieseberg, Head, 812-855-7614, *E-mail:* lriesebe@bio.indiana.edu.
Application contact: Gretchen Clearwater, Advisor for Graduate Affairs, 812-855-1861, *Fax:* 812-855-6705, *E-mail:* biograd@bio.indiana.edu.

Find an in-depth description at www.petersons.com/graduate.

■ IOWA STATE UNIVERSITY OF SCIENCE AND TECHNOLOGY

Graduate College, College of Liberal Arts and Sciences, Department of Botany, Ames, IA 50011

AWARDS MS, PhD.

Faculty: 17 full-time.
Students: 38 full-time (18 women), 10 part-time (6 women); includes 1 minority (Asian American or Pacific Islander), 11 international. *28 applicants, 21% accepted.*
Degree requirements: For master's, thesis or alternative required; for doctorate, dissertation required.
Entrance requirements: For master's and doctorate, GRE General Test, TOEFL. *Application fee:* $20 ($50 for international students). Electronic applications accepted.
Expenses: Tuition, state resident: full-time $3,308. Tuition, nonresident: full-time $9,744. Part-time tuition and fees vary according to course load, campus/location and program.
Financial aid: In 1999–00, 23 research assistantships with partial tuition reimbursements (averaging $11,339 per year), 18 teaching assistantships with partial tuition reimbursements (averaging $11,646 per year) were awarded; fellowships, scholarships also available.
Faculty research: Aquatic and wetland ecology, cytology, ecology, physiology and molecular biology, systematics and evolution.
Dr. David J. Oliver, Chair, 515-294-3522, *Fax:* 515-294-1337, *E-mail:* bot@iastate.edu.
Application contact: Dr. Robert S. Wallace, Director of Graduate Education, 515-294-0367, *E-mail:* bot@iastate.edu.

■ KENT STATE UNIVERSITY

College of Arts and Sciences, Department of Biological Sciences, Program in Botany, Kent, OH 44242-0001

AWARDS MA, MS, PhD.

Degree requirements: For master's and doctorate, thesis/dissertation required, foreign language not required.
Entrance requirements: For master's, GRE General Test, minimum GPA of 2.75; for doctorate, GRE General Test,

minimum GPA of 3.0. *Application deadline:* For fall admission, 7/12; for spring admission, 11/29. Applications are processed on a rolling basis. *Application fee:* $30.
Expenses: Tuition, state resident: full-time $5,334; part-time $243 per hour. Tuition, nonresident: full-time $10,238; part-time $466 per hour.
Financial aid: Fellowships, research assistantships, teaching assistantships, Federal Work-Study, institutionally sponsored loans, and tuition waivers (full) available. Financial aid application deadline: 2/1.
Application contact: Dr. John R. D. Stalvey, Coordinator of Graduate Studies, 330-672-2819.

■ LEHMAN COLLEGE OF THE CITY UNIVERSITY OF NEW YORK

Division of Natural and Social Sciences, Department of Biological Sciences, Program in Plant Sciences, Bronx, NY 10468-1589

AWARDS PhD.

Degree requirements: For doctorate, dissertation required.
Entrance requirements: For doctorate, GRE General Test. *Application deadline:* For fall admission, 4/1; for spring admission, 11/1. Applications are processed on a rolling basis. *Application fee:* $40.
Expenses: Tuition, state resident: full-time $4,350; part-time $185 per credit. Tuition, nonresident: full-time $7,600; part-time $320 per credit.
Financial aid: Fellowships, research assistantships, teaching assistantships, tuition waivers (full and partial) available. Financial aid application deadline: 5/15.
Jack Valdovinos, Adviser, 718-960-8235.

■ LOUISIANA STATE UNIVERSITY AND AGRICULTURAL AND MECHANICAL COLLEGE

Graduate School, College of Basic Sciences, Department of Biological Sciences, Baton Rouge, LA 70803

AWARDS Biochemistry (MS, PhD); microbiology (MS, PhD); plant biology (MS, PhD); zoology (MS, PhD). Part-time programs available.

Faculty: 79 full-time (7 women), 2 part-time/adjunct (0 women).
Students: 91 full-time (44 women), 22 part-time (10 women); includes 9 minority (3 African Americans, 1 Asian American or Pacific Islander, 5 Hispanic Americans), 29 international. Average age 28. *98 applicants, 28% accepted.* In 1999, 13 master's, 7 doctorates awarded. Terminal master's awarded for partial completion of doctoral program.

Degree requirements: For master's, foreign language not required; for doctorate, dissertation required.
Entrance requirements: For master's and doctorate, GRE General Test, minimum GPA of 3.0. *Application deadline:* Applications are processed on a rolling basis. *Application fee:* $25.
Expenses: Tuition, state resident: full-time $2,881. Tuition, nonresident: full-time $7,081. Part-time tuition and fees vary according to course load and program.
Financial aid: In 1999–00, 12 fellowships, 20 research assistantships with partial tuition reimbursements, 62 teaching assistantships with partial tuition reimbursements were awarded; Federal Work-Study, institutionally sponsored loans, and unspecified assistantships also available. Aid available to part-time students. *Total annual research expenditures:* $2.2 million.
Dr. Harold Silverman, Chairman, 225-388-2601, *Fax:* 225-388-2597, *E-mail:* cxsiv@unix1.sncc.lsu.edu.

Find an in-depth description at www.petersons.com/graduate.

■ MIAMI UNIVERSITY

Graduate School, College of Arts and Sciences, Department of Botany, Oxford, OH 45056

AWARDS Biological sciences (MAT); botany (MA, MS, PhD). Part-time programs available.

Faculty: 12 full-time (3 women).
Students: 12 full-time (7 women), 25 part-time (10 women), 7 international. *27 applicants, 96% accepted.* In 1999, 20 master's, 3 doctorates awarded.
Degree requirements: For master's, thesis (for some programs), final exam required; for doctorate, dissertation, comprehensive and final exams required.
Entrance requirements: For master's, GRE General Test, GRE Subject Test, minimum undergraduate GPA of 3.0 during previous 2 years or 2.75 overall; for doctorate, GRE General Test, GRE Subject Test, minimum GPA of 2.75 (undergraduate) or 3.0 (graduate). *Application deadline:* For fall admission, 3/1. *Application fee:* $35.
Expenses: Tuition, state resident: part-time $260 per hour. Tuition, nonresident: full-time $3,125; part-time $538 per hour. International tuition: $6,452 full-time. Required fees: $18 per semester. Tuition and fees vary according to campus/location.
Financial aid: In 1999–00, 22 fellowships, 3 research assistantships, 11 teaching assistantships were awarded; Federal Work-Study and tuition waivers (full) also available. Financial aid application deadline: 3/1.

Dr. David Francko, Chair, 513-529-4200. **Application contact:** Dr. James Hickey, Director of Graduate Studies, 513-529-4200, *Fax:* 513-529-4243, *E-mail:* botany@muohio.edu.

Find an in-depth description at www.petersons.com/graduate.

■ MICHIGAN STATE UNIVERSITY

Graduate School, College of Natural Science, Department of Botany and Plant Pathology, East Lansing, MI 48824

AWARDS Botany and plant pathology (MS, PhD); botany-environmental toxicology (PhD).

Faculty: 26.
Students: 42 full-time (15 women), 19 part-time (12 women); includes 2 minority (1 African American, 1 Hispanic American), 13 international. Average age 30. *30 applicants, 37% accepted.* In 1999, 5 master's, 4 doctorates awarded. Terminal master's awarded for partial completion of doctoral program.
Degree requirements: For master's, thesis required (for some programs), foreign language not required; for doctorate, dissertation required, foreign language not required.
Entrance requirements: For master's, GRE, TOEFL; for doctorate, GRE, TOEFL, master's degree, minimum GPA of 3.0. *Application deadline:* Applications are processed on a rolling basis. *Application fee:* $30 ($40 for international students).
Expenses: Tuition, state resident: part-time $229 per credit. Tuition, nonresident: part-time $464 per credit. Required fees: $241 per semester. Tuition and fees vary according to course load, degree level and program.
Financial aid: In 1999–00, 23 research assistantships (averaging $10,666 per year), 29 teaching assistantships (averaging $10,468 per year) were awarded; fellowships, institutionally sponsored loans also available. Financial aid application deadline: 1/15; financial aid applicants required to submit FAFSA.
Faculty research: Ecology and systematics, plant physiology and biochemistry, plant genetics and molecular biology, plant anatomy and morphology. *Total annual research expenditures:* $4.2 million.
Dr. R. Hammerschmidt, Acting Chair, 517-355-4683, *Fax:* 517-353-1926.
Application contact: Richard Allison, Director, 517-432-1548, *E-mail:* allison@msu.edu.

Find an in-depth description at www.petersons.com/graduate.

■ MICHIGAN STATE UNIVERSITY

Graduate School, College of Natural Science, Plant Research Laboratory, East Lansing, MI 48824

AWARDS Biochemistry (PhD); botany and plant pathology (PhD); cellular and molecular biology (PhD); genetics (PhD); horticulture (PhD); microbiology (PhD). Offered jointly with the Department of Energy.

Faculty: 10 full-time (2 women).
Students: Average age 28.
Degree requirements: For doctorate, dissertation, comprehensive exam required.
Entrance requirements: For doctorate, GRE General Test, TOEFL, minimum GPA of 3.0. *Application deadline:* For fall admission, 2/1 (priority date); for spring admission, 9/30. *Application fee:* $30 ($40 for international students).
Expenses: All students are awarded annual assistantships of $16,200 and tuition is waived for 6 credits and fees.
Financial aid: In 1999–00, 25 research assistantships with tuition reimbursements (averaging $12,852 per year) were awarded. Financial aid applicants required to submit FAFSA.
Faculty research: Plant-microbe interactions, regulation of plant growth and development, protein targeting, energy transduction, phytohormones. *Total annual research expenditures:* $5.3 million.
Dr. Kenneth Keegstra, Director, 517-353-2270, *Fax:* 517-353-9168.

Find an in-depth description at www.petersons.com/graduate.

■ MISSISSIPPI STATE UNIVERSITY

College of Agriculture and Life Sciences, Department of Plant and Soil Sciences, Mississippi State, MS 39762

AWARDS Agronomy (MS, PhD); horticulture (MS, PhD); weed science (MS, PhD). Part-time programs available.

Students: 57 full-time (10 women), 23 part-time (4 women); includes 1 minority (African American), 20 international. Average age 29. *37 applicants, 51% accepted.* In 1999, 11 master's, 19 doctorates awarded.
Degree requirements: For master's, thesis, comprehensive oral or written exam required, foreign language not required; for doctorate, computer language, dissertation, comprehensive oral or written exam required, foreign language not required.
Entrance requirements: For master's and doctorate, TOEFL. *Application deadline:* For fall admission, 7/1; for spring admission, 11/1. Applications are processed on a rolling basis. *Application fee:* $25 for international students.

Expenses: Tuition, state resident: full-time $3,017; part-time $168 per credit. Tuition, nonresident: full-time $6,119; part-time $340 per credit. Part-time tuition and fees vary according to course load and program.
Financial aid: In 1999–00, 4 research assistantships with full tuition reimbursements (averaging $13,098 per year), 3 teaching assistantships with full tuition reimbursements (averaging $15,424 per year) were awarded; career-related internships or fieldwork, Federal Work-Study, institutionally sponsored loans, and unspecified assistantships also available. Financial aid applicants required to submit FAFSA.
Faculty research: Metabolism, morphology, growth regulators, biotechnology, stress physiology, weed physiology, soil sciences/conservation, breeding genetics. *Total annual research expenditures:* $4 million.
Dr. Richard Mullenax, Head, 662-325-2311, *Fax:* 662-325-8742, *E-mail:* mullenax@pss.msstate.edu.
Application contact: Jerry B. Inmon, Director of Admissions, 662-325-2224, *Fax:* 662-325-7360, *E-mail:* admit@admissions.msstate.edu.

■ MONTANA STATE UNIVERSITY–BOZEMAN

College of Graduate Studies, College of Agriculture, Department of Plant Sciences, Bozeman, MT 59717

AWARDS Plant pathology (MS); plant science (MS, PhD). Part-time programs available.

Students: 10 full-time (6 women), 9 part-time (4 women); includes 1 minority (Asian American or Pacific Islander). Average age 34. *4 applicants, 75% accepted.* In 1999, 5 master's, 5 doctorates awarded.
Degree requirements: For master's and doctorate, thesis/dissertation required, foreign language not required.
Entrance requirements: For master's, GRE General Test, TOEFL; for doctorate, GRE General Test, TOEFL, minimum GPA of 3.0. *Application deadline:* For fall admission, 6/1; for spring admission, 11/1. Applications are processed on a rolling basis. *Application fee:* $50. Electronic applications accepted.
Expenses: Tuition, state resident: full-time $2,674. Tuition, nonresident: full-time $6,986. International tuition: $7,136 full-time. Tuition and fees vary according to course load and program.
Financial aid: In 1999–00, research assistantships with full tuition reimbursements (averaging $15,000 per year); fellowships with full tuition reimbursements, career-related internships or fieldwork, Federal Work-Study, scholarships, and

Montana State University–Bozeman (continued)

tuition waivers (full and partial) also available. Financial aid application deadline: 3/1; financial aid applicants required to submit FAFSA.

Faculty research: Virology, genetics, crop breeding, plant pathology. *Total annual research expenditures:* $3.1 million.

Dr. Norman F. Weedon, Head, 406-994-4832, *Fax:* 406-994-1848, *E-mail:* plantsciences@montana.edu.

■ NORTH CAROLINA AGRICULTURAL AND TECHNICAL STATE UNIVERSITY

Graduate School, School of Agriculture, Department of Natural Resource and Environmental Design, Greensboro, NC 27411

AWARDS Plant science (MS). Part-time and evening/weekend programs available.

Degree requirements: For master's, comprehensive exam, qualifying exam required, thesis optional, foreign language not required.

Entrance requirements: For master's, GRE General Test, minimum GPA of 3.0.

Expenses: Tuition, state resident: full-time $982; part-time $368 per semester. Tuition, nonresident: full-time $8,252; part-time $3,095 per semester. Required fees: $464 per semester.

Faculty research: Soil parameters and compaction of forest site, controlled traffic effects on soil, improving soybean and vegetable crops.

■ NORTH CAROLINA STATE UNIVERSITY

Graduate School, College of Agriculture and Life Sciences, Department of Botany, Raleigh, NC 27695

AWARDS MLS, MS, PhD. Part-time programs available.

Faculty: 18 full-time (6 women), 8 part-time/adjunct (0 women).

Students: 22 full-time (11 women), 7 part-time (4 women); includes 2 minority (1 Asian American or Pacific Islander, 1 Native American), 5 international. Average age 30. *20 applicants, 50% accepted.* In 1999, 2 master's, 3 doctorates awarded. Terminal master's awarded for partial completion of doctoral program.

Degree requirements: For master's, thesis required (for some programs); for doctorate, dissertation required.

Entrance requirements: For master's and doctorate, TOEFL. *Application deadline:* For fall admission, 6/25; for spring admission, 11/25. Applications are processed on a rolling basis. *Application fee:* $45.

Expenses: Tuition, state resident: full-time $1,578. Tuition, nonresident: full-time $10,744. Required fees: $892. Full-time tuition and fees vary according to program.

Financial aid: In 1999–00, 1 fellowship (averaging $3,838 per year), 22 research assistantships (averaging $4,886 per year), 1 teaching assistantship (averaging $5,041 per year) were awarded. Financial aid application deadline: 6/25.

Faculty research: Molecular biology, cell biology, and physiology of development; calcium, the cytoskeleton, and signal transduction; biochemistry of crown gall; physiological ecology of freshwater, marine, and terrestrial plants; community ecology; wetland plants. *Total annual research expenditures:* $6.6 million.

Dr. Margaret Daub, Head, 919-515-3807, *Fax:* 919-515-3436, *E-mail:* margaret_daub@ncsu.edu.

Application contact: Dr. Nina S. Allen, Director of Graduate Programs, 919-515-8382, *Fax:* 919-515-3436, *E-mail:* nina_allen@ncsu.edu.

■ NORTH DAKOTA STATE UNIVERSITY

Graduate Studies and Research, College of Agriculture, Department of Plant Sciences, Fargo, ND 58105

AWARDS Crop and weed sciences (MS); horticulture (MS); natural resources management (MS); plant sciences (PhD). Part-time programs available.

Faculty: 35 full-time (1 woman), 15 part-time/adjunct (2 women).

Students: 29 full-time (8 women), 4 part-time (1 woman), 13 international. Average age 26. *40 applicants, 50% accepted.* In 1999, 3 master's, 4 doctorates awarded.

Degree requirements: For master's and doctorate, thesis/dissertation required, foreign language not required.

Entrance requirements: For master's and doctorate, TOEFL. *Application deadline:* Applications are processed on a rolling basis. *Application fee:* $25. Electronic applications accepted.

Expenses: Tuition, state resident: full-time $3,096; part-time $112 per credit hour. Tuition, nonresident: full-time $7,588; part-time $299 per credit hour. Tuition and fees vary according to course load, campus/location and reciprocity agreements.

Financial aid: In 1999–00, 2 fellowships, 23 research assistantships (averaging $13,200 per year) were awarded; teaching assistantships, Federal Work-Study and institutionally sponsored loans also available. Financial aid application deadline: 4/15.

Faculty research: Biotechnology, weed control science, plant breeding, plant genetics, crop physiology. *Total annual research expenditures:* $880,000.

Dr. Al Schneiter, Chair, 701-231-7971, *Fax:* 701-231-8474, *E-mail:* aschneit@plains.nodak.edu.

■ NORTH DAKOTA STATE UNIVERSITY

Graduate Studies and Research, College of Science and Mathematics, Department of Botany/Biology, Fargo, ND 58105

AWARDS Botany (MS, PhD); cellular and molecular biology (PhD); natural resources management (MS). Part-time programs available.

Faculty: 5 full-time (0 women), 4 part-time/adjunct (0 women).

Students: 6 full-time (4 women), 4 part-time (3 women), 1 international. Average age 28. *7 applicants, 0% accepted.* In 1999, 1 master's awarded.

Degree requirements: For master's, foreign language not required; for doctorate, dissertation required. *Average time to degree:* Master's–4 years full-time; doctorate–3 years full-time.

Entrance requirements: For master's and doctorate, TOEFL. *Application deadline:* For fall admission, 3/1 (priority date); for spring admission, 10/1 (priority date). Applications are processed on a rolling basis. *Application fee:* $25.

Expenses: Tuition, state resident: full-time $3,096; part-time $112 per credit hour. Tuition, nonresident: full-time $7,588; part-time $299 per credit hour. Tuition and fees vary according to course load, campus/location and reciprocity agreements.

Financial aid: In 1999–00, 2 research assistantships with full tuition reimbursements (averaging $8,800 per year), teaching assistantships with full tuition reimbursements (averaging $8,800 per year) were awarded; Federal Work-Study and institutionally sponsored loans also available. Financial aid application deadline: 3/15.

Faculty research: Biosynthesis of plant cell walls; taxonomy, chemosystematis, and ecology of lichen fungi; structure-function relations in natural ecosystems, evolution of algae, plant stress physiology. *Total annual research expenditures:* $194,000.

Dr. Alan R. White, Chair, 701-231-8679, *Fax:* 701-231-7149, *E-mail:* alwhite@plains.nodak.edu.

Application contact: Marvin W. Fawley, Associate Professor, 701-231-7353, *Fax:* 701-231-7149, *E-mail:* fawley@plains.nodak.edu.

■ THE OHIO STATE UNIVERSITY

Graduate School, College of Biological Sciences, Department of Plant Biology, Columbus, OH 43210

AWARDS MS, PhD.

Faculty: 11 full-time, 6 part-time/adjunct. **Students:** 16 full-time (5 women), 4 part-time (all women); includes 1 minority (African American), 14 international. *24 applicants, 38% accepted.* In 1999, 2 master's, 2 doctorates awarded.

Degree requirements: For master's, thesis optional, foreign language not required; for doctorate, dissertation required.

Entrance requirements: For master's and doctorate, GRE General Test, GRE Subject Test. *Application deadline:* For fall admission, 8/15. Applications are processed on a rolling basis. *Application fee:* $30 ($40 for international students).

Expenses: Tuition, state resident: full-time $5,400. Tuition, nonresident: full-time $14,535. Part-time tuition and fees vary according to course load and program.

Financial aid: Fellowships, research assistantships, teaching assistantships, Federal Work-Study and institutionally sponsored loans available. Aid available to part-time students.

Faculty research: Regulatory, environmental, structural, systematic, and evolutionary botany.

Dr. Richard T. Sayre, Chairman, 614-292-8952, *Fax:* 614-292-6345, *E-mail:* sayre.2@osu.edu.

■ OHIO UNIVERSITY

Graduate Studies, College of Arts and Sciences, Department of Environmental and Plant Biology, Athens, OH 45701-2979

AWARDS MS, PhD. Part-time programs available.

Faculty: 13 full-time (2 women), 1 (woman) part-time/adjunct.

Students: 19 full-time (13 women), 3 part-time (all women), 5 international. *14 applicants, 79% accepted.* In 1999, 3 master's awarded (100% found work related to degree); 2 doctorates awarded.

Degree requirements: For master's, thesis (for some programs), 2 quarters of teaching experience required, foreign language not required; for doctorate, dissertation, 2 quarters of teaching experience required, foreign language not required.

Entrance requirements: For master's, GRE General Test, minimum GPA of 3.0; for doctorate, GRE General Test, minimum GPA of 3.2. *Application deadline:* For fall admission, 3/31 (priority date). Applications are processed on a rolling basis. *Application fee:* $30.

Expenses: Tuition, state resident: full-time $5,754; part-time $238 per credit hour. Tuition, nonresident: full-time $11,055; part-time $457 per credit hour. Tuition and fees vary according to course load, degree level and campus/location.

Financial aid: In 1999–00, 2 research assistantships with tuition reimbursements, 21 teaching assistantships with tuition reimbursements were awarded; fellowships, Federal Work-Study, institutionally sponsored loans, scholarships, and tuition waivers (full and partial) also available. Financial aid application deadline: 2/15.

Faculty research: Cellular and molecular biology, ecology, paleobotany, physiology, ethnobotany.

Dr. James Braselton, Chair, 740-593-1126.

Application contact: Dr. Irwin Ungar, Graduate Chair, 740-593-1120.

■ OKLAHOMA STATE UNIVERSITY

Graduate College, College of Arts and Sciences, Department of Botany, Stillwater, OK 74078

AWARDS MS, PhD.

Faculty: 3 full-time (0 women). **Students:** 2 full-time (0 women), 3 part-time, 1 international. Average age 30. In 1999, 1 master's awarded.

Degree requirements: For master's, foreign language not required; for doctorate, dissertation required.

Entrance requirements: For master's and doctorate, GRE General Test, TOEFL. *Application deadline:* For fall admission, 6/1 (priority date). *Application fee:* $25.

Expenses: Tuition, state resident: part-time $86 per credit hour. Tuition, nonresident: part-time $275 per credit hour. Required fees: $17 per credit hour. $14 per semester. One-time fee: $20 full-time. Tuition and fees vary according to course load.

Financial aid: In 1999–00, 1 research assistantship (averaging $9,009 per year), teaching assistantships (averaging $14,675 per year) were awarded; career-related internships or fieldwork, Federal Work-Study, and tuition waivers (partial) also available. Aid available to part-time students. Financial aid application deadline: 3/1.

Faculty research: Ethnobotany, developmental genetics of Arabidopsis, biological roles of Plasmodesmata, community ecology and biodiversity, nutrient cycling in grassland ecosystems.

Dr. Becky L. Johnson, Head, 405-744-5559.

■ OREGON STATE UNIVERSITY

Graduate School, College of Science, Department of Botany and Plant Pathology, Corvallis, OR 97331

AWARDS Ecology (MA, MAIS, MS, PhD); genetics (MA, MAIS, MS, PhD); molecular and cellular biology (MA, MAIS, MS, PhD); mycology (MA, MAIS, MS, PhD); plant pathology (MA, MAIS, MS, PhD); plant physiology (MA, MAIS, MS, PhD); structural botany (MA, MAIS, MS, PhD); systematics (MA, MAIS, MS, PhD). Part-time programs available.

Faculty: 45 full-time (10 women). **Students:** 39 full-time (25 women), 1 part-time; includes 1 minority (Hispanic American), 4 international. Average age 30. In 1999, 7 master's, 3 doctorates awarded.

Degree requirements: For master's, variable foreign language requirement required, thesis optional; for doctorate, dissertation required, foreign language not required.

Entrance requirements: For master's and doctorate, GRE General Test, TOEFL, minimum GPA of 3.0 in last 90 hours. *Application deadline:* For fall admission, 2/1 (priority date). Applications are processed on a rolling basis. *Application fee:* $50.

Expenses: Tuition, state resident: full-time $6,489. Tuition, nonresident: full-time $11,061. Tuition and fees vary according to program.

Financial aid: Fellowships, research assistantships, teaching assistantships, career-related internships or fieldwork, Federal Work-Study, institutionally sponsored loans, and scholarships available. Aid available to part-time students. Financial aid application deadline: 2/1.

Faculty research: Plant ecology, plant molecular biology, systematic botany, epidemiology, host-pathogen interaction.

Dr. Stella M. Coakley, Chair, 541-737-5264, *Fax:* 541-737-3573, *E-mail:* coakleys@bcc.orst.edu.

Application contact: Dr. Donald B. Zobel, Professor, 541-737-3451, *Fax:* 541-737-3573, *E-mail:* zobeld@bcc.orst.edu.

■ PURDUE UNIVERSITY

Graduate School, Interdisciplinary Program in Plant Biology, West Lafayette, IN 47907

AWARDS PhD.

Faculty: 47 full-time (7 women). **Students:** 11 full-time (6 women), 7 international. Average age 25. *25 applicants, 48% accepted.*

Degree requirements: For doctorate, dissertation, preliminary and qualifying exams required, foreign language not required. *Average time to degree:* Doctorate–6 years full-time.

Purdue University (continued)

Entrance requirements: For doctorate, GRE General Test, TOEFL. *Application deadline:* For fall admission, 2/15 (priority date). *Application fee:* $30. Electronic applications accepted.
Expenses: Tuition, state resident: full-time $4,530; part-time $130 per credit hour. Tuition, nonresident: full-time $15,310; part-time $404 per credit hour. Tuition and fees vary according to campus/location and program.
Financial aid: In 1999–00, 3 fellowships with partial tuition reimbursements (averaging $16,800 per year), 7 research assistantships with partial tuition reimbursements (averaging $16,800 per year), 1 teaching assistantship with partial tuition reimbursement were awarded; tuition waivers (full) also available. Financial aid application deadline: 4/15.
Faculty research: Chloroplast bioenergetics and membrane biochemistry, plant molecular biology, plant hormones, physiological stress.
Dr. C. C. Chapple, Chair, 765-494-0494, *Fax:* 765-496-1475.
Application contact: Elizabeth A. Chandler, Coordinator, 765-494-1634, *Fax:* 765-496-1475.

Find an in-depth description at www.petersons.com/graduate.

■ PURDUE UNIVERSITY

Graduate School, School of Agriculture, Department of Botany and Plant Pathology, West Lafayette, IN 47907

AWARDS MS, PhD. Part-time programs available.

Faculty: 26 full-time (4 women).
Students: 19 full-time (6 women), 6 part-time (2 women); includes 1 minority (African American), 11 international. Average age 27. *52 applicants, 29% accepted.* In 1999, 6 master's, 1 doctorate awarded. Terminal master's awarded for partial completion of doctoral program.
Degree requirements: For master's and doctorate, thesis/dissertation required, foreign language not required.
Entrance requirements: For master's and doctorate, GRE, TOEFL, TWE. *Application deadline:* For fall admission, 5/1 (priority date); for spring admission, 11/1. Applications are processed on a rolling basis. *Application fee:* $30. Electronic applications accepted.
Expenses: Tuition, state resident: full-time $4,530; part-time $130 per credit hour. Tuition, nonresident: full-time $15,310; part-time $404 per credit hour. Tuition and fees vary according to campus/location and program.

Financial aid: In 1999–00, 8 research assistantships, 6 teaching assistantships were awarded; fellowships, career-related internships or fieldwork also available. Aid available to part-time students. Financial aid application deadline: 7/1; financial aid applicants required to submit FAFSA.
Faculty research: Biotechnology, plant growth, weed control, crop improvement, plant physiology.
Dr. R. D. Martyn, Head, 765-494-4614, *Fax:* 765-494-0363, *E-mail:* martyn@btny.purdue.edu.
Application contact: Pam Mow, Graduate Admissions Office, 765-494-0352, *Fax:* 765-494-0363, *E-mail:* mow@btny.purdue.edu.

Find an in-depth description at www.petersons.com/graduate.

■ RENSSELAER POLYTECHNIC INSTITUTE

Graduate School, School of Science, Department of Biology, Troy, NY 12180-3590

AWARDS Biochemistry (MS, PhD); biophysics (MS, PhD); cell biology (MS, PhD); developmental biology (MS, PhD); microbiology (MS, PhD); molecular biology (MS, PhD); plant science (MS, PhD). Part-time programs available.

Faculty: 14 full-time (5 women).
Students: 16 full-time (8 women), 2 part-time (1 woman); includes 1 minority (Asian American or Pacific Islander), 4 international. *37 applicants, 41% accepted.* In 1999, 6 master's, 3 doctorates awarded. Terminal master's awarded for partial completion of doctoral program.
Degree requirements: For master's and doctorate, thesis/dissertation required, foreign language not required.
Entrance requirements: For master's and doctorate, GRE General Test, TOEFL. *Application deadline:* For fall admission, 2/1 (priority date). Applications are processed on a rolling basis. *Application fee:* $35.
Expenses: Tuition: Part-time $665 per credit hour. Required fees: $980.
Financial aid: In 1999–00, 8 research assistantships with partial tuition reimbursements (averaging $15,000 per year), 11 teaching assistantships with full tuition reimbursements (averaging $15,000 per year) were awarded; fellowships, career-related internships or fieldwork and institutionally sponsored loans also available. Financial aid application deadline: 2/1.
Faculty research: Applied environmental biology, genetics, environmental science, fresh water ecology, microbial ecology.
Dr. John Salerno, Chair, 518-276-2699, *Fax:* 518-276-2344.

Application contact: Dr. Jackie L. Collier, Assistant Professor, 518-276-6446, *Fax:* 518-276-2344, *E-mail:* collij3@rpi.edu.

Find an in-depth description at www.petersons.com/graduate.

■ RUTGERS, THE STATE UNIVERSITY OF NEW JERSEY, NEW BRUNSWICK

Graduate School, Program in Plant Biology, New Brunswick, NJ 08901-1281

AWARDS Horticulture (MS, PhD); molecular biology and biochemistry (MS, PhD); pathology (MS, PhD); plant ecology (MS, PhD); plant genetics (PhD); plant physiology (MS, PhD); production and management (MS); structure and plant groups (MS, PhD). Part-time programs available.

Faculty: 64 full-time (8 women), 1 part-time/adjunct (0 women).
Students: 42 full-time (17 women), 13 part-time (6 women); includes 4 minority (1 African American, 2 Asian Americans or Pacific Islanders, 1 Hispanic American), 28 international. Average age 27. *58 applicants, 38% accepted.* In 1999, 4 master's, 9 doctorates awarded. Terminal master's awarded for partial completion of doctoral program.
Degree requirements: For master's, thesis or alternative required, foreign language not required; for doctorate, dissertation required, foreign language not required. *Average time to degree:* Master's–2.5 years full-time; doctorate–5 years full-time.
Entrance requirements: For master's and doctorate, GRE General Test, GRE Subject Test (recommended). *Application deadline:* For fall admission, 6/1. *Application fee:* $50. Electronic applications accepted.
Expenses: Tuition, state resident: full-time $6,776; part-time $279 per credit. Tuition, nonresident: full-time $9,936; part-time $412 per credit. Required fees: $20 per credit. $89 per semester. Tuition and fees vary according to course load, campus/location and program.
Financial aid: In 1999–00, 42 students received aid, including 12 fellowships with full tuition reimbursements available (averaging $16,000 per year), 21 research assistantships with full tuition reimbursements available (averaging $14,000 per year), 9 teaching assistantships with full tuition reimbursements available (averaging $13,100 per year). Financial aid application deadline: 2/15; financial aid applicants required to submit FAFSA.
Faculty research: Molecular genetics, disease resistance, molecular biology of plant viruses.

Dr. Thomas Leustek, Director, 732-932-8165 Ext. 326, *Fax:* 732-932-9377, *E-mail:* leustek@aesop.rutgers.edu.
Application contact: Barbara Mulder, Administrative Assistant, 732-932-9375 Ext. 358, *Fax:* 732-932-9377, *E-mail:* plantbio@aesop.rutgers.edu.
Find an in-depth description at www.petersons.com/graduate.

■ SOUTH DAKOTA STATE UNIVERSITY

Graduate School, College of Agriculture and Biological Sciences, Department of Plant Science, Brookings, SD 57007
AWARDS Agronomy (MS, PhD); biological sciences (PhD); entomology (MS); plant pathology (MS).

Degree requirements: For master's, thesis, oral exam required, foreign language not required; for doctorate, dissertation, preliminary oral and written exams required.
Entrance requirements: For master's and doctorate, GRE General Test, TOEFL.

■ SOUTHERN ILLINOIS UNIVERSITY CARBONDALE

Graduate School, College of Agriculture, Department of Plant, Soil, and General Agriculture, Carbondale, IL 62901-6806
AWARDS Horticultural science (MS); plant and soil science (MS).

Faculty: 20 full-time (1 woman).
Students: 36 full-time (15 women), 17 part-time (6 women), 3 international. *13 applicants, 54% accepted.* In 1999, 10 degrees awarded.
Degree requirements: For master's, thesis required, foreign language not required.
Entrance requirements: For master's, TOEFL, minimum GPA of 2.7. *Application deadline:* Applications are processed on a rolling basis. *Application fee:* $0.
Expenses: Tuition, state resident: full-time $2,902. Tuition, nonresident: full-time $5,810. Tuition and fees vary according to course load.
Financial aid: In 1999-00, 22 students received aid, including 16 research assistantships with full tuition reimbursements available, 6 teaching assistantships with full tuition reimbursements available; fellowships with full tuition reimbursements available, Federal Work-Study, institutionally sponsored loans, and tuition waivers (full) also available. Aid available to part-time students.
Faculty research: Herbicides, fertilizers, agriculture education, landscape design,

plant breeding. *Total annual research expenditures:* $2 million.
Donald Stucky, Chairperson, 618-453-2496.

■ SOUTHERN ILLINOIS UNIVERSITY CARBONDALE

Graduate School, College of Science, Department of Plant Biology, Carbondale, IL 62901-6806
AWARDS MS, PhD.

Faculty: 13 full-time (2 women).
Students: 24 full-time (14 women), 8 part-time (3 women); includes 2 minority (1 Hispanic American, 1 Native American), 6 international. Average age 25. *4 applicants, 75% accepted.* In 1999, 4 master's, 3 doctorates awarded.
Degree requirements: For master's, thesis required, foreign language not required; for doctorate, one foreign language, dissertation required.
Entrance requirements: For master's, GRE General Test, TOEFL, minimum GPA of 2.7; for doctorate, GRE General Test, TOEFL, minimum GPA of 3.25. *Application deadline:* Applications are processed on a rolling basis. *Application fee:* $20.
Expenses: Tuition, state resident: full-time $2,902. Tuition, nonresident: full-time $5,810. Tuition and fees vary according to course load.
Financial aid: In 1999-00, 24 students received aid, including 4 fellowships with full tuition reimbursements available, 6 research assistantships with full tuition reimbursements available, 13 teaching assistantships with full tuition reimbursements available; Federal Work-Study, institutionally sponsored loans, and tuition waivers (full) also available. Aid available to part-time students.
Faculty research: Algal toxins, ethnobotany, community and wetland ecology, morphogenesis, systematics and evolution. *Total annual research expenditures:* $524,140.
Donald R. Tindall, Chairperson, 618-453-3210.
Application contact: Dr. Walter Schmid, Graduate Coordinator, 618-536-2331.

■ SOUTHWEST MISSOURI STATE UNIVERSITY

Graduate College, College of Natural and Applied Sciences, Program in Plant Science, Springfield, MO 65804-0094
AWARDS MS.

Faculty: 38 full-time (6 women).
Students: 3 full-time (1 woman), 5 part-time (2 women), 1 international. In 1999, 1 degree awarded.

Degree requirements: For master's, thesis, comprehensive exams required, foreign language not required.
Entrance requirements: For master's, GRE General Test, minimum undergraduate GPA of 3.0. *Application deadline:* For fall admission, 8/2 (priority date); for spring admission, 12/28 (priority date). Applications are processed on a rolling basis. *Application fee:* $25. Electronic applications accepted.
Expenses: Tuition, state resident: full-time $2,070; part-time $115 per credit. Tuition, nonresident: full-time $4,140; part-time $230 per credit. Required fees: $91 per credit. Tuition and fees vary according to course level, course load and program.
Financial aid: In 1999-00, research assistantships with full tuition reimbursements (averaging $6,150 per year), teaching assistantships with full tuition reimbursements (averaging $6,150 per year) were awarded; Federal Work-Study and unspecified assistantships also available. Financial aid application deadline: 3/31.
Dr. Russell G. Rhodes, Director, 417-836-6887, *E-mail:* rgr592f@mail.smsu.edu.

■ TEXAS A&M UNIVERSITY

College of Agriculture and Life Sciences, Department of Soil and Crop Sciences, Intercollegiate Program in Molecular and Environmental Plant Sciences, College Station, TX 77843
AWARDS MS, PhD.

Faculty: 40 full-time (5 women), 8 part-time/adjunct (1 woman).
Students: 26 full-time (12 women), 1 part-time, 17 international. Average age 29. *31 applicants, 16% accepted.* In 1999, 4 doctorates awarded (50% found work related to degree, 50% continued full-time study).
Degree requirements: For master's and doctorate, thesis/dissertation required, foreign language not required. *Average time to degree:* Doctorate–5 years full-time, 8 years part-time.
Entrance requirements: For master's and doctorate, GRE General Test, TOEFL. *Application deadline:* For fall admission, 3/1 (priority date); for spring admission, 8/1. Applications are processed on a rolling basis. *Application fee:* $50 ($75 for international students).
Expenses: Tuition, state resident: part-time $76 per semester hour. Tuition, nonresident: part-time $292 per semester hour. Required fees: $11 per semester hour. Tuition and fees vary according to program.
Financial aid: In 1999-00, 1 fellowship, 4 research assistantships were awarded; teaching assistantships Financial aid

Texas A&M University (continued)
application deadline: 3/1; financial aid applicants required to submit FAFSA.
Faculty research: Fuctional genomics, bioremediation, physiological ecology, transformation systems, fruit ripening. *Total annual research expenditures:* $6.2 million.
Dr. Wayne R. Jordan, Chair, 979-845-1851, *Fax:* 979-845-8554, *E-mail:* wjordan@tamu.edu.
Application contact: Dr. Jean Gould, Admissions Chair, 979-845-5078, *Fax:* 979-845-6049, *E-mail:* jean@silva.tamu.edu.

Find an in-depth description at www.petersons.com/graduate.

■ TEXAS A&M UNIVERSITY

College of Science, Department of Biology, Program in Botany, College Station, TX 77843
AWARDS MS, PhD.
Students: 5 full-time (4 women). Average age 28. *9 applicants, 33% accepted.* In 1999, 1 master's, 6 doctorates awarded.
Degree requirements: For master's, foreign language not required; for doctorate, dissertation required, foreign language not required.
Entrance requirements: For master's and doctorate, GRE General Test, TOEFL. *Application fee:* $50 ($75 for international students).
Expenses: Tuition, state resident: part-time $76 per semester hour. Tuition, nonresident: part-time $292 per semester hour. Required fees: $11 per semester hour. Tuition and fees vary according to program.
Financial aid: Application deadline: 4/1.
Dr. Mark Zoran, Head, 979-845-7755.

■ TEXAS A&M UNIVERSITY– KINGSVILLE

College of Graduate Studies, College of Agriculture and Home Economics, Program in Plant and Soil Sciences, Kingsville, TX 78363
AWARDS MS.
Faculty: 2 full-time (0 women), 3 part-time/adjunct (0 women).
Students: 10 full-time (3 women), 10 part-time (5 women); includes 4 minority (1 Asian American or Pacific Islander, 3 Hispanic Americans), 8 international. Average age 32. In 1999, 1 degree awarded.
Degree requirements: For master's, thesis or alternative, comprehensive exam required, foreign language not required.
Entrance requirements: For master's, GRE General Test, TOEFL, minimum GPA of 3.0. *Application deadline:* For fall admission, 6/1; for spring admission,

11/15. Applications are processed on a rolling basis. *Application fee:* $15 ($25 for international students).
Expenses: Tuition, state resident: full-time $2,062; part-time $102 per hour. Tuition, nonresident: full-time $7,246; part-time $316 per hour. Tuition and fees vary according to course load.
Financial aid: Fellowships, research assistantships, teaching assistantships available. Financial aid application deadline: 5/15.

■ TEXAS TECH UNIVERSITY

Graduate School, College of Agricultural Sciences and Natural Resources, Department of Plant and Soil Science, Lubbock, TX 79409
AWARDS Agronomy (PhD); crop science (MS); entomology (MS); horticulture (MS); plant and soil science (M Agr); soil science (MS). Part-time programs available.
Faculty: 14 full-time (2 women), 1 part-time/adjunct (0 women).
Students: 36 full-time (12 women), 17 part-time (4 women); includes 1 minority (Asian American or Pacific Islander), 7 international. Average age 31. *12 applicants, 50% accepted.* In 2000, 10 master's, 5 doctorates awarded.
Degree requirements: For master's, foreign language and thesis not required; for doctorate, dissertation required.
Entrance requirements: For master's and doctorate, GRE General Test. *Application deadline:* For fall admission, 4/15 (priority date); for spring admission, 11/1 (priority date). Applications are processed on a rolling basis. *Application fee:* $25 ($50 for international students). Electronic applications accepted.
Expenses: Tuition, state resident: full-time $2,736; part-time $114 per hour. Tuition, nonresident: full-time $7,920; part-time $330 per hour. Required fees: $464 per semester. Tuition and fees vary according to course level, course load and program.
Financial aid: In 2000–01, 25 students received aid, including 21 research assistantships (averaging $10,882 per year), 3 teaching assistantships (averaging $11,250 per year); fellowships, Federal Work-Study and institutionally sponsored loans also available. Aid available to part-time students. Financial aid application deadline: 5/15; financial aid applicants required to submit FAFSA.
Faculty research: Molecular and cellular biology of plant stress, physiology/genetics of cotton and sorghum production in semiarid conditions, biology of red fire ants. *Total annual research expenditures:* $1.6 million.
Dr. Dick L. Auld, Chair, 806-742-2837, *Fax:* 806-742-0775.

Application contact: Graduate Adviser, 806-742-2837.

■ UNIVERSITY OF ALASKA FAIRBANKS

Graduate School, College of Science, Engineering and Mathematics, Department of Biology and Wildlife, Program in Biological Sciences, Fairbanks, AK 99775
AWARDS Biology (MAT, MS, PhD); botany (MS, PhD); zoology (MS, PhD). Part-time programs available.
Faculty: 24 full-time (2 women), 2 part-time/adjunct (0 women).
Students: 39 full-time (19 women), 11 part-time (6 women); includes 1 minority (Native American), 6 international. Average age 31. *32 applicants, 38% accepted.* In 1999, 5 master's, 5 doctorates awarded.
Degree requirements: For master's, thesis, comprehensive exam required, foreign language not required; for doctorate, one foreign language (computer language can substitute), dissertation, comprehensive exam required.
Entrance requirements: For master's and doctorate, GRE General Test, GRE Subject Test, TOEFL. *Application deadline:* For fall admission, 8/1. Applications are processed on a rolling basis. *Application fee:* $35.
Expenses: Tuition, state resident: full-time $3,006; part-time $167 per credit. Tuition, nonresident: full-time $5,868; part-time $326 per credit. Required fees: $370; $10 per credit. $140 per semester.
Financial aid: Research assistantships, teaching assistantships, career-related internships or fieldwork available. Financial aid application deadline: 6/1.
Faculty research: Plant insect interactions, wildlife ecology, adaptations to winter/cold, cell/molecular biology, ecology.
Dr. Ed Murphy, Acting Dean, College of Science, Engineering and Mathematics, 907-474-7941.

■ THE UNIVERSITY OF ARIZONA

Graduate College, College of Agriculture, Department of Plant Sciences, Tucson, AZ 85721
AWARDS MS, PhD. Part-time programs available.
Degree requirements: For master's, thesis or alternative required, foreign language not required; for doctorate, dissertation required, foreign language not required.
Entrance requirements: For master's and doctorate, GRE General Test, GRE Subject Test (biology or chemistry), TOEFL, minimum GPA of 3.0.

Expenses: Tuition, nonresident: full-time $4,814; part-time $274 per unit. Required fees: $1,094; $115 per unit. Tuition and fees vary according to course load and program.
Faculty research: Molecular/cell biology, plant genetics and physiology, agronomic and horticultural production (including turf and ornamentals).
Find an in-depth description at www.petersons.com/graduate.

■ THE UNIVERSITY OF ARIZONA

Graduate College, College of Science, Department of Ecology and Evolutionary Biology, Program in Botany, Tucson, AZ 85721
AWARDS MS, PhD.

Degree requirements: For master's, foreign language and thesis not required; for doctorate, one foreign language, computer language, dissertation required.
Entrance requirements: For master's and doctorate, GRE General Test, GRE Subject Test, TOEFL.
Expenses: Tuition, nonresident: full-time $4,814; part-time $274 per unit. Required fees: $1,094; $115 per unit. Tuition and fees vary according to course load and program.
Faculty research: Systematics, plant population, biology, plant ecology, algabiology.

■ UNIVERSITY OF ARKANSAS

Graduate School, Dale Bumpers College of Agricultural, Food and Life Sciences, Interdepartmental Program in Plant Science, Fayetteville, AR 72701-1201
AWARDS PhD.

Students: 7 full-time (3 women), 3 part-time (all women); includes 2 minority (both Asian Americans or Pacific Islanders), 6 international. *2 applicants, 50% accepted.* In 1999, 6 degrees awarded.
Degree requirements: For doctorate, dissertation required, foreign language not required.
Application fee: $40 ($50 for international students).
Expenses: Tuition, state resident: full-time $3,186; part-time $177 per credit. Tuition, nonresident: full-time $7,560; part-time $420 per credit. Required fees: $756; $21 per credit. One-time fee: $22 part-time. Tuition and fees vary according to course load and program.
Financial aid: Research assistantships, teaching assistantships, career-related internships or fieldwork and Federal Work-Study available. Aid available to part-time students. Financial aid application deadline: 4/1; financial aid applicants required to submit FAFSA.

Dr. David Tebeest, Chair, 501-575-2678, *E-mail:* dtebeest@comp.uark.edu.

■ UNIVERSITY OF CALIFORNIA, BERKELEY

Graduate Division, College of Natural Resources, Department of Plant and Microbial Biology, Berkeley, CA 94720-1500
AWARDS Plant biology (PhD).

Degree requirements: For doctorate, dissertation, qualifying exam, seminar presentation required, foreign language not required.
Entrance requirements: For doctorate, GRE General Test, minimum GPA of 3.0.
Expenses: Tuition, nonresident: full-time $9,804. Required fees: $4,268. Tuition and fees vary according to program.
Faculty research: Development, molecular biology, genetics, microbial biology, mycology.

■ UNIVERSITY OF CALIFORNIA, DAVIS

Graduate Studies, Program in Plant Protection and Pest Management, Davis, CA 95616
AWARDS MS.

Faculty: 34 full-time (1 woman).
Students: 12 full-time (5 women), 2 international. Average age 28. *5 applicants, 100% accepted.* In 1999, 3 degrees awarded.
Degree requirements: For master's, thesis optional.
Entrance requirements: For master's, GRE General Test, GRE Subject Test (biology), minimum GPA of 3.0. *Application deadline:* For fall admission, 1/15. *Application fee:* $40. Electronic applications accepted.
Expenses: Tuition, nonresident: full-time $9,804. Tuition and fees vary according to program and student level.
Financial aid: In 1999–00, 7 students received aid, including 2 fellowships with full and partial tuition reimbursements available, 5 research assistantships with full and partial tuition reimbursements available; teaching assistantships with partial tuition reimbursements available, career-related internships or fieldwork, Federal Work-Study, institutionally sponsored loans, and scholarships also available. Financial aid application deadline: 1/15; financial aid applicants required to submit FAFSA.
Ken Giles, Chair, 530-752-0687, *E-mail:* dkgiles@ucdavis.edu.
Application contact: Brenda Nakamoto, Administrative Assistant, 530-752-0492, *Fax:* 530-752-1537, *E-mail:* bunakamota@ucdavis.edu.

■ UNIVERSITY OF CALIFORNIA, DAVIS

Graduate Studies, Programs in the Biological Sciences, Program in Plant Biology, Davis, CA 95616
AWARDS MS, PhD.

Faculty: 91 full-time (21 women).
Students: 75 full-time (39 women); includes 10 minority (1 African American, 6 Asian Americans or Pacific Islanders, 3 Hispanic Americans), 22 international. Average age 28. *74 applicants, 34% accepted.* In 1999, 5 master's, 12 doctorates awarded.
Degree requirements: For master's, thesis optional; for doctorate, dissertation required.
Entrance requirements: For master's, GRE General Test, GRE Subject Test (biology), minimum GPA of 3.0; for doctorate, GRE General Test, GRE Subject Test (biology). *Application deadline:* For fall admission, 3/15. *Application fee:* $40. Electronic applications accepted.
Expenses: Tuition, nonresident: full-time $9,804. Tuition and fees vary according to program and student level.
Financial aid: In 1999–00, 70 students received aid, including 19 fellowships with full and partial tuition reimbursements available, 21 research assistantships with full and partial tuition reimbursements available, 21 teaching assistantships with partial tuition reimbursements available; tuition waivers (full and partial) also available. Financial aid application deadline: 1/15; financial aid applicants required to submit FAFSA.
Faculty research: Cell and molecular biology, ecology, systematics and evolution, integrative plant and crop physiology, plant development and structure.
John Labavitch, Graduate Adviser, 530-752-0920, *Fax:* 530-752-5410, *E-mail:* jmlabavitch@ucdavis.edu.
Application contact: Cheryl Taylor, Administrative Assistant, 530-752-8131, *Fax:* 530-752-6635, *E-mail:* gradcord@ucdmath.ucdavis.edu.

■ UNIVERSITY OF CALIFORNIA, RIVERSIDE

Graduate Division, College of Natural and Agricultural Sciences, Department of Botany and Plant Sciences, Riverside, CA 92521-0102
AWARDS Botany (MS, PhD); botany (plant genetics) (PhD); plant science (MS). Part-time programs available.

Students: 46 full-time (25 women); includes 4 minority (2 Asian Americans or Pacific Islanders, 2 Hispanic Americans), 23 international. In 1999, 5 master's, 3

University of California, Riverside
(continued)
doctorates awarded. Terminal master's awarded for partial completion of doctoral program.
Degree requirements: For master's, comprehensive exams or thesis required; for doctorate, dissertation, qualifying exams required, foreign language not required. *Average time to degree:* Master's–2.3 years full-time; doctorate–5.5 years full-time.
Entrance requirements: For master's and doctorate, GRE General Test, TOEFL, minimum GPA of 3.2. *Application deadline:* For fall admission, 5/1; for spring admission, 12/1. Applications are processed on a rolling basis. *Application fee:* $40.
Expenses: Tuition, nonresident: full-time $9,804. Required fees: $4,758. Full-time tuition and fees vary according to program.
Financial aid: Fellowships, research assistantships, teaching assistantships, career-related internships or fieldwork, Federal Work-Study, grants, institutionally sponsored loans, and tuition waivers (full and partial) available. Financial aid application deadline: 2/1; financial aid applicants required to submit FAFSA.
Faculty research: Plant molecular biology, plant cell and developmental biology, plant systematics and evolution, ecology and natural resources, agriculture and crop physiology. *Total annual research expenditures:* $7.6 million.
Dr. Elizabeth Lord, Chair.
Application contact: 909-787-3424, *Fax:* 909-787-4437, *E-mail:* bpscmari@ucrac1.ucr.edu.

Find an in-depth description at www.petersons.com/graduate.

■ UNIVERSITY OF COLORADO AT BOULDER

Graduate School, College of Arts and Sciences, Department of Environmental, Population, and Organic Biology, Boulder, CO 80309

AWARDS Animal behavior (MA, PhD); aquatic biology (MA, PhD); behavioral genetics (MA, PhD); ecology (MA, PhD); microbiology (MA, PhD); neurobiology (MA, PhD); plant and animal physiology (MA, PhD); plant and animal systematics (MA, PhD); population biology (MA, PhD); population genetics (MA, PhD).

Faculty: 39 full-time (9 women).
Students: 84 full-time (36 women), 14 part-time (7 women); includes 17 minority (6 Asian Americans or Pacific Islanders, 10 Hispanic Americans, 1 Native American). Average age 29. *147 applicants, 14% accepted.* In 1999, 7 master's, 13 doctorates

awarded. Terminal master's awarded for partial completion of doctoral program.
Degree requirements: For master's, thesis or alternative, comprehensive exam required, foreign language not required; for doctorate, dissertation, comprehensive exam required, foreign language not required. *Average time to degree:* Master's–3 years full-time; doctorate–5 years full-time.
Entrance requirements: For master's, GRE General Test, GRE Subject Test, minimum undergraduate GPA of 3.0; for doctorate, GRE General Test, GRE Subject Test. *Application deadline:* For fall admission, 1/15 (priority date). *Application fee:* $40 ($60 for international students).
Expenses: Tuition, state resident: part-time $181 per credit hour. Tuition, nonresident: part-time $542 per credit hour. Required fees: $99 per term. Tuition and fees vary according to course load and program.
Financial aid: Fellowships, research assistantships, teaching assistantships, Federal Work-Study, institutionally sponsored loans, and tuition waivers (full) available. Financial aid application deadline: 3/1.
Faculty research: Evolution, developmental biology, behavior and neurobiology. *Total annual research expenditures:* $1.8 million.
Michael Breed, Chair, 303-492-8981, *Fax:* 303-492-8699, *E-mail:* michael.breed@colorado.edu.
Application contact: Jill Skarstadt, Graduate Secretary, 303-492-7654, *Fax:* 303-492-8699, *E-mail:* jill.skarstadt@colorado.edu.

■ UNIVERSITY OF CONNECTICUT

Graduate School, College of Agriculture and Natural Resources, Field of Plant Science, Storrs, CT 06269

AWARDS Plant and soil sciences (MS, PhD).

Degree requirements: For doctorate, dissertation required.
Entrance requirements: For master's and doctorate, GRE General Test, GRE Subject Test.
Expenses: Tuition, state resident: full-time $5,118. Tuition, nonresident: full-time $13,298. Required fees: $1,022.

■ UNIVERSITY OF CONNECTICUT

Graduate School, College of Liberal Arts and Sciences, Biological Sciences Group, Storrs, CT 06269

AWARDS Ecology and evolutionary biology (MS, PhD), including botany, ecology, entomology, systematics, zoology; molecular and cell biology (MS, PhD), including biochemistry, biophysics, biotechnology (MS); cell and developmental biology, genetics,

microbiology, plant molecular and cell biology; physiology and neurobiology (MS, PhD), including neurobiology, physiology.

Degree requirements: For doctorate, dissertation required.
Entrance requirements: For master's and doctorate, GRE General Test, GRE Subject Test, TOEFL.
Expenses: Tuition, state resident: full-time $5,118. Tuition, nonresident: full-time $13,298. Required fees: $1,022.

■ UNIVERSITY OF CONNECTICUT

Graduate School, College of Liberal Arts and Sciences, Biological Sciences Group, Department of Ecology and Evolutionary Biology, Field of Botany, Storrs, CT 06269

AWARDS MS, PhD.

Degree requirements: For doctorate, dissertation required.
Entrance requirements: For master's and doctorate, GRE General Test, GRE Subject Test, TOEFL.
Expenses: Tuition, state resident: full-time $5,118. Tuition, nonresident: full-time $13,298. Required fees: $1,022.

Find an in-depth description at www.petersons.com/graduate.

■ UNIVERSITY OF CONNECTICUT

Graduate School, College of Liberal Arts and Sciences, Biological Sciences Group, Department of Molecular and Cell Biology, Field of Plant Molecular and Cell Biology, Storrs, CT 06269

AWARDS MS, PhD.

Degree requirements: For doctorate, dissertation required.
Entrance requirements: For master's and doctorate, GRE General Test, GRE Subject Test, TOEFL.
Expenses: Tuition, state resident: full-time $5,118. Tuition, nonresident: full-time $13,298. Required fees: $1,022.

Find an in-depth description at www.petersons.com/graduate.

■ UNIVERSITY OF DELAWARE

College of Agriculture and Natural Resources, Department of Plant and Soil Sciences, Newark, DE 19716

AWARDS MS, PhD. Part-time programs available.

Faculty: 21 full-time (4 women).
Students: 32 full-time (13 women), 12 part-time (6 women); includes 1 minority (Hispanic American), 16 international. Average age 25. *35 applicants, 40% accepted.* In 1999, 2 master's, 2 doctorates awarded. Terminal master's awarded for partial completion of doctoral program.

Degree requirements: For master's and doctorate, thesis/dissertation required, foreign language not required. *Average time to degree:* Master's–3.7 years full-time; doctorate–4 years full-time.

Entrance requirements: For master's and doctorate, GRE General Test. *Application deadline:* For fall admission, 7/1. *Application fee:* $45. Electronic applications accepted.

Expenses: Tuition, state resident: full-time $4,380; part-time $243 per credit. Tuition, nonresident: full-time $12,750; part-time $708 per credit. Required fees: $15 per term. Tuition and fees vary according to program.

Financial aid: In 1999–00, 26 students received aid, including 1 fellowship with full tuition reimbursement available (averaging $15,000 per year), 22 research assistantships with full tuition reimbursements available (averaging $13,500 per year), 3 teaching assistantships with full tuition reimbursements available (averaging $11,000 per year); career-related internships or fieldwork also available. Financial aid application deadline: 3/1.

Faculty research: Soil chemistry, plant and cell tissue culture, plant breeding and genetics, soil physics, soil biochemistry, plant molecular biology, soil microbiology. *Total annual research expenditures:* $2.8 million.

Dr. Donald L. Sparks, Chairman, 302-831-2532, *Fax:* 302-831-3651, *E-mail:* dlsparks@udel.edu.

Application contact: Dr. Sherry Kitto, Graduate Coordinator, 302-831-2534, *E-mail:* kitto@udel.edu.

■ **UNIVERSITY OF FLORIDA**

Graduate School, College of Agriculture and Life Sciences and College of Liberal Arts and Sciences, Program in Plant Molecular and Cellular Biology, Gainesville, FL 32611

AWARDS MS, PhD.

Faculty: 27.

Students: 28 full-time (14 women), 1 part-time; includes 3 minority (1 African American, 2 Asian Americans or Pacific Islanders), 15 international. *15 applicants, 53% accepted.* In 1999, 2 master's, 2 doctorates awarded.

Degree requirements: For master's and doctorate, thesis/dissertation required, foreign language not required.

Entrance requirements: For master's and doctorate, GRE General Test, minimum GPA of 3.0. *Application deadline:* For fall admission, 2/1 (priority date). Applications are processed on a rolling basis. *Application fee:* $20. Electronic applications accepted.

Expenses: Tuition, state resident: part-time $144 per credit hour. Tuition, nonresident: part-time $505 per credit

hour. Tuition and fees vary according to course level, course load and program.

Financial aid: In 1999–00, 23 research assistantships, 1 teaching assistantship were awarded; fellowships, unspecified assistantships also available.

Faculty research: Plant pathology, genetics, biochemistry, microbiology.

Application contact: Dr. Kenneth Cline, Graduate Coordinator, 352-392-1928, *Fax:* 352-392-6479, *E-mail:* kcline@ufl.edu.

■ **UNIVERSITY OF FLORIDA**

Graduate School, College of Liberal Arts and Sciences and College of Agriculture and Life Sciences, Department of Botany, Gainesville, FL 32611

AWARDS Botany (M Ag, MS, PhD); botany education (MST). Part-time programs available.

Faculty: 25.

Students: 28 full-time (18 women), 3 part-time (2 women); includes 4 minority (1 African American, 2 Asian Americans or Pacific Islanders, 1 Hispanic American), 9 international. *41 applicants, 44% accepted.* In 1999, 5 master's, 2 doctorates awarded.

Degree requirements: For doctorate, dissertation required.

Entrance requirements: For master's and doctorate, GRE General Test, minimum GPA of 3.0. *Application deadline:* For fall admission, 6/1 (priority date). Applications are processed on a rolling basis. *Application fee:* $20. Electronic applications accepted.

Expenses: Tuition, state resident: part-time $144 per credit hour. Tuition, nonresident: part-time $505 per credit hour. Tuition and fees vary according to course level, course load and program.

Financial aid: In 1999–00, 19 students received aid, including 3 fellowships, 3 research assistantships, 17 teaching assistantships; Federal Work-Study and institutionally sponsored loans also available. Aid available to part-time students. Financial aid application deadline: 2/15.

Faculty research: Ecology, physiology, systematics, biochemistry, ecological genetics.

Dr. George Bowes, Chair, 352-392-1175, *Fax:* 352-392-3993, *E-mail:* gbowes@botany.ufl.edu.

Application contact: Dr. J. T. Mullins, Graduate Coordinator, 352-392-1095, *Fax:* 352-392-3993, *E-mail:* tmullins@botany.ufl.edu.

■ **UNIVERSITY OF GEORGIA**

Graduate School, College of Arts and Sciences, Department of Botany, Athens, GA 30602

AWARDS MS, PhD.

Degree requirements: For master's, thesis required, foreign language not required; for doctorate, one foreign language (computer language can substitute), dissertation required.

Entrance requirements: For master's and doctorate, GRE General Test. Electronic applications accepted.

Expenses: Tuition, state resident: full-time $7,516; part-time $431 per credit hour. Tuition, nonresident: full-time $12,204; part-time $793 per credit hour. Tuition and fees vary according to program.

■ **UNIVERSITY OF HAWAII AT MANOA**

Graduate Division, College of Arts and Sciences, College of Natural Sciences, Department of Botany, Honolulu, HI 96822

AWARDS MS, PhD.

Faculty: 24 full-time (4 women), 8 part-time/adjunct (2 women).

Students: 19 full-time (2 women), 5 part-time (2 women). Average age 29. *39 applicants, 54% accepted.* In 1999, 4 master's, 3 doctorates awarded. Terminal master's awarded for partial completion of doctoral program.

Degree requirements: For master's, thesis (for some programs), presentation required, foreign language not required; for doctorate, one foreign language (computer language can substitute), dissertation, presentation required.

Entrance requirements: For master's and doctorate, GRE General Test, GRE Subject Test (biology). *Application deadline:* For fall admission, 2/1; for spring admission, 8/1. *Application fee:* $25 ($50 for international students).

Expenses: Tuition, state resident: full-time $4,032; part-time $168 per credit. Tuition, nonresident: full-time $9,960; part-time $415 per credit. Required fees: $51 per semester. Part-time tuition and fees vary according to course load and program.

Financial aid: In 1999–00, 14 research assistantships (averaging $15,354 per year), 18 teaching assistantships (averaging $13,049 per year) were awarded; tuition waivers (full and partial) also available.

Faculty research: Plant ecology, evolution, systematics, conservation biology, ethnobotany.

Sterling Keeley, Chairperson, 808-956-8369, *Fax:* 808-956-3923, *E-mail:* sterling@hawaii.edu.

Application contact: George Wong, Graduate Field Chair, 808-956-3940, *Fax:* 808-956-3923, *E-mail:* gwong@hawaii.edu.

■ UNIVERSITY OF HAWAII AT MANOA

Graduate Division, College of Tropical Agriculture and Human Resources, Department of Plant Pathology, Honolulu, HI 96822

AWARDS Botanical sciences (MS, PhD); plant pathology (MS, PhD). Part-time programs available.

Faculty: 17 full-time (3 women), 1 part-time/adjunct (0 women). **Students:** 6 full-time (2 women), 2 part-time (both women). Average age 32. *10 applicants, 70% accepted.* Terminal master's awarded for partial completion of doctoral program.
Degree requirements: For master's, thesis optional, foreign language not required; for doctorate, dissertation required, foreign language not required. *Average time to degree:* Doctorate–4 years full-time.
Entrance requirements: For master's, GRE General Test (average 477 verbal, 629 quantitative, 562 analytical); for doctorate, GRE General Test (average 485 verbal, 570 quantitative). *Application deadline:* For fall admission, 3/1; for spring admission, 9/1. *Application fee:* $25 ($50 for international students).
Expenses: Tuition, state resident: full-time $4,032; part-time $168 per credit. Tuition, nonresident: full-time $9,960; part-time $415 per credit. Required fees: $51 per semester. Part-time tuition and fees vary according to course load and program.
Financial aid: In 1999–00, 8 research assistantships (averaging $15,606 per year) were awarded; teaching assistantships, tuition waivers (full) also available.
Faculty research: Nematology, virology, mycology, bacteriology, epidemiology. *Total annual research expenditures:* $1.9 million.
Dr. Donald P. Schmitt, Chairperson, 808-956-8329, *Fax:* 808-956-2832.
Application contact: Dr. John Hu, Graduate Chairperson, 808-956-8329, *Fax:* 808-956-2832.

■ UNIVERSITY OF IDAHO

College of Graduate Studies, College of Agriculture, Department of Plant, Soil, and Entomological Sciences, Program in Plant Science, Moscow, ID 83844-4140

AWARDS MS, PhD.

Students: 16 full-time (5 women), 11 part-time (4 women), 10 international. In 1999, 4 master's, 3 doctorates awarded.
Degree requirements: For master's, foreign language not required; for doctorate, dissertation required.
Entrance requirements: For master's and doctorate, GRE General Test, minimum

GPA of 3.0. *Application deadline:* For fall admission, 8/1; for spring admission, 12/15. *Application fee:* $35 ($45 for international students).
Expenses: Tuition, nonresident: full-time $6,000; part-time $239 per credit hour. Required fees: $2,888; $144 per credit hour. Tuition and fees vary according to program.
Financial aid: Application deadline: 2/15. Dr. Robert B. Dwelle, Chair, 208-885-7775.

■ UNIVERSITY OF IDAHO

College of Graduate Studies, College of Letters and Science, Department of Biological Sciences, Programs in Botany, Moscow, ID 83844-4140

AWARDS MS, PhD.

Students: 2 full-time (1 woman), 1 part-time. In 1999, 1 master's awarded.
Degree requirements: For master's, foreign language not required; for doctorate, dissertation required.
Entrance requirements: For master's, GRE General Test, minimum GPA of 2.8; for doctorate, GRE General Test, minimum GPA of 2.8 (undergraduate), 3.0 (graduate). *Application deadline:* For fall admission, 8/1; for spring admission, 12/15. *Application fee:* $35 ($45 for international students).
Expenses: Tuition, nonresident: full-time $6,000; part-time $239 per credit hour. Required fees: $2,888; $144 per credit hour. Tuition and fees vary according to program.
Financial aid: Application deadline: 2/15. Dr. Rolf L. Ingermann, Interim Chair, Department of Biological Sciences, 208-885-7764.

■ UNIVERSITY OF ILLINOIS AT CHICAGO

Graduate College, College of Liberal Arts and Sciences, Department of Biological Sciences, Chicago, IL 60607-7128

AWARDS Cell and developmental biology (PhD); ecology and evolution (MS, DA, PhD); genetics and development (PhD); molecular biology (MS, PhD); neurobiology (MS, PhD); plant biology (MS, DA, PhD).

Faculty: 40 full-time (5 women). **Students:** 100 full-time (47 women), 14 part-time (10 women); includes 13 minority (11 Asian Americans or Pacific Islanders, 2 Hispanic Americans), 42 international. Average age 29. *99 applicants, 36% accepted.* In 1999, 3 master's, 9 doctorates awarded.
Degree requirements: For master's, thesis required, foreign language not required; for doctorate, dissertation,

preliminary exam required, foreign language not required.
Entrance requirements: For master's and doctorate, GRE General Test, GRE Subject Test, TOEFL, previous course work in physics, calculus, and organic chemistry; minimum GPA of 3.75 on a 5.0 scale. *Application deadline:* For fall admission, 6/1. Applications are processed on a rolling basis. *Application fee:* $40 ($50 for international students). Electronic applications accepted.
Expenses: Tuition, state resident: full-time $3,750; part-time $1,250 per semester. Tuition, nonresident: full-time $10,588; part-time $3,530 per semester. Required fees: $507 per semester. Tuition and fees vary according to course load and program.
Financial aid: In 1999–00, 87 students received aid; fellowships, research assistantships, teaching assistantships, career-related internships or fieldwork, Federal Work-Study, traineeships, and tuition waivers (full) available. Financial aid application deadline: 3/1; financial aid applicants required to submit FAFSA. Dr. Lon Kaufman, Head, 312-996-2213.
Application contact: Dr. Leo Miller, Director of Graduate Studies, 312-996-2220.

Find an in-depth description at www.petersons.com/graduate.

■ UNIVERSITY OF ILLINOIS AT URBANA–CHAMPAIGN

Graduate College, College of Liberal Arts and Sciences, School of Life Sciences, Department of Plant Biology, Urbana, IL 61801

AWARDS MS, PhD.

Faculty: 11 full-time (2 women), 3 part-time/adjunct (0 women). **Students:** 28 full-time (15 women); includes 2 minority (1 African American, 1 Asian American or Pacific Islander), 10 international. *31 applicants, 13% accepted.* In 1999, 7 master's, 4 doctorates awarded.
Degree requirements: For master's, foreign language and thesis not required; for doctorate, dissertation required.
Entrance requirements: For master's, GRE General Test, GRE Subject Test, minimum GPA of 4.0 on a 5.0 scale. *Application deadline:* Applications are processed on a rolling basis. *Application fee:* $40 ($50 for international students).
Expenses: Tuition, state resident: full-time $4,616. Tuition, nonresident: full-time $11,768. Full-time tuition and fees vary according to course load.
Financial aid: In 1999–00, 4 fellowships, 11 research assistantships, 19 teaching assistantships were awarded. Financial aid application deadline: 2/15.

John Cheesman, Head, 217-333-3261, *Fax:* 217-244-7246, *E-mail:* j-cheese@uiuc.edu. **Application contact:** Lisa Boise, Director of Graduate Studies, 217-333-3261, *Fax:* 217-244-7246, *E-mail:* j-boise@uiuc.edu.

■ UNIVERSITY OF KENTUCKY

Graduate School, Graduate School Programs from the College of Agriculture, Program in Plant and Soil Science, Lexington, KY 40506-0032

AWARDS MS.

Degree requirements: For master's, comprehensive exam required, thesis optional, foreign language not required. **Entrance requirements:** For master's, GRE General Test, minimum GPA of 2.5 (undergraduate), 3.0 (graduate). **Expenses:** Tuition, state resident: full-time $3,596; part-time $188 per credit hour. Tuition, nonresident: full-time $10,116; part-time $550 per credit hour.

■ UNIVERSITY OF MAINE

Graduate School, College of Natural Sciences, Forestry, and Agriculture, Department of Biological Sciences, Program in Botany and Plant Pathology, Orono, ME 04469

AWARDS MS. Part-time programs available.

Degree requirements: For master's, thesis required, foreign language not required. **Entrance requirements:** For master's, GRE General Test, TOEFL. **Expenses:** Tuition, state resident: full-time $3,564. Tuition, nonresident: full-time $10,116. Required fees: $378. Tuition and fees vary according to course load. **Faculty research:** Molecular biology of viral and fungal pathogens, marine ecology, paleoecology and acid systematics and evolution.

■ UNIVERSITY OF MAINE

Graduate School, College of Natural Sciences, Forestry, and Agriculture, Department of Biological Sciences, Program in Plant Science, Orono, ME 04469

AWARDS PhD. Part-time programs available.

Degree requirements: For doctorate, dissertation required. **Entrance requirements:** For doctorate, GRE General Test, TOEFL. **Expenses:** Tuition, state resident: full-time $3,564. Tuition, nonresident: full-time $10,116. Required fees: $378. Tuition and fees vary according to course load.

■ UNIVERSITY OF MAINE

Graduate School, College of Natural Sciences, Forestry, and Agriculture, Department of Plant, Soil, and Environmental Sciences, Orono, ME 04469

AWARDS Biological sciences (PhD); ecology and environmental sciences (MS, PhD); forest resources (PhD); plant science (PhD); plant, soil, and environmental sciences (MS); resource utilization (MS).

Entrance requirements: For master's and doctorate, GRE General Test, TOEFL. **Expenses:** Tuition, state resident: full-time $3,564. Tuition, nonresident: full-time $10,116. Required fees: $378. Tuition and fees vary according to course load.

■ UNIVERSITY OF MARYLAND, COLLEGE PARK

Graduate Studies and Research, College of Life Sciences, Department of Cell Biology and Molecular Genetics, Program in Plant Biology, College Park, MD 20742

AWARDS MS, PhD. Part-time and evening/weekend programs available.

Students: 22 full-time (7 women), 2 part-time (1 woman), 8 international. *32 applicants, 13% accepted.* In 1999, 1 master's, 4 doctorates awarded. Terminal master's awarded for partial completion of doctoral program. **Degree requirements:** For master's and doctorate, thesis/dissertation required. **Entrance requirements:** For master's and doctorate, GRE General Test, minimum GPA of 3.0. *Application deadline:* For fall admission, 1/25. Applications are processed on a rolling basis. *Application fee:* $50 ($70 for international students). Electronic applications accepted. **Expenses:** Tuition, state resident: part-time $272 per credit hour. Tuition, nonresident: part-time $415 per credit hour. Required fees: $632; $379 per year. **Financial aid:** Fellowships, research assistantships, teaching assistantships available. Financial aid applicants required to submit FAFSA. **Faculty research:** Genetics and molecular biology, virology, plant pathology, mycology, nematology. **Application contact:** Trudy Lindsey, Director, Graduate Admissions and Records, 301-405-4198, *Fax:* 301-314-9305, *E-mail:* grschool@deans.umd.edu.

■ UNIVERSITY OF MASSACHUSETTS AMHERST

Graduate School, College of Food and Natural Resources, Department of Plant and Soil Sciences, Amherst, MA 01003

AWARDS Plant science (PhD); soil science (MS, PhD).

Faculty: 21 full-time (3 women). **Students:** 16 full-time (9 women), 25 part-time (11 women); includes 1 minority (Hispanic American), 16 international. Average age 35. *28 applicants, 61% accepted.* In 1999, 8 master's awarded. Terminal master's awarded for partial completion of doctoral program. **Degree requirements:** For master's, thesis optional, foreign language not required; for doctorate, dissertation required, foreign language not required. **Entrance requirements:** For master's and doctorate, GRE General Test. *Application deadline:* For fall admission, 2/1 (priority date); for spring admission, 10/1. Applications are processed on a rolling basis. *Application fee:* $40. **Expenses:** Tuition, state resident: full-time $2,640; part-time $165 per credit. Tuition, nonresident: full-time $9,756; part-time $407 per credit. Required fees: $1,221 per term. One-time fee: $110. Full-time tuition and fees vary according to course load, campus/location and reciprocity agreements. **Financial aid:** In 1999–00, 28 research assistantships with full tuition reimbursements (averaging $7,039 per year); 17 teaching assistantships with full tuition reimbursements (averaging $4,882 per year) were awarded; fellowships with full tuition reimbursements, career-related internships or fieldwork, Federal Work-Study, grants, scholarships, traineeships, and unspecified assistantships also available. Aid available to part-time students. Financial aid application deadline: 2/1. Dr. William J. Bramlage, Director, 413-545-2242, *Fax:* 413-545-3075, *E-mail:* bramlage@pssci.umass.edu.

■ UNIVERSITY OF MASSACHUSETTS AMHERST

Graduate School, Interdisciplinary Programs, Program in Plant Biology, Amherst, MA 01003

AWARDS MS, PhD.

Students: 7 full-time (5 women), 7 part-time (2 women); includes 1 minority (Asian American or Pacific Islander), 7 international. Average age 32. *32 applicants, 16% accepted.* In 2000, 1 master's, 2 doctorates awarded. **Degree requirements:** For master's and doctorate, thesis/dissertation required.

University of Massachusetts Amherst (continued)

Entrance requirements: For master's and doctorate, GRE General Test. *Application deadline:* For fall admission, 2/1 (priority date); for spring admission, 10/1. Applications are processed on a rolling basis. *Application fee:* $40.

Expenses: Tuition, state resident: full-time $2,640; part-time $165 per credit. Tuition, nonresident: full-time $9,756; part-time $407 per credit. Required fees: $1,221 per term. One-time fee: $110. Full-time tuition and fees vary according to course load, campus/location and reciprocity agreements.

Financial aid: Fellowships with full tuition reimbursements, research assistantships with full tuition reimbursements, teaching assistantships with full tuition reimbursements, career-related internships or fieldwork, Federal Work-Study, grants, scholarships, traineeships, and unspecified assistantships available. Aid available to part-time students. Financial aid application deadline: 1/15.

Dr. Peter Hepler, Head, 413-577-3217, *Fax:* 413-545-3243, *E-mail:* hepler@bio.umass.edu.

Application contact: Information Contact, 413-577-3217, *Fax:* 413-545-3243.

Find an in-depth description at www.petersons.com/graduate.

■ **UNIVERSITY OF MICHIGAN**

Horace H. Rackham School of Graduate Studies, College of Literature, Science, and the Arts, Department of Biology, Ann Arbor, MI 48109

AWARDS Biology (MS, PhD); plant biology (MS).

Faculty: 72 full-time (14 women).
Students: 148 full-time (74 women); includes 9 minority (2 African Americans, 5 Asian Americans or Pacific Islanders, 2 Hispanic Americans), 33 international. Average age 29. *231 applicants, 29% accepted.* In 1999, 19 master's, 10 doctorates awarded.

Degree requirements: For master's, thesis not required; for doctorate, oral defense of dissertation, preliminary exam required.

Entrance requirements: For master's and doctorate, GRE General Test, TOEFL. *Application deadline:* Applications are processed on a rolling basis. *Application fee:* $55.

Expenses: Tuition, state resident: full-time $10,316. Tuition, nonresident: full-time $20,922. Required fees: $185. Part-time tuition and fees vary according to course load and program.

Financial aid: In 1999–00, 30 fellowships with full tuition reimbursements (averaging $10,750 per year), 33 research assistantships with full tuition reimbursements (averaging $11,810 per year), 129 teaching assistantships with full tuition reimbursements (averaging $11,810 per year) were awarded; career-related internships or fieldwork, grants, traineeships, and unspecified assistantships also available.

Faculty research: Evolution, ecology and organismal biology; development and genetics; neurobiology and animal physiology; molecular plant biology; microbiology and gene action. *Total annual research expenditures:* $5.1 million.

Dr. Julian Adams, Chair, 734-764-7427, *Fax:* 734-747-0884, *E-mail:* chair@biology.lsa.umich.edu.

Application contact: Lisa Herring, Graduate Coordinator, 734-764-1443, *Fax:* 734-764-0884, *E-mail:* gradcord@biology.lsa.umich.edu.

Find an in-depth description at www.petersons.com/graduate.

■ **UNIVERSITY OF MINNESOTA, TWIN CITIES CAMPUS**

Graduate School, College of Agricultural, Food, and Environmental Sciences, Department of Applied Plant Sciences, Minneapolis, MN 55455-0213

AWARDS MS, PhD.

Faculty: 23 full-time (2 women), 7 part-time/adjunct (0 women).
Students: 33 full-time (16 women), 3 part-time (1 woman); includes 2 minority (1 African American, 1 Asian American or Pacific Islander), 10 international. Average age 24. In 1999, 10 master's, 9 doctorates awarded.

Degree requirements: For master's and doctorate, thesis/dissertation required, foreign language not required. *Average time to degree:* Master's–2.5 years full-time; doctorate–3 years full-time.

Entrance requirements: For master's and doctorate, GRE General Test, TOEFL. *Application fee:* $50 ($55 for international students).

Expenses: Tuition, state resident: full-time $5,040; part-time $420 per credit. Tuition, nonresident: full-time $9,900; part-time $825 per credit. Full-time tuition and fees vary according to course load, program and reciprocity agreements.

Financial aid: Fellowships, research assistantships available.

Faculty research: Weed science, crop management, sustainable agriculture, biotechnology, plant breeding.

Dr. Burle G. Gengenbach, Head, 612-625-8761, *Fax:* 612-625-1268.

Application contact: Dr. Nancy J. Ehlke, Professor, 612-625-1791, *Fax:* 612-625-1268, *E-mail:* ehlke001@tc.umn.edu.

■ **UNIVERSITY OF MINNESOTA, TWIN CITIES CAMPUS**

Graduate School, College of Biological Sciences, Department of Plant Biology, Minneapolis, MN 55455-0213

AWARDS MS, PhD. Part-time programs available.

Faculty: 41 full-time (10 women).
Students: 41 full-time (22 women); includes 4 minority (1 African American, 2 Asian Americans or Pacific Islanders, 1 Hispanic American), 4 international. Average age 26. In 1999, 3 master's awarded (67% found work related to degree, 33% continued full-time study). Terminal master's awarded for partial completion of doctoral program.

Degree requirements: For master's, thesis or alternative required, foreign language not required; for doctorate, dissertation, written and oral preliminary exams required, foreign language not required. *Average time to degree:* Master's–3 years full-time; doctorate–5 years full-time.

Entrance requirements: For master's and doctorate, GRE General Test, TOEFL. *Application deadline:* For fall admission, 12/15 (priority date). Applications are processed on a rolling basis. *Application fee:* $50 ($55 for international students). Electronic applications accepted.

Expenses: Tuition, state resident: full-time $5,040; part-time $420 per credit. Tuition, nonresident: full-time $9,900; part-time $825 per credit. Full-time tuition and fees vary according to course load, program and reciprocity agreements.

Financial aid: In 1999–00, 35 students received aid, including 2 fellowships with full tuition reimbursements available (averaging $16,000 per year), 2 research assistantships with full tuition reimbursements available (averaging $14,550 per year), 3 teaching assistantships with full tuition reimbursements available (averaging $11,115 per year); career-related internships or fieldwork, Federal Work-Study, grants, scholarships, traineeships, and tuition waivers (full and partial) also available. Financial aid application deadline: 12/15; financial aid applicants required to submit FAFSA.

Faculty research: Cell and molecular biology; plant physiology; plant structure, diversity, and development; ecology; systematics and evolution.

Prof. Steve Gantt, Head, 612-625-1234.

Application contact: Prof. M. David Marks, Director of Graduate Studies, 612-625-4222, *Fax:* 612-625-1738.

Find an in-depth description at www.petersons.com/graduate.

■ UNIVERSITY OF MISSOURI–ST. LOUIS

Graduate School, College of Arts and Sciences, Department of Biology, St. Louis, MO 63121-4499

AWARDS Biology (MS, PhD), including animal behavior (MS), biochemistry, biotechnology (MS), conservation biology (MS), development (MS), ecology (MS), environmental studies (PhD), evolution (MS), genetics (MS), molecular biology and biotechnology (PhD), molecular/cellular biology (MS), physiology (MS), plant systematics, population biology (MS), tropical biology (MS); biotechnology (Certificate); tropical biology and conservation (Certificate). Part-time programs available.

Faculty: 46.
Students: 21 full-time (11 women), 75 part-time (44 women); includes 13 minority (2 African Americans, 2 Asian Americans or Pacific Islanders, 8 Hispanic Americans, 1 Native American), 23 international. In 1999, 14 master's, 4 doctorates awarded.
Degree requirements: For master's, thesis or alternative required, foreign language not required; for doctorate, one foreign language, dissertation, 1 semester of teaching experience required.
Entrance requirements: For doctorate, GRE General Test. *Application deadline:* For fall admission, 7/1 (priority date); for spring admission, 11/1 (priority date). Applications are processed on a rolling basis. *Application fee:* $25 ($40 for international students). Electronic applications accepted.
Expenses: Tuition, state resident: full-time $4,932; part-time $173 per credit hour. Tuition, nonresident: full-time $13,279; part-time $521 per credit hour. Required fees: $775; $33 per credit hour. Tuition and fees vary according to degree level and program.
Financial aid: In 1999–00, 8 research assistantships with partial tuition reimbursements (averaging $10,635 per year), 14 teaching assistantships with partial tuition reimbursements (averaging $11,488 per year) were awarded; career-related internships or fieldwork and Federal Work-Study also available. Aid available to part-time students. Financial aid application deadline: 2/1. *Total annual research expenditures:* $908,828.
Application contact: Graduate Admissions, 314-516-5458, *Fax:* 314-516-6759, *E-mail:* gradadm@umsl.edu.

■ UNIVERSITY OF NEW HAMPSHIRE

Graduate School, College of Life Sciences and Agriculture, Graduate Programs in the Biological Sciences and Natural Resources, Department of Plant Biology, Durham, NH 03824

AWARDS MS, PhD. Part-time programs available.

Faculty: 23 full-time.
Students: 12 full-time (10 women), 14 part-time (6 women), 6 international. Average age 30. *14 applicants, 64% accepted.* In 1999, 7 master's awarded. Terminal master's awarded for partial completion of doctoral program.
Degree requirements: For master's and doctorate, thesis/dissertation required, foreign language not required.
Entrance requirements: For master's and doctorate, GRE General Test, GRE Subject Test. *Application deadline:* For fall admission, 4/1 (priority date). Applications are processed on a rolling basis. *Application fee:* $50.
Expenses: Tuition, area resident: Full-time $5,750; part-time $319 per credit. Tuition, state resident: full-time $8,625; part-time $478. Tuition, nonresident: full-time $14,640; part-time $598 per credit. Required fees: $224 per semester. Tuition and fees vary according to course load, degree level and program.
Financial aid: In 1999–00, 8 research assistantships, 10 teaching assistantships were awarded; fellowships, career-related internships or fieldwork, Federal Work-Study, scholarships, and tuition waivers (full and partial) also available. Aid available to part-time students. Financial aid application deadline: 2/15.
Dr. Robert Blanchard, Chairperson, 603-862-3697, *E-mail:* rb@cisunix.unh.edu.
Application contact: Dr. Thomas Davis, Contact, 603-862-3217, *E-mail:* tmdavis@cisunix.unh.edu.

■ UNIVERSITY OF NEW MEXICO

Graduate School, College of Arts and Sciences, Department of Biology, Albuquerque, NM 87131-2039

AWARDS Biology (MS, PhD), including air land ecology, behavioral ecology, botany, cellular and molecular biology, community ecology, comparative immunology, comparative physiology, conservation biology, ecology, ecosystem ecology, evolutionary biology, evolutionary genetics, microbiology, molecular genetics, parasitology, physiological ecology, physiology, population biology, vertebrate and invertebrate zoology. Part-time programs available.

Faculty: 35 full-time (5 women), 18 part-time/adjunct (11 women).

Students: 71 full-time (37 women), 28 part-time (11 women); includes 8 minority (2 Asian Americans or Pacific Islanders, 5 Hispanic Americans, 1 Native American), 11 international. Average age 33. *93 applicants, 30% accepted.* In 1999, 11 master's, 12 doctorates awarded. Terminal master's awarded for partial completion of doctoral program.
Degree requirements: For master's, one foreign language (computer language can substitute), thesis required (for some programs); for doctorate, 2 foreign languages (computer language can substitute for one), dissertation required.
Entrance requirements: For master's and doctorate, GRE General Test, GRE Subject Test, minimum GPA of 3.2. *Application deadline:* For fall admission, 1/15. Application fee: $25.
Expenses: Tuition, state resident: full-time $2,514; part-time $105 per credit hour. Tuition, nonresident: full-time $10,304; part-time $417 per credit hour. International tuition: $10,304 full-time. Required fees: $516; $22 per credit hour. Tuition and fees vary according to program.
Financial aid: In 1999–00, 58 students received aid, including 24 fellowships (averaging $1,645 per year), 26 research assistantships with tuition reimbursements available (averaging $8,921 per year), 40 teaching assistantships with tuition reimbursements available (averaging $11,066 per year); career-related internships or fieldwork, Federal Work-Study, institutionally sponsored loans, and tuition waivers (full and partial) also available. Aid available to part-time students. Financial aid applicants required to submit FAFSA.
Faculty research: Developmental biology, immunobiology. *Total annual research expenditures:* $4.5 million.
Dr. Kathryn Vogel, Chair, 505-277-3411, *Fax:* 505-277-0304, *E-mail:* kgvogel@unm.edu.
Application contact: Vivian Kent, Information Contact, 505-277-1712, *Fax:* 505-277-0304, *E-mail:* vkent@unm.edu.

■ THE UNIVERSITY OF NORTH CAROLINA AT CHAPEL HILL

Graduate School, College of Arts and Sciences, Department of Biology, Program in Botany, Chapel Hill, NC 27599

AWARDS MA, MS, PhD.

Degree requirements: For master's, comprehensive exams required; for doctorate, dissertation, comprehensive exams required.
Entrance requirements: For master's and doctorate, GRE General Test, GRE Subject Test.

The University of North Carolina at Chapel Hill (continued)

Expenses: Tuition, state resident: full-time $1,578. Tuition, nonresident: full-time $10,744. Required fees: $827. One-time fee: $15 full-time. Tuition and fees vary according to program.

■ UNIVERSITY OF NORTH DAKOTA

Graduate School, College of Arts and Sciences, Department of Biology, Grand Forks, ND 58202

AWARDS Botany (MS, PhD); ecology (MS, PhD); entomology (MS, PhD); environmental biology (MS, PhD); fisheries/wildlife (MS, PhD); genetics (MS, PhD); zoology (MS, PhD).

Faculty: 18 full-time (3 women).
Students: 21 full-time (8 women). *13 applicants, 62% accepted.* In 1999, 3 master's awarded. Terminal master's awarded for partial completion of doctoral program.
Degree requirements: For master's, thesis, final exam required; for doctorate, dissertation, comprehensive exam, final exam required.
Entrance requirements: For master's, GRE General Test, GRE Subject Test, TOEFL, minimum GPA of 3.0; for doctorate, GRE General Test, GRE Subject Test, TOEFL, minimum GPA of 3.5. *Application deadline:* For fall admission, 3/1 (priority date). Applications are processed on a rolling basis. *Application fee:* $25.
Expenses: Tuition, state resident: full-time $3,166; part-time $158 per credit. Tuition, nonresident: full-time $7,658; part-time $345 per credit. International tuition: $7,658 full-time. Required fees: $46 per credit. Tuition and fees vary according to program and reciprocity agreements.
Financial aid: In 1999–00, 6 research assistantships with full tuition reimbursements (averaging $11,250 per year), 15 teaching assistantships with full tuition reimbursements (averaging $11,250 per year) were awarded; fellowships, Federal Work-Study, institutionally sponsored loans, scholarships, and tuition waivers (full and partial) also available. Aid available to part-time students. Financial aid application deadline: 3/15; financial aid applicants required to submit FAFSA.
Faculty research: Population biology, wildlife ecology, RNA processing, hormonal control of behavior.
Dr. Jeff Lang, Director, 701-777-2621, *Fax:* 701-777-2623, *E-mail:* jlang@badlands.nodak.edu.

■ UNIVERSITY OF OKLAHOMA

Graduate College, College of Arts and Sciences, Department of Botany and Microbiology, Program in Botany, Norman, OK 73019-0390

AWARDS M Nat Sci, MS, PhD. Part-time programs available.

Students: 10 full-time (6 women), 2 part-time (1 woman); includes 1 minority (Native American), 2 international. *12 applicants, 67% accepted.* In 1999, 3 master's awarded.
Degree requirements: For master's, thesis, oral exam required; for doctorate, one foreign language, dissertation, general exam required.
Entrance requirements: For master's and doctorate, TOEFL. *Application deadline:* For fall admission, 6/1; for spring admission, 12/1. Applications are processed on a rolling basis. *Application fee:* $25.
Expenses: Tuition, state resident: full-time $2,064; part-time $86 per credit hour. Tuition, nonresident: full-time $6,588; part-time $275 per credit hour. Required fees: $468; $12 per credit hour. $94 per semester. Tuition and fees vary according to course level, course load and program.
Financial aid: In 1999–00, 7 students received aid; research assistantships, teaching assistantships, Federal Work-Study and scholarships available. Aid available to part-time students.
Faculty research: Plant ecology, phytoremediation, cell biology, global change biology, plant physiology.
Application contact: Mary Mason, Secretary I, 405-325-4321, *Fax:* 405-325-7619.

■ UNIVERSITY OF PENNSYLVANIA

School of Arts and Sciences, Graduate Group in Biology, Program in Plant Science, Philadelphia, PA 19104

AWARDS PhD.

Students: *11 applicants, 45% accepted.*
Degree requirements: For doctorate, dissertation required, foreign language not required.
Entrance requirements: For doctorate, GRE General Test, GRE Subject Test, TOEFL. *Application fee:* $65.
Expenses: Tuition: Full-time $23,670. Required fees: $1,546. Full-time tuition and fees vary according to degree level and program.
Financial aid: Fellowships available. Financial aid application deadline: 1/2.
Faculty research: Plant-cell interaction, plant developmental genetics, cell fine structure, reproductive biology, tropical animal-plant interactions.

Application contact: Allan Aiken, Graduate Group Coordinator, 215-898-6786, *Fax:* 215-898-8780, *E-mail:* aaiken@sas.upenn.edu.

Find an in-depth description at www.petersons.com/graduate.

■ UNIVERSITY OF RHODE ISLAND

Graduate School, College of Arts and Sciences, Department of Botany, Kingston, RI 02881

AWARDS MS, PhD.

Degree requirements: For master's and doctorate, thesis/dissertation required. *Application deadline:* For fall admission, 4/15 (priority date). Applications are processed on a rolling basis. *Application fee:* $35.
Expenses: Tuition, state resident: full-time $3,540; part-time $197 per credit. Tuition, nonresident: full-time $10,116; part-time $197 per credit. Required fees: $1,352; $37 per credit. $65 per term.
Dr. Roger Goos, Chairperson, 401-874-2161.

■ UNIVERSITY OF RHODE ISLAND

Graduate School, College of Resource Development, Department of Plant Sciences, Program in Plant Science, Kingston, RI 02881

AWARDS MS, PhD.

Degree requirements: For master's, thesis, professional seminar required; for doctorate, one foreign language (computer language can substitute), dissertation, professional seminar required.
Entrance requirements: For master's, GRE General Test. *Application deadline:* For fall admission, 4/15. *Application fee:* $35.
Expenses: Tuition, state resident: full-time $3,540; part-time $197 per credit. Tuition, nonresident: full-time $10,116; part-time $197 per credit. Required fees: $1,352; $37 per credit. $65 per term.
Financial aid: Unspecified assistantships available.
Faculty research: Ecology, physiology, improvement of turf, ornamental and food-crop plants.
Dr. Richard Hull, Chairperson, 401-874-2791.

■ UNIVERSITY OF SOUTH FLORIDA

Graduate School, College of Arts and Sciences, Department of Biology, Program in Botany, Tampa, FL 33620-9951

AWARDS MS. Part-time programs available.

Degree requirements: For master's, thesis required, foreign language not required.

Entrance requirements: For master's, GRE General Test, minimum GPA of 3.0 in last 60 hours. Electronic applications accepted.

Expenses: Tuition, state resident: part-time $148 per credit hour. Tuition, nonresident: part-time $509 per credit hour.

■ THE UNIVERSITY OF TENNESSEE

Graduate School, College of Agricultural Sciences and Natural Resources, Department of Plant and Soil Sciences, Knoxville, TN 37996

AWARDS Crop physiology and ecology (MS, PhD); plant breeding and genetics (MS, PhD); soil science (MS, PhD). Part-time programs available.

Faculty: 19 full-time (1 woman).
Students: 11 full-time (3 women), 11 part-time (7 women), 2 international. 7 *applicants, 43% accepted.* In 1999, 7 master's, 5 doctorates awarded.
Degree requirements: For master's, thesis or alternative required, foreign language not required; for doctorate, dissertation required, foreign language not required.
Entrance requirements: For master's and doctorate, GRE General Test, TOEFL, minimum GPA of 2.7. *Application deadline:* For fall admission, 2/1 (priority date). Applications are processed on a rolling basis. *Application fee:* $35. Electronic applications accepted.
Expenses: Tuition, state resident: full-time $3,806; part-time $184 per credit hour. Tuition, nonresident: full-time $9,874; part-time $522 per credit hour. Tuition and fees vary according to program.
Financial aid: In 1999–00, 17 research assistantships, 2 teaching assistantships were awarded; fellowships, career-related internships or fieldwork, Federal Work-Study, and institutionally sponsored loans also available. Financial aid application deadline: 2/1; financial aid applicants required to submit FAFSA.
Dr. Fred L. Allen, Head, 865-974-7101, *Fax:* 865-974-7997, *E-mail:* allenf@utk.edu.
Application contact: Dr. Mike Essington, Graduate Representative, 865-974-8818, *E-mail:* messington@utk.edu.

■ THE UNIVERSITY OF TENNESSEE

Graduate School, College of Arts and Sciences, Department of Botany, Knoxville, TN 37996

AWARDS MS, PhD.

Faculty: 14 full-time (3 women), 1 part-time/adjunct (0 women).
Students: 20 full-time (11 women), 1 (woman) part-time; includes 1 minority (Hispanic American), 9 international. *10 applicants, 40% accepted.* In 1999, 1 master's awarded. Terminal master's awarded for partial completion of doctoral program.
Degree requirements: For master's, thesis or alternative required, foreign language not required; for doctorate, dissertation required.
Entrance requirements: For master's and doctorate, GRE General Test, TOEFL, minimum GPA of 2.7. *Application deadline:* For fall admission, 2/1 (priority date). Applications are processed on a rolling basis. *Application fee:* $35. Electronic applications accepted.
Expenses: Tuition, state resident: full-time $3,806; part-time $184 per credit hour. Tuition, nonresident: full-time $9,874; part-time $522 per credit hour. Tuition and fees vary according to program.
Financial aid: In 1999–00, 16 teaching assistantships were awarded; fellowships, research assistantships, Federal Work-Study, institutionally sponsored loans, and unspecified assistantships also available. Financial aid application deadline: 2/1; financial aid applicants required to submit FAFSA.
Dr. Edward Schilling, Head, 865-974-2256, *Fax:* 865-974-0978, *E-mail:* eschilling@utk.edu.
Application contact: Dr. Otto Schwarz, Graduate Representative, *E-mail:* oschwarz@utk.edu.

■ THE UNIVERSITY OF TEXAS AT AUSTIN

Graduate School, College of Natural Sciences, School of Biological Sciences, Program in Plant Biology, Austin, TX 78712-1111

AWARDS MA, PhD.

Entrance requirements: For master's and doctorate, GRE General Test, TOEFL, minimum GPA of 3.0. *Application deadline:* Applications are processed on a rolling basis. *Application fee:* $50 ($75 for international students).
Expenses: Tuition, state resident: part-time $114 per semester hour. Tuition, nonresident: part-time $330 per semester hour. Tuition and fees vary according to program.
Financial aid: Fellowships, teaching assistantships available. Financial aid application deadline: 1/15.
Dr. Guy A. Thompson, Chairman, 512-471-5858.
Application contact: Dr. Robert K. Jansen, Graduate Admissions Chair, 512-471-8827.

■ UNIVERSITY OF VERMONT

Graduate College, College of Agriculture and Life Sciences, Department of Botany, Burlington, VT 05405

AWARDS Biology (MST); botany (MAT, MS, PhD); field naturalist (MS).

Degree requirements: For master's, thesis required, foreign language not required; for doctorate, dissertation required.
Entrance requirements: For master's, GRE General Test, TOEFL; for doctorate, GRE General Test, GRE Subject Test, TOEFL.
Expenses: Tuition, state resident: full-time $7,464; part-time $311 per credit. Tuition, nonresident: full-time $18,672; part-time $778 per credit. Full-time tuition and fees vary according to degree level and program.
Faculty research: Ecology, physiology, cytology, genetics.

■ UNIVERSITY OF VERMONT

Graduate College, College of Agriculture and Life Sciences, Department of Plant and Soil Science, Burlington, VT 05405

AWARDS MS, PhD.

Degree requirements: For master's, thesis required, foreign language not required; for doctorate, dissertation required.
Entrance requirements: For master's and doctorate, GRE General Test, TOEFL.
Expenses: Tuition, state resident: full-time $7,464; part-time $311 per credit. Tuition, nonresident: full-time $18,672; part-time $778 per credit. Full-time tuition and fees vary according to degree level and program.
Faculty research: Soil chemistry, plant nutrition.

■ UNIVERSITY OF WASHINGTON

Graduate School, College of Arts and Sciences, Department of Botany, Seattle, WA 98195

AWARDS MS, PhD.

Faculty: 21 full-time (4 women), 6 part-time/adjunct (0 women).
Students: 26 full-time (15 women); includes 1 minority (Hispanic American), 5 international. Average age 30. *85 applicants, 7% accepted.* In 1999, 3 master's awarded (100% found work related to degree); 5 doctorates awarded (60% entered university research/teaching, 40% found other work related to degree).
Degree requirements: For master's, thesis optional, foreign language not required; for doctorate, dissertation required, foreign language not required.

University of Washington (continued)
Average time to degree: Master's–3 years full-time; doctorate–6 years full-time.
Entrance requirements: For master's and doctorate, GRE General Test, GRE Subject Test, TOEFL, minimum GPA of 3.0. *Application deadline:* For fall admission, 1/15. *Application fee:* $50. Electronic applications accepted.
Expenses: Tuition, state resident: full-time $5,196; part-time $495 per credit. Tuition, nonresident: full-time $13,485; part-time $1,285 per credit. Required fees: $387; $36 per credit. Tuition and fees vary according to course load and program.
Financial aid: In 1999–00, 18 fellowships with full tuition reimbursements, 6 teaching assistantships with full tuition reimbursements were awarded; research assistantships with full tuition reimbursements, career-related internships or fieldwork, grants, institutionally sponsored loans, and traineeships also available. Aid available to part-time students. Financial aid application deadline: 1/15; financial aid applicants required to submit FAFSA.
Faculty research: Molecular and cellular botany, plant physiology, systematics, plant population biology, ecosystems and environmental studies.
Prof. Joseph F. Ammirati, Chair, 206-543-1942, *Fax:* 206-685-1728.
Application contact: Jerry Pangilinan, Graduate Program Assistant, 206-543-1972, *Fax:* 206-685-1728, *E-mail:* jlpang@u.washington.edu.

■ UNIVERSITY OF WISCONSIN–MADISON

Graduate School, College of Agricultural and Life Sciences, Department of Horticulture, Plant Breeding and Plant Genetics Program, Madison, WI 53706-1380

AWARDS MS, PhD. Part-time programs available.

Faculty: 34 full-time (4 women).
Students: 48 full-time (14 women); includes 1 minority (Hispanic American), 30 international. Average age 24. *45 applicants, 11% accepted.* In 1999, 6 master's awarded (100% continued full-time study); 7 doctorates awarded (50% entered university research/teaching, 50% found other work related to degree). Terminal master's awarded for partial completion of doctoral program.
Degree requirements: For master's and doctorate, foreign language and thesis not required. *Average time to degree:* Master's–2 years full-time; doctorate–4 years full-time.
Entrance requirements: For master's and doctorate, GRE, minimum GPA of 3.0. *Application deadline:* Applications are

processed on a rolling basis. *Application fee:* $45. Electronic applications accepted.
Expenses: Tuition, state resident: full-time $5,406; part-time $339 per credit. Tuition, nonresident: full-time $17,110; part-time $1,071 per credit. Full-time tuition and fees vary according to program and reciprocity agreements. Part-time tuition and fees vary according to course load and program.
Financial aid: In 1999–00, 22 students received aid, including 5 research assistantships; career-related internships or fieldwork, Federal Work-Study, and tuition waivers (partial) also available.
Faculty research: Classical and molecular genetics.
Thomas C. Osborn, Chair, 608-262-2330, *Fax:* 608-262-4743, *E-mail:* tcosborn@fastaff.wisc.edu.

■ UNIVERSITY OF WISCONSIN–MADISON

Graduate School, College of Letters and Science, Department of Botany, Madison, WI 53706-1380

AWARDS MS, PhD. Part-time programs available.

Faculty: 16 full-time (5 women), 3 part-time/adjunct (0 women).
Students: 39 full-time (20 women); includes 2 minority (1 African American, 1 Hispanic American), 7 international. Average age 28. *70 applicants, 19% accepted.* In 1999, 1 master's awarded (100% continued full-time study); 4 doctorates awarded (100% entered university research/teaching). Terminal master's awarded for partial completion of doctoral program.
Degree requirements: For master's, thesis required, foreign language not required; for doctorate, one foreign language, dissertation required. *Average time to degree:* Master's–3 years full-time; doctorate–5 years full-time.
Entrance requirements: For master's and doctorate, GRE General Test. *Application deadline:* For fall admission, 1/2 (priority date). Applications are processed on a rolling basis. *Application fee:* $45. Electronic applications accepted.
Expenses: Tuition, state resident: full-time $5,406; part-time $339 per credit. Tuition, nonresident: full-time $17,110; part-time $1,071 per credit. Full-time tuition and fees vary according to program and reciprocity agreements. Part-time tuition and fees vary according to course load and program.
Financial aid: In 1999–00, 8 fellowships with full tuition reimbursements (averaging $12,700 per year), 7 research assistantships with full tuition reimbursements (averaging $14,500 per year), 10 teaching assistantships with full tuition reimbursements

(averaging $11,000 per year) were awarded; Federal Work-Study and institutionally sponsored loans also available.
Faculty research: Taxonomy and systematics; ecology; structural botany; physiological, cellular, and molecular biology.
Kenneth J. Sytsma, Chair, 608-262-1057, *Fax:* 608-262-7509, *E-mail:* botany@ls.wisc.edu.
Application contact: Barbara J. Schaack, Student Status Examiner, 608-262-0476, *Fax:* 608-262-7509, *E-mail:* botgrad@ls.wisc.edu.

■ UNIVERSITY OF WISCONSIN–OSHKOSH

Graduate School, College of Letters and Science, Department of Biology and Microbiology, Oshkosh, WI 54901

AWARDS Biology (MS), including botany, microbiology, zoology.

Degree requirements: For master's, thesis required, foreign language not required.
Entrance requirements: For master's, GRE General Test, minimum GPA of 3.0, BS in biology.
Expenses: Tuition, state resident: full-time $3,917; part-time $219 per credit. Tuition, nonresident: full-time $12,375; part-time $684 per credit. Part-time tuition and fees vary according to course load and program.

■ UNIVERSITY OF WYOMING

Graduate School, College of Arts and Sciences, Department of Botany, Laramie, WY 82071

AWARDS Botany (MS, PhD); botany/water resources (MS). Part-time programs available.

Faculty: 10 full-time (1 woman), 2 part-time/adjunct (1 woman).
Students: 18 full-time (6 women), 11 part-time (5 women). *12 applicants, 50% accepted.* In 1999, 3 master's awarded (25% entered university research/teaching, 25% found other work related to degree, 50% continued full-time study); 2 doctorates awarded (34% entered university research/teaching, 66% found other work related to degree). Terminal master's awarded for partial completion of doctoral program.
Degree requirements: For master's and doctorate, thesis/dissertation required, foreign language not required.
Entrance requirements: For master's and doctorate, GRE General Test, minimum GPA of 3.0. *Application deadline:* For fall admission, 2/1 (priority date). *Application fee:* $40. Electronic applications accepted.
Expenses: Tuition, state resident: full-time $2,520; part-time $140 per credit hour.

Tuition, nonresident: full-time $7,790; part-time $433 per credit hour. Required fees: $440; $7 per credit hour. Full-time tuition and fees vary according to course load and program.

Financial aid: In 1999–00, 14 research assistantships with full tuition reimbursements (averaging $8,667 per year), 9 teaching assistantships with full tuition reimbursements (averaging $8,667 per year) were awarded; Graduate School scholarships also available. Financial aid application deadline: 3/1.

Faculty research: Ecology, systematics, physiology, cell biology, mycology, biogeochemistry, paleoecology. *Total annual research expenditures:* $1.8 million.
Gregory K. Brown, Head, 307-766-2214, *Fax:* 307-766-2851, *E-mail:* gkbrown@uwyo.edu.

Application contact: Steve L. Miller, Associate Professor, 307-766-2834, *Fax:* 307-766-2851, *E-mail:* fungi@uwyo.edu.

■ UTAH STATE UNIVERSITY

School of Graduate Studies, College of Agriculture, Department of Plants, Soils, and Biometeorology, Logan, UT 84322

AWARDS Biometeorology (MS, PhD); ecology (MS, PhD); plant science (MS, PhD); soil science (MS, PhD). Part-time programs available.

Faculty: 28 full-time (3 women), 15 part-time/adjunct (0 women).

Students: 21 full-time (7 women), 7 part-time (2 women); includes 1 minority (African American), 4 international. *18 applicants, 61% accepted.* In 1999, 4 master's, 5 doctorates awarded. Terminal master's awarded for partial completion of doctoral program.

Degree requirements: For master's and doctorate, thesis/dissertation required, foreign language not required.

Entrance requirements: For master's, GRE General Test, TOEFL, BS in plant science, biological science, or related field, minimum GPA of 3.0; for doctorate, GRE General Test, TOEFL, minimum GPA of 3.0. *Application deadline:* For fall admission, 6/15 (priority date); for spring admission, 10/15. Applications are processed on a rolling basis. *Application fee:* $40.

Expenses: Tuition, state resident: full-time $1,553. Tuition, nonresident: full-time $5,436. International fee: $5,526 full-time. Required fees: $447. Tuition and fees vary according to course load and program.

Financial aid: In 1999–00, 1 fellowship with partial tuition reimbursement (averaging $12,000 per year), 25 research assistantships with partial tuition reimbursements (averaging $12,000 per year), 4 teaching assistantships with partial

tuition reimbursements were awarded; Federal Work-Study, institutionally sponsored loans, and tuition waivers (full and partial) also available. Aid available to part-time students. Financial aid application deadline: 4/15.

Faculty research: Plant physiology and developmental biology; plant improvement through breeding and biotechnology, soil, plant, water, and nutrient relationships; agronomic and landscape management; adaptations to weather. *Total annual research expenditures:* $4.3 million.
James H. Thomas, Head, 435-797-3394, *Fax:* 435-797-0464.

Application contact: Dr. John G. Carman, Graduate Student Coordinator, 435-797-2238, *Fax:* 435-797-3376, *E-mail:* jcarm@mendel.usu.edu.

■ VIRGINIA POLYTECHNIC INSTITUTE AND STATE UNIVERSITY

Graduate School, College of Arts and Sciences, Department of Biology, Program in Botany, Blacksburg, VA 24061

AWARDS MS, PhD.

Faculty: 11 full-time (0 women).

Degree requirements: For master's and doctorate, thesis/dissertation required, foreign language not required.

Entrance requirements: For master's, GRE General Test, TOEFL; for doctorate, GRE General Test, GRE Subject Test, TOEFL. *Application deadline:* For fall admission, 12/1 (priority date). Applications are processed on a rolling basis. *Application fee:* $25.

Expenses: Tuition, state resident: full-time $4,122; part-time $229 per credit hour. Tuition, nonresident: full-time $6,930; part-time $385 per credit hour. Required fees: $828; $107 per semester. Part-time tuition and fees vary according to course load.

Financial aid: Fellowships, research assistantships, teaching assistantships, unspecified assistantships available. Financial aid application deadline: 4/1.

Faculty research: Ecology, evolution, molecular biology, systematics, microbiology.
Dr. Joe R. Cowles, Chairman, Department of Biology, 540-231-8928, *E-mail:* cowlesjr@vt.edu.

■ WASHINGTON STATE UNIVERSITY

Graduate School, College of Sciences, School of Biological Sciences, Department of Botany, Pullman, WA 99164

AWARDS MS, PhD.

Faculty: 12 full-time (1 woman).

Students: 17 full-time (5 women), 2 part-time (both women); includes 2 minority (both Asian Americans or Pacific Islanders). Average age 28. In 1999, 1 master's, 2 doctorates awarded.

Degree requirements: For master's, oral exam required, thesis optional, foreign language not required; for doctorate, dissertation, oral exam required, foreign language not required. *Average time to degree:* Master's–2 years full-time; doctorate–4 years full-time.

Entrance requirements: For master's and doctorate, GRE General Test, GRE Subject Test (recommended), minimum GPA of 3.0. *Application deadline:* For fall admission, 2/1 (priority date); for spring admission, 9/15. Applications are processed on a rolling basis. *Application fee:* $35.

Expenses: Tuition, state resident: full-time $5,654. Tuition, nonresident: full-time $13,850. International tuition: $13,850 full-time. Tuition and fees vary according to program.

Financial aid: In 1999–00, 3 research assistantships with full and partial tuition reimbursements, 16 teaching assistantships with full and partial tuition reimbursements were awarded; fellowships, career-related internships or fieldwork, Federal Work-Study, institutionally sponsored loans, and tuition waivers (partial) also available. Financial aid application deadline: 4/1; financial aid applicants required to submit FAFSA.

Faculty research: Molecular biology, plant physiology, systematics, ecology-evolution. *Total annual research expenditures:* $647,177.
Dr. Richard N. Mack, Chair, 509-335-3066, *Fax:* 509-335-3517.

Application contact: Linda M. Thompson, Graduate Secretary, 509-335-1666, *Fax:* 509-335-3517, *E-mail:* linder@mail.wsu.edu.

■ WASHINGTON UNIVERSITY IN ST. LOUIS

Graduate School of Arts and Sciences, Division of Biology and Biomedical Sciences, Program in Plant Biology, St. Louis, MO 63130-4899

AWARDS PhD.

Degree requirements: For doctorate, dissertation required, foreign language not required.

Entrance requirements: For doctorate, GRE General Test, GRE Subject Test. *Application deadline:* For fall admission, 1/1 (priority date). Applications are processed on a rolling basis. *Application fee:* $0.

Expenses: Tuition: Full-time $23,400; part-time $975 per credit. Tuition and fees vary according to program.

Washington University in St. Louis (continued)
Financial aid: Fellowships, research assistantships available. Financial aid application deadline: 1/1.
Dr. Himadri Pakrasi, Co-Coordinator, 314-362-6853.
Application contact: Rosemary Garagneni, Director of Admissions, 800-852-9074, *E-mail:* admissions@ dbbs.wustl.edu.

■ WEST TEXAS A&M UNIVERSITY

College of Agriculture, Nursing, and Natural Sciences, Division of Agriculture, Emphasis in Plant Science, Canyon, TX 79016-0001

AWARDS MS. Part-time programs available.

Degree requirements: For master's, comprehensive exam required, thesis optional, foreign language not required.
Entrance requirements: For master's, GRE General Test. Electronic applications accepted.
Expenses: Tuition, state resident: full-time $1,152; part-time $48 per credit. Tuition, nonresident: full-time $6,336; part-time $264 per credit. Required fees: $1,063; $531 per semester.
Faculty research: Crop and soil disciplines.

■ WEST VIRGINIA UNIVERSITY

College of Agriculture, Forestry and Consumer Sciences, Division of Plant and Soil Sciences, Morgantown, WV 26506

AWARDS Agronomy (MS); entomology (MS); environmental microbiology (MS); horticulture (MS); plant and soil sciences (PhD); plant pathology (MS).

Faculty: 19 full-time (1 woman), 2 part-time/adjunct (1 woman).
Students: 18 full-time (10 women), 7 part-time (5 women); includes 1 minority (Native American), 2 international. Average age 26. In 1999, 1 master's awarded.
Degree requirements: For master's, thesis required, foreign language not required; for doctorate, dissertation, comprehensive exam required, foreign language not required.
Entrance requirements: For master's, GRE, TOEFL, minimum GPA of 2.5; for doctorate, GRE General Test, TOEFL, minimum GPA of 3.0. *Application fee:* $45.
Expenses: Tuition, state resident: full-time $2,910; part-time $154 per credit hour. Tuition, nonresident: full-time $8,368; part-time $457 per credit hour.
Financial aid: In 1999–00, 13 research assistantships, 3 teaching assistantships were awarded; Federal Work-Study,

institutionally sponsored loans, and tuition waivers (full and partial) also available. Financial aid application deadline: 2/1; financial aid applicants required to submit FAFSA.
Faculty research: Water quality, reclamation of disturbed land, crop production, pest control, environmental protection.
Dr. Barton Baker, Chair, 304-293-4817, *Fax:* 304-293-3740, *E-mail:* bbaker2@ wvu.edu.

■ WEST VIRGINIA UNIVERSITY

Eberly College of Arts and Sciences, Department of Biology, Morgantown, WV 26506

AWARDS Animal behavior (MS); cellular and molecular biology (MS, PhD); environmental plant biology (MS, PhD); plant systematics (MS); population genetics (MS).

Faculty: 18 full-time (4 women), 2 part-time/adjunct (1 woman).
Students: 19 full-time (10 women), 3 part-time (2 women); includes 2 minority (1 Asian American or Pacific Islander, 1 Hispanic American), 4 international. Average age 27. *50 applicants, 10% accepted.* In 1999, 4 master's, 1 doctorate awarded. Terminal master's awarded for partial completion of doctoral program.
Degree requirements: For master's, thesis, final exam required, foreign language not required; for doctorate, dissertation, preliminary and final exams required, foreign language not required. *Average time to degree:* Master's–2.5 years full-time; doctorate–5 years full-time.
Entrance requirements: For master's, GRE General Test, GRE Subject Test, TOEFL, minimum GPA of 3.0; for doctorate, GRE General Test, TOEFL, minimum GPA of 3.0. *Application deadline:* For fall admission, 4/1; for spring admission, 10/1. Applications are processed on a rolling basis. *Application fee:* $45.
Expenses: Tuition, state resident: full-time $2,910; part-time $154 per credit hour. Tuition, nonresident: full-time $8,368; part-time $457 per credit hour.
Financial aid: In 1999–00, 3 research assistantships, 13 teaching assistantships were awarded; Federal Work-Study and institutionally sponsored loans also available. Financial aid application deadline: 4/1; financial aid applicants required to submit FAFSA.
Faculty research: Environmental biology, genetic engineering, developmental biology, global change, biodiversity.
Dr. Keith Garbutt, Chair, 304-293-5394.
Application contact: Dr. James B. McGraw, Director of Graduate Studies, 304-293-5201, *Fax:* 304-293-6363.
Find an in-depth description at www.petersons.com/graduate.

■ YALE UNIVERSITY

Graduate School of Arts and Sciences, Department of Molecular, Cellular, and Developmental Biology, Program in Plant Sciences, New Haven, CT 06520

AWARDS PhD.

Degree requirements: For doctorate, dissertation required.
Entrance requirements: For doctorate, GRE General Test, GRE Subject Test. *Application deadline:* For fall admission, 1/4. *Application fee:* $65.
Expenses: Tuition: Full-time $22,300. Full-time tuition and fees vary according to program.
Financial aid: Fellowships, research assistantships, teaching assistantships available.
Application contact: Anne Scott, Graduate Registrar, 203-432-3538, *Fax:* 203-432-3597, *E-mail:* anne.scott@yale.edu.
Find an in-depth description at www.petersons.com/graduate.

PLANT MOLECULAR BIOLOGY

■ CORNELL UNIVERSITY

Graduate School, Graduate Fields of Agriculture and Life Sciences, Field of Plant Biology, Ithaca, NY 14853-0001

AWARDS Paleobotany (PhD); plant cell biology (PhD); plant ecology (PhD); plant molecular biology (PhD); plant morphology, anatomy and biomechanics (PhD); plant physiology (PhD); systematic botany (PhD).

Faculty: 38 full-time.
Students: 51 full-time (26 women); includes 10 minority (1 African American, 5 Asian Americans or Pacific Islanders, 4 Hispanic Americans), 16 international. *94 applicants, 24% accepted.* In 1999, 11 doctorates awarded.
Degree requirements: For doctorate, dissertation required, foreign language not required.
Entrance requirements: For doctorate, GRE General Test, GRE Subject Test (biology) (recommended), TOEFL. *Application deadline:* For fall admission, 1/15. *Application fee:* $65. Electronic applications accepted.
Expenses: Tuition: Full-time $12,100.
Financial aid: In 1999–00, 47 students received aid, including 12 fellowships with full tuition reimbursements available, 19 research assistantships with full tuition reimbursements available, 16 teaching assistantships with full tuition reimbursements available; institutionally sponsored loans, scholarships, tuition waivers (full

and partial), and unspecified assistantships also available. Financial aid applicants required to submit FAFSA.

Faculty research: Plant cell biology/cytology, plant molecular biology plant morphology/anatomy/biomechanics, plant physiology, systematic botany, paleobotany, plant ecology, ethnobotany, plant biochemistry, photosynthesis.

Application contact: Graduate Field Assistant, 607-255-8507, *E-mail:* plbio@cornell.edu.

■ UNIVERSITY OF CALIFORNIA, LOS ANGELES

Graduate Division, College of Letters and Science, Department of Organic Biology, Ecology and Evolution, Plant Molecular Biology Program, Los Angeles, CA 90095

AWARDS PhD.

Students: 66 full-time (26 women); includes 24 minority (21 Asian Americans or Pacific Islanders, 3 Hispanic Americans), 5 international. *8 applicants, 0% accepted.*

Degree requirements: For doctorate, dissertation, oral and written qualifying exams required, foreign language not required.

Entrance requirements: For doctorate, GRE General Test, GRE Subject Test, minimum undergraduate GPA of 3.0. *Application deadline:* For fall admission, 1/1. *Application fee:* $40. Electronic applications accepted.

Expenses: Tuition, nonresident: full-time $9,804. Required fees: $4,405. Full-time tuition and fees vary according to program and student level.

Financial aid: Fellowships, research assistantships, teaching assistantships available. Financial aid application deadline: 3/1.

Dr. Arnold Berk, Director, 310-206-6298.

Application contact: UCLA Access Coordinator, 800-284-8252, *Fax:* 310-206-5280.

■ UNIVERSITY OF CALIFORNIA, SAN DIEGO

Graduate Studies and Research, Department of Biology, Program in Plant Molecular Biology, La Jolla, CA 92093

AWARDS PhD.

Degree requirements: For doctorate, dissertation required.

Entrance requirements: For doctorate, GRE General Test, pre-application beginning in September. *Application deadline:* For fall admission, 1/8. *Application fee:* $40.

Expenses: Tuition, nonresident: full-time $14,691. Required fees: $4,697. Full-time tuition and fees vary according to program.

Application contact: Biology Graduate Admissions Committee, 858-534-3835.

■ UNIVERSITY OF CONNECTICUT

Graduate School, College of Liberal Arts and Sciences, Biological Sciences Group, Storrs, CT 06269

AWARDS Ecology and evolutionary biology (MS, PhD), including botany, ecology, entomology, systematics, zoology; molecular and cell biology (MS, PhD), including biochemistry, biophysics, biotechnology (MS), cell and developmental biology, genetics, microbiology, plant molecular and cell biology; physiology and neurobiology (MS, PhD), including neurobiology, physiology.

Degree requirements: For doctorate, dissertation required.

Entrance requirements: For master's and doctorate, GRE General Test, GRE Subject Test, TOEFL.

Expenses: Tuition, state resident: full-time $5,118. Tuition, nonresident: full-time $13,298. Required fees: $1,022.

■ UNIVERSITY OF CONNECTICUT

Graduate School, College of Liberal Arts and Sciences, Biological Sciences Group, Department of Molecular and Cell Biology, Field of Plant Molecular and Cell Biology, Storrs, CT 06269

AWARDS MS, PhD.

Degree requirements: For doctorate, dissertation required.

Entrance requirements: For master's and doctorate, GRE General Test, GRE Subject Test, TOEFL.

Expenses: Tuition, state resident: full-time $5,118. Tuition, nonresident: full-time $13,298. Required fees: $1,022.

Find an in-depth description at www.petersons.com/graduate.

■ UNIVERSITY OF DELAWARE

College of Arts and Science, Department of Biological Sciences, Newark, DE 19716

AWARDS Biotechnology (MS, PhD); cell and extracellular matrix biology (MS, PhD); cell and systems physiology (MS, PhD); ecology and evolution (MS, PhD); microbiology (MS, PhD); molecular biology and genetics (MS, PhD); plant biology (MS, PhD).

Faculty: 37 full-time (10 women).

Students: 22 full-time (11 women), 1 part-time; includes 2 minority (both African Americans), 9 international. Average age

---

25. *37 applicants, 27% accepted.* In 2000, 9 doctorates awarded.

Degree requirements: For master's and doctorate, thesis/dissertation required, foreign language not required. *Average time to degree:* Master's–2.5 years full-time; doctorate–6 years full-time.

Entrance requirements: For master's and doctorate, GRE General Test, GRE Subject Test (advanced biology). *Application deadline:* For fall admission, 6/15. Applications are processed on a rolling basis. *Application fee:* $50. Electronic applications accepted.

Expenses: Tuition, state resident: full-time $4,380; part-time $243 per credit. Tuition, nonresident: full-time $12,750; part-time $708 per credit. Required fees: $15 per term. Tuition and fees vary according to program.

Financial aid: In 2000–01, 18 students received aid, including 2 fellowships with full tuition reimbursements available (averaging $18,000 per year), 4 research assistantships with full tuition reimbursements available (averaging $18,000 per year), 11 teaching assistantships with full tuition reimbursements available (averaging $18,000 per year); tuition waivers (partial) also available. Financial aid application deadline: 6/15.

Faculty research: Cell interactions, molecular mechanisms, microorganisms, embryo implantation. *Total annual research expenditures:* $1.8 million.

Dr. Daniel D. Carson, Chair, 302-831-6977, *Fax:* 302-831-2281, *E-mail:* dcarson@udel.edu.

Application contact: Norman Karin, Graduate Coordinator, 302-831-1841, *Fax:* 302-831-2281, *E-mail:* ccoletta@udel.edu.

Find an in-depth description at www.petersons.com/graduate.

■ UNIVERSITY OF FLORIDA

Graduate School, College of Agriculture and Life Sciences and College of Liberal Arts and Sciences, Program in Plant Molecular and Cellular Biology, Gainesville, FL 32611

AWARDS MS, PhD.

Faculty: 27.

Students: 28 full-time (14 women), 1 part-time; includes 3 minority (1 African American, 2 Asian Americans or Pacific Islanders), 15 international. *15 applicants, 53% accepted.* In 1999, 2 master's, 2 doctorates awarded.

Degree requirements: For master's and doctorate, thesis/dissertation required, foreign language not required.

Entrance requirements: For master's and doctorate, GRE General Test, minimum GPA of 3.0. *Application deadline:* For fall admission, 2/1 (priority date). Applications

---

University of Florida (continued)
are processed on a rolling basis. *Application fee:* $20. Electronic applications accepted.
Expenses: Tuition, state resident: part-time $144 per credit hour. Tuition, nonresident: part-time $505 per credit hour. Tuition and fees vary according to course level, course load and program.
Financial aid: In 1999–00, 23 research assistantships, 1 teaching assistantship were awarded; fellowships, unspecified assistantships also available.
Faculty research: Plant pathology, genetics, biochemistry, microbiology.
Application contact: Dr. Kenneth Cline, Graduate Coordinator, 352-392-1928, *Fax:* 352-392-6479, *E-mail:* kcline@ufl.edu.

PLANT PATHOLOGY

■ AUBURN UNIVERSITY

Graduate School, College of Agriculture, Department of Entomology and Plant Pathology, Auburn, Auburn University, AL 36849-0002

AWARDS Entomology (M Ag, MS, PhD); plant pathology (M Ag, MS, PhD). Part-time programs available.

Faculty: 25 full-time (4 women).
Students: 11 full-time (3 women), 13 part-time (6 women); includes 4 minority (2 African Americans, 2 Asian Americans or Pacific Islanders), 9 international. *11 applicants, 45% accepted.* In 2000, 2 master's, 2 doctorates awarded.
Degree requirements: For master's, thesis (MS) required; for doctorate, dissertation required.
Entrance requirements: For master's, GRE General Test; for doctorate, GRE General Test, GRE Subject Test, master's degree with thesis. *Application deadline:* For fall admission, 7/7; for spring admission, 11/24. Applications are processed on a rolling basis. *Application fee:* $25 ($50 for international students). Electronic applications accepted.
Expenses: Tuition, state resident: full-time $2,895; part-time $80 per credit hour. Tuition, nonresident: full-time $8,685; part-time $240 per credit hour.
Financial aid: Research assistantships, teaching assistantships, Federal Work-Study available. Aid available to part-time students. Financial aid application deadline: 3/15.
Faculty research: Pest management, biological control, systematics, medical entomology.
Dr. Michael L. Williams, Chair, 334-844-5006.

Application contact: Dr. John F. Pritchett, Dean of the Graduate School, 334-844-4700.

■ CLEMSON UNIVERSITY

Graduate School, College of Agriculture, Forestry and Life Sciences, School of Plant, Statistical and Ecological Sciences, Department of Plant Pathology and Physiology and Department of Biological Sciences, Program in Plant Pathology, Clemson, SC 29634

AWARDS MS, PhD.

Students: 12 full-time (2 women), 2 part-time; includes 1 minority (African American), 7 international. Average age 28. *6 applicants, 33% accepted.* In 1999, 4 master's, 2 doctorates awarded.
Degree requirements: For master's and doctorate, thesis/dissertation required.
Entrance requirements: For master's, GRE General Test, TOEFL, minimum undergraduate GPA of 3.0; for doctorate, GRE General Test, TOEFL. *Application deadline:* For fall admission, 6/1; for spring admission, 11/1. Applications are processed on a rolling basis. *Application fee:* $40.
Expenses: Tuition, state resident: full-time $3,480; part-time $174 per credit hour. Tuition, nonresident: full-time $9,256; part-time $388 per credit hour. Required fees: $5 per term. Full-time tuition and fees vary according to course level, course load and campus/location.
Financial aid: Applicants required to submit FAFSA.
Dr. Stephen A. Lewis, Chair, 864-656-5741, *Fax:* 864-565-0274, *E-mail:* slewis@clemson.edu.

■ COLORADO STATE UNIVERSITY

Graduate School, College of Agricultural Sciences, Department of Bioagricultural Sciences and Pest Management, Plant Pathology and Weed Science, Fort Collins, CO 80523-0015

AWARDS Plant pathology (MS, PhD); weed science (MS, PhD).

Faculty: 10 full-time (1 woman), 2 part-time/adjunct (0 women).
Students: 12 full-time (4 women), 9 part-time (5 women), 3 international. Average age 31. *24 applicants, 29% accepted.* In 1999, 5 master's, 3 doctorates awarded.
Degree requirements: For master's, thesis required (for some programs), foreign language not required; for doctorate, dissertation required, foreign language not required.
Entrance requirements: For master's and doctorate, GRE General Test, TOEFL,

minimum GPA of 3.0. *Application deadline:* For fall admission, 2/1 (priority date). Applications are processed on a rolling basis. *Application fee:* $30. Electronic applications accepted.
Expenses: Tuition, state resident: full-time $2,694; part-time $150 per credit. Tuition, nonresident: full-time $10,460; part-time $581 per credit. Required fees: $32 per semester. Tuition and fees vary according to program.
Financial aid: In 1999–00, 10 research assistantships, 2 teaching assistantships were awarded; fellowships, traineeships also available.
Faculty research: Biological control of plant pathogens and weeds, integrated pest management, weed ecology/biology, seed physiology/technology, molecular biology of plant stress. *Total annual research expenditures:* $2 million.
Thomas O. Holtzer, Head, 970-491-5261, *Fax:* 970-491-3862, *E-mail:* tholtzer@lamar.agsci.colostate.edu.

■ CORNELL UNIVERSITY

Graduate School, Graduate Fields of Agriculture and Life Sciences, Field of Plant Pathology, Ithaca, NY 14853-0001

AWARDS Ecological and environmental plant pathology (MPS, MS, PhD); molecular plant pathology (MPS, MS, PhD); mycology (MPS, MS, PhD); plant disease epidemiology (MPS, MS, PhD); plant pathology (MPS, MS, PhD).

Faculty: 47 full-time.
Students: 59 full-time (29 women); includes 8 minority (7 Asian Americans or Pacific Islanders, 1 Hispanic American), 27 international. *43 applicants, 44% accepted.* In 1999, 4 master's, 19 doctorates awarded.
Degree requirements: For master's, thesis (MS), project paper (MPS) required; for doctorate, dissertation required, foreign language not required.
Entrance requirements: For master's, GRE General Test, GRE Subject Test (biology) (recommended), TOEFL; for doctorate, GRE General Test, GRE Subject Test (biology) (recommended), TOEFL. *Application deadline:* Applications are processed on a rolling basis. *Application fee:* $65. Electronic applications accepted.
Expenses: Tuition: Full-time $12,100.
Financial aid: In 1999–00, 36 students received aid, including 6 fellowships with full tuition reimbursements available, 27 research assistantships with full tuition reimbursements available, 3 teaching assistantships with full tuition reimbursements available; institutionally sponsored loans, scholarships, tuition waivers (full and partial), and unspecified assistantships also available. Financial aid applicants required to submit FAFSA.

Faculty research: Plant pathology; mycology; molecular plant pathology; plant disease epidemiology, ecological and environmental plant pathology; plant disease epidemiology and simulation modeling; ecology and environmental quality; biology, genetics and control of plant pathogens.

Application contact: Graduate Field Assistant, 607-255-3259, *Fax:* 607-255-4471, *E-mail:* plpathology@cornell.edu.

■ IOWA STATE UNIVERSITY OF SCIENCE AND TECHNOLOGY

Graduate College, College of Agriculture, Department of Plant Pathology, Ames, IA 50011

AWARDS MS, PhD.

Faculty: 14 full-time.

Students: 38 full-time (17 women), 1 part-time, 16 international. *26 applicants, 35% accepted.* In 1999, 5 master's awarded.

Degree requirements: For master's, thesis or alternative required, foreign language not required; for doctorate, dissertation required, foreign language not required.

Entrance requirements: For master's and doctorate, GRE General Test, TOEFL. *Application deadline:* For fall admission, 3/1 (priority date). Applications are processed on a rolling basis. *Application fee:* $20 ($50 for international students). Electronic applications accepted.

Expenses: Tuition, state resident: full-time $3,308. Tuition, nonresident: full-time $9,744. Part-time tuition and fees vary according to course load, campus/location and program.

Financial aid: In 1999–00, 29 research assistantships with partial tuition reimbursements (averaging $11,794 per year) were awarded; fellowships, teaching assistantships, scholarships also available. Dr. Edward Braun, Chair, 515-294-1741, *Fax:* 515-294-9420, *E-mail:* plantpath@iastate.edu.

■ KANSAS STATE UNIVERSITY

Graduate School, College of Agriculture, Department of Plant Pathology, Manhattan, KS 66506

AWARDS MS, PhD. Terminal master's awarded for partial completion of doctoral program.

Degree requirements: For master's and doctorate, thesis/dissertation required, foreign language not required.

Expenses: Tuition, state resident: part-time $103 per credit hour. Tuition, nonresident: part-time $338 per credit hour. Required fees: $17 per credit hour. One-time fee: $64 part-time.

Faculty research: Biological control, microbial ecology, epidemiology, disease management, cell and tissue culture, plant transformation and regeneration, disease physiology, host-parasite genetics and cytogenetics, molecular genetics and plant biotechnology.

■ LOUISIANA STATE UNIVERSITY AND AGRICULTURAL AND MECHANICAL COLLEGE

Graduate School, College of Agriculture, Department of Plant Pathology and Crop Physiology, Baton Rouge, LA 70803

AWARDS Plant health (MS, PhD).

Faculty: 20 full-time (1 woman).

Students: 18 full-time (2 women), 2 part-time; includes 2 minority (1 Asian American or Pacific Islander, 1 Hispanic American), 7 international. Average age 28. *11 applicants, 27% accepted.* In 1999, 1 master's, 6 doctorates awarded. Terminal master's awarded for partial completion of doctoral program.

Degree requirements: For master's and doctorate, thesis/dissertation required, foreign language not required.

Entrance requirements: For master's and doctorate, GRE General Test, TOEFL, minimum GPA of 3.0. *Application deadline:* For fall admission, 1/25 (priority date). Applications are processed on a rolling basis. *Application fee:* $25.

Expenses: Tuition, state resident: full-time $2,881. Tuition, nonresident: full-time $7,081. Part-time tuition and fees vary according to course load and program.

Financial aid: In 1999–00, 2 fellowships, 19 research assistantships with partial tuition reimbursements (averaging $13,392 per year) were awarded; teaching assistantships with partial tuition reimbursements, career-related internships or fieldwork, Federal Work-Study, and tuition waivers (full) also available. Aid available to part-time students.

Faculty research: Plant health and protection, weed biology and management, crop physiology and biotechnology. *Total annual research expenditures:* $31,464. Dr. Johnnie Snow, Head, 225-388-1464, *Fax:* 225-388-1415, *E-mail:* jsnow@unix1.sncc.lsu.edu.

Application contact: Dr. Rodrigo Valverde, Admissions Committee Chair, 225-388-1384, *E-mail:* rvalver@lsuvm.snc.lsu.edu.

■ MICHIGAN STATE UNIVERSITY

Graduate School, College of Natural Science, Department of Botany and Plant Pathology, East Lansing, MI 48824

AWARDS Botany and plant pathology (MS, PhD); botany-environmental toxicology (PhD).

Faculty: 26.

Students: 42 full-time (15 women), 19 part-time (12 women); includes 2 minority (1 African American, 1 Hispanic American), 13 international. Average age 30. *30 applicants, 37% accepted.* In 1999, 5 master's, 4 doctorates awarded. Terminal master's awarded for partial completion of doctoral program.

Degree requirements: For master's, thesis required (for some programs), foreign language not required; for doctorate, dissertation required, foreign language not required.

Entrance requirements: For master's, GRE, TOEFL; for doctorate, GRE, TOEFL, master's degree, minimum GPA of 3.0. *Application deadline:* Applications are processed on a rolling basis. *Application fee:* $30 ($40 for international students).

Expenses: Tuition, state resident: part-time $229 per credit. Tuition, nonresident: part-time $464 per credit. Required fees: $241 per semester. Tuition and fees vary according to course load, degree level and program.

Financial aid: In 1999–00, 23 research assistantships (averaging $10,666 per year), 29 teaching assistantships (averaging $10,468 per year) were awarded; fellowships, institutionally sponsored loans also available. Financial aid application deadline: 1/15; financial aid applicants required to submit FAFSA.

Faculty research: Ecology and systematics, plant physiology and biochemistry, plant genetics and molecular biology, plant anatomy and morphology. *Total annual research expenditures:* $4.2 million. Dr. R. Hammerschmidt, Acting Chair, 517-355-4683, *Fax:* 517-353-1926.

Application contact: Richard Allison, Director, 517-432-1548, *E-mail:* allison@msu.edu.

Find an in-depth description at www.petersons.com/graduate.

■ MISSISSIPPI STATE UNIVERSITY

College of Agriculture and Life Sciences, Department of Entomology and Plant Pathology, Mississippi State, MS 39762

AWARDS Agricultural pest management (MS); entomology (MS, PhD); plant pathology (MS, PhD).

Mississippi State University (continued)
Faculty: 26 full-time (1 woman), 27 part-time/adjunct (2 women).
Students: 12 full-time (2 women), 6 part-time (1 woman); includes 1 minority (Hispanic American), 7 international. Average age 30. *9 applicants, 22% accepted.* In 1999, 3 master's, 2 doctorates awarded.
Degree requirements: For master's and doctorate, thesis/dissertation, comprehensive oral or written exam required, foreign language not required.
Entrance requirements: For master's, GRE General Test, TOEFL, minimum GPA of 2.75; for doctorate, GRE General Test, TOEFL. *Application deadline:* For fall admission, 7/1; for spring admission, 11/1. Applications are processed on a rolling basis. *Application fee:* $25 for international students. Electronic applications accepted.
Expenses: Tuition, state resident: full-time $3,017; part-time $168 per credit. Tuition, nonresident: full-time $6,119; part-time $340 per credit. Part-time tuition and fees vary according to course load and program.
Financial aid: Research assistantships, teaching assistantships, Federal Work-Study, institutionally sponsored loans, and unspecified assistantships available. Financial aid applicants required to submit FAFSA.
Faculty research: Ecology and population dynamics, physiology, biochemistry and behavior, systematics. *Total annual research expenditures:* $2.1 million.
Dr. Clarence Collison, Head, 662-325-2085, *Fax:* 662-325-8837, *E-mail:* chc2@ra.msstate.edu.
Application contact: Jerry B. Inmon, Director of Admissions, 662-325-2224, *Fax:* 662-325-7360, *E-mail:* admit@admissions.msstate.edu.

■ MONTANA STATE UNIVERSITY–BOZEMAN

College of Graduate Studies, College of Agriculture, Department of Plant Sciences, Bozeman, MT 59717
AWARDS Plant pathology (MS); plant science (MS, PhD). Part-time programs available.
Students: 10 full-time (6 women), 9 part-time (4 women); includes 1 minority (Asian American or Pacific Islander). Average age 34. *4 applicants, 75% accepted.* In 1999, 5 master's, 5 doctorates awarded.
Degree requirements: For master's and doctorate, thesis/dissertation required, foreign language not required.
Entrance requirements: For master's, GRE General Test, TOEFL; for doctorate, GRE General Test, TOEFL, minimum GPA of 3.0. *Application deadline:* For fall admission, 6/1; for spring admission, 11/1. Applications are processed on a

rolling basis. *Application fee:* $50. Electronic applications accepted.
Expenses: Tuition, state resident: full-time $2,674. Tuition, nonresident: full-time $6,986. International tuition: $7,136 full-time. Tuition and fees vary according to course load and program.
Financial aid: In 1999–00, research assistantships with full tuition reimbursements (averaging $15,000 per year); fellowships with full tuition reimbursements, career-related internships or fieldwork, Federal Work-Study, scholarships, and tuition waivers (full and partial) also available. Financial aid application deadline: 3/1; financial aid applicants required to submit FAFSA.
Faculty research: Virology, genetics, crop breeding, plant pathology. *Total annual research expenditures:* $3.1 million.
Dr. Norman F. Weedon, Head, 406-994-4832, *Fax:* 406-994-1848, *E-mail:* plantsciences@montana.edu.

■ NEW MEXICO STATE UNIVERSITY

Graduate School, College of Agriculture and Home Economics, Department of Entomology, Plant Pathology and Weed Science, Las Cruces, NM 88003-8001
AWARDS MS. Part-time programs available.
Faculty: 12.
Students: 5 full-time (3 women); includes 1 minority (Hispanic American), 1 international. Average age 36. *1 applicant, 100% accepted.* In 1999, 3 degrees awarded.
Degree requirements: For master's, thesis required, foreign language not required.
Entrance requirements: For master's, GRE General Test. *Application deadline:* For fall admission, 7/1 (priority date); for spring admission, 11/1. Applications are processed on a rolling basis. *Application fee:* $15 ($35 for international students). Electronic applications accepted.
Expenses: Tuition, state resident: full-time $2,682; part-time $112 per credit. Tuition, nonresident: full-time $8,376; part-time $349 per credit. Tuition and fees vary according to course load.
Financial aid: Research assistantships, teaching assistantships, career-related internships or fieldwork available. Financial aid application deadline: 3/1.
Faculty research: Integrated pest management, pesticide application and safety, livestock ectoparasite research, biotechnology, nematology.
Dr. Grant Kinzer, Head, 505-646-3225, *Fax:* 505-646-8087, *E-mail:* gkinzer@nmsu.edu.

■ NORTH CAROLINA STATE UNIVERSITY

Graduate School, College of Agriculture and Life Sciences, Department of Plant Pathology, Raleigh, NC 27695
AWARDS M Ag, MLS, MS, PhD.
Faculty: 32 full-time (4 women), 20 part-time/adjunct (1 woman).
Students: 17 full-time (11 women), 12 part-time (7 women); includes 1 minority (Native American), 10 international. Average age 30. *21 applicants, 24% accepted.* In 1999, 7 master's, 1 doctorate awarded. Terminal master's awarded for partial completion of doctoral program.
Degree requirements: For master's and doctorate, thesis/dissertation required, foreign language not required.
Entrance requirements: For master's and doctorate, TOEFL. *Application deadline:* For fall admission, 6/25; for spring admission, 11/25. Applications are processed on a rolling basis. *Application fee:* $45.
Expenses: Tuition, state resident: full-time $1,578. Tuition, nonresident: full-time $10,744. Required fees: $892. Full-time tuition and fees vary according to program.
Financial aid: In 1999–00, 1 fellowship (averaging $1,289 per year), 23 research assistantships (averaging $5,050 per year) were awarded; teaching assistantships, career-related internships or fieldwork and institutionally sponsored loans also available. Aid available to part-time students.
Faculty research: Management of plant diseases and nematodes by cultural, biological, and chemical methods; epidemiology; host resistance; basic biology of plant pathogens and nematodes through use of new biotechnologies. *Total annual research expenditures:* $8 million.
Dr. O. W. Barnett, Head, 919-515-2730, *Fax:* 919-515-7716, *E-mail:* ow_barnett@ncsu.edu.
Application contact: Dr. David F. Ritchie, Director of Graduate Programs, 919-515-6809, *Fax:* 919-515-7716, *E-mail:* david_ritchie@ncsu.edu.

■ NORTH DAKOTA STATE UNIVERSITY

Graduate Studies and Research, College of Agriculture, Department of Plant Pathology, Fargo, ND 58105
AWARDS MS, PhD. Part-time programs available.
Faculty: 10 full-time (1 woman), 4 part-time/adjunct (0 women).
Students: 12 full-time (5 women), 1 part-time, 6 international. Average age 23. *3 applicants, 100% accepted.* In 1999, 1 master's awarded (100% found work

related to degree); 1 doctorate awarded (100% entered university research/teaching).

Degree requirements: For master's and doctorate, thesis/dissertation required, foreign language not required. *Average time to degree:* Master's–3 years full-time, 4.5 years part-time; doctorate–5 years full-time, 7 years part-time.

Entrance requirements: For master's and doctorate, TOEFL. *Application deadline:* Applications are processed on a rolling basis. *Application fee:* $25.

Expenses: Tuition, state resident: full-time $3,096; part-time $112 per credit hour. Tuition, nonresident: full-time $7,588; part-time $299 per credit hour. Tuition and fees vary according to course load, campus/location and reciprocity agreements.

Financial aid: In 1999–00, 9 research assistantships with full tuition reimbursements were awarded; Federal Work-Study and institutionally sponsored loans also available. Financial aid application deadline: 4/15.

Faculty research: Electron microscopy, disease physiology, molecular biology, genetic resistance, tissue culture.
Dr. Glen Statler, Chair, 701-231-7058, *Fax:* 701-231-7851, *E-mail:* ppth@ndsuext.nodak.edu.

■ THE OHIO STATE UNIVERSITY

Graduate School, College of Food, Agricultural, and Environmental Sciences, Department of Plant Pathology, Columbus, OH 43210
AWARDS MS, PhD.

Faculty: 17 full-time, 7 part-time/adjunct.
Students: 18 full-time (9 women), 6 international. *41 applicants, 37% accepted.* In 1999, 5 master's, 5 doctorates awarded.
Degree requirements: For master's, thesis optional, foreign language not required; for doctorate, dissertation required, foreign language not required.
Entrance requirements: For master's and doctorate, GRE General Test. *Application deadline:* For fall admission, 8/15. Applications are processed on a rolling basis. *Application fee:* $30 ($40 for international students).
Expenses: Tuition, state resident: full-time $5,400. Tuition, nonresident: full-time $14,535. Part-time tuition and fees vary according to course load and program.
Financial aid: Fellowships, research assistantships, teaching assistantships, Federal Work-Study and institutionally sponsored loans available. Aid available to part-time students.
Randall C. Rowe, Chairman, 614-292-1375, *Fax:* 614-292-4455, *E-mail:* rowe.4@osu.edu.

■ OKLAHOMA STATE UNIVERSITY

Graduate College, College of Agricultural Sciences and Natural Resources, Department of Entomology and Plant Pathology, Program in Plant Pathology, Stillwater, OK 74078
AWARDS M Ag, MS, PhD.

Degree requirements: For master's and doctorate, thesis/dissertation required, foreign language not required.
Entrance requirements: For master's and doctorate, GRE, TOEFL. *Application deadline:* For fall admission, 6/1 (priority date). *Application fee:* $25.
Expenses: Tuition, state resident: part-time $86 per credit hour. Tuition, nonresident: part-time $275 per credit hour. Required fees: $17 per credit hour. $14 per semester. One-time fee: $20 full-time. Tuition and fees vary according to course load.
Financial aid: Career-related internships or fieldwork, Federal Work-Study, and tuition waivers (partial) available. Aid available to part-time students. Financial aid application deadline: 3/1.
Russell Wright, Interim Head, Department of Entomology and Plant Pathology, 405-744-5527.

■ OREGON STATE UNIVERSITY

Graduate School, College of Science, Department of Botany and Plant Pathology, Corvallis, OR 97331
AWARDS Ecology (MA, MAIS, MS, PhD); genetics (MA, MAIS, MS, PhD); molecular and cellular biology (MA, MAIS, MS, PhD); mycology (MA, MAIS, MS, PhD); plant pathology (MA, MAIS, MS, PhD); plant physiology (MA, MAIS, MS, PhD); structural botany (MA, MAIS, MS, PhD); systematics (MA, MAIS, MS, PhD). Part-time programs available.

Faculty: 45 full-time (10 women).
Students: 39 full-time (25 women), 1 part-time; includes 1 minority (Hispanic American), 4 international. Average age 30. In 1999, 7 master's, 3 doctorates awarded.
Degree requirements: For master's, variable foreign language requirement required, thesis optional; for doctorate, dissertation required, foreign language not required.
Entrance requirements: For master's and doctorate, GRE General Test, TOEFL, minimum GPA of 3.0 in last 90 hours. *Application deadline:* For fall admission, 2/1 (priority date). Applications are processed on a rolling basis. *Application fee:* $50.
Expenses: Tuition, state resident: full-time $6,489. Tuition, nonresident: full-time $11,061. Tuition and fees vary according to program.

Financial aid: Fellowships, research assistantships, teaching assistantships, career-related internships or fieldwork, Federal Work-Study, institutionally sponsored loans, and scholarships available. Aid available to part-time students. Financial aid application deadline: 2/1.
Faculty research: Plant ecology, plant molecular biology, systematic botany, epidemiology, host-pathogen interaction.
Dr. Stella M. Coakley, Chair, 541-737-5264, *Fax:* 541-737-3573, *E-mail:* coakleys@bcc.orst.edu.
Application contact: Dr. Donald B. Zobel, Professor, 541-737-3451, *Fax:* 541-737-3573, *E-mail:* zobeld@bcc.orst.edu.

■ THE PENNSYLVANIA STATE UNIVERSITY UNIVERSITY PARK CAMPUS

Graduate School, College of Agricultural Sciences, Department of Plant Pathology, State College, University Park, PA 16802-1503
AWARDS M Agr, MS, PhD.

Students: 11 full-time (7 women), 12 part-time (5 women).
Entrance requirements: For master's and doctorate, GRE General Test. *Application fee:* $50.
Expenses: Tuition, state resident: full-time $6,886; part-time $291 per credit. Tuition, nonresident: full-time $14,118; part-time $588 per credit. Required fees: $46 per semester. Part-time tuition and fees vary according to course load and program.
Dr. Elwin L. Stewart, Head, 814-865-7069.
Application contact: Dr. John E. Ayers, Professor in Charge, 814-865-7069.

■ PURDUE UNIVERSITY

Graduate School, School of Agriculture, Department of Botany and Plant Pathology, West Lafayette, IN 47907
AWARDS MS, PhD. Part-time programs available.

Faculty: 26 full-time (4 women).
Students: 19 full-time (6 women), 6 part-time (2 women); includes 1 minority (African American), 11 international. Average age 27. *52 applicants, 29% accepted.* In 1999, 6 master's, 1 doctorate awarded. Terminal master's awarded for partial completion of doctoral program.
Degree requirements: For master's and doctorate, thesis/dissertation required, foreign language not required.
Entrance requirements: For master's and doctorate, GRE, TOEFL, TWE. *Application deadline:* For fall admission, 5/1 (priority date); for spring admission, 11/1. Applications are processed on a rolling

Purdue University (continued)
basis. *Application fee:* $30. Electronic applications accepted.

Expenses: Tuition, state resident: full-time $4,530; part-time $130 per credit hour. Tuition, nonresident: full-time $15,310; part-time $404 per credit hour. Tuition and fees vary according to campus/location and program.

Financial aid: In 1999–00, 8 research assistantships, 6 teaching assistantships were awarded; fellowships, career-related internships or fieldwork also available. Aid available to part-time students. Financial aid application deadline: 7/1; financial aid applicants required to submit FAFSA.

Faculty research: Biotechnology, plant growth, weed control, crop improvement, plant physiology.

Dr. R. D. Martyn, Head, 765-494-4614, *Fax:* 765-494-0363, *E-mail:* martyn@btny.purdue.edu.

Application contact: Pam Mow, Graduate Admissions Office, 765-494-0352, *Fax:* 765-494-0363, *E-mail:* mow@btny.purdue.edu.

Find an in-depth description at www.petersons.com/graduate.

■ **RUTGERS, THE STATE UNIVERSITY OF NEW JERSEY, NEW BRUNSWICK**

Graduate School, Program in Plant Biology, New Brunswick, NJ 08901-1281

AWARDS Horticulture (MS, PhD); molecular biology and biochemistry (MS, PhD); pathology (MS, PhD); plant ecology (MS, PhD); plant genetics (PhD); plant physiology (MS, PhD); production and management (MS); structure and plant groups (MS, PhD). Part-time programs available.

Faculty: 64 full-time (8 women), 1 part-time/adjunct (0 women).

Students: 42 full-time (17 women), 13 part-time (6 women); includes 4 minority (1 African American, 2 Asian Americans or Pacific Islanders, 1 Hispanic American), 28 international. Average age 27. *58 applicants, 38% accepted.* In 1999, 4 master's, 9 doctorates awarded. Terminal master's awarded for partial completion of doctoral program.

Degree requirements: For master's, thesis or alternative required, foreign language not required; for doctorate, dissertation required, foreign language not required. *Average time to degree:* Master's–2.5 years full-time; doctorate–5 years full-time.

Entrance requirements: For master's and doctorate, GRE General Test, GRE Subject Test (recommended). *Application deadline:* For fall admission, 6/1. *Application fee:* $50. Electronic applications accepted.

Expenses: Tuition, state resident: full-time $6,776; part-time $279 per credit. Tuition, nonresident: full-time $9,936; part-time $412 per credit. Required fees: $20 per credit. $89 per semester. Tuition and fees vary according to course load, campus/location and program.

Financial aid: In 1999–00, 42 students received aid, including 12 fellowships with full tuition reimbursements available (averaging $16,000 per year), 21 research assistantships with full tuition reimbursements available (averaging $14,000 per year), 9 teaching assistantships with full tuition reimbursements available (averaging $13,100 per year). Financial aid application deadline: 2/15; financial aid applicants required to submit FAFSA.

Faculty research: Molecular genetics, disease resistance, molecular biology of plant viruses.

Dr. Thomas Leustek, Director, 732-932-8165 Ext. 326, *Fax:* 732-932-9377, *E-mail:* leustek@aesop.rutgers.edu.

Application contact: Barbara Mulder, Administrative Assistant, 732-932-9375 Ext. 358, *Fax:* 732-932-9377, *E-mail:* plantbio@aesop.rutgers.edu.

Find an in-depth description at www.petersons.com/graduate.

■ **SOUTH DAKOTA STATE UNIVERSITY**

Graduate School, College of Agriculture and Biological Sciences, Department of Plant Science, Program in Plant Pathology, Brookings, SD 57007

AWARDS MS.

Degree requirements: For master's, thesis, oral exam required, foreign language not required.

Entrance requirements: For master's, GRE General Test, TOEFL.

Faculty research: Small grain and row crop pathology, virology, nematology, molecular biology, host/pathogen systems, epidemiology and disease management.

■ **TEXAS A&M UNIVERSITY**

College of Agriculture and Life Sciences, Department of Plant Pathology and Microbiology, College Station, TX 77843

AWARDS Plant pathology (MS, PhD); plant protection (M Agr). Part-time programs available. Postbaccalaureate distance learning degree programs offered.

Faculty: 20 full-time (5 women).

Students: 49 full-time (18 women); includes 2 minority (1 Asian American or Pacific Islander, 1 Hispanic American), 22 international. Average age 30. In 1999, 4 master's, 6 doctorates awarded.

Degree requirements: For master's, thesis required (for some programs), foreign language not required; for doctorate, dissertation required, foreign language not required.

Entrance requirements: For master's and doctorate, GRE General Test, TOEFL. *Application deadline:* Applications are processed on a rolling basis. *Application fee:* $50 ($75 for international students).

Expenses: Tuition, state resident: part-time $76 per semester hour. Tuition, nonresident: part-time $292 per semester hour. Required fees: $11 per semester hour. Tuition and fees vary according to program.

Financial aid: In 1999–00, 28 research assistantships with partial tuition reimbursements (averaging $15,000 per year), 7 teaching assistantships with partial tuition reimbursements (averaging $15,000 per year) were awarded; fellowships, career-related internships or fieldwork, Federal Work-Study, and institutionally sponsored loans also available. Aid available to part-time students. Financial aid application deadline: 4/1; financial aid applicants required to submit FAFSA.

Dr. David N. Appel, Interim Head, 979-845-7311.

Application contact: Douglas R. Cook, Chair, Graduate Program Committee, 979-845-7311.

■ **THE UNIVERSITY OF ARIZONA**

Graduate College, College of Agriculture, Department of Plant Pathology, Tucson, AZ 85721

AWARDS MS, PhD. Part-time programs available.

Degree requirements: For master's, thesis optional, foreign language not required; for doctorate, computer language, dissertation required, foreign language not required.

Entrance requirements: For master's and doctorate, GRE (recommended), TOEFL, minimum GPA of 3.0.

Expenses: Tuition, nonresident: full-time $4,814; part-time $274 per unit. Required fees: $1,094; $115 per unit. Tuition and fees vary according to course load and program.

Faculty research: Fungal molecular biology, ecology of soil-borne plant pathogens, plant virology, plant bacteriology, plant/pathogen interactions.

■ **UNIVERSITY OF ARKANSAS**

Graduate School, Dale Bumpers College of Agricultural, Food and Life Sciences, Department of Plant Pathology, Fayetteville, AR 72701-1201

AWARDS MS.

Faculty: 16 full-time (1 woman).

Students: 4 full-time (1 woman), 1 international. *2 applicants, 0% accepted.* In 1999, 2 degrees awarded.
Degree requirements: For master's, thesis required, foreign language not required.
Application fee: $40 ($50 for international students).
Expenses: Tuition, state resident: full-time $3,186; part-time $177 per credit. Tuition, nonresident: full-time $7,560; part-time $420 per credit. Required fees: $756; $21 per credit. One-time fee: $22 part-time. Tuition and fees vary according to course load and program.
Financial aid: Research assistantships, teaching assistantships, career-related internships or fieldwork and Federal Work-Study available. Aid available to part-time students. Financial aid application deadline: 4/1; financial aid applicants required to submit FAFSA.
Dr. Sung M. Lim, Chair, 501-575-2446.
Application contact: Craig Rothrock, Graduate Coordinator, *E-mail:* rothrock@comp.uark.edu.

■ UNIVERSITY OF CALIFORNIA, DAVIS

Graduate Studies, Programs in the Biological Sciences, Program in Plant Pathology, Davis, CA 95616

AWARDS MS, PhD.

Faculty: 15 full-time (2 women).
Students: 30 full-time (16 women); includes 5 minority (1 African American, 2 Asian Americans or Pacific Islanders, 2 Hispanic Americans), 11 international. Average age 29. *36 applicants, 25% accepted.* In 1999, 6 master's, 4 doctorates awarded.
Degree requirements: For master's, thesis optional; for doctorate, dissertation required.
Entrance requirements: For master's and doctorate, GRE General Test. *Application deadline:* For fall admission, 1/15. *Application fee:* $40. Electronic applications accepted.
Expenses: Tuition, nonresident: full-time $9,804. Tuition and fees vary according to program and student level.
Financial aid: In 1999–00, 27 students received aid, including 20 research assistantships with full and partial tuition reimbursements available; fellowships with full and partial tuition reimbursements available, teaching assistantships with partial tuition reimbursements available, Federal Work-Study, grants, institutionally sponsored loans, scholarships, and tuition waivers (full and partial) also available. Financial aid application deadline: 1/15; financial aid applicants required to submit FAFSA.

Faculty research: Soil microbiology, diagnosis and control, molecular biology, biotechnology, ecology and epidemiology. Richard Bostock, Graduate Chair, 530-752-4269, *Fax:* 530-752-5674, *E-mail:* rmbostock@ucdavis.edu.
Application contact: Elizabeth Jeffery, Administrative Assistant, 530-752-0300, *Fax:* 530-752-5674, *E-mail:* emjeffery@ucdavis.edu.

■ UNIVERSITY OF CALIFORNIA, RIVERSIDE

Graduate Division, College of Natural and Agricultural Sciences, Department of Plant Pathology, Riverside, CA 92521-0102

AWARDS MS, PhD. Part-time programs available.

Faculty: 26 full-time (2 women).
Students: 19 full-time (11 women); includes 2 minority (1 Asian American or Pacific Islander, 1 Hispanic American), 9 international. Average age 32. In 1999, 1 master's, 6 doctorates awarded. Terminal master's awarded for partial completion of doctoral program.
Degree requirements: For master's, comprehensive exams or thesis required; for doctorate, dissertation, qualifying exams required, foreign language not required. *Average time to degree:* Doctorate–5 years full-time.
Entrance requirements: For master's and doctorate, GRE General Test, TOEFL, minimum GPA of 3.2. *Application deadline:* For fall admission, 2/1 (priority date); for winter admission, 9/1; for spring admission, 12/1. Applications are processed on a rolling basis. *Application fee:* $40. Electronic applications accepted.
Expenses: Tuition, nonresident: full-time $9,804. Required fees: $4,758. Full-time tuition and fees vary according to program.
Financial aid: Fellowships, research assistantships, teaching assistantships, career-related internships or fieldwork, Federal Work-Study, institutionally sponsored loans, and tuition waivers (full and partial) available. Financial aid application deadline: 2/1; financial aid applicants required to submit FAFSA.
Faculty research: Host-pathogen interactions, biological control and integrated approaches to disease management, fungicide behavior, molecular genetics.
Dr. Don Cooksey, Chair, 909-787-4115.
Application contact: Dr. John Menge, Graduate Adviser, 909-787-4130, *Fax:* 909-787-4294, *E-mail:* plpath@ucracl.ucr.edu.

Find an in-depth description at www.petersons.com/graduate.

■ UNIVERSITY OF FLORIDA

Graduate School, College of Agriculture and Life Sciences, Department of Plant Pathology, Gainesville, FL 32611

AWARDS MS, PhD.

Faculty: 51.
Students: 24 full-time (11 women), 2 part-time (1 woman); includes 2 minority (both Hispanic Americans), 12 international. *26 applicants, 58% accepted.* In 1999, 2 master's, 2 doctorates awarded.
Degree requirements: For master's, thesis optional; for doctorate, dissertation required.
Entrance requirements: For master's and doctorate, GRE General Test, minimum GPA of 3.0. *Application deadline:* For fall admission, 6/1 (priority date). Applications are processed on a rolling basis. *Application fee:* $20. Electronic applications accepted.
Expenses: Tuition, state resident: part-time $144 per credit hour. Tuition, nonresident: part-time $505 per credit hour. Tuition and fees vary according to course level, course load and program.
Financial aid: In 1999–00, 20 students received aid, including 15 research assistantships, 1 teaching assistantship; fellowships, career-related internships or fieldwork also available.
Faculty research: Causes of development of disease in plants, molecular and biochemical aspects of disease, biological control of pathogens and weeds, genetic engineering of resistant plants.
Dr. G. N. Agrios, Chair, 352-392-3631, *Fax:* 352-392-6532, *E-mail:* gna@gnv.ifas.ufl.edu.
Application contact: Dr. F. W. Zettler, Graduate Coordinator, 352-392-7245, *Fax:* 352-392-6532, *E-mail:* fwz@gnv.ifas.ufl.edu.

■ UNIVERSITY OF GEORGIA

Graduate School, College of Agricultural and Environmental Sciences, Department of Plant Pathology, Athens, GA 30602

AWARDS Plant pathology (MS, PhD); plant protection and pest management (MPPPM).

Degree requirements: For master's, thesis (MS) required; for doctorate, one foreign language (computer language can substitute), dissertation required.
Entrance requirements: For master's and doctorate, GRE General Test. Electronic applications accepted.
Expenses: Tuition, state resident: full-time $7,516; part-time $431 per credit hour. Tuition, nonresident: full-time $12,204; part-time $793 per credit hour. Tuition and fees vary according to program.

■ UNIVERSITY OF HAWAII AT MANOA

Graduate Division, College of Tropical Agriculture and Human Resources, Department of Plant Pathology, Honolulu, HI 96822

AWARDS Botanical sciences (MS, PhD); plant pathology (MS, PhD). Part-time programs available.

Faculty: 17 full-time (3 women), 1 part-time/adjunct (0 women).
Students: 6 full-time (2 women), 2 part-time (both women). Average age 32. *10 applicants, 70% accepted.* Terminal master's awarded for partial completion of doctoral program.
Degree requirements: For master's, thesis optional, foreign language not required; for doctorate, dissertation required, foreign language not required. *Average time to degree:* Doctorate–4 years full-time.
Entrance requirements: For master's, GRE General Test (average 477 verbal, 629 quantitative, 562 analytical); for doctorate, GRE General Test (average 485 verbal, 570 quantitative). *Application deadline:* For fall admission, 3/1; for spring admission, 9/1. *Application fee:* $25 ($50 for international students).
Expenses: Tuition, state resident: full-time $4,032; part-time $168 per credit. Tuition, nonresident: full-time $9,960; part-time $415 per credit. Required fees: $51 per semester. Part-time tuition and fees vary according to course load and program.
Financial aid: In 1999–00, 8 research assistantships (averaging $15,606 per year) were awarded; teaching assistantships, tuition waivers (full) also available.
Faculty research: Nematology, virology, mycology, bacteriology, epidemiology. *Total annual research expenditures:* $1.9 million.
Dr. Donald P. Schmitt, Chairperson, 808-956-8329, *Fax:* 808-956-2832.
Application contact: Dr. John Hu, Graduate Chairperson, 808-956-8329, *Fax:* 808-956-2832.

■ UNIVERSITY OF IDAHO

College of Graduate Studies, College of Agriculture, Department of Plant, Soil, and Entomological Sciences, Program in Plant Pathology, Moscow, ID 83844-4140

AWARDS MS, PhD.

Degree requirements: For master's, thesis required (for some programs), foreign language not required; for doctorate, dissertation required.
Entrance requirements: For master's and doctorate, GRE General Test, minimum GPA of 3.0. *Application deadline:* For fall admission, 8/1; for spring admission,

12/15. *Application fee:* $35 ($45 for international students).
Expenses: Tuition, nonresident: full-time $6,000; part-time $239 per credit hour. Required fees: $2,888; $144 per credit hour. Tuition and fees vary according to program.
Financial aid: Application deadline: 2/15.
Dr. Maurice V. Wiese, Chairman, 208-885-6650.

■ UNIVERSITY OF KENTUCKY

Graduate School, Graduate School Programs from the College of Agriculture, Department of Plant Pathology, Lexington, KY 40506-0032

AWARDS MS, PhD.

Degree requirements: For master's, comprehensive exam required, thesis optional, foreign language not required; for doctorate, dissertation, comprehensive exam required, foreign language not required.
Entrance requirements: For master's, GRE General Test, minimum undergraduate GPA of 2.5; for doctorate, GRE General Test, minimum graduate GPA of 3.0.
Expenses: Tuition, state resident: full-time $3,596; part-time $188 per credit hour. Tuition, nonresident: full-time $10,116; part-time $550 per credit hour.
Faculty research: Molecular biology of viruses and fungi, biochemistry and physiology of disease resistance, plant transformation, disease ecology, forest pathology.

■ UNIVERSITY OF MAINE

Graduate School, College of Natural Sciences, Forestry, and Agriculture, Department of Biological Sciences, Program in Botany and Plant Pathology, Orono, ME 04469

AWARDS MS. Part-time programs available.

Degree requirements: For master's, thesis required, foreign language not required.
Entrance requirements: For master's, GRE General Test, TOEFL.
Expenses: Tuition, state resident: full-time $3,564. Tuition, nonresident: full-time $10,116. Required fees: $378. Tuition and fees vary according to course load.
Faculty research: Molecular biology of viral and fungal pathogens, marine ecology, paleoecology and acid systematics and evolution.

■ UNIVERSITY OF MINNESOTA, TWIN CITIES CAMPUS

Graduate School, College of Agricultural, Food, and Environmental Sciences, Department of Plant Pathology, Minneapolis, MN 55455-0213

AWARDS MS, PhD. Part-time programs available.

Faculty: 24 full-time (3 women), 6 part-time/adjunct (2 women).
Students: 25 full-time (16 women); includes 1 minority (Asian American or Pacific Islander), 14 international. Average age 25. *14 applicants, 71% accepted.* In 1999, 6 master's, 4 doctorates awarded. Terminal master's awarded for partial completion of doctoral program.
Degree requirements: For master's, thesis required (for some programs); for doctorate, dissertation required. *Average time to degree:* Master's–2.5 years full-time; doctorate–3.5 years full-time.
Entrance requirements: For master's and doctorate, GRE General Test, TOEFL. *Application deadline:* For fall admission, 5/1 (priority date); for spring admission, 12/1. Applications are processed on a rolling basis. *Application fee:* $40 ($50 for international students). Electronic applications accepted.
Expenses: Tuition, state resident: full-time $5,040; part-time $420 per credit. Tuition, nonresident: full-time $9,900; part-time $825 per credit. Full-time tuition and fees vary according to course load, program and reciprocity agreements.
Financial aid: In 1999–00, 4 students received aid, including research assistantships with full tuition reimbursements available (averaging $14,375 per year); teaching assistantships. Aid available to part-time students.
Faculty research: Plant disease management, biological control, product deterioration, international agriculture, molecular biology. *Total annual research expenditures:* $1.4 million.
Dr. Francis L. Pfleger, Head, 612-625-9736, *Fax:* 612-625-9728, *E-mail:* francisp@puccini.crl.umn.edu.
Application contact: James Groth, Professor, 612-625-0299, *Fax:* 612-625-9728, *E-mail:* jamesg@puccini-crl.umn.edu.

■ UNIVERSITY OF MISSOURI–COLUMBIA

Graduate School, College of Agriculture, Program in Plant Pathology, Columbia, MO 65211

AWARDS MS, PhD. Terminal master's awarded for partial completion of doctoral program.

Degree requirements: For master's and doctorate, thesis/dissertation required, foreign language not required.
Entrance requirements: For master's and doctorate, GRE General Test, minimum GPA of 3.0.
Expenses: Tuition, state resident: full-time $3,020; part-time $168 per hour. Tuition, nonresident: full-time $6,066; part-time $505 per hour. Required fees: $445; $18 per hour. Tuition and fees vary according to course load and program.

■ UNIVERSITY OF RHODE ISLAND

Graduate School, College of Resource Development, Department of Plant Sciences, Program in Plant Pathology-Entomology, Kingston, RI 02881

AWARDS Entomology (MS, PhD); plant pathology (MS, PhD).

Degree requirements: For master's, thesis, comprehensive exam, professional seminar required; for doctorate, dissertation, professional seminar required, foreign language not required.
Entrance requirements: For master's and doctorate, GRE General Test. *Application deadline:* For fall admission, 4/15 (priority date). Applications are processed on a rolling basis. *Application fee:* $35.
Expenses: Tuition, state resident: full-time $3,540; part-time $197 per credit. Tuition, nonresident: full-time $10,116; part-time $197 per credit. Required fees: $1,352; $37 per term.
Faculty research: Physiology and fine structure of host-parasite relations, etiology of plant disease, biological control of insects, integrated pest management.
Dr. Richard Hull, Chairman, Department of Plant Sciences, 401-874-5995.

■ THE UNIVERSITY OF TENNESSEE

Graduate School, College of Agricultural Sciences and Natural Resources, Department of Entomology and Plant Pathology, Knoxville, TN 37996

AWARDS Entomology (MS); plant pathology (MS). Part-time programs available.

Faculty: 13 full-time (2 women).
Students: 11 full-time (6 women), 8 part-time (4 women); includes 2 minority (1 African American, 1 Asian American or Pacific Islander), 1 international. *14 applicants, 79% accepted.* In 1999, 10 degrees awarded.
Degree requirements: For master's, thesis, seminar required, foreign language not required.
Entrance requirements: For master's, TOEFL, minimum GPA of 2.7. *Application*

deadline: For fall admission, 3/15 (priority date). Applications are processed on a rolling basis. *Application fee:* $35. Electronic applications accepted.
Expenses: Tuition, state resident: full-time $3,806; part-time $184 per credit hour. Tuition, nonresident: full-time $9,874; part-time $522 per credit hour. Tuition and fees vary according to program.
Financial aid: In 1999–00, 8 research assistantships, 1 teaching assistantship were awarded; fellowships, career-related internships or fieldwork, Federal Work-Study, institutionally sponsored loans, and unspecified assistantships also available. Financial aid application deadline: 3/15; financial aid applicants required to submit FAFSA.
Dr. Gene Burgess, Interim Head, 865-974-7135, *Fax:* 865-974-4744, *E-mail:* gburgess1@utk.edu.
Application contact: Dr. Reid R. Gerhardt, Graduate Representative, 865-974-7135, *Fax:* 865-974-4744, *E-mail:* rgerhard@utk.edu.

■ UNIVERSITY OF WISCONSIN–MADISON

Graduate School, College of Agricultural and Life Sciences, Department of Plant Pathology, Madison, WI 53706-1380

AWARDS). Part-time programs available.

Faculty: 21 full-time (5 women).
Students: 29 full-time (15 women), 11 international. Average age 24. *50 applicants, 36% accepted.* In 1999, 3 master's awarded (100% found work related to degree); 8 doctorates awarded (99% entered university research/teaching, 1% found other work related to degree). Terminal master's awarded for partial completion of doctoral program.
Degree requirements: For master's and doctorate, thesis/dissertation required, foreign language not required. *Average time to degree:* Master's–2 years full-time; doctorate–4.7 years full-time.
Entrance requirements: For master's and doctorate, GRE. *Application deadline:* For fall admission, 1/15 (priority date). *Application fee:* $45. Electronic applications accepted.
Expenses: Tuition, state resident: full-time $5,406; part-time $339 per credit. Tuition, nonresident: full-time $17,110; part-time $1,071 per credit. Full-time tuition and fees vary according to program and reciprocity agreements. Part-time tuition and fees vary according to course load and program.
Financial aid: In 1999–00, 24 students received aid, including 4 fellowships with tuition reimbursements available (averaging $15,960 per year), 24 research

assistantships with tuition reimbursements available (averaging $15,960 per year); career-related internships or fieldwork and institutionally sponsored loans also available. Financial aid application deadline: 1/15.
Faculty research: Plant disease, plant health, plant-microbe interactions, plant disease management, biological control. *Total annual research expenditures:* $3 million.
John H. Andrews, Chair, 608-262-1410, *Fax:* 608-263-2626.
Application contact: Cathy Davis Gram, Admissions Coordinator, 608-262-9926, *Fax:* 608-263-2626, *E-mail:* admissions@plantpath.wisc.edu.

■ VIRGINIA POLYTECHNIC INSTITUTE AND STATE UNIVERSITY

Graduate School, College of Agriculture and Life Sciences, Department of Plant Pathology, Physiology and Weed Science, Blacksburg, VA 24061

AWARDS Plant pathology (MS, PhD); plant physiology (MS, PhD).

Students: 21 full-time (9 women), 3 part-time (1 woman); includes 3 minority (2 African Americans, 1 Asian American or Pacific Islander), 6 international. *19 applicants, 58% accepted.* In 1999, 3 master's, 1 doctorate awarded.
Entrance requirements: For master's and doctorate, TOEFL. *Application deadline:* For fall admission, 12/1 (priority date). Applications are processed on a rolling basis. *Application fee:* $25.
Expenses: Tuition, state resident: full-time $4,122; part-time $229 per credit hour. Tuition, nonresident: full-time $6,930; part-time $385 per credit hour. Required fees: $828; $107 per semester. Part-time tuition and fees vary according to course load.
Financial aid: In 1999–00, 5 research assistantships with full tuition reimbursements (averaging $11,390 per year) were awarded; fellowships, teaching assistantships, unspecified assistantships also available. Financial aid application deadline: 4/1.
Dr. Kriton K. Hatzios, Head, 540-231-6361, *E-mail:* hatzios@vt.edu.

■ WASHINGTON STATE UNIVERSITY

Graduate School, College of Agriculture and Home Economics, Department of Plant Pathology, Pullman, WA 99164
AWARDS MS, PhD.

Washington State University (continued)
Faculty: 25 full-time (3 women), 4 part-time/adjunct (0 women).
Students: 15 full-time (5 women); includes 1 minority (Asian American or Pacific Islander), 5 international. Average age 29. In 1999, 4 master's, 1 doctorate awarded. Terminal master's awarded for partial completion of doctoral program.
Degree requirements: For master's and doctorate, thesis/dissertation, oral exam required, foreign language not required. *Average time to degree:* Master's–2.4 years full-time, 3 years part-time; doctorate–3.6 years full-time.
Entrance requirements: For master's and doctorate, minimum GPA of 3.0. *Application deadline:* For fall admission, 3/1 (priority date); for spring admission, 8/1. Applications are processed on a rolling basis. *Application fee:* $35. Electronic applications accepted.
Expenses: Tuition, state resident: full-time $5,654. Tuition, nonresident: full-time $13,850. International tuition: $13,850 full-time. Tuition and fees vary according to program.
Financial aid: In 1999–00, 10 research assistantships with full and partial tuition reimbursements, 2 teaching assistantships with full and partial tuition reimbursements were awarded; career-related internships or fieldwork, Federal Work-Study, institutionally sponsored loans, and teaching associateships also available. Financial aid application deadline: 4/1; financial aid applicants required to submit FAFSA.
Faculty research: Biology of fungi, bacteria, and viruses; diseases of plants; genetics of fungi, bacteria, and viruses. *Total annual research expenditures:* $1.6 million.
Dr. Dennis Gross, Chair, 509-335-9541.

■ WEST VIRGINIA UNIVERSITY

College of Agriculture, Forestry and Consumer Sciences, Division of Plant and Soil Sciences, Morgantown, WV 26506

AWARDS Agronomy (MS); entomology (MS); environmental microbiology (MS); horticulture (MS); plant and soil sciences (PhD); plant pathology (MS).

Faculty: 19 full-time (1 woman), 2 part-time/adjunct (1 woman).
Students: 18 full-time (10 women), 7 part-time (5 women); includes 1 minority (Native American), 2 international. Average age 26. In 1999, 1 master's awarded.
Degree requirements: For master's, thesis required, foreign language not required; for doctorate, dissertation, comprehensive exam required, foreign language not required.
Entrance requirements: For master's, GRE, TOEFL, minimum GPA of 2.5; for

doctorate, GRE General Test, TOEFL, minimum GPA of 3.0. *Application fee:* $45.
Expenses: Tuition, state resident: full-time $2,910; part-time $154 per credit hour. Tuition, nonresident: full-time $8,368; part-time $457 per credit hour.
Financial aid: In 1999–00, 13 research assistantships, 3 teaching assistantships were awarded; Federal Work-Study, institutionally sponsored loans, and tuition waivers (full and partial) also available. Financial aid application deadline: 2/1; financial aid applicants required to submit FAFSA.
Faculty research: Water quality, reclamation of disturbed land, crop production, pest control, environmental protection.
Dr. Barton Baker, Chair, 304-293-4817, *Fax:* 304-293-3740, *E-mail:* bbaker2@wvu.edu.

PLANT PHYSIOLOGY

■ CLEMSON UNIVERSITY

Graduate School, College of Agriculture, Forestry and Life Sciences, School of Plant, Statistical and Ecological Sciences, Department of Horticulture, Clemson, SC 29634

AWARDS Genetics (PhD); horticulture (MS); plant physiology (PhD). Part-time programs available.

Students: 7 full-time (2 women), 4 part-time (2 women), 2 international. Average age 30. *11 applicants, 45% accepted.* In 1999, 4 degrees awarded.
Degree requirements: For master's and doctorate, thesis/dissertation required, foreign language not required.
Entrance requirements: For master's, GRE General Test, TOEFL, minimum undergraduate GPA of 3.0; for doctorate, GRE General Test, TOEFL, minimum graduate GPA of 3.0. *Application deadline:* For fall admission, 6/1. *Application fee:* $40.
Expenses: Tuition, state resident: full-time $3,480; part-time $174 per credit hour. Tuition, nonresident: full-time $9,256; part-time $388 per credit hour. Required fees: $5 per term. Full-time tuition and fees vary according to course level, course load and campus/location.
Financial aid: Fellowships, research assistantships, teaching assistantships available. Financial aid applicants required to submit FAFSA.
Faculty research: Molecular biology, tissue culture, plant breeding, plant cultural systems.
Dr. Calvin L. Shoulties, Chair, 864-656-7592, *Fax:* 864-656-4960.
Application contact: Dr. W. Vance Baird, Coordinator, 864-656-4953, *Fax:* 864-656-4960, *E-mail:* vbaird@clemson.edu.

■ CLEMSON UNIVERSITY

Graduate School, College of Agriculture, Forestry and Life Sciences, School of Plant, Statistical and Ecological Sciences, Department of Plant Pathology and Physiology, Program in Plant Physiology, Clemson, SC 29634

AWARDS PhD. Offered in cooperation with the Departments of Biological Sciences and Horticulture. Part-time programs available.

Students: 7 full-time (5 women), 7 part-time (4 women); includes 1 minority (Hispanic American), 3 international. *5 applicants, 60% accepted.* In 1999, 1 degree awarded.
Degree requirements: For doctorate, dissertation required.
Entrance requirements: For doctorate, GRE General Test, TOEFL. *Application deadline:* For fall admission, 6/1; for spring admission, 11/1. Applications are processed on a rolling basis. *Application fee:* $40.
Expenses: Tuition, state resident: full-time $3,480; part-time $174 per credit hour. Tuition, nonresident: full-time $9,256; part-time $388 per credit hour. Required fees: $5 per term. Full-time tuition and fees vary according to course level, course load and campus/location.
Financial aid: Fellowships, research assistantships, teaching assistantships, institutionally sponsored loans available. Financial aid applicants required to submit FAFSA.
Dr. N. D. Camper, Coordinator, 864-656-5743, *Fax:* 864-656-0274, *E-mail:* dcamper@clemson.edu.

■ COLORADO STATE UNIVERSITY

Graduate School, College of Agricultural Sciences, Department of Horticulture and Landscape Architecture, Fort Collins, CO 80523-0015

AWARDS Floriculture (M Agr, MS, PhD); horticultural food crops (M Agr, MS, PhD); nursery and landscape management (M Agr, MS, PhD); plant genetics (MS, PhD); plant physiology (MS, PhD); turf management (M Agr, MS, PhD). Part-time programs available.

Faculty: 9 full-time (2 women), 2 part-time/adjunct (0 women).
Students: 13 full-time (8 women), 12 part-time (5 women), 11 international. Average age 33. *26 applicants, 46% accepted.* In 1999, 7 master's, 2 doctorates awarded.
Degree requirements: For master's, thesis optional, foreign language not required; for doctorate, dissertation required, foreign language not required.

Entrance requirements: For master's and doctorate, GRE General Test, TOEFL, minimum GPA of 3.0. *Application deadline:* For fall admission, 2/1 (priority date). Applications are processed on a rolling basis. *Application fee:* $30. Electronic applications accepted.

Expenses: Tuition, state resident: full-time $2,694; part-time $150 per credit. Tuition, nonresident: full-time $10,460; part-time $581 per credit. Required fees: $32 per semester. Tuition and fees vary according to program.

Financial aid: In 1999–00, 3 research assistantships with full tuition reimbursements, 5 teaching assistantships with full tuition reimbursements were awarded; fellowships with full tuition reimbursements, career-related internships or fieldwork, Federal Work-Study, institutionally sponsored loans, and traineeships also available. *Total annual research expenditures:* $600,000.

Dr. Stephen J. Wallner, Head, 970-491-7018, *Fax:* 970-491-7745.

Application contact: Judith A. Croissant, Administrative Assistant III, 970-491-7018, *Fax:* 970-491-7745, *E-mail:* jcroissa@ceres.agsci.colostate.edu.

■ CORNELL UNIVERSITY

Graduate School, Graduate Fields of Agriculture and Life Sciences, Field of Plant Biology, Ithaca, NY 14853-0001

AWARDS Paleobotany (PhD); plant cell biology (PhD); plant ecology (PhD); plant molecular biology (PhD); plant morphology, anatomy and biomechanics (PhD); plant physiology (PhD); systematic botany (PhD).

Faculty: 38 full-time.

Students: 51 full-time (26 women); includes 10 minority (1 African American, 5 Asian Americans or Pacific Islanders, 4 Hispanic Americans), 16 international. *94 applicants, 24% accepted.* In 1999, 11 doctorates awarded.

Degree requirements: For doctorate, dissertation required, foreign language not required.

Entrance requirements: For doctorate, GRE General Test, GRE Subject Test (biology) (recommended), TOEFL. *Application deadline:* For fall admission, 1/15. *Application fee:* $65. Electronic applications accepted.

Expenses: Tuition: Full-time $12,100.

Financial aid: In 1999–00, 47 students received aid, including 12 fellowships with full tuition reimbursements available, 19 research assistantships with full tuition reimbursements, 16 teaching assistantships with full tuition reimbursements available; institutionally sponsored loans, scholarships, tuition waivers (full and partial), and unspecified assistantships

also available. Financial aid applicants required to submit FAFSA.

Faculty research: Plant cell biology/cytology, plant molecular biology plant morphology/anatomy/biomechanics, plant physiology, systematic botany, paleobotany, plant ecology, ethnobotany, plant biochemistry, photosynthesis.

Application contact: Graduate Field Assistant, 607-255-8507, *E-mail:* plbio@cornell.edu.

■ IOWA STATE UNIVERSITY OF SCIENCE AND TECHNOLOGY

Graduate College, Interdisciplinary Programs, Program in Plant Physiology, Ames, IA 50011

AWARDS MS, PhD.

Students: 6 full-time (4 women), 3 international. *13 applicants, 46% accepted.* In 1999, 4 master's, 1 doctorate awarded.

Degree requirements: For master's, thesis or alternative required; for doctorate, dissertation required.

Entrance requirements: For master's and doctorate, GRE General Test, TOEFL. *Application deadline:* For fall admission, 6/1 (priority date); for spring admission, 11/1. Applications are processed on a rolling basis. *Application fee:* $20 ($50 for international students). Electronic applications accepted.

Expenses: Tuition, state resident: full-time $3,308. Tuition, nonresident: full-time $9,744. Part-time tuition and fees vary according to course load, campus/location and program.

Financial aid: In 1999–00, 3 research assistantships with partial tuition reimbursements (averaging $10,913 per year), 1 teaching assistantship (averaging $11,250 per year) were awarded; scholarships also available.

Dr. Martin H. Spalding, Supervisory Committee Chair, 515-294-0132, *E-mail:* ippm@iastate.edu.

Find an in-depth description at www.petersons.com/graduate.

■ OREGON STATE UNIVERSITY

Graduate School, Program in Plant Physiology, Corvallis, OR 97331

AWARDS MS, PhD.

Faculty: 38 full-time (5 women).

Students: 3 full-time (2 women), (all international). Average age 31.

Degree requirements: For master's and doctorate, thesis/dissertation required, foreign language not required.

Entrance requirements: For master's, TOEFL, BS in related area; for doctorate, TOEFL, BS or MS in related area, minimum GPA of 3.0 in last 90 hours. *Application deadline:* For fall admission, 3/1.

Applications are processed on a rolling basis. *Application fee:* $50.

Expenses: Tuition, state resident: full-time $6,489. Tuition, nonresident: full-time $11,061. Tuition and fees vary according to program.

Financial aid: Fellowships, research assistantships, teaching assistantships, career-related internships or fieldwork, Federal Work-Study, and institutionally sponsored loans available. Aid available to part-time students. Financial aid application deadline: 2/1.

Faculty research: Nitrogen metabolism, physiological ecology, phloem transport, mineral nutrition, plant hormones.

Dr. Patrick J. Breen, Director, 541-737-5469, *Fax:* 541-737-3479, *E-mail:* breenp@bcc.orst.edu.

■ THE PENNSYLVANIA STATE UNIVERSITY UNIVERSITY PARK CAMPUS

Graduate School, Intercollege Graduate Programs, Intercollege Graduate Program in Integrative Biosciences, State College, University Park, PA 16802-1503

AWARDS Integrative biosciences (MS, PhD), including biomolecular transport dynamics, cell and developmental biology, cellular and molecular mechanisms of toxicity, chemical biology, ecological and molecular plant physiology, immunobiology, molecular medicine, neuroscience, nutrition science.

Students: 39 full-time (24 women), 1 part-time.

Entrance requirements: For master's and doctorate, GRE General Test. *Application fee:* $50.

Expenses: Tuition, state resident: full-time $6,886; part-time $291 per credit. Tuition, nonresident: full-time $14,118; part-time $588 per credit. Required fees: $46 per semester. Part-time tuition and fees vary according to course load and program.

Financial aid: Fellowships available.

Dr. C. R. Matthews, Co-Director, 814-863-3650.

Application contact: Admissions Committee, 814-865-3155, *Fax:* 814-863-1357, *E-mail:* lscgradadm@mail.biotec.psu.edu.

Find an in-depth description at www.petersons.com/graduate.

■ THE PENNSYLVANIA STATE UNIVERSITY UNIVERSITY PARK CAMPUS

Graduate School, Intercollege Graduate Programs, Intercollege Graduate Program in Plant Physiology, State College, University Park, PA 16802-1503

AWARDS MS, PhD.

Students: 34 full-time (20 women), 3 part-time (1 woman).
Entrance requirements: For master's and doctorate, GRE General Test. *Application fee:* $50.
Expenses: Tuition, state resident: full-time $6,886; part-time $291 per credit. Tuition, nonresident: full-time $14,118; part-time $588 per credit. Required fees: $46 per semester. Part-time tuition and fees vary according to course load and program.
Dr. Teh-hui Kao, Chair, 814-865-2626.
Find an in-depth description at www.petersons.com/graduate.

■ PURDUE UNIVERSITY

Graduate School, School of Science, Department of Biological Sciences, Program in Plant Physiology, West Lafayette, IN 47907

AWARDS PhD.

Faculty: 10 full-time (0 women).
Degree requirements: For doctorate, dissertation, seminars, teaching experience required, foreign language not required.
Entrance requirements: For doctorate, GRE General Test, TOEFL, TSE. *Application deadline:* For fall admission, 2/15. *Application fee:* $30. Electronic applications accepted.
Expenses: Tuition, state resident: full-time $4,530; part-time $130 per credit hour. Tuition, nonresident: full-time $15,310; part-time $404 per credit hour. Tuition and fees vary according to campus/location and program.
Financial aid: Fellowships, research assistantships, teaching assistantships available. Aid available to part-time students. Financial aid application deadline: 2/15; financial aid applicants required to submit FAFSA.
Application contact: Nancy Konopka, Graduate Studies Office Manager, 765-494-8142, *Fax:* 765-494-0876, *E-mail:* njk@bilbo.bio.purdue.edu.

■ RUTGERS, THE STATE UNIVERSITY OF NEW JERSEY, NEW BRUNSWICK

Graduate School, Program in Plant Biology, New Brunswick, NJ 08901-1281

AWARDS Horticulture (MS, PhD); molecular biology and biochemistry (MS, PhD); pathology (MS, PhD); plant ecology (MS, PhD); plant genetics (PhD); plant physiology (MS, PhD); production and management (MS); structure and plant groups (MS, PhD). Part-time programs available.

Faculty: 64 full-time (8 women), 1 part-time/adjunct (0 women).
Students: 42 full-time (17 women), 13 part-time (6 women); includes 4 minority (1 African American, 2 Asian Americans or Pacific Islanders, 1 Hispanic American), 28 international. Average age 27. *58 applicants, 38% accepted.* In 1999, 4 master's, 9 doctorates awarded. Terminal master's awarded for partial completion of doctoral program.
Degree requirements: For master's, thesis or alternative required, foreign language not required; for doctorate, dissertation required, foreign language not required. *Average time to degree:* Master's–2.5 years full-time; doctorate–5 years full-time.
Entrance requirements: For master's and doctorate, GRE General Test, GRE Subject Test (recommended). *Application deadline:* For fall admission, 6/1. *Application fee:* $50. Electronic applications accepted.
Expenses: Tuition, state resident: full-time $6,776; part-time $279 per credit. Tuition, nonresident: full-time $9,936; part-time $412 per credit. Required fees: $20 per credit. $89 per semester. Tuition and fees vary according to course load, campus/location and program.
Financial aid: In 1999–00, 42 students received aid, including 12 fellowships with full tuition reimbursements available (averaging $16,000 per year), 21 research assistantships with full tuition reimbursements available (averaging $14,000 per year), 9 teaching assistantships with full tuition reimbursements available (averaging $13,100 per year). Financial aid application deadline: 2/15; financial aid applicants required to submit FAFSA.
Faculty research: Molecular genetics, disease resistance, molecular biology of plant viruses.
Dr. Thomas Leustek, Director, 732-932-8165 Ext. 326, *Fax:* 732-932-9377, *E-mail:* leustek@aesop.rutgers.edu.
Application contact: Barbara Mulder, Administrative Assistant, 732-932-9375

Ext. 358, *Fax:* 732-932-9377, *E-mail:* plantbio@aesop.rutgers.edu.
Find an in-depth description at www.petersons.com/graduate.

■ UNIVERSITY OF COLORADO AT BOULDER

Graduate School, College of Arts and Sciences, Department of Environmental, Population, and Organic Biology, Boulder, CO 80309

AWARDS Animal behavior (MA, PhD); aquatic biology (MA, PhD); behavioral genetics (MA, PhD); ecology (MA, PhD); microbiology (MA, PhD); neurobiology (MA, PhD); plant and animal physiology (MA, PhD); plant and animal systematics (MA, PhD); population biology (MA, PhD); population genetics (MA, PhD).

Faculty: 39 full-time (9 women).
Students: 84 full-time (36 women), 14 part-time (7 women); includes 17 minority (6 Asian Americans or Pacific Islanders, 10 Hispanic Americans, 1 Native American). Average age 29. *147 applicants, 14% accepted.* In 1999, 7 master's, 13 doctorates awarded. Terminal master's awarded for partial completion of doctoral program.
Degree requirements: For master's, thesis or alternative, comprehensive exam required, foreign language not required; for doctorate, dissertation, comprehensive exam required, foreign language not required. *Average time to degree:* Master's–3 years full-time; doctorate–5 years full-time.
Entrance requirements: For master's, GRE General Test, GRE Subject Test, minimum undergraduate GPA of 3.0; for doctorate, GRE General Test, GRE Subject Test. *Application deadline:* For fall admission, 1/15 (priority date). *Application fee:* $40 ($60 for international students).
Expenses: Tuition, state resident: part-time $181 per credit hour. Tuition, nonresident: part-time $542 per credit hour. Required fees: $99 per term. Tuition and fees vary according to course load and program.
Financial aid: Fellowships, research assistantships, teaching assistantships, Federal Work-Study, institutionally sponsored loans, and tuition waivers (full) available. Financial aid application deadline: 3/1.
Faculty research: Evolution, developmental biology, behavior and neurobiology. *Total annual research expenditures:* $1.8 million.
Michael Breed, Chair, 303-492-8981, *Fax:* 303-492-8699, *E-mail:* michael.breed@colorado.edu.
Application contact: Jill Skarstadt, Graduate Secretary, 303-492-7654, *Fax:* 303-492-8699, *E-mail:* jill.skarstadt@colorado.edu.

■ UNIVERSITY OF HAWAII AT MANOA

Graduate Division, College of Tropical Agriculture and Human Resources, Department of Plant Molecular Physiology, Honolulu, HI 96822

AWARDS Botanical sciences (MS, PhD), including plant physiology.

Faculty: 12 full-time (0 women), 4 part-time/adjunct (1 woman).
Students: 6 full-time (0 women), 1 part-time. Average age 34. *17 applicants, 59% accepted.* In 1999, 2 doctorates awarded (100% entered university research/teaching).
Degree requirements: For master's, foreign language and thesis not required; for doctorate, dissertation required, foreign language not required. *Average time to degree:* Master's–3 years part-time; doctorate–4 years part-time.
Entrance requirements: For master's and doctorate, GRE General Test. *Application deadline:* For fall admission, 3/1; for spring admission, 9/1. *Application fee:* $25 ($50 for international students).
Expenses: Tuition, state resident: full-time $4,032; part-time $168 per credit. Tuition, nonresident: full-time $9,960; part-time $415 per credit. Required fees: $51 per semester. Part-time tuition and fees vary according to course load and program.
Financial aid: In 1999–00, research assistantships (averaging $16,024 per year), 2 teaching assistantships (averaging $13,584 per year) were awarded.
Faculty research: Plant molecular and cellular biology, biochemistry, photosynthesis and postharvest physiology. *Total annual research expenditures:* $650,000.
Robert E. Paull, Chairman, 808-956-5900, *Fax:* 808-956-3894, *E-mail:* paull@hawaii.edu.
Application contact: Dulal Borthukur, Graduate Chairperson, 808-956-6600, *Fax:* 808-956-3542, *E-mail:* dulal@hawaii.edu.

■ UNIVERSITY OF KENTUCKY

Graduate School, Graduate School Programs from the College of Agriculture, Program in Plant Physiology, Lexington, KY 40506-0032

AWARDS PhD.

Degree requirements: For doctorate, dissertation, comprehensive exam required, foreign language not required.
Entrance requirements: For doctorate, GRE General Test, minimum graduate GPA of 3.0.
Expenses: Tuition, state resident: full-time $3,596; part-time $188 per credit hour. Tuition, nonresident: full-time $10,116; part-time $550 per credit hour.

Faculty research: Biochemistry and biophysics of photosynthesis, biochemical and molecular basis for resistance of plants to pathogens, plant gene expression, physiological aspects of crop production.

■ THE UNIVERSITY OF TENNESSEE

Graduate School, College of Agricultural Sciences and Natural Resources, Department of Plant and Soil Sciences, Knoxville, TN 37996

AWARDS Crop physiology and ecology (MS, PhD); plant breeding and genetics (MS, PhD); soil science (MS, PhD). Part-time programs available.

Faculty: 19 full-time (1 woman).
Students: 11 full-time (3 women), 11 part-time (7 women), 2 international. *7 applicants, 43% accepted.* In 1999, 7 master's, 5 doctorates awarded.
Degree requirements: For master's, thesis or alternative required, foreign language not required; for doctorate, dissertation required, foreign language not required.
Entrance requirements: For master's and doctorate, GRE General Test, TOEFL, minimum GPA of 2.7. *Application deadline:* For fall admission, 2/1 (priority date). Applications are processed on a rolling basis. *Application fee:* $35. Electronic applications accepted.
Expenses: Tuition, state resident: full-time $3,806; part-time $184 per credit hour. Tuition, nonresident: full-time $9,874; part-time $522 per credit hour. Tuition and fees vary according to program.
Financial aid: In 1999–00, 17 research assistantships, 2 teaching assistantships were awarded; fellowships, career-related internships or fieldwork, Federal Work-Study, and institutionally sponsored loans also available. Financial aid application deadline: 2/1; financial aid applicants required to submit FAFSA.
Dr. Fred L. Allen, Head, 865-974-7101, *Fax:* 865-974-7997, *E-mail:* allenf@utk.edu.
Application contact: Dr. Mike Essington, Graduate Representative, 865-974-8818, *E-mail:* messington@utk.edu.

■ THE UNIVERSITY OF TENNESSEE

Graduate School, College of Arts and Sciences, Program in Life Sciences, Knoxville, TN 37996

AWARDS Genome science and technology (MS, PhD); plant physiology and genetics (MS, PhD).

Faculty: 4 full-time (1 woman).
Students: 12 full-time (7 women), 5 part-time (2 women); includes 1 minority (Hispanic American), 9 international. *4 applicants, 0% accepted.* In 1999, 2 master's, 1 doctorate awarded.
Degree requirements: For doctorate, dissertation required.
Entrance requirements: For master's and doctorate, GRE General Test, TOEFL, minimum GPA of 2.7. *Application deadline:* For fall admission, 2/1 (priority date). Applications are processed on a rolling basis. *Application fee:* $35. Electronic applications accepted.
Expenses: Tuition, state resident: full-time $3,806; part-time $184 per credit hour. Tuition, nonresident: full-time $9,874; part-time $522 per credit hour. Tuition and fees vary according to program.
Financial aid: Fellowships, unspecified assistantships available. Financial aid application deadline: 2/1; financial aid applicants required to submit FAFSA.
Dr. Jeffrey M. Becker, Chairperson, 865-974-6841, *Fax:* 865-974-4057, *E-mail:* russellg@utk.edu.

■ VIRGINIA POLYTECHNIC INSTITUTE AND STATE UNIVERSITY

Graduate School, College of Agriculture and Life Sciences, Department of Plant Pathology, Physiology and Weed Science, Blacksburg, VA 24061

AWARDS Plant pathology (MS, PhD); plant physiology (MS, PhD).

Students: 21 full-time (9 women), 3 part-time (1 woman); includes 3 minority (2 African Americans, 1 Asian American or Pacific Islander), 6 international. *19 applicants, 58% accepted.* In 1999, 3 master's, 1 doctorate awarded.
Entrance requirements: For master's and doctorate, TOEFL. *Application deadline:* For fall admission, 12/1 (priority date). Applications are processed on a rolling basis. *Application fee:* $25.
Expenses: Tuition, state resident: full-time $4,122; part-time $229 per credit hour. Tuition, nonresident: full-time $6,930; part-time $385 per credit hour. Required fees: $828; $107 per semester. Part-time tuition and fees vary according to course load.
Financial aid: In 1999–00, 5 research assistantships with full tuition reimbursements (averaging $11,390 per year) were awarded; fellowships, teaching assistantships, unspecified assistantships also available. Financial aid application deadline: 4/1.
Dr. Kriton K. Hatzios, Head, 540-231-6361, *E-mail:* hatzios@vt.edu.

■ WASHINGTON STATE UNIVERSITY

Graduate School, College of Agriculture and Home Economics, Program in Plant Physiology, Pullman, WA 99164

AWARDS MS, PhD.

Faculty: 20 full-time (2 women), 2 part-time/adjunct (0 women).
Students: 25 full-time (17 women), 1 (woman) part-time; includes 1 minority (Asian American or Pacific Islander), 8 international. In 1999, 2 master's, 1 doctorate awarded. Terminal master's awarded for partial completion of doctoral program.
Degree requirements: For master's and doctorate, thesis/dissertation, oral exam, written exam required, foreign language not required.
Entrance requirements: For master's and doctorate, GRE General Test, minimum GPA of 3.0. *Application deadline:* For fall admission, 3/1 (priority date). Applications are processed on a rolling basis. *Application fee:* $35.
Expenses: Tuition, state resident: full-time $5,654. Tuition, nonresident: full-time $13,850. International tuition: $13,850 full-time. Tuition and fees vary according to program.
Financial aid: In 1999–00, 18 research assistantships with full and partial tuition reimbursements, 1 teaching assistantship with full and partial tuition reimbursement were awarded; career-related internships or fieldwork, Federal Work-Study, institutionally sponsored loans, and tuition waivers (partial) also available. Financial aid application deadline: 4/1; financial aid applicants required to submit FAFSA. Dr. Howard Grimes, Co-Chair, 509-335-2333.

Cell, Molecular, and Structural Biology

CELL BIOLOGY

■ ALBANY MEDICAL COLLEGE

Graduate Programs in the Biological Sciences, Program in Cell Biology and Cancer Research, Albany, NY 12208-3479

AWARDS MS, PhD. Part-time programs available.

Students: 21 full-time (12 women); includes 6 minority (2 African Americans, 4 Asian Americans or Pacific Islanders), 1 international. Average age 26. Terminal master's awarded for partial completion of doctoral program.
Degree requirements: For master's, thesis required, foreign language not required; for doctorate, dissertation, comprehensive written exam, oral qualifying exam required, foreign language not required.
Entrance requirements: For master's, GRE General Test, TOEFL; for doctorate, GRE General Test, TOEFL. *Application deadline:* Applications are processed on a rolling basis.
Expenses: Tuition: Full-time $13,367; part-time $446 per credit hour.
Financial aid: Federal Work-Study, grants, scholarships, and tuition waivers (full) available.
Dr. Kevin Pumiglia, Graduate Director, 518-262-6587, *Fax:* 518-262-5696, *E-mail:* pumiglk@mail.amc.edu.

Find an in-depth description at www.petersons.com/graduate.

■ ARIZONA STATE UNIVERSITY

Graduate College, College of Liberal Arts and Sciences, Department of Biology, Cell and Developmental Biology Group, Tempe, AZ 85287

AWARDS MS, PhD. Terminal master's awarded for partial completion of doctoral program.

Degree requirements: For master's, thesis required, foreign language not required; for doctorate, dissertation, oral exam required, foreign language not required.
Entrance requirements: For master's and doctorate, GRE General Test, GRE Subject Test. *Application deadline:* For fall admission, 12/15. *Application fee:* $45.
Expenses: Tuition, state resident: part-time $115 per credit hour. Tuition, nonresident: part-time $389 per credit hour. Required fees: $18 per semester. Tuition and fees vary according to program.
Financial aid: Application deadline: 12/15.
Faculty research: Cytoskeleton assembly, exocytosis, cyclic nucleotides, membrane fusion, chromosome distribution.

■ ARIZONA STATE UNIVERSITY

Graduate College, College of Liberal Arts and Sciences, Molecular and Cellular Biology Program, Tempe, AZ 85287

AWARDS MS, PhD.

Students: *54 applicants, 33% accepted.*
Degree requirements: For master's and doctorate, thesis/dissertation required.
Entrance requirements: For master's and doctorate, GRE. *Application deadline:* For fall admission, 3/1 (priority date); for spring admission, 10/1. Applications are processed on a rolling basis. *Application fee:* $45.
Expenses: Tuition, state resident: part-time $115 per credit hour. Tuition, nonresident: part-time $389 per credit hour. Required fees: $18 per semester. Tuition and fees vary according to program.
Financial aid: Fellowships, research assistantships, teaching assistantships available. Financial aid application deadline: 3/1.
Faculty research: Biochemistry, organometallic chemistry, genetics of outer membrane proteins, history and philosophy of biology.
Dr. Bertram Jacobs, Director, 480-965-0743.

Find an in-depth description at www.petersons.com/graduate.

■ BAYLOR COLLEGE OF MEDICINE

Graduate School of Biomedical Sciences, Department of Molecular and Cellular Biology, Houston, TX 77030-3498

AWARDS PhD, MD/PhD.

Faculty: 73 full-time (18 women).
Students: 72 full-time (34 women); includes 12 minority (6 African Americans, 3 Asian Americans or Pacific Islanders, 3 Hispanic Americans), 14 international. Average age 27. *60 applicants, 35% accepted.* In 2000, 13 doctorates awarded (62% entered university research/teaching, 8%

found other work related to degree, 30% continued full-time study).

Degree requirements: For doctorate, dissertation, public defense, qualifying exam required, foreign language not required. *Average time to degree:* Doctorate–6.6 years full-time.

Entrance requirements: For doctorate, GRE General Test (average 80th percentile), GRE Subject Test (strongly recommended), TOEFL, minimum GPA of 3.0. *Application deadline:* For fall admission, 2/1 (priority date). *Application fee:* $30. Electronic applications accepted.

Expenses: Tuition: Full-time $8,200. Required fees: $175. Full-time tuition and fees vary according to student level.

Financial aid: In 2000–01, 72 students received aid, including 32 fellowships (averaging $16,000 per year), 40 research assistantships (averaging $16,000 per year); Federal Work-Study, institutionally sponsored loans, and tuition waivers (full) also available. Financial aid applicants required to submit FAFSA.

Faculty research: Gene regulation, cell structure/function, developmental biology, neurobiology, reproductive endocrinology. Dr. JoAnne Richards, Director, 713-798-4598.

Application contact: Caroline Kosnik, Graduate Program Administrator, 713-798-4598, *Fax:* 713-790-0545, *E-mail:* ckosnik@bcm.tmc.edu.

Find an in-depth description at www.petersons.com/graduate.

◼ BAYLOR COLLEGE OF MEDICINE

Graduate School of Biomedical Sciences, Program in Cell and Molecular Biology, Houston, TX 77030-3498

AWARDS PhD, MD/PhD.

Faculty: 91 full-time (24 women).

Students: 41 full-time (23 women); includes 7 minority (1 African American, 3 Asian Americans or Pacific Islanders, 3 Hispanic Americans), 4 international. Average age 26. *114 applicants, 17% accepted.* In 1999, 3 doctorates awarded (100% entered university research/teaching).

Degree requirements: For doctorate, dissertation, public defense, qualifying exam required, foreign language not required. *Average time to degree:* Doctorate–5.79 years full-time.

Entrance requirements: For doctorate, GRE General Test (average 80th percentile), GRE Subject Test (strongly recommended), TOEFL, minimum GPA of 3.0. *Application deadline:* For fall admission, 2/1 (priority date). Applications are processed on a rolling basis. *Application fee:* $30. Electronic applications accepted.

Expenses: Tuition: Full-time $8,200. Required fees: $175. Full-time tuition and fees vary according to student level.

Financial aid: In 1999–00, 41 students received aid, including 29 fellowships (averaging $16,000 per year), 12 research assistantships (averaging $16,000 per year); teaching assistantships, career-related internships or fieldwork, Federal Work-Study, institutionally sponsored loans, and tuition waivers (full) also available. Financial aid applicants required to submit FAFSA.

Faculty research: Gene expression and regulation, developmental biology and genetics, signal transduction and membrane biology, aging process, molecular virology. Dr. Susan M. Berget, Director, 713-798-6557.

Application contact: Lourdes Fernandez, Graduate Program Administrator, 713-798-6557, *Fax:* 713-798-6325, *E-mail:* lourdesf@bcm.tmc.edu.

Find an in-depth description at www.petersons.com/graduate.

◼ BOSTON UNIVERSITY

Graduate School of Arts and Sciences, Department of Biology, Boston, MA 02215

AWARDS Botany (MA, PhD); cell and molecular biology (MA, PhD); cell biology (MA, PhD); ecology (PhD); ecology, behavior, and evolution (MA, PhD); ecology/physiology, endocrinology and reproduction (MA); marine biology (MA, PhD); molecular biology, cell biology and biochemistry (MA, PhD); neurobiology, neuroendocrinology and reproduction (MA, PhD); physiology, endocrinology, and neurobiology (MA, PhD); zoology (MA, PhD). Part-time programs available.

Faculty: 41 full-time (8 women).

Students: 131 full-time (74 women), 11 part-time (7 women); includes 10 minority (7 Asian Americans or Pacific Islanders, 3 Hispanic Americans), 33 international. Average age 27. *238 applicants, 39% accepted.* In 1999, 61 master's, 45 doctorates awarded. Terminal master's awarded for partial completion of doctoral program.

Degree requirements: For master's, one foreign language, thesis not required; for doctorate, one foreign language, dissertation, qualifying exam required. *Average time to degree:* Master's–1 year full-time, 3 years part-time; doctorate–5.75 years full-time.

Entrance requirements: For master's and doctorate, GRE General Test, GRE Subject Test, TOEFL. *Application deadline:* For fall admission, 1/1 (priority date); for spring admission, 11/1. *Application fee:* $50.

Expenses: Tuition: Full-time $23,770; part-time $743 per credit. Required fees: $220. Tuition and fees vary according to class time, course level, campus/location and program.

Financial aid: In 1999–00, 82 students received aid, including 1 fellowship with full tuition reimbursement available (averaging $12,000 per year), 28 research assistantships with full tuition reimbursements available (averaging $11,500 per year), 43 teaching assistantships with full tuition reimbursements available (averaging $11,500 per year); Federal Work-Study, grants, institutionally sponsored loans, scholarships, and traineeships also available. Financial aid application deadline: 1/15; financial aid applicants required to submit FAFSA.

Faculty research: Marine science, endocrinology, behavior. *Total annual research expenditures:* $5 million. Geoffrey M. Cooper, Chairman, 617-353-2432, *Fax:* 617-353-6340, *E-mail:* gmcooper@bu.edu.

Application contact: Yolanta Kovalko, Senior Staff Assistant, 617-353-2432, *Fax:* 617-353-6340, *E-mail:* yolanta@bu.edu.

Find an in-depth description at www.petersons.com/graduate.

◼ BOSTON UNIVERSITY

Graduate School of Arts and Sciences, Molecular Biology, Cell Biology, and Biochemistry Program (MCBB), Boston, MA 02215

AWARDS MA, PhD. Part-time programs available.

Faculty: 45 full-time (10 women).

Students: 31 full-time (16 women), 2 part-time (1 woman); includes 7 minority (5 Asian Americans or Pacific Islanders, 2 Hispanic Americans), 10 international. Average age 26. *87 applicants, 30% accepted.* Terminal master's awarded for partial completion of doctoral program.

Degree requirements: For master's, one foreign language, thesis required (for some programs); for doctorate, one foreign language, dissertation, qualifying exam required. *Average time to degree:* Master's–2 years full-time.

Entrance requirements: For master's and doctorate, GRE General Test, GRE Subject Test, TOEFL. *Application deadline:* For fall admission, 1/1 (priority date). *Application fee:* $50.

Expenses: Tuition: Full-time $23,770; part-time $743 per credit. Required fees: $220. Tuition and fees vary according to class time, course level, campus/location and program.

Financial aid: In 1999–00, 16 students received aid, including 9 research assistantships with full tuition reimbursements

Boston University (continued)
available (averaging $11,500 per year), 7 teaching assistantships with full tuition reimbursements available (averaging $11,500 per year); fellowships, Federal Work-Study, scholarships, and traineeships also available. Financial aid application deadline: 1/15; financial aid applicants required to submit FAFSA.

Faculty research: Signal transduction, gene expression, protein and nucleic acid biochemistry, genomics, modular physiology and development. *Total annual research expenditures:* $4.5 million.
Gary R. Jacobson, Director, 617-353-2432, *Fax:* 617-353-6340, *E-mail:* jacobson@bio.bu.edu.

Application contact: Michelle Brodkowitz, Academic Administrator, 617-353-2432, *Fax:* 617-353-6340, *E-mail:* mcbb@bio.bu.edu.

Find an in-depth description at www.petersons.com/graduate.

■ BOSTON UNIVERSITY

School of Medicine, Division of Graduate Medical Sciences, Department of Pathology and Laboratory Medicine, Boston, MA 02118

AWARDS Cell and molecular biology (PhD); experimental pathology (PhD); immunology (PhD). Part-time programs available.

Faculty: 12 full-time (4 women), 13 part-time/adjunct (2 women).
Students: 33 full-time (18 women), 1 part-time; includes 4 minority (2 African Americans, 2 Asian Americans or Pacific Islanders), 14 international. Average age 29. In 1999, 3 degrees awarded.
Degree requirements: For doctorate, dissertation, qualifying exam required, foreign language not required.
Entrance requirements: For doctorate, GRE General Test, GRE Subject Test, TOEFL. *Application deadline:* For fall admission, 1/15 (priority date); for spring admission, 10/15 (priority date). *Application fee:* $50. Electronic applications accepted.
Expenses: Tuition: Full-time $24,700; part-time $772 per credit. Required fees: $220.
Financial aid: In 1999–00, 10 fellowships, 3 research assistantships were awarded; Federal Work-Study, scholarships, and traineeships also available.
Faculty research: Toxicology, carcinogenesis, endocytosis, cytogenetics.
Leonard Gottlieb, Chairman, 617-638-4500, *Fax:* 617-638-4085.
Application contact: Dr. Adrianne Rogers, Associate Chairman, 617-638-4500, *Fax:* 617-638-4085, *E-mail:* aerogers@bu.edu.

■ BOSTON UNIVERSITY

School of Medicine, Division of Graduate Medical Sciences, Program in Cell and Molecular Biology, Boston, MA 02118

AWARDS MA, PhD, MD/PhD.

Degree requirements: For doctorate, dissertation required, foreign language not required.
Entrance requirements: For master's and doctorate, GRE General Test, GRE Subject Test, TOEFL. *Application deadline:* For fall admission, 1/15 (priority date); for spring admission, 10/15 (priority date). *Application fee:* $45. Electronic applications accepted.
Expenses: Tuition: Full-time $24,700; part-time $772 per credit. Required fees: $220.
Financial aid: Fellowships, research assistantships, Federal Work-Study, scholarships, and traineeships available.
Dr. Vickery Trinkaus Randall, Director, 617-638-6099, *Fax:* 617-638-5337, *E-mail:* fickery@biochem.bumc.bu.edu.
Application contact: Dr. Mary Jo Murnanl, Admissions Director, 617-638-4926, *Fax:* 617-638-4085, *E-mail:* mmuranl@bu.edu.

Find an in-depth description at www.petersons.com/graduate.

■ BRANDEIS UNIVERSITY

Graduate School of Arts and Sciences, Programs in Biological Sciences, Program in Molecular and Cell Biology, Waltham, MA 02454-9110

AWARDS Cell biology (PhD); developmental biology (PhD); genetics (PhD); microbiology (PhD); molecular and cell biology (MS); molecular biology (PhD); neurobiology (PhD).

Faculty: 19 full-time (8 women).
Students: 58 full-time (31 women), 1 (woman) part-time; includes 13 minority (1 African American, 8 Asian Americans or Pacific Islanders, 4 Hispanic Americans), 8 international. Average age 27. *83 applicants, 11% accepted.* In 1999, 1 master's, 8 doctorates awarded (100% entered university research/teaching). Terminal master's awarded for partial completion of doctoral program.
Degree requirements: For master's, thesis not required; for doctorate, dissertation required. *Average time to degree:* Master's–1 year full-time; doctorate–6 years full-time.
Entrance requirements: For doctorate, GRE General Test. *Application deadline:* For fall admission, 1/15 (priority date). Applications are processed on a rolling basis. *Application fee:* $60. Electronic applications accepted.

Expenses: Tuition: Full-time $25,392; part-time $3,174 per course. Required fees: $509. Tuition and fees vary according to class time, degree level, program and student level.
Financial aid: In 1999–00, 10 fellowships with tuition reimbursements (averaging $17,000 per year), 49 research assistantships with tuition reimbursements (averaging $17,000 per year), 9 teaching assistantships (averaging $2,312 per year) were awarded; scholarships and tuition waivers (full and partial) also available. Financial aid application deadline: 4/15; financial aid applicants required to submit CSS PROFILE or FAFSA.
Faculty research: Regulation of gene expression by transcription factors, molecular neurobiology, immunology, molecular mechanisms of genetic recombination, and cell differentiation.
Dr. Ranjan Sen, Chair, 781-736-2455, *Fax:* 781-736-3107.
Application contact: Marcia Cabral, Information Officer, 781-736-3100, *Fax:* 781-736-3107, *E-mail:* cabral@brandeis.edu.

■ BROWN UNIVERSITY

Graduate School, Division of Biology and Medicine, Program in Molecular Biology, Cell Biology, and Biochemistry, Providence, RI 02912

AWARDS Biochemistry (M Med Sc, Sc M, PhD), including biochemistry (Sc M, PhD), biology (Sc M, PhD), medical science (M Med Sc, PhD); biology (MA); cell biology (M Med Sc, Sc M, PhD), including biochemistry (Sc M, PhD), biology (Sc M, PhD), medical science (M Med Sc, PhD); developmental biology (M Med Sc, Sc M, PhD), including biochemistry (Sc M, PhD), biology (Sc M, PhD), medical science (M Med Sc, PhD); immunology (M Med Sc, Sc M, PhD), including biochemistry (Sc M, PhD), biology (Sc M, PhD), medical science (M Med Sc, PhD); molecular microbiology (M Med Sc, Sc M, PhD), including biochemistry (Sc M, PhD), biology (Sc M, PhD), medical science (M Med Sc, PhD). Part-time programs available.

Faculty: 50 full-time (14 women).
Students: 61 full-time (34 women); includes 4 minority (1 African American, 3 Asian Americans or Pacific Islanders), 21 international. Average age 25. *106 applicants, 28% accepted.* In 1999, 1 master's, 3 doctorates awarded. Terminal master's awarded for partial completion of doctoral program.
Degree requirements: For master's, thesis required (for some programs), foreign language not required; for doctorate, one foreign language, dissertation, preliminary exam required. *Average time to degree:* Doctorate–5 years full-time.

Entrance requirements: For master's and doctorate, GRE General Test, GRE Subject Test. *Application deadline:* For fall admission, 1/2 (priority date). Applications are processed on a rolling basis. *Application fee:* $60. Electronic applications accepted.
Financial aid: In 1999–00, 58 students received aid, including 11 fellowships (averaging $18,916 per year), 9 research assistantships (averaging $18,916 per year), 13 teaching assistantships (averaging $12,690 per year); institutionally sponsored loans and traineeships also available. Financial aid application deadline: 1/2.
Faculty research: Molecular genetics, gene regulation.
Dr. Gary Wessel, Director, 401-863-1051, *E-mail:* chet@brown.edu.
Application contact: Mary C. Esser, Graduate Program Coordinator, 401-863-1661, *Fax:* 401-863-1348, *E-mail:* mary_esser@brown.edu.

Find an in-depth description at www.petersons.com/graduate.

■ **CALIFORNIA INSTITUTE OF TECHNOLOGY**

Division of Biology, Program in Cell Biology and Biophysics, Pasadena, CA 91125-0001

AWARDS PhD.

Degree requirements: For doctorate, dissertation, qualifying exam required, foreign language not required.
Entrance requirements: For doctorate, GRE General Test. *Application deadline:* For fall admission, 1/1. *Application fee:* $0.
Expenses: Tuition: Full-time $19,260. Required fees: $24. One-time fee: $100 full-time.
Financial aid: Application deadline: 1/1.
Application contact: Elizabeth Ayala, Graduate Option Coordinator, 626-395-4497, *Fax:* 626-449-0756, *E-mail:* biograd@cco.caltech.edu.

■ **CARNEGIE MELLON UNIVERSITY**

Mellon College of Science, Department of Biological Sciences, Pittsburgh, PA 15213-3891

AWARDS Biochemistry (PhD); biophysics (PhD); cell biology (PhD); developmental biology (PhD); genetics (PhD); molecular biology (PhD).
Faculty: 38 full-time (17 women), 1 (woman) part-time/adjunct.
Students: 35 full-time (24 women); includes 2 minority (1 African American, 1 Asian American or Pacific Islander), 18 international. Average age 27. In 1999, 4 degrees awarded.

Degree requirements: For doctorate, dissertation required, foreign language not required.
Entrance requirements: For doctorate, GRE General Test, GRE Subject Test, interview. *Application deadline:* For fall admission, 2/1 (priority date). Applications are processed on a rolling basis. *Application fee:* $0.
Expenses: Tuition: Full-time $22,100; part-time $307 per unit. Required fees: $200. Tuition and fees vary according to program.
Financial aid: Fellowships, research assistantships, teaching assistantships, traineeships available.
Faculty research: Genetic structure, function, and regulation; protein structure and function; biological membranes; biological spectroscopy. *Total annual research expenditures:* $4.4 million.
Dr. William E. Brown, Head, 412-268-3416, *Fax:* 412-268-7129, *E-mail:* wb02@andrew.cmu.edu.
Application contact: Stacey L. Young, Assistant Head, 412-268-7372, *Fax:* 412-268-7129, *E-mail:* sf38+@andrew.cmu.edu.

Find an in-depth description at www.petersons.com/graduate.

■ **CASE WESTERN RESERVE UNIVERSITY**

School of Medicine and School of Graduate Studies, Graduate Programs in Medicine, Department of Anatomy, Cleveland, OH 44106

AWARDS Applied anatomy (MS); biological anthropology (MS, PhD); cellular biology (MS, PhD); developmental biology (PhD); molecular biology (PhD). Part-time programs available.
Faculty: 14 full-time (5 women), 14 part-time/adjunct (4 women).
Students: 17 full-time (9 women), 12 part-time (1 woman); includes 11 minority (1 African American, 9 Asian Americans or Pacific Islanders, 1 Hispanic American), 2 international. Average age 26. *16 applicants, 50% accepted.* In 1999, 5 master's awarded (50% found work related to degree, 50% continued full-time study); 1 doctorate awarded (100% entered university research/teaching).
Degree requirements: For master's, thesis required (for some programs); for doctorate, dissertation required. *Average time to degree:* Master's–5 years full-time; doctorate–1 year full-time.
Entrance requirements: For master's, GRE General Test, TOEFL; for doctorate, GRE General Test, GRE Subject Test, TOEFL. *Application deadline:* For fall admission, 5/1 (priority date); for spring admission, 8/1 (priority date). Applications are processed on a rolling basis. *Application fee:* $25.
Financial aid: In 1999–00, 7 research assistantships with full tuition reimbursements (averaging $16,000 per year) were awarded; fellowships, grants also available.
Faculty research: Hypoxia, cell injury, biochemical aberration occurrences in ischemic tissue, human functional morphology, evolutionary morphology. *Total annual research expenditures:* $562,716.
Joseph C. LaManna, Acting Chairman, 216-368-1100, *Fax:* 216-368-8669, *E-mail:* jcl4@po.cwru.edu.
Application contact: Laila Boesinger, Administrator, 216-368-3430, *Fax:* 216-368-8669, *E-mail:* lvb2@po.cwru.edu.

■ **CASE WESTERN RESERVE UNIVERSITY**

School of Medicine and School of Graduate Studies, Graduate Programs in Medicine, Department of Molecular Biology and Microbiology, Cleveland, OH 44106-4960

AWARDS Cellular biology (PhD); microbiology (PhD); molecular biology (PhD). Students are admitted to an integrated Biomedical Sciences Training Program involving 11 basic science programs at Case Western Reserve University.

Faculty: 14 full-time (4 women).
Students: 24 full-time (9 women); includes 2 minority (both Asian Americans or Pacific Islanders), 5 international. In 1999, 1 doctorate awarded (100% found work related to degree).
Degree requirements: For doctorate, dissertation required, foreign language not required. *Average time to degree:* Doctorate–5.3 years full-time.
Entrance requirements: For doctorate, GRE General Test, GRE Subject Test, TOEFL. *Application fee:* $25.
Financial aid: In 1999–00, 1 fellowship with full tuition reimbursement (averaging $16,000 per year), 23 research assistantships with full tuition reimbursements (averaging $16,000 per year) were awarded; Federal Work-Study, scholarships, traineeships, and tuition waivers (full) also available.
Faculty research: Gene expression in eukaryotic and prokaryotic systems; microbial physiology; intracellular transport and signaling; mechanisms of oncogenesis; molecular mechanisms of RNA processing, editing, and catalysis. *Total annual research expenditures:* $4.6 million.
Lloyd A. Culp, Acting Chair, 216-368-3420, *Fax:* 216-368-3055, *E-mail:* lac7@po.cwru.edu.

Case Western Reserve University (continued)
Application contact: Dr. Michael E. Harris, Admissions Coordinator, 216-368-3347, *E-mail:* mbio@po.cwru.edu.
Find an in-depth description at www.petersons.com/graduate.

■ CASE WESTERN RESERVE UNIVERSITY

School of Medicine and School of Graduate Studies, Graduate Programs in Medicine, Program in Cell Biology, Cleveland, OH 44106

AWARDS PhD.

Faculty: 20 full-time (10 women).
Students: 3 full-time (2 women). Average age 26. In 1999, 1 degree awarded.
Degree requirements: For doctorate, dissertation required, foreign language not required. *Average time to degree:* Doctorate–5 years full-time.
Entrance requirements: For doctorate, GRE General Test, GRE Subject Test, TOEFL, previous course work in biochemistry. *Application deadline:* Applications are processed on a rolling basis. *Application fee:* $25.
Financial aid: In 1999–00, fellowships (averaging $16,000 per year); tuition waivers (full) also available.
Faculty research: Macromolecular transport, membrane traffic, signal transduction, nuclear organization.
Dr. Alan Tartakoff, Head, 216-368-5544, *Fax:* 216-368-5484.

Find an in-depth description at www.petersons.com/graduate.

■ CASE WESTERN RESERVE UNIVERSITY

School of Medicine and School of Graduate Studies, Graduate Programs in Medicine, Programs in Molecular and Cellular Basis of Disease, Cleveland, OH 44106

AWARDS Cell biology (MS, PhD); immunology (MS, PhD); pathology (MS, PhD). Terminal master's awarded for partial completion of doctoral program.

Degree requirements: For master's and doctorate, thesis/dissertation required, foreign language not required.
Entrance requirements: For master's and doctorate, GRE General Test, GRE Subject Test, TOEFL.
Faculty research: Neurobiology, molecular biology, cancer biology, biomaterials, biocompatibility.

Find an in-depth description at www.petersons.com/graduate.

■ THE CATHOLIC UNIVERSITY OF AMERICA

School of Arts and Sciences, Department of Biology, Program in Cell and Microbial Biology, Washington, DC 20064

AWARDS Cell biology (MS, PhD); microbiology (MS, PhD). Part-time programs available.

Students: 5 full-time (4 women), 4 part-time (2 women); includes 2 minority (1 African American, 1 Hispanic American), 2 international. Average age 26. *23 applicants, 57% accepted.* Terminal master's awarded for partial completion of doctoral program.
Degree requirements: For master's, thesis or alternative, comprehensive exam required, foreign language not required; for doctorate, dissertation, comprehensive exam required, foreign language not required.
Entrance requirements: For master's, GRE General Test, GRE Subject Test, TOEFL; for doctorate, GRE General Test, GRE Subject Test. *Application deadline:* For fall admission, 8/1 (priority date); for spring admission, 12/1. Applications are processed on a rolling basis. *Application fee:* $55. Electronic applications accepted.
Expenses: Tuition: Full-time $18,200; part-time $700 per credit hour. Required fees: $378 per semester. Part-time tuition and fees vary according to campus/location and program.
Financial aid: Fellowships, research assistantships, teaching assistantships, career-related internships or fieldwork, institutionally sponsored loans, and tuition waivers (full and partial) available. Aid available to part-time students. Financial aid application deadline: 2/1.
Faculty research: Cell differentiation, regulation of cell growth, drug resistance, gene cloning and sequencing, developmental biology and neurobiology. *Total annual research expenditures:* $230,000.
Dr. John Golin, Chair, Department of Biology, 202-319-5279, *Fax:* 202-319-5721.

■ COLORADO STATE UNIVERSITY

College of Veterinary Medicine and Biomedical Sciences and Graduate School, Graduate Programs in Veterinary Medicine and Biomedical Sciences, Fort Collins, CO 80523-0015

AWARDS Anatomy and neurobiology (MS, PhD); cell and molecular biology (MS, PhD); clinical sciences (MS, PhD); environmental health (MS, PhD); microbiology (MS, PhD), including immunology, microbiology; pathology (MS, PhD); physiology (MS, PhD);

radiological health sciences (MS, PhD), including cellular and molecular biology, health physics, mammalian radiobiology, nuclear-waste management (MS), radiobiology (MS), radioecology, radiology, veterinary radiology (MS). Part-time programs available.

Faculty: 134 full-time (17 women).
Students: 221 full-time (119 women), 61 part-time (34 women); includes 21 minority (4 African Americans, 7 Asian Americans or Pacific Islanders, 9 Hispanic Americans, 1 Native American), 37 international. Average age 29. *303 applicants, 42% accepted.* In 1999, 81 master's, 23 doctorates awarded. Terminal master's awarded for partial completion of doctoral program.
Degree requirements: For master's, foreign language not required; for doctorate, dissertation required.
Entrance requirements: For master's and doctorate, GRE General Test, TOEFL. *Application deadline:* Applications are processed on a rolling basis. *Application fee:* $30. Electronic applications accepted.
Expenses: Tuition, state resident: full-time $2,694; part-time $150 per credit. Tuition, nonresident: full-time $10,460; part-time $581 per credit. Required fees: $32 per semester.
Financial aid: In 1999–00, 93 research assistantships, 11 teaching assistantships were awarded; fellowships, career-related internships or fieldwork, Federal Work-Study, institutionally sponsored loans, traineeships, and tuition waivers (partial) also available. Aid available to part-time students.
Faculty research: Reproductive physiology, infectious diseases, comparative oncology, environmental toxicology.
Application contact: Dr. Alan Tucker, Assistant Dean, 970-491-6106, *Fax:* 970-491-2250.

■ COLORADO STATE UNIVERSITY

Graduate School, Program in Cell and Molecular Biology, Fort Collins, CO 80523-0015

AWARDS MS, PhD.

Faculty: 46 full-time (13 women).
Students: 15 full-time (11 women), 6 part-time (4 women), 6 international. Average age 28. *147 applicants, 12% accepted.* In 1999, 3 degrees awarded. Terminal master's awarded for partial completion of doctoral program.
Degree requirements: For master's, thesis required (for some programs), foreign language not required; for doctorate, dissertation required, foreign language not required.

Entrance requirements: For master's and doctorate, GRE General Test, GRE Subject Test, TOEFL, minimum GPA of 3.0. *Application deadline:* For fall admission, 1/15 (priority date). Applications are processed on a rolling basis. *Application fee:* $30. Electronic applications accepted.
Expenses: Tuition, state resident: full-time $2,694; part-time $150 per credit. Tuition, nonresident: full-time $10,460; part-time $581 per credit. Required fees: $32 per semester. Tuition and fees vary according to program.
Financial aid: In 1999–00, 8 research assistantships, 4 teaching assistantships were awarded; fellowships, Federal Work-Study and institutionally sponsored loans also available. Financial aid application deadline: 2/1.
Faculty research: Regulation of gene expression, cancer biology, plant molecular genetics, reproductive physiology, environmental toxicology. *Total annual research expenditures:* $10 million.
Dr. Michael H. Fox, Chairman, 970-491-7618, *Fax:* 970-491-0623, *E-mail:* mfox@cvmbs.colostate.edu.
Application contact: Linda Jones, Administrative Assistant, 970-491-0241, *Fax:* 970-491-0623, *E-mail:* ljones@cvmbs.colostate.edu.
Find an in-depth description at www.petersons.com/graduate.

■ **COLUMBIA UNIVERSITY**
College of Physicians and Surgeons and Graduate School of Arts and Sciences, Graduate School of Arts and Sciences at the College of Physicians and Surgeons, Department of Anatomy and Cell Biology, New York, NY 10032

AWARDS Anatomy (M Phil, MA, PhD); anatomy and cell biology (PhD). Only candidates for the PhD are admitted. Terminal master's awarded for partial completion of doctoral program.

Degree requirements: For master's, foreign language and thesis not required; for doctorate, dissertation, oral exam required, foreign language not required.
Entrance requirements: For master's and doctorate, GRE General Test, TOEFL.
Expenses: Tuition: Full-time $25,072.
Faculty research: Protein sorting, membrane biophysics, muscle energetics, neuroendocrinology, developmental biology, cytoskeleton, transcription factors.

Find an in-depth description at www.petersons.com/graduate.

■ **COLUMBIA UNIVERSITY**
College of Physicians and Surgeons and Graduate School of Arts and Sciences, Graduate School of Arts and Sciences at the College of Physicians and Surgeons, Integrated Program in Cellular, Molecular and Biophysical Studies, New York, NY 10032
AWARDS M Phil, MA, PhD, MD/PhD. Only candidates for the PhD are admitted. Terminal master's awarded for partial completion of doctoral program.

Degree requirements: For master's, foreign language and thesis not required; for doctorate, dissertation required, foreign language not required.
Entrance requirements: For master's, GRE General Test, TOEFL; for doctorate, GRE General Test, GRE Subject Test, TOEFL.
Expenses: Tuition: Full-time $25,072.
Faculty research: Transcription, macromolecular sorting, gene expression during development, cellular interaction.

■ **CORNELL UNIVERSITY**
Graduate School, Graduate Fields of Agriculture and Life Sciences, Field of Biochemistry, Molecular and Cell Biology, Ithaca, NY 14853-0001
AWARDS Biochemistry (PhD); biophysics (PhD); cell biology (PhD); molecular and cell biology (PhD); molecular biology (PhD).
Faculty: 43 full-time.
Students: 75 full-time (37 women); includes 15 minority (1 African American, 6 Asian Americans or Pacific Islanders, 8 Hispanic Americans), 19 international. *288 applicants, 21% accepted.* In 1999, 9 doctorates awarded.
Degree requirements: For doctorate, dissertation, 2 semesters of teaching experience required.
Entrance requirements: For doctorate, GRE General Test, GRE Subject Test (biology, chemistry, physics, biochemistry or, cell and molecular biology), TOEFL. *Application deadline:* For fall admission, 1/5. *Application fee:* $65. Electronic applications accepted.
Expenses: Tuition: Full-time $12,100.
Financial aid: In 1999–00, 74 students received aid, including 35 fellowships with full tuition reimbursements available, 32 research assistantships with full tuition reimbursements available, 7 teaching assistantships with full tuition reimbursements available; institutionally sponsored loans, scholarships, tuition waivers (full and partial), and unspecified assistantships also available. Financial aid applicants required to submit FAFSA.
Faculty research: Biophysics, structural biology.

Application contact: Graduate Field Assistant, 607-255-2317, *E-mail:* bmcb@cornell.edu.

Find an in-depth description at www.petersons.com/graduate.

■ **DUKE UNIVERSITY**
Graduate School, Department of Biological Anthropology and Anatomy, Durham, NC 27708-0586
AWARDS Cellular and molecular biology (PhD); gross anatomy and physical anthropology (PhD), including comparative morphology of human and non-human primates, primate social behavior, vertebrate paleontology; neuroanatomy (PhD).

Faculty: 14 full-time, 1 part-time/adjunct.
Students: 20 full-time (12 women); includes 3 minority (2 African Americans, 1 Hispanic American), 1 international. *62 applicants, 6% accepted.* In 1999, 2 doctorates awarded.
Degree requirements: For doctorate, dissertation required.
Entrance requirements: For doctorate, GRE General Test. *Application deadline:* For fall admission, 12/31. *Application fee:* $75.
Expenses: Tuition: Full-time $21,406; part-time $760 per unit. Required fees: $3,136; $3,136 per year. One-time fee: $30. Tuition and fees vary according to program.
Financial aid: Fellowships, teaching assistantships, Federal Work-Study available. Financial aid application deadline: 12/31.
Kathleen Smith, Director of Graduate Studies, 919-684-4124, *Fax:* 919-684-8034, *E-mail:* rachel_hougom@baa.mc.duke.edu.

■ **DUKE UNIVERSITY**
Graduate School, Department of Cell Biology, Durham, NC 27708-0586
AWARDS Cell biology (PhD); physiology and cellular biophysics (PhD).

Faculty: 72 full-time, 13 part-time/adjunct.
Students: 43 full-time (21 women); includes 4 minority (1 African American, 2 Asian Americans or Pacific Islanders, 1 Hispanic American), 8 international. *20 applicants, 25% accepted.* In 1999, 8 doctorates awarded.
Degree requirements: For doctorate, dissertation required, foreign language not required.
Entrance requirements: For doctorate, GRE General Test, GRE Subject Test (recommended). *Application deadline:* For fall admission, 12/31. *Application fee:* $75.
Expenses: Tuition: Full-time $21,406; part-time $760 per unit. Required fees:

Duke University (continued)
$3,136; $3,136 per year. One-time fee: $30. Tuition and fees vary according to program.
Financial aid: Fellowships, research assistantships, teaching assistantships, Federal Work-Study available. Financial aid application deadline: 12/31.
G. Vann Bennett, Director of Graduate Studies, 919-684-3538, *Fax:* 919-684-3590, *E-mail:* b.sampson@cellbio.duke.edu.

■ DUKE UNIVERSITY

Graduate School, Program in Cellular and Molecular Biology, Durham, NC 27708-0586

AWARDS Certificate. Students must be enrolled in a participating PhD program.

Faculty: 115 full-time.
Students: 23 full-time (14 women); includes 1 minority (Asian American or Pacific Islander), 2 international. *177 applicants, 29% accepted.*
Entrance requirements: For degree, GRE General Test, GRE Subject Test (recommended). *Application deadline:* For fall admission, 12/31. *Application fee:* $75.
Expenses: Tuition: Full-time $21,406; part-time $760 per unit. Required fees: $3,136; $3,136 per year. One-time fee: $30. Tuition and fees vary according to program.
Financial aid: Fellowships available. Financial aid application deadline: 12/31.
Dr. Kenneth Kreuzer, Director of Graduate Studies, 919-684-6559, *Fax:* 919-681-8911, *E-mail:* cmbtgp@acpub.duke.edu.
Find an in-depth description at www.petersons.com/graduate.

■ EAST CAROLINA UNIVERSITY

School of Medicine, Department of Anatomy and Cell Biology, Greenville, NC 27858-4353

AWARDS PhD.

Faculty: 13 full-time (2 women).
Students: 2 full-time (1 woman), 4 part-time (2 women); includes 1 minority (Asian American or Pacific Islander), 1 international. Average age 32. *2 applicants, 50% accepted.* In 1999, 1 degree awarded.
Degree requirements: For doctorate, one foreign language (computer language can substitute), dissertation required.
Entrance requirements: For doctorate, GRE General Test, GRE Subject Test, TOEFL. *Application deadline:* For fall admission, 6/1 (priority date). Applications are processed on a rolling basis. *Application fee:* $40.
Expenses: Tuition, state resident: full-time $1,012. Tuition, nonresident: full-time $8,578. Required fees: $1,006. Full-time tuition and fees vary according to degree

level. Part-time tuition and fees vary according to course load.
Financial aid: Fellowships available. Financial aid application deadline: 6/1.
Faculty research: Diabetes mellitus, cellular biology of pancreas, autonomic nervous system, neural cytoarchitecture, CNS neuropeptides.
Dr. Jack Brinn, Chairman, 252-816-2851, *Fax:* 252-816-2850, *E-mail:* brinnj@mail.ecu.edu.
Application contact: Dr. David Terrian, Senior Director of Graduate Studies, 252-816-3284, *Fax:* 252-816-2850, *E-mail:* grad@mail.ecu.edu.

■ EAST TENNESSEE STATE UNIVERSITY

James H. Quillen College of Medicine and School of Graduate Studies, Biomedical Science Graduate Program, Johnson City, TN 37614

AWARDS Anatomy and cell biology (MS, PhD); biochemistry and molecular biology (MS, PhD); biophysics (MS, PhD); microbiology (MS, PhD); pharmacology (MS, PhD); physiology (MS, PhD). Part-time programs available.

Faculty: 40 full-time (9 women).
Students: 23 full-time (10 women), 2 part-time (1 woman); includes 1 minority (Asian American or Pacific Islander), 4 international. Average age 30. *83 applicants, 20% accepted.* In 1999, 5 master's, 4 doctorates awarded. Terminal master's awarded for partial completion of doctoral program.
Degree requirements: For master's, one foreign language (computer language can substitute), thesis, comprehensive qualifying exam required; for doctorate, 2 foreign languages (computer language can substitute for one), dissertation required.
Entrance requirements: For master's, GRE General Test, minimum GPA of 3.0, bachelor's degree in biological or related science; for doctorate, GRE General Test, GRE Subject Test. *Application deadline:* For fall admission, 3/15 (priority date); for spring admission, 3/1. *Application fee:* $25 ($35 for international students).
Expenses: Tuition, state resident: full-time $10,342. Tuition, nonresident: full-time $21,080. Required fees: $532.
Financial aid: In 1999–00, 16 research assistantships, 5 teaching assistantships were awarded; fellowships, career-related internships or fieldwork, Federal Work-Study, grants, institutionally sponsored loans, and tuition waivers (full) also available.
Dr. Mitchell Robinson, Assistant Dean for Graduate Studies, 423-439-4658, *E-mail:* robinson@etsu.edu.

■ EMORY UNIVERSITY

Graduate School of Arts and Sciences, Division of Biological and Biomedical Sciences, Program in Biochemistry, Cell and Developmental Biology, Atlanta, GA 30322-1100

AWARDS PhD.

Faculty: 66 full-time (12 women).
Students: 52 full-time (33 women); includes 4 African Americans, 1 Hispanic American, 13 international. In 1999, 12 degrees awarded.
Degree requirements: For doctorate, dissertation required, foreign language not required.
Entrance requirements: For doctorate, GRE General Test, TOEFL, minimum GPA of 3.0 in science course work. *Application deadline:* For fall admission, 1/20 (priority date). *Application fee:* $45.
Expenses: Tuition: Full-time $22,770. Tuition and fees vary according to program.
Financial aid: In 1999–00, fellowships with full tuition reimbursements (averaging $18,000 per year).
Faculty research: Signal transduction, molecular biology, enzymes and cofactors, receptor and ion channel function, membrane biology.
Steven L'Hernault, Director, 404-727-3924, *Fax:* 404-727-3949, *E-mail:* jfridov@emory.edu.
Application contact: 404-727-2547, *Fax:* 404-727-3322, *E-mail:* gdbbs@gsas.emory.edu.
Find an in-depth description at www.petersons.com/graduate.

■ EMPORIA STATE UNIVERSITY

School of Graduate Studies, College of Liberal Arts and Sciences, Division of Biological Sciences, Emporia, KS 66801-5087

AWARDS Botany (MS); environmental biology (MS); general biology (MS); microbial and cellular biology (MS); zoology (MS). Part-time programs available.

Faculty: 15 full-time (3 women), 4 part-time/adjunct (0 women).
Students: 20 full-time (10 women), 8 part-time (3 women), 2 international. *8 applicants, 75% accepted.* In 1999, 12 degrees awarded.
Degree requirements: For master's, comprehensive exam or thesis required.
Entrance requirements: For master's, TOEFL, written exam. *Application deadline:* For fall admission, 8/15 (priority date). Applications are processed on a rolling basis. *Application fee:* $30 ($75 for international students). Electronic applications accepted.

Expenses: Tuition, state resident: full-time $2,410; part-time $108 per credit hour. Tuition, nonresident: full-time $6,212; part-time $266 per credit hour.

Financial aid: In 1999–00, 2 fellowships (averaging $1,396 per year), 1 research assistantship (averaging $5,390 per year), 11 teaching assistantships with full tuition reimbursements (averaging $5,047 per year) were awarded; career-related internships or fieldwork, Federal Work-Study, and institutionally sponsored loans also available. Financial aid application deadline: 3/15; financial aid applicants required to submit FAFSA.

Faculty research: Fisheries, range, and wildlife management; aquatic, plant, grassland, vertebrate, and invertebrate ecology; mammalian and plant systematics, taxonomy, and evolution; immunology, virology, and molecular biology.

Dr. Marshall Sundberg, Chair, 316-341-5311, *Fax:* 316-341-5607, *E-mail:* sundberm@emporia.edu.

■ FINCH UNIVERSITY OF HEALTH SCIENCES/THE CHICAGO MEDICAL SCHOOL

School of Graduate and Postdoctoral Studies, Department of Cell Biology and Anatomy, Program in Cell Biology, North Chicago, IL 60064-3095

AWARDS MS, PhD, MD/MS, MD/PhD.

Faculty: 4 full-time.

Degree requirements: For master's, computer language, thesis, qualifying exam required, foreign language not required; for doctorate, computer language, dissertation, comprehensive exam, original research project required, foreign language not required.

Entrance requirements: For master's and doctorate, GRE General Test, TOEFL, TWE, minimum GPA of 3.0. *Application deadline:* For fall admission, 6/1 (priority date). Applications are processed on a rolling basis. *Application fee:* $25.

Expenses: Tuition: Full-time $14,054; part-time $391 per credit hour. Tuition and fees vary according to program.

Financial aid: Application deadline: 6/9.

Application contact: Dr. Monica M. Oblinger, Graduate Student Coordinator, 847-578-3440.

Find an in-depth description at www.petersons.com/graduate.

■ FLORIDA INSTITUTE OF TECHNOLOGY

Graduate School, College of Science and Liberal Arts, Department of Biological Sciences, Program in Cell and Molecular Biology, Melbourne, FL 32901-6975

AWARDS MS. Part-time programs available.

Students: 4 full-time (3 women), 4 part-time; includes 1 minority (Asian American or Pacific Islander), 2 international. Average age 34. *10 applicants, 40% accepted.* In 1999, 1 degree awarded.

Degree requirements: For master's, thesis required, foreign language not required.

Entrance requirements: For master's, GRE General Test, minimum GPA of 3.0. *Application deadline:* Applications are processed on a rolling basis. *Application fee:* $50. Electronic applications accepted.

Expenses: Tuition: Part-time $575 per credit hour. Required fees: $50. Tuition and fees vary according to campus/location and program.

Financial aid: In 1999–00, 3 students received aid, including 1 research assistantship with full and partial tuition reimbursement available (averaging $3,150 per year), 2 teaching assistantships with full and partial tuition reimbursements available (averaging $4,180 per year); career-related internships or fieldwork and tuition remissions also available. Financial aid application deadline: 3/1; financial aid applicants required to submit FAFSA.

Faculty research: Changes in DNA molecule and differential expression of genetic information during aging.

Dr. Gary N. Wells, Head, 321-674-8034, *Fax:* 321-674-7238, *E-mail:* gwells@fit.edu.

Application contact: Carolyn P. Farrior, Associate Dean of Graduate Admissions, 321-674-7118, *Fax:* 321-674-9468, *E-mail:* cfarrior@fit.edu.

Find an in-depth description at www.petersons.com/graduate.

■ FLORIDA STATE UNIVERSITY

Graduate Studies, College of Arts and Sciences, Department of Biological Science, Program in Cell Biology, Tallahassee, FL 32306

AWARDS MS, PhD.

Faculty: 11 full-time (1 woman).
Students: 24 full-time (11 women); includes 4 minority (1 Asian American or Pacific Islander, 3 Hispanic Americans), 8 international.
Degree requirements: For master's and doctorate, thesis/dissertation, teaching experience required.

Entrance requirements: For master's, GRE General Test, TOEFL; for doctorate, GRE General Test, GRE Subject Test, TOEFL. *Application deadline:* For fall admission, 1/15; for spring admission, 10/15. *Application fee:* $20.

Expenses: Tuition, state resident: full-time $3,504; part-time $146 per credit hour. Tuition, nonresident: full-time $12,162; part-time $507 per credit hour. Tuition and fees vary according to program.

Financial aid: In 1999–00, fellowships with full tuition reimbursements (averaging $13,740 per year), research assistantships with full tuition reimbursements (averaging $13,740 per year), teaching assistantships with full tuition reimbursements (averaging $13,740 per year) were awarded. Financial aid application deadline: 1/15; financial aid applicants required to submit FAFSA.

Faculty research: Cell-to-cell interactions, nitrogen metabolism, carbon metabolism, cell motility, DNA methylation.

Dr. Thomas C. S. Keller, Associate Professor and Associate Chairman, 850-644-3023, *Fax:* 850-644-9829.

Application contact: Judy Bowers, Coordinator, Graduate Affairs, 850-644-3023, *Fax:* 850-644-9829, *E-mail:* bowers@bio.fsu.edu.

■ FORDHAM UNIVERSITY

Graduate School of Arts and Sciences, Department of Biological Sciences, New York, NY 10458

AWARDS Biological sciences (MS, PhD), including cell and molecular biology, ecology. Part-time and evening/weekend programs available.

Faculty: 17 full-time (1 woman).
Students: 29 full-time (17 women), 10 part-time (6 women); includes 2 minority (both Asian Americans or Pacific Islanders), 6 international. *57 applicants, 35% accepted.* In 1999, 11 master's, 2 doctorates awarded. Terminal master's awarded for partial completion of doctoral program.
Degree requirements: For master's, comprehensive exam required, thesis optional; for doctorate, 2 foreign languages (computer language can substitute for one), dissertation, comprehensive exam required.

Entrance requirements: For master's and doctorate, GRE General Test, GRE Subject Test (recommended). *Application deadline:* For fall admission, 1/16 (priority date); for spring admission, 12/1. *Application fee:* $60. Electronic applications accepted.

Expenses: Tuition: Full-time $14,400; part-time $600 per credit. Required fees: $125 per semester. Tuition and fees vary according to program.

Financial aid: In 1999–00, 29 students received aid, including 4 fellowships

Fordham University (continued)
(averaging $15,000 per year), 3 research assistantships (averaging $15,000 per year), teaching assistantships (averaging $15,000 per year); institutionally sponsored loans, tuition waivers (full and partial), and unspecified assistantships also available. Aid available to part-time students. Financial aid application deadline: 1/15. *Total annual research expenditures:* $365,196. Dr. Berish Rubin, Chair, 718-817-3641, *Fax:* 718-817-3645, *E-mail:* rubin@fordham.edu.

Application contact: Dr. Craig W. Pilant, Assistant Dean, 718-817-4420, *Fax:* 718-817-3566, *E-mail:* pilant@fordham.edu.

Find an in-depth description at www.petersons.com/graduate.

■ **GEORGE MASON UNIVERSITY**

College of Arts and Sciences, Department of Biology, Master's Program in Biology, Fairfax, VA 22030-4444

AWARDS Bioinformatics (MS); ecology, systematics and evolution (MS); environmental science and public policy (MS); interpretive biology (MS); molecular, microbial, and cellular biology (MS); organismal biology (MS). Part-time programs available.

Faculty: 30 full-time (11 women), 32 part-time/adjunct (20 women).

Students: 5 full-time (3 women), 55 part-time (37 women); includes 7 minority (1 African American, 4 Asian Americans or Pacific Islanders, 2 Hispanic Americans), 8 international. Average age 34. *36 applicants, 44% accepted.* In 1999, 18 degrees awarded.

Degree requirements: For master's, thesis or alternative required, foreign language not required.

Entrance requirements: For master's, GRE General Test, GRE Subject Test, bachelor's degree in biology or equivalent. *Application deadline:* For fall admission, 5/1; for spring admission, 11/1. *Application fee:* $30. Electronic applications accepted.

Expenses: Tuition, state resident: full-time $4,416; part-time $184 per credit hour. Tuition, nonresident: full-time $12,516; part-time $522 per credit hour. Tuition and fees vary according to program.

Financial aid: Available to part-time students. Application deadline: 3/1. Dr. George E. Andrykovitch, Director, 703-993-1027, *Fax:* 703-993-1046.

■ **GEORGETOWN UNIVERSITY**

Graduate School of Arts and Sciences, Programs in Biomedical Sciences, Department of Cell Biology, Washington, DC 20057

AWARDS PhD, MD/PhD.

Degree requirements: For doctorate, dissertation, comprehensive exam required, foreign language not required.

Entrance requirements: For doctorate, GRE General Test, TOEFL.

Find an in-depth description at www.petersons.com/graduate.

■ **GEORGIA STATE UNIVERSITY**

College of Arts and Sciences, Department of Biology, Program in Cell Biology and Physiology, Atlanta, GA 30303-3083

AWARDS MS, PhD.

Degree requirements: For master's, one foreign language (computer language can substitute), thesis or alternative, exam required; for doctorate, dissertation required.

Entrance requirements: For master's and doctorate, GRE General Test, TOEFL, minimum GPA of 3.0. *Application deadline:* For fall admission, 7/18; for spring admission, 2/13. Applications are processed on a rolling basis. *Application fee:* $25.

Expenses: Tuition, state resident: full-time $2,896; part-time $121 per credit hour. Tuition, nonresident: full-time $11,584; part-time $483 per credit hour. Required fees: $228. Full-time tuition and fees vary according to course load and program.

Financial aid: Application deadline: 2/6.

Application contact: Latesha Morrison, Graduate Administrative Coordinator, 404-651-2759, *Fax:* 404-651-2509, *E-mail:* biolxm@langate.gsu.edu.

Find an in-depth description at www.petersons.com/graduate.

■ **HARVARD UNIVERSITY**

Graduate School of Arts and Sciences, Department of Molecular and Cellular Biology, Cambridge, MA 02138

AWARDS PhD.

Students: 81 full-time (37 women). *325 applicants, 16% accepted.* In 1999, 16 degrees awarded.

Degree requirements: For doctorate, dissertation, oral exam required.

Entrance requirements: For doctorate, GRE General Test, GRE Subject Test (recommended), TOEFL. *Application deadline:* For fall admission, 1/2. *Application fee:* $60.

Expenses: Tuition: Full-time $22,054. Required fees: $711. Tuition and fees vary according to program.

Financial aid: Fellowships, research assistantships, teaching assistantships, career-related internships or fieldwork, Federal Work-Study, and institutionally sponsored loans available. Financial aid application deadline: 12/30.

Josephine Ferraro, Officer, 617-495-5396. **Application contact:** James Wang, Assistant Director, 617-495-1901.

Find an in-depth description at www.petersons.com/graduate.

■ **HARVARD UNIVERSITY**

Graduate School of Arts and Sciences, Program in Biological and Biomedical Sciences, Department of Cell Biology, Boston, MA 02115

AWARDS PhD. Applications through the Program in Biological and Biomedical Sciences (BBS).

Degree requirements: For doctorate, dissertation, qualifying exam required, foreign language not required.

Entrance requirements: For doctorate, GRE General Test, GRE Subject Test, TOEFL. *Application deadline:* For fall admission, 1/1. *Application fee:* $60.

Expenses: Tuition: Full-time $22,054. Required fees: $711. Tuition and fees vary according to program.

Financial aid: Fellowships, research assistantships, teaching assistantships, institutionally sponsored loans and tuition waivers (full) available. Financial aid application deadline: 1/1. Dr. Marc Kirschner, Chairman, 617-432-2250.

Application contact: Leah Simons, Manager of Student Affairs, 617-432-0162.

Find an in-depth description at www.petersons.com/graduate.

■ **ILLINOIS INSTITUTE OF TECHNOLOGY**

Graduate College, Armour College of Engineering and Sciences, Department of Biological, Chemical and Physical Sciences, Biology Division, Chicago, IL 60616-3793

AWARDS Biochemistry (MS); biology (PhD); biotechnology (MS); cell biology (MS); microbiology (MS). Part-time and evening/weekend programs available.

Faculty: 8 full-time (0 women), 2 part-time/adjunct (0 women).

Students: 38 full-time (15 women), 80 part-time (52 women); includes 34 minority (27 African Americans, 5 Asian Americans or Pacific Islanders, 2 Hispanic Americans), 39 international. *155 applicants, 36% accepted.* In 1999, 10 master's, 2 doctorates awarded. Terminal master's awarded for partial completion of doctoral program.

Degree requirements: For master's, thesis (for some programs), comprehensive exam required, foreign language not required; for doctorate, dissertation, comprehensive exam required, foreign language not required.

Entrance requirements: For master's and doctorate, GRE General Test, TOEFL, minimum undergraduate GPA of 3.0. *Application deadline:* For fall admission, 7/1; for spring admission, 11/1. Applications are processed on a rolling basis. *Application fee:* $30. Electronic applications accepted.
Expenses: Tuition: Part-time $590 per credit hour. Required fees: $100. Tuition and fees vary according to course load and program.
Financial aid: In 1999–00, 7 fellowships, 1 research assistantship, 5 teaching assistantships were awarded; Federal Work-Study, institutionally sponsored loans, scholarships, and unspecified assistantships also available. Aid available to part-time students. Financial aid application deadline: 3/1; financial aid applicants required to submit FAFSA.
Faculty research: Genetics, molecular biology.
Dr. Benjamin Stark, Associate Chair, 312-567-3980, *Fax:* 312-567-3494, *E-mail:* starkb@iit.edu.
Application contact: Dr. S. Mohammad Shahidehpour, Dean of Graduate College, 312-567-3024, *Fax:* 312-567-7517, *E-mail:* gradstu@alpha1.ais.iit.edu.

■ INDIANA UNIVERSITY BLOOMINGTON

Graduate School, College of Arts and Sciences, Department of Biology, Program in Molecular, Cellular, and Developmental Biology, Bloomington, IN 47405

AWARDS PhD. Offered through the University Graduate School.

Students: 54 full-time (21 women); includes 4 minority (2 Asian Americans or Pacific Islanders, 2 Hispanic Americans), 9 international. In 1999, 9 degrees awarded.
Degree requirements: For doctorate, dissertation required.
Entrance requirements: For doctorate, GRE General Test, TOEFL. *Application deadline:* For fall admission, 1/15 (priority date); for spring admission, 9/1 (priority date). Applications are processed on a rolling basis. *Application fee:* $45. Electronic applications accepted.
Expenses: Tuition, state resident: full-time $3,853; part-time $161 per credit hour. Tuition, nonresident: full-time $11,226; part-time $468 per credit hour. Required fees: $360 per year. Tuition and fees vary according to course load and program.
Financial aid: Fellowships with tuition reimbursements, research assistantships with tuition reimbursements, teaching assistantships with tuition reimbursements available. Financial aid application deadline: 2/15.

Faculty research: Developmental genetics, molecular evolution, macromolecular structure and function, cell biology and microbial molecular genetics.
Dr. Susan Strome, Head, 812-855-5450, *Fax:* 812-855-6705, *E-mail:* sstrome@bio.indiana.edu.
Application contact: Gretchen Clearwater, Advisor for Graduate Affairs, 812-855-1861, *Fax:* 812-855-6705, *E-mail:* biograd@bio.indiana.edu.
Find an in-depth description at www.petersons.com/graduate.

■ INDIANA UNIVERSITY BLOOMINGTON

Medical Sciences Program, Bloomington, IN 47405

AWARDS Anatomy and cell biology (MA, PhD); pharmacology (MS, PhD); physiology (MA, PhD).

Faculty: 12 full-time (3 women).
Students: 10 full-time (6 women), 6 part-time (2 women); includes 1 minority (Asian American or Pacific Islander), 3 international. In 1999, 1 master's, 1 doctorate awarded.
Entrance requirements: For master's, GRE, TOEFL, minimum GPA of 3.0; for doctorate, GRE, TOEFL. *Application deadline:* For fall admission, 1/15. *Application fee:* $45.
Expenses: Tuition, state resident: full-time $3,853; part-time $161 per credit hour. Tuition, nonresident: full-time $11,226; part-time $468 per credit hour. Required fees: $360 per year. Tuition and fees vary according to course load and program.
Dr. Talmage Bosin, Assistant Dean, 812-855-8118, *E-mail:* bosin@indiana.edu.
Application contact: Kimberly Bunch, Director of Graduate Admissions, 812-855-1119, *E-mail:* kbunch@indiana.edu.

■ IOWA STATE UNIVERSITY OF SCIENCE AND TECHNOLOGY

Graduate College, College of Agriculture and College of Liberal Arts and Sciences, Department of Biochemistry, Biophysics, and Molecular Biology, Ames, IA 50011

AWARDS Biochemistry (MS, PhD); biophysics (MS, PhD); genetics (MS, PhD); molecular, cellular, and developmental biology (MS, PhD); toxicology (MS, PhD).

Faculty: 19 full-time, 1 part-time/adjunct.
Students: 57 full-time (24 women), 6 part-time (1 woman); includes 1 minority (African American), 20 international. *22 applicants, 59% accepted.* In 2000, 4 master's, 4 doctorates awarded.
Degree requirements: For master's and doctorate, thesis/dissertation required.

Entrance requirements: For master's and doctorate, GRE General Test, TOEFL. *Application deadline:* For fall admission, 6/15 (priority date); for spring admission, 11/15 (priority date). *Application fee:* $20 ($50 for international students). Electronic applications accepted.
Expenses: Tuition, state resident: full-time $3,308. Tuition, nonresident: full-time $9,744. Part-time tuition and fees vary according to course load, campus/location and program.
Financial aid: In 2000–01, 44 research assistantships with partial tuition reimbursements (averaging $12,314 per year), 2 teaching assistantships with partial tuition reimbursements (averaging $12,375 per year) were awarded; scholarships also available.
Dr. Marit Nilsen-Hamilton, Chair, 515-294-2231, *E-mail:* biochem@iastate.edu.
Find an in-depth description at www.petersons.com/graduate.

■ IOWA STATE UNIVERSITY OF SCIENCE AND TECHNOLOGY

Graduate College, Interdisciplinary Programs, Program in Molecular, Cellular, and Developmental Biology, Ames, IA 50011

AWARDS MS, PhD.

Students: 10 full-time (6 women), 9 international. *89 applicants, 20% accepted.* In 1999, 1 master's, 1 doctorate awarded.
Degree requirements: For master's, thesis or alternative required; for doctorate, dissertation required.
Entrance requirements: For master's and doctorate, GRE General Test, TOEFL. *Application deadline:* For fall admission, 2/1 (priority date). *Application fee:* $20 ($50 for international students). Electronic applications accepted.
Expenses: Tuition, state resident: full-time $3,308. Tuition, nonresident: full-time $9,744. Part-time tuition and fees vary according to course load, campus/location and program.
Financial aid: In 1999–00, 8 research assistantships with partial tuition reimbursements (averaging $11,748 per year) were awarded; teaching assistantships, scholarships also available.
Dr. Jorgen Johansen, Supervisory Committee Chair, 515-294-7252, *E-mail:* idgp@iastate.edu.
Application contact: 515-294-7252, *Fax:* 515-924-6790, *E-mail:* idgp@iastate.edu.
Find an in-depth description at www.petersons.com/graduate.

■ JOAN AND SANFORD I. WEILL MEDICAL COLLEGE AND GRADUATE SCHOOL OF MEDICAL SCIENCES OF CORNELL UNIVERSITY

Graduate School of Medical Sciences, Cell Biology and Genetics Graduate Program, New York, NY 10021

AWARDS PhD, MD/PhD.

Faculty: 41 full-time (7 women).
Students: 44 full-time (24 women); includes 12 minority (3 African Americans, 8 Asian Americans or Pacific Islanders, 1 Hispanic American), 9 international. *91 applicants, 23% accepted.* In 2000, 6 doctorates awarded.
Degree requirements: For doctorate, dissertation, final exam required.
Entrance requirements: For doctorate, GRE General Test, GRE Subject Test, MCAT (MD/PhD). *Application deadline:* For fall admission, 1/15. *Application fee:* $50.
Expenses: All students in good standing receive an annual stipend of $22,880.
Financial aid: Fellowships, stipends available.
Dr. Leonard Freedman, Director, 212-639-2976.

■ JOHNS HOPKINS UNIVERSITY

School of Medicine, Graduate Programs in Medicine, Department of Cell Biology and Anatomy, Baltimore, MD 21218-2699

AWARDS PhD.

Faculty: 5 full-time (1 woman).
Students: 10 full-time (7 women), 4 international. Average age 25. In 1999, 2 degrees awarded.
Degree requirements: For doctorate, one foreign language, dissertation, comprehensive and oral exams required.
Entrance requirements: For doctorate, GRE. *Application deadline:* For fall admission, 1/15. *Application fee:* $50.
Expenses: Tuition: Full-time $23,660.
Financial aid: In 1999–00, 7 students received aid, including 7 teaching assistantships; career-related internships or fieldwork and institutionally sponsored loans also available.
Faculty research: Vertebrate evolution, functional anatomy, primate evolution, vertebrate paleobiology.
Dr. Joan T. Richtsmeier, Director, 410-955-7892, *Fax:* 410-955-4129, *E-mail:* jtr@welchlink.welch.jhu.edu.
Application contact: Catherine L. Will, Coordinator, Graduate Student Affairs, 410-614-3385, *E-mail:* grad_study@som.adm.jhu.edu.

■ JOHNS HOPKINS UNIVERSITY

School of Medicine, Graduate Programs in Medicine, Graduate Program in Cellular and Molecular Medicine, Baltimore, MD 21218-2699

AWARDS PhD.

Faculty: 112 full-time (23 women).
Students: 49 full-time (15 women); includes 25 minority (1 African American, 22 Asian Americans or Pacific Islanders, 2 Hispanic Americans). Average age 28. *95 applicants, 24% accepted.* In 1999, 3 degrees awarded (100% entered university research/teaching).
Degree requirements: For doctorate, dissertation, oral exam required. *Average time to degree:* Doctorate–4.4 years full-time.
Entrance requirements: For doctorate, GRE General Test (average 600 verbal, 750 quantitative, 700 analytical), GRE Subject Test (average 600). *Application deadline:* For fall admission, 2/1 (priority date). Applications are processed on a rolling basis. *Application fee:* $50.
Expenses: Tuition: Full-time $23,660.
Financial aid: In 1999–00, 11 fellowships (averaging $8,128 per year) were awarded; tuition waivers (full) also available. Financial aid application deadline: 2/15.
Faculty research: Cellular and molecular basis of disease. *Total annual research expenditures:* $24 million.
Dr. John T. Isaacs, Director, 410-955-7777, *Fax:* 410-955-0840, *E-mail:* jtisaacs@jhmi.edu.
Application contact: Theo M. Karpovich, Admissions Coordinator, 410-614-0391, *Fax:* 410-614-7294, *E-mail:* karpoo@jhmi.edu.

Find an in-depth description at www.petersons.com/graduate.

■ JOHNS HOPKINS UNIVERSITY

School of Medicine, Graduate Programs in Medicine, Program in Biochemistry, Cellular and Molecular Biology, Baltimore, MD 21205

AWARDS PhD.

Faculty: 83 full-time (23 women).
Students: 173 full-time (83 women); includes 29 minority (2 African Americans, 24 Asian Americans or Pacific Islanders, 2 Hispanic Americans, 1 Native American), 46 international. Average age 25. *296 applicants, 23% accepted.* In 1999, 17 doctorates awarded.
Degree requirements: For doctorate, dissertation, comprehensive oral exams required, foreign language not required.
Entrance requirements: For doctorate, GRE General Test, GRE Subject Test. *Application deadline:* For fall admission, 1/15. Applications are processed on a rolling basis. *Application fee:* $50.

Expenses: Tuition: Full-time $23,660.
Financial aid: In 1999–00, 49 fellowships with full tuition reimbursements (averaging $16,760 per year), 124 research assistantships with full tuition reimbursements (averaging $16,760 per year) were awarded. Financial aid application deadline: 1/15.
Faculty research: DNA topology, protein folding, enzyme mechanisms and glycoproteins, bioenergetics, gene transcription.
Dr. Craig Montell, Director, 410-955-3225, *Fax:* 410-614-8842.
Application contact: Dr. Daniel Raben, Admissions Chairman, 410-955-3506, *Fax:* 410-614-8842, *E-mail:* pantol@jhmi.edu.

Find an in-depth description at www.petersons.com/graduate.

■ JOHNS HOPKINS UNIVERSITY

Zanvyl Krieger School of Arts and Sciences, Department of Biology, Baltimore, MD 21218-2699

AWARDS Biochemistry (PhD); cell biology (PhD); developmental biology (PhD); genetic biology (PhD); molecular biology (PhD).

Faculty: 23 full-time (4 women).
Students: 85 full-time (50 women); includes 23 minority (4 African Americans, 14 Asian Americans or Pacific Islanders, 5 Hispanic Americans), 18 international. Average age 24. *202 applicants, 27% accepted.* In 1999, 17 doctorates awarded.
Degree requirements: For doctorate, dissertation required. *Average time to degree:* Doctorate–6 years full-time.
Entrance requirements: For doctorate, GRE General Test, GRE Subject Test. *Application deadline:* For fall admission, 12/15 (priority date). Applications are processed on a rolling basis. *Application fee:* $55.
Expenses: Tuition: Full-time $24,930. Tuition and fees vary according to program.
Financial aid: In 1999–00, 73 students received aid, including 12 fellowships, 61 research assistantships, 12 teaching assistantships; Federal Work-Study and institutionally sponsored loans also available. Financial aid application deadline: 12/15; financial aid applicants required to submit FAFSA.
Faculty research: Protein and nucleic acid biochemistry and biophysical chemistry, molecular biology and development. *Total annual research expenditures:* $9.3 million.
Dr. Victor G. Corces, Chair, 410-516-4693, *Fax:* 410-516-5213, *E-mail:* corces_v@jhuvms.hct.jhu.edu.

Application contact: Joan Miller, Graduate Admissions Coordinator, 410-516-5502, *Fax:* 410-516-5213, *E-mail:* joan@jhu.edu.

Find an in-depth description at www.petersons.com/graduate.

■ KANSAS STATE UNIVERSITY

Graduate School, College of Arts and Sciences, Division of Biology, Manhattan, KS 66506

AWARDS Cell biology (MS, PhD); developmental biology and physiology (MS, PhD); microbiology and immunology (MS, PhD); molecular biology and genetics (MS, PhD); systematics and ecology (MS, PhD); virology and oncology (MS, PhD). Terminal master's awarded for partial completion of doctoral program.

Degree requirements: For master's and doctorate, thesis/dissertation required, foreign language not required.
Entrance requirements: For master's and doctorate, GRE General Test. Electronic applications accepted.
Expenses: Tuition, state resident: part-time $103 per credit hour. Tuition, nonresident: part-time $338 per credit hour. Required fees: $17 per credit hour. One-time fee: $64 part-time.
Faculty research: Immune cell function, prairie ecology.

■ KENT STATE UNIVERSITY

School of Biomedical Sciences, Program in Cellular and Molecular Biology, Kent, OH 44242-0001

AWARDS MS, PhD. Offered in cooperation with Northeastern Ohio Universities College of Medicine. Terminal master's awarded for partial completion of doctoral program.

Degree requirements: For master's and doctorate, thesis/dissertation required, foreign language not required.
Entrance requirements: For master's and doctorate, GRE General Test.
Expenses: Tuition, state resident: full-time $5,334; part-time $243 per hour. Tuition, nonresident: full-time $10,238; part-time $466 per hour.
Faculty research: Molecular genetics, molecular endocrinology, virology and tumor biology, P450 enzymology and catalysis, membrane structure and function.

■ LOUISIANA STATE UNIVERSITY HEALTH SCIENCES CENTER

School of Graduate Studies in New Orleans, Department of Cell Biology and Anatomy, New Orleans, LA 70112-2223

AWARDS Cell biology and anatomy (MS, PhD), including cell biology, developmental biology, neurobiology and anatomy.

Faculty: 18 full-time (3 women), 1 part-time/adjunct (0 women).
Students: 9 full-time (2 women); includes 1 minority (African American), 4 international. Average age 26. *6 applicants, 50% accepted.* In 1999, 3 doctorates awarded (34% entered university research/teaching, 33% found other work related to degree, 33% continued full-time study).
Degree requirements: For master's and doctorate, thesis/dissertation required, foreign language not required. *Average time to degree:* Doctorate–4.5 years full-time.
Entrance requirements: For master's and doctorate, GRE General Test, GRE Subject Test, TOEFL, minimum undergraduate GPA of 3.0. *Application deadline:* For fall admission, 3/1 (priority date); for spring admission, 10/15. Applications are processed on a rolling basis. *Application fee:* $30.
Expenses: Tuition, state resident: full-time $2,878; part-time $126 per hour. Tuition, nonresident: full-time $6,003; part-time $265 per hour. Required fees: $2,272. Tuition and fees vary according to course load, degree level and program.
Financial aid: In 1999–00, 1 fellowship with full tuition reimbursement (averaging $15,000 per year), teaching assistantships with full tuition reimbursements (averaging $14,000 per year) were awarded; research assistantships, career-related internships or fieldwork, Federal Work-Study, grants, institutionally sponsored loans, tuition waivers (full), and unspecified assistantships also available. Aid available to part-time students. Financial aid application deadline: 4/1.
Faculty research: Visual system organization, neural development, plasticity of sensory systems, information processing through the nervous system, visuomotor integration. *Total annual research expenditures:* $772,550.
Dr. R. Ranney Mize, Head, 504-599-1458, *Fax:* 504-568-4392, *E-mail:* rmize@lsumc.edu.
Application contact: Dr. Mark C. Alliegro, Director of Graduate Studies, 504-568-7618, *Fax:* 504-568-4392, *E-mail:* mallie@lsumc.edu.

Find an in-depth description at www.petersons.com/graduate.

■ LOUISIANA STATE UNIVERSITY HEALTH SCIENCES CENTER

School of Graduate Studies in Shreveport, Department of Cellular Biology and Anatomy, Shreveport, LA 71130-3932

AWARDS MS, PhD, MD/PhD. MD/PhD open only to residents of Louisiana. Terminal master's awarded for partial completion of doctoral program.

Degree requirements: For master's and doctorate, thesis/dissertation required, foreign language not required.
Entrance requirements: For master's and doctorate, GRE General Test, TOEFL.
Expenses: Tuition, state resident: full-time $2,878; part-time $126 per hour. Tuition, nonresident: full-time $6,003; part-time $265 per hour. Required fees: $2,272. Tuition and fees vary according to course load, degree level and program.
Faculty research: Alcohol and immunity, neuroscience, olfactory physiology, extracellular matrix, cancer cell biology and gene therapy.

■ LOYOLA UNIVERSITY CHICAGO

Graduate School, Department of Cell Biology, Neurobiology and Anatomy, Maywood, IL 60153

AWARDS MS, PhD, MD/PhD. Part-time programs available.

Faculty: 13 full-time, 13 part-time/adjunct.
Students: 26 full-time (12 women), 3 part-time (all women); includes 4 minority (all Asian Americans or Pacific Islanders), 1 international. Average age 26. *42 applicants, 14% accepted.* In 1999, 1 master's awarded (100% continued full-time study); 3 doctorates awarded (100% entered university research/teaching).
Degree requirements: For master's, thesis or alternative, comprehensive exams required, foreign language not required; for doctorate, dissertation, comprehensive exams required, foreign language not required. *Average time to degree:* Master's–2 years full-time; doctorate–5 years full-time.
Entrance requirements: For master's and doctorate, GRE General Test, GRE Subject Test (biology), minimum GPA of 3.0. *Application deadline:* For fall admission, 5/1 (priority date). Applications are processed on a rolling basis. *Application fee:* $35.
Expenses: Tuition: Part-time $500 per credit hour. Required fees: $42 per term.
Financial aid: In 1999–00, 8 fellowships with full tuition reimbursements (averaging $17,500 per year), 8 research assistantships with full tuition reimbursements (averaging $17,500 per year) were awarded; Federal

Loyola University Chicago (continued)
Work-Study and institutionally sponsored loans also available. Aid available to part-time students. Financial aid application deadline: 5/1; financial aid applicants required to submit FAFSA.
Faculty research: Brain steroids, immunology, neuroregeneration, cytokines. *Total annual research expenditures:* $1.4 million.
Dr. John Clancy, Chair, 708-216-3352.
Application contact: Thackery S. Gray, Graduate Program Director, 708-216-3352.

Find an in-depth description at www.petersons.com/graduate.

■ MAHARISHI UNIVERSITY OF MANAGEMENT

Graduate Studies, Program in Physiology, Molecular, and Cell Biology, Fairfield, IA 52557

AWARDS MS, PhD. Program admits applicants every other year. Terminal master's awarded for partial completion of doctoral program.

Degree requirements: For master's, thesis or alternative required, foreign language not required; for doctorate, dissertation required, foreign language not required.
Entrance requirements: For master's, GMAT or GRE, minimum GPA of 3.0; for doctorate, GRE General Test, GRE Subject Test, bachelor's degree in biology, chemistry, or related quantitative science; minimum GPA of 3.0.
Faculty research: Developmental neurobiology, aging, neurochemistry.

■ MARQUETTE UNIVERSITY

Graduate School, College of Arts and Sciences, Department of Biology, Milwaukee, WI 53201-1881

AWARDS Cell biology (MS, PhD); developmental biology (MS, PhD); ecology (MS, PhD); endocrinology (MS, PhD); evolutionary biology (MS, PhD); genetics (MS, PhD); microbiology (MS, PhD); molecular biology (MS, PhD); muscle and exercise physiology (MS, PhD); neurobiology (MS, PhD); reproductive physiology (MS, PhD).

Faculty: 16 full-time (4 women), 2 part-time/adjunct (0 women).
Students: 34 full-time (20 women), 3 part-time; includes 3 minority (all Asian Americans or Pacific Islanders), 2 international. Average age 31. *42 applicants, 29% accepted.* In 1999, 1 master's, 4 doctorates awarded. Terminal master's awarded for partial completion of doctoral program.

Degree requirements: For master's, thesis, 1 year of teaching experience or equivalent, comprehensive exam required, foreign language not required; for doctorate, dissertation, 1 year of teaching experience or equivalent, qualifying exam required, foreign language not required.
Entrance requirements: For master's and doctorate, GRE General Test, GRE Subject Test, TOEFL. *Application fee:* $40.
Expenses: Tuition: Part-time $510 per credit hour. Tuition and fees vary according to program.
Financial aid: In 1999–00, 4 fellowships, 22 teaching assistantships were awarded; research assistantships, Federal Work-Study, institutionally sponsored loans, scholarships, and tuition waivers (full and partial) also available. Aid available to part-time students. Financial aid application deadline: 2/15.
Faculty research: Microbial and invertebrate ecology, evolution of gene function, DNA methylation, DNA arrangement. *Total annual research expenditures:* $1.5 million.
Dr. Brian Unsworth, Chairman, 414-288-7355, *Fax:* 414-288-7357.
Application contact: Barbara DeNoyer, Graduate Studies Coordinator, 414-288-7355, *Fax:* 414-288-7357.

Find an in-depth description at www.petersons.com/graduate.

■ MASSACHUSETTS INSTITUTE OF TECHNOLOGY

School of Science, Department of Biology, Program in Cellular and Developmental Biology, Cambridge, MA 02139-4307

AWARDS PhD.

Degree requirements: For doctorate, dissertation, general exam required, foreign language not required.
Entrance requirements: For doctorate, GRE General Test. *Application deadline:* For fall admission, 1/1. *Application fee:* $55.
Expenses: Tuition: Full-time $25,000. Full-time tuition and fees vary according to degree level, program and student level.
Financial aid: Application deadline: 1/1.
Application contact: Dr. Janice D. Chang, Educational Administrator, 617-253-3717, *Fax:* 617-258-9329, *E-mail:* gradbio@mit.edu.

■ MCP HAHNEMANN UNIVERSITY

School of Medicine, Biomedical Graduate Programs, Program in Molecular and Cell Biology, Philadelphia, PA 19102-1192

AWARDS MS, PhD, MD/PhD. Terminal master's awarded for partial completion of doctoral program.

Degree requirements: For master's, thesis, comprehensive exam required, foreign language not required; for doctorate, dissertation, qualifying exam required, foreign language not required.
Entrance requirements: For master's, GRE General Test, TOEFL, minimum GPA of 2.75; for doctorate, GRE General Test, TOEFL, minimum GPA of 3.0.
Faculty research: Molecular anatomy, biochemistry, medical biotechnology, molecular pathology, microbiology and immunology.

Find an in-depth description at www.petersons.com/graduate.

■ MEDICAL COLLEGE OF GEORGIA

School of Graduate Studies, Department of Cellular Biology and Anatomy, Augusta, GA 30912-1500

AWARDS MS, PhD.

Faculty: 18 full-time (2 women).
Students: 11 full-time (6 women); includes 1 minority (African American), 3 international. *10 applicants, 20% accepted.* In 1999, 5 doctorates awarded. Terminal master's awarded for partial completion of doctoral program.

Degree requirements: For master's and doctorate, thesis/dissertation required, foreign language not required.
Entrance requirements: For master's and doctorate, GRE General Test, TOEFL. *Application deadline:* For fall admission, 6/30 (priority date). Applications are processed on a rolling basis. *Application fee:* $25.
Expenses: Tuition, state resident: full-time $2,896; part-time $121 per hour. Tuition, nonresident: full-time $11,584; part-time $483 per hour. Required fees: $286; $143 per semester. Tuition and fees vary according to program.
Financial aid: In 1999–00, 11 research assistantships with partial tuition reimbursements (averaging $15,500 per year) were awarded; teaching assistantships, grants and institutionally sponsored loans also available. Aid available to part-time students. Financial aid application deadline: 3/31; financial aid applicants required to submit FAFSA.
Faculty research: Stereocilia, biomedicine, neuroanatomy, gross anatomy, histology.
Dr. Dale E. Bockman, Chair, 706-721-3731, *E-mail:* dbockman@mail.mcg.edu.
Application contact: Dr. Dale Sickles, Director, 706-721-3913, *E-mail:* dsickles@mail.mcg.edu.

■ MEDICAL COLLEGE OF OHIO

Graduate School, Program in Molecular and Cellular Biology, Toledo, OH 43614-5805

AWARDS PhD, MD/PhD.

Students: 23 full-time (12 women), 1 part-time; includes 2 minority (1 Asian American or Pacific Islander, 1 Hispanic American), 12 international. Average age 26.

Degree requirements: For doctorate, dissertation, qualifying exam required, foreign language not required.

Entrance requirements: For doctorate, GRE General Test, minimum undergraduate GPA of 3.0. *Application deadline:* Applications are processed on a rolling basis. *Application fee:* $30.

Expenses: Tuition, state resident: part-time $193 per hour. Tuition, nonresident: part-time $445 per hour. Tuition and fees vary according to degree level.

Financial aid: Fellowships, Federal Work-Study and institutionally sponsored loans available. Financial aid applicants required to submit FAFSA.

Dr. Dorothea Sawicki, Director, 419-383-4117, *Fax:* 419-383-6140, *E-mail:* mcogradschool@mco.edu.

Application contact: Joann Braatz, Clerk, 419-383-4117, *Fax:* 419-383-6140, *E-mail:* mcogradschool@mco.edu.

■ MEDICAL COLLEGE OF WISCONSIN

Graduate School of Biomedical Sciences, Program in Cell and Developmental Biology, Milwaukee, WI 53226-0509

AWARDS MS, PhD. Terminal master's awarded for partial completion of doctoral program.

Degree requirements: For master's and doctorate, thesis/dissertation required, foreign language not required.

Entrance requirements: For master's and doctorate, GRE General Test, TOEFL.

Expenses: Tuition, state resident: full-time $9,318. Tuition, nonresident: full-time $9,318. Required fees: $115.

Faculty research: Neurobiology, development, neuroscience, teratology.

■ MEDICAL UNIVERSITY OF SOUTH CAROLINA

College of Graduate Studies, Program in Molecular and Cellular Biology and Pathobiology, Charleston, SC 29425-0002

AWARDS Cell biology and anatomy (PhD); marine biomedicine (PhD).

Faculty: 72 part-time/adjunct (8 women).

Students: 84 full-time (36 women). Average age 29. *56 applicants, 64% accepted.* In 1999, 9 degrees awarded.

Degree requirements: For doctorate, dissertation, teaching and research seminar, oral and written exams required.

Entrance requirements: For doctorate, GRE General Test, TOEFL, interview, minimum GPA of 3.2. *Application deadline:* Applications are processed on a rolling basis. *Application fee:* $55. Electronic applications accepted.

Expenses: Tuition, state resident: full-time $3,470; part-time $160 per semester hour. Tuition, nonresident: full-time $4,426; part-time $213 per semester hour. Required fees: $408 per semester. One-time fee: $160. Tuition and fees vary according to program.

Financial aid: In 1999–00, 27 fellowships (averaging $16,000 per year) were awarded; research assistantships, teaching assistantships, Federal Work-Study and tuition waivers (partial) also available. Financial aid application deadline: 4/1; financial aid applicants required to submit FAFSA.

Faculty research: Structural biology, marine biology, neurobiology.

Dr. Barry E. Ledford, Interim Dean, 843-792-3391, *Fax:* 843-792-6590.

Application contact: Julie Johnston, Director of Admissions, 843-792-8710, *Fax:* 843-792-3764.

■ MICHIGAN STATE UNIVERSITY

Graduate School, College of Natural Science, Program in Cell and Molecular Biology, East Lansing, MI 48824

AWARDS PhD.

Faculty: 71 full-time (21 women).

Students: 19 full-time (14 women); includes 3 minority (1 Asian American or Pacific Islander, 2 Hispanic Americans), 10 international. Average age 26. *309 applicants, 2% accepted.* In 1999, 1 degree awarded (100% found work related to degree).

Degree requirements: For doctorate, dissertation required, foreign language not required. *Average time to degree:* Doctorate–5 years full-time.

Entrance requirements: For doctorate, GRE General Test, TOEFL. *Application deadline:* For fall admission, 2/1. *Application fee:* $30 ($40 for international students). Electronic applications accepted.

Expenses: Tuition, state resident: part-time $229 per credit. Tuition, nonresident: part-time $464 per credit. Required fees: $241 per semester. Tuition and fees vary according to course load, degree level and program.

Financial aid: In 1999–00, 19 students received aid, including fellowships with full

tuition reimbursements available (averaging $17,000 per year), 19 research assistantships with full tuition reimbursements available (averaging $16,680 per year); teaching assistantships with full tuition reimbursements available, career-related internships or fieldwork, grants, institutionally sponsored loans, and unspecified assistantships also available. Financial aid application deadline: 1/15.

Faculty research: Molecular genetics, molecular basis of plant and animal diseases, plant molecular biology, protein structure and function, signal transduction.

Dr. Susan E. Conrad, Director, 517-353-5161, *Fax:* 517-353-8957, *E-mail:* conrad@pilot.msu.edu.

Application contact: Angela Zell, Secretary, 517-353-8916, *Fax:* 517-353-8957, *E-mail:* cmb@pilot.msu.edu.

Find an in-depth description at www.petersons.com/graduate.

■ MOUNT SINAI SCHOOL OF MEDICINE OF NEW YORK UNIVERSITY

Graduate School of Biological Sciences, Molecular, Cellular, Biochemical and Developmental Sciences (MCBDS) Training Area, New York, NY 10029-6504

AWARDS PhD, MD/PhD.

Students: 48 full-time (18 women).

Degree requirements: For doctorate, dissertation required, foreign language not required.

Entrance requirements: For doctorate, GRE General Test, GRE Subject Test, MCAT, TOEFL. *Application deadline:* For fall admission, 4/15. *Application fee:* $60.

Expenses: Tuition: Full-time $21,750; part-time $725 per credit. Required fees: $750; $25 per credit. Full-time tuition and fees vary according to student level.

Financial aid: Fellowships with full tuition reimbursements, grants available.

Dr. Gillian Small, Program Director, *E-mail:* small@msvax.mssm.edu.

Application contact: C. Gita Bosch, Administrative Manager and Assistant Dean, 212-241-6546, *Fax:* 212-241-0651, *E-mail:* grads@mssm.edu.

■ NEW YORK MEDICAL COLLEGE

Graduate School of Basic Medical Sciences, Department of Cell Biology and Anatomy, Valhalla, NY 10595-1691

AWARDS Cell biology and neuroscience (MS, PhD). Part-time and evening/weekend programs available.

Faculty: 18 full-time (6 women).

New York Medical College (continued)
Students: 8 full-time (5 women), 1 (woman) part-time; includes 1 minority (Asian American or Pacific Islander), 3 international. Average age 30. *8 applicants, 38% accepted.* In 1999, 2 master's, 2 doctorates awarded. Terminal master's awarded for partial completion of doctoral program.
Degree requirements: For master's and doctorate, computer language, thesis/dissertation required, foreign language not required.
Entrance requirements: For master's, GRE General Test, TOEFL; for doctorate, GRE General Test, GRE Subject Test, TOEFL. *Application deadline:* For fall admission, 7/1 (priority date); for spring admission, 12/1 (priority date). Applications are processed on a rolling basis. *Application fee:* $35 ($60 for international students).
Expenses: Tuition: Part-time $430 per credit. Required fees: $15 per semester. One-time fee: $100.
Financial aid: In 1999–00, 7 research assistantships with full tuition reimbursements were awarded; career-related internships or fieldwork, Federal Work-Study, grants, institutionally sponsored loans, and tuition waivers (full) also available. Aid available to part-time students. Financial aid applicants required to submit FAFSA.
Faculty research: Mechanisms of growth control in skeletal muscle, cartilage differentiation, cytoskeletal functions, signal transduction pathways, neuronal development and plasticity.
Dr. Anna B. Drakontides, Director, 914-594-4036.

Find an in-depth description at www.petersons.com/graduate.

■ **NEW YORK UNIVERSITY**

Graduate School of Arts and Science, Department of Basic Medical Sciences, New York, NY 10012-1019

AWARDS Biochemistry (MS, PhD); cell biology (MS, PhD); microbiology (MS, PhD); parasitology (PhD); pathology (MS, PhD); pharmacology (PhD); physiology (MS, PhD). Part-time programs available.

Faculty: 23 full-time (2 women), 3 part-time/adjunct (1 woman).
Students: 182 full-time (73 women), 4 part-time (1 woman); includes 52 minority (11 African Americans, 34 Asian Americans or Pacific Islanders, 7 Hispanic Americans), 48 international. Average age 26. *671 applicants, 11% accepted.* In 1999, 15 master's, 32 doctorates awarded. Terminal master's awarded for partial completion of doctoral program.
Degree requirements: For master's, thesis or alternative, written

comprehensive exam required, foreign language not required; for doctorate, one foreign language, dissertation, oral and written comprehensive exams required.
Entrance requirements: For master's and doctorate, GRE General Test, GRE Subject Test, TOEFL. *Application deadline:* For fall admission, 2/1 (priority date). *Application fee:* $60.
Expenses: Tuition: Full-time $17,880; part-time $745 per credit. Required fees: $1,140; $35 per credit. Tuition and fees vary according to course load and program.
Financial aid: Fellowships with tuition reimbursements, research assistantships with tuition reimbursements, teaching assistantships with tuition reimbursements, career-related internships or fieldwork, Federal Work-Study, institutionally sponsored loans, and tuition waivers (full and partial) available. Financial aid application deadline: 2/1; financial aid applicants required to submit FAFSA.
Dr. Joel D. Oppenheim, Director, 212-263-5648, *Fax:* 212-263-7600, *E-mail:* sackler-info@nyumed.med.nyu.edu.

■ **NEW YORK UNIVERSITY**

Graduate School of Arts and Science, Department of Biology, New York, NY 10012-1019

AWARDS Applied recombinant DNA technology (MS); biochemistry (PhD); biomedical journalism (MA); cell biology (PhD); computers in biological research (MS); environmental biology (PhD); general biology (MS); neural sciences and physiology (PhD); oral biology (MS); population and evolutionary biology (PhD). Part-time programs available.

Faculty: 22 full-time (5 women), 8 part-time/adjunct.
Students: 99 full-time (48 women), 61 part-time (31 women); includes 30 minority (4 African Americans, 22 Asian Americans or Pacific Islanders, 4 Hispanic Americans), 52 international. Average age 24. *371 applicants, 41% accepted.* In 1999, 54 master's, 4 doctorates awarded. Terminal master's awarded for partial completion of doctoral program.
Degree requirements: For master's, thesis or alternative, qualifying paper required, foreign language not required; for doctorate, dissertation, oral and written comprehensive exams required, foreign language not required.
Entrance requirements: For master's, GRE General Test, TOEFL; for doctorate, GRE General Test, GRE Subject Test, TOEFL. *Application deadline:* For fall admission, 1/4 (priority date). *Application fee:* $60.
Expenses: Tuition: Full-time $17,880; part-time $745 per credit. Required fees:

$1,140; $35 per credit. Tuition and fees vary according to course load and program.
Financial aid: Fellowships with tuition reimbursements, research assistantships with tuition reimbursements, teaching assistantships with tuition reimbursements, career-related internships or fieldwork, Federal Work-Study, institutionally sponsored loans, and tuition waivers (full and partial) available. Financial aid application deadline: 1/4; financial aid applicants required to submit FAFSA.
Faculty research: Development and genetics, neurobiology, plant sciences, molecular and cell biology.
Philip Furmanski, Chairman, 212-998-8200.

Application contact: Gloria Coruzzi, Director of Graduate Studies, 212-998-8200, *Fax:* 212-995-4015, *E-mail:* biology@nyu.edu.

Find an in-depth description at www.petersons.com/graduate.

■ **NEW YORK UNIVERSITY**

School of Medicine and Graduate School of Arts and Science, Medical Scientist Training Program, New York, NY 10012-1019

AWARDS Biochemistry (MD/PhD); cell biology (MD/PhD); environmental health sciences (MD/PhD); microbiology (MD/PhD); parasitology (MD/PhD); pathology (MD/PhD); pharmacology (MD/PhD). Students must be accepted by both the School of Medicine and the Graduate School of Arts and Science.

Faculty: 150 full-time (35 women).
Students: 80 full-time (18 women); includes 37 minority (3 African Americans, 32 Asian Americans or Pacific Islanders, 2 Hispanic Americans), 1 international. Average age 25. *195 applicants, 18% accepted.*
Degree requirements: One foreign language.
Application deadline: For fall admission, 11/15. Applications are processed on a rolling basis. *Application fee:* $60.
Expenses: Students receive full tuition support and an annual stipend of $17,500.
Financial aid: Application deadline: 5/1.
Faculty research: Genetics, tumor biology, cardiovascular biology, neuroscience, host defense mechanisms.
Dr. James Salzer, Director, 212-263-0758.
Application contact: Arlene Kohler, Administrative Officer, 212-263-5649.

■ NEW YORK UNIVERSITY

School of Medicine and Graduate School of Arts and Science, Sackler Institute of Graduate Biomedical Sciences, Program in Cellular and Molecular Biology, New York, NY 10012-1019

AWARDS Biochemistry (PhD); cell biology (PhD).

Faculty: 29 full-time (8 women).
Students: 50 full-time; includes 20 minority (5 African Americans, 12 Asian Americans or Pacific Islanders, 3 Hispanic Americans), 12 international. Average age 25. In 1999, 8 degrees awarded (100% entered university research/teaching).
Degree requirements: For doctorate, one foreign language, dissertation, qualifying exams required. *Average time to degree:* Doctorate–5.5 years full-time.
Entrance requirements: For doctorate, GRE General Test, GRE Subject Test, TOEFL. *Application deadline:* For fall admission, 2/1 (priority date). Applications are processed on a rolling basis. *Application fee:* $60.
Expenses: Tuition: Full-time $17,880; part-time $745 per credit. Required fees: $1,140; $35 per credit. Tuition and fees vary according to course load and program.
Financial aid: Fellowships, research assistantships, teaching assistantships available. Financial aid application deadline: 1/15.
Faculty research: Membrane and organelle structure and biogenesis, intracellular transport and processing of proteins, cellular recognition and cell adhesion, oncogene structure and function, action of growth factors. *Total annual research expenditures:* $1.9 million.
Dr. Daniel Rifkin, Director, 212-263-5109, *E-mail:* rifkind01@popmail.med.nyu.edu.
Application contact: Alan Frey, 212-263-8129, *E-mail:* freya01@popmail.med.nyu.edu.

■ NORTH CAROLINA STATE UNIVERSITY

College of Veterinary Medicine and Graduate School, Graduate Programs in Comparative Biomedical Sciences, Raleigh, NC 27695

AWARDS Cell biology and morphology (MS, PhD); epidemiology and population medicine (MS, PhD); immunology (MS, PhD); microbiology and immunology (MS, PhD); pathology (MS, PhD); pharmacology (MS, PhD); specialized veterinary medicine (MS). Part-time programs available.

Students: 40 full-time (23 women), 22 part-time (12 women); includes 11 minority (8 African Americans, 2 Asian Americans or Pacific Islanders, 1 Hispanic American), 14 international. Average age 34. *33 applicants, 33% accepted.* In 1999, 2 master's, 2 doctorates awarded.
Degree requirements: For master's and doctorate, thesis/dissertation required.
Entrance requirements: For master's and doctorate, GRE General Test. *Application deadline:* For fall admission, 6/25; for spring admission, 11/25. Applications are processed on a rolling basis. *Application fee:* $45.
Expenses: Tuition, state resident: full-time $1,578. Tuition, nonresident: full-time $10,744. Required fees: $892. Full-time tuition and fees vary according to program.
Financial aid: Fellowships, research assistantships, teaching assistantships available. Financial aid application deadline: 2/15.
Faculty research: Infectious diseases, immunology and virology, tumor biology, toxicological pathology, food safety.
Dr. Neil C. Olson, Associate Dean, 919-513-6213, *Fax:* 919-513-6222, *E-mail:* neil_olson@ncsu.edu.

■ NORTH DAKOTA STATE UNIVERSITY

Graduate Studies and Research, College of Agriculture and College of Science and Mathematics, Cellular and Molecular Biology Program, Fargo, ND 58105

AWARDS PhD. Offered in cooperation with eleven departments in the university.

Faculty: 33 full-time (4 women), 4 part-time/adjunct (1 woman).
Students: 15 full-time (7 women); includes 1 minority (Hispanic American), 5 international. Average age 27. *8 applicants, 63% accepted.*
Degree requirements: For doctorate, dissertation required, foreign language not required.
Entrance requirements: For doctorate, GRE, TOEFL. *Application deadline:* Applications are processed on a rolling basis. *Application fee:* $25.
Expenses: Tuition, state resident: full-time $3,096; part-time $112 per credit hour. Tuition, nonresident: full-time $7,588; part-time $299 per credit hour. Tuition and fees vary according to course load, campus/location and reciprocity agreements.
Financial aid: Fellowships, research assistantships, teaching assistantships available. Financial aid application deadline: 3/15.
Faculty research: Plant and animal cell biology, gene regulation, molecular genetics, plant and animal virology.

Dr. Zong-Ming Cheng, Director, 701-231-7405, *Fax:* 701-231-8474, *E-mail:* zcheng@plains.nodak.edu.

■ NORTHWESTERN UNIVERSITY

The Graduate School, Division of Interdepartmental Programs and Medical School, Integrated Graduate Programs in the Life Sciences, Chicago, IL 60611

AWARDS Cancer biology (PhD); cell biology (PhD); developmental biology (PhD); evolutionary biology (PhD); immunology and microbial pathogenesis (PhD); molecular biology and genetics (PhD); neurobiology (PhD); pharmacology and toxicology (PhD); structural biology and biochemistry (PhD).

Degree requirements: For doctorate, dissertation, written and oral qualifying exams required, foreign language not required.
Entrance requirements: For doctorate, GRE General Test, TOEFL.
Expenses: Tuition: Full-time $23,301. Full-time tuition and fees vary according to program.

Find an in-depth description at www.petersons.com/graduate.

■ NORTHWESTERN UNIVERSITY

The Graduate School, Judd A. and Marjorie Weinberg College of Arts and Sciences, Interdepartmental Biological Sciences Program (IBiS), Concentration in Cell and Molecular Biology, Evanston, IL 60208

AWARDS PhD.

Faculty: 59 full-time (11 women).
Students: 79 full-time (46 women); includes 4 minority (2 African Americans, 2 Hispanic Americans), 14 international. *236 applicants, 19% accepted.*
Degree requirements: For doctorate, dissertation, 2 quarters of teaching experience required, foreign language not required.
Entrance requirements: For doctorate, GRE General Test, TOEFL. *Application deadline:* For fall admission, 1/15. Applications are processed on a rolling basis. *Application fee:* $50 ($55 for international students).
Expenses: Tuition: Full-time $23,301. Full-time tuition and fees vary according to program.
Financial aid: In 1999–00, 15 fellowships with full tuition reimbursements (averaging $12,078 per year), 64 research assistantships (averaging $17,000 per year), 15 teaching assistantships (averaging $16,620 per year) were awarded; Federal Work-Study, institutionally sponsored loans, and traineeships also available. Financial aid application deadline: 12/31; financial aid applicants required to submit FAFSA.

Northwestern University (continued)
Application contact: Latonia Trimuel, Program Assistant, 800-546-1761, *E-mail:* ibis@northwestern.edu.

Find an in-depth description at www.petersons.com/graduate.

■ NORTHWESTERN UNIVERSITY

The Graduate School, Judd A. and Marjorie Weinberg College of Arts and Sciences, Interdepartmental Biological Sciences Program (IBiS), Department of Biochemistry, Molecular Biology, and Cell Biology, Evanston, IL 60208

AWARDS Cell and molecular biology (PhD); molecular biophysics (PhD). Department participates in the Interdepartmental Biological Sciences Program (IBiS).

Faculty: 59 full-time (11 women).
Students: 79 full-time (46 women); includes 5 minority (2 African Americans, 3 Hispanic Americans), 14 international. *Application fee:* $50 ($55 for international students).
Expenses: Tuition: Full-time $23,301. Full-time tuition and fees vary according to program.
Financial aid: In 1999–00, 15 fellowships (averaging $12,078 per year), 64 research assistantships (averaging $17,000 per year), 16 teaching assistantships (averaging $16,620 per year) were awarded.
Application contact: Latonia Trimuel, Program Assistant, 800-546-1761, *E-mail:* ibis@northwestern.edu.

Find an in-depth description at www.petersons.com/graduate.

■ OAKLAND UNIVERSITY

Graduate Studies, College of Arts and Sciences, Department of Biological Sciences, Rochester, MI 48309-4401

AWARDS Biological sciences (MS); cellular biology of aging (MS).

Faculty: 19 full-time (4 women), 1 (woman) part-time/adjunct.
Students: 8 full-time (all women), 4 part-time (3 women); includes 1 minority (Asian American or Pacific Islander), 3 international. Average age 29. In 1999, 2 degrees awarded.
Degree requirements: For master's, thesis required, foreign language not required.
Entrance requirements: For master's, GRE Subject Test, minimum GPA of 3.0 for unconditional admission. *Application deadline:* For fall admission, 7/15; for spring admission, 3/15. *Application fee:* $30.
Expenses: Tuition, state resident: full-time $5,294; part-time $221 per credit hour. Tuition, nonresident: full-time $11,720; part-time $488 per credit hour. Required

fees: $214 per semester. Tuition and fees vary according to campus/location and program.
Financial aid: Federal Work-Study, institutionally sponsored loans, and tuition waivers (full) available. Financial aid application deadline: 3/1; financial aid applicants required to submit FAFSA. Dr. Virinder K. Moudgil, Chair, 248-370-3553.
Application contact: Dr. George Gamboa, Coordinator, 248-370-3550.

■ THE OHIO STATE UNIVERSITY

College of Medicine and Public Health and Graduate School, Graduate Programs in the Basic Medical Sciences, Department of Physiology and Cell Biology, Columbus, OH 43210

AWARDS PhD.

Faculty: 17 full-time (3 women), 7 part-time/adjunct (1 woman).
Students: 14 full-time (7 women); includes 1 minority (Hispanic American), 9 international. Average age 27. *27 applicants, 19% accepted.* In 1999, 3 doctorates awarded.
Degree requirements: For doctorate, dissertation required, foreign language not required.
Entrance requirements: For doctorate, GRE General Test. *Application deadline:* For fall admission, 3/15 (priority date). Applications are processed on a rolling basis. *Application fee:* $30 ($40 for international students).
Expenses: Tuition, state resident: full-time $5,400. Tuition, nonresident: full-time $14,535. Part-time tuition and fees vary according to course load.
Financial aid: In 1999–00, 13 students received aid, including fellowships with full tuition reimbursements available (averaging $12,224 per year), research assistantships with full tuition reimbursements available (averaging $15,504 per year); institutionally sponsored loans also available. Financial aid application deadline: 4/1.
Faculty research: Neurobiology of cell and muscle, intestinal mucosal control, autonomic neurophysiology, cardiovascular regulation, cell biology. *Total annual research expenditures:* $2 million.
Jack A. Rall, Chair, 614-292-5448, *Fax:* 614-292-4888, *E-mail:* rall.1@osu.edu.
Application contact: P. E. Ward, Graduate Committee Chair, 614-292-5448, *Fax:* 614-292-4888, *E-mail:* ward.10@osu.edu.

■ THE OHIO STATE UNIVERSITY

College of Veterinary Medicine and Graduate School, Graduate Programs in Veterinary Medicine, Department of Veterinary Biosciences, Columbus, OH 43210

AWARDS Anatomy and cellular biology (MS, PhD); pathobiology (MS, PhD); pharmacology (MS, PhD); toxicology (MS, PhD); veterinary physiology (MS, PhD).

Faculty: 28 full-time (8 women).
Students: 49 full-time (20 women); includes 2 minority (1 African American, 1 Asian American or Pacific Islander), 18 international.
Degree requirements: For master's and doctorate, thesis/dissertation, final exam required.
Entrance requirements: For master's, GRE General Test; for doctorate, GRE General Test, master's degree. *Application fee:* $25.
Expenses: Tuition, state resident: full-time $5,757. Tuition, nonresident: full-time $14,892.
Financial aid: Fellowships, research assistantships, teaching assistantships available.
Faculty research: Microvasculature, muscle biology, neonatal lung and bone development.
Charles C. Capen, Interim Chair, 614-292-4489.
Application contact: Graduate Admission Committee, 614-292-4489.

■ THE OHIO STATE UNIVERSITY

Graduate School, College of Biological Sciences, Department of Molecular Genetics, Columbus, OH 43210

AWARDS Cell and developmental biology (MS, PhD); genetics (MS, PhD); molecular biology (MS, PhD).

Faculty: 18 full-time, 6 part-time/adjunct.
Students: 46 full-time (17 women), 3 part-time (1 woman); includes 3 minority (2 African Americans, 1 Asian American or Pacific Islander), 21 international. *68 applicants, 28% accepted.* In 1999, 6 master's, 8 doctorates awarded.
Degree requirements: For master's and doctorate, thesis/dissertation required, foreign language not required.
Entrance requirements: For master's and doctorate, GRE General Test, GRE Subject Test (biology). *Application deadline:* For fall admission, 8/15. Applications are processed on a rolling basis. *Application fee:* $30 ($40 for international students).
Expenses: Tuition, state resident: full-time $5,400. Tuition, nonresident: full-time $14,535. Part-time tuition and fees vary according to course load and program.
Financial aid: Fellowships, research assistantships, teaching assistantships,

Federal Work-Study and institutionally sponsored loans available. Aid available to part-time students.
Dr. Lee F. Johnson, Chairman, 614-292-8084, *Fax:* 614-292-4466, *E-mail:* johnson.6@osu.edu.

Find an in-depth description at www.petersons.com/graduate.

■ THE OHIO STATE UNIVERSITY

Graduate School, College of Biological Sciences, Program in Molecular, Cellular and Developmental Biology, Columbus, OH 43210

AWARDS MS, PhD.

Faculty: 14 full-time, 73 part-time/adjunct.
Students: 65 full-time (38 women); includes 7 minority (2 African Americans, 4 Asian Americans or Pacific Islanders, 1 Hispanic American), 33 international. *114 applicants, 27% accepted.* In 1999, 4 master's, 7 doctorates awarded.
Degree requirements: For master's and doctorate, thesis/dissertation required, foreign language not required.
Entrance requirements: For master's and doctorate, GRE General Test, GRE Subject Test. *Application deadline:* For fall admission, 8/15. Applications are processed on a rolling basis. *Application fee:* $30 ($40 for international students).
Expenses: Tuition, state resident: full-time $5,400. Tuition, nonresident: full-time $14,535. Part-time tuition and fees vary according to course load and program.
Financial aid: Fellowships, research assistantships, teaching assistantships, Federal Work-Study and institutionally sponsored loans available. Aid available to part-time students.
Dr. David Bisaro, Director, 614-292-2804, *Fax:* 614-292-1538, *E-mail:* bisaro.1@osu.edu.

Find an in-depth description at www.petersons.com/graduate.

■ OHIO UNIVERSITY

Graduate Studies, College of Arts and Sciences, Interdisciplinary Program in Molecular and Cellular Biology, Athens, OH 45701-2979

AWARDS MS, PhD.

Faculty: 33 full-time (3 women).
Students: 20 full-time (8 women), 10 international. *10 applicants, 30% accepted.*
Degree requirements: For master's, thesis required; for doctorate, one foreign language (computer language can substitute), dissertation, exam, research proposal, teaching experience required.
Entrance requirements: For master's and doctorate, GRE General Test, GRE Subject Test, TOEFL. *Application deadline:*

For fall admission, 3/1 (priority date). Applications are processed on a rolling basis. *Application fee:* $30.
Expenses: Tuition, state resident: full-time $5,754; part-time $238 per credit hour. Tuition, nonresident: full-time $11,055; part-time $457 per credit hour. Tuition and fees vary according to course load, degree level and campus/location.
Financial aid: In 1999–00, research assistantships with tuition reimbursements (averaging $14,500 per year), 10 teaching assistantships with tuition reimbursements (averaging $14,500 per year) were awarded; Federal Work-Study, institutionally sponsored loans, and tuition waivers (full and partial) also available.
Faculty research: Animal biotechnology, plant molecular biology, immunology, cellular genetics, biochemistry of signal transduction.
Dr. Martin Tuck, Chair, 740-593-1747, *E-mail:* tuck@ohiou.edu.
Application contact: Dr. Robert A. Colvin, Graduate Chair, 740-593-0193, *Fax:* 740-593-0300, *E-mail:* rcolvin1@ohio.edu.

Find an in-depth description at www.petersons.com/graduate.

■ OREGON HEALTH SCIENCES UNIVERSITY

School of Medicine, Graduate Programs in Medicine, Department of Cell and Developmental Biology, Portland, OR 97201-3098

AWARDS PhD, MD/PhD.

Faculty: 11 full-time (5 women).
Students: 29 full-time (12 women); includes 1 minority (Asian American or Pacific Islander), 8 international. Average age 30. In 1999, 5 doctorates awarded (100% entered university research/teaching).
Degree requirements: For doctorate, dissertation required, foreign language not required. *Average time to degree:* Doctorate–7 years full-time.
Entrance requirements: For doctorate, GRE General Test, GRE Subject Test, MCAT. *Application deadline:* For fall admission, 1/15. Applications are processed on a rolling basis. *Application fee:* $60.
Expenses: Tuition, state resident: full-time $3,132; part-time $174 per credit hour. Tuition, nonresident: full-time $5,256; part-time $292 per credit hour. Required fees: $8.5 per credit hour. $146 per term. Part-time tuition and fees vary according to course load.
Financial aid: In 1999–00, research assistantships (averaging $16,500 per year); Federal Work-Study, institutionally sponsored loans, scholarships, and tuition waivers (full) also available. Financial aid

application deadline: 3/1; financial aid applicants required to submit FAFSA.
Faculty research: Developmental mechanisms, molecular biology of cancer, molecular neurobiology, intracellular signalling, growth factors and development. *Total annual research expenditures:* $3.4 million.
Dr. Bruce Magun, Chair, 503-494-7811, *Fax:* 503-494-4253, *E-mail:* magunb@ohsu.edu.
Application contact: Dr. Philip Copenhaver, Associate Professor, 503-494-4646, *Fax:* 503-494-4253, *E-mail:* copenhav@ohsu.edu.

■ OREGON STATE UNIVERSITY

Graduate School, College of Science, Program in Molecular and Cellular Biology, Corvallis, OR 97331

AWARDS PhD.

Degree requirements: For doctorate, dissertation, oral and written qualifying exams required, foreign language not required.
Entrance requirements: For doctorate, TOEFL, minimum GPA of 3.0 in last 90 hours. *Application deadline:* For fall admission, 2/15. Applications are processed on a rolling basis. *Application fee:* $50.
Expenses: Tuition, state resident: full-time $6,489. Tuition, nonresident: full-time $11,061. Tuition and fees vary according to program.
Financial aid: Fellowships, career-related internships or fieldwork, Federal Work-Study, and institutionally sponsored loans available. Aid available to part-time students.
Dr. Daniel J. Arp, Director, 541-737-3799, *Fax:* 541-737-3045, *E-mail:* arpd@bcc.orst.edu.

■ THE PENNSYLVANIA STATE UNIVERSITY MILTON S. HERSHEY MEDICAL CENTER

Graduate School, Interdepartmental Graduate Program in Cell and Molecular Biology, Hershey, PA 17033-2360

AWARDS MS, PhD, MD/PhD.

Students: 22 full-time (16 women), 1 (woman) part-time. Average age 27. In 1999, 3 doctorates awarded.
Entrance requirements: For master's and doctorate, GRE General Test. *Application deadline:* For fall admission, 7/26. *Application fee:* $50.
Expenses: Tuition, state resident: full-time $6,886; part-time $291 per credit. Tuition, nonresident: full-time $14,118; part-time $588 per credit. Required fees: $43 per semester. Part-time tuition and fees vary according to course load.

The Pennsylvania State University Milton S. Hershey Medical Center (continued)
Financial aid: Research assistantships available.
Dr. Robert Levenson, Director, 717-531-1045.

■ THE PENNSYLVANIA STATE UNIVERSITY UNIVERSITY PARK CAMPUS

Graduate School, Eberly College of Science, Department of Biochemistry and Molecular Biology, Program in Cell and Developmental Biology, State College, University Park, PA 16802-1503

AWARDS PhD.

Degree requirements: For doctorate, dissertation, comprehensive exam required.
Entrance requirements: For doctorate, GRE General Test, GRE Subject Test. *Application fee:* $50.
Expenses: Tuition, state resident: full-time $6,886; part-time $291 per credit. Tuition, nonresident: full-time $14,118; part-time $588 per credit. Required fees: $46 per semester. Part-time tuition and fees vary according to course load and program.
Application contact: Dr. Ronald D. Porter, Director, 814-865-2538.

Find an in-depth description at www.petersons.com/graduate.

■ THE PENNSYLVANIA STATE UNIVERSITY UNIVERSITY PARK CAMPUS

Graduate School, Intercollege Graduate Programs, Intercollege Graduate Program in Integrative Biosciences, State College, University Park, PA 16802-1503

AWARDS Integrative biosciences (MS, PhD), including biomolecular transport dynamics, cell and developmental biology, cellular and molecular mechanisms of toxicity, chemical biology, ecological and molecular plant physiology, immunobiology, molecular medicine, neuroscience, nutrition science.

Students: 39 full-time (24 women), 1 part-time.
Entrance requirements: For master's and doctorate, GRE General Test. *Application fee:* $50.
Expenses: Tuition, state resident: full-time $6,886; part-time $291 per credit. Tuition, nonresident: full-time $14,118; part-time $588 per credit. Required fees: $46 per semester. Part-time tuition and fees vary according to course load and program.
Financial aid: Fellowships available.
Dr. C. R. Matthews, Co-Director, 814-863-3650.

■ PRINCETON UNIVERSITY

Graduate School, Department of Molecular Biology, Princeton, NJ 08544-1019

AWARDS Cell biology (PhD); developmental biology (PhD); molecular biology (PhD); neuroscience (PhD).

Degree requirements: For doctorate, dissertation required.
Entrance requirements: For doctorate, GRE General Test.
Expenses: Tuition: Full-time $25,050.
Faculty research: Genetics, virology, biochemistry.

Find an in-depth description at www.petersons.com/graduate.

■ PURDUE UNIVERSITY

Graduate School, School of Science, Department of Biological Sciences, Program in Cell and Developmental Biology, West Lafayette, IN 47907

AWARDS PhD.

Faculty: 13 full-time (3 women).
Degree requirements: For doctorate, dissertation, seminars, teaching experience required.
Entrance requirements: For doctorate, GRE General Test, TOEFL, TSE. *Application deadline:* For fall admission, 2/15. *Application fee:* $30. Electronic applications accepted.
Expenses: Tuition, state resident: full-time $4,530; part-time $130 per credit hour. Tuition, nonresident: full-time $15,310; part-time $404 per credit hour. Tuition and fees vary according to campus/location and program.
Financial aid: Fellowships, research assistantships, teaching assistantships available. Aid available to part-time students. Financial aid application deadline: 2/15; financial aid applicants required to submit FAFSA.
Faculty research: Cellular signalling, patterns of development in eukaryotes, genomics, gene expression, cell structure and function.
Application contact: Nancy Konopka, Graduate Studies Office Manager, 765-494-8142, *Fax:* 765-494-0876, *E-mail:* njk@bilbo.bio.purdue.edu.

■ QUINNIPIAC UNIVERSITY

School of Health Sciences, Program in Molecular and Cell Biology, Hamden, CT 06518-1940

AWARDS MS. Part-time and evening/weekend programs available.

Faculty: 6 full-time (1 woman), 2 part-time/adjunct (1 woman).
Students: 3 full-time (2 women), 12 part-time (8 women); includes 1 minority (Hispanic American). Average age 29. *8 applicants, 88% accepted.*
Degree requirements: For master's, thesis optional.
Entrance requirements: For master's, minimum GPA of 2.5. *Application deadline:* For fall admission, 8/1 (priority date); for spring admission, 12/15 (priority date). Applications are processed on a rolling basis. *Application fee:* $45. Electronic applications accepted.
Expenses: Tuition: Part-time $410 per credit hour. Required fees: $20 per term. Tuition and fees vary according to program.
Financial aid: Available to part-time students.
Dr. Charlotte Hammond, Director, 203-582-8058 Ext. 8058, *E-mail:* charlotte.hammond@quinnipiac.edu.
Application contact: Scott Farber, Director of Graduate Admissions, 800-462-1944, *Fax:* 203-582-3443, *E-mail:* graduate@quinnipiac.edu.

Find an in-depth description at www.petersons.com/graduate.

■ RENSSELAER POLYTECHNIC INSTITUTE

Graduate School, School of Science, Department of Biology, Troy, NY 12180-3590

AWARDS Biochemistry (MS, PhD); biophysics (MS, PhD); cell biology (MS, PhD); developmental biology (MS, PhD); microbiology (MS, PhD); molecular biology (MS, PhD); plant science (MS, PhD). Part-time programs available.

Faculty: 14 full-time (5 women).
Students: 16 full-time (8 women), 2 part-time (1 woman); includes 1 minority (Asian American or Pacific Islander), 4 international. *37 applicants, 41% accepted.* In 1999, 6 master's, 3 doctorates awarded. Terminal master's awarded for partial completion of doctoral program.
Degree requirements: For master's and doctorate, thesis/dissertation required, foreign language not required.
Entrance requirements: For master's and doctorate, GRE General Test, TOEFL. *Application deadline:* For fall admission, 2/1 (priority date). Applications are processed on a rolling basis. *Application fee:* $35.

Application contact: Admissions Committee, 814-865-3155, *Fax:* 814-863-1357, *E-mail:* lscgradadm@mail.biotec.psu.edu.

Find an in-depth description at www.petersons.com/graduate.

Expenses: Tuition: Part-time $665 per credit hour. Required fees: $980.
Financial aid: In 1999–00, 8 research assistantships with partial tuition reimbursements (averaging $15,000 per year), 11 teaching assistantships with full tuition reimbursements (averaging $15,000 per year) were awarded; fellowships, career-related internships or fieldwork and institutionally sponsored loans also available. Financial aid application deadline: 2/1.
Faculty research: Applied environmental biology, genetics, environmental science, fresh water ecology, microbial ecology. Dr. John Salerno, Chair, 518-276-2699, *Fax:* 518-276-2344.
Application contact: Dr. Jackie L. Collier, Assistant Professor, 518-276-6446, *Fax:* 518-276-2344, *E-mail:* collij3@rpi.edu.
Find an in-depth description at www.petersons.com/graduate.

■ **RICE UNIVERSITY**
Graduate Programs, Wiess School of Natural Sciences, Department of Biochemistry and Cell Biology, Houston, TX 77251-1892
AWARDS MA, PhD.

Faculty: 21 full-time (5 women), 1 part-time/adjunct (0 women).
Students: 54 full-time (29 women); includes 3 minority (all Hispanic Americans), 15 international. Average age 22. *295 applicants, 5% accepted.* In 1999, 2 master's awarded (50% found work related to degree, 50% continued full-time study); 12 doctorates awarded (90% found work related to degree, 10% continued full-time study).
Degree requirements: For master's and doctorate, thesis/dissertation required, foreign language not required. *Average time to degree:* Master's–3 years full-time; doctorate–5 years full-time.
Entrance requirements: For master's and doctorate, GRE. *Application deadline:* For fall admission, 2/1 (priority date). Applications are processed on a rolling basis. *Application fee:* $0. Electronic applications accepted.
Expenses: Tuition: Full-time $16,100. Required fees: $300.
Financial aid: In 1999–00, 21 fellowships, 32 research assistantships were awarded; traineeships and tuition waivers (full) also available.
Faculty research: Steroid metabolism, protein structure NMR, biophysics, cell growth and movement. *Total annual research expenditures:* $5 million.
Dr. F. B. Rudolph, Chair, 713-348-4015, *Fax:* 713-348-5154, *E-mail:* fbr@rice.edu.

Application contact: Dolores M. Schwartz, Office Manager, 713-348-4230, *Fax:* 713-348-5154, *E-mail:* dolores@rice.edu.
Find an in-depth description at www.petersons.com/graduate.

■ **RUTGERS, THE STATE UNIVERSITY OF NEW JERSEY, NEW BRUNSWICK**
Graduate School, Core Curriculum in Molecular and Cell Biology, New Brunswick, NJ 08901-1281
AWARDS PhD.

Students: Average age 26.
Degree requirements: For doctorate, dissertation required.
Entrance requirements: For doctorate, GRE General Test, GRE Subject Test, TOEFL. *Application fee:* $40.
Expenses: Tuition, state resident: full-time $6,776; part-time $279 per credit. Tuition, nonresident: full-time $9,936; part-time $412 per credit. Required fees: $20 per credit. $89 per semester. Tuition and fees vary according to course load, campus/location and program.
Financial aid: Fellowships available. Michael Leibowitz, Program Director, 732-235-4795.

■ **RUTGERS, THE STATE UNIVERSITY OF NEW JERSEY, NEW BRUNSWICK**
Graduate School, Program in Cell and Developmental Biology, New Brunswick, NJ 08901-1281
AWARDS Cell biology (MS, PhD); developmental biology (MS, PhD); immunology (MS). Part-time programs available.

Faculty: 84 full-time (15 women).
Students: 30 full-time (16 women), 42 part-time (22 women); includes 24 minority (1 African American, 15 Asian Americans or Pacific Islanders, 8 Hispanic Americans), 17 international. Average age 24. *109 applicants, 34% accepted.* In 1999, 8 master's, 8 doctorates awarded. Terminal master's awarded for partial completion of doctoral program.
Degree requirements: For master's, thesis or alternative required; for doctorate, dissertation, written qualifying exam required.
Entrance requirements: For master's, GRE General Test, TOEFL; for doctorate, GRE General Test, GRE Subject Test, TOEFL, minimum GPA of 3.0. *Application deadline:* For fall admission, 2/15. Applications are processed on a rolling basis. *Application fee:* $50.
Expenses: Tuition, state resident: full-time $6,776; part-time $279 per credit. Tuition,

nonresident: full-time $9,936; part-time $412 per credit. Required fees: $20 per credit. $89 per semester. Tuition and fees vary according to course load, campus/location and program.
Financial aid: In 1999–00, 6 fellowships with full tuition reimbursements (averaging $15,000 per year), 17 research assistantships with full tuition reimbursements (averaging $15,000 per year), 10 teaching assistantships with full tuition reimbursements (averaging $14,000 per year) were awarded. Financial aid application deadline: 2/15; financial aid applicants required to submit FAFSA.
Faculty research: Genetics, cell matrix interactions, signal transduction, gene expression, cytoskeletal proteins, neurobiology.
Dr. Alice Y. Liu, Director, 732-445-2730, *Fax:* 732-445-6370, *E-mail:* liu@biology.rutgers.edu.
Application contact: Carolyn Ambrose, Administrative Assistant, 732-445-3430, *Fax:* 732-445-6370, *E-mail:* ambrose@biology.rutgers.edu.

■ **SAINT LOUIS UNIVERSITY**
School of Medicine and Graduate School, Graduate Programs in Biomedical Sciences, Cell and Molecular Biology Training Program, St. Louis, MO 63103-2097
AWARDS PhD.

Students: Average age 31. In 1999, 8 degrees awarded.
Degree requirements: For doctorate, dissertation, departmental qualifying exams required.
Entrance requirements: For doctorate, GRE General Test, GRE Subject Test. *Application deadline:* For fall admission, 4/15 (priority date). Applications are processed on a rolling basis. *Application fee:* $40.
Expenses: Tuition: Part-time $507 per credit hour. Required fees: $38 per term.
Financial aid: In 1999–00, 8 students received aid, including 1 research assistantship; fellowships, teaching assistantships, traineeships also available. Aid available to part-time students. Financial aid application deadline: 8/1; financial aid applicants required to submit FAFSA.
Dr. Duane P. Grandgenette, Director, 314-577-8418.
Application contact: Sheila Trimble, Administrative Assistant, 314-577-8418, *E-mail:* trimble@slu.edu.

■ SAN DIEGO STATE UNIVERSITY

Graduate and Research Affairs, College of Sciences, Department of Biological Sciences, San Diego, CA 92182

AWARDS Biology (MA, MS), including ecology (MS), molecular biology (MS), physiology (MS), systematics/evolution (MS); cell and molecular biology (PhD); ecology (PhD); microbiology (MS).

Students: 37 full-time (23 women), 86 part-time (46 women); includes 13 minority (1 African American, 10 Asian Americans or Pacific Islanders, 2 Hispanic Americans), 4 international. Average age 26. *135 applicants, 33% accepted.* In 1999, 20 master's, 4 doctorates awarded. Terminal master's awarded for partial completion of doctoral program.

Degree requirements: For master's, thesis required; for doctorate, dissertation required.

Entrance requirements: For master's, GRE General Test, GRE Subject Test, TOEFL. *Application deadline:* For fall admission, 7/1 (priority date); for spring admission, 12/1. Applications are processed on a rolling basis. *Application fee:* $55.

Expenses: Tuition, nonresident: part-time $246 per unit. Required fees: $1,932; $633 per semester. Tuition and fees vary according to course load.

Financial aid: Fellowships, research assistantships, teaching assistantships, career-related internships or fieldwork and unspecified assistantships available. *Total annual research expenditures:* $11.5 million. Sanford Bernstein, Chair, 619-594-5629, *Fax:* 619-594-5676, *E-mail:* sanford.bernstein@sdsu.edu.

Application contact: Ken Johnson, Graduate Coordinator, 619-594-6919, *Fax:* 619-594-5676, *E-mail:* kjohnson@sunstroke.sdsu.edu.

■ SAN DIEGO STATE UNIVERSITY

Graduate and Research Affairs, College of Sciences, Molecular Biology Institute and Department of Chemistry and Department of Biological Sciences, Program in Cell and Molecular Biology, San Diego, CA 92182

AWARDS PhD.

Students: 7 full-time (5 women), 24 part-time (14 women); includes 6 minority (5 Asian Americans or Pacific Islanders, 1 Hispanic American), 1 international.

Degree requirements: For doctorate, dissertation, oral comprehensive qualifying exam required. *Average time to degree:* Doctorate–5 years full-time.

Entrance requirements: For doctorate, GRE General Test, GRE Subject Test. *Application deadline:* For fall admission, 7/1 (priority date); for spring admission, 12/1. Applications are processed on a rolling basis. *Application fee:* $55.

Expenses: Tuition, nonresident: part-time $246 per unit. Required fees: $1,932; $633 per semester. Tuition and fees vary according to course load.

Financial aid: In 1999–00, 5 fellowships were awarded; institutionally sponsored loans also available. Financial aid applicants required to submit CSS PROFILE or FAFSA.

Faculty research: Structure/dynamics of protein kinesis, chromatin structure and DNA methylation membrane biochemistry, secretory protein targeting, molecular biology of cardiac myocytes.
Dr. Skai Krisans, Head, 619-594-5368, *Fax:* 619-594-5676, *E-mail:* skrisans@sunstroke.sdsu.edu.

Find an in-depth description at www.petersons.com/graduate.

■ SAN FRANCISCO STATE UNIVERSITY

Graduate Division, College of Science and Engineering, Department of Biology, Program in Cell and Molecular Biology, San Francisco, CA 94132-1722

AWARDS MA.

Entrance requirements: For master's, minimum GPA of 2.5 in last 60 units.

Expenses: Tuition, nonresident: full-time $5,904; part-time $246 per unit. Required fees: $1,904; $637 per semester. Tuition and fees vary according to course load.

■ THE SCRIPPS RESEARCH INSTITUTE

Office of Graduate Studies, Macromolecular and Cellular Structure and Chemistry Graduate Program, La Jolla, CA 92037

AWARDS PhD.

Faculty: 84 full-time (12 women).
Students: 73 full-time (29 women). *121 applicants, 26% accepted.* In 1999, 10 degrees awarded.

Degree requirements: For doctorate, dissertation required, foreign language not required. *Average time to degree:* Doctorate–5 years full-time.

Entrance requirements: For doctorate, GRE General Test (average 90th percentile), GRE Subject Test (average 72nd percentile), TOEFL. *Application deadline:* For fall admission, 1/1. *Application fee:* $0.

Expenses: All students are fully funded.

Financial aid: Institutionally sponsored loans and stipends available.

Faculty research: Biocatalysis and enzyme engineering, molecular structure and function, neurosciences, immunology, plant biology.

Application contact: Marylyn Rinaldi, Graduate Program Administrator, 858-784-8469, *Fax:* 858-784-2802, *E-mail:* mrinaldi@scripps.edu.

■ SOUTHWEST MISSOURI STATE UNIVERSITY

Graduate College, College of Health and Human Services, Department of Biomedical Sciences, Program in Cell and Molecular Biology, Springfield, MO 65804-0094

AWARDS MS.

Degree requirements: For master's, thesis or alternative, oral and written exams required, foreign language not required.

Entrance requirements: For master's, GRE General Test, 2 semesters of organic chemistry and physics, 1 semester of calculus, minimum GPA of 3.0 during last 60 hours of course work.

Expenses: Tuition, state resident: full-time $2,070; part-time $115 per credit. Tuition, nonresident: full-time $4,140; part-time $230 per credit. Required fees: $91 per credit. Tuition and fees vary according to course level, course load and program.

■ STATE UNIVERSITY OF NEW YORK AT ALBANY

College of Arts and Sciences, Department of Biological Sciences, Specialization in Molecular, Cellular, Developmental, and Neural Biology, Albany, NY 12222-0001

AWARDS MS, PhD.

Degree requirements: For master's, one foreign language required; for doctorate, dissertation required.

Entrance requirements: For master's and doctorate, GRE General Test. *Application fee:* $50.

Expenses: Tuition, state resident: full-time $5,100; part-time $214 per credit. Tuition, nonresident: full-time $8,416; part-time $352 per credit. Required fees: $31 per credit.

Financial aid: Minority assistantships available.
Dr. David Shub, Chair, Department of Biological Sciences, 518-442-4300.

■ STATE UNIVERSITY OF NEW YORK AT ALBANY

School of Public Health, Department of Biomedical Sciences, Program in Cell and Molecular Structure, Albany, NY 12222-0001

AWARDS MS, PhD.

Degree requirements: For master's and doctorate, thesis/dissertation required.
Entrance requirements: For master's and doctorate, GRE General Test, GRE Subject Test. *Application deadline:* For fall admission, 1/15 (priority date); for spring admission, 11/1 (priority date). *Application fee:* $50.
Expenses: Tuition, state resident: full-time $5,100; part-time $214 per credit. Tuition, nonresident: full-time $8,416; part-time $352 per credit. Required fees: $31 per credit.
Financial aid: Application deadline: 2/1. Dr. Harry Taber, Chair, Department of Biomedical Sciences, 518-474-2662.

■ STATE UNIVERSITY OF NEW YORK AT BUFFALO

Graduate School, Graduate Programs in Biomedical Sciences at Roswell Park Cancer Institute, Department of Cellular and Molecular Biology at Roswell Park Cancer Institute, Buffalo, NY 14263

AWARDS PhD.

Faculty: 19 full-time (1 woman), 1 part-time/adjunct (0 women).
Students: 11 full-time (6 women), 7 part-time (4 women); includes 3 minority (all Asian Americans or Pacific Islanders), 8 international. Average age 25. *40 applicants, 33% accepted.* In 1999, 4 doctorates awarded (100% continued full-time study).
Degree requirements: For doctorate, dissertation, exam project required. *Average time to degree:* Doctorate–6.4 years full-time.
Entrance requirements: For doctorate, GRE General Test, GRE Subject Test, TOEFL, TSE, TWE, minimum B average in undergraduate coursework. *Application deadline:* For fall admission, 2/1 (priority date). Applications are processed on a rolling basis. *Application fee:* $35. Electronic applications accepted.
Expenses: Tuition, state resident: full-time $5,100; part-time $213 per credit hour. Tuition, nonresident: full-time $8,416; part-time $351 per credit hour. Required fees: $935; $75 per semester. Tuition and fees vary according to course load and program.
Financial aid: In 1999–00, 4 fellowships with full tuition reimbursements (averaging $15,000 per year), 17 research assistantships with full tuition reimbursements (averaging $15,000 per year) were awarded; unspecified assistantships also available. Financial aid application deadline: 2/1; financial aid applicants required to submit FAFSA.
Faculty research: Cancer genetics, chromatin structure and replication, regulation of transcription, human gene mapping, genetic and structural approaches to regulation of gene expression, mouse genetics, gene libraries. *Total annual research expenditures:* $4.5 million. Dr. John Yates, Chair, 716-845-8964, *Fax:* 716-845-8449, *E-mail:* yates@sc3101.med.buffalo.edu.
Application contact: Craig R. Johnson, Director of Graduate Studies, 716-845-2339, *Fax:* 716-845-8178, *E-mail:* rpgradapp@sc3103.med.buffalo.edu.

■ STATE UNIVERSITY OF NEW YORK AT BUFFALO

Graduate School, Interdisciplinary and Interdepartmental Graduate Program in Molecular Cell Biology, Buffalo, NY 14260

AWARDS PhD, Certificate. PhD awarded through participating departments.

Faculty: 53.
Students: *80 applicants, 6% accepted.*
Degree requirements: For doctorate, dissertation required.
Entrance requirements: For doctorate, GRE General Test, GRE Subject Test (recommended), TOEFL, minimum GPA of 3.0. *Application deadline:* For fall admission, 2/1 (priority date). Applications are processed on a rolling basis. *Application fee:* $35.
Expenses: All students receive a full tuition waiver, health coverage, and an annual stipend of $15,000.
Financial aid: In 1999–00, 16 students received aid, including 4 research assistantships with full tuition reimbursements available (averaging $15,000 per year); institutionally sponsored loans and stipends also available. Financial aid applicants required to submit FAFSA.
Faculty research: Nucleic acid-protein interactions, protein structure, membrane proteins and channels, microbiology and parasitology, virology. *Total annual research expenditures:* $28 million. Dr. Stephen J. Free, Director, 716-645-2865, *Fax:* 716-645-3776, *E-mail:* free@acsu.buffalo.edu.
Application contact: Xochitl E. Nicholson, Program Administrator, 716-645-2164, *Fax:* 716-645-3776, *E-mail:* ubmcbp@hopper.bio.buffalo.edu.

Find an in-depth description at www.petersons.com/graduate.

■ STATE UNIVERSITY OF NEW YORK AT BUFFALO

Graduate School, School of Medicine and Biomedical Sciences, Graduate Programs in Medicine and Biomedical Sciences, Department of Anatomy and Cell Biology, Buffalo, NY 14214

AWARDS MA, PhD.

Faculty: 10 full-time (2 women), 2 part-time/adjunct (1 woman).
Students: 3 full-time (1 woman), 2 part-time, 3 international. Average age 28. 7 *applicants, 14% accepted.* In 1999, 1 master's, 1 doctorate awarded.
Degree requirements: For master's, thesis, exam required; for doctorate, one foreign language (computer language can substitute), dissertation, exam required.
Entrance requirements: For master's, GRE General Test, TOEFL, previous course work in biology, chemistry, or physics; for doctorate, GRE General Test, TOEFL, previous course work in biology, chemistry, and physics. *Application deadline:* For fall admission, 2/1 (priority date). Applications are processed on a rolling basis. *Application fee:* $35. Electronic applications accepted.
Expenses: Tuition, state resident: full-time $5,100. Tuition, nonresident: full-time $8,416. Required fees: $935.
Financial aid: In 1999–00, 4 students received aid, including 3 teaching assistantships with full tuition reimbursements available (averaging $14,700 per year); fellowships with full tuition reimbursements available, research assistantships with full tuition reimbursements available, Federal Work-Study and institutionally sponsored loans also available. Financial aid application deadline: 2/28; financial aid applicants required to submit FAFSA.
Faculty research: Developmental biology, neuroscience, digital imaging, cell biology. *Total annual research expenditures:* $350,000. Dr. Frank Mendel, Interim Chairperson, 716-829-2912, *Fax:* 716-829-2915, *E-mail:* fcmendel@buffalo.edu.
Application contact: Debbie Tomasulo, Keyboard Specialist, 716-829-2912, *Fax:* 716-829-2915.

■ STATE UNIVERSITY OF NEW YORK AT STONY BROOK

Graduate School, College of Arts and Sciences, Molecular and Cellular Biology Program, Stony Brook, NY 11794

AWARDS Biochemistry and molecular biology (PhD); cellular and developmental biology (PhD); immunology and pathology (PhD).

Degree requirements: For doctorate, dissertation, comprehensive exam, teaching experience required.

State University of New York at Stony Brook (continued)

Entrance requirements: For doctorate, GRE General Test, GRE Subject Test, TOEFL. *Application deadline:* For fall admission, 1/15. *Application fee:* $50.

Expenses: Tuition, state resident: full-time $5,100; part-time $213 per credit hour. Tuition, nonresident: full-time $8,416; part-time $351 per credit hour. Required fees: $492. Tuition and fees vary according to program.

Financial aid: Fellowships, research assistantships, teaching assistantships, Federal Work-Study available.

Application contact: Information Contact, 631-632-8533, *Fax:* 631-632-9730.

Find an in-depth description at www.petersons.com/graduate.

■ **STATE UNIVERSITY OF NEW YORK HEALTH SCIENCE CENTER AT BROOKLYN**

School of Graduate Studies, Core Program in Molecular and Cellular Biology, Brooklyn, NY 11203-2098

AWARDS Anatomy and cell biology (PhD); biochemistry (PhD); biophysics (PhD); microbiology and immunology (PhD); neural and behavioral sciences (PhD); pharmacology (PhD); physiology (PhD). Affiliation with a particular PhD degree-granting program is deferred to the second year.

Degree requirements: For doctorate, one foreign language, dissertation required.

Entrance requirements: For doctorate, GRE General Test.

Expenses: Tuition, state resident: full-time $5,100; part-time $213 per credit. Tuition, nonresident: full-time $8,416; part-time $351 per credit. Required fees: $200. Full-time tuition and fees vary according to program and student level.

Faculty research: Mechanism of gene regulation, molecular virology.

■ **STATE UNIVERSITY OF NEW YORK HEALTH SCIENCE CENTER AT BROOKLYN**

School of Graduate Studies, Department of Anatomy and Cell Biology, Brooklyn, NY 11203-2098

AWARDS PhD, MD/PhD.

Degree requirements: For doctorate, one foreign language, dissertation required.

Entrance requirements: For doctorate, GRE.

Expenses: Tuition, state resident: full-time $5,100; part-time $213 per credit. Tuition, nonresident: full-time $8,416; part-time

$351 per credit. Required fees: $200. Full-time tuition and fees vary according to program and student level.

Faculty research: Role of oncogenes in early cardiogenesis, transepithelial migration of human neutrophils, biochemical pathways of platelet activation, mechanism of action of interferon-T, nerve growth factor's role in inner ear development.

■ **STATE UNIVERSITY OF NEW YORK UPSTATE MEDICAL UNIVERSITY**

College of Graduate Studies, Department of Anatomy and Cell Biology, Syracuse, NY 13210-2334

AWARDS MS, PhD, MD/PhD.

Faculty: 33.

Students: 12 full-time (6 women); includes 3 minority (1 African American, 2 Asian Americans or Pacific Islanders), 2 international. *22 applicants, 41% accepted.* In 1999, 2 doctorates awarded (50% found work related to degree, 50% continued full-time study). Terminal master's awarded for partial completion of doctoral program.

Degree requirements: For master's, thesis required, foreign language not required; for doctorate, dissertation, comprehensive exam required, foreign language not required.

Entrance requirements: For master's and doctorate, GRE General Test, GRE Subject Test, TSE. *Application deadline:* For fall admission, 4/1 (priority date). Applications are processed on a rolling basis. *Application fee:* $40.

Expenses: Tuition, state resident: full-time $5,100; part-time $213 per credit. Tuition, nonresident: full-time $8,416; part-time $351 per credit. Required fees: $410; $25 per credit. Part-time tuition and fees vary according to course load and program.

Financial aid: Fellowships, research assistantships, Federal Work-Study and institutionally sponsored loans available. Aid available to part-time students. Dr. James Schwob, Interim Chair, 315-464-5120.

Application contact: Dr. Ira Ames, Professor, 315-464-5120.

■ **STATE UNIVERSITY OF NEW YORK UPSTATE MEDICAL UNIVERSITY**

College of Graduate Studies, Department of Cell and Developmental Biology, Syracuse, NY 13210-2334

AWARDS PhD, MD/PhD.

Degree requirements: For doctorate, dissertation, comprehensive exam required, foreign language not required.

Entrance requirements: For doctorate, GRE General Test, GRE Subject Test, TSE. *Application deadline:* For fall admission, 4/1 (priority date). Applications are processed on a rolling basis. *Application fee:* $40.

Expenses: Tuition, state resident: full-time $5,100; part-time $213 per credit. Tuition, nonresident: full-time $8,416; part-time $351 per credit. Required fees: $410; $25 per credit. Part-time tuition and fees vary according to course load and program. Dr. David R. Mitchell, Interim Chair.

Find an in-depth description at www.petersons.com/graduate.

■ **STATE UNIVERSITY OF NEW YORK UPSTATE MEDICAL UNIVERSITY**

College of Graduate Studies, Program in Cell and Molecular Biology, Syracuse, NY 13210-2334

AWARDS PhD.

Faculty: 42.

Students: 22 full-time (5 women), 1 (woman) part-time; includes 2 minority (both Asian Americans or Pacific Islanders), 4 international. *17 applicants, 24% accepted.*

Degree requirements: For doctorate, dissertation, comprehensive exam required, foreign language not required.

Entrance requirements: For doctorate, GRE General Test, GRE Subject Test, TSE. *Application deadline:* For fall admission, 4/1 (priority date). Applications are processed on a rolling basis. *Application fee:* $40.

Expenses: Tuition, state resident: full-time $5,100; part-time $213 per credit. Tuition, nonresident: full-time $8,416; part-time $351 per credit. Required fees: $410; $25 per credit. Part-time tuition and fees vary according to course load and program.

Application contact: Dr. Richard Wojcikiewicz, Associate Professor, 315-464-7655.

Find an in-depth description at www.petersons.com/graduate.

■ **TEMPLE UNIVERSITY**

Health Sciences Center, School of Medicine and Graduate School, Graduate Programs in Medicine, Department of Anatomy and Cell Biology, Philadelphia, PA 19140

AWARDS PhD.

Faculty: 18 full-time (6 women).

Students: 4 full-time (2 women); all minorities (1 African American, 3 Asian Americans or Pacific Islanders). *8 applicants, 50% accepted.*

Degree requirements: For doctorate, dissertation, research seminars required, foreign language not required.

Entrance requirements: For doctorate, GRE General Test, GRE Subject Test, minimum GPA of 3.0. *Application deadline:* For fall admission, 9/1 (priority date); for spring admission, 2/1. *Application fee:* $40. Electronic applications accepted.

Expenses: Tuition, state resident: full-time $6,030. Tuition, nonresident: full-time $8,298. Required fees: $230. One-time fee: $10 full-time.

Financial aid: Fellowships, Federal Work-Study available.

Faculty research: Neurobiology, reproductive biology, cardiovascular system, musculoskeletal biology, developmental biology. *Total annual research expenditures:* $632,373.
Dr. Steve Popoff, Chair, 215-707-3161, *E-mail:* spopoff@vm.temple.edu.

Application contact: Dr. Judith Litvin, Admissions Chair, 215-707-2070, *Fax:* 215-707-2966, *E-mail:* jl1@astro.ocis.temple.edu.

■ TEXAS A&M UNIVERSITY

College of Science, Program in Molecular and Cell Biology, College Station, TX 77843

AWARDS PhD. Program composed of members from 4 colleges and 11 departments.

Degree requirements: For doctorate, dissertation required, foreign language not required.

Entrance requirements: For doctorate, GRE General Test, TOEFL. *Application fee:* $50 ($75 for international students).

Expenses: Tuition, state resident: part-time $76 per semester hour. Tuition, nonresident: part-time $292 per semester hour. Required fees: $11 per semester hour. Tuition and fees vary according to program.

Financial aid: Fellowships available. Financial aid application deadline: 4/1; financial aid applicants required to submit FAFSA.
Dr. Terry L. Thomas, Head, 979-845-0184, *Fax:* 979-845-2891.

Application contact: Graduate Adviser, 979-826-2465, *Fax:* 979-845-2891.

■ TEXAS TECH UNIVERSITY HEALTH SCIENCES CENTER

Graduate School of Biomedical Sciences, Department of Cell Biology and Biochemistry, Program in Anatomy/Cell Biology, Lubbock, TX 79430

AWARDS MS, PhD, MD/PhD.

Faculty: 12 full-time (4 women).

Students: 8 full-time (6 women), 4 international. Average age 27. *17 applicants, 12% accepted.* In 1999, 1 doctorate awarded (100% entered university research/teaching). Terminal master's awarded for partial completion of doctoral program.

Degree requirements: For master's and doctorate, thesis/dissertation required, foreign language not required. *Average time to degree:* Doctorate–6 years full-time.

Entrance requirements: For master's and doctorate, GRE General Test, TOEFL, minimum GPA of 3.0. *Application deadline:* For fall admission, 4/15 (priority date). *Application fee:* $30 ($55 for international students).

Expenses: Tuition, state resident: part-time $38 per credit hour. Tuition, nonresident: part-time $254 per credit hour. Part-time tuition and fees vary according to program.

Financial aid: In 1999–00, 2 fellowships with full tuition reimbursements (averaging $14,688 per year), 6 research assistantships (averaging $14,500 per year) were awarded.

Faculty research: Biochemical endocrinology, neurobiology, molecular biology, reproductive biology, biology of developing systems.
Dr. James Hutson, Graduate Director, 806-743-2700, *Fax:* 806-743-2990, *E-mail:* james.hutson@ttmc.ttuhsc.edu.

Find an in-depth description at www.petersons.com/graduate.

■ THOMAS JEFFERSON UNIVERSITY

College of Graduate Studies, Program in Pathology and Cell Biology, Philadelphia, PA 19107

AWARDS PhD.

Faculty: 48 full-time.

Students: 19 full-time (6 women); includes 4 minority (1 African American, 1 Asian American or Pacific Islander, 1 Hispanic American, 1 Native American), 2 international. *31 applicants, 6% accepted.* In 1999, 3 degrees awarded.

Degree requirements: For doctorate, dissertation required, foreign language not required.

Entrance requirements: For doctorate, GRE General Test, TOEFL, minimum GPA of 3.2. *Application deadline:* For fall admission, 3/1 (priority date). Applications are processed on a rolling basis. *Application fee:* $40.

Expenses: Tuition: Full-time $12,670. Tuition and fees vary according to degree level and program.

Financial aid: In 1999–00, 17 fellowships with full tuition reimbursements were awarded; research assistantships, Federal Work-Study, institutionally sponsored

loans, traineeships, and training grants also available. Aid available to part-time students. Financial aid application deadline: 5/1; financial aid applicants required to submit FAFSA.

Faculty research: Liver diseases, alcohol metabolism, structure-function relationships in biological membranes, chemical carcinogenesis. *Total annual research expenditures:* $8 million.
Dr. Jan B. Hoek, Chairman, Graduate Committee, 215-503-5016, *Fax:* 215-923-2218, *E-mail:* jan.hoek@mail.tju.edu.

Application contact: Jessie F. Pervall, Director of Admissions, 215-503-4400, *Fax:* 215-503-3433, *E-mail:* cgs-info@mail.tju.edu.

Find an in-depth description at www.petersons.com/graduate.

■ TUFTS UNIVERSITY

Sackler School of Graduate Biomedical Sciences, Program in Cell, Molecular and Developmental Biology, Medford, MA 02155

AWARDS PhD.

Faculty: 31 full-time (6 women), 6 part-time/adjunct (3 women).

Students: 25 full-time (14 women); includes 1 minority (Asian American or Pacific Islander), 3 international. Average age 25. *143 applicants, 3% accepted.* In 1999, 4 degrees awarded.

Degree requirements: For doctorate, dissertation required, foreign language not required. *Average time to degree:* Doctorate–5 years full-time.

Entrance requirements: For doctorate, GRE General Test, GRE Subject Test, TOEFL. *Application deadline:* For fall admission, 1/15 (priority date). Applications are processed on a rolling basis. *Application fee:* $45.

Expenses: Tuition: Full-time $19,325.

Financial aid: In 1999–00, 2 fellowships, 23 research assistantships with full tuition reimbursements (averaging $18,805 per year) were awarded; grants, scholarships, traineeships, and tuition waivers (full) also available. Financial aid application deadline: 2/1.

Faculty research: Reproduction and hormone action, control of gene expression, cell-matrix and cell-cell interactions, growth control and tumorigenesis, cytoskeleton and contractile proteins.
Dr. John J. Castellot, Director, 617-636-0303, *Fax:* 617-636-6536, *E-mail:* jcastellot@infonet.tufts.edu.

Application contact: Secretary, 617-636-6767, *Fax:* 617-636-0375, *E-mail:* sackler-school@tufts.edu.

Find an in-depth description at www.petersons.com/graduate.

■ TULANE UNIVERSITY

Graduate School, Department of Cell and Molecular Biology, New Orleans, LA 70118-5669

AWARDS Biology (MS, PhD). Part-time programs available.

Students: 10 full-time (4 women); includes 1 minority (African American), 3 international. *3 applicants, 67% accepted.* In 1999, 5 master's, 11 doctorates awarded (100% found work related to degree). Terminal master's awarded for partial completion of doctoral program.
Degree requirements: For master's, foreign language and thesis not required; for doctorate, dissertation required, foreign language not required. *Average time to degree:* Master's–2 years full-time; doctorate–5 years full-time.
Entrance requirements: For master's, GRE General Test, TSE, minimum B average in undergraduate course work; for doctorate, GRE General Test, TSE. *Application deadline:* For fall admission, 2/1. *Application fee:* $45.
Expenses: Tuition: Full-time $23,500. Tuition and fees vary according to program.
Financial aid: Fellowships, research assistantships, teaching assistantships available. Financial aid application deadline: 2/1.
Dr. Leonard Thein, Chair, 504-865-5546, *Fax:* 504-865-0785.
Application contact: Dr. Jeffrey Tasker, Graduate Director, 504-865-5546, *Fax:* 504-865-6785, *E-mail:* tasker@mailhost.tcs.tulane.edu.

■ TULANE UNIVERSITY

School of Medicine and Graduate School, Graduate Programs in Medicine, Interdisciplinary Graduate Program in Molecular and Cellular Biology, New Orleans, LA 70118-5669

AWARDS PhD, MD/PhD. PhD offered through the Graduate School.

Students: 41 full-time (18 women); includes 6 minority (3 African Americans, 3 Asian Americans or Pacific Islanders), 9 international. *117 applicants, 15% accepted.* In 1999, 10 doctorates awarded.
Degree requirements: For doctorate, dissertation required.
Entrance requirements: For doctorate, GRE General Test, GRE Subject Test, TOEFL. *Application deadline:* For fall admission, 2/1. *Application fee:* $45.
Expenses: Tuition: Full-time $23,030.
Financial aid: Fellowships, research assistantships, teaching assistantships available. Financial aid application deadline: 2/1.

Dr. Barbara Beckman, Director, 504-588-5226, *Fax:* 504-584-3779, *E-mail:* mcbpgm@mailhost.tcs.tulane.edu.

Find an in-depth description at www.petersons.com/graduate.

■ UNIFORMED SERVICES UNIVERSITY OF THE HEALTH SCIENCES

School of Medicine, Division of Basic Medical Sciences, Department of Anatomy and Cell Biology, Bethesda, MD 20814-4799

AWARDS Cell biology, developmental biology, and neurobiology (PhD).

Faculty: 20 full-time (9 women), 12 part-time/adjunct (2 women).
Students: 8 full-time (1 woman); includes 2 minority (both Asian Americans or Pacific Islanders). Average age 28. *3 applicants, 33% accepted.* In 1999, 1 degree awarded (100% entered university research/teaching).
Degree requirements: For doctorate, dissertation, qualifying exam required, foreign language not required. *Average time to degree:* Doctorate–5 years full-time.
Entrance requirements: For doctorate, GRE General Test, GRE Subject Test, minimum GPA of 3.0, U.S. citizenship. *Application deadline:* For fall admission, 1/15 (priority date). Applications are processed on a rolling basis. *Application fee:* $0.
Financial aid: In 1999–00, 3 fellowships with full tuition reimbursements (averaging $15,000 per year) were awarded; tuition waivers (full) also available.
Faculty research: Neuroanatomy and neurocytology, molecular biology.
Dr. Harvey Pollard, Chair, 301-295-3200, *E-mail:* hpollard@usuhs.mil.
Application contact: Janet M. Anastasi, Graduate Program Coordinator, 301-295-9474, *Fax:* 301-295-6772, *E-mail:* janastasi@usuhs.mil.

■ UNIFORMED SERVICES UNIVERSITY OF THE HEALTH SCIENCES

School of Medicine, Division of Basic Medical Sciences, Program in Molecular and Cell Biology, Bethesda, MD 20814-4799

AWARDS PhD.

Faculty: 25 part-time/adjunct (7 women).
Students: 13 full-time (7 women), 1 part-time; includes 3 minority (1 African American, 2 Asian Americans or Pacific Islanders). Average age 26. *19 applicants, 21% accepted.*
Degree requirements: For doctorate, dissertation, qualifying exam required.

Entrance requirements: For doctorate, GRE General Test, GRE Subject Test, minimum GPA of 3.0, U.S. citizenship. *Application deadline:* For fall admission, 1/15 (priority date). Applications are processed on a rolling basis. *Application fee:* $0.
Financial aid: In 1999–00, 8 fellowships with full tuition reimbursements (averaging $15,000 per year) were awarded; career-related internships or fieldwork and tuition waivers (full) also available.
Faculty research: Immunology, biochemistry.
Dr. William C. Gause, Director, 301-295-1958, *Fax:* 301-295-3220.
Application contact: Janet M. Anastasi, Graduate Program Coordinator, 301-295-9474, *Fax:* 301-295-6772, *E-mail:* janastasi@usuhs.mil.

Find an in-depth description at www.petersons.com/graduate.

■ THE UNIVERSITY OF ALABAMA AT BIRMINGHAM

Graduate School and School of Medicine and School of Dentistry, Graduate Programs in Joint Health Sciences, Department of Cell Biology, Birmingham, AL 35294

AWARDS Anatomy (PhD). The department participates in the Cellular and Molecular Biology Graduate Program.

Students: 49 full-time (19 women), 1 part-time; includes 2 minority (1 Asian American or Pacific Islander, 1 Hispanic American), 28 international. *6 applicants, 100% accepted.* In 1999, 10 degrees awarded.
Degree requirements: For doctorate, dissertation, qualifying exam required.
Entrance requirements: For doctorate, GRE General Test, interview. *Application deadline:* Applications are processed on a rolling basis. Electronic applications accepted.
Expenses: Tuition, state resident: part-time $104 per semester hour. Tuition, nonresident: part-time $208 per semester hour. Required fees: $17 per semester hour. $57 per quarter. Tuition and fees vary according to program.
Financial aid: In 1999–00, 4 fellowships were awarded.
Faculty research: Neuroscience, immunology.
Dr. Richard B. Marchase, Chair, 205-934-9672, *Fax:* 205-934-0950, *E-mail:* marchase@uab.edu.
Application contact: Injformation Contact, 205-975-7145, *Fax:* 205-975-6748.

Find an in-depth description at www.petersons.com/graduate.

■ THE UNIVERSITY OF ALABAMA AT BIRMINGHAM

Graduate School, School of Natural Sciences and Mathematics, Department of Biology, Birmingham, AL 35294

AWARDS Comparative and cellular biology (PhD); comparative and cellular physiology (MS); marine science (MS, PhD); microbial ecology and physiology (MS, PhD); reproduction and development (MS, PhD).

Students: 34 full-time (19 women), 6 international. *61 applicants, 38% accepted.* In 1999, 1 master's, 3 doctorates awarded. Terminal master's awarded for partial completion of doctoral program.
Degree requirements: For master's and doctorate, thesis/dissertation required.
Entrance requirements: For master's and doctorate, GRE General Test, TOEFL, previous course work in biology, calculus, organic chemistry, physics. *Application deadline:* Applications are processed on a rolling basis. *Application fee:* $35 ($60 for international students). Electronic applications accepted.
Expenses: Tuition, state resident: part-time $104 per semester hour. Tuition, nonresident: part-time $208 per semester hour. Required fees: $17 per semester hour; $57 per quarter. Tuition and fees vary according to program.
Financial aid: In 1999–00, 22 students received aid, including 3 fellowships with full tuition reimbursements available (averaging $14,000 per year), 19 teaching assistantships with full tuition reimbursements available (averaging $14,000 per year); research assistantships, career-related internships or fieldwork, Federal Work-Study, institutionally sponsored loans, and tuition waivers (full) also available. Aid available to part-time students.
Faculty research: Invertebrate physiology, marine biology, environmental biology. Dr. Daniel D. Jones, Chairman, 205-934-4290, *Fax:* 205-975-6097, *E-mail:* ddjones@uab.edu.

Find an in-depth description at www.petersons.com/graduate.

■ THE UNIVERSITY OF ARIZONA

College of Medicine, Graduate Programs in Medicine, Department of Cell Biology and Anatomy, Tucson, AZ 85721

AWARDS PhD.

Degree requirements: For doctorate, dissertation required, foreign language not required.
Entrance requirements: For doctorate, GRE General Test, GRE Subject Test (optional).

Expenses: Tuition, nonresident: full-time $4,814; part-time $274 per unit. Required fees: $1,094; $115 per unit.
Faculty research: Heart development, neural development, cellular toxicology and microcirculation; membrane traffic and cytoskeleton; cell-surface receptors.

Find an in-depth description at www.petersons.com/graduate.

■ THE UNIVERSITY OF ARIZONA

Graduate College, College of Science, Department of Molecular and Cellular Biology, Tucson, AZ 85721

AWARDS MS, PhD. Terminal master's awarded for partial completion of doctoral program.

Degree requirements: For master's and doctorate, thesis/dissertation required, foreign language not required.
Entrance requirements: For master's, GRE General Test, GRE Subject Test, TOEFL; for doctorate, GRE General Test, GRE Subject Test, TOEFL, undergraduate research experience.
Expenses: Tuition, nonresident: full-time $4,814; part-time $274 per unit. Required fees: $1,094; $115 per unit. Tuition and fees vary according to course load and program.
Faculty research: Plant molecular biology, cellular and molecular aspects of development, genetics of bacteria and lower eukaryotes.

Find an in-depth description at www.petersons.com/graduate.

■ UNIVERSITY OF CALIFORNIA, BERKELEY

Graduate Division, College of Letters and Science, Department of Molecular and Cell Biology, Berkeley, CA 94720-1500

AWARDS PhD.

Degree requirements: For doctorate, dissertation, qualifying exam, 2 semesters of teaching required, foreign language not required.
Entrance requirements: For doctorate, GRE General Test, GRE Subject Test, minimum GPA of 3.0. Electronic applications accepted.
Expenses: Tuition, nonresident: full-time $9,804. Required fees: $4,268. Tuition and fees vary according to program.

Find an in-depth description at www.petersons.com/graduate.

■ UNIVERSITY OF CALIFORNIA, DAVIS

Graduate Studies, Programs in the Biological Sciences, Program in Cell and Developmental Biology, Davis, CA 95616

AWARDS PhD.

Faculty: 36.
Students: 25 full-time (16 women); includes 7 minority (2 Asian Americans or Pacific Islanders, 3 Hispanic Americans, 2 Native Americans), 7 international. Average age 29. *30 applicants, 53% accepted.* In 1999, 1 degree awarded.
Degree requirements: For doctorate, dissertation required.
Entrance requirements: For doctorate, GRE General Test, GRE Subject Test. *Application deadline:* For fall admission, 1/15. *Application fee:* $40. Electronic applications accepted.
Expenses: Tuition, nonresident: full-time $9,804. Tuition and fees vary according to program and student level.
Financial aid: In 1999–00, 23 students received aid, including 5 fellowships with full and partial tuition reimbursements available, 13 research assistantships with full and partial tuition reimbursements available, 3 teaching assistantships with partial tuition reimbursements available; Federal Work-Study, grants, institutionally sponsored loans, scholarships, and tuition waivers (full and partial) also available. Financial aid application deadline: 1/15; financial aid applicants required to submit FAFSA.
Faculty research: Molecular basis of cell function and development. Paul FitzGerald, Chair, Cell Biology and Human Anatomy, 530-752-7130, *E-mail:* pfitzgerald@ucdavis.edu.
Application contact: Dawne Shell, Graduate Program Assistant, 530-752-9091, *Fax:* 530-752-8822, *E-mail:* drshell@ucdavis.edu.

Find an in-depth description at www.petersons.com/graduate.

■ UNIVERSITY OF CALIFORNIA, IRVINE

Office of Research and Graduate Studies, School of Biological Sciences, Department of Developmental and Cell Biology, Irvine, CA 92697

AWARDS Biological sciences (MS, PhD). Students apply through the Graduate Program in Molecular Biology, Genetics, and Biochemistry.

Faculty: 16 full-time (3 women).
Students: 7 full-time (4 women), 3 part-time; includes 2 minority (1 Asian American or Pacific Islander, 1 Hispanic

University of California, Irvine (continued) American), 2 international. *292 applicants, 25% accepted.* In 1999, 1 master's awarded. Terminal master's awarded for partial completion of doctoral program.
Degree requirements: For master's, one foreign language required; for doctorate, dissertation required.
Entrance requirements: For master's, GRE General Test, GRE Subject Test, minimum GPA of 3.0; for doctorate, GRE General Test, GRE Subject Test. *Application deadline:* For fall admission, 2/1 (priority date). *Application fee:* $40. Electronic applications accepted.
Expenses: Tuition, nonresident: full-time $10,244; part-time $1,720 per quarter. Required fees: $5,252; $1,300 per quarter. Tuition and fees vary according to course load and program.
Financial aid: Fellowships, research assistantships, teaching assistantships, institutionally sponsored loans and tuition waivers (full and partial) available. Financial aid application deadline: 3/2; financial aid applicants required to submit FAFSA.
Faculty research: Genetics and development, oncogene signaling pathways, gene regulation, tissue regeneration and molecular genetics.
J. Lawrence Marsh, Chair, 949-824-6677, *Fax:* 949-824-4709, *E-mail:* jlmarsh@uci.edu.
Application contact: Administrator, 949-824-8145, *Fax:* 949-824-7407, *E-mail:* gp-mbgb@uci.edu.

■ UNIVERSITY OF CALIFORNIA, LOS ANGELES

Graduate Division, College of Letters and Science and School of Medicine, UCLA ACCESS to Programs in the Molecular and Cellular Life Sciences, Los Angeles, CA 90095

AWARDS PhD.

Students: 60 full-time (36 women); includes 22 minority (1 African American, 13 Asian Americans or Pacific Islanders, 8 Hispanic Americans), 8 international. *375 applicants, 26% accepted.*
Degree requirements: For doctorate, dissertation, oral and written qualifying exams required.
Entrance requirements: For doctorate, GRE General Test, minimum undergraduate GPA of 3.0. *Application deadline:* For fall admission, 12/15. *Application fee:* $40. Electronic applications accepted.
Expenses: Tuition, nonresident: full-time $9,804. Required fees: $4,405. Full-time tuition and fees vary according to program and student level.
Financial aid: Fellowships, research assistantships, teaching assistantships,

scholarships available. Financial aid application deadline: 3/1.
Faculty research: Molecular, cellular, and developmental biology; immunology; microbiology; integrative biology.
David Meyer, Director, 800-284-8252.
Application contact: UCLA Access Coordinator, 800-284-8252, *Fax:* 310-206-5280, *E-mail:* uclaaccess@ibes.medsch.ucla.edu.
Find an in-depth description at www.petersons.com/graduate.

■ UNIVERSITY OF CALIFORNIA, LOS ANGELES

School of Medicine and Graduate Division, Graduate Programs in Medicine, Department of Molecular, Cell and Developmental Biology, Los Angeles, CA 90095

AWARDS MA, PhD.

Students: 35 full-time (15 women); includes 15 minority (13 Asian Americans or Pacific Islanders, 2 Hispanic Americans), 5 international. *66 applicants, 18% accepted.*
Degree requirements: For doctorate, dissertation, qualifying exams required, foreign language not required.
Entrance requirements: For doctorate, GRE General Test, GRE Subject Test, TOEFL. *Application fee:* $40.
Expenses: Tuition, nonresident: full-time $9,804. Required fees: $4,405.
Financial aid: Fellowships, research assistantships, teaching assistantships, scholarships available. Financial aid application deadline: 3/1.
Dr. Lutz Birnbaumer, Chair, 310-825-7109.
Application contact: UCLA Access Coordinator, 800-284-8252, *Fax:* 310-206-5280, *E-mail:* uclaaccess@ibes.medsch.ucla.edu.
Find an in-depth description at www.petersons.com/graduate.

■ UNIVERSITY OF CALIFORNIA, LOS ANGELES

School of Medicine and Graduate Division, Graduate Programs in Medicine, Department of Neurobiology, Los Angeles, CA 90095

AWARDS Anatomy and cell biology (PhD).

Students: 16 full-time (8 women); includes 7 minority (1 African American, 2 Asian Americans or Pacific Islanders, 3 Hispanic Americans, 1 Native American), 1 international. *8 applicants, 25% accepted.*
Degree requirements: For doctorate, dissertation, oral and written qualifying exams required, foreign language not required.
Entrance requirements: For doctorate, GRE General Test, GRE Subject Test,

bachelor's degree in physical or biological science. *Application fee:* $40.
Expenses: Tuition, nonresident: full-time $9,804. Required fees: $4,405.
Financial aid: In 1999–00, 6 fellowships, 8 research assistantships, 1 teaching assistantship were awarded; Federal Work-Study, institutionally sponsored loans, scholarships, and tuition waivers (full and partial) also available. Financial aid application deadline: 3/1.
Faculty research: Neuroendocrinology, neurophysiology.
Dr. Jack Feldman, Chair, 310-825-9558.
Application contact: UCLA Access Coordinator, 800-284-8252, *Fax:* 310-206-5280, *E-mail:* uclaaccess@ibes.medsch.ucla.edu.

■ UNIVERSITY OF CALIFORNIA, RIVERSIDE

Graduate Division, College of Natural and Agricultural Sciences, Program in Cell, Molecular, and Developmental Biology, Riverside, CA 92521-0102

AWARDS MS, PhD.

Faculty: 49 full-time (15 women).
Students: 3 full-time (2 women); includes 1 minority (Asian American or Pacific Islander).
Degree requirements: For master's, thesis, oral defense of thesis required; for doctorate, dissertation, oral defense of thesis, qualifying exams, 2 quarters of teaching experience required. *Average time to degree:* Master's–2 years full-time; doctorate–5 years full-time.
Entrance requirements: For master's and doctorate, GRE General Test, TOEFL, minimum GPA of 3.2. *Application deadline:* For fall admission, 5/1; for winter admission, 9/1; for spring admission, 12/1. *Application fee:* $40.
Expenses: Tuition, nonresident: full-time $9,804. Required fees: $4,758. Full-time tuition and fees vary according to program.
Financial aid: Fellowships, research assistantships, teaching assistantships available. Financial aid application deadline: 1/5.
Dr. Prudence Talbot, Graduate Advisor, 909-787-5913, *Fax:* 909-787-4286, *E-mail:* talbot@pop.ucr.edu.
Application contact: Helene Serewis, Graduate Student Affairs Officer, 800-735-0717, *Fax:* 909-787-5517, *E-mail:* biopgrad@pep.ycr.edu.

■ UNIVERSITY OF CALIFORNIA, SAN DIEGO

Graduate Studies and Research, Department of Biology, Program in Cell and Developmental Biology, La Jolla, CA 92093-0348

AWARDS PhD. Offered in association with the Salk Institute.

Degree requirements: For doctorate, dissertation required.
Entrance requirements: For doctorate, GRE General Test, pre-application beginning in September. *Application deadline:* For fall admission, 1/7. *Application fee:* $40.
Expenses: Tuition, nonresident: full-time $14,691. Required fees: $4,697. Full-time tuition and fees vary according to program.
Financial aid: Tuition waivers (full) and stipends available.
Application contact: 858-534-3835.
Find an in-depth description at www.petersons.com/graduate.

■ UNIVERSITY OF CALIFORNIA, SAN DIEGO

Graduate Studies and Research, Department of Biology, Program in Molecular and Cellular Biology, La Jolla, CA 92093

AWARDS PhD.

Degree requirements: For doctorate, dissertation required.
Entrance requirements: For doctorate, GRE General Test, GRE Subject Test (optional), pre-application beginning in September. *Application deadline:* For fall admission, 1/8. *Application fee:* $40.
Expenses: Tuition, nonresident: full-time $14,691. Required fees: $4,697. Full-time tuition and fees vary according to program.
Application contact: Biology Graduate Admissions Committee, 858-534-3835.

■ UNIVERSITY OF CALIFORNIA, SAN DIEGO

School of Medicine and Graduate Studies and Research, Graduate Studies in Biomedical Sciences, Cell and Molecular Biology Program, La Jolla, CA 92093-0685

AWARDS PhD.

Faculty: 106.
Students: 219.
Degree requirements: For doctorate, dissertation, qualifying exam required, foreign language not required.
Entrance requirements: For doctorate, GRE General Test, TOEFL. *Application deadline:* For fall admission, 1/5. *Application fee:* $40.

Expenses: Program pays tuition, fees, health insurance, and stipend for all students in good standing.
Faculty research: Molecular and cellular pharmacology, cell and organ physiology. Kim Barrett, Chair, 868-543-3726.
Application contact: Gina Butcher, Graduate Program Representative, 858-534-3982.
Find an in-depth description at www.petersons.com/graduate.

■ UNIVERSITY OF CALIFORNIA, SAN DIEGO

School of Medicine and Graduate Studies and Research, Graduate Studies in Biomedical Sciences, Regulatory Biology Program, La Jolla, CA 92093

AWARDS PhD.

Faculty: 71.
Degree requirements: For doctorate, dissertation, 2 qualifying exams required, foreign language not required.
Entrance requirements: For doctorate, GRE General Test, GRE Subject Test, TOEFL. *Application deadline:* For fall admission, 1/5. *Application fee:* $40.
Expenses: Program pays tuition, fees, health insurance, and stipend for all students in good standing.
Financial aid: Tuition waivers (full) available.
Faculty research: Eukaryotic regulatory and molecular biology, molecular and cellular pharmacology, cell and organ physiology.
M. Geoffrey Rosenfeld, Director.
Application contact: Gina Butcher, Graduate Program Representative, 858-534-3982.

■ UNIVERSITY OF CALIFORNIA, SAN FRANCISCO

Graduate Division and School of Medicine, Department of Biochemistry and Biophysics, Program in Cell Biology, San Francisco, CA 94143

AWARDS PhD, MD/PhD.

Students: In 1999, 2 degrees awarded.
Degree requirements: For doctorate, dissertation required, foreign language not required.
Entrance requirements: For doctorate, GRE General Test, GRE Subject Test, TOEFL. *Application deadline:* For fall admission, 1/5. *Application fee:* $40.
Expenses: All students guaranteed an annual stipend of $17,600.
Financial aid: Application deadline: 2/1.
Dr. Peter Walter, Director, 415-476-5017.

Application contact: Sue Adams, Program Assistant, 415-476-1495.
Find an in-depth description at www.petersons.com/graduate.

■ UNIVERSITY OF CALIFORNIA, SANTA BARBARA

Graduate Division, College of Letters and Sciences, Division of Mathematics, Life, and Physical Sciences, Department of Molecular, Cellular, and Developmental Biology, Santa Barbara, CA 93106

AWARDS MA, PhD.

Faculty: 25 full-time (6 women), 1 part-time/adjunct (0 women).
Students: 59 full-time (34 women); includes 7 minority (all Asian Americans or Pacific Islanders), 1 international. *91 applicants, 34% accepted.* In 1999, 7 master's, 6 doctorates awarded.
Degree requirements: For doctorate, dissertation required. *Average time to degree:* Master's–2 years full-time; doctorate–6 years full-time.
Entrance requirements: For master's and doctorate, GRE General Test, GRE Subject Test, TOEFL. *Application deadline:* For fall admission, 1/1 (priority date). *Application fee:* $40. Electronic applications accepted.
Expenses: Tuition, state resident: full-time $14,637. Tuition, nonresident: full-time $24,441.
Financial aid: Fellowships, research assistantships, teaching assistantships available. Financial aid application deadline: 1/15; financial aid applicants required to submit FAFSA.
Faculty research: Regulation of pap operon, recombinant composite protein, interferons, microtubules, embryonic development, bacterial pathogens.
Stu Feinstein, Chair, 805-893-2659, *Fax:* 805-893-4724, *E-mail:* feinstei@lifesci.ucsb.edu.
Application contact: Shari Profant, Graduate Program Assistant, 805-893-8499, *Fax:* 805-893-4724, *E-mail:* profant@lifesci.ucsb.edu.

■ UNIVERSITY OF CALIFORNIA, SANTA CRUZ

Graduate Division, Division of Natural Sciences, Department of Biology, Santa Cruz, CA 95064

AWARDS Molecular, cellular, and developmental biology (PhD), including biology.

Faculty: 38 full-time.
Students: 90 full-time (47 women); includes 18 minority (1 African American, 13 Asian Americans or Pacific Islanders, 4 Hispanic Americans), 5 international. *196*

University of California, Santa Cruz (continued)
applicants, 20% accepted. In 1999, 11 doctorates awarded.

Degree requirements: For doctorate, one foreign language (computer language can substitute), dissertation, oral and written qualifying exams required.
Entrance requirements: For doctorate, GRE General Test, GRE Subject Test. *Application deadline:* For fall admission, 1/1. *Application fee:* $40.
Expenses: Tuition, state resident: full-time $4,925. Tuition, nonresident: full-time $14,919.
Financial aid: Fellowships, research assistantships, teaching assistantships, career-related internships or fieldwork, Federal Work-Study, and institutionally sponsored loans available. Financial aid application deadline: 1/1.
Faculty research: Neurophysiology and psychophysiology; plant sciences; population, environmental, and evolutionary biology.
Barry Bowman, Chairperson, 831-459-2385.
Application contact: Graduate Admissions, 831-459-2301.

■ UNIVERSITY OF CHICAGO

Division of the Biological Sciences, Molecular Biosciences: Biochemistry, Genetics, Cell and Developmental Biology, Department of Molecular Genetics and Cell Biology, Chicago, IL 60637-1513
AWARDS PhD.

Faculty: 36 full-time (12 women).
Students: 41 full-time (17 women); includes 9 minority (1 African American, 5 Asian Americans or Pacific Islanders, 3 Hispanic Americans), 4 international. Average age 25. In 1999, 7 doctorates awarded.
Degree requirements: For doctorate, dissertation required. *Average time to degree:* Doctorate–5.5 years full-time.
Entrance requirements: For doctorate, GRE General Test, TOEFL. *Application deadline:* For fall admission, 1/5 (priority date). *Application fee:* $55.
Expenses: Tuition: Full-time $24,804; part-time $3,422 per course. Required fees: $390. Tuition and fees vary according to program.
Financial aid: In 1999–00, 36 students received aid; fellowships, research assistantships, institutionally sponsored loans available. Financial aid application deadline: 6/1.
Faculty research: Gene expression, chromosome structure, animal viruses, plant molecular genetics. *Total annual research expenditures:* $8 million.
Dr. Anthony P. Mahowald, Chairman, 773-702-1620.

Application contact: Kristine Gaston, Graduate Administrative Director, 773-702-8037, *Fax:* 773-702-3172, *E-mail:* kristine@cummings.uchicago.edu.

■ UNIVERSITY OF CINCINNATI

Division of Research and Advanced Studies, College of Medicine, Graduate Programs in Medicine, Department of Cell Biology, Neurobiology and Anatomy, Cincinnati, OH 45221-0091
AWARDS Anatomy (PhD); cell biology (PhD); neurobiology (PhD).

Faculty: 14 full-time.
Students: 31 full-time (17 women), 2 part-time (both women); includes 3 minority (1 African American, 2 Asian Americans or Pacific Islanders), 12 international. 77 *applicants, 18% accepted.* In 1999, 2 degrees awarded.
Degree requirements: For doctorate, dissertation, qualifying exam required. *Average time to degree:* Doctorate–5.6 years full-time.
Entrance requirements: For doctorate, GRE General Test, TOEFL. *Application deadline:* For fall admission, 2/1 (priority date). Applications are processed on a rolling basis. *Application fee:* $30.
Expenses: Tuition, state resident: full-time $5,139; part-time $196 per credit hour. Tuition, nonresident: full-time $10,326; part-time $369 per credit hour. Required fees: $561; $187 per quarter.
Financial aid: Tuition waivers (full) and unspecified assistantships available. Financial aid application deadline: 5/1.
Faculty research: Cell structure, molecular genetics. *Total annual research expenditures:* $5.3 million.
Dr. Peter Stambrook, Head, 513-558-5685, *Fax:* 513-556-4454, *E-mail:* peter.stambrook@uc.edu.
Application contact: Robert Brackenbury, Graduate Program Director, 513-558-6080, *Fax:* 513-556-4454, *E-mail:* robert.brackenbury@uc.edu.

■ UNIVERSITY OF CINCINNATI

Division of Research and Advanced Studies, College of Medicine, Graduate Programs in Medicine, Graduate Program in Cell and Molecular Biology, Cincinnati, OH 45221-0091
AWARDS PhD.

Degree requirements: For doctorate, dissertation, qualifying exam required.
Entrance requirements: For doctorate, GRE General Test, TOEFL, TWE. *Application deadline:* For fall admission, 2/1 (priority date). Applications are processed on a rolling basis. *Application fee:* $30.

Expenses: Tuition, state resident: full-time $5,139; part-time $196 per credit hour. Tuition, nonresident: full-time $10,326; part-time $369 per credit hour. Required fees: $561; $187 per quarter.
Financial aid: Unspecified assistantships available. Financial aid application deadline: 5/1.

Find an in-depth description at www.petersons.com/graduate.

■ UNIVERSITY OF COLORADO AT BOULDER

Graduate School, College of Arts and Sciences, Department of Molecular, Cellular, and Developmental Biology, Boulder, CO 80309
AWARDS Cellular structure and function (MA, PhD); developmental biology (MA, PhD); molecular biology (MA, PhD).

Faculty: 25 full-time (7 women).
Students: 54 full-time (28 women), 1 (woman) part-time; includes 6 minority (1 African American, 3 Asian Americans or Pacific Islanders, 2 Hispanic Americans), 4 international. Average age 27. 174 *applicants, 9% accepted.* In 1999, 2 master's, 15 doctorates awarded. Terminal master's awarded for partial completion of doctoral program.
Degree requirements: For master's, thesis or alternative, comprehensive exam required, foreign language not required; for doctorate, dissertation, comprehensive exam required, foreign language not required. *Average time to degree:* Doctorate–6 years full-time.
Entrance requirements: For master's, GRE General Test, GRE Subject Test, minimum undergraduate GPA of 2.75; for doctorate, GRE General Test, GRE Subject Test. *Application deadline:* For fall admission, 1/1. *Application fee:* $40 ($60 for international students).
Expenses: Tuition, state resident: part-time $181 per credit hour. Tuition, nonresident: part-time $542 per credit hour. Required fees: $99 per term. Tuition and fees vary according to course load and program.
Financial aid: Fellowships, research assistantships, teaching assistantships, tuition waivers (full) available. Financial aid application deadline: 3/1. *Total annual research expenditures:* $9.7 million.
Leslie Leinward, Chair, 303-492-7606, *Fax:* 303-492-7744, *E-mail:* leinward@stripe.colorado.edu.
Application contact: Karen Brown, Student Affairs Office, 303-492-7230, *Fax:* 303-492-7744, *E-mail:* mcdbgradinfo@beagle.colorado.edu.

Find an in-depth description at www.petersons.com/graduate.

■ UNIVERSITY OF COLORADO HEALTH SCIENCES CENTER

Graduate School, Programs in Biological and Medical Sciences, Program in Cell and Developmental Biology, Denver, CO 80262

AWARDS PhD.

Degree requirements: For doctorate, dissertation required, foreign language not required.

Entrance requirements: For doctorate, GRE, minimum GPA of 2.75.

Expenses: Tuition, state resident: full-time $1,512; part-time $56 per hour. Tuition, nonresident: full-time $7,209; part-time $267 per hour. Full-time tuition and fees vary according to course load and program.

Find an in-depth description at www.petersons.com/graduate.

■ UNIVERSITY OF CONNECTICUT

Graduate School, College of Liberal Arts and Sciences, Biological Sciences Group, Storrs, CT 06269

AWARDS Ecology and evolutionary biology (MS, PhD), including botany, ecology, entomology, systematics, zoology; molecular and cell biology (MS, PhD), including biochemistry, biophysics, biotechnology (MS), cell and developmental biology, genetics, microbiology, plant molecular and cell biology; physiology and neurobiology (MS, PhD), including neurobiology, physiology.

Degree requirements: For doctorate, dissertation required.

Entrance requirements: For master's and doctorate, GRE General Test, GRE Subject Test, TOEFL.

Expenses: Tuition, state resident: full-time $5,118. Tuition, nonresident: full-time $13,298. Required fees: $1,022.

■ UNIVERSITY OF CONNECTICUT

Graduate School, College of Liberal Arts and Sciences, Biological Sciences Group, Department of Molecular and Cell Biology, Field of Cell and Developmental Biology, Storrs, CT 06269

AWARDS MS, PhD.

Degree requirements: For doctorate, dissertation required.

Entrance requirements: For master's and doctorate, GRE General Test, GRE Subject Test, TOEFL.

Expenses: Tuition, state resident: full-time $5,118. Tuition, nonresident: full-time $13,298. Required fees: $1,022.

Find an in-depth description at www.petersons.com/graduate.

■ UNIVERSITY OF CONNECTICUT HEALTH CENTER

Graduate School, Programs in Biomedical Sciences, Program in Cell Biology, Farmington, CT 06030

AWARDS PhD, DMD/PhD, MD/PhD.

Faculty: 36.
Students: 11 full-time (6 women), 5 international.

Degree requirements: For doctorate, one foreign language (computer language can substitute), dissertation required.

Entrance requirements: For doctorate, GRE General Test, TOEFL. *Application deadline:* For fall admission, 2/1 (priority date); for spring admission, 10/1. *Application fee:* $40 ($45 for international students).

Expenses: Tuition, state resident: full-time $5,272; part-time $293 per credit. Tuition, nonresident: full-time $13,696; part-time $761 per credit. Required fees: $320; $198 per semester. One-time fee: $50 full-time. Full-time tuition and fees vary according to course load, program and reciprocity agreements.

Financial aid: In 1999–00, research assistantships (averaging $17,000 per year); fellowships. Financial aid application deadline: 3/1.

Faculty research: Membrane structure and function, cell regulation, electrophysiology, gene expression and regulation, intracellular messengers.
Dr. Tim Hla, Director, 860-679-4128.

Application contact: Marizta Barta, Information Contact, 860-679-4306, *Fax:* 860-679-1282, *E-mail:* barta@adp.uchc.edu.

Find an in-depth description at www.petersons.com/graduate.

■ UNIVERSITY OF DELAWARE

College of Arts and Science, Department of Biological Sciences, Newark, DE 19716

AWARDS Biotechnology (MS, PhD); cell and extracellular matrix biology (MS, PhD); cell and systems physiology (MS, PhD); ecology and evolution (MS, PhD); microbiology (MS, PhD); molecular biology and genetics (MS, PhD); plant biology (MS, PhD).

Faculty: 37 full-time (10 women).
Students: 22 full-time (11 women), 1 part-time; includes 2 minority (both African Americans), 9 international. Average age 25. 37 *applicants,* 27% *accepted.* In 2000, 9 doctorates awarded.

Degree requirements: For master's and doctorate, thesis/dissertation required, foreign language not required. *Average time to degree:* Master's–2.5 years full-time; doctorate–6 years full-time.

Entrance requirements: For master's and doctorate, GRE General Test, GRE Subject Test (advanced biology). *Application deadline:* For fall admission, 6/15. Applications are processed on a rolling basis. *Application fee:* $50. Electronic applications accepted.

Expenses: Tuition, state resident: full-time $4,380; part-time $243 per credit. Tuition, nonresident: full-time $12,750; part-time $708 per credit. Required fees: $15 per term. Tuition and fees vary according to program.

Financial aid: In 2000–01, 18 students received aid, including 2 fellowships with full tuition reimbursements available (averaging $18,000 per year), 4 research assistantships with full tuition reimbursements available (averaging $18,000 per year), 11 teaching assistantships with full tuition reimbursements available (averaging $18,000 per year); tuition waivers (partial) also available. Financial aid application deadline: 6/15.

Faculty research: Cell interactions, molecular mechanisms, microorganisms, embryo implantation. *Total annual research expenditures:* $1.8 million.
Dr. Daniel D. Carson, Chair, 302-831-6977, *Fax:* 302-831-2281, *E-mail:* dcarson@udel.edu.

Application contact: Norman Karin, Graduate Coordinator, 302-831-1841, *Fax:* 302-831-2281, *E-mail:* ccoletta@udel.edu.

Find an in-depth description at www.petersons.com/graduate.

■ UNIVERSITY OF FLORIDA

College of Medicine and Graduate School, Interdisciplinary Program in Biomedical Sciences, Concentration in Molecular Cell Biology, Gainesville, FL 32611

AWARDS PhD.

Degree requirements: For doctorate, dissertation required, foreign language not required.

Entrance requirements: For doctorate, GRE General Test, minimum GPA of 3.0. Electronic applications accepted.

Expenses: Tuition, state resident: part-time $144 per credit hour. Tuition, nonresident: part-time $505 per credit hour. Tuition and fees vary according to course level, course load and program.

Find an in-depth description at www.petersons.com/graduate.

■ UNIVERSITY OF FLORIDA

College of Medicine and Graduate School, Interdisciplinary Program in Biomedical Sciences, Department of Anatomy and Cell Biology, Gainesville, FL 32610

AWARDS PhD.

University of Florida (continued)

Degree requirements: For doctorate, dissertation required, foreign language not required.

Entrance requirements: For doctorate, GRE General Test, TOEFL, minimum GPA of 3.0. Electronic applications accepted.

Expenses: Tuition, state resident: part-time $144 per credit hour. Tuition, nonresident: part-time $505 per credit hour. Tuition and fees vary according to course level, course load and program.

Faculty research: Structure and function of intracellular organelles, cell adhesion, differentiation.

■ **UNIVERSITY OF FLORIDA**

Graduate School, College of Agriculture and Life Sciences, Department of Microbiology and Cell Science, Gainesville, FL 32611

AWARDS Cell biology (MS, PhD); microbiology (MS, PhD); microbiology and cell science (M Ag).

Faculty: 19.
Students: 21 full-time (11 women), 2 part-time; includes 7 minority (3 African Americans, 1 Asian American or Pacific Islander, 3 Hispanic Americans), 1 international. *31 applicants, 26% accepted.* In 1999, 5 doctorates awarded.
Degree requirements: For master's, computer language required; for doctorate, dissertation required.
Entrance requirements: For master's and doctorate, GRE General Test, minimum GPA of 3.0. *Application deadline:* For fall admission, 6/1 (priority date). Applications are processed on a rolling basis. *Application fee:* $20. Electronic applications accepted.
Expenses: Tuition, state resident: part-time $144 per credit hour. Tuition, nonresident: part-time $505 per credit hour. Tuition and fees vary according to course level, course load and program.
Financial aid: In 1999–00, 20 students received aid, including 1 fellowship, 11 research assistantships, 7 teaching assistantships
Faculty research: Biomass conversion, membrane and cell wall chemistry, plant biochemistry and genetics.
Dr. Edward M. Hoffman, Chair, 352-392-1906, *Fax:* 352-392-5922.
Application contact: Dr. Francis Davis, Graduate Coordinator, 352-392-1179, *Fax:* 352-392-5922, *E-mail:* fdavis@micro.ifas.ufl.edu.

Find an in-depth description at www.petersons.com/graduate.

■ **UNIVERSITY OF GEORGIA**

Graduate School, College of Arts and Sciences, Department of Cellular Biology, Athens, GA 30602

AWARDS MS, PhD.

Degree requirements: For master's, thesis required, foreign language not required; for doctorate, one foreign language (computer language can substitute), dissertation required.
Entrance requirements: For master's and doctorate, GRE General Test. Electronic applications accepted.
Expenses: Tuition, state resident: full-time $7,516; part-time $431 per credit hour. Tuition, nonresident: full-time $12,204; part-time $793 per credit hour. Tuition and fees vary according to program.

Find an in-depth description at www.petersons.com/graduate.

■ **UNIVERSITY OF HAWAII AT MANOA**

Graduate Division, Specialization in Cell, Molecular, and Neuro Sciences, Honolulu, HI 96822

AWARDS MS, PhD. Program is interdisciplinary; degree is in specific discipline with the specialization in Cell, Molecular, and Neuro Sciences. Part-time programs available.

Faculty: 55 full-time (11 women).
Students: 60 full-time (34 women); includes 24 minority (22 Asian Americans or Pacific Islanders, 2 Hispanic Americans), 13 international. Average age 30. *72 applicants, 36% accepted.*
Degree requirements: For doctorate, dissertation required, foreign language not required.
Entrance requirements: For doctorate, GRE Subject Test (recommended). *Application deadline:* For fall admission, 3/1; for spring admission, 9/1. Applications are processed on a rolling basis. *Application fee:* $25 ($50 for international students).
Expenses: Tuition, state resident: full-time $4,032; part-time $168 per credit. Tuition, nonresident: full-time $9,960; part-time $415 per credit. Required fees: $51 per semester. Part-time tuition and fees vary according to course load and program.
Financial aid: In 1999–00, 22 research assistantships with tuition reimbursements (averaging $15,000 per year), 5 teaching assistantships with tuition reimbursements (averaging $13,000 per year) were awarded; tuition waivers (full and partial) and unspecified assistantships also available. Financial aid application deadline: 2/1.
Faculty research: Anatomy, cellular neurobiology, genetics,. *Total annual research expenditures:* $500,000.

Dr. Martin D. Rayner, Chair, 808-956-7269, *Fax:* 808-956-6984, *E-mail:* martin@pbre.hawaii.edu.

■ **UNIVERSITY OF ILLINOIS AT CHICAGO**

College of Medicine and Graduate College, Graduate Programs in Medicine, Department of Anatomy and Cell Biology, Chicago, IL 60607-7128

AWARDS MS, PhD, MD/PhD.

Faculty: 13 full-time (5 women).
Students: 10 full-time (5 women); includes 1 minority (Asian American or Pacific Islander), 7 international. Average age 29. *26 applicants, 15% accepted.* In 1999, 1 doctorate awarded.
Degree requirements: For master's and doctorate, thesis/dissertation required, foreign language not required.
Entrance requirements: For master's and doctorate, GRE General Test, TOEFL. *Application deadline:* For fall admission, 6/1; for spring admission, 11/1. *Application fee:* $40 ($50 for international students).
Expenses: Tuition, state resident: full-time $3,750. Tuition, nonresident: full-time $10,588. Tuition and fees vary according to course load.
Financial aid: In 1999–00, 7 students received aid; fellowships, research assistantships, teaching assistantships, Federal Work-Study, scholarships, traineeships, and tuition waivers (full) available. Financial aid application deadline: 3/1; financial aid applicants required to submit FAFSA.
Faculty research: Neuroanatomy, functional morphology, cytoskeleton, synapses, neural transplants.
Rochelle Cohen, Acting Head.
Application contact: Conwell Anderson, Director of Graduate Studies, 312-996-3360, *E-mail:* conwell@uic.edu.

Find an in-depth description at www.petersons.com/graduate.

■ **UNIVERSITY OF ILLINOIS AT CHICAGO**

Graduate College, College of Liberal Arts and Sciences, Department of Biological Sciences, Chicago, IL 60607-7128

AWARDS Cell and developmental biology (PhD); ecology and evolution (MS, DA, PhD); genetics and development (PhD); molecular biology (MS, PhD); neurobiology (MS, PhD); plant biology (MS, DA, PhD).

Faculty: 40 full-time (5 women).
Students: 100 full-time (47 women), 14 part-time (10 women); includes 13 minority (11 Asian Americans or Pacific Islanders, 2 Hispanic Americans), 42 international. Average age 29. *99 applicants,*

36% accepted. In 1999, 3 master's, 9 doctorates awarded.

Degree requirements: For master's, thesis required, foreign language not required; for doctorate, dissertation, preliminary exam required, foreign language not required.

Entrance requirements: For master's and doctorate, GRE General Test, GRE Subject Test, TOEFL, previous course work in physics, calculus, and organic chemistry; minimum GPA of 3.75 on a 5.0 scale. *Application deadline:* For fall admission, 6/1. Applications are processed on a rolling basis. *Application fee:* $40 ($50 for international students). Electronic applications accepted.

Expenses: Tuition, state resident: full-time $3,750; part-time $1,250 per semester. Tuition, nonresident: full-time $10,588; part-time $3,530 per semester. Required fees: $507 per semester. Tuition and fees vary according to course load and program.

Financial aid: In 1999–00, 87 students received aid; fellowships, research assistantships, teaching assistantships, career-related internships or fieldwork, Federal Work-Study, traineeships, and tuition waivers (full) available. Financial aid application deadline: 3/1; financial aid applicants required to submit FAFSA. Dr. Lon Kaufman, Head, 312-996-2213.

Application contact: Dr. Leo Miller, Director of Graduate Studies, 312-996-2220.

Find an in-depth description at www.petersons.com/graduate.

■ UNIVERSITY OF ILLINOIS AT URBANA–CHAMPAIGN

Graduate College, College of Liberal Arts and Sciences, School of Life Sciences, Department of Cell and Structural Biology, Urbana, IL 61801

AWARDS PhD.

Faculty: 9 full-time (2 women), 3 part-time/adjunct (1 woman).

Students: 51 full-time (20 women); includes 7 minority (1 African American, 5 Asian Americans or Pacific Islanders, 1 Hispanic American), 9 international. *60 applicants, 10% accepted.*

Degree requirements: For doctorate, dissertation required.

Application deadline: Applications are processed on a rolling basis. *Application fee:* $40 ($50 for international students).

Expenses: Tuition, state resident: full-time $4,616. Tuition, nonresident: full-time $11,768. Full-time tuition and fees vary according to course load.

Financial aid: Fellowships, research assistantships, teaching assistantships available. Financial aid application deadline: 2/15.

Dr. Martha U. Gillette, Acting Head, 217-333-6118, *Fax:* 217-244-1648, *E-mail:* mgillett@uiuc.edu.

Application contact: Dr. Lorie Hatfield, Director of Graduate Studies, 217-244-8116, *Fax:* 217-244-1648, *E-mail:* loriehat@uiuc.edu.

Find an in-depth description at www.petersons.com/graduate.

■ THE UNIVERSITY OF IOWA

College of Medicine and Graduate College, Graduate Programs in Medicine, Department of Anatomy and Cell Biology, Iowa City, IA 52242-1316

AWARDS PhD.

Faculty: 20 full-time (7 women).

Students: 12 full-time (5 women); includes 1 minority (Hispanic American), 8 international. Average age 28. *23 applicants, 26% accepted.* In 1999, 2 degrees awarded (100% entered university research/teaching).

Degree requirements: For doctorate, dissertation required, foreign language not required. *Average time to degree:* Doctorate–5 years full-time.

Entrance requirements: For doctorate, GRE General Test, TOEFL, minimum GPA of 3.0. *Application deadline:* For fall admission, 2/1 (priority date). Applications are processed on a rolling basis. *Application fee:* $30 ($50 for international students). Electronic applications accepted.

Expenses: Tuition, state resident: full-time $3,308. Tuition, nonresident: full-time $10,662. Tuition and fees vary according to course load and program.

Financial aid: In 1999–00, 11 students received aid, including 11 teaching assistantships with full tuition reimbursements available (averaging $16,277 per year); fellowships with full tuition reimbursements available, research assistantships with full tuition reimbursements available, Federal Work-Study, grants, institutionally sponsored loans, and scholarships also available. Financial aid application deadline: 3/1.

Faculty research: Biology of differentiation and transformation, developmental and vascular cell biology, neurobiology. *Total annual research expenditures:* $4.1 million.

Dr. Mary J. C. Hendrix, Head, 319-335-7755, *Fax:* 319-335-7198, *E-mail:* mary-hendrix@uiowa.edu.

Application contact: Tracy M. Middleton, Program Assistant, 319-335-7744, *Fax:* 319-335-7198, *E-mail:* tracy-middleton@uiowa.edu.

Find an in-depth description at www.petersons.com/graduate.

■ UNIVERSITY OF KANSAS

Graduate School, College of Liberal Arts and Sciences, Division of Biological Sciences, Department of Molecular Biosciences, Lawrence, KS 66045

AWARDS Biochemistry and biophysics (MA, MS, PhD); microbiology (MA, MS, PhD); molecular, cellular, and developmental biology (MA, MS, PhD).

Students: 13 full-time (7 women), 8 part-time (2 women); includes 1 minority (Asian American or Pacific Islander), 5 international. *57 applicants, 21% accepted.* In 2000, 11 master's, 10 doctorates awarded.

Degree requirements: For master's, thesis required, foreign language not required; for doctorate, dissertation required.

Entrance requirements: For master's and doctorate, GRE General Test, GRE Subject Test, TOEFL. *Application deadline:* For fall admission, 1/15 (priority date). *Application fee:* $25.

Expenses: Tuition, state resident: full-time $2,482; part-time $103 per credit hour. Tuition, nonresident: full-time $8,104; part-time $338 per credit hour. Required fees: $428; $31 per credit hour. Tuition and fees vary according to program.

Financial aid: In 2000–01, research assistantships (averaging $11,588 per year), teaching assistantships (averaging $11,588 per year) were awarded; fellowships. Financial aid application deadline: 3/1. Paul Kelly, Chair, 785-864-4311, *Fax:* 785-864-5294.

Application contact: Information Contact, 785-864-4311, *Fax:* 785-864-5294.

Find an in-depth description at www.petersons.com/graduate.

■ UNIVERSITY OF KANSAS

Graduate Studies Medical Center, Graduate Programs in Biomedical and Basic Sciences, Department of Anatomy and Cell Biology, Lawrence, KS 66045

AWARDS MA, PhD, MD/PhD. Part-time programs available.

Faculty: 14 full-time (3 women).

Students: 1 full-time (0 women), 9 part-time (4 women), 2 international. Average age 31. *0 applicants, 0% accepted.* In 1999, 2

University of Kansas (continued) doctorates awarded. Terminal master's awarded for partial completion of doctoral program.

Degree requirements: For master's, comprehensive oral exam, oral defense of thesis required; for doctorate, dissertation, comprehensive oral exam required.

Entrance requirements: For master's and doctorate, GRE General Test, GRE Subject Test, TOEFL, TSE. *Application deadline:* For fall admission, 1/31 (priority date). Applications are processed on a rolling basis. *Application fee:* $0. Electronic applications accepted.

Expenses: Tuition, state resident: full-time $2,482; part-time $103 per credit hour. Tuition, nonresident: full-time $8,104; part-time $338 per credit hour. Required fees: $428; $31 per credit hour. Tuition and fees vary according to program.

Financial aid: Fellowships, research assistantships, teaching assistantships, Federal Work-Study, institutionally sponsored loans, and scholarships available. Aid available to part-time students. Financial aid application deadline: 3/31; financial aid applicants required to submit FAFSA.

Faculty research: Neuropeptides, exocrine cell biology, immunobiology of pregnancy, platelet activating facor, signal transduction across nuclear membrane. *Total annual research expenditures:* $2.1 million.

Dr. Dale R. Abrahamson, Chairman, 913-588-7000, *Fax:* 913-588-2710, *E-mail:* dabrahamson@kume.edu.

Application contact: Dr. Robert C. De Lisle, Graduate Adviser, 913-588-2742, *Fax:* 913-588-2710, *E-mail:* rdelisle@ kumc.edu.

■ UNIVERSITY OF MARYLAND

Graduate School, Graduate Programs in Medicine, Program in Molecular and Cell Biology, Baltimore, MD 21201-1627

AWARDS PhD, MD/PhD.

Degree requirements: For doctorate, dissertation required.

Entrance requirements: For doctorate, GRE General Test, TOEFL, minimum GPA of 3.0.

Expenses: Tuition, state resident: part-time $261 per credit hour. Tuition, nonresident: part-time $468 per credit hour. Tuition and fees vary according to program.

Find an in-depth description at www.petersons.com/graduate.

■ UNIVERSITY OF MARYLAND, BALTIMORE COUNTY

Graduate School, Department of Biological Sciences, Program in Molecular and Cell Biology, Baltimore, MD 21250-5398

AWARDS PhD.

Students: 22 full-time (9 women), 3 part-time (2 women); includes 2 minority (1 African American, 1 Asian American or Pacific Islander), 9 international. *37 applicants, 46% accepted.* In 1999, 3 degrees awarded.

Degree requirements: For doctorate, dissertation required, foreign language not required.

Entrance requirements: For doctorate, GRE General Test, GRE Subject Test, TOEFL, minimum GPA of 3.0. *Application deadline:* For fall admission, 2/1. Applications are processed on a rolling basis. *Application fee:* $45.

Expenses: Tuition, state resident: part-time $268 per credit hour. Tuition, nonresident: part-time $470 per credit hour. Required fees: $38 per credit hour. $557 per semester.

Financial aid: Fellowships, research assistantships, teaching assistantships available.

Richard E. Wolf, Director, Graduate Program, 410-455-3669, *Fax:* 410-455-3875, *E-mail:* biograd@umbc.edu.

Find an in-depth description at www.petersons.com/graduate.

■ UNIVERSITY OF MARYLAND, COLLEGE PARK

Graduate Studies and Research, College of Life Sciences, Department of Cell Biology and Molecular Genetics, Program in Molecular and Cell Biology, College Park, MD 20742

AWARDS PhD. Part-time and evening/weekend programs available.

Students: 35 full-time (22 women), 11 part-time (8 women); includes 5 minority (2 African Americans, 1 Asian American or Pacific Islander, 2 Hispanic Americans), 22 international. *116 applicants, 22% accepted.* In 1999, 3 degrees awarded.

Degree requirements: For doctorate, dissertation, exam required.

Entrance requirements: For doctorate, GRE General Test, TOEFL. *Application deadline:* For fall admission, 2/1; for spring admission, 11/1. Applications are processed on a rolling basis. *Application fee:* $50 ($70 for international students). Electronic applications accepted.

Expenses: Tuition, state resident: part-time $272 per credit hour. Tuition, nonresident: part-time $415 per credit hour. Required fees: $632; $379 per year.

Financial aid: Fellowships, research assistantships, teaching assistantships available. Financial aid applicants required to submit FAFSA.

Faculty research: Genetics, transduction, photoregulation, parasitic interactions, cell cultures, monoclonal antibody production, oligonucleotide synthesis.

Application contact: Trudy Lindsey, Director, Graduate Admissions and Records, 301-405-4198, *Fax:* 301-314-9305, *E-mail:* grschool@deans.umd.edu.

■ UNIVERSITY OF MASSACHUSETTS AMHERST

Graduate School, Interdisciplinary Programs, Molecular and Cellular Biology Graduate Program, Amherst, MA 01003

AWARDS Biological chemistry (PhD); cell and developmental biology (PhD). Part-time programs available.

Students: 29 full-time (19 women), 33 part-time (21 women); includes 4 minority (1 African American, 3 Asian Americans or Pacific Islanders), 28 international. Average age 28. *288 applicants, 15% accepted.* In 2000, 8 degrees awarded.

Degree requirements: For doctorate, one foreign language, dissertation required.

Entrance requirements: For doctorate, GRE General Test. *Application deadline:* For fall admission, 1/15 (priority date). Applications are processed on a rolling basis. *Application fee:* $40.

Expenses: Tuition, state resident: full-time $2,640; part-time $165 per credit. Tuition, nonresident: full-time $9,756; part-time $407 per credit. Required fees: $1,221 per term. One-time fee: $110. Full-time tuition and fees vary according to course load, campus/location and reciprocity agreements.

Financial aid: In 2000–01, 17 research assistantships with full tuition reimbursements (averaging $3,796 per year), 3 teaching assistantships with full tuition reimbursements (averaging $5,510 per year) were awarded; fellowships with full tuition reimbursements, career-related internships or fieldwork, Federal Work-Study, grants, scholarships, traineeships, and unspecified assistantships also available. Aid available to part-time students. Financial aid application deadline: 1/15.

Dr. Rodney K. Murphey, Acting Head, 413-545-3246, *Fax:* 413-545-1812, *E-mail:* rmurphey@bio.umass.edu.

Find an in-depth description at www.petersons.com/graduate.

■ UNIVERSITY OF MASSACHUSETTS WORCESTER

Graduate School of Biomedical Sciences, Department of Cell Biology, Worcester, MA 01655-0115

AWARDS PhD.

Faculty: 61 full-time (11 women).
Students: 34 full-time (24 women); includes 2 minority (1 Asian American or Pacific Islander, 1 Hispanic American), 11 international. In 1999, 2 degrees awarded.
Degree requirements: For doctorate, dissertation required. *Average time to degree:* Doctorate–8.1 years full-time.
Entrance requirements: For doctorate, GRE General Test, GRE Subject Test, 1 year of calculus, physics, organic chemistry and biology. *Application deadline:* For fall admission, 1/15 (priority date). Applications are processed on a rolling basis. *Application fee:* $25 ($50 for international students).
Expenses: Tuition, state resident: full-time $2,640. Tuition, nonresident: full-time $9,756. Required fees: $825. Full-time tuition and fees vary according to program.
Financial aid: In 1999–00, research assistantships with full tuition reimbursements (averaging $17,500 per year); unspecified assistantships also available.
Faculty research: Cellular development and function, growth, development, differentiation, molecular mechanisms.
Dr. Gary Stein, Chair, 508-856-5625.
Application contact: Dr. Janet Stein, Graduate Director, 508-856-4996.
Find an in-depth description at www.petersons.com/graduate.

■ UNIVERSITY OF MEDICINE AND DENTISTRY OF NEW JERSEY

Graduate School of Biomedical Sciences, Graduate Programs in Biomedical Sciences, Department of Anatomy, Cell Biology, and Injury Sciences, Newark, NJ 07107

AWARDS MS, PhD. Terminal master's awarded for partial completion of doctoral program.

Degree requirements: For master's, thesis required, foreign language not required; for doctorate, dissertation, qualifying exam required, foreign language not required.
Entrance requirements: For master's and doctorate, GRE General Test, TOEFL. *Application deadline:* For fall admission, 2/1; for spring admission, 10/1. *Application fee:* $40.
Expenses: Tuition, state resident: part-time $270 per credit hour. Tuition, nonresident: part-time $407 per credit

hour. Part-time tuition and fees vary according to campus/location and program.
Financial aid: Fellowships, research assistantships, Federal Work-Study, institutionally sponsored loans, and tuition waivers (full and partial) available. Financial aid application deadline: 5/1.
Dr. John Siegel, Chairperson, 973-972-4414.
Application contact: Dr. Henry E. Brezenoff, Dean, Graduate School of Biomedical Sciences, 973-972-5333, *Fax:* 973-972-7148, *E-mail:* hbrezeno@ umdnj.edu.
Find an in-depth description at www.petersons.com/graduate.

■ UNIVERSITY OF MEDICINE AND DENTISTRY OF NEW JERSEY

Graduate School of Biomedical Sciences, Graduate Programs in Biomedical Sciences, Department of Neuroscience and Cell Biology, Piscataway, NJ 08854-5635

AWARDS MS, PhD.

Degree requirements: For master's and doctorate, thesis/dissertation, qualifying exam required, foreign language not required.
Entrance requirements: For master's, GRE General Test, GRE Subject Test (biology or chemistry), TOEFL; for doctorate, GRE General Test, GRE Subject Test (biology or chemistry), TOEFL, minimum undergraduate GPA of 3.0. *Application deadline:* For fall admission, 3/1 (priority date); for spring admission, 10/1. Applications are processed on a rolling basis. *Application fee:* $40.
Expenses: Tuition, state resident: part-time $270 per credit hour. Tuition, nonresident: part-time $407 per credit hour. Part-time tuition and fees vary according to campus/location and program.
Financial aid: Fellowships, research assistantships, teaching assistantships available. Financial aid application deadline: 5/1.
Faculty research: Neuronal growth factors, neuronal gene expression, neurogenetics, circulation controls, reproduction.
Dr. Ira B. Black, Chair, 732-235-5388, *E-mail:* black@mcbl.rutgers.edu.
Application contact: David Egger, Director of Admissions, 732-235-4522, *E-mail:* egger@umdnj.edu.

■ UNIVERSITY OF MEDICINE AND DENTISTRY OF NEW JERSEY

Graduate School of Biomedical Sciences, Graduate Programs in Biomedical Sciences, Program in Cell and Developmental Biology, Piscataway, NJ 08854-5635

AWARDS Cell biology (MS, PhD); developmental biology (MS, PhD); immunology (MS).

Degree requirements: For master's, thesis, qualifying exam required; for doctorate, dissertation, qualifying exam required, foreign language not required.
Entrance requirements: For master's, GRE General Test, TOEFL; for doctorate, GRE General Test, GRE Subject Test, TOEFL. *Application deadline:* For fall admission, 2/1; for spring admission, 10/1. *Application fee:* $40.
Expenses: Tuition, state resident: part-time $270 per credit hour. Tuition, nonresident: part-time $407 per credit hour. Part-time tuition and fees vary according to campus/location and program.
Financial aid: Fellowships, research assistantships, teaching assistantships available. Financial aid application deadline: 5/1.
Faculty research: Genetics, cell matrix interactions, signal transduction, gene expression, cytoskeletal proteins.
Dr. Alice Y. Liu, Director, 732-445-3430, *Fax:* 732-445-2730.

■ UNIVERSITY OF MEDICINE AND DENTISTRY OF NEW JERSEY

Graduate School of Biomedical Sciences, Graduate Programs in Biomedical Sciences, Program in Cell and Molecular Biology, Stratford, NJ 08084-5634

AWARDS MS, PhD, DO/PhD.

Students: 11 full-time (5 women), 7 part-time (5 women); includes 10 minority (1 African American, 9 Asian Americans or Pacific Islanders).
Degree requirements: For master's, thesis required; for doctorate, dissertation, qualifying exam required, foreign language not required.
Entrance requirements: For master's and doctorate, GRE General Test, TOEFL. *Application deadline:* For fall admission, 2/1; for spring admission, 10/1. Applications are processed on a rolling basis. *Application fee:* $40.
Expenses: Tuition, state resident: part-time $270 per credit hour. Tuition, nonresident: part-time $407 per credit hour. Part-time tuition and fees vary according to campus/location and program.

University of Medicine and Dentistry of New Jersey (continued)
Financial aid: Application deadline: 5/1. Dr. Salvatore J. Caradonna, Co-Chairperson, 856-566-6056.

■ THE UNIVERSITY OF MEMPHIS

Graduate School, College of Arts and Sciences, Department of Microbiology and Molecular Cell Sciences, Memphis, TN 38152

AWARDS Biology (MS, PhD). Part-time programs available.

Faculty: 10 full-time (2 women).
Students: Average age 29. *33 applicants, 79% accepted.* In 1999, 4 master's awarded (100% found work related to degree); 1 doctorate awarded (100% found work related to degree). Terminal master's awarded for partial completion of doctoral program.
Degree requirements: For master's, thesis or alternative, oral and/or written comprehensive exam required, foreign language not required; for doctorate, one foreign language (computer language can substitute), dissertation, comprehensive exam required. *Average time to degree:* Master's–2 years full-time; doctorate–5.5 years full-time.
Entrance requirements: For master's and doctorate, GRE General Test, GRE Subject Test (biochemistry), minimum GPA of 2.5. *Application deadline:* For fall admission, 8/1; for spring admission, 12/1. Applications are processed on a rolling basis. *Application fee:* $25 ($50 for international students).
Expenses: Tuition, state resident: full-time $3,410; part-time $178 per credit hour. Tuition, nonresident: full-time $8,670; part-time $408 per credit hour. Tuition and fees vary according to program.
Financial aid: In 1999–00, 2 research assistantships with full tuition reimbursements (averaging $12,000 per year), 16 teaching assistantships with full tuition reimbursements (averaging $8,000 per year) were awarded; fellowships with full tuition reimbursements
Faculty research: Molecular genetics, cell biology, biochemistry, microbiology and toxicology.
Dr. H. Delano Black, Interim Chair, 901-678-2955, *Fax:* 901-678-4457, *E-mail:* dblack@cc.memphis.edu.
Application contact: Dr. Barbara J. Taller, Graduate Coordinator, 901-678-2955, *Fax:* 901-678-4457, *E-mail:* bjtaller@cc.memphis.edu.
Find an in-depth description at www.petersons.com/graduate.

■ UNIVERSITY OF MIAMI

School of Medicine and Graduate School, Graduate Programs in Medicine, Department of Cell Biology and Anatomy, Coral Gables, FL 33124

AWARDS Molecular, cell and developmental biology (PhD).

Faculty: 13 full-time (4 women).
Students: 11 full-time (5 women); includes 5 minority (1 African American, 3 Asian Americans or Pacific Islanders, 1 Hispanic American). In 1999, 1 degree awarded (100% entered university research/teaching).
Degree requirements: For doctorate, one foreign language (computer language can substitute), dissertation required.
Entrance requirements: For doctorate, GRE General Test, GRE Subject Test, TOEFL. *Application deadline:* For fall admission, 3/1 (priority date). Applications are processed on a rolling basis. *Application fee:* $35.
Expenses: Tuition, area resident: Part-time $899 per credit.
Financial aid: Fellowships available.
Faculty research: Signal transduction cell-matrix interactions, gene expression during development, cell polarity, protein sorting. *Total annual research expenditures:* $1.8 million.
Dr. Robert Warren, Interim Chairman, 305-243-6691, *Fax:* 305-545-7166.
Application contact: Dr. Richard Rotundo, Professor, 305-243-6691, *Fax:* 305-545-7166.

■ UNIVERSITY OF MIAMI

School of Medicine and Graduate School, Graduate Programs in Medicine, Program in Molecular Cell and Developmental Biology, Coral Gables, FL 33124

AWARDS PhD.

Faculty: 13 full-time (4 women).
Students: 14 full-time (8 women); includes 6 minority (4 Asian Americans or Pacific Islanders, 1 Hispanic American, 1 Native American), 2 international. In 1999, 2 degrees awarded.
Entrance requirements: For doctorate, GRE General Test, GRE Subject Test, TOEFL. *Application deadline:* For fall admission, 3/1 (priority date). Applications are processed on a rolling basis. *Application fee:* $35.
Expenses: Tuition, area resident: Part-time $899 per credit.
Financial aid: Fellowships available.
Dr. Robert Warren, Interim Chair, 305-243-6691, *Fax:* 305-545-7166.

Application contact: Dr. Richard Rotundo, Professor, 305-243-6691, *Fax:* 305-545-7166.
Find an in-depth description at www.petersons.com/graduate.

■ UNIVERSITY OF MICHIGAN

Medical School and Horace H. Rackham School of Graduate Studies, Program in Biomedical Sciences (PIBS), Department of Cell and Developmental Biology, Ann Arbor, MI 48109

AWARDS PhD.

Faculty: 30 full-time (8 women), 5 part-time/adjunct (0 women).
Students: 9 full-time (5 women), 6 international. Average age 33.
Degree requirements: For doctorate, oral defense of dissertation, preliminary exam required.
Entrance requirements: For doctorate, GRE General Test, GRE Subject Test, master's degree. *Application deadline:* For fall admission, 1/5. Applications are processed on a rolling basis. *Application fee:* $55. Electronic applications accepted.
Expenses: Tuition, state resident: full-time $10,316. Tuition, nonresident: full-time $20,922.
Financial aid: In 1999–00, 9 students received aid, including 9 research assistantships with full tuition reimbursements available (averaging $17,000 per year); fellowships, teaching assistantships, grants, scholarships, traineeships, tuition waivers (full), and unspecified assistantships also available.
Faculty research: Small stress proteins, cellular stress response, muscle, male reproductive, toxicology, cell cytoskeleton. *Total annual research expenditures:* $3.8 million.
Dr. Bruce M. Carlson, Chair, 734-763-2538, *Fax:* 734-763-1166, *E-mail:* brcarl@umich.edu.
Application contact: Program in Biomedical Sciences (PIBS), 734-647-7005, *Fax:* 734-647-7022, *E-mail:* pibs@umich.edu.
Find an in-depth description at www.petersons.com/graduate.

■ UNIVERSITY OF MICHIGAN

Medical School and Horace H. Rackham School of Graduate Studies, Program in Biomedical Sciences (PIBS), Interdepartmental Program in Cellular and Molecular Biology, Ann Arbor, MI 48109

AWARDS PhD.

Degree requirements: For doctorate, oral defense of dissertation, preliminary exam required.

Entrance requirements: For doctorate, GRE General Test, GRE Subject Test.
Expenses: Tuition, state resident: full-time $10,316. Tuition, nonresident: full-time $20,922.

Find an in-depth description at www.petersons.com/graduate.

■ UNIVERSITY OF MINNESOTA, DULUTH

School of Medicine, Department of Anatomy and Cell Biology, Duluth, MN 55812-2496

AWARDS MS, PhD. Part-time programs available.

Faculty: 4 full-time (1 woman).
Students: 1 full-time (0 women), 1 international. Average age 25. *10 applicants, 10% accepted.* In 1999, 1 master's awarded (100% found work related to degree). Terminal master's awarded for partial completion of doctoral program.
Degree requirements: For master's and doctorate, thesis/dissertation required, foreign language not required. *Average time to degree:* Master's–2 years full-time.
Entrance requirements: For master's and doctorate, GRE, biology/chemistry background. *Application deadline:* For fall admission, 7/15. *Application fee:* $30.
Expenses: Tuition, state resident: full-time $5,040; part-time $420 per credit. Tuition, nonresident: full-time $9,900; part-time $825 per credit. Required fees: $509. Tuition and fees vary according to course load and program.
Financial aid: Research assistantships, teaching assistantships, Federal Work-Study available. Financial aid application deadline: 4/1.
Faculty research: Cytoskeleton, neurobiology, cancer biology, cardiovascular biology, developmental biology. *Total annual research expenditures:* $45,000.
Dr. Arlen R. Severson, Head, 218-726-7903, *Fax:* 218-726-6235, *E-mail:* aseverso@d.umn.edu.
Application contact: Dr. David J. Schimpf, Director, Graduate Studies in Biology, 218-726-7265, *Fax:* 218-726-8142, *E-mail:* dschimpf@d.umn.edu.

■ UNIVERSITY OF MINNESOTA, TWIN CITIES CAMPUS

Graduate School, Program in Molecular, Cellular, Developmental Biology and Genetics, Minneapolis, MN 55455-0213

AWARDS Genetic counseling (MS); molecular, cellular, developmental biology and genetics (PhD). Part-time programs available.

Faculty: 80 full-time (24 women), 17 part-time/adjunct (16 women).

Students: 62 full-time (43 women), 2 part-time (both women). Average age 24. *188 applicants, 19% accepted.* In 1999, 9 master's awarded (100% found work related to degree); 5 doctorates awarded (20% entered university research/teaching, 80% found other work related to degree). Terminal master's awarded for partial completion of doctoral program.
Degree requirements: For master's, thesis optional, foreign language not required; for doctorate, dissertation required, foreign language not required. *Average time to degree:* Master's–2 years full-time; doctorate–5 years full-time.
Entrance requirements: For master's and doctorate, GRE General Test, TOEFL. *Application deadline:* For fall admission, 1/15 (priority date). Applications are processed on a rolling basis. *Application fee:* $40 ($50 for international students).
Expenses: Tuition, state resident: full-time $5,040; part-time $420 per credit. Tuition, nonresident: full-time $9,900; part-time $825 per credit. Full-time tuition and fees vary according to course load, program and reciprocity agreements.
Financial aid: In 1999–00, 53 students received aid, including 9 fellowships with full tuition reimbursements available (averaging $16,504 per year), 39 research assistantships with full tuition reimbursements available (averaging $16,504 per year), 5 teaching assistantships with full tuition reimbursements available (averaging $16,504 per year); Federal Work-Study, grants, institutionally sponsored loans, and traineeships also available. Financial aid application deadline: 1/15.
Faculty research: Membrane receptors and membrane transport, cell interactions, cytoskeleton and cell mobility, regulation of gene expression, plant cell and molecular biology.
Dr. Perry B. Hackett, Director of Graduate Studies, 612-624-3053, *Fax:* 612-625-1700, *E-mail:* dgsmcdbg@biosci.cbs.umn.edu.
Application contact: Sue Knoblauch, Principal Secretary, 612-624-7470, *Fax:* 612-625-5754, *E-mail:* mcdbg@biosci.cbs.umn.edu.

Find an in-depth description at www.petersons.com/graduate.

■ UNIVERSITY OF MISSOURI–KANSAS CITY

School of Biological Sciences, Program in Cell Biology and Biophysics, Kansas City, MO 64110-2499

AWARDS PhD. PhD offered through School of Graduate Studies.

Degree requirements: For doctorate, dissertation required.

Entrance requirements: For doctorate, GRE General Test, TOEFL, minimum GPA of 3.0, bachelor's degree in chemistry, biology or related field. *Application deadline:* For fall admission, 3/1 (priority date). Applications are processed on a rolling basis. *Application fee:* $25. Electronic applications accepted.
Expenses: Tuition, state resident: part-time $173 per hour. Tuition, nonresident: part-time $348 per hour. Required fees: $22 per hour. $15 per term. Part-time tuition and fees vary according to course load and program.
Financial aid: In 1999–00, research assistantships with full tuition reimbursements (averaging $16,000 per year), teaching assistantships with full and partial tuition reimbursements (averaging $16,000 per year) were awarded; grants, scholarships, and unspecified assistantships also available.
Application contact: Graduate Adviser, 816-235-2580.

Find an in-depth description at www.petersons.com/graduate.

■ UNIVERSITY OF MISSOURI–KANSAS CITY

School of Biological Sciences, Program in Cellular and Molecular Biology, Kansas City, MO 64110-2499

AWARDS MS, PhD. PhD offered through the School of Graduate Studies. Part-time programs available.

Students: 2 full-time (both women), 1 international. Average age 24. In 1999, 2 master's, 3 doctorates awarded. Terminal master's awarded for partial completion of doctoral program.
Degree requirements: For master's, thesis required (for some programs), foreign language not required; for doctorate, dissertation required. *Average time to degree:* Master's–2 years full-time.
Entrance requirements: For master's, GRE General Test, TOEFL, bachelor's degree in chemistry, biology, or a related discipline; minimum GPA of 3.0. *Application deadline:* For fall admission, 3/1 (priority date). Applications are processed on a rolling basis. *Application fee:* $25. Electronic applications accepted.
Expenses: Tuition, state resident: part-time $173 per hour. Tuition, nonresident: part-time $348 per hour. Required fees: $22 per hour. $15 per term. Part-time tuition and fees vary according to course load and program.
Financial aid: In 1999–00, research assistantships with full tuition reimbursements (averaging $16,000 per year), teaching assistantships with full and partial tuition reimbursements (averaging $16,000

University of Missouri–Kansas City (continued)

per year) were awarded; grants, scholarships, and unspecified assistantships also available.

Faculty research: Structural biology and molecular genetics; protein and gene regulation; structure-function studies; molecular, cell, and neurobiology.

Application contact: Graduate Adviser, 816-235-2580.

■ UNIVERSITY OF MISSOURI–ST. LOUIS

Graduate School, College of Arts and Sciences, Department of Biology, St. Louis, MO 63121-4499

AWARDS Biology (MS, PhD), including animal behavior (MS), biochemistry, biotechnology (MS), conservation biology (MS), development (MS), ecology (MS), environmental studies (PhD), evolution (MS), genetics (MS), molecular biology and biotechnology (PhD), molecular/cellular biology (MS), physiology (MS), plant systematics, population biology (MS), tropical biology (MS); biotechnology (Certificate); tropical biology and conservation (Certificate). Part-time programs available.

Faculty: 46.

Students: 21 full-time (11 women), 75 part-time (44 women); includes 13 minority (2 African Americans, 2 Asian Americans or Pacific Islanders, 8 Hispanic Americans, 1 Native American), 23 international. In 1999, 14 master's, 4 doctorates awarded.

Degree requirements: For master's, thesis or alternative required, foreign language not required; for doctorate, one foreign language, dissertation, 1 semester of teaching experience required.

Entrance requirements: For doctorate, GRE General Test. *Application deadline:* For fall admission, 7/1 (priority date); for spring admission, 11/1 (priority date). Applications are processed on a rolling basis. *Application fee:* $25 ($40 for international students). Electronic applications accepted.

Expenses: Tuition, state resident: full-time $4,932; part-time $173 per credit hour. Tuition, nonresident: full-time $13,279; part-time $521 per credit hour. Required fees: $775; $33 per credit hour. Tuition and fees vary according to degree level and program.

Financial aid: In 1999–00, 8 research assistantships with partial tuition reimbursements (averaging $10,635 per year), 14 teaching assistantships with partial tuition reimbursements (averaging $11,488 per year) were awarded; career-related internships or fieldwork and Federal Work-Study also available. Aid available to part-time students. Financial

aid application deadline: 2/1. *Total annual research expenditures:* $908,828.

Application contact: Graduate Admissions, 314-516-5458, *Fax:* 314-516-6759, *E-mail:* gradadm@umsl.edu.

■ UNIVERSITY OF NEBRASKA MEDICAL CENTER

Graduate College, Department of Cell Biology and Anatomy, Omaha, NE 68198

AWARDS MS, PhD. Part-time programs available.

Faculty: 16 full-time (1 woman), 2 part-time/adjunct (1 woman).

Students: 5 full-time (4 women), 1 (woman) part-time. Average age 28. *4 applicants, 0% accepted.* In 1999, 4 doctorates awarded. Terminal master's awarded for partial completion of doctoral program.

Degree requirements: For master's, thesis required, foreign language not required; for doctorate, dissertation required.

Entrance requirements: For master's and doctorate, GRE General Test. *Application deadline:* For fall admission, 3/1 (priority date). Applications are processed on a rolling basis. *Application fee:* $35.

Expenses: Tuition, state resident: part-time $116 per semester hour. Tuition, nonresident: part-time $270 per semester hour. Tuition and fees vary according to program.

Financial aid: In 1999–00, 1 fellowship, 1 research assistantship, 4 teaching assistantships were awarded; institutionally sponsored loans also available. Aid available to part-time students. Financial aid application deadline: 3/1.

Faculty research: Hematology, immunology, developmental biology, cardiovascular biology, neuroscience.

Dr. J. B. Turpen, Graduate Committee Chair, 402-559-4388, *Fax:* 402-559-7328, *E-mail:* jturpen@unmc.edu.

Application contact: Jo Wagner, Associate Director of Admissions, 402-559-6468.

Find an in-depth description at www.petersons.com/graduate.

■ UNIVERSITY OF NEVADA, RENO

Graduate School, Interdisciplinary Program in Cell and Molecular Biology, Reno, NV 89557

AWARDS MS, PhD. Offered through the College of Arts and Science, the M. C. Fleischmann College of Agriculture, and the School of Medicine.

Students: 25 full-time (18 women), 1 part-time; includes 2 minority (both Asian Americans or Pacific Islanders), 3 international. Average age 33. *28 applicants,*

36% accepted. In 1999, 4 master's, 6 doctorates awarded. Terminal master's awarded for partial completion of doctoral program.

Degree requirements: For master's and doctorate, thesis/dissertation required, foreign language not required.

Entrance requirements: For master's, GRE, TOEFL, minimum GPA of 2.75; for doctorate, GRE, TOEFL, minimum GPA of 3.0. *Application deadline:* For fall admission, 3/1. Applications are processed on a rolling basis. *Application fee:* $40.

Expenses: Tuition, area resident: Part-time $3,173 per semester. Tuition, nonresident: full-time $6,347. Required fees: $101 per credit. $101 per credit.

Financial aid: In 1999–00, 17 research assistantships were awarded; teaching assistantships, Federal Work-Study also available. Financial aid application deadline: 3/1.

Faculty research: Cancer biology, plant virology.

Dr. Stephen St. Jeor, Chair, 775-784-4113, *E-mail:* stjeor@unr.edu.

Find an in-depth description at www.petersons.com/graduate.

■ UNIVERSITY OF NEW HAVEN

Graduate School, College of Arts and Sciences, Program in Cellular and Molecular Biology, West Haven, CT 06516-1916

AWARDS MS.

Students: 7 full-time (3 women), 33 part-time (23 women); includes 5 minority (2 African Americans, 2 Asian Americans or Pacific Islanders, 1 Hispanic American), 5 international. *24 applicants, 67% accepted.* In 1999, 11 degrees awarded.

Degree requirements: For master's, thesis or alternative required, foreign language not required.

Application deadline: Applications are processed on a rolling basis. *Application fee:* $50.

Expenses: Tuition: Part-time $1,170 per course. Part-time tuition and fees vary according to course level, degree level and program.

Financial aid: Career-related internships or fieldwork and Federal Work-Study available. Financial aid application deadline: 5/1; financial aid applicants required to submit FAFSA.

Dr. Michael J. Rossi, Coordinator, 203-932-7125.

■ UNIVERSITY OF NEW MEXICO

Graduate School, College of Arts and Sciences, Department of Biology, Albuquerque, NM 87131-2039

AWARDS Biology (MS, PhD), including air land ecology, behavioral ecology, botany, cellular and molecular biology, community ecology, comparative immunology, comparative physiology, conservation biology, ecology, ecosystem ecology, evolutionary biology, evolutionary genetics, microbiology, molecular genetics, parasitology, physiological ecology, physiology, population biology, vertebrate and invertebrate zoology. Part-time programs available.

Faculty: 35 full-time (5 women), 18 part-time/adjunct (11 women).
Students: 71 full-time (37 women), 28 part-time (11 women); includes 8 minority (2 Asian Americans or Pacific Islanders, 5 Hispanic Americans, 1 Native American), 11 international. Average age 33. *93 applicants, 30% accepted.* In 1999, 11 master's, 12 doctorates awarded. Terminal master's awarded for partial completion of doctoral program.
Degree requirements: For master's, one foreign language (computer language can substitute), thesis required (for some programs); for doctorate, 2 foreign languages (computer language can substitute for one), dissertation required.
Entrance requirements: For master's and doctorate, GRE General Test, GRE Subject Test, minimum GPA of 3.2. *Application deadline:* For fall admission, 1/15. *Application fee:* $25.
Expenses: Tuition, state resident: full-time $2,514; part-time $105 per credit hour. Tuition, nonresident: full-time $10,304; part-time $417 per credit hour. International tuition: $10,304 full-time. Required fees: $516; $22 per credit hour. Tuition and fees vary according to program.
Financial aid: In 1999–00, 58 students received aid, including 24 fellowships (averaging $1,645 per year), 26 research assistantships with tuition reimbursements available (averaging $8,921 per year), 40 teaching assistantships with tuition reimbursements available (averaging $11,066 per year); career-related internships or fieldwork, Federal Work-Study, institutionally sponsored loans, and tuition waivers (full and partial) also available. Aid available to part-time students. Financial aid applicants required to submit FAFSA.
Faculty research: Developmental biology, immunobiology. *Total annual research expenditures:* $4.5 million.
Dr. Kathryn Vogel, Chair, 505-277-3411, *Fax:* 505-277-0304, *E-mail:* kgvogel@unm.edu.

Application contact: Vivian Kent, Information Contact, 505-277-1712, *Fax:* 505-277-0304, *E-mail:* vkent@unm.edu.

■ THE UNIVERSITY OF NORTH CAROLINA AT CHAPEL HILL

Graduate School, College of Arts and Sciences, Department of Biology, Chapel Hill, NC 27599

AWARDS Botany (MA, MS, PhD); cell biology, development, and physiology (MA, MS, PhD); ecology and behavior (MA, MS, PhD); genetics and molecular biology (MA, MS, PhD); morphology, systematics, and evolution (MA, MS, PhD).

Degree requirements: For master's, thesis (for some programs), comprehensive exams required; for doctorate, dissertation, comprehensive exams required.
Entrance requirements: For master's and doctorate, GRE General Test, GRE Subject Test. Electronic applications accepted.
Expenses: Tuition, state resident: full-time $1,578. Tuition, nonresident: full-time $10,744. Required fees: $827. One-time fee: $15 full-time. Tuition and fees vary according to program.
Faculty research: Gene expression, biomechanics, yeast genetics, plant ecology, plant molecular biology.

Find an in-depth description at www.petersons.com/graduate.

■ THE UNIVERSITY OF NORTH CAROLINA AT CHAPEL HILL

School of Medicine and Graduate School, Graduate Programs in Medicine, Department of Cell Biology and Anatomy, Chapel Hill, NC 27599

AWARDS PhD.

Faculty: 27 full-time (7 women).
Students: 25 full-time (14 women); includes 6 minority (all Asian Americans or Pacific Islanders). Average age 24. *36 applicants, 19% accepted.* In 1999, 5 degrees awarded (100% entered university research/teaching).
Degree requirements: For doctorate, dissertation, comprehensive exams required. *Average time to degree:* Doctorate–5 years full-time.
Entrance requirements: For doctorate, GRE General Test, GRE Subject Test. *Application deadline:* For fall admission, 1/1 (priority date). Applications are processed on a rolling basis. *Application fee:* $55. Electronic applications accepted.
Expenses: Tuition, state resident: full-time $1,966. Tuition, nonresident: full-time $11,026. Required fees: $8,940. One-time fee: $15 full-time. Part-time tuition and fees vary according to course load.

Financial aid: In 1999–00, 25 students received aid, including 13 research assistantships with full tuition reimbursements available (averaging $17,000 per year), 12 teaching assistantships with full tuition reimbursements available (averaging $17,000 per year); fellowships, tuition waivers (full) and unspecified assistantships also available. Financial aid application deadline: 2/1; financial aid applicants required to submit FAFSA.
Faculty research: Cell adhesion, motility and cytoskeleton; molecular analysis of signal transduction; development biology and toxicology; reproductive biology; cell and molecular imaging. *Total annual research expenditures:* $4.5 million.
Dr. Michael G. O'Rand, Interim Chair, 919-966-3026.

Application contact: Dr. Ellen Weiss, Director of Graduate Studies, 919-966-7683, *E-mail:* erweiss@med.unc.edu.

Find an in-depth description at www.petersons.com/graduate.

■ UNIVERSITY OF NOTRE DAME

Graduate School, College of Science, Department of Biological Sciences, Program in Cellular and Molecular Biology, Notre Dame, IN 46556

AWARDS MS, PhD. Terminal master's awarded for partial completion of doctoral program.

Degree requirements: For master's and doctorate, thesis/dissertation required, foreign language not required.
Entrance requirements: For master's and doctorate, GRE General Test, GRE Subject Test, TOEFL. *Application deadline:* For fall admission, 2/1 (priority date); for spring admission, 11/1. Applications are processed on a rolling basis. *Application fee:* $50.
Expenses: Tuition: Full-time $21,930; part-time $1,218 per credit. Required fees: $95. Tuition and fees vary according to program.
Financial aid: Fellowships with full tuition reimbursements, research assistantships with full tuition reimbursements, teaching assistantships with full tuition reimbursements, traineeships and tuition waivers (full) available. Financial aid application deadline: 2/1.
Faculty research: Gene expression, developmental biology, neurobiology, genetics, cell biology of membranes and cytoskeleton.
Application contact: Dr. Terrence J. Akai, Director of Graduate Admissions, 219-631-7706, *Fax:* 219-631-4183, *E-mail:* gradad@nd.edu.

■ UNIVERSITY OF OKLAHOMA HEALTH SCIENCES CENTER

College of Medicine and Graduate College, Graduate Programs in Medicine, Department of Cell Biology, Oklahoma City, OK 73190

AWARDS MS, PhD.

Faculty: 24 full-time (7 women), 3 part-time/adjunct (0 women).
Students: 2 full-time (both women), 1 part-time; includes 2 minority (both Asian Americans or Pacific Islanders), 1 international. Average age 25. *21 applicants, 10% accepted.*
Degree requirements: For master's and doctorate, thesis/dissertation required, foreign language not required.
Entrance requirements: For master's and doctorate, GRE General Test, GRE Subject Test, TOEFL, minimum GPA of 3.0. *Application deadline:* For fall admission, 7/1; for spring admission, 12/1. *Application fee:* $25 ($50 for international students).
Expenses: Tuition, state resident: part-time $90 per semester hour. Tuition, nonresident: part-time $264 per semester hour. Tuition and fees vary according to program.
Financial aid: Research assistantships, teaching assistantships, career-related internships or fieldwork and institutionally sponsored loans available. Aid available to part-time students. Financial aid application deadline: 7/1.
Faculty research: Neurobiology, reproductive, neuronal plasticity, extracellular matrix, neuroendocrinology. Dr. Robert Eugene Anderson, Chair, 405-271-2377.
Application contact: Dr. Allan Wiechmann, Director of Graduate Studies, 405-271-2377.

■ UNIVERSITY OF PENNSYLVANIA

School of Arts and Sciences, Graduate Group in Biology, Program in Cell, Molecular, and Developmental Biology, Philadelphia, PA 19104

AWARDS PhD.

Students: 27 full-time (8 women); includes 9 minority (2 African Americans, 6 Asian Americans or Pacific Islanders, 1 Hispanic American), 2 international. In 1999, 5 degrees awarded.
Degree requirements: For doctorate, dissertation required, foreign language not required.
Entrance requirements: For doctorate, GRE General Test, GRE Subject Test, TOEFL. *Application fee:* $65.

Expenses: Tuition: Full-time $23,670. Required fees: $1,546. Full-time tuition and fees vary according to degree level and program.
Financial aid: Fellowships, research assistantships, teaching assistantships, tuition waivers (full and partial) available. Financial aid application deadline: 1/2.
Application contact: Allan Aiken, Graduate Group Coordinator, 215-898-6786, *Fax:* 215-898-8780, *E-mail:* aaiken@ sas.upenn.edu.

■ UNIVERSITY OF PENNSYLVANIA

School of Medicine, Biomedical Graduate Studies, Graduate Group in Cell and Molecular Biology, Philadelphia, PA 19104

AWARDS Cell growth and cancer (PhD); cell structure and function (PhD); developmental biology (PhD); gene therapy (PhD); genetics and gene regulation (PhD); microbiology and virology (PhD).

Faculty: 297 full-time (62 women).
Students: 203 full-time (92 women); includes 47 minority (11 African Americans, 30 Asian Americans or Pacific Islanders, 6 Hispanic Americans), 18 international. *407 applicants, 15% accepted.* In 1999, 11 doctorates awarded.
Degree requirements: For doctorate, dissertation required, foreign language not required.
Entrance requirements: For doctorate, GRE General Test, TOEFL. *Application deadline:* For fall admission, 1/2 (priority date). Applications are processed on a rolling basis. *Application fee:* $65. Electronic applications accepted.
Expenses: Tuition: Full-time $17,256; part-time $2,991 per course. Required fees: $2,588; $363 per course. $726 per term.
Financial aid: In 1999–00, 154 students received aid, including 123 fellowships, 67 research assistantships; teaching assistantships, grants and institutionally sponsored loans also available. Financial aid application deadline: 1/2.
Dr. James C. Alwine, Chair, 215-898-3256.
Application contact: Mary Webster, Coordinator, 215-898-4360, *E-mail:* camb@mail.med.upenn.edu.

Find an in-depth description at www.petersons.com/graduate.

■ UNIVERSITY OF PITTSBURGH

Faculty of Arts and Sciences, Department of Biological Sciences, Program in Molecular, Cellular, and Developmental Biology, Pittsburgh, PA 15260

AWARDS PhD.

Degree requirements: For doctorate, dissertation required, foreign language not required.
Entrance requirements: For doctorate, GRE General Test, GRE Subject Test, TOEFL. *Application deadline:* For fall admission, 2/1 (priority date). Applications are processed on a rolling basis. *Application fee:* $40.
Expenses: Tuition, state resident: full-time $8,338; part-time $342 per credit. Tuition, nonresident: full-time $17,168; part-time $707 per credit. Required fees: $480; $90 per semester. Tuition and fees vary according to program.

Find an in-depth description at www.petersons.com/graduate.

■ UNIVERSITY OF PITTSBURGH

School of Medicine, Graduate Programs in Medicine, Program in Cell Biology and Molecular Physiology, Pittsburgh, PA 15260

AWARDS MS, PhD.

Faculty: 26 full-time (6 women).
Students: 9 full-time (3 women); includes 3 minority (all Asian Americans or Pacific Islanders), 1 international. *325 applicants, 22% accepted.* In 1999, 1 master's, 1 doctorate awarded.
Degree requirements: For doctorate, dissertation required, foreign language not required. *Average time to degree:* Doctorate–5.3 years full-time.
Entrance requirements: For doctorate, GRE General Test, GRE Subject Test, TOEFL, minimum QPA of 3.0. *Application deadline:* For fall admission, 1/15 (priority date). Applications are processed on a rolling basis. *Application fee:* $30 ($40 for international students).
Expenses: Tuition, state resident: full-time $9,778; part-time $403 per credit. Tuition, nonresident: full-time $20,146; part-time $830 per credit. Required fees: $480; $90 per semester.
Financial aid: Research assistantships with tuition reimbursements, teaching assistantships with tuition reimbursements, Federal Work-Study, institutionally sponsored loans, scholarships, traineeships, and unspecified assistantships available.
Faculty research: Epithelial cell biology, developmental biology of muscle, reproductive physiology and neuroendocrinology. *Total annual research expenditures:* $76.3 million.
Application contact: Graduate Studies Administrator, 412-648-8957, *Fax:* 412-648-1236, *E-mail:* biomed_phd@fs1.dean-med.pitt.edu.

Find an in-depth description at www.petersons.com/graduate.

■ UNIVERSITY OF ROCHESTER

The College, Arts and Sciences, Department of Biology, Program in Cellular and Molecular Biology, Rochester, NY 14627-0250

AWARDS MS, PhD.

Degree requirements: For master's, thesis not required; for doctorate, dissertation, qualifying exam required, foreign language not required.
Entrance requirements: For master's and doctorate, GRE General Test, GRE Subject Test, TOEFL. *Application deadline:* For fall admission, 2/1 (priority date). *Application fee:* $25.
Expenses: Tuition: Part-time $697 per credit hour. Tuition and fees vary according to program.
Financial aid: Application deadline: 2/1.
Application contact: Cindy Landry, Graduate Program Secretary, 716-275-7991.

■ UNIVERSITY OF SOUTH ALABAMA

College of Medicine and Graduate School, Program in Basic Medical Sciences, Specialization in Cellular Biology and Neuroscience, Mobile, AL 36688-0002

AWARDS PhD.

Faculty: 13 full-time (2 women), 1 (woman) part-time/adjunct.
Degree requirements: For doctorate, dissertation, oral and written preliminary exams, research proposal, qualifying exam required, foreign language not required.
Entrance requirements: For doctorate, GRE General Test, TOEFL, minimum GPA of 3.0. *Application deadline:* For fall admission, 4/30. Applications are processed on a rolling basis. *Application fee:* $25.
Expenses: Tuition, state resident: part-time $116 per semester hour. Tuition, nonresident: part-time $230 per semester hour. Required fees: $121 per semester.
Financial aid: Fellowships, institutionally sponsored loans available. Financial aid application deadline: 4/1.
Faculty research: Cytoskeleton-membrane interactions, neural basis of oral motor behavior, microtubule organizing centers, molecular biology of human chromosomes, mechanisms of synaptic transmissions.
Dr. Steven R. Goodman, Chair, 334-460-6490.
Application contact: Lanette Flagge, Coordinator, 334-460-6153.

Find an in-depth description at www.petersons.com/graduate.

■ UNIVERSITY OF SOUTH CAROLINA

Graduate School, College of Science and Mathematics, Department of Biological Sciences, Graduate Training Program in Molecular, Cellular, and Developmental Biology, Columbia, SC 29208

AWARDS MS, PhD.

Degree requirements: For master's, one foreign language required; for doctorate, dissertation required.
Entrance requirements: For master's and doctorate, GRE General Test, minimum GPA of 3.0 in science. *Application deadline:* For fall admission, 2/15 (priority date). *Application fee:* $35.
Expenses: Tuition, state resident: full-time $4,014; part-time $202 per credit hour. Tuition, nonresident: full-time $8,528; part-time $428 per credit hour. Required fees: $100; $4 per credit hour. Tuition and fees vary according to program.
Application contact: Dr. Franklyn F. Bolander, Director of Graduate Studies, 803-777-2755, *Fax:* 803-777-4002, *E-mail:* bolander@sc.edu.

Find an in-depth description at www.petersons.com/graduate.

■ UNIVERSITY OF SOUTH DAKOTA

School of Medicine and Graduate School, Biomedical Sciences Graduate Program, Cellular and Molecular Biology Group, Vermillion, SD 57069-2390

AWARDS MA, PhD.

Faculty: 10 full-time (1 woman), 1 part-time/adjunct (0 women).
Students: 11 full-time (2 women), 4 international. Average age 28. *72 applicants, 1% accepted.* In 1999, 1 degree awarded (100% continued full-time study). Terminal master's awarded for partial completion of doctoral program.
Degree requirements: For master's and doctorate, thesis/dissertation required, foreign language not required. *Average time to degree:* Doctorate–4 years full-time.
Entrance requirements: For master's and doctorate, GRE General Test, GRE Subject Test, TOEFL, minimum GPA of 3.0. *Application deadline:* For fall admission, 3/15 (priority date). Applications are processed on a rolling basis. *Application fee:* $15.
Expenses: Tuition, state resident: full-time $2,126; part-time $89 per credit. Tuition, nonresident: full-time $6,270; part-time $261 per credit. Required fees: $1,194; $50 per credit. Tuition and fees vary according to course load and reciprocity agreements.

Financial aid: In 1999–00, 11 students received aid, including 3 research assistantships with full tuition reimbursements available (averaging $13,500 per year); fellowships, tuition waivers (partial) also available.
Faculty research: Molecular aspects of protein and DNA, neurochemistry and energy transduction, gene regulation, cellular development.
W. Keith Miskimins, Coordinator, 605-677-5132, *Fax:* 605-677-6381, *E-mail:* kmiskimi@usd.edu.
Application contact: Robin Miskimins, Graduate Program Director, 605-677-5237, *Fax:* 605-677-6381, *E-mail:* rmiskim@usd.edu.

■ UNIVERSITY OF SOUTHERN CALIFORNIA

Keck School of Medicine and Graduate School, Graduate Programs in Medicine, Department of Cell and Neurobiology, Los Angeles, CA 90089

AWARDS Anatomy and cell biology (MS, PhD), including anatomy (PhD), cell biology (PhD); cell and neurobiology (MS, PhD); pharmacology and nutrition (MS, PhD); preventive nutrition (MS).

Faculty: 21 full-time (3 women), 8 part-time/adjunct (2 women).
Students: 19 full-time (9 women); includes 10 minority (all Asian Americans or Pacific Islanders). Average age 23. *54 applicants, 22% accepted.* In 1999, 5 master's awarded (40% found work related to degree, 60% continued full-time study). Terminal master's awarded for partial completion of doctoral program.
Degree requirements: For master's, thesis or alternative required, foreign language not required; for doctorate, dissertation required. *Average time to degree:* Master's–2 years full-time; doctorate–5 years full-time.
Entrance requirements: For master's, GRE General Test, TOEFL, minimum GPA of 3.0; for doctorate, GRE General Test, TOEFL. *Application fee:* $55. Electronic applications accepted.
Expenses: Tuition: Full-time $22,198; part-time $748 per unit. Required fees: $406.
Financial aid: In 1999–00, 13 students received aid; fellowships, research assistantships, teaching assistantships, Federal Work-Study, institutionally sponsored loans, and tuition waivers (partial) available. Aid available to part-time students.
Faculty research: Neurobiology and development, circaulian rhythm, gene therapy in vision, lacrimal glands, neuroendocrinology, signal transduction mechanisms.

*University of Southern California
(continued)*

Dr. Cheryl Craft, Chair, 323-442-1881, *Fax:* 323-442-2709, *E-mail:* ccraft@hsc.usc.edu.
Application contact: Darlene Marie Campbell, Administrative Assistant, 323-442-1881, *Fax:* 323-442-0466, *E-mail:* dmc@hsc.usc.edu.

■ UNIVERSITY OF SOUTH FLORIDA

Graduate School, College of Arts and Sciences, Interdisciplinary Program in Cellular and Molecular Biology, Tampa, FL 33620-9951

AWARDS PhD.

Degree requirements: For doctorate, dissertation, 2 tools of research required, foreign language not required.
Entrance requirements: For doctorate, GRE General Test, minimum GPA of 3.4.
Expenses: Tuition, state resident: part-time $148 per credit hour. Tuition, nonresident: part-time $509 per credit hour.
Faculty research: Protein structure and function, gene regulation, signal transduction, molecular immunology, cancer.

Find an in-depth description at www.petersons.com/graduate.

■ THE UNIVERSITY OF TENNESSEE HEALTH SCIENCE CENTER

College of Graduate Health Sciences, Department of Microbiology and Immunology, Concentration in Molecular and Cell Biology, Memphis, TN 38163-0002

AWARDS PhD.

Degree requirements: For doctorate, dissertation, oral and written preliminary and comprehensive exams required, foreign language not required.
Entrance requirements: For doctorate, GRE General Test, GRE Subject Test, TOEFL, minimum GPA of 3.0.

Find an in-depth description at www.petersons.com/graduate.

■ THE UNIVERSITY OF TEXAS AT DALLAS

School of Natural Sciences and Mathematics, Program in Molecular and Cell Biology, Richardson, TX 75083-0688

AWARDS Biology (MS, PhD); molecular and cell biology (MS, PhD). Part-time and evening/weekend programs available.

Faculty: 15 full-time (2 women).

Students: 43 full-time (26 women), 11 part-time (7 women); includes 7 minority (5 Asian Americans or Pacific Islanders, 2 Hispanic Americans), 26 international. Average age 29. In 1999, 7 master's, 8 doctorates awarded.
Degree requirements: For master's, thesis optional, foreign language not required; for doctorate, dissertation, publishable paper required, foreign language not required.
Entrance requirements: For master's and doctorate, GRE General Test, TOEFL. *Application deadline:* For fall admission, 7/15; for spring admission, 11/15. Applications are processed on a rolling basis. *Application fee:* $25 ($75 for international students). Electronic applications accepted.
Expenses: Tuition, state resident: full-time $2,052; part-time $76 per semester hour. Tuition, nonresident: full-time $5,256; part-time $292 per semester hour. Required fees: $1,504; $656 per year. One-time fee: $10. Full-time tuition and fees vary according to course level, course load, degree level and program.
Financial aid: In 1999–00, 9 research assistantships (averaging $5,115 per year), 18 teaching assistantships (averaging $5,154 per year) were awarded; fellowships, career-related internships or fieldwork, Federal Work-Study, grants, institutionally sponsored loans, and scholarships also available. Aid available to part-time students. Financial aid application deadline: 4/30; financial aid applicants required to submit FAFSA.
Faculty research: DNA replication, regulation of gene expression, subcellular organelles, physical chemistry of macromolecules, damage and repair of cellular DNA. *Total annual research expenditures:* $1.5 million.
Dr. Lawrence Reitzer, Head, 972-883-2502, *Fax:* 972-883-2502.
Application contact: Ricky Robichaud, Assistant Coordinator, 972-883-2500, *E-mail:* rickyr@utdallas.edu.

■ THE UNIVERSITY OF TEXAS HEALTH SCIENCE CENTER AT SAN ANTONIO

Graduate School of Biomedical Sciences, Department of Cellular and Structural Biology, San Antonio, TX 78229-3900

AWARDS PhD.

Degree requirements: For doctorate, dissertation, oral qualifying exam required, foreign language not required.
Entrance requirements: For doctorate, GRE General Test, previous course work in biology, chemistry, physics, and calculus.

Expenses: Tuition, state resident: part-time $38 per credit hour. Tuition, nonresident: part-time $249 per credit hour.
Faculty research: Human/molecular genetics, endocrinology and neurobiology, cell biology, cancer biology, biology of aging.

Find an in-depth description at www.petersons.com/graduate.

■ THE UNIVERSITY OF TEXAS–HOUSTON HEALTH SCIENCE CENTER

Graduate School of Biomedical Sciences, Program in Integrative Biology, Houston, TX 77225-0036

AWARDS Cell biology (MS, PhD); physiology (MS, PhD).

Faculty: 25 full-time (8 women).
Students: 4 full-time (1 woman); includes 1 minority (Hispanic American), 1 international. Average age 27. *12 applicants, 50% accepted.* In 1999, 3 master's awarded. Terminal master's awarded for partial completion of doctoral program.
Degree requirements: For master's and doctorate, thesis/dissertation required, foreign language not required.
Entrance requirements: For master's and doctorate, GRE General Test, TOEFL, TWE. *Application deadline:* For fall admission, 1/15 (priority date); for spring admission, 11/1. Applications are processed on a rolling basis. *Application fee:* $10. Electronic applications accepted.
Financial aid: Fellowships, research assistantships, teaching assistantships, institutionally sponsored loans available. Financial aid application deadline: 1/15.
Faculty research: Cell signaling, regulation of gene expression, cell growth and adaptation, cell injury and repair/protection, membrane trafficking and ion channels.
Dr. Roger G. O'Neil, Director, 713-500-6316, *Fax:* 713-500-7444, *E-mail:* roneil@girch1.med.uth.tmc.edu.
Application contact: Anne Baronitis, Director of Admissions, 713-500-9860, *Fax:* 713-500-9877, *E-mail:* abaron@gsbs.gs.uth.tmc.edu.

Find an in-depth description at www.petersons.com/graduate.

■ THE UNIVERSITY OF TEXAS–HOUSTON HEALTH SCIENCE CENTER

Graduate School of Biomedical Sciences, Program in Regulatory Biology, Houston, TX 77225-0036

AWARDS MS, PhD, MD/PhD.

Faculty: 18 full-time (5 women).

Students: *1 applicant, 100% accepted.*
Degree requirements: For master's and doctorate, thesis/dissertation required, foreign language not required.
Entrance requirements: For master's and doctorate, GRE General Test, TOEFL, TWE. *Application deadline:* For fall admission, 1/15 (priority date); for spring admission, 11/1. Applications are processed on a rolling basis. *Application fee:* $10. Electronic applications accepted.
Financial aid: Fellowships, research assistantships, institutionally sponsored loans available. Financial aid application deadline: 1/15.
Faculty research: Mechanisms of action of hormones and neurotransmitters, cancer chemotherapy, molecular biology of regulatory processes, cell signaling.
Dr. Fernando R. Cabral, Director, 713-500-7485, *Fax:* 713-500-7455, *E-mail:* fcabral@farmr1.med.uth.tmc.edu.
Application contact: Anne Baronitis, Director of Admissions, 713-500-9860, *Fax:* 713-500-9877, *E-mail:* abaron@gsbs.gs.uth.tmc.edu.

Find an in-depth description at www.petersons.com/graduate.

■ THE UNIVERSITY OF TEXAS MEDICAL BRANCH AT GALVESTON

Graduate School of Biomedical Sciences, Program in Cell Biology, Galveston, TX 77555
AWARDS PhD.

Faculty: 39 full-time (14 women).
Students: 12 full-time (8 women), 1 (woman) part-time; includes 3 minority (1 Asian American or Pacific Islander, 2 Hispanic Americans), 5 international. Average age 29. *27 applicants, 56% accepted.* In 1999, 3 degrees awarded.
Degree requirements: For doctorate, dissertation required, foreign language not required.
Entrance requirements: For doctorate, GRE General Test. *Application deadline:* For fall admission, 3/31. Applications are processed on a rolling basis. *Application fee:* $25 ($50 for international students). Electronic applications accepted.
Expenses: Tuition, state resident: full-time $684; part-time $38 per credit hour. Tuition, nonresident: full-time $4,572; part-time $254 per credit hour. Required fees: $29; $7.5 per credit hour. One-time fee: $55. Tuition and fees vary according to program.
Financial aid: In 1999–00, 2 research assistantships, 2 teaching assistantships were awarded; Federal Work-Study and institutionally sponsored loans also available. Financial aid applicants required to submit FAFSA.

Faculty research: Neurosciences, neuroendocrinology, developmental biology, aging, marine biology, stress, reproductive biology, toxicology, oncology.
Dr. Gwendolyn Childs, Director, 409-772-1942, *Fax:* 409-772-3222, *E-mail:* gvchilds@utmb.edu.
Application contact: Dr. Eric Smith, Chairman, Admissions Committee, 409-772-2729, *Fax:* 409-772-3222.

Find an in-depth description at www.petersons.com/graduate.

■ THE UNIVERSITY OF TEXAS MEDICAL BRANCH AT GALVESTON

Graduate School of Biomedical Sciences, Program in Human Biological Chemistry and Genetics, Galveston, TX 77555

AWARDS Biochemistry (MS, PhD); cell biology (MS, PhD); genetics (MS, PhD); molecular biology (PhD); structural biology (PhD).

Faculty: 61 full-time (4 women).
Students: 39 full-time (12 women); includes 2 minority (both Hispanic Americans), 12 international. Average age 28. *41 applicants, 66% accepted.* In 1999, 1 master's awarded (100% found work related to degree); 4 doctorates awarded (100% found work related to degree).
Degree requirements: For master's and doctorate, thesis/dissertation required, foreign language not required.
Entrance requirements: For master's and doctorate, GRE General Test. *Application deadline:* For fall admission, 2/1 (priority date). Applications are processed on a rolling basis. *Application fee:* $25 ($50 for international students). Electronic applications accepted.
Expenses: Tuition, state resident: full-time $684; part-time $38 per credit hour. Tuition, nonresident: full-time $4,572; part-time $254 per credit hour. Required fees: $29; $7.5 per credit hour. One-time fee: $55. Tuition and fees vary according to program.
Financial aid: Fellowships, research assistantships, Federal Work-Study and institutionally sponsored loans available. Financial aid application deadline: 2/1; financial aid applicants required to submit FAFSA.
Faculty research: Cell surfaces and transmembrane signaling, structural biology, growth factors, DNA repair and aging.
Dr. James C. Lee, Director, 409-772-2769, *Fax:* 409-747-0552, *E-mail:* gradprog.hbcg@utmb.edu.

Application contact: Becky L. Hansen, Senior Administrative Secretary, 409-772-2769, *Fax:* 409-747-0552, *E-mail:* bhansen@utmb.edu.

Find an in-depth description at www.petersons.com/graduate.

■ THE UNIVERSITY OF TEXAS SOUTHWESTERN MEDICAL CENTER AT DALLAS

Southwestern Graduate School of Biomedical Sciences, Division of Cell and Molecular Biology, Program in Cell Regulation, Dallas, TX 75390
AWARDS PhD.

Faculty: 49 full-time (8 women), 1 part-time/adjunct (0 women).
Students: 49 full-time (18 women); includes 6 minority (4 Asian Americans or Pacific Islanders, 2 Hispanic Americans), 13 international. Average age 25. In 1999, 14 doctorates awarded (100% entered university research/teaching).
Degree requirements: For doctorate, dissertation required, foreign language not required.
Entrance requirements: For doctorate, GRE General Test, minimum GPA of 3.0. *Application deadline:* For fall admission, 1/5 (priority date). *Application fee:* $0. Electronic applications accepted.
Expenses: Tuition, state resident: full-time $912. Tuition, nonresident: full-time $6,096. Required fees: $216. Full-time tuition and fees vary according to course load and program.
Financial aid: Fellowships, research assistantships, institutionally sponsored loans and traineeships available. Financial aid application deadline: 3/15; financial aid applicants required to submit FAFSA.
Faculty research: Molecular and cellular approaches to regulatory biology; receptor-effector coupling; membrane structure, function, and assembly.
Dr. Mark Lehrman, Chair, 214-648-2323, *Fax:* 214-648-8812, *E-mail:* lehrman@utsw.swmed.edu.
Application contact: Nancy McKinney, Education Coordinator, 214-648-8099, *Fax:* 214-648-2978, *E-mail:* dcmbinfo@utsouthwestern.edu.

Find an in-depth description at www.petersons.com/graduate.

■ UNIVERSITY OF THE SCIENCES IN PHILADELPHIA

College of Graduate Studies, Program in Cell Biology and Biotechnology, Philadelphia, PA 19104-4495
AWARDS MS. Part-time and evening/weekend programs available.

Faculty: 8 full-time (2 women).

University of the Sciences in Philadelphia (continued)

Students: Average age 29. *8 applicants, 88% accepted.*

Degree requirements: For master's, thesis required (for some programs), foreign language not required.

Entrance requirements: For master's, GRE General Test, TOEFL. *Application deadline:* For fall admission, 5/1; for spring admission, 10/1. Applications are processed on a rolling basis. *Application fee:* $45.

Expenses: Tuition: Part-time $464 per credit.

Financial aid: In 1999–00, 1 student received aid. Grants available. Financial aid application deadline: 5/1.

Faculty research: Invertebrate cell adhesion, plant-microbe interactions, natural product mechanisms, cell signal transduction, gene regulation and organization. *Total annual research expenditures:* $30,000.

Dr. John R. Porter, Director, 215-596-8917, *Fax:* 215-596-8710, *E-mail:* j.porter@usip.edu.

■ **UNIVERSITY OF UTAH**

Graduate School, College of Science, Program in Cell Biology, Salt Lake City, UT 84132

AWARDS PhD.

Degree requirements: For doctorate, dissertation required.

Entrance requirements: For doctorate, GRE General Test, GRE Subject Test, TOEFL.

Expenses: Tuition, state resident: full-time $1,663. Tuition, nonresident: full-time $5,201. Tuition and fees vary according to course load and program.

■ **UNIVERSITY OF VERMONT**

Graduate College, Cell and Molecular Biology Program, Burlington, VT 05405

AWARDS MS, PhD.

Degree requirements: For master's and doctorate, thesis/dissertation required, foreign language not required.

Entrance requirements: For master's and doctorate, GRE General Test, TOEFL.

Expenses: Tuition, state resident: full-time $7,464; part-time $311 per credit. Tuition, nonresident: full-time $18,672; part-time $778 per credit. Full-time tuition and fees vary according to degree level and program.

Find an in-depth description at www.petersons.com/graduate.

■ **UNIVERSITY OF VIRGINIA**

College and Graduate School of Arts and Sciences, Department of Cell Biology, Charlottesville, VA 22903

AWARDS PhD.

Faculty: 24 full-time (4 women), 1 part-time/adjunct (0 women).

Students: 23 full-time (16 women); includes 4 minority (2 African Americans, 2 Asian Americans or Pacific Islanders), 5 international. Average age 26. *19 applicants, 26% accepted.* In 1999, 1 degree awarded.

Degree requirements: For doctorate, dissertation required.

Entrance requirements: For doctorate, GRE General Test. *Application deadline:* For fall admission, 7/15; for spring admission, 12/1. Applications are processed on a rolling basis. *Application fee:* $40. Electronic applications accepted.

Expenses: Tuition, state resident: full-time $3,832. Tuition, nonresident: full-time $15,519. Required fees: $1,084. Tuition and fees vary according to course load and program.

Financial aid: Application deadline: 2/1.

Dr. Charles J. Flickinger, Chairman, 804-924-9979.

Application contact: Duane J. Osheim, Associate Dean, 804-924-7184, *E-mail:* microbiology@virginia.edu.

Find an in-depth description at www.petersons.com/graduate.

■ **UNIVERSITY OF VIRGINIA**

College and Graduate School of Arts and Sciences, Interdisciplinary Program in Cell and Molecular Biology, Charlottesville, VA 22903

AWARDS PhD. PhD awarded by related department after one year of general course work and research within CMB program.

Students: 4 full-time (1 woman). Average age 24. *87 applicants, 15% accepted.*

Degree requirements: For doctorate, dissertation required.

Entrance requirements: For doctorate, GRE General Test. *Application fee:* $40. Electronic applications accepted.

Expenses: Tuition, state resident: full-time $3,832. Tuition, nonresident: full-time $15,519. Required fees: $1,084. Tuition and fees vary according to course load and program.

Financial aid: Applicants required to submit FAFSA.

Find an in-depth description at www.petersons.com/graduate.

■ **UNIVERSITY OF WASHINGTON**

School of Medicine and Graduate School, Graduate Programs in Medicine, Program in Molecular and Cellular Biology, Seattle, WA 98195

AWARDS PhD. Offered jointly with Fred Hutchinson Cancer Research Center.

Faculty: 178 full-time (42 women).

Students: 98 full-time (42 women); includes 23 minority (3 African Americans, 16 Asian Americans or Pacific Islanders, 3 Hispanic Americans, 1 Native American). *244 applicants, 31% accepted.* In 1999, 6 degrees awarded (100% entered university research/teaching).

Degree requirements: For doctorate, dissertation required. *Average time to degree:* Doctorate–5 years full-time.

Entrance requirements: For doctorate, GRE General Test, GRE Subject Test. *Application deadline:* For fall admission, 1/3. *Application fee:* $45. Electronic applications accepted.

Expenses: Tuition, state resident: full-time $9,210; part-time $236 per credit. Tuition, nonresident: full-time $23,256; part-time $596 per credit.

Financial aid: In 1999–00, 33 fellowships with tuition reimbursements (averaging $14,688 per year), 20 research assistantships with tuition reimbursements (averaging $17,304 per year) were awarded; career-related internships or fieldwork, Federal Work-Study, and traineeships also available.

Dr. Randall T. Moon, Co-Director, 206-543-1722, *Fax:* 206-616-4230, *E-mail:* rtmoon@ul.washington.edu.

Application contact: MaryEllin Robinson, Program Specialist, 206-685-3155, *Fax:* 206-685-8174, *E-mail:* mcb@u.washington.edu.

Find an in-depth description at www.petersons.com/graduate.

■ **UNIVERSITY OF WISCONSIN–MADISON**

Graduate School, College of Agricultural and Life Sciences, Department of Animal Health and Biomedical Sciences, Program in Comparative Biosciences, Madison, WI 53706-1380

AWARDS Anatomy (MS, PhD); biochemistry (MS, PhD); cellular and molecular biology (MS, PhD); environmental toxicology (MS, PhD); neurosciences (MS, PhD); pharmacology (MS, PhD); physiology (MS, PhD).

Degree requirements: For doctorate, dissertation required.

Expenses: Tuition, state resident: full-time $5,406; part-time $339 per credit. Tuition, nonresident: full-time $17,110; part-time $1,071 per credit. Full-time tuition and fees vary according to program and reciprocity agreements. Part-time tuition and fees vary according to course load and program.

■ UNIVERSITY OF WISCONSIN–MADISON

Graduate School, Program in Cellular and Molecular Biology, Madison, WI 53706-1380

AWARDS Cellular and molecular biology (PhD); developmental biology (PhD).

Degree requirements: For doctorate, dissertation required, foreign language not required.

Entrance requirements: For doctorate, GRE General Test, GRE Subject Test. Electronic applications accepted.

Expenses: Tuition, state resident: full-time $5,406; part-time $339 per credit. Tuition, nonresident: full-time $17,110; part-time $1,071 per credit. Full-time tuition and fees vary according to program and reciprocity agreements. Part-time tuition and fees vary according to course load and program.

Faculty research: Virology, carcinogensis, cell structure, hormone regulation, developmental biology.

Find an in-depth description at www.petersons.com/graduate.

■ VANDERBILT UNIVERSITY

Graduate School and School of Medicine, Department of Cell Biology, Nashville, TN 37240-1001

AWARDS MS, PhD, MD/PhD.

Faculty: 25 full-time (9 women).
Students: 54 full-time (31 women); includes 6 minority (3 African Americans, 2 Asian Americans or Pacific Islanders, 1 Hispanic American), 9 international. Average age 28. In 1999, 2 master's, 7 doctorates awarded.

Degree requirements: For master's, thesis required, foreign language not required; for doctorate, dissertation, preliminary, qualifying, and final exams required, foreign language not required.

Entrance requirements: For master's, GRE General Test; for doctorate, GRE General Test, GRE Subject Test (recommended). *Application deadline:* For fall admission, 1/15. *Application fee:* $40.

Expenses: Tuition: Full-time $17,244; part-time $958 per hour. Required fees: $242; $121 per semester. Tuition and fees vary according to program.

Financial aid: In 1999–00, fellowships with full tuition reimbursements (averaging $17,000 per year), research assistantships (averaging $17,000 per year) were awarded; career-related internships or fieldwork, institutionally sponsored loans,

traineeships, and tuition waivers (partial) also available. Financial aid application deadline: 1/15.

Faculty research: Reproductive biology, neurobiology, cancer biology, molecular biology, developmental biology.
Lynn M. Matrisian, Interim Chair, 615-322-2134, *Fax:* 615-343-4539, *E-mail:* dot.blue@vanderbilt.edu.

Application contact: Steven K. Hanks, Director of Graduate Studies, 615-322-2134, *Fax:* 615-343-4539, *E-mail:* steven.k.hanks@vanderbilt.edu.

Find an in-depth description at www.petersons.com/graduate.

■ VIRGINIA POLYTECHNIC INSTITUTE AND STATE UNIVERSITY

Graduate School, College of Agriculture and Life Sciences, Department of Biochemistry, Blacksburg, VA 24061

AWARDS Biochemistry (MS, PhD); cell and molecular biology (PhD).

Students: 17 full-time (8 women), 4 part-time (all women), 9 international. *40 applicants, 35% accepted.* In 1999, 2 master's, 2 doctorates awarded.

Degree requirements: For doctorate, dissertation required, foreign language not required.

Entrance requirements: For master's, TOEFL; for doctorate, GRE General Test, TOEFL. *Application deadline:* For fall admission, 12/1 (priority date). Applications are processed on a rolling basis. *Application fee:* $25.

Expenses: Tuition, state resident: full-time $4,122; part-time $229 per credit hour. Tuition, nonresident: full-time $6,930; part-time $385 per credit hour. Required fees: $828; $107 per semester. Part-time tuition and fees vary according to course load.

Financial aid: In 1999–00, research assistantships with full tuition reimbursements (averaging $15,815 per year), 8 teaching assistantships with full tuition reimbursements (averaging $15,815 per year) were awarded; fellowships. Financial aid application deadline: 4/1.
Dr. John Hess, Head, 540-231-6315, *E-mail:* jlhess@vt.edu.

Find an in-depth description at www.petersons.com/graduate.

■ WASHINGTON STATE UNIVERSITY

Graduate School, College of Sciences, School of Molecular Biosciences, Department of Genetics and Cell Biology, Pullman, WA 99164

AWARDS MS, PhD.

Faculty: 14 full-time (4 women), 2 part-time/adjunct (both women).
Students: 28 full-time (11 women), 2 part-time (both women); includes 2 minority (both Asian Americans or Pacific Islanders), 5 international. Average age 24. In 1999, 4 master's awarded (100% continued full-time study); 1 doctorate awarded (100% entered university research/teaching). Terminal master's awarded for partial completion of doctoral program.

Degree requirements: For master's, thesis or alternative, oral exam required, foreign language not required; for doctorate, dissertation, oral exam required, foreign language not required. *Average time to degree:* Master's–2.5 years full-time; doctorate–4.67 years full-time.

Entrance requirements: For master's and doctorate, GRE General Test, minimum GPA of 3.0. *Application deadline:* For fall admission, 3/1 (priority date). Applications are processed on a rolling basis. *Application fee:* $35.

Expenses: Tuition, state resident: full-time $5,654. Tuition, nonresident: full-time $13,850. International tuition: $13,850 full-time. Tuition and fees vary according to program.

Financial aid: In 1999–00, 25 students received aid, including 13 research assistantships with full and partial tuition reimbursements available, 11 teaching assistantships with full and partial tuition reimbursements available; fellowships, Federal Work-Study, institutionally sponsored loans, tuition waivers (partial), and teaching associateships also available. Financial aid application deadline: 4/1; financial aid applicants required to submit FAFSA.

Faculty research: Plant molecular biology, population genetics, growth factors, cancer induction and DNA repair, gene regulation and genetic engineering. *Total annual research expenditures:* $426,134.
Dr. Paul Lurquin, Chair, 509-335-5733, *Fax:* 509-335-1907.

Application contact: Mary Ann Storms, Program Assistant, 509-335-4566, *Fax:* 509-335-1907, *E-mail:* storms@mail.wsu.edu.

Find an in-depth description at www.petersons.com/graduate.

■ WASHINGTON UNIVERSITY IN ST. LOUIS

Graduate School of Arts and Sciences, Division of Biology and Biomedical Sciences, Program in Molecular Cell Biology, St. Louis, MO 63110

AWARDS PhD.

Washington University in St. Louis (continued)

Degree requirements: For doctorate, dissertation required, foreign language not required.
Entrance requirements: For doctorate, GRE General Test, GRE Subject Test. *Application deadline:* For fall admission, 1/1 (priority date). Applications are processed on a rolling basis. *Application fee:* $0.
Expenses: Tuition: Full-time $23,400; part-time $975 per credit. Tuition and fees vary according to program.
Financial aid: Application deadline: 1/1. Dr. Robert Mercer, Coordinator, 314-362-8849.
Application contact: Rosemary Garagneni, Director of Admissions, 800-852-9074, *E-mail:* admissions@dbbs.wustl.edu.

■ WAYNE STATE UNIVERSITY

School of Medicine and Graduate School, Graduate Programs in Medicine, Department of Anatomy and Cell Biology, Detroit, MI 48202

AWARDS MS, PhD, MD/PhD. Terminal master's awarded for partial completion of doctoral program.

Degree requirements: For master's, thesis required (for some programs), foreign language not required; for doctorate, dissertation required, foreign language not required.
Entrance requirements: For master's and doctorate, GRE General Test, minimum GPA of 3.0.
Faculty research: Cytoskeletal proteins, neuronal plasticity, neural connections, glial cells, receptor interaction.

Find an in-depth description at www.petersons.com/graduate.

■ WESLEYAN UNIVERSITY

Graduate Programs, Department of Biology, Middletown, CT 06459-0260

AWARDS Cell biology (PhD); comparative physiology (PhD); developmental biology (PhD); genetics (PhD); neurophysiology (PhD); population biology (PhD).

Faculty: 12 full-time (3 women).
Students: 24 full-time (12 women); includes 1 minority (African American), 11 international. Average age 28. *125 applicants, 4% accepted.* In 1999, 2 doctorates awarded.
Degree requirements: For doctorate, one foreign language (computer language can substitute), dissertation required.
Entrance requirements: For doctorate, GRE Subject Test. *Application deadline:* For fall admission, 1/15. Applications are processed on a rolling basis. *Application fee:* $0.

Expenses: Tuition: Full-time $24,876. Required fees: $650. Tuition and fees vary according to program.
Financial aid: Research assistantships, teaching assistantships, stipends available.
Faculty research: Microbial population genetics, genetic basis of evolutionary adaptation, genetic regulation of differentiation and pattern formation in *drosophila.*
Dr. Fred Cohan, Chairman, 860-685-3489.
Application contact: Marina J. Melendez, Director of Graduate Student Services, 860-685-2390, *Fax:* 860-685-2439, *E-mail:* mmelendez@wesleyan.edu.

Find an in-depth description at www.petersons.com/graduate.

■ WEST VIRGINIA UNIVERSITY

Eberly College of Arts and Sciences, Department of Biology, Morgantown, WV 26506

AWARDS Animal behavior (MS); cellular and molecular biology (MS, PhD); environmental plant biology (MS, PhD); plant systematics (MS); population genetics (MS).

Faculty: 18 full-time (4 women), 2 part-time/adjunct (1 woman).
Students: 19 full-time (10 women), 3 part-time (2 women); includes 2 minority (1 Asian American or Pacific Islander, 1 Hispanic American), 4 international. Average age 27. *50 applicants, 10% accepted.* In 1999, 4 master's, 1 doctorate awarded. Terminal master's awarded for partial completion of doctoral program.
Degree requirements: For master's, thesis, final exam required, foreign language not required; for doctorate, dissertation, preliminary and final exams required, foreign language not required. *Average time to degree:* Master's–2.5 years full-time; doctorate–5 years full-time.
Entrance requirements: For master's, GRE General Test, GRE Subject Test, TOEFL, minimum GPA of 3.0; for doctorate, GRE General Test, TOEFL, minimum GPA of 3.0. *Application deadline:* For fall admission, 4/1; for spring admission, 10/1. Applications are processed on a rolling basis. *Application fee:* $45.
Expenses: Tuition, state resident: full-time $2,910; part-time $154 per credit hour. Tuition, nonresident: full-time $8,368; part-time $457 per credit hour.
Financial aid: In 1999–00, 3 research assistantships, 13 teaching assistantships were awarded; Federal Work-Study and institutionally sponsored loans also available. Financial aid application deadline: 4/1; financial aid applicants required to submit FAFSA.
Faculty research: Environmental biology, genetic engineering, developmental biology, global change, biodiversity.

Dr. Keith Garbutt, Chair, 304-293-5394.
Application contact: Dr. James B. McGraw, Director of Graduate Studies, 304-293-5201, *Fax:* 304-293-6363.
Find an in-depth description at www.petersons.com/graduate.

■ YALE UNIVERSITY

Graduate School of Arts and Sciences, Department of Cell Biology, New Haven, CT 06520

AWARDS PhD.

Faculty: 30 full-time (7 women), 2 part-time/adjunct (1 woman).
Students: 36 full-time (26 women); includes 8 minority (1 African American, 3 Asian Americans or Pacific Islanders, 4 Hispanic Americans), 4 international. *68 applicants, 26% accepted.* In 1999, 8 degrees awarded.
Degree requirements: For doctorate, dissertation required, foreign language not required. *Average time to degree:* Doctorate–5.9 years full-time.
Entrance requirements: For doctorate, GRE General Test. *Application deadline:* For fall admission, 1/4. *Application fee:* $65.
Expenses: Tuition: Full-time $22,300. Full-time tuition and fees vary according to program.
Financial aid: Fellowships, research assistantships, teaching assistantships, Federal Work-Study, institutionally sponsored loans, and traineeships available. Aid available to part-time students.
Application contact: Admissions Information, 203-432-2770.

Find an in-depth description at www.petersons.com/graduate.

■ YALE UNIVERSITY

Graduate School of Arts and Sciences, Department of Molecular, Cellular, and Developmental Biology, Program in Cell Biology, New Haven, CT 06520

AWARDS PhD.

Degree requirements: For doctorate, dissertation required.
Entrance requirements: For doctorate, GRE General Test, GRE Subject Test. *Application deadline:* For fall admission, 1/4. *Application fee:* $65.
Expenses: Tuition: Full-time $22,300. Full-time tuition and fees vary according to program.
Financial aid: Fellowships, research assistantships, teaching assistantships available.
Application contact: Admissions Information, 203-432-2770.

■ YALE UNIVERSITY

School of Medicine and Graduate School of Arts and Sciences, Combined Program in Biological and Biomedical Sciences (BBS), Cell Biology and Molecular Physiology Track, New Haven, CT 06520

AWARDS PhD, MD/PhD.

Degree requirements: For doctorate, dissertation required.
Entrance requirements: For doctorate, GRE General Test, TOEFL. *Application deadline:* For fall admission, 1/2. *Application fee:* $65. Electronic applications accepted.
Expenses: All students receive full tuition of $22,330 and an annual stipend of $17,600 .
Financial aid: Fellowships, research assistantships available.
Dr. Susan Ferro-Novick, Co-Director of Graduate Studies, 203-787-5207, *E-mail:* bbs.cbmp@yale.edu.

Find an in-depth description at www.petersons.com/graduate.

■ YESHIVA UNIVERSITY

Albert Einstein College of Medicine, Sue Golding Graduate Division of Medical Sciences, Department of Anatomy and Structural Biology, Bronx, NY 10461

AWARDS Anatomy (PhD); cell and developmental biology (PhD).

Faculty: 13 full-time.
Students: 14 full-time (8 women); includes 3 minority (1 African American, 2 Asian Americans or Pacific Islanders), 8 international. In 1999, 1 degree awarded.
Degree requirements: For doctorate, dissertation required, foreign language not required.
Entrance requirements: For doctorate, GRE General Test, TOEFL. *Application deadline:* For fall admission, 1/15. *Application fee:* $0. Electronic applications accepted.
Expenses: Tuition: Part-time $525 per credit. Tuition and fees vary according to degree level and program.
Financial aid: In 1999–00, 14 fellowships were awarded.
Faculty research: Cell motility, cell membranes and membrane-cytoskeletal interactions as applied to processing of pancreatic hormones, mechanisms of secretion.
Dr. Peter Satir, Chairperson, 718-430-2836.
Application contact: Sheila Cleeton, Assistant Director, 718-430-2128, *Fax:* 718-430-8655, *E-mail:* phd@aecom.yu.edu.

■ YESHIVA UNIVERSITY

Albert Einstein College of Medicine, Sue Golding Graduate Division of Medical Sciences, Division of Biological Sciences, Department of Cell Biology, Bronx, NY 10461

AWARDS PhD, MD/PhD.

Faculty: 12 full-time.
Students: 44 full-time (18 women); includes 9 minority (1 African American, 6 Asian Americans or Pacific Islanders, 2 Hispanic Americans), 21 international. In 1999, 6 degrees awarded.
Degree requirements: For doctorate, dissertation required, foreign language not required.
Entrance requirements: For doctorate, GRE General Test, TOEFL. *Application deadline:* For fall admission, 1/15. *Application fee:* $0.
Expenses: Tuition: Part-time $525 per credit. Tuition and fees vary according to degree level and program.
Financial aid: In 1999–00, 44 fellowships were awarded.
Faculty research: Molecular and genetic basis of gene expression in animal cells; expression of differentiated traits of albumin, hemoglobin, myosin, and immunoglobin.
Dr. Art Skoultchi, Director, 718-430-2169.
Application contact: Sheila Cleeton, Assistant Director, 718-430-2128, *Fax:* 718-430-8655, *E-mail:* phd@aecom.yu.edu.

MOLECULAR BIOLOGY

■ ALBANY MEDICAL COLLEGE

Graduate Programs in the Biological Sciences, Program in Cell Biology and Cancer Research, Albany, NY 12208-3479

AWARDS MS, PhD. Part-time programs available.

Students: 21 full-time (12 women); includes 6 minority (2 African Americans, 4 Asian Americans or Pacific Islanders), 1 international. Average age 26. Terminal master's awarded for partial completion of doctoral program.
Degree requirements: For master's, thesis required, foreign language not required; for doctorate, dissertation, comprehensive written exam, oral qualifying exam required, foreign language not required.
Entrance requirements: For master's, GRE General Test, TOEFL; for doctorate, GRE General Test ,TOEFL. *Application deadline:* Applications are processed on a rolling basis.
Expenses: Tuition: Full-time $13,367; part-time $446 per credit hour.

Financial aid: Federal Work-Study, grants, scholarships, and tuition waivers (full) available.
Dr. Kevin Pumiglia, Graduate Director, 518-262-6587, *Fax:* 518-262-5696, *E-mail:* pumiglk@mail.amc.edu.

Find an in-depth description at www.petersons.com/graduate.

■ ARIZONA STATE UNIVERSITY

Graduate College, College of Liberal Arts and Sciences, Molecular and Cellular Biology Program, Tempe, AZ 85287

AWARDS MS, PhD.

Students: *54 applicants, 33% accepted.*
Degree requirements: For master's and doctorate, thesis/dissertation required.
Entrance requirements: For master's and doctorate, GRE. *Application deadline:* For fall admission, 3/1 (priority date); for spring admission, 10/1. Applications are processed on a rolling basis. *Application fee:* $45.
Expenses: Tuition, state resident: part-time $115 per credit hour. Tuition, nonresident: part-time $389 per credit hour. Required fees: $18 per semester. Tuition and fees vary according to program.
Financial aid: Fellowships, research assistantships, teaching assistantships available. Financial aid application deadline: 3/1.
Faculty research: Biochemistry, organometallic chemistry, genetics of outer membrane proteins, history and philosophy of biology.
Dr. Bertram Jacobs, Director, 480-965-0743.

Find an in-depth description at www.petersons.com/graduate.

■ BAYLOR COLLEGE OF MEDICINE

Graduate School of Biomedical Sciences, Department of Molecular and Cellular Biology, Houston, TX 77030-3498

AWARDS PhD, MD/PhD.

Faculty: 73 full-time (18 women).
Students: 72 full-time (34 women); includes 12 minority (6 African Americans, 3 Asian Americans or Pacific Islanders, 3 Hispanic Americans), 14 international. Average age 27. *60 applicants, 35% accepted.* In 2000, 13 doctorates awarded (62% entered university research/teaching, 8% found other work related to degree, 30% continued full-time study).
Degree requirements: For doctorate, dissertation, public defense, qualifying exam

www.petersons.com · *Graduate Programs in Biology 2002* **343**

Baylor College of Medicine (continued)
required, foreign language not required. *Average time to degree:* Doctorate–6.6 years full-time.

Entrance requirements: For doctorate, GRE General Test (average 80th percentile), GRE Subject Test (strongly recommended), TOEFL, minimum GPA of 3.0. *Application deadline:* For fall admission, 2/1 (priority date). *Application fee:* $30. Electronic applications accepted.

Expenses: Tuition: Full-time $8,200. Required fees: $175. Full-time tuition and fees vary according to student level.

Financial aid: In 2000–01, 72 students received aid, including 32 fellowships (averaging $16,000 per year), 40 research assistantships (averaging $16,000 per year); Federal Work-Study, institutionally sponsored loans, and tuition waivers (full) also available. Financial aid applicants required to submit FAFSA.

Faculty research: Gene regulation, cell structure/function, developmental biology, neurobiology, reproductive endocrinology. Dr. JoAnne Richards, Director, 713-798-4598.

Application contact: Caroline Kosnik, Graduate Program Administrator, 713-798-4598, *Fax:* 713-790-0545, *E-mail:* ckosnik@bcm.tmc.edu.

Find an in-depth description at www.petersons.com/graduate.

■ BAYLOR COLLEGE OF MEDICINE

Graduate School of Biomedical Sciences, Program in Cell and Molecular Biology, Houston, TX 77030-3498

AWARDS PhD, MD/PhD.

Faculty: 91 full-time (24 women).

Students: 41 full-time (23 women); includes 7 minority (1 African American, 3 Asian Americans or Pacific Islanders, 3 Hispanic Americans), 4 international. Average age 26. *114 applicants, 17% accepted.* In 1999, 3 doctorates awarded (100% entered university research/teaching).

Degree requirements: For doctorate, dissertation, public defense, qualifying exam required, foreign language not required. *Average time to degree:* Doctorate–5.79 years full-time.

Entrance requirements: For doctorate, GRE General Test (average 80th percentile), GRE Subject Test (strongly recommended), TOEFL, minimum GPA of 3.0. *Application deadline:* For fall admission, 2/1 (priority date). Applications are processed on a rolling basis. *Application fee:* $30. Electronic applications accepted.

Expenses: Tuition: Full-time $8,200. Required fees: $175. Full-time tuition and fees vary according to student level.

Financial aid: In 1999–00, 41 students received aid, including 29 fellowships (averaging $16,000 per year), 12 research assistantships (averaging $16,000 per year); teaching assistantships, career-related internships or fieldwork, Federal Work-Study, institutionally sponsored loans, and tuition waivers (full) also available. Financial aid applicants required to submit FAFSA.

Faculty research: Gene expression and regulation, developmental biology and genetics, signal transduction and membrane biology, aging process, molecular virology. Dr. Susan M. Berget, Director, 713-798-6557.

Application contact: Lourdes Fernandez, Graduate Program Administrator, 713-798-6557, *Fax:* 713-798-6325, *E-mail:* lourdesf@bcm.tmc.edu.

Find an in-depth description at www.petersons.com/graduate.

■ BOSTON UNIVERSITY

Graduate School of Arts and Sciences, Department of Biology, Boston, MA 02215

AWARDS Botany (MA, PhD); cell and molecular biology (MA, PhD); cell biology (MA, PhD); ecology (PhD); ecology, behavior, and evolution (MA, PhD); ecology/physiology, endocrinology and reproduction (MA); marine biology (MA, PhD); molecular biology, cell biology and biochemistry (MA, PhD); neurobiology, neuroendocrinology and reproduction (MA, PhD); physiology, endocrinology, and neurobiology (MA, PhD); zoology (MA, PhD). Part-time programs available.

Faculty: 41 full-time (8 women).

Students: 131 full-time (74 women), 11 part-time (7 women); includes 10 minority (7 Asian Americans or Pacific Islanders, 3 Hispanic Americans), 33 international. Average age 27. *238 applicants, 39% accepted.* In 1999, 61 master's, 45 doctorates awarded. Terminal master's awarded for partial completion of doctoral program.

Degree requirements: For master's, one foreign language, thesis not required; for doctorate, one foreign language, dissertation, qualifying exam required. *Average time to degree:* Master's–1 year full-time, 3 years part-time; doctorate–5.75 years full-time.

Entrance requirements: For master's and doctorate, GRE General Test, GRE Subject Test, TOEFL. *Application deadline:* For fall admission, 1/1 (priority date); for spring admission, 11/1. *Application fee:* $50.

Expenses: Tuition: Full-time $23,770; part-time $743 per credit. Required fees: $220. Tuition and fees vary according to

class time, course level, campus/location and program.

Financial aid: In 1999–00, 82 students received aid, including 1 fellowship with full tuition reimbursement available (averaging $12,000 per year), 28 research assistantships with full tuition reimbursements available (averaging $11,500 per year), 43 teaching assistantships with full tuition reimbursements available (averaging $11,500 per year); Federal Work-Study, grants, institutionally sponsored loans, scholarships, and traineeships also available. Financial aid application deadline: 1/15; financial aid applicants required to submit FAFSA.

Faculty research: Marine science, endocrinology, behavior. *Total annual research expenditures:* $5 million. Geoffrey M. Cooper, Chairman, 617-353-2432, *Fax:* 617-353-6340, *E-mail:* gmcooper@bu.edu.

Application contact: Yolanta Kovalko, Senior Staff Assistant, 617-353-2432, *Fax:* 617-353-6340, *E-mail:* yolanta@bu.edu.

Find an in-depth description at www.petersons.com/graduate.

■ BOSTON UNIVERSITY

Graduate School of Arts and Sciences, Molecular Biology, Cell Biology, and Biochemistry Program (MCBB), Boston, MA 02215

AWARDS MA, PhD. Part-time programs available.

Faculty: 45 full-time (10 women).

Students: 31 full-time (16 women), 2 part-time (1 woman); includes 7 minority (5 Asian Americans or Pacific Islanders, 2 Hispanic Americans), 10 international. Average age 26. *87 applicants, 30% accepted.* Terminal master's awarded for partial completion of doctoral program.

Degree requirements: For master's, one foreign language, thesis required (for some programs); for doctorate, one foreign language, dissertation, qualifying exam required. *Average time to degree:* Master's–2 years full-time.

Entrance requirements: For master's and doctorate, GRE General Test, GRE Subject Test, TOEFL. *Application deadline:* For fall admission, 1/1 (priority date). *Application fee:* $50.

Expenses: Tuition: Full-time $23,770; part-time $743 per credit. Required fees: $220. Tuition and fees vary according to class time, course level, campus/location and program.

Financial aid: In 1999–00, 16 students received aid, including 9 research assistantships with full tuition reimbursements available (averaging $11,500 per year), 7 teaching assistantships with full tuition reimbursements available (averaging

$11,500 per year); fellowships, Federal Work-Study, scholarships, and traineeships also available. Financial aid application deadline: 1/15; financial aid applicants required to submit FAFSA.

Faculty research: Signal transduction, gene expression, protein and nucleic acid biochemistry, genomics, modular physiology and development. *Total annual research expenditures:* $4.5 million.
Gary R. Jacobson, Director, 617-353-2432, *Fax:* 617-353-6340, *E-mail:* jacobson@bio.bu.edu.

Application contact: Michelle Brodkowitz, Academic Administrator, 617-353-2432, *Fax:* 617-353-6340, *E-mail:* mcbb@bio.bu.edu.

Find an in-depth description at www.petersons.com/graduate.

■ **BOSTON UNIVERSITY**

School of Medicine, Division of Graduate Medical Sciences, Department of Pathology and Laboratory Medicine, Boston, MA 02118

AWARDS Cell and molecular biology (PhD); experimental pathology (PhD); immunology (PhD). Part-time programs available.

Faculty: 12 full-time (4 women), 13 part-time/adjunct (2 women).
Students: 33 full-time (18 women), 1 part-time; includes 4 minority (2 African Americans, 2 Asian Americans or Pacific Islanders), 14 international. Average age 29. In 1999, 3 degrees awarded.
Degree requirements: For doctorate, dissertation, qualifying exam required, foreign language not required.
Entrance requirements: For doctorate, GRE General Test, GRE Subject Test, TOEFL. *Application deadline:* For fall admission, 1/15 (priority date); for spring admission, 10/15 (priority date). *Application fee:* $50. Electronic applications accepted.
Expenses: Tuition: Full-time $24,700; part-time $772 per credit. Required fees: $220.
Financial aid: In 1999–00, 10 fellowships, 3 research assistantships were awarded; Federal Work-Study, scholarships, and traineeships also available.
Faculty research: Toxicology, carcinogenesis, endocytosis, cytogenetics.
Leonard Gottlieb, Chairman, 617-638-4500, *Fax:* 617-638-4085.

Application contact: Dr. Adrianne Rogers, Associate Chairman, 617-638-4500, *Fax:* 617-638-4085, *E-mail:* aerogers@bu.edu.

■ **BOSTON UNIVERSITY**

School of Medicine, Division of Graduate Medical Sciences, Program in Cell and Molecular Biology, Boston, MA 02118

AWARDS MA, PhD, MD/PhD.

Degree requirements: For doctorate, dissertation required, foreign language not required.
Entrance requirements: For master's and doctorate, GRE General Test, GRE Subject Test, TOEFL. *Application deadline:* For fall admission, 1/15 (priority date); for spring admission, 10/15 (priority date). *Application fee:* $45. Electronic applications accepted.
Expenses: Tuition: Full-time $24,700; part-time $772 per credit. Required fees: $220.
Financial aid: Fellowships, research assistantships, Federal Work-Study, scholarships, and traineeships available.
Dr. Vickery Trinkaus Randall, Director, 617-638-6099, *Fax:* 617-638-5337; *E-mail:* fickery@biochem.bumc.bu.edu.
Application contact: Dr. Mary Jo Murnanl, Admissions Director, 617-638-4926, *Fax:* 617-638-4085, *E-mail:* mmuranl@bu.edu.

Find an in-depth description at www.petersons.com/graduate.

■ **BRANDEIS UNIVERSITY**

Graduate School of Arts and Sciences, Programs in Biological Sciences, Program in Molecular and Cell Biology, Waltham, MA 02454-9110

AWARDS Cell biology (PhD); developmental biology (PhD); genetics (PhD); microbiology (PhD); molecular and cell biology (MS); molecular biology (PhD); neurobiology (PhD).

Faculty: 19 full-time (8 women).
Students: 58 full-time (31 women), 1 (woman) part-time; includes 13 minority (1 African American, 8 Asian Americans or Pacific Islanders, 4 Hispanic Americans), 8 international. Average age 27. *83 applicants, 11% accepted.* In 1999, 1 master's, 8 doctorates awarded (100% entered university research/teaching). Terminal master's awarded for partial completion of doctoral program.
Degree requirements: For master's, thesis not required; for doctorate, dissertation required. *Average time to degree:* Master's–1 year full-time; doctorate–6 years full-time.
Entrance requirements: For doctorate, GRE General Test. *Application deadline:* For fall admission, 1/15 (priority date). Applications are processed on a rolling basis. *Application fee:* $60. Electronic applications accepted.

Expenses: Tuition: Full-time $25,392; part-time $3,174 per course. Required fees: $509. Tuition and fees vary according to class time, degree level, program and student level.
Financial aid: In 1999–00, 10 fellowships with tuition reimbursements (averaging $17,000 per year), 49 research assistantships with tuition reimbursements (averaging $17,000 per year), 9 teaching assistantships (averaging $2,312 per year) were awarded; scholarships and tuition waivers (full and partial) also available. Financial aid application deadline: 4/15; financial aid applicants required to submit CSS PROFILE or FAFSA.
Faculty research: Regulation of gene expression by transcription factors, molecular neurobiology, immunology, molecular mechanisms of genetic recombination, and cell differentiation.
Dr. Ranjan Sen, Chair, 781-736-2455, *Fax:* 781-736-3107.
Application contact: Marcia Cabral, Information Officer, 781-736-3100, *Fax:* 781-736-3107, *E-mail:* cabral@brandeis.edu.

■ **BRIGHAM YOUNG UNIVERSITY**

Graduate Studies, College of Biological and Agricultural Sciences, Department of Zoology, Provo, UT 84602-1001

AWARDS Biological science education (MS); molecular biology (MS, PhD); wildlife and range resources (MS, PhD); zoology (MS, PhD). Part-time programs available.

Faculty: 32 full-time (2 women), 3 part-time/adjunct (0 women).
Students: 54 full-time (20 women), 8 part-time (3 women); includes 6 minority (2 Asian Americans or Pacific Islanders, 4 Hispanic Americans), 6 international. Average age 25. *32 applicants, 63% accepted.* In 1999, 17 master's awarded (23% continued full-time study); 5 doctorates awarded (80% entered university research/teaching, 20% found other work related to degree).
Degree requirements: For master's and doctorate, thesis/dissertation required, foreign language not required. *Average time to degree:* Master's–2.5 years full-time; doctorate–6 years full-time.
Entrance requirements: For master's, GRE General Test, minimum GPA of 3.0 during previous 2 years; for doctorate, GRE General Test, minimum GPA of 3.0 overall. *Application deadline:* For fall admission, 2/1 (priority date). *Application fee:* $30. Electronic applications accepted.
Expenses: Tuition: Full-time $3,330; part-time $185 per credit hour. Tuition and fees vary according to program and student's religious affiliation.
Financial aid: In 1999–00, 54 students received aid, including fellowships with

Brigham Young University (continued)
partial tuition reimbursements available (averaging $11,000 per year), 12 research assistantships with full tuition reimbursements available (averaging $11,000 per year), 50 teaching assistantships with partial tuition reimbursements available (averaging $11,000 per year); career-related internships or fieldwork, institutionally sponsored loans, scholarships, tuition waivers (partial), unspecified assistantships, and tuition awards also available. Financial aid application deadline: 2/1.
Faculty research: Sex differentiation of brain, exercise physiology, toxicology, phylogenetic systematics, population biology.
Dr. John D. Bell, Chair, 801-378-2006, *Fax:* 801-378-7423, *E-mail:* jdb32@ email.byu.edu.
Application contact: Dr. R. Ward Rhees, Graduate Coordinator, 801-378-2158, *Fax:* 801-378-7423, *E-mail:* ward_rhees@ byu.edu.

■ BROWN UNIVERSITY

Graduate School, Division of Biology and Medicine, Program in Molecular Biology, Cell Biology, and Biochemistry, Providence, RI 02912

AWARDS Biochemistry (M Med Sc, Sc M, PhD), including biochemistry (Sc M, PhD), biology (Sc M, PhD), medical science (M Med Sc, PhD); biology (MA); cell biology (M Med Sc, Sc M, PhD), including biochemistry (Sc M, PhD), biology (Sc M, PhD), medical science (M Med Sc, PhD); developmental biology (M Med Sc, Sc M, PhD), including biochemistry (Sc M, PhD), biology (Sc M, PhD), medical science (M Med Sc, PhD); immunology (M Med Sc, Sc M, PhD), including biochemistry (Sc M, PhD), biology (Sc M, PhD), medical science (M Med Sc, PhD); molecular microbiology (M Med Sc, Sc M, PhD), including biochemistry (Sc M, PhD), biology (Sc M, PhD), medical science (M Med Sc, PhD). Part-time programs available.

Faculty: 50 full-time (14 women).
Students: 61 full-time (34 women); includes 4 minority (1 African American, 3 Asian Americans or Pacific Islanders), 21 international. Average age 25. *106 applicants, 28% accepted.* In 1999, 1 master's, 3 doctorates awarded. Terminal master's awarded for partial completion of doctoral program.
Degree requirements: For master's, thesis required (for some programs), foreign language not required; for doctorate, one foreign language, dissertation, preliminary exam required. *Average time to degree:* Doctorate–5 years full-time.

Entrance requirements: For master's and doctorate, GRE General Test, GRE Subject Test. *Application deadline:* For fall admission, 1/2 (priority date). Applications are processed on a rolling basis. *Application fee:* $60. Electronic applications accepted.
Financial aid: In 1999–00, 58 students received aid, including 11 fellowships (averaging $18,916 per year), 9 research assistantships (averaging $18,916 per year), 13 teaching assistantships (averaging $12,690 per year); institutionally sponsored loans and traineeships also available. Financial aid application deadline: 1/2.
Faculty research: Molecular genetics, gene regulation.
Dr. Gary Wessel, Director, 401-863-1051, *E-mail:* chet@brown.edu.
Application contact: Mary C. Esser, Graduate Program Coordinator, 401-863-1661, *Fax:* 401-863-1348, *E-mail:* mary_esser@brown.edu.

Find an in-depth description at www.petersons.com/graduate.

■ CALIFORNIA INSTITUTE OF TECHNOLOGY

Division of Biology, Program in Molecular Biology, Pasadena, CA 91125-0001

AWARDS PhD.

Degree requirements: For doctorate, dissertation, qualifying exam required, foreign language not required.
Entrance requirements: For doctorate, GRE General Test. *Application deadline:* For fall admission, 1/1. *Application fee:* $0.
Expenses: Tuition: Full-time $19,260. Required fees: $24. One-time fee: $100 full-time.
Financial aid: Application deadline: 1/1.
Application contact: Elizabeth Ayala, Graduate Option Coordinator, 626-395-4497, *Fax:* 626-449-0756, *E-mail:* biograd@cco.caltech.edu.

■ CARNEGIE MELLON UNIVERSITY

Mellon College of Science, Department of Biological Sciences, Pittsburgh, PA 15213-3891

AWARDS Biochemistry (PhD); biophysics (PhD); cell biology (PhD); developmental biology (PhD); genetics (PhD); molecular biology (PhD).

Faculty: 38 full-time (17 women), 1 (woman) part-time/adjunct.
Students: 35 full-time (24 women); includes 2 minority (1 African American, 1 Asian American or Pacific Islander), 18 international. Average age 27. In 1999, 4 degrees awarded.

Degree requirements: For doctorate, dissertation required, foreign language not required.
Entrance requirements: For doctorate, GRE General Test, GRE Subject Test, interview. *Application deadline:* For fall admission, 2/1 (priority date). Applications are processed on a rolling basis. *Application fee:* $0.
Expenses: Tuition: Full-time $22,100; part-time $307 per unit. Required fees: $200. Tuition and fees vary according to program.
Financial aid: Fellowships, research assistantships, teaching assistantships, traineeships available.
Faculty research: Genetic structure, function, and regulation; protein structure and function; biological membranes; biological spectroscopy. *Total annual research expenditures:* $4.4 million.
Dr. William E. Brown, Head, 412-268-3416, *Fax:* 412-268-7129, *E-mail:* wb02@ andrew.cmu.edu.
Application contact: Stacey L. Young, Assistant Head, 412-268-7372, *Fax:* 412-268-7129, *E-mail:* sf38+@andrew.cmu.edu.

Find an in-depth description at www.petersons.com/graduate.

■ CASE WESTERN RESERVE UNIVERSITY

School of Medicine and School of Graduate Studies, Graduate Programs in Medicine, Department of Anatomy, Cleveland, OH 44106

AWARDS Applied anatomy (MS); biological anthropology (MS, PhD); cellular biology (MS, PhD); developmental biology (PhD); molecular biology (PhD). Part-time programs available.

Faculty: 14 full-time (5 women), 14 part-time/adjunct (4 women).
Students: 17 full-time (9 women), 12 part-time (1 woman); includes 11 minority (1 African American, 9 Asian Americans or Pacific Islanders, 1 Hispanic American), 2 international. Average age 26. *16 applicants, 50% accepted.* In 1999, 5 master's awarded (50% found work related to degree, 50% continued full-time study); 1 doctorate awarded (100% entered university research/teaching).
Degree requirements: For master's, thesis required (for some programs); for doctorate, dissertation required. *Average time to degree:* Master's–5 years full-time; doctorate–1 year full-time.
Entrance requirements: For master's, GRE General Test, TOEFL; for doctorate, GRE General Test, GRE Subject Test, TOEFL. *Application deadline:* For fall admission, 5/1 (priority date); for spring

admission, 8/1 (priority date). Applications are processed on a rolling basis. *Application fee:* $25.

Financial aid: In 1999–00, 7 research assistantships with full tuition reimbursements (averaging $16,000 per year) were awarded; fellowships, grants also available.

Faculty research: Hypoxia, cell injury, biochemical aberration occurrences in ischemic tissue, human functional morphology, evolutionary morphology. *Total annual research expenditures:* $562,716. Joseph C. LaManna, Acting Chairman, 216-368-1100, *Fax:* 216-368-8669, *E-mail:* jcl4@po.cwru.edu.

Application contact: Laila Boesinger, Administrator, 216-368-3430, *Fax:* 216-368-8669, *E-mail:* lvb2@po.cwru.edu.

■ CASE WESTERN RESERVE UNIVERSITY

School of Medicine and School of Graduate Studies, Graduate Programs in Medicine, Department of Molecular Biology and Microbiology, Cleveland, OH 44106-4960

AWARDS Cellular biology (PhD); microbiology (PhD); molecular biology (PhD). Students are admitted to an integrated Biomedical Sciences Training Program involving 11 basic science programs at Case Western Reserve University.

Faculty: 14 full-time (4 women).

Students: 24 full-time (9 women); includes 2 minority (both Asian Americans or Pacific Islanders), 5 international. In 1999, 1 doctorate awarded (100% found work related to degree).

Degree requirements: For doctorate, dissertation required, foreign language not required. *Average time to degree:* Doctorate–5.3 years full-time.

Entrance requirements: For doctorate, GRE General Test, GRE Subject Test, TOEFL. *Application fee:* $25.

Financial aid: In 1999–00, 1 fellowship with full tuition reimbursement (averaging $16,000 per year), 23 research assistantships with full tuition reimbursements (averaging $16,000 per year) were awarded; Federal Work-Study, scholarships, traineeships, and tuition waivers (full) also available.

Faculty research: Gene expression in eukaryotic and prokaryotic systems; microbial physiology; intracellular transport and signaling; mechanisms of oncogenesis; molecular mechanisms of RNA processing, editing, and catalysis. *Total annual research expenditures:* $4.6 million.

Lloyd A. Culp, Acting Chair, 216-368-3420, *Fax:* 216-368-3055, *E-mail:* lac7@po.cwru.edu.

Application contact: Dr. Michael E. Harris, Admissions Coordinator, 216-368-3347, *E-mail:* mbio@po.cwru.edu.

Find an in-depth description at www.petersons.com/graduate.

■ CLARK UNIVERSITY

Graduate School, Concentration in Biochemistry/Molecular Biology, Worcester, MA 01610-1477

AWARDS MA, PhD.

Students: 16 full-time (7 women), 8 international. *24 applicants, 17% accepted.* In 1999, 2 master's, 3 doctorates awarded.

Degree requirements: For doctorate, dissertation required.

Entrance requirements: For master's and doctorate, GRE General Test, TOEFL. *Application deadline:* For fall admission, 2/15 (priority date). Applications are processed on a rolling basis. *Application fee:* $40.

Expenses: Tuition: Full-time $22,400; part-time $2,800 per course.

Faculty research: Molecular genetics, neurochemistry, protein chemistry. Dr. David Thurlow, Head, 508-793-7173.

Application contact: Rene Baril, Department Secretary, 528-793-7173.

■ COLORADO STATE UNIVERSITY

College of Veterinary Medicine and Biomedical Sciences and Graduate School, Graduate Programs in Veterinary Medicine and Biomedical Sciences, Fort Collins, CO 80523-0015

AWARDS Anatomy and neurobiology (MS, PhD); cell and molecular biology (MS, PhD); clinical sciences (MS, PhD); environmental health (MS, PhD); microbiology (MS, PhD), including immunology, microbiology; pathology (MS, PhD); physiology (MS, PhD); radiological health sciences (MS, PhD), including cellular and molecular biology, health physics, mammalian radiobiology, nuclear-waste management (MS), radiobiology (MS), radioecology, radiology, veterinary radiology (MS). Part-time programs available.

Faculty: 134 full-time (17 women).

Students: 221 full-time (119 women), 61 part-time (34 women); includes 21 minority (4 African Americans, 7 Asian Americans or Pacific Islanders, 9 Hispanic Americans, 1 Native American), 37 international. Average age 29. *303 applicants, 42% accepted.* In 1999, 81 master's, 23 doctorates awarded. Terminal master's awarded for partial completion of doctoral program.

Degree requirements: For master's, foreign language not required; for doctorate, dissertation required.

Entrance requirements: For master's and doctorate, GRE General Test, TOEFL. *Application deadline:* Applications are processed on a rolling basis. *Application fee:* $30. Electronic applications accepted.

Expenses: Tuition, state resident: full-time $2,694; part-time $150 per credit. Tuition, nonresident: full-time $10,460; part-time $581 per credit. Required fees: $32 per semester.

Financial aid: In 1999–00, 93 research assistantships, 11 teaching assistantships were awarded; fellowships, career-related internships or fieldwork, Federal Work-Study, institutionally sponsored loans, traineeships, and tuition waivers (partial) also available. Aid available to part-time students.

Faculty research: Reproductive physiology, infectious diseases, comparative oncology, environmental toxicology.

Application contact: Dr. Alan Tucker, Assistant Dean, 970-491-6106, *Fax:* 970-491-2250.

■ COLORADO STATE UNIVERSITY

Graduate School, College of Natural Sciences, Department of Biochemistry and Molecular Biology, Fort Collins, CO 80523-0015

AWARDS MS, PhD.

Faculty: 20 full-time (6 women), 2 part-time/adjunct (0 women).

Students: 18 full-time (9 women), 12 part-time (7 women); includes 1 minority (Asian American or Pacific Islander), 8 international. Average age 29. *181 applicants, 5% accepted.* In 1999, 14 master's, 4 doctorates awarded (100% entered university research/teaching). Terminal master's awarded for partial completion of doctoral program.

Degree requirements: For master's, thesis or alternative required, foreign language not required; for doctorate, dissertation required, foreign language not required.

Entrance requirements: For master's, GRE General Test, TOEFL, minimum GPA of 3.0; for doctorate, GRE General Test, TOEFL, minimum GPA of 3.2. *Application deadline:* For fall admission, 1/15 (priority date). Applications are processed on a rolling basis. *Application fee:* $30. Electronic applications accepted.

Expenses: Tuition, state resident: full-time $2,694; part-time $150 per credit. Tuition, nonresident: full-time $10,460; part-time $581 per credit. Required fees: $32 per semester. Tuition and fees vary according to program.

Financial aid: In 1999–00, 1 fellowship, 21 research assistantships, 9 teaching assistantships were awarded; Federal

Colorado State University (continued)
Work-Study, institutionally sponsored loans, traineeships, and tuition waivers (partial) also available. Financial aid application deadline: 1/15; financial aid applicants required to submit FAFSA.
Faculty research: Cellular biology, molecular gene expression, structure and function of macromolecules, neurobiology, transcriptional control mechanisms. Norman P. Curthoys, Chair, 970-491-5566, *Fax:* 970-491-0494, *E-mail:* ncurth@lamar.colostate.edu.
Application contact: Diane Keith, Graduate Recruitment Committee, 970-491-6841, *Fax:* 970-491-0494, *E-mail:* dkeith@vines.colostate.edu.

Find an in-depth description at www.petersons.com/graduate.

■ COLORADO STATE UNIVERSITY

Graduate School, Program in Cell and Molecular Biology, Fort Collins, CO 80523-0015
AWARDS MS, PhD.

Faculty: 46 full-time (13 women).
Students: 15 full-time (11 women), 6 part-time (4 women), 6 international. Average age 28. *147 applicants, 12% accepted.* In 1999, 3 degrees awarded. Terminal master's awarded for partial completion of doctoral program.
Degree requirements: For master's, thesis required (for some programs), foreign language not required; for doctorate, dissertation required, foreign language not required.
Entrance requirements: For master's and doctorate, GRE General Test, GRE Subject Test, TOEFL, minimum GPA of 3.0. *Application deadline:* For fall admission, 1/15 (priority date). Applications are processed on a rolling basis. *Application fee:* $30. Electronic applications accepted.
Expenses: Tuition, state resident: full-time $2,694; part-time $150 per credit. Tuition, nonresident: full-time $10,460; part-time $581 per credit. Required fees: $32 per semester. Tuition and fees vary according to program.
Financial aid: In 1999–00, 8 research assistantships, 4 teaching assistantships were awarded; fellowships, Federal Work-Study and institutionally sponsored loans also available. Financial aid application deadline: 2/1.
Faculty research: Regulation of gene expression, cancer biology, plant molecular genetics, reproductive physiology, environmental toxicology. *Total annual research expenditures:* $10 million. Dr. Michael H. Fox, Chairman, 970-491-7618, *Fax:* 970-491-0623, *E-mail:* mfox@cvmbs.colostate.edu.

Application contact: Linda Jones, Administrative Assistant, 970-491-0241, *Fax:* 970-491-0623, *E-mail:* ljones@cvmbs.colostate.edu.

Find an in-depth description at www.petersons.com/graduate.

■ COLUMBIA UNIVERSITY

College of Physicians and Surgeons and Graduate School of Arts and Sciences, Graduate School of Arts and Sciences at the College of Physicians and Surgeons, Integrated Program in Cellular, Molecular and Biophysical Studies, New York, NY 10032
AWARDS M Phil, MA, PhD, MD/PhD. Only candidates for the PhD are admitted. Terminal master's awarded for partial completion of doctoral program.

Degree requirements: For master's, foreign language and thesis not required; for doctorate, dissertation required, foreign language not required.
Entrance requirements: For master's, GRE General Test, TOEFL; for doctorate, GRE General Test, GRE Subject Test, TOEFL.
Expenses: Tuition: Full-time $25,072.
Faculty research: Transcription, macromolecular sorting, gene expression during development, cellular interaction.

■ CORNELL UNIVERSITY

Graduate School, Graduate Fields of Agriculture and Life Sciences, Field of Biochemistry, Molecular and Cell Biology, Ithaca, NY 14853-0001
AWARDS Biochemistry (PhD); biophysics (PhD); cell biology (PhD); molecular and cell biology (PhD); molecular biology (PhD).

Faculty: 43 full-time.
Students: 75 full-time (37 women); includes 15 minority (1 African American, 6 Asian Americans or Pacific Islanders, 8 Hispanic Americans), 19 international. *288 applicants, 21% accepted.* In 1999, 9 doctorates awarded.
Degree requirements: For doctorate, dissertation, 2 semesters of teaching experience required.
Entrance requirements: For doctorate, GRE General Test, GRE Subject Test (biology, chemistry, physics, biochemistry or, cell and molecular biology), TOEFL. *Application deadline:* For fall admission, 1/5. *Application fee:* $65. Electronic applications accepted.
Expenses: Tuition: Full-time $12,100.
Financial aid: In 1999–00, 74 students received aid, including 35 fellowships with full tuition reimbursements available, 32 research assistantships with full tuition reimbursements available, 7 teaching

assistantships with full tuition reimbursements available; institutionally sponsored loans, scholarships, tuition waivers (full and partial), and unspecified assistantships also available. Financial aid applicants required to submit FAFSA.
Faculty research: Biophysics, structural biology.
Application contact: Graduate Field Assistant, 607-255-2317, *E-mail:* bmcb@cornell.edu.

Find an in-depth description at www.petersons.com/graduate.

■ DUKE UNIVERSITY

Graduate School, Department of Biological Anthropology and Anatomy, Durham, NC 27708-0586
AWARDS Cellular and molecular biology (PhD); gross anatomy and physical anthropology (PhD), including comparative morphology of human and non-human primates, primate social behavior, vertebrate paleontology; neuroanatomy (PhD).

Faculty: 14 full-time, 1 part-time/adjunct.
Students: 20 full-time (12 women); includes 3 minority (2 African Americans, 1 Hispanic American), 1 international. *62 applicants, 6% accepted.* In 1999, 2 doctorates awarded.
Degree requirements: For doctorate, dissertation required.
Entrance requirements: For doctorate, GRE General Test. *Application deadline:* For fall admission, 12/31. *Application fee:* $75.
Expenses: Tuition: Full-time $21,406; part-time $760 per unit. Required fees: $3,136; $3,136 per year. One-time fee: $30. Tuition and fees vary according to program.
Financial aid: Fellowships, teaching assistantships, Federal Work-Study available. Financial aid application deadline: 12/31. Kathleen Smith, Director of Graduate Studies, 919-684-4124, *Fax:* 919-684-8034, *E-mail:* rachel_hougom@baa.mc.duke.edu.

■ DUKE UNIVERSITY

Graduate School, Program in Cellular and Molecular Biology, Durham, NC 27708-0586
AWARDS Certificate. Students must be enrolled in a participating PhD program.

Faculty: 115 full-time.
Students: 23 full-time (14 women); includes 1 minority (Asian American or Pacific Islander), 2 international. *177 applicants, 29% accepted.*
Entrance requirements: For degree, GRE General Test, GRE Subject Test (recommended). *Application deadline:* For fall admission, 12/31. *Application fee:* $75.

Expenses: Tuition: Full-time $21,406; part-time $760 per unit. Required fees: $3,136; $3,136 per year. One-time fee: $30. Tuition and fees vary according to program.
Financial aid: Fellowships available. Financial aid application deadline: 12/31. Dr. Kenneth Kreuzer, Director of Graduate Studies, 919-684-6559, *Fax:* 919-681-8911, *E-mail:* cmbtgp@acpub.duke.edu.

Find an in-depth description at www.petersons.com/graduate.

■ EAST CAROLINA UNIVERSITY

Graduate School, College of Arts and Sciences, Department of Biology, Greenville, NC 27858-4353

AWARDS Biology (MS); molecular biology/biotechnology (MS). Part-time programs available.

Faculty: 19 full-time (5 women).
Students: 29 full-time (14 women), 50 part-time (22 women); includes 9 minority (5 African Americans, 2 Asian Americans or Pacific Islanders, 2 Native Americans), 1 international. Average age 28. *52 applicants, 65% accepted.* In 1999, 17 degrees awarded.
Degree requirements: For master's, one foreign language (computer language can substitute), thesis, comprehensive exams required.
Entrance requirements: For master's, GRE General Test, GRE Subject Test, TOEFL. *Application deadline:* For fall admission, 6/1 (priority date); for spring admission, 10/15. Applications are processed on a rolling basis. *Application fee:* $40.
Expenses: Tuition, state resident: full-time $1,012. Tuition, nonresident: full-time $8,578. Required fees: $1,006. Full-time tuition and fees vary according to degree level. Part-time tuition and fees vary according to course load.
Financial aid: Fellowships with partial tuition reimbursements, research assistantships with partial tuition reimbursements, teaching assistantships with partial tuition reimbursements, career-related internships or fieldwork, Federal Work-Study, scholarships, and unspecified assistantships available. Aid available to part-time students. Financial aid application deadline: 6/1.
Faculty research: Biochemistry, microbiology, cell biology.
Dr. Gerhard W. Kalmus, Director of Graduate Studies, 252-328-6722, *Fax:* 252-328-4178, *E-mail:* kalmusg@mail.ecu.edu.
Application contact: Dr. Paul D. Tschetter, Senior Associate Dean, 252-328-6012, *Fax:* 252-328-6071, *E-mail:* grad@mail.ecu.edu.

Find an in-depth description at www.petersons.com/graduate.

■ EMORY UNIVERSITY

Graduate School of Arts and Sciences, Division of Biological and Biomedical Sciences, Program in Genetics and Molecular Biology, Atlanta, GA 30322-1100

AWARDS PhD.

Faculty: 41 full-time (10 women).
Students: 44 full-time (26 women); includes 2 African Americans, 2 Hispanic Americans, 2 international. In 1999, 10 degrees awarded.
Degree requirements: For doctorate, dissertation required, foreign language not required.
Entrance requirements: For doctorate, GRE General Test, TOEFL, minimum GPA of 3.0 in science course work. *Application deadline:* For fall admission, 1/20 (priority date). *Application fee:* $45.
Expenses: Tuition: Full-time $22,770. Tuition and fees vary according to program.
Financial aid: In 1999–00, fellowships with full tuition reimbursements (averaging $18,000 per year).
Faculty research: Gene regulation, genetic combination, developmental regulation.
Dr. Jeremy Boss, Director, 404-727-5973, *Fax:* 404-727-3659, *E-mail:* boss@microbio.emory.edu.
Application contact: 404-727-2547, *Fax:* 404-727-3322, *E-mail:* gdbbs@gsas.emory.edu.

Find an in-depth description at www.petersons.com/graduate.

■ FLORIDA INSTITUTE OF TECHNOLOGY

Graduate School, College of Science and Liberal Arts, Department of Biological Sciences, Program in Cell and Molecular Biology, Melbourne, FL 32901-6975

AWARDS MS. Part-time programs available.

Students: 4 full-time (3 women), 4 part-time; includes 1 minority (Asian American or Pacific Islander), 2 international. Average age 34. *10 applicants, 40% accepted.* In 1999, 1 degree awarded.
Degree requirements: For master's, thesis required, foreign language not required.
Entrance requirements: For master's, GRE General Test, minimum GPA of 3.0. *Application deadline:* Applications are processed on a rolling basis. *Application fee:* $50. Electronic applications accepted.
Expenses: Tuition: Part-time $575 per credit hour. Required fees: $50. Tuition and fees vary according to campus/location and program.

Financial aid: In 1999–00, 3 students received aid, including 1 research assistantship with full and partial tuition reimbursement available (averaging $3,150 per year), 2 teaching assistantships with full and partial tuition reimbursements available (averaging $4,180 per year); career-related internships or fieldwork and tuition remissions also available. Financial aid application deadline: 3/1; financial aid applicants required to submit FAFSA.
Faculty research: Changes in DNA molecule and differential expression of genetic information during aging.
Dr. Gary N. Wells, Head, 321-674-8034, *Fax:* 321-674-7238, *E-mail:* gwells@fit.edu.
Application contact: Carolyn P. Farrior, Associate Dean of Graduate Admissions, 321-674-7118, *Fax:* 321-674-9468, *E-mail:* cfarrior@fit.edu.

Find an in-depth description at www.petersons.com/graduate.

■ FLORIDA STATE UNIVERSITY

Graduate Studies, College of Arts and Sciences, Department of Biological Science, Program in Molecular Biology, Tallahassee, FL 32306

AWARDS MS, PhD.

Faculty: 23 full-time (4 women).
Students: 40 full-time (21 women); includes 6 minority (2 Asian Americans or Pacific Islanders, 4 Hispanic Americans), 12 international.
Degree requirements: For master's and doctorate, thesis/dissertation, teaching experience required.
Entrance requirements: For master's, GRE General Test, TOEFL; for doctorate, GRE General Test, GRE Subject Test, TOEFL. *Application deadline:* For fall admission, 1/15; for spring admission, 10/15. *Application fee:* $20.
Expenses: Tuition, state resident: full-time $3,504; part-time $146 per credit hour. Tuition, nonresident: full-time $12,162; part-time $507 per credit hour. Tuition and fees vary according to program.
Financial aid: In 1999–00, fellowships with full tuition reimbursements (averaging $13,740 per year), research assistantships with full tuition reimbursements (averaging $13,740 per year), teaching assistantships with full tuition reimbursements (averaging $13,740 per year) were awarded. Financial aid application deadline: 1/15; financial aid applicants required to submit FAFSA.
Faculty research: Development, fertilization, photosynthesis, nuclei, chromosomes, motility, cell division.
Dr. Thomas C. S. Keller, Associate Professor and Associate Chairman, 850-644-3023, *Fax:* 850-644-9829.

Florida State University (continued)
Application contact: Judy Bowers, Coordinator, Graduate Affairs, 850-644-3023, *Fax:* 850-644-9829, *E-mail:* bowers@bio.fsu.edu.

■ **FORDHAM UNIVERSITY**

Graduate School of Arts and Sciences, Department of Biological Sciences, New York, NY 10458

AWARDS Biological sciences (MS, PhD), including cell and molecular biology, ecology. Part-time and evening/weekend programs available.

Faculty: 17 full-time (1 woman).
Students: 29 full-time (17 women), 10 part-time (6 women); includes 2 minority (both Asian Americans or Pacific Islanders), 6 international. *57 applicants, 35% accepted.* In 1999, 11 master's, 2 doctorates awarded. Terminal master's awarded for partial completion of doctoral program.
Degree requirements: For master's, comprehensive exam required, thesis optional; for doctorate, 2 foreign languages (computer language can substitute for one), dissertation, comprehensive exam required.
Entrance requirements: For master's and doctorate, GRE General Test, GRE Subject Test (recommended). *Application deadline:* For fall admission, 1/16 (priority date); for spring admission, 12/1. *Application fee:* $60. Electronic applications accepted.
Expenses: Tuition: Full-time $14,400; part-time $600 per credit. Required fees: $125 per semester. Tuition and fees vary according to program.
Financial aid: In 1999–00, 29 students received aid, including 4 fellowships (averaging $15,000 per year), 3 research assistantships (averaging $15,000 per year), teaching assistantships (averaging $15,000 per year); institutionally sponsored loans, tuition waivers (full and partial), and unspecified assistantships also available. Aid available to part-time students. Financial aid application deadline: 1/15. *Total annual research expenditures:* $365,196.
Dr. Berish Rubin, Chair, 718-817-3641, *Fax:* 718-817-3645, *E-mail:* rubin@fordham.edu.
Application contact: Dr. Craig W. Pilant, Assistant Dean, 718-817-4420, *Fax:* 718-817-3566, *E-mail:* pilant@fordham.edu.
Find an in-depth description at www.petersons.com/graduate.

■ **GEORGE MASON UNIVERSITY**

College of Arts and Sciences, Department of Biology, Master's Program in Biology, Fairfax, VA 22030-4444

AWARDS Bioinformatics (MS); ecology, systematics and evolution (MS); environmental science and public policy (MS); interpretive biology (MS); molecular, microbial, and cellular biology (MS); organismal biology (MS). Part-time programs available.

Faculty: 30 full-time (11 women), 32 part-time/adjunct (20 women).
Students: 5 full-time (3 women), 55 part-time (37 women); includes 7 minority (1 African American, 4 Asian Americans or Pacific Islanders, 2 Hispanic Americans), 8 international. Average age 34. *36 applicants, 44% accepted.* In 1999, 18 degrees awarded.
Degree requirements: For master's, thesis or alternative required, foreign language not required.
Entrance requirements: For master's, GRE General Test, GRE Subject Test, bachelor's degree in biology or equivalent. *Application deadline:* For fall admission, 5/1; for spring admission, 11/1. *Application fee:* $30. Electronic applications accepted.
Expenses: Tuition, state resident: full-time $4,416; part-time $184 per credit hour. Tuition, nonresident: full-time $12,516; part-time $522 per credit hour. Tuition and fees vary according to program.
Financial aid: Available to part-time students. Application deadline: 3/1.
Dr. George E. Andrykovitch, Director, 703-993-1027, *Fax:* 703-993-1046.

■ **GEORGETOWN UNIVERSITY**

Graduate School of Arts and Sciences, Programs in Biomedical Sciences, Department of Biochemistry and Molecular Biology, Washington, DC 20057

AWARDS PhD, MD/PhD.

Degree requirements: For doctorate, dissertation, comprehensive exam required.
Entrance requirements: For doctorate, GRE General Test, TOEFL.
Find an in-depth description at www.petersons.com/graduate.

■ **THE GEORGE WASHINGTON UNIVERSITY**

Columbian School of Arts and Sciences, Department of Biochemistry and Molecular Biology, Washington, DC 20037

AWARDS MS, PhD, MD/PhD.

Faculty: 11 full-time (4 women).
Students: 3 full-time (2 women), 11 part-time (3 women); includes 4 minority (3 Asian Americans or Pacific Islanders, 1 Hispanic American), 4 international. Average age 29. *7 applicants, 100% accepted.* In 1999, 5 master's, 1 doctorate awarded.
Degree requirements: For master's, thesis (for some programs), comprehensive exam required; for doctorate, dissertation, general exam required.
Entrance requirements: For master's, GRE General Test, interview, minimum GPA of 3.0; for doctorate, GRE General Test, minimum GPA of 3.0. *Application fee:* $55.
Expenses: Tuition: Full-time $16,836; part-time $702 per credit hour. Required fees: $828; $35 per credit hour. Tuition and fees vary according to campus/location and program.
Financial aid: In 1999–00, 6 students received aid, including 6 fellowships. Financial aid application deadline: 2/1.
Dr. Allan L. Goldstein, Chair, 202-994-3517.
Application contact: Dr. Glenn Walker, Director of Graduate Studies, 202-994-2919.
Find an in-depth description at www.petersons.com/graduate.

■ **HARVARD UNIVERSITY**

Graduate School of Arts and Sciences, Department of Molecular and Cellular Biology, Cambridge, MA 02138

AWARDS PhD.

Students: 81 full-time (37 women). *325 applicants, 16% accepted.* In 1999, 16 degrees awarded.
Degree requirements: For doctorate, dissertation, oral exam required.
Entrance requirements: For doctorate, GRE General Test, GRE Subject Test (recommended), TOEFL. *Application deadline:* For fall admission, 1/2. *Application fee:* $60.
Expenses: Tuition: Full-time $22,054. Required fees: $711. Tuition and fees vary according to program.
Financial aid: Fellowships, research assistantships, teaching assistantships, career-related internships or fieldwork, Federal Work-Study, and institutionally sponsored loans available. Financial aid application deadline: 12/30.
Josephine Ferraro, Officer, 617-495-5396.
Application contact: James Wang, Assistant Director, 617-495-1901.
Find an in-depth description at www.petersons.com/graduate.

■ HOWARD UNIVERSITY

College of Medicine, Department of Biochemistry and Molecular Biology, Washington, DC 20059-0002

AWARDS Biochemistry and molecular biology (PhD); biotechnology (MS). Part-time programs available.

Faculty: 16.

Students: 10; includes 8 minority (6 African Americans, 2 Asian Americans or Pacific Islanders), 2 international. *15 applicants, 33% accepted.* In 1999, 1 doctorate awarded.

Degree requirements: For master's, externship required, foreign language and thesis not required; for doctorate, dissertation, oral and written comprehensive exams required, foreign language not required. *Average time to degree:* Doctorate–4 years full-time.

Entrance requirements: For master's and doctorate, GRE General Test, minimum GPA of 3.0. *Application deadline:* For fall admission, 4/1; for spring admission, 11/1. Applications are processed on a rolling basis. *Application fee:* $45.

Financial aid: In 1999–00, research assistantships with full tuition reimbursements (averaging $10,000 per year), teaching assistantships with full tuition reimbursements (averaging $10,000 per year) were awarded; fellowships.

Faculty research: Cellular and molecular biology of olfaction, gene regulation and expression, enzymology, NMR spectroscopy of molecular structure, hormone regulation/metabolism. *Total annual research expenditures:* $3.4 million.

Dr. Matthew George, Interim Chair, 202-806-6289, *Fax:* 202-806-5784, *E-mail:* mgeorge@howard.edu.

Application contact: Dr. Cynthia K. Abrams, Director of Graduate Studies, 202-806-6289, *Fax:* 202-806-5784, *E-mail:* cabrams@fac.howard.edu.

Find an in-depth description at www.petersons.com/graduate.

■ INDIANA UNIVERSITY BLOOMINGTON

Graduate School, College of Arts and Sciences, Department of Biology, Program in Molecular, Cellular, and Developmental Biology, Bloomington, IN 47405

AWARDS PhD. Offered through the University Graduate School.

Students: 54 full-time (21 women); includes 4 minority (2 Asian Americans or Pacific Islanders, 2 Hispanic Americans), 9 international. In 1999, 9 degrees awarded.

Degree requirements: For doctorate, dissertation required.

Entrance requirements: For doctorate, GRE General Test, TOEFL. *Application deadline:* For fall admission, 1/15 (priority date); for spring admission, 9/1 (priority date). Applications are processed on a rolling basis. *Application fee:* $45. Electronic applications accepted.

Expenses: Tuition, state resident: full-time $3,853; part-time $161 per credit hour. Tuition, nonresident: full-time $11,226; part-time $468 per credit hour. Required fees: $360 per year. Tuition and fees vary according to course load and program.

Financial aid: Fellowships with tuition reimbursements, research assistantships with tuition reimbursements, teaching assistantships with tuition reimbursements available. Financial aid application deadline: 2/15.

Faculty research: Developmental genetics, molecular evolution, macromolecular structure and function, cell biology and microbial molecular genetics.

Dr. Susan Strome, Head, 812-855-5450, *Fax:* 812-855-6705, *E-mail:* sstrome@bio.indiana.edu.

Application contact: Gretchen Clearwater, Advisor for Graduate Affairs, 812-855-1861, *Fax:* 812-855-6705, *E-mail:* biograd@bio.indiana.edu.

Find an in-depth description at www.petersons.com/graduate.

■ INDIANA UNIVERSITY BLOOMINGTON

Graduate School, College of Arts and Sciences, Interdepartmental Program in Biochemistry and Molecular Biology, Bloomington, IN 47405

AWARDS MS, PhD. PhD offered through the University Graduate School.

Students: 1 full-time (0 women), 1 (woman) part-time; includes 1 minority (Asian American or Pacific Islander). Terminal master's awarded for partial completion of doctoral program.

Degree requirements: For master's and doctorate, thesis/dissertation required, foreign language not required. *Average time to degree:* Doctorate–6.7 years full-time.

Entrance requirements: For doctorate, GRE General Test, GRE Subject Test (biochemistry or chemistry), TOEFL, BA or BE in biochemistry or chemistry. *Application deadline:* For fall admission, 1/15 (priority date); for spring admission, 9/1 (priority date). *Application fee:* $45.

Expenses: Tuition, state resident: full-time $3,853; part-time $161 per credit hour. Tuition, nonresident: full-time $11,226; part-time $468 per credit hour. Required fees: $360 per year. Tuition and fees vary according to course load and program.

Financial aid: In 1999–00, fellowships with full tuition reimbursements (averaging $15,750 per year), research assistantships with full tuition reimbursements (averaging $15,696 per year), teaching assistantships with full tuition reimbursements (averaging $15,720 per year) were awarded.

Faculty research: Biological membranes, enzymology, bioanalytical chemistry, photosynthesis.

Dr. John P. Richardson, Director, 812-855-1520, *E-mail:* richardj@indiana.edu.

Application contact: Dr. Jack K. Crandall, Chairperson of Admissions, 812-855-2068, *Fax:* 812-855-8300, *E-mail:* chemgrad@indiana.edu.

■ INDIANA UNIVERSITY–PURDUE UNIVERSITY INDIANAPOLIS

School of Medicine, Graduate Programs in Medicine, Department of Biochemistry and Molecular Biology, Indianapolis, IN 46202-2896

AWARDS MS, PhD, MD/MS, MD/PhD.

Students: 19 full-time (9 women), 19 part-time (8 women); includes 2 minority (1 African American, 1 Asian American or Pacific Islander), 25 international. Average age 28. *173 applicants, 16% accepted.* In 1999, 3 master's, 7 doctorates awarded. Terminal master's awarded for partial completion of doctoral program.

Degree requirements: For master's and doctorate, thesis/dissertation required, foreign language not required. *Average time to degree:* Master's–3.3 years full-time; doctorate–7 years full-time.

Entrance requirements: For master's and doctorate, GRE General Test, GRE Subject Test (recommended), previous course work in organic chemistry. *Application deadline:* For fall admission, 1/15 (priority date). Applications are processed on a rolling basis. *Application fee:* $35 ($55 for international students).

Expenses: Tuition, state resident: full-time $13,245; part-time $158 per credit hour. Tuition, nonresident: full-time $30,330; part-time $455 per credit hour. Required fees: $121 per year. Tuition and fees vary according to course load and degree level.

Financial aid: In 1999–00, 15 fellowships with tuition reimbursements (averaging $18,000 per year), 28 research assistantships with tuition reimbursements (averaging $18,000 per year) were awarded; teaching assistantships, Federal Work-Study, grants, institutionally sponsored loans, and tuition waivers (partial) also available. Aid available to part-time students. Financial aid application deadline: 2/1.

Faculty research: Metabolic regulation, enzymology, peptide and protein

Indiana University–Purdue University Indianapolis (continued)

chemistry, cell biology, signal transduction, cancer, diabetes, structural biology.
Dr. Robert A. Harris, Chairman, 317-274-7151, *E-mail:* rharris@iupui.edu.
Application contact: Dr. David W. Allmann, Chairperson, Admissions Committee, 317-274-4096, *Fax:* 317-274-4686, *E-mail:* dallman@iupui.edu.

Find an in-depth description at www.petersons.com/graduate.

■ **IOWA STATE UNIVERSITY OF SCIENCE AND TECHNOLOGY**

Graduate College, College of Agriculture and College of Liberal Arts and Sciences, Department of Biochemistry, Biophysics, and Molecular Biology, Ames, IA 50011

AWARDS Biochemistry (MS, PhD); biophysics (MS, PhD); genetics (MS, PhD); molecular, cellular, and developmental biology (MS, PhD); toxicology (MS, PhD).

Faculty: 19 full-time, 1 part-time/adjunct.
Students: 57 full-time (24 women), 6 part-time (1 woman); includes 1 minority (African American), 20 international. *22 applicants, 59% accepted.* In 2000, 4 master's, 4 doctorates awarded.
Degree requirements: For master's and doctorate, thesis/dissertation required.
Entrance requirements: For master's and doctorate, GRE General Test, TOEFL. *Application deadline:* For fall admission, 6/15 (priority date); for spring admission, 11/15 (priority date). *Application fee:* $20 ($50 for international students). Electronic applications accepted.
Expenses: Tuition, state resident: full-time $3,308. Tuition, nonresident: full-time $9,744. Part-time tuition and fees vary according to course load, campus/location and program.
Financial aid: In 2000–01, 44 research assistantships with partial tuition reimbursements (averaging $12,314 per year), 2 teaching assistantships with partial tuition reimbursements (averaging $12,375 per year) were awarded; scholarships also available.
Dr. Marit Nilsen-Hamilton, Chair, 515-294-2231, *E-mail:* biochem@iastate.edu.

Find an in-depth description at www.petersons.com/graduate.

■ **IOWA STATE UNIVERSITY OF SCIENCE AND TECHNOLOGY**

Graduate College, Interdisciplinary Programs, Bioinformatics and Computational Biology Program, Ames, IA 50011-3260

AWARDS MS, PhD.

Faculty: 46 full-time (9 women).
Students: 40.
Degree requirements: For master's and doctorate, thesis/dissertation required, foreign language not required.
Entrance requirements: For master's and doctorate, GRE General Test (waived for applicants with PhD), TOEFL. *Application deadline:* For fall admission, 2/1 (priority date). Electronic applications accepted.
Expenses: Tuition, state resident: full-time $3,308. Tuition, nonresident: full-time $9,744. Part-time tuition and fees vary according to course load, campus/location and program.
Financial aid: In 1999–00, 7 fellowships with full tuition reimbursements (averaging $23,000 per year), 8 research assistantships with full tuition reimbursements (averaging $16,800 per year) were awarded.
Faculty research: Functional and structural genomics, genome evoultion, macromolecular structure and function, mathematical biology and computational modeling, metabolic and developmental networks.
Drena L. Dobbs, Chair, 515-294-1112, *Fax:* 515-294-6790, *E-mail:* ddobbs@iastate.edu.
Application contact: Kathy Wiederin, Program Assistant, 888-569-8509, *Fax:* 515-294-6790, *E-mail:* bioinformatics@iastate.edu.

Find an in-depth description at www.petersons.com/graduate.

■ **IOWA STATE UNIVERSITY OF SCIENCE AND TECHNOLOGY**

Graduate College, Interdisciplinary Programs, Program in Molecular, Cellular, and Developmental Biology, Ames, IA 50011

AWARDS MS, PhD.

Students: 10 full-time (6 women), 9 international. *89 applicants, 20% accepted.* In 1999, 1 master's, 1 doctorate awarded.
Degree requirements: For master's, thesis or alternative required; for doctorate, dissertation required.
Entrance requirements: For master's and doctorate, GRE General Test, TOEFL. *Application deadline:* For fall admission, 2/1 (priority date). *Application fee:* $20 ($50 for international students). Electronic applications accepted.
Expenses: Tuition, state resident: full-time $3,308. Tuition, nonresident: full-time $9,744. Part-time tuition and fees vary according to course load, campus/location and program.
Financial aid: In 1999–00, 8 research assistantships with partial tuition reimbursements (averaging $11,748 per year) were awarded; teaching assistantships, scholarships also available.

Dr. Jorgen Johansen, Supervisory Committee Chair, 515-294-7252, *E-mail:* idgp@iastate.edu.
Application contact: 515-294-7252, *Fax:* 515-924-6790, *E-mail:* idgp@iastate.edu.

Find an in-depth description at www.petersons.com/graduate.

■ **JOAN AND SANFORD I. WEILL MEDICAL COLLEGE AND GRADUATE SCHOOL OF MEDICAL SCIENCES OF CORNELL UNIVERSITY**

Graduate School of Medical Sciences, Molecular Biology Graduate Program, New York, NY 10021-4896

AWARDS MS, PhD, MD/PhD.

Faculty: 22 full-time (7 women).
Students: 32 full-time (13 women); includes 10 minority (9 Asian Americans or Pacific Islanders, 1 Hispanic American), 10 international. *57 applicants, 14% accepted.* In 2000, 5 degrees awarded.
Degree requirements: For doctorate, dissertation, final exam required.
Entrance requirements: For doctorate, GRE General Test, GRE Subject Test, MCAT (MD/PhD), background in genetics, molecular biology, chemistry, or biochemistry. *Application deadline:* For fall admission, 1/15. *Application fee:* $50.
Expenses: All students in good standing receive an annual stipend of $22,880.
Financial aid: Fellowships, stipends available.
Dr. Andrew Koff, Director, 212-639-2354.

■ **JOHNS HOPKINS UNIVERSITY**

School of Hygiene and Public Health, Department of Biochemistry and Molecular Biology, Baltimore, MD 21205

AWARDS Biochemistry (PhD); biophysics (PhD), including basic mechanisms of carcinogenesis, interferon induction and herpes simplex viral transformation of mammalian cells, nucleic acid structure, function, and interaction, structure and function of mammalian genetic apparatus, structure and interaction of biopolymers and cells; reproductive biology (MHS, Sc M, PhD).

Faculty: 25 full-time, 3 part-time/adjunct.
Students: 6 full-time (3 women), 42 part-time (20 women); includes 7 minority (1 African American, 6 Asian Americans or Pacific Islanders), 15 international. *24 applicants, 50% accepted.* In 1999, 7 master's, 2 doctorates awarded.
Degree requirements: For master's, thesis required, foreign language not required; for doctorate, dissertation, 1 year full-time residency, oral and written exams required, foreign language not required.

Entrance requirements: For master's and doctorate, GRE General Test, TOEFL. *Application deadline:* For fall admission, 2/1 (priority date). Applications are processed on a rolling basis. *Application fee:* $60. Electronic applications accepted.
Expenses: Tuition: Full-time $23,660; part-time $493 per unit. Full-time tuition and fees vary according to degree level, campus/location and program.
Financial aid: Federal Work-Study, institutionally sponsored loans, scholarships, and stipends available. Aid available to part-time students. Financial aid application deadline: 4/15.
Faculty research: DNA replication, recombination, repair, mutation, and structure; glycoprotein synthesis, enzyme catalysis, carcinogenesis, and protein structure. *Total annual research expenditures:* $2.9 million.
Dr. Roger McMacken, Chairman, 410-955-3671, *E-mail:* rmcmacke@jhsph.edu.
Application contact: Gerry Graziano, Admissions Coordinator, 410-955-3671, *Fax:* 410-955-2926, *E-mail:* ggrazian@jhsph.edu.
Find an in-depth description at www.petersons.com/graduate.

■ **JOHNS HOPKINS UNIVERSITY**
School of Medicine, Graduate Programs in Medicine, Department of Molecular Biology and Genetics, Baltimore, MD 21218-2699
AWARDS PhD. PhD awarded through the Program in Biochemistry, Cellular and Molecular Biology.

Faculty: 25 full-time (6 women).
Students: 48 full-time (23 women); includes 6 minority (1 African American, 5 Asian Americans or Pacific Islanders), 14 international. In 1999, 9 doctorates awarded.
Degree requirements: For doctorate, dissertation, oral comprehensive exams required, foreign language not required.
Entrance requirements: For doctorate, GRE General Test, GRE Subject Test. *Application deadline:* For fall admission, 2/1. *Application fee:* $50.
Expenses: Tuition: Full-time $23,660.
Financial aid: Traineeships available.
Faculty research: Molecular genetics of animal cells, yeasts, *Drosophila*, and viruses; segregation.
Dr. Thomas J. Kelly, Chairman, 410-955-3292.
Application contact: Dr. Daniel Raben, Admissions Chairman, 410-955-3506, *Fax:* 410-614-8842, *E-mail:* pantol@jhmi.edu.

■ **JOHNS HOPKINS UNIVERSITY**
School of Medicine, Graduate Programs in Medicine, Department of Pharmacology and Molecular Sciences, Baltimore, MD 21205
AWARDS PhD.

Faculty: 28 full-time (4 women).
Students: 32 full-time (18 women); includes 7 minority (1 African American, 5 Asian Americans or Pacific Islanders, 1 Hispanic American), 10 international. In 1999, 6 degrees awarded (67% entered university research/teaching, 33% found other work related to degree).
Degree requirements: For doctorate. *Average time to degree:* Doctorate–6 years full-time.
Entrance requirements: For doctorate, GRE General Test, GRE Subject Test. *Application deadline:* For fall admission, 2/1. *Application fee:* $50.
Expenses: Tuition: Full-time $23,660.
Dr. Philip A. Cole, Chairman, 410-614-0540, *E-mail:* pcde@jhmi.edu.
Application contact: Dr. Wade Gibson, Director of Admissions, 410-955-7117, *Fax:* 410-955-3023, *E-mail:* wgibson@jhmi.edu.
Find an in-depth description at www.petersons.com/graduate.

■ **JOHNS HOPKINS UNIVERSITY**
School of Medicine, Graduate Programs in Medicine, Predoctoral Training Program in Human Genetics and Molecular Biology, Baltimore, MD 21218-2699
AWARDS PhD, MD/PhD.

Faculty: 55 full-time (17 women).
Students: 58 full-time (31 women); includes 10 minority (9 Asian Americans or Pacific Islanders, 1 Hispanic American), 2 international. Average age 24. *70 applicants, 14% accepted.* In 1999, 5 degrees awarded (100% entered university research/teaching).
Degree requirements: For doctorate, dissertation required, foreign language not required. *Average time to degree:* Doctorate–6 years full-time.
Entrance requirements: For doctorate, GRE General Test, GRE Subject Test. *Application deadline:* For fall admission, 1/15 (priority date). *Application fee:* $50.
Expenses: Tuition: Full-time $23,660.
Financial aid: Fellowships, teaching assistantships available.
Faculty research: Human, mammalian, and molecular genetics.
Dr. David Valle, Director, 410-955-4260, *Fax:* 410-955-7397, *E-mail:* human.genetics@jhmi.edu.

Application contact: Sandy Muscelli, Program Administrator, 410-955-4260, *Fax:* 410-955-7397, *E-mail:* muscelli@jhmi.edu.
Find an in-depth description at www.petersons.com/graduate.

■ **JOHNS HOPKINS UNIVERSITY**
School of Medicine, Graduate Programs in Medicine, Program in Biochemistry, Cellular and Molecular Biology, Baltimore, MD 21205
AWARDS PhD.

Faculty: 83 full-time (23 women).
Students: 173 full-time (83 women); includes 29 minority (2 African Americans, 24 Asian Americans or Pacific Islanders, 2 Hispanic Americans, 1 Native American), 46 international. Average age 25. *296 applicants, 23% accepted.* In 1999, 17 doctorates awarded.
Degree requirements: For doctorate, dissertation, comprehensive oral exams required, foreign language not required.
Entrance requirements: For doctorate, GRE General Test, GRE Subject Test. *Application deadline:* For fall admission, 1/15. Applications are processed on a rolling basis. *Application fee:* $50.
Expenses: Tuition: Full-time $23,660.
Financial aid: In 1999–00, 49 fellowships with full tuition reimbursements (averaging $16,760 per year), 124 research assistantships with full tuition reimbursements (averaging $16,760 per year) were awarded. Financial aid application deadline: 1/15.
Faculty research: DNA topology, protein folding, enzyme mechanisms and glycoproteins, bioenergetics, gene transcription.
Dr. Craig Montell, Director, 410-955-3225, *Fax:* 410-614-8842.
Application contact: Dr. Daniel Raben, Admissions Chairman, 410-955-3506, *Fax:* 410-614-8842, *E-mail:* pantol@jhmi.edu.
Find an in-depth description at www.petersons.com/graduate.

■ **JOHNS HOPKINS UNIVERSITY**
Zanvyl Krieger School of Arts and Sciences, Department of Biology, Baltimore, MD 21218-2699
AWARDS Biochemistry (PhD); cell biology (PhD); developmental biology (PhD); genetic biology (PhD); molecular biology (PhD).

Faculty: 23 full-time (4 women).
Students: 85 full-time (50 women); includes 23 minority (4 African Americans, 14 Asian Americans or Pacific Islanders, 5 Hispanic Americans), 18 international. Average age 24. *202 applicants, 27% accepted.* In 1999, 17 doctorates awarded.

Johns Hopkins University (continued)

Degree requirements: For doctorate, dissertation required. *Average time to degree:* Doctorate–6 years full-time.

Entrance requirements: For doctorate, GRE General Test, GRE Subject Test. *Application deadline:* For fall admission, 12/15 (priority date). Applications are processed on a rolling basis. *Application fee:* $55.

Expenses: Tuition: Full-time $24,930. Tuition and fees vary according to program.

Financial aid: In 1999–00, 73 students received aid, including 12 fellowships, 61 research assistantships, 12 teaching assistantships; Federal Work-Study and institutionally sponsored loans also available. Financial aid application deadline: 12/15; financial aid applicants required to submit FAFSA.

Faculty research: Protein and nucleic acid biochemistry and biophysical chemistry, molecular biology and development. *Total annual research expenditures:* $9.3 million. Dr. Victor G. Corces, Chair, 410-516-4693, *Fax:* 410-516-5213, *E-mail:* corces_v@jhuvms.hct.jhu.edu.

Application contact: Joan Miller, Graduate Admissions Coordinator, 410-516-5502, *Fax:* 410-516-5213, *E-mail:* joan@jhu.edu.

Find an in-depth description at www.petersons.com/graduate.

■ KANSAS STATE UNIVERSITY

Graduate School, College of Arts and Sciences, Division of Biology, Manhattan, KS 66506

AWARDS Cell biology (MS, PhD); developmental biology and physiology (MS, PhD); microbiology and immunology (MS, PhD); molecular biology and genetics (MS, PhD); systematics and ecology (MS, PhD); virology and oncology (MS, PhD). Terminal master's awarded for partial completion of doctoral program.

Degree requirements: For master's and doctorate, thesis/dissertation required, foreign language not required.

Entrance requirements: For master's and doctorate, GRE General Test. Electronic applications accepted.

Expenses: Tuition, state resident: part-time $103 per credit hour. Tuition, nonresident: part-time $338 per credit hour. Required fees: $17 per credit hour. One-time fee: $64 part-time.

Faculty research: Immune cell function, prairie ecology.

■ KENT STATE UNIVERSITY

School of Biomedical Sciences, Program in Cellular and Molecular Biology, Kent, OH 44242-0001

AWARDS MS, PhD. Offered in cooperation with Northeastern Ohio Universities College of Medicine. Terminal master's awarded for partial completion of doctoral program.

Degree requirements: For master's and doctorate, thesis/dissertation required, foreign language not required.

Entrance requirements: For master's and doctorate, GRE General Test.

Expenses: Tuition, state resident: full-time $5,334; part-time $243 per hour. Tuition, nonresident: full-time $10,238; part-time $466 per hour.

Faculty research: Molecular genetics, molecular endocrinology, virology and tumor biology, P450 enzymology and catalysis, membrane structure and function.

■ LEHIGH UNIVERSITY

College of Arts and Sciences, Department of Biological Sciences, Bethlehem, PA 18015-3094

AWARDS Behavioral and evolutionary bioscience (PhD); behavioral neuroscience (PhD); biochemistry (PhD); biology (PhD); molecular biology (PhD). Part-time programs available. Postbaccalaureate distance learning degree programs offered (no on-campus study).

Students: 31 full-time (22 women), 77 part-time (47 women); includes 11 minority (3 African Americans, 5 Asian Americans or Pacific Islanders, 3 Hispanic Americans), 8 international. *142 applicants, 22% accepted.* In 1999, 5 doctorates awarded.

Degree requirements: For doctorate, dissertation, comprehensive exam required, foreign language not required. *Average time to degree:* Doctorate–6.3 years full-time.

Entrance requirements: For doctorate, GRE General Test, GRE Subject Test, TOEFL. *Application deadline:* For fall admission, 7/15; for spring admission, 12/1. Applications are processed on a rolling basis. *Application fee:* $40. Electronic applications accepted.

Expenses: Tuition: Part-time $860 per credit. Required fees: $6 per term. Tuition and fees vary according to program.

Financial aid: In 1999–00, 30 students received aid, including 4 fellowships, 6 research assistantships, 15 teaching assistantships; career-related internships or fieldwork, institutionally sponsored loans, and tuition waivers (full and partial) also available. Financial aid application deadline: 1/15.

Faculty research: Gene expression, cell biology, virology, bacteriology, developmental biology.
Dr. Neal G. Simon, Chairperson, 610-758-3680, *Fax:* 610-758-4004.

Application contact: Dr. Jennifer J. Swann, Graduate Coordinator, 610-758-5884, *Fax:* 610-758-4004, *E-mail:* jms5@lehigh.edu.

■ LOUISIANA STATE UNIVERSITY HEALTH SCIENCES CENTER

School of Graduate Studies in New Orleans, Department of Biochemistry and Molecular Biology, New Orleans, LA 70112-2223

AWARDS MS, PhD, MD/PhD.

Faculty: 12 full-time (2 women), 8 part-time/adjunct (2 women).

Students: 13 full-time (5 women); includes 5 minority (2 African Americans, 2 Asian Americans or Pacific Islanders, 1 Hispanic American), 2 international. Average age 29. *69 applicants, 10% accepted.* Terminal master's awarded for partial completion of doctoral program.

Degree requirements: For master's and doctorate, thesis/dissertation required, foreign language not required.

Entrance requirements: For master's and doctorate, GRE General Test, TOEFL, previous course work in physics, chemistry, and calculus. *Application deadline:* For fall admission, 6/1 (priority date); for spring admission, 11/1. Applications are processed on a rolling basis. *Application fee:* $30.

Expenses: Tuition, state resident: full-time $2,878; part-time $126 per hour. Tuition, nonresident: full-time $6,003; part-time $265 per hour. Required fees: $2,272. Tuition and fees vary according to course load, degree level and program.

Financial aid: In 1999–00, 1 fellowship with full tuition reimbursement, research assistantships with full tuition reimbursements (averaging $16,000 per year), 10 teaching assistantships with full tuition reimbursements (averaging $14,500 per year) were awarded; scholarships and tuition waivers (full) also available. Financial aid application deadline: 4/1.

Faculty research: Signal transduction; enzymology; gene structure, regulation, and cloning; cancer biology. *Total annual research expenditures:* $2.2 million. Dr. Robert Roskoski, Head, 504-619-8568, *Fax:* 504-619-8775, *E-mail:* biocrr@lsumc.edu.

Application contact: Mildred Williams, Administrative Assistant, 504-619-8568, *Fax:* 504-619-8775, *E-mail:* biocmcw@lsumc.edu.

Find an in-depth description at www.petersons.com/graduate.

■ LOUISIANA STATE UNIVERSITY HEALTH SCIENCES CENTER

School of Graduate Studies in Shreveport, Department of Biochemistry and Molecular Biology, Shreveport, LA 71130-3932

AWARDS MS, PhD.

Degree requirements: For master's and doctorate, thesis/dissertation required, foreign language not required.
Entrance requirements: For master's and doctorate, GRE General Test, TOEFL.
Expenses: Tuition, state resident: full-time $2,878; part-time $126 per hour. Tuition, nonresident: full-time $6,003; part-time $265 per hour. Required fees: $2,272. Tuition and fees vary according to course load, degree level and program.
Faculty research: Metabolite transport, regulation of translation and transcription, procaryotic molecular genetics, cell matrix biochemistry, yeast molecular genetics, oncogenes.

■ LOYOLA UNIVERSITY CHICAGO

Graduate School, Department of Molecular and Cellular Biochemistry, Chicago, IL 60611-2196

AWARDS Biochemistry (MS, PhD); molecular biology (PhD); neurochemistry (PhD).

Faculty: 12 full-time (2 women), 1 (woman) part-time/adjunct.
Students: 16 full-time (9 women); includes 3 minority (1 African American, 2 Asian Americans or Pacific Islanders), 3 international. Average age 29. *40 applicants, 33% accepted.* In 1999, 1 master's awarded (100% found work related to degree); 2 doctorates awarded (100% entered university research/teaching).
Degree requirements: For master's, oral and written reports required, foreign language and thesis not required; for doctorate, dissertation, oral and written comprehensive exams required, foreign language not required. *Average time to degree:* Master's–2 years full-time; doctorate–4.5 years full-time.
Entrance requirements: For master's and doctorate, GRE General Test. *Application deadline:* For fall admission, 2/15 (priority date). Applications are processed on a rolling basis. *Application fee:* $35. Electronic applications accepted.
Expenses: Tuition: Part-time $500 per credit hour. Required fees: $42 per term.
Financial aid: In 1999–00, 11 students received aid, including 10 fellowships with full tuition reimbursements available, 4 research assistantships with full tuition reimbursements available; Federal Work-Study, institutionally sponsored loans, and

scholarships also available. Financial aid application deadline: 3/15.
Faculty research: Molecular oncology; molecular neurochemical mechanisms of brain development and alcohol addiction; biochemistry of RNA and protein synthesis and intracellular protein degradation; developmentally regulated genes; cell membranes, neurotransmitters, and cell-cell interactions.
Dr. Richard M Schultz, Chairman, 708-216-3360, *Fax:* 708-216-8523.
Application contact: Dr. Michael A. Collins, Admissions Committee, 708-216-3361, *Fax:* 708-216-8523, *E-mail:* mcollin@luc.edu.
Find an in-depth description at www.petersons.com/graduate.

■ LOYOLA UNIVERSITY CHICAGO

Graduate School, Program in Molecular Biology, Chicago, IL 60626

AWARDS PhD, MD/PhD. MD/PhD offered in cooperation with the Stritch School of Medicine.

Degree requirements: For doctorate, dissertation, written comprehensive exam required, foreign language not required.
Entrance requirements: For doctorate, GRE General Test, TOEFL.
Expenses: Tuition: Part-time $500 per credit hour. Required fees: $42 per term.
Faculty research: Cell cycle regulation, molecular immunology, molecular genetics, molecular oncology, molecular virology.
Find an in-depth description at www.petersons.com/graduate.

■ MAHARISHI UNIVERSITY OF MANAGEMENT

Graduate Studies, Program in Physiology, Molecular, and Cell Biology, Fairfield, IA 52557

AWARDS MS, PhD. Program admits applicants every other year. Terminal master's awarded for partial completion of doctoral program.

Degree requirements: For master's, thesis or alternative required, foreign language not required; for doctorate, dissertation required, foreign language not required.
Entrance requirements: For master's, GMAT or GRE, minimum GPA of 3.0; for doctorate, GRE General Test, GRE Subject Test, bachelor's degree in biology, chemistry, or related quantitative science; minimum GPA of 3.0.
Faculty research: Developmental neurobiology, aging, neurochemistry.

■ MARQUETTE UNIVERSITY

Graduate School, College of Arts and Sciences, Department of Biology, Milwaukee, WI 53201-1881

AWARDS Cell biology (MS, PhD); developmental biology (MS, PhD); ecology (MS, PhD); endocrinology (MS, PhD); evolutionary biology (MS, PhD); genetics (MS, PhD); microbiology (MS, PhD); molecular biology (MS, PhD); muscle and exercise physiology (MS, PhD); neurobiology (MS, PhD); reproductive physiology (MS, PhD).

Faculty: 16 full-time (4 women), 2 part-time/adjunct (0 women).
Students: 34 full-time (20 women), 3 part-time; includes 3 minority (all Asian Americans or Pacific Islanders), 2 international. Average age 31. *42 applicants, 29% accepted.* In 1999, 1 master's, 4 doctorates awarded. Terminal master's awarded for partial completion of doctoral program.
Degree requirements: For master's, thesis, 1 year of teaching experience or equivalent, comprehensive exam required, foreign language not required; for doctorate, dissertation, 1 year of teaching experience or equivalent, qualifying exam required, foreign language not required.
Entrance requirements: For master's and doctorate, GRE General Test, GRE Subject Test, TOEFL. *Application fee:* $40.
Expenses: Tuition: Part-time $510 per credit hour. Tuition and fees vary according to program.
Financial aid: In 1999–00, 4 fellowships, 22 teaching assistantships were awarded; research assistantships, Federal Work-Study, institutionally sponsored loans, scholarships, and tuition waivers (full and partial) also available. Aid available to part-time students. Financial aid application deadline: 2/15.
Faculty research: Microbial and invertebrate ecology, evolution of gene function, DNA methylation, DNA arrangement. *Total annual research expenditures:* $1.5 million.
Dr. Brian Unsworth, Chairman, 414-288-7355, *Fax:* 414-288-7357.
Application contact: Barbara DeNoyer, Graduate Studies Coordinator, 414-288-7355, *Fax:* 414-288-7357.
Find an in-depth description at www.petersons.com/graduate.

■ MAYO GRADUATE SCHOOL

Graduate Programs in Biomedical Sciences, Program in Biochemistry and Molecular Biology, Rochester, MN 55905

AWARDS Biochemistry (PhD); molecular biology (PhD).

Mayo Graduate School (continued)
Faculty: 70 full-time (10 women).
Students: 33 full-time (19 women); includes 3 minority (1 Asian American or Pacific Islander, 2 Hispanic Americans), 7 international. In 1999, 6 degrees awarded.
Degree requirements: For doctorate, oral defense of dissertation, qualifying oral and written exam required.
Entrance requirements: For doctorate, GRE, TOEFL, 2 years of chemistry; 1 year of biology, calculus, and physics. *Application deadline:* For fall admission, 12/31 (priority date). Applications are processed on a rolling basis. *Application fee:* $0.
Expenses: Tuition: Full-time $17,900.
Financial aid: In 1999–00, 28 students received aid, including 28 fellowships with full tuition reimbursements available; tuition waivers (full) also available.
Faculty research: Gene structure and function, membranes and receptors/cytoskeleton, oncogenes and growth factors, protein structure and function, steroid hormonal action.
Dr. Edward B. Leof, Education Coordinator, 507-284-5717, *E-mail:* leof.edward@mayo.edu.
Application contact: Sherry Kallies, Information Contact, 507-266-0122, *Fax:* 507-284-0999, *E-mail:* phd.training@mayo.edu.
Find an in-depth description at www.petersons.com/graduate.

■ MCP HAHNEMANN UNIVERSITY

School of Medicine, Biomedical Graduate Programs, Program in Molecular and Cell Biology, Philadelphia, PA 19102-1192

AWARDS MS, PhD, MD/PhD. Terminal master's awarded for partial completion of doctoral program.

Degree requirements: For master's, thesis, comprehensive exam required, foreign language not required; for doctorate, dissertation, qualifying exam required, foreign language not required.
Entrance requirements: For master's, GRE General Test, TOEFL, minimum GPA of 2.75; for doctorate, GRE General Test, TOEFL, minimum GPA of 3.0.
Faculty research: Molecular anatomy, biochemistry, medical biotechnology, molecular pathology, microbiology and immunology.

Find an in-depth description at www.petersons.com/graduate.

■ MEDICAL COLLEGE OF GEORGIA

School of Graduate Studies, Department of Biochemistry and Molecular Biology, Augusta, GA 30912-1500

AWARDS MS, PhD.

Faculty: 14 full-time (1 woman).
Students: 9 full-time (2 women), 1 (woman) part-time, 5 international. *9 applicants, 33% accepted.* In 1999, 6 degrees awarded. Terminal master's awarded for partial completion of doctoral program.
Degree requirements: For master's and doctorate, thesis/dissertation required, foreign language not required.
Entrance requirements: For master's and doctorate, GRE General Test, TOEFL. *Application deadline:* For fall admission, 6/30 (priority date). Applications are processed on a rolling basis. *Application fee:* $25.
Expenses: Tuition, state resident: full-time $2,896; part-time $121 per hour. Tuition, nonresident: full-time $11,584; part-time $483 per hour. Required fees: $286; $143 per semester. Tuition and fees vary according to program.
Financial aid: In 1999–00, 9 research assistantships with partial tuition reimbursements (averaging $15,500 per year) were awarded; grants and institutionally sponsored loans also available. Aid available to part-time students. Financial aid application deadline: 3/31; financial aid applicants required to submit FAFSA.
Faculty research: Cancer biology, chemical biology, chemical kinetics, physical methods of structure determination, physical biochemistry.
Dr. Frederick H. Leibach, Chair, 706-721-7661, *Fax:* 706-721-9947, *E-mail:* fleibach@mail.mcg.edu.
Application contact: Dr. Eugene Howard, Director, 706-721-7647, *Fax:* 706-721-9947, *E-mail:* ehoward@mail.mcg.edu.

■ MEDICAL COLLEGE OF OHIO

Graduate School, Department of Biochemistry and Molecular Biology, Toledo, OH 43614-5805

AWARDS Medical sciences (MS), including biochemistry, genetics, molecular biology. Part-time programs available.

Faculty: 8 full-time (2 women).
Students: 4 full-time (2 women), 8 part-time (1 woman), 9 international. Average age 30. In 1999, 3 degrees awarded (100% found work related to degree).
Degree requirements: For master's, thesis, qualifying exam required, foreign language not required. *Average time to degree:* Master's–3 years part-time.

Entrance requirements: For master's, GRE General Test, TOEFL, minimum undergraduate GPA of 3.0. *Application fee:* $30.
Expenses: Tuition, state resident: part-time $193 per hour. Tuition, nonresident: part-time $445 per hour. Tuition and fees vary according to degree level.
Financial aid: Fellowships, Federal Work-Study and institutionally sponsored loans available. Financial aid applicants required to submit FAFSA.
Faculty research: Gene regulation, protein structure, receptors, protein phosphorylation, peptides. *Total annual research expenditures:* $81,820.
James P. Trempe, Interim Chairman, 419-383-4117, *Fax:* 419-383-6140, *E-mail:* mcograd school@mco.edu.
Application contact: Joann Braatz, Clerk, 419-383-4117, *Fax:* 419-383-6140, *E-mail:* mcograd school@mco.edu.

■ MEDICAL COLLEGE OF OHIO

Graduate School, Program in Molecular and Cellular Biology, Toledo, OH 43614-5805

AWARDS PhD, MD/PhD.

Students: 23 full-time (12 women), 1 part-time; includes 2 minority (1 Asian American or Pacific Islander, 1 Hispanic American), 12 international. Average age 26.
Degree requirements: For doctorate, dissertation, qualifying exam required, foreign language not required.
Entrance requirements: For doctorate, GRE General Test, minimum undergraduate GPA of 3.0. *Application deadline:* Applications are processed on a rolling basis. *Application fee:* $30.
Expenses: Tuition, state resident: part-time $193 per hour. Tuition, nonresident: part-time $445 per hour. Tuition and fees vary according to degree level.
Financial aid: Fellowships, Federal Work-Study and institutionally sponsored loans available. Financial aid applicants required to submit FAFSA.
Dr. Dorothea Sawicki, Director, 419-383-4117, *Fax:* 419-383-6140, *E-mail:* mcograd school@mco.edu.
Application contact: Joann Braatz, Clerk, 419-383-4117, *Fax:* 419-383-6140, *E-mail:* mcograd school@mco.edu.

■ MEDICAL UNIVERSITY OF SOUTH CAROLINA

College of Graduate Studies, Department of Biochemistry and Molecular Biology, Charleston, SC 29425-0002

AWARDS MS, PhD, MD/PhD.

Faculty: 14 part-time/adjunct (4 women).

Students: 17 full-time (3 women); includes 3 minority (2 African Americans, 1 Hispanic American), 7 international. Average age 31. *15 applicants, 67% accepted.* In 1999, 1 degree awarded. Terminal master's awarded for partial completion of doctoral program.
Degree requirements: For master's, thesis, research seminar required, foreign language not required; for doctorate, dissertation, teaching and research seminar, oral and written exams required, foreign language not required.
Entrance requirements: For master's and doctorate, GRE General Test, TOEFL, interview. *Application deadline:* Applications are processed on a rolling basis. *Application fee:* $55. Electronic applications accepted.
Expenses: Tuition, state resident: full-time $3,470; part-time $160 per semester hour. Tuition, nonresident: full-time $4,426; part-time $213 per semester hour. Required fees: $408 per semester. One-time fee: $160. Tuition and fees vary according to program.
Financial aid: In 1999–00, 5 students received aid, including 2 fellowships (averaging $16,000 per year), 3 research assistantships; teaching assistantships, Federal Work-Study and tuition waivers (partial) also available. Financial aid application deadline: 4/1; financial aid applicants required to submit FAFSA.
Faculty research: Protein chemistry, polypeptide hormones, gene expression, development, antineoplastic agents. *Total annual research expenditures:* $2.5 million.
Dr. Y. A. Hannun, Chairman, 843-792-2331.
Application contact: Julie Johnston, Director of Admissions, 843-792-8710, *Fax:* 843-792-3764.

■ MEDICAL UNIVERSITY OF SOUTH CAROLINA

College of Graduate Studies, Program in Molecular and Cellular Biology and Pathobiology, Charleston, SC 29425-0002

AWARDS Cell biology and anatomy (PhD); marine biomedicine (PhD).

Faculty: 72 part-time/adjunct (8 women).
Students: 84 full-time (36 women). Average age 29. *56 applicants, 64% accepted.* In 1999, 9 degrees awarded.
Degree requirements: For doctorate, dissertation, teaching and research seminar, oral and written exams required.
Entrance requirements: For doctorate, GRE General Test, TOEFL, interview, minimum GPA of 3.2. *Application deadline:* Applications are processed on a rolling basis. *Application fee:* $55. Electronic applications accepted.

Expenses: Tuition, state resident: full-time $3,470; part-time $160 per semester hour. Tuition, nonresident: full-time $4,426; part-time $213 per semester hour. Required fees: $408 per semester. One-time fee: $160. Tuition and fees vary according to program.
Financial aid: In 1999–00, 27 fellowships (averaging $16,000 per year) were awarded; research assistantships, teaching assistantships, Federal Work-Study and tuition waivers (partial) also available. Financial aid application deadline: 4/1; financial aid applicants required to submit FAFSA.
Faculty research: Structural biology, marine biology, neurobiology.
Dr. Barry E. Ledford, Interim Dean, 843-792-3391, *Fax:* 843-792-6590.
Application contact: Julie Johnston, Director of Admissions, 843-792-8710, *Fax:* 843-792-3764.

■ MICHIGAN STATE UNIVERSITY

Graduate School, College of Natural Science, Program in Cell and Molecular Biology, East Lansing, MI 48824

AWARDS PhD.

Faculty: 71 full-time (21 women).
Students: 19 full-time (14 women); includes 3 minority (1 Asian American or Pacific Islander, 2 Hispanic Americans), 10 international. Average age 26. *309 applicants, 2% accepted.* In 1999, 1 degree awarded (100% found work related to degree).
Degree requirements: For doctorate, dissertation required, foreign language not required. *Average time to degree:* Doctorate–5 years full-time.
Entrance requirements: For doctorate, GRE General Test, TOEFL. *Application deadline:* For fall admission, 2/1. *Application fee:* $30 ($40 for international students). Electronic applications accepted.
Expenses: Tuition, state resident: part-time $229 per credit. Tuition, nonresident: part-time $464 per credit. Required fees: $241 per semester. Tuition and fees vary according to course load, degree level and program.
Financial aid: In 1999–00, 19 students received aid, including fellowships with full tuition reimbursements available (averaging $17,000 per year), 19 research assistantships with full tuition reimbursements available (averaging $16,680 per year); teaching assistantships with full tuition reimbursements available, career-related internships or fieldwork, grants, institutionally sponsored loans, and unspecified assistantships also available. Financial aid application deadline: 1/15.
Faculty research: Molecular genetics, molecular basis of plant and animal

diseases, plant molecular biology, protein structure and function, signal transduction.
Dr. Susan E. Conrad, Director, 517-353-5161, *Fax:* 517-353-8957, *E-mail:* conrad@pilot.msu.edu.
Application contact: Angela Zell, Secretary, 517-353-8916, *Fax:* 517-353-8957, *E-mail:* cmb@pilot.msu.edu.

Find an in-depth description at www.petersons.com/graduate.

■ MISSISSIPPI STATE UNIVERSITY

College of Agriculture and Life Sciences, Department of Biochemistry and Molecular Biology, Mississippi State, MS 39762

AWARDS Biochemistry (MS); molecular biology (PhD).

Faculty: 8 full-time (1 woman), 3 part-time/adjunct (all women).
Students: 10 full-time (5 women), 3 part-time (1 woman); includes 1 minority (African American), 8 international. Average age 27. *42 applicants, 17% accepted.* In 1999, 1 master's, 2 doctorates awarded. Terminal master's awarded for partial completion of doctoral program.
Degree requirements: For master's, thesis, comprehensive oral or written exam required, foreign language not required; for doctorate, dissertation, comprehensive oral and written exam required, foreign language not required.
Entrance requirements: For master's, GRE General Test, TOEFL, minimum GPA of 2.75; for doctorate, GRE, TOEFL. *Application deadline:* For fall admission, 7/1; for spring admission, 11/1. Applications are processed on a rolling basis. *Application fee:* $25 for international students.
Expenses: Tuition, state resident: full-time $3,017; part-time $168 per credit. Tuition, nonresident: full-time $6,119; part-time $340 per credit. Part-time tuition and fees vary according to course load and program.
Financial aid: Federal Work-Study, institutionally sponsored loans, and unspecified assistantships available. Financial aid applicants required to submit FAFSA.
Faculty research: Fish nutrition, plant and animal molecular biology, plant biochemistry, enzymology, lipid metabolism. *Total annual research expenditures:* $50,000.
Dr. John A. Boyle, Head, 662-325-2640, *Fax:* 662-325-8664, *E-mail:* jab@ra.msstate.edu.
Application contact: Jerry B. Inmon, Director of Admissions, 662-325-2224, *Fax:* 662-325-7360, *E-mail:* admit@admissions.msstate.edu.

■ MONTANA STATE UNIVERSITY–BOZEMAN

College of Graduate Studies, College of Agriculture, Department of Veterinary Molecular Biology, Bozeman, MT 59717

AWARDS MS, PhD. Part-time programs available.

Students: 3 full-time (1 woman), 10 part-time (7 women). Average age 29. *5 applicants, 80% accepted.* In 1999, 1 master's awarded.
Degree requirements: For master's, thesis required, foreign language not required.
Entrance requirements: For master's and doctorate, GRE General Test, TOEFL, minimum GPA of 3.0. *Application deadline:* For fall admission, 6/1; for spring admission, 11/1. Applications are processed on a rolling basis. *Application fee:* $50. Electronic applications accepted.
Expenses: Tuition, state resident: full-time $2,674. Tuition, nonresident: full-time $6,986. International tuition: $7,136 full-time. Tuition and fees vary according to course load and program.
Financial aid: Fellowships, research assistantships, teaching assistantships, career-related internships or fieldwork, Federal Work-Study, and scholarships available. Financial aid application deadline: 3/1; financial aid applicants required to submit FAFSA.
Faculty research: Inflammatory disassociate immunology, parasitic liver flukes, targeted gene delivery to mucosal tissues for vaccines and gene therapy. *Total annual research expenditures:* $1.7 million.
Dr. Mark A. Jutila, Head, 406-994-4705, *Fax:* 406-994-4303, *E-mail:* uvsmj@ montana.edu.

■ MOUNT SINAI SCHOOL OF MEDICINE OF NEW YORK UNIVERSITY

Graduate School of Biological Sciences, Molecular, Cellular, Biochemical and Developmental Sciences (MCBDS) Training Area, New York, NY 10029-6504

AWARDS PhD, MD/PhD.

Students: 48 full-time (18 women).
Degree requirements: For doctorate, dissertation required, foreign language not required.
Entrance requirements: For doctorate, GRE General Test, GRE Subject Test, MCAT, TOEFL. *Application deadline:* For fall admission, 4/15. *Application fee:* $60.
Expenses: Tuition: Full-time $21,750; part-time $725 per credit. Required fees: $750; $25 per credit. Full-time tuition and fees vary according to student level.

Financial aid: Fellowships with full tuition reimbursements, grants available.
Dr. Gillian Small, Program Director, *E-mail:* small@msvax.mssm.edu.
Application contact: C. Gita Bosch, Administrative Manager and Assistant Dean, 212-241-6546, *Fax:* 212-241-0651, *E-mail:* grads@mssm.edu.

■ NEW MEXICO STATE UNIVERSITY

Graduate School, Graduate Program in Molecular Biology, Las Cruces, NM 88003-8001

AWARDS MS, PhD.

Students: 24 full-time (14 women), 3 part-time (2 women). *18 applicants, 6% accepted.* In 1999, 3 doctorates awarded.
Degree requirements: For master's and doctorate, computer language, thesis/ dissertation, oral seminars required.
Entrance requirements: For master's, GRE General Test, TOEFL, minimum GPA of 3.0; for doctorate, GRE General Test, TOEFL, minimum GPA of 3.3. *Application deadline:* For fall admission, 9/1 (priority date); for spring admission, 2/1 (priority date). Applications are processed on a rolling basis. *Application fee:* $15 ($35 for international students). Electronic applications accepted.
Expenses: Tuition, state resident: full-time $2,682; part-time $112 per credit. Tuition, nonresident: full-time $8,376; part-time $349 per credit. Tuition and fees vary according to course load.
Financial aid: Research assistantships, teaching assistantships, career-related internships or fieldwork available. Financial aid application deadline: 3/1.
Faculty research: Biochemistry, microbiology, plant cell biology, molecular genetics, genomics.
Dr. Peter Lammers, Director, 505-646-3437, *Fax:* 505-646-6846, *E-mail:* plammers@nmsu.edu.

■ NEW YORK MEDICAL COLLEGE

Graduate School of Basic Medical Sciences, Program in Biochemistry and Molecular Biology, Valhalla, NY 10595-1691

AWARDS MS, PhD, MD/PhD. Part-time and evening/weekend programs available.

Faculty: 9 full-time (4 women).
Students: 17 full-time (10 women), 2 part-time (both women); includes 3 minority (1 African American, 2 Asian Americans or Pacific Islanders), 7 international. Average age 30. *22 applicants, 36% accepted.* In 1999, 1 doctorate awarded. Terminal master's awarded for partial completion of doctoral program.

Degree requirements: For master's and doctorate, computer language, thesis/ dissertation required, foreign language not required.
Entrance requirements: For master's, GRE General Test, TOEFL; for doctorate, GRE General Test, GRE Subject Test, TOEFL. *Application deadline:* For fall admission, 7/1 (priority date); for spring admission, 12/1 (priority date). Applications are processed on a rolling basis. *Application fee:* $35 ($60 for international students).
Expenses: Tuition: Part-time $430 per credit. Required fees: $15 per semester. One-time fee: $100.
Financial aid: In 1999–00, 14 research assistantships with full tuition reimbursements were awarded; career-related internships or fieldwork, Federal Work-Study, grants, institutionally sponsored loans, and tuition waivers (full) also available. Aid available to part-time students. Financial aid applicants required to submit FAFSA.
Faculty research: Mechanisms of control of blood coagulation, molecular neurobiology, molecular probes for infectious disease, protein-DNA interactions, molecular biology and biochemistry of double-stranded RNA-dependent enzymes.
Dr. Joseph Wu, Director, 914-594-4062.

■ NEW YORK UNIVERSITY

School of Medicine and Graduate School of Arts and Science, Sackler Institute of Graduate Biomedical Sciences, Program in Cellular and Molecular Biology, New York, NY 10012-1019

AWARDS Biochemistry (PhD); cell biology (PhD).

Faculty: 29 full-time (8 women).
Students: 50 full-time; includes 20 minority (5 African Americans, 12 Asian Americans or Pacific Islanders, 3 Hispanic Americans), 12 international. Average age 25. In 1999, 8 degrees awarded (100% entered university research/teaching).
Degree requirements: For doctorate, one foreign language, dissertation, qualifying exams required. *Average time to degree:* Doctorate–5.5 years full-time.
Entrance requirements: For doctorate, GRE General Test, GRE Subject Test, TOEFL. *Application deadline:* For fall admission, 2/1 (priority date). Applications are processed on a rolling basis. *Application fee:* $60.
Expenses: Tuition: Full-time $17,880; part-time $745 per credit. Required fees: $1,140; $35 per credit. Tuition and fees vary according to course load and program.

Financial aid: Fellowships, research assistantships, teaching assistantships available. Financial aid application deadline: 1/15.

Faculty research: Membrane and organelle structure and biogenesis, intracellular transport and processing of proteins, cellular recognition and cell adhesion, oncogene structure and function, action of growth factors. *Total annual research expenditures:* $1.9 million.
Dr. Daniel Rifkin, Director, 212-263-5109, *E-mail:* rifkind01@popmail.med.nyu.edu.
Application contact: Alan Frey, 212-263-8129, *E-mail:* freya01@popmail.med.nyu.edu.

■ NORTH DAKOTA STATE UNIVERSITY

Graduate Studies and Research, College of Agriculture and College of Science and Mathematics, Cellular and Molecular Biology Program, Fargo, ND 58105

AWARDS PhD. Offered in cooperation with eleven departments in the university.

Faculty: 33 full-time (4 women), 4 part-time/adjunct (1 woman).
Students: 15 full-time (7 women); includes 1 minority (Hispanic American), 5 international. Average age 27. *8 applicants, 63% accepted.*
Degree requirements: For doctorate, dissertation required, foreign language not required.
Entrance requirements: For doctorate, GRE, TOEFL. *Application deadline:* Applications are processed on a rolling basis. *Application fee:* $25.
Expenses: Tuition, state resident: full-time $3,096; part-time $112 per credit hour. Tuition, nonresident: full-time $7,588; part-time $299 per credit hour. Tuition and fees vary according to course load, campus/location and reciprocity agreements.
Financial aid: Fellowships, research assistantships, teaching assistantships available. Financial aid application deadline: 3/15.
Faculty research: Plant and animal cell biology, gene regulation, molecular genetics, plant and animal virology.
Dr. Zong-Ming Cheng, Director, 701-231-7405, *Fax:* 701-231-8474, *E-mail:* zcheng@plains.nodak.edu.

■ NORTH DAKOTA STATE UNIVERSITY

Graduate Studies and Research, College of Science and Mathematics, Department of Biochemistry and Molecular Biology, Fargo, ND 58105

AWARDS MS, PhD. Part-time programs available.

Faculty: 6 full-time (0 women), 4 part-time/adjunct (0 women).
Students: 9 full-time (4 women), 3 international. Average age 26. *10 applicants, 0% accepted.* Terminal master's awarded for partial completion of doctoral program.
Degree requirements: For master's and doctorate, thesis/dissertation required, foreign language not required. *Average time to degree:* Master's–2.5 years full-time; doctorate–5 years full-time.
Entrance requirements: For master's and doctorate, TOEFL. *Application deadline:* For fall admission, 4/15 (priority date). Applications are processed on a rolling basis. *Application fee:* $25.
Expenses: Tuition, state resident: full-time $3,096; part-time $112 per credit hour. Tuition, nonresident: full-time $7,588; part-time $299 per credit hour. Tuition and fees vary according to course load, campus/location and reciprocity agreements.
Financial aid: Research assistantships with full tuition reimbursements, teaching assistantships with full tuition reimbursements, career-related internships or fieldwork, Federal Work-Study, and institutionally sponsored loans available. Financial aid application deadline: 4/15.
Faculty research: Biotechnology, molecular biology, metabolism and enzymology, pesticide metabolism, protein structures.
Dr. S. Derek Killilea, Chair, 701-231-7946, *Fax:* 701-231-8324, *E-mail:* skillile@badlands.nodak.edu.

■ NORTHWESTERN UNIVERSITY

The Graduate School, Division of Interdepartmental Programs and Medical School, Integrated Graduate Programs in the Life Sciences, Chicago, IL 60611

AWARDS Cancer biology (PhD); cell biology (PhD); developmental biology (PhD); evolutionary biology (PhD); immunology and microbial pathogenesis (PhD); molecular biology and genetics (PhD); neurobiology (PhD); pharmacology and toxicology (PhD); structural biology and biochemistry (PhD).

Degree requirements: For doctorate, dissertation, written and oral qualifying exams required, foreign language not required.
Entrance requirements: For doctorate, GRE General Test, TOEFL.
Expenses: Tuition: Full-time $23,301. Full-time tuition and fees vary according to program.

Find an in-depth description at www.petersons.com/graduate.

■ NORTHWESTERN UNIVERSITY

The Graduate School, Judd A. and Marjorie Weinberg College of Arts and Sciences, Interdepartmental Biological Sciences Program (IBiS), Concentration in Cell and Molecular Biology, Evanston, IL 60208
AWARDS PhD.

Faculty: 59 full-time (11 women).
Students: 79 full-time (46 women); includes 4 minority (2 African Americans, 2 Hispanic Americans), 14 international. *236 applicants, 19% accepted.*
Degree requirements: For doctorate, dissertation, 2 quarters of teaching experience required, foreign language not required.
Entrance requirements: For doctorate, GRE General Test, TOEFL. *Application deadline:* For fall admission, 1/15. Applications are processed on a rolling basis. *Application fee:* $50 ($55 for international students).
Expenses: Tuition: Full-time $23,301. Full-time tuition and fees vary according to program.
Financial aid: In 1999–00, 15 fellowships with full tuition reimbursements (averaging $12,078 per year), 64 research assistantships (averaging $17,000 per year), 15 teaching assistantships (averaging $16,620 per year) were awarded; Federal Work-Study, institutionally sponsored loans, and traineeships also available. Financial aid application deadline: 12/31; financial aid applicants required to submit FAFSA.
Application contact: Latonia Trimuel, Program Assistant, 800-546-1761, *E-mail:* ibis@northwestern.edu.

Find an in-depth description at www.petersons.com/graduate.

■ NORTHWESTERN UNIVERSITY

The Graduate School, Judd A. and Marjorie Weinberg College of Arts and Sciences, Interdepartmental Biological Sciences Program (IBiS), Department of Biochemistry, Molecular Biology, and Cell Biology, Evanston, IL 60208

AWARDS Cell and molecular biology (PhD); molecular biophysics (PhD). Department participates in the Interdepartmental Biological Sciences Program (IBiS).

Faculty: 59 full-time (11 women).
Students: 79 full-time (46 women); includes 5 minority (2 African Americans, 3 Hispanic Americans), 14 international. *Application fee:* $50 ($55 for international students).
Expenses: Tuition: Full-time $23,301. Full-time tuition and fees vary according to program.
Financial aid: In 1999–00, 15 fellowships (averaging $12,078 per year), 64 research assistantships (averaging $17,000 per year),

Northwestern University (continued)
16 teaching assistantships (averaging $16,620 per year) were awarded.
Application contact: Latonia Trimuel, Program Assistant, 800-546-1761, *E-mail:* ibis@northwestern.edu.

Find an in-depth description at www.petersons.com/graduate.

■ THE OHIO STATE UNIVERSITY

Graduate School, College of Biological Sciences, Department of Molecular Genetics, Columbus, OH 43210

AWARDS Cell and developmental biology (MS, PhD); genetics (MS, PhD); molecular biology (MS, PhD).

Faculty: 18 full-time, 6 part-time/adjunct.
Students: 46 full-time (17 women), 3 part-time (1 woman); includes 3 minority (2 African Americans, 1 Asian American or Pacific Islander), 21 international. *68 applicants, 28% accepted.* In 1999, 6 master's, 8 doctorates awarded.
Degree requirements: For master's and doctorate, thesis/dissertation required, foreign language not required.
Entrance requirements: For master's and doctorate, GRE General Test, GRE Subject Test (biology). *Application deadline:* For fall admission, 8/15. Applications are processed on a rolling basis. *Application fee:* $30 ($40 for international students).
Expenses: Tuition, state resident: full-time $5,400. Tuition, nonresident: full-time $14,535. Part-time tuition and fees vary according to course load and program.
Financial aid: Fellowships, research assistantships, teaching assistantships, Federal Work-Study and institutionally sponsored loans available. Aid available to part-time students.
Dr. Lee F. Johnson, Chairman, 614-292-8084, *Fax:* 614-292-4466, *E-mail:* johnson.6@osu.edu.

Find an in-depth description at www.petersons.com/graduate.

■ THE OHIO STATE UNIVERSITY

Graduate School, College of Biological Sciences, Program in Molecular, Cellular and Developmental Biology, Columbus, OH 43210

AWARDS MS, PhD.

Faculty: 14 full-time, 73 part-time/adjunct.
Students: 65 full-time (38 women); includes 7 minority (2 African Americans, 4 Asian Americans or Pacific Islanders, 1 Hispanic American), 33 international. *114 applicants, 27% accepted.* In 1999, 4 master's, 7 doctorates awarded.
Degree requirements: For master's and doctorate, thesis/dissertation required, foreign language not required.

Entrance requirements: For master's and doctorate, GRE General Test, GRE Subject Test. *Application deadline:* For fall admission, 8/15. Applications are processed on a rolling basis. *Application fee:* $30 ($40 for international students).
Expenses: Tuition, state resident: full-time $5,400. Tuition, nonresident: full-time $14,535. Part-time tuition and fees vary according to course load and program.
Financial aid: Fellowships, research assistantships, teaching assistantships, Federal Work-Study and institutionally sponsored loans available. Aid available to part-time students.
Dr. David Bisaro, Director, 614-292-2804, *Fax:* 614-292-1538, *E-mail:* bisaro.1@osu.edu.

Find an in-depth description at www.petersons.com/graduate.

■ OHIO UNIVERSITY

Graduate Studies, College of Arts and Sciences, Interdisciplinary Program in Molecular and Cellular Biology, Athens, OH 45701-2979

AWARDS MS, PhD.

Faculty: 33 full-time (3 women).
Students: 20 full-time (8 women), 10 international. *10 applicants, 30% accepted.*
Degree requirements: For master's, thesis required; for doctorate, one foreign language (computer language can substitute), dissertation, exam, research proposal, teaching experience required.
Entrance requirements: For master's and doctorate, GRE General Test, GRE Subject Test, TOEFL. *Application deadline:* For fall admission, 3/1 (priority date). Applications are processed on a rolling basis. *Application fee:* $30.
Expenses: Tuition, state resident: full-time $5,754; part-time $238 per credit hour. Tuition, nonresident: full-time $11,055; part-time $457 per credit hour. Tuition and fees vary according to course load, degree level and campus/location.
Financial aid: In 1999–00, research assistantships with tuition reimbursements (averaging $14,500 per year), 10 teaching assistantships with tuition reimbursements (averaging $14,500 per year) were awarded; Federal Work-Study, institutionally sponsored loans, and tuition waivers (full and partial) also available.
Faculty research: Animal biotechnology, plant molecular biology, immunology, cellular genetics, biochemistry of signal transduction.
Dr. Martin Tuck, Chair, 740-593-1747, *E-mail:* tuck@ohiou.edu.

Application contact: Dr. Robert A. Colvin, Graduate Chair, 740-593-0193, *Fax:* 740-593-0300, *E-mail:* rcolvin1@ohio.edu.

Find an in-depth description at www.petersons.com/graduate.

■ OKLAHOMA STATE UNIVERSITY

Graduate College, College of Agricultural Sciences and Natural Resources, Department of Biochemistry and Molecular Biology, Stillwater, OK 74078

AWARDS MS, PhD.

Faculty: 5 full-time (1 woman).
Students: 27 full-time (11 women), 20 part-time (8 women); includes 4 minority (3 African Americans, 1 Hispanic American), 31 international. Average age 28. In 1999, 4 master's, 4 doctorates awarded.
Degree requirements: For master's, thesis required, foreign language not required; for doctorate, dissertation required.
Entrance requirements: For master's and doctorate, TOEFL. *Application deadline:* For fall admission, 6/1 (priority date). *Application fee:* $25.
Expenses: Tuition, state resident: part-time $86 per credit hour. Tuition, nonresident: part-time $275 per credit hour. Required fees: $17 per credit hour. $14 per semester. One-time fee: $20 full-time. Tuition and fees vary according to course load.
Financial aid: In 1999–00, 36 students received aid, including 5 research assistantships (averaging $12,810 per year), 33 teaching assistantships (averaging $13,300 per year); career-related internships or fieldwork, Federal Work-Study, and tuition waivers (partial) also available. Aid available to part-time students. Financial aid application deadline: 3/1.
Dr. James Blair, Head, 405-744-6189.

Find an in-depth description at www.petersons.com/graduate.

■ OREGON GRADUATE INSTITUTE OF SCIENCE AND TECHNOLOGY

Graduate Studies, Department of Biochemistry and Molecular Biology, Beaverton, OR 97006-8921

AWARDS MS, PhD. Part-time programs available. Terminal master's awarded for partial completion of doctoral program.

Degree requirements: For master's, thesis optional, foreign language not required; for doctorate, comprehensive exam, oral defense of dissertation required.

Entrance requirements: For master's, TOEFL; for doctorate, GRE General Test, GRE Subject Test, TOEFL. Electronic applications accepted.
Expenses: Tuition: Full-time $17,860; part-time $425 per credit. Full-time tuition and fees vary according to degree level, program and reciprocity agreements.
Faculty research: Biotechnology, membrane transport, metallobiochemistry.

Find an in-depth description at www.petersons.com/graduate.

■ **OREGON HEALTH SCIENCES UNIVERSITY**

School of Medicine, Graduate Programs in Medicine, Department of Biochemistry and Molecular Biology, Portland, OR 97201-3098

AWARDS PhD, MD/PhD. Part-time programs available.
Faculty: 9 full-time (2 women), 18 part-time/adjunct (8 women).
Students: 18 full-time (8 women); includes 2 minority (both Asian Americans or Pacific Islanders), 6 international. Average age 29. In 1999, 2 degrees awarded (100% entered university research/teaching).
Degree requirements: For doctorate, dissertation required, foreign language not required. *Average time to degree:* Doctorate–5 years full-time.
Entrance requirements: For doctorate, GRE General Test, MCAT (MD/PhD). *Application deadline:* For fall admission, 1/15. Applications are processed on a rolling basis. *Application fee:* $0.
Expenses: Tuition, state resident: full-time $3,132; part-time $174 per credit hour. Tuition, nonresident: full-time $5,256; part-time $292 per credit hour. Required fees: $8.5 per credit hour. $146 per term. Part-time tuition and fees vary according to course load.
Financial aid: In 1999–00, 6 fellowships with tuition reimbursements (averaging $16,500 per year), 12 research assistantships with tuition reimbursements (averaging $16,500 per year) were awarded; Federal Work-Study, institutionally sponsored loans, and scholarships also available. Financial aid application deadline: 3/1; financial aid applicants required to submit FAFSA.
Faculty research: Protein structure and function, enzymology, metabolism, membranes transport.
Dr. Jack H. Kaplan, Chairperson, 503-494-1655.
Application contact: Dr. Richard G. Brennan, Director of Graduate Admissions, 503-494-7781, *E-mail:* brennanr@ohsu.edu.

Find an in-depth description at www.petersons.com/graduate.

■ **OREGON STATE UNIVERSITY**

Graduate School, College of Science, Program in Molecular and Cellular Biology, Corvallis, OR 97331
AWARDS PhD.

Degree requirements: For doctorate, dissertation, oral and written qualifying exams required, foreign language not required.
Entrance requirements: For doctorate, TOEFL, minimum GPA of 3.0 in last 90 hours. *Application deadline:* For fall admission, 2/15. Applications are processed on a rolling basis. *Application fee:* $50.
Expenses: Tuition, state resident: full-time $6,489. Tuition, nonresident: full-time $11,061. Tuition and fees vary according to program.
Financial aid: Fellowships, career-related internships or fieldwork, Federal Work-Study, and institutionally sponsored loans available. Aid available to part-time students.
Dr. Daniel J. Arp, Director, 541-737-3799, *Fax:* 541-737-3045, *E-mail:* arpd@bcc.orst.edu.

■ **THE PENNSYLVANIA STATE UNIVERSITY MILTON S. HERSHEY MEDICAL CENTER**

Graduate School, Department of Biochemistry and Molecular Biology, Hershey, PA 17033-2360
AWARDS MS, PhD, MD/PhD.

Students: 22 full-time (9 women), 1 (woman) part-time. Average age 27. In 1999, 4 degrees awarded.
Degree requirements: For master's, foreign language and thesis not required; for doctorate, dissertation required.
Entrance requirements: For master's and doctorate, GRE General Test, TOEFL. *Application deadline:* For fall admission, 7/26. *Application fee:* $50.
Expenses: Tuition, state resident: full-time $6,886; part-time $291 per credit. Tuition, nonresident: full-time $14,118; part-time $588 per credit. Required fees: $43 per semester. Part-time tuition and fees vary according to course load.
Dr. Judith S. Bond, Assistant Dean, 717-531-8585.

■ **THE PENNSYLVANIA STATE UNIVERSITY MILTON S. HERSHEY MEDICAL CENTER**

Graduate School, Department of Microbiology and Immunology, Hershey, PA 17033-2360
AWARDS Genetics (PhD); immunology (MS, PhD); microbiology (MS); microbiology/virology (PhD); molecular biology (PhD).

Students: 19 full-time (6 women). Average age 27. In 1999, 2 master's, 3 doctorates awarded.
Degree requirements: For doctorate, dissertation required.
Entrance requirements: For doctorate, GRE General Test, TOEFL, minimum GPA of 3.0. *Application deadline:* For fall admission, 7/26. *Application fee:* $50.
Expenses: Tuition, state resident: full-time $6,886; part-time $291 per credit. Tuition, nonresident: full-time $14,118; part-time $588 per credit. Required fees: $43 per semester. Part-time tuition and fees vary according to course load.
Financial aid: Unspecified assistantships available.
Dr. Richard J. Courtney, Chair, 717-531-7659.
Application contact: Dr. Brian Wigdahl, Chairman, Graduate Program Committee, 717-531-6682, *Fax:* 717-531-6522, *E-mail:* eneidigh@cor-mail.biochem.hmc.psu.edu.

■ **THE PENNSYLVANIA STATE UNIVERSITY MILTON S. HERSHEY MEDICAL CENTER**

Graduate School, Interdepartmental Graduate Program in Cell and Molecular Biology, Hershey, PA 17033-2360
AWARDS MS, PhD, MD/PhD.

Students: 22 full-time (16 women), 1 (woman) part-time. Average age 27. In 1999, 3 doctorates awarded.
Entrance requirements: For master's and doctorate, GRE General Test. *Application deadline:* For fall admission, 7/26. *Application fee:* $50.
Expenses: Tuition, state resident: full-time $6,886; part-time $291 per credit. Tuition, nonresident: full-time $14,118; part-time $588 per credit. Required fees: $43 per semester. Part-time tuition and fees vary according to course load.
Financial aid: Research assistantships available.
Dr. Robert Levenson, Director, 717-531-1045.

■ **THE PENNSYLVANIA STATE UNIVERSITY UNIVERSITY PARK CAMPUS**

Graduate School, Eberly College of Science, Department of Biochemistry and Molecular Biology, Program in Biochemistry, Microbiology, and Molecular Biology, State College, University Park, PA 16802-1503
AWARDS MS, PhD.

Students: 83 full-time (45 women), 12 part-time (6 women). In 1999, 4 master's, 12 doctorates awarded.

The Pennsylvania State University University Park Campus (continued)

Degree requirements: For master's, thesis required; for doctorate, dissertation, comprehensive exam required.

Entrance requirements: For master's and doctorate, GRE General Test, GRE Subject Test. *Application fee:* $50.

Expenses: Tuition, state resident: full-time $6,886; part-time $291 per credit. Tuition, nonresident: full-time $14,118; part-time $588 per credit. Required fees: $46 per semester. Part-time tuition and fees vary according to course load and program.

Financial aid: Fellowships, unspecified assistantships available.

Ronald D. Porter, Director, 814-865-2538.

Find an in-depth description at www.petersons.com/graduate.

■ THE PENNSYLVANIA STATE UNIVERSITY UNIVERSITY PARK CAMPUS

Graduate School, Intercollege Graduate Programs, Intercollege Graduate Program in Integrative Biosciences, State College, University Park, PA 16802-1503

AWARDS Integrative biosciences (MS, PhD), including biomolecular transport dynamics, cell and developmental biology, cellular and molecular mechanisms of toxicity, chemical biology, ecological and molecular plant physiology, immunobiology, molecular medicine, neuroscience, nutrition science.

Students: 39 full-time (24 women), 1 part-time.

Entrance requirements: For master's and doctorate, GRE General Test. *Application fee:* $50.

Expenses: Tuition, state resident: full-time $6,886; part-time $291 per credit. Tuition, nonresident: full-time $14,118; part-time $588 per credit. Required fees: $46 per semester. Part-time tuition and fees vary according to course load and program.

Financial aid: Fellowships available.

Dr. C. R. Matthews, Co-Director, 814-863-3650.

Application contact: Admissions Committee, 814-865-3155, *Fax:* 814-863-1357, *E-mail:* lscgradadm@mail.biotec.psu.edu.

Find an in-depth description at www.petersons.com/graduate.

■ PRINCETON UNIVERSITY

Graduate School, Department of Molecular Biology, Princeton, NJ 08544-1019

AWARDS Cell biology (PhD); developmental biology (PhD); molecular biology (PhD); neuroscience (PhD).

Degree requirements: For doctorate, dissertation required.

Entrance requirements: For doctorate, GRE General Test.

Expenses: Tuition: Full-time $25,050.

Faculty research: Genetics, virology, biochemistry.

Find an in-depth description at www.petersons.com/graduate.

■ PURDUE UNIVERSITY

Graduate School, Interdisciplinary Biochemistry and Molecular Biology Program, West Lafayette, IN 47907

AWARDS PhD.

Faculty: 60 full-time (11 women).

Students: 58 full-time (16 women); includes 8 minority (1 African American, 3 Asian Americans or Pacific Islanders, 4 Hispanic Americans), 19 international. Average age 25. *90 applicants, 30% accepted.*

Degree requirements: For doctorate, dissertation, preliminary and qualifying exams required, foreign language not required. *Average time to degree:* Doctorate–6.25 years full-time.

Entrance requirements: For doctorate, GRE General Test, TOEFL, TWE. *Application deadline:* For fall admission, 2/15 (priority date). *Application fee:* $30. Electronic applications accepted.

Expenses: Tuition, state resident: full-time $4,530; part-time $130 per credit hour. Tuition, nonresident: full-time $15,310; part-time $404 per credit hour. Tuition and fees vary according to campus/location and program.

Financial aid: In 1999–00, 8 fellowships with partial tuition reimbursements (averaging $16,800 per year), 50 research assistantships with partial tuition reimbursements (averaging $16,800 per year), 11 teaching assistantships with partial tuition reimbursements (averaging $16,800 per year) were awarded; grants also available. Financial aid application deadline: 4/15.

Faculty research: Structure of macromolecules, regulatory mechanisms in metabolism, gene expression and DNA/RNA processing, drug design, cell development and differentiation.

Dr. S. S. Broyles, Chair of the Executive Committee, 765-494-0745, *Fax:* 765-496-1475.

Application contact: Elizabeth A. Chandler, Coordinator, 765-494-1634, *Fax:* 765-496-1475.

Find an in-depth description at www.petersons.com/graduate.

■ PURDUE UNIVERSITY

Graduate School, School of Science, Department of Biological Sciences, Program in Molecular Biology, West Lafayette, IN 47907

AWARDS PhD.

Faculty: 12 full-time (2 women).

Degree requirements: For doctorate, dissertation, seminars, teaching experience required, foreign language not required.

Entrance requirements: For doctorate, GRE General Test, TOEFL, TSE. *Application deadline:* For fall admission, 2/15. *Application fee:* $30. Electronic applications accepted.

Expenses: Tuition, state resident: full-time $4,530; part-time $130 per credit hour. Tuition, nonresident: full-time $15,310; part-time $404 per credit hour. Tuition and fees vary according to campus/location and program.

Financial aid: Fellowships, research assistantships, teaching assistantships available. Aid available to part-time students. Financial aid application deadline: 2/15; financial aid applicants required to submit FAFSA.

Application contact: Nancy Konopka, Graduate Studies Office Manager, 765-494-8142, *Fax:* 765-494-0876, *E-mail:* njk@bilbo.bio.purdue.edu.

■ PURDUE UNIVERSITY

School of Pharmacy and Pharmacal Sciences and Graduate School, Graduate Programs in Pharmacy and Pharmacal Sciences, Department of Medicinal Chemistry and Molecular Pharmacology, West Lafayette, IN 47907

AWARDS Analytical medicinal chemistry (PhD); computational and biophysical medicinal chemistry (PhD); medicinal and bioorganic chemistry (PhD); medicinal biochemistry and molecular biology (PhD); molecular pharmacology and toxicology (PhD); natural products and pharmacognosy (PhD); nuclear pharmacy (MS); radiopharmaceutical chemistry and nuclear pharmacy (PhD).

Faculty: 24 full-time (2 women).

Students: 48 full-time (26 women), 3 part-time (1 woman); includes 4 minority (1 African American, 1 Asian American or Pacific Islander, 2 Hispanic Americans), 13 international. Average age 29. *139 applicants, 19% accepted.* In 1999, 3 master's, 11 doctorates awarded. Terminal master's awarded for partial completion of doctoral program.

Degree requirements: For master's and doctorate, thesis/dissertation required, foreign language not required.

Entrance requirements: For master's, GRE General Test, TOEFL, minimum B average; BS in biology, chemistry, or pharmacy; for doctorate, GRE General Test, TOEFL, minimum B average; BS in biology, chemistry, or pharmacology. *Application deadline:* Applications are processed on a rolling basis. *Application fee:* $30. Electronic applications accepted.

Expenses: Tuition, state resident: full-time $4,530; part-time $130 per credit hour. Tuition, nonresident: full-time $15,310; part-time $404 per credit hour. Tuition and fees vary according to campus/location and program.
Financial aid: Fellowships, research assistantships, teaching assistantships, traineeships available. Aid available to part-time students. Financial aid applicants required to submit FAFSA.
Faculty research: Drug design and development, cancer research, drug synthesis and analysis, chemical pharmacology, environmental toxicology. Dr. R. F. Borch, Graduate Head, 765-494-1403.
Application contact: Dr. D. E. Bergstrom, Graduate Committee, 765-494-6275, *E-mail:* bergstrom@ pharmacy.purdue.edu.

Find an in-depth description at www.petersons.com/graduate.

■ **QUINNIPIAC UNIVERSITY**

School of Health Sciences, Program in Molecular and Cell Biology, Hamden, CT 06518-1940

AWARDS MS. Part-time and evening/weekend programs available.

Faculty: 6 full-time (1 woman), 2 part-time/adjunct (1 woman).
Students: 3 full-time (2 women), 12 part-time (8 women); includes 1 minority (Hispanic American). Average age 29. *8 applicants, 88% accepted.*
Degree requirements: For master's, thesis optional.
Entrance requirements: For master's, minimum GPA of 2.5. *Application deadline:* For fall admission, 8/1 (priority date); for spring admission, 12/15 (priority date). Applications are processed on a rolling basis. *Application fee:* $45. Electronic applications accepted.
Expenses: Tuition: Part-time $410 per credit hour. Required fees: $20 per term. Tuition and fees vary according to program.
Financial aid: Available to part-time students.
Dr. Charlotte Hammond, Director, 203-582-8058 Ext. 8058, *E-mail:* charlotte.hammond@quinnipiac.edu.
Application contact: Scott Farber, Director of Graduate Admissions, 800-462-1944, *Fax:* 203-582-3443, *E-mail:* graduate@quinnipiac.edu.

Find an in-depth description at www.petersons.com/graduate.

■ **RENSSELAER POLYTECHNIC INSTITUTE**

Graduate School, School of Science, Department of Biology, Troy, NY 12180-3590

AWARDS Biochemistry (MS, PhD); biophysics (MS, PhD); cell biology (MS, PhD); developmental biology (MS, PhD); microbiology (MS, PhD); molecular biology (MS, PhD); plant science (MS, PhD). Part-time programs available.

Faculty: 14 full-time (5 women).
Students: 16 full-time (8 women), 2 part-time (1 woman); includes 1 minority (Asian American or Pacific Islander), 4 international. *37 applicants, 41% accepted.* In 1999, 6 master's, 3 doctorates awarded. Terminal master's awarded for partial completion of doctoral program.
Degree requirements: For master's and doctorate, thesis/dissertation required, foreign language not required.
Entrance requirements: For master's and doctorate, GRE General Test, TOEFL. *Application deadline:* For fall admission, 2/1 (priority date). Applications are processed on a rolling basis. *Application fee:* $35.
Expenses: Tuition: Part-time $665 per credit hour. Required fees: $980.
Financial aid: In 1999–00, 8 research assistantships with partial tuition reimbursements (averaging $15,000 per year), 11 teaching assistantships with full tuition reimbursements (averaging $15,000 per year) were awarded; fellowships, career-related internships or fieldwork and institutionally sponsored loans also available. Financial aid application deadline: 2/1.
Faculty research: Applied environmental biology, genetics, environmental science, fresh water ecology, microbial ecology. Dr. John Salerno, Chair, 518-276-2699, *Fax:* 518-276-2344.
Application contact: Dr. Jackie L. Collier, Assistant Professor, 518-276-6446, *Fax:* 518-276-2344, *E-mail:* collij3@rpi.edu.

Find an in-depth description at www.petersons.com/graduate.

■ **RUTGERS, THE STATE UNIVERSITY OF NEW JERSEY, NEW BRUNSWICK**

Graduate School, Core Curriculum in Molecular and Cell Biology, New Brunswick, NJ 08901-1281

AWARDS PhD.

Students: Average age 26.
Degree requirements: For doctorate, dissertation required.
Entrance requirements: For doctorate, GRE General Test, GRE Subject Test, TOEFL. *Application fee:* $40.

Expenses: Tuition, state resident: full-time $6,776; part-time $279 per credit. Tuition, nonresident: full-time $9,936; part-time $412 per credit. Required fees: $20 per credit. $89 per semester. Tuition and fees vary according to course load, campus/ location and program.
Financial aid: Fellowships available. Michael Leibowitz, Program Director, 732-235-4795.

■ **RUTGERS, THE STATE UNIVERSITY OF NEW JERSEY, NEW BRUNSWICK**

Graduate School, Program in Biochemistry, New Brunswick, NJ 08901-1281

AWARDS Biochemistry (MS, PhD); molecular biology (MS, PhD).

Faculty: 100 full-time (26 women).
Students: 47 full-time (14 women), 13 part-time (8 women); includes 5 minority (all Asian Americans or Pacific Islanders), 34 international. Average age 25. *101 applicants, 19% accepted.* In 1999, 2 master's, 7 doctorates awarded. Terminal master's awarded for partial completion of doctoral program.
Degree requirements: For master's, thesis, qualifying exam required, foreign language not required; for doctorate, dissertation, written qualifying exam required, foreign language not required.
Entrance requirements: For master's, GRE General Test, GRE Subject Test; for doctorate, GRE General Test, GRE Subject Test, TOEFL, minimum GPA of 3.0. *Application deadline:* For fall admission, 2/15. Applications are processed on a rolling basis. *Application fee:* $50.
Expenses: Tuition, state resident: full-time $6,776; part-time $279 per credit. Tuition, nonresident: full-time $9,936; part-time $412 per credit. Required fees: $20 per credit. $89 per semester. Tuition and fees vary according to course load, campus/ location and program.
Financial aid: In 1999–00, 21 fellowships with full tuition reimbursements (averaging $15,000 per year), 24 research assistantships with full tuition reimbursements (averaging $15,000 per year), 3 teaching assistantships with full tuition reimbursements (averaging $14,000 per year) were awarded. Financial aid application deadline: 3/1; financial aid applicants required to submit FAFSA.
Faculty research: Chemistry of biological systems, mechanisms of enzyme action. Dr. Alice Y. Liu, Director, 732-445-2730, *Fax:* 732-445-6370, *E-mail:* liu@ biology.rutgers.edu.
Application contact: Carolyn Ambrose, Administrative Assistant, 732-445-3430,

Rutgers, The State University of New Jersey, New Brunswick (continued)
Fax: 732-445-6370, *E-mail:* ambrose@biology.rutgers.edu.

■ RUTGERS, THE STATE UNIVERSITY OF NEW JERSEY, NEW BRUNSWICK

Graduate School, Program in Microbiology and Molecular Genetics, New Brunswick, NJ 08901-1281

AWARDS Applied microbiology (MS, PhD); clinical microbiology (MS, PhD); computational molecular biology (PhD); immunology (MS, PhD); microbial biochemistry (MS, PhD); molecular genetics (MS, PhD); virology (MS, PhD). Part-time programs available.

Faculty: 122 full-time (25 women), 4 part-time/adjunct (1 woman).
Students: 65 full-time (31 women), 51 part-time (24 women); includes 24 minority (14 Asian Americans or Pacific Islanders, 10 Hispanic Americans), 29 international. Average age 27. *200 applicants, 26% accepted.* In 1999, 15 master's, 10 doctorates awarded. Terminal master's awarded for partial completion of doctoral program.
Degree requirements: For master's, thesis or alternative required, foreign language not required; for doctorate, dissertation, written qualifying exam required, foreign language not required.
Entrance requirements: For master's, GRE General Test, GRE Subject Test; for doctorate, GRE General Test, GRE Subject Test, TOEFL, minimum GPA of 3.0. *Application deadline:* For fall admission, 2/15 (priority date). Applications are processed on a rolling basis. *Application fee:* $50.
Expenses: Tuition, state resident: full-time $6,776; part-time $279 per credit. Tuition, nonresident: full-time $9,936; part-time $412 per credit. Required fees: $20 per credit. $89 per semester. Tuition and fees vary according to course load, campus/location and program.
Financial aid: In 1999–00, 70 students received aid, including 32 fellowships with full tuition reimbursements available, 26 research assistantships with full tuition reimbursements available, 12 teaching assistantships with full tuition reimbursements available; institutionally sponsored loans also available. Financial aid application deadline: 2/15; financial aid applicants required to submit FAFSA.
Faculty research: Biochemistry of microbes, medical microbiology.
Dr. Howard Passmore, Director, 732-445-2812, *E-mail:* passmore@biology.rutgers.edu.

Application contact: Betty Green, Administrative Assistant, 732-445-5086, *Fax:* 732-445-5735, *E-mail:* green@mbcl.rutgers.edu.

■ SAINT LOUIS UNIVERSITY

School of Medicine and Graduate School, Graduate Programs in Biomedical Sciences, Cell and Molecular Biology Training Program, St. Louis, MO 63103-2097

AWARDS PhD.

Students: Average age 31. In 1999, 8 degrees awarded.
Degree requirements: For doctorate, dissertation, departmental qualifying exams required.
Entrance requirements: For doctorate, GRE General Test, GRE Subject Test. *Application deadline:* For fall admission, 4/15 (priority date). Applications are processed on a rolling basis. *Application fee:* $40.
Expenses: Tuition: Part-time $507 per credit hour. Required fees: $38 per term.
Financial aid: In 1999–00, 8 students received aid, including 1 research assistantship; fellowships, teaching assistantships, traineeships also available. Aid available to part-time students. Financial aid application deadline: 8/1; financial aid applicants required to submit FAFSA.
Dr. Duane P. Grandgenette, Director, 314-577-8418.
Application contact: Sheila Trimble, Administrative Assistant, 314-577-8418, *E-mail:* trimble@slu.edu.

■ SAINT LOUIS UNIVERSITY

School of Medicine and Graduate School, Graduate Programs in Biomedical Sciences, Department of Biochemistry and Molecular Biology, St. Louis, MO 63103-2097

AWARDS PhD.

Faculty: 26 full-time (9 women), 1 part-time/adjunct (0 women).
Students: 2 full-time (both women), 8 part-time, 1 international. Average age 26. In 1999, 6 degrees awarded.
Degree requirements: For doctorate, dissertation, departmental qualifying exams required, foreign language not required.
Entrance requirements: For doctorate, GRE General Test. *Application deadline:* Applications are processed on a rolling basis. *Application fee:* $40.
Expenses: Tuition: Part-time $507 per credit hour. Required fees: $38 per term.
Financial aid: In 1999–00, 9 students received aid, including 8 fellowships, 2 research assistantships; teaching assistantships. Aid available to part-time students.

Financial aid application deadline: 8/1; financial aid applicants required to submit FAFSA.
Faculty research: Biochemical genetics, lipid metabolism, cell biology, structure and function of proteins, gene expression. Dr. William S. Sly, Chairman, 314-577-8131.

Find an in-depth description at www.petersons.com/graduate.

■ SALEM-TEIKYO UNIVERSITY

Graduate School, Department of Bioscience, Salem, WV 26426-0500

AWARDS Biotechnology/molecular biology (MS).

Degree requirements: For master's, thesis required, foreign language not required.
Entrance requirements: For master's, GRE, minimum undergraduate GPA of 3.0. Electronic applications accepted.
Expenses: Tuition: Full-time $10,000; part-time $165 per credit hour. Required fees: $55; $55 per year.
Faculty research: Genetic engineering of seed storage proteins, virus replication and infection, gene therapy, programmed cell death, cell protocols for creation of gene transfer.

■ SAN DIEGO STATE UNIVERSITY

Graduate and Research Affairs, College of Sciences, Department of Biological Sciences, San Diego, CA 92182

AWARDS Biology (MA, MS), including ecology (MS), molecular biology (MS), physiology (MS), systematics/evolution (MS); cell and molecular biology (PhD); ecology (PhD); microbiology (MS).

Students: 37 full-time (23 women), 86 part-time (46 women); includes 13 minority (1 African American, 10 Asian Americans or Pacific Islanders, 2 Hispanic Americans), 4 international. Average age 26. *135 applicants, 33% accepted.* In 1999, 20 master's, 4 doctorates awarded. Terminal master's awarded for partial completion of doctoral program.
Degree requirements: For master's, thesis required; for doctorate, dissertation required.
Entrance requirements: For master's, GRE General Test, GRE Subject Test, TOEFL. *Application deadline:* For fall admission, 7/1 (priority date); for spring admission, 12/1. Applications are processed on a rolling basis. *Application fee:* $55.
Expenses: Tuition, nonresident: part-time $246 per unit. Required fees: $1,932; $633 per semester. Tuition and fees vary according to course load.

Financial aid: Fellowships, research assistantships, teaching assistantships, career-related internships or fieldwork and unspecified assistantships available. *Total annual research expenditures:* $11.5 million. Sanford Bernstein, Chair, 619-594-5629, *Fax:* 619-594-5676, *E-mail:* sanford.bernstcin@sdsu.edu.

Application contact: Ken Johnson, Graduate Coordinator, 619-594-6919, *Fax:* 619-594-5676, *E-mail:* kjohnson@ sunstroke.sdsu.edu.

■ SAN DIEGO STATE UNIVERSITY

Graduate and Research Affairs, College of Sciences, Molecular Biology Institute and Department of Chemistry and Department of Biological Sciences, Program in Cell and Molecular Biology, San Diego, CA 92182

AWARDS PhD.

Students: 7 full-time (5 women), 24 part-time (14 women); includes 6 minority (5 Asian Americans or Pacific Islanders, 1 Hispanic American), 1 international.
Degree requirements: For doctorate, dissertation, oral comprehensive qualifying exam required. *Average time to degree:* Doctorate–5 years full-time.
Entrance requirements: For doctorate, GRE General Test, GRE Subject Test. *Application deadline:* For fall admission, 7/1 (priority date); for spring admission, 12/1. Applications are processed on a rolling basis. *Application fee:* $55.
Expenses: Tuition, nonresident: part-time $246 per unit. Required fees: $1,932; $633 per semester. Tuition and fees vary according to course load.
Financial aid: In 1999–00, 5 fellowships were awarded; institutionally sponsored loans also available. Financial aid applicants required to submit CSS PROFILE or FAFSA.
Faculty research: Structure/dynamics of protein kinesis, chromatin structure and DNA methylation membrane biochemistry, secretory protein targeting, molecular biology of cardiac myocytes.
Dr. Skai Krisans, Head, 619-594-5368, *Fax:* 619-594-5676, *E-mail:* skrisans@ sunstroke.sdsu.edu.

Find an in-depth description at www.petersons.com/graduate.

■ SAN FRANCISCO STATE UNIVERSITY

Graduate Division, College of Science and Engineering, Department of Biology, Program in Cell and Molecular Biology, San Francisco, CA 94132-1722

AWARDS MA.

Entrance requirements: For master's, minimum GPA of 2.5 in last 60 units.
Expenses: Tuition, nonresident: full-time $5,904; part-time $246 per unit. Required fees: $1,904; $637 per semester. Tuition and fees vary according to course load.

■ THE SCRIPPS RESEARCH INSTITUTE

Office of Graduate Studies, Macromolecular and Cellular Structure and Chemistry Graduate Program, La Jolla, CA 92037

AWARDS PhD.

Faculty: 84 full-time (12 women).
Students: 73 full-time (29 women). *121 applicants, 26% accepted.* In 1999, 10 degrees awarded.
Degree requirements: For doctorate, dissertation required, foreign language not required. *Average time to degree:* Doctorate–5 years full-time.
Entrance requirements: For doctorate, GRE General Test (average 90th percentile), GRE Subject Test (average 72nd percentile), TOEFL. *Application deadline:* For fall admission, 1/1. *Application fee:* $0.
Expenses: All students are fully funded.
Financial aid: Institutionally sponsored loans and stipends available.
Faculty research: Biocatalysis and enzyme engineering, molecular structure and function, neurosciences, immunology, plant biology.
Application contact: Marylyn Rinaldi, Graduate Program Administrator, 858-784-8469, *Fax:* 858-784-2802, *E-mail:* mrinaldi@scripps.edu.

■ SOUTHERN ILLINOIS UNIVERSITY CARBONDALE

Graduate School, College of Science, Program in Molecular Biology, Microbiology, and Biochemistry, Carbondale, IL 62901-6806

AWARDS MS, PhD.

Faculty: 16 full-time (2 women).
Students: 51 full-time (24 women), 11 part-time (8 women); includes 4 minority (2 African Americans, 1 Asian American or Pacific Islander, 1 Hispanic American), 24 international. Average age 25. *34 applicants, 59% accepted.* In 1999, 7 master's, 5 doctorates awarded.

Degree requirements: For master's and doctorate, thesis/dissertation required, foreign language not required.
Entrance requirements: For master's, GRE, TOEFL, minimum GPA of 2.7; for doctorate, GRE, TOEFL, minimum GPA of 3.25. *Application deadline:* Applications are processed on a rolling basis. *Application fee:* $20.
Expenses: Tuition, state resident: full-time $2,902. Tuition, nonresident: full-time $5,810. Tuition and fees vary according to course load.
Financial aid: In 1999–00, 40 students received aid, including 3 fellowships with full tuition reimbursements available, 24 research assistantships with full tuition reimbursements available, 12 teaching assistantships with full tuition reimbursements available; Federal Work-Study and institutionally sponsored loans also available. Aid available to part-time students. Financial aid application deadline: 3/1.
Faculty research: Prokaryotic gene regulation and expression; eukaryotic gene regulation; microbial, phylogenetic, and metabolic diversity; immune responses to tumors, pathogens, and autoantigens; protein folding and structure.
Dr. John Martinko, Director, 618-536-2349, *Fax:* 618-453-8036, *E-mail:* martinko.mbmb@science.siu.edu.
Application contact: Donna Mueller, Office Systems Specialist II, 618-536-2349, *Fax:* 618-453-8036, *E-mail:* mueller@ micro.siu.edu.

■ SOUTHWEST MISSOURI STATE UNIVERSITY

Graduate College, College of Health and Human Services, Department of Biomedical Sciences, Program in Cell and Molecular Biology, Springfield, MO 65804-0094

AWARDS MS.

Degree requirements: For master's, thesis or alternative, oral and written exams required, foreign language not required.
Entrance requirements: For master's, GRE General Test, 2 semesters of organic chemistry and physics, 1 semester of calculus, minimum GPA of 3.0 during last 60 hours of course work.
Expenses: Tuition, state resident: full-time $2,070; part-time $115 per credit. Tuition, nonresident: full-time $4,140; part-time $230 per credit. Required fees: $91 per credit. Tuition and fees vary according to course level, course load and program.

■ STATE UNIVERSITY OF NEW YORK AT ALBANY

College of Arts and Sciences, Department of Biological Sciences, Specialization in Molecular, Cellular, Developmental, and Neural Biology, Albany, NY 12222-0001

AWARDS MS, PhD.

Degree requirements: For master's, one foreign language required; for doctorate, dissertation required.

Entrance requirements: For master's and doctorate, GRE General Test. *Application fee:* $50.

Expenses: Tuition, state resident: full-time $5,100; part-time $214 per credit. Tuition, nonresident: full-time $8,416; part-time $352 per credit. Required fees: $31 per credit.

Financial aid: Minority assistantships available.
Dr. David Shub, Chair, Department of Biological Sciences, 518-442-4300.

■ STATE UNIVERSITY OF NEW YORK AT ALBANY

School of Public Health, Department of Biomedical Sciences, Program in Biochemistry, Molecular Biology, and Genetics, Albany, NY 12222-0001

AWARDS MS, PhD.

Degree requirements: For master's and doctorate, thesis/dissertation required.

Entrance requirements: For master's and doctorate, GRE General Test, GRE Subject Test. *Application deadline:* For fall admission, 1/15 (priority date); for spring admission, 11/1 (priority date). *Application fee:* $50.

Expenses: Tuition, state resident: full-time $5,100; part-time $214 per credit. Tuition, nonresident: full-time $8,416; part-time $352 per credit. Required fees: $31 per credit.

Financial aid: Application deadline: 2/1.
Dr. Harry Taber, Chair, Department of Biomedical Sciences, 518-474-2662.

■ STATE UNIVERSITY OF NEW YORK AT BUFFALO

Graduate School, Graduate Programs in Biomedical Sciences at Roswell Park Cancer Institute, Department of Cellular and Molecular Biology at Roswell Park Cancer Institute, Buffalo, NY 14263

AWARDS PhD.

Faculty: 19 full-time (1 woman), 1 part-time/adjunct (0 women).

Students: 11 full-time (6 women), 7 part-time (4 women); includes 3 minority (all Asian Americans or Pacific Islanders), 8 international. Average age 25. *40 applicants,*

33% accepted. In 1999, 4 doctorates awarded (100% continued full-time study).

Degree requirements: For doctorate, dissertation, exam project required. *Average time to degree:* Doctorate–6.4 years full-time.

Entrance requirements: For doctorate, GRE General Test, GRE Subject Test, TOEFL, TSE, TWE, minimum B average in undergraduate coursework. *Application deadline:* For fall admission, 2/1 (priority date). Applications are processed on a rolling basis. *Application fee:* $35. Electronic applications accepted.

Expenses: Tuition, state resident: full-time $5,100; part-time $213 per credit hour. Tuition, nonresident: full-time $8,416; part-time $351 per credit hour. Required fees: $935; $75 per semester. Tuition and fees vary according to course load and program.

Financial aid: In 1999–00, 4 fellowships with full tuition reimbursements (averaging $15,000 per year), 17 research assistantships with full tuition reimbursements (averaging $15,000 per year) were awarded; unspecified assistantships also available. Financial aid application deadline: 2/1; financial aid applicants required to submit FAFSA.

Faculty research: Cancer genetics, chromatin structure and replication, regulation of transcription, human gene mapping, genetic and structural approaches to regulation of gene expression, mouse genetics, gene libraries. *Total annual research expenditures:* $4.5 million.
Dr. John Yates, Chair, 716-845-8964, *Fax:* 716-845-8449, *E-mail:* yates@sc3101.med.buffalo.edu.

Application contact: Craig R. Johnson, Director of Graduate Studies, 716-845-2339, *Fax:* 716-845-8178, *E-mail:* rpgradapp@sc3103.med.buffalo.edu.

■ STATE UNIVERSITY OF NEW YORK AT BUFFALO

Graduate School, Interdisciplinary and Interdepartmental Graduate Program in Molecular Cell Biology, Buffalo, NY 14260

AWARDS PhD, Certificate. PhD awarded through participating departments.

Faculty: 53.

Students: *80 applicants, 6% accepted.*

Degree requirements: For doctorate, dissertation required.

Entrance requirements: For doctorate, GRE General Test, GRE Subject Test (recommended), TOEFL, minimum GPA of 3.0. *Application deadline:* For fall admission, 2/1 (priority date). Applications are processed on a rolling basis. *Application fee:* $35.

Expenses: All students receive a full tuition waiver, health coverage, and an annual stipend of $15,000.

Financial aid: In 1999–00, 16 students received aid, including 4 research assistantships with full tuition reimbursements available (averaging $15,000 per year); institutionally sponsored loans and stipends also available. Financial aid applicants required to submit FAFSA.

Faculty research: Nucleic acid-protein interactions, protein structure, membrane proteins and channels, microbiology and parasitology, virology. *Total annual research expenditures:* $28 million.
Dr. Stephen J. Free, Director, 716-645-2865, *Fax:* 716-645-3776, *E-mail:* free@acsu.buffalo.edu.

Application contact: Xochitl E. Nicholson, Program Administrator, 716-645-2164, *Fax:* 716-645-3776, *E-mail:* ubmcbp@hopper.bio.buffalo.edu.

Find an in-depth description at www.petersons.com/graduate.

■ STATE UNIVERSITY OF NEW YORK AT STONY BROOK

Graduate School, College of Arts and Sciences, Molecular and Cellular Biology Program, Stony Brook, NY 11794

AWARDS Biochemistry and molecular biology (PhD); cellular and developmental biology (PhD); immunology and pathology (PhD).

Degree requirements: For doctorate, dissertation, comprehensive exam, teaching experience required.

Entrance requirements: For doctorate, GRE General Test, GRE Subject Test, TOEFL. *Application deadline:* For fall admission, 1/15. *Application fee:* $50.

Expenses: Tuition, state resident: full-time $5,100; part-time $213 per credit hour. Tuition, nonresident: full-time $8,416; part-time $351 per credit hour. Required fees: $492. Tuition and fees vary according to program.

Financial aid: Fellowships, research assistantships, teaching assistantships, Federal Work-Study available.

Application contact: Information Contact, 631-632-8533, *Fax:* 631-632-9730.

Find an in-depth description at www.petersons.com/graduate.

■ STATE UNIVERSITY OF NEW YORK HEALTH SCIENCE CENTER AT BROOKLYN

School of Graduate Studies, Core Program in Molecular and Cellular Biology, Brooklyn, NY 11203-2098

AWARDS Anatomy and cell biology (PhD); biochemistry (PhD); biophysics (PhD);

microbiology and immunology (PhD); neural and behavioral sciences (PhD); pharmacology (PhD); physiology (PhD). Affiliation with a particular PhD degree-granting program is deferred to the second year.

Degree requirements: For doctorate, one foreign language, dissertation required.
Entrance requirements: For doctorate, GRE General Test.
Expenses: Tuition, state resident: full-time $5,100; part-time $213 per credit. Tuition, nonresident: full-time $8,416; part-time $351 per credit. Required fees: $200. Full-time tuition and fees vary according to program and student level.
Faculty research: Mechanism of gene regulation, molecular virology.

■ STATE UNIVERSITY OF NEW YORK UPSTATE MEDICAL UNIVERSITY

College of Graduate Studies, Department of Biochemistry and Molecular Biology, Syracuse, NY 13210-2334

AWARDS MS, PhD, MD/PhD.

Faculty: 10 full-time.
Students: 13 full-time (9 women), 1 part-time; includes 3 minority (all Asian Americans or Pacific Islanders), 6 international. *48 applicants, 10% accepted.* In 1999, 1 master's awarded (100% found work related to degree). Terminal master's awarded for partial completion of doctoral program.
Degree requirements: For master's, thesis required, foreign language not required; for doctorate, dissertation, comprehensive exam required, foreign language not required.
Entrance requirements: For master's and doctorate, GRE General Test, GRE Subject Test, TSE. *Application deadline:* For fall admission, 4/1 (priority date). Applications are processed on a rolling basis. *Application fee:* $40.
Expenses: Tuition, state resident: full-time $5,100; part-time $213 per credit. Tuition, nonresident: full-time $8,416; part-time $351 per credit. Required fees: $410; $25 per credit. Part-time tuition and fees vary according to course load and program.
Financial aid: Fellowships, research assistantships, Federal Work-Study and institutionally sponsored loans available. Aid available to part-time students. Financial aid application deadline: 4/15.
Faculty research: Enzymology, membrane structure and functions, developmental biochemistry.
Dr. Richard L. Cross, Chairperson, 315-464-5127.
Application contact: Dr. David Gilbert, Associate Professor, 315-464-8723.

■ STATE UNIVERSITY OF NEW YORK UPSTATE MEDICAL UNIVERSITY

College of Graduate Studies, Program in Cell and Molecular Biology, Syracuse, NY 13210-2334

AWARDS PhD.

Faculty: 42.
Students: 22 full-time (5 women), 1 (woman) part-time; includes 2 minority (both Asian Americans or Pacific Islanders), 4 international. *17 applicants, 24% accepted.*
Degree requirements: For doctorate, dissertation, comprehensive exam required, foreign language not required.
Entrance requirements: For doctorate, GRE General Test, GRE Subject Test, TSE. *Application deadline:* For fall admission, 4/1 (priority date). Applications are processed on a rolling basis. *Application fee:* $40.
Expenses: Tuition, state resident: full-time $5,100; part-time $213 per credit. Tuition, nonresident: full-time $8,416; part-time $351 per credit. Required fees: $410; $25 per credit. Part-time tuition and fees vary according to course load and program.
Application contact: Dr. Richard Wojcikiewicz, Associate Professor, 315-464-7655.

Find an in-depth description at www.petersons.com/graduate.

■ TEMPLE UNIVERSITY

Health Sciences Center, School of Medicine and Graduate School, Graduate Programs in Medicine, Program in Molecular Biology and Genetics, Philadelphia, PA 19140

AWARDS PhD, MD/PhD.

Faculty: 7 full-time (2 women).
Students: 27 full-time (15 women), 2 international.
Degree requirements: For doctorate, dissertation, presentation of research and literature seminars, including a research proposal distinct from the student's area of concentration required.
Entrance requirements: For doctorate, GRE General Test, GRE Subject Test (biology, chemistry, or molecular cell biology recommended). *Application deadline:* For fall admission, 3/1. *Application fee:* $40. Electronic applications accepted.
Expenses: Tuition, state resident: full-time $6,030. Tuition, nonresident: full-time $8,298. Required fees: $230. One-time fee: $10 full-time.
Financial aid: Fellowships, research assistantships, Federal Work-Study, institutionally sponsored loans, and tuition waivers (full) available. Financial aid application deadline: 3/1.

Faculty research: Molecular genetics of normal and malignant cell growth, regulation of gene expression, DNA repair systems and carcinogenesis, hormone-receptor interactions and signal transduction systems, structural biology.
Dr. E. Premkumar Reddy, Chair, 215-707-4307, *Fax:* 215-707-4588, *E-mail:* reddy@sgi1.fels.temple.edu.
Application contact: Scott K. Shore, Graduate Chair, 215-707-4302, *Fax:* 215-707-4588, *E-mail:* sks@unix.temple.edu.

Find an in-depth description at www.petersons.com/graduate.

■ TEXAS A&M UNIVERSITY

College of Science, Program in Molecular and Cell Biology, College Station, TX 77843

AWARDS PhD. Program composed of members from 4 colleges and 11 departments.

Degree requirements: For doctorate, dissertation required, foreign language not required.
Entrance requirements: For doctorate, GRE General Test, TOEFL. *Application fee:* $50 ($75 for international students).
Expenses: Tuition, state resident: part-time $76 per semester hour. Tuition, nonresident: part-time $292 per semester hour. Required fees: $11 per semester hour. Tuition and fees vary according to program.
Financial aid: Fellowships available. Financial aid application deadline: 4/1; financial aid applicants required to submit FAFSA.
Dr. Terry L. Thomas, Head, 979-845-0184, *Fax:* 979-845-2891.
Application contact: Graduate Adviser, 979-826-2465, *Fax:* 979-845-2891.

■ TEXAS A&M UNIVERSITY SYSTEM HEALTH SCIENCE CENTER

College of Medicine, Graduate School of Biomedical Sciences, Department of Medical Microbiology and Immunology, College Station, TX 77840-7896

AWARDS Immunology (PhD); microbiology (PhD); molecular biology (PhD); virology (PhD).

Faculty: 7 full-time (0 women), 1 part-time/adjunct (0 women).
Students: 16 full-time (10 women); includes 2 minority (1 African American, 1 Asian American or Pacific Islander).
Degree requirements: For doctorate, dissertation required, foreign language not required.
Entrance requirements: For doctorate, GRE General Test, minimum GPA of 3.0.

Texas A&M University System Health Science Center (continued)
Application deadline: For fall admission, 2/1 (priority date). *Application fee:* $50 ($75 for international students).
Expenses: Tuition, area resident: Full-time $1,368. Tuition, state resident: part-time $76 per credit. Tuition, nonresident: full-time $5,256; part-time $292 per credit. International tuition: $5,256 full-time. Required fees: $678; $38 per credit. Full-time tuition and fees vary according to course load and student level.
Financial aid: In 1999–00, 16 students received aid, including 1 fellowship (averaging $17,000 per year), 15 research assistantships (averaging $17,000 per year). Financial aid application deadline: 4/1; financial aid applicants required to submit FAFSA.
Faculty research: Molecular pathogenesis, microbial therapeutics. *Total annual research expenditures:* $1.2 million.
Dr. John M. Quarles, Interim Head, 979-845-1313, *Fax:* 979-845-3479, *E-mail:* jmquarles@medicine.tamu.edu.
Application contact: Dr. James Samuel, Graduate Adviser, 979-845-1313, *Fax:* 979-845-3479, *E-mail:* jsamuel@medicine.tamu.edu.
Find an in-depth description at www.petersons.com/graduate.

■ **TEXAS A&M UNIVERSITY SYSTEM HEALTH SCIENCE CENTER**

College of Medicine, Graduate School of Biomedical Sciences, Department of Medical Physiology, College Station, TX 77840-7896
AWARDS PhD.
Faculty: 10 full-time (1 woman).
Students: 11 full-time (3 women); includes 2 minority (both Asian Americans or Pacific Islanders), 2 international. Average age 28. *32 applicants, 13% accepted.*
Degree requirements: For doctorate, dissertation required, foreign language not required. *Average time to degree:* Doctorate–4 years full-time.
Entrance requirements: For doctorate, GRE General Test. *Application deadline:* For fall admission, 2/1 (priority date). *Application fee:* $50 ($75 for international students).
Expenses: Tuition, area resident: Full-time $1,368. Tuition, state resident: part-time $76 per credit. Tuition, nonresident: full-time $5,256; part-time $292 per credit. International tuition: $5,256 full-time. Required fees: $678; $38 per credit. Full-time tuition and fees vary according to course load and student level.
Financial aid: In 1999–00, 10 students received aid; fellowships, research

assistantships, institutionally sponsored loans available. Financial aid application deadline: 4/1; financial aid applicants required to submit FAFSA.
Faculty research: Cardiovascular physiology, vascular cell and molecular biology. *Total annual research expenditures:* $1.5 million.
Dr. Harris J. Granger, Head, 409-845-7816, *Fax:* 409-847-8635.
Application contact: Dr. David C. Zawieja, Assistant Professor, 409-845-7816, *Fax:* 409-847-8635, *E-mail:* dcz@tamu.edu.
Find an in-depth description at www.petersons.com/graduate.

■ **TEXAS WOMAN'S UNIVERSITY**

Graduate School, College of Arts and Sciences, Department of Biology, Denton, TX 76204
AWARDS Biology (MS); biology teaching (MS); molecular biology (PhD). Part-time programs available.
Faculty: 11 full-time (4 women), 2 part-time/adjunct (1 woman).
Students: 27 full-time (23 women), 35 part-time (26 women); includes 20 minority (7 African Americans, 8 Asian Americans or Pacific Islanders, 5 Hispanic Americans), 6 international. Average age 32. *33 applicants, 70% accepted.* In 1999, 7 master's, 1 doctorate awarded (100% continued full-time study). Terminal master's awarded for partial completion of doctoral program.
Degree requirements: For master's, thesis required (for some programs), foreign language not required; for doctorate, variable foreign language requirement (computer language can substitute for one), dissertation, residency required. *Average time to degree:* Master's–2.5 years full-time, 4 years part-time; doctorate–5 years full-time, 8 years part-time.
Entrance requirements: For master's and doctorate, GRE General Test, minimum GPA of 3.0. *Application deadline:* For fall admission, 4/1 (priority date); for spring admission, 8/1. Applications are processed on a rolling basis. *Application fee:* $30.
Expenses: Tuition, state resident: full-time $2,045; part-time $83 per semester hour. Tuition, nonresident: full-time $5,933; part-time $279 per semester hour. Required fees: $500 per semester. Tuition and fees vary according to course load.
Financial aid: In 1999–00, 8 research assistantships with partial tuition reimbursements (averaging $10,000 per year), 18 teaching assistantships with partial tuition reimbursements (averaging $9,000 per year) were awarded; career-related internships or fieldwork, Federal Work-Study, institutionally sponsored

loans, and tuition waivers (partial) also available. Aid available to part-time students. Financial aid application deadline: 4/1. *Total annual research expenditures:* $2 million.
Dr. Fritz E. Schwalm, Chair, 940-898-2352, *Fax:* 940-898-2382, *E-mail:* d_schwalm@venus.twu.edu.
Find an in-depth description at www.petersons.com/graduate.

■ **THOMAS JEFFERSON UNIVERSITY**

College of Graduate Studies, Program in Biochemistry and Molecular Biology, Philadelphia, PA 19107
AWARDS PhD.
Faculty: 18 full-time (5 women), 17 part-time/adjunct (2 women).
Students: 21 full-time (14 women); includes 4 minority (2 African Americans, 1 Asian American or Pacific Islander, 1 Hispanic American). Average age 24. *82 applicants, 18% accepted.* In 1999, 9 degrees awarded.
Degree requirements: For doctorate, dissertation, preliminary exam required, foreign language not required. *Average time to degree:* Doctorate–4.5 years full-time.
Entrance requirements: For doctorate, GRE General Test, TOEFL, minimum GPA of 3.2. *Application deadline:* For fall admission, 3/1 (priority date). Applications are processed on a rolling basis. *Application fee:* $40.
Expenses: Tuition: Full-time $12,670. Tuition and fees vary according to degree level and program.
Financial aid: Fellowships with full tuition reimbursements, research assistantships, Federal Work-Study, institutionally sponsored loans, traineeships, and training grants available. Aid available to part-time students. Financial aid application deadline: 5/1; financial aid applicants required to submit FAFSA.
Faculty research: Human mitochondrial genetics, molecular biology of protein-RNA interaction, mammalian mitochondrial biogenesis and function, glucocorticoid-induced programmed cell death, molecular cytogenetics of aneuploidy syndromes, cri-du-chat syndrome. *Total annual research expenditures:* $1.9 million.
Dr. Gerald Litwack, Chairman, Graduate Committee, 215-503-4634, *Fax:* 215-503-5393, *E-mail:* gerry.litwack@mail.tju.edu.
Application contact: Jessie F. Pervall, Director of Admissions, 215-503-4400, *Fax:* 215-503-3433, *E-mail:* cgs-info@mail.tju.edu.
Find an in-depth description at www.petersons.com/graduate.

■ TUFTS UNIVERSITY

Sackler School of Graduate Biomedical Sciences, Department of Molecular Biology and Microbiology, Medford, MA 02155

AWARDS Molecular microbiology (PhD).

Faculty: 16 full-time (5 women).
Students: 42 full-time (20 women). *116 applicants, 7% accepted.* In 1999, 10 degrees awarded (8% entered university research/teaching, 1% found other work related to degree, 1% continued full-time study).
Degree requirements: For doctorate, dissertation required, foreign language not required. *Average time to degree:* Doctorate–6 years full-time.
Entrance requirements: For doctorate, GRE General Test, TOEFL. *Application deadline:* For fall admission, 1/15 (priority date). Applications are processed on a rolling basis. *Application fee:* $45.
Expenses: Tuition: Full-time $19,325.
Financial aid: In 1999–00, 42 research assistantships with full tuition reimbursements (averaging $18,805 per year) were awarded; scholarships and tuition waivers (full) also available. Financial aid application deadline: 2/1.
Faculty research: Fundamental problems of molecular biology of prokaryotes, eukaryotes, and their viruses. *Total annual research expenditures:* $5 million.
Dr. Catherine L. Squires, Chair, 617-636-6947, *Fax:* 617-636-0337, *E-mail:* csquires_rib@opal.tufts.edu.
Application contact: Secretary, 617-636-6767, *Fax:* 617-636-0375, *E-mail:* sackler-school@tufts.edu.

■ TUFTS UNIVERSITY

Sackler School of Graduate Biomedical Sciences, Program in Cell, Molecular and Developmental Biology, Medford, MA 02155

AWARDS PhD.

Faculty: 31 full-time (6 women), 6 part-time/adjunct (3 women).
Students: 25 full-time (14 women); includes 1 minority (Asian American or Pacific Islander), 3 international. Average age 25. *143 applicants, 3% accepted.* In 1999, 4 degrees awarded.
Degree requirements: For doctorate, dissertation required, foreign language not required. *Average time to degree:* Doctorate–5 years full-time.
Entrance requirements: For doctorate, GRE General Test, GRE Subject Test, TOEFL. *Application deadline:* For fall admission, 1/15 (priority date). Applications are processed on a rolling basis. *Application fee:* $45.
Expenses: Tuition: Full-time $19,325.
Financial aid: In 1999–00, 2 fellowships, 23 research assistantships with full tuition reimbursements (averaging $18,805 per year) were awarded; grants, scholarships, traineeships, and tuition waivers (full) also available. Financial aid application deadline: 2/1.
Faculty research: Reproduction and hormone action, control of gene expression, cell-matrix and cell-cell interactions, growth control and tumorigenesis, cytoskeleton and contractile proteins.
Dr. John J. Castellot, Director, 617-636-0303, *Fax:* 617-636-6536, *E-mail:* jcastellot@infonet.tufts.edu.
Application contact: Secretary, 617-636-6767, *Fax:* 617-636-0375, *E-mail:* sackler-school@tufts.edu.

Find an in-depth description at www.petersons.com/graduate.

■ TULANE UNIVERSITY

Graduate School, Department of Cell and Molecular Biology, New Orleans, LA 70118-5669

AWARDS Biology (MS, PhD). Part-time programs available.

Students: 10 full-time (4 women); includes 1 minority (African American), 3 international. *3 applicants, 67% accepted.* In 1999, 5 master's, 11 doctorates awarded (100% found work related to degree). Terminal master's awarded for partial completion of doctoral program.
Degree requirements: For master's, foreign language and thesis not required; for doctorate, dissertation required, foreign language not required. *Average time to degree:* Master's–2 years full-time; doctorate–5 years full-time.
Entrance requirements: For master's, GRE General Test, TSE, minimum B average in undergraduate course work; for doctorate, GRE General Test, TSE. *Application deadline:* For fall admission, 2/1. *Application fee:* $45.
Expenses: Tuition: Full-time $23,500. Tuition and fees vary according to program.
Financial aid: Fellowships, research assistantships, teaching assistantships available. Financial aid application deadline: 2/1.
Dr. Leonard Thein, Chair, 504-865-5546, *Fax:* 504-865-0785.
Application contact: Dr. Jeffrey Tasker, Graduate Director, 504-865-5546, *Fax:* 504-865-6785, *E-mail:* tasker@mailhost.tcs.tulane.edu.

■ TULANE UNIVERSITY

School of Medicine and Graduate School, Graduate Programs in Medicine, Interdisciplinary Graduate Program in Molecular and Cellular Biology, New Orleans, LA 70118-5669

AWARDS PhD, MD/PhD. PhD offered through the Graduate School.

Students: 41 full-time (18 women); includes 6 minority (3 African Americans, 3 Asian Americans or Pacific Islanders), 9 international. *117 applicants, 15% accepted.* In 1999, 10 doctorates awarded.
Degree requirements: For doctorate, dissertation required.
Entrance requirements: For doctorate, GRE General Test, GRE Subject Test, TOEFL. *Application deadline:* For fall admission, 2/1. *Application fee:* $45.
Expenses: Tuition: Full-time $23,030.
Financial aid: Fellowships, research assistantships, teaching assistantships available. Financial aid application deadline: 2/1.
Dr. Barbara Beckman, Director, 504-588-5226, *Fax:* 504-584-3779, *E-mail:* mcbpgm@mailhost.tcs.tulane.edu.

Find an in-depth description at www.petersons.com/graduate.

■ UNIFORMED SERVICES UNIVERSITY OF THE HEALTH SCIENCES

School of Medicine, Division of Basic Medical Sciences, Program in Molecular and Cell Biology, Bethesda, MD 20814-4799

AWARDS PhD.

Faculty: 25 part-time/adjunct (7 women).
Students: 13 full-time (7 women), 1 part-time; includes 3 minority (1 African American, 2 Asian Americans or Pacific Islanders). Average age 26. *19 applicants, 21% accepted.*
Degree requirements: For doctorate, dissertation, qualifying exam required.
Entrance requirements: For doctorate, GRE General Test, GRE Subject Test, minimum GPA of 3.0, U.S. citizenship. *Application deadline:* For fall admission, 1/15 (priority date). Applications are processed on a rolling basis. *Application fee:* $0.
Financial aid: In 1999–00, 8 fellowships with full tuition reimbursements (averaging $15,000 per year) were awarded; career-related internships or fieldwork and tuition waivers (full) also available.
Faculty research: Immunology, biochemistry.
Dr. William C. Gause, Director, 301-295-1958, *Fax:* 301-295-3220.

Uniformed Services University of the Health Sciences (continued)
Application contact: Janet M. Anastasi, Graduate Program Coordinator, 301-295-9474, *Fax:* 301-295-6772, *E-mail:* janastasi@usuhs.mil.

Find an in-depth description at www.petersons.com/graduate.

■ THE UNIVERSITY OF ARIZONA

Graduate College, College of Science, Department of Molecular and Cellular Biology, Tucson, AZ 85721

AWARDS MS, PhD. Terminal master's awarded for partial completion of doctoral program.

Degree requirements: For master's and doctorate, thesis/dissertation required, foreign language not required.
Entrance requirements: For master's, GRE General Test, GRE Subject Test, TOEFL; for doctorate, GRE General Test, GRE Subject Test, TOEFL, undergraduate research experience.
Expenses: Tuition, nonresident: full-time $4,814; part-time $274 per unit. Required fees: $1,094; $115 per unit. Tuition and fees vary according to course load and program.
Faculty research: Plant molecular biology, cellular and molecular aspects of development, genetics of bacteria and lower eukaryotes.

Find an in-depth description at www.petersons.com/graduate.

■ UNIVERSITY OF ARKANSAS FOR MEDICAL SCIENCES

College of Medicine and Graduate School, Graduate Programs in Medicine, Department of Biochemistry and Molecular Biology, Little Rock, AR 72205-7199

AWARDS MS, PhD, MD/PhD.

Faculty: 25 full-time (5 women), 7 part-time/adjunct (3 women).
Students: 28 full-time (11 women), 5 part-time (3 women); includes 6 minority (all Asian Americans or Pacific Islanders), 5 international. In 1999, 2 master's, 5 doctorates awarded.
Degree requirements: For master's, thesis, comprehensive exam required, foreign language not required; for doctorate, dissertation, qualifying exam required, foreign language not required.
Entrance requirements: For master's, GRE General Test, TOEFL, bachelor's degree in biology, chemistry, or related field; for doctorate, GRE General Test. *Application fee:* $0.
Expenses: Tuition: Full-time $8,928.
Financial aid: In 1999–00, 27 research assistantships were awarded; unspecified

assistantships also available. Aid available to part-time students.
Faculty research: Gene regulation, growth factors, oncogenes, metabolic diseases, hormone regulation.
Dr. Alan D. Elbein, Chairman, 501-686-5185.
Application contact: Dr. Richard Drake, Graduate Coordinator, 501-686-5185.

■ UNIVERSITY OF CALIFORNIA, BERKELEY

Graduate Division, College of Letters and Science, Department of Molecular and Cell Biology, Berkeley, CA 94720-1500

AWARDS PhD.

Degree requirements: For doctorate, dissertation, qualifying exam, 2 semesters of teaching required, foreign language not required.
Entrance requirements: For doctorate, GRE General Test, GRE Subject Test, minimum GPA of 3.0. Electronic applications accepted.
Expenses: Tuition, nonresident: full-time $9,804. Required fees: $4,268. Tuition and fees vary according to program.

Find an in-depth description at www.petersons.com/graduate.

■ UNIVERSITY OF CALIFORNIA, DAVIS

Graduate Studies, Programs in the Biological Sciences, Program in Biochemistry and Molecular Biology, Davis, CA 95616

AWARDS MS, PhD.

Faculty: 80 full-time (14 women).
Students: 59 full-time (28 women); includes 20 minority (2 African Americans, 16 Asian Americans or Pacific Islanders, 2 Hispanic Americans), 2 international. Average age 28. *105 applicants, 38% accepted.* In 1999, 8 doctorates awarded.
Degree requirements: For master's and doctorate, thesis/dissertation required.
Entrance requirements: For master's and doctorate, GRE General Test, GRE Subject Test. *Application deadline:* For fall admission, 4/1. *Application fee:* $40. Electronic applications accepted.
Expenses: Tuition, nonresident: full-time $9,804. Tuition and fees vary according to program and student level.
Financial aid: In 1999–00, 55 students received aid, including 13 fellowships with full and partial tuition reimbursements available, 28 research assistantships with full and partial tuition reimbursements available, 9 teaching assistantships with partial tuition reimbursements available; Federal Work-Study, institutionally sponsored loans, and scholarships also

available. Financial aid application deadline: 1/15; financial aid applicants required to submit FAFSA.
Faculty research: Gene expression, protein structure, molecular virology, protein synthesis, enzymology, membrane transport and structural biology.
J. Clark Lagarias, Chair, 530-752-1865, *E-mail:* jclagarias@ucdavis.edu.
Application contact: Dawne Shell, Graduate Program Assistant, 530-752-9091, *Fax:* 530-752-8822, *E-mail:* drshell@ucdavis.edu.

Find an in-depth description at www.petersons.com/graduate.

■ UNIVERSITY OF CALIFORNIA, IRVINE

Office of Research and Graduate Studies, School of Biological Sciences, Department of Molecular Biology and Biochemistry, Irvine, CA 92697

AWARDS Biological sciences (PhD).

Faculty: 22 full-time (4 women).
Students: 12 full-time (6 women), 2 part-time. *252 applicants, 34% accepted.* In 1999, 2 doctorates awarded.
Degree requirements: For doctorate, dissertation required.
Entrance requirements: For doctorate, GRE General Test, GRE Subject Test. *Application deadline:* For fall admission, 2/1 (priority date). Applications are processed on a rolling basis. *Application fee:* $40. Electronic applications accepted.
Expenses: Tuition, nonresident: full-time $10,244; part-time $1,720 per quarter. Required fees: $5,252; $1,300 per quarter. Tuition and fees vary according to course load and program.
Financial aid: Fellowships, research assistantships, teaching assistantships, institutionally sponsored loans and tuition waivers (full and partial) available. Financial aid application deadline: 3/2; financial aid applicants required to submit FAFSA.
Faculty research: Structure and synthesis of nucleic acids and proteins, regulation, virology, biochemical genetics, gene organization.
Jerry Manning, Chair, 949-824-5578, *Fax:* 949-824-8551, *E-mail:* gp-mbgb@uci.edu.
Application contact: Administrator, 949-824-8145, *Fax:* 949-824-7407, *E-mail:* gp-mbgb@uci.edu.

Find an in-depth description at www.petersons.com/graduate.

■ UNIVERSITY OF CALIFORNIA, IRVINE

Office of Research and Graduate Studies, School of Biological Sciences and College of Medicine, Graduate Program in Molecular Biology, Genetics, and Biochemistry, Irvine, CA 92697-1450

AWARDS Biological sciences (PhD).

Faculty: 102 full-time (20 women).
Students: 30 full-time (14 women); includes 7 minority (4 Asian Americans or Pacific Islanders, 3 Hispanic Americans), 2 international. Average age 25. 287 applicants, 37% accepted.

Degree requirements: For doctorate, dissertation, teaching assignment required. *Average time to degree:* Doctorate–5 years full-time.

Entrance requirements: For doctorate, GRE General Test, GRE Subject Test (biochemistry, cell and molecular biology; biology; or chemistry), TOEFL, TSE, minimum GPA of 3.0, research experience. *Application deadline:* For fall admission, 1/1. *Application fee:* $40. Electronic applications accepted.

Expenses: Tuition, nonresident: full-time $9,384. Required fees: $5,178. Full-time tuition and fees vary according to program.

Financial aid: In 1999–00, 30 students received aid, including 30 fellowships with full tuition reimbursements available (averaging $33,995 per year); grants, institutionally sponsored loans, scholarships, and tuition waivers (full) also available. Financial aid application deadline: 3/2; financial aid applicants required to submit FAFSA.

Faculty research: Cellular biochemistry; gene structure and expression; protein structure, function, and design; molecular genetics; pathogenesis and inherited disease; molecular virology and immunology.

Dr. Rozanne Sandri-Goldin, Director, 949-824-7570, *Fax:* 949-824-7407, *E-mail:* gp-mbgb@uci.edu.

Application contact: Kimberly McKinney, Administrator, 949-824-8145, *Fax:* 949-824-7407, *E-mail:* gp-mbgb@uci.edu.

Find an in-depth description at www.petersons.com/graduate.

■ UNIVERSITY OF CALIFORNIA, LOS ANGELES

Graduate Division, College of Letters and Science, Department of Chemistry and Biochemistry, Program in Biochemistry and Molecular Biology, Los Angeles, CA 90095

AWARDS MS, PhD. MS admission to program only under exceptional circumstances.

Students: 94 full-time (29 women); includes 28 minority (2 African Americans, 22 Asian Americans or Pacific Islanders, 3 Hispanic Americans, 1 Native American), 19 international. *111 applicants, 29% accepted.*
Entrance requirements: For master's, GRE General Test, GRE Subject Test, minimum GPA of 3.0; for doctorate, GRE General Test, GRE Subject Test, minimum undergraduate GPA of 3.0. *Application deadline:* For fall admission, 1/15. *Application fee:* $40. Electronic applications accepted.
Expenses: Tuition, nonresident: full-time $9,804. Required fees: $4,405. Full-time tuition and fees vary according to program and student level.
Financial aid: In 1999–00, 67 fellowships, 82 research assistantships, 32 teaching assistantships were awarded; scholarships also available.
Application contact: Departmental Office, 310-825-3150, *E-mail:* grad@chem.ucla.edu.

■ UNIVERSITY OF CALIFORNIA, LOS ANGELES

Graduate Division, College of Letters and Science, Program in Molecular Biology, Los Angeles, CA 90095

AWARDS PhD.

Students: 66 full-time (26 women); includes 24 minority (21 Asian Americans or Pacific Islanders, 3 Hispanic Americans), 5 international. *8 applicants, 0% accepted.*
Degree requirements: For doctorate, dissertation, oral and written qualifying exams required, foreign language not required.
Entrance requirements: For doctorate, GRE General Test. *Application deadline:* For fall admission, 12/15. *Application fee:* $40. Electronic applications accepted.
Expenses: Tuition, nonresident: full-time $9,804. Required fees: $4,405. Full-time tuition and fees vary according to program and student level.
Financial aid: In 1999–00, 55 fellowships, 60 research assistantships, 18 teaching assistantships were awarded; Federal Work-Study, institutionally sponsored loans, scholarships, and tuition waivers

(full and partial) also available. Financial aid application deadline: 3/1.
Dr. Arnold Berk, Director, 310-206-6298.
Application contact: UCLA Access Coordinator, 800-284-8252, *Fax:* 310-206-5280, *E-mail:* uclaaccess@ibes.medsch.ucla.edu.

■ UNIVERSITY OF CALIFORNIA, LOS ANGELES

Graduate Division, College of Letters and Science and School of Medicine, UCLA ACCESS to Programs in the Molecular and Cellular Life Sciences, Los Angeles, CA 90095

AWARDS PhD.

Students: 60 full-time (36 women); includes 22 minority (1 African American, 13 Asian Americans or Pacific Islanders, 8 Hispanic Americans), 8 international. *375 applicants, 26% accepted.*
Degree requirements: For doctorate, dissertation, oral and written qualifying exams required.
Entrance requirements: For doctorate, GRE General Test, minimum undergraduate GPA of 3.0. *Application deadline:* For fall admission, 12/15. *Application fee:* $40. Electronic applications accepted.
Expenses: Tuition, nonresident: full-time $9,804. Required fees: $4,405. Full-time tuition and fees vary according to program and student level.
Financial aid: Fellowships, research assistantships, teaching assistantships, scholarships available. Financial aid application deadline: 3/1.
Faculty research: Molecular, cellular, and developmental biology; immunology; microbiology; integrative biology.
David Meyer, Director, 800-284-8252.
Application contact: UCLA Access Coordinator, 800-284-8252, *Fax:* 310-206-5280, *E-mail:* uclaaccess@ibes.medsch.ucla.edu.

Find an in-depth description at www.petersons.com/graduate.

■ UNIVERSITY OF CALIFORNIA, LOS ANGELES

School of Medicine and Graduate Division, Graduate Programs in Medicine, Department of Molecular, Cell and Developmental Biology, Los Angeles, CA 90095

AWARDS MA, PhD.

Students: 35 full-time (15 women); includes 15 minority (13 Asian Americans or Pacific Islanders, 2 Hispanic Americans), 5 international. *66 applicants, 18% accepted.*
Degree requirements: For doctorate, dissertation, qualifying exams required, foreign language not required.

University of California, Los Angeles (continued)

Entrance requirements: For doctorate, GRE General Test, GRE Subject Test, TOEFL. *Application fee:* $40.

Expenses: Tuition, nonresident: full-time $9,804. Required fees: $4,405.

Financial aid: Fellowships, research assistantships, teaching assistantships, scholarships available. Financial aid application deadline: 3/1.

Dr. Lutz Birnbaumer, Chair, 310-825-7109.

Application contact: UCLA Access Coordinator, 800-284-8252, *Fax:* 310-206-5280, *E-mail:* uclaaccess@ ibes.medsch.ucla.edu.

Find an in-depth description at www.petersons.com/graduate.

■ UNIVERSITY OF CALIFORNIA, RIVERSIDE

Graduate Division, College of Natural and Agricultural Sciences, Program in Cell, Molecular, and Developmental Biology, Riverside, CA 92521-0102

AWARDS MS, PhD.

Faculty: 49 full-time (15 women).
Students: 3 full-time (2 women); includes 1 minority (Asian American or Pacific Islander).
Degree requirements: For master's, thesis, oral defense of thesis required; for doctorate, dissertation, oral defense of thesis, qualifying exams, 2 quarters of teaching experience required. *Average time to degree:* Master's–2 years full-time; doctorate–5 years full-time.
Entrance requirements: For master's and doctorate, GRE General Test, TOEFL, minimum GPA of 3.2. *Application deadline:* For fall admission, 5/1; for winter admission, 9/1; for spring admission, 12/1. *Application fee:* $40.
Expenses: Tuition, nonresident: full-time $9,804. Required fees: $4,758. Full-time tuition and fees vary according to program.
Financial aid: Fellowships, research assistantships, teaching assistantships available. Financial aid application deadline: 1/5.
Dr. Prudence Talbot, Graduate Advisor, 909-787-5913, *Fax:* 909-787-4286, *E-mail:* talbot@pop.ucr.edu.
Application contact: Helene Serewis, Graduate Student Affairs Officer, 800-735-0717, *Fax:* 909-787-5517, *E-mail:* biopgrad@pep.ycr.edu.

■ UNIVERSITY OF CALIFORNIA, SAN DIEGO

Graduate Studies and Research, Department of Biology, Program in Genetics and Molecular Biology, La Jolla, CA 92093-0348

AWARDS PhD. Offered in association with the Salk Institute.

Degree requirements: For doctorate, dissertation required.
Entrance requirements: For doctorate, GRE General Test, pre-application beginning in September. *Application deadline:* For fall admission, 1/7. *Application fee:* $40.
Expenses: Tuition, nonresident: full-time $14,691. Required fees: $4,697. Full-time tuition and fees vary according to program.
Financial aid: Tuition waivers (full) and stipends available.
Application contact: 858-534-3835.

Find an in-depth description at www.petersons.com/graduate.

■ UNIVERSITY OF CALIFORNIA, SAN DIEGO

Graduate Studies and Research, Department of Biology, Program in Molecular and Cellular Biology, La Jolla, CA 92093

AWARDS PhD.

Degree requirements: For doctorate, dissertation required.
Entrance requirements: For doctorate, GRE General Test, GRE Subject Test (optional), pre-application beginning in September. *Application deadline:* For fall admission, 1/8. *Application fee:* $40.
Expenses: Tuition, nonresident: full-time $14,691. Required fees: $4,697. Full-time tuition and fees vary according to program.
Application contact: Biology Graduate Admissions Committee, 858-534-3835.

■ UNIVERSITY OF CALIFORNIA, SAN DIEGO

School of Medicine and Graduate Studies and Research, Graduate Studies in Biomedical Sciences, Cell and Molecular Biology Program, La Jolla, CA 92093-0685

AWARDS PhD.

Faculty: 106.
Students: 219.
Degree requirements: For doctorate, dissertation, qualifying exam required, foreign language not required.
Entrance requirements: For doctorate, GRE General Test, TOEFL. *Application deadline:* For fall admission, 1/5. *Application fee:* $40.

Expenses: Program pays tuition, fees, health insurance, and stipend for all students in good standing.
Faculty research: Molecular and cellular pharmacology, cell and organ physiology. Kim Barrett, Chair, 868-543-3726.
Application contact: Gina Butcher, Graduate Program Representative, 858-534-3982.

Find an in-depth description at www.petersons.com/graduate.

■ UNIVERSITY OF CALIFORNIA, SAN DIEGO

School of Medicine and Graduate Studies and Research, Graduate Studies in Biomedical Sciences, Regulatory Biology Program, La Jolla, CA 92093

AWARDS PhD.

Faculty: 71.
Degree requirements: For doctorate, dissertation, 2 qualifying exams required, foreign language not required.
Entrance requirements: For doctorate, GRE General Test, GRE Subject Test, TOEFL. *Application deadline:* For fall admission, 1/5. *Application fee:* $40.
Expenses: Program pays tuition, fees, health insurance, and stipend for all students in good standing.
Financial aid: Tuition waivers (full) available.
Faculty research: Eukaryotic regulatory and molecular biology, molecular and cellular pharmacology, cell and organ physiology.
M. Geoffrey Rosenfeld, Director.
Application contact: Gina Butcher, Graduate Program Representative, 858-534-3982.

■ UNIVERSITY OF CALIFORNIA, SAN FRANCISCO

Graduate Division and School of Medicine, Department of Biochemistry and Biophysics, Program in Biochemistry and Molecular Biology, San Francisco, CA 94143

AWARDS PhD, MD/PhD.

Students: In 1999, 22 degrees awarded.
Degree requirements: For doctorate, dissertation required, foreign language not required.
Entrance requirements: For doctorate, GRE General Test, GRE Subject Test, TOEFL. *Application deadline:* For fall admission, 1/5. *Application fee:* $40.
Expenses: All students guaranteed an annual stipend of $17,600.
Financial aid: Fellowships available. Financial aid application deadline: 2/1.

Faculty research: Structural biology, genetics, cell biology, cell physiology, metabolism.
Keith R. Yamamoto, Director, 415-476-3941.
Application contact: Ray Herrman, Graduate Admissions Assistant, 415-476-3941, *E-mail:* admissions@ biochem.ucsf.edu.

Find an in-depth description at www.petersons.com/graduate.

■ UNIVERSITY OF CALIFORNIA, SANTA BARBARA

Graduate Division, College of Letters and Sciences, Division of Mathematics, Life, and Physical Sciences, Department of Molecular, Cellular, and Developmental Biology, Santa Barbara, CA 93106

AWARDS MA, PhD.

Faculty: 25 full-time (6 women), 1 part-time/adjunct (0 women).
Students: 59 full-time (34 women); includes 7 minority (all Asian Americans or Pacific Islanders), 1 international. *91 applicants, 34% accepted.* In 1999, 7 master's, 6 doctorates awarded.
Degree requirements: For doctorate, dissertation required. *Average time to degree:* Master's–2 years full-time; doctorate–6 years full-time.
Entrance requirements: For master's and doctorate, GRE General Test, GRE Subject Test, TOEFL. *Application deadline:* For fall admission, 1/1 (priority date). *Application fee:* $40. Electronic applications accepted.
Expenses: Tuition, state resident: full-time $14,637. Tuition, nonresident: full-time $24,441.
Financial aid: Fellowships, research assistantships, teaching assistantships available. Financial aid application deadline: 1/15; financial aid applicants required to submit FAFSA.
Faculty research: Regulation of pap operon, recombinant composite protein, interferons, microtubules, embryonic development, bacterial pathogens.
Stu Feinstein, Chair, 805-893-2659, *Fax:* 805-893-4724, *E-mail:* feinstei@ lifesci.ucsb.edu.
Application contact: Shari Profant, Graduate Program Assistant, 805-893-8499, *Fax:* 805-893-4724, *E-mail:* profant@ lifesci.ucsb.edu.

■ UNIVERSITY OF CALIFORNIA, SANTA BARBARA

Graduate Division, College of Letters and Sciences, Division of Mathematics, Life, and Physical Sciences, Interdepartmental Program in Biochemistry and Molecular Biology, Santa Barbara, CA 93106

AWARDS PhD.

Faculty: 21 part-time/adjunct (3 women).
Students: 15 full-time (7 women). Average age 27. *55 applicants, 27% accepted.* In 1999, 1 doctorate awarded.
Degree requirements: For doctorate, dissertation required, foreign language not required. *Average time to degree:* Doctorate–6 years full-time.
Entrance requirements: For doctorate, GRE General Test, TOEFL. *Application deadline:* For fall admission, 12/15 (priority date). *Application fee:* $40. Electronic applications accepted.
Expenses: Tuition, state resident: full-time $14,637. Tuition, nonresident: full-time $24,441.
Financial aid: In 1999–00, 15 students received aid, including 5 fellowships with full tuition reimbursements available (averaging $17,500 per year); research assistantships, teaching assistantships with partial tuition reimbursements available, career-related internships or fieldwork, Federal Work-Study, institutionally sponsored loans, and tuition waivers (full and partial) also available. Financial aid application deadline: 12/15; financial aid applicants required to submit FAFSA.
Faculty research: Genetics and biochemistry of bacterial gene expression; structure-function relationships in proteins and nucleic acids, protein chemistry; biochemistsry and biophysics of marine adhesion.
James Cooper, Director, 805-893-8028, *Fax:* 805-893-7558.
Application contact: *E-mail:* bmbprog@ lifesci.ucsb.edu.

Find an in-depth description at www.petersons.com/graduate.

■ UNIVERSITY OF CALIFORNIA, SANTA CRUZ

Graduate Division, Division of Natural Sciences, Department of Biology, Santa Cruz, CA 95064

AWARDS Molecular, cellular, and developmental biology (PhD), including biology.

Faculty: 38 full-time.
Students: 90 full-time (47 women); includes 18 minority (1 African American, 13 Asian Americans or Pacific Islanders, 4 Hispanic Americans), 5 international. *196*

applicants, 20% accepted. In 1999, 11 doctorates awarded.
Degree requirements: For doctorate, one foreign language (computer language can substitute), dissertation, oral and written qualifying exams required.
Entrance requirements: For doctorate, GRE General Test, GRE Subject Test. *Application deadline:* For fall admission, 1/1. *Application fee:* $40.
Expenses: Tuition, state resident: full-time $4,925. Tuition, nonresident: full-time $14,919.
Financial aid: Fellowships, research assistantships, teaching assistantships, career-related internships or fieldwork, Federal Work-Study, and institutionally sponsored loans available. Financial aid application deadline: 1/1.
Faculty research: Neurophysiology and psychophysiology; plant sciences; population, environmental, and evolutionary biology.
Barry Bowman, Chairperson, 831-459-2385.
Application contact: Graduate Admissions, 831-459-2301.

■ UNIVERSITY OF CENTRAL FLORIDA

College of Health and Public Affairs, Department of Molecular and Microbiology, Orlando, FL 32816

AWARDS Microbiology (MS); molecular biology (MS). Part-time and evening/weekend programs available.

Faculty: 10 full-time, 6 part-time/adjunct.
Students: 27 full-time (18 women), 11 part-time (8 women); includes 8 minority (1 African American, 7 Hispanic Americans), 5 international. Average age 30. *26 applicants, 69% accepted.* In 1999, 6 degrees awarded.
Degree requirements: For master's, thesis, comprehensive exam required, foreign language not required.
Entrance requirements: For master's, GRE General Test, TOEFL, minimum GPA of 3.0 in last 60 hours. *Application deadline:* For fall admission, 3/15 (priority date); for spring admission, 12/1. *Application fee:* $20.
Expenses: Tuition, state resident: full-time $2,054; part-time $137 per credit. Tuition, nonresident: full-time $7,207; part-time $480 per credit. Required fees: $47 per term.
Financial aid: In 1999–00, 10 fellowships with partial tuition reimbursements (averaging $3,100 per year), 31 research assistantships with partial tuition reimbursements (averaging $2,974 per year), 35 teaching assistantships with partial tuition reimbursements (averaging

University of Central Florida (continued)
$3,690 per year) were awarded; career-related internships or fieldwork, Federal Work-Study, institutionally sponsored loans, tuition waivers (partial), and unspecified assistantships also available. Financial aid application deadline: 3/1; financial aid applicants required to submit FAFSA.
Dr. Robert Gennaro, Chair, 407-823-5932, *E-mail:* gennaro@mail.ucf.edu.

■ UNIVERSITY OF CHICAGO

Division of the Biological Sciences, Molecular Biosciences: Biochemistry, Genetics, Cell and Developmental Biology, Department of Biochemistry and Molecular Biology, Chicago, IL 60637-1513

AWARDS PhD, MD/PhD.

Faculty: 32 full-time (3 women).
Students: 32 full-time (10 women); includes 12 minority (1 African American, 8 Asian Americans or Pacific Islanders, 3 Hispanic Americans). Average age 26. In 1999, 5 doctorates awarded (60% entered university research/teaching, 40% found other work related to degree).
Degree requirements: For doctorate, one foreign language, dissertation, qualifying exam required. *Average time to degree:* Doctorate–6 years full-time.
Entrance requirements: For doctorate, GRE General Test, GRE Subject Test, TOEFL. *Application deadline:* For fall admission, 1/5 (priority date). *Application fee:* $55.
Expenses: Tuition: Full-time $24,804; part-time $3,422 per course. Required fees: $390. Tuition and fees vary according to program.
Financial aid: In 1999–00, 32 students received aid; fellowships, research assistantships, institutionally sponsored loans and traineeships available. Financial aid application deadline: 6/1.
Faculty research: Molecular biology, gene expression, and DNA-protein interactions; membrane biochemistry, molecular endocrinology, and transmembrane signaling; enzyme mechanisms, physical biochemistry. and structural biology. *Total annual research expenditures:* $5 million.
Dr. Anthony A. Kossiakoff, Chairman, 773-702-9297, *Fax:* 773-702-0439, *E-mail:* koss@cummings.uchicago.edu.
Application contact: Anna P. Wright, Graduate Student Administrator, 773-702-0571, *Fax:* 773-702-0439, *E-mail:* anna@cummings.uchicago.edu.

■ UNIVERSITY OF CINCINNATI

Division of Research and Advanced Studies, College of Medicine, Graduate Programs in Medicine, Department of Molecular Genetics, Biochemistry, Microbiology and Immunology, Cincinnati, OH 45267

AWARDS PhD.

Faculty: 22 full-time.
Students: 42 full-time (21 women), 3 part-time (2 women); includes 2 minority (both Asian Americans or Pacific Islanders), 3 international. *109 applicants, 17% accepted.* In 1999, 8 doctorates awarded.
Degree requirements: For doctorate, dissertation, qualifying exam required, foreign language not required. *Average time to degree:* Doctorate–7.3 years full-time.
Entrance requirements: For doctorate, GRE General Test, GRE Subject Test. *Application deadline:* For fall admission, 2/1 (priority date). Applications are processed on a rolling basis. *Application fee:* $30.
Expenses: Tuition, state resident: full-time $5,139; part-time $196 per credit hour. Tuition, nonresident: full-time $10,326; part-time $369 per credit hour. Required fees: $561; $187 per quarter.
Financial aid: Tuition waivers (full) and unspecified assistantships available. Financial aid application deadline: 5/1. *Total annual research expenditures:* $6.8 million.
Dr. Jerry Lingrel, Head, 513-558-5324, *Fax:* 513-558-1190, *E-mail:* jerry.lingrel@uc.edu.
Application contact: Iain Cartwright, Graduate Program Director, 513-558-5532, *Fax:* 513-558-8474, *E-mail:* iain.cartwright@uc.edu.
Find an in-depth description at www.petersons.com/graduate.

■ UNIVERSITY OF CINCINNATI

Division of Research and Advanced Studies, College of Medicine, Graduate Programs in Medicine, Department of Pediatrics, Program in Molecular and Developmental Biology, Cincinnati, OH 45221-0091

AWARDS Molecular and developmental biology (MS, PhD); teratology (MS, PhD).

Faculty: 12 full-time.
Students: 29 full-time (16 women), 4 part-time (3 women); includes 5 minority (2 African Americans, 3 Asian Americans or Pacific Islanders), 10 international. *38 applicants, 32% accepted.* In 1999, 3 master's, 6 doctorates awarded. Terminal master's awarded for partial completion of doctoral program.
Degree requirements: For master's, thesis required, foreign language not required; for doctorate, dissertation,

qualifying exam required, foreign language not required. *Average time to degree:* Master's–3.8 years full-time; doctorate–6.4 years full-time.
Entrance requirements: For master's and doctorate, GRE General Test, GRE Subject Test (biology or chemistry), TOEFL. *Application deadline:* For fall admission, 2/1 (priority date). Applications are processed on a rolling basis. *Application fee:* $30.
Expenses: Tuition, state resident: full-time $5,139; part-time $196 per credit hour. Tuition, nonresident: full-time $10,326; part-time $369 per credit hour. Required fees: $561; $187 per quarter.
Financial aid: Tuition waivers (full) and unspecified assistantships available. Financial aid application deadline: 5/1. *Total annual research expenditures:* $2.5 million.
Dr. James Lessard, Interim Head, 513-636-4549, *Fax:* 513-636-4317, *E-mail:* james.lessard@uc.edu.
Application contact: Dan Wigginton, Graduate Program Director, 513-636-4547, *Fax:* 513-559-4317, *E-mail:* dan.wigginton@chmcc.org.

Find an in-depth description at www.petersons.com/graduate.

■ UNIVERSITY OF CINCINNATI

Division of Research and Advanced Studies, College of Medicine, Graduate Programs in Medicine, Graduate Program in Cell and Molecular Biology, Cincinnati, OH 45221-0091

AWARDS PhD.

Degree requirements: For doctorate, dissertation, qualifying exam required.
Entrance requirements: For doctorate, GRE General Test, TOEFL, TWE. *Application deadline:* For fall admission, 2/1 (priority date). Applications are processed on a rolling basis. *Application fee:* $30.
Expenses: Tuition, state resident: full-time $5,139; part-time $196 per credit hour. Tuition, nonresident: full-time $10,326; part-time $369 per credit hour. Required fees: $561; $187 per quarter.
Financial aid: Unspecified assistantships available. Financial aid application deadline: 5/1.

Find an in-depth description at www.petersons.com/graduate.

■ UNIVERSITY OF COLORADO AT BOULDER

Graduate School, College of Arts and Sciences, Department of Molecular, Cellular, and Developmental Biology, Boulder, CO 80309

AWARDS Cellular structure and function (MA, PhD); developmental biology (MA, PhD); molecular biology (MA, PhD).

Faculty: 25 full-time (7 women).
Students: 54 full-time (28 women), 1 (woman) part-time; includes 6 minority (1 African American, 3 Asian Americans or Pacific Islanders, 2 Hispanic Americans), 4 international. Average age 27. *174 applicants, 9% accepted.* In 1999, 2 master's, 15 doctorates awarded. Terminal master's awarded for partial completion of doctoral program.
Degree requirements: For master's, thesis or alternative, comprehensive exam required, foreign language not required; for doctorate, dissertation, comprehensive exam required, foreign language not required. *Average time to degree:* Doctorate–6 years full-time.
Entrance requirements: For master's, GRE General Test, GRE Subject Test, minimum undergraduate GPA of 2.75; for doctorate, GRE General Test, GRE Subject Test. *Application deadline:* For fall admission, 1/1. *Application fee:* $40 ($60 for international students).
Expenses: Tuition, state resident: part-time $181 per credit hour. Tuition, nonresident: part-time $542 per credit hour. Required fees: $99 per term. Tuition and fees vary according to course load and program.
Financial aid: Fellowships, research assistantships, teaching assistantships, tuition waivers (full) available. Financial aid application deadline: 3/1. *Total annual research expenditures:* $9.7 million.
Leslie Leinward, Chair, 303-492-7606, *Fax:* 303-492-7744, *E-mail:* leinward@ stripe.colorado.edu.
Application contact: Karen Brown, Student Affairs Office, 303-492-7230, *Fax:* 303-492-7744, *E-mail:* mcdbgradinfo@ beagle.colorado.edu.

Find an in-depth description at www.petersons.com/graduate.

■ UNIVERSITY OF COLORADO HEALTH SCIENCES CENTER

Graduate School, Programs in Biological and Medical Sciences, Program in Molecular Biology, Denver, CO 80262

AWARDS PhD.

Degree requirements: For doctorate, dissertation required, foreign language not required.

Entrance requirements: For doctorate, GRE, minimum GPA of 2.75.
Expenses: Tuition, state resident: full-time $1,512; part-time $56 per hour. Tuition, nonresident: full-time $7,209; part-time $267 per hour. Full-time tuition and fees vary according to course load and program.

Find an in-depth description at www.petersons.com/graduate.

■ UNIVERSITY OF CONNECTICUT HEALTH CENTER

Graduate School, Programs in Biomedical Sciences, Program in Genetics, Molecular Biology, and Biochemistry, Farmington, CT 06030

AWARDS PhD, DMD/PhD, MD/PhD.

Faculty: 29.
Students: 31 full-time (18 women); includes 2 minority (1 African American, 1 Asian American or Pacific Islander), 10 international. In 1999, 3 degrees awarded.
Degree requirements: For doctorate, one foreign language (computer language can substitute), dissertation required.
Entrance requirements: For doctorate, GRE General Test, GRE Subject Test, TOEFL. *Application deadline:* For fall admission, 2/1 (priority date); for spring admission, 10/1. Applications are processed on a rolling basis. *Application fee:* $40 ($45 for international students).
Expenses: Tuition, state resident: full-time $5,272; part-time $293 per credit. Tuition, nonresident: full-time $13,696; part-time $761 per credit. Required fees: $320; $198 per semester. One-time fee: $50 full-time. Full-time tuition and fees vary according to course load, program and reciprocity agreements.
Financial aid: In 1999–00, research assistantships (averaging $17,000 per year); fellowships also available.
Dr. Steve Pfeiffer, Director, 860-679-3395.
Application contact: Marizta Barta, Information Contact, 860-679-4306, *Fax:* 860-679-1282, *E-mail:* barta@ adp.uchc.edu.

Find an in-depth description at www.petersons.com/graduate.

■ UNIVERSITY OF DELAWARE

College of Arts and Science, Department of Biological Sciences, Newark, DE 19716

AWARDS Biotechnology (MS, PhD); cell and extracellular matrix biology (MS, PhD); cell and systems physiology (MS, PhD); ecology and evolution (MS, PhD); microbiology (MS, PhD); molecular biology and genetics (MS, PhD); plant biology (MS, PhD).

Faculty: 37 full-time (10 women).

Students: 22 full-time (11 women), 1 part-time; includes 2 minority (both African Americans), 9 international. Average age 25. *37 applicants, 27% accepted.* In 2000, 9 doctorates awarded.
Degree requirements: For master's and doctorate, thesis/dissertation required, foreign language not required. *Average time to degree:* Master's–2.5 years full-time; doctorate–6 years full-time.
Entrance requirements: For master's and doctorate, GRE General Test, GRE Subject Test (advanced biology). *Application deadline:* For fall admission, 6/15. Applications are processed on a rolling basis. *Application fee:* $50. Electronic applications accepted.
Expenses: Tuition, state resident: full-time $4,380; part-time $243 per credit. Tuition, nonresident: full-time $12,750; part-time $708 per credit. Required fees: $15 per term. Tuition and fees vary according to program.
Financial aid: In 2000–01, 18 students received aid, including 2 fellowships with full tuition reimbursements available (averaging $18,000 per year), 4 research assistantships with full tuition reimbursements available (averaging $18,000 per year), 11 teaching assistantships with full tuition reimbursements available (averaging $18,000 per year); tuition waivers (partial) also available. Financial aid application deadline: 6/15.
Faculty research: Cell interactions, molecular mechanisms, microorganisms, embryo implantation. *Total annual research expenditures:* $1.8 million.
Dr. Daniel D. Carson, Chair, 302-831-6977, *Fax:* 302-831-2281, *E-mail:* dcarson@udel.edu.
Application contact: Norman Karin, Graduate Coordinator, 302-831-1841, *Fax:* 302-831-2281, *E-mail:* ccoletta@udel.edu.

Find an in-depth description at www.petersons.com/graduate.

■ UNIVERSITY OF FLORIDA

College of Medicine and Graduate School, Interdisciplinary Program in Biomedical Sciences, Concentration in Biochemistry and Molecular Biology, Gainesville, FL 32611

AWARDS PhD.

Degree requirements: For doctorate, dissertation required, foreign language not required.
Entrance requirements: For doctorate, GRE General Test, TOEFL (average 550), minimum GPA of 3.0. Electronic applications accepted.
Expenses: Tuition, state resident: part-time $144 per credit hour. Tuition, nonresident: part-time $505 per credit

University of Florida (continued)
hour. Tuition and fees vary according to course level, course load and program.
Faculty research: Gene expression, metabolic regulation, structural biology, enzyme mechanism, membrane transporters.

Find an in-depth description at www.petersons.com/graduate.

■ **UNIVERSITY OF FLORIDA**

College of Medicine and Graduate School, Interdisciplinary Program in Biomedical Sciences, Department of Biochemistry and Molecular Biology, Gainesville, FL 32610

AWARDS PhD.

Degree requirements: For doctorate, dissertation required, foreign language not required.
Entrance requirements: For doctorate, GRE General Test, TOEFL, minimum GPA of 3.0. Electronic applications accepted.
Expenses: Tuition, state resident: part-time $144 per credit hour. Tuition, nonresident: part-time $505 per credit hour. Tuition and fees vary according to course level, course load and program.
Faculty research: Gene expression, metabolic regulation, structural biology, enzyme mechanism, membrane transporters.

Find an in-depth description at www.petersons.com/graduate.

■ **UNIVERSITY OF GEORGIA**

Graduate School, College of Arts and Sciences, Department of Biochemistry and Molecular Biology, Athens, GA 30602

AWARDS MS, PhD.

Degree requirements: For master's, one foreign language, thesis required; for doctorate, one foreign language (computer language can substitute), dissertation required.
Entrance requirements: For master's and doctorate, GRE General Test. Electronic applications accepted.
Expenses: Tuition, state resident: full-time $7,516; part-time $431 per credit hour. Tuition, nonresident: full-time $12,204; part-time $793 per credit hour. Tuition and fees vary according to program.

■ **UNIVERSITY OF HAWAII AT MANOA**

Graduate Division, Specialization in Cell, Molecular, and Neuro Sciences, Honolulu, HI 96822

AWARDS MS, PhD. Program is interdisciplinary; degree is in specific

discipline with the specialization in Cell, Molecular, and Neuro Sciences. Part-time programs available.

Faculty: 55 full-time (11 women).
Students: 60 full-time (34 women); includes 24 minority (22 Asian Americans or Pacific Islanders, 2 Hispanic Americans), 13 international. Average age 30. *72 applicants, 36% accepted.*
Degree requirements: For doctorate, dissertation required, foreign language not required.
Entrance requirements: For doctorate, GRE Subject Test (recommended). *Application deadline:* For fall admission, 3/1; for spring admission, 9/1. Applications are processed on a rolling basis. *Application fee:* $25 ($50 for international students).
Expenses: Tuition, state resident: full-time $4,032; part-time $168 per credit. Tuition, nonresident: full-time $9,960; part-time $415 per credit. Required fees: $51 per semester. Part-time tuition and fees vary according to course load and program.
Financial aid: In 1999–00, 22 research assistantships with tuition reimbursements (averaging $15,000 per year), 5 teaching assistantships with tuition reimbursements (averaging $13,000 per year) were awarded; tuition waivers (full and partial) and unspecified assistantships also available. Financial aid application deadline: 2/1.
Faculty research: Anatomy, cellular neurobiology, genetics,. *Total annual research expenditures:* $500,000.
Dr. Martin D. Rayner, Chair, 808-956-7269, *Fax:* 808-956-6984, *E-mail:* martin@pbre.hawaii.edu.

■ **UNIVERSITY OF HAWAII AT MANOA**

John A. Burns School of Medicine and Graduate Division, Graduate Programs in Biomedical Sciences, Department of Genetics and Molecular Biology, Honolulu, HI 96822

AWARDS MS, PhD.

Faculty: 18 full-time (3 women), 1 part-time/adjunct (0 women).
Students: 4 full-time (1 woman), 2 part-time (1 woman). Average age 29. *10 applicants, 90% accepted.* In 1999, 1 master's, 4 doctorates awarded (100% found work related to degree). Terminal master's awarded for partial completion of doctoral program.
Degree requirements: For master's, thesis required (for some programs), foreign language not required; for doctorate, dissertation required, foreign language not required. *Average time to degree:* Doctorate–4 years full-time.
Entrance requirements: For master's and doctorate, GRE, minimum GPA of 3.0.

Application deadline: For fall admission, 3/1; for spring admission, 5/1. Applications are processed on a rolling basis. *Application fee:* $25 ($50 for international students).
Expenses: Tuition, state resident: part-time $168 per credit. Tuition, nonresident: part-time $415 per credit. Required fees: $51 per semester. Part-time tuition and fees vary according to course load.
Financial aid: In 1999–00, 6 research assistantships (averaging $16,628 per year), 2 teaching assistantships (averaging $13,041 per year) were awarded; Federal Work-Study and institutionally sponsored loans also available. Financial aid application deadline: 2/1.
Dr. John A. Hunt, Chairperson, 808-956-8552, *Fax:* 808-956-5506, *E-mail:* jahunt@hawaii.edu.
Application contact: Dr. David S. Haymer, Chairman, Admissions Committee, 808-956-5517.

■ **UNIVERSITY OF IDAHO**

College of Graduate Studies, College of Agriculture, Department of Microbiology, Molecular Biology and Biochemistry, Program in Microbiology, Molecular Biology and Biochemistry, Moscow, ID 83844-4140

AWARDS MS, PhD.

Students: 28 full-time (7 women), 5 part-time (1 woman); includes 1 minority (Asian American or Pacific Islander), 14 international.
Degree requirements: For master's, thesis required, foreign language not required; for doctorate, dissertation required.
Entrance requirements: For master's, minimum GPA of 2.8; for doctorate, minimum undergraduate GPA of 2.8, 3.0 graduate. *Application deadline:* For fall admission, 8/1; for spring admission, 12/15. *Application fee:* $35 ($45 for international students).
Expenses: Tuition, nonresident: full-time $6,000; part-time $239 per credit hour. Required fees: $2,888; $144 per credit hour. Tuition and fees vary according to program.
Financial aid: Application deadline: 2/15.
Dr. Greg Bohach, Head, Department of Microbiology, Molecular Biology and Biochemistry, 208-885-6666.

■ **UNIVERSITY OF ILLINOIS AT CHICAGO**

College of Medicine and Graduate College, Graduate Programs in Medicine, Department of Biochemistry and Molecular Biology, Chicago, IL 60607-7128

AWARDS MS, PhD.

Faculty: 17 full-time (5 women).
Students: 19 full-time (13 women), 1 (woman) part-time; includes 4 minority (3 Asian Americans or Pacific Islanders, 1 Hispanic American), 10 international. Average age 27. *107 applicants, 13% accepted.* In 1999, 4 master's, 2 doctorates awarded. Terminal master's awarded for partial completion of doctoral program.
Degree requirements: For master's and doctorate, thesis/dissertation required, foreign language not required.
Entrance requirements: For master's and doctorate, GRE General Test. *Application deadline:* For fall admission, 4/1. *Application fee:* $40 ($50 for international students).
Expenses: Tuition, state resident: full-time $3,750. Tuition, nonresident: full-time $10,588. Tuition and fees vary according to course load.
Financial aid: In 1999–00, 18 students received aid; fellowships, research assistantships, teaching assistantships, career-related internships or fieldwork, Federal Work-Study, institutionally sponsored loans, and traineeships available. Financial aid application deadline: 3/1; financial aid applicants required to submit FAFSA.
Faculty research: Nature of cellular components, control of metabolic processes, regulation of gene expression. Dr. Donald Chambers, Head, 312-996-7670.
Application contact: Dr. James Vary, Contact, 312-996-8444.
Find an in-depth description at www.petersons.com/graduate.

■ UNIVERSITY OF ILLINOIS AT CHICAGO

Graduate College, College of Liberal Arts and Sciences, Department of Biological Sciences, Laboratory for Molecular Biology, Chicago, IL 60607-7128

AWARDS MS, PhD.

Degree requirements: For master's, thesis required, foreign language not required; for doctorate, dissertation, preliminary exam required, foreign language not required.
Entrance requirements: For master's and doctorate, GRE General Test, TOEFL, minimum GPA of 4.0 on a 5.0 scale during final 2 years of undergraduate study.
Expenses: Tuition, state resident: full-time $3,750; part-time $1,250 per semester. Tuition, nonresident: full-time $10,588; part-time $3,530 per semester. Required fees: $507 per semester. Tuition and fees vary according to course load and program.

Financial aid: Fellowships, research assistantships, teaching assistantships available.
Application contact: Dr. Leo Miller, Director of Graduate Studies, 312-996-2220.
Find an in-depth description at www.petersons.com/graduate.

■ THE UNIVERSITY OF IOWA

College of Medicine and Graduate College, Graduate Programs in Medicine, Molecular Biology PhD Program, Iowa City, IA 52242-1316

AWARDS PhD, MD/PhD.

Faculty: 62 full-time (16 women).
Students: 24 full-time (10 women); includes 1 minority (Hispanic American), 8 international. Average age 25. *64 applicants, 14% accepted.* In 1999, 4 degrees awarded (50% found work related to degree).
Degree requirements: For doctorate, dissertation, comprehensive exam required, foreign language not required. *Average time to degree:* Doctorate–5 years full-time.
Entrance requirements: For doctorate, GRE General Test, TOEFL, minimum GPA of 3.0. *Application deadline:* Applications are processed on a rolling basis. *Application fee:* $30 ($50 for international students).
Expenses: Tuition, state resident: full-time $3,308. Tuition, nonresident: full-time $10,662. Tuition and fees vary according to course load and program.
Financial aid: In 1999–00, 3 fellowships with full tuition reimbursements (averaging $17,000 per year), 23 research assistantships with full tuition reimbursements (averaging $16,770 per year) were awarded; scholarships also available.
Faculty research: Regulation of gene expression, inherited human genetic diseases, signal transduction mechanisms, structural biology and function. Dr. Curt Sigmund, Director, 319-335-7604, *Fax:* 319-335-7656, *E-mail:* curt-sigmund@uiouwa.edu.
Application contact: Paulette Scheler, Program Assistant, 800-551-6787, *Fax:* 319-335-7656, *E-mail:* interdis@blue.weeg.uiowa.edu.

■ UNIVERSITY OF KANSAS

Graduate School, College of Liberal Arts and Sciences, Division of Biological Sciences, Department of Molecular Biosciences, Lawrence, KS 66045

AWARDS Biochemistry and biophysics (MA, MS, PhD); microbiology (MA, MS, PhD); molecular, cellular, and developmental biology (MA, MS, PhD).

Students: 13 full-time (7 women), 8 part-time (2 women); includes 1 minority (Asian American or Pacific Islander), 5 international. *57 applicants, 21% accepted.* In 2000, 11 master's, 10 doctorates awarded.
Degree requirements: For master's, thesis required, foreign language not required; for doctorate, dissertation required.
Entrance requirements: For master's and doctorate, GRE General Test, GRE Subject Test, TOEFL. *Application deadline:* For fall admission, 1/15 (priority date). *Application fee:* $25.
Expenses: Tuition, state resident: full-time $2,482; part-time $103 per credit hour. Tuition, nonresident: full-time $8,104; part-time $338 per credit hour. Required fees: $428; $31 per credit hour. Tuition and fees vary according to program.
Financial aid: In 2000–01, research assistantships (averaging $11,588 per year), teaching assistantships (averaging $11,588 per year) were awarded; fellowships. Financial aid application deadline: 3/1. Paul Kelly, Chair, 785-864-4311, *Fax:* 785-864-5294.
Application contact: Information Contact, 785-864-4311, *Fax:* 785-864-5294.

Find an in-depth description at www.petersons.com/graduate.

■ UNIVERSITY OF KANSAS

Graduate Studies Medical Center, Graduate Programs in Biomedical and Basic Sciences, Department of Biochemistry and Molecular Biology, Lawrence, KS 66045

AWARDS MS, PhD, MD/PhD. Part-time programs available.

Faculty: 12 full-time (0 women).
Students: 1 full-time (0 women), 10 part-time (5 women), 5 international. Average age 28. *0 applicants, 0% accepted.* Terminal master's awarded for partial completion of doctoral program.
Degree requirements: For master's, oral defense of thesis required; for doctorate, one foreign language (computer language can substitute), dissertation, comprehensive oral and written exam required.
Entrance requirements: For master's and doctorate, GRE, TOEFL, TSE, minimum GPA of 3.0. *Application deadline:* For fall admission, 4/15 (priority date). Applications are processed on a rolling basis. *Application fee:* $0. Electronic applications accepted.
Expenses: Tuition, state resident: full-time $2,482; part-time $103 per credit hour. Tuition, nonresident: full-time $8,104; part-time $338 per credit hour. Required fees: $428; $31 per credit hour. Tuition and fees vary according to program.

University of Kansas (continued)

Financial aid: Fellowships, research assistantships, teaching assistantships, Federal Work-Study, institutionally sponsored loans, scholarships, and unspecified assistantships available. Aid available to part-time students. Financial aid application deadline: 3/31; financial aid applicants required to submit FAFSA.

Faculty research: Regulation of gene expression, molecular genetics of kidney disease, molecular chaperones, glycoproteins, cell cycle control. *Total annual research expenditures:* $3 million.

Dr. Billy G. Hudson, Chairman, 913-588-7008, *Fax:* 913-588-7440, *E-mail:* bhudson@kumc.edu.

Application contact: Dr. Glen Andrews, Director of Graduate Studies, 913-588-6935, *Fax:* 913-588-7440, *E-mail:* gandrews@kumc.edu.

■ UNIVERSITY OF LOUISVILLE

School of Medicine and Graduate School, Integrated Programs in Biomedical Sciences, Department of Biochemistry and Molecular Biology, Louisville, KY 40292-0001

AWARDS MS, PhD, MD/MS, MD/PhD.

Degree requirements: For master's, thesis required; for doctorate, dissertation, comprehensive exams required.

Entrance requirements: For master's and doctorate, GRE General Test, TOEFL, minimum GPA of 3.0. Electronic applications accepted.

Expenses: Tuition, state resident: full-time $3,260; part-time $182 per hour. Tuition, nonresident: full-time $9,780; part-time $544 per hour. Required fees: $143; $28 per hour.

Faculty research: Protein structure and function, genetic and metabolic regulation, molecular endocrinology.

■ UNIVERSITY OF MAINE

Graduate School, College of Natural Sciences, Forestry, and Agriculture, Department of Biochemistry, Molecular Biology, and Microbiology, Orono, ME 04469

AWARDS Biochemistry (MPS, MS); biochemistry and molecular biology (PhD); microbiology (MPS, MS, PhD).

Degree requirements: For doctorate, dissertation required.

Entrance requirements: For master's and doctorate, GRE General Test, TOEFL.

Expenses: Tuition, state resident: full-time $3,564. Tuition, nonresident: full-time $10,116. Required fees: $378. Tuition and fees vary according to course load.

■ UNIVERSITY OF MARYLAND

Graduate School, Graduate Programs in Medicine, Program in Molecular and Cell Biology, Baltimore, MD 21201-1627

AWARDS PhD, MD/PhD.

Degree requirements: For doctorate, dissertation required.

Entrance requirements: For doctorate, GRE General Test, TOEFL, minimum GPA of 3.0.

Expenses: Tuition, state resident: part-time $261 per credit hour. Tuition, nonresident: part-time $468 per credit hour. Tuition and fees vary according to program.

Find an in-depth description at www.petersons.com/graduate.

■ UNIVERSITY OF MARYLAND, BALTIMORE COUNTY

Graduate School, Department of Biological Sciences, Program in Applied Molecular Biology, Baltimore, MD 21250-5398

AWARDS MS.

Faculty: 19 full-time.

Students: 8 full-time (3 women); includes 1 minority (Asian American or Pacific Islander), 1 international. *31 applicants, 48% accepted.* In 1999, 12 degrees awarded.

Degree requirements: For master's, foreign language and thesis not required.

Entrance requirements: For master's, GRE General Test, GRE Subject Test, TOEFL, minimum GPA of 3.0. *Application deadline:* For fall admission, 7/1. Applications are processed on a rolling basis. *Application fee:* $45.

Expenses: Tuition, state resident: part-time $268 per credit hour. Tuition, nonresident: part-time $470 per credit hour. Required fees: $38 per credit hour. $557 per semester.

Financial aid: In 1999–00, 1 teaching assistantship with tuition reimbursement (averaging $12,193 per year) was awarded; fellowships, research assistantships, tuition waivers (partial) also available.

Faculty research: Structure-function of RNA, genetics and molecular biology, biological chemistry.

Application contact: Dr. Philip Farabaugh, Director, Graduate Program, 410-455-3669, *Fax:* 410-455-3875, *E-mail:* biograd@umbc.edu.

■ UNIVERSITY OF MARYLAND, BALTIMORE COUNTY

Graduate School, Department of Biological Sciences, Program in Molecular and Cell Biology, Baltimore, MD 21250-5398

AWARDS PhD.

Students: 22 full-time (9 women), 3 part-time (2 women); includes 2 minority (1 African American, 1 Asian American or Pacific Islander), 9 international. *37 applicants, 46% accepted.* In 1999, 3 degrees awarded.

Degree requirements: For doctorate, dissertation required, foreign language not required.

Entrance requirements: For doctorate, GRE General Test, GRE Subject Test, TOEFL, minimum GPA of 3.0. *Application deadline:* For fall admission, 2/1. Applications are processed on a rolling basis. *Application fee:* $45.

Expenses: Tuition, state resident: part-time $268 per credit hour. Tuition, nonresident: part-time $470 per credit hour. Required fees: $38 per credit hour. $557 per semester.

Financial aid: Fellowships, research assistantships, teaching assistantships available.

Richard E. Wolf, Director, Graduate Program, 410-455-3669, *Fax:* 410-455-3875, *E-mail:* biograd@umbc.edu.

Find an in-depth description at www.petersons.com/graduate.

■ UNIVERSITY OF MARYLAND, COLLEGE PARK

Graduate Studies and Research, College of Life Sciences, Department of Cell Biology and Molecular Genetics, Program in Molecular and Cell Biology, College Park, MD 20742

AWARDS PhD. Part-time and evening/weekend programs available.

Students: 35 full-time (22 women), 11 part-time (8 women); includes 5 minority (2 African Americans, 1 Asian American or Pacific Islander, 2 Hispanic Americans), 22 international. *116 applicants, 22% accepted.* In 1999, 3 degrees awarded.

Degree requirements: For doctorate, dissertation, exam required.

Entrance requirements: For doctorate, GRE General Test, TOEFL. *Application deadline:* For fall admission, 2/1; for spring admission, 11/1. Applications are processed on a rolling basis. *Application fee:* $50 ($70 for international students). Electronic applications accepted.

Expenses: Tuition, state resident: part-time $272 per credit hour. Tuition, nonresident: part-time $415 per credit hour. Required fees: $632; $379 per year.

Financial aid: Fellowships, research assistantships, teaching assistantships available. Financial aid applicants required to submit FAFSA.

Faculty research: Genetics, transduction, photoregulation, parasitic interactions, cell cultures, monoclonal antibody production, oligonucleotide synthesis.

Application contact: Trudy Lindsey, Director, Graduate Admissions and Records, 301-405-4198, *Fax:* 301-314-9305, *E-mail:* grschool@deans.umd.edu.

■ **UNIVERSITY OF MASSACHUSETTS AMHERST**

Graduate School, Interdisciplinary Programs, Molecular and Cellular Biology Graduate Program, Amherst, MA 01003

AWARDS Biological chemistry (PhD); cell and developmental biology (PhD). Part-time programs available.

Students: 29 full-time (19 women), 33 part-time (21 women); includes 4 minority (1 African American, 3 Asian Americans or Pacific Islanders), 28 international. Average age 28. *288 applicants, 15% accepted.* In 2000, 8 degrees awarded.

Degree requirements: For doctorate, one foreign language, dissertation required.

Entrance requirements: For doctorate, GRE General Test. *Application deadline:* For fall admission, 1/15 (priority date). Applications are processed on a rolling basis. *Application fee:* $40.

Expenses: Tuition, state resident: full-time $2,640; part-time $165 per credit. Tuition, nonresident: full-time $9,756; part-time $407 per credit. Required fees: $1,221 per term. One-time fee: $110. Full-time tuition and fees vary according to course load, campus/location and reciprocity agreements.

Financial aid: In 2000–01, 17 research assistantships with full tuition reimbursements (averaging $3,796 per year), 3 teaching assistantships with full tuition reimbursements (averaging $5,510 per year) were awarded; fellowships with full tuition reimbursements, career-related internships or fieldwork, Federal Work-Study, grants, scholarships, traineeships, and unspecified assistantships also available. Aid available to part-time students. Financial aid application deadline: 1/15. Dr. Rodney K. Murphey, Acting Head, 413-545-3246, *Fax:* 413-545-1812, *E-mail:* rmurphey@bio.umass.edu.

Find an in-depth description at www.petersons.com/graduate.

■ **UNIVERSITY OF MASSACHUSETTS WORCESTER**

Graduate School of Biomedical Sciences, Department of Biochemistry and Molecular Biology, Worcester, MA 01655-0115

AWARDS PhD.

Faculty: 29 full-time (1 woman).
Students: 39 full-time (16 women); includes 1 minority (Asian American or Pacific Islander), 14 international. In 1999, 3 degrees awarded. *Average time to degree:* Doctorate–5.9 years full-time.
Entrance requirements: For doctorate, GRE General Test, 1 year of calculus, physics, organic chemistry and biology. *Application deadline:* 1/15 (priority date). Applications are processed on a rolling basis. *Application fee:* $25 ($50 for international students).
Expenses: Tuition, state resident: full-time $2,640. Tuition, nonresident: full-time $9,756. Required fees: $825. Full-time tuition and fees vary according to program.
Financial aid: In 1999–00, research assistantships with full tuition reimbursements (averaging $17,500 per year); unspecified assistantships also available.
Faculty research: Metabolic regulation, membrane structure, nutrition, hormone action, cloning.
Dr. Anthony Carruthers, Acting Chair, 508-856-2254, *Fax:* 508-856-6231, *E-mail:* anthony.carruthers@umassmed.edu.
Application contact: Dr. William Royer, Graduate Director, 508-856-6912.

■ **UNIVERSITY OF MASSACHUSETTS WORCESTER**

Graduate School of Biomedical Sciences, Department of Biochemistry and Molecular Pharmacology, Worcester, MA 01655-0115

AWARDS Biochemistry and molecular biology (PhD); pharmacology and molecular toxicology (PhD).

Faculty: 25 full-time (2 women).
Students: 12 full-time (6 women); includes 2 minority (both Asian Americans or Pacific Islanders), 3 international.
Degree requirements: For doctorate, dissertation required.
Entrance requirements: For doctorate, GRE General Test, GRE Subject Test, 1 year of calculus, physics, organic chemistry and biology. *Application deadline:* For fall admission, 1/15 (priority date). Applications are processed on a rolling basis. *Application fee:* $25 ($50 for international students).
Expenses: Tuition, state resident: full-time $2,640. Tuition, nonresident: full-time

$9,756. Required fees: $825. Full-time tuition and fees vary according to program.
Financial aid: In 1999–00, research assistantships with full tuition reimbursements (averaging $17,500 per year); unspecified assistantships also available.
Faculty research: Neuropharmacology, control of DNA metabolism, clinical pharmacology and toxicology.
Dr. C. Robert Matthews, Chair, 508-856-2251, *Fax:* 508-856-2151.
Application contact: Dr. Alonzo Ross, Graduate Director, 508-856-8016, *Fax:* 508-856-2151, *E-mail:* alonzo.ross@umassmed.edu.

Find an in-depth description at www.petersons.com/graduate.

■ **UNIVERSITY OF MEDICINE AND DENTISTRY OF NEW JERSEY**

Graduate School of Biomedical Sciences, Graduate Programs in Biomedical Sciences, Department of Biochemistry and Molecular Biology, Newark, NJ 07107

AWARDS MS, PhD.

Degree requirements: For master's, thesis required; for doctorate, dissertation, qualifying exam required, foreign language not required.
Entrance requirements: For master's and doctorate, GRE General Test, TOEFL. *Application deadline:* For fall admission, 2/1; for spring admission, 10/1. *Application fee:* $40.
Expenses: Tuition, state resident: part-time $270 per credit hour. Tuition, nonresident: part-time $407 per credit hour. Part-time tuition and fees vary according to campus/location and program.
Financial aid: Fellowships, research assistantships, Federal Work-Study, institutionally sponsored loans, and tuition waivers (full and partial) available. Financial aid application deadline: 5/1. Dr. Michael B. Mathews, Chairperson, 973-972-4411.
Application contact: Dr. Henry E. Brezenoff, Dean, Graduate School of Biomedical Sciences, 973-972-5333, *Fax:* 973-972-7148, *E-mail:* hbrezeno@umdnj.edu.

Find an in-depth description at www.petersons.com/graduate.

■ UNIVERSITY OF MEDICINE AND DENTISTRY OF NEW JERSEY

Graduate School of Biomedical Sciences, Graduate Programs in Biomedical Sciences, Program in Biochemistry and Molecular Biology, Piscataway, NJ 08854-5635

AWARDS MS, PhD. Terminal master's awarded for partial completion of doctoral program.

Degree requirements: For master's, thesis, qualifying exam required; for doctorate, dissertation, qualifying exam required, foreign language not required.
Entrance requirements: For master's and doctorate, GRE General Test, TOEFL. *Application deadline:* For fall admission, 2/1; for spring admission, 10/1. *Application fee:* $40.
Expenses: Tuition, state resident: part-time $270 per credit hour. Tuition, nonresident: part-time $407 per credit hour. Part-time tuition and fees vary according to campus/location and program.
Financial aid: Fellowships, research assistantships, teaching assistantships available. Financial aid application deadline: 5/1.
Faculty research: Signal transduction, regulation of RNA, polymerase II transcribed genes, developmental gene expression.
Dr. Masayori Inouye, Director, 732-235-4115.
Application contact: Dr. Michael J. Leibowitz, Associate Dean, Graduate School, 732-235-5016, *Fax:* 732-235-4720, *E-mail:* gsbspisc@umdnj.edu.

■ UNIVERSITY OF MEDICINE AND DENTISTRY OF NEW JERSEY

Graduate School of Biomedical Sciences, Graduate Programs in Biomedical Sciences, Program in Cell and Molecular Biology, Stratford, NJ 08084-5634

AWARDS MS, PhD, DO/PhD.

Students: 11 full-time (5 women), 7 part-time (5 women); includes 10 minority (1 African American, 9 Asian Americans or Pacific Islanders).
Degree requirements: For master's, thesis required; for doctorate, dissertation, qualifying exam required, foreign language not required.
Entrance requirements: For master's and doctorate, GRE General Test, TOEFL. *Application deadline:* For fall admission, 2/1; for spring admission, 10/1. Applications are processed on a rolling basis. *Application fee:* $40.
Expenses: Tuition, state resident: part-time $270 per credit hour. Tuition, nonresident: part-time $407 per credit hour. Part-time tuition and fees vary according to campus/location and program.
Financial aid: Application deadline: 5/1.
Dr. Salvatore J. Caradonna, Co-Chairperson, 856-566-6056.

■ THE UNIVERSITY OF MEMPHIS

Graduate School, College of Arts and Sciences, Department of Microbiology and Molecular Cell Sciences, Memphis, TN 38152

AWARDS Biology (MS, PhD). Part-time programs available.

Faculty: 10 full-time (2 women).
Students: Average age 29. *33 applicants, 79% accepted.* In 1999, 4 master's awarded (100% found work related to degree); 1 doctorate awarded (100% found work related to degree). Terminal master's awarded for partial completion of doctoral program.
Degree requirements: For master's, thesis or alternative, oral and/or written comprehensive exam required, foreign language not required; for doctorate, one foreign language (computer language can substitute), dissertation, comprehensive exam required. *Average time to degree:* Master's–2 years full-time; doctorate–5.5 years full-time.
Entrance requirements: For master's and doctorate, GRE General Test, GRE Subject Test (biochemistry), minimum GPA of 2.5. *Application deadline:* For fall admission, 8/1; for spring admission, 12/1. Applications are processed on a rolling basis. *Application fee:* $25 ($50 for international students).
Expenses: Tuition, state resident: full-time $3,410; part-time $178 per credit hour. Tuition, nonresident: full-time $8,670; part-time $408 per credit hour. Tuition and fees vary according to program.
Financial aid: In 1999–00, 2 research assistantships with full tuition reimbursements (averaging $12,000 per year), 16 teaching assistantships with full tuition reimbursements (averaging $8,000 per year) were awarded; fellowships with full tuition reimbursements
Faculty research: Molecular genetics, cell biology, biochemistry, microbiology and toxicology.
Dr. H. Delano Black, Interim Chair, 901-678-2955, *Fax:* 901-678-4457, *E-mail:* dblack@cc.memphis.edu.
Application contact: Dr. Barbara J. Taller, Graduate Coordinator, 901-678-2955, *Fax:* 901-678-4457, *E-mail:* bjtaller@cc.memphis.edu.
Find an in-depth description at www.petersons.com/graduate.

■ UNIVERSITY OF MIAMI

School of Medicine and Graduate School, Graduate Programs in Medicine, Department of Biochemistry and Molecular Biology, Coral Gables, FL 33124

AWARDS PhD, MD/PhD. Part-time programs available.

Faculty: 24 full-time (5 women).
Students: 31 full-time (15 women); includes 3 minority (all Hispanic Americans), 22 international. Average age 28. *70 applicants, 11% accepted.* In 1999, 2 doctorates awarded.
Degree requirements: For doctorate, dissertation, comprehensive and proposition exams required, foreign language not required.
Entrance requirements: For doctorate, GRE General Test, TOEFL. *Application deadline:* For fall admission, 4/15. Applications are processed on a rolling basis. *Application fee:* $35.
Expenses: Tuition, area resident: Part-time $899 per credit.
Financial aid: In 1999–00, 29 students received aid; fellowships, research assistantships, Federal Work-Study and scholarships available.
Faculty research: Macromolecule metabolism, molecular genetics, protein folding and 3-D structure, regulation of gene expression and enzyme function, signal transduction and developmental biology. *Total annual research expenditures:* $2.9 million.
Dr. Murray P. Deutscher, Chairman, 305-243-3150, *Fax:* 305-243-3955, *E-mail:* mdeutsch@mednet.med.miami.edu.
Application contact: Dr. Rudolf Werner, Director, 305-243-6998, *Fax:* 305-547-3955, *E-mail:* rwerner@mednet.med.miami.edu.
Find an in-depth description at www.petersons.com/graduate.

■ UNIVERSITY OF MIAMI

School of Medicine and Graduate School, Graduate Programs in Medicine, Program in Molecular Cell and Developmental Biology, Coral Gables, FL 33124

AWARDS PhD.

Faculty: 13 full-time (4 women).
Students: 14 full-time (8 women); includes 6 minority (4 Asian Americans or Pacific Islanders, 1 Hispanic American, 1 Native American), 2 international. In 1999, 2 degrees awarded.
Entrance requirements: For doctorate, GRE General Test, GRE Subject Test, TOEFL. *Application deadline:* For fall admission, 3/1 (priority date). Applications are processed on a rolling basis. *Application fee:* $35.

Expenses: Tuition, area resident: Part-time $899 per credit.
Financial aid: Fellowships available.
Dr. Robert Warren, Interim Chair, 305-243-6691, *Fax:* 305-545-7166.
Application contact: Dr. Richard Rotundo, Professor, 305-243-6691, *Fax:* 305-545-7166.

Find an in-depth description at www.petersons.com/graduate.

■ UNIVERSITY OF MICHIGAN

Medical School and Horace H. Rackham School of Graduate Studies, Program in Biomedical Sciences (PIBS), Interdepartmental Program in Cellular and Molecular Biology, Ann Arbor, MI 48109

AWARDS PhD.

Degree requirements: For doctorate, oral defense of dissertation, preliminary exam required.
Entrance requirements: For doctorate, GRE General Test, GRE Subject Test.
Expenses: Tuition, state resident: full-time $10,316. Tuition, nonresident: full-time $20,922.

Find an in-depth description at www.petersons.com/graduate.

■ UNIVERSITY OF MINNESOTA, DULUTH

School of Medicine, Department of Biochemistry and Molecular Biology, Duluth, MN 55812-2496

AWARDS MS, PhD.

Faculty: 6 full-time (1 woman), 3 part-time/adjunct (2 women).
Students: 4 full-time (2 women), 1 (woman) part-time; includes 2 minority (1 Asian American or Pacific Islander, 1 Native American). Average age 24. *22 applicants, 9% accepted.* In 1999, 1 master's awarded (100% found work related to degree). Terminal master's awarded for partial completion of doctoral program.
Degree requirements: For master's and doctorate, thesis/dissertation required, foreign language not required. *Average time to degree:* Master's–2 years full-time.
Entrance requirements: For master's and doctorate, GRE General Test, TOEFL. *Application deadline:* For fall admission, 2/15 (priority date). *Application fee:* $50 ($55 for international students).
Expenses: Tuition, state resident: full-time $5,040; part-time $420 per credit. Tuition, nonresident: full-time $9,900; part-time $825 per credit. Required fees: $509. Tuition and fees vary according to course load and program.
Financial aid: In 1999–00, 5 students received aid, including 4 research assistantships with tuition reimbursements available

(averaging $12,487 per year), 1 teaching assistantship with tuition reimbursement available (averaging $13,947 per year); fellowships with tuition reimbursements available, institutionally sponsored loans also available. Financial aid application deadline: 4/1.
Faculty research: Fish liver metabolism, yeast molecular genetics. *Total annual research expenditures:* $750,000.
Dr. Lester R. Drewes, Head, 218-726-7925, *Fax:* 218-726-8014, *E-mail:* ldrewes@d.umn.edu.

■ UNIVERSITY OF MINNESOTA, TWIN CITIES CAMPUS

Graduate School, Program in Molecular, Cellular, Developmental Biology and Genetics, Minneapolis, MN 55455-0213

AWARDS Genetic counseling (MS); molecular, cellular, developmental biology and genetics (PhD). Part-time programs available.

Faculty: 80 full-time (24 women), 17 part-time/adjunct (16 women).
Students: 62 full-time (43 women), 2 part-time (both women). Average age 24. *188 applicants, 19% accepted.* In 1999, 9 master's awarded (100% found work related to degree); 5 doctorates awarded (20% entered university research/teaching, 80% found other work related to degree). Terminal master's awarded for partial completion of doctoral program.
Degree requirements: For master's, thesis optional, foreign language not required; for doctorate, dissertation required, foreign language not required. *Average time to degree:* Master's–2 years full-time; doctorate–5 years full-time.
Entrance requirements: For master's and doctorate, GRE General Test, TOEFL. *Application deadline:* For fall admission, 1/15 (priority date). Applications are processed on a rolling basis. *Application fee:* $40 ($50 for international students).
Expenses: Tuition, state resident: full-time $5,040; part-time $420 per credit. Tuition, nonresident: full-time $9,900; part-time $825 per credit. Full-time tuition and fees vary according to course load, program and reciprocity agreements.
Financial aid: In 1999–00, 53 students received aid, including 9 fellowships with full tuition reimbursements available (averaging $16,504 per year), 39 research assistantships with full tuition reimbursements available (averaging $16,504 per year), 5 teaching assistantships with full tuition reimbursements available (averaging $16,504 per year); Federal Work-Study, grants, institutionally sponsored loans, and traineeships also available. Financial aid application deadline: 1/15.

Faculty research: Membrane receptors and membrane transport, cell interactions, cytoskeleton and cell mobility, regulation of gene expression, plant cell and molecular biology.
Dr. Perry B. Hackett, Director of Graduate Studies, 612-624-3053, *Fax:* 612-625-1700, *E-mail:* dgsmcdbg@biosci.cbs.umn.edu.
Application contact: Sue Knoblauch, Principal Secretary, 612-624-7470, *Fax:* 612-625-5754, *E-mail:* mcdbg@biosci.cbs.umn.edu.

Find an in-depth description at www.petersons.com/graduate.

■ UNIVERSITY OF MINNESOTA, TWIN CITIES CAMPUS

Medical School and Graduate School, Graduate Programs in Medicine, Department of Biochemistry, Molecular Biology and Biophysics, Minneapolis, MN 55455-0213

AWARDS PhD.

Faculty: 67 full-time (13 women).
Students: 77 full-time (39 women); includes 1 African American, 1 Hispanic American, 5 international. *189 applicants, 22% accepted.* In 1999, 16 degrees awarded (94% entered university research/teaching, 6% found other work related to degree).
Degree requirements: For doctorate, dissertation required, foreign language not required. *Average time to degree:* Doctorate–5 years full-time.
Entrance requirements: For doctorate, GRE General Test. *Application deadline:* For fall admission, 1/15 (priority date). *Application fee:* $50 ($55 for international students).
Expenses: Program provides tuition for all students.
Financial aid: In 1999–00, 12 fellowships with full tuition reimbursements (averaging $16,500 per year), research assistantships with full tuition reimbursements (averaging $16,500 per year) were awarded.
Faculty research: Physical biochemistry, enzymology, physiological chemistry. *Total annual research expenditures:* $10.9 million.
Dr. Douglas Ohlendorf, Director of Admissions, 612-625-6100.
Application contact: Mary F. Petrie-Terry, Student Support Assistant, 612-625-5179, *Fax:* 612-625-2163, *E-mail:* petri008@umn.edu.

Find an in-depth description at www.petersons.com/graduate.

■ UNIVERSITY OF MISSOURI–KANSAS CITY

School of Biological Sciences, Program in Cellular and Molecular Biology, Kansas City, MO 64110-2499

AWARDS MS, PhD. PhD offered through the School of Graduate Studies. Part-time programs available.

Students: 2 full-time (both women), 1 international. Average age 24. In 1999, 2 master's, 3 doctorates awarded. Terminal master's awarded for partial completion of doctoral program.

Degree requirements: For master's, thesis required (for some programs), foreign language not required; for doctorate, dissertation required. *Average time to degree:* Master's–2 years full-time.

Entrance requirements: For master's, GRE General Test, TOEFL, bachelor's degree in chemistry, biology, or a related discipline; minimum GPA of 3.0. *Application deadline:* For fall admission, 3/1 (priority date). Applications are processed on a rolling basis. *Application fee:* $25. Electronic applications accepted.

Expenses: Tuition, state resident: part-time $173 per hour. Tuition, nonresident: part-time $348 per hour. Required fees: $22 per hour. $15 per term. Part-time tuition and fees vary according to course load and program.

Financial aid: In 1999–00, research assistantships with full tuition reimbursements (averaging $16,000 per year), teaching assistantships with full and partial tuition reimbursements (averaging $16,000 per year) were awarded; grants, scholarships, and unspecified assistantships also available.

Faculty research: Structural biology and molecular genetics; protein and gene regulation; structure-function studies; molecular, cell, and neurobiology.

Application contact: Graduate Adviser, 816-235-2580.

■ UNIVERSITY OF MISSOURI–KANSAS CITY

School of Biological Sciences, Program in Molecular Biology and Biochemistry, Kansas City, MO 64110-2499

AWARDS PhD. PhD offered through the School of Graduate Studies.

Students: In 2000, 1 degree awarded.

Degree requirements: For doctorate, dissertation required.

Entrance requirements: For doctorate, GRE General Test, TOEFL, bachelor's degree in chemistry, biology, or a related discipline; minimum GPA of 3.0. *Application deadline:* For fall admission, 3/1. *Application fee:* $25.

Expenses: Tuition, state resident: part-time $173 per hour. Tuition, nonresident: part-time $348 per hour. Required fees: $22 per hour. $15 per term. Part-time tuition and fees vary according to course load and program.

Financial aid: Fellowships with tuition reimbursements, research assistantships, teaching assistantships, grants, scholarships, tuition waivers (full and partial), and unspecified assistantships available.

Application contact: Graduate Adviser, 816-235-2352, *Fax:* 816-235-5158.

Find an in-depth description at www.petersons.com/graduate.

■ UNIVERSITY OF MISSOURI–ST. LOUIS

Graduate School, College of Arts and Sciences, Department of Biology, St. Louis, MO 63121-4499

AWARDS Biology (MS, PhD), including animal behavior (MS), biochemistry, biotechnology (MS), conservation biology (MS), development (MS), ecology (MS), environmental studies (PhD), evolution (MS), genetics (MS), molecular biology and biotechnology (PhD), molecular/cellular biology (MS), physiology (MS), plant systematics, population biology (MS), tropical biology (MS); biotechnology (Certificate); tropical biology and conservation (Certificate). Part-time programs available.

Faculty: 46.

Students: 21 full-time (11 women), 75 part-time (44 women); includes 13 minority (2 African Americans, 2 Asian Americans or Pacific Islanders, 8 Hispanic Americans, 1 Native American), 23 international. In 1999, 14 master's, 4 doctorates awarded.

Degree requirements: For master's, thesis or alternative required, foreign language not required; for doctorate, one foreign language, dissertation, 1 semester of teaching experience required.

Entrance requirements: For doctorate, GRE General Test. *Application deadline:* For fall admission, 7/1 (priority date); for spring admission, 11/1 (priority date). Applications are processed on a rolling basis. *Application fee:* $25 ($40 for international students). Electronic applications accepted.

Expenses: Tuition, state resident: full-time $4,932; part-time $173 per credit hour. Tuition, nonresident: full-time $13,279; part-time $521 per credit hour. Required fees: $775; $33 per credit hour. Tuition and fees vary according to degree level and program.

Financial aid: In 1999–00, 8 research assistantships with partial tuition reimbursements (averaging $10,635 per year), 14 teaching assistantships with partial tuition reimbursements (averaging $11,488 per year) were awarded; career-related internships or fieldwork and Federal Work-Study also available. Aid available to part-time students. Financial aid application deadline: 2/1. *Total annual research expenditures:* $908,828.

Application contact: Graduate Admissions, 314-516-5458, *Fax:* 314-516-6759, *E-mail:* gradadm@umsl.edu.

■ UNIVERSITY OF NEBRASKA MEDICAL CENTER

Graduate College, Department of Biochemistry and Molecular Biology, Omaha, NE 68198

AWARDS MS, PhD.

Faculty: 13 full-time, 23 part-time/adjunct.

Students: 21 full-time (11 women), 3 part-time (2 women); includes 2 minority (1 Asian American or Pacific Islander, 1 Native American), 11 international. Average age 30. *25 applicants, 32% accepted.* In 1999, 1 master's, 4 doctorates awarded (100% entered university research/teaching).

Degree requirements: For master's and doctorate, thesis/dissertation required, foreign language not required. *Average time to degree:* Master's–1.5 years part-time.

Entrance requirements: For master's and doctorate, GRE General Test. *Application fee:* $35.

Expenses: Tuition, state resident: part-time $116 per semester hour. Tuition, nonresident: part-time $270 per semester hour. Tuition and fees vary according to program.

Financial aid: In 1999–00, 24 students received aid, including 4 fellowships with tuition reimbursements available (averaging $15,000 per year), 18 research assistantships with tuition reimbursements available (averaging $15,000 per year); institutionally sponsored loans also available. Aid available to part-time students. Financial aid application deadline: 3/1.

Faculty research: Recombinant DNA, cancer biology, diabetes and drug metabolism, biochemical endocrinology. Dr. Pi-Wan Cheng, Chairman, Graduate Committee, 402-559-4419, *Fax:* 402-559-6650, *E-mail:* broberts@unmc.edu.

Application contact: Jo Wagner, Associate Director of Admissions, 402-559-6468.

■ UNIVERSITY OF NEVADA, RENO

Graduate School, Interdisciplinary Program in Cell and Molecular Biology, Reno, NV 89557

AWARDS MS, PhD. Offered through the College of Arts and Science, the M. C.

Fleischmann College of Agriculture, and the School of Medicine.

Students: 25 full-time (18 women), 1 part-time; includes 2 minority (both Asian Americans or Pacific Islanders), 3 international. Average age 33. *28 applicants, 36% accepted.* In 1999, 4 master's, 6 doctorates awarded. Terminal master's awarded for partial completion of doctoral program.

Degree requirements: For master's and doctorate, thesis/dissertation required, foreign language not required.

Entrance requirements: For master's, GRE, TOEFL, minimum GPA of 2.75; for doctorate, GRE, TOEFL, minimum GPA of 3.0. *Application deadline:* For fall admission, 3/1. Applications are processed on a rolling basis. *Application fee:* $40.

Expenses: Tuition, area resident: Part-time $3,173 per semester. Tuition, nonresident: full-time $6,347. Required fees: $101 per credit. $101 per credit.

Financial aid: In 1999–00, 17 research assistantships were awarded; teaching assistantships, Federal Work-Study also available. Financial aid application deadline: 3/1.

Faculty research: Cancer biology, plant virology.

Dr. Stephen St. Jeor, Chair, 775-784-4113, *E-mail:* stjeor@unr.edu.

Find an in-depth description at www.petersons.com/graduate.

■ UNIVERSITY OF NEW HAMPSHIRE

Graduate School, College of Life Sciences and Agriculture, Graduate Programs in the Biological Sciences and Natural Resources, Department of Biochemistry and Molecular Biology, Durham, NH 03824

AWARDS MS, PhD. Part-time programs available.

Faculty: 12 full-time.

Students: 13 full-time (8 women), 11 part-time (6 women), 8 international. Average age 28. *37 applicants, 30% accepted.* In 1999, 1 master's, 1 doctorate awarded. Terminal master's awarded for partial completion of doctoral program.

Degree requirements: For master's, thesis required, foreign language not required; for doctorate, dissertation required.

Entrance requirements: For master's and doctorate, GRE General Test. *Application deadline:* For fall admission, 4/1 (priority date). Applications are processed on a rolling basis. *Application fee:* $50.

Expenses: Tuition, area resident: Full-time $5,750; part-time $319 per credit. Tuition, state resident: full-time $8,625; part-time $478. Tuition, nonresident: full-time

$14,640; part-time $598 per credit. Required fees: $224 per semester. Tuition and fees vary according to course load, degree level and program.

Financial aid: In 1999–00, 2 fellowships, 8 research assistantships, 8 teaching assistantships were awarded; career-related internships or fieldwork, Federal Work-Study, scholarships, and tuition waivers (full and partial) also available. Aid available to part-time students. Financial aid application deadline: 2/15.

Faculty research: Developmental biochemistry, biochemistry of natural products, physical biochemistry, biochemical genetics, structure and metabolism of macromolecules.

Dr. Clyde Denis, Chairperson, 603-862-2427, *E-mail:* cdenis@cisunix.unh.edu.

Application contact: Dr. Stacia Sower, Graduate Coordinator, 603-862-2103, *E-mail:* sasower@cisunix.unh.edu.

■ UNIVERSITY OF NEW HAVEN

Graduate School, College of Arts and Sciences, Program in Cellular and Molecular Biology, West Haven, CT 06516-1916

AWARDS MS.

Students: 7 full-time (3 women), 33 part-time (23 women); includes 5 minority (2 African Americans, 2 Asian Americans or Pacific Islanders, 1 Hispanic American), 5 international. *24 applicants, 67% accepted.* In 1999, 11 degrees awarded.

Degree requirements: For master's, thesis or alternative required, foreign language not required.

Application deadline: Applications are processed on a rolling basis. *Application fee:* $50.

Expenses: Tuition: Part-time $1,170 per course. Part-time tuition and fees vary according to course level, degree level and program.

Financial aid: Career-related internships or fieldwork and Federal Work-Study available. Financial aid application deadline: 5/1; financial aid applicants required to submit FAFSA.

Dr. Michael J. Rossi, Coordinator, 203-932-7125.

■ UNIVERSITY OF NEW MEXICO

Graduate School, College of Arts and Sciences, Department of Biology, Albuquerque, NM 87131-2039

AWARDS Biology (MS, PhD), including air land ecology, behavioral ecology, botany, cellular and molecular biology, community ecology, comparative immunology, comparative physiology, conservation biology, ecology, ecosystem ecology, evolutionary biology, evolutionary genetics, microbiology, molecular genetics, parasitology, physiological ecology,

physiology, population biology, vertebrate and invertebrate zoology. Part-time programs available.

Faculty: 35 full-time (5 women), 18 part-time/adjunct (11 women).

Students: 71 full-time (37 women), 28 part-time (11 women); includes 8 minority (2 Asian Americans or Pacific Islanders, 5 Hispanic Americans, 1 Native American), 11 international. Average age 33. *93 applicants, 30% accepted.* In 1999, 11 master's, 12 doctorates awarded. Terminal master's awarded for partial completion of doctoral program.

Degree requirements: For master's, one foreign language (computer language can substitute), thesis required (for some programs); for doctorate, 2 foreign languages (computer language can substitute for one), dissertation required.

Entrance requirements: For master's and doctorate, GRE General Test, GRE Subject Test, minimum GPA of 3.2. *Application deadline:* For fall admission, 1/15. *Application fee:* $25.

Expenses: Tuition, state resident: full-time $2,514; part-time $105 per credit hour. Tuition, nonresident: full-time $10,304; part-time $417 per credit hour. International tuition: $10,304 full-time. Required fees: $516; $22 per credit hour. Tuition and fees vary according to program.

Financial aid: In 1999–00, 58 students received aid, including 24 fellowships (averaging $1,645 per year), 26 research assistantships with tuition reimbursements available (averaging $8,921 per year), 40 teaching assistantships with tuition reimbursements available (averaging $11,066 per year); career-related internships or fieldwork, Federal Work-Study, institutionally sponsored loans, and tuition waivers (full and partial) also available. Aid available to part-time students. Financial aid applicants required to submit FAFSA.

Faculty research: Developmental biology, immunobiology. *Total annual research expenditures:* $4.5 million.

Dr. Kathryn Vogel, Chair, 505-277-3411, *Fax:* 505-277-0304, *E-mail:* kgvogel@unm.edu.

Application contact: Vivian Kent, Information Contact, 505-277-1712, *Fax:* 505-277-0304, *E-mail:* vkent@unm.edu.

■ THE UNIVERSITY OF NORTH CAROLINA AT CHAPEL HILL

Graduate School, College of Arts and Sciences, Department of Biology, Chapel Hill, NC 27599

AWARDS Botany (MA, MS, PhD); cell biology, development, and physiology (MA, MS, PhD);

The University of North Carolina at Chapel Hill (continued)
ecology and behavior (MA, MS, PhD); genetics and molecular biology (MA, MS, PhD); morphology, systematics, and evolution (MA, MS, PhD).

Degree requirements: For master's, thesis (for some programs), comprehensive exams required; for doctorate, dissertation, comprehensive exams required.
Entrance requirements: For master's and doctorate, GRE General Test, GRE Subject Test. Electronic applications accepted.
Expenses: Tuition, state resident: full-time $1,578. Tuition, nonresident: full-time $10,744. Required fees: $827. One-time fee: $15 full-time. Tuition and fees vary according to program.
Faculty research: Gene expression, biomechanics, yeast genetics, plant ecology, plant molecular biology.
Find an in-depth description at www.petersons.com/graduate.

■ **THE UNIVERSITY OF NORTH CAROLINA AT CHAPEL HILL**

School of Medicine and Graduate School, Graduate Programs in Medicine, Curriculum in Genetics and Molecular Biology, Chapel Hill, NC 27599

AWARDS MS, PhD.

Faculty: 66 full-time (22 women), 2 part-time/adjunct (0 women).
Students: 56 full-time (33 women); includes 5 minority (3 African Americans, 1 Asian American or Pacific Islander, 1 Hispanic American), 2 international. *68 applicants, 35% accepted.* In 1999, 9 doctorates awarded (45% entered university research/teaching, 33% found other work related to degree, 22% continued full-time study).
Degree requirements: For master's, thesis, comprehensive exam required, foreign language not required; for doctorate, dissertation, comprehensive exams required, foreign language not required. *Average time to degree:* Doctorate–5 years full-time.
Entrance requirements: For master's and doctorate, GRE General Test, minimum GPA of 3.0. *Application deadline:* For fall admission, 1/1 (priority date). Applications are processed on a rolling basis. Electronic applications accepted.
Expenses: Tuition, state resident: full-time $1,966. Tuition, nonresident: full-time $11,026. Required fees: $8,940. One-time fee: $15 full-time. Part-time tuition and fees vary according to course load.
Financial aid: In 1999–00, 1 fellowship with tuition reimbursement (averaging

$16,500 per year), 30 research assistantships with tuition reimbursements (averaging $16,500 per year), 1 teaching assistantship with tuition reimbursement (averaging $16,500 per year) were awarded; traineeships also available.
Dr. Susan T. Lord, Director, 919-966-3548, *E-mail:* stl@med.unc.edu.
Application contact: Janet Collins, Administrative Assistant, 919-966-2681, *Fax:* 919-966-0401, *E-mail:* lcollins@med.unc.edu.

Find an in-depth description at www.petersons.com/graduate.

■ **UNIVERSITY OF NORTH TEXAS**

Robert B. Toulouse School of Graduate Studies, College of Arts and Sciences, Department of Biological Sciences, Program in Molecular Biology, Denton, TX 76203

AWARDS MA, MS, PhD.

Degree requirements: For master's, oral defense of thesis required; for doctorate, one foreign language (computer language can substitute), dissertation, oral and written comprehensive exams required.
Entrance requirements: For master's and doctorate, GRE General Test, minimum GPA of 3.0.
Expenses: Tuition, state resident: full-time $2,865; part-time $600 per semester. Tuition, nonresident: full-time $8,049; part-time $1,896 per semester. Required fees: $26 per hour.
Faculty research: Adenosine diphosphate ribosylation reactions, mammalian gene structure, pyrimidine metabolism, enzymology.

■ **UNIVERSITY OF NORTH TEXAS HEALTH SCIENCE CENTER AT FORT WORTH**

Graduate School of Biomedical Sciences, Fort Worth, TX 76107-2699

AWARDS Anatomy and cell biology (MS, PhD); biochemistry and molecular biology (MS, PhD); biotechnology (MS); integrative physiology (MS, PhD); microbiology and immunology (MS, PhD); pharmacology (MS, PhD).

Faculty: 65 full-time (9 women), 11 part-time/adjunct (1 woman).
Students: 59 full-time (27 women), 44 part-time (20 women); includes 30 minority (10 African Americans, 9 Asian Americans or Pacific Islanders, 11 Hispanic Americans), 23 international. *70 applicants, 70% accepted.* In 1999, 5 master's awarded (40% found work related to degree, 60% continued full-time study); 5 doctorates awarded (80% entered university research/teaching, 20% found other work related to degree).

Degree requirements: For doctorate, computer language, dissertation required. *Average time to degree:* Master's–2.5 years full-time, 4 years part-time; doctorate–5 years full-time.
Entrance requirements: For master's and doctorate, GRE General Test, TOEFL. *Application deadline:* For fall admission, 5/1; for spring admission, 11/1. Applications are processed on a rolling basis. *Application fee:* $25 ($50 for international students).
Expenses: Tuition, state resident: full-time $1,188; part-time $66 per credit. Tuition, nonresident: full-time $5,058; part-time $281 per credit. Required fees: $366; $183 per semester.
Financial aid: In 1999–00, 11 fellowships, 70 research assistantships (averaging $16,500 per year) were awarded; teaching assistantships, career-related internships or fieldwork, Federal Work-Study, grants, institutionally sponsored loans, and traineeships also available. Aid available to part-time students. Financial aid application deadline: 4/1; financial aid applicants required to submit FAFSA.
Faculty research: Alzheimer's disease, diabetes, eye diseases, cancer, cardiovascular physiology.
Dr. Thomas Yorio, Dean, 817-735-2560, *Fax:* 817-735-0243, *E-mail:* yoriot@hsc.unt.edu.
Application contact: Jan Sharp, Administrative Assistant, 817-735-0258, *Fax:* 817-735-0243, *E-mail:* gsbs@hsc.unt.edu.

Find an in-depth description at www.petersons.com/graduate.

■ **UNIVERSITY OF NOTRE DAME**

Graduate School, College of Science, Department of Biological Sciences, Program in Cellular and Molecular Biology, Notre Dame, IN 46556

AWARDS MS, PhD. Terminal master's awarded for partial completion of doctoral program.

Degree requirements: For master's and doctorate, thesis/dissertation required, foreign language not required.
Entrance requirements: For master's and doctorate, GRE General Test, GRE Subject Test, TOEFL. *Application deadline:* For fall admission, 2/1 (priority date); for spring admission, 11/1. Applications are processed on a rolling basis. *Application fee:* $50.
Expenses: Tuition: Full-time $21,930; part-time $1,218 per credit. Required fees: $95. Tuition and fees vary according to program.
Financial aid: Fellowships with full tuition reimbursements, research assistantships with full tuition reimbursements, teaching

assistantships with full tuition reimbursements, traineeships and tuition waivers (full) available. Financial aid application deadline: 2/1.

Faculty research: Gene expression, developmental biology, neurobiology, genetics, cell biology of membranes and cytoskeleton.

Application contact: Dr. Terrence J. Akai, Director of Graduate Admissions, 219-631-7706, *Fax:* 219-631-4183, *E-mail:* gradad@nd.edu.

■ UNIVERSITY OF OKLAHOMA HEALTH SCIENCES CENTER

College of Medicine and Graduate College, Graduate Programs in Medicine, Department of Biochemistry and Molecular Biology, Oklahoma City, OK 73190

AWARDS Biochemistry (MS, PhD); molecular biology (MS, PhD). Part-time programs available.

Faculty: 22 full-time (4 women), 8 part-time/adjunct (3 women).
Students: 7 full-time (4 women), 23 part-time (10 women); includes 4 minority (2 Asian Americans or Pacific Islanders, 2 Native Americans), 9 international. Average age 25. *53 applicants, 21% accepted.* In 1999, 5 master's, 4 doctorates awarded. Terminal master's awarded for partial completion of doctoral program.
Degree requirements: For master's and doctorate, thesis/dissertation required, foreign language not required.
Entrance requirements: For master's and doctorate, GRE General Test, GRE Subject Test, TOEFL. *Application deadline:* For fall admission, 3/31. *Application fee:* $25 ($50 for international students).
Expenses: Tuition, state resident: part-time $90 per semester hour. Tuition, nonresident: part-time $264 per semester hour. Tuition and fees vary according to program.
Financial aid: Research assistantships, institutionally sponsored loans available. Aid available to part-time students. Financial aid application deadline: 3/31; financial aid applicants required to submit FAFSA.
Faculty research: Gene expression, regulation of transcription, enzyme evolution, melanogenesis, signal transduction. Dr. Paul H. Weigel, Chairman, 405-271-2227.
Application contact: Dr. A. C. Cox, Committee Chairperson, Biochemistry Admissions, 405-271-2227, *Fax:* 405-271-3092, *E-mail:* biochemistry@uokhsc.edu.

Find an in-depth description at www.petersons.com/graduate.

■ UNIVERSITY OF OREGON

Graduate School, College of Arts and Sciences, Department of Biology, Eugene, OR 97403

AWARDS Ecology and evolution (MA, MS, PhD); marine biology (MA, MS, PhD); molecular, cellular and genetic biology (PhD); neuroscience and development (PhD).

Faculty: 42 full-time (14 women), 5 part-time/adjunct (2 women).
Students: 69 full-time (33 women), 8 part-time (5 women); includes 6 minority (2 Asian Americans or Pacific Islanders, 3 Hispanic Americans, 1 Native American), 5 international. *18 applicants, 83% accepted.* In 1999, 11 master's, 7 doctorates awarded. Terminal master's awarded for partial completion of doctoral program.
Degree requirements: For master's, thesis required (for some programs); for doctorate, dissertation required, foreign language not required.
Entrance requirements: For master's and doctorate, GRE General Test, TOEFL, minimum GPA of 3.2. *Application deadline:* For fall admission, 1/10. *Application fee:* $50.
Expenses: Tuition, state resident: full-time $6,750. Tuition, nonresident: full-time $11,409. Part-time tuition and fees vary according to course load.
Financial aid: In 1999–00, 36 teaching assistantships were awarded; research assistantships, Federal Work-Study, grants, and institutionally sponsored loans also available. Financial aid application deadline: 2/1.
Faculty research: Developmental neurobiology; evolution, population biology, and quantitative genetics; regulation of gene expression; biochemistry of marine organisms.
Janis C. Weeks, Head, 541-346-4502, *Fax:* 541-346-6056.
Application contact: Donna Overall, Graduate Program Coordinator, 541-346-4503, *Fax:* 541-346-6056, *E-mail:* doverall@oregon.uoregon.edu.

Find an in-depth description at www.petersons.com/graduate.

■ UNIVERSITY OF PENNSYLVANIA

School of Arts and Sciences, Graduate Group in Biology, Program in Cell, Molecular, and Developmental Biology, Philadelphia, PA 19104

AWARDS PhD.

Students: 27 full-time (8 women); includes 9 minority (2 African Americans, 6 Asian Americans or Pacific Islanders, 1 Hispanic American), 2 international. In 1999, 5 degrees awarded.

Degree requirements: For doctorate, dissertation required, foreign language not required.
Entrance requirements: For doctorate, GRE General Test, GRE Subject Test, TOEFL. *Application fee:* $65.
Expenses: Tuition: Full-time $23,670. Required fees: $1,546. Full-time tuition and fees vary according to degree level and program.
Financial aid: Fellowships, research assistantships, teaching assistantships, tuition waivers (full and partial) available. Financial aid application deadline: 1/2.
Application contact: Allan Aiken, Graduate Group Coordinator, 215-898-6786, *Fax:* 215-898-8780, *E-mail:* aaiken@sas.upenn.edu.

■ UNIVERSITY OF PENNSYLVANIA

School of Medicine, Biomedical Graduate Studies, Graduate Group in Cell and Molecular Biology, Philadelphia, PA 19104

AWARDS Cell growth and cancer (PhD); cell structure and function (PhD); developmental biology (PhD); gene therapy (PhD); genetics and gene regulation (PhD); microbiology and virology (PhD).

Faculty: 297 full-time (62 women).
Students: 203 full-time (92 women); includes 47 minority (11 African Americans, 30 Asian Americans or Pacific Islanders, 6 Hispanic Americans), 18 international. *407 applicants, 15% accepted.* In 1999, 11 doctorates awarded.
Degree requirements: For doctorate, dissertation required, foreign language not required.
Entrance requirements: For doctorate, GRE General Test, TOEFL. *Application deadline:* For fall admission, 1/2 (priority date). Applications are processed on a rolling basis. *Application fee:* $65. Electronic applications accepted.
Expenses: Tuition: Full-time $17,256; part-time $2,991 per course. Required fees: $2,588; $363 per course. $726 per term.
Financial aid: In 1999–00, 154 students received aid, including 123 fellowships, 67 research assistantships; teaching assistantships, grants and institutionally sponsored loans also available. Financial aid application deadline: 1/2.
Dr. James C. Alwine, Chair, 215-898-3256.
Application contact: Mary Webster, Coordinator, 215-898-4360, *E-mail:* camb@mail.med.upenn.edu.

Find an in-depth description at www.petersons.com/graduate.

■ UNIVERSITY OF PITTSBURGH

Faculty of Arts and Sciences, Department of Biological Sciences, Program in Molecular, Cellular, and Developmental Biology, Pittsburgh, PA 15260

AWARDS PhD.

Degree requirements: For doctorate, dissertation required, foreign language not required.

Entrance requirements: For doctorate, GRE General Test, GRE Subject Test, TOEFL. *Application deadline:* For fall admission, 2/1 (priority date). Applications are processed on a rolling basis. *Application fee:* $40.

Expenses: Tuition, state resident: full-time $8,338; part-time $342 per credit. Tuition, nonresident: full-time $17,168; part-time $707 per credit. Required fees: $480; $90 per semester. Tuition and fees vary according to program.

Find an in-depth description at www.petersons.com/graduate.

■ UNIVERSITY OF ROCHESTER

The College, Arts and Sciences, Department of Biology, Program in Cellular and Molecular Biology, Rochester, NY 14627-0250

AWARDS MS, PhD.

Degree requirements: For master's, thesis not required; for doctorate, dissertation, qualifying exam required, foreign language not required.

Entrance requirements: For master's and doctorate, GRE General Test, GRE Subject Test, TOEFL. *Application deadline:* For fall admission, 2/1 (priority date). *Application fee:* $25.

Expenses: Tuition: Part-time $697 per credit hour. Tuition and fees vary according to program.

Financial aid: Application deadline: 2/1.

Application contact: Cindy Landry, Graduate Program Secretary, 716-275-7991.

■ UNIVERSITY OF SOUTH ALABAMA

College of Medicine and Graduate School, Program in Basic Medical Sciences, Specialization in Biochemistry and Molecular Biology, Mobile, AL 36688-0002

AWARDS PhD.

Faculty: 8 full-time (1 woman).
Students: In 1999, 5 degrees awarded.
Degree requirements: For doctorate, dissertation required, foreign language not required.
Entrance requirements: For doctorate, GRE General Test or MCAT. *Application*

deadline: For fall admission, 4/1. Applications are processed on a rolling basis.
Application fee: $25.

Expenses: Tuition, state resident: part-time $116 per semester hour. Tuition, nonresident: part-time $230 per semester hour. Required fees: $121 per semester.

Financial aid: Fellowships, institutionally sponsored loans available. Financial aid application deadline: 4/1.

Faculty research: Biochemistry of aging, mechanisms of metabolic regulation and oxygen metabolism.
Dr. Nathan N. Aronson, Chair, 334-460-6402.

Application contact: Lanette Flagge, Coordinator, 334-460-6153.

■ UNIVERSITY OF SOUTH CAROLINA

Graduate School, College of Science and Mathematics, Department of Biological Sciences, Graduate Training Program in Molecular, Cellular, and Developmental Biology, Columbia, SC 29208

AWARDS MS, PhD.

Degree requirements: For master's, one foreign language required; for doctorate, dissertation required.

Entrance requirements: For master's and doctorate, GRE General Test, minimum GPA of 3.0 in science. *Application deadline:* For fall admission, 2/15 (priority date). *Application fee:* $35.

Expenses: Tuition, state resident: full-time $4,014; part-time $202 per credit hour. Tuition, nonresident: full-time $8,528; part-time $428 per credit hour. Required fees: $100; $4 per credit hour. Tuition and fees vary according to program.

Application contact: Dr. Franklyn F. Bolander, Director of Graduate Studies, 803-777-2755, *Fax:* 803-777-4002, *E-mail:* bolander@sc.edu.

Find an in-depth description at www.petersons.com/graduate.

■ UNIVERSITY OF SOUTH DAKOTA

School of Medicine and Graduate School, Biomedical Sciences Graduate Program, Cellular and Molecular Biology Group, Vermillion, SD 57069-2390

AWARDS MA, PhD.

Faculty: 10 full-time (1 woman), 1 part-time/adjunct (0 women).
Students: 11 full-time (2 women), 4 international. Average age 28. *72 applicants, 1% accepted.* In 1999, 1 degree awarded (100% continued full-time study). Terminal master's awarded for partial completion of doctoral program.

Degree requirements: For master's and doctorate, thesis/dissertation required, foreign language not required. *Average time to degree:* Doctorate–4 years full-time.

Entrance requirements: For master's and doctorate, GRE General Test, GRE Subject Test, TOEFL, minimum GPA of 3.0. *Application deadline:* For fall admission, 3/15 (priority date). Applications are processed on a rolling basis. *Application fee:* $15.

Expenses: Tuition, state resident: full-time $2,126; part-time $89 per credit. Tuition, nonresident: full-time $6,270; part-time $261 per credit. Required fees: $1,194; $50 per credit. Tuition and fees vary according to course load and reciprocity agreements.

Financial aid: In 1999–00, 11 students received aid, including 3 research assistantships with full tuition reimbursements available (averaging $13,500 per year); fellowships, tuition waivers (partial) also available.

Faculty research: Molecular aspects of protein and DNA, neurochemistry and energy transduction, gene regulation, cellular development.
W. Keith Miskimins, Coordinator, 605-677-5132, *Fax:* 605-677-6381, *E-mail:* kmiskimi@usd.edu.

Application contact: Robin Miskimins, Graduate Program Director, 605-677-5237, *Fax:* 605-677-6381, *E-mail:* rmiskim@usd.edu.

■ UNIVERSITY OF SOUTHERN CALIFORNIA

Graduate School, College of Letters, Arts and Sciences, Department of Biological Sciences, Program in Molecular Biology, Los Angeles, CA 90089

AWARDS MS, PhD.

Faculty: 21 full-time (2 women).
Students: 36 full-time (15 women); includes 1 minority (Asian American or Pacific Islander), 26 international. Average age 27. *81 applicants, 23% accepted.* In 1999, 6 degrees awarded.

Degree requirements: For doctorate, dissertation required.

Entrance requirements: For master's and doctorate, GRE General Test. *Application deadline:* For fall admission, 1/15 (priority date). *Application fee:* $55.

Expenses: Tuition: Full-time $17,952; part-time $748 per unit. Required fees: $406; $203 per unit. Tuition and fees vary according to program.

Financial aid: In 1999–00, 7 fellowships with full tuition reimbursements (averaging $18,000 per year), 14 research assistantships with full tuition reimbursements (averaging $18,000 per year), 13 teaching

assistantships with full tuition reimbursements (averaging $18,000 per year) were awarded; Federal Work-Study, institutionally sponsored loans, and scholarships also available. Aid available to part-time students. Financial aid application deadline: 2/15; financial aid applicants required to submit FAFSA.
Dr. Myron F. Goodman, Director, 213-740-1109.
Application contact: Admissions Office, 213-740-5557.

Find an in-depth description at www.petersons.com/graduate.

■ UNIVERSITY OF SOUTHERN CALIFORNIA

Keck School of Medicine and Graduate School, Graduate Programs in Medicine, Department of Biochemistry and Molecular Biology, Los Angeles, CA 90089

AWARDS MS, PhD.

Faculty: 39 full-time (7 women).
Students: 84 full-time (39 women); includes 23 minority (1 African American, 20 Asian Americans or Pacific Islanders, 2 Hispanic Americans), 44 international. Average age 23. *205 applicants, 13% accepted.* In 1999, 8 master's awarded (75% entered university research/teaching, 25% found other work related to degree); 8 doctorates awarded (100% entered university research/teaching). Terminal master's awarded for partial completion of doctoral program.
Degree requirements: For master's and doctorate, thesis/dissertation required, foreign language not required.
Entrance requirements: For master's and doctorate, GRE General Test, GRE Subject Test, TOEFL, minimum GPA of 3.0. *Application deadline:* For fall admission, 1/15 (priority date). *Application fee:* $55. Electronic applications accepted.
Expenses: Tuition: Full-time $22,198; part-time $748 per unit. Required fees: $406.
Financial aid: In 1999–00, 57 students received aid, including 5 fellowships with tuition reimbursements available, 51 research assistantships with tuition reimbursements available (averaging $1,465 per year), 1 teaching assistantship with tuition reimbursement available; career-related internships or fieldwork, grants, institutionally sponsored loans, and tuition waivers (full) also available. Aid available to part-time students. Financial aid application deadline: 4/15.
Faculty research: Molecular genetics, gene expression, membrane biochemistry, metabolic regulation, cancer biology.

Dr. Laurence H. Kedes, Chairman, 323-442-1144, *Fax:* 323-442-2764, *E-mail:* kedes@zygote.hsc.usc.edu.
Application contact: Anne L. Vazquez, Admissions Counselor, 323-442-1145, *Fax:* 323-442-1224, *E-mail:* annvazqu@hsc.usc.edu.

■ UNIVERSITY OF SOUTHERN CALIFORNIA

Keck School of Medicine and Graduate School, Graduate Programs in Medicine, Department of Preventive Medicine, Division of Biometry, Los Angeles, CA 90089

AWARDS Applied biometry/epidemiology (MS); biometry (MS, PhD); epidemiology (PhD); molecular biology and epidemiological methods (MS).

Faculty: 53 full-time (21 women), 2 part-time/adjunct (1 woman).
Students: 78 full-time (48 women); includes 20 minority (1 African American, 17 Asian Americans or Pacific Islanders, 2 Hispanic Americans), 31 international. Average age 30. *53 applicants, 42% accepted.* In 1999, 10 master's, 3 doctorates awarded (100% entered university research/teaching). Terminal master's awarded for partial completion of doctoral program.
Degree requirements: For master's and doctorate, computer language, thesis/dissertation required, foreign language not required. *Average time to degree:* Master's–2 years full-time; doctorate–5 years full-time.
Entrance requirements: For master's and doctorate, GRE General Test, GRE Subject Test, TOEFL, minimum GPA of 3.0. *Application deadline:* For fall admission, 1/15 (priority date). Applications are processed on a rolling basis. *Application fee:* $55.
Expenses: Tuition: Full-time $22,198; part-time $748 per unit. Required fees: $406.
Financial aid: In 1999–00, 32 students received aid, including 19 research assistantships with tuition reimbursements available (averaging $15,000 per year), 13 teaching assistantships with tuition reimbursements available (averaging $15,000 per year); fellowships, career-related internships or fieldwork, Federal Work-Study, and institutionally sponsored loans also available. Financial aid application deadline: 4/1.
Faculty research: Clinical trials in ophthalmology and cancer research, methods of analysis for epidemiological studies, genetic epidemiology. *Total annual research expenditures:* $1.3 million.
Dr. Stanley P. Azen, Director, 323-442-1810, *Fax:* 323-442-2993, *E-mail:* mtrujill@hsc.usc.edu.

Application contact: Mary L. Trujillo, Student Advisor, 323-442-1810, *Fax:* 323-442-2993, *E-mail:* mtrujill@hsc.usc.edu.

■ UNIVERSITY OF SOUTHERN MAINE

School of Applied Science, Program in Applied Immunology and Molecular Biology, Portland, ME 04104-9300

AWARDS MS. Part-time programs available.

Degree requirements: For master's, thesis required, foreign language not required.
Entrance requirements: For master's, GRE General Test, GRE Subject Test, minimum GPA 3.0.
Expenses: Tuition, state resident: full-time $5,944. Tuition, nonresident: full-time $15,634.
Faculty research: Flow cytometry, cancer, epidemiology, monoclonal antibodies, DNA diagnostics.

■ UNIVERSITY OF SOUTHERN MISSISSIPPI

Graduate School, College of Science and Technology, Department of Biological Sciences, Hattiesburg, MS 39406

AWARDS Environmental biology (MS, PhD); marine biology (MS, PhD); microbiology (MS, PhD); molecular biology (MS, PhD).

Degree requirements: For master's, thesis required, foreign language not required; for doctorate, 2 foreign languages (computer language can substitute for one), dissertation required.
Entrance requirements: For master's, GRE General Test, TOEFL, minimum GPA of 3.0; for doctorate, GRE General Test, TOEFL, minimum GPA of 3.5.
Expenses: Tuition, state resident: full-time $2,250; part-time $137 per semester hour. Tuition, nonresident: full-time $3,102; part-time $172 per semester hour. Required fees: $602.

■ UNIVERSITY OF SOUTH FLORIDA

College of Medicine and Graduate School, Graduate Programs in Medical Sciences, Department of Biochemistry and Molecular Biology, Tampa, FL 33620-9951

AWARDS PhD.

Degree requirements: For doctorate, dissertation required, foreign language not required.
Entrance requirements: For doctorate, GRE General Test, minimum GPA of 3.0.

University of South Florida (continued)

Expenses: Tuition, state resident: part-time $148 per credit hour. Tuition, nonresident: part-time $509 per credit hour.

Faculty research: Eukaryotic systems.

Find an in-depth description at www.petersons.com/graduate.

■ UNIVERSITY OF SOUTH FLORIDA

Graduate School, College of Arts and Sciences, Interdisciplinary Program in Cellular and Molecular Biology, Tampa, FL 33620-9951

AWARDS PhD.

Degree requirements: For doctorate, dissertation, 2 tools of research required, foreign language not required.

Entrance requirements: For doctorate, GRE General Test, minimum GPA of 3.4.

Expenses: Tuition, state resident: part-time $148 per credit hour. Tuition, nonresident: part-time $509 per credit hour.

Faculty research: Protein structure and function, gene regulation, signal transduction, molecular immunology, cancer.

Find an in-depth description at www.petersons.com/graduate.

■ THE UNIVERSITY OF TENNESSEE HEALTH SCIENCE CENTER

College of Graduate Health Sciences, Department of Microbiology and Immunology, Concentration in Molecular and Cell Biology, Memphis, TN 38163-0002

AWARDS PhD.

Degree requirements: For doctorate, dissertation, oral and written preliminary and comprehensive exams required, foreign language not required.

Entrance requirements: For doctorate, GRE General Test, GRE Subject Test, TOEFL, minimum GPA of 3.0.

Find an in-depth description at www.petersons.com/graduate.

■ THE UNIVERSITY OF TEXAS AT AUSTIN

Graduate School, College of Natural Sciences, Program in Molecular Biology, Austin, TX 78712-1111

AWARDS MA, PhD.

Students: Average age 25.

Degree requirements: For doctorate, dissertation, qualifying exam required. *Average time to degree:* Master's–3 years full-time.

Entrance requirements: For master's and doctorate, GRE General Test. *Application deadline:* For fall admission, 1/15 (priority date); for spring admission, 10/1. Applications are processed on a rolling basis. *Application fee:* $50 ($75 for international students).

Expenses: Tuition, state resident: part-time $114 per semester hour. Tuition, nonresident: part-time $330 per semester hour. Tuition and fees vary according to program.

Financial aid: In 1999–00, 39 students received aid; fellowships, research assistantships, teaching assistantships, institutionally sponsored loans and traineeships available. Financial aid application deadline: 2/1.

Faculty research: Virology and immunology, RNA, molecular genetics, developmental biology, plant science, biochemistry evolution.

Dr. Alan M. Lambowitz, Director, 512-471-1156, *Fax:* 512-471-2149, *E-mail:* lambowitz@mail.utexas.edu.

Application contact: Dean R. Appling, Graduate Adviser, 512-471-2150, *Fax:* 512-471-2149, *E-mail:* molbio@hpcf.cc.utexas.edu.

Find an in-depth description at www.petersons.com/graduate.

■ THE UNIVERSITY OF TEXAS AT DALLAS

School of Natural Sciences and Mathematics, Program in Molecular and Cell Biology, Richardson, TX 75083-0688

AWARDS Biology (MS, PhD); molecular and cell biology (MS, PhD). Part-time and evening/weekend programs available.

Faculty: 15 full-time (2 women).

Students: 43 full-time (26 women), 11 part-time (7 women); includes 7 minority (5 Asian Americans or Pacific Islanders, 2 Hispanic Americans), 26 international. Average age 29. In 1999, 7 master's, 8 doctorates awarded.

Degree requirements: For master's, thesis optional, foreign language not required; for doctorate, dissertation, publishable paper required, foreign language not required.

Entrance requirements: For master's and doctorate, GRE General Test, TOEFL. *Application deadline:* For fall admission, 7/15; for spring admission, 11/15. Applications are processed on a rolling basis. *Application fee:* $25 ($75 for international students). Electronic applications accepted.

Expenses: Tuition, state resident: full-time $2,052; part-time $76 per semester hour. Tuition, nonresident: full-time $5,256; part-time $292 per semester hour.

Required fees: $1,504; $656 per year. One-time fee: $10. Full-time tuition and fees vary according to course level, course load, degree level and program.

Financial aid: In 1999–00, 9 research assistantships (averaging $5,115 per year), 18 teaching assistantships (averaging $5,154 per year) were awarded; fellowships, career-related internships or fieldwork, Federal Work-Study, grants, institutionally sponsored loans, and scholarships also available. Aid available to part-time students. Financial aid application deadline: 4/30; financial aid applicants required to submit FAFSA.

Faculty research: DNA replication, regulation of gene expression, subcellular organelles, physical chemistry of macromolecules, damage and repair of cellular DNA. *Total annual research expenditures:* $1.5 million.

Dr. Lawrence Reitzer, Head, 972-883-2502, *Fax:* 972-883-2502.

Application contact: Ricky Robichaud, Assistant Coordinator, 972-883-2500, *E-mail:* rickyr@utdallas.edu.

■ THE UNIVERSITY OF TEXAS–HOUSTON HEALTH SCIENCE CENTER

Graduate School of Biomedical Sciences, Program in Biochemistry and Molecular Biology, Houston, TX 77225-0036

AWARDS MS, PhD, MD/PhD.

Faculty: 17 full-time (3 women).

Students: 8 full-time (3 women), 6 international. Average age 27. *38 applicants, 61% accepted.* In 1999, 1 master's, 6 doctorates awarded. Terminal master's awarded for partial completion of doctoral program.

Degree requirements: For master's and doctorate, thesis/dissertation required, foreign language not required.

Entrance requirements: For master's and doctorate, GRE General Test, TOEFL, TWE. *Application deadline:* For fall admission, 1/15 (priority date); for spring admission, 11/1. Applications are processed on a rolling basis. *Application fee:* $10. Electronic applications accepted.

Financial aid: Fellowships, research assistantships, teaching assistantships, institutionally sponsored loans available. Financial aid application deadline: 1/15.

Faculty research: Protein structure and function, regulation of gene expression, cellular signal transduction, regulation of cell growth, membrane protein assembly and function.

Dr. John Putkey, Director, 713-500-6061, *Fax:* 713-500-0652, *E-mail:* john.putkey@uth.tmc.edu.

Application contact: Anne Baronitis, Director of Admissions, 713-500-9860, *Fax:* 713-500-9877, *E-mail:* abaron@gsbs.gs.uth.tmc.edu.

Find an in-depth description at www.petersons.com/graduate.

■ THE UNIVERSITY OF TEXAS–HOUSTON HEALTH SCIENCE CENTER

Graduate School of Biomedical Sciences, Program in Regulatory Biology, Houston, TX 77225-0036

AWARDS MS, PhD, MD/PhD.

Faculty: 18 full-time (5 women).
Students: *1 applicant, 100% accepted.*
Degree requirements: For master's and doctorate, thesis/dissertation required, foreign language not required.
Entrance requirements: For master's and doctorate, GRE General Test, TOEFL, TWE. *Application deadline:* For fall admission, 1/15 (priority date); for spring admission, 11/1. Applications are processed on a rolling basis. *Application fee:* $10. Electronic applications accepted.
Financial aid: Fellowships, research assistantships, institutionally sponsored loans available. Financial aid application deadline: 1/15.
Faculty research: Mechanisms of action of hormones and neurotransmitters, cancer chemotherapy, molecular biology of regulatory processes, cell signaling.
Dr. Fernando R. Cabral, Director, 713-500-7485, *Fax:* 713-500-7455, *E-mail:* fcabral@farmr1.med.uth.tmc.edu.
Application contact: Anne Baronitis, Director of Admissions, 713-500-9860, *Fax:* 713-500-9877, *E-mail:* abaron@gsbs.gs.uth.tmc.edu.

Find an in-depth description at www.petersons.com/graduate.

■ THE UNIVERSITY OF TEXAS MEDICAL BRANCH AT GALVESTON

Graduate School of Biomedical Sciences, Program in Human Biological Chemistry and Genetics, Galveston, TX 77555

AWARDS Biochemistry (MS, PhD); cell biology (MS, PhD); genetics (MS, PhD); molecular biology (PhD); structural biology (PhD).

Faculty: 61 full-time (4 women).
Students: 39 full-time (12 women); includes 2 minority (both Hispanic Americans), 12 international. Average age 28. *41 applicants, 66% accepted.* In 1999, 1 master's awarded (100% found work related to degree); 4 doctorates awarded (100% found work related to degree).

Degree requirements: For master's and doctorate, thesis/dissertation required, foreign language not required.
Entrance requirements: For master's and doctorate, GRE General Test. *Application deadline:* For fall admission, 2/1 (priority date). Applications are processed on a rolling basis. *Application fee:* $25 ($50 for international students). Electronic applications accepted.
Expenses: Tuition, state resident: full-time $684; part-time $38 per credit hour. Tuition, nonresident: full-time $4,572; part-time $254 per credit hour. Required fees: $29; $7.5 per credit hour. One-time fee: $55. Tuition and fees vary according to program.
Financial aid: Fellowships, research assistantships, Federal Work-Study and institutionally sponsored loans available. Financial aid application deadline: 2/1; financial aid applicants required to submit FAFSA.
Faculty research: Cell surfaces and transmembrane signaling, structural biology, growth factors, DNA repair and aging.
Dr. James C. Lee, Director, 409-772-2769, *Fax:* 409-747-0552, *E-mail:* gradprog.hbcg@utmb.edu.
Application contact: Becky L. Hansen, Senior Administrative Secretary, 409-772-2769, *Fax:* 409-747-0552, *E-mail:* bhansen@utmb.edu.

Find an in-depth description at www.petersons.com/graduate.

■ THE UNIVERSITY OF TEXAS SOUTHWESTERN MEDICAL CENTER AT DALLAS

Southwestern Graduate School of Biomedical Sciences, Division of Cell and Molecular Biology, Dallas, TX 75390

AWARDS Biological chemistry (PhD); cell regulation (PhD); genetics and development (PhD); immunology (PhD); integrative biology (PhD); medical science (PhD); molecular biophysics (PhD); molecular microbiology (PhD); neuroscience (PhD).

Faculty: 184 full-time (26 women), 10 part-time/adjunct (0 women).
Students: 93 full-time (31 women); includes 8 minority (1 African American, 4 Asian Americans or Pacific Islanders, 3 Hispanic Americans), 16 international. Average age 27. *699 applicants, 16% accepted.*
Degree requirements: For doctorate, dissertation required, foreign language not required.
Entrance requirements: For doctorate, GRE General Test, minimum GPA of 3.0. *Application deadline:* For fall admission, 1/5

(priority date). *Application fee:* $0. Electronic applications accepted.
Expenses: Tuition, state resident: full-time $912. Tuition, nonresident: full-time $6,096. Required fees: $216. Full-time tuition and fees vary according to course load and program.
Financial aid: Fellowships, research assistantships, institutionally sponsored loans and traineeships available. Financial aid application deadline: 3/15; financial aid applicants required to submit FAFSA.
Dr. Philip S. Perlman, Associate Dean, 214-648-1464, *Fax:* 214-648-1488, *E-mail:* perlman@utsw.swmed.edu.
Application contact: Nancy McKinney, Education Coordinator, 214-648-8099, *Fax:* 214-648-2978, *E-mail:* dcmbinfo@utsouthwestern.edu.

Find an in-depth description at www.petersons.com/graduate.

■ UNIVERSITY OF UTAH

Graduate School, College of Science and Graduate Programs in Medicine, Program in Molecular Biology, Salt Lake City, UT 84132

AWARDS PhD.

Degree requirements: For doctorate, dissertation, preliminary exams required, foreign language not required.
Entrance requirements: For doctorate, GRE General Test, GRE Subject Test, TOEFL, minimum GPA of 3.0.
Expenses: Tuition, state resident: full-time $1,663. Tuition, nonresident: full-time $5,201. Tuition and fees vary according to course load and program.
Faculty research: Biochemistry; cellular, viral, and molecular biology; human genetics; pathology; procaryotic development.

Find an in-depth description at www.petersons.com/graduate.

■ UNIVERSITY OF VERMONT

Graduate College, Cell and Molecular Biology Program, Burlington, VT 05405

AWARDS MS, PhD.

Degree requirements: For master's and doctorate, thesis/dissertation required, foreign language not required.
Entrance requirements: For master's and doctorate, GRE General Test, TOEFL.
Expenses: Tuition, state resident: full-time $7,464; part-time $311 per credit. Tuition, nonresident: full-time $18,672; part-time $778 per credit. Full-time tuition and fees vary according to degree level and program.

Find an in-depth description at www.petersons.com/graduate.

■ UNIVERSITY OF VIRGINIA

College and Graduate School of Arts and Sciences, Interdisciplinary Program in Cell and Molecular Biology, Charlottesville, VA 22903

AWARDS PhD. PhD awarded by related department after one year of general course work and research within CMB program.

Students: 4 full-time (1 woman). Average age 24. *87 applicants, 15% accepted.*
Degree requirements: For doctorate, dissertation required.
Entrance requirements: For doctorate, GRE General Test. *Application fee:* $40. Electronic applications accepted.
Expenses: Tuition, state resident: full-time $3,832. Tuition, nonresident: full-time $15,519. Required fees: $1,084. Tuition and fees vary according to course load and program.
Financial aid: Applicants required to submit FAFSA.

Find an in-depth description at www.petersons.com/graduate.

■ UNIVERSITY OF WASHINGTON

School of Medicine and Graduate School, Graduate Programs in Medicine, Program in Molecular and Cellular Biology, Seattle, WA 98195

AWARDS PhD. Offered jointly with Fred Hutchinson Cancer Research Center.

Faculty: 178 full-time (42 women).
Students: 98 full-time (42 women); includes 23 minority (3 African Americans, 16 Asian Americans or Pacific Islanders, 3 Hispanic Americans, 1 Native American). *244 applicants, 31% accepted.* In 1999, 6 degrees awarded (100% entered university research/teaching).
Degree requirements: For doctorate, dissertation required. *Average time to degree:* Doctorate–5 years full-time.
Entrance requirements: For doctorate, GRE General Test, GRE Subject Test. *Application deadline:* For fall admission, 1/3. *Application fee:* $45. Electronic applications accepted.
Expenses: Tuition, state resident: full-time $9,210; part-time $236 per credit. Tuition, nonresident: full-time $23,256; part-time $596 per credit.
Financial aid: In 1999–00, 33 fellowships with tuition reimbursements (averaging $14,688 per year), 20 research assistantships with tuition reimbursements (averaging $17,304 per year) were awarded; career-related internships or fieldwork, Federal Work-Study, and traineeships also available.
Dr. Randall T. Moon, Co-Director, 206-543-1722, *Fax:* 206-616-4230, *E-mail:* rtmoon@ul.washington.edu.

Application contact: MaryEllin Robinson, Program Specialist, 206-685-3155, *Fax:* 206-685-8174, *E-mail:* mcb@u.washington.edu.

Find an in-depth description at www.petersons.com/graduate.

■ UNIVERSITY OF WISCONSIN–MADISON

Graduate School, College of Agricultural and Life Sciences, Department of Animal Health and Biomedical Sciences, Program in Comparative Biosciences, Madison, WI 53706-1380

AWARDS Anatomy (MS, PhD); biochemistry (MS, PhD); cellular and molecular biology (MS, PhD); environmental toxicology (MS, PhD); neurosciences (MS, PhD); pharmacology (MS, PhD); physiology (MS, PhD).

Degree requirements: For doctorate, dissertation required.
Expenses: Tuition, state resident: full-time $5,406; part-time $339 per credit. Tuition, nonresident: full-time $17,110; part-time $1,071 per credit. Full-time tuition and fees vary according to program and reciprocity agreements. Part-time tuition and fees vary according to course load and program.

■ UNIVERSITY OF WISCONSIN–MADISON

Graduate School, Program in Cellular and Molecular Biology, Madison, WI 53706-1380

AWARDS Cellular and molecular biology (PhD); developmental biology (PhD).

Degree requirements: For doctorate, dissertation required, foreign language not required.
Entrance requirements: For doctorate, GRE General Test, GRE Subject Test. Electronic applications accepted.
Expenses: Tuition, state resident: full-time $5,406; part-time $339 per credit. Tuition, nonresident: full-time $17,110; part-time $1,071 per credit. Full-time tuition and fees vary according to program and reciprocity agreements. Part-time tuition and fees vary according to course load and program.
Faculty research: Virology, carcinogensis, cell structure, hormone regulation, developmental biology.

Find an in-depth description at www.petersons.com/graduate.

■ UNIVERSITY OF WISCONSIN–PARKSIDE

College of Arts and Sciences, Program in Applied Molecular Biology, Kenosha, WI 53141-2000

AWARDS MAMB.

Faculty: 12 full-time.
Students: 10 full-time (2 women); includes 3 minority (all Asian Americans or Pacific Islanders). Average age 25. *21 applicants, 33% accepted.* In 1999, 5 degrees awarded.
Degree requirements: For master's, thesis, internship, oral exam required, foreign language not required. *Average time to degree:* Master's–2 years full-time.
Entrance requirements: For master's, GRE General Test, minimum GPA of 3.0. *Application deadline:* For fall admission, 7/1 (priority date). Applications are processed on a rolling basis. *Application fee:* $45.
Expenses: Tuition, state resident: full-time $3,960; part-time $233 per credit. Tuition, nonresident: full-time $13,118; part-time $698 per credit. Tuition and fees vary according to program.
Financial aid: In 1999–00, 6 research assistantships were awarded; career-related internships or fieldwork, Federal Work-Study, and unspecified assistantships also available. Financial aid application deadline: 7/1.
Faculty research: Gene cloning, genome structure, cell cycle effects on gene expression, molecular biology of plant hormones, laboratory toxin production and resistance. Dr. Gary Wood, Chair, 414-595-2434, *Fax:* 414-595-2056.

■ UNIVERSITY OF WYOMING

Graduate School, College of Agriculture, Department of Molecular Biology, Laramie, WY 82071

AWARDS MS, PhD.

Faculty: 13 full-time (2 women).
Students: 17 full-time (7 women), 7 part-time (4 women), 7 international. Average age 31. *18 applicants, 72% accepted.* In 1999, 5 master's awarded (60% found work related to degree, 40% continued full-time study); 4 doctorates awarded (43% entered university research/teaching, 57% found other work related to degree). Terminal master's awarded for partial completion of doctoral program.
Degree requirements: For master's and doctorate, thesis/dissertation required, foreign language not required. *Average time to degree:* Master's–1.7 years full-time; doctorate–4.5 years full-time.
Entrance requirements: For master's and doctorate, GRE General Test, GRE Subject Test (recommended), minimum GPA of 3.0. *Application deadline:* For fall

admission, 3/1 (priority date). Applications are processed on a rolling basis. *Application fee:* $40.

Expenses: Tuition, state resident: full-time $2,520; part-time $140 per credit hour. Tuition, nonresident: full-time $7,790; part-time $433 per credit hour. Required fees: $440; $7 per credit hour. Full-time tuition and fees vary according to course load and program.

Financial aid: In 1999–00, 10 research assistantships with full tuition reimbursements (averaging $13,700 per year) were awarded; institutionally sponsored loans also available. Financial aid application deadline: 3/1.

Faculty research: Protein structure/function, molecular parasitology, hormone receptors, developmental regulation, yeast genetics. *Total annual research expenditures:* $1.5 million.

Dr. Jerry Johnson, Chairman, 307-766-3300, *Fax:* 307-766-5098, *E-mail:* uwmbio@uwyo.edu.

Application contact: Dr. Pamela J. Langer, Graduate Program Chairperson, 307-766-3300 Ext. 307, *Fax:* 307-766-5098, *E-mail:* uwmbio.edu.

■ **UTAH STATE UNIVERSITY**

School of Graduate Studies, College of Agriculture, Program in Toxicology, Logan, UT 84322

AWARDS Molecular biology (MS, PhD); toxicology (MS, PhD).

Faculty: 14 full-time (2 women), 3 part-time/adjunct (0 women).

Students: 9 full-time (1 woman), 1 (woman) part-time. Average age 25. *9 applicants, 33% accepted.* In 1999, 2 doctorates awarded. Terminal master's awarded for partial completion of doctoral program.

Degree requirements: For master's and doctorate, thesis/dissertation required, foreign language not required.

Entrance requirements: For master's and doctorate, GRE General Test, TOEFL, minimum GPA of 3.0. *Application deadline:* For fall admission, 6/15 (priority date); for spring admission, 10/15. Applications are processed on a rolling basis. *Application fee:* $40.

Expenses: Tuition, state resident: full-time $1,553. Tuition, nonresident: full-time $5,436. International tuition: $5,526 full-time. Required fees: $447. Tuition and fees vary according to course load and program.

Financial aid: In 1999–00, 1 fellowship with partial tuition reimbursement (averaging $15,000 per year), 5 research assistantships with partial tuition reimbursements (averaging $15,000 per year) were awarded; teaching assistantships with partial tuition reimbursements, Federal Work-Study, institutionally sponsored

loans, and tuition waivers (partial) also available. Aid available to part-time students.

Faculty research: Free-radical mechanisms, toxicity of iron, carcinogenesis of natural compounds, molecular mechanisms of retinoid toxicity, aflatoxins.

Roger A. Coulombe, Director, 435-797-1598, *Fax:* 435-797-1601, *E-mail:* rogerc@cc.usu.edu.

■ **UTAH STATE UNIVERSITY**

School of Graduate Studies, College of Family Life and College of Agriculture, Department of Nutrition and Food Sciences, Logan, UT 84322

AWARDS Food microbiology and safety (MFMS); molecular biology (MS, PhD); nutrition and food sciences (MS, PhD).

Faculty: 17 full-time (8 women), 2 part-time/adjunct (0 women).

Students: 18 full-time (11 women), 8 part-time (4 women); includes 2 minority (1 Asian American or Pacific Islander, 1 Hispanic American), 12 international. Average age 27. *34 applicants, 21% accepted.* In 1999, 10 master's, 3 doctorates awarded.

Degree requirements: For master's, thesis required, foreign language not required; for doctorate, computer language, dissertation, teaching experience required, foreign language not required.

Entrance requirements: For master's and doctorate, GRE General Test, TOEFL, minimum GPA of 3.0, previous course work in chemistry. *Application deadline:* For fall admission, 2/1 (priority date); for spring admission, 10/15 (priority date). Applications are processed on a rolling basis. *Application fee:* $40.

Expenses: Tuition, state resident: full-time $1,553. Tuition, nonresident: full-time $5,436. International tuition: $5,526 full-time. Required fees: $447. Tuition and fees vary according to course load and program.

Financial aid: In 1999–00, 22 research assistantships with partial tuition reimbursements were awarded; fellowships with partial tuition reimbursements, teaching assistantships with partial tuition reimbursements, Federal Work-Study, institutionally sponsored loans, and tuition waivers (full and partial) also available.

Faculty research: Mineral balance, meat microbiology and nitrate interactions, milk ultrafiltration, lactic culture, milk coagulation.

Ann Sorenson, Head, 435-797-2102, *Fax:* 435-797-2379.

Application contact: Carolyn Glover, Receptionist, 435-797-2126, *Fax:* 435-797-2379, *E-mail:* nfs@cc.usu.edu.

■ **VANDERBILT UNIVERSITY**

Graduate School and School of Medicine, Department of Molecular Biology, Nashville, TN 37240-1001

AWARDS MS, PhD, MD/PhD.

Faculty: 16 full-time (4 women), 2 part-time/adjunct (0 women).

Students: 26 full-time (13 women), 1 part-time; includes 1 minority (African American), 8 international. Average age 26. *22 applicants, 23% accepted.* In 1999, 2 master's, 3 doctorates awarded.

Degree requirements: For master's, thesis required, foreign language not required; for doctorate, dissertation, final and qualifying exams required, foreign language not required.

Entrance requirements: For master's and doctorate, GRE General Test, GRE Subject Test (recommended). *Application deadline:* For fall admission, 1/15. *Application fee:* $40.

Expenses: Tuition: Full-time $17,244; part-time $958 per hour. Required fees: $242; $121 per semester. Tuition and fees vary according to program.

Financial aid: In 1999–00, 9 fellowships with full tuition reimbursements (averaging $17,270 per year) were awarded; research assistantships, Federal Work-Study, institutionally sponsored loans, and traineeships also available. Financial aid application deadline: 1/15.

Faculty research: Macromolecular structure, molecular genetics, developmental biology, cell ultrastructure, molecular biophysics.

Ellen H. Fanning, Chair, 615-322-2008, *Fax:* 615-343-6707, *E-mail:* fannine@ctrvax.vanderbilt.edu.

Application contact: Charles K. Singleton, Director of Graduate Studies, 615-322-6516, *Fax:* 615-343-6707, *E-mail:* charles.k.singleton@vanderbilt.edu.

■ **VANDERBILT UNIVERSITY**

Graduate School and School of Medicine, Department of Molecular Physiology and Biophysics, Nashville, TN 37240-1001

AWARDS PhD, MD/PhD.

Faculty: 20 full-time (2 women).

Students: 23 full-time (11 women), 1 part-time; includes 3 minority (1 Asian American or Pacific Islander, 2 Hispanic Americans), 4 international. Average age 29. In 1999, 11 degrees awarded.

Degree requirements: For doctorate, dissertation, preliminary, qualifying, and final exams required, foreign language not required.

Entrance requirements: For doctorate, GRE General Test, GRE Subject Test (recommended). *Application deadline:* For fall admission, 1/15. *Application fee:* $40.

Vanderbilt University (continued)

Expenses: Tuition: Full-time $17,244; part-time $958 per hour. Required fees: $242; $121 per semester. Tuition and fees vary according to program.

Financial aid: In 1999–00, fellowships with full tuition reimbursements (averaging $17,000 per year), research assistantships with full tuition reimbursements (averaging $17,000 per year) were awarded; Federal Work-Study, institutionally sponsored loans, traineeships, and tuition waivers (partial) also available. Financial aid application deadline: 1/15.

Faculty research: Molecular endocrinology, membrane transport biophysics, metabolic regulation, neurobiology. Alan D. Cherrington, Chair, 615-322-7000, *Fax:* 615-322-7236, *E-mail:* alan.cherrington@mcmail.vanderbilt.edu.

Application contact: Roger J. Colbran, Director of Graduate Studies, 615-322-7000, *Fax:* 615-322-7236, *E-mail:* roger.j.colbran@vanderbilt.edu.

Find an in-depth description at www.petersons.com/graduate.

■ **VIRGINIA COMMONWEALTH UNIVERSITY**

School of Graduate Studies and School of Medicine, School of Medicine Graduate Programs, Department of Biochemistry and Molecular Biophysics, Richmond, VA 23284-9005

AWARDS Biochemistry (PhD); biochemistry and molecular biophysics (MS, CBHS); molecular biology and genetics (PhD); neurosciences (PhD).

Students: 5 full-time (3 women), 21 part-time (10 women); includes 12 minority (9 Asian Americans or Pacific Islanders, 1 Hispanic American, 2 Native Americans). In 1999, 1 master's, 10 doctorates awarded.

Degree requirements: For master's, thesis required, foreign language not required; for doctorate, dissertation, comprehensive oral and written exams required, foreign language not required.

Entrance requirements: For master's and doctorate, GRE General Test. *Application deadline:* For fall admission, 5/1. *Application fee:* $30.

Expenses: Tuition, state resident: full-time $4,031; part-time $224 per credit hour. Tuition, nonresident: full-time $11,946; part-time $664 per credit hour. Required fees: $1,081; $40 per credit hour. Tuition and fees vary according to campus/location and program.

Financial aid: Fellowships, research assistantships available.

Faculty research: Molecular biology, peptide/protein chemistry, neurochemistry,

enzyme mechanisms, macromolecular structure determination. *Total annual research expenditures:* $3.5 million. Chair, 804-828-9762, *Fax:* 804-828-1473.

Application contact: Dr. Zendra E. Zehner, Program Director, 804-828-8753, *Fax:* 804-828-1473, *E-mail:* zezehner@vcu.edu.

■ **VIRGINIA COMMONWEALTH UNIVERSITY**

School of Graduate Studies and School of Medicine, School of Medicine Graduate Programs, Department of Human Genetics, Richmond, VA 23284-9005

AWARDS Genetic counseling (MS); human genetics (PhD, CBHS); molecular biology and genetics (PhD).

Students: In 1999, 3 master's, 3 doctorates awarded.

Degree requirements: For master's, computer language, thesis required, foreign language not required; for doctorate, computer language, dissertation, comprehensive oral and written exams required, foreign language not required.

Entrance requirements: For master's, DAT, GRE General Test, or MCAT; for doctorate, GRE General Test. *Application deadline:* For fall admission, 2/1. *Application fee:* $30.

Expenses: Tuition, state resident: full-time $4,031; part-time $224 per credit hour. Tuition, nonresident: full-time $11,946; part-time $664 per credit hour. Required fees: $1,081; $40 per credit hour. Tuition and fees vary according to campus/location and program.

Financial aid: Fellowships available.

Faculty research: Genetic epidemiology, biochemical genetics, quantitative genetics, human cytogenetics, molecular genetics. Dr. Walter E. Nance, Chair, 804-828-9632, *Fax:* 804-828-3760.

Application contact: Dr. Linda A. Corey, Graduate Program Director, 804-828-9632, *Fax:* 804-828-3760, *E-mail:* lacorey@vcu.edu.

Find an in-depth description at www.petersons.com/graduate.

■ **VIRGINIA COMMONWEALTH UNIVERSITY**

School of Graduate Studies and School of Medicine, School of Medicine Graduate Programs, Department of Microbiology and Immunology, Richmond, VA 23284-9005

AWARDS Microbiology (PhD); microbiology and immunology (MS, CBHS); molecular biology and genetics (PhD).

Students: 13 full-time (6 women), 46 part-time (26 women); includes 22 minority (2 African Americans, 17 Asian Americans or Pacific Islanders, 2 Hispanic Americans, 1 Native American). In 1999, 2 master's, 9 doctorates, 2 other advanced degrees awarded.

Degree requirements: For master's, thesis required, foreign language not required; for doctorate, dissertation, comprehensive oral and written exams required, foreign language not required.

Entrance requirements: For master's, DAT, GRE General Test, or MCAT; for doctorate, GRE General Test. *Application deadline:* For fall admission, 4/15. *Application fee:* $30.

Expenses: Tuition, state resident: full-time $4,031; part-time $224 per credit hour. Tuition, nonresident: full-time $11,946; part-time $664 per credit hour. Required fees: $1,081; $40 per credit hour. Tuition and fees vary according to campus/location and program.

Financial aid: Fellowships, research assistantships, teaching assistantships available.

Faculty research: Microbial physiology and genetics, molecular biology, crystallography of biological molecules, antibiotics and chemotherapy, membrane transport. Dr. Dennis E. Ohman, Chair, 804-828-9728, *Fax:* 804-828-9946.

Find an in-depth description at www.petersons.com/graduate.

■ **VIRGINIA COMMONWEALTH UNIVERSITY**

School of Graduate Studies and School of Medicine, School of Medicine Graduate Programs, Department of Pharmacology and Toxicology, Richmond, VA 23284-9005

AWARDS Molecular biology and genetics (PhD); neurosciences (PhD); pharmacology (PhD, CBHS); pharmacology and toxicology (MS).

Students: 14 full-time (5 women), 51 part-time (27 women); includes 27 minority (4 African Americans, 22 Asian Americans or Pacific Islanders, 1 Hispanic American). In 1999, 3 master's, 10 doctorates, 1 other advanced degree awarded. Terminal master's awarded for partial completion of doctoral program.

Degree requirements: For master's, thesis required, foreign language not required; for doctorate, dissertation, comprehensive oral and written exams required, foreign language not required.

Entrance requirements: For master's, DAT, GRE General Test or MCAT; for doctorate, GRE General Test. *Application* *deadline:* For fall admission, 2/1 (priority date). *Application fee:* $30.

Expenses: Tuition, state resident: full-time $4,031; part-time $224 per credit hour. Tuition, nonresident: full-time $11,946; part-time $664 per credit hour. Required fees: $1,081; $40 per credit hour. Tuition and fees vary according to campus/location and program.

Financial aid: Fellowships, teaching assistantships available.

Faculty research: Drug abuse, drug metabolism, pharmacodynamics, peptide synthesis, receptor mechanisms.

Dr. George Kunos, Chair, 804-828-2073, *Fax:* 804-828-2117.

Application contact: Sheryol Cox, Graduate Program Coordinator, 804-828-8400, *Fax:* 804-828-2117, *E-mail:* swcox@ vcu.edu.

Find an in-depth description at www.petersons.com/graduate.

■ VIRGINIA POLYTECHNIC INSTITUTE AND STATE UNIVERSITY

Graduate School, College of Agriculture and Life Sciences, Department of Biochemistry, Blacksburg, VA 24061

AWARDS Biochemistry (MS, PhD); cell and molecular biology (PhD).

Students: 17 full-time (8 women), 4 part-time (all women), 9 international. *40 applicants, 35% accepted.* In 1999, 2 master's, 2 doctorates awarded.

Degree requirements: For doctorate, dissertation required, foreign language not required.

Entrance requirements: For master's, TOEFL; for doctorate, GRE General Test, TOEFL. *Application deadline:* For fall admission, 12/1 (priority date). Applications are processed on a rolling basis. *Application fee:* $25.

Expenses: Tuition, state resident: full-time $4,122; part-time $229 per credit hour. Tuition, nonresident: full-time $6,930; part-time $385 per credit hour. Required fees: $828; $107 per semester. Part-time tuition and fees vary according to course load.

Financial aid: In 1999–00, research assistantships with full tuition reimbursements (averaging $15,815 per year), 8 teaching assistantships with full tuition reimbursements (averaging $15,815 per year) were awarded; fellowships. Financial aid application deadline: 4/1.

Dr. John Hess, Head, 540-231-6315, *E-mail:* jlhess@vt.edu.

Find an in-depth description at www.petersons.com/graduate.

■ WAKE FOREST UNIVERSITY

School of Medicine and Graduate School, Graduate Programs in Medicine, Molecular Genetics Program, Winston-Salem, NC 27109

AWARDS PhD.

Degree requirements: For doctorate, dissertation required.

Entrance requirements: For doctorate, GRE General Test, GRE Subject Test.

Expenses: Tuition: Full-time $18,300.

Find an in-depth description at www.petersons.com/graduate.

■ WASHINGTON STATE UNIVERSITY

Graduate School, College of Sciences, School of Molecular Biosciences, Pullman, WA 99164

AWARDS Biochemistry and biophysics (MS, PhD); genetics and cell biology (MS, PhD); microbiology (MS, PhD).

Students: 94 full-time (37 women), 6 part-time (4 women); includes 7 minority (6 Asian Americans or Pacific Islanders, 1 Hispanic American), 19 international. Terminal master's awarded for partial completion of doctoral program.

Degree requirements: For master's and doctorate, oral exam required.

Entrance requirements: For master's and doctorate, GRE General Test, minimum GPA of 3.0. *Application deadline:* Applications are processed on a rolling basis. *Application fee:* $35.

Expenses: Tuition, state resident: full-time $5,654. Tuition, nonresident: full-time $13,850. International tuition: $13,850 full-time. Tuition and fees vary according to program.

Financial aid: Application deadline: 4/1.

Dr. Michael D. Griswold, Director, 509-335-1276, *Fax:* 509-335-9688.

■ WASHINGTON UNIVERSITY IN ST. LOUIS

Graduate School of Arts and Sciences, Division of Biology and Biomedical Sciences, Program in Molecular Cell Biology, St. Louis, MO 63110

AWARDS PhD.

Degree requirements: For doctorate, dissertation required, foreign language not required.

Entrance requirements: For doctorate, GRE General Test, GRE Subject Test. *Application deadline:* For fall admission, 1/1 (priority date). Applications are processed on a rolling basis. *Application fee:* $0.

Expenses: Tuition: Full-time $23,400; part-time $975 per credit. Tuition and fees vary according to program.

Financial aid: Application deadline: 1/1.

Dr. Robert Mercer, Coordinator, 314-362-8849.

Application contact: Rosemary Garagneni, Director of Admissions, 800-852-9074, *E-mail:* admissions@ dbbs.wustl.edu.

■ WAYNE STATE UNIVERSITY

School of Medicine and Graduate School, Graduate Programs in Medicine, Program in Molecular Biology and Genetics, Detroit, MI 48202

AWARDS MS, PhD. Terminal master's awarded for partial completion of doctoral program.

Degree requirements: For master's and doctorate, thesis/dissertation required, foreign language not required.

Entrance requirements: For master's and doctorate, GRE General Test.

Faculty research: Human gene mapping, genome organization and sequencing, gene regulation, molecular evolution.

Find an in-depth description at www.petersons.com/graduate.

■ WESLEYAN UNIVERSITY

Graduate Programs, Department of Molecular Biology and Biochemistry, Middletown, CT 06459-0260

AWARDS Biochemistry (PhD); molecular biology (PhD).

Faculty: 9 full-time (2 women).

Students: 22 full-time (11 women); includes 2 minority (both Hispanic Americans), 16 international. Average age 28. *89 applicants, 3% accepted.* In 1999, 1 doctorate awarded.

Degree requirements: For doctorate, computer language, dissertation required.

Entrance requirements: For doctorate, GRE General Test, GRE Subject Test. *Application deadline:* For fall admission, 3/15. Applications are processed on a rolling basis. *Application fee:* $0.

Expenses: Tuition: Full-time $24,876. Required fees: $650. Tuition and fees vary according to program.

Financial aid: Research assistantships, teaching assistantships, institutionally sponsored loans available.

Faculty research: Genome organization, regulation of gene expression, molecular biology of development, physical biochemistry.

Dr. Irina Russu, Chair, 860-685-3556.

Application contact: Marina J. Melendez, Director of Graduate Student Services, 860-685-2390, *Fax:* 860-685-2439, *E-mail:* mmelendez@wesleyan.edu.

Find an in-depth description at www.petersons.com/graduate.

■ WEST VIRGINIA UNIVERSITY

Eberly College of Arts and Sciences, Department of Biology, Morgantown, WV 26506

AWARDS Animal behavior (MS); cellular and molecular biology (MS, PhD); environmental plant biology (MS, PhD); plant systematics (MS); population genetics (MS).

Faculty: 18 full-time (4 women), 2 part-time/adjunct (1 woman).

Students: 19 full-time (10 women), 3 part-time (2 women); includes 2 minority (1 Asian American or Pacific Islander, 1 Hispanic American), 4 international. Average age 27. *50 applicants, 10% accepted.* In 1999, 4 master's, 1 doctorate awarded. Terminal master's awarded for partial completion of doctoral program.

Degree requirements: For master's, thesis, final exam required, foreign language not required; for doctorate, dissertation, preliminary and final exams required, foreign language not required. *Average time to degree:* Master's–2.5 years full-time; doctorate–5 years full-time.

Entrance requirements: For master's, GRE General Test, GRE Subject Test, TOEFL, minimum GPA of 3.0; for doctorate, GRE General Test, TOEFL, minimum GPA of 3.0. *Application deadline:* For fall admission, 4/1; for spring admission, 10/1. Applications are processed on a rolling basis. *Application fee:* $45.

Expenses: Tuition, state resident: full-time $2,910; part-time $154 per credit hour. Tuition, nonresident: full-time $8,368; part-time $457 per credit hour.

Financial aid: In 1999–00, 3 research assistantships, 13 teaching assistantships were awarded; Federal Work-Study and institutionally sponsored loans also available. Financial aid application deadline: 4/1; financial aid applicants required to submit FAFSA.

Faculty research: Environmental biology, genetic engineering, developmental biology, global change, biodiversity.

Dr. Keith Garbutt, Chair, 304-293-5394.

Application contact: Dr. James B. McGraw, Director of Graduate Studies, 304-293-5201, *Fax:* 304-293-6363.

Find an in-depth description at www.petersons.com/graduate.

■ WILLIAM PATERSON UNIVERSITY OF NEW JERSEY

College of Science and Health, Department of Biology, General Biology Program, Wayne, NJ 07470-8420

AWARDS General biology (MA); limnology and terrestrial ecology (MA); molecular biology (MA); physiology (MA). Part-time and evening/weekend programs available.

Students: 1 (woman) full-time, 6 part-time (3 women); includes 1 minority (Hispanic American). Average age 25. *9 applicants, 44% accepted.* In 1999, 1 degree awarded.

Degree requirements: For master's, comprehensive exam, independent study or thesis required.

Entrance requirements: For master's, GRE General Test, minimum GPA of 2.75. *Application deadline:* For fall admission, 4/1; for spring admission, 10/15. Applications are processed on a rolling basis. *Application fee:* $35. Electronic applications accepted.

Expenses: Tuition, state resident: part-time $244 per credit. Tuition, nonresident: part-time $350 per credit.

Financial aid: In 1999–00, 2 teaching assistantships (averaging $9,000 per year) were awarded; career-related internships or fieldwork and unspecified assistantships also available. Financial aid application deadline: 4/1; financial aid applicants required to submit FAFSA.

Application contact: Office of Graduate Studies, 973-720-2237, *Fax:* 973-720-2035.

Find an in-depth description at www.petersons.com/graduate.

■ WRIGHT STATE UNIVERSITY

School of Graduate Studies, College of Science and Mathematics, Department of Biochemistry and Molecular Biology, Dayton, OH 45435

AWARDS MS.

Students: 7 full-time (1 woman), 1 (woman) part-time; includes 1 minority (African American), 1 international. *10 applicants, 40% accepted.* In 1999, 3 degrees awarded.

Degree requirements: For master's, thesis required, foreign language not required.

Entrance requirements: For master's, TOEFL. *Application deadline:* For fall admission, 8/20. *Application fee:* $25.

Expenses: Tuition, state resident: full-time $5,568; part-time $175 per quarter hour. Tuition, nonresident: full-time $9,696; part-time $302 per quarter hour. Full-time tuition and fees vary according to course load, campus/location and program.

Financial aid: Fellowships, research assistantships, teaching assistantships, unspecified assistantships available. Aid available to part-time students. Financial aid application deadline: 6/15; financial aid applicants required to submit FAFSA.

Faculty research: Regulation of gene expression, macromolecular structural function, NMR imaging, visual biochemistry.

Dr. Daniel T. Organisciak, Chair, 937-775-3041, *Fax:* 937-775-3730, *E-mail:* dto@wright.edu.

Application contact: Dr. John J. Turchi, Director, 937-775-2853, *Fax:* 937-775-3730, *E-mail:* john.turchi@wright.edu.

Find an in-depth description at www.petersons.com/graduate.

■ YALE UNIVERSITY

Graduate School of Arts and Sciences, Department of Molecular, Cellular, and Developmental Biology, Program in Molecular Biology, New Haven, CT 06520

AWARDS PhD.

Degree requirements: For doctorate, dissertation required.

Entrance requirements: For doctorate, GRE General Test, GRE Subject Test. *Application deadline:* For fall admission, 1/4. *Application fee:* $65.

Expenses: Tuition: Full-time $22,300. Full-time tuition and fees vary according to program.

Financial aid: Fellowships, research assistantships, teaching assistantships available.

Application contact: Admissions Information, 203-432-2770.

■ YESHIVA UNIVERSITY

Albert Einstein College of Medicine, Sue Golding Graduate Division of Medical Sciences, Division of Biological Sciences, Department of Developmental and Molecular Biology, Bronx, NY 10461

AWARDS PhD, MD/PhD.

Faculty: 12 full-time.

Students: 32 full-time (12 women); includes 6 minority (all Asian Americans or Pacific Islanders), 17 international. In 1999, 4 degrees awarded.

Degree requirements: For doctorate, dissertation required, foreign language not required.

Entrance requirements: For doctorate, GRE General Test, TOEFL. *Application deadline:* For fall admission, 1/15. *Application fee:* $0.

Expenses: Tuition: Part-time $525 per credit. Tuition and fees vary according to degree level and program.

Financial aid: In 1999–00, 32 fellowships were awarded.

Faculty research: DNA, RNA, and protein synthesis in prokaryotes and eukaryotes; chemical and enzymatic alteration of RNA; glycoproteins.

Dr. R. E. Stanley, Chairperson, 718-430-2094.

Application contact: Sheila Cleeton, Assistant Director, 718-430-2128, *Fax:* 718-430-8655, *E-mail:* phd@aecom.yu.edu.

MOLECULAR MEDICINE

■ BOSTON UNIVERSITY

School of Medicine, Division of Graduate Medical Sciences, Program in Molecular Medicine, Boston, MA 02215

AWARDS PhD, MD/PhD.

Degree requirements: For doctorate, dissertation, qualifying exam required, foreign language not required.
Application deadline: For fall admission, 1/15 (priority date); for spring admission, 10/15 (priority date). *Application fee:* $50. Electronic applications accepted.
Expenses: Tuition: Full-time $24,700; part-time $772 per credit. Required fees: $220.
Financial aid: Fellowships, research assistantships, Federal Work-Study, scholarships, and traineeships available.
Dr. Joseph Loscalzo, Director, 617-414-1519, *Fax:* 617-414-1515, *E-mail:* gpmm@med-med1.bu.edu.

Find an in-depth description at www.petersons.com/graduate.

■ JOAN AND SANFORD I. WEILL MEDICAL COLLEGE AND GRADUATE SCHOOL OF MEDICAL SCIENCES OF CORNELL UNIVERSITY

Graduate School of Medical Sciences, Interdisciplinary Training Program in Molecular Medicine, New York, NY 10021-4896

AWARDS PhD, MD/PhD.

Degree requirements: For doctorate, dissertation, final exam required.
Entrance requirements: For doctorate, GRE General Test, GRE Subject Test, MCAT (MD/PhD). *Application deadline:* For fall admission, 1/15. *Application fee:* $50.
Expenses: All students in good standing receive an annual stipend of $22,880.

■ JOHNS HOPKINS UNIVERSITY

School of Medicine, Graduate Programs in Medicine, Graduate Program in Cellular and Molecular Medicine, Baltimore, MD 21218-2699

AWARDS PhD.

Faculty: 112 full-time (23 women).
Students: 49 full-time (15 women); includes 25 minority (1 African American, 22 Asian Americans or Pacific Islanders, 2 Hispanic Americans). Average age 28. *95 applicants, 24% accepted.* In 1999, 3 degrees awarded (100% entered university research/teaching).

Degree requirements: For doctorate, dissertation, oral exam required. *Average time to degree:* Doctorate–4.4 years full-time.
Entrance requirements: For doctorate, GRE General Test (average 600 verbal, 750 quantitative, 700 analytical), GRE Subject Test (average 600). *Application deadline:* For fall admission, 2/1 (priority date). Applications are processed on a rolling basis. *Application fee:* $50.
Expenses: Tuition: Full-time $23,660.
Financial aid: In 1999–00, 11 fellowships (averaging $8,128 per year) were awarded; tuition waivers (full) also available. Financial aid application deadline: 2/15.
Faculty research: Cellular and molecular basis of disease. *Total annual research expenditures:* $24 million.
Dr. John T. Isaacs, Director, 410-955-7777, *Fax:* 410-955-0840, *E-mail:* jtisaacs@jhmi.edu.
Application contact: Theo M. Karpovich, Admissions Coordinator, 410-614-0391, *Fax:* 410-614-7294, *E-mail:* karpoo@jhmi.edu.

Find an in-depth description at www.petersons.com/graduate.

■ MEDICAL COLLEGE OF GEORGIA

School of Graduate Studies, Program in Molecular Medicine, Augusta, GA 30912

AWARDS PhD.

Faculty: 1 full-time (0 women).
Students: 21 full-time (5 women); includes 1 minority (Hispanic American), 4 international. *12 applicants, 83% accepted.*
Degree requirements: For doctorate, dissertation required, foreign language not required.
Entrance requirements: For doctorate, GRE General Test, TOEFL. *Application deadline:* For fall admission, 6/30 (priority date). Applications are processed on a rolling basis. *Application fee:* $25.
Expenses: Tuition, state resident: full-time $2,896; part-time $121 per hour. Tuition, nonresident: full-time $11,584; part-time $483 per hour. Required fees: $286; $143 per semester. Tuition and fees vary according to program.
Financial aid: In 1999–00, 14 research assistantships with partial tuition reimbursements (averaging $15,500 per year) were awarded; fellowships, teaching assistantships, Federal Work-Study, grants, and institutionally sponsored loans also available. Aid available to part-time students. Financial aid application deadline: 3/31; financial aid applicants required to submit FAFSA.
Dr. Andrew Mellor, Acting Director, 706-721-8735, *E-mail:* amellor@mail.mcg.edu.

Application contact: Dr. William Dynan, Graduate Director, 706-721-8756, *E-mail:* wdynan@mail.mcg.edu.

■ THE PENNSYLVANIA STATE UNIVERSITY UNIVERSITY PARK CAMPUS

Graduate School, Intercollege Graduate Programs, Intercollege Graduate Program in Integrative Biosciences, State College, University Park, PA 16802-1503

AWARDS Integrative biosciences (MS, PhD), including biomolecular transport dynamics, cell and developmental biology, cellular and molecular mechanisms of toxicity, chemical biology, ecological and molecular plant physiology, immunobiology, molecular medicine, neuroscience, nutrition science.

Students: 39 full-time (24 women), 1 part-time.
Entrance requirements: For master's and doctorate, GRE General Test. *Application fee:* $50.
Expenses: Tuition, state resident: full-time $6,886; part-time $291 per credit. Tuition, nonresident: full-time $14,118; part-time $588 per credit. Required fees: $46 per semester. Part-time tuition and fees vary according to course load and program.
Financial aid: Fellowships available.
Dr. C. R. Matthews, Co-Director, 814-863-3650.
Application contact: Admissions Committee, 814-865-3155, *Fax:* 814-863-1357, *E-mail:* lscgradadm@mail.biotec.psu.edu.

Find an in-depth description at www.petersons.com/graduate.

■ PICOWER GRADUATE SCHOOL OF MOLECULAR MEDICINE

Program in Molecular Medicine, Manhasset, NY 11030

AWARDS PhD.

Faculty: 10 full-time (3 women).
Students: 3 full-time (0 women). *25 applicants, 12% accepted.*
Degree requirements: For doctorate, dissertation required. *Average time to degree:* Doctorate–3 years full-time.
Entrance requirements: For doctorate, MD. *Application deadline:* For fall admission, 2/1 (priority date); for spring admission, 9/1 (priority date). Applications are processed on a rolling basis. *Application fee:* $25.
Financial aid: In 1999–00, fellowships (averaging $35,000 per year); tuition waivers (full) also available.
Faculty research: Diabetes, cardiovascular disease, HIV disease, cancer, cytokine biology. *Total annual research expenditures:* $8.8 million.

Picower Graduate School of Molecular Medicine (continued)

Annette T. Lee, Dean, 516-365-4200, *Fax:* 516-365-5090.

Application contact: Lydia M. Moser, Administrative Assistant, 516-562-9442, *Fax:* 516-365-5090, *E-mail:* lmoser@picower.edu.

■ UNIVERSITY OF CINCINNATI

Division of Research and Advanced Studies, College of Medicine, Graduate Programs in Medicine, Department of Pathobiology and Molecular Medicine, Cincinnati, OH 45267

AWARDS Pathology (PhD), including anatomic pathology, laboratory medicine, pathobiology and molecular medicine.

Faculty: 10 full-time.
Students: 20 full-time (11 women); includes 1 minority (Native American), 5 international. *116 applicants, 3% accepted.* In 1999, 1 doctorate awarded.
Degree requirements: For doctorate, dissertation, qualifying exam required, foreign language not required. *Average time to degree:* Doctorate–5.3 years full-time.
Entrance requirements: For doctorate, GRE General Test, TOEFL. *Application deadline:* For fall admission, 2/1 (priority date). Applications are processed on a rolling basis. *Application fee:* $30.
Expenses: Tuition, state resident: full-time $5,139; part-time $196 per credit hour. Tuition, nonresident: full-time $10,326; part-time $369 per credit hour. Required fees: $561; $187 per quarter.
Financial aid: Tuition waivers (full) and unspecified assistantships available. Financial aid application deadline: 5/1.
Faculty research: Carcinogenesis; cardiovascular disease; inflammation, immunology, infectious disease; toxicology/environmental health. *Total annual research expenditures:* $3.8 million.
Dr. Cecilia Fenoglio-Preiser, Head, 513-558-4500, *Fax:* 513-558-2289, *E-mail:* cecilia.fenogliopreiser@uc.edu.
Application contact: Dr. Thomas Clemens, Acting Director of Graduate Studies, 513-558-4444, *Fax:* 513-558-2289, *E-mail:* thomas.clemens@uc.edu.

Find an in-depth description at www.petersons.com/graduate.

■ UNIVERSITY OF OKLAHOMA HEALTH SCIENCES CENTER

College of Medicine and Graduate College, Graduate Programs in Medicine, Oklahoma City, OK 73190

AWARDS Biochemistry and molecular biology (MS, PhD), including biochemistry, molecular

biology; cell biology (MS, PhD); medical sciences (MS); microbiology and immunology (MS, PhD), including immunology, microbiology; molecular medicine (PhD); neuroscience (MS, PhD); pathology (PhD); physiology (MS, PhD); psychiatry and behavioral sciences (MS, PhD), including biological psychology; radiological sciences (MS, PhD), including medical radiation physics. Part-time programs available.

Faculty: 132 full-time (25 women), 46 part-time/adjunct (14 women).
Students: 44 full-time (24 women), 56 part-time (28 women); includes 13 minority (1 African American, 8 Asian Americans or Pacific Islanders, 1 Hispanic American, 3 Native Americans), 20 international. Average age 27. *186 applicants, 25% accepted.* In 1999, 15 master's, 15 doctorates awarded. Terminal master's awarded for partial completion of doctoral program.
Degree requirements: For master's, foreign language not required; for doctorate, dissertation required.
Entrance requirements: For master's and doctorate, GRE General Test, TOEFL. *Application fee:* $25 ($50 for international students).
Expenses: Tuition, state resident: part-time $90 per semester hour. Tuition, nonresident: part-time $264 per semester hour. Tuition and fees vary according to program.
Financial aid: Fellowships, research assistantships, teaching assistantships, career-related internships or fieldwork, Federal Work-Study, institutionally sponsored loans, and tuition waivers (full and partial) available. Aid available to part-time students.
Faculty research: Behavior and drugs, structure and function of endothelium, genetics and behavior, gene structure and function, action of antibiotics.
Dr. O. Ray Kling, Dean, 405-271-2085, *Fax:* 405-271-1155, *E-mail:* ray-kling@uokhsc.edu.

Find an in-depth description at www.petersons.com/graduate.

■ THE UNIVERSITY OF TEXAS HEALTH SCIENCE CENTER AT SAN ANTONIO

Graduate School of Biomedical Sciences, Program in Molecular Medicine, San Antonio, TX 78229-3900

AWARDS MS, PhD.

Degree requirements: For doctorate, oral qualifying exam required.
Entrance requirements: For master's and doctorate, GRE General Test, TOEFL.

Expenses: Tuition, state resident: part-time $38 per credit hour. Tuition, nonresident: part-time $249 per credit hour.
Faculty research: DNA repair, tumor suppressor genes, vision in drosophila, *gene expression (nervous system), cell-type specific gene regulation and development.*

Find an in-depth description at www.petersons.com/graduate.

■ UNIVERSITY OF VIRGINIA

College and Graduate School of Arts and Sciences, Program in Molecular Medicine, Charlottesville, VA 22903

AWARDS Biochemistry (PhD); biophysics (PhD); cell biology/anatomy (PhD); microbiology (PhD); pharmacology (PhD); physiology (PhD). Offered through the Departments of Biochemistry and Molecular Genetics, Cell Biology, Microbiology, Molecular Physiology and Biological Physics, and Pharmacology, the Interdisciplinary Program in Biophysics, and the School of Medicine.

Faculty: 77 full-time (6 women).
Students: 6 full-time (5 women); includes 1 minority (Asian American or Pacific Islander), 1 international. Average age 25. *58 applicants, 10% accepted.*
Degree requirements: For doctorate, dissertation required.
Entrance requirements: For doctorate, GRE, TOEFL. *Application deadline:* For fall admission, 2/15 (priority date). Applications are processed on a rolling basis. *Application fee:* $40. Electronic applications accepted.
Expenses: Tuition, state resident: full-time $3,832. Tuition, nonresident: full-time $15,519. Required fees: $1,084. Tuition and fees vary according to course load and program.
Financial aid: Fellowships, tuition waivers (full) available. Financial aid application deadline: 2/15; financial aid applicants required to submit FAFSA.
Faculty research: Molecular biology of human disease.
Dr. Michael Weber, Director, 804-924-2181, *Fax:* 804-982-0689.
Application contact: Graduate Program Office, 804-924-2181.

Find an in-depth description at www.petersons.com/graduate.

■ UNIVERSITY OF WASHINGTON

Graduate School, School of Public Health and Community Medicine, Department of Pathobiology, Seattle, WA 98195

AWARDS MS, PhD.

Faculty: 33 full-time (11 women), 9 part-time/adjunct (0 women).

Students: 24 full-time (18 women), 3 part-time (1 woman); includes 11 minority (1 African American, 8 Asian Americans or Pacific Islanders, 2 Hispanic Americans), 4 international. Average age 28. *45 applicants, 20% accepted.* In 1999, 3 master's awarded (33% found work related to degree, 67% continued full-time study); 8 doctorates awarded (100% entered university research/teaching). Terminal master's awarded for partial completion of doctoral program.
Degree requirements: For master's and doctorate, thesis/dissertation required, foreign language not required. *Average time to degree:* Master's–2 years full-time; doctorate–8 years full-time.
Entrance requirements: For master's and doctorate, GRE General Test, TOEFL, minimum GPA of 3.0. *Application deadline:* For fall admission, 1/1. *Application fee:* $50.
Expenses: Tuition, state resident: full-time $5,196; part-time $495 per credit. Tuition, nonresident: full-time $13,485; part-time $1,285 per credit. Required fees: $387; $36 per credit. Tuition and fees vary according to course load and program.
Financial aid: In 1999–00, 2 fellowships with tuition reimbursements (averaging $16,068 per year), 18 research assistantships with tuition reimbursements (averaging $16,068 per year) were awarded; career-related internships or fieldwork, Federal Work-Study, institutionally sponsored loans, traineeships, and tuition waivers (full and partial) also available. Financial aid application deadline: 3/1; financial aid applicants required to submit FAFSA.
Faculty research: Pathogenesis of chlamydiae, molecular biology of parasites, signal transduction, antigenic analysis, molecular biology of tumor viruses.
Dr. Kenneth Stuart, Chair, 206-543-8350, *Fax:* 206-543-3873, *E-mail:* kstuart@u.washington.edu.
Application contact: Leslie G. Miller, Coordinator, 206-543-4338, *Fax:* 206-543-3873, *E-mail:* pathiobio@u.washington.edu.
Find an in-depth description at www.petersons.com/graduate.

■ **WAKE FOREST UNIVERSITY**
School of Medicine and Graduate School, Graduate Programs in Medicine, Program in Molecular Medicine, Winston-Salem, NC 27109
AWARDS PhD.

Degree requirements: For doctorate, dissertation required.
Entrance requirements: For doctorate, GRE General Test, GRE Subject Test.
Expenses: Tuition: Full-time $18,300.
Find an in-depth description at www.petersons.com/graduate.

■ **YALE UNIVERSITY**
School of Medicine and Graduate School of Arts and Sciences, Combined Program in Biological and Biomedical Sciences (BBS), Pharmacological Sciences and Molecular Medicine Track, New Haven, CT 06520
AWARDS PhD, MD/PhD.

Degree requirements: For doctorate, dissertation required.
Entrance requirements: For doctorate, GRE General Test, TOEFL. *Application deadline:* For fall admission, 1/2. *Application fee:* $65. Electronic applications accepted.
Expenses: All students receive full tuition of $22,330 and an annual stipend of $17,600 .
Financial aid: Fellowships, research assistantships available.
Dr. David F. Stern, Co-Director, 203-785-4832, *Fax:* 203-785-7467, *E-mail:* bbs.pharm@yale.edu.

ONCOLOGY

■ **BROWN UNIVERSITY**
Graduate School, Division of Biology and Medicine, Pathobiology Graduate Program, Providence, RI 02912
AWARDS Biology (PhD); cancer biology (PhD); immunology and infection (PhD); medical science (PhD); pathobiology (Sc M); toxicology and environmental pathology (PhD).

Faculty: 36 full-time (9 women), 1 part-time/adjunct (0 women).
Students: 18 full-time (11 women); includes 4 minority (2 African Americans, 2 Asian Americans or Pacific Islanders), 1 international. Average age 25. *42 applicants, 17% accepted.* In 1999, 2 doctorates awarded (100% entered university research/teaching). Terminal master's awarded for partial completion of doctoral program.
Degree requirements: For master's, foreign language not required; for doctorate, dissertation, preliminary exam required, foreign language not required. *Average time to degree:* Doctorate–6 years full-time.
Entrance requirements: For master's, GRE General Test, GRE Subject Test; for doctorate, GRE General Test, GRE Subject Test, TOEFL. *Application deadline:* For fall admission, 1/2 (priority date). Applications are processed on a rolling basis. *Application fee:* $60. Electronic applications accepted.
Financial aid: In 1999–00, 11 fellowships with tuition reimbursements, 1 research assistantship with tuition reimbursement, 4

teaching assistantships with tuition reimbursements were awarded; institutionally sponsored loans and traineeships also available. Financial aid application deadline: 1/2.
Faculty research: Environmental pathology, carcinogenesis, immunopathology, signal transduction.
Dr. Nancy Thompson, Director, 401-444-8860, *E-mail:* nancy_thompson@brown.edu.
Application contact: Marilyn May, Program Coordinator, 401-863-2913, *Fax:* 401-863-9008, *E-mail:* marilyn_may@brown.edu.
Find an in-depth description at www.petersons.com/graduate.

■ **CORNELL UNIVERSITY**
Graduate School, Graduate Fields of Veterinary Medicine, Field of Veterinary Medicine, Ithaca, NY 14853-0001
AWARDS Anatomy (MS, PhD); cancer biology (MS, PhD); clinical sciences (MS, PhD); infectious diseases (MS, PhD); pathology (MS, PhD); pharmacology (MS, PhD); veterinary physiology (MS, PhD); virology (MS, PhD).

Faculty: 83 full-time.
Students: 25 full-time (13 women); includes 2 minority (1 African American, 1 Asian American or Pacific Islander), 12 international. *28 applicants, 18% accepted.* In 1999, 1 master's, 12 doctorates awarded.
Degree requirements: For master's and doctorate, thesis/dissertation required, foreign language not required.
Entrance requirements: For master's and doctorate, GRE General Test, TOEFL. *Application deadline:* For fall admission, 1/15; for spring admission, 10/1. Applications are processed on a rolling basis. *Application fee:* $65. Electronic applications accepted.
Expenses: Tuition: Full-time $12,400.
Financial aid: In 1999–00, 24 students received aid, including 7 fellowships with full tuition reimbursements available, 17 research assistantships with full tuition reimbursements available; teaching assistantships with full tuition reimbursements available, institutionally sponsored loans, scholarships, tuition waivers (full and partial), and unspecified assistantships also available. Financial aid applicants required to submit FAFSA.
Faculty research: Receptors and signal transduction, viral and bacterial infectious diseases, tumor metastasis, clinical sciences/nutritional disease, development/neurologic disorders.
Application contact: Graduate Field Assistant, 607-253-3276, *E-mail:* vetgradpgms@cornell.edu.

■ DUKE UNIVERSITY

Graduate School, Department of Molecular Cancer Biology, Durham, NC 27708-0586

AWARDS PhD.

Faculty: 28 full-time.
Students: 29 full-time (15 women); includes 5 minority (2 Asian Americans or Pacific Islanders, 3 Hispanic Americans), 11 international. *51 applicants, 18% accepted.* In 1999, 6 doctorates awarded.
Degree requirements: For doctorate, dissertation required.
Entrance requirements: For doctorate, GRE General Test, GRE Subject Test (recommended). *Application deadline:* For fall admission, 12/31. *Application fee:* $75.
Expenses: Tuition: Full-time $21,406; part-time $760 per unit. Required fees: $3,136; $3,136 per year. One-time fee: $30. Tuition and fees vary according to program.
Financial aid: Fellowships, research assistantships available. Financial aid application deadline: 12/31.
Shirish Shenolikar, Director of Graduate Studies, 919-613-8601, *Fax:* 919-681-7767, *E-mail:* means003@mc.duke.edu.

■ EMORY UNIVERSITY

School of Medicine, Programs in Allied Health Professions, Program in Radiation Oncology Physics, Atlanta, GA 30322-1100

AWARDS MM Sc.

Faculty: 2 full-time (0 women).
Students: 5 full-time (0 women). Average age 30. *7 applicants, 57% accepted.* In 1999, 4 degrees awarded (100% found work related to degree). *Average time to degree:* Master's–1.8 years full-time.
Entrance requirements: For master's, GRE General Test. *Application deadline:* For fall admission, 6/1 (priority date). *Application fee:* $35.
Expenses: Tuition: Full-time $12,516; part-time $348 per credit hour. Required fees: $380; $5 per credit hour. Tuition and fees vary according to program.
Financial aid: In 1999–00, 3 students received aid. Federal Work-Study, institutionally sponsored loans, and scholarships available. Aid available to part-time students. Financial aid application deadline: 3/15; financial aid applicants required to submit CSS PROFILE or FAFSA.
Dr. Pat McGinley, Director, 404-778-3535.

■ THE GEORGE WASHINGTON UNIVERSITY

Columbian School of Arts and Sciences, Institute for Biomedical Sciences, Program in Molecular and Cellular Oncology, Washington, DC 20052

AWARDS PhD.

Faculty: 43 full-time (8 women).
Students: 3 full-time (all women), 8 part-time (6 women); includes 3 minority (1 African American, 2 Asian Americans or Pacific Islanders), 2 international. Average age 30. In 1999, 3 degrees awarded.
Degree requirements: For doctorate, dissertation, comprehensive and general exams required.
Entrance requirements: For doctorate, GRE General Test, interview, minimum GPA of 3.0. *Application fee:* $55.
Expenses: Tuition: Full-time $16,836; part-time $702 per credit hour. Required fees: $828; $35 per credit hour. Tuition and fees vary according to campus/location and program.
Financial aid: In 1999–00, 1 student received aid, including 1 fellowship; Federal Work-Study and institutionally sponsored loans also available. Financial aid application deadline: 2/1.
Dr. Steven R. Patierno, Director, 202-994-3286.
Application contact: 202-994-2179.

Find an in-depth description at www.petersons.com/graduate.

■ HARVARD UNIVERSITY

School of Public Health, Department of Cancer Cell Biology, Boston, MA 02115-6096

AWARDS PhD.

Faculty: 7 full-time (1 woman), 5 part-time/adjunct (0 women).
Students: 2 full-time (both women), 2 part-time, 1 international. Average age 29. *1 applicant, 0% accepted.* In 1999, 2 doctorates awarded.
Degree requirements: For doctorate, dissertation, qualifying exam required.
Entrance requirements: For doctorate, GRE, TOEFL. *Application deadline:* For fall admission, 1/3. *Application fee:* $60.
Expenses: Tuition: Full-time $22,950; part-time $574 per credit.
Financial aid: Fellowships, research assistantships, Federal Work-Study, grants, scholarships, traineeships, tuition waivers (partial), and unspecified assistantships available. Financial aid application deadline: 2/12; financial aid applicants required to submit FAFSA.
Faculty research: Toxicology, radiation biology.

Dr. John Little, Chairman, 617-432-0054, *Fax:* 617-432-0107, *E-mail:* gbraga@hsph.harvard.edu.

■ KANSAS STATE UNIVERSITY

Graduate School, College of Arts and Sciences, Division of Biology, Manhattan, KS 66506

AWARDS Cell biology (MS, PhD); developmental biology and physiology (MS, PhD); microbiology and immunology (MS, PhD); molecular biology and genetics (MS, PhD); systematics and ecology (MS, PhD); virology and oncology (MS, PhD). Terminal master's awarded for partial completion of doctoral program.

Degree requirements: For master's and doctorate, thesis/dissertation required, foreign language not required.
Entrance requirements: For master's and doctorate, GRE General Test. Electronic applications accepted.
Expenses: Tuition, state resident: part-time $103 per credit hour. Tuition, nonresident: part-time $338 per credit hour. Required fees: $17 per credit hour. One-time fee: $64 part-time.
Faculty research: Immune cell function, prairie ecology.

■ MAYO GRADUATE SCHOOL

Graduate Programs in Biomedical Sciences, Program in Tumor Biology, Rochester, MN 55905

AWARDS PhD.

Faculty: 66 full-time (13 women).
Students: 16 full-time (11 women); includes 3 minority (2 Hispanic Americans, 1 Native American), 3 international.
Degree requirements: For doctorate, oral defense of dissertation, qualifying oral and written exam required.
Entrance requirements: For doctorate, GRE, TOEFL, 2 years of chemistry; 1 year of biology, calculus, and physics. *Application deadline:* For fall admission, 12/31 (priority date). Applications are processed on a rolling basis. *Application fee:* $0.
Expenses: Tuition: Full-time $17,900.
Financial aid: In 1999–00, 14 students received aid, including 14 fellowships with full tuition reimbursements available; tuition waivers (full) also available. Financial aid application deadline: 12/31.
Dr. Jeffrey L. Salisbury, Education Coordinator, 507-284-3326, *E-mail:* salisbury@mayo.edu.
Application contact: Sherry Kallies, Information Contact, 507-266-0122, *Fax:* 507-284-0999, *E-mail:* phd.training@mayo.edu.

Find an in-depth description at www.petersons.com/graduate.

■ MCP HAHNEMANN UNIVERSITY

School of Medicine, Biomedical Graduate Programs, Department of Radiation Oncology, Philadelphia, PA 19102-1192

AWARDS Medical physics (MS, PhD); radiation science (MS, PhD). Part-time programs available. Terminal master's awarded for partial completion of doctoral program.

Degree requirements: For master's, thesis, comprehensive exam required, foreign language not required; for doctorate, one foreign language (computer language can substitute), dissertation, qualifying exam required.

Entrance requirements: For master's, GRE General Test, TOEFL, minimum GPA of 2.75; for doctorate, GRE General Test, TOEFL, minimum GPA of 3.0.

Faculty research: Mechanisms for improving use of radiolabeled antibodies for cancer diagnosis and therapy; improved cancer therapy by linear accelerators and internal, sealed radiation sources; algorithms for improved superminicomputer-assisted radiation therapy simulation and tumor imaging; molecular and cellular mechanisms of radiation damage.

Find an in-depth description at www.petersons.com/graduate.

■ NEW YORK UNIVERSITY

School of Medicine and Graduate School of Arts and Science, Sackler Institute of Graduate Biomedical Sciences, Graduate Programs in Molecular Oncology and Immunology, New York, NY 10012-1019

AWARDS Immunology (PhD); molecular oncology (PhD).

Degree requirements: For doctorate, one foreign language, dissertation, qualifying exam required.

Entrance requirements: For doctorate, GRE General Test, GRE Subject Test, TOEFL. *Application deadline:* For fall admission, 2/1. Applications are processed on a rolling basis. *Application fee:* $60. Electronic applications accepted.

Expenses: Tuition: Full-time $17,880; part-time $745 per credit. Required fees: $1,140; $35 per credit. Tuition and fees vary according to course load and program.

Financial aid: Tuition waivers (full) available. Financial aid application deadline: 1/15.

Application contact: Dr. Robert B. Carroll, Graduate Adviser, 212-263-5347, *Fax:* 212-263-8211, *E-mail:* carror01@mcrcr.med.nyu.edu.

Find an in-depth description at www.petersons.com/graduate.

■ NORTHWESTERN UNIVERSITY

The Graduate School, Division of Interdepartmental Programs and Medical School, Integrated Graduate Programs in the Life Sciences, Chicago, IL 60611

AWARDS Cancer biology (PhD); cell biology (PhD); developmental biology (PhD); evolutionary biology (PhD); immunology and microbial pathogenesis (PhD); molecular biology and genetics (PhD); neurobiology (PhD); pharmacology and toxicology (PhD); structural biology and biochemistry (PhD).

Degree requirements: For doctorate, dissertation, written and oral qualifying exams required, foreign language not required.

Entrance requirements: For doctorate, GRE General Test, TOEFL.

Expenses: Tuition: Full-time $23,301. Full-time tuition and fees vary according to program.

Find an in-depth description at www.petersons.com/graduate.

■ STANFORD UNIVERSITY

School of Medicine, Graduate Programs in Medicine, Program in Cancer Biology, Stanford, CA 94305-9991

AWARDS PhD.

Students: 29 full-time (15 women), 14 part-time (5 women); includes 13 minority (8 Asian Americans or Pacific Islanders, 3 Hispanic Americans, 2 Native Americans), 6 international. Average age 27. *113 applicants, 12% accepted.* In 1999, 9 doctorates awarded.

Degree requirements: For doctorate, dissertation required.

Entrance requirements: For doctorate, GRE General Test, GRE Subject Test, TOEFL. *Application deadline:* For fall admission, 12/15. *Application fee:* $65 ($80 for international students). Electronic applications accepted.

Expenses: Tuition: Full-time $23,058. Required fees: $152. Part-time tuition and fees vary according to course load.

Financial aid: Research assistantships available. Financial aid application deadline: 12/15.

Application contact: Graduate Admissions Coordinator, 650-723-6198, *Fax:* 650-725-7855.

■ STATE UNIVERSITY OF NEW YORK AT BUFFALO

Graduate School, Graduate Programs in Biomedical Sciences at Roswell Park Cancer Institute, Department of Molecular Pharmacology and Cancer Therapeutics at Roswell Park Cancer Institute, Buffalo, NY 14263

AWARDS PhD.

Faculty: 19 full-time (6 women).

Students: 17 full-time (8 women); includes 2 minority (both Asian Americans or Pacific Islanders), 2 international. Average age 26. *14 applicants, 64% accepted.* In 1999, 2 doctorates awarded (100% entered university research/teaching).

Degree requirements: For doctorate, dissertation, departmental qualifying exam, grant proposal required, foreign language not required. *Average time to degree:* Doctorate–7 years full-time.

Entrance requirements: For doctorate, GRE General Test, TOEFL, TWE (recommended). *Application deadline:* For fall admission, 2/1 (priority date). Applications are processed on a rolling basis. *Application fee:* $35. Electronic applications accepted.

Expenses: Tuition, state resident: full-time $5,100; part-time $213 per credit hour. Tuition, nonresident: full-time $8,416; part-time $351 per credit hour. Required fees: $935; $75 per semester. Tuition and fees vary according to course load and program.

Financial aid: In 1999–00, 17 students received aid, including 8 fellowships with full tuition reimbursements available (averaging $15,500 per year), 9 research assistantships with full tuition reimbursements available (averaging $15,500 per year).

Faculty research: Molecular pharmacology, cancer cell biology, molecular biology, biochemistry, chemotherapy. *Total annual research expenditures:* $6.5 million.

Dr. Enrico Mihich, Chair, 716-845-8223, *Fax:* 716-845-8857.

Application contact: Dr. Jennifer D. Black, Director of Graduate Studies, 716-845-5766, *Fax:* 716-845-8857, *E-mail:* jblack@sc3103.med.buffalo.edu.

Find an in-depth description at www.petersons.com/graduate.

■ THOMAS JEFFERSON UNIVERSITY

Kimmel Cancer Institute, Philadelphia, PA 19107

AWARDS Genetics (PhD); immunology (PhD); microbiology and molecular virology (PhD); molecular pharmacology and structural biology (PhD).

Faculty: 57 full-time (8 women), 1 part-time/adjunct (0 women).

Degree requirements: For doctorate, dissertation required, foreign language not required.

Entrance requirements: For doctorate, GRE General Test, GRE Subject Test (recommended), TOEFL. *Application deadline:* For fall admission, 3/1 (priority date). Applications are processed on a rolling basis.

Thomas Jefferson University (continued)
Expenses: Tuition: Full-time $12,670. Tuition and fees vary according to degree level and program.
Financial aid: Fellowships available. Financial aid application deadline: 3/1.
Faculty research: Genetics of normal and malignant cells, cellular and molecular immunology, virology and parasitology, signal transduction. *Total annual research expenditures:* $17 million.
Dr. Carlo M. Croce, Director, 215-503-4645.
Application contact: Joanne Balitzky, Graduate Coordinator, 215-503-6687, *Fax:* 215-503-0622, *E-mail:* joanne.balitzky@mail.tju.edu.

Find an in-depth description at www.petersons.com/graduate.

■ THE UNIVERSITY OF ARIZONA

Graduate College, Graduate Interdisciplinary Programs, Graduate Interdisciplinary Program in Cancer Biology, Tucson, AZ 85721

AWARDS PhD.

Degree requirements: For doctorate, dissertation required.
Entrance requirements: For doctorate, GRE General Test, GRE Subject Test, TOEFL.
Expenses: Tuition, nonresident: full-time $4,814; part-time $274 per unit. Required fees: $1,094; $115 per unit. Tuition and fees vary according to course load and program.
Faculty research: Differential gene expression, DNA-protein cross linking, cell growth regulation steroid, receptor proteins.

Find an in-depth description at www.petersons.com/graduate.

■ UNIVERSITY OF CALIFORNIA, SAN DIEGO

Graduate Studies and Research, Department of Biology, Program in Immunology, Virology, and Cancer Biology, La Jolla, CA 92093

AWARDS PhD.

Degree requirements: For doctorate, dissertation required.
Entrance requirements: For doctorate, GRE General Test, pre-application beginning in September. *Application deadline:* For fall admission, 1/8. *Application fee:* $40.
Expenses: Tuition, nonresident: full-time $14,691. Required fees: $4,697. Full-time tuition and fees vary according to program.
Application contact: Biology Graduate Admissions Committee, 858-534-3835.

■ UNIVERSITY OF CHICAGO

Division of the Biological Sciences, Biomedical Sciences: Cancer, Immunology, Nutrition, Pathology, and Virology, Committee on Cancer Biology, Chicago, IL 60637-1513

AWARDS PhD.

Faculty: 55 full-time (11 women).
Students: 14 full-time (6 women); includes 2 minority (both Asian Americans or Pacific Islanders), 1 international. Average age 25. *94 applicants, 6% accepted.*
Degree requirements: For doctorate, dissertation required.
Entrance requirements: For doctorate, GRE General Test, TOEFL. *Application deadline:* For fall admission, 1/5 (priority date). *Application fee:* $55.
Expenses: Tuition: Full-time $24,804; part-time $3,422 per course. Required fees: $390. Tuition and fees vary according to program.
Financial aid: In 1999–00, 14 students received aid, including 13 fellowships with full tuition reimbursements available (averaging $16,500 per year), 1 research assistantship with full tuition reimbursement available (averaging $16,500 per year); institutionally sponsored loans and traineeships also available. Financial aid application deadline: 4/1.
Faculty research: Cancer genetics, apoptosis, signal transduction, tumor biology, cell cycle regulation.
Dr. Michelle LeBeau, Chair, 773-702-0795, *E-mail:* mrosner@huggins.bsd.uchicago.edu.
Application contact: Rebecca Levine, Administrative Assistant, Student Services, 773-834-3899, *Fax:* 773-702-4634, *E-mail:* rlevine@huggins.bsd.uchicago.edu.

■ UNIVERSITY OF PENNSYLVANIA

School of Medicine, Biomedical Graduate Studies, Graduate Group in Cell and Molecular Biology, Program in Cell Growth and Cancer, Philadelphia, PA 19104

AWARDS PhD, MD/PhD, VMD/PhD.

Degree requirements: For doctorate, dissertation required, foreign language not required.
Entrance requirements: For doctorate, GRE General Test, TOEFL. *Application deadline:* For fall admission, 1/2 (priority date). Applications are processed on a rolling basis. *Application fee:* $65. Electronic applications accepted.
Expenses: Tuition: Full-time $17,256; part-time $2,991 per course. Required fees: $2,588; $363 per course. $726 per term.

Financial aid: Fellowships, research assistantships, teaching assistantships, grants and institutionally sponsored loans available. Financial aid application deadline: 1/2.
Dr. Craig Thompson, Chair, 215-746-5514.
Application contact: Mary Webster, Coordinator, 215-898-4360, *E-mail:* camb@mail.med.upenn.edu.

■ UNIVERSITY OF SOUTH ALABAMA

College of Medicine and Graduate School, Program in Basic Medical Sciences, Specialization in Cancer Biology, Mobile, AL 36688-0002

AWARDS PhD.

Degree requirements: For doctorate, dissertation required, foreign language not required.
Application deadline: For fall admission, 4/1. *Application fee:* $25.
Expenses: Tuition, state resident: part-time $116 per semester hour. Tuition, nonresident: part-time $230 per semester hour. Required fees: $121 per semester. Lanette Flagge, Coordinator, Program in Basic Medical Sciences, 334-460-6153.

Find an in-depth description at www.petersons.com/graduate.

■ UNIVERSITY OF SOUTH FLORIDA

H. Lee Moffitt Cancer Center and Research Institute, Tampa, FL 33612-9497

AWARDS Oncology (PhD).

Expenses: Tuition, state resident: part-time $148 per credit hour. Tuition, nonresident: part-time $509 per credit hour.

■ THE UNIVERSITY OF TEXAS– HOUSTON HEALTH SCIENCE CENTER

Graduate School of Biomedical Sciences, Program in Cancer Biology, Houston, TX 77225-0036

AWARDS MS, PhD, MD/PhD.

Faculty: 51 full-time (12 women).
Students: 26 full-time (17 women); includes 9 minority (1 African American, 4 Asian Americans or Pacific Islanders, 3 Hispanic Americans, 1 Native American), 7 international. Average age 27. *65 applicants, 54% accepted.* In 1999, 3 master's, 5 doctorates awarded. Terminal master's awarded for partial completion of doctoral program.

Degree requirements: For master's and doctorate, thesis/dissertation required, foreign language not required.
Entrance requirements: For master's and doctorate, GRE General Test, TOEFL, TWE. *Application deadline:* For fall admission, 1/15 (priority date); for spring admission, 11/1. Applications are processed on a rolling basis. *Application fee:* $10. Electronic applications accepted.
Financial aid: Fellowships, research assistantships, institutionally sponsored loans available. Financial aid application deadline: 1/15.
Faculty research: Mechanisms of carcinogenesis, mechanisms of tumor metastasis, activation of host antitumor defense, molecular biology of oncogenes and tumor suppressor genes, molecular biology of gene expression.
Dr. Gary E. Gallick, Director, 713-792-3657, *Fax:* 713-745-1927, *E-mail:* ggallick@mdanderson.org.
Application contact: Anne Baronitis, Director of Admissions, 713-500-9860, *Fax:* 713-500-9877, *E-mail:* abaron@ gsbs.gs.uth.tmc.edu.
Find an in-depth description at www.petersons.com/graduate.

■ **THE UNIVERSITY OF TEXAS– HOUSTON HEALTH SCIENCE CENTER**

Graduate School of Biomedical Sciences, Program in Environmental and Molecular Carcinogenesis, Houston, TX 77225-0036
AWARDS MS, PhD, MD/PhD.
Faculty: 34 full-time (11 women).
Students: 5 full-time (1 woman); includes 1 minority (Asian American or Pacific Islander). Average age 27.
Degree requirements: For doctorate, dissertation required, foreign language not required.
Entrance requirements: For doctorate, GRE General Test, TOEFL, TWE. *Application deadline:* For fall admission, 1/15 (priority date); for spring admission, 11/1. Applications are processed on a rolling basis. *Application fee:* $10. Electronic applications accepted.
Financial aid: Fellowships, research assistantships, institutionally sponsored loans available. Financial aid application deadline: 1/15.
Faculty research: Cellular and molecular biology of tumor development.
Dr. Ellen R. Richie, Director, 512-237-9435, *Fax:* 512-237-2444, *E-mail:* erichie@ odin.mdacc.tmc.edu.
Application contact: Anne Baronitis, Director of Admissions, 713-500-9860,

Fax: 713-500-9877, *E-mail:* abaron@ gsbs.gs.uth.tmc.edu.
Find an in-depth description at www.petersons.com/graduate.

■ **UNIVERSITY OF UTAH**

School of Medicine and Graduate School, Graduate Programs in Medicine, Department of Oncological Sciences, Salt Lake City, UT 84112-1107
AWARDS MS, PhD.
Faculty: 20 full-time (5 women), 11 part-time/adjunct (4 women).
Students: 38 full-time (11 women). Average age 24. *634 applicants, 9% accepted.* In 1999, 2 master's awarded (100% entered university research/teaching); 1 doctorate awarded (100% entered university research/teaching). Terminal master's awarded for partial completion of doctoral program.
Degree requirements: For master's, thesis required (for some programs), foreign language not required; for doctorate, dissertation required, foreign language not required. *Average time to degree:* Master's–3 years full-time; doctorate–6 years full-time.
Entrance requirements: For master's and doctorate, GRE General Test, GRE Subject Test, minimum GPA of 3.0. *Application deadline:* For winter admission, 1/15 (priority date). Electronic applications accepted.
Expenses: Tuition, state resident: full-time $2,105. Tuition, nonresident: full-time $6,312.
Financial aid: In 1999–00, 3 fellowships with full tuition reimbursements (averaging $17,000 per year), 35 research assistantships with full tuition reimbursements (averaging $17,000 per year) were awarded; grants, institutionally sponsored loans, scholarships, traineeships, and tuition waivers (full) also available.
Faculty research: Molecular basis of cell growth and differences, regulation of gene expression, biochemical mechanics of DNA replication, molecular biology and biochemistry of signal transduction, somatic cell genetics, molecular genetics of human cancer. *Total annual research expenditures:* $1.4 million.
Dr. Raymond L. White, Chair, 801-581-4330, *Fax:* 801-585-3833, *E-mail:* ray.white@hci.utah.edu.
Application contact: Tami Sue Brunson, Academic Coordinator, 801-581-5207, *Fax:* 801-585-2465, *E-mail:* tbrunson@ medschool.med.utah.edu.

■ **UNIVERSITY OF WISCONSIN– MADISON**

Medical School and Graduate School, Graduate Programs in Medicine, Cancer Biology Program, Madison, WI 53706-1380
AWARDS PhD.
Application fee: $45.
Expenses: Tuition, state resident: full-time $5,406; part-time $339 per credit. Tuition, nonresident: full-time $17,110; part-time $1,071 per credit.

■ **UNIVERSITY OF WISCONSIN– MADISON**

Medical School and Graduate School, Graduate Programs in Medicine, Department of Oncology, Madison, WI 53706-1380
AWARDS PhD.
Faculty: 21 full-time (4 women).
Students: 55 full-time (27 women); includes 7 minority (3 Asian Americans or Pacific Islanders, 4 Hispanic Americans), 10 international. *Average time to degree:* Doctorate–5.5 years full-time.
Entrance requirements: For doctorate, GRE General Test, GRE Subject Test. *Application deadline:* For fall admission, 1/15 (priority date). *Application fee:* $45. Electronic applications accepted.
Expenses: Tuition, state resident: full-time $5,406; part-time $339 per credit. Tuition, nonresident: full-time $17,110; part-time $1,071 per credit.
Financial aid: In 1999–00, 27 students received aid, including fellowships with full tuition reimbursements available (averaging $15,500 per year), research assistantships with tuition reimbursements available (averaging $15,500 per year); traineeships also available.
Faculty research: Cancer genetics, tumor virology, chemical carcinogenesis, signal transduction, cell cycle.
Dr. Norman R. Drinkwater, Director, 608-262-2177, *Fax:* 608-262-2824, *E-mail:* drinkwater@oncology.wisc.edu.
Application contact: Bette Sheehan, Administrative Program Manager, 608-262-8651, *Fax:* 608-262-2824, *E-mail:* bsheehan@oncology.wisc.edu.
Find an in-depth description at www.petersons.com/graduate.

■ **WAKE FOREST UNIVERSITY**

School of Medicine and Graduate School, Graduate Programs in Medicine, Department of Cancer Biology, Winston-Salem, NC 27109
AWARDS PhD.
Degree requirements: For doctorate, dissertation required.

Wake Forest University (continued)
Entrance requirements: For doctorate, GRE General Test, GRE Subject Test. Electronic applications accepted.
Expenses: Tuition: Full-time $18,300.

Find an in-depth description at www.petersons.com/graduate.

■ WAYNE STATE UNIVERSITY

School of Medicine and Graduate School, Graduate Programs in Medicine, Department of Radiation Oncology, Detroit, MI 48202

AWARDS Medical physics (PhD); radiological physics (MS). Part-time and evening/weekend programs available. Terminal master's awarded for partial completion of doctoral program.

Degree requirements: For master's, thesis, essay, exit exam required; for doctorate, dissertation, qualifying exam required.
Entrance requirements: For master's, GRE General Test, BS in physics or related area; for doctorate, GRE General Test, GRE Subject Test, BS in physics or related area.
Faculty research: Radiotherapy physics, hyperthermia, magnetic resonance imaging and spectroscopy, clinical ultrasound, x-ray physics.

■ WAYNE STATE UNIVERSITY

School of Medicine and Graduate School, Graduate Programs in Medicine, Program in Cancer Biology, Detroit, MI 48202

AWARDS PhD.

Degree requirements: For doctorate, dissertation required, foreign language not required.
Entrance requirements: For doctorate, GRE General Test.
Faculty research: Cell regulation, molecular biology, carcinogenesis, virology, immunological modulation.

Find an in-depth description at www.petersons.com/graduate.

■ WEST VIRGINIA UNIVERSITY

College of Agriculture, Forestry and Consumer Sciences, Interdisciplinary Program in Genetics and Developmental Biology, Morgantown, WV 26506

AWARDS Animal breeding (MS, PhD); biochemical and molecular genetics (MS, PhD); cytogenetics (MS, PhD); descriptive embryology (MS, PhD); developmental genetics (MS); experimental morphogenesis (MS); human genetics (MS, PhD); immunogenetics (MS, PhD); life cycles of animals and plants (MS, PhD); molecular aspects of development (MS, PhD); mutagenesis (PhD); mutagenetics (MS); oncology (MS, PhD); plant genetics (MS, PhD); population and quantitative genetics (PhD); population and quantitative genetics (MS); regeneration (MS, PhD); teratology (MS, PhD); toxicology (MS, PhD).

Students: 18 full-time (8 women), 5 part-time (2 women); includes 1 minority (Asian American or Pacific Islander), 8 international. Average age 27. In 1999, 4 doctorates awarded.
Degree requirements: For master's, thesis required, foreign language not required; for doctorate, dissertation, comprehensive exam required, foreign language not required.
Entrance requirements: For master's, GRE or MCAT, TOEFL, minimum GPA of 2.75; for doctorate, TOEFL. *Application fee:* $45.
Expenses: Tuition, state resident: full-time $2,910; part-time $154 per credit hour. Tuition, nonresident: full-time $8,368; part-time $457 per credit hour.
Financial aid: In 1999–00, 11 research assistantships, 3 teaching assistantships were awarded; fellowships, Federal Work-Study, institutionally sponsored loans, and tuition waivers (full and partial) also available. Financial aid application deadline: 2/1; financial aid applicants required to submit FAFSA.
Dr. J. Nath, Chairman, 304-293-6256 Ext. 4333, *E-mail:* jnath@wvu.edu.

STRUCTURAL BIOLOGY

■ BAYLOR COLLEGE OF MEDICINE

Graduate School of Biomedical Sciences, Program in Structural and Computational Biology and Molecular Biophysics, Houston, TX 77030-3498

AWARDS PhD, MD/PhD.

Faculty: 44 full-time (4 women).
Students: 15 full-time (1 woman); includes 2 minority (1 African American, 1 Asian American or Pacific Islander), 7 international. Average age 27. *55 applicants, 20% accepted.*
Degree requirements: For doctorate, dissertation, public defense, qualifying exam required, foreign language not required.
Entrance requirements: For doctorate, GRE General Test (average 80th percentile), GRE Subject Test (strongly recommended), TOEFL, minimum GPA of 3.0. *Application deadline:* For fall admission, 2/1. *Application fee:* $30. Electronic applications accepted.

Expenses: Tuition: Full-time $8,200. Required fees: $175. Full-time tuition and fees vary according to student level.
Financial aid: In 1999–00, 15 students received aid, including 13 fellowships (averaging $16,000 per year), 2 research assistantships (averaging $16,000 per year); Federal Work-Study, institutionally sponsored loans, and tuition waivers (full) also available. Financial aid applicants required to submit FAFSA.
Faculty research: X-ray and electron crystallography, light and electron microscopy, computer image reconstruction, molecular spectroscopy.
Dr. Wah Chiu, Director, 713-798-6985.
Application contact: Wanda Waguespack, Graduate Program Administrator, 713-798-5197, *Fax:* 713-798-6325, *E-mail:* wandaw@bcm.tmc.edu.

Find an in-depth description at www.petersons.com/graduate.

■ BRANDEIS UNIVERSITY

Graduate School of Arts and Sciences, Programs in Biological Sciences, Program in Biophysics and Structural Biology, Waltham, MA 02454-9110

AWARDS PhD.

Faculty: 16 full-time (4 women).
Students: 25 full-time (10 women); includes 8 minority (all Asian Americans or Pacific Islanders), 9 international. Average age 24. *15 applicants, 7% accepted.* In 1999, 1 degree awarded (100% entered university research/teaching).
Degree requirements: For doctorate, one foreign language (computer language can substitute), dissertation required. *Average time to degree:* Doctorate–6 years full-time.
Entrance requirements: For doctorate, GRE General Test. *Application deadline:* For fall admission, 1/15 (priority date). Applications are processed on a rolling basis. *Application fee:* $60. Electronic applications accepted.
Expenses: Tuition: Full-time $25,392; part-time $3,174 per course. Required fees: $509. Tuition and fees vary according to class time, degree level, program and student level.
Financial aid: In 1999–00, 3 fellowships with tuition reimbursements (averaging $17,000 per year), 20 research assistantships with tuition reimbursements (averaging $17,000 per year), 1 teaching assistantship with tuition reimbursement (averaging $2,000 per year) were awarded; scholarships and tuition waivers (full and partial) also available. Aid available to part-time students. Financial aid application deadline: 4/15; financial aid applicants required to submit CSS PROFILE or FAFSA.

Faculty research: Biophysical chemistry, structural biology, protein crystallography, neuroscience, photobiology.
Dr. Jeff Gelles, Chair, 781-736-2377, *Fax:* 781-736-3107.

Application contact: Marcia Cabral, Information Officer, 781-736-3100, *Fax:* 781-736-3107, *E-mail:* cabral@ brandeis.edu.

■ IOWA STATE UNIVERSITY OF SCIENCE AND TECHNOLOGY

Graduate College, Interdisciplinary Programs, Bioinformatics and Computational Biology Program, Ames, IA 50011-3260

AWARDS MS, PhD.

Faculty: 46 full-time (9 women).
Students: 40.
Degree requirements: For master's and doctorate, thesis/dissertation required, foreign language not required.
Entrance requirements: For master's and doctorate, GRE General Test (waived for applicants with PhD), TOEFL. *Application deadline:* For fall admission, 2/1 (priority date). Electronic applications accepted.
Expenses: Tuition, state resident: full-time $3,308. Tuition, nonresident: full-time $9,744. Part-time tuition and fees vary according to course load, campus/location and program.
Financial aid: In 1999–00, 7 fellowships with full tuition reimbursements (averaging $23,000 per year), 8 research assistantships with full tuition reimbursements (averaging $16,800 per year) were awarded.
Faculty research: Functional and structural genomics, genome evoultion, macromolecular structure and function, mathematical biology and computational modeling, metabolic and developmental networks.
Drena L. Dobbs, Chair, 515-294-1112, *Fax:* 515-294-6790, *E-mail:* ddobbs@ iastate.edu.
Application contact: Kathy Wiederin, Program Assistant, 888-569-8509, *Fax:* 515-294-6790, *E-mail:* bioinformatics@ iastate.edu.

Find an in-depth description at www.petersons.com/graduate.

■ JOAN AND SANFORD I. WEILL MEDICAL COLLEGE AND GRADUATE SCHOOL OF MEDICAL SCIENCES OF CORNELL UNIVERSITY

Graduate School of Medical Sciences, Department of Biochemistry and Structural Biology, New York, NY 10021

AWARDS PhD, MD/PhD.

Faculty: 19 full-time (3 women).
Students: 14 full-time (9 women); includes 2 minority (1 African American, 1 Asian American or Pacific Islander), 5 international. *41 applicants, 15% accepted.* In 2000, 1 degree awarded.
Degree requirements: For doctorate, dissertation, final exam required.
Entrance requirements: For doctorate, GRE General Test, GRE Subject Test, MCAT (MD/PhD). *Application deadline:* For fall admission, 1/15. *Application fee:* $50.
Expenses: All students in good standing receive an annual stipend of $22,880.
Financial aid: Fellowships, stipends available.
Dr. Timothy McGraw, Director, 212-746-4982.

■ MOUNT SINAI SCHOOL OF MEDICINE OF NEW YORK UNIVERSITY

Graduate School of Biological Sciences, Biophysics, Structural Biology and Biomathematics (BSBB) Training Area, New York, NY 10029-6504

AWARDS PhD, MD/PhD.

Students: 12 full-time (3 women).
Degree requirements: For doctorate, dissertation required, foreign language not required.
Entrance requirements: For doctorate, GRE General Test, GRE Subject Test, MCAT, TOEFL. *Application deadline:* For fall admission, 4/15. *Application fee:* $60.
Expenses: Tuition: Full-time $21,750; part-time $725 per credit. Required fees: $750; $25 per credit. Full-time tuition and fees vary according to student level.
Financial aid: Fellowships with full tuition reimbursements, grants available.
Dr. Rami Osman, Program Director, *E-mail:* osman@msvax.mssm.edu.
Application contact: C. Gita Bosch, Administrative Manager and Assistant Dean, 212-241-6546, *Fax:* 212-241-0651, *E-mail:* grads@mssm.edu.

■ NORTHWESTERN UNIVERSITY

The Graduate School, Division of Interdepartmental Programs and Medical School, Integrated Graduate Programs in the Life Sciences, Chicago, IL 60611

AWARDS Cancer biology (PhD); cell biology (PhD); developmental biology (PhD); evolutionary biology (PhD); immunology and microbial pathogenesis (PhD); molecular biology and genetics (PhD); neurobiology (PhD); pharmacology and toxicology (PhD); structural biology and biochemistry (PhD).

Degree requirements: For doctorate, dissertation, written and oral qualifying exams required, foreign language not required.
Entrance requirements: For doctorate, GRE General Test, TOEFL.
Expenses: Tuition: Full-time $23,301. Full-time tuition and fees vary according to program.

Find an in-depth description at www.petersons.com/graduate.

■ NORTHWESTERN UNIVERSITY

The Graduate School, Judd A. and Marjorie Weinberg College of Arts and Sciences, Interdepartmental Biological Sciences Program (IBiS), Concentration in Structural Biology, Evanston, IL 60208

AWARDS PhD.

Faculty: 59 full-time (11 women).
Students: 79 full-time (46 women); includes 5 minority (2 African Americans, 3 Hispanic Americans), 14 international. *236 applicants, 19% accepted.*
Degree requirements: For doctorate, dissertation, 2 quarters of teaching experience required, foreign language not required.
Entrance requirements: For doctorate, GRE General Test, TOEFL. *Application deadline:* For fall admission, 1/15. Applications are processed on a rolling basis. *Application fee:* $50 ($55 for international students).
Expenses: Tuition: Full-time $23,301. Full-time tuition and fees vary according to program.
Financial aid: In 1999–00, 15 fellowships (averaging $12,078 per year), 64 research assistantships (averaging $17,000 per year), 15 teaching assistantships (averaging $16,620 per year) were awarded. Financial aid application deadline: 12/31; financial aid applicants required to submit FAFSA.

Find an in-depth description at www.petersons.com/graduate.

■ STANFORD UNIVERSITY

School of Medicine, Graduate Programs in Medicine, Department of Structural Biology, Stanford, CA 94305-9991

AWARDS PhD. Students are admitted to the Department of Biochemistry and can transfer to the Department of Structural Biology after one year of study.

Faculty: 7 full-time (1 woman).
Students: 2 full-time (1 woman), 1 international. Average age 27. *28 applicants, 18% accepted.*
Degree requirements: For doctorate, dissertation required.
Entrance requirements: For doctorate, GRE General Test, GRE Subject Test

Stanford University (continued)
(biology or chemistry), TOEFL. *Application deadline:* For fall admission, 12/15. *Application fee:* $65 ($80 for international students). Electronic applications accepted.
Expenses: Tuition: Full-time $23,058. Required fees: $152. Part-time tuition and fees vary according to course load.
Financial aid: Research assistantships available. Financial aid application deadline: 1/1.
Dr. Michael Levitt, Chairman, 650-723-6800, *Fax:* 650-723-8464, *E-mail:* levitt@stanford.edu.
Application contact: Admissions Office, 650-723-2460.

■ STATE UNIVERSITY OF NEW YORK AT ALBANY

School of Public Health, Department of Biomedical Sciences, Program in Cell and Molecular Structure, Albany, NY 12222-0001

AWARDS MS, PhD.

Degree requirements: For master's and doctorate, thesis/dissertation required.
Entrance requirements: For master's and doctorate, GRE General Test, GRE Subject Test. *Application deadline:* For fall admission, 1/15 (priority date); for spring admission, 11/1 (priority date). *Application fee:* $50.
Expenses: Tuition, state resident: full-time $5,100; part-time $214 per credit. Tuition, nonresident: full-time $8,416; part-time $352 per credit. Required fees: $31 per credit.
Financial aid: Application deadline: 2/1.
Dr. Harry Taber, Chair, Department of Biomedical Sciences, 518-474-2662.

■ STATE UNIVERSITY OF NEW YORK AT STONY BROOK

Graduate School, College of Arts and Sciences, Department of Biochemistry and Cell Biology, Program in Biochemistry and Structural Biology, Stony Brook, NY 11794

AWARDS PhD.

Expenses: Tuition, state resident: full-time $5,100; part-time $213 per credit hour. Tuition, nonresident: full-time $8,416; part-time $351 per credit hour. Required fees: $492. Tuition and fees vary according to program.

Find an in-depth description at www.petersons.com/graduate.

■ THOMAS JEFFERSON UNIVERSITY

College of Graduate Studies, Program in Molecular Pharmacology and Structural Biology, Philadelphia, PA 19107

AWARDS PhD.

Faculty: 48 full-time.
Students: 13 full-time (2 women), 2 part-time; includes 1 minority (Asian American or Pacific Islander), 2 international. *45 applicants, 7% accepted.* In 1999, 4 degrees awarded.
Degree requirements: For doctorate, dissertation required, foreign language not required.
Entrance requirements: For doctorate, GRE General Test, TOEFL, minimum GPA of 3.2. *Application deadline:* For fall admission, 3/1 (priority date). Applications are processed on a rolling basis. *Application fee:* $40.
Expenses: Tuition: Full-time $12,670. Tuition and fees vary according to degree level and program.
Financial aid: In 1999–00, 14 fellowships with full tuition reimbursements were awarded; research assistantships, Federal Work-Study, institutionally sponsored loans, traineeships, and training grants also available. Aid available to part-time students. Financial aid application deadline: 5/1; financial aid applicants required to submit FAFSA.
Faculty research: Biochemistry and cell, molecular and structural biology of cell-surface and intracellular receptors, molecular modeling, signal transduction.
Dr. Jeffrey L. Benovic, Chair, Graduate Committee, 215-503-4607, *Fax:* 215-923-1098.
Application contact: Jessie F. Pervall, Director of Admissions, 215-503-4400, *Fax:* 215-503-3433, *E-mail:* cgs-info@mail.tju.edu.

Find an in-depth description at www.petersons.com/graduate.

■ TULANE UNIVERSITY

School of Medicine and Graduate School, Graduate Programs in Medicine, Department of Structural and Cellular Biology, New Orleans, LA 70118-5669

AWARDS MS, PhD, MD/PhD. MS and PhD offered through the Graduate School.
Students: 8 full-time (6 women), 4 international. *9 applicants, 33% accepted.* In 1999, 2 doctorates awarded.
Degree requirements: For master's, one foreign language, thesis required; for doctorate, 2 foreign languages (computer language can substitute for one), dissertation required.

Entrance requirements: For master's, GRE General Test, TOEFL, or TSE, minimum B average in undergraduate course work; for doctorate, GRE General Test, TOEFL, or TSE. *Application deadline:* For fall admission, 2/1. *Application fee:* $45.
Expenses: Tuition: Full-time $23,030.
Financial aid: Fellowships, research assistantships, career-related internships or fieldwork, Federal Work-Study, and institutionally sponsored loans available. Financial aid application deadline: 2/1.
Dr. R. Yates, Chairman, 504-588-5258.

■ UNIVERSITY OF ILLINOIS AT URBANA–CHAMPAIGN

Graduate College, College of Liberal Arts and Sciences, School of Life Sciences, Department of Cell and Structural Biology, Urbana, IL 61801

AWARDS PhD.

Faculty: 9 full-time (2 women), 3 part-time/adjunct (1 woman).
Students: 51 full-time (20 women); includes 7 minority (1 African American, 5 Asian Americans or Pacific Islanders, 1 Hispanic American), 9 international. *60 applicants, 10% accepted.*
Degree requirements: For doctorate, dissertation required.
Application deadline: Applications are processed on a rolling basis. *Application fee:* $40 ($50 for international students).
Expenses: Tuition, state resident: full-time $4,616. Tuition, nonresident: full-time $11,768. Full-time tuition and fees vary according to course load.
Financial aid: Fellowships, research assistantships, teaching assistantships available. Financial aid application deadline: 2/15.
Dr. Martha U. Gillette, Acting Head, 217-333-6118, *Fax:* 217-244-1648, *E-mail:* mgillett@uiuc.edu.
Application contact: Dr. Lorie Hatfield, Director of Graduate Studies, 217-244-8116, *Fax:* 217-244-1648, *E-mail:* loriehat@uiuc.edu.

Find an in-depth description at www.petersons.com/graduate.

■ UNIVERSITY OF PENNSYLVANIA

School of Medicine, Biomedical Graduate Studies, Graduate Group in Cell and Molecular Biology, Program in Cell Biology and Physiology, Philadelphia, PA 19104

AWARDS PhD, MD/PhD, VMD/PhD.

Degree requirements: For doctorate, dissertation required, foreign language not required.
Entrance requirements: For doctorate, GRE General Test, TOEFL. *Application*

deadline: For fall admission, 1/2 (priority date). Applications are processed on a rolling basis. *Application fee:* $65.
Expenses: Tuition: Full-time $17,256; part-time $2,991 per course. Required fees: $2,588; $363 per course. $726 per term.
Financial aid: Fellowships, research assistantships, teaching assistantships, grants and institutionally sponsored loans available. Financial aid application deadline: 1/2.
Dr. Morris Birnbaum, Head, 215-898-5001.
Application contact: Mary Webster, Coordinator, 215-898-4360, *E-mail:* camb@mail.med.upenn.edu.

■ THE UNIVERSITY OF TEXAS HEALTH SCIENCE CENTER AT SAN ANTONIO

Graduate School of Biomedical Sciences, Department of Cellular and Structural Biology, San Antonio, TX 78229-3900

AWARDS PhD.

Degree requirements: For doctorate, dissertation, oral qualifying exam required, foreign language not required.
Entrance requirements: For doctorate, GRE General Test, previous course work in biology, chemistry, physics, and calculus.
Expenses: Tuition, state resident: part-time $38 per credit hour. Tuition, nonresident: part-time $249 per credit hour.
Faculty research: Human/molecular genetics, endocrinology and neurobiology, cell biology, cancer biology, biology of aging.

Find an in-depth description at www.petersons.com/graduate.

■ THE UNIVERSITY OF TEXAS MEDICAL BRANCH AT GALVESTON

Graduate School of Biomedical Sciences, Program in Human Biological Chemistry and Genetics, Galveston, TX 77555

AWARDS Biochemistry (MS, PhD); cell biology (MS, PhD); genetics (MS, PhD); molecular biology (PhD); structural biology (PhD).

Faculty: 61 full-time (4 women).
Students: 39 full-time (12 women); includes 2 minority (both Hispanic Americans), 12 international. Average age 28. *41 applicants, 66% accepted.* In 1999, 1 master's awarded (100% found work related to degree); 4 doctorates awarded (100% found work related to degree).
Degree requirements: For master's and doctorate, thesis/dissertation required, foreign language not required.
Entrance requirements: For master's and doctorate, GRE General Test. *Application deadline:* For fall admission, 2/1 (priority date). Applications are processed on a rolling basis. *Application fee:* $25 ($50 for international students). Electronic applications accepted.
Expenses: Tuition, state resident: full-time $684; part-time $38 per credit hour. Tuition, nonresident: full-time $4,572; part-time $254 per credit hour. Required fees: $29; $7.5 per credit hour. One-time fee: $55. Tuition and fees vary according to program.
Financial aid: Fellowships, research assistantships, Federal Work-Study and institutionally sponsored loans available. Financial aid application deadline: 2/1; financial aid applicants required to submit FAFSA.
Faculty research: Cell surfaces and transmembrane signaling, structural biology, growth factors, DNA repair and aging.
Dr. James C. Lee, Director, 409-772-2769, *Fax:* 409-747-0552, *E-mail:* gradprog.hbcg@utmb.edu.
Application contact: Becky L. Hansen, Senior Administrative Secretary, 409-772-2769, *Fax:* 409-747-0552, *E-mail:* bhansen@utmb.edu.

Find an in-depth description at www.petersons.com/graduate.

■ UNIVERSITY OF WASHINGTON

School of Medicine and Graduate School, Graduate Programs in Medicine, Department of Biological Structure, Seattle, WA 98195

AWARDS PhD.

Faculty: 19 full-time (3 women), 2 part-time/adjunct (0 women).
Students: 2 full-time (both women); includes 5 minority (all Asian Americans or

Pacific Islanders). Average age 32. In 1999, 1 degree awarded (100% found work related to degree).
Degree requirements: For doctorate, dissertation required. *Average time to degree:* Doctorate–5 years full-time.
Application fee: $45.
Expenses: Tuition, state resident: full-time $9,210; part-time $236 per credit. Tuition, nonresident: full-time $23,256; part-time $596 per credit.
Financial aid: In 1999–00, 2 students received aid, including 2 fellowships (averaging $17,304 per year), research assistantships (averaging $17,304 per year), teaching assistantships (averaging $17,304 per year); grants, scholarships, and traineeships also available.
Faculty research: Cellular and developmental biology, experimental immunology and hematology, molecular structure and molecular biology, neurobiology, x-rays.
Ronald E. Stenkamp, Graduate Program Advisor, 206-685-1721, *Fax:* 206-543-1524, *E-mail:* stenkamp@u.washington.edu.
Application contact: Pat Breen Fern, Graduate Program Assistant, 206-543-1860, *Fax:* 206-543-1524, *E-mail:* killian@u.washington.edu.

Ecology, Environmental Biology, and Evolutionary Biology

CONSERVATION BIOLOGY

■ ARIZONA STATE UNIVERSITY

Graduate College, College of Liberal Arts and Sciences, Department of Biology, Program in Conservation, Tempe, AZ 85287

AWARDS MS, PhD. Terminal master's awarded for partial completion of doctoral program.

Degree requirements: For master's, thesis required, foreign language not required; for doctorate, dissertation, oral exam required, foreign language not required.
Entrance requirements: For master's and doctorate, GRE General Test, GRE Subject Test. *Application deadline:* For fall admission, 12/15. *Application fee:* $45.
Expenses: Tuition, state resident: part-time $115 per credit hour. Tuition, nonresident: part-time $389 per credit hour. Required fees: $18 per semester. Tuition and fees vary according to program.
Financial aid: Application deadline: 12/15.
Application contact: Dr. Michael C. Moore, Director, 480-965-0386, *Fax:* 480-965-2519.

■ CENTRAL MICHIGAN UNIVERSITY

College of Graduate Studies, College of Science and Technology, Department of Biology, Mount Pleasant, MI 48859

AWARDS Biology (MS); conservation biology (MS).

Faculty: 39 full-time (6 women).
Students: 20 full-time (10 women), 38 part-time (17 women). Average age 27. In 1999, 18 degrees awarded.
Degree requirements: For master's, thesis or alternative required, foreign language not required.
Entrance requirements: For master's, bachelor's degree in biology, minimum GPA of 3.0. *Application deadline:* Applications are processed on a rolling basis. *Application fee:* $30.
Expenses: Tuition, state resident: part-time $144 per credit hour. Tuition, nonresident: part-time $285 per credit hour. Required fees: $240 per semester.

Tuition and fees vary according to degree level and program.
Financial aid: In 1999–00, 2 fellowships with tuition reimbursements, 11 research assistantships with tuition reimbursements, 26 teaching assistantships with tuition reimbursements were awarded; career-related internships or fieldwork and Federal Work-Study also available. Financial aid application deadline: 3/7.
Faculty research: Vertebrates, morphology and taxonomy of aquatic plants, molecular biology and genetics, microbials and invertebrate ecology.
Dr. John Scheide, Chairperson, 517-774-3227, *Fax:* 517-774-3462, *E-mail:* john.iver.scheide@cmich.edu.

■ COLUMBIA UNIVERSITY

Graduate School of Arts and Sciences, Graduate Program in Ecology and Evolutionary Biology, New York, NY 10027

AWARDS Conservation biology (Certificate); ecology and evolutionary biology (PhD); environmental policy (Certificate).

Degree requirements: For doctorate, dissertation, teaching experience required.
Entrance requirements: For doctorate, GRE General Test, TOEFL, previous course work in biology. Electronic applications accepted.
Expenses: Tuition: Full-time $25,072. Full-time tuition and fees vary according to course load and program.
Faculty research: Tropical ecology, ethnobotany, global change, systematics.
Find an in-depth description at www.petersons.com/graduate.

■ COLUMBIA UNIVERSITY

Graduate School of Arts and Sciences, Program in Conservation Biology, New York, NY 10027

AWARDS MA.

Degree requirements: For master's, thesis required, foreign language not required.
Expenses: Tuition: Full-time $25,072. Full-time tuition and fees vary according to course load and program.

■ FROSTBURG STATE UNIVERSITY

Graduate School, College of Liberal Arts and Sciences, Department of Biology, Program in Applied Ecology and Conservation Biology, Frostburg, MD 21532-1099

AWARDS MS.

Faculty: 8.
Students: 13 full-time (2 women), 14 part-time (7 women), 2 international. Average age 30. In 1999, 3 degrees awarded.
Degree requirements: For master's, thesis required, foreign language not required.
Entrance requirements: For master's, GRE General Test, resume. *Application deadline:* For fall admission, 7/15 (priority date). Applications are processed on a rolling basis. *Application fee:* $30.
Expenses: Tuition, state resident: full-time $3,132; part-time $174 per credit hour. Tuition, nonresident: full-time $3,636; part-time $202 per credit hour. Required fees: $31 per credit hour. $8 per semester.
Financial aid: In 1999–00, 14 research assistantships with full tuition reimbursements (averaging $5,000 per year) were awarded; career-related internships or fieldwork and Federal Work-Study also available. Financial aid application deadline: 4/1; financial aid applicants required to submit FAFSA.
Faculty research: Forest ecology, microbiology of man-made wetlands, invertebrate zoology and entomology, wildlife and carnivore ecology, aquatic pollution ecology.
Dr. Gwen Brewer, Coordinator, 301-687-4166.
Application contact: Robert E. Smith, Assistant Dean for Graduate Services, 301-687-7053, *Fax:* 301-687-4597, *E-mail:* rsmith@frostburg.edu.

■ SAN FRANCISCO STATE UNIVERSITY

Graduate Division, College of Science and Engineering, Department of Biology, Program in Conservation Biology, San Francisco, CA 94132-1722

AWARDS MA.

Entrance requirements: For master's, minimum GPA of 2.5 in last 60 units.

Expenses: Tuition, nonresident: full-time $5,904; part-time $246 per unit. Required fees: $1,904; $637 per semester. Tuition and fees vary according to course load.

■ STATE UNIVERSITY OF NEW YORK AT ALBANY

College of Arts and Sciences, Department of Biological Sciences, Program in Biodiversity, Conservation, and Policy, Albany, NY 12222-0001

AWARDS MS.

Degree requirements: For master's, one foreign language required.
Entrance requirements: For master's, GRE General Test. *Application fee:* $50.
Expenses: Tuition, state resident: full-time $5,100; part-time $214 per credit. Tuition, nonresident: full-time $8,416; part-time $352 per credit. Required fees: $31 per credit.
Dr. David Shub, Chair, Department of Biological Sciences, 518-442-4300.

■ UNIVERSITY OF ARKANSAS

Graduate School, J. William Fulbright College of Arts and Sciences, Interdisciplinary Program in Environmental Dynamics, Fayetteville, AR 72701-1201

AWARDS PhD.

Students: 21 full-time (11 women), 1 part-time; includes 1 minority (Asian American or Pacific Islander), 3 international. *12 applicants, 92% accepted.*
Degree requirements: For doctorate, dissertation required.
Application fee: $40 ($50 for international students).
Expenses: Tuition, state resident: full-time $3,186; part-time $177 per credit. Tuition, nonresident: full-time $7,560; part-time $420 per credit. Required fees: $756; $21 per credit. One-time fee: $22 part-time. Tuition and fees vary according to course load and program.
Financial aid: In 1999–00, 5 teaching assistantships were awarded. Financial aid application deadline: 4/1.
Allen McCartney, Head, 501-575-2508, *E-mail:* endy@comp.uark.edu.

■ UNIVERSITY OF CENTRAL FLORIDA

College of Arts and Sciences, Program in Biological Sciences, Orlando, FL 32816

AWARDS Biological sciences (MS); conservation biology (Certificate). Part-time and evening/weekend programs available.

Faculty: 15 full-time, 2 part-time/adjunct.
Students: 17 full-time (7 women), 36 part-time (22 women); includes 3 minority (2

Asian Americans or Pacific Islanders, 1 Hispanic American), 4 international. Average age 30. *38 applicants, 55% accepted.* In 1999, 7 degrees awarded.
Degree requirements: For master's, thesis or alternative, comprehensive exam, biology field exam required, foreign language not required.
Entrance requirements: For master's, GRE General Test, TOEFL, minimum GPA of 3.0 in last 60 hours. *Application deadline:* For fall admission, 3/1 (priority date); for spring admission, 10/15. *Application fee:* $20.
Expenses: Tuition, state resident: full-time $2,054; part-time $137 per credit. Tuition, nonresident: full-time $7,207; part-time $480 per term. Required fees: $47 per term.
Financial aid: In 1999–00, 17 fellowships with partial tuition reimbursements (averaging $2,676 per year), 50 research assistantships with partial tuition reimbursements (averaging $2,316 per year), 48 teaching assistantships with partial tuition reimbursements (averaging $4,028 per year) were awarded; career-related internships or fieldwork, Federal Work-Study, institutionally sponsored loans, tuition waivers (partial), and unspecified assistantships also available. Financial aid application deadline: 3/1; financial aid applicants required to submit FAFSA.
Dr. D. H. Vickers, Chair, 407-823-2141, *Fax:* 407-823-5769, *E-mail:* dvickers@pegasus.cc.ucf.edu.
Application contact: Dr. David Kuhn, Coordinator, 407-823-2141, *Fax:* 407-823-5769, *E-mail:* dkuhn@pegasus.cc.ucf.edu.

■ UNIVERSITY OF HAWAII AT MANOA

Graduate Division, Specialization in Ecology, Evolution and Conservation Biology, Honolulu, HI 96822

AWARDS MS, PhD. Program is interdisciplinary; degree is in specific discipline with the specialization in Ecology, Evolution and Conservation Biology.

Faculty: 35 full-time (8 women), 11 part-time/adjunct (0 women).
Students: 52 full-time (34 women); includes 12 minority (8 Asian Americans or Pacific Islanders, 3 Hispanic Americans, 1 Native American), 2 international.
Degree requirements: For doctorate, dissertation required.
Expenses: Tuition, state resident: full-time $4,032; part-time $168 per credit. Tuition, nonresident: full-time $9,960; part-time $415 per credit. Required fees: $51 per semester. Part-time tuition and fees vary according to course load and program.

Financial aid: In 1999–00, 15 students received aid, including 3 fellowships, 5 research assistantships, 2 teaching assistantships; career-related internships or fieldwork and tuition waivers (full) also available.
Faculty research: Agronomy and soil science, zoology, entomology, genetics and molecular biology, botanical sciences.
Dr. Sheila Conant, Chair, 808-956-4602, *Fax:* 808-956-9608, *E-mail:* sconant@zoogatc.zoo.hawaii.edu.

■ UNIVERSITY OF MARYLAND, COLLEGE PARK

Graduate Studies and Research, College of Life Sciences, Department of Biology, Program in Sustainable Development and Conservation Biology, College Park, MD 20742

AWARDS MS. Part-time and evening/weekend programs available.

Students: 23 full-time (18 women), 14 part-time (10 women); includes 4 minority (3 Hispanic Americans, 1 Native American), 5 international. *73 applicants, 32% accepted.* In 1999, 6 degrees awarded.
Degree requirements: For master's, internship, scholarly paper required, foreign language and thesis not required.
Entrance requirements: For master's, GRE General Test, minimum GPA of 3.0. *Application deadline:* For fall admission, 4/1; for spring admission, 12/1. Applications are processed on a rolling basis. *Application fee:* $50 ($70 for international students). Electronic applications accepted.
Expenses: Tuition, state resident: part-time $272 per credit hour. Tuition, nonresident: part-time $415 per credit hour. Required fees: $632; $379 per year.
Financial aid: Fellowships, research assistantships, teaching assistantships available. Financial aid application deadline: 2/1; financial aid applicants required to submit FAFSA.
Faculty research: Biodiversity, global change, conservation.
Dr. David W. Inouye, Director, 301-405-7409, *Fax:* 301-314-9358.
Application contact: Trudy Lindsey, Director, Graduate Admissions and Records, 301-405-4198, *Fax:* 301-314-9305, *E-mail:* grschool@deans.umd.edu.

■ UNIVERSITY OF MINNESOTA, TWIN CITIES CAMPUS

Graduate School, College of Natural Resources, Department of Fisheries and Wildlife, Minneapolis, MN 55455-0213

AWARDS Conservation biology (MS, PhD); fisheries (MS, PhD); wildlife conservation (MS, PhD).

University of Minnesota, Twin Cities Campus (continued)

Faculty: 13 full-time (2 women), 11 part-time/adjunct (0 women).

Students: 34 full-time (14 women), 1 (woman) part-time; includes 5 minority (3 Asian Americans or Pacific Islanders, 1 Hispanic American, 1 Native American), 2 international. *73 applicants, 12% accepted.* In 1999, 5 master's, 2 doctorates awarded (50% entered university research/teaching, 50% found other work related to degree). Terminal master's awarded for partial completion of doctoral program.

Degree requirements: For master's, thesis optional; for doctorate, dissertation required.

Entrance requirements: For master's and doctorate, GRE. *Application deadline:* For fall admission, 1/1 (priority date). *Application fee:* $50 ($55 for international students).

Expenses: Tuition, state resident: full-time $5,040; part-time $420 per credit. Tuition, nonresident: full-time $9,900; part-time $825 per credit. Full-time tuition and fees vary according to course load, program and reciprocity agreements.

Financial aid: In 1999–00, 30 students received aid, including 5 fellowships with full tuition reimbursements available (averaging $12,000 per year), 25 research assistantships with full and partial tuition reimbursements available, 1 teaching assistantship with full and partial tuition reimbursement available Financial aid application deadline: 1/1.

Faculty research: Management, ecology, physiology, genetics, and computer modeling of fish and wildlife.

Dr. Ira R. Adelman, Head, 612-624-3600.

Application contact: Kathleen Walter, Assistant for Graduate Studies, 612-624-2748, *Fax:* 612-624-8701, *E-mail:* kwalter@forestry.umn.edu.

Find an in-depth description at www.petersons.com/graduate.

■ UNIVERSITY OF MISSOURI–ST. LOUIS

Graduate School, College of Arts and Sciences, Department of Biology, St. Louis, MO 63121-4499

AWARDS Biology (MS, PhD), including animal behavior (MS), biochemistry, biotechnology (MS), conservation biology (MS), development (MS), ecology (MS), environmental studies (PhD), evolution (MS), genetics (MS), molecular biology and biotechnology (PhD), molecular/cellular biology (MS), physiology (MS), plant systematics, population biology (MS), tropical biology (MS); biotechnology (Certificate); tropical biology and conservation (Certificate). Part-time programs available.

Faculty: 46.

Students: 21 full-time (11 women), 75 part-time (44 women); includes 13 minority (2 African Americans, 2 Asian Americans or Pacific Islanders, 8 Hispanic Americans, 1 Native American), 23 international. In 1999, 14 master's, 4 doctorates awarded.

Degree requirements: For master's, thesis or alternative required, foreign language not required; for doctorate, one foreign language, dissertation, 1 semester of teaching experience required.

Entrance requirements: For doctorate, GRE General Test. *Application deadline:* For fall admission, 7/1 (priority date); for spring admission, 11/1 (priority date). Applications are processed on a rolling basis. *Application fee:* $25 ($40 for international students). Electronic applications accepted.

Expenses: Tuition, state resident: full-time $4,932; part-time $173 per credit hour. Tuition, nonresident: full-time $13,279; part-time $521 per credit hour. Required fees: $775; $33 per credit hour. Tuition and fees vary according to degree level and program.

Financial aid: In 1999–00, 8 research assistantships with partial tuition reimbursements (averaging $10,635 per year), 14 teaching assistantships with partial tuition reimbursements (averaging $11,488 per year) were awarded; career-related internships or fieldwork and Federal Work-Study also available. Aid available to part-time students. Financial aid application deadline: 2/1. *Total annual research expenditures:* $908,828.

Application contact: Graduate Admissions, 314-516-5458, *Fax:* 314-516-6759, *E-mail:* gradadm@umsl.edu.

■ UNIVERSITY OF NEVADA, RENO

Graduate School, Interdisciplinary Program in Ecology, Evolution, and Conservation Biology, Reno, NV 89557

AWARDS PhD. Offered through the College of Arts and Science, the M. C. Fleischmann College of Agriculture, and the Desert Research Institute.

Faculty: 29.

Students: 27 full-time (16 women), 9 part-time (3 women); includes 2 minority (1 Asian American or Pacific Islander, 1 Hispanic American), 2 international. Average age 33. *36 applicants, 14% accepted.* In 1999, 7 degrees awarded.

Degree requirements: For doctorate, dissertation required, foreign language not required.

Entrance requirements: For doctorate, GRE General Test, GRE Subject Test, TOEFL, minimum GPA of 3.0. *Application*

deadline: For fall admission, 2/15. *Application fee:* $40.

Expenses: Tuition, area resident: Part-time $3,173 per semester. Tuition, nonresident: full-time $6,347. Required fees: $101 per credit. $101 per credit.

Financial aid: In 1999–00, 1 research assistantship was awarded; teaching assistantships Financial aid application deadline: 3/1.

Faculty research: Population biology, behavioral ecology, plant response to climate change, conservation of endangered species, restoration of natural ecosystems.

Dr. C. Richard Tracy, Director, 775-784-4419, *Fax:* 775-784-4583.

Application contact: Larry Hillerman, Program Manager, 775-784-4439, *Fax:* 775-784-1306, *E-mail:* eecb@biodiversity.unr.edu.

■ UNIVERSITY OF NEW MEXICO

Graduate School, College of Arts and Sciences, Department of Biology, Albuquerque, NM 87131-2039

AWARDS Biology (MS, PhD), including air land ecology, behavioral ecology, botany, cellular and molecular biology, community ecology, comparative immunology, comparative physiology, conservation biology, ecology, ecosystem ecology, evolutionary biology, evolutionary genetics, microbiology, molecular genetics, parasitology, physiological ecology, physiology, population biology, vertebrate and invertebrate zoology. Part-time programs available.

Faculty: 35 full-time (5 women), 18 part-time/adjunct (11 women).

Students: 71 full-time (37 women), 28 part-time (11 women); includes 8 minority (2 Asian Americans or Pacific Islanders, 5 Hispanic Americans, 1 Native American), 11 international. Average age 33. *93 applicants, 30% accepted.* In 1999, 11 master's, 12 doctorates awarded. Terminal master's awarded for partial completion of doctoral program.

Degree requirements: For master's, one foreign language (computer language can substitute), thesis required (for some programs); for doctorate, 2 foreign languages (computer language can substitute for one), dissertation required.

Entrance requirements: For master's and doctorate, GRE General Test, GRE Subject Test, minimum GPA of 3.2. *Application deadline:* For fall admission, 1/15. *Application fee:* $25.

Expenses: Tuition, state resident: full-time $2,514; part-time $105 per credit hour. Tuition, nonresident: full-time $10,304; part-time $417 per credit hour. International tuition: $10,304 full-time.

Required fees: $516; $22 per credit hour. Tuition and fees vary according to program.

Financial aid: In 1999–00, 58 students received aid, including 24 fellowships (averaging $1,645 per year), 26 research assistantships with tuition reimbursements available (averaging $8,921 per year), 40 teaching assistantships with tuition reimbursements available (averaging $11,066 per year); career-related internships or fieldwork, Federal Work-Study, institutionally sponsored loans, and tuition waivers (full and partial) also available. Aid available to part-time students. Financial aid applicants required to submit FAFSA.

Faculty research: Developmental biology, immunobiology. *Total annual research expenditures:* $4.5 million.

Dr. Kathryn Vogel, Chair, 505-277-3411, *Fax:* 505-277-0304, *E-mail:* kgvogel@unm.edu.

Application contact: Vivian Kent, Information Contact, 505-277-1712, *Fax:* 505-277-0304, *E-mail:* vkent@unm.edu.

■ UNIVERSITY OF NEW ORLEANS

Graduate School, College of Sciences, Department of Biological Sciences, New Orleans, LA 70148

AWARDS Biological sciences (MS); conservation biology (PhD).

Faculty: 28 full-time (10 women), 3 part-time/adjunct (all women).

Students: 14 full-time (7 women), 6 part-time (2 women); includes 3 minority (1 African American, 2 Hispanic Americans), 2 international. Average age 29. *33 applicants, 21% accepted.* In 1999, 7 degrees awarded.

Degree requirements: For master's, one foreign language (computer language can substitute), thesis required.

Entrance requirements: For master's, GRE General Test. *Application deadline:* For fall admission, 7/1 (priority date). Applications are processed on a rolling basis. *Application fee:* $20.

Expenses: Tuition, state resident: full-time $2,362. Tuition, nonresident: full-time $7,888. Part-time tuition and fees vary according to course load.

Faculty research: Biochemistry, genetics, vertebrate and invertebrate systematics and ecology, cell and mammalian physiology, morphology.

Dr. Sam Rogers, Chairman, 504-280-6307, *Fax:* 504-280-6121, *E-mail:* jsrogers@uno.edu.

Application contact: Dr. Jerry Howard, Graduate Coordinator, 504-280-7059, *Fax:* 504-280-6121, *E-mail:* jjhoward@uno.edu.

Find an in-depth description at www.petersons.com/graduate.

■ UNIVERSITY OF WISCONSIN–MADISON

Graduate School, Institute for Environmental Studies, Conservation Biology and Sustainable Development Program, Madison, WI 53706-1380

AWARDS MS. Part-time programs available.

Students: 19 full-time (13 women), 10 part-time (7 women); includes 3 minority (1 African American, 2 Hispanic Americans), 1 international. *81 applicants, 25% accepted.* In 1999, 5 degrees awarded.

Degree requirements: For master's, thesis optional, foreign language not required.

Entrance requirements: For master's, GRE General Test. *Application deadline:* For fall admission, 2/15; for spring admission, 10/15. *Application fee:* $45. Electronic applications accepted.

Expenses: Tuition, state resident: full-time $5,406; part-time $339 per credit. Tuition, nonresident: full-time $17,110; part-time $1,071 per credit. Full-time tuition and fees vary according to program and reciprocity agreements. Part-time tuition and fees vary according to course load and program.

Financial aid: In 1999–00, 3 fellowships (averaging $12,430 per year), 3 research assistantships (averaging $15,042 per year), 11 teaching assistantships (averaging $8,712 per year) were awarded; career-related internships or fieldwork, Federal Work-Study, grants, and unspecified assistantships also available.

Faculty research: Ornithology, forestry, sociology, rural sociology, plant ecology, wildlife conservation, resource economics, education, land use.

Stanley A. Temple, Chair, 608-263-1796, *Fax:* 608-262-2273, *E-mail:* iesgrad@mail.ies.wisc.edu.

Application contact: Jim E. Miller, Clerical Assistant, 608-263-1796, *Fax:* 608-262-2273, *E-mail:* jemiller@facstaff.wisc.edu.

■ VIRGINIA POLYTECHNIC INSTITUTE AND STATE UNIVERSITY

Graduate School, College of Natural Resources, Department of Fisheries and Wildlife Sciences, Blacksburg, VA 24061

AWARDS Aquaculture (MS, PhD); conservation biology (MS, PhD); fisheries science (MS, PhD); wildlife science (MS, PhD).

Faculty: 21 full-time (2 women).

Students: 40 full-time (13 women), 5 part-time (4 women); includes 2 minority (both African Americans), 8 international. *91 applicants, 19% accepted.* In 1999, 12 master's, 2 doctorates awarded.

Degree requirements: For master's and doctorate, thesis/dissertation required.

Entrance requirements: For master's and doctorate, GRE General Test, TOEFL, minimum GPA of 3.0. *Application deadline:* For fall admission, 12/1 (priority date). Applications are processed on a rolling basis. *Application fee:* $25.

Financial aid: Fellowships, research assistantships, teaching assistantships, Federal Work-Study and tuition waivers (full) available. Financial aid application deadline: 4/1.

Faculty research: Fisheries management, wildlife management, wildlife toxicology and physiology, endangered species, computer applications.

Dr. Donald Orth, Head, 540-231-5573.

Find an in-depth description at www.petersons.com/graduate.

ECOLOGY

■ ARIZONA STATE UNIVERSITY

Graduate College, College of Liberal Arts and Sciences, Department of Biology, Program in Ecology, Tempe, AZ 85287

AWARDS MS, PhD. Terminal master's awarded for partial completion of doctoral program.

Degree requirements: For master's, thesis required, foreign language not required; for doctorate, dissertation, oral exam required, foreign language not required.

Entrance requirements: For master's and doctorate, GRE General Test, GRE Subject Test. *Application deadline:* For fall admission, 12/15. *Application fee:* $45.

Expenses: Tuition, state resident: part-time $115 per credit hour. Tuition, nonresident: part-time $389 per credit hour. Required fees: $18 per semester. Tuition and fees vary according to program.

Financial aid: Application deadline: 12/15.

Application contact: Dr. Nancy Grimm, Associate Professor, 480-965-3571, *Fax:* 480-965-2519.

■ BOSTON UNIVERSITY

Graduate School of Arts and Sciences, Department of Biology, Boston, MA 02215

AWARDS Botany (MA, PhD); cell and molecular biology (MA, PhD); cell biology (MA, PhD); ecology (PhD); ecology, behavior, and evolution (MA, PhD); ecology/physiology, endocrinology and reproduction (MA); marine biology (MA, PhD); molecular biology, cell biology and biochemistry (MA, PhD); neurobiology, neuroendocrinology and

Boston University (continued)
reproduction (MA, PhD); physiology, endocrinology, and neurobiology (MA, PhD); zoology (MA, PhD). Part-time programs available.

Faculty: 41 full-time (8 women).
Students: 131 full-time (74 women), 11 part-time (7 women); includes 10 minority (7 Asian Americans or Pacific Islanders, 3 Hispanic Americans), 33 international. Average age 27. *238 applicants, 39% accepted.* In 1999, 61 master's, 45 doctorates awarded. Terminal master's awarded for partial completion of doctoral program.
Degree requirements: For master's, one foreign language, thesis not required; for doctorate, one foreign language, dissertation, qualifying exam required. *Average time to degree:* Master's–1 year full-time, 3 years part-time; doctorate–5.75 years full-time.
Entrance requirements: For master's and doctorate, GRE General Test, GRE Subject Test, TOEFL. *Application deadline:* For fall admission, 1/1 (priority date); for spring admission, 11/1. *Application fee:* $50.
Expenses: Tuition: Full-time $23,770; part-time $743 per credit. Required fees: $220. Tuition and fees vary according to class time, course level, campus/location and program.
Financial aid: In 1999–00, 82 students received aid, including 1 fellowship with full tuition reimbursement available (averaging $12,000 per year), 28 research assistantships with full tuition reimbursements available (averaging $11,500 per year), 43 teaching assistantships with full tuition reimbursements available (averaging $11,500 per year); Federal Work-Study, grants, institutionally sponsored loans, scholarships, and traineeships also available. Financial aid application deadline: 1/15; financial aid applicants required to submit FAFSA.
Faculty research: Marine science, endocrinology, behavior. *Total annual research expenditures:* $5 million.
Geoffrey M. Cooper, Chairman, 617-353-2432, *Fax:* 617-353-6340, *E-mail:* gmcooper@bu.edu.
Application contact: Yolanta Kovalko, Senior Staff Assistant, 617-353-2432, *Fax:* 617-353-6340, *E-mail:* yolanta@bu.edu.
Find an in-depth description at www.petersons.com/graduate.

■ BROWN UNIVERSITY

Graduate School, Division of Biology and Medicine, Program in Ecology and Evolutionary Biology, Providence, RI 02912

AWARDS PhD.

Faculty: 11 full-time (4 women).

Students: 13 full-time (4 women), 1 international. Average age 24. *33 applicants, 9% accepted.* In 1999, 3 degrees awarded.
Degree requirements: For doctorate, dissertation, preliminary exam required.
Entrance requirements: For doctorate, GRE General Test, GRE Subject Test. *Application deadline:* For fall admission, 1/2 (priority date). Applications are processed on a rolling basis. *Application fee:* $60.
Financial aid: In 1999–00, 3 fellowships, 1 research assistantship, 7 teaching assistantships were awarded. Financial aid application deadline: 1/2.
Faculty research: Marine ecology, behavioral ecology, population genetics, evolutionary morphology, plant ecology. Dr. Mark Bertness, Director, 401-863-2280, *E-mail:* mark_bertness@brown.edu.

■ COLORADO STATE UNIVERSITY

Graduate School, College of Natural Resources, Program in Ecology, Fort Collins, CO 80523-0015

AWARDS MS, PhD.

Students: 38 full-time (26 women), 32 part-time (13 women); includes 1 minority (Asian American or Pacific Islander), 5 international. Average age 30. *114 applicants, 12% accepted.* In 1999, 8 master's, 11 doctorates awarded.
Degree requirements: For master's, thesis, oral or written exam required, foreign language not required; for doctorate, dissertation, oral and written exams required, foreign language not required.
Entrance requirements: For master's and doctorate, GRE General Test (average 75th percentile), minimum GPA of 3.0. *Application deadline:* For fall admission, 2/1 (priority date); for spring admission, 9/1. Applications are processed on a rolling basis. *Application fee:* $30. Electronic applications accepted.
Expenses: Tuition, state resident: full-time $2,694; part-time $150 per credit. Tuition, nonresident: full-time $10,460; part-time $581 per credit. Required fees: $32 per semester. Tuition and fees vary according to program.
Financial aid: In 1999–00, 4 fellowships, 29 research assistantships, 14 teaching assistantships were awarded; career-related internships or fieldwork, institutionally sponsored loans, and traineeships also available.
Faculty research: Plant and animal ecology at organismal, population, community, and ecosystem levels.
Dr. Dan Binkley, Director, 970-491-4373, *Fax:* 970-491-2796.
Application contact: Sally Dunphy, Program Assistant, 970-491-4373, *Fax:*

970-491-2796, *E-mail:* ecology@picea.cnr.colostate.edu.

■ COLUMBIA UNIVERSITY

Graduate School of Arts and Sciences, Graduate Program in Ecology and Evolutionary Biology, New York, NY 10027

AWARDS Conservation biology (Certificate); ecology and evolutionary biology (PhD); environmental policy (Certificate).

Degree requirements: For doctorate, dissertation, teaching experience required.
Entrance requirements: For doctorate, GRE General Test, TOEFL, previous course work in biology. Electronic applications accepted.
Expenses: Tuition: Full-time $25,072. Full-time tuition and fees vary according to course load and program.
Faculty research: Tropical ecology, ethnobotany, global change, systematics.
Find an in-depth description at www.petersons.com/graduate.

■ CORNELL UNIVERSITY

Graduate School, Graduate Fields of Agriculture and Life Sciences, Field of Ecology and Evolutionary Biology, Ithaca, NY 14853-0001

AWARDS Ecology (PhD), including animal ecology, applied ecology, biogeochemistry, community and ecosystem ecology, limnology, oceanography, physiological ecology, plant ecology, population ecology, theoretical ecology, vertebrate zoology; evolutionary biology (PhD), including ecological genetics, paleobiology, population biology, systematics.

Faculty: 41 full-time.
Students: 55 full-time (27 women); includes 4 minority (3 Asian Americans or Pacific Islanders, 1 Hispanic American), 2 international. *115 applicants, 11% accepted.* In 1999, 12 doctorates awarded.
Degree requirements: For doctorate, dissertation, 2 semesters of teaching experience required, foreign language not required.
Entrance requirements: For doctorate, GRE General Test, GRE Subject Test (biology), TOEFL. *Application deadline:* For fall admission, 12/15. *Application fee:* $65. Electronic applications accepted.
Expenses: Tuition: Full-time $12,100.
Financial aid: In 1999–00, 54 students received aid, including 28 fellowships with full tuition reimbursements available, 4 research assistantships with full tuition reimbursements available, 22 teaching assistantships with full tuition reimbursements available; institutionally sponsored loans, scholarships, tuition waivers (full and partial), and unspecified assistantships

also available. Financial aid applicants required to submit FAFSA.

Faculty research: Population and organismal biology, population and evolutionary genetics, systematics and macroevolution, biochemistry, conservation biology.

Application contact: Graduate Field Assistant, 607-254-4230, *E-mail:* eeb_grad_req@cornell.edu.

■ DUKE UNIVERSITY

Graduate School, Department of Ecology, Durham, NC 27708-0342

AWARDS PhD, Certificate.

Faculty: 26 full-time.
Entrance requirements: For doctorate, GRE General Test. *Application fee:* $75.
Expenses: Tuition: Full-time $21,406; part-time $760 per unit. Required fees: $3,136; $3,136 per year. One-time fee: $30. Tuition and fees vary according to program.
Jim Clark, Director of Graduate Studies, 919-660-7339, *Fax:* 919-684-5412, *E-mail:* susan.gillispie@duke.edu.

■ DUKE UNIVERSITY

Graduate School, Department of Environment, Durham, NC 27708-0586

AWARDS Natural resource economics/policy (AM, PhD); natural resource science/ecology (AM, PhD); natural resource systems science (AM, PhD). Part-time programs available.

Faculty: 37 full-time, 11 part-time/adjunct.
Students: 67 full-time, 2 part-time; includes 7 minority (3 African Americans, 1 Asian American or Pacific Islander, 3 Hispanic Americans), 16 international. *155 applicants, 14% accepted.* In 1999, 4 master's, 8 doctorates awarded. Terminal master's awarded for partial completion of doctoral program.
Degree requirements: For master's, foreign language not required; for doctorate, dissertation required.
Entrance requirements: For master's and doctorate, GRE General Test. *Application deadline:* For fall admission, 12/31. *Application fee:* $75.
Expenses: Tuition: Full-time $21,406; part-time $760 per unit. Required fees: $3,136; $3,136 per year. One-time fee: $30. Tuition and fees vary according to program.
Financial aid: Fellowships, research assistantships, teaching assistantships, Federal Work-Study available. Financial aid application deadline: 12/31.
Kenneth Knoerr, Director of Graduate Studies, 919-613-8002, *Fax:* 919-684-8741, *E-mail:* nettleto@acpub.duke.edu.

■ DUKE UNIVERSITY

Nicholas School of the Environment, Durham, NC 27708-0328

AWARDS Coastal environmental management (MEM); environmental science and policy (PhD); environmental toxicology, chemistry, and risk assessment (MEM); forest resource management (MF); resource ecology (MEM); resource economics and policy (MEM); water and air resources (MEM). PhD offered through the Graduate School. Part-time programs available.

Faculty: 61 full-time (10 women), 23 part-time/adjunct (3 women).
Students: 225 full-time (131 women), 1 part-time. Average age 25. *400 applicants, 63% accepted.* In 1999, 113 master's, 10 doctorates awarded. Terminal master's awarded for partial completion of doctoral program.
Degree requirements: For master's, thesis required (for some programs), foreign language not required; for doctorate, dissertation required, foreign language not required. *Average time to degree:* Master's–2 years full-time, 3 years part-time; doctorate–5 years full-time, 8 years part-time.
Entrance requirements: For master's, GRE General Test, TOEFL, previous course work in biology or ecology, calculus, statistics, and microeconomics; computer familiarity with word processing and data analysis; for doctorate, GRE General Test, TOEFL. *Application deadline:* For fall admission, 2/1; for spring admission, 10/15. Applications are processed on a rolling basis. *Application fee:* $75. Electronic applications accepted.
Expenses: Tuition: Full-time $18,900; part-time $850 per credit. Required fees: $493; $231 per semester.
Financial aid: In 1999–00, 163 students received aid, including 152 fellowships (averaging $10,000 per year), 40 research assistantships (averaging $2,700 per year), 15 teaching assistantships (averaging $6,000 per year); career-related internships or fieldwork, Federal Work-Study, institutionally sponsored loans, scholarships, and unspecified assistantships also available. Financial aid application deadline: 2/1; financial aid applicants required to submit FAFSA.
Faculty research: Ecosystem management, conservation ecology, earth systems, risk assessment.
Dr. Norman L. Christensen, Dean, 919-613-8004, *Fax:* 919-684-8741.
Application contact: Bertie S. Belvin, Associate Dean for Academic Services, 919-613-8070, *Fax:* 919-684-8741, *E-mail:* envadm@duke.edu.

Find an in-depth description at www.petersons.com/graduate.

■ EASTERN KENTUCKY UNIVERSITY

The Graduate School, College of Natural and Mathematical Sciences, Department of Biological Sciences, Richmond, KY 40475-3102

AWARDS Biological sciences (MS); ecology (MS). Part-time programs available.

Faculty: 17 full-time (3 women).
Students: 32; includes 1 minority (African American), 1 international. In 1999, 10 degrees awarded.
Degree requirements: For master's, thesis required.
Entrance requirements: For master's, GRE General Test, minimum GPA of 2.5. *Application deadline:* For fall admission, 8/1; for spring admission, 12/1. Applications are processed on a rolling basis. *Application fee:* $0.
Expenses: Tuition, state resident: full-time $2,390; part-time $145 per credit hour. Tuition, nonresident: full-time $6,430; part-time $391 per credit hour.
Financial aid: Research assistantships, teaching assistantships, career-related internships or fieldwork and Federal Work-Study available. Aid available to part-time students. Financial aid applicants required to submit FAFSA.
Faculty research: Systematics, ecology, and biodiversity; animal behavior; protein structure and molecular genetics; biomonitoring and aquatic toxicology; pathogenesis of microbes and parasites. *Total annual research expenditures:* $45,000.
Dr. Ross Clark, Chair, 606-622-1531, *Fax:* 606-622-1020, *E-mail:* bioclark@acs.eku.edu.

■ EMORY UNIVERSITY

Graduate School of Arts and Sciences, Division of Biological and Biomedical Sciences, Program in Population Biology, Ecology, and Evolution, Atlanta, GA 30322-1100

AWARDS PhD.

Faculty: 20 full-time (3 women).
Students: 11 full-time (5 women), 2 international.
Degree requirements: For doctorate, dissertation required, foreign language not required.
Entrance requirements: For doctorate, GRE General Test, TOEFL, minimum GPA of 3.0 in science course work. *Application deadline:* For fall admission, 1/20 (priority date). *Application fee:* $45.
Expenses: Tuition: Full-time $22,770. Tuition and fees vary according to program.
Financial aid: In 1999–00, fellowships with full tuition reimbursements (averaging $18,000 per year)

Emory University (continued)
Dr. Leslie Real, Director, 404-727-4099, *Fax:* 404-727-2880, *E-mail:* lreal@biology.emory.edu.
Application contact: 404-727-2547, *Fax:* 404-727-3322, *E-mail:* gdbbs@gsas.emory.edu.

Find an in-depth description at www.petersons.com/graduate.

■ FLORIDA INSTITUTE OF TECHNOLOGY

Graduate School, College of Science and Liberal Arts, Department of Biological Sciences, Program in Ecology, Melbourne, FL 32901-6975

AWARDS MS. Part-time programs available.

Students: 2 full-time (1 woman), 4 part-time (1 woman), 1 international. Average age 26. *11 applicants, 18% accepted.*
Degree requirements: For master's, thesis required, foreign language not required.
Entrance requirements: For master's, GRE General Test, minimum GPA of 3.0. *Application deadline:* Applications are processed on a rolling basis. *Application fee:* $50. Electronic applications accepted.
Expenses: Tuition: Part-time $575 per credit hour. Required fees: $50. Tuition and fees vary according to campus/location and program.
Financial aid: In 1999–00, 3 students received aid, including 1 research assistantship with full and partial tuition reimbursement available (averaging $5,844 per year), 2 teaching assistantships (averaging $4,267 per year); career-related internships or fieldwork and tuition remissions also available. Financial aid application deadline: 3/1; financial aid applicants required to submit FAFSA.
Faculty research: Endangered or threatened avian and mammalian species, hydroacoustics and feeding preference of the West Indian manatee, habitat preference of the Florida scrub jay.
Dr. Gary N. Wells, Head, 321-674-8034, *Fax:* 321-674-7238, *E-mail:* gwells@fit.edu.
Application contact: Carolyn P. Farrior, Associate Dean of Graduate Admissions, 321-674-7118, *Fax:* 321-674-9468, *E-mail:* cfarrior@fit.edu.

Find an in-depth description at www.petersons.com/graduate.

■ FLORIDA STATE UNIVERSITY

Graduate Studies, College of Arts and Sciences, Department of Biological Science, Program in Ecology, Tallahassee, FL 32306

AWARDS MS, PhD.

Faculty: 13 full-time (2 women).

Students: 30 full-time (13 women); includes 2 minority (1 African American, 1 Asian American or Pacific Islander), 1 international.
Degree requirements: For master's and doctorate, thesis/dissertation, teaching experience required.
Entrance requirements: For master's, GRE General Test, TOEFL; for doctorate, GRE General Test, GRE Subject Test, TOEFL. *Application deadline:* For fall admission, 1/15; for spring admission, 10/15. *Application fee:* $20.
Expenses: Tuition, state resident: full-time $3,504; part-time $146 per credit hour. Tuition, nonresident: full-time $12,162; part-time $507 per credit hour. Tuition and fees vary according to program.
Financial aid: In 1999–00, fellowships with full tuition reimbursements (averaging $13,740 per year), research assistantships with full tuition reimbursements (averaging $13,740 per year), teaching assistantships with full tuition reimbursements (averaging $13,740 per year) were awarded. Financial aid application deadline: 1/15; financial aid applicants required to submit FAFSA.
Faculty research: Community ecology, biogeography, functional morphology, adaptation groundwater-bearing environments, ecophysiology.
Dr. Thomas C. S. Keller, Associate Professor and Associate Chairman, 850-644-3023, *Fax:* 850-644-9829.
Application contact: Judy Bowers, Coordinator, Graduate Affairs, 850-644-3023, *Fax:* 850-644-9829, *E-mail:* bowers@bio.fsu.edu.

■ FORDHAM UNIVERSITY

Graduate School of Arts and Sciences, Department of Biological Sciences, New York, NY 10458

AWARDS Biological sciences (MS, PhD), including cell and molecular biology, ecology. Part-time and evening/weekend programs available.

Faculty: 17 full-time (1 woman).
Students: 29 full-time (17 women), 10 part-time (6 women); includes 2 minority (both Asian Americans or Pacific Islanders), 6 international. *57 applicants, 35% accepted.* In 1999, 11 master's, 2 doctorates awarded. Terminal master's awarded for partial completion of doctoral program.
Degree requirements: For master's, comprehensive exam required, thesis optional; for doctorate, 2 foreign languages (computer language can substitute for one), dissertation, comprehensive exam required.
Entrance requirements: For master's and doctorate, GRE General Test, GRE Subject Test (recommended). *Application deadline:* For fall admission, 1/16 (priority date); for spring admission, 12/1. *Application fee:* $60. Electronic applications accepted.
Expenses: Tuition: Full-time $14,400; part-time $600 per credit. Required fees: $125 per semester. Tuition and fees vary according to program.
Financial aid: In 1999–00, 29 students received aid, including 4 fellowships (averaging $15,000 per year), 3 research assistantships (averaging $15,000 per year), teaching assistantships (averaging $15,000 per year); institutionally sponsored loans, tuition waivers (full and partial), and unspecified assistantships also available. Aid available to part-time students. Financial aid application deadline: 1/15. *Total annual research expenditures:* $365,196.
Dr. Berish Rubin, Chair, 718-817-3641, *Fax:* 718-817-3645, *E-mail:* rubin@fordham.edu.
Application contact: Dr. Craig W. Pilant, Assistant Dean, 718-817-4420, *Fax:* 718-817-3566, *E-mail:* pilant@fordham.edu.

Find an in-depth description at www.petersons.com/graduate.

■ FROSTBURG STATE UNIVERSITY

Graduate School, College of Liberal Arts and Sciences, Department of Biology, Program in Applied Ecology and Conservation Biology, Frostburg, MD 21532-1099

AWARDS MS.

Faculty: 8.
Students: 13 full-time (2 women), 14 part-time (7 women), 2 international. Average age 30. In 1999, 3 degrees awarded.
Degree requirements: For master's, thesis required, foreign language not required.
Entrance requirements: For master's, GRE General Test, resume. *Application deadline:* For fall admission, 7/15 (priority date). Applications are processed on a rolling basis. *Application fee:* $30.
Expenses: Tuition, state resident: full-time $3,132; part-time $174 per credit hour. Tuition, nonresident: full-time $3,636; part-time $202 per credit hour. Required fees: $31 per credit hour. $8 per semester.
Financial aid: In 1999–00, 14 research assistantships with full tuition reimbursements (averaging $5,000 per year) were awarded; career-related internships or fieldwork and Federal Work-Study also available. Financial aid application deadline: 4/1; financial aid applicants required to submit FAFSA.
Faculty research: Forest ecology, microbiology of man-made wetlands, invertebrate zoology and entomology, wildlife and carnivore ecology, aquatic pollution ecology.

Dr. Gwen Brewer, Coordinator, 301-687-4166.

Application contact: Robert E. Smith, Assistant Dean for Graduate Services, 301-687-7053, *Fax:* 301-687-4597, *E-mail:* rsmith@frostburg.edu.

■ GEORGE MASON UNIVERSITY

College of Arts and Sciences, Department of Biology, Master's Program in Biology, Fairfax, VA 22030-4444

AWARDS Bioinformatics (MS); ecology, systematics and evolution (MS); environmental science and public policy (MS); interpretive biology (MS); molecular, microbial, and cellular biology (MS); organismal biology (MS). Part-time programs available.

Faculty: 30 full-time (11 women), 32 part-time/adjunct (20 women).
Students: 5 full-time (3 women), 55 part-time (37 women); includes 7 minority (1 African American, 4 Asian Americans or Pacific Islanders, 2 Hispanic Americans), 8 international. Average age 34. *36 applicants, 44% accepted.* In 1999, 18 degrees awarded.
Degree requirements: For master's, thesis or alternative required, foreign language not required.
Entrance requirements: For master's, GRE General Test, GRE Subject Test, bachelor's degree in biology or equivalent. *Application deadline:* For fall admission, 5/1; for spring admission, 11/1. *Application fee:* $30. Electronic applications accepted.
Expenses: Tuition, state resident: full-time $4,416; part-time $184 per credit hour. Tuition, nonresident: full-time $12,516; part-time $522 per credit hour. Tuition and fees vary according to program.
Financial aid: Available to part-time students. Application deadline: 3/1.
Dr. George E. Andrykovitch, Director, 703-993-1027, *Fax:* 703-993-1046.

■ GODDARD COLLEGE

Graduate Programs, Program in Social Ecology, Plainfield, VT 05667-9432

AWARDS MA. Offered jointly with the Institute for Social Ecology. Postbaccalaureate distance learning degree programs offered (minimal on-campus study).

Degree requirements: For master's, thesis required.
Electronic applications accepted.
Expenses: Tuition: Full-time $9,650.
Faculty research: Biology, alternative technology, computer science.

■ ILLINOIS STATE UNIVERSITY

Graduate School, College of Arts and Sciences, Department of Biological Sciences, Normal, IL 61790-2200

AWARDS Biological sciences (MS); biology (PhD); botany (PhD); ecology (PhD); genetics (PhD); microbiology (PhD); physiology (PhD); zoology (PhD). Part-time programs available.

Faculty: 24 full-time (4 women).
Students: 57 full-time (25 women), 27 part-time (10 women); includes 6 minority (1 African American, 3 Asian Americans or Pacific Islanders, 2 Hispanic Americans), 13 international. *58 applicants, 59% accepted.* In 1999, 12 master's, 3 doctorates awarded.
Degree requirements: For master's, thesis or alternative required; for doctorate, variable foreign language requirement (computer language can substitute for one), dissertation, 2 terms of residency required.
Entrance requirements: For master's, GRE General Test, minimum GPA of 2.6 in last 60 hours; for doctorate, GRE General Test. *Application deadline:* Applications are processed on a rolling basis. *Application fee:* $0.
Expenses: Tuition, state resident: full-time $2,526; part-time $105 per credit hour. Tuition, nonresident: full-time $7,578; part-time $316 per credit hour. Required fees: $1,082; $38 per credit hour. Tuition and fees vary according to course load and program.
Financial aid: In 1999–00, 8 research assistantships, 63 teaching assistantships were awarded; Federal Work-Study, tuition waivers (full), and unspecified assistantships also available. Financial aid application deadline: 4/1.
Faculty research: Phenotypic plasticity in reproduction: molecular mechanisms, physiological control, adaptive significance; molecular stress physiology of *Listeria monocytogenes*; enzymology of eggshell formation in *Schistosoma mansoni*, compensatory adaptation in the rat midbrain after neurodegeneration, analysis of staphylococcal virulence germs. *Total annual research expenditures:* $586,648.
Dr. Hou Cheung, Chairperson, 309-438-3669.

Application contact: Derek A. McCracken, Graduate Adviser, 309-438-3664.

Find an in-depth description at www.petersons.com/graduate.

■ INDIANA STATE UNIVERSITY

School of Graduate Studies, College of Arts and Sciences, Department of Life Sciences, Terre Haute, IN 47809-1401

AWARDS Clinical laboratory sciences (MS); ecology (MA, MS, PhD); microbiology (MA, MS, PhD); physiology (MA, MS, PhD).

Degree requirements: For doctorate, computer language, dissertation required.
Entrance requirements: For master's and doctorate, GRE General Test. Electronic applications accepted.
Expenses: Tuition, state resident: full-time $3,552; part-time $148 per hour. Tuition, nonresident: full-time $8,088; part-time $337 per hour.
Find an in-depth description at www.petersons.com/graduate.

■ INDIANA UNIVERSITY BLOOMINGTON

Graduate School, College of Arts and Sciences, Department of Biology, Program in Evolution, Ecology, and Behavior, Bloomington, IN 47405

AWARDS Ecology (MA, PhD); evolutionary biology (MA, PhD); zoology (MA, PhD). PhD offered through the University Graduate School. Part-time programs available.

Students: 66 full-time (35 women), 2 part-time (1 woman); includes 8 minority (2 African Americans, 2 Asian Americans or Pacific Islanders, 4 Hispanic Americans), 3 international. In 1999, 2 master's, 7 doctorates awarded. Terminal master's awarded for partial completion of doctoral program.
Degree requirements: For master's, thesis or alternative required; for doctorate, dissertation required.
Entrance requirements: For master's and doctorate, GRE General Test, TOEFL. *Application deadline:* For fall admission, 1/5 (priority date); for spring admission, 9/1. Applications are processed on a rolling basis. *Application fee:* $45. Electronic applications accepted.
Expenses: Tuition, state resident: full-time $3,853; part-time $161 per credit hour. Tuition, nonresident: full-time $11,226; part-time $468 per credit hour. Required fees: $360 per year. Tuition and fees vary according to course load and program.
Financial aid: Fellowships with tuition reimbursements, research assistantships with tuition reimbursements, teaching assistantships with tuition reimbursements, scholarships available. Financial aid application deadline: 1/15.
Faculty research: Ecosystem of community, plant and animal population biology, avian sociobiology, fish ethology, evolutionary genetics.

Indiana University Bloomington (continued)
Dr. Gerald Gastony, Head, 812-855-3333, *Fax:* 812-855-6705, *E-mail:* gastony@indiana.edu.

Application contact: Gretchen Clearwater, Advisor for Graduate Affairs, 812-855-1861, *Fax:* 812-855-6705, *E-mail:* biograd@bio.indiana.edu.

Find an in-depth description at www.petersons.com/graduate.

■ **IOWA STATE UNIVERSITY OF SCIENCE AND TECHNOLOGY**

Graduate College, College of Agriculture, Department of Animal Ecology, Ames, IA 50011

AWARDS Animal ecology (MS, PhD); fisheries biology (MS, PhD); wildlife biology (MS, PhD).

Faculty: 14 full-time, 3 part-time/adjunct.
Students: 28 full-time (10 women), 5 part-time (1 woman); includes 2 minority (1 African American, 1 Asian American or Pacific Islander), 1 international. *25 applicants, 16% accepted.* In 1999, 3 master's awarded.
Degree requirements: For master's and doctorate, thesis/dissertation required.
Entrance requirements: For master's and doctorate, GRE General Test, TOEFL. *Application deadline:* For fall admission, 6/1 (priority date); for spring admission, 11/1 (priority date). *Application fee:* $20 ($50 for international students). Electronic applications accepted.
Expenses: Tuition, state resident: full-time $3,308. Tuition, nonresident: full-time $9,744. Part-time tuition and fees vary according to course load, campus/location and program.
Financial aid: In 1999–00, 19 research assistantships with partial tuition reimbursements (averaging $11,752 per year), 7 teaching assistantships with partial tuition reimbursements (averaging $11,571 per year) were awarded; fellowships, scholarships also available.
Faculty research: Animal behavior, ecology, limnology.
Dr. Bruce W. Menzel, Chair, 515-294-6148, *Fax:* 515-294-7874, *E-mail:* aecgradm@iastate.edu.

■ **IOWA STATE UNIVERSITY OF SCIENCE AND TECHNOLOGY**

Graduate College, Interdisciplinary Programs, Program in Ecology and Evolutionary Biology, Ames, IA 50011
AWARDS MS, PhD.

Students: In 1999, 3 master's, 2 doctorates awarded.

Degree requirements: For master's, thesis or alternative required; for doctorate, dissertation required.
Entrance requirements: For master's and doctorate, GRE General Test, TOEFL. *Application deadline:* For fall admission, 2/1. *Application fee:* $20 ($50 for international students). Electronic applications accepted.
Expenses: Tuition, state resident: full-time $3,308. Tuition, nonresident: full-time $9,744. Part-time tuition and fees vary according to course load, campus/location and program.
Financial aid: Scholarships available.
Faculty research: Landscape ecology, aquatic and method ecology, physiological ecology, population genetics and evolution, systematics.
Dr. William R. Clark, Supervisory Committee Chair, 515-294-7252, *E-mail:* eebadm@iastate.edu.

■ **KANSAS STATE UNIVERSITY**

Graduate School, College of Arts and Sciences, Division of Biology, Manhattan, KS 66506

AWARDS Cell biology (MS, PhD); developmental biology and physiology (MS, PhD); microbiology and immunology (MS, PhD); molecular biology and genetics (MS, PhD); systematics and ecology (MS, PhD); virology and oncology (MS, PhD). Terminal master's awarded for partial completion of doctoral program.

Degree requirements: For master's and doctorate, thesis/dissertation required, foreign language not required.
Entrance requirements: For master's and doctorate, GRE General Test. Electronic applications accepted.
Expenses: Tuition, state resident: part-time $103 per credit hour. Tuition, nonresident: part-time $338 per credit hour. Required fees: $17 per credit hour. One-time fee: $64 part-time.
Faculty research: Immune cell function, prairie ecology.

■ **KENT STATE UNIVERSITY**

College of Arts and Sciences, Department of Biological Sciences, Program in Ecology, Kent, OH 44242-0001
AWARDS MS, PhD.

Degree requirements: For master's and doctorate, thesis/dissertation required, foreign language not required.
Entrance requirements: For master's, GRE General Test, minimum GPA of 2.75; for doctorate, GRE General Test, minimum GPA of 3.0. *Application deadline:* For fall admission, 7/12; for spring admission, 11/29. Applications are processed on a rolling basis. *Application fee:* $30.

Expenses: Tuition, state resident: full-time $5,334; part-time $243 per hour. Tuition, nonresident: full-time $10,238; part-time $466 per hour.
Financial aid: Fellowships, research assistantships, teaching assistantships, Federal Work-Study, institutionally sponsored loans, and tuition waivers (full) available. Financial aid application deadline: 2/1.
Application contact: Dr. John R. D. Stalvey, Coordinator of Graduate Studies, 330-672-2819.

Find an in-depth description at www.petersons.com/graduate.

■ **LESLEY UNIVERSITY**

Graduate School of Arts and Social Sciences, Cambridge, MA 02138-2790

AWARDS Clinical mental health counseling (MA), including expressive therapies counseling, holistic counseling, school and community counseling; counseling psychology (MA, CAGS), including school counseling (MA); creative arts in learning (M Ed, CAGS), including individually designed (M Ed, MA), multicultural education (M Ed, MA), storytelling (M Ed), theater studies (M Ed); ecological literacy (MS); environmental education (MS); expressive therapies (MA, CAGS), including art therapy (MA), dance therapy (MA), individually designed (M Ed, MA), mental health counseling (MA), music therapy (MA); independent studies (M Ed); independent study (MA); intercultural relations (MA, CAGS), including development project administration (MA), individually designed (M Ed, MA), intercultural conflict resolution (MA), intercultural health and human services (MA), intercultural relations (CAGS), intercultural training and consulting (MA), international education exchange (MA), international student advising (MA), managing culturally diverse human resources (MA), multicultural education (M Ed, MA); interdisciplinary studies (MA). MS (environmental education) offered jointly with the Audubon Society Expedition Institute. Part-time and evening/weekend programs available. Postbaccalaureate distance learning degree programs offered (minimal on-campus study).

Faculty: 32 full-time (22 women), 389 part-time/adjunct (260 women).
Students: 168 full-time (150 women), 1,755 part-time (1,607 women); includes 145 minority (76 African Americans, 27 Asian Americans or Pacific Islanders, 36 Hispanic Americans, 6 Native Americans), 150 international. Average age 35. *634 applicants, 75% accepted.* In 1999, 855 degrees awarded.
Degree requirements: For master's, internship, practicum, thesis (expressive therapies) required; for CAGS, thesis,

internship (counseling psychology, expressive therapies) required.

Entrance requirements: For master's, MAT (counseling psychology), TOEFL, interview; for CAGS, interview, master's degree. *Application deadline:* Applications are processed on a rolling basis. *Application fee:* $45.

Expenses: Tuition: Full-time $10,200; part-time $425 per credit. One-time fee: $15.

Financial aid: Research assistantships, teaching assistantships, career-related internships or fieldwork, Federal Work-Study, and unspecified assistantships available. Aid available to part-time students. Financial aid application deadline: 4/1; financial aid applicants required to submit FAFSA.

Faculty research: Developmental psychology, women's issues, health psychology, limited supervision group psychology. Dr. Martha B. McKenna, Dean, 617-349-8467, *Fax:* 617-349-8366.

Application contact: Maxine Lentz, Dean of Admissions and Enrollment Planning, 800-999-1959, *Fax:* 617-349-8366.

■ MARQUETTE UNIVERSITY

Graduate School, College of Arts and Sciences, Department of Biology, Milwaukee, WI 53201-1881

AWARDS Cell biology (MS, PhD); developmental biology (MS, PhD); ecology (MS, PhD); endocrinology (MS, PhD); evolutionary biology (MS, PhD); genetics (MS, PhD); microbiology (MS, PhD); molecular biology (MS, PhD); muscle and exercise physiology (MS, PhD); neurobiology (MS, PhD); reproductive physiology (MS, PhD).

Faculty: 16 full-time (4 women), 2 part-time/adjunct (0 women).

Students: 34 full-time (20 women), 3 part-time; includes 3 minority (all Asian Americans or Pacific Islanders), 2 international. Average age 31. *42 applicants, 29% accepted.* In 1999, 1 master's, 4 doctorates awarded. Terminal master's awarded for partial completion of doctoral program.

Degree requirements: For master's, thesis, 1 year of teaching experience or equivalent, comprehensive exam required, foreign language not required; for doctorate, dissertation, 1 year of teaching experience or equivalent, qualifying exam required, foreign language not required.

Entrance requirements: For master's and doctorate, GRE General Test, GRE Subject Test, TOEFL. *Application fee:* $40.

Expenses: Tuition: Part-time $510 per credit hour. Tuition and fees vary according to program.

Financial aid: In 1999–00, 4 fellowships, 22 teaching assistantships were awarded; research assistantships, Federal Work-Study, institutionally sponsored loans, scholarships, and tuition waivers (full and partial) also available. Aid available to part-time students. Financial aid application deadline: 2/15.

Faculty research: Microbial and invertebrate ecology, evolution of gene function, DNA methylation, DNA arrangement. *Total annual research expenditures:* $1.5 million. Dr. Brian Unsworth, Chairman, 414-288-7355, *Fax:* 414-288-7357.

Application contact: Barbara DeNoyer, Graduate Studies Coordinator, 414-288-7355, *Fax:* 414-288-7357.

Find an in-depth description at www.petersons.com/graduate.

■ MINNESOTA STATE UNIVERSITY, MANKATO

College of Graduate Studies, College of Science, Engineering and Technology, Department of Biological Sciences, Program in Environmental Science, Mankato, MN 56001

AWARDS Ecology (MS); economic and political systems (MS); human ecosystems (MS); physical science (MS); technology (MS).

Faculty: 1 (woman) full-time.

Students: 9 full-time (5 women), 1 part-time; includes 1 minority (Asian American or Pacific Islander). Average age 31. In 1999, 5 degrees awarded.

Degree requirements: For master's, one foreign language, thesis or alternative, comprehensive exam required.

Entrance requirements: For master's, minimum GPA of 3.0 during previous 2 years. *Application deadline:* For fall admission, 7/9 (priority date); for spring admission, 11/27. Applications are processed on a rolling basis. *Application fee:* $3.

Expenses: Tuition, state resident: part-time $152 per credit hour. Tuition, nonresident: part-time $228 per credit hour.

Financial aid: Research assistantships with partial tuition reimbursements, teaching assistantships with partial tuition reimbursements, career-related internships or fieldwork, Federal Work-Study, and institutionally sponsored loans available. Financial aid application deadline: 3/15; financial aid applicants required to submit FAFSA. Dr. Beth Proctor, Graduate Coordinator, 507-389-5697.

Application contact: Joni Roberts, Admissions Coordinator, 507-389-2321, *Fax:* 507-389-5974, *E-mail:* grad@mankato.msus.edu.

■ NORTH CAROLINA STATE UNIVERSITY

Graduate School, College of Physical and Mathematical Sciences, Program in Biomathematics, Raleigh, NC 27695

AWARDS Biomathematics (M Biomath, MS, PhD); ecology (PhD). Part-time programs available.

Faculty: 12 full-time (2 women), 3 part-time/adjunct (0 women).

Students: 18 full-time (9 women), 10 part-time (3 women); includes 1 minority (Hispanic American), 6 international. Average age 30. *13 applicants, 62% accepted.* In 1999, 4 master's, 2 doctorates awarded. Terminal master's awarded for partial completion of doctoral program.

Degree requirements: For master's, thesis or alternative required; for doctorate, dissertation required.

Entrance requirements: For master's and doctorate, GRE General Test, TOEFL. *Application deadline:* For fall admission, 3/1 (priority date); for spring admission, 9/15. *Application fee:* $45.

Expenses: Tuition, state resident: full-time $1,578. Tuition, nonresident: full-time $10,744. Required fees: $892. Full-time tuition and fees vary according to program.

Financial aid: Fellowships, research assistantships, teaching assistantships, career-related internships or fieldwork available. Financial aid application deadline: 3/1.

Faculty research: Theory and methods of biological modeling, theoretical biology (genetics), applied biology (wildlife). Dr. Thomas B. Kepler, Director of Graduate Programs, 919-515-1911, *Fax:* 919-515-1909, *E-mail:* kepler@stat.ncsu.edu.

■ THE OHIO STATE UNIVERSITY

Graduate School, College of Biological Sciences, Department of Evolution, Ecology, and Organismal Biology, Columbus, OH 43210

AWARDS MS, PhD.

Faculty: 41 full-time, 12 part-time/adjunct.

Students: 72 full-time (37 women), 7 part-time (4 women); includes 5 minority (2 Asian Americans or Pacific Islanders, 2 Hispanic Americans, 1 Native American), 8 international. *50 applicants, 46% accepted.* In 1999, 7 master's, 5 doctorates awarded.

Degree requirements: For master's, thesis optional, foreign language not required; for doctorate, dissertation required, foreign language not required.

Entrance requirements: For master's and doctorate, GRE General Test, GRE Subject Test (biology), TOEFL, TSE. *Application deadline:* For fall admission,

The Ohio State University (continued)
8/15. Applications are processed on a rolling basis. *Application fee:* $30 ($40 for international students).
Expenses: Tuition, state resident: full-time $5,400. Tuition, nonresident: full-time $14,535. Part-time tuition and fees vary according to course load and program.
Financial aid: Fellowships, research assistantships, teaching assistantships, Federal Work-Study and institutionally sponsored loans available. Aid available to part-time students.
Dr. Ralph E. J. Boerner, Chairperson, 614-292-8088, *Fax:* 614-292-2030, *E-mail:* boerner.1@osu.edu.

■ OKLAHOMA STATE UNIVERSITY

Graduate College, College of Arts and Sciences, Department of Zoology, Program in Wildlife and Fisheries Ecology, Stillwater, OK 74078

AWARDS MS, PhD.

Degree requirements: For master's, thesis required, foreign language not required; for doctorate, dissertation required.
Entrance requirements: For master's and doctorate, GRE General Test, GRE Subject Test, TOEFL. *Application deadline:* For fall admission, 7/1 (priority date). *Application fee:* $25.
Expenses: Tuition, state resident: part-time $86 per credit hour. Tuition, nonresident: part-time $275 per credit hour. Required fees: $17 per credit hour. $14 per semester. One-time fee: $20 full-time. Tuition and fees vary according to course load.
Financial aid: Career-related internships or fieldwork, Federal Work-Study, and tuition waivers (partial) available. Aid available to part-time students. Financial aid application deadline: 3/1.
Dr. Jim Shaw, Head, Department of Zoology, 405-744-5555.

■ OLD DOMINION UNIVERSITY

College of Sciences, Department of Biological Sciences, Program in Ecological Sciences, Norfolk, VA 23529

AWARDS PhD.

Faculty: 27 full-time (5 women).
Students: 17 full-time (5 women), 9 part-time (2 women); includes 3 minority (1 African American, 1 Hispanic American, 1 Native American), 5 international. Average age 35. *6 applicants, 100% accepted.* In 1999, 3 degrees awarded.
Degree requirements: For doctorate, one foreign language (computer language can

substitute), dissertation, comprehensive exam, internships required.
Entrance requirements: For doctorate, GRE General Test, TOEFL, master's degree or GRE Subject Test (minimum score of 600 required), minimum GPA of 3.0. *Application deadline:* For fall admission, 2/15 (priority date). Applications are processed on a rolling basis. *Application fee:* $30.
Expenses: Tuition, state resident: full-time $4,440; part-time $185 per credit. Tuition, nonresident: full-time $11,784; part-time $477 per credit. Required fees: $1,612. Tuition and fees vary according to program.
Financial aid: In 1999–00, 25 students received aid, including 2 fellowships (averaging $6,575 per year), 14 research assistantships with tuition reimbursements available (averaging $9,527 per year), 1 teaching assistantship with tuition reimbursement available (averaging $10,000 per year); career-related internships or fieldwork, grants, and tuition waivers (partial) also available. Aid available to part-time students. Financial aid application deadline: 2/15; financial aid applicants required to submit FAFSA.
Faculty research: Spiny lobsters in Florida, pollution in marine environment, wetlands and barrier islands. *Total annual research expenditures:* $1.8 million.
Dr. Frank P. Day, Director, 757-683-3595, *Fax:* 757-683-5283, *E-mail:* ecolgpd@odu.edu.

■ THE PENNSYLVANIA STATE UNIVERSITY UNIVERSITY PARK CAMPUS

Graduate School, Intercollege Graduate Programs, Intercollege Graduate Program in Ecology, State College, University Park, PA 16802-1503

AWARDS MS, PhD.

Students: 21 full-time (11 women), 11 part-time (4 women). In 1999, 11 master's, 2 doctorates awarded.
Entrance requirements: For master's and doctorate, GRE General Test, GRE Subject Test (biology). *Application fee:* $50.
Expenses: Tuition, state resident: full-time $6,886; part-time $291 per credit. Tuition, nonresident: full-time $14,118; part-time $588 per credit. Required fees: $46 per semester. Part-time tuition and fees vary according to course load and program.
Christopher Uhl, Chair, 814-863-3201.

■ PRINCETON UNIVERSITY

Graduate School, Department of Ecology and Evolutionary Biology, Princeton, NJ 08544-1019

AWARDS Biology (PhD); neuroscience (PhD).

Degree requirements: For doctorate, dissertation required, foreign language not required.
Entrance requirements: For doctorate, GRE General Test, GRE Subject Test.
Expenses: Tuition: Full-time $25,050.
Find an in-depth description at www.petersons.com/graduate.

■ PURDUE UNIVERSITY

Graduate School, School of Science, Department of Biological Sciences, Program in Ecology, Evolutionary and Population Biology, West Lafayette, IN 47907

AWARDS Ecology (MS, PhD); evolutionary biology (MS, PhD); population biology (MS, PhD).

Faculty: 9 full-time (1 woman).
Degree requirements: For master's, thesis required (for some programs), foreign language not required; for doctorate, dissertation, seminars, teaching experience required, foreign language not required.
Entrance requirements: For master's and doctorate, GRE General Test, TOEFL, TSE. *Application deadline:* For fall admission, 2/15. *Application fee:* $30. Electronic applications accepted.
Expenses: Tuition, state resident: full-time $4,530; part-time $130 per credit hour. Tuition, nonresident: full-time $15,310; part-time $404 per credit hour. Tuition and fees vary according to campus/location and program.
Financial aid: Fellowships, research assistantships, teaching assistantships available. Aid available to part-time students. Financial aid application deadline: 2/15; financial aid applicants required to submit FAFSA.
Faculty research: Host-parasite coevolution, social behavior, foraging ecology, conservation biology and patterns of evolution in speciation.
Application contact: Nancy Konopka, Graduate Studies Office Manager, 765-494-8142, *Fax:* 765-494-0876, *E-mail:* njk@bilbo.bio.purdue.edu.

■ RICE UNIVERSITY

Graduate Programs, Wiess School of Natural Sciences, Department of Ecology and Evolutionary Biology, Houston, TX 77251-1892

AWARDS MA, PhD.

Degree requirements: For master's and doctorate, thesis/dissertation required.
Entrance requirements: For master's and doctorate, GRE General Test, GRE Subject Test, TOEFL, minimum GPA of 3.0.

Expenses: Tuition: Full-time $16,700. Required fees: $250. Tuition and fees vary according to program.

■ RUTGERS, THE STATE UNIVERSITY OF NEW JERSEY, NEW BRUNSWICK

Graduate School, Program in Ecology and Evolution, New Brunswick, NJ 08901-1281

AWARDS MS, PhD. Part-time programs available.

Faculty: 65 full-time (11 women).
Students: 53 full-time (26 women), 41 part-time (24 women); includes 7 minority (1 African American, 2 Asian Americans or Pacific Islanders, 3 Hispanic Americans, 1 Native American), 13 international. Average age 27. *86 applicants, 29% accepted.* In 1999, 8 master's awarded (75% found work related to degree, 25% continued full-time study); 6 doctorates awarded.
Degree requirements: For master's, foreign language and thesis not required; for doctorate, dissertation required, foreign language not required. *Average time to degree:* Master's–4 years full-time; doctorate–6 years full-time, 8 years part-time.
Entrance requirements: For master's and doctorate, GRE General Test, GRE Subject Test, TOEFL, minimum GPA of 3.0. *Application deadline:* For fall admission, 3/1 (priority date); for spring admission, 12/1. Applications are processed on a rolling basis. *Application fee:* $50.
Expenses: Tuition, state resident: full-time $6,776; part-time $279 per credit. Tuition, nonresident: full-time $9,936; part-time $412 per credit. Required fees: $20 per credit. $89 per semester. Tuition and fees vary according to course load, campus/location and program.
Financial aid: In 1999–00, 5 fellowships with full tuition reimbursements (averaging $12,000 per year), 17 research assistantships with full tuition reimbursements (averaging $13,100 per year), 23 teaching assistantships with full tuition reimbursements (averaging $13,100 per year) were awarded. Federal Work-Study also available. Financial aid application deadline: 3/1; financial aid applicants required to submit FAFSA.
Faculty research: Population and community ecology, population genetics, evolutionary biology, conservation biology, ecosystem ecology.
Dr. Michael Sukhdeo, Director, Graduate Program, 732-932-2971, *Fax:* 732-932-2972.
Application contact: Nancy Tiedge, Secretary, 732-932-2971, *Fax:* 732-932-2972, *E-mail:* tiedge@rci.rutgers.edu.

■ SAN DIEGO STATE UNIVERSITY

Graduate and Research Affairs, College of Sciences, Department of Biological Sciences, Program in Ecology, San Diego, CA 92182

AWARDS PhD.

Students: 7 full-time (6 women), 25 part-time (17 women); includes 1 minority (Asian American or Pacific Islander). Average age 30. In 1999, 2 degrees awarded (50% entered university research/teaching, 50% found other work related to degree).
Degree requirements: For doctorate, dissertation required, foreign language not required. *Average time to degree:* Doctorate–6 years full-time.
Entrance requirements: For doctorate, GRE General Test, GRE Subject Test. *Application deadline:* For fall admission, 7/1 (priority date); for spring admission, 12/1. Applications are processed on a rolling basis. *Application fee:* $55.
Expenses: Tuition, nonresident: part-time $246 per unit. Required fees: $1,932; $633 per semester. Tuition and fees vary according to course load.
Financial aid: Research assistantships, teaching assistantships, career-related internships or fieldwork available.
Faculty research: Conservation and restoration ecology, coastal and marine ecology, global change and ecosystem ecology. *Total annual research expenditures:* $4 million.
Mike Allen, Coordinator, 619-594-4460, *E-mail:* michael.allen@sdsu.edu.

■ SAN FRANCISCO STATE UNIVERSITY

Graduate Division, College of Science and Engineering, Department of Biology, Program in Ecology and Systematic Biology, San Francisco, CA 94132-1722

AWARDS MA.

Entrance requirements: For master's, minimum GPA of 2.5 in last 60 units.
Expenses: Tuition, nonresident: full-time $5,904; part-time $246 per unit. Required fees: $1,904; $637 per semester. Tuition and fees vary according to course load.

■ STATE UNIVERSITY OF NEW YORK AT ALBANY

College of Arts and Sciences, Department of Biological Sciences, Specialization in Ecology, Evolution, and Behavior, Albany, NY 12222-0001

AWARDS MS, PhD.

Degree requirements: For master's, one foreign language required; for doctorate, dissertation required.

Entrance requirements: For master's and doctorate, GRE General Test. *Application fee:* $50.
Expenses: Tuition, state resident: full-time $5,100; part-time $214 per credit. Tuition, nonresident: full-time $8,416; part-time $352 per credit. Required fees: $31 per credit.
Financial aid: Minority assistantships available.
Dr. David Shub, Chair, Department of Biological Sciences, 518-442-4300.

■ STATE UNIVERSITY OF NEW YORK AT STONY BROOK

Graduate School, College of Arts and Sciences, Department of Ecology and Evolution, Stony Brook, NY 11794

AWARDS Ecology and evolution (PhD).

Faculty: 18 full-time (3 women), 2 part-time/adjunct (0 women).
Students: 28 full-time (15 women), 18 part-time (8 women); includes 2 minority (1 African American, 1 Hispanic American), 12 international. *71 applicants, 28% accepted.* In 1999, 2 doctorates awarded.
Degree requirements: For doctorate, one foreign language (computer language can substitute), dissertation, comprehensive exam, teaching experience required.
Entrance requirements: For doctorate, GRE General Test, GRE Subject Test, TOEFL. *Application deadline:* For fall admission, 1/15. *Application fee:* $50.
Expenses: Tuition, state resident: full-time $5,100; part-time $213 per credit hour. Tuition, nonresident: full-time $8,416; part-time $351 per credit hour. Required fees: $492. Tuition and fees vary according to program.
Financial aid: In 1999–00, 12 fellowships, 8 research assistantships, 22 teaching assistantships were awarded; Federal Work-Study also available.
Faculty research: Theoretical and experimental population genetics, numerical taxonomy, biostatistics, population and community ecology, plant ecology. *Total annual research expenditures:* $1.2 million.
Dr. Walter Danes, Chairman, 631-632-8600.
Application contact: Dr. Dan Dykhuizen, Director, 631-246-8604, *E-mail:* dandyk@life.bio.sunysb.edu.

■ THE UNIVERSITY OF ALABAMA AT BIRMINGHAM

Graduate School, School of Natural Sciences and Mathematics, Department of Biology, Birmingham, AL 35294

AWARDS Comparative and cellular biology (PhD); comparative and cellular physiology

The University of Alabama at Birmingham (continued)
(MS); marine science (MS, PhD); microbial ecology and physiology (MS, PhD); reproduction and development (MS, PhD).

Students: 34 full-time (19 women), 6 international. *61 applicants, 38% accepted.* In 1999, 1 master's, 3 doctorates awarded. Terminal master's awarded for partial completion of doctoral program.
Degree requirements: For master's and doctorate, thesis/dissertation required.
Entrance requirements: For master's and doctorate, GRE General Test, TOEFL, previous course work in biology, calculus, organic chemistry, physics. *Application deadline:* Applications are processed on a rolling basis. *Application fee:* $35 ($60 for international students). Electronic applications accepted.
Expenses: Tuition, state resident: part-time $104 per semester hour. Tuition, nonresident: part-time $208 per semester hour. Required fees: $17 per semester hour. $57 per quarter. Tuition and fees vary according to program.
Financial aid: In 1999–00, 22 students received aid, including 3 fellowships with full tuition reimbursements available (averaging $14,000 per year), 19 teaching assistantships with full tuition reimbursements available (averaging $14,000 per year); research assistantships, career-related internships or fieldwork, Federal Work-Study, institutionally sponsored loans, and tuition waivers (full) also available. Aid available to part-time students.
Faculty research: Invertebrate physiology, marine biology, environmental biology. Dr. Daniel D. Jones, Chairman, 205-934-4290, *Fax:* 205-975-6097, *E-mail:* ddjones@uab.edu.
Find an in-depth description at www.petersons.com/graduate.

■ THE UNIVERSITY OF ARIZONA

Graduate College, College of Science, Department of Ecology and Evolutionary Biology, Tucson, AZ 85721

AWARDS Botany (MS, PhD); ecology and evolutionary biology (MS, PhD).

Degree requirements: For master's, foreign language and thesis not required; for doctorate, one foreign language, computer language, dissertation required.
Entrance requirements: For master's and doctorate, GRE General Test, GRE Subject Test, TOEFL.
Expenses: Tuition, nonresident: full-time $4,814; part-time $274 per unit. Required fees: $1,094; $115 per unit. Tuition and fees vary according to course load and program.

Faculty research: Biological diversity, evolutionary history, evolutionary mechanisms, community structure.

■ UNIVERSITY OF CALIFORNIA, DAVIS

Graduate Studies, Programs in the Biological Sciences, Program in Ecology, Davis, CA 95616

AWARDS MS, PhD. Part-time programs available.

Faculty: 104.
Students: 173 full-time (91 women), 1 (woman) part-time; includes 16 minority (9 Asian Americans or Pacific Islanders, 5 Hispanic Americans, 2 Native Americans), 12 international. Average age 31. *235 applicants, 23% accepted.* In 1999, 10 master's, 27 doctorates awarded.
Degree requirements: For master's, thesis optional; for doctorate, dissertation required.
Entrance requirements: For master's and doctorate, GRE General Test. *Application deadline:* For fall admission, 1/15. *Application fee:* $40. Electronic applications accepted.
Expenses: Tuition, nonresident: full-time $9,804. Tuition and fees vary according to program and student level.
Financial aid: In 1999–00, 150 students received aid, including 51 fellowships with full and partial tuition reimbursements available, 51 research assistantships with full and partial tuition reimbursements available, 29 teaching assistantships with partial tuition reimbursements available; Federal Work-Study, grants, institutionally sponsored loans, scholarships, and tuition waivers (full and partial) also available. Financial aid application deadline: 1/15; financial aid applicants required to submit FAFSA.
Faculty research: Agricultural conservation, physiological restoration, environmental policy, ecotoxicology. Kevin Rice, Graduate Chair, 530-752-8529, *Fax:* 530-752-3350, *E-mail:* kjrice@ucdavis.edu.
Application contact: Silvia Hillyer, Graduate Coordinator, 530-752-6752, *Fax:* 530-752-3350, *E-mail:* schillyer@ucdavis.edu.

■ UNIVERSITY OF CALIFORNIA, IRVINE

Office of Research and Graduate Studies, School of Biological Sciences, Department of Ecology and Evolutionary Biology, Irvine, CA 92697

AWARDS Biological sciences (MS, PhD).

Faculty: 21 full-time (4 women).

Students: 32 full-time (20 women); includes 1 minority (African American), 4 international. *45 applicants, 31% accepted.*
Degree requirements: For master's, one foreign language (computer language can substitute), thesis required; for doctorate, dissertation required.
Entrance requirements: For master's, GRE General Test, GRE Subject Test, minimum GPA of 3.0; for doctorate, GRE General Test, GRE Subject Test. *Application deadline:* For fall admission, 1/15 (priority date). Applications are processed on a rolling basis. *Application fee:* $40. Electronic applications accepted.
Expenses: Tuition, nonresident: full-time $10,244; part-time $1,720 per quarter. Required fees: $5,252; $1,300 per quarter. Tuition and fees vary according to course load and program.
Financial aid: Fellowships, research assistantships, teaching assistantships, career-related internships or fieldwork, institutionally sponsored loans, and tuition waivers (full and partial) available. Financial aid application deadline: 3/2; financial aid applicants required to submit FAFSA.
Faculty research: Ecological energetics, quantitative genetics, life history evolution, plant-herbivore and plant-pollinator interactions, molecular evolution. Albert F. Bennett, Chair, 949-824-6930.
Application contact: Pam McDonald, Administrative Assistant, 949-824-4743.

■ UNIVERSITY OF CALIFORNIA, SAN DIEGO

Graduate Studies and Research, Department of Biology, Program in Ecology, Behavior, and Evolution, La Jolla, CA 92093

AWARDS PhD.

Degree requirements: For doctorate, dissertation required.
Entrance requirements: For doctorate, GRE General Test, pre-application beginning in September. *Application deadline:* For fall admission, 1/8. *Application fee:* $40.
Expenses: Tuition, nonresident: full-time $14,691. Required fees: $4,697. Full-time tuition and fees vary according to program.
Application contact: Biology Graduate Admissions Committee, 858-534-3835.

Find an in-depth description at www.petersons.com/graduate.

■ UNIVERSITY OF CALIFORNIA, SANTA BARBARA

Graduate Division, College of Letters and Sciences, Division of Mathematics, Life, and Physical Sciences, Department of Ecology, Evolution, and Marine Biology, Santa Barbara, CA 93106

AWARDS MA, PhD.

Faculty: 28 full-time (4 women), 16 part-time/adjunct (5 women).
Students: 85 full-time (38 women); includes 10 minority (2 Asian Americans or Pacific Islanders, 7 Hispanic Americans, 1 Native American), 6 international. *122 applicants, 16% accepted.* In 1999, 14 master's, 12 doctorates awarded. Terminal master's awarded for partial completion of doctoral program.
Degree requirements: For master's, thesis or alternative required, foreign language not required; for doctorate, dissertation required, foreign language not required. *Average time to degree:* Master's–2 years full-time; doctorate–6 years full-time.
Entrance requirements: For master's and doctorate, GRE General Test, TOEFL. *Application deadline:* For fall admission, 12/15. *Application fee:* $40. Electronic applications accepted.
Expenses: Tuition, state resident: full-time $14,637. Tuition, nonresident: full-time $24,441.
Financial aid: Fellowships, research assistantships, teaching assistantships, career-related internships or fieldwork, Federal Work-Study, institutionally sponsored loans, and tuition waivers (full and partial) available. Financial aid application deadline: 12/15; financial aid applicants required to submit FAFSA.
Scott Cooper, Chair, 805-893-2979, *Fax:* 805-893-4724, *E-mail:* scooper@ lifesci.ucsb.edu.
Application contact: Stephanie Slosser, Graduate Program Assistant, 805-893-3023, *Fax:* 805-893-4724, *E-mail:* slosser@ lifesci.ucsb.edu.

■ UNIVERSITY OF CHICAGO

Division of the Biological Sciences, Darwinian Sciences: Ecological, Integrative and Evolutionary Biology, Department of Ecology and Evolution, Chicago, IL 60637-1513

AWARDS PhD.

Faculty: 14 full-time (2 women).
Students: 24 full-time (8 women); includes 4 minority (2 Asian Americans or Pacific Islanders, 2 Hispanic Americans), 6 international. *41 applicants, 24% accepted.* In 1999, 3 doctorates awarded (67% entered university research/teaching, 33% found other work related to degree).

Degree requirements: For doctorate, dissertation required, foreign language not required. *Average time to degree:* Doctorate–5.5 years full-time.
Entrance requirements: For doctorate, GRE General Test, TOEFL. *Application deadline:* For fall admission, 1/5 (priority date). *Application fee:* $55.
Expenses: Tuition: Full-time $24,804; part-time $3,422 per course. Required fees: $390. Tuition and fees vary according to program.
Financial aid: In 1999–00, 22 students received aid, including fellowships with tuition reimbursements available (averaging $17,050 per year). Financial aid application deadline: 6/1.
Faculty research: Population genetics, molecular evolution, behavior.
Dr. Chung-I Wu, Chairman, 773-702-2565, *Fax:* 773-702-9740.
Application contact: Carolyn Johnson, Graduate Administrative Director, 773-702-9474, *Fax:* 773-702-4699, *E-mail:* cs-johnson@uchicago.edu.

■ UNIVERSITY OF COLORADO AT BOULDER

Graduate School, College of Arts and Sciences, Department of Environmental, Population, and Organic Biology, Boulder, CO 80309

AWARDS Animal behavior (MA, PhD); aquatic biology (MA, PhD); behavioral genetics (MA, PhD); ecology (MA, PhD); microbiology (MA, PhD); neurobiology (MA, PhD); plant and animal physiology (MA, PhD); plant and animal systematics (MA, PhD); population biology (MA, PhD); population genetics (MA, PhD).

Faculty: 39 full-time (9 women).
Students: 84 full-time (36 women), 14 part-time (7 women); includes 17 minority (6 Asian Americans or Pacific Islanders, 10 Hispanic Americans, 1 Native American). Average age 29. *147 applicants, 14% accepted.* In 1999, 7 master's, 13 doctorates awarded. Terminal master's awarded for partial completion of doctoral program.
Degree requirements: For master's, thesis or alternative, comprehensive exam required, foreign language not required; for doctorate, dissertation, comprehensive exam required, foreign language not required. *Average time to degree:* Master's–3 years full-time; doctorate–5 years full-time.
Entrance requirements: For master's, GRE General Test, GRE Subject Test, minimum undergraduate GPA of 3.0; for doctorate, GRE General Test, GRE Subject Test. *Application deadline:* For fall admission, 1/15 (priority date). *Application fee:* $40 ($60 for international students).
Expenses: Tuition, state resident: part-time $181 per credit hour. Tuition,

nonresident: part-time $542 per credit hour. Required fees: $99 per term. Tuition and fees vary according to course load and program.
Financial aid: Fellowships, research assistantships, teaching assistantships, Federal Work-Study, institutionally sponsored loans, and tuition waivers (full) available. Financial aid application deadline: 3/1.
Faculty research: Evolution, developmental biology, behavior and neurobiology. *Total annual research expenditures:* $1.8 million.
Michael Breed, Chair, 303-492-8981, *Fax:* 303-492-8699, *E-mail:* michael.breed@ colorado.edu.
Application contact: Jill Skarstadt, Graduate Secretary, 303-492-7654, *Fax:* 303-492-8699, *E-mail:* jill.skarstadt@ colorado.edu.

■ UNIVERSITY OF CONNECTICUT

Graduate School, College of Liberal Arts and Sciences, Biological Sciences Group, Storrs, CT 06269

AWARDS Ecology and evolutionary biology (MS, PhD), including botany, ecology, entomology, systematics, zoology; molecular and cell biology (MS, PhD), including biochemistry, biophysics, biotechnology (MS), cell and developmental biology, genetics, microbiology, plant molecular and cell biology; physiology and neurobiology (MS, PhD), including neurobiology, physiology.

Degree requirements: For doctorate, dissertation required.
Entrance requirements: For master's and doctorate, GRE General Test, GRE Subject Test, TOEFL.
Expenses: Tuition, state resident: full-time $5,118. Tuition, nonresident: full-time $13,298. Required fees: $1,022.

■ UNIVERSITY OF CONNECTICUT

Graduate School, College of Liberal Arts and Sciences, Biological Sciences Group, Department of Ecology and Evolutionary Biology, Field of Ecology, Storrs, CT 06269

AWARDS MS, PhD.

Degree requirements: For doctorate, dissertation required.
Entrance requirements: For master's and doctorate, GRE General Test, GRE Subject Test, TOEFL.
Expenses: Tuition, state resident: full-time $5,118. Tuition, nonresident: full-time $13,298. Required fees: $1,022.

Find an in-depth description at www.petersons.com/graduate.

■ UNIVERSITY OF DELAWARE

College of Agriculture and Natural Resources, Department of Entomology and Applied Ecology, Newark, DE 19716

AWARDS Biology (PhD); entomology and applied ecology (MS, PhD), including avian ecology (MS), evolution and taxonomy (MS), insect biological control (MS), insect ecology and behavior (MS), insect genetics (MS), pest management (MS), plant-insect interactions (MS); plant science (PhD); turtle ecology (PhD). PhD offered in cooperation with the Department of Biological Sciences or the Department of Plant and Soil Sciences. Part-time programs available.

Faculty: 6 full-time (0 women).
Students: 26 full-time (11 women), 1 part-time; includes 1 minority (Asian American or Pacific Islander), 2 international. Average age 24. *10 applicants, 80% accepted.* In 1999, 5 degrees awarded.
Degree requirements: For master's, thesis, oral exam, seminar required, foreign language not required; for doctorate, dissertation, qualifying exam required, foreign language not required. *Average time to degree:* Master's–3.4 years full-time.
Entrance requirements: For master's and doctorate, GRE General Test, TOEFL, minimum GPA of 3.0 in field, 2.8 overall. *Application deadline:* For fall admission, 3/1 (priority date); for spring admission, 11/1. Applications are processed on a rolling basis. *Application fee:* $45. Electronic applications accepted.
Expenses: Tuition, state resident: full-time $4,380; part-time $243 per credit. Tuition, nonresident: full-time $12,750; part-time $708 per credit. Required fees: $15 per term. Tuition and fees vary according to program.
Financial aid: In 1999–00, 18 students received aid, including 1 fellowship with full tuition reimbursement available (averaging $15,000 per year), 9 research assistantships with full tuition reimbursements available (averaging $12,400 per year), 9 teaching assistantships with full tuition reimbursements available (averaging $12,000 per year); career-related internships or fieldwork, institutionally sponsored loans, scholarships, and tuition waivers (full) also available. Financial aid application deadline: 3/1.
Faculty research: Genetics and resistance, biological control, chemically mediated behavioral ecology, ecology and evolution of plant-insect interactions, ecology of species conservation. *Total annual research expenditures:* $885,313.
Dr. Judith A. Hough-Goldstein, Chairperson, 302-831-2526, *Fax:* 302-831-3651, *E-mail:* jhough@udel.edu.

Application contact: Dr. Charles E. Mason, Graduate Coordinator, 302-831-8888, *Fax:* 302-831-3651, *E-mail:* mason@udel.edu.

■ UNIVERSITY OF DELAWARE

College of Arts and Science, Department of Biological Sciences, Newark, DE 19716

AWARDS Biotechnology (MS, PhD); cell and extracellular matrix biology (MS, PhD); cell and systems physiology (MS, PhD); ecology and evolution (MS, PhD); microbiology (MS, PhD); molecular biology and genetics (MS, PhD); plant biology (MS, PhD).

Faculty: 37 full-time (10 women).
Students: 22 full-time (11 women), 1 part-time; includes 2 minority (both African Americans), 9 international. Average age 25. *37 applicants, 27% accepted.* In 2000, 9 doctorates awarded.
Degree requirements: For master's and doctorate, thesis/dissertation required, foreign language not required. *Average time to degree:* Master's–2.5 years full-time; doctorate–6 years full-time.
Entrance requirements: For master's and doctorate, GRE General Test, GRE Subject Test (advanced biology). *Application deadline:* For fall admission, 6/15. Applications are processed on a rolling basis. *Application fee:* $50. Electronic applications accepted.
Expenses: Tuition, state resident: full-time $4,380; part-time $243 per credit. Tuition, nonresident: full-time $12,750; part-time $708 per credit. Required fees: $15 per term. Tuition and fees vary according to program.
Financial aid: In 2000–01, 18 students received aid, including 2 fellowships with full tuition reimbursements available (averaging $18,000 per year), 4 research assistantships with full tuition reimbursements available (averaging $18,000 per year), 11 teaching assistantships with full tuition reimbursements available (averaging $18,000 per year); tuition waivers (partial) also available. Financial aid application deadline: 6/15.
Faculty research: Cell interactions, molecular mechanisms, microorganisms, embryo implantation. *Total annual research expenditures:* $1.8 million.
Dr. Daniel D. Carson, Chair, 302-831-6977, *Fax:* 302-831-2281, *E-mail:* dcarson@udel.edu.

Application contact: Norman Karin, Graduate Coordinator, 302-831-1841, *Fax:* 302-831-2281, *E-mail:* ccoletta@udel.edu.

Find an in-depth description at www.petersons.com/graduate.

■ UNIVERSITY OF FLORIDA

Graduate School, College of Agriculture and Life Sciences, Department of Wildlife Ecology, Gainesville, FL 32611

AWARDS MS, PhD.

Faculty: 34.
Students: 54 full-time (26 women), 32 part-time (13 women); includes 8 minority (2 African Americans, 6 Hispanic Americans), 16 international. *65 applicants, 29% accepted.* In 1999, 14 master's, 5 doctorates awarded.
Degree requirements: For master's, thesis optional; for doctorate, dissertation required.
Entrance requirements: For master's and doctorate, GRE General Test, minimum GPA of 3.3. *Application deadline:* For fall admission, 6/1 (priority date); for spring admission, 12/1. Applications are processed on a rolling basis. *Application fee:* $20. Electronic applications accepted.
Expenses: Tuition, state resident: part-time $144 per credit hour. Tuition, nonresident: part-time $505 per credit hour. Tuition and fees vary according to course level, course load and program.
Financial aid: In 1999–00, 46 students received aid, including 6 fellowships, 23 research assistantships, 17 teaching assistantships; institutionally sponsored loans also available.
Faculty research: Wildlife biology and management, tropical ecology and conservation, conservation biology, landscape ecology and restoration, conservation education.
Dr. Nat Frazer, Chair, 352-846-0552, *Fax:* 352-392-6984, *E-mail:* gwt@gnv.ifas.ufl.edu.

Application contact: Dr. George Tanner, Graduate Coordinator, 352-896-0552, *Fax:* 352-392-6984, *E-mail:* gwt@gnv.ifas.ufl.edu.

■ UNIVERSITY OF FLORIDA

Graduate School, College of Natural Resources and Environment, Program in Interdisciplinary Ecology, Gainesville, FL 32611

AWARDS MS, PhD.

Faculty: 280 full-time.
Students: 19 full-time (10 women), 1 (woman) part-time; includes 3 minority (2 African Americans, 1 Hispanic American), 9 international. *42 applicants, 57% accepted.*
Degree requirements: For master's, thesis optional, foreign language not required; for doctorate, dissertation required, foreign language not required.
Entrance requirements: For master's and doctorate, GRE General Test, TOEFL, minimum GPA of 3.0. *Application deadline:*

For fall admission, 1/15. Applications are processed on a rolling basis. *Application fee:* $20.

Expenses: Tuition, state resident: part-time $144 per credit hour. Tuition, nonresident: part-time $505 per credit hour. Tuition and fees vary according to course level, course load and program.

Financial aid: In 1999–00, 6 research assistantships, 1 teaching assistantship were awarded.

Dr. Stephen R. Humphrey, Dean, 352-392-9230, *Fax:* 352-392-9748, *E-mail:* humphrey@ufl.edu.

Application contact: Meisha Wade, Professional Advisor, 352-846-1634, *Fax:* 352-392-9748, *E-mail:* mwade@ufl.edu.

Find an in-depth description at www.petersons.com/graduate.

■ UNIVERSITY OF GEORGIA

Graduate School, College of Arts and Sciences, Program in Ecology, Athens, GA 30602

AWARDS Conservation ecology and sustainable development (MS); ecology (MS, PhD).

Degree requirements: For master's, thesis required; for doctorate, one foreign language (computer language can substitute), dissertation required.

Entrance requirements: For master's and doctorate, GRE General Test. Electronic applications accepted.

Expenses: Tuition, state resident: full-time $7,516; part-time $431 per credit hour. Tuition, nonresident: full-time $12,204; part-time $793 per credit hour. Tuition and fees vary according to program.

■ UNIVERSITY OF HAWAII AT MANOA

Graduate Division, Specialization in Ecology, Evolution and Conservation Biology, Honolulu, HI 96822

AWARDS MS, PhD. Program is interdisciplinary; degree is in specific discipline with the specialization in Ecology, Evolution and Conservation Biology.

Faculty: 35 full-time (8 women), 11 part-time/adjunct (0 women).

Students: 52 full-time (34 women); includes 12 minority (8 Asian Americans or Pacific Islanders, 3 Hispanic Americans, 1 Native American), 2 international.

Degree requirements: For doctorate, dissertation required.

Expenses: Tuition, state resident: full-time $4,032; part-time $168 per credit. Tuition, nonresident: full-time $9,960; part-time $415 per credit. Required fees: $51 per semester. Part-time tuition and fees vary according to course load and program.

Financial aid: In 1999–00, 15 students received aid, including 3 fellowships, 5

research assistantships, 2 teaching assistantships; career-related internships or fieldwork and tuition waivers (full) also available.

Faculty research: Agronomy and soil science, zoology, entomology, genetics and molecular biology, botanical sciences.

Dr. Sheila Conant, Chair, 808-956-4602, *Fax:* 808-956-9608, *E-mail:* sconant@zoogatc.zoo.hawaii.edu.

■ UNIVERSITY OF ILLINOIS AT CHICAGO

Graduate College, College of Liberal Arts and Sciences, Department of Biological Sciences, Chicago, IL 60607-7128

AWARDS Cell and developmental biology (PhD); ecology and evolution (MS, DA, PhD); genetics and development (PhD); molecular biology (MS, PhD); neurobiology (MS, PhD); plant biology (MS, DA, PhD).

Faculty: 40 full-time (5 women).

Students: 100 full-time (47 women), 14 part-time (10 women); includes 13 minority (11 Asian Americans or Pacific Islanders, 2 Hispanic Americans), 42 international. Average age 29. *99 applicants, 36% accepted.* In 1999, 3 master's, 9 doctorates awarded.

Degree requirements: For master's, thesis required, foreign language not required; for doctorate, dissertation, preliminary exam required, foreign language not required.

Entrance requirements: For master's and doctorate, GRE General Test, GRE Subject Test, TOEFL, previous course work in physics, calculus, and organic chemistry; minimum GPA of 3.75 on a 5.0 scale. *Application deadline:* For fall admission, 6/1. Applications are processed on a rolling basis. *Application fee:* $40 ($50 for international students). Electronic applications accepted.

Expenses: Tuition, state resident: full-time $3,750; part-time $1,250 per semester. Tuition, nonresident: full-time $10,588; part-time $3,530 per semester. Required fees: $507 per semester. Tuition and fees vary according to course load and program.

Financial aid: In 1999–00, 87 students received aid; fellowships, research assistantships, teaching assistantships, career-related internships or fieldwork, Federal Work-Study, traineeships, and tuition waivers (full) available. Financial aid application deadline: 3/1; financial aid applicants required to submit FAFSA.

Dr. Lon Kaufman, Head, 312-996-2213.

Application contact: Dr. Leo Miller, Director of Graduate Studies, 312-996-2220.

Find an in-depth description at www.petersons.com/graduate.

■ UNIVERSITY OF ILLINOIS AT URBANA–CHAMPAIGN

Graduate College, College of Liberal Arts and Sciences, School of Life Sciences, Department of Ecology, Ethnology, and Evolution, Urbana, IL 61801

AWARDS PhD.

Faculty: 9 full-time (0 women), 1 part-time/adjunct (0 women).

Students: 29 full-time (9 women); includes 1 minority (Native American), 3 international.

Degree requirements: For doctorate, dissertation required.

Application deadline: Applications are processed on a rolling basis. *Application fee:* $30 ($50 for international students).

Expenses: Tuition, state resident: full-time $4,616. Tuition, nonresident: full-time $11,768. Full-time tuition and fees vary according to course load.

Financial aid: In 1999–00, 2 fellowships, 9 research assistantships, 16 teaching assistantships were awarded; tuition waivers (full and partial) also available. Financial aid application deadline: 2/15.

Scott K. Robinson, Head, 217-333-6857, *Fax:* 217-244-4565, *E-mail:* skrobins@uiuc.edu.

Application contact: Linda Thorman, Director of Graduate Studies, 217-333-7802, *Fax:* 217-244-4565, *E-mail:* thorman@uiuc.edu.

■ UNIVERSITY OF KANSAS

Graduate School, College of Liberal Arts and Sciences, Division of Biological Sciences, Department of Ecology and Evolutionary Biology, Lawrence, KS 66045

AWARDS MA, PhD. Part-time programs available.

Faculty: 34.

Students: 19 full-time (10 women), 30 part-time (11 women); includes 2 minority (1 African American, 1 Asian American or Pacific Islander), 8 international. *65 applicants, 14% accepted.*

Degree requirements: For master's, thesis or alternative, general exam required, foreign language not required; for doctorate, dissertation, oral comprehensive exam required.

Entrance requirements: For master's, GRE General Test, GRE Subject Test, TOEFL, minimum GPA of 3.1; for doctorate, GRE General Test, GRE

University of Kansas (continued)
Subject Test, TOEFL. *Application deadline:* For fall admission, 1/1 (priority date). *Application fee:* $25.

Expenses: Tuition, state resident: full-time $2,482; part-time $103 per credit hour. Tuition, nonresident: full-time $8,104; part-time $338 per credit hour. Required fees: $428; $31 per credit hour. Tuition and fees vary according to program.

Financial aid: In 1999–00, 9 fellowships, teaching assistantships (averaging $11,588 per year) were awarded; research assistantships Financial aid application deadline: 3/1.

Faculty research: Aquatic ecology, environmental studies, population and evolutionary ecology, tropical biology. Thomas N. Taylor, Chair, 785-864-3625.

Application contact: Linda Trueb, Graduate Director, 785-864-3342, *Fax:* 785-864-5335, *E-mail:* trueb@kuhub.cc.ukans.edu.

■ UNIVERSITY OF MAINE

Graduate School, College of Natural Sciences, Forestry, and Agriculture, Department of Biological Sciences, Program in Ecology and Environmental Science, Orono, ME 04469

AWARDS MS, PhD. Part-time programs available.

Degree requirements: For doctorate, dissertation required.

Entrance requirements: For master's and doctorate, GRE General Test, TOEFL.

Expenses: Tuition, state resident: full-time $3,564. Tuition, nonresident: full-time $10,116. Required fees: $378. Tuition and fees vary according to course load.

■ UNIVERSITY OF MAINE

Graduate School, College of Natural Sciences, Forestry, and Agriculture, Department of Plant, Soil, and Environmental Sciences, Orono, ME 04469

AWARDS Biological sciences (PhD); ecology and environmental sciences (MS, PhD); forest resources (PhD); plant science (PhD); plant, soil, and environmental sciences (MS); resource utilization (MS).

Entrance requirements: For master's and doctorate, GRE General Test, TOEFL.

Expenses: Tuition, state resident: full-time $3,564. Tuition, nonresident: full-time $10,116. Required fees: $378. Tuition and fees vary according to course load.

■ UNIVERSITY OF MIAMI

Graduate School, College of Arts and Sciences, Department of Biology, Coral Gables, FL 33124

AWARDS Biology (MS, PhD); genetics and evolution (MS, PhD); tropical biology, ecology, and behavior (MS, PhD).

Faculty: 24 full-time (3 women), 5 part-time/adjunct (1 woman).

Students: 42 full-time (23 women); includes 7 minority (1 African American, 3 Asian Americans or Pacific Islanders, 3 Hispanic Americans), 2 international. Average age 26. *56 applicants, 13% accepted.* In 1999, 3 doctorates awarded (34% entered university research/teaching, 66% found other work related to degree). Terminal master's awarded for partial completion of doctoral program.

Degree requirements: For master's, oral defense required, thesis optional, foreign language not required; for doctorate, dissertation, oral and written qualifying exam required, foreign language not required. *Average time to degree:* Doctorate–5.5 years full-time.

Entrance requirements: For master's and doctorate, GRE General Test, GRE Subject Test, TOEFL. *Application deadline:* For fall admission, 3/15. Applications are processed on a rolling basis. *Application fee:* $50. Electronic applications accepted.

Expenses: Tuition: Full-time $15,336; part-time $852 per credit. Required fees: $174. Tuition and fees vary according to program.

Financial aid: In 1999–00, 37 students received aid, including 6 fellowships with tuition reimbursements available (averaging $15,000 per year), 10 research assistantships with tuition reimbursements available (averaging $13,000 per year), 21 teaching assistantships with tuition reimbursements available (averaging $13,202 per year); career-related internships or fieldwork and institutionally sponsored loans also available. Financial aid application deadline: 3/15.

Faculty research: Population biology, behavioral ecology, plant-animal and plant-environment interactions and genetic co-evolution, biogeography, conservation biology. Dr. Julian C. Lee, Director, 305-284-6420, *Fax:* 305-284-3039, *E-mail:* jlee@fig.cox.miami.edu.

Application contact: 305-284-3973, *Fax:* 305-284-3039, *E-mail:* gaac@fig.cox.miami.edu.

Find an in-depth description at www.petersons.com/graduate.

■ UNIVERSITY OF MIAMI

Graduate School, College of Arts and Sciences, Program in Tropical Biology, Ecology, and Behavior, Coral Gables, FL 33124

AWARDS MS, PhD.

Faculty: 21 full-time (4 women), 2 part-time/adjunct (1 woman).

Students: 52 full-time (20 women); includes 9 minority (2 African Americans, 3 Asian Americans or Pacific Islanders, 4 Hispanic Americans), 14 international. Average age 27. *80 applicants, 14% accepted.* Terminal master's awarded for partial completion of doctoral program.

Degree requirements: For master's, thesis optional, foreign language not required; for doctorate, dissertation, oral and written qualifying exam required, foreign language not required. *Average time to degree:* Master's–3 years full-time; doctorate–6 years full-time.

Entrance requirements: For master's and doctorate, GRE General Test, GRE Subject Test, TOEFL. *Application deadline:* For fall admission, 2/1. Applications are processed on a rolling basis. *Application fee:* $35.

Expenses: Tuition: Full-time $15,336; part-time $852 per credit. Required fees: $174. Tuition and fees vary according to program.

Financial aid: In 1999–00, 8 fellowships, 12 research assistantships, 23 teaching assistantships were awarded; career-related internships or fieldwork and institutionally sponsored loans also available. Financial aid application deadline: 3/15.

Faculty research: Behavioral ecology, plant-animal and plant-environmental interactions and coevolution, biogeography, conservation biology, genetics.
Jean Crawford, Graduate Coordinator, 305-284-3973, *Fax:* 305-284-3039, *E-mail:* jean@fig.cox.miami.edu.

■ UNIVERSITY OF MINNESOTA, TWIN CITIES CAMPUS

Graduate School, College of Agricultural, Food, and Environmental Sciences, Microbial Ecology Program, Minneapolis, MN 55455-0213

AWARDS MS, PhD.

Faculty: 15 full-time (1 woman).

Students: 3 full-time. Average age 30. *1 applicant, 100% accepted.* In 1999, 1 master's awarded (100% entered university research/teaching); 1 doctorate awarded (100% entered university research/teaching). Terminal master's awarded for partial completion of doctoral program.

Degree requirements: For master's and doctorate, thesis/dissertation required, foreign language not required. *Average time to degree:* Master's–2 years full-time; doctorate–5 years full-time. *Application fee:* $0.

Expenses: Tuition, state resident: full-time $5,040; part-time $420 per credit. Tuition, nonresident: full-time $9,900; part-time $825 per credit. Full-time tuition and fees vary according to course load, program and reciprocity agreements.

Faculty research: Arthropod-microbe interactions, plant-microbe interactions, resource competition in plants, systematics and ecology of mushrooms, physiology and ecology of oral bacteria, biodegradation. Dr. Michael J. Sadowsky, Professor, 612-624-2706, *Fax:* 612-625-2208, *E-mail:* sadowsky@soils.umn.edu.

■ UNIVERSITY OF MINNESOTA, TWIN CITIES CAMPUS

Graduate School, College of Biological Sciences, Department of Ecology, Evolution, and Behavior, Minneapolis, MN 55455-0213

AWARDS Ecology, animal behavior, and evolution (MS, PhD).

Faculty: 55 full-time (11 women).
Students: 45 full-time (24 women), 1 (woman) part-time; includes 2 minority (1 Asian American or Pacific Islander, 1 Hispanic American). Average age 26. 71 *applicants, 17% accepted.* In 1999, 2 master's awarded (50% found work related to degree, 50% continued full-time study); 3 doctorates awarded (33% entered university research/teaching, 67% found other work related to degree). Terminal master's awarded for partial completion of doctoral program.
Degree requirements: For master's, thesis or projects, comprehensive exam required; for doctorate, dissertation, comprehensive exam required, foreign language not required. *Average time to degree:* Master's–6.5 years full-time; doctorate–9 years full-time.
Entrance requirements: For master's and doctorate, GRE General Test, minimum GPA of 3.2. *Application deadline:* For fall admission, 1/4. *Application fee:* $50 ($55 for international students). Electronic applications accepted.
Expenses: Tuition, state resident: full-time $5,040; part-time $420 per credit. Tuition, nonresident: full-time $9,900; part-time $825 per credit. Full-time tuition and fees vary according to course load, program and reciprocity agreements.
Financial aid: In 1999–00, 44 students received aid, including 5 fellowships with tuition reimbursements available (averaging $12,000 per year), 12 research assistantships with tuition reimbursements available (averaging $10,920 per year), 14

teaching assistantships with tuition reimbursements available (averaging $10,920 per year); Federal Work-Study, institutionally sponsored loans, and tuition waivers (partial) also available. Financial aid application deadline: 1/4. *Total annual research expenditures:* $2.7 million.
Prof. Robert W. Sterner, Head, 612-625-5700, *Fax:* 612-624-6777, *E-mail:* stern007@tc.umn.edu.
Application contact: Prof. Elmer C. Birney, Director of Graduate Studies, 612-625-5713, *Fax:* 612-624-6777, *E-mail:* ecbirney@cbs.umn.edu.

■ UNIVERSITY OF MISSOURI–ST. LOUIS

Graduate School, College of Arts and Sciences, Department of Biology, St. Louis, MO 63121-4499

AWARDS Biology (MS, PhD), including animal behavior (MS), biochemistry, biotechnology (MS), conservation biology (MS), development (MS), ecology (MS), environmental studies (PhD), evolution (MS), genetics (MS), molecular biology and biotechnology (PhD), molecular/cellular biology (MS), physiology (MS), plant systematics, population biology (MS), tropical biology (MS); biotechnology (Certificate); tropical biology and conservation (Certificate). Part-time programs available.

Faculty: 46.
Students: 21 full-time (11 women), 75 part-time (44 women); includes 13 minority (2 African Americans, 2 Asian Americans or Pacific Islanders, 8 Hispanic Americans, 1 Native American), 23 international. In 1999, 14 master's, 4 doctorates awarded.
Degree requirements: For master's, thesis or alternative required, foreign language not required; for doctorate, one foreign language, dissertation, 1 semester of teaching experience required.
Entrance requirements: For doctorate, GRE General Test. *Application deadline:* For fall admission, 7/1 (priority date); for spring admission, 11/1 (priority date). Applications are processed on a rolling basis. *Application fee:* $25 ($40 for international students). Electronic applications accepted.
Expenses: Tuition, state resident: full-time $4,932; part-time $173 per credit hour. Tuition, nonresident: full-time $13,279; part-time $521 per credit hour. Required fees: $775; $33 per credit hour. Tuition and fees vary according to degree level and program.
Financial aid: In 1999–00, 8 research assistantships with partial tuition reimbursements (averaging $10,635 per year), 14 teaching assistantships with partial tuition reimbursements (averaging

$11,488 per year) were awarded; career-related internships or fieldwork and Federal Work-Study also available. Aid available to part-time students. Financial aid application deadline: 2/1. *Total annual research expenditures:* $908,828.
Application contact: Graduate Admissions, 314-516-5458, *Fax:* 314-516-6759, *E-mail:* gradadm@umsl.edu.

■ THE UNIVERSITY OF MONTANA–MISSOULA

Graduate School, Division of Biological Sciences, Program in Organismal Biology and Ecology, Missoula, MT 59812-0002

AWARDS MS, PhD.

Students: 24 full-time (8 women), 16 part-time (9 women); includes 4 minority (1 Asian American or Pacific Islander, 3 Native Americans). 77 *applicants, 16% accepted.* In 1999, 8 master's, 3 doctorates awarded. Terminal master's awarded for partial completion of doctoral program.
Degree requirements: For master's, thesis required; for doctorate, dissertation required.
Entrance requirements: For master's and doctorate, GRE General Test. *Application deadline:* For fall admission, 2/1. *Application fee:* $45.
Expenses: Tuition, state resident: full-time $2,484; part-time $151 per credit. Tuition, nonresident: full-time $8,000; part-time $305 per credit. Required fees: $1,600. Full-time tuition and fees vary according to degree level and program.
Financial aid: In 1999–00, research assistantships with full tuition reimbursements (averaging $9,400 per year), teaching assistantships with full tuition reimbursements (averaging $9,400 per year) were awarded; Federal Work-Study also available. Financial aid application deadline: 3/1.
Faculty research: Conservation biology, ecology and behavior, evolutionary genetics, avian biology.
Application contact: Janean Clark, Graduate Programs Secretary, 406-243-5222, *Fax:* 406-243-4184, *E-mail:* jmclark@selway.umt.edu.

■ UNIVERSITY OF NEVADA, RENO

Graduate School, Interdisciplinary Program in Ecology, Evolution, and Conservation Biology, Reno, NV 89557

AWARDS PhD. Offered through the College of Arts and Science, the M. C. Fleischmann College of Agriculture, and the Desert Research Institute.
Faculty: 29.

University of Nevada, Reno (continued)
Students: 27 full-time (16 women), 9 part-time (3 women); includes 2 minority (1 Asian American or Pacific Islander, 1 Hispanic American), 2 international. Average age 33. *36 applicants, 14% accepted.* In 1999, 7 degrees awarded.
Degree requirements: For doctorate, dissertation required, foreign language not required.
Entrance requirements: For doctorate, GRE General Test, GRE Subject Test, TOEFL, minimum GPA of 3.0. *Application deadline:* For fall admission, 2/15. *Application fee:* $40.
Expenses: Tuition, area resident: Part-time $3,173 per semester. Tuition, nonresident: full-time $6,347. Required fees: $101 per credit. $101 per credit.
Financial aid: In 1999–00, 1 research assistantship was awarded; teaching assistantships Financial aid application deadline: 3/1.
Faculty research: Population biology, behavioral ecology, plant response to climate change, conservation of endangered species, restoration of natural ecosystems.
Dr. C. Richard Tracy, Director, 775-784-4419, *Fax:* 775-784-4583.
Application contact: Larry Hillerman, Program Manager, 775-784-4439, *Fax:* 775-784-1306, *E-mail:* eecb@ biodiversity.unr.edu.

■ UNIVERSITY OF NEW MEXICO

Graduate School, College of Arts and Sciences, Department of Biology, Albuquerque, NM 87131-2039

AWARDS Biology (MS, PhD), including air land ecology, behavioral ecology, botany, cellular and molecular biology, community ecology, comparative immunology, comparative physiology, conservation biology, ecology, ecosystem ecology, evolutionary biology, evolutionary genetics, microbiology, molecular genetics, parasitology, physiological ecology, physiology, population biology, vertebrate and invertebrate zoology. Part-time programs available.

Faculty: 35 full-time (5 women), 18 part-time/adjunct (11 women).
Students: 71 full-time (37 women), 28 part-time (11 women); includes 8 minority (2 Asian Americans or Pacific Islanders, 5 Hispanic Americans, 1 Native American), 11 international. Average age 33. *93 applicants, 30% accepted.* In 1999, 11 master's, 12 doctorates awarded. Terminal master's awarded for partial completion of doctoral program.
Degree requirements: For master's, one foreign language (computer language can substitute), thesis required (for some programs); for doctorate, 2 foreign

languages (computer language can substitute for one), dissertation required.
Entrance requirements: For master's and doctorate, GRE General Test, GRE Subject Test, minimum GPA of 3.2. *Application deadline:* For fall admission, 1/15. *Application fee:* $25.
Expenses: Tuition, state resident: full-time $2,514; part-time $105 per credit hour. Tuition, nonresident: full-time $10,304; part-time $417 per credit hour. International tuition: $10,304 full-time. Required fees: $516; $22 per credit hour. Tuition and fees vary according to program.
Financial aid: In 1999–00, 58 students received aid, including 24 fellowships (averaging $1,645 per year), 26 research assistantships with tuition reimbursements available (averaging $8,921 per year), 40 teaching assistantships with tuition reimbursements available (averaging $11,066 per year); career-related internships or fieldwork, Federal Work-Study, institutionally sponsored loans, and tuition waivers (full and partial) also available. Aid available to part-time students. Financial aid applicants required to submit FAFSA.
Faculty research: Developmental biology, immunobiology. *Total annual research expenditures:* $4.5 million.
Dr. Kathryn Vogel, Chair, 505-277-3411, *Fax:* 505-277-0304, *E-mail:* kgvogel@ unm.edu.
Application contact: Vivian Kent, Information Contact, 505-277-1712, *Fax:* 505-277-0304, *E-mail:* vkent@unm.edu.

■ THE UNIVERSITY OF NORTH CAROLINA AT CHAPEL HILL

Graduate School, College of Arts and Sciences, Curriculum in Ecology, Chapel Hill, NC 27599

AWARDS MA, MS, PhD. Part-time programs available.

Faculty: 25 full-time (1 woman).
Students: 20 full-time (14 women). Average age 28. *52 applicants, 15% accepted.* In 1999, 4 master's awarded (100% found work related to degree); 3 doctorates awarded (100% found work related to degree).
Degree requirements: For master's, comprehensive exam, oral defense of thesis required; for doctorate, comprehensive and oral exams, oral defense of dissertation required. *Average time to degree:* Master's–3.5 years full-time; doctorate–6.7 years full-time.
Entrance requirements: For master's and doctorate, GRE General Test. *Application deadline:* For fall admission, 1/1 (priority date). Applications are processed on a rolling basis. *Application fee:* $55. Electronic applications accepted.

Expenses: Tuition, state resident: full-time $1,578. Tuition, nonresident: full-time $10,744. Required fees: $827. One-time fee: $15 full-time. Tuition and fees vary according to program.
Financial aid: In 1999–00, 6 fellowships with tuition reimbursements (averaging $12,480 per year), 2 research assistantships (averaging $8,740 per year), 12 teaching assistantships with tuition reimbursements (averaging $5,690 per year) were awarded; grants also available. Financial aid application deadline: 3/1.
Dr. Bruce Winterhalder, Chairman, 919-962-1270.

■ THE UNIVERSITY OF NORTH CAROLINA AT CHAPEL HILL

Graduate School, College of Arts and Sciences, Department of Biology, Chapel Hill, NC 27599

AWARDS Botany (MA, MS, PhD); cell biology, development, and physiology (MA, MS, PhD); ecology and behavior (MA, MS, PhD); genetics and molecular biology (MA, MS, PhD); morphology, systematics, and evolution (MA, MS, PhD).

Degree requirements: For master's, thesis (for some programs), comprehensive exams required; for doctorate, dissertation, comprehensive exams required.
Entrance requirements: For master's and doctorate, GRE General Test, GRE Subject Test. Electronic applications accepted.
Expenses: Tuition, state resident: full-time $1,578. Tuition, nonresident: full-time $10,744. Required fees: $827. One-time fee: $15 full-time. Tuition and fees vary according to program.
Faculty research: Gene expression, biomechanics, yeast genetics, plant ecology, plant molecular biology.
Find an in-depth description at www.petersons.com/graduate.

■ UNIVERSITY OF NORTH DAKOTA

Graduate School, College of Arts and Sciences, Department of Biology, Grand Forks, ND 58202

AWARDS Botany (MS, PhD); ecology (MS, PhD); entomology (MS, PhD); environmental biology (MS, PhD); fisheries/wildlife (MS, PhD); genetics (MS, PhD); zoology (MS, PhD).

Faculty: 18 full-time (3 women).
Students: 21 full-time (8 women). *13 applicants, 62% accepted.* In 1999, 3 master's awarded. Terminal master's awarded for partial completion of doctoral program.
Degree requirements: For master's, thesis, final exam required; for doctorate,

dissertation, comprehensive exam, final exam required.

Entrance requirements: For master's, GRE General Test, GRE Subject Test, TOEFL, minimum GPA of 3.0; for doctorate, GRE General Test, GRE Subject Test, TOEFL, minimum GPA of 3.5. *Application deadline:* For fall admission, 3/1 (priority date). Applications are processed on a rolling basis. *Application fee:* $25.

Expenses: Tuition, state resident: full-time $3,166; part-time $158 per credit. Tuition, nonresident: full-time $7,658; part-time $345 per credit. International tuition: $7,658 full-time. Required fees: $46 per credit. Tuition and fees vary according to program and reciprocity agreements.

Financial aid: In 1999–00, 6 research assistantships with full tuition reimbursements (averaging $11,250 per year), 15 teaching assistantships with full tuition reimbursements (averaging $11,250 per year) were awarded; fellowships, Federal Work-Study, institutionally sponsored loans, scholarships, and tuition waivers (full and partial) also available. Aid available to part-time students. Financial aid application deadline: 3/15; financial aid applicants required to submit FAFSA.

Faculty research: Population biology, wildlife ecology, RNA processing, hormonal control of behavior.
Dr. Jeff Lang, Director, 701-777-2621, *Fax:* 701-777-2623, *E-mail:* jlang@ badlands.nodak.edu.

■ UNIVERSITY OF NOTRE DAME

Graduate School, College of Science, Department of Biological Sciences, Program in Aquatic Ecology, Evolution and Environmental Biology, Notre Dame, IN 46556

AWARDS MS, PhD. Terminal master's awarded for partial completion of doctoral program.

Degree requirements: For master's and doctorate, thesis/dissertation required, foreign language not required.

Entrance requirements: For master's and doctorate, GRE General Test, GRE Subject Test, TOEFL. *Application deadline:* For fall admission, 2/1 (priority date); for spring admission, 11/1. Applications are processed on a rolling basis. *Application fee:* $50.

Expenses: Tuition: Full-time $21,930; part-time $1,218 per credit. Required fees: $95. Tuition and fees vary according to program.

Financial aid: Fellowships with full tuition reimbursements, research assistantships with full tuition reimbursements, teaching assistantships with full tuition reimbursements, traineeships and tuition waivers

(full) available. Financial aid application deadline: 2/1.

Faculty research: Aquatic benthic ecology, biogeochemistry, plant-animal interactions, species invasions, ecological genetics and speciation, aquatic community studies.

Application contact: Dr. Terrence J. Akai, Director of Graduate Admissions, 219-631-7706, *Fax:* 219-631-4183, *E-mail:* gradad@nd.edu.

■ UNIVERSITY OF OREGON

Graduate School, College of Arts and Sciences, Department of Biology, Eugene, OR 97403

AWARDS Ecology and evolution (MA, MS, PhD); marine biology (MA, MS, PhD); molecular, cellular and genetic biology (PhD); neuroscience and development (PhD).

Faculty: 42 full-time (14 women), 5 part-time/adjunct (2 women).
Students: 69 full-time (33 women), 8 part-time (5 women); includes 6 minority (2 Asian Americans or Pacific Islanders, 3 Hispanic Americans, 1 Native American), 5 international. *18 applicants, 83% accepted.* In 1999, 11 master's, 7 doctorates awarded. Terminal master's awarded for partial completion of doctoral program.

Degree requirements: For master's, thesis required (for some programs); for doctorate, dissertation required, foreign language not required.

Entrance requirements: For master's and doctorate, GRE General Test, TOEFL, minimum GPA of 3.2. *Application deadline:* For fall admission, 1/10. *Application fee:* $50.

Expenses: Tuition, state resident: full-time $6,750. Tuition, nonresident: full-time $11,409. Part-time tuition and fees vary according to course load.

Financial aid: In 1999–00, 36 teaching assistantships were awarded; research assistantships, Federal Work-Study, grants, and institutionally sponsored loans also available. Financial aid application deadline: 2/1.

Faculty research: Developmental neurobiology; evolution, population biology, and quantitative genetics; regulation of gene expression; biochemistry of marine organisms.
Janis C. Weeks, Head, 541-346-4502, *Fax:* 541-346-6056.

Application contact: Donna Overall, Graduate Program Coordinator, 541-346-4503, *Fax:* 541-346-6056, *E-mail:* doverall@oregon.uoregon.edu.

Find an in-depth description at www.petersons.com/graduate.

■ UNIVERSITY OF PENNSYLVANIA

School of Arts and Sciences, Graduate Group in Biology, Program in Ecology and Population Biology, Philadelphia, PA 19104

AWARDS PhD.

Degree requirements: For doctorate, dissertation required, foreign language not required.

Entrance requirements: For doctorate, GRE General Test, GRE Subject Test, TOEFL. *Application fee:* $65.

Expenses: Tuition: Full-time $23,670. Required fees: $1,546. Full-time tuition and fees vary according to degree level and program.

Financial aid: Fellowships, research assistantships, teaching assistantships available. Financial aid application deadline: 1/2.

Application contact: Allan Aiken, Graduate Group Coordinator, 215-898-6786, *Fax:* 215-898-8780, *E-mail:* aaiken@ sas.upenn.edu.

■ UNIVERSITY OF PITTSBURGH

Faculty of Arts and Sciences, Department of Biological Sciences, Program in Ecology and Evolution, Pittsburgh, PA 15260

AWARDS MS, PhD.

Degree requirements: For master's and doctorate, thesis/dissertation required, foreign language not required.

Entrance requirements: For master's and doctorate, GRE General Test, GRE Subject Test, TOEFL. *Application deadline:* For fall admission, 2/1 (priority date). Applications are processed on a rolling basis. *Application fee:* $40.

Expenses: Tuition, state resident: full-time $8,338; part-time $342 per credit. Tuition, nonresident: full-time $17,168; part-time $707 per credit. Required fees: $480; $90 per semester. Tuition and fees vary according to program.

Find an in-depth description at www.petersons.com/graduate.

■ UNIVERSITY OF ROCHESTER

The College, Arts and Sciences, Department of Biology, Program in Ecology and Evolutionary Biology, Rochester, NY 14627-0250

AWARDS MS, PhD.

Degree requirements: For master's, thesis not required; for doctorate, dissertation, qualifying exam required, foreign language not required.

Entrance requirements: For master's and doctorate, GRE General Test, GRE Subject Test, TOEFL. *Application deadline:*

University of Rochester (continued)
For fall admission, 2/1 (priority date). *Application fee:* $25.
Expenses: Tuition: Part-time $697 per credit hour. Tuition and fees vary according to program.
Financial aid: Application deadline: 2/1.
Application contact: Cindy Landry, Graduate Program Secretary, 716-275-7991.

■ UNIVERSITY OF SOUTH CAROLINA

Graduate School, College of Science and Mathematics, Department of Biological Sciences, Graduate Training Program in Ecology, Evolution, and Organismal Biology, Columbia, SC 29208

AWARDS MS, PhD.

Degree requirements: For master's, one foreign language required; for doctorate, dissertation required.
Entrance requirements: For master's and doctorate, GRE General Test, minimum GPA of 3.0 in science. *Application deadline:* For fall admission, 2/15 (priority date). *Application fee:* $35.
Expenses: Tuition, state resident: full-time $4,014; part-time $202 per credit hour. Tuition, nonresident: full-time $8,528; part-time $428 per credit hour. Required fees: $100; $4 per credit hour. Tuition and fees vary according to program.
Application contact: Dr. Franklyn F. Bolander, Director of Graduate Studies, 803-777-2755, *Fax:* 803-777-4002, *E-mail:* bolander@sc.edu.

Find an in-depth description at www.petersons.com/graduate.

■ UNIVERSITY OF SOUTH FLORIDA

Graduate School, College of Arts and Sciences, Department of Biology, Tampa, FL 33620-9951

AWARDS Biology (PhD); botany (MS); ecology (PhD); marine biology (MS, PhD); microbiology (MS); physiology (PhD); zoology (MS). Part-time programs available.

Degree requirements: For master's, foreign language not required; for doctorate, 2 foreign languages (computer language can substitute for one), dissertation required.
Entrance requirements: For master's, GRE General Test, minimum GPA of 3.0 in last 60 hours; for doctorate, GRE General Test, GRE Subject Test in biology. Electronic applications accepted.
Expenses: Tuition, state resident: part-time $148 per credit hour. Tuition, nonresident: part-time $509 per credit hour.

■ THE UNIVERSITY OF TENNESSEE

Graduate School, College of Agricultural Sciences and Natural Resources, Department of Plant and Soil Sciences, Knoxville, TN 37996

AWARDS Crop physiology and ecology (MS, PhD); plant breeding and genetics (MS, PhD); soil science (MS, PhD). Part-time programs available.

Faculty: 19 full-time (1 woman).
Students: 11 full-time (3 women), 11 part-time (7 women), 2 international. 7 *applicants, 43% accepted.* In 1999, 7 master's, 5 doctorates awarded.
Degree requirements: For master's, thesis or alternative required, foreign language not required; for doctorate, dissertation required, foreign language not required.
Entrance requirements: For master's and doctorate, GRE General Test, TOEFL, minimum GPA of 2.7. *Application deadline:* For fall admission, 2/1 (priority date). Applications are processed on a rolling basis. *Application fee:* $35. Electronic applications accepted.
Expenses: Tuition, state resident: full-time $3,806; part-time $184 per credit hour. Tuition, nonresident: full-time $9,874; part-time $522 per credit hour. Tuition and fees vary according to program.
Financial aid: In 1999–00, 17 research assistantships, 2 teaching assistantships were awarded; fellowships, career-related internships or fieldwork, Federal Work-Study, and institutionally sponsored loans also available. Financial aid application deadline: 2/1; financial aid applicants required to submit FAFSA.
Dr. Fred L. Allen, Head, 865-974-7101, *Fax:* 865-974-7997, *E-mail:* allenf@utk.edu.
Application contact: Dr. Mike Essington, Graduate Representative, 865-974-8818, *E-mail:* messington@utk.edu.

■ THE UNIVERSITY OF TENNESSEE

Graduate School, College of Arts and Sciences, Department of Ecology and Evolutionary Biology, Knoxville, TN 37996

AWARDS Behavior (MS, PhD); ecology (MS, PhD); evolutionary biology (MS, PhD). Part-time programs available.

Faculty: 21 full-time (4 women), 1 part-time/adjunct (0 women).
Students: 30 full-time (18 women), 26 part-time (10 women); includes 1 minority (Asian American or Pacific Islander), 9 international. *54 applicants, 17% accepted.* In 1999, 6 master's, 9 doctorates awarded.

Degree requirements: For master's and doctorate, thesis/dissertation required, foreign language not required.
Entrance requirements: For master's and doctorate, GRE General Test, TOEFL, minimum GPA of 2.7. *Application deadline:* For fall admission, 2/1 (priority date). Applications are processed on a rolling basis. *Application fee:* $35. Electronic applications accepted.
Expenses: Tuition, state resident: full-time $3,806; part-time $184 per credit hour. Tuition, nonresident: full-time $9,874; part-time $522 per credit hour. Tuition and fees vary according to program.
Financial aid: In 1999–00, 4 fellowships, 9 research assistantships, 26 teaching assistantships were awarded; Federal Work-Study, institutionally sponsored loans, and unspecified assistantships also available. Financial aid application deadline: 2/1; financial aid applicants required to submit FAFSA.
Dr. Tom Hallam, Head, 865-974-3065, *Fax:* 865-974-3067, *E-mail:* thallam@utk.edu.
Application contact: Dr. C. R. Boake, Graduate Representative, *E-mail:* cboake@utk.edu.

■ THE UNIVERSITY OF TENNESSEE

Graduate School, College of Arts and Sciences, Department of Mathematics, Knoxville, TN 37996

AWARDS Applied mathematics (MS); mathematical ecology (PhD); mathematics (M Math, MS, PhD). Part-time programs available.

Faculty: 49 full-time (4 women), 3 part-time/adjunct (1 woman).
Students: 54 full-time (18 women), 30 part-time (14 women); includes 9 minority (6 African Americans, 2 Asian Americans or Pacific Islanders, 1 Hispanic American), 20 international. *80 applicants, 66% accepted.* In 1999, 14 master's, 8 doctorates awarded.
Degree requirements: For master's, thesis or alternative required, foreign language not required; for doctorate, dissertation required.
Entrance requirements: For master's and doctorate, TOEFL, minimum GPA of 2.7. *Application deadline:* For fall admission, 2/1 (priority date). Applications are processed on a rolling basis. *Application fee:* $35. Electronic applications accepted.
Expenses: Tuition, state resident: full-time $3,806; part-time $184 per credit hour. Tuition, nonresident: full-time $9,874; part-time $522 per credit hour. Tuition and fees vary according to program.
Financial aid: In 1999–00, 15 fellowships, 61 teaching assistantships were awarded;

research assistantships, Federal Work-Study, institutionally sponsored loans, and unspecified assistantships also available. Financial aid application deadline: 2/1; financial aid applicants required to submit FAFSA.

Dr. John B. Conway, Head, 865-974-2464, *Fax:* 865-974-6576, *E-mail:* gradprogram@ novell.math.utk.edu.

■ THE UNIVERSITY OF TEXAS AT AUSTIN

Graduate School, College of Natural Sciences, School of Biological Sciences, Program in Ecology, Evolution and Behavior, Austin, TX 78712-1111

AWARDS MA, PhD.

Entrance requirements: For master's and doctorate, GRE General Test. *Application deadline:* Applications are processed on a rolling basis. *Application fee:* $50 ($75 for international students). Electronic applications accepted.

Expenses: Tuition, state resident: part-time $114 per semester hour. Tuition, nonresident: part-time $330 per semester hour. Tuition and fees vary according to program.

Financial aid: Fellowships, research assistantships, teaching assistantships available. Financial aid application deadline: 1/5.

Dr. Klaus O. Kalthoff, Chairman, 512-471-1152.

Application contact: Michael J. Ryan, Graduate Adviser, 512-471-7131, *E-mail:* mryan@mail.utexas.edu.

■ UNIVERSITY OF UTAH

Graduate School, College of Science, Department of Biology, Salt Lake City, UT 84112-1107

AWARDS Biology (M Phil); ecology and evolutionary biology (MS, PhD); genetics (MS, PhD); molecular biology (PhD). Part-time programs available. Terminal master's awarded for partial completion of doctoral program.

Degree requirements: For master's and doctorate, thesis/dissertation required, foreign language not required.

Entrance requirements: For master's and doctorate, GRE General Test, GRE Subject Test, TOEFL, minimum GPA of 3.0.

Expenses: Tuition, state resident: full-time $1,663. Tuition, nonresident: full-time $5,201. Tuition and fees vary according to course load and program.

Faculty research: Behavioral ecology, cellular neurobiology, DNA replication, ecological genetics, herpetology.

Find an in-depth description at www.petersons.com/graduate.

■ UNIVERSITY OF WISCONSIN–MADISON

Graduate School, College of Agricultural and Life Sciences, Department of Wildlife Ecology, Madison, WI 53706-1380

AWARDS MS, PhD.

Faculty: 9 full-time (2 women).
Students: 29 full-time (11 women), 2 part-time (both women); includes 1 minority (African American), 5 international. *500 applicants, 1% accepted.* In 1999, 5 master's awarded (20% entered university research/teaching, 40% continued full-time study); 3 doctorates awarded (100% found work related to degree).
Degree requirements: For master's and doctorate, thesis/dissertation required. *Average time to degree:* Master's–3 years full-time; doctorate–4 years full-time.
Entrance requirements: For master's and doctorate, GRE General Test. *Application deadline:* For fall admission, 1/15 (priority date); for spring admission, 10/15 (priority date). *Application fee:* $45. Electronic applications accepted.
Expenses: Tuition, state resident: full-time $5,406; part-time $339 per credit. Tuition, nonresident: full-time $17,110; part-time $1,071 per credit. Full-time tuition and fees vary according to program and reciprocity agreements. Part-time tuition and fees vary according to course load and program.
Financial aid: In 1999–00, 31 students received aid, including 1 fellowship with tuition reimbursement available (averaging $15,800 per year), 29 research assistantships with tuition reimbursements available (averaging $15,800 per year), 1 teaching assistantship with tuition reimbursement available (averaging $15,800 per year); career-related internships or fieldwork, Federal Work-Study, and institutionally sponsored loans also available.
Faculty research: Agroecology, ecosystem management, physiological ecology, waterfowl ecology, endangered species. *Total annual research expenditures:* $900,000.
Robert L. Ruff, Chair, 608-263-2071, *Fax:* 608-262-6099, *E-mail:* rlruff@ facstaff.wisc.edu.

■ UTAH STATE UNIVERSITY

School of Graduate Studies, College of Natural Resources, Department of Fisheries and Wildlife, Logan, UT 84322

AWARDS Ecology (MS, PhD); fisheries biology (MS, PhD); wildlife biology (MS, PhD).

Faculty: 19 full-time (0 women).
Students: 49 full-time (21 women), 19 part-time (7 women); includes 2 minority (1 Asian American or Pacific Islander, 1 Native American), 3 international. Average age 27. *36 applicants, 44% accepted.* In 1999, 11 master's, 4 doctorates awarded.
Degree requirements: For master's, thesis required (for some programs), foreign language not required; for doctorate, dissertation required, foreign language not required. *Average time to degree:* Master's–2.5 years full-time; doctorate–6 years full-time.
Entrance requirements: For master's and doctorate, GRE General Test, TOEFL, minimum GPA of 3.2. *Application deadline:* For fall admission, 2/15 (priority date); for spring admission, 10/15. Applications are processed on a rolling basis. *Application fee:* $40.
Expenses: Tuition, state resident: full-time $1,553. Tuition, nonresident: full-time $5,436. International tuition: $5,526 full-time. Required fees: $447. Tuition and fees vary according to course load and program.
Financial aid: In 1999–00, 5 fellowships with partial tuition reimbursements, 47 research assistantships with partial tuition reimbursements, 6 teaching assistantships with partial tuition reimbursements were awarded; career-related internships or fieldwork, Federal Work-Study, and institutionally sponsored loans also available. Aid available to part-time students. Financial aid application deadline: 2/1.
Faculty research: Behavior, population ecology, habitat, conservation biology, restoration. *Total annual research expenditures:* $3 million.
Chris Luecke, Interim Head, 435-797-2463, *Fax:* 435-797-1871, *E-mail:* fishnwlf@cc.usu.edu.

Application contact: Suzanne S. Stoker, Senior Secretary, 435-797-2459, *Fax:* 435-797-1871, *E-mail:* fishnwlf@cc.usu.edu.

■ UTAH STATE UNIVERSITY

School of Graduate Studies, College of Natural Resources, Department of Forest Resources, Logan, UT 84322

AWARDS Ecology (MS, PhD); forestry (MS, PhD); recreation resources management (MS, PhD).

Faculty: 13 full-time (2 women), 8 part-time/adjunct (3 women).

Utah State University (continued)
Students: 18 full-time (9 women), 15 part-time (10 women), 3 international. Average age 33. *9 applicants, 67% accepted.* In 1999, 7 master's awarded. Terminal master's awarded for partial completion of doctoral program.
Degree requirements: For master's, computer language, thesis required (for some programs), foreign language not required; for doctorate, one foreign language, computer language, dissertation required.
Entrance requirements: For master's and doctorate, GRE General Test, TOEFL, minimum GPA of 3.0. *Application deadline:* For fall admission, 6/15 (priority date); for spring admission, 10/15. Applications are processed on a rolling basis. *Application fee:* $40.
Expenses: Tuition, state resident: full-time $1,553. Tuition, nonresident: full-time $5,436. International tuition: $5,526 full-time. Required fees: $447. Tuition and fees vary according to course load and program.
Financial aid: In 1999–00, 1 fellowship with partial tuition reimbursement (averaging $15,000 per year), 16 research assistantships with partial tuition reimbursements (averaging $7,455 per year), 4 teaching assistantships with partial tuition reimbursements (averaging $6,344 per year) were awarded; Federal Work-Study, institutionally sponsored loans, and tuition waivers (full and partial) also available. Financial aid application deadline: 2/15.
Faculty research: Disturbance ecology, natural resource policy, outdoor recreation, ecological modeling. *Total annual research expenditures:* $592,844.
Dr. Terry L. Sharik, Head, 435-797-3219, *Fax:* 435-797-4040, *E-mail:* forestry@cc.usu.edu.
Application contact: B. J. Tueller, Staff Assistant, 435-797-3488, *Fax:* 435-797-4040, *E-mail:* bjtuell@cnr.usu.edu.

■ UTAH STATE UNIVERSITY

School of Graduate Studies, College of Natural Resources, Department of Rangeland Resources, Logan, UT 84322

AWARDS Ecology (MS, PhD); range science (MS, PhD). Part-time programs available.

Faculty: 12 full-time (1 woman).
Students: 16 full-time (3 women), 4 part-time (1 woman); includes 1 minority (African American), 7 international. Average age 24. *10 applicants, 50% accepted.* In 1999, 2 master's, 7 doctorates awarded.
Degree requirements: For master's, thesis required; for doctorate, computer language, dissertation required.

Entrance requirements: For master's and doctorate, GRE General Test, TOEFL, minimum GPA of 3.0. *Application deadline:* For fall admission, 6/15 (priority date); for spring admission, 10/15. Applications are processed on a rolling basis. *Application fee:* $40.
Expenses: Tuition, state resident: full-time $1,553. Tuition, nonresident: full-time $5,436. International tuition: $5,526 full-time. Required fees: $447. Tuition and fees vary according to course load and program.
Financial aid: In 1999–00, research assistantships with partial tuition reimbursements (averaging $12,000 per year); fellowships, teaching assistantships, career-related internships or fieldwork, Federal Work-Study, and institutionally sponsored loans also available.
Faculty research: Range plant ecophysiology, plant community ecology, ruminant nutrition, population ecology. *Total annual research expenditures:* $3.5 million.
Dr. John C. Malechek, Head, 435-797-2471, *Fax:* 435-797-3796, *E-mail:* rangesci@cc.usu.edu.

■ UTAH STATE UNIVERSITY

School of Graduate Studies, College of Science, Department of Biology, Logan, UT 84322

AWARDS Biology (MS, PhD); ecology (MS, PhD). Part-time programs available.

Faculty: 39 full-time (7 women).
Students: 39 full-time (15 women), 25 part-time (11 women); includes 2 minority (both Asian Americans or Pacific Islanders), 8 international. Average age 26. *47 applicants, 23% accepted.* In 1999, 9 master's, 3 doctorates awarded.
Degree requirements: For master's and doctorate, thesis/dissertation required, foreign language not required.
Entrance requirements: For master's and doctorate, GRE General Test, TOEFL, minimum GPA of 3.0. *Application deadline:* For fall admission, 6/15 (priority date); for spring admission, 10/15. Applications are processed on a rolling basis. *Application fee:* $40.
Expenses: Tuition, state resident: full-time $1,553. Tuition, nonresident: full-time $5,436. International tuition: $5,526 full-time. Required fees: $447. Tuition and fees vary according to course load and program.
Financial aid: In 1999–00, 3 fellowships with partial tuition reimbursements, 27 research assistantships with partial tuition reimbursements (averaging $10,250 per year), 44 teaching assistantships with partial tuition reimbursements (averaging $10,250 per year) were awarded; career-related internships or fieldwork, Federal

Work-Study, and institutionally sponsored loans also available. Aid available to part-time students. Financial aid application deadline: 3/1.
Faculty research: Plant, insect, microbial, and animal biology.
Dr. Edmund D. Brodie, Head, 435-797-2483.
Application contact: Nancy Kay Harrison, Coordinator of Graduate Studies, 435-797-1770, *Fax:* 435-797-1575, *E-mail:* nancykay@biology.usu.edu.

■ VIRGINIA POLYTECHNIC INSTITUTE AND STATE UNIVERSITY

Graduate School, College of Arts and Sciences, Department of Biology, Program in Ecology, Blacksburg, VA 24061

AWARDS MS, PhD.

Degree requirements: For master's and doctorate, thesis/dissertation required, foreign language not required.
Entrance requirements: For master's, GRE General Test, TOEFL; for doctorate, GRE General Test, GRE Subject Test, TOEFL. *Application fee:* $25.
Expenses: Tuition, state resident: full-time $4,122; part-time $229 per credit hour. Tuition, nonresident: full-time $6,930; part-time $385 per credit hour. Required fees: $828; $107 per semester. Part-time tuition and fees vary according to course load.
Dr. Joe R. Cowles, Chairman, Department of Biology, 540-231-8928, *E-mail:* cowlesjr@vt.edu.

■ WASHINGTON UNIVERSITY IN ST. LOUIS

Graduate School of Arts and Sciences, Division of Biology and Biomedical Sciences, Program in Evolutionary and Population Biology, St. Louis, MO 63130-4899

AWARDS Ecology (PhD); environmental biology (PhD); evolutionary biology (PhD); genetics (PhD).

Degree requirements: For doctorate, dissertation required, foreign language not required.
Entrance requirements: For doctorate, GRE General Test, GRE Subject Test. *Application deadline:* For fall admission, 1/1 (priority date). Applications are processed on a rolling basis. *Application fee:* $0.
Expenses: Tuition: Full-time $23,400; part-time $975 per credit. Tuition and fees vary according to program.
Financial aid: Fellowships, research assistantships available. Financial aid application deadline: 1/1.

Dr. Jonathan Losos, Coordinator, 314-362-4188.
Application contact: Rosemary Garagneni, Director of Admissions, 800-852-9074, *E-mail:* admissions@ dbbs.wustl.edu.

■ WILLIAM PATERSON UNIVERSITY OF NEW JERSEY

College of Science and Health, Department of Biology, General Biology Program, Wayne, NJ 07470-8420

AWARDS General biology (MA); limnology and terrestrial ecology (MA); molecular biology (MA); physiology (MA). Part-time and evening/weekend programs available.

Students: 1 (woman) full-time, 6 part-time (3 women); includes 1 minority (Hispanic American). Average age 25. *9 applicants, 44% accepted.* In 1999, 1 degree awarded.
Degree requirements: For master's, comprehensive exam, independent study or thesis required.
Entrance requirements: For master's, GRE General Test, minimum GPA of 2.75. *Application deadline:* For fall admission, 4/1; for spring admission, 10/15. Applications are processed on a rolling basis. *Application fee:* $35. Electronic applications accepted.
Expenses: Tuition, state resident: part-time $244 per credit. Tuition, nonresident: part-time $350 per credit.
Financial aid: In 1999–00, 2 teaching assistantships (averaging $9,000 per year) were awarded; career-related internships or fieldwork and unspecified assistantships also available. Financial aid application deadline: 4/1; financial aid applicants required to submit FAFSA.
Application contact: Office of Graduate Studies, 973-720-2237, *Fax:* 973-720-2035.
Find an in-depth description at www.petersons.com/graduate.

■ YALE UNIVERSITY

Graduate School of Arts and Sciences, Department of Ecology and Evolutionary Biology, New Haven, CT 06520

AWARDS PhD.

Faculty: 15 full-time (5 women), 2 part-time/adjunct (both women).
Students: 10 full-time (5 women), 1 international. Average age 28. *28 applicants, 43% accepted.*
Entrance requirements: For doctorate, GRE General Test, GRE Subject Test (biology). *Application deadline:* For fall admission, 1/4. *Application fee:* $65.
Expenses: Tuition: Full-time $22,300. Full-time tuition and fees vary according to program.

Financial aid: In 1999–00, fellowships with full tuition reimbursements (averaging $17,600 per year); grants also available. Dr. Gunter Wagner, Chair, 203-432-6138, *Fax:* 203-432-5176.
Application contact: Maureen F. Cunningham, Registrar, 203-432-3837, *Fax:* 203-432-5176, *E-mail:* maureen.cunningham@yale.edu.
Find an in-depth description at www.petersons.com/graduate.

ENVIRONMENTAL BIOLOGY

■ ANTIOCH NEW ENGLAND GRADUATE SCHOOL

Graduate School, Department of Environmental Studies, Program in Environmental Studies, Keene, NH 03431-3516

AWARDS Environmental biology (MS); environmental education (MS); teaching certification in high school biology (9th-12th) (MS); teaching certification in middle school general science (5th-9th) (MS).

Faculty: 4 full-time (1 woman), 20 part-time/adjunct (8 women).
Students: 91 full-time (65 women), 31 part-time (19 women). Average age 31. *92 applicants, 92% accepted.* In 1999, 62 degrees awarded.
Degree requirements: For master's, practicum required, thesis not required.
Entrance requirements: For master's, previous undergraduate course work in biology, chemistry, mathematics (environmental biology). *Application deadline:* For fall admission, 8/1; for spring admission, 12/1. Applications are processed on a rolling basis. *Application fee:* $40.
Expenses: Tuition: Full-time $12,600. Full-time tuition and fees vary according to course load, degree level, program and student level.
Financial aid: In 1999–00, 70 students received aid, including 2 fellowships (averaging $1,000 per year); career-related internships or fieldwork and Federal Work-Study also available. Financial aid applicants required to submit FAFSA.
Faculty research: Sustainability, natural resources inventory.
Application contact: Robbie P. Hertneky, Director of Admissions, 603-357-6265 Ext. 287, *Fax:* 603-357-0718, *E-mail:* rhertneky@antiochne.edu.

■ BAYLOR UNIVERSITY

Graduate School, College of Arts and Sciences, Department of Biology, Waco, TX 76798

AWARDS Biology (MA, MS, PhD); environmental biology (MS); limnology (MSL). Part-time programs available.

Faculty: 13 full-time (3 women).
Students: 10 full-time (3 women), 4 part-time (3 women); includes 2 minority (both Asian Americans or Pacific Islanders), 2 international. In 1999, 7 degrees awarded.
Degree requirements: For master's, thesis required (for some programs); for doctorate, dissertation required.
Entrance requirements: For master's and doctorate, GRE General Test. *Application deadline:* For fall admission, 1/31 (priority date). Applications are processed on a rolling basis. *Application fee:* $25.
Expenses: Tuition: Part-time $329 per semester hour. Tuition and fees vary according to program.
Financial aid: Teaching assistantships, career-related internships or fieldwork, Federal Work-Study, institutionally sponsored loans, and tuition waivers (full and partial) available. Aid available to part-time students. Financial aid application deadline: 2/28.
Faculty research: Terrestrial ecology, aquatic ecology, genetics.
Dr. Richard E. Duhrkopf, Director of Graduate Studies, 254-710-2911, *Fax:* 254-710-2969, *E-mail:* rick_duhrkopf@ baylor.edu.
Application contact: Sandy Tighe, Administrative Assistant, 254-710-2911, *Fax:* 254-710-2969, *E-mail:* sandy_tighe@ baylor.edu.

■ EASTERN ILLINOIS UNIVERSITY

Graduate School, College of Sciences, Program in Biological Sciences, Charleston, IL 61920-3099

AWARDS Biological sciences (MS); botany (MS); environmental biology (MS); zoology (MS).

Degree requirements: For master's, exam required, foreign language and thesis not required.

■ EMPORIA STATE UNIVERSITY

School of Graduate Studies, College of Liberal Arts and Sciences, Division of Biological Sciences, Emporia, KS 66801-5087

AWARDS Botany (MS); environmental biology (MS); general biology (MS); microbial and cellular biology (MS); zoology (MS). Part-time programs available.

Emporia State University (continued)
Faculty: 15 full-time (3 women), 4 part-time/adjunct (0 women).
Students: 20 full-time (10 women), 8 part-time (3 women), 2 international. *8 applicants, 75% accepted.* In 1999, 12 degrees awarded.
Degree requirements: For master's, comprehensive exam or thesis required.
Entrance requirements: For master's, TOEFL, written exam. *Application deadline:* For fall admission, 8/15 (priority date). Applications are processed on a rolling basis. *Application fee:* $30 ($75 for international students). Electronic applications accepted.
Expenses: Tuition, state resident: full-time $2,410; part-time $108 per credit hour. Tuition, nonresident: full-time $6,212; part-time $266 per credit hour.
Financial aid: In 1999–00, 2 fellowships (averaging $1,396 per year), 1 research assistantship (averaging $5,390 per year), 11 teaching assistantships with full tuition reimbursements (averaging $5,047 per year) were awarded; career-related internships or fieldwork, Federal Work-Study, and institutionally sponsored loans also available. Financial aid application deadline: 3/15; financial aid applicants required to submit FAFSA.
Faculty research: Fisheries, range, and wildlife management; aquatic, plant, grassland, vertebrate, and invertebrate ecology; mammalian and plant systematics, taxonomy, and evolution; immunology, virology, and molecular biology.
Dr. Marshall Sundberg, Chair, 316-341-5311, *Fax:* 316-341-5607, *E-mail:* sundberm@emporia.edu.

■ GEORGIA STATE UNIVERSITY

College of Arts and Sciences, Department of Biology, Program in Applied and Environmental Microbiology, Atlanta, GA 30303-3083
AWARDS MS, PhD.

Degree requirements: For master's, one foreign language (computer language can substitute), thesis or alternative, exam required; for doctorate, dissertation required.
Entrance requirements: For master's and doctorate, GRE General Test, TOEFL, minimum GPA of 3.0. *Application deadline:* For fall admission, 7/18; for spring admission, 2/13. Applications are processed on a rolling basis. *Application fee:* $25.
Expenses: Tuition, state resident: full-time $2,896; part-time $121 per credit hour. Tuition, nonresident: full-time $11,584; part-time $483 per credit hour. Required fees: $228. Full-time tuition and fees vary according to course load and program.
Financial aid: Application deadline: 2/6.

Application contact: Latesha Morrison, Graduate Administrative Coordinator, 404-651-2759, *Fax:* 404-651-2509, *E-mail:* biolxm@langate.gsu.edu.

Find an in-depth description at www.petersons.com/graduate.

■ GOVERNORS STATE UNIVERSITY

College of Arts and Sciences, Division of Science, Program in Environmental Biology, University Park, IL 60466-0975

AWARDS MS. Part-time and evening/weekend programs available.

Faculty: 5 full-time (2 women), 2 part-time/adjunct (1 woman).
Students: 39. In 1999, 4 degrees awarded.
Degree requirements: For master's, computer language, thesis or alternative required, foreign language not required. *Application deadline:* For fall admission, 7/15 (priority date); for spring admission, 11/10. Applications are processed on a rolling basis. *Application fee:* $0.
Expenses: Tuition, state resident: full-time $2,352; part-time $98 per semester hour. Tuition, nonresident: full-time $7,056; part-time $294 per semester hour. Required fees: $220; $110 per semester.
Financial aid: Research assistantships, career-related internships or fieldwork, Federal Work-Study, institutionally sponsored loans, and scholarships available. Aid available to part-time students. Financial aid application deadline: 5/1.
Faculty research: Animal physiology, cell biology, animal behavior, plant physiology, plant populations.
Dr. Edwin Cehelnik, Chairperson, Division of Science, 708-534-4520.

■ HOOD COLLEGE

Graduate School, Program in Environmental Biology, Frederick, MD 21701-8575

AWARDS MS. Part-time and evening/weekend programs available.

Students: 5 full-time (3 women), 53 part-time (34 women); includes 4 minority (2 African Americans, 1 Asian American or Pacific Islander, 1 Hispanic American). Average age 33. In 1999, 10 degrees awarded.
Degree requirements: For master's, thesis required, foreign language not required.
Entrance requirements: For master's, minimum GPA of 2.5, 1 year of undergraduate biology and chemistry. *Application deadline:* Applications are processed on a rolling basis. *Application fee:* $30.

Expenses: Tuition: Full-time $5,310; part-time $295 per credit hour. Required fees: $5 per credit hour.
Financial aid: Career-related internships or fieldwork, institutionally sponsored loans, and tuition waivers (partial) available. Aid available to part-time students. Financial aid applicants required to submit FAFSA.
Faculty research: Epidemiology and risk assessment, forest and plant ecology, marine and freshwater ecology, behavioral ecology and population biology, symbiotic relationships.
Dr. Drew Ferrier, Director, 301-696-3649, *Fax:* 301-694-3597, *E-mail:* dferrier@hood.edu.
Application contact: 301-696-3600, *Fax:* 301-696-3597, *E-mail:* hoodgrad@hood.edu.

■ NEW YORK UNIVERSITY

Graduate School of Arts and Science, Department of Biology, New York, NY 10012-1019

AWARDS Applied recombinant DNA technology (MS); biochemistry (PhD); biomedical journalism (MA); cell biology (PhD); computers in biological research (MS); environmental biology (PhD); general biology (MS); neural sciences and physiology (PhD); oral biology (MS); population and evolutionary biology (PhD). Part-time programs available.

Faculty: 22 full-time (5 women), 8 part-time/adjunct.
Students: 99 full-time (48 women), 61 part-time (31 women); includes 30 minority (4 African Americans, 22 Asian Americans or Pacific Islanders, 4 Hispanic Americans), 52 international. Average age 24. *371 applicants, 41% accepted.* In 1999, 54 master's, 4 doctorates awarded. Terminal master's awarded for partial completion of doctoral program.
Degree requirements: For master's, thesis or alternative, qualifying paper required, foreign language not required; for doctorate, dissertation, oral and written comprehensive exams required, foreign language not required.
Entrance requirements: For master's, GRE General Test, TOEFL; for doctorate, GRE General Test, GRE Subject Test, TOEFL. *Application deadline:* For fall admission, 1/4 (priority date). *Application fee:* $60.
Expenses: Tuition: Full-time $17,880; part-time $745 per credit. Required fees: $1,140; $35 per credit. Tuition and fees vary according to course load and program.
Financial aid: Fellowships with tuition reimbursements, research assistantships with tuition reimbursements, teaching assistantships with tuition reimbursements, career-related internships or fieldwork,

Federal Work-Study, institutionally sponsored loans, and tuition waivers (full and partial) available. Financial aid application deadline: 1/4; financial aid applicants required to submit FAFSA.

Faculty research: Development and genetics, neurobiology, plant sciences, molecular and cell biology.

Philip Furmanski, Chairman, 212-998-8200.

Application contact: Gloria Coruzzi, Director of Graduate Studies, 212-998-8200, *Fax:* 212-995-4015, *E-mail:* biology@nyu.edu.

Find an in-depth description at www.petersons.com/graduate.

■ OHIO UNIVERSITY

Graduate Studies, College of Arts and Sciences, Department of Environmental and Plant Biology, Athens, OH 45701-2979

AWARDS MS, PhD. Part-time programs available.

Faculty: 13 full-time (2 women), 1 (woman) part-time/adjunct.

Students: 19 full-time (13 women), 3 part-time (all women), 5 international. *14 applicants, 79% accepted.* In 1999, 3 master's awarded (100% found work related to degree); 2 doctorates awarded.

Degree requirements: For master's, thesis (for some programs), 2 quarters of teaching experience required, foreign language not required; for doctorate, dissertation, 2 quarters of teaching experience required, foreign language not required.

Entrance requirements: For master's, GRE General Test, minimum GPA of 3.0; for doctorate, GRE General Test, minimum GPA of 3.2. *Application deadline:* For fall admission, 3/31 (priority date). Applications are processed on a rolling basis. *Application fee:* $30.

Expenses: Tuition, state resident: full-time $5,754; part-time $238 per credit hour. Tuition, nonresident: full-time $11,055; part-time $457 per credit hour. Tuition and fees vary according to course load, degree level and campus/location.

Financial aid: In 1999–00, 2 research assistantships with tuition reimbursements, 21 teaching assistantships with tuition reimbursements were awarded; fellowships, Federal Work-Study, institutionally sponsored loans, scholarships, and tuition waivers (full and partial) also available. Financial aid application deadline: 2/15.

Faculty research: Cellular and molecular biology, ecology, paleobotany, physiology, ethnobotany.

Dr. James Braselton, Chair, 740-593-1126.

Application contact: Dr. Irwin Ungar, Graduate Chair, 740-593-1120.

■ RUTGERS, THE STATE UNIVERSITY OF NEW JERSEY, NEW BRUNSWICK

Graduate School, Program in Environmental Sciences, New Brunswick, NJ 08901-1281

AWARDS Air resources (MS, PhD); aquatic biology (MS, PhD); aquatic chemistry (MS, PhD); chemistry and physics of aerosol and hydrosol systems (MS, PhD); environmental chemistry (MS, PhD); environmental microbiology (MS, PhD); environmental toxicology (MS, PhD); exposure assessment (PhD); water and wastewater treatment (MS, PhD); water resources (MS, PhD). Part-time and evening/weekend programs available.

Faculty: 33 full-time (7 women), 36 part-time/adjunct (6 women).

Students: 68 full-time (25 women), 58 part-time (26 women); includes 10 minority (8 Asian Americans or Pacific Islanders, 2 Hispanic Americans), 44 international. Average age 26. *128 applicants, 40% accepted.* In 1999, 15 master's, 18 doctorates awarded. Terminal master's awarded for partial completion of doctoral program.

Degree requirements: For master's, thesis or alternative, oral final exam required, foreign language not required; for doctorate, thesis defense, qualifying exam required.

Entrance requirements: For master's and doctorate, GRE General Test, TOEFL. *Application deadline:* For fall admission, 3/1; for spring admission, 11/1. Applications are processed on a rolling basis. *Application fee:* $50.

Expenses: Tuition, state resident: full-time $6,776; part-time $279 per credit. Tuition, nonresident: full-time $9,936; part-time $412 per credit. Required fees: $20 per credit. $89 per semester. Tuition and fees vary according to course load, campus/location and program.

Financial aid: In 1999–00, 1 fellowship (averaging $3,000 per year), 30 research assistantships with full tuition reimbursements (averaging $13,956 per year), 10 teaching assistantships with full tuition reimbursements (averaging $13,100 per year) were awarded; career-related internships or fieldwork and Federal Work-Study also available. Financial aid application deadline: 3/1; financial aid applicants required to submit FAFSA.

Faculty research: Atmospheric sciences; biological waste treatment; contaminant fate and transport; exposure assessment; air, soil and water quality.

Dr. Peter F. Strom, Director, 732-932-8078, *Fax:* 732-932-8644, *E-mail:* strom@aesop.rutgers.edu.

Application contact: Paul J. Lioy, Graduate Admissions Committee, 732-932-0150,

Fax: 732-445-0116, *E-mail:* plioy@eohsi.rutgers.edu.

■ SONOMA STATE UNIVERSITY

School of Natural Sciences, Department of Biology, Rohnert Park, CA 94928-3609

AWARDS Environmental biology (MA); general biology (MA). Part-time programs available.

Faculty: 13 full-time (3 women), 16 part-time/adjunct (8 women).

Students: 12 full-time (8 women), 16 part-time (13 women); includes 1 minority (Hispanic American). Average age 28. *18 applicants, 39% accepted.* In 1999, 4 degrees awarded.

Degree requirements: For master's, thesis or alternative, oral exam required, foreign language not required.

Entrance requirements: For master's, GRE General Test, GRE Subject Test, minimum GPA of 3.0. *Application deadline:* For fall admission, 11/30. Applications are processed on a rolling basis. *Application fee:* $55.

Expenses: Tuition, nonresident: part-time $246 per unit. Required fees: $2,064; $715 per semester. Tuition and fees vary according to course load.

Financial aid: In 1999–00, 11 students received aid, including 2 research assistantships, 9 teaching assistantships; career-related internships or fieldwork and Federal Work-Study also available. Financial aid application deadline: 3/2.

Faculty research: Molecular biology, genetics, riparian and wetland ecology, plant ecology, feeding mechanisms of invertebrates, ichthyology, microbiology. *Total annual research expenditures:* $32,000.

Dr. Philip Northen, Chairperson, 707-664-2189, *E-mail:* philip.northen@sonoma.edu.

Application contact: John Hopkirk, Graduate Adviser, 707-664-2180.

■ STATE UNIVERSITY OF NEW YORK COLLEGE OF ENVIRONMENTAL SCIENCE AND FORESTRY

Faculty of Environmental and Forest Biology, Syracuse, NY 13210-2779

AWARDS MPS, MS, PhD.

Faculty: 31 full-time (4 women), 3 part-time/adjunct (1 woman).

Students: 82 full-time (49 women), 57 part-time (18 women); includes 1 minority (Hispanic American), 16 international. Average age 29. *82 applicants, 52% accepted.* In 1999, 28 master's, 6 doctorates awarded. Terminal master's awarded for partial completion of doctoral program.

State University of New York College of Environmental Science and Forestry (continued)

Degree requirements: For master's, thesis or alternative required, foreign language not required; for doctorate, dissertation required. *Average time to degree:* Master's–2 years full-time, 3 years part-time; doctorate–6 years full-time, 9 years part-time.

Entrance requirements: For master's and doctorate, GRE General Test, GRE Subject Test, minimum GPA of 3.0. *Application deadline:* For fall admission, 4/15 (priority date); for spring admission, 11/15. Applications are processed on a rolling basis. *Application fee:* $50.

Expenses: Tuition, area resident: Part-time $213 per credit hour. Tuition, state resident: full-time $5,100; part-time $351 per credit hour. Tuition, nonresident: full-time $8,416. Required fees: $286.

Financial aid: In 1999–00, 7 fellowships with full and partial tuition reimbursements, 38 research assistantships with full and partial tuition reimbursements (averaging $9,000 per year), 35 teaching assistantships with full and partial tuition reimbursements (averaging $8,700 per year) were awarded; Federal Work-Study also available.

Faculty research: Ecology, fish and wildlife biology and management, plant science, entomology. *Total annual research expenditures:* $1.9 million.

Dr. Neil H. Ringler, Chairperson, 315-470-6770, *Fax:* 315-470-6934, *E-mail:* neilringler@csf.edu.

Application contact: Dr. Robert H. Frey, Dean, Instruction and Graduate Studies, 315-470-6599, *Fax:* 315-470-6978, *E-mail:* esfgrad@esf.edu.

Find an in-depth description at www.petersons.com/graduate.

■ TENNESSEE TECHNOLOGICAL UNIVERSITY

Graduate School, College of Arts and Sciences, Department of Biology, Cookeville, TN 38505

AWARDS Environmental biology (MS); fish, game, and wildlife management (MS). Part-time programs available.

Faculty: 22 full-time (2 women).
Students: 23 full-time (8 women), 7 part-time (3 women); includes 3 minority (2 Asian Americans or Pacific Islanders, 1 Hispanic American). Average age 25. *16 applicants, 13% accepted.* In 1999, 15 degrees awarded.

Degree requirements: For master's, thesis required, foreign language not required.

Entrance requirements: For master's, GRE General Test, TOEFL. *Application*

deadline: For fall admission, 3/1 (priority date); for spring admission, 8/1. *Application fee:* $25 ($30 for international students).

Expenses: Tuition, state resident: full-time $3,082; part-time $154 per hour. Tuition, nonresident: full-time $7,908; part-time $365 per hour. Required fees: $1,541; $154 per hour. Tuition and fees vary according to course load.

Financial aid: In 1999–00, 24 students received aid, including 17 research assistantships (averaging $7,000 per year), 7 teaching assistantships (averaging $5,320 per year). Financial aid application deadline: 4/1.

Faculty research: Aquatics, environmental studies.

Dr. Daniel Combs, Interim Chairperson, 931-372-3134, *Fax:* 931-372-6257, *E-mail:* dcombs@tntech.edu.

Application contact: Dr. Rebecca F. Quattlebaum, Dean of the Graduate School, 931-372-3233, *Fax:* 931-372-3497, *E-mail:* rquattlebaum@tntech.edu.

■ UNIVERSITY OF CALIFORNIA, BERKELEY

Graduate Division, School of Public Health, Program in Maternal and Child Health, Berkeley, CA 94720-1500
AWARDS MPH.

Entrance requirements: For master's, GRE General Test, minimum GPA of 3.0.
Expenses: Tuition, nonresident: full-time $9,804. Required fees: $4,268. Tuition and fees vary according to program.

■ UNIVERSITY OF COLORADO AT BOULDER

Graduate School, College of Arts and Sciences, Department of Environmental, Population, and Organic Biology, Boulder, CO 80309

AWARDS Animal behavior (MA, PhD); aquatic biology (MA, PhD); behavioral genetics (MA, PhD); ecology (MA, PhD); microbiology (MA, PhD); neurobiology (MA, PhD); plant and animal physiology (MA, PhD); plant and animal systematics (MA, PhD); population biology (MA, PhD); population genetics (MA, PhD).

Faculty: 39 full-time (9 women).
Students: 84 full-time (36 women), 14 part-time (7 women); includes 17 minority (6 Asian Americans or Pacific Islanders, 10 Hispanic Americans, 1 Native American). Average age 29. *147 applicants, 14% accepted.* In 1999, 7 master's, 13 doctorates awarded. Terminal master's awarded for partial completion of doctoral program.

Degree requirements: For master's, thesis or alternative, comprehensive exam required, foreign language not required; for doctorate, dissertation, comprehensive

exam required, foreign language not required. *Average time to degree:* Master's–3 years full-time; doctorate–5 years full-time.

Entrance requirements: For master's, GRE General Test, GRE Subject Test, minimum undergraduate GPA of 3.0; for doctorate, GRE General Test, GRE Subject Test. *Application deadline:* For fall admission, 1/15 (priority date). *Application fee:* $40 ($60 for international students).

Expenses: Tuition, state resident: part-time $181 per credit hour. Tuition, nonresident: part-time $542 per credit hour. Required fees: $99 per term. Tuition and fees vary according to course load and program.

Financial aid: Fellowships, research assistantships, teaching assistantships, Federal Work-Study, institutionally sponsored loans, and tuition waivers (full) available. Financial aid application deadline: 3/1.

Faculty research: Evolution, developmental biology, behavior and neurobiology. *Total annual research expenditures:* $1.8 million.

Michael Breed, Chair, 303-492-8981, *Fax:* 303-492-8699, *E-mail:* michael.breed@colorado.edu.

Application contact: Jill Skarstadt, Graduate Secretary, 303-492-7654, *Fax:* 303-492-8699, *E-mail:* jill.skarstadt@colorado.edu.

■ UNIVERSITY OF LOUISIANA AT LAFAYETTE

Graduate School, College of Sciences, Department of Biology, Lafayette, LA 70504

AWARDS Biology (MS); environmental and evolutionary biology (PhD).

Faculty: 40 full-time (7 women).
Students: 59 full-time (25 women), 8 part-time (4 women); includes 2 minority (1 African American, 1 Asian American or Pacific Islander), 17 international. *62 applicants, 52% accepted.* In 1999, 7 master's, 6 doctorates awarded. Terminal master's awarded for partial completion of doctoral program.

Degree requirements: For master's, thesis required, foreign language not required; for doctorate, 2 foreign languages (computer language can substitute for one), dissertation required.

Entrance requirements: For master's, GRE General Test, minimum GPA of 2.75; for doctorate, GRE General Test, GRE Subject Test, minimum GPA of 3.0. *Application deadline:* For fall admission, 5/15. *Application fee:* $20 ($30 for international students).

Expenses: Tuition, state resident: full-time $2,021; part-time $287 per credit. Tuition, nonresident: full-time $7,253; part-time

$287 per credit. Part-time tuition and fees vary according to course load.

Financial aid: In 1999–00, 14 fellowships with full tuition reimbursements (averaging $14,572 per year), 7 research assistantships with full tuition reimbursements (averaging $4,565 per year), 24 teaching assistantships with full tuition reimbursements (averaging $5,377 per year) were awarded; Federal Work-Study and institutionally sponsored loans also available. Financial aid application deadline: 5/1.

Faculty research: Structure and ultrastructure, system biology, ecology, processes, environmental physiology.
Dr. Darryl L. Felder, Head, 337-482-6748.
Application contact: Dr. Karl Hasenstein, Graduate Coordinator, 337-482-6750.
Find an in-depth description at www.petersons.com/graduate.

■ **UNIVERSITY OF LOUISVILLE**

Graduate School, College of Arts and Sciences, Department of Biology, Program in Environmental Biology, Louisville, KY 40292-0001

AWARDS PhD.

Degree requirements: For doctorate, dissertation required.

Entrance requirements: For doctorate, GRE General Test.

Expenses: Tuition, state resident: full-time $3,260; part-time $182 per hour. Tuition, nonresident: full-time $9,780; part-time $544 per hour. Required fees: $143; $28 per hour. Tuition and fees vary according to program.

■ **UNIVERSITY OF MASSACHUSETTS AMHERST**

Graduate School, College of Food and Natural Resources, Department of Natural Resources Conservation, Program in Wildlife and Fisheries Conservation, Amherst, MA 01003

AWARDS MS, PhD. Part-time programs available.

Faculty: 25 full-time (3 women).
Students: 33 full-time (16 women), 31 part-time (16 women); includes 1 minority (Asian American or Pacific Islander), 2 international. Average age 31. *79 applicants, 23% accepted.* In 1999, 19 master's, 2 doctorates awarded. Terminal master's awarded for partial completion of doctoral program.
Degree requirements: For master's, thesis optional, foreign language not required; for doctorate, variable foreign language requirement, dissertation required.
Entrance requirements: For master's and doctorate, GRE General Test. *Application deadline:* For fall admission, 2/1 (priority

date); for spring admission, 10/1. Applications are processed on a rolling basis.
Application fee: $40.

Expenses: Tuition, state resident: full-time $2,640; part-time $165 per credit. Tuition, nonresident: full-time $9,756; part-time $407 per credit. Required fees: $1,221 per term. One-time fee: $110. Full-time tuition and fees vary according to course load, campus/location and reciprocity agreements.

Financial aid: Fellowships with full tuition reimbursements, research assistantships with full tuition reimbursements, teaching assistantships with full tuition reimbursements, career-related internships or fieldwork, Federal Work-Study, grants, scholarships, traineeships, and unspecified assistantships available. Aid available to part-time students. Financial aid application deadline: 2/1.
Dr. Todd K. Fuller, Director, 413-545-2665, *Fax:* 413-545-4358, *E-mail:* tkfuller@forwild.umass.edu.

■ **UNIVERSITY OF MASSACHUSETTS BOSTON**

Office of Graduate Studies and Research, College of Arts and Sciences, Faculty of Sciences, Program in Environmental, Coastal and Ocean Sciences, Track in Environmental Biology, Boston, MA 02125-3393

AWARDS PhD. Part-time and evening/weekend programs available.

Students: 18 full-time (15 women), 15 part-time (8 women); includes 2 minority (1 African American, 1 Asian American or Pacific Islander), 6 international. In 1999, 1 degree awarded.
Degree requirements: For doctorate, dissertation, comprehensive exams, oral exams required, foreign language not required.
Entrance requirements: For doctorate, GRE General Test, minimum GPA of 2.75. *Application deadline:* For fall admission, 2/1; for spring admission, 10/15. *Application fee:* $25 ($35 for international students).
Expenses: Tuition, state resident: full-time $2,590; part-time $108 per credit. Tuition, nonresident: full-time $4,758; part-time $407 per credit. Required fees: $150; $159 per term.
Financial aid: In 1999–00, research assistantships with full tuition reimbursements (averaging $8,000 per year), teaching assistantships with full tuition reimbursements (averaging $8,000 per year) were awarded; career-related internships or fieldwork, Federal Work-Study, and unspecified assistantships also available. Aid available to part-time students.

Financial aid application deadline: 3/1; financial aid applicants required to submit FAFSA.

Faculty research: Polychoets biology, predator and prey relationships, population and evolutionary biology, neurobiology, biodiversity.
Dr. Michael Shiaris, Director, 617-287-6600.
Application contact: Lisa Lavely, Director of Graduate Admissions and Records, 617-287-6400, *Fax:* 617-287-6236, *E-mail:* bos.gadm@dpc.umassp.edu.

■ **UNIVERSITY OF NEVADA, LAS VEGAS**

Graduate College, College of Science, Department of Biological Sciences, Program in Environmental Biology, Las Vegas, NV 89154-9900

AWARDS PhD.

Students: 11 full-time (7 women), 9 part-time (5 women); includes 2 minority (both Asian Americans or Pacific Islanders). *9 applicants, 22% accepted.* In 1999, 1 degree awarded.
Degree requirements: For doctorate, one foreign language, computer language, dissertation required.
Entrance requirements: For doctorate, GRE General Test, GRE Subject Test, minimum GPA of 3.5. *Application deadline:* For fall admission, 6/15. *Application fee:* $40 ($95 for international students).
Expenses: Tuition, state resident: part-time $97 per credit. Tuition, nonresident: full-time $6,347; part-time $198 per credit. Required fees: $62; $31 per semester.
Financial aid: In 1999–00, 4 research assistantships with full tuition reimbursements (averaging $12,800 per year), 7 teaching assistantships with partial tuition reimbursements (averaging $9,800 per year) were awarded. Financial aid application deadline: 3/1.
Dr. Peter Starkweather, Chair, 702-895-3399.

■ **UNIVERSITY OF NORTH DAKOTA**

Graduate School, College of Arts and Sciences, Department of Biology, Grand Forks, ND 58202

AWARDS Botany (MS, PhD); ecology (MS, PhD); entomology (MS, PhD); environmental biology (MS, PhD); fisheries/wildlife (MS, PhD); genetics (MS, PhD); zoology (MS, PhD).

Faculty: 18 full-time (3 women).
Students: 21 full-time (8 women). *13 applicants, 62% accepted.* In 1999, 3 master's awarded. Terminal master's awarded for partial completion of doctoral program.

University of North Dakota (continued)

Degree requirements: For master's, thesis, final exam required; for doctorate, dissertation, comprehensive exam, final exam required.

Entrance requirements: For master's, GRE General Test, GRE Subject Test, TOEFL, minimum GPA of 3.0; for doctorate, GRE General Test, GRE Subject Test, TOEFL, minimum GPA of 3.5. *Application deadline:* For fall admission, 3/1 (priority date). Applications are processed on a rolling basis. *Application fee:* $25.

Expenses: Tuition, state resident: full-time $3,166; part-time $158 per credit. Tuition, nonresident: full-time $7,658; part-time $345 per credit. International tuition: $7,658 full-time. Required fees: $46 per credit. Tuition and fees vary according to program and reciprocity agreements.

Financial aid: In 1999–00, 6 research assistantships with full tuition reimbursements (averaging $11,250 per year), 15 teaching assistantships with full tuition reimbursements (averaging $11,250 per year) were awarded; fellowships, Federal Work-Study, institutionally sponsored loans, scholarships, and tuition waivers (full and partial) also available. Aid available to part-time students. Financial aid application deadline: 3/15; financial aid applicants required to submit FAFSA.

Faculty research: Population biology, wildlife ecology, RNA processing, hormonal control of behavior.

Dr. Jeff Lang, Director, 701-777-2621, *Fax:* 701-777-2623, *E-mail:* jlang@ badlands.nodak.edu.

■ UNIVERSITY OF NOTRE DAME

Graduate School, College of Science, Department of Biological Sciences, Program in Aquatic Ecology, Evolution and Environmental Biology, Notre Dame, IN 46556

AWARDS MS, PhD. Terminal master's awarded for partial completion of doctoral program.

Degree requirements: For master's and doctorate, thesis/dissertation required, foreign language not required.

Entrance requirements: For master's and doctorate, GRE General Test, GRE Subject Test, TOEFL. *Application deadline:* For fall admission, 2/1 (priority date); for spring admission, 11/1. Applications are processed on a rolling basis. *Application fee:* $50.

Expenses: Tuition: Full-time $21,930; part-time $1,218 per credit. Required fees: $95. Tuition and fees vary according to program.

Financial aid: Fellowships with full tuition reimbursements, research assistantships

with full tuition reimbursements, teaching assistantships with full tuition reimbursements, traineeships and tuition waivers (full) available. Financial aid application deadline: 2/1.

Faculty research: Aquatic benthic ecology, biogeochemistry, plant-animal interactions, species invasions, ecological genetics and speciation, aquatic community studies.

Application contact: Dr. Terrence J. Akai, Director of Graduate Admissions, 219-631-7706, *Fax:* 219-631-4183, *E-mail:* gradad@nd.edu.

■ UNIVERSITY OF SOUTHERN MISSISSIPPI

Graduate School, College of Science and Technology, Department of Biological Sciences, Hattiesburg, MS 39406

AWARDS Environmental biology (MS, PhD); marine biology (MS, PhD); microbiology (MS, PhD); molecular biology (MS, PhD).

Degree requirements: For master's, thesis required, foreign language not required; for doctorate, 2 foreign languages (computer language can substitute for one), dissertation required.

Entrance requirements: For master's, GRE General Test, TOEFL, minimum GPA of 3.0; for doctorate, GRE General Test, TOEFL, minimum GPA of 3.5.

Expenses: Tuition, state resident: full-time $2,250; part-time $137 per semester hour. Tuition, nonresident: full-time $3,102; part-time $172 per semester hour. Required fees: $602.

■ UNIVERSITY OF WISCONSIN–MADISON

Graduate School, College of Agricultural and Life Sciences, Environmental Toxicology Center, Madison, WI 53706-1380

AWARDS Molecular and environmental toxicology (MS, PhD).

Degree requirements: For doctorate, dissertation required.

Expenses: Tuition, state resident: full-time $5,406; part-time $339 per credit. Tuition, nonresident: full-time $17,110; part-time $1,071 per credit. Full-time tuition and fees vary according to program and reciprocity agreements. Part-time tuition and fees vary according to course load and program.

Find an in-depth description at www.petersons.com/graduate.

■ WASHINGTON UNIVERSITY IN ST. LOUIS

Graduate School of Arts and Sciences, Division of Biology and Biomedical Sciences, Program in Evolutionary and Population Biology, St. Louis, MO 63130-4899

AWARDS Ecology (PhD); environmental biology (PhD); evolutionary biology (PhD); genetics (PhD).

Degree requirements: For doctorate, dissertation required, foreign language not required.

Entrance requirements: For doctorate, GRE General Test, GRE Subject Test. *Application deadline:* For fall admission, 1/1 (priority date). Applications are processed on a rolling basis. *Application fee:* $0.

Expenses: Tuition: Full-time $23,400; part-time $975 per credit. Tuition and fees vary according to program.

Financial aid: Fellowships, research assistantships available. Financial aid application deadline: 1/1.

Dr. Jonathan Losos, Coordinator, 314-362-4188.

Application contact: Rosemary Garagneni, Director of Admissions, 800-852-9074, *E-mail:* admissions@ dbbs.wustl.edu.

■ WEST VIRGINIA UNIVERSITY

College of Agriculture, Forestry and Consumer Sciences, Division of Plant and Soil Sciences, Morgantown, WV 26506

AWARDS Agronomy (MS); entomology (MS); environmental microbiology (MS); horticulture (MS); plant and soil sciences (PhD); plant pathology (MS).

Faculty: 19 full-time (1 woman), 2 part-time/adjunct (1 woman).

Students: 18 full-time (10 women), 7 part-time (5 women); includes 1 minority (Native American), 2 international. Average age 26. In 1999, 1 master's awarded.

Degree requirements: For master's, thesis required, foreign language not required; for doctorate, dissertation, comprehensive exam required, foreign language not required.

Entrance requirements: For master's, GRE, TOEFL, minimum GPA of 2.5; for doctorate, GRE General Test, TOEFL, minimum GPA of 3.0. *Application fee:* $45.

Expenses: Tuition, state resident: full-time $2,910; part-time $154 per credit hour. Tuition, nonresident: full-time $8,368; part-time $457 per credit hour.

Financial aid: In 1999–00, 13 research assistantships, 3 teaching assistantships were awarded; Federal Work-Study, institutionally sponsored loans, and tuition waivers (full and partial) also available.

Financial aid application deadline: 2/1; financial aid applicants required to submit FAFSA.

Faculty research: Water quality, reclamation of disturbed land, crop production, pest control, environmental protection. Dr. Barton Baker, Chair, 304-293-4817, *Fax:* 304-293-3740, *E-mail:* bbaker2@ wvu.edu.

■ WEST VIRGINIA UNIVERSITY

Eberly College of Arts and Sciences, Department of Biology, Morgantown, WV 26506

AWARDS Animal behavior (MS); cellular and molecular biology (MS, PhD); environmental plant biology (MS, PhD); plant systematics (MS); population genetics (MS).

Faculty: 18 full-time (4 women), 2 part-time/adjunct (1 woman).
Students: 19 full-time (10 women), 3 part-time (2 women); includes 2 minority (1 Asian American or Pacific Islander, 1 Hispanic American), 4 international. Average age 27. *50 applicants, 10% accepted.* In 1999, 4 master's, 1 doctorate awarded. Terminal master's awarded for partial completion of doctoral program.
Degree requirements: For master's, thesis, final exam required, foreign language not required; for doctorate, dissertation, preliminary and final exams required, foreign language not required. *Average time to degree:* Master's–2.5 years full-time; doctorate–5 years full-time.
Entrance requirements: For master's, GRE General Test, GRE Subject Test, TOEFL, minimum GPA of 3.0; for doctorate, GRE General Test, TOEFL, minimum GPA of 3.0. *Application deadline:* For fall admission, 4/1; for spring admission, 10/1. Applications are processed on a rolling basis. *Application fee:* $45.
Expenses: Tuition, state resident: full-time $2,910; part-time $154 per credit hour. Tuition, nonresident: full-time $8,368; part-time $457 per credit hour.
Financial aid: In 1999–00, 3 research assistantships, 13 teaching assistantships were awarded; Federal Work-Study and institutionally sponsored loans also available. Financial aid application deadline: 4/1; financial aid applicants required to submit FAFSA.
Faculty research: Environmental biology, genetic engineering, developmental biology, global change, biodiversity. Dr. Keith Garbutt, Chair, 304-293-5394.
Application contact: Dr. James B. McGraw, Director of Graduate Studies, 304-293-5201, *Fax:* 304-293-6363.

Find an in-depth description at www.petersons.com/graduate.

EVOLUTIONARY BIOLOGY

■ ARIZONA STATE UNIVERSITY

Graduate College, College of Liberal Arts and Sciences, Department of Biology, Program in Evolution, Tempe, AZ 85287

AWARDS MS, PhD. Terminal master's awarded for partial completion of doctoral program.

Degree requirements: For master's, thesis required, foreign language not required; for doctorate, dissertation, oral exam required, foreign language not required.
Entrance requirements: For master's and doctorate, GRE General Test, GRE Subject Test. *Application deadline:* For fall admission, 12/15. *Application fee:* $45.
Expenses: Tuition, state resident: part-time $115 per credit hour. Tuition, nonresident: part-time $389 per credit hour. Required fees: $18 per semester. Tuition and fees vary according to program.
Financial aid: Application deadline: 12/15.

■ BOSTON UNIVERSITY

Graduate School of Arts and Sciences, Department of Biology, Boston, MA 02215

AWARDS Botany (MA, PhD); cell and molecular biology (MA, PhD); cell biology (MA, PhD); ecology (PhD); ecology, behavior, and evolution (MA, PhD); ecology/physiology, endocrinology and reproduction (MA); marine biology (MA, PhD); molecular biology, cell biology and biochemistry (MA, PhD); neurobiology, neuroendocrinology and reproduction (MA, PhD); physiology, endocrinology, and neurobiology (MA, PhD); zoology (MA, PhD). Part-time programs available.

Faculty: 41 full-time (8 women).
Students: 131 full-time (74 women), 11 part-time (7 women); includes 10 minority (7 Asian Americans or Pacific Islanders, 3 Hispanic Americans), 33 international. Average age 27. *238 applicants, 39% accepted.* In 1999, 61 master's, 45 doctorates awarded. Terminal master's awarded for partial completion of doctoral program.
Degree requirements: For master's, one foreign language, thesis not required; for doctorate, one foreign language, dissertation, qualifying exam required. *Average time to degree:* Master's–1 year full-time, 3 years part-time; doctorate–5.75 years full-time.
Entrance requirements: For master's and doctorate, GRE General Test, GRE Subject Test, TOEFL. *Application deadline:*

For fall admission, 1/1 (priority date); for spring admission, 11/1. *Application fee:* $50.
Expenses: Tuition: Full-time $23,770; part-time $743 per credit. Required fees: $220. Tuition and fees vary according to class time, course level, campus/location and program.
Financial aid: In 1999–00, 82 students received aid, including 1 fellowship with full tuition reimbursement available (averaging $12,000 per year), 28 research assistantships with full tuition reimbursements available (averaging $11,500 per year), 43 teaching assistantships with full tuition reimbursements available (averaging $11,500 per year); Federal Work-Study, grants, institutionally sponsored loans, scholarships, and traineeships also available. Financial aid application deadline: 1/15; financial aid applicants required to submit FAFSA.
Faculty research: Marine science, endocrinology, behavior. *Total annual research expenditures:* $5 million. Geoffrey M. Cooper, Chairman, 617-353-2432, *Fax:* 617-353-6340, *E-mail:* gmcooper@bu.edu.
Application contact: Yolanta Kovalko, Senior Staff Assistant, 617-353-2432, *Fax:* 617-353-6340, *E-mail:* yolanta@bu.edu.

Find an in-depth description at www.petersons.com/graduate.

■ BROWN UNIVERSITY

Graduate School, Division of Biology and Medicine, Program in Ecology and Evolutionary Biology, Providence, RI 02912

AWARDS PhD.

Faculty: 11 full-time (4 women).
Students: 13 full-time (4 women), 1 international. Average age 24. *33 applicants, 9% accepted.* In 1999, 3 degrees awarded.
Degree requirements: For doctorate, dissertation, preliminary exam required.
Entrance requirements: For doctorate, GRE General Test, GRE Subject Test. *Application deadline:* For fall admission, 1/2 (priority date). Applications are processed on a rolling basis. *Application fee:* $60.
Financial aid: In 1999–00, 3 fellowships, 1 research assistantship, 7 teaching assistantships were awarded. Financial aid application deadline: 1/2.
Faculty research: Marine ecology, behavioral ecology, population genetics, evolutionary morphology, plant ecology. Dr. Mark Bertness, Director, 401-863-2280, *E-mail:* mark_bertness@brown.edu.

■ COLUMBIA UNIVERSITY

Graduate School of Arts and Sciences, Graduate Program in Ecology and Evolutionary Biology, New York, NY 10027

AWARDS Conservation biology (Certificate); ecology and evolutionary biology (PhD); environmental policy (Certificate).

Degree requirements: For doctorate, dissertation, teaching experience required.
Entrance requirements: For doctorate, GRE General Test, TOEFL, previous course work in biology. Electronic applications accepted.
Expenses: Tuition: Full-time $25,072. Full-time tuition and fees vary according to course load and program.
Faculty research: Tropical ecology, ethnobotany, global change, systematics.

Find an in-depth description at www.petersons.com/graduate.

■ CORNELL UNIVERSITY

Graduate School, Graduate Fields of Agriculture and Life Sciences, Field of Ecology and Evolutionary Biology, Ithaca, NY 14853-0001

AWARDS Ecology (PhD), including animal ecology, applied ecology, biogeochemistry, community and ecosystem ecology, limnology, oceanography, physiological ecology, plant ecology, population ecology, theoretical ecology, vertebrate zoology; evolutionary biology (PhD), including ecological genetics, paleobiology, population biology, systematics.

Faculty: 41 full-time.
Students: 55 full-time (27 women); includes 4 minority (3 Asian Americans or Pacific Islanders, 1 Hispanic American), 2 international. *115 applicants, 11% accepted.* In 1999, 12 doctorates awarded.
Degree requirements: For doctorate, dissertation, 2 semesters of teaching experience required, foreign language not required.
Entrance requirements: For doctorate, GRE General Test, GRE Subject Test (biology), TOEFL. *Application deadline:* For fall admission, 12/15. *Application fee:* $65. Electronic applications accepted.
Expenses: Tuition: Full-time $12,100.
Financial aid: In 1999–00, 54 students received aid, including 28 fellowships with full tuition reimbursements available, 4 research assistantships with full tuition reimbursements available, 22 teaching assistantships with full tuition reimbursements available; institutionally sponsored loans, scholarships, tuition waivers (full and partial), and unspecified assistantships also available. Financial aid applicants required to submit FAFSA.

Faculty research: Population and organismal biology, population and evolutionary genetics, systematics and macroevolution, biochemistry, conservation biology.
Application contact: Graduate Field Assistant, 607-254-4230, *E-mail:* eeb_grad_req@cornell.edu.

■ EMORY UNIVERSITY

Graduate School of Arts and Sciences, Division of Biological and Biomedical Sciences, Program in Population Biology, Ecology, and Evolution, Atlanta, GA 30322-1100
AWARDS PhD.

Faculty: 20 full-time (3 women).
Students: 11 full-time (5 women), 2 international.
Degree requirements: For doctorate, dissertation required, foreign language not required.
Entrance requirements: For doctorate, GRE General Test, TOEFL, minimum GPA of 3.0 in science course work. *Application deadline:* For fall admission, 1/20 (priority date). *Application fee:* $45.
Expenses: Tuition: Full-time $22,770. Tuition and fees vary according to program.
Financial aid: In 1999–00, fellowships with full tuition reimbursements (averaging $18,000 per year)
Dr. Leslie Real, Director, 404-727-4099, *Fax:* 404-727-2880, *E-mail:* lreal@biology.emory.edu.
Application contact: 404-727-2547, *Fax:* 404-727-3322, *E-mail:* gdbbs@gsas.emory.edu.

Find an in-depth description at www.petersons.com/graduate.

■ FLORIDA STATE UNIVERSITY

Graduate Studies, College of Arts and Sciences, Department of Biological Science, Program in Evolutionary Biology, Tallahassee, FL 32306
AWARDS MS, PhD.

Faculty: 13 full-time (2 women).
Students: 30 full-time (12 women); includes 2 minority (1 African American, 1 Asian American or Pacific Islander), 1 international.
Degree requirements: For master's and doctorate, thesis/dissertation, teaching experience required.
Entrance requirements: For master's, GRE General Test, TOEFL; for doctorate, GRE General Test, GRE Subject Test, TOEFL. *Application deadline:* For fall admission, 1/15; for spring admission, 10/15. *Application fee:* $20.
Expenses: Tuition, state resident: full-time $3,504; part-time $146 per credit hour.

Tuition, nonresident: full-time $12,162; part-time $507 per credit hour. Tuition and fees vary according to program.
Financial aid: In 1999–00, fellowships with full tuition reimbursements (averaging $13,740 per year), research assistantships with full tuition reimbursements (averaging $13,740 per year), teaching assistantships with full tuition reimbursements (averaging $13,740 per year) were awarded. Financial aid application deadline: 1/15; financial aid applicants required to submit FAFSA.
Faculty research: Population biology, community ecology, genetics, evolution of protein structure.
Dr. Thomas C. S. Keller, Associate Professor and Associate Chairman, 850-644-3023, *Fax:* 850-644-9829.
Application contact: Judy Bowers, Coordinator, Graduate Affairs, 850-644-3023, *Fax:* 850-644-9829, *E-mail:* bowers@bio.fsu.edu.

■ GEORGE MASON UNIVERSITY

College of Arts and Sciences, Department of Biology, Master's Program in Biology, Fairfax, VA 22030-4444

AWARDS Bioinformatics (MS); ecology, systematics and evolution (MS); environmental science and public policy (MS); interpretive biology (MS); molecular, microbial, and cellular biology (MS); organismal biology (MS). Part-time programs available.

Faculty: 30 full-time (11 women), 32 part-time/adjunct (20 women).
Students: 5 full-time (3 women), 55 part-time (37 women); includes 7 minority (1 African American, 4 Asian Americans or Pacific Islanders, 2 Hispanic Americans), 8 international. Average age 34. *36 applicants, 44% accepted.* In 1999, 18 degrees awarded.
Degree requirements: For master's, thesis or alternative required, foreign language not required.
Entrance requirements: For master's, GRE General Test, GRE Subject Test, bachelor's degree in biology or equivalent. *Application deadline:* For fall admission, 5/1; for spring admission, 11/1. *Application fee:* $30. Electronic applications accepted.
Expenses: Tuition, state resident: full-time $4,416; part-time $184 per credit hour. Tuition, nonresident: full-time $12,516; part-time $522 per credit hour. Tuition and fees vary according to program.
Financial aid: Available to part-time students. Application deadline: 3/1.
Dr. George E. Andrykovitch, Director, 703-993-1027, *Fax:* 703-993-1046.

■ HARVARD UNIVERSITY

Graduate School of Arts and Sciences, Department of Organismic and Evolutionary Biology, Cambridge, MA 02138

AWARDS Biology (PhD).

Students: 57 full-time (22 women). *98 applicants, 13% accepted.* In 1999, 11 doctorates awarded.

Degree requirements: For doctorate, public presentation of thesis research, exam required.

Entrance requirements: For doctorate, GRE General Test, GRE Subject Test (recommended), TOEFL, 7 courses in biology, chemistry, physics, mathematics, computer science, or geology. *Application deadline:* For fall admission, 12/15. *Application fee:* $60.

Expenses: Tuition: Full-time $22,054. Required fees: $711. Tuition and fees vary according to program.

Financial aid: Fellowships, research assistantships, teaching assistantships, career-related internships or fieldwork, Federal Work-Study, and institutionally sponsored loans available. Financial aid application deadline: 12/30.
Josephine Ferraro, Officer, 617-495-5396.
Application contact: Departmental Office, 617-495-2305.

■ INDIANA UNIVERSITY BLOOMINGTON

Graduate School, College of Arts and Sciences, Department of Biology, Program in Evolution, Ecology, and Behavior, Bloomington, IN 47405

AWARDS Ecology (MA, PhD); evolutionary biology (MA, PhD); zoology (MA, PhD). PhD offered through the University Graduate School. Part-time programs available.

Students: 66 full-time (35 women), 2 part-time (1 woman); includes 8 minority (2 African Americans, 2 Asian Americans or Pacific Islanders, 4 Hispanic Americans), 3 international. In 1999, 2 master's, 7 doctorates awarded. Terminal master's awarded for partial completion of doctoral program.

Degree requirements: For master's, thesis or alternative required; for doctorate, dissertation required.

Entrance requirements: For master's and doctorate, GRE General Test, TOEFL. *Application deadline:* For fall admission, 1/5 (priority date); for spring admission, 9/1. Applications are processed on a rolling basis. *Application fee:* $45. Electronic applications accepted.

Expenses: Tuition, state resident: full-time $3,853; part-time $161 per credit hour. Tuition, nonresident: full-time $11,226; part-time $468 per credit hour. Required

fees: $360 per year. Tuition and fees vary according to course load and program.

Financial aid: Fellowships with tuition reimbursements, research assistantships with tuition reimbursements, teaching assistantships with tuition reimbursements, scholarships available. Financial aid application deadline: 1/15.

Faculty research: Ecosystem of community, plant and animal population biology, avian sociobiology, fish ethology, evolutionary genetics.
Dr. Gerald Gastony, Head, 812-855-3333, *Fax:* 812-855-6705, *E-mail:* gastony@indiana.edu.
Application contact: Gretchen Clearwater, Advisor for Graduate Affairs, 812-855-1861, *Fax:* 812-855-6705, *E-mail:* biograd@bio.indiana.edu.

Find an in-depth description at www.petersons.com/graduate.

■ IOWA STATE UNIVERSITY OF SCIENCE AND TECHNOLOGY

Graduate College, Interdisciplinary Programs, Program in Ecology and Evolutionary Biology, Ames, IA 50011

AWARDS MS, PhD.

Students: In 1999, 3 master's, 2 doctorates awarded.

Degree requirements: For master's, thesis or alternative required; for doctorate, dissertation required.

Entrance requirements: For master's and doctorate, GRE General Test, TOEFL. *Application deadline:* For fall admission, 2/1. *Application fee:* $20 ($50 for international students). Electronic applications accepted.

Expenses: Tuition, state resident: full-time $3,308. Tuition, nonresident: full-time $9,744. Part-time tuition and fees vary according to course load, campus/location and program.

Financial aid: Scholarships available.

Faculty research: Landscape ecology, aquatic and method ecology, physiological ecology, population genetics and evolution, systematics.
Dr. William R. Clark, Supervisory Committee Chair, 515-294-7252, *E-mail:* eebadm@iastate.edu.

■ LEHIGH UNIVERSITY

College of Arts and Sciences, Department of Biological Sciences, Bethlehem, PA 18015-3094

AWARDS Behavioral and evolutionary bioscience (PhD); behavioral neuroscience (PhD); biochemistry (PhD); biology (PhD); molecular biology (PhD). Part-time programs available. Postbaccalaureate distance learning degree programs offered (no on-campus study).

Students: 31 full-time (22 women), 77 part-time (47 women); includes 11 minority (3 African Americans, 5 Asian Americans or Pacific Islanders, 3 Hispanic Americans), 8 international. *142 applicants, 22% accepted.* In 1999, 5 doctorates awarded.

Degree requirements: For doctorate, dissertation, comprehensive exam required, foreign language not required. *Average time to degree:* Doctorate–6.3 years full-time.

Entrance requirements: For doctorate, GRE General Test, GRE Subject Test, TOEFL. *Application deadline:* For fall admission, 7/15; for spring admission, 12/1. Applications are processed on a rolling basis. *Application fee:* $40. Electronic applications accepted.

Expenses: Tuition: Part-time $860 per credit. Required fees: $6 per term. Tuition and fees vary according to program.

Financial aid: In 1999–00, 30 students received aid, including 4 fellowships, 6 research assistantships, 15 teaching assistantships; career-related internships or fieldwork, institutionally sponsored loans, and tuition waivers (full and partial) also available. Financial aid application deadline: 1/15.

Faculty research: Gene expression, cell biology, virology, bacteriology, developmental biology.
Dr. Neal G. Simon, Chairperson, 610-758-3680, *Fax:* 610-758-4004.
Application contact: Dr. Jennifer J. Swann, Graduate Coordinator, 610-758-5884, *Fax:* 610-758-4004, *E-mail:* jms5@lehigh.edu.

■ MARQUETTE UNIVERSITY

Graduate School, College of Arts and Sciences, Department of Biology, Milwaukee, WI 53201-1881

AWARDS Cell biology (MS, PhD); developmental biology (MS, PhD); ecology (MS, PhD); endocrinology (MS, PhD); evolutionary biology (MS, PhD); genetics (MS, PhD); microbiology (MS, PhD); molecular biology (MS, PhD); muscle and exercise physiology (MS, PhD); neurobiology (MS, PhD); reproductive physiology (MS, PhD).

Faculty: 16 full-time (4 women), 2 part-time/adjunct (0 women).

Students: 34 full-time (20 women), 3 part-time; includes 3 minority (all Asian Americans or Pacific Islanders), 2 international. Average age 31. *42 applicants, 29% accepted.* In 1999, 1 master's, 4 doctorates awarded. Terminal master's awarded for partial completion of doctoral program.

Degree requirements: For master's, thesis, 1 year of teaching experience or equivalent, comprehensive exam required,

Marquette University (continued)
foreign language not required; for doctorate, dissertation, 1 year of teaching experience or equivalent, qualifying exam required, foreign language not required.
Entrance requirements: For master's and doctorate, GRE General Test, GRE Subject Test, TOEFL. *Application fee:* $40.
Expenses: Tuition: Part-time $510 per credit hour. Tuition and fees vary according to program.
Financial aid: In 1999–00, 4 fellowships, 22 teaching assistantships were awarded; research assistantships, Federal Work-Study, institutionally sponsored loans, scholarships, and tuition waivers (full and partial) also available. Aid available to part-time students. Financial aid application deadline: 2/15.
Faculty research: Microbial and invertebrate ecology, evolution of gene function, DNA methylation, DNA arrangement. *Total annual research expenditures:* $1.5 million.
Dr. Brian Unsworth, Chairman, 414-288-7355, *Fax:* 414-288-7357.
Application contact: Barbara DeNoyer, Graduate Studies Coordinator, 414-288-7355, *Fax:* 414-288-7357.

Find an in-depth description at www.petersons.com/graduate.

■ NEW YORK UNIVERSITY

Graduate School of Arts and Science, Department of Biology, New York, NY 10012-1019

AWARDS Applied recombinant DNA technology (MS); biochemistry (PhD); biomedical journalism (MA); cell biology (PhD); computers in biological research (MS); environmental biology (PhD); general biology (MS); neural sciences and physiology (PhD); oral biology (MS); population and evolutionary biology (PhD). Part-time programs available.

Faculty: 22 full-time (5 women), 8 part-time/adjunct.
Students: 99 full-time (48 women), 61 part-time (31 women); includes 30 minority (4 African Americans, 22 Asian Americans or Pacific Islanders, 4 Hispanic Americans), 52 international. Average age 24. *371 applicants, 41% accepted.* In 1999, 54 master's, 4 doctorates awarded. Terminal master's awarded for partial completion of doctoral program.
Degree requirements: For master's, thesis or alternative, qualifying paper required, foreign language not required; for doctorate, dissertation, oral and written comprehensive exams required, foreign language not required.
Entrance requirements: For master's, GRE General Test, TOEFL; for doctorate, GRE General Test, GRE Subject Test,

TOEFL. *Application deadline:* For fall admission, 1/4 (priority date). *Application fee:* $60.
Expenses: Tuition: Full-time $17,880; part-time $745 per credit. Required fees: $1,140; $35 per credit. Tuition and fees vary according to course load and program.
Financial aid: Fellowships with tuition reimbursements, research assistantships with tuition reimbursements, teaching assistantships with tuition reimbursements, career-related internships or fieldwork, Federal Work-Study, institutionally sponsored loans, and tuition waivers (full and partial) available. Financial aid application deadline: 1/4; financial aid applicants required to submit FAFSA.
Faculty research: Development and genetics, neurobiology, plant sciences, molecular and cell biology.
Philip Furmanski, Chairman, 212-998-8200.
Application contact: Gloria Coruzzi, Director of Graduate Studies, 212-998-8200, *Fax:* 212-995-4015, *E-mail:* biology@nyu.edu.

Find an in-depth description at www.petersons.com/graduate.

■ NORTHWESTERN UNIVERSITY

The Graduate School, Division of Interdepartmental Programs and Medical School, Integrated Graduate Programs in the Life Sciences, Chicago, IL 60611

AWARDS Cancer biology (PhD); cell biology (PhD); developmental biology (PhD); evolutionary biology (PhD); immunology and microbial pathogenesis (PhD); molecular biology and genetics (PhD); neurobiology (PhD); pharmacology and toxicology (PhD); structural biology and biochemistry (PhD).

Degree requirements: For doctorate, dissertation, written and oral qualifying exams required, foreign language not required.
Entrance requirements: For doctorate, GRE General Test, TOEFL.
Expenses: Tuition: Full-time $23,301. Full-time tuition and fees vary according to program.

Find an in-depth description at www.petersons.com/graduate.

■ THE OHIO STATE UNIVERSITY

Graduate School, College of Biological Sciences, Department of Evolution, Ecology, and Organismal Biology, Columbus, OH 43210

AWARDS MS, PhD.

Faculty: 41 full-time, 12 part-time/adjunct.
Students: 72 full-time (37 women), 7 part-time (4 women); includes 5 minority (2

Asian Americans or Pacific Islanders, 2 Hispanic Americans, 1 Native American), 8 international. *50 applicants, 46% accepted.* In 1999, 7 master's, 5 doctorates awarded.
Degree requirements: For master's, thesis optional, foreign language not required; for doctorate, dissertation required, foreign language not required.
Entrance requirements: For master's and doctorate, GRE General Test, GRE Subject Test (biology), TOEFL, TSE. *Application deadline:* For fall admission, 8/15. Applications are processed on a rolling basis. *Application fee:* $30 ($40 for international students).
Expenses: Tuition, state resident: full-time $5,400. Tuition, nonresident: full-time $14,535. Part-time tuition and fees vary according to course load and program.
Financial aid: Fellowships, research assistantships, teaching assistantships, Federal Work-Study and institutionally sponsored loans available. Aid available to part-time students.
Dr. Ralph E. J. Boerner, Chairperson, 614-292-8088, *Fax:* 614-292-2030, *E-mail:* boerner.1@osu.edu.

■ THE PENNSYLVANIA STATE UNIVERSITY UNIVERSITY PARK CAMPUS

Graduate School, Eberly College of Science, Department of Biology, State College, University Park, PA 16802-1503

AWARDS Biology (MS, PhD); molecular evolutionary biology (MS, PhD).

Students: 28 full-time (12 women), 10 part-time (4 women). In 1999, 6 master's, 2 doctorates awarded.
Entrance requirements: For master's and doctorate, GRE General Test. *Application fee:* $50.
Expenses: Tuition, state resident: full-time $6,886; part-time $291 per credit. Tuition, nonresident: full-time $14,118; part-time $588 per credit. Required fees: $46 per semester. Part-time tuition and fees vary according to course load and program.
Financial aid: Fellowships, research assistantships, teaching assistantships available.
Dr. Charles Fisher, Head, 814-865-7034.
Application contact: Dr. Stephen Schaeffer, Chair, 814-863-7034.

Find an in-depth description at www.petersons.com/graduate.

■ PRINCETON UNIVERSITY

Graduate School, Department of Ecology and Evolutionary Biology, Princeton, NJ 08544-1019

AWARDS Biology (PhD); neuroscience (PhD).

Degree requirements: For doctorate, dissertation required, foreign language not required.

Entrance requirements: For doctorate, GRE General Test, GRE Subject Test.

Expenses: Tuition: Full-time $25,050.

Find an in-depth description at www.petersons.com/graduate.

■ PURDUE UNIVERSITY

Graduate School, School of Science, Department of Biological Sciences, Program in Ecology, Evolutionary and Population Biology, West Lafayette, IN 47907

AWARDS Ecology (MS, PhD); evolutionary biology (MS, PhD); population biology (MS, PhD).

Faculty: 9 full-time (1 woman).

Degree requirements: For master's, thesis required (for some programs), foreign language not required; for doctorate, dissertation, seminars, teaching experience required, foreign language not required.

Entrance requirements: For master's and doctorate, GRE General Test, TOEFL, TSE. *Application deadline:* For fall admission, 2/15. *Application fee:* $30. Electronic applications accepted.

Expenses: Tuition, state resident: full-time $4,530; part-time $130 per credit hour. Tuition, nonresident: full-time $15,310; part-time $404 per credit hour. Tuition and fees vary according to campus/location and program.

Financial aid: Fellowships, research assistantships, teaching assistantships available. Aid available to part-time students. Financial aid application deadline: 2/15; financial aid applicants required to submit FAFSA.

Faculty research: Host-parasite coevolution, social behavior, foraging ecology, conservation biology and patterns of evolution in speciation.

Application contact: Nancy Konopka, Graduate Studies Office Manager, 765-494-8142, *Fax:* 765-494-0876, *E-mail:* njk@bilbo.bio.purdue.edu.

■ RICE UNIVERSITY

Graduate Programs, Wiess School of Natural Sciences, Department of Ecology and Evolutionary Biology, Houston, TX 77251-1892

AWARDS MA, PhD.

Degree requirements: For master's and doctorate, thesis/dissertation required.

Entrance requirements: For master's and doctorate, GRE General Test, GRE Subject Test, TOEFL, minimum GPA of 3.0.

Expenses: Tuition: Full-time $16,700. Required fees: $250. Tuition and fees vary according to program.

■ RUTGERS, THE STATE UNIVERSITY OF NEW JERSEY, NEW BRUNSWICK

Graduate School, Program in Ecology and Evolution, New Brunswick, NJ 08901-1281

AWARDS MS, PhD. Part-time programs available.

Faculty: 65 full-time (11 women).

Students: 53 full-time (26 women), 41 part-time (24 women); includes 7 minority (1 African American, 2 Asian Americans or Pacific Islanders, 3 Hispanic Americans, 1 Native American), 13 international. Average age 27. *86 applicants, 29% accepted.* In 1999, 8 master's awarded (75% found work related to degree, 25% continued full-time study); 6 doctorates awarded.

Degree requirements: For master's, foreign language and thesis not required; for doctorate, dissertation required, foreign language not required. *Average time to degree:* Master's–4 years full-time; doctorate–6 years full-time, 8 years part-time.

Entrance requirements: For master's and doctorate, GRE General Test, GRE Subject Test, TOEFL, minimum GPA of 3.0. *Application deadline:* For fall admission, 3/1 (priority date); for spring admission, 12/1. Applications are processed on a rolling basis. *Application fee:* $50.

Expenses: Tuition, state resident: full-time $6,776; part-time $279 per credit. Tuition, nonresident: full-time $9,936; part-time $412 per credit. Required fees: $20 per credit. $89 per semester. Tuition and fees vary according to course load, campus/location and program.

Financial aid: In 1999–00, 5 fellowships with full tuition reimbursements (averaging $12,000 per year), 17 research assistantships with full tuition reimbursements (averaging $13,100 per year), 23 teaching assistantships with full tuition reimbursements (averaging $13,100 per year) were awarded; Federal Work-Study also available. Financial aid application deadline: 3/1; financial aid applicants required to submit FAFSA.

Faculty research: Population and community ecology, population genetics, evolutionary biology, conservation biology, ecosystem ecology.

Dr. Michael Sukhdeo, Director, Graduate Program, 732-932-2971, *Fax:* 732-932-2972.

Application contact: Nancy Tiedge, Secretary, 732-932-2971, *Fax:* 732-932-2972, *E-mail:* tiedge@rci.rutgers.edu.

■ STATE UNIVERSITY OF NEW YORK AT ALBANY

College of Arts and Sciences, Department of Biological Sciences, Specialization in Ecology, Evolution, and Behavior, Albany, NY 12222-0001

AWARDS MS, PhD.

Degree requirements: For master's, one foreign language required; for doctorate, dissertation required.

Entrance requirements: For master's and doctorate, GRE General Test. *Application fee:* $50.

Expenses: Tuition, state resident: full-time $5,100; part-time $214 per credit. Tuition, nonresident: full-time $8,416; part-time $352 per credit. Required fees: $31 per credit.

Financial aid: Minority assistantships available.

Dr. David Shub, Chair, Department of Biological Sciences, 518-442-4300.

■ STATE UNIVERSITY OF NEW YORK AT STONY BROOK

Graduate School, College of Arts and Sciences, Department of Ecology and Evolution, Stony Brook, NY 11794

AWARDS Ecology and evolution (PhD).

Faculty: 18 full-time (3 women), 2 part-time/adjunct (0 women).

Students: 28 full-time (15 women), 18 part-time (8 women); includes 2 minority (1 African American, 1 Hispanic American), 12 international. *71 applicants, 28% accepted.* In 1999, 2 doctorates awarded.

Degree requirements: For doctorate, one foreign language (computer language can substitute), dissertation, comprehensive exam, teaching experience required.

Entrance requirements: For doctorate, GRE General Test, GRE Subject Test, TOEFL. *Application deadline:* For fall admission, 1/15. *Application fee:* $50.

Expenses: Tuition, state resident: full-time $5,100; part-time $213 per credit hour. Tuition, nonresident: full-time $8,416; part-time $351 per credit hour. Required fees: $492. Tuition and fees vary according to program.

Financial aid: In 1999–00, 12 fellowships, 8 research assistantships, 22 teaching assistantships were awarded; Federal Work-Study also available.

Faculty research: Theoretical and experimental population genetics, numerical taxonomy, biostatistics, population and community ecology, plant ecology. *Total annual research expenditures:* $1.2 million.

Dr. Walter Danes, Chairman, 631-632-8600.

State University of New York at Stony Brook (continued)
Application contact: Dr. Dan Dykhuizen, Director, 631-246-8604, *E-mail:* dandyk@life.bio.sunysb.edu.

■ THE UNIVERSITY OF ARIZONA

Graduate College, College of Science, Department of Ecology and Evolutionary Biology, Tucson, AZ 85721

AWARDS Botany (MS, PhD); ecology and evolutionary biology (MS, PhD).

Degree requirements: For master's, foreign language and thesis not required; for doctorate, one foreign language, computer language, dissertation required.
Entrance requirements: For master's and doctorate, GRE General Test, GRE Subject Test, TOEFL.
Expenses: Tuition, nonresident: full-time $4,814; part-time $274 per unit. Required fees: $1,094; $115 per unit. Tuition and fees vary according to course load and program.
Faculty research: Biological diversity, evolutionary history, evolutionary mechanisms, community structure.

■ UNIVERSITY OF CALIFORNIA, DAVIS

Graduate Studies, Programs in the Biological Sciences, Program in Population Biology, Davis, CA 95616
AWARDS PhD.

Faculty: 33 full-time (7 women).
Students: 40 full-time (18 women); includes 2 minority (1 Asian American or Pacific Islander, 1 Hispanic American), 1 international. Average age 28. *76 applicants, 21% accepted.* In 1999, 5 doctorates awarded.
Degree requirements: For doctorate, computer language, dissertation required.
Entrance requirements: For doctorate, GRE General Test, GRE Subject Test. *Application deadline:* For fall admission, 1/15. *Application fee:* $40. Electronic applications accepted.
Expenses: Tuition, nonresident: full-time $9,804. Tuition and fees vary according to program and student level.
Financial aid: In 1999–00, 40 students received aid, including 38 fellowships with full and partial tuition reimbursements available, 8 research assistantships with full and partial tuition reimbursements available, 5 teaching assistantships with partial tuition reimbursements available; Federal Work-Study, grants, and scholarships also available. Financial aid application deadline: 1/15; financial aid applicants required to submit FAFSA.

Faculty research: Population ecology, population genetics, systematics, evolution, community ecology.
John Gillespie, Graduate Chair, 530-752-0605, *Fax:* 530-752-1449, *E-mail:* hjgillespie@ucdavis.edu.
Application contact: Janet Dillon, Administrative Assistant, 530-752-8523, *Fax:* 530-752-1449, *E-mail:* jldillon@ucdavis.edu.

■ UNIVERSITY OF CALIFORNIA, IRVINE

Office of Research and Graduate Studies, School of Biological Sciences, Department of Ecology and Evolutionary Biology, Irvine, CA 92697

AWARDS Biological sciences (MS, PhD).

Faculty: 21 full-time (4 women).
Students: 32 full-time (20 women); includes 1 minority (African American), 4 international. *45 applicants, 31% accepted.*
Degree requirements: For master's, one foreign language (computer language can substitute), thesis required; for doctorate, dissertation required.
Entrance requirements: For master's, GRE General Test, GRE Subject Test, minimum GPA of 3.0; for doctorate, GRE General Test, GRE Subject Test. *Application deadline:* For fall admission, 1/15 (priority date). Applications are processed on a rolling basis. *Application fee:* $40. Electronic applications accepted.
Expenses: Tuition, nonresident: full-time $10,244; part-time $1,720 per quarter. Required fees: $5,252; $1,300 per quarter. Tuition and fees vary according to course load and program.
Financial aid: Fellowships, research assistantships, teaching assistantships, career-related internships or fieldwork, institutionally sponsored loans, and tuition waivers (full and partial) available. Financial aid application deadline: 3/2; financial aid applicants required to submit FAFSA.
Faculty research: Ecological energetics, quantitative genetics, life history evolution, plant-herbivore and plant-pollinator interactions, molecular evolution.
Albert F. Bennett, Chair, 949-824-6930.
Application contact: Pam McDonald, Administrative Assistant, 949-824-4743.

■ UNIVERSITY OF CALIFORNIA, SAN DIEGO

Graduate Studies and Research, Department of Biology, Program in Ecology, Behavior, and Evolution, La Jolla, CA 92093
AWARDS PhD.

Degree requirements: For doctorate, dissertation required.

Entrance requirements: For doctorate, GRE General Test, pre-application beginning in September. *Application deadline:* For fall admission, 1/8. *Application fee:* $40.
Expenses: Tuition, nonresident: full-time $14,691. Required fees: $4,697. Full-time tuition and fees vary according to program.
Application contact: Biology Graduate Admissions Committee, 858-534-3835.

Find an in-depth description at www.petersons.com/graduate.

■ UNIVERSITY OF CALIFORNIA, SANTA BARBARA

Graduate Division, College of Letters and Sciences, Division of Mathematics, Life, and Physical Sciences, Department of Ecology, Evolution, and Marine Biology, Santa Barbara, CA 93106

AWARDS MA, PhD.

Faculty: 28 full-time (4 women), 16 part-time/adjunct (5 women).
Students: 85 full-time (38 women); includes 10 minority (2 Asian Americans or Pacific Islanders, 7 Hispanic Americans, 1 Native American), 6 international. *122 applicants, 16% accepted.* In 1999, 14 master's, 12 doctorates awarded. Terminal master's awarded for partial completion of doctoral program.
Degree requirements: For master's, thesis or alternative required, foreign language not required; for doctorate, dissertation required, foreign language not required. *Average time to degree:* Master's–2 years full-time; doctorate–6 years full-time.
Entrance requirements: For master's and doctorate, GRE General Test, TOEFL. *Application deadline:* For fall admission, 12/15. *Application fee:* $40. Electronic applications accepted.
Expenses: Tuition, state resident: full-time $14,637. Tuition, nonresident: full-time $24,441.
Financial aid: Fellowships, research assistantships, teaching assistantships, career-related internships or fieldwork, Federal Work-Study, institutionally sponsored loans, and tuition waivers (full and partial) available. Financial aid application deadline: 12/15; financial aid applicants required to submit FAFSA.
Scott Cooper, Chair, 805-893-2979, *Fax:* 805-893-4724, *E-mail:* scooper@lifesci.ucsb.edu.
Application contact: Stephanie Slosser, Graduate Program Assistant, 805-893-3023, *Fax:* 805-893-4724, *E-mail:* slosser@lifesci.ucsb.edu.

■ UNIVERSITY OF CHICAGO

Division of the Biological Sciences, Darwinian Sciences: Ecological, Integrative and Evolutionary Biology, Committee on Evolutionary Biology, Chicago, IL 60637-1513

AWARDS Functional and evolutionary biology (PhD).

Faculty: 49 full-time (7 women).
Students: 35 full-time (22 women); includes 4 minority (1 African American, 3 Hispanic Americans), 3 international. *53 applicants, 19% accepted.* In 1999, 6 doctorates awarded (33% entered university research/teaching, 67% found other work related to degree).
Degree requirements: For doctorate, dissertation required, foreign language not required. *Average time to degree:* Doctorate–5.8 years full-time.
Entrance requirements: For doctorate, GRE General Test, TOEFL. *Application deadline:* For fall admission, 1/5 (priority date). *Application fee:* $55.
Expenses: Tuition: Full-time $24,804; part-time $3,422 per course. Required fees: $390. Tuition and fees vary according to program.
Financial aid: In 1999–00, 31 students received aid, including fellowships with tuition reimbursements available (averaging $17,050 per year). Financial aid application deadline: 6/1.
Faculty research: Systematics and evolutionary theory, genetics, functional morphology and physiology, behavior, ecology and biogeography.
Dr. Mathew Leibold, Chair, 773-702-0953, *Fax:* 773-702-4699.
Application contact: Carolyn Johnson, Graduate Administrative Director, 773-702-9474, *Fax:* 773-702-4699, *E-mail:* cs-johnson@uchicago.edu.

■ UNIVERSITY OF COLORADO AT BOULDER

Graduate School, College of Arts and Sciences, Department of Environmental, Population, and Organic Biology, Boulder, CO 80309

AWARDS Animal behavior (MA, PhD); aquatic biology (MA, PhD); behavioral genetics (MA, PhD); ecology (MA, PhD); microbiology (MA, PhD); neurobiology (MA, PhD); plant and animal physiology (MA, PhD); plant and animal systematics (MA, PhD); population biology (MA, PhD); population genetics (MA, PhD).

Faculty: 39 full-time (9 women).
Students: 84 full-time (36 women), 14 part-time (7 women); includes 17 minority (6 Asian Americans or Pacific Islanders, 10 Hispanic Americans, 1 Native American). Average age 29. *147 applicants, 14%*

accepted. In 1999, 7 master's, 13 doctorates awarded. Terminal master's awarded for partial completion of doctoral program.
Degree requirements: For master's, thesis or alternative, comprehensive exam required, foreign language not required; for doctorate, dissertation, comprehensive exam required, foreign language not required. *Average time to degree:* Master's–3 years full-time; doctorate–5 years full-time.
Entrance requirements: For master's, GRE General Test, GRE Subject Test, minimum undergraduate GPA of 3.0; for doctorate, GRE General Test, GRE Subject Test. *Application deadline:* For fall admission, 1/15 (priority date). *Application fee:* $40 ($60 for international students).
Expenses: Tuition, state resident: part-time $181 per credit hour. Tuition, nonresident: part-time $542 per credit hour. Required fees: $99 per term. Tuition and fees vary according to course load and program.
Financial aid: Fellowships, research assistantships, teaching assistantships, Federal Work-Study, institutionally sponsored loans, and tuition waivers (full) available. Financial aid application deadline: 3/1.
Faculty research: Evolution, developmental biology, behavior and neurobiology. *Total annual research expenditures:* $1.8 million.
Michael Breed, Chair, 303-492-8981, *Fax:* 303-492-8699, *E-mail:* michael.breed@colorado.edu.
Application contact: Jill Skarstadt, Graduate Secretary, 303-492-7654, *Fax:* 303-492-8699, *E-mail:* jill.skarstadt@colorado.edu.

■ UNIVERSITY OF DELAWARE

College of Arts and Science, Department of Biological Sciences, Newark, DE 19716

AWARDS Biotechnology (MS, PhD); cell and extracellular matrix biology (MS, PhD); cell and systems physiology (MS, PhD); ecology and evolution (MS, PhD); microbiology (MS, PhD); molecular biology and genetics (MS, PhD); plant biology (MS, PhD).

Faculty: 37 full-time (10 women).
Students: 22 full-time (11 women), 1 part-time; includes 2 minority (both African Americans), 9 international. Average age 25. *37 applicants, 27% accepted.* In 2000, 9 doctorates awarded.
Degree requirements: For master's and doctorate, thesis/dissertation required, foreign language not required. *Average time to degree:* Master's–2.5 years full-time; doctorate–6 years full-time.
Entrance requirements: For master's and doctorate, GRE General Test, GRE Subject Test (advanced biology). *Application*

deadline: For fall admission, 6/15. Applications are processed on a rolling basis.
Application fee: $50. Electronic applications accepted.
Expenses: Tuition, state resident: full-time $4,380; part-time $243 per credit. Tuition, nonresident: full-time $12,750; part-time $708 per credit. Required fees: $15 per term. Tuition and fees vary according to program.
Financial aid: In 2000–01, 18 students received aid, including 2 fellowships with full tuition reimbursements available (averaging $18,000 per year), 4 research assistantships with full tuition reimbursements available (averaging $18,000 per year), 11 teaching assistantships with full tuition reimbursements available (averaging $18,000 per year); tuition waivers (partial) also available. Financial aid application deadline: 6/15.
Faculty research: Cell interactions, molecular mechanisms, microorganisms, embryo implantation. *Total annual research expenditures:* $1.8 million.
Dr. Daniel D. Carson, Chair, 302-831-6977, *Fax:* 302-831-2281, *E-mail:* dcarson@udel.edu.
Application contact: Norman Karin, Graduate Coordinator, 302-831-1841, *Fax:* 302-831-2281, *E-mail:* ccoletta@udel.edu.

Find an in-depth description at www.petersons.com/graduate.

■ UNIVERSITY OF HAWAII AT MANOA

Graduate Division, Specialization in Ecology, Evolution and Conservation Biology, Honolulu, HI 96822

AWARDS MS, PhD. Program is interdisciplinary; degree is in specific discipline with the specialization in Ecology, Evolution and Conservation Biology.

Faculty: 35 full-time (8 women), 11 part-time/adjunct (0 women).
Students: 52 full-time (34 women); includes 12 minority (8 Asian Americans or Pacific Islanders, 3 Hispanic Americans, 1 Native American), 2 international.
Degree requirements: For doctorate, dissertation required.
Expenses: Tuition, state resident: full-time $4,032; part-time $168 per credit. Tuition, nonresident: full-time $9,960; part-time $415 per credit. Required fees: $51 per semester. Part-time tuition and fees vary according to course load and program.
Financial aid: In 1999–00, 15 students received aid, including 3 fellowships, 5 research assistantships, 2 teaching assistantships; career-related internships or fieldwork and tuition waivers (full) also available.

University of Hawaii at Manoa (continued)

Faculty research: Agronomy and soil science, zoology, entomology, genetics and molecular biology, botanical sciences. Dr. Sheila Conant, Chair, 808-956-4602, *Fax:* 808-956-9608, *E-mail:* sconant@ zoogatc.zoo.hawaii.edu.

■ UNIVERSITY OF ILLINOIS AT CHICAGO

Graduate College, College of Liberal Arts and Sciences, Department of Biological Sciences, Chicago, IL 60607-7128

AWARDS Cell and developmental biology (PhD); ecology and evolution (MS, DA, PhD); genetics and development (PhD); molecular biology (MS, PhD); neurobiology (MS, PhD); plant biology (MS, DA, PhD).

Faculty: 40 full-time (5 women).
Students: 100 full-time (47 women), 14 part-time (10 women); includes 13 minority (11 Asian Americans or Pacific Islanders, 2 Hispanic Americans), 42 international. Average age 29. *99 applicants, 36% accepted.* In 1999, 3 master's, 9 doctorates awarded.
Degree requirements: For master's, thesis required, foreign language not required; for doctorate, dissertation, preliminary exam required, foreign language not required.
Entrance requirements: For master's and doctorate, GRE General Test, GRE Subject Test, TOEFL, previous course work in physics, calculus, and organic chemistry; minimum GPA of 3.75 on a 5.0 scale. *Application deadline:* For fall admission, 6/1. Applications are processed on a rolling basis. *Application fee:* $40 ($50 for international students). Electronic applications accepted.
Expenses: Tuition, state resident: full-time $3,750; part-time $1,250 per semester. Tuition, nonresident: full-time $10,588; part-time $3,530 per semester. Required fees: $507 per semester. Tuition and fees vary according to course load and program.
Financial aid: In 1999–00, 87 students received aid; fellowships, research assistantships, teaching assistantships, career-related internships or fieldwork, Federal Work-Study, traineeships, and tuition waivers (full) available. Financial aid application deadline: 3/1; financial aid applicants required to submit FAFSA. Dr. Lon Kaufman, Head, 312-996-2213.
Application contact: Dr. Leo Miller, Director of Graduate Studies, 312-996-2220.
Find an in-depth description at www.petersons.com/graduate.

■ UNIVERSITY OF ILLINOIS AT URBANA–CHAMPAIGN

Graduate College, College of Liberal Arts and Sciences, School of Life Sciences, Department of Ecology, Ethnology, and Evolution, Urbana, IL 61801

AWARDS PhD.

Faculty: 9 full-time (0 women), 1 part-time/adjunct (0 women).
Students: 29 full-time (9 women); includes 1 minority (Native American), 3 international.
Degree requirements: For doctorate, dissertation required.
Application deadline: Applications are processed on a rolling basis. *Application fee:* $30 ($50 for international students).
Expenses: Tuition, state resident: full-time $4,616. Tuition, nonresident: full-time $11,768. Full-time tuition and fees vary according to course load.
Financial aid: In 1999–00, 2 fellowships, 9 research assistantships, 16 teaching assistantships were awarded; tuition waivers (full and partial) also available. Financial aid application deadline: 2/15. Scott K. Robinson, Head, 217-333-6857, *Fax:* 217-244-4565, *E-mail:* skrobins@ uiuc.edu.
Application contact: Linda Thorman, Director of Graduate Studies, 217-333-7802, *Fax:* 217-244-4565, *E-mail:* thorman@uiuc.edu.

■ UNIVERSITY OF KANSAS

Graduate School, College of Liberal Arts and Sciences, Division of Biological Sciences, Department of Ecology and Evolutionary Biology, Lawrence, KS 66045

AWARDS MA, PhD. Part-time programs available.

Faculty: 34.
Students: 19 full-time (10 women), 30 part-time (11 women); includes 2 minority (1 African American, 1 Asian American or Pacific Islander), 8 international. *65 applicants, 14% accepted.*
Degree requirements: For master's, thesis or alternative, general exam required, foreign language not required; for doctorate, dissertation, oral comprehensive exam required.
Entrance requirements: For master's, GRE General Test, GRE Subject Test, TOEFL, minimum GPA of 3.1; for doctorate, GRE General Test, GRE Subject Test, TOEFL. *Application deadline:* For fall admission, 1/1 (priority date). *Application fee:* $25.
Expenses: Tuition, state resident: full-time $2,482; part-time $103 per credit hour. Tuition, nonresident: full-time $8,104;

part-time $338 per credit hour. Required fees: $428; $31 per credit hour. Tuition and fees vary according to program.
Financial aid: In 1999–00, 9 fellowships, teaching assistantships (averaging $11,588 per year) were awarded; research assistantships Financial aid application deadline: 3/1.
Faculty research: Aquatic ecology, environmental studies, population and evolutionary ecology, tropical biology. Thomas N. Taylor, Chair, 785-864-3625.
Application contact: Linda Trueb, Graduate Director, 785-864-3342, *Fax:* 785-864-5335, *E-mail:* trueb@kuhub.cc.ukans.edu.

■ UNIVERSITY OF LOUISIANA AT LAFAYETTE

Graduate School, College of Sciences, Department of Biology, Lafayette, LA 70504

AWARDS Biology (MS); environmental and evolutionary biology (PhD).

Faculty: 40 full-time (7 women).
Students: 59 full-time (25 women), 8 part-time (4 women); includes 2 minority (1 African American, 1 Asian American or Pacific Islander), 17 international. *62 applicants, 52% accepted.* In 1999, 7 master's, 6 doctorates awarded. Terminal master's awarded for partial completion of doctoral program.
Degree requirements: For master's, thesis required, foreign language not required; for doctorate, 2 foreign languages (computer language can substitute for one), dissertation required.
Entrance requirements: For master's, GRE General Test, minimum GPA of 2.75; for doctorate, GRE General Test, GRE Subject Test, minimum GPA of 3.0. *Application deadline:* For fall admission, 5/15. *Application fee:* $20 ($30 for international students).
Expenses: Tuition, state resident: full-time $2,021; part-time $287 per credit. Tuition, nonresident: full-time $7,253; part-time $287 per credit. Part-time tuition and fees vary according to course load.
Financial aid: In 1999–00, 14 fellowships with full tuition reimbursements (averaging $14,572 per year), 7 research assistantships with full tuition reimbursements (averaging $4,565 per year), 24 teaching assistantships with full tuition reimbursements (averaging $5,377 per year) were awarded; Federal Work-Study and institutionally sponsored loans also available. Financial aid application deadline: 5/1.
Faculty research: Structure and ultrastructure, system biology, ecology, processes, environmental physiology. Dr. Darryl L. Felder, Head, 337-482-6748.

Application contact: Dr. Karl Hasenstein, Graduate Coordinator, 337-482-6750.

Find an in-depth description at www.petersons.com/graduate.

■ UNIVERSITY OF MASSACHUSETTS AMHERST

Graduate School, Interdisciplinary Programs, Program in Organic and Evolutionary Biology, Amherst, MA 01003

AWARDS MS, PhD. Part-time programs available.

Students: 25 full-time (12 women), 10 part-time (3 women); includes 1 minority (Hispanic American), 6 international. Average age 33. *39 applicants, 44% accepted.* In 1999, 4 master's, 4 doctorates awarded. Terminal master's awarded for partial completion of doctoral program.
Degree requirements: For master's, thesis or alternative required, foreign language not required; for doctorate, 2 foreign languages, dissertation required.
Entrance requirements: For master's and doctorate, GRE General Test, GRE Subject Test. *Application deadline:* For fall admission, 1/15 (priority date). Applications are processed on a rolling basis. *Application fee:* $40.
Expenses: Tuition, state resident: full-time $2,640; part-time $165 per credit. Tuition, nonresident: full-time $9,756; part-time $407 per credit. Required fees: $1,221 per term. One-time fee: $110. Full-time tuition and fees vary according to course load, campus/location and reciprocity agreements.
Financial aid: In 1999–00, 22 fellowships with full tuition reimbursements (averaging $4,499 per year), 7 research assistantships with full tuition reimbursements (averaging $5,517 per year) were awarded; teaching assistantships with full tuition reimbursements, career-related internships or fieldwork, Federal Work-Study, grants, scholarships, traineeships, and unspecified assistantships also available. Aid available to part-time students. Financial aid application deadline: 1/15.
Dr. Joseph Elkinton, Director, 413-545-0928, *Fax:* 413-545-3243, *E-mail:* elkinton@ent.umass.edu.

Find an in-depth description at www.petersons.com/graduate.

■ UNIVERSITY OF MIAMI

Graduate School, College of Arts and Sciences, Department of Biology, Coral Gables, FL 33124

AWARDS Biology (MS, PhD); genetics and evolution (MS, PhD); tropical biology, ecology, and behavior (MS, PhD).

Faculty: 24 full-time (3 women), 5 part-time/adjunct (1 woman).
Students: 42 full-time (23 women); includes 7 minority (1 African American, 3 Asian Americans or Pacific Islanders, 3 Hispanic Americans), 2 international. Average age 26. *56 applicants, 13% accepted.* In 1999, 3 doctorates awarded (34% entered university research/teaching, 66% found other work related to degree). Terminal master's awarded for partial completion of doctoral program.
Degree requirements: For master's, oral defense required, thesis optional, foreign language not required; for doctorate, dissertation, oral and written qualifying exam required, foreign language not required. *Average time to degree:* Doctorate–5.5 years full-time.
Entrance requirements: For master's and doctorate, GRE General Test, GRE Subject Test, TOEFL. *Application deadline:* For fall admission, 3/15. Applications are processed on a rolling basis. *Application fee:* $50. Electronic applications accepted.
Expenses: Tuition: Full-time $15,336; part-time $852 per credit. Required fees: $174. Tuition and fees vary according to program.
Financial aid: In 1999–00, 37 students received aid, including 6 fellowships with tuition reimbursements available (averaging $15,000 per year), 10 research assistantships with tuition reimbursements available (averaging $13,000 per year), 21 teaching assistantships with tuition reimbursements available (averaging $13,202 per year); career-related internships or fieldwork and institutionally sponsored loans also available. Financial aid application deadline: 3/15.
Faculty research: Population biology, behavioral ecology, plant-animal and plant-environment interactions and genetic co-evolution, biogeography, conservation biology.
Dr. Julian C. Lee, Director, 305-284-6420, *Fax:* 305-284-3039, *E-mail:* jlee@fig.cox.miami.edu.
Application contact: 305-284-3973, *Fax:* 305-284-3039, *E-mail:* gaac@fig.cox.miami.edu.

Find an in-depth description at www.petersons.com/graduate.

■ UNIVERSITY OF MINNESOTA, TWIN CITIES CAMPUS

Graduate School, College of Biological Sciences, Department of Ecology, Evolution, and Behavior, Minneapolis, MN 55455-0213

AWARDS Ecology, animal behavior, and evolution (MS, PhD).

Faculty: 55 full-time (11 women).

Students: 45 full-time (24 women), 1 (woman) part-time; includes 2 minority (1 Asian American or Pacific Islander, 1 Hispanic American). Average age 26. *71 applicants, 17% accepted.* In 1999, 2 master's awarded (50% found work related to degree, 50% continued full-time study); 3 doctorates awarded (33% entered university research/teaching, 67% found other work related to degree). Terminal master's awarded for partial completion of doctoral program.
Degree requirements: For master's, thesis or projects, comprehensive exam required; for doctorate, dissertation, comprehensive exam required, foreign language not required. *Average time to degree:* Master's–6.5 years full-time; doctorate–9 years full-time.
Entrance requirements: For master's and doctorate, GRE General Test, minimum GPA of 3.2. *Application deadline:* For fall admission, 1/4. *Application fee:* $50 ($55 for international students). Electronic applications accepted.
Expenses: Tuition, state resident: full-time $5,040; part-time $420 per credit. Tuition, nonresident: full-time $9,900; part-time $825 per credit. Full-time tuition and fees vary according to course load, program and reciprocity agreements.
Financial aid: In 1999–00, 44 students received aid, including 5 fellowships with tuition reimbursements available (averaging $12,000 per year), 12 research assistantships with tuition reimbursements available (averaging $10,920 per year), 14 teaching assistantships with tuition reimbursements available (averaging $10,920 per year); Federal Work-Study, institutionally sponsored loans, and tuition waivers (partial) also available. Financial aid application deadline: 1/4. *Total annual research expenditures:* $2.7 million.
Prof. Robert W. Sterner, Head, 612-625-5700, *Fax:* 612-624-6777, *E-mail:* stern007@tc.umn.edu.
Application contact: Prof. Elmer C. Birney, Director of Graduate Studies, 612-625-5713, *Fax:* 612-624-6777, *E-mail:* ecbirney@cbs.umn.edu.

■ UNIVERSITY OF MISSOURI–ST. LOUIS

Graduate School, College of Arts and Sciences, Department of Biology, St. Louis, MO 63121-4499

AWARDS Biology (MS, PhD), including animal behavior (MS), biochemistry, biotechnology (MS), conservation biology (MS), development (MS), ecology (MS), environmental studies (PhD), evolution (MS), genetics (MS), molecular biology and biotechnology (PhD), molecular/cellular biology (MS), physiology (MS), plant systematics, population biology

University of Missouri–St. Louis (continued)

(MS), tropical biology (MS); biotechnology (Certificate); tropical biology and conservation (Certificate). Part-time programs available.

Faculty: 46.

Students: 21 full-time (11 women), 75 part-time (44 women); includes 13 minority (2 African Americans, 2 Asian Americans or Pacific Islanders, 8 Hispanic Americans, 1 Native American), 23 international. In 1999, 14 master's, 4 doctorates awarded.

Degree requirements: For master's, thesis or alternative required, foreign language not required; for doctorate, one foreign language, dissertation, 1 semester of teaching experience required.

Entrance requirements: For doctorate, GRE General Test. *Application deadline:* For fall admission, 7/1 (priority date); for spring admission, 11/1 (priority date). Applications are processed on a rolling basis. *Application fee:* $25 ($40 for international students). Electronic applications accepted.

Expenses: Tuition, state resident: full-time $4,932; part-time $173 per credit hour. Tuition, nonresident: full-time $13,279; part-time $521 per credit hour. Required fees: $775; $33 per credit hour. Tuition and fees vary according to degree level and program.

Financial aid: In 1999–00, 8 research assistantships with partial tuition reimbursements (averaging $10,635 per year), 14 teaching assistantships with partial tuition reimbursements (averaging $11,488 per year) were awarded; career-related internships or fieldwork and Federal Work-Study also available. Aid available to part-time students. Financial aid application deadline: 2/1. *Total annual research expenditures:* $908,828.

Application contact: Graduate Admissions, 314-516-5458, *Fax:* 314-516-6759, *E-mail:* gradadm@umsl.edu.

■ UNIVERSITY OF NEVADA, RENO

Graduate School, Interdisciplinary Program in Ecology, Evolution, and Conservation Biology, Reno, NV 89557

AWARDS PhD. Offered through the College of Arts and Science, the M. C. Fleischmann College of Agriculture, and the Desert Research Institute.

Faculty: 29.

Students: 27 full-time (16 women), 9 part-time (3 women); includes 2 minority (1 Asian American or Pacific Islander, 1 Hispanic American), 2 international. Average age 33. *36 applicants, 14% accepted.* In 1999, 7 degrees awarded.

Degree requirements: For doctorate, dissertation required, foreign language not required.

Entrance requirements: For doctorate, GRE General Test, GRE Subject Test, TOEFL, minimum GPA of 3.0. *Application deadline:* For fall admission, 2/15. *Application fee:* $40.

Expenses: Tuition, area resident: Part-time $3,173 per semester. Tuition, nonresident: full-time $6,347. Required fees: $101 per credit. $101 per credit.

Financial aid: In 1999–00, 1 research assistantship was awarded; teaching assistantships Financial aid application deadline: 3/1.

Faculty research: Population biology, behavioral ecology, plant response to climate change, conservation of endangered species, restoration of natural ecosystems.

Dr. C. Richard Tracy, Director, 775-784-4419, *Fax:* 775-784-4583.

Application contact: Larry Hillerman, Program Manager, 775-784-4439, *Fax:* 775-784-1306, *E-mail:* eecb@biodiversity.unr.edu.

■ UNIVERSITY OF NEW MEXICO

Graduate School, College of Arts and Sciences, Department of Biology, Albuquerque, NM 87131-2039

AWARDS Biology (MS, PhD), including air land ecology, behavioral ecology, botany, cellular and molecular biology, community ecology, comparative immunology, comparative physiology, conservation biology, ecology, ecosystem ecology, evolutionary biology, evolutionary genetics, microbiology, molecular genetics, parasitology, physiological ecology, physiology, population biology, vertebrate and invertebrate zoology. Part-time programs available.

Faculty: 35 full-time (5 women), 18 part-time/adjunct (11 women).

Students: 71 full-time (37 women), 28 part-time (11 women); includes 8 minority (2 Asian Americans or Pacific Islanders, 5 Hispanic Americans, 1 Native American), 11 international. Average age 33. *93 applicants, 30% accepted.* In 1999, 11 master's, 12 doctorates awarded. Terminal master's awarded for partial completion of doctoral program.

Degree requirements: For master's, one foreign language (computer language can substitute), thesis required (for some programs); for doctorate, 2 foreign languages (computer language can substitute for one), dissertation required.

Entrance requirements: For master's and doctorate, GRE General Test, GRE Subject Test, minimum GPA of 3.2. *Application deadline:* For fall admission, 1/15. *Application fee:* $25.

Expenses: Tuition, state resident: full-time $2,514; part-time $105 per credit hour. Tuition, nonresident: full-time $10,304; part-time $417 per credit hour. International tuition: $10,304 full-time. Required fees: $516; $22 per credit hour. Tuition and fees vary according to program.

Financial aid: In 1999–00, 58 students received aid, including 24 fellowships (averaging $1,645 per year), 26 research assistantships with tuition reimbursements available (averaging $8,921 per year), 40 teaching assistantships with tuition reimbursements available (averaging $11,066 per year); career-related internships or fieldwork, Federal Work-Study, institutionally sponsored loans, and tuition waivers (full and partial) also available. Aid available to part-time students. Financial aid applicants required to submit FAFSA.

Faculty research: Developmental biology, immunobiology. *Total annual research expenditures:* $4.5 million.

Dr. Kathryn Vogel, Chair, 505-277-3411, *Fax:* 505-277-0304, *E-mail:* kgvogel@unm.edu.

Application contact: Vivian Kent, Information Contact, 505-277-1712, *Fax:* 505-277-0304, *E-mail:* vkent@unm.edu.

■ THE UNIVERSITY OF NORTH CAROLINA AT CHAPEL HILL

Graduate School, College of Arts and Sciences, Department of Biology, Chapel Hill, NC 27599

AWARDS Botany (MA, MS, PhD); cell biology, development, and physiology (MA, MS, PhD); ecology and behavior (MA, MS, PhD); genetics and molecular biology (MA, MS, PhD); morphology, systematics, and evolution (MA, MS, PhD).

Degree requirements: For master's, thesis (for some programs), comprehensive exams required; for doctorate, dissertation, comprehensive exams required.

Entrance requirements: For master's and doctorate, GRE General Test, GRE Subject Test. Electronic applications accepted.

Expenses: Tuition, state resident: full-time $1,578. Tuition, nonresident: full-time $10,744. Required fees: $827. One-time fee: $15 full-time. Tuition and fees vary according to program.

Faculty research: Gene expression, biomechanics, yeast genetics, plant ecology, plant molecular biology.

Find an in-depth description at www.petersons.com/graduate.

■ UNIVERSITY OF NOTRE DAME

Graduate School, College of Science, Department of Biological Sciences, Program in Aquatic Ecology, Evolution and Environmental Biology, Notre Dame, IN 46556

AWARDS MS, PhD. Terminal master's awarded for partial completion of doctoral program.

Degree requirements: For master's and doctorate, thesis/dissertation required, foreign language not required.
Entrance requirements: For master's and doctorate, GRE General Test, GRE Subject Test, TOEFL. *Application deadline:* For fall admission, 2/1 (priority date); for spring admission, 11/1. Applications are processed on a rolling basis. *Application fee:* $50.
Expenses: Tuition: Full-time $21,930; part-time $1,218 per credit. Required fees: $95. Tuition and fees vary according to program.
Financial aid: Fellowships with full tuition reimbursements, research assistantships with full tuition reimbursements, teaching assistantships with full tuition reimbursements, traineeships and tuition waivers (full) available. Financial aid application deadline: 2/1.
Faculty research: Aquatic benthic ecology, biogeochemistry, plant-animal interactions, species invasions, ecological genetics and speciation, aquatic community studies.
Application contact: Dr. Terrence J. Akai, Director of Graduate Admissions, 219-631-7706, *Fax:* 219-631-4183, *E-mail:* gradad@nd.edu.

■ UNIVERSITY OF OREGON

Graduate School, College of Arts and Sciences, Department of Biology, Eugene, OR 97403

AWARDS Ecology and evolution (MA, MS, PhD); marine biology (MA, MS, PhD); molecular, cellular and genetic biology (PhD); neuroscience and development (PhD).

Faculty: 42 full-time (14 women), 5 part-time/adjunct (2 women).
Students: 69 full-time (33 women), 8 part-time (5 women); includes 6 minority (2 Asian Americans or Pacific Islanders, 3 Hispanic Americans, 1 Native American), 5 international. *18 applicants, 83% accepted.* In 1999, 11 master's, 7 doctorates awarded. Terminal master's awarded for partial completion of doctoral program.
Degree requirements: For master's, thesis required (for some programs); for doctorate, dissertation required, foreign language not required.
Entrance requirements: For master's and doctorate, GRE General Test, TOEFL,

minimum GPA of 3.2. *Application deadline:* For fall admission, 1/10. *Application fee:* $50.
Expenses: Tuition, state resident: full-time $6,750. Tuition, nonresident: full-time $11,409. Part-time tuition and fees vary according to course load.
Financial aid: In 1999–00, 36 teaching assistantships were awarded; research assistantships, Federal Work-Study, grants, and institutionally sponsored loans also available. Financial aid application deadline: 2/1.
Faculty research: Developmental neurobiology; evolution, population biology, and quantitative genetics; regulation of gene expression; biochemistry of marine organisms.
Janis C. Weeks, Head, 541-346-4502, *Fax:* 541-346-6056.
Application contact: Donna Overall, Graduate Program Coordinator, 541-346-4503, *Fax:* 541-346-6056, *E-mail:* doverall@oregon.uoregon.edu.

Find an in-depth description at www.petersons.com/graduate.

■ UNIVERSITY OF PITTSBURGH

Faculty of Arts and Sciences, Department of Biological Sciences, Program in Ecology and Evolution, Pittsburgh, PA 15260

AWARDS MS, PhD.

Degree requirements: For master's and doctorate, thesis/dissertation required, foreign language not required.
Entrance requirements: For master's and doctorate, GRE General Test, GRE Subject Test, TOEFL. *Application deadline:* For fall admission, 2/1 (priority date). Applications are processed on a rolling basis. *Application fee:* $40.
Expenses: Tuition, state resident: full-time $8,338; part-time $342 per credit. Tuition, nonresident: full-time $17,168; part-time $707 per credit. Required fees: $480; $90 per semester. Tuition and fees vary according to program.

Find an in-depth description at www.petersons.com/graduate.

■ UNIVERSITY OF ROCHESTER

The College, Arts and Sciences, Department of Biology, Program in Ecology and Evolutionary Biology, Rochester, NY 14627-0250

AWARDS MS, PhD.

Degree requirements: For master's, thesis not required; for doctorate, dissertation, qualifying exam required, foreign language not required.
Entrance requirements: For master's and doctorate, GRE General Test, GRE Subject Test, TOEFL. *Application deadline:*

For fall admission, 2/1 (priority date). *Application fee:* $25.
Expenses: Tuition: Part-time $697 per credit hour. Tuition and fees vary according to program.
Financial aid: Application deadline: 2/1..
Application contact: Cindy Landry, Graduate Program Secretary, 716-275-7991.

■ UNIVERSITY OF SOUTH CAROLINA

Graduate School, College of Science and Mathematics, Department of Biological Sciences, Graduate Training Program in Ecology, Evolution, and Organismal Biology, Columbia, SC 29208

AWARDS MS, PhD.

Degree requirements: For master's, one foreign language required; for doctorate, dissertation required.
Entrance requirements: For master's and doctorate, GRE General Test, minimum GPA of 3.0 in science. *Application deadline:* For fall admission, 2/15 (priority date). *Application fee:* $35.
Expenses: Tuition, state resident: full-time $4,014; part-time $202 per credit hour. Tuition, nonresident: full-time $8,528; part-time $428 per credit hour. Required fees: $100; $4 per credit hour. Tuition and fees vary according to program.
Application contact: Dr. Franklyn F. Bolander, Director of Graduate Studies, 803-777-2755, *Fax:* 803-777-4002, *E-mail:* bolander@sc.edu.

Find an in-depth description at www.petersons.com/graduate.

■ THE UNIVERSITY OF TENNESSEE

Graduate School, College of Arts and Sciences, Department of Ecology and Evolutionary Biology, Knoxville, TN 37996

AWARDS Behavior (MS, PhD); ecology (MS, PhD); evolutionary biology (MS, PhD). Part-time programs available.

Faculty: 21 full-time (4 women), 1 part-time/adjunct (0 women).
Students: 30 full-time (18 women), 26 part-time (10 women); includes 1 minority (Asian American or Pacific Islander), 9 international. *54 applicants, 17% accepted.* In 1999, 6 master's, 9 doctorates awarded.
Degree requirements: For master's and doctorate, thesis/dissertation required, foreign language not required.
Entrance requirements: For master's and doctorate, GRE General Test, TOEFL, minimum GPA of 2.7. *Application deadline:* For fall admission, 2/1 (priority date). Applications are processed on a rolling

The University of Tennessee (continued) basis. *Application fee:* $35. Electronic applications accepted.

Expenses: Tuition, state resident: full-time $3,806; part-time $184 per credit hour. Tuition, nonresident: full-time $9,874; part-time $522 per credit hour. Tuition and fees vary according to program.

Financial aid: In 1999–00, 4 fellowships, 9 research assistantships, 26 teaching assistantships were awarded; Federal Work-Study, institutionally sponsored loans, and unspecified assistantships also available. Financial aid application deadline: 2/1; financial aid applicants required to submit FAFSA.

Dr. Tom Hallam, Head, 865-974-3065, *Fax:* 865-974-3067, *E-mail:* thallam@ utk.edu.

Application contact: Dr. C. R. Boake, Graduate Representative, *E-mail:* cboake@ utk.edu.

■ THE UNIVERSITY OF TEXAS AT AUSTIN

Graduate School, College of Natural Sciences, School of Biological Sciences, Program in Ecology, Evolution and Behavior, Austin, TX 78712-1111

AWARDS MA, PhD.

Entrance requirements: For master's and doctorate, GRE General Test. *Application deadline:* Applications are processed on a rolling basis. *Application fee:* $50 ($75 for international students). Electronic applications accepted.

Expenses: Tuition, state resident: part-time $114 per semester hour. Tuition, nonresident: part-time $330 per semester hour. Tuition and fees vary according to program.

Financial aid: Fellowships, research assistantships, teaching assistantships available. Financial aid application deadline: 1/5.

Dr. Klaus O. Kalthoff, Chairman, 512-471-1152.

Application contact: Michael J. Ryan, Graduate Adviser, 512-471-7131, *E-mail:* mryan@mail.utexas.edu.

■ UNIVERSITY OF UTAH

Graduate School, College of Science, Department of Biology, Salt Lake City, UT 84112-1107

AWARDS Biology (M Phil); ecology and evolutionary biology (MS, PhD); genetics

(MS, PhD); molecular biology (PhD). Part-time programs available. Terminal master's awarded for partial completion of doctoral program.

Degree requirements: For master's and doctorate, thesis/dissertation required, foreign language not required.

Entrance requirements: For master's and doctorate, GRE General Test, GRE Subject Test, TOEFL, minimum GPA of 3.0.

Expenses: Tuition, state resident: full-time $1,663. Tuition, nonresident: full-time $5,201. Tuition and fees vary according to course load and program.

Faculty research: Behavioral ecology, cellular neurobiology, DNA replication, ecological genetics, herpetology.

Find an in-depth description at www.petersons.com/graduate.

■ WASHINGTON UNIVERSITY IN ST. LOUIS

Graduate School of Arts and Sciences, Division of Biology and Biomedical Sciences, Program in Evolutionary and Population Biology, St. Louis, MO 63130-4899

AWARDS Ecology (PhD); environmental biology (PhD); evolutionary biology (PhD); genetics (PhD).

Degree requirements: For doctorate, dissertation required, foreign language not required.

Entrance requirements: For doctorate, GRE General Test, GRE Subject Test. *Application deadline:* For fall admission, 1/1 (priority date). Applications are processed on a rolling basis. *Application fee:* $0.

Expenses: Tuition: Full-time $23,400; part-time $975 per credit. Tuition and fees vary according to program.

Financial aid: Fellowships, research assistantships available. Financial aid application deadline: 1/1.

Dr. Jonathan Losos, Coordinator, 314-362-4188.

Application contact: Rosemary Garagneni, Director of Admissions, 800-852-9074, *E-mail:* admissions@ dbbs.wustl.edu.

■ WESLEYAN UNIVERSITY

Graduate Programs, Department of Biology, Middletown, CT 06459-0260

AWARDS Cell biology (PhD); comparative physiology (PhD); developmental biology (PhD); genetics (PhD); neurophysiology (PhD); population biology (PhD).

Faculty: 12 full-time (3 women).

Students: 24 full-time (12 women); includes 1 minority (African American), 11 international. Average age 28. *125 applicants, 4% accepted.* In 1999, 2 doctorates awarded.

Degree requirements: For doctorate, one foreign language (computer language can substitute), dissertation required.

Entrance requirements: For doctorate, GRE Subject Test. *Application deadline:* For fall admission, 1/15. Applications are processed on a rolling basis. *Application fee:* $0.

Expenses: Tuition: Full-time $24,876. Required fees: $650. Tuition and fees vary according to program.

Financial aid: Research assistantships, teaching assistantships, stipends available.

Faculty research: Microbial population genetics, genetic basis of evolutionary adaptation, genetic regulation of differentiation and pattern formation in *drosophila*.

Dr. Fred Cohan, Chairman, 860-685-3489.

Application contact: Marina J. Melendez, Director of Graduate Student Services, 860-685-2390, *Fax:* 860-685-2439, *E-mail:* mmelendez@wesleyan.edu.

Find an in-depth description at www.petersons.com/graduate.

■ YALE UNIVERSITY

Graduate School of Arts and Sciences, Department of Ecology and Evolutionary Biology, New Haven, CT 06520

AWARDS PhD.

Faculty: 15 full-time (5 women), 2 part-time/adjunct (both women).

Students: 10 full-time (5 women), 1 international. Average age 28. *28 applicants, 43% accepted.*

Entrance requirements: For doctorate, GRE General Test, GRE Subject Test (biology). *Application deadline:* For fall admission, 1/4. *Application fee:* $65.

Expenses: Tuition: Full-time $22,300. Full-time tuition and fees vary according to program.

Financial aid: In 1999–00, fellowships with full tuition reimbursements (averaging $17,600 per year); grants also available.

Dr. Gunter Wagner, Chair, 203-432-6138, *Fax:* 203-432-5176.

Application contact: Maureen F. Cunningham, Registrar, 203-432-3837, *Fax:* 203-432-5176, *E-mail:* maureen.cunningham@yale.edu.

Find an in-depth description at www.petersons.com/graduate.

Entomology

ENTOMOLOGY

■ AUBURN UNIVERSITY

Graduate School, College of Agriculture, Department of Entomology and Plant Pathology, Auburn, Auburn University, AL 36849-0002

AWARDS Entomology (M Ag, MS, PhD); plant pathology (M Ag, MS, PhD). Part-time programs available.

Faculty: 25 full-time (4 women).
Students: 11 full-time (3 women), 13 part-time (6 women); includes 4 minority (2 African Americans, 2 Asian Americans or Pacific Islanders), 9 international. *11 applicants, 45% accepted.* In 2000, 2 master's, 2 doctorates awarded.
Degree requirements: For master's, thesis (MS) required; for doctorate, dissertation required.
Entrance requirements: For master's, GRE General Test; for doctorate, GRE General Test, GRE Subject Test, master's degree with thesis. *Application deadline:* For fall admission, 7/7; for spring admission, 11/24. Applications are processed on a rolling basis. *Application fee:* $25 ($50 for international students). Electronic applications accepted.
Expenses: Tuition, state resident: full-time $2,895; part-time $80 per credit hour. Tuition, nonresident: full-time $8,685; part-time $240 per credit hour.
Financial aid: Research assistantships, teaching assistantships, Federal Work-Study available. Aid available to part-time students. Financial aid application deadline: 3/15.
Faculty research: Pest management, biological control, systematics, medical entomology.
Dr. Michael L. Williams, Chair, 334-844-5006.
Application contact: Dr. John F. Pritchett, Dean of the Graduate School, 334-844-4700.

■ CLEMSON UNIVERSITY

Graduate School, College of Agriculture, Forestry and Life Sciences, School of Plant, Statistical and Ecological Sciences, Department of Entomology, Clemson, SC 29634

AWARDS MS, PhD.

Students: 31 full-time (12 women), 4 part-time; includes 1 minority (African American), 9 international. Average age 31.

19 applicants, 74% accepted. In 1999, 7 master's, 1 doctorate awarded.
Degree requirements: For master's, thesis required; for doctorate, dissertation required.
Entrance requirements: For master's and doctorate, GRE General Test, TOEFL, minimum GPA of 3.0. *Application deadline:* For fall admission, 7/1 (priority date). Applications are processed on a rolling basis. *Application fee:* $40.
Expenses: Tuition, state resident: full-time $3,480; part-time $174 per credit hour. Tuition, nonresident: full-time $9,256; part-time $388 per credit hour. Required fees: $5 per term. Full-time tuition and fees vary according to course level, course load and campus/location.
Financial aid: Fellowships, research assistantships, institutionally sponsored loans and unspecified assistantships available. Financial aid applicants required to submit FAFSA.
Faculty research: Aquatic arthropod diversity, crop insect management, medical and veterinary entomology, genetics and biotechnology, urban entomology.
Dr. Paul M. Horton, Interim Chair, 864-656-5051, *Fax:* 864-656-5065.
Application contact: John Morse, Coordinator, 864-656-5049, *Fax:* 864-656-5065, *E-mail:* jmorse@clemson.edu.

■ COLORADO STATE UNIVERSITY

Graduate School, College of Agricultural Sciences, Department of Bioagricultural Science and Pest Management/Entomology, Fort Collins, CO 80523-0015

AWARDS Entomology (MS, PhD). Part-time programs available.

Faculty: 8 full-time (1 woman).
Students: 8 full-time (2 women), 7 part-time (4 women); includes 1 minority (Hispanic American), 2 international. Average age 35. *16 applicants, 38% accepted.* In 1999, 3 master's, 1 doctorate awarded.
Degree requirements: For master's, thesis optional, foreign language not required; for doctorate, dissertation, comprehensive exams required, foreign language not required.
Entrance requirements: For master's and doctorate, GRE General Test, TOEFL, minimum GPA of 3.0. *Application deadline:* For fall admission, 2/1 (priority date). Applications are processed on a rolling basis. *Application fee:* $30. Electronic applications accepted.

Expenses: Tuition, state resident: full-time $2,694; part-time $150 per credit. Tuition, nonresident: full-time $10,460; part-time $581 per credit. Required fees: $32 per semester. Tuition and fees vary according to program.
Financial aid: In 1999–00, 2 research assistantships, 4 teaching assistantships were awarded; fellowships, career-related internships or fieldwork, Federal Work-Study, and traineeships also available. Aid available to part-time students.
Faculty research: Ecology, systematics, behavior, molecular genetics, integrated pest management. *Total annual research expenditures:* $1.8 million.
Thomas O. Holtzer, Head, 970-491-5261, *Fax:* 970-491-3862, *E-mail:* tholtzer@lamar.agsci.colostate.edu.

■ CORNELL UNIVERSITY

Graduate School, Graduate Fields of Agriculture and Life Sciences, Field of Entomology, Ithaca, NY 14853-0001

AWARDS Acarology (MS, PhD); apiculture (MS, PhD); applied entomology (MS, PhD); aquatic entomology (MS, PhD); biological control (MS, PhD); insect behavior (MS, PhD); insect biochemistry (MS, PhD); insect ecology (MS, PhD); insect genetics (MS, PhD); insect morphology (MS, PhD); insect pathology (MS, PhD); insect physiology (MS, PhD); insect systematics (MS, PhD); insect toxicology and insecticide chemistry (MS, PhD); integrated pest management (MS, PhD); medical and veterinary entomology (MS, PhD).

Faculty: 42 full-time.
Students: 39 full-time (18 women); includes 4 minority (1 African American, 1 Asian American or Pacific Islander, 2 Hispanic Americans), 14 international. *59 applicants, 14% accepted.* In 1999, 2 master's, 6 doctorates awarded.
Degree requirements: For master's and doctorate, thesis/dissertation required, foreign language not required.
Entrance requirements: For master's and doctorate, GRE General Test, GRE Subject Test (biology), TOEFL. *Application deadline:* For fall admission, 1/15. *Application fee:* $65. Electronic applications accepted.
Expenses: Tuition: Full-time $12,100.
Financial aid: In 1999–00, 31 students received aid, including 10 fellowships with full tuition reimbursements available, 13 research assistantships with full tuition reimbursements available, 8 teaching

Cornell University (continued)
assistantships with full tuition reimbursements available; institutionally sponsored loans, scholarships, tuition waivers (full and partial), and unspecified assistantships also available. Financial aid applicants required to submit FAFSA.
Faculty research: Biodiversity.
Application contact: Graduate Field Assistant, 607-255-3250, *E-mail:* fieldofent2@cornell.edu.

■ IOWA STATE UNIVERSITY OF SCIENCE AND TECHNOLOGY

Graduate College, College of Agriculture, Department of Entomology, Ames, IA 50011
AWARDS MS, PhD.

Faculty: 17 full-time, 1 part-time/adjunct.
Students: 32 full-time (8 women), 4 part-time (1 woman); includes 2 minority (1 African American, 1 Hispanic American), 10 international. *15 applicants, 53% accepted.* In 1999, 4 master's, 4 doctorates awarded.
Degree requirements: For master's and doctorate, thesis/dissertation required.
Entrance requirements: For master's and doctorate, GRE General Test, GRE Subject Test (biology), TOEFL. *Application deadline:* For fall admission, 6/1 (priority date); for spring admission, 11/1 (priority date). *Application fee:* $20 ($50 for international students). Electronic applications accepted.
Expenses: Tuition, state resident: full-time $3,308. Tuition, nonresident: full-time $9,744. Part-time tuition and fees vary according to course load, campus/location and program.
Financial aid: In 1999–00, research assistantships with partial tuition reimbursements (averaging $11,598 per year); fellowships, teaching assistantships with partial tuition reimbursements, scholarships also available.
Dr. Joel R. Coats, Chair, 515-294-7400, *Fax:* 515-294-2125, *E-mail:* entomology@ iastate.edu.

■ KANSAS STATE UNIVERSITY

Graduate School, College of Agriculture, Department of Entomology, Manhattan, KS 66506
AWARDS MS, PhD.

Degree requirements: For master's and doctorate, thesis/dissertation required, foreign language not required.
Expenses: Tuition, state resident: part-time $103 per credit hour. Tuition, nonresident: part-time $338 per credit hour. Required fees: $17 per credit hour. One-time fee: $64 part-time.

Faculty research: Plant resistance to insects, anthropod pest management, insect behavior and ecology, biological control, insect genetics, physiology, toxicology, stored products insects.

■ LOUISIANA STATE UNIVERSITY AND AGRICULTURAL AND MECHANICAL COLLEGE

Graduate School, College of Agriculture, Department of Entomology, Baton Rouge, LA 70803
AWARDS MS, PhD.

Faculty: 21 full-time (1 woman).
Students: 26 full-time (11 women), 6 part-time (1 woman), 8 international. Average age 29. *10 applicants, 50% accepted.* In 1999, 4 master's, 2 doctorates awarded.
Degree requirements: For master's and doctorate, thesis/dissertation required, foreign language not required.
Entrance requirements: For master's and doctorate, GRE General Test, minimum GPA of 3.0. *Application deadline:* For fall admission, 1/25 (priority date). Applications are processed on a rolling basis. *Application fee:* $25.
Expenses: Tuition, state resident: full-time $2,881. Tuition, nonresident: full-time $7,081. Part-time tuition and fees vary according to course load and program.
Financial aid: In 1999–00, 1 fellowship (averaging $15,307 per year), 22 research assistantships with partial tuition reimbursements (averaging $13,124 per year) were awarded; teaching assistantships with partial tuition reimbursements
Faculty research: Integrated pest management, ecology and biology, parasitoids and pathogens, host-pest resistance, molecular biology. *Total annual research expenditures:* $63,073.
Dr. Frank Guillot, Head, 225-388-1634, *Fax:* 225-388-1643, *E-mail:* fguillo@ unix1.sncc.lsu.edu.

■ MICHIGAN STATE UNIVERSITY

Graduate School, College of Natural Science and College of Agriculture and Natural Resources, Department of Entomology, East Lansing, MI 48824
AWARDS Entomology (MS, PhD); entomology-environmental toxicology (PhD); entomology-urban studies (MS, PhD). Part-time programs available.

Faculty: 25.
Students: 15 full-time (7 women), 26 part-time (10 women); includes 3 minority (1 Asian American or Pacific Islander, 2 Hispanic Americans), 11 international. Average age 29. *34 applicants, 41% accepted.* In 1999, 15 master's, 3 doctorates awarded.

Entrance requirements: For master's, GRE, minimum GPA of 3.0; for doctorate, GRE, MS. *Application deadline:* For fall admission, 4/1. Applications are processed on a rolling basis. *Application fee:* $30 ($40 for international students).
Expenses: Tuition, state resident: part-time $229 per credit. Tuition, nonresident: part-time $464 per credit. Required fees: $241 per semester. Tuition and fees vary according to course load, degree level and program.
Financial aid: In 1999–00, 27 research assistantships (averaging $11,743 per year), 6 teaching assistantships (averaging $12,126 per year) were awarded. Financial aid applicants required to submit FAFSA.
Faculty research: Agroaquatic and forest ecology, insect physiology, toxicology and behavior, agricultural entomology. *Total annual research expenditures:* $3.9 million.
Dr. Edward Grafius, Chairperson, 517-355-4662.
Application contact: Dr. Frederick Stehr, Assistant Chair, 517-353-8739.

■ MISSISSIPPI STATE UNIVERSITY

College of Agriculture and Life Sciences, Department of Entomology and Plant Pathology, Mississippi State, MS 39762

AWARDS Agricultural pest management (MS); entomology (MS, PhD); plant pathology (MS, PhD).

Faculty: 26 full-time (1 woman), 27 part-time/adjunct (2 women).
Students: 12 full-time (2 women), 6 part-time (1 woman); includes 1 minority (Hispanic American), 7 international. Average age 30. *9 applicants, 22% accepted.* In 1999, 3 master's, 2 doctorates awarded.
Degree requirements: For master's and doctorate, thesis/dissertation, comprehensive oral or written exam required, foreign language not required.
Entrance requirements: For master's, GRE General Test, TOEFL, minimum GPA of 2.75; for doctorate, GRE General Test, TOEFL. *Application deadline:* For fall admission, 7/1; for spring admission, 11/1. Applications are processed on a rolling basis. *Application fee:* $25 for international students. Electronic applications accepted.
Expenses: Tuition, state resident: full-time $3,017; part-time $168 per credit. Tuition, nonresident: full-time $6,119; part-time $340 per credit. Part-time tuition and fees vary according to course load and program.
Financial aid: Research assistantships, teaching assistantships, Federal Work-Study, institutionally sponsored loans, and

unspecified assistantships available. Financial aid applicants required to submit FAFSA.

Faculty research: Ecology and population dynamics, physiology, biochemistry and behavior, systematics. *Total annual research expenditures:* $2.1 million.

Dr. Clarence Collison, Head, 662-325-2085, *Fax:* 662-325-8837, *E-mail:* chc2@ ra.msstate.edu.

Application contact: Jerry B. Inmon, Director of Admissions, 662-325-2224, *Fax:* 662-325-7360, *E-mail:* admit@ admissions.msstate.edu.

■ **MONTANA STATE UNIVERSITY–BOZEMAN**

College of Graduate Studies, College of Agriculture, Entomology Research Laboratory, Bozeman, MT 59717

AWARDS MS. Part-time programs available.

Students: 5 full-time (1 woman), 7 part-time (4 women). Average age 28. *5 applicants, 100% accepted.* In 1999, 4 degrees awarded.

Degree requirements: For master's, thesis required, foreign language not required.

Entrance requirements: For master's, GRE General Test, TOEFL, minimum GPA of 3.0. *Application deadline:* For fall admission, 6/1; for spring admission, 11/1. Applications are processed on a rolling basis. *Application fee:* $50. Electronic applications accepted.

Expenses: Tuition, state resident: full-time $2,674. Tuition, nonresident: full-time $6,986. International tuition: $7,136 full-time. Tuition and fees vary according to course load and program.

Financial aid: In 1999–00, 1 student received aid, including research assistantships with full tuition reimbursements available (averaging $12,600 per year), teaching assistantships with full tuition reimbursements available (averaging $1,000 per year); fellowships, career-related internships or fieldwork, Federal Work-Study, and scholarships also available. Financial aid application deadline: 3/1.

Faculty research: Arthropods, biocontrol, insects, biosystems, ecology, behavior, cropping systems. *Total annual research expenditures:* $1.1 million.

Dr. Greg Johnson, Head, 406-994-3860, *Fax:* 406-994-6029, *E-mail:* entomology@ montana.edu.

■ **NEW MEXICO STATE UNIVERSITY**

Graduate School, College of Agriculture and Home Economics, Department of Entomology, Plant Pathology and Weed Science, Las Cruces, NM 88003-8001

AWARDS MS. Part-time programs available.

Faculty: 12.

Students: 5 full-time (3 women); includes 1 minority (Hispanic American), 1 international. Average age 36. *1 applicant, 100% accepted.* In 1999, 3 degrees awarded.

Degree requirements: For master's, thesis required, foreign language not required.

Entrance requirements: For master's, GRE General Test. *Application deadline:* For fall admission, 7/1 (priority date); for spring admission, 11/1. Applications are processed on a rolling basis. *Application fee:* $15 ($35 for international students). Electronic applications accepted.

Expenses: Tuition, state resident: full-time $2,682; part-time $112 per credit. Tuition, nonresident: full-time $8,376; part-time $349 per credit. Tuition and fees vary according to course load.

Financial aid: Research assistantships, teaching assistantships, career-related internships or fieldwork available. Financial aid application deadline: 3/1.

Faculty research: Integrated pest management, pesticide application and safety, livestock ectoparasite research, biotechnology, nematology.

Dr. Grant Kinzer, Head, 505-646-3225, *Fax:* 505-646-8087, *E-mail:* gkinzer@ nmsu.edu.

■ **NORTH CAROLINA STATE UNIVERSITY**

Graduate School, College of Agriculture and Life Sciences, Department of Entomology, Raleigh, NC 27695

AWARDS M Ag, MS, PhD.

Faculty: 29 full-time (2 women), 17 part-time/adjunct (1 woman).

Students: 30 full-time (7 women), 12 part-time (3 women); includes 4 minority (1 African American, 2 Asian Americans or Pacific Islanders, 1 Hispanic American), 9 international. Average age 31. *30 applicants, 60% accepted.* In 1999, 5 master's, 13 doctorates awarded. Terminal master's awarded for partial completion of doctoral program.

Degree requirements: For master's and doctorate, thesis/dissertation required, foreign language not required.

Entrance requirements: For master's and doctorate, GRE General Test. *Application deadline:* For fall admission, 6/25 (priority date); for spring admission, 11/25. Applications are processed on a rolling basis. *Application fee:* $45.

Expenses: Tuition, state resident: full-time $1,578. Tuition, nonresident: full-time $10,744. Required fees: $892. Full-time tuition and fees vary according to program.

Financial aid: In 1999–00, 27 research assistantships (averaging $5,174 per year), 1 teaching assistantship (averaging $5,293 per year) were awarded; fellowships, career-related internships or fieldwork, Federal Work-Study, and institutionally sponsored loans also available. Aid available to part-time students.

Faculty research: Agriculture, physiology, biocontrol, ecology, forest entomology, medical and veterinary entomology. *Total annual research expenditures:* $7.1 million.

Dr. James D. Harper, Head, 919-515-2746, *Fax:* 919-515-7746, *E-mail:* james_ harper@ncsu.edu.

Application contact: Dr. Wayne M. Brooks, Director of Graduate Programs, 919-515-3771, *Fax:* 919-515-7746, *E-mail:* wayne_brooks@ncsu.edu.

■ **NORTH DAKOTA STATE UNIVERSITY**

Graduate Studies and Research, College of Agriculture, Department of Entomology, Fargo, ND 58105

AWARDS MS, PhD. Part-time programs available.

Faculty: 6 full-time (1 woman), 2 part-time/adjunct (0 women).

Students: 7 full-time (3 women), 1 part-time. Average age 30. *3 applicants, 33% accepted.* In 1999, 2 degrees awarded.

Degree requirements: For master's, thesis required, foreign language not required; for doctorate, one foreign language, dissertation required. *Average time to degree:* Master's–2.5 years full-time.

Entrance requirements: For master's and doctorate, TOEFL, minimum GPA of 3.0. *Application deadline:* Applications are processed on a rolling basis. *Application fee:* $25.

Expenses: Tuition, state resident: full-time $3,096; part-time $112 per credit hour. Tuition, nonresident: full-time $7,588; part-time $299 per credit hour. Tuition and fees vary according to course load, campus/location and reciprocity agreements.

Financial aid: In 1999–00, research assistantships with full tuition reimbursements (averaging $9,360 per year); Federal Work-Study and institutionally sponsored loans also available. Financial aid application deadline: 4/15.

North Dakota State University (continued)
Faculty research: Insect systematics, conservation biology, integrated pest management, biological control.
Dr. Gary J. Brewer, Chair, 701-231-7908, *Fax:* 701-231-8557, *E-mail:* brewer@badlands.nodak.edu.

■ THE OHIO STATE UNIVERSITY

Graduate School, College of Biological Sciences, Department of Entomology, Columbus, OH 43210
AWARDS MS, PhD.

Faculty: 26 full-time, 8 part-time/adjunct.
Students: 32 full-time (7 women), 4 part-time (2 women); includes 2 minority (1 Asian American or Pacific Islander, 1 Hispanic American), 12 international. *14 applicants, 43% accepted.* In 1999, 6 master's, 2 doctorates awarded.
Degree requirements: For master's, thesis optional; for doctorate, dissertation required.
Entrance requirements: For master's and doctorate, GRE General Test, GRE Subject Test (biology), TOEFL. *Application deadline:* For fall admission, 8/15. Applications are processed on a rolling basis. *Application fee:* $30 ($40 for international students).
Expenses: Tuition, state resident: full-time $5,400. Tuition, nonresident: full-time $14,535. Part-time tuition and fees vary according to course load and program.
Financial aid: Fellowships, research assistantships, teaching assistantships, Federal Work-Study and institutionally sponsored loans available. Aid available to part-time students.
Faculty research: Acarology, insect systematics, soil ecology, integrated pest management, chemical ecology.
Dr. David L. Denlinger, Chairman, 614-292-8209, *Fax:* 614-292-2180, *E-mail:* denlinger.1@osu.edu.

■ OKLAHOMA STATE UNIVERSITY

Graduate College, College of Agricultural Sciences and Natural Resources, Department of Entomology and Plant Pathology, Program in Entomology, Stillwater, OK 74078
AWARDS MS, PhD.

Degree requirements: For master's, thesis or alternative required; for doctorate, dissertation required.
Entrance requirements: For master's and doctorate, TOEFL. *Application deadline:* For fall admission, 6/1 (priority date). *Application fee:* $25.
Expenses: Tuition, state resident: part-time $86 per credit hour. Tuition, nonresident: part-time $275 per credit

hour. Required fees: $17 per credit hour. $14 per semester. One-time fee: $20 full-time. Tuition and fees vary according to course load.
Financial aid: Research assistantships, teaching assistantships, career-related internships or fieldwork, Federal Work-Study, and tuition waivers (partial) available. Aid available to part-time students. Financial aid application deadline: 3/1.
Russell Wright, Interim Head, Department of Entomology and Plant Pathology, 405-744-5527.

■ OREGON STATE UNIVERSITY

Graduate School, College of Science, Department of Entomology, Corvallis, OR 97331
AWARDS M Agr, MA, MAIS, MS, PhD. Part-time programs available.

Faculty: 19 full-time (1 woman), 4 part-time/adjunct (1 woman).
Students: 24 full-time (7 women), 3 part-time, 6 international. Average age 30. In 1999, 1 master's, 1 doctorate awarded.
Degree requirements: For master's and doctorate, thesis/dissertation required, foreign language not required.
Entrance requirements: For master's and doctorate, GRE General Test, GRE Subject Test, TOEFL, minimum GPA of 3.0 in last 90 hours. *Application deadline:* For fall admission, 3/1 (priority date). Applications are processed on a rolling basis. *Application fee:* $50.
Expenses: Tuition, state resident: full-time $6,489. Tuition, nonresident: full-time $11,061. Tuition and fees vary according to program.
Financial aid: Research assistantships, teaching assistantships, career-related internships or fieldwork, Federal Work-Study, and institutionally sponsored loans available. Aid available to part-time students. Financial aid application deadline: 2/1.
Faculty research: Insect ecology, insect biosystematics, integrated pest management, insect biochemistry, insect physiology.
Dr. Paul C. Jepson, Chair, 541-737-4733, *Fax:* 541-737-3643, *E-mail:* jepsonp@bcc.orst.edu.
Application contact: Deanna Watkins, Office Specialist, 541-737-5488, *Fax:* 541-737-3643, *E-mail:* watkinsd@bcc.orst.edu.

■ THE PENNSYLVANIA STATE UNIVERSITY UNIVERSITY PARK CAMPUS

Graduate School, College of Agricultural Sciences, Department of Entomology, State College, University Park, PA 16802-1503
AWARDS M Agr, MS, PhD.

Students: 21 full-time (6 women), 4 part-time.
Entrance requirements: For master's and doctorate, GRE General Test. *Application fee:* $50.
Expenses: Tuition, state resident: full-time $6,886; part-time $291 per credit. Tuition, nonresident: full-time $14,118; part-time $588 per credit. Required fees: $46 per semester. Part-time tuition and fees vary according to course load and program.
Dr. James L. Frazier, Head, 814-863-7344.

■ PURDUE UNIVERSITY

Graduate School, School of Agriculture, Department of Entomology, West Lafayette, IN 47907
AWARDS MS, PhD. Part-time programs available.

Faculty: 18 full-time (2 women), 4 part-time/adjunct (1 woman).
Students: 25 full-time (5 women), 1 part-time, 8 international. Average age 29. *17 applicants, 47% accepted.* In 1999, 3 doctorates awarded.
Degree requirements: For master's, thesis (for some programs), seminar required; for doctorate, dissertation, seminar required. *Average time to degree:* Master's–3 years full-time; doctorate–5 years full-time.
Entrance requirements: For master's and doctorate, TOEFL. *Application deadline:* For fall admission, 4/15 (priority date); for spring admission, 9/15. Applications are processed on a rolling basis. *Application fee:* $30. Electronic applications accepted.
Expenses: Tuition, state resident: full-time $4,530; part-time $130 per credit hour. Tuition, nonresident: full-time $15,310; part-time $404 per credit hour. Tuition and fees vary according to campus/location and program.
Financial aid: In 1999–00, 1 fellowship with tuition reimbursement, 20 research assistantships with tuition reimbursements (averaging $14,000 per year), 3 teaching assistantships with tuition reimbursements (averaging $14,000 per year) were awarded; career-related internships or fieldwork also available. Aid available to part-time students. Financial aid applicants required to submit FAFSA.
Faculty research: Insect biochemistry, nematology, aquatic diptera, behavioral ecology, insect physiology.
Dr. C. R. Edwards, Head, 765-494-4562, *Fax:* 765-494-0535, *E-mail:* rich_edwards@entm.purdue.edu.
Application contact: Jenny Franklin, Graduate Admissions Office, 765-494-9061, *Fax:* 765-494-0535, *E-mail:* jenny_franklin@entm.purdue.edu.

■ RUTGERS, THE STATE UNIVERSITY OF NEW JERSEY, NEW BRUNSWICK

Graduate School, Program in Entomology, New Brunswick, NJ 08901-1281

AWARDS MS, PhD.

Faculty: 10 full-time (1 woman).
Students: 10 full-time (6 women), 4 part-time (2 women); includes 2 minority (1 Asian American or Pacific Islander, 1 Hispanic American), 6 international. *16 applicants, 31% accepted.* In 1999, 2 master's awarded.
Degree requirements: For master's, thesis or alternative required, foreign language not required; for doctorate, dissertation required, foreign language not required. *Average time to degree:* Master's–3 years full-time.
Entrance requirements: For master's and doctorate, GRE General Test. *Application fee:* $50.
Expenses: Tuition, state resident: full-time $6,776; part-time $279 per credit. Tuition, nonresident: full-time $9,936; part-time $412 per credit. Required fees: $20 per credit. $89 per semester. Tuition and fees vary according to course load, campus/location and program.
Financial aid: In 1999–00, 9 students received aid, including 4 research assistantships with full tuition reimbursements available, 5 teaching assistantships with full tuition reimbursements available; fellowships with full tuition reimbursements available Financial aid application deadline: 3/1; financial aid applicants required to submit FAFSA.
Faculty research: Insect toxicology, physiology, morphology, and ecology, insect systematics.
Dr. George C. Hamilton, Director, 732-932-9774, *Fax:* 732-932-7229.

■ SOUTH DAKOTA STATE UNIVERSITY

Graduate School, College of Agriculture and Biological Sciences, Department of Plant Science, Program in Entomology, Brookings, SD 57007

AWARDS MS.

Degree requirements: For master's, thesis, oral exam required, foreign language not required.
Entrance requirements: For master's, GRE General Test, TOEFL.
Faculty research: Integrated pest management, biological control, behavioral ecology, biodiversity, systematics.

■ TEXAS A&M UNIVERSITY

College of Agriculture and Life Sciences, Department of Entomology, College Station, TX 77843

AWARDS M Agr, MS, PhD.

Faculty: 48 full-time (3 women), 15 part-time/adjunct (2 women).
Students: 55 full-time (21 women). Average age 34. *22 applicants, 77% accepted.* In 1999, 7 master's, 6 doctorates awarded.
Entrance requirements: For master's and doctorate, GRE General Test, TOEFL. *Application deadline:* For fall admission, 2/1 (priority date); for spring admission, 10/1. Applications are processed on a rolling basis. *Application fee:* $50 ($75 for international students). Electronic applications accepted.
Expenses: Tuition, state resident: part-time $76 per semester hour. Tuition, nonresident: part-time $292 per semester hour. Required fees: $11 per semester hour. Tuition and fees vary according to program.
Financial aid: In 1999–00, 33 research assistantships with partial tuition reimbursements (averaging $14,700 per year), 5 teaching assistantships with partial tuition reimbursements (averaging $14,700 per year) were awarded; fellowships, Federal Work-Study also available. Financial aid application deadline: 3/1; financial aid applicants required to submit FAFSA.
Faculty research: Biology, biological control. biosystematics of parasitic hymenoptera, integrated management of corn and sorghum insects, integrated pest management.
Dr. Ray E. Frisbie, Head, 979-845-2516, *Fax:* 979-845-6305.
Application contact: Jim Woolley, Adviser, 979-845-9349, *Fax:* 979-845-9938, *E-mail:* jimwoooley@tamu.edu.

Find an in-depth description at www.petersons.com/graduate.

■ TEXAS TECH UNIVERSITY

Graduate School, College of Agricultural Sciences and Natural Resources, Department of Plant and Soil Science, Lubbock, TX 79409

AWARDS Agronomy (PhD); crop science (MS); entomology (MS); horticulture (MS); plant and soil science (M Agr); soil science (MS). Part-time programs available.

Faculty: 14 full-time (2 women), 1 part-time/adjunct (0 women).
Students: 36 full-time (12 women), 17 part-time (4 women); includes 1 minority (Asian American or Pacific Islander), 7 international. Average age 31. *12 applicants, 50% accepted.* In 2000, 10 master's, 5 doctorates awarded.

Degree requirements: For master's, foreign language and thesis not required; for doctorate, dissertation required.
Entrance requirements: For master's and doctorate, GRE General Test. *Application deadline:* For fall admission, 4/15 (priority date); for spring admission, 11/1 (priority date). Applications are processed on a rolling basis. *Application fee:* $25 ($50 for international students). Electronic applications accepted.
Expenses: Tuition, state resident: full-time $2,736; part-time $114 per hour. Tuition, nonresident: full-time $7,920; part-time $330 per hour. Required fees: $464 per semester. Tuition and fees vary according to course level, course load and program.
Financial aid: In 2000–01, 25 students received aid, including 21 research assistantships (averaging $10,882 per year), 3 teaching assistantships (averaging $11,250 per year); fellowships, Federal Work-Study and institutionally sponsored loans also available. Aid available to part-time students. Financial aid application deadline: 5/15; financial aid applicants required to submit FAFSA.
Faculty research: Molecular and cellular biology of plant stress, physiology/genetics of cotton and sorghum production in semiarid conditions, biology of red fire ants. *Total annual research expenditures:* $1.6 million.
Dr. Dick L. Auld, Chair, 806-742-2837, *Fax:* 806-742-0775.
Application contact: Graduate Adviser, 806-742-2837.

■ THE UNIVERSITY OF ARIZONA

Graduate College, College of Agriculture, Department of Entomology, Tucson, AZ 85721

AWARDS MS, PhD. Part-time programs available.

Degree requirements: For master's and doctorate, thesis/dissertation required, foreign language not required.
Entrance requirements: For master's and doctorate, GRE General Test, GRE Subject Test, TOEFL.
Expenses: Tuition, nonresident: full-time $4,814; part-time $274 per unit. Required fees: $1,094; $115 per unit. Tuition and fees vary according to course load and program.
Faculty research: Toxicology and physiology, plant/insect relations, vector biology, insect pest management, chemical ecology.

■ THE UNIVERSITY OF ARIZONA

Graduate College, Graduate Interdisciplinary Programs, Graduate Interdisciplinary Program in Insect Science, Tucson, AZ 85721

AWARDS PhD.

The University of Arizona (continued)
Degree requirements: For doctorate, dissertation required.
Entrance requirements: For doctorate, GRE General Test, TOEFL.
Expenses: Tuition, nonresident: full-time $4,814; part-time $274 per unit. Required fees: $1,094; $115 per unit. Tuition and fees vary according to course load and program.

■ UNIVERSITY OF ARKANSAS

Graduate School, Dale Bumpers College of Agricultural, Food and Life Sciences, Department of Entomology, Fayetteville, AR 72701-1201

AWARDS MS, PhD.

Faculty: 14 full-time (0 women).
Students: 19 full-time (3 women); includes 1 minority (African American), 5 international. *11 applicants, 73% accepted.* In 1999, 2 master's, 1 doctorate awarded.
Degree requirements: For master's, thesis required, foreign language not required; for doctorate, one foreign language, dissertation required.
Entrance requirements: For master's, GRE General Test, minimum GPA of 3.0; for doctorate, GRE General Test, GRE Subject Test, minimum GPA of 3.25. *Application fee:* $40 ($50 for international students).
Expenses: Tuition, state resident: full-time $3,186; part-time $177 per credit. Tuition, nonresident: full-time $7,560; part-time $420 per credit. Required fees: $756; $21 per credit. One-time fee: $22 part-time. Tuition and fees vary according to course load and program.
Financial aid: In 1999–00, 15 research assistantships, 1 teaching assistantship were awarded; career-related internships or fieldwork and Federal Work-Study also available. Aid available to part-time students. Financial aid application deadline: 4/1; financial aid applicants required to submit FAFSA.
Faculty research: Integrated pest management, insect virology, insect taxonomy.
Dr. W. C. Yearian, Chair, 501-575-2451, *E-mail:* wyearia@comp.uark.edu.

■ UNIVERSITY OF CALIFORNIA, DAVIS

Graduate Studies, Programs in the Biological Sciences, Program in Entomology, Davis, CA 95616

AWARDS MS, PhD.

Faculty: 27 full-time (4 women), 3 part-time/adjunct.
Students: 26 full-time (15 women); includes 7 minority (2 Asian Americans or Pacific Islanders, 5 Hispanic Americans), 1 international. Average age 29. *31 applicants, 32% accepted.* In 1999, 7 doctorates awarded.
Degree requirements: For master's, thesis optional; for doctorate, dissertation required. *Average time to degree:* Master's– 2.5 years full-time; doctorate–4 years full-time.
Entrance requirements: For master's and doctorate, GRE General Test, GRE Subject Test (biology). *Application deadline:* For fall admission, 4/1. *Application fee:* $40. Electronic applications accepted.
Expenses: Tuition, nonresident: full-time $9,804. Tuition and fees vary according to program and student level.
Financial aid: In 1999–00, 25 students received aid, including 3 fellowships with full and partial tuition reimbursements available, 7 research assistantships with full and partial tuition reimbursements available, 1 teaching assistantship with partial tuition reimbursement available; Federal Work-Study, grants, institutionally sponsored loans, scholarships, and tuition waivers (full and partial) also available. Financial aid application deadline: 1/15; financial aid applicants required to submit FAFSA.
Faculty research: Bee biology, biological control, systematics, medical/veterinary entomology, pest management, agricultural entomology, molecular biology, environmental toxicology, physiology and biochemistry, behavior, systematics, parasitology.
Robert Page, Graduate Chair, 530-752-0492, *E-mail:* repage@ucdavis.edu.
Application contact: Brenda Nakamoto, Administrative Assistant, 530-752-0492, *Fax:* 530-752-1537, *E-mail:* bunakamota@ucdavis.edu.

■ UNIVERSITY OF CALIFORNIA, RIVERSIDE

Graduate Division, College of Natural and Agricultural Sciences, Department of Entomology, Riverside, CA 92521-0102

AWARDS MS, PhD. Part-time programs available.

Faculty: 38.
Students: 37 full-time (14 women), 10 international. Average age 31. In 1999, 2 master's, 6 doctorates awarded. Terminal master's awarded for partial completion of doctoral program.
Degree requirements: For master's, thesis required, foreign language not required; for doctorate, dissertation, qualifying exams required, foreign language not required. *Average time to degree:* Master's–3 years full-time; doctorate–6 years full-time.

Entrance requirements: For master's and doctorate, GRE General Test, TOEFL, minimum GPA of 3.2. *Application deadline:* For fall admission, 5/1; for winter admission, 9/1; for spring admission, 12/1. Applications are processed on a rolling basis. *Application fee:* $40.
Expenses: Tuition, nonresident: full-time $9,804. Required fees: $4,758. Full-time tuition and fees vary according to program.
Financial aid: Fellowships, research assistantships, teaching assistantships, career-related internships or fieldwork, Federal Work-Study, institutionally sponsored loans, and tuition waivers (full and partial) available. Financial aid application deadline: 12/31; financial aid applicants required to submit FAFSA.
Faculty research: Agricultural, urban, medical, and veterinary entomology; biological control; chemical ecology; insect pathogens; novel toxicants.
Dr. Timothy Paine, Chair, 909-787-5831, *Fax:* 909-787-3086.
Application contact: Kathy Redd, Student Affairs Officer, 909-787-4716, *Fax:* 909-787-5517, *E-mail:* insects@ucr.edu.

Find an in-depth description at www.petersons.com/graduate.

■ UNIVERSITY OF CONNECTICUT

Graduate School, College of Liberal Arts and Sciences, Biological Sciences Group, Storrs, CT 06269

AWARDS Ecology and evolutionary biology (MS, PhD), including botany, ecology, entomology, systematics, zoology; molecular and cell biology (MS, PhD), including biochemistry, biophysics, biotechnology (MS), cell and developmental biology, genetics, microbiology, plant molecular and cell biology; physiology and neurobiology (MS, PhD), including neurobiology, physiology.

Degree requirements: For doctorate, dissertation required.
Entrance requirements: For master's and doctorate, GRE General Test, GRE Subject Test, TOEFL.
Expenses: Tuition, state resident: full-time $5,118. Tuition, nonresident: full-time $13,298. Required fees: $1,022.

■ UNIVERSITY OF CONNECTICUT

Graduate School, College of Liberal Arts and Sciences, Biological Sciences Group, Department of Ecology and Evolutionary Biology, Field of Entomology, Storrs, CT 06269

AWARDS MS, PhD.

Degree requirements: For doctorate, dissertation required.

Entrance requirements: For master's and doctorate, GRE General Test, GRE Subject Test, TOEFL.
Expenses: Tuition, state resident: full-time $5,118. Tuition, nonresident: full-time $13,298. Required fees: $1,022.
Find an in-depth description at www.petersons.com/graduate.

■ UNIVERSITY OF DELAWARE

College of Agriculture and Natural Resources, Department of Entomology and Applied Ecology, Newark, DE 19716

AWARDS Biology (PhD); entomology and applied ecology (MS, PhD), including avian ecology (MS), evolution and taxonomy (MS), insect biological control (MS), insect ecology and behavior (MS), insect genetics (MS), pest management (MS), plant-insect interactions (MS); plant science (PhD); turtle ecology (PhD). PhD offered in cooperation with the Department of Biological Sciences or the Department of Plant and Soil Sciences. Part-time programs available.

Faculty: 6 full-time (0 women).
Students: 26 full-time (11 women), 1 part-time; includes 1 minority (Asian American or Pacific Islander), 2 international. Average age 24. *10 applicants, 80% accepted.* In 1999, 5 degrees awarded.
Degree requirements: For master's, thesis, oral exam, seminar required, foreign language not required; for doctorate, dissertation, qualifying exam required, foreign language not required. *Average time to degree:* Master's–3.4 years full-time.
Entrance requirements: For master's and doctorate, GRE General Test, TOEFL, minimum GPA of 3.0 in field, 2.8 overall. *Application deadline:* For fall admission, 3/1 (priority date); for spring admission, 11/1. Applications are processed on a rolling basis. *Application fee:* $45. Electronic applications accepted.
Expenses: Tuition, state resident: full-time $4,380; part-time $243 per credit. Tuition, nonresident: full-time $12,750; part-time $708 per credit. Required fees: $15 per term. Tuition and fees vary according to program.
Financial aid: In 1999–00, 18 students received aid, including 1 fellowship with full tuition reimbursement available (averaging $15,000 per year), 9 research assistantships with full tuition reimbursements available (averaging $12,400 per year), 9 teaching assistantships with full tuition reimbursements available (averaging $12,000 per year); career-related internships or fieldwork, institutionally sponsored loans, scholarships, and tuition waivers (full) also available. Financial aid application deadline: 3/1.

Faculty research: Genetics and resistance, biological control, chemically mediated behavioral ecology, ecology and evolution of plant-insect interactions, ecology of species conservation. *Total annual research expenditures:* $885,313.
Dr. Judith A. Hough-Goldstein, Chairperson, 302-831-2526, *Fax:* 302-831-3651, *E-mail:* jhough@udel.edu.
Application contact: Dr. Charles E. Mason, Graduate Coordinator, 302-831-8888, *Fax:* 302-831-3651, *E-mail:* mason@udel.edu.

■ UNIVERSITY OF FLORIDA

Graduate School, College of Agriculture and Life Sciences, Department of Entomology and Nematology, Gainesville, FL 32611

AWARDS MS, PhD.

Faculty: 128.
Students: 69 full-time (29 women), 8 part-time (2 women); includes 8 minority (3 African Americans, 5 Hispanic Americans), 29 international. *36 applicants, 78% accepted.* In 1999, 9 master's, 9 doctorates awarded. Terminal master's awarded for partial completion of doctoral program.
Degree requirements: For master's, thesis optional; for doctorate, dissertation required.
Entrance requirements: For master's and doctorate, GRE General Test, GRE Subject Test (biology), minimum GPA of 3.0. *Application deadline:* For fall admission, 6/1 (priority date). Applications are processed on a rolling basis. *Application fee:* $20. Electronic applications accepted.
Expenses: Tuition, state resident: part-time $144 per credit hour. Tuition, nonresident: part-time $505 per credit hour. Tuition and fees vary according to course level, course load and program.
Financial aid: In 1999–00, 49 students received aid, including 3 fellowships, 42 research assistantships, 9 teaching assistantships; career-related internships or fieldwork also available.
Faculty research: Medical, veterinary, and urban entomology; genetics; biology and management; biocontrol; insect ecology.
Dr. J. L. Capinera, Chair, 352-392-1901 Ext. 111, *Fax:* 352-392-0190, *E-mail:* jlcap@gnv.ifas.ufl.edu.
Application contact: Dr. Grover C. Smart, Graduate Coordinator, 352-392-1901 Ext. 118, *Fax:* 352-392-0190, *E-mail:* gradc@gnv.ifas.ufl.edu.
Find an in-depth description at www.petersons.com/graduate.

■ UNIVERSITY OF GEORGIA

Graduate School, College of Agricultural and Environmental Sciences, Department of Entomology, Athens, GA 30602

AWARDS Entomology (MS, PhD); plant protection and pest management (MPPPM).

Degree requirements: For master's, thesis (MS) required; for doctorate, one foreign language (computer language can substitute), dissertation required.
Entrance requirements: For master's and doctorate, GRE General Test. Electronic applications accepted.
Expenses: Tuition, state resident: full-time $7,516; part-time $431 per credit hour. Tuition, nonresident: full-time $12,204; part-time $793 per credit hour. Tuition and fees vary according to program.
Faculty research: Apiculture, acarology, aquatic and soil biology, ecology, systematics.

Find an in-depth description at www.petersons.com/graduate.

■ UNIVERSITY OF HAWAII AT MANOA

Graduate Division, College of Tropical Agriculture and Human Resources, Department of Entomology, Honolulu, HI 96822

AWARDS MS, PhD. Part-time programs available.

Faculty: 17 full-time (3 women), 13 part-time/adjunct (0 women).
Students: 9 full-time (1 woman), 2 part-time (1 woman). Average age 35. *9 applicants, 89% accepted.* In 1999, 1 master's awarded (100% found work related to degree).
Degree requirements: For master's and doctorate, thesis/dissertation required, foreign language not required. *Average time to degree:* Master's–1 year full-time; doctorate–3 years full-time, 8 years part-time.
Entrance requirements: For master's and doctorate, GRE General Test, GRE Subject Test (biology). *Application deadline:* For fall admission, 3/1; for spring admission, 9/1. *Application fee:* $25 ($50 for international students).
Expenses: Tuition, state resident: full-time $4,032; part-time $168 per credit. Tuition, nonresident: full-time $9,960; part-time $415 per credit. Required fees: $51 per semester. Part-time tuition and fees vary according to course load and program.
Financial aid: In 1999–00, 8 research assistantships (averaging $16,010 per year) were awarded; teaching assistantships, tuition waivers (full) also available.

University of Hawaii at Manoa
(continued)

Faculty research: Integrated pest management, biological control, urban entomology, medical/forensic entomology resistance.

Dr. Marshall Johnson, Chair, 808-956-6747, *Fax:* 808-956-2428, *E-mail:* mjohnson@hawaii.edu.

Application contact: Dr. Stephen Saul, Graduate Chair, 808-956-7076, *Fax:* 808-956-2428, *E-mail:* saul@hawaii.edu.

■ UNIVERSITY OF IDAHO

College of Graduate Studies, College of Agriculture, Department of Plant, Soil, and Entomological Sciences, Program in Entomology, Moscow, ID 83844-4140

AWARDS MS, PhD.

Students: 7 full-time (3 women), 2 part-time; includes 1 minority (Hispanic American). In 1999, 4 master's awarded.

Degree requirements: For master's, thesis required (for some programs), foreign language not required; for doctorate, dissertation required.

Entrance requirements: For master's and doctorate, GRE General Test, minimum GPA of 3.0. *Application deadline:* For fall admission, 8/1; for spring admission, 12/15. *Application fee:* $35 ($45 for international students).

Expenses: Tuition, nonresident: full-time $6,000; part-time $239 per credit hour. Required fees: $2,888; $144 per credit hour. Tuition and fees vary according to program.

Financial aid: Application deadline: 2/15.

Dr. James B. Johnson, Chair, 208-885-7543.

■ UNIVERSITY OF ILLINOIS AT URBANA–CHAMPAIGN

Graduate College, College of Liberal Arts and Sciences, School of Life Sciences, Department of Entomology, Urbana, IL 61801

AWARDS Entomology (MS, PhD); insect pest management (MS).

Faculty: 6 full-time (2 women), 12 part-time/adjunct (1 woman).

Students: 32 full-time (11 women); includes 3 minority (2 African Americans, 1 Asian American or Pacific Islander), 4 international. *8 applicants, 0% accepted.* In 1999, 9 master's, 1 doctorate awarded. Terminal master's awarded for partial completion of doctoral program.

Degree requirements: For master's, thesis required, foreign language not required; for doctorate, one foreign language (computer language can substitute), dissertation required.

Entrance requirements: For master's and doctorate, GRE General Test, GRE Subject Test, TOEFL, minimum GPA of 4.0 on a 5.0 scale. *Application deadline:* Applications are processed on a rolling basis. *Application fee:* $40 ($50 for international students).

Expenses: Tuition, state resident: full-time $4,616. Tuition, nonresident: full-time $11,768. Full-time tuition and fees vary according to course load.

Financial aid: In 1999–00, 3 fellowships, 12 research assistantships, 16 teaching assistantships were awarded; career-related internships or fieldwork, Federal Work-Study, and institutionally sponsored loans also available. Financial aid application deadline: 2/15.

Dr. May R. Berenbaum, Head, 217-333-2910, *Fax:* 217-244-3499, *E-mail:* maybe@uiuc.edu.

Application contact: Dorothy Houchens, Director of Graduate Studies, 217-244-2888, *Fax:* 217-244-3799, *E-mail:* dorothyh@uiuc.edu.

Find an in-depth description at www.petersons.com/graduate.

■ UNIVERSITY OF KANSAS

Graduate School, College of Liberal Arts and Sciences, Department of Entomology, Lawrence, KS 66045

AWARDS MA, PhD.

Faculty: 9.

Students: 6 full-time (1 woman), 16 part-time (6 women); includes 2 minority (1 African American, 1 Hispanic American), 6 international. *13 applicants, 69% accepted.* In 1999, 1 master's, 4 doctorates awarded.

Entrance requirements: For master's and doctorate, GRE, TSE, TOEFL. *Application deadline:* For fall admission, 4/1; for spring admission, 8/15. *Application fee:* $25.

Expenses: Tuition, state resident: full-time $2,482; part-time $103 per credit hour. Tuition, nonresident: full-time $8,104; part-time $338 per credit hour. Required fees: $428; $31 per credit hour. Tuition and fees vary according to program.

Financial aid: In 1999–00, 1 research assistantship, 1 teaching assistantship were awarded.

Orley Taylor, Acting Chair, 785-864-4301.

■ UNIVERSITY OF KENTUCKY

Graduate School, Graduate School Programs from the College of Agriculture, Program in Entomology, Lexington, KY 40506-0032

AWARDS MS, PhD.

Degree requirements: For master's, comprehensive exam required, thesis optional, foreign language not required; for doctorate, dissertation, comprehensive exam required, foreign language not required.

Entrance requirements: For master's, GRE General Test, minimum undergraduate GPA of 2.75; for doctorate, GRE General Test, minimum graduate GPA of 3.25.

Expenses: Tuition, state resident: full-time $3,596; part-time $188 per credit hour. Tuition, nonresident: full-time $10,116; part-time $550 per credit hour.

Faculty research: Applied entomology, behavior, insect biology and ecology, biological control, insect physiology and molecular biology.

■ UNIVERSITY OF MAINE

Graduate School, College of Natural Sciences, Forestry, and Agriculture, Department of Biological Sciences, Program in Entomology, Orono, ME 04469

AWARDS MS. Part-time programs available.

Entrance requirements: For master's, GRE General Test, TOEFL.

Expenses: Tuition, state resident: full-time $3,564. Tuition, nonresident: full-time $10,116. Required fees: $378. Tuition and fees vary according to course load.

■ UNIVERSITY OF MARYLAND, COLLEGE PARK

Graduate Studies and Research, College of Life Sciences, Department of Entomology, College Park, MD 20742

AWARDS MS, PhD. Part-time and evening/weekend programs available.

Faculty: 32 full-time (11 women), 1 part-time/adjunct (0 women).

Students: 18 full-time (9 women), 6 part-time (4 women); includes 3 minority (1 African American, 2 Asian Americans or Pacific Islanders), 5 international. *31 applicants, 16% accepted.* In 1999, 4 master's, 3 doctorates awarded. Terminal master's awarded for partial completion of doctoral program.

Degree requirements: For master's, thesis required, foreign language not required; for doctorate, dissertation, oral qualifying exam required.

Entrance requirements: For master's and doctorate, GRE General Test, minimum GPA of 3.0. *Application deadline:* For fall admission, 7/1; for spring admission, 11/1. Applications are processed on a rolling basis. *Application fee:* $50 ($70 for international students). Electronic applications accepted.

Expenses: Tuition, state resident: part-time $272 per credit hour. Tuition,

nonresident: part-time $415 per credit hour. Required fees: $632; $379 per year.
Financial aid: In 1999–00, 19 teaching assistantships with tuition reimbursements (averaging $11,408 per year) were awarded; fellowships with full tuition reimbursements, research assistantships, career-related internships or fieldwork and Federal Work-Study also available. Aid available to part-time students. Financial aid applicants required to submit FAFSA.
Faculty research: Pest management, biosystematics, physiology and morphology, toxicology. *Total annual research expenditures:* $1.2 million.
Dr. Michael Raupp, Chair, 301-405-3912, *Fax:* 301-314-9290, *E-mail:* grschool@ deans.umd.edu.
Application contact: Trudy Lindsey, Director, Graduate Admissions and Records, 301-405-4198, *Fax:* 301-314-9305, *E-mail:* grschool@deans.umd.edu.

■ UNIVERSITY OF MASSACHUSETTS AMHERST

Graduate School, College of Food and Natural Resources, Department of Entomology, Amherst, MA 01003

AWARDS MS, PhD. Part-time programs available.

Faculty: 13 full-time (3 women).
Students: 14 full-time (4 women), 8 part-time (5 women); includes 1 minority (African American), 9 international. Average age 30. *13 applicants, 23% accepted.* In 1999, 3 master's, 6 doctorates awarded. Terminal master's awarded for partial completion of doctoral program.
Degree requirements: For master's, thesis or alternative required, foreign language not required; for doctorate, dissertation required, foreign language not required.
Entrance requirements: For master's and doctorate, GRE General Test, GRE Subject Test. *Application deadline:* For fall admission, 2/1 (priority date); for spring admission, 10/1. Applications are processed on a rolling basis. *Application fee:* $40.
Expenses: Tuition, state resident: full-time $2,640; part-time $165 per credit. Tuition, nonresident: full-time $9,756; part-time $407 per credit. Required fees: $1,221 per term. One-time fee: $110. Full-time tuition and fees vary according to course load, campus/location and reciprocity agreements.
Financial aid: In 1999–00, 1 fellowship with full tuition reimbursement (averaging $11,582 per year), research assistantships with full tuition reimbursements (averaging $7,760 per year), 6 teaching assistantships with full tuition reimbursements (averaging $2,304 per year) were awarded; career-related internships or fieldwork, Federal

Work-Study, grants, scholarships, traineeships, and unspecified assistantships also available. Aid available to part-time students. Financial aid application deadline: 2/1.
Dr. David N. Ferro, Chair, 413-545-1059, *Fax:* 413-545-2115 or 545-5858, *E-mail:* ferro@ent.umass.edu.

■ UNIVERSITY OF MINNESOTA, TWIN CITIES CAMPUS

Graduate School, College of Agricultural, Food, and Environmental Sciences, Department of Entomology, Minneapolis, MN 55455-0213

AWARDS MS, PhD. Part-time programs available.

Faculty: 19 full-time (5 women), 4 part-time/adjunct (1 woman).
Students: 29 full-time (8 women); includes 1 minority (Hispanic American), 10 international. Average age 33. *22 applicants, 23% accepted.* In 1999, 3 master's, 1 doctorate awarded.
Degree requirements: For master's and doctorate, thesis/dissertation required. *Average time to degree:* Master's–3 years full-time, 5 years part-time; doctorate–4 years full-time, 6 years part-time.
Entrance requirements: For master's, minimum undergraduate GPA of 3.0; for doctorate, minimum GPA of 3.0 (undergraduate), 3.5 (graduate). *Application deadline:* For fall admission, 7/15; for spring admission, 12/15. Applications are processed on a rolling basis. *Application fee:* $50 ($55 for international students). Electronic applications accepted.
Expenses: Tuition, state resident: full-time $5,040; part-time $420 per credit. Tuition, nonresident: full-time $9,900; part-time $825 per credit. Full-time tuition and fees vary according to course load, program and reciprocity agreements.
Financial aid: In 1999–00, 3 fellowships with full tuition reimbursements, 19 research assistantships with full tuition reimbursements, 1 teaching assistantship with full tuition reimbursement were awarded.
Faculty research: Behavior, ecology, molecular genetics, physiology, systematics and taxonomy.
Mark E. Ascerno, Head, 612-624-3278, *Fax:* 612-625-5299, *E-mail:* mascerno@ tc.umn.edu.
Application contact: Ann M. Fallon, Director of Graduate Studies, 612-625-3728, *Fax:* 612-625-5299, *E-mail:* fallo002@tc.umn.edu.

■ UNIVERSITY OF MISSOURI– COLUMBIA

Graduate School, College of Agriculture, Program in Entomology, Columbia, MO 65211

AWARDS MS, PhD.

Degree requirements: For master's, foreign language and thesis not required; for doctorate, dissertation required, foreign language not required.
Entrance requirements: For master's and doctorate, GRE General Test, minimum GPA of 3.0.
Expenses: Tuition, state resident: full-time $3,020; part-time $168 per hour. Tuition, nonresident: full-time $6,066; part-time $505 per hour. Required fees: $445; $18 per hour. Tuition and fees vary according to course load and program.

■ UNIVERSITY OF NEBRASKA– LINCOLN

Graduate College, College of Agricultural Sciences and Natural Resources, Department of Entomology, Lincoln, NE 68588

AWARDS MS, PhD.

Faculty: 14 full-time (0 women).
Students: 28 full-time (12 women), 15 part-time (4 women), 23 international. Average age 30. *18 applicants, 50% accepted.* In 1999, 1 master's, 5 doctorates awarded.
Degree requirements: For master's, thesis optional, foreign language not required; for doctorate, dissertation, comprehensive exams required.
Entrance requirements: For master's and doctorate, GRE General Test, TOEFL. *Application deadline:* For fall admission, 3/1 (priority date). Applications are processed on a rolling basis. *Application fee:* $35. Electronic applications accepted.
Expenses: Tuition, state resident: part-time $116 per credit hour. Tuition, nonresident: part-time $285 per credit hour. Required fees: $119 per semester. Tuition and fees vary according to course load and program.
Financial aid: In 1999–00, 5 fellowships, 27 research assistantships were awarded; teaching assistantships, Federal Work-Study also available. Aid available to part-time students. Financial aid application deadline: 2/15.
Faculty research: Ecology and behavior, insect-plant interactions, integrated pest management, genetics, urban entomology.
Dr. Z. B. Mayo, Head, 402-472-2123, *Fax:* 402-472-4687.

■ UNIVERSITY OF NORTH DAKOTA

Graduate School, College of Arts and Sciences, Department of Biology, Grand Forks, ND 58202

AWARDS Botany (MS, PhD); ecology (MS, PhD); entomology (MS, PhD); environmental biology (MS, PhD); fisheries/wildlife (MS, PhD); genetics (MS, PhD); zoology (MS, PhD).

Faculty: 18 full-time (3 women).
Students: 21 full-time (8 women). *13 applicants, 62% accepted.* In 1999, 3 master's awarded. Terminal master's awarded for partial completion of doctoral program.
Degree requirements: For master's, thesis, final exam required; for doctorate, dissertation, comprehensive exam, final exam required.
Entrance requirements: For master's, GRE General Test, GRE Subject Test, TOEFL, minimum GPA of 3.0; for doctorate, GRE General Test, GRE Subject Test, TOEFL, minimum GPA of 3.5. *Application deadline:* For fall admission, 3/1 (priority date). Applications are processed on a rolling basis. *Application fee:* $25.
Expenses: Tuition, state resident: full-time $3,166; part-time $158 per credit. Tuition, nonresident: full-time $7,658; part-time $345 per credit. International tuition: $7,658 full-time. Required fees: $46 per credit. Tuition and fees vary according to program and reciprocity agreements.
Financial aid: In 1999–00, 6 research assistantships with full tuition reimbursements (averaging $11,250 per year), 15 teaching assistantships with full tuition reimbursements (averaging $11,250 per year) were awarded; fellowships, Federal Work-Study, institutionally sponsored loans, scholarships, and tuition waivers (full and partial) also available. Aid available to part-time students. Financial aid application deadline: 3/15; financial aid applicants required to submit FAFSA.
Faculty research: Population biology, wildlife ecology, RNA processing, hormonal control of behavior.
Dr. Jeff Lang, Director, 701-777-2621, *Fax:* 701-777-2623, *E-mail:* jlang@badlands.nodak.edu.

■ UNIVERSITY OF RHODE ISLAND

Graduate School, College of Resource Development, Department of Plant Sciences, Program in Plant Pathology-Entomology, Kingston, RI 02881

AWARDS Entomology (MS, PhD); plant pathology (MS, PhD).

Degree requirements: For master's, thesis, comprehensive exam, professional seminar required; for doctorate, dissertation, professional seminar required, foreign language not required.
Entrance requirements: For master's and doctorate, GRE General Test. *Application deadline:* For fall admission, 4/15 (priority date). Applications are processed on a rolling basis. *Application fee:* $35.
Expenses: Tuition, state resident: full-time $3,540; part-time $197 per credit. Tuition, nonresident: full-time $10,116; part-time $197 per credit. Required fees: $1,352; $37 per credit. $65 per term.
Faculty research: Physiology and fine structure of host-parasite relations, etiology of plant disease, biological control of insects, integrated pest management.
Dr. Richard Hull, Chairman, Department of Plant Sciences, 401-874-5995.

■ THE UNIVERSITY OF TENNESSEE

Graduate School, College of Agricultural Sciences and Natural Resources, Department of Entomology and Plant Pathology, Knoxville, TN 37996

AWARDS Entomology (MS); plant pathology (MS). Part-time programs available.

Faculty: 13 full-time (2 women).
Students: 11 full-time (6 women), 8 part-time (4 women); includes 2 minority (1 African American, 1 Asian American or Pacific Islander), 1 international. *14 applicants, 79% accepted.* In 1999, 10 degrees awarded.
Degree requirements: For master's, thesis, seminar required, foreign language not required.
Entrance requirements: For master's, TOEFL, minimum GPA of 2.7. *Application deadline:* For fall admission, 3/15 (priority date). Applications are processed on a rolling basis. *Application fee:* $35. Electronic applications accepted.
Expenses: Tuition, state resident: full-time $3,806; part-time $184 per credit hour. Tuition, nonresident: full-time $9,874; part-time $522 per credit hour. Tuition and fees vary according to program.
Financial aid: In 1999–00, 8 research assistantships, 1 teaching assistantship were awarded; fellowships, career-related internships or fieldwork, Federal Work-Study, institutionally sponsored loans, and unspecified assistantships also available. Financial aid application deadline: 3/15; financial aid applicants required to submit FAFSA.
Dr. Gene Burgees, Interim Head, 865-974-7135, *Fax:* 865-974-4744, *E-mail:* gburgess1@utk.edu.

Application contact: Dr. Reid R. Gerhardt, Graduate Representative, 865-974-7135, *Fax:* 865-974-4744, *E-mail:* rgerhard@utk.edu.

■ UNIVERSITY OF WISCONSIN–MADISON

Graduate School, College of Agricultural and Life Sciences, Department of Entomology, Madison, WI 53706-1380

AWARDS MS, PhD.

Degree requirements: For doctorate, dissertation required.
Entrance requirements: For master's and doctorate, GRE General Test, minimum GPA of 3.0. Electronic applications accepted.
Expenses: Tuition, state resident: full-time $5,406; part-time $339 per credit. Tuition, nonresident: full-time $17,110; part-time $1,071 per credit. Full-time tuition and fees vary according to program and reciprocity agreements. Part-time tuition and fees vary according to course load and program.

■ UNIVERSITY OF WYOMING

Graduate School, College of Agriculture, Department of Renewable Resources, Program in Entomology, Laramie, WY 82071

AWARDS MS, PhD.

Faculty: 6 full-time (0 women).
Students: 8 full-time (4 women), 2 part-time (1 woman); includes 1 minority (Asian American or Pacific Islander), 2 international. *4 applicants, 25% accepted.* In 1999, 2 master's, 1 doctorate awarded.
Degree requirements: For master's, thesis required, foreign language not required; for doctorate, one foreign language (computer language can substitute), dissertation required.
Entrance requirements: For master's and doctorate, GRE General Test, minimum GPA of 3.0. *Application deadline:* For fall admission, 6/1 (priority date). Applications are processed on a rolling basis. *Application fee:* $40.
Expenses: Tuition, state resident: full-time $2,520; part-time $140 per credit hour. Tuition, nonresident: full-time $7,790; part-time $433 per credit hour. Required fees: $440; $7 per credit hour. Full-time tuition and fees vary according to course load and program.
Financial aid: In 1999–00, 10 research assistantships with full tuition reimbursements (averaging $8,667 per year) were awarded. Financial aid application deadline: 3/1.

Faculty research: Insect pest management, taxonomy, biocontrol of weeds, forest insects, insects affecting humans and animals.
Application contact: Kimm Mann-Malody, Office Assistant, Sr., 307-766-2263, *Fax:* 307-766-6403, *E-mail:* kimmmann@uwyo.edu.

■ VIRGINIA POLYTECHNIC INSTITUTE AND STATE UNIVERSITY

Graduate School, College of Agriculture and Life Sciences, Department of Entomology, Blacksburg, VA 24061

AWARDS MS, PhD. Part-time programs available.

Faculty: 22 full-time (1 woman).
Students: 21 full-time (12 women); includes 2 minority (1 Asian American or Pacific Islander, 1 Hispanic American), 5 international. Average age 29. *23 applicants, 70% accepted.* In 1999, 3 master's, 2 doctorates awarded.
Degree requirements: For master's and doctorate, thesis/dissertation required, foreign language not required.
Entrance requirements: For master's and doctorate, GRE, TOEFL. *Application deadline:* For fall admission, 12/1 (priority date). Applications are processed on a rolling basis. *Application fee:* $25.
Expenses: Tuition, state resident: full-time $4,122; part-time $229 per credit hour. Tuition, nonresident: full-time $6,930; part-time $385 per credit hour. Required fees: $828; $107 per semester. Part-time tuition and fees vary according to course load.
Financial aid: In 1999–00, 5 research assistantships with full tuition reimbursements (averaging $11,637 per year), 4 teaching assistantships with full tuition reimbursements (averaging $11,637 per year) were awarded; fellowships, career-related internships or fieldwork, Federal Work-Study, institutionally sponsored loans, and unspecified assistantships also available. Aid available to part-time

students. Financial aid application deadline: 4/1.
Faculty research: Physiology, ecology, biocontrol, genetics, taxonomy.
Dr. Tim P. Mack, Head, 540-231-6341, *E-mail:* tmack@vt.edu.

■ WASHINGTON STATE UNIVERSITY

Graduate School, College of Agriculture and Home Economics, Department of Entomology, Pullman, WA 99164

AWARDS MS, PhD.

Faculty: 18 full-time (2 women), 4 part-time/adjunct (1 woman).
Students: 16 full-time (6 women), 5 part-time (2 women); includes 1 minority (African American), 4 international. Average age 26. In 1999, 2 degrees awarded (100% found work related to degree).
Degree requirements: For master's, oral exam required, thesis optional, foreign language not required; for doctorate, dissertation, oral exam, written exam required, foreign language not required. *Average time to degree:* Doctorate–4 years full-time.
Entrance requirements: For master's and doctorate, GRE General Test, minimum GPA of 3.0. *Application deadline:* For fall admission, 3/1. Applications are processed on a rolling basis. *Application fee:* $35. Electronic applications accepted.
Expenses: Tuition, state resident: full-time $5,654. Tuition, nonresident: full-time $13,850. International tuition: $13,850 full-time. Tuition and fees vary according to program.
Financial aid: In 1999–00, 9 research assistantships with full and partial tuition reimbursements, 2 teaching assistantships with full and partial tuition reimbursements were awarded; career-related internships or fieldwork, Federal Work-Study, institutionally sponsored loans, tuition waivers (partial), and teaching associateships also available. Financial aid application deadline: 4/1; financial aid applicants required to submit FAFSA.

Faculty research: Apiculture, biological control of arthropods, integrated pest management, ecology, physiology and systematics of insects. *Total annual research expenditures:* $1.8 million.
Dr. John J. Brown, Chair, 509-335-5504, *Fax:* 509-335-1009.
Application contact: Doris Lohrey-Birch, Senior Secretary-Academic, 509-335-5422, *Fax:* 509-335-1009, *E-mail:* entom@wsu.edu.

■ WEST VIRGINIA UNIVERSITY

College of Agriculture, Forestry and Consumer Sciences, Division of Plant and Soil Sciences, Morgantown, WV 26506

AWARDS Agronomy (MS); entomology (MS); environmental microbiology (MS); horticulture (MS); plant and soil sciences (PhD); plant pathology (MS).

Faculty: 19 full-time (1 woman), 2 part-time/adjunct (1 woman).
Students: 18 full-time (10 women), 7 part-time (5 women); includes 1 minority (Native American), 2 international. Average age 26. In 1999, 1 master's awarded.
Degree requirements: For master's, thesis required, foreign language not required; for doctorate, dissertation, comprehensive exam required, foreign language not required.
Entrance requirements: For master's, GRE, TOEFL, minimum GPA of 2.5; for doctorate, GRE General Test, TOEFL, minimum GPA of 3.0. *Application fee:* $45.
Expenses: Tuition, state resident: full-time $2,910; part-time $154 per credit hour. Tuition, nonresident: full-time $8,368; part-time $457 per credit hour.
Financial aid: In 1999–00, 13 research assistantships, 3 teaching assistantships were awarded; Federal Work-Study, institutionally sponsored loans, and tuition waivers (full and partial) also available. Financial aid application deadline: 2/1; financial aid applicants required to submit FAFSA.
Faculty research: Water quality, reclamation of disturbed land, crop production, pest control, environmental protection.
Dr. Barton Baker, Chair, 304-293-4817, *Fax:* 304-293-3740, *E-mail:* bbaker2@wvu.edu.

Genetics, Developmental Biology, and Reproductive Biology

DEVELOPMENTAL BIOLOGY

■ ARIZONA STATE UNIVERSITY

Graduate College, College of Liberal Arts and Sciences, Department of Biology, Cell and Developmental Biology Group, Tempe, AZ 85287

AWARDS MS, PhD. Terminal master's awarded for partial completion of doctoral program.

Degree requirements: For master's, thesis required, foreign language not required; for doctorate, dissertation, oral exam required, foreign language not required.
Entrance requirements: For master's and doctorate, GRE General Test, GRE Subject Test. *Application deadline:* For fall admission, 12/15. *Application fee:* $45.
Expenses: Tuition, state resident: part-time $115 per credit hour. Tuition, nonresident: part-time $389 per credit hour. Required fees: $18 per semester. Tuition and fees vary according to program.
Financial aid: Application deadline: 12/15.
Faculty research: Cytoskeleton assembly, exocytosis, cyclic nucleotides, membrane fusion, chromosome distribution.

■ BAYLOR COLLEGE OF MEDICINE

Graduate School of Biomedical Sciences, Program in Developmental Biology, Houston, TX 77030-3498

AWARDS PhD, MD/PhD.

Faculty: 32 full-time (6 women).
Students: 25 full-time (10 women); includes 3 minority (all Asian Americans or Pacific Islanders, 18 international. Average age 27. *53 applicants, 21% accepted.* In 1999, 3 degrees awarded (100% entered university research/teaching).
Degree requirements: For doctorate, dissertation required, foreign language not required. *Average time to degree:* Doctorate–5.23 years full-time.
Entrance requirements: For doctorate, GRE General Test; GRE Subject Test (strongly recommended), TOEFL, minimum GPA of 3.0. *Application deadline:* For fall admission, 2/1 (priority date). *Application fee:* $30. Electronic applications accepted.

Expenses: Tuition: Full-time $8,200. Required fees: $175. Full-time tuition and fees vary according to student level.
Financial aid: In 1999–00, 25 students received aid, including 12 fellowships (averaging $16,000 per year), 13 research assistantships (averaging $16,000 per year); tuition waivers (full) and stipends also available.
Faculty research: Molecular and genetic approaches to study pattern formation in *Dictyostelium, Drosophila, C.elegans,* mouse, *Xenopus,* and zebrafish; cross-species approach.
Dr. Hugo Bellen, Director, 713-798-6410.
Application contact: Catherine Tasnier, Graduate Program Administrator, 713-798-6410, *Fax:* 713-798-5386, *E-mail:* cat@bcm.tmc.edu.

Find an in-depth description at www.petersons.com/graduate.

■ BRANDEIS UNIVERSITY

Graduate School of Arts and Sciences, Programs in Biological Sciences, Program in Molecular and Cell Biology, Waltham, MA 02454-9110

AWARDS Cell biology (PhD); developmental biology (PhD); genetics (PhD); microbiology (PhD); molecular and cell biology (MS); molecular biology (PhD); neurobiology (PhD).

Faculty: 19 full-time (8 women).
Students: 58 full-time (31 women), 1 (woman) part-time; includes 13 minority (1 African American, 8 Asian Americans or Pacific Islanders, 4 Hispanic Americans), 8 international. Average age 27. *83 applicants, 11% accepted.* In 1999, 1 master's, 8 doctorates awarded (100% entered university research/teaching). Terminal master's awarded for partial completion of doctoral program.
Degree requirements: For master's, thesis not required; for doctorate, dissertation required. *Average time to degree:* Master's–1 year full-time; doctorate–6 years full-time.
Entrance requirements: For doctorate, GRE General Test. *Application deadline:* For fall admission, 1/15 (priority date). Applications are processed on a rolling basis. *Application fee:* $60. Electronic applications accepted.
Expenses: Tuition: Full-time $25,392; part-time $3,174 per course. Required fees: $509. Tuition and fees vary according to class time, degree level, program and student level.

Financial aid: In 1999–00, 10 fellowships with tuition reimbursements (averaging $17,000 per year), 49 research assistantships with tuition reimbursements (averaging $17,000 per year), 9 teaching assistantships (averaging $2,312 per year) were awarded; scholarships and tuition waivers (full and partial) also available. Financial aid application deadline: 4/15; financial aid applicants required to submit CSS PROFILE or FAFSA.
Faculty research: Regulation of gene expression by transcription factors, molecular neurobiology, immunology, molecular mechanisms of genetic recombination, and cell differentiation.
Dr. Ranjan Sen, Chair, 781-736-2455, *Fax:* 781-736-3107.
Application contact: Marcia Cabral, Information Officer, 781-736-3100, *Fax:* 781-736-3107, *E-mail:* cabral@brandeis.edu.

■ BROWN UNIVERSITY

Graduate School, Division of Biology and Medicine, Program in Molecular Biology, Cell Biology, and Biochemistry, Providence, RI 02912

AWARDS Biochemistry (M Med Sc, Sc M, PhD), including biochemistry (Sc M, PhD), biology (Sc M, PhD), medical science (M Med Sc, PhD); biology (MA); cell biology (M Med Sc, Sc M, PhD), including biochemistry (Sc M, PhD), biology (Sc M, PhD), medical science (M Med Sc, PhD); developmental biology (M Med Sc, Sc M, PhD), including biochemistry (Sc M, PhD), biology (Sc M, PhD), medical science (M Med Sc, PhD); immunology (M Med Sc, Sc M, PhD), including biochemistry (Sc M, PhD), biology (Sc M, PhD), medical science (M Med Sc, PhD); molecular microbiology (M Med Sc, Sc M, PhD), including biochemistry (Sc M, PhD), biology (Sc M, PhD), medical science (M Med Sc, PhD). Part-time programs available.

Faculty: 50 full-time (14 women).
Students: 61 full-time (34 women); includes 4 minority (1 African American, 3 Asian Americans or Pacific Islanders), 21 international. Average age 25. *106 applicants, 28% accepted.* In 1999, 1 master's, 3 doctorates awarded. Terminal master's awarded for partial completion of doctoral program.
Degree requirements: For master's, thesis required (for some programs),

foreign language not required; for doctorate, one foreign language, dissertation, preliminary exam required. *Average time to degree:* Doctorate–5 years full-time. **Entrance requirements:** For master's and doctorate, GRE General Test, GRE Subject Test. *Application deadline:* For fall admission, 1/2 (priority date). Applications are processed on a rolling basis. *Application fee:* $60. Electronic applications accepted. **Financial aid:** In 1999–00, 58 students received aid, including 11 fellowships (averaging $18,916 per year), 9 research assistantships (averaging $18,916 per year), 13 teaching assistantships (averaging $12,690 per year); institutionally sponsored loans and traineeships also available. Financial aid application deadline: 1/2. **Faculty research:** Molecular genetics, gene regulation. Dr. Gary Wessel, Director, 401-863-1051, *E-mail:* chet@brown.edu. **Application contact:** Mary C. Esser, Graduate Program Coordinator, 401-863-1661, *Fax:* 401-863-1348, *E-mail:* mary_esser@brown.edu.

Find an in-depth description at www.petersons.com/graduate.

■ **CALIFORNIA INSTITUTE OF TECHNOLOGY**

Division of Biology, Program in Developmental Biology, Pasadena, CA 91125-0001
AWARDS PhD.

Degree requirements: For doctorate, dissertation, qualifying exam required, foreign language not required. **Entrance requirements:** For doctorate, GRE General Test. *Application deadline:* For fall admission, 1/1. *Application fee:* $0. **Expenses:** Tuition: Full-time $19,260. Required fees: $24. One-time fee: $100 full-time. **Financial aid:** Application deadline: 1/1. **Application contact:** Elizabeth Ayala, Graduate Option Coordinator, 626-395-4497, *Fax:* 626-449-0756, *E-mail:* biograd@cco.caltech.edu.

■ **CARNEGIE MELLON UNIVERSITY**

Mellon College of Science, Department of Biological Sciences, Pittsburgh, PA 15213-3891
AWARDS Biochemistry (PhD); biophysics (PhD); cell biology (PhD); developmental biology (PhD); genetics (PhD); molecular biology (PhD).

Faculty: 38 full-time (17 women), 1 (woman) part-time/adjunct. **Students:** 35 full-time (24 women); includes 2 minority (1 African American, 1 Asian American or Pacific Islander), 18

international. Average age 27. In 1999, 4 degrees awarded.
Degree requirements: For doctorate, dissertation required, foreign language not required. **Entrance requirements:** For doctorate, GRE General Test, GRE Subject Test, interview. *Application deadline:* For fall admission, 2/1 (priority date). Applications are processed on a rolling basis. *Application fee:* $0. **Expenses:** Tuition: Full-time $22,100; part-time $307 per unit. Required fees: $200. Tuition and fees vary according to program. **Financial aid:** Fellowships, research assistantships, teaching assistantships, traineeships available. **Faculty research:** Genetic structure, function, and regulation; protein structure and function; biological membranes; biological spectroscopy. *Total annual research expenditures:* $4.4 million. Dr. William E. Brown, Head, 412-268-3416, *Fax:* 412-268-7129, *E-mail:* wb02@andrew.cmu.edu. **Application contact:** Stacey L. Young, Assistant Head, 412-268-7372, *Fax:* 412-268-7129, *E-mail:* sf38+@andrew.cmu.edu.

Find an in-depth description at www.petersons.com/graduate.

■ **CASE WESTERN RESERVE UNIVERSITY**

School of Medicine and School of Graduate Studies, Graduate Programs in Medicine, Department of Anatomy, Cleveland, OH 44106
AWARDS Applied anatomy (MS); biological anthropology (MS, PhD); cellular biology (MS, PhD); developmental biology (PhD); molecular biology (PhD). Part-time programs available.

Faculty: 14 full-time (5 women), 14 part-time/adjunct (4 women). **Students:** 17 full-time (9 women), 12 part-time (1 woman); includes 11 minority (1 African American, 9 Asian Americans or Pacific Islanders, 1 Hispanic American), 2 international. Average age 26. *16 applicants, 50% accepted.* In 1999, 5 master's awarded (50% found work related to degree, 50% continued full-time study); 1 doctorate awarded (100% entered university research/teaching). **Degree requirements:** For master's, thesis required (for some programs); for doctorate, dissertation required. *Average time to degree:* Master's–5 years full-time; doctorate–1 year full-time. **Entrance requirements:** For master's, GRE General Test, TOEFL; for doctorate, GRE General Test, GRE Subject Test, TOEFL. *Application deadline:* For fall admission, 5/1 (priority date); for spring

admission, 8/1 (priority date). Applications are processed on a rolling basis. *Application fee:* $25. **Financial aid:** In 1999–00, 7 research assistantships with full tuition reimbursements (averaging $16,000 per year) were awarded; fellowships, grants also available. **Faculty research:** Hypoxia, cell injury, biochemical aberration occurrences in ischemic tissue, human functional morphology, evolutionary morphology. *Total annual research expenditures:* $562,716. Joseph C. LaManna, Acting Chairman, 216-368-1100, *Fax:* 216-368-8669, *E-mail:* jcl4@po.cwru.edu. **Application contact:** Laila Boesinger, Administrator, 216-368-3430, *Fax:* 216-368-8669, *E-mail:* lvb2@po.cwru.edu.

■ **COLUMBIA UNIVERSITY**

College of Physicians and Surgeons and Graduate School of Arts and Sciences, Graduate School of Arts and Sciences at the College of Physicians and Surgeons, Department of Genetics and Development, New York, NY 10032
AWARDS Genetics (M Phil, MA, PhD). Only candidates for the PhD are admitted. Terminal master's awarded for partial completion of doctoral program.

Degree requirements: For master's, foreign language and thesis not required; for doctorate, dissertation required, foreign language not required. **Entrance requirements:** For master's and doctorate, GRE General Test, TOEFL. **Expenses:** Tuition: Full-time $25,072. **Faculty research:** Mammalian cell differentiation and meiosis, developmental genetics, yeast and human genetics, chromosome structure, molecular and cellular biology.

■ **CORNELL UNIVERSITY**

Graduate School, Graduate Fields of Agriculture and Life Sciences, Field of Genetics and Development, Ithaca, NY 14853-0001
AWARDS Developmental biology (PhD); genetics (PhD).

Faculty: 31 full-time. **Students:** 34 full-time (17 women); includes 4 minority (1 African American, 1 Asian American or Pacific Islander, 1 Hispanic American, 1 Native American), 8 international. *95 applicants, 25% accepted.* In 1999, 9 doctorates awarded. **Degree requirements:** For doctorate, dissertation, 2 semesters of teaching experience required, foreign language not required. **Entrance requirements:** For doctorate, GRE General Test, GRE Subject Test (biology or biochemistry) (recommended).

Cornell University (continued)
Application deadline: For fall admission, 1/5.
Application fee: $65. Electronic applications accepted.
Expenses: Tuition: Full-time $12,100.
Financial aid: In 1999–00, 33 students received aid, including 17 fellowships with full tuition reimbursements available, 11 research assistantships with full tuition reimbursements available, 5 teaching assistantships with full tuition reimbursements available; institutionally sponsored loans, scholarships, tuition waivers (full and partial), and unspecified assistantships also available. Financial aid applicants required to submit FAFSA.
Faculty research: Molecular and general genetics, developmental genetics, evolution and population genetics, plant genetics, microbial genetics.
Application contact: Graduate Field Assistant, 607-254-4840, *E-mail:* gendev@cornell.edu.

Find an in-depth description at www.petersons.com/graduate.

■ EMORY UNIVERSITY

Graduate School of Arts and Sciences, Division of Biological and Biomedical Sciences, Program in Biochemistry, Cell and Developmental Biology, Atlanta, GA 30322-1100

AWARDS PhD.

Faculty: 66 full-time (12 women).
Students: 52 full-time (33 women); includes 4 African Americans, 1 Hispanic American, 13 international. In 1999, 12 degrees awarded.
Degree requirements: For doctorate, dissertation required, foreign language not required.
Entrance requirements: For doctorate, GRE General Test, TOEFL, minimum GPA of 3.0 in science course work. *Application deadline:* For fall admission, 1/20 (priority date). *Application fee:* $45.
Expenses: Tuition: Full-time $22,770. Tuition and fees vary according to program.
Financial aid: In 1999–00, fellowships with full tuition reimbursements (averaging $18,000 per year).
Faculty research: Signal transduction, molecular biology, enzymes and cofactors, receptor and ion channel function, membrane biology.
Steven L'Hernault, Director, 404-727-3924, *Fax:* 404-727-3949, *E-mail:* jfridov@emory.edu.
Application contact: 404-727-2547, *Fax:* 404-727-3322, *E-mail:* gdbbs@gsas.emory.edu.

Find an in-depth description at www.petersons.com/graduate.

■ FLORIDA STATE UNIVERSITY

Graduate Studies, College of Arts and Sciences, Department of Biological Science, Program in Developmental Biology, Tallahassee, FL 32306

AWARDS MS, PhD.

Faculty: 9 full-time (2 women).
Students: 22 full-time (11 women); includes 1 minority (Hispanic American), 11 international.
Degree requirements: For master's and doctorate, thesis/dissertation, teaching experience required.
Entrance requirements: For master's, GRE General Test, TOEFL; for doctorate, GRE General Test, GRE Subject Test, TOEFL. *Application deadline:* For fall admission, 1/15; for spring admission, 10/15. *Application fee:* $20.
Expenses: Tuition, state resident: full-time $3,504; part-time $146 per credit hour. Tuition, nonresident: full-time $12,162; part-time $507 per credit hour. Tuition and fees vary according to program.
Financial aid: In 1999–00, fellowships with full tuition reimbursements (averaging $13,740 per year), research assistantships with full tuition reimbursements (averaging $13,740 per year), teaching assistantships with full tuition reimbursements (averaging $13,740 per year) were awarded. Financial aid application deadline: 1/15; financial aid applicants required to submit FAFSA.
Faculty research: Cell fusion, fertilization of nucleic acids, meiotic mechanisms, mitosis, DNA repair.
Dr. Thomas C. S. Keller, Associate Professor and Associate Chairman, 850-644-3023, *Fax:* 850-644-9829.
Application contact: Judy Bowers, Coordinator, Graduate Affairs, 850-644-3023, *Fax:* 850-644-9829, *E-mail:* bowers@bio.fsu.edu.

■ INDIANA UNIVERSITY BLOOMINGTON

Graduate School, College of Arts and Sciences, Department of Biology, Program in Molecular, Cellular, and Developmental Biology, Bloomington, IN 47405

AWARDS PhD. Offered through the University Graduate School.

Students: 54 full-time (21 women); includes 4 minority (2 Asian Americans or Pacific Islanders, 2 Hispanic Americans), 9 international. In 1999, 9 degrees awarded.
Degree requirements: For doctorate, dissertation required.
Entrance requirements: For doctorate, GRE General Test, TOEFL. *Application deadline:* For fall admission, 1/15 (priority date); for spring admission, 9/1 (priority

date). Applications are processed on a rolling basis. *Application fee:* $45. Electronic applications accepted.
Expenses: Tuition, state resident: full-time $3,853; part-time $161 per credit hour. Tuition, nonresident: full-time $11,226; part-time $468 per credit hour. Required fees: $360 per year. Tuition and fees vary according to course load and program.
Financial aid: Fellowships with tuition reimbursements, research assistantships with tuition reimbursements, teaching assistantships with tuition reimbursements available. Financial aid application deadline: 2/15.
Faculty research: Developmental genetics, molecular evolution, macromolecular structure and function, cell biology and microbial molecular genetics.
Dr. Susan Strome, Head, 812-855-5450, *Fax:* 812-855-6705, *E-mail:* sstrome@bio.indiana.edu.
Application contact: Gretchen Clearwater, Advisor for Graduate Affairs, 812-855-1861, *Fax:* 812-855-6705, *E-mail:* biograd@bio.indiana.edu.

Find an in-depth description at www.petersons.com/graduate.

■ IOWA STATE UNIVERSITY OF SCIENCE AND TECHNOLOGY

Graduate College, Interdisciplinary Programs, Program in Molecular, Cellular, and Developmental Biology, Ames, IA 50011

AWARDS MS, PhD.

Students: 10 full-time (6 women), 9 international. *89 applicants, 20% accepted.* In 1999, 1 master's, 1 doctorate awarded.
Degree requirements: For master's, thesis or alternative required; for doctorate, dissertation required.
Entrance requirements: For master's and doctorate, GRE General Test, TOEFL. *Application deadline:* For fall admission, 2/1 (priority date). *Application fee:* $20 ($50 for international students). Electronic applications accepted.
Expenses: Tuition, state resident: full-time $3,308. Tuition, nonresident: full-time $9,744. Part-time tuition and fees vary according to course load, campus/location and program.
Financial aid: In 1999–00, 8 research assistantships with partial tuition reimbursements (averaging $11,748 per year) were awarded; teaching assistantships, scholarships also available.
Dr. Jorgen Johansen, Supervisory Committee Chair, 515-294-7252, *E-mail:* idgp@iastate.edu.
Application contact: 515-294-7252, *Fax:* 515-924-6790, *E-mail:* idgp@iastate.edu.

Find an in-depth description at www.petersons.com/graduate.

■ JOHNS HOPKINS UNIVERSITY

Zanvyl Krieger School of Arts and Sciences, Department of Biology, Baltimore, MD 21218-2699

AWARDS Biochemistry (PhD); cell biology (PhD); developmental biology (PhD); genetic biology (PhD); molecular biology (PhD).

Faculty: 23 full-time (4 women).
Students: 85 full-time (50 women); includes 23 minority (4 African Americans, 14 Asian Americans or Pacific Islanders, 5 Hispanic Americans), 18 international. Average age 24. *202 applicants, 27% accepted.* In 1999, 17 doctorates awarded.
Degree requirements: For doctorate, dissertation required. *Average time to degree:* Doctorate–6 years full-time.
Entrance requirements: For doctorate, GRE General Test, GRE Subject Test. *Application deadline:* For fall admission, 12/15 (priority date). Applications are processed on a rolling basis. *Application fee:* $55.
Expenses: Tuition: Full-time $24,930. Tuition and fees vary according to program.
Financial aid: In 1999–00, 73 students received aid, including 12 fellowships, 61 research assistantships, 12 teaching assistantships; Federal Work-Study and institutionally sponsored loans also available. Financial aid application deadline: 12/15; financial aid applicants required to submit FAFSA.
Faculty research: Protein and nucleic acid biochemistry and biophysical chemistry, molecular biology and development. *Total annual research expenditures:* $9.3 million.
Dr. Victor G. Corces, Chair, 410-516-4693, *Fax:* 410-516-5213, *E-mail:* corces_v@jhuvms.hct.jhu.edu.
Application contact: Joan Miller, Graduate Admissions Coordinator, 410-516-5502, *Fax:* 410-516-5213, *E-mail:* joan@jhu.edu.

Find an in-depth description at www.petersons.com/graduate.

■ KANSAS STATE UNIVERSITY

Graduate School, College of Arts and Sciences, Division of Biology, Manhattan, KS 66506

AWARDS Cell biology (MS, PhD); developmental biology and physiology (MS, PhD); microbiology and immunology (MS, PhD); molecular biology and genetics (MS, PhD); systematics and ecology (MS, PhD); virology and oncology (MS, PhD). Terminal master's awarded for partial completion of doctoral program.

Degree requirements: For master's and doctorate, thesis/dissertation required, foreign language not required.

Entrance requirements: For master's and doctorate, GRE General Test. Electronic applications accepted.
Expenses: Tuition, state resident: part-time $103 per credit hour. Tuition, nonresident: part-time $338 per credit hour. Required fees: $17 per credit hour. One-time fee: $64 part-time.
Faculty research: Immune cell function, prairie ecology.

■ LOUISIANA STATE UNIVERSITY HEALTH SCIENCES CENTER

School of Graduate Studies in New Orleans, Department of Cell Biology and Anatomy, New Orleans, LA 70112-2223

AWARDS Cell biology and anatomy (MS, PhD), including cell biology, developmental biology, neurobiology and anatomy.

Faculty: 18 full-time (3 women), 1 part-time/adjunct (0 women).
Students: 9 full-time (2 women); includes 1 minority (African American), 4 international. Average age 26. *6 applicants, 50% accepted.* In 1999, 3 doctorates awarded (34% entered university research/teaching, 33% found other work related to degree, 33% continued full-time study).
Degree requirements: For master's and doctorate, thesis/dissertation required, foreign language not required. *Average time to degree:* Doctorate–4.5 years full-time.
Entrance requirements: For master's and doctorate, GRE General Test, GRE Subject Test, TOEFL, minimum undergraduate GPA of 3.0. *Application deadline:* For fall admission, 3/1 (priority date); for spring admission, 10/15. Applications are processed on a rolling basis. *Application fee:* $30.
Expenses: Tuition, state resident: full-time $2,878; part-time $126 per hour. Tuition, nonresident: full-time $6,003; part-time $265 per hour. Required fees: $2,272. Tuition and fees vary according to course load, degree level and program.
Financial aid: In 1999–00, 1 fellowship with full tuition reimbursement (averaging $15,000 per year), teaching assistantships with full tuition reimbursements (averaging $14,000 per year) were awarded; research assistantships, career-related internships or fieldwork, Federal Work-Study, grants, institutionally sponsored loans, tuition waivers (full), and unspecified assistantships also available. Aid available to part-time students. Financial aid application deadline: 4/1.
Faculty research: Visual system organization, neural development, plasticity of sensory systems, information processing through the nervous system, visuomotor integration. *Total annual research expenditures:* $772,550.
Dr. R. Ranney Mize, Head, 504-599-1458, *Fax:* 504-568-4392, *E-mail:* rmize@lsumc.edu.
Application contact: Dr. Mark C. Alliegro, Director of Graduate Studies, 504-568-7618, *Fax:* 504-568-4392, *E-mail:* mallie@lsumc.edu.

Find an in-depth description at www.petersons.com/graduate.

■ MARQUETTE UNIVERSITY

Graduate School, College of Arts and Sciences, Department of Biology, Milwaukee, WI 53201-1881

AWARDS Cell biology (MS, PhD); developmental biology (MS, PhD); ecology (MS, PhD); endocrinology (MS, PhD); evolutionary biology (MS, PhD); genetics (MS, PhD); microbiology (MS, PhD); molecular biology (MS, PhD); muscle and exercise physiology (MS, PhD); neurobiology (MS, PhD); reproductive physiology (MS, PhD).

Faculty: 16 full-time (4 women), 2 part-time/adjunct (0 women).
Students: 34 full-time (20 women), 3 part-time; includes 3 minority (all Asian Americans or Pacific Islanders), 2 international. Average age 31. *42 applicants, 29% accepted.* In 1999, 1 master's, 4 doctorates awarded. Terminal master's awarded for partial completion of doctoral program.
Degree requirements: For master's, thesis, 1 year of teaching experience or equivalent, comprehensive exam required, foreign language not required; for doctorate, dissertation, 1 year of teaching experience or equivalent, qualifying exam required, foreign language not required.
Entrance requirements: For master's and doctorate, GRE General Test, GRE Subject Test, TOEFL. *Application fee:* $40.
Expenses: Tuition: Part-time $510 per credit hour. Tuition and fees vary according to program.
Financial aid: In 1999–00, 4 fellowships, 22 teaching assistantships were awarded; research assistantships, Federal Work-Study, institutionally sponsored loans, scholarships, and tuition waivers (full and partial) also available. Aid available to part-time students. Financial aid application deadline: 2/15.
Faculty research: Microbial and invertebrate ecology, evolution of gene function, DNA methylation, DNA arrangement. *Total annual research expenditures:* $1.5 million.
Dr. Brian Unsworth, Chairman, 414-288-7355, *Fax:* 414-288-7357.

Marquette University (continued)
Application contact: Barbara DeNoyer, Graduate Studies Coordinator, 414-288-7355, *Fax:* 414-288-7357.

Find an in-depth description at www.petersons.com/graduate.

■ MASSACHUSETTS INSTITUTE OF TECHNOLOGY

School of Science, Department of Biology, Program in Cellular and Developmental Biology, Cambridge, MA 02139-4307

AWARDS PhD.

Degree requirements: For doctorate, dissertation, general exam required, foreign language not required.
Entrance requirements: For doctorate, GRE General Test. *Application deadline:* For fall admission, 1/1. *Application fee:* $55.
Expenses: Tuition: Full-time $25,000. Full-time tuition and fees vary according to degree level, program and student level.
Financial aid: Application deadline: 1/1.
Application contact: Dr. Janice D. Chang, Educational Administrator, 617-253-3717, *Fax:* 617-258-9329, *E-mail:* gradbio@mit.edu.

■ MEDICAL COLLEGE OF WISCONSIN

Graduate School of Biomedical Sciences, Program in Cell and Developmental Biology, Milwaukee, WI 53226-0509

AWARDS MS, PhD. Terminal master's awarded for partial completion of doctoral program.

Degree requirements: For master's and doctorate, thesis/dissertation required, foreign language not required.
Entrance requirements: For master's and doctorate, GRE General Test, TOEFL.
Expenses: Tuition, state resident: full-time $9,318. Tuition, nonresident: full-time $9,318. Required fees: $115.
Faculty research: Neurobiology, development, neuroscience, teratology.

■ MOUNT SINAI SCHOOL OF MEDICINE OF NEW YORK UNIVERSITY

Graduate School of Biological Sciences, Molecular, Cellular, Biochemical and Developmental Sciences (MCBDS) Training Area, New York, NY 10029-6504

AWARDS PhD, MD/PhD.

Students: 48 full-time (18 women).
Degree requirements: For doctorate, dissertation required, foreign language not required.

Entrance requirements: For doctorate, GRE General Test, GRE Subject Test, MCAT, TOEFL. *Application deadline:* For fall admission, 4/15. *Application fee:* $60.
Expenses: Tuition: Full-time $21,750; part-time $725 per credit. Required fees: $750; $25 per credit. Full-time tuition and fees vary according to student level.
Financial aid: Fellowships with full tuition reimbursements, grants available.
Dr. Gillian Small, Program Director, *E-mail:* small@msvax.mssm.edu.
Application contact: C. Gita Bosch, Administrative Manager and Assistant Dean, 212-241-6546, *Fax:* 212-241-0651, *E-mail:* grads@mssm.edu.

■ NORTHWESTERN UNIVERSITY

The Graduate School, Division of Interdepartmental Programs and Medical School, Integrated Graduate Programs in the Life Sciences, Chicago, IL 60611

AWARDS Cancer biology (PhD); cell biology (PhD); developmental biology (PhD); evolutionary biology (PhD); immunology and microbial pathogenesis (PhD); molecular biology and genetics (PhD); neurobiology (PhD); pharmacology and toxicology (PhD); structural biology and biochemistry (PhD).

Degree requirements: For doctorate, dissertation, written and oral qualifying exams required, foreign language not required.
Entrance requirements: For doctorate, GRE General Test, TOEFL.
Expenses: Tuition: Full-time $23,301. Full-time tuition and fees vary according to program.

Find an in-depth description at www.petersons.com/graduate.

■ NORTHWESTERN UNIVERSITY

The Graduate School, Judd A. and Marjorie Weinberg College of Arts and Sciences, Interdepartmental Biological Sciences Program (IBiS), Concentration in Genetics and Developmental Biology, Evanston, IL 60208

AWARDS PhD.

Faculty: 59 full-time (11 women).
Students: 79 full-time (46 women); includes 5 minority (2 African Americans, 3 Hispanic Americans). *236 applicants, 19% accepted.*
Degree requirements: For doctorate, dissertation, 2 quarters of teaching experience required, foreign language not required.
Entrance requirements: For doctorate, GRE General Test, TOEFL. *Application deadline:* For fall admission, 1/15. Applications are processed on a rolling basis. *Application fee:* $50 ($55 for international students).

Expenses: Tuition: Full-time $23,301. Full-time tuition and fees vary according to program.
Financial aid: In 1999–00, 15 fellowships (averaging $12,078 per year), 64 research assistantships (averaging $17,000 per year), 15 teaching assistantships (averaging $16,620 per year) were awarded; Federal Work-Study, institutionally sponsored loans, traineeships, and Howard Hughes Fellowship also available. Financial aid applicants required to submit FAFSA.
Application contact: Latonia Trimuel, Program Assistant, 800-546-1761, *E-mail:* ibis@northwestern.edu.

■ THE OHIO STATE UNIVERSITY

Graduate School, College of Biological Sciences, Department of Molecular Genetics, Columbus, OH 43210

AWARDS Cell and developmental biology (MS, PhD); genetics (MS, PhD); molecular biology (MS, PhD).

Faculty: 18 full-time, 6 part-time/adjunct.
Students: 46 full-time (17 women), 3 part-time (1 woman); includes 3 minority (2 African Americans, 1 Asian American or Pacific Islander), 21 international. *68 applicants, 28% accepted.* In 1999, 6 master's, 8 doctorates awarded.
Degree requirements: For master's and doctorate, thesis/dissertation required, foreign language not required.
Entrance requirements: For master's and doctorate, GRE General Test, GRE Subject Test (biology). *Application deadline:* For fall admission, 8/15. Applications are processed on a rolling basis. *Application fee:* $30 ($40 for international students).
Expenses: Tuition, state resident: full-time $5,400. Tuition, nonresident: full-time $14,535. Part-time tuition and fees vary according to course load and program.
Financial aid: Fellowships, research assistantships, teaching assistantships, Federal Work-Study and institutionally sponsored loans available. Aid available to part-time students.
Dr. Lee F. Johnson, Chairman, 614-292-8084, *Fax:* 614-292-4466, *E-mail:* johnson.6@osu.edu.

Find an in-depth description at www.petersons.com/graduate.

■ THE OHIO STATE UNIVERSITY

Graduate School, College of Biological Sciences, Program in Molecular, Cellular and Developmental Biology, Columbus, OH 43210

AWARDS MS, PhD.

Faculty: 14 full-time, 73 part-time/adjunct.
Students: 65 full-time (38 women); includes 7 minority (2 African Americans,

4 Asian Americans or Pacific Islanders, 1 Hispanic American), 33 international. *114 applicants, 27% accepted.* In 1999, 4 master's, 7 doctorates awarded.
Degree requirements: For master's and doctorate, thesis/dissertation required, foreign language not required.
Entrance requirements: For master's and doctorate, GRE General Test, GRE Subject Test. *Application deadline:* For fall admission, 8/15. Applications are processed on a rolling basis. *Application fee:* $30 ($40 for international students).
Expenses: Tuition, state resident: full-time $5,400. Tuition, nonresident: full-time $14,535. Part-time tuition and fees vary according to course load and program.
Financial aid: Fellowships, research assistantships, teaching assistantships, Federal Work-Study and institutionally sponsored loans available. Aid available to part-time students.
Dr. David Bisaro, Director, 614-292-2804, *Fax:* 614-292-1538, *E-mail:* bisaro.1@osu.edu.

Find an in-depth description at www.petersons.com/graduate.

■ OREGON HEALTH SCIENCES UNIVERSITY

School of Medicine, Graduate Programs in Medicine, Department of Cell and Developmental Biology, Portland, OR 97201-3098
AWARDS PhD, MD/PhD.

Faculty: 11 full-time (5 women).
Students: 29 full-time (12 women); includes 1 minority (Asian American or Pacific Islander), 8 international. Average age 30. In 1999, 5 doctorates awarded (100% entered university research/teaching).
Degree requirements: For doctorate, dissertation required, foreign language not required. *Average time to degree:* Doctorate–7 years full-time.
Entrance requirements: For doctorate, GRE General Test, GRE Subject Test, MCAT. *Application deadline:* For fall admission, 1/15. Applications are processed on a rolling basis. *Application fee:* $60.
Expenses: Tuition, state resident: full-time $3,132; part-time $174 per credit hour. Tuition, nonresident: full-time $5,256; part-time $292 per credit hour. Required fees: $8.5 per credit hour. $146 per term. Part-time tuition and fees vary according to course load.
Financial aid: In 1999–00, research assistantships (averaging $16,500 per year); Federal Work-Study, institutionally sponsored loans, scholarships, and tuition waivers (full) also available. Financial aid application deadline: 3/1; financial aid applicants required to submit FAFSA.

Faculty research: Developmental mechanisms, molecular biology of cancer, molecular neurobiology, intracellular signalling, growth factors and development. *Total annual research expenditures:* $3.4 million.
Dr. Bruce Magun, Chair, 503-494-7811, *Fax:* 503-494-4253, *E-mail:* magunb@ohsu.edu.
Application contact: Dr. Philip Copenhaver, Associate Professor, 503-494-4646, *Fax:* 503-494-4253, *E-mail:* copenhav@ohsu.edu.

■ THE PENNSYLVANIA STATE UNIVERSITY UNIVERSITY PARK CAMPUS

Graduate School, Eberly College of Science, Department of Biochemistry and Molecular Biology, Program in Cell and Developmental Biology, State College, University Park, PA 16802-1503
AWARDS PhD.

Degree requirements: For doctorate, dissertation, comprehensive exam required.
Entrance requirements: For doctorate, GRE General Test, GRE Subject Test. *Application fee:* $50.
Expenses: Tuition, state resident: full-time $6,886; part-time $291 per credit. Tuition, nonresident: full-time $14,118; part-time $588 per credit. Required fees: $46 per semester. Part-time tuition and fees vary according to course load and program.
Application contact: Dr. Ronald D. Porter, Director, 814-865-2538.

Find an in-depth description at www.petersons.com/graduate.

■ THE PENNSYLVANIA STATE UNIVERSITY UNIVERSITY PARK CAMPUS

Graduate School, Intercollege Graduate Programs, Intercollege Graduate Program in Integrative Biosciences, State College, University Park, PA 16802-1503
AWARDS Integrative biosciences (MS, PhD), including biomolecular transport dynamics, cell and developmental biology, cellular and molecular mechanisms of toxicity, chemical biology, ecological and molecular plant physiology, immunobiology, molecular medicine, neuroscience, nutrition science.

Students: 39 full-time (24 women), 1 part-time.
Entrance requirements: For master's and doctorate, GRE General Test. *Application fee:* $50.
Expenses: Tuition, state resident: full-time $6,886; part-time $291 per credit. Tuition, nonresident: full-time $14,118; part-time

$588 per credit. Required fees: $46 per semester. Part-time tuition and fees vary according to course load and program.
Financial aid: Fellowships available.
Dr. C. R. Matthews, Co-Director, 814-863-3650.
Application contact: Admissions Committee, 814-865-3155, *Fax:* 814-863-1357, *E-mail:* lscgradadm@mail.biotec.psu.edu.

Find an in-depth description at www.petersons.com/graduate.

■ PRINCETON UNIVERSITY

Graduate School, Department of Molecular Biology, Princeton, NJ 08544-1019
AWARDS Cell biology (PhD); developmental biology (PhD); molecular biology (PhD); neuroscience (PhD).

Degree requirements: For doctorate, dissertation required.
Entrance requirements: For doctorate, GRE General Test.
Expenses: Tuition: Full-time $25,050.
Faculty research: Genetics, virology, biochemistry.

Find an in-depth description at www.petersons.com/graduate.

■ PURDUE UNIVERSITY

Graduate School, School of Science, Department of Biological Sciences, Program in Cell and Developmental Biology, West Lafayette, IN 47907
AWARDS PhD.

Faculty: 13 full-time (3 women).
Degree requirements: For doctorate, dissertation, seminars, teaching experience required.
Entrance requirements: For doctorate, GRE General Test, TOEFL, TSE. *Application deadline:* For fall admission, 2/15. *Application fee:* $30. Electronic applications accepted.
Expenses: Tuition, state resident: full-time $4,530; part-time $130 per credit hour. Tuition, nonresident: full-time $15,310; part-time $404 per credit hour. Tuition and fees vary according to campus/location and program.
Financial aid: Fellowships, research assistantships, teaching assistantships available. Aid available to part-time students. Financial aid application deadline: 2/15; financial aid applicants required to submit FAFSA.
Faculty research: Cellular signalling, patterns of development in eukaryotes, genomics, gene expression, cell structure and function.
Application contact: Nancy Konopka, Graduate Studies Office Manager, 765-494-8142, *Fax:* 765-494-0876, *E-mail:* njk@bilbo.bio.purdue.edu.

■ RENSSELAER POLYTECHNIC INSTITUTE

Graduate School, School of Science, Department of Biology, Troy, NY 12180-3590

AWARDS Biochemistry (MS, PhD); biophysics (MS, PhD); cell biology (MS, PhD); developmental biology (MS, PhD); microbiology (MS, PhD); molecular biology (MS, PhD); plant science (MS, PhD). Part-time programs available.

Faculty: 14 full-time (5 women).
Students: 16 full-time (8 women), 2 part-time (1 woman); includes 1 minority (Asian American or Pacific Islander), 4 international. *37 applicants, 41% accepted.* In 1999, 6 master's, 3 doctorates awarded. Terminal master's awarded for partial completion of doctoral program.
Degree requirements: For master's and doctorate, thesis/dissertation required, foreign language not required.
Entrance requirements: For master's and doctorate, GRE General Test, TOEFL. *Application deadline:* For fall admission, 2/1 (priority date). Applications are processed on a rolling basis. *Application fee:* $35.
Expenses: Tuition: Part-time $665 per credit hour. Required fees: $980.
Financial aid: In 1999–00, 8 research assistantships with partial tuition reimbursements (averaging $15,000 per year), 11 teaching assistantships with full tuition reimbursements (averaging $15,000 per year) were awarded; fellowships, career-related internships or fieldwork and institutionally sponsored loans also available. Financial aid application deadline: 2/1.
Faculty research: Applied environmental biology, genetics, environmental science, fresh water ecology, microbial ecology.
Dr. John Salerno, Chair, 518-276-2699, *Fax:* 518-276-2344.
Application contact: Dr. Jackie L. Collier, Assistant Professor, 518-276-6446, *Fax:* 518-276-2344, *E-mail:* collij3@rpi.edu.

Find an in-depth description at www.petersons.com/graduate.

■ RUTGERS, THE STATE UNIVERSITY OF NEW JERSEY, NEW BRUNSWICK

Graduate School, Program in Cell and Developmental Biology, New Brunswick, NJ 08901-1281

AWARDS Cell biology (MS, PhD); developmental biology (MS, PhD); immunology (MS). Part-time programs available.

Faculty: 84 full-time (15 women).
Students: 30 full-time (16 women), 42 part-time (22 women); includes 24 minority (1 African American, 15 Asian Americans or Pacific Islanders, 8 Hispanic Americans), 17 international. Average age 24. *109 applicants, 34% accepted.* In 1999, 8 master's, 8 doctorates awarded. Terminal master's awarded for partial completion of doctoral program.
Degree requirements: For master's, thesis or alternative required; for doctorate, dissertation, written qualifying exam required.
Entrance requirements: For master's, GRE General Test, TOEFL; for doctorate, GRE General Test, GRE Subject Test, TOEFL, minimum GPA of 3.0. *Application deadline:* For fall admission, 2/15. Applications are processed on a rolling basis. *Application fee:* $50.
Expenses: Tuition, state resident: full-time $6,776; part-time $279 per credit. Tuition, nonresident: full-time $9,936; part-time $412 per credit. Required fees: $20 per credit. $89 per semester. Tuition and fees vary according to course load, campus/location and program.
Financial aid: In 1999–00, 6 fellowships with full tuition reimbursements (averaging $15,000 per year), 17 research assistantships with full tuition reimbursements (averaging $15,000 per year), 10 teaching assistantships with full tuition reimbursements (averaging $14,000 per year) were awarded. Financial aid application deadline: 2/15; financial aid applicants required to submit FAFSA.
Faculty research: Genetics, cell matrix interactions, signal transduction, gene expression, cytoskeletal proteins, neurobiology.
Dr. Alice Y. Liu, Director, 732-445-2730, *Fax:* 732-445-6370, *E-mail:* liu@biology.rutgers.edu.
Application contact: Carolyn Ambrose, Administrative Assistant, 732-445-3430, *Fax:* 732-445-6370, *E-mail:* ambrose@biology.rutgers.edu.

■ STANFORD UNIVERSITY

School of Medicine, Graduate Programs in Medicine, Department of Developmental Biology, Stanford, CA 94305-9991

AWARDS PhD.

Faculty: 9 full-time (3 women).
Students: 20 full-time (9 women), 10 part-time (2 women); includes 10 minority (all Asian Americans or Pacific Islanders), 3 international. Average age 27. *63 applicants, 14% accepted.* In 1999, 6 doctorates awarded.
Degree requirements: For doctorate, dissertation required.
Entrance requirements: For doctorate, GRE, TOEFL. *Application deadline:* For fall admission, 12/15. *Application fee:* $65 ($80 for international students). Electronic applications accepted.
Expenses: Tuition: Full-time $23,058. Required fees: $152. Part-time tuition and fees vary according to course load.
Financial aid: Research assistantships available. Financial aid application deadline: 1/1.
Faculty research: Mammalian embryology, developmental genetics with particular emphasis on microbial systems, *Dictyostelium, Drosophila,* the nematode, and the mouse.
Roel Nusse, Co-Chair, 650-725-7662, *Fax:* 650-725-7739, *E-mail:* rnusse@cmgm.stanford.edu.
Application contact: Admissions Office, 650-723-2460.

■ STATE UNIVERSITY OF NEW YORK AT ALBANY

College of Arts and Sciences, Department of Biological Sciences, Specialization in Molecular, Cellular, Developmental, and Neural Biology, Albany, NY 12222-0001

AWARDS MS, PhD.

Degree requirements: For master's, one foreign language required; for doctorate, dissertation required.
Entrance requirements: For master's and doctorate, GRE General Test. *Application fee:* $50.
Expenses: Tuition, state resident: full-time $5,100; part-time $214 per credit. Tuition, nonresident: full-time $8,416; part-time $352 per credit. Required fees: $31 per credit.
Financial aid: Minority assistantships available.
Dr. David Shub, Chair, Department of Biological Sciences, 518-442-4300.

■ STATE UNIVERSITY OF NEW YORK AT STONY BROOK

Graduate School, College of Arts and Sciences, Molecular and Cellular Biology Program, Specialization in Cellular and Developmental Biology, Stony Brook, NY 11794

AWARDS PhD.

Faculty: 98.
Students: 27 full-time (16 women), 9 part-time (5 women); includes 6 minority (1 African American, 5 Asian Americans or Pacific Islanders), 8 international.
Degree requirements: For doctorate, one foreign language (computer language can substitute), dissertation, comprehensive exam, teaching experience required.
Entrance requirements: For doctorate, GRE General Test, GRE Subject Test, TOEFL. *Application deadline:* For fall admission, 1/15. *Application fee:* $50.
Expenses: Tuition, state resident: full-time $5,100; part-time $213 per credit hour.

Tuition, nonresident: full-time $8,416; part-time $351 per credit hour. Required fees: $492. Tuition and fees vary according to program.
Financial aid: Fellowships, research assistantships, teaching assistantships available. *Total annual research expenditures:* $14,402.
Dr. James Quigley, Director, 631-632-8533, *Fax:* 631-632-9730, *E-mail:* jquigley@path.som.sunysb.edu.
Find an in-depth description at www.petersons.com/graduate.

■ STATE UNIVERSITY OF NEW YORK UPSTATE MEDICAL UNIVERSITY

College of Graduate Studies, Department of Cell and Developmental Biology, Syracuse, NY 13210-2334
AWARDS PhD, MD/PhD.

Degree requirements: For doctorate, dissertation, comprehensive exam required, foreign language not required.
Entrance requirements: For doctorate, GRE General Test, GRE Subject Test, TSE. *Application deadline:* For fall admission, 4/1 (priority date). Applications are processed on a rolling basis. *Application fee:* $40.
Expenses: Tuition, state resident: full-time $5,100; part-time $213 per credit. Tuition, nonresident: full-time $8,416; part-time $351 per credit. Required fees: $410; $25 per credit. Part-time tuition and fees vary according to course load and program.
Dr. David R. Mitchell, Interim Chair.
Find an in-depth description at www.petersons.com/graduate.

■ THOMAS JEFFERSON UNIVERSITY

College of Graduate Studies, Program in Developmental Biology and Teratology, Philadelphia, PA 19107
AWARDS MS, PhD.

Faculty: 33 full-time (10 women).
Students: 12 full-time (5 women), 7 part-time (6 women); includes 7 minority (4 African Americans, 2 Asian Americans or Pacific Islanders, 1 Hispanic American), 2 international. 17 *applicants, 12% accepted.* In 1999, 1 master's, 2 doctorates awarded. Terminal master's awarded for partial completion of doctoral program.
Degree requirements: For master's and doctorate, thesis/dissertation required, foreign language not required.
Entrance requirements: For master's, GRE General Test, TOEFL, minimum GPA of 3.0; for doctorate, GRE General Test, TOEFL, minimum GPA of 3.2.
Application deadline: For fall admission, 3/1

(priority date). Applications are processed on a rolling basis. *Application fee:* $40.
Expenses: Tuition: Full-time $12,670. Tuition and fees vary according to degree level and program.
Financial aid: In 1999–00, 10 fellowships with full tuition reimbursements were awarded; research assistantships, Federal Work-Study, institutionally sponsored loans, traineeships, and training grants also available. Aid available to part-time students. Financial aid application deadline: 5/1; financial aid applicants required to submit FAFSA.
Faculty research: Developmental toxicology, cell signaling, cell adhesion, genetic regulation. *Total annual research expenditures:* $6.3 million.
Dr. Gerald B. Grunwald, Chairman, Doctoral Committee, 215-503-4191, *Fax:* 215-923-3808, *E-mail:* gerald.grunwald@mail.tju.edu.
Application contact: Jessie F. Pervall, Director of Admissions, 215-503-4400, *Fax:* 215-503-3433, *E-mail:* cgs-info@mail.tju.edu.
Find an in-depth description at www.petersons.com/graduate.

■ TUFTS UNIVERSITY

Sackler School of Graduate Biomedical Sciences, Program in Cell, Molecular and Developmental Biology, Medford, MA 02155
AWARDS PhD.

Faculty: 31 full-time (6 women), 6 part-time/adjunct (3 women).
Students: 25 full-time (14 women); includes 1 minority (Asian American or Pacific Islander), 3 international. Average age 25. 143 *applicants, 3% accepted.* In 1999, 4 degrees awarded.
Degree requirements: For doctorate, dissertation required, foreign language not required. *Average time to degree:* Doctorate–5 years full-time.
Entrance requirements: For doctorate, GRE General Test, GRE Subject Test, TOEFL. *Application deadline:* For fall admission, 1/15 (priority date). Applications are processed on a rolling basis. *Application fee:* $45.
Expenses: Tuition: Full-time $19,325.
Financial aid: In 1999–00, 2 fellowships, 23 research assistantships with full tuition reimbursements (averaging $18,805 per year) were awarded; grants, scholarships, traineeships, and tuition waivers (full) also available. Financial aid application deadline: 2/1.
Faculty research: Reproduction and hormone action, control of gene expression, cell-matrix and cell-cell interactions, growth control and tumorigenesis, cytoskeleton and contractile proteins.

Dr. John J. Castellot, Director, 617-636-0303, *Fax:* 617-636-6536, *E-mail:* jcastellot@infonet.tufts.edu.
Application contact: Secretary, 617-636-6767, *Fax:* 617-636-0375, *E-mail:* sackler-school@tufts.edu.
Find an in-depth description at www.petersons.com/graduate.

■ UNIFORMED SERVICES UNIVERSITY OF THE HEALTH SCIENCES

School of Medicine, Division of Basic Medical Sciences, Department of Anatomy and Cell Biology, Bethesda, MD 20814-4799
AWARDS Cell biology, developmental biology, and neurobiology (PhD).

Faculty: 20 full-time (9 women), 12 part-time/adjunct (2 women).
Students: 8 full-time (1 woman); includes 2 minority (both Asian Americans or Pacific Islanders). Average age 28. 3 *applicants, 33% accepted.* In 1999, 1 degree awarded (100% entered university research/teaching).
Degree requirements: For doctorate, dissertation, qualifying exam required, foreign language not required. *Average time to degree:* Doctorate–5 years full-time.
Entrance requirements: For doctorate, GRE General Test, GRE Subject Test, minimum GPA of 3.0, U.S. citizenship. *Application deadline:* For fall admission, 1/15 (priority date). Applications are processed on a rolling basis. *Application fee:* $0.
Financial aid: In 1999–00, 3 fellowships with full tuition reimbursements (averaging $15,000 per year) were awarded; tuition waivers (full) also available.
Faculty research: Neuroanatomy and neurocytology, molecular biology.
Dr. Harvey Pollard, Chair, 301-295-3200, *E-mail:* hpollard@usuhs.mil.
Application contact: Janet M. Anastasi, Graduate Program Coordinator, 301-295-9474, *Fax:* 301-295-6772, *E-mail:* janastasi@usuhs.mil.

■ THE UNIVERSITY OF ALABAMA AT BIRMINGHAM

Graduate School, School of Natural Sciences and Mathematics, Department of Biology, Birmingham, AL 35294
AWARDS Comparative and cellular biology (PhD); comparative and cellular physiology (MS); marine science (MS, PhD); microbial ecology and physiology (MS, PhD); reproduction and development (MS, PhD).

Students: 34 full-time (19 women), 6 international. 61 *applicants, 38% accepted.* In

The University of Alabama at Birmingham (continued)

1999, 1 master's, 3 doctorates awarded. Terminal master's awarded for partial completion of doctoral program.

Degree requirements: For master's and doctorate, thesis/dissertation required.

Entrance requirements: For master's and doctorate, GRE General Test, TOEFL, previous course work in biology, calculus, organic chemistry, physics. *Application deadline:* Applications are processed on a rolling basis. *Application fee:* $35 ($60 for international students). Electronic applications accepted.

Expenses: Tuition, state resident: part-time $104 per semester hour. Tuition, nonresident: part-time $208 per semester hour. Required fees: $17 per semester hour. $57 per quarter. Tuition and fees vary according to program.

Financial aid: In 1999–00, 22 students received aid, including 3 fellowships with full tuition reimbursements available (averaging $14,000 per year), 19 teaching assistantships with full tuition reimbursements available (averaging $14,000 per year); research assistantships, career-related internships or fieldwork, Federal Work-Study, institutionally sponsored loans, and tuition waivers (full) also available. Aid available to part-time students.

Faculty research: Invertebrate physiology, marine biology, environmental biology.

Dr. Daniel D. Jones, Chairman, 205-934-4290, *Fax:* 205-975-6097, *E-mail:* ddjones@uab.edu.

Find an in-depth description at www.petersons.com/graduate.

■ UNIVERSITY OF CALIFORNIA, DAVIS

Graduate Studies, Programs in the Biological Sciences, Program in Cell and Developmental Biology, Davis, CA 95616

AWARDS PhD.

Faculty: 36.

Students: 25 full-time (16 women); includes 7 minority (2 Asian Americans or Pacific Islanders, 3 Hispanic Americans, 2 Native Americans), 7 international. Average age 29. *30 applicants, 53% accepted.* In 1999, 1 degree awarded.

Degree requirements: For doctorate, dissertation required.

Entrance requirements: For doctorate, GRE General Test, GRE Subject Test. *Application deadline:* For fall admission, 1/15. *Application fee:* $40. Electronic applications accepted.

Expenses: Tuition, nonresident: full-time $9,804. Tuition and fees vary according to program and student level.

Financial aid: In 1999–00, 23 students received aid, including 5 fellowships with full and partial tuition reimbursements available, 13 research assistantships with full and partial tuition reimbursements available, 3 teaching assistantships with partial tuition reimbursements available; Federal Work-Study, grants, institutionally sponsored loans, scholarships, and tuition waivers (full and partial) also available. Financial aid application deadline: 1/15; financial aid applicants required to submit FAFSA.

Faculty research: Molecular basis of cell function and development.

Paul FitzGerald, Chair, Cell Biology and Human Anatomy, 530-752-7130, *E-mail:* pfitzgerald@ucdavis.edu.

Application contact: Dawne Shell, Graduate Program Assistant, 530-752-9091, *Fax:* 530-752-8822, *E-mail:* drshell@ucdavis.edu.

Find an in-depth description at www.petersons.com/graduate.

■ UNIVERSITY OF CALIFORNIA, IRVINE

Office of Research and Graduate Studies, School of Biological Sciences, Department of Developmental and Cell Biology, Irvine, CA 92697

AWARDS Biological sciences (MS, PhD). Students apply through the Graduate Program in Molecular Biology, Genetics, and Biochemistry.

Faculty: 16 full-time (3 women).

Students: 7 full-time (4 women), 3 part-time; includes 2 minority (1 Asian American or Pacific Islander, 1 Hispanic American), 2 international. *292 applicants, 25% accepted.* In 1999, 1 master's awarded. Terminal master's awarded for partial completion of doctoral program.

Degree requirements: For master's, one foreign language required; for doctorate, dissertation required.

Entrance requirements: For master's, GRE General Test, GRE Subject Test, minimum GPA of 3.0; for doctorate, GRE General Test, GRE Subject Test. *Application deadline:* For fall admission, 2/1 (priority date). *Application fee:* $40. Electronic applications accepted.

Expenses: Tuition, nonresident: full-time $10,244; part-time $1,720 per quarter. Required fees: $5,252; $1,300 per quarter. Tuition and fees vary according to course load and program.

Financial aid: Fellowships, research assistantships, teaching assistantships, institutionally sponsored loans and tuition waivers (full and partial) available.

Financial aid application deadline: 3/2; financial aid applicants required to submit FAFSA.

Faculty research: Genetics and development, oncogene signaling pathways, gene regulation, tissue regeneration and molecular genetics.

J. Lawrence Marsh, Chair, 949-824-6677, *Fax:* 949-824-4709, *E-mail:* jlmarsh@uci.edu.

Application contact: Administrator, 949-824-8145, *Fax:* 949-824-7407, *E-mail:* gp-mbgb@uci.edu.

■ UNIVERSITY OF CALIFORNIA, LOS ANGELES

School of Medicine and Graduate Division, Graduate Programs in Medicine, Department of Molecular, Cell and Developmental Biology, Los Angeles, CA 90095

AWARDS MA, PhD.

Students: 35 full-time (15 women); includes 15 minority (13 Asian Americans or Pacific Islanders, 2 Hispanic Americans), 5 international. *66 applicants, 18% accepted.*

Degree requirements: For doctorate, dissertation, qualifying exams required, foreign language not required.

Entrance requirements: For doctorate, GRE General Test, GRE Subject Test, TOEFL. *Application fee:* $40.

Expenses: Tuition, nonresident: full-time $9,804. Required fees: $4,405.

Financial aid: Fellowships, research assistantships, teaching assistantships, scholarships available. Financial aid application deadline: 3/1.

Dr. Lutz Birnbaumer, Chair, 310-825-7109.

Application contact: UCLA Access Coordinator, 800-284-8252, *Fax:* 310-206-5280, *E-mail:* uclaaccess@ibes.medsch.ucla.edu.

Find an in-depth description at www.petersons.com/graduate.

■ UNIVERSITY OF CALIFORNIA, RIVERSIDE

Graduate Division, College of Natural and Agricultural Sciences, Program in Cell, Molecular, and Developmental Biology, Riverside, CA 92521-0102

AWARDS MS, PhD.

Faculty: 49 full-time (15 women).

Students: 3 full-time (2 women); includes 1 minority (Asian American or Pacific Islander).

Degree requirements: For master's, thesis, oral defense of thesis required; for doctorate, dissertation, oral defense of thesis, qualifying exams, 2 quarters of teaching experience required. *Average time* *to degree:* Master's–2 years full-time; doctorate–5 years full-time.

Entrance requirements: For master's and doctorate, GRE General Test, TOEFL, minimum GPA of 3.2. *Application deadline:* For fall admission, 5/1; for winter admission, 9/1; for spring admission, 12/1. *Application fee:* $40.

Expenses: Tuition, nonresident: full-time $9,804. Required fees: $4,758. Full-time tuition and fees vary according to program.

Financial aid: Fellowships, research assistantships, teaching assistantships available. Financial aid application deadline: 1/5.

Dr. Prudence Talbot, Graduate Advisor, 909-787-5913, *Fax:* 909-787-4286, *E-mail:* talbot@pop.ucr.edu.

Application contact: Helene Serewis, Graduate Student Affairs Officer, 800-735-0717, *Fax:* 909-787-5517, *E-mail:* biopgrad@pep.ycr.edu.

■ UNIVERSITY OF CALIFORNIA, SAN DIEGO

Graduate Studies and Research, Department of Biology, Program in Cell and Developmental Biology, La Jolla, CA 92093-0348

AWARDS PhD. Offered in association with the Salk Institute.

Degree requirements: For doctorate, dissertation required.

Entrance requirements: For doctorate, GRE General Test, pre-application beginning in September. *Application deadline:* For fall admission, 1/7. *Application fee:* $40.

Expenses: Tuition, nonresident: full-time $14,691. Required fees: $4,697. Full-time tuition and fees vary according to program.

Financial aid: Tuition waivers (full) and stipends available.

Application contact: 858-534-3835.

Find an in-depth description at www.petersons.com/graduate.

■ UNIVERSITY OF CALIFORNIA, SANTA BARBARA

Graduate Division, College of Letters and Sciences, Division of Mathematics, Life, and Physical Sciences, Department of Molecular, Cellular, and Developmental Biology, Santa Barbara, CA 93106

AWARDS MA, PhD.

Faculty: 25 full-time (6 women), 1 part-time/adjunct (0 women).

Students: 59 full-time (34 women); includes 7 minority (all Asian Americans or Pacific Islanders), 1 international. *91 applicants, 34% accepted.* In 1999, 7 master's, 6 doctorates awarded.

Degree requirements: For doctorate, dissertation required. *Average time to degree:* Master's–2 years full-time; doctorate–6 years full-time.

Entrance requirements: For master's and doctorate, GRE General Test, GRE Subject Test, TOEFL. *Application deadline:* For fall admission, 1/1 (priority date). *Application fee:* $40. Electronic applications accepted.

Expenses: Tuition, state resident: full-time $14,637. Tuition, nonresident: full-time $24,441.

Financial aid: Fellowships, research assistantships, teaching assistantships available. Financial aid application deadline: 1/15; financial aid applicants required to submit FAFSA.

Faculty research: Regulation of pap operon, recombinant composite protein, interferons, microtubules, embryonic development, bacterial pathogens. Stu Feinstein, Chair, 805-893-2659, *Fax:* 805-893-4724, *E-mail:* feinstei@lifesci.ucsb.edu.

Application contact: Shari Profant, Graduate Program Assistant, 805-893-8499, *Fax:* 805-893-4724, *E-mail:* profant@lifesci.ucsb.edu.

■ UNIVERSITY OF CHICAGO

Division of the Biological Sciences, Molecular Biosciences: Biochemistry, Genetics, Cell and Developmental Biology, Committee on Developmental Biology, Chicago, IL 60637-1513

AWARDS Cellular differentiation (PhD); developmental endocrinology (PhD); developmental genetics (PhD); developmental neurobiology (PhD); gene expression (PhD).

Faculty: 30 full-time (11 women).

Students: 13 full-time (6 women); includes 3 minority (1 Asian American or Pacific Islander, 2 Hispanic Americans), 3 international. Average age 26.

Degree requirements: For doctorate, dissertation required, foreign language not required. *Average time to degree:* Doctorate–5.5 years full-time.

Entrance requirements: For doctorate, GRE General Test, TOEFL. *Application deadline:* For fall admission, 1/5 (priority date). *Application fee:* $55.

Expenses: Tuition: Full-time $24,804; part-time $3,422 per course. Required fees: $390. Tuition and fees vary according to program.

Financial aid: In 1999–00, 11 students received aid; fellowships, research assistantships, institutionally sponsored loans and traineeships available. Financial aid application deadline: 6/1.

Faculty research: Epidermal differentiation, neural lineages, pattern formation. Dr. Edwin Ferguson, Chairman, 773-702-8037.

Application contact: Kristine Gaston, Graduate Administrative Director, 773-702-8037, *Fax:* 773-702-3172, *E-mail:* kristine@cummings.uchicago.edu.

■ UNIVERSITY OF CINCINNATI

Division of Research and Advanced Studies, College of Medicine, Graduate Programs in Medicine, Department of Pediatrics, Program in Molecular and Developmental Biology, Cincinnati, OH 45221-0091

AWARDS Molecular and developmental biology (MS, PhD); teratology (MS, PhD).

Faculty: 12 full-time.

Students: 29 full-time (16 women), 4 part-time (3 women); includes 5 minority (2 African Americans, 3 Asian Americans or Pacific Islanders), 10 international. *38 applicants, 32% accepted.* In 1999, 3 master's, 6 doctorates awarded. Terminal master's awarded for partial completion of doctoral program.

Degree requirements: For master's, thesis required, foreign language not required; for doctorate, dissertation, qualifying exam required, foreign language not required. *Average time to degree:* Master's–3.8 years full-time; doctorate–6.4 years full-time.

Entrance requirements: For master's and doctorate, GRE General Test, GRE Subject Test (biology or chemistry), TOEFL. *Application deadline:* For fall admission, 2/1 (priority date). Applications are processed on a rolling basis. *Application fee:* $30.

Expenses: Tuition, state resident: full-time $5,139; part-time $196 per credit hour. Tuition, nonresident: full-time $10,326; part-time $369 per credit hour. Required fees: $561; $187 per quarter.

Financial aid: Tuition waivers (full) and unspecified assistantships available. Financial aid application deadline: 5/1. *Total annual research expenditures:* $2.5 million.

Dr. James Lessard, Interim Head, 513-636-4549, *Fax:* 513-636-4317, *E-mail:* james.lessard@uc.edu.

Application contact: Dan Wigginton, Graduate Program Director, 513-636-4547, *Fax:* 513-559-4317, *E-mail:* dan.wigginton@chmcc.org.

Find an in-depth description at www.petersons.com/graduate.

■ UNIVERSITY OF COLORADO AT BOULDER

Graduate School, College of Arts and Sciences, Department of Molecular, Cellular, and Developmental Biology, Boulder, CO 80309

AWARDS Cellular structure and function (MA, PhD); developmental biology (MA, PhD); molecular biology (MA, PhD).

Faculty: 25 full-time (7 women).
Students: 54 full-time (28 women), 1 (woman) part-time; includes 6 minority (1 African American, 3 Asian Americans or Pacific Islanders, 2 Hispanic Americans), 4 international. Average age 27. *174 applicants, 9% accepted.* In 1999, 2 master's, 15 doctorates awarded. Terminal master's awarded for partial completion of doctoral program.
Degree requirements: For master's, thesis or alternative, comprehensive exam required, foreign language not required; for doctorate, dissertation, comprehensive exam required, foreign language not required. *Average time to degree:* Doctorate–6 years full-time.
Entrance requirements: For master's, GRE General Test, GRE Subject Test, minimum undergraduate GPA of 2.75; for doctorate, GRE General Test, GRE Subject Test. *Application deadline:* For fall admission, 1/1. *Application fee:* $40 ($60 for international students).
Expenses: Tuition, state resident: part-time $181 per credit hour. Tuition, nonresident: part-time $542 per credit hour. Required fees: $99 per term. Tuition and fees vary according to course load and program.
Financial aid: Fellowships, research assistantships, teaching assistantships, tuition waivers (full) available. Financial aid application deadline: 3/1. *Total annual research expenditures:* $9.7 million.
Leslie Leinward, Chair, 303-492-7606, *Fax:* 303-492-7744, *E-mail:* leinward@ stripe.colorado.edu.
Application contact: Karen Brown, Student Affairs Office, 303-492-7230, *Fax:* 303-492-7744, *E-mail:* mcdbgradinfo@ beagle.colorado.edu.

Find an in-depth description at www.petersons.com/graduate.

■ UNIVERSITY OF COLORADO HEALTH SCIENCES CENTER

Graduate School, Programs in Biological and Medical Sciences, Program in Cell and Developmental Biology, Denver, CO 80262

AWARDS PhD.

Degree requirements: For doctorate, dissertation required, foreign language not required.

Entrance requirements: For doctorate, GRE, minimum GPA of 2.75.
Expenses: Tuition, state resident: full-time $1,512; part-time $56 per hour. Tuition, nonresident: full-time $7,209; part-time $267 per hour. Full-time tuition and fees vary according to course load and program.

Find an in-depth description at www.petersons.com/graduate.

■ UNIVERSITY OF CONNECTICUT

Graduate School, College of Liberal Arts and Sciences, Biological Sciences Group, Storrs, CT 06269

AWARDS Ecology and evolutionary biology (MS, PhD), including botany, ecology, entomology, systematics, zoology; molecular and cell biology (MS, PhD), including biochemistry, biophysics, biotechnology (MS), cell and developmental biology, genetics, microbiology, plant molecular and cell biology; physiology and neurobiology (MS, PhD), including neurobiology, physiology.

Degree requirements: For doctorate, dissertation required.
Entrance requirements: For master's and doctorate, GRE General Test, GRE Subject Test, TOEFL.
Expenses: Tuition, state resident: full-time $5,118. Tuition, nonresident: full-time $13,298. Required fees: $1,022.

■ UNIVERSITY OF CONNECTICUT

Graduate School, College of Liberal Arts and Sciences, Biological Sciences Group, Department of Molecular and Cell Biology, Field of Cell and Developmental Biology, Storrs, CT 06269

AWARDS MS, PhD.

Degree requirements: For doctorate, dissertation required.
Entrance requirements: For master's and doctorate, GRE General Test, GRE Subject Test, TOEFL.
Expenses: Tuition, state resident: full-time $5,118. Tuition, nonresident: full-time $13,298. Required fees: $1,022.

Find an in-depth description at www.petersons.com/graduate.

■ UNIVERSITY OF CONNECTICUT HEALTH CENTER

Graduate School, Programs in Biomedical Sciences, Program in Developmental Biology, Farmington, CT 06030

AWARDS PhD, DMD/PhD, MD/PhD.

Faculty: 28.
Students: 12 full-time (6 women); includes 1 minority (Hispanic American), 4 international. In 1999, 1 degree awarded.

Degree requirements: For doctorate, one foreign language (computer language can substitute), dissertation required.
Entrance requirements: For doctorate, GRE General Test, GRE Subject Test, TOEFL. *Application deadline:* For fall admission, 2/1 (priority date); for spring admission, 10/1. Applications are processed on a rolling basis. *Application fee:* $40 ($45 for international students).
Expenses: Tuition, state resident: full-time $5,272; part-time $293 per credit. Tuition, nonresident: full-time $13,696; part-time $761 per credit. Required fees: $320; $198 per semester. One-time fee: $50 full-time. Full-time tuition and fees vary according to course load, program and reciprocity agreements.
Financial aid: In 1999–00, research assistantships (averaging $17,000 per year); fellowships.
Faculty research: Limb development/ bone formation; genetics and human disease; apoptosis and development; signal transduction in development.
Dr. Steve L. Helfand, Director, 860-679-4200, *E-mail:* helfand@sun.uchc.edu.
Application contact: Marizta Barta, Information Contact, 860-679-4306, *Fax:* 860-679-1282, *E-mail:* barta@ adp.uchc.edu.

Find an in-depth description at www.petersons.com/graduate.

■ UNIVERSITY OF ILLINOIS AT CHICAGO

Graduate College, College of Liberal Arts and Sciences, Department of Biological Sciences, Chicago, IL 60607-7128

AWARDS Cell and developmental biology (PhD); ecology and evolution (MS, DA, PhD); genetics and development (PhD); molecular biology (MS, PhD); neurobiology (MS, PhD); plant biology (MS, DA, PhD).

Faculty: 40 full-time (5 women).
Students: 100 full-time (47 women), 14 part-time (10 women); includes 13 minority (11 Asian Americans or Pacific Islanders, 2 Hispanic Americans), 42 international. Average age 29. *99 applicants, 36% accepted.* In 1999, 3 master's, 9 doctorates awarded.
Degree requirements: For master's, thesis required, foreign language not required; for doctorate, dissertation, preliminary exam required, foreign language not required.
Entrance requirements: For master's and doctorate, GRE General Test, GRE Subject Test, TOEFL, previous course work in physics, calculus, and organic chemistry; minimum GPA of 3.75 on a 5.0 scale. *Application deadline:* For fall admission, 6/1. Applications are processed on a

rolling basis. *Application fee:* $40 ($50 for international students). Electronic applications accepted.

Expenses: Tuition, state resident: full-time $3,750; part-time $1,250 per semester. Tuition, nonresident: full-time $10,588; part-time $3,530 per semester. Required fees: $507 per semester. Tuition and fees vary according to course load and program.

Financial aid: In 1999–00, 87 students received aid; fellowships, research assistantships, teaching assistantships, career-related internships or fieldwork, Federal Work-Study, traineeships, and tuition waivers (full) available. Financial aid application deadline: 3/1; financial aid applicants required to submit FAFSA. Dr. Lon Kaufman, Head, 312-996-2213.

Application contact: Dr. Leo Miller, Director of Graduate Studies, 312-996-2220.

Find an in-depth description at www.petersons.com/graduate.

■ UNIVERSITY OF KANSAS

Graduate School, College of Liberal Arts and Sciences, Division of Biological Sciences, Department of Molecular Biosciences, Lawrence, KS 66045

AWARDS Biochemistry and biophysics (MA, MS, PhD); microbiology (MA, MS, PhD); molecular, cellular, and developmental biology (MA, MS, PhD).

Students: 13 full-time (7 women), 8 part-time (2 women); includes 1 minority (Asian American or Pacific Islander), 5 international. *57 applicants, 21% accepted.* In 2000, 11 master's, 10 doctorates awarded.

Degree requirements: For master's, thesis required, foreign language not required; for doctorate, dissertation required.

Entrance requirements: For master's and doctorate, GRE General Test, GRE Subject Test, TOEFL. *Application deadline:* For fall admission, 1/15 (priority date). *Application fee:* $25.

Expenses: Tuition, state resident: full-time $2,482; part-time $103 per credit hour. Tuition, nonresident: full-time $8,104; part-time $338 per credit hour. Required fees: $428; $31 per credit hour. Tuition and fees vary according to program.

Financial aid: In 2000–01, research assistantships (averaging $11,588 per year), teaching assistantships (averaging $11,588 per year) were awarded; fellowships. Financial aid application deadline: 3/1. Paul Kelly, Chair, 785-864-4311, *Fax:* 785-864-5294.

Application contact: Information Contact, 785-864-4311, *Fax:* 785-864-5294.

Find an in-depth description at www.petersons.com/graduate.

■ UNIVERSITY OF MASSACHUSETTS AMHERST

Graduate School, Interdisciplinary Programs, Molecular and Cellular Biology Graduate Program, Amherst, MA 01003

AWARDS Biological chemistry (PhD); cell and developmental biology (PhD). Part-time programs available.

Students: 29 full-time (19 women), 33 part-time (21 women); includes 4 minority (1 African American, 3 Asian Americans or Pacific Islanders), 28 international. Average age 28. *288 applicants, 15% accepted.* In 2000, 8 degrees awarded.

Degree requirements: For doctorate, one foreign language, dissertation required.

Entrance requirements: For doctorate, GRE General Test. *Application deadline:* For fall admission, 1/15 (priority date). Applications are processed on a rolling basis. *Application fee:* $40.

Expenses: Tuition, state resident: full-time $2,640; part-time $165 per credit. Tuition, nonresident: full-time $9,756; part-time $407 per credit. Required fees: $1,221 per term. One-time fee: $110. Full-time tuition and fees vary according to course load, campus/location and reciprocity agreements.

Financial aid: In 2000–01, 17 research assistantships with full tuition reimbursements (averaging $3,796 per year), 3 teaching assistantships with full tuition reimbursements (averaging $5,510 per year) were awarded; fellowships with full tuition reimbursements, career-related internships or fieldwork, Federal Work-Study, grants, scholarships, traineeships, and unspecified assistantships also available. Aid available to part-time students. Financial aid application deadline: 1/15. Dr. Rodney K. Murphey, Acting Head, 413-545-3246, *Fax:* 413-545-1812, *E-mail:* rmurphey@bio.umass.edu.

Find an in-depth description at www.petersons.com/graduate.

■ UNIVERSITY OF MEDICINE AND DENTISTRY OF NEW JERSEY

Graduate School of Biomedical Sciences, Graduate Programs in Biomedical Sciences, Program in Cell and Developmental Biology, Piscataway, NJ 08854-5635

AWARDS Cell biology (MS, PhD); developmental biology (MS, PhD); immunology (MS).

Degree requirements: For master's, thesis, qualifying exam required; for doctorate, dissertation, qualifying exam required, foreign language not required.

Entrance requirements: For master's, GRE General Test, TOEFL; for doctorate, GRE General Test, GRE Subject Test, TOEFL. *Application deadline:* For fall admission, 2/1; for spring admission, 10/1. *Application fee:* $40.

Expenses: Tuition, state resident: part-time $270 per credit hour. Tuition, nonresident: part-time $407 per credit hour. Part-time tuition and fees vary according to campus/location and program.

Financial aid: Fellowships, research assistantships, teaching assistantships available. Financial aid application deadline: 5/1.

Faculty research: Genetics, cell matrix interactions, signal transduction, gene expression, cytoskeletal proteins. Dr. Alice Y. Liu, Director, 732-445-3430, *Fax:* 732-445-2730.

■ UNIVERSITY OF MIAMI

School of Medicine and Graduate School, Graduate Programs in Medicine, Department of Cell Biology and Anatomy, Coral Gables, FL 33124

AWARDS Molecular, cell and developmental biology (PhD).

Faculty: 13 full-time (4 women).

Students: 11 full-time (5 women); includes 5 minority (1 African American, 3 Asian Americans or Pacific Islanders, 1 Hispanic American). In 1999, 1 degree awarded (100% entered university research/teaching).

Degree requirements: For doctorate, one foreign language (computer language can substitute), dissertation required.

Entrance requirements: For doctorate, GRE General Test, GRE Subject Test, TOEFL. *Application deadline:* For fall admission, 3/1 (priority date). Applications are processed on a rolling basis. *Application fee:* $35.

Expenses: Tuition, area resident: Part-time $899 per credit.

Financial aid: Fellowships available.

Faculty research: Signal transduction cell-matrix interactions, gene expression during development, cell polarity, protein sorting. *Total annual research expenditures:* $1.8 million. Dr. Robert Warren, Interim Chairman, 305-243-6691, *Fax:* 305-545-7166.

Application contact: Dr. Richard Rotundo, Professor, 305-243-6691, *Fax:* 305-545-7166.

■ UNIVERSITY OF MIAMI

School of Medicine and Graduate School, Graduate Programs in Medicine, Program in Molecular Cell and Developmental Biology, Coral Gables, FL 33124

AWARDS PhD.

Faculty: 13 full-time (4 women).
Students: 14 full-time (8 women); includes 6 minority (4 Asian Americans or Pacific Islanders, 1 Hispanic American, 1 Native American), 2 international. In 1999, 2 degrees awarded.
Entrance requirements: For doctorate, GRE General Test, GRE Subject Test, TOEFL. *Application deadline:* For fall admission, 3/1 (priority date). Applications are processed on a rolling basis. *Application fee:* $35.
Expenses: Tuition, area resident: Part-time $899 per credit.
Financial aid: Fellowships available.
Dr. Robert Warren, Interim Chair, 305-243-6691, *Fax:* 305-545-7166.
Application contact: Dr. Richard Rotundo, Professor, 305-243-6691, *Fax:* 305-545-7166.

Find an in-depth description at www.petersons.com/graduate.

■ UNIVERSITY OF MINNESOTA, TWIN CITIES CAMPUS

Graduate School, Program in Molecular, Cellular, Developmental Biology and Genetics, Minneapolis, MN 55455-0213

AWARDS Genetic counseling (MS); molecular, cellular, developmental biology and genetics (PhD). Part-time programs available.

Faculty: 80 full-time (24 women), 17 part-time/adjunct (16 women).
Students: 62 full-time (43 women), 2 part-time (both women). Average age 24. *188 applicants, 19% accepted.* In 1999, 9 master's awarded (100% found work related to degree); 5 doctorates awarded (20% entered university research/teaching, 80% found other work related to degree). Terminal master's awarded for partial completion of doctoral program.
Degree requirements: For master's, thesis optional, foreign language not required; for doctorate, dissertation required, foreign language not required. *Average time to degree:* Master's–2 years full-time; doctorate–5 years full-time.
Entrance requirements: For master's and doctorate, GRE General Test, TOEFL. *Application deadline:* For fall admission, 1/15 (priority date). Applications are processed on a rolling basis. *Application fee:* $40 ($50 for international students).
Expenses: Tuition, state resident: full-time $5,040; part-time $420 per credit. Tuition,

nonresident: full-time $9,900; part-time $825 per credit. Full-time tuition and fees vary according to course load, program and reciprocity agreements.
Financial aid: In 1999–00, 53 students received aid, including 9 fellowships with full tuition reimbursements available (averaging $16,504 per year), 39 research assistantships with full tuition reimbursements available (averaging $16,504 per year), 5 teaching assistantships with full tuition reimbursements available (averaging $16,504 per year); Federal Work-Study, grants, institutionally sponsored loans, and traineeships also available. Financial aid application deadline: 1/15.
Faculty research: Membrane receptors and membrane transport, cell interactions, cytoskeleton and cell mobility, regulation of gene expression, plant cell and molecular biology.
Dr. Perry B. Hackett, Director of Graduate Studies, 612-624-3053, *Fax:* 612-625-1700, *E-mail:* dgsmcdbg@biosci.cbs.umn.edu.
Application contact: Sue Knoblauch, Principal Secretary, 612-624-7470, *Fax:* 612-625-5754, *E-mail:* mcdbg@biosci.cbs.umn.edu.

Find an in-depth description at www.petersons.com/graduate.

■ UNIVERSITY OF MISSOURI–ST. LOUIS

Graduate School, College of Arts and Sciences, Department of Biology, St. Louis, MO 63121-4499

AWARDS Biology (MS, PhD), including animal behavior (MS); biochemistry, biotechnology (MS), conservation biology (MS), development (MS), ecology (MS), environmental studies (PhD), evolution (MS), genetics (MS), molecular biology and biotechnology (PhD), molecular/cellular biology (MS), physiology (MS), plant systematics, population biology (MS), tropical biology (MS); biotechnology (Certificate), tropical biology and conservation (Certificate). Part-time programs available.

Faculty: 46.
Students: 21 full-time (11 women), 75 part-time (44 women); includes 13 minority (2 African Americans, 2 Asian Americans or Pacific Islanders, 8 Hispanic Americans, 1 Native American), 23 international. In 1999, 14 master's, 4 doctorates awarded.
Degree requirements: For master's, thesis or alternative required, foreign language not required; for doctorate, one foreign language, dissertation, 1 semester of teaching experience required.
Entrance requirements: For doctorate, GRE General Test. *Application deadline:* For fall admission, 7/1 (priority date); for spring admission, 11/1 (priority date).

Applications are processed on a rolling basis. *Application fee:* $25 ($40 for international students). Electronic applications accepted.
Expenses: Tuition, state resident: full-time $4,932; part-time $173 per credit hour. Tuition, nonresident: full-time $13,279; part-time $521 per credit hour. Required fees: $775; $33 per credit hour. Tuition and fees vary according to degree level and program.
Financial aid: In 1999–00, 8 research assistantships with partial tuition reimbursements (averaging $10,635 per year), 14 teaching assistantships with partial tuition reimbursements (averaging $11,488 per year) were awarded; career-related internships or fieldwork and Federal Work-Study also available. Aid available to part-time students. Financial aid application deadline: 2/1. *Total annual research expenditures:* $908,828.
Application contact: Graduate Admissions, 314-516-5458, *Fax:* 314-516-6759, *E-mail:* gradadm@umsl.edu.

■ THE UNIVERSITY OF NORTH CAROLINA AT CHAPEL HILL

Graduate School, College of Arts and Sciences, Department of Biology, Chapel Hill, NC 27599

AWARDS Botany (MA, MS, PhD); cell biology, development, and physiology (MA, MS, PhD); ecology and behavior (MA, MS, PhD); genetics and molecular biology (MA, MS, PhD); morphology, systematics, and evolution (MA, MS, PhD).

Degree requirements: For master's, thesis (for some programs), comprehensive exams required; for doctorate, dissertation, comprehensive exams required.
Entrance requirements: For master's and doctorate, GRE General Test, GRE Subject Test. Electronic applications accepted.
Expenses: Tuition, state resident: full-time $1,578. Tuition, nonresident: full-time $10,744. Required fees: $827. One-time fee: $15 full-time. Tuition and fees vary according to program.
Faculty research: Gene expression, biomechanics, yeast genetics, plant ecology, plant molecular biology.

Find an in-depth description at www.petersons.com/graduate.

■ UNIVERSITY OF NOTRE DAME

Graduate School, College of Science, Department of Biological Sciences, Program in Developmental Biology, Notre Dame, IN 46556

AWARDS MS, PhD. Terminal master's awarded for partial completion of doctoral program.

Degree requirements: For master's and doctorate, thesis/dissertation required, foreign language not required.

Entrance requirements: For master's and doctorate, GRE General Test, GRE Subject Test, TOEFL. *Application deadline:* For fall admission, 2/1 (priority date); for spring admission, 11/1. Applications are processed on a rolling basis. *Application fee:* $50.

Expenses: Tuition: Full-time $21,930; part-time $1,218 per credit. Required fees: $95. Tuition and fees vary according to program.

Financial aid: Fellowships with full tuition reimbursements, research assistantships with full tuition reimbursements, teaching assistantships with full tuition reimbursements, traineeships and tuition waivers (full) available. Financial aid application deadline: 2/1.

Faculty research: Developmental biology of Drosophila and zebra fish.

Application contact: Dr. Terrence J. Akai, Director of Graduate Admissions, 219-631-7706, *Fax:* 219-631-4183, *E-mail:* gradad@nd.edu.

■ UNIVERSITY OF PENNSYLVANIA

School of Arts and Sciences, Graduate Group in Biology, Program in Cell, Molecular, and Developmental Biology, Philadelphia, PA 19104

AWARDS PhD.

Students: 27 full-time (8 women); includes 9 minority (2 African Americans, 6 Asian Americans or Pacific Islanders, 1 Hispanic American), 2 international. In 1999, 5 degrees awarded.

Degree requirements: For doctorate, dissertation required, foreign language not required.

Entrance requirements: For doctorate, GRE General Test, GRE Subject Test, TOEFL. *Application fee:* $65.

Expenses: Tuition: Full-time $23,670. Required fees: $1,546. Full-time tuition and fees vary according to degree level and program.

Financial aid: Fellowships, research assistantships, teaching assistantships, tuition waivers (full and partial) available. Financial aid application deadline: 1/2.

Application contact: Allan Aiken, Graduate Group Coordinator, 215-898-6786, *Fax:* 215-898-8780, *E-mail:* aaiken@sas.upenn.edu.

■ UNIVERSITY OF PENNSYLVANIA

School of Medicine, Biomedical Graduate Studies, Graduate Group in Cell and Molecular Biology, Program in Developmental Biology, Philadelphia, PA 19104

AWARDS PhD.

Degree requirements: For doctorate, dissertation required, foreign language not required.

Entrance requirements: For doctorate, GRE General Test, TOEFL. *Application deadline:* For fall admission, 1/2 (priority date). Applications are processed on a rolling basis. *Application fee:* $65. Electronic applications accepted.

Expenses: Tuition: Full-time $17,256; part-time $2,991 per course. Required fees: $2,588; $363 per course. $726 per term.

Financial aid: Fellowships, research assistantships, teaching assistantships, institutionally sponsored loans available. Financial aid application deadline: 1/2. Dr. Jonathan Raper, Chair, 215-898-2180.

Application contact: Mary Webster, Coordinator, 215-898-4360, *E-mail:* camb@mail.med.upenn.edu.

■ UNIVERSITY OF PITTSBURGH

Faculty of Arts and Sciences, Department of Biological Sciences, Program in Molecular, Cellular, and Developmental Biology, Pittsburgh, PA 15260

AWARDS PhD.

Degree requirements: For doctorate, dissertation required, foreign language not required.

Entrance requirements: For doctorate, GRE General Test, GRE Subject Test, TOEFL. *Application deadline:* For fall admission, 2/1 (priority date). Applications are processed on a rolling basis. *Application fee:* $40.

Expenses: Tuition, state resident: full-time $8,338; part-time $342 per credit. Tuition, nonresident: full-time $17,168; part-time $707 per credit. Required fees: $480; $90 per semester. Tuition and fees vary according to program.

Find an in-depth description at www.petersons.com/graduate.

■ UNIVERSITY OF ROCHESTER

The College, Arts and Sciences, Department of Biology, Program in Genetics and Developmental Biology, Rochester, NY 14627-0250

AWARDS MS, PhD.

Degree requirements: For master's, thesis not required; for doctorate, dissertation, qualifying exam required, foreign language not required.

Entrance requirements: For master's and doctorate, GRE General Test, GRE Subject Test, TOEFL. *Application deadline:* For fall admission, 2/1 (priority date). *Application fee:* $25.

Expenses: Tuition: Part-time $697 per credit hour. Tuition and fees vary according to program.

Financial aid: Application deadline: 2/1.

Application contact: Cindy Landry, Graduate Program Secretary, 716-275-7991.

■ UNIVERSITY OF SOUTH CAROLINA

Graduate School, College of Science and Mathematics, Department of Biological Sciences, Graduate Training Program in Molecular, Cellular, and Developmental Biology, Columbia, SC 29208

AWARDS MS, PhD.

Degree requirements: For master's, one foreign language required; for doctorate, dissertation required.

Entrance requirements: For master's and doctorate, GRE General Test, minimum GPA of 3.0 in science. *Application deadline:* For fall admission, 2/15 (priority date). *Application fee:* $35.

Expenses: Tuition, state resident: full-time $4,014; part-time $202 per credit hour. Tuition, nonresident: full-time $8,528; part-time $428 per credit hour. Required fees: $100; $4 per credit hour. Tuition and fees vary according to program.

Application contact: Dr. Franklyn F. Bolander, Director of Graduate Studies, 803-777-2755, *Fax:* 803-777-4002, *E-mail:* bolander@sc.edu.

Find an in-depth description at www.petersons.com/graduate.

■ THE UNIVERSITY OF TEXAS AT AUSTIN

Graduate School, College of Natural Sciences, Institute for Cellular and Molecular Biology, Program in Genetics and Developmental Biology, Austin, TX 78712-1111

AWARDS PhD.

Expenses: Tuition, state resident: part-time $114 per semester hour. Tuition, nonresident: part-time $330 per semester hour. Tuition and fees vary according to program.

The University of Texas at Austin (continued)

Application contact: Information Contact, 512-471-1156, *Fax:* 512-471-2149.

Find an in-depth description at www.petersons.com/graduate.

■ THE UNIVERSITY OF TEXAS–HOUSTON HEALTH SCIENCE CENTER

Graduate School of Biomedical Sciences, Program in Genes and Development, Houston, TX 77225-0036

AWARDS MS, PhD, MD/PhD.

Faculty: 25 full-time (7 women).
Students: 26 full-time (15 women); includes 4 minority (all Asian Americans or Pacific Islanders), 13 international. Average age 27. *9 applicants, 67% accepted.* In 1999, 3 master's, 9 doctorates awarded. Terminal master's awarded for partial completion of doctoral program.
Degree requirements: For master's and doctorate, thesis/dissertation required, foreign language not required.
Entrance requirements: For master's and doctorate, GRE General Test, TOEFL, TWE. *Application deadline:* For fall admission, 1/15 (priority date); for spring admission, 11/1. Applications are processed on a rolling basis. *Application fee:* $10. Electronic applications accepted.
Financial aid: Fellowships, research assistantships, teaching assistantships, institutionally sponsored loans available. Financial aid application deadline: 1/15.
Faculty research: Mammalian embryogenesis, genetics, transgenic mice, tissue differentiation, sex determination.
Dr. Richard R. Behringer, Director, 713-794-4618, *Fax:* 713-794-4394, *E-mail:* rrb@notes.mdacc.tmc.edu.
Application contact: Anne Baronitis, Director of Admissions, 713-500-9860, *Fax:* 713-500-9877, *E-mail:* abaron@gsbs.gs.uth.tmc.edu.

Find an in-depth description at www.petersons.com/graduate.

■ THE UNIVERSITY OF TEXAS SOUTHWESTERN MEDICAL CENTER AT DALLAS

Southwestern Graduate School of Biomedical Sciences, Division of Cell and Molecular Biology, Program in Genetics and Development, Dallas, TX 75390

AWARDS PhD.

Faculty: 38 full-time (4 women), 2 part-time/adjunct (0 women).
Students: 50 full-time (18 women); includes 4 minority (1 Asian American or Pacific Islander, 3 Hispanic Americans), 19 international. Average age 28. In 1999, 9 doctorates awarded.
Degree requirements: For doctorate, dissertation required, foreign language not required.
Entrance requirements: For doctorate, GRE General Test, minimum GPA of 3.0. *Application deadline:* For fall admission, 1/5 (priority date). *Application fee:* $0. Electronic applications accepted.
Expenses: Tuition, state resident: full-time $912. Tuition, nonresident: full-time $6,096. Required fees: $216. Full-time tuition and fees vary according to course load and program.
Financial aid: Fellowships, research assistantships, institutionally sponsored loans available. Financial aid application deadline: 3/15; financial aid applicants required to submit FAFSA.
Faculty research: Human molecular genetics, chromosome structure, gene regulation, molecular biology, gene expression.
Dr. Dennis M. McKearin, Chair, 214-648-4944, *Fax:* 214-648-1915, *E-mail:* mckearin@utsw.swmed.edu.
Application contact: Nancy McKinney, Education Coordinator, 214-648-8099, *Fax:* 214-648-2978, *E-mail:* dcmbinfo@utsouthwestern.edu.

Find an in-depth description at www.petersons.com/graduate.

■ UNIVERSITY OF WISCONSIN–MADISON

Graduate School, Program in Cellular and Molecular Biology, Madison, WI 53706-1380

AWARDS Cellular and molecular biology (PhD); developmental biology (PhD).

Degree requirements: For doctorate, dissertation required, foreign language not required.
Entrance requirements: For doctorate, GRE General Test, GRE Subject Test. Electronic applications accepted.
Expenses: Tuition, state resident: full-time $5,406; part-time $339 per credit. Tuition, nonresident: full-time $17,110; part-time $1,071 per credit. Full-time tuition and fees vary according to program and reciprocity agreements. Part-time tuition and fees vary according to course load and program.
Faculty research: Virology, carcinogensis, cell structure, hormone regulation, developmental biology.

Find an in-depth description at www.petersons.com/graduate.

■ WASHINGTON UNIVERSITY IN ST. LOUIS

Graduate School of Arts and Sciences, Division of Biology and Biomedical Sciences, Program in Developmental Biology, St. Louis, MO 63130-4899

AWARDS PhD.

Degree requirements: For doctorate, dissertation required, foreign language not required.
Entrance requirements: For doctorate, GRE General Test, GRE Subject Test. *Application deadline:* For fall admission, 1/1 (priority date). Applications are processed on a rolling basis. *Application fee:* $0.
Expenses: Tuition: Full-time $23,400; part-time $975 per credit. Tuition and fees vary according to program.
Financial aid: Application deadline: 1/1.
Dr. David Ornitz, Coordinator.
Application contact: Rosemary Garagneni, Director of Admissions, 800-852-9074, *E-mail:* admissions@dbbs.wustl.edu.

■ WESLEYAN UNIVERSITY

Graduate Programs, Department of Biology, Middletown, CT 06459-0260

AWARDS Cell biology (PhD); comparative physiology (PhD); developmental biology (PhD); genetics (PhD); neurophysiology (PhD); population biology (PhD).

Faculty: 12 full-time (3 women).
Students: 24 full-time (12 women); includes 1 minority (African American), 11 international. Average age 28. *125 applicants, 4% accepted.* In 1999, 2 doctorates awarded.
Degree requirements: For doctorate, one foreign language (computer language can substitute), dissertation required.
Entrance requirements: For doctorate, GRE Subject Test. *Application deadline:* For fall admission, 1/15. Applications are processed on a rolling basis. *Application fee:* $0.
Expenses: Tuition: Full-time $24,876. Required fees: $650. Tuition and fees vary according to program.
Financial aid: Research assistantships, teaching assistantships, stipends available.
Faculty research: Microbial population genetics, genetic basis of evolutionary adaptation, genetic regulation of differentiation and pattern formation in *drosophila*.
Dr. Fred Cohan, Chairman, 860-685-3489.
Application contact: Marina J. Melendez, Director of Graduate Student Services, 860-685-2390, *Fax:* 860-685-2439, *E-mail:* mmelendez@wesleyan.edu.

Find an in-depth description at www.petersons.com/graduate.

■ WEST VIRGINIA UNIVERSITY

College of Agriculture, Forestry and Consumer Sciences, Interdisciplinary Program in Genetics and Developmental Biology, Morgantown, WV 26506

AWARDS Animal breeding (MS, PhD); biochemical and molecular genetics (MS, PhD); cytogenetics (MS, PhD); descriptive embryology (MS, PhD); developmental genetics (MS); experimental morphogenesis (MS); human genetics (MS, PhD); immunogenetics (MS, PhD); life cycles of animals and plants (MS, PhD); molecular aspects of development (MS, PhD); mutagenesis (PhD); mutagenetics (MS); oncology (MS, PhD); plant genetics (MS, PhD); population and quantitative genetics (PhD); population and quantitative genetics (MS); regeneration (MS, PhD); teratology (MS, PhD); toxicology (MS, PhD).

Students: 18 full-time (8 women), 5 part-time (2 women); includes 1 minority (Asian American or Pacific Islander), 8 international. Average age 27. In 1999, 4 doctorates awarded.

Degree requirements: For master's, thesis required, foreign language not required; for doctorate, dissertation, comprehensive exam required, foreign language not required.

Entrance requirements: For master's, GRE or MCAT, TOEFL, minimum GPA of 2.75; for doctorate, TOEFL. *Application fee:* $45.

Expenses: Tuition, state resident: full-time $2,910; part-time $154 per credit hour. Tuition, nonresident: full-time $8,368; part-time $457 per credit hour.

Financial aid: In 1999–00, 11 research assistantships, 3 teaching assistantships were awarded; fellowships, Federal Work-Study, institutionally sponsored loans, and tuition waivers (full and partial) also available. Financial aid application deadline: 2/1; financial aid applicants required to submit FAFSA.

Dr. J. Nath, Chairman, 304-293-6256 Ext. 4333, *E-mail:* jnath@wvu.edu.

■ YALE UNIVERSITY

Graduate School of Arts and Sciences, Department of Molecular, Cellular, and Developmental Biology, Program in Developmental Biology, New Haven, CT 06520

AWARDS PhD.

Degree requirements: For doctorate, dissertation required.

Entrance requirements: For doctorate, GRE General Test, GRE Subject Test. *Application deadline:* For fall admission, 1/4. *Application fee:* $65.

Expenses: Tuition: Full-time $22,300. Full-time tuition and fees vary according to program.

Financial aid: Fellowships, research assistantships, teaching assistantships available.

Application contact: Admissions Information, 203-432-2770.

Find an in-depth description at www.petersons.com/graduate.

■ YALE UNIVERSITY

School of Medicine and Graduate School of Arts and Sciences, Combined Program in Biological and Biomedical Sciences (BBS), Genetics and Development Track, New Haven, CT 06520

AWARDS Genetics, development, and molecular biology (PhD).

Degree requirements: For doctorate, dissertation required.

Entrance requirements: For doctorate, GRE General Test, TOEFL. *Application deadline:* For fall admission, 1/2. *Application fee:* $65. Electronic applications accepted.

Expenses: All students receive full tuition of $22,330 and an annual stipend of $17,600 .

Financial aid: Fellowships, research assistantships available.

Dr. Lynn Cooley, Director of Graduate Studies, 203-785-5067, *E-mail:* bbs.gendev@yale.edu.

Application contact: Betsy Jasiorkowski, Graduate Registrar, 203-785-5846, *E-mail:* bbs.gendev@yale.edu.

Find an in-depth description at www.petersons.com/graduate.

■ YESHIVA UNIVERSITY

Albert Einstein College of Medicine, Sue Golding Graduate Division of Medical Sciences, Department of Anatomy and Structural Biology, Bronx, NY 10461

AWARDS Anatomy (PhD); cell and developmental biology (PhD).

Faculty: 13 full-time.

Students: 14 full-time (8 women); includes 3 minority (1 African American, 2 Asian Americans or Pacific Islanders), 8 international. In 1999, 1 degree awarded.

Degree requirements: For doctorate, dissertation required, foreign language not required.

Entrance requirements: For doctorate, GRE General Test, TOEFL. *Application deadline:* For fall admission, 1/15. *Application fee:* $0. Electronic applications accepted.

Expenses: Tuition: Part-time $525 per credit. Tuition and fees vary according to degree level and program.

Financial aid: In 1999–00, 14 fellowships were awarded.

Faculty research: Cell motility, cell membranes and membrane-cytoskeletal interactions as applied to processing of pancreatic hormones, mechanisms of secretion.

Dr. Peter Satir, Chairperson, 718-430-2836.

Application contact: Sheila Cleeton, Assistant Director, 718-430-2128, *Fax:* 718-430-8655, *E-mail:* phd@aecom.yu.edu.

■ YESHIVA UNIVERSITY

Albert Einstein College of Medicine, Sue Golding Graduate Division of Medical Sciences, Division of Biological Sciences, Department of Developmental and Molecular Biology, Bronx, NY 10461

AWARDS PhD, MD/PhD.

Faculty: 12 full-time.

Students: 32 full-time (12 women); includes 6 minority (all Asian Americans or Pacific Islanders), 17 international. In 1999, 4 degrees awarded.

Degree requirements: For doctorate, dissertation required, foreign language not required.

Entrance requirements: For doctorate, GRE General Test, TOEFL. *Application deadline:* For fall admission, 1/15. *Application fee:* $0.

Expenses: Tuition: Part-time $525 per credit. Tuition and fees vary according to degree level and program.

Financial aid: In 1999–00, 32 fellowships were awarded.

Faculty research: DNA, RNA, and protein synthesis in prokaryotes and eukaryotes; chemical and enzymatic alteration of RNA; glycoproteins.

Dr. R. E. Stanley, Chairperson, 718-430-2094.

Application contact: Sheila Cleeton, Assistant Director, 718-430-2128, *Fax:* 718-430-8655, *E-mail:* phd@aecom.yu.edu.

GENETICS

■ ARIZONA STATE UNIVERSITY

Graduate College, College of Liberal Arts and Sciences, Department of Biology, Program in Genetics, Tempe, AZ 85287

AWARDS MS, PhD. Terminal master's awarded for partial completion of doctoral program.

Degree requirements: For master's, thesis required, foreign language not required; for doctorate, dissertation, oral exam required, foreign language not required.

Entrance requirements: For master's and doctorate, GRE General Test, GRE

Arizona State University (continued)
Subject Test. *Application deadline:* For fall admission, 12/15. *Application fee:* $45.
Expenses: Tuition, state resident: part-time $115 per credit hour. Tuition, nonresident: part-time $389 per credit hour. Required fees: $18 per semester. Tuition and fees vary according to program.
Financial aid: Application deadline: 12/15.
Faculty research: Molecular, developmental, and behavioral genetics; cytogenetics; population studies.

■ BAYLOR COLLEGE OF MEDICINE

Graduate School of Biomedical Sciences, Department of Molecular and Human Genetics, Houston, TX 77030-3498

AWARDS PhD, MD/PhD.

Faculty: 49 full-time (10 women).
Students: 57 full-time (32 women); includes 9 minority (1 African American, 4 Asian Americans or Pacific Islanders, 4 Hispanic Americans), 18 international. Average age 27. *127 applicants, 33% accepted.* In 1999, 7 degrees awarded.
Degree requirements: For doctorate, dissertation, public defense, qualifying exam required, foreign language not required. *Average time to degree:* Doctorate–5.4 years full-time.
Entrance requirements: For doctorate, GRE General Test (average 80th percentile), GRE Subject Test (strongly recommended), TOEFL, minimum GPA of 3.0. *Application deadline:* For fall admission, 2/1 (priority date). *Application fee:* $30. Electronic applications accepted.
Expenses: Tuition: Full-time $8,200. Required fees: $175. Full-time tuition and fees vary according to student level.
Financial aid: In 1999–00, 57 students received aid, including 30 fellowships (averaging $16,000 per year), 27 research assistantships (averaging $16,000 per year); Federal Work-Study, institutionally sponsored loans, and tuition waivers (full) also available. Financial aid applicants required to submit FAFSA.
Faculty research: Cytogenetics, biochemical genetics, somatic cell genetics, gene therapy.
Dr. Craig Chinault, Director, 713-798-5056, *E-mail:* rtanagho@bcm.tmc.edu.
Application contact: Lillie Tanagho, Graduate Program Administrator, 713-798-5056, *Fax:* 713-798-8597, *E-mail:* rtanagho@bcm.tmc.edu.
Find an in-depth description at www.petersons.com/graduate.

■ BRANDEIS UNIVERSITY

Graduate School of Arts and Sciences, Programs in Biological Sciences, Program in Molecular and Cell Biology, Waltham, MA 02454-9110

AWARDS Cell biology (PhD); developmental biology (PhD); genetics (PhD); microbiology (PhD); molecular and cell biology (MS); molecular biology (PhD); neurobiology (PhD).

Faculty: 19 full-time (8 women).
Students: 58 full-time (31 women), 1 (woman) part-time; includes 13 minority (1 African American, 8 Asian Americans or Pacific Islanders, 4 Hispanic Americans), 8 international. Average age 27. *83 applicants, 11% accepted.* In 1999, 1 master's, 8 doctorates awarded (100% entered university research/teaching). Terminal master's awarded for partial completion of doctoral program.
Degree requirements: For master's, thesis not required; for doctorate, dissertation required. *Average time to degree:* Master's–1 year full-time; doctorate–6 years full-time.
Entrance requirements: For doctorate, GRE General Test. *Application deadline:* For fall admission, 1/15 (priority date). Applications are processed on a rolling basis. *Application fee:* $60. Electronic applications accepted.
Expenses: Tuition: Full-time $25,392; part-time $3,174 per course. Required fees: $509. Tuition and fees vary according to class time, degree level, program and student level.
Financial aid: In 1999–00, 10 fellowships with tuition reimbursements (averaging $17,000 per year), 49 research assistantships with tuition reimbursements (averaging $17,000 per year), 9 teaching assistantships (averaging $2,312 per year) were awarded; scholarships and tuition waivers (full and partial) also available. Financial aid application deadline: 4/15; financial aid applicants required to submit CSS PROFILE or FAFSA.
Faculty research: Regulation of gene expression by transcription factors, molecular neurobiology, immunology, molecular mechanisms of genetic recombination, and cell differentiation.
Dr. Ranjan Sen, Chair, 781-736-2455, *Fax:* 781-736-3107.
Application contact: Marcia Cabral, Information Officer, 781-736-3100, *Fax:* 781-736-3107, *E-mail:* cabral@brandeis.edu.

■ CALIFORNIA INSTITUTE OF TECHNOLOGY

Division of Biology, Program in Genetics, Pasadena, CA 91125-0001
AWARDS PhD.

Degree requirements: For doctorate, dissertation, qualifying exam required, foreign language not required.
Entrance requirements: For doctorate, GRE General Test. *Application deadline:* For fall admission, 1/1. *Application fee:* $0.
Expenses: Tuition: Full-time $19,260. Required fees: $24. One-time fee: $100 full-time.
Financial aid: Application deadline: 1/1.
Application contact: Elizabeth Ayala, Graduate Option Coordinator, 626-395-4497, *Fax:* 626-449-0756, *E-mail:* biograd@cco.caltech.edu.

■ CARNEGIE MELLON UNIVERSITY

Mellon College of Science, Department of Biological Sciences, Program in Genetics, Pittsburgh, PA 15213-3891
AWARDS PhD. Offered in cooperation with the University of Pittsburgh.

Degree requirements: For doctorate, dissertation required, foreign language not required.
Entrance requirements: For doctorate, GRE General Test, GRE Subject Test, interview. *Application deadline:* For fall admission, 2/1 (priority date). Applications are processed on a rolling basis. *Application fee:* $0.
Expenses: Tuition: Full-time $22,100; part-time $307 per unit. Required fees: $200. Tuition and fees vary according to program.
Financial aid: Fellowships, research assistantships, traineeships available.
Application contact: Stacey L. Young, Assistant Head, 412-268-7372, *Fax:* 412-268-7129, *E-mail:* sf38+@andrew.cmu.edu.

■ CASE WESTERN RESERVE UNIVERSITY

School of Medicine and School of Graduate Studies, Graduate Programs in Medicine, Department of Genetics, Program in Human, Molecular, and Developmental Genetics and Genomics, Cleveland, OH 44106
AWARDS PhD, MD/PhD.

Faculty: 26 full-time (9 women).
Students: 33 full-time (11 women). In 1999, 5 degrees awarded (80% entered university research/teaching, 20% continued full-time study).
Degree requirements: For doctorate, dissertation required. *Average time to degree:* Doctorate–6 years full-time.
Entrance requirements: For doctorate, GRE General Test, GRE Subject Test, TOEFL. *Application deadline:* For fall admission, 2/28 (priority date). Applications are processed on a rolling basis. *Application fee:* $25.

Financial aid: Fellowships, research assistantships available. Financial aid application deadline: 2/28.

Faculty research: Regulation of gene expression, molecular control of development, genomics. *Total annual research expenditures:* $8 million.

Application contact: Chair, Graduate Committee, 216-368-5847, *Fax:* 216-368-5857.

Find an in-depth description at www.petersons.com/graduate.

■ CLEMSON UNIVERSITY

Graduate School, College of Agriculture, Forestry and Life Sciences, School of Animal, Biomedical and Biological Sciences, Department of Biological Sciences, Program in Genetics, Clemson, SC 29634

AWARDS MS, PhD. Offered in cooperation with the Department of Animal, Dairy and Veterinary Sciences and the Department of Horticulture.

Students: 24 full-time (15 women), 4 part-time (1 woman); includes 3 minority (1 African American, 1 Asian American or Pacific Islander, 1 Hispanic American), 17 international. *29 applicants, 24% accepted.* In 1999, 2 master's, 3 doctorates awarded.

Degree requirements: For master's and doctorate, thesis/dissertation required, foreign language not required.

Entrance requirements: For master's and doctorate, GRE General Test, TOEFL, minimum GPA of 3.2. *Application deadline:* For fall admission, 6/1. Applications are processed on a rolling basis. *Application fee:* $40.

Expenses: Tuition, state resident: full-time $3,480; part-time $174 per credit hour. Tuition, nonresident: full-time $9,256; part-time $388 per credit hour. Required fees: $5 per term. Full-time tuition and fees vary according to course level, course load and campus/location.

Financial aid: Fellowships, research assistantships, teaching assistantships available. Financial aid application deadline: 3/15; financial aid applicants required to submit FAFSA.

Faculty research: Animal, plant, microbial, molecular, and biometrical genetics.
Dr. A. G. Abbott, Coordinator, 864-656-2328, *Fax:* 864-656-0234, *E-mail:* aalbert@clemson.edu.

■ COLORADO STATE UNIVERSITY

Graduate School, College of Agricultural Sciences, Department of Soil and Crop Sciences, Fort Collins, CO 80523-0015

AWARDS Crop science (MS, PhD); plant genetics (MS, PhD); soil science (MS, PhD). Part-time programs available.

Faculty: 20 full-time (3 women).
Students: 18 full-time (9 women), 12 part-time (3 women), 9 international. Average age 34. *25 applicants, 48% accepted.* In 1999, 4 master's, 6 doctorates awarded.

Degree requirements: For master's, computer language, thesis required (for some programs), foreign language not required; for doctorate, one foreign language, computer language, dissertation required.

Entrance requirements: For master's, GRE General Test, TOEFL, minimum GPA of 3.0, appropriate bachelor's degree; for doctorate, GRE General Test, TOEFL, minimum GPA of 3.0, appropriate master's degree. *Application deadline:* For fall admission, 2/1 (priority date). Applications are processed on a rolling basis. *Application fee:* $30. Electronic applications accepted.

Expenses: Tuition, state resident: full-time $2,694; part-time $150 per credit. Tuition, nonresident: full-time $10,460; part-time $581 per credit. Required fees: $32 per semester. Tuition and fees vary according to program.

Financial aid: In 1999–00, 1 fellowship, 10 research assistantships, 1 teaching assistantship were awarded; career-related internships or fieldwork and traineeships also available.

Faculty research: Water quality, soil fertility, soil/plant ecosystems, plant breeding and genetics, crop physiology. *Total annual research expenditures:* $3 million.
Dr. James S. Quick, Head, 970-491-6517, *Fax:* 970-491-0564.

Application contact: Dr. Dan H. Smith, Graduate Studies Coordinator, 970-491-6371, *Fax:* 970-491-0564, *E-mail:* dsmith@ceres.agsci.colostate.edu.

■ COLUMBIA UNIVERSITY

College of Physicians and Surgeons and Graduate School of Arts and Sciences, Graduate School of Arts and Sciences at the College of Physicians and Surgeons, Department of Genetics and Development, New York, NY 10032

AWARDS Genetics (M Phil, MA, PhD). Only candidates for the PhD are admitted. Terminal master's awarded for partial completion of doctoral program.

Degree requirements: For master's, foreign language and thesis not required; for doctorate, dissertation required, foreign language not required.

Entrance requirements: For master's and doctorate, GRE General Test, TOEFL.
Expenses: Tuition: Full-time $25,072.
Faculty research: Mammalian cell differentiation and meiosis, developmental genetics, yeast and human genetics, chromosome structure, molecular and cellular biology.

■ CORNELL UNIVERSITY

Graduate School, Graduate Fields of Agriculture and Life Sciences, Field of Genetics and Development, Ithaca, NY 14853-0001

AWARDS Developmental biology (PhD); genetics (PhD).

Faculty: 31 full-time.
Students: 34 full-time (17 women); includes 4 minority (1 African American, 1 Asian American or Pacific Islander, 1 Hispanic American, 1 Native American), 8 international. *95 applicants, 25% accepted.* In 1999, 9 doctorates awarded.

Degree requirements: For doctorate, dissertation, 2 semesters of teaching experience required, foreign language not required.

Entrance requirements: For doctorate, GRE General Test, GRE Subject Test (biology or biochemistry) (recommended). *Application deadline:* For fall admission, 1/5. *Application fee:* $65. Electronic applications accepted.

Expenses: Tuition: Full-time $12,100.
Financial aid: In 1999–00, 33 students received aid, including 17 fellowships with full tuition reimbursements available, 11 research assistantships with full tuition reimbursements available, 5 teaching assistantships with full tuition reimbursements available; institutionally sponsored loans, scholarships, tuition waivers (full and partial), and unspecified assistantships also available. Financial aid applicants required to submit FAFSA.

Faculty research: Molecular and general genetics, developmental genetics, evolution and population genetics, plant genetics, microbial genetics.

Application contact: Graduate Field Assistant, 607-254-4840, *E-mail:* gendev@cornell.edu.

Find an in-depth description at www.petersons.com/graduate.

■ DUKE UNIVERSITY

Graduate School, Department of Biochemistry, Durham, NC 27708-0586

AWARDS Crystallography of macromolecules (PhD); enzyme mechanisms (PhD); lipid biochemistry (PhD); membrane structure and

Duke University (continued)
function (PhD); molecular genetics (PhD); neurochemistry (PhD); nucleic acid structure and function (PhD); protein structure and function (PhD).

Faculty: 31 full-time, 10 part-time/adjunct.
Students: 51 full-time (17 women); includes 3 minority (2 Asian Americans or Pacific Islanders, 1 Hispanic American), 10 international. *106 applicants, 19% accepted.* In 1999, 6 doctorates awarded.
Degree requirements: For doctorate, dissertation required, foreign language not required.
Entrance requirements: For doctorate, GRE General Test, GRE Subject Test (recommended). *Application deadline:* For fall admission, 12/31. *Application fee:* $75.
Expenses: Tuition: Full-time $21,406; part-time $760 per unit. Required fees: $3,136; $3,136 per year. One-time fee: $30. Tuition and fees vary according to program.
Financial aid: Fellowships, research assistantships, teaching assistantships, Federal Work-Study available. Financial aid application deadline: 12/31.
Terry Oas, Director of Graduate Studies, 919-681-8770, *Fax:* 919-684-8885, *E-mail:* pwilkison@biochem.duke.edu.

Find an in-depth description at www.petersons.com/graduate.

■ DUKE UNIVERSITY

Graduate School, Program in Genetics, Durham, NC 27708-0586

AWARDS PhD.

Faculty: 56 full-time, 3 part-time/adjunct.
Students: 19 full-time (13 women); includes 4 minority (2 African Americans, 2 Hispanic Americans), 3 international. *104 applicants, 28% accepted.* In 1999, 2 degrees awarded.
Degree requirements: For doctorate, dissertation required.
Entrance requirements: For doctorate, GRE General Test, GRE Subject Test (recommended). *Application deadline:* For fall admission, 12/31. *Application fee:* $75.
Expenses: Tuition: Full-time $21,406; part-time $760 per unit. Required fees: $3,136; $3,136 per year. One-time fee: $30. Tuition and fees vary according to program.
Financial aid: Fellowships available. Financial aid application deadline: 12/31.
Dr. Mariano Garcia-Blanco, Director of Graduate Studies, 919-684-6629, *Fax:* 919-613-8646, *E-mail:* genetics@biochem.duke.edu.

Find an in-depth description at www.petersons.com/graduate.

■ EMORY UNIVERSITY

Graduate School of Arts and Sciences, Division of Biological and Biomedical Sciences, Program in Genetics and Molecular Biology, Atlanta, GA 30322-1100

AWARDS PhD.

Faculty: 41 full-time (10 women).
Students: 44 full-time (26 women); includes 2 African Americans, 2 Hispanic Americans, 2 international. In 1999, 10 degrees awarded.
Degree requirements: For doctorate, dissertation required, foreign language not required.
Entrance requirements: For doctorate, GRE General Test, TOEFL, minimum GPA of 3.0 in science course work. *Application deadline:* For fall admission, 1/20 (priority date). *Application fee:* $45.
Expenses: Tuition: Full-time $22,770. Tuition and fees vary according to program.
Financial aid: In 1999–00, fellowships with full tuition reimbursements (averaging $18,000 per year).
Faculty research: Gene regulation, genetic combination, developmental regulation.
Dr. Jeremy Boss, Director, 404-727-5973, *Fax:* 404-727-3659, *E-mail:* boss@microbio.emory.edu.
Application contact: 404-727-2547, *Fax:* 404-727-3322, *E-mail:* gdbbs@gsas.emory.edu.

Find an in-depth description at www.petersons.com/graduate.

■ EMORY UNIVERSITY

Graduate School of Arts and Sciences, Division of Biological and Biomedical Sciences, Program in Microbiology and Molecular Genetics, Atlanta, GA 30322-1100

AWARDS PhD.

Faculty: 20 full-time (3 women).
Students: 30 full-time (15 women); includes 1 Native American, 3 international. In 1999, 7 degrees awarded.
Degree requirements: For doctorate, dissertation required, foreign language not required.
Entrance requirements: For doctorate, GRE General Test, TOEFL, minimum GPA of 3.0 in science course work. *Application deadline:* For fall admission, 1/20 (priority date). *Application fee:* $45.
Expenses: Tuition: Full-time $22,770. Tuition and fees vary according to program.
Financial aid: In 1999–00, fellowships with full tuition reimbursements (averaging $18,000 per year)

Dr. Charles Moran, Director, 404-727-5969, *Fax:* 404-727-3659, *E-mail:* moran@microbio.emory.edu.
Application contact: 404-727-2547, *Fax:* 404-727-3322, *E-mail:* gdbbs@gsas.emory.edu.

Find an in-depth description at www.petersons.com/graduate.

■ FLORIDA STATE UNIVERSITY

Graduate Studies, College of Arts and Sciences, Department of Biological Science, Program in Genetics, Tallahassee, FL 32306

AWARDS MS, PhD.

Faculty: 16 full-time (3 women).
Students: 34 full-time (20 women); includes 1 minority (Hispanic American), 11 international.
Degree requirements: For master's and doctorate, thesis/dissertation, teaching experience required.
Entrance requirements: For master's, GRE General Test, TOEFL; for doctorate, GRE General Test, GRE Subject Test, TOEFL. *Application deadline:* For fall admission, 1/15; for spring admission, 10/15. *Application fee:* $20.
Expenses: Tuition, state resident: full-time $3,504; part-time $146 per credit hour. Tuition, nonresident: full-time $12,162; part-time $507 per credit hour. Tuition and fees vary according to program.
Financial aid: In 1999–00, fellowships with full tuition reimbursements (averaging $13,740 per year), research assistantships with full tuition reimbursements (averaging $13,740 per year), teaching assistantships with full tuition reimbursements (averaging $13,740 per year) were awarded. Financial aid application deadline: 1/15; financial aid applicants required to submit FAFSA.
Faculty research: Population genetics, fertilization, eukaryotic gene expression, biotechnology, molecular biology.
Dr. Thomas C. S. Keller, Associate Professor and Associate Chairman, 850-644-3023, *Fax:* 850-644-9829.
Application contact: Judy Bowers, Coordinator, Graduate Affairs, 850-644-3023, *Fax:* 850-644-9829, *E-mail:* bowers@bio.fsu.edu.

■ THE GEORGE WASHINGTON UNIVERSITY

Columbian School of Arts and Sciences, Institute for Biomedical Sciences, Program in Genetics, Washington, DC 20052

AWARDS MS, PhD. Part-time and evening/weekend programs available.

Faculty: 1 (woman) full-time, 1 part-time/adjunct (0 women).

Students: 18 full-time (8 women), 47 part-time (35 women); includes 10 minority (1 African American, 6 Asian Americans or Pacific Islanders, 3 Hispanic Americans), 8 international. Average age 31. *17 applicants, 59% accepted.* In 1999, 4 master's, 6 doctorates awarded. Terminal master's awarded for partial completion of doctoral program.
Degree requirements: For master's, thesis, comprehensive exam required; for doctorate, dissertation, general exam required.
Entrance requirements: For master's and doctorate, GRE General Test, interview, minimum GPA of 3.0. *Application fee:* $55.
Expenses: Tuition: Full-time $16,836; part-time $702 per credit hour. Required fees: $828; $35 per credit hour. Tuition and fees vary according to campus/location and program.
Financial aid: In 1999–00, 16 students received aid, including 12 fellowships; Federal Work-Study and institutionally sponsored loans also available. Financial aid application deadline: 2/1.
Dr. Diana Johnson, Director, 202-994-7120.

Find an in-depth description at www.petersons.com/graduate.

■ **GEORGIA STATE UNIVERSITY**

College of Arts and Sciences, Department of Biology, Program in Molecular Genetics and Biochemistry, Atlanta, GA 30303-3083

AWARDS MS, PhD.

Degree requirements: For master's, one foreign language (computer language can substitute), thesis or alternative, exam required; for doctorate, dissertation required.
Entrance requirements: For master's and doctorate, GRE General Test, TOEFL, minimum GPA of 3.0. *Application deadline:* For fall admission, 7/18; for spring admission, 2/13. Applications are processed on a rolling basis. *Application fee:* $25.
Expenses: Tuition, state resident: full-time $2,896; part-time $121 per credit hour. Tuition, nonresident: full-time $11,584; part-time $483 per credit hour. Required fees: $228. Full-time tuition and fees vary according to course load and program.
Financial aid: Application deadline: 2/6.
Application contact: Latesha Morrison, Graduate Administrative Coordinator, 404-651-2759, *Fax:* 404-651-2509, *E-mail:* biolxm@langate.gsu.edu.

Find an in-depth description at www.petersons.com/graduate.

■ **HARVARD UNIVERSITY**

Graduate School of Arts and Sciences, Program in Biological and Biomedical Sciences, Department of Genetics, Boston, MA 02115

AWARDS PhD. Applications through the Program in Biological and Biomedical Sciences (BBS).

Degree requirements: For doctorate, dissertation, qualifying exam required, foreign language not required.
Entrance requirements: For doctorate, GRE General Test, GRE Subject Test, TOEFL. *Application deadline:* For fall admission, 12/15. *Application fee:* $60.
Expenses: Tuition: Full-time $22,054. Required fees: $711. Tuition and fees vary according to program.
Financial aid: Fellowships, research assistantships, teaching assistantships, institutionally sponsored loans and tuition waivers (full) available. Financial aid application deadline: 1/1.
Faculty research: Complex immunoglobulin loci in humans, nitrogen-fixing genes of plant-associated bacteria, gene regulation.
Dr. Philip Leder, Chair, 617-432-7667.
Application contact: Leah Simons, Manager of Student Affairs, 617-432-0162.

Find an in-depth description at www.petersons.com/graduate.

■ **HARVARD UNIVERSITY**

Graduate School of Arts and Sciences, Program in Biological and Biomedical Sciences, Department of Microbiology and Molecular Genetics, Boston, MA 02115

AWARDS PhD. Applications through the Program in Biological and Biomedical Sciences (BBS).

Degree requirements: For doctorate, dissertation required, foreign language not required.
Entrance requirements: For doctorate, GRE General Test, GRE Subject Test, TOEFL. *Application fee:* $60.
Expenses: Tuition: Full-time $22,054. Required fees: $711. Tuition and fees vary according to program.
Financial aid: Fellowships, research assistantships, teaching assistantships available. Financial aid application deadline: 1/1.
Faculty research: Bacterial physiology, cell biology, embryology.
Dr. John J. Mekalanos, Chairman, 617-432-2263.
Application contact: Leah Simons, Manager of Student Affairs, 617-432-0162.

Find an in-depth description at www.petersons.com/graduate.

■ **HOWARD UNIVERSITY**

Graduate School of Arts and Sciences, Department of Genetics and Human Genetics, Washington, DC 20059

AWARDS MS, PhD. Part-time programs available.

Faculty: 13.
Students: 23; includes 16 minority (15 African Americans, 1 Asian American or Pacific Islander), 6 international. In 1999, 11 master's, 2 doctorates awarded.
Degree requirements: For master's, thesis, comprehensive exam required; for doctorate, dissertation, comprehensive exam required. *Average time to degree:* Master's–2 years full-time; doctorate–4 years full-time.
Entrance requirements: For master's, GRE General Test, TOEFL, minimum GPA of 3.0, previous course work in biology, chemistry, physics, mathematics, and genetics; for doctorate, GRE General Test, TOEFL, master's degree in genetics or related field, minimum GPA of 3.2. *Application deadline:* For fall admission, 4/1; for spring admission, 11/1. *Application fee:* $45.
Expenses: Tuition: Full-time $10,500; part-time $583 per credit hour. Required fees: $405; $203 per semester.
Financial aid: Fellowships, research assistantships, teaching assistantships available.
Dr. Robert F. Murray, Chairman, 202-806-6340.
Application contact: Dr. Verle E. Headings, Director of Graduate Studies, 202-806-6381.

■ **ILLINOIS STATE UNIVERSITY**

Graduate School, College of Arts and Sciences, Department of Biological Sciences, Normal, IL 61790-2200

AWARDS Biological sciences (MS); biology (PhD); botany (PhD); ecology (PhD); genetics (PhD); microbiology (PhD); physiology (PhD); zoology (PhD). Part-time programs available.

Faculty: 24 full-time (4 women).
Students: 57 full-time (25 women), 27 part-time (10 women); includes 6 minority (1 African American, 3 Asian Americans or Pacific Islanders, 2 Hispanic Americans), 13 international. *58 applicants, 59% accepted.* In 1999, 12 master's, 3 doctorates awarded.
Degree requirements: For master's, thesis or alternative required; for doctorate, variable foreign language requirement (computer language can substitute for one), dissertation, 2 terms of residency required.
Entrance requirements: For master's, GRE General Test, minimum GPA of 2.6 in last 60 hours; for doctorate, GRE

Illinois State University (continued)
General Test. *Application deadline:* Applications are processed on a rolling basis. *Application fee:* $0.

Expenses: Tuition, state resident: full-time $2,526; part-time $105 per credit hour. Tuition, nonresident: full-time $7,578; part-time $316 per credit hour. Required fees: $1,082; $38 per credit hour. Tuition and fees vary according to course load and program.

Financial aid: In 1999–00, 8 research assistantships, 63 teaching assistantships were awarded; Federal Work-Study, tuition waivers (full), and unspecified assistantships also available. Financial aid application deadline: 4/1.

Faculty research: Phenotypic plasticity in reproduction: molecular mechanisms, physiological control, adaptive significance; molecular stress physiology of *Listeria monocytogenes*; enzymology of eggshell formation in *Schistosoma mansoni*, compensatory adaptation in the rat midbrain after neurodegeneration, analysis of staphylococcal virulence germs. *Total annual research expenditures:* $586,648.
Dr. Hou Cheung, Chairperson, 309-438-3669.

Application contact: Derek A. McCracken, Graduate Adviser, 309-438-3664.

Find an in-depth description at www.petersons.com/graduate.

■ INDIANA UNIVERSITY BLOOMINGTON

Graduate School, College of Arts and Sciences, Department of Biology, Program in Genetics, Bloomington, IN 47405

AWARDS PhD. Offered through the University Graduate School.

Students: 54 full-time (21 women); includes 4 minority (2 Asian Americans or Pacific Islanders, 2 Hispanic Americans), 9 international. In 1999, 1 degree awarded.
Degree requirements: For doctorate, dissertation required.
Entrance requirements: For doctorate, GRE General Test, TOEFL. *Application deadline:* For fall admission, 1/5 (priority date); for spring admission, 9/1 (priority date). Applications are processed on a rolling basis. *Application fee:* $45. Electronic applications accepted.
Expenses: Tuition, state resident: full-time $3,853; part-time $161 per credit hour. Tuition, nonresident: full-time $11,226; part-time $468 per credit hour. Required fees: $360 per year. Tuition and fees vary according to course load and program.
Financial aid: Fellowships with tuition reimbursements, research assistantships with tuition reimbursements, teaching

assistantships with tuition reimbursements, grants available. Financial aid application deadline: 1/15.
Faculty research: Transmission genetics of *Drosophila* and maize, population genetics, viral hybrid DNA techniques, cytogenetics, molecular genetics.
Dr. Susan Strome, Head, 812-855-5450, *Fax:* 812-855-6705, *E-mail:* sstrome@bio.indiana.edu.
Application contact: Gretchen Clearwater, Advisor for Graduate Affairs, 812-855-1861, *Fax:* 812-855-6705, *E-mail:* biograd@bio.indiana.edu.

Find an in-depth description at www.petersons.com/graduate.

■ INDIANA UNIVERSITY–PURDUE UNIVERSITY INDIANAPOLIS

School of Medicine, Graduate Programs in Medicine, Department of Medical and Molecular Genetics, Indianapolis, IN 46202-2896

AWARDS MS, PhD, MD/MS, MD/PhD. Part-time programs available.

Students: 13 full-time (12 women), 7 part-time (3 women), 2 international. Average age 28. In 1999, 9 master's awarded (63% found work related to degree, 37% continued full-time study); 3 doctorates awarded. Terminal master's awarded for partial completion of doctoral program.
Degree requirements: For master's, thesis optional, foreign language not required; for doctorate, dissertation, research ethics required, foreign language not required. *Average time to degree:* Master's–2.2 years full-time.
Entrance requirements: For master's and doctorate, GRE General Test, minimum GPA of 3.2. *Application deadline:* For fall admission, 1/15 (priority date). *Application fee:* $35 ($55 for international students).
Expenses: Tuition, state resident: full-time $13,245; part-time $158 per credit hour. Tuition, nonresident: full-time $30,330; part-time $455 per credit hour. Required fees: $121 per year. Tuition and fees vary according to course load and degree level.
Financial aid: In 1999–00, 11 students received aid, including 4 fellowships with tuition reimbursements available (averaging $15,167 per year), 7 research assistantships with tuition reimbursements available (averaging $15,000 per year); Federal Work-Study and institutionally sponsored loans also available. Aid available to part-time students. Financial aid application deadline: 1/15.
Faculty research: Twins, human gene mapping, chromosomes and malignancy, clinical genetics. *Total annual research expenditures:* $2.1 million.
Dr. Gail Vance, Chairman, 317-274-2241, *E-mail:* ghvance@iupui.edu.

Application contact: Kathleen Wilhelm, Admissions Secretary, 317-274-2241, *Fax:* 317-274-2387, *E-mail:* medgen@iupui.edu.

Find an in-depth description at www.petersons.com/graduate.

■ IOWA STATE UNIVERSITY OF SCIENCE AND TECHNOLOGY

Graduate College, College of Agriculture and College of Liberal Arts and Sciences, Department of Biochemistry, Biophysics, and Molecular Biology, Ames, IA 50011

AWARDS Biochemistry (MS, PhD); biophysics (MS, PhD); genetics (MS, PhD); molecular, cellular, and developmental biology (MS, PhD); toxicology (MS, PhD).

Faculty: 19 full-time, 1 part-time/adjunct.
Students: 57 full-time (24 women), 6 part-time (1 woman); includes 1 minority (African American), 20 international. *22 applicants, 59% accepted.* In 2000, 4 master's, 4 doctorates awarded.
Degree requirements: For master's and doctorate, thesis/dissertation required.
Entrance requirements: For master's and doctorate, GRE General Test, TOEFL. *Application deadline:* For fall admission, 6/15 (priority date); for spring admission, 11/15 (priority date). *Application fee:* $20 ($50 for international students). Electronic applications accepted.
Expenses: Tuition, state resident: full-time $3,308. Tuition, nonresident: full-time $9,744. Part-time tuition and fees vary according to course load, campus/location and program.
Financial aid: In 2000–01, 44 research assistantships with partial tuition reimbursements (averaging $12,314 per year), 2 teaching assistantships with partial tuition reimbursements (averaging $12,375 per year) were awarded; scholarships also available.
Dr. Marit Nilsen-Hamilton, Chair, 515-294-2231, *E-mail:* biochem@iastate.edu.

Find an in-depth description at www.petersons.com/graduate.

■ IOWA STATE UNIVERSITY OF SCIENCE AND TECHNOLOGY

Graduate College, College of Liberal Arts and Sciences and College of Agriculture, Department of Zoology and Genetics, Ames, IA 50011

AWARDS MS, PhD.

Faculty: 31 full-time, 7 part-time/adjunct.
Students: 59 full-time (27 women), 6 part-time (2 women); includes 2 minority (1 African American, 1 Hispanic American), 30 international. *12 applicants, 50% accepted.* In 1999, 3 doctorates awarded.

Degree requirements: For master's and doctorate, thesis/dissertation required.
Entrance requirements: For master's and doctorate, GRE General Test, TOEFL. *Application deadline:* For fall admission, 2/1 (priority date). *Application fee:* $20 ($50 for international students). Electronic applications accepted.
Expenses: Tuition, state resident: full-time $3,308. Tuition, nonresident: full-time $9,744. Part-time tuition and fees vary according to course load, campus/location and program.
Financial aid: In 1999–00, 65 students received aid, including 1 fellowship with full tuition reimbursement available (averaging $20,000 per year), 34 research assistantships with partial tuition reimbursements available (averaging $15,000 per year), 18 teaching assistantships with partial tuition reimbursements available (averaging $13,000 per year); scholarships also available. Financial aid application deadline: 2/1.
Faculty research: Animal behavior, animal models of gene therapy, cell biology, comparative physiology, developmental biology.
Dr. M. Duane Enger, Chair, 515-294-3908, *Fax:* 515-294-8457, *E-mail:* zg@iastate.edu.
Application contact: Dr. Eileen Muff, Graduate Coordinator, 515-294-3909, *E-mail:* zg@iastate.edu.

Find an in-depth description at www.petersons.com/graduate.

■ **IOWA STATE UNIVERSITY OF SCIENCE AND TECHNOLOGY**
Graduate College, Interdisciplinary Programs, Bioinformatics and Computational Biology Program, Ames, IA 50011-3260
AWARDS MS, PhD.

Faculty: 46 full-time (9 women).
Students: 40.
Degree requirements: For master's and doctorate, thesis/dissertation required, foreign language not required.
Entrance requirements: For master's and doctorate, GRE General Test (waived for applicants with PhD), TOEFL. *Application deadline:* For fall admission, 2/1 (priority date). Electronic applications accepted.
Expenses: Tuition, state resident: full-time $3,308. Tuition, nonresident: full-time $9,744. Part-time tuition and fees vary according to course load, campus/location and program.
Financial aid: In 1999–00, 7 fellowships with full tuition reimbursements (averaging $23,000 per year), 8 research assistantships with full tuition reimbursements (averaging $16,800 per year) were awarded.

Faculty research: Functional and structural genomics, genome evoultion, macromolecular structure and function, mathematical biology and computational modeling, metabolic and developmental networks.
Drena L. Dobbs, Chair, 515-294-1112, *Fax:* 515-294-6790, *E-mail:* ddobbs@iastate.edu.
Application contact: Kathy Wiederin, Program Assistant, 888-569-8509, *Fax:* 515-294-6790, *E-mail:* bioinformatics@iastate.edu.

Find an in-depth description at www.petersons.com/graduate.

■ **IOWA STATE UNIVERSITY OF SCIENCE AND TECHNOLOGY**
Graduate College, Interdisciplinary Programs, Program in Genetics, Ames, IA 50011
AWARDS MS, PhD.

Students: 16 full-time (6 women), 1 part-time, 11 international. *76 applicants, 28% accepted.* In 1999, 7 master's, 4 doctorates awarded. Terminal master's awarded for partial completion of doctoral program.
Degree requirements: For master's and doctorate, thesis/dissertation required.
Entrance requirements: For master's and doctorate, GRE General Test, TOEFL. *Application deadline:* For fall admission, 2/1 (priority date). Applications are processed on a rolling basis. *Application fee:* $20 ($50 for international students).
Expenses: Tuition, state resident: full-time $3,308. Tuition, nonresident: full-time $9,744. Part-time tuition and fees vary according to course load, campus/location and program.
Financial aid: In 1999–00, 16 research assistantships with partial tuition reimbursements (averaging $11,990 per year) were awarded; fellowships, scholarships also available.
Dr. Steven Rodermel, Supervisory Committee Chair, 515-294-7697, *Fax:* 515-294-6669, *E-mail:* genetics@iastate.edu.

Find an in-depth description at www.petersons.com/graduate.

■ **JOAN AND SANFORD I. WEILL MEDICAL COLLEGE AND GRADUATE SCHOOL OF MEDICAL SCIENCES OF CORNELL UNIVERSITY**
Graduate School of Medical Sciences, Cell Biology and Genetics Graduate Program, New York, NY 10021
AWARDS PhD, MD/PhD.
Faculty: 41 full-time (7 women).

Students: 44 full-time (24 women); includes 12 minority (3 African Americans, 8 Asian Americans or Pacific Islanders, 1 Hispanic American), 9 international. *91 applicants, 23% accepted.* In 2000, 6 doctorates awarded.
Degree requirements: For doctorate, dissertation, final exam required.
Entrance requirements: For doctorate, GRE General Test, GRE Subject Test, MCAT (MD/PhD). *Application deadline:* For fall admission, 1/15. *Application fee:* $50.
Expenses: All students in good standing receive an annual stipend of $22,880.
Financial aid: Fellowships, stipends available.
Dr. Leonard Freedman, Director, 212-639-2976.

■ **JOHNS HOPKINS UNIVERSITY**
School of Hygiene and Public Health, Department of Epidemiology, Baltimore, MD 21203
AWARDS Chronic disease epidemiology (MHS, Sc M, Dr PH, PhD, Sc D); clinical epidemiology (MHS, Sc M, Dr PH, PhD, Sc D); epidemiology (MHS, Sc M, Dr PH, PhD, Sc D); genetics (MHS, Sc M, Dr PH, PhD, Sc D); infectious disease epidemiology (MHS, Sc M, Dr PH, PhD, Sc D); occupational/environmental epidemiology (MHS, Sc M, Dr PH, PhD, Sc D).

Faculty: 57 full-time, 61 part-time/adjunct.
Students: 169 (115 women); includes 42 minority (10 African Americans, 28 Asian Americans or Pacific Islanders, 4 Hispanic Americans) 37 international. *245 applicants, 33% accepted.* In 1999, 20 master's, 20 doctorates awarded.
Degree requirements: For master's, thesis required (for some programs), foreign language not required; for doctorate, dissertation, 1 year full-time residency, oral and written exams required, foreign language not required.
Entrance requirements: For master's and doctorate, GRE General Test, TOEFL. *Application deadline:* For fall admission, 2/1 (priority date). Applications are processed on a rolling basis. *Application fee:* $60. Electronic applications accepted.
Expenses: Tuition: Full-time $23,660; part-time $493 per unit. Full-time tuition and fees vary according to degree level, campus/location and program.
Financial aid: Federal Work-Study, institutionally sponsored loans, and scholarships available. Aid available to part-time students. Financial aid application deadline: 4/15.
Faculty research: Cancer and congenital malformations, nutritional epidemiology, AIDS, tuberculosis, cardiovascular disease,

Johns Hopkins University (continued)
risk assessment. *Total annual research expenditures:* $24.8 million.
Dr. Jonathan Samet, Chairman, 410-955-3286, *Fax:* 410-955-0863, *E-mail:* jsamet@jhsph.edu.
Application contact: Dr. Adolfo Correa, Chair, Admissions/Credentials Committee, 410-955-3483, *Fax:* 410-955-0863, *E-mail:* acorrea@jhsph.edu.

■ JOHNS HOPKINS UNIVERSITY

School of Medicine, Graduate Programs in Medicine, Department of Molecular Biology and Genetics, Baltimore, MD 21218-2699

AWARDS PhD. PhD awarded through the Program in Biochemistry, Cellular and Molecular Biology.

Faculty: 25 full-time (6 women).
Students: 48 full-time (23 women); includes 6 minority (1 African American, 5 Asian Americans or Pacific Islanders), 14 international. In 1999, 9 doctorates awarded.
Degree requirements: For doctorate, dissertation, oral comprehensive exams required, foreign language not required.
Entrance requirements: For doctorate, GRE General Test, GRE Subject Test. *Application deadline:* For fall admission, 2/1. *Application fee:* $50.
Expenses: Tuition: Full-time $23,660.
Financial aid: Traineeships available.
Faculty research: Molecular genetics of animal cells, yeasts, *Drosophila*, and viruses; segregation.
Dr. Thomas J. Kelly, Chairman, 410-955-3292.
Application contact: Dr. Daniel Raben, Admissions Chairman, 410-955-3506, *Fax:* 410-614-8842, *E-mail:* pantol@jhmi.edu.

■ JOHNS HOPKINS UNIVERSITY

Zanvyl Krieger School of Arts and Sciences, Department of Biology, Baltimore, MD 21218-2699

AWARDS Biochemistry (PhD); cell biology (PhD); developmental biology (PhD); genetic biology (PhD); molecular biology (PhD).

Faculty: 23 full-time (4 women).
Students: 85 full-time (50 women); includes 23 minority (4 African Americans, 14 Asian Americans or Pacific Islanders, 5 Hispanic Americans), 18 international. Average age 24. *202 applicants, 27% accepted.* In 1999, 17 doctorates awarded.
Degree requirements: For doctorate, dissertation required. *Average time to degree:* Doctorate–6 years full-time.
Entrance requirements: For doctorate, GRE General Test, GRE Subject Test. *Application deadline:* For fall admission,

12/15 (priority date). Applications are processed on a rolling basis. *Application fee:* $55.
Expenses: Tuition: Full-time $24,930. Tuition and fees vary according to program.
Financial aid: In 1999–00, 73 students received aid, including 12 fellowships, 61 research assistantships, 12 teaching assistantships; Federal Work-Study and institutionally sponsored loans also available. Financial aid application deadline: 12/15; financial aid applicants required to submit FAFSA.
Faculty research: Protein and nucleic acid biochemistry and biophysical chemistry, molecular biology and development. *Total annual research expenditures:* $9.3 million.
Dr. Victor G. Corces, Chair, 410-516-4693, *Fax:* 410-516-5213, *E-mail:* corces_v@jhuvms.hct.jhu.edu.
Application contact: Joan Miller, Graduate Admissions Coordinator, 410-516-5502, *Fax:* 410-516-5213, *E-mail:* joan@jhu.edu.

Find an in-depth description at www.petersons.com/graduate.

■ KANSAS STATE UNIVERSITY

Graduate School, College of Agriculture, Department of Animal Sciences and Industry, Manhattan, KS 66506

AWARDS Animal nutrition (MS, PhD); animal reproduction (MS, PhD); animal sciences and industry (MS, PhD); genetics (MS, PhD); meat science (MS, PhD).

Degree requirements: For master's and doctorate, thesis/dissertation required, foreign language not required.
Expenses: Tuition, state resident: part-time $103 per credit hour. Tuition, nonresident: part-time $338 per credit hour. Required fees: $17 per credit hour. One-time fee: $64 part-time.

■ KANSAS STATE UNIVERSITY

Graduate School, College of Arts and Sciences, Division of Biology, Manhattan, KS 66506

AWARDS Cell biology (MS, PhD); developmental biology and physiology (MS, PhD); microbiology and immunology (MS, PhD); molecular biology and genetics (MS, PhD); systematics and ecology (MS, PhD); virology and oncology (MS, PhD). Terminal master's awarded for partial completion of doctoral program.

Degree requirements: For master's and doctorate, thesis/dissertation required, foreign language not required.
Entrance requirements: For master's and doctorate, GRE General Test. Electronic applications accepted.

Expenses: Tuition, state resident: part-time $103 per credit hour. Tuition, nonresident: part-time $338 per credit hour. Required fees: $17 per credit hour. One-time fee: $64 part-time.
Faculty research: Immune cell function, prairie ecology.

■ KANSAS STATE UNIVERSITY

Graduate School, Program in Genetics, Manhattan, KS 66506

AWARDS MS, PhD. Terminal master's awarded for partial completion of doctoral program.

Degree requirements: For master's, computer language, thesis required, foreign language not required; for doctorate, one foreign language, computer language, dissertation required.
Entrance requirements: For master's and doctorate, GRE General Test.
Expenses: Tuition, state resident: part-time $103 per credit hour. Tuition, nonresident: part-time $338 per credit hour. Required fees: $17 per credit hour. One-time fee: $64 part-time.
Faculty research: Plant transformation, yeast genetics, molecular genetics of insects, molecular cytogenetics.

■ MARQUETTE UNIVERSITY

Graduate School, College of Arts and Sciences, Department of Biology, Milwaukee, WI 53201-1881

AWARDS Cell biology (MS, PhD); developmental biology (MS, PhD); ecology (MS, PhD); endocrinology (MS, PhD); evolutionary biology (MS, PhD); genetics (MS, PhD); microbiology (MS, PhD); molecular biology (MS, PhD); muscle and exercise physiology (MS, PhD); neurobiology (MS, PhD); reproductive physiology (MS, PhD).

Faculty: 16 full-time (4 women), 2 part-time/adjunct (0 women).
Students: 34 full-time (20 women), 3 part-time; includes 3 minority (all Asian Americans or Pacific Islanders), 2 international. Average age 31. *42 applicants, 29% accepted.* In 1999, 1 master's, 4 doctorates awarded. Terminal master's awarded for partial completion of doctoral program.
Degree requirements: For master's, thesis, 1 year of teaching experience or equivalent, comprehensive exam required, foreign language not required; for doctorate, dissertation, 1 year of teaching experience or equivalent, qualifying exam required, foreign language not required.
Entrance requirements: For master's and doctorate, GRE General Test, GRE Subject Test, TOEFL. *Application fee:* $40.

Expenses: Tuition: Part-time $510 per credit hour. Tuition and fees vary according to program.

Financial aid: In 1999–00, 4 fellowships, 22 teaching assistantships were awarded; research assistantships, Federal Work-Study, institutionally sponsored loans, scholarships, and tuition waivers (full and partial) also available. Aid available to part-time students. Financial aid application deadline: 2/15.

Faculty research: Microbial and invertebrate ecology, evolution of gene function, DNA methylation, DNA arrangement. *Total annual research expenditures:* $1.5 million.

Dr. Brian Unsworth, Chairman, 414-288-7355, *Fax:* 414-288-7357.

Application contact: Barbara DeNoyer, Graduate Studies Coordinator, 414-288-7355, *Fax:* 414-288-7357.

Find an in-depth description at www.petersons.com/graduate.

■ MASSACHUSETTS INSTITUTE OF TECHNOLOGY

School of Science, Department of Biology, Program in Genetics, Cambridge, MA 02139-4307

AWARDS PhD.

Degree requirements: For doctorate, dissertation, general exam required, foreign language not required.

Entrance requirements: For doctorate, GRE General Test. *Application deadline:* For fall admission, 1/1. *Application fee:* $55.

Expenses: Tuition: Full-time $25,000. Full-time tuition and fees vary according to degree level, program and student level.

Financial aid: Application deadline: 1/1.

Application contact: Dr. Janice D. Chang, Educational Administrator, 617-253-3717, *Fax:* 617-258-9329, *E-mail:* gradbio@mit.edu.

■ MCP HAHNEMANN UNIVERSITY

School of Medicine, Biomedical Graduate Programs, Interdisciplinary Program in Molecular and Human Genetics, Philadelphia, PA 19102-1192

AWARDS MS, PhD, MD/PhD.

Degree requirements: For master's, comprehensive exam required; for doctorate, dissertation, qualifying exam required, foreign language not required.

Entrance requirements: For master's and doctorate, GRE General Test, TOEFL.

Find an in-depth description at www.petersons.com/graduate.

■ MEDICAL COLLEGE OF OHIO

Graduate School, Department of Biochemistry and Molecular Biology, Toledo, OH 43614-5805

AWARDS Medical sciences (MS), including biochemistry, genetics, molecular biology. Part-time programs available.

Faculty: 8 full-time (2 women).

Students: 4 full-time (2 women), 8 part-time (1 woman), 9 international. Average age 30. In 1999, 3 degrees awarded (100% found work related to degree).

Degree requirements: For master's, thesis, qualifying exam required, foreign language not required. *Average time to degree:* Master's–3 years part-time.

Entrance requirements: For master's, GRE General Test, TOEFL, minimum undergraduate GPA of 3.0. *Application fee:* $30.

Expenses: Tuition, state resident: part-time $193 per hour. Tuition, nonresident: part-time $445 per hour. Tuition and fees vary according to degree level.

Financial aid: Fellowships, Federal Work-Study and institutionally sponsored loans available. Financial aid applicants required to submit FAFSA.

Faculty research: Gene regulation, protein structure, receptors, protein phosphorylation, peptides. *Total annual research expenditures:* $81,820.

James P. Trempe, Interim Chairman, 419-383-4117, *Fax:* 419-383-6140, *E-mail:* mcogradschool@mco.edu.

Application contact: Joann Braatz, Clerk, 419-383-4117, *Fax:* 419-383-6140, *E-mail:* mcogradschool@mco.edu.

■ MEDICAL COLLEGE OF WISCONSIN

Graduate School of Biomedical Sciences, Department of Microbiology and Molecular Genetics, Milwaukee, WI 53226-0509

AWARDS MS, PhD, MD/MS, MD/PhD. Terminal master's awarded for partial completion of doctoral program.

Degree requirements: For master's and doctorate, thesis/dissertation required, foreign language not required.

Entrance requirements: For master's and doctorate, GRE General Test, TOEFL.

Expenses: Tuition, state resident: full-time $9,318. Tuition, nonresident: full-time $9,318. Required fees: $115.

Faculty research: Virology, immunology, bacterial toxins, regulation of gene expression.

■ MICHIGAN STATE UNIVERSITY

Graduate School, College of Natural Science, Genetics Program, East Lansing, MI 48824

AWARDS Genetics (PhD); genetics-environmental toxicology (PhD).

Faculty: 73.

Students: 42 (25 women); includes 3 minority (2 African Americans, 1 Asian American or Pacific Islander) 25 international. *17 applicants, 47% accepted.*

Degree requirements: For doctorate, dissertation required, foreign language not required.

Entrance requirements: For doctorate, GRE General Test, GRE Subject Test. *Application deadline:* For fall admission, 2/1 (priority date). Applications are processed on a rolling basis. *Application fee:* $30 ($40 for international students).

Expenses: Tuition, state resident: part-time $229 per credit. Tuition, nonresident: part-time $464 per credit. Required fees: $241 per semester. Tuition and fees vary according to course load, degree level and program.

Financial aid: In 1999–00, 8 research assistantships (averaging $19,058 per year) were awarded; fellowships, teaching assistantships with tuition reimbursements Financial aid applicants required to submit FAFSA.

Faculty research: Gene expression, RNA structure, molecular endocrinology.

Dr. Rebecca Grumet, Director, 517-353-9845, *E-mail:* genetics@msu.edu.

Application contact: Jeannine Lee, Information Contact, 517-353-9845, *Fax:* 517-355-0112, *E-mail:* genetics@msu.edu.

Find an in-depth description at www.petersons.com/graduate.

■ MISSISSIPPI STATE UNIVERSITY

College of Agriculture and Life Sciences, Program in Genetics, Mississippi State, MS 39762

AWARDS MS. Part-time programs available.

Faculty: 24 full-time (2 women).

Students: 7 full-time (3 women), 4 international. Average age 25. *6 applicants, 50% accepted.* In 1999, 1 degree awarded.

Degree requirements: For master's, thesis required (for some programs), foreign language not required.

Entrance requirements: For master's, GRE General Test, TOEFL, minimum GPA of 2.75. *Application deadline:* For fall admission, 7/1; for spring admission, 11/1. Applications are processed on a rolling basis. *Application fee:* $0 ($25 for international students).

Expenses: Tuition, state resident: full-time $3,017; part-time $168 per credit. Tuition,

Mississippi State University (continued)
nonresident: full-time $6,119; part-time $340 per credit. Part-time tuition and fees vary according to course load and program.

Financial aid: In 1999–00, 5 students received aid, including 5 research assistantships; teaching assistantships, Federal Work-Study, institutionally sponsored loans, and unspecified assistantships also available. Financial aid application deadline: 6/1; financial aid applicants required to submit FAFSA.

Faculty research: Host plant resistance, disease resistance, catfish genetics, cytogenetics, genetic engineering.
Dr. E. David Peebles, Graduate Coordinator, 662-325-3379, *Fax:* 662-325-8292, *E-mail:* dpeebles@poultry.msstate.edu.

Application contact: Jerry B. Inmon, Director of Admissions, 662-325-2224, *Fax:* 662-325-7360, *E-mail:* admit@admissions.msstate.edu.

■ MOUNT SINAI SCHOOL OF MEDICINE OF NEW YORK UNIVERSITY

Graduate School of Biological Sciences, Genetics and Genomic Sciences (GGS) Training Area, New York, NY 10029-6504

AWARDS PhD, MD/PhD.

Students: 8 full-time (3 women).
Degree requirements: For doctorate, dissertation required, foreign language not required.
Entrance requirements: For doctorate, GRE General Test, GRE Subject Test, MCAT, TOEFL. *Application deadline:* For fall admission, 4/15. *Application fee:* $60.
Expenses: Tuition: Full-time $21,750; part-time $725 per credit. Required fees: $750; $25 per credit. Full-time tuition and fees vary according to student level.
Financial aid: Fellowships with full tuition reimbursements, grants available.
Dr. Edward Schuchman, Program Director, *E-mail:* schuchman@msuak.mssm.edu.
Application contact: C. Gita Bosch, Administrative Manager and Assistant Dean, 212-241-6546, *Fax:* 212-241-0651, *E-mail:* grads@mssm.edu.

■ NEW YORK UNIVERSITY

Graduate School of Arts and Science, Department of Biology, New York, NY 10012-1019

AWARDS Applied recombinant DNA technology (MS); biochemistry (PhD); biomedical journalism (MA); cell biology (PhD); computers in biological research (MS); environmental biology (PhD); general biology (MS); neural sciences and physiology (PhD); oral biology (MS); population and evolutionary biology (PhD). Part-time programs available.

Faculty: 22 full-time (5 women), 8 part-time/adjunct.
Students: 99 full-time (48 women), 61 part-time (31 women); includes 30 minority (4 African Americans, 22 Asian Americans or Pacific Islanders, 4 Hispanic Americans), 52 international. Average age 24. *371 applicants, 41% accepted.* In 1999, 54 master's, 4 doctorates awarded. Terminal master's awarded for partial completion of doctoral program.
Degree requirements: For master's, thesis or alternative, qualifying paper required, foreign language not required; for doctorate, dissertation, oral and written comprehensive exams required, foreign language not required.
Entrance requirements: For master's, GRE General Test, TOEFL; for doctorate, GRE General Test, GRE Subject Test, TOEFL. *Application deadline:* For fall admission, 1/4 (priority date). *Application fee:* $60.
Expenses: Tuition: Full-time $17,880; part-time $745 per credit. Required fees: $1,140; $35 per credit. Tuition and fees vary according to course load and program.
Financial aid: Fellowships with tuition reimbursements, research assistantships with tuition reimbursements, teaching assistantships with tuition reimbursements, career-related internships or fieldwork, Federal Work-Study, institutionally sponsored loans, and tuition waivers (full and partial) available. Financial aid application deadline: 1/4; financial aid applicants required to submit FAFSA.
Faculty research: Development and genetics, neurobiology, plant sciences, molecular and cell biology.
Philip Furmanski, Chairman, 212-998-8200.
Application contact: Gloria Coruzzi, Director of Graduate Studies, 212-998-8200, *Fax:* 212-995-4015, *E-mail:* biology@nyu.edu.

Find an in-depth description at www.petersons.com/graduate.

■ NORTH CAROLINA STATE UNIVERSITY

Graduate School, College of Agriculture and Life Sciences, Department of Genetics, Raleigh, NC 27695

AWARDS MS, PhD.

Faculty: 17 full-time (2 women), 8 part-time/adjunct (1 woman).
Students: 26 full-time (19 women), 4 part-time (2 women); includes 1 minority (Hispanic American), 7 international. Average age 29. *51 applicants, 22% accepted.* In 1999, 1 master's, 9 doctorates awarded. Terminal master's awarded for partial completion of doctoral program.
Degree requirements: For master's and doctorate, thesis/dissertation required, foreign language not required.
Entrance requirements: For master's and doctorate, GRE General Test, minimum GPA of 3.0. *Application deadline:* For fall admission, 6/25; for spring admission, 11/25. *Application fee:* $45.
Expenses: Tuition, state resident: full-time $1,578. Tuition, nonresident: full-time $10,744. Required fees: $892. Full-time tuition and fees vary according to program.
Financial aid: In 1999–00, 8 fellowships (averaging $6,173 per year), 21 research assistantships (averaging $5,316 per year) were awarded; teaching assistantships Financial aid application deadline: 6/25.
Faculty research: Population and quantitative genetics, plant molecular genetics, developmental genetics,. *Total annual research expenditures:* $4.5 million.
Dr. Stephanie E. Curtis, Head, 919-515-2291, *Fax:* 919-515-3355, *E-mail:* securtis@ncsu.edu.

Find an in-depth description at www.petersons.com/graduate.

■ NORTHWESTERN UNIVERSITY

The Graduate School, Division of Interdepartmental Programs and Medical School, Integrated Graduate Programs in the Life Sciences, Chicago, IL 60611

AWARDS Cancer biology (PhD); cell biology (PhD); developmental biology (PhD); evolutionary biology (PhD); immunology and microbial pathogenesis (PhD); molecular biology and genetics (PhD); neurobiology (PhD); pharmacology and toxicology (PhD); structural biology and biochemistry (PhD).

Degree requirements: For doctorate, dissertation, written and oral qualifying exams required, foreign language not required.
Entrance requirements: For doctorate, GRE General Test, TOEFL.
Expenses: Tuition: Full-time $23,301. Full-time tuition and fees vary according to program.

Find an in-depth description at www.petersons.com/graduate.

■ NORTHWESTERN UNIVERSITY

The Graduate School, Judd A. and Marjorie Weinberg College of Arts and Sciences, Interdepartmental Biological Sciences Program (IBiS), Concentration in Genetics and Developmental Biology, Evanston, IL 60208

AWARDS PhD.

Faculty: 59 full-time (11 women).
Students: 79 full-time (46 women); includes 5 minority (2 African Americans, 3 Hispanic Americans). *236 applicants, 19% accepted.*
Degree requirements: For doctorate, dissertation, 2 quarters of teaching experience required, foreign language not required.
Entrance requirements: For doctorate, GRE General Test, TOEFL. *Application deadline:* For fall admission, 1/15. Applications are processed on a rolling basis. *Application fee:* $50 ($55 for international students).
Expenses: Tuition: Full-time $23,301. Full-time tuition and fees vary according to program.
Financial aid: In 1999–00, 15 fellowships (averaging $12,078 per year), 64 research assistantships (averaging $17,000 per year), 15 teaching assistantships (averaging $16,620 per year) were awarded; Federal Work-Study, institutionally sponsored loans, traineeships, and Howard Hughes Fellowship also available. Financial aid applicants required to submit FAFSA.
Application contact: Latonia Trimuel, Program Assistant, 800-546-1761, *E-mail:* ibis@northwestern.edu.

■ THE OHIO STATE UNIVERSITY

College of Medicine and Public Health and Graduate School, Graduate Programs in the Basic Medical Sciences, Department of Molecular Virology, Immunology and Medical Genetics, Columbus, OH 43210
AWARDS MS, PhD, MD/PhD.

Faculty: 30 full-time (5 women).
Students: 24 full-time (10 women), 10 international. Average age 27. *160 applicants, 8% accepted.* In 1999, 1 master's, 3 doctorates awarded (100% entered university research/teaching). Terminal master's awarded for partial completion of doctoral program.
Degree requirements: For master's, thesis optional, foreign language not required; for doctorate, dissertation required, foreign language not required. *Average time to degree:* Master's–4 years part-time; doctorate–5.6 years full-time.
Entrance requirements: For master's and doctorate, GRE General Test. *Application deadline:* For fall admission, 1/15. Applications are processed on a rolling basis. *Application fee:* $30 ($40 for international students). Electronic applications accepted.
Expenses: Tuition, state resident: full-time $5,400. Tuition, nonresident: full-time $14,535. Part-time tuition and fees vary according to course load.
Financial aid: In 1999–00, 1 fellowship with full tuition reimbursement (averaging $19,260 per year), 8 research assistantships with full tuition reimbursements (averaging

$15,600 per year), 7 teaching assistantships with full tuition reimbursements (averaging $15,600 per year) were awarded; grants also available. Financial aid application deadline: 1/15.
Faculty research: Virology, bacteriology, molecular biology. *Total annual research expenditures:* $912,708.
Dr. Caroline C. Whitacre, Chairperson, 614-292-5889, *Fax:* 614-292-9805, *E-mail:* whitacre.3@osu.edu.
Application contact: Dr. Deborah S. Parris, Chairman, Graduate Studies Committee, 614-292-0735, *Fax:* 614-292-9805, *E-mail:* mvimg@osu.edu.

Find an in-depth description at www.petersons.com/graduate.

■ THE OHIO STATE UNIVERSITY

Graduate School, College of Biological Sciences, Department of Molecular Genetics, Columbus, OH 43210
AWARDS Cell and developmental biology (MS, PhD); genetics (MS, PhD); molecular biology (MS, PhD).

Faculty: 18 full-time, 6 part-time/adjunct.
Students: 46 full-time (17 women), 3 part-time (1 woman); includes 3 minority (2 African Americans, 1 Asian American or Pacific Islander), 21 international. *68 applicants, 28% accepted.* In 1999, 6 master's, 8 doctorates awarded.
Degree requirements: For master's and doctorate, thesis/dissertation required, foreign language not required.
Entrance requirements: For master's and doctorate, GRE General Test, GRE Subject Test (biology). *Application deadline:* For fall admission, 8/15. Applications are processed on a rolling basis. *Application fee:* $30 ($40 for international students).
Expenses: Tuition, state resident: full-time $5,400. Tuition, nonresident: full-time $14,535. Part-time tuition and fees vary according to course load and program.
Financial aid: Fellowships, research assistantships, teaching assistantships, Federal Work-Study and institutionally sponsored loans available. Aid available to part-time students.
Dr. Lee F. Johnson, Chairman, 614-292-8084, *Fax:* 614-292-4466, *E-mail:* johnson.6@osu.edu.

Find an in-depth description at www.petersons.com/graduate.

■ OKLAHOMA STATE UNIVERSITY

Graduate College, College of Arts and Sciences, Department of Microbiology and Molecular Genetics, Stillwater, OK 74078
AWARDS MS, PhD.

Faculty: 6 full-time (0 women).

Students: 9 full-time (4 women), 15 part-time (8 women); includes 2 minority (both Hispanic Americans), 11 international. Average age 25. In 1999, 1 master's, 2 doctorates awarded.
Degree requirements: For master's and doctorate, thesis/dissertation required.
Entrance requirements: For master's and doctorate, GRE General Test, TOEFL. *Application deadline:* For fall admission, 6/1. *Application fee:* $25.
Expenses: Tuition, state resident: part-time $86 per credit hour. Tuition, nonresident: part-time $275 per credit hour. Required fees: $17 per credit hour. $14 per semester. One-time fee: $20 full-time. Tuition and fees vary according to course load.
Financial aid: In 1999–00, 6 research assistantships (averaging $12,015 per year), 20 teaching assistantships (averaging $13,500 per year) were awarded; tuition waivers (full) also available. Financial aid application deadline: 3/1.
Faculty research: Bioinformatics, genomics-genetics, virology, environmental microbiology, development-molecular mechanisms.
Dr. James T. Blankemeyer, Head, 405-744-6243.

■ OREGON HEALTH SCIENCES UNIVERSITY

School of Medicine, Graduate Programs in Medicine, Department of Molecular and Medical Genetics, Portland, OR 97201-3098
AWARDS PhD.

Faculty: 27 full-time (11 women), 24 part-time/adjunct (9 women).
Degree requirements: For doctorate, dissertation required, foreign language not required.
Entrance requirements: For doctorate, GRE General Test, TOEFL. *Application deadline:* For fall admission, 9/1. *Application fee:* $60.
Expenses: Tuition, state resident: full-time $3,132; part-time $174 per credit hour. Tuition, nonresident: full-time $5,256; part-time $292 per credit hour. Required fees: $8.5 per credit hour. $146 per term. Part-time tuition and fees vary according to course load.
Financial aid: Fellowships, research assistantships, teaching assistantships, career-related internships or fieldwork, Federal Work-Study, institutionally sponsored loans, and scholarships available. Financial aid application deadline: 3/1; financial aid applicants required to submit FAFSA.
Faculty research: Molecular studies of metabolic diseases, gene therapy, control

Oregon Health Sciences University (continued)

of mycogenesis, regulation of gene expression, DNA replication and repair. *Total annual research expenditures:* $2.8 million.
Dr. Robb E. Moses, Chairperson, 503-494-6881, *Fax:* 503-494-6882.
Application contact: 503-494-7703, *Fax:* 503-494-6886.

Find an in-depth description at www.petersons.com/graduate.

■ OREGON STATE UNIVERSITY

Graduate School, College of Agricultural Sciences, Program in Genetics, Corvallis, OR 97331

AWARDS MA, MAIS, MS, PhD. Part-time programs available.

Faculty: 12 full-time (2 women).
Students: 11 full-time (4 women), 2 part-time (1 woman), 8 international. Average age 30. In 1999, 2 master's awarded. Terminal master's awarded for partial completion of doctoral program.
Degree requirements: For master's, variable foreign language requirement, thesis or alternative required; for doctorate, dissertation required, foreign language not required. *Average time to degree:* Master's–2 years full-time; doctorate–3 years full-time, 4 years part-time.
Entrance requirements: For master's and doctorate, GRE General Test, TOEFL, minimum GPA of 3.0 in last 90 hours. *Application deadline:* For fall admission, 3/1. Applications are processed on a rolling basis. *Application fee:* $50.
Expenses: Tuition, state resident: full-time $6,489. Tuition, nonresident: full-time $11,061. Tuition and fees vary according to program.
Financial aid: Fellowships, research assistantships, teaching assistantships, Federal Work-Study and institutionally sponsored loans available. Financial aid application deadline: 2/1.
Faculty research: Molecular genetics, cytogenetics, population and quantitative genetics, microbial genetics, plant genetics. Dr. Warren E. Kronstad, Co-Director, 541-737-3728, *Fax:* 541-737-3132, *E-mail:* warren.e.kronstad@orst.edu.

■ THE PENNSYLVANIA STATE UNIVERSITY MILTON S. HERSHEY MEDICAL CENTER

Graduate School, Department of Microbiology and Immunology, Hershey, PA 17033-2360

AWARDS Genetics (PhD); immunology (MS, PhD); microbiology (MS); microbiology/virology (PhD); molecular biology (PhD).

Students: 19 full-time (6 women). Average age 27. In 1999, 2 master's, 3 doctorates awarded.
Degree requirements: For doctorate, dissertation required.
Entrance requirements: For doctorate, GRE General Test, TOEFL, minimum GPA of 3.0. *Application deadline:* For fall admission, 7/26. *Application fee:* $50.
Expenses: Tuition, state resident: full-time $6,886; part-time $291 per credit. Tuition, nonresident: full-time $14,118; part-time $588 per credit. Required fees: $43 per semester. Part-time tuition and fees vary according to course load.
Financial aid: Unspecified assistantships available.
Dr. Richard J. Courtney, Chair, 717-531-7659.
Application contact: Dr. Brian Wigdahl, Chairman, Graduate Program Committee, 717-531-6682, *Fax:* 717-531-6522, *E-mail:* eneidigh@cor-mail.biochem.hmc.psu.edu.

■ THE PENNSYLVANIA STATE UNIVERSITY MILTON S. HERSHEY MEDICAL CENTER

Graduate School, Intercollege Graduate Program in Genetics, Hershey, PA 17033-2360

AWARDS MS, PhD.

Students: 1 (woman) full-time. Average age 27. In 1999, 1 doctorate awarded.
Entrance requirements: For master's and doctorate, GRE General Test. *Application deadline:* For fall admission, 7/26. *Application fee:* $50.
Expenses: Tuition, state resident: full-time $6,886; part-time $291 per credit. Tuition, nonresident: full-time $14,118; part-time $588 per credit. Required fees: $43 per semester. Part-time tuition and fees vary according to course load.
Dr. Guy Barbato, Chairman, 814-865-5202.

■ THE PENNSYLVANIA STATE UNIVERSITY UNIVERSITY PARK CAMPUS

Graduate School, Intercollege Graduate Programs, Intercollege Graduate Program in Genetics, State College, University Park, PA 16802-1503

AWARDS MS, PhD.

Students: 16 full-time (9 women), 5 part-time (3 women). In 1999, 4 master's, 6 doctorates awarded.
Entrance requirements: For master's and doctorate, GRE General Test. *Application fee:* $50.
Expenses: Tuition, state resident: full-time $6,886; part-time $291 per credit. Tuition,

nonresident: full-time $14,118; part-time $588 per credit. Required fees: $46 per semester. Part-time tuition and fees vary according to course load and program.
Dr. Guy F. Barbato, Head, 814-865-5202.

Find an in-depth description at www.petersons.com/graduate.

■ PURDUE UNIVERSITY

Graduate School, School of Science, Department of Biological Sciences, Program in Genetics, West Lafayette, IN 47907

AWARDS MS, PhD.

Faculty: 50 full-time (10 women).
Students: 8 full-time (2 women), 3 international. Average age 25. *17 applicants, 24% accepted.* Terminal master's awarded for partial completion of doctoral program.
Degree requirements: For master's, foreign language and thesis not required; for doctorate, dissertation, preliminary and qualifying exams, seminars, teaching experience required, foreign language not required. *Average time to degree:* Doctorate–5.2 years full-time.
Entrance requirements: For master's, GRE General Test, TOEFL; for doctorate, GRE General Test, TOEFL, TSE. *Application deadline:* For fall admission, 2/15 (priority date). *Application fee:* $30. Electronic applications accepted.
Expenses: Tuition, state resident: full-time $4,530; part-time $130 per credit hour. Tuition, nonresident: full-time $15,310; part-time $404 per credit hour. Tuition and fees vary according to campus/location and program.
Financial aid: In 1999–00, fellowships with partial tuition reimbursements (averaging $15,000 per year), 6 research assistantships with partial tuition reimbursements (averaging $16,800 per year), 2 teaching assistantships with partial tuition reimbursements (averaging $16,800 per year) were awarded. Financial aid application deadline: 2/15.
Faculty research: Molecular genetics, oncogenes, transgenics, plant genetics, mutagenesis, host-parasite interactions, gene regulatory mechanisms, signal transduction, functional genomics.
S. B. Gelvin, Head, 765-494-4939, *Fax:* 765-496-1475.
Application contact: Nancy Konopka, Graduate Studies Office Manager, 765-494-8142, *Fax:* 765-494-0876, *E-mail:* njk@bilbo.bio.purdue.edu.

Find an in-depth description at www.petersons.com/graduate.

■ RUTGERS, THE STATE UNIVERSITY OF NEW JERSEY, NEW BRUNSWICK

Graduate School, Program in Microbiology and Molecular Genetics, New Brunswick, NJ 08901-1281

AWARDS Applied microbiology (MS, PhD); clinical microbiology (MS, PhD); computational molecular biology (PhD); immunology (MS, PhD); microbial biochemistry (MS, PhD); molecular genetics (MS, PhD); virology (MS, PhD). Part-time programs available.

Faculty: 122 full-time (25 women), 4 part-time/adjunct (1 woman).
Students: 65 full-time (31 women), 51 part-time (24 women); includes 24 minority (14 Asian Americans or Pacific Islanders, 10 Hispanic Americans), 29 international. Average age 27. *200 applicants, 26% accepted.* In 1999, 15 master's, 10 doctorates awarded. Terminal master's awarded for partial completion of doctoral program.
Degree requirements: For master's, thesis or alternative required, foreign language not required; for doctorate, dissertation, written qualifying exam required, foreign language not required.
Entrance requirements: For master's, GRE General Test, GRE Subject Test; for doctorate, GRE General Test, GRE Subject Test, TOEFL, minimum GPA of 3.0. *Application deadline:* For fall admission, 2/15 (priority date). Applications are processed on a rolling basis. *Application fee:* $50.
Expenses: Tuition, state resident: full-time $6,776; part-time $279 per credit. Tuition, nonresident: full-time $9,936; part-time $412 per credit. Required fees: $20 per credit. $89 per semester. Tuition and fees vary according to course load, campus/location and program.
Financial aid: In 1999–00, 70 students received aid, including 32 fellowships with full tuition reimbursements available, 26 research assistantships with full tuition reimbursements available, 12 teaching assistantships with full tuition reimbursements available; institutionally sponsored loans also available. Financial aid application deadline: 2/15; financial aid applicants required to submit FAFSA.
Faculty research: Biochemistry of microbes, medical microbiology.
Dr. Howard Passmore, Director, 732-445-2812, *E-mail:* passmore@biology.rutgers.edu.
Application contact: Betty Green, Administrative Assistant, 732-445-5086, *Fax:* 732-445-5735, *E-mail:* green@mbcl.rutgers.edu.

■ STANFORD UNIVERSITY

School of Medicine, Graduate Programs in Medicine, Department of Genetics, Stanford, CA 94305-9991

AWARDS PhD.

Faculty: 10 full-time (3 women).
Students: 26 full-time (10 women), 12 part-time (6 women); includes 9 minority (1 African American, 4 Asian Americans or Pacific Islanders, 4 Hispanic Americans), 2 international. Average age 26. *84 applicants, 17% accepted.*
Degree requirements: For doctorate, dissertation required.
Entrance requirements: For doctorate, GRE General Test, GRE Subject Test (biology or chemistry), TOEFL. *Application deadline:* For fall admission, 12/15. *Application fee:* $65 ($80 for international students). Electronic applications accepted.
Expenses: Tuition: Full-time $23,058. Required fees: $152. Part-time tuition and fees vary according to course load.
Financial aid: Research assistantships available. Financial aid application deadline: 12/15.
Faculty research: Molecular biology of DNA replication in human cells, analysis of existing and search for new DNA polymorphisms in humans, molecular genetics of prokaryotic and eukaryotic genetic elements, proteins in DNA replication.
David Botstein, Chair, 650-723-3488, *Fax:* 650-723-7016, *E-mail:* botstein@genome.stanford.edu.
Application contact: Admissions Office, 650-723-2460.

Find an in-depth description at www.petersons.com/graduate.

■ STATE UNIVERSITY OF NEW YORK AT ALBANY

School of Public Health, Department of Biomedical Sciences, Program in Biochemistry, Molecular Biology, and Genetics, Albany, NY 12222-0001

AWARDS MS, PhD.

Degree requirements: For master's and doctorate, thesis/dissertation required.
Entrance requirements: For master's and doctorate, GRE General Test, GRE Subject Test. *Application deadline:* For fall admission, 1/15 (priority date); for spring admission, 11/1 (priority date). *Application fee:* $50.
Expenses: Tuition, state resident: full-time $5,100; part-time $214 per credit. Tuition, nonresident: full-time $8,416; part-time $352 per credit. Required fees: $31 per credit.
Financial aid: Application deadline: 2/1.
Dr. Harry Taber, Chair, Department of Biomedical Sciences, 518-474-2662.

■ STATE UNIVERSITY OF NEW YORK AT STONY BROOK

Graduate School, College of Arts and Sciences, Program in Genetics, Stony Brook, NY 11794

AWARDS PhD.

Faculty: 77.
Students: 32 full-time (12 women), 21 part-time (9 women); includes 13 minority (2 African Americans, 6 Asian Americans or Pacific Islanders, 4 Hispanic Americans, 1 Native American), 10 international. Average age 26. *118 applicants, 26% accepted.* In 1999, 4 degrees awarded.
Degree requirements: For doctorate, dissertation, comprehensive exam, teaching experience required, foreign language not required.
Entrance requirements: For doctorate, GRE General Test, GRE Subject Test, TOEFL. *Application deadline:* For fall admission, 1/15. *Application fee:* $50.
Expenses: Tuition, state resident: full-time $5,100; part-time $213 per credit hour. Tuition, nonresident: full-time $8,416; part-time $351 per credit hour. Required fees: $492. Tuition and fees vary according to program.
Financial aid: Fellowships, research assistantships, teaching assistantships, Federal Work-Study available.
Faculty research: Gene structure, gene regulation.
Dr. Sidney Strickland, Director, 631-632-8812, *Fax:* 631-632-9797, *E-mail:* sid@pharm.som.sunysb.edu.

Find an in-depth description at www.petersons.com/graduate.

■ STATE UNIVERSITY OF NEW YORK AT STONY BROOK

Health Sciences Center, School of Medicine and Graduate School, Graduate Programs in Medicine, Department of Molecular Genetics and Microbiology, Stony Brook, NY 11794

AWARDS Molecular microbiology (PhD).

Faculty: 28.
Students: 22 full-time (12 women), 11 part-time (4 women); includes 5 minority (2 African Americans, 2 Asian Americans or Pacific Islanders, 1 Hispanic American), 13 international. Average age 26. 77 *applicants, 19% accepted.* In 1999, 8 degrees awarded.
Degree requirements: For doctorate, dissertation, comprehensive exam required.
Entrance requirements: For doctorate, GRE General Test, GRE Subject Test, TOEFL. *Application deadline:* For fall admission, 1/15. *Application fee:* $50.
Expenses: Tuition, state resident: full-time $5,100. Tuition, nonresident: full-time $8,416. Required fees: $492.

State University of New York at Stony Brook (continued)

Financial aid: In 1999–00, 18 fellowships, 1 teaching assistantship were awarded; research assistantships, Federal Work-Study also available. Financial aid application deadline: 3/15.

Faculty research: Adenovirus molecular genetics, molecular biology of tumors, virus SV40, mechanism of tumor infection by SAV virus. *Total annual research expenditures:* $4.7 million.

Dr. Michael Hayman, Acting Chair, 631-632-8813, *Fax:* 631-632-9797.

Application contact: Dr. Patrick Hearing, Director, 631-632-8813, *Fax:* 631-632-9797, *E-mail:* hearing@asterix.bio.sunysb.edu.

Find an in-depth description at www.petersons.com/graduate.

■ TEMPLE UNIVERSITY

Health Sciences Center, School of Medicine and Graduate School, Graduate Programs in Medicine, Program in Molecular Biology and Genetics, Philadelphia, PA 19140

AWARDS PhD, MD/PhD.

Faculty: 7 full-time (2 women).
Students: 27 full-time (15 women), 2 international.
Degree requirements: For doctorate, dissertation, presentation of research and literature seminars, including a research proposal distinct from the student's area of concentration required.
Entrance requirements: For doctorate, GRE General Test, GRE Subject Test (biology, chemistry, or molecular cell biology recommended). *Application deadline:* For fall admission, 3/1. *Application fee:* $40. Electronic applications accepted.
Expenses: Tuition, state resident: full-time $6,030. Tuition, nonresident: full-time $8,298. Required fees: $230. One-time fee: $10 full-time.
Financial aid: Fellowships, research assistantships, Federal Work-Study, institutionally sponsored loans, and tuition waivers (full) available. Financial aid application deadline: 3/1.
Faculty research: Molecular genetics of normal and malignant cell growth, regulation of gene expression, DNA repair systems and carcinogenesis, hormone-receptor interactions and signal transduction systems, structural biology.
Dr. E. Premkumar Reddy, Chair, 215-707-4307, *Fax:* 215-707-4588, *E-mail:* reddy@sgi1.fels.temple.edu.
Application contact: Scott K. Shore, Graduate Chair, 215-707-4302, *Fax:* 215-707-4588, *E-mail:* sks@unix.temple.edu.

Find an in-depth description at www.petersons.com/graduate.

■ TEXAS A&M UNIVERSITY

College of Agriculture and Life Sciences, Department of Animal Science, College Station, TX 77843

AWARDS Animal breeding (MS, PhD); animal science (M Agr, MS, PhD); dairy science (M Agr, MS); food science and technology (MS, PhD); genetics (MS, PhD); nutrition (MS, PhD); physiology of reproduction (MS, PhD).

Faculty: 73 full-time (11 women), 18 part-time/adjunct (4 women).
Students: 188 full-time (108 women), 4 part-time (2 women); includes 9 minority (2 Asian Americans or Pacific Islanders, 7 Hispanic Americans), 36 international. Average age 26. *158 applicants, 25% accepted.* In 1999, 43 master's awarded (40% found work related to degree, 60% continued full-time study); 9 doctorates awarded (60% entered university research/teaching, 40% found other work related to degree).
Degree requirements: For master's and doctorate, thesis/dissertation required, foreign language not required. *Average time to degree:* Master's–2 years full-time; doctorate–4 years full-time.
Entrance requirements: For master's and doctorate, GRE General Test, TOEFL. *Application deadline:* For fall admission, 2/1 (priority date); for spring admission, 10/1 (priority date). Applications are processed on a rolling basis. *Application fee:* $50 ($75 for international students).
Expenses: Tuition, state resident: part-time $76 per semester hour. Tuition, nonresident: part-time $292 per semester hour. Required fees: $11 per semester hour. Tuition and fees vary according to program.
Financial aid: In 1999–00, 5 fellowships (averaging $12,000 per year), 39 research assistantships (averaging $11,250 per year), 33 teaching assistantships (averaging $11,250 per year) were awarded; career-related internships or fieldwork, Federal Work-Study, institutionally sponsored loans, and scholarships also available. Financial aid application deadline: 2/1; financial aid applicants required to submit FAFSA.
Faculty research: Genetic engineering/gene markers, dietary effects on colon cancer, biotechnology.
Dr. Bryan H. Johnson, Head, 979-845-1541.
Application contact: Ronnie Edwards, Graduate Adviser, 409-845-1542, *Fax:* 409-845-6433, *E-mail:* r-edwards@tamu.edu.

■ TEXAS A&M UNIVERSITY

College of Agriculture and Life Sciences, Faculty of Genetics, College Station, TX 77843

AWARDS MS, PhD. Program composed of members from 4 colleges and 14 departments.

Faculty: 61.
Students: 44 full-time (23 women), 5 part-time (2 women). Average age 29. 77 *applicants, 19% accepted.* In 1999, 1 master's awarded.
Degree requirements: For master's, thesis optional.
Entrance requirements: For master's and doctorate, GRE General Test, TOEFL. *Application deadline:* For fall admission, 4/15 (priority date). Applications are processed on a rolling basis. *Application fee:* $50 ($75 for international students).
Expenses: Tuition, state resident: part-time $76 per semester hour. Tuition, nonresident: part-time $292 per semester hour. Required fees: $11 per semester hour. Tuition and fees vary according to program.
Financial aid: Fellowships, research assistantships, teaching assistantships available. Financial aid application deadline: 4/15; financial aid applicants required to submit FAFSA.
Faculty research: Biochemical genetics, cytogenetics, developmental genetics, immunogenetics, molecular genetics.
Dr. Linda Guarino, Chair, 979-845-1013.
Application contact: Linda Fisher, Administrative Assistant, 979-845-6848, *Fax:* 979-845-9274.

Find an in-depth description at www.petersons.com/graduate.

■ TEXAS A&M UNIVERSITY

College of Veterinary Medicine, Department of Veterinary Pathobiology, College Station, TX 77843

AWARDS Genetics (MS, PhD); toxicology (MS, PhD); veterinary microbiology (MS, PhD); veterinary parasitology (MS); veterinary pathology (MS, PhD). Part-time programs available. Postbaccalaureate distance learning degree programs offered.

Faculty: 44 full-time (10 women), 17 part-time/adjunct (2 women).
Students: 33 full-time (13 women), 30 part-time (18 women); includes 11 minority (2 African Americans, 1 Asian American or Pacific Islander, 8 Hispanic Americans), 11 international. Average age 34. *28 applicants, 46% accepted.* In 1999, 4 master's awarded (25% found work related to degree, 75% continued full-time study); 10 doctorates awarded (40% entered university research/teaching, 60% found

other work related to degree). Terminal master's awarded for partial completion of doctoral program.

Degree requirements: For master's and doctorate, thesis/dissertation required, foreign language not required. *Average time to degree:* Master's–2 years full-time, 4.5 years part-time; doctorate–4.5 years full-time, 7 years part-time.

Entrance requirements: For master's and doctorate, GRE General Test, TOEFL. *Application deadline:* For fall admission, 3/1 (priority date); for spring admission, 8/1 (priority date). Applications are processed on a rolling basis. *Application fee:* $50 ($75 for international students). Electronic applications accepted.

Expenses: Tuition, state resident: full-time $5,400. Tuition, nonresident: full-time $16,200. Required fees: $2,936. Full-time tuition and fees vary according to student level.

Financial aid: In 1999–00, 7 fellowships with partial tuition reimbursements (averaging $16,000 per year), 13 research assistantships with partial tuition reimbursements (averaging $14,400 per year), 5 teaching assistantships with partial tuition reimbursements (averaging $16,000 per year) were awarded; career-related internships or fieldwork, Federal Work-Study, and institutionally sponsored loans also available. Aid available to part-time students. Financial aid applicants required to submit FAFSA.

Faculty research: Infectious and noninfectious diseases of animals and birds, animal genetics, molecular biology, immunology, virology. *Total annual research expenditures:* $2.6 million.

Dr. Ann B. Kier, Head, 979-845-5941, *Fax:* 979-845-9231, *E-mail:* akier@cvm.tamu.edu.

Application contact: Dr. G. G. Wagner, Graduate Adviser, 979-845-5941, *Fax:* 979-862-1147, *E-mail:* gwagner@cvm.tamus.edu.

■ TEXAS A&M UNIVERSITY

College of Veterinary Medicine and Office of Graduate Studies, Graduate Programs in Veterinary Medicine, Department of Veterinary Anatomy and Public Health, College Station, TX 77843

AWARDS Anatomy (MS, PhD); epidemiology (MS); genetics (PhD); toxicology (PhD); veterinary public health (MS).

Faculty: 23 full-time (6 women), 13 part-time/adjunct (6 women).

Students: 36 full-time (21 women), 2 part-time (both women); includes 13 minority (2 African Americans, 9 Asian Americans or Pacific Islanders, 2 Hispanic Americans). Average age 27. *19 applicants, 47% accepted.* In 1999, 4 master's, 10

doctorates awarded. Terminal master's awarded for partial completion of doctoral program.

Degree requirements: For master's and doctorate, thesis/dissertation required, foreign language not required.

Entrance requirements: For master's and doctorate, GRE General Test, TOEFL. *Application deadline:* For fall admission, 7/15 (priority date); for spring admission, 10/1. Applications are processed on a rolling basis. *Application fee:* $50 ($75 for international students).

Expenses: Tuition, state resident: part-time $76 per semester hour. Tuition, nonresident: part-time $292 per semester hour. Required fees: $11 per semester hour.

Financial aid: In 1999–00, 2 fellowships (averaging $12,000 per year), 18 research assistantships (averaging $13,500 per year), 5 teaching assistantships (averaging $14,000 per year) were awarded; Federal Work-Study, institutionally sponsored loans, and clinical associateships also available. Financial aid application deadline: 7/15; financial aid applicants required to submit FAFSA.

Faculty research: Metal toxicology, reproductive biology, genetics of neural development, developmental biology, environmental toxicology. *Total annual research expenditures:* $3.4 million.

Dr. Evelyn Tiffany-Castiglioni, Head, 979-845-2828, *Fax:* 979-847-8981, *E-mail:* ecastiglioni@cvm.tamu.edu.

■ TEXAS A&M UNIVERSITY SYSTEM HEALTH SCIENCE CENTER

College of Medicine, Graduate School of Biomedical Sciences, Department of Medical Biochemistry and Genetics, College Station, TX 77840-7896

AWARDS PhD.

Faculty: 13 full-time (1 woman).

Students: 24 full-time (9 women), 4 international. In 1999, 1 degree awarded.

Degree requirements: For doctorate, dissertation required, foreign language not required.

Entrance requirements: For doctorate, GRE General Test. *Application deadline:* For fall admission, 2/1 (priority date). Applications are processed on a rolling basis. *Application fee:* $50 ($75 for international students).

Expenses: Tuition, area resident: Full-time $1,368. Tuition, state resident: part-time $76 per credit. Tuition, nonresident: full-time $5,256; part-time $292 per credit. International tuition: $5,256 full-time. Required fees: $678; $38 per credit. Full-time tuition and fees vary according to course load and student level.

Financial aid: Fellowships, research assistantships available. Financial aid applicants required to submit FAFSA.

Faculty research: Immunology, cell and membrane biology, protein biochemistry, molecular genetics, parasitology, vertebrate embryogenesis and microbiology.

Dr. Hagan Bayley, Head, 979-845-2726, *Fax:* 979-847-9481.

Application contact: Janis Chmiel, Graduate Secretary, 800-298-2260, *Fax:* 979-847-9481, *E-mail:* jchmiel@tamu.edu.

Find an in-depth description at www.petersons.com/graduate.

■ THOMAS JEFFERSON UNIVERSITY

College of Graduate Studies, Program in Genetics, Philadelphia, PA 19107

AWARDS PhD.

Faculty: 36 full-time (6 women).

Students: 19 full-time (9 women); includes 3 minority (2 Asian Americans or Pacific Islanders, 1 Hispanic American), 3 international. *46 applicants, 26% accepted.* In 1999, 5 doctorates awarded.

Degree requirements: For doctorate, dissertation required, foreign language not required.

Entrance requirements: For doctorate, GRE General Test, TOEFL, minimum GPA of 3.2. *Application deadline:* For fall admission, 3/1 (priority date). Applications are processed on a rolling basis. *Application fee:* $40.

Expenses: Tuition: Full-time $12,670. Tuition and fees vary according to degree level and program.

Financial aid: In 1999–00, 19 fellowships with full tuition reimbursements were awarded; research assistantships, Federal Work-Study, institutionally sponsored loans, traineeships, and training grants also available. Aid available to part-time students. Financial aid application deadline: 5/1; financial aid applicants required to submit FAFSA.

Faculty research: Functional genomics, cancer susceptibility, cell cycle, regulation oncogenes and tumor suppressor genes, genetics of neoplastic disease. *Total annual research expenditures:* $17 million.

Dr. Arthur M. Buchberg, Chairman, Graduate Committee, 215-503-4533, *E-mail:* a_buchberg@lac.jci.tju.edu.

Application contact: Jessie F. Pervall, Director of Admissions, 215-503-4400, *Fax:* 215-503-3433, *E-mail:* cgs-info@mail.tju.edu.

Find an in-depth description at www.petersons.com/graduate.

■ TUFTS UNIVERSITY

Sackler School of Graduate Biomedical Sciences, Graduate Program in Genetics, Boston, MA 02111

AWARDS PhD.

Faculty: 38 full-time (15 women).
Students: 13 full-time (11 women); includes 5 minority (all Asian Americans or Pacific Islanders). Average age 25. 78 *applicants, 5% accepted.* In 1999, 1 doctorate awarded (100% continued full-time study).
Degree requirements: For doctorate, dissertation, qualifying exam required, foreign language not required. *Average time to degree:* Doctorate–4 years full-time.
Entrance requirements: For doctorate, TOEFL. *Application deadline:* For fall admission, 1/15 (priority date). *Application fee:* $45.
Expenses: Tuition: Full-time $19,325.
Financial aid: In 1999–00, 13 research assistantships with full tuition reimbursements (averaging $18,805 per year) were awarded; scholarships also available.
Faculty research: Abl-mediated leukemogenesis, retrovirus pathogenesis, ras signlaing, cell signaling and B cell growth, human fetal genetics. *Total annual research expenditures:* $6.9 million.
Dr. Naomi Rosenberg, Director, 617-636-6906, *Fax:* 617-636-0337, *E-mail:* nrosenbe@opal.tufts.edu.
Application contact: Dean Dawson, Associate Professor, 617-636-0393, *Fax:* 617-363-0337, *E-mail:* ddawson@opal.tufts.edu.

■ THE UNIVERSITY OF ALABAMA AT BIRMINGHAM

Graduate School and School of Medicine and School of Dentistry, Graduate Programs in Joint Health Sciences, Program in Medical Genetics, Birmingham, AL 35294

AWARDS PhD.

Students: 15 full-time (10 women). *25 applicants, 24% accepted.* In 1999, 3 degrees awarded.
Degree requirements: For doctorate, dissertation required.
Entrance requirements: For doctorate, GRE, interview. *Application deadline:* Applications are processed on a rolling basis. *Application fee:* $35 ($60 for international students). Electronic applications accepted.
Expenses: Tuition, state resident: part-time $104 per semester hour. Tuition, nonresident: part-time $208 per semester hour. Required fees: $17 per semester hour. $57 per quarter. Tuition and fees vary according to program.

Financial aid: In 1999–00, 2 fellowships were awarded.
Faculty research: Clinical cytogenetics, cancer cytogenetics, prenatal diagnosis.
Dr. Jerry N. Thompson, Director, 205-934-1081, *Fax:* 205-934-1078.

■ THE UNIVERSITY OF ARIZONA

Graduate College, Graduate Interdisciplinary Programs, Graduate Interdisciplinary Program in Genetics, Tucson, AZ 85721

AWARDS MS, PhD.

Degree requirements: For master's, thesis required, foreign language not required; for doctorate, one foreign language (computer language can substitute), dissertation required.
Entrance requirements: For master's, GRE General Test, TOEFL, minimum GPA of 3.5; for doctorate, GRE General Test, TOEFL, master's degree or equivalent.
Expenses: Tuition, nonresident: full-time $4,814; part-time $274 per unit. Required fees: $1,094; $115 per unit. Tuition and fees vary according to course load and program.
Faculty research: Cancer research; DNA repair; plant and animal cytogenetics; molecular, population, and ecological genetics.

■ UNIVERSITY OF CALIFORNIA, DAVIS

Graduate Studies, Programs in the Biological Sciences, Program in Genetics, Davis, CA 95616

AWARDS MS, PhD.

Faculty: 76 full-time (15 women).
Students: 70 full-time (39 women), 1 (woman) part-time; includes 13 minority (1 African American, 7 Asian Americans or Pacific Islanders, 5 Hispanic Americans), 22 international. Average age 29. 76 *applicants, 47% accepted.* In 1999, 8 master's, 6 doctorates awarded.
Degree requirements: For master's, thesis optional; for doctorate, dissertation required.
Entrance requirements: For master's and doctorate, GRE General Test, GRE Subject Test. *Application deadline:* For fall admission, 3/15. Applications are processed on a rolling basis. *Application fee:* $40. Electronic applications accepted.
Expenses: Tuition, nonresident: full-time $9,804. Tuition and fees vary according to program and student level.
Financial aid: In 1999–00, 55 students received aid, including 15 fellowships with full and partial tuition reimbursements available, 28 research assistantships with full and partial tuition reimbursements

available, 11 teaching assistantships with partial tuition reimbursements available; Federal Work-Study, grants, institutionally sponsored loans, scholarships, and tuition waivers (full and partial) also available. Financial aid application deadline: 1/15; financial aid applicants required to submit FAFSA.
Faculty research: Molecular, quantitative, and developmental genetics; cytogenetics; plant breeding.
Richard Michelmore, Graduate Chair, 530-752-1799, *E-mail:* rwmichelmore@ucdavis.edu.
Application contact: Tori Hollowell, Administrative Assistant, 530-752-9092, *Fax:* 530-752-8822, *E-mail:* trhollowell@ucdavis.edu.

Find an in-depth description at www.petersons.com/graduate.

■ UNIVERSITY OF CALIFORNIA, IRVINE

College of Medicine and Office of Research and Graduate Studies, Graduate Programs in Medicine and School of Biological Sciences, Department of Microbiology and Molecular Genetics, Irvine, CA 92697

AWARDS Biological sciences (PhD). Students apply through the Graduate Program in Molecular Biology, Genetics, and Biochemistry.

Faculty: 11 full-time (2 women).
Students: 26 full-time (13 women); includes 7 minority (all Asian Americans or Pacific Islanders), 3 international. *252 applicants, 34% accepted.*
Degree requirements: For doctorate, dissertation required, foreign language not required.
Entrance requirements: For doctorate, GRE General Test, GRE Subject Test. *Application deadline:* For fall admission, 2/1 (priority date). *Application fee:* $40. Electronic applications accepted.
Expenses: Tuition, nonresident: full-time $10,322; part-time $1,720 per quarter. Required fees: $5,354; $1,300 per quarter. Tuition and fees vary according to program.
Financial aid: Fellowships, research assistantships, teaching assistantships, institutionally sponsored loans and tuition waivers (full and partial) available. Financial aid application deadline: 3/2; financial aid applicants required to submit FAFSA.
Faculty research: Molecular biology and genetics of viruses, bacteria, and yeast; immune response; molecular biology of cultured animal cells; genetic basis of cancer; genetics and physiology of infectious agents.

Bert L. Semler, Chair, 949-824-7573 Ext. 6058, *Fax:* 949-824-8598, *E-mail:* blsemler@uci.edu.
Application contact: 949-824-5261, *Fax:* 949-824-8598, *E-mail:* gp-mbgb@uci.edu.
Find an in-depth description at www.petersons.com/graduate.

■ UNIVERSITY OF CALIFORNIA, IRVINE

Office of Research and Graduate Studies, School of Biological Sciences and College of Medicine, Graduate Program in Molecular Biology, Genetics, and Biochemistry, Irvine, CA 92697-1450

AWARDS Biological sciences (PhD).

Faculty: 102 full-time (20 women).
Students: 30 full-time (14 women); includes 7 minority (4 Asian Americans or Pacific Islanders, 3 Hispanic Americans), 2 international. Average age 25. *287 applicants, 37% accepted.*
Degree requirements: For doctorate, dissertation, teaching assignment required. *Average time to degree:* Doctorate–5 years full-time.
Entrance requirements: For doctorate, GRE General Test, GRE Subject Test (biochemistry, cell and molecular biology; biology; or chemistry), TOEFL, TSE, minimum GPA of 3.0, research experience. *Application deadline:* For fall admission, 1/1. *Application fee:* $40. Electronic applications accepted.
Expenses: Tuition, nonresident: full-time $9,384. Required fees: $5,178. Full-time tuition and fees vary according to program.
Financial aid: In 1999–00, 30 students received aid, including 30 fellowships with full tuition reimbursements available (averaging $33,995 per year); grants, institutionally sponsored loans, scholarships, and tuition waivers (full) also available. Financial aid application deadline: 3/2; financial aid applicants required to submit FAFSA.
Faculty research: Cellular biochemistry; gene structure and expression; protein structure, function, and design; molecular genetics; pathogenesis and inherited disease; molecular virology and immunology.
Dr. Rozanne Sandri-Goldin, Director, 949-824-7570, *Fax:* 949-824-7407, *E-mail:* gp-mbgb@uci.edu.
Application contact: Kimberly McKinney, Administrator, 949-824-8145, *Fax:* 949-824-7407, *E-mail:* gp-mbgb@uci.edu.
Find an in-depth description at www.petersons.com/graduate.

■ UNIVERSITY OF CALIFORNIA, LOS ANGELES

Graduate Division, College of Letters and Science, Department of Microbiology and Molecular Genetics, Los Angeles, CA 90095

AWARDS PhD.

Students: 27 full-time (17 women); includes 10 minority (9 Asian Americans or Pacific Islanders, 1 Hispanic American), 2 international. *3 applicants, 67% accepted.*
Degree requirements: For doctorate, dissertation, oral and written qualifying exams required, foreign language not required.
Entrance requirements: For doctorate, GRE General Test, GRE Subject Test (biology or chemistry recommended), minimum undergraduate GPA of 3.0. *Application deadline:* For fall admission, 12/15. *Application fee:* $40. Electronic applications accepted.
Expenses: Tuition, nonresident: full-time $9,804. Required fees: $4,405. Full-time tuition and fees vary according to program and student level.
Financial aid: Fellowships, research assistantships, teaching assistantships, Federal Work-Study, institutionally sponsored loans, scholarships, and tuition waivers (full and partial) available. Financial aid application deadline: 3/1.
Dr. Sherie Morrison, Chair, 310-825-3578.
Application contact: UCLA Access Coordinator, 800-284-8252, *Fax:* 310-206-5280, *E-mail:* uclaaccess@ibes.medsch.ucla.edu.
Find an in-depth description at www.petersons.com/graduate.

■ UNIVERSITY OF CALIFORNIA, RIVERSIDE

Graduate Division, College of Natural and Agricultural Sciences, Department of Botany and Plant Sciences, Riverside, CA 92521-0102

AWARDS Botany (MS, PhD); botany (plant genetics) (PhD); plant science (MS). Part-time programs available.

Students: 46 full-time (25 women); includes 4 minority (2 Asian Americans or Pacific Islanders, 2 Hispanic Americans), 23 international. In 1999, 5 master's, 3 doctorates awarded. Terminal master's awarded for partial completion of doctoral program.
Degree requirements: For master's, comprehensive exams or thesis required; for doctorate, dissertation, qualifying exams required, foreign language not required. *Average time to degree:* Master's–2.3 years full-time; doctorate–5.5 years full-time.
Entrance requirements: For master's and doctorate, GRE General Test, TOEFL,

minimum GPA of 3.2. *Application deadline:* For fall admission, 5/1; for spring admission, 12/1. Applications are processed on a rolling basis. *Application fee:* $40.
Expenses: Tuition, nonresident: full-time $9,804. Required fees: $4,758. Full-time tuition and fees vary according to program.
Financial aid: Fellowships, research assistantships, teaching assistantships, career-related internships or fieldwork, Federal Work-Study, grants, institutionally sponsored loans, and tuition waivers (full and partial) available. Financial aid application deadline: 2/1; financial aid applicants required to submit FAFSA.
Faculty research: Plant molecular biology, plant cell and developmental biology, plant systematics and evolution, ecology and natural resources, agriculture and crop physiology. *Total annual research expenditures:* $7.6 million.
Dr. Elizabeth Lord, Chair.
Application contact: 909-787-3424, *Fax:* 909-787-4437, *E-mail:* bpscmari@ucrac1.ucr.edu.
Find an in-depth description at www.petersons.com/graduate.

■ UNIVERSITY OF CALIFORNIA, RIVERSIDE

Graduate Division, College of Natural and Agricultural Sciences, Program in Genetics, Riverside, CA 92521-0102

AWARDS PhD.

Students: 7 full-time (5 women); includes 2 minority (both Asian Americans or Pacific Islanders), 1 international. In 1999, 5 degrees awarded.
Degree requirements: For doctorate, dissertation, qualifying exams, teaching experience required, foreign language not required. *Average time to degree:* Doctorate–5 years full-time.
Entrance requirements: For doctorate, GRE General Test, TOEFL, minimum GPA of 3.2. *Application deadline:* For fall admission, 5/1; for spring admission, 12/1. Applications are processed on a rolling basis. *Application fee:* $40.
Expenses: Tuition, nonresident: full-time $9,804. Required fees: $4,758. Full-time tuition and fees vary according to program.
Financial aid: Fellowships, research assistantships, teaching assistantships, career-related internships or fieldwork, Federal Work-Study, institutionally sponsored loans, and tuition waivers (full and partial) available. Financial aid application deadline: 2/1; financial aid applicants required to submit FAFSA.
Faculty research: Molecular genetics, microbial genetics, plant genetics, population genetics, evolutionary genetics.

University of California, Riverside (continued)
Dr. Bradley Hyman, Chair.
Application contact: 909-787-5688, *Fax:* 909-787-5517, *E-mail:* genetics@ucrac1.ucr.edu.
Find an in-depth description at www.petersons.com/graduate.

■ UNIVERSITY OF CALIFORNIA, SAN DIEGO

Graduate Studies and Research, Department of Biology, Program in Genetics and Molecular Biology, La Jolla, CA 92093-0348

AWARDS PhD. Offered in association with the Salk Institute.

Degree requirements: For doctorate, dissertation required.
Entrance requirements: For doctorate, GRE General Test, pre-application beginning in September. *Application deadline:* For fall admission, 1/7. *Application fee:* $40.
Expenses: Tuition, nonresident: full-time $14,691. Required fees: $4,697. Full-time tuition and fees vary according to program.
Financial aid: Tuition waivers (full) and stipends available.
Application contact: 858-534-3835.
Find an in-depth description at www.petersons.com/graduate.

■ UNIVERSITY OF CALIFORNIA, SAN FRANCISCO

Graduate Division and School of Medicine, Department of Biochemistry and Biophysics, Program in Genetics, San Francisco, CA 94143

AWARDS PhD, MD/PhD.

Students: In 1999, 2 degrees awarded.
Degree requirements: For doctorate, dissertation required, foreign language not required.
Entrance requirements: For doctorate, GRE General Test, GRE Subject Test, TOEFL. *Application deadline:* For fall admission, 1/5. *Application fee:* $40.
Expenses: All students guaranteed an annual stipend of $17,600.
Financial aid: Application deadline: 2/1.
Faculty research: Gene expression; chromosome structure and mechanics; medical, somatic cell, and radiation genetics.
Ira Herskowitz, Director, 415-476-3941.
Application contact: Sue Adams, Program Assistant, 415-476-1495.
Find an in-depth description at www.petersons.com/graduate.

■ UNIVERSITY OF CHICAGO

Division of the Biological Sciences, Molecular Biosciences: Biochemistry, Genetics, Cell and Developmental Biology, Committee on Genetics, Chicago, IL 60637-1513

AWARDS PhD.

Faculty: 58 full-time (24 women).
Students: 19 full-time (7 women); includes 2 minority (both Asian Americans or Pacific Islanders). Average age 26. In 1999, 2 degrees awarded (100% entered university research/teaching).
Degree requirements: For doctorate, dissertation required, foreign language not required. *Average time to degree:* Doctorate–6 years full-time.
Entrance requirements: For doctorate, GRE General Test, TOEFL, minimum GPA of 3.0. *Application deadline:* For fall admission, 1/5 (priority date). *Application fee:* $55.
Expenses: Tuition: Full-time $24,804; part-time $3,422 per course. Required fees: $390. Tuition and fees vary according to program.
Financial aid: In 1999–00, 17 students received aid; fellowships, research assistantships, institutionally sponsored loans available. Financial aid application deadline: 6/1; financial aid applicants required to submit FAFSA.
Faculty research: Molecular genetics, developmental genetics, population genetics, human genetics.
Dr. Rochelle Esposito, Chair, 773-702-8046, *Fax:* 773-702-8093, *E-mail:* committee-on-genetics@uchicago.edu.
Application contact: Elizabeth Hope Adams, Administrator, 773-702-2464, *Fax:* 773-702-3172, *E-mail:* committee-on-genetics@uchicago.edu.

■ UNIVERSITY OF CHICAGO

Division of the Biological Sciences, Molecular Biosciences: Biochemistry, Genetics, Cell and Developmental Biology, Department of Molecular Genetics and Cell Biology, Chicago, IL 60637-1513

AWARDS PhD.

Faculty: 36 full-time (12 women).
Students: 41 full-time (17 women); includes 9 minority (1 African American, 5 Asian Americans or Pacific Islanders, 3 Hispanic Americans), 4 international. Average age 25. In 1999, 7 doctorates awarded.
Degree requirements: For doctorate, dissertation required. *Average time to degree:* Doctorate–5.5 years full-time.
Entrance requirements: For doctorate, GRE General Test, TOEFL. *Application deadline:* For fall admission, 1/5 (priority date). *Application fee:* $55.

Expenses: Tuition: Full-time $24,804; part-time $3,422 per course. Required fees: $390. Tuition and fees vary according to program.
Financial aid: In 1999–00, 36 students received aid; fellowships, research assistantships, institutionally sponsored loans available. Financial aid application deadline: 6/1.
Faculty research: Gene expression, chromosome structure, animal viruses, plant molecular genetics. *Total annual research expenditures:* $8 million.
Dr. Anthony P. Mahowald, Chairman, 773-702-1620.
Application contact: Kristine Gaston, Graduate Administrative Director, 773-702-8037, *Fax:* 773-702-3172, *E-mail:* kristine@cummings.uchicago.edu.

■ UNIVERSITY OF CINCINNATI

Division of Research and Advanced Studies, College of Medicine, Graduate Programs in Medicine, Department of Molecular Genetics, Biochemistry, Microbiology and Immunology, Cincinnati, OH 45267

AWARDS PhD.

Faculty: 22 full-time.
Students: 42 full-time (21 women), 3 part-time (2 women); includes 2 minority (both Asian Americans or Pacific Islanders), 3 international. *109 applicants, 17% accepted.* In 1999, 8 doctorates awarded.
Degree requirements: For doctorate, dissertation, qualifying exam required, foreign language not required. *Average time to degree:* Doctorate–7.3 years full-time.
Entrance requirements: For doctorate, GRE General Test, GRE Subject Test. *Application deadline:* For fall admission, 2/1 (priority date). Applications are processed on a rolling basis. *Application fee:* $30.
Expenses: Tuition, state resident: full-time $5,139; part-time $196 per credit hour. Tuition, nonresident: full-time $10,326; part-time $369 per credit hour. Required fees: $561; $187 per quarter.
Financial aid: Tuition waivers (full) and unspecified assistantships available. Financial aid application deadline: 5/1. *Total annual research expenditures:* $6.8 million.
Dr. Jerry Lingrel, Head, 513-558-5324, *Fax:* 513-558-1190, *E-mail:* jerry.lingrel@uc.edu.
Application contact: Iain Cartwright, Graduate Program Director, 513-558-5532, *Fax:* 513-558-8474, *E-mail:* iain.cartwright@uc.edu.
Find an in-depth description at www.petersons.com/graduate.

■ UNIVERSITY OF COLORADO AT BOULDER

Graduate School, College of Arts and Sciences, Department of Environmental, Population, and Organic Biology, Boulder, CO 80309

AWARDS Animal behavior (MA, PhD); aquatic biology (MA, PhD); behavioral genetics (MA, PhD); ecology (MA, PhD); microbiology (MA, PhD); neurobiology (MA, PhD); plant and animal physiology (MA, PhD); plant and animal systematics (MA, PhD); population biology (MA, PhD); population genetics (MA, PhD).

Faculty: 39 full-time (9 women).
Students: 84 full-time (36 women), 14 part-time (7 women); includes 17 minority (6 Asian Americans or Pacific Islanders, 10 Hispanic Americans, 1 Native American). Average age 29. *147 applicants, 14% accepted.* In 1999, 7 master's, 13 doctorates awarded. Terminal master's awarded for partial completion of doctoral program.
Degree requirements: For master's, thesis or alternative, comprehensive exam required, foreign language not required; for doctorate, dissertation, comprehensive exam required, foreign language not required. *Average time to degree:* Master's–3 years full-time; doctorate–5 years full-time.
Entrance requirements: For master's, GRE General Test, GRE Subject Test, minimum undergraduate GPA of 3.0; for doctorate, GRE General Test, GRE Subject Test. *Application deadline:* For fall admission, 1/15 (priority date). *Application fee:* $40 ($60 for international students).
Expenses: Tuition, state resident: part-time $181 per credit hour. Tuition, nonresident: part-time $542 per credit hour. Required fees: $99 per term. Tuition and fees vary according to course load and program.
Financial aid: Fellowships, research assistantships, teaching assistantships, Federal Work-Study, institutionally sponsored loans, and tuition waivers (full) available. Financial aid application deadline: 3/1.
Faculty research: Evolution, developmental biology, behavior and neurobiology. *Total annual research expenditures:* $1.8 million.
Michael Breed, Chair, 303-492-8981, *Fax:* 303-492-8699, *E-mail:* michael.breed@colorado.edu.
Application contact: Jill Skarstadt, Graduate Secretary, 303-492-7654, *Fax:* 303-492-8699, *E-mail:* jill.skarstadt@colorado.edu.

■ UNIVERSITY OF COLORADO HEALTH SCIENCES CENTER

Graduate School, Programs in Biological and Medical Sciences, Programs in Biophysics and Genetics, Denver, CO 80262

AWARDS Biophysics and genetics (PhD); genetic counseling (MS). Terminal master's awarded for partial completion of doctoral program.

Degree requirements: For master's, thesis optional, foreign language not required; for doctorate, dissertation required, foreign language not required.
Entrance requirements: For master's, GRE General Test, minimum GPA of 2.75; for doctorate, GRE General Test, TOEFL, minimum GPA of 2.75.
Expenses: Tuition, state resident: full-time $1,512; part-time $56 per hour. Tuition, nonresident: full-time $7,209; part-time $267 per hour. Full-time tuition and fees vary according to course load and program.

■ UNIVERSITY OF CONNECTICUT

Graduate School, College of Liberal Arts and Sciences, Biological Sciences Group, Storrs, CT 06269

AWARDS Ecology and evolutionary biology (MS, PhD), including botany, ecology, entomology, systematics, zoology; molecular and cell biology (MS, PhD), including biochemistry, biophysics, biotechnology (MS), cell and developmental biology, genetics, microbiology, plant molecular and cell biology; physiology and neurobiology (MS, PhD), including neurobiology, physiology.

Degree requirements: For doctorate, dissertation required.
Entrance requirements: For master's and doctorate, GRE General Test, GRE Subject Test, TOEFL.
Expenses: Tuition, state resident: full-time $5,118. Tuition, nonresident: full-time $13,298. Required fees: $1,022.

■ UNIVERSITY OF CONNECTICUT

Graduate School, College of Liberal Arts and Sciences, Biological Sciences Group, Department of Molecular and Cell Biology, Field of Genetics, Storrs, CT 06269

AWARDS MS, PhD.

Degree requirements: For doctorate, dissertation required.
Entrance requirements: For master's and doctorate, GRE General Test, GRE Subject Test, TOEFL.

Expenses: Tuition, state resident: full-time $5,118. Tuition, nonresident: full-time $13,298. Required fees: $1,022.

Find an in-depth description at www.petersons.com/graduate.

■ UNIVERSITY OF CONNECTICUT HEALTH CENTER

Graduate School, Programs in Biomedical Sciences, Program in Genetics, Molecular Biology, and Biochemistry, Farmington, CT 06030
AWARDS PhD, DMD/PhD, MD/PhD.

Faculty: 29.
Students: 31 full-time (18 women); includes 2 minority (1 African American, 1 Asian American or Pacific Islander), 10 international. In 1999, 3 degrees awarded.
Degree requirements: For doctorate, one foreign language (computer language can substitute), dissertation required.
Entrance requirements: For doctorate, GRE General Test, GRE Subject Test, TOEFL. *Application deadline:* For fall admission, 2/1 (priority date); for spring admission, 10/1. Applications are processed on a rolling basis. *Application fee:* $40 ($45 for international students).
Expenses: Tuition, state resident: full-time $5,272; part-time $293 per credit. Tuition, nonresident: full-time $13,696; part-time $761 per credit. Required fees: $320; $198 per semester. One-time fee: $50 full-time. Full-time tuition and fees vary according to course load, program and reciprocity agreements.
Financial aid: In 1999–00, research assistantships (averaging $17,000 per year); fellowships also available.
Dr. Steve Pfeiffer, Director, 860-679-3395.
Application contact: Marizta Barta, Information Contact, 860-679-4306, *Fax:* 860-679-1282, *E-mail:* barta@adp.uchc.edu.

Find an in-depth description at www.petersons.com/graduate.

■ UNIVERSITY OF DELAWARE

College of Arts and Science, Department of Biological Sciences, Newark, DE 19716

AWARDS Biotechnology (MS, PhD); cell and extracellular matrix biology (MS, PhD); cell and systems physiology (MS, PhD); ecology and evolution (MS, PhD); microbiology (MS, PhD); molecular biology and genetics (MS, PhD); plant biology (MS, PhD).

Faculty: 37 full-time (10 women).
Students: 22 full-time (11 women), 1 part-time; includes 2 minority (both African Americans), 9 international. Average age 25. *37 applicants, 27% accepted.* In 2000, 9 doctorates awarded.

University of Delaware (continued)

Degree requirements: For master's and doctorate, thesis/dissertation required, foreign language not required. *Average time to degree:* Master's–2.5 years full-time; doctorate–6 years full-time.
Entrance requirements: For master's and doctorate, GRE General Test, GRE Subject Test (advanced biology). *Application deadline:* For fall admission, 6/15. Applications are processed on a rolling basis. *Application fee:* $50. Electronic applications accepted.
Expenses: Tuition, state resident: full-time $4,380; part-time $243 per credit. Tuition, nonresident: full-time $12,750; part-time $708 per credit. Required fees: $15 per term. Tuition and fees vary according to program.
Financial aid: In 2000–01, 18 students received aid, including 2 fellowships with full tuition reimbursements available (averaging $18,000 per year), 4 research assistantships with full tuition reimbursements available (averaging $18,000 per year), 11 teaching assistantships with full tuition reimbursements available (averaging $18,000 per year); tuition waivers (partial) also available. Financial aid application deadline: 6/15.
Faculty research: Cell interactions, molecular mechanisms, microorganisms, embryo implantation. *Total annual research expenditures:* $1.8 million.
Dr. Daniel D. Carson, Chair, 302-831-6977, *Fax:* 302-831-2281, *E-mail:* dcarson@udel.edu.
Application contact: Norman Karin, Graduate Coordinator, 302-831-1841, *Fax:* 302-831-2281, *E-mail:* ccoletta@udel.edu.
Find an in-depth description at www.petersons.com/graduate.

■ **UNIVERSITY OF FLORIDA**
College of Medicine and Graduate School, Interdisciplinary Program in Biomedical Sciences, Concentration in Genetics, Gainesville, FL 32611
AWARDS PhD.

Degree requirements: For doctorate, dissertation required, foreign language not required.
Entrance requirements: For doctorate, GRE General Test, minimum GPA of 3.0. Electronic applications accepted.
Expenses: Tuition, state resident: part-time $144 per credit hour. Tuition, nonresident: part-time $505 per credit hour. Tuition and fees vary according to course level, course load and program.
Find an in-depth description at www.petersons.com/graduate.

■ **UNIVERSITY OF FLORIDA**
College of Medicine and Graduate School, Interdisciplinary Program in Biomedical Sciences, Department of Molecular Genetics and Microbiology, Gainesville, FL 32611
AWARDS MS, PhD. Terminal master's awarded for partial completion of doctoral program.

Degree requirements: For master's and doctorate, thesis/dissertation required, foreign language not required.
Entrance requirements: For master's and doctorate, GRE General Test, TOEFL, minimum GPA of 3.0. Electronic applications accepted.
Expenses: Tuition, state resident: part-time $144 per credit hour. Tuition, nonresident: part-time $505 per credit hour. Tuition and fees vary according to course level, course load and program.

■ **UNIVERSITY OF GEORGIA**
Graduate School, College of Arts and Sciences, Department of Genetics, Athens, GA 30602
AWARDS MS, PhD.

Degree requirements: For master's, thesis required, foreign language not required; for doctorate, one foreign language (computer language can substitute), dissertation required.
Entrance requirements: For master's and doctorate, GRE General Test. Electronic applications accepted.
Expenses: Tuition, state resident: full-time $7,516; part-time $431 per credit hour. Tuition, nonresident: full-time $12,204; part-time $793 per credit hour. Tuition and fees vary according to program.
Find an in-depth description at www.petersons.com/graduate.

■ **UNIVERSITY OF HAWAII AT MANOA**
John A. Burns School of Medicine and Graduate Division, Graduate Programs in Biomedical Sciences, Department of Genetics and Molecular Biology, Honolulu, HI 96822
AWARDS MS, PhD.

Faculty: 18 full-time (3 women), 1 part-time/adjunct (0 women).
Students: 4 full-time (1 woman), 2 part-time (1 woman). Average age 29. *10 applicants, 90% accepted.* In 1999, 1 master's, 4 doctorates awarded (100% found work related to degree). Terminal master's awarded for partial completion of doctoral program.
Degree requirements: For master's, thesis required (for some programs),

foreign language not required; for doctorate, dissertation required, foreign language not required. *Average time to degree:* Doctorate–4 years full-time.
Entrance requirements: For master's and doctorate, GRE, minimum GPA of 3.0. *Application deadline:* For fall admission, 3/1; for spring admission, 5/1. Applications are processed on a rolling basis. *Application fee:* $25 ($50 for international students).
Expenses: Tuition, state resident: part-time $168 per credit. Tuition, nonresident: part-time $415 per credit. Required fees: $51 per semester. Part-time tuition and fees vary according to course load.
Financial aid: In 1999–00, 6 research assistantships (averaging $16,628 per year), 2 teaching assistantships (averaging $13,041 per year) were awarded; Federal Work-Study and institutionally sponsored loans also available. Financial aid application deadline: 2/1.
Dr. John A. Hunt, Chairperson, 808-956-8552, *Fax:* 808-956-5506, *E-mail:* jahunt@hawaii.edu.
Application contact: Dr. David S. Haymer, Chairman, Admissions Committee, 808-956-5517.

■ **UNIVERSITY OF ILLINOIS AT CHICAGO**
College of Medicine and Graduate College, Graduate Programs in Medicine, Chicago, IL 60607-7128
AWARDS Anatomy and cell biology (MS, PhD); biochemistry and molecular biology (MS, PhD); genetics (PhD), including molecular genetics; health professions education (MHPE); microbiology and immunology (PhD); pathology (MS, PhD); pharmacology (PhD), including pharmacology; physiology and biophysics (MS, PhD); surgery (MS). Part-time programs available.

Students: 169 full-time (98 women), 40 part-time (22 women); includes 29 minority (4 African Americans, 21 Asian Americans or Pacific Islanders, 3 Hispanic Americans, 1 Native American), 100 international. Average age 29. *648 applicants, 15% accepted.* In 1999, 22 master's, 22 doctorates awarded. Terminal master's awarded for partial completion of doctoral program.
Degree requirements: For master's and doctorate, thesis/dissertation required, foreign language not required.
Entrance requirements: For master's and doctorate, GRE General Test. *Application deadline:* For fall admission, 6/1; for spring admission, 11/1. *Application fee:* $40 ($50 for international students).
Expenses: Tuition, state resident: full-time $3,750. Tuition, nonresident: full-time $10,588. Tuition and fees vary according to course load.

Financial aid: In 1999–00, 119 students received aid; fellowships, research assistantships, teaching assistantships, career-related internships or fieldwork, Federal Work-Study, institutionally sponsored loans, scholarships, traineeships, and tuition waivers (full) available. Financial aid application deadline: 3/1; financial aid applicants required to submit FAFSA.
Gerald S. Moss, Dean, College of Medicine, 312-996-3500.

Find an in-depth description at www.petersons.com/graduate.

■ UNIVERSITY OF ILLINOIS AT CHICAGO

College of Medicine and Graduate College, Graduate Programs in Medicine, Department of Genetics, Program in Molecular Genetics, Chicago, IL 60607-7128

AWARDS PhD.

Faculty: 12 full-time (2 women).
Students: 42 full-time (18 women), 2 part-time (1 woman); includes 3 minority (2 Asian Americans or Pacific Islanders, 1 Hispanic American), 28 international. Average age 25. *128 applicants, 20% accepted.* In 1999, 6 degrees awarded.
Degree requirements: For doctorate, dissertation required, foreign language not required.
Entrance requirements: For doctorate, GRE General Test, TOEFL. *Application deadline:* For fall admission, 2/15. *Application fee:* $40 ($50 for international students).
Expenses: Tuition, state resident: full-time $3,750. Tuition, nonresident: full-time $10,588. Tuition and fees vary according to course load.
Financial aid: In 1999–00, 25 students received aid; fellowships, research assistantships, institutionally sponsored loans and tuition waivers (full) available. Financial aid application deadline: 2/15.
Dr. Lester F. Lau, Director of Graduate Studies.
Application contact: Phyllis Gallegos, Graduate Studies Assistant, 312-996-6984, *E-mail:* prgalleg@uic.edu.

Find an in-depth description at www.petersons.com/graduate.

■ UNIVERSITY OF ILLINOIS AT CHICAGO

Graduate College, College of Liberal Arts and Sciences, Department of Biological Sciences, Chicago, IL 60607-7128

AWARDS Cell and developmental biology (PhD); ecology and evolution (MS, DA, PhD); genetics and development (PhD); molecular biology (MS, PhD); neurobiology (MS, PhD); plant biology (MS, DA, PhD).

Faculty: 40 full-time (5 women).
Students: 100 full-time (47 women), 14 part-time (10 women); includes 13 minority (11 Asian Americans or Pacific Islanders, 2 Hispanic Americans), 42 international. Average age 29. *99 applicants, 36% accepted.* In 1999, 3 master's, 9 doctorates awarded.
Degree requirements: For master's, thesis required, foreign language not required; for doctorate, dissertation, preliminary exam required, foreign language not required.
Entrance requirements: For master's and doctorate, GRE General Test, GRE Subject Test, TOEFL, previous course work in physics, calculus, and organic chemistry; minimum GPA of 3.75 on a 5.0 scale. *Application deadline:* For fall admission, 6/1. Applications are processed on a rolling basis. *Application fee:* $40 ($50 for international students). Electronic applications accepted.
Expenses: Tuition, state resident: full-time $3,750; part-time $1,250 per semester. Tuition, nonresident: full-time $10,588; part-time $3,530 per semester. Required fees: $507 per semester. Tuition and fees vary according to course load and program.
Financial aid: In 1999–00, 87 students received aid; fellowships, research assistantships, teaching assistantships, career-related internships or fieldwork, Federal Work-Study, traineeships, and tuition waivers (full) available. Financial aid application deadline: 3/1; financial aid applicants required to submit FAFSA.
Dr. Lon Kaufman, Head, 312-996-2213.
Application contact: Dr. Leo Miller, Director of Graduate Studies, 312-996-2220.

Find an in-depth description at www.petersons.com/graduate.

■ THE UNIVERSITY OF IOWA

College of Medicine and Graduate College, Graduate Programs in Medicine, Department of Microbiology, Iowa City, IA 52242-1316

AWARDS General microbiology and microbial physiology (MS, PhD); immunology (MS, PhD); microbial genetics (MS, PhD); pathogenic bacteriology (MS, PhD); virology (MS, PhD).

Faculty: 21 full-time (4 women), 10 part-time/adjunct (1 woman).
Students: 50 full-time (21 women); includes 7 minority (2 African Americans, 4 Asian Americans or Pacific Islanders, 1 Hispanic American), 5 international. *144*

applicants, 8% accepted. In 1999, 7 doctorates awarded (100% entered university research/teaching).
Degree requirements: For master's, thesis required; for doctorate, dissertation, comprehensive exam required. *Average time to degree:* Doctorate–6 years full-time.
Entrance requirements: For master's and doctorate, GRE General Test. *Application deadline:* For fall admission, 2/1. *Application fee:* $30 ($50 for international students). Electronic applications accepted.
Expenses: Tuition, state resident: full-time $3,308. Tuition, nonresident: full-time $10,662. Tuition and fees vary according to course load and program.
Financial aid: In 1999–00, 50 students received aid, including fellowships with full tuition reimbursements available (averaging $16,787 per year), research assistantships with full tuition reimbursements available (averaging $16,787 per year); grants, institutionally sponsored loans, and traineeships also available.
Faculty research: Biocatalysis and blue jeans, gene regulation, processing and transport of HIV, retroviral pathogenesis, biodegradation. *Total annual research expenditures:* $8.2 million.
Dr. Michael A. Apicella, Head, 319-335-7810, *E-mail:* grad-micro-info@uiowa.edu.

Find an in-depth description at www.petersons.com/graduate.

■ THE UNIVERSITY OF IOWA

College of Medicine and Graduate College, Graduate Programs in Medicine, Graduate Program in Genetics, Iowa City, IA 52242-1316

AWARDS PhD.

Faculty: 44 full-time (10 women).
Students: 36 full-time (13 women); includes 3 minority (1 African American, 1 Hispanic American, 1 Native American), 8 international. *136 applicants, 5% accepted.* In 1999, 6 degrees awarded (67% entered university research/teaching, 33% continued full-time study).
Degree requirements: For doctorate, dissertation, comprehensive exam required. *Average time to degree:* Doctorate–5 years full-time.
Entrance requirements: For doctorate, GRE General Test. *Application deadline:* For fall admission, 1/15 (priority date). Applications are processed on a rolling basis. *Application fee:* $30 ($50 for international students).
Expenses: Tuition, state resident: full-time $3,308. Tuition, nonresident: full-time $10,662. Tuition and fees vary according to course load and program.
Financial aid: In 1999–00, 36 students received aid, including 2 fellowships with full tuition reimbursements available

The University of Iowa (continued) (averaging $18,150 per year), 34 research assistantships with full tuition reimbursements available (averaging $16,787 per year); traineeships also available.

Faculty research: Developmental genetics, eukaryotic gene expression, human genetics, molecular and biochemical genetics, evolutionary genetics.

Dr. Robert Deschenes, Director, 319-335-9968, *Fax:* 319-353-5330, *E-mail:* phd@genetics.uiowa.edu.

Application contact: Karen J. Kriege, Program Assistant, 319-335-9968, *Fax:* 319-353-5330, *E-mail:* phd@genetics.uiowa.edu.

Find an in-depth description at www.petersons.com/graduate.

■ UNIVERSITY OF KANSAS

Graduate Studies Medical Center, Graduate Programs in Biomedical and Basic Sciences, Department of Microbiology, Molecular Genetics and Immunology, Lawrence, KS 66045

AWARDS PhD, MD/PhD.

Faculty: 10 full-time (1 woman), 1 part-time/adjunct (0 women).

Students: Average age 27. *0 applicants, 0% accepted.* In 1999, 4 doctorates awarded.

Degree requirements: For doctorate, dissertation, comprehensive oral exam required.

Entrance requirements: For doctorate, GRE General Test, TOEFL, TSE. *Application deadline:* For fall admission, 1/31 (priority date). Applications are processed on a rolling basis. *Application fee:* $0.

Expenses: Tuition, state resident: full-time $2,482; part-time $103 per credit hour. Tuition, nonresident: full-time $8,104; part-time $338 per credit hour. Required fees: $428; $31 per credit hour. Tuition and fees vary according to program.

Financial aid: Fellowships, research assistantships, teaching assistantships, Federal Work-Study, institutionally sponsored loans, scholarships, and unspecified assistantships available. Aid available to part-time students. Financial aid application deadline: 3/31; financial aid applicants required to submit FAFSA.

Faculty research: Bacterial and viral pathogenesis, molecular virology, microbial molecular biology and genetics, immunochemistry. *Total annual research expenditures:* $4.2 million.

Dr. Opendra Narayan, Chairman, 913-5887010, *Fax:* 913-588-7295, *E-mail:* bnarayan@kume.edu.

Application contact: Dr. Joe Lutkenhaus, Director of Graduate Studies, 913-588-7010, *Fax:* 913-588-7295, *E-mail:* jlutkenh@kumc.edu.

■ UNIVERSITY OF MASSACHUSETTS WORCESTER

Graduate School of Biomedical Sciences, Department of Molecular Genetics and Microbiology, Worcester, MA 01655-0115

AWARDS Medical sciences (PhD).

Faculty: 42 full-time (6 women).

Students: 28 full-time (11 women); includes 1 minority (Asian American or Pacific Islander), 11 international.

Degree requirements: For doctorate, dissertation required.

Entrance requirements: For doctorate, GRE General Test, GRE Subject Test, 1 year of calculus, organic chemistry, physics, and biology. *Application deadline:* For fall admission, 1/15 (priority date). Applications are processed on a rolling basis. *Application fee:* $25 ($50 for international students).

Expenses: Tuition, state resident: full-time $2,640. Tuition, nonresident: full-time $9,756. Required fees: $825. Full-time tuition and fees vary according to program.

Financial aid: In 1999–00, research assistantships with full tuition reimbursements (averaging $17,500 per year); unspecified assistantships also available.

Faculty research: Gene structure, regulation of gene expression.

Dr. Allan Jacobson, Chair.

Application contact: Dr. Janet Stavnezer, Professor, 508-856-4100, *E-mail:* janet.stavnezer@umassmed.edu.

Find an in-depth description at www.petersons.com/graduate.

■ UNIVERSITY OF MEDICINE AND DENTISTRY OF NEW JERSEY

Graduate School of Biomedical Sciences, Graduate Programs in Biomedical Sciences, Department of Molecular Genetics and Microbiology, Newark, NJ 07107

AWARDS MS, PhD.

Degree requirements: For master's, thesis required; for doctorate, dissertation, qualifying exam required, foreign language not required.

Entrance requirements: For master's and doctorate, GRE General Test, TOEFL. *Application deadline:* For fall admission, 2/1; for spring admission, 10/1. *Application fee:* $40.

Expenses: Tuition, state resident: part-time $270 per credit hour. Tuition, nonresident: part-time $407 per credit hour. Part-time tuition and fees vary according to campus/location and program.

Financial aid: Fellowships, research assistantships, Federal Work-Study,

institutionally sponsored loans, and tuition waivers (full and partial) available. Financial aid application deadline: 5/1.

Faculty research: Molecular genetics of yeast, mutagenesis and carcinogenesis of DNA, bacterial protein synthesis, mammalian cell genetics, adenovirus gene expression.

Dr. Harvey Ozer, Chairperson, 973-972-4483, *Fax:* 973-972-1286, *E-mail:* ozerhl@umdnj.edu.

Application contact: Dr. Henry E. Brezenoff, Dean, Graduate School of Biomedical Sciences, 973-972-5333, *Fax:* 973-972-7148, *E-mail:* hbrezeno@umdnj.edu.

Find an in-depth description at www.petersons.com/graduate.

■ UNIVERSITY OF MEDICINE AND DENTISTRY OF NEW JERSEY

Graduate School of Biomedical Sciences, Graduate Programs in Biomedical Sciences, Program in Molecular Genetics and Microbiology, Piscataway, NJ 08854-5635

AWARDS MS, PhD. Terminal master's awarded for partial completion of doctoral program.

Degree requirements: For master's, thesis, qualifying exam required; for doctorate, dissertation, qualifying exam required, foreign language not required.

Entrance requirements: For master's and doctorate, GRE General Test, TOEFL. *Application deadline:* For fall admission, 2/1; for spring admission, 10/1. Applications are processed on a rolling basis. *Application fee:* $40.

Expenses: Tuition, state resident: part-time $270 per credit hour. Tuition, nonresident: part-time $407 per credit hour. Part-time tuition and fees vary according to campus/location and program.

Financial aid: Fellowships, research assistantships, teaching assistantships available. Financial aid application deadline: 5/1.

Faculty research: Interferon, receptors, retrovirus evolution, Arbo virus/host cell interactions.

Dr. Sidney Pestka, Director, 732-235-4567.

Application contact: Dr. Michael J. Leibowitz, Associate Dean, Graduate School, 732-235-5016, *Fax:* 732-235-4720, *E-mail:* gsbspisc@umdnj.edu.

■ UNIVERSITY OF MIAMI

Graduate School, College of Arts and Sciences, Department of Biology, Coral Gables, FL 33124

AWARDS Biology (MS, PhD); genetics and evolution (MS, PhD); tropical biology, ecology, and behavior (MS, PhD).

Faculty: 24 full-time (3 women), 5 part-time/adjunct (1 woman).
Students: 42 full-time (23 women); includes 7 minority (1 African American, 3 Asian Americans or Pacific Islanders, 3 Hispanic Americans), 2 international. Average age 26. *56 applicants, 13% accepted.* In 1999, 3 doctorates awarded (34% entered university research/teaching, 66% found other work related to degree). Terminal master's awarded for partial completion of doctoral program.
Degree requirements: For master's, oral defense required, thesis optional, foreign language not required; for doctorate, dissertation, oral and written qualifying exam required, foreign language not required. *Average time to degree:* Doctorate–5.5 years full-time.
Entrance requirements: For master's and doctorate, GRE General Test, GRE Subject Test, TOEFL. *Application deadline:* For fall admission, 3/15. Applications are processed on a rolling basis. *Application fee:* $50. Electronic applications accepted.
Expenses: Tuition: Full-time $15,336; part-time $852 per credit. Required fees: $174. Tuition and fees vary according to program.
Financial aid: In 1999–00, 37 students received aid, including 6 fellowships with tuition reimbursements available (averaging $15,000 per year), 10 research assistantships with tuition reimbursements available (averaging $13,000 per year), 21 teaching assistantships with tuition reimbursements available (averaging $13,202 per year); career-related internships or fieldwork and institutionally sponsored loans also available. Financial aid application deadline: 3/15.
Faculty research: Population biology, behavioral ecology, plant-animal and plant-environment interactions and genetic co-evolution, biogeography, conservation biology.
Dr. Julian C. Lee, Director, 305-284-6420, *Fax:* 305-284-3039, *E-mail:* jlee@fig.cox.miami.edu.
Application contact: 305-284-3973, *Fax:* 305-284-3039, *E-mail:* gaac@fig.cox.miami.edu.

Find an in-depth description at www.petersons.com/graduate.

■ UNIVERSITY OF MINNESOTA, TWIN CITIES CAMPUS

Graduate School, Program in Molecular, Cellular, Developmental Biology and Genetics, Minneapolis, MN 55455-0213

AWARDS Genetic counseling (MS); molecular, cellular, developmental biology and genetics (PhD). Part-time programs available.

Faculty: 80 full-time (24 women), 17 part-time/adjunct (16 women).
Students: 62 full-time (43 women), 2 part-time (both women). Average age 24. *188 applicants, 19% accepted.* In 1999, 9 master's awarded (100% found work related to degree); 5 doctorates awarded (20% entered university research/teaching, 80% found other work related to degree). Terminal master's awarded for partial completion of doctoral program.
Degree requirements: For master's, thesis optional, foreign language not required; for doctorate, dissertation required, foreign language not required. *Average time to degree:* Master's–2 years full-time; doctorate–5 years full-time.
Entrance requirements: For master's and doctorate, GRE General Test, TOEFL. *Application deadline:* For fall admission, 1/15 (priority date). Applications are processed on a rolling basis. *Application fee:* $40 ($50 for international students).
Expenses: Tuition, state resident: full-time $5,040; part-time $420 per credit. Tuition, nonresident: full-time $9,900; part-time $825 per credit. Full-time tuition and fees vary according to course load, program and reciprocity agreements.
Financial aid: In 1999–00, 53 students received aid, including 9 fellowships with full tuition reimbursements available (averaging $16,504 per year), 39 research assistantships with full tuition reimbursements available (averaging $16,504 per year), 5 teaching assistantships with full tuition reimbursements available (averaging $16,504 per year); Federal Work-Study, grants, institutionally sponsored loans, and traineeships also available. Financial aid application deadline: 1/15.
Faculty research: Membrane receptors and membrane transport, cell interactions, cytoskeleton and cell mobility, regulation of gene expression, plant cell and molecular biology.
Dr. Perry B. Hackett, Director of Graduate Studies, 612-624-3053, *Fax:* 612-625-1700, *E-mail:* dgsmcdbg@biosci.cbs.umn.edu.
Application contact: Sue Knoblauch, Principal Secretary, 612-624-7470, *Fax:* 612-625-5754, *E-mail:* mcdbg@biosci.cbs.umn.edu.

Find an in-depth description at www.petersons.com/graduate.

■ UNIVERSITY OF MISSOURI–COLUMBIA

Graduate School, College of Arts and Sciences, Division of Biological Sciences, Program in Genetics, Columbia, MO 65211
AWARDS MA, PhD.

Degree requirements: For master's, thesis required; for doctorate, dissertation, comprehensive exam required.
Entrance requirements: For master's and doctorate, GRE General Test, minimum GPA of 3.0.
Expenses: Tuition, state resident: full-time $3,020; part-time $168 per hour. Tuition, nonresident: full-time $6,066; part-time $505 per hour. Required fees: $445; $18 per hour. Tuition and fees vary according to course load and program.

■ UNIVERSITY OF MISSOURI–ST. LOUIS

Graduate School, College of Arts and Sciences, Department of Biology, St. Louis, MO 63121-4499

AWARDS Biology (MS, PhD), including animal behavior (MS), biochemistry, biotechnology (MS), conservation biology (MS), development (MS), ecology (MS), environmental studies (PhD), evolution (MS), genetics (MS), molecular biology and biotechnology (PhD), molecular/cellular biology (MS), physiology (MS), plant systematics, population biology (MS), tropical biology (MS); biotechnology (Certificate); tropical biology and conservation (Certificate). Part-time programs available.

Faculty: 46.
Students: 21 full-time (11 women), 75 part-time (44 women); includes 13 minority (2 African Americans, 2 Asian Americans or Pacific Islanders, 8 Hispanic Americans, 1 Native American), 23 international. In 1999, 14 master's, 4 doctorates awarded.
Degree requirements: For master's, thesis or alternative required, foreign language not required; for doctorate, one foreign language, dissertation, 1 semester of teaching experience required.
Entrance requirements: For doctorate, GRE General Test. *Application deadline:* For fall admission, 7/1 (priority date); for spring admission, 11/1 (priority date). Applications are processed on a rolling basis. *Application fee:* $25 ($40 for international students). Electronic applications accepted.
Expenses: Tuition, state resident: full-time $4,932; part-time $173 per credit hour. Tuition, nonresident: full-time $13,279; part-time $521 per credit hour. Required fees: $775; $33 per credit hour. Tuition and fees vary according to degree level and program.

University of Missouri–St. Louis (continued)

Financial aid: In 1999–00, 8 research assistantships with partial tuition reimbursements (averaging $10,635 per year), 14 teaching assistantships with partial tuition reimbursements (averaging $11,488 per year) were awarded; career-related internships or fieldwork and Federal Work-Study also available. Aid available to part-time students. Financial aid application deadline: 2/1. *Total annual research expenditures:* $908,828.

Application contact: Graduate Admissions, 314-516-5458, *Fax:* 314-516-6759, *E-mail:* gradadm@umsl.edu.

■ UNIVERSITY OF NEW HAMPSHIRE

Graduate School, College of Life Sciences and Agriculture, Graduate Programs in the Biological Sciences and Natural Resources, Program in Genetics, Durham, NH 03824

AWARDS MS, PhD.

Faculty: 14 full-time.
Students: 5 full-time (2 women), 8 part-time (5 women), 3 international. Average age 28. *12 applicants, 83% accepted.* In 1999, 1 master's awarded.
Degree requirements: For master's, thesis required, foreign language not required; for doctorate, dissertation required.
Entrance requirements: For master's and doctorate, GRE General Test, GRE Subject Test. *Application deadline:* For fall admission, 4/1 (priority date); for winter admission, 12/1. Applications are processed on a rolling basis. *Application fee:* $50.
Expenses: Tuition, area resident: Full-time $5,750; part-time $319 per credit. Tuition, state resident: full-time $8,625; part-time $478. Tuition, nonresident: full-time $14,640; part-time $598 per credit. Required fees: $224 per semester. Tuition and fees vary according to course load, degree level and program.
Financial aid: In 1999–00, 1 research assistantship, 8 teaching assistantships were awarded; fellowships, scholarships also available. Financial aid application deadline: 2/15.
Dr. Thomas Davis, Head, 603-862-3217, *E-mail:* tmdavis@cisunix.unh.edu.
Application contact: John Collins, Coordinator, 603-862-2462, *E-mail:* john.collins@unh.edu.

■ UNIVERSITY OF NEW MEXICO

Graduate School, College of Arts and Sciences, Department of Biology, Albuquerque, NM 87131-2039

AWARDS Biology (MS, PhD), including air land ecology, behavioral ecology, botany, cellular and molecular biology, community ecology, comparative immunology, comparative physiology, conservation biology, ecology, ecosystem ecology, evolutionary biology, evolutionary genetics, microbiology, molecular genetics, parasitology, physiological ecology, physiology, population biology, vertebrate and invertebrate zoology. Part-time programs available.

Faculty: 35 full-time (5 women), 18 part-time/adjunct (11 women).
Students: 71 full-time (37 women), 28 part-time (11 women); includes 8 minority (2 Asian Americans or Pacific Islanders, 5 Hispanic Americans, 1 Native American), 11 international. Average age 33. *93 applicants, 30% accepted.* In 1999, 11 master's, 12 doctorates awarded. Terminal master's awarded for partial completion of doctoral program.
Degree requirements: For master's, one foreign language (computer language can substitute), thesis required (for some programs); for doctorate, 2 foreign languages (computer language can substitute for one), dissertation required.
Entrance requirements: For master's and doctorate, GRE General Test, GRE Subject Test, minimum GPA of 3.2. *Application deadline:* For fall admission, 1/15. *Application fee:* $25.
Expenses: Tuition, state resident: full-time $2,514; part-time $105 per credit hour. Tuition, nonresident: full-time $10,304; part-time $417 per credit hour. International tuition: $10,304 full-time. Required fees: $516; $22 per credit hour. Tuition and fees vary according to program.
Financial aid: In 1999–00, 58 students received aid, including 24 fellowships (averaging $1,645 per year), 26 research assistantships with tuition reimbursements available (averaging $8,921 per year), 40 teaching assistantships with tuition reimbursements available (averaging $11,066 per year); career-related internships or fieldwork, Federal Work-Study, institutionally sponsored loans, and tuition waivers (full and partial) also available. Aid available to part-time students. Financial aid applicants required to submit FAFSA.
Faculty research: Developmental biology, immunobiology. *Total annual research expenditures:* $4.5 million.
Dr. Kathryn Vogel, Chair, 505-277-3411, *Fax:* 505-277-0304, *E-mail:* kgvogel@unm.edu.

Application contact: Vivian Kent, Information Contact, 505-277-1712, *Fax:* 505-277-0304, *E-mail:* vkent@unm.edu.

■ THE UNIVERSITY OF NORTH CAROLINA AT CHAPEL HILL

Graduate School, College of Arts and Sciences, Department of Biology, Chapel Hill, NC 27599

AWARDS Botany (MA, MS, PhD); cell biology, development, and physiology (MA, MS, PhD); ecology and behavior (MA, MS, PhD); genetics and molecular biology (MA, MS, PhD); morphology, systematics, and evolution (MA, MS, PhD).

Degree requirements: For master's, thesis (for some programs), comprehensive exams required; for doctorate, dissertation, comprehensive exams required.
Entrance requirements: For master's and doctorate, GRE General Test, GRE Subject Test. Electronic applications accepted.
Expenses: Tuition, state resident: full-time $1,578. Tuition, nonresident: full-time $10,744. Required fees: $827. One-time fee: $15 full-time. Tuition and fees vary according to program.
Faculty research: Gene expression, biomechanics, yeast genetics, plant ecology, plant molecular biology.

Find an in-depth description at www.petersons.com/graduate.

■ THE UNIVERSITY OF NORTH CAROLINA AT CHAPEL HILL

School of Medicine and Graduate School, Graduate Programs in Medicine, Curriculum in Genetics and Molecular Biology, Chapel Hill, NC 27599

AWARDS MS, PhD.

Faculty: 66 full-time (22 women), 2 part-time/adjunct (0 women).
Students: 56 full-time (33 women); includes 5 minority (3 African Americans, 1 Asian American or Pacific Islander, 1 Hispanic American), 2 international. *68 applicants, 35% accepted.* In 1999, 9 doctorates awarded (45% entered university research/teaching, 33% found other work related to degree, 22% continued full-time study).
Degree requirements: For master's, thesis, comprehensive exam required, foreign language not required; for doctorate, dissertation, comprehensive exams required, foreign language not required. *Average time to degree:* Doctorate–5 years full-time.
Entrance requirements: For master's and doctorate, GRE General Test, minimum GPA of 3.0. *Application deadline:* For fall admission, 1/1 (priority date). Applications

are processed on a rolling basis. Electronic applications accepted.

Expenses: Tuition, state resident: full-time $1,966. Tuition, nonresident: full-time $11,026. Required fees: $8,940. One-time fee: $15 full-time. Part-time tuition and fees vary according to course load.

Financial aid: In 1999–00, 1 fellowship with tuition reimbursement (averaging $16,500 per year), 30 research assistantships with tuition reimbursements (averaging $16,500 per year), 1 teaching assistantship with tuition reimbursement (averaging $16,500 per year) were awarded; traineeships also available. Dr. Susan T. Lord, Director, 919-966-3548, *E-mail:* stl@med.unc.edu.

Application contact: Janet Collins, Administrative Assistant, 919-966-2681, *Fax:* 919-966-0401, *E-mail:* lcollins@med.unc.edu.

Find an in-depth description at www.petersons.com/graduate.

■ UNIVERSITY OF NORTH DAKOTA

Graduate School, College of Arts and Sciences, Department of Biology, Grand Forks, ND 58202

AWARDS Botany (MS, PhD); ecology (MS, PhD); entomology (MS, PhD); environmental biology (MS, PhD); fisheries/wildlife (MS, PhD); genetics (MS, PhD); zoology (MS, PhD).

Faculty: 18 full-time (3 women).

Students: 21 full-time (8 women). *13 applicants, 62% accepted.* In 1999, 3 master's awarded. Terminal master's awarded for partial completion of doctoral program.

Degree requirements: For master's, thesis, final exam required; for doctorate, dissertation, comprehensive exam, final exam required.

Entrance requirements: For master's, GRE General Test, GRE Subject Test, TOEFL, minimum GPA of 3.0; for doctorate, GRE General Test, GRE Subject Test, TOEFL, minimum GPA of 3.5. *Application deadline:* For fall admission, 3/1 (priority date). Applications are processed on a rolling basis. *Application fee:* $25.

Expenses: Tuition, state resident: full-time $3,166; part-time $158 per credit. Tuition, nonresident: full-time $7,658; part-time $345 per credit. International tuition: $7,658 full-time. Required fees: $46 per credit. Tuition and fees vary according to program and reciprocity agreements.

Financial aid: In 1999–00, 6 research assistantships with full tuition reimbursements (averaging $11,250 per year), 15 teaching assistantships with full tuition reimbursements (averaging $11,250 per year) were awarded; fellowships, Federal

Work-Study, institutionally sponsored loans, scholarships, and tuition waivers (full and partial) also available. Aid available to part-time students. Financial aid application deadline: 3/15; financial aid applicants required to submit FAFSA.

Faculty research: Population biology, wildlife ecology, RNA processing, hormonal control of behavior. Dr. Jeff Lang, Director, 701-777-2621, *Fax:* 701-777-2623, *E-mail:* jlang@badlands.nodak.edu.

■ UNIVERSITY OF NOTRE DAME

Graduate School, College of Science, Department of Biological Sciences, Program in Genetics, Notre Dame, IN 46556

AWARDS MS, PhD. Terminal master's awarded for partial completion of doctoral program.

Degree requirements: For master's and doctorate, thesis/dissertation required, foreign language not required.

Entrance requirements: For master's and doctorate, GRE General Test, GRE Subject Test, TOEFL. *Application deadline:* For fall admission, 2/1 (priority date); for spring admission, 11/1. Applications are processed on a rolling basis. *Application fee:* $50.

Expenses: Tuition: Full-time $21,930; part-time $1,218 per credit. Required fees: $95. Tuition and fees vary according to program.

Financial aid: Fellowships with full tuition reimbursements, research assistantships with full tuition reimbursements, teaching assistantships with full tuition reimbursements, tuition waivers (full) available. Financial aid application deadline: 2/1.

Faculty research: Genetics of Drosophila and zebrafish, genetic models of human retinal diseases, genetics of speciation, transgenic model systems.

Application contact: Dr. Terrence J. Akai, Director of Graduate Admissions, 219-631-7706, *Fax:* 219-631-4183, *E-mail:* gradad@nd.edu.

■ UNIVERSITY OF OREGON

Graduate School, College of Arts and Sciences, Department of Biology, Eugene, OR 97403

AWARDS Ecology and evolution (MA, MS, PhD); marine biology (MA, MS, PhD); molecular, cellular and genetic biology (PhD); neuroscience and development (PhD).

Faculty: 42 full-time (14 women), 5 part-time/adjunct (2 women).

Students: 69 full-time (33 women), 8 part-time (5 women); includes 6 minority (2 Asian Americans or Pacific Islanders, 3 Hispanic Americans, 1 Native American), 5 international. *18 applicants, 83% accepted.*

In 1999, 11 master's, 7 doctorates awarded. Terminal master's awarded for partial completion of doctoral program.

Degree requirements: For master's, thesis required (for some programs); for doctorate, dissertation required, foreign language not required.

Entrance requirements: For master's and doctorate, GRE General Test, TOEFL, minimum GPA of 3.2. *Application deadline:* For fall admission, 1/10. *Application fee:* $50.

Expenses: Tuition, state resident: full-time $6,750. Tuition, nonresident: full-time $11,409. Part-time tuition and fees vary according to course load.

Financial aid: In 1999–00, 36 teaching assistantships were awarded; research assistantships, Federal Work-Study, grants, and institutionally sponsored loans also available. Financial aid application deadline: 2/1.

Faculty research: Developmental neurobiology; evolution, population biology, and quantitative genetics; regulation of gene expression; biochemistry of marine organisms. Janis C. Weeks, Head, 541-346-4502, *Fax:* 541-346-6056.

Application contact: Donna Overall, Graduate Program Coordinator, 541-346-4503, *Fax:* 541-346-6056, *E-mail:* doverall@oregon.uoregon.edu.

Find an in-depth description at www.petersons.com/graduate.

■ UNIVERSITY OF PENNSYLVANIA

School of Medicine, Biomedical Graduate Studies, Graduate Group in Cell and Molecular Biology, Program in Gene Therapy, Philadelphia, PA 19104

AWARDS PhD, MD/PhD, VMD/PhD.

Degree requirements: For doctorate, dissertation required, foreign language not required.

Entrance requirements: For doctorate, GRE General Test, TOEFL. *Application deadline:* For fall admission, 1/2 (priority date). *Application fee:* $65. Electronic applications accepted.

Expenses: Tuition: Full-time $17,256; part-time $2,991 per course. Required fees: $2,588; $363 per course. $726 per term.

Financial aid: Fellowships, research assistantships, teaching assistantships, grants and institutionally sponsored loans available. Financial aid application deadline: 1/2. Dr. James Wilson, Chair, 215-898-1979.

Application contact: Mary Webster, Coordinator, 215-898-4360, *E-mail:* camb@mail.med.upenn.edu.

■ UNIVERSITY OF PENNSYLVANIA

School of Medicine, Biomedical Graduate Studies, Graduate Group in Cell and Molecular Biology, Program in Genetics and Gene Regulation, Philadelphia, PA 19104

AWARDS PhD, MD/PhD, VMD/PhD.

Degree requirements: For doctorate, dissertation required, foreign language not required.

Entrance requirements: For doctorate, GRE General Test, TOEFL. *Application deadline:* For fall admission, 1/2 (priority date). Applications are processed on a rolling basis. *Application fee:* $65. Electronic applications accepted.

Expenses: Tuition: Full-time $17,256; part-time $2,991 per course. Required fees: $2,588; $363 per course. $726 per term.

Financial aid: Fellowships, research assistantships, teaching assistantships, grants and institutionally sponsored loans available. Financial aid application deadline: 1/2.

Dr. Mark Fortini, Chair, 215-573-6446. **Application contact:** Mary Webster, Coordinator, 215-898-4360, *E-mail:* camb@mail.med.upenn.edu.

■ UNIVERSITY OF PITTSBURGH

School of Medicine, Graduate Programs in Medicine, Program in Biochemistry and Molecular Genetics, Pittsburgh, PA 15260

AWARDS MS, PhD.

Faculty: 29 full-time (9 women), 1 part-time/adjunct (0 women).

Students: 16 full-time (5 women); includes 1 minority (Asian American or Pacific Islander), 1 international. *325 applicants, 22% accepted.* In 1999, 3 master's, 4 doctorates awarded.

Degree requirements: For doctorate, dissertation required, foreign language not required. *Average time to degree:* Doctorate–5.3 years full-time.

Entrance requirements: For doctorate, GRE General Test, GRE Subject Test, TOEFL, minimum QPA of 3.0. *Application deadline:* For fall admission, 1/15 (priority date). Applications are processed on a rolling basis. *Application fee:* $30 ($40 for international students).

Expenses: Tuition, state resident: full-time $9,778; part-time $403 per credit. Tuition, nonresident: full-time $20,146; part-time $830 per credit. Required fees: $480; $90 per semester.

Financial aid: Research assistantships with full tuition reimbursements, teaching assistantships with full tuition reimbursements, Federal Work-Study, institutionally

sponsored loans, traineeships, and unspecified assistantships available.

Faculty research: Cell cycle control and DNA replication, gene expression, human gene therapy, molecular genetics of inherited diseases, protein structure and function. *Total annual research expenditures:* $76.3 million.

Application contact: Graduate Studies Administrator, 412-648-8957, *Fax:* 412-648-1236, *E-mail:* biomed_phd@fs1.deanmed.pitt.edu.

Find an in-depth description at www.petersons.com/graduate.

■ UNIVERSITY OF RHODE ISLAND

Graduate School, College of Arts and Sciences, Department of Biochemistry, Microbiology, and Molecular Genetics, Kingston, RI 02881

AWARDS Biochemistry (MS, PhD); microbiology (MS, PhD), including biodegradation (MS), cellular development (MS), electron microscopy and ultrastructure (MS), genetics and molecular biology (MS), immunology (MS), marine and freshwater ecosystems (MS), microbial pathogenesis (MS), microbial physiology (MS), protozoology (MS), virology (MS), water-pollution microbiology (MS).

Degree requirements: For master's and doctorate, thesis/dissertation required.

Entrance requirements: For master's and doctorate, GRE General Test, TOEFL.

Expenses: Tuition, state resident: full-time $3,540; part-time $197 per credit. Tuition, nonresident: full-time $10,116; part-time $197 per credit. Required fees: $1,352; $37 per credit. $65 per term.

Financial aid: Fellowships, research assistantships, teaching assistantships available.

Dr. David Laux, Chairperson, 401-874-2201.

■ UNIVERSITY OF ROCHESTER

The College, Arts and Sciences, Department of Biology, Program in Genetics and Developmental Biology, Rochester, NY 14627-0250

AWARDS MS, PhD.

Degree requirements: For master's, thesis not required; for doctorate, dissertation, qualifying exam required, foreign language not required.

Entrance requirements: For master's and doctorate, GRE General Test, GRE Subject Test, TOEFL. *Application deadline:* For fall admission, 2/1 (priority date). *Application fee:* $25.

Expenses: Tuition: Part-time $697 per credit hour. Tuition and fees vary according to program.

Financial aid: Application deadline: 2/1. **Application contact:** Cindy Landry, Graduate Program Secretary, 716-275-7991.

■ UNIVERSITY OF ROCHESTER

School of Medicine and Dentistry, Graduate Programs in Medicine and Dentistry, Department of Biochemistry and Biophysics, Program in Genetics, Rochester, NY 14627-0250

AWARDS MS, PhD.

Students: 5 full-time (2 women); includes 2 minority (1 African American, 1 Hispanic American), 2 international. Terminal master's awarded for partial completion of doctoral program.

Degree requirements: For master's, foreign language not required; for doctorate, dissertation, qualifying exam required, foreign language not required.

Entrance requirements: For doctorate, GRE General Test. *Application deadline:* For fall admission, 2/1.

Expenses: Tuition: Part-time $697 per credit hour. Tuition and fees vary according to program.

Financial aid: Fellowships, research assistantships, teaching assistantships, tuition waivers (full and partial) available. Financial aid application deadline: 2/1.

Application contact: Rose Burgholzer, Graduate Program Secretary, 716-275-3417.

■ THE UNIVERSITY OF TENNESSEE

Graduate School, College of Agricultural Sciences and Natural Resources, Department of Plant and Soil Sciences, Knoxville, TN 37996

AWARDS Crop physiology and ecology (MS, PhD); plant breeding and genetics (MS, PhD); soil science (MS, PhD). Part-time programs available.

Faculty: 19 full-time (1 woman).

Students: 11 full-time (3 women), 11 part-time (7 women), 2 international. 7 *applicants, 43% accepted.* In 1999, 7 master's, 5 doctorates awarded.

Degree requirements: For master's, thesis or alternative required, foreign language not required; for doctorate, dissertation required, foreign language not required.

Entrance requirements: For master's and doctorate, GRE General Test, TOEFL, minimum GPA of 2.7. *Application deadline:* For fall admission, 2/1 (priority date). Applications are processed on a rolling basis. *Application fee:* $35. Electronic applications accepted.

Expenses: Tuition, state resident: full-time $3,806; part-time $184 per credit hour. Tuition, nonresident: full-time $9,874; part-time $522 per credit hour. Tuition and fees vary according to program.

Financial aid: In 1999–00, 17 research assistantships, 2 teaching assistantships were awarded; fellowships, career-related internships or fieldwork, Federal Work-Study, and institutionally sponsored loans also available. Financial aid application deadline: 2/1; financial aid applicants required to submit FAFSA.

Dr. Fred L. Allen, Head, 865-974-7101, *Fax:* 865-974-7997, *E-mail:* allenf@utk.edu.

Application contact: Dr. Mike Essington, Graduate Representative, 865-974-8818, *E-mail:* messington@utk.edu.

■ THE UNIVERSITY OF TEXAS AT AUSTIN

Graduate School, College of Natural Sciences, Institute for Cellular and Molecular Biology, Program in Genetics and Developmental Biology, Austin, TX 78712-1111

AWARDS PhD.

Expenses: Tuition, state resident: part-time $114 per semester hour. Tuition, nonresident: part-time $330 per semester hour. Tuition and fees vary according to program.

Application contact: Information Contact, 512-471-1156, *Fax:* 512-471-2149.

Find an in-depth description at www.petersons.com/graduate.

■ THE UNIVERSITY OF TEXAS–HOUSTON HEALTH SCIENCE CENTER

Graduate School of Biomedical Sciences, Program in Genes and Development, Houston, TX 77225-0036

AWARDS MS, PhD, MD/PhD.

Faculty: 25 full-time (7 women).
Students: 26 full-time (15 women); includes 4 minority (all Asian Americans or Pacific Islanders), 13 international. Average age 27. *9 applicants, 67% accepted.* In 1999, 3 master's, 9 doctorates awarded. Terminal master's awarded for partial completion of doctoral program.
Degree requirements: For master's and doctorate, thesis/dissertation required, foreign language not required.
Entrance requirements: For master's and doctorate, GRE General Test, TOEFL, TWE. *Application deadline:* For fall admission, 1/15 (priority date); for spring admission, 11/1. Applications are processed on a

rolling basis. *Application fee:* $10. Electronic applications accepted.

Financial aid: Fellowships, research assistantships, teaching assistantships, institutionally sponsored loans available. Financial aid application deadline: 1/15.

Faculty research: Mammalian embryogenesis, genetics, transgenic mice, tissue differentiation, sex determination.

Dr. Richard R. Behringer, Director, 713-794-4618, *Fax:* 713-794-4394, *E-mail:* rrb@notes.mdacc.tmc.edu.

Application contact: Anne Baronitis, Director of Admissions, 713-500-9860, *Fax:* 713-500-9877, *E-mail:* abaron@gsbs.gs.uth.tmc.edu.

Find an in-depth description at www.petersons.com/graduate.

■ THE UNIVERSITY OF TEXAS–HOUSTON HEALTH SCIENCE CENTER

Graduate School of Biomedical Sciences, Program in Human and Molecular Genetics, Houston, TX 77225-0036

AWARDS MS, PhD, MD/PhD.

Faculty: 38 full-time (10 women).
Students: 19 full-time (12 women); includes 2 minority (both Asian Americans or Pacific Islanders), 5 international. Average age 27. *47 applicants, 49% accepted.* In 1999, 2 master's, 5 doctorates awarded. Terminal master's awarded for partial completion of doctoral program.
Degree requirements: For master's and doctorate, thesis/dissertation required, foreign language not required.
Entrance requirements: For master's and doctorate, GRE General Test, TOEFL, TWE. *Application deadline:* For fall admission, 1/15 (priority date); for spring admission, 11/1. Applications are processed on a rolling basis. *Application fee:* $10. Electronic applications accepted.
Financial aid: Fellowships, research assistantships, institutionally sponsored loans available. Financial aid application deadline: 1/15.
Faculty research: Cancer genetics, genetic epidemiology, genome evolution, cytogenetics.

Dr. Michael J. Siciliano, Director, 713-792-2910, *Fax:* 713-794-4394, *E-mail:* mjs@mdanderson.org.

Application contact: Anne Baronitis, Director of Admissions, 713-500-9860, *Fax:* 713-500-9877, *E-mail:* abaron@gsbs.gs.uth.tmc.edu.

Find an in-depth description at www.petersons.com/graduate.

■ THE UNIVERSITY OF TEXAS–HOUSTON HEALTH SCIENCE CENTER

Graduate School of Biomedical Sciences, Program in Microbiology and Molecular Genetics, Houston, TX 77225-0036

AWARDS MS, PhD, MD/PhD.

Faculty: 21 full-time (11 women).
Students: 19 full-time (10 women); includes 3 minority (1 African American, 1 Asian American or Pacific Islander, 1 Hispanic American), 8 international. Average age 27. *35 applicants, 43% accepted.* In 1999, 4 master's, 7 doctorates awarded. Terminal master's awarded for partial completion of doctoral program.
Degree requirements: For master's and doctorate, thesis/dissertation required, foreign language not required.
Entrance requirements: For master's and doctorate, GRE General Test, TOEFL, TWE. *Application deadline:* For fall admission, 1/15 (priority date); for spring admission, 11/1. Applications are processed on a rolling basis. *Application fee:* $10. Electronic applications accepted.
Financial aid: Fellowships, research assistantships, institutionally sponsored loans available. Financial aid application deadline: 1/15.
Faculty research: Microbial genetics, microbial diversity, gene regulation, molecular pathogenesis, sensory transduction.

Dr. Samuel Kaplan, Chairman and Director, 713-500-5502, *Fax:* 713-500-5499, *E-mail:* skaplan@utmmg.med.uth.tmc.edu.

Application contact: Anne Baronitis, Director of Admissions, 713-500-9860, *Fax:* 713-500-9877, *E-mail:* abaron@gsbs.gs.uth.tmc.edu.

Find an in-depth description at www.petersons.com/graduate.

■ THE UNIVERSITY OF TEXAS MEDICAL BRANCH AT GALVESTON

Graduate School of Biomedical Sciences, Program in Human Biological Chemistry and Genetics, Galveston, TX 77555

AWARDS Biochemistry (MS, PhD); cell biology (MS, PhD); genetics (MS, PhD); molecular biology (PhD); structural biology (PhD).

Faculty: 61 full-time (4 women).
Students: 39 full-time (12 women); includes 2 minority (both Hispanic Americans), 12 international. Average age 28. *41 applicants, 66% accepted.* In 1999, 1 master's awarded (100% found work

The University of Texas Medical Branch at Galveston (continued)

related to degree); 4 doctorates awarded (100% found work related to degree).

Degree requirements: For master's and doctorate, thesis/dissertation required, foreign language not required.

Entrance requirements: For master's and doctorate, GRE General Test. *Application deadline:* For fall admission, 2/1 (priority date). Applications are processed on a rolling basis. *Application fee:* $25 ($50 for international students). Electronic applications accepted.

Expenses: Tuition, state resident: full-time $684; part-time $38 per credit hour. Tuition, nonresident: full-time $4,572; part-time $254 per credit hour. Required fees: $29; $7.5 per credit hour. One-time fee: $55. Tuition and fees vary according to program.

Financial aid: Fellowships, research assistantships, Federal Work-Study and institutionally sponsored loans available. Financial aid application deadline: 2/1; financial aid applicants required to submit FAFSA.

Faculty research: Cell surfaces and transmembrane signaling, structural biology, growth factors, DNA repair and aging.

Dr. James C. Lee, Director, 409-772-2769, *Fax:* 409-747-0552, *E-mail:* gradprog.hbcg@utmb.edu.

Application contact: Becky L. Hansen, Senior Administrative Secretary, 409-772-2769, *Fax:* 409-747-0552, *E-mail:* bhansen@utmb.edu.

Find an in-depth description at www.petersons.com/graduate.

■ THE UNIVERSITY OF TEXAS SOUTHWESTERN MEDICAL CENTER AT DALLAS

Southwestern Graduate School of Biomedical Sciences, Division of Cell and Molecular Biology, Program in Genetics and Development, Dallas, TX 75390

AWARDS PhD.

Faculty: 38 full-time (4 women), 2 part-time/adjunct (0 women).

Students: 50 full-time (18 women); includes 4 minority (1 Asian American or Pacific Islander, 3 Hispanic Americans), 19 international. Average age 28. In 1999, 9 doctorates awarded.

Degree requirements: For doctorate, dissertation required, foreign language not required.

Entrance requirements: For doctorate, GRE General Test, minimum GPA of 3.0. *Application deadline:* For fall admission, 1/5 (priority date). *Application fee:* $0. Electronic applications accepted.

Expenses: Tuition, state resident: full-time $912. Tuition, nonresident: full-time $6,096. Required fees: $216. Full-time tuition and fees vary according to course load and program.

Financial aid: Fellowships, research assistantships, institutionally sponsored loans available. Financial aid application deadline: 3/15; financial aid applicants required to submit FAFSA.

Faculty research: Human molecular genetics, chromosome structure, gene regulation, molecular biology, gene expression.

Dr. Dennis M. McKearin, Chair, 214-648-4944, *Fax:* 214-648-1915, *E-mail:* mckearin@utsw.swmed.edu.

Application contact: Nancy McKinney, Education Coordinator, 214-648-8099, *Fax:* 214-648-2978, *E-mail:* dcmbinfo@ utsouthwestern.edu.

Find an in-depth description at www.petersons.com/graduate.

■ UNIVERSITY OF UTAH

Graduate School, College of Science, Department of Biology, Salt Lake City, UT 84112-1107

AWARDS Biology (M Phil); ecology and evolutionary biology (MS, PhD); genetics (MS, PhD); molecular biology (PhD). Part-time programs available. Terminal master's awarded for partial completion of doctoral program.

Degree requirements: For master's and doctorate, thesis/dissertation required, foreign language not required.

Entrance requirements: For master's and doctorate, GRE General Test, GRE Subject Test, TOEFL, minimum GPA of 3.0.

Expenses: Tuition, state resident: full-time $1,663. Tuition, nonresident: full-time $5,201. Tuition and fees vary according to course load and program.

Faculty research: Behavioral ecology, cellular neurobiology, DNA replication, ecological genetics, herpetology.

Find an in-depth description at www.petersons.com/graduate.

■ UNIVERSITY OF VERMONT

College of Medicine and Graduate College, Graduate Programs in Medicine, Department of Microbiology and Molecular Genetics, Burlington, VT 05405

AWARDS MS, PhD, MD/MS, MD/PhD.

Degree requirements: For master's and doctorate, thesis/dissertation required.

Entrance requirements: For master's and doctorate, GRE General Test, TOEFL.

Expenses: Tuition, state resident: full-time $7,464; part-time $311 per credit. Tuition, nonresident: full-time $18,672; part-time $778 per credit. Full-time tuition and fees vary according to degree level and program.

Find an in-depth description at www.petersons.com/graduate.

■ UNIVERSITY OF VIRGINIA

College and Graduate School of Arts and Sciences, Department of Biochemistry and Molecular Genetics, Charlottesville, VA 22903

AWARDS Biochemistry (PhD); molecular genetics (PhD).

Faculty: 21 full-time (4 women), 1 part-time/adjunct (0 women).

Students: 24 full-time (11 women); includes 3 minority (2 African Americans, 1 Asian American or Pacific Islander), 9 international. Average age 26. *43 applicants, 21% accepted.* In 1999, 1 degree awarded.

Degree requirements: For doctorate, dissertation required, foreign language not required.

Entrance requirements: For doctorate, GRE General Test, GRE Subject Test. *Application fee:* $40. Electronic applications accepted.

Expenses: Tuition, state resident: full-time $3,832. Tuition, nonresident: full-time $15,519. Required fees: $1,084. Tuition and fees vary according to course load and program.

Financial aid: Application deadline: 2/1. Joyce L. Hamlin, Chairman, 804-924-5139.

Application contact: Duane J. Osheim, Associate Dean, 804-924-7184, *E-mail:* microbiology@virginia.edu.

Find an in-depth description at www.petersons.com/graduate.

■ UNIVERSITY OF WASHINGTON

Graduate School, College of Arts and Sciences, Department of Genetics, Seattle, WA 98195

AWARDS PhD.

Faculty: 18 full-time (4 women), 9 part-time/adjunct (3 women).

Students: 37 full-time (23 women); includes 7 minority (4 African Americans, 2 Asian Americans or Pacific Islanders, 1 Hispanic American), 1 international. *150 applicants, 20% accepted.* In 1999, 7 doctorates awarded (100% entered university research/teaching).

Degree requirements: For doctorate, dissertation, general exam required, foreign language not required. *Average time to degree:* Doctorate–6 years full-time.

Entrance requirements: For doctorate, GRE General Test, TOEFL, minimum GPA of 3.0. *Application deadline:* For fall

admission, 1/1. *Application fee:* $50.
Electronic applications accepted.
Expenses: Tuition, state resident: full-time
$5,196; part-time $495 per credit. Tuition,
nonresident: full-time $13,485; part-time
$1,285 per credit. Required fees: $387; $36
per credit. Tuition and fees vary according
to course load and program.
Financial aid: In 1999–00, 37 students
received aid, including 5 fellowships with
full tuition reimbursements available
(averaging $17,304 per year); traineeships
also available. Financial aid application
deadline: 1/1.
Faculty research: Genetics of bacteria,
yeast, plants, mammals; development. *Total
annual research expenditures:* $3 million.
Breck E. Byers, Chairman, 206-543-9068,
Fax: 206-543-0754, *E-mail:* byers@
genetics.u.washington.edu.
Application contact: Carol H. Sibley,
Graduate Program Coordinator, 206-685-
9378, *Fax:* 206-543-0754, *E-mail:* sibley@
genetics.u.washington.edu.
**Find an in-depth description at
www.petersons.com/graduate.**

■ **UNIVERSITY OF WASHINGTON**

**Graduate School, School of Public
Health and Community Medicine,
Department of Epidemiology, Program
in Public Health Genetics, Seattle, WA
98195**

AWARDS Genetic epidemiology (MS); public
health genetics (MPH). Part-time programs
available.

Faculty: 15 full-time (10 women).
Students: 8 full-time (6 women); includes
1 minority (Asian American or Pacific
Islander). Average age 29. *13 applicants,
62% accepted.* In 1999, 1 degree awarded
(100% continued full-time study).
Degree requirements: For master's,
thesis, practicum required, foreign
language not required. *Average time to
degree:* Master's–2 years full-time.
Entrance requirements: For master's,
GRE General Test, TOEFL, experience in
health sciences preferred, minimum GPA
of 3.0. *Application deadline:* For fall admission, 2/1. *Application fee:* $50.
Expenses: Tuition, state resident: full-time
$5,196; part-time $495 per credit. Tuition,
nonresident: full-time $13,485; part-time
$1,285 per credit. Required fees: $387; $36
per credit. Tuition and fees vary according
to course load and program.
Financial aid: In 1999–00, 1 student
received aid, including 1 research assistantship with tuition reimbursement available
(averaging $10,440 per year). Financial aid
application deadline: 2/28.
Faculty research: Genetic epidemiology;
ethical, legal, social issues of genetics;
ecogenetics; health policy.

Dr. Melissa A. Austin, Director, 206-543-
0709, *Fax:* 206-685-9651, *E-mail:*
maustin@u.washington.edu.
Application contact: Kevin M. Schuda,
Program Manager, 206-616-9286, *Fax:*
206-685-9651, *E-mail:* phgen@
u.washington.edu.

■ **UNIVERSITY OF WISCONSIN–
MADISON**

**Graduate School, College of
Agricultural and Life Sciences and
Graduate Programs in Medicine,
Departments of Genetics and Medical
Genetics, Program in Genetics,
Madison, WI 53706-1380**

AWARDS PhD.

Degree requirements: For doctorate, dissertation required.
Expenses: Tuition, state resident: full-time
$5,406; part-time $339 per credit. Tuition,
nonresident: full-time $17,110; part-time
$1,071 per credit. Full-time tuition and
fees vary according to program and
reciprocity agreements. Part-time tuition
and fees vary according to course load and
program.
**Find an in-depth description at
www.petersons.com/graduate.**

■ **UNIVERSITY OF WISCONSIN–
MADISON**

**Graduate School, College of
Agricultural and Life Sciences and
Graduate Programs in Medicine,
Departments of Genetics and Medical
Genetics, Program in Medical
Genetics, Madison, WI 53706-1380**

AWARDS MS.

Expenses: Tuition, state resident: full-time
$5,406; part-time $339 per credit. Tuition,
nonresident: full-time $17,110; part-time
$1,071 per credit. Full-time tuition and
fees vary according to program and
reciprocity agreements. Part-time tuition
and fees vary according to course load and
program.
**Find an in-depth description at
www.petersons.com/graduate.**

■ **UNIVERSITY OF WISCONSIN–
MADISON**

**Medical School and Graduate School,
Graduate Programs in Medicine,
Madison, WI 53706-1380**

AWARDS Biomolecular chemistry (MS, PhD);
cancer biology (PhD); genetics and medical
genetics (MS, PhD), including genetics (PhD),
medical genetics (MS); medical physics (MS,
PhD), including health physics (MS), medical
physics; microbiology (PhD); molecular and
cellular pharmacology (PhD); oncology (PhD);

pathology and laboratory medicine (PhD);
physiology (PhD), including neurophysiology,
physiology; population health (MS, PhD).
Part-time programs available.
Postbaccalaureate distance learning degree
programs offered (minimal on-campus study).
Terminal master's awarded for partial completion of doctoral program.
Degree requirements: For master's,
foreign language not required.
Application fee: $45. Electronic applications
accepted.
Expenses: Tuition, state resident: full-time
$5,406; part-time $339 per credit. Tuition,
nonresident: full-time $17,110; part-time
$1,071 per credit.
Financial aid: Fellowships with full tuition
reimbursements, research assistantships
with full tuition reimbursements, teaching
assistantships with full tuition reimbursements, grants, scholarships, traineeships,
and tuition waivers (full) available.
Dr. Paul M. DeLuca, Associate Dean of
Research and Graduate Studies, 608-265-
0524, *Fax:* 608-265-0522, *E-mail:*
pmdeluca@facstaff.wisc.edu.

■ **VIRGINIA COMMONWEALTH
UNIVERSITY**

**School of Graduate Studies and
School of Medicine, School of
Medicine Graduate Programs,
Department of Biochemistry and
Molecular Biophysics, Richmond, VA
23284-9005**

AWARDS Biochemistry (PhD); biochemistry
and molecular biophysics (MS, CBHS);
molecular biology and genetics (PhD);
neurosciences (PhD).

Students: 5 full-time (3 women), 21 part-
time (10 women); includes 12 minority (9
Asian Americans or Pacific Islanders, 1
Hispanic American, 2 Native Americans).
In 1999, 1 master's, 10 doctorates
awarded.
Degree requirements: For master's,
thesis required, foreign language not
required; for doctorate, dissertation,
comprehensive oral and written exams
required, foreign language not required.
Entrance requirements: For master's and
doctorate, GRE General Test. *Application
deadline:* For fall admission, 5/1. *Application
fee:* $30.
Expenses: Tuition, state resident: full-time
$4,031; part-time $224 per credit hour.
Tuition, nonresident: full-time $11,946;
part-time $664 per credit hour. Required
fees: $1,081; $40 per credit hour. Tuition
and fees vary according to campus/location
and program.
Financial aid: Fellowships, research
assistantships available.
Faculty research: Molecular biology,
peptide/protein chemistry, neurochemistry,

Virginia Commonwealth University (continued)

enzyme mechanisms, macromolecular structure determination. *Total annual research expenditures:* $3.5 million.
Chair, 804-828-9762, *Fax:* 804-828-1473.
Application contact: Dr. Zendra E. Zehner, Program Director, 804-828-8753, *Fax:* 804-828-1473, *E-mail:* zezehner@vcu.edu.

■ VIRGINIA COMMONWEALTH UNIVERSITY

School of Graduate Studies and School of Medicine, School of Medicine Graduate Programs, Department of Human Genetics, Richmond, VA 23284-9005

AWARDS Genetic counseling (MS); human genetics (PhD, CBHS); molecular biology and genetics (PhD).

Students: In 1999, 3 master's, 3 doctorates awarded.
Degree requirements: For master's, computer language, thesis required, foreign language not required; for doctorate, computer language, dissertation, comprehensive oral and written exams required, foreign language not required.
Entrance requirements: For master's, DAT, GRE General Test, or MCAT; for doctorate, GRE General Test. *Application deadline:* For fall admission, 2/1. *Application fee:* $30.
Expenses: Tuition, state resident: full-time $4,031; part-time $224 per credit hour. Tuition, nonresident: full-time $11,946; part-time $664 per credit hour. Required fees: $1,081; $40 per credit hour. Tuition and fees vary according to campus/location and program.
Financial aid: Fellowships available.
Faculty research: Genetic epidemiology, biochemical genetics, quantitative genetics, human cytogenetics, molecular genetics.
Dr. Walter E. Nance, Chair, 804-828-9632, *Fax:* 804-828-3760.
Application contact: Dr. Linda A. Corey, Graduate Program Director, 804-828-9632, *Fax:* 804-828-3760, *E-mail:* lacorey@vcu.edu.

Find an in-depth description at www.petersons.com/graduate.

■ VIRGINIA COMMONWEALTH UNIVERSITY

School of Graduate Studies and School of Medicine, School of Medicine Graduate Programs, Department of Microbiology and Immunology, Richmond, VA 23284-9005

AWARDS Microbiology (PhD); microbiology and immunology (MS, CBHS); molecular biology and genetics (PhD).

Students: 13 full-time (6 women), 46 part-time (26 women); includes 22 minority (2 African Americans, 17 Asian Americans or Pacific Islanders, 2 Hispanic Americans, 1 Native American). In 1999, 2 master's, 9 doctorates, 2 other advanced degrees awarded.
Degree requirements: For master's, thesis required, foreign language not required; for doctorate, dissertation, comprehensive oral and written exams required, foreign language not required.
Entrance requirements: For master's, DAT, GRE General Test, or MCAT; for doctorate, GRE General Test. *Application deadline:* For fall admission, 4/15. *Application fee:* $30.
Expenses: Tuition, state resident: full-time $4,031; part-time $224 per credit hour. Tuition, nonresident: full-time $11,946; part-time $664 per credit hour. Required fees: $1,081; $40 per credit hour. Tuition and fees vary according to campus/location and program.
Financial aid: Fellowships, research assistantships, teaching assistantships available.
Faculty research: Microbial physiology and genetics, molecular biology, crystallography of biological molecules, antibiotics and chemotherapy, membrane transport.
Dr. Dennis E. Ohman, Chair, 804-828-9728, *Fax:* 804-828-9946.

Find an in-depth description at www.petersons.com/graduate.

■ VIRGINIA COMMONWEALTH UNIVERSITY

School of Graduate Studies and School of Medicine, School of Medicine Graduate Programs, Department of Pharmacology and Toxicology, Richmond, VA 23284-9005

AWARDS Molecular biology and genetics (PhD); neurosciences (PhD); pharmacology (PhD, CBHS); pharmacology and toxicology (MS).

Students: 14 full-time (5 women), 51 part-time (27 women); includes 27 minority (4 African Americans, 22 Asian Americans or Pacific Islanders, 1 Hispanic American). In 1999, 3 master's, 10 doctorates, 1 other

advanced degree awarded. Terminal master's awarded for partial completion of doctoral program.
Degree requirements: For master's, thesis required, foreign language not required; for doctorate, dissertation, comprehensive oral and written exams required, foreign language not required.
Entrance requirements: For master's, DAT, GRE General Test or MCAT; for doctorate, GRE General Test. *Application deadline:* For fall admission, 2/1 (priority date). *Application fee:* $30.
Expenses: Tuition, state resident: full-time $4,031; part-time $224 per credit hour. Tuition, nonresident: full-time $11,946; part-time $664 per credit hour. Required fees: $1,081; $40 per credit hour. Tuition and fees vary according to campus/location and program.
Financial aid: Fellowships, teaching assistantships available.
Faculty research: Drug abuse, drug metabolism, pharmacodynamics, peptide synthesis, receptor mechanisms.
Dr. George Kunos, Chair, 804-828-2073, *Fax:* 804-828-2117.
Application contact: Sheryol Cox, Graduate Program Coordinator, 804-828-8400, *Fax:* 804-828-2117, *E-mail:* swcox@vcu.edu.

Find an in-depth description at www.petersons.com/graduate.

■ VIRGINIA POLYTECHNIC INSTITUTE AND STATE UNIVERSITY

Graduate School, College of Agriculture and Life Sciences and Department of Biology, Program in Genetics, Blacksburg, VA 24061

AWARDS PhD.

Faculty: 30 full-time (3 women).
Students: 3 full-time (1 woman), 2 international. *4 applicants, 25% accepted.* In 1999, 1 degree awarded.
Degree requirements: For doctorate, dissertation required.
Entrance requirements: For doctorate, GRE General Test, TOEFL, master's degree in related field. *Application deadline:* For fall admission, 12/1 (priority date). Applications are processed on a rolling basis. *Application fee:* $25.
Expenses: Tuition, state resident: full-time $4,122; part-time $229 per credit hour. Tuition, nonresident: full-time $6,930; part-time $385 per credit hour. Required fees: $828; $107 per semester. Part-time tuition and fees vary according to course load.
Financial aid: Application deadline: 4/1.
Faculty research: Behavioral genetics, evolutionary genetics, plant and animal breeding, plant molecular genetics.

Dr. Ina Hoeschele, Chairman, 540-231-8944, *E-mail:* inah@vt.edu.

■ VIRGINIA POLYTECHNIC INSTITUTE AND STATE UNIVERSITY

Graduate School, College of Arts and Sciences, Department of Biology, Blacksburg, VA 24061

AWARDS Botany (MS, PhD); ecology (MS, PhD); genetics (PhD); microbiology (MS, PhD); zoology (MS, PhD).

Faculty: 44 full-time (3 women).
Students: 67 full-time (30 women), 12 part-time (8 women); includes 5 minority (2 African Americans, 2 Asian Americans or Pacific Islanders, 1 Hispanic American), 14 international. Average age 25. *98 applicants, 26% accepted.* In 1999, 14 master's, 10 doctorates awarded.
Degree requirements: For master's and doctorate, thesis/dissertation required, foreign language not required.
Entrance requirements: For master's, GRE General Test, TOEFL; for doctorate, GRE General Test, GRE Subject Test, TOEFL. *Application fee:* $25.
Expenses: Tuition, state resident: full-time $4,122; part-time $229 per credit hour. Tuition, nonresident: full-time $6,930; part-time $385 per credit hour. Required fees: $828; $107 per semester. Part-time tuition and fees vary according to course load.
Financial aid: In 1999–00, 19 research assistantships, 39 teaching assistantships were awarded; fellowships, unspecified assistantships also available.
Faculty research: Evolution, molecular biology, systematics.
Dr. Joe R. Cowles, Chairman, 540-231-8928, *E-mail:* cowlesjr@vt.edu.
Find an in-depth description at www.petersons.com/graduate.

■ WAKE FOREST UNIVERSITY

School of Medicine and Graduate School, Graduate Programs in Medicine, Department of Pediatrics, Winston-Salem, NC 27109

AWARDS Medical genetics (MS).

Degree requirements: For master's, thesis required.
Entrance requirements: For master's, GRE General Test, GRE Subject Test.
Expenses: Tuition: Full-time $18,300.
Faculty research: Morphological and biochemical aspects of Down's syndrome.

■ WASHINGTON STATE UNIVERSITY

Graduate School, College of Sciences, School of Molecular Biosciences, Department of Genetics and Cell Biology, Pullman, WA 99164

AWARDS MS, PhD.

Faculty: 14 full-time (4 women), 2 part-time/adjunct (both women).
Students: 28 full-time (11 women), 2 part-time (both women); includes 2 minority (both Asian Americans or Pacific Islanders), 5 international. Average age 24. In 1999, 4 master's awarded (100% continued full-time study); 1 doctorate awarded (100% entered university research/teaching). Terminal master's awarded for partial completion of doctoral program.
Degree requirements: For master's, thesis or alternative, oral exam required, foreign language not required; for doctorate, dissertation, oral exam required, foreign language not required. *Average time to degree:* Master's–2.5 years full-time; doctorate–4.67 years full-time.
Entrance requirements: For master's and doctorate, GRE General Test, minimum GPA of 3.0. *Application deadline:* For fall admission, 3/1 (priority date). Applications are processed on a rolling basis. *Application fee:* $35.
Expenses: Tuition, state resident: full-time $5,654. Tuition, nonresident: full-time $13,850. International tuition: $13,850 full-time. Tuition and fees vary according to program.
Financial aid: In 1999–00, 25 students received aid, including 13 research assistantships with full and partial tuition reimbursements available, 11 teaching assistantships with full and partial tuition reimbursements available; fellowships, Federal Work-Study, institutionally sponsored loans, tuition waivers (partial), and teaching associateships also available. Financial aid application deadline: 4/1; financial aid applicants required to submit FAFSA.
Faculty research: Plant molecular biology, population genetics, growth factors, cancer induction and DNA repair, gene regulation and genetic engineering. *Total annual research expenditures:* $426,134.
Dr. Paul Lurquin, Chair, 509-335-5733, *Fax:* 509-335-1907.
Application contact: Mary Ann Storms, Program Assistant, 509-335-4566, *Fax:* 509-335-1907, *E-mail:* storms@mail.wsu.edu.
Find an in-depth description at www.petersons.com/graduate.

■ WASHINGTON UNIVERSITY IN ST. LOUIS

Graduate School of Arts and Sciences, Division of Biology and Biomedical Sciences, Program in Evolutionary and Population Biology, St. Louis, MO 63130-4899

AWARDS Ecology (PhD); environmental biology (PhD); evolutionary biology (PhD); genetics (PhD).

Degree requirements: For doctorate, dissertation required, foreign language not required.
Entrance requirements: For doctorate, GRE General Test, GRE Subject Test. *Application deadline:* For fall admission, 1/1 (priority date). Applications are processed on a rolling basis. *Application fee:* $0.
Expenses: Tuition: Full-time $23,400; part-time $975 per credit. Tuition and fees vary according to program.
Financial aid: Fellowships, research assistantships available. Financial aid application deadline: 1/1.
Dr. Jonathan Losos, Coordinator, 314-362-4188.
Application contact: Rosemary Garagneni, Director of Admissions, 800-852-9074, *E-mail:* admissions@dbbs.wustl.edu.

■ WASHINGTON UNIVERSITY IN ST. LOUIS

Graduate School of Arts and Sciences, Division of Biology and Biomedical Sciences, Program in Molecular Genetics, St. Louis, MO 63130-4899

AWARDS PhD.

Degree requirements: For doctorate, dissertation required, foreign language not required.
Entrance requirements: For doctorate, GRE General Test, GRE Subject Test. *Application deadline:* For fall admission, 1/1 (priority date). Applications are processed on a rolling basis. *Application fee:* $0.
Expenses: Tuition: Full-time $23,400; part-time $975 per credit. Tuition and fees vary according to program.
Financial aid: Application deadline: 1/1.
Dr. Mark Johnston, Coordinator, 314-362-9060.
Application contact: Rosemary Garagneni, Director of Admissions, 800-852-9074, *E-mail:* admissions@dbbs.wustl.edu.

■ WAYNE STATE UNIVERSITY

School of Medicine and Graduate School, Graduate Programs in Medicine, Program in Molecular Biology and Genetics, Detroit, MI 48202

AWARDS MS, PhD. Terminal master's awarded for partial completion of doctoral program.

Degree requirements: For master's and doctorate, thesis/dissertation required, foreign language not required.
Entrance requirements: For master's and doctorate, GRE General Test.
Faculty research: Human gene mapping, genome organization and sequencing, gene regulation, molecular evolution.

Find an in-depth description at www.petersons.com/graduate.

■ WESLEYAN UNIVERSITY

Graduate Programs, Department of Biology, Middletown, CT 06459-0260

AWARDS Cell biology (PhD); comparative physiology (PhD); developmental biology (PhD); genetics (PhD); neurophysiology (PhD); population biology (PhD).

Faculty: 12 full-time (3 women).
Students: 24 full-time (12 women); includes 1 minority (African American), 11 international. Average age 28. *125 applicants, 4% accepted.* In 1999, 2 doctorates awarded.
Degree requirements: For doctorate, one foreign language (computer language can substitute), dissertation required.
Entrance requirements: For doctorate, GRE Subject Test. *Application deadline:* For fall admission, 1/15. Applications are processed on a rolling basis. *Application fee:* $0.
Expenses: Tuition: Full-time $24,876. Required fees: $650. Tuition and fees vary according to program.
Financial aid: Research assistantships, teaching assistantships, stipends available.
Faculty research: Microbial population genetics, genetic basis of evolutionary adaptation, genetic regulation of differentiation and pattern formation in *drosophila.*
Dr. Fred Cohan, Chairman, 860-685-3489.
Application contact: Marina J. Melendez, Director of Graduate Student Services, 860-685-2390, *Fax:* 860-685-2439, *E-mail:* mmelendez@wesleyan.edu.

Find an in-depth description at www.petersons.com/graduate.

■ WEST VIRGINIA UNIVERSITY

College of Agriculture, Forestry and Consumer Sciences, Interdisciplinary Program in Genetics and Developmental Biology, Morgantown, WV 26506

AWARDS Animal breeding (MS, PhD); biochemical and molecular genetics (MS, PhD); cytogenetics (MS, PhD); descriptive embryology (MS, PhD); developmental genetics (MS); experimental morphogenesis (MS); human genetics (MS, PhD); immunogenetics (MS, PhD); life cycles of animals and plants (MS, PhD); molecular aspects of development (MS, PhD); mutagenesis (PhD); mutagenetics (MS); oncology (MS, PhD); plant genetics (MS, PhD); population and quantitative genetics (PhD); population and quantitative genetics (MS); regeneration (MS, PhD); teratology (MS, PhD); toxicology (MS, PhD).

Students: 18 full-time (8 women), 5 part-time (2 women); includes 1 minority (Asian American or Pacific Islander), 8 international. Average age 27. In 1999, 4 doctorates awarded.
Degree requirements: For master's, thesis required, foreign language not required; for doctorate, dissertation, comprehensive exam required, foreign language not required.
Entrance requirements: For master's, GRE or MCAT, TOEFL, minimum GPA of 2.75; for doctorate, TOEFL. *Application fee:* $45.
Expenses: Tuition, state resident: full-time $2,910; part-time $154 per credit hour. Tuition, nonresident: full-time $8,368; part-time $457 per credit hour.
Financial aid: In 1999–00, 11 research assistantships, 3 teaching assistantships were awarded; fellowships, Federal Work-Study, institutionally sponsored loans, and tuition waivers (full and partial) also available. Financial aid application deadline: 2/1; financial aid applicants required to submit FAFSA.
Dr. J. Nath, Chairman, 304-293-6256 Ext. 4333, *E-mail:* jnath@wvu.edu.

■ WEST VIRGINIA UNIVERSITY

Eberly College of Arts and Sciences, Department of Biology, Morgantown, WV 26506

AWARDS Animal behavior (MS); cellular and molecular biology (MS, PhD); environmental plant biology (MS, PhD); plant systematics (MS); population genetics (MS).

Faculty: 18 full-time (4 women), 2 part-time/adjunct (1 woman).
Students: 19 full-time (10 women), 3 part-time (2 women); includes 2 minority (1 Asian American or Pacific Islander, 1 Hispanic American), 4 international. Average age 27. *50 applicants, 10% accepted.* In

1999, 4 master's, 1 doctorate awarded. Terminal master's awarded for partial completion of doctoral program.
Degree requirements: For master's, thesis, final exam required, foreign language not required; for doctorate, dissertation, preliminary and final exams required, foreign language not required. *Average time to degree:* Master's–2.5 years full-time; doctorate–5 years full-time.
Entrance requirements: For master's, GRE General Test, GRE Subject Test, TOEFL, minimum GPA of 3.0; for doctorate, GRE General Test, TOEFL, minimum GPA of 3.0. *Application deadline:* For fall admission, 4/1; for spring admission, 10/1. Applications are processed on a rolling basis. *Application fee:* $45.
Expenses: Tuition, state resident: full-time $2,910; part-time $154 per credit hour. Tuition, nonresident: full-time $8,368; part-time $457 per credit hour.
Financial aid: In 1999–00, 3 research assistantships, 13 teaching assistantships were awarded; Federal Work-Study and institutionally sponsored loans also available. Financial aid application deadline: 4/1; financial aid applicants required to submit FAFSA.
Faculty research: Environmental biology, genetic engineering, developmental biology, global change, biodiversity.
Dr. Keith Garbutt, Chair, 304-293-5394.
Application contact: Dr. James B. McGraw, Director of Graduate Studies, 304-293-5201, *Fax:* 304-293-6363.

Find an in-depth description at www.petersons.com/graduate.

■ WEST VIRGINIA UNIVERSITY

School of Medicine, Graduate Programs in Health Sciences, Department of Microbiology and Immunology, Morgantown, WV 26506

AWARDS Genetics (MS, PhD); immunology (MS, PhD); mycology (PhD); parasitology (MS, PhD); pathogenic bacteriology (MS, PhD); physiology (PhD); psysiology (MS); virology (MS, PhD).

Students: 25 full-time (11 women), 1 (woman) part-time; includes 2 minority (both Asian Americans or Pacific Islanders), 4 international. Average age 27. *62 applicants, 8% accepted.* In 1999, 1 master's, 5 doctorates awarded. Terminal master's awarded for partial completion of doctoral program.
Degree requirements: For master's, thesis required, foreign language not required; for doctorate, dissertation, comprehensive exam required, foreign language not required. *Average time to degree:* Doctorate–4.5 years full-time.
Entrance requirements: For master's and doctorate, GRE General Test, GRE

Subject Test, TOEFL, minimum GPA of 3.0. *Application deadline:* For fall admission, 3/1 (priority date). Applications are processed on a rolling basis. *Application fee:* $45. Electronic applications accepted.
Expenses: Tuition, state resident: full-time $3,564. Tuition, nonresident: full-time $10,230.
Financial aid: In 1999–00, 7 research assistantships, 14 teaching assistantships were awarded; fellowships, Federal Work-Study and institutionally sponsored loans also available. Financial aid application deadline: 3/1; financial aid applicants required to submit FAFSA.
Faculty research: Mechanisms of pathogenesis, microbial genetics, molecular virology, immunotoxicology, oncogenes. *Total annual research expenditures:* $2.6 million.
John Barnett, Chair, 304-293-2649.
Application contact: James Sheil, Graduate Coordinator, 304-293-3559, *Fax:* 304-293-7823, *E-mail:* jsheil@wvu.edu.

■ YALE UNIVERSITY

Graduate School of Arts and Sciences, Department of Genetics, New Haven, CT 06520

AWARDS PhD, MD/PhD.

Faculty: 66 full-time (24 women), 4 part-time/adjunct (3 women).
Students: 39 full-time (20 women); includes 3 minority (1 Asian American or Pacific Islander, 2 Hispanic Americans), 7 international. *87 applicants, 32% accepted.* In 1999, 6 degrees awarded.
Degree requirements: For doctorate, dissertation required, foreign language not required. *Average time to degree:* Doctorate–5.5 years full-time.
Entrance requirements: For doctorate, GRE General Test, GRE Subject Test. *Application deadline:* For fall admission, 1/4. *Application fee:* $65.
Expenses: Tuition: Full-time $22,300. Full-time tuition and fees vary according to program.
Financial aid: Fellowships, research assistantships, teaching assistantships, Federal Work-Study and institutionally sponsored loans available. Aid available to part-time students.
Application contact: Admissions Information, 203-432-2770.

■ YALE UNIVERSITY

Graduate School of Arts and Sciences, Department of Molecular, Cellular, and Developmental Biology, Program in Genetics, New Haven, CT 06520

AWARDS PhD.

Degree requirements: For doctorate, dissertation required.

Entrance requirements: For doctorate, GRE General Test, GRE Subject Test. *Application deadline:* For fall admission, 1/4. *Application fee:* $65.
Expenses: Tuition: Full-time $22,300. Full-time tuition and fees vary according to program.
Application contact: Admissions Information, 203-432-2770.

Find an in-depth description at www.petersons.com/graduate.

■ YALE UNIVERSITY

School of Medicine and Graduate School of Arts and Sciences, Combined Program in Biological and Biomedical Sciences (BBS), Genetics and Development Track, New Haven, CT 06520

AWARDS Genetics, development, and molecular biology (PhD).

Degree requirements: For doctorate, dissertation required.
Entrance requirements: For doctorate, GRE General Test, TOEFL. *Application deadline:* For fall admission, 1/2. *Application fee:* $65. Electronic applications accepted.
Expenses: All students receive full tuition of $22,330 and an annual stipend of $17,600 .
Financial aid: Fellowships, research assistantships available.
Dr. Lynn Cooley, Director of Graduate Studies, 203-785-5067, *E-mail:* bbs.gendev@yale.edu.
Application contact: Betsy Jasiorkowski, Graduate Registrar, 203-785-5846, *E-mail:* bbs.gendev@yale.edu.

Find an in-depth description at www.petersons.com/graduate.

■ YESHIVA UNIVERSITY

Albert Einstein College of Medicine, Sue Golding Graduate Division of Medical Sciences, Division of Biological Sciences, Department of Molecular Genetics, Bronx, NY 10461

AWARDS PhD, MD/PhD.

Faculty: 10 full-time.
Students: 32 full-time (18 women); includes 8 minority (1 African American, 5 Asian Americans or Pacific Islanders, 2 Hispanic Americans), 14 international. In 1999, 2 degrees awarded.
Degree requirements: For doctorate, dissertation required, foreign language not required.
Entrance requirements: For doctorate, GRE General Test, TOEFL. *Application deadline:* For fall admission, 1/15. *Application fee:* $0.
Expenses: Tuition: Part-time $525 per credit. Tuition and fees vary according to degree level and program.

Financial aid: In 1999–00, 32 fellowships were awarded.
Faculty research: Neurologic genetics in *Drosophila*, biochemical genetics of yeast, developmental genetics in the mouse.
Dr. R. Kucherlapati, Chairperson, 718-430-2824.

HUMAN GENETICS

■ BAYLOR COLLEGE OF MEDICINE

Graduate School of Biomedical Sciences, Department of Molecular and Human Genetics, Houston, TX 77030-3498

AWARDS PhD, MD/PhD.

Faculty: 49 full-time (10 women).
Students: 57 full-time (32 women); includes 9 minority (1 African American, 4 Asian Americans or Pacific Islanders, 4 Hispanic Americans), 18 international. Average age 27. *127 applicants, 33% accepted.* In 1999, 7 degrees awarded.
Degree requirements: For doctorate, dissertation, public defense, qualifying exam required, foreign language not required. *Average time to degree:* Doctorate–5.4 years full-time.
Entrance requirements: For doctorate, GRE General Test (average 80th percentile), GRE Subject Test (strongly recommended), TOEFL, minimum GPA of 3.0. *Application deadline:* For fall admission, 2/1 (priority date). *Application fee:* $30. Electronic applications accepted.
Expenses: Tuition: Full-time $8,200. Required fees: $175. Full-time tuition and fees vary according to student level.
Financial aid: In 1999–00, 57 students received aid, including 30 fellowships (averaging $16,000 per year), 27 research assistantships (averaging $16,000 per year); Federal Work-Study, institutionally sponsored loans, and tuition waivers (full) also available. Financial aid applicants required to submit FAFSA.
Faculty research: Cytogenetics, biochemical genetics, somatic cell genetics, gene therapy.
Dr. Craig Chinault, Director, 713-798-5056, *E-mail:* rtanagho@bcm.tmc.edu.
Application contact: Lillie Tanagho, Graduate Program Administrator, 713-798-5056, *Fax:* 713-798-8597, *E-mail:* rtanagho@bcm.tmc.edu.

Find an in-depth description at www.petersons.com/graduate.

■ CASE WESTERN RESERVE UNIVERSITY

School of Medicine and School of Graduate Studies, Graduate Programs in Medicine, Department of Genetics, Program in Human, Molecular, and Developmental Genetics and Genomics, Cleveland, OH 44106

AWARDS PhD, MD/PhD.

Faculty: 26 full-time (9 women).
Students: 33 full-time (11 women). In 1999, 5 degrees awarded (80% entered university research/teaching, 20% continued full-time study).
Degree requirements: For doctorate, dissertation required. *Average time to degree:* Doctorate–6 years full-time.
Entrance requirements: For doctorate, GRE General Test, GRE Subject Test, TOEFL. *Application deadline:* For fall admission, 2/28 (priority date). Applications are processed on a rolling basis. *Application fee:* $25.
Financial aid: Fellowships, research assistantships available. Financial aid application deadline: 2/28.
Faculty research: Regulation of gene expression, molecular control of development, genomics. *Total annual research expenditures:* $8 million.
Application contact: Chair, Graduate Committee, 216-368-5847, *Fax:* 216-368-5857.

Find an in-depth description at www.petersons.com/graduate.

■ HOFSTRA UNIVERSITY

College of Liberal Arts and Sciences, Division of Natural Sciences, Mathematics, Engineering, and Computer Science, Department of Biology, Hempstead, NY 11549

AWARDS Biology (MA); human cytogenetics (MS). Part-time and evening/weekend programs available.

Degree requirements: For master's, thesis, internship (MS) required.
Entrance requirements: For master's, bachelor's degree in biology or equivalent.
Expenses: Tuition: Full-time $11,400. Required fees: $670. Tuition and fees vary according to course load and program.
Faculty research: Evolution, genetics, aquaculture, molecular and cellular biology, marine sciences.

Find an in-depth description at www.petersons.com/graduate.

■ HOWARD UNIVERSITY

Graduate School of Arts and Sciences, Department of Genetics and Human Genetics, Washington, DC 20059

AWARDS MS, PhD. Part-time programs available.

Faculty: 13.
Students: 23; includes 16 minority (15 African Americans, 1 Asian American or Pacific Islander), 6 international. In 1999, 11 master's, 2 doctorates awarded.
Degree requirements: For master's, thesis, comprehensive exam required; for doctorate, dissertation, comprehensive exam required. *Average time to degree:* Master's–2 years full-time; doctorate–4 years full-time.
Entrance requirements: For master's, GRE General Test, TOEFL, minimum GPA of 3.0, previous course work in biology, chemistry, physics, mathematics, and genetics; for doctorate, GRE General Test, TOEFL, master's degree in genetics or related field, minimum GPA of 3.2. *Application deadline:* For fall admission, 4/1; for spring admission, 11/1. *Application fee:* $45.
Expenses: Tuition: Full-time $10,500; part-time $583 per credit hour. Required fees: $405; $203 per semester.
Financial aid: Fellowships, research assistantships, teaching assistantships available.
Dr. Robert F. Murray, Chairman, 202-806-6340.
Application contact: Dr. Verle E. Headings, Director of Graduate Studies, 202-806-6381.

■ JOHNS HOPKINS UNIVERSITY

School of Medicine, Graduate Programs in Medicine, Predoctoral Training Program in Human Genetics and Molecular Biology, Baltimore, MD 21218-2699

AWARDS PhD, MD/PhD.

Faculty: 55 full-time (17 women).
Students: 58 full-time (31 women); includes 10 minority (9 Asian Americans or Pacific Islanders, 1 Hispanic American), 2 international. Average age 24. *70 applicants, 14% accepted.* In 1999, 5 degrees awarded (100% entered university research/teaching).
Degree requirements: For doctorate, dissertation required, foreign language not required. *Average time to degree:* Doctorate–6 years full-time.
Entrance requirements: For doctorate, GRE General Test, GRE Subject Test. *Application deadline:* For fall admission, 1/15 (priority date). *Application fee:* $50.
Expenses: Tuition: Full-time $23,660.

Financial aid: Fellowships, teaching assistantships available.
Faculty research: Human, mammalian, and molecular genetics.
Dr. David Valle, Director, 410-955-4260, *Fax:* 410-955-7397, *E-mail:* human.genetics@jhmi.edu.
Application contact: Sandy Muscelli, Program Administrator, 410-955-4260, *Fax:* 410-955-7397, *E-mail:* muscelli@jhmi.edu.

Find an in-depth description at www.petersons.com/graduate.

■ LOUISIANA STATE UNIVERSITY HEALTH SCIENCES CENTER

School of Graduate Studies in New Orleans, Department of Biometry and Genetics, Program in Human Genetics, New Orleans, LA 70112-2223

AWARDS MS, PhD, MD/PhD. Part-time programs available.

Faculty: 2 full-time (both women), 7 part-time/adjunct (2 women).
Students: 4 full-time (2 women); includes 1 minority (African American). Average age 25. *3 applicants, 67% accepted.* Terminal master's awarded for partial completion of doctoral program.
Degree requirements: For master's and doctorate, computer language, thesis/dissertation required, foreign language not required. *Average time to degree:* Master's–3 years full-time.
Entrance requirements: For master's and doctorate, GRE General Test, TOEFL. *Application deadline:* For fall admission, 4/1 (priority date). Applications are processed on a rolling basis. *Application fee:* $30.
Expenses: Tuition, state resident: full-time $2,878; part-time $126 per hour. Tuition, nonresident: full-time $6,003; part-time $265 per hour. Required fees: $2,272. Tuition and fees vary according to course load, degree level and program.
Financial aid: In 1999–00, 2 research assistantships with tuition reimbursements (averaging $11,000 per year) were awarded; tuition waivers (full) also available.
Faculty research: Genetic epidemiology, segregation and linkage analysis, gene mapping.
Application contact: Dr. Bronya J. Keats, Acting Head, 504-568-6150, *Fax:* 504-568-8500, *E-mail:* biombjk@lsumc.edu.

■ MCP HAHNEMANN UNIVERSITY

School of Medicine, Biomedical Graduate Programs, Interdisciplinary Program in Molecular and Human Genetics, Philadelphia, PA 19102-1192

AWARDS MS, PhD, MD/PhD.

Degree requirements: For master's, comprehensive exam required; for doctorate, dissertation, qualifying exam required, foreign language not required.
Entrance requirements: For master's and doctorate, GRE General Test, TOEFL.
Find an in-depth description at www.petersons.com/graduate.

■ SARAH LAWRENCE COLLEGE
Graduate Studies, Program in Genetic Counseling, Bronxville, NY 10708
AWARDS Human genetics (MPS, MS).

Faculty: 1 full-time, 8 part-time/adjunct.
Students: 39 full-time (38 women), 5 part-time (all women); includes 4 minority (all Asian Americans or Pacific Islanders), 3 international. *155 applicants, 31% accepted.* In 1999, 22 degrees awarded.
Degree requirements: For master's, thesis, fieldwork required, foreign language not required. *Average time to degree:* Master's–2 years full-time, 3 years part-time.
Entrance requirements: For master's, previous course work in biology, chemistry, developmental biology, genetics, probability and statistics. *Application deadline:* For fall admission, 4/1. *Application fee:* $45.
Expenses: Tuition: Full-time $12,400; part-time $628. Required fees: $300; $150 per semester.
Financial aid: Career-related internships or fieldwork, Federal Work-Study, grants, and unspecified assistantships available. Aid available to part-time students. Financial aid application deadline: 3/1.
Faculty research: Genetics.
Caroline Lieber, Director, 914-395-2371.
Application contact: Susan P. Guma, Director of Graduate Studies, 914-395-2373.

■ TULANE UNIVERSITY
School of Medicine and Graduate School, Graduate Programs in Medicine, Program in Human Genetics/Genetic Counseling, New Orleans, LA 70118-5669
AWARDS Human genetics (MS, PhD). MS and PhD offered through the Graduate School.

Students: 2 full-time (both women), 2 part-time (both women); includes 1 minority (Asian American or Pacific Islander), 1 international. *25 applicants, 8% accepted.*
Degree requirements: For doctorate, dissertation required.
Entrance requirements: For doctorate, GRE General Test, TOEFL, or TSE. *Application deadline:* For fall admission, 2/1. *Application fee:* $45.
Expenses: Tuition: Full-time $23,030.

Financial aid: Fellowships, research assistantships available. Financial aid application deadline: 2/1.
Dr. Jess Thuene, Director, 504-588-5229.

■ UNIVERSITY OF CALIFORNIA, LOS ANGELES
School of Medicine and Graduate Division, Graduate Programs in Medicine, Department of Human Genetics, Los Angeles, CA 90095
AWARDS MS, PhD.

Students: 9 full-time (5 women); includes 4 minority (3 Asian Americans or Pacific Islanders, 1 Hispanic American), 1 international. *9 applicants, 11% accepted.*
Entrance requirements: For master's and doctorate, GRE General Test. *Application fee:* $40.
Expenses: Tuition, nonresident: full-time $9,804. Required fees: $4,405.
Dr. Leena Peltonen, Chair, 310-794-7505.
Application contact: School of Medicine Admissions Office, 310-825-6081.

Find an in-depth description at www.petersons.com/graduate.

■ UNIVERSITY OF CHICAGO
Division of the Biological Sciences, Molecular Biosciences: Biochemistry, Genetics, Cell and Developmental Biology, Department of Human Genetics, Chicago, IL 60637-1513
AWARDS PhD.

Expenses: Tuition: Full-time $24,804; part-time $3,422 per course. Required fees: $390. Tuition and fees vary according to program.
Dr. David Ledbetter, Chairman, 773-834-0575.

■ UNIVERSITY OF MARYLAND
Graduate School, Graduate Programs in Medicine, Program in Human Genetics, Baltimore, MD 21201-1627
AWARDS MS, PhD, MD/PhD. Part-time programs available.

Degree requirements: For master's, thesis or alternative required, foreign language not required; for doctorate, dissertation, qualifying exam required, foreign language not required.
Entrance requirements: For master's and doctorate, GRE General Test, TOEFL, minimum GPA of 3.0.
Expenses: Tuition, state resident: part-time $261 per credit hour. Tuition, nonresident: part-time $468 per credit hour. Tuition and fees vary according to program.
Faculty research: Cytogenetics, biochemical genetics, metabolic diseases, population genetics, family data analysis.

■ UNIVERSITY OF MICHIGAN
Medical School and Horace H. Rackham School of Graduate Studies, Program in Biomedical Sciences (PIBS), Department of Human Genetics, Ann Arbor, MI 48109
AWARDS MS, PhD.

Faculty: 21 full-time (5 women), 3 part-time/adjunct (all women).
Students: 32 full-time (24 women), 1 part-time; includes 1 minority (African American), 1 international. Average age 27. *113 applicants, 7% accepted.* In 1999, 4 master's, 5 doctorates awarded (100% entered university research/teaching). Terminal master's awarded for partial completion of doctoral program.
Degree requirements: For master's, foreign language and thesis not required; for doctorate, oral defense of dissertation, oral preliminary exam required. *Average time to degree:* Master's–1.5 years full-time, 2 years part-time; doctorate–6 years full-time.
Entrance requirements: For master's and doctorate, GRE General Test, GRE Subject Test (biology, biochemistry). *Application deadline:* For fall admission, 1/15. *Application fee:* $55. Electronic applications accepted.
Expenses: Tuition, state resident: full-time $10,316. Tuition, nonresident: full-time $20,922.
Financial aid: In 1999–00, 14 fellowships with full tuition reimbursements (averaging $14,688 per year), 10 research assistantships with full tuition reimbursements (averaging $17,000 per year), 6 teaching assistantships with full tuition reimbursements were awarded; grants, scholarships, traineeships, and unspecified assistantships also available.
Faculty research: Molecular, cellular, and population genetics.
Dr. Thomas Gelehrter, Chair, 734-764-5491.
Application contact: Janet Miller, Student Services Associate I, 734-764-5490, *Fax:* 734-763-3784, *E-mail:* janmil@umich.edu.

Find an in-depth description at www.petersons.com/graduate.

■ UNIVERSITY OF PITTSBURGH
Graduate School of Public Health, Department of Human Genetics, Program in Human Genetics, Pittsburgh, PA 15260
AWARDS MS, PhD.

Faculty: 17 full-time (9 women).
Students: 19 full-time (13 women), 5 part-time (4 women), 10 international. In 1999, 5 master's, 6 doctorates awarded.
Degree requirements: For master's and doctorate, thesis/dissertation required,

University of Pittsburgh (continued) foreign language not required. *Average time to degree:* Master's–2 years full-time; doctorate–4 years full-time.

Entrance requirements: For master's, GRE General Test, TOEFL, previous course work in biochemistry, calculus, and genetics; for doctorate, GRE General Test, TOEFL. *Application deadline:* For fall admission, 4/1 (priority date); for spring admission, 11/1 (priority date). Applications are processed on a rolling basis. *Application fee:* $50 ($60 for international students). Electronic applications accepted. **Expenses:** Tuition, state resident: full-time $9,778; part-time $403 per credit. Tuition, nonresident: full-time $20,146; part-time $830 per credit. Required fees: $480; $90 per term.

Financial aid: In 1999–00, 17 students received aid, including 17 research assistantships with full tuition reimbursements available (averaging $1,082 per year); fellowships, grants also available. **Faculty research:** Statistical genetics, molecular genetics, cytogenetics, gene therapy, genetic counseling. *Total annual research expenditures:* $4 million.

Application contact: Jeanette Norbut, Administrative Secretary, 412-624-9951, *Fax:* 412-624-3020, *E-mail:* jnorbut@ helix.hgen.pitt.edu.

■ **THE UNIVERSITY OF TEXAS– HOUSTON HEALTH SCIENCE CENTER**

Graduate School of Biomedical Sciences, Program in Human and Molecular Genetics, Houston, TX 77225-0036

AWARDS MS, PhD, MD/PhD.

Faculty: 38 full-time (10 women). **Students:** 19 full-time (12 women); includes 2 minority (both Asian Americans or Pacific Islanders), 5 international. Average age 27. *47 applicants, 49% accepted.* In 1999, 2 master's, 5 doctorates awarded. Terminal master's awarded for partial completion of doctoral program. **Degree requirements:** For master's and doctorate, thesis/dissertation required, foreign language not required. **Entrance requirements:** For master's and doctorate, GRE General Test, TOEFL, TWE. *Application deadline:* For fall admission, 1/15 (priority date); for spring admission, 11/1. Applications are processed on a rolling basis. *Application fee:* $10. Electronic applications accepted. **Financial aid:** Fellowships, research assistantships, institutionally sponsored loans available. Financial aid application deadline: 1/15.

Faculty research: Cancer genetics, genetic epidemiology, genome evolution, cytogenetics.
Dr. Michael J. Siciliano, Director, 713-792-2910, *Fax:* 713-794-4394, *E-mail:* mjs@mdanderson.org.
Application contact: Anne Baronitis, Director of Admissions, 713-500-9860, *Fax:* 713-500-9877, *E-mail:* abaron@ gsbs.gs.uth.tmc.edu.
Find an in-depth description at www.petersons.com/graduate.

■ **UNIVERSITY OF UTAH**

School of Medicine and Graduate School, Graduate Programs in Medicine, Department of Human Genetics, Salt Lake City, UT 84112-1107

AWARDS MS, PhD.

Faculty: 27 full-time (5 women). **Students:** 29 full-time (12 women); includes 9 minority (all Asian Americans or Pacific Islanders), 14 international. Average age 26. In 1999, 4 master's, 14 doctorates awarded. *Average time to degree:* Master's–3 years full-time; doctorate–7.5 years full-time.

Expenses: Tuition, state resident: full-time $2,105. Tuition, nonresident: full-time $6,312.

Financial aid: In 1999–00, 2 fellowships with full tuition reimbursements (averaging $17,000 per year), 4 research assistantships with full tuition reimbursements (averaging $14,688 per year), teaching assistantships with full tuition reimbursements (averaging $17,000 per year) were awarded; tuition waivers (full) also available.
Raymond Gresteland, MD, Chair, 801-581-5190.

■ **VIRGINIA COMMONWEALTH UNIVERSITY**

School of Graduate Studies and School of Medicine, School of Medicine Graduate Programs, Department of Human Genetics, Richmond, VA 23284-9005

AWARDS Genetic counseling (MS); human genetics (PhD, CBHS); molecular biology and genetics (PhD).

Students: In 1999, 3 master's, 3 doctorates awarded.
Degree requirements: For master's, computer language, thesis required, foreign language not required; for doctorate, computer language, dissertation, comprehensive oral and written exams required, foreign language not required.
Entrance requirements: For master's, DAT, GRE General Test, or MCAT; for

doctorate, GRE General Test. *Application deadline:* For fall admission, 2/1. *Application fee:* $30.
Expenses: Tuition, state resident: full-time $4,031; part-time $224 per credit hour. Tuition, nonresident: full-time $11,946; part-time $664 per credit hour. Required fees: $1,081; $40 per credit hour. Tuition and fees vary according to campus/location and program.
Financial aid: Fellowships available. **Faculty research:** Genetic epidemiology, biochemical genetics, quantitative genetics, human cytogenetics, molecular genetics. Dr. Walter E. Nance, Chair, 804-828-9632, *Fax:* 804-828-3760.
Application contact: Dr. Linda A. Corey, Graduate Program Director, 804-828-9632, *Fax:* 804-828-3760, *E-mail:* lacorey@ vcu.edu.
Find an in-depth description at www.petersons.com/graduate.

■ **WEST VIRGINIA UNIVERSITY**

College of Agriculture, Forestry and Consumer Sciences, Interdisciplinary Program in Genetics and Developmental Biology, Morgantown, WV 26506

AWARDS Animal breeding (MS, PhD); biochemical and molecular genetics (MS, PhD); cytogenetics (MS, PhD); descriptive embryology (MS, PhD); developmental genetics (MS); experimental morphogenesis (MS); human genetics (MS, PhD); immunogenetics (MS, PhD); life cycles of animals and plants (MS, PhD); molecular aspects of development (MS, PhD); mutagenesis (PhD); mutagenetics (MS); oncology (MS, PhD); plant genetics (MS, PhD); population and quantitative genetics (PhD); population and quantitative genetics (MS); regeneration (MS, PhD); teratology (MS, PhD); toxicology (MS, PhD).

Students: 18 full-time (8 women), 5 part-time (2 women); includes 1 minority (Asian American or Pacific Islander), 8 international. Average age 27. In 1999, 4 doctorates awarded.
Degree requirements: For master's, thesis required, foreign language not required; for doctorate, dissertation, comprehensive exam required, foreign language not required.
Entrance requirements: For master's, GRE or MCAT, TOEFL, minimum GPA of 2.75; for doctorate, TOEFL. *Application fee:* $45.
Expenses: Tuition, state resident: full-time $2,910; part-time $154 per credit hour. Tuition, nonresident: full-time $8,368; part-time $457 per credit hour.
Financial aid: In 1999–00, 11 research assistantships, 3 teaching assistantships were awarded; fellowships, Federal Work-Study, institutionally sponsored loans, and

tuition waivers (full and partial) also available. Financial aid application deadline: 2/1; financial aid applicants required to submit FAFSA.

Dr. J. Nath, Chairman, 304-293-6256 Ext. 4333, *E-mail:* jnath@wvu.edu.

REPRODUCTIVE BIOLOGY

■ JOHNS HOPKINS UNIVERSITY

School of Hygiene and Public Health, Department of Biochemistry and Molecular Biology, Division of Reproductive Biology, Baltimore, MD 21218-2699

AWARDS MHS, Sc M, PhD.

Degree requirements: For master's, thesis required, foreign language not required; for doctorate, dissertation, 1 year full-time residency, oral and written exams required, foreign language not required.
Entrance requirements: For master's and doctorate, GRE General Test, TOEFL. *Application deadline:* For fall admission, 2/1 (priority date). Applications are processed on a rolling basis. *Application fee:* $60. Electronic applications accepted.
Expenses: Tuition: Full-time $23,660; part-time $493 per unit. Full-time tuition and fees vary according to degree level, campus/location and program.
Financial aid: Federal Work-Study, institutionally sponsored loans, and scholarships available. Aid available to part-time students. Financial aid application deadline: 4/15.
Faculty research: Reproductive endocrinology, gene regulation, regulation of cell development, male contraception, aging.
Dr. Barry R. Zirkin, Director, 410-955-7827, *Fax:* 410-955-0792, *E-mail:* brzirkin@jhsph.edu.
Application contact: Gerry Graziano, Academic Coordinator, 410-955-1763, *Fax:* 410-955-0792, *E-mail:* ggrazian@jhsph.edu.

■ NORTHWESTERN UNIVERSITY

The Graduate School, Judd A. and Marjorie Weinberg College of Arts and Sciences, Interdepartmental Biological Sciences Program (IBiS), Concentration in Reproductive Biology, Evanston, IL 60208

AWARDS PhD.

Faculty: 59 full-time (11 women).
Students: 79 full-time (46 women); includes 5 minority (2 African Americans, 3 Hispanic Americans), 14 international. *236 applicants, 19% accepted.*

Degree requirements: For doctorate, dissertation, 2 quarters of teaching experience required, foreign language not required.
Entrance requirements: For doctorate, GRE General Test, TOEFL. *Application deadline:* For fall admission, 1/15. Applications are processed on a rolling basis. *Application fee:* $50 ($55 for international students).
Expenses: Tuition: Full-time $23,301. Full-time tuition and fees vary according to program.
Financial aid: In 1999–00, 15 fellowships (averaging $12,078 per year), 64 research assistantships (averaging $17,000 per year), 15 teaching assistantships (averaging $16,620 per year) were awarded. Financial aid application deadline: 12/31; financial aid applicants required to submit FAFSA.

Find an in-depth description at www.petersons.com/graduate.

■ UNIVERSITY OF HAWAII AT MANOA

John A. Burns School of Medicine and Graduate Division, Graduate Programs in Biomedical Sciences, Department of Anatomy and Reproductive Biology, Honolulu, HI 96822

AWARDS Reproductive biology (PhD). Part-time programs available.

Faculty: 10 full-time (0 women).
Students: 1 full-time (0 women). Average age 25. *4 applicants, 50% accepted.* In 1999, 1 doctorate awarded.
Degree requirements: For doctorate, dissertation required.
Entrance requirements: For doctorate, GRE General Test, GRE Subject Test. *Application fee:* $25 ($50 for international students).
Expenses: Tuition, state resident: part-time $168 per credit. Tuition, nonresident: part-time $415 per credit. Required fees: $51 per semester. Part-time tuition and fees vary according to course load.
Financial aid: Fellowships, research assistantships, teaching assistantships available.
Faculty research: Biology of gametes and fertilization, reproductive endocrinology.
Dr. Gillian D. Bryant-Greenwood, Head, 808-956-3388, *Fax:* 808-956-9481, *E-mail:* gbg@ahi.pbrc.hawaii.edu.
Application contact: Scott Lozanoff, Graduate Chairperson, 808-956-7131, *Fax:* 808-956-9481, *E-mail:* lozanoffs@jabsom.biomed.hawaii.edu.

■ THE UNIVERSITY OF TEXAS–HOUSTON HEALTH SCIENCE CENTER

Graduate School of Biomedical Sciences, Program in Reproductive Biology, Houston, TX 77225-0036

AWARDS MS, PhD, MD/PhD.

Faculty: 20 full-time (5 women).
Students: 3 full-time (2 women); includes 1 minority (African American), 1 international. *3 applicants, 33% accepted.* In 1999, 1 doctorate awarded. Terminal master's awarded for partial completion of doctoral program.
Degree requirements: For master's and doctorate, thesis/dissertation required, foreign language not required.
Entrance requirements: For master's and doctorate, GRE General Test, TOEFL, TWE. *Application deadline:* For fall admission, 1/15 (priority date); for spring admission, 11/1. Applications are processed on a rolling basis. *Application fee:* $10. Electronic applications accepted.
Financial aid: Fellowships, research assistantships, teaching assistantships, institutionally sponsored loans available. Financial aid application deadline: 1/15.
Faculty research: Mechanisms of hormone action in reproductive tissues, reproductive tract and gonadal cell cancers, molecular genetics of sex determination, gametogenesis and gonadal cells, embryonic and placental development.
Dr. Marvin L. Meistrich, Director, 713-792-3424, *Fax:* 713-794-5369, *E-mail:* meistrich@mdanderson.org.
Application contact: Anne Baronitis, Director of Admissions, 713-500-9860, *Fax:* 713-500-9877, *E-mail:* abaron@gsbs.gs.uth.tmc.edu.

Find an in-depth description at www.petersons.com/graduate.

■ UNIVERSITY OF WYOMING

Graduate School, College of Agriculture, Department of Animal Sciences, Program in Reproductive Biology, Laramie, WY 82071

AWARDS MS, PhD.

Students: 3 full-time (2 women). *0 applicants, 0% accepted.*
Degree requirements: For master's, thesis required, foreign language not required; for doctorate, dissertation required.
Entrance requirements: For master's and doctorate, GRE General Test, minimum GPA of 3.0. *Application deadline:* For fall admission, 6/1 (priority date). Applications are processed on a rolling basis. *Application fee:* $40.
Expenses: Tuition, state resident: full-time $2,520; part-time $140 per credit hour.

University of Wyoming (continued)
Tuition, nonresident: full-time $7,790; part-time $433 per credit hour. Required fees: $440; $7 per credit hour. Full-time tuition and fees vary according to course load and program.
Financial aid: Application deadline: 3/1. Melvin Riley, Head, Department of Animal Sciences, 307-766-2224, *Fax:* 307-766-2355, *E-mail:* mriley@uwyo.edu.

■ **WEST VIRGINIA UNIVERSITY**
College of Agriculture, Forestry and Consumer Sciences, Interdisciplinary Program in Genetics and Developmental Biology, Morgantown, WV 26506
AWARDS Animal breeding (MS, PhD); biochemical and molecular genetics (MS,

PhD); cytogenetics (MS, PhD); descriptive embryology (MS, PhD); developmental genetics (MS); experimental morphogenesis (MS); human genetics (MS, PhD); immunogenetics (MS, PhD); life cycles of animals and plants (MS, PhD); molecular aspects of development (MS, PhD); mutagenesis (PhD); mutagenetics (MS); oncology (MS, PhD); plant genetics (MS, PhD); population and quantitative genetics (PhD); population and quantitative genetics (MS); regeneration (MS, PhD); teratology (MS, PhD); toxicology (MS, PhD).

Students: 18 full-time (8 women), 5 part-time (2 women); includes 1 minority (Asian American or Pacific Islander), 8 international. Average age 27. In 1999, 4 doctorates awarded.
Degree requirements: For master's, thesis required, foreign language not

required; for doctorate, dissertation, comprehensive exam required, foreign language not required.
Entrance requirements: For master's, GRE or MCAT, TOEFL, minimum GPA of 2.75; for doctorate, TOEFL. *Application fee:* $45.
Expenses: Tuition, state resident: full-time $2,910; part-time $154 per credit hour. Tuition, nonresident: full-time $8,368; part-time $457 per credit hour.
Financial aid: In 1999–00, 11 research assistantships, 3 teaching assistantships were awarded; fellowships, Federal Work-Study, institutionally sponsored loans, and tuition waivers (full and partial) also available. Financial aid application deadline: 2/1; financial aid applicants required to submit FAFSA.
Dr. J. Nath, Chairman, 304-293-6256 Ext. 4333, *E-mail:* jnath@wvu.edu.

Marine Biology

MARINE BIOLOGY

■ **BOSTON UNIVERSITY**
Graduate School of Arts and Sciences, Department of Biology, Boston, MA 02215
AWARDS Botany (MA, PhD); cell and molecular biology (MA, PhD); cell biology (MA, PhD); ecology (PhD); ecology, behavior, and evolution (MA, PhD); ecology/physiology, endocrinology and reproduction (MA); marine biology (MA, PhD); molecular biology, cell biology and biochemistry (MA, PhD); neurobiology, neuroendocrinology and reproduction (MA, PhD); physiology, endocrinology, and neurobiology (MA, PhD); zoology (MA, PhD). Part-time programs available.

Faculty: 41 full-time (8 women).
Students: 131 full-time (74 women), 11 part-time (7 women); includes 10 minority (7 Asian Americans or Pacific Islanders, 3 Hispanic Americans), 33 international. Average age 27. *238 applicants, 39% accepted.* In 1999, 61 master's, 45 doctorates awarded. Terminal master's awarded for partial completion of doctoral program.
Degree requirements: For master's, one foreign language, thesis not required; for doctorate, one foreign language, dissertation, qualifying exam required. *Average time to degree:* Master's–1 year full-time, 3 years part-time; doctorate–5.75 years full-time.
Entrance requirements: For master's and doctorate, GRE General Test, GRE

Subject Test, TOEFL. *Application deadline:* For fall admission, 1/1 (priority date); for spring admission, 11/1. *Application fee:* $50.
Expenses: Tuition: Full-time $23,770; part-time $743 per credit. Required fees: $220. Tuition and fees vary according to class time, course level, campus/location and program.
Financial aid: In 1999–00, 82 students received aid, including 1 fellowship with full tuition reimbursement available (averaging $12,000 per year), 28 research assistantships with full tuition reimbursements available (averaging $11,500 per year), 43 teaching assistantships with full tuition reimbursements available (averaging $11,500 per year); Federal Work-Study, grants, institutionally sponsored loans, scholarships, and traineeships also available. Financial aid application deadline: 1/15; financial aid applicants required to submit FAFSA.
Faculty research: Marine science, endocrinology, behavior. *Total annual research expenditures:* $5 million.
Geoffrey M. Cooper, Chairman, 617-353-2432, *Fax:* 617-353-6340, *E-mail:* gmcooper@bu.edu.
Application contact: Yolanta Kovalko, Senior Staff Assistant, 617-353-2432, *Fax:* 617-353-6340, *E-mail:* yolanta@bu.edu.
Find an in-depth description at www.petersons.com/graduate.

■ **CALIFORNIA STATE UNIVERSITY, STANISLAUS**
Graduate Programs, College of Arts, Letters, and Sciences, Department of Biological Sciences, Turlock, CA 95382
AWARDS Marine science (MS). Part-time programs available.

Faculty: 9 full-time (1 woman), 6 part-time/adjunct (1 woman).
Students: 7 (4 women); includes 1 minority (Hispanic American). *9 applicants, 22% accepted.* In 1999, 5 degrees awarded.
Degree requirements: For master's, thesis required, foreign language not required.
Entrance requirements: For master's, GRE General Test, GRE Subject Test, minimum GPA of 3.0. *Application deadline:* For fall admission, 2/15; for spring admission, 9/15. *Application fee:* $55. Electronic applications accepted.
Expenses: Tuition, nonresident: part-time $246 per unit. Required fees: $1,955; $1,291 per year.
Financial aid: In 1999–00, 1 fellowship (averaging $2,500 per year) was awarded; career-related internships or fieldwork, Federal Work-Study, and scholarships also available. Aid available to part-time students. Financial aid application deadline: 3/2; financial aid applicants required to submit FAFSA.
Faculty research: Marine mollusks, kelp forest ecology, marine fish, trace metals, paleoceanography and sedimentology.
Dr. Wayne Pierce, Chair, 209-667-3476.

Application contact: Dr. Pamela Roe, Director, 209-667-3484.

■ FLORIDA INSTITUTE OF TECHNOLOGY

Graduate School, College of Science and Liberal Arts, Department of Biological Sciences, Program in Marine Biology, Melbourne, FL 32901-6975

AWARDS MS. Part-time programs available.

Students: 3 full-time (1 woman), 20 part-time (10 women); includes 1 minority (Asian American or Pacific Islander), 3 international. Average age 28. *60 applicants, 15% accepted.* In 1999, 5 degrees awarded.
Degree requirements: For master's, thesis required, foreign language not required.
Entrance requirements: For master's, GRE General Test, minimum GPA of 3.0. *Application deadline:* Applications are processed on a rolling basis. *Application fee:* $50. Electronic applications accepted.
Expenses: Tuition: Part-time $575 per credit hour. Required fees: $50. Tuition and fees vary according to campus/location and program.
Financial aid: In 1999–00, 8 students received aid, including 1 research assistantship with full and partial tuition reimbursement available (averaging $3,230 per year), 6 teaching assistantships with full and partial tuition reimbursements available (averaging $3,927 per year); career-related internships or fieldwork and tuition remissions also available. Financial aid application deadline: 3/1; financial aid applicants required to submit FAFSA.
Faculty research: Ecology of coral reef fish communities and biology of *Foraminiferida*; ecology, physiology, reproduction, and morphology of sea stars, sea urchins, and other echinoderms.
Dr. Gary N. Wells, Head, 321-674-8034, *Fax:* 321-674-7238, *E-mail:* gwells@fit.edu.
Application contact: Carolyn P. Farrior, Associate Dean of Graduate Admissions, 321-674-7118, *Fax:* 321-674-9468, *E-mail:* cfarrior@fit.edu.

Find an in-depth description at www.petersons.com/graduate.

■ FLORIDA STATE UNIVERSITY

Graduate Studies, College of Arts and Sciences, Department of Biological Science, Program in Marine Biology, Tallahassee, FL 32306

AWARDS MS, PhD.

Faculty: 7 full-time (0 women).
Students: 14 full-time (7 women); includes 2 minority (1 Asian American or Pacific Islander, 1 Hispanic American), 1 international.

Degree requirements: For master's and doctorate, thesis/dissertation, teaching experience required.
Entrance requirements: For master's, GRE General Test, TOEFL; for doctorate, GRE General Test, GRE Subject Test, TOEFL. *Application deadline:* For fall admission, 1/15; for spring admission, 10/15. *Application fee:* $20.
Expenses: Tuition, state resident: full-time $3,504; part-time $146 per credit hour. Tuition, nonresident: full-time $12,162; part-time $507 per credit hour. Tuition and fees vary according to program.
Financial aid: In 1999–00, fellowships with full tuition reimbursements (averaging $13,740 per year), research assistantships with full tuition reimbursements (averaging $13,740 per year), teaching assistantships with full tuition reimbursements (averaging $13,740 per year) were awarded. Financial aid application deadline: 1/15; financial aid applicants required to submit FAFSA.
Faculty research: Community ecology, physiological and biochemical adaptation, marine algae, behavioral ecology, marine ecology.
Dr. Thomas C. S. Keller, Associate Professor and Associate Chairman, 850-644-3023, *Fax:* 850-644-9829.
Application contact: Judy Bowers, Coordinator, Graduate Affairs, 850-644-3023, *Fax:* 850-644-9829, *E-mail:* bowers@bio.fsu.edu.

■ MASSACHUSETTS INSTITUTE OF TECHNOLOGY

Program in Oceanography/Applied Ocean Science and Engineering, Cambridge, MA 02139-4307

AWARDS Biological oceanography (PhD, Sc D); chemical oceanography (PhD, Sc D); marine geochemistry (PhD, Sc D); marine geology (PhD, Sc D); oceanographic engineering (M Eng, MS, PhD, Sc D, Eng); physical oceanography (PhD, Sc D). MS, PhD, and Sc D offered jointly with Woods Hole Oceanographic Institution.

Faculty: 170 full-time.
Students: 133 full-time, 43 international. Average age 27. *158 applicants, 25% accepted.* In 1999, 8 master's, 21 doctorates, 2 other advanced degrees awarded. Terminal master's awarded for partial completion of doctoral program.
Degree requirements: For master's and Eng, thesis required (for some programs), foreign language not required; for doctorate, dissertation required, foreign language not required. *Average time to degree:* Master's–2.5 years full-time; doctorate–6 years full-time; Eng–2 years full-time.
Entrance requirements: For master's, GRE General Test; for doctorate, GRE

General Test, GRE Subject Test. *Application deadline:* For fall admission, 1/15 (priority date). *Application fee:* $55.
Expenses: Tuition: Full-time $25,000. Full-time tuition and fees vary according to degree level, program and student level.
Financial aid: Fellowships, research assistantships, teaching assistantships available.
Paola Rizzoli, Director, 617-253-2451, *E-mail:* rizzoli@mit.edu.
Application contact: Ronni Schwartz, Administrator, 617-253-7544, *Fax:* 617-253-9784, *E-mail:* mspiggy@mit.edu.

Find an in-depth description at www.petersons.com/graduate.

■ MURRAY STATE UNIVERSITY

College of Science, Department of Water Science, Murray, KY 42071-0009

AWARDS MS. Part-time programs available.

Students: 1 (woman) full-time, 4 part-time (1 woman). *1 applicant, 100% accepted.* In 1999, 1 degree awarded.
Application deadline: Applications are processed on a rolling basis. *Application fee:* $20.
Expenses: Tuition, state resident: full-time $2,600; part-time $130 per hour. Tuition, nonresident: full-time $7,040; part-time $374 per hour. Required fees: $90 per semester. Part-time tuition and fees vary according to course load and program.
Financial aid: Application deadline: 4/1.
Dr. David White, Graduate Coordinator, 270-762-3194, *E-mail:* david.white@murraystate.edu.

■ NOVA SOUTHEASTERN UNIVERSITY

Oceanographic Center, Fort Lauderdale, FL 33314-7721

AWARDS Coastal-zone management (MS); marine biology (MS); marine environmental science (MS); oceanography (PhD). Part-time and evening/weekend programs available.

Faculty: 14 full-time (1 woman), 5 part-time/adjunct (0 women).
Students: 23 full-time (13 women), 90 part-time (50 women). Average age 30. *67 applicants, 75% accepted.* In 1999, 4 master's, 1 doctorate awarded.
Degree requirements: For master's, thesis required, foreign language not required; for doctorate, dissertation, departmental qualifying exam required. *Average time to degree:* Master's–2 years full-time, 3 years part-time; doctorate–4 years full-time, 6 years part-time.
Entrance requirements: For master's, GRE General Test; for doctorate, GRE General Test, master's degree. *Application deadline:* Applications are processed on a rolling basis. *Application fee:* $50.

Nova Southeastern University (continued)
Expenses: Tuition: Part-time $417 per credit hour. Required fees: $50 per semester. Tuition and fees vary according to degree level and program.
Financial aid: In 1999–00, 6 research assistantships (averaging $4,000 per year), 3 teaching assistantships (averaging $3,500 per year) were awarded; career-related internships or fieldwork, Federal Work-Study, tuition waivers (partial), and unspecified assistantships also available. Aid available to part-time students.
Faculty research: Physical, geological, chemical, and biological oceanography.
Dr. Richard Dodge, Dean, 954-262-3600, *Fax:* 954-262-4020, *E-mail:* dodge@ ocean.nova.edu.
Application contact: Dr. Andrew Rogerson, Director, Graduate Programs, 954-262-3600, *Fax:* 954-262-4020, *E-mail:* arogers@ocean.nova.edu.
Find an in-depth description at www.petersons.com/graduate.

■ RUTGERS, THE STATE UNIVERSITY OF NEW JERSEY, NEW BRUNSWICK

Graduate School, Program in Environmental Sciences, New Brunswick, NJ 08901-1281

AWARDS Air resources (MS, PhD); aquatic biology (MS, PhD); aquatic chemistry (MS, PhD); chemistry and physics of aerosol and hydrosol systems (MS, PhD); environmental chemistry (MS, PhD); environmental microbiology (MS, PhD); environmental toxicology (MS, PhD); exposure assessment (PhD); water and wastewater treatment (MS, PhD); water resources (MS, PhD). Part-time and evening/weekend programs available.

Faculty: 33 full-time (7 women), 36 part-time/adjunct (6 women).
Students: 68 full-time (25 women), 58 part-time (26 women); includes 10 minority (8 Asian Americans or Pacific Islanders, 2 Hispanic Americans), 44 international. Average age 26. *128 applicants, 40% accepted.* In 1999, 15 master's, 18 doctorates awarded. Terminal master's awarded for partial completion of doctoral program.
Degree requirements: For master's, thesis or alternative, oral final exam required, foreign language not required; for doctorate, thesis defense, qualifying exam required.
Entrance requirements: For master's and doctorate, GRE General Test, TOEFL. *Application deadline:* For fall admission, 3/1; for spring admission, 11/1. Applications are processed on a rolling basis. *Application fee:* $50.
Expenses: Tuition, state resident: full-time $6,776; part-time $279 per credit. Tuition,

nonresident: full-time $9,936; part-time $412 per credit. Required fees: $20 per credit. $89 per semester. Tuition and fees vary according to course load, campus/ location and program.
Financial aid: In 1999–00, 1 fellowship (averaging $3,000 per year), 30 research assistantships with full tuition reimbursements (averaging $13,956 per year), 10 teaching assistantships with full tuition reimbursements (averaging $13,100 per year) were awarded; career-related internships or fieldwork and Federal Work-Study also available. Financial aid application deadline: 3/1; financial aid applicants required to submit FAFSA.
Faculty research: Atmospheric sciences; biological waste treatment; contaminant fate and transport; exposure assessment; air, soil and water quality.
Dr. Peter F. Strom, Director, 732-932-8078, *Fax:* 732-932-8644, *E-mail:* strom@ aesop.rutgers.edu.
Application contact: Paul J. Lioy, Graduate Admissions Committee, 732-932-0150, *Fax:* 732-445-0116, *E-mail:* plioy@ eohsi.rutgers.edu.

■ SAN FRANCISCO STATE UNIVERSITY

Graduate Division, College of Science and Engineering, Department of Biology, Program in Marine Biology, San Francisco, CA 94132-1722
AWARDS MA.

Entrance requirements: For master's, minimum GPA of 2.5 in last 60 units.
Expenses: Tuition, nonresident: full-time $5,904; part-time $246 per unit. Required fees: $1,904; $637 per semester. Tuition and fees vary according to course load.

■ SOUTHWEST TEXAS STATE UNIVERSITY

Graduate School, College of Science, Department of Biology, Program in Aquatic Biology, San Marcos, TX 78666
AWARDS MS.

Students: 10 full-time (6 women), 14 part-time (2 women); includes 2 minority (both Native Americans). Average age 30. In 1999, 2 degrees awarded.
Degree requirements: For master's, thesis, 3 seminars, comprehensive exam required, foreign language not required.
Entrance requirements: For master's, GRE General Test, TOEFL, previous course work in biology, minimum GPA of 2.75 in last 60 hours. *Application deadline:* For fall admission, 6/15 (priority date); for spring admission, 10/15 (priority date). Applications are processed on a rolling

basis. *Application fee:* $25 ($75 for international students).
Expenses: Tuition, state resident: full-time $720; part-time $40 per semester hour. Tuition, nonresident: full-time $4,608; part-time $256 per semester hour. Required fees: $1,470; $122.
Financial aid: Teaching assistantships available. Financial aid application deadline: 4/1; financial aid applicants required to submit FAFSA.
Dr. Thomas L. Arsuffi, Director, Aquatic Station, 512-245-2284, *Fax:* 512-245-7919, *E-mail:* ta04@swt.edu.

■ THE UNIVERSITY OF ALABAMA AT BIRMINGHAM

Graduate School, School of Natural Sciences and Mathematics, Department of Biology, Birmingham, AL 35294

AWARDS Comparative and cellular biology (PhD); comparative and cellular physiology (MS); marine science (MS, PhD); microbial ecology and physiology (MS, PhD); reproduction and development (MS, PhD).

Students: 34 full-time (19 women), 6 international. *61 applicants, 38% accepted.* In 1999, 1 master's, 3 doctorates awarded. Terminal master's awarded for partial completion of doctoral program.
Degree requirements: For master's and doctorate, thesis/dissertation required.
Entrance requirements: For master's and doctorate, GRE General Test, TOEFL, previous course work in biology, calculus, organic chemistry, physics. *Application deadline:* Applications are processed on a rolling basis. *Application fee:* $35 ($60 for international students). Electronic applications accepted.
Expenses: Tuition, state resident: part-time $104 per semester hour. Tuition, nonresident: part-time $208 per semester hour. Required fees: $17 per semester hour. $57 per quarter. Tuition and fees vary according to program.
Financial aid: In 1999–00, 22 students received aid, including 3 fellowships with full tuition reimbursements available (averaging $14,000 per year), 19 teaching assistantships with full tuition reimbursements available (averaging $14,000 per year); research assistantships, career-related internships or fieldwork, Federal Work-Study, institutionally sponsored loans, and tuition waivers (full) also available. Aid available to part-time students.
Faculty research: Invertebrate physiology, marine biology, environmental biology.
Dr. Daniel D. Jones, Chairman, 205-934-4290, *Fax:* 205-975-6097, *E-mail:* ddjones@uab.edu.
Find an in-depth description at www.petersons.com/graduate.

■ UNIVERSITY OF ALASKA FAIRBANKS

Graduate School, School of Fisheries and Ocean Sciences, Program in Marine Sciences and Limnology, Fairbanks, AK 99775

AWARDS Marine biology (MS); oceanography (MS, PhD), including biological oceanography (PhD), chemical oceanography (PhD), fisheries (PhD), physical oceanography (PhD).

Faculty: 9 full-time (1 woman), 2 part-time/adjunct (0 women).
Students: 37 full-time (27 women), 9 part-time (2 women); includes 2 minority (1 African American, 1 Asian American or Pacific Islander), 15 international. Average age 29. *38 applicants, 47% accepted.* In 1999, 2 master's, 1 doctorate awarded.
Degree requirements: For master's, thesis, comprehensive exam required, foreign language not required; for doctorate, one foreign language (computer language can substitute), dissertation, comprehensive exam required.
Entrance requirements: For master's and doctorate, GRE General Test, TOEFL. *Application deadline:* For fall admission, 3/1 (priority date). Applications are processed on a rolling basis. *Application fee:* $35.
Expenses: Tuition, state resident: full-time $3,006; part-time $167 per credit. Tuition, nonresident: full-time $5,868; part-time $326 per credit. Required fees: $370; $10 per credit. $140 per semester.
Financial aid: Research assistantships, teaching assistantships available.
Dr. Susan Henrichs, Head, 907-474-7289.

■ UNIVERSITY OF CALIFORNIA, SAN DIEGO

Graduate Studies and Research, Scripps Institution of Oceanography, La Jolla, CA 92093

AWARDS Biological oceanography (MS, PhD); geochemistry and marine chemistry (MS, PhD); marine biology (MS, PhD); physical oceanography and geological sciences (MS, PhD).

Faculty: 73.
Students: 167 (69 women). *288 applicants, 25% accepted.* In 1999, 8 master's, 23 doctorates awarded.
Entrance requirements: For master's and doctorate, GRE General Test, GRE Subject Test (marine biology). *Application fee:* $40.
Expenses: Tuition, nonresident: full-time $14,691. Required fees: $4,697. Full-time tuition and fees vary according to program.
Financial aid: Fellowships, research assistantships available.
W. Kendall Melville, Chair.

Application contact: Graduate Coordinator, 858-534-3206.

■ UNIVERSITY OF CALIFORNIA, SANTA BARBARA

Graduate Division, College of Letters and Sciences, Division of Mathematics, Life, and Physical Sciences, Department of Ecology, Evolution, and Marine Biology, Santa Barbara, CA 93106

AWARDS MA, PhD.

Faculty: 28 full-time (4 women), 16 part-time/adjunct (5 women).
Students: 85 full-time (38 women); includes 10 minority (2 Asian Americans or Pacific Islanders, 7 Hispanic Americans, 1 Native American), 6 international. *122 applicants, 16% accepted.* In 1999, 14 master's, 12 doctorates awarded. Terminal master's awarded for partial completion of doctoral program.
Degree requirements: For master's, thesis or alternative required, foreign language not required; for doctorate, dissertation required, foreign language not required. *Average time to degree:* Master's–2 years full-time; doctorate–6 years full-time.
Entrance requirements: For master's and doctorate, GRE General Test, TOEFL. *Application deadline:* For fall admission, 12/15. *Application fee:* $40. Electronic applications accepted.
Expenses: Tuition, state resident: full-time $14,637. Tuition, nonresident: full-time $24,441.
Financial aid: Fellowships, research assistantships, teaching assistantships, career-related internships or fieldwork, Federal Work-Study, institutionally sponsored loans, and tuition waivers (full and partial) available. Financial aid application deadline: 12/15; financial aid applicants required to submit FAFSA.
Scott Cooper, Chair, 805-893-2979, *Fax:* 805-893-4724, *E-mail:* scooper@lifesci.ucsb.edu.
Application contact: Stephanie Slosser, Graduate Program Assistant, 805-893-3023, *Fax:* 805-893-4724, *E-mail:* slosser@lifesci.ucsb.edu.

■ UNIVERSITY OF CHARLESTON, SOUTH CAROLINA

Graduate School, School of Sciences and Mathematics, Program in Marine Biology, Charleston, SC 29424-0001

AWARDS MS. Includes cooperative program faculty and laboratories from the Charleston Laboratory of the National Ocean Service; South Carolina Marine Resources Research Institute (MRRI); Medical University of South Carolina; and The Citadel, The Military College of South Carolina.

Faculty: 99.
Students: 55 full-time (35 women); includes 3 minority (1 African American, 1 Asian American or Pacific Islander, 1 Hispanic American), 4 international. Average age 25. *59 applicants, 22% accepted.* In 1999, 8 degrees awarded.
Degree requirements: For master's, thesis, oral comprehensive exam required. *Average time to degree:* Master's–3 years full-time.
Entrance requirements: For master's, GRE General Test, GRE Subject Test, TOEFL. *Application deadline:* For fall admission, 2/1; for spring admission, 11/15. *Application fee:* $35.
Expenses: Tuition, state resident: part-time $152 per hour. Tuition, nonresident: part-time $305 per hour. Required fees: $2 per hour. $15 per semester. One-time fee: $45 part-time.
Financial aid: In 1999–00, 3 fellowships, 20 research assistantships (averaging $11,000 per year), 24 teaching assistantships (averaging $8,100 per year) were awarded; career-related internships or fieldwork, Federal Work-Study, and institutionally sponsored loans also available. Financial aid application deadline: 4/1.
Faculty research: Ecology, biological oceanography, environmental physiology, mariculture, cellular-molecular developmental biology.
Dr. David W. Owens, Director, 843-406-4000, *Fax:* 843-406-4001, *E-mail:* owensd@cofc.edu.
Application contact: Suyapa Germain, Administrative Specialist, 843-406-4000, *Fax:* 843-406-4001, *E-mail:* germains@cofc.edu.

Find an in-depth description at www.petersons.com/graduate.

■ UNIVERSITY OF COLORADO AT BOULDER

Graduate School, College of Arts and Sciences, Department of Environmental, Population, and Organic Biology, Boulder, CO 80309

AWARDS Animal behavior (MA, PhD); aquatic biology (MA, PhD); behavioral genetics (MA, PhD); ecology (MA, PhD); microbiology (MA, PhD); neurobiology (MA, PhD); plant and animal physiology (MA, PhD); plant and animal systematics (MA, PhD); population biology (MA, PhD); population genetics (MA, PhD).

Faculty: 39 full-time (9 women).
Students: 84 full-time (36 women), 14 part-time (7 women); includes 17 minority (6 Asian Americans or Pacific Islanders, 10 Hispanic Americans, 1 Native American). Average age 29. *147 applicants, 14% accepted.* In 1999, 7 master's, 13 doctorates

University of Colorado at Boulder (continued)

awarded. Terminal master's awarded for partial completion of doctoral program. **Degree requirements:** For master's, thesis or alternative, comprehensive exam required, foreign language not required; for doctorate, dissertation, comprehensive exam required, foreign language not required. *Average time to degree:* Master's–3 years full-time; doctorate–5 years full-time. **Entrance requirements:** For master's, GRE General Test, GRE Subject Test, minimum undergraduate GPA of 3.0; for doctorate, GRE General Test, GRE Subject Test. *Application deadline:* For fall admission, 1/15 (priority date). *Application fee:* $40 ($60 for international students). **Expenses:** Tuition, state resident: part-time $181 per credit hour. Tuition, nonresident: part-time $542 per credit hour. Required fees: $99 per term. Tuition and fees vary according to course load and program. **Financial aid:** Fellowships, research assistantships, teaching assistantships, Federal Work-Study, institutionally sponsored loans, and tuition waivers (full) available. Financial aid application deadline: 3/1. **Faculty research:** Evolution, developmental biology, behavior and neurobiology. *Total annual research expenditures:* $1.8 million. Michael Breed, Chair, 303-492-8981, *Fax:* 303-492-8699, *E-mail:* michael.breed@ colorado.edu.

Application contact: Jill Skarstadt, Graduate Secretary, 303-492-7654, *Fax:* 303-492-8699, *E-mail:* jill.skarstadt@ colorado.edu.

■ UNIVERSITY OF GUAM

Graduate School and Research, College of Arts and Sciences, Program in Biology, Mangilao, GU 96923

AWARDS Tropical marine biology (MS).

Degree requirements: For master's, thesis, oral comprehensive exam required, foreign language not required. **Entrance requirements:** For master's, GRE General Test, GRE Subject Test, TOEFL. **Expenses:** Tuition, state resident: part-time $99 per credit hour. Tuition, nonresident: part-time $246 per credit hour. Required fees: $170 per semester. Tuition and fees vary according to course load. **Faculty research:** Maintenance and ecology of coral reefs.

■ UNIVERSITY OF HAWAII AT MANOA

Graduate Division, Specialization in Marine Biology, Honolulu, HI 96822

AWARDS MS, PhD. Program is interdisciplinary; degree is in specific discipline with the specialization in Marine Biology.

Faculty: 37 full-time (7 women). **Students:** 20 full-time (10 women); includes 2 minority (1 Asian American or Pacific Islander, 1 Hispanic American). **Degree requirements:** For master's and doctorate, thesis/dissertation, research project required. **Entrance requirements:** For master's, GRE Subject Test, TOEFL. *Application deadline:* For fall admission, 2/1. **Expenses:** Tuition, state resident: full-time $4,032; part-time $168 per credit. Tuition, nonresident: full-time $9,960; part-time $415 per credit. Required fees: $51 per semester. Part-time tuition and fees vary according to course load and program. **Financial aid:** Research assistantships, teaching assistantships available. **Faculty research:** Ecology, ichthyology, behavior of marine animals, developmental biology. Dr. Julie H. Bailey-Brock, Chair, 808-956-6149, *Fax:* 808-956-9812, *E-mail:* brock@ zoogate.zoo.hawaii.edu.

■ UNIVERSITY OF MAINE

Graduate School, College of Natural Sciences, Forestry, and Agriculture, School of Marine Sciences, Program in Marine Biology, Orono, ME 04469

AWARDS MS, PhD.

Degree requirements: For master's and doctorate, thesis/dissertation required, foreign language not required. **Entrance requirements:** For master's and doctorate, GRE General Test, GRE Subject Test, TOEFL. **Expenses:** Tuition, state resident: full-time $3,564. Tuition, nonresident: full-time $10,116. Required fees: $378. Tuition and fees vary according to course load.

■ UNIVERSITY OF MASSACHUSETTS DARTMOUTH

Graduate School, College of Arts and Sciences, Department of Biology, North Dartmouth, MA 02747-2300

AWARDS Biology (MS); marine biology (MS). Part-time programs available.

Faculty: 18 full-time (5 women). **Students:** 5 full-time (3 women), 17 part-time (10 women); includes 4 minority (all Asian Americans or Pacific Islanders). Average age 26. *29 applicants, 34% accepted.* In 1999, 10 degrees awarded.

Degree requirements: For master's, thesis or alternative required, foreign language not required. **Entrance requirements:** For master's, GRE General Test, GRE Subject Test, TOEFL. *Application deadline:* For fall admission, 5/7; for spring admission, 11/15 (priority date). *Application fee:* $40 for international students. **Expenses:** Tuition, area resident: Full-time $2,071; part-time $86 per credit. Tuition, state resident: full-time $2,071; part-time $86 per credit. Tuition, nonresident: full-time $7,845; part-time $327 per credit. Required fees: $127 per credit. Full-time tuition and fees vary according to program and reciprocity agreements. Part-time tuition and fees vary according to course load and reciprocity agreements. **Financial aid:** In 1999–00, 2 research assistantships with full tuition reimbursements (averaging $5,615 per year), 13 teaching assistantships with full tuition reimbursements (averaging $7,158 per year) were awarded; Federal Work-Study also available. Aid available to part-time students. Financial aid application deadline: 3/1; financial aid applicants required to submit FAFSA. **Faculty research:** Bacterial regulatery, gene, phytoplankton analysis, estimating age from teeth, microbial community structure, cranberry production and pesticicles. *Total annual research expenditures:* $944,000. Dr. Nancy O'Connor, Director, 508-999-8217, *Fax:* 508-999-8196, *E-mail:* noconnor@unmassd.edu.

Application contact: Carol A. Novo, Graduate Admissions Office, 508-999-8026, *Fax:* 508-999-8183, *E-mail:* graduate@umassd.edu.

■ UNIVERSITY OF MIAMI

Graduate School, Rosenstiel School of Marine and Atmospheric Science, Division of Marine Biology and Fisheries, Miami, FL 33149

AWARDS MA, MS, PhD.

Faculty: 19 full-time (4 women), 1 (woman) part-time/adjunct. **Students:** 42 full-time (16 women), 2 part-time (1 woman); includes 4 minority (1 African American, 2 Hispanic Americans, 1 Native American), 12 international. Average age 28. *135 applicants, 7% accepted.* In 1999, 3 master's, 5 doctorates awarded. Terminal master's awarded for partial completion of doctoral program. **Degree requirements:** For master's and doctorate, thesis/dissertation required. *Average time to degree:* Master's–3 years full-time; doctorate–6 years full-time. **Entrance requirements:** For master's and doctorate, GRE General Test, GRE Subject Test, TOEFL. *Application deadline:*

For fall admission, 1/1 (priority date). Applications are processed on a rolling basis. *Application fee:* $50. Electronic applications accepted.

Expenses: Tuition: Full-time $15,336; part-time $852 per credit. Required fees: $174. Tuition and fees vary according to program.

Financial aid: In 1999–00, 43 students received aid, including 6 fellowships with tuition reimbursements available, 30 research assistantships with tuition reimbursements available (averaging $17,000 per year), 4 teaching assistantships with tuition reimbursements available (averaging $17,000 per year); institutionally sponsored loans also available. Financial aid application deadline: 3/1; financial aid applicants required to submit FAFSA.

Faculty research: Biochemistry, physiology, plankton, coral, fisheries biology. *Total annual research expenditures:* $5 million.
Dr. Samuel Snedaker, Chairman, 305-361-4176.

Application contact: Dr. Frank Millero, Associate Dean, 305-361-4155, *Fax:* 305-361-4771, *E-mail:* gso@rsmas.miami.edu.

■ THE UNIVERSITY OF NORTH CAROLINA AT WILMINGTON

College of Arts and Sciences, Department of Biological Sciences, Wilmington, NC 28403-3201

AWARDS Biology (MS); marine biology (MS). Part-time programs available.

Faculty: 19 full-time (2 women).
Students: 8 full-time (all women), 56 part-time (30 women); includes 3 minority (all Asian Americans or Pacific Islanders). Average age 29. *100 applicants, 23% accepted.* In 1999, 25 degrees awarded.
Degree requirements: For master's, thesis, oral and written comprehensive exams required.
Entrance requirements: For master's, GRE General Test, GRE Subject Test, minimum B average in undergraduate major. *Application deadline:* For fall admission, 3/15. Applications are processed on a rolling basis. *Application fee:* $45.
Expenses: Tuition, state resident: full-time $982. Tuition, nonresident: full-time $2,252. Required fees: $1,106. Part-time tuition and fees vary according to course load.
Financial aid: In 1999–00, 5 research assistantships, 31 teaching assistantships were awarded; career-related internships or fieldwork and Federal Work-Study also available. Aid available to part-time students. Financial aid application deadline: 3/15.

Faculty research: Marine processes, estuaries studies, biotechnology, underwater research, acid rain.
Dr. L. Scott Quackenbush, Chairman, 910-962-3470.

Application contact: Dr. Neil F. Hadley, Dean, Graduate School, 910-962-4117, *Fax:* 910-962-3787, *E-mail:* hadlcyn@uncwil.edu.

■ UNIVERSITY OF OREGON

Graduate School, College of Arts and Sciences, Department of Biology, Eugene, OR 97403

AWARDS Ecology and evolution (MA, MS, PhD); marine biology (MA, MS, PhD); molecular, cellular and genetic biology (PhD); neuroscience and development (PhD).

Faculty: 42 full-time (14 women), 5 part-time/adjunct (2 women).
Students: 69 full-time (33 women), 8 part-time (5 women); includes 6 minority (2 Asian Americans or Pacific Islanders, 3 Hispanic Americans, 1 Native American), 5 international. *18 applicants, 83% accepted.* In 1999, 11 master's, 7 doctorates awarded. Terminal master's awarded for partial completion of doctoral program.
Degree requirements: For master's, thesis required (for some programs); for doctorate, dissertation required, foreign language not required.
Entrance requirements: For master's and doctorate, GRE General Test, TOEFL, minimum GPA of 3.2. *Application deadline:* For fall admission, 1/10. *Application fee:* $50.
Expenses: Tuition, state resident: full-time $6,750. Tuition, nonresident: full-time $11,409. Part-time tuition and fees vary according to course load.
Financial aid: In 1999–00, 36 teaching assistantships were awarded; research assistantships, Federal Work-Study, grants, and institutionally sponsored loans also available. Financial aid application deadline: 2/1.
Faculty research: Developmental neurobiology; evolution, population biology, and quantitative genetics; regulation of gene expression; biochemistry of marine organisms.
Janis C. Weeks, Head, 541-346-4502, *Fax:* 541-346-6056.

Application contact: Donna Overall, Graduate Program Coordinator, 541-346-4503, *Fax:* 541-346-6056, *E-mail:* doverall@oregon.uoregon.edu.

Find an in-depth description at www.petersons.com/graduate.

■ UNIVERSITY OF SOUTHERN CALIFORNIA

Graduate School, College of Letters, Arts and Sciences, Department of Biological Sciences, Program in Marine Biology and Biological Oceanography, Los Angeles, CA 90089

AWARDS MS, PhD.

Faculty: 10 full-time (1 woman).
Students: 1 full-time (0 women). Average age 36. *0 applicants, 0% accepted.*
Degree requirements: For master's and doctorate, thesis/dissertation required.
Entrance requirements: For master's and doctorate, GRE General Test. *Application deadline:* For fall admission, 2/1; for spring admission, 10/15. *Application fee:* $55.
Expenses: Tuition: Full-time $17,952; part-time $748 per unit. Required fees: $406; $203 per unit. Tuition and fees vary according to program.
Financial aid: In 1999–00, 3 fellowships with full tuition reimbursements (averaging $18,000 per year), 6 research assistantships with full tuition reimbursements (averaging $18,000 per year), 10 teaching assistantships with full tuition reimbursements (averaging $18,000 per year) were awarded; Federal Work-Study, institutionally sponsored loans, and scholarships also available. Aid available to part-time students. Financial aid application deadline: 2/15; financial aid applicants required to submit FAFSA.
Dr. Donal Manahan, Chair, 213-740-1109.

Find an in-depth description at www.petersons.com/graduate.

■ UNIVERSITY OF SOUTHERN MISSISSIPPI

Graduate School, College of Science and Technology, Department of Biological Sciences, Hattiesburg, MS 39406

AWARDS Environmental biology (MS, PhD); marine biology (MS, PhD); microbiology (MS, PhD); molecular biology (MS, PhD).

Degree requirements: For master's, thesis required, foreign language not required; for doctorate, 2 foreign languages (computer language can substitute for one), dissertation required.
Entrance requirements: For master's, GRE General Test, TOEFL, minimum GPA of 3.0; for doctorate, GRE General Test, TOEFL, minimum GPA of 3.5.
Expenses: Tuition, state resident: full-time $2,250; part-time $137 per semester hour. Tuition, nonresident: full-time $3,102; part-time $172 per semester hour. Required fees: $602.

■ UNIVERSITY OF SOUTH FLORIDA

Graduate School, College of Arts and Sciences, Department of Biology, Tampa, FL 33620-9951

AWARDS Biology (PhD); botany (MS); ecology (PhD); marine biology (MS, PhD); microbiology (MS); physiology (PhD); zoology (MS). Part-time programs available.

Degree requirements: For master's, foreign language not required; for doctorate, 2 foreign languages (computer language can substitute for one), dissertation required.

Entrance requirements: For master's, GRE General Test, minimum GPA of 3.0 in last 60 hours; for doctorate, GRE General Test, GRE Subject Test in biology. Electronic applications accepted.

Expenses: Tuition, state resident: part-time $148 per credit hour. Tuition, nonresident: part-time $509 per credit hour.

Microbiological Sciences

BACTERIOLOGY

■ PURDUE UNIVERSITY

School of Veterinary Medicine and Graduate School, Graduate Programs in Veterinary Medicine, Department of Veterinary Pathobiology, West Lafayette, IN 47907

AWARDS Bacteriology (MS, PhD); epidemiology (MS, PhD); immunology (MS, PhD); infectious diseases (MS, PhD); microbiology (MS, PhD); parasitology (MS, PhD); pathology (MS, PhD); toxicology (MS, PhD); virology (MS, PhD).

Faculty: 26 full-time (5 women).
Students: 42 full-time (24 women), 2 part-time (1 woman); includes 5 minority (1 African American, 3 Asian Americans or Pacific Islanders, 1 Hispanic American), 18 international. Average age 32. In 1999, 2 master's, 6 doctorates awarded (100% found work related to degree). Terminal master's awarded for partial completion of doctoral program.
Degree requirements: For master's, thesis required (for some programs), foreign language not required; for doctorate, dissertation required, foreign language not required.
Entrance requirements: For master's and doctorate, GRE General Test, TOEFL. *Application fee:* $30.
Expenses: Tuition, state resident: full-time $3,732. Tuition, nonresident: full-time $8,732.
Financial aid: Fellowships, research assistantships, teaching assistantships available. Financial aid application deadline: 3/1; financial aid applicants required to submit FAFSA.
Dr. H. L. Thacker, Head, 765-494-7543.

■ THE UNIVERSITY OF IOWA

College of Medicine and Graduate College, Graduate Programs in Medicine, Department of Microbiology, Iowa City, IA 52242-1316

AWARDS General microbiology and microbial physiology (MS, PhD); immunology (MS, PhD); microbial genetics (MS, PhD); pathogenic bacteriology (MS, PhD); virology (MS, PhD).

Faculty: 21 full-time (4 women), 10 part-time/adjunct (1 woman).
Students: 50 full-time (21 women); includes 7 minority (2 African Americans, 4 Asian Americans or Pacific Islanders, 1 Hispanic American), 5 international. *144 applicants, 8% accepted.* In 1999, 7 doctorates awarded (100% entered university research/teaching).
Degree requirements: For master's, thesis required; for doctorate, dissertation, comprehensive exam required. *Average time to degree:* Doctorate–6 years full-time.
Entrance requirements: For master's and doctorate, GRE General Test. *Application deadline:* For fall admission, 2/1. *Application fee:* $30 ($50 for international students). Electronic applications accepted.
Expenses: Tuition, state resident: full-time $3,308. Tuition, nonresident: full-time $10,662. Tuition and fees vary according to course load and program.
Financial aid: In 1999–00, 50 students received aid, including fellowships with full tuition reimbursements available (averaging $16,787 per year), research assistantships with full tuition reimbursements available (averaging $16,787 per year); grants, institutionally sponsored loans, and traineeships also available.
Faculty research: Biocatalysis and blue jeans, gene regulation, processing and transport of HIV, retroviral pathogenesis, biodegradation. *Total annual research expenditures:* $8.2 million.

Dr. Michael A. Apicella, Head, 319-335-7810, *E-mail:* grad-micro-info@uiowa.edu.
Find an in-depth description at www.petersons.com/graduate.

■ THE UNIVERSITY OF TENNESSEE HEALTH SCIENCE CENTER

College of Graduate Health Sciences, Department of Microbiology and Immunology, Memphis, TN 38163-0002

AWARDS Bacteriology (MS, PhD); immunology (MS, PhD); medical microbiology (MS, PhD); microbiology (MS, PhD); molecular and cell biology (PhD); molecular biology (MS, PhD); virology (MS, PhD). Terminal master's awarded for partial completion of doctoral program.

Degree requirements: For master's, thesis, oral and written comprehensive exams required, foreign language not required; for doctorate, dissertation, oral and written preliminary and comprehensive exams required, foreign language not required.
Entrance requirements: For master's, GRE General Test, TOEFL, minimum GPA of 3.0; for doctorate, GRE General Test, GRE Subject Test, minimum GPA of 3.0.

Find an in-depth description at www.petersons.com/graduate.

■ THE UNIVERSITY OF TEXAS MEDICAL BRANCH AT GALVESTON

Graduate School of Biomedical Sciences, Program in Emerging and Tropical Infectious Diseases, Galveston, TX 77555

AWARDS PhD.

Faculty: 39 full-time (6 women).
Students: 4 full-time (3 women).

Degree requirements: For doctorate, dissertation required, foreign language not required.

Entrance requirements: For doctorate, GRE General Test. *Application deadline:* Applications are processed on a rolling basis. *Application fee:* $25 ($50 for international students).

Expenses: Tuition, state resident: full-time $684; part-time $38 per credit hour. Tuition, nonresident: full-time $4,572; part-time $254 per credit hour. Required fees: $29; $7.5 per credit hour. One-time fee: $55. Tuition and fees vary according to program.

Financial aid: In 1999–00, 4 fellowships with tuition reimbursements (averaging $15,060 per year) were awarded; traineeships also available.

Faculty research: Emerging diseases, tropical diseases, parasitology, vitology and bacteriology.

Dr. Alan D.T. Barrett, Director, 409-772-2521, *Fax:* 409-747-2400.

Find an in-depth description at www.petersons.com/graduate.

■ UNIVERSITY OF WISCONSIN–MADISON

Graduate School, College of Agricultural and Life Sciences and Medical School, Department of Bacteriology, Madison, WI 53706-1380

AWARDS MS.

Students: In 1999, 13 degrees awarded. *Average time to degree:* Master's–2 years full-time.

Entrance requirements: For master's, GRE. *Application deadline:* For fall admission, 1/15 (priority date). *Application fee:* $45. Electronic applications accepted.

Expenses: Tuition, state resident: full-time $5,406; part-time $339 per credit. Tuition, nonresident: full-time $17,110; part-time $1,071 per credit. Full-time tuition and fees vary according to program and reciprocity agreements. Part-time tuition and fees vary according to course load and program.

Financial aid: In 1999–00, 26 students received aid, including 6 fellowships, 12 research assistantships; grants, scholarships, traineeships, and tuition waivers (full) also available.

Faculty research: Cytochrome biosynthesis, metabolic regulation, anaerobic metabolism, transcriptional regulation, nitrogen fixation.

■ WEST VIRGINIA UNIVERSITY

School of Medicine, Graduate Programs in Health Sciences, Department of Microbiology and Immunology, Morgantown, WV 26506

AWARDS Genetics (MS, PhD); immunology (MS, PhD); mycology (PhD); parasitology (MS, PhD); pathogenic bacteriology (MS, PhD); physiology (PhD); psysiology (MS); virology (MS, PhD).

Students: 25 full-time (11 women), 1 (woman) part-time; includes 2 minority (both Asian Americans or Pacific Islanders), 4 international. Average age 27. *62 applicants, 8% accepted.* In 1999, 1 master's, 5 doctorates awarded. Terminal master's awarded for partial completion of doctoral program.

Degree requirements: For master's, thesis required, foreign language not required; for doctorate, dissertation, comprehensive exam required, foreign language not required. *Average time to degree:* Doctorate–4.5 years full-time.

Entrance requirements: For master's and doctorate, GRE General Test, GRE Subject Test, TOEFL, minimum GPA of 3.0. *Application deadline:* For fall admission, 3/1 (priority date). Applications are processed on a rolling basis. *Application fee:* $45. Electronic applications accepted.

Expenses: Tuition, state resident: full-time $3,564. Tuition, nonresident: full-time $10,230.

Financial aid: In 1999–00, 7 research assistantships, 14 teaching assistantships were awarded; fellowships, Federal Work-Study and institutionally sponsored loans also available. Financial aid application deadline: 3/1; financial aid applicants required to submit FAFSA.

Faculty research: Mechanisms of pathogenesis, microbial genetics, molecular virology, immunotoxicology, oncogenes. *Total annual research expenditures:* $2.6 million.

John Barnett, Chair, 304-293-2649.

Application contact: James Sheil, Graduate Coordinator, 304-293-3559, *Fax:* 304-293-7823, *E-mail:* jsheil@wvu.edu.

IMMUNOLOGY

■ ALBANY MEDICAL COLLEGE

Graduate Programs in the Biological Sciences, Program in Immunology and Microbial Disease, Albany, NY 12208-3479

AWARDS MS, PhD. Part-time programs available.

Students: 30 full-time (17 women); includes 2 minority (1 African American, 1 Asian American or Pacific Islander), 11

international. Terminal master's awarded for partial completion of doctoral program.

Degree requirements: For master's, thesis required, foreign language not required; for doctorate, dissertation, oral qualifying exam, written preliminary exam required, foreign language not required.

Entrance requirements: For master's and doctorate, GRE General Test, TOEFL. *Application deadline:* Applications are processed on a rolling basis. *Application fee:* $0 ($60 for international students).

Expenses: Tuition: Full-time $13,367; part-time $446 per credit hour.

Financial aid: Federal Work-Study, grants, scholarships, and tuition waivers (full) available.

Dr. Jeffrey Banas, Graduate Director, 518-262-6286, *Fax:* 518-262-5748, *E-mail:* banasj@mail.amc.edu.

Find an in-depth description at www.petersons.com/graduate.

■ BAYLOR COLLEGE OF MEDICINE

Graduate School of Biomedical Sciences, Department of Immunology, Houston, TX 77030-3498

AWARDS PhD, MD/PhD.

Faculty: 24 full-time (3 women).

Students: 32 full-time (16 women); includes 6 minority (1 African American, 3 Asian Americans or Pacific Islanders, 2 Hispanic Americans), 8 international. Average age 28. *86 applicants, 10% accepted.* In 1999, 5 degrees awarded (80% entered university research/teaching, 20% continued full-time study).

Degree requirements: For doctorate, dissertation, public defense, qualifying exam required, foreign language not required. *Average time to degree:* Doctorate–5.54 years full-time.

Entrance requirements: For doctorate, GRE General Test (average 80th percentile), GRE Subject Test (strongly recommended), TOEFL, minimum GPA of 3.0. *Application deadline:* For fall admission, 2/1 (priority date). *Application fee:* $30. Electronic applications accepted.

Expenses: Tuition: Full-time $8,200. Required fees: $175. Full-time tuition and fees vary according to student level.

Financial aid: In 1999–00, 32 students received aid, including 15 fellowships (averaging $20,000 per year), 17 research assistantships (averaging $20,000 per year); teaching assistantships, Federal Work-Study, institutionally sponsored loans, and tuition waivers (full) also available. Financial aid applicants required to submit FAFSA.

Faculty research: Structure and function of major histocompatibility antigens, induction and regulation of T-cell immune

Baylor College of Medicine (continued) responses, microbial genetics, pathophysiology of bacterial and viral infections, control and epidemiology of respiratory viruses.

Dr. Sarah Highlander, Director, 713-798-6311, *Fax:* 713-798-7375.

Application contact: Beatrice Torres, Graduate Program Administrator, 713-798-4472, *Fax:* 713-798-7375, *E-mail:* bft@bcm.tmc.edu.

Find an in-depth description at www.petersons.com/graduate.

■ BOSTON UNIVERSITY

School of Medicine, Division of Graduate Medical Sciences, Department of Pathology and Laboratory Medicine, Immunology Training Program, Boston, MA 02215

AWARDS PhD.

Degree requirements: For doctorate, dissertation, qualifying exam required, foreign language not required.

Entrance requirements: For doctorate, GRE General Test, GRE Subject Test, TOEFL. *Application deadline:* For fall admission, 1/15 (priority date); for spring admission, 10/15 (priority date). *Application fee:* $50. Electronic applications accepted.

Expenses: Tuition: Full-time $24,700; part-time $772 per credit. Required fees: $220.

Financial aid: Fellowships with tuition reimbursements, research assistantships with tuition reimbursements, Federal Work-Study, scholarships, and traineeships available.

Dr. Ann Marshak-Rothstein, Program Director, 617-638-4284, *Fax:* 617-638-4286, *E-mail:* itp@bu.edu.

Find an in-depth description at www.petersons.com/graduate.

■ BROWN UNIVERSITY

Graduate School, Division of Biology and Medicine, Pathobiology Graduate Program, Providence, RI 02912

AWARDS Biology (PhD); cancer biology (PhD); immunology and infection (PhD); medical science (PhD); pathobiology (Sc M); toxicology and environmental pathology (PhD).

Faculty: 36 full-time (9 women), 1 part-time/adjunct (0 women).

Students: 18 full-time (11 women); includes 4 minority (2 African Americans, 2 Asian Americans or Pacific Islanders), 1 international. Average age 25. *42 applicants, 17% accepted.* In 1999, 2 doctorates awarded (100% entered university research/teaching). Terminal master's awarded for partial completion of doctoral program.

Degree requirements: For master's, foreign language not required; for doctorate, dissertation, preliminary exam required, foreign language not required. *Average time to degree:* Doctorate–6 years full-time.

Entrance requirements: For master's, GRE General Test, GRE Subject Test; for doctorate, GRE General Test, GRE Subject Test, TOEFL. *Application deadline:* For fall admission, 1/2 (priority date). Applications are processed on a rolling basis. *Application fee:* $60. Electronic applications accepted.

Financial aid: In 1999–00, 11 fellowships with tuition reimbursements, 1 research assistantship with tuition reimbursement, 4 teaching assistantships with tuition reimbursements were awarded; institutionally sponsored loans and traineeships also available. Financial aid application deadline: 1/2.

Faculty research: Environmental pathology, carcinogenesis, immunopathology, signal transduction.

Dr. Nancy Thompson, Director, 401-444-8860, *E-mail:* nancy_thompson@ brown.edu.

Application contact: Marilyn May, Program Coordinator, 401-863-2913, *Fax:* 401-863-9008, *E-mail:* marilyn_may@ brown.edu.

Find an in-depth description at www.petersons.com/graduate.

■ BROWN UNIVERSITY

Graduate School, Division of Biology and Medicine, Program in Molecular Biology, Cell Biology, and Biochemistry, Providence, RI 02912

AWARDS Biochemistry (M Med Sc, Sc M, PhD), including biochemistry (Sc M, PhD), biology (Sc M, PhD), medical science (M Med Sc, PhD); biology (MA); cell biology (M Med Sc, Sc M, PhD), including biochemistry (Sc M, PhD), biology (Sc M, PhD), medical science (M Med Sc, PhD); developmental biology (M Med Sc, Sc M, PhD), including biochemistry (Sc M, PhD), biology (Sc M, PhD), medical science (M Med Sc, PhD); immunology (M Med Sc, Sc M, PhD), including biochemistry (Sc M, PhD), biology (Sc M, PhD), medical science (M Med Sc, PhD); molecular microbiology (M Med Sc, Sc M, PhD), including biochemistry (Sc M, PhD), biology (Sc M, PhD), medical science (M Med Sc, PhD). Part-time programs available.

Faculty: 50 full-time (14 women).

Students: 61 full-time (34 women); includes 4 minority (1 African American, 3 Asian Americans or Pacific Islanders), 21 international. Average age 25. *106 applicants, 28% accepted.* In 1999, 1 master's, 3 doctorates awarded. Terminal master's awarded for partial completion of doctoral program.

Degree requirements: For master's, thesis required (for some programs), foreign language not required; for doctorate, one foreign language, dissertation, preliminary exam required. *Average time to degree:* Doctorate–5 years full-time.

Entrance requirements: For master's and doctorate, GRE General Test, GRE Subject Test. *Application deadline:* For fall admission, 1/2 (priority date). Applications are processed on a rolling basis. *Application fee:* $60. Electronic applications accepted.

Financial aid: In 1999–00, 58 students received aid, including 11 fellowships (averaging $18,916 per year), 9 research assistantships (averaging $18,916 per year), 13 teaching assistantships (averaging $12,690 per year); institutionally sponsored loans and traineeships also available. Financial aid application deadline: 1/2.

Faculty research: Molecular genetics, gene regulation.

Dr. Gary Wessel, Director, 401-863-1051, *E-mail:* chet@brown.edu.

Application contact: Mary C. Esser, Graduate Program Coordinator, 401-863-1661, *Fax:* 401-863-1348, *E-mail:* mary_esser@brown.edu.

Find an in-depth description at www.petersons.com/graduate.

■ CALIFORNIA INSTITUTE OF TECHNOLOGY

Division of Biology, Program in Immunology, Pasadena, CA 91125-0001

AWARDS PhD.

Degree requirements: For doctorate, dissertation, qualifying exam required, foreign language not required.

Entrance requirements: For doctorate, GRE General Test. *Application deadline:* For fall admission, 1/1. *Application fee:* $0.

Expenses: Tuition: Full-time $19,260. Required fees: $24. One-time fee: $100 full-time.

Financial aid: Application deadline: 1/1.

Application contact: Elizabeth Ayala, Graduate Option Coordinator, 626-395-4497, *Fax:* 626-449-0756, *E-mail:* biograd@cco.caltech.edu.

■ CASE WESTERN RESERVE UNIVERSITY

School of Medicine and School of Graduate Studies, Graduate Programs in Medicine, Programs in Molecular and Cellular Basis of Disease, Program in Immunology, Cleveland, OH 44106

AWARDS MS, PhD, MD/PhD.

Degree requirements: For doctorate, dissertation required, foreign language not required.

Entrance requirements: For doctorate, GRE General Test, GRE Subject Test, TOEFL.

Faculty research: Immunopathology, immunochemistry.

■ COLORADO STATE UNIVERSITY

College of Veterinary Medicine and Biomedical Sciences and Graduate School, Graduate Programs in Veterinary Medicine and Biomedical Sciences, Department of Microbiology, Fort Collins, CO 80523-0015

AWARDS Immunology (MS, PhD); microbiology (MS, PhD).

Faculty: 27 full-time (8 women).
Students: 42 full-time (26 women), 6 part-time (5 women); includes 6 minority (1 African American, 2 Asian Americans or Pacific Islanders, 3 Hispanic Americans), 7 international. Average age 28. *103 applicants, 21% accepted.* In 1999, 8 master's, 5 doctorates awarded.
Degree requirements: For master's, thesis required, foreign language not required; for doctorate, dissertation required.
Entrance requirements: For master's and doctorate, GRE General Test, TOEFL, minimum GPA of 3.0. *Application deadline:* For fall admission, 2/1 (priority date); for spring admission, 10/1 (priority date). Applications are processed on a rolling basis. *Application fee:* $30. Electronic applications accepted.
Expenses: Tuition, state resident: full-time $2,694; part-time $150 per credit. Tuition, nonresident: full-time $10,460; part-time $581 per credit. Required fees: $32 per semester.
Financial aid: In 1999–00, 3 fellowships, 35 research assistantships, 4 teaching assistantships were awarded; traineeships also available.
Faculty research: Medical and veterinary bacteriology, virology, industrial and environmental microbiology, vector-borne disease. *Total annual research expenditures:* $5 million.
Ralph E. Smith, Head.
Application contact: Graduate Coordinator, 970-491-6136, *Fax:* 970-491-1815, *E-mail:* microbio@cvmbs.colostate.edu.

Find an in-depth description at www.petersons.com/graduate.

■ CORNELL UNIVERSITY

Graduate School, Graduate Fields of Veterinary Medicine, Field of Immunology, Ithaca, NY 14853-0001
AWARDS Cellular immunology (MS, PhD); immunochemistry (MS, PhD); immunogenetics (MS, PhD); immunopathology (MS, PhD); infection and immunity (MS, PhD).

Faculty: 18 full-time.
Students: 14 full-time (12 women), 4 international. *38 applicants, 8% accepted.* In 1999, 2 master's, 1 doctorate awarded.
Degree requirements: For master's and doctorate, thesis/dissertation required, foreign language not required.
Entrance requirements: For master's and doctorate, GRE General Test, TOEFL. *Application deadline:* For fall admission, 1/15; for spring admission, 10/1. *Application fee:* $65. Electronic applications accepted.
Expenses: Tuition: Full-time $12,400.
Financial aid: In 1999–00, 13 students received aid, including 5 fellowships with full tuition reimbursements available, 8 research assistantships with full tuition reimbursements available; teaching assistantships with full tuition reimbursements available, institutionally sponsored loans, scholarships, tuition waivers (full and partial), and unspecified assistantships also available. Financial aid applicants required to submit FAFSA.
Faculty research: T-cell subset function, avian immunology, mucosal immunity, receptor signaling, anti-parasite and anti-viral immunity.
Application contact: Graduate Field Assistant, 607-253-3276, *E-mail:* vetgradpgms@cornell.edu.

■ CREIGHTON UNIVERSITY

School of Medicine and Graduate School, Graduate Programs in Medicine and College of Arts and Sciences, Department of Medical Microbiology and Immunology, Omaha, NE 68178-0001
AWARDS MS, PhD.

Faculty: 20 full-time (3 women), 5 part-time/adjunct (0 women).
Students: 15 full-time (7 women), 2 part-time (both women), 5 international. Average age 24. *40 applicants, 7% accepted.* In 1999, 3 master's, 1 doctorate awarded. Terminal master's awarded for partial completion of doctoral program.
Degree requirements: For master's and doctorate, thesis/dissertation required, foreign language not required. *Average time to degree:* Master's–3 years full-time; doctorate–5 years full-time.
Entrance requirements: For master's and doctorate, GRE General Test. *Application deadline:* For fall admission, 2/1 (priority date). Applications are processed on a rolling basis. *Application fee:* $30.
Expenses: Tuition: Full-time $8,940; part-time $447 per credit hour. Required fees: $598; $50 per semester.
Financial aid: In 1999–00, 17 students received aid, including 6 fellowships with tuition reimbursements available; research assistantships with tuition reimbursements available, career-related internships or fieldwork and institutionally sponsored loans also available.
Faculty research: Infectious diseases, molecular biology, genetics, antimicrobial agents and chemotherapy, virology, microbial physiology, prions, neuropathology. *Total annual research expenditures:* $770,612.
Dr. Roderick Nairn, Chair, 402-280-2921, *Fax:* 402-280-1875, *E-mail:* rnairn@creighton.edu.
Application contact: Dr. Philip D. Lister, Graduate Director, 402-280-2921, *Fax:* 402-280-1875, *E-mail:* pdlister@creighton.edu.

Find an in-depth description at www.petersons.com/graduate.

■ DUKE UNIVERSITY

Graduate School, Department of Immunology, Durham, NC 27708-0586
AWARDS PhD.

Faculty: 21 full-time, 2 part-time/adjunct.
Students: 22 full-time (10 women), 5 international.
In 2000, 4 degrees awarded.
Degree requirements: For doctorate, dissertation required.
Entrance requirements: For doctorate, GRE General Test, GRE Subject Test (recommended). *Application deadline:* For fall admission, 12/31. *Application fee:* $75.
Expenses: Tuition: Full-time $21,406; part-time $760 per unit. Required fees: $3,136; $3,136 per year. One-time fee: $30. Tuition and fees vary according to program.
Financial aid: Fellowships, research assistantships available. Financial aid application deadline: 12/31.
Dr. Garnett Kesloe, Director of Graduate Studies, 919-613-7815, *Fax:* 919-613-7878, *E-mail:* pinkn001@mc.duke.edu.
Application contact: Mel Glazebrook, Assistant to the Director of Graduate Studies, 919-613-7815, *Fax:* 919-613-7878, *E-mail:* glaze002@mc.duke.edu.

Find an in-depth description at www.petersons.com/graduate.

■ EAST CAROLINA UNIVERSITY

School of Medicine, Department of Microbiology and Immunology, Greenville, NC 27858-4353

AWARDS PhD.

Faculty: 12 full-time (1 woman), 1 part-time/adjunct (0 women).
Students: 6 full-time (2 women), 7 part-time (2 women); includes 1 minority (Asian American or Pacific Islander), 3 international. Average age 26. 7 *applicants, 86% accepted.* In 1999, 1 degree awarded.
Degree requirements: For doctorate, dissertation required, foreign language not required.
Entrance requirements: For doctorate, GRE General Test, GRE Subject Test, TOEFL. *Application deadline:* For fall admission, 4/15 (priority date). Applications are processed on a rolling basis. *Application fee:* $40.
Expenses: Tuition, state resident: full-time $1,012. Tuition, nonresident: full-time $8,578. Required fees: $1,006. Full-time tuition and fees vary according to degree level. Part-time tuition and fees vary according to course load.
Financial aid: Fellowships available. Financial aid application deadline: 6/1.
Faculty research: Molecular virology, genetics of bacteria, yeast and somatic cells, bacterial physiology and metabolism. Dr. Paul V. Phibbs, Chairman, 252-816-2700, *Fax:* 252-816-3104, *E-mail:* phibbs@brody.med.ecu.edu.
Application contact: Dr. Henry Stone, Director of Graduate Studies, 252-816-2700, *Fax:* 252-816-3104, *E-mail:* stone@brody.med.ecu.edu.

■ EMORY UNIVERSITY

Graduate School of Arts and Sciences, Division of Biological and Biomedical Sciences, Program in Immunology and Molecular Pathogenesis, Atlanta, GA 30322-1100

AWARDS PhD.

Faculty: 37 full-time (8 women).
Students: 44 full-time (26 women); includes 3 African Americans, 5 international. In 1999, 7 degrees awarded.
Degree requirements: For doctorate, dissertation required, foreign language not required.
Entrance requirements: For doctorate, GRE General Test, TOEFL, minimum GPA of 3.0 in science course work. *Application deadline:* For fall admission, 1/20 (priority date). *Application fee:* $45.
Expenses: Tuition: Full-time $22,770. Tuition and fees vary according to program.
Financial aid: In 1999–00, fellowships with full tuition reimbursements (averaging $18,000 per year).

Faculty research: Transplantation immunology, autoimmunity, microbial pathogenesis.
Dr. Peter Jensen, Director, 404-727-5763, *Fax:* 404-727-8540, *E-mail:* bimcore@emory.edu.
Application contact: 404-727-2547, *Fax:* 404-727-3322, *E-mail:* gdbbs@gsas.emory.edu.

Find an in-depth description at www.petersons.com/graduate.

■ FINCH UNIVERSITY OF HEALTH SCIENCES/THE CHICAGO MEDICAL SCHOOL

School of Graduate and Postdoctoral Studies, Department of Microbiology and Immunology, North Chicago, IL 60064-3095

AWARDS Medical microbiology (MS, PhD); microbiology and immunology (MS, PhD). Part-time programs available.

Faculty: 10 full-time, 1 part-time/adjunct.
Students: 16 full-time (3 women); includes 3 minority (all Asian Americans or Pacific Islanders), 8 international. In 1999, 5 doctorates awarded. Terminal master's awarded for partial completion of doctoral program.
Degree requirements: For master's and doctorate, thesis/dissertation required, foreign language not required.
Entrance requirements: For master's and doctorate, GRE General Test, TOEFL, TWE. *Application deadline:* For fall admission, 6/1 (priority date). Applications are processed on a rolling basis. *Application fee:* $25.
Expenses: Tuition: Full-time $14,054; part-time $391 per credit hour. Tuition and fees vary according to program.
Financial aid: In 1999–00, fellowships (averaging $15,500 per year); tuition waivers (full) also available. Financial aid application deadline: 6/9; financial aid applicants required to submit FAFSA.
Faculty research: Molecular biology, parasitology, virology.
Dr. Yoon B. Kim, Chairman, 847-578-3230.
Application contact: Dana Frederick, Admissions Officer, 847-578-3209.

Find an in-depth description at www.petersons.com/graduate.

■ FLORIDA STATE UNIVERSITY

Graduate Studies, College of Arts and Sciences, Department of Biological Science, Program in Immunology, Tallahassee, FL 32306

AWARDS MS, PhD.

Faculty: 1 full-time (0 women).
Students: 1 full-time (0 women).

Degree requirements: For master's and doctorate, thesis/dissertation, teaching experience required.
Entrance requirements: For master's, GRE General Test, TOEFL; for doctorate, GRE General Test, GRE Subject Test, TOEFL. *Application deadline:* For fall admission, 1/15; for spring admission, 10/15. *Application fee:* $20.
Expenses: Tuition, state resident: full-time $3,504; part-time $146 per credit hour. Tuition, nonresident: full-time $12,162; part-time $507 per credit hour. Tuition and fees vary according to program.
Financial aid: In 1999–00, fellowships with full tuition reimbursements (averaging $13,740 per year), research assistantships with full tuition reimbursements (averaging $13,740 per year), teaching assistantships with full tuition reimbursements (averaging $13,740 per year) were awarded. Financial aid application deadline: 1/15; financial aid applicants required to submit FAFSA.
Faculty research: Immunogenetics.
Dr. Thomas C. S. Keller, Associate Professor and Associate Chairman, 850-644-3023, *Fax:* 850-644-9829.
Application contact: Judy Bowers, Coordinator, Graduate Affairs, 850-644-3023, *Fax:* 850-644-9829, *E-mail:* bowers@bio.fsu.edu.

■ GEORGETOWN UNIVERSITY

Graduate School of Arts and Sciences, Programs in Biomedical Sciences, Department of Microbiology and Immunology, Washington, DC 20057

AWARDS PhD, MD/PhD.

Degree requirements: For doctorate, dissertation, comprehensive exam required, foreign language not required.
Entrance requirements: For doctorate, GRE General Test, TOEFL.

■ THE GEORGE WASHINGTON UNIVERSITY

Columbian School of Arts and Sciences, Institute for Biomedical Sciences, Program in Immunology, Washington, DC 20052

AWARDS PhD.

Degree requirements: For doctorate, dissertation required.
Entrance requirements: For doctorate, GRE General Test, minimum GPA of 3.0.
Expenses: Tuition: Full-time $16,836; part-time $702 per credit hour. Required fees: $828; $35 per credit hour. Tuition and fees vary according to campus/location and program.

Find an in-depth description at www.petersons.com/graduate.

■ HARVARD UNIVERSITY

Graduate School of Arts and Sciences, Program in Immunology, Boston, MA 02115

AWARDS PhD.

Degree requirements: For doctorate, dissertation, qualifying exam required, foreign language not required.

Entrance requirements: For doctorate, GRE General Test, GRE Subject Test, TOEFL. *Application deadline:* For fall admission, 12/15. *Application fee:* $60.

Expenses: Tuition: Full-time $22,054. Required fees: $711. Tuition and fees vary according to program.

Financial aid: Fellowships, research assistantships, teaching assistantships, institutionally sponsored loans and tuition waivers (full) available. Financial aid application deadline: 1/1.

Faculty research: Immunochemistry, cellular immunology, tumor immunology, allergic inflammation, immunogenetics. Dr. Martin Dorf, Chair, 617-432-1983.

Application contact: Leah Simons, Manager of Student Affairs, 617-432-0162.

Find an in-depth description at www.petersons.com/graduate.

■ HARVARD UNIVERSITY

School of Public Health, Department of Immunology and Infectious Diseases, Boston, MA 02115-6096

AWARDS DPH, SD. Part-time programs available.

Faculty: 13 full-time (4 women), 12 part-time/adjunct (1 woman).

Students: 7 full-time (5 women), 3 part-time (2 women), 4 international. Average age 29. *33 applicants, 15% accepted.*

Degree requirements: For doctorate, dissertation, qualifying exam required.

Entrance requirements: For doctorate, GRE, TOEFL. *Application deadline:* For fall admission, 1/3. *Application fee:* $60.

Expenses: Tuition: Full-time $22,950; part-time $574 per credit.

Financial aid: Fellowships, research assistantships, Federal Work-Study, grants, scholarships, traineeships, tuition waivers (partial), and unspecified assistantships available. Financial aid application deadline: 2/12; financial aid applicants required to submit FAFSA.

Faculty research: Infectious disease epidemiology and tropical public health, vector biology, ecology and control, virology.

Dr. Myron Essex, Chairman, 617-432-1023.

Application contact: Sara Thomford, Administrative Assistant, 617-432-0975.

Find an in-depth description at www.petersons.com/graduate.

■ INDIANA UNIVERSITY–PURDUE UNIVERSITY INDIANAPOLIS

School of Medicine, Graduate Programs in Medicine, Department of Microbiology and Immunology, Indianapolis, IN 46202-2896

AWARDS MS, PhD, MD/MS, MD/PhD.

Students: 18 full-time (12 women), 7 part-time (3 women); includes 3 minority (2 African Americans, 1 Hispanic American), 6 international. Average age 25. *72 applicants, 21% accepted.* In 1999, 1 master's awarded (100% found work related to degree); 5 doctorates awarded. Terminal master's awarded for partial completion of doctoral program.

Degree requirements: For master's and doctorate, thesis/dissertation required. *Average time to degree:* Master's–3 years full-time; doctorate–5 years full-time.

Entrance requirements: For master's and doctorate, GRE General Test, previous course work in calculus, cell biology, chemistry, genetics, physics, and biochemistry. *Application deadline:* For fall admission, 3/1. Applications are processed on a rolling basis. *Application fee:* $35 ($55 for international students).

Expenses: Tuition, state resident: full-time $13,245; part-time $158 per credit hour. Tuition, nonresident: full-time $30,330; part-time $455 per credit hour. Required fees: $121 per year. Tuition and fees vary according to course load and degree level.

Financial aid: In 1999–00, 20 research assistantships with full tuition reimbursements (averaging $17,000 per year) were awarded; fellowships with full tuition reimbursements, teaching assistantships with full tuition reimbursements, Federal Work-Study, grants, institutionally sponsored loans, traineeships, and tuition waivers (partial) also available. Financial aid application deadline: 2/1.

Faculty research: Host-parasite interactions, molecular biology, cellular and molecular immunology and hematology, viral and bacterial pathogenesis, cancer research. *Total annual research expenditures:* $4.2 million.

Dr. Hal E. Broxmeyer, Chairman, 317-274-7672, *Fax:* 317-274-4090, *E-mail:* hbroxmey@iupui.edu.

Application contact: 317-274-7671, *Fax:* 317-274-4090, *E-mail:* rhaak@iupui.edu.

Find an in-depth description at www.petersons.com/graduate.

■ IOWA STATE UNIVERSITY OF SCIENCE AND TECHNOLOGY

Graduate College, Interdisciplinary Programs, Program in Immunobiology, Ames, IA 50011

AWARDS MS, PhD.

Students: 6 full-time (2 women), 2 part-time (both women); includes 1 minority (African American), 3 international. *25 applicants, 36% accepted.* In 1999, 1 master's, 5 doctorates awarded.

Degree requirements: For master's and doctorate, thesis/dissertation required.

Entrance requirements: For master's and doctorate, GRE General Test, TOEFL. *Application deadline:* For fall admission, 2/1 (priority date). Applications are processed on a rolling basis. *Application fee:* $20 ($50 for international students). Electronic applications accepted.

Expenses: Tuition, state resident: full-time $3,308. Tuition, nonresident: full-time $9,744. Part-time tuition and fees vary according to course load, campus/location and program.

Financial aid: In 1999–00, 4 research assistantships with partial tuition reimbursements (averaging $12,015 per year) were awarded; fellowships, teaching assistantships, scholarships also available.

Faculty research: Immunogenetics, cellular and molecular immunology, infectious disease, neuroimmunology.

Dr. Ricardo Rosenbusch, Supervisory Committee Chair, 515-294-7252, *E-mail:* idgp@iastate.edu.

■ JOAN AND SANFORD I. WEILL MEDICAL COLLEGE AND GRADUATE SCHOOL OF MEDICAL SCIENCES OF CORNELL UNIVERSITY

Graduate School of Medical Sciences, Graduate Program in Immunology, New York, NY 10021

AWARDS Immunology (MS, PhD), including immunology, microbiology, pathology.

Faculty: 26 full-time (6 women).

Students: 23 full-time (16 women); includes 3 minority (all Asian Americans or Pacific Islanders), 14 international. *54 applicants, 20% accepted.* In 2000, 5 doctorates awarded.

Degree requirements: For doctorate, dissertation, final exam required.

Entrance requirements: For doctorate, GRE General Test, GRE Subject Test, MCAT (MD/PhD), previous course work in biological sciences, laboratory research experience. *Application deadline:* For fall admission, 1/15. *Application fee:* $50.

Expenses: All students in good standing receive an annual stipend of $22,880.

Financial aid: Fellowships, stipends available.

Paolo Casali, Director, 212-746-6454.

■ JOHNS HOPKINS UNIVERSITY

School of Hygiene and Public Health, Department of Molecular Microbiology and Immunology, Baltimore, MD 21218-2699

AWARDS MHS, Sc M, PhD, Sc D.

Faculty: 26 full-time, 18 part-time/adjunct.

Students: 56 (35 women); includes 8 minority (1 African American, 7 Asian Americans or Pacific Islanders) 10 international. *138 applicants, 34% accepted.* In 1999, 15 master's, 9 doctorates awarded.

Degree requirements: For master's, thesis required (for some programs), foreign language not required; for doctorate, dissertation, 1 year full-time residency, oral and written exams required, foreign language not required.

Entrance requirements: For master's and doctorate, GRE General Test, GRE Subject Test (biology or chemistry recommended), TOEFL. *Application deadline:* For fall admission, 2/1 (priority date). Applications are processed on a rolling basis. *Application fee:* $60. Electronic applications accepted.

Expenses: Tuition: Full-time $23,660; part-time $493 per unit. Full-time tuition and fees vary according to degree level, campus/location and program.

Financial aid: Federal Work-Study, institutionally sponsored loans, and scholarships available. Aid available to part-time students. Financial aid application deadline: 4/15.

Faculty research: Immunopathology, immunodiagnosis, pathogenesis of enteric infections, parasite immunology, biochemistry of parasitic protozoa. *Total annual research expenditures:* $10.3 million.
Dr. Diane Griffin, Chair, 410-955-3459, *Fax:* 410-955-0105, *E-mail:* dgriffin@jhsph.edu.

Application contact: Denise Patrick, Academic Coordinator, 410-614-4232, *Fax:* 410-955-0105, *E-mail:* dpatrick@jhsph.edu.

Find an in-depth description at www.petersons.com/graduate.

■ JOHNS HOPKINS UNIVERSITY

School of Hygiene and Public Health, Program in Public Health, Baltimore, MD 21218-2699

AWARDS Biochemistry (MPH); biostatistics (MPH); environmental health sciences (MPH); epidemiology (MPH); health policy and management (MPH); international health (MPH); mental hygiene (MPH); molecular microbiology and immunology (MPH); population and family health sciences (MPH). Part-time and evening/weekend programs available. Postbaccalaureate distance learning degree programs offered.

Students: *694 applicants, 73% accepted.* In 1999, 196 degrees awarded.

Degree requirements: For master's, foreign language and thesis not required.

Entrance requirements: For master's, GRE General Test, TOEFL, 2 years of work related experience. *Application deadline:* For fall admission, 12/1 (priority date). Applications are processed on a rolling basis. *Application fee:* $60. Electronic applications accepted.

Expenses: Tuition: Full-time $23,660; part-time $493 per unit. Full-time tuition and fees vary according to degree level, campus/location and program.

Financial aid: Federal Work-Study, institutionally sponsored loans, and scholarships available. Aid available to part-time students. Financial aid application deadline: 4/15.
Dr. Miriam Alexander, Director, 410-955-1291, *Fax:* 410-955-4749.

Application contact: Lenora Davis, Administrator, 410-955-1291, *Fax:* 410-955-4749, *E-mail:* lrdavis@jhsph.edu.

■ JOHNS HOPKINS UNIVERSITY

School of Medicine, Graduate Programs in Medicine, Program in Immunology, Baltimore, MD 21218-2699

AWARDS PhD.

Faculty: 21 full-time (3 women).
Students: 26 full-time (17 women); includes 14 minority (12 Asian Americans or Pacific Islanders, 2 Hispanic Americans), 2 international. Average age 23. *71 applicants, 10% accepted.* In 1999, 5 degrees awarded.

Degree requirements: For doctorate, dissertation, oral and written comprehensive exams, seminar required, foreign language not required. *Average time to degree:* Doctorate–5 years full-time.

Entrance requirements: For doctorate, GRE General Test, TOEFL. *Application deadline:* For fall admission, 1/10. *Application fee:* $50.

Expenses: Tuition: Full-time $23,660.
Financial aid: In 1999–00, 24 students received aid, including 5 fellowships with tuition reimbursements available (averaging $17,642 per year); tuition waivers (full) and research grants, institutional funds also available. Financial aid application deadline: 1/10.

Faculty research: HIV immunity, tumor immunity, major histocompatibility complex, transplantation, genetics of antibodies and T-cell receptors. *Total annual research expenditures:* $19 million.
Dr. Mark J. Soloski, Director, 410-955-3344, *Fax:* 410-955-0964, *E-mail:* mski@welch.jhu.edu.

Application contact: Angela James, Academic Program Coordinator II, 410-955-2709, *Fax:* 410-955-0964, *E-mail:* ajames@welch.jhu.edu.

Find an in-depth description at www.petersons.com/graduate.

■ KANSAS STATE UNIVERSITY

Graduate School, College of Arts and Sciences, Division of Biology, Program in Microbiology and Immunology, Manhattan, KS 66506

AWARDS MS, PhD.

Degree requirements: For master's and doctorate, thesis/dissertation required, foreign language not required.

Expenses: Tuition, state resident: part-time $103 per credit hour. Tuition, nonresident: part-time $338 per credit hour. Required fees: $17 per credit hour. One-time fee: $64 part-time.

Faculty research: Immune cell function, virology, cell viability, water quality.

■ LONG ISLAND UNIVERSITY, C.W. POST CAMPUS

School of Health Professions, Department of Biomedical Sciences, Program in Medical Biology, Brookville, NY 11548-1300

AWARDS Hematology (MS); immunology (MS); medical chemistry (MS); microbiology (MS). Part-time and evening/weekend programs available.

Faculty: 3 full-time (1 woman), 12 part-time/adjunct (6 women).
Students: 4 full-time (2 women), 41 part-time (21 women); includes 25 minority (10 African Americans, 10 Asian Americans or Pacific Islanders, 5 Hispanic Americans). *37 applicants, 89% accepted.* In 1999, 11 degrees awarded.

Degree requirements: For master's, computer language, thesis required, foreign language not required. *Average time to degree:* Master's–2 years full-time, 4 years part-time.

Entrance requirements: For master's, TOEFL, minimum GPA of 2.75 in major. *Application deadline:* For fall admission, 9/1 (priority date); for spring admission, 1/20 (priority date). Applications are processed on a rolling basis. *Application fee:* $30. Electronic applications accepted.

Expenses: Tuition: Part-time $405 per credit. Required fees: $310; $65 per year. Tuition and fees vary according to course load and program.

Financial aid: In 1999–00, 20 students received aid; fellowships with partial tuition reimbursements available, teaching assistantships with partial tuition reimbursements available, career-related

internships or fieldwork, Federal Work-Study, and institutionally sponsored loans available. Aid available to part-time students. Financial aid application deadline: 5/15; financial aid applicants required to submit FAFSA.
Faculty research: Hematopoiesis, growth factors in cancer, interleukins in allergy, PCR techniques. *Total annual research expenditures:* $15,000.
Application contact: Robin Steadman, Graduate Adviser, 516-299-2337, *Fax:* 516-299-2527, *E-mail:* rsteadman@phoenix.liu.edu.

■ LOUISIANA STATE UNIVERSITY HEALTH SCIENCES CENTER

School of Graduate Studies in New Orleans, Department of Microbiology, Immunology, and Parasitology, New Orleans, LA 70112-1393

AWARDS Microbiology and immunology (MS, PhD).

Faculty: 13 full-time (3 women).
Students: 19 full-time (13 women); includes 5 minority (3 African Americans, 1 Asian American or Pacific Islander, 1 Hispanic American), 4 international. Average age 23. *30 applicants, 50% accepted.* In 1999, 3 master's awarded. Terminal master's awarded for partial completion of doctoral program.
Degree requirements: For master's, thesis required, foreign language not required; for doctorate, dissertation, preliminary exam, qualifying exam required, foreign language not required. *Average time to degree:* Master's–3.5 years full-time; doctorate–7 years full-time.
Entrance requirements: For master's and doctorate, GRE General Test, TOEFL. *Application deadline:* For fall admission, 4/15 (priority date). Applications are processed on a rolling basis. *Application fee:* $30.
Expenses: Tuition, state resident: full-time $2,878; part-time $126 per hour. Tuition, nonresident: full-time $6,003; part-time $265 per hour. Required fees: $2,272. Tuition and fees vary according to course load, degree level and program.
Financial aid: In 1999–00, 16 students received aid, including 16 research assistantships (averaging $13,000 per year); Federal Work-Study and tuition waivers (full) also available. Financial aid application deadline: 4/15.
Faculty research: Microbial physiology, animal virology, vaccine development, AIDS drug studies, pathogenic mechanisms, molecular immunology. Dr. Ronald B. Luftig, Head, 504-568-4062.

Application contact: Jeanine M. Campbell, Asst. Business Manager, 504-568-4061, *Fax:* 504-568-2918, *E-mail:* jcampb@lsumc.edu.
Find an in-depth description at www.petersons.com/graduate.

■ LOUISIANA STATE UNIVERSITY HEALTH SCIENCES CENTER

School of Graduate Studies in Shreveport, Department of Microbiology, Shreveport, LA 71130-3932

AWARDS Microbiology/immunology (MS, PhD). Terminal master's awarded for partial completion of doctoral program.

Degree requirements: For master's and doctorate, thesis/dissertation required, foreign language not required.
Entrance requirements: For master's and doctorate, GRE General Test, TOEFL.
Expenses: Tuition, state resident: full-time $2,878; part-time $126 per hour. Tuition, nonresident: full-time $6,003; part-time $265 per hour. Required fees: $2,272. Tuition and fees vary according to course load, degree level and program.
Faculty research: Infectious disease, pathogenesis, molecular virology and biology.

■ LOYOLA UNIVERSITY CHICAGO

Graduate School, Department of Microbiology and Immunology, Maywood, IL 60153

AWARDS Immunology (MS, PhD); microbiology (MS, PhD); virology (MS, PhD).

Faculty: 10 full-time (3 women).
Students: 26 full-time (18 women). Average age 27. *138 applicants, 9% accepted.* Terminal master's awarded for partial completion of doctoral program.
Degree requirements: For master's, thesis required, foreign language not required; for doctorate, computer language, dissertation, oral comprehensive exams required, foreign language not required.
Entrance requirements: For master's and doctorate, GRE General Test, TOEFL. *Application deadline:* Applications are processed on a rolling basis. *Application fee:* $35. Electronic applications accepted.
Expenses: Tuition: Part-time $500 per credit hour. Required fees: $42 per term.
Financial aid: In 1999–00, 10 fellowships with tuition reimbursements, 12 research assistantships with tuition reimbursements were awarded; Federal Work-Study, institutionally sponsored loans, and scholarships also available. Financial aid application deadline: 2/15.

Faculty research: Viral pathogenesis, microbial physiology and genetics, immunoglobulin genetics and differentiation of the immune response, signal transduction and host-parasite interactions. Dr. Katherine L. Knight, Chair, 708-216-3385, *Fax:* 708-216-9574, *E-mail:* kknight@luc.edu.
Application contact: Dr. Thomas Gallagher, Head, 708-216-3385, *Fax:* 708-216-9574, *E-mail:* tgallag@luc.edu.
Find an in-depth description at www.petersons.com/graduate.

■ MASSACHUSETTS INSTITUTE OF TECHNOLOGY

School of Science, Department of Biology, Program in Immunology, Cambridge, MA 02139-4307

AWARDS PhD.

Degree requirements: For doctorate, dissertation, general exam required, foreign language not required.
Entrance requirements: For doctorate, GRE General Test. *Application deadline:* For fall admission, 1/1. *Application fee:* $55.
Expenses: Tuition: Full-time $25,000. Full-time tuition and fees vary according to degree level, program and student level.
Financial aid: Application deadline: 1/1.
Application contact: Dr. Janice D. Chang, Educational Administrator, 617-253-3717, *Fax:* 617-258-9329, *E-mail:* gradbio@mit.edu.

■ MAYO GRADUATE SCHOOL

Graduate Programs in Biomedical Sciences, Program in Immunology, Rochester, MN 55905

AWARDS PhD.

Faculty: 20 full-time (5 women).
Students: 24 full-time (10 women); includes 6 minority (4 Asian Americans or Pacific Islanders, 2 Hispanic Americans), 2 international. In 1999, 5 degrees awarded.
Degree requirements: For doctorate, oral defense of dissertation, qualifying oral and written exam required.
Entrance requirements: For doctorate, GRE, TOEFL, 2 years of chemistry; 1 year of biology, calculus, and physics. *Application deadline:* For fall admission, 12/31 (priority date). Applications are processed on a rolling basis. *Application fee:* $0.
Expenses: Tuition: Full-time $17,900.
Financial aid: In 1999–00, 24 students received aid, including 24 fellowships with full tuition reimbursements available
Faculty research: Immunogenetics, autoimmunity, receptor signal transduction, T lymphocyte activation, transplantation.

Mayo Graduate School (continued)
Dr. Larry R. Pease, Education Coordinator, 507-284-9891, *E-mail:* pease@mayo.edu.
Application contact: Sherry Kallies, Information Contact, 507-266-0122, *Fax:* 507-284-0999, *E-mail:* phd.training@mayo.edu.

■ MCP HAHNEMANN UNIVERSITY

School of Medicine, Biomedical Graduate Programs, Department of Microbiology and Immunology, Philadelphia, PA 19102-1192

AWARDS MS, PhD, MD/MS, MD/PhD. Terminal master's awarded for partial completion of doctoral program.

Degree requirements: For master's, thesis, comprehensive exam required, foreign language not required; for doctorate, dissertation, qualifying exam required, foreign language not required.
Entrance requirements: For master's, GRE General Test, TOEFL, minimum GPA of 2.75; for doctorate, GRE General Test, TOEFL, minimum GPA of 3.0.
Faculty research: Immunology of malarial parasites, virology, bacteriology, molecular biology, parasitology.

Find an in-depth description at www.petersons.com/graduate.

■ MEDICAL UNIVERSITY OF SOUTH CAROLINA

College of Graduate Studies, Department of Microbiology and Immunology, Charleston, SC 29425-0002

AWARDS MS, PhD, DMD/PhD, MD/PhD.

Faculty: 18 part-time/adjunct (3 women).
Students: 34 full-time (21 women); includes 5 minority (2 African Americans, 3 Asian Americans or Pacific Islanders), 3 international. Average age 27. *27 applicants, 41% accepted.* In 1999, 1 degree awarded. Terminal master's awarded for partial completion of doctoral program.
Degree requirements: For master's, thesis, research seminar required; for doctorate, dissertation, teaching and research seminar, oral and written exams required, foreign language not required.
Entrance requirements: For master's and doctorate, GRE General Test, TOEFL, interview. *Application deadline:* Applications are processed on a rolling basis. *Application fee:* $55. Electronic applications accepted.
Expenses: Tuition, state resident: full-time $3,470; part-time $160 per semester hour. Tuition, nonresident: full-time $4,426; part-time $213 per semester hour. Required fees: $408 per semester. One-time fee: $160. Tuition and fees vary according to program.

Financial aid: In 1999–00, 14 students received aid, including 10 fellowships (averaging $16,000 per year), 12 research assistantships; teaching assistantships, Federal Work-Study and tuition waivers (partial) also available. Financial aid application deadline: 4/1; financial aid applicants required to submit FAFSA.
Faculty research: Cellular immunology, immunochemistry, immunogenetics, molecular genetics, tumor immunology. *Total annual research expenditures:* $3.3 million.
Dr. E. C. LeRoy, Chairman, 843-792-4421.
Application contact: Julie Johnston, Director of Admissions, 843-792-8710, *Fax:* 843-792-3764.

■ NEW YORK MEDICAL COLLEGE

Graduate School of Basic Medical Sciences, Program in Microbiology and Immunology, Valhalla, NY 10595-1691

AWARDS MS, PhD, MD/PhD. Part-time and evening/weekend programs available.

Faculty: 15 full-time (1 woman).
Students: 14 full-time (8 women), 16 part-time (12 women); includes 4 minority (3 Asian Americans or Pacific Islanders, 1 Hispanic American), 5 international. Average age 30. *27 applicants, 11% accepted.* In 1999, 8 master's awarded (100% found work related to degree); 3 doctorates awarded. Terminal master's awarded for partial completion of doctoral program.
Degree requirements: For master's and doctorate, computer language, thesis/dissertation required, foreign language not required.
Entrance requirements: For master's, GRE General Test, TOEFL; for doctorate, GRE General Test, GRE Subject Test, TOEFL. *Application deadline:* For fall admission, 7/1 (priority date); for spring admission, 12/1 (priority date). Applications are processed on a rolling basis. *Application fee:* $35 ($60 for international students).
Expenses: Tuition: Part-time $430 per credit. Required fees: $15 per semester. One-time fee: $100.
Financial aid: In 1999–00, 6 research assistantships with full tuition reimbursements were awarded; career-related internships or fieldwork, Federal Work-Study, grants, institutionally sponsored loans, and tuition waivers (full) also available. Aid available to part-time students. Financial aid applicants required to submit FAFSA.
Faculty research: Tumor and transplantation immunology, molecular mechanisms of DNA repair, virus-host interactions.
Dr. Raj Tiwari, Director, 914-594-4192.

■ NEW YORK UNIVERSITY

School of Medicine and Graduate School of Arts and Science, Sackler Institute of Graduate Biomedical Sciences, Graduate Programs in Molecular Oncology and Immunology, New York, NY 10012-1019

AWARDS Immunology (PhD); molecular oncology (PhD).

Degree requirements: For doctorate, one foreign language, dissertation, qualifying exam required.
Entrance requirements: For doctorate, GRE General Test, GRE Subject Test, TOEFL. *Application deadline:* For fall admission, 2/1. Applications are processed on a rolling basis. *Application fee:* $60. Electronic applications accepted.
Expenses: Tuition: Full-time $17,880; part-time $745 per credit. Required fees: $1,140; $35 per credit. Tuition and fees vary according to course load and program.
Financial aid: Tuition waivers (full) available. Financial aid application deadline: 1/15.
Application contact: Dr. Robert B. Carroll, Graduate Adviser, 212-263-5347, *Fax:* 212-263-8211, *E-mail:* carror01@mcrcr.med.nyu.edu.

Find an in-depth description at www.petersons.com/graduate.

■ NORTH CAROLINA STATE UNIVERSITY

College of Veterinary Medicine and Graduate School, Graduate Programs in Comparative Biomedical Sciences, Graduate Program in Immunology, Raleigh, NC 27695

AWARDS MS, PhD.

Students: 17 full-time (13 women), 4 part-time (all women); includes 3 minority (2 Asian Americans or Pacific Islanders, 1 Hispanic American), 3 international. Average age 35. *31 applicants, 16% accepted.* In 1999, 1 master's, 2 doctorates awarded.
Entrance requirements: For master's and doctorate, GRE General Test. *Application fee:* $45.
Expenses: Tuition, state resident: full-time $1,578. Tuition, nonresident: full-time $10,744. Required fees: $892. Full-time tuition and fees vary according to program.
Financial aid: Fellowships, research assistantships, teaching assistantships available.
Faculty research: Molecular and cellular pathogenesis of AIDS, mucosal immunology, immunopathogenesis of aquatic pathogens, cytokine regulation in intestinal inflammation, cytokine regulation of the immune response to parasitic infections.

Dr. Wayne Tompkins, Director of Graduate Programs, 919-513-6262, *Fax:* 919-513-6452, *E-mail:* wayne_tompkins@ncsu.edu.

■ NORTHWESTERN UNIVERSITY

The Graduate School, Division of Interdepartmental Programs and Medical School, Integrated Graduate Programs in the Life Sciences, Chicago, IL 60611

AWARDS Cancer biology (PhD); cell biology (PhD); developmental biology (PhD); evolutionary biology (PhD); immunology (PhD); microbial pathogenesis (PhD); molecular biology and genetics (PhD); neurobiology (PhD); pharmacology and toxicology (PhD); structural biology and biochemistry (PhD).

Degree requirements: For doctorate, dissertation, written and oral qualifying exams required, foreign language not required.
Entrance requirements: For doctorate, GRE General Test, TOEFL.
Expenses: Tuition: Full-time $23,301. Full-time tuition and fees vary according to program.

Find an in-depth description at www.petersons.com/graduate.

■ THE OHIO STATE UNIVERSITY

College of Medicine and Public Health and Graduate School, Graduate Programs in the Basic Medical Sciences, Department of Molecular Virology, Immunology and Medical Genetics, Columbus, OH 43210

AWARDS MS, PhD, MD/PhD.

Faculty: 30 full-time (5 women).
Students: 24 full-time (10 women), 10 international. Average age 27. *160 applicants, 8% accepted.* In 1999, 1 master's, 3 doctorates awarded (100% entered university research/teaching). Terminal master's awarded for partial completion of doctoral program.
Degree requirements: For master's, thesis optional, foreign language not required; for doctorate, dissertation required, foreign language not required. *Average time to degree:* Master's–4 years part-time; doctorate–5.6 years full-time.
Entrance requirements: For master's and doctorate, GRE General Test. *Application deadline:* For fall admission, 1/15. Applications are processed on a rolling basis. *Application fee:* $30 ($40 for international students). Electronic applications accepted.
Expenses: Tuition, state resident: full-time $5,400. Tuition, nonresident: full-time $14,535. Part-time tuition and fees vary according to course load.
Financial aid: In 1999–00, 1 fellowship with full tuition reimbursement (averaging $19,260 per year), 8 research assistantships

with full tuition reimbursements (averaging $15,600 per year), 7 teaching assistantships with full tuition reimbursements (averaging $15,600 per year) were awarded; grants also available. Financial aid application deadline: 1/15.
Faculty research: Virology, bacteriology, molecular biology. *Total annual research expenditures:* $912,708.
Dr. Caroline C. Whitacre, Chairperson, 614-292-5889, *Fax:* 614-292-9805, *E-mail:* whitacre.3@osu.edu.
Application contact: Dr. Deborah S. Parris, Chairman, Graduate Studies Committee, 614-292-0735, *Fax:* 614-292-9805, *E-mail:* mvimg@osu.edu.

Find an in-depth description at www.petersons.com/graduate.

■ OREGON HEALTH SCIENCES UNIVERSITY

School of Medicine, Graduate Programs in Medicine, Department of Molecular Microbiology and Immunology, Portland, OR 97201-3098

AWARDS PhD.

Degree requirements: For doctorate, dissertation required, foreign language not required.
Entrance requirements: For doctorate, GRE General Test.
Expenses: Tuition, state resident: full-time $3,132; part-time $174 per credit hour. Tuition, nonresident: full-time $5,256; part-time $292 per credit hour. Required fees: $8.5 per credit hour. $146 per term. Part-time tuition and fees vary according to course load.
Faculty research: Molecular biology of bacterial and viral pathogens, cellular and humoral immunology, molecular biology of microbes.

Find an in-depth description at www.petersons.com/graduate.

■ THE PENNSYLVANIA STATE UNIVERSITY MILTON S. HERSHEY MEDICAL CENTER

Graduate School, Department of Microbiology and Immunology, Hershey, PA 17033-2360

AWARDS Genetics (PhD); immunology (MS, PhD); microbiology (MS); microbiology/virology (PhD); molecular biology (PhD).

Students: 19 full-time (6 women). Average age 27. In 1999, 2 master's, 3 doctorates awarded.
Degree requirements: For doctorate, dissertation required.
Entrance requirements: For doctorate, GRE General Test, TOEFL, minimum GPA of 3.0. *Application deadline:* For fall admission, 7/26. *Application fee:* $50.

Expenses: Tuition, state resident: full-time $6,886; part-time $291 per credit. Tuition, nonresident: full-time $14,118; part-time $588 per credit. Required fees: $43 per semester. Part-time tuition and fees vary according to course load.
Financial aid: Unspecified assistantships available.
Dr. Richard J. Courtney, Chair, 717-531-7659.
Application contact: Dr. Brian Wigdahl, Chairman, Graduate Program Committee, 717-531-6682, *Fax:* 717-531-6522, *E-mail:* eneidigh@cor-mail.biochem.hmc.psu.edu.

■ PURDUE UNIVERSITY

School of Veterinary Medicine and Graduate School, Graduate Programs in Veterinary Medicine, Department of Veterinary Pathobiology, West Lafayette, IN 47907

AWARDS Bacteriology (MS, PhD); epidemiology (MS, PhD); immunology (MS, PhD); infectious diseases (MS, PhD); microbiology (MS, PhD); parasitology (MS, PhD); pathology (MS, PhD); toxicology (MS, PhD); virology (MS, PhD).

Faculty: 26 full-time (5 women).
Students: 42 full-time (24 women), 2 part-time (1 woman); includes 5 minority (1 African American, 3 Asian Americans or Pacific Islanders, 1 Hispanic American), 18 international. Average age 32. In 1999, 2 master's, 6 doctorates awarded (100% found work related to degree). Terminal master's awarded for partial completion of doctoral program.
Degree requirements: For master's, thesis required (for some programs), foreign language not required; for doctorate, dissertation required, foreign language not required.
Entrance requirements: For master's and doctorate, GRE General Test, TOEFL. *Application fee:* $30.
Expenses: Tuition, state resident: full-time $3,732. Tuition, nonresident: full-time $8,732.
Financial aid: Fellowships, research assistantships, teaching assistantships available. Financial aid application deadline: 3/1; financial aid applicants required to submit FAFSA.
Dr. H. L. Thacker, Head, 765-494-7543.

■ RUSH UNIVERSITY

Graduate College, Division of Immunology, Chicago, IL 60612-3832

AWARDS Immunology (PhD); microbiology (PhD); virology (PhD).

Degree requirements: For doctorate, dissertation, comprehensive preliminary exam required, foreign language not required.

Rush University (continued)

Entrance requirements: For doctorate, GRE General Test, TOEFL, interview, minimum GPA of 3.0.
Expenses: Tuition: Full-time $13,020; part-time $390 per credit. Tuition and fees vary according to program.
Faculty research: Immune interactions of cells and membranes.
Find an in-depth description at www.petersons.com/graduate.

■ RUTGERS, THE STATE UNIVERSITY OF NEW JERSEY, NEW BRUNSWICK

Graduate School, Program in Cell and Developmental Biology, New Brunswick, NJ 08901-1281

AWARDS Cell biology (MS, PhD); developmental biology (MS, PhD); immunology (MS). Part-time programs available.

Faculty: 84 full-time (15 women).
Students: 30 full-time (16 women), 42 part-time (22 women); includes 24 minority (1 African American, 15 Asian Americans or Pacific Islanders, 8 Hispanic Americans), 17 international. Average age 24. *109 applicants, 34% accepted.* In 1999, 8 master's, 8 doctorates awarded. Terminal master's awarded for partial completion of doctoral program.
Degree requirements: For master's, thesis or alternative required; for doctorate, dissertation, written qualifying exam required.
Entrance requirements: For master's, GRE General Test, TOEFL; for doctorate, GRE General Test, GRE Subject Test, TOEFL, minimum GPA of 3.0. *Application deadline:* For fall admission, 2/15. Applications are processed on a rolling basis. *Application fee:* $50.
Expenses: Tuition, state resident: full-time $6,776; part-time $279 per credit. Tuition, nonresident: full-time $9,936; part-time $412 per credit. Required fees: $20 per credit. $89 per semester. Tuition and fees vary according to course load, campus/location and program.
Financial aid: In 1999–00, 6 fellowships with full tuition reimbursements (averaging $15,000 per year), 17 research assistantships with full tuition reimbursements (averaging $15,000 per year), 10 teaching assistantships with full tuition reimbursements (averaging $14,000 per year) were awarded. Financial aid application deadline: 2/15; financial aid applicants required to submit FAFSA.
Faculty research: Genetics, cell matrix interactions, signal transduction, gene expression, cytoskeletal proteins, neurobiology.

Dr. Alice Y. Liu, Director, 732-445-2730, *Fax:* 732-445-6370, *E-mail:* liu@biology.rutgers.edu.
Application contact: Carolyn Ambrose, Administrative Assistant, 732-445-3430, *Fax:* 732-445-6370, *E-mail:* ambrose@biology.rutgers.edu.

■ RUTGERS, THE STATE UNIVERSITY OF NEW JERSEY, NEW BRUNSWICK

Graduate School, Program in Microbiology and Molecular Genetics, New Brunswick, NJ 08901-1281

AWARDS Applied microbiology (MS, PhD); clinical microbiology (MS, PhD); computational molecular biology (PhD); immunology (MS, PhD); microbial biochemistry (MS, PhD); molecular genetics (MS, PhD); virology (MS, PhD). Part-time programs available.

Faculty: 122 full-time (25 women), 4 part-time/adjunct (1 woman).
Students: 65 full-time (31 women), 51 part-time (24 women); includes 24 minority (14 Asian Americans or Pacific Islanders, 10 Hispanic Americans), 29 international. Average age 27. *200 applicants, 26% accepted.* In 1999, 15 master's, 10 doctorates awarded. Terminal master's awarded for partial completion of doctoral program.
Degree requirements: For master's, thesis or alternative required, foreign language not required; for doctorate, dissertation, written qualifying exam required, foreign language not required.
Entrance requirements: For master's, GRE General Test, GRE Subject Test; for doctorate, GRE General Test, GRE Subject Test, TOEFL, minimum GPA of 3.0. *Application deadline:* For fall admission, 2/15 (priority date). Applications are processed on a rolling basis. *Application fee:* $50.
Expenses: Tuition, state resident: full-time $6,776; part-time $279 per credit. Tuition, nonresident: full-time $9,936; part-time $412 per credit. Required fees: $20 per credit. $89 per semester. Tuition and fees vary according to course load, campus/location and program.
Financial aid: In 1999–00, 70 students received aid, including 32 fellowships with full tuition reimbursements available, 26 research assistantships with full tuition reimbursements available, 12 teaching assistantships with full tuition reimbursements available; institutionally sponsored loans also available. Financial aid application deadline: 2/15; financial aid applicants required to submit FAFSA.
Faculty research: Biochemistry of microbes, medical microbiology.

Dr. Howard Passmore, Director, 732-445-2812, *E-mail:* passmore@biology.rutgers.edu.
Application contact: Betty Green, Administrative Assistant, 732-445-5086, *Fax:* 732-445-5735, *E-mail:* green@mbcl.rutgers.edu.

■ SAINT LOUIS UNIVERSITY

School of Medicine and Graduate School, Graduate Programs in Biomedical Sciences, Department of Molecular Microbiology and Immunology, St. Louis, MO 63103-2097

AWARDS PhD.

Faculty: 16 full-time (4 women), 1 part-time/adjunct (0 women).
Students: Average age 30. In 1999, 1 degree awarded.
Degree requirements: For doctorate, dissertation, qualifying exams required.
Entrance requirements: For doctorate, GRE General Test, GRE Subject Test. *Application deadline:* For fall admission, 5/1. Applications are processed on a rolling basis. *Application fee:* $40.
Expenses: Tuition: Part-time $507 per credit hour. Required fees: $38 per term.
Financial aid: In 1999–00, 7 students received aid, including 4 research assistantships; fellowships, teaching assistantships. Aid available to part-time students. Financial aid application deadline: 8/1; financial aid applicants required to submit FAFSA.
Faculty research: Helper and suppressor T cells, cellular immunology, lymphocyte cloning, T and B cell collaboration, genetic control of immune response.
Dr. William Wold, Chairman, 314-577-8435, *Fax:* 314-773-3403.
Application contact: Dr. H. Peter Zassenhaus, Director of Graduate Studies, 314-577-8432, *Fax:* 314-773-3403, *E-mail:* zassenp@slu.edu.

Find an in-depth description at www.petersons.com/graduate.

■ STANFORD UNIVERSITY

School of Medicine, Graduate Programs in Medicine, Department of Microbiology and Immunology, Stanford, CA 94305-9991

AWARDS PhD.

Faculty: 11 full-time (3 women).
Students: 20 full-time (9 women), 9 part-time (4 women); includes 7 minority (1 African American, 5 Asian Americans or Pacific Islanders, 1 Hispanic American), 3 international. Average age 27. *77 applicants, 17% accepted.* In 1999, 1 doctorate awarded.

Degree requirements: For doctorate, dissertation required, foreign language not required.

Entrance requirements: For doctorate, GRE General Test, GRE Subject Test (biology or chemistry), TOEFL. *Application deadline:* For fall admission, 12/15. *Application fee:* $65 ($80 for international students). Electronic applications accepted.

Expenses: Tuition: Full-time $23,058. Required fees: $152. Part-time tuition and fees vary according to course load.

Financial aid: Research assistantships available. Financial aid application deadline: 12/15.

Faculty research: Molecular pathogenesis of bacteria viruses and parasites, immune system function, autoimmunity, molecular biology.

Edward Mocarski, Chair, 650-723-6435, *Fax:* 650-725-6757, *E-mail:* mocarski@stanford.edu.

Application contact: Student Services Coordinator, 650-725-8541.

Find an in-depth description at www.petersons.com/graduate.

■ STANFORD UNIVERSITY

School of Medicine, Graduate Programs in Medicine, Program in Immunology, Stanford, CA 94305-9991

AWARDS PhD.

Students: 22 full-time (10 women), 17 part-time (7 women); includes 12 minority (3 African Americans, 6 Asian Americans or Pacific Islanders, 3 Hispanic Americans), 4 international. Average age 28. *68 applicants, 18% accepted.* In 1999, 14 doctorates awarded.

Degree requirements: For doctorate, dissertation required.

Entrance requirements: For doctorate, GRE General Test, GRE Subject Test, TOEFL. *Application deadline:* For fall admission, 12/15. *Application fee:* $65 ($80 for international students). Electronic applications accepted.

Expenses: Tuition: Full-time $23,058. Required fees: $152. Part-time tuition and fees vary according to course load.

Financial aid: Application deadline: 12/15.

Application contact: Graduate Admissions Coordinator, 650-723-6120, *Fax:* 650-725-7855.

Find an in-depth description at www.petersons.com/graduate.

■ STATE UNIVERSITY OF NEW YORK AT ALBANY

School of Public Health, Department of Biomedical Sciences, Program in Immunobiology and Immunochemistry, Albany, NY 12222-0001

AWARDS MS, PhD.

Degree requirements: For master's and doctorate, thesis/dissertation required.

Entrance requirements: For master's and doctorate, GRE General Test, GRE Subject Test. *Application deadline:* For fall admission, 1/15 (priority date); for spring admission, 11/1 (priority date). *Application fee:* $50.

Expenses: Tuition, state resident: full-time $5,100; part-time $214 per credit. Tuition, nonresident: full-time $8,416; part-time $352 per credit. Required fees: $31 per credit.

Financial aid: Application deadline: 2/1. Dr. Harry Taber, Chair, Department of Biomedical Sciences, 518-474-2662.

■ STATE UNIVERSITY OF NEW YORK AT BUFFALO

Graduate School, Graduate Programs in Biomedical Sciences at Roswell Park Cancer Institute, Department of Immunology at Roswell Park Cancer Institute, Buffalo, NY 14263

AWARDS PhD.

Faculty: 15 full-time (4 women).

Students: 12 full-time (6 women), 3 part-time (1 woman), 9 international. Average age 27. *32 applicants, 38% accepted.* In 2000, 1 doctorate awarded.

Degree requirements: For doctorate, dissertation required.

Entrance requirements: For doctorate, GRE General Test, GRE Subject Test, TOEFL, TSE, TWE. *Application deadline:* For fall admission, 2/1 (priority date). Applications are processed on a rolling basis. *Application fee:* $35. Electronic applications accepted.

Expenses: Tuition, state resident: full-time $5,100; part-time $213 per credit hour. Tuition, nonresident: full-time $8,416; part-time $351 per credit hour. Required fees: $935; $75 per semester. Tuition and fees vary according to course load and program.

Financial aid: In 2000–01, 10 students received aid, including 4 fellowships with full tuition reimbursements available (averaging $15,000 per year), 6 research assistantships with full tuition reimbursements available (averaging $15,000 per year); Federal Work-Study also available. Financial aid application deadline: 6/1; financial aid applicants required to submit FAFSA.

Faculty research: Immunochemistry, immunobiology, molecular immunology, hybridoma studies, recombinant DNA studies, vaccine research, diagnostics. *Total annual research expenditures:* $2 million. Dr. Soldano Ferrone, Chair, 716-845-8534.

Application contact: Dr. Elizabeth Repasky, Graduate Program Director, 716-845-8534.

Find an in-depth description at www.petersons.com/graduate.

■ STATE UNIVERSITY OF NEW YORK AT STONY BROOK

Graduate School, College of Arts and Sciences, Molecular and Cellular Biology Program, Specialization in Immunology and Pathology, Stony Brook, NY 11794

AWARDS PhD.

Faculty: 23 full-time (6 women).

Students: 13 full-time (10 women), 6 part-time (1 woman); includes 5 minority (3 Asian Americans or Pacific Islanders, 2 Hispanic Americans), 3 international. Average age 27.

Degree requirements: For doctorate, one foreign language (computer language can substitute), dissertation, exam, teaching experience required.

Entrance requirements: For doctorate, GRE General Test, GRE Subject Test, TOEFL. *Application deadline:* For fall admission, 1/15. *Application fee:* $50.

Expenses: Tuition, state resident: full-time $5,100; part-time $213 per credit hour. Tuition, nonresident: full-time $8,416; part-time $351 per credit hour. Required fees: $492. Tuition and fees vary according to program.

Financial aid: Fellowships, research assistantships, teaching assistantships, career-related internships or fieldwork available. Financial aid application deadline: 3/15.

Faculty research: Environmental pathology of respiratory tract, immunology, bone disease, platelet physiology. *Total annual research expenditures:* $2.2 million. Dr. Frederick Miller, Chairman, Department of Pathology, 631-444-3000, *Fax:* 631-444-3424.

Application contact: Dr. Nancy Reich, Director, 631-444-7503, *Fax:* 631-444-3424, *E-mail:* nreich@path.som.sunysb.edu.

Find an in-depth description at www.petersons.com/graduate.

■ STATE UNIVERSITY OF NEW YORK HEALTH SCIENCE CENTER AT BROOKLYN

School of Graduate Studies, Department of Microbiology and Immunology, Brooklyn, NY 11203-2098
AWARDS PhD, MD/PhD.

Degree requirements: For doctorate, one foreign language, dissertation, comprehensive qualifying exam required.
Entrance requirements: For doctorate, GRE General Test, GRE Subject Test (strongly recommended).
Expenses: Tuition, state resident: full-time $5,100; part-time $213 per credit. Tuition, nonresident: full-time $8,416; part-time $351 per credit. Required fees: $200. Full-time tuition and fees vary according to program and student level.
Faculty research: Yeast molecular genetics, regulation of gene expression, retroviral-mediated gene transfer, nuclear acid structure: regulation and change within RNA genomes.

■ STATE UNIVERSITY OF NEW YORK UPSTATE MEDICAL UNIVERSITY

College of Graduate Studies, Department of Microbiology and Immunology, Syracuse, NY 13210-2334
AWARDS MS, PhD, MD/PhD.

Faculty: 17.
Students: 16 full-time (7 women); includes 2 minority (1 Asian American or Pacific Islander, 1 Hispanic American), 3 international. *50 applicants, 36% accepted.* In 1999, 3 doctorates awarded (33% found work related to degree, 67% continued full-time study). Terminal master's awarded for partial completion of doctoral program.
Degree requirements: For master's, thesis required, foreign language not required; for doctorate, dissertation, comprehensive exam required, foreign language not required.
Entrance requirements: For master's and doctorate, GRE General Test, GRE Subject Test, TSE, interview. *Application deadline:* For fall admission, 4/1 (priority date). Applications are processed on a rolling basis. *Application fee:* $40.
Expenses: Tuition, state resident: full-time $5,100; part-time $213 per credit. Tuition, nonresident: full-time $8,416; part-time $351 per credit. Required fees: $410; $25 per credit. Part-time tuition and fees vary according to course load and program.
Financial aid: Research assistantships, teaching assistantships available.
Faculty research: Virology, molecular biology, microbial genetics, parasitology.

Dr. Edward Shillitoe, Chairperson, 315-464-5453.
Application contact: Dr. Allen E. Silverstone, Graduate Program Director, 315-464-5453.

■ TEMPLE UNIVERSITY

Health Sciences Center, School of Medicine and Graduate School, Graduate Programs in Medicine, Department of Microbiology and Immunology, Philadelphia, PA 19122-6096
AWARDS MS, PhD, MD/PhD.

Faculty: 19 full-time (4 women), 17 part-time/adjunct (7 women).
Students: 51 full-time. In 1999, 2 master's awarded (100% found work related to degree); 6 doctorates awarded (100% found work related to degree). Terminal master's awarded for partial completion of doctoral program.
Degree requirements: For master's, thesis required; for doctorate, dissertation, research seminars required. *Average time to degree:* Master's–3 years full-time; doctorate–5 years full-time.
Entrance requirements: For master's and doctorate, GRE General Test, TOEFL, minimum GPA of 3.0. *Application deadline:* For fall admission, 7/1; for spring admission, 11/1. Applications are processed on a rolling basis. *Application fee:* $40.
Expenses: Tuition, state resident: full-time $6,030. Tuition, nonresident: full-time $8,298. Required fees: $230. One-time fee: $10 full-time.
Financial aid: Fellowships, research assistantships, career-related internships or fieldwork, Federal Work-Study, grants, institutionally sponsored loans, and tuition waivers (full and partial) available. Aid available to part-time students. Financial aid application deadline: 3/15; financial aid applicants required to submit FAFSA.
Faculty research: Molecular and cellular immunology, molecular and biochemical microbiology, molecular genetics. *Total annual research expenditures:* $3.4 million.
Dr. Chris D. Platsoucas, Chair, 215-707-7929, *Fax:* 215-707-7788, *E-mail:* microimm@vm.temple.edu.

Find an in-depth description at www.petersons.com/graduate.

■ TEXAS A&M UNIVERSITY SYSTEM HEALTH SCIENCE CENTER

College of Medicine, Graduate School of Biomedical Sciences, Department of Medical Microbiology and Immunology, College Station, TX 77840-7896
AWARDS Immunology (PhD); microbiology (PhD); molecular biology (PhD); virology (PhD).

Faculty: 7 full-time (0 women), 1 part-time/adjunct (0 women).
Students: 16 full-time (10 women); includes 2 minority (1 African American, 1 Asian American or Pacific Islander).
Degree requirements: For doctorate, dissertation required, foreign language not required.
Entrance requirements: For doctorate, GRE General Test, minimum GPA of 3.0. *Application deadline:* For fall admission, 2/1 (priority date). *Application fee:* $50 ($75 for international students).
Expenses: Tuition, area resident: Full-time $1,368. Tuition, state resident: part-time $76 per credit. Tuition, nonresident: full-time $5,256; part-time $292 per credit. International tuition: $5,256 full-time. Required fees: $678; $38 per credit. Full-time tuition and fees vary according to course load and student level.
Financial aid: In 1999–00, 16 students received aid, including 1 fellowship (averaging $17,000 per year), 15 research assistantships (averaging $17,000 per year). Financial aid application deadline: 4/1; financial aid applicants required to submit FAFSA.
Faculty research: Molecular pathogenesis, microbial therapeutics. *Total annual research expenditures:* $1.2 million.
Dr. John M. Quarles, Interim Head, 979-845-1313, *Fax:* 979-845-3479, *E-mail:* jmquarles@medicine.tamu.edu.
Application contact: Dr. James Samuel, Graduate Adviser, 979-845-1313, *Fax:* 979-845-3479, *E-mail:* jsamuel@medicine.tamu.edu.

Find an in-depth description at www.petersons.com/graduate.

■ THOMAS JEFFERSON UNIVERSITY

College of Graduate Studies, Program in Immunology, Philadelphia, PA 19107
AWARDS PhD.

Faculty: 19 full-time (3 women).
Students: 16 full-time (6 women); includes 2 minority (1 African American, 1 Asian American or Pacific Islander), 1 international. *40 applicants, 18% accepted.*

Degree requirements: For doctorate, dissertation required, foreign language not required.

Entrance requirements: For doctorate, GRE General Test, TOEFL, minimum GPA of 3.2. *Application deadline:* For fall admission, 3/1 (priority date). Applications are processed on a rolling basis. *Application fee:* $40.

Expenses: Tuition: Full-time $12,670. Tuition and fees vary according to degree level and program.

Financial aid: In 1999–00, 15 fellowships with full tuition reimbursements were awarded; research assistantships, Federal Work-Study, institutionally sponsored loans, traineeships, and training grants also available. Aid available to part-time students. Financial aid application deadline: 5/1; financial aid applicants required to submit FAFSA.

Faculty research: Autoimmunity, cancer immunology, cellular immunology, cytokines, immunochemistry.

Dr. Robert Korngold, Chairman, Graduate Committee, 215-503-4552, *Fax:* 215-923-4153, *E-mail:* r_korngold@lac.jci.tju.edu.

Application contact: Jessie F. Pervall, Director of Admissions, 215-503-4400, *Fax:* 215-503-3433, *E-mail:* cgs-info@mail.tju.edu.

Find an in-depth description at www.petersons.com/graduate.

■ **TUFTS UNIVERSITY**

Sackler School of Graduate Biomedical Sciences, Program in Immunology, Medford, MA 02155

AWARDS PhD.

Faculty: 24 full-time (5 women).

Students: 33 full-time (17 women); includes 5 minority (all Asian Americans or Pacific Islanders), 7 international. Average age 26. *116 applicants, 4% accepted.* In 1999, 4 degrees awarded (100% entered university research/teaching).

Degree requirements: For doctorate, dissertation required, foreign language not required. *Average time to degree:* Doctorate–6 years full-time.

Entrance requirements: For doctorate, GRE General Test, TOEFL. *Application deadline:* For fall admission, 1/15. Applications are processed on a rolling basis. *Application fee:* $45.

Expenses: Tuition: Full-time $19,325.

Financial aid: In 1999–00, 33 students received aid, including 6 fellowships, 27 research assistantships with full tuition reimbursements available (averaging $18,805 per year); scholarships and tuition waivers (full) also available. Financial aid application deadline: 1/15.

Faculty research: Genetic analysis of lymphocyte function, ontogeny and activation, transformation of hematopoietic cells, autoimmunity, the immune response to infection.

Dr. Henry H. Wortis, Director, 617-636-6836, *Fax:* 617-636-2990, *E-mail:* hwortis_sup@opal.tufts.edu.

Application contact: Secretary, 617-636-6767, *Fax:* 617-636-0375, *E-mail:* sackler-school@tufts.edu.

■ **TULANE UNIVERSITY**

School of Medicine and Graduate School, Graduate Programs in Medicine, Department of Microbiology and Immunology, New Orleans, LA 70118-5669

AWARDS MS, PhD, MD/PhD. MS and PhD offered through the Graduate School.

Students: 41 full-time (18 women); includes 6 minority (3 African Americans, 3 Asian Americans or Pacific Islanders), 9 international. *87 applicants, 9% accepted.* In 1999, 2 master's awarded.

Degree requirements: For master's, thesis required; for doctorate, 2 foreign languages (computer language can substitute for one), dissertation required.

Entrance requirements: For master's, GRE General Test, TOEFL, or TSE, minimum B average in undergraduate course work; for doctorate, GRE General Test, GRE Subject Test, TOEFL, or TSE. *Application deadline:* For fall admission, 2/1. *Application fee:* $45.

Expenses: Tuition: Full-time $23,030.

Financial aid: Fellowships, research assistantships, career-related internships or fieldwork, Federal Work-Study, and institutionally sponsored loans available. Financial aid application deadline: 2/1.

Dr. John Clements, Chairman, 504-588-5159.

Find an in-depth description at www.petersons.com/graduate.

■ **UNIFORMED SERVICES UNIVERSITY OF THE HEALTH SCIENCES**

School of Medicine, Division of Basic Medical Sciences, Department of Microbiology and Immunology, Bethesda, MD 20814-4799

AWARDS PhD.

Faculty: 13 full-time (7 women), 8 part-time/adjunct (4 women).

Students: 9 full-time (7 women); includes 2 minority (1 African American, 1 Hispanic American). Average age 28. *20 applicants, 5% accepted.*

Degree requirements: For doctorate, one foreign language (computer language can substitute), dissertation, qualifying exam required.

Entrance requirements: For doctorate, GRE General Test, GRE Subject Test, minimum GPA of 3.0, U.S. citizenship. *Application deadline:* For fall admission, 1/15 (priority date). Applications are processed on a rolling basis. *Application fee:* $0.

Financial aid: In 1999–00, 6 fellowships with full tuition reimbursements (averaging $15,000 per year) were awarded; tuition waivers (full) also available.

Faculty research: Infections diseases, virology, bacteriology, microbial pathogenesis.

Dr. Alison O'Brien, Chair, 301-295-3400.

Application contact: Janet M. Anastasi, Graduate Program Coordinator, 301-295-9474, *Fax:* 301-295-6772, *E-mail:* janastasi@usuhs.mil.

Find an in-depth description at www.petersons.com/graduate.

■ **UNIFORMED SERVICES UNIVERSITY OF THE HEALTH SCIENCES**

School of Medicine, Division of Basic Medical Sciences, Program in Emerging Infectious Diseases, Bethesda, MD 20814-4799

AWARDS PhD.

Degree requirements: For doctorate, dissertation, qualifying exam required.

Entrance requirements: For doctorate, GRE General Test, U.S. citizenship. *Application deadline:* For fall admission, 1/15 (priority date). Applications are processed on a rolling basis. *Application fee:* $0.

Dr. Eleanor Metcalf, Director, 301-295-3413, *E-mail:* emetcalf@usuhs.mil.

Application contact: Janet M. Anastasi, Graduate Program Coordinator, 301-295-9474, *Fax:* 301-295-6772, *E-mail:* janastasi@usuhs.mil.

Find an in-depth description at www.petersons.com/graduate.

■ **THE UNIVERSITY OF ARIZONA**

College of Medicine, Graduate Programs in Medicine, Department of Microbiology and Immunology, Tucson, AZ 85721

AWARDS MS, PhD.

Degree requirements: For master's and doctorate, thesis/dissertation required, foreign language not required.

Entrance requirements: For master's and doctorate, GRE General Test, minimum GPA of 3.0.

The University of Arizona (continued)
Expenses: Tuition, nonresident: full-time $4,814; part-time $274 per unit. Required fees: $1,094; $115 per unit.
Faculty research: Environmental and pathogenic microbiology, molecular biology.
Find an in-depth description at www.petersons.com/graduate.

■ UNIVERSITY OF ARKANSAS FOR MEDICAL SCIENCES

College of Medicine and Graduate School, Graduate Programs in Medicine, Department of Microbiology and Immunology, Little Rock, AR 72205-7199

AWARDS MS, PhD, MD/PhD.

Faculty: 15 full-time (2 women).
Students: 17 full-time (10 women), 1 part-time; includes 1 minority (African American), 5 international. In 1999, 2 master's, 2 doctorates awarded.
Degree requirements: For master's and doctorate, thesis/dissertation required, foreign language not required.
Entrance requirements: For master's and doctorate, GRE General Test. *Application fee:* $0.
Expenses: Tuition: Full-time $8,928.
Financial aid: Research assistantships available. Aid available to part-time students.
Dr. Roger G. Rank, Chairman, 501-686-5144.
Application contact: Dr. Marie Chow, Co-Coordinator of Graduate Studies, 501-686-5144.
Find an in-depth description at www.petersons.com/graduate.

■ UNIVERSITY OF CALIFORNIA, BERKELEY

Graduate Division, School of Public Health, Group in Infectious Diseases and Immunity, Berkeley, CA 94720-1500

AWARDS PhD.

Entrance requirements: For doctorate, GRE General Test, minimum GPA of 3.0.
Expenses: Tuition, nonresident: full-time $9,804. Required fees: $4,268. Tuition and fees vary according to program.

■ UNIVERSITY OF CALIFORNIA, BERKELEY

Graduate Division, School of Public Health, Program in Infectious Diseases, Berkeley, CA 94720-1500

AWARDS MPH, MS, PhD.

Entrance requirements: For master's and doctorate, GRE General Test, minimum GPA of 3.0.
Expenses: Tuition, nonresident: full-time $9,804. Required fees: $4,268. Tuition and fees vary according to program.

■ UNIVERSITY OF CALIFORNIA, DAVIS

Graduate Studies, Programs in the Biological Sciences, Program in Immunology, Davis, CA 95616

AWARDS MS, PhD.

Faculty: 37.
Students: 16 full-time (3 women); includes 4 minority (2 Asian Americans or Pacific Islanders, 1 Hispanic American, 1 Native American), 3 international. Average age 27. *34 applicants, 24% accepted.* In 1999, 6 master's, 2 doctorates awarded.
Degree requirements: For master's, thesis optional; for doctorate, dissertation required.
Entrance requirements: For master's and doctorate, GRE General Test. *Application deadline:* For fall admission, 1/15. *Application fee:* $40. Electronic applications accepted.
Expenses: Tuition, nonresident: full-time $9,804. Tuition and fees vary according to program and student level.
Financial aid: In 1999–00, 16 students received aid, including 5 fellowships with full and partial tuition reimbursements available, 8 research assistantships with full and partial tuition reimbursements available, 2 teaching assistantships with partial tuition reimbursements available; Federal Work-Study, institutionally sponsored loans, scholarships, and tuition waivers (full and partial) also available. Financial aid application deadline: 1/15; financial aid applicants required to submit FAFSA.
Faculty research: Immune regulation in autoimmunity, immunopathology, immunotoxicology, tumor immunology, avian immunology.
Eric Gershwin, Graduate Adviser, 530-752-2884, *E-mail:* megershwin@ucdavis.edu.
Application contact: Marlene Belz, Administrative Assistant, 530-754-7684, *Fax:* 530-752-8966, *E-mail:* mhbelz@ucdavis.edu.

■ UNIVERSITY OF CALIFORNIA, LOS ANGELES

School of Medicine and Graduate Division, Graduate Programs in Medicine, Department of Microbiology and Immunology, Los Angeles, CA 90095

AWARDS MS, PhD.

Students: 47 full-time (25 women); includes 19 minority (2 African Americans, 15 Asian Americans or Pacific Islanders, 2 Hispanic Americans), 3 international. *1 applicant, 0% accepted.*
Degree requirements: For doctorate, dissertation, oral and written qualifying exams required, foreign language not required.
Entrance requirements: For doctorate, GRE General Test, GRE Subject Test, TOEFL. *Application fee:* $40.
Expenses: Tuition, nonresident: full-time $9,804. Required fees: $4,405.
Financial aid: In 1999–00, 44 students received aid, including 44 fellowships, 9 teaching assistantships; research assistantships, Federal Work-Study, institutionally sponsored loans, and tuition waivers (full and partial) also available. Financial aid application deadline: 3/1.
Dr. Sherie Morrison, Chair, 310-206-5148.
Application contact: UCLA Access Coordinator, 800-284-8252, *Fax:* 310-206-3280, *E-mail:* uclaaccess@ibes.medsch.ucla.edu.
Find an in-depth description at www.petersons.com/graduate.

■ UNIVERSITY OF CALIFORNIA, SAN DIEGO

Graduate Studies and Research, Department of Biology, Program in Immunology, Virology, and Cancer Biology, La Jolla, CA 92093

AWARDS PhD.

Degree requirements: For doctorate, dissertation required.
Entrance requirements: For doctorate, GRE General Test, pre-application beginning in September. *Application deadline:* For fall admission, 1/8. *Application fee:* $40.
Expenses: Tuition, nonresident: full-time $14,691. Required fees: $4,697. Full-time tuition and fees vary according to program.
Application contact: Biology Graduate Admissions Committee, 858-534-3835.

■ UNIVERSITY OF CALIFORNIA, SAN FRANCISCO

Graduate Division, Department of Microbiology and Immunology, San Francisco, CA 94143

AWARDS PhD.

Faculty: 13 full-time (3 women).
Students: 15 full-time (7 women). *66 applicants, 15% accepted.* In 1999, 4 doctorates awarded.
Degree requirements: For doctorate, dissertation required, foreign language not required.

Entrance requirements: For doctorate, GRE General Test. *Application deadline:* For fall admission, 1/15. *Application fee:* $40.
Financial aid: Application deadline: 1/10. Liz Blackburn, Chair, 415-476-1212.
Application contact: Grace Stauffer, Program Assistant, 415-476-8204.

■ UNIVERSITY OF CHICAGO

Division of the Biological Sciences, Biomedical Sciences: Cancer, Immunology, Nutrition, Pathology, and Virology, Committee on Immunology, Chicago, IL 60637-1513

AWARDS PhD.

Faculty: 41 full-time (10 women).
Students: 25 full-time (9 women); includes 6 minority (5 Asian Americans or Pacific Islanders, 1 Hispanic American). Average age 23. *88 applicants, 7% accepted.* In 1999, 5 degrees awarded (80% entered university research/teaching, 20% continued full-time study).
Degree requirements: For doctorate, dissertation required, foreign language not required. *Average time to degree:* Doctorate–5.5 years full-time.
Entrance requirements: For doctorate, GRE General Test, TOEFL. *Application deadline:* For fall admission, 1/5 (priority date). *Application fee:* $55.
Expenses: Tuition: Full-time $24,804; part-time $3,422 per course. Required fees: $390. Tuition and fees vary according to program.
Financial aid: In 1999–00, 25 students received aid, including 9 fellowships with full tuition reimbursements available (averaging $16,500 per year), 16 research assistantships with full tuition reimbursements available (averaging $16,500 per year); institutionally sponsored loans and traineeships also available. Financial aid application deadline: 6/1.
Faculty research: Molecular immunology, transplantation, autoimmunology, neuroimmunology, tumor immunology, regulation of B-Cell development.
Dr. Jim Miller, Chairman, 773-702-0981, *Fax:* 773-702-4634, *E-mail:* jbluest@ flowcity.bsd.uchicago.edu.
Application contact: Rebecca Levine, Administrative Assistant, Student Services, 773-834-3899, *Fax:* 773-702-4634, *E-mail:* rlevine@huggins.bsd.uchicago.edu.

■ UNIVERSITY OF COLORADO HEALTH SCIENCES CENTER

Graduate School, Programs in Biological and Medical Sciences, Department of Immunology, Denver, CO 80262

AWARDS PhD.

Degree requirements: For doctorate, dissertation required, foreign language not required.
Entrance requirements: For doctorate, GRE.
Expenses: Tuition, state resident: full-time $1,512; part-time $56 per hour. Tuition, nonresident: full-time $7,209; part-time $267 per hour. Full-time tuition and fees vary according to course load and program.
Find an in-depth description at www.petersons.com/graduate.

■ UNIVERSITY OF CONNECTICUT HEALTH CENTER

Graduate School, Programs in Biomedical Sciences, Program in Immunology, Farmington, CT 06030

AWARDS PhD, DMD/PhD, MD/PhD.

Faculty: 21.
Students: 18 full-time (9 women), 12 international. In 1999, 3 degrees awarded.
Degree requirements: For doctorate, one foreign language (computer language can substitute), dissertation required.
Entrance requirements: For doctorate, GRE General Test, GRE Subject Test (recommended), TOEFL. *Application deadline:* For fall admission, 2/1 (priority date); for spring admission, 10/1. Applications are processed on a rolling basis. *Application fee:* $40 ($45 for international students).
Expenses: Tuition, state resident: full-time $5,272; part-time $293 per credit. Tuition, nonresident: full-time $13,696; part-time $761 per credit. Required fees: $320; $198 per semester. One-time fee: $50 full-time. Full-time tuition and fees vary according to course load, program and reciprocity agreements.
Financial aid: In 1999–00, research assistantships (averaging $17,000 per year); fellowships also available.
Dr. Leo Lefrancois, Co-Director, 860-679-3242, *E-mail:* llefranc@neuton.uchc.edu.
Application contact: Marizta Barta, Information Contact, 860-679-4306, *Fax:* 860-679-1282, *E-mail:* barta@ adp.uchc.edu.
Find an in-depth description at www.petersons.com/graduate.

■ UNIVERSITY OF FLORIDA

College of Medicine and Graduate School, Interdisciplinary Program in Biomedical Sciences, Concentration in Immunology and Microbiology, Gainesville, FL 32611

AWARDS PhD.

Degree requirements: For doctorate, dissertation required, foreign language not required.

Entrance requirements: For doctorate, GRE General Test, TOEFL, minimum GPA of 3.0. Electronic applications accepted.
Expenses: Tuition, state resident: part-time $144 per credit hour. Tuition, nonresident: part-time $505 per credit hour. Tuition and fees vary according to course level, course load and program.
Find an in-depth description at www.petersons.com/graduate.

■ UNIVERSITY OF FLORIDA

College of Medicine and Graduate School, Interdisciplinary Program in Biomedical Sciences, Department of Pathology, Gainesville, FL 32611

AWARDS Clinical chemistry (MS); immunology and molecular pathology (PhD). Terminal master's awarded for partial completion of doctoral program.

Degree requirements: For master's and doctorate, thesis/dissertation required, foreign language not required.
Entrance requirements: For master's and doctorate, GRE General Test, TOEFL, minimum GPA of 3.0. Electronic applications accepted.
Expenses: Tuition, state resident: part-time $144 per credit hour. Tuition, nonresident: part-time $505 per credit hour. Tuition and fees vary according to course level, course load and program.
Faculty research: Molecular immunology, autoimmunity and transplantation, tumor biology, oncogenic viruses, human immunodeficiency viruses.

■ UNIVERSITY OF ILLINOIS AT CHICAGO

College of Medicine and Graduate College, Graduate Programs in Medicine, Department of Microbiology and Immunology, Chicago, IL 60607-7128

AWARDS PhD, MD/PhD.

Faculty: 16 full-time (3 women).
Students: 28 full-time (20 women), 1 (woman) part-time; includes 3 minority (1 African American, 2 Asian Americans or Pacific Islanders), 18 international. Average age 27. *171 applicants, 10% accepted.* In 1999, 2 doctorates awarded.
Degree requirements: For doctorate, dissertation required, foreign language not required.
Entrance requirements: For doctorate, GRE General Test, TOEFL, minimum GPA of 3.75 on a 5.0 scale. *Application deadline:* For fall admission, 6/1; for spring admission, 11/1. *Application fee:* $40 ($50 for international students).
Expenses: Tuition, state resident: full-time $3,750. Tuition, nonresident: full-time

University of Illinois at Chicago (continued)

$10,588. Tuition and fees vary according to course load.

Financial aid: In 1999–00, 15 students received aid; fellowships, research assistantships, teaching assistantships, Federal Work-Study, grants, and tuition waivers (full) available. Financial aid application deadline: 3/1; financial aid applicants required to submit FAFSA. Dr. Bellur Prabhakar, Head.

Application contact: Mia Johnson, Admissions and Records Officer, 312-996-9477, *Fax:* 312-996-6415, *E-mail:* ui8735@uicvm.uic.edu.

Find an in-depth description at www.petersons.com/graduate.

■ **THE UNIVERSITY OF IOWA**

College of Medicine and Graduate College, Graduate Programs in Medicine, Department of Microbiology, Iowa City, IA 52242-1316

AWARDS General microbiology and microbial physiology (MS, PhD); immunology (MS, PhD); microbial genetics (MS, PhD); pathogenic bacteriology (MS, PhD); virology (MS, PhD).

Faculty: 21 full-time (4 women), 10 part-time/adjunct (1 woman).

Students: 50 full-time (21 women); includes 7 minority (2 African Americans, 4 Asian Americans or Pacific Islanders, 1 Hispanic American), 5 international. *144 applicants, 8% accepted.* In 1999, 7 doctorates awarded (100% entered university research/teaching).

Degree requirements: For master's, thesis required; for doctorate, dissertation, comprehensive exam required. *Average time to degree:* Doctorate–6 years full-time.

Entrance requirements: For master's and doctorate, GRE General Test. *Application deadline:* For fall admission, 2/1. *Application fee:* $30 ($50 for international students). Electronic applications accepted.

Expenses: Tuition, state resident: full-time $3,308. Tuition, nonresident: full-time $10,662. Tuition and fees vary according to course load and program.

Financial aid: In 1999–00, 50 students received aid, including fellowships with full tuition reimbursements available (averaging $16,787 per year), research assistantships with full tuition reimbursements available (averaging $16,787 per year); grants, institutionally sponsored loans, and traineeships also available.

Faculty research: Biocatalysis and blue jeans, gene regulation, processing and transport of HIV, retroviral pathogenesis, biodegradation. *Total annual research expenditures:* $8.2 million.

Dr. Michael A. Apicella, Head, 319-335-7810, *E-mail:* grad-micro-info@uiowa.edu.

Find an in-depth description at www.petersons.com/graduate.

■ **THE UNIVERSITY OF IOWA**

College of Medicine and Graduate College, Graduate Programs in Medicine, Interdisciplinary Program in Immunology, Iowa City, IA 52242-1316

AWARDS PhD, MD/PhD.

Faculty: 31 full-time (3 women).

Students: 20 full-time (7 women); includes 2 minority (1 Asian American or Pacific Islander, 1 Hispanic American), 6 international. Average age 25. *87 applicants, 10% accepted.* In 1999, 3 degrees awarded (50% found work related to degree).

Degree requirements: For doctorate, dissertation, comprehensive exam required, foreign language not required. *Average time to degree:* Doctorate–5 years full-time.

Entrance requirements: For doctorate, GRE General Test, minimum GPA of 3.0. *Application fee:* $30 ($50 for international students).

Expenses: Tuition, state resident: full-time $3,308. Tuition, nonresident: full-time $10,662. Tuition and fees vary according to course load and program.

Financial aid: In 1999–00, fellowships with full tuition reimbursements (averaging $17,000 per year), 18 research assistantships with full tuition reimbursements (averaging $16,770 per year) were awarded; teaching assistantships with full tuition reimbursements, grants also available.

Dr. Gail Bishop, Director, 319-335-7945, *Fax:* 319-335-7656, *E-mail:* gail-bishop@uiowa.edu.

Application contact: Dr. Morris Dailey, Chairman, Graduate Admissions Committee, 800-551-6787, *Fax:* 319-335-7656, *E-mail:* interdis@blue.weeg.uiowa.edu.

Find an in-depth description at www.petersons.com/graduate.

■ **UNIVERSITY OF KANSAS**

Graduate Studies Medical Center, Graduate Programs in Biomedical and Basic Sciences, Department of Microbiology, Molecular Genetics and Immunology, Lawrence, KS 66045

AWARDS PhD, MD/PhD.

Faculty: 10 full-time (1 woman), 1 part-time/adjunct (0 women).

Students: Average age 27. *0 applicants, 0% accepted.* In 1999, 4 doctorates awarded.

Degree requirements: For doctorate, dissertation, comprehensive oral exam required.

Entrance requirements: For doctorate, GRE General Test, TOEFL, TSE.

Application deadline: For fall admission, 1/31 (priority date). Applications are processed on a rolling basis. *Application fee:* $0.

Expenses: Tuition, state resident: full-time $2,482; part-time $103 per credit hour. Tuition, nonresident: full-time $8,104; part-time $338 per credit hour. Required fees: $428; $31 per credit hour. Tuition and fees vary according to program.

Financial aid: Fellowships, research assistantships, teaching assistantships, Federal Work-Study, institutionally sponsored loans, scholarships, and unspecified assistantships available. Aid available to part-time students. Financial aid application deadline: 3/31; financial aid applicants required to submit FAFSA.

Faculty research: Bacterial and viral pathogenesis, molecular virology, microbial molecular biology and genetics, immunochemistry. *Total annual research expenditures:* $4.2 million.

Dr. Opendra Narayan, Chairman, 913-5887010, *Fax:* 913-588-7295, *E-mail:* bnarayan@kume.edu.

Application contact: Dr. Joe Lutkenhaus, Director of Graduate Studies, 913-588-7010, *Fax:* 913-588-7295, *E-mail:* jlutkenh@kumc.edu.

■ **UNIVERSITY OF KENTUCKY**

Graduate School and College of Medicine, Graduate Programs in Medicine, Department of Microbiology and Immunology, Lexington, KY 40506-0032

AWARDS PhD.

Degree requirements: For doctorate, dissertation, comprehensive exam required, foreign language not required.

Entrance requirements: For doctorate, GRE General Test, minimum GPA of 3.0.

Expenses: Tuition, state resident: full-time $3,596; part-time $188 per credit hour. Tuition, nonresident: full-time $10,116; part-time $550 per credit hour.

Faculty research: Immunology transplantation, cellular antibody reactions, immunobiology of tumors.

Find an in-depth description at www.petersons.com/graduate.

■ **UNIVERSITY OF LOUISVILLE**

School of Medicine and Graduate School, Integrated Programs in Biomedical Sciences, Department of Microbiology and Immunology, Louisville, KY 40292-0001

AWARDS MS, PhD.

Degree requirements: For master's and doctorate, thesis/dissertation required.

Entrance requirements: For master's and doctorate, GRE General Test, 1 year of

course work in biology, organic chemistry, physics; 1 semester of course work in calculus and quantitative analysis, biochemistry, or molecular biology. Electronic applications accepted.
Expenses: Tuition, state resident: full-time $3,260; part-time $182 per hour. Tuition, nonresident: full-time $9,780; part-time $544 per hour. Required fees: $143; $28 per hour.
Faculty research: Virology, molecular pathogenesis, microbial genetics, cell biology.

Find an in-depth description at www.petersons.com/graduate.

■ UNIVERSITY OF MARYLAND

Graduate School, Graduate Programs in Medicine, Department of Microbiology and Immunology, Baltimore, MD 21201-1627
AWARDS MS, PhD, MD/PhD. Part-time programs available.

Degree requirements: For master's, thesis required, foreign language not required; for doctorate, dissertation, oral exam required, foreign language not required.
Entrance requirements: For master's and doctorate, GRE General Test, TOEFL, minimum GPA of 3.0.
Expenses: Tuition, state resident: part-time $261 per credit hour. Tuition, nonresident: part-time $468 per credit hour. Tuition and fees vary according to program.
Faculty research: Epidemiology, ecology of infectious microorganisms, electron microscopy, medical microbiology, molecular biology.

Find an in-depth description at www.petersons.com/graduate.

■ UNIVERSITY OF MASSACHUSETTS WORCESTER

Graduate School of Biomedical Sciences, Program in Immunology-Virology, Worcester, MA 01655-0115
AWARDS Medical sciences (PhD).

Faculty: 41 full-time (10 women).
Students: 41 full-time (20 women); includes 4 minority (2 Asian Americans or Pacific Islanders, 1 Hispanic American, 1 Native American), 10 international. In 1999, 5 degrees awarded.
Degree requirements: For doctorate, dissertation required, foreign language not required. *Average time to degree:* Doctorate–5.7 years full-time.
Entrance requirements: For doctorate, GRE General Test, GRE Subject Test, 1 year of calculus, organic chemistry, physics, and biology. *Application deadline:* For fall admission, 1/15 (priority date).

Applications are processed on a rolling basis. *Application fee:* $25 ($50 for international students).
Expenses: Tuition, state resident: full-time $2,640. Tuition, nonresident: full-time $9,756. Required fees: $825. Full-time tuition and fees vary according to program.
Financial aid: In 1999–00, research assistantships with full tuition reimbursements (averaging $17,500 per year); unspecified assistantships also available.
Faculty research: Molecular and immunological studies on viral pathogenesis and oncology, AIDS viruses and other retroviruses, herpes viruses, influenza, Dengue and Newcastle disease viruses.
Dr. Carol Miller-Graziano, Director, 508-856-2387.

Find an in-depth description at www.petersons.com/graduate.

■ UNIVERSITY OF MEDICINE AND DENTISTRY OF NEW JERSEY

Graduate School of Biomedical Sciences, Graduate Programs in Biomedical Sciences, Program in Cell and Developmental Biology, Piscataway, NJ 08854-5635
AWARDS Cell biology (MS, PhD); developmental biology (MS, PhD); immunology (MS).

Degree requirements: For master's, thesis, qualifying exam required; for doctorate, dissertation, qualifying exam required, foreign language not required.
Entrance requirements: For master's, GRE General Test, TOEFL; for doctorate, GRE General Test, GRE Subject Test, TOEFL. *Application deadline:* For fall admission, 2/1; for spring admission, 10/1. *Application fee:* $40.
Expenses: Tuition, state resident: part-time $270 per credit hour. Tuition, nonresident: part-time $407 per credit hour. Part-time tuition and fees vary according to campus/location and program.
Financial aid: Fellowships, research assistantships, teaching assistantships available. Financial aid application deadline: 5/1.
Faculty research: Genetics, cell matrix interactions, signal transduction, gene expression, cytoskeletal proteins.
Dr. Alice Y. Liu, Director, 732-445-3430, *Fax:* 732-445-2730.

■ UNIVERSITY OF MIAMI

School of Medicine and Graduate School, Graduate Programs in Medicine, Department of Microbiology and Immunology, Coral Gables, FL 33124
AWARDS PhD, MD/PhD.

Faculty: 24 full-time (5 women).
Students: 22 full-time (9 women); includes 6 minority (1 African American, 2 Asian Americans or Pacific Islanders, 3 Hispanic Americans), 7 international. Average age 24. *105 applicants, 8% accepted.* In 1999, 4 degrees awarded (50% entered university research/teaching, 50% found other work related to degree).
Degree requirements: For doctorate, dissertation, oral and written qualifying exams required, foreign language not required. *Average time to degree:* Doctorate–6 years full-time.
Entrance requirements: For doctorate, GRE General Test, TOEFL. *Application deadline:* For fall admission, 3/15. Applications are processed on a rolling basis. *Application fee:* $35.
Expenses: Tuition, area resident: Part-time $899 per credit.
Financial aid: In 1999–00, 22 students received aid, including 4 fellowships, 12 research assistantships; institutionally sponsored loans also available.
Faculty research: Cellular and molecular immunology, molecular and pathogenic virology, pathogenic bacteriology and gene therapy of cancer. *Total annual research expenditures:* $4.5 million.
Dr. Eckhard R. Podack, Chairman, 305-243-6694, *Fax:* 305-243-5522.
Application contact: Dr. Robert B. Levy, Director, Graduate Program, 305-243-5682, *Fax:* 305-243-6903, *E-mail:* kdelrio@mednet.med.miami.edu.

Find an in-depth description at www.petersons.com/graduate.

■ UNIVERSITY OF MICHIGAN

Medical School and Horace H. Rackham School of Graduate Studies, Program in Biomedical Sciences (PIBS), Department of Microbiology and Immunology, Ann Arbor, MI 48109
AWARDS PhD.

Degree requirements: For doctorate, computer language, oral defense of dissertation, preliminary exam required.
Entrance requirements: For doctorate, GRE General Test, TOEFL, TWE.
Expenses: Tuition, state resident: full-time $10,316. Tuition, nonresident: full-time $20,922.
Faculty research: Gene regulation, molecular biology of animal and bacterial

University of Michigan (continued)
viruses, molecular and cellular networks, pathogenesis and microbial genetics.

Find an in-depth description at www.petersons.com/graduate.

■ **UNIVERSITY OF MICHIGAN**

Medical School and Horace H. Rackham School of Graduate Studies, Program in Biomedical Sciences (PIBS), Program in Immunology, Ann Arbor, MI 48109-0619

AWARDS PhD.

Faculty: 37 part-time/adjunct (6 women).
Students: 2 full-time (both women); includes 1 minority (Asian American or Pacific Islander).
Degree requirements: For doctorate, dissertation required.
Expenses: Tuition, state resident: full-time $10,316. Tuition, nonresident: full-time $20,922.
Financial aid: In 1999–00, 2 students received aid, including 2 research assistantships with tuition reimbursements available (averaging $17,000 per year); fellowships with tuition reimbursements available
Faculty research: Tumor immunology, aging, tolerance, autoimmunity, antigen presentation. *Total annual research expenditures:* $2 million.
Dr. James Mule, Director, 734-647-2779, *Fax:* 734-936-9715, *E-mail:* jimmule@umich.edu.

Application contact: Ellen Elkin, Student Services Representative, 734-615-4846, *Fax:* 734-936-9715, *E-mail:* eelkin@umich.edu.

Find an in-depth description at www.petersons.com/graduate.

■ **UNIVERSITY OF MINNESOTA, DULUTH**

School of Medicine, Department of Medical Microbiology and Immunology, Duluth, MN 55812-2496

AWARDS MS, PhD.

Faculty: 4 full-time (1 woman), 3 part-time/adjunct (1 woman).
Students: 1 (woman) full-time; minority (Asian American or Pacific Islander). *18 applicants, 0% accepted.* Terminal master's awarded for partial completion of doctoral program.
Degree requirements: For master's, thesis, final oral exam required, foreign language not required; for doctorate, dissertation, final exam, oral and written preliminary exams required, foreign language not required.
Entrance requirements: For master's and doctorate, GRE General Test, TOEFL.
Application deadline: For fall admission, 5/1

(priority date). Applications are processed on a rolling basis. *Application fee:* $30.
Expenses: Tuition, state resident: full-time $5,040; part-time $420 per credit. Tuition, nonresident: full-time $9,900; part-time $825 per credit. Required fees: $509. Tuition and fees vary according to course load and program.
Financial aid: Research assistantships available.
Faculty research: Immunomodulation, molecular diagnosis of rabies, cytokines, cancer immunology, cytomegalovirus infection. *Total annual research expenditures:* $60,000.
Dr. Arthur Johnson, Head, 218-726-7561, *Fax:* 218-726-6235, *E-mail:* ajohnso1@d.umn.edu.

■ **UNIVERSITY OF MINNESOTA, TWIN CITIES CAMPUS**

Medical School and Graduate School, Graduate Programs in Medicine, Department of Microbiology, Minneapolis, MN 55455-0213

AWARDS Microbiology (MS); microbiology, immunology, and molecular pathobiology (PhD). Terminal master's awarded for partial completion of doctoral program.

Degree requirements: For master's and doctorate, thesis/dissertation required.
Entrance requirements: For master's, GRE General Test.
Expenses: Tuition, state resident: full-time $11,984; part-time $1,498 per semester. Tuition, nonresident: full-time $22,264; part-time $2,783 per semester. Full-time tuition and fees vary according to program and student level. Part-time tuition and fees vary according to course load and program.
Faculty research: Virology.

Find an in-depth description at www.petersons.com/graduate.

■ **UNIVERSITY OF MISSOURI–COLUMBIA**

School of Medicine and Graduate School, Graduate Programs in Medicine, Department of Molecular Microbiology and Immunology, Columbia, MO 65211

AWARDS MS, PhD. Terminal master's awarded for partial completion of doctoral program.

Degree requirements: For master's and doctorate, thesis/dissertation required.
Entrance requirements: For master's and doctorate, GRE General Test, minimum GPA of 3.0.
Expenses: Tuition, state resident: full-time $3,020; part-time $168 per hour. Tuition,

nonresident: full-time $6,066; part-time $505 per hour. Required fees: $445; $18 per hour.
Faculty research: Molecular biology, host-parasite interactions.

Find an in-depth description at www.petersons.com/graduate.

■ **UNIVERSITY OF NEW MEXICO**

Graduate School, College of Arts and Sciences, Department of Biology, Albuquerque, NM 87131-2039

AWARDS Biology (MS, PhD), including air land ecology, behavioral ecology, botany, cellular and molecular biology, community ecology, comparative immunology, comparative physiology, conservation biology, ecology, ecosystem ecology, evolutionary biology, evolutionary genetics, microbiology, molecular genetics, parasitology, physiological ecology, physiology, population biology, vertebrate and invertebrate zoology. Part-time programs available.

Faculty: 35 full-time (5 women), 18 part-time/adjunct (11 women).
Students: 71 full-time (37 women), 28 part-time (11 women); includes 8 minority (2 Asian Americans or Pacific Islanders, 5 Hispanic Americans, 1 Native American), 11 international. Average age 33. *93 applicants, 30% accepted.* In 1999, 11 master's, 12 doctorates awarded. Terminal master's awarded for partial completion of doctoral program.
Degree requirements: For master's, one foreign language (computer language can substitute), thesis required (for some programs); for doctorate, 2 foreign languages (computer language can substitute for one), dissertation required.
Entrance requirements: For master's and doctorate, GRE General Test, GRE Subject Test, minimum GPA of 3.2. *Application deadline:* For fall admission, 1/15. *Application fee:* $25.
Expenses: Tuition, state resident: full-time $2,514; part-time $105 per credit hour. Tuition, nonresident: full-time $10,304; part-time $417 per credit hour. International tuition: $10,304 full-time. Required fees: $516; $22 per credit hour. Tuition and fees vary according to program.
Financial aid: In 1999–00, 58 students received aid, including 24 fellowships (averaging $1,645 per year), 26 research assistantships with tuition reimbursements available (averaging $8,921 per year), 40 teaching assistantships with tuition reimbursements available (averaging $11,066 per year); career-related internships or fieldwork, Federal Work-Study, institutionally sponsored loans, and tuition waivers (full and partial) also available. Aid

available to part-time students. Financial aid applicants required to submit FAFSA. **Faculty research:** Developmental biology, immunobiology. *Total annual research expenditures:* $4.5 million.
Dr. Kathryn Vogel, Chair, 505-277-3411, *Fax:* 505-277-0304, *E-mail:* kgvogel@unm.edu.

Application contact: Vivian Kent, Information Contact, 505-277-1712, *Fax:* 505-277-0304, *E-mail:* vkent@unm.edu.

■ THE UNIVERSITY OF NORTH CAROLINA AT CHAPEL HILL

School of Medicine and Graduate School, Graduate Programs in Medicine, Department of Microbiology and Immunology, Chapel Hill, NC 27599

AWARDS Immunology (MS, PhD); microbiology (MS, PhD).

Faculty: 24 full-time (6 women), 17 part-time/adjunct (2 women).
Students: 45 full-time (23 women), 3 part-time (all women); includes 7 minority (1 African American, 4 Asian Americans or Pacific Islanders, 2 Hispanic Americans), 3 international. Average age 27. *112 applicants, 23% accepted.* In 1999, 2 master's awarded (100% entered university research/teaching); 6 doctorates awarded (100% entered university research/teaching). Terminal master's awarded for partial completion of doctoral program.
Degree requirements: For master's, thesis, comprehensive exam required, foreign language not required; for doctorate, dissertation, comprehensive exams required, foreign language not required. *Average time to degree:* Master's–4 years full-time, 6 years part-time; doctorate–5.83 years full-time.
Entrance requirements: For master's and doctorate, GRE General Test, minimum GPA of 3.0. *Application deadline:* For fall admission, 1/15. *Application fee:* $55. Electronic applications accepted.
Expenses: Tuition, state resident: full-time $1,966. Tuition, nonresident: full-time $11,026. Required fees: $8,940. One-time fee: $15 full-time. Part-time tuition and fees vary according to course load.
Financial aid: In 1999–00, 1 fellowship with full tuition reimbursement (averaging $16,500 per year), 38 research assistantships with full tuition reimbursements (averaging $16,500 per year) were awarded; teaching assistantships with full tuition reimbursements, grants, traineeships, tuition waivers (full), and unspecified assistantships also available. *Total annual research expenditures:* $10.6 million.

Jeffrey Frelinger, Chairman, 919-966-1191, *Fax:* 919-962-8103, *E-mail:* microimm@med.unc.edu.
Application contact: Dixie Flannery, Administrator, 919-966-1191, *Fax:* 919-962-8103, *E-mail:* microimm@med.unc.edu.

Find an in-depth description at www.petersons.com/graduate.

■ UNIVERSITY OF NORTH DAKOTA

School of Medicine and Graduate School, Graduate Programs in Medicine, Department of Microbiology and Immunology, Grand Forks, ND 58202

AWARDS MS, PhD.

Faculty: 6 full-time (1 woman).
Students: 8 full-time (5 women), 3 part-time (1 woman). *13 applicants, 31% accepted.* In 1999, 2 master's, 1 doctorate awarded.
Degree requirements: For master's, thesis or alternative, comprehensive final examination required, foreign language not required; for doctorate, computer language, dissertation, comprehensive examination, final examination required.
Entrance requirements: For master's, GRE General Test, TOEFL, minimum GPA of 3.0; for doctorate, GRE General Test, TOEFL, minimum GPA of 3.5. *Application deadline:* For fall admission, 3/1 (priority date). Applications are processed on a rolling basis. *Application fee:* $25.
Expenses: Tuition, state resident: full-time $2,690; part-time $112 per credit. Tuition, nonresident: full-time $7,182; part-time $299 per credit. Required fees: $46 per semester.
Financial aid: In 1999–00, 8 students received aid, including 6 teaching assistantships with full tuition reimbursements available (averaging $10,586 per year); fellowships, research assistantships, Federal Work-Study, institutionally sponsored loans, scholarships, and tuition waivers (full and partial) also available. Aid available to part-time students. Financial aid application deadline: 3/15; financial aid applicants required to submit FAFSA.
Faculty research: Genetic and immunological aspects of a murine model of human multiple sclerosis, termination of DNA replication, cell division in bacteria, yersinia pestis.
Dr. Kevin Young, Director, 701-777-2214, *Fax:* 701-777-3527, *E-mail:* kyoung@medicine.nodak.edu.

■ UNIVERSITY OF NORTH TEXAS HEALTH SCIENCE CENTER AT FORT WORTH

Graduate School of Biomedical Sciences, Fort Worth, TX 76107-2699

AWARDS Anatomy and cell biology (MS, PhD); biochemistry and molecular biology (MS, PhD); biotechnology (MS); integrative physiology (MS, PhD); microbiology and immunology (MS, PhD); pharmacology (MS, PhD).

Faculty: 65 full-time (9 women), 11 part-time/adjunct (1 woman).
Students: 59 full-time (27 women), 44 part-time (20 women); includes 30 minority (10 African Americans, 9 Asian Americans or Pacific Islanders, 11 Hispanic Americans), 23 international. *70 applicants, 70% accepted.* In 1999, 5 master's awarded (40% found work related to degree, 60% continued full-time study); 5 doctorates awarded (80% entered university research/teaching, 20% found other work related to degree).
Degree requirements: For doctorate, computer language, dissertation required. *Average time to degree:* Master's–2.5 years full-time, 4 years part-time; doctorate–5 years full-time.
Entrance requirements: For master's and doctorate, GRE General Test, TOEFL. *Application deadline:* For fall admission, 5/1; for spring admission, 11/1. Applications are processed on a rolling basis. *Application fee:* $25 ($50 for international students).
Expenses: Tuition, state resident: full-time $1,188; part-time $66 per credit. Tuition, nonresident: full-time $5,058; part-time $281 per credit. Required fees: $366; $183 per semester.
Financial aid: In 1999–00, 11 fellowships, 70 research assistantships (averaging $16,500 per year) were awarded; teaching assistantships, career-related internships or fieldwork, Federal Work-Study, grants, institutionally sponsored loans, and traineeships also available. Aid available to part-time students. Financial aid application deadline: 4/1; financial aid applicants required to submit FAFSA.
Faculty research: Alzheimer's disease, diabetes, eye diseases, cancer, cardiovascular physiology.
Dr. Thomas Yorio, Dean, 817-735-2560, *Fax:* 817-735-0243, *E-mail:* yoriot@hsc.unt.edu.
Application contact: Jan Sharp, Administrative Assistant, 817-735-0258, *Fax:* 817-735-0243, *E-mail:* gsbs@hsc.unt.edu.

Find an in-depth description at www.petersons.com/graduate.

■ UNIVERSITY OF OKLAHOMA HEALTH SCIENCES CENTER

College of Medicine and Graduate College, Graduate Programs in Medicine, Department of Microbiology and Immunology, Oklahoma City, OK 73190

AWARDS Immunology (MS, PhD); microbiology (MS, PhD). Part-time programs available.

Faculty: 21 full-time (3 women), 11 part-time/adjunct (4 women).
Students: 8 full-time (5 women), 25 part-time (14 women); includes 3 minority (1 African American, 1 Asian American or Pacific Islander, 1 Hispanic American), 6 international. Average age 24. *44 applicants, 23% accepted.* In 1999, 5 master's, 5 doctorates awarded. Terminal master's awarded for partial completion of doctoral program.
Degree requirements: For master's, thesis or alternative required, foreign language not required; for doctorate, one foreign language (computer language can substitute), dissertation required.
Entrance requirements: For master's and doctorate, GRE General Test, TOEFL. *Application deadline:* For fall admission, 3/31. *Application fee:* $25 ($50 for international students).
Expenses: Tuition, state resident: part-time $90 per semester hour. Tuition, nonresident: part-time $264 per semester hour. Tuition and fees vary according to program.
Financial aid: Fellowships, teaching assistantships available. Financial aid applicants required to submit FAFSA.
Faculty research: Molecular genetics, pathogenesis, streptococcal infections, gram-positive virulence, monoclonal antibodies.
Dr. John Iandolo, Chairman, 405-271-2133.
Application contact: Dr. Don Graves, Graduate Liaison, 405-271-2133.

■ UNIVERSITY OF PENNSYLVANIA

School of Medicine, Biomedical Graduate Studies, Graduate Group in Immunology, Philadelphia, PA 19104

AWARDS PhD, MD/PhD, VMD/PhD.

Faculty: 84 full-time (14 women).
Students: 63 full-time (29 women), 3 part-time (1 woman); includes 18 minority (2 African Americans, 11 Asian Americans or Pacific Islanders, 5 Hispanic Americans), 7 international. *79 applicants, 18% accepted.* In 1999, 12 doctorates awarded (100% entered university research/teaching).
Degree requirements: For doctorate, dissertation, 2 preliminary exams required, foreign language not required.
Entrance requirements: For doctorate, GRE General Test, GRE Subject Test, TOEFL, undergraduate major in natural or physical science. *Application deadline:* For fall admission, 1/2 (priority date). Applications are processed on a rolling basis. *Application fee:* $65. Electronic applications accepted.
Expenses: Tuition: Full-time $17,256; part-time $2,991 per course. Required fees: $2,588; $363 per course. $726 per term.
Financial aid: In 1999–00, 32 fellowships, 30 research assistantships were awarded; teaching assistantships, grants and institutionally sponsored loans also available. Financial aid application deadline: 1/2.
Faculty research: Immunoglobulin structure and function, cell surface receptors, lymphocyte functional transplantation immunology, cellular immunology, molecular biology of immunoglobulins.
Dr. John G. Monroe, Chairman, 215-898-2317.
Application contact: Suzanne Hakanen, Graduate Coordinator, 215-573-9450, *Fax:* 215-573-3934, *E-mail:* shakanen@mail.med.upenn.edu.

Find an in-depth description at www.petersons.com/graduate.

■ UNIVERSITY OF PITTSBURGH

School of Medicine, Graduate Programs in Medicine, Program in Immunology, Pittsburgh, PA 15260

AWARDS MS, PhD.

Students: 8 full-time (5 women); includes 1 minority (Asian American or Pacific Islander), 6 international. *325 applicants, 22% accepted.* In 1999, 1 degree awarded.
Degree requirements: For doctorate, dissertation required, foreign language not required. *Average time to degree:* Doctorate–5.3 years full-time.
Entrance requirements: For doctorate, GRE General Test, GRE Subject Test, TOEFL, minimum QPA of 3.0. *Application deadline:* For fall admission, 1/15 (priority date). Applications are processed on a rolling basis. *Application fee:* $30 ($40 for international students).
Expenses: Tuition, state resident: full-time $9,778; part-time $403 per credit. Tuition, nonresident: full-time $20,146; part-time $830 per credit. Required fees: $480; $90 per semester.
Financial aid: Research assistantships with full tuition reimbursements, teaching assistantships with full tuition reimbursements, Federal Work-Study, institutionally sponsored loans, scholarships, traineeships, and unspecified assistantships available.
Faculty research: Cancer immunology, organ transplantation, autoimmunity and immunology of infectious disease. *Total annual research expenditures:* $76.3 million.
Application contact: Graduate Studies Administrator, 412-648-8957, *Fax:* 412-648-1236, *E-mail:* biomed_phd@fs1.dean-med.pitt.edu.

Find an in-depth description at www.petersons.com/graduate.

■ UNIVERSITY OF ROCHESTER

School of Medicine and Dentistry, Graduate Programs in Medicine and Dentistry, Department of Microbiology and Immunology, Rochester, NY 14627-0250

AWARDS Microbiology (MS, PhD).

Faculty: 21.
Students: 49 full-time (27 women), 17 part-time (8 women); includes 12 minority (2 African Americans, 7 Asian Americans or Pacific Islanders, 2 Hispanic Americans, 1 Native American), 8 international. In 1999, 17 master's, 7 doctorates awarded.
Degree requirements: For doctorate, dissertation, qualifying exam required, foreign language not required.
Entrance requirements: For master's and doctorate, GRE General Test. *Application deadline:* For fall admission, 2/1. *Application fee:* $25.
Expenses: Tuition: Part-time $697 per credit hour. Tuition and fees vary according to program.
Financial aid: Fellowships, research assistantships, teaching assistantships, tuition waivers (full and partial) available. Financial aid application deadline: 2/1.
Dr. Barbara Iglewski, Chair, 716-275-3402.
Application contact: Brenda Khorr, Graduate Program Secretary, 716-275-3402.

Find an in-depth description at www.petersons.com/graduate.

■ UNIVERSITY OF SOUTH ALABAMA

College of Medicine and Graduate School, Program in Basic Medical Sciences, Specialization in Microbiology and Immunology, Mobile, AL 36688-0002

AWARDS PhD.

Faculty: 11 full-time (0 women).
Students: In 1999, 1 degree awarded.
Degree requirements: For doctorate, dissertation required, foreign language not required.
Entrance requirements: For doctorate, GRE General Test or MCAT. *Application*

deadline: For fall admission, 4/1. Applications are processed on a rolling basis. *Application fee:* $25.
Expenses: Tuition, state resident: part-time $116 per semester hour. Tuition, nonresident: part-time $230 per semester hour. Required fees: $121 per semester.
Financial aid: Fellowships, research assistantships, institutionally sponsored loans available. Financial aid application deadline: 4/1.
Faculty research: Mechanisms of tumor immunity, host response to infectious agents, virus replication, immune regulation, mechanisms of resistance to viruses and bacteria.
Dr. Joseph H. Coggin, Chair, 334-460-6339.
Application contact: Lanette Flagge, Coordinator, 334-460-6153.

■ UNIVERSITY OF SOUTH DAKOTA

School of Medicine and Graduate School, Biomedical Sciences Graduate Program, Molecular Microbiology and Immunology Group, Vermillion, SD 57069-2390

AWARDS MA, PhD.

Faculty: 4 full-time (1 woman).
Students: 4 full-time (1 woman). Average age 25. *15 applicants, 13% accepted.* Terminal master's awarded for partial completion of doctoral program.
Degree requirements: For master's and doctorate, thesis/dissertation required, foreign language not required. *Average time to degree:* Doctorate–4 years full-time.
Entrance requirements: For master's and doctorate, GRE General Test, TOEFL, minimum GPA of 3.0. *Application deadline:* For fall admission, 3/15 (priority date). Applications are processed on a rolling basis. *Application fee:* $15.
Expenses: Tuition, state resident: full-time $2,126; part-time $89 per credit. Tuition, nonresident: full-time $6,270; part-time $261 per credit. Required fees: $1,194; $50 per credit. Tuition and fees vary according to course load and reciprocity agreements.
Financial aid: In 1999–00, 4 students received aid; fellowships, research assistantships, teaching assistantships, Federal Work-Study available.
Faculty research: Structure-function membranes, plasmids, immunology, virology, pathogenesis. *Total annual research expenditures:* $280,000.
Dr. Leigh Washburn, Coordinator, 605-677-5170, *Fax:* 605-677-5658, *E-mail:* lwashbur@usd.edu.

■ UNIVERSITY OF SOUTHERN CALIFORNIA

Keck School of Medicine and Graduate School, Graduate Programs in Medicine, Department of Molecular Microbiology and Immunology, Los Angeles, CA 90033

AWARDS MS, PhD.

Faculty: 8 full-time (1 woman), 1 (woman) part-time/adjunct.
Students: 40 full-time (23 women), 1 part-time; includes 2 minority (1 Asian American or Pacific Islander, 1 Hispanic American), 25 international. Average age 28. *140 applicants, 9% accepted.* In 1999, 5 master's, 5 doctorates awarded (60% entered university research/teaching, 40% continued full-time study).
Degree requirements: For master's, thesis optional, foreign language not required; for doctorate, dissertation required, foreign language not required. *Average time to degree:* Master's–3 years full-time; doctorate–4.75 years full-time.
Entrance requirements: For master's, GRE General Test, TOEFL, minimum GPA of 3.0; for doctorate, GRE General Test, GRE Subject Test, TOEFL, minimum GPA of 3.0. *Application deadline:* For fall admission, 3/15 (priority date); for spring admission, 7/15. Applications are processed on a rolling basis. *Application fee:* $55. Electronic applications accepted.
Expenses: Tuition: Full-time $22,198; part-time $748 per unit. Required fees: $406.
Financial aid: In 1999–00, 28 students received aid, including 1 fellowship with full tuition reimbursement available (averaging $17,575 per year), 26 research assistantships with full tuition reimbursement available (averaging $17,575 per year), 1 teaching assistantship with full tuition reimbursement available (averaging $17,575 per year); institutionally sponsored loans, traineeships, and tuition waivers (partial) also available. Financial aid application deadline: 6/1.
Faculty research: Animal virology, microbial genetics, molecular and cellular immunology, cellular differentiation control of protein synthesis. *Total annual research expenditures:* $3 million.
Dr. Stanley M. Tahara, Director, Graduate Studies, 323-442-1722, *Fax:* 323-442-1721, *E-mail:* stahara@hsc.usc.edu.
Application contact: Laura C. Steel, Graduate Administrator, 323-442-2337, *Fax:* 323-442-1721, *E-mail:* lsteel@hsc.usc.edu.

■ UNIVERSITY OF SOUTHERN MAINE

School of Applied Science, Program in Applied Immunology and Molecular Biology, Portland, ME 04104-9300
AWARDS MS. Part-time programs available.

Degree requirements: For master's, thesis required, foreign language not required.
Entrance requirements: For master's, GRE General Test, GRE Subject Test, minimum GPA of 3.0.
Expenses: Tuition, state resident: full-time $5,944. Tuition, nonresident: full-time $15,634.
Faculty research: Flow cytometry, cancer, epidemiology, monoclonal antibodies, DNA diagnostics.

■ UNIVERSITY OF SOUTH FLORIDA

College of Medicine and Graduate School, Graduate Programs in Medical Sciences, Department of Medical Microbiology and Immunology, Tampa, FL 33620-9951
AWARDS PhD.

Degree requirements: For doctorate, dissertation required, foreign language not required.
Entrance requirements: For doctorate, GRE General Test, TOEFL, interview, minimum GPA of 3.0.
Expenses: Tuition, state resident: part-time $148 per credit hour. Tuition, nonresident: part-time $509 per credit hour.
Faculty research: Molecular genetics and pathogenesis of bacteria and viruses, molecular biology of DNA tumor viruses, immunomodulation by drugs and viruses, retrovirus vaccine development.

■ THE UNIVERSITY OF TENNESSEE HEALTH SCIENCE CENTER

College of Graduate Health Sciences, Department of Microbiology and Immunology, Memphis, TN 38163-0002
AWARDS Bacteriology (MS, PhD); immunology (MS, PhD); medical microbiology (MS, PhD); microbiology (MS, PhD); molecular and cell biology (PhD); molecular biology (MS, PhD); virology (MS, PhD). Terminal master's awarded for partial completion of doctoral program.

Degree requirements: For master's, thesis, oral and written comprehensive exams required, foreign language not required; for doctorate, dissertation, oral and written preliminary and

The University of Tennessee Health Science Center (continued)
comprehensive exams required, foreign language not required.
Entrance requirements: For master's, GRE General Test, TOEFL, minimum GPA of 3.0; for doctorate, GRE General Test, GRE Subject Test, TOEFL, minimum GPA of 3.0.

Find an in-depth description at www.petersons.com/graduate.

■ **THE UNIVERSITY OF TEXAS AT AUSTIN**

Graduate School, College of Natural Sciences, Institute for Cellular and Molecular Biology, Program in Microbiology and Immunology, Austin, TX 78712-1111

AWARDS PhD.

Entrance requirements: For doctorate, GRE General Test. *Application fee:* $50 ($75 for international students). Electronic applications accepted.
Expenses: Tuition, state resident: part-time $114 per semester hour. Tuition, nonresident: part-time $330 per semester hour. Tuition and fees vary according to program.
Financial aid: In 1999–00, fellowships with full tuition reimbursements (averaging $19,000 per year), research assistantships with full and partial tuition reimbursements (averaging $18,000 per year), teaching assistantships with full tuition reimbursements (averaging $17,300 per year) were awarded.
Paul O. Gottlieb, Chairman, 512-471-5105.
Application contact: Dr. Clarence Chan, Graduate Adviser, 512-471-6860, *E-mail:* clarence_chan@mail.utexas.edu.

Find an in-depth description at www.petersons.com/graduate.

■ **THE UNIVERSITY OF TEXAS– HOUSTON HEALTH SCIENCE CENTER**

Graduate School of Biomedical Sciences, Program in Immunology, Houston, TX 77225-0036

AWARDS MS, PhD, MD/PhD.

Faculty: 42 full-time (8 women).
Students: 17 full-time (9 women); includes 3 minority (2 Asian Americans or Pacific Islanders, 1 Hispanic American), 3 international. Average age 27. *58 applicants, 43% accepted.* In 1999, 1 master's, 5 doctorates awarded. Terminal master's awarded for partial completion of doctoral program.

Degree requirements: For master's and doctorate, thesis/dissertation required, foreign language not required.
Entrance requirements: For master's and doctorate, GRE General Test, TOEFL, TWE. *Application deadline:* For fall admission, 1/15 (priority date); for spring admission, 11/1. Applications are processed on a rolling basis. *Application fee:* $10. Electronic applications accepted.
Financial aid: Fellowships, research assistantships, teaching assistantships, institutionally sponsored loans available. Financial aid application deadline: 1/15.
Faculty research: Tumor immunobiology, lymphocyte activation and differentiation, cytokine and lymphocyte biology, intracellular signaling in lymphocyte activation, photoimmunology.
Dr. Stephen E. Ullrich, Director, 713-792-8593, *Fax:* 713-745-1633, *E-mail:* sullrich@notes.mdacc.tmc.edu.
Application contact: Anne Baronitis, Director of Admissions, 713-500-9860, *Fax:* 713-500-9877, *E-mail:* abaron@gsbs.gs.uth.tmc.edu.

Find an in-depth description at www.petersons.com/graduate.

■ **THE UNIVERSITY OF TEXAS MEDICAL BRANCH AT GALVESTON**

Graduate School of Biomedical Sciences, Program in Microbiology and Immunology, Galveston, TX 77555

AWARDS MS, PhD.

Faculty: 52 full-time (6 women).
Students: 30 full-time (16 women); includes 8 minority (1 African American, 4 Asian Americans or Pacific Islanders, 3 Hispanic Americans), 12 international. Average age 29. *53 applicants, 30% accepted.* In 1999, 6 degrees awarded (100% entered university research/teaching).
Degree requirements: For master's, thesis or alternative required, foreign language not required; for doctorate, dissertation required, foreign language not required.
Entrance requirements: For master's and doctorate, GRE General Test, minimum GPA of 3.0. *Application deadline:* For fall admission, 3/15. Applications are processed on a rolling basis. *Application fee:* $25 ($50 for international students). Electronic applications accepted.
Expenses: Tuition, state resident: full-time $684; part-time $38 per credit hour. Tuition, nonresident: full-time $4,572; part-time $254 per credit hour. Required fees: $29; $7.5 per credit hour. One-time fee: $55. Tuition and fees vary according to program.
Financial aid: Fellowships, research assistantships, Federal Work-Study and

institutionally sponsored loans available. Financial aid applicants required to submit FAFSA.
Faculty research: Bacterial and viral pathogenesis, microbial genetics, molecular virology, host defense, neuroimmunology, autoimmunity, and structural and molecular biology.
Dr. Thomas K. Hughes, Director, 409-772-2322, *Fax:* 409-772-2366, *E-mail:* thughes@utmb.edu.
Application contact: Martha Lewis, Secretary, 409-772-2322, *Fax:* 409-772-5065, *E-mail:* mlewis@utmb.edu.

Find an in-depth description at www.petersons.com/graduate.

■ **THE UNIVERSITY OF TEXAS SOUTHWESTERN MEDICAL CENTER AT DALLAS**

Southwestern Graduate School of Biomedical Sciences, Division of Cell and Molecular Biology, Program in Immunology, Dallas, TX 75390

AWARDS PhD.

Faculty: 43 full-time (11 women), 8 part-time/adjunct (0 women).
Students: 30 full-time (20 women); includes 9 minority (1 African American, 6 Asian Americans or Pacific Islanders, 2 Hispanic Americans), 7 international. Average age 30. In 1999, 8 doctorates awarded (100% continued full-time study).
Degree requirements: For doctorate, dissertation required, foreign language not required. *Average time to degree:* Doctorate–5 years full-time.
Entrance requirements: For doctorate, GRE General Test, minimum GPA of 3.0. *Application deadline:* For fall admission, 1/5 (priority date). *Application fee:* $0. Electronic applications accepted.
Expenses: Tuition, state resident: full-time $912. Tuition, nonresident: full-time $6,096. Required fees: $216. Full-time tuition and fees vary according to course load and program.
Financial aid: Fellowships, research assistantships available. Financial aid application deadline: 3/15; financial aid applicants required to submit FAFSA.
Faculty research: Antibody diversity and idiotype, cytotoxic effector mechanisms, natural killer cells, biology of immunoglobulins, oncogenes.
Dr. Richard H. Scheuermann, Chair, 214-648-4115, *Fax:* 214-648-1902, *E-mail:* scheuerm@utsw.swmed.edu.
Application contact: Nancy McKinney, Education Coordinator, 214-648-8099, *Fax:* 214-648-2978, *E-mail:* dcmbinfo@utsouthwestern.edu.

Find an in-depth description at www.petersons.com/graduate.

■ UNIVERSITY OF VIRGINIA

College and Graduate School of Arts and Sciences, Immunology Training Program, Charlottesville, VA 22903

AWARDS PhD, MD/PhD.

Students: 55 full-time (25 women); includes 8 minority (3 African Americans, 5 Asian Americans or Pacific Islanders). **Entrance requirements:** For doctorate, GRE General Test, GRE Subject Test, TOEFL. *Application deadline:* For fall admission, 2/15. *Application fee:* $60. **Expenses:** Tuition, state resident: full-time $3,832. Tuition, nonresident: full-time $15,519. Required fees: $1,084. Tuition and fees vary according to course load and program. **Application contact:** Glenn Glover, Graduate Program Administrator, 804-924-2412, *Fax:* 804-924-1221, *E-mail:* gmg6n@virginia.edu.

Find an in-depth description at www.petersons.com/graduate.

■ UNIVERSITY OF WASHINGTON

School of Medicine and Graduate School, Graduate Programs in Medicine, Department of Immunology, Seattle, WA 98195

AWARDS PhD.

Faculty: 21 full-time (2 women). **Students:** 29 full-time (9 women); includes 9 minority (7 Asian Americans or Pacific Islanders, 1 Hispanic American, 1 Native American), 1 international. Average age 29. *75 applicants, 17% accepted.* In 1999, 2 doctorates awarded (100% continued full-time study). **Degree requirements:** For doctorate, dissertation required. *Average time to degree:* Doctorate–6 years full-time. **Entrance requirements:** For doctorate, GRE General Test, TOEFL, BA or BS in related field. *Application deadline:* For fall admission, 1/15. *Application fee:* $50. **Expenses:** Tuition, state resident: full-time $9,210; part-time $236 per credit. Tuition, nonresident: full-time $23,256; part-time $596 per credit. **Financial aid:** In 1999–00, 7 fellowships with full tuition reimbursements (averaging $17,304 per year), 1 research assistantship with full tuition reimbursement (averaging $12,600 per year) were awarded; tuition waivers (full) and stipends also available. **Faculty research:** Molecular and cellular immunology, regulation of lymphocyte differentiation and responses, genetics of immune recognition genetics and pathogenesis of autoimmune diseases, signal transduction. Dr. Christopher B. Wilson, Chairman, 206-543-1010, *Fax:* 206-543-1013.

Application contact: Peggy A. McCune, Training Program Manager, 206-685-3955, *Fax:* 206-543-1013, *E-mail:* pmccune@u.washington.edu.

Find an in-depth description at www.petersons.com/graduate.

■ VANDERBILT UNIVERSITY

Graduate School and School of Medicine, Department of Microbiology and Immunology, Nashville, TN 37232-2363

AWARDS MS, PhD, MD/PhD.

Faculty: 16 full-time (2 women), 17 part-time/adjunct (1 woman). **Students:** 35 full-time (14 women), 8 part-time (3 women); includes 3 minority (1 African American, 1 Asian American or Pacific Islander, 1 Hispanic American), 13 international. Average age 29. In 1999, 4 degrees awarded. **Degree requirements:** For master's, thesis required, foreign language not required; for doctorate, dissertation, final and qualifying exams required, foreign language not required. **Entrance requirements:** For master's and doctorate, GRE General Test, GRE Subject Test (recommended). *Application deadline:* For fall admission, 1/15. *Application fee:* $40. **Expenses:** Tuition: Full-time $17,244; part-time $958 per hour. Required fees: $242; $121 per semester. Tuition and fees vary according to program. **Financial aid:** In 1999–00, fellowships with full tuition reimbursements (averaging $17,000 per year), research assistantships with full tuition reimbursements (averaging $17,000 per year) were awarded; Federal Work-Study, institutionally sponsored loans, traineeships, and tuition waivers (partial) also available. Financial aid application deadline: 1/15. **Faculty research:** Molecular and cellular immunology, molecular genetics and immunogenetics, cellular microbiology of pathogen-host interaction, virology, biotechnology. Jacek Hawiger, Chair, 615-343-8280, *Fax:* 615-343-7392, *E-mail:* jacek.hawiger@mcmail.vanderbilt.edu. **Application contact:** G. Neil Green, Director of Graduate Studies, 615-343-0453, *Fax:* 615-343-7392, *E-mail:* neil.green@mcmail.vanderbilt.edu.

Find an in-depth description at www.petersons.com/graduate.

■ VIRGINIA COMMONWEALTH UNIVERSITY

School of Graduate Studies and School of Medicine, School of Medicine Graduate Programs, Department of Microbiology and Immunology, Richmond, VA 23284-9005

AWARDS Microbiology (PhD); microbiology and immunology (MS, CBHS); molecular biology and genetics (PhD).

Students: 13 full-time (6 women), 46 part-time (26 women); includes 22 minority (2 African Americans, 17 Asian Americans or Pacific Islanders, 2 Hispanic Americans, 1 Native American). In 1999, 2 master's, 9 doctorates, 2 other advanced degrees awarded. **Degree requirements:** For master's, thesis required, foreign language not required; for doctorate, dissertation, comprehensive oral and written exams required, foreign language not required. **Entrance requirements:** For master's, DAT, GRE General Test, or MCAT; for doctorate, GRE General Test. *Application deadline:* For fall admission, 4/15. *Application fee:* $30. **Expenses:** Tuition, state resident: full-time $4,031; part-time $224 per credit hour. Tuition, nonresident: full-time $11,946; part-time $664 per credit hour. Required fees: $1,081; $40 per credit hour. Tuition and fees vary according to campus/location and program. **Financial aid:** Fellowships, research assistantships, teaching assistantships available. **Faculty research:** Microbial physiology and genetics, molecular biology, crystallography of biological molecules, antibiotics and chemotherapy, membrane transport. Dr. Dennis E. Ohman, Chair, 804-828-9728, *Fax:* 804-828-9946.

Find an in-depth description at www.petersons.com/graduate.

■ WAKE FOREST UNIVERSITY

School of Medicine and Graduate School, Graduate Programs in Medicine, Department of Microbiology and Immunology, Winston-Salem, NC 27109

AWARDS PhD.

Faculty: 13. **Students:** 22 full-time (12 women); includes 2 African Americans. *54 applicants, 22% accepted.* In 1999, 2 degrees awarded. **Degree requirements:** For doctorate, dissertation required. **Entrance requirements:** For doctorate, GRE General Test, GRE Subject Test. *Application deadline:* For fall admission,

Wake Forest University (continued)
2/15 (priority date). Applications are processed on a rolling basis. *Application fee:* $25. Electronic applications accepted.
Expenses: Tuition: Full-time $18,300.
Financial aid: In 1999–00, 14 fellowships, 8 research assistantships were awarded; grants, scholarships, and traineeships also available. Financial aid application deadline: 2/15.
Faculty research: Molecular immunology, bacterial pathogenesis and molecular genetics, viral pathogenesis, regulation of mRNA metabolism, leukocyte biology. *Total annual research expenditures:* $1.5 million.
Dr. Steven B. Mizel, Chair, 336-716-4471.
Application contact: Dr. Griffith D. Parks, Graduate Recruitment Director, 336-716-9083, *Fax:* 336-716-9928, *E-mail:* gparks@wfubmc.edu.

Find an in-depth description at www.petersons.com/graduate.

■ **WASHINGTON UNIVERSITY IN ST. LOUIS**

Graduate School of Arts and Sciences, Division of Biology and Biomedical Sciences, Program in Immunology, St. Louis, MO 63130-4899

AWARDS PhD.

Degree requirements: For doctorate, dissertation required, foreign language not required.
Entrance requirements: For doctorate, GRE General Test, GRE Subject Test. *Application deadline:* For fall admission, 1/1 (priority date). Applications are processed on a rolling basis. *Application fee:* $0.
Expenses: Tuition: Full-time $23,400; part-time $975 per credit. Tuition and fees vary according to program.
Financial aid: Fellowships, research assistantships available. Financial aid application deadline: 1/1.
Dr. Robert Schreiber, Coordinator, 314-362-8747.
Application contact: Rosemary Garagneni, Director of Admissions, 800-852-9074, *E-mail:* admissions@dbbs.wustl.edu.

■ **WAYNE STATE UNIVERSITY**

School of Medicine and Graduate School, Graduate Programs in Medicine, Department of Immunology and Microbiology, Detroit, MI 48202

AWARDS MS, PhD, MD/PhD. Terminal master's awarded for partial completion of doctoral program.

Degree requirements: For master's and doctorate, thesis/dissertation required, foreign language not required.

Entrance requirements: For master's, GRE, minimum GPA of 2.5; for doctorate, GRE, minimum GPA of 3.0.
Faculty research: Cellular immunity, immune regulation, bacterial pathophysiology, microbial physiology, molecular biology/viruses/bacteria.

Find an in-depth description at www.petersons.com/graduate.

■ **WEST VIRGINIA UNIVERSITY**

College of Agriculture, Forestry and Consumer Sciences, Interdisciplinary Program in Genetics and Developmental Biology, Morgantown, WV 26506

AWARDS Animal breeding (MS, PhD); biochemical and molecular genetics (MS, PhD); cytogenetics (MS, PhD); descriptive embryology (MS, PhD); developmental genetics (MS); experimental morphogenesis (MS); human genetics (MS, PhD); immunogenetics (MS, PhD); life cycles of animals and plants (MS, PhD); molecular aspects of development (MS, PhD); mutagenesis (PhD); mutagenetics (MS); oncology (MS, PhD); plant genetics (MS, PhD); population and quantitative genetics (PhD); population and quantitative genetics (MS); regeneration (MS, PhD); teratology (MS, PhD); toxicology (MS, PhD).

Students: 18 full-time (8 women), 5 part-time (2 women); includes 1 minority (Asian American or Pacific Islander), 8 international. Average age 27. In 1999, 4 doctorates awarded.
Degree requirements: For master's, thesis required, foreign language not required; for doctorate, dissertation, comprehensive exam required, foreign language not required.
Entrance requirements: For master's, GRE or MCAT, TOEFL, minimum GPA of 2.75; for doctorate, TOEFL. *Application fee:* $45.
Expenses: Tuition, state resident: full-time $2,910; part-time $154 per credit hour. Tuition, nonresident: full-time $8,368; part-time $457 per credit hour.
Financial aid: In 1999–00, 11 research assistantships, 3 teaching assistantships were awarded; fellowships, Federal Work-Study, institutionally sponsored loans, and tuition waivers (full and partial) also available. Financial aid application deadline: 2/1; financial aid applicants required to submit FAFSA.
Dr. J. Nath, Chairman, 304-293-6256 Ext. 4333, *E-mail:* jnath@wvu.edu.

■ **WEST VIRGINIA UNIVERSITY**

School of Medicine, Graduate Programs in Health Sciences, Department of Microbiology and Immunology, Morgantown, WV 26506

AWARDS Genetics (MS, PhD); immunology (MS, PhD); mycology (PhD); parasitology (MS, PhD); pathogenic bacteriology (MS, PhD); physiology (PhD); psysiology (MS); virology (MS, PhD).

Students: 25 full-time (11 women), 1 (woman) part-time; includes 2 minority (both Asian Americans or Pacific Islanders), 4 international. Average age 27. *62 applicants, 8% accepted.* In 1999, 1 master's, 5 doctorates awarded. Terminal master's awarded for partial completion of doctoral program.
Degree requirements: For master's, thesis required, foreign language not required; for doctorate, dissertation, comprehensive exam required, foreign language not required. *Average time to degree:* Doctorate–4.5 years full-time.
Entrance requirements: For master's and doctorate, GRE General Test, GRE Subject Test, TOEFL, minimum GPA of 3.0. *Application deadline:* For fall admission, 3/1 (priority date). Applications are processed on a rolling basis. *Application fee:* $45. Electronic applications accepted.
Expenses: Tuition, state resident: full-time $3,564. Tuition, nonresident: full-time $10,230.
Financial aid: In 1999–00, 7 research assistantships, 14 teaching assistantships were awarded; fellowships, Federal Work-Study and institutionally sponsored loans also available. Financial aid application deadline: 3/1; financial aid applicants required to submit FAFSA.
Faculty research: Mechanisms of pathogenesis, microbial genetics, molecular virology, immunotoxicology, oncogenes. *Total annual research expenditures:* $2.6 million.
John Barnett, Chair, 304-293-2649.
Application contact: James Sheil, Graduate Coordinator, 304-293-3559, *Fax:* 304-293-7823, *E-mail:* jsheil@wvu.edu.

■ **WRIGHT STATE UNIVERSITY**

School of Graduate Studies, College of Science and Mathematics, Department of Microbiology and Immunology, Dayton, OH 45435

AWARDS MS. Part-time programs available.

Students: 13 full-time (9 women), 7 part-time (4 women); includes 4 minority (3 African Americans, 1 Hispanic American), 2 international. Average age 28. *17 applicants, 35% accepted.* In 1999, 4 degrees awarded.

Degree requirements: For master's, thesis required, foreign language not required.
Entrance requirements: For master's, TOEFL. *Application fee:* $25.
Expenses: Tuition, state resident: full-time $5,568; part-time $175 per quarter hour. Tuition, nonresident: full-time $9,696; part-time $302 per quarter hour. Full-time tuition and fees vary according to course load, campus/location and program.
Financial aid: Fellowships, research assistantships, teaching assistantships, Federal Work-Study, institutionally sponsored loans, and unspecified assistantships available. Aid available to part-time students. Financial aid applicants required to submit FAFSA.
Faculty research: Reproductive immunology, viral pathogenesis, virus-host cell interactions.
Dr. Neal S. Rote, Chair, 937-775-2996, *Fax:* 937-775-2012, *E-mail:* neal.rote@wright.edu.

■ YALE UNIVERSITY

Graduate School of Arts and Sciences, Department of Immunobiology, New Haven, CT 06520
AWARDS PhD.

Faculty: 26 full-time (13 women), 2 part-time/adjunct (1 woman).
Students: 22 full-time (18 women), 1 part-time; includes 4 minority (1 African American, 2 Asian Americans or Pacific Islanders, 1 Hispanic American), 7 international. 73 *applicants, 18% accepted.* In 1999, 6 degrees awarded.
Degree requirements: For doctorate, dissertation required. *Average time to degree:* Doctorate–6 years full-time.
Entrance requirements: For doctorate, GRE General Test. *Application deadline:* For fall admission, 1/4. *Application fee:* $65.
Expenses: Tuition: Full-time $22,300. Full-time tuition and fees vary according to program.
Financial aid: Fellowships, research assistantships, teaching assistantships, Federal Work-Study, institutionally sponsored loans, and traineeships available. Aid available to part-time students.
Application contact: Admissions Information, 203-432-2770.

Find an in-depth description at www.petersons.com/graduate.

■ YALE UNIVERSITY

School of Medicine and Graduate School of Arts and Sciences, Combined Program in Biological and Biomedical Sciences (BBS), Immunology Track, New Haven, CT 06520
AWARDS PhD, MD/PhD.

Degree requirements: For doctorate, dissertation required.
Entrance requirements: For doctorate, GRE General Test, TOEFL. *Application deadline:* For fall admission, 1/2. *Application fee:* $65. Electronic applications accepted.
Expenses: All students receive full tuition of $22,330 and an annual stipend of $17,600 .
Financial aid: Fellowships, research assistantships available.
Dr. David Schatz, Director of Graduate Studies, 203-785-3857, *Fax:* 203-737-1764, *E-mail:* immunol.bbs@yale.edu.

Find an in-depth description at www.petersons.com/graduate.

■ YESHIVA UNIVERSITY

Albert Einstein College of Medicine, Sue Golding Graduate Division of Medical Sciences, Department of Microbiology and Immunology, Bronx, NY 10461
AWARDS PhD, MD/PhD.

Faculty: 10 full-time.
Students: 51 full-time (20 women); includes 19 minority (5 African Americans, 6 Asian Americans or Pacific Islanders, 8 Hispanic Americans), 6 international. In 1999, 7 degrees awarded.
Degree requirements: For doctorate, dissertation required, foreign language not required.
Entrance requirements: For doctorate, GRE General Test, TOEFL. *Application deadline:* For fall admission, 1/15. *Application fee:* $0.
Expenses: Tuition: Part-time $525 per credit. Tuition and fees vary according to degree level and program.
Financial aid: In 1999–00, 51 fellowships were awarded.
Faculty research: Nature of histocompatibility antigens, lymphoid cell receptors, regulation of immune responses and mechanisms of resistance to infection.
Dr. Marshall Horwitz, Chairperson, 718-430-2811.

Find an in-depth description at www.petersons.com/graduate.

MEDICAL MICROBIOLOGY

■ CREIGHTON UNIVERSITY

School of Medicine and Graduate School, Graduate Programs in Medicine and College of Arts and Sciences, Department of Medical Microbiology and Immunology, Omaha, NE 68178-0001
AWARDS MS, PhD.

Faculty: 20 full-time (3 women), 5 part-time/adjunct (0 women).
Students: 15 full-time (7 women), 2 part-time (both women), 5 international. Average age 24. *40 applicants, 7% accepted.* In 1999, 3 master's, 1 doctorate awarded. Terminal master's awarded for partial completion of doctoral program.
Degree requirements: For master's and doctorate, thesis/dissertation required, foreign language not required. *Average time to degree:* Master's–3 years full-time; doctorate–5 years full-time.
Entrance requirements: For master's and doctorate, GRE General Test. *Application deadline:* For fall admission, 2/1 (priority date). Applications are processed on a rolling basis. *Application fee:* $30.
Expenses: Tuition: Full-time $8,940; part-time $447 per credit hour. Required fees: $598; $50 per semester.
Financial aid: In 1999–00, 17 students received aid, including 6 fellowships with tuition reimbursements available; research assistantships with tuition reimbursements available, career-related internships or fieldwork and institutionally sponsored loans also available.
Faculty research: Infectious diseases, molecular biology, genetics, antimicrobial agents and chemotherapy, virology, microbial physiology, prions, neuropathology. *Total annual research expenditures:* $770,612.
Dr. Roderick Nairn, Chair, 402-280-2921, *Fax:* 402-280-1875, *E-mail:* rnairn@creighton.edu.
Application contact: Dr. Philip D. Lister, Graduate Director, 402-280-2921, *Fax:* 402-280-1875, *E-mail:* pdlister@creighton.edu.

Find an in-depth description at www.petersons.com/graduate.

■ FINCH UNIVERSITY OF HEALTH SCIENCES/THE CHICAGO MEDICAL SCHOOL

School of Graduate and Postdoctoral Studies, Department of Microbiology and Immunology, Program in Medical Microbiology, North Chicago, IL 60064-3095
AWARDS MS, PhD.

Students: 1 full-time (0 women).
Degree requirements: For master's and doctorate, thesis/dissertation required, foreign language not required.
Entrance requirements: For master's and doctorate, GRE General Test, TOEFL, TWE. *Application deadline:* For fall admission, 6/1 (priority date). Applications are processed on a rolling basis. *Application fee:* $25.

Finch University of Health Sciences/The Chicago Medical School (continued)

Expenses: Tuition: Full-time $14,054; part-time $391 per credit hour. Tuition and fees vary according to program. **Financial aid:** Application deadline: 6/9. **Application contact:** Dana Frederick, Admissions Officer, 847-578-3209.

■ **MCP HAHNEMANN UNIVERSITY**

School of Medicine, Biomedical Graduate Programs, Department of Pathology and Laboratory Medicine, Program in Clinical Microbiology, Philadelphia, PA 19102-1192

AWARDS MS, MD/MS. Part-time programs available.

Degree requirements: For master's, thesis, comprehensive exam required, foreign language not required. **Entrance requirements:** For master's, GRE General Test, TOEFL, interview, minimum GPA of 3.0. **Faculty research:** Rapid diagnostic methods.

■ **THE OHIO STATE UNIVERSITY**

College of Medicine and Public Health and Graduate School, Graduate Programs in the Basic Medical Sciences, Department of Molecular Virology, Immunology and Medical Genetics, Columbus, OH 43210

AWARDS MS, PhD, MD/PhD.

Faculty: 30 full-time (5 women). **Students:** 24 full-time (10 women), 10 international. Average age 27. *160 applicants, 8% accepted.* In 1999, 1 master's, 3 doctorates awarded (100% entered university research/teaching). Terminal master's awarded for partial completion of doctoral program. **Degree requirements:** For master's, thesis optional, foreign language not required; for doctorate, dissertation required, foreign language not required. *Average time to degree:* Master's–4 years part-time; doctorate–5.6 years full-time. **Entrance requirements:** For master's and doctorate, GRE General Test. *Application deadline:* For fall admission, 1/15. Applications are processed on a rolling basis. *Application fee:* $30 ($40 for international students). Electronic applications accepted. **Expenses:** Tuition, state resident: full-time $5,400. Tuition, nonresident: full-time $14,535. Part-time tuition and fees vary according to course load. **Financial aid:** In 1999–00, 1 fellowship with full tuition reimbursement (averaging $19,260 per year), 8 research assistantships with full tuition reimbursements (averaging $15,600 per year), 7 teaching assistantships with full tuition reimbursements (averaging $15,600 per year) were awarded; grants

also available. Financial aid application deadline: 1/15. **Faculty research:** Virology, bacteriology, molecular biology. *Total annual research expenditures:* $912,708. Dr. Caroline C. Whitacre, Chairperson, 614-292-5889, *Fax:* 614-292-9805, *E-mail:* whitacre.3@osu.edu. **Application contact:** Dr. Deborah S. Parris, Chairman, Graduate Studies Committee, 614-292-0735, *Fax:* 614-292-9805, *E-mail:* mvimg@osu.edu.

Find an in-depth description at www.petersons.com/graduate.

■ **RUTGERS, THE STATE UNIVERSITY OF NEW JERSEY, NEW BRUNSWICK**

Graduate School, Program in Microbiology and Molecular Genetics, New Brunswick, NJ 08901-1281

AWARDS Applied microbiology (MS, PhD); clinical microbiology (MS, PhD); computational molecular biology (PhD); immunology (MS, PhD); microbial biochemistry (MS, PhD); molecular genetics (MS, PhD); virology (MS, PhD). Part-time programs available.

Faculty: 122 full-time (25 women), 4 part-time/adjunct (1 woman). **Students:** 65 full-time (31 women), 51 part-time (24 women); includes 24 minority (14 Asian Americans or Pacific Islanders, 10 Hispanic Americans), 29 international. Average age 27. *200 applicants, 26% accepted.* In 1999, 15 master's, 10 doctorates awarded. Terminal master's awarded for partial completion of doctoral program. **Degree requirements:** For master's, thesis or alternative required, foreign language not required; for doctorate, dissertation, written qualifying exam required, foreign language not required. **Entrance requirements:** For master's, GRE General Test, GRE Subject Test; for doctorate, GRE General Test, GRE Subject Test, TOEFL, minimum GPA of 3.0. *Application deadline:* For fall admission, 2/15 (priority date). Applications are processed on a rolling basis. *Application fee:* $50. **Expenses:** Tuition, state resident: full-time $6,776; part-time $279 per credit. Tuition, nonresident: full-time $9,936; part-time $412 per credit. Required fees: $20 per credit. $89 per semester. Tuition and fees vary according to course load, campus/location and program. **Financial aid:** In 1999–00, 70 students received aid, including 32 fellowships with full tuition reimbursements available, 26 research assistantships with full tuition reimbursements available, 12 teaching

assistantships with full tuition reimbursements available; institutionally sponsored loans also available. Financial aid application deadline: 2/15; financial aid applicants required to submit FAFSA. **Faculty research:** Biochemistry of microbes, medical microbiology. Dr. Howard Passmore, Director, 732-445-2812, *E-mail:* passmore@biology.rutgers.edu. **Application contact:** Betty Green, Administrative Assistant, 732-445-5086, *Fax:* 732-445-5735, *E-mail:* green@mbcl.rutgers.edu.

■ **TEXAS A&M UNIVERSITY SYSTEM HEALTH SCIENCE CENTER**

College of Medicine, Graduate School of Biomedical Sciences, Department of Medical Microbiology and Immunology, College Station, TX 77840-7896

AWARDS Immunology (PhD); microbiology (PhD); molecular biology (PhD); virology (PhD).

Faculty: 7 full-time (0 women), 1 part-time/adjunct (0 women). **Students:** 16 full-time (10 women); includes 2 minority (1 African American, 1 Asian American or Pacific Islander). **Degree requirements:** For doctorate, dissertation required, foreign language not required. **Entrance requirements:** For doctorate, GRE General Test, minimum GPA of 3.0. *Application deadline:* For fall admission, 2/1 (priority date). *Application fee:* $50 ($75 for international students). **Expenses:** Tuition, area resident: Full-time $1,368. Tuition, state resident: part-time $76 per credit. Tuition, nonresident: full-time $5,256; part-time $292 per credit. International tuition: $5,256 full-time. Required fees: $678; $38 per credit. Full-time tuition and fees vary according to course load and student level. **Financial aid:** In 1999–00, 16 students received aid, including 1 fellowship (averaging $17,000 per year), 15 research assistantships (averaging $17,000 per year). Financial aid application deadline: 4/1; financial aid applicants required to submit FAFSA. **Faculty research:** Molecular pathogenesis, microbial therapeutics. *Total annual research expenditures:* $1.2 million. Dr. John M. Quarles, Interim Head, 979-845-1313, *Fax:* 979-845-3479, *E-mail:* jmquarles@medicine.tamu.edu.

Application contact: Dr. James Samuel, Graduate Adviser, 979-845-1313, *Fax:* 979-845-3479, *E-mail:* jsamuel@medicine.tamu.edu.

Find an in-depth description at www.petersons.com/graduate.

■ TEXAS TECH UNIVERSITY HEALTH SCIENCES CENTER

Graduate School of Biomedical Sciences, Department of Microbiology and Immunology, Lubbock, TX 79430

AWARDS Medical microbiology (MS, PhD).

Faculty: 7 full-time (1 woman), 4 part-time/adjunct (0 women).
Students: 12 full-time (8 women); includes 2 minority (both Asian Americans or Pacific Islanders), 1 international. Average age 30. *13 applicants, 31% accepted.* In 1999, 2 doctorates awarded (100% entered university research/teaching). Terminal master's awarded for partial completion of doctoral program.
Degree requirements: For master's and doctorate, thesis/dissertation required, foreign language not required. *Average time to degree:* Doctorate–5 years full-time.
Entrance requirements: For master's and doctorate, GRE General Test, TOEFL, minimum GPA of 3.0. *Application deadline:* For fall admission, 4/15 (priority date). Applications are processed on a rolling basis. *Application fee:* $30 ($55 for international students). Electronic applications accepted.
Expenses: Tuition, state resident: part-time $38 per credit hour. Tuition, nonresident: part-time $254 per credit hour. Part-time tuition and fees vary according to program.
Financial aid: In 1999–00, 2 fellowships (averaging $25,222 per year), 7 research assistantships (averaging $14,500 per year) were awarded.
Faculty research: Genetics, pathogenic bacteriology, molecular biology, virology, medical mycology.
Dr. W. La Jean Chaffin, Graduate Adviser, 806-743-2545, *Fax:* 806-743-2334, *E-mail:* micwlc@ttuhsc.edu.
Application contact: Dr. Joe A. Fralick, Graduate Adviser, 806-743-2555, *Fax:* 806-743-2334, *E-mail:* micjaf@ttuhsc.edu.

■ UNIVERSITY OF GEORGIA

College of Veterinary Medicine and Graduate School, Graduate Programs in Veterinary Medicine, Athens, GA 30602

AWARDS Avian medicine (MAM); medical microbiology and parasitology (MS, PhD), including medical microbiology, parasitology; pathology (MS, PhD); physiology and pharmacology (MS, PhD), including

pharmacology, physiology; toxicology (MS, PhD). Evening/weekend programs available.

Faculty: 66 full-time (20 women).
Students: 63 full-time, 11 part-time. *70 applicants, 30% accepted.* In 1999, 13 master's, 8 doctorates awarded.
Degree requirements: For master's, foreign language not required; for doctorate, one foreign language (computer language can substitute), dissertation required.
Entrance requirements: For master's and doctorate, GRE General Test. *Application deadline:* For fall admission, 7/1 (priority date); for spring admission, 11/15. *Application fee:* $30. Electronic applications accepted.
Expenses: Tuition, state resident: full-time $7,516; part-time $431 per credit hour. Tuition, nonresident: full-time $12,204; part-time $793 per credit hour.
Financial aid: Fellowships, research assistantships, teaching assistantships, unspecified assistantships available.
Dr. Harry W. Dickerson, Associate Dean, 706-542-5734, *Fax:* 706-542-8254, *E-mail:* hwd@calc.vet.uga.edu.

■ UNIVERSITY OF GEORGIA

College of Veterinary Medicine and Graduate School, Graduate Programs in Veterinary Medicine, Department of Medical Microbiology and Parasitology, Program in Medical Microbiology, Athens, GA 30602

AWARDS MS, PhD.

Degree requirements: For master's, thesis required, foreign language not required; for doctorate, one foreign language (computer language can substitute), dissertation required.
Entrance requirements: For master's and doctorate, GRE General Test. Electronic applications accepted.
Expenses: Tuition, state resident: full-time $7,516; part-time $431 per credit hour. Tuition, nonresident: full-time $12,204; part-time $793 per credit hour.

■ UNIVERSITY OF HAWAII AT MANOA

John A. Burns School of Medicine and Graduate Division, Graduate Programs in Biomedical Sciences, Department of Tropical Medicine and Medical Microbiology, Honolulu, HI 96822

AWARDS Tropical medicine (MS, PhD). Part-time programs available.

Faculty: 13 full-time (3 women), 1 (woman) part-time/adjunct.
Students: 3 full-time (2 women). Average age 29. *5 applicants, 40% accepted.* In 1999,

1 doctorate awarded. Terminal master's awarded for partial completion of doctoral program.
Degree requirements: For master's, thesis required; for doctorate, dissertation required, foreign language not required. *Average time to degree:* Master's–1.5 years full-time.
Entrance requirements: For master's and doctorate, GRE. *Application deadline:* For fall admission, 3/1; for spring admission, 9/1. *Application fee:* $25 ($50 for international students).
Expenses: Tuition, state resident: part-time $168 per credit. Tuition, nonresident: part-time $415 per credit. Required fees: $51 per semester. Part-time tuition and fees vary according to course load.
Financial aid: In 1999–00, 1 research assistantship was awarded; fellowships, teaching assistantships, tuition waivers (full) also available.
Faculty research: Immunological studies of dengue, malaria, Kawasaki's disease, lupus erythematosus, and rheumatoid disease.
Dr. Karen Yamaga, Acting Chair, 808-732-1477, *Fax:* 808-732-1483, *E-mail:* yamaga@hawaii.edu.

■ UNIVERSITY OF MINNESOTA, DULUTH

School of Medicine, Department of Medical Microbiology and Immunology, Duluth, MN 55812-2496

AWARDS MS, PhD.

Faculty: 4 full-time (1 woman), 3 part-time/adjunct (1 woman).
Students: 1 (woman) full-time; minority (Asian American or Pacific Islander). *18 applicants, 0% accepted.* Terminal master's awarded for partial completion of doctoral program.
Degree requirements: For master's, thesis, final oral exam required, foreign language not required; for doctorate, dissertation, final exam, oral and written preliminary exams required, foreign language not required.
Entrance requirements: For master's and doctorate, GRE General Test, TOEFL. *Application deadline:* For fall admission, 5/1 (priority date). Applications are processed on a rolling basis. *Application fee:* $30.
Expenses: Tuition, state resident: full-time $5,040; part-time $420 per credit. Tuition, nonresident: full-time $9,900; part-time $825 per credit. Required fees: $509. Tuition and fees vary according to course load and program.
Financial aid: Research assistantships available.
Faculty research: Immunomodulation, molecular diagnosis of rabies, cytokines,

University of Minnesota, Duluth (continued)

cancer immunology, cytomegalovirus infection. *Total annual research expenditures:* $60,000.

Dr. Arthur Johnson, Head, 218-726-7561, *Fax:* 218-726-6235, *E-mail:* ajohnso1@d.umn.edu.

■ UNIVERSITY OF SOUTH FLORIDA

College of Medicine and Graduate School, Graduate Programs in Medical Sciences, Department of Medical Microbiology and Immunology, Tampa, FL 33620-9951

AWARDS PhD.

Degree requirements: For doctorate, dissertation required, foreign language not required.
Entrance requirements: For doctorate, GRE General Test, TOEFL, interview, minimum GPA of 3.0.
Expenses: Tuition, state resident: part-time $148 per credit hour. Tuition, nonresident: part-time $509 per credit hour.
Faculty research: Molecular genetics and pathogenesis of bacteria and viruses, molecular biology of DNA tumor viruses, immunomodulation by drugs and viruses, retrovirus vaccine development.

■ THE UNIVERSITY OF TENNESSEE HEALTH SCIENCE CENTER

College of Graduate Health Sciences, Department of Microbiology and Immunology, Memphis, TN 38163-0002

AWARDS Bacteriology (MS, PhD); immunology (MS, PhD); medical microbiology (MS, PhD); microbiology (MS, PhD); molecular and cell biology (PhD); molecular biology (MS, PhD); virology (MS, PhD). Terminal master's awarded for partial completion of doctoral program.

Degree requirements: For master's, thesis, oral and written comprehensive exams required, foreign language not required; for doctorate, dissertation, oral and written preliminary and comprehensive exams required, foreign language not required.
Entrance requirements: For master's, GRE General Test, TOEFL, minimum GPA of 3.0; for doctorate, GRE General Test, GRE Subject Test, TOEFL, minimum GPA of 3.0.

Find an in-depth description at www.petersons.com/graduate.

■ UNIVERSITY OF WASHINGTON

Graduate School, School of Public Health and Community Medicine, Department of Pathobiology, Seattle, WA 98195

AWARDS MS, PhD.

Faculty: 33 full-time (11 women), 9 part-time/adjunct (0 women).
Students: 24 full-time (18 women), 3 part-time (1 woman); includes 11 minority (1 African American, 8 Asian Americans or Pacific Islanders, 2 Hispanic Americans), 4 international. Average age 28. *45 applicants, 20% accepted.* In 1999, 3 master's awarded (33% found work related to degree, 67% continued full-time study); 8 doctorates awarded (100% entered university research/teaching). Terminal master's awarded for partial completion of doctoral program.
Degree requirements: For master's and doctorate, thesis/dissertation required, foreign language not required. *Average time to degree:* Master's–2 years full-time; doctorate–8 years full-time.
Entrance requirements: For master's and doctorate, GRE General Test, TOEFL, minimum GPA of 3.0. *Application deadline:* For fall admission, 1/1. *Application fee:* $50.
Expenses: Tuition, state resident: full-time $5,196; part-time $495 per credit. Tuition, nonresident: full-time $13,485; part-time $1,285 per credit. Required fees: $387; $36 per credit. Tuition and fees vary according to course load and program.
Financial aid: In 1999–00, 2 fellowships with tuition reimbursements (averaging $16,068 per year), 18 research assistantships with tuition reimbursements (averaging $16,068 per year) were awarded; career-related internships or fieldwork, Federal Work-Study, institutionally sponsored loans, traineeships, and tuition waivers (full and partial) also available. Financial aid application deadline: 3/1; financial aid applicants required to submit FAFSA.
Faculty research: Pathogenesis of chlamydiae, molecular biology of parasites, signal transduction, antigenic analysis, molecular biology of tumor viruses.

Dr. Kenneth Stuart, Chair, 206-543-8350, *Fax:* 206-543-3873, *E-mail:* kstuart@u.washington.edu.
Application contact: Leslie G. Miller, Coordinator, 206-543-4338, *Fax:* 206-543-3873, *E-mail:* pathiobio@u.washington.edu.

Find an in-depth description at www.petersons.com/graduate.

■ UNIVERSITY OF WISCONSIN–MADISON

Medical School and Graduate School, Graduate Programs in Medicine and College of Agricultural and Life Sciences, Microbiology Doctoral Training Program, Madison, WI 53706-1380

AWARDS PhD.

Faculty: 80 full-time (15 women).
Students: 44 full-time (38 women); includes 8 minority (2 African Americans, 2 Asian Americans or Pacific Islanders, 4 Hispanic Americans), 4 international. Average age 24. *269 applicants, 15% accepted.* In 1999, 6 degrees awarded.
Degree requirements: For doctorate, dissertation required, foreign language not required. *Average time to degree:* Doctorate–6 years full-time.
Entrance requirements: For doctorate, GRE, TOEFL. *Application deadline:* For fall admission, 1/1 (priority date). *Application fee:* $45. Electronic applications accepted.
Expenses: Tuition, state resident: full-time $5,406; part-time $339 per credit. Tuition, nonresident: full-time $17,110; part-time $1,071 per credit.
Financial aid: In 1999–00, 26 students received aid, including 15 fellowships with tuition reimbursements available, 12 research assistantships with tuition reimbursements available; grants, scholarships, traineeships, and tuition waivers (full) also available. *Total annual research expenditures:* $6.5 million.

Dr. Robert C. Landick, Director, 608-265-8475, *Fax:* 608-262-9865, *E-mail:* landick@bact.wisc.edu.
Application contact: Kathryn A. Holtgraver, Program Coordinator, 608-262-0689, *Fax:* 608-262-9865, *E-mail:* kathyh@bact.wisc.edu.

Find an in-depth description at www.petersons.com/graduate.

MICROBIOLOGY

■ ALBANY MEDICAL COLLEGE

Graduate Programs in the Biological Sciences, Program in Immunology and Microbial Disease, Albany, NY 12208-3479

AWARDS MS, PhD. Part-time programs available.

Students: 30 full-time (17 women); includes 2 minority (1 African American, 1 Asian American or Pacific Islander), 11 international. Terminal master's awarded for partial completion of doctoral program.

Degree requirements: For master's, thesis required, foreign language not required; for doctorate, dissertation, oral qualifying exam, written preliminary exam required, foreign language not required.
Entrance requirements: For master's and doctorate, GRE General Test, TOEFL. *Application deadline:* Applications are processed on a rolling basis. *Application fee:* $0 ($60 for international students). *
Expenses: Tuition: Full-time $13,367; part-time $446 per credit hour.
Financial aid: Federal Work-Study, grants, scholarships, and tuition waivers (full) available.
Dr. Jeffrey Banas, Graduate Director, 518-262-6286, *Fax:* 518-262-5748, *E-mail:* banasj@mail.amc.edu.

Find an in-depth description at www.petersons.com/graduate.

■ **ARIZONA STATE UNIVERSITY**

Graduate College, College of Liberal Arts and Sciences, Department of Microbiology, Tempe, AZ 85287

AWARDS Biology (MNS); cell and developmental biology (MS, PhD); microbiology (MNS, MS, PhD).

Faculty: 14 full-time (8 women), 1 part-time/adjunct (0 women).
Students: 23 full-time (14 women), 8 part-time (4 women); includes 4 minority (2 African Americans, 1 Hispanic American, 1 Native American), 6 international. Average age 28. *34 applicants, 47% accepted.* In 1999, 3 master's, 1 doctorate awarded.
Degree requirements: For doctorate, one foreign language, dissertation required.
Entrance requirements: For master's and doctorate, GRE. *Application fee:* $45.
Expenses: Tuition, state resident: part-time $115 per credit hour. Tuition, nonresident: part-time $389 per credit hour. Required fees: $18 per semester. Tuition and fees vary according to program.
Faculty research: Bacterial enzymology, bacterial genetics, immunology, host-parasite relationships, medical molecular biology, physiology.
Dr. Edward A. Birge, Chair, 480-965-1457.

■ **AUBURN UNIVERSITY**

Graduate School, College of Sciences and Mathematics, Department of Biological Sciences, Auburn, Auburn University, AL 36849-0002

AWARDS Botany (MS, PhD); microbiology (MS, PhD); zoology (MS, PhD).

Faculty: 31 full-time (6 women).
Students: 38 full-time (23 women), 35 part-time (18 women); includes 6 minority (5 African Americans, 1 Hispanic

American), 11 international. *66 applicants, 48% accepted.* In 1999, 12 master's, 1 doctorate awarded.
Entrance requirements: For master's and doctorate, GRE General Test, TOEFL. *Application deadline:* For fall admission, 7/7; for spring admission, 11/24. Electronic applications accepted.
Expenses: Tuition, state resident: full-time $2,895; part-time $80 per credit hour. Tuition, nonresident: full-time $8,685; part-time $240 per credit hour.
Financial aid: Research assistantships, teaching assistantships available.
Dr. Alfred E. Brown, Interim Chair, 334-844-4830, *Fax:* 334-844-1645.

Find an in-depth description at www.petersons.com/graduate.

■ **BAYLOR COLLEGE OF MEDICINE**

Graduate School of Biomedical Sciences, Department of Molecular Virology and Microbiology, Houston, TX 77030-3498

AWARDS PhD, MD/PhD.

Faculty: 42 full-time (15 women).
Students: 31 full-time (22 women), 10 international. Average age 28. *64 applicants, 19% accepted.* In 1999, 3 doctorates awarded (100% entered university research/teaching).
Degree requirements: For doctorate, dissertation, public defense, qualifying exam required, foreign language not required. *Average time to degree:* Doctorate–5.92 years full-time.
Entrance requirements: For doctorate, GRE General Test (average 80th percentile), GRE Subject Test (strongly recommended), TOEFL, minimum GPA of 3.0. *Application deadline:* For fall admission, 2/1 (priority date). Applications are processed on a rolling basis. *Application fee:* $30. Electronic applications accepted.
Expenses: Tuition: Full-time $8,200. Required fees: $175. Full-time tuition and fees vary according to student level.
Financial aid: In 1999–00, 31 students received aid, including 15 fellowships (averaging $16,000 per year), 14 research assistantships (averaging $16,000 per year), 2 teaching assistantships (averaging $16,000 per year); Federal Work-Study, institutionally sponsored loans, and tuition waivers (full) also available. Financial aid applicants required to submit FAFSA.
Faculty research: Molecular biology of virus replication, viruses and cancer, viral genetics, viral infectious diseases, environmental virology.

Application contact: Beatrice Torres, Graduate Program Administrator, 713-798-4472, *Fax:* 713-798-5075, *E-mail:* bft@bmc.tmc.edu.

Find an in-depth description at www.petersons.com/graduate.

■ **BOSTON UNIVERSITY**

School of Medicine, Division of Graduate Medical Sciences, Department of Microbiology, Boston, MA 02118

AWARDS MA, PhD, MD/PhD.

Faculty: 12 full-time (3 women), 12 part-time/adjunct (2 women).
Students: 22 full-time (14 women); includes 2 minority (1 African American, 1 Asian American or Pacific Islander), 5 international. Average age 28. In 1999, 1 master's, 3 doctorates awarded. Terminal master's awarded for partial completion of doctoral program.
Degree requirements: For master's, thesis required, foreign language not required; for doctorate, dissertation, comprehensive exam required, foreign language not required.
Entrance requirements: For master's and doctorate, GRE General Test, GRE Subject Test, TOEFL. *Application fee:* $45. Electronic applications accepted.
Expenses: Tuition: Full-time $24,700; part-time $772 per credit. Required fees: $220.
Financial aid: Fellowships, research assistantships, Federal Work-Study, scholarships, and traineeships available.
Faculty research: Eukaryotic cell biology, tumor cell biology, nutrition and cancer, experimental tumor therapy, photobiology.
Dr. Ronald B. Corley, Chairman, 617-638-4284, *Fax:* 617-638-4286, *E-mail:* rbcorley@bu.edu.
Application contact: Dr. Gregory Viglianti, Graduate Director, 617-638-7790, *Fax:* 617-638-4286, *E-mail:* gviglian@bu.edu.

Find an in-depth description at www.petersons.com/graduate.

■ **BRANDEIS UNIVERSITY**

Graduate School of Arts and Sciences, Programs in Biological Sciences, Program in Molecular and Cell Biology, Waltham, MA 02454-9110

AWARDS Cell biology (PhD); developmental biology (PhD); genetics (PhD); microbiology (PhD); molecular and cell biology (MS); molecular biology (PhD); neurobiology (PhD).

Faculty: 19 full-time (8 women).
Students: 58 full-time (31 women), 1 (woman) part-time; includes 13 minority (1 African American, 8 Asian Americans or Pacific Islanders, 4 Hispanic Americans), 8

Brandeis University (continued)
international. Average age 27. *83 applicants, 11% accepted.* In 1999, 1 master's, 8 doctorates awarded (100% entered university research/teaching). Terminal master's awarded for partial completion of doctoral program.

Degree requirements: For master's, thesis not required; for doctorate, dissertation required. *Average time to degree:* Master's–1 year full-time; doctorate–6 years full-time.

Entrance requirements: For doctorate, GRE General Test. *Application deadline:* For fall admission, 1/15 (priority date). Applications are processed on a rolling basis. *Application fee:* $60. Electronic applications accepted.

Expenses: Tuition: Full-time $25,392; part-time $3,174 per course. Required fees: $509. Tuition and fees vary according to class time, degree level, program and student level.

Financial aid: In 1999–00, 10 fellowships with tuition reimbursements (averaging $17,000 per year), 49 research assistantships with tuition reimbursements (averaging $17,000 per year), 9 teaching assistantships (averaging $2,312 per year) were awarded; scholarships and tuition waivers (full and partial) also available. Financial aid application deadline: 4/15; financial aid applicants required to submit CSS PROFILE or FAFSA.

Faculty research: Regulation of gene expression by transcription factors, molecular neurobiology, immunology, molecular mechanisms of genetic recombination, and cell differentiation.
Dr. Ranjan Sen, Chair, 781-736-2455, *Fax:* 781-736-3107.

Application contact: Marcia Cabral, Information Officer, 781-736-3100, *Fax:* 781-736-3107, *E-mail:* cabral@brandeis.edu.

■ BRIGHAM YOUNG UNIVERSITY

Graduate Studies, College of Biological and Agricultural Sciences, Department of Microbiology, Provo, UT 84602-1001

AWARDS MS, PhD.

Faculty: 12 full-time (1 woman).
Students: 27 full-time (8 women); includes 2 minority (both Asian Americans or Pacific Islanders), 1 international. Average age 26. *25 applicants, 56% accepted.* In 1999, 6 master's awarded (100% found work related to degree); 2 doctorates awarded (100% continued full-time study). Terminal master's awarded for partial completion of doctoral program.

Degree requirements: For master's and doctorate, thesis/dissertation required, foreign language not required. *Average*

time to degree: Master's–2 years full-time; doctorate–5 years full-time.

Entrance requirements: For master's, GRE General Test, GRE Subject Test, minimum GPA of 3.0 during previous 2 years; for doctorate, GRE General Test, GRE Subject Test. *Application deadline:* For fall admission, 2/1 (priority date). *Application fee:* $30.

Expenses: Tuition: Full-time $3,330; part-time $185 per credit hour. Tuition and fees vary according to program and student's religious affiliation.

Financial aid: In 1999–00, 18 students received aid, including 18 research assistantships with tuition reimbursements available (averaging $11,300 per year), 2 teaching assistantships; institutionally sponsored loans and tuition waivers (partial) also available. Financial aid application deadline: 2/1.

Faculty research: Immunobiology, environmental microbiology, molecular genetics, molecular virology, tumor biology.
Dr. Shauna C. Anderson, Chair, 801-378-2889, *Fax:* 801-378-9197, *E-mail:* shauna_anderson@byu.edu.

Application contact: Dr. Kim L. O'Neill, Graduate Coordinator, 801-378-2449, *Fax:* 801-378-9197, *E-mail:* kim_oneill@byu.edu.

■ BROWN UNIVERSITY

Graduate School, Division of Biology and Medicine, Program in Molecular Biology, Cell Biology, and Biochemistry, Providence, RI 02912

AWARDS Biochemistry (M Med Sc, Sc M, PhD), including biochemistry (Sc M, PhD), biology (Sc M, PhD), medical science (M Med Sc, PhD); biology (MA); cell biology (M Med Sc, Sc M, PhD), including biochemistry (Sc M, PhD), biology (Sc M, PhD), medical science (M Med Sc, PhD); developmental biology (M Med Sc, Sc M, PhD), including biochemistry (Sc M, PhD), biology (Sc M, PhD), medical science (M Med Sc, PhD); immunology (M Med Sc, Sc M, PhD), including biochemistry (Sc M, PhD), biology (Sc M, PhD), medical science (M Med Sc, PhD); molecular microbiology (M Med Sc, Sc M, PhD), including biochemistry (Sc M, PhD), biology (Sc M, PhD), medical science (M Med Sc, PhD). Part-time programs available.

Faculty: 50 full-time (14 women).
Students: 61 full-time (34 women); includes 4 minority (1 African American, 3 Asian Americans or Pacific Islanders), 21 international. Average age 25. *106 applicants, 28% accepted.* In 1999, 1 master's, 3 doctorates awarded. Terminal master's awarded for partial completion of doctoral program.

Degree requirements: For master's, thesis required (for some programs), foreign language not required; for doctorate, one foreign language, dissertation, preliminary exam required. *Average time to degree:* Doctorate–5 years full-time.

Entrance requirements: For master's and doctorate, GRE General Test, GRE Subject Test. *Application deadline:* For fall admission, 1/2 (priority date). Applications are processed on a rolling basis. *Application fee:* $60. Electronic applications accepted.

Financial aid: In 1999–00, 58 students received aid, including 11 fellowships (averaging $18,916 per year), 9 research assistantships (averaging $18,916 per year), 13 teaching assistantships (averaging $12,690 per year); institutionally sponsored loans and traineeships also available. Financial aid application deadline: 1/2.

Faculty research: Molecular genetics, gene regulation.
Dr. Gary Wessel, Director, 401-863-1051, *E-mail:* chet@brown.edu.

Application contact: Mary C. Esser, Graduate Program Coordinator, 401-863-1661, *Fax:* 401-863-1348, *E-mail:* mary_esser@brown.edu.

Find an in-depth description at www.petersons.com/graduate.

■ CALIFORNIA STATE UNIVERSITY, FULLERTON

Graduate Studies, College of Natural Science and Mathematics, Department of Biological Science, Fullerton, CA 92834-9480

AWARDS Biological science (MA); botany (MA); microbiology (MA). Part-time programs available.

Faculty: 23 full-time (7 women), 41 part-time/adjunct.
Students: 4 full-time (all women), 60 part-time (35 women); includes 20 minority (1 African American, 8 Asian Americans or Pacific Islanders, 10 Hispanic Americans, 1 Native American), 3 international. Average age 29. *51 applicants, 45% accepted.* In 1999, 13 degrees awarded.

Degree requirements: For master's, thesis required, foreign language not required.

Entrance requirements: For master's, DAT, GRE General Test and GRE Subject Test, or MCAT, minimum GPA of 3.0 in biology. *Application fee:* $55.

Expenses: Tuition, nonresident: part-time $264 per unit. Required fees: $1,887; $629 per year.

Financial aid: Teaching assistantships, career-related internships or fieldwork, Federal Work-Study, grants, and institutionally sponsored loans available. Aid available to part-time students. Financial aid application deadline: 3/1.

Faculty research: Glycosidase release and the block to polyspermy in ascidian eggs. Dr. Eugene Jones, Chair, 714-278-3614. **Application contact:** Dr. Michael Horn, Adviser, 714-278-3707.

■ CALIFORNIA STATE UNIVERSITY, LONG BEACH

Graduate Studies, College of Natural Sciences, Department of Biological Sciences, Program in Microbiology, Long Beach, CA 90840

AWARDS Medical technology (MPH); microbiology (MS); nurse epidemiology (MPH). Part-time programs available.

Faculty: 11 full-time.

Students: 1 full-time (0 women), 7 part-time (3 women); includes 4 minority (3 Asian Americans or Pacific Islanders, 1 Hispanic American). Average age 28. *18 applicants, 17% accepted.* In 1999, 1 degree awarded.

Degree requirements: For master's, thesis required (for some programs), foreign language not required.

Entrance requirements: For master's, GRE Subject Test, minimum GPA of 3.0. *Application deadline:* For fall admission, 8/1; for spring admission, 12/1. Applications are processed on a rolling basis. *Application fee:* $55. Electronic applications accepted.

Expenses: Tuition, nonresident: part-time $246 per credit. Required fees: $569 per semester. Tuition and fees vary according to course load.

Financial aid: Teaching assistantships, Federal Work-Study, grants, institutionally sponsored loans, and unspecified assistantships available. Financial aid application deadline: 3/2.

Dr. Terry Shuster, Coordinator, 562-985-4820, *Fax:* 562-985-8878, *E-mail:* tshuster@csulb.edu.

■ CALIFORNIA STATE UNIVERSITY, LOS ANGELES

Graduate Studies, School of Natural and Social Sciences, Department of Biology and Microbiology, Los Angeles, CA 90032-8530

AWARDS Biology (MS). Part-time and evening/weekend programs available.

Faculty: 19 full-time, 24 part-time/adjunct.

Students: 22 full-time (10 women), 54 part-time (25 women); includes 36 minority (4 African Americans, 14 Asian Americans or Pacific Islanders, 18 Hispanic Americans), 12 international. In 1999, 6 degrees awarded.

Degree requirements: For master's, comprehensive exam or thesis required.

Entrance requirements: For master's, TOEFL. *Application deadline:* For fall

admission, 6/30; for spring admission, 2/1. Applications are processed on a rolling basis. *Application fee:* $55.

Expenses: Tuition, nonresident: full-time $7,703; part-time $164 per unit. Required fees: $1,799; $387 per quarter.

Financial aid: In 1999–00, 37 students received aid. Federal Work-Study available. Aid available to part-time students. Financial aid application deadline: 3/1.

Faculty research: Ecology, environmental biology, cell and molecular biology, physiology, medical microbiology.

Dr. Alan Muchlinski, Chair, 323-343-2050.

■ CASE WESTERN RESERVE UNIVERSITY

School of Medicine and School of Graduate Studies, Graduate Programs in Medicine, Department of Molecular Biology and Microbiology, Cleveland, OH 44106-4960

AWARDS Cellular biology (PhD); microbiology (PhD); molecular biology (PhD). Students are admitted to an integrated Biomedical Sciences Training Program involving 11 basic science programs at Case Western Reserve University.

Faculty: 14 full-time (4 women).

Students: 24 full-time (9 women); includes 2 minority (both Asian Americans or Pacific Islanders), 5 international. In 1999, 1 doctorate awarded (100% found work related to degree).

Degree requirements: For doctorate, dissertation required, foreign language not required. *Average time to degree:* Doctorate–5.3 years full-time.

Entrance requirements: For doctorate, GRE General Test, GRE Subject Test, TOEFL. *Application fee:* $25.

Financial aid: In 1999–00, 1 fellowship with full tuition reimbursement (averaging $16,000 per year), 23 research assistantships with full tuition reimbursements (averaging $16,000 per year) were awarded; Federal Work-Study, scholarships, traineeships, and tuition waivers (full) also available.

Faculty research: Gene expression in eukaryotic and prokaryotic systems; microbial physiology; intracellular transport and signaling; mechanisms of oncogenesis; molecular mechanisms of RNA processing, editing, and catalysis. *Total annual research expenditures:* $4.6 million.

Lloyd A. Culp, Acting Chair, 216-368-3420, *Fax:* 216-368-3055, *E-mail:* lac7@po.cwru.edu.

Application contact: Dr. Michael E. Harris, Admissions Coordinator, 216-368-3347, *E-mail:* mbio@po.cwru.edu.

Find an in-depth description at www.petersons.com/graduate.

■ THE CATHOLIC UNIVERSITY OF AMERICA

School of Arts and Sciences, Department of Biology, Program in Cell and Microbial Biology, Washington, DC 20064

AWARDS Cell biology (MS, PhD); microbiology (MS, PhD). Part-time programs available.

Students: 5 full-time (4 women), 4 part-time (2 women); includes 2 minority (1 African American, 1 Hispanic American), 2 international. Average age 26. *23 applicants, 57% accepted.* Terminal master's awarded for partial completion of doctoral program.

Degree requirements: For master's, thesis or alternative, comprehensive exam required, foreign language not required; for doctorate, dissertation, comprehensive exam required, foreign language not required.

Entrance requirements: For master's, GRE General Test, GRE Subject Test, TOEFL; for doctorate, GRE General Test, GRE Subject Test. *Application deadline:* For fall admission, 8/1 (priority date); for spring admission, 12/1. Applications are processed on a rolling basis. *Application fee:* $55. Electronic applications accepted.

Expenses: Tuition: Full-time $18,200; part-time $700 per credit hour. Required fees: $378 per semester. Part-time tuition and fees vary according to campus/location and program.

Financial aid: Fellowships, research assistantships, teaching assistantships, career-related internships or fieldwork, institutionally sponsored loans, and tuition waivers (full and partial) available. Aid available to part-time students. Financial aid application deadline: 2/1.

Faculty research: Cell differentiation, regulation of cell growth, drug resistance, gene cloning and sequencing, developmental biology and neurobiology. *Total annual research expenditures:* $230,000.

Dr. John Golin, Chair, Department of Biology, 202-319-5279, *Fax:* 202-319-5721.

■ CLEMSON UNIVERSITY

Graduate School, College of Agriculture, Forestry and Life Sciences, School of Animal, Biomedical and Biological Sciences, Department of Microbiology and Molecular Medicine, Clemson, SC 29634

AWARDS Microbiology (MS, PhD).

Students: 28 full-time (12 women), 6 part-time (3 women), 7 international. Average age 26. *62 applicants, 34% accepted.* In 1999, 7 master's, 5 doctorates awarded.

Clemson University (continued)

Degree requirements: For master's and doctorate, thesis/dissertation required, foreign language not required.

Entrance requirements: For master's and doctorate, GRE General Test, TOEFL. *Application deadline:* For fall admission, 6/1. *Application fee:* $40.

Expenses: Tuition, state resident: full-time $3,480; part-time $174 per credit hour. Tuition, nonresident: full-time $9,256; part-time $388 per credit hour. Required fees: $5 per term. Full-time tuition and fees vary according to course level, course load and campus/location.

Financial aid: Research assistantships, teaching assistantships available. Financial aid application deadline: 3/1; financial aid applicants required to submit FAFSA.

Faculty research: Anaerobic microbiology, microbiology and ecology of soil and aquatic systems, genetic engineering, monoclonal antibodies and immunomodulation.

Dr. Steven S. Hayasaka, Chair, 864-656-5432, *Fax:* 864-656-1127, *E-mail:* hayasas@clemson.edu.

Application contact: Dr. Malcolm J. B. Paynter, Graduate Coordinator, 864-656-3058, *Fax:* 864-656-0245, *E-mail:* pmalcol@clemson.edu.

■ COLORADO STATE UNIVERSITY

College of Veterinary Medicine and Biomedical Sciences and Graduate School, Graduate Programs in Veterinary Medicine and Biomedical Sciences, Department of Microbiology, Fort Collins, CO 80523-0015

AWARDS Immunology (MS, PhD); microbiology (MS, PhD).

Faculty: 27 full-time (8 women).
Students: 42 full-time (26 women), 6 part-time (5 women); includes 6 minority (1 African American, 2 Asian Americans or Pacific Islanders, 3 Hispanic Americans), 7 international. Average age 28. *103 applicants, 21% accepted.* In 1999, 8 master's, 5 doctorates awarded.

Degree requirements: For master's, thesis required, foreign language not required; for doctorate, dissertation required.

Entrance requirements: For master's and doctorate, GRE General Test, TOEFL, minimum GPA of 3.0. *Application deadline:* For fall admission, 2/1 (priority date); for spring admission, 10/1 (priority date). Applications are processed on a rolling basis. *Application fee:* $30. Electronic applications accepted.

Expenses: Tuition, state resident: full-time $2,694; part-time $150 per credit. Tuition,

nonresident: full-time $10,460; part-time $581 per credit. Required fees: $32 per semester.

Financial aid: In 1999–00, 3 fellowships, 35 research assistantships, 4 teaching assistantships were awarded; traineeships also available.

Faculty research: Medical and veterinary bacteriology, virology, industrial and environmental microbiology, vector-borne disease. *Total annual research expenditures:* $5 million.

Ralph E. Smith, Head.

Application contact: Graduate Coordinator, 970-491-6136, *Fax:* 970-491-1815, *E-mail:* microbio@cvmbs.colostate.edu.

Find an in-depth description at www.petersons.com/graduate.

■ COLUMBIA UNIVERSITY

College of Physicians and Surgeons and Graduate School of Arts and Sciences, Graduate School of Arts and Sciences at the College of Physicians and Surgeons, Department of Microbiology, New York, NY 10032

AWARDS Biomedical sciences (M Phil, MA, PhD). Only candidates for the PhD are admitted. Terminal master's awarded for partial completion of doctoral program.

Degree requirements: For master's, foreign language and thesis not required; for doctorate, dissertation required, foreign language not required.

Entrance requirements: For master's, GRE General Test, TOEFL; for doctorate, GRE, TOEFL.

Expenses: Tuition: Full-time $25,072.

Faculty research: Prokaryotic molecular biology, immunology, virology, yeast molecular genetics, regulation of gene expression.

Find an in-depth description at www.petersons.com/graduate.

■ CORNELL UNIVERSITY

Graduate School, Graduate Fields of Agriculture and Life Sciences, Field of Microbiology, Ithaca, NY 14853-0001

AWARDS PhD.

Faculty: 30 full-time.
Students: 32 full-time (20 women); includes 2 minority (1 Asian American or Pacific Islander, 1 Hispanic American), 14 international. *91 applicants, 21% accepted.* In 1999, 3 doctorates awarded.

Degree requirements: For doctorate, dissertation, 2 semesters of teaching experience required, foreign language not required.

Entrance requirements: For doctorate, GRE General Test, TOEFL. *Application*

deadline: For fall admission, 1/15. *Application fee:* $65. Electronic applications accepted.

Expenses: Tuition: Full-time $12,100.

Financial aid: In 1999–00, 31 students received aid, including 7 fellowships with full tuition reimbursements available, 18 research assistantships with full tuition reimbursements available, 6 teaching assistantships with full tuition reimbursements available; institutionally sponsored loans, scholarships, tuition waivers (full and partial), and unspecified assistantships also available. Financial aid applicants required to submit FAFSA.

Faculty research: Microbial diversity, molecular biology, biotechnology, microbial ecology, phytobacteriology.

Application contact: Graduate Field Assistant, 607-255-3088, *E-mail:* microfield@cornell.edu.

■ DUKE UNIVERSITY

Graduate School, Department of Microbiology, Durham, NC 27708-0586

AWARDS PhD.

Faculty: 22 full-time, 4 part-time/adjunct.
Students: 16 full-time (6 women); includes 2 minority (both African Americans), 3 international. *44 applicants, 27% accepted.* In 1999, 3 doctorates awarded.

Degree requirements: For doctorate, dissertation required, foreign language not required.

Entrance requirements: For doctorate, GRE General Test, GRE Subject Test (recommended). *Application deadline:* For fall admission, 12/31. *Application fee:* $75.

Expenses: Tuition: Full-time $21,406; part-time $760 per unit. Required fees: $3,136; $3,136 per year. One-time fee: $30. Tuition and fees vary according to program.

Financial aid: Fellowships, research assistantships, Federal Work-Study available. Financial aid application deadline: 12/31.

Dr. David J. Pickup, Director of Graduate Studies, 919-684-4657, *Fax:* 919-684-8735, *E-mail:* akuehn@abacus.mc.duke.edu.

Find an in-depth description at www.petersons.com/graduate.

■ EAST CAROLINA UNIVERSITY

School of Medicine, Department of Microbiology and Immunology, Greenville, NC 27858-4353

AWARDS PhD.

Faculty: 12 full-time (1 woman), 1 part-time/adjunct (0 women).
Students: 6 full-time (2 women), 7 part-time (2 women); includes 1 minority (Asian American or Pacific Islander), 3 international. Average age 26. *7 applicants, 86% accepted.* In 1999, 1 degree awarded.

Degree requirements: For doctorate, dissertation required, foreign language not required.

Entrance requirements: For doctorate, GRE General Test, GRE Subject Test, TOEFL. *Application deadline:* For fall admission, 4/15 (priority date). Applications are processed on a rolling basis. *Application fee:* $40.

Expenses: Tuition, state resident: full-time $1,012. Tuition, nonresident: full-time $8,578. Required fees: $1,006. Full-time tuition and fees vary according to degree level. Part-time tuition and fees vary according to course load.

Financial aid: Fellowships available. Financial aid application deadline: 6/1.

Faculty research: Molecular virology, genetics of bacteria, yeast and somatic cells, bacterial physiology and metabolism. Dr. Paul V. Phibbs, Chairman, 252-816-2700, *Fax:* 252-816-3104, *E-mail:* phibbs@brody.med.ecu.edu.

Application contact: Dr. Henry Stone, Director of Graduate Studies, 252-816-2700, *Fax:* 252-816-3104, *E-mail:* stone@brody.med.ecu.edu.

■ EAST TENNESSEE STATE UNIVERSITY

James H. Quillen College of Medicine and School of Graduate Studies, Biomedical Science Graduate Program, Johnson City, TN 37614

AWARDS Anatomy and cell biology (MS, PhD); biochemistry and molecular biology (MS, PhD); biophysics (MS, PhD); microbiology (MS, PhD); pharmacology (MS, PhD); physiology (MS, PhD). Part-time programs available.

Faculty: 40 full-time (9 women).
Students: 23 full-time (10 women), 2 part-time (1 woman); includes 1 minority (Asian American or Pacific Islander), 4 international. Average age 30. *83 applicants, 20% accepted.* In 1999, 5 master's, 4 doctorates awarded. Terminal master's awarded for partial completion of doctoral program.
Degree requirements: For master's, one foreign language (computer language can substitute), thesis, comprehensive qualifying exam required; for doctorate, 2 foreign languages (computer language can substitute for one), dissertation required.
Entrance requirements: For master's, GRE General Test, minimum GPA of 3.0, bachelor's degree in biological or related science; for doctorate, GRE General Test, GRE Subject Test. *Application deadline:* For fall admission, 3/15 (priority date); for spring admission, 3/1. *Application fee:* $25 ($35 for international students).

Expenses: Tuition, state resident: full-time $10,342. Tuition, nonresident: full-time $21,080. Required fees: $532.
Financial aid: In 1999–00, 16 research assistantships, 5 teaching assistantships were awarded; fellowships, career-related internships or fieldwork, Federal Work-Study, grants, institutionally sponsored loans, and tuition waivers (full) also available.
Dr. Mitchell Robinson, Assistant Dean for Graduate Studies, 423-439-4658, *E-mail:* robinson@etsu.edu.

■ EMORY UNIVERSITY

Graduate School of Arts and Sciences, Division of Biological and Biomedical Sciences, Program in Microbiology and Molecular Genetics, Atlanta, GA 30322-1100

AWARDS PhD.

Faculty: 20 full-time (3 women).
Students: 30 full-time (15 women); includes 1 Native American, 3 international. In 1999, 7 degrees awarded.
Degree requirements: For doctorate, dissertation required, foreign language not required.
Entrance requirements: For doctorate, GRE General Test, TOEFL, minimum GPA of 3.0 in science course work. *Application deadline:* For fall admission, 1/20 (priority date). *Application fee:* $45.
Expenses: Tuition: Full-time $22,770. Tuition and fees vary according to program.
Financial aid: In 1999–00, fellowships with full tuition reimbursements (averaging $18,000 per year)
Dr. Charles Moran, Director, 404-727-5969, *Fax:* 404-727-3659, *E-mail:* moran@microbio.emory.edu.
Application contact: 404-727-2547, *Fax:* 404-727-3322, *E-mail:* gdbbs@gsas.emory.edu.

Find an in-depth description at www.petersons.com/graduate.

■ EMPORIA STATE UNIVERSITY

School of Graduate Studies, College of Liberal Arts and Sciences, Division of Biological Sciences, Emporia, KS 66801-5087

AWARDS Botany (MS); environmental biology (MS); general biology (MS); microbial and cellular biology (MS); zoology (MS). Part-time programs available.

Faculty: 15 full-time (3 women), 4 part-time/adjunct (0 women).
Students: 20 full-time (10 women), 8 part-time (3 women), 2 international. *8 applicants, 75% accepted.* In 1999, 12 degrees awarded.

Degree requirements: For master's, comprehensive exam or thesis required.
Entrance requirements: For master's, TOEFL, written exam. *Application deadline:* For fall admission, 8/15 (priority date). Applications are processed on a rolling basis. *Application fee:* $30 ($75 for international students). Electronic applications accepted.
Expenses: Tuition, state resident: full-time $2,410; part-time $108 per credit hour. Tuition, nonresident: full-time $6,212; part-time $266 per credit hour.
Financial aid: In 1999–00, 2 fellowships (averaging $1,396 per year), 1 research assistantship (averaging $5,390 per year), 11 teaching assistantships with full tuition reimbursements (averaging $5,047 per year) were awarded; career-related internships or fieldwork, Federal Work-Study, and institutionally sponsored loans also available. Financial aid application deadline: 3/15; financial aid applicants required to submit FAFSA.
Faculty research: Fisheries, range, and wildlife management; aquatic, plant, grassland, vertebrate, and invertebrate ecology; mammalian and plant systematics, taxonomy, and evolution; immunology, virology, and molecular biology.
Dr. Marshall Sundberg, Chair, 316-341-5311, *Fax:* 316-341-5607, *E-mail:* sundberm@emporia.edu.

■ FINCH UNIVERSITY OF HEALTH SCIENCES/THE CHICAGO MEDICAL SCHOOL

School of Graduate and Postdoctoral Studies, Department of Microbiology and Immunology, North Chicago, IL 60064-3095

AWARDS Medical microbiology (MS, PhD); microbiology and immunology (MS, PhD). Part-time programs available.

Faculty: 10 full-time, 1 part-time/adjunct.
Students: 16 full-time (3 women); includes 3 minority (all Asian Americans or Pacific Islanders), 8 international. In 1999, 5 doctorates awarded. Terminal master's awarded for partial completion of doctoral program.
Degree requirements: For master's and doctorate, thesis/dissertation required, foreign language not required.
Entrance requirements: For master's and doctorate, GRE General Test, TOEFL, TWE. *Application deadline:* For fall admission, 6/1 (priority date). Applications are processed on a rolling basis. *Application fee:* $25.
Expenses: Tuition: Full-time $14,054; part-time $391 per credit hour. Tuition and fees vary according to program.

Finch University of Health Sciences/The Chicago Medical School (continued)
Financial aid: In 1999–00, fellowships (averaging $15,500 per year); tuition waivers (full) also available. Financial aid application deadline: 6/9; financial aid applicants required to submit FAFSA.
Faculty research: Molecular biology, parasitology, virology.
Dr. Yoon B. Kim, Chairman, 847-578-3230.
Application contact: Dana Frederick, Admissions Officer, 847-578-3209.

Find an in-depth description at www.petersons.com/graduate.

■ FLORIDA STATE UNIVERSITY

Graduate Studies, College of Arts and Sciences, Department of Biological Science, Program in Microbiology, Tallahassee, FL 32306

AWARDS MS, PhD.

Faculty: 6 full-time (0 women).
Students: 8 full-time (6 women), 1 international.
Degree requirements: For master's and doctorate, thesis/dissertation, teaching experience required.
Entrance requirements: For master's, GRE General Test, TOEFL; for doctorate, GRE General Test, GRE Subject Test, TOEFL. *Application deadline:* For fall admission, 1/15; for spring admission, 10/15. *Application fee:* $20.
Expenses: Tuition, state resident: full-time $3,504; part-time $146 per credit hour. Tuition, nonresident: full-time $12,162; part-time $507 per credit hour. Tuition and fees vary according to program.
Financial aid: In 1999–00, fellowships with full tuition reimbursements (averaging $13,740 per year), research assistantships with full tuition reimbursements (averaging $13,740 per year), teaching assistantships with full tuition reimbursements (averaging $13,740 per year) were awarded. Financial aid application deadline: 1/15; financial aid applicants required to submit FAFSA.
Faculty research: Prokaryotes and eukaryotes in biotechnology, recombinant DNA technology, microbioecology, biochemistry.
Dr. Thomas C. S. Keller, Associate Professor and Associate Chairman, 850-644-3023, *Fax:* 850-644-9829.
Application contact: Judy Bowers, Coordinator, Graduate Affairs, 850-644-3023, *Fax:* 850-644-9829, *E-mail:* bowers@bio.fsu.edu.

■ GEORGETOWN UNIVERSITY

Graduate School of Arts and Sciences, Programs in Biomedical Sciences, Department of Microbiology and Immunology, Washington, DC 20057

AWARDS PhD, MD/PhD.

Degree requirements: For doctorate, dissertation, comprehensive exam required, foreign language not required.
Entrance requirements: For doctorate, GRE General Test, TOEFL.

■ GEORGIA STATE UNIVERSITY

College of Arts and Sciences, Department of Biology, Program in Applied and Environmental Microbiology, Atlanta, GA 30303-3083

AWARDS MS, PhD.

Degree requirements: For master's, one foreign language (computer language can substitute), thesis or alternative, exam required; for doctorate, dissertation required.
Entrance requirements: For master's and doctorate, GRE General Test, TOEFL, minimum GPA of 3.0. *Application deadline:* For fall admission, 7/18; for spring admission, 2/13. Applications are processed on a rolling basis. *Application fee:* $25.
Expenses: Tuition, state resident: full-time $2,896; part-time $121 per credit hour. Tuition, nonresident: full-time $11,584; part-time $483 per credit hour. Required fees: $228. Full-time tuition and fees vary according to course load and program.
Financial aid: Application deadline: 2/6.
Application contact: Latesha Morrison, Graduate Administrative Coordinator, 404-651-2759, *Fax:* 404-651-2509, *E-mail:* biolxm@langate.gsu.edu.

Find an in-depth description at www.petersons.com/graduate.

■ HARVARD UNIVERSITY

Graduate School of Arts and Sciences, Program in Biological and Biomedical Sciences, Department of Microbiology and Molecular Genetics, Boston, MA 02115

AWARDS PhD. Applications through the Program in Biological and Biomedical Sciences (BBS).

Degree requirements: For doctorate, dissertation required, foreign language not required.
Entrance requirements: For doctorate, GRE General Test, GRE Subject Test, TOEFL. *Application fee:* $60.
Expenses: Tuition: Full-time $22,054. Required fees: $711. Tuition and fees vary according to program.

Financial aid: Fellowships, research assistantships, teaching assistantships available. Financial aid application deadline: 1/1.
Faculty research: Bacterial physiology, cell biology, embryology.
Dr. John J. Mekalanos, Chairman, 617-432-2263.
Application contact: Leah Simons, Manager of Student Affairs, 617-432-0162.

Find an in-depth description at www.petersons.com/graduate.

■ HOWARD UNIVERSITY

Graduate School of Arts and Sciences, Department of Microbiology, Washington, DC 20059-0002

AWARDS PhD.

Faculty: 16.
Students: 18; includes 16 minority (15 African Americans, 1 Asian American or Pacific Islander), 2 international. In 1999, 3 degrees awarded.
Degree requirements: For doctorate, one foreign language (computer language can substitute), dissertation, comprehensive exam, qualifying exam, teaching experience required. *Average time to degree:* Doctorate–4 years full-time.
Entrance requirements: For doctorate, GRE General Test, TOEFL, minimum GPA of 3.0 in sciences. *Application deadline:* For fall admission, 4/1; for spring admission, 11/1. *Application fee:* $45.
Expenses: Tuition: Full-time $10,500; part-time $583 per credit hour. Required fees: $405; $203 per semester.
Financial aid: Fellowships, research assistantships, teaching assistantships, grants and institutionally sponsored loans available. Financial aid application deadline: 4/1.
Faculty research: Immunology, molecular and cellular microbiology, microbial genetics, microbial physiology, pathogenic bacteriology, medical mycology, medical parasitology, virology.
Dr. Georgia Dunston, Chair, 202-806-6284.
Application contact: Secretary, 202-806-6290, *Fax:* 202-806-4508.

Find an in-depth description at www.petersons.com/graduate.

■ IDAHO STATE UNIVERSITY

Office of Graduate Studies, College of Arts and Sciences, Department of Biological Sciences, Pocatello, ID 83209

AWARDS Biology (MS, DA, PhD); microbiology (MS); natural science (MNS).

Faculty: 37 full-time (10 women), 6 part-time/adjunct (2 women).

Students: 61 full-time (19 women), 10 part-time (5 women), 4 international. Average age 30. In 1999, 7 master's, 4 doctorates awarded.

Degree requirements: For master's, one foreign language (computer language can substitute), thesis required; for doctorate, 2 foreign languages (computer language can substitute for one), dissertation required.

Entrance requirements: For master's, GRE General Test; for doctorate, GRE General Test, GRE Subject Test. *Application deadline:* For fall admission, 7/1; for spring admission, 12/1. Applications are processed on a rolling basis. *Application fee:* $30.

Expenses: Tuition, nonresident: full-time $6,240; part-time $90 per credit. Required fees: $3,384; $147 per credit.

Financial aid: In 1999–00, 8 fellowships, 20 research assistantships, 19 teaching assistantships were awarded; Federal Work-Study and institutionally sponsored loans also available.

Faculty research: Ecology and evolutionary biology, plant and animal physiology, plant and animal developmental biology, immunology, molecular biology. *Total annual research expenditures:* $800,000.

Dr. Rod R. Seeley, Chairman, 208-282-3765, *Fax:* 208-282-4570.

Find an in-depth description at www.petersons.com/graduate.

■ ILLINOIS INSTITUTE OF TECHNOLOGY

Graduate College, Armour College of Engineering and Sciences, Department of Biological, Chemical and Physical Sciences, Biology Division, Chicago, IL 60616-3793

AWARDS Biochemistry (MS); biology (PhD); biotechnology (MS); cell biology (MS); microbiology (MS). Part-time and evening/weekend programs available.

Faculty: 8 full-time (0 women), 2 part-time/adjunct (0 women).

Students: 38 full-time (15 women), 80 part-time (52 women); includes 34 minority (27 African Americans, 5 Asian Americans or Pacific Islanders, 2 Hispanic Americans), 39 international. *155 applicants, 36% accepted.* In 1999, 10 master's, 2 doctorates awarded. Terminal master's awarded for partial completion of doctoral program.

Degree requirements: For master's, thesis (for some programs), comprehensive exam required, foreign language not required; for doctorate, dissertation, comprehensive exam required, foreign language not required.

Entrance requirements: For master's and doctorate, GRE General Test, TOEFL, minimum undergraduate GPA of 3.0.

Application deadline: For fall admission, 7/1; for spring admission, 11/1. Applications are processed on a rolling basis. *Application fee:* $30. Electronic applications accepted.

Expenses: Tuition: Part-time $590 per credit hour. Required fees: $100. Tuition and fees vary according to course load and program.

Financial aid: In 1999–00, 7 fellowships, 1 research assistantship, 5 teaching assistantships were awarded; Federal Work-Study, institutionally sponsored loans, scholarships, and unspecified assistantships also available. Aid available to part-time students. Financial aid application deadline: 3/1; financial aid applicants required to submit FAFSA.

Faculty research: Genetics, molecular biology.

Dr. Benjamin Stark, Associate Chair, 312-567-3980, *Fax:* 312-567-3494, *E-mail:* starkb@iit.edu.

Application contact: Dr. S. Mohammad Shahidehpour, Dean of Graduate College, 312-567-3024, *Fax:* 312-567-7517, *E-mail:* gradstu@alpha1.ais.iit.edu.

■ ILLINOIS STATE UNIVERSITY

Graduate School, College of Arts and Sciences, Department of Biological Sciences, Normal, IL 61790-2200

AWARDS Biological sciences (MS); biology (PhD); botany (PhD); ecology (PhD); genetics (PhD); microbiology (PhD); physiology (PhD); zoology (PhD). Part-time programs available.

Faculty: 24 full-time (4 women).

Students: 57 full-time (25 women), 27 part-time (10 women); includes 6 minority (1 African American, 3 Asian Americans or Pacific Islanders, 2 Hispanic Americans), 13 international. *58 applicants, 59% accepted.* In 1999, 12 master's, 3 doctorates awarded.

Degree requirements: For master's, thesis or alternative required; for doctorate, variable foreign language requirement (computer language can substitute for one), dissertation, 2 terms of residency required.

Entrance requirements: For master's, GRE General Test, minimum GPA of 2.6 in last 60 hours; for doctorate, GRE General Test. *Application deadline:* Applications are processed on a rolling basis. *Application fee:* $0.

Expenses: Tuition, state resident: full-time $2,526; part-time $105 per credit hour. Tuition, nonresident: full-time $7,578; part-time $316 per credit hour. Required fees: $1,082; $38 per credit hour. Tuition and fees vary according to course load and program.

Financial aid: In 1999–00, 8 research assistantships, 63 teaching assistantships were awarded; Federal Work-Study, tuition waivers (full), and unspecified assistantships also available. Financial aid application deadline: 4/1.

Faculty research: Phenotypic plasticity in reproduction: molecular mechanisms, physiological control, adaptive significance; molecular stress physiology of *Listeria monocytogenes*; enzymology of eggshell formation in *Schistosoma mansoni*, compensatory adaptation in the rat midbrain after neurodegeneration, analysis of staphylococcal virulence germs. *Total annual research expenditures:* $586,648.

Dr. Hou Cheung, Chairperson, 309-438-3669.

Application contact: Derek A. McCracken, Graduate Adviser, 309-438-3664.

Find an in-depth description at www.petersons.com/graduate.

■ INDIANA STATE UNIVERSITY

School of Graduate Studies, College of Arts and Sciences, Department of Life Sciences, Terre Haute, IN 47809-1401

AWARDS Clinical laboratory sciences (MS); ecology (MA, MS, PhD); microbiology (MA, MS, PhD); physiology (MA, MS, PhD).

Degree requirements: For doctorate, computer language, dissertation required.

Entrance requirements: For master's and doctorate, GRE General Test. Electronic applications accepted.

Expenses: Tuition, state resident: full-time $3,552; part-time $148 per hour. Tuition, nonresident: full-time $8,088; part-time $337 per hour.

Find an in-depth description at www.petersons.com/graduate.

■ INDIANA UNIVERSITY BLOOMINGTON

Graduate School, College of Arts and Sciences, Department of Biology, Program in Microbiology, Bloomington, IN 47405

AWARDS MA, PhD. PhD offered through the University Graduate School. Part-time programs available.

Students: 12 full-time (6 women); includes 3 minority (1 African American, 1 Asian American or Pacific Islander, 1 Hispanic American). In 1999, 3 master's, 2 doctorates awarded.

Degree requirements: For master's, thesis required, foreign language not required; for doctorate, dissertation required.

Entrance requirements: For master's and doctorate, GRE General Test, TOEFL. *Application deadline:* For fall admission, 1/5 (priority date); for spring admission, 9/1 (priority date). Applications are processed

Indiana University Bloomington (continued)

on a rolling basis. *Application fee:* $45. Electronic applications accepted.

Expenses: Tuition, state resident: full-time $3,853; part-time $161 per credit hour. Tuition, nonresident: full-time $11,226; part-time $468 per credit hour. Required fees: $360 per year. Tuition and fees vary according to course load and program.

Financial aid: Fellowships with tuition reimbursements, research assistantships with tuition reimbursements, teaching assistantships with tuition reimbursements available. Financial aid application deadline: 1/15.

Faculty research: Fungal ecology, bacterial photogenesis, microbial genetics.

Dr. Carl Bauer, Head, 812-855-6595, *Fax:* 812-855-6705, *E-mail:* cbauer@bio.indiana.edu.

Application contact: Gretchen Clearwater, Advisor for Graduate Affairs, 812-855-1861, *Fax:* 812-855-6705, *E-mail:* biograd@bio.indiana.edu.

Find an in-depth description at www.petersons.com/graduate.

■ INDIANA UNIVERSITY–PURDUE UNIVERSITY INDIANAPOLIS

School of Medicine, Graduate Programs in Medicine, Department of Microbiology and Immunology, Indianapolis, IN 46202-2896

AWARDS MS, PhD, MD/MS, MD/PhD.

Students: 18 full-time (12 women), 7 part-time (3 women); includes 3 minority (2 African Americans, 1 Hispanic American), 6 international. Average age 25. *72 applicants, 21% accepted.* In 1999, 1 master's awarded (100% found work related to degree); 5 doctorates awarded. Terminal master's awarded for partial completion of doctoral program.

Degree requirements: For master's and doctorate, thesis/dissertation required. *Average time to degree:* Master's–3 years full-time; doctorate–5 years full-time.

Entrance requirements: For master's and doctorate, GRE General Test, previous course work in calculus, cell biology, chemistry, genetics, physics, and biochemistry. *Application deadline:* For fall admission, 3/1. Applications are processed on a rolling basis. *Application fee:* $35 ($55 for international students).

Expenses: Tuition, state resident: full-time $13,245; part-time $158 per credit hour. Tuition, nonresident: full-time $30,330; part-time $455 per credit hour. Required fees: $121 per year. Tuition and fees vary according to course load and degree level.

Financial aid: In 1999–00, 20 research assistantships with full tuition reimbursements (averaging $17,000 per year) were

awarded; fellowships with full tuition reimbursements, teaching assistantships with full tuition reimbursements, Federal Work-Study, grants, institutionally sponsored loans, traineeships, and tuition waivers (partial) also available. Financial aid application deadline: 2/1.

Faculty research: Host-parasite interactions, molecular biology, cellular and molecular immunology and hematology, viral and bacterial pathogenesis, cancer research. *Total annual research expenditures:* $4.2 million.

Dr. Hal E. Broxmeyer, Chairman, 317-274-7672, *Fax:* 317-274-4090, *E-mail:* hbroxmey@iupui.edu.

Application contact: 317-274-7671, *Fax:* 317-274-4090, *E-mail:* rhaak@iupui.edu.

Find an in-depth description at www.petersons.com/graduate.

■ IOWA STATE UNIVERSITY OF SCIENCE AND TECHNOLOGY

College of Veterinary Medicine and Graduate College, Graduate Programs in Veterinary Medicine, Department of Veterinary Microbiology and Preventive Medicine, Ames, IA 50011

AWARDS Veterinary microbiology (MS, PhD); veterinary preventive medicine (MS).

Faculty: 18 full-time, 10 part-time/adjunct.

Students: 18 full-time (6 women), 16 part-time (8 women); includes 4 minority (1 African American, 2 Asian Americans or Pacific Islanders, 1 Hispanic American), 9 international. *15 applicants, 33% accepted.* In 1999, 6 master's, 3 doctorates awarded.

Degree requirements: For master's, thesis or alternative required; for doctorate, dissertation required.

Entrance requirements: For master's and doctorate, GRE General Test, TOEFL. *Application deadline:* For fall admission, 2/1 (priority date). Applications are processed on a rolling basis. *Application fee:* $20 ($50 for international students). Electronic applications accepted.

Expenses: Tuition, state resident: full-time $3,308. Tuition, nonresident: full-time $9,744. Tuition and fees vary according to course load and program.

Financial aid: In 1999–00, 20 research assistantships with partial tuition reimbursements (averaging $11,205 per year), 1 teaching assistantship with partial tuition reimbursement (averaging $11,250 per year) were awarded; fellowships, scholarships also available.

Faculty research: Food microbiology, infectious diseases, medical microbiology, microbial genetics, mycology.

Dr. Charles Thoen, Chair, 515-294-5776, *E-mail:* vetmicro@iastate.edu.

Application contact: Dr. F. Christopher Minion, Director of Graduate Education, 515-294-6347, *E-mail:* vetmicro@instate.edu.

Find an in-depth description at www.petersons.com/graduate.

■ IOWA STATE UNIVERSITY OF SCIENCE AND TECHNOLOGY

Graduate College, College of Agriculture, Department of Microbiology, Ames, IA 50011

AWARDS MS, PhD.

Faculty: 10 full-time, 2 part-time/adjunct.

Students: 15 full-time (7 women), 14 part-time (7 women); includes 3 minority (2 African Americans, 1 Hispanic American), 7 international. *68 applicants, 9% accepted.* In 1999, 5 master's awarded.

Degree requirements: For master's, thesis or alternative required; for doctorate, dissertation required.

Entrance requirements: For master's and doctorate, GRE General Test, TOEFL. *Application deadline:* For fall admission, 2/1 (priority date). *Application fee:* $20 ($50 for international students). Electronic applications accepted.

Expenses: Tuition, state resident: full-time $3,308. Tuition, nonresident: full-time $9,744. Part-time tuition and fees vary according to course load, campus/location and program.

Financial aid: In 1999–00, 9 research assistantships with partial tuition reimbursements (averaging $10,968 per year), 7 teaching assistantships with partial tuition reimbursements (averaging $10,619 per year) were awarded; scholarships also available.

Dr. James Dickson, Chair, 515-294-1630.

Application contact: Dr. Alan Di Spirito, Director of Graduate Education, 515-294-2944.

Find an in-depth description at www.petersons.com/graduate.

■ JOHNS HOPKINS UNIVERSITY

School of Hygiene and Public Health, Department of Molecular Microbiology and Immunology, Baltimore, MD 21218-2699

AWARDS MHS, Sc M, PhD, Sc D.

Faculty: 26 full-time, 18 part-time/adjunct.

Students: 56 (35 women); includes 8 minority (1 African American, 7 Asian Americans or Pacific Islanders) 10 international. *138 applicants, 34% accepted.* In 1999, 15 master's, 9 doctorates awarded.

Degree requirements: For master's, thesis required (for some programs), foreign language not required; for doctorate, dissertation, 1 year full-time residency, oral and written exams required, foreign language not required.

Entrance requirements: For master's and doctorate, GRE General Test, GRE Subject Test (biology or chemistry recommended), TOEFL. *Application deadline:* For fall admission, 2/1 (priority date). Applications are processed on a rolling basis. *Application fee:* $60. Electronic applications accepted.

Expenses: Tuition: Full-time $23,660; part-time $493 per unit. Full-time tuition and fees vary according to degree level, campus/location and program.

Financial aid: Federal Work-Study, institutionally sponsored loans, and scholarships available. Aid available to part-time students. Financial aid application deadline: 4/15.

Faculty research: Immunopathology, immunodiagnosis, pathogenesis of enteric infections, parasite immunology, biochemistry of parasitic protozoa. *Total annual research expenditures:* $10.3 million. Dr. Diane Griffin, Chair, 410-955-3459, *Fax:* 410-955-0105, *E-mail:* dgriffin@jhsph.edu.

Application contact: Denise Patrick, Academic Coordinator, 410-614-4232, *Fax:* 410-955-0105, *E-mail:* dpatrick@jhsph.edu.

Find an in-depth description at www.petersons.com/graduate.

■ JOHNS HOPKINS UNIVERSITY

School of Hygiene and Public Health, Program in Public Health, Baltimore, MD 21218-2699

AWARDS Biochemistry (MPH); biostatistics (MPH); environmental health sciences (MPH); epidemiology (MPH); health policy and management (MPH); international health (MPH); mental hygiene (MPH); molecular microbiology and immunology (MPH); population and family health sciences (MPH). Part-time and evening/weekend programs available. Postbaccalaureate distance learning degree programs offered.

Students: *694 applicants, 73% accepted.* In 1999, 196 degrees awarded.

Degree requirements: For master's, foreign language and thesis not required.

Entrance requirements: For master's, GRE General Test, TOEFL, 2 years of work related experience. *Application deadline:* For fall admission, 12/1 (priority date). Applications are processed on a rolling basis. *Application fee:* $60. Electronic applications accepted.

Expenses: Tuition: Full-time $23,660; part-time $493 per unit. Full-time tuition and fees vary according to degree level, campus/location and program.

Financial aid: Federal Work-Study, institutionally sponsored loans, and scholarships available. Aid available to part-time students. Financial aid application deadline: 4/15.

Dr. Miriam Alexander, Director, 410-955-1291, *Fax:* 410-955-4749.

Application contact: Lenora Davis, Administrator, 410-955-1291, *Fax:* 410-955-4749, *E-mail:* lrdavis@jhsph.edu.

■ KANSAS STATE UNIVERSITY

Graduate School, College of Arts and Sciences, Division of Biology, Program in Microbiology and Immunology, Manhattan, KS 66506

AWARDS MS, PhD.

Degree requirements: For master's and doctorate, thesis/dissertation required, foreign language not required.

Expenses: Tuition, state resident: part-time $103 per credit hour. Tuition, nonresident: part-time $338 per credit hour. Required fees: $17 per credit hour. One-time fee: $64 part-time.

Faculty research: Immune cell function, virology, cell viability, water quality.

■ LOMA LINDA UNIVERSITY

Graduate School, Graduate Programs in Medicine, Department of Microbiology and Molecular Genetics, Loma Linda, CA 92350

AWARDS Microbiology (MS, PhD). Part-time programs available.

Degree requirements: For master's and doctorate, thesis/dissertation required, foreign language not required.

Entrance requirements: For master's and doctorate, GRE General Test, TOEFL. *Application deadline:* Applications are processed on a rolling basis. *Application fee:* $40.

Expenses: Tuition: Part-time $395 per unit.

Financial aid: Tuition waivers (full and partial) available. Aid available to part-time students. Financial aid application deadline: 2/1.

Faculty research: Bacterial chemotaxis, tumor immunology.

Dr. James D. Kettering, Graduate Adviser, 909-824-4480, *Fax:* 909-824-4035, *E-mail:* jkettering@ccmail.llu.edu.

Find an in-depth description at www.petersons.com/graduate.

■ LONG ISLAND UNIVERSITY, C.W. POST CAMPUS

School of Health Professions, Department of Biomedical Sciences, Program in Medical Biology, Brookville, NY 11548-1300

AWARDS Hematology (MS); immunology (MS); medical chemistry (MS); microbiology (MS). Part-time and evening/weekend programs available.

Faculty: 3 full-time (1 woman), 12 part-time/adjunct (6 women).

Students: 4 full-time (2 women), 41 part-time (21 women); includes 25 minority (10 African Americans, 10 Asian Americans or Pacific Islanders, 5 Hispanic Americans). *37 applicants, 89% accepted.* In 1999, 11 degrees awarded.

Degree requirements: For master's, computer language, thesis required, foreign language not required. *Average time to degree:* Master's–2 years full-time, 4 years part-time.

Entrance requirements: For master's, TOEFL, minimum GPA of 2.75 in major. *Application deadline:* For fall admission, 9/1 (priority date); for spring admission, 1/20 (priority date). Applications are processed on a rolling basis. *Application fee:* $30. Electronic applications accepted.

Expenses: Tuition: Part-time $405 per credit. Required fees: $310; $65 per year. Tuition and fees vary according to course load and program.

Financial aid: In 1999–00, 20 students received aid; fellowships with partial tuition reimbursements available, teaching assistantships with partial tuition reimbursements available, career-related internships or fieldwork, Federal Work-Study, and institutionally sponsored loans available. Aid available to part-time students. Financial aid application deadline: 5/15; financial aid applicants required to submit FAFSA.

Faculty research: Hematopoiesis, growth factors in cancer, interleukins in allergy, PCR techniques. *Total annual research expenditures:* $15,000.

Application contact: Robin Steadman, Graduate Adviser, 516-299-2337, *Fax:* 516-299-2527, *E-mail:* rsteadman@phoenix.liu.edu.

■ LOUISIANA STATE UNIVERSITY AND AGRICULTURAL AND MECHANICAL COLLEGE

Graduate School, College of Basic Sciences, Department of Biological Sciences, Baton Rouge, LA 70803

AWARDS Biochemistry (MS, PhD); microbiology (MS, PhD); plant biology (MS, PhD); zoology (MS, PhD). Part-time programs available.

Louisiana State University and Agricultural and Mechanical College (continued)

Faculty: 79 full-time (7 women), 2 part-time/adjunct (0 women).
Students: 91 full-time (44 women), 22 part-time (10 women); includes 9 minority (3 African Americans, 1 Asian American or Pacific Islander, 5 Hispanic Americans), 29 international. Average age 28. *98 applicants, 28% accepted.* In 1999, 13 master's, 7 doctorates awarded. Terminal master's awarded for partial completion of doctoral program.
Degree requirements: For master's, foreign language not required; for doctorate, dissertation required.
Entrance requirements: For master's and doctorate, GRE General Test, minimum GPA of 3.0. *Application deadline:* Applications are processed on a rolling basis. *Application fee:* $25.
Expenses: Tuition, state resident: full-time $2,881. Tuition, nonresident: full-time $7,081. Part-time tuition and fees vary according to course load and program.
Financial aid: In 1999–00, 12 fellowships, 20 research assistantships with partial tuition reimbursements, 62 teaching assistantships with partial tuition reimbursements were awarded; Federal Work-Study, institutionally sponsored loans, and unspecified assistantships also available. Aid available to part-time students. *Total annual research expenditures:* $2.2 million.
Dr. Harold Silverman, Chairman, 225-388-2601, *Fax:* 225-388-2597, *E-mail:* cxsiv@unix1.sncc.lsu.edu.
Find an in-depth description at www.petersons.com/graduate.

■ **LOUISIANA STATE UNIVERSITY HEALTH SCIENCES CENTER**

School of Graduate Studies in New Orleans, Department of Microbiology, Immunology, and Parasitology, New Orleans, LA 70112-1393

AWARDS Microbiology and immunology (MS, PhD).

Faculty: 13 full-time (3 women).
Students: 19 full-time (13 women); includes 5 minority (3 African Americans, 1 Asian American or Pacific Islander, 1 Hispanic American), 4 international. Average age 23. *30 applicants, 50% accepted.* In 1999, 3 master's awarded. Terminal master's awarded for partial completion of doctoral program.
Degree requirements: For master's, thesis required, foreign language not required; for doctorate, dissertation, preliminary exam, qualifying exam required, foreign language not required.

Average time to degree: Master's–3.5 years full-time; doctorate–7 years full-time.
Entrance requirements: For master's and doctorate, GRE General Test, TOEFL.
Application deadline: For fall admission, 4/15 (priority date). Applications are processed on a rolling basis. *Application fee:* $30.
Expenses: Tuition, state resident: full-time $2,878; part-time $126 per hour. Tuition, nonresident: full-time $6,003; part-time $265 per hour. Required fees: $2,272. Tuition and fees vary according to course load, degree level and program.
Financial aid: In 1999–00, 16 students received aid, including 16 research assistantships (averaging $13,000 per year); Federal Work-Study and tuition waivers (full) also available. Financial aid application deadline: 4/15.
Faculty research: Microbial physiology, animal virology, vaccine development, AIDS drug studies, pathogenic mechanisms, molecular immunology.
Dr. Ronald B. Luftig, Head, 504-568-4062.
Application contact: Jeanine M. Campbell, Asst. Business Manager, 504-568-4061, *Fax:* 504-568-2918, *E-mail:* jcampb@lsumc.edu.
Find an in-depth description at www.petersons.com/graduate.

■ **LOUISIANA STATE UNIVERSITY HEALTH SCIENCES CENTER**

School of Graduate Studies in Shreveport, Department of Microbiology, Shreveport, LA 71130-3932

AWARDS Microbiology/immunology (MS, PhD). Terminal master's awarded for partial completion of doctoral program.

Degree requirements: For master's and doctorate, thesis/dissertation required, foreign language not required.
Entrance requirements: For master's and doctorate, GRE General Test, TOEFL.
Expenses: Tuition, state resident: full-time $2,878; part-time $126 per hour. Tuition, nonresident: full-time $6,003; part-time $265 per hour. Required fees: $2,272. Tuition and fees vary according to course load, degree level and program.
Faculty research: Infectious disease, pathogenesis, molecular virology and biology.

■ **LOYOLA UNIVERSITY CHICAGO**

Graduate School, Department of Microbiology and Immunology, Maywood, IL 60153

AWARDS Immunology (MS, PhD); microbiology (MS, PhD); virology (MS, PhD).

Faculty: 10 full-time (3 women).
Students: 26 full-time (18 women). Average age 27. *138 applicants, 9% accepted.* Terminal master's awarded for partial completion of doctoral program.
Degree requirements: For master's, thesis required, foreign language not required; for doctorate, computer language, dissertation, oral comprehensive exams required, foreign language not required.
Entrance requirements: For master's and doctorate, GRE General Test, TOEFL.
Application deadline: Applications are processed on a rolling basis. *Application fee:* $35. Electronic applications accepted.
Expenses: Tuition: Part-time $500 per credit hour. Required fees: $42 per term.
Financial aid: In 1999–00, 10 fellowships with tuition reimbursements, 12 research assistantships with tuition reimbursements were awarded; Federal Work-Study, institutionally sponsored loans, and scholarships also available. Financial aid application deadline: 2/15.
Faculty research: Viral pathogenesis, microbial physiology and genetics, immunoglobulin genetics and differentiation of the immune response, signal transduction and host-parasite interactions.
Dr. Katherine L. Knight, Chair, 708-216-3385, *Fax:* 708-216-9574, *E-mail:* kknight@luc.edu.
Application contact: Dr. Thomas Gallagher, Head, 708-216-3385, *Fax:* 708-216-9574, *E-mail:* tgallag@luc.edu.
Find an in-depth description at www.petersons.com/graduate.

■ **MARQUETTE UNIVERSITY**

Graduate School, College of Arts and Sciences, Department of Biology, Milwaukee, WI 53201-1881

AWARDS Cell biology (MS, PhD); developmental biology (MS, PhD); ecology (MS, PhD); endocrinology (MS, PhD); evolutionary biology (MS, PhD); genetics (MS, PhD); microbiology (MS, PhD); molecular biology (MS, PhD); muscle and exercise physiology (MS, PhD); neurobiology (MS, PhD); reproductive physiology (MS, PhD).

Faculty: 16 full-time (4 women), 2 part-time/adjunct (0 women).
Students: 34 full-time (20 women), 3 part-time; includes 3 minority (all Asian Americans or Pacific Islanders), 2 international. Average age 31. *42 applicants, 29% accepted.* In 1999, 1 master's, 4 doctorates awarded. Terminal master's awarded for partial completion of doctoral program.
Degree requirements: For master's, thesis, 1 year of teaching experience or equivalent, comprehensive exam required,

foreign language not required; for doctorate, dissertation, 1 year of teaching experience or equivalent, qualifying exam required, foreign language not required.
Entrance requirements: For master's and doctorate, GRE General Test, GRE Subject Test, TOEFL. *Application fee:* $40.
Expenses: Tuition: Part-time $510 per credit hour. Tuition and fees vary according to program.
Financial aid: In 1999–00, 4 fellowships, 22 teaching assistantships were awarded; research assistantships, Federal Work-Study, institutionally sponsored loans, scholarships, and tuition waivers (full and partial) also available. Aid available to part-time students. Financial aid application deadline: 2/15.
Faculty research: Microbial and invertebrate ecology, evolution of gene function, DNA methylation, DNA arrangement. *Total annual research expenditures:* $1.5 million.
Dr. Brian Unsworth, Chairman, 414-288-7355, *Fax:* 414-288-7357.
Application contact: Barbara DeNoyer, Graduate Studies Coordinator, 414-288-7355, *Fax:* 414-288-7357.

Find an in-depth description at www.petersons.com/graduate.

■ MASSACHUSETTS INSTITUTE OF TECHNOLOGY

School of Science, Department of Biology, Program in Microbiology, Cambridge, MA 02139-4307

AWARDS PhD.

Degree requirements: For doctorate, dissertation, general exam required, foreign language not required.
Entrance requirements: For doctorate, GRE General Test. *Application deadline:* For fall admission, 1/1. *Application fee:* $55.
Expenses: Tuition: Full-time $25,000. Full-time tuition and fees vary according to degree level, program and student level.
Financial aid: Application deadline: 1/1.
Application contact: Dr. Janice D. Chang, Educational Administrator, 617-253-3717, *Fax:* 617-258-9329, *E-mail:* gradbio@mit.edu.

■ MCP HAHNEMANN UNIVERSITY

School of Medicine, Biomedical Graduate Programs, Department of Microbiology and Immunology, Philadelphia, PA 19102-1192

AWARDS MS, PhD, MD/MS, MD/PhD. Terminal master's awarded for partial completion of doctoral program.

Degree requirements: For master's, thesis, comprehensive exam required,

foreign language not required; for doctorate, dissertation, qualifying exam required, foreign language not required.
Entrance requirements: For master's, GRE General Test, TOEFL, minimum GPA of 2.75; for doctorate, GRE General Test, TOEFL, minimum GPA of 3.0.
Faculty research: Immunology of malarial parasites, virology, bacteriology, molecular biology, parasitology.

Find an in-depth description at www.petersons.com/graduate.

■ MEDICAL COLLEGE OF OHIO

Graduate School, Department of Microbiology, Toledo, OH 43614-5805

AWARDS MS. Part-time programs available.

Faculty: 9 full-time (2 women), 1 part-time/adjunct (0 women).
Students: 1 full-time (0 women), 1 (woman) part-time. Average age 25.
Degree requirements: For master's, thesis, qualifying exam required, foreign language not required.
Entrance requirements: For master's, GRE General Test, minimum undergraduate GPA of 3.0. *Application fee:* $30.
Expenses: Tuition, state resident: part-time $193 per hour. Tuition, nonresident: part-time $445 per hour. Tuition and fees vary according to degree level.
Financial aid: Federal Work-Study and institutionally sponsored loans available. Financial aid applicants required to submit FAFSA.
Faculty research: Gene regulation, bacterial and fungal genetics, viral replication, immunology, microbial ecology. *Total annual research expenditures:* $1.9 million.
Dr. Garry Cole, Chairman, 419-383-4117, *Fax:* 419-383-6140, *E-mail:* mcograduate school@mco.edu.
Application contact: Joann Braatz, Clerk, 419-383-4117, *Fax:* 419-383-6140, *E-mail:* mcograduate school@mco.edu.

■ MEDICAL COLLEGE OF WISCONSIN

Graduate School of Biomedical Sciences, Department of Microbiology and Molecular Genetics, Milwaukee, WI 53226-0509

AWARDS MS, PhD, MD/MS, MD/PhD. Terminal master's awarded for partial completion of doctoral program.

Degree requirements: For master's and doctorate, thesis/dissertation required, foreign language not required.
Entrance requirements: For master's and doctorate, GRE General Test, TOEFL.
Expenses: Tuition, state resident: full-time $9,318. Tuition, nonresident: full-time $9,318. Required fees: $115.

Faculty research: Virology, immunology, bacterial toxins, regulation of gene expression.

■ MEDICAL UNIVERSITY OF SOUTH CAROLINA

College of Graduate Studies, Department of Microbiology and Immunology, Charleston, SC 29425-0002

AWARDS MS, PhD, DMD/PhD, MD/PhD.

Faculty: 18 part-time/adjunct (3 women).
Students: 34 full-time (21 women); includes 5 minority (2 African Americans, 3 Asian Americans or Pacific Islanders), 3 international. Average age 27. *27 applicants, 41% accepted.* In 1999, 1 degree awarded. Terminal master's awarded for partial completion of doctoral program.
Degree requirements: For master's, thesis, research seminar required; for doctorate, dissertation, teaching and research seminar, oral and written exams required, foreign language not required.
Entrance requirements: For master's and doctorate, GRE General Test, TOEFL, interview. *Application deadline:* Applications are processed on a rolling basis. *Application fee:* $55. Electronic applications accepted.
Expenses: Tuition, state resident: full-time $3,470; part-time $160 per semester hour. Tuition, nonresident: full-time $4,426; part-time $213 per semester hour. Required fees: $408 per semester. One-time fee: $160. Tuition and fees vary according to program.
Financial aid: In 1999–00, 14 students received aid, including 10 fellowships (averaging $16,000 per year), 12 research assistantships; teaching assistantships, Federal Work-Study and tuition waivers (partial) also available. Financial aid application deadline: 4/1; financial aid applicants required to submit FAFSA.
Faculty research: Cellular immunology, immunochemistry, immunogenetics, molecular genetics, tumor immunology. *Total annual research expenditures:* $3.3 million.
Dr. E. C. LeRoy, Chairman, 843-792-4421.
Application contact: Julie Johnston, Director of Admissions, 843-792-8710, *Fax:* 843-792-3764.

■ MEHARRY MEDICAL COLLEGE

School of Graduate Studies, Department of Microbiology, Nashville, TN 37208-9989

AWARDS PhD.

Faculty: 10 full-time (4 women).
Students: 13 full-time (9 women); includes 11 minority (10 African Americans, 1

Meharry Medical College (continued)
Asian American or Pacific Islander). Average age 30. *10 applicants, 60% accepted.* In 1999, 3 degrees awarded.
Degree requirements: For doctorate, dissertation, oral and written comprehensive exams required, foreign language not required.
Entrance requirements: For doctorate, GRE General Test, GRE Subject Test, undergraduate degree in related science. *Application deadline:* For fall admission, 6/1. Applications are processed on a rolling basis. *Application fee:* $45.
Expenses: Tuition: Full-time $8,732. Required fees: $2,133.
Financial aid: Fellowships, research assistantships, Federal Work-Study available. Financial aid application deadline: 4/15.
Faculty research: Microbial and bacterial pathogenesis, viral transcription, immune response to viruses and parasites.
Dr. Shirley Russell, Chair, 615-327-6281, *Fax:* 615-327-6072, *E-mail:* russel29@ccvax.mmc.edu.

■ MIAMI UNIVERSITY

Graduate School, College of Arts and Sciences, Department of Microbiology, Oxford, OH 45056

AWARDS MS, PhD. Part-time programs available.

Faculty: 9 full-time (2 women).
Students: 4 full-time (all women), 18 part-time (9 women); includes 3 minority (1 African American, 1 Asian American or Pacific Islander, 1 Hispanic American). *27 applicants, 100% accepted.* In 1999, 3 master's, 3 doctorates awarded.
Degree requirements: For master's, thesis, final exam required; for doctorate, dissertation, comprehensive and final exams required.
Entrance requirements: For master's, GRE General Test, minimum undergraduate GPA of 3.0 during previous 2 years or 2.75 overall; for doctorate, GRE General Test, minimum GPA of 2.75 (undergraduate), 3.0 (graduate). *Application deadline:* For fall admission, 3/1 (priority date); for spring admission, 12/15. Applications are processed on a rolling basis. *Application fee:* $35.
Expenses: Tuition, state resident: part-time $260 per hour. Tuition, nonresident: full-time $3,125; part-time $538 per hour. International tuition: $6,452 full-time. Required fees: $18 per semester. Tuition and fees vary according to campus/location.
Financial aid: In 1999–00, 15 fellowships, 5 teaching assistantships were awarded;

research assistantships, Federal Work-Study and tuition waivers (full) also available. Financial aid application deadline: 3/1.
Dr. Gary Janssen, Director of Graduate Studies, 513-529-5422, *Fax:* 513-529-2431, *E-mail:* micro@muohio.edu.

■ MICHIGAN STATE UNIVERSITY

Graduate School, College of Natural Science, Department of Microbiology, East Lansing, MI 48824

AWARDS PhD.

Faculty: 64 full-time (19 women).
Students: 73 full-time (30 women); includes 14 minority (7 African Americans, 4 Asian Americans or Pacific Islanders, 3 Hispanic Americans), 16 international. Average age 26. *256 applicants, 4% accepted.* In 1999, 5 doctorates awarded.
Degree requirements: For doctorate, dissertation required, foreign language not required. *Average time to degree:* Doctorate–5 years full-time.
Entrance requirements: For doctorate, GRE General Test, TOEFL. *Application deadline:* For fall admission, 2/1. *Application fee:* $30 ($40 for international students). Electronic applications accepted.
Expenses: Tuition, state resident: part-time $229 per credit. Tuition, nonresident: part-time $464 per credit. Required fees: $241 per semester. Tuition and fees vary according to course load, degree level and program.
Financial aid: In 1999–00, 73 students received aid, including 2 fellowships with full tuition reimbursements available (averaging $17,000 per year), 65 research assistantships with full tuition reimbursements available (averaging $16,560 per year), 6 teaching assistantships with full tuition reimbursements available (averaging $16,560 per year); career-related internships or fieldwork, grants, institutionally sponsored loans, and unspecified assistantships also available. Financial aid application deadline: 1/15.
Faculty research: Microbial diversity and ecology, cell growth and differentiation, molecular pathogenesis, and functional and comparative genomics.
Dr. Jerry B. Dodgson, Chairperson, 517-355-6464, *Fax:* 517-353-8957, *E-mail:* dodgson@pilot.msu.edu.
Application contact: Dr. Thomas Schmidt, Graduate Recruiting Officer, 517-432-2288, *Fax:* 517-353-8957, *E-mail:* micgrad@pilot.msu.edu.
Find an in-depth description at www.petersons.com/graduate.

■ MONTANA STATE UNIVERSITY–BOZEMAN

College of Graduate Studies, College of Letters and Science, Department of Microbiology, Bozeman, MT 59717

AWARDS MS, PhD. Part-time programs available.

Students: 14 full-time (6 women), 12 part-time (6 women); includes 1 minority (Asian American or Pacific Islander). Average age 30. *7 applicants, 100% accepted.* In 1999, 3 master's, 3 doctorates awarded.
Degree requirements: For master's, thesis (for some programs), comprehensive exam required, foreign language not required; for doctorate, dissertation, comprehensive exam required, foreign language not required.
Entrance requirements: For master's and doctorate, GRE General Test, TOEFL, minimum GPA of 3.0. *Application deadline:* For fall admission, 6/1; for spring admission, 11/1. Applications are processed on a rolling basis. *Application fee:* $50. Electronic applications accepted.
Expenses: Tuition, state resident: full-time $2,674. Tuition, nonresident: full-time $6,986. International tuition: $7,136 full-time. Tuition and fees vary according to course load and program.
Financial aid: In 1999–00, research assistantships with full tuition reimbursements (averaging $12,000 per year), teaching assistantships with full tuition reimbursements (averaging $9,000 per year) were awarded; career-related internships or fieldwork, Federal Work-Study, scholarships, and tuition waivers (full and partial) also available. Financial aid application deadline: 3/1; financial aid applicants required to submit FAFSA.
Faculty research: Medical mycology, environmental ecology, molecular genetics, biofilms, immunology. *Total annual research expenditures:* $2.8 million.
Dr. Seth Pincus, Head, 406-994-2903, *Fax:* 406-994-4926, *E-mail:* spincus@montana.edu.

■ MOUNT SINAI SCHOOL OF MEDICINE OF NEW YORK UNIVERSITY

Graduate School of Biological Sciences, Department of Microbiology, New York, NY 10029-6504

AWARDS PhD, MD/PhD.

Faculty: 13 full-time.
Students: 18 full-time (9 women); includes 1 minority (African American), 13 international. *38 applicants, 21% accepted.* In 1999, 2 degrees awarded.
Degree requirements: For doctorate, dissertation required, foreign language not required.

Entrance requirements: For doctorate, GRE General Test, GRE Subject Test, TOEFL. *Application deadline:* For fall admission, 4/15. *Application fee:* $35.
Expenses: Tuition: Full-time $21,750; part-time $725 per credit. Required fees: $750; $25 per credit. Full-time tuition and fees vary according to student level.
Financial aid: Grants available.
Dr. Peter Palese, Chairman, 212-241-7318.
Application contact: C. Gita Bosch, Administrative Manager and Assistant Dean, 212-241-6546, *Fax:* 212-241-0651, *E-mail:* grads@mssm.edu.

Find an in-depth description at www.petersons.com/graduate.

■ MOUNT SINAI SCHOOL OF MEDICINE OF NEW YORK UNIVERSITY

Graduate School of Biological Sciences, Mount Sinai Microbiology (MSM) Training Area, New York, NY 10029-6504

AWARDS PhD, MD/PhD.

Students: 11 full-time (4 women).
Degree requirements: For doctorate, dissertation required, foreign language not required.
Entrance requirements: For doctorate, GRE General Test, GRE Subject Test, MCAT, TOEFL. *Application deadline:* For fall admission, 4/15. *Application fee:* $60.
Expenses: Tuition: Full-time $21,750; part-time $725 per credit. Required fees: $750; $25 per credit. Full-time tuition and fees vary according to student level.
Financial aid: Fellowships with full tuition reimbursements, grants available.
Dr. John Blaho, Program Director, *E-mail:* blaho@msvax.mssm.edu.
Application contact: C. Gita Bosch, Administrative Manager and Assistant Dean, 212-241-6546, *Fax:* 212-241-0651, *E-mail:* grads@mssm.edu.

■ NEW YORK MEDICAL COLLEGE

Graduate School of Basic Medical Sciences, Program in Microbiology and Immunology, Valhalla, NY 10595-1691

AWARDS MS, PhD, MD/PhD. Part-time and evening/weekend programs available.

Faculty: 15 full-time (1 woman).
Students: 14 full-time (8 women), 16 part-time (12 women); includes 4 minority (3 Asian Americans or Pacific Islanders, 1 Hispanic American), 5 international. Average age 30. *27 applicants, 11% accepted.* In 1999, 8 master's awarded (100% found work related to degree); 3 doctorates

awarded. Terminal master's awarded for partial completion of doctoral program.
Degree requirements: For master's and doctorate, computer language, thesis/dissertation required, foreign language not required.
Entrance requirements: For master's, GRE General Test, TOEFL; for doctorate, GRE General Test, GRE Subject Test, TOEFL. *Application deadline:* For fall admission, 7/1 (priority date); for spring admission, 12/1 (priority date). Applications are processed on a rolling basis. *Application fee:* $35 ($60 for international students).
Expenses: Tuition: Part-time $430 per credit. Required fees: $15 per semester. One-time fee: $100.
Financial aid: In 1999–00, 6 research assistantships with full tuition reimbursements were awarded; career-related internships or fieldwork, Federal Work-Study, grants, institutionally sponsored loans, and tuition waivers (full) also available. Aid available to part-time students. Financial aid applicants required to submit FAFSA.
Faculty research: Tumor and transplantation immunology, molecular mechanisms of DNA repair, virus-host interactions.
Dr. Raj Tiwari, Director, 914-594-4192.

■ NEW YORK UNIVERSITY

Graduate School of Arts and Science, Department of Basic Medical Sciences, New York, NY 10012-1019

AWARDS Biochemistry (MS, PhD); cell biology (MS, PhD); microbiology (MS, PhD); parasitology (PhD); pathology (MS, PhD); pharmacology (PhD); physiology (MS, PhD). Part-time programs available.

Faculty: 23 full-time (2 women), 3 part-time/adjunct (1 woman).
Students: 182 full-time (73 women), 4 part-time (1 woman); includes 52 minority (11 African Americans, 34 Asian Americans or Pacific Islanders, 7 Hispanic Americans), 48 international. Average age 26. *671 applicants, 11% accepted.* In 1999, 15 master's, 32 doctorates awarded. Terminal master's awarded for partial completion of doctoral program.
Degree requirements: For master's, thesis or alternative, written comprehensive exam required, foreign language not required; for doctorate, one foreign language, dissertation, oral and written comprehensive exams required.
Entrance requirements: For master's and doctorate, GRE General Test, GRE Subject Test, TOEFL. *Application deadline:* For fall admission, 2/1 (priority date). *Application fee:* $60.
Expenses: Tuition: Full-time $17,880; part-time $745 per credit. Required fees:

$1,140; $35 per credit. Tuition and fees vary according to course load and program.
Financial aid: Fellowships with tuition reimbursements, research assistantships with tuition reimbursements, teaching assistantships with tuition reimbursements, career-related internships or fieldwork, Federal Work-Study, institutionally sponsored loans, and tuition waivers (full and partial) available. Financial aid application deadline: 2/1; financial aid applicants required to submit FAFSA.
Dr. Joel D. Oppenheim, Director, 212-263-5648, *Fax:* 212-263-7600, *E-mail:* sackler-info@nyumed.med.nyu.edu.

■ NEW YORK UNIVERSITY

School of Medicine and Graduate School of Arts and Science, Medical Scientist Training Program, New York, NY 10012-1019

AWARDS Biochemistry (MD/PhD); cell biology (MD/PhD); environmental health sciences (MD/PhD); microbiology (MD/PhD); parasitology (MD/PhD); pathology (MD/PhD); pharmacology (MD/PhD). Students must be accepted by both the School of Medicine and the Graduate School of Arts and Science.

Faculty: 150 full-time (35 women).
Students: 80 full-time (18 women); includes 37 minority (3 African Americans, 32 Asian Americans or Pacific Islanders, 2 Hispanic Americans), 1 international. Average age 25. *195 applicants, 18% accepted.*
Degree requirements: One foreign language.
Application deadline: For fall admission, 11/15. Applications are processed on a rolling basis. *Application fee:* $60.
Expenses: Students receive full tuition support and an annual stipend of $17,500.
Financial aid: Application deadline: 5/1.
Faculty research: Genetics, tumor biology, cardiovascular biology, neuroscience, host defense mechanisms.
Dr. James Salzer, Director, 212-263-0758.
Application contact: Arlene Kohler, Administrative Officer, 212-263-5649.

■ NEW YORK UNIVERSITY

School of Medicine and Graduate School of Arts and Science, Sackler Institute of Graduate Biomedical Sciences, Department of Microbiology, New York, NY 10012-1019

AWARDS PhD, MD/PhD.

Faculty: 15 full-time (2 women).
Students: 30 full-time (23 women); includes 4 minority (2 African Americans, 2 Asian Americans or Pacific Islanders), 7 international. Average age 25. In 1999, 12 degrees awarded (100% entered university research/teaching).

New York University (continued)

Degree requirements: For doctorate, one foreign language, dissertation, qualifying exam required. *Average time to degree:* Doctorate–5.5 years full-time.

Entrance requirements: For doctorate, GRE General Test, GRE Subject Test, TOEFL. *Application deadline:* For fall admission, 2/1 (priority date). Applications are processed on a rolling basis. *Application fee:* $60.

Expenses: Tuition: Full-time $17,880; part-time $745 per credit. Required fees: $1,140; $35 per credit. Tuition and fees vary according to course load and program.

Financial aid: Fellowships, research assistantships, teaching assistantships available. Financial aid application deadline: 1/15.

Faculty research: Aspects of microbiology, parasitology, and genetics; virology. *Total annual research expenditures:* $1.8 million.

Dr. Claudio Basilico, Chairman, 212-263-5341, *Fax:* 212-263-8276.

Application contact: Dr. Arturo Zychlinsky, Graduate Adviser, 212-263-7058, *Fax:* 212-263-8276, *E-mail:* zychlins@saturn.med.nyu.edu.

Find an in-depth description at www.petersons.com/graduate.

■ **NORTH CAROLINA STATE UNIVERSITY**

College of Veterinary Medicine and Graduate School, Graduate Programs in Comparative Biomedical Sciences, Raleigh, NC 27695

AWARDS Cell biology and morphology (MS, PhD); epidemiology and population medicine (MS, PhD); immunology (MS, PhD); microbiology and immunology (MS, PhD); pathology (MS, PhD); pharmacology (MS, PhD); specialized veterinary medicine (MS). Part-time programs available.

Students: 40 full-time (23 women), 22 part-time (12 women); includes 11 minority (8 African Americans, 2 Asian Americans or Pacific Islanders, 1 Hispanic American), 14 international. Average age 34. *33 applicants, 33% accepted.* In 1999, 2 master's, 2 doctorates awarded.

Degree requirements: For master's and doctorate, thesis/dissertation required.

Entrance requirements: For master's and doctorate, GRE General Test. *Application deadline:* For fall admission, 6/25; for spring admission, 11/25. Applications are processed on a rolling basis. *Application fee:* $45.

Expenses: Tuition, state resident: full-time $1,578. Tuition, nonresident: full-time $10,744. Required fees: $892. Full-time tuition and fees vary according to program.

Financial aid: Fellowships, research assistantships, teaching assistantships available. Financial aid application deadline: 2/15.

Faculty research: Infectious diseases, immunology and virology, tumor biology, toxicological pathology, food safety.

Dr. Neil C. Olson, Associate Dean, 919-513-6213, *Fax:* 919-513-6222, *E-mail:* neil_olson@ncsu.edu.

■ **NORTH CAROLINA STATE UNIVERSITY**

Graduate School, College of Agriculture and Life Sciences, Department of Microbiology, Raleigh, NC 27695

AWARDS MLS, MS, PhD.

Faculty: 12 full-time (1 woman), 10 part-time/adjunct (1 woman).

Students: 27 full-time (14 women), 12 part-time (8 women); includes 7 minority (5 African Americans, 1 Asian American or Pacific Islander, 1 Hispanic American), 3 international. Average age 28. *66 applicants, 15% accepted.* In 1999, 6 master's, 2 doctorates awarded.

Degree requirements: For master's and doctorate, thesis/dissertation required.

Entrance requirements: For master's and doctorate, GRE General Test. *Application deadline:* For fall admission, 3/1. *Application fee:* $45.

Expenses: Tuition, state resident: full-time $1,578. Tuition, nonresident: full-time $10,744. Required fees: $892. Full-time tuition and fees vary according to program.

Financial aid: In 1999–00, 3 fellowships (averaging $7,119 per year), 27 research assistantships (averaging $5,637 per year) were awarded; teaching assistantships, career-related internships or fieldwork and institutionally sponsored loans also available.

Faculty research: Molecular pathogenesis, gene regulation, yeast genetics and molecular biology, tumor immunology, microbial ecology. *Total annual research expenditures:* $2.8 million.

Dr. Hosni M. Hassan, Head, 919-515-6663, *Fax:* 919-515-2047, *E-mail:* hmhassan@mbio.ncsu.edu.

Application contact: Dr. Scott M. Laster, Director of Graduate Programs, 919-515-7958, *Fax:* 919-515-7867, *E-mail:* scott_laster@ncsu.edu.

■ **NORTH DAKOTA STATE UNIVERSITY**

Graduate Studies and Research, College of Agriculture, Department of Veterinary and Microbiological Sciences, Fargo, ND 58105

AWARDS Cellular and molecular biology (PhD); microbiology (MS); natural resources management (MS); veterinary sciences (MS). Part-time programs available.

Faculty: 9 full-time (2 women).

Students: 11 full-time (9 women), 3 part-time (all women); includes 1 minority (Native American). Average age 25. *18 applicants, 17% accepted.* In 1999, 5 master's awarded (100% found work related to degree).

Degree requirements: For master's, thesis required, foreign language not required; for doctorate, dissertation, oral and written preliminary exams required, foreign language not required. *Average time to degree:* Master's–2 years full-time.

Entrance requirements: For master's and doctorate, GRE, TOEFL. *Application deadline:* For fall admission, 3/15 (priority date). Applications are processed on a rolling basis. *Application fee:* $25.

Expenses: Tuition, state resident: full-time $3,096; part-time $112 per credit hour. Tuition, nonresident: full-time $7,588; part-time $299 per credit hour. Tuition and fees vary according to course load, campus/location and reciprocity agreements.

Financial aid: In 1999–00, 2 fellowships with full tuition reimbursements, 9 research assistantships with full tuition reimbursements, 7 teaching assistantships with full tuition reimbursements were awarded; Federal Work-Study and institutionally sponsored loans also available. Financial aid application deadline: 4/15.

Faculty research: Bacterial gene regulation, antibiotic resistance, molecular virology, mechanisms of bacterial pathogenesis, immunology of animals.

Dr. D. L. Berryhill, Interim Chair, 701-231-7511, *Fax:* 701-231-7514, *E-mail:* berryhil@badlands.nodak.edu.

Application contact: Dr. E. S. Berry, Graduate Program Director, 701-231-7520, *Fax:* 701-231-7514, *E-mail:* berry@plains.nodak.edu.

■ **NORTHWESTERN UNIVERSITY**

The Graduate School, Division of Interdepartmental Programs and Medical School, Integrated Graduate Programs in the Life Sciences, Chicago, IL 60611

AWARDS Cancer biology (PhD); cell biology (PhD); developmental biology (PhD);

evolutionary biology (PhD); immunology and microbial pathogenesis (PhD); molecular biology and genetics (PhD); neurobiology (PhD); pharmacology and toxicology (PhD); structural biology and biochemistry (PhD).

Degree requirements: For doctorate, dissertation, written and oral qualifying exams required, foreign language not required.

Entrance requirements: For doctorate, GRE General Test, TOEFL.

Expenses: Tuition: Full-time $23,301. Full-time tuition and fees vary according to program.

Find an in-depth description at www.petersons.com/graduate.

■ THE OHIO STATE UNIVERSITY

Graduate School, College of Biological Sciences, Department of Microbiology, Columbus, OH 43210

AWARDS MS, PhD.

Faculty: 14 full-time, 5 part-time/adjunct.
Students: 53 full-time (28 women), 1 part-time; includes 6 minority (3 African Americans, 2 Asian Americans or Pacific Islanders, 1 Hispanic American), 17 international. *110 applicants, 24% accepted.* In 1999, 5 doctorates awarded.
Degree requirements: For master's, thesis optional, foreign language not required; for doctorate, dissertation required, foreign language not required.
Entrance requirements: For master's and doctorate, GRE General Test. *Application deadline:* For fall admission, 8/15. Applications are processed on a rolling basis. *Application fee:* $30 ($40 for international students).
Expenses: Tuition, state resident: full-time $5,400. Tuition, nonresident: full-time $14,535. Part-time tuition and fees vary according to course load and program.
Financial aid: Fellowships, research assistantships, teaching assistantships, Federal Work-Study and institutionally sponsored loans available. Aid available to part-time students.
Dr. John N. Reeve, Chairman, 614-292-2301, *Fax:* 614-292-8120, *E-mail:* reeve.2@osu.edu.

Find an in-depth description at www.petersons.com/graduate.

■ OHIO UNIVERSITY

Graduate Studies, College of Arts and Sciences, Department of Biological Sciences, Program in Microbiology, Athens, OH 45701-2979

AWARDS MS, PhD, DO/PhD.

Students: 2 full-time (1 woman). *6 applicants, 33% accepted.* In 1999, 1 doctorate awarded.
Degree requirements: For master's, thesis, 1 quarter of teaching experience

required, foreign language not required; for doctorate, dissertation, 2 quarters of teaching experience required, foreign language not required.
Entrance requirements: For master's and doctorate, GRE General Test. *Application deadline:* For fall admission, 1/15. *Application fee:* $30.
Expenses: Tuition, state resident: full-time $5,754; part-time $238 per credit hour. Tuition, nonresident: full-time $11,055; part-time $457 per credit hour. Tuition and fees vary according to course load, degree level and campus/location.
Financial aid: Fellowships with full tuition reimbursements, research assistantships with full tuition reimbursements, teaching assistantships with full tuition reimbursements, Federal Work-Study, institutionally sponsored loans, and tuition waivers (full) available. Financial aid application deadline: 1/15.
Faculty research: Bacteriology, virology, microbial genetics, cell and molecular biology.
Application contact: Dr. William R. Holmes, Graduate Chair, 740-593-2334, *Fax:* 740-593-0300, *E-mail:* holmes@ohiou.edu.

■ OKLAHOMA STATE UNIVERSITY

Graduate College, College of Arts and Sciences, Department of Microbiology and Molecular Genetics, Stillwater, OK 74078

AWARDS MS, PhD.

Faculty: 6 full-time (0 women).
Students: 9 full-time (4 women), 15 part-time (8 women); includes 2 minority (both Hispanic Americans), 11 international. Average age 25. In 1999, 1 master's, 2 doctorates awarded.
Degree requirements: For master's and doctorate, thesis/dissertation required.
Entrance requirements: For master's and doctorate, GRE General Test, TOEFL. *Application deadline:* For fall admission, 6/1. *Application fee:* $25.
Expenses: Tuition, state resident: part-time $86 per credit hour. Tuition, nonresident: part-time $275 per credit hour. Required fees: $17 per credit hour. $14 per semester. One-time fee: $20 full-time. Tuition and fees vary according to course load.
Financial aid: In 1999–00, 6 research assistantships (averaging $12,015 per year), 20 teaching assistantships (averaging $13,500 per year) were awarded; tuition waivers (full) also available. Financial aid application deadline: 3/1.
Faculty research: Bioinformatics, genomics-genetics, virology, environmental

microbiology, development-molecular mechanisms.
Dr. James T. Blankemeyer, Head, 405-744-6243.

■ OREGON HEALTH SCIENCES UNIVERSITY

School of Medicine, Graduate Programs in Medicine, Department of Molecular Microbiology and Immunology, Portland, OR 97201-3098

AWARDS PhD.

Degree requirements: For doctorate, dissertation required, foreign language not required.
Entrance requirements: For doctorate, GRE General Test.
Expenses: Tuition, state resident: full-time $3,132; part-time $174 per credit hour. Tuition, nonresident: full-time $5,256; part-time $292 per credit hour. Required fees: $8.5 per credit hour. $146 per term. Part-time tuition and fees vary according to course load.
Faculty research: Molecular biology of bacterial and viral pathogens, cellular and humoral immunology, molecular biology of microbes.

Find an in-depth description at www.petersons.com/graduate.

■ OREGON STATE UNIVERSITY

College of Veterinary Medicine, Program in Veterinary Sciences, Corvallis, OR 97331

AWARDS Microbiology (MS); pathology (MS); toxicology (MS). Part-time programs available.

Faculty: 4 full-time (0 women).
Students: 5 full-time (4 women). Average age 34. In 1999, 1 degree awarded (100% found work related to degree).
Degree requirements: For master's, thesis required, foreign language not required.
Entrance requirements: For master's, TOEFL, minimum GPA of 3.0 in last 90 hours. *Application deadline:* For fall admission, 11/1. *Application fee:* $50.
Expenses: Tuition, state resident: full-time $4,334. Tuition, nonresident: full-time $7,382.
Financial aid: Research assistantships, Federal Work-Study, institutionally sponsored loans, and scholarships available. Aid available to part-time students. Financial aid application deadline: 2/1.
Faculty research: Calf diseases, bovine foot rot, caliciviruses, effects of toxic agents on immune systems.
Linda L. Blythe, Associate Dean, 541-737-2098, *Fax:* 541-737-4245, *E-mail:* linda.blythe@orst.edu.

■ OREGON STATE UNIVERSITY

Graduate School, College of Science, Department of Microbiology, Corvallis, OR 97331

AWARDS M Agr, MA, MAIS, MS, PhD. Part-time programs available.

Students: 26 full-time (14 women), 1 (woman) part-time; includes 2 minority (both Asian Americans or Pacific Islanders), 5 international. Average age 27. In 1999, 6 master's awarded (100% entered university research/teaching); 1 doctorate awarded. Terminal master's awarded for partial completion of doctoral program.
Degree requirements: For master's, thesis required, foreign language not required; for doctorate, one foreign language, dissertation required.
Entrance requirements: For master's and doctorate, GRE General Test, TOEFL, minimum GPA of 3.0 in last 90 hours. *Application deadline:* For fall admission, 3/1. Applications are processed on a rolling basis. *Application fee:* $50.
Expenses: Tuition, state resident: full-time $6,489. Tuition, nonresident: full-time $11,061. Tuition and fees vary according to program.
Financial aid: Fellowships, research assistantships, teaching assistantships, career-related internships or fieldwork, Federal Work-Study, and institutionally sponsored loans available. Aid available to part-time students. Financial aid application deadline: 2/1.
Faculty research: Genetics, physiology, biotechnology, pathogenic microbiology, plant virology.
Dr. Jo-Ann C. Leong, Chair, 541-737-1834, *Fax:* 541-737-0496, *E-mail:* leongj@bcc.orst.edu.

Application contact: Sharon Jansen, Graduate Admissions Clerk, 541-737-4441, *Fax:* 541-737-0496, *E-mail:* jansens@orst.edu.

■ THE PENNSYLVANIA STATE UNIVERSITY MILTON S. HERSHEY MEDICAL CENTER

Graduate School, Department of Microbiology and Immunology, Hershey, PA 17033-2360

AWARDS Genetics (PhD); immunology (MS, PhD); microbiology (MS); microbiology/virology (PhD); molecular biology (PhD).

Students: 19 full-time (6 women). Average age 27. In 1999, 2 master's, 3 doctorates awarded.
Degree requirements: For doctorate, dissertation required.
Entrance requirements: For doctorate, GRE General Test, TOEFL, minimum GPA of 3.0. *Application deadline:* For fall admission, 7/26. *Application fee:* $50.

Expenses: Tuition, state resident: full-time $6,886; part-time $291 per credit. Tuition, nonresident: full-time $14,118; part-time $588 per credit. Required fees: $43 per semester. Part-time tuition and fees vary according to course load.
Financial aid: Unspecified assistantships available.
Dr. Richard J. Courtney, Chair, 717-531-7659.

Application contact: Dr. Brian Wigdahl, Chairman, Graduate Program Committee, 717-531-6682, *Fax:* 717-531-6522, *E-mail:* eneidigh@cor-mail.biochem.hmc.psu.edu.

■ THE PENNSYLVANIA STATE UNIVERSITY UNIVERSITY PARK CAMPUS

Graduate School, Eberly College of Science, Department of Biochemistry and Molecular Biology, Program in Biochemistry, Microbiology, and Molecular Biology, State College, University Park, PA 16802-1503

AWARDS MS, PhD.

Students: 83 full-time (45 women), 12 part-time (6 women). In 1999, 4 master's, 12 doctorates awarded.
Degree requirements: For master's, thesis required; for doctorate, dissertation, comprehensive exam required.
Entrance requirements: For master's and doctorate, GRE General Test, GRE Subject Test. *Application fee:* $50.
Expenses: Tuition, state resident: full-time $6,886; part-time $291 per credit. Tuition, nonresident: full-time $14,118; part-time $588 per credit. Required fees: $46 per semester. Part-time tuition and fees vary according to course load and program.
Financial aid: Fellowships, unspecified assistantships available.
Ronald D. Porter, Director, 814-865-2538.

Find an in-depth description at www.petersons.com/graduate.

■ PURDUE UNIVERSITY

Graduate School, School of Science, Department of Biological Sciences, Program in Microbiology, West Lafayette, IN 47907

AWARDS MS, PhD.

Faculty: 8 full-time (0 women). Terminal master's awarded for partial completion of doctoral program.
Degree requirements: For master's, thesis required, foreign language not required; for doctorate, dissertation, seminars, teaching experience required, foreign language not required.
Entrance requirements: For master's and doctorate, GRE General Test, TOEFL,

TSE. *Application deadline:* For fall admission, 2/15. *Application fee:* $30. Electronic applications accepted.
Expenses: Tuition, state resident: full-time $4,530; part-time $130 per credit hour. Tuition, nonresident: full-time $15,310; part-time $404 per credit hour. Tuition and fees vary according to campus/location and program.
Financial aid: Fellowships, research assistantships, teaching assistantships available. Aid available to part-time students. Financial aid application deadline: 2/15; financial aid applicants required to submit FAFSA.
Faculty research: Microbial ecology, physiological ecology, biodegradation, mechanisms of response to osmotic stress, insecticidal protoxin synthesis and structure.
Application contact: Nancy Konopka, Graduate Studies Office Manager, 765-494-8142, *Fax:* 765-494-0876, *E-mail:* njk@bilbo.bio.purdue.edu.

■ PURDUE UNIVERSITY

School of Veterinary Medicine and Graduate School, Graduate Programs in Veterinary Medicine, Department of Veterinary Pathobiology, West Lafayette, IN 47907

AWARDS Bacteriology (MS, PhD); epidemiology (MS, PhD); immunology (MS, PhD); infectious diseases (MS, PhD); microbiology (MS, PhD); parasitology (MS, PhD); pathology (MS, PhD); toxicology (MS, PhD); virology (MS, PhD).

Faculty: 26 full-time (5 women).
Students: 42 full-time (24 women), 2 part-time (1 woman); includes 5 minority (1 African American, 3 Asian Americans or Pacific Islanders, 1 Hispanic American), 18 international. Average age 32. In 1999, 2 master's, 6 doctorates awarded (100% found work related to degree). Terminal master's awarded for partial completion of doctoral program.
Degree requirements: For master's, thesis required (for some programs), foreign language not required; for doctorate, dissertation required, foreign language not required.
Entrance requirements: For master's and doctorate, GRE General Test, TOEFL. *Application fee:* $30.
Expenses: Tuition, state resident: full-time $3,732. Tuition, nonresident: full-time $8,732.
Financial aid: Fellowships, research assistantships, teaching assistantships available. Financial aid application deadline: 3/1; financial aid applicants required to submit FAFSA.
Dr. H. L. Thacker, Head, 765-494-7543.

■ QUINNIPIAC UNIVERSITY

School of Health Sciences, Programs in Medical Laboratory Sciences, Hamden, CT 06518-1940

AWARDS Biomedical sciences (MHS); laboratory management (MHS); microbiology (MHS). Part-time and evening/weekend programs available.

Faculty: 6 full-time (1 woman), 2 part-time/adjunct (1 woman).
Students: 3 full-time (all women), 27 part-time (22 women); includes 2 minority (both Asian Americans or Pacific Islanders). Average age 28. *10 applicants, 70% accepted.* In 1999, 10 degrees awarded (100% found work related to degree).
Degree requirements: For master's, comprehensive exam required, thesis optional. *Average time to degree:* Master's–2 years full-time, 4 years part-time.
Entrance requirements: For master's, minimum GPA of 2.5. *Application deadline:* For fall admission, 8/1 (priority date); for spring admission, 12/15 (priority date). Applications are processed on a rolling basis. *Application fee:* $45. Electronic applications accepted.
Expenses: Tuition: Part-time $410 per credit hour. Required fees: $20 per term. Tuition and fees vary according to program.
Financial aid: Available to part-time students. Applicants required to submit FAFSA.
Faculty research: Microbial physiology, fermentation technology.
Dr. Kenneth Kaloustian, Director, 203-582-8676, *Fax:* 203-582-3443, *E-mail:* ken.kaloustian@quinnipiac.edu.
Application contact: Scott Farber, Director of Graduate Admissions, 800-462-1944, *Fax:* 203-582-3443, *E-mail:* graduate@quinnipiac.edu.

Find an in-depth description at www.petersons.com/graduate.

■ RENSSELAER POLYTECHNIC INSTITUTE

Graduate School, School of Science, Department of Biology, Troy, NY 12180-3590

AWARDS Biochemistry (MS, PhD); biophysics (MS, PhD); cell biology (MS, PhD); developmental biology (MS, PhD); microbiology (MS, PhD); molecular biology (MS, PhD); plant science (MS, PhD). Part-time programs available.

Faculty: 14 full-time (5 women).
Students: 16 full-time (8 women), 2 part-time (1 woman); includes 1 minority (Asian American or Pacific Islander), 4 international. *37 applicants, 41% accepted.* In 1999, 6 master's, 3 doctorates awarded.

Terminal master's awarded for partial completion of doctoral program.
Degree requirements: For master's and doctorate, thesis/dissertation required, foreign language not required.
Entrance requirements: For master's and doctorate, GRE General Test, TOEFL. *Application deadline:* For fall admission, 2/1 (priority date). Applications are processed on a rolling basis. *Application fee:* $35.
Expenses: Tuition: Part-time $665 per credit hour. Required fees: $980.
Financial aid: In 1999–00, 8 research assistantships with partial tuition reimbursements (averaging $15,000 per year), 11 teaching assistantships with full tuition reimbursements (averaging $15,000 per year) were awarded; fellowships, career-related internships or fieldwork and institutionally sponsored loans also available. Financial aid application deadline: 2/1.
Faculty research: Applied environmental biology, genetics, environmental science, fresh water ecology, microbial ecology.
Dr. John Salerno, Chair, 518-276-2699, *Fax:* 518-276-2344.
Application contact: Dr. Jackie L. Collier, Assistant Professor, 518-276-6446, *Fax:* 518-276-2344, *E-mail:* collij3@rpi.edu.

Find an in-depth description at www.petersons.com/graduate.

■ RUSH UNIVERSITY

Graduate College, Division of Immunology, Chicago, IL 60612-3832

AWARDS Immunology (PhD); microbiology (PhD); virology (PhD).

Degree requirements: For doctorate, dissertation, comprehensive preliminary exam required, foreign language not required.
Entrance requirements: For doctorate, GRE General Test, TOEFL, interview, minimum GPA of 3.0.
Expenses: Tuition: Full-time $13,020; part-time $390 per credit. Tuition and fees vary according to program.
Faculty research: Immune interactions of cells and membranes.

Find an in-depth description at www.petersons.com/graduate.

■ RUTGERS, THE STATE UNIVERSITY OF NEW JERSEY, NEW BRUNSWICK

Graduate School, Program in Microbiology and Molecular Genetics, New Brunswick, NJ 08901-1281

AWARDS Applied microbiology (MS, PhD); clinical microbiology (MS, PhD); computational molecular biology (PhD); immunology (MS, PhD); microbial biochemistry (MS, PhD); molecular genetics

(MS, PhD); virology (MS, PhD). Part-time programs available.

Faculty: 122 full-time (25 women), 4 part-time/adjunct (1 woman).
Students: 65 full-time (31 women), 51 part-time (24 women); includes 24 minority (14 Asian Americans or Pacific Islanders, 10 Hispanic Americans), 29 international. Average age 27. *200 applicants, 26% accepted.* In 1999, 15 master's, 10 doctorates awarded. Terminal master's awarded for partial completion of doctoral program.
Degree requirements: For master's, thesis or alternative required, foreign language not required; for doctorate, dissertation, written qualifying exam required, foreign language not required.
Entrance requirements: For master's, GRE General Test, GRE Subject Test; for doctorate, GRE General Test, GRE Subject Test, TOEFL, minimum GPA of 3.0. *Application deadline:* For fall admission, 2/15 (priority date). Applications are processed on a rolling basis. *Application fee:* $50.
Expenses: Tuition, state resident: full-time $6,776; part-time $279 per credit. Tuition, nonresident: full-time $9,936; part-time $412 per credit. Required fees: $20 per credit. $89 per semester. Tuition and fees vary according to course load, campus/location and program.
Financial aid: In 1999–00, 70 students received aid, including 32 fellowships with full tuition reimbursements available, 26 research assistantships with full tuition reimbursements available, 12 teaching assistantships with full tuition reimbursements available; institutionally sponsored loans also available. Financial aid application deadline: 2/15; financial aid applicants required to submit FAFSA.
Faculty research: Biochemistry of microbes, medical microbiology.
Dr. Howard Passmore, Director, 732-445-2812, *E-mail:* passmore@biology.rutgers.edu.
Application contact: Betty Green, Administrative Assistant, 732-445-5086, *Fax:* 732-445-5735, *E-mail:* green@mbcl.rutgers.edu.

■ SAINT LOUIS UNIVERSITY

School of Medicine and Graduate School, Graduate Programs in Biomedical Sciences, Department of Molecular Microbiology and Immunology, St. Louis, MO 63103-2097

AWARDS PhD.

Faculty: 16 full-time (4 women), 1 part-time/adjunct (0 women).
Students: Average age 30. In 1999, 1 degree awarded.

Saint Louis University (continued)

Degree requirements: For doctorate, dissertation, qualifying exams required.
Entrance requirements: For doctorate, GRE General Test, GRE Subject Test. *Application deadline:* For fall admission, 5/1. Applications are processed on a rolling basis. *Application fee:* $40.
Expenses: Tuition: Part-time $507 per credit hour. Required fees: $38 per term.
Financial aid: In 1999–00, 7 students received aid, including 4 research assistantships; fellowships, teaching assistantships. Aid available to part-time students. Financial aid application deadline: 8/1; financial aid applicants required to submit FAFSA.
Faculty research: Helper and suppressor T cells, cellular immunology, lymphocyte cloning, T and B cell collaboration, genetic control of immune response.
Dr. William Wold, Chairman, 314-577-8435, *Fax:* 314-773-3403.

Application contact: Dr. H. Peter Zassenhaus, Director of Graduate Studies, 314-577-8432, *Fax:* 314-773-3403, *E-mail:* zassenp@slu.edu.

Find an in-depth description at www.petersons.com/graduate.

■ **SAN DIEGO STATE UNIVERSITY**

Graduate and Research Affairs, College of Sciences, Department of Biological Sciences, Program in Microbiology, San Diego, CA 92182

AWARDS MS.

Students: Average age 30.
Degree requirements: For master's, thesis, oral exam required, foreign language not required.
Entrance requirements: For master's, GRE General Test, GRE Subject Test, TOEFL. *Application deadline:* For fall admission, 7/1 (priority date); for spring admission, 12/1. Applications are processed on a rolling basis. *Application fee:* $55.
Expenses: Tuition, nonresident: part-time $246 per unit. Required fees: $1,932; $633 per semester. Tuition and fees vary according to course load.
Financial aid: Fellowships, research assistantships, teaching assistantships available.
Application contact: Ken Johnson, Graduate Coordinator, 619-594-7029, *Fax:* 619-594-5676, *E-mail:* kjohnson@sunstroke.sdsu.edu.

■ **SAN FRANCISCO STATE UNIVERSITY**

Graduate Division, College of Science and Engineering, Department of Biology, Program in Microbiology, San Francisco, CA 94132-1722

AWARDS MA.

Entrance requirements: For master's, minimum GPA of 2.5 in last 60 units.
Expenses: Tuition, nonresident: full-time $5,904; part-time $246 per unit. Required fees: $1,904; $637 per semester. Tuition and fees vary according to course load.

■ **SETON HALL UNIVERSITY**

College of Arts and Sciences, Department of Biology, South Orange, NJ 07079-2697

AWARDS Biology (MS); microbiology (MS). Part-time and evening/weekend programs available.

Faculty: 11 full-time (5 women).
Students: 16 full-time (7 women), 61 part-time (36 women); includes 10 minority (1 African American, 6 Asian Americans or Pacific Islanders, 3 Hispanic Americans), 3 international. In 1999, 10 degrees awarded (90% found work related to degree, 10% continued full-time study).
Degree requirements: For master's, research paper or thesis, seminar required.
Entrance requirements: For master's, minimum GPA of 3.0. *Application deadline:* For fall admission, 7/15; for spring admission, 12/15. Applications are processed on a rolling basis. *Application fee:* $50.
Expenses: Tuition: Full-time $10,404; part-time $578 per credit. Required fees: $185 per year. Tuition and fees vary according to course load, campus/location, program and student's religious affiliation.
Financial aid: Teaching assistantships, career-related internships or fieldwork available.
Faculty research: Neurobiology, genetics, immunology, molecular biology, cellular physiology.
Dr. Suli Chang, Chairperson, 973-761-9044, *Fax:* 973-761-9596.

Application contact: Dr. Eliot Krause, Graduate Adviser, 973-761-9532, *E-mail:* krauseel@lanmail.shu.edu.

■ **SOUTH DAKOTA STATE UNIVERSITY**

Graduate School, College of Agriculture and Biological Sciences, Department of Biology/Microbiology, Brookings, SD 57007

AWARDS Biology (MS); microbiology (MS).

Degree requirements: For master's, thesis, oral exam required, foreign language not required.

Entrance requirements: For master's, GRE, TOEFL.
Faculty research: Plant tissue culture, molecular biology studies of metabolic regulation in plants, mechanisms of mammalian gene expression, aquatic-wetland ecosystem ecology, stress-induced immunosuppression on parasite-induced pathology.

■ **SOUTHERN ILLINOIS UNIVERSITY CARBONDALE**

Graduate School, College of Science, Program in Molecular Biology, Microbiology, and Biochemistry, Carbondale, IL 62901-6806

AWARDS MS, PhD.

Faculty: 16 full-time (2 women).
Students: 51 full-time (24 women), 11 part-time (8 women); includes 4 minority (2 African Americans, 1 Asian American or Pacific Islander, 1 Hispanic American), 24 international. Average age 25. *34 applicants, 59% accepted.* In 1999, 7 master's, 5 doctorates awarded.
Degree requirements: For master's and doctorate, thesis/dissertation required, foreign language not required.
Entrance requirements: For master's, GRE, TOEFL, minimum GPA of 2.7; for doctorate, GRE, TOEFL, minimum GPA of 3.25. *Application deadline:* Applications are processed on a rolling basis. *Application fee:* $20.
Expenses: Tuition, state resident: full-time $2,902. Tuition, nonresident: full-time $5,810. Tuition and fees vary according to course load.
Financial aid: In 1999–00, 40 students received aid, including 3 fellowships with full tuition reimbursements available, 24 research assistantships with full tuition reimbursements available, 12 teaching assistantships with full tuition reimbursements available; Federal Work-Study and institutionally sponsored loans also available. Aid available to part-time students. Financial aid application deadline: 3/1.
Faculty research: Prokaryotic gene regulation and expression; eukaryotic gene regulation; microbial, phylogenetic, and metabolic diversity; immune responses to tumors, pathogens, and autoantigens; protein folding and structure.
Dr. John Martinko, Director, 618-536-2349, *Fax:* 618-453-8036, *E-mail:* martinko.mbmb@science.siu.edu.

Application contact: Donna Mueller, Office Systems Specialist II, 618-536-2349, *Fax:* 618-453-8036, *E-mail:* mueller@micro.siu.edu.

■ STANFORD UNIVERSITY

School of Medicine, Graduate Programs in Medicine, Department of Microbiology and Immunology, Stanford, CA 94305-9991
AWARDS PhD.

Faculty: 11 full-time (3 women).
Students: 20 full-time (9 women), 9 part-time (4 women); includes 7 minority (1 African American, 5 Asian Americans or Pacific Islanders, 1 Hispanic American), 3 international. Average age 27. 77 *applicants, 17% accepted.* In 1999, 1 doctorate awarded.
Degree requirements: For doctorate, dissertation required, foreign language not required.
Entrance requirements: For doctorate, GRE General Test, GRE Subject Test (biology or chemistry), TOEFL. *Application deadline:* For fall admission, 12/15. *Application fee:* $65 ($80 for international students). Electronic applications accepted.
Expenses: Tuition: Full-time $23,058. Required fees: $152. Part-time tuition and fees vary according to course load.
Financial aid: Research assistantships available. Financial aid application deadline: 12/15.
Faculty research: Molecular pathogenesis of bacteria viruses and parasites, immune system function, autoimmunity, molecular biology.
Edward Mocarski, Chair, 650-723-6435, *Fax:* 650-725-6757, *E-mail:* mocarski@stanford.edu.
Application contact: Student Services Coordinator, 650-725-8541.

Find an in-depth description at www.petersons.com/graduate.

■ STATE UNIVERSITY OF NEW YORK AT BUFFALO

Graduate School, School of Medicine and Biomedical Sciences, Graduate Programs in Medicine and Biomedical Sciences, Department of Microbiology, Buffalo, NY 14260
AWARDS MA, PhD.

Faculty: 21 full-time (6 women).
Students: 12 full-time (10 women), 14 part-time (8 women); includes 3 minority (1 African American, 2 Hispanic Americans), 9 international. Average age 29. 7 *applicants, 100% accepted.* In 1999, 4 master's, 4 doctorates awarded. Terminal master's awarded for partial completion of doctoral program.
Degree requirements: For master's, comprehensive exam required, foreign language and thesis not required; for doctorate, dissertation, departmental qualifying exam required, foreign language not required.

Entrance requirements: For master's and doctorate, GRE General Test, TOEFL. *Application deadline:* For fall admission, 2/1 (priority date). *Application fee:* $35.
Expenses: Tuition, state resident: full-time $5,100. Tuition, nonresident: full-time $8,416. Required fees: $935.
Financial aid: In 1999–00, 4 fellowships with tuition reimbursements, 18 research assistantships with tuition reimbursements (averaging $15,000 per year) were awarded; teaching assistantships, Federal Work-Study, institutionally sponsored loans, tuition waivers (full and partial), and unspecified assistantships also available. Financial aid application deadline: 2/1; financial aid applicants required to submit FAFSA.
Faculty research: Bacteriology, immunology, parasitology, virology, microbial pathogenesis. *Total annual research expenditures:* $5.6 million.
Dr. John Hay, Chairman, 716-829-2907, *Fax:* 716-829-2158.
Application contact: Dr. Noreen Williams, Director of Graduate Studies, 716-829-2176, *Fax:* 716-829-2158, *E-mail:* cumming@buffalo.edu.

Find an in-depth description at www.petersons.com/graduate.

■ STATE UNIVERSITY OF NEW YORK AT STONY BROOK

Health Sciences Center, School of Medicine and Graduate School, Graduate Programs in Medicine, Department of Molecular Genetics and Microbiology, Stony Brook, NY 11794
AWARDS Molecular microbiology (PhD).

Faculty: 28.
Students: 22 full-time (12 women), 11 part-time (4 women); includes 5 minority (2 African Americans, 2 Asian Americans or Pacific Islanders, 1 Hispanic American), 13 international. Average age 26. 77 *applicants, 19% accepted.* In 1999, 8 degrees awarded.
Degree requirements: For doctorate, dissertation, comprehensive exam required.
Entrance requirements: For doctorate, GRE General Test, GRE Subject Test, TOEFL. *Application deadline:* For fall admission, 1/15. *Application fee:* $50.
Expenses: Tuition, state resident: full-time $5,100. Tuition, nonresident: full-time $8,416. Required fees: $492.
Financial aid: In 1999–00, 18 fellowships, 1 teaching assistantship were awarded; research assistantships, Federal Work-Study also available. Financial aid application deadline: 3/15.
Faculty research: Adenovirus molecular genetics, molecular biology of tumors, virus SV40, mechanism of tumor infection

by SAV virus. *Total annual research expenditures:* $4.7 million.
Dr. Michael Hayman, Acting Chair, 631-632-8813, *Fax:* 631-632-9797.
Application contact: Dr. Patrick Hearing, Director, 631-632-8813, *Fax:* 631-632-9797, *E-mail:* hearing@asterix.bio.sunysb.edu.

Find an in-depth description at www.petersons.com/graduate.

■ STATE UNIVERSITY OF NEW YORK HEALTH SCIENCE CENTER AT BROOKLYN

School of Graduate Studies, Department of Microbiology and Immunology, Brooklyn, NY 11203-2098
AWARDS PhD, MD/PhD.

Degree requirements: For doctorate, one foreign language, dissertation, comprehensive qualifying exam required.
Entrance requirements: For doctorate, GRE General Test, GRE Subject Test (strongly recommended).
Expenses: Tuition, state resident: full-time $5,100; part-time $213 per credit. Tuition, nonresident: full-time $8,416; part-time $351 per credit. Required fees: $200. Full-time tuition and fees vary according to program and student level.
Faculty research: Yeast molecular genetics, regulation of gene expression, retroviral-mediated gene transfer, nuclear acid structure: regulation and change within RNA genomes.

■ STATE UNIVERSITY OF NEW YORK UPSTATE MEDICAL UNIVERSITY

College of Graduate Studies, Department of Microbiology and Immunology, Syracuse, NY 13210-2334
AWARDS MS, PhD, MD/PhD.

Faculty: 17.
Students: 16 full-time (7 women); includes 2 minority (1 Asian American or Pacific Islander, 1 Hispanic American), 3 international. *50 applicants, 36% accepted.* In 1999, 3 doctorates awarded (33% found work related to degree, 67% continued full-time study). Terminal master's awarded for partial completion of doctoral program.
Degree requirements: For master's, thesis required, foreign language not required; for doctorate, dissertation, comprehensive exam required, foreign language not required.
Entrance requirements: For master's and doctorate, GRE General Test, GRE Subject Test, TSE, interview. *Application deadline:* For fall admission, 4/1 (priority

State University of New York Upstate Medical University (continued)
date). Applications are processed on a rolling basis. *Application fee:* $40.

Expenses: Tuition, state resident: full-time $5,100; part-time $213 per credit. Tuition, nonresident: full-time $8,416; part-time $351 per credit. Required fees: $410; $25 per credit. Part-time tuition and fees vary according to course load and program.

Financial aid: Research assistantships, teaching assistantships available.

Faculty research: Virology, molecular biology, microbial genetics, parasitology. Dr. Edward Shillitoe, Chairperson, 315-464-5453.

Application contact: Dr. Allen E. Silverstone, Graduate Program Director, 315-464-5453.

■ TEMPLE UNIVERSITY

Health Sciences Center, School of Medicine and Graduate School, Graduate Programs in Medicine, Department of Microbiology and Immunology, Philadelphia, PA 19122-6096

AWARDS MS, PhD, MD/PhD.

Faculty: 19 full-time (4 women), 17 part-time/adjunct (7 women).

Students: 51 full-time. In 1999, 2 master's awarded (100% found work related to degree); 6 doctorates awarded (100% found work related to degree). Terminal master's awarded for partial completion of doctoral program.

Degree requirements: For master's, thesis required; for doctorate, dissertation, research seminars required. *Average time to degree:* Master's–3 years full-time; doctorate–5 years full-time.

Entrance requirements: For master's and doctorate, GRE General Test, TOEFL, minimum GPA of 3.0. *Application deadline:* For fall admission, 7/1; for spring admission, 11/1. Applications are processed on a rolling basis. *Application fee:* $40.

Expenses: Tuition, state resident: full-time $6,030. Tuition, nonresident: full-time $8,298. Required fees: $230. One-time fee: $10 full-time.

Financial aid: Fellowships, research assistantships, career-related internships or fieldwork, Federal Work-Study, grants, institutionally sponsored loans, and tuition waivers (full and partial) available. Aid available to part-time students. Financial aid application deadline: 3/15; financial aid applicants required to submit FAFSA.

Faculty research: Molecular and cellular immunology, molecular and biochemical microbiology, molecular genetics. *Total annual research expenditures:* $3.4 million.

Dr. Chris D. Platsoucas, Chair, 215-707-7929, *Fax:* 215-707-7788, *E-mail:* microimm@vm.temple.edu.

Find an in-depth description at www.petersons.com/graduate.

■ TEXAS A&M UNIVERSITY

College of Science, Department of Biology, Program in Microbiology, College Station, TX 77843

AWARDS MS, PhD.

Students: 20 full-time (9 women), 3 part-time (2 women). Average age 28. *55 applicants, 9% accepted.* In 1999, 2 doctorates awarded.

Degree requirements: For master's, foreign language not required; for doctorate, dissertation required, foreign language not required.

Entrance requirements: For master's and doctorate, GRE General Test, TOEFL. *Application fee:* $50 ($75 for international students).

Expenses: Tuition, state resident: part-time $76 per semester hour. Tuition, nonresident: part-time $292 per semester hour. Required fees: $11 per semester hour. Tuition and fees vary according to program.

Financial aid: Application deadline: 4/1. Dr. Mark Zoran, Head, 979-845-7755.

■ TEXAS A&M UNIVERSITY

College of Veterinary Medicine, Department of Veterinary Pathobiology, College Station, TX 77843

AWARDS Genetics (MS, PhD); toxicology (MS, PhD); veterinary microbiology (MS, PhD); veterinary parasitology (MS); veterinary pathology (MS, PhD). Part-time programs available. Postbaccalaureate distance learning degree programs offered.

Faculty: 44 full-time (10 women), 17 part-time/adjunct (2 women).

Students: 33 full-time (13 women), 30 part-time (18 women); includes 11 minority (2 African Americans, 1 Asian American or Pacific Islander, 8 Hispanic Americans), 11 international. Average age 34. *28 applicants, 46% accepted.* In 1999, 4 master's awarded (25% found work related to degree, 75% continued full-time study); 10 doctorates awarded (40% entered university research/teaching, 60% found other work related to degree). Terminal master's awarded for partial completion of doctoral program.

Degree requirements: For master's and doctorate, thesis/dissertation required, foreign language not required. *Average time to degree:* Master's–2 years full-time, 4.5 years part-time; doctorate–4.5 years full-time, 7 years part-time.

Entrance requirements: For master's and doctorate, GRE General Test, TOEFL. *Application deadline:* For fall admission, 3/1 (priority date); for spring admission, 8/1 (priority date). Applications are processed on a rolling basis. *Application fee:* $50 ($75 for international students). Electronic applications accepted.

Expenses: Tuition, state resident: full-time $5,400. Tuition, nonresident: full-time $16,200. Required fees: $2,936. Full-time tuition and fees vary according to student level.

Financial aid: In 1999–00, 7 fellowships with partial tuition reimbursements (averaging $16,000 per year), 13 research assistantships with partial tuition reimbursements (averaging $14,400 per year), 5 teaching assistantships with partial tuition reimbursements (averaging $16,000 per year) were awarded; career-related internships or fieldwork, Federal Work-Study, and institutionally sponsored loans also available. Aid available to part-time students. Financial aid applicants required to submit FAFSA.

Faculty research: Infectious and noninfectious diseases of animals and birds, animal genetics, molecular biology, immunology, virology. *Total annual research expenditures:* $2.6 million.

Dr. Ann B. Kier, Head, 979-845-5941, *Fax:* 979-845-9231, *E-mail:* akier@cvm.tamu.edu.

Application contact: Dr. G. G. Wagner, Graduate Adviser, 979-845-5941, *Fax:* 979-862-1147, *E-mail:* gwagner@cvm.tamus.edu.

■ TEXAS A&M UNIVERSITY SYSTEM HEALTH SCIENCE CENTER

College of Medicine, Graduate School of Biomedical Sciences, Department of Medical Microbiology and Immunology, College Station, TX 77840-7896

AWARDS Immunology (PhD); microbiology (PhD); molecular biology (PhD); virology (PhD).

Faculty: 7 full-time (0 women), 1 part-time/adjunct (0 women).

Students: 16 full-time (10 women); includes 2 minority (1 African American, 1 Asian American or Pacific Islander).

Degree requirements: For doctorate, dissertation required, foreign language not required.

Entrance requirements: For doctorate, GRE General Test, minimum GPA of 3.0. *Application deadline:* For fall admission, 2/1 (priority date). *Application fee:* $50 ($75 for international students).

Expenses: Tuition, area resident: Full-time $1,368. Tuition, state resident: part-time

$76 per credit. Tuition, nonresident: full-time $5,256; part-time $292 per credit. International tuition: $5,256 full-time. Required fees: $678; $38 per credit. Full-time tuition and fees vary according to course load and student level.

Financial aid: In 1999–00, 16 students received aid, including 1 fellowship (averaging $17,000 per year), 15 research assistantships (averaging $17,000 per year). Financial aid application deadline: 4/1; financial aid applicants required to submit FAFSA.

Faculty research: Molecular pathogenesis, microbial therapeutics. *Total annual research expenditures:* $1.2 million.

Dr. John M. Quarles, Interim Head, 979-845-1313, *Fax:* 979-845-3479, *E-mail:* jmquarles@medicine.tamu.edu.

Application contact: Dr. James Samuel, Graduate Adviser, 979-845-1313, *Fax:* 979-845-3479, *E-mail:* jsamuel@medicine.tamu.edu.

Find an in-depth description at www.petersons.com/graduate.

■ **TEXAS TECH UNIVERSITY**

Graduate School, College of Arts and Sciences, Department of Biological Sciences, Lubbock, TX 79409

AWARDS Biology (MS, PhD); environmental toxicology (MS); microbiology (MS); zoology (MS, PhD). Part-time programs available.

Faculty: 36 full-time (4 women), 1 part-time/adjunct (0 women).

Students: 86 full-time (41 women), 26 part-time (10 women); includes 5 minority (1 Asian American or Pacific Islander, 4 Hispanic Americans), 28 international. Average age 30. *54 applicants, 46% accepted.* In 2000, 26 master's, 11 doctorates awarded.

Degree requirements: For master's, thesis required (for some programs), foreign language not required; for doctorate, dissertation required, foreign language not required.

Entrance requirements: For master's and doctorate, GRE General Test. *Application deadline:* For fall admission, 4/15 (priority date); for spring admission, 11/1 (priority date). Applications are processed on a rolling basis. *Application fee:* $25 ($50 for international students). Electronic applications accepted.

Expenses: Tuition, state resident: full-time $2,376; part-time $99 per credit hour. Tuition, nonresident: full-time $7,560; part-time $315 per credit hour. Required fees: $464 per semester. Part-time tuition and fees vary according to course load, program and reciprocity agreements.

Financial aid: In 2000–01, 60 students received aid, including 40 research assistantships (averaging $10,405 per year),

53 teaching assistantships (averaging $11,067 per year); fellowships, career-related internships or fieldwork, Federal Work-Study, and institutionally sponsored loans also available. Aid available to part-time students. Financial aid application deadline: 5/15; financial aid applicants required to submit FAFSA.

Faculty research: Development of strains of transgenic plants, ecological studies of Arctic tundra and Puerto Rican rain forests, genome organization and evolution. *Total annual research expenditures:* $2.1 million.

Dr. Carleton Phillips, Chairman, 806-742-2715, *Fax:* 806-742-2963.

Application contact: Graduate Adviser, 806-742-2715, *Fax:* 806-742-2963.

■ **THOMAS JEFFERSON UNIVERSITY**

College of Graduate Studies, Program in Microbiology, Philadelphia, PA 19107

AWARDS MS. Part-time and evening/weekend programs available.

Faculty: 11 full-time (2 women), 2 part-time/adjunct (0 women).

Students: 6 full-time (4 women), 54 part-time (32 women); includes 10 minority (5 African Americans, 4 Asian Americans or Pacific Islanders, 1 Hispanic American), 2 international. Average age 30. *46 applicants, 80% accepted.* In 1999, 28 degrees awarded.

Degree requirements: For master's, thesis required, foreign language not required.

Entrance requirements: For master's, GRE General Test, minimum GPA of 3.0. *Application deadline:* Applications are processed on a rolling basis. *Application fee:* $40.

Expenses: Tuition: Full-time $17,625; part-time $610 per credit.

Financial aid: In 1999–00, 22 students received aid. Federal Work-Study and institutionally sponsored loans available. Aid available to part-time students. Financial aid application deadline: 5/1; financial aid applicants required to submit FAFSA.

Faculty research: Virological procedures for characterization of HIV isolates, epidemiological studies of nasocomial infections, immunogenetics, molecular microbiology for identification of infectious diseases.

Dr. Georganne K. Buescher, Associate Dean, 215-503-5799, *Fax:* 215-503-3433, *E-mail:* georganne.buescher@mail.tju.edu.

Application contact: Jessie F. Pervall, Director of Admissions, 215-503-4400, *Fax:* 215-503-3433, *E-mail:* cgs-info@mail.tju.edu.

■ **THOMAS JEFFERSON UNIVERSITY**

College of Graduate Studies, Program in Microbiology and Molecular Virology, Philadelphia, PA 19107

AWARDS PhD.

Faculty: 24 full-time (1 woman).

Students: 13 full-time (7 women), 2 part-time; includes 2 minority (1 African American, 1 Asian American or Pacific Islander), 4 international. *56 applicants, 13% accepted.* In 1999, 2 degrees awarded.

Degree requirements: For doctorate, dissertation required, foreign language not required.

Entrance requirements: For doctorate, GRE General Test, TOEFL, minimum GPA of 3.2. *Application deadline:* For fall admission, 3/1 (priority date). Applications are processed on a rolling basis. *Application fee:* $40.

Expenses: Tuition: Full-time $12,670. Tuition and fees vary according to degree level and program.

Financial aid: In 1999–00, 14 fellowships with full tuition reimbursements were awarded; research assistantships, Federal Work-Study, institutionally sponsored loans, traineeships, and training grants also available. Aid available to part-time students. Financial aid application deadline: 5/1; financial aid applicants required to submit FAFSA.

Faculty research: Chemical and antigenic structure of virus particles, viral replication, viral gene expression, viral oncogenes, parasite immunology. *Total annual research expenditures:* $17 million.

Dr. David Abraham, Chairman, Graduate Committee, 215-503-8917, *Fax:* 215-923-9248, *E-mail:* abrahamd@jeflin.tju.edu.

Application contact: Jessie F. Pervall, Director of Admissions, 215-503-4400, *Fax:* 215-503-3433, *E-mail:* cgs-info@mail.tju.edu.

Find an in-depth description at www.petersons.com/graduate.

■ **TUFTS UNIVERSITY**

Sackler School of Graduate Biomedical Sciences, Department of Molecular Biology and Microbiology, Medford, MA 02155

AWARDS Molecular microbiology (PhD).

Faculty: 16 full-time (5 women).

Students: 42 full-time (20 women). *116 applicants, 7% accepted.* In 1999, 10 degrees awarded (8% entered university research/teaching, 1% found other work related to degree, 1% continued full-time study).

Degree requirements: For doctorate, dissertation required, foreign language not required. *Average time to degree:* Doctorate–6 years full-time.

Tufts University (continued)

Entrance requirements: For doctorate, GRE General Test, TOEFL. *Application deadline:* For fall admission, 1/15 (priority date). Applications are processed on a rolling basis. *Application fee:* $45.

Expenses: Tuition: Full-time $19,325.

Financial aid: In 1999–00, 42 research assistantships with full tuition reimbursements (averaging $18,805 per year) were awarded; scholarships and tuition waivers (full) also available. Financial aid application deadline: 2/1.

Faculty research: Fundamental problems of molecular biology of prokaryotes, eukaryotes, and their viruses. *Total annual research expenditures:* $5 million.

Dr. Catherine L. Squires, Chair, 617-636-6947, *Fax:* 617-636-0337, *E-mail:* csquires_rib@opal.tufts.edu.

Application contact: Secretary, 617-636-6767, *Fax:* 617-636-0375, *E-mail:* sackler-school@tufts.edu.

■ **TULANE UNIVERSITY**

School of Medicine and Graduate School, Graduate Programs in Medicine, Department of Microbiology and Immunology, New Orleans, LA 70118-5669

AWARDS MS, PhD, MD/PhD. MS and PhD offered through the Graduate School.

Students: 41 full-time (18 women); includes 6 minority (3 African Americans, 3 Asian Americans or Pacific Islanders), 9 international. 87 *applicants,* 9% *accepted.* In 1999, 2 master's awarded.

Degree requirements: For master's, thesis required; for doctorate, 2 foreign languages (computer language can substitute for one), dissertation required.

Entrance requirements: For master's, GRE General Test, TOEFL, or TSE, minimum B average in undergraduate course work; for doctorate, GRE General Test, GRE Subject Test, TOEFL, or TSE. *Application deadline:* For fall admission, 2/1. *Application fee:* $45.

Expenses: Tuition: Full-time $23,030.

Financial aid: Fellowships, research assistantships, career-related internships or fieldwork, Federal Work-Study, and institutionally sponsored loans available. Financial aid application deadline: 2/1.

Dr. John Clements, Chairman, 504-588-5159.

Find an in-depth description at www.petersons.com/graduate.

■ **UNIFORMED SERVICES UNIVERSITY OF THE HEALTH SCIENCES**

School of Medicine, Division of Basic Medical Sciences, Department of Microbiology and Immunology, Bethesda, MD 20814-4799

AWARDS PhD.

Faculty: 13 full-time (7 women), 8 part-time/adjunct (4 women).

Students: 9 full-time (7 women); includes 2 minority (1 African American, 1 Hispanic American). Average age 28. *20 applicants,* 5% *accepted.*

Degree requirements: For doctorate, one foreign language (computer language can substitute), dissertation, qualifying exam required.

Entrance requirements: For doctorate, GRE General Test, GRE Subject Test, minimum GPA of 3.0, U.S. citizenship. *Application deadline:* For fall admission, 1/15 (priority date). Applications are processed on a rolling basis. *Application fee:* $0.

Financial aid: In 1999–00, 6 fellowships with full tuition reimbursements (averaging $15,000 per year) were awarded; tuition waivers (full) also available.

Faculty research: Infections diseases, virology, bacteriology, microbial pathogenesis.

Dr. Alison O'Brien, Chair, 301-295-3400.

Application contact: Janet M. Anastasi, Graduate Program Coordinator, 301-295-9474, *Fax:* 301-295-6772, *E-mail:* janastasi@usuhs.mil.

Find an in-depth description at www.petersons.com/graduate.

■ **THE UNIVERSITY OF ALABAMA AT BIRMINGHAM**

Graduate School and School of Medicine and School of Dentistry, Graduate Programs in Joint Health Sciences, Department of Microbiology, Birmingham, AL 35294

AWARDS PhD. The department participates in the Cellular and Molecular Biology Graduate Program.

Students: In 1999, 13 degrees awarded.

Degree requirements: For doctorate, dissertation required, foreign language not required.

Entrance requirements: For doctorate, GRE General Test, interview. *Application deadline:* Applications are processed on a rolling basis. Electronic applications accepted.

Expenses: Tuition, state resident: part-time $104 per semester hour. Tuition, nonresident: part-time $208 per semester hour. Required fees: $17 per semester

hour. $57 per quarter. Tuition and fees vary according to program.

Financial aid: Fellowships available.

Dr. Suzanne M. Michalek, Interim Chair, 205-934-3470, *Fax:* 205-934-1426, *E-mail:* suemich@uab.edu.

Application contact: Information Contact, 205-934-0621, *Fax:* 205-975-2536.

Find an in-depth description at www.petersons.com/graduate.

■ **THE UNIVERSITY OF ARIZONA**

College of Medicine, Graduate Programs in Medicine, Department of Microbiology and Immunology, Tucson, AZ 85721

AWARDS MS, PhD.

Degree requirements: For master's and doctorate, thesis/dissertation required, foreign language not required.

Entrance requirements: For master's and doctorate, GRE General Test, minimum GPA of 3.0.

Expenses: Tuition, nonresident: full-time $4,814; part-time $274 per unit. Required fees: $1,094; $115 per unit.

Faculty research: Environmental and pathogenic microbiology, molecular biology.

Find an in-depth description at www.petersons.com/graduate.

■ **UNIVERSITY OF ARKANSAS FOR MEDICAL SCIENCES**

College of Medicine and Graduate School, Graduate Programs in Medicine, Department of Microbiology and Immunology, Little Rock, AR 72205-7199

AWARDS MS, PhD, MD/PhD.

Faculty: 15 full-time (2 women).

Students: 17 full-time (10 women), 1 part-time; includes 1 minority (African American), 5 international. In 1999, 2 master's, 2 doctorates awarded.

Degree requirements: For master's and doctorate, thesis/dissertation required, foreign language not required.

Entrance requirements: For master's and doctorate, GRE General Test. *Application fee:* $0.

Expenses: Tuition: Full-time $8,928.

Financial aid: Research assistantships available. Aid available to part-time students.

Dr. Roger G. Rank, Chairman, 501-686-5144.

Application contact: Dr. Marie Chow, Co-Coordinator of Graduate Studies, 501-686-5144.

Find an in-depth description at www.petersons.com/graduate.

■ UNIVERSITY OF CALIFORNIA, BERKELEY

Graduate Division, Group in Microbiology, Berkeley, CA 94720-1500

AWARDS PhD.

Degree requirements: For doctorate, dissertation required.
Entrance requirements: For doctorate, GRE General Test, minimum GPA of 3.0.
Expenses: Tuition, nonresident: full-time $9,804. Required fees: $4,268. Tuition and fees vary according to program.

■ UNIVERSITY OF CALIFORNIA, DAVIS

Graduate Studies, Programs in the Biological Sciences, Program in Microbiology, Davis, CA 95616

AWARDS MS, PhD.

Faculty: 76.
Students: 56 full-time (30 women); includes 15 minority (1 African American, 12 Asian Americans or Pacific Islanders, 2 Hispanic Americans), 15 international. Average age 29. *86 applicants, 31% accepted.* In 1999, 2 master's, 11 doctorates awarded. Terminal master's awarded for partial completion of doctoral program.
Degree requirements: For master's and doctorate, thesis/dissertation required.
Entrance requirements: For master's and doctorate, GRE General Test. *Application deadline:* For fall admission, 1/15. *Application fee:* $40. Electronic applications accepted.
Expenses: Tuition, nonresident: full-time $9,804. Tuition and fees vary according to program and student level.
Financial aid: In 1999–00, 51 students received aid, including 10 fellowships with full and partial tuition reimbursements available, 32 research assistantships with full and partial tuition reimbursements available, 6 teaching assistantships with full and partial tuition reimbursements available; Federal Work-Study, institutionally sponsored loans, scholarships, and tuition waivers (full and partial) also available. Financial aid application deadline: 1/15; financial aid applicants required to submit FAFSA.
Faculty research: Microbial physiology and genetics, microbial molecular and cellular biology, microbial ecology, microbial pathogenesis and immunology, urology.
Linda Bisson, Graduate Chair, 530-752-3835, *E-mail:* lfbisson@ucdavis.edu.
Application contact: Lewanna Archen, Administrative Assistant, 530-752-0262, *Fax:* 530-752-0914, *E-mail:* leachen@ucdavis.edu.

Find an in-depth description at www.petersons.com/graduate.

■ UNIVERSITY OF CALIFORNIA, IRVINE

College of Medicine and Office of Research and Graduate Studies, Graduate Programs in Medicine and School of Biological Sciences, Department of Microbiology and Molecular Genetics, Irvine, CA 92697

AWARDS Biological sciences (PhD). Students apply through the Graduate Program in Molecular Biology, Genetics, and Biochemistry.

Faculty: 11 full-time (2 women).
Students: 26 full-time (13 women); includes 7 minority (all Asian Americans or Pacific Islanders), 3 international. *252 applicants, 34% accepted.*
Degree requirements: For doctorate, dissertation required, foreign language not required.
Entrance requirements: For doctorate, GRE General Test, GRE Subject Test. *Application deadline:* For fall admission, 2/1 (priority date). *Application fee:* $40. Electronic applications accepted.
Expenses: Tuition, nonresident: full-time $10,322; part-time $1,720 per quarter. Required fees: $5,354; $1,300 per quarter. Tuition and fees vary according to program.
Financial aid: Fellowships, research assistantships, teaching assistantships, institutionally sponsored loans and tuition waivers (full and partial) available. Financial aid application deadline: 3/2; financial aid applicants required to submit FAFSA.
Faculty research: Molecular biology and genetics of viruses, bacteria, and yeast; immune response; molecular biology of cultured animal cells; genetic basis of cancer; genetics and physiology of infectious agents.
Bert L. Semler, Chair, 949-824-7573 Ext. 6058, *Fax:* 949-824-8598, *E-mail:* blsemler@uci.edu.
Application contact: 949-824-5261, *Fax:* 949-824-8598, *E-mail:* gp-mbgb@uci.edu.

Find an in-depth description at www.petersons.com/graduate.

■ UNIVERSITY OF CALIFORNIA, LOS ANGELES

Graduate Division, College of Letters and Science, Department of Microbiology and Molecular Genetics, Los Angeles, CA 90095

AWARDS PhD.

Students: 27 full-time (17 women); includes 10 minority (9 Asian Americans or Pacific Islanders, 1 Hispanic American), 2 international. *3 applicants, 67% accepted.*

Degree requirements: For doctorate, dissertation, oral and written qualifying exams required, foreign language not required.
Entrance requirements: For doctorate, GRE General Test, GRE Subject Test (biology or chemistry recommended), minimum undergraduate GPA of 3.0. *Application deadline:* For fall admission, 12/15. *Application fee:* $40. Electronic applications accepted.
Expenses: Tuition, nonresident: full-time $9,804. Required fees: $4,405. Full-time tuition and fees vary according to program and student level.
Financial aid: Fellowships, research assistantships, teaching assistantships, Federal Work-Study, institutionally sponsored loans, scholarships, and tuition waivers (full and partial) available. Financial aid application deadline: 3/1.
Dr. Sherie Morrison, Chair, 310-825-3578.
Application contact: UCLA Access Coordinator, 800-284-8252, *Fax:* 310-206-5280, *E-mail:* uclaaccess@ibes.medsch.ucla.edu.

Find an in-depth description at www.petersons.com/graduate.

■ UNIVERSITY OF CALIFORNIA, LOS ANGELES

School of Medicine and Graduate Division, Graduate Programs in Medicine, Department of Microbiology and Immunology, Los Angeles, CA 90095

AWARDS MS, PhD.

Students: 47 full-time (25 women); includes 19 minority (2 African Americans, 15 Asian Americans or Pacific Islanders, 2 Hispanic Americans), 3 international. *1 applicant, 0% accepted.*
Degree requirements: For doctorate, dissertation, oral and written qualifying exams required, foreign language not required.
Entrance requirements: For doctorate, GRE General Test, GRE Subject Test, TOEFL. *Application fee:* $40.
Expenses: Tuition, nonresident: full-time $9,804. Required fees: $4,405.
Financial aid: In 1999–00, 44 students received aid, including 44 fellowships, 9 teaching assistantships; research assistantships, Federal Work-Study, institutionally sponsored loans, and tuition waivers (full and partial) also available. Financial aid application deadline: 3/1.
Dr. Sherie Morrison, Chair, 310-206-5148.
Application contact: UCLA Access Coordinator, 800-284-8252, *Fax:* 310-206-3280, *E-mail:* uclaaccess@ibes.medsch.ucla.edu.

Find an in-depth description at www.petersons.com/graduate.

■ UNIVERSITY OF CALIFORNIA, RIVERSIDE

Graduate Division, College of Natural and Agricultural Sciences, Program in Microbiology, Riverside, CA 92521-0102

AWARDS MS, PhD. Part-time programs available.

Faculty: 32 full-time (5 women).
Students: 13 full-time (7 women); includes 5 minority (3 Asian Americans or Pacific Islanders, 1 Hispanic American, 1 Native American), 3 international. In 1999, 3 master's awarded. Terminal master's awarded for partial completion of doctoral program.
Degree requirements: For master's, thesis required; for doctorate, dissertation, qualifying exams required.
Entrance requirements: For master's and doctorate, GRE General Test, TOEFL, minimum GPA of 3.2. *Application deadline:* For fall admission, 2/1 (priority date); for winter admission, 9/1; for spring admission, 12/1. Applications are processed on a rolling basis. *Application fee:* $40. Electronic applications accepted.
Expenses: Tuition, nonresident: full-time $9,804. Required fees: $4,758. Full-time tuition and fees vary according to program.
Financial aid: Fellowships, research assistantships, teaching assistantships, career-related internships or fieldwork, Federal Work-Study, institutionally sponsored loans, and tuition waivers (full and partial) available. Financial aid application deadline: 2/1; financial aid applicants required to submit FAFSA.
Faculty research: Host-pathogen interactions; environmental microbiology; bioremediation; molecular microbiology; microbial genetics, physiology, and pathogenesis.
Dr. David Crowley, Director, 909-787-4716, *Fax:* 909-787-3719, *E-mail:* microbio@ucrac1.ucr.edu.
Application contact: 909-787-4716, *Fax:* 909-787-5517, *E-mail:* microbio@ucrac1.ucr.edu.

■ UNIVERSITY OF CALIFORNIA, SAN FRANCISCO

Graduate Division, Department of Microbiology and Immunology, San Francisco, CA 94143

AWARDS PhD.

Faculty: 13 full-time (3 women).
Students: 15 full-time (7 women). *66 applicants, 15% accepted.* In 1999, 4 doctorates awarded.
Degree requirements: For doctorate, dissertation required, foreign language not required.

Entrance requirements: For doctorate, GRE General Test. *Application deadline:* For fall admission, 1/15. *Application fee:* $40.
Financial aid: Application deadline: 1/10. Liz Blackburn, Chair, 415-476-1212.
Application contact: Grace Stauffer, Program Assistant, 415-476-8204.

■ UNIVERSITY OF CENTRAL FLORIDA

College of Health and Public Affairs, Department of Molecular and Microbiology, Orlando, FL 32816

AWARDS Microbiology (MS); molecular biology (MS). Part-time and evening/weekend programs available.

Faculty: 10 full-time, 6 part-time/adjunct.
Students: 27 full-time (18 women), 11 part-time (8 women); includes 8 minority (1 African American, 7 Hispanic Americans), 5 international. Average age 30. *26 applicants, 69% accepted.* In 1999, 6 degrees awarded.
Degree requirements: For master's, thesis, comprehensive exam required, foreign language not required.
Entrance requirements: For master's, GRE General Test, TOEFL, minimum GPA of 3.0 in last 60 hours. *Application deadline:* For fall admission, 3/15 (priority date); for spring admission, 12/1. *Application fee:* $20.
Expenses: Tuition, state resident: full-time $2,054; part-time $137 per credit. Tuition, nonresident: full-time $7,207; part-time $480 per credit. Required fees: $47 per term.
Financial aid: In 1999–00, 10 fellowships with partial tuition reimbursements (averaging $3,100 per year), 31 research assistantships with partial tuition reimbursements (averaging $2,974 per year), 35 teaching assistantships with partial tuition reimbursements (averaging $3,690 per year) were awarded; career-related internships or fieldwork, Federal Work-Study, institutionally sponsored loans, tuition waivers (partial), and unspecified assistantships also available. Financial aid application deadline: 3/1; financial aid applicants required to submit FAFSA.
Dr. Robert Gennaro, Chair, 407-823-5932, *E-mail:* gennaro@mail.ucf.edu.

■ UNIVERSITY OF CINCINNATI

Division of Research and Advanced Studies, College of Medicine, Graduate Programs in Medicine, Department of Molecular Genetics, Biochemistry, Microbiology and Immunology, Cincinnati, OH 45267

AWARDS PhD.

Faculty: 22 full-time.
Students: 42 full-time (21 women), 3 part-time (2 women); includes 2 minority (both Asian Americans or Pacific Islanders), 3 international. *109 applicants, 17% accepted.* In 1999, 8 doctorates awarded.
Degree requirements: For doctorate, dissertation, qualifying exam required, foreign language not required. *Average time to degree:* Doctorate–7.3 years full-time.
Entrance requirements: For doctorate, GRE General Test, GRE Subject Test. *Application deadline:* For fall admission, 2/1 (priority date). Applications are processed on a rolling basis. *Application fee:* $30.
Expenses: Tuition, state resident: full-time $5,139; part-time $196 per credit hour. Tuition, nonresident: full-time $10,326; part-time $369 per credit hour. Required fees: $561; $187 per quarter.
Financial aid: Tuition waivers (full) and unspecified assistantships available. Financial aid application deadline: 5/1. *Total annual research expenditures:* $6.8 million.
Dr. Jerry Lingrel, Head, 513-558-5324, *Fax:* 513-558-1190, *E-mail:* jerry.lingrel@uc.edu.
Application contact: Iain Cartwright, Graduate Program Director, 513-558-5532, *Fax:* 513-558-8474, *E-mail:* iain.cartwright@uc.edu.

Find an in-depth description at www.petersons.com/graduate.

■ UNIVERSITY OF COLORADO AT BOULDER

Graduate School, College of Arts and Sciences, Department of Environmental, Population, and Organic Biology, Boulder, CO 80309

AWARDS Animal behavior (MA, PhD); aquatic biology (MA, PhD); behavioral genetics (MA, PhD); ecology (MA, PhD); microbiology (MA, PhD); neurobiology (MA, PhD); plant and animal physiology (MA, PhD); plant and animal systematics (MA, PhD); population biology (MA, PhD); population genetics (MA, PhD).

Faculty: 39 full-time (9 women).
Students: 84 full-time (36 women), 14 part-time (7 women); includes 17 minority (6 Asian Americans or Pacific Islanders, 10 Hispanic Americans, 1 Native American). Average age 29. *147 applicants, 14% accepted.* In 1999, 7 master's, 13 doctorates awarded. Terminal master's awarded for partial completion of doctoral program.
Degree requirements: For master's, thesis or alternative, comprehensive exam required, foreign language not required; for doctorate, dissertation, comprehensive exam required, foreign language not required. *Average time to degree:* Master's–3 years full-time; doctorate–5 years full-time.

Entrance requirements: For master's, GRE General Test, GRE Subject Test, minimum undergraduate GPA of 3.0; for doctorate, GRE General Test, GRE Subject Test. *Application deadline:* For fall admission, 1/15 (priority date). *Application fee:* $40 ($60 for international students).
Expenses: Tuition, state resident: part-time $181 per credit hour. Tuition, nonresident: part-time $542 per credit hour. Required fees: $99 per term. Tuition and fees vary according to course load and program.
Financial aid: Fellowships, research assistantships, teaching assistantships, Federal Work-Study, institutionally sponsored loans, and tuition waivers (full) available. Financial aid application deadline: 3/1.
Faculty research: Evolution, developmental biology, behavior and neurobiology. *Total annual research expenditures:* $1.8 million.
Michael Breed, Chair, 303-492-8981, *Fax:* 303-492-8699, *E-mail:* michael.breed@colorado.edu.
Application contact: Jill Skarstadt, Graduate Secretary, 303-492-7654, *Fax:* 303-492-8699, *E-mail:* jill.skarstadt@colorado.edu.

■ UNIVERSITY OF COLORADO HEALTH SCIENCES CENTER

Graduate School, Programs in Biological and Medical Sciences, Program in Microbiology, Denver, CO 80262

AWARDS PhD.

Degree requirements: For doctorate, dissertation required, foreign language not required.
Entrance requirements: For doctorate, GRE, minimum GPA of 2.75.
Expenses: Tuition, state resident: full-time $1,512; part-time $56 per hour. Tuition, nonresident: full-time $7,209; part-time $267 per hour. Full-time tuition and fees vary according to course load and program.
Find an in-depth description at www.petersons.com/graduate.

■ UNIVERSITY OF CONNECTICUT

Graduate School, College of Liberal Arts and Sciences, Biological Sciences Group, Storrs, CT 06269

AWARDS Ecology and evolutionary biology (MS, PhD), including botany, ecology, entomology, systematics, zoology; molecular and cell biology (MS, PhD), including biochemistry, biophysics, biotechnology (MS), cell and developmental biology, genetics,

microbiology, plant molecular and cell biology; physiology and neurobiology (MS, PhD), including neurobiology, physiology.

Degree requirements: For doctorate, dissertation required.
Entrance requirements: For master's and doctorate, GRE General Test, GRE Subject Test, TOEFL.
Expenses: Tuition, state resident: full-time $5,118. Tuition, nonresident: full-time $13,298. Required fees: $1,022.

■ UNIVERSITY OF CONNECTICUT

Graduate School, College of Liberal Arts and Sciences, Biological Sciences Group, Department of Molecular and Cell Biology, Field of Microbiology, Storrs, CT 06269

AWARDS MS, PhD.

Degree requirements: For doctorate, dissertation required.
Entrance requirements: For master's and doctorate, GRE General Test, GRE Subject Test, TOEFL.
Expenses: Tuition, state resident: full-time $5,118. Tuition, nonresident: full-time $13,298. Required fees: $1,022.
Find an in-depth description at www.petersons.com/graduate.

■ UNIVERSITY OF DELAWARE

College of Arts and Science, Department of Biological Sciences, Newark, DE 19716

AWARDS Biotechnology (MS, PhD); cell and extracellular matrix biology (MS, PhD); cell and systems physiology (MS, PhD); ecology and evolution (MS, PhD); microbiology (MS, PhD); molecular biology and genetics (MS, PhD); plant biology (MS, PhD).
Faculty: 37 full-time (10 women).
Students: 22 full-time (11 women), 1 part-time; includes 2 minority (both African Americans), 9 international. Average age 25. *37 applicants, 27% accepted.* In 2000, 9 doctorates awarded.
Degree requirements: For master's and doctorate, thesis/dissertation required, foreign language not required. *Average time to degree:* Master's–2.5 years full-time; doctorate–6 years full-time.
Entrance requirements: For master's and doctorate, GRE General Test, GRE Subject Test (advanced biology). *Application deadline:* For fall admission, 6/15. Applications are processed on a rolling basis. *Application fee:* $50. Electronic applications accepted.
Expenses: Tuition, state resident: full-time $4,380; part-time $243 per credit. Tuition, nonresident: full-time $12,750; part-time $708 per credit. Required fees: $15 per term. Tuition and fees vary according to program.

Financial aid: In 2000–01, 18 students received aid, including 2 fellowships with full tuition reimbursements available (averaging $18,000 per year), 4 research assistantships with full tuition reimbursements available (averaging $18,000 per year), 11 teaching assistantships with full tuition reimbursements available (averaging $18,000 per year); tuition waivers (partial) also available. Financial aid application deadline: 6/15.
Faculty research: Cell interactions, molecular mechanisms, microorganisms, embryo implantation. *Total annual research expenditures:* $1.8 million.
Dr. Daniel D. Carson, Chair, 302-831-6977, *Fax:* 302-831-2281, *E-mail:* dcarson@udel.edu.
Application contact: Norman Karin, Graduate Coordinator, 302-831-1841, *Fax:* 302-831-2281, *E-mail:* ccoletta@udel.edu.
Find an in-depth description at www.petersons.com/graduate.

■ UNIVERSITY OF FLORIDA

College of Medicine and Graduate School, Interdisciplinary Program in Biomedical Sciences, Concentration in Immunology and Microbiology, Gainesville, FL 32611

AWARDS PhD.

Degree requirements: For doctorate, dissertation required, foreign language not required.
Entrance requirements: For doctorate, GRE General Test, TOEFL, minimum GPA of 3.0. Electronic applications accepted.
Expenses: Tuition, state resident: part-time $144 per credit hour. Tuition, nonresident: part-time $505 per credit hour. Tuition and fees vary according to course level, course load and program.
Find an in-depth description at www.petersons.com/graduate.

■ UNIVERSITY OF FLORIDA

College of Medicine and Graduate School, Interdisciplinary Program in Biomedical Sciences, Department of Molecular Genetics and Microbiology, Gainesville, FL 32611

AWARDS MS, PhD. Terminal master's awarded for partial completion of doctoral program.

Degree requirements: For master's and doctorate, thesis/dissertation required, foreign language not required.
Entrance requirements: For master's and doctorate, GRE General Test, TOEFL, minimum GPA of 3.0. Electronic applications accepted.
Expenses: Tuition, state resident: part-time $144 per credit hour. Tuition,

University of Florida (continued)
nonresident: part-time $505 per credit hour. Tuition and fees vary according to course level, course load and program.

■ UNIVERSITY OF FLORIDA

Graduate School, College of Agriculture and Life Sciences, Department of Microbiology and Cell Science, Gainesville, FL 32611

AWARDS Cell biology (MS, PhD); microbiology (MS, PhD); microbiology and cell science (M Ag).

Faculty: 19.
Students: 21 full-time (11 women), 2 part-time; includes 7 minority (3 African Americans, 1 Asian American or Pacific Islander, 3 Hispanic Americans), 1 international. *31 applicants, 26% accepted.* In 1999, 5 doctorates awarded.
Degree requirements: For master's, computer language required; for doctorate, dissertation required.
Entrance requirements: For master's and doctorate, GRE General Test, minimum GPA of 3.0. *Application deadline:* For fall admission, 6/1 (priority date). Applications are processed on a rolling basis. *Application fee:* $20. Electronic applications accepted.
Expenses: Tuition, state resident: part-time $144 per credit hour. Tuition, nonresident: part-time $505 per credit hour. Tuition and fees vary according to course level, course load and program.
Financial aid: In 1999–00, 20 students received aid, including 1 fellowship, 11 research assistantships, 7 teaching assistantships
Faculty research: Biomass conversion, membrane and cell wall chemistry, plant biochemistry and genetics.
Dr. Edward M. Hoffman, Chair, 352-392-1906, *Fax:* 352-392-5922.
Application contact: Dr. Francis Davis, Graduate Coordinator, 352-392-1179, *Fax:* 352-392-5922, *E-mail:* fdavis@micro.ifas.ufl.edu.
Find an in-depth description at www.petersons.com/graduate.

■ UNIVERSITY OF GEORGIA

Graduate School, College of Arts and Sciences, Department of Microbiology, Athens, GA 30602

AWARDS MS, PhD.

Degree requirements: For master's, thesis required, foreign language not required; for doctorate, one foreign language (computer language can substitute), dissertation required.
Entrance requirements: For master's and doctorate, GRE General Test. Electronic applications accepted.

Expenses: Tuition, state resident: full-time $7,516; part-time $431 per credit hour. Tuition, nonresident: full-time $12,204; part-time $793 per credit hour. Tuition and fees vary according to program.

Find an in-depth description at www.petersons.com/graduate.

■ UNIVERSITY OF HAWAII AT MANOA

Graduate Division, College of Arts and Sciences, College of Natural Sciences, Department of Microbiology, Honolulu, HI 96822

AWARDS MS, PhD.

Faculty: 14 full-time (4 women).
Students: 17 full-time (2 women), 5 part-time (2 women). Average age 29. *31 applicants, 74% accepted.* In 1999, 3 master's awarded.
Degree requirements: For master's, thesis required (for some programs); for doctorate, dissertation required.
Entrance requirements: For master's and doctorate, GRE. *Application deadline:* For fall admission, 3/1; for spring admission, 9/1. *Application fee:* $25 ($50 for international students).
Expenses: Tuition, state resident: full-time $4,032; part-time $168 per credit. Tuition, nonresident: full-time $9,960; part-time $415 per credit. Required fees: $51 per semester. Part-time tuition and fees vary according to course load and program.
Financial aid: In 1999–00, 11 research assistantships (averaging $15,622 per year), 11 teaching assistantships (averaging $13,363 per year) were awarded.
Faculty research: Virology, immunology, microbial physiology, medical microbiology, bacterial genetics.
Dr. Paul Q. Patek, Chairperson, 808-956-8603.
Application contact: Philip Loh, Graduate Chair, 808-956-8553, *Fax:* 808-956-5339, *E-mail:* kkamiya@hawaii.edu.

■ UNIVERSITY OF IDAHO

College of Graduate Studies, College of Agriculture, Department of Microbiology, Molecular Biology and Biochemistry, Program in Microbiology, Moscow, ID 83844-4140

AWARDS MS, PhD.

Degree requirements: For master's, thesis required, foreign language not required; for doctorate, dissertation required.
Entrance requirements: For master's, minimum GPA of 2.8; for doctorate, minimum undergraduate GPA of 2.8, 3.0 graduate. *Application deadline:* For fall admission, 8/1; for spring admission,

12/15. *Application fee:* $35 ($45 for international students).
Expenses: Tuition, nonresident: full-time $6,000; part-time $239 per credit hour. Required fees: $2,888; $144 per credit hour. Tuition and fees vary according to program.
Financial aid: Application deadline: 2/15.
Dr. Greg Bohach, Head, Department of Microbiology, Molecular Biology and Biochemistry, 208-885-6666.

■ UNIVERSITY OF IDAHO

College of Graduate Studies, College of Agriculture, Department of Microbiology, Molecular Biology and Biochemistry, Program in Microbiology, Molecular Biology and Biochemistry, Moscow, ID 83844-4140

AWARDS MS, PhD.

Students: 28 full-time (7 women), 5 part-time (1 woman); includes 1 minority (Asian American or Pacific Islander), 14 international.
Degree requirements: For master's, thesis required, foreign language not required; for doctorate, dissertation required.
Entrance requirements: For master's, minimum GPA of 2.8; for doctorate, minimum undergraduate GPA of 2.8, 3.0 graduate. *Application deadline:* For fall admission, 8/1; for spring admission, 12/15. *Application fee:* $35 ($45 for international students).
Expenses: Tuition, nonresident: full-time $6,000; part-time $239 per credit hour. Required fees: $2,888; $144 per credit hour. Tuition and fees vary according to program.
Financial aid: Application deadline: 2/15.
Dr. Greg Bohach, Head, Department of Microbiology, Molecular Biology and Biochemistry, 208-885-6666.

■ UNIVERSITY OF ILLINOIS AT CHICAGO

College of Medicine and Graduate College, Graduate Programs in Medicine, Department of Microbiology and Immunology, Chicago, IL 60607-7128

AWARDS PhD, MD/PhD.

Faculty: 16 full-time (3 women).
Students: 28 full-time (20 women), 1 (woman) part-time; includes 3 minority (1 African American, 2 Asian Americans or Pacific Islanders), 18 international. Average age 27. *171 applicants, 10% accepted.* In 1999, 2 doctorates awarded.
Degree requirements: For doctorate, dissertation required, foreign language not required.

Entrance requirements: For doctorate, GRE General Test, TOEFL, minimum GPA of 3.75 on a 5.0 scale. *Application deadline:* For fall admission, 6/1; for spring admission, 11/1. *Application fee:* $40 ($50 for international students).
Expenses: Tuition, state resident: full-time $3,750. Tuition, nonresident: full-time $10,588. Tuition and fees vary according to course load.
Financial aid: In 1999–00, 15 students received aid; fellowships, research assistantships, teaching assistantships, Federal Work-Study, grants, and tuition waivers (full) available. Financial aid application deadline: 3/1; financial aid applicants required to submit FAFSA. Dr. Bellur Prabhakar, Head.
Application contact: Mia Johnson, Admissions and Records Officer, 312-996-9477, *Fax:* 312-996-6415, *E-mail:* ui8735@uicvm.uic.edu.

Find an in-depth description at www.petersons.com/graduate.

■ **UNIVERSITY OF ILLINOIS AT URBANA–CHAMPAIGN**

Graduate College, College of Liberal Arts and Sciences, School of Life Sciences, Department of Microbiology, Urbana, IL 61801

AWARDS MS, PhD, MD/PhD.

Faculty: 8 full-time (2 women), 3 part-time/adjunct (1 woman).
Students: 55 full-time (29 women); includes 10 minority (1 African American, 5 Asian Americans or Pacific Islanders, 4 Hispanic Americans), 19 international. *108 applicants, 7% accepted.* In 1999, 12 master's, 9 doctorates awarded.
Degree requirements: For master's, foreign language and thesis not required; for doctorate, dissertation required, foreign language not required.
Entrance requirements: For master's, minimum GPA of 4.0 on a 5.0 scale. *Application deadline:* Applications are processed on a rolling basis. *Application fee:* $40 ($50 for international students).
Expenses: Tuition, state resident: full-time $4,616. Tuition, nonresident: full-time $11,768. Full-time tuition and fees vary according to course load.
Financial aid: Fellowships, research assistantships, teaching assistantships available. Financial aid application deadline: 2/15.
Faculty research: Bacterial physiology and genetics, bacterial pathogenesis, host-pathogen interaction, molecular immunology.
Dr. John E. Cronan, Head, 217-333-1737, *Fax:* 217-244-6697, *E-mail:* microinfo@life.uiuc.edu.

Application contact: Dr. Diane Combs, Director of Graduate Studies, 217-333-9765, *Fax:* 217-244-6697, *E-mail:* dcombs@uiuc.edu.

Find an in-depth description at www.petersons.com/graduate.

■ **THE UNIVERSITY OF IOWA**

College of Medicine and Graduate College, Graduate Programs in Medicine, Department of Microbiology, Iowa City, IA 52242-1316

AWARDS General microbiology and microbial physiology (MS, PhD); immunology (MS, PhD); microbial genetics (MS, PhD); pathogenic bacteriology (MS, PhD); virology (MS, PhD).
Faculty: 21 full-time (4 women), 10 part-time/adjunct (1 woman).
Students: 50 full-time (21 women); includes 7 minority (2 African Americans, 4 Asian Americans or Pacific Islanders, 1 Hispanic American), 5 international. *144 applicants, 8% accepted.* In 1999, 7 doctorates awarded (100% entered university research/teaching).
Degree requirements: For master's, thesis required; for doctorate, dissertation, comprehensive exam required. *Average time to degree:* Doctorate–6 years full-time.
Entrance requirements: For master's and doctorate, GRE General Test. *Application deadline:* For fall admission, 2/1. *Application fee:* $30 ($50 for international students). Electronic applications accepted.
Expenses: Tuition, state resident: full-time $3,308. Tuition, nonresident: full-time $10,662. Tuition and fees vary according to course load and program.
Financial aid: In 1999–00, 50 students received aid, including fellowships with full tuition reimbursements available (averaging $16,787 per year), research assistantships with full tuition reimbursements available (averaging $16,787 per year); grants, institutionally sponsored loans, and traineeships also available.
Faculty research: Biocatalysis and blue jeans, gene regulation, processing and transport of HIV, retroviral pathogenesis, biodegradation. *Total annual research expenditures:* $8.2 million.
Dr. Michael A. Apicella, Head, 319-335-7810, *E-mail:* grad-micro-info@uiowa.edu.

Find an in-depth description at www.petersons.com/graduate.

■ **UNIVERSITY OF KANSAS**

Graduate School, College of Liberal Arts and Sciences, Division of Biological Sciences, Department of Molecular Biosciences, Lawrence, KS 66045

AWARDS Biochemistry and biophysics (MA, MS, PhD); microbiology (MA, MS, PhD);

molecular, cellular, and developmental biology (MA, MS, PhD).
Students: 13 full-time (7 women), 8 part-time (2 women); includes 1 minority (Asian American or Pacific Islander), 5 international. *57 applicants, 21% accepted.* In 2000, 11 master's, 10 doctorates awarded.
Degree requirements: For master's, thesis required, foreign language not required; for doctorate, dissertation required.
Entrance requirements: For master's and doctorate, GRE General Test, GRE Subject Test, TOEFL. *Application deadline:* For fall admission, 1/15 (priority date). *Application fee:* $25.
Expenses: Tuition, state resident: full-time $2,482; part-time $103 per credit hour. Tuition, nonresident: full-time $8,104; part-time $338 per credit hour. Required fees: $428; $31 per credit hour. Tuition and fees vary according to program.
Financial aid: In 2000–01, research assistantships (averaging $11,588 per year), teaching assistantships (averaging $11,588 per year) were awarded; fellowships. Financial aid application deadline: 3/1. Paul Kelly, Chair, 785-864-4311, *Fax:* 785-864-5294.
Application contact: Information Contact, 785-864-4311, *Fax:* 785-864-5294.

Find an in-depth description at www.petersons.com/graduate.

■ **UNIVERSITY OF KANSAS**

Graduate Studies Medical Center, Graduate Programs in Biomedical and Basic Sciences, Department of Microbiology, Molecular Genetics and Immunology, Lawrence, KS 66045

AWARDS PhD, MD/PhD.

Faculty: 10 full-time (1 woman), 1 part-time/adjunct (0 women).
Students: Average age 27. *0 applicants, 0% accepted.* In 1999, 4 doctorates awarded.
Degree requirements: For doctorate, dissertation, comprehensive oral exam required.
Entrance requirements: For doctorate, GRE General Test, TOEFL, TSE. *Application deadline:* For fall admission, 1/31 (priority date). Applications are processed on a rolling basis. *Application fee:* $0.
Expenses: Tuition, state resident: full-time $2,482; part-time $103 per credit hour. Tuition, nonresident: full-time $8,104; part-time $338 per credit hour. Required fees: $428; $31 per credit hour. Tuition and fees vary according to program.
Financial aid: Fellowships, research assistantships, teaching assistantships, Federal Work-Study, institutionally

University of Kansas (continued)
sponsored loans, scholarships, and unspecified assistantships available. Aid available to part-time students. Financial aid application deadline: 3/31; financial aid applicants required to submit FAFSA.
Faculty research: Bacterial and viral pathogenesis, molecular virology, microbial molecular biology and genetics, immunochemistry. *Total annual research expenditures:* $4.2 million.
Dr. Opendra Narayan, Chairman, 913-5887010, *Fax:* 913-588-7295, *E-mail:* bnarayan@kume.edu.
Application contact: Dr. Joe Lutkenhaus, Director of Graduate Studies, 913-588-7010, *Fax:* 913-588-7295, *E-mail:* jlutkenh@kumc.edu.

■ UNIVERSITY OF KENTUCKY

Graduate School and College of Medicine, Graduate Programs in Medicine, Department of Microbiology and Immunology, Lexington, KY 40506-0032

AWARDS PhD.

Degree requirements: For doctorate, dissertation, comprehensive exam required, foreign language not required.
Entrance requirements: For doctorate, GRE General Test, minimum GPA of 3.0.
Expenses: Tuition, state resident: full-time $3,596; part-time $188 per credit hour. Tuition, nonresident: full-time $10,116; part-time $550 per credit hour.
Faculty research: Immunology transplantation, cellular antibody reactions, immunobiology of tumors.
Find an in-depth description at www.petersons.com/graduate.

■ UNIVERSITY OF LOUISVILLE

School of Medicine and Graduate School, Integrated Programs in Biomedical Sciences, Department of Microbiology and Immunology, Louisville, KY 40292-0001

AWARDS MS, PhD.

Degree requirements: For master's and doctorate, thesis/dissertation required.
Entrance requirements: For master's and doctorate, GRE General Test, 1 year of course work in biology, organic chemistry, physics; 1 semester of course work in calculus and quantitative analysis, biochemistry, or molecular biology. Electronic applications accepted.
Expenses: Tuition, state resident: full-time $3,260; part-time $182 per hour. Tuition, nonresident: full-time $9,780; part-time $544 per hour. Required fees: $143; $28 per hour.

Faculty research: Virology, molecular pathogenesis, microbial genetics, cell biology.
Find an in-depth description at www.petersons.com/graduate.

■ UNIVERSITY OF MAINE

Graduate School, College of Natural Sciences, Forestry, and Agriculture, Department of Biochemistry, Molecular Biology, and Microbiology, Orono, ME 04469

AWARDS Biochemistry (MPS, MS); biochemistry and molecular biology (PhD); microbiology (MPS, MS, PhD).

Degree requirements: For doctorate, dissertation required.
Entrance requirements: For master's and doctorate, GRE General Test, TOEFL.
Expenses: Tuition, state resident: full-time $3,564. Tuition, nonresident: full-time $10,116. Required fees: $378. Tuition and fees vary according to course load.

■ UNIVERSITY OF MARYLAND

Graduate School, Graduate Programs in Medicine, Department of Microbiology and Immunology, Baltimore, MD 21201-1627

AWARDS MS, PhD, MD/PhD. Part-time programs available.

Degree requirements: For master's, thesis required, foreign language not required; for doctorate, dissertation, oral exam required, foreign language not required.
Entrance requirements: For master's and doctorate, GRE General Test, TOEFL, minimum GPA of 3.0.
Expenses: Tuition, state resident: part-time $261 per credit hour. Tuition, nonresident: part-time $468 per credit hour. Tuition and fees vary according to program.
Faculty research: Epidemiology, ecology of infectious microorganisms, electron microscopy, medical microbiology, molecular biology.
Find an in-depth description at www.petersons.com/graduate.

■ UNIVERSITY OF MARYLAND, COLLEGE PARK

Graduate Studies and Research, College of Life Sciences, Department of Cell Biology and Molecular Genetics, Program in Microbiology, College Park, MD 20742

AWARDS MS, PhD. Part-time and evening/weekend programs available.

Students: 29 full-time (14 women), 6 part-time (3 women); includes 3 minority (1 African American, 1 Asian American or

Pacific Islander, 1 Hispanic American), 16 international. *85 applicants, 29% accepted.* In 1999, 4 master's awarded. Terminal master's awarded for partial completion of doctoral program.
Degree requirements: For master's, thesis, oral exam required, foreign language not required; for doctorate, dissertation required.
Entrance requirements: For master's, GRE General Test, minimum GPA of 3.0; for doctorate, GRE Subject Test. *Application deadline:* For fall admission, 1/25. Applications are processed on a rolling basis. *Application fee:* $50 ($70 for international students). Electronic applications accepted.
Expenses: Tuition, state resident: part-time $272 per credit hour. Tuition, nonresident: part-time $415 per credit hour. Required fees: $632; $379 per year.
Financial aid: Fellowships, research assistantships, teaching assistantships, career-related internships or fieldwork available. Financial aid applicants required to submit FAFSA.
Faculty research: Biomedical, marine, and environmental microbiology, electron microscopy.
Application contact: Trudy Lindsey, Director, Graduate Admissions and Records, 301-405-4198, *Fax:* 301-314-9305, *E-mail:* grschool@deans.umd.edu.

■ UNIVERSITY OF MASSACHUSETTS AMHERST

Graduate School, College of Food and Natural Resources, Department of Microbiology, Amherst, MA 01003

AWARDS MS, PhD. Part-time programs available.

Faculty: 15 full-time (3 women).
Students: 24 full-time (9 women), 13 part-time (8 women); includes 1 minority (Asian American or Pacific Islander), 13 international. Average age 29. *103 applicants, 17% accepted.* In 1999, 8 master's, 3 doctorates awarded. Terminal master's awarded for partial completion of doctoral program.
Degree requirements: For master's, thesis or alternative required, foreign language not required; for doctorate, dissertation required, foreign language not required.
Entrance requirements: For master's and doctorate, GRE General Test. *Application deadline:* For fall admission, 2/1 (priority date); for spring admission, 10/1. Applications are processed on a rolling basis. *Application fee:* $40.
Expenses: Tuition, state resident: full-time $2,640; part-time $165 per credit. Tuition, nonresident: full-time $9,756; part-time $407 per credit. Required fees: $1,221 per

term. One-time fee: $110. Full-time tuition and fees vary according to course load, campus/location and reciprocity agreements.

Financial aid: In 1999–00, 31 research assistantships with full tuition reimbursements (averaging $7,859 per year), 20 teaching assistantships with full tuition reimbursements (averaging $8,060 per year) were awarded; fellowships with full tuition reimbursements, career-related internships or fieldwork, Federal Work-Study, grants, scholarships, traineeships, and unspecified assistantships also available. Aid available to part-time students. Financial aid application deadline: 2/1. Dr. Derek Lovely, Director, 413-545-6663, *Fax:* 413-545-1578, *E-mail:* dlovely@microbio.umass.edu.

Find an in-depth description at www.petersons.com/graduate.

■ UNIVERSITY OF MASSACHUSETTS WORCESTER

Graduate School of Biomedical Sciences, Department of Molecular Genetics and Microbiology, Worcester, MA 01655-0115

AWARDS Medical sciences (PhD).

Faculty: 42 full-time (6 women).
Students: 28 full-time (11 women); includes 1 minority (Asian American or Pacific Islander), 11 international.
Degree requirements: For doctorate, dissertation required.
Entrance requirements: For doctorate, GRE General Test, GRE Subject Test, 1 year of calculus, organic chemistry, physics, and biology. *Application deadline:* For fall admission, 1/15 (priority date). Applications are processed on a rolling basis. *Application fee:* $25 ($50 for international students).
Expenses: Tuition, state resident: full-time $2,640. Tuition, nonresident: full-time $9,756. Required fees: $825. Full-time tuition and fees vary according to program.
Financial aid: In 1999–00, research assistantships with full tuition reimbursements (averaging $17,500 per year); unspecified assistantships also available.
Faculty research: Gene structure, regulation of gene expression.
Dr. Allan Jacobson, Chair.
Application contact: Dr. Janet Stavnezer, Professor, 508-856-4100, *E-mail:* janet.stavnezer@umassmed.edu.

Find an in-depth description at www.petersons.com/graduate.

■ UNIVERSITY OF MEDICINE AND DENTISTRY OF NEW JERSEY

Graduate School of Biomedical Sciences, Graduate Programs in Biomedical Sciences, Department of Molecular Genetics and Microbiology, Newark, NJ 07107

AWARDS MS, PhD.

Degree requirements: For master's, thesis required; for doctorate, dissertation, qualifying exam required, foreign language not required.
Entrance requirements: For master's and doctorate, GRE General Test, TOEFL. *Application deadline:* For fall admission, 2/1; for spring admission, 10/1. *Application fee:* $40.
Expenses: Tuition, state resident: part-time $270 per credit hour. Tuition, nonresident: part-time $407 per credit hour. Part-time tuition and fees vary according to campus/location and program.
Financial aid: Fellowships, research assistantships, Federal Work-Study, institutionally sponsored loans, and tuition waivers (full and partial) available. Financial aid application deadline: 5/1.
Faculty research: Molecular genetics of yeast, mutagenesis and carcinogenesis of DNA, bacterial protein synthesis, mammalian cell genetics, adenovirus gene expression.
Dr. Harvey Ozer, Chairperson, 973-972-4483, *Fax:* 973-972-1286, *E-mail:* ozerhl@umdnj.edu.
Application contact: Dr. Henry E. Brezenoff, Dean, Graduate School of Biomedical Sciences, 973-972-5333, *Fax:* 973-972-7148, *E-mail:* hbrezeno@umdnj.edu.

Find an in-depth description at www.petersons.com/graduate.

■ UNIVERSITY OF MEDICINE AND DENTISTRY OF NEW JERSEY

Graduate School of Biomedical Sciences, Graduate Programs in Biomedical Sciences, Program in Molecular Genetics and Microbiology, Piscataway, NJ 08854-5635

AWARDS MS, PhD. Terminal master's awarded for partial completion of doctoral program.

Degree requirements: For master's, thesis, qualifying exam required; for doctorate, dissertation, qualifying exam required, foreign language not required.
Entrance requirements: For master's and doctorate, GRE General Test, TOEFL. *Application deadline:* For fall admission, 2/1; for spring admission, 10/1. Applications are processed on a rolling basis. *Application fee:* $40.

Expenses: Tuition, state resident: part-time $270 per credit hour. Tuition, nonresident: part-time $407 per credit hour. Part-time tuition and fees vary according to campus/location and program.
Financial aid: Fellowships, research assistantships, teaching assistantships available. Financial aid application deadline: 5/1.
Faculty research: Interferon, receptors, retrovirus evolution, Arbo virus/host cell interactions.
Dr. Sidney Pestka, Director, 732-235-4567.
Application contact: Dr. Michael J. Leibowitz, Associate Dean, Graduate School, 732-235-5016, *Fax:* 732-235-4720, *E-mail:* gsbspisc@umdnj.edu.

■ THE UNIVERSITY OF MEMPHIS

Graduate School, College of Arts and Sciences, Department of Microbiology and Molecular Cell Sciences, Memphis, TN 38152

AWARDS Biology (MS, PhD). Part-time programs available.

Faculty: 10 full-time (2 women).
Students: Average age 29. *33 applicants, 79% accepted.* In 1999, 4 master's awarded (100% found work related to degree); 1 doctorate awarded (100% found work related to degree). Terminal master's awarded for partial completion of doctoral program.
Degree requirements: For master's, thesis or alternative, oral and/or written comprehensive exam required, foreign language not required; for doctorate, one foreign language (computer language can substitute), dissertation, comprehensive exam required. *Average time to degree:* Master's–2 years full-time; doctorate–5.5 years full-time.
Entrance requirements: For master's and doctorate, GRE General Test, GRE Subject Test (biochemistry), minimum GPA of 2.5. *Application deadline:* For fall admission, 8/1; for spring admission, 12/1. Applications are processed on a rolling basis. *Application fee:* $25 ($50 for international students).
Expenses: Tuition, state resident: full-time $3,410; part-time $178 per credit hour. Tuition, nonresident: full-time $8,670; part-time $408 per credit hour. Tuition and fees vary according to program.
Financial aid: In 1999–00, 2 research assistantships with full tuition reimbursements (averaging $12,000 per year), 16 teaching assistantships with full tuition reimbursements (averaging $8,000 per year) were awarded; fellowships with full tuition reimbursements

The University of Memphis (continued)
Faculty research: Molecular genetics, cell biology, biochemistry, microbiology and toxicology.
Dr. H. Delano Black, Interim Chair, 901-678-2955, *Fax:* 901-678-4457, *E-mail:* dblack@cc.memphis.edu.
Application contact: Dr. Barbara J. Taller, Graduate Coordinator, 901-678-2955, *Fax:* 901-678-4457, *E-mail:* bjtaller@cc.memphis.edu.

Find an in-depth description at www.petersons.com/graduate.

■ **UNIVERSITY OF MIAMI**

School of Medicine and Graduate School, Graduate Programs in Medicine, Department of Microbiology and Immunology, Coral Gables, FL 33124

AWARDS PhD, MD/PhD.

Faculty: 24 full-time (5 women).
Students: 22 full-time (9 women); includes 6 minority (1 African American, 2 Asian Americans or Pacific Islanders, 3 Hispanic Americans), 7 international. Average age 24. *105 applicants, 8% accepted.* In 1999, 4 degrees awarded (50% entered university research/teaching, 50% found other work related to degree).
Degree requirements: For doctorate, dissertation, oral and written qualifying exams required, foreign language not required. *Average time to degree:* Doctorate–6 years full-time.
Entrance requirements: For doctorate, GRE General Test, TOEFL. *Application deadline:* For fall admission, 3/15. Applications are processed on a rolling basis. *Application fee:* $35.
Expenses: Tuition, area resident: Part-time $899 per credit.
Financial aid: In 1999–00, 22 students received aid, including 4 fellowships, 12 research assistantships; institutionally sponsored loans also available.
Faculty research: Cellular and molecular immunology, molecular and pathogenic virology, pathogenic bacteriology and gene therapy of cancer. *Total annual research expenditures:* $4.5 million.
Dr. Eckhard R. Podack, Chairman, 305-243-6694, *Fax:* 305-243-5522.
Application contact: Dr. Robert B. Levy, Director, Graduate Program, 305-243-5682, *Fax:* 305-243-6903, *E-mail:* kdelrio@mednet.med.miami.edu.

Find an in-depth description at www.petersons.com/graduate.

■ **UNIVERSITY OF MICHIGAN**

Medical School and Horace H. Rackham School of Graduate Studies, Program in Biomedical Sciences (PIBS), Department of Microbiology and Immunology, Ann Arbor, MI 48109

AWARDS PhD.

Degree requirements: For doctorate, computer language, oral defense of dissertation, preliminary exam required.
Entrance requirements: For doctorate, GRE General Test, TOEFL, TWE.
Expenses: Tuition, state resident: full-time $10,316. Tuition, nonresident: full-time $20,922.
Faculty research: Gene regulation, molecular biology of animal and bacterial viruses, molecular and cellular networks, pathogenesis and microbial genetics.

Find an in-depth description at www.petersons.com/graduate.

■ **UNIVERSITY OF MINNESOTA, TWIN CITIES CAMPUS**

Medical School and Graduate School, Graduate Programs in Medicine, Department of Microbiology, Minneapolis, MN 55455-0213

AWARDS Microbiology (MS); microbiology, immunology, and molecular pathobiology (PhD). Terminal master's awarded for partial completion of doctoral program.

Degree requirements: For master's and doctorate, thesis/dissertation required.
Entrance requirements: For master's, GRE General Test.
Expenses: Tuition, state resident: full-time $11,984; part-time $1,498 per semester. Tuition, nonresident: full-time $22,264; part-time $2,783 per semester. Full-time tuition and fees vary according to program and student level. Part-time tuition and fees vary according to course load and program.
Faculty research: Virology.

Find an in-depth description at www.petersons.com/graduate.

■ **UNIVERSITY OF MINNESOTA, TWIN CITIES CAMPUS**

School of Public Health, Division of Environmental and Occupational Health, Area in Environmental Microbiology, Minneapolis, MN 55455-0213

AWARDS MPH, MS, PhD.

Degree requirements: For master's, foreign language not required; for doctorate, dissertation required, foreign language not required.
Entrance requirements: For master's and doctorate, GRE General Test, minimum

GPA of 3.0. *Application deadline:* For fall admission, 3/1 (priority date). Applications are processed on a rolling basis. *Application fee:* $50 ($75 for international students).
Expenses: Tuition, state resident: full-time $4,270; part-time $267 per credit. Tuition, nonresident: full-time $8,400; part-time $525 per credit. Tuition and fees vary according to program.
Financial aid: Application deadline: 3/1.
Application contact: Kathy Soupir, Student Coordinator, 612-625-0622, *Fax:* 612-626-4837, *E-mail:* ksoupir@mail.eoh.umn.edu.

■ **UNIVERSITY OF MISSISSIPPI MEDICAL CENTER**

Graduate Programs in Biomedical Sciences, Department of Microbiology, Jackson, MS 39216-4505

AWARDS MS, PhD, MD/PhD.

Faculty: 13 full-time (3 women).
Students: 25 full-time (11 women); includes 1 minority (African American), 9 international. Average age 27. *51 applicants, 16% accepted.* In 1999, 1 master's awarded (100% found work related to degree); 3 doctorates awarded (100% entered university research/teaching).
Degree requirements: For master's, thesis required, foreign language not required; for doctorate, dissertation, first authored publication required, foreign language not required. *Average time to degree:* Master's–2 years full-time; doctorate–6 years full-time.
Entrance requirements: For master's and doctorate, GRE General Test, minimum GPA of 3.0. *Application deadline:* For fall admission, 7/1. Applications are processed on a rolling basis. *Application fee:* $10.
Expenses: Tuition, state resident: full-time $2,378; part-time $132 per hour. Tuition, nonresident: full-time $4,697; part-time $261 per hour. Tuition and fees vary according to program.
Financial aid: In 1999–00, 19 students received aid, including 19 research assistantships (averaging $16,234 per year); grants also available. Financial aid application deadline: 4/1.
Faculty research: Immunology, virology, microbial physiology/genetics, parasitology.
Dr. L. W. Clem, Director, 601-984-1700, *Fax:* 601-984-1708, *E-mail:* lclem@microbio.umsmed.edu.
Application contact: Dr. Billy M. Bishop, Director, Student Services and Records, 601-984-1080, *Fax:* 601-984-1079, *E-mail:* bbishop@registrar.umsmed.edu.

■ UNIVERSITY OF MISSOURI–COLUMBIA

School of Medicine and Graduate School, Graduate Programs in Medicine, Department of Molecular Microbiology and Immunology, Columbia, MO 65211

AWARDS MS, PhD. Terminal master's awarded for partial completion of doctoral program.

Degree requirements: For master's and doctorate, thesis/dissertation required.
Entrance requirements: For master's and doctorate, GRE General Test, minimum GPA of 3.0.
Expenses: Tuition, state resident: full-time $3,020; part-time $168 per hour. Tuition, nonresident: full-time $6,066; part-time $505 per hour. Required fees: $445; $18 per hour.
Faculty research: Molecular biology, host-parasite interactions.
Find an in-depth description at www.petersons.com/graduate.

■ THE UNIVERSITY OF MONTANA–MISSOULA

Graduate School, Division of Biological Sciences, Program in Biochemistry and Microbiology, Missoula, MT 59812-0002

AWARDS MS, PhD.

Students: 17 full-time (5 women), 12 part-time (5 women); includes 3 minority (2 Asian Americans or Pacific Islanders, 1 Native American). *41 applicants, 27% accepted.* In 1999, 1 master's, 2 doctorates awarded. Terminal master's awarded for partial completion of doctoral program.
Degree requirements: For master's, thesis required (for some programs); for doctorate, dissertation required.
Entrance requirements: For master's and doctorate, GRE General Test. *Application deadline:* For fall admission, 2/1. *Application fee:* $45.
Expenses: Tuition, state resident: full-time $2,484; part-time $151 per credit. Tuition, nonresident: full-time $8,000; part-time $305 per credit. Required fees: $1,600. Full-time tuition and fees vary according to degree level and program.
Financial aid: In 1999–00, research assistantships with tuition reimbursements (averaging $9,400 per year), teaching assistantships with full tuition reimbursements (averaging $9,400 per year) were awarded; Federal Work-Study and tuition waivers (full and partial) also available. Financial aid application deadline: 3/1.
Faculty research: Ribosome structure, medical microbiology/pathogenesis, microbial ecology/environmental microbiology.

Application contact: Janean Clark, Graduate Programs Secretary, 406-243-5222, *Fax:* 406-243-4184, *E-mail:* jmclark@selway.umt.edu.

■ UNIVERSITY OF NEBRASKA MEDICAL CENTER

Graduate College, Department of Pathology and Microbiology, Omaha, NE 68198

AWARDS MS, PhD. Part-time programs available.

Faculty: 43 full-time (7 women), 29 part-time/adjunct (3 women).
Students: 27 full-time (11 women), 3 part-time (1 woman); includes 2 minority (both African Americans), 2 international. Average age 27. *54 applicants, 19% accepted.* In 1999, 2 master's awarded (100% entered university research/teaching); 5 doctorates awarded. Terminal master's awarded for partial completion of doctoral program.
Degree requirements: For master's and doctorate, thesis/dissertation required, foreign language not required.
Entrance requirements: For master's, previous course work in biology, chemistry, mathematics, and physics; for doctorate, GRE General Test, previous course work in biology, chemistry, mathematics, and physics. *Application deadline:* Applications are processed on a rolling basis. *Application fee:* $35.
Expenses: Tuition, state resident: part-time $116 per semester hour. Tuition, nonresident: part-time $270 per semester hour. Tuition and fees vary according to program.
Financial aid: In 1999–00, 2 fellowships with tuition reimbursements (averaging $14,500 per year), 21 research assistantships with tuition reimbursements (averaging $14,500 per year) were awarded; teaching assistantships, institutionally sponsored loans and tuition waivers (full) also available. Aid available to part-time students. Financial aid application deadline: 3/1.
Faculty research: Carcinogenesis, cancer biology, immunobiology, molecular virology, molecular genetics.
Dr. Donald R. Johnson, Chairman, Graduate Committee, 402-559-4042, *Fax:* 402-559-4077.
Application contact: Jo Wagner, Associate Director of Admissions, 402-559-6468.

■ UNIVERSITY OF NEW HAMPSHIRE

Graduate School, College of Life Sciences and Agriculture, Graduate Programs in the Biological Sciences and Natural Resources, Department of Microbiology, Durham, NH 03824

AWARDS MS, PhD. Part-time programs available.

Faculty: 7 full-time.
Students: 12 full-time (7 women), 14 part-time (9 women), 4 international. Average age 28. *27 applicants, 56% accepted.* In 1999, 3 master's, 2 doctorates awarded. Terminal master's awarded for partial completion of doctoral program.
Degree requirements: For master's and doctorate, thesis/dissertation required, foreign language not required.
Entrance requirements: For master's and doctorate, GRE General Test. *Application deadline:* For fall admission, 4/1 (priority date); for winter admission, 12/1. Applications are processed on a rolling basis. *Application fee:* $50.
Expenses: Tuition, area resident: Full-time $5,750; part-time $319 per credit. Tuition, state resident: full-time $8,625; part-time $478. Tuition, nonresident: full-time $14,640; part-time $598 per credit. Required fees: $224 per semester. Tuition and fees vary according to course load, degree level and program.
Financial aid: In 1999–00, 1 fellowship, 5 research assistantships, 12 teaching assistantships were awarded; career-related internships or fieldwork, Federal Work-Study, scholarships, and tuition waivers (full and partial) also available. Aid available to part-time students. Financial aid application deadline: 2/15.
Faculty research: Bacterial host-parasite interactions, immunology, microbial structures, bacterial and bacteriophage genetics, virology.
Dr. Robert Zsigary, Chairperson, 603-862-4095, *E-mail:* robert.zsigary@unh.edu.
Application contact: Dr. Lou Tisa, Graduate Coordinator, 603-862-2442, *E-mail:* lst@cisunix.unh.edu.

■ UNIVERSITY OF NEW MEXICO

Graduate School, College of Arts and Sciences, Department of Biology, Albuquerque, NM 87131-2039

AWARDS Biology (MS, PhD), including air land ecology, behavioral ecology, botany, cellular and molecular biology, community ecology, comparative immunology, comparative physiology, conservation biology, ecology, ecosystem ecology, evolutionary biology, evolutionary genetics, microbiology, molecular genetics, parasitology, physiological ecology,

University of New Mexico (continued)
physiology, population biology, vertebrate and invertebrate zoology. Part-time programs available.

Faculty: 35 full-time (5 women), 18 part-time/adjunct (11 women).
Students: 71 full-time (37 women), 28 part-time (11 women); includes 8 minority (2 Asian Americans or Pacific Islanders, 5 Hispanic Americans, 1 Native American), 11 international. Average age 33. *93 applicants, 30% accepted.* In 1999, 11 master's, 12 doctorates awarded. Terminal master's awarded for partial completion of doctoral program.
Degree requirements: For master's, one foreign language (computer language can substitute), thesis required (for some programs); for doctorate, 2 foreign languages (computer language can substitute for one), dissertation required.
Entrance requirements: For master's and doctorate, GRE General Test, GRE Subject Test, minimum GPA of 3.2. *Application deadline:* For fall admission, 1/15. *Application fee:* $25.
Expenses: Tuition, state resident: full-time $2,514; part-time $105 per credit hour. Tuition, nonresident: full-time $10,304; part-time $417 per credit hour. International tuition: $10,304 full-time. Required fees: $516; $22 per credit hour. Tuition and fees vary according to program.
Financial aid: In 1999–00, 58 students received aid, including 24 fellowships (averaging $1,645 per year), 26 research assistantships with tuition reimbursements available (averaging $8,921 per year), 40 teaching assistantships with tuition reimbursements available (averaging $11,066 per year); career-related internships or fieldwork, Federal Work-Study, institutionally sponsored loans, and tuition waivers (full and partial) also available. Aid available to part-time students. Financial aid applicants required to submit FAFSA.
Faculty research: Developmental biology, immunobiology. *Total annual research expenditures:* $4.5 million.
Dr. Kathryn Vogel, Chair, 505-277-3411, *Fax:* 505-277-0304, *E-mail:* kgvogel@unm.edu.
Application contact: Vivian Kent, Information Contact, 505-277-1712, *Fax:* 505-277-0304, *E-mail:* vkent@unm.edu.

■ THE UNIVERSITY OF NORTH CAROLINA AT CHAPEL HILL

School of Medicine and Graduate School, Graduate Programs in Medicine, Department of Microbiology and Immunology, Chapel Hill, NC 27599

AWARDS Immunology (MS, PhD); microbiology (MS, PhD).

Faculty: 24 full-time (6 women), 17 part-time/adjunct (2 women).
Students: 45 full-time (23 women), 3 part-time (all women); includes 7 minority (1 African American, 4 Asian Americans or Pacific Islanders, 2 Hispanic Americans), 3 international. Average age 27. *112 applicants, 23% accepted.* In 1999, 2 master's awarded (100% entered university research/teaching); 6 doctorates awarded (100% entered university research/teaching). Terminal master's awarded for partial completion of doctoral program.
Degree requirements: For master's, thesis, comprehensive exam required, foreign language not required; for doctorate, dissertation, comprehensive exams required, foreign language not required. *Average time to degree:* Master's–4 years full-time, 6 years part-time; doctorate–5.83 years full-time.
Entrance requirements: For master's and doctorate, GRE General Test, minimum GPA of 3.0. *Application deadline:* For fall admission, 1/15. *Application fee:* $55. Electronic applications accepted.
Expenses: Tuition, state resident: full-time $1,966. Tuition, nonresident: full-time $11,026. Required fees: $8,940. One-time fee: $15 full-time. Part-time tuition and fees vary according to course load.
Financial aid: In 1999–00, 1 fellowship with full tuition reimbursement (averaging $16,500 per year), 38 research assistantships with full tuition reimbursements (averaging $16,500 per year) were awarded; teaching assistantships with full tuition reimbursements, grants, traineeships, tuition waivers (full), and unspecified assistantships also available. *Total annual research expenditures:* $10.6 million.
Jeffrey Frelinger, Chairman, 919-966-1191, *Fax:* 919-962-8103, *E-mail:* microimm@med.unc.edu.
Application contact: Dixie Flannery, Administrator, 919-966-1191, *Fax:* 919-962-8103, *E-mail:* microimm@med.unc.edu.

Find an in-depth description at www.petersons.com/graduate.

■ UNIVERSITY OF NORTH DAKOTA

School of Medicine and Graduate School, Graduate Programs in Medicine, Department of Microbiology and Immunology, Grand Forks, ND 58202

AWARDS MS, PhD.

Faculty: 6 full-time (1 woman).
Students: 8 full-time (5 women), 3 part-time (1 woman). *13 applicants, 31% accepted.* In 1999, 2 master's, 1 doctorate awarded.
Degree requirements: For master's, thesis or alternative, comprehensive final examination required, foreign language not required; for doctorate, computer language, dissertation, comprehensive examination, final examination required.
Entrance requirements: For master's, GRE General Test, TOEFL, minimum GPA of 3.0; for doctorate, GRE General Test, TOEFL, minimum GPA of 3.5. *Application deadline:* For fall admission, 3/1 (priority date). Applications are processed on a rolling basis. *Application fee:* $25.
Expenses: Tuition, state resident: full-time $2,690; part-time $112 per credit. Tuition, nonresident: full-time $7,182; part-time $299 per credit. Required fees: $46 per semester.
Financial aid: In 1999–00, 8 students received aid, including 6 teaching assistantships with full tuition reimbursements available (averaging $10,586 per year); fellowships, research assistantships, Federal Work-Study, institutionally sponsored loans, scholarships, and tuition waivers (full and partial) also available. Aid available to part-time students. Financial aid application deadline: 3/15; financial aid applicants required to submit FAFSA.
Faculty research: Genetic and immunological aspects of a murine model of human multiple sclerosis, termination of DNA replication, cell division in bacteria, yersinia pestis.
Dr. Kevin Young, Director, 701-777-2214, *Fax:* 701-777-3527, *E-mail:* kyoung@medicine.nodak.edu.

■ UNIVERSITY OF NORTH TEXAS HEALTH SCIENCE CENTER AT FORT WORTH

Graduate School of Biomedical Sciences, Fort Worth, TX 76107-2699

AWARDS Anatomy and cell biology (MS, PhD); biochemistry and molecular biology (MS, PhD); biotechnology (MS); integrative physiology (MS, PhD); microbiology and immunology (MS, PhD); pharmacology (MS, PhD).

Faculty: 65 full-time (9 women), 11 part-time/adjunct (1 woman).

Students: 59 full-time (27 women), 44 part-time (20 women); includes 30 minority (10 African Americans, 9 Asian Americans or Pacific Islanders, 11 Hispanic Americans), 23 international. *70 applicants, 70% accepted.* In 1999, 5 master's awarded (40% found work related to degree, 60% continued full-time study); 5 doctorates awarded (80% entered university research/teaching, 20% found other work related to degree).
Degree requirements: For doctorate, computer language, dissertation required. *Average time to degree:* Master's–2.5 years full-time, 4 years part-time; doctorate–5 years full-time.
Entrance requirements: For master's and doctorate, GRE General Test, TOEFL. *Application deadline:* For fall admission, 5/1; for spring admission, 11/1. Applications are processed on a rolling basis. *Application fee:* $25 ($50 for international students).
Expenses: Tuition, state resident: full-time $1,188; part-time $66 per credit. Tuition, nonresident: full-time $5,058; part-time $281 per credit. Required fees: $366; $183 per semester.
Financial aid: In 1999–00, 11 fellowships, 70 research assistantships (averaging $16,500 per year) were awarded; teaching assistantships, career-related internships or fieldwork, Federal Work-Study, grants, institutionally sponsored loans, and traineeships also available. Aid available to part-time students. Financial aid application deadline: 4/1; financial aid applicants required to submit FAFSA.
Faculty research: Alzheimer's disease, diabetes, eye diseases, cancer, cardiovascular physiology.
Dr. Thomas Yorio, Dean, 817-735-2560, *Fax:* 817-735-0243, *E-mail:* yoriot@hsc.unt.edu.
Application contact: Jan Sharp, Administrative Assistant, 817-735-0258, *Fax:* 817-735-0243, *E-mail:* gsbs@hsc.unt.edu.

Find an in-depth description at www.petersons.com/graduate.

■ UNIVERSITY OF OKLAHOMA

Graduate College, College of Arts and Sciences, Department of Botany and Microbiology, Program in Microbiology, Norman, OK 73019-0390
AWARDS M Nat Sci, MS, PhD. Part-time programs available.

Students: 25 full-time (11 women), 4 part-time (3 women); includes 3 minority (1 Asian American or Pacific Islander, 1 Hispanic American, 1 Native American), 12 international. *21 applicants, 24% accepted.* In 1999, 4 master's, 2 doctorates awarded.

Degree requirements: For master's, thesis, oral exam required; for doctorate, one foreign language, dissertation, general exam required.
Entrance requirements: For master's and doctorate, TOEFL. *Application deadline:* For fall admission, 6/1; for spring admission, 12/1. Applications are processed on a rolling basis. *Application fee:* $25.
Expenses: Tuition, state resident: full-time $2,064; part-time $86 per credit hour. Tuition, nonresident: full-time $6,588; part-time $275 per credit hour. Required fees: $468; $12 per credit hour. $94 per semester. Tuition and fees vary according to course level, course load and program.
Financial aid: In 1999–00, 10 students received aid; research assistantships, teaching assistantships, Federal Work-Study and scholarships available. Aid available to part-time students.
Faculty research: Bioremediation, functional genomics, pathogenic microbiology and geomicrobiology, microbial pathology, anaerobic microbiology, molecular biology.
Application contact: Mary Mason, Secretary I, 405-325-4321, *Fax:* 405-325-7619.

■ UNIVERSITY OF OKLAHOMA HEALTH SCIENCES CENTER

College of Medicine and Graduate College, Graduate Programs in Medicine, Department of Microbiology and Immunology, Oklahoma City, OK 73190

AWARDS Immunology (MS, PhD); microbiology (MS, PhD). Part-time programs available.

Faculty: 21 full-time (3 women), 11 part-time/adjunct (4 women).
Students: 8 full-time (5 women), 25 part-time (14 women); includes 3 minority (1 African American, 1 Asian American or Pacific Islander, 1 Hispanic American), 6 international. Average age 24. *44 applicants, 23% accepted.* In 1999, 5 master's, 5 doctorates awarded. Terminal master's awarded for partial completion of doctoral program.
Degree requirements: For master's, thesis or alternative required, foreign language not required; for doctorate, one foreign language (computer language can substitute), dissertation required.
Entrance requirements: For master's and doctorate, GRE General Test, TOEFL. *Application deadline:* For fall admission, 3/31. *Application fee:* $25 ($50 for international students).
Expenses: Tuition, state resident: part-time $90 per semester hour. Tuition,

nonresident: part-time $264 per semester hour. Tuition and fees vary according to program.
Financial aid: Fellowships, teaching assistantships available. Financial aid applicants required to submit FAFSA.
Faculty research: Molecular genetics, pathogenesis, streptococcal infections, gram-positive virulence, monoclonal antibodies.
Dr. John Iandolo, Chairman, 405-271-2133.
Application contact: Dr. Don Graves, Graduate Liaison, 405-271-2133.

■ UNIVERSITY OF PENNSYLVANIA

School of Medicine, Biomedical Graduate Studies, Graduate Group in Cell and Molecular Biology, Program in Microbiology and Virology, Philadelphia, PA 19104
AWARDS PhD, MD/PhD, VMD/PhD.

Degree requirements: For doctorate, dissertation required, foreign language not required.
Entrance requirements: For doctorate, GRE General Test, GRE Subject Test (biology), TOEFL, previous course work in science. *Application deadline:* For fall admission, 1/2 (priority date). Applications are processed on a rolling basis. *Application fee:* $65. Electronic applications accepted.
Expenses: Tuition: Full-time $17,256; part-time $2,991 per course. Required fees: $2,588; $363 per course. $726 per term.
Financial aid: Fellowships, research assistantships, teaching assistantships, grants and institutionally sponsored loans available. Financial aid application deadline: 1/2.
Dr. Michael Malim, Chair, 215-573-3493.
Application contact: Mary Webster, Coordinator, 215-898-4360, *E-mail:* camb@mail.med.upenn.edu.

Find an in-depth description at www.petersons.com/graduate.

■ UNIVERSITY OF PITTSBURGH

Graduate School of Public Health, Department of Infectious Diseases and Microbiology, Pittsburgh, PA 15260
AWARDS MPH, MS, Dr PH, PhD.

Faculty: 13 full-time (4 women), 1 (woman) part-time/adjunct.
Students: 23 full-time (13 women), 2 part-time (both women); includes 3 minority (1 African American, 2 Asian Americans or Pacific Islanders), 5 international. *47 applicants, 40% accepted.* In 1999, 1 master's, 3 doctorates awarded. Terminal master's awarded for partial completion of doctoral program.

University of Pittsburgh (continued)

Degree requirements: For master's and doctorate, one foreign language, thesis not required. *Average time to degree:* Master's–2 years full-time, 4 years part-time; doctorate–4 years full-time.

Entrance requirements: For master's and doctorate, GRE General Test. *Application deadline:* For fall admission, 4/1 (priority date). Applications are processed on a rolling basis. *Application fee:* $50 ($60 for international students).

Expenses: Tuition, state resident: full-time $9,778; part-time $403 per credit. Tuition, nonresident: full-time $20,146; part-time $830 per credit. Required fees: $480; $90 per term.

Financial aid: In 1999–00, 18 students received aid, including 17 research assistantships with tuition reimbursements available (averaging $15,000 per year), 1 teaching assistantship with tuition reimbursement available (averaging $15,515 per year).

Faculty research: AIDS, Epstein-Barr virus, virology, immunology, HIV. *Total annual research expenditures:* $4.6 million. Dr. Charles Rinaldo, Chairman, 412-624-3928, *Fax:* 412-624-4953, *E-mail:* rinaldo@vms.cis.pitt.edu.

Application contact: Dr. Phalguni Gupta, Associate Professor, 412-624-7998, *Fax:* 412-624-4953, *E-mail:* pguptal@vms.cis.pitt.edu.

Find an in-depth description at www.petersons.com/graduate.

■ **UNIVERSITY OF PITTSBURGH**

School of Medicine, Graduate Programs in Medicine, Program in Molecular Virology and Microbiology, Pittsburgh, PA 15260

AWARDS MS, PhD.

Students: 6 full-time (0 women); includes 2 minority (1 African American, 1 Hispanic American). *325 applicants, 22% accepted.* In 1999, 2 master's, 2 doctorates awarded.

Degree requirements: For doctorate, dissertation required, foreign language not required. *Average time to degree:* Doctorate–5.3 years full-time.

Entrance requirements: For doctorate, GRE General Test, GRE Subject Test, TOEFL, minimum QPA of 3.0. *Application deadline:* For fall admission, 1/15 (priority date). Applications are processed on a rolling basis. *Application fee:* $30 ($40 for international students).

Expenses: Tuition, state resident: full-time $9,778; part-time $403 per credit. Tuition, nonresident: full-time $20,146; part-time $830 per credit. Required fees: $480; $90 per semester.

Financial aid: Research assistantships with full tuition reimbursements, teaching assistantships with full tuition reimbursements, Federal Work-Study, institutionally sponsored loans, scholarships, traineeships, and unspecified assistantships available.

Faculty research: Herpes, HIV-1, sexually transmitted diseases, molecular basis of infectious disease. *Total annual research expenditures:* $76.3 million.

Application contact: Graduate Studies Administrator, 412-648-8957, *Fax:* 412-648-1236, *E-mail:* biomed_phd@fs1.dean-med.pitt.edu.

Find an in-depth description at www.petersons.com/graduate.

■ **UNIVERSITY OF PUERTO RICO, MEDICAL SCIENCES CAMPUS**

School of Medicine, Division of Graduate Studies, Department of Microbiology and Medical Zoology, San Juan, PR 00936-5067

AWARDS Medical zoology (MS, PhD); microbiology and medical zoology (MS, PhD).

Faculty: 16 full-time (8 women).

Students: 30 full-time (23 women); all minorities (1 Asian American or Pacific Islander, 29 Hispanic Americans). Average age 23. *15 applicants, 40% accepted.* In 1999, 4 master's awarded.

Degree requirements: For master's and doctorate, one foreign language, thesis/dissertation required.

Entrance requirements: For master's and doctorate, GRE General Test, GRE Subject Test, interview, minimum GPA or GPS of 3.0. *Application deadline:* For fall admission, 2/15; for spring admission, 9/15. *Application fee:* $15.

Expenses: Tuition, state resident: full-time $5,500. Tuition, nonresident: full-time $8,400. Required fees: $600. Tuition and fees vary according to class time, course load, degree level and program.

Financial aid: Fellowships, research assistantships, teaching assistantships, institutionally sponsored loans and tuition waivers (full) available. Financial aid application deadline: 4/30.

Faculty research: Molecular and general parasitology, immunology, development of viral vaccines and antiviral agents, fungal dimorphism, AIDS, pathogenesis. Dr. Guillermo Vázquez, Director, 787-758-2525 Ext. 1309, *Fax:* 787-758-4808.

Application contact: Dr. Adelfa Serrano, Coordinator, 787-758-2525 Ext. 1313, *Fax:* 787-758-4808.

■ **UNIVERSITY OF RHODE ISLAND**

Graduate School, College of Arts and Sciences, Department of Biochemistry, Microbiology, and Molecular Genetics, Kingston, RI 02881

AWARDS Biochemistry (MS, PhD); microbiology (MS, PhD), including biodegradation (MS), cellular development (MS), electron microscopy and ultrastructure (MS), genetics and molecular biology (MS), immunology (MS), marine and freshwater ecosystems (MS), microbial pathogenesis (MS), microbial physiology (MS), protozoology (MS), virology (MS), water-pollution microbiology (MS).

Degree requirements: For master's and doctorate, thesis/dissertation required.

Entrance requirements: For master's and doctorate, GRE General Test, TOEFL.

Expenses: Tuition, state resident: full-time $3,540; part-time $197 per credit. Tuition, nonresident: full-time $10,116; part-time $197 per credit. Required fees: $1,352; $37 per credit. $65 per term.

Financial aid: Fellowships, research assistantships, teaching assistantships available.

Dr. David Laux, Chairperson, 401-874-2201.

■ **UNIVERSITY OF ROCHESTER**

School of Medicine and Dentistry, Graduate Programs in Medicine and Dentistry, Department of Microbiology and Immunology, Rochester, NY 14627-0250

AWARDS Microbiology (MS, PhD).

Faculty: 21.

Students: 49 full-time (27 women), 17 part-time (8 women); includes 12 minority (2 African Americans, 7 Asian Americans or Pacific Islanders, 2 Hispanic Americans, 1 Native American), 8 international. In 1999, 17 master's, 7 doctorates awarded.

Degree requirements: For doctorate, dissertation, qualifying exam required, foreign language not required.

Entrance requirements: For master's and doctorate, GRE General Test. *Application deadline:* For fall admission, 2/1. *Application fee:* $25.

Expenses: Tuition: Part-time $697 per credit hour. Tuition and fees vary according to program.

Financial aid: Fellowships, research assistantships, teaching assistantships, tuition waivers (full and partial) available. Financial aid application deadline: 2/1.

Dr. Barbara Iglewski, Chair, 716-275-3402.

Application contact: Brenda Khorr, Graduate Program Secretary, 716-275-3402.

Find an in-depth description at www.petersons.com/graduate.

■ **UNIVERSITY OF SOUTH ALABAMA**

College of Medicine and Graduate School, Program in Basic Medical Sciences, Specialization in Microbiology and Immunology, Mobile, AL 36688-0002

AWARDS PhD.

Faculty: 11 full-time (0 women).
Students: In 1999, 1 degree awarded.
Degree requirements: For doctorate, dissertation required, foreign language not required.
Entrance requirements: For doctorate, GRE General Test or MCAT. *Application deadline:* For fall admission, 4/1. Applications are processed on a rolling basis. *Application fee:* $25.
Expenses: Tuition, state resident: part-time $116 per semester hour. Tuition, nonresident: part-time $230 per semester hour. Required fees: $121 per semester.
Financial aid: Fellowships, research assistantships, institutionally sponsored loans available. Financial aid application deadline: 4/1.
Faculty research: Mechanisms of tumor immunity, host response to infectious agents, virus replication, immune regulation, mechanisms of resistance to viruses and bacteria.
Dr. Joseph H. Coggin, Chair, 334-460-6339.
Application contact: Lanette Flagge, Coordinator, 334-460-6153.

■ **UNIVERSITY OF SOUTH DAKOTA**

School of Medicine and Graduate School, Biomedical Sciences Graduate Program, Molecular Microbiology and Immunology Group, Vermillion, SD 57069-2390

AWARDS MA, PhD.

Faculty: 4 full-time (1 woman).
Students: 4 full-time (1 woman). Average age 25. *15 applicants, 13% accepted.* Terminal master's awarded for partial completion of doctoral program.
Degree requirements: For master's and doctorate, thesis/dissertation required, foreign language not required. *Average time to degree:* Doctorate–4 years full-time.
Entrance requirements: For master's and doctorate, GRE General Test, TOEFL, minimum GPA of 3.0. *Application deadline:* For fall admission, 3/15 (priority date).

Applications are processed on a rolling basis. *Application fee:* $15.
Expenses: Tuition, state resident: full-time $2,126; part-time $89 per credit. Tuition, nonresident: full-time $6,270; part-time $261 per credit. Required fees: $1,194; $50 per credit. Tuition and fees vary according to course load and reciprocity agreements.
Financial aid: In 1999–00, 4 students received aid; fellowships, research assistantships, teaching assistantships, Federal Work-Study available.
Faculty research: Structure-function membranes, plasmids, immunology, virology, pathogenesis. *Total annual research expenditures:* $280,000.
Dr. Leigh Washburn, Coordinator, 605-677-5170, *Fax:* 605-677-5658, *E-mail:* lwashbur@usd.edu.

■ **UNIVERSITY OF SOUTHERN CALIFORNIA**

Keck School of Medicine and Graduate School, Graduate Programs in Medicine, Department of Molecular Microbiology and Immunology, Los Angeles, CA 90033

AWARDS MS, PhD.

Faculty: 8 full-time (1 woman), 1 (woman) part-time/adjunct.
Students: 40 full-time (23 women), 1 part-time; includes 2 minority (1 Asian American or Pacific Islander, 1 Hispanic American), 25 international. Average age 28. *140 applicants, 9% accepted.* In 1999, 5 master's, 5 doctorates awarded (60% entered university research/teaching, 40% continued full-time study).
Degree requirements: For master's, thesis optional, foreign language not required; for doctorate, dissertation required, foreign language not required. *Average time to degree:* Master's–3 years full-time; doctorate–4.75 years full-time.
Entrance requirements: For master's, GRE General Test, TOEFL, minimum GPA of 3.0; for doctorate, GRE General Test, GRE Subject Test, TOEFL, minimum GPA of 3.0. *Application deadline:* For fall admission, 3/15 (priority date); for spring admission, 7/15. Applications are processed on a rolling basis. *Application fee:* $55. Electronic applications accepted.
Expenses: Tuition: Full-time $22,198; part-time $748 per unit. Required fees: $406.
Financial aid: In 1999–00, 28 students received aid, including 1 fellowship with full tuition reimbursement available (averaging $17,575 per year), 26 research assistantships with full tuition reimbursements available (averaging $17,575 per year), 1 teaching assistantship with full tuition reimbursement available (averaging $17,575 per year); institutionally sponsored

loans, traineeships, and tuition waivers (partial) also available. Financial aid application deadline: 6/1.
Faculty research: Animal virology, microbial genetics, molecular and cellular immunology, cellular differentiation control of protein synthesis. *Total annual research expenditures:* $3 million.
Dr. Stanley M. Tahara, Director, Graduate Studies, 323-442-1722, *Fax:* 323-442-1721, *E-mail:* stahara@hsc.usc.edu.
Application contact: Laura C. Steel, Graduate Administrator, 323-442-2337, *Fax:* 323-442-1721, *E-mail:* lsteel@hsc.usc.edu.

■ **UNIVERSITY OF SOUTHERN MISSISSIPPI**

Graduate School, College of Science and Technology, Department of Biological Sciences, Hattiesburg, MS 39406

AWARDS Environmental biology (MS, PhD); marine biology (MS, PhD); microbiology (MS, PhD); molecular biology (MS, PhD).

Degree requirements: For master's, thesis required, foreign language not required; for doctorate, 2 foreign languages (computer language can substitute for one), dissertation required.
Entrance requirements: For master's, GRE General Test, TOEFL, minimum GPA of 3.0; for doctorate, GRE General Test, TOEFL, minimum GPA of 3.5.
Expenses: Tuition, state resident: full-time $2,250; part-time $137 per semester hour. Tuition, nonresident: full-time $3,102; part-time $172 per semester hour. Required fees: $602.

■ **UNIVERSITY OF SOUTH FLORIDA**

Graduate School, College of Arts and Sciences, Department of Biology, Program in Microbiology, Tampa, FL 33620-9951

AWARDS MS. Part-time programs available.

Degree requirements: For master's, thesis required, foreign language not required.
Entrance requirements: For master's, GRE General Test, minimum GPA of 3.0 in last 60 hours. Electronic applications accepted.
Expenses: Tuition, state resident: part-time $148 per credit hour. Tuition, nonresident: part-time $509 per credit hour.

■ THE UNIVERSITY OF TENNESSEE

Graduate School, College of Arts and Sciences, Department of Microbiology, Knoxville, TN 37996

AWARDS MS, PhD. Part-time programs available.

Faculty: 11 full-time (0 women).
Students: 31 full-time (14 women), 2 part-time (1 woman); includes 1 minority (Asian American or Pacific Islander), 5 international. *60 applicants, 17% accepted.* In 1999, 3 master's, 3 doctorates awarded.
Degree requirements: For master's and doctorate, thesis/dissertation required, foreign language not required.
Entrance requirements: For master's and doctorate, GRE General Test, TOEFL, minimum GPA of 2.7. *Application deadline:* For fall admission, 2/1 (priority date). Applications are processed on a rolling basis. *Application fee:* $35. Electronic applications accepted.
Expenses: Tuition, state resident: full-time $3,806; part-time $184 per credit hour. Tuition, nonresident: full-time $9,874; part-time $522 per credit hour. Tuition and fees vary according to program.
Financial aid: In 1999–00, 10 research assistantships, 14 teaching assistantships were awarded; fellowships, Federal Work-Study, institutionally sponsored loans, and unspecified assistantships also available. Financial aid application deadline: 2/1; financial aid applicants required to submit FAFSA.
Dr. Robert Moore, Head, 865-974-3441, *Fax:* 865-974-4007, *E-mail:* rmoore1@utk.edu.
Application contact: Dr. Gary Stacey, Graduate Representative, *E-mail:* gstacey@utk.edu.

Find an in-depth description at www.petersons.com/graduate.

■ THE UNIVERSITY OF TENNESSEE HEALTH SCIENCE CENTER

College of Graduate Health Sciences, Department of Microbiology and Immunology, Memphis, TN 38163-0002

AWARDS Bacteriology (MS, PhD); immunology (MS, PhD); medical microbiology (MS, PhD); microbiology (MS, PhD); molecular and cell biology (PhD); molecular biology (MS, PhD); virology (MS, PhD). Terminal master's awarded for partial completion of doctoral program.

Degree requirements: For master's, thesis, oral and written comprehensive exams required, foreign language not required; for doctorate, dissertation, oral and written preliminary and comprehensive exams required, foreign language not required.
Entrance requirements: For master's, GRE General Test, TOEFL, minimum GPA of 3.0; for doctorate, GRE General Test, GRE Subject Test, TOEFL, minimum GPA of 3.0.

Find an in-depth description at www.petersons.com/graduate.

■ THE UNIVERSITY OF TEXAS AT AUSTIN

Graduate School, College of Natural Sciences, Institute for Cellular and Molecular Biology, Program in Microbiology and Immunology, Austin, TX 78712-1111

AWARDS PhD.

Entrance requirements: For doctorate, GRE General Test. *Application fee:* $50 ($75 for international students). Electronic applications accepted.
Expenses: Tuition, state resident: part-time $114 per semester hour. Tuition, nonresident: part-time $330 per semester hour. Tuition and fees vary according to program.
Financial aid: In 1999–00, fellowships with full tuition reimbursements (averaging $19,000 per year), research assistantships with full and partial tuition reimbursements (averaging $18,000 per year), teaching assistantships with full tuition reimbursements (averaging $17,300 per year) were awarded.
Paul O. Gottlieb, Chairman, 512-471-5105.
Application contact: Dr. Clarence Chan, Graduate Adviser, 512-471-6860, *E-mail:* clarence_chan@mail.utexas.edu.

Find an in-depth description at www.petersons.com/graduate.

■ THE UNIVERSITY OF TEXAS AT AUSTIN

Graduate School, College of Natural Sciences, School of Biological Sciences, Program in Microbiology, Austin, TX 78712-1111

AWARDS MA, PhD.

Entrance requirements: For master's and doctorate, GRE General Test. *Application fee:* $50 ($75 for international students). Electronic applications accepted.
Expenses: Tuition, state resident: part-time $114 per semester hour. Tuition, nonresident: part-time $330 per semester hour. Tuition and fees vary according to program.
Financial aid: In 1999–00, fellowships with full tuition reimbursements (averaging $19,000 per year), research assistantships

with full and partial tuition reimbursements (averaging $18,000 per year), teaching assistantships with full tuition reimbursements (averaging $17,300 per year) were awarded.
Paul O. Gottlieb, Chairman, 512-471-5105.
Application contact: Dr. Clarence Chan, Graduate Adviser, 512-471-6860, *E-mail:* clarence_chan@mail.utexas.edu.

Find an in-depth description at www.petersons.com/graduate.

■ THE UNIVERSITY OF TEXAS HEALTH SCIENCE CENTER AT SAN ANTONIO

Graduate School of Biomedical Sciences, Department of Microbiology, San Antonio, TX 78229-3900

AWARDS MS, PhD. Terminal master's awarded for partial completion of doctoral program.

Degree requirements: For master's and doctorate, thesis/dissertation required.
Entrance requirements: For master's and doctorate, GRE General Test, minimum GPA of 3.0.
Expenses: Tuition, state resident: part-time $38 per credit hour. Tuition, nonresident: part-time $249 per credit hour.
Faculty research: Molecular immunology, mechanisms of pathogenesis, molecular genetics, vaccine and immunodiagnostic development.

Find an in-depth description at www.petersons.com/graduate.

■ THE UNIVERSITY OF TEXAS–HOUSTON HEALTH SCIENCE CENTER

Graduate School of Biomedical Sciences, Program in Microbiology and Molecular Genetics, Houston, TX 77225-0036

AWARDS MS, PhD, MD/PhD.

Faculty: 21 full-time (11 women).
Students: 19 full-time (10 women); includes 3 minority (1 African American, 1 Asian American or Pacific Islander, 1 Hispanic American), 8 international. Average age 27. *35 applicants, 43% accepted.* In 1999, 4 master's, 7 doctorates awarded. Terminal master's awarded for partial completion of doctoral program.
Degree requirements: For master's and doctorate, thesis/dissertation required, foreign language not required.
Entrance requirements: For master's and doctorate, GRE General Test, TOEFL, TWE. *Application deadline:* For fall admission, 1/15 (priority date); for spring admission, 11/1. Applications are processed on a

rolling basis. *Application fee:* $10. Electronic applications accepted.

Financial aid: Fellowships, research assistantships, institutionally sponsored loans available. Financial aid application deadline: 1/15.

Faculty research: Microbial genetics, microbial diversity, gene regulation, molecular pathogenesis, sensory transduction.

Dr. Samuel Kaplan, Chairman and Director, 713-500-5502, *Fax:* 713-500-5499, *E-mail:* skaplan@utmmg.med.uth.tmc.edu.

Application contact: Anne Baronitis, Director of Admissions, 713-500-9860, *Fax:* 713-500-9877, *E-mail:* abaron@ gsbs.gs.uth.tmc.edu.

Find an in-depth description at www.petersons.com/graduate.

■ THE UNIVERSITY OF TEXAS MEDICAL BRANCH AT GALVESTON

Graduate School of Biomedical Sciences, Program in Microbiology and Immunology, Galveston, TX 77555
AWARDS MS, PhD.

Faculty: 52 full-time (6 women).
Students: 30 full-time (16 women); includes 8 minority (1 African American, 4 Asian Americans or Pacific Islanders, 3 Hispanic Americans), 12 international. Average age 29. *53 applicants, 30% accepted.* In 1999, 6 degrees awarded (100% entered university research/teaching).
Degree requirements: For master's, thesis or alternative required, foreign language not required; for doctorate, dissertation required, foreign language not required.
Entrance requirements: For master's and doctorate, GRE General Test, minimum GPA of 3.0. *Application deadline:* For fall admission, 3/15. Applications are processed on a rolling basis. *Application fee:* $25 ($50 for international students). Electronic applications accepted.
Expenses: Tuition, state resident: full-time $684; part-time $38 per credit hour. Tuition, nonresident: full-time $4,572; part-time $254 per credit hour. Required fees: $29; $7.5 per credit hour. One-time fee: $55. Tuition and fees vary according to program.
Financial aid: Fellowships, research assistantships, Federal Work-Study and institutionally sponsored loans available. Financial aid applicants required to submit FAFSA.
Faculty research: Bacterial and viral pathogenesis, microbial genetics, molecular virology, host defense, neuroimmunology, autoimmunity, and structural and molecular biology.

Dr. Thomas K. Hughes, Director, 409-772-2322, *Fax:* 409-772-2366, *E-mail:* thughes@utmb.edu.
Application contact: Martha Lewis, Secretary, 409-772-2322, *Fax:* 409-772-5065, *E-mail:* mlewis@utmb.edu.

Find an in-depth description at www.petersons.com/graduate.

■ THE UNIVERSITY OF TEXAS SOUTHWESTERN MEDICAL CENTER AT DALLAS

Southwestern Graduate School of Biomedical Sciences, Division of Cell and Molecular Biology, Program in Molecular Microbiology, Dallas, TX 75390
AWARDS PhD.

Faculty: 28 full-time (5 women), 1 part-time/adjunct (0 women).
Students: 12 full-time (7 women), 1 international. Average age 30.
Degree requirements: For doctorate, computer language, dissertation, oral and written exams required, foreign language not required.
Entrance requirements: For doctorate, GRE General Test, minimum GPA of 3.0. *Application deadline:* For fall admission, 1/5 (priority date). *Application fee:* $0. Electronic applications accepted.
Expenses: Tuition, state resident: full-time $912. Tuition, nonresident: full-time $6,096. Required fees: $216. Full-time tuition and fees vary according to course load and program.
Financial aid: In 1999–00, 5 students received aid; fellowships, research assistantships, institutionally sponsored loans available. Financial aid application deadline: 3/15; financial aid applicants required to submit FAFSA.
Faculty research: Cell and molecular immunology, molecular pathogenesis of infectious disease, virology.

Dr. Michael Norgard, Chair, 214-648-5904, *Fax:* 214-648-5906, *E-mail:* norgard@utsw.swmed.edu.
Application contact: Nancy McKinney, Education Coordinator, 214-648-8099, *Fax:* 214-648-2978, *E-mail:* dcmbinfo@ utsouthwestern.edu.

Find an in-depth description at www.petersons.com/graduate.

■ UNIVERSITY OF VERMONT

College of Medicine and Graduate College, Graduate Programs in Medicine, Department of Microbiology and Molecular Genetics, Burlington, VT 05405
AWARDS MS, PhD, MD/MS, MD/PhD.

Degree requirements: For master's and doctorate, thesis/dissertation required.

Entrance requirements: For master's and doctorate, GRE General Test, TOEFL.
Expenses: Tuition, state resident: full-time $7,464; part-time $311 per credit. Tuition, nonresident: full-time $18,672; part-time $778 per credit. Full-time tuition and fees vary according to degree level and program.

Find an in-depth description at www.petersons.com/graduate.

■ UNIVERSITY OF VIRGINIA

College and Graduate School of Arts and Sciences, Department of Microbiology, Charlottesville, VA 22903
AWARDS PhD.

Faculty: 38 full-time (12 women), 1 part-time/adjunct (0 women).
Students: 80 full-time (42 women); includes 11 minority (4 African Americans, 4 Asian Americans or Pacific Islanders, 3 Hispanic Americans), 6 international. Average age 27. *94 applicants, 21% accepted.* In 1999, 8 doctorates awarded.
Degree requirements: For doctorate, dissertation required.
Entrance requirements: For doctorate, GRE. *Application deadline:* For fall admission, 3/15; for spring admission, 12/1. Applications are processed on a rolling basis. *Application fee:* $40. Electronic applications accepted.
Expenses: Tuition, state resident: full-time $3,832. Tuition, nonresident: full-time $15,519. Required fees: $1,084. Tuition and fees vary according to course load and program.
Financial aid: Fellowships, traineeships and unspecified assistantships available. Financial aid application deadline: 2/1; financial aid applicants required to submit FAFSA.
Faculty research: Virology, membrane biology and molecular genetics.

J. Thomas Parsons, Chairman, 804-924-5111.
Application contact: Duane J. Osheim, Associate Dean, 804-924-7184, *E-mail:* microbiology@virginia.edu.

Find an in-depth description at www.petersons.com/graduate.

■ UNIVERSITY OF WASHINGTON

School of Medicine and Graduate School, Graduate Programs in Medicine, Department of Microbiology, Seattle, WA 98195
AWARDS PhD.

Faculty: 18 full-time (3 women), 5 part-time/adjunct (3 women).
Students: 28 full-time (15 women); includes 4 minority (2 Asian Americans or Pacific Islanders, 2 Hispanic Americans), 3

University of Washington (continued) international. Average age 28. *124 applicants, 4% accepted.* In 1999, 8 degrees awarded.

Degree requirements: For doctorate, dissertation required.

Entrance requirements: For doctorate, GRE General Test, GRE Subject Test (recommended). *Application deadline:* For fall admission, 12/31 (priority date). *Application fee:* $50. Electronic applications accepted.

Expenses: Tuition, state resident: full-time $9,210; part-time $236 per credit. Tuition, nonresident: full-time $23,256; part-time $596 per credit.

Financial aid: In 1999–00, 28 students received aid, including 3 fellowships (averaging $17,300 per year), 20 research assistantships (averaging $17,300 per year), 5 teaching assistantships (averaging $17,300 per year); tuition waivers (full) also available. Financial aid application deadline: 12/31.

Faculty research: Bacterial genetics and physiology, mechanisms of bacterial and viral pathogenesis, bacterial-plant interaction.

Dr. James I. Mullins, Chairman, 206-543-5824.

Application contact: Sarah Mears, Coordinator, 206-543-2572, *Fax:* 206-543-8297, *E-mail:* mears@u.washington.edu.

Find an in-depth description at www.petersons.com/graduate.

■ UNIVERSITY OF WISCONSIN– LA CROSSE

Graduate Studies, College of Science and Allied Health, Department of Biology, La Crosse, WI 54601-3742

AWARDS Biology (MS); clinical microbiology (MS); nurse anesthetist (MS). Part-time programs available.

Degree requirements: For master's, thesis (for some programs), oral comprehensive exam required, foreign language not required.

Entrance requirements: For master's, GRE General Test, minimum GPA of 3.0 during previous 2 years or 2.85 overall.

Expenses: Tuition, state resident: full-time $3,737; part-time $208 per credit. Tuition, nonresident: full-time $11,380; part-time $633 per credit. Tuition and fees vary according to course load, program and reciprocity agreements.

Faculty research: Ecology, river studies, aquatic toxicology, aquatic microbiology, molecular biology, physiology.

■ UNIVERSITY OF WISCONSIN– MADISON

Medical School and Graduate School, Graduate Programs in Medicine and College of Agricultural and Life Sciences, Microbiology Doctoral Training Program, Madison, WI 53706-1380

AWARDS PhD.

Faculty: 80 full-time (15 women).
Students: 44 full-time (38 women); includes 8 minority (2 African Americans, 2 Asian Americans or Pacific Islanders, 4 Hispanic Americans), 4 international. Average age 24. *269 applicants, 15% accepted.* In 1999, 6 degrees awarded.
Degree requirements: For doctorate, dissertation required, foreign language not required. *Average time to degree:* Doctorate–6 years full-time.
Entrance requirements: For doctorate, GRE, TOEFL. *Application deadline:* For fall admission, 1/1 (priority date). *Application fee:* $45. Electronic applications accepted.
Expenses: Tuition, state resident: full-time $5,406; part-time $339 per credit. Tuition, nonresident: full-time $17,110; part-time $1,071 per credit.
Financial aid: In 1999–00, 26 students received aid, including 15 fellowships with tuition reimbursements available, 12 research assistantships with tuition reimbursements available; grants, scholarships, traineeships, and tuition waivers (full) also available. *Total annual research expenditures:* $6.5 million.
Dr. Robert C. Landick, Director, 608-265-8475, *Fax:* 608-262-9865, *E-mail:* landick@bact.wisc.edu.
Application contact: Kathryn A. Holtgraver, Program Coordinator, 608-262-0689, *Fax:* 608-262-9865, *E-mail:* kathyh@bact.wisc.edu.

Find an in-depth description at www.petersons.com/graduate.

■ UNIVERSITY OF WISCONSIN– OSHKOSH

Graduate School, College of Letters and Science, Department of Biology and Microbiology, Oshkosh, WI 54901

AWARDS Biology (MS), including botany, microbiology, zoology.

Degree requirements: For master's, thesis required, foreign language not required.
Entrance requirements: For master's, GRE General Test, minimum GPA of 3.0, BS in biology.
Expenses: Tuition, state resident: full-time $3,917; part-time $219 per credit. Tuition, nonresident: full-time $12,375; part-time

$684 per credit. Part-time tuition and fees vary according to course load and program.

■ UTAH STATE UNIVERSITY

School of Graduate Studies, College of Family Life and College of Agriculture, Department of Nutrition and Food Sciences, Logan, UT 84322

AWARDS Food microbiology and safety (MFMS); molecular biology (MS, PhD); nutrition and food sciences (MS, PhD).

Faculty: 17 full-time (8 women), 2 part-time/adjunct (0 women).
Students: 18 full-time (11 women), 8 part-time (4 women); includes 2 minority (1 Asian American or Pacific Islander, 1 Hispanic American), 12 international. Average age 27. *34 applicants, 21% accepted.* In 1999, 10 master's, 3 doctorates awarded.
Degree requirements: For master's, thesis required, foreign language not required; for doctorate, computer language, dissertation, teaching experience required, foreign language not required.
Entrance requirements: For master's and doctorate, GRE General Test, TOEFL, minimum GPA of 3.0, previous course work in chemistry. *Application deadline:* For fall admission, 2/1 (priority date); for spring admission, 10/15 (priority date). Applications are processed on a rolling basis. *Application fee:* $40.
Expenses: Tuition, state resident: full-time $1,553. Tuition, nonresident: full-time $5,436. International tuition: $5,526 full-time. Required fees: $447. Tuition and fees vary according to course load and program.
Financial aid: In 1999–00, 22 research assistantships with partial tuition reimbursements were awarded; fellowships with partial tuition reimbursements, teaching assistantships with partial tuition reimbursements, Federal Work-Study, institutionally sponsored loans, and tuition waivers (full and partial) also available.
Faculty research: Mineral balance, meat microbiology and nitrate interactions, milk ultrafiltration, lactic culture, milk coagulation.
Ann Sorenson, Head, 435-797-2102, *Fax:* 435-797-2379.
Application contact: Carolyn Glover, Receptionist, 435-797-2126, *Fax:* 435-797-2379, *E-mail:* nfs@cc.usu.edu.

■ VANDERBILT UNIVERSITY

Graduate School and School of Medicine, Department of Microbiology and Immunology, Nashville, TN 37232-2363

AWARDS MS, PhD, MD/PhD.

Faculty: 16 full-time (2 women), 17 part-time/adjunct (1 woman).

Students: 35 full-time (14 women), 8 part-time (3 women); includes 3 minority (1 African American, 1 Asian American or Pacific Islander, 1 Hispanic American), 13 international. Average age 29. In 1999, 4 degrees awarded.

Degree requirements: For master's, thesis required, foreign language not required; for doctorate, dissertation, final and qualifying exams required, foreign language not required.

Entrance requirements: For master's and doctorate, GRE General Test, GRE Subject Test (recommended). *Application deadline:* For fall admission, 1/15. *Application fee:* $40.

Expenses: Tuition: Full-time $17,244; part-time $958 per hour. Required fees: $242; $121 per semester. Tuition and fees vary according to program.

Financial aid: In 1999–00, fellowships with full tuition reimbursements (averaging $17,000 per year), research assistantships with full tuition reimbursements (averaging $17,000 per year) were awarded; Federal Work-Study, institutionally sponsored loans, traineeships, and tuition waivers (partial) also available. Financial aid application deadline: 1/15.

Faculty research: Molecular and cellular immunology, molecular genetics and immunogenetics, cellular microbiology of pathogen-host interaction, virology, biotechnology.
Jacek Hawiger, Chair, 615-343-8280, *Fax:* 615-343-7392, *E-mail:* jacek.hawiger@ mcmail.vanderbilt.edu.

Application contact: G. Neil Green, Director of Graduate Studies, 615-343-0453, *Fax:* 615-343-7392, *E-mail:* neil.green@mcmail.vanderbilt.edu.

Find an in-depth description at www.petersons.com/graduate.

■ VIRGINIA COMMONWEALTH UNIVERSITY

School of Graduate Studies and School of Medicine, School of Medicine Graduate Programs, Department of Microbiology and Immunology, Richmond, VA 23284-9005

AWARDS Microbiology (PhD); microbiology and immunology (MS, CBHS); molecular biology and genetics (PhD).

Students: 13 full-time (6 women), 46 part-time (26 women); includes 22 minority (2 African Americans, 17 Asian Americans or Pacific Islanders, 2 Hispanic Americans, 1 Native American). In 1999, 2 master's, 9 doctorates, 2 other advanced degrees awarded.

Degree requirements: For master's, thesis required, foreign language not required; for doctorate, dissertation, comprehensive oral and written exams required, foreign language not required.

Entrance requirements: For master's, DAT, GRE General Test, or MCAT; for doctorate, GRE General Test. *Application deadline:* For fall admission, 4/15. *Application fee:* $30.

Expenses: Tuition, state resident: full-time $4,031; part-time $224 per credit hour. Tuition, nonresident: full-time $11,946; part-time $664 per credit hour. Required fees: $1,081; $40 per credit hour. Tuition and fees vary according to campus/location and program.

Financial aid: Fellowships, research assistantships, teaching assistantships available.

Faculty research: Microbial physiology and genetics, molecular biology, crystallography of biological molecules, antibiotics and chemotherapy, membrane transport.
Dr. Dennis E. Ohman, Chair, 804-828-9728, *Fax:* 804-828-9946.

Find an in-depth description at www.petersons.com/graduate.

■ VIRGINIA POLYTECHNIC INSTITUTE AND STATE UNIVERSITY

Graduate School, College of Arts and Sciences, Department of Biology, Program in Microbiology, Blacksburg, VA 24061-0406

AWARDS MS, PhD.

Faculty: 14 full-time (2 women).

Degree requirements: For master's and doctorate, thesis/dissertation required, foreign language not required.

Entrance requirements: For master's, GRE General Test, TOEFL; for doctorate, GRE General Test, GRE Subject Test, TOEFL. *Application deadline:* For fall admission, 1/31 (priority date). Applications are processed on a rolling basis. *Application fee:* $25.

Financial aid: Fellowships, research assistantships, teaching assistantships available. Financial aid application deadline: 1/31.

Faculty research: Ecology, evolution, molecular biology, systematics, physiology.
Dr. Joe R. Cowles, Chairman, Department of Biology, 540-231-8928, *E-mail:* cowlesjr@vt.edu.

■ WAGNER COLLEGE

Division of Graduate Studies, Department of Biological Sciences, Program in Microbiology, Staten Island, NY 10301-4495

AWARDS MS.

Faculty: 7 full-time (4 women), 2 part-time/adjunct (1 woman).

Students: 9 full-time (7 women), 6 part-time (4 women); includes 1 minority (Hispanic American). *10 applicants, 30% accepted.* In 1999, 7 degrees awarded.

Degree requirements: For master's, T required.

Application deadline: For fall admission, 8/1 (priority date); for spring admission, 12/10. Applications are processed on a rolling basis. *Application fee:* $50 ($65 for international students).

Expenses: Tuition: Part-time $580 per credit.

Financial aid: In 1999–00, 1 fellowship, 4 teaching assistantships with full tuition reimbursements (averaging $2,400 per year) were awarded; tuition waivers (partial) also available.

Application contact: Admissions Office, 718-390-3411.

■ WAKE FOREST UNIVERSITY

School of Medicine and Graduate School, Graduate Programs in Medicine, Department of Microbiology and Immunology, Winston-Salem, NC 27109

AWARDS PhD.

Faculty: 13.

Students: 22 full-time (12 women); includes 2 African Americans. *54 applicants, 22% accepted.* In 1999, 2 degrees awarded.

Degree requirements: For doctorate, dissertation required.

Entrance requirements: For doctorate, GRE General Test, GRE Subject Test. *Application deadline:* For fall admission, 2/15 (priority date). Applications are processed on a rolling basis. *Application fee:* $25. Electronic applications accepted.

Expenses: Tuition: Full-time $18,300.

Financial aid: In 1999–00, 14 fellowships, 8 research assistantships were awarded; grants, scholarships, and traineeships also available. Financial aid application deadline: 2/15.

Faculty research: Molecular immunology, bacterial pathogenesis and molecular genetics, viral pathogenesis, regulation of mRNA metabolism, leukocyte biology. *Total annual research expenditures:* $1.5 million.
Dr. Steven B. Mizel, Chair, 336-716-4471.

Application contact: Dr. Griffith D. Parks, Graduate Recruitment Director,

Wake Forest University (continued)
336-716-9083, *Fax:* 336-716-9928, *E-mail:* gparks@wfubmc.edu.

Find an in-depth description at www.petersons.com/graduate.

■ WASHINGTON STATE UNIVERSITY

College of Veterinary Medicine and Graduate School, Graduate Programs in Veterinary Science, Department of Veterinary Microbiology and Pathology, Pullman, WA 99164

AWARDS Veterinary science (MS, PhD).

Faculty: 23 full-time (4 women).
Students: 32 full-time (13 women); includes 14 minority (2 African Americans, 6 Asian Americans or Pacific Islanders, 6 Hispanic Americans). Average age 33. *10 applicants, 40% accepted.* In 1999, 2 doctorates awarded (100% found work related to degree). Terminal master's awarded for partial completion of doctoral program.
Degree requirements: For master's and doctorate, computer language, thesis/dissertation, oral exam required. *Average time to degree:* Master's–3 years full-time; doctorate–5.6 years full-time.
Entrance requirements: For master's and doctorate, GRE General Test, minimum GPA of 3.0. *Application deadline:* For fall admission, 1/31 (priority date). Applications are processed on a rolling basis. *Application fee:* $35. Electronic applications accepted.
Expenses: Tuition, state resident: full-time $5,494. Tuition, nonresident: full-time $13,390.
Financial aid: In 1999–00, 17 students received aid, including 17 research assistantships (averaging $4,887 per year); grants and traineeships also available. Financial aid application deadline: 3/1.
Faculty research: Microbial pathogenesis, veterinary and wildlife parasitology, laboratory animal pathology, immune responses to infectious diseases.
Dr. David J. Prieur, Chair, 509-335-6030, *Fax:* 509-335-8529, *E-mail:* dprieur@vetmed.wsu.edu.
Application contact: Dr. Guy Palmer, Professor, 509-335-6033, *Fax:* 509-335-8529, *E-mail:* gpalmer@vetmed.wsu.edu.

■ WASHINGTON STATE UNIVERSITY

Graduate School, College of Sciences, School of Molecular Biosciences, Department of Microbiology, Pullman, WA 99164

AWARDS MS, PhD.

Faculty: 10.
Students: 23 full-time (12 women), 1 part-time; includes 2 minority (1 Asian

American or Pacific Islander, 1 Hispanic American), 5 international. Average age 29. In 1999, 3 master's, 2 doctorates awarded.
Degree requirements: For master's and doctorate, thesis/dissertation, oral exam required, foreign language not required. *Average time to degree:* Master's–4 years full-time; doctorate–5 years full-time.
Entrance requirements: For master's and doctorate, GRE General Test, GRE Subject Test, minimum GPA of 3.0. *Application deadline:* For fall admission, 2/1 (priority date). Applications are processed on a rolling basis. *Application fee:* $35.
Expenses: Tuition, state resident: full-time $5,654. Tuition, nonresident: full-time $13,850. International tuition: $13,850 full-time. Tuition and fees vary according to program.
Financial aid: In 1999–00, 1 fellowship, 15 research assistantships with full and partial tuition reimbursements, 9 teaching assistantships with full and partial tuition reimbursements were awarded; Federal Work-Study, institutionally sponsored loans, tuition waivers (partial), and teaching associateships also available. Financial aid application deadline: 4/1; financial aid applicants required to submit FAFSA. *Total annual research expenditures:* $1.2 million.
Dr. Michael Kahn, Interim Chair, 509-335-3323, *Fax:* 509-335-1907, *E-mail:* kahn@wsu.edu.
Application contact: Kathleen Postle, Graduate Admissions Coordinator, 509-335-5614, *Fax:* 509-335-1907, *E-mail:* postle@mail.wsu.edu.

Find an in-depth description at www.petersons.com/graduate.

■ WASHINGTON UNIVERSITY IN ST. LOUIS

Graduate School of Arts and Sciences, Division of Biology and Biomedical Sciences, Program in Molecular Microbiology and Microbial Pathogenesis, St. Louis, MO 63110

AWARDS PhD.

Degree requirements: For doctorate, dissertation required, foreign language not required.
Entrance requirements: For doctorate, GRE General Test, GRE Subject Test. *Application deadline:* For fall admission, 1/1 (priority date). Applications are processed on a rolling basis. *Application fee:* $0.
Expenses: Tuition: Full-time $23,400; part-time $975 per credit. Tuition and fees vary according to program.
Financial aid: Application deadline: 1/1.
Dr. William E. Goldman, Coordinator.
Application contact: Rosemary Garagneni, Director of Admissions, 800-852-9074, *E-mail:* admissions@dbbs.wustl.edu.

■ WAYNE STATE UNIVERSITY

School of Medicine and Graduate School, Graduate Programs in Medicine, Department of Immunology and Microbiology, Detroit, MI 48202

AWARDS MS, PhD, MD/PhD. Terminal master's awarded for partial completion of doctoral program.

Degree requirements: For master's and doctorate, thesis/dissertation required, foreign language not required.
Entrance requirements: For master's, GRE, minimum GPA of 2.5; for doctorate, GRE, minimum GPA of 3.0.
Faculty research: Cellular immunity, immune regulation, bacterial pathophysiology, microbial physiology, molecular biology/viruses/bacteria.

Find an in-depth description at www.petersons.com/graduate.

■ WEST VIRGINIA UNIVERSITY

School of Medicine, Graduate Programs in Health Sciences, Department of Microbiology and Immunology, Morgantown, WV 26506

AWARDS Genetics (MS, PhD); immunology (MS, PhD); mycology (PhD); parasitology (MS, PhD); pathogenic bacteriology (MS, PhD); physiology (PhD); psysiology (MS); virology (MS, PhD).

Students: 25 full-time (11 women), 1 (woman) part-time; includes 2 minority (both Asian Americans or Pacific Islanders), 4 international. Average age 27. *62 applicants, 8% accepted.* In 1999, 1 master's, 5 doctorates awarded. Terminal master's awarded for partial completion of doctoral program.
Degree requirements: For master's, thesis required, foreign language not required; for doctorate, dissertation, comprehensive exam required, foreign language not required. *Average time to degree:* Doctorate–4.5 years full-time.
Entrance requirements: For master's and doctorate, GRE General Test, GRE Subject Test, TOEFL, minimum GPA of 3.0. *Application deadline:* For fall admission, 3/1 (priority date). Applications are processed on a rolling basis. *Application fee:* $45. Electronic applications accepted.
Expenses: Tuition, state resident: full-time $3,564. Tuition, nonresident: full-time $10,230.
Financial aid: In 1999–00, 7 research assistantships, 14 teaching assistantships were awarded; fellowships, Federal Work-Study and institutionally sponsored loans also available. Financial aid application deadline: 3/1; financial aid applicants required to submit FAFSA.
Faculty research: Mechanisms of pathogenesis, microbial genetics, molecular

virology, immunotoxicology, oncogenes. *Total annual research expenditures:* $2.6 million.
John Barnett, Chair, 304-293-2649.
Application contact: James Sheil, Graduate Coordinator, 304-293-3559, *Fax:* 304-293-7823, *E-mail:* jsheil@wvu.edu.

■ WEST VIRGINIA UNIVERSITY

School of Medicine, Graduate Programs in Health Sciences, Medical Technology Program, Morgantown, WV 26506

AWARDS Clinical chemistry (MS); hematology (MS); immunohematology (MS); microbiology (MS). Part-time programs available.

Students: Average age 37. *2 applicants, 100% accepted.*
Degree requirements: For master's, thesis, written exam required, foreign language not required. *Average time to degree:* Master's–2 years full-time, 3 years part-time.
Entrance requirements: For master's, GRE, TOEFL, minimum GPA of 2.5, interview. *Application deadline:* Applications are processed on a rolling basis. *Application fee:* $45.
Expenses: Tuition, state resident: full-time $3,564. Tuition, nonresident: full-time $10,230.
Financial aid: In 1999–00, 1 teaching assistantship was awarded; Federal Work-Study and institutionally sponsored loans also available. Financial aid application deadline: 2/1; financial aid applicants required to submit FAFSA.
Faculty research: Management, lipids, coagulation, blood bank, molecular pathology, hematology.
Dr. Jean Holter, Director, 304-293-2069.

■ WRIGHT STATE UNIVERSITY

School of Graduate Studies, College of Science and Mathematics, Department of Microbiology and Immunology, Dayton, OH 45435

AWARDS MS. Part-time programs available.

Students: 13 full-time (9 women), 7 part-time (4 women); includes 4 minority (3 African Americans, 1 Hispanic American), 2 international. Average age 28. *17 applicants, 35% accepted.* In 1999, 4 degrees awarded.
Degree requirements: For master's, thesis required, foreign language not required.
Entrance requirements: For master's, TOEFL. *Application fee:* $25.
Expenses: Tuition, state resident: full-time $5,568; part-time $175 per quarter hour. Tuition, nonresident: full-time $9,696; part-time $302 per quarter hour. Full-time tuition and fees vary according to course load, campus/location and program.

Financial aid: Fellowships, research assistantships, teaching assistantships, Federal Work-Study, institutionally sponsored loans, and unspecified assistantships available. Aid available to part-time students. Financial aid applicants required to submit FAFSA.
Faculty research: Reproductive immunology, viral pathogenesis, virus-host cell interactions.
Dr. Neal S. Rote, Chair, 937-775-2996, *Fax:* 937-775-2012, *E-mail:* neal.rote@wright.edu.

■ YALE UNIVERSITY

School of Medicine and Graduate School of Arts and Sciences, Combined Program in Biological and Biomedical Sciences (BBS), Microbiology Track, New Haven, CT 06520

AWARDS PhD, MD/PhD.

Degree requirements: For doctorate, dissertation required.
Entrance requirements: For doctorate, GRE General Test, GRE Subject Test, TOEFL. *Application deadline:* For fall admission, 1/2. *Application fee:* $65. Electronic applications accepted.
Expenses: All students receive full tuition of $22,330 and an annual stipend of $17,600 .
Financial aid: Fellowships, research assistantships available.
Dr. Peter Tattersall, Director of Graduate Studies, 203-785-4586, *E-mail:* bbs@yale.edu.

Find an in-depth description at www.petersons.com/graduate.

■ YESHIVA UNIVERSITY

Albert Einstein College of Medicine, Sue Golding Graduate Division of Medical Sciences, Department of Microbiology and Immunology, Bronx, NY 10461

AWARDS PhD, MD/PhD.

Faculty: 10 full-time.
Students: 51 full-time (20 women); includes 19 minority (5 African Americans, 6 Asian Americans or Pacific Islanders, 8 Hispanic Americans), 6 international. In 1999, 7 degrees awarded.
Degree requirements: For doctorate, dissertation required, foreign language not required.
Entrance requirements: For doctorate, GRE General Test, TOEFL. *Application deadline:* For fall admission, 1/15. *Application fee:* $0.
Expenses: Tuition: Part-time $525 per credit. Tuition and fees vary according to degree level and program.

Financial aid: In 1999–00, 51 fellowships were awarded.
Faculty research: Nature of histocompatibility antigens, lymphoid cell receptors, regulation of immune responses and mechanisms of resistance to infection.
Dr. Marshall Horwitz, Chairperson, 718-430-2811.

Find an in-depth description at www.petersons.com/graduate.

VIROLOGY

■ BAYLOR COLLEGE OF MEDICINE

Graduate School of Biomedical Sciences, Department of Molecular Virology and Microbiology, Houston, TX 77030-3498

AWARDS PhD, MD/PhD.

Faculty: 42 full-time (15 women).
Students: 31 full-time (22 women), 10 international. Average age 28. *64 applicants, 19% accepted.* In 1999, 3 doctorates awarded (100% entered university research/teaching).
Degree requirements: For doctorate, dissertation, public defense, qualifying exam required, foreign language not required. *Average time to degree:* Doctorate–5.92 years full-time.
Entrance requirements: For doctorate, GRE General Test (average 80th percentile), GRE Subject Test (strongly recommended), TOEFL, minimum GPA of 3.0. *Application deadline:* For fall admission, 2/1 (priority date). Applications are processed on a rolling basis. *Application fee:* $30. Electronic applications accepted.
Expenses: Tuition: Full-time $8,200. Required fees: $175. Full-time tuition and fees vary according to student level.
Financial aid: In 1999–00, 31 students received aid, including 15 fellowships (averaging $16,000 per year), 14 research assistantships (averaging $16,000 per year), 2 teaching assistantships (averaging $16,000 per year), Federal Work-Study, institutionally sponsored loans, and tuition waivers (full) also available. Financial aid applicants required to submit FAFSA.
Faculty research: Molecular biology of virus replication, viruses and cancer, viral genetics, viral infectious diseases, environmental virology.
Application contact: Beatrice Torres, Graduate Program Administrator, 713-798-4472, *Fax:* 713-798-5075, *E-mail:* bft@bmc.tmc.edu.

Find an in-depth description at www.petersons.com/graduate.

■ CORNELL UNIVERSITY

Graduate School, Graduate Fields of Veterinary Medicine, Field of Veterinary Medicine, Ithaca, NY 14853-0001

AWARDS Anatomy (MS, PhD); cancer biology (MS, PhD); clinical sciences (MS, PhD); infectious diseases (MS, PhD); pathology (MS, PhD); pharmacology (MS, PhD); veterinary physiology (MS, PhD); virology (MS, PhD).

Faculty: 83 full-time.
Students: 25 full-time (13 women); includes 2 minority (1 African American, 1 Asian American or Pacific Islander), 12 international. *28 applicants, 18% accepted.* In 1999, 1 master's, 12 doctorates awarded.
Degree requirements: For master's and doctorate, thesis/dissertation required, foreign language not required.
Entrance requirements: For master's and doctorate, GRE General Test, TOEFL. *Application deadline:* For fall admission, 1/15; for spring admission, 10/1. Applications are processed on a rolling basis. *Application fee:* $65. Electronic applications accepted.
Expenses: Tuition: Full-time $12,400.
Financial aid: In 1999–00, 24 students received aid, including 7 fellowships with full tuition reimbursements available, 17 research assistantships with full tuition reimbursements available; teaching assistantships with full tuition reimbursements available, institutionally sponsored loans, scholarships, tuition waivers (full and partial), and unspecified assistantships also available. Financial aid applicants required to submit FAFSA.
Faculty research: Receptors and signal transduction, viral and bacterial infectious diseases, tumor metastasis, clinical sciences/nutritional disease, development/ neurologic disorders.
Application contact: Graduate Field Assistant, 607-253-3276, *E-mail:* vetgradpgms@cornell.edu.

■ HARVARD UNIVERSITY

Graduate School of Arts and Sciences, Program in Virology, Boston, MA 02115

AWARDS PhD.

Degree requirements: For doctorate, dissertation, qualifying exam required, foreign language not required.
Entrance requirements: For doctorate, GRE General Test, GRE Subject Test, TOEFL. *Application deadline:* For fall admission, 12/15. *Application fee:* $60.
Expenses: Tuition: Full-time $22,054. Required fees: $711. Tuition and fees vary according to program.
Financial aid: Fellowships, research assistantships, teaching assistantships, tuition waivers (full) available. Financial aid application deadline: 1/1.
Faculty research: Molecular genetics of viral oncogene systems, molecular biology of viral replication, antiviral immune response.
Dr. Elliott Kieff, Chair, 617-432-1977.
Application contact: Leah Simons, Manager of Student Affairs, 617-432-0162.
Find an in-depth description at www.petersons.com/graduate.

■ JOHNS HOPKINS UNIVERSITY

School of Hygiene and Public Health, Department of International Health, Baltimore, MD 21218-2699

AWARDS Community health and health systems (MHS, PhD, Sc D); disease control (MHS, PhD, Sc D); human nutrition (MHS, PhD, Sc D); international health (Dr PH); social sciences and public health (MHS, Sc D); vaccine development (Sc M, PhD, Sc D). Part-time programs available.

Faculty: 92 full-time, 145 part-time/ adjunct.
Students: 145 (96 women); includes 24 minority (3 African Americans, 15 Asian Americans or Pacific Islanders, 5 Hispanic Americans, 1 Native American) 46 international. *248 applicants, 58% accepted.* In 1999, 26 master's, 13 doctorates awarded.
Degree requirements: For master's, thesis (for some programs), internship required, foreign language not required; for doctorate, dissertation, 1 year full-time residency, oral and written exams required, foreign language not required.
Entrance requirements: For master's and doctorate, GRE General Test, TOEFL. *Application deadline:* For fall admission, 2/1 (priority date). Applications are processed on a rolling basis. *Application fee:* $60. Electronic applications accepted.
Expenses: Tuition: Full-time $23,660; part-time $493 per unit. Full-time tuition and fees vary according to degree level, campus/location and program.
Financial aid: Federal Work-Study, institutionally sponsored loans, and scholarships available. Aid available to part-time students. Financial aid application deadline: 4/15.
Faculty research: Respiratory diseases, oral rehydration programs, vitamin A deficiency, infant malnutrition, health manpower planning. *Total annual research expenditures:* $21.1 million.
Dr. Robert E. Black, Chairman, 410-955-3934, *E-mail:* rblack@jhsph.edu.
Application contact: Nancy Stephens, Student Coordinator, 410-955-3734, *Fax:* 410-955-8734, *E-mail:* nstephen@ jhsph.edu.

■ KANSAS STATE UNIVERSITY

Graduate School, College of Arts and Sciences, Division of Biology, Manhattan, KS 66506

AWARDS Cell biology (MS, PhD); developmental biology and physiology (MS, PhD); microbiology and immunology (MS, PhD); molecular biology and genetics (MS, PhD); systematics and ecology (MS, PhD); virology and oncology (MS, PhD). Terminal master's awarded for partial completion of doctoral program.

Degree requirements: For master's and doctorate, thesis/dissertation required, foreign language not required.
Entrance requirements: For master's and doctorate, GRE General Test. Electronic applications accepted.
Expenses: Tuition, state resident: part-time $103 per credit hour. Tuition, nonresident: part-time $338 per credit hour. Required fees: $17 per credit hour. One-time fee: $64 part-time.
Faculty research: Immune cell function, prairie ecology.

■ LOYOLA UNIVERSITY CHICAGO

Graduate School, Department of Microbiology and Immunology, Maywood, IL 60153

AWARDS Immunology (MS, PhD); microbiology (MS, PhD); virology (MS, PhD).

Faculty: 10 full-time (3 women).
Students: 26 full-time (18 women). Average age 27. *138 applicants, 9% accepted.* Terminal master's awarded for partial completion of doctoral program.
Degree requirements: For master's, thesis required, foreign language not required; for doctorate, computer language, dissertation, oral comprehensive exams required, foreign language not required.
Entrance requirements: For master's and doctorate, GRE General Test, TOEFL. *Application deadline:* Applications are processed on a rolling basis. *Application fee:* $35. Electronic applications accepted.
Expenses: Tuition: Part-time $500 per credit hour. Required fees: $42 per term.
Financial aid: In 1999–00, 10 fellowships with tuition reimbursements, 12 research assistantships with tuition reimbursements were awarded; Federal Work-Study, institutionally sponsored loans, and scholarships also available. Financial aid application deadline: 2/15.
Faculty research: Viral pathogenesis, microbial physiology and genetics, immunoglobulin genetics and differentiation of the immune response, signal transduction and host-parasite interactions.

Dr. Katherine L. Knight, Chair, 708-216-3385, *Fax:* 708-216-9574, *E-mail:* kknight@luc.edu.
Application contact: Dr. Thomas Gallagher, Head, 708-216-3385, *Fax:* 708-216-9574, *E-mail:* tgallag@luc.edu.
Find an in-depth description at www.petersons.com/graduate.

■ THE PENNSYLVANIA STATE UNIVERSITY MILTON S. HERSHEY MEDICAL CENTER

Graduate School, Department of Microbiology and Immunology, Hershey, PA 17033-2360

AWARDS Genetics (PhD); immunology (MS, PhD); microbiology (MS); microbiology/virology (PhD); molecular biology (PhD).

Students: 19 full-time (6 women). Average age 27. In 1999, 2 master's, 3 doctorates awarded.
Degree requirements: For doctorate, dissertation required.
Entrance requirements: For doctorate, GRE General Test, TOEFL, minimum GPA of 3.0. *Application deadline:* For fall admission, 7/26. *Application fee:* $50.
Expenses: Tuition, state resident: full-time $6,886; part-time $291 per credit. Tuition, nonresident: full-time $14,118; part-time $588 per credit. Required fees: $43 per semester. Part-time tuition and fees vary according to course load.
Financial aid: Unspecified assistantships available.
Dr. Richard J. Courtney, Chair, 717-531-7659.
Application contact: Dr. Brian Wigdahl, Chairman, Graduate Program Committee, 717-531-6682, *Fax:* 717-531-6522, *E-mail:* eneidigh@cor-mail.biochem.hmc.psu.edu.

■ PURDUE UNIVERSITY

School of Veterinary Medicine and Graduate School, Graduate Programs in Veterinary Medicine, Department of Veterinary Pathobiology, West Lafayette, IN 47907

AWARDS Bacteriology (MS, PhD); epidemiology (MS, PhD); immunology (MS, PhD); infectious diseases (MS, PhD); microbiology (MS, PhD); parasitology (MS, PhD); pathology (MS, PhD); toxicology (MS, PhD); virology (MS, PhD).

Faculty: 26 full-time (5 women).
Students: 42 full-time (24 women), 2 part-time (1 woman); includes 5 minority (1 African American, 3 Asian Americans or Pacific Islanders, 1 Hispanic American), 18 international. Average age 32. In 1999, 2 master's, 6 doctorates awarded (100% found work related to degree). Terminal master's awarded for partial completion of doctoral program.

Degree requirements: For master's, thesis required (for some programs), foreign language not required; for doctorate, dissertation required, foreign language not required.
Entrance requirements: For master's and doctorate, GRE General Test, TOEFL. *Application fee:* $30.
Expenses: Tuition, state resident: full-time $3,732. Tuition, nonresident: full-time $8,732.
Financial aid: Fellowships, research assistantships, teaching assistantships available. Financial aid application deadline: 3/1; financial aid applicants required to submit FAFSA.
Dr. H. L. Thacker, Head, 765-494-7543.

■ RUSH UNIVERSITY

Graduate College, Division of Immunology, Program in Virology, Chicago, IL 60612-3832

AWARDS PhD, MD/PhD.

Degree requirements: For doctorate, dissertation, comprehensive preliminary exam required, foreign language not required.
Entrance requirements: For doctorate, GRE General Test, TOEFL, interview, minimum GPA of 3.0.
Expenses: Tuition: Full-time $13,020; part-time $390 per credit. Tuition and fees vary according to program.
Faculty research: HIV, hepatitis B, negative strand RNA viruses, respiratory syncytial virus.
Find an in-depth description at www.petersons.com/graduate.

■ RUTGERS, THE STATE UNIVERSITY OF NEW JERSEY, NEW BRUNSWICK

Graduate School, Program in Microbiology and Molecular Genetics, New Brunswick, NJ 08901-1281

AWARDS Applied microbiology (MS, PhD); clinical microbiology (MS, PhD); computational molecular biology (PhD); immunology (MS, PhD); microbial biochemistry (MS, PhD); molecular genetics (MS, PhD); virology (MS, PhD). Part-time programs available.

Faculty: 122 full-time (25 women), 4 part-time/adjunct (1 woman).
Students: 65 full-time (31 women), 51 part-time (24 women); includes 24 minority (14 Asian Americans or Pacific Islanders, 10 Hispanic Americans), 29 international. Average age 27. *200 applicants, 26% accepted.* In 1999, 15 master's, 10 doctorates awarded. Terminal master's awarded for partial completion of doctoral program.
Degree requirements: For master's, thesis or alternative required, foreign

language not required; for doctorate, dissertation, written qualifying exam required, foreign language not required.
Entrance requirements: For master's, GRE General Test, GRE Subject Test; for doctorate, GRE General Test, GRE Subject Test, TOEFL, minimum GPA of 3.0. *Application deadline:* For fall admission, 2/15 (priority date). Applications are processed on a rolling basis. *Application fee:* $50.
Expenses: Tuition, state resident: full-time $6,776; part-time $279 per credit. Tuition, nonresident: full-time $9,936; part-time $412 per credit. Required fees: $20 per credit. $89 per semester. Tuition and fees vary according to course load, campus/location and program.
Financial aid: In 1999–00, 70 students received aid, including 32 fellowships with full tuition reimbursements available, 26 research assistantships with full tuition reimbursements available, 12 teaching assistantships with full tuition reimbursements available; institutionally sponsored loans also available. Financial aid application deadline: 2/15; financial aid applicants required to submit FAFSA.
Faculty research: Biochemistry of microbes, medical microbiology.
Dr. Howard Passmore, Director, 732-445-2812, *E-mail:* passmore@biology.rutgers.edu.
Application contact: Betty Green, Administrative Assistant, 732-445-5086, *Fax:* 732-445-5735, *E-mail:* green@mbcl.rutgers.edu.

■ TEXAS A&M UNIVERSITY SYSTEM HEALTH SCIENCE CENTER

College of Medicine, Graduate School of Biomedical Sciences, Department of Medical Microbiology and Immunology, College Station, TX 77840-7896

AWARDS Immunology (PhD); microbiology (PhD); molecular biology (PhD); virology (PhD).

Faculty: 7 full-time (0 women), 1 part-time/adjunct (0 women).
Students: 16 full-time (10 women); includes 2 minority (1 African American, 1 Asian American or Pacific Islander).
Degree requirements: For doctorate, dissertation required, foreign language not required.
Entrance requirements: For doctorate, GRE General Test, minimum GPA of 3.0. *Application deadline:* For fall admission, 2/1 (priority date). *Application fee:* $50 ($75 for international students).
Expenses: Tuition, area resident: Full-time $1,368. Tuition, state resident: part-time

Texas A&M University System Health Science Center (continued)
$76 per credit. Tuition, nonresident: full-time $5,256; part-time $292 per credit. International tuition: $5,256 full-time. Required fees: $678; $38 per credit. Full-time tuition and fees vary according to course load and student level.
Financial aid: In 1999–00, 16 students received aid, including 1 fellowship (averaging $17,000 per year), 15 research assistantships (averaging $17,000 per year). Financial aid application deadline: 4/1; financial aid applicants required to submit FAFSA.
Faculty research: Molecular pathogenesis, microbial therapeutics. *Total annual research expenditures:* $1.2 million.
Dr. John M. Quarles, Interim Head, 979-845-1313, *Fax:* 979-845-3479, *E-mail:* jmquarles@medicine.tamu.edu.
Application contact: Dr. James Samuel, Graduate Adviser, 979-845-1313, *Fax:* 979-845-3479, *E-mail:* jsamuel@medicine.tamu.edu.
Find an in-depth description at www.petersons.com/graduate.

■ **THOMAS JEFFERSON UNIVERSITY**
College of Graduate Studies, Program in Microbiology and Molecular Virology, Philadelphia, PA 19107
AWARDS PhD.

Faculty: 24 full-time (1 woman).
Students: 13 full-time (7 women), 2 part-time; includes 2 minority (1 African American, 1 Asian American or Pacific Islander), 4 international. *56 applicants, 13% accepted.* In 1999, 2 degrees awarded.
Degree requirements: For doctorate, dissertation required, foreign language not required.
Entrance requirements: For doctorate, GRE General Test, TOEFL, minimum GPA of 3.2. *Application deadline:* For fall admission, 3/1 (priority date). Applications are processed on a rolling basis. *Application fee:* $40.
Expenses: Tuition: Full-time $12,670. Tuition and fees vary according to degree level and program.
Financial aid: In 1999–00, 14 fellowships with full tuition reimbursements were awarded; research assistantships, Federal Work-Study, institutionally sponsored loans, traineeships, and training grants also available. Aid available to part-time students. Financial aid application deadline: 5/1; financial aid applicants required to submit FAFSA.
Faculty research: Chemical and antigenic structure of virus particles, viral replication, viral gene expression, viral

oncogenes, parasite immunology. *Total annual research expenditures:* $17 million.
Dr. David Abraham, Chairman, Graduate Committee, 215-503-8917, *Fax:* 215-923-9248, *E-mail:* abrahamd@jeflin.tju.edu.
Application contact: Jessie F. Pervall, Director of Admissions, 215-503-4400, *Fax:* 215-503-3433, *E-mail:* cgs-info@mail.tju.edu.
Find an in-depth description at www.petersons.com/graduate.

■ **UNIVERSITY OF CALIFORNIA, SAN DIEGO**
Graduate Studies and Research, Department of Biology, Program in Immunology, Virology, and Cancer Biology, La Jolla, CA 92093
AWARDS PhD.

Degree requirements: For doctorate, dissertation required.
Entrance requirements: For doctorate, GRE General Test, pre-application beginning in September. *Application deadline:* For fall admission, 1/8. *Application fee:* $40.
Expenses: Tuition, nonresident: full-time $14,691. Required fees: $4,697. Full-time tuition and fees vary according to program.
Application contact: Biology Graduate Admissions Committee, 858-534-3835.

■ **UNIVERSITY OF CHICAGO**
Division of the Biological Sciences, Biomedical Sciences: Cancer, Immunology, Nutrition, Pathology, and Virology, Committee on Virology, Chicago, IL 60637-1513
AWARDS PhD.

Faculty: 5 full-time (1 woman).
Students: 8 full-time (2 women); includes 1 minority (Asian American or Pacific Islander). Average age 29. *53 applicants, 6% accepted.*
Degree requirements: For doctorate, dissertation required, foreign language not required. *Average time to degree:* Doctorate–5.5 years full-time.
Entrance requirements: For doctorate, GRE General Test, TOEFL. *Application deadline:* For fall admission, 1/5 (priority date). *Application fee:* $55.
Expenses: Tuition: Full-time $24,804; part-time $3,422 per course. Required fees: $390. Tuition and fees vary according to program.
Financial aid: In 1999–00, 8 students received aid, including 3 fellowships with full tuition reimbursements available (averaging $16,500 per year), 5 research assistantships with full tuition reimbursements available (averaging $16,500 per year); institutionally sponsored loans and

traineeships also available. Financial aid application deadline: 1/5.
Faculty research: Molecular genetics, herpes virus, adipoviruses, Picarna viruses, ENS viruses.
Dr. Bernard Roizman, Chairman, 773-702-1898, *Fax:* 773-702-1631, *E-mail:* bernard@cummings.uchicago.edu.
Application contact: Rebecca Levine, Administrative Assistant, Student Services, 773-834-3899, *Fax:* 773-702-4634, *E-mail:* rlevine@huggins.bsd.uchicago.edu.

■ **THE UNIVERSITY OF IOWA**
College of Medicine and Graduate College, Graduate Programs in Medicine, Department of Microbiology, Iowa City, IA 52242-1316

AWARDS General microbiology and microbial physiology (MS, PhD); immunology (MS, PhD); microbial genetics (MS, PhD); pathogenic bacteriology (MS, PhD); virology (MS, PhD).

Faculty: 21 full-time (4 women), 10 part-time/adjunct (1 woman).
Students: 50 full-time (21 women); includes 7 minority (2 African Americans, 4 Asian Americans or Pacific Islanders, 1 Hispanic American), 5 international. *144 applicants, 8% accepted.* In 1999, 7 doctorates awarded (100% entered university research/teaching).
Degree requirements: For master's, thesis required; for doctorate, dissertation, comprehensive exam required. *Average time to degree:* Doctorate–6 years full-time.
Entrance requirements: For master's and doctorate, GRE General Test. *Application deadline:* For fall admission, 2/1. *Application fee:* $30 ($50 for international students). Electronic applications accepted.
Expenses: Tuition, state resident: full-time $3,308. Tuition, nonresident: full-time $10,662. Tuition and fees vary according to course load and program.
Financial aid: In 1999–00, 50 students received aid, including fellowships with full tuition reimbursements available (averaging $16,787 per year), research assistantships with full tuition reimbursements available (averaging $16,787 per year); grants, institutionally sponsored loans, and traineeships also available.
Faculty research: Biocatalysis and blue jeans, gene regulation, processing and transport of HIV, retroviral pathogenesis, biodegradation. *Total annual research expenditures:* $8.2 million.
Dr. Michael A. Apicella, Head, 319-335-7810, *E-mail:* grad-micro-info@uiowa.edu.
Find an in-depth description at www.petersons.com/graduate.

■ UNIVERSITY OF MASSACHUSETTS WORCESTER

Graduate School of Biomedical Sciences, Program in Immunology-Virology, Worcester, MA 01655-0115

AWARDS Medical sciences (PhD).

Faculty: 41 full-time (10 women).
Students: 41 full-time (20 women); includes 4 minority (2 Asian Americans or Pacific Islanders, 1 Hispanic American, 1 Native American), 10 international. In 1999, 5 degrees awarded.
Degree requirements: For doctorate, dissertation required, foreign language not required. *Average time to degree:* Doctorate–5.7 years full-time.
Entrance requirements: For doctorate, GRE General Test, GRE Subject Test, 1 year of calculus, organic chemistry, physics, and biology. *Application deadline:* For fall admission, 1/15 (priority date). Applications are processed on a rolling basis. *Application fee:* $25 ($50 for international students).
Expenses: Tuition, state resident: full-time $2,640. Tuition, nonresident: full-time $9,756. Required fees: $825. Full-time tuition and fees vary according to program.
Financial aid: In 1999–00, research assistantships with full tuition reimbursements (averaging $17,500 per year); unspecified assistantships also available.
Faculty research: Molecular and immunological studies on viral pathogenesis and oncology, AIDS viruses and other retroviruses, herpes viruses, influenza, Dengue and Newcastle disease viruses.
Dr. Carol Miller-Graziano, Director, 508-856-2387.

Find an in-depth description at www.petersons.com/graduate.

■ UNIVERSITY OF PENNSYLVANIA

School of Medicine, Biomedical Graduate Studies, Graduate Group in Cell and Molecular Biology, Program in Microbiology and Virology, Philadelphia, PA 19104

AWARDS PhD, MD/PhD, VMD/PhD.

Degree requirements: For doctorate, dissertation required, foreign language not required.
Entrance requirements: For doctorate, GRE General Test, GRE Subject Test (biology), TOEFL, previous course work in science. *Application deadline:* For fall admission, 1/2 (priority date). Applications are processed on a rolling basis. *Application fee:* $65. Electronic applications accepted.

Expenses: Tuition: Full-time $17,256; part-time $2,991 per course. Required fees: $2,588; $363 per course. $726 per term.
Financial aid: Fellowships, research assistantships, teaching assistantships, grants and institutionally sponsored loans available. Financial aid application deadline: 1/2.
Dr. Michael Malim, Chair, 215-573-3493.
Application contact: Mary Webster, Coordinator, 215-898-4360, *E-mail:* camb@mail.med.upenn.edu.

Find an in-depth description at www.petersons.com/graduate.

■ UNIVERSITY OF PITTSBURGH

School of Medicine, Graduate Programs in Medicine, Program in Molecular Virology and Microbiology, Pittsburgh, PA 15260

AWARDS MS, PhD.

Students: 6 full-time (0 women); includes 2 minority (1 African American, 1 Hispanic American). *325 applicants, 22% accepted.* In 1999, 2 master's, 2 doctorates awarded.
Degree requirements: For doctorate, dissertation required, foreign language not required. *Average time to degree:* Doctorate–5.3 years full-time.
Entrance requirements: For doctorate, GRE General Test, GRE Subject Test, TOEFL, minimum QPA of 3.0. *Application deadline:* For fall admission, 1/15 (priority date). Applications are processed on a rolling basis. *Application fee:* $30 ($40 for international students).
Expenses: Tuition, state resident: full-time $9,778; part-time $403 per credit. Tuition, nonresident: full-time $20,146; part-time $830 per credit. Required fees: $480; $90 per semester.
Financial aid: Research assistantships with full tuition reimbursements, teaching assistantships with full tuition reimbursements, Federal Work-Study, institutionally sponsored loans, scholarships, traineeships, and unspecified assistantships available.
Faculty research: Herpes, HIV-1, sexually transmitted diseases, molecular basis of infectious disease. *Total annual research expenditures:* $76.3 million.
Application contact: Graduate Studies Administrator, 412-648-8957, *Fax:* 412-648-1236, *E-mail:* biomed_phd@fs1.dean-med.pitt.edu.

Find an in-depth description at www.petersons.com/graduate.

■ THE UNIVERSITY OF TENNESSEE HEALTH SCIENCE CENTER

College of Graduate Health Sciences, Department of Microbiology and Immunology, Memphis, TN 38163-0002

AWARDS Bacteriology (MS, PhD); immunology (MS, PhD); medical microbiology (MS, PhD); microbiology (MS, PhD); molecular and cell biology (PhD); molecular biology (MS, PhD); virology (MS, PhD). Terminal master's awarded for partial completion of doctoral program.

Degree requirements: For master's, thesis, oral and written comprehensive exams required, foreign language not required; for doctorate, dissertation, oral and written preliminary and comprehensive exams required, foreign language not required.
Entrance requirements: For master's, GRE General Test, TOEFL, minimum GPA of 3.0; for doctorate, GRE General Test, GRE Subject Test, TOEFL, minimum GPA of 3.0.

Find an in-depth description at www.petersons.com/graduate.

■ THE UNIVERSITY OF TEXAS–HOUSTON HEALTH SCIENCE CENTER

Graduate School of Biomedical Sciences, Program in Virology and Gene Therapy, Houston, TX 77225-0036

AWARDS MS, PhD, MD/PhD.

Faculty: 19 full-time (3 women).
Students: 4 full-time (3 women); includes 1 minority (Asian American or Pacific Islander). Average age 27. *24 applicants, 54% accepted.* In 1999, 1 master's awarded. Terminal master's awarded for partial completion of doctoral program.
Degree requirements: For master's and doctorate, thesis/dissertation required, foreign language not required.
Entrance requirements: For master's and doctorate, GRE General Test, TOEFL, TWE. *Application deadline:* For fall admission, 1/15 (priority date); for spring admission, 11/1. Applications are processed on a rolling basis. *Application fee:* $10. Electronic applications accepted.
Financial aid: Fellowships, research assistantships, institutionally sponsored loans available. Financial aid application deadline: 1/15.
Faculty research: Molecular virology and biology, viral carcinogenesis, viral immunology, viral pathogenesis, viral nucleic acid protein biochemistry, gene therapy.

The University of Texas–Houston Health Science Center (continued)
Dr. Paul Gershon, Director, 713-677-7665, *Fax:* 713-677-7689, *E-mail:* pgershon@ibt03.tamu.edu.
Application contact: Anne Baronitis, Director of Admissions, 713-500-9860, *Fax:* 713-500-9877, *E-mail:* abaron@gsbs.gs.uth.tmc.edu.

Find an in-depth description at www.petersons.com/graduate.

■ **THE UNIVERSITY OF TEXAS MEDICAL BRANCH AT GALVESTON**

Graduate School of Biomedical Sciences, Program in Emerging and Tropical Infectious Diseases, Galveston, TX 77555

AWARDS PhD.

Faculty: 39 full-time (6 women).
Students: 4 full-time (3 women).
Degree requirements: For doctorate, dissertation required, foreign language not required.
Entrance requirements: For doctorate, GRE General Test. *Application deadline:* Applications are processed on a rolling basis. *Application fee:* $25 ($50 for international students).
Expenses: Tuition, state resident: full-time $684; part-time $38 per credit hour.

Tuition, nonresident: full-time $4,572; part-time $254 per credit hour. Required fees: $29; $7.5 per credit hour. One-time fee: $55. Tuition and fees vary according to program.
Financial aid: In 1999–00, 4 fellowships with tuition reimbursements (averaging $15,060 per year) were awarded; traineeships also available.
Faculty research: Emerging diseases, tropical diseases, parasitology, vitology and bacteriology.
Dr. Alan D.T. Barrett, Director, 409-772-2521, *Fax:* 409-747-2400.

Find an in-depth description at www.petersons.com/graduate.

■ **WEST VIRGINIA UNIVERSITY**

School of Medicine, Graduate Programs in Health Sciences, Department of Microbiology and Immunology, Morgantown, WV 26506

AWARDS Genetics (MS, PhD); immunology (MS, PhD); mycology (PhD); parasitology (MS, PhD); pathogenic bacteriology (MS, PhD); physiology (PhD); psysiology (MS); virology (MS, PhD).

Students: 25 full-time (11 women), 1 (woman) part-time; includes 2 minority (both Asian Americans or Pacific Islanders), 4 international. Average age 27. *62 applicants, 8% accepted.* In 1999, 1 master's,

5 doctorates awarded. Terminal master's awarded for partial completion of doctoral program.
Degree requirements: For master's, thesis required, foreign language not required; for doctorate, dissertation, comprehensive exam required, foreign language not required. *Average time to degree:* Doctorate–4.5 years full-time.
Entrance requirements: For master's and doctorate, GRE General Test, GRE Subject Test, TOEFL, minimum GPA of 3.0. *Application deadline:* For fall admission, 3/1 (priority date). Applications are processed on a rolling basis. *Application fee:* $45. Electronic applications accepted.
Expenses: Tuition, state resident: full-time $3,564. Tuition, nonresident: full-time $10,230.
Financial aid: In 1999–00, 7 research assistantships, 14 teaching assistantships were awarded; fellowships, Federal Work-Study and institutionally sponsored loans also available. Financial aid application deadline: 3/1; financial aid applicants required to submit FAFSA.
Faculty research: Mechanisms of pathogenesis, microbial genetics, molecular virology, immunotoxicology, oncogenes. *Total annual research expenditures:* $2.6 million.
John Barnett, Chair, 304-293-2649.
Application contact: James Sheil, Graduate Coordinator, 304-293-3559, *Fax:* 304-293-7823, *E-mail:* jsheil@wvu.edu.

Neuroscience

BIOPSYCHOLOGY

■ **ALLIANT UNIVERSITY–SAN DIEGO**

Graduate Programs, California School of Professional Psychology, Program in Clinical Psychophysiology and Biofeedback, San Diego, CA 92121-3725

AWARDS MS.

Faculty: 3 full-time (1 woman), 6 part-time/adjunct (1 woman).
Students: 7 full-time (6 women), 2 part-time (both women); includes 3 minority (2 African Americans, 1 Asian American or Pacific Islander). Average age 31. *14 applicants, 64% accepted.* In 1999, 3 degrees awarded.
Degree requirements: For master's, foreign language and thesis not required.
Entrance requirements: For master's, TOEFL, minimum GPA of 3.0 in both

psychology and overall. *Application deadline:* For fall admission, 4/1 (priority date). *Application fee:* $50.
Expenses: Tuition: Full-time $14,330; part-time $341 per semester hour.
Financial aid: In 1999–00, 9 students received aid. Federal Work-Study available. Financial aid application deadline: 4/1.
Faculty research: Relationship between stress and health, coping with chronic illness, acculturation and health.
Dr. Perry Nicassio, Director, 858-623-2777, *Fax:* 858-552-1974.
Application contact: Patricia J. Mullen, Vice President, Enrollment and Student Services, 800-457-1273 Ext. 303, *Fax:* 415-931-8322, *E-mail:* admissions@mail.cspp.edu.

Find an in-depth description at www.petersons.com/graduate.

■ **AMERICAN UNIVERSITY**

College of Arts and Sciences, Department of Psychology, Program in Psychology, Washington, DC 20016-8001

AWARDS Experimental/biological psychology (MA); general psychology (MA); personality/social psychology (MA). Part-time programs available.

Faculty: 14 full-time (3 women).
Students: In 1999, 11 degrees awarded.
Degree requirements: For master's, thesis required (for some programs), foreign language not required.
Entrance requirements: For master's, GRE General Test, GRE Subject Test. *Application deadline:* For fall admission, 4/30. Applications are processed on a rolling basis. *Application fee:* $50.
Expenses: Tuition: Part-time $721 per credit hour. Required fees: $90 per semester. Tuition and fees vary according to program.

Financial aid: In 1999–00, 2 students received aid; teaching assistantships available. Financial aid application deadline: 2/1.

Faculty research: Behavior therapy, cognitive behavior modification, pro-social behavior, conditioning and learning, olfaction.

Dr. Michele Carter, Director, 202-885-1712.

■ BRYN MAWR COLLEGE

Graduate School of Arts and Sciences, Department of Biology, Program in Neural and Behavioral Science, Bryn Mawr, PA 19010-2899

AWARDS PhD.

Degree requirements: For doctorate, dissertation required.

Entrance requirements: For doctorate, GRE General Test, GRE Subject Test. *Application deadline:* For fall admission, 6/30; for spring admission, 12/7. *Application fee:* $40.

Expenses: Tuition: Full-time $20,790; part-time $3,530 per course.

Financial aid: Application deadline: 1/2.

Application contact: Graduate School of Arts and Sciences, 610-526-5075.

■ CARNEGIE MELLON UNIVERSITY

College of Humanities and Social Sciences, Department of Psychology, Area of Cognitive Neuropsychology, Pittsburgh, PA 15213-3891

AWARDS PhD.

Degree requirements: For doctorate, dissertation, oral and written comprehensive exams required.

Entrance requirements: For doctorate, GRE General Test, TOEFL, TSE. *Application deadline:* For fall admission, 1/15. *Application fee:* $45.

Expenses: Tuition: Full-time $22,100; part-time $307 per unit. Required fees: $200. Tuition and fees vary according to program.

Application contact: Dr. Margaret Clark, Graduate Director, 412-268-5690.

Find an in-depth description at www.petersons.com/graduate.

■ COLUMBIA UNIVERSITY

Graduate School of Arts and Sciences, Division of Natural Sciences, Department of Psychology, New York, NY 10027

AWARDS Experimental psychology (M Phil, MA, PhD); psychobiology (M Phil, MA, PhD); social psychology (M Phil, MA, PhD).

Degree requirements: For master's and doctorate, thesis/dissertation required, foreign language not required.

Entrance requirements: For master's and doctorate, GRE General Test, TOEFL.

Expenses: Tuition: Full-time $25,072. Full-time tuition and fees vary according to course load and program.

■ CORNELL UNIVERSITY

Graduate School, Graduate Fields of Arts and Sciences, Field of Psychology, Ithaca, NY 14853-0001

AWARDS Biopsychology (PhD); general psychology (PhD); human experimental psychology (PhD); social and personality psychology (PhD).

Faculty: 35 full-time.

Students: 34 full-time (17 women); includes 2 minority (1 Asian American or Pacific Islander, 1 Hispanic American), 8 international. *137 applicants, 4% accepted.* In 1999, 5 doctorates awarded.

Degree requirements: For doctorate, dissertation, 2 semesters of teaching experience required, foreign language not required.

Entrance requirements: For doctorate, GRE General Test, TOEFL. *Application deadline:* For fall admission, 1/15. *Application fee:* $65. Electronic applications accepted.

Expenses: Tuition: Full-time $23,760. Required fees: $48. Full-time tuition and fees vary according to program.

Financial aid: In 1999–00, 32 students received aid, including 12 fellowships with full tuition reimbursements available, 20 teaching assistantships with full tuition reimbursements available; research assistantships with full tuition reimbursements available, institutionally sponsored loans, scholarships, tuition waivers (full and partial), and unspecified assistantships also available. Financial aid applicants required to submit FAFSA.

Faculty research: Sensory and perceptual systems, social cognition, cognitive development, neuroscience, quantitative and computational modeling.

Application contact: Graduate Field Assistant, 607-255-6364, *E-mail:* lap5@cornell.edu.

■ DREXEL UNIVERSITY

Graduate School, College of Arts and Sciences, Program in Clinical Psychology, Philadelphia, PA 19104-2875

AWARDS Clinical neuropsychology (MS, PhD).

Faculty: 19 full-time (6 women), 16 part-time/adjunct (11 women).

Students: 6 full-time (5 women), 19 part-time (13 women); includes 6 minority (2 African Americans, 3 Asian Americans or Pacific Islanders, 1 Hispanic American). Average age 27. *114 applicants, 5% accepted.* In 1999, 4 master's, 10 doctorates awarded.

Degree requirements: For doctorate, dissertation, internship required.

Entrance requirements: For master's, GRE, TOEFL; for doctorate, GRE General Test, TOEFL. *Application deadline:* For fall admission, 2/1. Applications are processed on a rolling basis. *Application fee:* $35. Electronic applications accepted.

Expenses: Tuition: Part-time $511 per credit. Required fees: $67 per term.

Financial aid: Research assistantships, teaching assistantships, career-related internships or fieldwork, Federal Work-Study, institutionally sponsored loans, tuition waivers (full and partial), unspecified assistantships, and practicum positions available. Financial aid application deadline: 2/1.

Faculty research: Neurosciences, rehabilitation psychology, cognitive science, neurological assessment.

Dr. Eric Zillmer, Director, 215-895-2402, *Fax:* 215-895-1333.

Application contact: Director of Graduate Admissions, 215-895-6700, *Fax:* 215-895-5939, *E-mail:* enroll@drexel.edu.

Find an in-depth description at www.petersons.com/graduate.

■ DUKE UNIVERSITY

Graduate School, Department of Psychology, Durham, NC 27708-0586

AWARDS Biological psychology (PhD); clinical psychology (PhD); cognitive psychology (PhD); developmental psychology (PhD); experimental psychology (PhD); health psychology (PhD); human social development (PhD).

Faculty: 67 full-time, 14 part-time/adjunct.

Students: 62 full-time (37 women); includes 9 minority (5 African Americans, 2 Asian Americans or Pacific Islanders, 2 Hispanic Americans), 14 international. *379 applicants, 8% accepted.* In 1999, 20 doctorates awarded.

Degree requirements: For doctorate, dissertation required, foreign language not required.

Entrance requirements: For doctorate, GRE General Test. *Application deadline:* For fall admission, 12/31. *Application fee:* $75.

Expenses: Tuition: Full-time $21,406; part-time $760 per unit. Required fees: $3,136; $3,136 per year. One-time fee: $30. Tuition and fees vary according to program.

Financial aid: Fellowships, research assistantships, teaching assistantships,

Duke University (continued)
career-related internships or fieldwork and Federal Work-Study available. Financial aid application deadline: 12/31.
Steven Asher, Co-Director of Graduate Studies, 919-660-5716, *Fax:* 919-660-5726, *E-mail:* bseymore@acpub.duke.edu.

■ EMORY UNIVERSITY

Graduate School of Arts and Sciences, Department of Psychology, Atlanta, GA 30322-1100

AWARDS Clinical psychology (PhD); cognition and development (PhD); psychobiology (PhD).

Faculty: 25 full-time (6 women), 2 part-time/adjunct (0 women).
Students: 81 full-time (56 women); includes 12 minority (3 African Americans, 6 Asian Americans or Pacific Islanders, 2 Hispanic Americans, 1 Native American), 6 international. *250 applicants, 8% accepted.* In 1999, 10 doctorates awarded.
Degree requirements: For doctorate, computer language, dissertation, comprehensive exams required, foreign language not required.
Entrance requirements: For doctorate, GRE General Test, TOEFL, minimum GPA of 3.0. *Application deadline:* For fall admission, 1/15. *Application fee:* $45.
Expenses: Tuition: Full-time $22,770. Tuition and fees vary according to program.
Financial aid: In 1999–00, 56 fellowships were awarded; research assistantships, teaching assistantships, career-related internships or fieldwork, Federal Work-Study, institutionally sponsored loans, scholarships, and tuition waivers (full and partial) also available. Financial aid application deadline: 1/20.
Faculty research: Neurophysiology, drugs and behavior, hormones and behavior, nonverbal behavior.
Dr. Darryl Neill, Chair, 404-727-7437.
Application contact: Katherine Gaddie, Graduate Coordinator, 404-727-7456.

■ FLORIDA STATE UNIVERSITY

Graduate Studies, College of Arts and Sciences, Department of Psychology, Interdisciplinary Program in Psychology and Neuroscience, Tallahassee, FL 32306

AWARDS Neuroscience (PhD).

Faculty: 12 full-time (3 women).
Students: 18 full-time (7 women), 1 part-time; includes 2 minority (both African Americans). Average age 23. *47 applicants, 9% accepted.*
Degree requirements: For doctorate, dissertation, preliminary exam required, foreign language not required.

Entrance requirements: For doctorate, GRE General Test, minimum GPA of 3.0. *Application deadline:* For fall admission, 1/15. *Application fee:* $20. Electronic applications accepted.
Expenses: Tuition, state resident: full-time $3,504; part-time $146 per credit hour. Tuition, nonresident: full-time $12,162; part-time $507 per credit hour. Tuition and fees vary according to program.
Financial aid: In 1999–00, 4 fellowships with full tuition reimbursements (averaging $13,500 per year), 6 research assistantships with full tuition reimbursements (averaging $13,500 per year), 8 teaching assistantships with full tuition reimbursements (averaging $13,000 per year) were awarded; Federal Work-Study, institutionally sponsored loans, and traineeships also available. Financial aid applicants required to submit FAFSA.
Faculty research: Sensory processes, biophysiology and electrophysiology, neuroanatomy, circadian rhythms, genetics. *Total annual research expenditures:* $1.4 million.
Dr. Michael Rasholte, Director, 850-644-3511.
Application contact: Cherie P. Dilworth, Graduate Program Assistant, 850-644-2499, *Fax:* 850-644-7739, *E-mail:* dilworth@psy.fsu.edu.

■ GRADUATE SCHOOL AND UNIVERSITY CENTER OF THE CITY UNIVERSITY OF NEW YORK

Graduate Studies, Program in Psychology, New York, NY 10016-4039

AWARDS Basic applied neurocognition (PhD); biopsychology (PhD); clinical psychology (PhD); developmental psychology (PhD); environmental psychology (PhD); experimental psychology (PhD); industrial psychology (PhD); learning processes (PhD); neuropsychology (PhD); psychology (PhD); social personality (PhD).

Degree requirements: For doctorate, dissertation required.
Entrance requirements: For doctorate, GRE General Test.
Expenses: Tuition, state resident: full-time $4,350; part-time $245 per credit hour. Tuition, nonresident: full-time $7,600; part-time $425 per credit hour.

■ HARVARD UNIVERSITY

Graduate School of Arts and Sciences, Department of Psychology, Cambridge, MA 02138

AWARDS Psychology (AM, PhD), including behavior and decision analysis, cognition, developmental psychology, experimental

psychology, personality, psychobiology, psychopathology; social psychology (AM, PhD).

Faculty: 25 full-time, 4 part-time/adjunct.
Students: 71 full-time (37 women). *166 applicants, 10% accepted.* In 1999, 19 master's, 7 doctorates awarded.
Degree requirements: For doctorate, dissertation, general exams required, foreign language not required.
Entrance requirements: For master's and doctorate, GRE General Test, TOEFL. *Application deadline:* For fall admission, 12/30. *Application fee:* $60.
Expenses: Tuition: Full-time $22,054. Required fees: $711. Tuition and fees vary according to program.
Financial aid: Fellowships, research assistantships, teaching assistantships, career-related internships or fieldwork, Federal Work-Study, and institutionally sponsored loans available. Financial aid application deadline: 12/30.
Robert LaPointe, Officer, 617-495-5396.
Application contact: Office of Admissions and Financial Aid, 617-495-5315.

■ HOWARD UNIVERSITY

Graduate School of Arts and Sciences, Department of Psychology, Washington, DC 20059-0002

AWARDS Clinical psychology (PhD); developmental psychology (PhD); experimental psychology (PhD); neuropsychology (PhD); personality psychology (PhD); psychology (MS); social psychology (PhD). Part-time programs available.

Faculty: 13.
Students: 100; includes 86 minority (85 African Americans, 1 Asian American or Pacific Islander), 13 international. In 1999, 13 master's, 15 doctorates awarded.
Degree requirements: For master's, thesis, comprehensive exam required, foreign language not required; for doctorate, dissertation, comprehensive exam, qualifying exam required, foreign language not required. *Average time to degree:* Master's–2 years full-time; doctorate–4 years full-time.
Entrance requirements: For master's, GRE General Test, minimum GPA of 2.5, bachelor's degree in psychology or related field; for doctorate, GRE General Test, minimum GPA of 3.0. *Application deadline:* For fall admission, 4/1; for spring admission, 11/1. Applications are processed on a rolling basis. *Application fee:* $45.
Expenses: Tuition: Full-time $10,500; part-time $583 per credit hour. Required fees: $405; $203 per semester.
Financial aid: Fellowships, research assistantships, teaching assistantships, career-related internships or fieldwork, grants, and institutionally sponsored loans

available. Financial aid application deadline: 4/1.
Faculty research: Personality and psychophysiology, educational and social development of African-American children, child and adult psychopathology.
Dr. Leslie Hicks, Chair, 202-806-6805.
Application contact: Dr. Alfonso Campbell, Chairman, Graduate Admissions Committee, 202-806-6805, *Fax:* 202-806-4873.

■ HUNTER COLLEGE OF THE CITY UNIVERSITY OF NEW YORK

Graduate School, School of Arts and Sciences, Department of Psychology, New York, NY 10021-5085

AWARDS Applied and evaluative psychology (MA); biopsychology and comparative psychology (MA); social, cognitive, and developmental psychology (MA). Part-time and evening/weekend programs available.

Faculty: 26 full-time (15 women), 19 part-time/adjunct (10 women).
Students: 3 full-time (all women), 14 part-time (11 women), 3 international. Average age 30. *56 applicants, 55% accepted.* In 1999, 15 degrees awarded (67% continued full-time study).
Degree requirements: For master's, thesis, comprehensive exam required, foreign language not required. *Average time to degree:* Master's–2 years full-time, 3 years part-time.
Entrance requirements: For master's, GRE General Test, TOEFL, minimum 12 credits in psychology, including statistics and experimental psychology. *Application deadline:* For fall admission, 4/7; for spring admission, 11/7. Applications are processed on a rolling basis. *Application fee:* $40.
Expenses: Tuition, state resident: full-time $4,350; part-time $185 per credit. Tuition, nonresident: full-time $7,600; part-time $320 per credit. Required fees: $8 per term.
Faculty research: Personality, cognitive and linguistic development, hormonal and neural control of behavior, gender and culture, social cognition of health and attitudes.
Vita C. Rabinowitz, Chair, 212-772-5550.
Application contact: Michael Goldstein, Assistant Director for Graduate Admissions, 212-772-4288, *Fax:* 212-650-3336, *E-mail:* admissions@hunter.cuny.edu.

■ LOUISIANA STATE UNIVERSITY AND AGRICULTURAL AND MECHANICAL COLLEGE

Graduate School, College of Arts and Sciences, Department of Psychology, Baton Rouge, LA 70803

AWARDS Biological psychology (MA, PhD); clinical psychology (MA, PhD); cognitive psychology (MA, PhD); developmental psychology (MA, PhD); industrial/organizational psychology (MA, PhD); school psychology (MA, PhD).

Faculty: 27 full-time (8 women).
Students: 80 full-time (51 women), 40 part-time (21 women). Average age 29. *257 applicants, 6% accepted.* In 1999, 15 master's, 21 doctorates awarded. Terminal master's awarded for partial completion of doctoral program.
Degree requirements: For master's, thesis required, foreign language not required; for doctorate, dissertation, 1 year internship required, foreign language not required.
Entrance requirements: For master's and doctorate, GRE General Test, minimum GPA of 3.0. *Application deadline:* For fall admission, 1/25 (priority date). Applications are processed on a rolling basis. *Application fee:* $25.
Expenses: Tuition, state resident: full-time $2,881. Tuition, nonresident: full-time $7,081. Part-time tuition and fees vary according to course load and program.
Financial aid: In 1999–00, 4 fellowships (averaging $14,000 per year), 12 research assistantships with partial tuition reimbursements (averaging $10,257 per year), 39 teaching assistantships with partial tuition reimbursements (averaging $8,507 per year) were awarded; career-related internships or fieldwork, institutionally sponsored loans, and contracts also available. *Total annual research expenditures:* $69,217.
Dr. Irving Lane, Chair, 225-388-8745, *Fax:* 225-388-4125, *E-mail:* irvlane@unix1.sncc.lsu.edu.
Application contact: Dr. Janet McDonald, Coordinator of Graduate Studies, 225-388-8745, *E-mail:* psmcdo@lsuvm.sncc.lsu.edu.

■ NORTHWESTERN UNIVERSITY

The Graduate School, Judd A. and Marjorie Weinberg College of Arts and Sciences, Department of Psychology, Evanston, IL 60208

AWARDS Clinical psychology (PhD); cognitive psychology (PhD); personality (PhD); psychobiology (PhD); social psychology (PhD). Admissions and degrees offered through The Graduate School. Part-time programs available.

Faculty: 23 full-time (7 women), 4 part-time/adjunct (3 women).
Students: 53 full-time (30 women); includes 8 minority (1 African American, 4 Asian Americans or Pacific Islanders, 2 Hispanic Americans, 1 Native American), 10 international. *282 applicants, 10% accepted.* In 1999, 10 doctorates awarded.
Degree requirements: For doctorate, dissertation required, foreign language not required.
Entrance requirements: For doctorate, GRE General Test, GRE Subject Test (average 720), TOEFL. *Application deadline:* For fall admission, 8/30. *Application fee:* $50 ($55 for international students).
Expenses: Tuition: Full-time $23,301. Full-time tuition and fees vary according to program.
Financial aid: In 1999–00, 18 fellowships with full tuition reimbursements (averaging $15,600 per year), 3 research assistantships with partial tuition reimbursements (averaging $12,465 per year), 17 teaching assistantships with full tuition reimbursements (averaging $12,465 per year) were awarded; career-related internships or fieldwork, Federal Work-Study, institutionally sponsored loans, and tuition waivers (full and partial) also available. Financial aid application deadline: 1/15; financial aid applicants required to submit FAFSA.
Faculty research: Memory and higher order cognition, anxiety and depression, effectiveness of psychotherapy, social cognition, molecular basis of memory. *Total annual research expenditures:* $2.6 million.
William Revelle, Chair, 847-491-5190.
Application contact: Florence Sales, Secretary, 847-491-5190, *Fax:* 847-491-7859, *E-mail:* f-sales@northwestern.edu.

■ NORTHWESTERN UNIVERSITY

The Graduate School and Medical School, Program in Clinical Psychology, Evanston, IL 60208

AWARDS Clinical psychology (PhD), including clinical health, clinical neuropsychology, general clinical. PhD admissions and degree offered through The Graduate School.

Faculty: 16 full-time (8 women), 61 part-time/adjunct (35 women).
Students: 25 full-time (18 women); includes 3 minority (all African Americans), 1 international. *139 applicants, 4% accepted.* In 1999, 9 degrees awarded (11% entered university research/teaching, 44% found other work related to degree, 44% continued full-time study).
Degree requirements: For doctorate, computer language, dissertation, clinical internship required, foreign language not required. *Average time to degree:* Doctorate–6 years full-time.

Northwestern University (continued)

Entrance requirements: For doctorate, GRE General Test, GRE Subject Test; TOEFL, minimum GPA of 3.2, previous course work in psychology. *Application deadline:* For fall admission, 1/5. *Application fee:* $50 ($55 for international students).
Expenses: Tuition: Full-time $23,301. Full-time tuition and fees vary according to program.
Financial aid: In 1999–00, 1 fellowship with full and partial tuition reimbursement (averaging $11,673 per year), 6 research assistantships with partial tuition reimbursements (averaging $16,285 per year) were awarded; career-related internships or fieldwork, grants, institutionally sponsored loans, scholarships, and tuition waivers (full) also available. Financial aid application deadline: 1/5; financial aid applicants required to submit FAFSA.
Faculty research: Cancer and cardiovascular risk reduction, evaluation of mental health services and policy, neuropsychological assessment, outcome of psychotherapy and HIV risk behavior in adolescents.
Walter Burke, Director, 312-908-8262, *Fax:* 312-503-0466.
Application contact: Chelsea Kelley, Admission Contact, 312-908-8262, *Fax:* 312-503-0466.

■ THE OHIO STATE UNIVERSITY

Graduate School, College of Social and Behavioral Sciences, Department of Psychology, Columbus, OH 43210
AWARDS Clinical psychology (PhD); cognitive/experimental psychology (PhD); counseling psychology (PhD); developmental psychology (PhD); mental retardation and developmental disabilities (PhD); psychobiology (PhD); quantitative psychology (PhD); social psychology (PhD).

Faculty: 83 full-time, 20 part-time/adjunct.
Students: 168 full-time (105 women), 15 part-time (7 women); includes 39 minority (14 African Americans, 17 Asian Americans or Pacific Islanders, 7 Hispanic Americans, 1 Native American), 15 international. *567 applicants, 17% accepted.* In 2000, 19 degrees awarded.
Degree requirements: For doctorate, dissertation required, foreign language not required.
Entrance requirements: For doctorate, GRE General Test, TOEFL. *Application deadline:* For fall admission, 12/15. *Application fee:* $30 ($40 for international students).
Expenses: Tuition, state resident: full-time $5,400. Tuition, nonresident: full-time $14,535. Part-time tuition and fees vary according to course load and program.

Financial aid: Fellowships, research assistantships, teaching assistantships, Federal Work-Study and institutionally sponsored loans available. Aid available to part-time students.
Richard E. Petty, Chairman, 614-292-4112, *Fax:* 614-292-4537, *E-mail:* petty.1@osu.edu.
Application contact: Graduate Program Coordinator, 614-292-4112, *Fax:* 614-292-4537, *E-mail:* psygrad@osu.edu.

■ OREGON HEALTH SCIENCES UNIVERSITY

School of Medicine, Graduate Programs in Medicine, Department of Behavioral Neuroscience, Portland, OR 97201-3098
AWARDS MS, PhD, MD/PhD.

Degree requirements: For master's, thesis required, foreign language not required; for doctorate, dissertation, written exam required, foreign language not required.
Entrance requirements: For master's and doctorate, GRE General Test.
Expenses: Tuition, state resident: full-time $3,132; part-time $174 per credit hour. Tuition, nonresident: full-time $5,256; part-time $292 per credit hour. Required fees: $8.5 per credit hour. $146 per term. Part-time tuition and fees vary according to course load.
Faculty research: Neural basis of behavior, behavioral pharmacology, behavioral genetics, neuropharmacology and neuroendocrinology, biological basis of drug seeking and addiction.
Find an in-depth description at www.petersons.com/graduate.

■ THE PENNSYLVANIA STATE UNIVERSITY UNIVERSITY PARK CAMPUS

Graduate School, College of Health and Human Development, Department of Biobehavioral Health, State College, University Park, PA 16802-1503
AWARDS MS, PhD.

Students: 25 full-time (19 women), 3 part-time (2 women).
Entrance requirements: For doctorate, GRE General Test or MCAT. *Application fee:* $50.
Expenses: Tuition, state resident: full-time $6,886; part-time $291 per credit. Tuition, nonresident: full-time $14,118; part-time $588 per credit. Required fees: $46 per semester. Part-time tuition and fees vary according to course load and program.
Dr. Lynn T. Kozlowski, Head, 814-863-7256.

■ THE PENNSYLVANIA STATE UNIVERSITY UNIVERSITY PARK CAMPUS

Graduate School, College of Liberal Arts, Department of Psychology, State College, University Park, PA 16802-1503
AWARDS Clinical psychology (MS, PhD); cognitive psychology (MS, PhD); developmental psychology (MS, PhD); industrial/organizational psychology (MS, PhD); psychobiology (MS, PhD); social psychology (MS, PhD).

Students: 117 full-time (72 women), 15 part-time (9 women). In 2000, 17 master's, 16 doctorates awarded.
Degree requirements: For master's and doctorate, thesis/dissertation required.
Entrance requirements: For master's and doctorate, GRE General Test, GRE Subject Test. *Application fee:* $50.
Expenses: Tuition, state resident: full-time $6,886; part-time $291 per credit. Tuition, nonresident: full-time $14,118; part-time $588 per credit. Required fees: $46 per semester. Part-time tuition and fees vary according to course load and program.
Financial aid: Fellowships, research assistantships, teaching assistantships available.
Dr. Keith Crnic, Head, 814-863-1721.

■ RUTGERS, THE STATE UNIVERSITY OF NEW JERSEY, NEWARK

Graduate School, Department of Psychology, Newark, NJ 07102
AWARDS Cognitive science (PhD); perception (PhD); psychobiology (PhD); social cognition (PhD). Part-time and evening/weekend programs available.

Faculty: 15 full-time (5 women), 3 part-time/adjunct (1 woman).
Students: 16 full-time (6 women), 14 part-time (9 women); includes 8 minority (2 African Americans, 4 Asian Americans or Pacific Islanders, 2 Hispanic Americans). *54 applicants, 44% accepted.* In 1999, 1 doctorate awarded.
Degree requirements: For doctorate, dissertation, comprehensive exam required, foreign language not required.
Entrance requirements: For doctorate, GRE Subject Test, minimum undergraduate B average. *Application deadline:* For fall admission, 2/1 (priority date); for spring admission, 11/1. Applications are processed on a rolling basis. *Application fee:* $50. Electronic applications accepted.
Expenses: Tuition, state resident: full-time $6,776; part-time $279 per credit hour. Tuition, nonresident: full-time $9,936; part-time $412 per credit hour. Required

fees: $201 per semester. Tuition and fees vary according to course load and program.

Financial aid: In 1999–00, 12 students received aid, including 9 fellowships with full tuition reimbursements available (averaging $13,000 per year), 8 teaching assistantships with full tuition reimbursements available (averaging $13,350 per year); research assistantships, career-related internships or fieldwork, Federal Work-Study, and minority scholarships also available. Aid available to part-time students. Financial aid application deadline: 3/1.

Faculty research: Visual perception (luminance, motion), neuroendocrine mechanisms in behavior (reproduction, pain), attachment theory, connectionist modeling of cognition. *Total annual research expenditures:* $300,000.

Dr. Maggie Shiffrar, Director, 973-353-5971, *Fax:* 973-353-1171, *E-mail:* mag@psychology.rutgers.edu.

Find an in-depth description at www.petersons.com/graduate.

■ RUTGERS, THE STATE UNIVERSITY OF NEW JERSEY, NEWARK

Graduate School, Program in Behavioral and Neural Sciences, Newark, NJ 07102

AWARDS PhD. Part-time programs available.

Faculty: 17 full-time (10 women), 2 part-time/adjunct (0 women).

Students: 25 full-time (12 women), 5 part-time (2 women); includes 11 minority (1 African American, 10 Hispanic Americans). *68 applicants, 26% accepted.* In 1999, 6 doctorates awarded (100% found work related to degree).

Degree requirements: For doctorate, dissertation required, foreign language not required. *Average time to degree:* Doctorate–5 years full-time.

Entrance requirements: For doctorate, GRE, minimum GPA of 3.0. *Application deadline:* For fall admission, 2/1 (priority date). Applications are processed on a rolling basis. *Application fee:* $50. Electronic applications accepted.

Expenses: Tuition, state resident: full-time $6,776; part-time $279 per credit hour. Tuition, nonresident: full-time $9,936; part-time $412 per credit hour. Required fees: $201 per semester. Tuition and fees vary according to course load and program.

Financial aid: In 1999–00, 1 fellowship with full tuition reimbursement (averaging $15,000 per year), 21 teaching assistantships with full tuition reimbursements (averaging $15,352 per year) were awarded; research assistantships Financial aid application deadline: 3/1.

Faculty research: Systems neuroscience, cognitive neuroscience, molecular neuroscience, behavioral neuroscience. *Total annual research expenditures:* $3 million.

Dr. Ian Creese, Co-Director, 973-353-1080 Ext. 3300, *Fax:* 973-353-1272, *E-mail:* creese@axon.rutgers.edu.

Application contact: Dr. Howard Poizner, First Year Advisor, 973-353-1080 Ext. 3231, *Fax:* 973-353-1272, *E-mail:* poizner@axon.rutgers.edu.

■ RUTGERS, THE STATE UNIVERSITY OF NEW JERSEY, NEW BRUNSWICK

Graduate School, Program in Psychology, New Brunswick, NJ 08901-1281

AWARDS Biopsychology and behavioral neuroscience (PhD); clinical psychology (PhD); cognitive psychology (PhD); interdisciplinary developmental psychology (PhD); interdisciplinary health psychology (PhD); social psychology (PhD).

Faculty: 77 full-time (16 women), 32 part-time/adjunct (6 women).

Students: 68 full-time (50 women), 27 part-time (20 women); includes 11 minority (4 African Americans, 4 Asian Americans or Pacific Islanders, 3 Hispanic Americans), 12 international. Average age 24. *430 applicants, 11% accepted.* In 1999, 18 degrees awarded.

Degree requirements: For doctorate, dissertation required, foreign language not required. *Average time to degree:* Doctorate–6.5 years full-time.

Entrance requirements: For doctorate, GRE General Test. *Application deadline:* For fall admission, 12/15. *Application fee:* $50.

Expenses: Tuition, state resident: full-time $6,776; part-time $279 per credit. Tuition, nonresident: full-time $9,936; part-time $412 per credit. Required fees: $20 per credit. $89 per semester. Tuition and fees vary according to course load, campus/location and program.

Financial aid: In 1999–00, 68 students received aid, including 15 fellowships with full tuition reimbursements available (averaging $13,000 per year), 9 research assistantships with full tuition reimbursements available (averaging $13,100 per year), 44 teaching assistantships with full tuition reimbursements available (averaging $13,100 per year); career-related internships or fieldwork, Federal Work-Study, grants, scholarships, and traineeships also available. Financial aid application deadline: 3/1.

Dr. G. Terence Wilson, Director, 732-445-2556, *Fax:* 732-445-2263.

Application contact: Joan Olmizzi, Administrative Assistant, 732-445-2555, *Fax:* 732-445-2263, *E-mail:* joan@psych-b.rutgers.edu.

■ STATE UNIVERSITY OF NEW YORK AT ALBANY

College of Arts and Sciences, Department of Psychology, Albany, NY 12222-0001

AWARDS Biopsychology (PhD); clinical psychology (PhD); general/experimental psychology (PhD); industrial/organizational psychology (PhD); psychology (MA); social/personality psychology (PhD).

Students: 76 full-time (44 women), 68 part-time (49 women); includes 10 minority (2 African Americans, 3 Asian Americans or Pacific Islanders, 5 Hispanic Americans), 7 international. Average age 30. *290 applicants, 33% accepted.* In 1999, 6 master's awarded (100% continued full-time study); 15 doctorates awarded.

Degree requirements: For doctorate, dissertation required, foreign language not required.

Entrance requirements: For doctorate, GRE General Test, GRE Subject Test. *Application deadline:* For fall admission, 1/15. *Application fee:* $50.

Expenses: Tuition, state resident: full-time $5,100; part-time $214 per credit. Tuition, nonresident: full-time $8,416; part-time $352 per credit. Required fees: $31 per credit.

Financial aid: Fellowships, research assistantships, teaching assistantships, career-related internships or fieldwork available. Financial aid application deadline: 2/1.

Robert Rosellini, Chair, 518-442-4820.

Application contact: Glenn Sanders, Graduate Director, 518-442-4853.

■ STATE UNIVERSITY OF NEW YORK AT BINGHAMTON

Graduate School, School of Arts and Sciences, Department of Psychology, Specialization in Behavioral Neuroscience, Binghamton, NY 13902-6000

AWARDS MA, PhD.

Students: 18 full-time (11 women), 2 part-time (both women), 3 international. Average age 28. *26 applicants, 46% accepted.* In 1999, 3 master's, 1 doctorate awarded.

Degree requirements: For master's, thesis required; for doctorate, dissertation, departmental qualifying exam required.

Entrance requirements: For master's and doctorate, GRE General Test, GRE Subject Test, TOEFL. *Application deadline:* For fall admission, 4/15 (priority date); for spring admission, 11/1. Applications are

State University of New York at Binghamton (continued)

processed on a rolling basis. *Application fee:* $50. Electronic applications accepted.

Expenses: Tuition, state resident: full-time $5,100; part-time $213 per credit. Tuition, nonresident: full-time $8,416; part-time $351 per credit. Required fees: $77 per credit. Part-time tuition and fees vary according to course load.

Financial aid: In 1999–00, 18 students received aid, including 1 fellowship (averaging $12,000 per year), 6 research assistantships with full tuition reimbursements available (averaging $5,382 per year), 11 teaching assistantships with full tuition reimbursements available (averaging $7,639 per year); career-related internships or fieldwork, Federal Work-Study, institutionally sponsored loans, and unspecified assistantships also available. Aid available to part-time students. Financial aid application deadline: 2/15. Dr. Paul Gold, Graduate Coordinator, 607-777-4644.

■ STATE UNIVERSITY OF NEW YORK AT BUFFALO

Graduate School, College of Arts and Sciences, Department of Psychology, Buffalo, NY 14260

AWARDS Biopsychology (PhD); clinical psychology (PhD); cognitive psychology (PhD); general psychology (MA).

Faculty: 31 full-time (12 women), 1 part-time/adjunct (0 women).
Students: 51 full-time (27 women), 24 part-time (14 women); includes 2 African Americans, 3 Asian Americans or Pacific Islanders, 4 Hispanic Americans, 5 international. Average age 25. *233 applicants, 18% accepted.* In 1999, 6 master's, 26 doctorates awarded. Terminal master's awarded for partial completion of doctoral program.
Degree requirements: For master's, project required, foreign language and thesis not required; for doctorate, dissertation required, foreign language not required. *Average time to degree:* Master's–3 years full-time; doctorate–7 years full-time.
Entrance requirements: For master's and doctorate, GRE General Test, TOEFL. *Application deadline:* For fall admission, 1/15. *Application fee:* $35.
Expenses: Tuition, state resident: full-time $5,100; part-time $213 per credit hour. Tuition, nonresident: full-time $8,416; part-time $351 per credit hour. Required fees: $935; $75 per semester. Tuition and fees vary according to course load and program.
Financial aid: In 1999–00, 47 students received aid, including 3 fellowships with full tuition reimbursements available

(averaging $14,000 per year), 6 research assistantships with full tuition reimbursements available (averaging $9,866 per year), 41 teaching assistantships with full tuition reimbursements available (averaging $8,400 per year); career-related internships or fieldwork, Federal Work-Study, institutionally sponsored loans, and minority fellowships also available. Financial aid application deadline: 1/15; financial aid applicants required to submit FAFSA.
Faculty research: Spoken language processing, close relationships, opiates in behavior, depression and anxiety, attention deficit disorder and hyperactivity in children. *Total annual research expenditures:* $2.4 million.
Dr. Jack R. Meacham, Chair, 716-645-3650 Ext. 203, *Fax:* 716-645-3801, *E-mail:* meacham@acsu.buffalo.edu.
Application contact: Michele Nowacki, Coordinator of Admissions, 716-645-3650 Ext. 209, *Fax:* 716-695-3801.

■ STATE UNIVERSITY OF NEW YORK AT STONY BROOK

Graduate School, College of Arts and Sciences, Department of Psychology, Program in Biopsychology, Stony Brook, NY 11794

AWARDS PhD.

Students: 13 full-time (5 women), 5 part-time (3 women); includes 3 minority (2 African Americans, 1 Hispanic American), 5 international. Average age 29. *28 applicants, 14% accepted.* In 1999, 1 degree awarded.
Degree requirements: For doctorate, dissertation required, foreign language not required.
Entrance requirements: For doctorate, GRE General Test, GRE Subject Test, TOEFL. *Application deadline:* For fall admission, 1/15. *Application fee:* $50.
Expenses: Tuition, state resident: full-time $5,100; part-time $213 per credit hour. Tuition, nonresident: full-time $8,416; part-time $351 per credit hour. Required fees: $492. Tuition and fees vary according to program.
Application contact: Dr. Harriet Waters, Director, 631-632-7855, *Fax:* 631-632-7876, *E-mail:* hwaters@ccvm.sunysb.edu.

■ UNIVERSITY OF COLORADO AT BOULDER

Graduate School, College of Arts and Sciences, Department of Environmental, Population, and Organic Biology, Boulder, CO 80309

AWARDS Animal behavior (MA, PhD); aquatic biology (MA, PhD); behavioral genetics (MA, PhD); ecology (MA, PhD); microbiology (MA, PhD); neurobiology (MA, PhD); plant and

animal physiology (MA, PhD); plant and animal systematics (MA, PhD); population biology (MA, PhD); population genetics (MA, PhD).

Faculty: 39 full-time (9 women).
Students: 84 full-time (36 women), 14 part-time (7 women); includes 17 minority (6 Asian Americans or Pacific Islanders, 10 Hispanic Americans, 1 Native American). Average age 29. *147 applicants, 14% accepted.* In 1999, 7 master's, 13 doctorates awarded. Terminal master's awarded for partial completion of doctoral program.
Degree requirements: For master's, thesis or alternative, comprehensive exam required, foreign language not required; for doctorate, dissertation, comprehensive exam required, foreign language not required. *Average time to degree:* Master's–3 years full-time; doctorate–5 years full-time.
Entrance requirements: For master's, GRE General Test, GRE Subject Test, minimum undergraduate GPA of 3.0; for doctorate, GRE General Test, GRE Subject Test. *Application deadline:* For fall admission, 1/15 (priority date). *Application fee:* $40 ($60 for international students).
Expenses: Tuition, state resident: part-time $181 per credit hour. Tuition, nonresident: part-time $542 per credit hour. Required fees: $99 per term. Tuition and fees vary according to course load and program.
Financial aid: Fellowships, research assistantships, teaching assistantships, Federal Work-Study, institutionally sponsored loans, and tuition waivers (full) available. Financial aid application deadline: 3/1.
Faculty research: Evolution, developmental biology, behavior and neurobiology. *Total annual research expenditures:* $1.8 million.
Michael Breed, Chair, 303-492-8981, *Fax:* 303-492-8699, *E-mail:* michael.breed@colorado.edu.
Application contact: Jill Skarstadt, Graduate Secretary, 303-492-7654, *Fax:* 303-492-8699, *E-mail:* jill.skarstadt@colorado.edu.

■ UNIVERSITY OF CONNECTICUT

Graduate School, College of Liberal Arts and Sciences, Field of Biobehavioral Sciences, Storrs, CT 06269

AWARDS Developmental psychobiology (MS, PhD).

Degree requirements: For doctorate, dissertation required.
Entrance requirements: For master's and doctorate, GRE General Test, TOEFL.
Expenses: Tuition, state resident: full-time $5,118. Tuition, nonresident: full-time $13,298. Required fees: $1,022.

■ UNIVERSITY OF CONNECTICUT

Graduate School, College of Liberal Arts and Sciences, Field of Psychology, Storrs, CT 06269

AWARDS Behavioral neuroscience (PhD); biopsychology (MA, PhD); clinical psychology (PhD); cognition/instruction psychology (PhD); developmental psychology (PhD); ecological psychology (PhD); general experimental psychology (PhD); industrial and organizational psychology (PhD); language psychology (PhD); neuroscience (MA); social psychology (PhD).

Faculty: 48 full-time (12 women), 3 part-time/adjunct (1 woman).
Students: 132 full-time (79 women), 18 part-time (9 women); includes 16 minority (6 African Americans, 3 Asian Americans or Pacific Islanders, 6 Hispanic Americans, 1 Native American), 8 international. Average age 28. *415 applicants, 7% accepted.* In 1999, 18 master's awarded (15% found work related to degree, 85% continued full-time study); 30 doctorates awarded (37% entered university research/teaching, 53% found other work related to degree). Terminal master's awarded for partial completion of doctoral program.
Degree requirements: For master's, computer language, thesis or alternative required; for doctorate, computer language, dissertation, internship required. *Average time to degree:* Master's–2 years full-time; doctorate–5 years full-time.
Entrance requirements: For master's and doctorate, GRE General Test, GRE Subject Test. *Application deadline:* For fall admission, 1/1 (priority date). *Application fee:* $40 ($45 for international students).
Expenses: Tuition, state resident: full-time $5,118. Tuition, nonresident: full-time $13,298. Required fees: $1,022.
Financial aid: In 1999–00, 45 fellowships (averaging $4,500 per year), 23 research assistantships with tuition reimbursements (averaging $18,145 per year), 79 teaching assistantships with tuition reimbursements (averaging $18,145 per year) were awarded; career-related internships or fieldwork, Federal Work-Study, grants, and scholarships also available. Financial aid application deadline: 1/1; financial aid applicants required to submit FAFSA.
Faculty research: Behavioral neuroscience; physiology/chemistry of memory, sensation, motor control/motivation; cognitive, social, personality and language development; infancy/person perceptions and intergroup relations; anxiety, expectancies, health, and neuropsychology; speech, reading, perceiving.
Charles A. Lowe, Head.
Application contact: Nicole Dolat, Graduate Admissions, 860-486-3528, *Fax:* 860-486-2760, *E-mail:* futuregr@psych.psy.uconn.edu.

■ UNIVERSITY OF DELAWARE

College of Arts and Science, Department of Psychology, Newark, DE 19716

AWARDS Biopsychology (PhD); clinical psychology (PhD); cognitive psychology (PhD); social psychology (PhD).

Faculty: 25 full-time (6 women), 3 part-time/adjunct (0 women).
Students: 46 full-time (29 women), 3 part-time (all women). Average age 28. *199 applicants, 10% accepted.* In 1999, 9 degrees awarded.
Degree requirements: For doctorate, dissertation required, foreign language not required. *Average time to degree:* Doctorate–6.6 years full-time.
Entrance requirements: For doctorate, GRE General Test. *Application deadline:* For fall admission, 1/7 (priority date). *Application fee:* $45. Electronic applications accepted.
Expenses: Tuition, state resident: full-time $4,380; part-time $243 per credit. Tuition, nonresident: full-time $12,750; part-time $708 per credit. Required fees: $15 per term. Tuition and fees vary according to program.
Financial aid: In 1999–00, 32 students received aid, including 5 fellowships with full tuition reimbursements available (averaging $10,200 per year), 6 research assistantships with full tuition reimbursements available (averaging $10,400 per year), 21 teaching assistantships with full tuition reimbursements available (averaging $10,900 per year); career-related internships or fieldwork, Federal Work-Study, institutionally sponsored loans, scholarships, tuition waivers (full and partial), and minority fellowships also available. Financial aid application deadline: 1/7.
Faculty research: Emotion development, neural and cognitive aspects of memory, neural control of feeding, intergroup relations, social cognition and communication. *Total annual research expenditures:* $1.3 million.
Dr. Evelyn Satinoff, Chairperson, 302-831-2271, *Fax:* 302-831-3645, *E-mail:* satinoff@bach.udel.edu.
Application contact: Dr. Brian Ackerman, Chairperson, Graduate Committee, 302-831-2271, *Fax:* 302-831-3645, *E-mail:* bpa@udel.edu.

■ UNIVERSITY OF ILLINOIS AT URBANA–CHAMPAIGN

Graduate College, College of Liberal Arts and Sciences, Department of Psychology, Urbana, IL 61801

AWARDS Applied measurement (MS); biological psychology (AM, PhD); clinical psychology (AM, PhD); cognitive psychology (AM, PhD); developmental psychology (AM, PhD); engineering psychology (MS); perception and performance psychology (AM, PhD); personality-social-organizational (AM, PhD); personnel psychology (MS); quantitative psychology (AM, PhD).

Faculty: 51 full-time (17 women), 10 part-time/adjunct (4 women).
Students: 160 full-time (85 women); includes 27 minority (6 African Americans, 14 Asian Americans or Pacific Islanders, 7 Hispanic Americans), 28 international. *422 applicants, 19% accepted.* In 1999, 4 master's, 41 doctorates awarded.
Degree requirements: For master's, foreign language and thesis not required; for doctorate, dissertation required, foreign language not required. *Average time to degree:* Master's–1.5 years full-time; doctorate–6 years full-time.
Entrance requirements: For master's, GRE General Test, GRE Subject Test (recommended), minimum GPA of 3.0; for doctorate, GRE General Test, GRE Subject Test (recommended). *Application deadline:* For fall admission, 1/1. *Application fee:* $40 ($50 for international students).
Expenses: Tuition, state resident: full-time $4,040. Tuition, nonresident: full-time $11,192. Full-time tuition and fees vary according to program.
Financial aid: In 1999–00, 18 fellowships with full tuition reimbursements (averaging $14,000 per year), 67 research assistantships with full tuition reimbursements (averaging $9,640 per year), 57 teaching assistantships with full tuition reimbursements (averaging $9,640 per year) were awarded; career-related internships or fieldwork, traineeships, and tuition waivers (full and partial) also available. Financial aid application deadline: 1/1. *Total annual research expenditures:* $4.4 million.
Dr. Edward J. Shoben, Head.
Application contact: Cheryl Berger, Assistant Head for Graduate Affairs, 217-333-3429, *Fax:* 217-244-5876, *E-mail:* gradstdy@s.psych.uiuc.edu.

■ UNIVERSITY OF MICHIGAN

Horace H. Rackham School of Graduate Studies, College of Literature, Science, and the Arts, Department of Psychology, Ann Arbor, MI 48109

AWARDS Biopsychology (PhD); clinical psychology (PhD); cognition and perception

University of Michigan (continued)
(PhD); developmental psychology (PhD);
organizational psychology (PhD); personality
psychology (PhD); social psychology (PhD).
Faculty: 68 full-time, 23 part-time/
adjunct.
Students: 176 full-time (115 women);
includes 72 minority (45 African
Americans, 10 Asian Americans or Pacific
Islanders, 15 Hispanic Americans, 2 Native
Americans), 14 international. *586
applicants, 10% accepted.* In 1999, 26
degrees awarded.
Degree requirements: For doctorate, oral
defense of dissertation, preliminary exam
required. *Average time to degree:*
Doctorate–5.8 years full-time.
Entrance requirements: For doctorate,
GRE General Test, GRE Subject Test,
TOEFL. *Application deadline:* For fall
admission, 12/15. *Application fee:* $55.
Expenses: Tuition, state resident: full-time
$10,316. Tuition, nonresident: full-time
$20,922. Required fees: $185. Part-time
tuition and fees vary according to course
load and program.
Financial aid: Fellowships, research
assistantships, teaching assistantships,
career-related internships or fieldwork
available. Financial aid application
deadline: 3/15.
Patricia Gurin, Chair, 734-764-7429.
Application contact: Jill Becker, Graduate
Chair, 734-764-6316.

■ UNIVERSITY OF MINNESOTA, TWIN CITIES CAMPUS

**Graduate School, College of Liberal
Arts, Department of Psychology,
Program in Cognitive and Biological
Psychology, Minneapolis, MN 55455-
0213**
AWARDS PhD.
Students: 31 full-time (12 women);
includes 3 minority (1 Asian American or
Pacific Islander, 2 Hispanic Americans), 3
international. *22 applicants, 50% accepted.* In
1999, 2 degrees awarded.
Degree requirements: For doctorate, dissertation, comprehensive exams required,
foreign language not required.
Entrance requirements: For doctorate,
GRE General Test, GRE Subject Test
(recommended), TOEFL, 12 credits of
upper-level psychology courses, including a
course in statistics or psychological
measurement. *Application deadline:* For fall
admission, 1/5. *Application fee:* $50 ($55 for
international students).
Expenses: Tuition, state resident: full-time
$5,040; part-time $420 per credit. Tuition,
nonresident: full-time $9,900; part-time
$825 per credit. Full-time tuition and fees
vary according to course load, program
and reciprocity agreements.

Financial aid: Fellowships, research
assistantships, teaching assistantships,
career-related internships or fieldwork and
tuition waivers (partial) available. Financial
aid application deadline: 12/15.
Application contact: Susan Prahl, Graduate Admissions, 612-624-4181, *Fax:* 612-
626-2079, *E-mail:* psyapply@tc.umn.edu.

■ UNIVERSITY OF NEW ORLEANS

**Graduate School, College of Sciences,
Department of Psychology, New
Orleans, LA 70148**
AWARDS Applied psychology (PhD), including
applied biopsychology, applied developmental
psychology; psychology (MS).
Faculty: 22 full-time (8 women), 2 part-
time/adjunct (1 woman).
Students: 17 full-time (11 women), 1
(woman) part-time; includes 3 minority (2
Asian Americans or Pacific Islanders, 1
Hispanic American). Average age 28. *41
applicants, 12% accepted.* In 1999, 7
master's, 3 doctorates awarded (100%
found work related to degree).
Degree requirements: For doctorate, dissertation required, foreign language not
required.
Entrance requirements: For doctorate,
GRE General Test, minimum GPA of 3.0,
21 hours of course work in psychology.
Application deadline: For fall admission,
2/15 (priority date). Applications are
processed on a rolling basis. *Application fee:*
$20.
Expenses: Tuition, state resident: full-time
$2,362. Tuition, nonresident: full-time
$7,888. Part-time tuition and fees vary
according to course load.
Financial aid: Fellowships, research
assistantships, teaching assistantships,
career-related internships or fieldwork and
unspecified assistantships available.
Financial aid application deadline: 2/15.
Faculty research: Biofeedback, visual and
auditory perception, psychopharmacology,
neuropeptides.
Dr. Richard David Olson, Chair, 504-280-
6778, *Fax:* 504-280-6049, *E-mail:* rolson@
uno.edu.
Application contact: Dr. Matthew
Stanford, Graduate Coordinator, 504-280-
6185, *Fax:* 504-280-6049, *E-mail:*
mstanfor@uno.edu.

■ UNIVERSITY OF OKLAHOMA HEALTH SCIENCES CENTER

**College of Medicine and Graduate
College, Graduate Programs in
Medicine, Department of Psychiatry
and Behavioral Sciences, Oklahoma
City, OK 73190**
AWARDS Biological psychology (MS, PhD).

Faculty: 17 full-time (6 women), 18 part-
time/adjunct (6 women).
Students: 13 full-time (6 women), 4 part-
time (2 women), 1 international. Average
age 25. *17 applicants, 29% accepted.* In
1999, 2 master's, 2 doctorates awarded.
Degree requirements: For master's and
doctorate, thesis/dissertation required,
foreign language not required.
Entrance requirements: For master's and
doctorate, GRE General Test, TOEFL.
Application deadline: For fall admission, 3/1.
Application fee: $25 ($50 for international
students).
Expenses: Tuition, state resident: part-
time $90 per semester hour. Tuition,
nonresident: part-time $264 per semester
hour. Tuition and fees vary according to
program.
Financial aid: Fellowships, research
assistantships, career-related internships or
fieldwork, institutionally sponsored loans,
and tuition waivers (full and partial) available. Aid available to part-time students.
Faculty research: Behavioral
neuroscience, human neuropsychology,
psychophysiology, behavioral medicine,
health psychology.
Dr. Frank Holloway, Director, 405-271-
2011.
Application contact: Dr. Bill Beatty,
Graduate Liaison, 405-271-2474.

■ UNIVERSITY OF OREGON

**Graduate School, College of Arts and
Sciences, Department of Psychology,
Eugene, OR 97403**
AWARDS Clinical psychology (PhD); cognitive
psychology (MA, MS, PhD); developmental
psychology (MA, MS, PhD); physiological
psychology (MA, MS, PhD); psychology (MA,
MS, PhD); social/personality psychology (MA,
MS, PhD).
Faculty: 27 full-time (11 women), 11 part-
time/adjunct (8 women).
Students: 68 full-time (48 women), 10
part-time (8 women); includes 9 minority
(1 African American, 6 Asian Americans or
Pacific Islanders, 1 Hispanic American, 1
Native American), 6 international. *176
applicants, 11% accepted.* In 1999, 21
master's, 11 doctorates awarded. Terminal
master's awarded for partial completion of
doctoral program.
Degree requirements: For master's,
foreign language and thesis not required;
for doctorate, dissertation required, foreign
language not required. *Average time to
degree:* Master's–2 years full-time;
doctorate–4.5 years full-time.
Entrance requirements: For master's,
GRE General Test, TOEFL, median GPA
of most recent class: 3.67, minimum GPA
of 3.00; for doctorate, GRE General Test,
TOEFL. *Application deadline:* For fall
admission, 12/1. *Application fee:* $50.

Expenses: Tuition, state resident: full-time $6,750. Tuition, nonresident: full-time $11,409. Part-time tuition and fees vary according to course load.
Financial aid: In 1999–00, 48 teaching assistantships were awarded; research assistantships, career-related internships or fieldwork also available.
Dr. Robert Mauro, Head, 541-346-4921, *Fax:* 541-346-4911.
Application contact: Lori Olsen, Graduate Secretary, 541-346-5060, *Fax:* 541-346-4911.

■ UNIVERSITY OF WISCONSIN–MADISON

Graduate School, College of Letters and Science, Department of Psychology, Program in Biological Psychology, Madison, WI 53706-1380

AWARDS PhD.

Faculty: 4 full-time (1 woman), 1 part-time/adjunct (0 women).
Students: 16 full-time (6 women); includes 3 minority (1 African American, 1 Asian American or Pacific Islander, 1 Hispanic American), 1 international. Average age 23. *29 applicants, 21% accepted.* In 1999, 1 doctorate awarded (100% entered university research/teaching).
Degree requirements: For doctorate, dissertation required, foreign language not required. *Average time to degree:* Doctorate–5 years full-time.
Entrance requirements: For doctorate, GRE General Test, minimum undergraduate GPA of 3.0. *Application deadline:* For fall admission, 1/5 (priority date). Applications are processed on a rolling basis. *Application fee:* $45. Electronic applications accepted.
Expenses: Tuition, state resident: full-time $5,406; part-time $339 per credit. Tuition, nonresident: full-time $17,110; part-time $1,071 per credit. Full-time tuition and fees vary according to program and reciprocity agreements. Part-time tuition and fees vary according to course load and program.
Financial aid: In 1999–00, 16 students received aid, including 6 fellowships with full tuition reimbursements available (averaging $12,420 per year), 5 research assistantships with full tuition reimbursements available (averaging $12,307 per year), 2 teaching assistantships with full tuition reimbursements available (averaging $12,000 per year); institutionally sponsored loans also available. Financial aid application deadline: 1/5.
Chris Coe, Chair, 608-262-3550, *Fax:* 608-262-6020, *E-mail:* ccoe@facstaff.wisc.edu.

Application contact: Jane Fox-Anderson, Graduate Admissions Secretary, 608-262-2079, *Fax:* 608-262-4029, *E-mail:* jefoxand@facstaff.wisc.edu.

NEUROSCIENCE

■ ALBANY MEDICAL COLLEGE

Graduate Programs in the Biological Sciences, Program in Neuropharmacology and Neuroscience, Albany, NY 12208-3479

AWARDS MS, PhD. Part-time programs available.

Students: 14 full-time (8 women); includes 2 minority (1 African American, 1 Hispanic American), 6 international. Terminal master's awarded for partial completion of doctoral program.
Degree requirements: For master's, thesis required, foreign language not required; for doctorate, dissertation, comprehensive written exam, oral qualifying exam required, foreign language not required.
Entrance requirements: For master's and doctorate, GRE General Test, TOEFL. *Application deadline:* Applications are processed on a rolling basis. *Application fee:* $0 ($60 for international students).
Expenses: Tuition: Full-time $13,367; part-time $446 per credit hour.
Financial aid: Federal Work-Study, grants, scholarships, and tuition waivers (full) available.
Dr. Stanley D. Glick, Co-Director, 518-262-5303, *Fax:* 518-262-5799, *E-mail:* pharmneuroinfo@mail.amc.edu.

Find an in-depth description at www.petersons.com/graduate.

■ ARIZONA STATE UNIVERSITY

Graduate College, College of Liberal Arts and Sciences, Department of Biology, Program in Neuroscience, Tempe, AZ 85287

AWARDS MS, PhD. Terminal master's awarded for partial completion of doctoral program.

Degree requirements: For master's, thesis required, foreign language not required; for doctorate, dissertation, oral exam required, foreign language not required.
Entrance requirements: For master's and doctorate, GRE General Test, GRE Subject Test. *Application deadline:* For fall admission, 12/15. *Application fee:* $45.
Expenses: Tuition, state resident: part-time $115 per credit hour. Tuition, nonresident: part-time $389 per credit

hour. Required fees: $18 per semester. Tuition and fees vary according to program.
Financial aid: Application deadline: 12/15.
Application contact: Dr. Michael C. Moore, Director, 480-965-0386, *Fax:* 480-965-2519.

■ ARIZONA STATE UNIVERSITY

Graduate College, College of Liberal Arts and Sciences, Department of Psychology, Tempe, AZ 85287

AWARDS Behavioral neuroscience (PhD); clinical psychology (PhD); cognitive/behavioral systems (PhD); developmental psychology (PhD); environmental psychology (PhD); quantitative research methods (PhD); social psychology (PhD).

Faculty: 71 full-time (27 women), 49 part-time/adjunct (35 women).
Students: 89 full-time (60 women), 20 part-time (16 women); includes 21 minority (1 African American, 4 Asian Americans or Pacific Islanders, 14 Hispanic Americans, 2 Native Americans), 7 international. Average age 29. *449 applicants, 10% accepted.* In 1999, 8 doctorates awarded.
Degree requirements: For doctorate, dissertation required.
Entrance requirements: For doctorate, GRE General Test, GRE Subject Test. *Application fee:* $45.
Expenses: Tuition, state resident: part-time $115 per credit hour. Tuition, nonresident: part-time $389 per credit hour. Required fees: $18 per semester. Tuition and fees vary according to program.
Faculty research: Reduction of personal stress, cognitive aspects of motor skills training, behavior analysis and treatment of depression.
Dr. Darwin Linder, Chair, 480-965-3326.
Application contact: Dr. Laurie Chassin, 480-965-3326.

■ BAYLOR COLLEGE OF MEDICINE

Graduate School of Biomedical Sciences, Division of Neuroscience, Houston, TX 77030-3498

AWARDS PhD, MD/PhD.

Faculty: 43 full-time (7 women).
Students: 31 full-time (10 women); includes 4 minority (1 African American, 1 Asian American or Pacific Islander, 2 Hispanic Americans), 6 international. Average age 27. *102 applicants, 13% accepted.* In 1999, 3 degrees awarded (67% entered university research/teaching, 33% continued full-time study).
Degree requirements: For doctorate, dissertation, public defense, qualifying exam

Baylor College of Medicine (continued) required, foreign language not required. *Average time to degree:* Doctorate–5.12 years full-time.

Entrance requirements: For doctorate, GRE General Test (average 80th percentile), GRE Subject Test (strongly recommended), TOEFL, minimum GPA of 3.0. *Application deadline:* For fall admission, 2/1 (priority date). *Application fee:* $30. Electronic applications accepted.

Expenses: Tuition: Full-time $8,200. Required fees: $175. Full-time tuition and fees vary according to student level.

Financial aid: In 1999–00, 31 students received aid, including 18 fellowships (averaging $16,000 per year), 13 research assistantships (averaging $16,000 per year); Federal Work-Study, institutionally sponsored loans, and tuition waivers (full) also available. Financial aid applicants required to submit FAFSA.

Faculty research: Molecular and developmental neurobiology, neurobiology of disease, neuroanatomy, neurophysiology, neural systems analysis.

Dr. Daniel Johnston, Director, 713-798-5984.

Application contact: Lori Christman, Graduate Program Administrator, 713-798-7270, *Fax:* 713-798-3946, *E-mail:* loric@bcm.tmc.edu.

Find an in-depth description at www.petersons.com/graduate.

■ BAYLOR UNIVERSITY

Graduate School, College of Arts and Sciences, Department of Psychology and Neuroscience, Program in Neuroscience, Waco, TX 76798

AWARDS MA, PhD.

Students: 7 full-time (4 women), 1 (woman) part-time. In 1999, 1 doctorate awarded.

Degree requirements: For doctorate, comprehensive exam required.

Entrance requirements: For master's and doctorate, GRE General Test. *Application deadline:* Applications are processed on a rolling basis. *Application fee:* $25.

Expenses: Tuition: Part-time $329 per semester hour. Tuition and fees vary according to program.

Dr. Charles Weaver, Director, 254-710-2961, *Fax:* 254-710-3033, *E-mail:* charles_weaver@baylor.edu.

Application contact: Bettye Keel, Graduate Coordinator, 254-710-2811, *Fax:* 254-710-3033, *E-mail:* bettye_keel@baylor.edu.

■ BOSTON UNIVERSITY

Graduate School of Arts and Sciences, Department of Biology, Boston, MA 02215

AWARDS Botany (MA, PhD); cell and molecular biology (MA, PhD); cell biology (MA, PhD); ecology (PhD); ecology, behavior, and evolution (MA, PhD); ecology/physiology, endocrinology and reproduction (MA); marine biology (MA, PhD); molecular biology, cell biology and biochemistry (MA, PhD); neurobiology, neuroendocrinology and reproduction (MA, PhD); physiology, endocrinology, and neurobiology (MA, PhD); zoology (MA, PhD). Part-time programs available.

Faculty: 41 full-time (8 women).

Students: 131 full-time (74 women), 11 part-time (7 women); includes 10 minority (7 Asian Americans or Pacific Islanders, 3 Hispanic Americans), 33 international. Average age 27. *238 applicants, 39% accepted.* In 1999, 61 master's, 45 doctorates awarded. Terminal master's awarded for partial completion of doctoral program.

Degree requirements: For master's, one foreign language, thesis not required; for doctorate, one foreign language, dissertation, qualifying exam required. *Average time to degree:* Master's–1 year full-time, 3 years part-time; doctorate–5.75 years full-time.

Entrance requirements: For master's and doctorate, GRE General Test, GRE Subject Test, TOEFL. *Application deadline:* For fall admission, 1/1 (priority date); for spring admission, 11/1. *Application fee:* $50.

Expenses: Tuition: Full-time $23,770; part-time $743 per credit. Required fees: $220. Tuition and fees vary according to class time, course level, campus/location and program.

Financial aid: In 1999–00, 82 students received aid, including 1 fellowship with full tuition reimbursement available (averaging $12,000 per year), 28 research assistantships with full tuition reimbursements available (averaging $11,500 per year), 43 teaching assistantships with full tuition reimbursements available (averaging $11,500 per year); Federal Work-Study, grants, institutionally sponsored loans, scholarships, and traineeships also available. Financial aid application deadline: 1/15; financial aid applicants required to submit FAFSA.

Faculty research: Marine science, endocrinology, behavior. *Total annual research expenditures:* $5 million.

Geoffrey M. Cooper, Chairman, 617-353-2432, *Fax:* 617-353-6340, *E-mail:* gmcooper@bu.edu.

Application contact: Yolanta Kovalko, Senior Staff Assistant, 617-353-2432, *Fax:* 617-353-6340, *E-mail:* yolanta@bu.edu.

Find an in-depth description at www.petersons.com/graduate.

■ BOSTON UNIVERSITY

Graduate School of Arts and Sciences, Department of Cognitive and Neural Systems, Boston, MA 02215

AWARDS MA, PhD. Part-time programs available.

Faculty: 12 full-time (2 women), 6 part-time/adjunct (0 women).

Students: 39 full-time (3 women), 11 part-time (2 women), 24 international. Average age 29. *74 applicants, 41% accepted.* Terminal master's awarded for partial completion of doctoral program.

Degree requirements: For master's, one foreign language, comprehensive exam (MA) required, thesis not required; for doctorate, one foreign language, dissertation, qualifying exam required. *Average time to degree:* Master's–2 years full-time, 3 years part-time; doctorate–4 years full-time, 5 years part-time.

Entrance requirements: For master's and doctorate, GRE General Test, GRE Subject Test (recommended), TOEFL. *Application deadline:* For fall admission, 7/1; for spring admission, 11/15. Applications are processed on a rolling basis. *Application fee:* $50.

Expenses: Tuition: Full-time $23,770; part-time $743 per credit. Required fees: $220. Tuition and fees vary according to class time, course level, campus/location and program.

Financial aid: In 1999–00, 4 fellowships, 25 research assistantships, 2 teaching assistantships were awarded; Federal Work-Study and unspecified assistantships also available. Aid available to part-time students. Financial aid application deadline: 1/15; financial aid applicants required to submit FAFSA.

Faculty research: Brain models, intelligent technology, vision, recognition, motor control.

Stephen Grossberg, Chairman, 617-353-7858, *Fax:* 617-353-7755, *E-mail:* steve@cns.bu.edu.

Application contact: Carol Jefferson, Administrative Assistant, 617-353-9481, *Fax:* 617-353-7755, *E-mail:* caroly@cns.bu.edu.

■ BOSTON UNIVERSITY

Graduate School of Arts and Sciences, Interdepartmental Program in Experimental and Computational Neuroscience, Boston, MA 02215

AWARDS Neuroscience (MA, PhD).

Entrance requirements: For master's and doctorate, GRE General Test, TOEFL.
Expenses: Tuition: Full-time $23,770; part-time $743 per credit. Required fees: $220. Tuition and fees vary according to class time, course level, campus/location and program.

Find an in-depth description at www.petersons.com/graduate.

■ BOSTON UNIVERSITY

School of Medicine, Division of Graduate Medical Sciences, Department of Anatomy and Neurobiology, Boston, MA 02118
AWARDS MA, PhD, MD/PhD. Part-time programs available.

Faculty: 13 full-time (3 women), 6 part-time/adjunct (0 women).
Students: 20 full-time (10 women); includes 7 minority (1 African American, 4 Asian Americans or Pacific Islanders, 2 Hispanic Americans), 1 international. Average age 27. Terminal master's awarded for partial completion of doctoral program.
Degree requirements: For master's and doctorate, thesis/dissertation, qualifying exam required, foreign language not required.
Entrance requirements: For master's and doctorate, GRE General Test, GRE Subject Test, TOEFL. *Application deadline:* For fall admission, 1/15 (priority date); for spring admission, 10/15 (priority date). *Application fee:* $50. Electronic applications accepted.
Expenses: Tuition: Full-time $24,700; part-time $772 per credit. Required fees: $220.
Financial aid: Fellowships with tuition reimbursements, research assistantships with tuition reimbursements, Federal Work-Study, scholarships, and traineeships available.
Faculty research: Neuroanatomy, development of the nervous system, aging, respiratory system, reproductive system. Mark Moss, Chairman, 617-638-4200, *Fax:* 617-638-4216 Ext. -, *E-mail:* mmos@cajal-1.bu.edu.

■ BOSTON UNIVERSITY

School of Medicine, Division of Graduate Medical Sciences, Program in Behavioral Neurosciences, Boston, MA 02215
AWARDS PhD, MD/PhD. Part-time programs available.

Faculty: 14 part-time/adjunct (5 women).
Students: 11 full-time (7 women), 4 part-time (3 women); includes 2 minority (1 Asian American or Pacific Islander, 1 Native American), 1 international. Average age 35. In 1999, 13 doctorates awarded.

Degree requirements: For doctorate, dissertation required, foreign language not required.
Entrance requirements: For doctorate, GRE General Test, GRE Subject Test, TOEFL. *Application deadline:* For fall admission, 1/15 (priority date); for spring admission, 10/15 (priority date). *Application fee:* $50. Electronic applications accepted.
Expenses: Tuition: Full-time $24,700; part-time $772 per credit. Required fees: $220.
Financial aid: In 1999–00, 4 fellowships, 5 research assistantships were awarded; Federal Work-Study, scholarships, and traineeships also available.
Faculty research: Human brain dysfunction, language disorders, disorders of purposeful movement, path of learning. Dr. Marlene Oscar Berman, Director, 617-638-4803, *Fax:* 617-638-4806, *E-mail:* oscar@bu.edu.

■ BRANDEIS UNIVERSITY

Graduate School of Arts and Sciences, Program in Psychology, Waltham, MA 02454-9110
AWARDS Cognitive neuroscience (PhD); general psychology (MA); social/developmental psychology (PhD). Part-time programs available.

Faculty: 17 full-time (3 women), 6 part-time/adjunct (5 women).
Students: 28 full-time (23 women), 1 part-time; includes 5 minority (all Asian Americans or Pacific Islanders). Average age 24. *83 applicants, 26% accepted.* In 1999, 6 master's awarded (50% found work related to degree, 50% continued full-time study); 6 doctorates awarded (50% entered university research/teaching, 50% found other work related to degree). Terminal master's awarded for partial completion of doctoral program.
Degree requirements: For master's and doctorate, thesis/dissertation required, foreign language not required. *Average time to degree:* Master's–1 year full-time, 1.5 years part-time; doctorate–5 years full-time.
Entrance requirements: For master's, GRE General Test; for doctorate, GRE General Test, GRE Subject Test. *Application fee:* $60. Electronic applications accepted.
Expenses: Tuition: Full-time $25,392; part-time $3,174 per course. Required fees: $509. Tuition and fees vary according to class time, degree level, program and student level.
Financial aid: In 1999–00, 24 students received aid, including 17 fellowships with full tuition reimbursements available (averaging $10,100 per year), 4 research assistantships (averaging $1,500 per year), teaching assistantships (averaging $2,000

per year); traineeships, tuition waivers (full and partial), and tuition remissions also available. Aid available to part-time students. Financial aid application deadline: 4/15; financial aid applicants required to submit FAFSA.
Faculty research: Development, cognition, social aging, perception. *Total annual research expenditures:* $2.2 million. Margie Lachman, Director of Graduate Studies, 781-736-3300, *Fax:* 781-736-3291.
Application contact: Janice Steinberg, Secretary, 781-736-3300, *Fax:* 781-736-3291, *E-mail:* jsteinberg@brandeis.edu.

■ BRANDEIS UNIVERSITY

Graduate School of Arts and Sciences, Programs in Biological Sciences, Program in Molecular and Cell Biology, Waltham, MA 02454-9110
AWARDS Cell biology (PhD); developmental biology (PhD); genetics (PhD); microbiology (PhD); molecular and cell biology (MS); molecular biology (PhD); neurobiology (PhD).

Faculty: 19 full-time (8 women).
Students: 58 full-time (31 women), 1 (woman) part-time; includes 13 minority (1 African American, 8 Asian Americans or Pacific Islanders, 4 Hispanic Americans), 8 international. Average age 27. *83 applicants, 11% accepted.* In 1999, 1 master's, 8 doctorates awarded (100% entered university research/teaching). Terminal master's awarded for partial completion of doctoral program.
Degree requirements: For master's, thesis not required; for doctorate, dissertation required. *Average time to degree:* Master's–1 year full-time; doctorate–6 years full-time.
Entrance requirements: For doctorate, GRE General Test. *Application deadline:* For fall admission, 1/15 (priority date). Applications are processed on a rolling basis. *Application fee:* $60. Electronic applications accepted.
Expenses: Tuition: Full-time $25,392; part-time $3,174 per course. Required fees: $509. Tuition and fees vary according to class time, degree level, program and student level.
Financial aid: In 1999–00, 10 fellowships with tuition reimbursements (averaging $17,000 per year), 49 research assistantships with tuition reimbursements (averaging $17,000 per year), 9 teaching assistantships (averaging $2,312 per year) were awarded; scholarships and tuition waivers (full and partial) also available. Financial aid application deadline: 4/15; financial aid applicants required to submit CSS PROFILE or FAFSA.
Faculty research: Regulation of gene expression by transcription factors, molecular neurobiology, immunology,

Brandeis University (continued)
molecular mechanisms of genetic recombination, and cell differentiation.
Dr. Ranjan Sen, Chair, 781-736-2455, *Fax:* 781-736-3107.
Application contact: Marcia Cabral, Information Officer, 781-736-3100, *Fax:* 781-736-3107, *E-mail:* cabral@ brandeis.edu.

■ BRANDEIS UNIVERSITY

Graduate School of Arts and Sciences, Programs in Biological Sciences, Program in Neuroscience, Waltham, MA 02454-9110

AWARDS MS, PhD.

Faculty: 22 full-time (6 women), 1 part-time/adjunct (0 women).
Students: 40 full-time (14 women), 1 part-time; includes 8 minority (2 African Americans, 5 Asian Americans or Pacific Islanders, 1 Hispanic American), 5 international. Average age 24. *47 applicants, 26% accepted.* In 1999, 2 master's, 2 doctorates awarded. Terminal master's awarded for partial completion of doctoral program.
Degree requirements: For master's, research project required, thesis not required; for doctorate, dissertation, qualifying exams, teaching experience, journal club required. *Average time to degree:* Master's–1 year full-time; doctorate–6 years full-time.
Entrance requirements: For master's and doctorate, GRE General Test. *Application deadline:* For fall admission, 1/15 (priority date). Applications are processed on a rolling basis. *Application fee:* $60. Electronic applications accepted.
Expenses: Tuition: Full-time $25,392; part-time $3,174 per course. Required fees: $509. Tuition and fees vary according to class time, degree level, program and student level.
Financial aid: In 1999–00, 5 fellowships with tuition reimbursements (averaging $17,000 per year), 36 research assistantships with tuition reimbursements (averaging $17,000 per year), 8 teaching assistantships with tuition reimbursements (averaging $2,312 per year) were awarded; scholarships and tuition waivers (full and partial) also available. Aid available to part-time students. Financial aid application deadline: 4/15; financial aid applicants required to submit CSS PROFILE or FAFSA.
Faculty research: Behavioral neuroscience, cellular and molecular neuroscience, computational and integrative neuroscience.
Dr. Leslie Griffith, Chair, 781-736-3125, *Fax:* 781-736-3107, *E-mail:* griffith@ brandeis.edu.

Application contact: Marcia Cabral, Information Officer, 781-736-3100, *Fax:* 781-736-3107, *E-mail:* cabral@ brandeis.edu.

■ BROWN UNIVERSITY

Graduate School, Division of Biology and Medicine, Department of Neuroscience, Providence, RI 02912

AWARDS Sc M, PhD, MD/PhD.

Faculty: 17 full-time (8 women).
Students: 17 full-time (8 women); includes 1 minority (African American). Average age 25. *94 applicants, 5% accepted.* In 1999, 3 master's, 2 doctorates awarded. Terminal master's awarded for partial completion of doctoral program.
Degree requirements: For master's, thesis required (for some programs), foreign language not required; for doctorate, dissertation, preliminary exam required.
Entrance requirements: For master's, GRE General Test, GRE Subject Test; for doctorate, GRE General Test, GRE Subject Test, TOEFL. *Application deadline:* For fall admission, 1/2 (priority date). Applications are processed on a rolling basis. *Application fee:* $60.
Financial aid: In 1999–00, 3 fellowships, 2 research assistantships, 6 teaching assistantships were awarded; institutionally sponsored loans and traineeships also available. Financial aid application deadline: 1/2.
Faculty research: Neurophysiology, systems neuroscience, membrane biophysics, neuropharmacology, sensory systems.
Dr. Justin Fallon, Director, 401-863-9308.
Find an in-depth description at www.petersons.com/graduate.

■ BRYN MAWR COLLEGE

Graduate School of Arts and Sciences, Department of Biology, Program in Neural and Behavioral Science, Bryn Mawr, PA 19010-2899

AWARDS PhD.

Degree requirements: For doctorate, dissertation required.
Entrance requirements: For doctorate, GRE General Test, GRE Subject Test. *Application deadline:* For fall admission, 6/30; for spring admission, 12/7. *Application fee:* $40.
Expenses: Tuition: Full-time $20,790; part-time $3,530 per course.
Financial aid: Application deadline: 1/2.
Application contact: Graduate School of Arts and Sciences, 610-526-5075.

■ CALIFORNIA INSTITUTE OF TECHNOLOGY

Division of Biology, Program in Neurobiology, Pasadena, CA 91125-0001

AWARDS PhD.

Degree requirements: For doctorate, dissertation, qualifying exam required, foreign language not required.
Entrance requirements: For doctorate, GRE General Test. *Application deadline:* For fall admission, 1/1. *Application fee:* $0.
Expenses: Tuition: Full-time $19,260. Required fees: $24. One-time fee: $100 full-time.
Financial aid: Application deadline: 1/1.
Application contact: Elizabeth Ayala, Graduate Option Coordinator, 626-395-4497, *Fax:* 626-449-0756, *E-mail:* biograd@cco.caltech.edu.

■ CALIFORNIA INSTITUTE OF TECHNOLOGY

Division of Engineering and Applied Science, Option in Computation and Neural Systems, Pasadena, CA 91125-0001

AWARDS MS, PhD.

Faculty: 2 full-time (0 women).
Students: 35 full-time (4 women), 10 international. *76 applicants, 1% accepted.* In 1999, 2 master's, 7 doctorates awarded. Terminal master's awarded for partial completion of doctoral program.
Degree requirements: For master's, foreign language not required; for doctorate, dissertation, qualifying exam required, foreign language not required.
Entrance requirements: For doctorate, GRE General Test. *Application deadline:* For fall admission, 1/15. *Application fee:* $0.
Expenses: Tuition: Full-time $19,260. Required fees: $24. One-time fee: $100 full-time.
Financial aid: Fellowships, research assistantships, teaching assistantships, Federal Work-Study and institutionally sponsored loans available. Financial aid application deadline: 1/15.
Faculty research: Biological and artificial computational devices, modeling of sensory processes and learning, theory of collective computation.
Dr. Christof Koch, Executive Officer, 626-395-6855.

■ CASE WESTERN RESERVE UNIVERSITY

School of Medicine and School of Graduate Studies, Graduate Programs in Medicine, Department of Neurosciences, Cleveland, OH 44106

AWARDS Neurobiology (PhD); neuroscience (PhD).

Faculty: 15 full-time (3 women).
Students: 26 full-time (15 women). *80 applicants, 8% accepted.* In 1999, 8 degrees awarded (100% entered university research/teaching).
Degree requirements: For doctorate, dissertation required. *Average time to degree:* Doctorate–5 years full-time.
Entrance requirements: For doctorate, GRE General Test, GRE Subject Test, TOEFL. *Application deadline:* For fall admission, 3/15 (priority date). Applications are processed on a rolling basis. *Application fee:* $25.
Financial aid: In 1999–00, 34 students received aid; fellowships, research assistantships available. Financial aid application deadline: 4/1.
Faculty research: Neurotropic factors, synapse formation, regeneration, determination of cell fate, cellular neuroscience. *Total annual research expenditures:* $2.2 million.
Dr. Lynn Landmesser, Chair, 216-368-3996, *Fax:* 216-368-4650, *E-mail:* ltl@po.cwru.edu.
Application contact: Narlene Brown, Administrator, 216-368-6253, *Fax:* 216-368-4650, *E-mail:* nrb@po.cwru.edu.

Find an in-depth description at www.petersons.com/graduate.

■ COLLEGE OF STATEN ISLAND OF THE CITY UNIVERSITY OF NEW YORK

Graduate Programs, Center for Developmental Neuroscience and Developmental Disabilities, Program in Neuroscience, Staten Island, NY 10314-6600

AWARDS Biology (PhD).

Faculty: 5 full-time (0 women).
Students: Average age 28.
Degree requirements: For doctorate, one foreign language (computer language can substitute), dissertation, comprehensive exam required. *Average time to degree:* Doctorate–6 years full-time.
Entrance requirements: For doctorate, GRE, TOEFL. *Application deadline:* For fall admission, 4/1; for spring admission, 11/1. Applications are processed on a rolling basis. *Application fee:* $40.
Expenses: Tuition, state resident: full-time $4,350; part-time $185 per credit. Tuition,

nonresident: full-time $7,600; part-time $320 per credit. Required fees: $53; $27 per term.
Financial aid: Fellowships, research assistantships, teaching assistantships, career-related internships or fieldwork available.
Faculty research: Neuronal development, synaptic plasticity.
Dr. Fred Naider, Head, 718-982-3896, *Fax:* 718-982-3944, *E-mail:* naider@postbox.csi.cuny.edu.

■ COLORADO STATE UNIVERSITY

College of Veterinary Medicine and Biomedical Sciences and Graduate School, Graduate Programs in Veterinary Medicine and Biomedical Sciences, Department of Anatomy and Neurobiology, Fort Collins, CO 80523-0015

AWARDS MS, PhD.

Faculty: 19 full-time (7 women).
Students: 42 full-time (25 women), 5 part-time (3 women); includes 6 minority (2 Asian Americans or Pacific Islanders, 3 Hispanic Americans, 1 Native American), 1 international. Average age 28. *47 applicants, 68% accepted.* In 1999, 33 master's, 4 doctorates awarded.
Degree requirements: For master's, thesis required (for some programs), foreign language not required; for doctorate, dissertation required, foreign language not required.
Entrance requirements: For master's and doctorate, GRE General Test, GRE Subject Test, TOEFL. *Application deadline:* For fall admission, 6/1 (priority date). Applications are processed on a rolling basis. *Application fee:* $30. Electronic applications accepted.
Expenses: Tuition, state resident: full-time $2,694; part-time $150 per credit. Tuition, nonresident: full-time $10,460; part-time $581 per credit. Required fees: $32 per semester.
Financial aid: In 1999–00, 6 research assistantships, 3 teaching assistantships were awarded; fellowships, traineeships also available.
Faculty research: Structural biology, integrative neuroscience, developmental neurobiology. *Total annual research expenditures:* $4 million.
F. Edward Dudek, Chair, 970-491-5847.
Application contact: Dr. Robert Handa, Graduate Coordinator, 970-491-7130, *Fax:* 970-491-7907, *E-mail:* dadams@cvmbs.colostate.edu.

■ COLUMBIA UNIVERSITY

College of Physicians and Surgeons and Graduate School of Arts and Sciences, Graduate School of Arts and Sciences at the College of Physicians and Surgeons, Program in Neurobiology and Behavior, New York, NY 10032

AWARDS M Phil, PhD, MD/PhD. Only candidates for the PhD are admitted.

Degree requirements: For master's, foreign language and thesis not required; for doctorate, dissertation required, foreign language not required.
Entrance requirements: For master's and doctorate, GRE General Test, TOEFL.
Expenses: Accepted students receive full tuition, fees, and an annual stipend of $21,000.
Faculty research: Cellular and molecular mechanisms of neural development, neuropathology, neuropharmacology.

Find an in-depth description at www.petersons.com/graduate.

■ CORNELL UNIVERSITY

Graduate School, Graduate Fields of Agriculture and Life Sciences, Field of Neurobiology and Behavior, Ithaca, NY 14853-0001

AWARDS Behavioral biology (PhD), including behavioral ecology, chemical ecology, ethology, neuroethology, sociobiology; neurobiology (PhD), including cellular and molecular neurobiology, neuroanatomy, neurochemistry, neuropharmacology, neurophysiology, sensory physiology.

Faculty: 33 full-time.
Students: 35 full-time (12 women); includes 5 minority (2 African Americans, 1 Asian American or Pacific Islander, 2 Hispanic Americans), 8 international. *65 applicants, 14% accepted.* In 1999, 6 doctorates awarded.
Degree requirements: For doctorate, dissertation, 1 year of teaching experience, seminar presentation required, foreign language not required.
Entrance requirements: For doctorate, GRE General Test, GRE Subject Test (biology), TOEFL. *Application deadline:* For fall admission, 12/1. *Application fee:* $65. Electronic applications accepted.
Expenses: Tuition: Full-time $12,100.
Financial aid: In 1999–00, 33 students received aid, including 20 fellowships with full tuition reimbursements available, 6 research assistantships with full tuition reimbursements available, 7 teaching assistantships with full tuition reimbursements available; institutionally sponsored loans, scholarships, tuition waivers (full and partial), and unspecified assistantships

Cornell University (continued)
also available. Financial aid applicants required to submit FAFSA.

Faculty research: Cellular neurobiology and neuropharmacology, integrative neurobiology, social behavior, chemical ecology.

Application contact: Graduate Field Assistant, 607-254-4340, *E-mail:* tmn3@cornell.edu.

Find an in-depth description at www.petersons.com/graduate.

■ DARTMOUTH COLLEGE

School of Arts and Sciences, Department of Psychological and Brain Sciences, Hanover, NH 03755

AWARDS Cognitive neuroscience (PhD); psychology (PhD).

Faculty: 16 full-time (3 women), 5 part-time/adjunct (1 woman).

Students: 18 full-time (8 women); includes 1 minority (Native American), 1 international. Average age 26. *66 applicants, 9% accepted.* In 1999, 3 doctorates awarded (66% entered university research/teaching).

Degree requirements: For doctorate, dissertation required.

Entrance requirements: For doctorate, GRE General Test, GRE Subject Test. *Application deadline:* For fall admission, 1/15 (priority date). *Application fee:* $40.

Expenses: Tuition: Full-time $24,624. Required fees: $916. One-time fee: $15 full-time. Full-time tuition and fees vary according to program.

Financial aid: In 1999–00, 17 students received aid, including 12 fellowships with full tuition reimbursements available (averaging $15,338 per year), 1 research assistantship with full tuition reimbursement available (averaging $15,338 per year); Federal Work-Study, institutionally sponsored loans, and tuition waivers (full) also available.

Dr. Howard C. Hughes, Chair, 603-646-3181, *Fax:* 603-646-1419, *E-mail:* howard.hughes@dartmouth.edu.

Application contact: Tina Mason, Administrative Assistant, 603-646-3181.

Find an in-depth description at www.petersons.com/graduate.

■ DUKE UNIVERSITY

Graduate School, Department of Biological Anthropology and Anatomy, Durham, NC 27708-0586

AWARDS Cellular and molecular biology (PhD); gross anatomy and physical anthropology (PhD), including comparative morphology of human and non-human primates, primate social behavior, vertebrate paleontology; neuroanatomy (PhD).

Faculty: 14 full-time, 1 part-time/adjunct.

Students: 20 full-time (12 women); includes 3 minority (2 African Americans, 1 Hispanic American), 1 international. *62 applicants, 6% accepted.* In 1999, 2 doctorates awarded.

Degree requirements: For doctorate, dissertation required.

Entrance requirements: For doctorate, GRE General Test. *Application deadline:* For fall admission, 12/31. *Application fee:* $75.

Expenses: Tuition: Full-time $21,406; part-time $760 per unit. Required fees: $3,136; $3,136 per year. One-time fee: $30. Tuition and fees vary according to program.

Financial aid: Fellowships, teaching assistantships, Federal Work-Study available. Financial aid application deadline: 12/31.

Kathleen Smith, Director of Graduate Studies, 919-684-4124, *Fax:* 919-684-8034, *E-mail:* rachel_hougom@baa.mc.duke.edu.

■ DUKE UNIVERSITY

Graduate School, Department of Neurobiology, Durham, NC 27708-0586

AWARDS PhD.

Faculty: 44 full-time.

Students: 34 full-time (19 women); includes 1 minority (African American), 5 international. *88 applicants, 19% accepted.* In 1999, 1 degree awarded.

Degree requirements: For doctorate, dissertation required.

Entrance requirements: For doctorate, GRE General Test, GRE Subject Test (biology recommended). *Application deadline:* For fall admission, 12/31. *Application fee:* $75.

Expenses: Tuition: Full-time $21,406; part-time $760 per unit. Required fees: $3,136; $3,136 per year. One-time fee: $30. Tuition and fees vary according to program.

Financial aid: Fellowships, research assistantships, teaching assistantships, Federal Work-Study available. Financial aid application deadline: 12/31.

Dona Chikaraishi, Director, 919-681-4243, *Fax:* 919-684-4431, *E-mail:* susand@neuro.duke.edu.

Find an in-depth description at www.petersons.com/graduate.

■ EMORY UNIVERSITY

Graduate School of Arts and Sciences, Division of Biological and Biomedical Sciences, Program in Neuroscience, Atlanta, GA 30322-1100

AWARDS PhD.

Faculty: 79 full-time (5 women).

Students: 54 full-time (30 women), 8 international. In 1999, 5 degrees awarded.

Degree requirements: For doctorate, dissertation required, foreign language not required.

Entrance requirements: For doctorate, GRE General Test, TOEFL, minimum GPA of 3.0 in science course work. *Application deadline:* For fall admission, 1/20 (priority date). *Application fee:* $45.

Expenses: Tuition: Full-time $22,770. Tuition and fees vary according to program.

Financial aid: In 1999–00, fellowships with full tuition reimbursements (averaging $18,000 per year).

Faculty research: Cell and molecular biology, development, behavior, neurodegenerative disease.

Dr. Ronald Calabrese, Director, 404-727-0319, *Fax:* 404-727-2880, *E-mail:* rcalabre@biology.emory.edu.

Application contact: 404-727-2547, *Fax:* 404-727-3322, *E-mail:* gdbbs@gsas.emory.edu.

Find an in-depth description at www.petersons.com/graduate.

■ FINCH UNIVERSITY OF HEALTH SCIENCES/THE CHICAGO MEDICAL SCHOOL

School of Graduate and Postdoctoral Studies, Department of Neuroscience, North Chicago, IL 60064-3095

AWARDS PhD, MD/PhD.

Faculty: 5 full-time.

Students: 9 full-time (3 women); includes 1 minority (Asian American or Pacific Islander), 2 international. In 1999, 3 doctorates awarded.

Degree requirements: For doctorate, dissertation, comprehensive exam, original research project required.

Entrance requirements: For doctorate, GRE General Test, TOEFL, TWE. *Application deadline:* For fall admission, 6/1 (priority date). Applications are processed on a rolling basis. *Application fee:* $25.

Expenses: Tuition: Full-time $14,054; part-time $391 per credit hour. Tuition and fees vary according to program.

Financial aid: In 1999–00, fellowships (averaging $15,500 per year); grants and tuition waivers (full) also available. Financial aid application deadline: 6/9; financial aid applicants required to submit FAFSA.

Dr. John Sladek, Chairman, 847-578-3429.

Application contact: Dana Frederick, Admissions Officer, 847-578-3209.

Find an in-depth description at www.petersons.com/graduate.

■ FLORIDA STATE UNIVERSITY

Graduate Studies, College of Arts and Sciences, Department of Biological Science and Department of Psychology, Program in Neuroscience, Tallahassee, FL 32306

AWARDS PhD.

Faculty: 21 full-time (5 women).
Students: 31 full-time (16 women).
Degree requirements: For doctorate, variable foreign language requirement (computer language can substitute for one), dissertation, teaching experience required.
Entrance requirements: For doctorate, GRE General Test, TOEFL. *Application deadline:* For fall admission, 1/15; for spring admission, 10/15. *Application fee:* $20.
Expenses: Tuition, state resident: full-time $3,504; part-time $146 per credit hour. Tuition, nonresident: full-time $12,162; part-time $507 per credit hour. Tuition and fees vary according to program.
Financial aid: Fellowships with full tuition reimbursements, research assistantships with full tuition reimbursements, teaching assistantships with full tuition reimbursements available. Financial aid application deadline: 1/15; financial aid applicants required to submit FAFSA.
Faculty research: Behavioral ecology, neurophysiology, sensory physiology, neuroendocrinology, neurotransmission.
Dr. Michael Meredith, Co-Director, 850-644-3076, *Fax:* 850-644-0989, *E-mail:* brooks@neuro.fsu.edu.
Application contact: Judy Bowers, Coordinator, Graduate Affairs, 850-644-3023, *Fax:* 850-644-9829, *E-mail:* bowers@bio.fsu.edu.

■ FLORIDA STATE UNIVERSITY

Graduate Studies, College of Arts and Sciences, Department of Psychology, Interdisciplinary Program in Psychology and Neuroscience, Tallahassee, FL 32306

AWARDS Neuroscience (PhD).

Faculty: 12 full-time (3 women).
Students: 18 full-time (7 women), 1 part-time; includes 2 minority (both African Americans). Average age 23. *47 applicants, 9% accepted.*
Degree requirements: For doctorate, dissertation, preliminary exam required, foreign language not required.
Entrance requirements: For doctorate, GRE General Test, minimum GPA of 3.0. *Application deadline:* For fall admission, 1/15. *Application fee:* $20. Electronic applications accepted.
Expenses: Tuition, state resident: full-time $3,504; part-time $146 per credit hour. Tuition, nonresident: full-time $12,162;

part-time $507 per credit hour. Tuition and fees vary according to program.
Financial aid: In 1999–00, 4 fellowships with full tuition reimbursements (averaging $13,500 per year), 6 research assistantships with full tuition reimbursements (averaging $13,500 per year), 8 teaching assistantships with full tuition reimbursements (averaging $13,000 per year) were awarded; Federal Work-Study, institutionally sponsored loans, and traineeships also available. Financial aid applicants required to submit FAFSA.
Faculty research: Sensory processes, biophysiology and electrophysiology, neuroanatomy, circadian rhythms, genetics. *Total annual research expenditures:* $1.4 million.
Dr. Michael Rasholte, Director, 850-644-3511.
Application contact: Cherie P. Dilworth, Graduate Program Assistant, 850-644-2499, *Fax:* 850-644-7739, *E-mail:* dilworth@psy.fsu.edu.

■ GEORGETOWN UNIVERSITY

Graduate School of Arts and Sciences, Programs in Biomedical Sciences, Program in Neuroscience, Washington, DC 20057

AWARDS PhD, MD/PhD.

Degree requirements: For doctorate, dissertation required.
Entrance requirements: For doctorate, GRE General Test, TOEFL.

Find an in-depth description at www.petersons.com/graduate.

■ THE GEORGE WASHINGTON UNIVERSITY

Columbian School of Arts and Sciences, Institute for Biomedical Sciences, Program in Neuroscience, Washington, DC 20052

AWARDS PhD.

Faculty: 8 part-time/adjunct (2 women).
Students: Average age 32. In 1999, 1 degree awarded.
Degree requirements: For doctorate, dissertation, general exam required.
Entrance requirements: For doctorate, GRE General Test, interview, minimum GPA of 3.0. *Application fee:* $55.
Expenses: Tuition: Full-time $16,836; part-time $702 per credit hour. Required fees: $828; $35 per credit hour. Tuition and fees vary according to campus/location and program.
Financial aid: In 1999–00, 2 students received aid; fellowships, Federal Work-Study and institutionally sponsored loans available. Financial aid application deadline: 2/1.

Faculty research: Cerebral ischemia, neural transplantation, molecular mechanisms of action, neurotransmitter systems, psychobiology of learning and memory.
Dr. Vincent A. Chiappinelli, Director, 202-994-3541.
Application contact: 202-994-2179.

Find an in-depth description at www.petersons.com/graduate.

■ GEORGIA STATE UNIVERSITY

College of Arts and Sciences, Department of Biology, Program in Neurobiology, Atlanta, GA 30303-3083

AWARDS MS, PhD.

Degree requirements: For master's, one foreign language (computer language can substitute), thesis or alternative, exam required; for doctorate, dissertation required.
Entrance requirements: For master's and doctorate, GRE General Test, TOEFL, minimum GPA of 3.0. *Application deadline:* For fall admission, 7/18; for spring admission, 2/13. Applications are processed on a rolling basis. *Application fee:* $25.
Expenses: Tuition, state resident: full-time $2,896; part-time $121 per credit hour. Tuition, nonresident: full-time $11,584; part-time $483 per credit hour. Required fees: $228. Full-time tuition and fees vary according to course load and program.
Financial aid: Application deadline: 2/6.
Application contact: Latesha Morrison, Graduate Administrative Coordinator, 404-651-2759, *Fax:* 404-651-2509, *E-mail:* biolxm@langate.gsu.edu.

Find an in-depth description at www.petersons.com/graduate.

■ GRADUATE SCHOOL AND UNIVERSITY CENTER OF THE CITY UNIVERSITY OF NEW YORK

Graduate Studies, Program in Psychology, New York, NY 10016-4039

AWARDS Basic applied neurocognition (PhD); biopsychology (PhD); clinical psychology (PhD); developmental psychology (PhD); environmental psychology (PhD); experimental psychology (PhD); industrial psychology (PhD); learning processes (PhD); neuropsychology (PhD); psychology (PhD); social personality (PhD).

Degree requirements: For doctorate, dissertation required.
Entrance requirements: For doctorate, GRE General Test.
Expenses: Tuition, state resident: full-time $4,350; part-time $245 per credit hour. Tuition, nonresident: full-time $7,600; part-time $425 per credit hour.

■ HARVARD UNIVERSITY

Graduate School of Arts and Sciences, Program in Neuroscience, Boston, MA 02115

AWARDS Neurobiology (PhD).

Degree requirements: For doctorate, dissertation, qualifying exam required, foreign language not required.

Entrance requirements: For doctorate, GRE General Test, GRE Subject Test, TOEFL. *Application deadline:* For fall admission, 12/15. *Application fee:* $60.

Expenses: Tuition: Full-time $22,054. Required fees: $711. Tuition and fees vary according to program.

Financial aid: Fellowships, research assistantships, teaching assistantships, grants, institutionally sponsored loans, tuition waivers (full), and stipends available. Financial aid application deadline: 1/1.

Faculty research: Relationship between diseases of the nervous system and basic science.

Dr. Jonathan Cohen, Chair, 617-432-1728.
Application contact: Leah Simons, Manager of Student Affairs, 617-432-0162.

Find an in-depth description at www.petersons.com/graduate.

■ INDIANA UNIVERSITY BLOOMINGTON

Graduate School, College of Arts and Sciences, Program in Neural Sciences, Bloomington, IN 47405

AWARDS PhD. Offered through the University Graduate School.

Students: 7 full-time (2 women), 7 part-time (2 women), 1 international. In 1999, 3 degrees awarded.

Degree requirements: For doctorate, dissertation, oral exam required. *Average time to degree:* Doctorate–5 years full-time.

Entrance requirements: For doctorate, GRE General Test, TOEFL. *Application deadline:* For fall admission, 1/15 (priority date); for spring admission, 9/1 (priority date). Applications are processed on a rolling basis. *Application fee:* $45. Electronic applications accepted.

Expenses: Tuition, state resident: full-time $3,853; part-time $161 per credit hour. Tuition, nonresident: full-time $11,226; part-time $468 per credit hour. Required fees: $360 per year. Tuition and fees vary according to course load and program.

Financial aid: In 1999–00, fellowships with tuition reimbursements (averaging $18,000 per year), teaching assistantships with full tuition reimbursements (averaging $13,000 per year) were awarded; institutionally sponsored loans also available. Financial aid application deadline: 1/15.

Faculty research: Synaptic transmitter systems, neurophysiology, neuropharmacology, neurochemistry, somatosensory and sensorimotor functions.
Dr. George V. Rebec, Director, 812-855-7756, *Fax:* 812-855-4520, *E-mail:* rebec@indiana.edu.

Application contact: Faye Caylor, Administrative Assistant, 812-855-7756, *Fax:* 812-855-4520, *E-mail:* fcaylor@indiana.edu.

■ INDIANA UNIVERSITY–PURDUE UNIVERSITY INDIANAPOLIS

School of Medicine, Graduate Programs in Medicine, Program in Medical Neurobiology, Indianapolis, IN 46202-2896

AWARDS MS, PhD, MD/MS, MD/PhD.

Students: 7 full-time (4 women), 10 part-time (6 women). Average age 30. *26 applicants, 38% accepted.* In 1999, 1 master's, 2 doctorates awarded (50% entered university research/teaching, 50% found other work related to degree). Terminal master's awarded for partial completion of doctoral program.

Degree requirements: For master's and doctorate, thesis/dissertation required, foreign language not required. *Average time to degree:* Doctorate–6 years full-time.

Entrance requirements: For master's and doctorate, GRE General Test, previous course work in calculus, organic chemistry, and physics. *Application deadline:* For fall admission, 2/1 (priority date); for spring admission, 9/15 (priority date). Applications are processed on a rolling basis. *Application fee:* $35 ($55 for international students).

Expenses: Tuition, state resident: full-time $13,245; part-time $158 per credit hour. Tuition, nonresident: full-time $30,330; part-time $455 per credit hour. Required fees: $121 per year. Tuition and fees vary according to course load and degree level.

Financial aid: In 1999–00, 4 fellowships with full tuition reimbursements (averaging $14,000 per year), research assistantships with full tuition reimbursements (averaging $14,000 per year), 2 teaching assistantships with full tuition reimbursements (averaging $14,000 per year) were awarded; career-related internships or fieldwork, Federal Work-Study, grants, institutionally sponsored loans, scholarships, traineeships, tuition waivers (partial), and unspecified assistantships also available. Financial aid application deadline: 2/1.

Faculty research: Neurobiology from molecular level to complex behavioral interactions. *Total annual research expenditures:* $5.3 million.

Dr. J. N. Hingtgen, Director, 317-274-7397, *Fax:* 317-274-1365, *E-mail:* jhingtge@iupui.edu.

■ IOWA STATE UNIVERSITY OF SCIENCE AND TECHNOLOGY

Graduate College, Interdisciplinary Programs, Program in Neuroscience, Ames, IA 50011

AWARDS MS, PhD.

Students: 8 full-time (5 women), 1 (woman) part-time; includes 1 minority (Asian American or Pacific Islander), 8 international. *29 applicants, 21% accepted.* In 1999, 2 doctorates awarded. Terminal master's awarded for partial completion of doctoral program.

Degree requirements: For master's and doctorate, thesis/dissertation required. *Average time to degree:* Master's–3 years full-time; doctorate–4 years full-time.

Entrance requirements: For master's and doctorate, GRE General Test, TOEFL. *Application deadline:* For fall admission, 2/1 (priority date). *Application fee:* $20 ($50 for international students).

Expenses: Tuition, state resident: full-time $3,308. Tuition, nonresident: full-time $9,744. Part-time tuition and fees vary according to course load, campus/location and program.

Financial aid: In 1999–00, 5 research assistantships with partial tuition reimbursements (averaging $11,448 per year), 5 teaching assistantships with partial tuition reimbursements (averaging $11,394 per year) were awarded; scholarships also available.

Faculty research: Behavioral pharmacology and immunology, developmental neurobiology, neuroendocrinology, neuroregulatory mechanisms at the cellular level, signal transduction in neurons.
Dr. Donald S. Sakaguchi, Supervisory Committee Chair, 515-294-7252, *E-mail:* idgp@iastate.edu.

Application contact: 515-294-7252, *Fax:* 515-294-6790, *E-mail:* idgp@iastate.edu.

■ JOAN AND SANFORD I. WEILL MEDICAL COLLEGE AND GRADUATE SCHOOL OF MEDICAL SCIENCES OF CORNELL UNIVERSITY

Graduate School of Medical Sciences, Graduate Program in Neuroscience, New York, NY 10021

AWARDS PhD, MD/PhD.

Faculty: 32 full-time (7 women).
Students: 38 full-time (19 women); includes 12 minority (1 African American, 8 Asian Americans or Pacific Islanders, 3

Hispanic Americans), 5 international. *78 applicants, 22% accepted.* In 2000, 2 degrees awarded.

Degree requirements: For doctorate, dissertation, final exam required.

Entrance requirements: For doctorate, GRE General Test, GRE Subject Test, MCAT (MD/PhD), undergraduate training in biology, organic chemistry, physics, and mathematics. *Application deadline:* For fall admission, 1/15. *Application fee:* $50.

Expenses: All students in good standing receive an annual stipend of $22,880.

Financial aid: Fellowships, stipends available.

John Wagner, Co-Director, 212-746-6586.

■ JOHNS HOPKINS UNIVERSITY

School of Medicine, Graduate Programs in Medicine, Department of Neuroscience, Baltimore, MD 21218-2699

AWARDS PhD.

Faculty: 67 full-time (7 women), 1 part-time/adjunct (0 women).

Students: 56 full-time (22 women); includes 8 minority (2 African Americans, 6 Asian Americans or Pacific Islanders), 22 international. Average age 28. *140 applicants, 15% accepted.* In 1999, 4 doctorates awarded.

Degree requirements: For doctorate, oral comprehensive exam, thesis defense required. *Average time to degree:* Doctorate–5 years full-time.

Entrance requirements: For doctorate, GRE General Test, GRE Subject Test, bachelor's degree in science or mathematics. *Application deadline:* For fall admission, 1/10. *Application fee:* $50.

Expenses: Tuition: Full-time $23,660.

Financial aid: In 1999–00, 56 students received aid, including 53 fellowships (averaging $17,642 per year). Financial aid application deadline: 1/1.

Faculty research: Neurophysiology, neurochemistry, neuroanatomy, pharmacology, development.

Solomon H. Snyder, Chairman, 410-955-3024, *Fax:* 410-955-3623.

Application contact: Kenneth Johnson, Director of Graduate Studies, 410-955-7947, *Fax:* 410-955-3623, *E-mail:* johnson@bard.mb.jhu.edu.

Find an in-depth description at www.petersons.com/graduate.

■ KENT STATE UNIVERSITY

School of Biomedical Sciences, Program in Neuroscience, Kent, OH 44242-0001

AWARDS MS, PhD. Offered in cooperation with Northeastern Ohio Universities College of Medicine. Terminal master's awarded for partial completion of doctoral program.

Degree requirements: For master's and doctorate, thesis/dissertation required, foreign language not required.

Entrance requirements: For master's and doctorate, GRE General Test.

Expenses: Tuition, state resident: full-time $5,334; part-time $243 per hour. Tuition, nonresident: full-time $10,238; part-time $466 per hour.

Faculty research: Plasticity of the nervous system, learning and memory processes–neural correlates, neuroendocrinology of cyclic behavior, synaptic neurochemistry.

■ LEHIGH UNIVERSITY

College of Arts and Sciences, Department of Biological Sciences, Bethlehem, PA 18015-3094

AWARDS Behavioral and evolutionary bioscience (PhD); behavioral neuroscience (PhD); biochemistry (PhD); biology (PhD); molecular biology (PhD). Part-time programs available. Postbaccalaureate distance learning degree programs offered (no on-campus study).

Students: 31 full-time (22 women), 77 part-time (47 women); includes 11 minority (3 African Americans, 5 Asian Americans or Pacific Islanders, 3 Hispanic Americans), 8 international. *142 applicants, 22% accepted.* In 1999, 5 doctorates awarded.

Degree requirements: For doctorate, dissertation, comprehensive exam required, foreign language not required. *Average time to degree:* Doctorate–6.3 years full-time.

Entrance requirements: For doctorate, GRE General Test, GRE Subject Test, TOEFL. *Application deadline:* For fall admission, 7/15; for spring admission, 12/1. Applications are processed on a rolling basis. *Application fee:* $40. Electronic applications accepted.

Expenses: Tuition: Part-time $860 per credit. Required fees: $6 per term. Tuition and fees vary according to program.

Financial aid: In 1999–00, 30 students received aid, including 4 fellowships, 6 research assistantships, 15 teaching assistantships; career-related internships or fieldwork, institutionally sponsored loans, and tuition waivers (full and partial) also available. Financial aid application deadline: 1/15.

Faculty research: Gene expression, cell biology, virology, bacteriology, developmental biology.

Dr. Neal G. Simon, Chairperson, 610-758-3680, *Fax:* 610-758-4004.

Application contact: Dr. Jennifer J. Swann, Graduate Coordinator, 610-758-5884, *Fax:* 610-758-4004, *E-mail:* jms5@lehigh.edu.

■ LOUISIANA STATE UNIVERSITY HEALTH SCIENCES CENTER

School of Graduate Studies in New Orleans, Department of Cell Biology and Anatomy, New Orleans, LA 70112-2223

AWARDS Cell biology and anatomy (MS, PhD), including cell biology, developmental biology, neurobiology and anatomy.

Faculty: 18 full-time (3 women), 1 part-time/adjunct (0 women).

Students: 9 full-time (2 women); includes 1 minority (African American), 4 international. Average age 26. *6 applicants, 50% accepted.* In 1999, 3 doctorates awarded (34% entered university research/teaching, 33% found other work related to degree, 33% continued full-time study).

Degree requirements: For master's and doctorate, thesis/dissertation required, foreign language not required. *Average time to degree:* Doctorate–4.5 years full-time.

Entrance requirements: For master's and doctorate, GRE General Test, GRE Subject Test, TOEFL, minimum undergraduate GPA of 3.0. *Application deadline:* For fall admission, 3/1 (priority date); for spring admission, 10/15. Applications are processed on a rolling basis. *Application fee:* $30.

Expenses: Tuition, state resident: full-time $2,878; part-time $126 per hour. Tuition, nonresident: full-time $6,003; part-time $265 per hour. Required fees: $2,272. Tuition and fees vary according to course load, degree level and program.

Financial aid: In 1999–00, 1 fellowship with full tuition reimbursement (averaging $15,000 per year), teaching assistantships with full tuition reimbursements (averaging $14,000 per year) were awarded; research assistantships, career-related internships or fieldwork, Federal Work-Study, grants, institutionally sponsored loans, tuition waivers (full), and unspecified assistantships also available. Aid available to part-time students. Financial aid application deadline: 4/1.

Faculty research: Visual system organization, neural development, plasticity of sensory systems, information processing through the nervous system, visuomotor integration. *Total annual research expenditures:* $772,550.

Dr. R. Ranney Mize, Head, 504-599-1458, *Fax:* 504-568-4392, *E-mail:* rmize@lsumc.edu.

Application contact: Dr. Mark C. Alliegro, Director of Graduate Studies, 504-568-7618, *Fax:* 504-568-4392, *E-mail:* mallie@lsumc.edu.

Find an in-depth description at www.petersons.com/graduate.

■ LOUISIANA STATE UNIVERSITY HEALTH SCIENCES CENTER

School of Graduate Studies in New Orleans, Interdisciplinary Neuroscience PhD Training Program, New Orleans, LA 70112-2223

AWARDS PhD, MD/PhD.

Faculty: 48 full-time (11 women).
Students: 15 full-time (3 women); includes 1 minority (Asian American or Pacific Islander), 2 international. Average age 26. *25 applicants, 24% accepted.* In 1999, 1 degree awarded (100% entered university research/teaching).
Degree requirements: For doctorate, dissertation required, foreign language not required. *Average time to degree:* Doctorate–5 years full-time.
Entrance requirements: For doctorate, GRE General Test, GRE Subject Test, TOEFL, previous course work in chemistry, mathematics, physics, and computer science. *Application deadline:* For fall admission, 2/15 (priority date). Applications are processed on a rolling basis. *Application fee:* $30.
Expenses: Tuition, state resident: full-time $2,878; part-time $126 per hour. Tuition, nonresident: full-time $6,003; part-time $265 per hour. Required fees: $2,272. Tuition and fees vary according to course load, degree level and program.
Financial aid: In 1999–00, 13 students received aid; fellowships, tuition waivers (full) available. Financial aid application deadline: 8/1.
Faculty research: Visual system, second messengers, drugs and behavior, signal transduction, plasticity and development.
Dr. Nicholas G. Bazan, Co-Director, 504-599-0909 Ext. 320, *Fax:* 504-568-5801.

Find an in-depth description at www.petersons.com/graduate.

■ LOYOLA UNIVERSITY CHICAGO

Graduate School, Department of Cell Biology, Neurobiology and Anatomy, Maywood, IL 60153

AWARDS MS, PhD, MD/PhD. Part-time programs available.

Faculty: 13 full-time, 13 part-time/adjunct.
Students: 26 full-time (12 women), 3 part-time (all women); includes 4 minority (all Asian Americans or Pacific Islanders), 1 international. Average age 26. *42 applicants, 14% accepted.* In 1999, 1 master's awarded (100% continued full-time study); 3 doctorates awarded (100% entered university research/teaching).
Degree requirements: For master's, thesis or alternative, comprehensive exams required, foreign language not required; for doctorate, dissertation, comprehensive exams required, foreign language not required. *Average time to degree:* Master's–2 years full-time; doctorate–5 years full-time.
Entrance requirements: For master's and doctorate, GRE General Test, GRE Subject Test (biology), minimum GPA of 3.0. *Application deadline:* For fall admission, 5/1 (priority date). Applications are processed on a rolling basis. *Application fee:* $35.
Expenses: Tuition: Part-time $500 per credit hour. Required fees: $42 per term.
Financial aid: In 1999–00, 8 fellowships with full tuition reimbursements (averaging $17,500 per year), 8 research assistantships with full tuition reimbursements (averaging $17,500 per year) were awarded; Federal Work-Study and institutionally sponsored loans also available. Aid available to part-time students. Financial aid application deadline: 5/1; financial aid applicants required to submit FAFSA.
Faculty research: Brain steroids, immunology, neuroregeneration, cytokines. *Total annual research expenditures:* $1.4 million.
Dr. John Clancy, Chair, 708-216-3352.
Application contact: Thackery S. Gray, Graduate Program Director, 708-216-3352.

Find an in-depth description at www.petersons.com/graduate.

■ LOYOLA UNIVERSITY CHICAGO

Graduate School, Department of Molecular and Cellular Biochemistry, Chicago, IL 60611-2196

AWARDS Biochemistry (MS, PhD); molecular biology (PhD); neurochemistry (PhD).

Faculty: 12 full-time (2 women), 1 (woman) part-time/adjunct.
Students: 16 full-time (9 women); includes 3 minority (1 African American, 2 Asian Americans or Pacific Islanders), 3 international. Average age 29. *40 applicants, 33% accepted.* In 1999, 1 master's awarded (100% found work related to degree); 2 doctorates awarded (100% entered university research/teaching).
Degree requirements: For master's, oral and written reports required, foreign language and thesis not required; for doctorate, dissertation, oral and written comprehensive exams required, foreign language not required. *Average time to degree:* Master's–2 years full-time; doctorate–4.5 years full-time.
Entrance requirements: For master's and doctorate, GRE General Test. *Application deadline:* For fall admission, 2/15 (priority date). Applications are processed on a rolling basis. *Application fee:* $35. Electronic applications accepted.

Expenses: Tuition: Part-time $500 per credit hour. Required fees: $42 per term.
Financial aid: In 1999–00, 11 students received aid, including 10 fellowships with full tuition reimbursements available, 4 research assistantships with full tuition reimbursements available; Federal Work-Study, institutionally sponsored loans, and scholarships also available. Financial aid application deadline: 3/15.
Faculty research: Molecular oncology; molecular neurochemical mechanisms of brain development and alcohol addiction; biochemistry of RNA and protein synthesis and intracellular protein degradation; developmentally regulated genes; cell membranes, neurotransmitters, and cell-cell interactions.
Dr. Richard M Schultz, Chairman, 708-216-3360, *Fax:* 708-216-8523.
Application contact: Dr. Michael A. Collins, Admissions Committee, 708-216-3361, *Fax:* 708-216-8523, *E-mail:* mcollin@luc.edu.

Find an in-depth description at www.petersons.com/graduate.

■ LOYOLA UNIVERSITY CHICAGO

Graduate School, Program in Neuroscience, Maywood, IL 60153

AWARDS MS, PhD, MD/PhD.

Degree requirements: For doctorate, dissertation, written comprehensive exams required, foreign language not required.
Entrance requirements: For doctorate, GRE General Test, TOEFL.
Expenses: Tuition: Part-time $500 per credit hour. Required fees: $42 per term.
Faculty research: Alzheimer's and Parkinson's disease, drugs of abuse, neuroendocrinology, neuroimmunology, brain cancer.

Find an in-depth description at www.petersons.com/graduate.

■ MAHARISHI UNIVERSITY OF MANAGEMENT

Graduate Studies, Program in the Neuroscience of Human Consciousness, Fairfield, IA 52557

AWARDS MS, PhD. Program admits applicants every other year. Terminal master's awarded for partial completion of doctoral program.

Degree requirements: For master's, foreign language and thesis not required; for doctorate, dissertation required, foreign language not required.
Entrance requirements: For master's, minimum GPA of 3.0; for doctorate, GRE General Test, bachelor's degree in biology or related quantitative science, minimum GPA of 3.0.

Faculty research: Aging and immortality, gene structure and gene regulation, developmental neurobiology, EEG coherence and information processing, evoked potentials and brain development.

■ MARQUETTE UNIVERSITY

Graduate School, College of Arts and Sciences, Department of Biology, Milwaukee, WI 53201-1881

AWARDS Cell biology (MS, PhD); developmental biology (MS, PhD); ecology (MS, PhD); endocrinology (MS, PhD); evolutionary biology (MS, PhD); genetics (MS, PhD); microbiology (MS, PhD); molecular biology (MS, PhD); muscle and exercise physiology (MS, PhD); neurobiology (MS, PhD); reproductive physiology (MS, PhD).

Faculty: 16 full-time (4 women), 2 part-time/adjunct (0 women).
Students: 34 full-time (20 women), 3 part-time; includes 3 minority (all Asian Americans or Pacific Islanders), 2 international. Average age 31. *42 applicants, 29% accepted.* In 1999, 1 master's, 4 doctorates awarded. Terminal master's awarded for partial completion of doctoral program.
Degree requirements: For master's, thesis, 1 year of teaching experience or equivalent, comprehensive exam required, foreign language not required; for doctorate, dissertation, 1 year of teaching experience or equivalent, qualifying exam required, foreign language not required.
Entrance requirements: For master's and doctorate, GRE General Test, GRE Subject Test, TOEFL. *Application fee:* $40.
Expenses: Tuition: Part-time $510 per credit hour. Tuition and fees vary according to program.
Financial aid: In 1999–00, 4 fellowships, 22 teaching assistantships were awarded; research assistantships, Federal Work-Study, institutionally sponsored loans, scholarships, and tuition waivers (full and partial) also available. Aid available to part-time students. Financial aid application deadline: 2/15.
Faculty research: Microbial and invertebrate ecology, evolution of gene function, DNA methylation, DNA arrangement. *Total annual research expenditures:* $1.5 million.
Dr. Brian Unsworth, Chairman, 414-288-7355, *Fax:* 414-288-7357.
Application contact: Barbara DeNoyer, Graduate Studies Coordinator, 414-288-7355, *Fax:* 414-288-7357.

Find an in-depth description at www.petersons.com/graduate.

■ MASSACHUSETTS INSTITUTE OF TECHNOLOGY

School of Science, Department of Biology, Program in Neurobiology, Cambridge, MA 02139-4307

AWARDS PhD.

Degree requirements: For doctorate, dissertation, general exam required, foreign language not required.
Entrance requirements: For doctorate, GRE General Test. *Application deadline:* For fall admission, 1/1. *Application fee:* $55.
Expenses: Tuition: Full-time $25,000. Full-time tuition and fees vary according to degree level, program and student level.
Financial aid: Application deadline: 1/1.
Application contact: Dr. Janice D. Chang, Educational Administrator, 617-253-3717, *Fax:* 617-258-9329, *E-mail:* gradbio@mit.edu.

■ MASSACHUSETTS INSTITUTE OF TECHNOLOGY

School of Science, Department of Brain and Cognitive Sciences, Cambridge, MA 02139-4307

AWARDS Cellular/molecular neuroscience (PhD); cognitive neuroscience (PhD); cognitive science (PhD); computational neuroscience (PhD); systems neuroscience (PhD).

Faculty: 27 full-time (7 women).
Students: 45 full-time (13 women); includes 1 minority (Asian American or Pacific Islander), 19 international. Average age 27. *210 applicants, 5% accepted.* In 1999, 8 degrees awarded (100% entered university research/teaching).
Degree requirements: For doctorate, dissertation required, foreign language not required. *Average time to degree:* Doctorate–5.13 years full-time.
Entrance requirements: For doctorate, GRE General Test. *Application deadline:* For fall admission, 1/1. *Application fee:* $55.
Expenses: Tuition: Full-time $25,000. Full-time tuition and fees vary according to degree level, program and student level.
Financial aid: In 1999–00, 32 fellowships, 8 research assistantships, 8 teaching assistantships were awarded; grants, scholarships, and traineeships also available. Financial aid application deadline: 1/1.
Mriganka Sur, Head, 617-253-8784.
Application contact: Denise Heintze, Graduate Administrator, 617-253-5742, *Fax:* 617-253-9767, *E-mail:* bcsadmiss@wccf.mit.edu.

Find an in-depth description at www.petersons.com/graduate.

■ MAYO GRADUATE SCHOOL

Graduate Programs in Biomedical Sciences, Program in Molecular Neuroscience, Rochester, MN 55905

AWARDS PhD.

Faculty: 58 full-time (6 women).
Students: 28 full-time (15 women); includes 5 minority (1 African American, 2 Asian Americans or Pacific Islanders, 2 Hispanic Americans), 5 international. In 1999, 1 degree awarded.
Degree requirements: For doctorate, oral defense of dissertation, qualifying oral and written exam required.
Entrance requirements: For doctorate, GRE, TOEFL, 2 years of chemistry; 1 year of biology, calculus, and physics. *Application deadline:* For fall admission, 12/31 (priority date). Applications are processed on a rolling basis. *Application fee:* $0.
Expenses: Tuition: Full-time $17,900.
Financial aid: In 1999–00, 26 students received aid, including 26 fellowships with full tuition reimbursements available; tuition waivers (full) also available.
Faculty research: Cholinergic receptor/Alzheimer's; molecular biology, channels, receptors, and mental disease; neuronal cytoskeleton; growth factors; gene regulation.
Education Coordinator, 507-284-4717, *Fax:* 507-284-0999.
Application contact: Sherry Kallies, Information Contact, 507-266-0122, *Fax:* 507-284-0999, *E-mail:* phd.training@mayo.edu.

Find an in-depth description at www.petersons.com/graduate.

■ MCP HAHNEMANN UNIVERSITY

School of Medicine, Biomedical Graduate Programs, Department of Anatomy and Neurobiology, Philadelphia, PA 19102-1192

AWARDS MS, PhD, MD/PhD. Terminal master's awarded for partial completion of doctoral program.

Degree requirements: For master's, thesis, comprehensive exam required, foreign language not required; for doctorate, one foreign language (computer language can substitute), dissertation, qualifying exam required.
Entrance requirements: For master's, GRE General Test, TOEFL, minimum GPA of 2.75; for doctorate, GRE General Test, TOEFL, minimum GPA of 3.0.
Faculty research: Cell biology, anatomy of brain tumors, membrane excitability.

■ MCP HAHNEMANN UNIVERSITY

School of Medicine, Biomedical Graduate Programs, Program in Neuroscience, Philadelphia, PA 19102-1192

AWARDS PhD.

Degree requirements: For doctorate, dissertation, qualifying exam required, foreign language not required.
Entrance requirements: For doctorate, GRE General Test, TOEFL, minimum GPA of 2.75.
Faculty research: Central monoamine systems, drugs of abuse, anatomy/physiology of sensory systems, neurodegenerative disorders and recovery of function, neuromodulation and synaptic plasticity.

Find an in-depth description at www.petersons.com/graduate.

■ MEDICAL COLLEGE OF OHIO

Graduate School, Program in Cellular and Molecular Neurobiology, Toledo, OH 43614-5805

AWARDS MS, PhD. Part-time programs available.

Students: 15 full-time (6 women), 4 part-time (1 woman); includes 13 minority (all Asian Americans or Pacific Islanders). Average age 30. In 1999, 1 master's awarded. Terminal master's awarded for partial completion of doctoral program.
Degree requirements: For master's and doctorate, thesis/dissertation, qualifying exam required, foreign language not required. *Average time to degree:* Master's–2 years part-time; doctorate–5 years full-time.
Entrance requirements: For master's and doctorate, GRE General Test, minimum undergraduate GPA of 3.0. *Application fee:* $30.
Expenses: Tuition, state resident: part-time $193 per hour. Tuition, nonresident: part-time $445 per hour. Tuition and fees vary according to degree level.
Financial aid: Federal Work-Study and institutionally sponsored loans available. Financial aid applicants required to submit FAFSA.
Faculty research: Developmental neuroscience, sensory systems, neuropharmacology, neurophysiology, substances. *Total annual research expenditures:* $457,984.
Dr. Donald Godfrey, Program Director, 419-383-3571, *Fax:* 419-383-6140, *E-mail:* mcogradschool@mco.edu.
Application contact: Dr. Linda Dokas, Coordinator, 419-383-3890, *Fax:* 419-383-6140, *E-mail:* mcogradschool@mco.edu.

■ MEDICAL UNIVERSITY OF SOUTH CAROLINA

College of Graduate Studies, Department of Physiology, Charleston, SC 29425-0002

AWARDS MS, PhD, MD/PhD.

Faculty: 8 part-time/adjunct (0 women).
Students: 13 full-time (4 women); includes 4 minority (1 African American, 3 Asian Americans or Pacific Islanders). Average age 32. *15 applicants, 73% accepted.* In 1999, 2 doctorates awarded. Terminal master's awarded for partial completion of doctoral program.
Degree requirements: For master's, thesis, research seminar required; for doctorate, dissertation, teaching and research seminar, oral and written exams required.
Entrance requirements: For master's and doctorate, GRE General Test, TOEFL, interview. *Application deadline:* Applications are processed on a rolling basis. *Application fee:* $55. Electronic applications accepted.
Expenses: Tuition, state resident: full-time $3,470; part-time $160 per semester hour. Tuition, nonresident: full-time $4,426; part-time $213 per semester hour. Required fees: $408 per semester. One-time fee: $160. Tuition and fees vary according to program.
Financial aid: In 1999–00, 1 fellowship (averaging $16,000 per year), 2 research assistantships were awarded; teaching assistantships, Federal Work-Study and tuition waivers (partial) also available. Financial aid application deadline: 4/1; financial aid applicants required to submit FAFSA.
Faculty research: Molecular and cellular physiology, circulatory shock, neuroendocrinology, mammalian myocardium, spinal cord injury. *Total annual research expenditures:* $905,507.
Peter Kalivas, Chairman, 843-792-2005.
Application contact: Julie Johnston, Director of Admissions, 843-792-8710, *Fax:* 843-792-3764.

Find an in-depth description at www.petersons.com/graduate.

■ MICHIGAN STATE UNIVERSITY

Graduate School, College of Natural Science, Neuroscience Program, East Lansing, MI 48824

AWARDS PhD.

Faculty: 39 full-time (11 women).
Students: 5 (2 women) 2 international. *47 applicants, 19% accepted.*
Degree requirements: For doctorate, dissertation required, foreign language not required.
Entrance requirements: For doctorate, GRE General Test. *Application deadline:*

For fall admission, 1/5. *Application fee:* $30 ($40 for international students).
Expenses: Tuition, state resident: part-time $229 per credit. Tuition, nonresident: part-time $464 per credit. Required fees: $241 per semester. Tuition and fees vary according to course load, degree level and program.
Financial aid: Fellowships, research assistantships, teaching assistantships available.
Faculty research: Neurobiology and behavior, neuroendocrinology.
Cheryl Sisk, Director, 517-353-8947, *Fax:* 517-432-2744.
Application contact: Information Contact, 513-353-8947.

Find an in-depth description at www.petersons.com/graduate.

■ MOUNT SINAI SCHOOL OF MEDICINE OF NEW YORK UNIVERSITY

Graduate School of Biological Sciences, Neurosciences Training Area, New York, NY 10029-6504

AWARDS PhD, MD/PhD.

Students: 17 full-time (9 women).
Degree requirements: For doctorate, dissertation required, foreign language not required.
Entrance requirements: For doctorate, GRE General Test, GRE Subject Test, MCAT, TOEFL. *Application deadline:* For fall admission, 4/15. *Application fee:* $60.
Expenses: Tuition: Full-time $21,750; part-time $725 per credit. Required fees: $750; $25 per credit. Full-time tuition and fees vary according to student level.
Financial aid: Fellowships with full tuition reimbursements, grants available.
Dr. James Roberts, Program Director, 212-241-7368, *E-mail:* roberts@msvax.mssm.edu.
Application contact: C. Gita Bosch, Administrative Manager and Assistant Dean, 212-241-6546, *Fax:* 212-241-0651, *E-mail:* grads@mssm.edu.

■ NEW YORK MEDICAL COLLEGE

Graduate School of Basic Medical Sciences, Department of Cell Biology and Anatomy, Valhalla, NY 10595-1691

AWARDS Cell biology and neuroscience (MS, PhD). Part-time and evening/weekend programs available.

Faculty: 18 full-time (6 women).
Students: 8 full-time (5 women), 1 (woman) part-time; includes 1 minority (Asian American or Pacific Islander), 3 international. Average age 30. *8 applicants, 38% accepted.* In 1999, 2 master's, 2

doctorates awarded. Terminal master's awarded for partial completion of doctoral program.
Degree requirements: For master's and doctorate, computer language, thesis/dissertation required, foreign language not required.
Entrance requirements: For master's, GRE General Test, TOEFL; for doctorate, GRE General Test, GRE Subject Test, TOEFL. *Application deadline:* For fall admission, 7/1 (priority date); for spring admission, 12/1 (priority date). Applications are processed on a rolling basis. *Application fee:* $35 ($60 for international students).
Expenses: Tuition: Part-time $430 per credit. Required fees: $15 per semester. One-time fee: $100.
Financial aid: In 1999–00, 7 research assistantships with full tuition reimbursements were awarded; career-related internships or fieldwork, Federal Work-Study, grants, institutionally sponsored loans, and tuition waivers (full) also available. Aid available to part-time students. Financial aid applicants required to submit FAFSA.
Faculty research: Mechanisms of growth control in skeletal muscle, cartilage differentiation, cytoskeletal functions, signal transduction pathways, neuronal development and plasticity.
Dr. Anna B. Drakontides, Director, 914-594-4036.

Find an in-depth description at www.petersons.com/graduate.

■ **NEW YORK UNIVERSITY**
Graduate School of Arts and Science, Center for Neural Science, New York, NY 10012-1019
AWARDS PhD.
Faculty: 15 full-time (2 women), 4 part-time/adjunct.
Students: 20 full-time (8 women), 15 part-time (8 women); includes 4 minority (1 African American, 2 Hispanic Americans, 1 Native American), 6 international. Average age 26. *78 applicants, 15% accepted.* In 1999, 3 degrees awarded.
Degree requirements: For doctorate, one foreign language, dissertation required.
Entrance requirements: For doctorate, GRE, TOEFL, interview. *Application deadline:* For fall admission, 1/4 (priority date). *Application fee:* $60.
Expenses: Tuition: Full-time $17,880; part-time $745 per credit. Required fees: $1,140; $35 per credit. Tuition and fees vary according to course load and program.
Financial aid: Fellowships with tuition reimbursements, research assistantships with tuition reimbursements, career-related

internships or fieldwork, Federal Work-Study, and tuition waivers (full and partial) available. Financial aid application deadline: 1/4; financial aid applicants required to submit FAFSA.
Faculty research: Systems and integrative neuroscience; combining biology, cognition, computation, and theory.
Daniel Sanes, Chairman, 212-998-7780.
Application contact: Samuel Feldman, Director of Graduate Studies, 212-998-7780, *Fax:* 212-995-4011, *E-mail:* cns@nyu.edu.

Find an in-depth description at www.petersons.com/graduate.

■ **NEW YORK UNIVERSITY**
Graduate School of Arts and Science, Department of Biology, New York, NY 10012-1019
AWARDS Applied recombinant DNA technology (MS); biochemistry (PhD); biomedical journalism (MA); cell biology (PhD); computers in biological research (MS); environmental biology (PhD); general biology (MS); neural sciences and physiology (PhD); oral biology (MS); population and evolutionary biology (PhD). Part-time programs available.
Faculty: 22 full-time (5 women), 8 part-time/adjunct.
Students: 99 full-time (48 women), 61 part-time (31 women); includes 30 minority (4 African Americans, 22 Asian Americans or Pacific Islanders, 4 Hispanic Americans), 52 international. Average age 24. *371 applicants, 41% accepted.* In 1999, 54 master's, 4 doctorates awarded. Terminal master's awarded for partial completion of doctoral program.
Degree requirements: For master's, thesis or alternative, qualifying paper required, foreign language not required; for doctorate, dissertation, oral and written comprehensive exams required, foreign language not required.
Entrance requirements: For master's, GRE General Test, TOEFL; for doctorate, GRE General Test, GRE Subject Test, TOEFL. *Application deadline:* For fall admission, 1/4 (priority date). *Application fee:* $60.
Expenses: Tuition: Full-time $17,880; part-time $745 per credit. Required fees: $1,140; $35 per credit. Tuition and fees vary according to course load and program.
Financial aid: Fellowships with tuition reimbursements, research assistantships with tuition reimbursements, teaching assistantships with tuition reimbursements, career-related internships or fieldwork, Federal Work-Study, institutionally sponsored loans, and tuition waivers (full

and partial) available. Financial aid application deadline: 1/4; financial aid applicants required to submit FAFSA.
Faculty research: Development and genetics, neurobiology, plant sciences, molecular and cell biology.
Philip Furmanski, Chairman, 212-998-8200.
Application contact: Gloria Coruzzi, Director of Graduate Studies, 212-998-8200, *Fax:* 212-995-4015, *E-mail:* biology@nyu.edu.

Find an in-depth description at www.petersons.com/graduate.

■ **NEW YORK UNIVERSITY**
School of Medicine and Graduate School of Arts and Science, Sackler Institute of Graduate Biomedical Sciences, Department of Neuroscience and Physiology, New York, NY 10012-1019
AWARDS Neuroscience (PhD); physiology (PhD).
Faculty: 22 full-time (4 women).
Students: 16 full-time (2 women); includes 3 minority (2 Asian Americans or Pacific Islanders, 1 Hispanic American), 4 international. Average age 25. In 1999, 2 degrees awarded.
Degree requirements: For doctorate, one foreign language, dissertation, qualifying exam required. *Average time to degree:* Doctorate–5.5 years full-time.
Entrance requirements: For doctorate, GRE General Test, GRE Subject Test, TOEFL. *Application deadline:* For fall admission, 2/1 (priority date). Applications are processed on a rolling basis. *Application fee:* $60.
Expenses: Tuition: Full-time $17,880; part-time $745 per credit. Required fees: $1,140; $35 per credit. Tuition and fees vary according to course load and program.
Financial aid: In 1999–00, 6 research assistantships were awarded; fellowships, teaching assistantships Financial aid application deadline: 1/15.
Faculty research: Synaptic transmission, retinal physiology, signal transduction, CNS intrinsic properties, cerebellar function. *Total annual research expenditures:* $2.3 million.
Dr. Rodolfo R. Llinás, Chairman, 212-263-5415.
Application contact: Dr. Stewart A. Bloomfield, Graduate Adviser, 212-263-5770, *Fax:* 212-263-8072, *E-mail:* blooms01@med.nyu.edu.

Find an in-depth description at www.petersons.com/graduate.

■ NORTHWESTERN UNIVERSITY

The Graduate School, Division of Interdepartmental Programs and Medical School, Integrated Graduate Programs in the Life Sciences, Chicago, IL 60611

AWARDS Cancer biology (PhD); cell biology (PhD); developmental biology (PhD); evolutionary biology (PhD); immunology and microbial pathogenesis (PhD); molecular biology and genetics (PhD); neurobiology (PhD); pharmacology and toxicology (PhD); structural biology and biochemistry (PhD).

Degree requirements: For doctorate, dissertation, written and oral qualifying exams required, foreign language not required.
Entrance requirements: For doctorate, GRE General Test, TOEFL.
Expenses: Tuition: Full-time $23,301. Full-time tuition and fees vary according to program.

Find an in-depth description at www.petersons.com/graduate.

■ NORTHWESTERN UNIVERSITY

The Graduate School, Division of Interdepartmental Programs, Neuroscience Institute Graduate Program, Evanston, IL 60208

AWARDS PhD. Admissions and degree offered through The Graduate School.

Faculty: 118 full-time (26 women).
Students: 76 full-time (39 women); includes 10 minority (2 African Americans, 6 Asian Americans or Pacific Islanders, 2 Hispanic Americans), 15 international. *135 applicants, 27% accepted.* In 1999, 9 degrees awarded.
Degree requirements: For doctorate, dissertation required, foreign language not required.
Entrance requirements: For doctorate, GRE General Test, TOEFL. *Application deadline:* For fall admission, 1/15 (priority date). *Application fee:* $50 ($55 for international students).
Expenses: Tuition: Full-time $23,301. Full-time tuition and fees vary according to program.
Financial aid: In 1999–00, 7 fellowships with full tuition reimbursements (averaging $17,000 per year), 35 research assistantships with partial tuition reimbursements (averaging $17,000 per year), 10 teaching assistantships with full tuition reimbursements (averaging $12,843 per year) were awarded. Financial aid application deadline: 1/15; financial aid applicants required to submit FAFSA.
Faculty research: Circadian rhythms, synaptic neurotransmissions, cognitive neuroscience, sensory/motor systems, cell biology and structure/function, neurobiology of disease.

Enrico Mugnaini, Director, 847-491-2862.
Application contact: Robert Harper-Mangels, Assistant Director, NUIN, 847-467-4246, *Fax:* 847-491-5211, *E-mail:* r_mangels@northwestern.edu.

Find an in-depth description at www.petersons.com/graduate.

■ NORTHWESTERN UNIVERSITY

The Graduate School, Judd A. and Marjorie Weinberg College of Arts and Sciences, Department of Neurobiology and Physiology, Evanston, IL 60208

AWARDS MS. Admissions and degrees offered through The Graduate School. Part-time programs available.

Faculty: 12 full-time (3 women), 11 part-time/adjunct (3 women).
Students: 7 full-time (4 women), 5 part-time (2 women). *31 applicants, 52% accepted.* In 1999, 8 degrees awarded.
Degree requirements: For master's, thesis required, foreign language not required.
Entrance requirements: For master's, GRE General Test and MCAT (strongly recommended), TOEFL. *Application deadline:* For fall admission, 8/1. Applications are processed on a rolling basis. *Application fee:* $50 ($55 for international students).
Expenses: Tuition: Full-time $23,301. Full-time tuition and fees vary according to program.
Financial aid: Career-related internships or fieldwork, Federal Work-Study, and institutionally sponsored loans available. Financial aid application deadline: 1/15; financial aid applicants required to submit FAFSA.
Faculty research: Sensory neurobiology and neuroendocrinology, reproductive biology, vision physiology and psychophysics, cell and developmental biology.

Lawrence Pinto, Chair, 847-491-5521.
Application contact: Michael Kennedy, Assistant Chair, 847-491-5521, *Fax:* 847-491-5211, *E-mail:* m-kennedy@northwestern.edu.

Find an in-depth description at www.petersons.com/graduate.

■ THE OHIO STATE UNIVERSITY

Graduate School, Program in Neuroscience, Columbus, OH 43210

AWARDS PhD.

Faculty: 61 full-time.
Students: 19 full-time (11 women), 8 international. *39 applicants, 21% accepted.* In 1999, 5 degrees awarded.
Degree requirements: For doctorate, dissertation required, foreign language not required.

Entrance requirements: For doctorate, GRE General Test. *Application deadline:* For fall admission, 8/15. Applications are processed on a rolling basis. *Application fee:* $30 ($40 for international students).
Expenses: Tuition, state resident: full-time $5,400. Tuition, nonresident: full-time $14,535. Part-time tuition and fees vary according to course load and program.
Financial aid: Fellowships, research assistantships, teaching assistantships, Federal Work-Study, institutionally sponsored loans, and administrative associateships available. Aid available to part-time students.

Gregory J. Cole, Graduate Studies Committee Chair, 614-292-2379, *Fax:* 614-292-0490, *E-mail:* cole.115@osu.edu.

■ OREGON HEALTH SCIENCES UNIVERSITY

School of Medicine, Graduate Programs in Medicine, Department of Behavioral Neuroscience, Portland, OR 97201-3098

AWARDS MS, PhD, MD/PhD.

Degree requirements: For master's, thesis required, foreign language not required; for doctorate, dissertation, written exam required, foreign language not required.
Entrance requirements: For master's and doctorate, GRE General Test.
Expenses: Tuition, state resident: full-time $3,132; part-time $174 per credit hour. Tuition, nonresident: full-time $5,256; part-time $292 per credit hour. Required fees: $8.5 per credit hour. $146 per term. Part-time tuition and fees vary according to course load.
Faculty research: Neural basis of behavior, behavioral pharmacology, behavioral genetics, neuropharmacology and neuroendocrinology, biological basis of drug seeking and addiction.

Find an in-depth description at www.petersons.com/graduate.

■ OREGON HEALTH SCIENCES UNIVERSITY

School of Medicine, Graduate Programs in Medicine, Neuroscience Graduate Program, Portland, OR 97201-3098

AWARDS PhD, MD/PhD.

Faculty: 107 full-time (24 women).
Students: 43 full-time (22 women); includes 7 minority (1 African American, 2 Asian Americans or Pacific Islanders, 1 Hispanic American, 3 Native Americans), 2 international. *70 applicants, 30% accepted.* In 1999, 2 degrees awarded (50% entered university research/teaching, 50% continued full-time study).

Degree requirements: For doctorate, dissertation required, foreign language not required. *Average time to degree:* Doctorate–6 years full-time.

Entrance requirements: For doctorate, GRE General Test, TOEFL. *Application deadline:* For fall admission, 1/15 (priority date).

Expenses: Tuition, state resident: full-time $3,132; part-time $174 per credit hour. Tuition, nonresident: full-time $5,256; part-time $292 per credit hour. Required fees: $8.5 per credit hour. $146 per term. Part-time tuition and fees vary according to course load.

Financial aid: Fellowships, research assistantships, institutionally sponsored loans, scholarships, and tuition waivers (full) available. Aid available to part-time students. Financial aid application deadline: 3/1; financial aid applicants required to submit FAFSA.

Faculty research: Signal transduction, receptors and ion channels, transcriptional regulation, drug abuse, systems, behavioral neuroscience. *Total annual research expenditures:* $15 million.

Dr. Edwin W. McCleskey, Director, 503-494-6933.

Application contact: 503-494-6932, *Fax:* 503-464-5518, *E-mail:* ngp@ohsu.edu.

Find an in-depth description at www.petersons.com/graduate.

■ **THE PENNSYLVANIA STATE UNIVERSITY MILTON S. HERSHEY MEDICAL CENTER**

Graduate School, Department of Neuroscience and Anatomy, Interdepartmental Graduate Program in Neuroscience, Hershey, PA 17033-2360

AWARDS MS, PhD, MD/PhD.

Students: 9 full-time (8 women). Average age 27. In 1999, 1 doctorate awarded.

Degree requirements: For doctorate, dissertation required.

Entrance requirements: For master's and doctorate, GRE General Test. *Application deadline:* For fall admission, 7/26. *Application fee:* $50.

Expenses: Tuition, state resident: full-time $6,886; part-time $291 per credit. Tuition, nonresident: full-time $14,118; part-time $588 per credit. Required fees: $43 per semester. Part-time tuition and fees vary according to course load.

Dr. Robert J. Milner, Head, Department of Neuroscience and Anatomy, 717-531-8650.

■ **THE PENNSYLVANIA STATE UNIVERSITY UNIVERSITY PARK CAMPUS**

Graduate School, Intercollege Graduate Programs, Intercollege Graduate Program in Integrative Biosciences, State College, University Park, PA 16802-1503

AWARDS Integrative biosciences (MS, PhD), including biomolecular transport dynamics, cell and developmental biology, cellular and molecular mechanisms of toxicity, chemical biology, ecological and molecular plant physiology, immunobiology, molecular medicine, neuroscience, nutrition science.

Students: 39 full-time (24 women), 1 part-time.

Entrance requirements: For master's and doctorate, GRE General Test. *Application fee:* $50.

Expenses: Tuition, state resident: full-time $6,886; part-time $291 per credit. Tuition, nonresident: full-time $14,118; part-time $588 per credit. Required fees: $46 per semester. Part-time tuition and fees vary according to course load and program.

Financial aid: Fellowships available.

Dr. C. R. Matthews, Co-Director, 814-863-3650.

Application contact: Admissions Committee, 814-865-3155, *Fax:* 814-863-1357, *E-mail:* lscgradadm@mail.biotec.psu.edu.

Find an in-depth description at www.petersons.com/graduate.

■ **PRINCETON UNIVERSITY**

Graduate School, Department of Ecology and Evolutionary Biology, Princeton, NJ 08544-1019

AWARDS Biology (PhD); neuroscience (PhD).

Degree requirements: For doctorate, dissertation required, foreign language not required.

Entrance requirements: For doctorate, GRE General Test, GRE Subject Test.

Expenses: Tuition: Full-time $25,050.

Find an in-depth description at www.petersons.com/graduate.

■ **PRINCETON UNIVERSITY**

Graduate School, Department of Molecular Biology, Princeton, NJ 08544-1019

AWARDS Cell biology (PhD); developmental biology (PhD); molecular biology (PhD); neuroscience (PhD).

Degree requirements: For doctorate, dissertation required.

Entrance requirements: For doctorate, GRE General Test.

Expenses: Tuition: Full-time $25,050.

Faculty research: Genetics, virology, biochemistry.

Find an in-depth description at www.petersons.com/graduate.

■ **PRINCETON UNIVERSITY**

Graduate School, Department of Psychology and Department of Ecology and Evolutionary Biology and Department of Molecular Biology, Program in Neuroscience, Princeton, NJ 08544-1019

AWARDS PhD.

Degree requirements: For doctorate, dissertation required.

Entrance requirements: For doctorate, GRE General Test, GRE Subject Test.

Expenses: Tuition: Full-time $25,050.

Faculty research: CNS neurochemistry, visual system, invertebrate neurophysiology, neural bases of behavior, hormones and behavior.

■ **PURDUE UNIVERSITY**

Graduate School, Interdisciplinary Neuroscience Program, West Lafayette, IN 47907

AWARDS PhD. Degree awarded through one of 11 participating departments.

Faculty: 48 full-time (15 women), 1 part-time/adjunct (0 women).

Students: 13 full-time (9 women); includes 1 minority (African American), 5 international. Average age 25. *29 applicants, 24% accepted.*

Degree requirements: For doctorate, dissertation required, foreign language not required.

Entrance requirements: For doctorate, TOEFL. *Application deadline:* For fall admission, 2/15 (priority date). *Application fee:* $30. Electronic applications accepted.

Expenses: Tuition, state resident: full-time $4,530; part-time $130 per credit hour. Tuition, nonresident: full-time $15,310; part-time $404 per credit hour. Tuition and fees vary according to campus/location and program.

Financial aid: In 1999–00, 3 fellowships with full tuition reimbursements (averaging $15,000 per year), 3 research assistantships with full tuition reimbursements (averaging $15,000 per year), teaching assistantships (averaging $15,000 per year) were awarded. Aid available to part-time students. Financial aid applicants required to submit FAFSA.

Faculty research: Molecular signalling, plasticity and development, behavior and cognition.

Dr. R. L. Meisel, Chair, 765-494-7669, *Fax:* 765-496-1264, *E-mail:* meisel@psych.purdue.edu.

■ PURDUE UNIVERSITY

Graduate School, School of Science, Department of Biological Sciences, Program in Neurobiology, West Lafayette, IN 47907

AWARDS MS, PhD.

Faculty: 7 full-time (2 women). Terminal master's awarded for partial completion of doctoral program.

Degree requirements: For master's, foreign language and thesis not required; for doctorate, dissertation, seminars, teaching experience required, foreign language not required.

Entrance requirements: For master's and doctorate, GRE General Test, TOEFL, TSE. *Application deadline:* For fall admission, 2/15. *Application fee:* $30. Electronic applications accepted.

Expenses: Tuition, state resident: full-time $4,530; part-time $130 per credit hour. Tuition, nonresident: full-time $15,310; part-time $404 per credit hour. Tuition and fees vary according to campus/location and program.

Financial aid: Fellowships, research assistantships, teaching assistantships available. Aid available to part-time students. Financial aid application deadline: 2/15; financial aid applicants required to submit FAFSA.

Faculty research: Molecular and cellular neuroscience, signal transduction, mechanisms of learning and memory, neuronal morphogenesis, molecular genetics of development.

Application contact: Nancy Konopka, Graduate Studies Office Manager, 765-494-8142, *Fax:* 765-494-0876, *E-mail:* njk@bilbo.bio.purdue.edu.

■ RUTGERS, THE STATE UNIVERSITY OF NEW JERSEY, NEWARK

Graduate School, Program in Behavioral and Neural Sciences, Newark, NJ 07102

AWARDS PhD. Part-time programs available.

Faculty: 17 full-time (10 women), 2 part-time/adjunct (0 women).

Students: 25 full-time (12 women), 5 part-time (2 women); includes 11 minority (1 African American, 10 Hispanic Americans). *68 applicants, 26% accepted.* In 1999, 6 doctorates awarded (100% found work related to degree).

Degree requirements: For doctorate, dissertation required, foreign language not required. *Average time to degree:* Doctorate–5 years full-time.

Entrance requirements: For doctorate, GRE, minimum GPA of 3.0. *Application deadline:* For fall admission, 2/1 (priority

date). Applications are processed on a rolling basis. *Application fee:* $50. Electronic applications accepted.

Expenses: Tuition, state resident: full-time $6,776; part-time $279 per credit hour. Tuition, nonresident: full-time $9,936; part-time $412 per credit hour. Required fees: $201 per semester. Tuition and fees vary according to course load and program.

Financial aid: In 1999–00, 1 fellowship with full tuition reimbursement (averaging $15,000 per year), 21 teaching assistantships with full tuition reimbursements (averaging $15,352 per year) were awarded; research assistantships Financial aid application deadline: 3/1.

Faculty research: Systems neuroscience, cognitive neuroscience, molecular neuroscience, behavioral neuroscience. *Total annual research expenditures:* $3 million.

Dr. Ian Creese, Co-Director, 973-353-1080 Ext. 3300, *Fax:* 973-353-1272, *E-mail:* creese@axon.rutgers.edu.

Application contact: Dr. Howard Poizner, First Year Advisor, 973-353-1080 Ext. 3231, *Fax:* 973-353-1272, *E-mail:* poizner@axon.rutgers.edu.

■ RUTGERS, THE STATE UNIVERSITY OF NEW JERSEY, NEW BRUNSWICK

Graduate School, Program in Physiology and Neurobiology, New Brunswick, NJ 08901-1281

AWARDS PhD.

Faculty: 62 full-time (10 women).

Students: 18 full-time (9 women), 18 part-time (14 women); includes 8 minority (3 African Americans, 1 Asian American or Pacific Islander, 4 Hispanic Americans), 13 international. *44 applicants, 20% accepted.* In 1999, 1 degree awarded.

Degree requirements: For doctorate, dissertation, qualifying exam required, foreign language not required. *Average time to degree:* Doctorate–4.9 years full-time.

Entrance requirements: For doctorate, GRE General Test, GRE Subject Test (biology or chemistry), TOEFL, minimum undergraduate GPA of 3.0. *Application deadline:* For fall admission, 3/1 (priority date). Applications are processed on a rolling basis. *Application fee:* $50.

Expenses: Tuition, state resident: full-time $6,776; part-time $279 per credit. Tuition, nonresident: full-time $9,936; part-time $412 per credit. Required fees: $20 per credit. $89 per semester. Tuition and fees vary according to course load, campus/location and program.

Financial aid: In 1999–00, 7 students received aid, including 1 fellowship, 5 research assistantships with full tuition

reimbursements available, 1 teaching assistantship with full tuition reimbursement available; tuition waivers (full) also available. Financial aid applicants required to submit FAFSA.

Faculty research: Neuronal growth factors, neuronal gene expression, neurogenetics, circulation controls, reproduction. *Total annual research expenditures:* $3.4 million.

Dr. Ira B. Black, Director, 732-235-5388, *Fax:* 732-235-5885.

Application contact: David Egger, Director of Admissions, 732-235-4522, *Fax:* 732-235-4029.

■ SAINT LOUIS UNIVERSITY

School of Medicine and Graduate School, Graduate Programs in Biomedical Sciences, Department of Anatomy and Neurobiology, St. Louis, MO 63103-2097

AWARDS Anatomy (MS(R), PhD); neurobiology (PhD).

Faculty: 11 full-time (1 woman).

Students: Average age 29. *13 applicants, 100% accepted.* In 1999, 2 degrees awarded.

Degree requirements: For master's, thesis, comprehensive oral exam required, foreign language not required; for doctorate, dissertation, departmental qualifying exams required.

Entrance requirements: For master's and doctorate, GRE General Test, GRE Subject Test. *Application deadline:* For fall admission, 4/15 (priority date). Applications are processed on a rolling basis. *Application fee:* $40.

Expenses: Tuition: Part-time $507 per credit hour. Required fees: $38 per term.

Financial aid: In 1999–00, 11 students received aid, including 2 fellowships, 2 teaching assistantships; research assistantships, Federal Work-Study and institutionally sponsored loans also available. Aid available to part-time students. Financial aid application deadline: 8/1; financial aid applicants required to submit FAFSA.

Faculty research: Systems neurobiology.

Dr. Paul Young, Chairman, 314-577-8274, *Fax:* 314-268-5127.

Application contact: Kris Sherman, Director, 314-577-8275, *Fax:* 314-268-5127, *E-mail:* shermankb@slu.edu.

■ STANFORD UNIVERSITY

School of Medicine, Graduate Programs in Medicine, Neurosciences Program, Stanford, CA 94305-9991

AWARDS PhD.

Faculty: 8 full-time (0 women).

Students: 35 full-time (17 women), 18 part-time (2 women); includes 15 minority (2 African Americans, 12 Asian Americans or Pacific Islanders, 1 Hispanic American),

4 international. *Average age 27. 98 applicants, 13% accepted.* In 1999, 3 doctorates awarded.

Degree requirements: For doctorate, dissertation required.

Entrance requirements: For doctorate, GRE General Test, GRE Subject Test, TOEFL. *Application deadline:* For fall admission, 12/15. *Application fee:* $65 ($80 for international students). Electronic applications accepted.

Expenses: Tuition: Full-time $23,058. Required fees: $152. Part-time tuition and fees vary according to course load.

Financial aid: Research assistantships, teaching assistantships available. Financial aid application deadline: 12/15.

Eric I. Knudsen, Director, 650-723-9855, *Fax:* 650-725-7855, *E-mail:* eric@brio.stanford.edu.

Application contact: Graduate Admissions Coordinator, 650-723-9855.

■ **STATE UNIVERSITY OF NEW YORK AT ALBANY**

College of Arts and Sciences, Department of Biological Sciences, Specialization in Molecular, Cellular, Developmental, and Neural Biology, Albany, NY 12222-0001

AWARDS MS, PhD.

Degree requirements: For master's, one foreign language required; for doctorate, dissertation required.

Entrance requirements: For master's and doctorate, GRE General Test. *Application fee:* $50.

Expenses: Tuition, state resident: full-time $5,100; part-time $214 per credit. Tuition, nonresident: full-time $8,416; part-time $352 per credit. Required fees: $31 per credit.

Financial aid: Minority assistantships available.

Dr. David Shub, Chair, Department of Biological Sciences, 518-442-4300.

■ **STATE UNIVERSITY OF NEW YORK AT ALBANY**

School of Public Health, Department of Biomedical Sciences, Program in Neuroscience, Albany, NY 12222-0001

AWARDS MS, PhD.

Degree requirements: For master's and doctorate, thesis/dissertation required.

Entrance requirements: For master's and doctorate, GRE General Test, GRE Subject Test. *Application deadline:* For fall admission, 1/15 (priority date); for spring admission, 11/1 (priority date). *Application fee:* $50.

Expenses: Tuition, state resident: full-time $5,100; part-time $214 per credit. Tuition,

nonresident: full-time $8,416; part-time $352 per credit. Required fees: $31 per credit.

Financial aid: Application deadline: 2/1.

Dr. Harry Taber, Chair, Department of Biomedical Sciences, 518-474-2662.

■ **STATE UNIVERSITY OF NEW YORK AT STONY BROOK**

Graduate School, College of Arts and Sciences, Department of Neurobiology and Behavior, Stony Brook, NY 11794

AWARDS PhD.

Faculty: 34.

Students: 14 full-time (9 women), 19 part-time (9 women); includes 7 minority (5 Asian Americans or Pacific Islanders, 2 Hispanic Americans), 14 international. *Average age 28. 60 applicants, 30% accepted.* In 1999, 2 degrees awarded.

Degree requirements: For doctorate, dissertation, comprehensive exam, teaching experience required.

Entrance requirements: For doctorate, GRE General Test, GRE Subject Test, TOEFL, minimum GPA of 3.0. *Application deadline:* For fall admission, 1/15. *Application fee:* $50.

Expenses: Tuition, state resident: full-time $5,100; part-time $213 per credit hour. Tuition, nonresident: full-time $8,416; part-time $351 per credit hour. Required fees: $492. Tuition and fees vary according to program.

Financial aid: In 1999–00, 5 fellowships, 14 research assistantships, 13 teaching assistantships were awarded; Federal Work-Study also available.

Faculty research: Biophysics; neurochemistry; cellular, developmental, and integrative neurobiology. *Total annual research expenditures:* $6 million.

Dr. Lorne Mendell, Chairperson, 631-632-8616.

Application contact: Dr. Jim Gnadt, Director, 631-632-8616.

Find an in-depth description at www.petersons.com/graduate.

■ **STATE UNIVERSITY OF NEW YORK HEALTH SCIENCE CENTER AT BROOKLYN**

School of Graduate Studies, Program in Neural and Behavioral Science, Brooklyn, NY 11203-2098

AWARDS PhD, MD/PhD.

Degree requirements: For doctorate, one foreign language, dissertation required.

Entrance requirements: For doctorate, GRE.

Expenses: Tuition, state resident: full-time $5,100; part-time $213 per credit. Tuition, nonresident: full-time $8,416; part-time

$351 per credit. Required fees: $200. Full-time tuition and fees vary according to program and student level.

Faculty research: Molecular neuroscience, cellular neuroscience, systems neuroscience, behavioral neuroscience, behavior.

■ **STATE UNIVERSITY OF NEW YORK UPSTATE MEDICAL UNIVERSITY**

College of Graduate Studies, Program in Neuroscience, Syracuse, NY 13210-2334

AWARDS PhD.

Faculty: 28.

Students: 17 full-time (9 women), 1 part-time; includes 4 minority (3 Asian Americans or Pacific Islanders, 1 Hispanic American), 3 international. *24 applicants, 42% accepted.* In 1999, 2 degrees awarded.

Degree requirements: For doctorate, dissertation, comprehensive exam required, foreign language not required.

Entrance requirements: For doctorate, GRE General Test, GRE Subject Test, TSE. *Application deadline:* For fall admission, 4/1 (priority date). Applications are processed on a rolling basis. *Application fee:* $40.

Expenses: Tuition, state resident: full-time $5,100; part-time $213 per credit. Tuition, nonresident: full-time $8,416; part-time $351 per credit. Required fees: $410; $25 per credit. Part-time tuition and fees vary according to course load and program.

Dr. James McCasland, Director, 315-464-7655.

Application contact: Dr. Robert Barlow, Professor, 315-464-7770.

Find an in-depth description at www.petersons.com/graduate.

■ **SYRACUSE UNIVERSITY**

Graduate School, L. C. Smith College of Engineering and Computer Science, Institute for Sensory Research, Syracuse, NY 13244-0003

AWARDS Neuroscience (MS, PhD).

Faculty: 13.

Students: 21 full-time (5 women), 6 part-time (1 woman); includes 1 minority (African American), 5 international. *Average age 26. 16 applicants, 44% accepted.* In 1999, 4 master's awarded.

Degree requirements: For master's, foreign language not required; for doctorate, computer language, dissertation, 4 formal qualifying exams required, foreign language not required.

Entrance requirements: For master's and doctorate, GRE General Test, GRE

Syracuse University (continued)
Subject Test. *Application deadline:* Applications are processed on a rolling basis. *Application fee:* $40.
Expenses: Tuition: Full-time $13,992; part-time $583 per credit hour.
Financial aid: Fellowships, research assistantships, teaching assistantships, Federal Work-Study, tuition waivers (partial), and full support awards available. Financial aid application deadline: 3/1.
Faculty research: Psychophysics of auditory, somatosensory, and visual systems; psychophysics; neuroanatomy; neurochemistry.
Gustav Engbretson, Director, 315-443-1931.
Application contact: Norma Slepecky, Contact, 315-443-4164.

■ TEACHERS COLLEGE, COLUMBIA UNIVERSITY

Graduate Faculty of Education, Department of Organization and Leadership, Program in Neuroscience and Education, New York, NY 10027-6696

AWARDS Ed M.

Students: 1 (woman) full-time, 6 part-time (5 women), 1 international. Average age 34.
Degree requirements: For master's, foreign language not required. *Application deadline:* For fall admission, 5/15. *Application fee:* $50.
Expenses: Tuition: Part-time $670 per credit. Required fees: $161 per semester. Part-time tuition and fees vary according to program.
Financial aid: Career-related internships or fieldwork, Federal Work-Study, institutionally sponsored loans, and tuition waivers (full and partial) available. Aid available to part-time students. Financial aid application deadline: 2/1.
Faculty research: Neuropsychological diagnosis and intervention.
Application contact: Ursula Felton, Office of Admissions, 212-678-3710, *Fax:* 212-678-4171.

■ TEXAS A&M UNIVERSITY SYSTEM HEALTH SCIENCE CENTER

College of Medicine, Graduate School of Biomedical Sciences, Department of Human Anatomy and Medical Neurobiology, College Station, TX 77840-7896

AWARDS PhD.

Faculty: 11 full-time (2 women).
Students: 5 full-time (2 women); includes 1 minority (Hispanic American). Average

age 26. *30 applicants, 7% accepted.* In 1999, 2 degrees awarded (100% found work related to degree).
Degree requirements: For doctorate, dissertation required, foreign language not required. *Average time to degree:* Doctorate–4.5 years full-time.
Entrance requirements: For doctorate, GRE General Test. *Application deadline:* For fall admission, 2/1 (priority date). Applications are processed on a rolling basis. *Application fee:* $50 ($75 for international students). Electronic applications accepted.
Expenses: Tuition, area resident: Full-time $1,368. Tuition, state resident: part-time $76 per credit. Tuition, nonresident: full-time $5,256; part-time $292 per credit. International tuition: $5,256 full-time. Required fees: $678; $38 per credit. Full-time tuition and fees vary according to course load and student level.
Financial aid: In 1999–00, 4 research assistantships (averaging $17,400 per year) were awarded; fellowships, teaching assistantships Financial aid application deadline: 3/1; financial aid applicants required to submit FAFSA.
Faculty research: Fetal alcohol syndrome, circadian rhythms, neuroendocrinology, bone and joint diseases, molecular neurobiology. *Total annual research expenditures:* $850,000.
Dr. James R. West, Head, 979-845-4915, *Fax:* 979-845-0790, *E-mail:* jrwest@tamu.edu.
Application contact: Dr. Farida Sohrabji, Assistant Professor, 979-845-4072, *Fax:* 979-845-0790, *E-mail:* sohrabji@medicine.tamu.edu.

Find an in-depth description at www.petersons.com/graduate.

■ TEXAS A&M UNIVERSITY SYSTEM HEALTH SCIENCE CENTER

College of Medicine, Graduate School of Biomedical Sciences, Faculty of Neuroscience, College Station, TX 77840-7896

AWARDS PhD, MD/PhD.

Faculty: 31 full-time (9 women).
Degree requirements: For doctorate, dissertation, qualifying exam required, foreign language not required.
Entrance requirements: For doctorate, GRE General Test, TOEFL. *Application deadline:* For fall admission, 2/1 (priority date). Applications are processed on a rolling basis. *Application fee:* $50 ($75 for international students).
Expenses: Tuition, area resident: Full-time $1,368. Tuition, state resident: part-time $76 per credit. Tuition, nonresident: full-time $5,256; part-time $292 per credit.

International tuition: $5,256 full-time. Required fees: $678; $38 per credit. Full-time tuition and fees vary according to course load and student level.
Financial aid: Fellowships, teaching assistantships available. Financial aid application deadline: 4/1.
Faculty research: Learning and memory, circadian rhythms, neural development and function, neuroendocrinology, neuropharmacology and neurotoxicology. *Total annual research expenditures:* $1.6 million.
Application contact: Dr. Vincent M. Cassone, Chair, 409-845-2888, *E-mail:* vmc@bio.tamu.edu.

Find an in-depth description at www.petersons.com/graduate.

■ TUFTS UNIVERSITY

Sackler School of Graduate Biomedical Sciences, Program in Neuroscience, Medford, MA 02155

AWARDS PhD.

Faculty: 8 full-time (3 women), 10 part-time/adjunct (1 woman).
Students: 11 full-time (7 women); includes 3 minority (all Asian Americans or Pacific Islanders), 2 international. Average age 27. *66 applicants, 6% accepted.* In 1999, 4 degrees awarded (75% entered university research/teaching, 25% found other work related to degree).
Degree requirements: For doctorate, dissertation required, foreign language not required. *Average time to degree:* Doctorate–6 years full-time.
Entrance requirements: For doctorate, GRE General Test, TOEFL. *Application deadline:* For fall admission, 1/15 (priority date). *Application fee:* $45.
Expenses: Tuition: Full-time $19,325.
Financial aid: In 1999–00, research assistantships with full tuition reimbursements (averaging $18,805 per year); fellowships, tuition waivers (full) also available. Financial aid application deadline: 1/15.
Faculty research: Electrophysiology, molecular neurobiology, structure and function of sensory systems, neural regulation of development. *Total annual research expenditures:* $1.2 million.
Dr. Barbara R. L. Talamo, Director, 617-636-3642, *Fax:* 617-636-2413.
Application contact: Secretary, 617-636-6767, *Fax:* 617-636-0375, *E-mail:* sackler-school@tufts.edu.

■ TULANE UNIVERSITY

School of Medicine and Graduate School, Graduate Programs in Medicine, Program in Neuroscience, New Orleans, LA 70118-5669

AWARDS MS, PhD, MD/PhD. MS and PhD offered through the Graduate School.

Students: 15 full-time (7 women), 1 (woman) part-time; includes 2 minority (both African Americans), 1 international. *33 applicants, 30% accepted.* In 1999, 7 doctorates awarded.
Degree requirements: For doctorate, dissertation, qualifying exam required.
Entrance requirements: For doctorate, GRE General Test, TOEFL, or TSE. *Application deadline:* For fall admission, 2/1. *Application fee:* $45.
Expenses: Tuition: Full-time $23,030.
Financial aid: Fellowships available. Financial aid application deadline: 2/1. Dr. Richard Harlan, Director, 504-584-2744.

■ UNIFORMED SERVICES UNIVERSITY OF THE HEALTH SCIENCES

School of Medicine, Division of Basic Medical Sciences, Department of Anatomy and Cell Biology, Bethesda, MD 20814-4799

AWARDS Cell biology, developmental biology, and neurobiology (PhD).

Faculty: 20 full-time (9 women), 12 part-time/adjunct (2 women).
Students: 8 full-time (1 woman); includes 2 minority (both Asian Americans or Pacific Islanders). Average age 28. *3 applicants, 33% accepted.* In 1999, 1 degree awarded (100% entered university research/teaching).
Degree requirements: For doctorate, dissertation, qualifying exam required, foreign language not required. *Average time to degree:* Doctorate–5 years full-time.
Entrance requirements: For doctorate, GRE General Test, GRE Subject Test, minimum GPA of 3.0, U.S. citizenship. *Application deadline:* For fall admission, 1/15 (priority date). Applications are processed on a rolling basis. *Application fee:* $0.
Financial aid: In 1999–00, 3 fellowships with full tuition reimbursements (averaging $15,000 per year) were awarded; tuition waivers (full) also available.
Faculty research: Neuroanatomy and neurocytology, molecular biology. Dr. Harvey Pollard, Chair, 301-295-3200, *E-mail:* hpollard@usuhs.mil.
Application contact: Janet M. Anastasi, Graduate Program Coordinator, 301-295-9474, *Fax:* 301-295-6772, *E-mail:* janastasi@usuhs.mil.

■ UNIFORMED SERVICES UNIVERSITY OF THE HEALTH SCIENCES

School of Medicine, Division of Basic Medical Sciences, Program in Neuroscience, Bethesda, MD 20814-4799

AWARDS PhD.

Faculty: 36 part-time/adjunct (14 women).
Students: 11 full-time (5 women), 1 part-time. Average age 26. *13 applicants, 23% accepted.* In 1999, 1 doctorate awarded (100% found work related to degree).
Degree requirements: For doctorate, dissertation, comprehensive and qualifying exams required.
Entrance requirements: For doctorate, GRE General Test, GRE Subject Test, minimum GPA of 3.0; previous course work in biology, general chemistry, organic chemistry; U.S. citizenship. *Application deadline:* For fall admission, 1/15 (priority date). Applications are processed on a rolling basis. *Application fee:* $0.
Financial aid: In 1999–00, 5 fellowships with full tuition reimbursements (averaging $15,000 per year) were awarded; tuition waivers (full) also available.
Faculty research: Neuronal development and plasticity, molecular neurobiology, environmental adaptations, stress and injury. Dr. Cinda Helke, Director, 301-295-3238, *E-mail:* chelke@usuhs.mil.
Application contact: Janet M. Anastasi, Graduate Program Coordinator, 301-295-9474, *Fax:* 301-295-6772, *E-mail:* janastasi@usuhs.mil.

Find an in-depth description at www.petersons.com/graduate.

■ THE UNIVERSITY OF ALABAMA AT BIRMINGHAM

Graduate School and School of Medicine and School of Dentistry, Graduate Programs in Joint Health Sciences, Department of Neurobiology, Birmingham, AL 35294

AWARDS PhD. The department participates in the Cellular and Molecular Biology Graduate Program.

Students: 21 full-time (10 women), 1 part-time, 6 international. *2 applicants, 50% accepted.*
Degree requirements: For doctorate, dissertation required.
Entrance requirements: For doctorate, GRE, interview. *Application deadline:* Applications are processed on a rolling basis. Electronic applications accepted.
Expenses: Tuition, state resident: part-time $104 per semester hour. Tuition, nonresident: part-time $208 per semester hour. Required fees: $17 per semester hour. $57 per quarter. Tuition and fees vary according to program. Dr. Michael J. Friedlauder, Chair, 205-934-0100, *Fax:* 205-934-6571, *E-mail:* mjf@uab.edu.
Application contact: Information Contact, 205-975-5573, *Fax:* 205-934-6571.

Find an in-depth description at www.petersons.com/graduate.

■ THE UNIVERSITY OF ALABAMA AT BIRMINGHAM

Graduate School, School of Social and Behavioral Sciences, Department of Psychology, Birmingham, AL 35294

AWARDS Behavioral neuroscience (PhD); developmental psychology (PhD); medical psychology (PhD); psychology (MS).

Students: 59 full-time (35 women), 3 part-time (all women); includes 6 minority (3 African Americans, 2 Asian Americans or Pacific Islanders, 1 Hispanic American), 2 international. *179 applicants, 12% accepted.* In 1999, 8 master's, 12 doctorates awarded.
Application deadline: Applications are processed on a rolling basis. *Application fee:* $35 ($60 for international students). Electronic applications accepted.
Expenses: Tuition, state resident: part-time $104 per semester hour. Tuition, nonresident: part-time $208 per semester hour. Required fees: $17 per semester hour. $57 per quarter. Tuition and fees vary according to program.
Financial aid: Career-related internships or fieldwork available.
Faculty research: Biological basis of behavior structure, function of the nervous system. Dr. Carl E. McFarland, Chairman, 205-934-3850, *E-mail:* cmcfarla@uab.edu.

■ THE UNIVERSITY OF ARIZONA

Graduate College, Graduate Interdisciplinary Programs, Graduate Interdisciplinary Program in Neuroscience, Tucson, AZ 85721

AWARDS PhD.

Degree requirements: For doctorate, computer language, dissertation required.
Entrance requirements: For doctorate, GRE, TOEFL.
Expenses: Tuition, nonresident: full-time $4,814; part-time $274 per unit. Required fees: $1,094; $115 per unit. Tuition and fees vary according to course load and program.
Faculty research: Cognitive neuroscience, developmental neurobiology, speech and hearing, motor control, insect neurobiology.

■ UNIVERSITY OF CALIFORNIA, BERKELEY

Graduate Division, Neuroscience Graduate Program, Berkeley, CA 94720-3190

AWARDS PhD.

Degree requirements: For doctorate, dissertation, qualifying exam required.
Entrance requirements: For doctorate, GRE General Test, GRE Subject Test (Biochemistry and Cell Biology; Biology, Chemistry, Computer Science, Physics, or Psychology), TOEFL, minimum GPA of 3.0.
Expenses: Tuition, nonresident: full-time $9,804. Required fees: $4,268. Tuition and fees vary according to program.
Faculty research: Ion channels, signal transduction, plasticity, neural networks, neural cell fate and pattern formation.
Find an in-depth description at www.petersons.com/graduate.

■ UNIVERSITY OF CALIFORNIA, DAVIS

Graduate Studies, Programs in the Biological Sciences, Program in Neuroscience, Davis, CA 95616

AWARDS PhD.

Faculty: 34 full-time (8 women).
Students: 30 full-time (18 women); includes 4 minority (3 Asian Americans or Pacific Islanders, 1 Native American), 5 international. Average age 27. *47 applicants, 21% accepted.* In 1999, 2 degrees awarded.
Degree requirements: For doctorate, dissertation required. *Average time to degree:* Doctorate–5 years full-time.
Entrance requirements: For doctorate, GRE General Test, GRE Subject Test. *Application deadline:* For fall admission, 1/15. *Application fee:* $40. Electronic applications accepted.
Expenses: Tuition, nonresident: full-time $9,804. Tuition and fees vary according to program and student level.
Financial aid: In 1999–00, 30 students received aid, including 10 fellowships with full and partial tuition reimbursements available, 15 research assistantships with full and partial tuition reimbursements available, 2 teaching assistantships with partial tuition reimbursements available; Federal Work-Study, grants, institutionally sponsored loans, scholarships, and tuition waivers (full) also available. Financial aid application deadline: 1/15; financial aid applicants required to submit FAFSA.
Faculty research: Neuroethology, cognitive neurosciences, cortical neurophysics, cellular and molecular neurobiology.
David Amaral, Professor of Psychiatry, 530-757-8813, *E-mail:* dgamaral@ucdavis.edu.

Application contact: Vincent Heckerd, Gradaute Program Assistant, 530-757-8845, *Fax:* 530-752-8832, *E-mail:* vlheckand@ucdavis.edu.

■ UNIVERSITY OF CALIFORNIA, IRVINE

College of Medicine and Office of Research and Graduate Studies, Graduate Programs in Medicine and School of Biological Sciences, Department of Anatomy and Neurobiology, Irvine, CA 92697

AWARDS Biological sciences (MS, PhD).

Faculty: 14 full-time (3 women), 8 part-time/adjunct (4 women).
Students: 9 full-time (2 women); includes 1 minority (Asian American or Pacific Islander), 1 international. *292 applicants, 25% accepted.*
Degree requirements: For doctorate, dissertation required.
Entrance requirements: For master's and doctorate, GRE General Test, GRE Subject Test. *Application deadline:* For fall admission, 1/15 (priority date). Applications are processed on a rolling basis. *Application fee:* $40. Electronic applications accepted.
Expenses: Tuition, nonresident: full-time $10,322; part-time $1,720 per quarter. Required fees: $5,354; $1,300 per quarter. Tuition and fees vary according to program.
Financial aid: Fellowships, research assistantships, teaching assistantships, institutionally sponsored loans and tuition waivers (full and partial) available. Financial aid application deadline: 3/2; financial aid applicants required to submit FAFSA.
Faculty research: Neurotransmitter immunocytochemistry, intracellular physiology, molecular neurobiology, forebrain organization and development, structure and function of sensory and motor systems.
Dr. Richard T. Robertson, Professor and Chair, 949-824-6553, *Fax:* 949-824-1105, *E-mail:* rtrobert@uci.edu.
Application contact: Kimberly McKinney, Biological Sciences Contact, 949-824-8145, *Fax:* 949-824-7407, *E-mail:* kamckinn@uci.edu.

■ UNIVERSITY OF CALIFORNIA, IRVINE

Office of Research and Graduate Studies, School of Biological Sciences, Department of Neurobiology and Behavior, Irvine, CA 92697

AWARDS Biological sciences (PhD).

Faculty: 21 full-time (2 women).

Students: 25 full-time (15 women), 4 part-time (1 woman); includes 7 minority (6 Asian Americans or Pacific Islanders, 1 Hispanic American). *54 applicants, 31% accepted.*
Degree requirements: For doctorate, dissertation required, foreign language not required.
Entrance requirements: For doctorate, GRE General Test, GRE Subject Test. *Application deadline:* For fall admission, 1/7 (priority date). Applications are processed on a rolling basis. *Application fee:* $40. Electronic applications accepted.
Expenses: Tuition, nonresident: full-time $10,244; part-time $1,720 per quarter. Required fees: $5,252; $1,300 per quarter. Tuition and fees vary according to course load and program.
Financial aid: Fellowships, research assistantships, teaching assistantships, institutionally sponsored loans and tuition waivers (full and partial) available. Financial aid application deadline: 3/2; financial aid applicants required to submit FAFSA.
Faculty research: Synaptic processes, neurophysiology, neuroendocrinology, neuroanatomy, molecular neurobiology. Pauline Yahr, Acting Chair, 949-824-7050, *Fax:* 949-824-2447, *E-mail:* piyahr@uci.edu.
Application contact: Lee Johnson, Graduate Admissions Assistant, 949-824-8519, *Fax:* 949-824-2447, *E-mail:* jljohnso@uci.edu.

■ UNIVERSITY OF CALIFORNIA, LOS ANGELES

School of Medicine and Graduate Division, Graduate Programs in Medicine, Department of Neurobiology, Los Angeles, CA 90095

AWARDS Anatomy and cell biology (PhD).

Students: 16 full-time (8 women); includes 7 minority (1 African American, 2 Asian Americans or Pacific Islanders, 3 Hispanic Americans, 1 Native American), 1 international. *8 applicants, 25% accepted.*
Degree requirements: For doctorate, dissertation, oral and written qualifying exams required, foreign language not required.
Entrance requirements: For doctorate, GRE General Test, GRE Subject Test, bachelor's degree in physical or biological science. *Application fee:* $40.
Expenses: Tuition, nonresident: full-time $9,804. Required fees: $4,405.
Financial aid: In 1999–00, 6 fellowships, 8 research assistantships, 1 teaching assistantship were awarded; Federal Work-Study, institutionally sponsored loans, scholarships, and tuition waivers (full and partial) also available. Financial aid application deadline: 3/1.

Faculty research: Neuroendocrinology, neurophysiology.
Dr. Jack Feldman, Chair, 310-825-9558.
Application contact: UCLA Access Coordinator, 800-284-8252, *Fax:* 310-206-5280, *E-mail:* uclaaccess@ibes.medsch.ucla.edu.

■ **UNIVERSITY OF CALIFORNIA, LOS ANGELES**

School of Medicine and Graduate Division, Graduate Programs in Medicine, Neuroscience Program, Los Angeles, CA 90095

AWARDS PhD.

Students: 71 full-time (27 women); includes 12 minority (7 Asian Americans or Pacific Islanders, 4 Hispanic Americans, 1 Native American), 2 international. *132 applicants, 27% accepted.*
Degree requirements: For doctorate, dissertation, oral and written qualifying exams required, foreign language not required.
Entrance requirements: For doctorate, GRE General Test. *Application deadline:* For fall admission, 12/15. *Application fee:* $40.
Expenses: Tuition, nonresident: full-time $9,804. Required fees: $4,405.
Financial aid: In 1999–00, 64 students received aid, including 58 fellowships, 36 research assistantships, 15 teaching assistantships; Federal Work-Study, institutionally sponsored loans, scholarships, and tuition waivers (full and partial) also available. Financial aid application deadline: 3/1.
Dr. Marie-Francoise Chesselet, Chair, 310-825-8153.
Application contact: School of Medicine Admissions, 310-825-6081, *E-mail:* neurophd@bri.medsch.ucla.edu.

■ **UNIVERSITY OF CALIFORNIA, RIVERSIDE**

Graduate Division, College of Natural and Agricultural Sciences, Program in Neuroscience, Riverside, CA 92521-0102

AWARDS PhD.

Faculty: 18 full-time (3 women).
Students: 3 full-time (1 woman); includes 1 minority (Asian American or Pacific Islander). Average age 27.
Degree requirements: For doctorate, dissertation, qualifying exams required, foreign language not required. *Average time to degree:* Doctorate–5 years full-time.
Entrance requirements: For doctorate, GRE General Test, TOEFL, minimum GPA of 3.2. *Application deadline:* For fall admission, 5/1; for winter admission, 9/1; for spring admission, 12/1. Applications

are processed on a rolling basis. *Application fee:* $40. Electronic applications accepted.
Expenses: Tuition, nonresident: full-time $9,804. Required fees: $4,758. Full-time tuition and fees vary according to program.
Financial aid: Fellowships, research assistantships, teaching assistantships, tuition waivers (full and partial) available. Financial aid application deadline: 1/5.
Faculty research: Modulation of ion channels, synaptic transmission, sensor-motor processing, plasticity in adult CNS, and neural control of eating behaviors.
Dr. Glenn I. Hatton, Chair, 909-787-4419, *Fax:* 909-787-2966, *E-mail:* ghatton@pop.ucr.edu.
Application contact: Helene Serewis, Graduate Student Affairs Officer, 800-735-0717, *Fax:* 909-787-5517, *E-mail:* biopgrad@pep.ycr.edu.

■ **UNIVERSITY OF CALIFORNIA, SAN DIEGO**

Graduate Studies and Research, Department of Biology, Program in Computational Neurobiology, La Jolla, CA 92093-0348

AWARDS PhD. Offered in association with the Salk Institute.

Degree requirements: For doctorate, dissertation required.
Entrance requirements: For doctorate, GRE General Test, pre-application beginning in September. *Application deadline:* For fall admission, 1/7. *Application fee:* $40.
Expenses: Tuition, nonresident: full-time $14,691. Required fees: $4,697. Full-time tuition and fees vary according to program.
Financial aid: Tuition waivers (full) and stipends available.
Application contact: 858-534-3835.

■ **UNIVERSITY OF CALIFORNIA, SAN DIEGO**

Graduate Studies and Research, Department of Biology, Program in Neurobiology, La Jolla, CA 92093

AWARDS PhD.

Degree requirements: For doctorate, dissertation required.
Entrance requirements: For doctorate, GRE General Test, GRE Subject Test, pre-application beginning in September. *Application deadline:* For fall admission, 1/8. *Application fee:* $40.
Expenses: Tuition, nonresident: full-time $14,691. Required fees: $4,697. Full-time tuition and fees vary according to program.

Application contact: Biology Graduate Admissions Committee, 858-534-3835.
Find an in-depth description at www.petersons.com/graduate.

■ **UNIVERSITY OF CALIFORNIA, SAN DIEGO**

Graduate Studies and Research, Interdisciplinary Program in Cognitive Science, La Jolla, CA 92093

AWARDS Cognitive science/anthropology (PhD); cognitive science/communication (PhD); cognitive science/computer science and engineering (PhD); cognitive science/linguistics (PhD); cognitive science/neuroscience (PhD); cognitive science/philosophy (PhD); cognitive science/psychology (PhD); cognitive science/sociology (PhD). Admissions through affiliated departments.

Faculty: 51 full-time (6 women).
Students: 12 full-time (4 women). Average age 26. *2 applicants, 100% accepted.* In 1999, 2 degrees awarded (100% entered university research/teaching).
Degree requirements: For doctorate, dissertation required. *Average time to degree:* Doctorate–6 years full-time.
Entrance requirements: For doctorate, GRE General Test. *Application deadline:* Applications are processed on a rolling basis. *Application fee:* $40.
Expenses: Tuition, nonresident: full-time $14,691. Required fees: $4,697. Full-time tuition and fees vary according to program.
Faculty research: Cognition, neurobiology of cognition, artificial intelligence, neural networks, psycholinguistics.
Walter J. Savitch, Director, 858-534-7141, *Fax:* 858-534-1128, *E-mail:* wsavitch@ucsd.edu.
Application contact: Graduate Coordinator, 858-534-7141, *Fax:* 858-534-1128, *E-mail:* gradinfo@cogsci.ucsd.edu.

■ **UNIVERSITY OF CALIFORNIA, SAN DIEGO**

School of Medicine and Graduate Studies and Research, Graduate Studies in Biomedical Sciences, La Jolla, CA 92093-0685

AWARDS Cell and molecular biology (PhD); molecular pathology (PhD); neuroscience (PhD); pharmacology (PhD); physiology (PhD); regulatory biology (PhD).

Faculty: 106.
Students: 219. *241 applicants, 23% accepted.* In 1999, 16 doctorates awarded.
Degree requirements: For doctorate, dissertation, qualifying exam required, foreign language not required.

University of California, San Diego *(continued)*

Entrance requirements: For doctorate, GRE General Test, TOEFL. *Application deadline:* For fall admission, 1/5. *Application fee:* $40.

Expenses: Program pays tuition, fees, health insurance, and stipend for all students in good standing.

Financial aid: Fellowships, research assistantships, career-related internships or fieldwork, tuition waivers (full), and stipends available.

Faculty research: Molecular and cellular biology, molecular and cellular pharmacology, cell and organ physiology.

Kim Barrett, Chair, 858-543-3726.

Application contact: Gina Butcher, Graduate Program Representative, 858-534-3982.

Find an in-depth description at www.petersons.com/graduate.

■ UNIVERSITY OF CALIFORNIA, SAN DIEGO

School of Medicine and Graduate Studies and Research, Neurosciences Program, La Jolla, CA 92093

AWARDS PhD.

Faculty: 32.

Students: 63 (29 women). *166 applicants, 17% accepted.* In 1999, 9 doctorates awarded.

Degree requirements: For doctorate, dissertation, qualifying exam required.

Entrance requirements: For doctorate, GRE General Test, GRE Subject Test, TOEFL. *Application deadline:* For fall admission, 1/5. *Application fee:* $40.

Expenses: Tuition, nonresident: full-time $9,804. Full-time tuition and fees vary according to program and student level.

Financial aid: Application deadline: 1/15.

Faculty research: Neurophysiology, neuropharmacology, neurochemistry.

William Kristan, Chair.

Application contact: Graduate Coordinator, 858-534-3377.

Find an in-depth description at www.petersons.com/graduate.

■ UNIVERSITY OF CALIFORNIA, SAN FRANCISCO

Graduate Division, Program in Neuroscience, San Francisco, CA 94143

AWARDS PhD.

Faculty: 24.

Students: In 1999, 5 doctorates awarded.

Degree requirements: For doctorate, dissertation required.

Entrance requirements: For doctorate, GRE General Test, GRE Subject Test.

Application deadline: For fall admission, 1/15. *Application fee:* $40.

Financial aid: Application deadline: 1/10.

Faculty research: Molecular neurobiology, synaptic plasticity, mechanisms of motor learning.

Louis Reichardt, Director, 415-476-2248.

Application contact: Patricia Arrandale, Program Assistant, 415-476-2248.

■ UNIVERSITY OF CHICAGO

Division of the Biological Sciences, Neurobiology, Pharmacology, and Cell Physiology, Committee on Neurobiology, Chicago, IL 60637-1513

AWARDS PhD.

Faculty: 45 full-time (14 women).

Students: 28 full-time (9 women). In 1999, 7 degrees awarded.

Degree requirements: For doctorate, dissertation, preliminary exam required.

Entrance requirements: For doctorate, GRE General Test, TOEFL. *Application deadline:* For fall admission, 1/5 (priority date). Applications are processed on a rolling basis. *Application fee:* $55.

Expenses: Tuition: Full-time $24,804; part-time $3,422 per course. Required fees: $390. Tuition and fees vary according to program.

Financial aid: In 1999–00, 28 students received aid; fellowships, institutionally sponsored loans available. Financial aid application deadline: 6/1.

Faculty research: Immunogenetic aspects of neurologic disease.

Application contact: Chairman, 773-705-3849, *E-mail:* neurobiology@chicago.edu.

■ UNIVERSITY OF CINCINNATI

Division of Research and Advanced Studies, College of Medicine, Graduate Programs in Medicine, Department of Cell Biology, Neurobiology and Anatomy, Cincinnati, OH 45221-0091

AWARDS Anatomy (PhD); cell biology (PhD); neurobiology (PhD).

Faculty: 14 full-time.

Students: 31 full-time (17 women), 2 part-time (both women); includes 3 minority (1 African American, 2 Asian Americans or Pacific Islanders), 12 international. *77 applicants, 18% accepted.* In 1999, 2 degrees awarded.

Degree requirements: For doctorate, dissertation, qualifying exam required. *Average time to degree:* Doctorate–5.6 years full-time.

Entrance requirements: For doctorate, GRE General Test, TOEFL. *Application deadline:* For fall admission, 2/1 (priority date). Applications are processed on a rolling basis. *Application fee:* $30.

Expenses: Tuition, state resident: full-time $5,139; part-time $196 per credit hour. Tuition, nonresident: full-time $10,326; part-time $369 per credit hour. Required fees: $561; $187 per quarter.

Financial aid: Tuition waivers (full) and unspecified assistantships available. Financial aid application deadline: 5/1.

Faculty research: Cell structure, molecular genetics. *Total annual research expenditures:* $5.3 million.

Dr. Peter Stambrook, Head, 513-558-5685, *Fax:* 513-556-4454, *E-mail:* peter.stambrook@uc.edu.

Application contact: Robert Brackenbury, Graduate Program Director, 513-558-6080, *Fax:* 513-556-4454, *E-mail:* robert.brackenbury@uc.edu.

■ UNIVERSITY OF CINCINNATI

Division of Research and Advanced Studies, College of Medicine, Graduate Programs in Medicine, Interdisciplinary PhD Study Program in Neuroscience, Cincinnati, OH 45221-0091

AWARDS PhD.

Students: *88 applicants, 9% accepted.*

Degree requirements: For doctorate, dissertation, qualifying exam required, foreign language not required.

Entrance requirements: For doctorate, GRE General Test, TOEFL. *Application deadline:* For fall admission, 2/1 (priority date). Applications are processed on a rolling basis. *Application fee:* $30.

Expenses: Tuition, state resident: full-time $5,139; part-time $196 per credit hour. Tuition, nonresident: full-time $10,326; part-time $369 per credit hour. Required fees: $561; $187 per quarter.

Financial aid: Tuition waivers (full) and unspecified assistantships available. Financial aid application deadline: 5/1.

Peter Stambrook, Department Chair, 513-558-5685, *Fax:* 513-558-4454, *E-mail:* peter.stambrook@uc.edu.

Application contact: Michael Lehman, Graduate Program Director, 513-558-7626, *Fax:* 513-558-4454, *E-mail:* michael.lehman@uc.edu.

Find an in-depth description at www.petersons.com/graduate.

■ UNIVERSITY OF COLORADO AT BOULDER

Graduate School, College of Arts and Sciences, Department of Environmental, Population, and Organic Biology, Boulder, CO 80309

AWARDS Animal behavior (MA, PhD); aquatic biology (MA, PhD); behavioral genetics (MA, PhD); ecology (MA, PhD); microbiology (MA, PhD); neurobiology (MA, PhD); plant and animal physiology (MA, PhD); plant and animal systematics (MA, PhD); population biology (MA, PhD); population genetics (MA, PhD).

Faculty: 39 full-time (9 women).
Students: 84 full-time (36 women), 14 part-time (7 women); includes 17 minority (6 Asian Americans or Pacific Islanders, 10 Hispanic Americans, 1 Native American). Average age 29. *147 applicants, 14% accepted.* In 1999, 7 master's, 13 doctorates awarded. Terminal master's awarded for partial completion of doctoral program.
Degree requirements: For master's, thesis or alternative, comprehensive exam required, foreign language not required; for doctorate, dissertation, comprehensive exam required, foreign language not required. *Average time to degree:* Master's–3 years full-time; doctorate–5 years full-time.
Entrance requirements: For master's, GRE General Test, GRE Subject Test, minimum undergraduate GPA of 3.0; for doctorate, GRE General Test, GRE Subject Test. *Application deadline:* For fall admission, 1/15 (priority date). *Application fee:* $40 ($60 for international students).
Expenses: Tuition, state resident: part-time $181 per credit hour. Tuition, nonresident: part-time $542 per credit hour. Required fees: $99 per term. Tuition and fees vary according to course load and program.
Financial aid: Fellowships, research assistantships, teaching assistantships, Federal Work-Study, institutionally sponsored loans, and tuition waivers (full) available. Financial aid application deadline: 3/1.
Faculty research: Evolution, developmental biology, behavior and neurobiology. *Total annual research expenditures:* $1.8 million.
Michael Breed, Chair, 303-492-8981, *Fax:* 303-492-8699, *E-mail:* michael.breed@colorado.edu.
Application contact: Jill Skarstadt, Graduate Secretary, 303-492-7654, *Fax:* 303-492-8699, *E-mail:* jill.skarstadt@colorado.edu.

■ UNIVERSITY OF COLORADO HEALTH SCIENCES CENTER

Graduate School, Programs in Biological and Medical Sciences, Department of Physiology and Biophysics, Denver, CO 80262
AWARDS Cellular and molecular physiology (PhD), including cellular, molecular, and developmental neuroscience.
Degree requirements: For doctorate, dissertation required, foreign language not required.
Entrance requirements: For doctorate, GRE, minimum GPA of 2.75.

Expenses: Tuition, state resident: full-time $1,512; part-time $56 per hour. Tuition, nonresident: full-time $7,209; part-time $267 per hour. Full-time tuition and fees vary according to course load and program.
Find an in-depth description at www.petersons.com/graduate.

■ UNIVERSITY OF COLORADO HEALTH SCIENCES CENTER

Graduate School, Programs in Biological and Medical Sciences, Program in Neuroscience, Denver, CO 80262
AWARDS PhD.
Degree requirements: For doctorate, dissertation required, foreign language not required.
Entrance requirements: For doctorate, GRE, minimum GPA of 2.75.
Expenses: Tuition, state resident: full-time $1,512; part-time $56 per hour. Tuition, nonresident: full-time $7,209; part-time $267 per hour. Full-time tuition and fees vary according to course load and program.

■ UNIVERSITY OF CONNECTICUT

Graduate School, College of Liberal Arts and Sciences, Biological Sciences Group, Department of Physiology and Neurobiology, Storrs, CT 06269
AWARDS Neurobiology (MS, PhD); physiology (MS, PhD).
Degree requirements: For doctorate, dissertation required.
Entrance requirements: For master's and doctorate, GRE General Test, GRE Subject Test, TOEFL.
Expenses: Tuition, state resident: full-time $5,118. Tuition, nonresident: full-time $13,298. Required fees: $1,022.
Find an in-depth description at www.petersons.com/graduate.

■ UNIVERSITY OF CONNECTICUT

Graduate School, College of Liberal Arts and Sciences, Field of Psychology, Storrs, CT 06269
AWARDS Behavioral neuroscience (PhD); biopsychology (MA, PhD); clinical psychology (PhD); cognition/instruction psychology (PhD); developmental psychology (PhD); ecological psychology (PhD); general experimental psychology (PhD); industrial and organizational psychology (PhD); language psychology (PhD); neuroscience (MA); social psychology (PhD).
Faculty: 48 full-time (12 women), 3 part-time/adjunct (1 woman).

Students: 132 full-time (79 women), 18 part-time (9 women); includes 16 minority (6 African Americans, 3 Asian Americans or Pacific Islanders, 6 Hispanic Americans, 1 Native American), 8 international. Average age 28. *415 applicants, 7% accepted.* In 1999, 18 master's awarded (15% found work related to degree, 85% continued full-time study); 30 doctorates awarded (37% entered university research/teaching, 53% found other work related to degree). Terminal master's awarded for partial completion of doctoral program.
Degree requirements: For master's, computer language, thesis or alternative required; for doctorate, computer language, dissertation, internship required. *Average time to degree:* Master's–2 years full-time; doctorate–5 years full-time.
Entrance requirements: For master's and doctorate, GRE General Test, GRE Subject Test. *Application deadline:* For fall admission, 1/1 (priority date). *Application fee:* $40 ($45 for international students).
Expenses: Tuition, state resident: full-time $5,118. Tuition, nonresident: full-time $13,298. Required fees: $1,022.
Financial aid: In 1999–00, 45 fellowships (averaging $4,500 per year), 23 research assistantships with tuition reimbursements (averaging $18,145 per year), 79 teaching assistantships with tuition reimbursements (averaging $18,145 per year) were awarded; career-related internships or fieldwork, Federal Work-Study, grants, and scholarships also available. Financial aid application deadline: 1/1; financial aid applicants required to submit FAFSA.
Faculty research: Behavioral neuroscience; physiology/chemistry of memory, sensation, motor control/motivation; cognitive, social, personality and language development; infancy/person perceptions and intergroup relations; anxiety, expectancies, health, and neuropsychology; speech, reading, perceiving.
Charles A. Lowe, Head.
Application contact: Nicole Dolat, Graduate Admissions, 860-486-3528, *Fax:* 860-486-2760, *E-mail:* futuregr@psych.psy.uconn.edu.

■ UNIVERSITY OF CONNECTICUT HEALTH CENTER

Graduate School, Programs in Biomedical Sciences, Program in Neuroscience, Farmington, CT 06030
AWARDS PhD.
Faculty: 38.
Students: 14 full-time (7 women), 5 international. In 1999, 2 degrees awarded.
Degree requirements: For doctorate, one foreign language (computer language can substitute), dissertation required.

University of Connecticut Health Center (continued)

Entrance requirements: For doctorate, GRE General Test, TOEFL, interview (recommended). *Application deadline:* For fall admission, 2/1 (priority date); for spring admission, 10/1. Applications are processed on a rolling basis. *Application fee:* $40 ($45 for international students).
Expenses: Tuition, state resident: full-time $5,272; part-time $293 per credit. Tuition, nonresident: full-time $13,696; part-time $761 per credit. Required fees: $320; $198 per semester. One-time fee: $50 full-time. Full-time tuition and fees vary according to course load, program and reciprocity agreements.
Financial aid: Fellowships available.
Faculty research: Biology of degeneration, regeneration, plasticity, and transplantation; biology of neurotransmission in the nervous system; cell structure of neural tissue and neural development.
Dr. Steven Potashner, Director, 860-679-4904, *Fax:* 860-679-3693.
Application contact: Marizta Barta, Information Contact, 860-679-4306, *Fax:* 860-679-1282, *E-mail:* barta@adp.uchc.edu.
Find an in-depth description at www.petersons.com/graduate.

■ **UNIVERSITY OF DELAWARE**

College of Arts and Science, Program in Neuroscience, Newark, DE 19716
AWARDS Neuroscience and biology (PhD); neuroscience and psychology (PhD).
Faculty: 16 full-time (1 woman), 2 part-time/adjunct (0 women).
Students: 7 full-time (6 women); includes 1 minority (Asian American or Pacific Islander), 3 international. *13 applicants, 31% accepted.* In 1999, 1 degree awarded.
Degree requirements: For doctorate, dissertation required, foreign language not required. *Average time to degree:* Doctorate–3 years full-time.
Entrance requirements: For doctorate, GRE General Test, GRE Subject Test (advanced biology). *Application deadline:* For fall admission, 2/1 (priority date). *Application fee:* $45. Electronic applications accepted.
Expenses: Tuition, state resident: full-time $4,380; part-time $243 per credit. Tuition, nonresident: full-time $12,750; part-time $708 per credit. Required fees: $15 per term. Tuition and fees vary according to program.
Financial aid: In 1999–00, 7 students received aid, including 2 research assistantships with full tuition reimbursements available (averaging $10,400 per year), 5 teaching assistantships with full tuition

reimbursements available (averaging $10,300 per year); fellowships, tuition waivers (full) also available. Financial aid application deadline: 2/1.
Seymour Levine, Director, 302-831-1191, *Fax:* 302-831-3645, *E-mail:* glevine@udel.edu.

■ **UNIVERSITY OF FLORIDA**

College of Medicine and Graduate School, Interdisciplinary Program in Biomedical Sciences, Concentration in Neuroscience, Gainesville, FL 32611
AWARDS PhD.

Degree requirements: For doctorate, dissertation required, foreign language not required.
Entrance requirements: For doctorate, GRE General Test, TOEFL, minimum GPA of 3.0. Electronic applications accepted.
Expenses: Tuition, state resident: part-time $144 per credit hour. Tuition, nonresident: part-time $505 per credit hour. Tuition and fees vary according to course level, course load and program.
Faculty research: Neural injury and repair, neurophysiology, neurotoxicology, cellular and molecular neurobiology, neuroimmunology and endocrinology.
Find an in-depth description at www.petersons.com/graduate.

■ **UNIVERSITY OF FLORIDA**

College of Medicine and Graduate School, Interdisciplinary Program in Biomedical Sciences, Department of Neuroscience, Gainesville, FL 32610
AWARDS MS, PhD. Terminal master's awarded for partial completion of doctoral program.

Degree requirements: For master's and doctorate, thesis/dissertation required, foreign language not required.
Entrance requirements: For master's and doctorate, GRE General Test, TOEFL, minimum GPA of 3.0. Electronic applications accepted.
Expenses: Tuition, state resident: part-time $144 per credit hour. Tuition, nonresident: part-time $505 per credit hour. Tuition and fees vary according to course level, course load and program.
Faculty research: Neural injury and repair, neuroimmunology and endocrinology, neurophysiology, neurotoxicology, cellular and molecular neurobiology.
Find an in-depth description at www.petersons.com/graduate.

■ **UNIVERSITY OF HARTFORD**

College of Arts and Sciences, Program in Neuroscience, West Hartford, CT 06117-1599
AWARDS MA. Part-time and evening/weekend programs available.

Faculty: 5 full-time (1 woman), 1 part-time/adjunct (0 women).
Students: 3 full-time (1 woman), 8 part-time (5 women); includes 1 minority (Hispanic American), 1 international. Average age 33. *2 applicants, 50% accepted.* In 1999, 6 degrees awarded.
Degree requirements: For master's, comprehensive and oral exams required, thesis optional, foreign language not required.
Entrance requirements: For master's, GRE General Test, GRE Subject Test, MCAT, TOEFL. *Application deadline:* Applications are processed on a rolling basis. *Application fee:* $40 ($55 for international students). Electronic applications accepted.
Expenses: Tuition: Full-time $6,570; part-time $365 per hour. Required fees: $50 per term. Full-time tuition and fees vary according to degree level, program and student level.
Financial aid: In 1999–00, 2 research assistantships (averaging $8,000 per year), 5 teaching assistantships with partial tuition reimbursements (averaging $5,000 per year) were awarded. Financial aid application deadline: 6/1.
Faculty research: Neurobiology of aging, central actions of neural steroids, neuroendocrine control of reproduction, retinopatheis in sharks, plasticity in the central nervous system. *Total annual research expenditures:* $15,900.
Dr. Jacob Harney, Head, 860-768-5372, *Fax:* 860-768-5002.
Application contact: Nancy Clubb-Lazzerini, Coordinator of Graduate Applications, 860-768-4373, *Fax:* 860-768-5160, *E-mail:* gettoknow@mail.hartford.edu.
Find an in-depth description at www.petersons.com/graduate.

■ **UNIVERSITY OF HAWAII AT MANOA**

Graduate Division, Specialization in Cell, Molecular, and Neuro Sciences, Honolulu, HI 96822
AWARDS MS, PhD. Program is interdisciplinary; degree is in specific discipline with the specialization in Cell, Molecular, and Neuro Sciences. Part-time programs available.

Faculty: 55 full-time (11 women).
Students: 60 full-time (34 women); includes 24 minority (22 Asian Americans

or Pacific Islanders, 2 Hispanic Americans), 13 international. Average age 30. *72 applicants, 36% accepted.*

Degree requirements: For doctorate, dissertation required, foreign language not required.

Entrance requirements: For doctorate, GRE Subject Test (recommended). *Application deadline:* For fall admission, 3/1; for spring admission, 9/1. Applications are processed on a rolling basis. *Application fee:* $25 ($50 for international students).

Expenses: Tuition, state resident: full-time $4,032; part-time $168 per credit. Tuition, nonresident: full-time $9,960; part-time $415 per credit. Required fees: $51 per semester. Part-time tuition and fees vary according to course load and program.

Financial aid: In 1999–00, 22 research assistantships with tuition reimbursements (averaging $15,000 per year), 5 teaching assistantships with tuition reimbursements (averaging $13,000 per year) were awarded; tuition waivers (full and partial) and unspecified assistantships also available. Financial aid application deadline: 2/1.

Faculty research: Anatomy, cellular neurobiology, genetics,. *Total annual research expenditures:* $500,000.

Dr. Martin D. Rayner, Chair, 808-956-7269, *Fax:* 808-956-6984, *E-mail:* martin@pbre.hawaii.edu.

■ UNIVERSITY OF ILLINOIS AT CHICAGO

Graduate College, College of Liberal Arts and Sciences, Department of Biological Sciences, Chicago, IL 60607-7128

AWARDS Cell and developmental biology (PhD); ecology and evolution (MS, DA, PhD); genetics and development (PhD); molecular biology (MS, PhD); neurobiology (MS, PhD); plant biology (MS, DA, PhD).

Faculty: 40 full-time (5 women).

Students: 100 full-time (47 women), 14 part-time (10 women); includes 13 minority (11 Asian Americans or Pacific Islanders, 2 Hispanic Americans), 42 international. Average age 29. *99 applicants, 36% accepted.* In 1999, 3 master's, 9 doctorates awarded.

Degree requirements: For master's, thesis required, foreign language not required; for doctorate, dissertation, preliminary exam required, foreign language not required.

Entrance requirements: For master's and doctorate, GRE General Test, GRE Subject Test, TOEFL, previous course work in physics, calculus, and organic chemistry; minimum GPA of 3.75 on a 5.0 scale. *Application deadline:* For fall admission, 6/1. Applications are processed on a

rolling basis. *Application fee:* $40 ($50 for international students). Electronic applications accepted.

Expenses: Tuition, state resident: full-time $3,750; part-time $1,250 per semester. Tuition, nonresident: full-time $10,588; part-time $3,530 per semester. Required fees: $507 per semester. Tuition and fees vary according to course load and program.

Financial aid: In 1999–00, 87 students received aid; fellowships, research assistantships, teaching assistantships, career-related internships or fieldwork, Federal Work-Study, traineeships, and tuition waivers (full) available. Financial aid application deadline: 3/1; financial aid applicants required to submit FAFSA.

Dr. Lon Kaufman, Head, 312-996-2213.

Application contact: Dr. Leo Miller, Director of Graduate Studies, 312-996-2220.

Find an in-depth description at www.petersons.com/graduate.

■ UNIVERSITY OF ILLINOIS AT CHICAGO

Graduate College, Program in Neuroscience, Chicago, IL 60607-7128

AWARDS PhD, MD/PhD. Admissions and degrees offered through participating Departments of Anatomy and Cell Biology, Biochemistry, Bioengineering, Biological Sciences, Chemistry, Pathology, Pharmacology, Physiology and Biophysics, and Psychology.

Faculty: 77 full-time (12 women).

Degree requirements: For doctorate, dissertation required, foreign language not required.

Entrance requirements: For doctorate, GRE General Test, TOEFL, minimum GPA of 3.75 on a 5.0 scale. *Application fee:* $40 ($50 for international students).

Expenses: Tuition, state resident: full-time $3,750; part-time $1,250 per semester. Tuition, nonresident: full-time $10,588; part-time $3,530 per semester. Required fees: $507 per semester. Tuition and fees vary according to course load and program.

Faculty research: Neurobiology and behavior.

Dr. Emanuel D. Pollack, Chairperson, 312-413-1552.

■ UNIVERSITY OF ILLINOIS AT URBANA–CHAMPAIGN

Graduate College, College of Liberal Arts and Sciences, School of Life Sciences, Neuroscience Program, Urbana, IL 61801

AWARDS PhD, MD/PhD.

Students: 40 full-time (9 women); includes 5 minority (1 African American, 2 Asian

Americans or Pacific Islanders, 2 Hispanic Americans), 11 international. *66 applicants, 9% accepted.* In 1999, 2 doctorates awarded.

Degree requirements: For doctorate, dissertation required.

Application deadline: Applications are processed on a rolling basis. *Application fee:* $40 ($50 for international students).

Expenses: Tuition, state resident: full-time $4,616. Tuition, nonresident: full-time $11,768. Full-time tuition and fees vary according to course load.

Financial aid: In 1999–00, 10 fellowships, 14 research assistantships, 8 teaching assistantships were awarded. Financial aid application deadline: 2/15.

William T. Greenough, Chairman, 217-333-4472, *Fax:* 217-244-1224, *E-mail:* wgreenou@uiuc.edu.

Application contact: Joan Cornell, Director of Graduate Studies, 217-333-3166, *Fax:* 217-244-1224, *E-mail:* jcornell@uiuc.edu.

■ THE UNIVERSITY OF IOWA

College of Medicine and Graduate College, Graduate Programs in Medicine, Neuroscience Graduate Program, Iowa City, IA 52242-1316

AWARDS PhD, MD/PhD.

Faculty: 84 full-time (14 women).

Students: 21 full-time (9 women), 2 international. Average age 27. *39 applicants, 33% accepted.* In 1999, 2 degrees awarded.

Degree requirements: For doctorate, dissertation, comprehensive exam required, foreign language not required. *Average time to degree:* Doctorate–5 years full-time.

Entrance requirements: For doctorate, GRE General Test, minimum GPA of 3.0. *Application deadline:* Applications are processed on a rolling basis. *Application fee:* $30 ($50 for international students). Electronic applications accepted.

Expenses: Tuition, state resident: full-time $3,308. Tuition, nonresident: full-time $10,662. Tuition and fees vary according to course load and program.

Financial aid: In 1999–00, 3 fellowships with full tuition reimbursements (averaging $17,000 per year), 18 research assistantships with full tuition reimbursements (averaging $16,770 per year) were awarded.

Faculty research: Molecular, cellular, and developmental systems; behavioral neurosciences.

Dr. Chun-Fang Wu, Interim Director, 319-335-6512, *Fax:* 319-335-7656, *E-mail:* chun-fang-wu@uiowa.edu.

Application contact: Paulette Scheler, Program Assistant, 800-551-6787, *Fax:* 319-335-7656, *E-mail:* interdis@blue.weeg.uiowa.edu.

■ THE UNIVERSITY OF IOWA

Graduate College, College of Liberal Arts, Department of Biological Sciences, Iowa City, IA 52242-1316
AWARDS MS, PhD.

Expenses: Tuition, state resident: full-time $3,308; part-time $184 per semester hour. Tuition, nonresident: full-time $10,662; part-time $184 per semester hour. Required fees: $93 per semester. Tuition and fees vary according to course load and program.
Linda Maxson, Dean, College of Liberal Arts, 319-335-2611, *Fax:* 319-335-3755.

■ THE UNIVERSITY OF IOWA

Graduate College, College of Liberal Arts, Department of Psychology, Iowa City, IA 52242-1316
AWARDS Neural and behavioral sciences (PhD); psychology (MA, PhD).

Faculty: 30 full-time.
Students: 45 full-time (32 women), 29 part-time (18 women); includes 8 minority (2 African Americans, 4 Asian Americans or Pacific Islanders, 1 Hispanic American, 1 Native American), 5 international. *220 applicants, 12% accepted.* In 1999, 7 master's, 9 doctorates awarded.
Degree requirements: For master's, exam required, thesis optional; for doctorate, dissertation, comprehensive exam required.
Entrance requirements: For master's and doctorate, GRE General Test. *Application deadline:* For fall admission, 12/15 (priority date). *Application fee:* $30 ($50 for international students). Electronic applications accepted.
Expenses: Tuition, state resident: full-time $3,308; part-time $184 per semester hour. Tuition, nonresident: full-time $10,662; part-time $184 per semester hour. Required fees: $93 per semester. Tuition and fees vary according to course load and program.
Financial aid: In 1999–00, 3 fellowships, 28 research assistantships, 31 teaching assistantships were awarded. Financial aid applicants required to submit FAFSA.
Michael W. O'Hara, Chair, 319-335-2406, *Fax:* 319-335-0191.

■ UNIVERSITY OF KENTUCKY

Graduate School and College of Medicine, Graduate Programs in Medicine, Program in Anatomy and Neurobiology, Lexington, KY 40506-0032
AWARDS PhD, MD/PhD.

Degree requirements: For doctorate, dissertation, comprehensive exam required, foreign language not required.

Entrance requirements: For doctorate, GRE General Test, minimum undergraduate GPA of 3.0.
Expenses: Tuition, state resident: full-time $3,596; part-time $188 per credit hour. Tuition, nonresident: full-time $10,116; part-time $550 per credit hour.
Faculty research: Neuroendocrinology, developmental neurobiology, neurotrophic substances, neural plasticity and trauma, neurobiology of aging.

■ UNIVERSITY OF LOUISVILLE

School of Medicine and Graduate School, Integrated Programs in Biomedical Sciences, Department of Anatomical Sciences and Neurobiology, Louisville, KY 40292-0001
AWARDS MS, PhD.

Degree requirements: For master's and doctorate, thesis/dissertation required.
Entrance requirements: For master's and doctorate, GRE General Test, TOEFL. Electronic applications accepted.
Expenses: Tuition, state resident: full-time $3,260; part-time $182 per hour. Tuition, nonresident: full-time $9,780; part-time $544 per hour. Required fees: $143; $28 per hour.

■ UNIVERSITY OF MARYLAND

Graduate School, Graduate Programs in Medicine, Department of Anatomy and Neurobiology, Baltimore, MD 21201-1627
AWARDS MS, PhD, MD/PhD. Part-time and evening/weekend programs available.

Degree requirements: For master's, thesis optional; for doctorate, one foreign language (computer language can substitute), dissertation required.
Entrance requirements: For master's, GRE General Test, TOEFL, minimum GPA of 3.0; for doctorate, GRE General Test, GRE Subject Test (recommended), TOEFL, minimum GPA of 3.0.
Expenses: Tuition, state resident: part-time $261 per credit hour. Tuition, nonresident: part-time $468 per credit hour. Tuition and fees vary according to program.
Faculty research: Neural networks, chemical sensory pathways, electrophysiology, developmental neurobiology.

■ UNIVERSITY OF MARYLAND

Graduate School, Graduate Programs in Medicine, Department of Physiology (Medicine), Baltimore, MD 21201-1627
AWARDS Neuroscience (PhD); physiology (PhD); reproductive endocrinology (PhD).

Degree requirements: For doctorate, dissertation required, foreign language not required.
Entrance requirements: For doctorate, GRE General Test, GRE Subject Test, TOEFL, minimum GPA of 3.0.
Expenses: Tuition, state resident: part-time $261 per credit hour. Tuition, nonresident: part-time $468 per credit hour. Tuition and fees vary according to program.
Faculty research: Membrane physiology, biophysics and morphology, central nervous system physiology, EEG analysis, information theory.

■ UNIVERSITY OF MARYLAND

Graduate School, Graduate Programs in Medicine, Program in Neuroscience and Cognitive Sciences, Baltimore, MD 21201-1627
AWARDS MS, PhD, MD/PhD. Part-time programs available. Terminal master's awarded for partial completion of doctoral program.

Degree requirements: For master's and doctorate, thesis/dissertation required, foreign language not required.
Entrance requirements: For master's and doctorate, GRE General Test, TOEFL, minimum GPA of 3.0.
Expenses: Tuition, state resident: part-time $261 per credit hour. Tuition, nonresident: part-time $468 per credit hour. Tuition and fees vary according to program.
Faculty research: Molecular, biochemical, and cellular pharmacology; membrane biophysics; synaptology; developmental neurobiology.

■ UNIVERSITY OF MARYLAND, BALTIMORE COUNTY

Graduate School, Department of Biological Sciences and Department of Psychology, Program in Neurosciences and Cognitive Sciences, Baltimore, MD 21250-5398
AWARDS MS, PhD.

Students: 1 full-time (0 women); minority (Asian American or Pacific Islander). *5 applicants, 20% accepted.*
Degree requirements: For master's, foreign language not required; for doctorate, dissertation required, foreign language not required.
Entrance requirements: For master's and doctorate, GRE General Test, TOEFL, minimum GPA of 3.0. *Application deadline:* For fall admission, 2/1. Applications are processed on a rolling basis. *Application fee:* $45.
Expenses: Tuition, state resident: part-time $268 per credit hour. Tuition,

nonresident: part-time $470 per credit hour. Required fees: $38 per credit hour. $557 per semester.

Financial aid: In 1999–00, 1 fellowship with tuition reimbursement (averaging $18,122 per year) was awarded.

Dr. Phyllis Robinson, Director, 410-455-3669, *Fax:* 410-455-3875, *E-mail:* biograd@umbc.edu.

Find an in-depth description at www.petersons.com/graduate.

■ UNIVERSITY OF MARYLAND, BALTIMORE COUNTY

Graduate School, Department of Chemistry and Biochemistry, Program in Biochemistry, Baltimore, MD 21250-5398

AWARDS Biochemistry (PhD); neuroscience (PhD).

Faculty: 6 full-time, 1 part-time/adjunct.

Students: 6 full-time (4 women), 1 part-time; includes 1 minority (Asian American or Pacific Islander), 2 international. *24 applicants, 25% accepted.* In 1999, 1 degree awarded.

Degree requirements: For doctorate, dissertation, comprehensive exams required, foreign language not required.

Entrance requirements: For doctorate, GRE General Test, GRE Subject Test, TOEFL, minimum GPA of 3.0. *Application deadline:* For fall admission, 4/15. Applications are processed on a rolling basis. *Application fee:* $45.

Expenses: Tuition, state resident: part-time $268 per credit hour. Tuition, nonresident: part-time $470 per credit hour. Required fees: $38 per credit hour. $557 per semester.

Financial aid: In 1999–00, research assistantships with full tuition reimbursements (averaging $17,000 per year), teaching assistantships with tuition reimbursements (averaging $17,000 per year) were awarded; fellowships.

Faculty research: Biochemical genetics, chemistry of proteins.

Dr. Michael Summers, Graduate Coordinator, 410-455-2527.

Find an in-depth description at www.petersons.com/graduate.

■ UNIVERSITY OF MARYLAND, BALTIMORE COUNTY

Graduate School, Department of Psychology, Baltimore, MD 21250-5398

AWARDS Applied developmental psychology (PhD); neurosciences and cognitive sciences (MS, PhD); psychology/human services (PhD), including psychology.

Degree requirements: For doctorate, dissertation required, foreign language not required.

Entrance requirements: For doctorate, GRE General Test, GRE Subject Test, TOEFL, minimum GPA of 3.0.

Expenses: Tuition, state resident: part-time $268 per credit hour. Tuition, nonresident: part-time $470 per credit hour. Required fees: $38 per credit hour. $557 per semester.

Faculty research: Health services, interviewing and speech pattern analysis.

■ UNIVERSITY OF MARYLAND, COLLEGE PARK

Graduate Studies and Research, Interdepartmental Programs, Program in Neurosciences and Cognitive Sciences, College Park, MD 20742

AWARDS PhD.

Students: 13 full-time (6 women), 3 part-time (all women); includes 1 minority (Asian American or Pacific Islander), 6 international. *47 applicants, 28% accepted.*

Degree requirements: For doctorate, dissertation, comprehensive exams required.

Entrance requirements: For doctorate, GRE General Test. *Application deadline:* For fall admission, 2/1. Applications are processed on a rolling basis. *Application fee:* $50 ($70 for international students). Electronic applications accepted.

Expenses: Tuition, state resident: part-time $272 per credit hour. Tuition, nonresident: part-time $415 per credit hour. Required fees: $632; $379 per year.

Financial aid: Fellowships, teaching assistantships, Federal Work-Study and grants available. Aid available to part-time students. Financial aid applicants required to submit FAFSA.

Faculty research: Molecular neurobiology, cognition, neural and behavioral systems language.

Dr. Arthur Popper, Director, 301-405-8910, *Fax:* 301-314-9358.

■ UNIVERSITY OF MASSACHUSETTS AMHERST

Graduate School, Interdisciplinary Programs, Program in Neuroscience and Behavior, Amherst, MA 01003-5810

AWARDS MS, PhD.

Students: 8 full-time (5 women), 22 part-time (10 women); includes 8 minority (2 African Americans, 3 Asian Americans or Pacific Islanders, 3 Hispanic Americans), 4 international. Average age 27. *48 applicants, 31% accepted.* In 1999, 6 doctorates awarded. Terminal master's awarded for partial completion of doctoral program.

Degree requirements: For master's and doctorate, thesis/dissertation required, foreign language not required.

Entrance requirements: For master's and doctorate, GRE General Test, GRE Subject Test. *Application deadline:* For fall admission, 1/15 (priority date); for spring admission, 10/1. Applications are processed on a rolling basis. *Application fee:* $40.

Expenses: Tuition, state resident: full-time $2,640; part-time $165 per credit. Tuition, nonresident: full-time $9,756; part-time $407 per credit. Required fees: $1,221 per term. One-time fee: $110. Full-time tuition and fees vary according to course load, campus/location and reciprocity agreements.

Financial aid: In 1999–00, 11 fellowships with full tuition reimbursements (averaging $1,515 per year), 2 research assistantships with full tuition reimbursements (averaging $3,111 per year) were awarded; teaching assistantships with full tuition reimbursements, career-related internships or fieldwork, Federal Work-Study, grants, scholarships, traineeships, and unspecified assistantships also available. Aid available to part-time students. Financial aid application deadline: 1/15.

Dr. Katherine Fite, Acting Director, 413-545-2046, *Fax:* 413-545-3243, *E-mail:* kfite@psych.umass.edu.

Application contact: 413-545-0721.

Find an in-depth description at www.petersons.com/graduate.

■ UNIVERSITY OF MASSACHUSETTS WORCESTER

Graduate School of Biomedical Sciences, Program in Neuroscience, Worcester, MA 01655-0115

AWARDS PhD.

Faculty: 24 full-time (2 women).

Students: 10 full-time (3 women); includes 1 minority (Hispanic American), 2 international.

Degree requirements: For doctorate, dissertation required, foreign language not required.

Entrance requirements: For doctorate, GRE General Test, 1 year of calculus, physics, organic chemistry and biology. *Application deadline:* For fall admission, 1/15 (priority date). *Application fee:* $25 ($50 for international students).

Expenses: Tuition, state resident: full-time $2,640. Tuition, nonresident: full-time $9,756. Required fees: $825. Full-time tuition and fees vary according to program.

Financial aid: In 1999–00, research assistantships with full tuition reimbursements (averaging $17,500 per year); unspecified assistantships also available.

Faculty research: Molecular, biophysical, and cellular techniques to investigate cell function with emphasis on signal

University of Massachusetts Worcester *(continued)*

transduction processes, cell growth, and proliferation.

Dr. Steven Treistman, Director, 508-856-6985.

■ UNIVERSITY OF MEDICINE AND DENTISTRY OF NEW JERSEY

Graduate School of Biomedical Sciences, Graduate Programs in Biomedical Sciences, Department of Neuroscience and Cell Biology, Piscataway, NJ 08854-5635

AWARDS MS, PhD.

Degree requirements: For master's and doctorate, thesis/dissertation, qualifying exam required, foreign language not required.

Entrance requirements: For master's, GRE General Test, GRE Subject Test (biology or chemistry), TOEFL; for doctorate, GRE General Test, GRE Subject Test (biology or chemistry), TOEFL, minimum undergraduate GPA of 3.0. *Application deadline:* For fall admission, 3/1 (priority date); for spring admission, 10/1. Applications are processed on a rolling basis. *Application fee:* $40.

Expenses: Tuition, state resident: part-time $270 per credit hour. Tuition, nonresident: part-time $407 per credit hour. Part-time tuition and fees vary according to campus/location and program.

Financial aid: Fellowships, research assistantships, teaching assistantships available. Financial aid application deadline: 5/1.

Faculty research: Neuronal growth factors, neuronal gene expression, neurogenetics, circulation controls, reproduction.

Dr. Ira B. Black, Chair, 732-235-5388, *E-mail:* black@mcbl.rutgers.edu.

Application contact: David Egger, Director of Admissions, 732-235-4522, *E-mail:* egger@umdnj.edu.

■ UNIVERSITY OF MEDICINE AND DENTISTRY OF NEW JERSEY

Graduate School of Biomedical Sciences, Graduate Programs in Biomedical Sciences, Department of Neurosciences, Newark, NJ 07107

AWARDS MS, PhD. Terminal master's awarded for partial completion of doctoral program.

Degree requirements: For master's, thesis required; for doctorate, dissertation, qualifying exam required, foreign language not required.

Entrance requirements: For master's, GRE General Test, TOEFL; for doctorate, GRE General Test, TOEFL, minimum GPA of 3.5. *Application deadline:* For fall admission, 2/1; for spring admission, 10/1. *Application fee:* $40.

Expenses: Tuition, state resident: part-time $270 per credit hour. Tuition, nonresident: part-time $407 per credit hour. Part-time tuition and fees vary according to campus/location and program.

Financial aid: Fellowships, research assistantships available. Financial aid application deadline: 5/1.

Dr. Barry Levin, Acting Chairperson, 973-676-1000, *Fax:* 973-972-5059, *E-mail:* levin@umdnj.edu.

Application contact: Dr. Henry E. Brezenoff, Dean, Graduate School of Biomedical Sciences, 973-972-5333, *Fax:* 973-972-7148, *E-mail:* hbrezeno@umdnj.edu.

Find an in-depth description at www.petersons.com/graduate.

■ UNIVERSITY OF MIAMI

Graduate School, College of Arts and Sciences, Department of Psychology, Coral Gables, FL 33124

AWARDS Applied developmental psychology (PhD); behavioral neuroscience (PhD); clinical psychology (PhD); health psychology (PhD).

Faculty: 31 full-time (11 women), 2 part-time/adjunct (1 woman).

Students: 66 full-time (48 women); includes 18 minority (4 African Americans, 5 Asian Americans or Pacific Islanders, 9 Hispanic Americans). Average age 25. *370 applicants, 5% accepted.* In 1999, 32 degrees awarded.

Degree requirements: For doctorate, computer language, dissertation, comprehensive exam required, foreign language not required. *Average time to degree:* Doctorate–5 years full-time.

Entrance requirements: For doctorate, GRE General Test, TOEFL, minimum GPA of 3.5. *Application deadline:* For fall admission, 12/1. *Application fee:* $50. Electronic applications accepted.

Expenses: Tuition: Full-time $15,336; part-time $852 per credit. Required fees: $174. Tuition and fees vary according to program.

Financial aid: In 1999–00, 8 fellowships with full tuition reimbursements (averaging $15,000 per year), 36 research assistantships with full tuition reimbursements (averaging $14,600 per year), 22 teaching assistantships with full tuition reimbursements (averaging $11,290 per year) were awarded; career-related internships or

fieldwork, institutionally sponsored loans, scholarships, and traineeships also available.

Faculty research: Depression, social and emotional development, stress and coping, AIDS and psychoneuroimmunology, children's peer relations. *Total annual research expenditures:* $10 million.

Dr. A. Rodney Wellens, Chairman, 305-284-2814, *Fax:* 305-284-3402.

Application contact: Patricia L. Perreira, Staff Associate, 305-284-2814, *Fax:* 305-284-3402, *E-mail:* inquire@mail.psy.miami.edu.

■ UNIVERSITY OF MIAMI

School of Medicine and Graduate School, Graduate Programs in Medicine, Neuroscience Program, Miami, FL 33101

AWARDS PhD, MD/PhD.

Faculty: 56 full-time (10 women).

Students: 15 full-time (6 women); includes 4 minority (1 Asian American or Pacific Islander, 3 Hispanic Americans), 3 international. Average age 25. *90 applicants, 7% accepted.* In 1999, 1 degree awarded (100% entered university research/teaching).

Degree requirements: For doctorate, dissertation, qualifying exam required, foreign language not required. *Average time to degree:* Doctorate–5 years full-time.

Entrance requirements: For doctorate, GRE General Test, TOEFL. *Application deadline:* For fall admission, 3/1; for spring admission, 12/1. Applications are processed on a rolling basis. *Application fee:* $0.

Expenses: Tuition, area resident: Part-time $899 per credit.

Financial aid: In 1999–00, 11 fellowships, 4 research assistantships were awarded; institutionally sponsored loans also available.

Faculty research: Cellular and molecular biology, transduction, nerve regeneration and embryonic development, membrane biophysics. *Total annual research expenditures:* $6 million.

Dr. John L. Bixby, Chairman, Neuroscience Program Steering Committee, 305-243-3368, *Fax:* 305-243-3368, *E-mail:* neurosci@mednet.med.miami.edu.

Find an in-depth description at www.petersons.com/graduate.

■ UNIVERSITY OF MICHIGAN

Horace H. Rackham School of Graduate Studies, Interdepartmental Program in Neuroscience, Ann Arbor, MI 48109

AWARDS PhD.

Faculty: 83 full-time (21 women).

Students: 40 full-time (21 women); includes 10 minority (3 African Americans,

5 Asian Americans or Pacific Islanders, 1 Hispanic American, 1 Native American). Average age 37. *47 applicants, 34% accepted.* In 1999, 1 doctorate awarded (100% found work related to degree).

Degree requirements: For doctorate, oral defense of dissertation, preliminary exam required. *Average time to degree:* Doctorate–5 years full-time.

Entrance requirements: For doctorate, GRE General Test, GRE Subject Test. *Application deadline:* For fall admission, 1/1. *Application fee:* $55.

Expenses: Tuition, state resident: full-time $10,316. Tuition, nonresident: full-time $20,922. Required fees: $185. Part-time tuition and fees vary according to course load and program.

Financial aid: In 1999–00, 22 fellowships with full tuition reimbursements (averaging $17,500 per year), 18 research assistantships with full tuition reimbursements (averaging $17,500 per year), teaching assistantships with full tuition reimbursements (averaging $17,500 per year) were awarded; traineeships also available.

Faculty research: Developmental neurobiology, molecular neurobiology, systems neurobiology, hearing and chemical senses.

Dr. Richard I. Hume, Director, 734-763-9638, *Fax:* 734-936-2690, *E-mail:* neuroscience.program@umich.edu.

Find an in-depth description at www.petersons.com/graduate.

■ **UNIVERSITY OF MINNESOTA, TWIN CITIES CAMPUS**

Graduate School, Graduate Program in Neuroscience, Minneapolis, MN 55455-0213

AWARDS PhD.

Entrance requirements: For doctorate, GRE General Test, TOEFL.

Expenses: Tuition, state resident: full-time $5,040; part-time $420 per credit. Tuition, nonresident: full-time $9,900; part-time $825 per credit. Full-time tuition and fees vary according to course load, program and reciprocity agreements.

Faculty research: Molecular, cellular, systems, developmental and behavioral neurobiology.

Find an in-depth description at www.petersons.com/graduate.

■ **THE UNIVERSITY OF NORTH CAROLINA AT CHAPEL HILL**

School of Medicine and Graduate School, Graduate Programs in Medicine, Curriculum in Neurobiology, Chapel Hill, NC 27599

AWARDS PhD.

Faculty: 74 full-time.

Students: 31 full-time (21 women); includes 2 minority (1 African American, 1 Hispanic American). *49 applicants, 24% accepted.* In 1999, 2 degrees awarded.

Degree requirements: For doctorate, dissertation, comprehensive exams required, foreign language not required. *Average time to degree:* Doctorate–5 years full-time.

Entrance requirements: For doctorate, GRE General Test, minimum GPA of 3.0. *Application deadline:* For fall admission, 3/1. *Application fee:* $55. Electronic applications accepted.

Expenses: Tuition, state resident: full-time $1,966. Tuition, nonresident: full-time $11,026. Required fees: $8,940. One-time fee: $15 full-time. Part-time tuition and fees vary according to course load.

Financial aid: In 1999–00, 19 fellowships with full tuition reimbursements (averaging $16,500 per year), 17 research assistantships with full tuition reimbursements (averaging $16,500 per year) were awarded; teaching assistantships

Dr. Gerry S. Oxford, Director, 919-962-7157, *Fax:* 919-966-6927, *E-mail:* gsox@med.unc.edu.

Application contact: Ann Marie Gray, Administrative Assistant, 919-966-1260, *Fax:* 919-966-4348, *E-mail:* annmarie@med.unc.edu.

■ **UNIVERSITY OF OKLAHOMA HEALTH SCIENCES CENTER**

College of Medicine and Graduate College, Graduate Programs in Medicine, Department of Neuroscience, Oklahoma City, OK 73190

AWARDS MS, PhD.

Students: 3 full-time (1 woman); includes 1 minority (Asian American or Pacific Islander), 1 international. *11 applicants, 27% accepted.*

Degree requirements: For master's, foreign language not required; for doctorate, dissertation required.

Entrance requirements: For master's and doctorate, GRE General Test, TOEFL. *Application fee:* $25 ($50 for international students).

Expenses: Tuition, state resident: part-time $90 per semester hour. Tuition, nonresident: part-time $264 per semester hour. Tuition and fees vary according to program.

Dr. O. Ray Kling, Dean, Graduate Programs in Medicine, 405-271-2085, *Fax:* 405-271-1155, *E-mail:* ray-kling@uokhsc.edu.

■ **UNIVERSITY OF OREGON**

Graduate School, College of Arts and Sciences, Department of Biology, Eugene, OR 97403

AWARDS Ecology and evolution (MA, MS, PhD); marine biology (MA, MS, PhD); molecular, cellular and genetic biology (PhD); neuroscience and development (PhD).

Faculty: 42 full-time (14 women), 5 part-time/adjunct (2 women).

Students: 69 full-time (33 women), 8 part-time (5 women); includes 6 minority (2 Asian Americans or Pacific Islanders, 3 Hispanic Americans, 1 Native American), 5 international. *18 applicants, 83% accepted.* In 1999, 11 master's, 7 doctorates awarded. Terminal master's awarded for partial completion of doctoral program.

Degree requirements: For master's, thesis required (for some programs); for doctorate, dissertation required, foreign language not required.

Entrance requirements: For master's and doctorate, GRE General Test, TOEFL, minimum GPA of 3.2. *Application deadline:* For fall admission, 1/10. *Application fee:* $50.

Expenses: Tuition, state resident: full-time $6,750. Tuition, nonresident: full-time $11,409. Part-time tuition and fees vary according to course load.

Financial aid: In 1999–00, 36 teaching assistantships were awarded; research assistantships, Federal Work-Study, grants, and institutionally sponsored loans also available. Financial aid application deadline: 2/1.

Faculty research: Developmental neurobiology; evolution, population biology, and quantitative genetics; regulation of gene expression; biochemistry of marine organisms.

Janis C. Weeks, Head, 541-346-4502, *Fax:* 541-346-6056.

Application contact: Donna Overall, Graduate Program Coordinator, 541-346-4503, *Fax:* 541-346-6056, *E-mail:* doverall@oregon.uoregon.edu.

Find an in-depth description at www.petersons.com/graduate.

■ **UNIVERSITY OF PENNSYLVANIA**

School of Arts and Sciences, Graduate Group in Biology, Program in Neurobiology/Physiology and Behavior, Philadelphia, PA 19104

AWARDS PhD.

Degree requirements: For doctorate, dissertation required, foreign language not required.

Entrance requirements: For doctorate, GRE General Test, GRE Subject Test, TOEFL. *Application fee:* $65.

University of Pennsylvania (continued)
Expenses: Tuition: Full-time $23,670. Required fees: $1,546. Full-time tuition and fees vary according to degree level and program.
Financial aid: Fellowships, research assistantships, teaching assistantships available. Financial aid application deadline: 1/2.
Application contact: Allan Aiken, Graduate Group Coordinator, 215-898-6786, *Fax:* 215-898-8780, *E-mail:* aaiken@sas.upenn.edu.

■ UNIVERSITY OF PENNSYLVANIA

School of Medicine, Biomedical Graduate Studies, Graduate Group in Neuroscience, Philadelphia, PA 19104
AWARDS PhD, MD/PhD.

Students: 99 full-time (41 women); includes 24 minority (2 African Americans, 19 Asian Americans or Pacific Islanders, 2 Hispanic Americans, 1 Native American), 6 international. *149 applicants, 17% accepted.* In 1999, 12 doctorates awarded.
Degree requirements: For doctorate, dissertation, research project required, foreign language not required.
Entrance requirements: For doctorate, GRE General Test, TOEFL. *Application deadline:* For fall admission, 1/2 (priority date). Applications are processed on a rolling basis. *Application fee:* $65. Electronic applications accepted.
Expenses: Tuition: Full-time $17,256; part-time $2,991 per course. Required fees: $2,588; $363 per course. $726 per term.
Financial aid: In 1999–00, 58 students received aid, including 33 research assistantships, 2 teaching assistantships; fellowships, grants and institutionally sponsored loans also available. Financial aid application deadline: 1/2.
Faculty research: Molecular and cellular neuroscience, behavioral neuroscience, developmental neurobiology, systems neuroscience and neurophysiology, neurochemistry.
Dr. Michael Nusbaum, Chairperson, 215-898-8048, *Fax:* 215-573-2248.
Application contact: Fiona Cowan, Coordinator, 215-898-8048, *Fax:* 215-573-2248, *E-mail:* fiona@mail.med.upenn.edu.
Find an in-depth description at www.petersons.com/graduate.

■ UNIVERSITY OF PITTSBURGH

Center for Neuroscience, Pittsburgh, PA 15260
AWARDS Neurobiology (PhD); neuroscience (PhD).

Degree requirements: For doctorate, dissertation required.
Entrance requirements: For doctorate, GRE, TOEFL, interview.
Expenses: Tuition, state resident: full-time $8,338; part-time $342 per credit. Tuition, nonresident: full-time $17,168; part-time $707 per credit. Required fees: $480; $90 per semester. Tuition and fees vary according to program.
Faculty research: Molecular basis of cellular communication, neural development, psychiatric and neurological disorders, cognitive neuroscience, information processing in brain circuits, homeostatic regulatory systems.
Find an in-depth description at www.petersons.com/graduate.

■ UNIVERSITY OF PITTSBURGH

School of Medicine, Graduate Programs in Medicine, Program in Neurobiology, Pittsburgh, PA 15260
AWARDS MS, PhD.

Students: In 1999, 2 master's, 1 doctorate awarded.
Degree requirements: For doctorate, dissertation required, foreign language not required.
Entrance requirements: For doctorate, GRE General Test, GRE Subject Test, TOEFL, minimum QPA of 3.0. *Application fee:* $30 ($40 for international students).
Expenses: Tuition, state resident: full-time $9,778; part-time $403 per credit. Tuition, nonresident: full-time $20,146; part-time $830 per credit. Required fees: $480; $90 per semester.
Faculty research: Development and plasticity, biophysics and signal transduction, neural systems and computational modeling.
Application contact: Graduate Studies Administrator, 412-648-8957, *Fax:* 412-648-1236, *E-mail:* biomed_phd@fs1.deanmed.pitt.edu.

■ UNIVERSITY OF ROCHESTER

School of Medicine and Dentistry, Graduate Programs in Medicine and Dentistry, Department of Neurobiology and Anatomy, Program in Neuroscience, Rochester, NY 14627-0250
AWARDS MS, PhD.

Students: 20 full-time (10 women); includes 4 minority (3 Asian Americans or Pacific Islanders, 1 Hispanic American), 1 international. In 1999, 3 master's, 1 doctorate awarded. Terminal master's awarded for partial completion of doctoral program.
Degree requirements: For doctorate, one foreign language, dissertation, qualifying exam required.

Entrance requirements: For master's and doctorate, GRE General Test. *Application deadline:* For fall admission, 2/1. *Application fee:* $25.
Expenses: Tuition: Part-time $697 per credit hour. Tuition and fees vary according to program.
Financial aid: Fellowships, research assistantships, teaching assistantships, tuition waivers (full and partial) available. Financial aid application deadline: 2/1.
Dr. Howard Federoff, Director, 716-275-5788.
Application contact: Jennifer Dwyer, Graduate Program Secretary, 716-275-5788.
Find an in-depth description at www.petersons.com/graduate.

■ UNIVERSITY OF SOUTH ALABAMA

College of Medicine and Graduate School, Program in Basic Medical Sciences, Specialization in Cellular Biology and Neuroscience, Mobile, AL 36688-0002
AWARDS PhD.

Faculty: 13 full-time (2 women), 1 (woman) part-time/adjunct.
Degree requirements: For doctorate, dissertation, oral and written preliminary exams, research proposal, qualifying exam required, foreign language not required.
Entrance requirements: For doctorate, GRE General Test, TOEFL, minimum GPA of 3.0. *Application deadline:* For fall admission, 4/30. Applications are processed on a rolling basis. *Application fee:* $25.
Expenses: Tuition, state resident: part-time $116 per semester hour. Tuition, nonresident: part-time $230 per semester hour. Required fees: $121 per semester.
Financial aid: Fellowships, institutionally sponsored loans available. Financial aid application deadline: 4/1.
Faculty research: Cytoskeleton-membrane interactions, neural basis of oral motor behavior, microtubule organizing centers, molecular biology of human chromosomes, mechanisms of synaptic transmissions.
Dr. Steven R. Goodman, Chair, 334-460-6490.
Application contact: Lanette Flagge, Coordinator, 334-460-6153.
Find an in-depth description at www.petersons.com/graduate.

■ UNIVERSITY OF SOUTH DAKOTA

School of Medicine and Graduate School, Biomedical Sciences Graduate Program, Program in Neuroscience, Vermillion, SD 57069-2390

AWARDS MA, PhD.

Faculty: 6 full-time (4 women), 2 part-time/adjunct (0 women).
Students: 3 full-time (2 women). Average age 24. *3 applicants, 33% accepted.* Terminal master's awarded for partial completion of doctoral program.
Degree requirements: For master's and doctorate, thesis/dissertation required, foreign language not required.
Entrance requirements: For master's and doctorate, GRE General Test, TOEFL, minimum GPA of 3.0 required. *Application deadline:* For fall admission, 3/15. Applications are processed on a rolling basis. *Application fee:* $15.
Expenses: Tuition, state resident: full-time $2,126; part-time $89 per credit. Tuition, nonresident: full-time $6,270; part-time $261 per credit. Required fees: $1,194; $50 per credit. Tuition and fees vary according to course load and reciprocity agreements.
Financial aid: In 1999–00, 2 research assistantships with full tuition reimbursements (averaging $13,500 per year) were awarded; tuition waivers (partial) also available.
Faculty research: Central nervous system learning, neural plasticity, respiratory control.
Eevlyn Schlenker, Coordinator, 677-677-5160, *Fax:* 605-677-6381, *E-mail:* eschlenk@usd.edu.

■ UNIVERSITY OF SOUTHERN CALIFORNIA

Graduate School, College of Letters, Arts and Sciences, Department of Biological Sciences, Program in Neurobiology, Los Angeles, CA 90089

AWARDS PhD.

Students: 64 full-time (22 women), 1 part-time; includes 13 minority (2 African Americans, 9 Asian Americans or Pacific Islanders, 2 Hispanic Americans), 21 international. Average age 30. *87 applicants, 21% accepted.* In 1999, 11 degrees awarded.
Degree requirements: For doctorate, dissertation required.
Entrance requirements: For doctorate, GRE General Test, GRE Subject Test. *Application deadline:* For fall admission, 1/15 (priority date). *Application fee:* $55.
Expenses: Tuition: Full-time $17,952; part-time $748 per unit. Required fees: $406; $203 per unit. Tuition and fees vary according to program.

Financial aid: In 1999–00, 3 fellowships, 15 research assistantships, 9 teaching assistantships were awarded; Federal Work-Study, institutionally sponsored loans, and scholarships also available. Aid available to part-time students. Financial aid application deadline: 2/15; financial aid applicants required to submit FAFSA. Dr. Lou Byerly, Director, 213-740-1109.
Find an in-depth description at www.petersons.com/graduate.

■ UNIVERSITY OF SOUTHERN CALIFORNIA

Graduate School, College of Letters, Arts and Sciences, Program in Neuroscience, Los Angeles, CA 90089

AWARDS PhD.

Students: 26 full-time (13 women), 1 part-time; includes 8 minority (2 African Americans, 6 Asian Americans or Pacific Islanders), 7 international. Average age 28. *64 applicants, 19% accepted.*
Degree requirements: For doctorate, dissertation required.
Entrance requirements: For doctorate, GRE General Test. *Application deadline:* For fall admission, 1/15 (priority date). *Application fee:* $55.
Expenses: Tuition: Full-time $17,952; part-time $748 per unit. Required fees: $406; $203 per unit. Tuition and fees vary according to program.
Financial aid: In 1999–00, 4 fellowships, 9 research assistantships, 5 teaching assistantships were awarded; Federal Work-Study, institutionally sponsored loans, and scholarships also available. Aid available to part-time students. Financial aid application deadline: 2/15; financial aid applicants required to submit FAFSA. Richard F. Thompson, Director.

■ UNIVERSITY OF SOUTHERN CALIFORNIA

Keck School of Medicine and Graduate School, Graduate Programs in Medicine, Department of Cell and Neurobiology, Los Angeles, CA 90089

AWARDS Anatomy and cell biology (MS, PhD), including anatomy (PhD); cell biology (PhD); cell and neurobiology (MS, PhD); pharmacology and nutrition (MS, PhD); preventive nutrition (MS).

Faculty: 21 full-time (3 women), 8 part-time/adjunct (2 women).
Students: 19 full-time (9 women); includes 10 minority (all Asian Americans or Pacific Islanders). Average age 23. *54 applicants, 22% accepted.* In 1999, 5 master's awarded (40% found work related to degree, 60% continued full-time study). Terminal master's awarded for partial completion of doctoral program.

Degree requirements: For master's, thesis or alternative required, foreign language not required; for doctorate, dissertation required. *Average time to degree:* Master's–2 years full-time; doctorate–5 years full-time.
Entrance requirements: For master's, GRE General Test, TOEFL, minimum GPA of 3.0; for doctorate, GRE General Test, TOEFL. *Application fee:* $55. Electronic applications accepted.
Expenses: Tuition: Full-time $22,198; part-time $748 per unit. Required fees: $406.
Financial aid: In 1999–00, 13 students received aid; fellowships, research assistantships, teaching assistantships, Federal Work-Study, institutionally sponsored loans, and tuition waivers (partial) available. Aid available to part-time students.
Faculty research: Neurobiology and development, circaulian rhythm, gene therapy in vision, lacrimal glands, neuroendocrinology, signal transduction mechanisms.
Dr. Cheryl Craft, Chair, 323-442-1881, *Fax:* 323-442-2709, *E-mail:* ccraft@hsc.usc.edu.
Application contact: Darlene Marie Campbell, Administrative Assistant, 323-442-1881, *Fax:* 323-442-0466, *E-mail:* dmc@hsc.usc.edu.

■ THE UNIVERSITY OF TENNESSEE HEALTH SCIENCE CENTER

College of Graduate Health Sciences, Department of Anatomy and Neurobiology, Memphis, TN 38163-0002

AWARDS PhD.

Degree requirements: For doctorate, dissertation, oral and written preliminary and comprehensive exams required, foreign language not required.
Entrance requirements: For doctorate, GRE General Test, minimum GPA of 3.0.
Find an in-depth description at www.petersons.com/graduate.

■ THE UNIVERSITY OF TEXAS AT AUSTIN

Graduate School, The Institute for Neuroscience, Austin, TX 78712-1111

AWARDS MA, PhD.

Faculty: 53 full-time.
Students: 11 full-time (6 women).
Entrance requirements: For doctorate, GRE. *Application deadline:* For fall admission, 1/15 (priority date). *Application fee:* $50 ($75 for international students). Electronic applications accepted.

The University of Texas at Austin (continued)

Expenses: Tuition, state resident: part-time $114 per semester hour. Tuition, nonresident: part-time $330 per semester hour. Tuition and fees vary according to program.

Financial aid: In 1999–00, 11 students received aid, including teaching assistantships with tuition reimbursements available (averaging $12,500 per year); fellowships with tuition reimbursements available, research assistantships with tuition reimbursements available Financial aid application deadline: 2/1.

Faculty research: Cellular/molecular biology, neurobiology, pharmacology, behavioral neuroscience.

Dr. Creed W. Abell, Director, 512-471-3640, *Fax:* 512-471-2181, *E-mail:* dirins@uts.cc.utexas.edu.

Application contact: Dr. Timothy Schallert, Graduate Adviser, 512-471-3640, *Fax:* 512-471-0390, *E-mail:* ins_uta@psy.utexas.edu.

■ **THE UNIVERSITY OF TEXAS AT DALLAS**

School of Human Development, Program in Applied Cognition and Neuroscience, Richardson, TX 75083-0688

AWARDS MS. Part-time and evening/weekend programs available.

Faculty: 22 full-time (6 women).

Students: 8 full-time (7 women), 7 part-time (4 women); includes 1 minority (African American), 1 international. Average age 32. In 1999, 7 degrees awarded.

Degree requirements: For master's, internship required, foreign language and thesis not required.

Entrance requirements: For master's, GRE General Test, TOEFL, minimum GPA of 3.0 in upper-level course work in field. *Application deadline:* For fall admission, 7/15; for spring admission, 11/15. Applications are processed on a rolling basis. *Application fee:* $25 ($75 for international students). Electronic applications accepted.

Expenses: Tuition, state resident: full-time $2,052; part-time $76 per semester hour. Tuition, nonresident: full-time $5,256; part-time $292 per semester hour. Required fees: $1,504; $656 per year. One-time fee: $10. Full-time tuition and fees vary according to course level, course load, degree level and program.

Financial aid: Fellowships, research assistantships, teaching assistantships, Federal Work-Study available. Aid available to part-time students. Financial aid application deadline: 4/30; financial aid applicants required to submit FAFSA.

Faculty research: Combination of biological, behavioral, and computational approaches for evaluating biological and artificial information processing systems. Dr. W. Jay Dowling, Head, 972-883-2059, *Fax:* 972-883-2491, *E-mail:* jdowling@utdallas.edu.

Application contact: Dr. Robert D. Stillman, Associate Dean, 972-883-3106, *Fax:* 972-883-3022, *E-mail:* stillman@utdallas.edu.

■ **THE UNIVERSITY OF TEXAS AT SAN ANTONIO**

College of Sciences and Engineering, Division of Life Sciences, Program in Biology, San Antonio, TX 78249-0617

AWARDS Neurobiology (PhD).

Degree requirements: For doctorate, dissertation, comprehensive exam required.

Entrance requirements: For doctorate, GRE General Test, TOEFL, minimum GPA of 3.0.

Expenses: Program offers an annual stipend of $15,000 and provides tuition and fees.

Faculty research: Neurophysiology, neurotoxicology, neural circuit analysis, neuroendocrinology, development of biosensors for the detection of toxins, protein sorting, signal transduction.

■ **THE UNIVERSITY OF TEXAS– HOUSTON HEALTH SCIENCE CENTER**

Graduate School of Biomedical Sciences, Program in Neuroscience, Houston, TX 77225-0036

AWARDS MS, PhD, MD/PhD.

Faculty: 35 full-time (6 women).

Students: 17 full-time (12 women); includes 3 minority (2 Asian Americans or Pacific Islanders, 1 Native American), 2 international. Average age 27. *28 applicants, 54% accepted.* In 1999, 1 master's, 3 doctorates awarded. Terminal master's awarded for partial completion of doctoral program.

Degree requirements: For master's and doctorate, thesis/dissertation required, foreign language not required.

Entrance requirements: For master's and doctorate, GRE General Test, TOEFL, TWE. *Application deadline:* For fall admission, 1/15 (priority date); for spring admission, 11/1. Applications are processed on a rolling basis. *Application fee:* $10. Electronic applications accepted.

Financial aid: Research assistantships, teaching assistantships, institutionally sponsored loans available. Financial aid application deadline: 1/15.

Faculty research: Molecular and cellular neurobiology, neuroplasticity and memory, computer modeling of neural circuits, visual sciences, neurotrauma and stroke. Dr. Jack C. Waymire, Director, 713-500-5620, *Fax:* 713-500-0621, *E-mail:* jwaymire@nba19.med.uth.tmc.edu.

Application contact: Anne Baronitis, Director of Admissions, 713-500-9860, *Fax:* 713-500-9877, *E-mail:* abaron@gsbs.gs.uth.tmc.edu.

Find an in-depth description at www.petersons.com/graduate.

■ **THE UNIVERSITY OF TEXAS MEDICAL BRANCH AT GALVESTON**

Graduate School of Biomedical Sciences, Program in Neuroscience, Galveston, TX 77555

AWARDS PhD.

Faculty: 46 full-time (15 women).

Students: 24 full-time (7 women), 3 part-time (1 woman); includes 5 minority (1 Asian American or Pacific Islander, 3 Hispanic Americans, 1 Native American), 7 international. Average age 29. *39 applicants, 18% accepted.* In 1999, 1 degree awarded.

Degree requirements: For doctorate, dissertation required, foreign language not required.

Entrance requirements: For doctorate, GRE General Test. *Application deadline:* For fall admission, 3/1. Applications are processed on a rolling basis. *Application fee:* $25 ($50 for international students). Electronic applications accepted.

Expenses: Tuition, state resident: full-time $684; part-time $38 per credit hour. Tuition, nonresident: full-time $4,572; part-time $254 per credit hour. Required fees: $29; $7.5 per credit hour. One-time fee: $55. Tuition and fees vary according to program.

Financial aid: Research assistantships, Federal Work-Study and institutionally sponsored loans available. Financial aid applicants required to submit FAFSA.

Faculty research: Pain mechanisms, repair/regeneration, synaptic physiology/pharmacology, peptides/neuroendocrinology, drug abuse, oculomotor control, behavioral science. Dr. James E. Blankenship, Director, 409-772-2267, *Fax:* 409-762-9382, *E-mail:* jeblanke@utmb.edu.

Application contact: Lonnell Simmons, Coordinator, 409-772-2267, *Fax:* 409-762-9382, *E-mail:* losimmon@utmb.edu.

Find an in-depth description at www.petersons.com/graduate.

■ THE UNIVERSITY OF TEXAS SOUTHWESTERN MEDICAL CENTER AT DALLAS

Southwestern Graduate School of Biomedical Sciences, Division of Cell and Molecular Biology, Program in Neuroscience, Dallas, TX 75390

AWARDS PhD.

Faculty: 16 full-time (2 women), 2 part-time/adjunct (0 women).
Students: 8 full-time (4 women), 2 international. Average age 28. In 1999, 1 doctorate awarded.
Degree requirements: For doctorate, dissertation required, foreign language not required.
Entrance requirements: For doctorate, GRE General Test, minimum GPA of 3.0. *Application deadline:* For fall admission, 1/5 (priority date). *Application fee:* $0. Electronic applications accepted.
Expenses: Tuition, state resident: full-time $912. Tuition, nonresident: full-time $6,096. Required fees: $216. Full-time tuition and fees vary according to course load and program.
Financial aid: Fellowships, research assistantships available. Financial aid application deadline: 3/15; financial aid applicants required to submit FAFSA.
Faculty research: Ion channels, sensory transduction, membrane excitability and biophysics, synaptic transmission, developmental neurogenetics.
Dr. Thomas C. Südhof, Chair, 214-648-1802, *Fax:* 214-648-1801, *E-mail:* tsudho@mednet.swmed.edu.
Application contact: Nancy McKinney, Education Coordinator, 214-648-8099, *Fax:* 214-648-2978, *E-mail:* dcmbinfo@utsouthwestern.edu.

Find an in-depth description at www.petersons.com/graduate.

■ UNIVERSITY OF UTAH

School of Medicine and Graduate School, Graduate Programs in Medicine, Department of Neurology and Anatomy, Salt Lake City, UT 84112-1107

AWARDS M Phil, MS, PhD. Part-time programs available.

Faculty: 10 full-time (3 women), 3 part-time/adjunct (1 woman).
Students: 8 full-time (4 women), 1 (woman) part-time. Average age 26. *2 applicants, 100% accepted.* Terminal master's awarded for partial completion of doctoral program.
Degree requirements: For master's and doctorate, one foreign language, computer language, thesis/dissertation required.
Entrance requirements: For master's and doctorate, GRE. *Application deadline:* For

fall admission, 2/15. Applications are processed on a rolling basis. *Application fee:* $30 ($50 for international students).
Expenses: Tuition, state resident: full-time $2,105. Tuition, nonresident: full-time $6,312.
Financial aid: In 1999–00, 9 students received aid, including 9 research assistantships with tuition reimbursements available (averaging $17,000 per year); fellowships, teaching assistantships Financial aid application deadline: 2/15.
Faculty research: Neuroscience, neuroanatomy, developmental neurobiology, neurogenetics. *Total annual research expenditures:* $1.5 million.
Tom Parks, Chair, 801-581-5494, *Fax:* 801-585-9736, *E-mail:* tom.parks@hsc.utah.edu.
Application contact: Kathleen A. Kjaglien, Administrative Officer, 801-581-5494, *Fax:* 801-585-9736, *E-mail:* kathleen.kjaglien@usc.utah.edu.

■ UNIVERSITY OF UTAH

School of Medicine and Graduate School, Graduate Programs in Medicine, Program in Neuroscience, Salt Lake City, UT 84112-1107

AWARDS PhD.

Faculty: 66 full-time (14 women).
Students: 30 full-time (9 women). Average age 25. *48 applicants, 29% accepted.* In 1999, 1 degree awarded (100% entered university research/teaching).
Degree requirements: For doctorate, dissertation required, foreign language not required. *Average time to degree:* Doctorate–6 years full-time.
Entrance requirements: For doctorate, GRE General Test, minimum GPA of 3.0. *Application deadline:* For fall admission, 1/15 (priority date). Electronic applications accepted.
Expenses: Tuition, state resident: full-time $2,105. Tuition, nonresident: full-time $6,312.
Financial aid: In 1999–00, 2 students received aid, including 1 fellowship (averaging $10,000 per year); tuition waivers (full) also available. Financial aid application deadline: 2/15.
Faculty research: Brain and behavioral neuroscience, cellular neuroscience, molecular neuroscience, neurobiology of disease, developmental neuroscience.
Dr. Eric M. Lasater, Director, 801-585-6503, *Fax:* 801-581-3357, *E-mail:* eric.lasater@hsc.utah.edu.
Application contact: Dr. Monica Vetter, Chair, Admissions and Recruiting, 801-581-4984, *Fax:* 801-581-4233, *E-mail:* monica.vetter@hsc.utah.edu.

Find an in-depth description at www.petersons.com/graduate.

■ UNIVERSITY OF VERMONT

College of Medicine and Graduate College, Graduate Programs in Medicine, Department of Anatomy and Neurobiology, Burlington, VT 05405

AWARDS PhD, MD/PhD.

Degree requirements: For doctorate, dissertation required.
Entrance requirements: For doctorate, GRE General Test, TOEFL.
Expenses: Tuition, state resident: full-time $7,464; part-time $311 per credit. Tuition, nonresident: full-time $18,672; part-time $778 per credit. Full-time tuition and fees vary according to degree level and program.
Faculty research: Autonomic neurobiology, developmental neurobiology, neurotransmitter expression and release, plasticity and regeneration.

Find an in-depth description at www.petersons.com/graduate.

■ UNIVERSITY OF VIRGINIA

College and Graduate School of Arts and Sciences, Program in Neuroscience, Charlottesville, VA 22903

AWARDS PhD.

Faculty: 5 full-time (2 women).
Students: 20 full-time (10 women); includes 3 minority (1 African American, 1 Asian American or Pacific Islander, 1 Hispanic American), 1 international. Average age 27. *59 applicants, 17% accepted.* In 1999, 6 degrees awarded.
Degree requirements: For doctorate, dissertation required.
Entrance requirements: For doctorate, GRE General Test, GRE Subject Test. *Application deadline:* For fall admission, 7/15; for spring admission, 12/1. Applications are processed on a rolling basis. *Application fee:* $40. Electronic applications accepted.
Expenses: Tuition, state resident: full-time $3,832. Tuition, nonresident: full-time $15,519. Required fees: $1,084. Tuition and fees vary according to course load and program.
Financial aid: Application deadline: 2/1. Oswald Steward, Director, 804-924-9111.
Application contact: Information Contact, 804-982-4285, *Fax:* 804-982-4380, *E-mail:* neurograd@virginia.edu.

Find an in-depth description at www.petersons.com/graduate.

■ UNIVERSITY OF WASHINGTON

School of Medicine and Graduate School, Graduate Programs in Medicine, Program in Neurobiology and Behavior, Seattle, WA 98195

AWARDS PhD.

University of Washington (continued)
Faculty: 73 full-time (19 women), 2 part-time/adjunct (0 women).
Students: 40 full-time (18 women); includes 5 minority (2 Asian Americans or Pacific Islanders, 3 Hispanic Americans). Average age 26. *161 applicants, 6% accepted.* In 1999, 3 doctorates awarded (100% entered university research/teaching).
Degree requirements: For doctorate, dissertation required.
Entrance requirements: For doctorate, GRE, TOEFL. *Application deadline:* For fall admission, 1/15. *Application fee:* $49. Electronic applications accepted.
Expenses: Tuition, state resident: full-time $9,210; part-time $236 per credit. Tuition, nonresident: full-time $23,256; part-time $596 per credit.
Financial aid: In 1999–00, research assistantships with tuition reimbursements (averaging $17,280 per year), teaching assistantships (averaging $17,280 per year) were awarded; fellowships, grants and stipends also available.
Faculty research: Motor systems, sensory systems, neuroplasticity, animal behavior, neuroendocrinology.
Dr. Neil M. Nathanson, Co-Director, 206-543-9457, *Fax:* 206-685-3822.
Application contact: Lucia Wisdom, Administrator, 206-685-1647, *Fax:* 206-616-4230, *E-mail:* neubehav@u.washington.edu.

Find an in-depth description at www.petersons.com/graduate.

■ UNIVERSITY OF WISCONSIN–MADISON

Graduate School, College of Agricultural and Life Sciences, Department of Animal Health and Biomedical Sciences, Program in Comparative Biosciences, Madison, WI 53706-1380

AWARDS Anatomy (MS, PhD); biochemistry (MS, PhD); cellular and molecular biology (MS, PhD); environmental toxicology (MS, PhD); neurosciences (MS, PhD); pharmacology (MS, PhD); physiology (MS, PhD).

Degree requirements: For doctorate, dissertation required.
Expenses: Tuition, state resident: full-time $5,406; part-time $339 per credit. Tuition, nonresident: full-time $17,110; part-time $1,071 per credit. Full-time tuition and fees vary according to program and reciprocity agreements. Part-time tuition and fees vary according to course load and program.

■ UNIVERSITY OF WISCONSIN–MADISON

Graduate School, Neuroscience Training Program, Madison, WI 53706-1380

AWARDS PhD.

Faculty: 67 full-time (15 women).
Students: 17 full-time (4 women); includes 2 minority (1 Asian American or Pacific Islander, 1 Native American), 1 international.
Degree requirements: For doctorate, dissertation required, foreign language not required. *Average time to degree:* Doctorate–5 years full-time.
Entrance requirements: For doctorate, GRE General Test. *Application deadline:* For spring admission, 1/1. *Application fee:* $45. Electronic applications accepted.
Expenses: Tuition, state resident: full-time $5,406; part-time $339 per credit. Tuition, nonresident: full-time $17,110; part-time $1,071 per credit. Full-time tuition and fees vary according to program and reciprocity agreements. Part-time tuition and fees vary according to course load and program.
Financial aid: In 1999–00, fellowships with full tuition reimbursements (averaging $15,042 per year), research assistantships with tuition reimbursements (averaging $15,240 per year) were awarded; traineeships also available.
Ronald E. Kalil, Director, 608-262-4932, *Fax:* 608-265-2267, *E-mail:* rekalil@facstaff.wisc.edu.
Application contact: Heather M. Daniels, Program Administrator, 608-262-4932, *Fax:* 608-265-2267, *E-mail:* hdaniels@facstaff.wisc.edu.

Find an in-depth description at www.petersons.com/graduate.

■ UNIVERSITY OF WISCONSIN–MADISON

Medical School and Graduate School, Graduate Programs in Medicine, Department of Physiology, Madison, WI 53706-1380

AWARDS Neurophysiology (PhD); physiology (PhD).

Faculty: 24 full-time (6 women).
Students: 26 full-time (8 women); includes 1 minority (Hispanic American), 8 international.
Degree requirements: For doctorate, dissertation, written exams required, foreign language not required. *Average time to degree:* Doctorate–5 years full-time.
Entrance requirements: For doctorate, GRE, TOEFL. *Application deadline:* For fall admission, 1/15 (priority date). Applications are processed on a rolling basis. *Application fee:* $45. Electronic applications accepted.
Expenses: Tuition, state resident: full-time $5,406; part-time $339 per credit. Tuition, nonresident: full-time $17,110; part-time $1,071 per credit.
Financial aid: In 1999–00, fellowships with tuition reimbursements (averaging $16,400 per year), research assistantships with tuition reimbursements (averaging $16,400 per year), teaching assistantships with tuition reimbursements (averaging $16,400 per year) were awarded.
Faculty research: Studies in molecular cellular systems, cardiovascular, neuroscience.
Dr. Richard Moss, Chair, 608-262-1939, *Fax:* 609-265-5072, *E-mail:* rlmoss@facstaff.wisc.edu.
Application contact: Sue Krey, Program Assistant, 608-262-9114, *Fax:* 608-265-5512, *E-mail:* sskrey@facstaff.wisc.edu.

■ VANDERBILT UNIVERSITY

Graduate School and School of Medicine, Program in Neuroscience, Nashville, TN 37240-1001

AWARDS PhD.

Faculty: 32 full-time (8 women).
Students: 14 full-time (8 women); includes 2 minority (1 African American, 1 Asian American or Pacific Islander), 2 international.
Entrance requirements: For doctorate, GRE General Test. *Application deadline:* For fall admission, 1/15. *Application fee:* $40.
Expenses: Tuition: Full-time $17,244; part-time $958 per hour. Required fees: $242; $121 per semester. Tuition and fees vary according to program.
Financial aid: In 1999–00, fellowships with full tuition reimbursements (averaging $17,000 per year); research assistantships with full tuition reimbursements, institutionally sponsored loans, traineeships, and tuition waivers (partial) also available. Financial aid application deadline: 1/15.
Faculty research: Molecular neuroscience, neural development, synaptic and systems plasticity, neuropharmacology, synaptic transmission.
Marcie W. Pospichal, Director, 615-936-3037, *Fax:* 615-936-3040, *E-mail:* marcie.w.pospichal@vanderbilt.edu.
Application contact: David M. Lovinger, Director of Graduate Studies, 615-936-3037, *Fax:* 615-936-3040, *E-mail:* david.lovinger@mc.mail.vanderbilt.edu.

Find an in-depth description at www.petersons.com/graduate.

■ VIRGINIA COMMONWEALTH UNIVERSITY

School of Graduate Studies and School of Medicine, School of Medicine Graduate Programs, Department of Anatomy, Richmond, VA 23284-9005

AWARDS Anatomy (MS, PhD, CBHS), including neuroscience (MS, PhD); anatomy and physical therapy (PhD).

Students: 10 full-time (4 women), 35 part-time (11 women); includes 12 minority (1 African American, 10 Asian Americans or Pacific Islanders, 1 Hispanic American). In 1999, 8 master's, 4 doctorates, 7 other advanced degrees awarded.

Degree requirements: For master's, thesis required, foreign language not required; for doctorate, dissertation, comprehensive oral and written exams required, foreign language not required.

Entrance requirements: For master's, DAT, GRE General Test, or MCAT; for doctorate, GRE General Test. *Application deadline:* For fall admission, 5/15. *Application fee:* $30.

Expenses: Tuition, state resident: full-time $4,031; part-time $224 per credit hour. Tuition, nonresident: full-time $11,946; part-time $664 per credit hour. Required fees: $1,081; $40 per credit hour. Tuition and fees vary according to campus/location and program.

Financial aid: Fellowships available.
Dr. John T. Povlishock, Chair, 804-828-9535, *Fax:* 804-828-9477.

Application contact: Dr. George R. Leichnetz, Graduate Program Director, 804-828-9512, *Fax:* 804-828-9477, *E-mail:* grleichn@vcu.edu.

■ VIRGINIA COMMONWEALTH UNIVERSITY

School of Graduate Studies and School of Medicine, School of Medicine Graduate Programs, Department of Biochemistry and Molecular Biophysics, Richmond, VA 23284-9005

AWARDS Biochemistry (PhD); biochemistry and molecular biophysics (MS, CBHS); molecular biology and genetics (PhD); neurosciences (PhD).

Students: 5 full-time (3 women), 21 part-time (10 women); includes 12 minority (9 Asian Americans or Pacific Islanders, 1 Hispanic American, 2 Native Americans). In 1999, 1 master's, 10 doctorates awarded.

Degree requirements: For master's, thesis required, foreign language not required; for doctorate, dissertation, comprehensive oral and written exams required, foreign language not required.

Entrance requirements: For master's and doctorate, GRE General Test. *Application deadline:* For fall admission, 5/1. *Application fee:* $30.

Expenses: Tuition, state resident: full-time $4,031; part-time $224 per credit hour. Tuition, nonresident: full-time $11,946; part-time $664 per credit hour. Required fees: $1,081; $40 per credit hour. Tuition and fees vary according to campus/location and program.

Financial aid: Fellowships, research assistantships available.

Faculty research: Molecular biology, peptide/protein chemistry, neurochemistry, enzyme mechanisms, macromolecular structure determination. *Total annual research expenditures:* $3.5 million.
Chair, 804-828-9762, *Fax:* 804-828-1473.

Application contact: Dr. Zendra E. Zehner, Program Director, 804-828-8753, *Fax:* 804-828-1473, *E-mail:* zezehner@vcu.edu.

■ VIRGINIA COMMONWEALTH UNIVERSITY

School of Graduate Studies and School of Medicine, School of Medicine Graduate Programs, Department of Pharmacology and Toxicology, Richmond, VA 23284-9005

AWARDS Molecular biology and genetics (PhD); neurosciences (PhD); pharmacology (PhD, CBHS); pharmacology and toxicology (MS).

Students: 14 full-time (5 women), 51 part-time (27 women); includes 27 minority (4 African Americans, 22 Asian Americans or Pacific Islanders, 1 Hispanic American). In 1999, 3 master's, 10 doctorates, 1 other advanced degree awarded. Terminal master's awarded for partial completion of doctoral program.

Degree requirements: For master's, thesis required, foreign language not required; for doctorate, dissertation, comprehensive oral and written exams required, foreign language not required.

Entrance requirements: For master's, DAT, GRE General Test or MCAT; for doctorate, GRE General Test. *Application deadline:* For fall admission, 2/1 (priority date). *Application fee:* $30.

Expenses: Tuition, state resident: full-time $4,031; part-time $224 per credit hour. Tuition, nonresident: full-time $11,946; part-time $664 per credit hour. Required fees: $1,081; $40 per credit hour. Tuition and fees vary according to campus/location and program.

Financial aid: Fellowships, teaching assistantships available.

Faculty research: Drug abuse, drug metabolism, pharmacodynamics, peptide synthesis, receptor mechanisms.

Dr. George Kunos, Chair, 804-828-2073, *Fax:* 804-828-2117.

Application contact: Sheryol Cox, Graduate Program Coordinator, 804-828-8400, *Fax:* 804-828-2117, *E-mail:* swcox@vcu.edu.

Find an in-depth description at www.petersons.com/graduate.

■ VIRGINIA COMMONWEALTH UNIVERSITY

School of Graduate Studies and School of Medicine, School of Medicine Graduate Programs, Department of Physiology, Richmond, VA 23284-9005

AWARDS Neurosciences (PhD); physiology (MS, PhD, CBHS).

Students: 2 full-time (0 women), 6 part-time; includes 3 Asian Americans or Pacific Islanders. In 1999, 4 master's, 4 doctorates, 20 other advanced degrees awarded. Terminal master's awarded for partial completion of doctoral program.

Degree requirements: For master's, thesis required, foreign language not required; for doctorate, dissertation, comprehensive oral and written exams required, foreign language not required.

Entrance requirements: For master's, DAT, GRE General Test, or MCAT; for doctorate, GRE General Test. *Application fee:* $30.

Expenses: Tuition, state resident: full-time $4,031; part-time $224 per credit hour. Tuition, nonresident: full-time $11,946; part-time $664 per credit hour. Required fees: $1,081; $40 per credit hour. Tuition and fees vary according to campus/location and program.

Financial aid: Fellowships, research assistantships, teaching assistantships, career-related internships or fieldwork and tuition waivers (full) available.
Dr. Margaret C. Biber, Chair, 804-828-9756, *Fax:* 804-828-7382.

Application contact: Dr. James L. Poland, Graduate Program Director, 804-828-9557, *Fax:* 804-828-7382, *E-mail:* jlpoland@vcu.edu.

Find an in-depth description at www.petersons.com/graduate.

■ WAKE FOREST UNIVERSITY

School of Medicine and Graduate School, Graduate Programs in Medicine, Department of Neurobiology and Anatomy, Winston-Salem, NC 27109

AWARDS PhD.

Degree requirements: For doctorate, one foreign language (computer language can substitute), dissertation required.

Wake Forest University (continued)
Entrance requirements: For doctorate, GRE General Test, GRE Subject Test. Electronic applications accepted.
Expenses: Tuition: Full-time $18,300.
Faculty research: Sensory neurobiology, reproductive endocrinology, regulatory processes in cell biology.

Find an in-depth description at www.petersons.com/graduate.

■ WAKE FOREST UNIVERSITY

School of Medicine and Graduate School, Graduate Programs in Medicine, Interdisciplinary Program in Neuroscience, Winston-Salem, NC 27109

AWARDS PhD.

Degree requirements: For doctorate, one foreign language (computer language can substitute), dissertation, oral and written qualifying exams required.
Entrance requirements: For doctorate, GRE General Test, GRE Subject Test. Electronic applications accepted.
Expenses: Tuition: Full-time $18,300.
Faculty research: Neurobiology of substance abuse, learning and memory, aging, sensory neurobiology, nervous system development.

■ WASHINGTON STATE UNIVERSITY

College of Veterinary Medicine and Graduate School, Graduate Programs in Veterinary Science, Pullman, WA 99164

AWARDS Veterinary clinical sciences (MS, PhD); veterinary comparative anatomy, pharmacology, and physiology (MS, PhD), including neuroscience, veterinary science; veterinary microbiology and pathology (MS, PhD), including veterinary science. Part-time programs available.

Faculty: 72 full-time (13 women), 6 part-time/adjunct (4 women).
Students: 50 full-time (24 women). Average age 30. In 1999, 3 master's, 4 doctorates awarded. Terminal master's awarded for partial completion of doctoral program.
Degree requirements: For master's and doctorate, thesis/dissertation, oral exam required. *Average time to degree:* Master's–3 years full-time; doctorate–5 years full-time.
Entrance requirements: For master's and doctorate, GRE General Test, minimum GPA of 3.0. *Application deadline:* For fall admission, 12/31 (priority date). Applications are processed on a rolling basis. *Application fee:* $35. Electronic applications accepted.
Expenses: Tuition, state resident: full-time $5,494. Tuition, nonresident: full-time $13,390.

Financial aid: In 1999–00, 21 research assistantships with partial tuition reimbursements, 8 teaching assistantships with partial tuition reimbursements were awarded; fellowships, career-related internships or fieldwork, Federal Work-Study, grants, institutionally sponsored loans, traineeships, tuition waivers (partial), and teaching associateships also available. Financial aid application deadline: 12/1; financial aid applicants required to submit FAFSA.

■ WASHINGTON STATE UNIVERSITY

College of Veterinary Medicine and Graduate School, Graduate Programs in Veterinary Science, Department of Veterinary Comparative Anatomy, Pharmacology, and Physiology, Program in Neuroscience, Pullman, WA 99164

AWARDS MS, PhD. Part-time programs available.

Faculty: 23 full-time (6 women), 6 part-time/adjunct (4 women).
Students: 18 full-time (7 women); includes 1 minority (Hispanic American), 4 international. Average age 31. *13 applicants, 54% accepted.* In 1999, 1 master's awarded (100% entered university research/teaching); 1 doctorate awarded (100% entered university research/teaching). Terminal master's awarded for partial completion of doctoral program.
Degree requirements: For master's and doctorate, thesis/dissertation, oral exam required. *Average time to degree:* Master's–3 years full-time; doctorate–5 years full-time.
Entrance requirements: For master's and doctorate, GRE General Test, minimum GPA of 3.0. *Application deadline:* For fall admission, 12/31 (priority date). Applications are processed on a rolling basis. *Application fee:* $35. Electronic applications accepted.
Expenses: Tuition, state resident: full-time $5,494. Tuition, nonresident: full-time $13,390.
Financial aid: In 1999–00, 4 research assistantships with partial tuition reimbursements (averaging $13,566 per year), 8 teaching assistantships with partial tuition reimbursements (averaging $13,566 per year) were awarded; career-related internships or fieldwork, Federal Work-Study, and institutionally sponsored loans also available. Financial aid application deadline: 12/31.
Faculty research: Neural mechanisms of substance abuse, molecular mechanisms of sleep regulations, neuroendocrinology of pituitary cells, behavioral aspects of sexual differentiation. *Total annual research expenditures:* $3.5 million.

Dr. Bryan K. Slinker, Chair, 509-335-6624, *Fax:* 509-335-4650, *E-mail:* slinker@vetmed.wsu.edu.
Application contact: Pam Colbert, Coordinator, 509-335-0986, *Fax:* 509-335-4650, *E-mail:* colbertp@vetmed.wsu.edu.

■ WASHINGTON UNIVERSITY IN ST. LOUIS

Graduate School of Arts and Sciences, Department of Philosophy, Program in Philosophy/Neuroscience/Psychology, St. Louis, MO 63130-4899

AWARDS PhD.

Degree requirements: For doctorate, dissertation required.
Entrance requirements: For doctorate, GRE General Test, sample of written work. *Application deadline:* For fall admission, 1/15 (priority date). Applications are processed on a rolling basis. *Application fee:* $35.
Expenses: Tuition: Full-time $23,400; part-time $975 per credit. Tuition and fees vary according to program.
Financial aid: Fellowships, Federal Work-Study and tuition waivers (full and partial) available. Financial aid application deadline: 1/15.

Dr. Andy Clark, Coordinator, 314-935-7147, *Fax:* 314-935-7349.

■ WASHINGTON UNIVERSITY IN ST. LOUIS

Graduate School of Arts and Sciences, Division of Biology and Biomedical Sciences, Program in Neurosciences, St. Louis, MO 63130-4899

AWARDS PhD.

Degree requirements: For doctorate, dissertation required, foreign language not required.
Entrance requirements: For doctorate, GRE General Test, GRE Subject Test. *Application deadline:* For fall admission, 1/1 (priority date). Applications are processed on a rolling basis. *Application fee:* $0.
Expenses: Tuition: Full-time $23,400; part-time $975 per credit. Tuition and fees vary according to program.
Financial aid: Fellowships, research assistantships, tuition waivers (full) available. Financial aid application deadline: 1/1.

Dr. Jeanne Nerbonne, Coordinator, 314-362-2504.
Application contact: Rosemary Garagneni, Director of Admissions, 800-852-9074, *E-mail:* admissions@dbbs.wustl.edu.

■ WAYNE STATE UNIVERSITY

School of Medicine and Graduate School, Graduate Programs in Medicine, Cellular and Clinical Neurobiology Program, Detroit, MI 48202

AWARDS PhD.

Degree requirements: For doctorate, dissertation required, foreign language not required.
Entrance requirements: For doctorate, GRE General Test, GRE Subject Test, minimum B average.
Faculty research: Neurochemistry and neurophysiology of monoamine neurons, molecular neurobiology, signal transduction, neuropsychiatric disorders, cellular neurobiology.

Find an in-depth description at www.petersons.com/graduate.

■ WESLEYAN UNIVERSITY

Graduate Programs, Department of Biology, Middletown, CT 06459-0260

AWARDS Cell biology (PhD); comparative physiology (PhD); developmental biology (PhD); genetics (PhD); neurophysiology (PhD); population biology (PhD).

Faculty: 12 full-time (3 women).
Students: 24 full-time (12 women); includes 1 minority (African American), 11 international. Average age 28. *125 applicants, 4% accepted.* In 1999, 2 doctorates awarded.
Degree requirements: For doctorate, one foreign language (computer language can substitute), dissertation required.
Entrance requirements: For doctorate, GRE Subject Test. *Application deadline:* For fall admission, 1/15. Applications are processed on a rolling basis. *Application fee:* $0.
Expenses: Tuition: Full-time $24,876. Required fees: $650. Tuition and fees vary according to program.
Financial aid: Research assistantships, teaching assistantships, stipends available.
Faculty research: Microbial population genetics, genetic basis of evolutionary adaptation, genetic regulation of differentiation and pattern formation in *drosophila*.
Dr. Fred Cohan, Chairman, 860-685-3489.
Application contact: Marina J. Melendez, Director of Graduate Student Services, 860-685-2390, *Fax:* 860-685-2439, *E-mail:* mmelendez@wesleyan.edu.

Find an in-depth description at www.petersons.com/graduate.

■ YALE UNIVERSITY

Graduate School of Arts and Sciences, Department of Molecular, Cellular, and Developmental Biology, Program in Neurobiology, New Haven, CT 06520

AWARDS PhD.

Degree requirements: For doctorate, dissertation required.
Entrance requirements: For doctorate, GRE General Test, GRE Subject Test. *Application deadline:* For fall admission, 1/4. *Application fee:* $65.
Expenses: Tuition: Full-time $22,300. Full-time tuition and fees vary according to program.
Financial aid: Fellowships, research assistantships, teaching assistantships available.
Application contact: Admissions Information, 203-432-2770.

Find an in-depth description at www.petersons.com/graduate.

■ YALE UNIVERSITY

Graduate School of Arts and Sciences, Department of Neurobiology, New Haven, CT 06520
AWARDS PhD.

Faculty: 36 full-time (12 women), 2 part-time/adjunct (both women).
Students: 8 full-time (2 women); includes 1 minority (Asian American or Pacific Islander).
In 1999, 2 degrees awarded.
Degree requirements: For doctorate, dissertation required, foreign language not required. *Average time to degree:* Doctorate–6.8 years full-time.
Entrance requirements: For doctorate, GRE General Test, GRE Subject Test. *Application deadline:* For fall admission, 1/4. *Application fee:* $65.
Expenses: Tuition: Full-time $22,300. Full-time tuition and fees vary according to program.
Financial aid: Fellowships, research assistantships, Federal Work-Study and institutionally sponsored loans available. Aid available to part-time students.
Application contact: Admissions Information, 203-432-2770.

■ YALE UNIVERSITY

Graduate School of Arts and Sciences, Interdepartmental Neuroscience Program, New Haven, CT 06520

AWARDS PhD.

Faculty: 48 full-time (11 women).
Students: 25 full-time (9 women); includes 3 minority (2 Asian Americans or Pacific Islanders, 1 Hispanic American), 8

international. *133 applicants, 17% accepted.* In 1999, 6 degrees awarded.
Degree requirements: For doctorate, dissertation required. *Average time to degree:* Doctorate–6 years full-time.
Entrance requirements: For doctorate, GRE General Test. *Application deadline:* For fall admission, 1/4. *Application fee:* $65.
Expenses: Tuition: Full-time $22,300. Full-time tuition and fees vary according to program.
Financial aid: Fellowships, research assistantships, teaching assistantships, Federal Work-Study and institutionally sponsored loans available. Aid available to part-time students.
Application contact: Admissions Information, 203-432-2770.

■ YALE UNIVERSITY

School of Medicine and Graduate School of Arts and Sciences, Combined Program in Biological and Biomedical Sciences (BBS), Neuroscience Track, New Haven, CT 06520

AWARDS PhD, MD/PhD.

Degree requirements: For doctorate, dissertation required.
Entrance requirements: For doctorate, GRE General Test, TOEFL. *Application deadline:* For fall admission, 1/2. *Application fee:* $65. Electronic applications accepted.
Expenses: All students receive full tuition of $22,330 and an annual stipend of $17,600 .
Financial aid: Fellowships, research assistantships available.
Dr. Charles Greer, Co-Director, 203-785-4034.
Application contact: Carol Russo, Graduate Registrar, 203-785-5932, *Fax:* 203-785-5971, *E-mail:* carol.russo@yale.edu.

Find an in-depth description at www.petersons.com/graduate.

■ YESHIVA UNIVERSITY

Albert Einstein College of Medicine, Sue Golding Graduate Division of Medical Sciences, Department of Neuroscience, Bronx, NY 10461
AWARDS PhD, MD/PhD.

Faculty: 22 full-time.
Students: 42 full-time (16 women); includes 11 minority (5 African Americans, 3 Asian Americans or Pacific Islanders, 3 Hispanic Americans), 7 international. In 1999, 5 degrees awarded (100% entered university research/teaching).
Degree requirements: For doctorate, dissertation required, foreign language not required.
Entrance requirements: For doctorate, GRE General Test, TOEFL. *Application*

Yeshiva University (continued)
deadline: For fall admission, 1/15. *Application fee:* $0.
Expenses: Tuition: Part-time $525 per credit. Tuition and fees vary according to degree level and program.
Financial aid: In 1999–00, 42 fellowships were awarded.

Faculty research: Structure-function relations at chemical and electrical synapses, mechanisms of electrogenesis, analysis of neuronal subsystems.
Dr. Donald Faber, Chairperson, 718-430-2409.

Application contact: Sheila Cleeton, Assistant Director, 718-430-2128, *Fax:* 718-430-8655, *E-mail:* phd@aecom.yu.edu.
Find an in-depth description at www.petersons.com/graduate.

Nutrition

NUTRITION

■ ANDREWS UNIVERSITY

School of Graduate Studies, College of Arts and Sciences, Department of Nutrition, Berrien Springs, MI 49104
AWARDS MS. Part-time programs available.
Students: 6 full-time (4 women), 11 part-time (all women); includes 4 minority (3 African Americans, 1 Hispanic American), 3 international. In 1999, 3 degrees awarded.
Application deadline: Applications are processed on a rolling basis. *Application fee:* $40.
Expenses: Tuition: Full-time $11,040; part-time $300 per credit. Required fees: $80 per quarter. Tuition and fees vary according to degree level, campus/location and program.
Dr. Winston Craig, Chairperson, 616-471-3370.

■ AUBURN UNIVERSITY

Graduate School, College of Human Sciences, Department of Nutrition and Food Science, Auburn, Auburn University, AL 36849-0002
AWARDS MS, PhD. Part-time programs available.
Faculty: 14 full-time (9 women).
Students: 21 full-time (16 women), 14 part-time (10 women); includes 7 minority (4 African Americans, 3 Asian Americans or Pacific Islanders), 10 international. *26 applicants, 54% accepted.* In 1999, 7 master's, 1 doctorate awarded.
Degree requirements: For master's, thesis (MS) required; for doctorate, dissertation required.
Entrance requirements: For master's and doctorate, GRE General Test. *Application deadline:* For fall admission, 7/7; for spring admission, 11/24. Applications are processed on a rolling basis. *Application fee:* $25 ($50 for international students). Electronic applications accepted.

Expenses: Tuition, state resident: full-time $2,895; part-time $80 per credit hour. Tuition, nonresident: full-time $8,685; part-time $240 per credit hour.
Financial aid: Research assistantships, teaching assistantships, career-related internships or fieldwork and Federal Work-Study available. Aid available to part-time students. Financial aid application deadline: 3/15.
Faculty research: Food quality and safety, diet, food supply, physical activity in maintenance of health, prevention of selected chronic disease states.
Dr. Cheng-i Wei, Head, 334-844-4261.
Application contact: Dr. John F. Pritchett, Dean of the Graduate School, 334-844-4700.
Find an in-depth description at www.petersons.com/graduate.

■ BASTYR UNIVERSITY

Graduate and Professional Programs, Program in Nutrition, Kenmore, WA 98028-4966
AWARDS MS. Part-time programs available.
Students: 44 full-time (43 women), 10 part-time (6 women). Average age 31. *53 applicants, 49% accepted.* In 1999, 12 degrees awarded.
Degree requirements: For master's, thesis required, foreign language not required. *Average time to degree:* Master's–2 years full-time, 3 years part-time.
Entrance requirements: For master's, BS with 1 year of course work in biochemistry, anatomy, physics, nutrition, and developmental psychology. *Application deadline:* For fall admission, 4/1 (priority date). Applications are processed on a rolling basis. *Application fee:* $60.
Expenses: Tuition: Part-time $220 per credit.
Financial aid: Career-related internships or fieldwork and Federal Work-Study available. Aid available to part-time students. Financial aid application deadline: 4/15; financial aid applicants required to submit FAFSA.

Dr. Mark Kestin, Chair, 425-823-1300, *Fax:* 425-823-6222.
Application contact: Richard Dent, Director, Enrollment Services, 425-602-3080, *Fax:* 425-602-3090.
Find an in-depth description at www.petersons.com/graduate.

■ BOSTON UNIVERSITY

Henry M. Goldman School of Dental Medicine, Graduate Programs in Dentistry, Boston, MA 02215
AWARDS Advanced general dentistry (CAGS); dental public health (MS, MSD, D Sc D, CAGS); endodontics (MSD, D Sc D, CAGS); implantology (CAGS); nutritional science (MS, D Sc); operative dentistry (MSD, D.Sc D, CAGS); oral and maxillofacial surgery (MSD, D Sc D, CAGS); oral biology (MSD, D Sc, D Sc D, PhD); orthodontics (MSD, D Sc D, CAGS); pediatric dentistry (MSD, D Sc D, CAGS); periodontology (MSD, D Sc D, CAGS); prosthodontics (MSD, D Sc D, CAGS).
Faculty: 99 full-time (30 women), 213 part-time/adjunct (29 women).
Students: 175 full-time (61 women); includes 26 minority (19 Asian Americans or Pacific Islanders, 7 Hispanic Americans), 85 international. Average age 29. *751 applicants, 10% accepted.* In 1999, 14 master's, 18 doctorates, 63 other advanced degrees awarded.
Degree requirements: For master's and doctorate, thesis/dissertation required; for CAGS, thesis required (for some programs). *Average time to degree:* Master's–1 year full-time; doctorate–2 years full-time; CAGS–2.5 years full-time.
Entrance requirements: For degree, dental degree. *Application deadline:* Applications are processed on a rolling basis. *Application fee:* $50.
Expenses: Tuition: Full-time $32,540.
Financial aid: In 1999–00, 100 students received aid. Application deadline: 4/15.
Application contact: Postdoctoral Admissions, 617-638-4708.
Find an in-depth description at www.petersons.com/graduate.

■ BOSTON UNIVERSITY

Sargent College of Health and Rehabilitation Sciences, Department of Health Sciences, Boston, MA 02215

AWARDS Applied anatomy and physiology (MS, D Sc); nutrition (MS). Part-time programs available.

Faculty: 8 full-time (4 women), 2 part-time/adjunct (1 woman).

Students: 49 full-time (41 women), 7 part-time (5 women); includes 4 minority (3 Asian Americans or Pacific Islanders, 1 Hispanic American), 6 international. Average age 26. *86 applicants, 66% accepted.* In 1999, 8 master's, 3 doctorates awarded.

Degree requirements: For master's, thesis or alternative required, foreign language not required; for doctorate, computer language, dissertation required, foreign language not required.

Entrance requirements: For master's, GRE General Test, minimum GPA of 3.0; for doctorate, GRE General Test, master's degree. *Application deadline:* For fall admission, 4/1 (priority date); for spring admission, 10/1. Applications are processed on a rolling basis. *Application fee:* $60.

Expenses: Tuition: Full-time $23,770; part-time $743 per credit. Required fees: $220. Tuition and fees vary according to class time, course level, campus/location and program.

Financial aid: In 1999–00, 20 fellowships, 7 research assistantships, 6 teaching assistantships were awarded; career-related internships or fieldwork, Federal Work-Study, institutionally sponsored loans, and scholarships also available. Aid available to part-time students. Financial aid application deadline: 4/15.

Faculty research: Muscle metabolism, body acid-base balance, human performance, physical conditioning, diabetes.

Dr. Gary Skrinar, Chairman, 617-353-2717.

Application contact: Judy Skeffington, Senior Admissions Coordinator, 617-353-2713, *Fax:* 617-353-7500, *E-mail:* jaskeff@bu.edu.

■ BOWLING GREEN STATE UNIVERSITY

Graduate College, College of Education and Human Development, School of Family and Consumer Sciences, Bowling Green, OH 43403

AWARDS Food and nutrition (MFCS); human development and family studies (MFCS). Part-time and evening/weekend programs available.

Degree requirements: For master's, thesis required, foreign language not required.

Entrance requirements: For master's, GRE General Test, TOEFL, minimum GPA of 3.0.

Expenses: Tuition, state resident: full-time $6,362. Tuition, nonresident: full-time $11,910. Tuition and fees vary according to course load.

Faculty research: Public health, wellness, social issues and policies, ethnic foods, nutrition and aging, child development, abuse and neglect, gender roles and human sexuality.

■ BRIGHAM YOUNG UNIVERSITY

Graduate Studies, College of Biological and Agricultural Sciences, Department of Food Science and Nutrition, Provo, UT 84602-1001

AWARDS Food science (MS); nutrition (MS). Part-time programs available.

Faculty: 12 full-time (4 women), 4 part-time/adjunct (3 women).

Students: 17 full-time (10 women), 1 part-time. Average age 27. *10 applicants, 80% accepted.* In 1999, 3 degrees awarded.

Degree requirements: For master's, thesis required, foreign language not required. *Average time to degree:* Master's–2 years full-time.

Entrance requirements: For master's, GRE General Test, minimum GPA of 3.0 during previous 2 years. *Application deadline:* For fall admission, 2/1. *Application fee:* $30.

Expenses: Tuition: Full-time $3,330; part-time $185 per credit hour. Tuition and fees vary according to program and student's religious affiliation.

Financial aid: In 1999–00, 10 students received aid, including 5 research assistantships (averaging $4,300 per year), 4 teaching assistantships (averaging $4,300 per year); career-related internships or fieldwork, institutionally sponsored loans, and scholarships also available. Financial aid application deadline: 3/23.

Faculty research: Dairy foods, lipid oxidation, food processes, magnesium and selenium nutrition, nutrient effect on gene expression. *Total annual research expenditures:* $460,000.

Dr. Lynn V. Ogden, Chair, 801-378-3912, *Fax:* 801-378-8714, *E-mail:* lynn_ogden@byu.edu.

Application contact: Dr. Clayton S. Huber, Graduate Coordinator, 801-378-6038, *Fax:* 801-378-8714, *E-mail:* clayton_huber@byu.edu.

■ BROOKLYN COLLEGE OF THE CITY UNIVERSITY OF NEW YORK

Division of Graduate Studies, Department of Health and Nutrition Science, Program in Nutrition Sciences, Brooklyn, NY 11210-2889

AWARDS Nutrition (MS). Part-time programs available.

Students: 33 full-time (30 women), 326 part-time (253 women); includes 98 minority (66 African Americans, 14 Asian Americans or Pacific Islanders, 18 Hispanic Americans), 11 international. *128 applicants, 60% accepted.* In 1999, 81 degrees awarded.

Degree requirements: For master's, thesis or alternative required, foreign language not required.

Entrance requirements: For master's, TOEFL, 18 credits in health-related areas. *Application deadline:* For fall admission, 3/1; for spring admission, 11/1. *Application fee:* $40.

Expenses: Tuition, state resident: full-time $4,350; part-time $185 per credit. Tuition, nonresident: full-time $7,600; part-time $320 per credit.

Financial aid: Federal Work-Study, institutionally sponsored loans, and scholarships available. Aid available to part-time students. Financial aid application deadline: 5/1; financial aid applicants required to submit FAFSA.

Faculty research: Medical ethics, AIDS, history of public health, diet restriction, palliative care, risk reduction/disease prevention, metabolism, diabetes.

Dr. Erika Friedmann, Chair, 718-951-5026, *Fax:* 718-951-4670, *E-mail:* erikaf@brooklyn.cuny.edu.

Application contact: Kathleen Axen, Deputy Chair, 718-951-5909, *Fax:* 718-951-4670, *E-mail:* kaxen@brooklyn.cuny.edu.

■ CALIFORNIA STATE POLYTECHNIC UNIVERSITY, POMONA

Academic Affairs, College of Agriculture, Pomona, CA 91768-2557

AWARDS Agricultural science (MS); animal science (MS); foods and nutrition (MS). Part-time programs available.

Faculty: 24.

Students: 34 full-time (24 women), 31 part-time (21 women); includes 24 minority (3 African Americans, 13 Asian Americans or Pacific Islanders, 8 Hispanic Americans), 10 international. Average age 31. *63 applicants, 67% accepted.* In 1999, 16 degrees awarded.

Degree requirements: For master's, thesis or alternative required.

California State Polytechnic University, Pomona (continued)
Application deadline: Applications are processed on a rolling basis. *Application fee:* $55.
Expenses: Tuition, nonresident: part-time $164 per unit. Required fees: $306 per quarter.
Financial aid: Career-related internships or fieldwork, Federal Work-Study, and institutionally sponsored loans available. Aid available to part-time students. Financial aid application deadline: 3/2; financial aid applicants required to submit FAFSA.
Faculty research: Equine nutrition, physiology, and reproduction; leadership development; bioartificial pancreas; plant science; ruminant and human nutrition.
Dr. Wayne R. Bidlack, Dean, 909-869-2200, *E-mail:* wrbidlack@csupomona.edu.

■ **CALIFORNIA STATE UNIVERSITY, CHICO**

Graduate School, College of Natural Sciences, Department of Biological Sciences, Program in Nutritional Science, Chico, CA 95929-0722

AWARDS Nutrition education (MS).

Degree requirements: For master's, thesis, seminar presentation required, foreign language not required.
Entrance requirements: For master's, GRE General Test.
Expenses: Tuition, nonresident: part-time $246 per credit. Required fees: $2,108; $1,442 per year.

■ **CALIFORNIA STATE UNIVERSITY, FRESNO**

Division of Graduate Studies, College of Agricultural Sciences and Technology, Department of Food Science and Nutritional Services, Fresno, CA 93740

AWARDS Agriculture (MS), including agricultural chemistry, food science and nutrition. Part-time programs available.

Faculty: 12 full-time (4 women).
Students: 2 full-time (1 woman), 9 part-time (7 women); includes 1 minority (Asian American or Pacific Islander). Average age 31. *9 applicants, 56% accepted.* In 1999, 1 degree awarded.
Degree requirements: For master's, thesis required, foreign language not required. *Average time to degree:* Master's–3.5 years full-time.
Entrance requirements: For master's, GRE General Test, TOEFL, minimum GPA of 3.0 in last 60 hours. *Application deadline:* For fall admission, 6/1 (priority

date); for spring admission, 11/1. Applications are processed on a rolling basis.
Application fee: $55. Electronic applications accepted.
Expenses: Tuition, nonresident: part-time $246 per unit. Required fees: $1,906; $620 per semester.
Financial aid: Fellowships, career-related internships or fieldwork, Federal Work-Study, and scholarships available. Financial aid application deadline: 3/1; financial aid applicants required to submit FAFSA.
Faculty research: Liquid foods, agro-ecosystems, pruning evaluations, characterization of juice concentrates, evaluation of root systems.
Dr. Sandra Witte, Chair, 559-278-2164, *Fax:* 559-278-7623, *E-mail:* sandra_witte@csufresno.edu.

■ **CALIFORNIA STATE UNIVERSITY, LONG BEACH**

Graduate Studies, College of Health and Human Services, Department of Family and Consumer Sciences, Program in Nutritional Sciences, Long Beach, CA 90840

AWARDS MS. Part-time programs available.

Faculty: 8 full-time (6 women).
Students: 17 full-time (all women), 17 part-time (all women); includes 8 minority (7 Asian Americans or Pacific Islanders, 1 Hispanic American), 2 international. Average age 29. *32 applicants, 50% accepted.* In 1999, 1 degree awarded.
Degree requirements: For master's, comprehensive exam or thesis required.
Entrance requirements: For master's, minimum GPA of 3.0. *Application deadline:* For fall admission, 8/1; for spring admission, 12/1. Applications are processed on a rolling basis. *Application fee:* $55. Electronic applications accepted.
Expenses: Tuition, nonresident: part-time $246 per credit. Required fees: $569 per semester. Tuition and fees vary according to course load.
Financial aid: Federal Work-Study, grants, and institutionally sponsored loans available. Financial aid application deadline: 3/2.
Faculty research: Protein and water-soluble vitamins, sensory evaluation of foods, mineral deficiencies in humans, child nutrition, minerals and blood pressure.
Dr. Mary Jacob, Graduate Coordinator, 562-985-4516, *Fax:* 562-985-4414, *E-mail:* mjacob@csulb.edu.

■ **CALIFORNIA STATE UNIVERSITY, LOS ANGELES**

Graduate Studies, School of Health and Human Services, Department of Health and Nutritional Sciences, Major in Nutritional Science, Los Angeles, CA 90032-8530

AWARDS MS. Part-time and evening/weekend programs available.

Students: 35 full-time (32 women), 24 part-time (22 women); includes 28 minority (1 African American, 23 Asian Americans or Pacific Islanders, 4 Hispanic Americans), 5 international. In 1999, 12 degrees awarded.
Degree requirements: For master's, comprehensive exam, project, or thesis required.
Entrance requirements: For master's, TOEFL, minimum GPA of 3.0. *Application deadline:* For fall admission, 6/30; for spring admission, 2/1. Applications are processed on a rolling basis. *Application fee:* $55.
Expenses: Tuition, nonresident: full-time $7,703; part-time $164 per unit. Required fees: $1,799; $387 per quarter.
Financial aid: In 1999–00, 23 students received aid. Career-related internships or fieldwork and Federal Work-Study available. Aid available to part-time students. Financial aid application deadline: 3/1.
Faculty research: Human nutrition, nutrition education, nutrition and fitness for the elderly.
Dr. Chick Tam, Chair, Department of Health and Nutritional Sciences, 323-343-4740.

■ **CASE WESTERN RESERVE UNIVERSITY**

School of Medicine and School of Graduate Studies, Graduate Programs in Medicine, Department of Nutrition, Cleveland, OH 44106

AWARDS Dietetics (MS); nutrition (MS, PhD), including nutrition and biochemistry (PhD); public health nutrition (MS). Part-time programs available. Terminal master's awarded for partial completion of doctoral program.

Degree requirements: For master's, thesis required (for some programs), foreign language not required; for doctorate, dissertation required, foreign language not required.
Entrance requirements: For master's and doctorate, GRE General Test, GRE Subject Test, TOEFL.
Faculty research: Fatty acid metabolism, application of gene therapy to nutritional problems, dietary intake methodology,

nutrition and physical fitness, metabolism during infancy and pregnancy.

Find an in-depth description at www.petersons.com/graduate.

■ CENTRAL MICHIGAN UNIVERSITY

College of Graduate Studies, College of Education and Human Services, Department of Human Environmental Studies, Mount Pleasant, MI 48859

AWARDS Human development and family studies (MA); nutrition and dietetics (MS).

Faculty: 22 full-time (14 women).
Students: 1 (woman) full-time, 24 part-time (all women). Average age 26. In 1999, 3 degrees awarded.
Degree requirements: For master's, thesis or alternative required, foreign language not required.
Entrance requirements: For master's, GRE (MA), minimum GPA of 3.0 in last 60 hours, 15 credits in human development and family studies or related area (MA). *Application deadline:* For fall admission, 2/1; for spring admission, 9/15. *Application fee:* $30.
Expenses: Tuition, state resident: part-time $144 per credit hour. Tuition, nonresident: part-time $285 per credit hour. Required fees: $240 per semester. Tuition and fees vary according to degree level and program.
Financial aid: In 1999–00, 2 research assistantships were awarded; fellowships with tuition reimbursements, career-related internships or fieldwork and Federal Work-Study also available. Financial aid application deadline: 3/7.
Faculty research: Human growth and development, family studies and human sexuality, nutritional food science/food services, apparel and textile retailing, computer-aided design for apparel and interior design.
Dr. Kathryn Koch, Chairperson, 517-774-3218, *Fax:* 517-774-2435, *E-mail:* kathryn.e.koch@cmich.edu.

■ CENTRAL WASHINGTON UNIVERSITY

Graduate Studies and Research, College of Education and Professional Studies, Department of Family and Consumer Sciences, Ellensburg, WA 98926

AWARDS Apparel design (MS); family and consumer sciences education (MS); family studies (MS); nutrition (MS). Part-time programs available.

Faculty: 7 full-time (5 women).
Students: 6 full-time (all women), 2 part-time (both women); includes 1 minority (Hispanic American). 7 *applicants, 43% accepted.* In 1999, 3 degrees awarded.
Degree requirements: For master's, thesis or alternative required, foreign language not required.
Entrance requirements: For master's, GRE General Test (nutrition), minimum GPA of 3.0. *Application deadline:* For fall admission, 4/1 (priority date); for winter admission, 10/1; for spring admission, 1/1. Applications are processed on a rolling basis. *Application fee:* $35.
Expenses: Tuition, state resident: full-time $4,389; part-time $146 per credit. Tuition, nonresident: full-time $13,365; part-time $446 per credit. Tuition and fees vary according to course load.
Financial aid: In 1999–00, 2 teaching assistantships with partial tuition reimbursements (averaging $6,664 per year) were awarded; research assistantships, Federal Work-Study also available. Financial aid application deadline: 3/1; financial aid applicants required to submit FAFSA.
Dr. Jan Bowers, Chair, 509-963-2766.
Application contact: Barbara Sisko, Office Assistant, Graduate Studies and Research, 509-963-3103, *Fax:* 509-963-1799, *E-mail:* masters@cwu.edu.

■ CHAPMAN UNIVERSITY

Graduate Studies, Department of Food Science and Nutrition, Orange, CA 92866

AWARDS MS.

Faculty: 2 full-time (1 woman).
Students: In 1999, 6 degrees awarded.
Degree requirements: For master's, thesis or alternative required, foreign language not required.
Entrance requirements: For master's, GRE General Test, minimum undergraduate GPA of 3.0. *Application deadline:* Applications are processed on a rolling basis. *Application fee:* $40.
Expenses: Tuition: Part-time $475 per credit. Required fees: $140 per year. Tuition and fees vary according to program.
Financial aid: Application deadline: 3/1.
Dr. Fredric Caporaso, Chair, 714-997-6638.

■ CLEMSON UNIVERSITY

Graduate School, College of Agriculture, Forestry and Life Sciences, School of Animal, Biomedical and Biological Sciences, Department of Animal, Dairy and Veterinary Sciences, Program in Nutrition, Clemson, SC 29634

AWARDS MS, PhD. Offered in cooperation with the Departments of Food Science and Poultry Science.

Students: 11 full-time (9 women), 2 part-time, 2 international. *23 applicants, 48% accepted.* In 1999, 5 master's, 2 doctorates awarded.
Degree requirements: For master's, thesis optional; for doctorate, dissertation required.
Entrance requirements: For master's and doctorate, GRE General Test, TOEFL. *Application deadline:* For fall admission, 6/1; for spring admission, 11/1. *Application fee:* $40.
Expenses: Tuition, state resident: full-time $3,480; part-time $174 per credit hour. Tuition, nonresident: full-time $9,256; part-time $388 per credit hour. Required fees: $5 per term. Full-time tuition and fees vary according to course level, course load and campus/location.
Financial aid: Fellowships, research assistantships, Federal Work-Study available. Financial aid applicants required to submit FAFSA.
Faculty research: Availability, absorption, and metabolism of nutrients in humans and lab/farm animals; nutrition education.
Dr. D. V. Maurice, Coordinator, 864-656-4023, *Fax:* 864-656-1033, *E-mail:* dmrc@clemson.edu.

■ COLLEGE OF SAINT ELIZABETH

Department of Foods and Nutrition, Morristown, NJ 07960-6989

AWARDS Nutrition (MS). Part-time and evening/weekend programs available.

Faculty: 3 full-time (all women), 10 part-time/adjunct (8 women).
Students: Average age 30. *4 applicants, 100% accepted.* In 1999, 5 degrees awarded (100% found work related to degree).
Degree requirements: For master's, thesis or alternative, portfolio required, foreign language not required. *Average time to degree:* Master's–3 years part-time.
Entrance requirements: For master's, minimum GPA of 3.0. *Application deadline:* For fall admission, 6/30 (priority date); for spring admission, 11/30. Applications are processed on a rolling basis. *Application fee:* $35. Electronic applications accepted.
Expenses: Tuition: Full-time $6,930; part-time $385 per credit. Required fees: $500; $160 per course.
Financial aid: In 1999–00, 1 student received aid, including 1 teaching assistantship (averaging $3,850 per year); Federal Work-Study, tuition waivers (partial), and unspecified assistantships also available. Aid available to part-time students. Financial aid application deadline: 3/15; financial aid applicants required to submit FAFSA.

College of Saint Elizabeth (continued)
Faculty research: Medical nutrition intervention, health care ethics, public policy.
Dr. Mary Hager, Director of Graduate Program, 973-290-4122, *Fax:* 973-290-4167, *E-mail:* nutrition@liza.st-elizabeth.edu.

■ COLORADO STATE UNIVERSITY

Graduate School, College of Applied Human Sciences, Department of Food Science and Human Nutrition, Fort Collins, CO 80523-0015

AWARDS Food science (MS, PhD); nutrition (MS, PhD). Part-time programs available.

Faculty: 14 full-time (8 women).
Students: 39 full-time (35 women), 17 part-time (13 women); includes 4 minority (1 African American, 1 Asian American or Pacific Islander, 2 Hispanic Americans), 10 international. Average age 31. *113 applicants, 46% accepted.* In 1999, 10 master's, 1 doctorate awarded.
Degree requirements: For master's, thesis optional, foreign language not required; for doctorate, dissertation required, foreign language not required.
Entrance requirements: For master's and doctorate, GRE General Test, TOEFL, minimum GPA of 3.0. *Application deadline:* For fall admission, 2/1 (priority date); for spring admission, 9/1 (priority date). Applications are processed on a rolling basis. *Application fee:* $30. Electronic applications accepted.
Expenses: Tuition, state resident: full-time $2,694; part-time $150 per credit. Tuition, nonresident: full-time $10,460; part-time $581 per credit. Required fees: $32 per semester. Tuition and fees vary according to program.
Financial aid: In 1999–00, 3 fellowships, 13 research assistantships, 12 teaching assistantships were awarded; career-related internships or fieldwork, Federal Work-Study, and traineeships also available.
Faculty research: Exercise and energy metabolism, nutrition education, lipid metabolism, eicosanoids, tool product development. *Total annual research expenditures:* $700,000.
Pat Kendall, Graduate Coordinator, 970-491-5093, *Fax:* 970-491-3875, *E-mail:* kendall@cahs.colostate.edu.
Application contact: Irene Lewus, Staff Assistant, 970-491-6535, *Fax:* 970-491-3875, *E-mail:* lewus@cahs.colostate.edu.

■ COLUMBIA UNIVERSITY

College of Physicians and Surgeons, Institute of Human Nutrition, MS Program in Nutrition, New York, NY 10032

AWARDS MS, MPH/MS. Part-time and evening/weekend programs available.

Faculty: 62 full-time (32 women).
Students: 21 full-time (18 women), 22 part-time (9 women); includes 13 minority (2 African Americans, 10 Asian Americans or Pacific Islanders, 1 Hispanic American), 3 international. Average age 22. *52 applicants,* 77% accepted. In 1999, 19 degrees awarded.
Degree requirements: For master's, thesis required, foreign language not required. *Average time to degree:* Master's–1 year full-time, 2 years part-time.
Entrance requirements: For master's, GRE General Test, TOEFL. *Application deadline:* For fall admission, 4/15 (priority date). Applications are processed on a rolling basis. *Application fee:* $75.
Financial aid: In 1999–00, 15 students received aid; fellowships, research assistantships, Federal Work-Study, institutionally sponsored loans, and traineeships available. Aid available to part-time students. Financial aid application deadline: 5/1; financial aid applicants required to submit FAFSA.
Application contact: Dr. David Talmage, Student Adviser, 212-305-4808, *Fax:* 212-305-3079, *E-mail:* dat1@columbia.edu.
Find an in-depth description at www.petersons.com/graduate.

■ COLUMBIA UNIVERSITY

College of Physicians and Surgeons, Institute of Human Nutrition and Graduate School of Arts and Sciences at the College of Physicians and Surgeons, PhD Program in Nutrition, New York, NY 10032

AWARDS M Phil, MA, PhD, MD/PhD. Only candidates for the PhD are admitted.

Faculty: 41 full-time (15 women).
Students: 31 full-time (26 women); includes 7 minority (2 African Americans, 5 Asian Americans or Pacific Islanders), 12 international. Average age 24. *48 applicants,* 21% accepted. In 1999, 2 degrees awarded.
Degree requirements: For master's and doctorate, thesis/dissertation required, foreign language not required. *Average time to degree:* Doctorate–7 years full-time.
Entrance requirements: For doctorate, GRE General Test, TOEFL. *Application deadline:* For fall admission, 1/4. *Application fee:* $75.
Financial aid: In 1999–00, 14 fellowships (averaging $22,488 per year), 17 research assistantships (averaging $22,488 per year)

were awarded; institutionally sponsored loans and traineeships also available. Financial aid application deadline: 1/4.
Faculty research: Growth and development, nutrition and metabolism.
Application contact: Dr. William S. Blaner, Director, 212-305-4808, *Fax:* 212-305-3079, *E-mail:* wsb2@columbia.edu.
Find an in-depth description at www.petersons.com/graduate.

■ CORNELL UNIVERSITY

Graduate School, Graduate Fields of Agriculture and Life Sciences and Graduate Fields of Human Ecology, Field of Nutrition, Ithaca, NY 14853-0001

AWARDS Animal nutrition (MPS, MS, PhD); community nutrition (MPS, MS, PhD); human nutrition (MPS, MS, PhD); international nutrition (MPS, MS, PhD); nutritional biochemistry (MPS, MS, PhD).

Faculty: 48 full-time.
Students: 73 full-time (57 women); includes 8 minority (3 African Americans, 3 Asian Americans or Pacific Islanders, 2 Hispanic Americans), 30 international. *113 applicants,* 22% accepted. In 1999, 3 master's, 7 doctorates awarded.
Degree requirements: For master's, thesis, thesis (MS), project papers (MPS) required, foreign language not required; for doctorate, dissertation required, foreign language not required.
Entrance requirements: For master's and doctorate, GRE General Test, TOEFL. *Application deadline:* For fall admission, 1/10. *Application fee:* $65. Electronic applications accepted.
Expenses: Tuition: Full-time $12,100.
Financial aid: In 1999–00, 63 students received aid, including 29 fellowships with full tuition reimbursements available, 9 research assistantships with full tuition reimbursements available, 25 teaching assistantships with full tuition reimbursements available; institutionally sponsored loans, scholarships, tuition waivers (full and partial), and unspecified assistantships also available. Financial aid applicants required to submit FAFSA.
Faculty research: Nutritional biochemistry, experimental human and animal nutrition, international nutrition, community nutrition.
Application contact: Graduate Field Assistant, 607-255-4410, *Fax:* 607-255-0178, *E-mail:* nutrition_gfr@cornell.edu.
Find an in-depth description at www.petersons.com/graduate.

■ DREXEL UNIVERSITY

Graduate School, College of Arts and Sciences, Department of Bioscience and Biotechnology, Program in Nutrition and Food Sciences, Philadelphia, PA 19104-2875

AWARDS Food science (MS); nutrition science (PhD). Part-time programs available.

Faculty: 13 full-time (7 women), 1 part-time/adjunct (0 women).
Students: 1 (woman) full-time, 20 part-time (17 women); includes 1 minority (Hispanic American), 8 international. Average age 29. *48 applicants, 52% accepted.* In 1999, 2 degrees awarded. Terminal master's awarded for partial completion of doctoral program.
Degree requirements: For master's and doctorate, thesis/dissertation required, foreign language not required.
Entrance requirements: For master's and doctorate, GRE General Test, TOEFL. *Application deadline:* For fall admission, 8/21. Applications are processed on a rolling basis. *Application fee:* $35. Electronic applications accepted.
Expenses: Tuition: Part-time $585 per credit.
Financial aid: Research assistantships, teaching assistantships, Federal Work-Study and unspecified assistantships available. Financial aid application deadline: 2/1.
Faculty research: Metabolism of lipids, W-3 fatty acids, obesity, diabetes and heart disease, mineral metabolism.
Dr. Stanley Segall, Head, 215-895-2416, *Fax:* 215-895-2421.
Application contact: Director of Graduate Admissions, 215-895-6700, *Fax:* 215-895-5939, *E-mail:* enroll@drexel.edu.

■ EAST CAROLINA UNIVERSITY

Graduate School, School of Human Environmental Sciences, Department of Nutrition and Hospitality Management, Greenville, NC 27858-4353

AWARDS Nutrition (MS).

Faculty: 5 full-time (all women).
Students: 14 full-time (13 women), 11 part-time (all women); includes 2 minority (1 African American, 1 Asian American or Pacific Islander). Average age 27. *22 applicants, 73% accepted.* In 1999, 10 degrees awarded.
Degree requirements: For master's, thesis, comprehensive exams required, foreign language not required.
Entrance requirements: For master's, GRE or MAT, TOEFL. *Application deadline:* For fall admission, 6/1 (priority date). Applications are processed on a rolling basis. *Application fee:* $40.

Expenses: Tuition, state resident: full-time $1,012. Tuition, nonresident: full-time $8,578. Required fees: $1,006. Full-time tuition and fees vary according to degree level. Part-time tuition and fees vary according to course load.
Financial aid: Fellowships, teaching assistantships, Federal Work-Study available. Aid available to part-time students. Financial aid application deadline: 6/1.
Faculty research: Lifecycle nutrition, nutrition and disease, nutrition for fish species, food service management. *Total annual research expenditures:* $60,000.
Dr. Dori Finley, Chairperson, 252-328-6917, *Fax:* 252-328-4276, *E-mail:* finleyd@mail.ecu.edu.
Application contact: Dr. Paul D. Tschetter, Senior Associate Dean, 252-328-6012, *Fax:* 252-328-6071, *E-mail:* grad@mail.ecu.edu.

■ EASTERN ILLINOIS UNIVERSITY

Graduate School, Lumpkin College of Business and Applied Sciences, School of Family and Consumer Sciences, Charleston, IL 61920-3099

AWARDS Dietetics (MS); home economics (MS). Part-time programs available.

Degree requirements: For master's, comprehensive exams required, foreign language and thesis not required.

■ EASTERN KENTUCKY UNIVERSITY

The Graduate School, College of Applied Arts and Technology, Department of Human Environmental Sciences, Richmond, KY 40475-3102

AWARDS Community nutrition (MS). Part-time programs available.

Faculty: 3 full-time (all women).
Students: 17 full-time (16 women), 12 part-time (10 women). In 1999, 10 degrees awarded.
Degree requirements: For master's, thesis not required.
Entrance requirements: For master's, GRE General Test, minimum GPA of 2.5. *Application fee:* $0.
Expenses: Tuition, state resident: full-time $2,390; part-time $145 per credit hour. Tuition, nonresident: full-time $6,430; part-time $391 per credit hour.
Financial aid: Research assistantships, teaching assistantships, Federal Work-Study available. Aid available to part-time students.
Cherilyn N. Nelson, Chair, 606-622-3445.

■ EAST TENNESSEE STATE UNIVERSITY

School of Graduate Studies, College of Applied Science and Technology, Department of Applied Human Sciences, Johnson City, TN 37614

AWARDS Clinical nutrition (MS).

Faculty: 6 full-time (5 women).
Students: 13 full-time (all women), 9 part-time (all women); includes 1 minority (African American), 1 international. Average age 28. *20 applicants, 40% accepted.* In 1999, 6 degrees awarded.
Degree requirements: For master's, thesis, oral exam required, foreign language not required.
Entrance requirements: For master's, TOEFL, completion of ADA-approved undergraduate didactic program in dietetics. *Application deadline:* For fall admission, 2/15 (priority date). Applications are processed on a rolling basis. *Application fee:* $25 ($35 for international students).
Expenses: Tuition, state resident: full-time $2,404; part-time $123 per semester hour. Tuition, nonresident: full-time $2,558; part-time $224 per semester hour. International tuition: $7,400 full-time. Required fees: $172 per hour.
Financial aid: In 1999–00, 9 students received aid, including 6 research assistantships, 4 teaching assistantships; career-related internships or fieldwork, Federal Work-Study, and scholarships also available. Financial aid application deadline: 7/1; financial aid applicants required to submit FAFSA.
Faculty research: Kindergarten students, measures of percent body fat during pregnancy, writing to read in preschool children, computer research in food systems management, students' knowledge of aging.
Dr. Robert Acuff, Interim Chair, 423-439-7538, *Fax:* 423-439-5324, *E-mail:* acuffr@etsu.edu.

■ EMORY UNIVERSITY

Graduate School of Arts and Sciences, Division of Biological and Biomedical Sciences, Program in Nutrition and Health Sciences, Atlanta, GA 30322-1100

AWARDS PhD.

Faculty: 45 full-time (23 women).
Students: 18 full-time (16 women); includes 1 African American, 1 Native American, 3 international. In 1999, 1 degree awarded.
Degree requirements: For doctorate, dissertation required, foreign language not required.
Entrance requirements: For doctorate, GRE General Test, TOEFL, minimum

Emory University (continued)
GPA of 3.0 in science course work. *Application deadline:* For fall admission, 1/20 (priority date). *Application fee:* $45.
Expenses: Tuition: Full-time $22,770. Tuition and fees vary according to program.
Financial aid: In 1999–00, fellowships with full tuition reimbursements (averaging $18,000 per year).
Faculty research: Biochemistry, molecular and cell biology, clinical nutrition, community and preventive health, nutritional epidemiology.
Dr. Dean Jones, Director, 404-727-5970, *Fax:* 404-727-3231, *E-mail:* dpjones@ sph.emory.edu.
Application contact: 404-727-2547, *Fax:* 404-727-3322, *E-mail:* gdbbs@ gsas.emory.edu.
Find an in-depth description at www.petersons.com/graduate.

■ FINCH UNIVERSITY OF HEALTH SCIENCES/THE CHICAGO MEDICAL SCHOOL

School of Related Health Sciences, Department of Nutrition, North Chicago, IL 60064-3095

AWARDS MS. Part-time and evening/weekend programs available. Postbaccalaureate distance learning degree programs offered (no on-campus study).

Faculty: 3 full-time (all women), 2 part-time/adjunct (both women).
Students: Average age 34. *20 applicants, 100% accepted.* In 1999, 18 degrees awarded (100% found work related to degree).
Degree requirements: For master's, thesis optional, foreign language not required. *Average time to degree:* Master's–2.5 years part-time.
Entrance requirements: For master's, minimum GPA of 2.75, registered dietitian (RD), professional certificate or license. *Application deadline:* For fall admission, 6/15. Applications are processed on a rolling basis. *Application fee:* $25.
Financial aid: In 1999–00, 8 students received aid. Institutionally sponsored loans available. Aid available to part-time students. Financial aid application deadline: 6/9; financial aid applicants required to submit FAFSA.
Faculty research: Nutrition education, distance learning, computer-based graduate education, childhood obesity, nutrition medical education.
Application contact: Dr. Virginia Hammarlund, Chair, 847-578-3415, *Fax:* 847-578-8623, *E-mail:* hammarlundv@ finchcms.edu.

■ FLORIDA INTERNATIONAL UNIVERSITY

College of Health, Department of Dietetics and Nutrition, Miami, FL 33199

AWARDS MS, PhD. Part-time programs available.

Faculty: 8 full-time (7 women).
Students: 36 full-time (31 women), 36 part-time (31 women); includes 22 minority (6 African Americans, 3 Asian Americans or Pacific Islanders, 13 Hispanic Americans), 16 international. Average age 31. *42 applicants, 60% accepted.* In 1999, 6 master's awarded.
Degree requirements: For master's, thesis required, foreign language not required; for doctorate, dissertation required.
Entrance requirements: For master's, GRE General Test, TOEFL, minimum GPA of 3.0; for doctorate, GRE General Test, TOEFL. *Application deadline:* For fall admission, 4/1 (priority date); for spring admission, 10/1. Applications are processed on a rolling basis. *Application fee:* $20.
Expenses: Tuition, state resident: full-time $3,479; part-time $145 per credit hour. Tuition, nonresident: full-time $12,137; part-time $506 per credit hour. Required fees: $158; $158 per year.
Financial aid: In 1999–00, 1 fellowship, 1 research assistantship, 1 teaching assistantship were awarded; career-related internships or fieldwork, Federal Work-Study, and institutionally sponsored loans also available.
Faculty research: Clinical nutrition, cultural food habits, pediatric nutrition, diabetes, dietetic education.
Dr. Michelle W. Ciccazzo, Chairperson, 305-348-2878, *Fax:* 305-348-1996, *E-mail:* ciccazzo@fiu.edu.
Find an in-depth description at www.petersons.com/graduate.

■ FLORIDA STATE UNIVERSITY

Graduate Studies, College of Human Sciences, Department of Nutrition, Food, and Exercise Sciences, Tallahassee, FL 32306

AWARDS Exercise science (MS, PhD), including exercise physiology, motor learning and control; nutrition and food science (PhD); nutrition and food sciences (MS), including clinical nutrition, food science, nutrition and sport, nutrition science, nutrition, education and health promotion.

Faculty: 13 full-time (9 women).
Students: 67 full-time (37 women), 16 part-time (10 women); includes 14 minority (6 African Americans, 2 Asian Americans or Pacific Islanders, 6 Hispanic Americans), 8 international. *93 applicants,*

70% accepted. In 1999, 21 master's awarded (100% found work related to degree); 4 doctorates awarded.
Degree requirements: For master's, thesis optional, foreign language not required; for doctorate, dissertation required, foreign language not required.
Entrance requirements: For master's and doctorate, GRE General Test, minimum GPA of 3.0. *Application fee:* $20. Electronic applications accepted.
Expenses: Tuition, state resident: full-time $3,504; part-time $146 per credit hour. Tuition, nonresident: full-time $12,162; part-time $507 per credit hour. Tuition and fees vary according to program.
Financial aid: In 1999–00, 33 students received aid, including 2 fellowships (averaging $10,000 per year), 3 research assistantships with partial tuition reimbursements available (averaging $8,000 per year), 28 teaching assistantships with partial tuition reimbursements available (averaging $8,000 per year); career-related internships or fieldwork, Federal Work-Study, institutionally sponsored loans, scholarships, and unspecified assistantships also available. Financial aid applicants required to submit FAFSA.
Faculty research: Nutrition and exercise, vitamin A deficiency, protein biochemistry, cardiovascular responses to exercises, physiological effects of cigarette smoking related to health and wellness. *Total annual research expenditures:* $320,386.
Dr. Robert Moffatt, Chair, 850-644-1828, *Fax:* 850-644-0700, *E-mail:* rmoffatt@ mailer.fsu.edu.
Application contact: Dr. Natholyn Harris, Graduate Coordinator, 850-644-4800, *Fax:* 850-644-0700, *E-mail:* nharris@ mailer.fsu.edu.

■ FRAMINGHAM STATE COLLEGE

Graduate Programs, Department of Chemistry and Food Science, Framingham, MA 01701-9101

AWARDS Food science and nutrition science (MS). Part-time and evening/weekend programs available.

Degree requirements: For master's, foreign language and thesis not required.
Entrance requirements: For master's, GRE General Test.
Expenses: Tuition, state resident: part-time $523 per course. Tuition, nonresident: part-time $523 per course.

■ FRAMINGHAM STATE COLLEGE

Graduate Programs, Program in Nutrition Education, Framingham, MA 01701-9101
AWARDS M Ed.

Degree requirements: For master's, foreign language and thesis not required.
Expenses: Tuition, state resident: part-time $523 per course. Tuition, nonresident: part-time $523 per course.

■ HARVARD UNIVERSITY

School of Public Health, Department of Nutrition, Boston, MA 02115-6096

AWARDS Epidemiology/international nutrition (DPH, SD).

Faculty: 8 full-time (2 women), 10 part-time/adjunct (1 woman).
Students: 15 full-time (9 women), 5 part-time (all women); includes 3 minority (all Asian Americans or Pacific Islanders), 13 international. Average age 39. *28 applicants, 37% accepted.* In 1999, 3 degrees awarded.
Degree requirements: For doctorate, dissertation, qualifying exam required.
Entrance requirements: For doctorate, GRE, TOEFL. *Application deadline:* For fall admission, 1/3. *Application fee:* $60.
Expenses: Tuition: Full-time $22,950; part-time $574 per credit.
Financial aid: Fellowships, research assistantships, teaching assistantships, Federal Work-Study, grants, scholarships, traineeships, tuition waivers (partial), and unspecified assistantships available. Aid available to part-time students. Financial aid application deadline: 2/12; financial aid applicants required to submit FAFSA.
Dr. Walter Willett, Chairman, 617-432-1333.
Application contact: Stanley Hudson, Assistant Dean for Enrollment Services, 617-432-1031, *Fax:* 617-432-2009, *E-mail:* admisofc@hsph.harvard.edu.
Find an in-depth description at www.petersons.com/graduate.

■ HOWARD UNIVERSITY

Graduate School of Arts and Sciences, Department of Nutritional Sciences, Washington, DC 20059-0002

AWARDS Nutrition (MS, PhD). Part-time programs available.

Degree requirements: For master's and doctorate, thesis/dissertation, comprehensive exam required, foreign language not required.
Entrance requirements: For master's and doctorate, GRE General Test, minimum GPA of 3.0.
Expenses: Tuition: Full-time $10,500; part-time $583 per credit hour. Required fees: $405; $203 per semester.

■ HUNTER COLLEGE OF THE CITY UNIVERSITY OF NEW YORK

Graduate School, School of Health Sciences, Program in Urban Public Health, New York, NY 10021-5085

AWARDS Community health education (MPH); environmental and occupational health (MPH); public heatlh nutrition (MPH).

Faculty: 6 full-time (3 women), 12 part-time/adjunct (6 women).
Students: 24 full-time (22 women), 151 part-time (122 women); includes 84 minority (47 African Americans, 16 Asian Americans or Pacific Islanders, 19 Hispanic Americans, 2 Native Americans), 4 international. Average age 33. *129 applicants, 61% accepted.* In 1999, 31 degrees awarded.
Degree requirements: For master's, comprehensive exam required, thesis optional, foreign language not required. *Average time to degree:* Master's–2 years full-time, 4 years part-time.
Entrance requirements: For master's, GRE General Test, TOEFL, 1 year of health-related experience. *Application deadline:* For fall admission, 4/15; for spring admission, 11/15. Applications are processed on a rolling basis. *Application fee:* $40.
Expenses: Tuition, state resident: full-time $4,350; part-time $185 per credit. Tuition, nonresident: full-time $7,600; part-time $320 per credit. Required fees: $8 per term.
Financial aid: In 1999–00, 10 fellowships with partial tuition reimbursements, 3 research assistantships (averaging $2,000 per year) were awarded; career-related internships or fieldwork, Federal Work-Study, grants, scholarships, traineeships, and unspecified assistantships also available. Financial aid application deadline: 3/1.
Faculty research: Gerontology, epidemiology, AIDS policy, sexuality, urban health, women's health. *Total annual research expenditures:* $2.9 million.
Nicholas Freudenberg, Director, 212-481-5111, *Fax:* 212-481-5260, *E-mail:* nfreuden@hunter.cuny.edu.
Application contact: William Zlata, Director, Admissions, 212-772-4486, *Fax:* 212-650-3336, *E-mail:* admissions@hunter.cuny.edu.

■ IMMACULATA COLLEGE

Graduate Division, Program in Nutrition Education, Immaculata, PA 19345-0500

AWARDS Nutrition education (MA); nutrition education/approved pre-professional practice program (MA). Part-time and evening/weekend programs available.

Students: 2 full-time (both women), 53 part-time (51 women). Average age 34. *15 applicants, 87% accepted.* In 1999, 18 degrees awarded (100% found work related to degree).
Degree requirements: For master's, comprehensive exam required, thesis optional, foreign language not required.
Entrance requirements: For master's, GRE or MAT, minimum GPA of 3.0. *Application deadline:* Applications are processed on a rolling basis. *Application fee:* $25. Electronic applications accepted.
Expenses: Tuition: Part-time $355 per credit. Required fees: $40 per semester. Tuition and fees vary according to degree level.
Financial aid: Application deadline: 5/1.
Faculty research: Sports nutrition, pediatric nutrition, changes in food consumption patterns in weight loss, nutritional counseling.
Dr. Laura Frank, Chair, 610-647-4400 Ext. 3482, *E-mail:* lfrank@immaculate.edu.
Application contact: Office of Graduate Admission, 610-647-4400 Ext. 3211.

■ INDIANA STATE UNIVERSITY

School of Graduate Studies, College of Arts and Sciences, Department of Family and Consumer Sciences, Terre Haute, IN 47809-1401

AWARDS Child and family relations (MS); clothing and textiles (MS); dietetics (MS); home management (MS); nutrition and foods (MS). Part-time programs available.

Degree requirements: For master's, foreign language not required. Electronic applications accepted.
Expenses: Tuition, state resident: full-time $3,552; part-time $148 per hour. Tuition, nonresident: full-time $8,088; part-time $337 per hour.

■ INDIANA UNIVERSITY OF PENNSYLVANIA

Graduate School and Research, College of Health and Human Services, Program in Food and Nutrition, Indiana, PA 15705-1087

AWARDS MS. Part-time programs available.

Students: 9 full-time (all women), 11 part-time (10 women), 2 international. Average age 28. *14 applicants, 57% accepted.* In 1999, 3 degrees awarded.
Degree requirements: For master's, thesis optional, foreign language not required.
Entrance requirements: For master's, GRE General Test, TOEFL. *Application deadline:* For fall admission, 7/1 (priority date); for spring admission, 11/1. Applications are processed on a rolling basis. *Application fee:* $30.

Indiana University of Pennsylvania (continued)

Expenses: Tuition, state resident: full-time $3,780; part-time $210 per credit hour. Tuition, nonresident: full-time $6,610; part-time $367 per credit hour. Required fees: $705; $138 per semester.

Financial aid: Research assistantships, Federal Work-Study available. Aid available to part-time students. Financial aid application deadline: 3/15.

Dr. Joanne Steiner, Chairperson and Graduate Coordinator, 724-357-4440, *E-mail:* jsteiner@grove.iup.edu.

■ INDIANA UNIVERSITY–PURDUE UNIVERSITY INDIANAPOLIS

School of Medicine, Graduate Programs in Allied Health, Program in Nutrition and Dietetics, Indianapolis, IN 46202-2896

AWARDS MS.

Students: Average age 26.

Degree requirements: For master's, thesis required.

Entrance requirements: For master's, GRE General Test, RD exam (25 scaled required), minimum GPA of 3.0. *Application deadline:* For fall admission, 1/15 (priority date); for spring admission, 10/15. Applications are processed on a rolling basis. *Application fee:* $35 ($55 for international students).

Expenses: Tuition, state resident: full-time $13,245. Tuition, nonresident: full-time $30,330. Required fees: $121.

Financial aid: In 1999–00, research assistantships (averaging $15,000 per year); fellowships, career-related internships or fieldwork also available. Aid available to part-time students.

Faculty research: Clinical nutrition–human applications.

Dr. Jacquelynn O'Palka, Director, 317-298-0934.

■ IOWA STATE UNIVERSITY OF SCIENCE AND TECHNOLOGY

Graduate College, College of Family and Consumer Sciences and College of Agriculture, Department of Food Science and Human Nutrition, Ames, IA 50011

AWARDS Food science and technology (MS, PhD); nutrition (MS, PhD).

Faculty: 29 full-time.

Students: 31 full-time (21 women), 8 part-time (6 women); includes 1 minority (Asian American or Pacific Islander), 15 international. *69 applicants, 20% accepted.* In 1999, 9 master's, 8 doctorates awarded.

Degree requirements: For master's and doctorate, thesis/dissertation required.

Entrance requirements: For master's and doctorate, GRE General Test, TOEFL. *Application deadline:* For fall admission, 12/31 (priority date); for spring admission, 8/1. *Application fee:* $20 ($50 for international students). Electronic applications accepted.

Expenses: Tuition, state resident: full-time $3,308. Tuition, nonresident: full-time $9,744. Part-time tuition and fees vary according to course load, campus/location and program.

Financial aid: In 1999–00, research assistantships with partial tuition reimbursements (averaging $11,410 per year); fellowships, teaching assistantships, scholarships also available.

Dr. Diane F. Birt, Chair, 515-294-3011, *Fax:* 515-294-8181, *E-mail:* fshn@iastate.edu.

Application contact: Dr. Murray L. Kaplan, Director of Graduate Education, 515-294-9304, *E-mail:* fshn@iastate.edu.

Find an in-depth description at www.petersons.com/graduate.

■ JOHNS HOPKINS UNIVERSITY

School of Hygiene and Public Health, Department of International Health, Division of Human Nutrition, Baltimore, MD 21218-2699

AWARDS MHS, PhD, Sc D.

Degree requirements: For master's, internship required; for doctorate, dissertation, 1 year full-time residency, oral and written exams required, foreign language not required.

Entrance requirements: For master's and doctorate, GRE General Test, TOEFL. *Application deadline:* For fall admission, 2/1 (priority date). Applications are processed on a rolling basis. *Application fee:* $60. Electronic applications accepted.

Expenses: Tuition: Full-time $23,660; part-time $493 per unit. Full-time tuition and fees vary according to degree level, campus/location and program.

Financial aid: Federal Work-Study, institutionally sponsored loans, and scholarships available. Aid available to part-time students. Financial aid application deadline: 4/15.

Faculty research: Nutritional anthropology, food hygiene, malnutrition in children, nutritional status and infection.

Dr. Benjamin Caballero, Director, 410-614-4070, *E-mail:* bcaballe@jhsph.edu.

Application contact: Nancy Stephens, Student Coordinator, 410-955-3734, *Fax:* 410-955-8734, *E-mail:* nstephen@jhsph.edu.

Find an in-depth description at www.petersons.com/graduate.

■ KANSAS STATE UNIVERSITY

Graduate School, College of Human Ecology, Department of Foods and Nutrition, Manhattan, KS 66506

AWARDS Food science (MS, PhD); nutrition (MS, PhD). Part-time programs available.

Degree requirements: For master's, thesis or alternative required, foreign language not required; for doctorate, dissertation required, foreign language not required.

Entrance requirements: For master's and doctorate, GRE General Test. Electronic applications accepted.

Expenses: Tuition, state resident: part-time $103 per credit hour. Tuition, nonresident: part-time $338 per credit hour. Required fees: $17 per credit hour. One-time fee: $64 part-time.

Faculty research: Lipid soluble vitamins, dietary fiber, nutrition for young adults and the elderly, nutrition and exercise, sensory analysis.

■ KENT STATE UNIVERSITY

College of Fine and Professional Arts, School of Family and Consumer Studies, Kent, OH 44242-0001

AWARDS Child and family relations (MA); nutrition (MS).

Faculty: 8 full-time.

Students: 2 full-time (both women), 5 part-time (all women), 1 international. *8 applicants, 75% accepted.* In 1999, 3 degrees awarded.

Degree requirements: For master's, thesis required (for some programs), foreign language not required.

Entrance requirements: For master's, minimum GPA of 2.75. *Application deadline:* For fall admission, 7/12; for spring admission, 11/29. Applications are processed on a rolling basis. *Application fee:* $30.

Expenses: Tuition, state resident: full-time $5,334; part-time $243 per hour. Tuition, nonresident: full-time $10,238; part-time $466 per hour.

Financial aid: Research assistantships, teaching assistantships, Federal Work-Study and tuition waivers (full) available. Financial aid application deadline: 2/1.

Dr. Jeannie D. Sneed, Director, 330-672-2197, *Fax:* 330-672-2194.

■ LEHMAN COLLEGE OF THE CITY UNIVERSITY OF NEW YORK

Division of Natural and Social Sciences, Department of Health Services, Program in Nutrition, Bronx, NY 10468-1589

AWARDS Approved preprofessional practice (MS); clinical nutrition (MS); community nutrition (MS).

Faculty: 2 full-time (both women), 1 (woman) part-time/adjunct.
Degree requirements: For master's, thesis or alternative required. *Application deadline:* For fall admission, 4/1; for spring admission, 11/1. Applications are processed on a rolling basis. *Application fee:* $40.
Expenses: Tuition, state resident: full-time $4,350; part-time $185 per credit. Tuition, nonresident: full-time $7,600; part-time $320 per credit.
Financial aid: Federal Work-Study and tuition waivers (full and partial) available. Aid available to part-time students. Financial aid application deadline: 5/15; financial aid applicants required to submit FAFSA.
Alice Tobias, Adviser, 718-960-8775.

■ LOMA LINDA UNIVERSITY

Graduate School, Department of Nutrition, Loma Linda, CA 92350
AWARDS Clinical nutrition (MS); nutrition care management (MS); nutritional science (MS). Part-time programs available.
Degree requirements: For master's, thesis optional, foreign language not required.
Entrance requirements: For master's, GRE General Test. *Application fee:* $55.
Expenses: Tuition: Part-time $395 per unit.
Dr. Joan Sabaté, Chair, 909-824-4598, *Fax:* 909-824-4087.

■ LOMA LINDA UNIVERSITY

School of Public Health, Department of Nutrition, Loma Linda, CA 92350
AWARDS Public health nutrition (MPH, Dr PH).
Degree requirements: For doctorate, dissertation required.
Entrance requirements: For master's, Michigan English Language Assessment Battery or TOEFL; for doctorate, GRE General Test. *Application deadline:* Applications are processed on a rolling basis. *Application fee:* $100.
Expenses: Tuition: Part-time $395 per unit.
Financial aid: Application deadline: 5/15.
Faculty research: Sports nutrition in minorities, dietary determinance of chronic disease, protein adequacy in vegetarian diets, relationship of dietary intake to hormone level.
Dr. Joan Sabaté, Chair, 909-824-4598, *Fax:* 909-824-4087.
Application contact: Teri Tamayose, Director of Admissions and Academic Records, 909-824-4694, *Fax:* 909-824-4087, *E-mail:* ttamayose@sph.llu.edu.

■ LONG ISLAND UNIVERSITY, C.W. POST CAMPUS

School of Health Professions, Department of Nutrition, Brookville, NY 11548-1300
AWARDS Dietetic internship (Certificate); nutrition (MS). Part-time and evening/weekend programs available.
Faculty: 4 full-time (3 women), 12 part-time/adjunct (11 women).
Students: 27 full-time (25 women), 28 part-time (26 women); includes 2 minority (both Asian Americans or Pacific Islanders). Average age 30. *15 applicants, 80% accepted.* In 1999, 8 master's awarded (100% found work related to degree); 20 other advanced degrees awarded.
Degree requirements: For master's, computer language, thesis required, foreign language not required. *Average time to degree:* Master's–2 years full-time, 4 years part-time.
Entrance requirements: For master's, TOEFL, minimum GPA of 2.75 in major. *Application deadline:* Applications are processed on a rolling basis. *Application fee:* $30. Electronic applications accepted.
Expenses: Tuition: Part-time $405 per credit. Required fees: $310; $65 per year. Tuition and fees vary according to course load and program.
Financial aid: In 1999–00, 50 students received aid, including 3 teaching assistantships; fellowships, career-related internships or fieldwork, Federal Work-Study, institutionally sponsored loans, tuition waivers (partial), and unspecified assistantships also available. Aid available to part-time students. Financial aid application deadline: 5/15; financial aid applicants required to submit FAFSA.
Faculty research: Hematopoiesis, interleukins in allergy, growth factors effect in metastasis affecting behavioral change for nutrition.
Dr. Frances C. Gizis, Chair, 516-299-2670, *Fax:* 516-299-3106, *E-mail:* fgizis@liu.edu.
Application contact: Robin Steadman, Graduate Adviser, 516-299-2337, *Fax:* 516-299-2527, *E-mail:* rsteadman@phoenix.liu.edu.

■ LOUISIANA TECH UNIVERSITY

Graduate School, College of Applied and Natural Sciences, School for Human Ecology, Ruston, LA 71272
AWARDS Dietetics (MS); human ecology (MS). Part-time programs available.
Degree requirements: For master's, computer language, thesis or alternative, Registered Dietician Exam eligibility required, foreign language not required.

Entrance requirements: For master's, GRE General Test.

■ MARYWOOD UNIVERSITY

Graduate School of Arts and Sciences, Department of Nutrition and Dietetics, Scranton, PA 18509-1598
AWARDS MS. Part-time and evening/weekend programs available.
Faculty: 5 full-time (3 women), 6 part-time/adjunct (5 women).
Students: 21 full-time (19 women), 22 part-time (21 women). Average age 29. *22 applicants, 64% accepted.* In 1999, 5 degrees awarded.
Degree requirements: For master's, thesis required, foreign language not required.
Entrance requirements: For master's, GRE General Test or MAT, TOEFL. *Application deadline:* For fall admission, 7/15 (priority date); for spring admission, 12/1. Applications are processed on a rolling basis. *Application fee:* $20.
Expenses: Tuition: Part-time $499 per credit. Required fees: $90 per semester. Tuition and fees vary according to degree level, campus/location and program.
Financial aid: Research assistantships, career-related internships or fieldwork, scholarships, and tuition waivers (partial) available. Aid available to part-time students. Financial aid application deadline: 2/15; financial aid applicants required to submit FAFSA.
Faculty research: Community nutrition and the environment, wellness, human performance and sports nutrition, dietary regimens, food systems management.
Dr. Lee Harrison, Chair, 570-348-6277.
Application contact: Deborah M. Flynn, Coordinator of Admissions, 570-340-6002, *Fax:* 570-961-4745, *E-mail:* gsas_adm@ac.marywood.edu.

■ MCP HAHNEMANN UNIVERSITY

School of Medicine, Biomedical Graduate Programs, Department of Biochemistry, Program in Biomedical Nutrition, Philadelphia, PA 19102-1192
AWARDS PhD, MD/PhD.
Degree requirements: For doctorate, computer language, dissertation, qualifying exam required, foreign language not required.
Entrance requirements: For doctorate, GRE General Test, TOEFL, interview, minimum GPA of 3.0.
Faculty research: Biomedical nutrition, atherosclerosis, lipid metabolism, protein metabolism, enzymology.

■ MICHIGAN STATE UNIVERSITY

Graduate School, College of Human Ecology and College of Agriculture and Natural Resources, Department of Food Science and Human Nutrition, East Lansing, MI 48824

AWARDS Food science (MS, PhD); human nutrition (MS, PhD).

Faculty: 11.
Students: 43 full-time (35 women), 48 part-time (31 women); includes 12 minority (4 African Americans, 8 Asian Americans or Pacific Islanders), 40 international. Average age 30. *144 applicants, 14% accepted.* In 1999, 9 master's, 4 doctorates awarded. Terminal master's awarded for partial completion of doctoral program.
Degree requirements: For master's, thesis optional, foreign language not required; for doctorate, dissertation required, foreign language not required.
Entrance requirements: For master's, GRE, minimum GPA of 3.0 in last 2 years of undergraduate course work; for doctorate, GRE, minimum GPA of 3.0 (MS). *Application deadline:* For fall admission, 1/15 (priority date). Applications are processed on a rolling basis. *Application fee:* $30 ($40 for international students). Electronic applications accepted.
Expenses: Tuition, state resident: part-time $229 per credit. Tuition, nonresident: part-time $464 per credit. Required fees: $241 per semester. Tuition and fees vary according to course load, degree level and program.
Financial aid: In 1999–00, 13 research assistantships with tuition reimbursements (averaging $10,764 per year), 5 teaching assistantships with tuition reimbursements (averaging $10,422 per year) were awarded; fellowships. Financial aid application deadline: 1/15; financial aid applicants required to submit FAFSA.
Faculty research: Vitamin A and embryonic heart development, diet effect on energy metabolism, food safety. *Total annual research expenditures:* $2.4 million.
Dr. Mark A. Uebersax, Interim Chairperson, 517-355-8474, *Fax:* 517-353-8963, *E-mail:* uebersax@pilot.msu.edu.
Application contact: Dr. John E. Linz, Director of Graduate Studies, 517-353-9624, *Fax:* 517-353-8963, *E-mail:* jlinz@pilot.msu.edu.

Find an in-depth description at www.petersons.com/graduate.

■ MIDDLE TENNESSEE STATE UNIVERSITY

College of Graduate Studies, College of Education and Behavioral Science, Department of Human Sciences, Murfreesboro, TN 37132

AWARDS Child development and family studies (MS); nutrition and food science (MS). Part-time programs available.

Faculty: 10 full-time (all women), 1 (woman) part-time/adjunct.
Students: 3 full-time (2 women), 23 part-time (21 women); includes 2 minority (1 African American, 1 Asian American or Pacific Islander). Average age 33. *18 applicants, 22% accepted.* In 1999, 10 degrees awarded.
Degree requirements: For master's, thesis, comprehensive exams required, foreign language not required.
Entrance requirements: For master's, GRE or MAT. *Application deadline:* For fall admission, 8/1 (priority date). Applications are processed on a rolling basis. *Application fee:* $25. Electronic applications accepted.
Expenses: Tuition, state resident: full-time $1,356; part-time $137 per semester hour. Tuition, nonresident: full-time $3,914; part-time $361 per semester hour.
Financial aid: Application deadline: 5/1.
Dr. Emily K. Hughes, Chair, 615-898-2884, *E-mail:* khughes@mtsu.edu.

■ MISSISSIPPI STATE UNIVERSITY

College of Agriculture and Life Sciences, Program in Nutrition, Mississippi State, MS 39762

AWARDS MS, PhD. Part-time programs available.

Faculty: 2 full-time (both women), 9 part-time/adjunct (2 women).
Students: 40 full-time (31 women), 23 part-time (17 women); includes 6 minority (4 African Americans, 2 Asian Americans or Pacific Islanders), 6 international. Average age 28. *35 applicants, 83% accepted.* In 1999, 3 master's awarded (100% found work related to degree); 1 doctorate awarded (100% entered university research/teaching).
Degree requirements: For master's, thesis required, foreign language not required; for doctorate, dissertation required. *Average time to degree:* Master's–6 years part-time; doctorate–4.5 years full-time.
Entrance requirements: For master's, TOEFL, minimum GPA of 2.75; for doctorate, TOEFL. *Application deadline:* For fall admission, 7/1; for spring admission, 11/1. Applications are processed on a rolling basis. *Application fee:* $25 for international students.

Expenses: Tuition, state resident: full-time $3,017; part-time $168 per credit. Tuition, nonresident: full-time $6,119; part-time $340 per credit. Part-time tuition and fees vary according to course load and program.
Financial aid: In 1999–00, 12 students received aid, including 3 research assistantships with tuition reimbursements available (averaging $10,000 per year); Federal Work-Study, institutionally sponsored loans, tuition waivers, and unspecified assistantships also available. Financial aid applicants required to submit FAFSA.
Faculty research: Human monogastric, ruminant, poultry, and aquatic animal nutrition.
Dr. Robert P. Wilson, Graduate Coordinator, 662-325-7740, *Fax:* 662-325-8664, *E-mail:* rpwl@ra.msstate.edu.
Application contact: Jerry B. Inmon, Director of Admissions, 662-325-2224, *Fax:* 662-325-7360, *E-mail:* admit@admissions.msstate.edu.

■ MOUNT MARY COLLEGE

Graduate Programs, Program in Dietetics, Milwaukee, WI 53222-4597

AWARDS Administrative dietetics (MS); clinical dietetics (MS); nutrition education (MS). Part-time and evening/weekend programs available.

Faculty: 1 (woman) full-time, 6 part-time/adjunct (4 women).
Students: 6 full-time (all women), 16 part-time (all women); includes 1 minority (Hispanic American). Average age 30. *11 applicants, 100% accepted.* In 1999, 2 degrees awarded (100% found work related to degree).
Degree requirements: For master's, thesis required. *Average time to degree:* Master's–2 years full-time, 4 years part-time.
Entrance requirements: For master's, TOEFL, minimum GPA of 2.75, completion of ADA and DPD requirements. *Application deadline:* For fall admission, 8/15 (priority date). *Application fee:* $35.
Expenses: Tuition: Full-time $6,840; part-time $380 per credit. Required fees: $150; $40 per semester. One-time fee: $10.
Financial aid: Career-related internships or fieldwork, Federal Work-Study, and unspecified assistantships available. Aid available to part-time students. Financial aid application deadline: 5/1.
Dr. Lisa Stark, Director, 414-258-4810 Ext. 398, *E-mail:* starkl@mtmary.edu.

■ NEW YORK INSTITUTE OF TECHNOLOGY

Graduate Division, School of Allied Health and Life Sciences, Program in Clinical Nutrition, Old Westbury, NY 11568-8000

AWARDS MS, DO/MS. Part-time and evening/weekend programs available.

Students: 17 full-time (14 women), 36 part-time (25 women); includes 5 minority (2 African Americans, 2 Asian Americans or Pacific Islanders, 1 Hispanic American), 7 international. Average age 29. *43 applicants, 53% accepted.* In 1999, 4 degrees awarded.
Degree requirements: For master's, comprehensive exam required, foreign language and thesis not required. *Average time to degree:* Master's–3 years full-time, 4 years part-time.
Entrance requirements: For master's, interview, minimum QPA of 2.85. *Application deadline:* Applications are processed on a rolling basis. *Application fee:* $50. Electronic applications accepted.
Expenses: Tuition: Part-time $450 per credit. Required fees: $135 per year. Tuition and fees vary according to course load and program.
Financial aid: Fellowships, research assistantships, career-related internships or fieldwork, institutionally sponsored loans, tuition waivers (full and partial), and unspecified assistantships available. Aid available to part-time students.
Faculty research: Medical nutrition training.
Dr. Susan Ettinger, Chair, 516-686-3803.
Application contact: Jacquelyn Nealon, Dean of Admissions and Financial Aid, 516-686-7925, *Fax:* 516-626-0419, *E-mail:* jnealon@nyit.edu.

■ NEW YORK UNIVERSITY

School of Education, Department of Nutrition and Food Studies, Program in Food, Nutrition, and Dietetics, New York, NY 10012-1019

AWARDS MS, PhD.

Students: 82 full-time (79 women), 110 part-time (105 women); includes 29 minority (8 African Americans, 13 Asian Americans or Pacific Islanders, 8 Hispanic Americans). *96 applicants, 65% accepted.* In 1999, 84 master's, 2 doctorates awarded.
Degree requirements: For master's, thesis required (for some programs), foreign language not required; for doctorate, dissertation required.
Entrance requirements: For master's, TOEFL; for doctorate, GRE General Test, TOEFL, interview. *Application deadline:* For fall admission, 2/1 (priority date); for spring admission, 12/1. Applications are processed on a rolling basis. *Application fee:* $40 ($60 for international students).
Expenses: Tuition: Full-time $16,536; part-time $689 per credit. Required fees: $1,104; $36 per credit. Tuition and fees vary according to course load.
Financial aid: Career-related internships or fieldwork, Federal Work-Study, institutionally sponsored loans, and tuition waivers (partial) available. Financial aid application deadline: 3/1; financial aid applicants required to submit FAFSA.
Faculty research: Nutrition policy, nutrition and health promotion, international nutrition, weight management, nutrition and disease.
Sharon Dalton, Director, 212-998-5593, *Fax:* 212-995-4194.
Application contact: 212-998-5030, *Fax:* 212-995-4328, *E-mail:* grad.admissions@nyu.edu.

■ NEW YORK UNIVERSITY

School of Education, Department of Nutrition and Food Studies, Program in Public Health Nutrition, New York, NY 10012-1019

AWARDS MPH. Part-time and evening/weekend programs available.

Students: 3 full-time (all women), 15 part-time (14 women); includes 6 minority (2 African Americans, 1 Asian American or Pacific Islander, 3 Hispanic Americans). *10 applicants, 60% accepted.* In 1999, 6 degrees awarded.
Degree requirements: For master's, thesis required (for some programs), foreign language not required.
Entrance requirements: For master's, TOEFL. *Application deadline:* For fall admission, 2/1 (priority date); for spring admission, 12/1. Applications are processed on a rolling basis. *Application fee:* $40 ($60 for international students).
Expenses: Tuition: Full-time $16,536; part-time $689 per credit. Required fees: $1,104; $36 per credit. Tuition and fees vary according to course load.
Financial aid: Federal Work-Study, institutionally sponsored loans, and tuition waivers (partial) available. Financial aid application deadline: 3/1; financial aid applicants required to submit FAFSA.
Faculty research: Poverty and malnutrition, epidemiology, sociocultural determinants of food consumption, nutritional anthropology.
Jeffrey Backstrand, Director, 212-998-5586.
Application contact: 212-998-5030, *Fax:* 212-995-4328, *E-mail:* grad.admissions@nyu.edu.

■ NORTH CAROLINA AGRICULTURAL AND TECHNICAL STATE UNIVERSITY

Graduate School, School of Agriculture, Department of Human Environment and Family Services, Greensboro, NC 27411

AWARDS Food and nutrition (MS). Part-time and evening/weekend programs available.

Degree requirements: For master's, thesis or alternative, comprehensive exam, qualifying exam required, foreign language not required.
Entrance requirements: For master's, GRE General Test, minimum GPA of 2.6.
Expenses: Tuition, state resident: full-time $982; part-time $368 per semester. Tuition, nonresident: full-time $8,252; part-time $3,095 per semester. Required fees: $464 per semester.

■ NORTH CAROLINA STATE UNIVERSITY

Graduate School, College of Agriculture and Life Sciences, Program in Nutrition, Raleigh, NC 27695

AWARDS MS, PhD. Part-time programs available.

Students: 19 full-time (11 women), 11 part-time (8 women); includes 4 minority (2 African Americans, 1 Asian American or Pacific Islander, 1 Hispanic American), 3 international. Average age 32. *17 applicants, 18% accepted.*
Degree requirements: For master's and doctorate, thesis/dissertation required, foreign language not required.
Entrance requirements: For master's and doctorate, GRE General Test, TOEFL. *Application deadline:* For fall admission, 6/25; for spring admission, 11/25. *Application fee:* $45.
Expenses: Tuition, state resident: full-time $1,578. Tuition, nonresident: full-time $10,744. Required fees: $892. Full-time tuition and fees vary according to program.
Financial aid: Research assistantships, teaching assistantships available.
Faculty research: Experimental animal nutrition and nutritional biochemistry, nutrition and health in animal agriculture and animal models, regulation of growth and development, food components that affect human nutrition, animal waste management through nutrition, nutrition education.
Dr. Jonathan C. Allen, Director of Graduate Programs, 919-513-2257, *Fax:* 919-515-7124, *E-mail:* j_allen@ncsu.edu.

■ NORTH DAKOTA STATE UNIVERSITY

Graduate Studies and Research, College of Human Development and Education, Department of Food and Nutrition, Fargo, ND 58105

AWARDS Cellular and molecular biology (PhD); food and nutrition (MS).

Faculty: 7 full-time (5 women).
Students: 7 full-time (4 women); includes 5 minority (4 Asian Americans or Pacific Islanders, 1 Hispanic American). Average age 26. In 1999, 2 degrees awarded.
Degree requirements: For master's, thesis or alternative required, foreign language not required.
Entrance requirements: For master's, TOEFL, minimum GPA of 3.0; for doctorate, TOEFL. *Application deadline:* Applications are processed on a rolling basis. *Application fee:* $25.
Expenses: Tuition, state resident: full-time $3,096; part-time $112 per credit hour. Tuition, nonresident: full-time $7,588; part-time $299 per credit hour. Tuition and fees vary according to course load, campus/location and reciprocity agreements.
Financial aid: In 1999–00, 6 research assistantships with full tuition reimbursements (averaging $8,666 per year), 3 teaching assistantships with full tuition reimbursements (averaging $4,666 per year) were awarded; fellowships, Federal Work-Study and institutionally sponsored loans also available. Aid available to part-time students. Financial aid application deadline: 4/15.
Faculty research: Lipids, fiber, protein, nutrition education, sensory evaluation.
Application contact: Vickie H. Grossnickle, Administrative Secretary, 701-231-7474, *Fax:* 701-231-7174, *E-mail:* grossnic@plains.nodak.edu.

■ NORTHERN ILLINOIS UNIVERSITY

Graduate School, College of Health and Human Sciences, School of Family, Consumer and Nutrition Sciences, Program in Nutrition and Dietetics, De Kalb, IL 60115-2854

AWARDS MS. Part-time programs available.

Faculty: 5 full-time (4 women).
Students: 18 full-time (17 women), 18 part-time (17 women); includes 2 minority (1 African American, 1 Asian American or Pacific Islander). Average age 32. *25 applicants, 52% accepted.* In 1999, 7 degrees awarded.
Degree requirements: For master's, thesis, comprehensive exam, thesis defense, internship required, foreign language not required.

Entrance requirements: For master's, GRE General Test, TOEFL, minimum GPA of 3.0 in field, 2.75 overall. *Application deadline:* For fall admission, 4/15 (priority date); for spring admission, 9/15 (priority date). Applications are processed on a rolling basis. *Application fee:* $30.
Expenses: Tuition, state resident: part-time $169 per credit hour. Tuition, nonresident: part-time $295 per credit hour. Tuition and fees vary according to campus/location and program.
Financial aid: Fellowships with full tuition reimbursements, research assistantships with full tuition reimbursements, teaching assistantships with full tuition reimbursements, career-related internships or fieldwork, Federal Work-Study, tuition waivers (full), and unspecified assistantships available. Aid available to part-time students.
Application contact: Dr. Sondra King, Chair of Graduate Faculty, 815-753-8702.

■ THE OHIO STATE UNIVERSITY

Graduate School, College of Food, Agricultural, and Environmental Sciences, Program in Food Science and Nutrition, Columbus, OH 43210

AWARDS MS, PhD.

Faculty: 17 full-time, 8 part-time/adjunct.
Students: 52 full-time (32 women), 9 part-time (7 women); includes 7 minority (5 African Americans, 2 Asian Americans or Pacific Islanders), 30 international. *116 applicants, 19% accepted.* In 1999, 9 master's, 4 doctorates awarded.
Degree requirements: For master's, thesis optional, foreign language not required; for doctorate, dissertation required, foreign language not required.
Entrance requirements: For master's and doctorate, GRE General Test. *Application deadline:* For fall admission, 8/15. Applications are processed on a rolling basis. *Application fee:* $30 ($40 for international students).
Expenses: Tuition, state resident: full-time $5,400. Tuition, nonresident: full-time $14,535. Part-time tuition and fees vary according to course load and program.
Financial aid: Fellowships, research assistantships, Federal Work-Study and institutionally sponsored loans available. Aid available to part-time students.
Grady Chism, Graduate Studies Committee Chair, 614-292-7719, *Fax:* 614-292-0218, *E-mail:* chism.2@osu.edu.

■ THE OHIO STATE UNIVERSITY

Graduate School, College of Human Ecology, Department of Human Nutrition and Food Management, Columbus, OH 43210

AWARDS Food service management (MS, PhD); foods (MS, PhD); nutrition (MS, PhD).

Faculty: 11 full-time, 6 part-time/adjunct.
Students: 15 full-time (10 women), 7 part-time (all women); includes 3 minority (2 Asian Americans or Pacific Islanders, 1 Hispanic American), 5 international. *54 applicants, 30% accepted.* In 1999, 9 master's, 1 doctorate awarded.
Degree requirements: For master's, thesis optional, foreign language not required; for doctorate, dissertation required, foreign language not required.
Entrance requirements: For master's and doctorate, GRE General Test. *Application deadline:* For fall admission, 4/1. Applications are processed on a rolling basis. *Application fee:* $30 ($40 for international students).
Expenses: Tuition, state resident: full-time $5,400. Tuition, nonresident: full-time $14,535. Part-time tuition and fees vary according to course load and program.
Financial aid: Fellowships, research assistantships, teaching assistantships, Federal Work-Study and institutionally sponsored loans available. Aid available to part-time students.
Tammy Bray, Chair, 614-292-4485, *Fax:* 614-292-8880, *E-mail:* bray.21@osu.edu.

■ THE OHIO STATE UNIVERSITY

Graduate School, Program in Nutrition, Columbus, OH 43210

AWARDS PhD.

Faculty: 30 full-time.
Students: 12 full-time (7 women), 5 part-time (3 women); includes 1 minority (Asian American or Pacific Islander), 8 international. *16 applicants, 19% accepted.* In 1999, 4 degrees awarded.
Degree requirements: For doctorate, dissertation required.
Application fee: $30 ($40 for international students).
Expenses: Tuition, state resident: full-time $5,400. Tuition, nonresident: full-time $14,535. Part-time tuition and fees vary according to course load and program.
Jean T. Snook, Graduate Studies Committee Chairperson, 614-292-1680, *E-mail:* snook.3@osu.edu.

■ OHIO UNIVERSITY

Graduate Studies, College of Health and Human Services, School of Human and Consumer Sciences, Athens, OH 45701-2979

AWARDS Child development and family life (MSHCS); food and nutrition (MSHCS).

Faculty: 13 full-time (9 women), 5 part-time/adjunct (all women).

Students: 8 full-time (7 women), 13 part-time (all women), 4 international. Average age 26. *16 applicants, 88% accepted.* In 1999, 7 degrees awarded.

Degree requirements: For master's, thesis required, foreign language not required.

Entrance requirements: For master's, GRE. *Application deadline:* For fall admission, 8/30 (priority date). Applications are processed on a rolling basis. *Application fee:* $30.

Expenses: Tuition, state resident: full-time $5,754; part-time $238 per credit hour. Tuition, nonresident: full-time $11,055; part-time $457 per credit hour. Tuition and fees vary according to course load, degree level and campus/location.

Financial aid: In 1999–00, 6 teaching assistantships were awarded; career-related internships or fieldwork, Federal Work-Study, and institutionally sponsored loans also available. Financial aid application deadline: 3/15.

Faculty research: Diversity, developmentally appropriate activities, death and dying, gerontology, sexuality education.
Dr. V. Ann Pullins, Director, 740-593-2880.
Application contact: Dr. June Varner, Graduate Chair, 740-593-2877.

■ OKLAHOMA STATE UNIVERSITY

Graduate College, College of Human Environmental Sciences, Department of Nutritional Sciences, Stillwater, OK 74078

AWARDS MS, PhD.

Faculty: 7 full-time (6 women).

Students: 53 full-time (45 women), 62 part-time (49 women); includes 9 minority (4 African Americans, 2 Asian Americans or Pacific Islanders, 3 Native Americans), 23 international. Average age 33. In 1999, 7 master's, 2 doctorates awarded.

Degree requirements: For master's and doctorate, thesis/dissertation required, foreign language not required.

Entrance requirements: For master's, GRE, TOEFL; for doctorate, GRE. *Application deadline:* For fall admission, 7/1 (priority date). *Application fee:* $25.

Expenses: Tuition, state resident: part-time $86 per credit hour. Tuition, nonresident: part-time $275 per credit hour. Required fees: $17 per credit hour. $14 per semester. One-time fee: $20 full-time. Tuition and fees vary according to course load.

Financial aid: In 1999–00, 25 students received aid, including 11 research assistantships (averaging $9,313 per year), 12 teaching assistantships (averaging $8,441 per year); career-related internships or fieldwork, Federal Work-Study, and tuition waivers (partial) also available. Aid available to part-time students. Financial aid application deadline: 3/1.

Faculty research: Nutritional Sciences, micronutrients and chronic disease, phytochemicals, nutrition education, osteoporosis, food service administration.
Dr. Barbara Stoecker, Head, 405-744-5039.

■ OKLAHOMA STATE UNIVERSITY

Graduate College, College of Human Environmental Sciences, Program in Human Environmental Sciences, Stillwater, OK 74078

AWARDS Design, housing, and merchandising (PhD); family relations and child development (PhD); hotel and restaurant administration (PhD); nutritional sciences (PhD).

Faculty: 1 (woman) full-time.

Students: 1 (woman) full-time, 1 part-time, 1 international. Average age 41. In 1999, 7 degrees awarded.

Degree requirements: For doctorate, dissertation required, foreign language not required.

Entrance requirements: For doctorate, TOEFL. *Application deadline:* For fall admission, 7/1 (priority date). *Application fee:* $25.

Expenses: Tuition, state resident: part-time $86 per credit hour. Tuition, nonresident: part-time $275 per credit hour. Required fees: $17 per credit hour. $14 per semester. One-time fee: $20 full-time. Tuition and fees vary according to course load.

Financial aid: Research assistantships, teaching assistantships, career-related internships or fieldwork, Federal Work-Study, and tuition waivers (partial) available. Aid available to part-time students. Financial aid application deadline: 3/1.
Dr. Patricia Knaub, Dean, College of Human Environmental Sciences, 405-744-5053.

■ OREGON STATE UNIVERSITY

Graduate School, College of Home Economics and Education, Department of Nutrition and Food Management, Corvallis, OR 97331

AWARDS MAIS, MS, PhD. Part-time programs available.

Faculty: 11 full-time (9 women).

Students: 15 full-time (14 women), 16 part-time (15 women); includes 1 minority (Asian American or Pacific Islander), 7 international. Average age 35. In 1999, 6 master's, 1 doctorate awarded.

Degree requirements: For master's and doctorate, thesis/dissertation required, foreign language not required.

Entrance requirements: For master's and doctorate, TOEFL, minimum GPA of 3.0 in last 90 hours. *Application deadline:* For fall admission, 3/1. Applications are processed on a rolling basis. *Application fee:* $50.

Expenses: Tuition, state resident: full-time $6,489. Tuition, nonresident: full-time $11,061. Tuition and fees vary according to program.

Financial aid: Fellowships, research assistantships, teaching assistantships, career-related internships or fieldwork, Federal Work-Study, and institutionally sponsored loans available. Aid available to part-time students. Financial aid application deadline: 2/1.

Faculty research: Human metabolic studies, trace minerals, food science, food management.
Dr. Ann M. Messersmith, Head, 541-737-3561, *Fax:* 541-737-6914, *E-mail:* messersa@orst.edu.
Application contact: Laura Reid, Office Manager, 541-737-3561, *Fax:* 541-737-6914, *E-mail:* reidl@orst.edu.

■ THE PENNSYLVANIA STATE UNIVERSITY UNIVERSITY PARK CAMPUS

Graduate School, Intercollege Graduate Programs, Intercollege Graduate Program in Integrative Biosciences, State College, University Park, PA 16802-1503

AWARDS Integrative biosciences (MS, PhD), including biomolecular transport dynamics, cell and developmental biology, cellular and molecular mechanisms of toxicity, chemical biology, ecological and molecular plant physiology, immunobiology, molecular medicine, neuroscience, nutrition science.

Students: 39 full-time (24 women), 1 part-time.

Entrance requirements: For master's and doctorate, GRE General Test. *Application fee:* $50.

The Pennsylvania State University University Park Campus (continued)

Expenses: Tuition, state resident: full-time $6,886; part-time $291 per credit. Tuition, nonresident: full-time $14,118; part-time $588 per credit. Required fees: $46 per semester. Part-time tuition and fees vary according to course load and program.
Financial aid: Fellowships available.
Dr. C. R. Matthews, Co-Director, 814-863-3650.
Application contact: Admissions Committee, 814-865-3155, *Fax:* 814-863-1357, *E-mail:* lscgradadm@mail.biotec.psu.edu.

Find an in-depth description at www.petersons.com/graduate.

■ THE PENNSYLVANIA STATE UNIVERSITY UNIVERSITY PARK CAMPUS

Graduate School, Intercollege Graduate Programs, Intercollege Graduate Program in Nutrition, State College, University Park, PA 16802-1503

AWARDS Human nutrition (M Ed), including nutrition and public health (M Ed, MS), nutrition education (M Ed, MS); nutrition (MS, PhD), including applied human nutrition (MS), nutrition (PhD), nutrition and public health (M Ed, MS), nutrition education (M Ed, MS), nutrition science (MS).

Students: 33 full-time (29 women), 13 part-time (11 women). In 1999, 10 master's, 7 doctorates awarded.
Entrance requirements: For master's and doctorate, GRE General Test or MCAT. *Application fee:* $50.
Expenses: Tuition, state resident: full-time $6,886; part-time $291 per credit. Tuition, nonresident: full-time $14,118; part-time $588 per credit. Required fees: $46 per semester. Part-time tuition and fees vary according to course load and program.
John A. Milner, Head, 814-865-3448.

■ PURDUE UNIVERSITY

Graduate School, School of Consumer and Family Sciences, Department of Foods and Nutrition, West Lafayette, IN 47907

AWARDS Food sciences (MS, PhD); nutrition (MS, PhD).

Faculty: 15 full-time (8 women), 2 part-time/adjunct (0 women).
Students: 30 full-time (26 women), 9 part-time (8 women); includes 4 minority (2 African Americans, 1 Hispanic American, 1 Native American, 15 international. *37 applicants, 38% accepted.* In 1999, 3 master's, 4 doctorates awarded.
Degree requirements: For master's and doctorate, thesis/dissertation required.

Entrance requirements: For master's and doctorate, GRE General Test, TOEFL. *Application deadline:* Applications are processed on a rolling basis. *Application fee:* $30. Electronic applications accepted.
Expenses: Tuition, state resident: full-time $4,530; part-time $130 per credit hour. Tuition, nonresident: full-time $15,310; part-time $404 per credit hour. Tuition and fees vary according to campus/location and program.
Financial aid: Fellowships, research assistantships, teaching assistantships available. Aid available to part-time students. Financial aid applicants required to submit FAFSA.
Faculty research: Nutrient requirements, nutrient metabolism, nutrition and disease prevention.
Dr. C. M. Weaver, Head, 765-494-8237, *Fax:* 765-494-0674.
Application contact: Dawn Haan, Graduate Secretary, 765-494-8231, *Fax:* 765-494-0674.

Find an in-depth description at www.petersons.com/graduate.

■ RUSH UNIVERSITY

College of Health Sciences, Department of Clinical Nutrition, Chicago, IL 60612-3832

AWARDS MS. Part-time programs available.

Faculty: 16 full-time (all women), 5 part-time/adjunct (all women).
Students: 13 full-time (all women), 10 part-time (all women). Average age 24. *32 applicants, 31% accepted.* In 1999, 6 degrees awarded (100% found work related to degree).
Degree requirements: For master's, thesis required, foreign language not required. *Average time to degree:* Master's–2 years full-time, 5 years part-time.
Entrance requirements: For master's, GRE General Test, previous course work in statistics, minimum GPA of 3.0, undergraduate didactic program approved by the American Dietetic Association. *Application deadline:* For winter admission, 2/15. *Application fee:* $40.
Expenses: Tuition: Full-time $14,900; part-time $380 per credit. Full-time tuition and fees vary according to program and student level.
Financial aid: In 1999–00, 16 students received aid. Career-related internships or fieldwork, Federal Work-Study, grants, institutionally sponsored loans, and stipends available. Aid available to part-time students. Financial aid application deadline: 4/15; financial aid applicants required to submit FAFSA.
Faculty research: Lipid metabolism, cardiovascular risk factors, calcium risk factors, food service management,

parenteral nutrition. *Total annual research expenditures:* $40,000.
Dr. Rebecca A. Dowling, Chairperson, 312-942-7075, *Fax:* 312-942-3075, *E-mail:* rebecca_dowling@rush.edu.
Application contact: Annalynn Skipper, Director, 312-942-3349, *Fax:* 312-942-5203, *E-mail:* askipper@rush.edu.

■ RUTGERS, THE STATE UNIVERSITY OF NEW JERSEY, NEW BRUNSWICK

Graduate School, Program in Nutritional Sciences, New Brunswick, NJ 08901-1281

AWARDS MS, PhD. Part-time programs available.

Faculty: 28 full-time (13 women).
Students: 10 full-time (8 women), 16 part-time (all women); includes 7 minority (2 African Americans, 3 Asian Americans or Pacific Islanders, 2 Hispanic Americans), 7 international. *64 applicants, 14% accepted.* In 1999, 3 master's awarded. Terminal master's awarded for partial completion of doctoral program.
Degree requirements: For master's and doctorate, thesis/dissertation required, foreign language not required. *Average time to degree:* Master's–2 years full-time; doctorate–5 years full-time.
Entrance requirements: For master's and doctorate, GRE General Test. *Application deadline:* For fall admission, 4/1. Applications are processed on a rolling basis. *Application fee:* $50.
Expenses: Tuition, state resident: full-time $6,776; part-time $279 per credit. Tuition, nonresident: full-time $9,936; part-time $412 per credit. Required fees: $20 per credit. $89 per semester. Tuition and fees vary according to course load, campus/location and program.
Financial aid: In 1999–00, 13 students received aid, including 5 research assistantships, 6 teaching assistantships; fellowships, Federal Work-Study also available. Financial aid application deadline: 3/1; financial aid applicants required to submit FAFSA.
Faculty research: Nutrition and gene expression, nutrition and disease (obesity, diabetes, cancer, osteoporosis, alcohol), community nutrition and nutrition education, cellular lipid transport and metabolism.
Dr. Susan K. Fried, Director, 732-932-9039, *Fax:* 732-932-6837, *E-mail:* sfried@rci.rutgers.edu.
Application contact: Barbara Hannon, Administrative Assistant, 732-932-9379, *Fax:* 732-932-6837, *E-mail:* hannon@aesop.rutgers.edu.

■ SAGE GRADUATE SCHOOL

Graduate School, Division of Education, Program in Health Education, Troy, NY 12180-4115

AWARDS Health education (MS); nutrition and dietetics (MS). Part-time and evening/weekend programs available.

Students: 5 full-time (3 women), 33 part-time (21 women). Average age 30. *15 applicants, 67% accepted.* In 1999, 19 degrees awarded.

Degree requirements: For master's, thesis optional, foreign language not required.

Entrance requirements: For master's, minimum GPA of 2.75. *Application deadline:* For fall admission, 8/1; for spring admission, 12/15. Applications are processed on a rolling basis. *Application fee:* $40.

Expenses: Tuition: Part-time $372 per credit hour. Required fees: $100 per year.

Financial aid: Research assistantships, career-related internships or fieldwork available. Aid available to part-time students. Financial aid application deadline: 7/1; financial aid applicants required to submit FAFSA.

Faculty research: Policy development in health education and health care.

Dr. John J. Pelizza, Adviser, 518-244-2326, *Fax:* 218-244-2334, *E-mail:* peliz@sage.edu.

Application contact: Melissa M. Robertson, Associate Director of Admissions, 518-244-6878, *Fax:* 518-244-6880, *E-mail:* sgsadm@sage.edu.

■ SAGE GRADUATE SCHOOL

Graduate School, Division of Management, Communications and Legal Studies, Program in Health Services Administration, Troy, NY 12180-4115

AWARDS Gerontology (MS); health education (MS); management (MS); nutrition and dietetics (MS). Part-time and evening/weekend programs available.

Students: 12 full-time (9 women), 93 part-time (75 women); includes 6 minority (2 African Americans, 1 Asian American or Pacific Islander, 2 Hispanic Americans, 1 Native American). Average age 33. *18 applicants, 78% accepted.* In 1999, 35 degrees awarded.

Degree requirements: For master's, foreign language and thesis not required.

Entrance requirements: For master's, minimum GPA of 2.75. *Application deadline:* For fall admission, 8/1; for spring admission, 12/15. Applications are processed on a rolling basis. *Application fee:* $40.

Expenses: Tuition: Part-time $372 per credit hour. Required fees: $100 per year.

Financial aid: Career-related internships or fieldwork available. Aid available to part-time students. Financial aid application deadline: 7/1; financial aid applicants required to submit FAFSA.

Dr. Joan Dacher, Director, 518-244-3150, *E-mail:* dachej@sage.edu.

Application contact: Melissa M. Robertson, Associate Director of Admissions, 518-244-6878, *Fax:* 518-244-6880, *E-mail:* sgsadm@sage.edu.

Find an in-depth description at www.petersons.com/graduate.

■ SAGE GRADUATE SCHOOL

Graduate School, Division of Management, Communications and Legal Studies, Program in Public Administration, Troy, NY 12180-4115

AWARDS Communications (MS); gerontology (MS); human services administration (MS); nutrition and dietetics (MS); public management (MS). Part-time and evening/weekend programs available.

Students: 5 full-time (3 women), 25 part-time (13 women); includes 1 minority (Hispanic American). *4 applicants, 75% accepted.* In 1999, 16 degrees awarded.

Degree requirements: For master's, foreign language and thesis not required.

Entrance requirements: For master's, minimum GPA of 2.75. *Application deadline:* For fall admission, 8/1; for spring admission, 12/15. Applications are processed on a rolling basis. *Application fee:* $40.

Expenses: Tuition: Part-time $372 per credit hour. Required fees: $100 per year.

Financial aid: Career-related internships or fieldwork available. Aid available to part-time students. Financial aid application deadline: 7/1; financial aid applicants required to submit FAFSA.

Jeffrey Rinehart, Head, 518-292-1770, *E-mail:* rinehj@sage.edu.

Application contact: Melissa M. Robertson, Associate Director of Admissions, 518-244-6878, *Fax:* 518-244-6880, *E-mail:* sgsadm@sage.edu.

■ SAINT JOSEPH COLLEGE

Graduate Division, Field of Social Sciences, Department of Nutrition and Resource Management, West Hartford, CT 06117-2700

AWARDS Nutrition (MS). Part-time and evening/weekend programs available.

Faculty: 3 full-time (all women).

Students: 1 (woman). Average age 33. In 1999, 4 degrees awarded.

Degree requirements: For master's, project required, foreign language and thesis not required.

Entrance requirements: For master's, GRE General Test or MAT. *Application deadline:* For fall admission, 7/15 (priority date); for spring admission, 12/1 (priority date). Applications are processed on a rolling basis. *Application fee:* $25.

Expenses: Tuition: Part-time $420 per credit hour. Required fees: $25 per course.

Financial aid: In 1999-00, 1 research assistantship with full tuition reimbursement was awarded. Financial aid application deadline: 7/15; financial aid applicants required to submit FAFSA.

Dr. Margery Lawrence, Head, 860-231-5388, *E-mail:* mlawrence@sjc.edu.

■ SAINT LOUIS UNIVERSITY

Graduate School, School of Allied Health Professions, Department of Nutrition and Dietetics, St. Louis, MO 63103-2097

AWARDS Medical dietetics (MS); nutrition and physical performance (MS).

Faculty: 7 full-time (6 women), 58 part-time/adjunct (57 women).

Students: 16 full-time (15 women), 50 part-time (47 women); includes 9 minority (4 African Americans, 2 Asian Americans or Pacific Islanders, 3 Hispanic Americans), 7 international. Average age 28. *36 applicants, 94% accepted.* In 1999, 3 degrees awarded.

Degree requirements: For master's, comprehensive oral exam required, foreign language and thesis not required.

Entrance requirements: For master's, GRE General Test. *Application deadline:* For fall admission, 7/1; for spring admission, 11/1. Applications are processed on a rolling basis. *Application fee:* $40.

Expenses: Tuition: Full-time $20,520; part-time $570 per credit hour. Required fees: $38 per term. Tuition and fees vary according to program.

Financial aid: In 1999-00, 31 students received aid, including 6 teaching assistantships Financial aid application deadline: 4/1; financial aid applicants required to submit FAFSA.

Faculty research: Hypertension prevention, weight management, geriatrics, sports nutrition, pediatrics.

Dr. Mildred Mattfeldt-Beman, Chairperson, 314-577-8523.

Application contact: Dr. Marcia Buresch, Assistant Dean of the Graduate School, 314-977-2240, *Fax:* 314-977-3943, *E-mail:* bureschm@slu.edu.

■ SAN DIEGO STATE UNIVERSITY

Graduate and Research Affairs, College of Professional Studies and Fine Arts, Department of Exercise and Nutritional Sciences, Program in Nutritional Science, San Diego, CA 92182

AWARDS MS.

San Diego State University (continued)
Students: 13 full-time (12 women), 9 part-time (8 women); includes 2 minority (1 Asian American or Pacific Islander, 1 Hispanic American), 1 international. *19 applicants, 68% accepted.* In 1999, 1 degree awarded.
Degree requirements: For master's, thesis required, foreign language not required.
Entrance requirements: For master's, GRE General Test, TOEFL. *Application deadline:* For fall admission, 4/1 (priority date). Applications are processed on a rolling basis. *Application fee:* $55.
Expenses: Tuition, nonresident: part-time $246 per unit. Required fees: $1,932; $633 per semester. Tuition and fees vary according to course load.
Financial aid: Career-related internships or fieldwork available.
Application contact: Patricia Patterson, Graduate Adviser, 619-594-5999, *Fax:* 619-594-6553, *E-mail:* ensgrad@mail.sdsu.edu.

■ SAN JOSE STATE UNIVERSITY
Graduate Studies, College of Applied Arts and Sciences, Department of Nutritional Science, San Jose, CA 95192-0001
AWARDS MS.

Expenses: Tuition, nonresident: part-time $246 per unit. Required fees: $1,939; $1,309 per year.

■ SIMMONS COLLEGE
Graduate School for Health Studies, Program in Nutrition and Health Promotion, Boston, MA 02115
AWARDS MS, Certificate. Part-time and evening/weekend programs available.

Faculty: 4 full-time (3 women).
Students: 11 full-time (all women), 14 part-time (all women), 2 international. *69 applicants, 58% accepted.* In 1999, 14 master's, 6 other advanced degrees awarded.
Degree requirements: For master's, research project required. *Average time to degree:* Master's–1 year full-time, 2.5 years part-time.
Entrance requirements: For master's, GRE General Test, TOEFL. *Application deadline:* Applications are processed on a rolling basis. *Application fee:* $50.
Expenses: Tuition: Full-time $14,460; part-time $610 per semester hour. Required fees: $10 per semester. Tuition and fees vary according to course load and program.
Financial aid: Federal Work-Study and unspecified assistantships available. Financial aid application deadline: 3/1.
Faculty research: Nutrition supplements of athletes' nutrition and health-related

behaviors of college-age women, glutamine supplementation in AIDS.
Dr. Nancie Herbold, Program Director, 617-521-2711, *Fax:* 617-521-3137, *E-mail:* nherbold@simmons.edu.
Application contact: Christine Keuleyan, Admission Coordinator, 617-521-2650, *Fax:* 617-521-3137, *E-mail:* keuleyan@simmons.edu.
Find an in-depth description at www.petersons.com/graduate.

■ SOUTH CAROLINA STATE UNIVERSITY
School of Graduate Studies, School of Applied Professional Sciences, Department of Family and Consumer Sciences, Orangeburg, SC 29117-0001
AWARDS Individual and family development (MS); nutritional sciences (MS). Part-time and evening/weekend programs available.

Degree requirements: For master's, departmental qualifying exam required, thesis optional, foreign language not required.
Entrance requirements: For master's, GRE, MAT, or NTE, minimum GPA of 2.7.
Expenses: Tuition, state resident: full-time $3,410; part-time $192 per credit hour. Tuition, nonresident: full-time $6,702; part-time $375 per credit hour.
Faculty research: Societal competence, relationship of parent-child interaction to adult, quality of well-being of rural elders.

■ SOUTHERN ILLINOIS UNIVERSITY CARBONDALE
Graduate School, College of Agriculture, Department of Animal Science, Food and Nutrition, Program in Food and Nutrition, Carbondale, IL 62901-6806
AWARDS MS.

Faculty: 15 full-time (6 women).
Students: 6 full-time (5 women), 20 part-time (19 women); includes 1 minority (Hispanic American), 1 international. Average age 29. *8 applicants, 50% accepted.* In 1999, 7 degrees awarded.
Degree requirements: For master's, thesis or alternative required, foreign language not required.
Entrance requirements: For master's, TOEFL, minimum GPA of 2.7. *Application deadline:* Applications are processed on a rolling basis. *Application fee:* $0.
Expenses: Tuition, state resident: full-time $2,902. Tuition, nonresident: full-time $5,810. Tuition and fees vary according to course load.
Financial aid: In 1999–00, 12 students received aid; fellowships, research assistantships, teaching assistantships,

career-related internships or fieldwork, Federal Work-Study, institutionally sponsored loans, and tuition waivers (full) available. Aid available to part-time students.
Faculty research: Public health nutrition, nutrition physiology, soybean utilization, nutrition education. *Total annual research expenditures:* $100,000.
Application contact: Dr. Carol Boushey, Director, Dietetic Internship Program, 618-453-7514, *Fax:* 618-453-7517.

■ STATE UNIVERSITY OF NEW YORK AT BUFFALO
Graduate School, School of Health Related Professions, Program in Nutrition, Buffalo, NY 14260
AWARDS MS. Part-time programs available.

Faculty: 5 full-time (3 women), 1 (woman) part-time/adjunct.
Students: 18 full-time (12 women), 8 part-time (all women); includes 1 minority (Asian American or Pacific Islander), 13 international. Average age 27. In 1999, 11 degrees awarded.
Degree requirements: For master's, thesis required (for some programs), foreign language not required.
Entrance requirements: For master's, GRE General Test, TOEFL. *Application deadline:* Applications are processed on a rolling basis. *Application fee:* $35.
Expenses: Tuition, state resident: full-time $5,100; part-time $213 per credit hour. Tuition, nonresident: full-time $8,416; part-time $351 per credit hour. Required fees: $935; $75 per semester. Tuition and fees vary according to course load and program.
Financial aid: In 1999–00, 4 research assistantships, 4 teaching assistantships were awarded; fellowships, Federal Work-Study, grants, institutionally sponsored loans, scholarships, and tuition waivers (full and partial) also available. Financial aid application deadline: 1/31.
Faculty research: Nutrition and signal transduction, immunology, extracellular matrix differentiation, nutrient transport, exercise nutrition.
Atif B. Awad, Director, 716-829-3680, *Fax:* 716-829-3700, *E-mail:* awad@ascu.buffalo.edu.

■ SYRACUSE UNIVERSITY
Graduate School, College for Human Development, Department of Nutrition Science and Food Management, Syracuse, NY 13244-0003
AWARDS MA, MS, PhD.

Faculty: 8.
Students: 16 full-time (13 women), 5 part-time (all women); includes 1 minority

(Asian American or Pacific Islander), 1 international. Average age 28. *25 applicants, 80% accepted.* In 1999, 9 master's awarded.
Degree requirements: For master's, thesis required (for some programs); for doctorate, dissertation required.
Entrance requirements: For master's and doctorate, GRE General Test. *Application deadline:* For fall admission, 1/15. *Application fee:* $40.
Expenses: Tuition: Full-time $13,992; part-time $583 per credit hour.
Financial aid: Fellowships, research assistantships, teaching assistantships, Federal Work-Study and tuition waivers (partial) available. Financial aid application deadline: 3/1.
Norman Faiola, Chair, 315-443-4550.
Application contact: Jean Bowering, Graduate Program Director, 315-443-2396.

■ TEXAS A&M UNIVERSITY

College of Agriculture and Life Sciences, Department of Animal Science, College Station, TX 77843

AWARDS Animal breeding (MS, PhD); animal science (M Agr, MS, PhD); dairy science (M Agr, MS); food science and technology (MS, PhD); genetics (MS, PhD); nutrition (MS, PhD); physiology of reproduction (MS, PhD).

Faculty: 73 full-time (11 women), 18 part-time/adjunct (4 women).
Students: 188 full-time (108 women), 4 part-time (2 women); includes 9 minority (2 Asian Americans or Pacific Islanders, 7 Hispanic Americans), 36 international. Average age 26. *158 applicants, 25% accepted.* In 1999, 43 master's awarded (40% found work related to degree, 60% continued full-time study); 9 doctorates awarded (60% entered university research/teaching, 40% found other work related to degree).
Degree requirements: For master's and doctorate, thesis/dissertation required, foreign language not required. *Average time to degree:* Master's–2 years full-time; doctorate–4 years full-time.
Entrance requirements: For master's and doctorate, GRE General Test, TOEFL. *Application deadline:* For fall admission, 2/1 (priority date); for spring admission, 10/1 (priority date). Applications are processed on a rolling basis. *Application fee:* $50 ($75 for international students).
Expenses: Tuition, state resident: part-time $76 per semester hour. Tuition, nonresident: part-time $292 per semester hour. Required fees: $11 per semester hour. Tuition and fees vary according to program.
Financial aid: In 1999–00, 5 fellowships (averaging $12,000 per year), 39 research assistantships (averaging $11,250 per year),

33 teaching assistantships (averaging $11,250 per year) were awarded; career-related internships or fieldwork, Federal Work-Study, institutionally sponsored loans, and scholarships also available. Financial aid application deadline: 2/1; financial aid applicants required to submit FAFSA.
Faculty research: Genetic engineering/gene markers, dietary effects on colon cancer, biotechnology.
Dr. Bryan H. Johnson, Head, 979-845-1541.
Application contact: Ronnie Edwards, Graduate Adviser, 409-845-1542, *Fax:* 409-845-6433, *E-mail:* r-edwards@tamu.edu.

■ TEXAS A&M UNIVERSITY

Intercollegiate Faculty in Nutrition, College Station, TX 77843

AWARDS MS, PhD. Part-time programs available.

Students: 41 full-time (35 women), 23 part-time (14 women); includes 4 minority (1 Asian American or Pacific Islander, 3 Hispanic Americans), 11 international. *7 applicants, 14% accepted.* In 1999, 8 master's, 4 doctorates awarded.
Entrance requirements: For master's and doctorate, GRE General Test, TOEFL. *Application deadline:* Applications are processed on a rolling basis. *Application fee:* $50 ($75 for international students).
Expenses: Tuition, state resident: part-time $76 per semester hour. Tuition, nonresident: part-time $292 per semester hour. Required fees: $11 per semester hour. Tuition and fees vary according to program.
Financial aid: Fellowships, research assistantships, teaching assistantships available. Financial aid application deadline: 4/1; financial aid applicants required to submit FAFSA.
Faculty research: Cellular and molecular nutrition, domestic animal nutrition, applied human nutrition, food quality and nutrient composition. *Total annual research expenditures:* $2 million.
Dr. Gary R. Acuff, Chair, 979-845-4425, *Fax:* 979-862-1864.
Application contact: Kathy Martinez, Secretary, 979-845-1735, *Fax:* 979-862-2378, *E-mail:* dmcmurray@tamu.edu.

■ TEXAS SOUTHERN UNIVERSITY

Graduate School, College of Arts and Sciences, Department of Home Economics, Houston, TX 77004-4584

AWARDS Home economics education (MA, MS), including child development (MS), foods and nutrition (MS), home economics education (MS). Part-time and evening/weekend programs available.

Faculty: 4 full-time (3 women), 1 (woman) part-time/adjunct.
Students: 22 full-time (all women), 1 part-time; all minorities (22 African Americans, 1 Asian American or Pacific Islander). *9 applicants, 78% accepted.* In 1999, 5 degrees awarded.
Degree requirements: For master's, thesis (for some programs), comprehensive exam required, foreign language not required. *Average time to degree:* Master's–2 years full-time, 3 years part-time.
Entrance requirements: For master's, GRE General Test, TOEFL, minimum GPA of 2.5. *Application deadline:* For fall admission, 7/15 (priority date). Applications are processed on a rolling basis. *Application fee:* $35 ($75 for international students).
Expenses: Tuition, area resident: Part-time $296 per credit hour. Tuition, nonresident: part-time $449 per credit hour.
Financial aid: Teaching assistantships, career-related internships or fieldwork and institutionally sponsored loans available. Financial aid application deadline: 5/1.
Dr. Oddis Turner, Head, 713-313-7699, *Fax:* 713-313-7228.

■ TEXAS TECH UNIVERSITY

Graduate School, College of Human Sciences, Department of Education, Nutrition, and Restaurant/Hotel Management, Program in Food and Nutrition, Lubbock, TX 79409

AWARDS MS, PhD. Part-time programs available.

Faculty: 6 full-time (4 women), 2 part-time/adjunct (both women).
Students: 13 full-time (11 women), 1 (woman) part-time; includes 1 minority (Hispanic American), 3 international. Average age 30. *48 applicants, 69% accepted.* In 2000, 8 master's awarded (88% found work related to degree, 12% continued full-time study); 1 doctorate awarded (100% entered university research/teaching).
Degree requirements: For master's, thesis optional, foreign language not required; for doctorate, dissertation required, foreign language not required. *Average time to degree:* Master's–1.5 years full-time; doctorate–3.5 years full-time.
Entrance requirements: For master's and doctorate, GRE General Test. *Application deadline:* For fall admission, 4/15 (priority date). Applications are processed on a rolling basis. *Application fee:* $25 ($50 for international students). Electronic applications accepted.
Expenses: Tuition, state resident: full-time $2,474. Tuition, nonresident: full-time $6,272.

Texas Tech University (continued)

Financial aid: In 2000–01, 5 teaching assistantships (averaging $12,000 per year) were awarded; fellowships, research assistantships, career-related internships or fieldwork, Federal Work-Study, grants, institutionally sponsored loans, and scholarships also available. Aid available to part-time students. Financial aid application deadline: 2/1.

Faculty research: Assessment of nutritional status, nutritional status and health effects of selenium, vitamin E, biotin, and iron; diabetes; obesity. *Total annual research expenditures:* $200,000.

Dr. Mallory Boylan, Graduate Adviser, 806-742-3068, *Fax:* 806-742-3042, *E-mail:* mboylan@hs.ttu.edu.

Application contact: Graduate Adviser, 806-742-3068, *Fax:* 806-742-1849.

■ TEXAS WOMAN'S UNIVERSITY

Graduate School, College of Health Sciences, Department of Nutrition and Food Sciences, Denton, TX 76204

AWARDS Exercise and sports nutrition (MS); institutional administration (MS); nutrition (MS, PhD). Part-time and evening/weekend programs available.

Faculty: 11 full-time (8 women), 3 part-time/adjunct (2 women).

Students: 53 full-time (51 women), 98 part-time (93 women); includes 27 minority (3 African Americans, 9 Asian Americans or Pacific Islanders, 15 Hispanic Americans), 8 international. Average age 30. *48 applicants, 94% accepted.* In 1999, 35 master's, 4 doctorates awarded (33% entered university research/teaching, 67% found other work related to degree).

Degree requirements: For master's, thesis required (for some programs); for doctorate, dissertation, qualifying exam required, dissertation, qualifying exam required.

Entrance requirements: For master's, GRE General Test, minimum GPA of 3.25; for doctorate, GRE General Test, minimum GPA of 3.5. *Application deadline:* Applications are processed on a rolling basis. *Application fee:* $30.

Expenses: Tuition, state resident: full-time $2,045; part-time $83 per semester hour. Tuition, nonresident: full-time $5,933; part-time $279 per semester hour. Required fees: $500 per semester. Tuition and fees vary according to course load.

Financial aid: In 1999–00, 12 research assistantships (averaging $10,000 per year), 11 teaching assistantships (averaging $10,000 per year) were awarded; career-related internships or fieldwork, Federal Work-Study, institutionally sponsored loans, and scholarships also available. Aid available to part-time students. Financial aid application deadline: 4/1.

Faculty research: Nutrition and aging, food safety, exercise and sports nutrition, nutrition and cancer, animal models of disease.

Dr. Carolyn Bednar, Chair, 940-898-2636, *Fax:* 940-898-2634, *E-mail:* cbednar@twu.edu.

■ TUFTS UNIVERSITY

School of Nutrition Science and Policy, Medford, MA 02155

AWARDS Nutrition (MS, PhD). Part-time programs available. Terminal master's awarded for partial completion of doctoral program.

Degree requirements: For master's, foreign language and thesis not required; for doctorate, dissertation required, foreign language not required.

Entrance requirements: For master's and doctorate, GRE General Test, TOEFL. Electronic applications accepted.

Expenses: Tuition: Full-time $17,674.

Faculty research: Nutritional biochemistry, nutrition development, epidemiology, policy/planning, applied nutrition.

Find an in-depth description at www.petersons.com/graduate.

■ TULANE UNIVERSITY

School of Public Health and Tropical Medicine, Department of Community Health Sciences, Program in Nutrition, New Orleans, LA 70118-5669

AWARDS MPH.

Students: 10 full-time (all women), 3 part-time (all women); includes 2 minority (1 African American, 1 Asian American or Pacific Islander), 1 international. Average age 26.

Degree requirements: For master's, one foreign language, thesis not required.

Entrance requirements: For master's, GRE General Test, TOEFL. *Application deadline:* For fall admission, 4/15 (priority date); for spring admission, 10/15. Applications are processed on a rolling basis. *Application fee:* $40.

Expenses: Tuition: Full-time $16,625. Required fees: $1,352.

Financial aid: Application deadline: 2/1. Dr. Thu Chin, Acting Chair, Department of Community Health Sciences, 504-588-5391.

■ TUSKEGEE UNIVERSITY

Graduate Programs, College of Agricultural, Environmental and Natural Sciences, Department of Food and Nutritional Sciences, Tuskegee, AL 36088

AWARDS MS.

Faculty: 4 full-time (3 women).

Students: 7 full-time (5 women), 5 part-time (all women); includes 11 minority (all African Americans), 1 international. Average age 24. In 1999, 3 degrees awarded.

Degree requirements: For master's, computer language, thesis required, foreign language not required.

Entrance requirements: For master's, GRE General Test. *Application deadline:* For fall admission, 7/15. Applications are processed on a rolling basis. *Application fee:* $25 ($35 for international students).

Expenses: Tuition: Full-time $9,500. Tuition and fees vary according to course load and degree level.

Financial aid: Application deadline: 4/15. Dr. Ralphenia Pace, Head, 334-727-8162.

■ THE UNIVERSITY OF AKRON

Graduate School, College of Fine and Applied Arts, School of Family and Consumer Sciences, Program in Nutrition and Dietetics, Akron, OH 44325-0001

AWARDS MS.

Degree requirements: For master's, thesis or alternative required, foreign language not required.

Entrance requirements: For master's, GRE General Test, minimum GPA of 2.75.

Expenses: Tuition, state resident: part-time $189 per credit. Tuition, nonresident: part-time $353 per credit. Required fees: $7.3 per credit.

■ THE UNIVERSITY OF ALABAMA

Graduate School, College of Human Environmental Sciences, Department of Human Nutrition and Hospitality Management, Tuscaloosa, AL 35487

AWARDS MSHES. Part-time programs available.

Faculty: 5 full-time (4 women).

Students: 12 full-time (11 women), 1 (woman) part-time. In 1999, 4 degrees awarded (100% found work related to degree).

Degree requirements: For master's, thesis optional, foreign language not required.

Entrance requirements: For master's, GRE General Test or MAT, minimum GPA of 3.0. *Application deadline:* For fall admission, 7/6. Applications are processed on a rolling basis. *Application fee:* $25.

Expenses: Tuition, state resident: full-time $2,872. Tuition, nonresident: full-time $7,722. Part-time tuition and fees vary according to course load and program.

Financial aid: In 1999–00, 4 students received aid, including 2 research assistantships (averaging $8,100 per year), 2 teaching assistantships (averaging $8,100 per

year); career-related internships or fieldwork also available. Financial aid application deadline: 3/15.
Faculty research: Fat determination of low-fat foods, maternal and child nutrition, obesity and eating disorders, community nutrition interventions.
Dr. Judy L. Bonner, Head, 205-348-6157, *E-mail:* jbonner@ches.ua.edu.
Application contact: Dr. Olivia Kendrick, Coordinator of Graduate Education, 205-348-9146, *Fax:* 205-348-3789, *E-mail:* okendric@ches.ua.edu.

■ THE UNIVERSITY OF ALABAMA AT BIRMINGHAM

Graduate School, School of Health Related Professions, Department of Nutrition Sciences, Program in Clinical Nutrition and Dietetics, Birmingham, AL 35294

AWARDS Clinical nutrition (MS); dietetic internship (Certificate). Part-time programs available.

Students: 39 full-time (37 women), 4 part-time (all women); includes 4 minority (1 African American, 1 Asian American or Pacific Islander, 1 Hispanic American, 1 Native American), 1 international. *43 applicants, 100% accepted.* In 1999, 7 master's, 21 other advanced degrees awarded.
Degree requirements: For master's, thesis required.
Entrance requirements: For master's, GRE General Test or MAT, bachelor's degree in dietetics or related field; for Certificate, GRE General Test or MAT, bachelor's degree in dietetics. *Application deadline:* For fall admission, 7/1 (priority date). Applications are processed on a rolling basis. *Application fee:* $35 ($60 for international students). Electronic applications accepted.
Expenses: Tuition, state resident: part-time $135 per credit hour. Tuition, nonresident: part-time $270 per credit hour. Required fees: $16 per credit hour. $40 per quarter.
Financial aid: In 1999–00, 5 students received aid, including 5 research assistantships; career-related internships or fieldwork also available.
Faculty research: Clinical assessment, folic acid, energy metabolism, nutrition and cancer, nutrition for children and adolescents with special health care needs.
Dr. Gayl J. Canfield, Director, 205-934-3006, *Fax:* 205-934-7049, *E-mail:* canfield@uab.edu.

■ THE UNIVERSITY OF ALABAMA AT BIRMINGHAM

Graduate School, School of Health Related Professions, Department of Nutrition Sciences, Program in Nutrition Sciences, Birmingham, AL 35294

AWARDS PhD.

Students: 9 full-time (4 women); includes 2 minority (both African Americans), 4 international. *25 applicants, 64% accepted.* In 1999, 2 degrees awarded.
Degree requirements: For doctorate, dissertation required.
Entrance requirements: For doctorate, GRE General Test. *Application deadline:* Applications are processed on a rolling basis. *Application fee:* $35 ($60 for international students). Electronic applications accepted.
Expenses: Tuition, state resident: part-time $135 per credit hour. Tuition, nonresident: part-time $270 per credit hour. Required fees: $16 per credit hour. $40 per quarter.
Financial aid: In 1999–00, 6 students received aid, including 6 fellowships with tuition reimbursements available, 1 research assistantship with tuition reimbursement available; career-related internships or fieldwork also available.
Faculty research: Energy metabolism, obesity, body composition, cancer prevention, bone metabolism. *Total annual research expenditures:* $2 million.
Dr. Charles W. Prince, Interim Chair, 205-934-7757, *Fax:* 205-934-7049, *E-mail:* princecw@uab.edu.

Find an in-depth description at www.petersons.com/graduate.

■ THE UNIVERSITY OF ARIZONA

Graduate College, College of Agriculture, Department of Nutritional Sciences, Tucson, AZ 85721

AWARDS Dietetics (MS); nutritional sciences (MS). Part-time programs available.

Degree requirements: For master's, thesis required, foreign language not required.
Entrance requirements: For master's, GRE, TOEFL, minimum GPA of 3.0.
Expenses: Tuition, nonresident: full-time $4,814; part-time $274 per unit. Required fees: $1,094; $115 per unit. Tuition and fees vary according to course load and program.
Faculty research: Lipids, lipoproteins, minerals, aging, mycotoxins.

■ THE UNIVERSITY OF ARIZONA

Graduate College, Graduate Interdisciplinary Programs, Graduate Interdisciplinary Program in Nutritional Sciences, Tucson, AZ 85721

AWARDS PhD. Part-time programs available.

Degree requirements: For doctorate, dissertation required, foreign language not required.
Entrance requirements: For doctorate, GRE, TOEFL.
Expenses: Tuition, nonresident: full-time $4,814; part-time $274 per unit. Required fees: $1,094; $115 per unit. Tuition and fees vary according to course load and program.
Faculty research: International nutrition, nutrition and aging, vitamin metabolism.

■ UNIVERSITY OF ARKANSAS FOR MEDICAL SCIENCES

Graduate School, Program in Clinical Nutrition, Little Rock, AR 72205-7199

AWARDS MS. Part-time programs available.

Faculty: 12 full-time (6 women), 1 (woman) part-time/adjunct.
Students: 6 full-time (all women), 13 part-time (all women); includes 3 minority (2 African Americans, 1 Native American), 1 international. In 1999, 2 degrees awarded.
Degree requirements: For master's, thesis required, foreign language not required.
Application fee: $0.
Expenses: Tuition, state resident: full-time $3,540; part-time $177 per hour. Tuition, nonresident: full-time $7,600; part-time $380 per hour. Required fees: $75; $40 per trimester. Part-time tuition and fees vary according to course load.
Financial aid: In 1999–00, 1 research assistantship was awarded. Aid available to part-time students.
Dr. Beverly J. McCabe, Director, 501-686-6166.

■ UNIVERSITY OF BRIDGEPORT

College of Graduate and Undergraduate Studies, Division of Allied Health Technology, Human Nutrition Institute, Bridgeport, CT 06601

AWARDS MS. Part-time and evening/weekend programs available. Postbaccalaureate distance learning degree programs offered (no on-campus study).

Faculty: 1 (woman) full-time, 6 part-time/adjunct (3 women).
Students: 92 (56 women); includes 9 minority (2 African Americans, 4 Asian Americans or Pacific Islanders, 3 Hispanic Americans) 4 international. Average age

University of Bridgeport (continued)
37. *140 applicants, 64% accepted.* In 1999, 21 degrees awarded.

Degree requirements: For master's, thesis, research project required, foreign language not required.

Entrance requirements: For master's, previous course work in anatomy, biochemistry, organic chemistry, or physiology. *Application deadline:* Applications are processed on a rolling basis. *Application fee:* $25 ($35 for international students). Electronic applications accepted.

Expenses: Tuition: Part-time $370 per credit. Required fees: $75 per semester.

Financial aid: In 1999–00, 33 students received aid. Available to part-time students. Application deadline: 6/1. Dr. Blonnie Y. Thompson, Director, Division of Allied Health Technology, 203-576-4667.

Find an in-depth description at www.petersons.com/graduate.

■ UNIVERSITY OF CALIFORNIA, BERKELEY

Graduate Division, Group in Nutrition, Berkeley, CA 94720-1500

AWARDS MS; PhD. Terminal master's awarded for partial completion of doctoral program.

Degree requirements: For master's, thesis required; for doctorate, dissertation, qualifying exam required.

Entrance requirements: For master's and doctorate, GRE General Test, TOEFL, minimum GPA of 3.0.

Expenses: Tuition, nonresident: full-time $9,804. Required fees: $4,268. Tuition and fees vary according to program.

Find an in-depth description at www.petersons.com/graduate.

■ UNIVERSITY OF CALIFORNIA, BERKELEY

Graduate Division, School of Public Health, Master Internationalist Program, Berkeley, CA 94720-1500

AWARDS Epidemiology (MPH); interdisciplinary (MPH); maternal and child health (MPH); public health nutrition (MPH).

Entrance requirements: For master's, GRE General Test, minimum GPA of 3.0.

Expenses: Tuition, nonresident: full-time $9,804. Required fees: $4,268. Tuition and fees vary according to program.

■ UNIVERSITY OF CALIFORNIA, DAVIS

Graduate Studies, Programs in the Biological Sciences, Program in Nutrition, Davis, CA 95616

AWARDS MS, PhD. Part-time programs available.

Faculty: 44 full-time (10 women).

Students: 88 full-time (70 women), 2 part-time (1 woman); includes 20 minority (1 African American, 12 Asian Americans or Pacific Islanders, 7 Hispanic Americans), 8 international. Average age 30. *79 applicants, 42% accepted.* In 1999, 11 master's, 8 doctorates awarded.

Degree requirements: For master's, thesis optional; for doctorate, dissertation required.

Entrance requirements: For master's and doctorate, GRE General Test. *Application deadline:* For fall admission, 1/15. Applications are processed on a rolling basis. *Application fee:* $40. Electronic applications accepted.

Expenses: Tuition, nonresident: full-time $9,804. Tuition and fees vary according to program and student level.

Financial aid: In 1999–00, 75 students received aid, including 20 fellowships with full and partial tuition reimbursements available, 19 research assistantships with full and partial tuition reimbursements available, 18 teaching assistantships with partial tuition reimbursements available; Federal Work-Study, grants, scholarships, and tuition waivers (full and partial) also available. Aid available to part-time students. Financial aid application deadline: 1/15; financial aid applicants required to submit FAFSA.

Faculty research: Human/animal nutrition.
Katherine Dewey, Graduate Chair, 530-752-0851, *E-mail:* kgdewey@ucdavis.edu.
Application contact: Marlene Belz, Administrative Assistant, 530-754-7684, *Fax:* 530-752-8966, *E-mail:* mhbelz@ucdavis.edu.

■ UNIVERSITY OF CENTRAL OKLAHOMA

Graduate College, College of Education, Department of Human Environmental Sciences, Edmond, OK 73034-5209

AWARDS Family and child studies (MS); family and consumer science education (MS); interior design (MS); nutrition-food management (MS). Part-time programs available.

Faculty: 10 full-time (8 women), 3 part-time/adjunct (all women).

Students: 7 full-time (all women), 86 part-time (82 women); includes 9 minority (5 African Americans, 1 Asian American or Pacific Islander, 1 Hispanic American, 2 Native Americans), 9 international. Average age 34. *21 applicants, 99% accepted.* In 1999, 30 degrees awarded.

Degree requirements: For master's, foreign language and thesis not required. *Application deadline:* Applications are processed on a rolling basis. *Application fee:* $15.

Expenses: Tuition, state resident: part-time $66 per hour. Tuition, nonresident: part-time $84 per hour. Full-time tuition and fees vary according to course level and course load.

Financial aid: Career-related internships or fieldwork and unspecified assistantships available. Financial aid application deadline: 3/31; financial aid applicants required to submit FAFSA.

Faculty research: Dietetics and food science.
Dr. Valerie Knotts, Chairperson, 405-974-5787.

■ UNIVERSITY OF CHICAGO

Division of the Biological Sciences, Biomedical Sciences: Cancer, Immunology, Nutrition, Pathology, and Virology, Committee on Human Nutrition and Nutritional Biology, Chicago, IL 60637-1513

AWARDS PhD.

Faculty: 21 full-time (5 women).

Students: 7 full-time (4 women); includes 1 minority (Asian American or Pacific Islander), 1 international. Average age 30. *32 applicants, 3% accepted.* In 1999, 3 degrees awarded (67% entered university research/teaching).

Degree requirements: For doctorate, dissertation required, foreign language not required. *Average time to degree:* Doctorate–5.5 years full-time.

Entrance requirements: For doctorate, GRE General Test, TOEFL. *Application deadline:* For fall admission, 1/5 (priority date). *Application fee:* $55.

Expenses: Tuition: Full-time $24,804; part-time $3,422 per course. Required fees: $390. Tuition and fees vary according to program.

Financial aid: In 1999–00, 7 students received aid, including 1 fellowship with full tuition reimbursement available (averaging $16,500 per year), 6 research assistantships with full tuition reimbursements available (averaging $16,500 per year); institutionally sponsored loans also available. Financial aid application deadline: 6/1.

Faculty research: Regulation of lipoprotein metabolism, cellular vitamin metabolism, obesity and body composition, adipocyte differentiation.

Dr. Reed Graves, Interim Chairman, 773-702-6921, *Fax:* 773-702-4634, *E-mail:* ragraves@midway.uchicago.edu.
Application contact: Rebecca Levine, Administrative Assistant, Student Services, 773-834-3899, *Fax:* 773-702-4634, *E-mail:* rlevine@huggins.bsd.uchicago.edu.

■ UNIVERSITY OF CINCINNATI

Division of Research and Advanced Studies, College of Allied Health Sciences, Department of Nutrition Science, Cincinnati, OH 45221-0091

AWARDS M Ed. Part-time programs available.

Faculty: 4 full-time.
Students: 13 full-time (11 women), 10 part-time (9 women); includes 4 minority (1 African American, 2 Asian Americans or Pacific Islanders, 1 Hispanic American), 1 international. *21 applicants, 90% accepted.* In 1999, 13 degrees awarded.
Degree requirements: For master's, thesis or alternative required, foreign language not required. *Average time to degree:* Master's–2.8 years full-time.
Entrance requirements: For master's, GRE General Test. *Application deadline:* For fall admission, 2/1. *Application fee:* $30.
Expenses: Tuition, state resident: full-time $5,880; part-time $196 per credit hour. Tuition, nonresident: full-time $11,067; part-time $369 per credit hour. Required fees: $741; $247 per quarter. Tuition and fees vary according to program.
Financial aid: Fellowships, tuition waivers (full) and unspecified assistantships available. Aid available to part-time students. Financial aid application deadline: 5/1.
Dr. Bonnie Brehm, Chair, 513-558-7502, *Fax:* 513-558-7500, *E-mail:* bonnie.brehm@uc.edu.
Application contact: Grace Falciglia, Graduate Program Director, 513-558-7505, *Fax:* 513-558-7500, *E-mail:* grace.falciglia@uc.edu.

■ UNIVERSITY OF CONNECTICUT

Graduate School, College of Agriculture and Natural Resources, Department of Nutritional Sciences, Storrs, CT 06269

AWARDS MS, PhD.

Degree requirements: For master's and doctorate, thesis/dissertation required.
Entrance requirements: For master's and doctorate, GRE General Test, TOEFL.
Expenses: Tuition, state resident: full-time $5,118. Tuition, nonresident: full-time $13,298. Required fees: $1,022.

■ UNIVERSITY OF DELAWARE

College of Health and Nursing Sciences, Department of Nutrition and Dietetics, Newark, DE 19716

AWARDS Applied nutrition (MS); general human nutrition (MS); nutrient metabolism and utilization (MS). Part-time programs available.

Faculty: 10 full-time (9 women).
Students: 9 full-time (8 women), 3 part-time (1 woman); includes 2 minority (both African Americans). Average age 26. *20 applicants, 75% accepted.* In 1999, 5 degrees awarded.
Degree requirements: For master's, thesis required (for some programs), foreign language not required. *Average time to degree:* Master's–3.5 years full-time, 4 years part-time.
Entrance requirements: For master's, GRE General Test, minimum GPA of 2.75. *Application deadline:* For fall admission, 7/1; for spring admission, 12/15. Applications are processed on a rolling basis. *Application fee:* $45. Electronic applications accepted.
Expenses: Tuition, state resident: full-time $4,380; part-time $243 per credit. Tuition, nonresident: full-time $12,750; part-time $708 per credit. Required fees: $15 per term. Tuition and fees vary according to program.
Financial aid: In 1999–00, 5 students received aid, including 5 teaching assistantships with full tuition reimbursements available (averaging $10,455 per year); fellowships, career-related internships or fieldwork, Federal Work-Study, institutionally sponsored loans, and tuition waivers (partial) also available. Financial aid application deadline: 3/1.
Faculty research: Lipids, food composition databases, gerontology, diet and cancer prevention.
Dr. Jack L. Smith, Chair, 302-831-8729, *Fax:* 302-831-4186, *E-mail:* jack.smith@mvs.udel.edu.
Application contact: Dr. Nancy Cotugna, Graduate Coordinator, 302-831-2937, *Fax:* 302-831-4186, *E-mail:* nancy.cotugna@mvs.udel.edu.

■ UNIVERSITY OF FLORIDA

Graduate School, College of Agriculture and Life Sciences, Department of Food Science and Human Nutrition, Gainesville, FL 32611

AWARDS MS, PhD.

Faculty: 38.
Students: 65 full-time (42 women), 9 part-time (5 women); includes 16 minority (4 African Americans, 3 Asian Americans or Pacific Islanders, 9 Hispanic Americans), 15 international. *84 applicants, 61% accepted.* In 1999, 22 master's, 5 doctorates awarded.
Degree requirements: For master's, thesis optional; for doctorate, dissertation required.
Entrance requirements: For master's and doctorate, GRE General Test, TOEFL, minimum GPA of 3.0. *Application deadline:* For fall admission, 6/1 (priority date). Applications are processed on a rolling basis. *Application fee:* $20. Electronic applications accepted.
Expenses: Tuition, state resident: part-time $144 per credit hour. Tuition, nonresident: part-time $505 per credit hour. Tuition and fees vary according to course level, course load and program.
Financial aid: In 1999–00, 37 students received aid, including 9 fellowships, 24 research assistantships, 7 teaching assistantships; career-related internships or fieldwork also available.
Faculty research: Pesticide research, nutritional biochemistry and microbiology, food safety and toxicology assessment and dietetics, food chemistry.
Dr. Douglas L. Archer, Chair, 352-392-1991 Ext. 102, *Fax:* 352-392-9467, *E-mail:* dlar@gnv.ifas.ufl.edu.
Application contact: Dr. Rachel Shireman, Assistant Dean for Academic Programs, 352-392-2251, *Fax:* 352-392-8988, *E-mail:* rbs@gnv.ifas.ufl.edu.

Find an in-depth description at www.petersons.com/graduate.

■ UNIVERSITY OF GEORGIA

Graduate School, College of Family and Consumer Sciences, Department of Foods and Nutrition, Athens, GA 30602

AWARDS MHE, MS, PhD.

Degree requirements: For master's, thesis (MS) required; for doctorate, dissertation required, foreign language not required.
Entrance requirements: For master's, GRE General Test, minimum GPA of 3.0, previous course work in biochemistry and physiology; for doctorate, GRE General Test, master's degree, minimum GPA of 3.0. Electronic applications accepted.
Expenses: Tuition, state resident: full-time $7,516; part-time $431 per credit hour. Tuition, nonresident: full-time $12,204; part-time $793 per credit hour. Tuition and fees vary according to program.

■ UNIVERSITY OF HAWAII AT MANOA

Graduate Division, College of Tropical Agriculture and Human Resources, Department of Food Science and Human Nutrition, Program in Nutritional Science, Honolulu, HI 96822

AWARDS MS.

Students: Average age 27. *6 applicants, 33% accepted. Average time to degree:* Master's–2 years full-time.
Entrance requirements: For master's, GRE General Test. *Application deadline:* For fall admission, 3/1; for spring admission, 9/1. *Application fee:* $25 ($50 for international students).
Expenses: Tuition, state resident: full-time $4,032; part-time $168 per credit. Tuition, nonresident: full-time $9,960; part-time $415 per credit. Required fees: $51 per semester. Part-time tuition and fees vary according to course load and program.
Financial aid: In 1999–00, 3 teaching assistantships were awarded; research assistantships, tuition waivers (full) also available. Financial aid application deadline: 3/1.
Faculty research: Nutritional biochemistry, human nutrition, nutrition education, international nutrition, nutritional epidemiology.
Dr. Rachel Novotny, Graduate Chairperson, 808-956-8236, *Fax:* 808-956-4024, *E-mail:* novotny@hawaii.edu.

■ UNIVERSITY OF ILLINOIS AT CHICAGO

Graduate College, College of Associated Health Professions, Program in Human Nutrition and Dietetics, Chicago, IL 60607-7128

AWARDS MS, PhD.

Faculty: 6 full-time (4 women).
Students: 11 full-time (9 women), 14 part-time (all women); includes 5 minority (1 African American, 2 Asian Americans or Pacific Islanders, 2 Hispanic Americans), 6 international. Average age 33. *21 applicants, 29% accepted.* In 1999, 10 degrees awarded.
Degree requirements: For master's and doctorate, thesis/dissertation required, foreign language not required.
Entrance requirements: For master's, GRE General Test, TOEFL, minimum GPA of 3.75 on a 5.0 scale; for doctorate, GRE General Test, TOEFL (average 550), minimum GPA of 3.75 on a 5.0 scale. *Application deadline:* For fall admission, 6/1; for spring admission, 11/1. *Application fee:* $40 ($50 for international students). Electronic applications accepted.
Expenses: Tuition, state resident: full-time $3,750; part-time $1,250 per semester.

Tuition, nonresident: full-time $10,588; part-time $3,530 per semester. Required fees: $507 per semester. Tuition and fees vary according to course load and program.
Financial aid: In 1999–00, 11 students received aid; fellowships, research assistantships, teaching assistantships, career-related internships or fieldwork, Federal Work-Study, institutionally sponsored loans, and tuition waivers (full) available. Financial aid application deadline: 3/1; financial aid applicants required to submit FAFSA.
Faculty research: Nutrition for the elderly, inborn errors of metabolism, nutrition and cancer, lipid metabolism, dietary fat markers.
Shiriki Kumanyika, Head, 312-996-2083.
Application contact: Alan Diamond, Director of Graduate Studies, 312-996-1207.

■ UNIVERSITY OF ILLINOIS AT URBANA–CHAMPAIGN

Graduate College, College of Agricultural, Consumer and Environmental Sciences, Department of Food Science and Human Nutrition, Urbana, IL 61801

AWARDS MS, PhD, MD/PhD.

Faculty: 24 full-time (8 women), 2 part-time/adjunct (both women).
Students: 45 full-time (24 women), 17 international. *90 applicants, 13% accepted.* In 1999, 14 master's, 6 doctorates awarded.
Degree requirements: For master's, foreign language and thesis not required; for doctorate, dissertation required.
Entrance requirements: For master's, minimum GPA of 4.0 on a 5.0 scale. *Application deadline:* Applications are processed on a rolling basis. *Application fee:* $40 ($50 for international students).
Expenses: Tuition, state resident: full-time $4,040. Tuition, nonresident: full-time $11,192. Full-time tuition and fees vary according to program.
Financial aid: Fellowships, research assistantships, teaching assistantships, tuition waivers (full and partial) available. Financial aid application deadline: 2/15. *Total annual research expenditures:* $4.4 million.
Dr. Bruce M. Chassy, Head, 217-244-4498, *Fax:* 217-244-2455, *E-mail:* b-chassy@uiuc.edu.
Application contact: Diane Pickert, Graduate Admissions Coordinator, 217-333-1326, *Fax:* 217-244-2455, *E-mail:* pickert@uiuc.edu.

Find an in-depth description at www.petersons.com/graduate.

■ UNIVERSITY OF ILLINOIS AT URBANA–CHAMPAIGN

Graduate College, College of Agricultural, Consumer and Environmental Sciences, Division of Nutritional Sciences, Urbana, IL 61801

AWARDS MS, PhD, MD/PhD.

Faculty: 49.
Students: 38 full-time (25 women); includes 7 minority (4 Asian Americans or Pacific Islanders, 2 Hispanic Americans, 1 Native American), 9 international. *33 applicants, 12% accepted.* In 1999, 7 degrees awarded.
Degree requirements: For master's, foreign language and thesis not required; for doctorate, dissertation required.
Entrance requirements: For master's, minimum GPA of 4.0 on a 5.0 scale. *Application deadline:* Applications are processed on a rolling basis. *Application fee:* $40 ($50 for international students).
Expenses: Tuition, state resident: full-time $4,040. Tuition, nonresident: full-time $11,192. Full-time tuition and fees vary according to program.
Financial aid: Fellowships, research assistantships, teaching assistantships available. Financial aid application deadline: 2/15.
Sharon M. Donovan, Director, 217-333-4177, *Fax:* 217-333-9368, *E-mail:* sdonovan@uiuc.edu.
Application contact: Linda L. Barenthin, Director of Graduate Studies, 217-333-4177, *Fax:* 217-333-9368, *E-mail:* lbarenthin@uiuc.edu.

■ UNIVERSITY OF KANSAS

Graduate Studies Medical Center, School of Allied Health, Department of Dietetics and Nutrition, Lawrence, KS 66045

AWARDS MS. Part-time programs available.

Faculty: 4 full-time (2 women).
Students: 16 full-time (all women), 6 part-time (all women); includes 1 minority (Hispanic American). Average age 25. *23 applicants, 48% accepted.* In 1999, 14 degrees awarded.
Degree requirements: For master's, thesis, oral exam required, foreign language not required.
Entrance requirements: For master's, GRE General Test, TOEFL, TSE, bachelor's degree in related field, minimum GPA of 3.0. *Application deadline:* For fall admission, 2/15; for spring admission, 10/31. Applications are processed on a rolling basis. *Application fee:* $25. Electronic applications accepted.
Expenses: Tuition, state resident: full-time $2,482; part-time $103 per credit hour. Tuition, nonresident: full-time $8,104;

part-time $338 per credit hour. Required fees: $428; $31 per credit hour. Tuition and fees vary according to program.

Financial aid: In 1999–00, 15 students received aid; teaching assistantships, career-related internships or fieldwork, Federal Work-Study, institutionally sponsored loans, and traineeships available. Aid available to part-time students. Financial aid application deadline: 3/31; financial aid applicants required to submit FAFSA.

James Halling, Chairman, 913-588-7682, *Fax:* 913-588-7685, *E-mail:* jhalling1@kumc.edu.

Application contact: Rachel Barkley, Associate Director of Education, 913-588-5359, *Fax:* 913-588-7685, *E-mail:* rbarkley@kumc.edu.

■ UNIVERSITY OF KENTUCKY

Graduate School, College of Human Environmental Sciences, Program in Nutrition and Food Science, Lexington, KY 40506-0032

AWARDS MS.

Degree requirements: For master's, comprehensive exam required, thesis optional, foreign language not required.
Entrance requirements: For master's, GRE General Test, minimum undergraduate GPA of 2.5.
Expenses: Tuition, state resident: full-time $3,596; part-time $188 per credit hour. Tuition, nonresident: full-time $10,116; part-time $550 per credit hour.

■ UNIVERSITY OF KENTUCKY

Graduate School, Graduate School Programs from the College of Allied Health, Program in Clinical Nutrition, Lexington, KY 40506-0032

AWARDS MSCNU.

Degree requirements: For master's, comprehensive exam required, foreign language and thesis not required.
Entrance requirements: For master's, GRE General Test, minimum undergraduate GPA of 2.5.
Expenses: Tuition, state resident: full-time $3,596; part-time $188 per credit hour. Tuition, nonresident: full-time $10,116; part-time $550 per credit hour.
Faculty research: Dietary lipids and microcirculation, niacin and cancer, cholesterol and atherosclerosis, zinc and trauma.

■ UNIVERSITY OF KENTUCKY

Graduate School, Multidisciplinary PhD Program in Nutritional Sciences, Lexington, KY 40506-0032

AWARDS PhD.

Degree requirements: For doctorate, dissertation required, foreign language not required.
Entrance requirements: For doctorate, GRE General Test, minimum graduate GPA of 3.0.
Expenses: Tuition, state resident: full-time $3,596; part-time $188 per credit hour. Tuition, nonresident: full-time $10,116; part-time $550 per credit hour.
Faculty research: Nutrition and AIDS, nutrition and alcoholism, nutrition and cardiovascular disease, nutrition and cancer, nutrition and diabetes.

Find an in-depth description at www.petersons.com/graduate.

■ UNIVERSITY OF MAINE

Graduate School, College of Natural Sciences, Forestry, and Agriculture, Department of Food Science and Human Nutrition, Orono, ME 04469

AWARDS Food and nutritional sciences (PhD); food science and human nutrition (MS). Part-time programs available.

Degree requirements: For master's and doctorate, thesis/dissertation required, foreign language not required.
Entrance requirements: For master's, GRE General Test, TOEFL, minimum GPA of 3.0; for doctorate, GRE General Test, TOEFL.
Expenses: Tuition, state resident: full-time $3,564. Tuition, nonresident: full-time $10,116. Required fees: $378. Tuition and fees vary according to course load.
Faculty research: Product development of fruit and vegetables, lipid oxidation in fish and meat, analytical methods development, metabolism of potato glycoalkaloids, seafood quality.

■ UNIVERSITY OF MARYLAND, COLLEGE PARK

Graduate Studies and Research, College of Agriculture and Natural Resources, Department of Nutrition and Food Science, Program in Nutrition, College Park, MD 20742

AWARDS MS, PhD.

Students: 15 full-time (14 women), 10 part-time (all women); includes 4 minority (2 African Americans, 2 Asian Americans or Pacific Islanders), 7 international. *46 applicants, 22% accepted.* In 1999, 1 master's, 1 doctorate awarded.
Degree requirements: For master's, comprehensive exam, research-based thesis or equivalent paper required; for doctorate, dissertation, candidacy exam required, foreign language not required.
Entrance requirements: For master's and doctorate, GRE General Test, TOEFL, minimum GPA of 3.0. *Application deadline:*

For fall admission, 5/1; for spring admission, 11/1. Applications are processed on a rolling basis. *Application fee:* $50 ($70 for international students). Electronic applications accepted.
Expenses: Tuition, state resident: part-time $272 per credit hour. Tuition, nonresident: part-time $415 per credit hour. Required fees: $632; $379 per year.
Financial aid: Fellowships, research assistantships, teaching assistantships with tuition reimbursements available. Financial aid applicants required to submit FAFSA.
Dr. Phylis Moser-Veillon, Director, 301-405-4521.
Application contact: Trudy Lindsey, Director, Graduate Admissions and Records, 301-405-4198, *Fax:* 301-314-9305, *E-mail:* grschool@deans.umd.edu.

■ UNIVERSITY OF MASSACHUSETTS AMHERST

Graduate School, School of Public Health and Health Sciences, Department of Nutrition, Amherst, MA 01003

AWARDS MS.

Faculty: 18 full-time (3 women).
Students: 19 full-time (18 women), 12 part-time (11 women); includes 2 minority (1 Asian American or Pacific Islander, 1 Hispanic American), 10 international. Average age 30. *37 applicants, 76% accepted.* In 1999, 2 degrees awarded.
Degree requirements: For master's, thesis or alternative required, foreign language not required.
Entrance requirements: For master's, GRE General Test. *Application deadline:* For fall admission, 2/1 (priority date); for spring admission, 10/1. Applications are processed on a rolling basis. *Application fee:* $40.
Expenses: Tuition, state resident: full-time $2,640; part-time $165 per credit. Tuition, nonresident: full-time $9,756; part-time $407 per credit. Required fees: $1,221 per term. One-time fee: $110. Full-time tuition and fees vary according to course load, campus/location and reciprocity agreements.
Financial aid: In 1999–00, 1 fellowship with full tuition reimbursement (averaging $5,093 per year), 17 research assistantships with full tuition reimbursements (averaging $5,991 per year), 8 teaching assistantships with full tuition reimbursements (averaging $7,503 per year) were awarded; career-related internships or fieldwork, Federal Work-Study, grants, scholarships, traineeships, and unspecified assistantships also available. Aid available to part-time students. Financial aid application deadline: 2/1.

University of Massachusetts Amherst (continued)
Dr. Nancy Cohen, Acting Head, 413-545-1067, *Fax:* 413-545-1074, *E-mail:* cohen@nutrition.umass.edu.

■ UNIVERSITY OF MEDICINE AND DENTISTRY OF NEW JERSEY

School of Health Related Professions, Department of Interdisciplinary Studies, Program in Health Sciences, Newark, NJ 07107-3001

AWARDS Cardiopulmonary sciences (PhD); clinical laboratory sciences (PhD); health sciences (MSHS); interdisciplinary studies (PhD); nutrition science (PhD); physical therapy/movement science (PhD).

Degree requirements: For doctorate, dissertation required.
Entrance requirements: For doctorate, TOEFL (minimum score of 550 required; 213 for computer-based), interview, writing sample.
Expenses: Tuition, state resident: part-time $270 per credit hour. Tuition, nonresident: part-time $407 per credit hour. Part-time tuition and fees vary according to campus/location and program.
Find an in-depth description at www.petersons.com/graduate.

■ UNIVERSITY OF MEDICINE AND DENTISTRY OF NEW JERSEY

School of Health Related Professions, Department of Primary Care, Dietetic Internship Program, Newark, NJ 07107-3001

AWARDS Certificate.

Entrance requirements: For degree, TOEFL (minimum score of 550 required; 213 for computer-based), bachelor's degree in dietetics, nutrition, or related field; interview; minimum GPA of 2.5.
Expenses: Tuition, state resident: part-time $270 per credit hour. Tuition, nonresident: part-time $407 per credit hour. Part-time tuition and fees vary according to campus/location and program.

■ UNIVERSITY OF MEDICINE AND DENTISTRY OF NEW JERSEY

School of Health Related Professions, Department of Primary Care, Program in Clinical Nutrition, Newark, NJ 07107-3001

AWARDS MS.

Entrance requirements: For master's, minimum GPA of 3.0, proof of registered dietician status.

Expenses: Tuition, state resident: part-time $270 per credit hour. Tuition, nonresident: part-time $407 per credit hour. Part-time tuition and fees vary according to campus/location and program.
Find an in-depth description at www.petersons.com/graduate.

■ THE UNIVERSITY OF MEMPHIS

Graduate School, College of Education, Department of Consumer Science and Education, Memphis, TN 38152

AWARDS Clinical nutrition (MS); consumer science and education (MS). Part-time programs available.

Faculty: 6 full-time (all women), 4 part-time/adjunct (all women).
Students: 21 full-time (19 women), 7 part-time (6 women); includes 9 minority (8 African Americans, 1 Hispanic American). Average age 29. *21 applicants, 76% accepted.* In 1999, 5 degrees awarded.
Degree requirements: For master's, computer language, thesis (for some programs), written comprehensive exam required, foreign language not required. *Average time to degree:* Master's–2 years full-time, 3 years part-time.
Entrance requirements: For master's, GRE General Test or MAT. *Application deadline:* For fall admission, 8/1; for spring admission, 2/15. *Application fee:* $25 ($50 for international students).
Expenses: Tuition, state resident: full-time $3,410; part-time $178 per credit hour. Tuition, nonresident: full-time $8,670; part-time $408 per credit hour. Tuition and fees vary according to program.
Financial aid: In 1999–00, 15 students received aid, including 13 research assistantships with tuition reimbursements available, 1 teaching assistantship with tuition reimbursement available; career-related internships or fieldwork and scholarships also available.
Faculty research: State vocation education services, marketing education, clinical nutrition outcomes. *Total annual research expenditures:* $186,999.
Dr. Dixie R. Crase, Chair and Coordinator of Graduate Studies, 901-678-2301, *Fax:* 901-678-5324, *E-mail:* crase.dixie@coe.memphis.edu.
Application contact: Dr. Linda Clemens, Director, Clinical Nutrition, 901-678-3108, *Fax:* 901-678-5324, *E-mail:* lhclemns@memphis.edu.

■ UNIVERSITY OF MICHIGAN

School of Public Health, Department of Environmental and Industrial Health, Program in Human Nutrition, Ann Arbor, MI 48109

AWARDS MPH, MS. MS offered through the Horace H. Rackham School of Graduate Studies.

Degree requirements: For master's, thesis required, foreign language not required.
Entrance requirements: For master's, GRE General Test.
Expenses: Tuition, state resident: full-time $10,520. Tuition, nonresident: full-time $21,344.

■ UNIVERSITY OF MINNESOTA, TWIN CITIES CAMPUS

Graduate School, College of Agricultural, Food, and Environmental Sciences and College of Human Ecology, Department of Food Science and Nutrition, Graduate Program in Nutrition, Minneapolis, MN 55455-0213

AWARDS MS, PhD. Part-time programs available.

Faculty: 31 full-time (17 women).
Students: 43; includes 2 African Americans, 3 Asian Americans or Pacific Islanders, 9 international. Average age 30. *70 applicants, 36% accepted.* In 1999, 8 master's, 3 doctorates awarded. Terminal master's awarded for partial completion of doctoral program.
Degree requirements: For master's, thesis required (for some programs), foreign language not required; for doctorate, dissertation required, foreign language not required. *Average time to degree:* Master's–2.5 years full-time; doctorate–5 years full-time.
Entrance requirements: For master's, GRE General Test, TOEFL, previous course work in general chemistry, organic chemistry, physiology, biology, biochemistry, statistics; minimum GPA of 3.0; for doctorate, GRE General Test, TOEFL, previous course work in general chemistry, organic chemistry, calculus, biology, physics, physiology, biochemistry, statistics; minimum GPA of 3.0. *Application deadline:* For fall admission, 3/15 (priority date); for spring admission, 9/15 (priority date). Applications are processed on a rolling basis. *Application fee:* $50 ($55 for international students). Electronic applications accepted.
Expenses: Tuition, state resident: full-time $5,040; part-time $420 per credit. Tuition, nonresident: full-time $9,900; part-time $825 per credit. Full-time tuition and fees vary according to course load, program and reciprocity agreements.

Financial aid: In 1999–00, 40 students received aid, including 3 fellowships with full tuition reimbursements available (averaging $13,750 per year), 25 research assistantships with full and partial tuition reimbursements available (averaging $13,750 per year), 4 teaching assistantships with full and partial tuition reimbursements available (averaging $13,750 per year); career-related internships or fieldwork, Federal Work-Study, institutionally sponsored loans, and scholarships also available. Aid available to part-time students.

Faculty research: Diet and chronic disease: from basic biological and molecular biology approaches to a public health/intervention/epidemiology perspective. *Total annual research expenditures:* $1.5 million.

Dr. Dan D. Gallaher, Director of Graduate Studies, 612-624-0746, *Fax:* 612-625-5272, *E-mail:* dgallaher@ che2.che.umn.edu.

Application contact: Susan J. A. Punchochar, Student Services Coordinator, 612-624-6753, *Fax:* 612-625-5272, *E-mail:* spuncoch@che2.che.umn.edu.

■ UNIVERSITY OF MINNESOTA, TWIN CITIES CAMPUS

School of Public Health, Major in Public Health Nutrition, Minneapolis, MN 55455-0213

AWARDS MPH. Part-time programs available.

Faculty: 6 full-time.

Students: 12 full-time, 7 part-time; includes 1 minority (African American), 2 international. *29 applicants, 62% accepted.* In 1999, 15 degrees awarded.

Degree requirements: For master's, foreign language not required.

Entrance requirements: For master's, GRE General Test, minimum GPA of 3.0. *Application deadline:* For fall admission, 3/1. Applications are processed on a rolling basis. *Application fee:* $50 ($75 for international students).

Expenses: Tuition, state resident: full-time $4,270; part-time $267 per credit. Tuition, nonresident: full-time $8,400; part-time $525 per credit. Tuition and fees vary according to program.

Financial aid: Research assistantships with partial tuition reimbursements, teaching assistantships with partial tuition reimbursements, career-related internships or fieldwork, Federal Work-Study, institutionally sponsored loans, and traineeships available.

Faculty research: Life cycle nutrition, nutrition/disease relationships.

Dr. John Himes, Chair, 612-624-8210, *Fax:* 612-624-0315, *E-mail:* himes@ epi.umn.edu.

Application contact: Shelley Cooksey, Student Coordinator, 612-626-8802, *Fax:* 612-624-0315, *E-mail:* cooksey@ epivax.epi.umn.edu.

■ UNIVERSITY OF MISSOURI–COLUMBIA

Graduate School, College of Agriculture, Department of Food Science and Human Nutrition, Columbia, MO 65211

AWARDS Food science (MS, PhD); foods and food systems management (MS); human nutrition (MS). Terminal master's awarded for partial completion of doctoral program.

Degree requirements: For master's, foreign language not required; for doctorate, dissertation required.

Entrance requirements: For master's and doctorate, GRE General Test, minimum GPA of 3.0.

Expenses: Tuition, state resident: full-time $3,020; part-time $168 per hour. Tuition, nonresident: full-time $6,066; part-time $505 per hour. Required fees: $445; $18 per hour. Tuition and fees vary according to course load and program.

■ UNIVERSITY OF MISSOURI–COLUMBIA

Graduate School, College of Agriculture, Nutritional Sciences Program, Columbia, MO 65211

AWARDS Nutrition (MS, PhD).

Degree requirements: For master's, thesis required; for doctorate, dissertation, comprehensive oral and written exam, qualifying exam required.

Entrance requirements: For master's and doctorate, GRE General Test, TOEFL. *Application deadline:* For fall admission, 3/1 (priority date). Applications are processed on a rolling basis. *Application fee:* $25 ($50 for international students).

Expenses: Tuition, state resident: full-time $3,020; part-time $168 per hour. Tuition, nonresident: full-time $6,066; part-time $505 per hour. Required fees: $445; $18 per hour. Tuition and fees vary according to course load and program.

Financial aid: Fellowships, research assistantships, teaching assistantships, institutionally sponsored loans available.

Dr. Roger A. Sunde, Chair, 573-882-4526, *Fax:* 573-882-0185, *E-mail:* sunder@ missouri.edu.

Application contact: Dr. Kevin Fritsche, Director of Graduate Studies, 573-882-7240, *Fax:* 573-882-0185.

Find an in-depth description at www.petersons.com/graduate.

■ UNIVERSITY OF MISSOURI–COLUMBIA

Graduate School, College of Human Environmental Science, Department of Human Nutrition, Foods, Food System Management, and Exercise Physiology, Columbia, MO 65211

AWARDS Exercise physiology (PhD); exercise science (MA); food science (MS, PhD); foods and food systems management (MS); human nutrition (MS).

Degree requirements: For doctorate, dissertation required.

Entrance requirements: For master's and doctorate, GRE General Test, TOEFL, minimum GPA of 3.0.

Expenses: Tuition, state resident: full-time $3,020; part-time $168 per hour. Tuition, nonresident: full-time $6,066; part-time $505 per hour. Required fees: $445; $18 per hour. Tuition and fees vary according to course load and program.

■ UNIVERSITY OF NEBRASKA–LINCOLN

Graduate College, College of Agricultural Sciences and Natural Resources, Interdepartmental Area of Nutrition, Lincoln, NE 68588

AWARDS MS, PhD.

Students: 6 full-time (all women), 6 part-time (3 women); includes 2 minority (both African Americans), 6 international. Average age 29. *7 applicants, 43% accepted.* In 1999, 3 master's, 3 doctorates awarded.

Degree requirements: For master's, thesis optional; for doctorate, dissertation, comprehensive exams required.

Entrance requirements: For master's and doctorate, GRE General Test, TOEFL. *Application deadline:* For fall admission, 3/1 (priority date). Applications are processed on a rolling basis. *Application fee:* $35. Electronic applications accepted.

Expenses: Tuition, state resident: part-time $116 per credit hour. Tuition, nonresident: part-time $285 per credit hour. Required fees: $119 per semester. Tuition and fees vary according to course load and program.

Financial aid: Fellowships, research assistantships, teaching assistantships, Federal Work-Study available. Aid available to part-time students. Financial aid application deadline: 2/15.

Faculty research: Human nutrition and metabolism, animal nutrition and metabolism, biochemistry, community and clinical nutrition.

Dr. Richard Grant, Graduate Committee Chair, 402-472-6442.

Application contact: Graduate Secretary, 402-472-3866.

■ UNIVERSITY OF NEBRASKA–LINCOLN

Graduate College, College of Human Resources and Family Sciences, Department of Nutritional Science and Dietetics, Lincoln, NE 68588

AWARDS Human resources and family sciences (PhD); nutritional science and dietetics (MS).

Faculty: 8 full-time (7 women).
Students: 18 full-time (17 women), 10 part-time (9 women), 1 international. Average age 31. *33 applicants, 76% accepted.* In 1999, 7 degrees awarded.
Degree requirements: For master's, thesis optional, foreign language not required.
Entrance requirements: For master's, GRE General Test, TOEFL. *Application deadline:* For fall admission, 4/15 (priority date). Applications are processed on a rolling basis. *Application fee:* $35. Electronic applications accepted.
Expenses: Tuition, state resident: part-time $116 per credit hour. Tuition, nonresident: part-time $285 per credit hour. Required fees: $119 per semester. Tuition and fees vary according to course load and program.
Financial aid: In 1999–00, 5 fellowships, 10 research assistantships, 12 teaching assistantships were awarded; Federal Work-Study also available. Aid available to part-time students. Financial aid application deadline: 2/15.
Faculty research: Foods/food service management, community nutrition, diet-health relationships, vitamin toxicities, food components in reducing cancer risk.
Dr. Marilyn Schnepf, Chair, 402-472-3716, *Fax:* 402-472-1587.

■ UNIVERSITY OF NEVADA, RENO

Graduate School, College of Human and Community Sciences, Department of Nutrition, Reno, NV 89557

AWARDS MS.

Faculty: 5 full-time (2 women).
Students: 4 full-time (3 women), 3 part-time (all women), 1 international. Average age 32. *14 applicants, 50% accepted.*
Degree requirements: For master's, thesis optional, foreign language not required.
Entrance requirements: For master's, GRE, TOEFL, minimum GPA of 3.0. *Application deadline:* For fall admission, 2/1 (priority date); for spring admission, 11/1. Applications are processed on a rolling basis. *Application fee:* $40.
Expenses: Tuition, area resident: Part-time $3,173 per semester. Tuition,

nonresident: full-time $6,347. Required fees: $101 per credit. $101 per credit.
Financial aid: In 1999–00, 5 research assistantships, 3 teaching assistantships were awarded; Federal Work-Study, institutionally sponsored loans, and unspecified assistantships also available. Financial aid application deadline: 3/1.
Faculty research: Nutritional education, food technology, therapeutic human nutrition, human nutritional requirements, diet and disease.
Dr. Marsha Read, Chair, 775-784-6440.

■ UNIVERSITY OF NEW HAMPSHIRE

Graduate School, College of Life Sciences and Agriculture, Graduate Programs in the Biological Sciences and Natural Resources, Department of Animal and Nutritional Sciences, Durham, NH 03824

AWARDS MS, PhD. Part-time programs available.

Faculty: 24 full-time.
Students: 8 full-time (all women), 10 part-time (9 women), 1 international. Average age 32. *15 applicants, 60% accepted.* In 1999, 6 degrees awarded. Terminal master's awarded for partial completion of doctoral program.
Degree requirements: For master's and doctorate, thesis/dissertation required, foreign language not required.
Entrance requirements: For master's and doctorate, GRE General Test. *Application deadline:* For fall admission, 4/1 (priority date); for winter admission, 12/1. Applications are processed on a rolling basis. *Application fee:* $50.
Expenses: Tuition, area resident: Full-time $5,750; part-time $319 per credit. Tuition, state resident: full-time $8,625; part-time $478. Tuition, nonresident: full-time $14,640; part-time $598 per credit. Required fees: $224 per semester. Tuition and fees vary according to course load, degree level and program.
Financial aid: In 1999–00, 6 research assistantships, 8 teaching assistantships were awarded; fellowships, career-related internships or fieldwork, Federal Work-Study, scholarships, and tuition waivers (full and partial) also available. Aid available to part-time students. Financial aid application deadline: 2/15.
Faculty research: Diseases, nutrition, parasites, cell biology, animal breeding.
Dr. William E. Berndtson, Chairperson, 603-862-2553, *E-mail:* williamb@cisunix.unh.edu.
Application contact: Dr. Robert Taylor, Graduate Studies, 603-862-2178, *E-mail:* bob.taylor@unh.edu.

■ UNIVERSITY OF NEW HAVEN

Graduate School, College of Arts and Sciences, Program in Human Nutrition, West Haven, CT 06516-1916

AWARDS MS.

Students: 2 full-time (1 woman), 93 part-time (82 women); includes 18 minority (2 African Americans, 11 Asian Americans or Pacific Islanders, 5 Hispanic Americans), 2 international. *63 applicants, 73% accepted.* In 1999, 43 degrees awarded.
Degree requirements: For master's, foreign language not required.
Application deadline: Applications are processed on a rolling basis. *Application fee:* $50.
Expenses: Tuition: Part-time $1,170 per course. Part-time tuition and fees vary according to course level, degree level and program.
Financial aid: Application deadline: 5/1.
Robert Fitzgerald, Director, 203-932-7352.

■ UNIVERSITY OF NEW MEXICO

Graduate School, College of Education, Program in Nutrition, Albuquerque, NM 87131-2039

AWARDS Nutrition/dietetics (MS). Part-time programs available.

Faculty: 3 full-time (all women), 1 (woman) part-time/adjunct.
Students: 11 full-time (10 women), 7 part-time (6 women); includes 6 minority (all Hispanic Americans). Average age 27. *13 applicants, 85% accepted.* In 1999, 2 degrees awarded.
Degree requirements: For master's, comprehensive exam or thesis required.
Application deadline: For fall admission, 6/1; for spring admission, 11/1. *Application fee:* $25.
Expenses: Tuition, state resident: full-time $2,514; part-time $105 per credit hour. Tuition, nonresident: full-time $10,304; part-time $417 per credit hour. International tuition: $10,304 full-time. Required fees: $516; $22 per credit hour. Tuition and fees vary according to program.
Financial aid: In 1999–00, 9 students received aid, including 2 research assistantships (averaging $11,544 per year), 2 teaching assistantships with tuition reimbursements available (averaging $6,075 per year); fellowships, institutionally sponsored loans also available. Financial aid applicants required to submit FAFSA.
Faculty research: Nutrition education, birth defects prevention: folate, obesity prevention.
Dr. Karen Heller, Graduate Adviser, 505-277-8183, *Fax:* 505-277-8427, *E-mail:* kheller@unm.edu.

Application contact: Mary Justus, Secretary, 505-277-8183, *Fax:* 505-277-8427.

■ THE UNIVERSITY OF NORTH CAROLINA AT CHAPEL HILL

Graduate School, School of Public Health, Department of Nutrition, Chapel Hill, NC 27599

AWARDS Nutrition (MPH, Dr PH, PhD); professional practice program (MPH).

Faculty: 21 full-time (15 women), 19 part-time/adjunct (11 women).
Students: 66 full-time (57 women), 36 part-time (31 women); includes 19 minority (12 African Americans, 4 Asian Americans or Pacific Islanders, 3 Hispanic Americans), 14 international. Average age 28. *132 applicants, 32% accepted.* In 1999, 20 master's, 5 doctorates awarded.
Degree requirements: For master's, thesis, major paper, comprehensive exam required, foreign language not required; for doctorate, dissertation, comprehensive exam required, foreign language not required. *Average time to degree:* Master's–2 years full-time; doctorate–5 years full-time.
Entrance requirements: For master's and doctorate, GRE General Test, minimum GPA of 3.0. *Application deadline:* For fall admission, 1/1 (priority date); for spring admission, 10/15. Applications are processed on a rolling basis. *Application fee:* $55.
Expenses: Tuition, state resident: full-time $2,220. Tuition, nonresident: full-time $10,794. Required fees: $822. One-time fee: $15 full-time. Part-time tuition and fees vary according to course load.
Financial aid: In 1999–00, 38 students received aid, including 1 fellowship, 21 research assistantships, 12 teaching assistantships; career-related internships or fieldwork, Federal Work-Study, institutionally sponsored loans, scholarships, traineeships, and unspecified assistantships also available. Financial aid application deadline: 3/1; financial aid applicants required to submit FAFSA.
Faculty research: Nutrition policy, management and leadership development, lipid and carbohydrate metabolism, dietary trends and determinants, transmembrane signal transduction and carcinogenesis, maternal and child nutrition. *Total annual research expenditures:* $2.3 million.
Dr. Steven H. Zeisel, Chair, 919-966-7218, *Fax:* 919-966-7216, *E-mail:* steven_zeisel@unc.edu.
Application contact: 919-966-7212, *Fax:* 919-966-7216.

■ THE UNIVERSITY OF NORTH CAROLINA AT GREENSBORO

Graduate School, School of Human Environmental Sciences, Department of Nutrition and Food Service Systems, Greensboro, NC 27412-5001

AWARDS Human nutrition (M Ed, MS, PhD).

Faculty: 17 full-time (9 women), 8 part-time/adjunct (3 women).
Students: 24 full-time (19 women), 22 part-time (19 women); includes 4 minority (2 African Americans, 1 Asian American or Pacific Islander, 1 Native American), 10 international. *17 applicants, 88% accepted.*
Degree requirements: For master's and doctorate, thesis/dissertation required.
Entrance requirements: For master's and doctorate, GRE General Test, TOEFL. *Application deadline:* For fall admission, 3/1; for spring admission, 11/1. *Application fee:* $35.
Expenses: Tuition, state resident: full-time $2,200; part-time $182 per semester. Tuition, nonresident: full-time $10,600; part-time $1,238 per semester. Tuition and fees vary according to course load and program.
Financial aid: In 1999–00, 3 fellowships with full tuition reimbursements (averaging $4,000 per year), 30 research assistantships with full tuition reimbursements (averaging $8,475 per year) were awarded; teaching assistantships with full tuition reimbursements, unspecified assistantships also available.
Dr. Rosemary C. Wander, Chair, 336-334-5313, *Fax:* 336-334-4129, *E-mail:* rcwander@uncg.edu.
Application contact: Dr. James Lynch, Director of Graduate Recruitment and Information Services, 336-334-4881, *Fax:* 336-334-4424, *E-mail:* jmlynch@office.uncg.edu.

Find an in-depth description at www.petersons.com/graduate.

■ UNIVERSITY OF NORTH FLORIDA

College of Health, Department of Health Sciences, Jacksonville, FL 32224-2645

AWARDS Addictions counseling (MSH); aging studies (Certificate); employee health services (MSH); health administration (MHA); health care administration (MSH); human ecology and nutrition (MSH); human performance (MSH); physical therapy (MSPT). Part-time and evening/weekend programs available.

Faculty: 29 full-time (18 women).
Students: 46 full-time (37 women), 82 part-time (68 women); includes 24 minority (13 African Americans, 4 Asian Americans or Pacific Islanders, 6 Hispanic Americans, 1 Native American), 3

international. Average age 33. *50 applicants, 94% accepted.* In 1999, 52 degrees awarded.
Degree requirements: For master's, thesis optional, foreign language not required.
Entrance requirements: For master's, GMAT (MHA), GRE General Test (MSH), TOEFL. *Application deadline:* For fall admission, 7/6 (priority date); for winter admission, 11/2 (priority date); for spring admission, 3/10 (priority date). Applications are processed on a rolling basis. *Application fee:* $20. Electronic applications accepted.
Expenses: Tuition, state resident: full-time $2,848; part-time $119 per credit. Tuition, nonresident: full-time $8,245; part-time $462 per credit. Required fees: $719; $30 per credit.
Financial aid: In 1999–00, 42 students received aid, including 1 research assistantship (averaging $4,092 per year); career-related internships or fieldwork, Federal Work-Study, scholarships, and tuition waivers (partial) also available. Aid available to part-time students. Financial aid application deadline: 4/1; financial aid applicants required to submit FAFSA.
Faculty research: Alcohol, tobacco, and other drug use prevention; forgotten memories of childhood physical and sexual abuse/post traumatic stress disorder; barriers to choice in the rehabilitation process; future of managed care; turnover among health professionals. *Total annual research expenditures:* $118,959.
Dr. Terry R. Tabor, Chair, 904-620-2840, *E-mail:* ttabor@unf.edu.
Application contact: Danielle Dacquisto, Academic Adviser, 904-620-2812, *E-mail:* ddacquis@unf.edu.

■ UNIVERSITY OF OKLAHOMA HEALTH SCIENCES CENTER

Graduate College, College of Allied Health, Department of Nutritional Sciences, Oklahoma City, OK 73190
AWARDS MS.

Faculty: 5 full-time (3 women).
Students: 8 full-time (7 women), 8 part-time (7 women); includes 4 minority (1 Asian American or Pacific Islander, 3 Native Americans). *13 applicants, 69% accepted.* In 1999, 6 degrees awarded.
Degree requirements: For master's, comprehensive exam required, thesis optional, foreign language not required. *Average time to degree:* Master's–2 years full-time.

Entrance requirements: For master's, GRE General Test, TOEFL. *Application deadline:* For fall admission, 7/1; for spring admission, 12/1. *Application fee:* $25 ($50 for international students).

University of Oklahoma Health Sciences Center (continued)

Expenses: Tuition, state resident: part-time $90 per semester hour. Tuition, nonresident: part-time $264 per semester hour. Tuition and fees vary according to program.

Dr. Kathy Onley, Chair, 405-271-2113.

■ UNIVERSITY OF PUERTO RICO, MEDICAL SCIENCES CAMPUS

Graduate School of Public Health, Department of Human Development, Program in Health Science Nutrition, San Juan, PR 00936-5067

AWARDS MS. Part-time programs available.

Students: 24 (22 women) 1 international. *17 applicants, 76% accepted.* In 1999, 4 degrees awarded.

Degree requirements: For master's, computer language, thesis required, foreign language not required.

Entrance requirements: For master's, GRE, previous course work in algebra, biochemistry, biology, chemistry, and social sciences. *Application deadline:* For fall admission, 3/15. *Application fee:* $15.

Expenses: Tuition, state resident: full-time $5,500. Tuition, nonresident: full-time $8,400. Required fees: $600. Tuition and fees vary according to class time, course load, degree level and program.

Financial aid: Research assistantships, teaching assistantships, Federal Work-Study and institutionally sponsored loans available. Financial aid application deadline: 4/30.

Dr. Jaime Ariza, Coordinator, 787-758-2525 Ext. 1433, *Fax:* 787-759-6719.

Application contact: Mayra E. Santiago-Vargas, Counselor, 787-756-5244, *Fax:* 787-759-6719, *E-mail:* m_santiago@ rcmaxp.upr.clu.edu.

■ UNIVERSITY OF PUERTO RICO, MEDICAL SCIENCES CAMPUS

School of Medicine, Division of Graduate Studies, Department of Biochemistry, San Juan, PR 00936-5067

AWARDS MS, PhD.

Faculty: 9 full-time (3 women), 3 part-time/adjunct (1 woman).

Students: 19 full-time (13 women); all minorities (all Hispanic Americans). Average age 23. *4 applicants, 50% accepted.* In 1999, 2 doctorates awarded (50% entered university research/teaching, 50% found other work related to degree).

Degree requirements: For master's, one foreign language, thesis required; for

doctorate, one foreign language, dissertation, comprehensive exam required. *Average time to degree:* Doctorate–5 years full-time.

Entrance requirements: For master's, GRE General Test, GRE Subject Test, interview, minimum GPA of 3.0; for doctorate, GRE General Test, GRE Subject Test, interview, minimum GPA of 3.0 required. *Application deadline:* For fall admission, 9/15; for spring admission, 2/15. *Application fee:* $15.

Expenses: Tuition, state resident: full-time $5,500. Tuition, nonresident: full-time $8,400. Required fees: $600. Tuition and fees vary according to class time, course load, degree level and program.

Financial aid: In 1999–00, 14 students received aid, including 1 fellowship (averaging $12,000 per year), 13 research assistantships (averaging $9,350 per year); Federal Work-Study, grants, and institutionally sponsored loans also available. Financial aid application deadline: 4/30.

Faculty research: Aging, plasma proteins, vitamins and minerals, protein structure/function, glycosilation of proteins, tumorigenesis. *Total annual research expenditures:* $747,913.

Dr. José R. Rodriguez-Medina, Director, 787-758-2525 Ext. 1633, *Fax:* 787-274-8724, *E-mail:* jo_rodriguez@ rcmaca.upr.clu.edu.

Application contact: Dr. Barulio D. Jimenez, Coordinator, 787-758-2525 Ext. 1235, *Fax:* 787-274-8724, *E-mail:* b_jimenez@rcmaca.upr.clu.edu.

■ UNIVERSITY OF RHODE ISLAND

Graduate School, College of Resource Development, Department of Food Science and Nutrition, Program in Food and Nutrition Science, Kingston, RI 02881

AWARDS MS, PhD.

Entrance requirements: For master's and doctorate, GRE General Test, TOEFL. *Application deadline:* For fall admission, 4/15 (priority date). Applications are processed on a rolling basis. *Application fee:* $35.

Expenses: Tuition, state resident: full-time $3,540; part-time $197 per credit. Tuition, nonresident: full-time $10,116; part-time $197 per credit. Required fees: $1,352; $37 per credit. $65 per term.

Dr. Richard Traxler, Graduate Coordinator, Department of Food Science and Nutrition, 401-874-4028.

■ UNIVERSITY OF SOUTHERN CALIFORNIA

Keck School of Medicine and Graduate School, Graduate Programs in Medicine, Department of Cell and Neurobiology, Los Angeles, CA 90089

AWARDS Anatomy and cell biology (MS, PhD), including anatomy (PhD), cell biology (PhD); cell and neurobiology (MS, PhD); pharmacology and nutrition (MS, PhD); preventive nutrition (MS).

Faculty: 21 full-time (3 women), 8 part-time/adjunct (2 women).

Students: 19 full-time (9 women); includes 10 minority (all Asian Americans or Pacific Islanders). Average age 23. *54 applicants, 22% accepted.* In 1999, 5 master's awarded (40% found work related to degree, 60% continued full-time study). Terminal master's awarded for partial completion of doctoral program.

Degree requirements: For master's, thesis or alternative required, foreign language not required; for doctorate, dissertation required. *Average time to degree:* Master's–2 years full-time; doctorate–5 years full-time.

Entrance requirements: For master's, GRE General Test, TOEFL, minimum GPA of 3.0; for doctorate, GRE General Test, TOEFL. *Application fee:* $55. Electronic applications accepted.

Expenses: Tuition: Full-time $22,198; part-time $748 per unit. Required fees: $406.

Financial aid: In 1999–00, 13 students received aid; fellowships, research assistantships, teaching assistantships, Federal Work-Study, institutionally sponsored loans, and tuition waivers (partial) available. Aid available to part-time students.

Faculty research: Neurobiology and development, circaulian rhythm, gene therapy in vision, lacrimal glands, neuroendocrinology, signal transduction mechanisms.

Dr. Cheryl Craft, Chair, 323-442-1881, *Fax:* 323-442-2709, *E-mail:* ccraft@ hsc.usc.edu.

Application contact: Darlene Marie Campbell, Administrative Assistant, 323-442-1881, *Fax:* 323-442-0466, *E-mail:* dmc@hsc.usc.edu.

■ UNIVERSITY OF SOUTHERN CALIFORNIA

Keck School of Medicine and Graduate School, Graduate Programs in Medicine, Department of Preventive Medicine, Program in Public Health, Los Angeles, CA 90089

AWARDS Biometry/epidemiology (MPH); health promotion (MPH); preventive nutrition (MPH).

Faculty: 28 full-time (14 women), 10 part-time/adjunct (7 women).
Students: 34 full-time (25 women), 7 part-time (3 women); includes 19 minority (1 African American, 13 Asian Americans or Pacific Islanders, 5 Hispanic Americans), 4 international. *34 applicants, 76% accepted.*
Degree requirements: For master's, computer language, practicum, final report required, foreign language and thesis not required.
Entrance requirements: For master's, GRE General Test, TOEFL, minimum GPA of 3.0. *Application deadline:* For fall admission, 3/1 (priority date). *Application fee:* $55. Electronic applications accepted.
Financial aid: In 1999–00, 23 students received aid, including 2 research assistantships (averaging $18,847 per year); institutionally sponsored loans, scholarships, and staff tuition remission also available. Aid available to part-time students. Financial aid application deadline: 2/1; financial aid applicants required to submit CSS PROFILE or FAFSA.
Faculty research: Substance abuse prevention, cancer and heart disease prevention, mass media and health communication research, health promotion, treatment compliance.
Dr. C. Anderson Johnson, Director, 323-442-2628, *Fax:* 323-442-2601, *E-mail:* carljohn@hsc.usc.edu.
Application contact: Nemesia P. Lockhart, Student Services Assistant, 323-442-2580, *Fax:* 323-442-2601, *E-mail:* lockhart@usc.edu.

■ UNIVERSITY OF SOUTHERN MISSISSIPPI

Graduate School, College of Health and Human Sciences, Center for Community Health, Hattiesburg, MS 39406

AWARDS Health education (MPH); health policy/administration (MPH); occupational/environmental health (MPH); public health nutrition (MPH). Part-time and evening/weekend programs available.

Degree requirements: For master's, comprehensive exam required, foreign language and thesis not required.

Entrance requirements: For master's, GRE General Test, minimum GPA of 2.75.
Expenses: Tuition, state resident: full-time $2,250; part-time $137 per semester hour. Tuition, nonresident: full-time $3,102; part-time $172 per semester hour. Required fees: $602.
Faculty research: Rural health care delivery, school health, nutrition of pregnant teens, risk factor reduction, sexually transmitted diseases.

■ UNIVERSITY OF SOUTHERN MISSISSIPPI

Graduate School, College of Health and Human Sciences, School of Family and Consumer Sciences, Hattiesburg, MS 39406

AWARDS Early intervention (MS); family and consumer studies (MS); human nutrition (MS); institution management (MS); marriage and family therapy (MS); nutrition and food systems (PhD). Part-time programs available.

Degree requirements: For master's, thesis optional, foreign language not required; for doctorate, dissertation required.
Entrance requirements: For master's, GRE General Test, minimum GPA of 2.75; for doctorate, GRE General Test, minimum GPA of 3.5. Electronic applications accepted.
Expenses: Tuition, state resident: full-time $2,250; part-time $137 per semester hour. Tuition, nonresident: full-time $3,102; part-time $172 per semester hour. Required fees: $602.
Faculty research: School food service, teen pregnancy, diet and cholesterol metabolism.

■ THE UNIVERSITY OF TENNESSEE

Graduate School, College of Human Ecology, Department of Nutrition, Knoxville, TN 37996

AWARDS Nutrition (MS), including nutrition science, public health nutrition. Part-time programs available.

Faculty: 12 full-time (6 women).
Students: 21 full-time (17 women), 6 part-time (4 women), 4 international. *33 applicants, 58% accepted.* In 1999, 7 degrees awarded.
Degree requirements: For master's, thesis or alternative required, foreign language not required.
Entrance requirements: For master's, GRE General Test, TOEFL, minimum GPA of 2.7. *Application deadline:* For fall admission, 2/1 (priority date). Applications are processed on a rolling basis. *Application fee:* $35. Electronic applications accepted.

Expenses: Tuition, state resident: full-time $3,806; part-time $184 per credit hour. Tuition, nonresident: full-time $9,874; part-time $522 per credit hour. Tuition and fees vary according to program.
Financial aid: In 1999–00, 2 fellowships, 6 teaching assistantships were awarded; research assistantships, Federal Work-Study, institutionally sponsored loans, and unspecified assistantships also available. Financial aid application deadline: 2/1; financial aid applicants required to submit FAFSA.
Dr. Michael Zemel, Head, 865-974-5445, *Fax:* 865-974-3491, *E-mail:* mzemel@utk.edu.

Find an in-depth description at www.petersons.com/graduate.

■ THE UNIVERSITY OF TENNESSEE AT MARTIN

Graduate Studies, School of Agriculture and Human Environment, Department of Family and Consumer Sciences, Martin, TN 38238-1000

AWARDS Child development and family relations (MSHES); food science and nutrition (MSHES). Part-time programs available.

Faculty: 6 full-time (5 women), 1 (woman) part-time/adjunct.
Students: 26 (25 women). *38 applicants, 71% accepted.* In 1999, 3 degrees awarded.
Degree requirements: For master's, thesis optional, foreign language not required.
Entrance requirements: For master's, GRE General Test, minimum GPA of 2.5. *Application deadline:* For fall admission, 7/1 (priority date). Applications are processed on a rolling basis. *Application fee:* $25 ($50 for international students).
Expenses: Tuition, state resident: full-time $3,332; part-time $187 per credit hour. Tuition, nonresident: full-time $8,592; part-time $480 per credit hour.
Financial aid: In 1999–00, 4 students received aid; fellowships, research assistantships, teaching assistantships, tuition waivers (partial) and unspecified assistantships available. Financial aid application deadline: 3/1.
Faculty research: Children with developmental disabilities, regional food product development and marketing, parent education.
Dr. Lisa LeBleu, Coordinator, 901-587-7116, *E-mail:* llebleu@utm.edu.

■ THE UNIVERSITY OF TEXAS AT AUSTIN

Graduate School, College of Natural Sciences, Department of Human Ecology, Program in Nutritional Sciences, Austin, TX 78712-1111

AWARDS Nutrition (MA); nutritional sciences (PhD).

Students: 29 full-time (26 women); includes 16 minority (13 Asian Americans or Pacific Islanders, 3 Hispanic Americans). *27 applicants, 30% accepted.* In 1999, 1 master's awarded (100% found work related to degree); 2 doctorates awarded (100% found work related to degree).
Degree requirements: For master's and doctorate, thesis/dissertation required. *Average time to degree:* Master's–2 years full-time; doctorate–5 years full-time.
Entrance requirements: For master's and doctorate, GRE General Test. *Application deadline:* For fall admission, 1/1 (priority date); for spring admission, 10/1. *Application fee:* $50 ($75 for international students).
Expenses: Tuition, state resident: part-time $114 per semester hour. Tuition, nonresident: part-time $330 per semester hour. Tuition and fees vary according to program.
Financial aid: Application deadline: 2/1. Kimberly Kline, Graduate Adviser, 512-471-8911, *Fax:* 512-471-5844, *E-mail:* k.kline@mail.utexas.edu.
Application contact: Elsa Villanueva, Graduate Coordinator, 512-471-0337, *Fax:* 512-471-5844, *E-mail:* hegrad@ uts.cc.utexas.edu.

■ UNIVERSITY OF THE INCARNATE WORD

School of Graduate Studies and Research, School of Nursing and Health Professions, Program in Nutrition, San Antonio, TX 78209-6397

AWARDS MS. Part-time and evening/weekend programs available.

Students: 23 full-time (21 women), 10 part-time (9 women); includes 14 minority (1 African American, 1 Asian American or Pacific Islander, 12 Hispanic Americans), 3 international. Average age 27. *28 applicants, 89% accepted.* In 1999, 4 degrees awarded.
Degree requirements: For master's, thesis optional, foreign language not required. *Average time to degree:* Master's–2 years full-time, 3 years part-time.
Entrance requirements: For master's, GRE General Test, TOEFL, minimum GPA of 3.0. *Application deadline:* For fall admission, 8/15 (priority date); for spring

admission, 12/31. Applications are processed on a rolling basis. *Application fee:* $20.
Expenses: Tuition: Part-time $395 per hour. Required fees: $15 per hour. One-time fee: $130 part-time. Tuition and fees vary according to degree level.
Financial aid: In 1999–00, 1 research assistantship, 1 teaching assistantship were awarded; career-related internships or fieldwork and Federal Work-Study also available.
Faculty research: Diabetes assessment, modified diets, minority health issues, health promotion, outcomes research, elderly nutrition, childhood nutrition. *Total annual research expenditures:* $1,000.
Dr. Beth Senne-Duff, Coordinator, 210-828-3165, *Fax:* 210-829-3165, *E-mail:* beths@universe.uiwtx.edu.
Application contact: Andrea Cyterski, Director of Admissions, 210-829-6005, *Fax:* 210-829-3921, *E-mail:* cyterski@ universe.uiwtx.edu.

■ UNIVERSITY OF UTAH

Graduate School, College of Health, Division of Foods and Nutrition, Salt Lake City, UT 84112-1107

AWARDS MS. Part-time programs available.

Degree requirements: For master's, thesis required, foreign language not required.
Entrance requirements: For master's, GRE General Test, TOEFL.
Expenses: Tuition, state resident: full-time $1,663. Tuition, nonresident: full-time $5,201. Tuition and fees vary according to course load and program.
Faculty research: Cholesterol metabolism, nutrition education, nutrition in marketing, mineral metabolism, nutrition in wellness.

■ UNIVERSITY OF VERMONT

Graduate College, College of Agriculture and Life Sciences, Department of Nutritional Sciences, Burlington, VT 05405

AWARDS Nutritional sciences (MS); occupational and practical arts (MAT).

Degree requirements: For master's, thesis required.
Entrance requirements: For master's, GRE General Test, TOEFL.
Expenses: Tuition, state resident: full-time $7,464; part-time $311 per credit. Tuition, nonresident: full-time $18,672; part-time $778 per credit. Full-time tuition and fees vary according to degree level and program.

■ UNIVERSITY OF WASHINGTON

Graduate School, School of Public Health and Community Medicine, Department of Epidemiology, Program in Nutritional Sciences, Seattle, WA 98195

AWARDS MPH, MS, PhD. Part-time programs available.

Faculty: 11 full-time (7 women), 5 part-time/adjunct (all women).
Students: 21 full-time (19 women), 8 part-time (all women); includes 6 minority (2 African Americans, 4 Asian Americans or Pacific Islanders), 4 international. *77 applicants, 31% accepted.* In 1999, 9 master's, 3 doctorates awarded. Terminal master's awarded for partial completion of doctoral program.
Degree requirements: For master's, thesis, practicum (MPH) required, foreign language not required; for doctorate, dissertation required, foreign language not required.
Entrance requirements: For master's and doctorate, GRE General Test, TOEFL, experience in health sciences preferred, minimum GPA of 3.0. *Application fee:* $50.
Expenses: Tuition, state resident: full-time $5,196; part-time $495 per credit. Tuition, nonresident: full-time $13,485; part-time $1,285 per credit. Required fees: $387; $36 per credit. Tuition and fees vary according to course load and program.
Financial aid: In 1999–00, research assistantships with full tuition reimbursements (averaging $10,000 per year), teaching assistantships with full tuition reimbursements (averaging $10,000 per year) were awarded; career-related internships or fieldwork also available. Financial aid application deadline: 3/1.
Faculty research: Lipids, trace elements, nutrition in disease prevention.
Adam Drewnowski, Director, 206-543-1730, *Fax:* 206-685-1696, *E-mail:* adamdrew@u.washington.edu.
Application contact: Carey Purnell, Graduate Secretary, 206-543-1730, *Fax:* 206-685-1696, *E-mail:* brochure@ u.washington.edu.

■ UNIVERSITY OF WISCONSIN–MADISON

Graduate School, College of Agricultural and Life Sciences, Department of Nutritional Sciences, Madison, WI 53706-1380

AWARDS MS, PhD. Part-time programs available.

Faculty: 13 full-time (6 women).
Students: 39 full-time (28 women), 1 (woman) part-time; includes 6 minority (1 African American, 1 Asian American or Pacific Islander, 3 Hispanic Americans, 1

Native American), 1 international. Average age 27. *47 applicants, 17% accepted.* In 1999, 4 master's awarded (100% found work related to degree); 8 doctorates awarded (50% entered university research/teaching, 50% found other work related to degree). Terminal master's awarded for partial completion of doctoral program.
Degree requirements: For master's, thesis or research report required; for doctorate, dissertation required, foreign language not required. *Average time to degree:* Master's–3 years full-time; doctorate–5 years full-time.
Entrance requirements: For master's and doctorate, GRE General Test, TOEFL. *Application deadline:* For fall admission, 1/15 (priority date). Applications are processed on a rolling basis. *Application fee:* $45. Electronic applications accepted.
Expenses: Tuition, state resident: full-time $5,406; part-time $339 per credit. Tuition, nonresident: full-time $17,110; part-time $1,071 per credit. Full-time tuition and fees vary according to program and reciprocity agreements. Part-time tuition and fees vary according to course load and program.
Financial aid: In 1999–00, 38 students received aid, including 4 fellowships, 23 research assistantships, 1 teaching assistantship; Federal Work-Study, institutionally sponsored loans, and traineeships also available. Aid available to part-time students. Financial aid applicants required to submit FAFSA.
Faculty research: Human and animal nutrition, nutrition epidemiology, nutrition education, biochemical and molecular nutrition.
Denise M. Ney, Chair, 608-262-2727, *Fax:* 608-262-5860.
Application contact: Cheri Bill-Mohoney, Program Assistant, 608-262-2513, *Fax:* 608-262-5860, *E-mail:* cmohoney@ nutrisci.wisc.edu.
Find an in-depth description at www.petersons.com/graduate.

■ UNIVERSITY OF WISCONSIN–STEVENS POINT

College of Professional Studies, School of Health Promotion and Human Development, Program in Nutritional Sciences, Stevens Point, WI 54481-3897

AWARDS MS. Part-time programs available.

Students: In 1999, 1 degree awarded.
Degree requirements: For master's, thesis or alternative required, foreign language not required.
Entrance requirements: For master's, minimum GPA of 2.75. *Application deadline:* For fall admission, 5/1 (priority date).

Applications are processed on a rolling basis. *Application fee:* $45.
Expenses: Tuition, state resident: full-time $3,966; part-time $242 per credit. Tuition, nonresident: full-time $12,324; part-time $706 per credit. Part-time tuition and fees vary according to course load.
Financial aid: Research assistantships, teaching assistantships, career-related internships or fieldwork and Federal Work-Study available. Aid available to part-time students. Financial aid application deadline: 5/1; financial aid applicants required to submit FAFSA.
John Munson, Associate Dean, School of Health Promotion and Human Development, 715-346-2830, *Fax:* 715-346-3751.

■ UNIVERSITY OF WISCONSIN–STOUT

Graduate School, College of Human Development, Program in Food Nutritional Sciences, Menomonie, WI 54751

AWARDS MS. Part-time programs available.

Students: 8 full-time (7 women), 10 part-time (7 women); includes 3 minority (2 Asian Americans or Pacific Islanders, 1 Hispanic American), 3 international. *17 applicants, 76% accepted.* In 1999, 8 degrees awarded.
Degree requirements: For master's, thesis required, foreign language not required.
Application deadline: Applications are processed on a rolling basis. *Application fee:* $45.
Expenses: Tuition, state resident: full-time $4,194; part-time $205 per credit. Tuition, nonresident: full-time $12,552; part-time $465 per credit. Tuition and fees vary according to course load.
Financial aid: In 1999–00, 6 research assistantships, 2 teaching assistantships were awarded; Federal Work-Study and tuition waivers (full and partial) also available. Aid available to part-time students. Financial aid application deadline: 4/1; financial aid applicants required to submit FAFSA.
Dr. Barbara Knous, Director, 715-232-1994.

■ UNIVERSITY OF WYOMING

Graduate School, College of Agriculture, Department of Animal Sciences, Program in Food Science and Human Nutrition, Laramie, WY 82071

AWARDS MS.

Faculty: 7 full-time (1 woman).
Students: 2 full-time (both women), 1 (woman) part-time. *1 applicant, 100% accepted.* In 1999, 3 degrees awarded.

Degree requirements: For master's, thesis required, foreign language not required.
Entrance requirements: For master's, GRE General Test, minimum GPA of 3.0. *Application deadline:* For fall admission, 6/1 (priority date). Applications are processed on a rolling basis. *Application fee:* $40.
Expenses: Tuition, state resident: full-time $2,520; part-time $140 per credit hour. Tuition, nonresident: full-time $7,790; part-time $433 per credit hour. Required fees: $440; $7 per credit hour. Full-time tuition and fees vary according to course load and program.
Financial aid: Research assistantships, career-related internships or fieldwork, Federal Work-Study, and institutionally sponsored loans available. Financial aid application deadline: 3/1.
Faculty research: Protein and lipid metabolism, food microbiology, food safety, meat science.
Daniel Rule, Coordinator, 307-766-3404, *Fax:* 307-766-2355.

■ UTAH STATE UNIVERSITY

School of Graduate Studies, College of Family Life and College of Agriculture, Department of Nutrition and Food Sciences, Logan, UT 84322

AWARDS Food microbiology and safety (MFMS); molecular biology (MS, PhD); nutrition and food sciences (MS, PhD).

Faculty: 17 full-time (8 women), 2 part-time/adjunct (0 women).
Students: 18 full-time (11 women), 8 part-time (4 women); includes 2 minority (1 Asian American or Pacific Islander, 1 Hispanic American), 12 international. Average age 27. *34 applicants, 21% accepted.* In 1999, 10 master's, 3 doctorates awarded.
Degree requirements: For master's, thesis required, foreign language not required; for doctorate, computer language, dissertation, teaching experience required, foreign language not required.
Entrance requirements: For master's and doctorate, GRE General Test, TOEFL, minimum GPA of 3.0, previous course work in chemistry. *Application deadline:* For fall admission, 2/1 (priority date); for spring admission, 10/15 (priority date). Applications are processed on a rolling basis. *Application fee:* $40.
Expenses: Tuition, state resident: full-time $1,553. Tuition, nonresident: full-time $5,436. International tuition: $5,526 full-time. Required fees: $447. Tuition and fees vary according to course load and program.
Financial aid: In 1999–00, 22 research assistantships with partial tuition reimbursements were awarded; fellowships

Utah State University (continued)
with partial tuition reimbursements, teaching assistantships with partial tuition reimbursements, Federal Work-Study, institutionally sponsored loans, and tuition waivers (full and partial) also available.
Faculty research: Mineral balance, meat microbiology and nitrate interactions, milk ultrafiltration, lactic culture, milk coagulation.
Ann Sorenson, Head, 435-797-2102, *Fax:* 435-797-2379.
Application contact: Carolyn Glover, Receptionist, 435-797-2126, *Fax:* 435-797-2379, *E-mail:* nfs@cc.usu.edu.

■ VIRGINIA POLYTECHNIC INSTITUTE AND STATE UNIVERSITY

Graduate School, College of Human Resources and Education, Department of Human Nutrition, Foods and Exercise, Blacksburg, VA 24061

AWARDS Clinical exercise physiology (MS, PhD); community and international nutrition (MS, PhD); foods (MS, PhD); muscle physiology and biochemistry (MS, PhD); nutrition (MS, PhD); nutrition in sports and chronic disease (MS, PhD).

Faculty: 16 full-time (12 women).
Students: 33 full-time (16 women), 5 part-time (all women); includes 4 minority (3 African Americans, 1 Hispanic American), 5 international. Average age 25. *73 applicants, 47% accepted.* In 1999, 8 master's, 3 doctorates awarded.
Degree requirements: For master's and doctorate, thesis/dissertation required.
Entrance requirements: For master's and doctorate, GRE, TOEFL. *Application deadline:* For fall admission, 12/1 (priority date). Applications are processed on a rolling basis. *Application fee:* $25.
Expenses: Tuition, state resident: full-time $4,122; part-time $229 per credit hour. Tuition, nonresident: full-time $6,930; part-time $385 per credit hour. Required fees: $828; $107 per semester. Part-time tuition and fees vary according to course load.
Financial aid: In 1999–00, 1 research assistantship with full tuition reimbursement (averaging $7,782 per year), 11 teaching assistantships with full tuition reimbursements (averaging $7,782 per year) were awarded; Federal Work-Study, tuition waivers (full), and unspecified assistantships also available. Financial aid application deadline: 4/1.
Faculty research: Nutrition and food science research.
Dr. Mike Houston, Head, 540-231-4672, *E-mail:* houstonm@vt.edu.

■ WASHINGTON STATE UNIVERSITY

Graduate School, College of Agriculture and Home Economics, Department of Animal Sciences, Pullman, WA 99164

AWARDS Animal sciences (MS, PhD); nutrition (PhD).

Faculty: 21 full-time (3 women).
Students: 24 full-time (14 women), 6 part-time (3 women); includes 4 minority (1 African American, 2 Asian Americans or Pacific Islanders, 1 Hispanic American), 6 international. In 1999, 5 master's, 4 doctorates awarded.
Degree requirements: For master's, thesis, oral exam required, foreign language not required; for doctorate, dissertation, oral and written exam required, foreign language not required. *Average time to degree:* Master's–2 years full-time; doctorate–4 years full-time.
Entrance requirements: For master's and doctorate, GRE General Test, minimum GPA of 3.0. *Application deadline:* For fall admission, 3/1 (priority date). Applications are processed on a rolling basis. *Application fee:* $35.
Expenses: Tuition, state resident: full-time $5,654. Tuition, nonresident: full-time $13,850. International tuition: $13,850 full-time. Tuition and fees vary according to program.
Financial aid: In 1999–00, 11 research assistantships with full and partial tuition reimbursements, 8 teaching assistantships with full and partial tuition reimbursements were awarded; fellowships, career-related internships or fieldwork, Federal Work-Study, institutionally sponsored loans, tuition waivers (partial), and teaching associateships also available. Financial aid application deadline: 4/1; financial aid applicants required to submit FAFSA.
Faculty research: Reproduction, genetics. *Total annual research expenditures:* $634,534.
Dr. Raymond Wright, Chair, 509-335-5523.

■ WASHINGTON STATE UNIVERSITY

Graduate School, College of Agriculture and Home Economics, Department of Food Science and Human Nutrition, Program in Human Nutrition, Pullman, WA 99164

AWARDS MS.

Students: 10 full-time (9 women), 3 part-time (all women); includes 2 minority (both Asian Americans or Pacific Islanders), 4 international. In 1999, 10 degrees awarded.

Degree requirements: For master's, thesis, oral exam required, foreign language not required.
Entrance requirements: For master's, minimum GPA of 3.0. *Application deadline:* For fall admission, 3/1 (priority date). Applications are processed on a rolling basis. *Application fee:* $35.
Expenses: Tuition, state resident: full-time $5,654. Tuition, nonresident: full-time $13,850. International tuition: $13,850 full-time. Tuition and fees vary according to program.
Financial aid: In 1999–00, 2 research assistantships with full and partial tuition reimbursements, 5 teaching assistantships with full and partial tuition reimbursements were awarded. Financial aid application deadline: 4/1.
Dr. Alan McCurdy, Chair, Department of Food Science and Human Nutrition, 509-335-9103, *Fax:* 509-335-4815.

■ WASHINGTON STATE UNIVERSITY

Graduate School, College of Agriculture and Home Economics, Department of Food Science and Human Nutrition and Department of Animal Sciences, Program in Nutrition, Pullman, WA 99164

AWARDS PhD.

Faculty: 19 full-time (12 women).
Students: 4 full-time (all women). In 1999, 1 degree awarded.
Degree requirements: For doctorate, dissertation, oral exam required, foreign language not required. *Average time to degree:* Doctorate–4 years full-time.
Entrance requirements: For doctorate, minimum GPA of 3.0. *Application deadline:* For fall admission, 3/1 (priority date). Applications are processed on a rolling basis. *Application fee:* $35.
Expenses: Tuition, state resident: full-time $5,654. Tuition, nonresident: full-time $13,850. International tuition: $13,850 full-time. Tuition and fees vary according to program.
Financial aid: In 1999–00, 1 research assistantship with full and partial tuition reimbursement, 1 teaching assistantship with full and partial tuition reimbursement were awarded; fellowships, career-related internships or fieldwork, Federal Work-Study, institutionally sponsored loans, tuition waivers (partial), and unspecified assistantships also available. Financial aid application deadline: 4/1; financial aid applicants required to submit FAFSA.
Dr. Alan McCurdy, Chair, Department of Food Science and Human Nutrition, 509-335-9103, *Fax:* 509-335-4815.

■ WASHINGTON STATE UNIVERSITY SPOKANE

Graduate Programs, Program in Human Nutrition, Spokane, WA 99201-3899

AWARDS MS.

■ WAYNE STATE UNIVERSITY

Graduate School, College of Science, Department of Nutrition and Food Science, Detroit, MI 48202

AWARDS MA, MS, PhD. Terminal master's awarded for partial completion of doctoral program.

Degree requirements: For master's, thesis required (for some programs), foreign language not required; for doctorate, dissertation required, foreign language not required.
Entrance requirements: For master's and doctorate, GRE General Test, minimum GPA of 3.0.
Faculty research: Nutrition and cancer, obesity and diabetes, mineral metabolism, food microbiology, lipoprotein metabolism.

■ WEST VIRGINIA UNIVERSITY

College of Agriculture, Forestry and Consumer Sciences, Division of Animal and Veterinary Sciences, Program in Animal and Food Sciences, Morgantown, WV 26506

AWARDS Agricultural biochemistry (PhD); animal nutrition (PhD); animal physiology (PhD); production management (PhD).

Students: 12 full-time (6 women), 5 part-time (2 women), 9 international. Average age 36. In 1999, 2 degrees awarded.
Degree requirements: For doctorate, dissertation, oral and written exams required, foreign language not required.
Entrance requirements: For doctorate, TOEFL. *Application fee:* $45.

Expenses: Tuition, state resident: full-time $2,910; part-time $154 per credit hour. Tuition, nonresident: full-time $8,368; part-time $457 per credit hour.
Financial aid: In 1999–00, 9 research assistantships, 1 teaching assistantship were awarded; Federal Work-Study, institutionally sponsored loans, and tuition waivers (full and partial) also available. Financial aid application deadline: 2/1; financial aid applicants required to submit FAFSA.
Faculty research: Ruminant nutrition, metabolism, forage utilization, physiology, reproduction.
Dr. Hillar Klandorf, Coordinator, 304-293-2631 Ext. 4436.

■ WEST VIRGINIA UNIVERSITY

College of Agriculture, Forestry and Consumer Sciences, Division of Animal and Veterinary Sciences, Program in Animal and Veterinary Sciences, Morgantown, WV 26506

AWARDS Animal sciences (MS); breeding (MS); food sciences (MS); nutrition (MS); physiology (MS); production (MS). Part-time programs available.

Students: 10 full-time (3 women), 1 (woman) part-time, 3 international. Average age 28. In 1999, 4 degrees awarded.
Degree requirements: For master's, thesis, oral and written exams required, foreign language not required.
Entrance requirements: For master's, TOEFL, minimum GPA of 2.5. *Application fee:* $45.
Expenses: Tuition, state resident: full-time $2,910; part-time $154 per credit hour. Tuition, nonresident: full-time $8,368; part-time $457 per credit hour.
Financial aid: In 1999–00, 9 research assistantships, 1 teaching assistantship were awarded; Federal Work-Study, institutionally sponsored loans, and tuition waivers

(full and partial) also available. Financial aid application deadline: 2/1; financial aid applicants required to submit FAFSA.
Faculty research: Animal nutrition, reproductive physiology, food science.
Dr. Hillar Klandorf, Coordinator, 304-293-2631 Ext. 4436.

■ WINTHROP UNIVERSITY

College of Arts and Sciences, Department of Human Nutrition, Rock Hill, SC 29733

AWARDS MS. Part-time programs available.

Faculty: 3 full-time (all women).
Students: 23 full-time (all women), 14 part-time (all women); includes 3 minority (all African Americans), 1 international. Average age 27. In 1999, 10 degrees awarded.
Degree requirements: For master's, thesis optional, foreign language not required.
Entrance requirements: For master's, GRE General Test, or NTE, interview, minimum GPA of 3.0. *Application deadline:* For fall admission, 7/15 (priority date); for spring admission, 12/1. Applications are processed on a rolling basis. *Application fee:* $35.
Expenses: Tuition, state resident: full-time $4,020; part-time $168 per semester hour. Tuition, nonresident: full-time $7,240; part-time $302 per semester hour.
Financial aid: Federal Work-Study, scholarships, and unspecified assistantships available. Aid available to part-time students. Financial aid application deadline: 2/1.
Dr. Patricia G. Wolman, Chair, 803-323-2101, *E-mail:* wolmanp@winthrop.edu.
Application contact: Sharon Johnson, Director of Graduate Studies, 803-323-2204, *Fax:* 803-323-2292, *E-mail:* johnsons@winthrop.edu.

Parasitology

PARASITOLOGY

■ LOUISIANA STATE UNIVERSITY HEALTH SCIENCES CENTER

School of Graduate Studies in New Orleans, Department of Microbiology, Immunology, and Parasitology, New Orleans, LA 70112-1393

AWARDS Microbiology and immunology (MS, PhD).

Faculty: 13 full-time (3 women).
Students: 19 full-time (13 women); includes 5 minority (3 African Americans, 1 Asian American or Pacific Islander, 1 Hispanic American), 4 international. Average age 23. 30 applicants, 50% accepted. In 1999, 3 master's awarded. Terminal master's awarded for partial completion of doctoral program.
Degree requirements: For master's, thesis required, foreign language not required; for doctorate, dissertation,

preliminary exam, qualifying exam required, foreign language not required. *Average time to degree:* Master's–3.5 years full-time; doctorate–7 years full-time.
Entrance requirements: For master's and doctorate, GRE General Test, TOEFL. *Application deadline:* For fall admission, 4/15 (priority date). Applications are processed on a rolling basis. *Application fee:* $30.
Expenses: Tuition, state resident: full-time $2,878; part-time $126 per hour. Tuition,

Louisiana State University Health Sciences Center *(continued)*
nonresident: full-time $6,003; part-time $265 per hour. Required fees: $2,272. Tuition and fees vary according to course load, degree level and program.
Financial aid: In 1999–00, 16 students received aid, including 16 research assistantships (averaging $13,000 per year); Federal Work-Study and tuition waivers (full) also available. Financial aid application deadline: 4/15.
Faculty research: Microbial physiology, animal virology, vaccine development, AIDS drug studies, pathogenic mechanisms, molecular immunology.
Dr. Ronald B. Luftig, Head, 504-568-4062.
Application contact: Jeanine M. Campbell, Asst. Business Manager, 504-568-4061, *Fax:* 504-568-2918, *E-mail:* jcampb@lsumc.edu.
Find an in-depth description at www.petersons.com/graduate.

■ **NEW YORK UNIVERSITY**

Graduate School of Arts and Science, Department of Basic Medical Sciences, New York, NY 10012-1019
AWARDS Biochemistry (MS, PhD); cell biology (MS, PhD); microbiology (MS, PhD); parasitology (PhD); pathology (MS, PhD); pharmacology (PhD); physiology (MS, PhD). Part-time programs available.
Faculty: 23 full-time (2 women), 3 part-time/adjunct (1 woman).
Students: 182 full-time (73 women), 4 part-time (1 woman); includes 52 minority (11 African Americans, 34 Asian Americans or Pacific Islanders, 7 Hispanic Americans), 48 international. Average age 26. *671 applicants, 11% accepted.* In 1999, 15 master's, 32 doctorates awarded. Terminal master's awarded for partial completion of doctoral program.
Degree requirements: For master's, thesis or alternative, written comprehensive exam required, foreign language not required; for doctorate, one foreign language, dissertation, oral and written comprehensive exams required.
Entrance requirements: For master's and doctorate, GRE General Test, GRE Subject Test, TOEFL. *Application deadline:* For fall admission, 2/1 (priority date). *Application fee:* $60.
Expenses: Tuition: Full-time $17,880; part-time $745 per credit. Required fees: $1,140; $35 per credit. Tuition and fees vary according to course load and program.
Financial aid: Fellowships with tuition reimbursements, research assistantships with tuition reimbursements, teaching assistantships with tuition reimbursements,

career-related internships or fieldwork, Federal Work-Study, institutionally sponsored loans, and tuition waivers (full and partial) available. Financial aid application deadline: 2/1; financial aid applicants required to submit FAFSA.
Dr. Joel D. Oppenheim, Director, 212-263-5648, *Fax:* 212-263-7600, *E-mail:* sackler-info@nyumed.med.nyu.edu.

■ **NEW YORK UNIVERSITY**

School of Medicine and Graduate School of Arts and Science, Medical Scientist Training Program, New York, NY 10012-1019
AWARDS Biochemistry (MD/PhD); cell biology (MD/PhD); environmental health sciences (MD/PhD); microbiology (MD/PhD); parasitology (MD/PhD); pathology (MD/PhD); pharmacology (MD/PhD). Students must be accepted by both the School of Medicine and the Graduate School of Arts and Science.
Faculty: 150 full-time (35 women).
Students: 80 full-time (18 women); includes 37 minority (3 African Americans, 32 Asian Americans or Pacific Islanders, 2 Hispanic Americans), 1 international. Average age 25. *195 applicants, 18% accepted.*
Degree requirements: One foreign language.
Application deadline: For fall admission, 11/15. Applications are processed on a rolling basis. *Application fee:* $60.
Expenses: Students receive full tuition support and an annual stipend of $17,500.
Financial aid: Application deadline: 5/1.
Faculty research: Genetics, tumor biology, cardiovascular biology, neuroscience, host defense mechanisms.
Dr. James Salzer, Director, 212-263-0758.
Application contact: Arlene Kohler, Administrative Officer, 212-263-5649.

■ **NEW YORK UNIVERSITY**

School of Medicine and Graduate School of Arts and Science, Sackler Institute of Graduate Biomedical Sciences, Department of Medical and Molecular Parasitology, New York, NY 10012-1019
AWARDS PhD, MD/PhD.
Faculty: 13 full-time (6 women).
Students: 4 full-time (3 women); includes 3 minority (all Asian Americans or Pacific Islanders). Average age 25. In 1999, 4 degrees awarded (100% entered university research/teaching).
Degree requirements: For doctorate, one foreign language, dissertation, qualifying exam required. *Average time to degree:* Doctorate–5.7 years full-time.
Entrance requirements: For doctorate, GRE General Test, GRE Subject Test, TOEFL. *Application deadline:* For fall

admission, 2/1 (priority date). Applications are processed on a rolling basis. *Application fee:* $60.
Expenses: Tuition: Full-time $17,880; part-time $745 per credit. Required fees: $1,140; $35 per credit. Tuition and fees vary according to course load and program.
Financial aid: Fellowships, research assistantships, teaching assistantships available. Financial aid application deadline: 1/15.
Faculty research: Immunoparasitology, cell biology of parasites, genetics of parasites, mode of action of antiparasitic drugs. *Total annual research expenditures:* $2 million.
Dr. Ruth S. Nussenzweig, Chairperson, 212-263-6817.
Application contact: Dr. Laura Pologe, Graduate Adviser, 212-263-6763, *E-mail:* pologl01@mcrcro.med.nyu.edu.

■ **PURDUE UNIVERSITY**

School of Veterinary Medicine and Graduate School, Graduate Programs in Veterinary Medicine, Department of Veterinary Pathobiology, West Lafayette, IN 47907
AWARDS Bacteriology (MS, PhD); epidemiology (MS, PhD); immunology (MS, PhD); infectious diseases (MS, PhD); microbiology (MS, PhD); parasitology (MS, PhD); pathology (MS, PhD); toxicology (MS, PhD); virology (MS, PhD).
Faculty: 26 full-time (5 women).
Students: 42 full-time (24 women), 2 part-time (1 woman); includes 5 minority (1 African American, 3 Asian Americans or Pacific Islanders, 1 Hispanic American), 18 international. Average age 32. In 1999, 2 master's, 6 doctorates awarded (100% found work related to degree). Terminal master's awarded for partial completion of doctoral program.
Degree requirements: For master's, thesis required (for some programs), foreign language not required; for doctorate, dissertation required, foreign language not required.
Entrance requirements: For master's and doctorate, GRE General Test, TOEFL. *Application fee:* $30.
Expenses: Tuition, state resident: full-time $3,732. Tuition, nonresident: full-time $8,732.
Financial aid: Fellowships, research assistantships, teaching assistantships available. Financial aid application deadline: 3/1; financial aid applicants required to submit FAFSA.
Dr. H. L. Thacker, Head, 765-494-7543.

■ TEXAS A&M UNIVERSITY

College of Veterinary Medicine, Department of Veterinary Pathobiology, College Station, TX 77843

AWARDS Genetics (MS, PhD); toxicology (MS, PhD); veterinary microbiology (MS, PhD); veterinary parasitology (MS); veterinary pathology (MS, PhD). Part-time programs available. Postbaccalaureate distance learning degree programs offered.

Faculty: 44 full-time (10 women), 17 part-time/adjunct (2 women).
Students: 33 full-time (13 women), 30 part-time (18 women); includes 11 minority (2 African Americans, 1 Asian American or Pacific Islander, 8 Hispanic Americans), 11 international. Average age 34. *28 applicants, 46% accepted.* In 1999, 4 master's awarded (25% found work related to degree, 75% continued full-time study); 10 doctorates awarded (40% entered university research/teaching, 60% found other work related to degree). Terminal master's awarded for partial completion of doctoral program.
Degree requirements: For master's and doctorate, thesis/dissertation required, foreign language not required. *Average time to degree:* Master's–2 years full-time, 4.5 years part-time; doctorate–4.5 years full-time, 7 years part-time.
Entrance requirements: For master's and doctorate, GRE General Test, TOEFL. *Application deadline:* For fall admission, 3/1 (priority date); for spring admission, 8/1 (priority date). Applications are processed on a rolling basis. *Application fee:* $50 ($75 for international students). Electronic applications accepted.
Expenses: Tuition, state resident: full-time $5,400. Tuition, nonresident: full-time $16,200. Required fees: $2,936. Full-time tuition and fees vary according to student level.
Financial aid: In 1999–00, 7 fellowships with partial tuition reimbursements (averaging $16,000 per year), 13 research assistantships with partial tuition reimbursements (averaging $14,400 per year), 5 teaching assistantships with partial tuition reimbursements (averaging $16,000 per year) were awarded; career-related internships or fieldwork, Federal Work-Study, and institutionally sponsored loans also available. Aid available to part-time students. Financial aid applicants required to submit FAFSA.
Faculty research: Infectious and noninfectious diseases of animals and birds, animal genetics, molecular biology, immunology, virology. *Total annual research expenditures:* $2.6 million.

Dr. Ann B. Kier, Head, 979-845-5941, *Fax:* 979-845-9231, *E-mail:* akier@cvm.tamu.edu.
Application contact: Dr. G. G. Wagner, Graduate Adviser, 979-845-5941, *Fax:* 979-862-1147, *E-mail:* gwagner@cvm.tamus.edu.

■ TULANE UNIVERSITY

School of Medicine and Graduate School, Graduate Programs in Medicine, Department of Parasitology, New Orleans, LA 70118-5669
AWARDS PhD.
Expenses: Tuition: Full-time $23,030.
Dr. Donald Krogstad, Chair, 504-587-7313, *E-mail:* tropmed@mailhost.tcs.tulane.edu.

■ TULANE UNIVERSITY

School of Medicine and Graduate School, Graduate Programs in Medicine and School of Public Health and Tropical Medicine, Program in Parasitology, New Orleans, LA 70118-5669
AWARDS MS, MSPH, PhD, MD/PhD. MS and PhD offered through the Graduate School.
Students: 4 full-time (1 woman), 2 part-time (1 woman), 5 international. *8 applicants, 13% accepted.* In 1999, 1 doctorate awarded.
Degree requirements: For master's, thesis required; for doctorate, dissertation required.
Entrance requirements: For master's, GRE General Test, TOEFL, or TSE, minimum B average in undergraduate course work; for doctorate, GRE General Test, TOEFL, or TSE. *Application deadline:* For fall admission, 2/1. *Application fee:* $45.
Expenses: Tuition: Full-time $23,030.
Financial aid: Application deadline: 2/1.
Dr. Donald Krogstad, Chairman, 504-588-5199, *E-mail:* tropmed@mailhost.tcs.tulane.edu.

■ UNIVERSITY OF GEORGIA

College of Veterinary Medicine and Graduate School, Graduate Programs in Veterinary Medicine, Athens, GA 30602
AWARDS Avian medicine (MAM); medical microbiology and parasitology (MS, PhD), including medical microbiology, parasitology; pathology (MS, PhD); physiology and pharmacology (MS, PhD), including pharmacology, physiology; toxicology (MS, PhD). Evening/weekend programs available.
Faculty: 66 full-time (20 women).
Students: 63 full-time, 11 part-time. *70 applicants, 30% accepted.* In 1999, 13 master's, 8 doctorates awarded.

Degree requirements: For master's, foreign language not required; for doctorate, one foreign language (computer language can substitute), dissertation required.
Entrance requirements: For master's and doctorate, GRE General Test. *Application deadline:* For fall admission, 7/1 (priority date); for spring admission, 11/15. *Application fee:* $30. Electronic applications accepted.
Expenses: Tuition, state resident: full-time $7,516; part-time $431 per credit hour. Tuition, nonresident: full-time $12,204; part-time $793 per credit hour.
Financial aid: Fellowships, research assistantships, teaching assistantships, unspecified assistantships available.
Dr. Harry W. Dickerson, Associate Dean, 706-542-5734, *Fax:* 706-542-8254, *E-mail:* hwd@calc.vet.uga.edu.

■ UNIVERSITY OF GEORGIA

College of Veterinary Medicine and Graduate School, Graduate Programs in Veterinary Medicine, Department of Medical Microbiology and Parasitology, Program in Parasitology, Athens, GA 30602
AWARDS MS, PhD.

Degree requirements: For master's, thesis required, foreign language not required; for doctorate, one foreign language (computer language can substitute), dissertation required.
Entrance requirements: For master's and doctorate, GRE General Test. Electronic applications accepted.
Expenses: Tuition, state resident: full-time $7,516; part-time $431 per credit hour. Tuition, nonresident: full-time $12,204; part-time $793 per credit hour.

■ UNIVERSITY OF NEW MEXICO

Graduate School, College of Arts and Sciences, Department of Biology, Albuquerque, NM 87131-2039
AWARDS Biology (MS, PhD), including air land ecology, behavioral ecology, botany, cellular and molecular biology, community ecology, comparative immunology, comparative physiology, conservation biology, ecology, ecosystem ecology, evolutionary biology, evolutionary genetics, microbiology, molecular genetics, parasitology, physiological ecology, physiology, population biology, vertebrate and invertebrate zoology. Part-time programs available.
Faculty: 35 full-time (5 women), 18 part-time/adjunct (11 women).
Students: 71 full-time (37 women), 28 part-time (11 women); includes 8 minority (2 Asian Americans or Pacific Islanders, 5 Hispanic Americans, 1 Native American),

University of New Mexico (continued)
11 international. Average age 33. *93 applicants, 30% accepted.* In 1999, 11 master's, 12 doctorates awarded. Terminal master's awarded for partial completion of doctoral program.

Degree requirements: For master's, one foreign language (computer language can substitute), thesis required (for some programs); for doctorate, 2 foreign languages (computer language can substitute for one), dissertation required.

Entrance requirements: For master's and doctorate, GRE General Test, GRE Subject Test, minimum GPA of 3.2. *Application deadline:* For fall admission, 1/15. *Application fee:* $25.

Expenses: Tuition, state resident: full-time $2,514; part-time $105 per credit hour. Tuition, nonresident: full-time $10,304; part-time $417 per credit hour. International tuition: $10,304 full-time. Required fees: $516; $22 per credit hour. Tuition and fees vary according to program.

Financial aid: In 1999–00, 58 students received aid, including 24 fellowships (averaging $1,645 per year), 26 research assistantships with tuition reimbursements available (averaging $8,921 per year), 40 teaching assistantships with tuition reimbursements available (averaging $11,066 per year); career-related internships or fieldwork, Federal Work-Study, institutionally sponsored loans, and tuition waivers (full and partial) also available. Aid available to part-time students. Financial aid applicants required to submit FAFSA.

Faculty research: Developmental biology, immunobiology. *Total annual research expenditures:* $4.5 million.
Dr. Kathryn Vogel, Chair, 505-277-3411, *Fax:* 505-277-0304, *E-mail:* kgvogel@unm.edu.

Application contact: Vivian Kent, Information Contact, 505-277-1712, *Fax:* 505-277-0304, *E-mail:* vkent@unm.edu.

■ **UNIVERSITY OF NOTRE DAME**
Graduate School, College of Science, Department of Biological Sciences, Program in Vector Biology and Parasitology, Notre Dame, IN 46556
AWARDS MS, PhD. Terminal master's awarded for partial completion of doctoral program.

Degree requirements: For master's and doctorate, thesis/dissertation required, foreign language not required.

Entrance requirements: For master's and doctorate, GRE General Test, GRE Subject Test, TOEFL. *Application deadline:* For fall admission, 2/1 (priority date); for spring admission, 11/1. Applications are processed on a rolling basis. *Application fee:* $50.

Expenses: Tuition: Full-time $21,930; part-time $1,218 per credit. Required fees: $95. Tuition and fees vary according to program.

Financial aid: Fellowships with full tuition reimbursements, research assistantships with full tuition reimbursements, teaching assistantships with full tuition reimbursements, traineeships and tuition waivers (full) available. Financial aid application deadline: 2/1.

Faculty research: Parasite genetics, pharmacology, and mode of action; immunoparasitology; cell biology; gene expression.

Application contact: Dr. Terrence J. Akai, Director of Graduate Admissions, 219-631-7706, *Fax:* 219-631-4183, *E-mail:* gradad@nd.edu.

■ **UNIVERSITY OF PENNSYLVANIA**
School of Medicine, Biomedical Graduate Studies, Graduate Group in Parasitology, Philadelphia, PA 19104
AWARDS PhD, MD/PhD, VMD/PhD. Part-time programs available.

Faculty: 19 full-time (0 women), 2 part-time/adjunct (0 women).

Students: 12 full-time (6 women); includes 3 minority (1 African American, 1 Asian American or Pacific Islander, 1 Hispanic American), 4 international. *10 applicants, 40% accepted.* In 1999, 1 doctorate awarded.

Degree requirements: For doctorate, dissertation required, foreign language not required.

Entrance requirements: For doctorate, GRE General Test, TOEFL, previous course work in biological sciences. *Application deadline:* For fall admission, 1/2 (priority date). Applications are processed on a rolling basis. *Application fee:* $65. Electronic applications accepted.

Expenses: Tuition: Full-time $17,256; part-time $2,991 per course. Required fees: $2,588; $363 per course. $726 per term.

Financial aid: In 1999–00, 7 students received aid, including 4 research assistantships; fellowships, grants and institutionally sponsored loans also available. Financial aid application deadline: 1/2.

Faculty research: Cell biology, molecular biology, immunology, ecology and epidemiology of parasitic infections.
Dr. Jay P. Farrell, Chairperson, 215-898-8561.

Application contact: Colleen Dunn, Graduate Coordinator, 215-898-1030, *Fax:* 215-898-2671, *E-mail:* dunncoll@mail.med.upenn.edu.

Find an in-depth description at www.petersons.com/graduate.

■ **UNIVERSITY OF WASHINGTON**
Graduate School, School of Public Health and Community Medicine, Department of Pathobiology, Seattle, WA 98195
AWARDS MS, PhD.

Faculty: 33 full-time (11 women), 9 part-time/adjunct (0 women).

Students: 24 full-time (18 women), 3 part-time (1 woman); includes 11 minority (1 African American, 8 Asian Americans or Pacific Islanders, 2 Hispanic Americans), 4 international. Average age 28. *45 applicants, 20% accepted.* In 1999, 3 master's awarded (33% found work related to degree, 67% continued full-time study); 8 doctorates awarded (100% entered university research/teaching). Terminal master's awarded for partial completion of doctoral program.

Degree requirements: For master's and doctorate, thesis/dissertation required, foreign language not required. *Average time to degree:* Master's–2 years full-time; doctorate–8 years full-time.

Entrance requirements: For master's and doctorate, GRE General Test, TOEFL, minimum GPA of 3.0. *Application deadline:* For fall admission, 1/1. *Application fee:* $50.

Expenses: Tuition, state resident: full-time $5,196; part-time $495 per credit. Tuition, nonresident: full-time $13,485; part-time $1,285 per credit. Required fees: $387; $36 per credit. Tuition and fees vary according to course load and program.

Financial aid: In 1999–00, 2 fellowships with tuition reimbursements (averaging $16,068 per year), 18 research assistantships with tuition reimbursements (averaging $16,068 per year) were awarded; career-related internships or fieldwork, Federal Work-Study, institutionally sponsored loans, traineeships, and tuition waivers (full and partial) also available. Financial aid application deadline: 3/1; financial aid applicants required to submit FAFSA.

Faculty research: Pathogenesis of chlamydiae, molecular biology of parasites, signal transduction, antigenic analysis, molecular biology of tumor viruses.
Dr. Kenneth Stuart, Chair, 206-543-8350, *Fax:* 206-543-3873, *E-mail:* kstuart@u.washington.edu.

Application contact: Leslie G. Miller, Coordinator, 206-543-4338, *Fax:* 206-543-3873, *E-mail:* pathiobio@u.washington.edu.

Find an in-depth description at www.petersons.com/graduate.

■ WEST VIRGINIA UNIVERSITY

School of Medicine, Graduate Programs in Health Sciences, Department of Microbiology and Immunology, Morgantown, WV 26506

AWARDS Genetics (MS, PhD); immunology (MS, PhD); mycology (PhD); parasitology (MS, PhD); pathogenic bacteriology (MS, PhD); physiology (PhD); psysiology (MS); virology (MS, PhD).

Students: 25 full-time (11 women), 1 (woman) part-time; includes 2 minority (both Asian Americans or Pacific Islanders), 4 international. Average age 27. *62 applicants, 8% accepted.* In 1999, 1 master's, 5 doctorates awarded. Terminal master's awarded for partial completion of doctoral program.
Degree requirements: For master's, thesis required, foreign language not required; for doctorate, dissertation, comprehensive exam required, foreign language not required. *Average time to degree:* Doctorate–4.5 years full-time.

Entrance requirements: For master's and doctorate, GRE General Test, GRE Subject Test, TOEFL, minimum GPA of 3.0. *Application deadline:* For fall admission, 3/1 (priority date). Applications are processed on a rolling basis. *Application fee:* $45. Electronic applications accepted.
Expenses: Tuition, state resident: full-time $3,564. Tuition, nonresident: full-time $10,230.
Financial aid: In 1999–00, 7 research assistantships, 14 teaching assistantships were awarded; fellowships, Federal Work-Study and institutionally sponsored loans also available. Financial aid application deadline: 3/1; financial aid applicants required to submit FAFSA.
Faculty research: Mechanisms of pathogenesis, microbial genetics, molecular virology, immunotoxicology, oncogenes. *Total annual research expenditures:* $2.6 million.
John Barnett, Chair, 304-293-2649.

Application contact: James Sheil, Graduate Coordinator, 304-293-3559, *Fax:* 304-293-7823, *E-mail:* jsheil@wvu.edu.

■ YALE UNIVERSITY

School of Medicine, Department of Epidemiology and Public Health, Program in Parasitology, New Haven, CT 06520

AWARDS PhD.

Degree requirements: For doctorate, dissertation, comprehensive exams, residency required, foreign language not required.
Entrance requirements: For doctorate, GRE General Test, TOEFL.
Expenses: Tuition: Full-time $20,100. Required fees: $125. Tuition and fees vary according to program.

Find an in-depth description at www.petersons.com/graduate.

Pathology

PATHOLOGY

■ BOSTON UNIVERSITY

School of Medicine, Division of Graduate Medical Sciences, Department of Pathology and Laboratory Medicine, Boston, MA 02118

AWARDS Cell and molecular biology (PhD); experimental pathology (PhD); immunology (PhD). Part-time programs available.

Faculty: 12 full-time (4 women), 13 part-time/adjunct (2 women).
Students: 33 full-time (18 women), 1 part-time; includes 4 minority (2 African Americans, 2 Asian Americans or Pacific Islanders), 14 international. Average age 29. In 1999, 3 degrees awarded.
Degree requirements: For doctorate, dissertation, qualifying exam required, foreign language not required.
Entrance requirements: For doctorate, GRE General Test, GRE Subject Test, TOEFL. *Application deadline:* For fall admission, 1/15 (priority date); for spring admission, 10/15 (priority date). *Application fee:* $50. Electronic applications accepted.
Expenses: Tuition: Full-time $24,700; part-time $772 per credit. Required fees: $220.

Financial aid: In 1999–00, 10 fellowships, 3 research assistantships were awarded; Federal Work-Study, scholarships, and traineeships also available.
Faculty research: Toxicology, carcinogenesis, endocytosis, cytogenetics.
Leonard Gottlieb, Chairman, 617-638-4500, *Fax:* 617-638-4085.
Application contact: Dr. Adrianne Rogers, Associate Chairman, 617-638-4500, *Fax:* 617-638-4085, *E-mail:* aerogers@bu.edu.

■ BROWN UNIVERSITY

Graduate School, Division of Biology and Medicine, Pathobiology Graduate Program, Providence, RI 02912

AWARDS Biology (PhD); cancer biology (PhD); immunology and infection (PhD); medical science (PhD); pathobiology (Sc M); toxicology and environmental pathology (PhD).

Faculty: 36 full-time (9 women), 1 part-time/adjunct (0 women).
Students: 18 full-time (11 women); includes 4 minority (2 African Americans, 2 Asian Americans or Pacific Islanders), 1 international. Average age 25. *42 applicants, 17% accepted.* In 1999, 2 doctorates awarded (100% entered university

research/teaching). Terminal master's awarded for partial completion of doctoral program.
Degree requirements: For master's, foreign language not required; for doctorate, dissertation, preliminary exam required, foreign language not required. *Average time to degree:* Doctorate–6 years full-time.
Entrance requirements: For master's, GRE General Test, GRE Subject Test; for doctorate, GRE General Test, GRE Subject Test, TOEFL. *Application deadline:* For fall admission, 1/2 (priority date). Applications are processed on a rolling basis. *Application fee:* $60. Electronic applications accepted.
Financial aid: In 1999–00, 11 fellowships with tuition reimbursements, 1 research assistantship with tuition reimbursement, 4 teaching assistantships with tuition reimbursements were awarded; institutionally sponsored loans and traineeships also available. Financial aid application deadline: 1/2.
Faculty research: Environmental pathology, carcinogenesis, immunopathology, signal transduction.
Dr. Nancy Thompson, Director, 401-444-8860, *E-mail:* nancy_thompson@brown.edu.
Application contact: Marilyn May, Program Coordinator, 401-863-2913, *Fax:*

Brown University (continued)
401-863-9008, *E-mail:* marilyn_may@
brown.edu.
**Find an in-depth description at
www.petersons.com/graduate.**

■ **CASE WESTERN RESERVE
UNIVERSITY**

**School of Medicine and School of
Graduate Studies, Graduate Programs
in Medicine, Programs in Molecular
and Cellular Basis of Disease,
Cleveland, OH 44106**

AWARDS Cell biology (MS, PhD); immunology (MS, PhD); pathology (MS, PhD).
Terminal master's awarded for partial completion of doctoral program.

Degree requirements: For master's and
doctorate, thesis/dissertation required,
foreign language not required.
Entrance requirements: For master's and
doctorate, GRE General Test, GRE
Subject Test, TOEFL.
Faculty research: Neurobiology,
molecular biology, cancer biology,
biomaterials, biocompatibility.
**Find an in-depth description at
www.petersons.com/graduate.**

■ **COLORADO STATE
UNIVERSITY**

**College of Veterinary Medicine and
Biomedical Sciences and Graduate
School, Graduate Programs in
Veterinary Medicine and Biomedical
Sciences, Department of Pathology,
Fort Collins, CO 80523-0015**

AWARDS MS, PhD.

Faculty: 23 full-time (6 women), 2 part-
time/adjunct (1 woman).
Students: 17 full-time (9 women); includes
1 minority (Native American), 1
international. Average age 31. *13 applicants,
23% accepted.* In 1999, 1 master's, 1
doctorate awarded (100% entered
university research/teaching).
Degree requirements: For master's,
thesis or alternative required, foreign
language not required; for doctorate, dissertation required, foreign language not
required.
Entrance requirements: For master's and
doctorate, GRE General Test, TOEFL,
minimum GPA of 3.0. *Application deadline:*
For fall admission, 1/1. Applications are
processed on a rolling basis. *Application fee:*
$30. Electronic applications accepted.
Expenses: Tuition, state resident: full-time
$2,694; part-time $150 per credit. Tuition,
nonresident: full-time $10,460; part-time
$581 per credit. Required fees: $32 per
semester.

Financial aid: In 1999–00, 2 fellowships, 1
research assistantship, 1 teaching assistantship were awarded; career-related internships or fieldwork and traineeships also
available.
Faculty research: Veterinary
pathobiology, immunology, parasitology,
viral pathogenesis, toxicology. *Total annual
research expenditures:* $3.1 million.
Anthony A. Frank, Head, 970-491-6144,
Fax: 970-491-0603, *E-mail:* tfrank@
lamar.colostate.edu.
Application contact: Esta Moutoux,
Administrative Assistant, 970-491-6239,
Fax: 970-491-0603, *E-mail:* emoutoux@
cvmbs.colostate.edu.

■ **COLUMBIA UNIVERSITY**

**College of Physicians and Surgeons
and Graduate School of Arts and
Sciences, Graduate School of Arts
and Sciences at the College of
Physicians and Surgeons, Department
of Pathology, New York, NY 10032**

AWARDS Pathobiology (M Phil, MA, PhD).
Only candidates for the PhD are admitted.
Terminal master's awarded for partial comple-
tion of doctoral program.

Degree requirements: For master's,
foreign language and thesis not required;
for doctorate, dissertation required, foreign
language not required.
Entrance requirements: For master's and
doctorate, GRE General Test, TOEFL.
Expenses: Tuition: Full-time $25,072.
Faculty research: Virology, molecular
biology, cell biology, neurobiology,
immunology.

■ **CORNELL UNIVERSITY**

**Graduate School, Graduate Fields of
Veterinary Medicine, Field of
Veterinary Medicine, Ithaca, NY
14853-0001**

AWARDS Anatomy (MS, PhD); cancer biology
(MS, PhD); clinical sciences (MS, PhD); infectious diseases (MS, PhD); pathology (MS,
PhD); pharmacology (MS, PhD); veterinary
physiology (MS, PhD); virology (MS, PhD).

Faculty: 83 full-time.
Students: 25 full-time (13 women);
includes 2 minority (1 African American, 1
Asian American or Pacific Islander), 12
international. *28 applicants, 18% accepted.* In
1999, 1 master's, 12 doctorates awarded.
Degree requirements: For master's and
doctorate, thesis/dissertation required,
foreign language not required.
Entrance requirements: For master's and
doctorate, GRE General Test, TOEFL.
Application deadline: For fall admission,
1/15; for spring admission, 10/1. Applications are processed on a rolling basis.
Application fee: $65. Electronic applications
accepted.

Expenses: Tuition: Full-time $12,400.
Financial aid: In 1999–00, 24 students
received aid, including 7 fellowships with
full tuition reimbursements available, 17
research assistantships with full tuition
reimbursements available; teaching
assistantships with full tuition reimbursements available, institutionally sponsored
loans, scholarships, tuition waivers (full
and partial), and unspecified assistantships
also available. Financial aid applicants
required to submit FAFSA.
Faculty research: Receptors and signal
transduction, viral and bacterial infectious
diseases, tumor metastasis, clinical
sciences/nutritional disease, development/
neurologic disorders.
Application contact: Graduate Field
Assistant, 607-253-3276, *E-mail:*
vetgradpgms@cornell.edu.

■ **DUKE UNIVERSITY**

**Graduate School, Department of
Pathology, Durham, NC 27708-0586**
AWARDS PhD.

Faculty: 24 full-time.
Students: 17 full-time (8 women); includes
7 minority (2 African Americans, 4 Asian
Americans or Pacific Islanders, 1 Hispanic
American), 4 international. *28 applicants,
14% accepted.* In 1999, 1 doctorate
awarded.
Degree requirements: For doctorate, dissertation required, foreign language not
required.
Entrance requirements: For doctorate,
GRE General Test, GRE Subject Test
(recommended). *Application deadline:* For
fall admission, 12/31. *Application fee:* $75.
Expenses: Tuition: Full-time $21,406;
part-time $760 per unit. Required fees:
$3,136; $3,136 per year. One-time fee:
$30. Tuition and fees vary according to
program.
Financial aid: Fellowships, research
assistantships, Federal Work-Study available. Financial aid application deadline:
12/31.
Soman Abraham, Director of Graduate
Studies, 919-684-5343, *Fax:* 919-684-8756,
E-mail: penny007@mc.duke.edu.
**Find an in-depth description at
www.petersons.com/graduate.**

■ **DUKE UNIVERSITY**

**School of Medicine, Pathologists'
Assistant Program, Durham, NC
27708-0586**
AWARDS MHS.

Students: 12 full-time (6 women). Average
age 28. *42 applicants, 14% accepted.* In
1999, 6 degrees awarded.
Degree requirements: For master's,
comprehensive exams required.

Entrance requirements: For master's, GRE. *Application deadline:* For fall admission, 2/28 (priority date). *Application fee:* $35.
Expenses: Tuition: Full-time $12,400. Required fees: $1,166.
Financial aid: In 1999–00, 9 students received aid. Application deadline: 5/1. Dr. James G. Lewis, Director, 919-684-2169, *Fax:* 919-684-3324, *E-mail:* lewis026@mc.duke.edu.

■ EAST CAROLINA UNIVERSITY
School of Medicine, Department of Pathology and Laboratory Medicine, Greenville, NC 27858-4353
AWARDS Interdisciplinary biological sciences (PhD).

Faculty: 10 full-time (1 woman).
Students: 2 full-time (0 women), 1 part-time. Average age 30. *5 applicants, 60% accepted.*
Degree requirements: For doctorate, dissertation required, foreign language not required.
Entrance requirements: For doctorate, GRE General Test, GRE Subject Test, TOEFL, bachelor's degree in biological chemistry or physical science. *Application deadline:* For fall admission, 6/1 (priority date). Applications are processed on a rolling basis. *Application fee:* $40.
Expenses: Tuition, state resident: full-time $1,012. Tuition, nonresident: full-time $8,578. Required fees: $1,006. Full-time tuition and fees vary according to degree level. Part-time tuition and fees vary according to course load.
Financial aid: Fellowships available. Financial aid application deadline: 6/1.
Faculty research: Immunochemistry and allergens, immunological disorders, cell biology of tumors, membrane antigens, microphage biology.
Dr. Peter Kragel, Chairperson, 252-816-2801, *Fax:* 252-816-3616, *E-mail:* pkragel@brody.med.ecu.edu.
Application contact: Dr. Donald Hoffman, Director of Graduate Studies, 252-816-2816, *Fax:* 252-816-3616, *E-mail:* mdhoffman@eastnet.ecu.edu.

■ FINCH UNIVERSITY OF HEALTH SCIENCES/THE CHICAGO MEDICAL SCHOOL
School of Graduate and Postdoctoral Studies, Department of Pathology, North Chicago, IL 60064-3095
AWARDS MS, PhD, MD/MS, MD/PhD. Part-time programs available.

Students: 29 full-time (7 women); includes 9 minority (2 African Americans, 7 Asian Americans or Pacific Islanders). In 1999, 8

master's awarded. Terminal master's awarded for partial completion of doctoral program.
Degree requirements: For master's and doctorate, computer language, thesis/dissertation required, foreign language not required.
Entrance requirements: For master's and doctorate, GRE General Test, TOEFL, TWE. *Application deadline:* For fall admission, 6/1 (priority date). Applications are processed on a rolling basis. *Application fee:* $25.
Expenses: Tuition: Full-time $14,054; part-time $391 per credit hour. Tuition and fees vary according to program.
Financial aid: Fellowships, career-related internships or fieldwork available. Financial aid application deadline: 8/15; financial aid applicants required to submit FAFSA.
Faculty research: Clinical chemistry, clinical microbiology, hepatopathology.
Dr. Arthur S. Schneider, Chairman, 847-578-3260.
Application contact: Dana Frederick, Admissions Officer, 847-578-3209.

■ FINCH UNIVERSITY OF HEALTH SCIENCES/THE CHICAGO MEDICAL SCHOOL
School of Related Health Sciences, Department of Clinical Laboratory Sciences, North Chicago, IL 60064-3095
AWARDS Clinical laboratory science (MS); pathologist assistant (MS). Part-time programs available.

Faculty: 3 full-time (all women), 1 (woman) part-time/adjunct.
Students: Average age 32. *4 applicants, 100% accepted.* In 1999, 2 degrees awarded (50% found work related to degree, 50% continued full-time study).
Degree requirements: For master's, computer language, thesis required, foreign language not required. *Average time to degree:* Master's–2 years part-time.
Entrance requirements: For master's, TOEFL, minimum GPA of 2.8 in science, 2.5 overall. *Application deadline:* For fall admission, 8/15 (priority date); for spring admission, 1/25. Applications are processed on a rolling basis. *Application fee:* $25.
Expenses: Tuition: Full-time $14,054; part-time $391 per credit hour. Tuition and fees vary according to program.
Financial aid: Institutionally sponsored loans, scholarships, and tuition waivers (partial) available. Aid available to part-time students. Financial aid application deadline: 2/1.
Faculty research: Clinical microbiology, hematology, and chemistry.

Janet M. DeRobertis, Chair, 847-578-3303, *Fax:* 847-578-8651, *E-mail:* deroberj@finchcms.edu.

■ GEORGETOWN UNIVERSITY
Graduate School of Arts and Sciences, Programs in Biomedical Sciences, Department of Pathology, Washington, DC 20057
AWARDS MS, PhD, MD/PhD, MS/PhD.

Degree requirements: For master's, thesis required, foreign language not required; for doctorate, dissertation, comprehensive exam required, foreign language not required.
Entrance requirements: For master's and doctorate, GRE General Test, TOEFL.
Faculty research: Virus-induced diabetes, viral oncology, renal pathophysiology.

■ HARVARD UNIVERSITY
Graduate School of Arts and Sciences, Program in Biological and Biomedical Sciences, Department of Pathology, Boston, MA 02115
AWARDS Experimental pathology (PhD). Applications through the Program in Biological and Biomedical Sciences (BBS).

Degree requirements: For doctorate, dissertation, qualifying exam required, foreign language not required.
Entrance requirements: For doctorate, GRE General Test, GRE Subject Test, TOEFL. *Application fee:* $60.
Expenses: Tuition: Full-time $22,054. Required fees: $711. Tuition and fees vary according to program.
Financial aid: Fellowships, research assistantships, teaching assistantships, institutionally sponsored loans and tuition waivers (full) available. Financial aid application deadline: 1/1.
Faculty research: Immunology, virology, cell-surface ultrastructure.
Dr. Peter M. Howley, Chair, 617-432-2884.
Application contact: Leah Simons, Manager of Student Affairs, 617-432-0162.
Find an in-depth description at www.petersons.com/graduate.

■ INDIANA UNIVERSITY–PURDUE UNIVERSITY INDIANAPOLIS
School of Medicine, Graduate Programs in Medicine, Department of Pathology and Laboratory Medicine, Indianapolis, IN 46202-2896
AWARDS MS, PhD, MD/MS, MD/PhD.

Students: 2 full-time (1 woman), 5 part-time (3 women); includes 2 minority (both Asian Americans or Pacific Islanders), 2 international. Average age 35. In 1999, 2 master's, 3 doctorates awarded.

Indiana University–Purdue University Indianapolis (continued)

Degree requirements: For master's and doctorate, thesis/dissertation required, foreign language not required. *Average time to degree:* Master's–2.5 years full-time, 5 years part-time; doctorate–3.5 years full-time, 7 years part-time.

Entrance requirements: For master's, GRE General Test, TOEFL; for doctorate, GRE General Tes, TOEFL. *Application deadline:* For fall admission, 1/15 (priority date). Applications are processed on a rolling basis. *Application fee:* $35 ($55 for international students).

Expenses: Tuition, state resident: full-time $13,245; part-time $158 per credit hour. Tuition, nonresident: full-time $30,330; part-time $455 per credit hour. Required fees: $121 per year. Tuition and fees vary according to course load and degree level.

Financial aid: In 1999–00, 5 research assistantships with full tuition reimbursements (averaging $12,000 per year), 3 teaching assistantships with full tuition reimbursements (averaging $10,000 per year) were awarded; fellowships, institutionally sponsored loans also available. Financial aid application deadline: 2/1.

Faculty research: Intestinal microecology and anaerobes, molecular pathogenesis of infectious diseases, AIDS pneumocystis, sports medicine toxicology, neuropathology of aging.
Dr. James W. Smith, Chairman, 317-274-4806.

Application contact: Dr. Diane S. Leland, Graduate Adviser, 317-274-0148.

■ **IOWA STATE UNIVERSITY OF SCIENCE AND TECHNOLOGY**

College of Veterinary Medicine and Graduate College, Graduate Programs in Veterinary Medicine, Department of Veterinary Pathology, Ames, IA 50011
AWARDS MS, PhD.

Faculty: 12 full-time.
Students: 15 full-time (5 women), 8 part-time (4 women), 5 international. 7 *applicants, 86% accepted.* In 1999, 1 master's, 1 doctorate awarded.
Degree requirements: For master's, thesis or alternative required; for doctorate, dissertation required.
Entrance requirements: For master's and doctorate, GRE (international applicants), TOEFL. *Application deadline:* For fall admission, 6/1 (priority date); for spring admission, 11/1 (priority date). Applications are processed on a rolling basis. *Application fee:* $20 ($50 for international students). Electronic applications accepted.
Expenses: Tuition, state resident: full-time $3,308. Tuition, nonresident: full-time

$9,744. Tuition and fees vary according to course load and program.
Financial aid: In 1999–00, 9 research assistantships with partial tuition reimbursements (averaging $12,087 per year) were awarded; scholarships also available.
Faculty research: Veterinary clinical pathology, veterinary parasitology, veterinary toxicology.
Dr. Norman F. Cheville, Chair, 515-294-3282, *E-mail:* nchevill@iastate.edu.

■ **JOHNS HOPKINS UNIVERSITY**

School of Medicine, Graduate Programs in Medicine, Department of Pathology, Baltimore, MD 21218-2699
AWARDS Pathobiology (PhD).

Faculty: 22 full-time (5 women).
Students: 4 full-time (2 women); includes 1 minority (Asian American or Pacific Islander). 8 applicants, 25% accepted.
Degree requirements: For doctorate, dissertation required.
Entrance requirements: For doctorate, GRE General Test, GRE Subject Test (biology or chemistry), previous course work with laboratory in organic and inorganic chemistry, general biology, calculus; interview. *Application deadline:* For winter admission, 1/10. *Application fee:* $50.
Expenses: Tuition: Full-time $23,660.
Financial aid: Grants available.
Faculty research: Role of mutant proteins in Alzheimer's disease, nuclear protein function in breast and prostate cancer, medically important fungi, glycoproteins in HIV pathogenesis.
Dr. Fred Sanfilippo, Chair.
Application contact: Terry Aman, Educational Programs Supervisor, 410-955-3439, *Fax:* 410-614-9011, *E-mail:* taman@jhmi.edu.

Find an in-depth description at www.petersons.com/graduate.

■ **LOUISIANA STATE UNIVERSITY HEALTH SCIENCES CENTER**

School of Graduate Studies in New Orleans, Department of Pathology, New Orleans, LA 70112-2223
AWARDS MS, PhD, MD/PhD. Part-time programs available.

Faculty: 28 full-time (3 women).
Students: 8 full-time (5 women), 5 part-time (1 woman); includes 2 minority (1 African American, 1 Asian American or Pacific Islander). Average age 27. *13 applicants, 31% accepted.* In 1999, 1 master's awarded (100% entered university research/teaching); 2 doctorates awarded. Terminal master's awarded for partial completion of doctoral program.

Degree requirements: For master's and doctorate, thesis/dissertation required, foreign language not required. *Average time to degree:* Master's–2.5 years full-time.
Entrance requirements: For master's and doctorate, GRE General Test, TOEFL. *Application deadline:* For fall admission, 2/28 (priority date); for spring admission, 9/30. Applications are processed on a rolling basis. *Application fee:* $30.
Expenses: Tuition, state resident: full-time $2,878; part-time $126 per hour. Tuition, nonresident: full-time $6,003; part-time $265 per hour. Required fees: $2,272. Tuition and fees vary according to course load, degree level and program.
Financial aid: In 1999–00, 4 students received aid, including 4 research assistantships (averaging $13,000 per year); tuition waivers (full) also available. Financial aid application deadline: 2/15.
Faculty research: Immunohematology, hematology, experimental and epidemiological studies in atherosclerosis, epidemiology of cancer.
Dr. Jack Strong, Head, 504-582-5487.
Application contact: Dr. Peter Lehmann, Graduate Coordinator, 504-568-6057, *Fax:* 504-568-6037, *E-mail:* hlehma@lsumc.edu.

■ **MEDICAL COLLEGE OF OHIO**

Graduate School, Department of Pathology, Toledo, OH 43614-5805
AWARDS MS. Part-time programs available.

Faculty: 14 full-time (3 women), 3 part-time/adjunct (0 women).
Students: 5 full-time (2 women), 4 part-time (2 women), 5 international. Average age 35. In 1999, 1 degree awarded (100% continued full-time study).
Degree requirements: For master's, thesis, qualifying exam required, foreign language not required. *Average time to degree:* Master's–3 years full-time.
Entrance requirements: For master's, GRE General Test, minimum undergraduate GPA of 3.0. *Application fee:* $30.
Expenses: Tuition, state resident: part-time $193 per hour. Tuition, nonresident: part-time $445 per hour. Tuition and fees vary according to degree level.
Financial aid: Federal Work-Study and institutionally sponsored loans available. Financial aid applicants required to submit FAFSA.
Faculty research: Cell injury, molecular and clinical molecular carcinogenesis, chemoprevention, hypertension. *Total annual research expenditures:* $1.7 million.
Herman Schot, Interim Chair, 419-383-4117, *Fax:* 419-383-6140, *E-mail:* mcogradschool@mco.edu.
Application contact: Joann Braatz, Clerk, 419-383-4117, *Fax:* 419-383-6140, *E-mail:* mcogradschool@mco.edu.

■ MEDICAL COLLEGE OF WISCONSIN

Graduate School of Biomedical Sciences, Department of Pathology, Milwaukee, WI 53226-0509

AWARDS MS, PhD, MD/MS, MD/PhD. Terminal master's awarded for partial completion of doctoral program.

Degree requirements: For master's and doctorate, thesis/dissertation required, foreign language not required.
Entrance requirements: For master's, GRE General Test, TOEFL, minimum GPA of 3.0; previous course work in biochemistry, biology, calculus, chemistry, and physics; for doctorate, GRE General Test, TOEFL.
Expenses: Tuition, state resident: full-time $9,318. Tuition, nonresident: full-time $9,318. Required fees: $115.
Faculty research: Virology, immunology, metabolism, molecular pathology.

■ MEDICAL UNIVERSITY OF SOUTH CAROLINA

College of Graduate Studies, Department of Pathology and Laboratory Medicine, Charleston, SC 29425-0002

AWARDS MS, PhD, DMD/PhD, MD/PhD.
Faculty: 13 part-time/adjunct (3 women).
Students: 8 full-time (3 women), 1 international. Average age 30. *14 applicants, 21% accepted.* In 1999, 3 doctorates awarded. Terminal master's awarded for partial completion of doctoral program.
Degree requirements: For master's, thesis, research seminar required; for doctorate, dissertation, teaching and research seminar, oral and written exams required, foreign language not required.
Entrance requirements: For master's and doctorate, GRE General Test, TOEFL, interview, minimum GPA of 3.2. *Application deadline:* Applications are processed on a rolling basis. *Application fee:* $55. Electronic applications accepted.
Expenses: Tuition, state resident: full-time $3,470; part-time $160 per semester hour. Tuition, nonresident: full-time $4,426; part-time $213 per semester hour. Required fees: $408 per semester. One-time fee: $160. Tuition and fees vary according to program.
Financial aid: In 1999–00, 3 fellowships (averaging $16,000 per year) were awarded; research assistantships, teaching assistantships, Federal Work-Study and tuition waivers (partial) also available. Financial aid application deadline: 4/1; financial aid applicants required to submit FAFSA.
Faculty research: Epithelial cell tissue culture, nephrotoxicity, diabetes, ion transport mediators, histochemistry and cytochemistry. *Total annual research expenditures:* $556,881.
Dr. Janice M. Lage, Acting Chairman, 843-792-3121.

Application contact: Julie Johnston, Director of Admissions, 843-792-8710, *Fax:* 843-792-3764.

■ MICHIGAN STATE UNIVERSITY

College of Human Medicine and Graduate School, Graduate Programs in Human Medicine and Graduate Programs in Osteopathic Medicine, Division of Human Pathology, East Lansing, MI 48824

AWARDS MS, PhD. Part-time programs available.
Faculty: 4.
Students: Average age 33. Terminal master's awarded for partial completion of doctoral program.
Degree requirements: For master's and doctorate, thesis optional.
Entrance requirements: For master's and doctorate, GRE General Test, minimum GPA of 3.0. *Application fee:* $30 ($40 for international students).
Expenses: Tuition, state resident: full-time $10,868; part-time $229 per credit. Tuition, nonresident: full-time $23,168; part-time $464 per credit.
Financial aid: Fellowships, research assistantships, teaching assistantships, institutionally sponsored loans and tuition waivers (full) available. Aid available to part-time students. Financial aid application deadline: 4/1; financial aid applicants required to submit FAFSA.
Dr. David Rovner, Interim Director, 517-353-9160.

■ MICHIGAN STATE UNIVERSITY

College of Osteopathic Medicine and Graduate School, Graduate Programs in Osteopathic Medicine, East Lansing, MI 48824

AWARDS Anatomy (MS, PhD); biochemistry (MS, PhD); microbiology (PhD); pathology (MS, PhD); pharmacology/toxicology (MS, PhD), including pharmacology; physiology (MS, PhD), including environmental toxicology (PhD), neuroscience (PhD). Part-time programs available.
Students: 17 full-time (9 women), 1 part-time; includes 5 minority (3 Asian Americans or Pacific Islanders, 1 Hispanic American, 1 Native American), 5 international. Average age 26. *33 applicants, 9% accepted.* In 1999, 2 doctorates awarded.
Degree requirements: For doctorate, dissertation required.
Entrance requirements: For master's and doctorate, GRE. *Application deadline:* For fall admission, 3/1 (priority date). Applications are processed on a rolling basis. *Application fee:* $30 ($40 for international students).
Expenses: Tuition, area resident: Full-time $15,879. Tuition, state resident: full-time $33,797.
Financial aid: Fellowships, research assistantships, teaching assistantships, career-related internships or fieldwork, Federal Work-Study, and institutionally sponsored loans available. Financial aid application deadline: 4/2. *Total annual research expenditures:* $5 million.
Application contact: Dr. Veronica M. Maher, Associate Dean for Graduate Studies, 517-353-7785, *Fax:* 517-353-9004, *E-mail:* maher@com.msu.edu.

■ MICHIGAN STATE UNIVERSITY

College of Veterinary Medicine and Graduate School, Graduate Programs in Veterinary Medicine, East Lansing, MI 48824

AWARDS Anatomy (MS, PhD); large animal clinical sciences (MS, PhD); microbiology (MS, PhD); pathology (MS, PhD); pharmacology/toxicology (MS, PhD), including pharmacology; small animal clinical sciences (MS).
Students: 52 full-time (28 women); includes 6 minority (4 African Americans, 2 Asian Americans or Pacific Islanders), 18 international. Average age 32. *45 applicants, 20% accepted.* In 1999, 3 master's, 3 doctorates awarded.
Degree requirements: For master's, thesis or alternative required, foreign language not required; for doctorate, dissertation required, foreign language not required.
Application deadline: Applications are processed on a rolling basis. *Application fee:* $30 ($40 for international students). Electronic applications accepted.
Expenses: Tuition, state resident: full-time $9,766. Tuition, nonresident: full-time $20,082. Tuition and fees vary according to program.
Financial aid: Fellowships, research assistantships available.
Faculty research: Molecular genetics, food safety/toxicology, comparative orthopedics, airway disease, population medicine.
Dr. John C. Baker, Associate Dean for Research and Graduate Studies, 517-432-2388, *Fax:* 517-432-1037, *E-mail:* baker@cvm.msu.edu.
Application contact: Victoria Hoelzer-Maddox, Administrative Assistant, 517-353-3118, *Fax:* 517-432-1037, *E-mail:* hoelzer-maddox@cvm.msu.edu.

■ MOUNT SINAI SCHOOL OF MEDICINE OF NEW YORK UNIVERSITY

Graduate School of Biological Sciences, Department of Pathology, New York, NY 10029-6504

AWARDS Molecular, cellular, and environmental pathology (PhD).

Degree requirements: For doctorate, dissertation required, foreign language not required.

Entrance requirements: For doctorate, GRE General Test, GRE Subject Test, TOEFL. *Application deadline:* For fall admission, 5/15. *Application fee:* $35.

Expenses: Tuition: Full-time $21,750; part-time $725 per credit. Required fees: $750; $25 per credit. Full-time tuition and fees vary according to student level.

Financial aid: Fellowships, grants available.

Dr. Alan L. Schiller, Chair, 212-241-6546.

Application contact: C. Gita Bosch, Administrative Manager and Assistant Dean, 212-241-6546, *Fax:* 212-241-0651, *E-mail:* grads@mssm.edu.

Find an in-depth description at www.petersons.com/graduate.

■ MOUNT SINAI SCHOOL OF MEDICINE OF NEW YORK UNIVERSITY

Graduate School of Biological Sciences, Mechanisms of Disease and Therapy Training Area, New York, NY 10029-6504

AWARDS PhD, MD/PhD.

Students: 39 full-time (15 women).

Degree requirements: For doctorate, dissertation required, foreign language not required.

Entrance requirements: For doctorate, GRE General Test, GRE Subject Test, MCAT, TOEFL. *Application deadline:* For fall admission, 4/15. *Application fee:* $60.

Expenses: Tuition: Full-time $21,750; part-time $725 per credit. Required fees: $750; $25 per credit. Full-time tuition and fees vary according to student level.

Financial aid: Fellowships with full tuition reimbursements, grants available.

Dr. Ravi Iyengar, Program Director, *E-mail:* iyengar@msvax.mssm.edu.

■ NEW YORK MEDICAL COLLEGE

Graduate School of Basic Medical Sciences, Program in Experimental Pathology, Valhalla, NY 10595-1691

AWARDS MS, PhD, MD/PhD. Part-time and evening/weekend programs available.

Faculty: 9 full-time (4 women).

Students: 9 full-time (5 women), 11 part-time (5 women); includes 2 minority (both African Americans), 5 international. Average age 30. *13 applicants, 31% accepted.* In 1999, 6 master's, 2 doctorates awarded. Terminal master's awarded for partial completion of doctoral program.

Degree requirements: For master's and doctorate, computer language, thesis/dissertation required, foreign language not required.

Entrance requirements: For master's, GRE General Test, TOEFL; for doctorate, GRE General Test, GRE Subject Test, TOEFL. *Application deadline:* For fall admission, 7/1 (priority date); for spring admission, 12/1 (priority date). Applications are processed on a rolling basis. *Application fee:* $35 ($60 for international students).

Expenses: Tuition: Part-time $430 per credit. Required fees: $15 per semester. One-time fee: $100.

Financial aid: In 1999–00, 7 research assistantships with full tuition reimbursements were awarded; career-related internships or fieldwork, Federal Work-Study, grants, institutionally sponsored loans, and tuition waivers (full) also available. Aid available to part-time students. Financial aid applicants required to submit FAFSA.

Faculty research: Atherogenesis and endothelial cell biology, immunology and inflammation, tumor biology and metastasis, mechanisms of chemical carcinogens, mechanisms of free radial tissue damage.

Dr. Henry P. Godfrey, Co-Director, 914-594-4160.

Find an in-depth description at www.petersons.com/graduate.

■ NEW YORK UNIVERSITY

Graduate School of Arts and Science, Department of Basic Medical Sciences, New York, NY 10012-1019

AWARDS Biochemistry (MS, PhD); cell biology (MS, PhD); microbiology (MS, PhD); parasitology (PhD); pathology (MS, PhD); pharmacology (PhD); physiology (MS, PhD). Part-time programs available.

Faculty: 23 full-time (2 women), 3 part-time/adjunct (1 woman).

Students: 182 full-time (73 women), 4 part-time (1 woman); includes 52 minority (11 African Americans, 34 Asian Americans or Pacific Islanders, 7 Hispanic Americans), 48 international. Average age 26. *671 applicants, 11% accepted.* In 1999, 15 master's, 32 doctorates awarded. Terminal master's awarded for partial completion of doctoral program.

Degree requirements: For master's, thesis or alternative, written comprehensive exam required, foreign language not required; for doctorate, one foreign language, dissertation, oral and written comprehensive exams required.

Entrance requirements: For master's and doctorate, GRE General Test, GRE Subject Test, TOEFL. *Application deadline:* For fall admission, 2/1 (priority date). *Application fee:* $60.

Expenses: Tuition: Full-time $17,880; part-time $745 per credit. Required fees: $1,140; $35 per credit. Tuition and fees vary according to course load and program.

Financial aid: Fellowships with tuition reimbursements, research assistantships with tuition reimbursements, teaching assistantships with tuition reimbursements, career-related internships or fieldwork, Federal Work-Study, institutionally sponsored loans, and tuition waivers (full and partial) available. Financial aid application deadline: 2/1; financial aid applicants required to submit FAFSA.

Dr. Joel D. Oppenheim, Director, 212-263-5648, *Fax:* 212-263-7600, *E-mail:* sackler-info@nyumed.med.nyu.edu.

■ NEW YORK UNIVERSITY

School of Medicine and Graduate School of Arts and Science, Medical Scientist Training Program, New York, NY 10012-1019

AWARDS Biochemistry (MD/PhD); cell biology (MD/PhD); environmental health sciences (MD/PhD); microbiology (MD/PhD); parasitology (MD/PhD); pathology (MD/PhD); pharmacology (MD/PhD). Students must be accepted by both the School of Medicine and the Graduate School of Arts and Science.

Faculty: 150 full-time (35 women).

Students: 80 full-time (18 women); includes 37 minority (3 African Americans, 32 Asian Americans or Pacific Islanders, 2 Hispanic Americans), 1 international. Average age 25. *195 applicants, 18% accepted.*

Degree requirements: One foreign language.

Application deadline: For fall admission, 11/15. Applications are processed on a rolling basis. *Application fee:* $60.

Expenses: Students receive full tuition support and an annual stipend of $17,500.

Financial aid: Application deadline: 5/1.

Faculty research: Genetics, tumor biology, cardiovascular biology, neuroscience, host defense mechanisms.

Dr. James Salzer, Director, 212-263-0758.

Application contact: Arlene Kohler, Administrative Officer, 212-263-5649.

■ NEW YORK UNIVERSITY

School of Medicine and Graduate School of Arts and Science, Sackler Institute of Graduate Biomedical Sciences, Department of Pathology, New York, NY 10012-1019

AWARDS PhD, MD/PhD.

Faculty: 29 full-time (7 women).
Students: 28 full-time (12 women); includes 12 minority (1 African American, 8 Asian Americans or Pacific Islanders, 3 Hispanic Americans), 2 international. Average age 25. In 1999, 6 degrees awarded (100% entered university research/teaching).
Degree requirements: For doctorate, one foreign language, dissertation, qualifying exam required. *Average time to degree:* Doctorate–5.5 years full-time.
Entrance requirements: For doctorate, GRE General Test, GRE Subject Test, TOEFL. *Application deadline:* For fall admission, 2/1 (priority date). Applications are processed on a rolling basis. *Application fee:* $60.
Expenses: Tuition: Full-time $17,880; part-time $745 per credit. Required fees: $1,140; $35 per credit. Tuition and fees vary according to course load and program.
Financial aid: Fellowships, research assistantships, teaching assistantships, tuition waivers (full) available. Financial aid application deadline: 1/15.
Faculty research: Tumor suppressors, oncogenes, leukemia, cell differentiation, molecular biology of the immunoglobulin genes. *Total annual research expenditures:* $5.2 million.
Dr. Vittorio Defendi, Chairman, 212-263-5927, *Fax:* 212-263-8211.
Application contact: Dr. Robert B. Carroll, Graduate Adviser, 212-263-5347, *Fax:* 212-263-8211, *E-mail:* carror01@mcrcr.med.nyu.edu.

Find an in-depth description at www.petersons.com/graduate.

■ NORTH CAROLINA STATE UNIVERSITY

College of Veterinary Medicine and Graduate School, Graduate Programs in Comparative Biomedical Sciences, Raleigh, NC 27695

AWARDS Cell biology and morphology (MS, PhD); epidemiology and population medicine (MS, PhD); immunology (MS, PhD); microbiology and immunology (MS, PhD); pathology (MS, PhD); pharmacology (MS, PhD); specialized veterinary medicine (MS). Part-time programs available.

Students: 40 full-time (23 women), 22 part-time (12 women); includes 11 minority (8 African Americans, 2 Asian

Americans or Pacific Islanders, 1 Hispanic American), 14 international. Average age 34. *33 applicants, 33% accepted.* In 1999, 2 master's, 2 doctorates awarded.
Degree requirements: For master's and doctorate, thesis/dissertation required.
Entrance requirements: For master's and doctorate, GRE General Test. *Application deadline:* For fall admission, 6/25; for spring admission, 11/25. Applications are processed on a rolling basis. *Application fee:* $45.
Expenses: Tuition, state resident: full-time $1,578. Tuition, nonresident: full-time $10,744. Required fees: $892. Full-time tuition and fees vary according to program.
Financial aid: Fellowships, research assistantships, teaching assistantships available. Financial aid application deadline: 2/15.
Faculty research: Infectious diseases, immunology and virology, tumor biology, toxicological pathology, food safety.
Dr. Neil C. Olson, Associate Dean, 919-513-6213, *Fax:* 919-513-6222, *E-mail:* neil_olson@ncsu.edu.

■ THE OHIO STATE UNIVERSITY

College of Medicine and Public Health and Graduate School, Graduate Programs in the Basic Medical Sciences, Department of Pathology, Columbus, OH 43210

AWARDS Experimental pathobiology (MS, PhD); pathology assistant (MS).

Faculty: 55 full-time (12 women).
Students: 24 full-time (15 women); includes 4 minority (all Asian Americans or Pacific Islanders). Average age 24. *32 applicants, 19% accepted.* In 1999, 7 master's, 4 doctorates awarded. Terminal master's awarded for partial completion of doctoral program.
Degree requirements: For master's and doctorate, thesis/dissertation required, foreign language not required. *Average time to degree:* Master's–7 years full-time; doctorate–4 years full-time.
Entrance requirements: For master's and doctorate, GRE General Test. *Application deadline:* For fall admission, 8/1 (priority date). Applications are processed on a rolling basis. *Application fee:* $40.
Expenses: Tuition, state resident: full-time $5,400. Tuition, nonresident: full-time $14,535. Part-time tuition and fees vary according to course load.
Financial aid: In 1999–00, 18 students received aid, including 8 fellowships with full tuition reimbursements available (averaging $13,500 per year), 10 research assistantships with full tuition reimbursements available (averaging $13,500 per year); Federal Work-Study and institutionally sponsored loans also available.

Faculty research: Clinical pathology, transplantation pathology, cancer research, neuropathology, vascular pathology. *Total annual research expenditures:* $4.2 million.
Dr. Daniel Sedmak, Interim Dean and Vice President for Health Sciences, 614-292-4330, *Fax:* 614-292-7072, *E-mail:* lankford-l@medetr.osu.edu.
Application contact: Dr. Allan Yates, Graduate Studies Chair, 614-292-8881, *Fax:* 614-292-5849, *E-mail:* ayates@magnus.acs.ohio_state.edu.

Find an in-depth description at www.petersons.com/graduate.

■ OREGON STATE UNIVERSITY

College of Veterinary Medicine, Program in Veterinary Sciences, Corvallis, OR 97331

AWARDS Microbiology (MS); pathology (MS); toxicology (MS). Part-time programs available.

Faculty: 4 full-time (0 women).
Students: 5 full-time (4 women). Average age 34. In 1999, 1 degree awarded (100% found work related to degree).
Degree requirements: For master's, thesis required, foreign language not required.
Entrance requirements: For master's, TOEFL, minimum GPA of 3.0 in last 90 hours. *Application deadline:* For fall admission, 11/1. *Application fee:* $50.
Expenses: Tuition, state resident: full-time $4,334. Tuition, nonresident: full-time $7,382.
Financial aid: Research assistantships, Federal Work-Study, institutionally sponsored loans, and scholarships available. Aid available to part-time students. Financial aid application deadline: 2/1.
Faculty research: Calf diseases, bovine foot rot, caliciviruses, effects of toxic agents on immune systems.
Linda L. Blythe, Associate Dean, 541-737-2098, *Fax:* 541-737-4245, *E-mail:* linda.blythe@orst.edu.

■ PURDUE UNIVERSITY

School of Veterinary Medicine and Graduate School, Graduate Programs in Veterinary Medicine, Department of Veterinary Pathobiology, West Lafayette, IN 47907

AWARDS Bacteriology (MS, PhD); epidemiology (MS, PhD); immunology (MS, PhD); infectious diseases (MS, PhD); microbiology (MS, PhD); parasitology (MS, PhD); pathology (MS, PhD); toxicology (MS, PhD); virology (MS, PhD).

Faculty: 26 full-time (5 women).
Students: 42 full-time (24 women), 2 part-time (1 woman); includes 5 minority (1 African American, 3 Asian Americans or

Purdue University (continued)
Pacific Islanders, 1 Hispanic American), 18 international. Average age 32. In 1999, 2 master's, 6 doctorates awarded (100% found work related to degree). Terminal master's awarded for partial completion of doctoral program.

Degree requirements: For master's, thesis required (for some programs), foreign language not required; for doctorate, dissertation required, foreign language not required.

Entrance requirements: For master's and doctorate, GRE General Test, TOEFL. *Application fee:* $30.

Expenses: Tuition, state resident: full-time $3,732. Tuition, nonresident: full-time $8,732.

Financial aid: Fellowships, research assistantships, teaching assistantships available. Financial aid application deadline: 3/1; financial aid applicants required to submit FAFSA.

Dr. H. L. Thacker, Head, 765-494-7543.

■ QUINNIPIAC UNIVERSITY

School of Health Sciences, Program for Pathologists' Assistant, Hamden, CT 06518-1940

AWARDS MHS.

Faculty: 4 full-time (0 women), 1 part-time/adjunct (0 women).
Students: 32 full-time (20 women); includes 1 minority (African American). Average age 26. *59 applicants, 29% accepted.* In 1999, 15 degrees awarded (100% found work related to degree).
Degree requirements: For master's, residency required, thesis optional. *Average time to degree:* Master's–2 years full-time.
Entrance requirements: For master's, interview, BS in biomedical science, minimum GPA of 2.8. *Application deadline:* For fall admission, 1/15. Applications are processed on a rolling basis. *Application fee:* $45. Electronic applications accepted.
Expenses: Tuition: Part-time $410 per credit hour. Required fees: $20 per term. Tuition and fees vary according to program.
Financial aid: Career-related internships or fieldwork available. Financial aid applicants required to submit FAFSA.
Dr. Kenneth Kaloustian, Director, 203-582-8676, *Fax:* 203-582-3443, *E-mail:* ken.kaloustian@quinnipiac.edu.
Application contact: Scott Farber, Director of Graduate Admissions, 800-462-1944, *Fax:* 203-582-3443, *E-mail:* graduate@quinnipiac.edu.
Find an in-depth description at www.petersons.com/graduate.

■ SAINT LOUIS UNIVERSITY

School of Medicine and Graduate School, Graduate Programs in Biomedical Sciences, Department of Pathology, St. Louis, MO 63103-2097

AWARDS MS(R), PhD.

Faculty: 46 full-time (20 women), 33 part-time/adjunct (8 women).
Students: Average age 27. In 1999, 1 degree awarded. Terminal master's awarded for partial completion of doctoral program.
Degree requirements: For master's, comprehensive written exam, oral defense of thesis required; for doctorate, oral and written preliminary exams, oral defense of dissertation required.
Entrance requirements: For master's and doctorate, GRE General Test, GRE Subject Test. *Application deadline:* For fall admission, 8/20 (priority date). Applications are processed on a rolling basis. *Application fee:* $50.
Expenses: Tuition: Part-time $507 per credit hour. Required fees: $38 per term.
Financial aid: In 1999–00, 2 students received aid, including 1 fellowship, 2 research assistantships; teaching assistantships, institutionally sponsored loans and tuition waivers (full) also available. Aid available to part-time students. Financial aid application deadline: 8/1; financial aid applicants required to submit FAFSA.
Faculty research: Gap junctions, mast cells, coagulation, pulmonary development, extracellular matrix, natural killer cell, oncogene in multiple myeloma.
Dr. Jacki Kornblugh, Chairperson, 314-268-5445.
Find an in-depth description at www.petersons.com/graduate.

■ STATE UNIVERSITY OF NEW YORK AT ALBANY

School of Public Health, Department of Biomedical Sciences, Program in Molecular Pathogenesis, Albany, NY 12222-0001

AWARDS MS, PhD.

Degree requirements: For master's and doctorate, thesis/dissertation required.
Entrance requirements: For master's and doctorate, GRE General Test, GRE Subject Test. *Application deadline:* For fall admission, 1/15 (priority date); for spring admission, 11/1 (priority date). *Application fee:* $50.
Expenses: Tuition, state resident: full-time $5,100; part-time $214 per credit. Tuition, nonresident: full-time $8,416; part-time $352 per credit. Required fees: $31 per credit.
Financial aid: Application deadline: 2/1.

Dr. Harry Taber, Chair, Department of Biomedical Sciences, 518-474-2662.

■ STATE UNIVERSITY OF NEW YORK AT BUFFALO

Graduate School, Graduate Programs in Biomedical Sciences at Roswell Park Cancer Institute, Department of Experimental Pathology at Roswell Park Cancer Institute, Buffalo, NY 14263

AWARDS PhD.

Faculty: 15 full-time (6 women).
Students: 3 full-time (1 woman), 3 part-time (2 women); includes 1 minority (Asian American or Pacific Islander). Average age 25. *28 applicants, 36% accepted.*
Degree requirements: For doctorate, dissertation, comprehensive exam, dissertation defense, project required.
Entrance requirements: For doctorate, GRE General Test, GRE Subject Test (biochemistry, physics), TOEFL, TSE, TWE. *Application deadline:* For fall admission, 2/1 (priority date). Applications are processed on a rolling basis. *Application fee:* $35. Electronic applications accepted.
Expenses: Tuition, state resident: full-time $5,100; part-time $213 per credit hour. Tuition, nonresident: full-time $8,416; part-time $351 per credit hour. Required fees: $935; $75 per semester. Tuition and fees vary according to course load and program.
Financial aid: In 1999–00, 4 fellowships with full tuition reimbursements (averaging $15,000 per year), 5 research assistantships with full tuition reimbursements (averaging $15,000 per year) were awarded; Federal Work-Study also available. Financial aid application deadline: 2/1; financial aid applicants required to submit FAFSA.
Faculty research: Immunohistochemistry, tumor biology, molecular diagnostics, hematopathology, free radicals, immunodiagnosis and experimental immunotherapy of cancer. *Total annual research expenditures:* $1.8 million.
Dr. Bonnie Asch, Chairman, 716-845-3504, *Fax:* 716-845-8806, *E-mail:* basch@sc3101.cc.buffalo.edu.
Application contact: Craig R. Johnson, Director of Graduate Studies, 716-845-2339, *Fax:* 716-845-8178, *E-mail:* rpgradapp@sc3103.med.buffalo.edu.

■ STATE UNIVERSITY OF NEW YORK AT BUFFALO

Graduate School, School of Medicine and Biomedical Sciences, Graduate Programs in Medicine and Biomedical Sciences, Department of Pathology, Buffalo, NY 14260

AWARDS MA, PhD.

Faculty: 39 full-time (16 women), 2 part-time/adjunct (0 women).
Students: 3 full-time (1 woman), 3 part-time (2 women). Average age 25. *5 applicants, 20% accepted.* In 1999, 2 master's, 1 doctorate awarded.
Degree requirements: For master's, thesis required; for doctorate, dissertation, departmental qualifying exam required, foreign language not required.
Entrance requirements: For master's and doctorate, GRE General Test, GRE Subject Test, TOEFL. *Application deadline:* For fall admission, 2/1 (priority date). Applications are processed on a rolling basis. *Application fee:* $35. Electronic applications accepted.
Expenses: Tuition, state resident: full-time $5,100. Tuition, nonresident: full-time $8,416. Required fees: $935.
Financial aid: In 1999–00, 5 students received aid, including 4 teaching assistantships with tuition reimbursements available (averaging $15,000 per year); research assistantships, Federal Work-Study, institutionally sponsored loans, and tuition waivers (full and partial) also available. Financial aid application deadline: 2/1.
Faculty research: Immunopathology-immunobiology, experimental hypertension, neuromuscular disease, molecular pathology. *Total annual research expenditures:* $356,240.
Dr. Reid Heffner, Chairman, 716-829-2846, *Fax:* 716-829-2086, *E-mail:* rheffner@acsu.buffalo.edu.
Application contact: Dr. Peter A. Nickerson, Director of Graduate Studies, 716-829-2971, *Fax:* 716-829-2086, *E-mail:* pnickers@acsu.buffalo.edu.

■ STATE UNIVERSITY OF NEW YORK AT STONY BROOK

Graduate School, College of Arts and Sciences, Molecular and Cellular Biology Program, Specialization in Immunology and Pathology, Stony Brook, NY 11794
AWARDS PhD.
Faculty: 23 full-time (6 women).
Students: 13 full-time (10 women), 6 part-time (1 woman); includes 5 minority (3 Asian Americans or Pacific Islanders, 2 Hispanic Americans), 3 international. Average age 27.
Degree requirements: For doctorate, one foreign language (computer language can substitute); dissertation, exam, teaching experience required.
Entrance requirements: For doctorate, GRE General Test, GRE Subject Test, TOEFL. *Application deadline:* For fall admission, 1/15. *Application fee:* $50.
Expenses: Tuition, state resident: full-time $5,100; part-time $213 per credit hour.

Tuition, nonresident: full-time $8,416; part-time $351 per credit hour. Required fees: $492. Tuition and fees vary according to program.
Financial aid: Fellowships, research assistantships, teaching assistantships, career-related internships or fieldwork available. Financial aid application deadline: 3/15.
Faculty research: Environmental pathology of respiratory tract, immunology, bone disease, platelet physiology. *Total annual research expenditures:* $2.2 million.
Dr. Frederick Miller, Chairman, Department of Pathology, 631-444-3000, *Fax:* 631-444-3424.
Application contact: Dr. Nancy Reich, Director, 631-444-7503, *Fax:* 631-444-3424, *E-mail:* nreich@path.som.sunysb.edu.
Find an in-depth description at www.petersons.com/graduate.

■ TEMPLE UNIVERSITY

Health Sciences Center, School of Medicine and Graduate School, Graduate Programs in Medicine, Department of Pathology and Laboratory Medicine, Philadelphia, PA 19140
AWARDS PhD.
Faculty: 9 full-time (1 woman).
Students: 9 full-time (3 women); includes 4 minority (all Asian Americans or Pacific Islanders), 2 international. *10 applicants, 30% accepted.*
Degree requirements: For doctorate, one foreign language (computer language can substitute), dissertation, research seminars required.
Entrance requirements: For doctorate, GRE General Test, TOEFL, minimum GPA of 3.0. *Application deadline:* For fall admission, 4/15 (priority date); for spring admission, 11/1. Applications are processed on a rolling basis. *Application fee:* $40. Electronic applications accepted.
Expenses: Tuition, state resident: full-time $6,030. Tuition, nonresident: full-time $8,298. Required fees: $230. One-time fee: $10 full-time.
Financial aid: Fellowships, research assistantships available.
Faculty research: Molecular cloning, cell proliferation, cell cycle regulation, DNA repair, cytogenetics.
Dr. Henry Simpkins, Chair, 215-707-4353, *Fax:* 215-204-2781.
Application contact: Dr. Peter Wong, Graduate Chair, 215-707-8361, *Fax:* 215-707-2781.
Find an in-depth description at www.petersons.com/graduate.

■ TEXAS A&M UNIVERSITY

College of Veterinary Medicine, Department of Veterinary Pathobiology, College Station, TX 77843
AWARDS Genetics (MS, PhD); toxicology (MS, PhD); veterinary microbiology (MS, PhD); veterinary parasitology (MS); veterinary pathology (MS, PhD). Part-time programs available. Postbaccalaureate distance learning degree programs offered.
Faculty: 44 full-time (10 women), 17 part-time/adjunct (2 women).
Students: 33 full-time (13 women), 30 part-time (18 women); includes 11 minority (2 African Americans, 1 Asian American or Pacific Islander, 8 Hispanic Americans), 11 international. Average age 34. *28 applicants, 46% accepted.* In 1999, 4 master's awarded (25% found work related to degree, 75% continued full-time study); 10 doctorates awarded (40% entered university research/teaching, 60% found other work related to degree). Terminal master's awarded for partial completion of doctoral program.
Degree requirements: For master's and doctorate, thesis/dissertation required, foreign language not required. *Average time to degree:* Master's–2 years full-time, 4.5 years part-time; doctorate–4.5 years full-time, 7 years part-time.
Entrance requirements: For master's and doctorate, GRE General Test, TOEFL. *Application deadline:* For fall admission, 3/1 (priority date); for spring admission, 8/1 (priority date). Applications are processed on a rolling basis. *Application fee:* $50 ($75 for international students). Electronic applications accepted.
Expenses: Tuition, state resident: full-time $5,400. Tuition, nonresident: full-time $16,200. Required fees: $2,936. Full-time tuition and fees vary according to student level.
Financial aid: In 1999–00, 7 fellowships with partial tuition reimbursements (averaging $16,000 per year), 13 research assistantships with partial tuition reimbursements (averaging $14,400 per year), 5 teaching assistantships with partial tuition reimbursements (averaging $16,000 per year) were awarded; career-related internships or fieldwork, Federal Work-Study, and institutionally sponsored loans also available. Aid available to part-time students. Financial aid applicants required to submit FAFSA.
Faculty research: Infectious and noninfectious diseases of animals and birds, animal genetics, molecular biology, immunology, virology. *Total annual research expenditures:* $2.6 million.

Texas A&M University (continued)
Dr. Ann B. Kier, Head, 979-845-5941, *Fax:* 979-845-9231, *E-mail:* akier@ cvm.tamu.edu.

Application contact: Dr. G. G. Wagner, Graduate Adviser, 979-845-5941, *Fax:* 979-862-1147, *E-mail:* gwagner@ cvm.tamus.edu.

■ TEXAS A&M UNIVERSITY SYSTEM HEALTH SCIENCE CENTER

College of Medicine, Graduate School of Biomedical Sciences, Department of Pathology and Laboratory Medicine, College Station, TX 77840-7896

AWARDS Molecular pathology (PhD).

Faculty: 7 full-time (1 woman).
Students: 5 full-time (3 women), 2 international. *14 applicants, 14% accepted.*
Degree requirements: For doctorate, dissertation required, foreign language not required.
Entrance requirements: For doctorate, GRE General Test. *Application deadline:* For fall admission, 2/1 (priority date); for spring admission, 3/1 (priority date). *Application fee:* $50 ($75 for international students). Electronic applications accepted.
Expenses: Tuition, area resident: Full-time $1,368. Tuition, state resident: part-time $76 per credit. Tuition, nonresident: full-time $5,256; part-time $292 per credit. International tuition: $5,256 full-time. Required fees: $678; $38 per credit. Full-time tuition and fees vary according to course load and student level.
Financial aid: In 1999–00, 4 research assistantships (averaging $17,000 per year) were awarded. Financial aid application deadline: 4/1; financial aid applicants required to submit FAFSA.
Faculty research: Oncogenes, tumor growth, angiogenesis, signal transduction, demyelinating, diseases.
Dr. Julius Gordon, Head, 979-845-7235, *Fax:* 979-862-1299, *E-mail:* gordon@ tamu.edu.

Application contact: Dr. Steve Allen Maxwell, Graduate Adviser, 979-845-7206, *Fax:* 979-862-1299, *E-mail:* s-maxwell@ tamu.edu.

Find an in-depth description at www.petersons.com/graduate.

■ THOMAS JEFFERSON UNIVERSITY

College of Graduate Studies, Program in Pathology and Cell Biology, Philadelphia, PA 19107

AWARDS PhD.

Faculty: 48 full-time.

Students: 19 full-time (6 women); includes 4 minority (1 African American, 1 Asian American or Pacific Islander, 1 Hispanic American, 1 Native American), 2 international. *31 applicants, 6% accepted.* In 1999, 3 degrees awarded.
Degree requirements: For doctorate, dissertation required, foreign language not required.
Entrance requirements: For doctorate, GRE General Test, TOEFL, minimum GPA of 3.2. *Application deadline:* For fall admission, 3/1 (priority date). Applications are processed on a rolling basis. *Application fee:* $40.
Expenses: Tuition: Full-time $12,670. Tuition and fees vary according to degree level and program.
Financial aid: In 1999–00, 17 fellowships with full tuition reimbursements were awarded; research assistantships, Federal Work-Study, institutionally sponsored loans, traineeships, and training grants also available. Aid available to part-time students. Financial aid application deadline: 5/1; financial aid applicants required to submit FAFSA.
Faculty research: Liver diseases, alcohol metabolism, structure-function relationships in biological membranes, chemical carcinogenesis. *Total annual research expenditures:* $8 million.
Dr. Jan B. Hoek, Chairman, Graduate Committee, 215-503-5016, *Fax:* 215-923-2218, *E-mail:* jan.hoek@mail.tju.edu.
Application contact: Jessie F. Pervall, Director of Admissions, 215-503-4400, *Fax:* 215-503-3433, *E-mail:* cgs-info@ mail.tju.edu.

Find an in-depth description at www.petersons.com/graduate.

■ UNIFORMED SERVICES UNIVERSITY OF THE HEALTH SCIENCES

School of Medicine, Division of Basic Medical Sciences, Department of Pathology, Bethesda, MD 20814-4799

AWARDS Molecular pathobiology (PhD).

Faculty: 18 full-time (5 women), 69 part-time/adjunct (18 women).
Students: 5 full-time (1 woman); includes 1 minority (African American). Average age 25. *6 applicants, 33% accepted.* In 1999, 1 degree awarded (100% found work related to degree).
Degree requirements: For doctorate, one foreign language (computer language can substitute), dissertation, qualifying exam required. *Average time to degree:* Doctorate–5 years full-time.
Entrance requirements: For doctorate, GRE General Test, GRE Subject Test, minimum GPA of 3.0, U.S. citizenship. *Application deadline:* For fall admission,

1/15 (priority date). Applications are processed on a rolling basis. *Application fee:* $0.
Financial aid: In 1999–00, 4 fellowships with full tuition reimbursements (averaging $15,000 per year) were awarded; tuition waivers (full) also available.
Faculty research: Molecular pathology, genetics, virology.
Dr. Robert M. Friedman, Chair, 301-295-3450.
Application contact: Janet M. Anastasi, Graduate Program Coordinator, 301-295-9474, *Fax:* 301-295-6772, *E-mail:* janastasi@usuhs.mil.

■ THE UNIVERSITY OF ALABAMA AT BIRMINGHAM

Graduate School and School of Medicine and School of Dentistry, Graduate Programs in Joint Health Sciences, Department of Pathology, Birmingham, AL 35294

AWARDS PhD.

Students: 37 full-time (18 women); includes 4 minority (3 African Americans, 1 Asian American or Pacific Islander), 12 international. *23 applicants, 39% accepted.* In 1999, 3 degrees awarded.
Degree requirements: For doctorate, dissertation required, foreign language not required.
Entrance requirements: For doctorate, GRE General Test, interview. *Application deadline:* Applications are processed on a rolling basis. *Application fee:* $35 ($60 for international students). Electronic applications accepted.
Expenses: Tuition, state resident: part-time $104 per semester hour. Tuition, nonresident: part-time $208 per semester hour. Required fees: $17 per semester hour. $57 per quarter. Tuition and fees vary according to program.
Financial aid: In 1999–00, 15 fellowships were awarded; career-related internships or fieldwork also available.
Dr. Jay M. McDonald, Chairman, 205-934-4303, *Fax:* 205-934-5499, *E-mail:* mcdonald@uab.edu.

Find an in-depth description at www.petersons.com/graduate.

■ UNIVERSITY OF ARKANSAS FOR MEDICAL SCIENCES

College of Medicine and Graduate School, Graduate Programs in Medicine, Department of Pathology, Little Rock, AR 72205-7199

AWARDS MS.

Faculty: 20 full-time (8 women), 1 part-time/adjunct (0 women).
Students: 1 full-time (0 women).

Degree requirements: For master's, thesis required, foreign language not required.
Entrance requirements: For master's, GRE General Test. *Application fee:* $0.
Expenses: Tuition: Full-time $8,928.
Financial aid: Research assistantships available. Aid available to part-time students.
Dr. Aubrey J. Hough, Chairman, 501-686-5170.
Application contact: Dr. Louis Chang, Information Contact, 501-686-5170.

■ UNIVERSITY OF CALIFORNIA, DAVIS

Graduate Studies, Programs in the Biological Sciences, Program in Comparative Pathology, Davis, CA 95616

AWARDS MS, PhD.

Faculty: 110 full-time (30 women).
Students: 72 full-time (41 women), 5 part-time (4 women); includes 15 minority (2 African Americans, 11 Asian Americans or Pacific Islanders, 1 Hispanic American, 1 Native American), 17 international. Average age 31. *51 applicants, 63% accepted.* In 1999, 6 master's, 8 doctorates awarded (50% entered university research/teaching, 50% found other work related to degree).
Degree requirements: For master's, thesis or alternative required; for doctorate, dissertation required. *Average time to degree:* Doctorate–5 years full-time.
Entrance requirements: For master's and doctorate, GRE General Test. *Application deadline:* For fall admission, 3/1. *Application fee:* $40. Electronic applications accepted.
Expenses: Tuition, nonresident: full-time $9,804. Tuition and fees vary according to program and student level.
Financial aid: In 1999–00, 65 students received aid, including 8 fellowships with full and partial tuition reimbursements available, 35 research assistantships with full and partial tuition reimbursements available, 3 teaching assistantships with partial tuition reimbursements available; Federal Work-Study, grants, institutionally sponsored loans, scholarships, and tuition waivers (full and partial) also available. Financial aid application deadline: 1/15; financial aid applicants required to submit FAFSA.
Faculty research: Immunopathology, toxicological and environmental pathology, reproductive pathology, pathology of infectious diseases.
Dennis Wilson, Graduate Chair, 530-752-0158, *E-mail:* dwwilson@ucdavis.edu.
Application contact: Darlene Flemming, Administrative Assistant, 530-752-2657, *Fax:* 530-754-8124, *E-mail:* dhflemming@ucdavis.edu.

■ UNIVERSITY OF CALIFORNIA, LOS ANGELES

School of Medicine and Graduate Division, Graduate Programs in Medicine, Program in Experimental Pathology, Los Angeles, CA 90095

AWARDS MS, PhD.

Students: 26 full-time (13 women); includes 16 minority (1 African American, 12 Asian Americans or Pacific Islanders, 3 Hispanic Americans), 3 international. *2 applicants, 100% accepted.*
Degree requirements: For master's, foreign language not required; for doctorate, dissertation, oral and written qualifying exams required, foreign language not required.
Entrance requirements: For master's, GRE General Test; for doctorate, GRE General Test, previous course work in physical chemistry and physics. *Application fee:* $40.
Expenses: Tuition, nonresident: full-time $9,804. Required fees: $4,405.
Financial aid: In 1999–00, 25 students received aid, including 24 fellowships, 24 research assistantships, 7 teaching assistantships; Federal Work-Study, institutionally sponsored loans, scholarships, and tuition waivers (full and partial) also available. Financial aid application deadline: 3/1.
Dr. Jonathan Braun, Chair, 310-825-5719.
Application contact: UCLA Access Coordinator, 800-284-8252, *Fax:* 310-206-5280, *E-mail:* uclaaccess@lbes.medsch.ucla.edu.

■ UNIVERSITY OF CALIFORNIA, SAN DIEGO

School of Medicine and Graduate Studies and Research, Graduate Studies in Biomedical Sciences, La Jolla, CA 92093-0685

AWARDS Cell and molecular biology (PhD); molecular pathology (PhD); neuroscience (PhD); pharmacology (PhD); physiology (PhD); regulatory biology (PhD).

Faculty: 106.
Students: 219. *241 applicants, 23% accepted.* In 1999, 16 doctorates awarded.
Degree requirements: For doctorate, dissertation, qualifying exam required, foreign language not required.
Entrance requirements: For doctorate, GRE General Test, TOEFL. *Application deadline:* For fall admission, 1/5. *Application fee:* $40.
Expenses: Program pays tuition, fees, health insurance, and stipend for all students in good standing.
Financial aid: Fellowships, research assistantships, career-related internships or fieldwork, tuition waivers (full), and stipends available.
Faculty research: Molecular and cellular biology, molecular and cellular pharmacology, cell and organ physiology.
Kim Barrett, Chair, 858-543-3726.
Application contact: Gina Butcher, Graduate Program Representative, 858-534-3982.

Find an in-depth description at www.petersons.com/graduate.

■ UNIVERSITY OF CALIFORNIA, SAN DIEGO

School of Medicine and Graduate Studies and Research, Molecular Pathology Program, La Jolla, CA 92093

AWARDS PhD.

Faculty: 33.
Students: 41 (20 women). *94 applicants, 13% accepted.* In 1999, 6 degrees awarded.
Degree requirements: For doctorate, dissertation (for some programs), qualifying exam required.
Entrance requirements: For doctorate, GRE General Test, GRE Subject Test, TOEFL. *Application deadline:* For fall admission, 1/5. *Application fee:* $40.
Expenses: Tuition, nonresident: full-time $9,804. Full-time tuition and fees vary according to program and student level.
Financial aid: Fellowships, research assistantships, career-related internships or fieldwork available.
Dr. Colin Bloor, Chair.
Application contact: Kim Ciero, Graduate Coordinator, 858-534-4324.

Find an in-depth description at www.petersons.com/graduate.

■ UNIVERSITY OF CALIFORNIA, SAN FRANCISCO

Graduate Division, Biomedical Sciences Graduate Group, Program in Experimental Pathology, San Francisco, CA 94143

AWARDS PhD.

Faculty: 34 full-time (9 women).
Degree requirements: For doctorate, dissertation required, foreign language not required.
Entrance requirements: For doctorate, GRE General Test. *Application fee:* $40.
Financial aid: Application deadline: 1/10.
Faculty research: Regulation of cell growth, mechanisms of immunological response, biology of parasitic diseases.
Application contact: Pamela Humphrey, Program Administrator, 415-476-8467.

■ UNIVERSITY OF CHICAGO

Division of the Biological Sciences, Biomedical Sciences: Cancer, Immunology, Nutrition, Pathology, and Virology, Department of Pathology, Chicago, IL 60637-1513

AWARDS PhD.

Faculty: 47 full-time (10 women).
Students: 16 full-time (2 women); includes 5 minority (1 African American, 4 Asian Americans or Pacific Islanders). *50 applicants, 10% accepted.* In 1999, 5 doctorates awarded (20% continued full-time study).
Degree requirements: For doctorate, dissertation required, foreign language not required. *Average time to degree:* Doctorate–5.5 years full-time.
Entrance requirements: For doctorate, GRE General Test, TOEFL. *Application deadline:* For fall admission, 1/5 (priority date). *Application fee:* $55.
Expenses: Tuition: Full-time $24,804; part-time $3,422 per course. Required fees: $390. Tuition and fees vary according to program.
Financial aid: In 1999–00, 16 students received aid, including 7 fellowships with full tuition reimbursements available (averaging $16,500 per year), 9 research assistantships with full tuition reimbursements available (averaging $16,500 per year); institutionally sponsored loans also available. Financial aid application deadline: 6/1.
Faculty research: Vascular biology, apolipoproteins, cardiovascular disease, immunopathology.
Dr. Vinay Kumar, Chairman, 773-702-0647, *Fax:* 773-702-4634, *E-mail:* yargon@midway.uchicago.edu.
Application contact: Rebecca Levine, Administrative Assistant, Student Services, 773-834-3899, *Fax:* 773-702-4634, *E-mail:* rlevine@huggins.bsd.uchicago.edu.

■ UNIVERSITY OF CINCINNATI

Division of Research and Advanced Studies, College of Medicine, Graduate Programs in Medicine, Department of Pathobiology and Molecular Medicine, Cincinnati, OH 45267

AWARDS Pathology (PhD), including anatomic pathology, laboratory medicine, pathobiology and molecular medicine.

Faculty: 10 full-time.
Students: 20 full-time (11 women); includes 1 minority (Native American), 5 international. *116 applicants, 3% accepted.* In 1999, 1 doctorate awarded.
Degree requirements: For doctorate, dissertation, qualifying exam required, foreign language not required. *Average time to degree:* Doctorate–5.3 years full-time.

Entrance requirements: For doctorate, GRE General Test, TOEFL. *Application deadline:* For fall admission, 2/1 (priority date). Applications are processed on a rolling basis. *Application fee:* $30.
Expenses: Tuition, state resident: full-time $5,139; part-time $196 per credit hour. Tuition, nonresident: full-time $10,326; part-time $369 per credit hour. Required fees: $561; $187 per quarter.
Financial aid: Tuition waivers (full) and unspecified assistantships available. Financial aid application deadline: 5/1.
Faculty research: Carcinogenesis; cardiovascular disease; inflammation, immunology, infectious disease; toxicology/environmental health. *Total annual research expenditures:* $3.8 million.
Dr. Cecilia Fenoglio-Preiser, Head, 513-558-4500, *Fax:* 513-558-2289, *E-mail:* cecilia.fenogliopreiser@uc.edu.
Application contact: Dr. Thomas Clemens, Acting Director of Graduate Studies, 513-558-4444, *Fax:* 513-558-2289, *E-mail:* thomas.clemens@uc.edu.

Find an in-depth description at www.petersons.com/graduate.

■ UNIVERSITY OF CINCINNATI

Division of Research and Advanced Studies, College of Medicine, Graduate Programs in Medicine, Program in Molecular and Cellular Pathophysiology, Cincinnati, OH 45267

AWARDS D Sc.

Faculty: 6 full-time.
Degree requirements: For doctorate, one foreign language, dissertation, qualifying exam required.
Entrance requirements: For doctorate, GRE General Test, surgical residency. *Application deadline:* For fall admission, 2/1 (priority date). Applications are processed on a rolling basis. *Application fee:* $30.
Expenses: Tuition, state resident: full-time $5,139; part-time $196 per credit hour. Tuition, nonresident: full-time $10,326; part-time $369 per credit hour. Required fees: $561; $187 per quarter.
Financial aid: Unspecified assistantships available. Financial aid application deadline: 5/1. *Total annual research expenditures:* $2.6 million.
Dr. Josef Fischer, Chairman, 513-558-4202, *Fax:* 513-872-6999, *E-mail:* josef.fischer@uc.edu.
Application contact: Dr. George Babcock, Director, 513-872-6231, *Fax:* 513-872-6999, *E-mail:* george.babcock@uc.edu.

■ UNIVERSITY OF COLORADO HEALTH SCIENCES CENTER

Graduate School, Programs in Biological and Medical Sciences, Program in Experimental Pathology, Denver, CO 80262

AWARDS PhD.

Degree requirements: For doctorate, dissertation required, foreign language not required.
Entrance requirements: For doctorate, GRE General Test, minimum GPA of 2.75.
Expenses: Tuition, state resident: full-time $1,512; part-time $56 per hour. Tuition, nonresident: full-time $7,209; part-time $267 per hour. Full-time tuition and fees vary according to course load and program.

■ UNIVERSITY OF FLORIDA

College of Medicine and Graduate School, Interdisciplinary Program in Biomedical Sciences, Department of Pathology, Gainesville, FL 32611

AWARDS Clinical chemistry (MS); immunology and molecular pathology (PhD). Terminal master's awarded for partial completion of doctoral program.

Degree requirements: For master's and doctorate, thesis/dissertation required, foreign language not required.
Entrance requirements: For master's and doctorate, GRE General Test, TOEFL, minimum GPA of 3.0. Electronic applications accepted.
Expenses: Tuition, state resident: part-time $144 per credit hour. Tuition, nonresident: part-time $505 per credit hour. Tuition and fees vary according to course level, course load and program.
Faculty research: Molecular immunology, autoimmunity and transplantation, tumor biology, oncogenic viruses, human immunodeficiency viruses.

■ UNIVERSITY OF GEORGIA

College of Veterinary Medicine and Graduate School, Graduate Programs in Veterinary Medicine, Department of Pathology, Athens, GA 30602

AWARDS MS, PhD.

Degree requirements: For master's, thesis required, foreign language not required; for doctorate, one foreign language (computer language can substitute), dissertation required.
Entrance requirements: For master's and doctorate, GRE General Test. Electronic applications accepted.
Expenses: Tuition, state resident: full-time $7,516; part-time $431 per credit hour.

Tuition, nonresident: full-time $12,204; part-time $793 per credit hour.

■ UNIVERSITY OF ILLINOIS AT CHICAGO

College of Medicine and Graduate College, Graduate Programs in Medicine, Department of Pathology, Chicago, IL 60607-7128

AWARDS MS, PhD, MD/PhD.

Faculty: 16 full-time (3 women).
Students: 13 full-time (9 women), 5 part-time (2 women); includes 5 minority (all Asian Americans or Pacific Islanders), 5 international. Average age 32. *25 applicants, 20% accepted.* In 1999, 4 master's, 1 doctorate awarded.
Degree requirements: For master's and doctorate, thesis/dissertation required, foreign language not required.
Entrance requirements: For master's and doctorate, GRE General Test, TOEFL. *Application deadline:* For fall admission, 6/1; for spring admission, 11/1. *Application fee:* $40 ($50 for international students).
Expenses: Tuition, state resident: full-time $3,750. Tuition, nonresident: full-time $10,588. Tuition and fees vary according to course load.
Financial aid: In 1999–00, 7 students received aid; fellowships, research assistantships, teaching assistantships, Federal Work-Study available. Financial aid application deadline: 3/1; financial aid applicants required to submit FAFSA.
Faculty research: Atherosclerosis, cellular immunology, oncology and tumor immunology, molecular and cardiovascular pathology.
Dr. Jose Manaligod, Acting Head, 312-996-7312.
Application contact: Dr. Rameshwar M. Prasad, Director, Graduate Program, 312-996-2954.

■ THE UNIVERSITY OF IOWA

College of Medicine and Graduate College, Graduate Programs in Medicine, Department of Pathology, Iowa City, IA 52242-1316

AWARDS MS.

Faculty: 5 full-time (0 women).
Students: 5 full-time (2 women), 2 international. Average age 31. *3 applicants, 33% accepted.* In 1999, 1 degree awarded (100% entered university research/teaching).
Degree requirements: For master's, thesis required, foreign language not required. *Average time to degree:* Master's–2 years full-time.
Entrance requirements: For master's, GRE, TOEFL. *Application deadline:* Applications are processed on a rolling

basis. *Application fee:* $30 ($50 for international students).
Expenses: Tuition, state resident: full-time $3,308. Tuition, nonresident: full-time $10,662. Tuition and fees vary according to course load and program.
Financial aid: In 1999–00, 4 research assistantships with full tuition reimbursements (averaging $16,277 per year) were awarded; Federal Work-Study, institutionally sponsored loans, and tuition waivers (partial) also available. Aid available to part-time students.
Faculty research: Cell and tumor immunopathology, neuropathology, immunohematology, microbiology, clinical chemistry. *Total annual research expenditures:* $3.6 million.
Michael Cohen, Head, 319-335-8232, *Fax:* 319-335-8348, *E-mail:* michael-cohen@uiowa.edu.
Application contact: Robert T. Cook, Graduate Committee Chair, 319-338-0581 Ext. 6516, *E-mail:* robert-cook@uiowa.edu.

■ UNIVERSITY OF KANSAS

Graduate Studies Medical Center, Graduate Programs in Biomedical and Basic Sciences, Department of Pathology and Laboratory Medicine, Lawrence, KS 66045

AWARDS MA, PhD, MD/PhD. Part-time programs available.

Faculty: 12 full-time (3 women), 1 part-time/adjunct (0 women).
Students: 2 full-time (1 woman), 1 international. Average age 26. *2 applicants, 100% accepted.* Terminal master's awarded for partial completion of doctoral program.
Degree requirements: For master's, thesis, comprehensive oral exam required, foreign language not required; for doctorate, dissertation, comprehensive oral exam required.
Entrance requirements: For master's and doctorate, GRE, TOEFL, TSE. *Application deadline:* For fall admission, 1/31 (priority date). Applications are processed on a rolling basis. *Application fee:* $0.
Expenses: Tuition, state resident: full-time $2,482; part-time $103 per credit hour. Tuition, nonresident: full-time $8,104; part-time $338 per credit hour. Required fees: $428; $31 per credit hour. Tuition and fees vary according to program.
Financial aid: Fellowships, research assistantships, teaching assistantships, Federal Work-Study, institutionally sponsored loans, and scholarships available. Aid available to part-time students. Financial aid application deadline: 3/31; financial aid applicants required to submit FAFSA.
Faculty research: Calcification, CA neoplasia, cell signaling, cytogenetics. *Total annual research expenditures:* $1.5 million.

Dr. Barbara Atkinson, Chairman, 913-588-7070, *Fax:* 913-588-7073, *E-mail:* batkinson@kumc.edu.
Application contact: Dr. Jill Pelling, Director of Graduate Studies, 913-588-7240, *Fax:* 913-588-7073, *E-mail:* jpelling@kumc.edu.

■ UNIVERSITY OF MARYLAND

Graduate School, Graduate Programs in Medicine, Department of Pathology, Baltimore, MD 21201-1627

AWARDS Medical pathology (PhD); pathology (MS). Part-time programs available.

Degree requirements: For master's, foreign language not required; for doctorate, dissertation required, foreign language not required.
Entrance requirements: For master's and doctorate, GRE General Test, TOEFL, minimum GPA of 3.0.
Expenses: Tuition, state resident: part-time $261 per credit hour. Tuition, nonresident: part-time $468 per credit hour. Tuition and fees vary according to program.
Faculty research: Carcinogenesis, immunopathology, cell injury.

Find an in-depth description at www.petersons.com/graduate.

■ UNIVERSITY OF MEDICINE AND DENTISTRY OF NEW JERSEY

Graduate School of Biomedical Sciences, Graduate Programs in Biomedical Sciences, Department of Pathology and Laboratory Medicine, Newark, NJ 07107

AWARDS MS, PhD.

Degree requirements: For master's, thesis required; for doctorate, dissertation, qualifying exam required, foreign language not required.
Entrance requirements: For master's and doctorate, GRE General Test, TOEFL. *Application deadline:* For fall admission, 2/1; for spring admission, 10/1. *Application fee:* $40.
Expenses: Tuition, state resident: part-time $270 per credit hour. Tuition, nonresident: part-time $407 per credit hour. Part-time tuition and fees vary according to campus/location and program.
Financial aid: Fellowships, research assistantships, Federal Work-Study, institutionally sponsored loans, and tuition waivers (full and partial) available. Financial aid application deadline: 5/1.
Dr. Stanley Cohen, Chairperson, 973-972-4520, *E-mail:* cohenst@umdnj.edu.
Application contact: Dr. Henry E. Brezenoff, Dean, Graduate School of Biomedical Sciences, 973-972-5333, *Fax:*

University of Medicine and Dentistry of New Jersey (continued)
973-972-7148, *E-mail:* hbrezeno@umdnj.edu.

Find an in-depth description at www.petersons.com/graduate.

■ UNIVERSITY OF MICHIGAN

Medical School and Horace H. Rackham School of Graduate Studies, Program in Biomedical Sciences (PIBS), Department of Pathology, Ann Arbor, MI 48109

AWARDS PhD, MD/PhD.

Faculty: 27 full-time (5 women).
Students: 7 full-time (4 women); includes 4 minority (all Asian Americans or Pacific Islanders). *20 applicants, 40% accepted.*
Degree requirements: For doctorate, oral defense of dissertation, preliminary exam required, foreign language not required. *Average time to degree:* Doctorate–4 years full-time.
Entrance requirements: For doctorate, GRE General Test, GRE Subject Test. *Application deadline:* For fall admission, 1/5. *Application fee:* $55.
Expenses: Tuition, state resident: full-time $10,316. Tuition, nonresident: full-time $20,922.
Financial aid: In 1999–00, fellowships with full tuition reimbursements (averaging $17,000 per year), research assistantships with full tuition reimbursements (averaging $17,000 per year) were awarded.
Faculty research: Regulation of cytokine gene expression, programmed cell death, pathogenesis of fibrosis, soluble mediators of inflammation. *Total annual research expenditures:* $9.5 million.
Dr. Peter A. Ward, Chairman, 734-763-6454.
Application contact: Graduate Programs in Biomedical Sciences, 734-647-7005, *Fax:* 734-647-7022, *E-mail:* pibs@umich.edu.

Find an in-depth description at www.petersons.com/graduate.

■ UNIVERSITY OF MISSISSIPPI MEDICAL CENTER

Graduate Programs in Biomedical Sciences, Department of Pathology, Jackson, MS 39216-4505

AWARDS MS, PhD.

Faculty: 10 full-time (1 woman).
Students: 1 (woman) full-time, 2 part-time (1 woman). Average age 36. *6 applicants, 0% accepted.* In 1999, 1 master's awarded (100% continued full-time study).
Degree requirements: For master's, thesis required, foreign language not required; for doctorate, dissertation, first authored publication required, foreign

language not required. *Average time to degree:* Master's–4 years full-time; doctorate–5 years full-time.
Entrance requirements: For master's, GRE General Test, minimum GPA of 3.0; for doctorate, GRE General Test, GRE Subject Test, minimum GPA of 3.0. *Application deadline:* For fall admission, 7/1. Applications are processed on a rolling basis. *Application fee:* $10.
Expenses: Tuition, state resident: full-time $2,378; part-time $132 per hour. Tuition, nonresident: full-time $4,697; part-time $261 per hour. Tuition and fees vary according to program.
Financial aid: In 1999–00, 1 student received aid, including 1 research assistantship (averaging $16,234 per year); Federal Work-Study also available. Financial aid application deadline: 4/1.
Faculty research: Effects of rehabilitation therapy on immune system/hypothalamic/pituitary adrenal axis interaction; HLA, GC, CM, KM, and/or genetic factors in the pathogenesis of AIDS.
Dr. J. M. Cruse, Director, 601-984-1565, *Fax:* 601-984-1835, *E-mail:* jcruse@pathology.umsmed.edu.
Application contact: Dr. Billy M. Bishop, Director, Student Services and Records, 601-984-1080, *Fax:* 601-984-1079, *E-mail:* bbishop@registrar.umsmed.edu.

■ UNIVERSITY OF MISSOURI–COLUMBIA

College of Veterinary Medicine and Graduate School, Graduate Programs in Veterinary Medicine, Program in Pathobiology, Columbia, MO 65211

AWARDS Laboratory animal medicine (MS); pathobiology (PhD); pathology (MS).

Degree requirements: For master's, thesis required; for doctorate, dissertation required.
Entrance requirements: For master's and doctorate, GRE General Test, minimum GPA of 3.0.
Expenses: Tuition, state resident: full-time $3,020; part-time $168 per hour. Tuition, nonresident: full-time $6,066; part-time $505 per hour. Required fees: $445; $18 per hour. Tuition and fees vary according to course load and program.

■ UNIVERSITY OF NEBRASKA MEDICAL CENTER

Graduate College, Department of Pathology and Microbiology, Omaha, NE 68198

AWARDS MS, PhD. Part-time programs available.

Faculty: 43 full-time (7 women), 29 part-time/adjunct (3 women).

Students: 27 full-time (11 women), 3 part-time (1 woman); includes 2 minority (both African Americans), 2 international. Average age 27. *54 applicants, 19% accepted.* In 1999, 2 master's awarded (100% entered university research/teaching); 5 doctorates awarded. Terminal master's awarded for partial completion of doctoral program.
Degree requirements: For master's and doctorate, thesis/dissertation required, foreign language not required.
Entrance requirements: For master's, previous course work in biology, chemistry, mathematics, and physics; for doctorate, GRE General Test, previous course work in biology, chemistry, mathematics, and physics. *Application deadline:* Applications are processed on a rolling basis. *Application fee:* $35.
Expenses: Tuition, state resident: part-time $116 per semester hour. Tuition, nonresident: part-time $270 per semester hour. Tuition and fees vary according to program.
Financial aid: In 1999–00, 2 fellowships with tuition reimbursements (averaging $14,500 per year), 21 research assistantships with tuition reimbursements (averaging $14,500 per year) were awarded; teaching assistantships, institutionally sponsored loans and tuition waivers (full) also available. Aid available to part-time students. Financial aid application deadline: 3/1.
Faculty research: Carcinogenesis, cancer biology, immunobiology, molecular virology, molecular genetics.
Dr. Donald R. Johnson, Chairman, Graduate Committee, 402-559-4042, *Fax:* 402-559-4077.
Application contact: Jo Wagner, Associate Director of Admissions, 402-559-6468.

■ THE UNIVERSITY OF NORTH CAROLINA AT CHAPEL HILL

School of Medicine and Graduate School, Graduate Programs in Medicine, Department of Pathology and Laboratory Medicine, Chapel Hill, NC 27599

AWARDS Experimental pathology (PhD).

Faculty: 52 full-time (21 women), 9 part-time/adjunct (0 women).
Students: 23 full-time (18 women); includes 4 minority (2 African Americans, 1 Asian American or Pacific Islander, 1 Hispanic American), 2 international. Average age 26. *28 applicants, 32% accepted.* In 1999, 5 doctorates awarded (100% found work related to degree).
Degree requirements: For doctorate, dissertation, comprehensive exam, proposal defense required, foreign language not required. *Average time to degree:* Doctorate–6.3 years full-time.

Entrance requirements: For doctorate, GRE General Test, GRE Subject Test (recommended). *Application deadline:* For fall admission, 3/31; for spring admission, 10/15. Applications are processed on a rolling basis. *Application fee:* $55. Electronic applications accepted.

Expenses: Tuition, state resident: full-time $1,966. Tuition, nonresident: full-time $11,026. Required fees: $8,940. One-time fee: $15 full-time. Part-time tuition and fees vary according to course load.

Financial aid: In 1999–00, 12 fellowships with full and partial tuition reimbursements (averaging $16,500 per year), 9 research assistantships with full and partial tuition reimbursements (averaging $16,500 per year) were awarded; Federal Work-Study, institutionally sponsored loans, traineeships, tuition waivers (full), and unspecified assistantships also available. Financial aid application deadline: 1/1; financial aid applicants required to submit FAFSA.

Faculty research: Carcinogenesis, blood coagulation, molecular biology and genetics and animal models of human disease, opportunistic infections in AIDS immunopathology. *Total annual research expenditures:* $9.6 million.

Dr. J. Charles Jennette, Brinkhous—Distinguished Professor and Chair, 919-966-4676, *Fax:* 919-966-6718, *E-mail:* jcj@med.unc.edu.

Application contact: Dr. Frank C. Church, Director of Graduate Admissions, 919-966-4676, *Fax:* 919-966-6718, *E-mail:* fchurch@email.unc.edu.

Find an in-depth description at www.petersons.com/graduate.

■ **UNIVERSITY OF OKLAHOMA HEALTH SCIENCES CENTER**

College of Medicine and Graduate College, Graduate Programs in Medicine, Department of Pathology, Oklahoma City, OK 73190

AWARDS PhD.

Faculty: 19 full-time (4 women), 4 part-time/adjunct (0 women).

Students: 5 full-time (3 women), 1 (woman) part-time; includes 1 minority (Native American), 2 international. *14 applicants, 21% accepted.* In 1999, 2 doctorates awarded.

Degree requirements: For doctorate, dissertation required.

Entrance requirements: For doctorate, GRE General Test, TOEFL. *Application deadline:* For fall admission, 7/1; for spring admission, 12/1. *Application fee:* $25 ($50 for international students).

Expenses: Tuition, state resident: part-time $90 per semester hour. Tuition,

nonresident: part-time $264 per semester hour. Tuition and fees vary according to program.

Financial aid: Federal Work-Study, institutionally sponsored loans, and tuition waivers (full) available.

Faculty research: Molecular pathology, tissue response in disease, anatomic pathology, immunopathology, histocytochemistry. Dr. Fred Silva, Head, 405-271-2422.

Application contact: Dr. Paula Grammas, Graduate Liaison, 405-271-2031.

■ **UNIVERSITY OF PITTSBURGH**

School of Medicine, Graduate Programs in Medicine, Program in Cellular and Molecular Pathology, Pittsburgh, PA 15260

AWARDS MS, PhD.

Faculty: 107 full-time (29 women), 10 part-time/adjunct (4 women).

Students: 4 full-time (1 woman), 2 international. *325 applicants, 22% accepted.* In 1999, 3 doctorates awarded.

Degree requirements: For doctorate, dissertation required, foreign language not required. *Average time to degree:* Doctorate–5.3 years full-time.

Entrance requirements: For doctorate, GRE General Test, GRE Subject Test, TOEFL, minimum QPA of 3.0. *Application deadline:* For fall admission, 1/15 (priority date). Applications are processed on a rolling basis. *Application fee:* $30 ($40 for international students).

Expenses: Tuition, state resident: full-time $9,778; part-time $403 per credit. Tuition, nonresident: full-time $20,146; part-time $830 per credit. Required fees: $480; $90 per semester.

Financial aid: Research assistantships with full tuition reimbursements, teaching assistantships with full tuition reimbursements, Federal Work-Study, institutionally sponsored loans, scholarships, traineeships, and unspecified assistantships available.

Faculty research: Liver growth and differentiation, pathogenesis of neurodegeneration, cancer research. *Total annual research expenditures:* $76.3 million.

Application contact: Graduate Studies Administrator, 412-648-8957, *Fax:* 412-648-1236, *E-mail:* biomed_phd@fs1.dean-med.pitt.edu.

Find an in-depth description at www.petersons.com/graduate.

■ **UNIVERSITY OF ROCHESTER**

School of Medicine and Dentistry, Graduate Programs in Medicine and Dentistry, Department of Pathology and Laboratory Medicine, Rochester, NY 14642

AWARDS Pathology (MS, PhD).

Faculty: 14.

Students: 30 full-time (13 women); includes 1 minority (Asian American or Pacific Islander), 17 international.

Degree requirements: For doctorate, variable foreign language requirement, dissertation, qualifying exam required.

Entrance requirements: For doctorate, GRE General Test, GRE Subject Test. *Application deadline:* For fall admission, 2/1. *Application fee:* $25.

Expenses: Tuition: Part-time $697 per credit hour. Tuition and fees vary according to program.

Financial aid: Fellowships, research assistantships, teaching assistantships, tuition waivers (full and partial) available. Financial aid application deadline: 2/1. Dr. Steven Spitalnik, Chair, 716-275-3181.

Application contact: Barb Collins, Graduate Program Secretary, 716-275-3083.

Find an in-depth description at www.petersons.com/graduate.

■ **UNIVERSITY OF SOUTHERN CALIFORNIA**

Keck School of Medicine and Graduate School, Graduate Programs in Medicine, Department of Pathology, Los Angeles, CA 90089

AWARDS Experimental and molecular pathology (MS); pathobiology (PhD).

Faculty: 40 full-time (8 women).

Students: 51 full-time (24 women); includes 16 minority (12 Asian Americans or Pacific Islanders, 4 Hispanic Americans), 29 international. Average age 29. *73 applicants, 34% accepted.* In 1999, 3 master's awarded (100% continued full-time study); 9 doctorates awarded (100% continued full-time study).

Degree requirements: For master's, thesis or alternative required, foreign language not required; for doctorate, dissertation required, foreign language not required. *Average time to degree:* Master's–2 years full-time; doctorate–6.25 years full-time.

Entrance requirements: For master's, GRE General Test, TOEFL, minimum GPA of 3.0; for doctorate, GRE General Test, TOEFL, minimum GPA of 3.0, BS in natural sciences. *Application deadline:* For fall admission, 4/1. *Application fee:* $55.

Expenses: Tuition: Full-time $22,198; part-time $748 per unit. Required fees: $406.

Financial aid: In 1999–00, 38 students received aid, including 7 fellowships with tuition reimbursements available (averaging $16,500 per year), 27 research assistantships with tuition reimbursements available (averaging $16,900 per year), 4 teaching assistantships with tuition reimbursements available (averaging

University of Southern California (continued)

$16,900 per year); grants also available. Financial aid application deadline: 4/1.
Faculty research: Immunology of lymphomas and leukemias, lung cancer, molecular basis of oncogenesis, central nervous system disease, organic chemical carcinogens and carcinogenic metal salts. Dr. Clive R. Taylor, Chairman, 323-442-1180, *Fax:* 323-442-3314, *E-mail:* taylor@pathfinder.usc.edu.
Application contact: Lisa A. Doumak, Office Assistant II, 323-442-1168, *Fax:* 323-442-3049, *E-mail:* doumak@pathfinder.usc.edu.

Find an in-depth description at www.petersons.com/graduate.

■ UNIVERSITY OF SOUTH FLORIDA

College of Medicine and Graduate School, Graduate Programs in Medical Sciences, Department of Pathology, Tampa, FL 33620-9951

AWARDS PhD.

Degree requirements: For doctorate, dissertation required, foreign language not required.
Entrance requirements: For doctorate, GRE General Test, minimum GPA of 3.0.
Expenses: Tuition, state resident: part-time $148 per credit hour. Tuition, nonresident: part-time $509 per credit hour.
Faculty research: Ovarian pathobiology, experimental oncology, molecular oncology, hematopathology.

■ THE UNIVERSITY OF TENNESSEE HEALTH SCIENCE CENTER

College of Graduate Health Sciences, Department of Pathology, Memphis, TN 38163-0002

AWARDS MS, PhD. Part-time programs available. Terminal master's awarded for partial completion of doctoral program.

Degree requirements: For master's, thesis, oral and written comprehensive exams required, foreign language not required; for doctorate, dissertation, oral and written preliminary and comprehensive exams required, foreign language not required.
Entrance requirements: For master's and doctorate, GRE General Test, TOEFL, minimum GPA of 3.0.

Find an in-depth description at www.petersons.com/graduate.

■ THE UNIVERSITY OF TEXAS–HOUSTON HEALTH SCIENCE CENTER

Graduate School of Biomedical Sciences, Program in Molecular Pathology, Houston, TX 77225-0036

AWARDS MS, PhD, MD/PhD.

Faculty: 37 full-time (8 women).
Students: 13 full-time (8 women); includes 4 minority (1 African American, 3 Asian Americans or Pacific Islanders), 1 international. Average age 27. *11 applicants, 45% accepted.* In 1999, 1 master's, 3 doctorates awarded. Terminal master's awarded for partial completion of doctoral program.
Degree requirements: For master's and doctorate, thesis/dissertation required, foreign language not required.
Entrance requirements: For master's and doctorate, GRE General Test, TOEFL, TWE. *Application deadline:* For fall admission, 1/15 (priority date); for spring admission, 11/1. Applications are processed on a rolling basis. *Application fee:* $10. Electronic applications accepted.
Financial aid: Fellowships, research assistantships, institutionally sponsored loans available. Financial aid application deadline: 1/15.
Faculty research: Cellular pathology, immunopathology, infectious diseases, carcinogenesis and chemical pathology. Dr. Diane L. Hickson-Bick, Director, 713-500-5328, *Fax:* 713-500-0730, *E-mail:* diane.l.bick@uth.tmc.edu.
Application contact: Anne Baronitis, Director of Admissions, 713-500-9860, *Fax:* 713-500-9877, *E-mail:* abaron@gsbs.gs.uth.tmc.edu.

Find an in-depth description at www.petersons.com/graduate.

■ THE UNIVERSITY OF TEXAS MEDICAL BRANCH AT GALVESTON

Graduate School of Biomedical Sciences, Program in Experimental Pathology, Galveston, TX 77555

AWARDS PhD.

Faculty: 40 full-time (10 women).
Students: 25 full-time (14 women), 1 (woman) part-time; includes 4 minority (1 African American, 2 Hispanic Americans, 1 Native American), 6 international. Average age 29. *23 applicants, 35% accepted.* In 1999, 9 degrees awarded.
Degree requirements: For doctorate, dissertation required, foreign language not required. *Average time to degree:* Doctorate–4.75 years full-time.
Entrance requirements: For doctorate, GRE General Test. *Application deadline:*

For fall admission, 3/1. Applications are processed on a rolling basis. *Application fee:* $25 ($50 for international students). Electronic applications accepted.
Expenses: Tuition, state resident: full-time $684; part-time $38 per credit hour. Tuition, nonresident: full-time $4,572; part-time $254 per credit hour. Required fees: $29; $7.5 per credit hour. One-time fee: $55. Tuition and fees vary according to program.
Financial aid: Research assistantships, Federal Work-Study and institutionally sponsored loans available. Financial aid applicants required to submit FAFSA.
Faculty research: Environmental toxicology, pathobiology, infectious diseases, tropical diseases, cardiovascular pathology. Dr. Norbert Herzog, Director, 409-772-3938, *Fax:* 409-747-2400, *E-mail:* nherzog@utmb.edu.
Application contact: Dr. Judith Aronson, Co-Director, 409-772-6547, *Fax:* 409-747-2415, *E-mail:* jaronson@utmb.edu.

Find an in-depth description at www.petersons.com/graduate.

■ UNIVERSITY OF UTAH

School of Medicine and Graduate School, Graduate Programs in Medicine, Department of Pathology, Salt Lake City, UT 84112-1107

AWARDS Experimental pathology (PhD). PhD offered after acceptance into the combined Program in Molecular Biology.

Faculty: 7 full-time (3 women), 5 part-time/adjunct (1 woman).
Students: 19 full-time (8 women). Average age 26. In 1999, 4 degrees awarded (100% continued full-time study).
Degree requirements: For doctorate, dissertation required, foreign language not required. *Average time to degree:* Doctorate–6 years full-time.
Entrance requirements: For doctorate, GRE, minimum GPA of 3.0.
Expenses: All students receive full tuition and insurance and an annual stipend of $17,000.
Financial aid: In 1999–00, 19 students received aid, including 19 research assistantships with full tuition reimbursements available (averaging $17,000 per year); institutionally sponsored loans also available.
Faculty research: Immunology, cell biology, signal transduction, gene regulation, receptor biology. *Total annual research expenditures:* $2.1 million. Dr. Raymond A. Daynes, Head, Division of Cell Biology and Immunology, 801-581-3013, *Fax:* 801-581-8946, *E-mail:* ray.daynes@path.med.utah.edu.
Application contact: Kim R. Cash, Project Coordinator, 801-581-3013, *Fax:*

801-581-8946, *E-mail:* kim.cash@path.med.utah.edu.

■ UNIVERSITY OF VERMONT

College of Medicine and Graduate College, Graduate Programs in Medicine, Department of Pathology, Burlington, VT 05405

AWARDS MS, MD/MS.

Degree requirements: For master's, thesis required, foreign language not required.
Entrance requirements: For master's, GRE General Test, TOEFL.
Expenses: Tuition, state resident: full-time $7,464; part-time $311 per credit. Tuition, nonresident: full-time $18,672; part-time $778 per credit. Full-time tuition and fees vary according to degree level and program.

■ UNIVERSITY OF WASHINGTON

School of Medicine and Graduate School, Graduate Programs in Medicine, Department of Pathology, Seattle, WA 98195

AWARDS Molecular basis of disease (PhD); pathology (MS).

Faculty: 40 full-time (7 women).
Students: 32 full-time (16 women); includes 10 minority (8 Asian Americans or Pacific Islanders, 1 Hispanic American, 1 Native American). Average age 29. *37 applicants, 14% accepted.* In 1999, 3 doctorates awarded.
Degree requirements: For doctorate, dissertation required, foreign language not required. *Average time to degree:* Doctorate–5 years full-time.
Entrance requirements: For doctorate, GRE General Test. *Application deadline:* For fall admission, 2/15. *Application fee:* $50.
Expenses: Tuition, state resident: full-time $9,210; part-time $236 per credit. Tuition, nonresident: full-time $23,256; part-time $596 per credit.
Financial aid: In 1999–00, 25 fellowships with tuition reimbursements, 7 research assistantships with tuition reimbursements were awarded; teaching assistantships with tuition reimbursements
Faculty research: Viral oncogenesis, aging, mutagenesis and repair, extracellular matrix biology, vascular biology. *Total annual research expenditures:* $12 million.
Dr. Dan Bowen-Pope, Professor, 206-685-2448, *Fax:* 206-543-3644, *E-mail:* bp@u.washington.edu.
Application contact: Kathy Burdick, Graduate Director, 206-616-7551, *Fax:* 206-543-3644, *E-mail:* kah@u.washington.edu.

Find an in-depth description at www.petersons.com/graduate.

■ UNIVERSITY OF WISCONSIN–MADISON

Medical School and Graduate School, Graduate Programs in Medicine, Department of Pathology and Laboratory Medicine, Madison, WI 53706-1380

AWARDS PhD.

Faculty: 15 full-time (4 women).
Students: 33 full-time (17 women); includes 4 minority (all Hispanic Americans), 4 international.
Degree requirements: For doctorate, dissertation required, foreign language not required. *Average time to degree:* Doctorate–6 years full-time.
Entrance requirements: For doctorate, GRE, minimum GPA of 3.0. *Application deadline:* For fall admission, 1/15 (priority date). Applications are processed on a rolling basis. *Application fee:* $45. Electronic applications accepted.
Expenses: Tuition, state resident: full-time $5,406; part-time $339 per credit. Tuition, nonresident: full-time $17,110; part-time $1,071 per credit.
Financial aid: In 1999–00, fellowships with full tuition reimbursements (averaging $17,000 per year), research assistantships with full tuition reimbursements (averaging $17,000 per year), teaching assistantships with full tuition reimbursements (averaging $17,000 per year) were awarded.
Faculty research: Molecular and cell biology, virology, extracellular matrix, signal transduction, human disease.
Dr. Michael Hart, Chair, 608-262-1189, *Fax:* 608-265-3301, *E-mail:* gharvey@facstaff.wisc.edu.
Application contact: Cameron Millard, Department Secretary, 608-262-1188, *Fax:* 608-265-3301, *E-mail:* pathology@ums2.macc.wisc.edu.

Find an in-depth description at www.petersons.com/graduate.

■ VANDERBILT UNIVERSITY

Graduate School and School of Medicine, Department of Pathology, Nashville, TN 37240-1001

AWARDS Cellular and molecular pathology (PhD).

Faculty: 28 full-time (5 women).
Students: 10 full-time (5 women); includes 2 minority (both Asian Americans or Pacific Islanders). Average age 29. In 1999, 2 degrees awarded.
Degree requirements: For doctorate, dissertation, preliminary, qualifying, and final exams required, foreign language not required.

Entrance requirements: For doctorate, GRE General Test. *Application deadline:* For fall admission, 1/15. *Application fee:* $40.
Expenses: Tuition: Full-time $17,244; part-time $958 per hour. Required fees: $242; $121 per semester. Tuition and fees vary according to program.
Financial aid: In 1999–00, fellowships with full tuition reimbursements (averaging $17,000 per year), research assistantships with full tuition reimbursements (averaging $17,000 per year) were awarded; Federal Work-Study, institutionally sponsored loans, traineeships, and tuition waivers (partial) also available. Financial aid application deadline: 1/15.
Faculty research: Regulation of cell growth, vascular biology, tumor biology, immunology and retroviral pathology, angiogenesis.
Doyle G. Graham, Chair, 615-322-2123, *Fax:* 615-343-7023, *E-mail:* doyle.graham@mcmail.vanderbilt.edu.
Application contact: Larry L. Swift, Director of Graduate Studies, 615-322-3028, *Fax:* 615-343-7023, *E-mail:* larry.swift@mcmail.vanderbilt.edu.

Find an in-depth description at www.petersons.com/graduate.

■ VIRGINIA COMMONWEALTH UNIVERSITY

School of Graduate Studies and School of Medicine, School of Medicine Graduate Programs, Department of Pathology, Richmond, VA 23284-9005

AWARDS MS, PhD, MD/PhD. Part-time programs available.

Students: 2 full-time (0 women), 9 part-time (4 women); includes 2 minority (1 Asian American or Pacific Islander, 1 Native American). Terminal master's awarded for partial completion of doctoral program.
Degree requirements: For master's and doctorate, thesis/dissertation, comprehensive oral and written exams required, foreign language not required.
Entrance requirements: For master's, DAT, GRE General Test, or MCAT; for doctorate, GRE General Test. *Application fee:* $30.
Expenses: Tuition, state resident: full-time $4,031; part-time $224 per credit hour. Tuition, nonresident: full-time $11,946; part-time $664 per credit hour. Required fees: $1,081; $40 per credit hour. Tuition and fees vary according to campus/location and program.
Financial aid: Fellowships, teaching assistantships, tuition waivers (full) available.

Virginia Commonwealth University (continued)

Faculty research: Biochemical and clinical applications of enzyme and protein immobilization, clinical enzymology. Dr. David S. Wilkinson, Chair, 804-828-0183, *Fax:* 804-828-9749.

Application contact: Dr. Joy L. Ware, Director, Graduate Studies, 804-828-9746, *Fax:* 804-828-9749, *E-mail:* jlware@vcu.edu.

Find an in-depth description at www.petersons.com/graduate.

■ WASHINGTON STATE UNIVERSITY

College of Veterinary Medicine and Graduate School, Graduate Programs in Veterinary Science, Department of Veterinary Microbiology and Pathology, Pullman, WA 99164

AWARDS Veterinary science (MS, PhD).

Faculty: 23 full-time (4 women).

Students: 32 full-time (13 women); includes 14 minority (2 African Americans, 6 Asian Americans or Pacific Islanders, 6 Hispanic Americans). Average age 33. *10 applicants, 40% accepted.* In 1999, 2 doctorates awarded (100% found work related to degree). Terminal master's awarded for partial completion of doctoral program.

Degree requirements: For master's and doctorate, computer language, thesis/dissertation, oral exam required. *Average time to degree:* Master's–3 years full-time; doctorate–5.6 years full-time.

Entrance requirements: For master's and doctorate, GRE General Test, minimum GPA of 3.0. *Application deadline:* For fall admission, 1/31 (priority date). Applications are processed on a rolling basis. *Application fee:* $35. Electronic applications accepted.

Expenses: Tuition, state resident: full-time $5,494. Tuition, nonresident: full-time $13,390.

Financial aid: In 1999–00, 17 students received aid, including 17 research assistantships (averaging $4,887 per year); grants and traineeships also available. Financial aid application deadline: 3/1.

Faculty research: Microbial pathogenesis, veterinary and wildlife parasitology, laboratory animal pathology, immune responses to infectious diseases. Dr. David J. Prieur, Chair, 509-335-6030, *Fax:* 509-335-8529, *E-mail:* dprieur@vetmed.wsu.edu.

Application contact: Dr. Guy Palmer, Professor, 509-335-6033, *Fax:* 509-335-8529, *E-mail:* gpalmer@vetmed.wsu.edu.

■ WAYNE STATE UNIVERSITY

School of Medicine and Graduate School, Graduate Programs in Medicine, Department of Pathology, Detroit, MI 48202

AWARDS PhD, MD/PhD.

Degree requirements: For doctorate, dissertation required.

Entrance requirements: For doctorate, GRE General Test.

Faculty research: Cardiovascular pathology, respiratory pathology, hematology, clinical laboratory science, cytogenetics cancer biology.

■ YALE UNIVERSITY

Graduate School of Arts and Sciences, Department of Experimental Pathology, New Haven, CT 06520

AWARDS PhD.

Faculty: 67 full-time (24 women), 3 part-time/adjunct (1 woman).

Students: 3 full-time (1 woman), 1 international.

Degree requirements: For doctorate, dissertation, qualifying exam required.

Entrance requirements: For doctorate, GRE General Test. *Application deadline:* For fall admission, 1/4. *Application fee:* $65.

Expenses: Tuition: Full-time $22,300. Full-time tuition and fees vary according to program.

Financial aid: Fellowships, research assistantships, Federal Work-Study, institutionally sponsored loans, and traineeships available. Aid available to part-time students.

Application contact: Admissions Information, 203-432-2770.

■ YESHIVA UNIVERSITY

Albert Einstein College of Medicine, Sue Golding Graduate Division of Medical Sciences, Department of Pathology, Bronx, NY 10467

AWARDS PhD, MD/PhD.

Faculty: 25 full-time.

Students: 22 full-time (13 women); includes 7 minority (2 African Americans, 4 Asian Americans or Pacific Islanders, 1 Hispanic American), 7 international. In 1999, 3 degrees awarded.

Degree requirements: For doctorate, dissertation required, foreign language not required.

Entrance requirements: For doctorate, GRE General Test, TOEFL. *Application deadline:* For fall admission, 1/15. *Application fee:* $0.

Expenses: Tuition: Part-time $525 per credit. Tuition and fees vary according to degree level and program.

Financial aid: In 1999–00, 22 fellowships were awarded.

Faculty research: Clinical and disease-related research at tissue, cellular, and subcellular levels; biochemistry and morphology of enzyme and lysosome disorders. Dr. Michael Prystowsky, Interim Chairperson, 718-430-2827.

Application contact: Sheila Cleeton, Assistant Director, 718-430-2128, *Fax:* 718-430-8655, *E-mail:* phd@aecom.yu.edu.

Pharmacology and Toxicology

PHARMACOLOGY

■ ALBANY MEDICAL COLLEGE

Graduate Programs in the Biological Sciences, Program in Neuropharmacology and Neuroscience, Albany, NY 12208-3479

AWARDS MS, PhD. Part-time programs available.

Students: 14 full-time (8 women); includes 2 minority (1 African American, 1 Hispanic American), 6 international. Terminal master's awarded for partial completion of doctoral program.

Degree requirements: For master's, thesis required, foreign language not required; for doctorate, dissertation, comprehensive written exam, oral qualifying exam required, foreign language not required.

Entrance requirements: For master's and doctorate, GRE General Test, TOEFL. *Application deadline:* Applications are processed on a rolling basis. *Application fee:* $0 ($60 for international students).

Expenses: Tuition: Full-time $13,367; part-time $446 per credit hour.

Financial aid: Federal Work-Study, grants, scholarships, and tuition waivers (full) available.

Dr. Stanley D. Glick, Co-Director, 518-262-5303, *Fax:* 518-262-5799, *E-mail:* pharmneuroinfo@mail.amc.edu.

Find an in-depth description at www.petersons.com/graduate.

■ ALLIANT UNIVERSITY

Graduate Programs, California School of Professional Psychology, Program in Psychopharmacology, Alameda, CA 94501-1148

AWARDS MS.

Students: 93. *117 applicants, 83% accepted.*
Degree requirements: For master's, foreign language and thesis not required.
Entrance requirements: For master's, TOEFL, doctorate in clinical psychology, minimum GPA of 3.0 in both psychology and overall. *Application fee:* $50.
Expenses: Tuition: Full-time $16,990; part-time $640 per semester hour. Tuition and fees vary according to degree level and student level.
Financial aid: Federal Work-Study available. Financial aid application deadline: 2/15; financial aid applicants required to submit FAFSA.
Dr. Steven Tulkin, Director, 510-523-2300, *Fax:* 510-521-3678, *E-mail:* admissions@mail.cspp.edu.
Application contact: Patricia J. Mullen, Vice President, Enrollment and Student Services, 800-457-1273 Ext. 303, *Fax:* 415-931-8322, *E-mail:* admissions@mail.cspp.edu.

■ AUBURN UNIVERSITY

College of Veterinary Medicine and Graduate School, Graduate Program in Veterinary Medicine, Department of Anatomy, Physiology and Pharmacology, Auburn, Auburn University, AL 36849-0002

AWARDS Anatomy and histology (MS); physiology and pharmacology (MS). Part-time programs available.

Students: *9 applicants, 44% accepted.*
Degree requirements: For master's, thesis required, foreign language not required.
Entrance requirements: For master's, GRE General Test. *Application deadline:* For fall admission, 7/1; for spring admission, 12/1. Applications are processed on a rolling basis. *Application fee:* $25 ($50 for international students).
Expenses: Tuition, state resident: full-time $2,895; part-time $80 per credit hour. Tuition, nonresident: full-time $8,685; part-time $240 per credit hour.
Financial aid: Research assistantships, teaching assistantships available. Aid available to part-time students. Financial aid application deadline: 3/15.

Faculty research: Chemosensory systems, embryo transfer, osteoarthritis, neurosenses, audiology, cardiovascular physiology, molecular endocrinology. *Total annual research expenditures:* $400,000.
Dr. Philip Posner, Head, 334-844-4427.
Application contact: Dr. John F. Pritchett, Dean of the Graduate School, 334-844-4700.

■ BAYLOR COLLEGE OF MEDICINE

Graduate School of Biomedical Sciences, Department of Pharmacology, Houston, TX 77030-3498

AWARDS PhD, MD/PhD.

Faculty: 11 full-time (2 women).
Students: 7 full-time (2 women), 5 international. Average age 27. *50 applicants, 8% accepted.*
Degree requirements: For doctorate, dissertation, public defense, qualifying exam required, foreign language not required.
Entrance requirements: For doctorate, GRE General Test (average 80th percentile), GRE Subject Test (strongly recommended), TOEFL, minimum GPA of 3.0. *Application deadline:* For fall admission, 2/1 (priority date). *Application fee:* $30. Electronic applications accepted.
Expenses: Tuition: Full-time $8,200. Required fees: $175. Full-time tuition and fees vary according to student level.
Financial aid: In 1999–00, 7 students received aid, including 7 fellowships (averaging $16,000 per year); Federal Work-Study, institutionally sponsored loans, and tuition waivers (full) also available. Financial aid applicants required to submit FAFSA.
Faculty research: Cancer research, neuropharmacology, molecular proteins and U-RNA, gene cleaning, tumor markers.
Dr. P. K. Chan, Director, 713-798-7915.
Application contact: Laura Henderson, Graduate Program Administrator, 713-798-7915, *Fax:* 713-798-3145, *E-mail:* laurah@bcm.tmc.edu.

■ BOSTON UNIVERSITY

School of Medicine, Division of Graduate Medical Sciences, Department of Pharmacology and Experimental Therapeutics, Boston, MA 02118

AWARDS MA, PhD, MD/PhD.

Faculty: 12 full-time (4 women).
Students: 24 full-time (9 women), 2 part-time (both women); includes 1 minority (Asian American or Pacific Islander), 4 international. Average age 27. Terminal

master's awarded for partial completion of doctoral program.
Degree requirements: For master's and doctorate, thesis/dissertation required, foreign language not required.
Entrance requirements: For master's and doctorate, GRE General Test, GRE Subject Test, TOEFL. *Application deadline:* For fall admission, 1/15 (priority date); for spring admission, 10/15 (priority date). *Application fee:* $50. Electronic applications accepted.
Expenses: Tuition: Full-time $24,700; part-time $772 per credit. Required fees: $220.
Financial aid: In 1999–00, fellowships with tuition reimbursements (averaging $19,000 per year), research assistantships with tuition reimbursements (averaging $19,000 per year) were awarded; Federal Work-Study, scholarships, traineeships, tuition waivers, and research stipends also available.
Faculty research: Molecular pharmacology, neuropharmacology, peptide receptors, psychopharmacology.
Dr. David H. Farb, Chairman, 617-638-4300, *Fax:* 617-638-4329, *E-mail:* dfarb@bu.edu.
Application contact: Dr. Carol T. Walsh, Graduate Director, 617-638-4326, *Fax:* 617-638-4329, *E-mail:* ctwalsh@bu.edu.

Find an in-depth description at www.petersons.com/graduate.

■ BROWN UNIVERSITY

Graduate School, Division of Biology and Medicine, Program in Molecular Pharmacology and Physiology, Providence, RI 02912

AWARDS MA, Sc M, PhD, MD/PhD.

Faculty: 16 full-time (6 women).
Students: 7 full-time (2 women); includes 2 minority (both Asian Americans or Pacific Islanders), 4 international. Average age 24. *21 applicants, 14% accepted.* In 1999, 1 degree awarded. Terminal master's awarded for partial completion of doctoral program.
Degree requirements: For master's, foreign language not required; for doctorate, dissertation, preliminary exam required.
Entrance requirements: For master's, GRE General Test, GRE Subject Test; for doctorate, GRE General Test, GRE Subject Test, TOEFL. *Application deadline:* For fall admission, 1/2 (priority date). Applications are processed on a rolling basis. *Application fee:* $60.
Financial aid: In 1999–00, 2 fellowships, 1 research assistantship, 5 teaching assistantships were awarded; traineeships also available. Financial aid application deadline: 1/2.

Brown University (continued)
Faculty research: Structural biology, antiplatelet drugs, nicotinic receptor structure/function.
Dr. John Marshall, Director, 401-863-1596.
Application contact: Cheryl Pariseau, Administrative Assistant, 401-863-1596, *Fax:* 401-863-1595, *E-mail:* cheryl_pariseau@brown.edu.
Find an in-depth description at www.petersons.com/graduate.

■ CASE WESTERN RESERVE UNIVERSITY

School of Medicine and School of Graduate Studies, Graduate Programs in Medicine, Department of Pharmacology, Cleveland, OH 44106

AWARDS PhD, MD/PhD.

Faculty: 13 full-time (5 women), 18 part-time/adjunct (4 women).
Students: 16 full-time (10 women), 2 part-time (1 woman). Average age 27. *42 applicants, 10% accepted.* In 1999, 2 doctorates awarded.
Degree requirements: For doctorate, dissertation required, foreign language not required. *Average time to degree:* Doctorate–5 years full-time.
Entrance requirements: For doctorate, GRE General Test, GRE Subject Test, TOEFL. *Application deadline:* For fall admission, 3/1 (priority date); for winter admission, 8/1 (priority date); for spring admission, 8/15 (priority date). Applications are processed on a rolling basis. *Application fee:* $25.
Financial aid: In 1999–00, 18 students received aid, including 16 fellowships with full tuition reimbursements available (averaging $16,000 per year); research assistantships
Faculty research: Aspects of cellular, molecular, and clinical pharmacology; neuroendocrine pharmacology; drug metabolism. *Total annual research expenditures:* $2.4 million.
Dr. John H. Nilson, Professor and Chair, 216-368-3394, *Fax:* 216-368-3395, *E-mail:* jhn@po.cwru.edu.
Application contact: Dr. Ruth A. Keri, Coordinator of Graduate Admissions in Pharmacology, 216-368-3495, *Fax:* 216-368-3395, *E-mail:* rak5@po.cwru.edu.
Find an in-depth description at www.petersons.com/graduate.

■ COLUMBIA UNIVERSITY

College of Physicians and Surgeons and Graduate School of Arts and Sciences, Graduate School of Arts and Sciences at the College of Physicians and Surgeons, Department of Pharmacology, New York, NY 10032

AWARDS Pharmacology (M Phil, MA, PhD); pharmacology-toxicology (M Phil, MA, PhD). Only candidates for the PhD are admitted. Terminal master's awarded for partial completion of doctoral program.

Degree requirements: For master's, foreign language and thesis not required; for doctorate, dissertation required, foreign language not required.
Entrance requirements: For master's and doctorate, GRE General Test, TOEFL.
Expenses: Tuition: Full-time $25,072.
Faculty research: Cardiovascular pharmacology, receptor pharmacology, neuropharmacology, membrane biophysics, eicosanoids.
Find an in-depth description at www.petersons.com/graduate.

■ CORNELL UNIVERSITY

Graduate School, Graduate Fields of Veterinary Medicine, Field of Pharmacology, Ithaca, NY 14853-0001

AWARDS MS, PhD.

Faculty: 14 full-time.
Students: 13 full-time (4 women); includes 3 minority (2 Asian Americans or Pacific Islanders, 1 Hispanic American), 4 international. *33 applicants, 12% accepted.* In 1999, 4 doctorates awarded.
Degree requirements: For master's and doctorate, thesis/dissertation required, foreign language not required.
Entrance requirements: For master's and doctorate, GRE General Test, TOEFL. *Application deadline:* For fall admission, 1/2; for spring admission, 10/1. *Application fee:* $65. Electronic applications accepted.
Expenses: Tuition: Full-time $12,400.
Financial aid: In 1999–00, 13 students received aid, including 9 fellowships with full tuition reimbursements available, 4 research assistantships with full tuition reimbursements available; teaching assistantships with full tuition reimbursements available, institutionally sponsored loans, scholarships, tuition waivers (full and partial), and unspecified assistantships also available. Financial aid applicants required to submit FAFSA.
Faculty research: Signal transduction, receptor structure and function, ion channels, calcium signaling, G proteins.

Application contact: Graduate Field Assistant, 607-253-3276, *E-mail:* vetgradpgms@cornell.edu.
Find an in-depth description at www.petersons.com/graduate.

■ CREIGHTON UNIVERSITY

School of Medicine and Graduate School, Graduate Programs in Medicine and College of Arts and Sciences, Department of Pharmacology, Omaha, NE 68178-0001

AWARDS Pharmaceutical sciences (MS); pharmacology (PhD).

Faculty: 6 full-time (1 woman), 15 part-time/adjunct (2 women).
Students: 18 full-time (8 women); includes 3 minority (1 African American, 1 Asian American or Pacific Islander, 1 Hispanic American), 8 international. Average age 25. *35 applicants, 6% accepted.* In 1999, 1 master's, 1 doctorate awarded (100% entered university research/teaching). Terminal master's awarded for partial completion of doctoral program.
Degree requirements: For master's, thesis required; for doctorate, dissertation, oral and written preliminary exams required. *Average time to degree:* Master's–2 years full-time; doctorate–5 years full-time.
Entrance requirements: For master's and doctorate, GRE General Test, minimum GPA of 3.0, undergraduate degree in sciences. *Application deadline:* For spring admission, 4/1 (priority date). Applications are processed on a rolling basis. *Application fee:* $30.
Expenses: Tuition: Full-time $8,940; part-time $447 per credit hour. Required fees: $598; $50 per semester.
Financial aid: In 1999–00, 18 students received aid, including 6 fellowships with full tuition reimbursements available (averaging $11,400 per year); institutionally sponsored loans and tuition waivers (full and partial) also available. Financial aid application deadline: 4/1.
Faculty research: Pharmacology secretion, cardiovascular-renal pharmacology, adrenergic receptors, signal transduction, genetic regulation of receptors. *Total annual research expenditures:* $94,437.
Dr. Frank J. Dowd, Chair, 402-280-2726.
Application contact: Dr. Barbara J. Braden, Dean, Graduate School, 402-280-2870, *Fax:* 402-280-5762.

Find an in-depth description at www.petersons.com/graduate.

■ DARTMOUTH COLLEGE

School of Arts and Sciences, Department of Pharmacology and Toxicology, Hanover, NH 03755
AWARDS PhD, MD/PhD.

Faculty: 19 full-time (5 women).
Students: 22 full-time (7 women); includes 1 minority (Asian American or Pacific Islander), 2 international. Average age 25. *44 applicants, 27% accepted.* In 1999, 4 degrees awarded (75% entered university research/teaching, 25% found other work related to degree).
Degree requirements: For doctorate, dissertation required.
Entrance requirements: For doctorate, GRE General Test, GRE Subject Test, bachelor's degree in biological, chemical, or physical science. *Application deadline:* For fall admission, 1/15. *Application fee:* $10 ($0 for international students).
Expenses: Tuition: Full-time $24,624. Required fees: $916. One-time fee: $15 full-time. Full-time tuition and fees vary according to program.
Financial aid: In 1999–00, 22 students received aid, including 8 fellowships with full tuition reimbursements available (averaging $15,460 per year), 14 research assistantships with full tuition reimbursements available (averaging $15,460 per year); institutionally sponsored loans and traineeships also available. Financial aid application deadline: 4/15.
Faculty research: Molecular biology of carcinogenesis, DNA repair and gene expression, biochemical and environmental toxicology, protein-receptor ligand interactions.
Dr. Ethan Dmitrovsky, Chairman, 603-650-1667, *Fax:* 603-650-1129, *E-mail:* pharmacology.and.toxicology@dartmouth.edu.
Application contact: Linda Conrad, Coordinator of Graduate Studies, 603-650-1667, *Fax:* 603-650-1129, *E-mail:* lvc@dartmouth.edu.

Find an in-depth description at www.petersons.com/graduate.

■ DUKE UNIVERSITY

Graduate School, Department of Pharmacology and Cancer Biology, Durham, NC 27708-0586

AWARDS PhD.

Faculty: 35 full-time (6 women).
Students: 50 full-time (31 women); includes 6 minority (3 African Americans, 1 Asian American or Pacific Islander, 2 Hispanic Americans), 6 international. Average age 24. *81 applicants, 22% accepted.* In 1999, 10 degrees awarded.
Degree requirements: For doctorate, dissertation required, foreign language not required. *Average time to degree:* Doctorate–5 years full-time.
Entrance requirements: For doctorate, GRE General Test. *Application deadline:* For fall admission, 12/31. *Application fee:* $75. Electronic applications accepted.

Expenses: Tuition: Full-time $21,406; part-time $760 per unit. Required fees: $3,136; $3,136 per year. One-time fee: $30. Tuition and fees vary according to program.
Financial aid: In 1999–00, 24 fellowships, 14 research assistantships, 2 teaching assistantships were awarded; Federal Work-Study also available. Financial aid application deadline: 12/31.
Faculty research: Developmental pharmacology, neuropharmacology, molecular pharmacology, toxicology, cell growth.
Dr. Anthony M. Means, Chairman, 919-681-6209, *Fax:* 919-681-8461, *E-mail:* means001@mc.duke.edu.
Application contact: Dr. Robert Abraham, Director of Graduate Studies, 919-681-8020, *Fax:* 919-613-8647, *E-mail:* means0003@mc.duke.edu.

Find an in-depth description at www.petersons.com/graduate.

■ DUQUESNE UNIVERSITY

School of Pharmacy, Graduate School of Pharmaceutical Sciences, Program in Pharmacology/Toxicology, Pittsburgh, PA 15282-0001

AWARDS MS, PhD. Part-time programs available.

Faculty: 5 full-time (1 woman).
Students: 3 full-time (all women), 11 part-time (5 women), 1 international. *45 applicants, 11% accepted.*
Degree requirements: For master's, thesis required, foreign language not required; for doctorate, one foreign language, computer language, dissertation required.
Entrance requirements: For master's and doctorate, GRE General Test, TOEFL, TSE. *Application deadline:* For fall admission, 3/1 (priority date). Applications are processed on a rolling basis. *Application fee:* $40.
Financial aid: In 1999–00, 1 research assistantship with full tuition reimbursement, 8 teaching assistantships with full tuition reimbursements were awarded; fellowships.
Faculty research: Analytical/clinical/forensic toxicology, drugs for memory enhancement, behavioral/cardiovascular pharmacology, antidotal agents for clinical use.
Dr. Frederick W. Fochtman, Head, 412-396-6373.

■ EAST CAROLINA UNIVERSITY

School of Medicine, Department of Pharmacology, Greenville, NC 27858-4353

AWARDS PhD, MD/PhD.

Faculty: 10 full-time (0 women).
Students: 2 full-time (0 women), 5 part-time (2 women), 1 international. Average age 25. *6 applicants, 17% accepted.*
Degree requirements: For doctorate, dissertation required, foreign language not required.
Entrance requirements: For doctorate, GRE General Test, GRE Subject Test, TOEFL. *Application deadline:* For fall admission, 6/15 (priority date). Applications are processed on a rolling basis. *Application fee:* $40.
Expenses: Tuition, state resident: full-time $1,012. Tuition, nonresident: full-time $8,578. Required fees: $1,006. Full-time tuition and fees vary according to degree level. Part-time tuition and fees vary according to course load.
Financial aid: Fellowships available. Financial aid application deadline: 6/1.
Dr. Wallace R. Wooles, Chairman, 252-816-2734, *Fax:* 252-816-3203.
Application contact: Dr. Saeed Dar, Director of Graduate Studies, 252-816-2758, *Fax:* 252-816-3203, *E-mail:* dar@brody.med.ecu.edu.

Find an in-depth description at www.petersons.com/graduate.

■ EAST TENNESSEE STATE UNIVERSITY

James H. Quillen College of Medicine and School of Graduate Studies, Biomedical Science Graduate Program, Johnson City, TN 37614

AWARDS Anatomy and cell biology (MS, PhD); biochemistry and molecular biology (MS, PhD); biophysics (MS, PhD); microbiology (MS, PhD); pharmacology (MS, PhD); physiology (MS, PhD). Part-time programs available.

Faculty: 40 full-time (9 women).
Students: 23 full-time (10 women), 2 part-time (1 woman); includes 1 minority (Asian American or Pacific Islander), 4 international. Average age 30. *83 applicants, 20% accepted.* In 1999, 5 master's, 4 doctorates awarded. Terminal master's awarded for partial completion of doctoral program.
Degree requirements: For master's, one foreign language (computer language can substitute), thesis, comprehensive qualifying exam required; for doctorate, 2 foreign languages (computer language can substitute for one), dissertation required.
Entrance requirements: For master's, GRE General Test, minimum GPA of 3.0, bachelor's degree in biological or related science; for doctorate, GRE General Test, GRE Subject Test. *Application deadline:* For fall admission, 3/15 (priority date); for spring admission, 3/1. *Application fee:* $25 ($35 for international students).

East Tennessee State University (continued)

Expenses: Tuition, state resident: full-time $10,342. Tuition, nonresident: full-time $21,080. Required fees: $532.

Financial aid: In 1999–00, 16 research assistantships, 5 teaching assistantships were awarded; fellowships, career-related internships or fieldwork, Federal Work-Study, grants, institutionally sponsored loans, and tuition waivers (full) also available.

Dr. Mitchell Robinson, Assistant Dean for Graduate Studies, 423-439-4658, *E-mail:* robinson@etsu.edu.

■ EMORY UNIVERSITY

Graduate School of Arts and Sciences, Division of Biological and Biomedical Sciences, Program in Molecular and Systems Pharmacology, Atlanta, GA 30322-1100

AWARDS PhD.

Faculty: 39 full-time (4 women).
Students: 22 full-time (10 women); includes 2 African Americans, 2 international. In 1999, 4 degrees awarded.
Degree requirements: For doctorate, dissertation required, foreign language not required.
Entrance requirements: For doctorate, GRE General Test, TOEFL, minimum GPA of 3.0 in science course work. *Application deadline:* For fall admission, 1/20 (priority date). *Application fee:* $45.
Expenses: Tuition: Full-time $22,770. Tuition and fees vary according to program.
Financial aid: In 1999–00, fellowships with full tuition reimbursements (averaging $18,000 per year).
Faculty research: Transmembrane signaling, neuropharmacology, neurophysiology and neurodegeneration, metabolism and molecular toxicology, cell and developmental biology.
Dr. Edward Morgan, Director, 404-727-5986, *Fax:* 404-727-0365, *E-mail:* etmorgan@bimcore.emory.edu.
Application contact: 404-727-2547, *Fax:* 404-727-3322, *E-mail:* gdbbs@gsas.emory.edu.

Find an in-depth description at www.petersons.com/graduate.

■ FINCH UNIVERSITY OF HEALTH SCIENCES/THE CHICAGO MEDICAL SCHOOL

School of Graduate and Postdoctoral Studies, Department of Cellular and Molecular Pharmacology, North Chicago, IL 60064-3095

AWARDS MS, PhD, MD/MS, MD/PhD.
Faculty: 8 full-time.

Students: 4 full-time (2 women), 1 international. Average age 23. In 1999, 3 doctorates awarded. Terminal master's awarded for partial completion of doctoral program.
Degree requirements: For master's and doctorate, computer language, thesis/dissertation required.
Entrance requirements: For master's and doctorate, GRE General Test, TOEFL, TWE. *Application deadline:* For fall admission, 6/1 (priority date). Applications are processed on a rolling basis. *Application fee:* $25.
Expenses: Tuition: Full-time $14,054; part-time $391 per credit hour. Tuition and fees vary according to program.
Financial aid: In 1999–00, fellowships (averaging $15,500 per year); research assistantships, tuition waivers (full) also available. Financial aid application deadline: 6/9; financial aid applicants required to submit FAFSA.
Faculty research: Control of gene expression in higher organisms, molecular mechanism of action of growth factors and hormones, hormonal regulation in brain neuropsychopharmacology.
Dr. Francis White, Chairman, 847-578-3271.
Application contact: Dana Frederick, Admissions Officer, 847-578-3209.

■ FLORIDA AGRICULTURAL AND MECHANICAL UNIVERSITY

Division of Graduate Studies, Research, and Continuing Education, College of Pharmacy and Pharmaceutical Sciences, Graduate Programs in Pharmaceutical Sciences and Public Health, Tallahassee, FL 32307-3200

AWARDS Environmental toxicology (PhD); medicinal chemistry (MS, PhD); pharmaceutics (MS, PhD); pharmacology/toxicology (MS, PhD); pharmacy administration (MS); public health (MPH).

Students: 67 (50 women); includes 62 minority (60 African Americans, 1 Asian American or Pacific Islander, 1 Hispanic American) 1 international. In 1999, 20 master's awarded.
Degree requirements: For master's and doctorate, thesis/dissertation, publishable paper required, foreign language not required.
Entrance requirements: For master's and doctorate, GRE General Test, minimum GPA of 3.0 in last 60 hours. *Application fee:* $20.
Expenses: Tuition, state resident: full-time $2,644; part-time $147 per credit hour. Tuition, nonresident: full-time $9,137; part-time $508 per credit hour. Required

fees: $52 per semester. Tuition and fees vary according to course load.
Financial aid: Fellowships, research assistantships, Federal Work-Study and grants available.
Faculty research: Anticancer agents, anti-inflammatory drugs, chronopharmacology, neuroendocrinology, microbiology.
Application contact: Dr. Thomas J. Fitzgerald, Chairman, Graduate Committee, 850-599-3301.

Find an in-depth description at www.petersons.com/graduate.

■ GEORGETOWN UNIVERSITY

Graduate School of Arts and Sciences, Programs in Biomedical Sciences, Department of Pharmacology, Washington, DC 20057

AWARDS PhD, MD/PhD.

Degree requirements: For doctorate, dissertation, comprehensive exams required, foreign language not required.
Entrance requirements: For doctorate, GRE General Test, TOEFL, previous course work in biology and chemistry.
Faculty research: Neuropharmacology, techniques in biochemistry and tissue culture.

Find an in-depth description at www.petersons.com/graduate.

■ THE GEORGE WASHINGTON UNIVERSITY

Columbian School of Arts and Sciences, Institute for Biomedical Sciences, Program in Pharmacology, Washington, DC 20052

AWARDS PhD, MD/PhD.

Faculty: 10 full-time (2 women).
Students: Average age 28. In 1999, 1 doctorate awarded.
Degree requirements: For doctorate, dissertation, general exam required.
Entrance requirements: For doctorate, GRE General Test, interview, minimum GPA of 3.0. *Application fee:* $55.
Expenses: Tuition: Full-time $16,836; part-time $702 per credit hour. Required fees: $828; $35 per credit hour. Tuition and fees vary according to campus/location and program.
Financial aid: Fellowships, Federal Work-Study available. Financial aid application deadline: 1/15.
Faculty research: Biochemical pharmacology and toxicology, cancer chemotherapy, carcinogenesis, drug metabolism and disposition, pharmacokinetics.
Dr. Vincent A. Chiappinelli, Chair, 202-994-3541.
Application contact: 202-994-2179.

Find an in-depth description at www.petersons.com/graduate.

■ HARVARD UNIVERSITY

Graduate School of Arts and Sciences, Program in Biological and Biomedical Sciences, Department of Biological Chemistry and Molecular Pharmacology, Boston, MA 02115

AWARDS PhD. Applications through the Program in Biological and Biomedical Sciences (BBS).

Degree requirements: For doctorate, dissertation, qualifying exam required, foreign language not required.
Entrance requirements: For doctorate, GRE General Test, GRE Subject Test, TOEFL. *Application deadline:* For fall admission, 12/15. *Application fee:* $60.
Expenses: Tuition: Full-time $22,054. Required fees: $711. Tuition and fees vary according to program.
Financial aid: Fellowships, research assistantships, teaching assistantships, institutionally sponsored loans and tuition waivers (full) available. Financial aid application deadline: 1/1.
Faculty research: Cellular and molecular mechanisms of drug action with emphasis on basic approaches to chemotherapy, neuropharmacology, membrane biology, endocrinology, and toxicology; molecular mechanisms of receptor and drug enzyme interactions; genetics and molecular biology of DNA replication and transcription; molecular aspects of membrane protein function.
Dr. Kevin Struhl, Acting Chairman, 617-432-2104.
Application contact: Leah Simons, Manager of Student Affairs, 617-432-0162.
Find an in-depth description at www.petersons.com/graduate.

■ HOWARD UNIVERSITY

Graduate School of Arts and Sciences, Department of Pharmacology, Washington, DC 20059-0002

AWARDS MS, PhD. Part-time programs available.

Faculty: 16.
Students: 10; includes 9 minority (all African Americans), 1 international. Average age 30. *20 applicants, 20% accepted.* In 1999, 1 doctorate awarded.
Degree requirements: For master's, thesis, comprehensive exam required; for doctorate, one foreign language (computer language can substitute), dissertation, comprehensive exam, qualifying exam required. *Average time to degree:* Master's–2 years full-time; doctorate–3.5 years full-time.
Entrance requirements: For master's, GRE General Test, minimum GPA of 3.2, BS in chemistry, biology, pharmacy, or psychology; for doctorate, GRE General Test, minimum graduate GPA of 3.0.
Application deadline: For fall admission, 4/1; for spring admission, 11/1. *Application fee:* $45.
Expenses: Tuition: Full-time $10,500; part-time $583 per credit hour. Required fees: $405; $203 per semester.
Financial aid: In 1999–00, 7 students received aid; fellowships, research assistantships, teaching assistantships, career-related internships or fieldwork, grants, and institutionally sponsored loans available. Financial aid application deadline: 4/1.
Faculty research: Biochemical pharmacology, molecular pharmacology, neuropharmacology, drug metabolism, cancer research, clinical pharmacology. *Total annual research expenditures:* $4 million.
Dr. Robert Taylor, Chairman, 202-806-6311.

■ IDAHO STATE UNIVERSITY

Office of Graduate Studies, College of Pharmacy, Department of Pharmaceutical Sciences, Pocatello, ID 83209

AWARDS Biopharmaceutical analysis (PhD); biopharmaceutics (PhD); pharmaceutical chemistry (MS); pharmaceutical science (PhD); pharmaceutics (MS); pharmacognosy (MS); pharmacokinetics (PhD); pharmacology (MS, PhD).

Faculty: 10 full-time (2 women).
Students: 6 full-time (all women), 1 (woman) part-time; includes 1 minority (Asian American or Pacific Islander), 2 international. Average age 30. In 1999, 2 master's, 1 doctorate awarded.
Degree requirements: For master's, thesis required.
Entrance requirements: For master's and doctorate, GRE General Test. *Application deadline:* For fall admission, 8/1 (priority date). *Application fee:* $30.
Expenses: Tuition, nonresident: full-time $6,240; part-time $90 per credit. Required fees: $3,384; $147 per credit.
Faculty research: Metabolic toxicity of heavy metals, neuroendicine pharmacology, cardiovascular pharmacology, cancer biology, immunopharmacology. *Total annual research expenditures:* $320,000.
Dr. Christopher Daniels, Chair, 208-282-2682.

■ INDIANA UNIVERSITY BLOOMINGTON

Medical Sciences Program, Bloomington, IN 47405

AWARDS Anatomy and cell biology (MA, PhD); pharmacology (MS, PhD); physiology (MA, PhD).

Faculty: 12 full-time (3 women).
Students: 10 full-time (6 women), 6 part-time (2 women); includes 1 minority (Asian American or Pacific Islander), 3 international. In 1999, 1 master's, 1 doctorate awarded.
Entrance requirements: For master's, GRE, TOEFL, minimum GPA of 3.0; for doctorate, GRE, TOEFL. *Application deadline:* For fall admission, 1/15. *Application fee:* $45.
Expenses: Tuition, state resident: full-time $3,853; part-time $161 per credit hour. Tuition, nonresident: full-time $11,226; part-time $468 per credit hour. Required fees: $360 per year. Tuition and fees vary according to course load and program.
Dr. Talmage Bosin, Assistant Dean, 812-855-8118, *E-mail:* bosin@indiana.edu.
Application contact: Kimberly Bunch, Director of Graduate Admissions, 812-855-1119, *E-mail:* kbunch@indiana.edu.

■ INDIANA UNIVERSITY–PURDUE UNIVERSITY INDIANAPOLIS

School of Medicine, Graduate Programs in Medicine, Department of Pharmacology and Toxicology, Program in Pharmacology, Indianapolis, IN 46202-2896

AWARDS MS, PhD, MD/MS, MD/PhD.

Students: 8 full-time (4 women), 6 part-time (1 woman), 7 international. Average age 26. In 1999, 1 master's awarded. Terminal master's awarded for partial completion of doctoral program.
Degree requirements: For master's and doctorate, thesis/dissertation required, foreign language not required. *Average time to degree:* Doctorate–5 years full-time.
Entrance requirements: For master's and doctorate, GRE General Test, GRE Subject Test, minimum GPA of 3.0. *Application deadline:* For fall admission, 1/15 (priority date). Applications are processed on a rolling basis. *Application fee:* $35 ($55 for international students).
Expenses: Tuition, state resident: full-time $13,245; part-time $158 per credit hour. Tuition, nonresident: full-time $30,330; part-time $455 per credit hour. Required fees: $121 per year. Tuition and fees vary according to course load and degree level.
Financial aid: In 1999–00, 9 students received aid, including 1 fellowship with partial tuition reimbursement available (averaging $14,500 per year), 8 research assistantships with partial tuition reimbursements available (averaging $14,500 per year); Federal Work-Study, institutionally sponsored loans, and tuition waivers (partial) also available. Financial aid application deadline: 1/15.

Indiana University–Purdue University Indianapolis (continued)

Faculty research: Cardiovascular pharmacology, chemotherapy, pharmacokinetics, neuropharmacology.
Application contact: Victor Elharrar, Director of Graduate Studies, 317-274-1564, *Fax:* 317-274-7714, *E-mail:* velharra@iupui.edu.

Find an in-depth description at www.petersons.com/graduate.

■ JOAN AND SANFORD I. WEILL MEDICAL COLLEGE AND GRADUATE SCHOOL OF MEDICAL SCIENCES OF CORNELL UNIVERSITY

Graduate School of Medical Sciences, Pharmacology Graduate Program, New York, NY 10021

AWARDS PhD, MD/PhD.

Faculty: 24 full-time (7 women).
Students: 34 full-time (19 women); includes 7 minority (6 Asian Americans or Pacific Islanders, 1 Hispanic American), 9 international. *72 applicants, 17% accepted.* In 2000, 4 doctorates awarded.
Degree requirements: For doctorate, dissertation, final exam required.
Entrance requirements: For doctorate, GRE General Test, GRE Subject Test, MCAT (MD/PhD), previous course work in natural and/or health sciences. *Application deadline:* For fall admission, 1/15. *Application fee:* $50.
Expenses: All students in good standing receive an annual stipend of $22,880.
Financial aid: Fellowships, stipends available.
Kathleen Scotto, Director, 212-639-8972.

■ JOHNS HOPKINS UNIVERSITY

School of Medicine, Graduate Programs in Medicine, Department of Pharmacology and Molecular Sciences, Baltimore, MD 21205

AWARDS PhD.

Faculty: 28 full-time (4 women).
Students: 32 full-time (18 women); includes 7 minority (1 African American, 5 Asian Americans or Pacific Islanders, 1 Hispanic American), 10 international. In 1999, 6 degrees awarded (67% entered university research/teaching, 33% found other work related to degree).
Degree requirements: For doctorate. *Average time to degree:* Doctorate–6 years full-time.
Entrance requirements: For doctorate, GRE General Test, GRE Subject Test. *Application deadline:* For fall admission, 2/1. *Application fee:* $50.
Expenses: Tuition: Full-time $23,660.

Dr. Philip A. Cole, Chairman, 410-614-0540, *E-mail:* pcde@jhmi.edu.
Application contact: Dr. Wade Gibson, Director of Admissions, 410-955-7117, *Fax:* 410-955-3023, *E-mail:* wgibson@jhmi.edu.

Find an in-depth description at www.petersons.com/graduate.

■ KENT STATE UNIVERSITY

School of Biomedical Sciences, Program in Pharmacology, Kent, OH 44242-0001

AWARDS MS, PhD. Offered in cooperation with Northeastern Ohio Universities College of Medicine. Terminal master's awarded for partial completion of doctoral program.

Degree requirements: For master's and doctorate, thesis/dissertation required.
Entrance requirements: For master's and doctorate, GRE General Test.
Expenses: Tuition, state resident: full-time $5,334; part-time $243 per hour. Tuition, nonresident: full-time $10,238; part-time $466 per hour.
Faculty research: Neuropharmacology, psychotherapeutics and substance abuse, molecular biology of substance abuse, toxicology.

■ LOMA LINDA UNIVERSITY

Graduate School, Graduate Programs in Medicine, Department of Pharmacology, Loma Linda, CA 92350

AWARDS MS, PhD. Part-time programs available.

Degree requirements: For master's, thesis or alternative required, foreign language not required; for doctorate, 2 foreign languages (computer language can substitute for one), dissertation required.
Entrance requirements: For master's and doctorate, GRE General Test. *Application deadline:* Applications are processed on a rolling basis. *Application fee:* $40.
Expenses: Tuition: Part-time $395 per unit.
Financial aid: Tuition waivers (full and partial) available. Aid available to part-time students.
Faculty research: Drug metabolism, biochemical pharmacology, structure and function of cell membranes, neuropharmacology.
Dr. Marvin Peters, Coordinator, 909-824-4564.

■ LONG ISLAND UNIVERSITY, BROOKLYN CAMPUS

Arnold and Marie Schwartz College of Pharmacy and Health Sciences, Graduate Programs in Pharmacy, Division of Pharmacology/Toxicology/Medicinal Chemistry, Brooklyn, NY 11201-8423

AWARDS Pharmacology/toxicology (MS); pharmacotherapeutics (MS). Part-time and evening/weekend programs available.

Faculty: 11 full-time (3 women).
Students: 10 full-time (5 women), 26 part-time (15 women); includes 15 minority (13 Asian Americans or Pacific Islanders, 2 Hispanic Americans), 5 international. Average age 30. In 1999, 14 degrees awarded.
Degree requirements: For master's, thesis optional, foreign language not required. *Average time to degree:* Master's–2 years full-time, 5 years part-time.
Entrance requirements: For master's, minimum GPA of 3.0. *Application deadline:* Applications are processed on a rolling basis. *Application fee:* $30.
Expenses: Tuition: Part-time $550 per credit.
Financial aid: In 1999–00, 3 teaching assistantships with full tuition reimbursements (averaging $6,000 per year) were awarded.
Dr. R. R. Raje, Director, 718-488-1062.
Application contact: Bernard W. Sullivan, Associate Director of Admissions, 718-488-1011, *Fax:* 718-797-2399, *E-mail:* attend@liu.edu.

■ LOUISIANA STATE UNIVERSITY HEALTH SCIENCES CENTER

School of Graduate Studies in New Orleans, Department of Pharmacology and Experimental Therapeutics, New Orleans, LA 70112-2223

AWARDS PhD, MD/PhD.

Faculty: 16 full-time (3 women).
Students: 10 full-time (3 women); includes 3 minority (all African Americans), 5 international. Average age 25. *25 applicants, 16% accepted.*
Degree requirements: For doctorate, dissertation required, foreign language not required.
Entrance requirements: For doctorate, GRE General Test, TOEFL. *Application deadline:* For fall admission, 2/15. *Application fee:* $30.
Expenses: Tuition, state resident: full-time $2,878; part-time $126 per hour. Tuition, nonresident: full-time $6,003; part-time $265 per hour. Required fees: $2,272. Tuition and fees vary according to course load, degree level and program.
Financial aid: In 1999–00, 6 research assistantships were awarded; teaching

assistantships, Federal Work-Study and tuition waivers (full) also available. Financial aid application deadline: 2/15.
Faculty research: Neuropharmacology, gastrointestinal pharmacology, drug metabolism, behavioral pharmacology, cardiovascular pharmacology.
Application contact: Dr. Emel Songu-Mize, Graduate Coordinator, 504-568-4740, *Fax:* 504-568-2361, *E-mail:* emize@lsumc.edu.

Find an in-depth description at www.petersons.com/graduate.

■ LOUISIANA STATE UNIVERSITY HEALTH SCIENCES CENTER

School of Graduate Studies in Shreveport, Department of Pharmacology and Therapeutics, Shreveport, LA 71130-3932

AWARDS Pharmacology (MS, PhD). Terminal master's awarded for partial completion of doctoral program.

Degree requirements: For master's and doctorate, thesis/dissertation required, foreign language not required.
Entrance requirements: For master's, GRE General Test, TOEFL; for doctorate, GRE General Test, TOEFL, minimum GPA of 3.0.
Expenses: Tuition, state resident: full-time $2,878; part-time $126 per hour. Tuition, nonresident: full-time $6,003; part-time $265 per hour. Required fees: $2,272. Tuition and fees vary according to course load, degree level and program.
Faculty research: Behavioral, cardiovascular, clinical, and gastrointestinal pharmacology; neuropharmacology; psychopharmacology; drug abuse; pharmacokinetics; neuroendocrinology, psychoneuroimmunology, and stress; toxicology.

Find an in-depth description at www.petersons.com/graduate.

■ LOYOLA UNIVERSITY CHICAGO

Graduate School, Department of Pharmacology and Experimental Therapeutics, Chicago, IL 60626

AWARDS MS, PhD, MBA/MS, MD/PhD, PhD/MBA.

Faculty: 10 full-time (3 women), 5 part-time/adjunct (0 women).
Students: 16 full-time (9 women); includes 2 minority (both Asian Americans or Pacific Islanders), 3 international. Average age 27. *42 applicants, 14% accepted.* In 1999, 3 doctorates awarded (67% entered university research/teaching, 33% found other work related to degree). Terminal master's awarded for partial completion of doctoral program.

Degree requirements: For master's and doctorate, thesis/dissertation, comprehensive exam required, foreign language not required. *Average time to degree:* Doctorate–6 years full-time.
Entrance requirements: For master's and doctorate, GRE General Test, TOEFL, minimum GPA of 3.0. *Application deadline:* For fall admission, 2/1. *Application fee:* $35. Electronic applications accepted.
Expenses: Tuition: Part-time $500 per credit hour. Required fees: $42 per term.
Financial aid: In 1999–00, 8 fellowships with full tuition reimbursements (averaging $17,500 per year), 6 research assistantships with full tuition reimbursements (averaging $17,500 per year) were awarded; career-related internships or fieldwork and Federal Work-Study also available. Financial aid application deadline: 2/1; financial aid applicants required to submit FAFSA.
Faculty research: Neuropharmacology, molecular pharmacology, neuroendocrinology, hematopharmacology, neurodegeneration.
Dr. Israel Hanin, Chair, 708-216-3261, *Fax:* 708-216-6596, *E-mail:* ihanin@luc.edu.
Application contact: Dr. William A. Wolf, Graduate Program Director, 708-202-8387, *Fax:* 708-216-6596, *E-mail:* wwolf@luc.edu.

Find an in-depth description at www.petersons.com/graduate.

■ MASSACHUSETTS COLLEGE OF PHARMACY AND HEALTH SCIENCES

Graduate Studies, Program in Pharmacology, Boston, MA 02115-5896

AWARDS MS, PhD.

Faculty: 5 full-time (2 women), 2 part-time/adjunct (0 women).
Students: 7 full-time (4 women), 3 part-time (1 woman); includes 2 minority (both Asian Americans or Pacific Islanders), 6 international. Average age 24. *25 applicants, 48% accepted.* In 1999, 1 master's awarded (100% found work related to degree); 2 doctorates awarded (100% entered university research/teaching). Terminal master's awarded for partial completion of doctoral program.

Degree requirements: For master's, oral defense of thesis required; for doctorate, one foreign language (computer language can substitute), oral defense of dissertation, qualifying exam required. *Average time to degree:* Master's–3 years full-time; doctorate–5 years full-time.
Entrance requirements: For master's and doctorate, GRE General Test, TOEFL, minimum QPA of 3.0. *Application deadline:*

For fall admission, 2/1 (priority date). *Application fee:* $60.
Expenses: Tuition: Full-time $8,250; part-time $550 per semester hour. Required fees: $400 per year.
Financial aid: In 1999–00, 8 students received aid, including 3 research assistantships (averaging $11,000 per year), 5 teaching assistantships (averaging $11,000 per year); fellowships, tuition waivers (full) and animal caretaker, library assistantships also available. Financial aid application deadline: 2/1.
Faculty research: Neuropharmacology, cardiovascular pharmacology, nutritional pharmacology, pulmonary physiology, drug metabolism. *Total annual research expenditures:* $30,000.
Dr. Timothy Maher, Director of Research, 617-732-2940, *Fax:* 617-732-2963, *E-mail:* tmaher@mcp.edu.
Application contact: Lovie Condrick, Coordinator of Graduate Admissions, 617-732-2986, *Fax:* 617-732-2801, *E-mail:* admissions@mcp.edu.

■ MAYO GRADUATE SCHOOL

Graduate Programs in Biomedical Sciences, Department of Molecular Pharmacology and Experimental Therapeutics, Rochester, MN 55905

AWARDS PhD.

Faculty: 25 full-time (2 women).
Students: 21 full-time (10 women); includes 3 minority (1 African American, 2 Hispanic Americans), 6 international. In 1999, 2 degrees awarded.
Degree requirements: For doctorate, oral defense of dissertation, qualifying oral and written exam required.
Entrance requirements: For doctorate, GRE, TOEFL, 2 years of chemistry; 1 year of biology, calculus, and physics. *Application deadline:* For fall admission, 12/31 (priority date). Applications are processed on a rolling basis. *Application fee:* $0.
Expenses: Tuition: Full-time $17,900.
Financial aid: In 1999–00, 19 students received aid, including 19 fellowships with full tuition reimbursements available (averaging $17,500 per year); tuition waivers (full) also available.
Faculty research: Patch clamping, G-proteins, pharmacogenetics, receptor-induced transcriptional events, cholinesterase biology.
Dr. Matthew M. Ames, Education Coordinator, 507-284-2424, *Fax:* 507-284-9111, *E-mail:* ames.matthew@mayo.edu.

Mayo Graduate School (continued)
Application contact: Sherry Kallies, Information Contact, 507-266-0122, *Fax:* 507-284-0999, *E-mail:* phd.training@mayo.edu.

Find an in-depth description at www.petersons.com/graduate.

■ MCP HAHNEMANN UNIVERSITY

School of Medicine, Biomedical Graduate Programs, Department of Pharmacology, Philadelphia, PA 19102-1192

AWARDS MS, PhD, MD/PhD. Part-time programs available. Terminal master's awarded for partial completion of doctoral program.

Degree requirements: For master's, comprehensive exam required; for doctorate, dissertation, qualifying exam required, foreign language not required.
Entrance requirements: For master's, GRE General Test, TOEFL, minimum GPA of 2.75; for doctorate, GRE General Test, TOEFL, minimum GPA of 3.0.
Faculty research: Cardiovascular pharmacology, drugs of abuse, neurotransmitter mechanisms.

Find an in-depth description at www.petersons.com/graduate.

■ MEDICAL COLLEGE OF GEORGIA

School of Graduate Studies, Department of Pharmacology and Toxicology, Augusta, GA 30912-1500

AWARDS MS, PhD.

Faculty: 10 full-time (2 women).
Students: 17 full-time (6 women); includes 4 minority (2 African Americans, 2 Asian Americans or Pacific Islanders), 8 international. *18 applicants, 61% accepted.* Terminal master's awarded for partial completion of doctoral program.
Degree requirements: For master's and doctorate, thesis/dissertation required, foreign language not required.
Entrance requirements: For master's and doctorate, GRE General Test, TOEFL. *Application deadline:* For fall admission, 6/30 (priority date). Applications are processed on a rolling basis. *Application fee:* $25.
Expenses: Tuition, state resident: full-time $2,896; part-time $121 per hour. Tuition, nonresident: full-time $11,584; part-time $483 per hour. Required fees: $286; $143 per semester. Tuition and fees vary according to program.
Financial aid: In 1999–00, 12 research assistantships with partial tuition reimbursements (averaging $15,500 per year) were awarded; fellowships, Federal Work-Study, grants, and institutionally

sponsored loans also available. Aid available to part-time students. Financial aid application deadline: 3/31; financial aid applicants required to submit FAFSA.
Faculty research: Kinins, bradykinins and inflammation, cholinergic neurotransmitters, neuropharmacology.
Dr. R. William Caldwell, Chairman, 706-721-3384, *Fax:* 706-721-2347, *E-mail:* wcaldwell@mail.mcg.edu.
Application contact: Dr. Nevin Lambert, Director, 706-721-6345, *Fax:* 706-721-2347, *E-mail:* dlewis@mail.mcg.edu.

■ MEDICAL COLLEGE OF OHIO

Graduate School, Department of Pharmacology, Toledo, OH 43614-5805

AWARDS Pharmacology (MS). Part-time programs available.

Faculty: 14 full-time (4 women), 1 part-time/adjunct (0 women).
Students: 3 full-time (2 women), 4 part-time (2 women), 6 international. Average age 30. In 1999, 1 degree awarded (100% entered university research/teaching).
Degree requirements: For master's, thesis, qualifying exam required, foreign language not required. *Average time to degree:* Master's–3 years full-time.
Entrance requirements: For master's, GRE General Test, minimum undergraduate GPA of 3.0. *Application fee:* $30.
Expenses: Tuition, state resident: part-time $193 per hour. Tuition, nonresident: part-time $445 per hour. Tuition and fees vary according to degree level.
Financial aid: Fellowships, Federal Work-Study and institutionally sponsored loans available. Financial aid applicants required to submit FAFSA.
Faculty research: Neuropharmacology of drug tolerance and dependence, biochemical pharmacology, cardiovascular pharmacology, molecular pharmacology and bioenergetics, clinical pharmacology. *Total annual research expenditures:* $1.6 million.
Howard Rosenberg, Chairman, 419-383-4117, *Fax:* 419-383-6140, *E-mail:* mcogradschool@mco.edu.
Application contact: Joann Braatz, Clerk, 419-383-4117, *Fax:* 419-383-6140, *E-mail:* mcogradschool@mco.edu.

■ MEDICAL COLLEGE OF WISCONSIN

Graduate School of Biomedical Sciences, Department of Pharmacology and Toxicology, Milwaukee, WI 53226-0509

AWARDS MS, PhD, MD/MS, MD/PhD. Terminal master's awarded for partial completion of doctoral program.

Degree requirements: For master's, thesis required, foreign language not required; for doctorate, dissertation, oral and written qualifying exams required, foreign language not required.
Entrance requirements: For master's and doctorate, GRE General Test, TOEFL, minimum B average.
Expenses: Tuition, state resident: full-time $9,318. Tuition, nonresident: full-time $9,318. Required fees: $115.
Faculty research: Cardiovascular physiology and pharmacology, drugs of abuse, environmental and aquatic toxicology, central nervous system and biochemical pharmacology, signal transduction.

■ MEDICAL UNIVERSITY OF SOUTH CAROLINA

College of Graduate Studies, Department of Cell and Molecular Pharmacology and Experimental Therapeutics, Charleston, SC 29425-0002

AWARDS MS, PhD, DMD/PhD, MD/PhD.

Faculty: 18 part-time/adjunct (2 women).
Students: 18 full-time (10 women). Average age 28. *15 applicants, 33% accepted.* In 1999, 5 degrees awarded. Terminal master's awarded for partial completion of doctoral program.
Degree requirements: For master's, thesis, research seminar required; for doctorate, dissertation, teaching and research seminar, oral and written exams required, foreign language not required.
Entrance requirements: For master's and doctorate, GRE General Test, TOEFL, interview. *Application deadline:* Applications are processed on a rolling basis. *Application fee:* $55. Electronic applications accepted.
Expenses: Tuition, state resident: full-time $3,470; part-time $160 per semester hour. Tuition, nonresident: full-time $4,426; part-time $213 per semester hour. Required fees: $408 per semester. One-time fee: $160. Tuition and fees vary according to program.
Financial aid: In 1999–00, 2 fellowships (averaging $16,000 per year) were awarded; research assistantships, teaching assistantships, Federal Work-Study and tuition waivers (partial) also available. Financial aid application deadline: 4/1; financial aid applicants required to submit FAFSA.
Faculty research: Hypertension, kallikrein-kinin, sodium/calcium exchange, thromboxane receptors, molecular toxicology. *Total annual research expenditures:* $3.5 million.
Dr. H. S. Margolius, Chairman, 843-792-2471.

Application contact: Julie Johnston, Director of Admissions, 843-792-8710, *Fax:* 843-792-3764.

■ MEHARRY MEDICAL COLLEGE

School of Graduate Studies, Department of Pharmacology, Nashville, TN 37208-9989

AWARDS MS, PhD.

Faculty: 10 full-time (2 women), 1 part-time/adjunct (0 women).

Students: 14 full-time (13 women); all minorities (all African Americans). Average age 29. 7 *applicants, 86% accepted.* In 1999, 2 doctorates awarded.

Degree requirements: For master's, thesis required, foreign language not required; for doctorate, dissertation, oral and written comprehensive exams required, foreign language not required.

Entrance requirements: For master's and doctorate, GRE. *Application deadline:* For fall admission, 6/1. Applications are processed on a rolling basis. *Application fee:* $45.

Expenses: Tuition: Full-time $8,732. Required fees: $2,133.

Financial aid: In 1999–00, 7 students received aid, including 1 fellowship, 4 research assistantships, 2 teaching assistantships; Federal Work-Study, institutionally sponsored loans, and tuition waivers (full) also available. Aid available to part-time students. Financial aid application deadline: 4/15.

Faculty research: Neuropharmacology, cardiovascular pharmacology, behavioral pharmacology, molecular pharmacology, drug metabolism, anticancer.

Dr. Delores Shockley, Chair, 615-327-6510, *Fax:* 615-327-6632, *E-mail:* dshockley@mmc.edu.

Find an in-depth description at www.petersons.com/graduate.

■ MICHIGAN STATE UNIVERSITY

College of Human Medicine and Graduate School, Graduate Programs in Human Medicine and Graduate Programs in Osteopathic Medicine and Graduate Programs in Veterinary Medicine, Department of Pharmacology/Toxicology, East Lansing, MI 48824

AWARDS MS, PhD. Part-time programs available.

Faculty: 18.

Students: 21 full-time (12 women), 4 part-time (2 women); includes 7 minority (2 African Americans, 3 Asian Americans or Pacific Islanders, 1 Hispanic American, 1 Native American), 4 international. Average age 28. 46 *applicants, 20% accepted.* In 1999, 3 degrees awarded.

Degree requirements: For master's, thesis required, foreign language not required; for doctorate, dissertation, comprehensive exams required, foreign language not required.

Entrance requirements: For master's and doctorate, GRE General Test, TOEFL, minimum GPA of 3.0. *Application fee:* $30 ($40 for international students). Electronic applications accepted.

Expenses: Tuition, state resident: full-time $10,868; part-time $229 per credit. Tuition, nonresident: full-time $23,168; part-time $464 per credit.

Financial aid: In 1999–00, 13 research assistantships (averaging $11,500 per year) were awarded; fellowships, teaching assistantships, grants and institutionally sponsored loans also available. Aid available to part-time students. Financial aid applicants required to submit FAFSA.

Faculty research: Central neural control of cardiovascular function, blood pressure control by forebrain and brainstem neurons, endothelian receptors, sensory nerve excitation in the intestine. *Total annual research expenditures:* $3.1 million.

Dr. Kenneth Moore, Chairperson, 517-353-7145, *Fax:* 517-353-8915.

Application contact: Dr. James J. Galligan, Graduate Committee, 517-353-7145, *Fax:* 517-353-8915, *E-mail:* hummeld@msu.edu.

■ MICHIGAN STATE UNIVERSITY

College of Osteopathic Medicine and Graduate School, Graduate Programs in Osteopathic Medicine, East Lansing, MI 48824

AWARDS Anatomy (MS, PhD); biochemistry (MS, PhD); microbiology (PhD); pathology (MS, PhD); pharmacology/toxicology (MS, PhD), including pharmacology; physiology (MS, PhD), including environmental toxicology (PhD), neuroscience (PhD). Part-time programs available.

Students: 17 full-time (9 women), 1 part-time; includes 5 minority (3 Asian Americans or Pacific Islanders, 1 Hispanic American, 1 Native American), 5 international. Average age 26. 33 *applicants, 9% accepted.* In 1999, 2 doctorates awarded.

Degree requirements: For doctorate, dissertation required.

Entrance requirements: For master's and doctorate, GRE. *Application deadline:* For fall admission, 3/1 (priority date). Applications are processed on a rolling basis. *Application fee:* $30 ($40 for international students).

Expenses: Tuition, area resident: Full-time $15,879. Tuition, state resident: full-time $33,797.

Financial aid: Fellowships, research assistantships, teaching assistantships, career-related internships or fieldwork, Federal Work-Study, and institutionally sponsored loans available. Financial aid application deadline: 4/2. *Total annual research expenditures:* $5 million.

Application contact: Dr. Veronica M. Maher, Associate Dean for Graduate Studies, 517-353-7785, *Fax:* 517-353-9004, *E-mail:* maher@com.msu.edu.

■ MICHIGAN STATE UNIVERSITY

College of Veterinary Medicine and Graduate School, Graduate Programs in Veterinary Medicine, East Lansing, MI 48824

AWARDS Anatomy (MS, PhD); large animal clinical sciences (MS, PhD); microbiology (MS, PhD); pathology (MS, PhD); pharmacology/toxicology (MS, PhD), including pharmacology; small animal clinical sciences (MS).

Students: 52 full-time (28 women); includes 6 minority (4 African Americans, 2 Asian Americans or Pacific Islanders), 18 international. Average age 32. 45 *applicants, 20% accepted.* In 1999, 3 master's, 3 doctorates awarded.

Degree requirements: For master's, thesis or alternative required, foreign language not required; for doctorate, dissertation required, foreign language not required.

Application deadline: Applications are processed on a rolling basis. *Application fee:* $30 ($40 for international students). Electronic applications accepted.

Expenses: Tuition, state resident: full-time $9,766. Tuition, nonresident: full-time $20,082. Tuition and fees vary according to program.

Financial aid: Fellowships, research assistantships available.

Faculty research: Molecular genetics, food safety/toxicology, comparative orthopedics, airway disease, population medicine.

Dr. John C. Baker, Associate Dean for Research and Graduate Studies, 517-432-2388, *Fax:* 517-432-1037, *E-mail:* baker@cvm.msu.edu.

Application contact: Victoria Hoelzer-Maddox, Administrative Assistant, 517-353-3118, *Fax:* 517-432-1037, *E-mail:* hoelzer-maddox@cvm.msu.edu.

■ MOUNT SINAI SCHOOL OF MEDICINE OF NEW YORK UNIVERSITY

Graduate School of Biological Sciences, Department of Pharmacology, New York, NY 10029-6504

AWARDS PhD, MD/PhD.

Faculty: 11 full-time.

Mount Sinai School of Medicine of New York University (continued)
Students: 9 full-time (3 women), 3 international. *25 applicants, 8% accepted.*
Degree requirements: For doctorate, dissertation required, foreign language not required.
Entrance requirements: For doctorate, GRE General Test, GRE Subject Test, TOEFL. *Application deadline:* For fall admission, 4/15. *Application fee:* $35.
Expenses: Tuition: Full-time $21,750; part-time $725 per credit. Required fees: $750; $25 per credit. Full-time tuition and fees vary according to student level.
Financial aid: Grants available.
Dr. Jack Peter Green, Chairman, 212-241-7014.
Application contact: C. Gita Bosch, Administrative Manager and Assistant Dean, 212-241-6546, *Fax:* 212-241-0651, *E-mail:* grads@mssm.edu.
Find an in-depth description at www.petersons.com/graduate.

■ NEW YORK MEDICAL COLLEGE

Graduate School of Basic Medical Sciences, Program in Pharmacology, Valhalla, NY 10595-1691
AWARDS MS, PhD, MD/PhD. Part-time and evening/weekend programs available.

Faculty: 15 full-time (3 women).
Students: 12 full-time (5 women), 7 part-time (5 women); includes 2 minority (1 African American, 1 Asian American or Pacific Islander), 7 international. Average age 30. *31 applicants, 26% accepted.* In 1999, 5 master's, 4 doctorates awarded. Terminal master's awarded for partial completion of doctoral program.
Degree requirements: For master's and doctorate, computer language, thesis/dissertation required, foreign language not required.
Entrance requirements: For master's, GRE General Test, TOEFL; for doctorate, GRE General Test, GRE Subject Test, TOEFL. *Application deadline:* For fall admission, 7/1 (priority date); for spring admission, 12/1 (priority date). Applications are processed on a rolling basis. *Application fee:* $35 ($60 for international students).
Expenses: Tuition: Part-time $430 per credit. Required fees: $15 per semester. One-time fee: $100.
Financial aid: In 1999–00, 11 research assistantships with full tuition reimbursements were awarded; career-related internships or fieldwork, Federal Work-Study, grants, institutionally sponsored loans, and tuition waivers also available. Aid available to part-time students. Financial aid applicants required to submit FAFSA.

Faculty research: Hypertension, neuroendocrine and renal physiology, metabolism of vasoactive peptides, neuroendocrine and hormonal control of circulation.
Dr. Michael Schwartzman, Co-Director, 914-594-4153.

Find an in-depth description at www.petersons.com/graduate.

■ NEW YORK UNIVERSITY

Graduate School of Arts and Science, Department of Basic Medical Sciences, New York, NY 10012-1019

AWARDS Biochemistry (MS, PhD); cell biology (MS, PhD); microbiology (MS, PhD); parasitology (PhD); pathology (MS, PhD); pharmacology (PhD); physiology (MS, PhD). Part-time programs available.

Faculty: 23 full-time (2 women), 3 part-time/adjunct (1 woman).
Students: 182 full-time (73 women), 4 part-time (1 woman); includes 52 minority (11 African Americans, 34 Asian Americans or Pacific Islanders, 7 Hispanic Americans), 48 international. Average age 26. *671 applicants, 11% accepted.* In 1999, 15 master's, 32 doctorates awarded. Terminal master's awarded for partial completion of doctoral program.
Degree requirements: For master's, thesis or alternative, written comprehensive exam required, foreign language not required; for doctorate, one foreign language, dissertation, oral and written comprehensive exams required.
Entrance requirements: For master's and doctorate, GRE General Test, GRE Subject Test, TOEFL. *Application deadline:* For fall admission, 2/1 (priority date). *Application fee:* $60.
Expenses: Tuition: Full-time $17,880; part-time $745 per credit. Required fees: $1,140; $35 per credit. Tuition and fees vary according to course load and program.
Financial aid: Fellowships with tuition reimbursements, research assistantships with tuition reimbursements, teaching assistantships with tuition reimbursements, career-related internships or fieldwork, Federal Work-Study, institutionally sponsored loans, and tuition waivers (full and partial) available. Financial aid application deadline: 2/1; financial aid applicants required to submit FAFSA.
Dr. Joel D. Oppenheim, Director, 212-263-5648, *Fax:* 212-263-7600, *E-mail:* sackler-info@nyumed.med.nyu.edu.

■ NEW YORK UNIVERSITY

School of Medicine and Graduate School of Arts and Science, Medical Scientist Training Program, New York, NY 10012-1019

AWARDS Biochemistry (MD/PhD); cell biology (MD/PhD); environmental health sciences (MD/PhD); microbiology (MD/PhD); parasitology (MD/PhD); pathology (MD/PhD); pharmacology (MD/PhD). Students must be accepted by both the School of Medicine and the Graduate School of Arts and Science.

Faculty: 150 full-time (35 women).
Students: 80 full-time (18 women); includes 37 minority (3 African Americans, 32 Asian Americans or Pacific Islanders, 2 Hispanic Americans), 1 international. Average age 25. *195 applicants, 18% accepted.*
Degree requirements: One foreign language.
Application deadline: For fall admission, 11/15. Applications are processed on a rolling basis. *Application fee:* $60.
Expenses: Students receive full tuition support and an annual stipend of $17,500.
Financial aid: Application deadline: 5/1.
Faculty research: Genetics, tumor biology, cardiovascular biology, neuroscience, host defense mechanisms.
Dr. James Salzer, Director, 212-263-0758.
Application contact: Arlene Kohler, Administrative Officer, 212-263-5649.

■ NEW YORK UNIVERSITY

School of Medicine and Graduate School of Arts and Science, Sackler Institute of Graduate Biomedical Sciences, Department of Pharmacology, New York, NY 10012-1019
AWARDS PhD, MD/PhD.

Faculty: 19 full-time (3 women).
Students: 13 full-time (3 women); includes 4 minority (all Asian Americans or Pacific Islanders), 5 international. Average age 25. In 1999, 7 degrees awarded (100% entered university research/teaching).
Degree requirements: For doctorate, one foreign language, dissertation, qualifying exam required. *Average time to degree:* Doctorate–5.5 years full-time.
Entrance requirements: For doctorate, GRE General Test, GRE Subject Test, TOEFL. *Application deadline:* For fall admission, 2/1 (priority date). Applications are processed on a rolling basis. *Application fee:* $60.
Expenses: Tuition: Full-time $17,880; part-time $745 per credit. Required fees: $1,140; $35 per credit. Tuition and fees vary according to course load and program.
Financial aid: Fellowships, research assistantships, teaching assistantships,

tuition waivers (full) available. Financial aid application deadline: 1/15.

Faculty research: Pharmacology and neurobiology, neuropeptides, receptor biochemistry, cytoskeleton, endocrinology. *Total annual research expenditures:* $1.9 million.

Dr. Joseph Schlessinger, Chairman, 212-263-7111.

Application contact: Dr. Lakshmi Devi, Graduate Student Adviser, 212-263-7119, *E-mail:* devil01@popmail.med.nyu.edu.

■ **NORTH CAROLINA STATE UNIVERSITY**

College of Veterinary Medicine and Graduate School, Graduate Programs in Comparative Biomedical Sciences, Raleigh, NC 27695

AWARDS Cell biology and morphology (MS, PhD); epidemiology and population medicine (MS, PhD); immunology (MS, PhD); microbiology and immunology (MS, PhD); pathology (MS, PhD); pharmacology (MS, PhD); specialized veterinary medicine (MS). Part-time programs available.

Students: 40 full-time (23 women), 22 part-time (12 women); includes 11 minority (8 African Americans, 2 Asian Americans or Pacific Islanders, 1 Hispanic American), 14 international. Average age 34. *33 applicants, 33% accepted.* In 1999, 2 master's, 2 doctorates awarded.

Degree requirements: For master's and doctorate, thesis/dissertation required.

Entrance requirements: For master's and doctorate, GRE General Test. *Application deadline:* for fall admission, 6/25; for spring admission, 11/25. Applications are processed on a rolling basis. *Application fee:* $45.

Expenses: Tuition, state resident: full-time $1,578. Tuition, nonresident: full-time $10,744. Required fees: $892. Full-time tuition and fees vary according to program.

Financial aid: Fellowships, research assistantships, teaching assistantships available. Financial aid application deadline: 2/15.

Faculty research: Infectious diseases, immunology and virology, tumor biology, toxicological pathology, food safety. Dr. Neil C. Olson, Associate Dean, 919-513-6213, *Fax:* 919-513-6222, *E-mail:* neil_olson@ncsu.edu.

■ **NORTHEASTERN UNIVERSITY**

Bouvé College of Health Sciences Graduate School, Program in Pharmacology, Boston, MA 02115-5096

AWARDS MS. Part-time and evening/weekend programs available.

Students: 13 full-time (8 women), 10 part-time (4 women). Average age 30. *26 applicants, 69% accepted.* In 1999, 4 degrees awarded.

Degree requirements: For master's, comprehensive exam required, thesis optional, foreign language not required.

Entrance requirements: For master's, bachelor's degree in science, minimum GPA of 3.0. *Application deadline:* Applications are processed on a rolling basis. *Application fee:* $50.

Expenses: Tuition: Full-time $16,560; part-time $460 per quarter hour. Required fees: $150; $25 per year. Tuition and fees vary according to course load and program.

Financial aid: Federal Work-Study and tuition waivers (partial) available. Aid available to part-time students. Financial aid application deadline: 3/1; financial aid applicants required to submit FAFSA.

Faculty research: Nicotinic receptor subtypes, G-protein coupled receptors, dopamine receptor pharmacology, pathclamping of ionchannel. *Total annual research expenditures:* $1.6 million. Dr. Ralph Loring, Director, 617-373-3216, *Fax:* 617-266-6756, *E-mail:* r.loring@nunet.neu.edu.

Application contact: Bill Purnell, Director of Graduate Admissions, 617-373-2708, *Fax:* 617-373-4701, *E-mail:* w.purnell@nunet.neu.edu.

Find an in-depth description at www.petersons.com/graduate.

■ **NORTHEASTERN UNIVERSITY**

Bouvé College of Health Sciences Graduate School, Programs in Biomedical Sciences, Boston, MA 02115-5096

AWARDS Biomedical sciences (MS); medical laboratory science (PhD); medicinal chemistry (PhD); pharmaceutics (PhD); pharmacology (PhD); toxicology (MS, PhD).

Faculty: 14 full-time (1 woman), 14 part-time/adjunct (5 women).

Students: 47 full-time (30 women), 6 part-time (4 women). Average age 31. *162 applicants, 47% accepted.* In 1999, 15 master's, 6 doctorates awarded. Terminal master's awarded for partial completion of doctoral program.

Degree requirements: For master's, comprehensive exam required, thesis optional, foreign language not required; for doctorate, dissertation, qualifying exam required, foreign language not required.

Entrance requirements: For master's and doctorate, GRE General Test, TOEFL. *Application deadline:* For fall admission, 3/15. *Application fee:* $50.

Expenses: Tuition: Full-time $16,560; part-time $460 per quarter hour. Required

fees: $150; $25 per year. Tuition and fees vary according to course load and program.

Financial aid: In 1999–00, 40 students received aid, including 10 research assistantships with full tuition reimbursements available, 12 teaching assistantships with full tuition reimbursements available (averaging $12,650 per year); tuition waivers (partial) also available. Financial aid applicants required to submit FAFSA.

Faculty research: Neuropharmacology, cardiovascular pharmacology, steroid chemistry, anti-infectives, behavioral pharmacology.

Dr. Roger W. Giese, Director, 617-373-3227, *Fax:* 617-266-6756, *E-mail:* rgiese@lynx.neu.edu.

Application contact: Bill Purnell, Director of Graduate Admissions, 617-373-2708, *Fax:* 617-373-4701, *E-mail:* w.purnell@nunet.neu.edu.

Find an in-depth description at www.petersons.com/graduate.

■ **NORTHWESTERN UNIVERSITY**

The Graduate School, Division of Interdepartmental Programs and Medical School, Integrated Graduate Programs in the Life Sciences, Chicago, IL 60611

AWARDS Cancer biology (PhD); cell biology (PhD); developmental biology (PhD); evolutionary biology (PhD); immunology and microbial pathogenesis (PhD); molecular biology and genetics (PhD); neurobiology (PhD); pharmacology and toxicology (PhD); structural biology and biochemistry (PhD).

Degree requirements: For doctorate, dissertation, written and oral qualifying exams required, foreign language not required.

Entrance requirements: For doctorate, GRE General Test, TOEFL.

Expenses: Tuition: Full-time $23,301. Full-time tuition and fees vary according to program.

Find an in-depth description at www.petersons.com/graduate.

■ **NOVA SOUTHEASTERN UNIVERSITY**

Center for Psychological Studies, Postdoctoral Master's Program in Psychopharmacology, Fort Lauderdale, FL 33314-7721

AWARDS MS. Part-time programs available.

Faculty: 2 full-time (0 women), 1 part-time/adjunct (0 women).

Degree requirements: For master's, two practica required, foreign language and thesis not required.

Entrance requirements: For master's, licensed psychologist. *Application fee:* $50.

Nova Southeastern University (continued)

Expenses: Tuition: Part-time $420 per credit hour. Required fees: $50 per semester. Tuition and fees vary according to program.

Financial aid: Application deadline: 4/1. Nancy L. Smith, Supervisor, 954-262-5760, *Fax:* 954-262-3893, *E-mail:* cpsinfo@cps.nova.edu.

Application contact: Supervisor, 954-262-5760.

■ THE OHIO STATE UNIVERSITY

College of Medicine and Public Health and Graduate School, Graduate Programs in the Basic Medical Sciences, Department of Pharmacology, Columbus, OH 43210

AWARDS Pharmacology (MS, PhD); toxicology (MS, PhD). Part-time programs available.

Faculty: 10 full-time (2 women), 3 part-time/adjunct (2 women).

Students: 12 full-time; includes 6 minority (all Asian Americans or Pacific Islanders), 1 international. Average age 28. *26 applicants, 8% accepted.* In 1999, 1 doctorate awarded (100% continued full-time study). Terminal master's awarded for partial completion of doctoral program.

Degree requirements: For master's, thesis optional, foreign language not required; for doctorate, dissertation required, foreign language not required. *Average time to degree:* Doctorate–4 years full-time.

Entrance requirements: For master's and doctorate, GRE General Test, GRE Subject Test, TOEFL. *Application deadline:* For fall admission, 8/1 (priority date); for spring admission, 1/15. Applications are processed on a rolling basis. *Application fee:* $30 ($40 for international students).

Expenses: Tuition, state resident: full-time $5,400. Tuition, nonresident: full-time $14,535. Part-time tuition and fees vary according to course load.

Financial aid: In 1999–00, 12 students received aid, including 1 fellowship with full tuition reimbursement available (averaging $14,688 per year), 11 research assistantships with full tuition reimbursements available (averaging $14,476 per year); Federal Work-Study, grants, and traineeships also available.

Faculty research: Biochemical, clinical, and cardiac pharmacology; neuropharmacology. *Total annual research expenditures:* $2.6 million.

Dr. Norton H. Neff, Chairperson, 614-292-8608, *Fax:* 614-292-7232, *E-mail:* neff.2@osu.edu.

Application contact: Dr. Andrej Rotter, Graduate Studies Chairperson, 614-292-7747, *Fax:* 614-292-7232, *E-mail:* rotter.1@osu.edu.

■ THE OHIO STATE UNIVERSITY

College of Pharmacy and Graduate School, Graduate Programs in Pharmacy, Division of Pharmacology, Columbus, OH 43210

AWARDS MS, PhD.

Faculty: 8 full-time (0 women).

Students: 12 full-time (7 women); includes 1 minority (African American), 7 international. Average age 25. *70 applicants, 10% accepted.* In 1999, 1 master's awarded (100% found work related to degree); 1 doctorate awarded (100% entered university research/teaching).

Degree requirements: For master's, foreign language not required; for doctorate, dissertation required, foreign language not required. *Average time to degree:* Master's–2 years full-time; doctorate–5 years full-time.

Entrance requirements: For master's, GRE General Test, TSE, minimum GPA of 3.0; for doctorate, GRE General Test, TOEFL, TSE, minimum GPA of 3.3. *Application deadline:* For fall admission, 2/1 (priority date); for winter admission, 9/1 (priority date); for spring admission, 11/1 (priority date). *Application fee:* $30 ($40 for international students). Electronic applications accepted.

Expenses: Tuition, state resident: full-time $5,400. Tuition, nonresident: full-time $14,535. Part-time tuition and fees vary according to course load.

Financial aid: In 1999–00, 12 students received aid, including 2 fellowships with full and partial tuition reimbursements available (averaging $16,000 per year), research assistantships with full and partial tuition reimbursements available (averaging $15,200 per year), 10 teaching assistantships with full and partial tuition reimbursements available (averaging $15,200 per year). Financial aid application deadline: 2/1.

Faculty research: Neuropharmacology, biochemical pharmacology, toxicology, drug receptor theory, molecular pharmacology.

Dr. Norman J. Uretsky, Chairman, 614-292-5433, *Fax:* 614-292-9083, *E-mail:* uretsky.1@osu.edu.

Application contact: Kathy I. Brooks, Graduate Program Coordinator, 614-292-6822, *Fax:* 614-292-2588, *E-mail:* gadmbrks@dendrite.pharmacy.ohio-state.edu.

■ THE OHIO STATE UNIVERSITY

College of Veterinary Medicine and Graduate School, Graduate Programs in Veterinary Medicine, Department of Veterinary Biosciences, Columbus, OH 43210

AWARDS Anatomy and cellular biology (MS, PhD); pathobiology (MS, PhD); pharmacology (MS, PhD); toxicology (MS, PhD); veterinary physiology (MS, PhD).

Faculty: 28 full-time (8 women).

Students: 49 full-time (20 women); includes 2 minority (1 African American, 1 Asian American or Pacific Islander), 18 international.

Degree requirements: For master's and doctorate, thesis/dissertation, final exam required.

Entrance requirements: For master's, GRE General Test; for doctorate, GRE General Test, master's degree. *Application fee:* $25.

Expenses: Tuition, state resident: full-time $5,757. Tuition, nonresident: full-time $14,892.

Financial aid: Fellowships, research assistantships, teaching assistantships available.

Faculty research: Microvasculature, muscle biology, neonatal lung and bone development.

Charles C. Capen, Interim Chair, 614-292-4489.

Application contact: Graduate Admission Committee, 614-292-4489.

■ OREGON HEALTH SCIENCES UNIVERSITY

School of Medicine, Graduate Programs in Medicine, Department of Physiology and Pharmacology, Portland, OR 97201-3098

AWARDS Pharmacology (PhD); physiology (PhD).

Degree requirements: For doctorate, dissertation required, foreign language not required.

Entrance requirements: For doctorate, GRE General Test, MCAT, TOEFL. *Application deadline:* For fall admission, 1/15. Applications are processed on a rolling basis. *Application fee:* $60.

Expenses: Tuition, state resident: full-time $3,132; part-time $174 per credit hour. Tuition, nonresident: full-time $5,256; part-time $292 per credit hour. Required fees: $8.5 per credit hour. $146 per term. Part-time tuition and fees vary according to course load.

Financial aid: Fellowships, research assistantships, Federal Work-Study, institutionally sponsored loans, scholarships, and tuition waivers (full) available.

Financial aid application deadline: 3/1; financial aid applicants required to submit FAFSA.
Dr. John A. Resko, Chairman, 503-494-8262, *Fax:* 503-494-4352.
Application contact: Dr. Charles Roselli, Graduate Program Director, 503-494-5837, *Fax:* 503-494-4352.

■ THE PENNSYLVANIA STATE UNIVERSITY MILTON S. HERSHEY MEDICAL CENTER

Graduate School, Department of Pharmacology, Hershey, PA 17033-2360

AWARDS MS, PhD, MD/PhD, PhD/MBA.

Students: 15 full-time (10 women). Average age 27. In 1999, 4 doctorates awarded.
Degree requirements: For master's, thesis required, foreign language not required; for doctorate, dissertation required.
Entrance requirements: For master's and doctorate, GRE General Test. *Application deadline:* For fall admission, 7/26. *Application fee:* $50.
Expenses: Tuition, state resident: full-time $6,886; part-time $291 per credit. Tuition, nonresident: full-time $14,118; part-time $588 per credit. Required fees: $43 per semester. Part-time tuition and fees vary according to course load.
Dr. Elliott S. Vesell, Chairman, 717-531-8285.

■ PURDUE UNIVERSITY

School of Pharmacy and Pharmacal Sciences and Graduate School, Graduate Programs in Pharmacy and Pharmacal Sciences, Department of Medicinal Chemistry and Molecular Pharmacology, West Lafayette, IN 47907

AWARDS Analytical medicinal chemistry (PhD); computational and biophysical medicinal chemistry (PhD); medicinal and bioorganic chemistry (PhD); medicinal biochemistry and molecular biology (PhD); molecular pharmacology and toxicology (PhD); natural products and pharmacognosy (PhD); nuclear pharmacy (MS); radiopharmaceutical chemistry and nuclear pharmacy (PhD).

Faculty: 24 full-time (2 women).
Students: 48 full-time (26 women), 3 part-time (1 woman); includes 4 minority (1 African American, 1 Asian American or Pacific Islander, 2 Hispanic Americans), 13 international. Average age 29. *139 applicants, 19% accepted.* In 1999, 3 master's, 11 doctorates awarded. Terminal master's awarded for partial completion of doctoral program.

Degree requirements: For master's and doctorate, thesis/dissertation required, foreign language not required.
Entrance requirements: For master's, GRE General Test, TOEFL, minimum B average; BS in biology, chemistry, or pharmacy; for doctorate, GRE General Test, TOEFL, minimum B average; BS in biology, chemistry, or pharmacology. *Application deadline:* Applications are processed on a rolling basis. *Application fee:* $30. Electronic applications accepted.
Expenses: Tuition, state resident: full-time $4,530; part-time $130 per credit hour. Tuition, nonresident: full-time $15,310; part-time $404 per credit hour. Tuition and fees vary according to campus/location and program.
Financial aid: Fellowships, research assistantships, teaching assistantships, traineeships available. Aid available to part-time students. Financial aid applicants required to submit FAFSA.
Faculty research: Drug design and development, cancer research, drug synthesis and analysis, chemical pharmacology, environmental toxicology.
Dr. R. F. Borch, Graduate Head, 765-494-1403.
Application contact: Dr. D. E. Bergstrom, Graduate Committee, 765-494-6275, *E-mail:* bergstrom@ pharmacy.purdue.edu.
Find an in-depth description at www.petersons.com/graduate.

■ PURDUE UNIVERSITY

School of Veterinary Medicine and Graduate School, Graduate Programs in Veterinary Medicine, Department of Basic Medical Sciences, West Lafayette, IN 47907

AWARDS Anatomy (MS, PhD); pharmacology (MS, PhD); physiology (MS, PhD). Part-time programs available.

Faculty: 19 full-time (3 women).
Students: 26 full-time (13 women). Average age 27. *26 applicants, 27% accepted.* In 1999, 4 master's, 5 doctorates awarded. Terminal master's awarded for partial completion of doctoral program.
Degree requirements: For master's and doctorate, thesis/dissertation required, foreign language not required. *Average time to degree:* Master's–3 years full-time; doctorate–3 years full-time.
Entrance requirements: For master's and doctorate, GRE General Test, TOEFL. *Application deadline:* For fall admission, 7/1 (priority date); for spring admission, 12/1 (priority date). *Application fee:* $30. Electronic applications accepted.
Expenses: Tuition, state resident: full-time $3,732. Tuition, nonresident: full-time $8,732.

Financial aid: In 1999–00, 6 fellowships, 14 research assistantships, 3 teaching assistantships were awarded. Financial aid application deadline: 3/1; financial aid applicants required to submit FAFSA.
Faculty research: Development and regeneration, tissue injury and shock, biomedical engineering, ovarian function, bone and cartilage biology, cell and molecular biology. *Total annual research expenditures:* $764,843.
Dr. Gordon L. Coppoc, Head, 765-494-8632, *Fax:* 765-494-0781, *E-mail:* coppoc@ vet.purdue.edu.
Application contact: Dr. Ronald Hullinger, Chairman, Graduate Committee, 765-494-8580, *Fax:* 765-494-0781, *E-mail:* bmsgrad@vet.purdue.edu.

■ RUSH UNIVERSITY

Graduate College, Division of Pharmacology, Chicago, IL 60612-3832

AWARDS MS, PhD, MD/PhD. Terminal master's awarded for partial completion of doctoral program.

Degree requirements: For master's, thesis required; for doctorate, dissertation required, foreign language not required.
Entrance requirements: For master's and doctorate, GRE, TOEFL, interview.
Expenses: Tuition: Full-time $13,020; part-time $390 per credit. Tuition and fees vary according to program.
Faculty research: Dopamine neurobiology and Parkinson's disease; cardiac electrophysiology and clinical pharmacology; neutrophil motility, apoptosis, and adhesion; angiogenesis; pulmonary vascular physiology.

■ RUTGERS, THE STATE UNIVERSITY OF NEW JERSEY, NEW BRUNSWICK

Graduate School, Program in Cellular and Molecular Pharmacology, Piscataway, NJ 08854

AWARDS PhD.

Faculty: 30 full-time (5 women).
Students: 7 full-time (3 women), 8 part-time (5 women); includes 1 minority (Asian American or Pacific Islander), 9 international. Average age 28. *52 applicants, 12% accepted.* In 1999, 13 degrees awarded (100% entered university research/ teaching).
Degree requirements: For doctorate, dissertation, qualifying exam required, foreign language not required. *Average time to degree:* Doctorate–5 years full-time.
Entrance requirements: For doctorate, GRE General Test, TOEFL, GRE Subject Test. *Application deadline:* For fall admission, 2/15 (priority date). *Application fee:* $50.

Rutgers, The State University of New Jersey, New Brunswick (continued)

Expenses: Tuition, state resident: full-time $6,776; part-time $279 per credit. Tuition, nonresident: full-time $9,936; part-time $412 per credit. Required fees: $20 per credit. $89 per semester. Tuition and fees vary according to course load, campus/location and program.
Financial aid: In 1999–00, 15 students received aid, including 5 research assistantships; fellowships, traineeships also available. Financial aid application deadline: 3/1; financial aid applicants required to submit FAFSA.
Faculty research: Cellular neuropharmacology.
Dr. Ronald Morris, Director, 732-235-4590, *Fax:* 732-235-4073, *E-mail:* vendulpa@umdnj.edu.

■ ST. JOHN'S UNIVERSITY

College of Pharmacy and Allied Health Professions, Graduate Programs in Pharmacy, Program in Pharmaceutical Sciences, Jamaica, NY 11439

AWARDS Clinical pharmacy (MS); cosmetic sciences (MS); industrial pharmacy (MS, PhD); medicinal chemistry (MS, PhD); pharmacology (MS, PhD); pharmacotherapeutics (MS); toxicology (PhD). Part-time and evening/weekend programs available.

Faculty: 19 full-time (4 women), 2 part-time/adjunct (both women).
Students: 14 full-time (4 women), 90 part-time (38 women); includes 19 minority (4 African Americans, 13 Asian Americans or Pacific Islanders, 2 Hispanic Americans), 55 international. Average age 30. *153 applicants, 53% accepted.* In 1999, 19 master's, 5 doctorates awarded. Terminal master's awarded for partial completion of doctoral program.
Degree requirements: For master's, comprehensive exam required, thesis optional, foreign language not required; for doctorate, one foreign language (computer language can substitute), dissertation, comprehensive and qualifying exams, residency required.
Entrance requirements: For master's, GRE General Test, minimum GPA of 3.0; for doctorate, GRE General Test, minimum GPA of 3.5 (undergraduate), 3.0 (graduate). *Application deadline:* For fall admission, 7/15 (priority date); for spring admission, 12/1. Applications are processed on a rolling basis. *Application fee:* $40.
Expenses: Tuition: Full-time $16,800; part-time $700 per credit. Required fees: $150; $75 per semester.
Financial aid: Fellowships, research assistantships, career-related internships or fieldwork and scholarships available. Aid available to part-time students. Financial

aid application deadline: 3/1; financial aid applicants required to submit FAFSA.
Faculty research: Neurotoxicology, biochemical toxicology, molecular pharmacology, neuropharmacology, intermediary metabolism.
Dr. Louis Trombetta, Chair, 718-990-5008, *E-mail:* trombetl@stjohns.edu.
Application contact: Shamus J. McGrenra, TOR, Associate Director, Graduate Admissions, 718-990-2000, *Fax:* 718-990-2096, *E-mail:* mcgrenrs@stjohns.edu.

■ SAINT LOUIS UNIVERSITY

School of Medicine and Graduate School, Graduate Programs in Biomedical Sciences, Department of Pharmacological and Physiological Science, St. Louis, MO 63103-2097

AWARDS MS(R), PhD.

Faculty: 27 full-time (9 women), 8 part-time/adjunct (2 women).
Students: 9 full-time (6 women), 9 part-time (3 women); includes 2 minority (1 African American, 1 Hispanic American). Average age 28. In 1999, 5 doctorates awarded.
Degree requirements: For master's, thesis, comprehensive oral exam required; for doctorate, dissertation, departmental qualifying exams required.
Entrance requirements: For master's and doctorate, GRE General Test, minimum B average in undergraduate course work; previous course work in biology, chemistry, mathematics, and physics. *Application deadline:* Applications are processed on a rolling basis. *Application fee:* $50. Electronic applications accepted.
Expenses: Tuition: Part-time $507 per credit hour. Required fees: $38 per term.
Financial aid: In 1999–00, 17 students received aid, including 8 fellowships, 1 teaching assistantship; research assistantships, traineeships and tuition waivers (partial) also available. Aid available to part-time students. Financial aid application deadline: 8/1; financial aid applicants required to submit FAFSA.
Faculty research: Molecular endocrinology, neuropharmacology, cardiovascular science, drug abuse, neurotransmitter and hormonal signaling mechanisms.
Dr. Thomas C. Westfall, Chairman, 314-577-8551, *Fax:* 314-577-8554.
Application contact: Director of Admissions, 314-577-8551, *Fax:* 314-577-8554, *E-mail:* inquiry@slu.edu.

Find an in-depth description at www.petersons.com/graduate.

■ SOUTHERN ILLINOIS UNIVERSITY CARBONDALE

School of Medicine and Graduate School, Graduate Program in Medicine, Program in Pharmacology, Carbondale, IL 62901-6806

AWARDS MS, PhD.

Faculty: 13 full-time (1 woman).
Students: 6 full-time (2 women), 1 part-time, 6 international. Average age 30. *13 applicants, 46% accepted.* In 1999, 4 doctorates awarded.
Degree requirements: For master's, thesis required, foreign language not required; for doctorate, dissertation required.
Entrance requirements: For master's, TOEFL, minimum GPA of 3.0; for doctorate, TOEFL, minimum GPA of 3.25. *Application deadline:* For fall admission, 2/15; for spring admission, 12/31. Applications are processed on a rolling basis. *Application fee:* $0.
Expenses: Tuition, state resident: full-time $2,604. Tuition, nonresident: full-time $5,208. Required fees: $380 per semester.
Financial aid: Fellowships with full tuition reimbursements, tuition waivers (full) available.
Faculty research: Autonomic nervous system pharmacology, biochemical pharmacology, neuropharmacology, toxicology, cardiovascular pharmacology.
Dr. Carl L. Faingold, Chairman, 217-785-2185, *Fax:* 217-524-0145.
Application contact: Satu M. Somani, Director, 217-785-2196.

■ STANFORD UNIVERSITY

School of Medicine, Graduate Programs in Medicine, Department of Molecular Pharmacology, Stanford, CA 94305-9991

AWARDS PhD.

Faculty: 9 full-time (3 women).
Students: 14 full-time (3 women), 5 part-time (3 women); includes 9 minority (1 African American, 6 Asian Americans or Pacific Islanders, 2 Hispanic Americans). Average age 27. *39 applicants, 31% accepted.* In 1999, 3 doctorates awarded.
Degree requirements: For doctorate, dissertation required, foreign language not required.
Entrance requirements: For doctorate, GRE General Test, GRE Subject Test (biology or chemistry), TOEFL. *Application deadline:* For fall admission, 12/15. *Application fee:* $65 ($80 for international students). Electronic applications accepted.
Expenses: Tuition: Full-time $23,058. Required fees: $152. Part-time tuition and fees vary according to course load.

Financial aid: Research assistantships, teaching assistantships available. Financial aid application deadline: 12/15.
Faculty research: Action of such drugs as epinephrine, cell differentiation and development, microsomal enzymes, neuropeptide gene expression.
Helen Blau, Chair, 650-723-6834, *Fax:* 650-725-2952, *E-mail:* hblau@cmem.stanford.edu.
Application contact: Margaret Tuggle, Administrative Associate, 650-723-6834, *E-mail:* mtuggle@stanford.edu.

Find an in-depth description at www.petersons.com/graduate.

■ STATE UNIVERSITY OF NEW YORK AT BUFFALO

Graduate School, Graduate Programs in Biomedical Sciences at Roswell Park Cancer Institute, Department of Molecular Pharmacology and Cancer Therapeutics at Roswell Park Cancer Institute, Buffalo, NY 14263

AWARDS PhD.

Faculty: 19 full-time (6 women).
Students: 17 full-time (8 women); includes 2 minority (both Asian Americans or Pacific Islanders), 2 international. Average age 26. *14 applicants, 64% accepted.* In 1999, 2 doctorates awarded (100% entered university research/teaching).
Degree requirements: For doctorate, dissertation, departmental qualifying exam, grant proposal required, foreign language not required. *Average time to degree:* Doctorate–7 years full-time.
Entrance requirements: For doctorate, GRE General Test, TOEFL, TWE (recommended). *Application deadline:* For fall admission, 2/1 (priority date). Applications are processed on a rolling basis. *Application fee:* $35. Electronic applications accepted.
Expenses: Tuition, state resident: full-time $5,100; part-time $213 per credit hour. Tuition, nonresident: full-time $8,416; part-time $351 per credit hour. Required fees: $935; $75 per semester. Tuition and fees vary according to course load and program.
Financial aid: In 1999–00, 17 students received aid, including 8 fellowships with full tuition reimbursements available (averaging $15,500 per year), 9 research assistantships with full tuition reimbursements available (averaging $15,500 per year).
Faculty research: Molecular pharmacology, cancer cell biology, molecular biology, biochemistry, chemotherapy. *Total annual research expenditures:* $6.5 million.
Dr. Enrico Mihich, Chair, 716-845-8223, *Fax:* 716-845-8857.

Application contact: Dr. Jennifer D. Black, Director of Graduate Studies, 716-845-5766, *Fax:* 716-845-8857, *E-mail:* jblack@sc3103.med.buffalo.edu.

Find an in-depth description at www.petersons.com/graduate.

■ STATE UNIVERSITY OF NEW YORK AT BUFFALO

Graduate School, School of Medicine and Biomedical Sciences, Graduate Programs in Medicine and Biomedical Sciences, Department of Pharmacology and Toxicology, Buffalo, NY 14260

AWARDS MA, PhD.

Faculty: 16 full-time (2 women).
Students: 9 full-time (3 women), 7 part-time (6 women), 8 international. Average age 25. *13 applicants, 54% accepted.* In 1999, 1 master's, 2 doctorates awarded. Terminal master's awarded for partial completion of doctoral program.
Degree requirements: For master's and doctorate, thesis/dissertation required.
Entrance requirements: For master's and doctorate, GRE General Test, TOEFL. *Application deadline:* For fall admission, 3/1 (priority date). Applications are processed on a rolling basis. *Application fee:* $35.
Expenses: Tuition, state resident: full-time $5,100. Tuition, nonresident: full-time $8,416. Required fees: $935.
Financial aid: In 1999–00, 15 students received aid, including 15 research assistantships (averaging $13,549 per year); fellowships, teaching assistantships, Federal Work-Study, grants, and unspecified assistantships also available. Financial aid application deadline: 2/28; financial aid applicants required to submit FAFSA.
Faculty research: Neuropharmacology, toxicology, signal transduction, molecular pharmacology, behavioral pharmacology. *Total annual research expenditures:* $1.5 million.
Dr. Ronald P. Rubin, Chairman, 716-829-2800, *Fax:* 716-829-2801.
Application contact: Noreen A. Harbison, Information Contact, 716-829-2800, *Fax:* 716-829-2801, *E-mail:* harbison@acsu.buffalo.edu.

Find an in-depth description at www.petersons.com/graduate.

■ STATE UNIVERSITY OF NEW YORK AT STONY BROOK

Health Sciences Center, School of Medicine and Graduate School, Graduate Programs in Medicine, Department of Pharmacological Sciences, Graduate Program in Molecular and Cellular Pharmacology, Stony Brook, NY 11794

AWARDS PhD.

Faculty: 37.
Students: 11 full-time (8 women), 12 part-time (5 women); includes 7 minority (1 African American, 4 Asian Americans or Pacific Islanders, 2 Hispanic Americans), 4 international. Average age 25. *61 applicants, 18% accepted.* In 1999, 5 degrees awarded.
Degree requirements: For doctorate, dissertation, departmental qualifying exam required, foreign language not required.
Entrance requirements: For doctorate, GRE General Test, TOEFL. *Application deadline:* For fall admission, 1/15 (priority date). Applications are processed on a rolling basis. *Application fee:* $50. Electronic applications accepted.
Expenses: Tuition, state resident: full-time $5,100. Tuition, nonresident: full-time $8,416. Required fees: $492.
Financial aid: In 1999–00, 20 fellowships, 20 research assistantships, 4 teaching assistantships were awarded; Federal Work-Study also available. Financial aid application deadline: 3/15; financial aid applicants required to submit FAFSA.
Faculty research: Toxicology, molecular and cellular biochemistry. *Total annual research expenditures:* $7.1 million.
Daniel Bogenhagen, Director, 516-444-3057, *Fax:* 516-444-3218.
Application contact: Beverly Ponte, Graduate Program Administrator, 631-444-3057, *Fax:* 631-444-3218, *E-mail:* bev@pharm.som.sunysb.edu.

Find an in-depth description at www.petersons.com/graduate.

■ STATE UNIVERSITY OF NEW YORK HEALTH SCIENCE CENTER AT BROOKLYN

School of Graduate Studies, Department of Pharmacology, Brooklyn, NY 11203-2098

AWARDS PhD, MD/PhD.

Degree requirements: For doctorate, one foreign language, dissertation required.
Entrance requirements: For doctorate, GRE.
Expenses: Tuition, state resident: full-time $5,100; part-time $213 per credit. Tuition, nonresident: full-time $8,416; part-time

State University of New York Health Science Center at Brooklyn (continued)
$351 per credit. Required fees: $200. Full-time tuition and fees vary according to program and student level.

Faculty research: Cellular neurobiology, molecular neurobiology, cell signalling process, edothelial relaxing factors, signal transduction process.

■ STATE UNIVERSITY OF NEW YORK UPSTATE MEDICAL UNIVERSITY

College of Graduate Studies, Department of Pharmacology, Syracuse, NY 13210-2334

AWARDS MS, PhD, MD/PhD.

Faculty: 20.
Students: 7 full-time (2 women), 1 part-time; includes 3 minority (all Asian Americans or Pacific Islanders), 1 international. *10 applicants, 10% accepted.* In 1999, 1 master's awarded (100% found work related to degree); 1 doctorate awarded (100% continued full-time study). Terminal master's awarded for partial completion of doctoral program.
Degree requirements: For master's, thesis required, foreign language not required; for doctorate, dissertation, comprehensive exam required, foreign language not required.
Entrance requirements: For master's and doctorate, GRE General Test, GRE Subject Test, TSE. *Application deadline:* For fall admission, 4/1 (priority date). Applications are processed on a rolling basis. *Application fee:* $40.
Expenses: Tuition, state resident: full-time $5,100; part-time $213 per credit. Tuition, nonresident: full-time $8,416; part-time $351 per credit. Required fees: $410; $25 per credit. Part-time tuition and fees vary according to course load and program.
Financial aid: Fellowships, research assistantships, Federal Work-Study and institutionally sponsored loans available. Aid available to part-time students. Financial aid application deadline: 4/15.
Dr. Jose Jalife, Chairperson, 315-464-5138.
Application contact: Dr. Richard Wojcikiewicz, Associate Professor, 315-464-7655.

■ TEMPLE UNIVERSITY

Health Sciences Center, School of Medicine and Graduate School, Graduate Programs in Medicine, Department of Pharmacology, Philadelphia, PA 19140

AWARDS MS, PhD, MD/PhD.

Faculty: 12 full-time (1 woman).
Students: 16 full-time (11 women), 1 part-time; includes 3 minority (all African

Americans). *18 applicants, 17% accepted.* Terminal master's awarded for partial completion of doctoral program.
Degree requirements: For master's, one foreign language, computer language, thesis required; for doctorate, one foreign language, computer language, dissertation, research seminars required.
Entrance requirements: For master's and doctorate, GRE General Test, GRE Subject Test, minimum GPA of 3.0 during previous 2 years, 2.8 overall. *Application deadline:* For fall admission, 7/1 (priority date); for spring admission, 11/1. Applications are processed on a rolling basis. *Application fee:* $40. Electronic applications accepted.
Expenses: Tuition, state resident: full-time $6,030. Tuition, nonresident: full-time $8,298. Required fees: $230. One-time fee: $10 full-time.
Financial aid: Fellowships, research assistantships, Federal Work-Study available.
Faculty research: Cardiovascular and central nervous systems, biochemical pharmacology.
Dr. J. Bryan Smith, Chair, 215-707-3237, *Fax:* 215-707-7068, *E-mail:* jsmith@nimbus.ocis.temple.edu.
Application contact: Dr. Barrie Ashby, Admissions Chair, 215-707-4404, *Fax:* 215-707-7068, *E-mail:* bashby@nimbus.temple.edu.

■ TEXAS A&M UNIVERSITY SYSTEM HEALTH SCIENCE CENTER

College of Medicine, Graduate School of Biomedical Sciences, Department of Medical Pharmacology and Toxicology, College Station, TX 77840-7896

AWARDS PhD.

Faculty: 8 full-time (0 women).
Students: 6 full-time (3 women), 1 part-time, 2 international. Average age 25. *6 applicants, 17% accepted.* In 1999, 1 degree awarded (100% found work related to degree).
Degree requirements: For doctorate, dissertation required, foreign language not required. *Average time to degree:* Doctorate–5 years full-time.
Entrance requirements: For doctorate, GRE General Test. *Application deadline:* For fall admission, 2/1 (priority date). Applications are processed on a rolling basis. *Application fee:* $35 ($75 for international students).
Expenses: Tuition, area resident: Full-time $1,368. Tuition, state resident: part-time $76 per credit. Tuition, nonresident: full-time $5,256; part-time $292 per credit. International tuition: $5,256 full-time.

Required fees: $678; $38 per credit. Full-time tuition and fees vary according to course load and student level.
Financial aid: In 1999–00, 7 students received aid, including fellowships (averaging $17,000 per year); research assistantships Financial aid application deadline: 4/1; financial aid applicants required to submit FAFSA.
Faculty research: Medical treatment of eye disease, fetal alcohol syndrome, Alzheimer's disease, glycme receptory steriods.
Dr. George C. Y. Chiou, Head, 409-845-2817, *Fax:* 409-845-0699, *E-mail:* gchiou@tamsun.tamu.edu.
Application contact: Dr. Gerald Frye, Graduate Adviser, 979-845-2860, *E-mail:* gdfrye@tamu.edu.

Find an in-depth description at www.petersons.com/graduate.

■ TEXAS TECH UNIVERSITY HEALTH SCIENCES CENTER

Graduate School of Biomedical Sciences, Department of Pharmacology, Lubbock, TX 79430

AWARDS MS, PhD, MD/PhD.

Faculty: 13 full-time (1 woman), 1 part-time/adjunct (0 women).
Students: 5 full-time (4 women), 4 international. Average age 28. *17 applicants, 24% accepted.* In 1999, 1 doctorate awarded (100% entered university research/teaching). Terminal master's awarded for partial completion of doctoral program.
Degree requirements: For master's and doctorate, thesis/dissertation required, foreign language not required. *Average time to degree:* Doctorate–5 years full-time.
Entrance requirements: For master's and doctorate, GRE General Test, TOEFL, minimum GPA of 3.0. *Application deadline:* For fall admission, 4/15 (priority date). Applications are processed on a rolling basis. *Application fee:* $30 ($55 for international students). Electronic applications accepted.
Expenses: Tuition, state resident: part-time $38 per credit hour. Tuition, nonresident: part-time $254 per credit hour. Part-time tuition and fees vary according to program.
Financial aid: In 1999–00, 5 research assistantships (averaging $14,500 per year) were awarded; institutionally sponsored loans and scholarships also available. Financial aid applicants required to submit FAFSA.
Faculty research:
Neuropsychopharmacology, autonomic pharmacology, cardiovascular pharmacology, molecular pharmacology, toxicology.
Dr. Reid Norman, Chair, 806-743-2425, *Fax:* 806-743-2744.

Application contact: Dr. Joseph D. Miller, Graduate Director, 806-743-2425, *Fax:* 806-743-2744, *E-mail:* phrjdm2@ttuhsc.edu.

■ THOMAS JEFFERSON UNIVERSITY

College of Graduate Studies, Program in Molecular Pharmacology and Structural Biology, Philadelphia, PA 19107

AWARDS PhD.

Faculty: 48 full-time.
Students: 13 full-time (2 women), 2 part-time; includes 1 minority (Asian American or Pacific Islander), 2 international. *45 applicants, 7% accepted.* In 1999, 4 degrees awarded.
Degree requirements: For doctorate, dissertation required, foreign language not required.
Entrance requirements: For doctorate, GRE General Test, TOEFL, minimum GPA of 3.2. *Application deadline:* For fall admission, 3/1 (priority date). Applications are processed on a rolling basis. *Application fee:* $40.
Expenses: Tuition: Full-time $12,670. Tuition and fees vary according to degree level and program.
Financial aid: In 1999–00, 14 fellowships with full tuition reimbursements were awarded; research assistantships, Federal Work-Study, institutionally sponsored loans, traineeships, and training grants also available. Aid available to part-time students. Financial aid application deadline: 5/1; financial aid applicants required to submit FAFSA.
Faculty research: Biochemistry and cell, molecular and structural biology of cell-surface and intracellular receptors, molecular modeling, signal transduction. Dr. Jeffrey L. Benovic, Chair, Graduate Committee, 215-503-4607, *Fax:* 215-923-1098.
Application contact: Jessie F. Pervall, Director of Admissions, 215-503-4400, *Fax:* 215-503-3433, *E-mail:* cgs-info@mail.tju.edu.
Find an in-depth description at www.petersons.com/graduate.

■ THOMAS JEFFERSON UNIVERSITY

College of Graduate Studies, Program in Pharmacology, Philadelphia, PA 19107

AWARDS MS. Part-time and evening/weekend programs available.

Faculty: 17 full-time (2 women), 4 part-time/adjunct (1 woman).
Students: 3 full-time (2 women), 34 part-time (21 women); includes 7 minority (2 African Americans, 4 Asian Americans or Pacific Islanders, 1 Hispanic American), 1 international. Average age 41. *28 applicants, 71% accepted.* In 1999, 12 degrees awarded.
Degree requirements: For master's, thesis required, foreign language not required.
Entrance requirements: For master's, GRE General Test, minimum GPA of 3.0. *Application deadline:* Applications are processed on a rolling basis. *Application fee:* $40.
Expenses: Tuition: Full-time $17,625; part-time $610 per credit.
Financial aid: In 1999–00, 9 students received aid. Federal Work-Study and institutionally sponsored loans available. Aid available to part-time students. Financial aid application deadline: 5/1; financial aid applicants required to submit FAFSA.
Faculty research: Receptors, forensic toxicology, thrombosis and atherosclerosis, peptide receptors, signal transduction. Dr. Georganne K. Buescher, Associate Dean, 215-503-5799, *Fax:* 215-503-3433, *E-mail:* georganne.buescher@mail.tju.edu.
Application contact: Jessie F. Pervall, Director of Admissions, 215-503-4400, *Fax:* 215-503-3433, *E-mail:* cgs-info@mail.tju.edu.
Find an in-depth description at www.petersons.com/graduate.

■ TUFTS UNIVERSITY

Sackler School of Graduate Biomedical Sciences, Program in Pharmacology and Experimental Therapeutics, Medford, MA 02155

AWARDS PhD.

Faculty: 20 full-time (3 women), 3 part-time/adjunct (1 woman).
Students: 18 full-time (10 women); includes 3 minority (all Asian Americans or Pacific Islanders), 3 international. Average age 28. *64 applicants, 3% accepted.* In 1999, 1 degree awarded (100% continued full-time study).
Degree requirements: For doctorate, dissertation, qualifying exam required, foreign language not required. *Average time to degree:* Doctorate–5 years full-time.
Entrance requirements: For doctorate, GRE General Test, TOEFL. *Application deadline:* For fall admission, 1/15. Applications are processed on a rolling basis. *Application fee:* $45.
Expenses: Tuition: Full-time $19,325.
Financial aid: In 1999–00, 18 research assistantships with full tuition reimbursements (averaging $18,805 per year) were awarded; career-related internships or fieldwork and tuition waivers (partial) also available. Financial aid application deadline: 2/1.

Faculty research: Biochemical mechanisms of narcotic addiction, clinical psychopharmacology, pharmacokinetics, neurotransmitter receptors, neuropeptides. *Total annual research expenditures:* $1.5 million.
Dr. Louis Shuster, Director, 617-636-6863, *Fax:* 617-636-6738, *E-mail:* louis.shuster@tufts.edu.
Application contact: Barbara W. Richard, Administrative Programs Director, 617-636-6703, *Fax:* 617-636-6738, *E-mail:* barbara.richard@tufts.edu.

■ TULANE UNIVERSITY

School of Medicine and Graduate School, Graduate Programs in Medicine, Department of Pharmacology, New Orleans, LA 70118-5669

AWARDS MS, PhD, MD/PhD. MS and PhD offered through the Graduate School.

Students: 9 full-time (3 women); includes 2 minority (1 African American, 1 Asian American or Pacific Islander), 1 international. *45 applicants, 16% accepted.* In 1999, 2 doctorates awarded.
Degree requirements: For master's, one foreign language, thesis required; for doctorate, 2 foreign languages (computer language can substitute for one), dissertation required.
Entrance requirements: For master's, GRE General Test, TOEFL, or TSE, minimum B average in undergraduate course work; for doctorate, GRE General Test, TOEFL, or TSE. *Application deadline:* For fall admission, 2/1. *Application fee:* $45.
Expenses: Tuition: Full-time $23,030.
Financial aid: Fellowships, research assistantships available. Financial aid application deadline: 2/1.
Dr. Krishna Agrawal, Chairman, 504-588-5444.
Find an in-depth description at www.petersons.com/graduate.

■ UNIFORMED SERVICES UNIVERSITY OF THE HEALTH SCIENCES

School of Medicine, Division of Basic Medical Sciences, Department of Pharmacology, Bethesda, MD 20814-4799

AWARDS PhD.

Faculty: 12 full-time (3 women), 14 part-time/adjunct (2 women).
Students: *2 applicants, 0% accepted.*
Degree requirements: For doctorate, computer language, dissertation, qualifying exam required, foreign language not required.
Entrance requirements: For doctorate, GRE General Test, GRE Subject Test,

Uniformed Services University of the Health Sciences (continued)
minimum GPA of 3.0, U.S. citizenship. *Application deadline:* For fall admission, 1/15 (priority date). Applications are processed on a rolling basis. *Application fee:* $0.

Financial aid: Fellowships, research assistantships available.

Faculty research: Molecular and cellular studies in signal transduction, drug metabolism and toxicology, neuropharmacology.
Dr. Brian M. Cox, Chair, 301-295-3223, *Fax:* 301-295-3220, *E-mail:* coxb@ usuhsb.usuhs.mil.edu.

Application contact: Janet M. Anastasi, Graduate Program Coordinator, 301-295-9474, *Fax:* 301-295-6772, *E-mail:* janastasi@usuhs.mil.

■ THE UNIVERSITY OF ALABAMA AT BIRMINGHAM

Graduate School and School of Medicine and School of Dentistry, Graduate Programs in Joint Health Sciences, Department of Pharmacology, Birmingham, AL 35294

AWARDS PhD.

Students: 15 full-time (7 women), 1 part-time; includes 2 minority (1 African American, 1 Asian American or Pacific Islander), 8 international. *54 applicants, 19% accepted.* In 1999, 4 degrees awarded.
Degree requirements: For doctorate, dissertation required, foreign language not required.
Entrance requirements: For doctorate, GRE General Test, interview. *Application deadline:* Applications are processed on a rolling basis. *Application fee:* $35 ($60 for international students). Electronic applications accepted.
Expenses: Tuition, state resident: part-time $104 per semester hour. Tuition, nonresident: part-time $208 per semester hour. Required fees: $17 per semester hour. $57 per quarter. Tuition and fees vary according to program.
Financial aid: In 1999–00, 6 fellowships were awarded.
Faculty research: Biochemical pharmacology, neuropharmacology, endocrine pharmacology.
Dr. Robert B. Diasio, Chair, 205-934-4578.
Application contact: Graduate Coordinator, 205-934-4584, *Fax:* 205-934-4209, *E-mail:* rdiasio@uab.edu.

Find an in-depth description at www.petersons.com/graduate.

■ THE UNIVERSITY OF ARIZONA

Graduate College, College of Pharmacy, Department of Pharmacology and Toxicology, Tucson, AZ 85721

AWARDS Pharmacology (MS); toxicology (MS).

Degree requirements: For master's, thesis required.
Entrance requirements: For master's, GRE General Test, TOEFL, minimum GPA of 3.0.
Expenses: Tuition, nonresident: full-time $4,814; part-time $274 per unit. Required fees: $1,094; $115 per unit. Tuition and fees vary according to course load and program.

■ THE UNIVERSITY OF ARIZONA

Graduate College, Graduate Interdisciplinary Programs, Graduate Interdisciplinary Program in Pharmacology and Toxicology, Tucson, AZ 85721

AWARDS PhD.

Degree requirements: For doctorate, dissertation required, foreign language not required.
Entrance requirements: For doctorate, TOEFL.
Expenses: Tuition, nonresident: full-time $4,814; part-time $274 per unit. Required fees: $1,094; $115 per unit. Tuition and fees vary according to course load and program.
Faculty research: Neuropharmacology, carcinogenesis and cancer chemotherapy, molecular pharmacology and toxicology, biochemical pharmacology and toxicology.

Find an in-depth description at www.petersons.com/graduate.

■ UNIVERSITY OF ARKANSAS FOR MEDICAL SCIENCES

College of Medicine and Graduate School, Graduate Programs in Medicine, Department of Pharmacology and Toxicology, Little Rock, AR 72205-7199

AWARDS Pharmacology (MS, PhD); toxicology (MS, PhD).

Faculty: 24 full-time (0 women), 4 part-time/adjunct (1 woman).
Students: 10 full-time (5 women), 5 international. In 1999, 1 master's, 2 doctorates awarded.
Degree requirements: For master's and doctorate, thesis/dissertation required, foreign language not required.
Entrance requirements: For master's and doctorate, GRE General Test. *Application fee:* $0.
Expenses: Tuition: Full-time $8,928.

Financial aid: Research assistantships, teaching assistantships available. Aid available to part-time students.
Dr. Donald E. McMillan, Chairman, 501-686-5510.
Application contact: Dr. Philip R. Mayeux, Graduate Coordinator, 501-686-5510, *E-mail:* mayeuxphilipr@ exchange.uams.edu.

Find an in-depth description at www.petersons.com/graduate.

■ UNIVERSITY OF CALIFORNIA, DAVIS

Graduate Studies, Programs in the Biological Sciences, Program in Pharmacology/Toxicology, Davis, CA 95616

AWARDS MS, PhD.

Faculty: 50 full-time (7 women).
Students: 39 full-time (22 women), 1 part-time; includes 9 minority (4 Asian Americans or Pacific Islanders, 4 Hispanic Americans, 1 Native American), 6 international. Average age 29. *58 applicants, 38% accepted.* In 1999, 4 master's, 8 doctorates awarded. Terminal master's awarded for partial completion of doctoral program.
Degree requirements: For master's, comprehensive exam or thesis required; for doctorate, dissertation, qualifying exam required, foreign language not required.
Entrance requirements: For master's and doctorate, GRE General Test, minimum GPA of 3.0, previous course work in biochemistry and/or physiology. *Application deadline:* For fall admission, 1/15. *Application fee:* $40. Electronic applications accepted.
Expenses: Tuition, nonresident: full-time $9,804. Tuition and fees vary according to program and student level.
Financial aid: In 1999–00, 40 students received aid, including 8 fellowships with full and partial tuition reimbursements available, 26 research assistantships with full and partial tuition reimbursements available; teaching assistantships, career-related internships or fieldwork, Federal Work-Study, grants, institutionally sponsored loans, scholarships, and tuition waivers (full and partial) also available. Financial aid application deadline: 1/15; financial aid applicants required to submit FAFSA.
Faculty research: Respiratory, neurochemical, molecular, genetic, and ecological toxicology.
Jerry Last, Graduate Chair, 530-752-6230, *E-mail:* jalast@ucdavis.edu.
Application contact: Carol Barnes, Graduate Administrative Assistant, 530-752-4521, *Fax:* 530-752-3394, *E-mail:* cbarnes@ucdavis.edu.

■ UNIVERSITY OF CALIFORNIA, IRVINE

College of Medicine and Office of Research and Graduate Studies, Graduate Programs in Medicine, Department of Pharmacology, Irvine, CA 92697

AWARDS Pharmacology and toxicology (MS, PhD).

Faculty: 9 full-time (2 women), 2 part-time/adjunct (1 woman).

Students: 12 full-time (6 women), 2 part-time (1 woman); includes 3 minority (2 Asian Americans or Pacific Islanders, 1 Hispanic American), 2 international. *33 applicants, 9% accepted.* In 1999, 1 doctorate awarded.

Degree requirements: For doctorate, dissertation required.

Entrance requirements: For master's, GRE; for doctorate, GRE General Test, GRE Subject Test. *Application deadline:* For fall admission, 1/15 (priority date). Applications are processed on a rolling basis. *Application fee:* $40. Electronic applications accepted.

Expenses: Tuition, nonresident: full-time $10,322; part-time $1,720 per quarter. Required fees: $5,354; $1,300 per quarter. Tuition and fees vary according to program.

Financial aid: Fellowships, research assistantships, institutionally sponsored loans and tuition waivers (full and partial) available. Financial aid application deadline: 3/2; financial aid applicants required to submit FAFSA.

Faculty research: Mechanisms of action and effects of drugs on the nervous system, behavior, skeletal muscle, heart, and blood vessels; basic processes in the nervous system, skeletal muscle, heart, and blood vessels.
Larry Stein, Chair, 949-824-6771, *Fax:* 949-824-4855.

Application contact: Graduate Coordinator, 949-824-7651, *Fax:* 949-824-4855, *E-mail:* pharm@uci.edu.

Find an in-depth description at www.petersons.com/graduate.

■ UNIVERSITY OF CALIFORNIA, LOS ANGELES

School of Medicine and Graduate Division, Graduate Programs in Medicine, Department of Molecular and Medical Pharmacology, Los Angeles, CA 90095

AWARDS MS, PhD.

Students: 37 full-time (21 women); includes 10 minority (1 African American, 7 Asian Americans or Pacific Islanders, 1 Hispanic American, 1 Native American), 7 international. *62 applicants, 10% accepted.*

Degree requirements: For doctorate, dissertation, qualifying exams required; foreign language not required.

Entrance requirements: For doctorate, GRE General Test. *Application fee:* $40.

Expenses: Tuition, nonresident: full-time $9,804. Required fees: $4,405.

Financial aid: In 1999–00, 32 students received aid, including 31 fellowships, 10 teaching assistantships; research assistantships, scholarships also available. Financial aid application deadline: 3/1.

Faculty research: Cardiovascular pharmacology, chemical pharmacology, neuropharmacology, clinical pharmacology.
Dr. Michael Phelps, Chair, 310-794-7726.

Application contact: Departmental Office, 310-794-7726, *E-mail:* gradinfo@ww.medsch.ucla.edu.

Find an in-depth description at www.petersons.com/graduate.

■ UNIVERSITY OF CALIFORNIA, SAN DIEGO

School of Medicine and Graduate Studies and Research, Graduate Studies in Biomedical Sciences, Department of Pharmacology, La Jolla, CA 92093-0685

AWARDS PhD.

Faculty: 106.

Students: 219.

Degree requirements: For doctorate, dissertation, qualifying exam required, foreign language not required.

Entrance requirements: For doctorate, GRE General Test, TOEFL. *Application deadline:* For fall admission, 1/5. *Application fee:* $40.

Expenses: Program pays tuition, fees, health insurance, and stipend for all students in good standing.

Financial aid: Tuition waivers (full) and stipends available.

Faculty research: Molecular and cellular pharmacology, cell and organ physiology, cellular and molecular biology.
Palmer Taylor, Director, 858-534-4028.

Application contact: Gina Butcher, Graduate Program Representative, 858-534-3982.

Find an in-depth description at www.petersons.com/graduate.

■ UNIVERSITY OF CALIFORNIA, SAN FRANCISCO

School of Pharmacy and Graduate Division, Pharmaceutical Sciences and Pharmacogenomics Graduate Group, San Francisco, CA 94143

AWARDS PhD.

Faculty: 33 full-time (8 women).

Students: 33 full-time (24 women); includes 11 minority (all Asian Americans or Pacific Islanders), 8 international. *69 applicants, 19% accepted.* In 1999, 3 doctorates awarded (100% found work related to degree).

Degree requirements: For doctorate, dissertation required, foreign language not required. *Average time to degree:* Doctorate–5.4 years full-time.

Entrance requirements: For doctorate, GRE General Test, TOEFL, minimum GPA of 3.0. *Application deadline:* For fall admission, 1/15. *Application fee:* $40.

Expenses: Tuition, nonresident: full-time $98,042. Required fees: $7,757. Full-time tuition and fees vary according to program and student level.

Financial aid: In 1999–00, 2 fellowships with full tuition reimbursements (averaging $10,000 per year), 8 research assistantships with full tuition reimbursements (averaging $19,600 per year), 1 teaching assistantship with full tuition reimbursement (averaging $19,600 per year) were awarded; career-related internships or fieldwork, grants, institutionally sponsored loans, scholarships, traineeships, and tuition waivers (full) also available.

Faculty research: Drug development, drug delivery, molecular pharmacology.
Francis C. Szoka, Program Director, 415-473-3895, *Fax:* 415-514-0502, *E-mail:* szoka@cgl.ucsf.edu.

Application contact: Barbara J. Paschke, Coordinator, 415-502-7788, *Fax:* 415-514-0502, *E-mail:* mis@cgl.ucsf.edu.

Find an in-depth description at www.petersons.com/graduate.

■ UNIVERSITY OF CHICAGO

Division of the Biological Sciences, Neurobiology, Pharmacology, and Cell Physiology, Department of Neurobiology, Pharmacology and Physiology, Chicago, IL 60637-1513

AWARDS Cell physiology (PhD); pharmacological and physiological sciences (PhD).

Faculty: 33 full-time (8 women).

Students: 11 full-time (6 women). In 1999, 2 degrees awarded.

Degree requirements: For doctorate, dissertation, preliminary exam required. *Average time to degree:* Doctorate–6 years full-time.

Entrance requirements: For doctorate, GRE General Test, TOEFL. *Application deadline:* For fall admission, 1/5 (priority date). Applications are processed on a rolling basis. *Application fee:* $55.

Expenses: Tuition: Full-time $24,804; part-time $3,422 per course. Required fees: $390. Tuition and fees vary according to program.

University of Chicago (continued)
Financial aid: Fellowships, grants and institutionally sponsored loans available. Financial aid application deadline: 6/1.
Faculty research: Psychopharmacology, neuropharmacology.
Application contact: Dr. Sangiam S. Sisodia, Chairman, 773-834-2900.

■ UNIVERSITY OF CINCINNATI

Division of Research and Advanced Studies, College of Medicine, Graduate Programs in Medicine, Department of Molecular, Cellular, and Biochemical Pharmacology, Cincinnati, OH 45267

AWARDS Cell biophysics (PhD); pharmacology (PhD).

Faculty: 7 full-time.
Students: 13 full-time (5 women), 1 (woman) part-time; includes 1 minority (African American), 2 international. *27 applicants, 15% accepted.* In 1999, 2 degrees awarded.
Degree requirements: For doctorate, dissertation, qualifying exam required, foreign language not required. *Average time to degree:* Doctorate–6.3 years full-time.
Entrance requirements: For doctorate, GRE General Test, GRE Subject Test. *Application deadline:* For fall admission, 2/1 (priority date). Applications are processed on a rolling basis. *Application fee:* $30.
Expenses: Tuition, state resident: full-time $5,139; part-time $196 per credit hour. Tuition, nonresident: full-time $10,326; part-time $369 per credit hour. Required fees: $561; $187 per quarter.
Financial aid: Tuition waivers (full) and unspecified assistantships available. Financial aid application deadline: 5/1.
Faculty research: Lipoprotein research, enzyme regulation, electrophysiology, gene actuation. *Total annual research expenditures:* $3.1 million.
Dr. John Maggio, Chair, 513-558-4723, *Fax:* 513-558-1190, *E-mail:* john.maggio@ uc.edu.
Application contact: Dr. Robert Rapoport, Director of Graduate Studies, 513-558-2376, *Fax:* 513-558-1169, *E-mail:* robert.rapoport@uc.edu.

Find an in-depth description at www.petersons.com/graduate.

■ UNIVERSITY OF COLORADO HEALTH SCIENCES CENTER

Graduate School, Programs in Biological and Medical Sciences, Department of Pharmacology, Denver, CO 80262

AWARDS PhD.

Degree requirements: For doctorate, dissertation required, foreign language not required.
Entrance requirements: For doctorate, GRE General Test, minimum GPA of 2.75.
Expenses: Tuition, state resident: full-time $1,512; part-time $56 per hour. Tuition, nonresident: full-time $7,209; part-time $267 per hour. Full-time tuition and fees vary according to course load and program.

Find an in-depth description at www.petersons.com/graduate.

■ UNIVERSITY OF CONNECTICUT

Graduate School, School of Pharmacy, Graduate Program in Pharmacology and Toxicology, Storrs, CT 06269

AWARDS MS, PhD. Terminal master's awarded for partial completion of doctoral program.

Degree requirements: For master's and doctorate, thesis/dissertation required.
Entrance requirements: For master's and doctorate, GRE General Test.
Expenses: Tuition, state resident: full-time $5,118. Tuition, nonresident: full-time $13,298. Required fees: $1,022.

Find an in-depth description at www.petersons.com/graduate.

■ UNIVERSITY OF CONNECTICUT HEALTH CENTER

Graduate School, Programs in Biomedical Sciences, Program in Cellular and Molecular Pharmacology, Farmington, CT 06030

AWARDS PhD, DMD/PhD, MD/PhD.

Faculty: 35.
Students: 3 full-time (1 woman), 1 international. In 1999, 1 degree awarded.
Degree requirements: For doctorate, one foreign language (computer language can substitute), dissertation required.
Entrance requirements: For doctorate, GRE General Test, TOEFL. *Application deadline:* For fall admission, 2/1; for spring admission, 10/1. Applications are processed on a rolling basis. *Application fee:* $40 ($45 for international students).
Expenses: Tuition, state resident: full-time $5,272; part-time $293 per credit. Tuition, nonresident: full-time $13,696; part-time $761 per credit. Required fees: $320; $198 per semester. One-time fee: $50 full-time. Full-time tuition and fees vary according to course load, program and reciprocity agreements.
Financial aid: In 1999–00, research assistantships (averaging $17,000 per year); fellowships.
Faculty research: Neuropharmacology; cardiovascular-pulmonary pharmacology;

endocrine-reproductive pharmacology; immunopharmacology and chemotherapy. Dr. Joel S. Pachter, Director, 860-679-3698, *E-mail:* pachter@nsol.uchc.edu.
Application contact: Marizta Barta, Information Contact, 860-679-4306, *Fax:* 860-679-1282, *E-mail:* barta@ adp.uchc.edu.

Find an in-depth description at www.petersons.com/graduate.

■ UNIVERSITY OF FLORIDA

College of Medicine and Graduate School, Interdisciplinary Program in Biomedical Sciences, Concentration in Physiology and Pharmacology, Gainesville, FL 32611

AWARDS PhD.

Degree requirements: For doctorate, dissertation required, foreign language not required.
Entrance requirements: For doctorate, GRE General Test, minimum GPA of 3.0. Electronic applications accepted.
Expenses: Tuition, state resident: part-time $144 per credit hour. Tuition, nonresident: part-time $505 per credit hour. Tuition and fees vary according to course level, course load and program.

Find an in-depth description at www.petersons.com/graduate.

■ UNIVERSITY OF FLORIDA

College of Medicine and Graduate School, Interdisciplinary Program in Biomedical Sciences, Department of Pharmacology and Therapeutics, Gainesville, FL 32610

AWARDS PhD.

Degree requirements: For doctorate, dissertation required, foreign language not required.
Entrance requirements: For doctorate, GRE General Test, TOEFL, minimum GPA of 3.0. Electronic applications accepted.
Expenses: Tuition, state resident: part-time $144 per credit hour. Tuition, nonresident: part-time $505 per credit hour. Tuition and fees vary according to course level, course load and program.
Faculty research: Receptor and membrane pharmacology, autonomics, tetralogy, enzymes, opioid peptides.

■ UNIVERSITY OF FLORIDA

College of Pharmacy and Graduate School, Graduate Programs in Pharmacy, Department of Pharmacodynamics, Gainesville, FL 32611

AWARDS MSP, PhD, Pharm D/PhD. Part-time programs available.

Faculty: 8.

Students: 15 full-time (9 women); includes 4 minority (1 African American, 2 Asian Americans or Pacific Islanders, 1 Hispanic American), 6 international. Average age 25. *19 applicants, 42% accepted.* In 1999, 2 degrees awarded. Terminal master's awarded for partial completion of doctoral program.

Degree requirements: For master's and doctorate, thesis/dissertation required, foreign language not required.

Entrance requirements: For master's and doctorate, GRE General Test, TOEFL, minimum GPA of 3.0. *Application deadline:* For fall admission, 6/1 (priority date). Applications are processed on a rolling basis. *Application fee:* $20. Electronic applications accepted.

Expenses: Tuition, state resident: part-time $144 per credit hour. Tuition, nonresident: part-time $505 per credit hour. Tuition and fees vary according to course level, course load and program.

Financial aid: In 1999–00, 12 students received aid, including 1 fellowship, 7 research assistantships, 2 teaching assistantships; institutionally sponsored loans and unspecified assistantships also available. Aid available to part-time students. Financial aid application deadline: 4/1.

Faculty research: Hypertension, aging, alcoholism, diabetes, toxicology. *Total annual research expenditures:* $641,489.
Dr. Michael Meldrum, Chair, 352-392-3408, *Fax:* 352-392-9187, *E-mail:* meldrum@cop.health.ufl.edu.

Application contact: Dr. Michael Katovich, Graduate Coordinator, 352-392-3490, *Fax:* 352-392-9187, *E-mail:* katovich@cop.health.ufl.edu.

■ UNIVERSITY OF GEORGIA

College of Pharmacy, Department of Pharmaceutical and Biomedical Sciences, Athens, GA 30602

AWARDS Medicinal chemistry (MS, PhD); pharmaceutics (MS, PhD); pharmacology (MS, PhD); toxicology (MS, PhD). Terminal master's awarded for partial completion of doctoral program.

Degree requirements: For master's, thesis required, foreign language not required; for doctorate, one foreign language (computer language can substitute), dissertation required.

Entrance requirements: For master's and doctorate, GRE General Test, minimum GPA of 3.0.

Expenses: Tuition, state resident: full-time $7,516; part-time $431 per credit hour. Tuition, nonresident: full-time $12,204;

part-time $793 per credit hour. Tuition and fees vary according to program.

Find an in-depth description at www.petersons.com/graduate.

■ UNIVERSITY OF GEORGIA

College of Veterinary Medicine and Graduate School, Graduate Programs in Veterinary Medicine, Department of Physiology and Pharmacology, Athens, GA 30602

AWARDS Pharmacology (MS, PhD); physiology (MS, PhD).

Degree requirements: For master's, thesis required, foreign language not required; for doctorate, one foreign language (computer language can substitute), dissertation required.

Entrance requirements: For master's and doctorate, GRE General Test. Electronic applications accepted.

Expenses: Tuition, state resident: full-time $7,516; part-time $431 per credit hour. Tuition, nonresident: full-time $12,204; part-time $793 per credit hour.

■ UNIVERSITY OF HAWAII AT MANOA

John A. Burns School of Medicine and Graduate Division, Graduate Programs in Biomedical Sciences, Department of Pharmacology, Honolulu, HI 96822

AWARDS MS, PhD.

Faculty: 5 full-time (1 woman), 1 (woman) part-time/adjunct.

Students: 2 full-time (1 woman), (both international). Average age 34. *2 applicants, 100% accepted.* In 1999, 1 master's awarded.

Degree requirements: For master's, thesis required (for some programs), foreign language not required; for doctorate, dissertation required, foreign language not required.

Entrance requirements: For doctorate, GRE. *Application deadline:* For fall admission, 3/1; for spring admission, 9/1. *Application fee:* $25 ($50 for international students).

Expenses: Tuition, state resident: part-time $168 per credit. Tuition, nonresident: part-time $415 per credit. Required fees: $51 per semester. Part-time tuition and fees vary according to course load.

Financial aid: In 1999–00, 1 research assistantship (averaging $16,176 per year) was awarded; tuition waivers (full and partial) also available.

Faculty research: Automatic and cardiovascular, viral chemotherapy, biochemical and smooth muscle pharmacology, neuropharmacology and receptor mechanisms.

Dr. G. Causey Whittow, Chairperson, 808-956-8936, *Fax:* 808-956-3165, *E-mail:* whittowg@jabsom.biomed.hawaii.edu.

■ UNIVERSITY OF HOUSTON

College of Pharmacy, Department of Pharmacological and Pharmaceutical Sciences, Houston, TX 77004

AWARDS Medical chemistry and pharmacology (MS); pharmaceutics (MS, PhD); pharmacology (MS, PhD).

Faculty: 15 full-time (4 women).

Students: 14 full-time (10 women); includes 1 minority (African American), 9 international. Average age 24. *65 applicants, 8% accepted.* In 1999, 2 degrees awarded. Terminal master's awarded for partial completion of doctoral program.

Degree requirements: For master's and doctorate, thesis/dissertation required, foreign language not required.

Entrance requirements: For master's and doctorate, GRE General Test, TOEFL. *Application deadline:* For fall admission, 3/1 (priority date); for spring admission, 10/1. *Application fee:* $25 ($100 for international students).

Expenses: Tuition, state resident: full-time $1,296; part-time $72 per credit. Tuition, nonresident: full-time $4,932; part-time $274 per credit. Required fees: $1,162. Tuition and fees vary according to program.

Financial aid: Research assistantships, teaching assistantships, institutionally sponsored loans available. Aid available to part-time students. Financial aid application deadline: 4/1.

Faculty research: Cardiovascular and renal pharmacology, cellular pharmacology, signal transduction, aging, drug delivery systems.

Dr. Douglas Eikenburg, Chair, 713-743-1217, *Fax:* 713-743-1229, *E-mail:* deikenburg@uh.edu.

Application contact: Shaki Commisariat, Graduate Programs Office Manager, 713-743-1227, *Fax:* 713-743-1229, *E-mail:* shaki@uh.edu.

■ UNIVERSITY OF ILLINOIS AT CHICAGO

College of Medicine and Graduate College, Graduate Programs in Medicine, Department of Pharmacology, Chicago, IL 60607-7128

AWARDS PhD, MD/PhD. Part-time programs available.

Faculty: 14 full-time (0 women).

Students: 23 full-time (15 women); includes 1 minority (Asian American or Pacific Islander), 11 international. Average age 29. *97 applicants, 6% accepted.* In 1999, 5 doctorates awarded.

University of Illinois at Chicago (continued)

Degree requirements: For doctorate, dissertation required, foreign language not required.

Entrance requirements: For doctorate, GRE General Test, TOEFL. *Application deadline:* For fall admission, 6/1; for spring admission, 11/1. *Application fee:* $40 ($50 for international students).

Expenses: Tuition, state resident: full-time $3,750. Tuition, nonresident: full-time $10,588. Tuition and fees vary according to course load.

Financial aid: In 1999–00, 14 students received aid; fellowships, research assistantships, teaching assistantships, Federal Work-Study and tuition waivers (full) available. Financial aid application deadline: 3/1; financial aid applicants required to submit FAFSA.

Faculty research: Neuropharmacology; platelet, molecular, and behavioral pharmacology.
Dr. Asrar B. Malik, Head, 312-996-7635.
Application contact: Dr. R. D. Green, Director, Graduate Studies, 312-996-7640.

Find an in-depth description at www.petersons.com/graduate.

■ **THE UNIVERSITY OF IOWA**

College of Medicine and Graduate College, Graduate Programs in Medicine, Department of Pharmacology, Iowa City, IA 52242-1316

AWARDS MS, PhD.

Faculty: 15 full-time (2 women), 9 part-time/adjunct (2 women).

Students: 21 full-time (8 women); includes 3 minority (all Asian Americans or Pacific Islanders), 6 international. *41 applicants, 22% accepted.* In 1999, 1 degree awarded (100% entered university research/teaching). Terminal master's awarded for partial completion of doctoral program.

Degree requirements: For master's, thesis required; for doctorate, dissertation, comprehensive exam required, foreign language not required. *Average time to degree:* Doctorate–6 years full-time.

Entrance requirements: For master's, GRE; for doctorate, GRE General Test. *Application deadline:* For fall admission, 2/15 (priority date). Applications are processed on a rolling basis. *Application fee:* $30 ($50 for international students). Electronic applications accepted.

Expenses: Tuition, state resident: full-time $3,308. Tuition, nonresident: full-time $10,662. Tuition and fees vary according to course load and program.

Financial aid: In 1999–00, research assistantships with full tuition reimbursements (averaging $16,787 per year); grants also available.

Faculty research: Cancer and cell cycle, hormones and growth factors, nervous system function and dysfunction, receptors and signal transduction, stroke and hypertension. *Total annual research expenditures:* $2.8 million.
Dr. Gerald F. Gebhart, Head, 319-335-7965, *Fax:* 319-335-8930.
Application contact: Dr. Minnetta V. Gardinier, Director, Graduate Admissions, 319-335-6735, *Fax:* 319-335-8930, *E-mail:* pharmacology@uiowa.edu.

Find an in-depth description at www.petersons.com/graduate.

■ **UNIVERSITY OF KANSAS**

Graduate School, School of Pharmacy, Department of Pharmacology and Toxicology, Lawrence, KS 66045

AWARDS Pharmacology (MS, PhD); toxicology (MS, PhD).

Faculty: 7.

Students: 12 full-time (7 women), 2 part-time (1 woman), 8 international. Average age 26. *10 applicants, 40% accepted.* In 1999, 1 master's awarded. Terminal master's awarded for partial completion of doctoral program.

Degree requirements: For master's, thesis required; for doctorate, dissertation, oral comprehensive exams required.

Entrance requirements: For master's, GRE General Test, TSE, bachelor's degree in related field; for doctorate, GRE General Test, MAT, TOEFL, bachelor's degree in related field. *Application deadline:* For fall admission, 2/1 (priority date). Applications are processed on a rolling basis. *Application fee:* $20.

Expenses: Tuition, state resident: full-time $2,482; part-time $103 per credit hour. Tuition, nonresident: full-time $8,104; part-time $338 per credit hour. Required fees: $428; $31 per credit hour. Tuition and fees vary according to program.

Financial aid: In 1999–00, 8 fellowships (averaging $12,500 per year), 11 research assistantships (averaging $15,000 per year), 1 teaching assistantship (averaging $15,000 per year) were awarded. Financial aid application deadline: 1/15.

Faculty research: Molecular neurobiology, gene regulation, neurotransmitter receptors. *Total annual research expenditures:* $1.8 million.
Elias Michaelis, Chair, 785-864-4001, *Fax:* 785-864-5219.
Application contact: Mary Michaelis, Graduate Director, 785-864-3905, *Fax:* 785-864-5219, *E-mail:* mlm@smissman.hbc.ukans.edu.

■ **UNIVERSITY OF KANSAS**

Graduate Studies Medical Center, Graduate Programs in Biomedical and Basic Sciences, Department of Pharmacology, Toxicology and Therapeutics, Curriculum in Pharmacology, Kansas City, KS 66160

AWARDS MS, PhD, MD/MS, MD/PhD. Part-time programs available.

Students: 6 full-time (4 women), 2 international. Average age 30. *0 applicants, 0% accepted.* In 1999, 1 master's awarded. Terminal master's awarded for partial completion of doctoral program.

Degree requirements: For master's, thesis, comprehensive oral exam required; for doctorate, one foreign language (computer language can substitute), dissertation, comprehensive oral exam required.

Entrance requirements: For master's and doctorate, GRE General Test, TOEFL, TSE. *Application deadline:* For fall admission, 1/31 (priority date). Applications are processed on a rolling basis. *Application fee:* $0. Electronic applications accepted.

Expenses: Tuition, state resident: full-time $2,482; part-time $103 per credit hour. Tuition, nonresident: full-time $8,104; part-time $338 per credit hour. Required fees: $428; $31 per credit hour. Tuition and fees vary according to program.

Financial aid: Fellowships, research assistantships, teaching assistantships, Federal Work-Study, institutionally sponsored loans, and scholarships available. Aid available to part-time students. Financial aid application deadline: 3/31; financial aid applicants required to submit FAFSA.

Application contact: Beth Levant, Director of Admissions, 913-588-7527, *Fax:* 913-588-7501, *E-mail:* blevant@kumc.edu.

Find an in-depth description at www.petersons.com/graduate.

■ **UNIVERSITY OF KENTUCKY**

Graduate School and College of Medicine, Graduate Programs in Medicine, Program in Pharmacology, Lexington, KY 40506-0032

AWARDS MS, PhD, MD/PhD.

Degree requirements: For master's, comprehensive exam required, thesis optional, foreign language not required; for doctorate, dissertation, comprehensive exam required, foreign language not required.

Entrance requirements: For master's, GRE General Test, minimum undergraduate GPA of 2.5; for doctorate, GRE General Test, minimum graduate GPA of 3.0.

Expenses: Tuition, state resident: full-time $3,596; part-time $188 per credit hour. Tuition, nonresident: full-time $10,116; part-time $550 per credit hour.
Faculty research: Neuropharmacology, molecular mechanisms of cancer, molecular pharmacology of ion channels, cardiovascular pharmacology, neurobiology of aging.

■ UNIVERSITY OF LOUISVILLE

School of Medicine and Graduate School, Integrated Programs in Biomedical Sciences, Department of Pharmacology and Toxicology, Louisville, KY 40292-0001

AWARDS MS, PhD.

Degree requirements: For master's and doctorate, thesis/dissertation required.
Entrance requirements: For master's and doctorate, GRE General Test.
Expenses: Tuition, state resident: full-time $3,260; part-time $182 per hour. Tuition, nonresident: full-time $9,780; part-time $544 per hour. Required fees: $143; $28 per hour.
Faculty research: Metabolic activation of chemical carcinogens, biochemical neuropharmacology, analytical toxicology and kinetics of chemical disposition, biochemical toxicology, clinical pharmacology.

■ UNIVERSITY OF MARYLAND

Graduate School, Graduate Programs in Medicine, Department of Pharmacology and Experimental Therapeutics, Baltimore, MD 21201-1627

AWARDS Pharmacology (PhD); pharmacology and experimental therapeutics (MS); toxicology (PhD). Part-time programs available. Terminal master's awarded for partial completion of doctoral program.

Degree requirements: For master's and doctorate, thesis/dissertation required, foreign language not required.
Entrance requirements: For master's and doctorate, GRE General Test, TOEFL, minimum GPA of 3.0.
Expenses: Tuition, state resident: part-time $261 per credit hour. Tuition, nonresident: part-time $468 per credit hour. Tuition and fees vary according to program.
Faculty research: Molecular and cellular neuropharmacology, drug and endocrine mechanisms in cancer, viral immunopharmacology.

Find an in-depth description at www.petersons.com/graduate.

■ UNIVERSITY OF MASSACHUSETTS WORCESTER

Graduate School of Biomedical Sciences, Department of Biochemistry and Molecular Pharmacology, Worcester, MA 01655-0115

AWARDS Biochemistry and molecular biology (PhD); pharmacology and molecular toxicology (PhD).

Faculty: 25 full-time (2 women).
Students: 12 full-time (6 women); includes 2 minority (both Asian Americans or Pacific Islanders), 3 international.
Degree requirements: For doctorate, dissertation required.
Entrance requirements: For doctorate, GRE General Test, GRE Subject Test, 1 year of calculus, physics, organic chemistry and biology. *Application deadline:* For fall admission, 1/15 (priority date). Applications are processed on a rolling basis. *Application fee:* $25 ($50 for international students).
Expenses: Tuition, state resident: full-time $2,640. Tuition, nonresident: full-time $9,756. Required fees: $825. Full-time tuition and fees vary according to program.
Financial aid: In 1999–00, research assistantships with full tuition reimbursements (averaging $17,500 per year); unspecified assistantships also available.
Faculty research: Neuropharmacology, control of DNA metabolism, clinical pharmacology and toxicology.
Dr. C. Robert Matthews, Chair, 508-856-2251, *Fax:* 508-856-2151.
Application contact: Dr. Alonzo Ross, Graduate Director, 508-856-8016, *Fax:* 508-856-2151, *E-mail:* alonzo.ross@umassmed.edu.

Find an in-depth description at www.petersons.com/graduate.

■ UNIVERSITY OF MEDICINE AND DENTISTRY OF NEW JERSEY

Graduate School of Biomedical Sciences, Graduate Programs in Biomedical Sciences, Department of Pharmacology, Piscataway, NJ 08854-5634

AWARDS Cellular and molecular pharmacology (PhD).

Degree requirements: For doctorate, dissertation, qualifying exam required, foreign language not required.
Entrance requirements: For doctorate, GRE General Test, TOEFL. *Application deadline:* For fall admission, 2/1; for spring admission, 10/1. *Application fee:* $40.
Expenses: Tuition, state resident: part-time $270 per credit hour. Tuition, nonresident: part-time $407 per credit

hour. Part-time tuition and fees vary according to campus/location and program.
Financial aid: Fellowships, research assistantships, traineeships available. Financial aid application deadline: 5/1.
Faculty research: Cellular neuropharmacology.
Dr. Tariq M. Rana, Director, 732-235-4082, *Fax:* 732-235-4073.

■ UNIVERSITY OF MEDICINE AND DENTISTRY OF NEW JERSEY

Graduate School of Biomedical Sciences, Graduate Programs in Biomedical Sciences, Department of Pharmacology and Physiology, Newark, NJ 07107

AWARDS MS, PhD. Terminal master's awarded for partial completion of doctoral program.

Degree requirements: For master's, thesis required; for doctorate, dissertation, qualifying exam required, foreign language not required.
Entrance requirements: For master's and doctorate, GRE General Test, TOEFL. *Application deadline:* For fall admission, 2/1; for spring admission, 10/1. Applications are processed on a rolling basis. *Application fee:* $40.
Expenses: Tuition, state resident: part-time $270 per credit hour. Tuition, nonresident: part-time $407 per credit hour. Part-time tuition and fees vary according to campus/location and program.
Financial aid: Fellowships, research assistantships, Federal Work-Study and institutionally sponsored loans available. Financial aid application deadline: 5/1.
Dr. Andrew Philip Thomas, Acting Chairman, 973-972-4660, *E-mail:* thomasap@umdnj.edu.
Application contact: Dr. Henry E. Brezenoff, Dean, Graduate School of Biomedical Sciences, 973-972-5333, *Fax:* 973-972-7148, *E-mail:* hbrezeno@umdnj.edu.

Find an in-depth description at www.petersons.com/graduate.

■ UNIVERSITY OF MIAMI

School of Medicine and Graduate School, Graduate Programs in Medicine, Department of Molecular and Cellular Pharmacology, Coral Gables, FL 33124

AWARDS PhD, MD/PhD.

Faculty: 15 full-time (4 women).
Students: 17 full-time (5 women); includes 3 minority (1 African American, 2 Hispanic Americans), 3 international. Average age 25. *55 applicants, 7% accepted.* In

University of Miami (continued)
1999, 3 degrees awarded (100% entered university research/teaching).
Degree requirements: For doctorate, dissertation defense, laboratory rotations, qualifying exam required.
Entrance requirements: For doctorate, GRE General Test, TOEFL. *Application deadline:* For fall admission, 3/1 (priority date). Applications are processed on a rolling basis. *Application fee:* $35.
Expenses: Tuition, area resident: Part-time $899 per credit.
Financial aid: In 1999–00, 17 students received aid, including 1 fellowship, 8 research assistantships; institutionally sponsored loans and traineeships also available. Financial aid application deadline: 3/1; financial aid applicants required to submit FAFSA.
Faculty research: Membrane and cardiovascular pharmacology, neuropharmacological cell signaling, calcium regulation, growth factor receptors. *Total annual research expenditures:* $4.6 million.
Dr. James D. Potter, Chairman, 305-243-5874, *Fax:* 305-243-6643, *E-mail:* pharm@mednet.med.miami.edu.
Application contact: Dr. Charles W. Luetje, Director of Graduate Studies, 305-243-6643, *Fax:* 305-243-3420, *E-mail:* cluetje@chroma.med.miami.edu.
Find an in-depth description at www.petersons.com/graduate.

■ **UNIVERSITY OF MICHIGAN**

Medical School and Horace H. Rackham School of Graduate Studies, Program in Biomedical Sciences (PIBS), Department of Pharmacology, Ann Arbor, MI 48109

AWARDS PhD, MD/PhD, Pharm D/PhD.

Faculty: 15 full-time (3 women), 15 part-time/adjunct (2 women).
Students: 32 full-time (13 women); includes 5 minority (3 African Americans, 1 Asian American or Pacific Islander, 1 Hispanic American), 3 international. Average age 25. *46 applicants, 30% accepted.* In 1999, 3 doctorates awarded (100% continued full-time study).
Degree requirements: For doctorate, oral defense of dissertation, oral preliminary exam required. *Average time to degree:* Doctorate–5 years full-time.
Entrance requirements: For doctorate, GRE General Test. *Application deadline:* For fall admission, 1/5. *Application fee:* $55. Electronic applications accepted.
Expenses: Tuition, state resident: full-time $10,316. Tuition, nonresident: full-time $20,922.
Financial aid: In 1999–00, 12 fellowships with full tuition reimbursements (averaging $17,000 per year), 21 research assistantships with full tuition reimbursements (averaging $17,000 per year) were awarded; teaching assistantships, grants, institutionally sponsored loans, and scholarships also available. Financial aid application deadline: 1/5.
Faculty research: Signal transduction, addiction research, cancer pharmacology, drug metabolism and pharmacogenetics. *Total annual research expenditures:* $7 million.
Dr. Paul F. Hollenberg, Chair, 734-764-8166, *Fax:* 734-763-5387, *E-mail:* phollen@umich.edu.
Application contact: Eileen A. Ferguson, Student Services Assistant, 734-764-8166, *Fax:* 734-763-5387, *E-mail:* effergie@umich.edu.
Find an in-depth description at www.petersons.com/graduate.

■ **UNIVERSITY OF MINNESOTA, DULUTH**

School of Medicine, Graduate Program in Pharmacology, Duluth, MN 55812-2496

AWARDS MS, PhD.

Faculty: 5 full-time (2 women).
Students: 1 full-time (0 women); minority (Native American). Average age 26. *4 applicants, 0% accepted.* Terminal master's awarded for partial completion of doctoral program.
Degree requirements: For master's, thesis, final oral exam required, foreign language not required; for doctorate, dissertation, final oral exam, oral and written preliminary exams required, foreign language not required.
Entrance requirements: For master's and doctorate, GRE General Test, TOEFL. *Application deadline:* For fall admission, 8/1. Applications are processed on a rolling basis. *Application fee:* $40 ($50 for international students).
Expenses: Tuition, state resident: full-time $5,040; part-time $420 per credit. Tuition, nonresident: full-time $9,900; part-time $825 per credit. Required fees: $509. Tuition and fees vary according to course load and program.
Financial aid: In 1999–00, 1 student received aid, including 1 research assistantship with full tuition reimbursement available (averaging $18,247 per year); fellowships, institutionally sponsored loans also available. Financial aid application deadline: 8/1.
Faculty research: Drug addiction, alcohol and hypertension, neurotransmission, allergic airway disease, auditory neuroscience. *Total annual research expenditures:* $345,759.

Jean F. Regal, Associate Director of Graduate Studies, 218-726-8950, *Fax:* 218-726-6235, *E-mail:* jregal@ub.d.umn.edu.

■ **UNIVERSITY OF MINNESOTA, TWIN CITIES CAMPUS**

Medical School and Graduate School, Graduate Programs in Medicine, Department of Pharmacology, Minneapolis, MN 55455-0213

AWARDS MS, PhD. Terminal master's awarded for partial completion of doctoral program.

Degree requirements: For master's and doctorate, thesis/dissertation required, foreign language not required.
Entrance requirements: For master's and doctorate, GRE General Test, TOEFL.
Expenses: Tuition, state resident: full-time $11,984; part-time $1,498 per semester. Tuition, nonresident: full-time $22,264; part-time $2,783 per semester. Full-time tuition and fees vary according to program and student level. Part-time tuition and fees vary according to course load and program.
Faculty research: Molecular pharmacology, gene therapy, molecular mechanisms of opiate and endorphin actions, cancer chemotherapy, neuropharmacology.

■ **UNIVERSITY OF MISSISSIPPI**

Graduate School, School of Pharmacy, Department of Pharmacology, Oxford, University, MS 38677

AWARDS Pharmacology (MS, PhD); toxicology (PhD).

Faculty: 7 full-time (1 woman).
Students: 15 full-time (7 women), 2 part-time (1 woman); includes 1 minority (African American), 9 international. In 1999, 3 degrees awarded.
Degree requirements: For master's and doctorate, thesis/dissertation required, foreign language not required.
Entrance requirements: For master's, GRE General Test, TOEFL, minimum GPA of 3.0; for doctorate, GRE General Test, TOEFL. *Application deadline:* For fall admission, 8/1. Applications are processed on a rolling basis. *Application fee:* $0 ($25 for international students).
Expenses: Tuition, state resident: full-time $3,053; part-time $170 per credit hour. Tuition, nonresident: full-time $6,155; part-time $342 per credit hour. Tuition and fees vary according to program.
Financial aid: Application deadline: 3/1.
Faculty research: Behavioral and biochemical pharmacology.
Dr. Dennis R. Feller, Chairman, 662-915-5148.

■ UNIVERSITY OF MISSISSIPPI MEDICAL CENTER

Graduate Programs in Biomedical Sciences, Department of Pharmacology and Toxicology, Jackson, MS 39216-4505

AWARDS Pharmacology (MS, PhD); toxicology (MS, PhD).

Faculty: 11 full-time (1 woman), 10 part-time/adjunct (0 women).

Students: 26 full-time (9 women), 2 part-time; includes 1 minority (African American), 22 international. Average age 30. *16 applicants, 50% accepted.* In 1999, 1 master's awarded (100% found work related to degree); 3 doctorates awarded (100% continued full-time study). Terminal master's awarded for partial completion of doctoral program.

Degree requirements: For master's, thesis required, foreign language not required; for doctorate, dissertation, first authored publication required, foreign language not required. *Average time to degree:* Master's–4 years full-time; doctorate–5 years full-time.

Entrance requirements: For master's and doctorate, GRE General Test, minimum GPA of 3.0. *Application deadline:* For fall admission, 6/1 (priority date). Applications are processed on a rolling basis. *Application fee:* $10.

Expenses: Tuition, state resident: full-time $2,378; part-time $132 per hour. Tuition, nonresident: full-time $4,697; part-time $261 per hour. Tuition and fees vary according to program.

Financial aid: In 1999–00, 22 students received aid, including 22 research assistantships (averaging $16,234 per year). Financial aid application deadline: 4/1.

Faculty research: Neuropharmacology, environmental toxicology, aging, immunopharmacology, cardiovascular pharmacology. *Total annual research expenditures:* $1.2 million.

Dr. I. K. Ho, Interim Associate Vice Chancellor, 601-984-1600, *Fax:* 601-984-1637.

Find an in-depth description at www.petersons.com/graduate.

■ UNIVERSITY OF MISSOURI–COLUMBIA

School of Medicine and Graduate School, Graduate Programs in Medicine, Department of Pharmacology, Columbia, MO 65211

AWARDS MS, PhD.

Degree requirements: For master's, thesis required, foreign language not required; for doctorate, dissertation required.

Entrance requirements: For master's and doctorate, GRE General Test, minimum GPA of 3.0.

Expenses: Tuition, state resident: full-time $3,020; part-time $168 per hour. Tuition, nonresident: full-time $6,066; part-time $505 per hour. Required fees: $445; $18 per hour.

Faculty research: Endocrine and metabolic pharmacology, biochemical pharmacology, neuropharmacology, receptors and transmembrane signalling.

■ THE UNIVERSITY OF MONTANA–MISSOULA

Graduate School, School of Pharmacy and Allied Health Sciences, Programs in Pharmaceutical Sciences, Missoula, MT 59812-0002

AWARDS Pharmaceutical sciences (MS); pharmacology (PhD).

Degree requirements: For master's, oral defense of thesis required; for doctorate, research dissertation defense required.

Entrance requirements: For master's and doctorate, GRE General Test, TOEFL. Electronic applications accepted.

Expenses: Tuition, state resident: full-time $2,484; part-time $151 per credit. Tuition, nonresident: full-time $8,000; part-time $305 per credit. Required fees: $1,600. Full-time tuition and fees vary according to degree level and program.

Faculty research: Neuropharmacology, molecular biochemistry, cardiovascular pharmacology, toxicology, medicinal chemistry.

Find an in-depth description at www.petersons.com/graduate.

■ UNIVERSITY OF NEBRASKA MEDICAL CENTER

Graduate College, Department of Pharmacology, Omaha, NE 68198

AWARDS MS, PhD.

Faculty: 10 full-time, 12 part-time/adjunct.

Students: 14 full-time (3 women), 1 part-time, 1 international. Average age 27. *40 applicants, 13% accepted.* In 1999, 1 doctorate awarded (100% entered university research/teaching). Terminal master's awarded for partial completion of doctoral program.

Degree requirements: For master's, thesis required, foreign language not required; for doctorate, dissertation required. *Average time to degree:* Doctorate–5.1 years full-time.

Entrance requirements: For master's and doctorate, GRE General Test. *Application deadline:* For fall admission, 3/1 (priority date). Applications are processed on a rolling basis. *Application fee:* $35.

Expenses: Tuition, state resident: part-time $116 per semester hour. Tuition, nonresident: part-time $270 per semester hour. Tuition and fees vary according to program.

Financial aid: In 1999–00, 1 fellowship (averaging $15,000 per year), 9 research assistantships with full tuition reimbursements (averaging $15,000 per year) were awarded; career-related internships or fieldwork and institutionally sponsored loans also available. Aid available to part-time students. Financial aid application deadline: 3/1.

Faculty research: Neuropharmacology, molecular pharmacology, toxicology, molecular biology, neuroscience. *Total annual research expenditures:* $1.8 million.

Dr. Daniel Monaghan, Graduate Committee Chair, 402-559-7196, *Fax:* 402-559-7495, *E-mail:* dtmonagh@mail.unmc.edu.

Application contact: Jo Wagner, Associate Director of Admissions, 402-559-6468.

■ UNIVERSITY OF NEVADA, RENO

School of Medicine and Graduate School, Graduate Programs in Medicine, Interdisciplinary Program in Cellular and Molecular Pharmacology and Physiology, Reno, NV 89557

AWARDS MS, PhD.

Faculty: 23.

Students: 8 full-time (5 women), 2 part-time (1 woman); includes 2 minority (both Asian Americans or Pacific Islanders), 2 international. Average age 27. *9 applicants, 33% accepted.* In 1999, 3 doctorates awarded (100% continued full-time study). Terminal master's awarded for partial completion of doctoral program.

Degree requirements: For master's, thesis required, foreign language not required; for doctorate, one foreign language (computer language can substitute), dissertation required.

Entrance requirements: For master's, GRE General Test, TOEFL, minimum GPA of 2.75; for doctorate, GRE General Test, TOEFL, minimum GPA of 3.0. *Application deadline:* For fall admission, 3/1 (priority date); for spring admission, 11/1. Applications are processed on a rolling basis. *Application fee:* $40.

Expenses: Tuition, state resident: full-time $7,782. Tuition, nonresident: full-time $22,808. Required fees: $1,918.

Financial aid: Research assistantships, teaching assistantships available. Aid available to part-time students. Financial aid application deadline: 3/1.

Faculty research: Neuropharmacology, toxicology, cardiovascular pharmacology, neuromuscular pharmacology.

University of Nevada, Reno (continued)
Dr. Burton Horowitz, Director, 775-784-1462.

Find an in-depth description at www.petersons.com/graduate.

■ THE UNIVERSITY OF NORTH CAROLINA AT CHAPEL HILL

School of Medicine and Graduate School, Graduate Programs in Medicine, Department of Pharmacology, Chapel Hill, NC 27599
AWARDS PhD.

Faculty: 30 full-time (3 women), 8 part-time/adjunct (0 women).
Students: 37 full-time (19 women); includes 4 African Americans, 1 Hispanic American, 1 international. Average age 25. *62 applicants, 18% accepted.* In 1999, 6 doctorates awarded (83% entered university research/teaching, 17% found other work related to degree).
Degree requirements: For doctorate, dissertation, comprehensive exams oral and written required. *Average time to degree:* Doctorate–5.5 years full-time.
Entrance requirements: For doctorate, GRE General Test, GRE Subject Test, minimum GPA of 3.0. *Application deadline:* For fall admission, 1/1 (priority date). *Application fee:* $55. Electronic applications accepted.
Expenses: Tuition, state resident: full-time $1,966. Tuition, nonresident: full-time $11,026. Required fees: $8,940. One-time fee: $15 full-time. Part-time tuition and fees vary according to course load.
Financial aid: In 1999–00, 5 fellowships with tuition reimbursements (averaging $17,000 per year), 32 research assistantships with tuition reimbursements (averaging $17,000 per year) were awarded; tuition waivers (full) and unspecified assistantships also available.
Faculty research: Signal transduction, cell adhesion, receptors, ion channels. *Total annual research expenditures:* $5 million.
Dr. Rudolph L. Juliano, Chairman, 919-966-4383, *E-mail:* arjay@med.unc.edu.
Application contact: Dr. Robert A. Nicholas, Director of Graduate Studies, 919-966-1153, *Fax:* 919-966-5640, *E-mail:* phcograd@med.unc.edu.

Find an in-depth description at www.petersons.com/graduate.

■ UNIVERSITY OF NORTH DAKOTA

School of Medicine and Graduate School, Graduate Programs in Medicine, Department of Pharmacology and Toxicology, Grand Forks, ND 58202
AWARDS MS, PhD.

Faculty: 8 full-time (2 women).
Students: 4 full-time (1 woman). *11 applicants, 55% accepted.* In 1999, 2 doctorates awarded.
Degree requirements: For master's, thesis, final examination required, foreign language not required; for doctorate, dissertation, comprehensive examination, final examination required, foreign language not required.
Entrance requirements: For master's, GRE General Test or MCAT, TOEFL, minimum GPA of 3.0; for doctorate, GRE General Test or MCAT, TOEFL, minimum GPA of 3.5. *Application deadline:* For fall admission, 3/1 (priority date). Applications are processed on a rolling basis. *Application fee:* $25.
Expenses: Tuition, state resident: full-time $2,690; part-time $112 per credit. Tuition, nonresident: full-time $7,182; part-time $299 per credit. Required fees: $46 per semester.
Financial aid: In 1999–00, 4 students received aid, including 4 research assistantships with full tuition reimbursements available (averaging $10,586 per year); fellowships, teaching assistantships with full tuition reimbursements available, Federal Work-Study, institutionally sponsored loans, scholarships, and tuition waivers (full and partial) also available. Aid available to part-time students. Financial aid application deadline: 3/15; financial aid applicants required to submit FAFSA.
Faculty research: Molecular and cellular aspects of pharmacology, endrocrinology, genetics, cancer.
Dr. Mike Ebadi, Chairperson, 701-777-4293, *Fax:* 701-777-6124, *E-mail:* mebadi@medicine.nodak.edu.

■ UNIVERSITY OF NORTH TEXAS HEALTH SCIENCE CENTER AT FORT WORTH

Graduate School of Biomedical Sciences, Fort Worth, TX 76107-2699
AWARDS Anatomy and cell biology (MS, PhD); biochemistry and molecular biology (MS, PhD); biotechnology (MS); integrative physiology (MS, PhD); microbiology and immunology (MS, PhD); pharmacology (MS, PhD).

Faculty: 65 full-time (9 women), 11 part-time/adjunct (1 woman).
Students: 59 full-time (27 women), 44 part-time (20 women); includes 30 minority (10 African Americans, 9 Asian Americans or Pacific Islanders, 11 Hispanic Americans), 23 international. *70 applicants, 70% accepted.* In 1999, 5 master's awarded (40% found work related to degree, 60% continued full-time study); 5 doctorates awarded (80% entered

university research/teaching, 20% found other work related to degree.
Degree requirements: For doctorate, computer language, dissertation required. *Average time to degree:* Master's–2.5 years full-time, 4 years part-time; doctorate–5 years full-time.
Entrance requirements: For master's and doctorate, GRE General Test, TOEFL. *Application deadline:* For fall admission, 5/1; for spring admission, 11/1. Applications are processed on a rolling basis. *Application fee:* $25 ($50 for international students).
Expenses: Tuition, state resident: full-time $1,188; part-time $66 per credit. Tuition, nonresident: full-time $5,058; part-time $281 per credit. Required fees: $366; $183 per semester.
Financial aid: In 1999–00, 11 fellowships, 70 research assistantships (averaging $16,500 per year) were awarded; teaching assistantships, career-related internships or fieldwork, Federal Work-Study, grants, institutionally sponsored loans, and traineeships also available. Aid available to part-time students. Financial aid application deadline: 4/1; financial aid applicants required to submit FAFSA.
Faculty research: Alzheimer's disease, diabetes, eye diseases, cancer, cardiovascular physiology.
Dr. Thomas Yorio, Dean, 817-735-2560, *Fax:* 817-735-0243, *E-mail:* yoriot@hsc.unt.edu.
Application contact: Jan Sharp, Administrative Assistant, 817-735-0258, *Fax:* 817-735-0243, *E-mail:* gsbs@hsc.unt.edu.

Find an in-depth description at www.petersons.com/graduate.

■ UNIVERSITY OF PENNSYLVANIA

School of Medicine, Biomedical Graduate Studies, Graduate Group in Pharmacological Sciences, Philadelphia, PA 19104
AWARDS Pharmacology (PhD).

Faculty: 55 full-time (7 women), 4 part-time/adjunct (2 women).
Students: 42 full-time (22 women); includes 8 minority (2 African Americans, 3 Asian Americans or Pacific Islanders, 3 Hispanic Americans), 10 international. Average age 27. *87 applicants, 17% accepted.* In 1999, 8 doctorates awarded (50% entered university research/teaching, 50% found other work related to degree).
Degree requirements: For doctorate, dissertation required, foreign language not required.
Entrance requirements: For doctorate, GRE General Test, TOEFL, previous course work in physical or natural science; BA, BS, or equivalent. *Application deadline:*

For fall admission, 1/1 (priority date). Applications are processed on a rolling basis. *Application fee:* $65. Electronic applications accepted.

Expenses: Tuition: Full-time $17,256; part-time $2,991 per course. Required fees: $2,588; $363 per course. $726 per term.

Financial aid: In 1999–00, 25 fellowships, 17 research assistantships were awarded; teaching assistantships, grants, institutionally sponsored loans, and departmental funds also available. Financial aid application deadline: 1/1.

Faculty research: Properties and regulation of receptors for biogenic amines, molecular aspects of transduction, mechanisms of biosynthesis, biological mechanisms of depression, developmental events in the nervous system, enzyme structure and function, molecular modeling. *Total annual research expenditures:* $6 million.

Dr. Randall N. Pittman, Graduate Group Chair, 215-898-9736, *E-mail:* coord@ pharm.med.upenn.edu.

Application contact: Dr. Linda Leroy, Graduate Group Coordinator, 215-898-1790, *E-mail:* leroy@ pharm.med.upenn.edu.

Find an in-depth description at www.petersons.com/graduate.

■ UNIVERSITY OF PITTSBURGH

School of Medicine, Graduate Programs in Medicine, Program in Molecular Pharmacology, Pittsburgh, PA 15260

AWARDS MS, PhD.

Faculty: 32 full-time (9 women), 2 part-time/adjunct (0 women).

Students: 3 full-time (0 women); includes 1 minority (Hispanic American), 2 international. *325 applicants, 22% accepted.* In 1999, 1 master's, 5 doctorates awarded.

Degree requirements: For doctorate, dissertation required, foreign language not required. *Average time to degree:* Doctorate–5.3 years full-time.

Entrance requirements: For doctorate, GRE General Test, GRE Subject Test, TOEFL, minimum QPA of 3.0. *Application deadline:* For fall admission, 1/15 (priority date). Applications are processed on a rolling basis. *Application fee:* $30 ($40 for international students).

Expenses: Tuition, state resident: full-time $9,778; part-time $403 per credit. Tuition, nonresident: full-time $20,146; part-time $830 per credit. Required fees: $480; $90 per semester.

Financial aid: Research assistantships with full tuition reimbursements, teaching assistantships with full tuition reimbursements, Federal Work-Study, institutionally

sponsored loans, scholarships, traineeships, and unspecified assistantships available.

Faculty research: Cancer pharmacology, aeuropharmacology, cardiovascular and pulmonary pharmacology. *Total annual research expenditures:* $76.3 million.

Application contact: Graduate Studies Administrator, 412-648-8957, *Fax:* 412-648-1236, *E-mail:* biomed_phd@fs1.dean-med.pitt.edu.

Find an in-depth description at www.petersons.com/graduate.

■ UNIVERSITY OF PUERTO RICO, MEDICAL SCIENCES CAMPUS

School of Medicine, Division of Graduate Studies, Department of Pharmacology and Toxicology, San Juan, PR 00936-5067

AWARDS MS, PhD.

Faculty: 8 full-time (4 women).

Students: 9 full-time (6 women); all minorities (all Hispanic Americans). Average age 30. *5 applicants, 60% accepted.* In 1999, 1 master's awarded (100% found work related to degree).

Degree requirements: For master's, one foreign language, thesis required; for doctorate, one foreign language, computer language, dissertation, qualifying exam required. *Average time to degree:* Master's–4 years full-time; doctorate–6 years full-time.

Entrance requirements: For master's and doctorate, GRE General Test, GRE Subject Test, interview. *Application deadline:* For fall admission, 2/15. *Application fee:* $15.

Expenses: Tuition, state resident: full-time $5,500. Tuition, nonresident: full-time $8,400. Required fees: $600. Tuition and fees vary according to class time, course load, degree level and program.

Financial aid: In 1999–00, 4 research assistantships were awarded; fellowships, Federal Work-Study, institutionally sponsored loans, and tuition waivers (full and partial) also available. Aid available to part-time students. Financial aid application deadline: 4/30.

Faculty research: Cardiovascular, central nervous system, and endocrine pharmacology; anti-cancer drugs; toxicology, sodium pump. *Total annual research expenditures:* $323,000.

Dr. Walmor C. De Mello, Director, 787-766-4441, *Fax:* 787-282-0568.

Application contact: Dr. Susan Corey, Coordinator, 787-724-1006, *Fax:* 787-725-3804, *E-mail:* s_corey@rcmaca.upr.clu.edu.

■ UNIVERSITY OF RHODE ISLAND

Graduate School, College of Pharmacy, Graduate Programs in Pharmacy, Department of Pharmacology and Toxicology, Kingston, RI 02881

AWARDS MS, PhD.

Application deadline: For fall admission, 4/15. *Application fee:* $35.

Expenses: Tuition, state resident: full-time $3,540; part-time $197 per credit. Tuition, nonresident: full-time $10,116; part-time $197 per credit. Required fees: $1,352; $37 per credit. $65 per term.

Dr. Zahir A. Shaikh, Chairperson, 401-874-2362.

■ UNIVERSITY OF ROCHESTER

School of Medicine and Dentistry, Graduate Programs in Medicine and Dentistry, Department of Pharmacology and Physiology, Program in Pharmacology, Rochester, NY 14627-0250

AWARDS MS, PhD.

Students: 26 full-time (9 women); includes 3 minority (2 African Americans, 1 Asian American or Pacific Islander), 7 international. In 1999, 6 master's, 2 doctorates awarded. Terminal master's awarded for partial completion of doctoral program.

Degree requirements: For master's, thesis required, foreign language not required; for doctorate, dissertation, qualifying exam required, foreign language not required.

Entrance requirements: For master's and doctorate, GRE General Test. *Application deadline:* For fall admission, 2/1. *Application fee:* $25.

Expenses: Tuition: Part-time $697 per credit hour. Tuition and fees vary according to program.

Financial aid: Fellowships, research assistantships, teaching assistantships, tuition waivers (full and partial) available. Financial aid application deadline: 2/1.

Application contact: Linda Fullington, Graduate Program Secretary, 716-275-0447.

Find an in-depth description at www.petersons.com/graduate.

■ UNIVERSITY OF SOUTH ALABAMA

College of Medicine and Graduate School, Program in Basic Medical Sciences, Specialization in Pharmacology, Mobile, AL 36688-0002

AWARDS PhD.

Faculty: 8 full-time (2 women).

University of South Alabama (continued)
Students: In 1999, 1 degree awarded.
Degree requirements: For doctorate, dissertation required, foreign language not required.
Entrance requirements: For doctorate, GRE General Test or MCAT. *Application deadline:* For fall admission, 4/1. Applications are processed on a rolling basis. *Application fee:* $25.
Expenses: Tuition, state resident: part-time $116 per semester hour. Tuition, nonresident: part-time $230 per semester hour. Required fees: $121 per semester.
Financial aid: Fellowships, research assistantships, institutionally sponsored loans available. Financial aid application deadline: 4/1.
Faculty research: Cardiovascular, clinical, and molecular pharmacology.
Dr. Mark N. Gillespie, Chair, 334-460-6497.
Application contact: Lanette Flagge, Coordinator, 334-460-6153.

■ UNIVERSITY OF SOUTH DAKOTA

School of Medicine and Graduate School, Biomedical Sciences Graduate Program, Physiology and Pharmacology Group, Vermillion, SD 57069-2390

AWARDS MA, PhD.

Faculty: 9 full-time (3 women).
Students: 6 full-time (2 women), 3 international. Average age 25. *65 applicants, 5% accepted.* In 1999, 1 doctorate awarded (100% entered university research/teaching). Terminal master's awarded for partial completion of doctoral program.
Degree requirements: For master's and doctorate, thesis/dissertation required, foreign language not required. *Average time to degree:* Doctorate–4.5 years full-time.
Entrance requirements: For master's and doctorate, GRE General Test, TOEFL, minimum GPA of 3.0. *Application deadline:* For fall admission, 3/15 (priority date). Applications are processed on a rolling basis. *Application fee:* $15.
Expenses: Tuition, state resident: full-time $2,126; part-time $89 per credit. Tuition, nonresident: full-time $6,270; part-time $261 per credit. Required fees: $1,194; $50 per credit. Tuition and fees vary according to course load and reciprocity agreements.
Financial aid: Fellowships, research assistantships available.
Faculty research: Pulmonary physiology and pharmacology, drug abuse, reproduction, signal transduction, cardiovascular physiology and pharmacology.

Dr. Steven B. Waller, Head, 605-677-5157, *Fax:* 605-677-6381, *E-mail:* swaller@usd.edu.

■ UNIVERSITY OF SOUTHERN CALIFORNIA

Keck School of Medicine and Graduate School, Graduate Programs in Medicine, Department of Cell and Neurobiology, Los Angeles, CA 90089

AWARDS Anatomy and cell biology (MS, PhD), including anatomy (PhD), cell biology (PhD); cell and neurobiology (MS, PhD); pharmacology and nutrition (MS, PhD); preventive nutrition (MS).

Faculty: 21 full-time (3 women), 8 part-time/adjunct (2 women).
Students: 19 full-time (9 women); includes 10 minority (all Asian Americans or Pacific Islanders). Average age 23. *54 applicants, 22% accepted.* In 1999, 5 master's awarded (40% found work related to degree, 60% continued full-time study). Terminal master's awarded for partial completion of doctoral program.
Degree requirements: For master's, thesis or alternative required, foreign language not required; for doctorate, dissertation required. *Average time to degree:* Master's–2 years full-time; doctorate–5 years full-time.
Entrance requirements: For master's, GRE General Test, TOEFL, minimum GPA of 3.0; for doctorate, GRE General Test, TOEFL. *Application fee:* $55. Electronic applications accepted.
Expenses: Tuition: Full-time $22,198; part-time $748 per unit. Required fees: $406.
Financial aid: In 1999–00, 13 students received aid; fellowships, research assistantships, teaching assistantships, Federal Work-Study, institutionally sponsored loans, and tuition waivers (partial) available. Aid available to part-time students.
Faculty research: Neurobiology and development, circaulian rhythm, gene therapy in vision, lacrimal glands, neuroendocrinology, signal transduction mechanisms.
Dr. Cheryl Craft, Chair, 323-442-1881, *Fax:* 323-442-2709, *E-mail:* ccraft@hsc.usc.edu.
Application contact: Darlene Marie Campbell, Administrative Assistant, 323-442-1881, *Fax:* 323-442-0466, *E-mail:* dmc@hsc.usc.edu.

■ UNIVERSITY OF SOUTHERN CALIFORNIA

School of Pharmacy and Graduate School, Graduate Programs in Pharmacy, Graduate Program in Molecular Pharmacology and Toxicology, Los Angeles, CA 90089

AWARDS MS, PhD, Pharm D/PhD.

Faculty: 9 full-time (3 women).
Students: 21 full-time (13 women), 2 part-time (1 woman); includes 6 minority (all Asian Americans or Pacific Islanders), 12 international. Average age 28. *44 applicants, 7% accepted.* In 1999, 1 master's, 5 doctorates awarded.
Degree requirements: For master's and doctorate, thesis/dissertation required.
Entrance requirements: For master's and doctorate, GRE General Test. *Application deadline:* For fall admission, 2/1 (priority date); for spring admission, 10/15. Applications are processed on a rolling basis. *Application fee:* $55.
Expenses: Tuition: Full-time $22,198; part-time $748 per unit. Required fees: $406.
Financial aid: In 1999–00, 9 fellowships with full tuition reimbursements (averaging $18,200 per year), 9 research assistantships with full tuition reimbursements (averaging $18,200 per year), 6 teaching assistantships with full tuition reimbursements (averaging $17,800 per year) were awarded; Federal Work-Study, institutionally sponsored loans, and scholarships also available. Aid available to part-time students. Financial aid application deadline: 2/15; financial aid applicants required to submit FAFSA.
Dr. Enrique Cadenas, Chair, 323-442-1551.
Application contact: 323-442-1474.

■ UNIVERSITY OF SOUTH FLORIDA

College of Medicine and Graduate School, Graduate Programs in Medical Sciences, Department of Pharmacology and Therapeutics, Tampa, FL 33620-9951

AWARDS Medical sciences (PhD).

Degree requirements: For doctorate, dissertation required, foreign language not required.
Entrance requirements: For doctorate, GRE General Test, TOEFL, minimum GPA of 3.0.
Expenses: Tuition, state resident: part-time $148 per credit hour. Tuition, nonresident: part-time $509 per credit hour.
Faculty research: Aging and neurodegenerative disease, neuropharmacology, genetics of brain

disorders, smooth muscle pharmacology, immunopharmacology.

Find an in-depth description at www.petersons.com/graduate.

■ THE UNIVERSITY OF TENNESSEE HEALTH SCIENCE CENTER

College of Graduate Health Sciences, Department of Pharmacology, Memphis, TN 38163-0002

AWARDS MS, PhD. Terminal master's awarded for partial completion of doctoral program.

Degree requirements: For master's, thesis, oral and written comprehensive exams required, foreign language not required; for doctorate, dissertation, oral and written preliminary and comprehensive exams required, foreign language not required.

Entrance requirements: For master's, GRE General Test, TOEFL, minimum GPA of 3.0; for doctorate, GRE General Test, GRE Subject Test, TOEFL, minimum GPA of 3.0.

■ THE UNIVERSITY OF TEXAS HEALTH SCIENCE CENTER AT SAN ANTONIO

Graduate School of Biomedical Sciences, Department of Pharmacology, San Antonio, TX 78229-3900

AWARDS PhD.

Degree requirements: For doctorate, dissertation required, foreign language not required.

Entrance requirements: For doctorate, GRE General Test, TOEFL.

Expenses: Tuition, state resident: part-time $38 per credit hour. Tuition, nonresident: part-time $249 per credit hour.

Faculty research: Receptor pharmacology, psychopharmacology, autonomic-cardiovascular pharmacology, neuropharmacology, molecular pharmacology.

■ THE UNIVERSITY OF TEXAS–HOUSTON HEALTH SCIENCE CENTER

Graduate School of Biomedical Sciences, Program in Pharmacology, Houston, TX 77225-0036

AWARDS MS, PhD, MD/PhD.

Faculty: 18 full-time (3 women).
Students: 3 full-time (all women); includes 1 minority (Asian American or Pacific Islander), 1 international. Average age 27.

24 applicants, 50% accepted. In 1999, 1 doctorate awarded. Terminal master's awarded for partial completion of doctoral program.

Degree requirements: For master's and doctorate, thesis/dissertation required, foreign language not required.

Entrance requirements: For master's and doctorate, GRE General Test, TOEFL, TWE. *Application deadline:* For fall admission, 1/15 (priority date); for spring admission, 11/1. Applications are processed on a rolling basis. *Application fee:* $10. Electronic applications accepted.

Financial aid: Fellowships, research assistantships, institutionally sponsored loans available. Financial aid application deadline: 1/15.

Faculty research: Mechanisms of action of hormones and neurotransmitters, cancer chemotherapy, molecular biology of regulatory processes, cell signaling.
Dr. Fernando R. Cabral, Director, 713-500-7485, *Fax:* 713-500-7455, *E-mail:* fcabral@farmr1.med.uth.tmc.edu.

Application contact: Anne Baronitis, Director of Admissions, 713-500-9860, *Fax:* 713-500-9877, *E-mail:* abaron@ gsbs.gs.uth.tmc.edu.

Find an in-depth description at www.petersons.com/graduate.

■ THE UNIVERSITY OF TEXAS MEDICAL BRANCH AT GALVESTON

Graduate School of Biomedical Sciences, Program in Pharmacology and Toxicology, Galveston, TX 77555

AWARDS Pharmacology (MS, PhD); toxicology (PhD).

Faculty: 21 full-time (7 women), 1 part-time/adjunct (0 women).
Students: 17 full-time (12 women); includes 1 minority (Hispanic American), 8 international. Average age 32. *20 applicants, 35% accepted.* In 1999, 2 master's, 5 doctorates awarded.

Degree requirements: For master's, thesis or alternative required, foreign language not required; for doctorate, dissertation required, foreign language not required.

Entrance requirements: For master's and doctorate, GRE General Test. *Application deadline:* For fall admission, 8/1. Applications are processed on a rolling basis. *Application fee:* $25 ($50 for international students).

Expenses: Tuition, state resident: full-time $684; part-time $38 per credit hour. Tuition, nonresident: full-time $4,572; part-time $254 per credit hour. Required fees: $29; $7.5 per credit hour. One-time fee: $55. Tuition and fees vary according to program.

Financial aid: Fellowships, research assistantships, Federal Work-Study and institutionally sponsored loans available. Financial aid applicants required to submit FAFSA.

Faculty research: Neuropharmacology; biochemical, endocrine, and cardiovascular pharmacology; molecular pharmacology and toxicology; cellular cell signaling, molecular design, cancer cell biology. *Total annual research expenditures:* $2.5 million.
Dr. Joel P. Gallagher, Director, 409-772-9639, *Fax:* 409-772-9642, *E-mail:* jpgallag@utmb.edu.

Application contact: Ray Kay Santa, Administrative Secretary, 409-772-9644, *Fax:* 409-772-9642, *E-mail:* rasanta@ utmb.edu.

Find an in-depth description at www.petersons.com/graduate.

■ UNIVERSITY OF THE PACIFIC

School of Pharmacy and Graduate School, Graduate Programs in Pharmaceutical Sciences, Department of Physiology/Pharmacology, Stockton, CA 95211-0197

AWARDS Pharmacology (MS, PhD); physiology (MS, PhD).

Faculty: 5 full-time (1 woman).
Students: 2 full-time (1 woman), 6 part-time (all women), 4 international. Terminal master's awarded for partial completion of doctoral program.

Degree requirements: For master's, thesis required, foreign language not required; for doctorate, dissertation, qualifying exam required, dissertation, qualifying exam required.

Entrance requirements: For master's and doctorate, GRE General Test, TOEFL, BS in biology, chemistry, or pharmacy. *Application deadline:* For fall admission, 3/1; for spring admission, 10/15. *Application fee:* $50.

Expenses: Tuition: Full-time $19,570; part-time $612 per unit. Required fees: $260. Tuition and fees vary according to program.

Financial aid: Teaching assistantships, institutionally sponsored loans available. Financial aid application deadline: 3/1.

Faculty research: Polyamines, cardiovascular pharmacology, pharmacology of naturally-occurring drugs.
James Blankenship, Chairman, 209-946-3167.

Application contact: Dr. Tim Smith, Graduate Studies Coordinator, 209-946-2487, *Fax:* 209-946-2410, *E-mail:* tsmith@ uop.edu.

■ UNIVERSITY OF THE SCIENCES IN PHILADELPHIA

College of Graduate Studies, Program in Pharmacology and Toxicology, Philadelphia, PA 19104-4495

AWARDS Pharmacology (MS, PhD); toxicology (MS, PhD). Part-time programs available.

Faculty: 6 full-time (3 women).
Students: 9 full-time (8 women), 1 part-time; includes 1 minority (Asian American or Pacific Islander), 4 international. *26 applicants, 12% accepted.* In 1999, 2 doctorates awarded. Terminal master's awarded for partial completion of doctoral program.
Degree requirements: For master's and doctorate, thesis/dissertation required, foreign language not required.
Entrance requirements: For master's and doctorate, GRE General Test, TOEFL. *Application deadline:* For fall admission, 5/1. Applications are processed on a rolling basis. *Application fee:* $45.
Expenses: Tuition: Full-time $29,600; part-time $612 per credit. Tuition and fees vary according to program.
Financial aid: In 1999–00, 9 teaching assistantships with tuition reimbursements (averaging $12,500 per year) were awarded; fellowships, research assistantships, institutionally sponsored loans and tuition waivers (full) also available. Financial aid application deadline: 5/1.
Faculty research: Autonomic, cardiovascular, cellular, and molecular pharmacology; mechanisms of carcinogenesis; drug metabolism. *Total annual research expenditures:* $460,206.
Dr. Raymond F. Orzechowski, Director, 215-596-8825, *E-mail:* r.orzech@usip.edu.
Application contact: Dr. Rodney J. Wigent, Dean, College of Graduate Studies, 215-596-8937, *Fax:* 215-596-8764, *E-mail:* graduate@usip.edu.

Find an in-depth description at www.petersons.com/graduate.

■ UNIVERSITY OF TOLEDO

Graduate School, College of Pharmacy, Graduate Programs in Pharmacy, Program in Pharmaceutical Science, Toledo, OH 43606-3398

AWARDS Administrative pharmacy (MSPS); industrial pharmacy (MSPS); pharmacology (MSPS).

Faculty: 11 full-time (3 women).
Students: 40 full-time (19 women), 38 international. Average age 25. *152 applicants, 19% accepted.* In 1999, 9 degrees awarded.
Degree requirements: For master's, thesis required, foreign language not required. *Average time to degree:* Master's–2 years full-time.

Entrance requirements: For master's, GRE General Test, TOEFL. *Application deadline:* For fall admission, 2/1 (priority date). *Application fee:* $30. Electronic applications accepted.
Expenses: Tuition, state resident: full-time $2,741; part-time $228 per credit hour. Tuition, nonresident: full-time $5,926; part-time $494 per credit hour. Required fees: $402; $34 per credit hour.
Financial aid: In 1999–00, 14 students received aid; teaching assistantships, tuition waivers (full) available.
Faculty research: Drug disposition, neuropharmacology, pharmacokinetics, product stability, pharmacy and health care administration.
Dr. Kenneth A. Bachmann, Graduate Coordinator, 419-530-1912, *Fax:* 419-530-1909, *E-mail:* kbachma@utnet.utoledo.edu.

■ UNIVERSITY OF UTAH

College of Pharmacy and Graduate School, Graduate Programs in Pharmacy, Department of Pharmacology and Toxicology, Salt Lake City, UT 84112-1107

AWARDS MS, PhD, MD/PhD. Terminal master's awarded for partial completion of doctoral program.

Degree requirements: For master's, thesis required; for doctorate, dissertation, final exam required, foreign language not required.
Entrance requirements: For master's, GRE, TOEFL; for doctorate, GRE Subject Test, TOEFL, BS in biology, chemistry, or pharmacy.
Expenses: Tuition, state resident: full-time $1,663. Tuition, nonresident: full-time $5,201. Tuition and fees vary according to course load and program.
Faculty research: Neuropharmacology, neurochemistry, biochemistry, molecular pharmacology, analytical chemistry.

Find an in-depth description at www.petersons.com/graduate.

■ UNIVERSITY OF VERMONT

College of Medicine and Graduate College, Graduate Programs in Medicine, Department of Pharmacology, Burlington, VT 05405

AWARDS MS, PhD, MD/MS, MD/PhD.

Degree requirements: For master's and doctorate, thesis/dissertation required.
Entrance requirements: For master's and doctorate, GRE General Test, GRE Subject Test, TOEFL.
Expenses: Tuition, state resident: full-time $7,464; part-time $311 per credit. Tuition, nonresident: full-time $18,672; part-time

$778 per credit. Full-time tuition and fees vary according to degree level and program.
Faculty research: Cardiovascular drugs, anticancer drugs.

Find an in-depth description at www.petersons.com/graduate.

■ UNIVERSITY OF VIRGINIA

College and Graduate School of Arts and Sciences, Department of Pharmacology, Charlottesville, VA 22903

AWARDS PhD, MD/PhD.

Faculty: 21 full-time (7 women), 1 part-time/adjunct (0 women).
Students: 23 full-time (7 women); includes 2 minority (both Asian Americans or Pacific Islanders), 3 international. Average age 25. *33 applicants, 24% accepted.* In 1999, 2 degrees awarded.
Degree requirements: For doctorate, dissertation required.
Entrance requirements: For doctorate, GRE General Test, GRE Subject Test. *Application deadline:* Applications are processed on a rolling basis. *Application fee:* $40. Electronic applications accepted.
Expenses: Tuition, state resident: full-time $3,832. Tuition, nonresident: full-time $15,519. Required fees: $1,084. Tuition and fees vary according to course load and program.
Financial aid: Fellowships, research assistantships, teaching assistantships available. Financial aid application deadline: 2/1; financial aid applicants required to submit FAFSA.
James C. Garrison, Chairman, 804-924-1919.
Application contact: Duane J. Osheim, Associate Dean, 804-924-7184, *E-mail:* microbiology@virginia.edu.

Find an in-depth description at www.petersons.com/graduate.

■ UNIVERSITY OF WASHINGTON

School of Medicine and Graduate School, Graduate Programs in Medicine, Department of Pharmacology, Seattle, WA 98195

AWARDS MS, PhD.

Faculty: 22 full-time (4 women).
Students: 37 full-time (19 women); includes 8 minority (1 African American, 6 Asian Americans or Pacific Islanders, 1 Hispanic American). Average age 27. *74 applicants, 22% accepted.* In 1999, 2 master's awarded (100% found work related to degree); 4 doctorates awarded (100% entered university research/teaching).
Degree requirements: For doctorate, dissertation required, foreign language not

required. *Average time to degree:* Master's–2 years full-time; doctorate–5 years full-time. **Entrance requirements:** For doctorate, GRE General Test, minimum GPA of 3.0. *Application deadline:* For fall admission, 2/1 (priority date). Applications are processed on a rolling basis. *Application fee:* $50. **Expenses:** Tuition, state resident: full-time $9,210; part-time $236 per credit. Tuition, nonresident: full-time $23,256; part-time $596 per credit.
Financial aid: In 1999–00, 20 fellowships with full tuition reimbursements (averaging $17,304 per year), 11 research assistantships with full tuition reimbursements (averaging $17,304 per year), 6 teaching assistantships with full tuition reimbursements (averaging $17,304 per year) were awarded; Federal Work-Study, institutionally sponsored loans, traineeships, and tuition waivers (full) also available. Financial aid application deadline: 3/15.
Faculty research: Neuroscience, cell physiology, molecular biology, regulation of metabolism, signal transduction, growth regulation, neuropharmacology. *Total annual research expenditures:* $5.1 million. Dr. William A. Catterall, Chairman, 206-543-1925, *Fax:* 206-685-3822.
Application contact: Diane L. Schulstad, Administrative Assistant, 206-685-7204, *Fax:* 206-685-3822, *E-mail:* diansch@u.washington.edu.

Find an in-depth description at www.petersons.com/graduate.

■ UNIVERSITY OF WISCONSIN–MADISON

Graduate School, College of Agricultural and Life Sciences, Department of Animal Health and Biomedical Sciences, Program in Comparative Biosciences, Madison, WI 53706-1380

AWARDS Anatomy (MS, PhD); biochemistry (MS, PhD); cellular and molecular biology (MS, PhD); environmental toxicology (MS, PhD); neurosciences (MS, PhD); pharmacology (MS, PhD); physiology (MS, PhD).

Degree requirements: For doctorate, dissertation required.
Expenses: Tuition, state resident: full-time $5,406; part-time $339 per credit. Tuition, nonresident: full-time $17,110; part-time $1,071 per credit. Full-time tuition and fees vary according to program and reciprocity agreements. Part-time tuition and fees vary according to course load and program.

■ UNIVERSITY OF WISCONSIN–MADISON

Medical School and Graduate School, Graduate Programs in Medicine, Molecular and Cellular Pharmacology Program, Madison, WI 53706-1380

AWARDS PhD.

Faculty: 36 full-time (7 women).
Students: 39 full-time (19 women), 11 international. Average age 26.
Application deadline: For fall admission, 1/15 (priority date). *Application fee:* $45.
Expenses: Tuition, state resident: full-time $5,406; part-time $339 per credit. Tuition, nonresident: full-time $17,110; part-time $1,071 per credit.
Financial aid: In 1999–00, fellowships with full tuition reimbursements (averaging $15,500 per year), research assistantships with full tuition reimbursements (averaging $15,500 per year) were awarded; grants, scholarships, and traineeships also available.
Faculty research: Protein kinases, signaling pathways, neurotransmitters, molecular recognition, receptors and transporters. Dr. Richard A. Anderson, Director, 608-262-3753, *Fax:* 608-262-1257, *E-mail:* raanders@facstaff.wisc.edu.
Application contact: Lynn Louise Squire, Student Services Coordinator, 608-262-9826, *Fax:* 608-262-1257, *E-mail:* lsquire@facstaff.wisc.edu.

Find an in-depth description at www.petersons.com/graduate.

■ VANDERBILT UNIVERSITY

Graduate School and School of Medicine, Department of Pharmacology, Nashville, TN 37240-1001

AWARDS PhD, MD/PhD.

Faculty: 19 full-time (3 women), 5 part-time/adjunct (0 women).
Students: 34 full-time (13 women); includes 1 minority (African American), 4 international. Average age 26. In 1999, 5 degrees awarded.
Degree requirements: For doctorate, dissertation, preliminary, qualifying, and final exams required, foreign language not required. *Average time to degree:* Doctorate–5 years full-time.
Entrance requirements: For doctorate, GRE General Test, GRE Subject Test (recommended). *Application deadline:* For fall admission, 1/15. *Application fee:* $40.
Expenses: Tuition: Full-time $17,244; part-time $958 per hour. Required fees: $242; $121 per semester. Tuition and fees vary according to program.
Financial aid: In 1999–00, fellowships with full tuition reimbursements (averaging $17,000 per year), research assistantships

with full tuition reimbursements (averaging $17,000 per year) were awarded; Federal Work-Study, institutionally sponsored loans, traineeships, and tuition waivers (partial) also available. Financial aid application deadline: 1/15.
Faculty research: Molecular pharmacology, neuropharmacology, drug disposition and toxicology, genetic mechanics, cell regulation, eicosanoids. *Total annual research expenditures:* $6 million. Elaine Sanders-Bush, Chair, 615-936-3037, *Fax:* 615-936-6532, *E-mail:* elaine.bush@mcmail.vanderbilt.edu.
Application contact: Ronald B. Emeson, Director of Graduate Studies, 615-322-2207, *Fax:* 615-343-6532, *E-mail:* ronald.b.emeson@vanderbilt.edu.

Find an in-depth description at www.petersons.com/graduate.

■ VIRGINIA COMMONWEALTH UNIVERSITY

School of Graduate Studies and School of Medicine, School of Medicine Graduate Programs, Department of Pharmacology and Toxicology, Richmond, VA 23284-9005

AWARDS Molecular biology and genetics (PhD); neurosciences (PhD); pharmacology (PhD, CBHS); pharmacology and toxicology (MS).

Students: 14 full-time (5 women), 51 part-time (27 women); includes 27 minority (4 African Americans, 22 Asian Americans or Pacific Islanders, 1 Hispanic American). In 1999, 3 master's, 10 doctorates, 1 other advanced degree awarded. Terminal master's awarded for partial completion of doctoral program.
Degree requirements: For master's, thesis required, foreign language not required; for doctorate, dissertation, comprehensive oral and written exams required, foreign language not required.
Entrance requirements: For master's, DAT, GRE General Test or MCAT; for doctorate, GRE General Test. *Application deadline:* For fall admission, 2/1 (priority date). *Application fee:* $30.
Expenses: Tuition, state resident: full-time $4,031; part-time $224 per credit hour. Tuition, nonresident: full-time $11,946; part-time $664 per credit hour. Required fees: $1,081; $40 per credit hour. Tuition and fees vary according to campus/location and program.
Financial aid: Fellowships, teaching assistantships available.
Faculty research: Drug abuse, drug metabolism, pharmacodynamics, peptide synthesis, receptor mechanisms. Dr. George Kunos, Chair, 804-828-2073, *Fax:* 804-828-2117.

Virginia Commonwealth University (continued)
Application contact: Sheryol Cox, Graduate Program Coordinator, 804-828-8400, *Fax:* 804-828-2117, *E-mail:* swcox@vcu.edu.

Find an in-depth description at www.petersons.com/graduate.

■ WAKE FOREST UNIVERSITY

School of Medicine and Graduate School, Graduate Programs in Medicine, Department of Physiology and Pharmacology, Program in Pharmacology, Winston-Salem, NC 27109

AWARDS PhD.

Degree requirements: For doctorate, dissertation required.
Entrance requirements: For doctorate, GRE General Test, GRE Subject Test.
Expenses: Tuition: Full-time $18,300.
Faculty research: Renal-cardiovascular system, endocrine system, neuropharmacology.

■ WASHINGTON STATE UNIVERSITY

Graduate School, College of Pharmacy, Program in Pharmacology and Toxicology, Pullman, WA 99164

AWARDS MS, PhD.

Faculty: 36.
Students: 9 full-time (6 women), 2 part-time (1 woman); includes 2 minority (1 Asian American or Pacific Islander, 1 Hispanic American), 4 international. Average age 30. In 1999, 5 doctorates awarded.
Degree requirements: For master's and doctorate, thesis/dissertation, oral exam required, foreign language not required. *Average time to degree:* Doctorate–5 years full-time.
Entrance requirements: For master's and doctorate, GRE General Test, TOEFL, minimum GPA of 3.0. *Application deadline:* For fall admission, 3/1. Applications are processed on a rolling basis. *Application fee:* $35.
Expenses: Tuition, state resident: full-time $5,654. Tuition, nonresident: full-time $13,850. International tuition: $13,850 full-time. Tuition and fees vary according to program.
Financial aid: In 1999–00, 3 fellowships, 8 research assistantships with full and partial tuition reimbursements, 2 teaching assistantships with full and partial tuition reimbursements were awarded; Federal Work-Study, institutionally sponsored loans, tuition waivers (partial), and staff assistantships, teaching associateships also available. Financial aid application

deadline: 4/1; financial aid applicants required to submit FAFSA.
Faculty research: Pharmacokinetics/pharmacodynamics, cancer pharmacology, immunology, drug metabolism, neuropharmacology.
Dr. Raymond Quock, Chair, 509-335-7598.
Application contact: Mary Stormo, Coordinator, 509-335-7598, *E-mail:* stormom@mail.wsu.edu.

■ WAYNE STATE UNIVERSITY

School of Medicine and Graduate School, Graduate Programs in Medicine, Department of Pharmacology, Detroit, MI 48202

AWARDS MS, PhD, MD/PhD.

Degree requirements: For doctorate, dissertation required, foreign language not required.
Entrance requirements: For master's and doctorate, GRE General Test.
Faculty research: Biochemical toxicology, ubiquitin-dependent proteolysis, regulation of neurotransmitter release, carcinogenesis and tumor progression.

Find an in-depth description at www.petersons.com/graduate.

■ WEST VIRGINIA UNIVERSITY

School of Medicine, Graduate Programs in Health Sciences, Department of Pharmacology and Toxicology, Morgantown, WV 26506

AWARDS Autonomic pharmacology (MS, PhD); biomedical pharmacology (MS, PhD); chemotherapy (MS, PhD); endocrine pharmacology (MS, PhD); neuropharmacology (MS, PhD); toxicology (MS, PhD).

Students: 12 full-time (7 women), 3 part-time, 5 international. Average age 28. *232 applicants, 2% accepted.* In 1999, 1 master's, 3 doctorates awarded. Terminal master's awarded for partial completion of doctoral program.
Degree requirements: For master's and doctorate, thesis/dissertation required, foreign language not required.
Entrance requirements: For master's, GRE, TOEFL, minimum GPA of 2.5; for doctorate, GRE, TOEFL. *Application deadline:* For fall admission, 2/1 (priority date). Applications are processed on a rolling basis. *Application fee:* $45.
Expenses: Tuition, state resident: full-time $3,564. Tuition, nonresident: full-time $10,230.
Financial aid: In 1999–00, 5 research assistantships, 7 teaching assistantships were awarded; Federal Work-Study, institutionally sponsored loans, traineeships, and tuition waivers (full) also available. Financial aid application

deadline: 2/1; financial aid applicants required to submit FAFSA.
Faculty research: Neuropharmacology and toxicology, cardiovascular pharmacology, molecular pharmacology, biochemical toxicology, biochemical pharmacology. Robert E. Stityel, Chair, 304-293-5761.
Application contact: Charles R. Craig, Director of Graduate Studies, 304-293-5795, *E-mail:* ccraig@wvu.edu.

Find an in-depth description at www.petersons.com/graduate.

■ YALE UNIVERSITY

Graduate School of Arts and Sciences, Department of Pharmacology, New Haven, CT 06520

AWARDS PhD.

Faculty: 49 full-time (12 women), 1 (woman) part-time/adjunct.
Students: 20 full-time (12 women); includes 6 minority (2 African Americans, 2 Asian Americans or Pacific Islanders, 2 Hispanic Americans), 2 international. *74 applicants, 24% accepted.* In 1999, 4 degrees awarded.
Degree requirements: For doctorate, dissertation required, foreign language not required. *Average time to degree:* Doctorate–6.2 years full-time.
Entrance requirements: For doctorate, GRE General Test. *Application deadline:* For fall admission, 1/4. *Application fee:* $65.
Expenses: Tuition: Full-time $22,300. Full-time tuition and fees vary according to program.
Financial aid: Fellowships, research assistantships, teaching assistantships, Federal Work-Study, institutionally sponsored loans, and traineeships available. Aid available to part-time students.
Application contact: Admissions Information, 203-432-2770.

■ YALE UNIVERSITY

School of Medicine and Graduate School of Arts and Sciences, Combined Program in Biological and Biomedical Sciences (BBS), Pharmacological Sciences and Molecular Medicine Track, New Haven, CT 06520

AWARDS PhD, MD/PhD.

Degree requirements: For doctorate, dissertation required.
Entrance requirements: For doctorate, GRE General Test, TOEFL. *Application deadline:* For fall admission, 1/2. *Application fee:* $65. Electronic applications accepted.
Expenses: All students receive full tuition of $22,330 and an annual stipend of $17,600 .
Financial aid: Fellowships, research assistantships available.

Dr. David F. Stern, Co-Director, 203-785-4832, *Fax:* 203-785-7467, *E-mail:* bbs.pharm@yale.edu.

■ YESHIVA UNIVERSITY

Albert Einstein College of Medicine, Sue Golding Graduate Division of Medical Sciences, Division of Biological Sciences, Department of Molecular Pharmacology, Bronx, NY 10461

AWARDS PhD, MD/PhD.

Faculty: 10 full-time.
Students: 30 full-time (15 women); includes 5 minority (3 African Americans, 2 Asian Americans or Pacific Islanders), 15 international. In 1999, 4 degrees awarded.
Degree requirements: For doctorate, dissertation required, foreign language not required.
Entrance requirements: For doctorate, GRE General Test, TOEFL. *Application deadline:* For fall admission, 1/15. *Application fee:* $0.
Expenses: Tuition: Part-time $525 per credit. Tuition and fees vary according to degree level and program.
Financial aid: In 1999–00, 30 fellowships were awarded.
Faculty research: Effects of drugs on macromolecules, enzyme systems, cell morphology and function.
Dr. Susan Horwitz, Co-Chairperson, 718-430-2163.

Find an in-depth description at www.petersons.com/graduate.

TOXICOLOGY

■ AMERICAN UNIVERSITY

College of Arts and Sciences, Department of Chemistry, Emphasis in Toxicology, Washington, DC 20016-8001

AWARDS MS, Certificate. Part-time programs available.

Faculty: 8 full-time (3 women), 7 part-time/adjunct (2 women).
Students: 6 full-time (all women), 3 international. *6 applicants, 83% accepted.* In 1999, 1 degree awarded.
Degree requirements: For master's, thesis or alternative, comprehensive written exam, tool of research exam required, foreign language not required.
Entrance requirements: For master's, minimum GPA of 3.0. *Application deadline:* For fall admission, 2/1 (priority date); for spring admission, 10/1 (priority date). Applications are processed on a rolling basis. *Application fee:* $50.
Expenses: Tuition: Part-time $721 per credit hour. Required fees: $90 per

semester. Tuition and fees vary according to program.
Financial aid: In 1999–00, 5 students received aid, including 1 research assistantship with full tuition reimbursement available (averaging $8,500 per year), 2 teaching assistantships with full tuition reimbursements available (averaging $8,500 per year); fellowships, career-related internships or fieldwork and institutionally sponsored loans also available. Financial aid application deadline: 2/1.
Faculty research: Carbohydrate chemistry, chromatography, enzyme mechanisms and model systems, environmental analysis with monoclonal antibodies, polymers.
Dr. Nina Roscher, Chair, Department of Chemistry, 202-885-1750, *Fax:* 202-885-1752, *E-mail:* nrosche@american.edu.

■ BROWN UNIVERSITY

Graduate School, Division of Biology and Medicine, Pathobiology Graduate Program, Providence, RI 02912

AWARDS Biology (PhD); cancer biology (PhD); immunology and infection (PhD); medical science (PhD); pathobiology (Sc M); toxicology and environmental pathology (PhD).

Faculty: 36 full-time (9 women), 1 part-time/adjunct (0 women).
Students: 18 full-time (11 women); includes 4 minority (2 African Americans, 2 Asian Americans or Pacific Islanders), 1 international. Average age 25. *42 applicants, 17% accepted.* In 1999, 2 doctorates awarded (100% entered university research/teaching). Terminal master's awarded for partial completion of doctoral program.
Degree requirements: For master's, foreign language not required; for doctorate, dissertation, preliminary exam required, foreign language not required. *Average time to degree:* Doctorate–6 years full-time.
Entrance requirements: For master's, GRE General Test, GRE Subject Test; for doctorate, GRE General Test, GRE Subject Test, TOEFL. *Application deadline:* For fall admission, 1/2 (priority date). Applications are processed on a rolling basis. *Application fee:* $60. Electronic applications accepted.
Financial aid: In 1999–00, 11 fellowships with tuition reimbursements, 1 research assistantship with tuition reimbursement, 4 teaching assistantships with tuition reimbursements were awarded; institutionally sponsored loans and traineeships also available. Financial aid application deadline: 1/2.

Faculty research: Environmental pathology, carcinogenesis, immunopathology, signal transduction.
Dr. Nancy Thompson, Director, 401-444-8860, *E-mail:* nancy_thompson@brown.edu.
Application contact: Marilyn May, Program Coordinator, 401-863-2913, *Fax:* 401-863-9008, *E-mail:* marilyn_may@brown.edu.

Find an in-depth description at www.petersons.com/graduate.

■ CASE WESTERN RESERVE UNIVERSITY

School of Medicine and School of Graduate Studies, Graduate Programs in Medicine, Department of Environmental Health Sciences, Cleveland, OH 44106

AWARDS Environmental toxicology (MS, PhD); molecular toxicology (MS, PhD). Part-time programs available.

Faculty: 2 full-time (1 woman), 5 part-time/adjunct (2 women).
Students: 6 full-time (3 women); includes 1 minority (Asian American or Pacific Islander), 2 international. In 1999, 1 degree awarded (100% entered university research/teaching). Terminal master's awarded for partial completion of doctoral program.
Degree requirements: For master's, thesis optional; for doctorate, dissertation, qualifying exams required. *Average time to degree:* Doctorate–5 years full-time.
Entrance requirements: For master's and doctorate, GRE General Test, GRE Subject Test, TOEFL. *Application deadline:* For fall admission, 2/15. *Application fee:* $25.
Financial aid: In 1999–00, 3 research assistantships (averaging $15,000 per year) were awarded; career-related internships or fieldwork and tuition waivers also available. Financial aid application deadline: 2/15.
Faculty research: Xenobiotic metabolism, microbial mutagenicity, DNA damage and repair, chromosome structure cytogenetics, carcinogenesis, photodynamic therapy.
Dr. G. David McCoy, Acting Chairman and Associate Professor, 216-368-5961, *Fax:* 216-368-3194, *E-mail:* gdm@po.cwru.edu.
Application contact: Dr. Martina L. Veigl, Graduate Program Director and Associate Professor, 216-844-7525, *Fax:* 216-844-8230, *E-mail:* mlv2@po.cwru.edu.

Find an in-depth description at www.petersons.com/graduate.

■ COLUMBIA UNIVERSITY

College of Physicians and Surgeons and Graduate School of Arts and Sciences, Graduate School of Arts and Sciences at the College of Physicians and Surgeons, Department of Pharmacology, New York, NY 10032

AWARDS Pharmacology (M Phil, MA, PhD); pharmacology-toxicology (M Phil, MA, PhD). Only candidates for the PhD are admitted. Terminal master's awarded for partial completion of doctoral program.

Degree requirements: For master's, foreign language and thesis not required; for doctorate, dissertation required, foreign language not required.
Entrance requirements: For master's and doctorate, GRE General Test, TOEFL.
Expenses: Tuition: Full-time $25,072.
Faculty research: Cardiovascular pharmacology, receptor pharmacology, neuropharmacology, membrane biophysics, eicosanoids.

Find an in-depth description at www.petersons.com/graduate.

■ CORNELL UNIVERSITY

Graduate School, Graduate Fields of Agriculture and Life Sciences, Field of Environmental Toxicology, Ithaca, NY 14853-0001

AWARDS Cellular and biochemical toxicology (MS, PhD); ecotoxicology and environmental chemistry (MS, PhD); food and nutritional toxicology (MS, PhD).

Faculty: 40 full-time.
Students: 17 full-time (8 women); includes 2 minority (both Hispanic Americans), 7 international. *71 applicants, 20% accepted.* In 1999, 2 master's, 5 doctorates awarded.
Degree requirements: For master's and doctorate, thesis/dissertation required, foreign language not required.
Entrance requirements: For master's and doctorate, GRE General Test, GRE Subject Test (biology or chemistry) (recommended). *Application deadline:* For fall admission, 1/15. *Application fee:* $65. Electronic applications accepted.
Expenses: Tuition: Full-time $12,100.
Financial aid: In 1999–00, 16 students received aid, including 8 fellowships with full tuition reimbursements available, 8 research assistantships with full tuition reimbursements available; teaching assistantships with full tuition reimbursements available, institutionally sponsored loans, scholarships, tuition waivers (full and partial), and unspecified assistantships also available. Financial aid applicants required to submit FAFSA.
Faculty research: Cellular and molecular toxicology, cancer toxicology, bioremediation.

Application contact: Graduate Field Assistant, 607-255-8008, *E-mail:* envtox@cornell.edu.

Find an in-depth description at www.petersons.com/graduate.

■ DARTMOUTH COLLEGE

School of Arts and Sciences, Department of Pharmacology and Toxicology, Hanover, NH 03755

AWARDS PhD, MD/PhD.

Faculty: 19 full-time (5 women).
Students: 22 full-time (7 women); includes 1 minority (Asian American or Pacific Islander), 2 international. Average age 25. *44 applicants, 27% accepted.* In 1999, 4 degrees awarded (75% entered university research/teaching, 25% found other work related to degree).
Degree requirements: For doctorate, dissertation required.
Entrance requirements: For doctorate, GRE General Test, GRE Subject Test, bachelor's degree in biological, chemical, or physical science. *Application deadline:* For fall admission, 1/15. *Application fee:* $10 ($0 for international students).
Expenses: Tuition: Full-time $24,624. Required fees: $916. One-time fee: $15 full-time. Full-time tuition and fees vary according to program.
Financial aid: In 1999–00, 22 students received aid, including 8 fellowships with full tuition reimbursements available (averaging $15,460 per year), 14 research assistantships with full tuition reimbursements available (averaging $15,460 per year); institutionally sponsored loans and traineeships also available. Financial aid application deadline: 4/15.
Faculty research: Molecular biology of carcinogenesis, DNA repair and gene expression, biochemical and environmental toxicology, protein receptor ligand interactions.
Dr. Ethan Dmitrovsky, Chairman, 603-650-1667, *Fax:* 603-650-1129, *E-mail:* pharmacology.and.toxicology@dartmouth.edu.
Application contact: Linda Conrad, Coordinator of Graduate Studies, 603-650-1667, *Fax:* 603-650-1129, *E-mail:* lvc@dartmouth.edu.

Find an in-depth description at www.petersons.com/graduate.

■ DUKE UNIVERSITY

Graduate School, Integrated Toxicology Program, Durham, NC 27708-0586

AWARDS Certificate.

Faculty: 1 full-time.
Application deadline: For fall admission, 12/31. *Application fee:* $75.

Expenses: Tuition: Full-time $21,406; part-time $760 per unit. Required fees: $3,136; $3,136 per year. One-time fee: $30. Tuition and fees vary according to program.
Financial aid: Fellowships available. Financial aid application deadline: 12/31. Edward Levin, Director, 919-681-6273, *E-mail:* rgross@duke.edu.

Find an in-depth description at www.petersons.com/graduate.

■ DUQUESNE UNIVERSITY

School of Pharmacy, Graduate School of Pharmaceutical Sciences, Program in Pharmacology/Toxicology, Pittsburgh, PA 15282-0001

AWARDS MS, PhD. Part-time programs available.

Faculty: 5 full-time (1 woman).
Students: 3 full-time (all women), 11 part-time (5 women), 1 international. *45 applicants, 11% accepted.*
Degree requirements: For master's, thesis required, foreign language not required; for doctorate, one foreign language, computer language, dissertation required.
Entrance requirements: For master's and doctorate, GRE General Test, TOEFL, TSE. *Application deadline:* For fall admission, 3/1 (priority date). Applications are processed on a rolling basis. *Application fee:* $40.
Financial aid: In 1999–00, 1 research assistantship with full tuition reimbursement, 8 teaching assistantships with full tuition reimbursements were awarded; fellowships.
Faculty research: Analytical/clinical/forensic toxicology, drugs for memory enhancement, behavioral/cardiovascular pharmacology, antidotal agents for clinical use.
Dr. Frederick W. Fochtman, Head, 412-396-6373.

■ FLORIDA AGRICULTURAL AND MECHANICAL UNIVERSITY

Division of Graduate Studies, Research, and Continuing Education, College of Pharmacy and Pharmaceutical Sciences, Graduate Programs in Pharmaceutical Sciences and Public Health, Tallahassee, FL 32307-3200

AWARDS Environmental toxicology (PhD); medicinal chemistry (MS, PhD); pharmaceutics (MS, PhD); pharmacology/toxicology (MS, PhD); pharmacy administration (MS); public health (MPH).

Students: 67 (50 women); includes 62 minority (60 African Americans, 1 Asian American or Pacific Islander, 1 Hispanic

American) 1 international. In 1999, 20 master's awarded.

Degree requirements: For master's and doctorate, thesis/dissertation, publishable paper required, foreign language not required.

Entrance requirements: For master's and doctorate, GRE General Test, minimum GPA of 3.0 in last 60 hours. *Application fee:* $20.

Expenses: Tuition, state resident: full-time $2,644; part-time $147 per credit hour. Tuition, nonresident: full-time $9,137; part-time $508 per credit hour. Required fees: $52 per semester. Tuition and fees vary according to course load.

Financial aid: Fellowships, research assistantships, Federal Work-Study and grants available.

Faculty research: Anticancer agents, anti-inflammatory drugs, chronopharmacology, neuroendocrinology, microbiology.

Application contact: Dr. Thomas J. Fitzgerald, Chairman, Graduate Committee, 850-599-3301.

Find an in-depth description at www.petersons.com/graduate.

■ THE GEORGE WASHINGTON UNIVERSITY

Columbian School of Arts and Sciences, Department of Forensic Sciences, Program in Chemical Toxicology, Washington, DC 20052

AWARDS MS. Part-time and evening/weekend programs available.

Students: Average age 25. *2 applicants, 50% accepted.* In 1999, 2 degrees awarded.

Degree requirements: For master's, thesis, comprehensive exam required, foreign language not required.

Entrance requirements: For master's, GRE General Test, minimum GPA of 3.0. *Application deadline:* For fall admission, 5/1. *Application fee:* $55.

Expenses: Tuition: Full-time $16,836; part-time $702 per credit hour. Required fees: $828; $35 per credit hour. Tuition and fees vary according to campus/location and program.

Financial aid: Federal Work-Study available. Financial aid application deadline: 2/1.

Dr. David Rowley, Chair, Department of Forensic Sciences, 202-994-7319.

■ INDIANA UNIVERSITY–PURDUE UNIVERSITY INDIANAPOLIS

School of Medicine, Graduate Programs in Medicine, Department of Pharmacology and Toxicology, Program in Toxicology, Indianapolis, IN 46202-2896

AWARDS MS, PhD, MD/MS, MD/PhD.

Students: 1 full-time (0 women), 3 part-time, 3 international. Average age 26. In 1999, 2 master's awarded (100% found work related to degree); 2 doctorates awarded. Terminal master's awarded for partial completion of doctoral program.

Degree requirements: For master's and doctorate, thesis/dissertation required. *Average time to degree:* Master's–2 years full-time; doctorate–5 years full-time.

Entrance requirements: For master's and doctorate, GRE General Test, GRE Subject Test, minimum GPA of 3.0. *Application deadline:* For fall admission, 1/15 (priority date). Applications are processed on a rolling basis. *Application fee:* $35 ($55 for international students).

Expenses: Tuition, state resident: full-time $13,245; part-time $158 per credit hour. Tuition, nonresident: full-time $30,330; part-time $455 per credit hour. Required fees: $121 per year. Tuition and fees vary according to course load and degree level.

Financial aid: In 1999–00, fellowships with partial tuition reimbursements (averaging $14,500 per year), research assistantships with partial tuition reimbursements (averaging $14,500 per year) were awarded; Federal Work-Study and institutionally sponsored loans also available. Financial aid application deadline: 1/15.

Faculty research: Oncogenesis, liver metabolism.

Dr. James Klaunig, Director, 317-274-7824, *Fax:* 317-274-7787.

Application contact: Victor Elharrar, Director of Graduate Studies, 317-274-1564, *Fax:* 317-274-7714, *E-mail:* velharra@iupui.edu.

Find an in-depth description at www.petersons.com/graduate.

■ IOWA STATE UNIVERSITY OF SCIENCE AND TECHNOLOGY

Graduate College, College of Agriculture and College of Liberal Arts and Sciences, Department of Biochemistry, Biophysics, and Molecular Biology, Ames, IA 50011

AWARDS Biochemistry (MS, PhD); biophysics (MS, PhD); genetics (MS, PhD); molecular, cellular, and developmental biology (MS, PhD); toxicology (MS, PhD).

Faculty: 19 full-time, 1 part-time/adjunct.

Students: 57 full-time (24 women), 6 part-time (1 woman); includes 1 minority (African American), 20 international. *22 applicants, 59% accepted.* In 2000, 4 master's, 4 doctorates awarded.

Degree requirements: For master's and doctorate, thesis/dissertation required.

Entrance requirements: For master's and doctorate, GRE General Test, TOEFL. *Application deadline:* For fall admission,

6/15 (priority date); for spring admission, 11/15 (priority date). *Application fee:* $20 ($50 for international students). Electronic applications accepted.

Expenses: Tuition, state resident: full-time $3,308. Tuition, nonresident: full-time $9,744. Part-time tuition and fees vary according to course load, campus/location and program.

Financial aid: In 2000–01, 44 research assistantships with partial tuition reimbursements (averaging $12,314 per year), 2 teaching assistantships with partial tuition reimbursements (averaging $12,375 per year) were awarded; scholarships also available.

Dr. Marit Nilsen-Hamilton, Chair, 515-294-2231, *E-mail:* biochem@iastate.edu.

Find an in-depth description at www.petersons.com/graduate.

■ IOWA STATE UNIVERSITY OF SCIENCE AND TECHNOLOGY

Graduate College, Interdisciplinary Programs, Program in Toxicology, Ames, IA 50011

AWARDS MS, PhD.

Students: *19 applicants, 0% accepted.* In 1999, 4 master's, 1 doctorate awarded.

Degree requirements: For master's and doctorate, thesis/dissertation required.

Entrance requirements: For master's and doctorate, GRE General Test, TOEFL. *Application deadline:* For fall admission, 2/1 (priority date). *Application fee:* $20 ($50 for international students). Electronic applications accepted.

Expenses: Tuition, state resident: full-time $3,308. Tuition, nonresident: full-time $9,744. Part-time tuition and fees vary according to course load, campus/location and program.

Dr. Franklin Ahrens, Supervisory Committee Chair, 515-294-3396, *Fax:* 515-294-6669, *E-mail:* toxic@iastate.edu.

■ JOHNS HOPKINS UNIVERSITY

School of Hygiene and Public Health, Department of Environmental Health Sciences, Division of Toxicological Sciences, Baltimore, MD 21218-2699

AWARDS PhD.

Degree requirements: For doctorate, dissertation, 1 year full-time residency, oral and written exams required, foreign language not required.

Entrance requirements: For doctorate, GRE General Test, TOEFL. *Application deadline:* For fall admission, 2/1 (priority date). Applications are processed on a rolling basis. *Application fee:* $60. Electronic applications accepted.

Expenses: Tuition: Full-time $23,660; part-time $493 per unit. Full-time tuition

Johns Hopkins University (continued)
and fees vary according to degree level, campus/location and program.

Financial aid: Federal Work-Study, institutionally sponsored loans, and scholarships available. Aid available to part-time students. Financial aid application deadline: 4/15.

Faculty research: Xenobiotic and stress-responsive genes, *in vitro* hepatotoxicology, toxicokinetics, T-cell function in immunotoxicology, bone marrow toxicology.

Dr. James Yager, Director, 410-955-4712, *E-mail:* jyager@jhsph.edu.

Find an in-depth description at www.petersons.com/graduate.

■ LONG ISLAND UNIVERSITY, BROOKLYN CAMPUS

Arnold and Marie Schwartz College of Pharmacy and Health Sciences, Graduate Programs in Pharmacy, Division of Pharmacology/Toxicology/Medicinal Chemistry, Brooklyn, NY 11201-8423

AWARDS Pharmacology/toxicology (MS); pharmacotherapeutics (MS). Part-time and evening/weekend programs available.

Faculty: 11 full-time (3 women).
Students: 10 full-time (5 women), 26 part-time (15 women); includes 15 minority (13 Asian Americans or Pacific Islanders, 2 Hispanic Americans), 5 international. Average age 30. In 1999, 14 degrees awarded.
Degree requirements: For master's, thesis optional, foreign language not required. *Average time to degree:* Master's–2 years full-time, 5 years part-time.
Entrance requirements: For master's, minimum GPA of 3.0. *Application deadline:* Applications are processed on a rolling basis. *Application fee:* $30.
Expenses: Tuition: Part-time $550 per credit.
Financial aid: In 1999–00, 3 teaching assistantships with full tuition reimbursements (averaging $6,000 per year) were awarded.
Dr. R. R. Raje, Director, 718-488-1062.
Application contact: Bernard W. Sullivan, Associate Director of Admissions, 718-488-1011, *Fax:* 718-797-2399, *E-mail:* attend@liu.edu.

■ LOUISIANA STATE UNIVERSITY AND AGRICULTURAL AND MECHANICAL COLLEGE

Graduate School, Center for Coastal, Energy and Environmental Resources, Institute for Environmental Studies, Baton Rouge, LA 70803

AWARDS Environmental planning and management (MS); environmental toxicology (MS).

Faculty: 14 full-time (2 women), 1 (woman) part-time/adjunct.
Students: 21 full-time (12 women), 25 part-time (8 women); includes 2 minority (1 African American, 1 Hispanic American), 9 international. Average age 29. *33 applicants, 55% accepted.* In 1999, 18 degrees awarded.
Degree requirements: For master's, thesis required (for some programs), foreign language not required.
Entrance requirements: For master's, GRE General Test, minimum GPA of 3.0. *Application deadline:* For fall admission, 1/25 (priority date). Applications are processed on a rolling basis. *Application fee:* $25.
Expenses: Tuition, state resident: full-time $2,881. Tuition, nonresident: full-time $7,081. Part-time tuition and fees vary according to course load and program.
Financial aid: In 1999–00, 1 fellowship, 10 research assistantships with partial tuition reimbursements (averaging $9,170 per year), 2 teaching assistantships with partial tuition reimbursements (averaging $13,350 per year) were awarded; career-related internships or fieldwork and unspecified assistantships also available.
Faculty research: Fates and movement of pollutants, neurobiotic metabolism, application of cellular toxicity/mutagenicity testing. *Total annual research expenditures:* $1.3 million.
Dr. Michael Wascom, Director, 225-388-8521, *Fax:* 225-388-4286, *E-mail:* coewas@lsu.edu.
Application contact: Dr. Ralph J. Portier, Graduate Adviser, 225-388-8522, *E-mail:* rportie@lsu.edu.

■ MASSACHUSETTS INSTITUTE OF TECHNOLOGY

School of Engineering, Division of Bioengineering and Environmental Health, Cambridge, MA 02139-4307

AWARDS Bioengineering (PhD); toxicology (SM, PhD, Sc D).

Faculty: 7 full-time (1 woman).
Students: 50 full-time (31 women); includes 8 minority (7 Asian Americans or Pacific Islanders, 1 Hispanic American), 15 international. Average age 26. *95 applicants, 39% accepted.* In 1999, 3 master's, 2

doctorates awarded. Terminal master's awarded for partial completion of doctoral program.
Degree requirements: For master's, thesis required, foreign language not required; for doctorate, dissertation, oral and written qualifying exams required, foreign language not required.
Entrance requirements: For master's and doctorate, GRE General Test, TOEFL. *Application deadline:* For fall admission, 1/15. *Application fee:* $60.
Expenses: Tuition: Full-time $25,000. Full-time tuition and fees vary according to degree level, program and student level.
Financial aid: In 1999–00, 34 students received aid, including 33 fellowships, 15 research assistantships, 5 teaching assistantships; Federal Work-Study, grants, institutionally sponsored loans, and scholarships also available. Financial aid application deadline: 1/15; financial aid applicants required to submit FAFSA.
Faculty research: Biological imaging, biological microanalytics, biological transport process, biomaterials, cell and tissue engineering. *Total annual research expenditures:* $7.7 million.
Douglas A. Lauffenburger, Co-Director, 617-252-1629, *E-mail:* lauffen@mit.edu.
Application contact: Debra A. Luchanin, Academic Administrator, 617-253-5804.

Find an in-depth description at www.petersons.com/graduate.

■ MEDICAL COLLEGE OF GEORGIA

School of Graduate Studies, Department of Pharmacology and Toxicology, Augusta, GA 30912-1500

AWARDS MS, PhD.

Faculty: 10 full-time (2 women).
Students: 17 full-time (6 women); includes 4 minority (2 African Americans, 2 Asian Americans or Pacific Islanders), 8 international. *18 applicants, 61% accepted.* Terminal master's awarded for partial completion of doctoral program.
Degree requirements: For master's and doctorate, thesis/dissertation required, foreign language not required.
Entrance requirements: For master's and doctorate, GRE General Test, TOEFL. *Application deadline:* For fall admission, 6/30 (priority date). Applications are processed on a rolling basis. *Application fee:* $25.
Expenses: Tuition, state resident: full-time $2,896; part-time $121 per hour. Tuition, nonresident: full-time $11,584; part-time $483 per hour. Required fees: $286; $143 per semester. Tuition and fees vary according to program.
Financial aid: In 1999–00, 12 research assistantships with partial tuition

reimbursements (averaging $15,500 per year) were awarded; fellowships, Federal Work-Study, grants, and institutionally sponsored loans also available. Aid available to part-time students. Financial aid application deadline: 3/31; financial aid applicants required to submit FAFSA.
Faculty research: Kinins, bradykinins and inflammation, cholinergic neurotransmitters, neuropharmacology.
Dr. R. William Caldwell, Chairman, 706-721-3384, *Fax:* 706-721-2347, *E-mail:* wcaldwell@mail.mcg.edu.
Application contact: Dr. Nevin Lambert, Director, 706-721-6345, *Fax:* 706-721-2347, *E-mail:* dlewis@mail.mcg.edu.

■ MEDICAL COLLEGE OF WISCONSIN

Graduate School of Biomedical Sciences, Department of Pharmacology and Toxicology, Milwaukee, WI 53226-0509
AWARDS MS, PhD, MD/MS, MD/PhD. Terminal master's awarded for partial completion of doctoral program.

Degree requirements: For master's, thesis required, foreign language not required; for doctorate, dissertation, oral and written qualifying exams required, foreign language not required.
Entrance requirements: For master's and doctorate, GRE General Test, TOEFL, minimum B average.
Expenses: Tuition, state resident: full-time $9,318. Tuition, nonresident: full-time $9,318. Required fees: $115.
Faculty research: Cardiovascular physiology and pharmacology, drugs of abuse, environmental and aquatic toxicology, central nervous system and biochemical pharmacology, signal transduction.

■ MICHIGAN STATE UNIVERSITY

College of Human Medicine and Graduate School, Graduate Programs in Human Medicine and Graduate Programs in Osteopathic Medicine and Graduate Programs in Veterinary Medicine, Department of Pharmacology/Toxicology, East Lansing, MI 48824
AWARDS MS, PhD. Part-time programs available.
Faculty: 18.
Students: 21 full-time (12 women), 4 part-time (2 women); includes 7 minority (2 African Americans, 3 Asian Americans or Pacific Islanders, 1 Hispanic American, 1 Native American), 4 international. Average age 28. *46 applicants, 20% accepted.* In 1999, 3 degrees awarded.
Degree requirements: For master's, thesis required, foreign language not

required; for doctorate, dissertation, comprehensive exams required, foreign language not required.
Entrance requirements: For master's and doctorate, GRE General Test, TOEFL, minimum GPA of 3.0. *Application fee:* $30 ($40 for international students). Electronic applications accepted.
Expenses: Tuition, state resident: full-time $10,868; part-time $229 per credit. Tuition, nonresident: full-time $23,168; part-time $464 per credit.
Financial aid: In 1999–00, 13 research assistantships (averaging $11,500 per year) were awarded; fellowships, teaching assistantships, grants and institutionally sponsored loans also available. Aid available to part-time students. Financial aid applicants required to submit FAFSA.
Faculty research: Central neural control of cardiovascular function, blood pressure control by forebrain and brainstem neurons, endothelian receptors, sensory nerve excitation in the intestine. *Total annual research expenditures:* $3.1 million.
Dr. Kenneth Moore, Chairperson, 517-353-7145, *Fax:* 517-353-8915.
Application contact: Dr. James J. Galligan, Graduate Committee, 517-353-7145, *Fax:* 517-353-8915, *E-mail:* hummeld@msu.edu.

■ MICHIGAN STATE UNIVERSITY

College of Osteopathic Medicine and Graduate School, Graduate Programs in Osteopathic Medicine, East Lansing, MI 48824
AWARDS Anatomy (MS, PhD); biochemistry (MS, PhD); microbiology (PhD); pathology (MS, PhD); pharmacology/toxicology (MS, PhD), including pharmacology; physiology (MS, PhD), including environmental toxicology (PhD), neuroscience (PhD). Part-time programs available.
Students: 17 full-time (9 women), 1 part-time; includes 5 minority (3 Asian Americans or Pacific Islanders, 1 Hispanic American, 1 Native American), 5 international. Average age 26. *33 applicants, 9% accepted.* In 1999, 2 doctorates awarded.
Degree requirements: For doctorate, dissertation required.
Entrance requirements: For master's and doctorate, GRE. *Application deadline:* For fall admission, 3/1 (priority date). Applications are processed on a rolling basis. *Application fee:* $30 ($40 for international students).
Expenses: Tuition, area resident: Full-time $15,879. Tuition, state resident: full-time $33,797.
Financial aid: Fellowships, research assistantships, teaching assistantships, career-related internships or fieldwork, Federal Work-Study, and institutionally

sponsored loans available. Financial aid application deadline: 4/2. *Total annual research expenditures:* $5 million.
Application contact: Dr. Veronica M. Maher, Associate Dean for Graduate Studies, 517-353-7785, *Fax:* 517-353-9004, *E-mail:* maher@com.msu.edu.

■ MICHIGAN STATE UNIVERSITY

College of Veterinary Medicine and Graduate School, Graduate Programs in Veterinary Medicine, East Lansing, MI 48824
AWARDS Anatomy (MS, PhD); large animal clinical sciences (MS, PhD); microbiology (MS, PhD); pathology (MS, PhD); pharmacology/toxicology (MS, PhD), including pharmacology; small animal clinical sciences (MS).

Students: 52 full-time (28 women); includes 6 minority (4 African Americans, 2 Asian Americans or Pacific Islanders), 18 international. Average age 32. *45 applicants, 20% accepted.* In 1999, 3 master's, 3 doctorates awarded.
Degree requirements: For master's, thesis or alternative required, foreign language not required; for doctorate, dissertation required, foreign language not required.
Application deadline: Applications are processed on a rolling basis. *Application fee:* $30 ($40 for international students). Electronic applications accepted.
Expenses: Tuition, state resident: full-time $9,766. Tuition, nonresident: full-time $20,082. Tuition and fees vary according to program.
Financial aid: Fellowships, research assistantships available.
Faculty research: Molecular genetics, food safety/toxicology, comparative orthopedics, airway disease, population medicine.
Dr. John C. Baker, Associate Dean for Research and Graduate Studies, 517-432-2388, *Fax:* 517-432-1037, *E-mail:* baker@cvm.msu.edu.
Application contact: Victoria Hoelzer-Maddox, Administrative Assistant, 517-353-3118, *Fax:* 517-432-1037, *E-mail:* hoelzer-maddox@cvm.msu.edu.

■ MICHIGAN STATE UNIVERSITY

Graduate School, Institute for Environmental Toxicology, East Lansing, MI 48824
AWARDS PhD.
Faculty: 2.
Students: 6 full-time (4 women), 1 international.
Degree requirements: For doctorate, dissertation required, foreign language not required.

Michigan State University (continued)
Application fee: $30 ($40 for international students).
Expenses: Tuition, state resident: part-time $229 per credit. Tuition, nonresident: part-time $464 per credit. Required fees: $241 per semester. Tuition and fees vary according to course load, degree level and program.
Financial aid: In 1999–00, research assistantships with tuition reimbursements (averaging $11,233 per year); teaching assistantships, career-related internships or fieldwork and Federal Work-Study also available. Financial aid applicants required to submit FAFSA.
Faculty research: Pathology; biochemical, aquatic, wildlife, food, and pesticide toxicology; hazardous material management; analytical chemistry. *Total annual research expenditures:* $2.7 million.
Dr. Lawrence J. Fischer, Director, 517-353-6469, *Fax:* 517-355-4603, *E-mail:* lfischer@pilot.msu.edu.

Find an in-depth description at www.petersons.com/graduate.

■ **NORTH CAROLINA STATE UNIVERSITY**

Graduate School, College of Agriculture and Life Sciences, Department of Toxicology, Raleigh, NC 27695

AWARDS M Tox, MS, PhD.

Faculty: 9 full-time (2 women), 14 part-time/adjunct (4 women).
Students: 31 full-time (19 women), 9 part-time (3 women); includes 2 minority (both African Americans), 6 international. Average age 32. *40 applicants, 30% accepted.* In 1999, 2 master's, 7 doctorates awarded. Terminal master's awarded for partial completion of doctoral program.
Degree requirements: For master's, thesis required (for some programs), foreign language not required; for doctorate, dissertation required, foreign language not required.
Entrance requirements: For master's and doctorate, GRE General Test, minimum GPA of 3.0. *Application deadline:* For fall admission, 5/25 (priority date); for spring admission, 10/25. *Application fee:* $45.
Expenses: Tuition, state resident: full-time $1,578. Tuition, nonresident: full-time $10,744. Required fees: $892. Full-time tuition and fees vary according to program.
Financial aid: In 1999–00, 10 fellowships (averaging $5,206 per year), 19 research assistantships (averaging $5,319 per year) were awarded; teaching assistantships, career-related internships or fieldwork, traineeships, and tuition waivers (partial) also available.

Faculty research: Biotransformation enzymes, oxygen radicals, chemical fate, carcinogenesis, developmental and endocrine toxicity. *Total annual research expenditures:* $2.7 million.
Dr. Hosni M. Hassan, Interim Department Head, 919-515-6663, *Fax:* 919-515-7169, *E-mail:* hmhassan@mbio.ncsu.edu.
Application contact: Dr. Gerald A. Leblanc, Director of Graduate Programs, 919-515-7404, *Fax:* 919-515-7169, *E-mail:* ga_leblanc@ncsu.edu.

Find an in-depth description at www.petersons.com/graduate.

■ **NORTHEASTERN UNIVERSITY**

Bouvé College of Health Sciences Graduate School, Program in Toxicology, Boston, MA 02115-5096

AWARDS MS.

Students: In 1999, 2 degrees awarded.
Entrance requirements: For master's, GRE General Test. *Application deadline:* Applications are processed on a rolling basis. *Application fee:* $50.
Expenses: Tuition: Full-time $16,560; part-time $460 per quarter hour. Required fees: $150; $25 per year. Tuition and fees vary according to course load and program.
Financial aid: Federal Work-Study available. Aid available to part-time students. Financial aid application deadline: 3/1; financial aid applicants required to submit FAFSA.
Robert Schatz, Director, 617-373-3214.
Application contact: Bill Purnell, Director of Graduate Admissions, 617-373-2708, *Fax:* 617-373-4701, *E-mail:* w.purnell@nunet.neu.edu.

Find an in-depth description at www.petersons.com/graduate.

■ **NORTHEASTERN UNIVERSITY**

Bouvé College of Health Sciences Graduate School, Programs in Biomedical Sciences, Boston, MA 02115-5096

AWARDS Biomedical sciences (MS); medical laboratory science (PhD); medicinal chemistry (PhD); pharmaceutics (PhD); pharmacology (PhD); toxicology (MS, PhD).

Faculty: 14 full-time (1 woman), 14 part-time/adjunct (5 women).
Students: 47 full-time (30 women), 6 part-time (4 women). Average age 31. *162 applicants, 47% accepted.* In 1999, 15 master's, 6 doctorates awarded. Terminal master's awarded for partial completion of doctoral program.
Degree requirements: For master's, comprehensive exam required, thesis optional, foreign language not required; ·

for doctorate, dissertation, qualifying exam required, foreign language not required.
Entrance requirements: For master's and doctorate, GRE General Test, TOEFL. *Application deadline:* For fall admission, 3/15. *Application fee:* $50.
Expenses: Tuition: Full-time $16,560; part-time $460 per quarter hour. Required fees: $150; $25 per year. Tuition and fees vary according to course load and program.
Financial aid: In 1999–00, 40 students received aid, including 10 research assistantships with full tuition reimbursements available, 12 teaching assistantships with full tuition reimbursements available (averaging $12,650 per year); tuition waivers (partial) also available. Financial aid applicants required to submit FAFSA.
Faculty research: Neuropharmacology, cardiovascular pharmacology, steroid chemistry, anti-infectives, behavioral pharmacology.
Dr. Roger W. Giese, Director, 617-373-3227, *Fax:* 617-266-6756, *E-mail:* rgiese@lynx.neu.edu.
Application contact: Bill Purnell, Director of Graduate Admissions, 617-373-2708, *Fax:* 617-373-4701, *E-mail:* w.purnell@nunet.neu.edu.

Find an in-depth description at www.petersons.com/graduate.

■ **NORTHEASTERN UNIVERSITY**

Bouvé College of Health Sciences Graduate School, Programs in Health Professions, Boston, MA 02115-5096

AWARDS General health professions (MHP); health policy (MHP); physician assistant (MS); regulatory toxicology (MHP). Part-time and evening/weekend programs available.

Faculty: 11 full-time (4 women).
Students: 63 full-time (37 women), 12 part-time (8 women). Average age 34. In 1999, 76 degrees awarded.
Degree requirements: For master's, comprehensive exam required.
Entrance requirements: For master's, minimum GPA of 3.0, bachelor's degree in science. *Application deadline:* For spring admission, 3/1. Applications are processed on a rolling basis. *Application fee:* $50.
Expenses: Tuition: Full-time $16,560; part-time $460 per quarter hour. Required fees: $150; $25 per year. Tuition and fees vary according to course load and program.
Financial aid: Federal Work-Study and tuition waivers (partial) available. Aid available to part-time students. Financial aid application deadline: 3/1; financial aid applicants required to submit FAFSA.
Faculty research: Statistical analysis, management.
Judith Barr, Director, 617-373-4188.

Application contact: Bill Purnell, Director of Graduate Admissions, 617-373-2708, *Fax:* 617-373-4701, *E-mail:* w.purnell@nunet.neu.edu.

Find an in-depth description at www.petersons.com/graduate.

■ NORTHWESTERN UNIVERSITY

The Graduate School, Division of Interdepartmental Programs and Medical School, Integrated Graduate Programs in the Life Sciences, Chicago, IL 60611

AWARDS Cancer biology (PhD); cell biology (PhD); developmental biology (PhD); evolutionary biology (PhD); immunology and microbial pathogenesis (PhD); molecular biology and genetics (PhD); neurobiology (PhD); pharmacology and toxicology (PhD); structural biology and biochemistry (PhD).

Degree requirements: For doctorate, dissertation, written and oral qualifying exams required, foreign language not required.
Entrance requirements: For doctorate, GRE General Test, TOEFL.
Expenses: Tuition: Full-time $23,301. Full-time tuition and fees vary according to program.

Find an in-depth description at www.petersons.com/graduate.

■ THE OHIO STATE UNIVERSITY

College of Veterinary Medicine and Graduate School, Graduate Programs in Veterinary Medicine, Department of Veterinary Biosciences, Columbus, OH 43210

AWARDS Anatomy and cellular biology (MS, PhD); pathobiology (MS, PhD); pharmacology (MS, PhD); toxicology (MS, PhD); veterinary physiology (MS, PhD).

Faculty: 28 full-time (8 women).
Students: 49 full-time (20 women); includes 2 minority (1 African American, 1 Asian American or Pacific Islander), 18 international.
Degree requirements: For master's and doctorate, thesis/dissertation, final exam required.
Entrance requirements: For master's, GRE General Test; for doctorate, GRE General Test, master's degree. *Application fee:* $25.
Expenses: Tuition, state resident: full-time $5,757. Tuition, nonresident: full-time $14,892.
Financial aid: Fellowships, research assistantships, teaching assistantships available.
Faculty research: Microvasculature, muscle biology, neonatal lung and bone development.
Charles C. Capen, Interim Chair, 614-292-4489.

Application contact: Graduate Admission Committee, 614-292-4489.

■ OREGON STATE UNIVERSITY

College of Veterinary Medicine, Program in Veterinary Sciences, Corvallis, OR 97331

AWARDS Microbiology (MS); pathology (MS); toxicology (MS). Part-time programs available.

Faculty: 4 full-time (0 women).
Students: 5 full-time (4 women). Average age 34. In 1999, 1 degree awarded (100% found work related to degree).
Degree requirements: For master's, thesis required, foreign language not required.
Entrance requirements: For master's, TOEFL, minimum GPA of 3.0 in last 90 hours. *Application deadline:* For fall admission, 11/1. *Application fee:* $50.
Expenses: Tuition, state resident: full-time $4,334. Tuition, nonresident: full-time $7,382.
Financial aid: Research assistantships, Federal Work-Study, institutionally sponsored loans, and scholarships available. Aid available to part-time students. Financial aid application deadline: 2/1.
Faculty research: Calf diseases, bovine foot rot, caliciviruses, effects of toxic agents on immune systems.
Linda L. Blythe, Associate Dean, 541-737-2098, *Fax:* 541-737-4245, *E-mail:* linda.blythe@orst.edu.

■ OREGON STATE UNIVERSITY

Graduate School, College of Agricultural Sciences, Program in Toxicology, Corvallis, OR 97331

AWARDS MS, PhD.

Faculty: 20 full-time (1 woman).
Students: 31 full-time (20 women), 2 part-time; includes 3 minority (1 African American, 1 Hispanic American, 1 Native American), 12 international. Average age 33. In 1999, 2 master's, 5 doctorates awarded.
Degree requirements: For master's and doctorate, thesis/dissertation required, foreign language not required.
Entrance requirements: For master's and doctorate, GRE, TOEFL, bachelor's degree in chemistry or biological sciences, minimum GPA of 3.0 in last 90 hours. *Application deadline:* For fall admission, 3/1. Applications are processed on a rolling basis. *Application fee:* $50.
Expenses: Tuition, state resident: full-time $6,489. Tuition, nonresident: full-time $11,061. Tuition and fees vary according to program.
Financial aid: Fellowships, research assistantships, Federal Work-Study and institutionally sponsored loans available.

Aid available to part-time students. Financial aid application deadline: 2/1.
Faculty research: Biochemical mechanisms for toxicology; analytical, comparative, aquatic, and food toxicology; aquaculture of salmonids; immunotoxicology; fish toxicology.
Dr. Lawrence R. Curtis, Director, 541-737-2363, *Fax:* 541-737-0497, *E-mail:* lamy.curtis@orst.edu.

■ PURDUE UNIVERSITY

Graduate School and School of Pharmacy and Pharmacal Sciences, School of Health Sciences, Program in Toxicology, West Lafayette, IN 47907

AWARDS MS, PhD.

Faculty: 1 full-time (0 women).
Students: 2 full-time (0 women), 1 international.
Degree requirements: For master's, thesis optional, foreign language not required; for doctorate, one foreign language (computer language can substitute), dissertation required.
Entrance requirements: For master's and doctorate, GRE General Test, TOEFL, minimum B average. *Application deadline:* Applications are processed on a rolling basis. *Application fee:* $30. Electronic applications accepted.
Expenses: Tuition, state resident: full-time $4,530; part-time $130 per credit hour. Tuition, nonresident: full-time $15,310; part-time $404 per credit hour. Tuition and fees vary according to campus/location and program.
Application contact: Dr. Gary Carlson, Graduate Chairperson, 765-494-1412, *Fax:* 765-496-1377, *E-mail:* gcarlson@purdue.edu.

■ PURDUE UNIVERSITY

School of Pharmacy and Pharmacal Sciences and Graduate School, Graduate Programs in Pharmacy and Pharmacal Sciences, Department of Medicinal Chemistry and Molecular Pharmacology, West Lafayette, IN 47907

AWARDS Analytical medicinal chemistry (PhD); computational and biophysical medicinal chemistry (PhD); medicinal and bioorganic chemistry (PhD); medicinal biochemistry and molecular biology (PhD); molecular pharmacology and toxicology (PhD); natural products and pharmacognosy (PhD); nuclear pharmacy (MS); radiopharmaceutical chemistry and nuclear pharmacy (PhD).

Faculty: 24 full-time (2 women).
Students: 48 full-time (26 women), 3 part-time (1 woman); includes 4 minority (1 African American, 1 Asian American or

Purdue University (continued)
Pacific Islander, 2 Hispanic Americans), 13 international. Average age 29. *139 applicants, 19% accepted.* In 1999, 3 master's, 11 doctorates awarded. Terminal master's awarded for partial completion of doctoral program.
Degree requirements: For master's and doctorate, thesis/dissertation required, foreign language not required.
Entrance requirements: For master's, GRE General Test, TOEFL, minimum B average; BS in biology, chemistry, or pharmacy; for doctorate, GRE General Test, TOEFL, minimum B average; BS in biology, chemistry, or pharmacology. *Application deadline:* Applications are processed on a rolling basis. *Application fee:* $30. Electronic applications accepted.
Expenses: Tuition, state resident: full-time $4,530; part-time $130 per credit hour. Tuition, nonresident: full-time $15,310; part-time $404 per credit hour. Tuition and fees vary according to campus/location and program.
Financial aid: Fellowships, research assistantships, teaching assistantships, traineeships available. Aid available to part-time students. Financial aid applicants required to submit FAFSA.
Faculty research: Drug design and development, cancer research, drug synthesis and analysis, chemical pharmacology, environmental toxicology.
Dr. R. F. Borch, Graduate Head, 765-494-1403.
Application contact: Dr. D. E. Bergstrom, Graduate Committee, 765-494-6275, *E-mail:* bergstrom@ pharmacy.purdue.edu.
Find an in-depth description at www.petersons.com/graduate.

■ **PURDUE UNIVERSITY**
School of Veterinary Medicine and Graduate School, Graduate Programs in Veterinary Medicine, Department of Veterinary Pathobiology, West Lafayette, IN 47907
AWARDS Bacteriology (MS, PhD); epidemiology (MS, PhD); immunology (MS, PhD); infectious diseases (MS, PhD); microbiology (MS, PhD); parasitology (MS, PhD); pathology (MS, PhD); toxicology (MS, PhD); virology (MS, PhD).
Faculty: 26 full-time (5 women).
Students: 42 full-time (24 women), 2 part-time (1 woman); includes 5 minority (1 African American, 3 Asian Americans or Pacific Islanders, 1 Hispanic American), 18 international. Average age 32. In 1999, 2 master's, 6 doctorates awarded (100% found work related to degree). Terminal master's awarded for partial completion of doctoral program.

Degree requirements: For master's, thesis required (for some programs), foreign language not required; for doctorate, dissertation required, foreign language not required.
Entrance requirements: For master's and doctorate, GRE General Test, TOEFL. *Application fee:* $30.
Expenses: Tuition, state resident: full-time $3,732. Tuition, nonresident: full-time $8,732.
Financial aid: Fellowships, research assistantships, teaching assistantships available. Financial aid application deadline: 3/1; financial aid applicants required to submit FAFSA.
Dr. H. L. Thacker, Head, 765-494-7543.

■ **RUTGERS, THE STATE UNIVERSITY OF NEW JERSEY, NEW BRUNSWICK**
Graduate School, Program in Environmental Sciences, New Brunswick, NJ 08901-1281
AWARDS Air resources (MS, PhD); aquatic biology (MS, PhD); aquatic chemistry (MS, PhD); chemistry and physics of aerosol and hydrosol systems (MS, PhD); environmental chemistry (MS, PhD); environmental microbiology (MS, PhD); environmental toxicology (MS, PhD); exposure assessment (PhD); water and wastewater treatment (MS, PhD); water resources (MS, PhD). Part-time and evening/weekend programs available.
Faculty: 33 full-time (7 women), 36 part-time/adjunct (6 women).
Students: 68 full-time (25 women), 58 part-time (26 women); includes 10 minority (8 Asian Americans or Pacific Islanders, 2 Hispanic Americans), 44 international. Average age 26. *128 applicants, 40% accepted.* In 1999, 15 master's, 18 doctorates awarded. Terminal master's awarded for partial completion of doctoral program.
Degree requirements: For master's, thesis or alternative, oral final exam required, foreign language not required; for doctorate, thesis defense, qualifying exam required.
Entrance requirements: For master's and doctorate, GRE General Test, TOEFL. *Application deadline:* For fall admission, 3/1; for spring admission, 11/1. Applications are processed on a rolling basis. *Application fee:* $50.
Expenses: Tuition, state resident: full-time $6,776; part-time $279 per credit. Tuition, nonresident: full-time $9,936; part-time $412 per credit. Required fees: $20 per credit. $89 per semester. Tuition and fees vary according to course load, campus/location and program.
Financial aid: In 1999–00, 1 fellowship (averaging $3,000 per year), 30 research

assistantships with full tuition reimbursements (averaging $13,956 per year), 10 teaching assistantships with full tuition reimbursements (averaging $13,100 per year) were awarded; career-related internships or fieldwork and Federal Work-Study also available. Financial aid application deadline: 3/1; financial aid applicants required to submit FAFSA.
Faculty research: Atmospheric sciences; biological waste treatment; contaminant fate and transport; exposure assessment; air, soil and water quality.
Dr. Peter F. Strom, Director, 732-932-8078, *Fax:* 732-932-8644, *E-mail:* strom@ aesop.rutgers.edu.
Application contact: Paul J. Lioy, Graduate Admissions Committee, 732-932-0150, *Fax:* 732-445-0116, *E-mail:* plioy@ eohsi.rutgers.edu.

■ **RUTGERS, THE STATE UNIVERSITY OF NEW JERSEY, NEW BRUNSWICK**
Graduate School, Program in Toxicology, New Brunswick, NJ 08901-1281
AWARDS Environmental toxicology (MS, PhD); industrial-occupational toxicology (MS, PhD); nutritional toxicology (MS, PhD); pharmaceutical toxicology (MS, PhD).
Faculty: 18 full-time (6 women).
Students: 16 full-time (10 women), 33 part-time (19 women); includes 8 minority (2 African Americans, 4 Asian Americans or Pacific Islanders, 2 Hispanic Americans), 8 international. Average age 25. *32 applicants, 22% accepted.* In 1999, 2 master's, 4 doctorates awarded.
Degree requirements: For master's, thesis, qualifying exam required, foreign language not required; for doctorate, dissertation, qualifying exam required. *Average time to degree:* Master's–3.5 years full-time; doctorate–5.5 years full-time.
Entrance requirements: For master's, GRE General Test, TOEFL; for doctorate, GRE General Test, TOEFL, GRE Subject Test. *Application deadline:* For fall admission, 5/1. Applications are processed on a rolling basis. *Application fee:* $50.
Expenses: Tuition, state resident: full-time $6,776; part-time $279 per credit. Tuition, nonresident: full-time $9,936; part-time $412 per credit. Required fees: $20 per credit. $89 per semester. Tuition and fees vary according to course load, campus/location and program.
Financial aid: In 1999–00, 10 fellowships (averaging $14,000 per year), 5 research assistantships with tuition reimbursements (averaging $14,000 per year), 4 teaching assistantships with tuition reimbursements (averaging $14,000 per year) were awarded; career-related internships or

fieldwork, grants, traineeships, and unspecified assistantships also available. Financial aid application deadline: 3/1; financial aid applicants required to submit FAFSA.
Faculty research: Carcinogenesis, phototoxins, food additives, neurotoxicants. *Total annual research expenditures:* $10 million.
Dr. Kenneth R. Reuhl, Director, 732-445-3720, *Fax:* 732-445-0119, *E-mail:* reuhl@eohsi.rutgers.edu.
Application contact: Dr. Debra L. Laskin, Chair, Admissions Committee, 732-445-5862, *Fax:* 732-445-0119, *E-mail:* laskin@eohsi.rutgers.edu.

■ ST. JOHN'S UNIVERSITY

College of Pharmacy and Allied Health Professions, Graduate Programs in Pharmacy, Program in Pharmaceutical Sciences, Jamaica, NY 11439
AWARDS Clinical pharmacy (MS); cosmetic sciences (MS); industrial pharmacy (MS, PhD); medicinal chemistry (MS, PhD); pharmacology (MS, PhD); pharmacotherapeutics (MS); toxicology (PhD). Part-time and evening/weekend programs available.
Faculty: 19 full-time (4 women), 2 part-time/adjunct (both women).
Students: 14 full-time (4 women), 90 part-time (38 women); includes 19 minority (4 African Americans, 13 Asian Americans or Pacific Islanders, 2 Hispanic Americans), 55 international. Average age 30. *153 applicants, 53% accepted.* In 1999, 19 master's, 5 doctorates awarded. Terminal master's awarded for partial completion of doctoral program.
Degree requirements: For master's, comprehensive exam required, thesis optional, foreign language not required; for doctorate, one foreign language (computer language can substitute), dissertation, comprehensive and qualifying exams, residency required.
Entrance requirements: For master's, GRE General Test, minimum GPA of 3.0; for doctorate, GRE General Test, minimum GPA of 3.5 (undergraduate), 3.0 (graduate). *Application deadline:* For fall admission, 7/15 (priority date); for spring admission, 12/1. Applications are processed on a rolling basis. *Application fee:* $40.
Expenses: Tuition: Full-time $16,800; part-time $700 per credit. Required fees: $150; $75 per semester.
Financial aid: Fellowships, research assistantships, career-related internships or fieldwork and scholarships available. Aid available to part-time students. Financial aid application deadline: 3/1; financial aid applicants required to submit FAFSA.
Faculty research: Neurotoxicology, biochemical toxicology, molecular

pharmacology, neuropharmacology, intermediary metabolism.
Dr. Louis Trombetta, Chair, 718-990-5008, *E-mail:* trombetl@stjohns.edu.
Application contact: Shamus J. McGrenra, TOR, Associate Director, Graduate Admissions, 718-990-2000, *Fax:* 718-990-2096, *E-mail:* mcgrenrs@stjohns.edu.

■ ST. JOHN'S UNIVERSITY

College of Pharmacy and Allied Health Professions, Graduate Programs in Pharmacy, Program in Toxicology, Jamaica, NY 11439
AWARDS MS, PhD. Part-time and evening/weekend programs available.
Students: 2 full-time (both women), 14 part-time (9 women); includes 5 minority (1 African American, 2 Asian Americans or Pacific Islanders, 2 Hispanic Americans), 3 international. Average age 30. *10 applicants, 50% accepted.* In 1999, 1 degree awarded.
Degree requirements: For master's, comprehensive exam required, thesis optional, foreign language not required.
Entrance requirements: For master's, GRE General Test, minimum GPA of 3.0; for doctorate, GRE General Test, minimum GPA of 3.5 (undergraduate), 3.0 (graduate). *Application deadline:* For fall admission, 7/15 (priority date); for spring admission, 12/1. Applications are processed on a rolling basis. *Application fee:* $40.
Expenses: Tuition: Full-time $16,800; part-time $700 per credit. Required fees: $150; $75 per semester.
Financial aid: Fellowships, research assistantships, career-related internships or fieldwork and scholarships available. Aid available to part-time students. Financial aid application deadline: 3/1; financial aid applicants required to submit FAFSA.
Faculty research: Neurotoxicology, renal toxicology, toxicology of metals, regulatory toxicology.
Dr. Louis Trombetta, Chair, 718-990-5008, *E-mail:* trombetl@stjohns.edu.
Application contact: Patricia G. Armstrong, Director, Office of Admission, 718-990-2000, *Fax:* 718-990-2096, *E-mail:* armstrop@stjohns.edu.

■ SAN DIEGO STATE UNIVERSITY

Graduate and Research Affairs, College of Health and Human Services, Graduate School of Public Health, San Diego, CA 92182
AWARDS Environmental health (MPH, MS); epidemiology (MPH, PhD), including biostatistics (MPH); health promotion (MPH); health services administration (MPH); industrial hygiene (MS); toxicology (MS). Part-time programs available.

Faculty: 24 full-time (9 women), 19 part-time/adjunct (11 women).
Students: 197 full-time (138 women), 95 part-time (71 women); includes 96 minority (6 African Americans, 49 Asian Americans or Pacific Islanders, 38 Hispanic Americans, 3 Native Americans), 8 international. *357 applicants, 54% accepted.* In 1999, 105 master's, 3 doctorates awarded.
Degree requirements: For master's, thesis required (for some programs), foreign language not required; for doctorate, dissertation required, foreign language not required.
Entrance requirements: For master's, GMAT (health services administration), GRE General Test; for doctorate, GRE General Test. *Application deadline:* For fall admission, 5/15 (priority date); for spring admission, 10/15 (priority date). Applications are processed on a rolling basis. *Application fee:* $55.
Expenses: Tuition, nonresident: part-time $246 per unit. Required fees: $1,932; $633 per semester. Tuition and fees vary according to course load.
Financial aid: Research assistantships, teaching assistantships, career-related internships or fieldwork, Federal Work-Study, and traineeships available. Financial aid applicants required to submit FAFSA.
Faculty research: Evaluation of tobacco, AIDS prevalence and prevention, mammography, infant death project, Alzheimer's in elderly Chinese.
Dr. Kenneth Bart, Director, 619-594-6317.
Application contact: Brenda Fass-Holmes, Coordinator, Admissions and Student Affairs, 619-594-6317, *E-mail:* bholmes@mail.sdsu.edu.

■ STATE UNIVERSITY OF NEW YORK AT ALBANY

School of Public Health, Department of Environmental Health and Toxicology, Albany, NY 12222-0001
AWARDS Environmental and occupational health (MS, PhD); environmental chemistry (MS, PhD); toxicology (MS, PhD).
Students: 13 full-time (6 women), 24 part-time (7 women); includes 1 minority (African American), 18 international. Average age 33. *37 applicants, 70% accepted.* In 1999, 3 master's, 4 doctorates awarded.
Degree requirements: For master's and doctorate, thesis/dissertation required.
Entrance requirements: For master's and doctorate, GRE General Test, GRE Subject Test. *Application deadline:* For fall admission, 2/1 (priority date); for spring admission, 11/1 (priority date). *Application fee:* $50.
Expenses: Tuition, state resident: full-time $5,100; part-time $214 per credit. Tuition,

State University of New York at Albany (continued)

nonresident: full-time $8,416; part-time $352 per credit. Required fees: $31 per credit.

Financial aid: Fellowships, research assistantships available. Financial aid application deadline: 2/1.
Dr. Kenneth Jackson, Chair, 518-473-7553.

Find an in-depth description at www.petersons.com/graduate.

■ STATE UNIVERSITY OF NEW YORK AT BUFFALO

Graduate School, School of Medicine and Biomedical Sciences, Graduate Programs in Medicine and Biomedical Sciences, Department of Pharmacology and Toxicology, Buffalo, NY 14260

AWARDS MA, PhD.

Faculty: 16 full-time (2 women).
Students: 9 full-time (3 women), 7 part-time (6 women), 8 international. Average age 25. *13 applicants, 54% accepted.* In 1999, 1 master's, 2 doctorates awarded. Terminal master's awarded for partial completion of doctoral program.
Degree requirements: For master's and doctorate, thesis/dissertation required.
Entrance requirements: For master's and doctorate, GRE General Test, TOEFL. *Application deadline:* For fall admission, 3/1 (priority date). Applications are processed on a rolling basis. *Application fee:* $35.
Expenses: Tuition, state resident: full-time $5,100. Tuition, nonresident: full-time $8,416. Required fees: $935.
Financial aid: In 1999–00, 15 students received aid, including 15 research assistantships (averaging $13,549 per year); fellowships, teaching assistantships, Federal Work-Study, grants, and unspecified assistantships also available. Financial aid application deadline: 2/28; financial aid applicants required to submit FAFSA.
Faculty research: Neuropharmacology, toxicology, signal transduction, molecular pharmacology, behavioral pharmacology. *Total annual research expenditures:* $1.5 million.
Dr. Ronald P. Rubin, Chairman, 716-829-2800, *Fax:* 716-829-2801.
Application contact: Noreen A. Harbison, Information Contact, 716-829-2800, *Fax:* 716-829-2801, *E-mail:* harbison@acsu.buffalo.edu.

Find an in-depth description at www.petersons.com/graduate.

■ TEXAS A&M UNIVERSITY

College of Veterinary Medicine, Department of Veterinary Pathobiology, College Station, TX 77843

AWARDS Genetics (MS, PhD); toxicology (MS, PhD); veterinary microbiology (MS, PhD); veterinary parasitology (MS); veterinary pathology (MS, PhD). Part-time programs available. Postbaccalaureate distance learning degree programs offered.

Faculty: 44 full-time (10 women), 17 part-time/adjunct (2 women).
Students: 33 full-time (13 women), 30 part-time (18 women); includes 11 minority (2 African Americans, 1 Asian American or Pacific Islander, 8 Hispanic Americans), 11 international. Average age 34. *28 applicants, 46% accepted.* In 1999, 4 master's awarded (25% found work related to degree, 75% continued full-time study); 10 doctorates awarded (40% entered university research/teaching, 60% found other work related to degree). Terminal master's awarded for partial completion of doctoral program.
Degree requirements: For master's and doctorate, thesis/dissertation required, foreign language not required. *Average time to degree:* Master's–2 years full-time, 4.5 years part-time; doctorate–4.5 years full-time, 7 years part-time.
Entrance requirements: For master's and doctorate, GRE General Test, TOEFL. *Application deadline:* For fall admission, 3/1 (priority date); for spring admission, 8/1 (priority date). Applications are processed on a rolling basis. *Application fee:* $50 ($75 for international students). Electronic applications accepted.
Expenses: Tuition, state resident: full-time $5,400. Tuition, nonresident: full-time $16,200. Required fees: $2,936. Full-time tuition and fees vary according to student level.
Financial aid: In 1999–00, 7 fellowships with partial tuition reimbursements (averaging $16,000 per year), 13 research assistantships with partial tuition reimbursements (averaging $14,400 per year), 5 teaching assistantships with partial tuition reimbursements (averaging $16,000 per year) were awarded; career-related internships or fieldwork, Federal Work-Study, and institutionally sponsored loans also available. Aid available to part-time students. Financial aid applicants required to submit FAFSA.
Faculty research: Infectious and noninfectious diseases of animals and birds, animal genetics, molecular biology, immunology, virology. *Total annual research expenditures:* $2.6 million.

Dr. Ann B. Kier, Head, 979-845-5941, *Fax:* 979-845-9231, *E-mail:* akier@cvm.tamu.edu.
Application contact: Dr. G. G. Wagner, Graduate Adviser, 979-845-5941, *Fax:* 979-862-1147, *E-mail:* gwagner@cvm.tamus.edu.

■ TEXAS A&M UNIVERSITY

College of Veterinary Medicine and Office of Graduate Studies, Graduate Programs in Veterinary Medicine, Department of Veterinary Anatomy and Public Health, College Station, TX 77843

AWARDS Anatomy (MS, PhD); epidemiology (MS); genetics (PhD); toxicology (PhD); veterinary public health (MS).

Faculty: 23 full-time (6 women), 13 part-time/adjunct (6 women).
Students: 36 full-time (21 women), 2 part-time (both women); includes 13 minority (2 African Americans, 9 Asian Americans or Pacific Islanders, 2 Hispanic Americans). Average age 27. *19 applicants, 47% accepted.* In 1999, 4 master's, 10 doctorates awarded. Terminal master's awarded for partial completion of doctoral program.
Degree requirements: For master's and doctorate, thesis/dissertation required, foreign language not required.
Entrance requirements: For master's and doctorate, GRE General Test, TOEFL. *Application deadline:* For fall admission, 7/15 (priority date); for spring admission, 10/1. Applications are processed on a rolling basis. *Application fee:* $50 ($75 for international students).
Expenses: Tuition, state resident: part-time $76 per semester hour. Tuition, nonresident: part-time $292 per semester hour. Required fees: $11 per semester hour.
Financial aid: In 1999–00, 2 fellowships (averaging $12,000 per year), 18 research assistantships (averaging $13,500 per year), 5 teaching assistantships (averaging $14,000 per year) were awarded; Federal Work-Study, institutionally sponsored loans, and clinical associateships also available. Financial aid application deadline: 7/15; financial aid applicants required to submit FAFSA.
Faculty research: Metal toxicology, reproductive biology, genetics of neural development, developmental biology, environmental toxicology. *Total annual research expenditures:* $3.4 million.
Dr. Evelyn Tiffany-Castiglioni, Head, 979-845-2828, *Fax:* 979-847-8981, *E-mail:* ecastiglioni@cvm.tamu.edu.

■ TEXAS A&M UNIVERSITY

College of Veterinary Medicine and Office of Graduate Studies, Graduate Programs in Veterinary Medicine, Department of Veterinary Physiology and Pharmacology, College Station, TX 77843

AWARDS Physiology (MS, PhD); toxicology (MS, PhD).

Faculty: 21 full-time (2 women), 3 part-time/adjunct (1 woman).

Students: 14 full-time (9 women), 6 part-time (2 women). Average age 32. *11 applicants, 27% accepted.* In 1999, 7 master's, 6 doctorates awarded.

Degree requirements: For master's and doctorate, foreign language and thesis not required.

Entrance requirements: For master's and doctorate, GRE General Test, TOEFL. *Application fee:* $50 ($75 for international students).

Expenses: Tuition, state resident: part-time $76 per semester hour. Tuition, nonresident: part-time $292 per semester hour. Required fees: $11 per semester hour.

Financial aid: Fellowships, research assistantships, teaching assistantships available. Financial aid application deadline: 4/1; financial aid applicants required to submit FAFSA.

Faculty research: Gamete and embryo physiology, endocrinology, equine laminitis.

Glen Laine, Head, 979-845-7261.

Find an in-depth description at www.petersons.com/graduate.

■ TEXAS A&M UNIVERSITY

Interdisciplinary Faculty in Toxicology, College Station, TX 77843

AWARDS MS, PhD. Program composed of members from 7 colleges and 15 departments.

Faculty: 57 full-time (10 women), 4 part-time/adjunct (1 woman).

Students: 43 full-time (21 women); includes 8 minority (4 Asian Americans or Pacific Islanders, 4 Hispanic Americans), 14 international. Average age 27. *33 applicants, 24% accepted.* In 1999, 1 master's awarded (100% entered university research/teaching); 6 doctorates awarded (50% entered university research/teaching, 33% found other work related to degree, 17% continued full-time study). Terminal master's awarded for partial completion of doctoral program.

Degree requirements: For master's and doctorate, thesis/dissertation required, foreign language not required. *Average time to degree:* Master's–4 years part-time; doctorate–4.9 years full-time.

Entrance requirements: For master's and doctorate, GRE General Test, TOEFL, minimum GPA of 3.0. *Application deadline:* For fall admission, 2/1 (priority date); for spring admission, 8/1. Applications are processed on a rolling basis. *Application fee:* $50 ($75 for international students). Electronic applications accepted.

Expenses: Tuition, state resident: part-time $76 per semester hour. Tuition, nonresident: part-time $292 per semester hour. Required fees: $11 per semester hour. Tuition and fees vary according to program.

Financial aid: In 1999–00, 7 fellowships with partial tuition reimbursements (averaging $16,200 per year), 36 research assistantships (averaging $1,200 per year) were awarded; Federal Work-Study, institutionally sponsored loans, and scholarships also available. Financial aid application deadline: 3/1; financial aid applicants required to submit FAFSA.

Faculty research: Behavioral toxicology and neurotoxicology, cellular toxicology, developmental and reproductive toxicology, applied veterinary and food toxicology. *Total annual research expenditures:* $1.8 million.

Dr. Stephen H. Safe, Chair, 979-845-5529, *Fax:* 979-862-4929, *E-mail:* ssafe@cvm.tamu.edu.

Application contact: Kimberly D. Daniel, Staff Assistant, 979-845-5529, *Fax:* 979-862-4929, *E-mail:* tox@cvm.tamu.edu.

■ TEXAS A&M UNIVERSITY SYSTEM HEALTH SCIENCE CENTER

College of Medicine, Graduate School of Biomedical Sciences, Department of Medical Pharmacology and Toxicology, College Station, TX 77840-7896

AWARDS PhD.

Faculty: 8 full-time (0 women).

Students: 6 full-time (3 women), 1 part-time, 2 international. Average age 25. *6 applicants, 17% accepted.* In 1999, 1 degree awarded (100% found work related to degree).

Degree requirements: For doctorate, dissertation required, foreign language not required. *Average time to degree:* Doctorate–5 years full-time.

Entrance requirements: For doctorate, GRE General Test. *Application deadline:* For fall admission, 2/1 (priority date). Applications are processed on a rolling basis. *Application fee:* $35 ($75 for international students).

Expenses: Tuition, area resident: Full-time $1,368. Tuition, state resident: part-time $76 per credit. Tuition, nonresident: full-time $5,256; part-time $292 per credit.

International tuition: $5,256 full-time. Required fees: $678; $38 per credit. Full-time tuition and fees vary according to course load and student level.

Financial aid: In 1999–00, 7 students received aid, including fellowships (averaging $17,000 per year); research assistantships Financial aid application deadline: 4/1; financial aid applicants required to submit FAFSA.

Faculty research: Medical treatment of eye disease, fetal alcohol syndrome, Alzheimer's disease, glycme receptory steriods.

Dr. George C. Y. Chiou, Head, 409-845-2817, *Fax:* 409-845-0699, *E-mail:* gchiou@tamsun.tamu.edu.

Application contact: Dr. Gerald Frye, Graduate Adviser, 979-845-2860, *E-mail:* gdfrye@tamu.edu.

Find an in-depth description at www.petersons.com/graduate.

■ TEXAS SOUTHERN UNIVERSITY

Graduate School, College of Arts and Sciences, Program in Environmental Toxicology, Houston, TX 77004-4584

AWARDS MS, PhD.

Application deadline: For fall admission, 7/15 (priority date). *Application fee:* $35 ($75 for international students).

Expenses: Tuition, area resident: Part-time $296 per credit hour. Tuition, nonresident: part-time $449 per credit hour.

Financial aid: Application deadline: 5/1.

Dr. John Sapp, Acting Dean, College of Arts and Sciences, 713-313-7210.

■ TEXAS TECH UNIVERSITY

Graduate School, College of Arts and Sciences, Department of Biological Sciences, Lubbock, TX 79409

AWARDS Biology (MS, PhD); environmental toxicology (MS); microbiology (MS); zoology (MS, PhD). Part-time programs available.

Faculty: 36 full-time (4 women), 1 part-time/adjunct (0 women).

Students: 86 full-time (41 women), 26 part-time (10 women); includes 5 minority (1 Asian American or Pacific Islander, 4 Hispanic Americans), 28 international. Average age 30. *54 applicants, 46% accepted.* In 2000, 26 master's, 11 doctorates awarded.

Degree requirements: For master's, thesis required (for some programs), foreign language not required; for doctorate, dissertation required, foreign language not required.

Entrance requirements: For master's and doctorate, GRE General Test. *Application deadline:* For fall admission, 4/15 (priority

Texas Tech University (continued)
date); for spring admission, 11/1 (priority date). Applications are processed on a rolling basis. *Application fee:* $25 ($50 for international students). Electronic applications accepted.
Expenses: Tuition, state resident: full-time $2,376; part-time $99 per credit hour. Tuition, nonresident: full-time $7,560; part-time $315 per credit hour. Required fees: $464 per semester. Part-time tuition and fees vary according to course load, program and reciprocity agreements.
Financial aid: In 2000–01, 60 students received aid, including 40 research assistantships (averaging $10,405 per year), 53 teaching assistantships (averaging $11,067 per year); fellowships, career-related internships or fieldwork, Federal Work-Study, and institutionally sponsored loans also available. Aid available to part-time students. Financial aid application deadline: 5/15; financial aid applicants required to submit FAFSA.
Faculty research: Development of strains of transgenic plants, ecological studies of Arctic tundra and Puerto Rican rain forests, genome organization and evolution. *Total annual research expenditures:* $2.1 million.
Dr. Carleton Phillips, Chairman, 806-742-2715, *Fax:* 806-742-2963.
Application contact: Graduate Adviser, 806-742-2715, *Fax:* 806-742-2963.

■ THE UNIVERSITY OF ARIZONA

Graduate College, College of Pharmacy, Department of Pharmacology and Toxicology, Tucson, AZ 85721

AWARDS Pharmacology (MS); toxicology (MS).

Degree requirements: For master's, thesis required.
Entrance requirements: For master's, GRE General Test, TOEFL, minimum GPA of 3.0.
Expenses: Tuition, nonresident: full-time $4,814; part-time $274 per unit. Required fees: $1,094; $115 per unit. Tuition and fees vary according to course load and program.

■ THE UNIVERSITY OF ARIZONA

Graduate College, Graduate Interdisciplinary Programs, Graduate Interdisciplinary Program in Pharmacology and Toxicology, Tucson, AZ 85721

AWARDS PhD.

Degree requirements: For doctorate, dissertation required, foreign language not required.
Entrance requirements: For doctorate, TOEFL.

Expenses: Tuition, nonresident: full-time $4,814; part-time $274 per unit. Required fees: $1,094; $115 per unit. Tuition and fees vary according to course load and program.
Faculty research: Neuropharmacology, carcinogenesis and cancer chemotherapy, molecular pharmacology and toxicology, biochemical pharmacology and toxicology.
Find an in-depth description at www.petersons.com/graduate.

■ UNIVERSITY OF ARKANSAS FOR MEDICAL SCIENCES

College of Medicine and Graduate School, Graduate Programs in Medicine, Department of Pharmacology and Toxicology, Little Rock, AR 72205-7199

AWARDS Pharmacology (MS, PhD); toxicology (MS, PhD).

Faculty: 24 full-time (0 women), 4 part-time/adjunct (1 woman).
Students: 10 full-time (5 women), 5 international. In 1999, 1 master's, 2 doctorates awarded.
Degree requirements: For master's and doctorate, thesis/dissertation required, foreign language not required.
Entrance requirements: For master's and doctorate, GRE General Test. *Application fee:* $0.
Expenses: Tuition: Full-time $8,928.
Financial aid: Research assistantships, teaching assistantships available. Aid available to part-time students.
Dr. Donald E. McMillan, Chairman, 501-686-5510.
Application contact: Dr. Philip R. Mayeux, Graduate Coordinator, 501-686-5510, *E-mail:* mayeuxphilipr@exchange.uams.edu.

Find an in-depth description at www.petersons.com/graduate.

■ UNIVERSITY OF ARKANSAS FOR MEDICAL SCIENCES

College of Medicine and Graduate School, Graduate Programs in Medicine, Interdisciplinary Toxicology Program, Little Rock, AR 72205-7199
AWARDS MS, PhD, MD/PhD.

Faculty: 10 full-time (1 woman), 7 part-time/adjunct (0 women).
Students: 13 full-time (3 women), 6 part-time (4 women); includes 4 minority (1 African American, 3 Asian Americans or Pacific Islanders), 3 international. In 1999, 2 doctorates awarded.
Degree requirements: For master's and doctorate, thesis/dissertation required, foreign language not required.

Entrance requirements: For master's and doctorate, GRE General Test. *Application fee:* $0.
Expenses: Tuition: Full-time $8,928.
Financial aid: In 1999–00, 11 research assistantships were awarded. Aid available to part-time students.
Dr. Jack A. Hinson, Director, 501-686-5766.

■ UNIVERSITY OF CALIFORNIA, DAVIS

Graduate Studies, Programs in the Biological Sciences, Program in Pharmacology/Toxicology, Davis, CA 95616

AWARDS MS, PhD.

Faculty: 50 full-time (7 women).
Students: 39 full-time (22 women), 1 part-time; includes 9 minority (4 Asian Americans or Pacific Islanders, 4 Hispanic Americans, 1 Native American), 6 international. Average age 29. *58 applicants, 38% accepted.* In 1999, 4 master's, 8 doctorates awarded. Terminal master's awarded for partial completion of doctoral program.
Degree requirements: For master's, comprehensive exam or thesis required; for doctorate, dissertation, qualifying exam required, foreign language not required.
Entrance requirements: For master's and doctorate, GRE General Test, minimum GPA of 3.0, previous course work in biochemistry and/or physiology. *Application deadline:* For fall admission, 1/15. *Application fee:* $40. Electronic applications accepted.
Expenses: Tuition, nonresident: full-time $9,804. Tuition and fees vary according to program and student level.
Financial aid: In 1999–00, 40 students received aid, including 8 fellowships with full and partial tuition reimbursements available, 26 research assistantships with full and partial tuition reimbursements available; teaching assistantships, career-related internships or fieldwork, Federal Work-Study, grants, institutionally sponsored loans, scholarships, and tuition waivers (full and partial) also available. Financial aid application deadline: 1/15; financial aid applicants required to submit FAFSA.
Faculty research: Respiratory, neurochemical, molecular, genetic, and ecological toxicology.
Jerry Last, Graduate Chair, 530-752-6230, *E-mail:* jalast@ucdavis.edu.
Application contact: Carol Barnes, Graduate Administrative Assistant, 530-752-4521, *Fax:* 530-752-3394, *E-mail:* cbarnes@ucdavis.edu.

■ UNIVERSITY OF CALIFORNIA, IRVINE

College of Medicine and Office of Research and Graduate Studies, Graduate Programs in Medicine, Department of Community and Environmental Medicine, Program in Environmental Toxicology, Irvine, CA 92697

AWARDS MS, PhD.

Faculty: 11 full-time (1 woman), 3 part-time/adjunct (0 women).
Students: 17 full-time (9 women); includes 4 minority (all Asian Americans or Pacific Islanders), 2 international. *12 applicants, 75% accepted.* Terminal master's awarded for partial completion of doctoral program.
Degree requirements: For master's, comprehensive exams required; for doctorate, dissertation, comprehensive exams required.
Entrance requirements: For master's and doctorate, GRE General Test, GRE Subject Test. *Application deadline:* For fall admission, 1/15 (priority date). Applications are processed on a rolling basis. *Application fee:* $40. Electronic applications accepted.
Expenses: Tuition, nonresident: full-time $10,322; part-time $1,720 per quarter. Required fees: $5,354; $1,300 per quarter. Tuition and fees vary according to program.
Financial aid: Fellowships, research assistantships, institutionally sponsored loans and tuition waivers (full and partial) available. Financial aid application deadline: 3/2; financial aid applicants required to submit FAFSA.
Faculty research: Inhalation/pulmonary toxicology, environmental carcinogenesis, biochemical neurotoxicology, toxicokinetics, chemical pathology.
Application contact: Jane Reimund, Administrative Assistant, 949-824-5186, *Fax:* 949-824-2793, *E-mail:* ljreimun@ uci.edu.

■ UNIVERSITY OF CALIFORNIA, LOS ANGELES

Graduate Division, School of Public Health, Department of Environmental Health Sciences, Interdepartmental Program in Molecular Toxicology, Los Angeles, CA 90095

AWARDS PhD.

Degree requirements: For doctorate, dissertation, oral and written qualifying exams required, foreign language not required.
Entrance requirements: For doctorate, GRE General Test. *Application deadline:* For fall admission, 12/15. *Application fee:* $40. Electronic applications accepted.

Expenses: Tuition, nonresident: full-time $9,804. Required fees: $4,405. Full-time tuition and fees vary according to program and student level.
Financial aid: Application deadline: 3/1. Dr. Oliver Hankinson, Chair, 310-794-9271.
Application contact: John Bulger, Student Affairs Officer, 310-206-1619, *Fax:* 310-794-2106.
Find an in-depth description at www.petersons.com/graduate.

■ UNIVERSITY OF CALIFORNIA, RIVERSIDE

Graduate Division, College of Natural and Agricultural Sciences, Program in Environmental Toxicology, Riverside, CA 92521-0102

AWARDS MS, PhD. Part-time programs available.

Faculty: 35 full-time (10 women).
Students: 40 full-time (22 women); includes 7 minority (6 Asian Americans or Pacific Islanders, 1 Hispanic American), 17 international. Average age 28. In 1999, 1 master's, 3 doctorates awarded. Terminal master's awarded for partial completion of doctoral program.
Degree requirements: For master's, thesis required, foreign language not required; for doctorate, dissertation, qualifying exams required, foreign language not required. *Average time to degree:* Master's–3 years full-time; doctorate–5 years full-time.
Entrance requirements: For master's and doctorate, GRE General Test, TOEFL, minimum GPA of 3.2. *Application deadline:* For fall admission, 2/1 (priority date); for winter admission, 9/1; for spring admission, 12/1. Applications are processed on a rolling basis. *Application fee:* $40. Electronic applications accepted.
Expenses: Tuition, nonresident: full-time $9,804. Required fees: $4,758. Full-time tuition and fees vary according to program.
Financial aid: Fellowships, research assistantships, teaching assistantships, career-related internships or fieldwork, Federal Work-Study, institutionally sponsored loans, and tuition waivers (full and partial) available. Financial aid application deadline: 2/1; financial aid applicants required to submit FAFSA.
Faculty research: Cellular/molecular toxicology, atmospheric chemistry, bioremediation, carcinogenesis, mechanism of toxicity, biochemical toxicology, chemical toxicology. *Total annual research expenditures:* $2 million.
Dr. David Eastmond, Director, 909-787-4497, *Fax:* 909-787-3087, *E-mail:* eastmond@ucrac1.ucr.edu.

Application contact: Gladis Berkowitz, Student Affairs Officer, 800-735-0717, *Fax:* 909-787-5517, *E-mail:* gladis@ mail.ucr.edu.

Find an in-depth description at www.petersons.com/graduate.

■ UNIVERSITY OF CINCINNATI

Division of Research and Advanced Studies, College of Medicine, Graduate Programs in Medicine, Department of Environmental Health, Cincinnati, OH 45267

AWARDS Environmental and industrial hygiene (MS); environmental and occupational medicine (MS); environmental health (PhD); environmental hygiene science and engineering (MS, PhD); epidemiology and biostatistics (MS); occupational safety (MS); toxicology (MS, PhD).

Faculty: 20 full-time.
Students: 69 full-time (34 women), 66 part-time (32 women); includes 29 minority (16 African Americans, 12 Asian Americans or Pacific Islanders, 1 Hispanic American), 31 international. *115 applicants, 40% accepted.* In 1999, 20 master's, 9 doctorates awarded. Terminal master's awarded for partial completion of doctoral program.
Degree requirements: For master's, thesis required, foreign language not required; for doctorate, one foreign language, dissertation, qualifying exam required. *Average time to degree:* Master's–3.7 years full-time; doctorate–6.7 years full-time.
Entrance requirements: For master's, GRE General Test, TOEFL, bachelor's degree in science; for doctorate, GRE General Test, TOEFL. *Application deadline:* For fall admission, 2/1 (priority date). Applications are processed on a rolling basis. *Application fee:* $30.
Expenses: Tuition, state resident: full-time $5,139; part-time $196 per credit hour. Tuition, nonresident: full-time $10,326; part-time $369 per credit hour. Required fees: $561; $187 per quarter.
Financial aid: Career-related internships or fieldwork, Federal Work-Study, tuition waivers (full), and unspecified assistantships available. Financial aid application deadline: 5/1.
Faculty research: Carcinogens and mutagenesis, pulmonary studies, reproduction and development. *Total annual research expenditures:* $15.1 million.
Dr. Marshall W. Anderson, Chairman, 513-558-5701, *Fax:* 513-558-4397, *E-mail:* marshall.anderson@uc.edu.

University of Cincinnati (continued)
Application contact: Judy Jarrell, Graduate Program Director, 513-558-1729, *Fax:* 513-558-4397, *E-mail:* judy.jarrell@uc.edu.

Find an in-depth description at www.petersons.com/graduate.

■ UNIVERSITY OF CINCINNATI

Division of Research and Advanced Studies, College of Medicine, Graduate Programs in Medicine, Department of Pediatrics, Cincinnati, OH 45221-0091

AWARDS Molecular and developmental biology (MS, PhD), including molecular and developmental biology, teratology.

Faculty: 6 full-time.
Students: 45 full-time (30 women), 5 part-time (4 women); includes 6 minority (3 African Americans, 3 Asian Americans or Pacific Islanders), 8 international. *213 applicants, 5% accepted.* In 1999, 9 master's, 9 doctorates awarded. Terminal master's awarded for partial completion of doctoral program.
Degree requirements: For master's, thesis required, foreign language not required; for doctorate, dissertation, qualifying exam required, foreign language not required. *Average time to degree:* Master's–2.7 years full-time; doctorate–5.9 years full-time.
Entrance requirements: For master's and doctorate, GRE General Test, TOEFL. *Application deadline:* For fall admission, 2/1 (priority date). Applications are processed on a rolling basis. *Application fee:* $30.
Expenses: Tuition, state resident: full-time $5,139; part-time $196 per credit hour. Tuition, nonresident: full-time $10,326; part-time $369 per credit hour. Required fees: $561; $187 per quarter.
Financial aid: Tuition waivers (full) and unspecified assistantships available. Financial aid application deadline: 5/1.
Dr. Thomas Boat, Chair, 513-636-4588, *Fax:* 513-636-8453, *E-mail:* thomas.boat@chmcc.org.

■ UNIVERSITY OF CINCINNATI

Division of Research and Advanced Studies, College of Medicine, Graduate Programs in Medicine, Department of Pediatrics, Program in Molecular and Developmental Biology, Program in Teratology, Cincinnati, OH 45221-0091

AWARDS MS, PhD.

Degree requirements: For master's, thesis required, foreign language not required; for doctorate, dissertation, qualifying exam required, foreign language not required.

Entrance requirements: For master's and doctorate, GRE General Test, GRE Subject Test (biology or chemistry), TOEFL. *Application deadline:* For fall admission, 2/1 (priority date). Applications are processed on a rolling basis. *Application fee:* $30.
Expenses: Tuition, state resident: full-time $5,139; part-time $196 per credit hour. Tuition, nonresident: full-time $10,326; part-time $369 per credit hour. Required fees: $561; $187 per quarter.
Financial aid: Application deadline: 5/1.
Application contact: Dan Wigginton, Graduate Program Director, 513-636-4547, *Fax:* 513-559-4317, *E-mail:* dan.wigginton@chmcc.org.

Find an in-depth description at www.petersons.com/graduate.

■ UNIVERSITY OF COLORADO HEALTH SCIENCES CENTER

Graduate School, Programs in Pharmacy, Denver, CO 80262

AWARDS Pharmaceutical sciences (PhD); toxicology (PhD).

Degree requirements: For doctorate, dissertation required, foreign language not required.

Entrance requirements: For doctorate, GRE General Test, minimum GPA of 2.75.
Expenses: Tuition, state resident: full-time $1,512; part-time $56 per hour. Tuition, nonresident: full-time $7,209; part-time $267 per hour. Full-time tuition and fees vary according to course load and program.

Find an in-depth description at www.petersons.com/graduate.

■ UNIVERSITY OF CONNECTICUT

Graduate School, School of Pharmacy, Graduate Program in Pharmacology and Toxicology, Storrs, CT 06269

AWARDS MS, PhD. Terminal master's awarded for partial completion of doctoral program.

Degree requirements: For master's and doctorate, thesis/dissertation required.
Entrance requirements: For master's and doctorate, GRE General Test.
Expenses: Tuition, state resident: full-time $5,118. Tuition, nonresident: full-time $13,298. Required fees: $1,022.

Find an in-depth description at www.petersons.com/graduate.

■ UNIVERSITY OF FLORIDA

College of Veterinary Medicine, Graduate Program in Veterinary Medical Sciences, Gainesville, FL 32611

AWARDS Forensic toxicology (Certificate); veterinary medical science (MS, PhD). Postbaccalaureate distance learning degree programs offered (no on-campus study).

Faculty: 99 full-time (22 women), 1 part-time/adjunct (0 women).
Students: 61 full-time (33 women), 5 part-time (3 women); includes 6 minority (1 African American, 1 Asian American or Pacific Islander, 4 Hispanic Americans), 21 international. Average age 29. *22 applicants, 32% accepted.* In 1999, 12 master's awarded (59% found work related to degree, 34% continued full-time study); 4 doctorates awarded (25% entered university research/teaching, 75% found other work related to degree). Terminal master's awarded for partial completion of doctoral program.
Degree requirements: For master's and doctorate, thesis/dissertation required, foreign language not required. *Average time to degree:* Master's–3 years full-time, 5 years part-time; doctorate–5 years full-time, 7 years part-time.
Entrance requirements: For master's and doctorate, GRE General Test, minimum GPA of 3.0. *Application deadline:* For fall admission, 6/9 (priority date); for spring admission, 10/1 (priority date). Applications are processed on a rolling basis. *Application fee:* $20. Electronic applications accepted.
Expenses: Tuition, state resident: part-time $119 per credit hour. Tuition, nonresident: part-time $462 per credit hour. Required fees: $26 per credit hour. Tuition and fees vary according to program.
Financial aid: In 1999–00, 57 students received aid, including 1 fellowship with partial tuition reimbursement available (averaging $14,000 per year), 32 research assistantships with partial tuition reimbursements available (averaging $15,623 per year), 14 teaching assistantships with partial tuition reimbursements available (averaging $15,921 per year); institutionally sponsored loans also available.
Faculty research: Infectious diseases, pathology, physiology, toxicology, clinical sciences. *Total annual research expenditures:* $4.5 million.
Dr. Charles H. Courtney, Associate Dean for Research and Graduate Studies, 352-392-4700 Ext. 5100, *Fax:* 352-392-8351, *E-mail:* courtneyc@mail.vetmed.ufl.edu.
Application contact: Sally O'Connell, Program Assistant, 352-392-4700 Ext.

5100, *Fax:* 352-392-8351, *E-mail:* oconnells@mail.vetmed.ufl.edu.

Find an in-depth description at www.petersons.com/graduate.

■ UNIVERSITY OF GEORGIA

College of Pharmacy, Department of Pharmaceutical and Biomedical Sciences, Athens, GA 30602

AWARDS Medicinal chemistry (MS, PhD); pharmaceutics (MS, PhD); pharmacology (MS, PhD); toxicology (MS, PhD). Terminal master's awarded for partial completion of doctoral program.

Degree requirements: For master's, thesis required, foreign language not required; for doctorate, one foreign language (computer language can substitute), dissertation required.

Entrance requirements: For master's and doctorate, GRE General Test, minimum GPA of 3.0.

Expenses: Tuition, state resident: full-time $7,516; part-time $431 per credit hour. Tuition, nonresident: full-time $12,204; part-time $793 per credit hour. Tuition and fees vary according to program.

Find an in-depth description at www.petersons.com/graduate.

■ UNIVERSITY OF GEORGIA

College of Veterinary Medicine and Graduate School, Graduate Programs in Veterinary Medicine, Interdisciplinary Graduate Program in Toxicology, Athens, GA 30602

AWARDS MS, PhD.

Degree requirements: For master's, thesis required, foreign language not required; for doctorate, one foreign language (computer language can substitute), dissertation required.

Entrance requirements: For master's and doctorate, GRE General Test. Electronic applications accepted.

Expenses: Tuition, state resident: full-time $7,516; part-time $431 per credit hour. Tuition, nonresident: full-time $12,204; part-time $793 per credit hour.

Find an in-depth description at www.petersons.com/graduate.

■ UNIVERSITY OF KANSAS

Graduate School, School of Pharmacy, Department of Pharmacology and Toxicology, Lawrence, KS 66045

AWARDS Pharmacology (MS, PhD); toxicology (MS, PhD).

Faculty: 7.

Students: 12 full-time (7 women), 2 part-time (1 woman), 8 international. Average age 26. *10 applicants, 40% accepted.* In 1999, 1 master's awarded. Terminal

master's awarded for partial completion of doctoral program.

Degree requirements: For master's, thesis required; for doctorate, dissertation, oral comprehensive exams required.

Entrance requirements: For master's, GRE General Test, TSE, bachelor's degree in related field; for doctorate, GRE General Test, MAT, TOEFL, bachelor's degree in related field. *Application deadline:* For fall admission, 2/1 (priority date). Applications are processed on a rolling basis. *Application fee:* $20.

Expenses: Tuition, state resident: full-time $2,482; part-time $103 per credit hour. Tuition, nonresident: full-time $8,104; part-time $338 per credit hour. Required fees: $428; $31 per credit hour. Tuition and fees vary according to program.

Financial aid: In 1999–00, 8 fellowships (averaging $12,500 per year), 11 research assistantships (averaging $15,000 per year), 1 teaching assistantship (averaging $15,000 per year) were awarded. Financial aid application deadline: 1/15.

Faculty research: Molecular neurobiology, gene regulation, neurotransmitter receptors. *Total annual research expenditures:* $1.8 million.

Elias Michaelis, Chair, 785-864-4001, *Fax:* 785-864-5219.

Application contact: Mary Michaelis, Graduate Director, 785-864-3905, *Fax:* 785-864-5219, *E-mail:* mlm@smissman.hbc.ukans.edu.

■ UNIVERSITY OF KANSAS

Graduate Studies Medical Center, Graduate Programs in Biomedical and Basic Sciences, Department of Pharmacology, Toxicology and Therapeutics, Curriculum in Toxicology, Kansas City, KS 66160

AWARDS MS, PhD, MD/MS, MD/PhD. Part-time programs available.

Students: 5 full-time (2 women), 3 part-time (2 women), 2 international. Average age 29. *0 applicants, 0% accepted.* In 1999, 2 doctorates awarded. Terminal master's awarded for partial completion of doctoral program.

Degree requirements: For master's, thesis, comprehensive oral exam required; for doctorate, one foreign language (computer language can substitute), dissertation, comprehensive oral and written exams required.

Entrance requirements: For master's, GRE General Test, TOEFL, TSE; for doctorate, GRE General Test, TOEFL, TSE (. *Application deadline:* For fall admission, 1/31 (priority date). Applications are processed on a rolling basis. *Application fee:* $0. Electronic applications accepted.

Expenses: Tuition, state resident: full-time $2,482; part-time $103 per credit hour.

Tuition, nonresident: full-time $8,104; part-time $338 per credit hour. Required fees: $428; $31 per credit hour. Tuition and fees vary according to program.

Financial aid: Fellowships, research assistantships, teaching assistantships, Federal Work-Study, institutionally sponsored loans, and scholarships available. Aid available to part-time students. Financial aid application deadline: 3/31; financial aid applicants required to submit FAFSA.

Application contact: Beth Levant, Director of Admissions, 913-588-7527, *Fax:* 913-588-7501, *E-mail:* blevant@kumc.edu.

Find an in-depth description at www.petersons.com/graduate.

■ UNIVERSITY OF KENTUCKY

Graduate School, Program in Toxicology, Lexington, KY 40506-0032

AWARDS MS, PhD. Terminal master's awarded for partial completion of doctoral program.

Degree requirements: For master's, comprehensive exam required, thesis optional, foreign language not required; for doctorate, dissertation, comprehensive exam required, foreign language not required.

Entrance requirements: For master's, GRE General Test, minimum undergraduate GPA of 3.0; for doctorate, GRE General Test, minimum graduate GPA of 3.0.

Expenses: Tuition, state resident: full-time $3,596; part-time $188 per credit hour. Tuition, nonresident: full-time $10,116; part-time $550 per credit hour.

Faculty research: Chemical carcinogenesis, immunotoxicology, neurotoxicology, metabolism and disposition, gene regulation.

Find an in-depth description at www.petersons.com/graduate.

■ UNIVERSITY OF LOUISVILLE

School of Medicine and Graduate School, Integrated Programs in Biomedical Sciences, Department of Pharmacology and Toxicology, Louisville, KY 40292-0001

AWARDS MS, PhD.

Degree requirements: For master's and doctorate, thesis/dissertation required.

Entrance requirements: For master's and doctorate, GRE General Test.

Expenses: Tuition, state resident: full-time $3,260; part-time $182 per hour. Tuition, nonresident: full-time $9,780; part-time $544 per hour. Required fees: $143; $28 per hour.

Faculty research: Metabolic activation of chemical carcinogens, biochemical

University of Louisville (continued)
neuropharmacology, analytical toxicology and kinetics of chemical disposition, biochemical toxicology, clinical pharmacology.

■ UNIVERSITY OF MARYLAND

Graduate School, Graduate Programs in Medicine, Department of Pharmacology and Experimental Therapeutics, Baltimore, MD 21201-1627

AWARDS Pharmacology (PhD); pharmacology and experimental therapeutics (MS); toxicology (PhD). Part-time programs available. Terminal master's awarded for partial completion of doctoral program.

Degree requirements: For master's and doctorate, thesis/dissertation required, foreign language not required.
Entrance requirements: For master's and doctorate, GRE General Test, TOEFL, minimum GPA of 3.0.
Expenses: Tuition, state resident: part-time $261 per credit hour. Tuition, nonresident: part-time $468 per credit hour. Tuition and fees vary according to program.
Faculty research: Molecular and cellular neuropharmacology, drug and endocrine mechanisms in cancer, viral immunopharmacology.

Find an in-depth description at www.petersons.com/graduate.

■ UNIVERSITY OF MARYLAND

Graduate School, Graduate Programs in Medicine, Department of Toxicology, Baltimore, MD 21201-1627
AWARDS MS, PhD, MD/PhD. Part-time programs available.

Degree requirements: For master's, foreign language and thesis not required; for doctorate, dissertation required, foreign language not required.
Entrance requirements: For master's and doctorate, GRE General Test, GRE Subject Test, TOEFL, minimum GPA of 3.0.
Expenses: Tuition, state resident: part-time $261 per credit hour. Tuition, nonresident: part-time $468 per credit hour. Tuition and fees vary according to program.

■ UNIVERSITY OF MARYLAND, COLLEGE PARK

Graduate Studies and Research, College of Life Sciences, Toxicology Program, College Park, MD 20742
AWARDS MS, PhD.

Students: *8 applicants, 0% accepted.*

Degree requirements: For master's, thesis required, foreign language not required; for doctorate, dissertation required.
Entrance requirements: For master's and doctorate, GRE General Test, minimum GPA of 3.0. *Application deadline:* For fall admission, 5/1; for spring admission, 11/1. Applications are processed on a rolling basis. *Application fee:* $50 ($70 for international students). Electronic applications accepted.
Expenses: Tuition, state resident: part-time $272 per credit hour. Tuition, nonresident: part-time $415 per credit hour. Required fees: $632; $379 per year.
Financial aid: Teaching assistantships available. Financial aid applicants required to submit FAFSA.
Faculty research: Aquatic, neurological, environmental, and drug toxicology.
Dr. Judd Nelson, Coordinator, 301-405-3919.

Application contact: Trudy Lindsey, Director, Graduate Admissions and Records, 301-405-4198, *Fax:* 301-314-9305, *E-mail:* grschool@deans.umd.edu.

■ UNIVERSITY OF MARYLAND EASTERN SHORE

Graduate Programs, Department of Natural Sciences, Program in Toxicology, Princess Anne, MD 21853-1299

AWARDS MS, PhD.

Faculty: 15 full-time (2 women).
Students: *1 applicant, 0% accepted.*
Degree requirements: For master's, thesis required, foreign language not required; for doctorate, dissertation, comprehensive exams required, foreign language not required. *Average time to degree:* Master's–2 years full-time, 3.5 years part-time; doctorate–4.5 years full-time, 9 years part-time.
Entrance requirements: For master's and doctorate, GRE, TOEFL, minimum GPA of 3.0. *Application deadline:* For fall admission, 4/15 (priority date); for spring admission, 10/15 (priority date). Applications are processed on a rolling basis. *Application fee:* $30.
Expenses: Tuition, state resident: part-time $145 per credit hour. Tuition, nonresident: part-time $261 per credit hour. Required fees: $25 per semester.
Financial aid: Fellowships, research assistantships, teaching assistantships, career-related internships or fieldwork, Federal Work-Study, and grants available. Aid available to part-time students. Financial aid application deadline: 3/1.

Faculty research: Bioremediation, wastewater management, aquatic toxicology, air pollution. *Total annual research expenditures:* $702,000.
Dr. Gian Gupta, Coordinator, 410-651-6030, *Fax:* 410-651-7739, *E-mail:* ggupta@umes-bird.umd.edu.

■ UNIVERSITY OF MASSACHUSETTS WORCESTER

Graduate School of Biomedical Sciences, Department of Biochemistry and Molecular Pharmacology, Worcester, MA 01655-0115

AWARDS Biochemistry and molecular biology (PhD); pharmacology and molecular toxicology (PhD).

Faculty: 25 full-time (2 women).
Students: 12 full-time (6 women); includes 2 minority (both Asian Americans or Pacific Islanders), 3 international.
Degree requirements: For doctorate, dissertation required.
Entrance requirements: For doctorate, GRE General Test, GRE Subject Test, 1 year of calculus, physics, organic chemistry and biology. *Application deadline:* For fall admission, 1/15 (priority date). Applications are processed on a rolling basis. *Application fee:* $25 ($50 for international students).
Expenses: Tuition, state resident: full-time $2,640. Tuition, nonresident: full-time $9,756. Required fees: $825. Full-time tuition and fees vary according to program.
Financial aid: In 1999–00, research assistantships with full tuition reimbursements (averaging $17,500 per year); unspecified assistantships also available.
Faculty research: Neuropharmacology, control of DNA metabolism, clinical pharmacology and toxicology.
Dr. C. Robert Matthews, Chair, 508-856-2251, *Fax:* 508-856-2151.

Application contact: Dr. Alonzo Ross, Graduate Director, 508-856-8016, *Fax:* 508-856-2151, *E-mail:* alonzo.ross@umassmed.edu.

Find an in-depth description at www.petersons.com/graduate.

■ UNIVERSITY OF MEDICINE AND DENTISTRY OF NEW JERSEY

Graduate School of Biomedical Sciences, Graduate Programs in Biomedical Sciences, Program in Toxicology, Piscataway, NJ 08854-5635

AWARDS Environmental toxicology (MS, PhD); industrial-occupational toxicology (MS, PhD); nutritional toxicology (MS, PhD); pharmaceutical toxicology (MS, PhD).

Degree requirements: For master's and doctorate, thesis/dissertation, qualifying exam required, foreign language not required.
Entrance requirements: For master's and doctorate, GRE General Test, TOEFL. *Application deadline:* For fall admission, 2/1; for spring admission, 10/1. *Application fee:* $40.
Expenses: Tuition, state resident: part-time $270 per credit hour. Tuition, nonresident: part-time $407 per credit hour. Part-time tuition and fees vary according to campus/location and program.
Financial aid: Fellowships, research assistantships, teaching assistantships, career-related internships or fieldwork available. Financial aid application deadline: 5/1.
Faculty research: Carcinogenesis, phototoxins, food additives, neurotoxicants. *Total annual research expenditures:* $10 million.
Director, 732-445-0176, *Fax:* 732-445-0119.
Application contact: Michael Iba, Chair, Admissions Committee, 732-445-2354, *Fax:* 732-445-0119.

■ UNIVERSITY OF MICHIGAN

School of Public Health, Department of Environmental and Industrial Health, Toxicology Training Program, Ann Arbor, MI 48109
AWARDS MPH, MS, PhD. MS and PhD offered through the Horace H. Rackham School of Graduate Studies.

Degree requirements: For master's, foreign language not required; for doctorate, oral defense of dissertation, preliminary exam required.
Entrance requirements: For master's and doctorate, GRE General Test.
Expenses: Tuition, state resident: full-time $10,520. Tuition, nonresident: full-time $21,344.

Find an in-depth description at www.petersons.com/graduate.

■ UNIVERSITY OF MINNESOTA, DULUTH

Graduate School, Program in Toxicology, Duluth, MN 55812-2496
AWARDS MS, PhD.

Faculty: 24 full-time (2 women), 3 part-time/adjunct (0 women).
Students: 3 full-time (0 women), 1 international. Average age 28. *15 applicants, 7% accepted.*
Degree requirements: For doctorate, dissertation, oral preliminary and final exams required, foreign language not required.

Entrance requirements: For master's, minimum GPA of 3.0; for doctorate, GRE General Test, TOEFL. *Application deadline:* For fall admission, 7/15; for spring admission, 11/15. Applications are processed on a rolling basis. *Application fee:* $50 ($55 for international students).
Expenses: Tuition, state resident: full-time $5,040; part-time $420 per credit. Tuition, nonresident: full-time $9,900; part-time $825 per credit. Required fees: $509. Tuition and fees vary according to course load and program.
Financial aid: In 1999–00, 3 research assistantships with full tuition reimbursements (averaging $12,000 per year), 1 teaching assistantship with full tuition reimbursement (averaging $12,000 per year) were awarded; fellowships, grants, institutionally sponsored loans, and scholarships also available. Aid available to part-time students. Financial aid application deadline: 3/15.
Faculty research: Structure activity correlations, neurotoxicity, aquatic toxicology, biochemical mechanisms, immunotoxicology. *Total annual research expenditures:* $300,000.
Dr. Kendall B. Wallace, Associate Director of Graduate Studies, 218-726-7922, *Fax:* 218-726-8014, *E-mail:* kwallace@d.umn.edu.
Application contact: Dr. Mike Murphy, Director of Graduate Studies, 612-624-2232, *E-mail:* murph005@maroon.tc.umn.edu.

■ UNIVERSITY OF MINNESOTA, TWIN CITIES CAMPUS

School of Public Health, Division of Environmental and Occupational Health, Area in Environmental Toxicology, Minneapolis, MN 55455-0213
AWARDS MPH, MS, PhD.

Degree requirements: For master's, foreign language not required; for doctorate, dissertation required, foreign language not required.
Entrance requirements: For master's and doctorate, GRE General Test, minimum GPA of 3.0. *Application deadline:* For fall admission, 3/1 (priority date). Applications are processed on a rolling basis. *Application fee:* $50 ($75 for international students).
Expenses: Tuition, state resident: full-time $4,270; part-time $267 per credit. Tuition, nonresident: full-time $8,400; part-time $525 per credit. Tuition and fees vary according to program.
Financial aid: Fellowships, research assistantships available. Financial aid application deadline: 3/1.
Application contact: Kathy Soupir, Student Coordinator, 612-625-0622, *Fax:*

612-626-4837, *E-mail:* ksoupir@mail.eoh.umn.edu.

■ UNIVERSITY OF MISSISSIPPI

Graduate School, School of Pharmacy, Department of Pharmacology, Oxford, University, MS 38677
AWARDS Pharmacology (MS, PhD); toxicology (PhD).

Faculty: 7 full-time (1 woman).
Students: 15 full-time (7 women), 2 part-time (1 woman); includes 1 minority (African American), 9 international. In 1999, 3 degrees awarded.
Degree requirements: For master's and doctorate, thesis/dissertation required, foreign language not required.
Entrance requirements: For master's, GRE General Test, TOEFL, minimum GPA of 3.0; for doctorate, GRE General Test, TOEFL. *Application deadline:* For fall admission, 8/1. Applications are processed on a rolling basis. *Application fee:* $0 ($25 for international students).
Expenses: Tuition, state resident: full-time $3,053; part-time $170 per credit hour. Tuition, nonresident: full-time $6,155; part-time $342 per credit hour. Tuition and fees vary according to program.
Financial aid: Application deadline: 3/1.
Faculty research: Behavioral and biochemical pharmacology.
Dr. Dennis R. Feller, Chairman, 662-915-5148.

■ UNIVERSITY OF MISSISSIPPI MEDICAL CENTER

Graduate Programs in Biomedical Sciences, Department of Pharmacology and Toxicology, Jackson, MS 39216-4505
AWARDS Pharmacology (MS, PhD); toxicology (MS, PhD).

Faculty: 11 full-time (1 woman), 10 part-time/adjunct (0 women).
Students: 26 full-time (9 women), 2 part-time; includes 1 minority (African American), 22 international. Average age 30. *16 applicants, 50% accepted.* In 1999, 1 master's awarded (100% found work related to degree); 3 doctorates awarded (100% continued full-time study). Terminal master's awarded for partial completion of doctoral program.
Degree requirements: For master's, thesis required, foreign language not required; for doctorate, dissertation, first authored publication required, foreign language not required. *Average time to degree:* Master's–4 years full-time; doctorate–5 years full-time.
Entrance requirements: For master's and doctorate, GRE General Test, minimum GPA of 3.0. *Application deadline:* For fall

University of Mississippi Medical Center (continued)

admission, 6/1 (priority date). Applications are processed on a rolling basis. *Application fee:* $10.

Expenses: Tuition, state resident: full-time $2,378; part-time $132 per hour. Tuition, nonresident: full-time $4,697; part-time $261 per hour. Tuition and fees vary according to program.

Financial aid: In 1999–00, 22 students received aid, including 22 research assistantships (averaging $16,234 per year). Financial aid application deadline: 4/1.

Faculty research: Neuropharmacology, environmental toxicology, aging, immunopharmacology, cardiovascular pharmacology. *Total annual research expenditures:* $1.2 million.

Dr. I. K. Ho, Interim Associate Vice Chancellor, 601-984-1600, *Fax:* 601-984-1637.

Find an in-depth description at www.petersons.com/graduate.

■ **UNIVERSITY OF NEW MEXICO**

Graduate School, College of Pharmacy, Graduate Programs in Pharmaceutical Sciences, Albuquerque, NM 87131-2039

AWARDS Pharmaceutical sciences (MS, PhD), including hospital pharmacy (MS), radiopharmacy (MS), toxicology; pharmacy administration (MS, PhD). Part-time programs available.

Students: 7 full-time (4 women), 6 part-time (2 women); includes 3 minority (1 Asian American or Pacific Islander, 2 Hispanic Americans), 1 international. Average age 37. *4 applicants, 75% accepted.* In 1999, 3 degrees awarded. Terminal master's awarded for partial completion of doctoral program.

Degree requirements: For master's, thesis required (for some programs), foreign language not required; for doctorate, dissertation required, foreign language not required.

Entrance requirements: For master's, GRE General Test, Pharm D (hospital pharmacy or radiopharmacy), BS in biology or chemistry (toxicology); for doctorate, GRE General Test. *Application deadline:* For fall admission, 7/16; for spring admission, 11/13. *Application fee:* $25.

Expenses: Tuition, state resident: full-time $2,169; part-time $90 per credit hour. Tuition, nonresident: full-time $8,943; part-time $373 per credit hour. Required fees: $21 per credit hour.

Financial aid: In 1999–00, 6 research assistantships with tuition reimbursements (averaging $9,612 per year) were awarded; fellowships, teaching assistantships with tuition reimbursements, career-related internships or fieldwork, Federal Work-Study, and residencies also available. Aid available to part-time students. Financial aid application deadline: 6/1; financial aid applicants required to submit FAFSA.

Faculty research: Cancer research, molecular toxicology, developmental toxicology, the role of pharmacy in the health care system.

Dr. Scott Burchiel, Associate Dean of Graduate Studies, 505-272-0920, *Fax:* 505-272-6749, *E-mail:* burchiel@unm.edu.

Application contact: Irma Montano, Administrative Assistant to the Dean, Graduate Committee, 505-272-3241, *Fax:* 505-272-6749, *E-mail:* toxinfo@unm.edu.

■ **THE UNIVERSITY OF NORTH CAROLINA AT CHAPEL HILL**

Graduate School, Curriculum in Toxicology, Chapel Hill, NC 27599

AWARDS MS, PhD.

Faculty: 41 full-time (8 women), 31 part-time/adjunct (4 women).

Students: 41 full-time (24 women); includes 11 minority (7 African Americans, 1 Asian American or Pacific Islander, 2 Hispanic Americans, 1 Native American). *30 applicants, 37% accepted.* In 1999, 5 degrees awarded (60% entered university research/teaching, 40% found other work related to degree).

Degree requirements: For master's, thesis, comprehensive exam required, foreign language not required; for doctorate, dissertation, comprehensive exams required, foreign language not required. *Average time to degree:* Doctorate–5 years full-time.

Application deadline: For fall admission, 2/1 (priority date); for spring admission, 10/1. Applications are processed on a rolling basis. *Application fee:* $55. Electronic applications accepted.

Expenses: Tuition, state resident: full-time $1,578. Tuition, nonresident: full-time $10,744. Required fees: $827. One-time fee: $15 full-time. Tuition and fees vary according to program.

Financial aid: In 1999–00, 31 fellowships with full tuition reimbursements, 10 research assistantships with full tuition reimbursements were awarded. Financial aid application deadline: 3/1.

Faculty research: Molecular and cellular toxicology, carcinogenesis, neurotoxicology, pulmonary toxicology.

Dr. James A. Swenberg, Director.

Application contact: Dr. Gary J. Smith, Director of Graduate Admissions, 919-966-2699, *Fax:* 919-966-6357, *E-mail:* toxicology@unc.edu.

Find an in-depth description at www.petersons.com/graduate.

■ **UNIVERSITY OF NORTH DAKOTA**

School of Medicine and Graduate School, Graduate Programs in Medicine, Department of Pharmacology and Toxicology, Grand Forks, ND 58202

AWARDS MS, PhD.

Faculty: 8 full-time (2 women).

Students: 4 full-time (1 woman). *11 applicants, 55% accepted.* In 1999, 2 doctorates awarded.

Degree requirements: For master's, thesis, final examination required, foreign language not required; for doctorate, dissertation, comprehensive examination, final examination required, foreign language not required.

Entrance requirements: For master's, GRE General Test or MCAT, TOEFL, minimum GPA of 3.0; for doctorate, GRE General Test or MCAT, TOEFL, minimum GPA of 3.5. *Application deadline:* For fall admission, 3/1 (priority date). Applications are processed on a rolling basis. *Application fee:* $25.

Expenses: Tuition, state resident: full-time $2,690; part-time $112 per credit. Tuition, nonresident: full-time $7,182; part-time $299 per credit. Required fees: $46 per semester.

Financial aid: In 1999–00, 4 students received aid, including 4 research assistantships with full tuition reimbursements available (averaging $10,586 per year); fellowships, teaching assistantships with full tuition reimbursements available, Federal Work-Study, institutionally sponsored loans, scholarships, and tuition waivers (full and partial) also available. Aid available to part-time students. Financial aid application deadline: 3/15; financial aid applicants required to submit FAFSA.

Faculty research: Molecular and cellular aspects of pharmacology, endrocrinology, genetics, cancer.

Dr. Mike Ebadi, Chairperson, 701-777-4293, *Fax:* 701-777-6124, *E-mail:* mebadi@medicine.nodak.edu.

■ **UNIVERSITY OF PUERTO RICO, MEDICAL SCIENCES CAMPUS**

School of Medicine, Division of Graduate Studies, Department of Pharmacology and Toxicology, San Juan, PR 00936-5067

AWARDS MS, PhD.

Faculty: 8 full-time (4 women).

Students: 9 full-time (6 women); all minorities (all Hispanic Americans). Average age 30. *5 applicants, 60% accepted.* In 1999, 1 master's awarded (100% found work related to degree).

Degree requirements: For master's, one foreign language, thesis required; for doctorate, one foreign language, computer language, dissertation, qualifying exam required. *Average time to degree:* Master's–4 years full-time; doctorate–6 years full-time.
Entrance requirements: For master's and doctorate, GRE General Test, GRE Subject Test, interview. *Application deadline:* For fall admission, 2/15. *Application fee:* $15.
Expenses: Tuition, state resident: full-time $5,500. Tuition, nonresident: full-time $8,400. Required fees: $600. Tuition and fees vary according to class time, course load, degree level and program.
Financial aid: In 1999–00, 4 research assistantships were awarded; fellowships, Federal Work-Study, institutionally sponsored loans, and tuition waivers (full and partial) also available. Aid available to part-time students. Financial aid application deadline: 4/30.
Faculty research: Cardiovascular, central nervous system, and endocrine pharmacology; anti-cancer drugs; toxicology, sodium pump. *Total annual research expenditures:* $323,000.
Dr. Walmor C. De Mello, Director, 787-766-4441, *Fax:* 787-282-0568.
Application contact: Dr. Susan Corey, Coordinator, 787-724-1006, *Fax:* 787-725-3804, *E-mail:* s_corey@rcmaca.upr.clu.edu.

■ UNIVERSITY OF RHODE ISLAND

Graduate School, College of Pharmacy, Graduate Programs in Pharmacy, Department of Pharmacology and Toxicology, Kingston, RI 02881
AWARDS MS, PhD.

Application deadline: For fall admission, 4/15. *Application fee:* $35.
Expenses: Tuition, state resident: full-time $3,540; part-time $197 per credit. Tuition, nonresident: full-time $10,116; part-time $197 per credit. Required fees: $1,352; $37 per credit. $65 per term.
Dr. Zahir A. Shaikh, Chairperson, 401-874-2362.

■ UNIVERSITY OF ROCHESTER

School of Medicine and Dentistry, Graduate Programs in Medicine and Dentistry, Department of Environmental Medicine, Program in Toxicology, Rochester, NY 14627-0250
AWARDS MS, PhD.

Students: 29 full-time (12 women); includes 3 minority (2 African Americans, 1 Hispanic American). In 1999, 4 master's,

4 doctorates awarded. Terminal master's awarded for partial completion of doctoral program.
Degree requirements: For master's, foreign language not required; for doctorate, dissertation, qualifying exam required, foreign language not required.
Entrance requirements: For master's and doctorate, GRE General Test. *Application deadline:* For fall admission, 2/1. *Application fee:* $25.
Expenses: Tuition: Part-time $697 per credit hour. Tuition and fees vary according to program.
Financial aid: Fellowships, research assistantships, teaching assistantships, tuition waivers (full and partial) and training grant fellowships available. Financial aid application deadline: 2/1.
Dr. Ned Ballatori, Director, 716-275-0262.
Application contact: Joyce Morgan, Graduate Program Secretary, 716-275-6702.

Find an in-depth description at www.petersons.com/graduate.

■ UNIVERSITY OF SOUTHERN CALIFORNIA

School of Pharmacy and Graduate School, Graduate Programs in Pharmacy, Graduate Program in Molecular Pharmacology and Toxicology, Los Angeles, CA 90089
AWARDS MS, PhD, Pharm D/PhD.

Faculty: 9 full-time (3 women).
Students: 21 full-time (13 women), 2 part-time (1 woman); includes 6 minority (all Asian Americans or Pacific Islanders), 12 international. Average age 28. *44 applicants, 7% accepted.* In 1999, 1 master's, 5 doctorates awarded.
Degree requirements: For master's and doctorate, thesis/dissertation required.
Entrance requirements: For master's and doctorate, GRE General Test. *Application deadline:* For fall admission, 2/1 (priority date); for spring admission, 10/15. Applications are processed on a rolling basis. *Application fee:* $55.
Expenses: Tuition: Full-time $22,198; part-time $748 per unit. Required fees: $406.
Financial aid: In 1999–00, 9 fellowships with full tuition reimbursements (averaging $18,200 per year), 9 research assistantships with full tuition reimbursements (averaging $18,200 per year), 6 teaching assistantships with full tuition reimbursements (averaging $17,800 per year) were awarded; Federal Work-Study, institutionally sponsored loans, and scholarships also available. Aid available to part-time students. Financial aid application deadline: 2/15; financial aid applicants required to submit FAFSA.

Dr. Enrique Cadenas, Chair, 323-442-1551.
Application contact: 323-442-1474.

■ THE UNIVERSITY OF TEXAS–HOUSTON HEALTH SCIENCE CENTER

Graduate School of Biomedical Sciences, Program in Toxicology, Houston, TX 77225-0036
AWARDS MS, PhD, MD/PhD.

Faculty: 23 full-time (6 women).
Students: 5 full-time (3 women); includes 1 minority (African American). Average age 27. *13 applicants, 38% accepted.* In 1999, 1 master's, 1 doctorate awarded. Terminal master's awarded for partial completion of doctoral program.
Degree requirements: For master's and doctorate, thesis/dissertation required, foreign language not required.
Entrance requirements: For master's and doctorate, GRE General Test, TOEFL, TWE. *Application deadline:* For fall admission, 1/15 (priority date); for spring admission, 11/1. Applications are processed on a rolling basis. *Application fee:* $10. Electronic applications accepted.
Financial aid: Fellowships, research assistantships, grants and institutionally sponsored loans available. Financial aid application deadline: 1/15.
Faculty research: Apoptosis, carcinogenesis, chemotherapy, immunotoxicology, molecular and cellular mechanisms of injury.
Dr. David J. McConkey, Director, 713-792-8591, *Fax:* 713-792-8747, *E-mail:* dmcconke@notes.mdacc.tmc.edu.
Application contact: Anne Baronitis, Director of Admissions, 713-500-9860, *Fax:* 713-500-9877, *E-mail:* abaron@gsbs.gs.uth.tmc.edu.

Find an in-depth description at www.petersons.com/graduate.

■ THE UNIVERSITY OF TEXAS MEDICAL BRANCH AT GALVESTON

Graduate School of Biomedical Sciences, Program in Pharmacology and Toxicology, Curriculum in Toxicology, Galveston, TX 77555
AWARDS PhD.

Faculty: 5 full-time (2 women).
Degree requirements: For doctorate, dissertation, qualifying exams required, foreign language not required.
Entrance requirements: For doctorate, GRE General Test, minimum GPA of 3.0. *Application deadline:* For fall admission, 4/15 (priority date). *Application fee:* $25

The University of Texas Medical Branch at Galveston (continued)

($50 for international students). Electronic applications accepted.

Expenses: Tuition, state resident: full-time $684; part-time $38 per credit hour. Tuition, nonresident: full-time $4,572; part-time $254 per credit hour. Required fees: $29; $7.5 per credit hour. One-time fee: $55. Tuition and fees vary according to program.

Financial aid: Federal Work-Study, institutionally sponsored loans, and unspecified assistantships available. Financial aid applicants required to submit FAFSA.

Faculty research: P450, estrogen, cancer, PKC.

Dr. Mary T. Moslen, Director, 409-772-2665, *E-mail:* mmoslen@utmb.edu.

Find an in-depth description at www.petersons.com/graduate.

■ UNIVERSITY OF THE SCIENCES IN PHILADELPHIA

College of Graduate Studies, Program in Pharmacology and Toxicology, Philadelphia, PA 19104-4495

AWARDS Pharmacology (MS, PhD); toxicology (MS, PhD). Part-time programs available.

Faculty: 6 full-time (3 women).
Students: 9 full-time (8 women), 1 part-time; includes 1 minority (Asian American or Pacific Islander), 4 international. *26 applicants, 12% accepted.* In 1999, 2 doctorates awarded. Terminal master's awarded for partial completion of doctoral program.
Degree requirements: For master's and doctorate, thesis/dissertation required, foreign language not required.
Entrance requirements: For master's and doctorate, GRE General Test, TOEFL. *Application deadline:* For fall admission, 5/1. Applications are processed on a rolling basis. *Application fee:* $45.
Expenses: Tuition: Full-time $29,600; part-time $612 per credit. Tuition and fees vary according to program.
Financial aid: In 1999–00, 9 teaching assistantships with tuition reimbursements (averaging $12,500 per year) were awarded; fellowships, research assistantships, institutionally sponsored loans and tuition waivers (full) also available. Financial aid application deadline: 5/1.
Faculty research: Autonomic, cardiovascular, cellular, and molecular pharmacology; mechanisms of carcinogenesis; drug metabolism. *Total annual research expenditures:* $460,206.
Dr. Raymond F. Orzechowski, Director, 215-596-8825, *E-mail:* r.orzech@usip.edu.

Application contact: Dr. Rodney J. Wigent, Dean, College of Graduate Studies, 215-596-8937, *Fax:* 215-596-8764, *E-mail:* graduate@usip.edu.

Find an in-depth description at www.petersons.com/graduate.

■ UNIVERSITY OF UTAH

College of Pharmacy and Graduate School, Graduate Programs in Pharmacy, Department of Pharmacology and Toxicology, Salt Lake City, UT 84112-1107

AWARDS MS, PhD, MD/PhD. Terminal master's awarded for partial completion of doctoral program.

Degree requirements: For master's, thesis required; for doctorate, dissertation, final exam required, foreign language not required.
Entrance requirements: For master's, GRE, TOEFL; for doctorate, GRE Subject Test, TOEFL, BS in biology, chemistry, or pharmacy.
Expenses: Tuition, state resident: full-time $1,663. Tuition, nonresident: full-time $5,201. Tuition and fees vary according to course load and program.
Faculty research: Neuropharmacology, neurochemistry, biochemistry, molecular pharmacology, analytical chemistry.

Find an in-depth description at www.petersons.com/graduate.

■ UNIVERSITY OF WASHINGTON

Graduate School, School of Public Health and Community Medicine, Department of Environmental Health, Seattle, WA 98195

AWARDS Industrial hygiene (PhD); industrial hygiene and safety (MS); occupational medicine (MPH); preventive medicine (MPH); technology (MS); toxicology (MS, PhD). Part-time programs available.

Faculty: 35 full-time (9 women), 10 part-time/adjunct (1 woman).
Students: 64 full-time (41 women), 4 part-time (3 women); includes 10 minority (6 Asian Americans or Pacific Islanders, 3 Hispanic Americans, 1 Native American), 12 international. *85 applicants, 47% accepted.* In 1999, 17 master's awarded (100% found work related to degree); 2 doctorates awarded (100% found work related to degree). Terminal master's awarded for partial completion of doctoral program.
Degree requirements: For master's and doctorate, thesis/dissertation required, foreign language not required. *Average time to degree:* Master's–2 years full-time; doctorate–5 years full-time.
Entrance requirements: For master's and doctorate, GRE General Test, TOEFL,

minimum GPA of 3.0. *Application deadline:* For fall admission, 2/1 (priority date). Applications are processed on a rolling basis. *Application fee:* $50. Electronic applications accepted.

Expenses: Tuition, state resident: full-time $5,196; part-time $495 per credit. Tuition, nonresident: full-time $13,485; part-time $1,285 per credit. Required fees: $387; $36 per credit. Tuition and fees vary according to course load and program.

Financial aid: In 1999–00, 25 fellowships with tuition reimbursements (averaging $14,688 per year), 30 research assistantships with tuition reimbursements (averaging $13,920 per year), 3 teaching assistantships with tuition reimbursements (averaging $13,920 per year) were awarded; career-related internships or fieldwork, Federal Work-Study, institutionally sponsored loans, and traineeships also available. Financial aid application deadline: 2/28.

Faculty research: Developmental toxicology, biochemical toxicology, exposure assessment, hazardous waste, industrial chemistry. *Total annual research expenditures:* $15.5 million.

Dr. David Kalman, Interim Chair, 206-543-8342, *Fax:* 206-616-0477.

Application contact: Rory A. Murphy, Manager, Student Services, 206-543-3199, *Fax:* 206-543-9616, *E-mail:* ehgrad@u.washington.edu.

Find an in-depth description at www.petersons.com/graduate.

■ UNIVERSITY OF WISCONSIN–MADISON

Graduate School, College of Agricultural and Life Sciences, Department of Animal Health and Biomedical Sciences, Program in Comparative Biosciences, Madison, WI 53706-1380

AWARDS Anatomy (MS, PhD); biochemistry (MS, PhD); cellular and molecular biology (MS, PhD); environmental toxicology (MS, PhD); neurosciences (MS, PhD); pharmacology (MS, PhD); physiology (MS, PhD).

Degree requirements: For doctorate, dissertation required.
Expenses: Tuition, state resident: full-time $5,406; part-time $339 per credit. Tuition, nonresident: full-time $17,110; part-time $1,071 per credit. Full-time tuition and fees vary according to program and reciprocity agreements. Part-time tuition and fees vary according to course load and program.

■ UNIVERSITY OF WISCONSIN–MADISON

Graduate School, College of Agricultural and Life Sciences, Environmental Toxicology Center, Madison, WI 53706-1380

AWARDS Molecular and environmental toxicology (MS, PhD).

Degree requirements: For doctorate, dissertation required.

Expenses: Tuition, state resident: full-time $5,406; part-time $339 per credit. Tuition, nonresident: full-time $17,110; part-time $1,071 per credit. Full-time tuition and fees vary according to program and reciprocity agreements. Part-time tuition and fees vary according to course load and program.

Find an in-depth description at www.petersons.com/graduate.

■ UTAH STATE UNIVERSITY

School of Graduate Studies, College of Agriculture, Program in Toxicology, Logan, UT 84322

AWARDS Molecular biology (MS, PhD); toxicology (MS, PhD).

Faculty: 14 full-time (2 women), 3 part-time/adjunct (0 women).

Students: 9 full-time (1 woman), 1 (woman) part-time. Average age 25. *9 applicants, 33% accepted.* In 1999, 2 doctorates awarded. Terminal master's awarded for partial completion of doctoral program.

Degree requirements: For master's and doctorate, thesis/dissertation required, foreign language not required.

Entrance requirements: For master's and doctorate, GRE General Test, TOEFL, minimum GPA of 3.0. *Application deadline:* For fall admission, 6/15 (priority date); for spring admission, 10/15. Applications are processed on a rolling basis. *Application fee:* $40.

Expenses: Tuition, state resident: full-time $1,553. Tuition, nonresident: full-time $5,436. International tuition: $5,526 full-time. Required fees: $447. Tuition and fees vary according to course load and program.

Financial aid: In 1999–00, 1 fellowship with partial tuition reimbursement (averaging $15,000 per year), 5 research assistantships with partial tuition reimbursements (averaging $15,000 per year) were awarded; teaching assistantships with partial tuition reimbursements, Federal Work-Study, institutionally sponsored loans, and tuition waivers (partial) also available. Aid available to part-time students.

Faculty research: Free-radical mechanisms, toxicity of iron, carcinogenesis of natural compounds, molecular mechanisms of retinoid toxicity, aflatoxins.

Roger A. Coulombe, Director, 435-797-1598, *Fax:* 435-797-1601, *E-mail:* rogerc@cc.usu.edu.

■ VIRGINIA COMMONWEALTH UNIVERSITY

School of Graduate Studies and School of Medicine, School of Medicine Graduate Programs, Department of Pharmacology and Toxicology, Richmond, VA 23284-9005

AWARDS Molecular biology and genetics (PhD); neurosciences (PhD); pharmacology (PhD, CBHS); pharmacology and toxicology (MS).

Students: 14 full-time (5 women), 51 part-time (27 women); includes 27 minority (4 African Americans, 22 Asian Americans or Pacific Islanders, 1 Hispanic American). In 1999, 3 master's, 10 doctorates, 1 other advanced degree awarded. Terminal master's awarded for partial completion of doctoral program.

Degree requirements: For master's, thesis required, foreign language not required; for doctorate, dissertation, comprehensive oral and written exams required, foreign language not required.

Entrance requirements: For master's, DAT, GRE General Test or MCAT; for doctorate, GRE General Test. *Application deadline:* For fall admission, 2/1 (priority date). *Application fee:* $30.

Expenses: Tuition, state resident: full-time $4,031; part-time $224 per credit hour. Tuition, nonresident: full-time $11,946; part-time $664 per credit hour. Required fees: $1,081; $40 per credit hour. Tuition and fees vary according to campus/location and program.

Financial aid: Fellowships, teaching assistantships available.

Faculty research: Drug abuse, drug metabolism, pharmacodynamics, peptide synthesis, receptor mechanisms.

Dr. George Kunos, Chair, 804-828-2073, *Fax:* 804-828-2117.

Application contact: Sheryol Cox, Graduate Program Coordinator, 804-828-8400, *Fax:* 804-828-2117, *E-mail:* swcox@vcu.edu.

Find an in-depth description at www.petersons.com/graduate.

■ WASHINGTON STATE UNIVERSITY

Graduate School, College of Pharmacy, Program in Pharmacology and Toxicology, Pullman, WA 99164

AWARDS MS, PhD.

Faculty: 36.

Students: 9 full-time (6 women), 2 part-time (1 woman); includes 2 minority (1 Asian American or Pacific Islander, 1 Hispanic American), 4 international. Average age 30. In 1999, 5 doctorates awarded.

Degree requirements: For master's and doctorate, thesis/dissertation, oral exam required, foreign language not required. *Average time to degree:* Doctorate–5 years full-time.

Entrance requirements: For master's and doctorate, GRE General Test, TOEFL, minimum GPA of 3.0. *Application deadline:* For fall admission, 3/1. Applications are processed on a rolling basis. *Application fee:* $35.

Expenses: Tuition, state resident: full-time $5,654. Tuition, nonresident: full-time $13,850. International tuition: $13,850 full-time. Tuition and fees vary according to program.

Financial aid: In 1999–00, 3 fellowships, 8 research assistantships with full and partial tuition reimbursements, 2 teaching assistantships with full and partial tuition reimbursements were awarded; Federal Work-Study, institutionally sponsored loans, tuition waivers (partial), and staff assistantships, teaching associateships also available. Financial aid application deadline: 4/1; financial aid applicants required to submit FAFSA.

Faculty research: Pharmacokinetics/pharmacodynamics, cancer pharmacology, immunology, drug metabolism, neuropharmacology.

Dr. Raymond Quock, Chair, 509-335-7598.

Application contact: Mary Stormo, Coordinator, 509-335-7598, *E-mail:* stormom@mail.wsu.edu.

■ WAYNE STATE UNIVERSITY

College of Pharmacy and Allied Health Professions, Faculty of Allied Health Professions, Department of Occupational and Environmental Health Sciences, Detroit, MI 48202

AWARDS Occupational health sciences (MS), including industrial hygiene, industrial toxicology, occupational medicine. Part-time and evening/weekend programs available.

Degree requirements: For master's, thesis optional, foreign language not required.

Entrance requirements: For master's, GRE General Test, 1 year of course work in biology, mathematics, and physics; 2 years of course work in chemistry.

Faculty research: Air sampling: methods and development, pulmonary, toxicity oxidant air pollutants, inflammatory reactions, DNA damage and repair.

■ WAYNE STATE UNIVERSITY

Graduate School, Interdisciplinary Program in Molecular and Cellular Toxicology, Detroit, MI 48202

AWARDS MS, PhD.

Degree requirements: For master's, thesis or alternative required; for doctorate, dissertation required.

Entrance requirements: For master's, GRE General Test, GRE Subject Test; for doctorate, GRE General Test, GRE Subject Test, minimum GPA of 3.0.

Faculty research: Molecular and cellular mechanisms of chemically-induced cell injury and cell death, effects of xenobiotics on cell growth and differentiation, gene expression, hormone signal transduction in animal and human cells.

Find an in-depth description at www.petersons.com/graduate.

■ WEST VIRGINIA UNIVERSITY

College of Agriculture, Forestry and Consumer Sciences, Interdisciplinary Program in Genetics and Developmental Biology, Morgantown, WV 26506

AWARDS Animal breeding (MS, PhD); biochemical and molecular genetics (MS, PhD); cytogenetics (MS, PhD); descriptive embryology (MS, PhD); developmental genetics (MS); experimental morphogenesis (MS); human genetics (MS, PhD); immunogenetics (MS, PhD); life cycles of animals and plants (MS, PhD); molecular aspects of development (MS, PhD); mutagenesis (PhD); mutagenetics (MS); oncology (MS, PhD); plant genetics (MS, PhD); population and quantitative genetics (PhD); population and quantitative genetics (MS); regeneration (MS, PhD); teratology (MS, PhD); toxicology (MS, PhD).

Students: 18 full-time (8 women), 5 part-time (2 women); includes 1 minority (Asian American or Pacific Islander), 8 international. Average age 27. In 1999, 4 doctorates awarded.

Degree requirements: For master's, thesis required, foreign language not required; for doctorate, dissertation, comprehensive exam required, foreign language not required.

Entrance requirements: For master's, GRE or MCAT, TOEFL, minimum GPA of 2.75; for doctorate, TOEFL. *Application fee:* $45.

Expenses: Tuition, state resident: full-time $2,910; part-time $154 per credit hour. Tuition, nonresident: full-time $8,368; part-time $457 per credit hour.

Financial aid: In 1999–00, 11 research assistantships, 3 teaching assistantships were awarded; fellowships, Federal Work-Study, institutionally sponsored loans, and tuition waivers (full and partial) also available. Financial aid application deadline: 2/1; financial aid applicants required to submit FAFSA.

Dr. J. Nath, Chairman, 304-293-6256 Ext. 4333, *E-mail:* jnath@wvu.edu.

■ WEST VIRGINIA UNIVERSITY

School of Medicine, Graduate Programs in Health Sciences, Department of Pharmacology and Toxicology, Morgantown, WV 26506

AWARDS Autonomic pharmacology (MS, PhD); biomedical pharmacology (MS, PhD); chemotherapy (MS, PhD); endocrine pharmacology (MS, PhD); neuropharmacology (MS, PhD); toxicology (MS, PhD).

Students: 12 full-time (7 women), 3 part-time, 5 international. Average age 28. *232 applicants, 2% accepted.* In 1999, 1 master's, 3 doctorates awarded. Terminal master's awarded for partial completion of doctoral program.

Degree requirements: For master's and doctorate, thesis/dissertation required, foreign language not required.

Entrance requirements: For master's, GRE, TOEFL, minimum GPA of 2.5; for doctorate, GRE, TOEFL. *Application deadline:* For fall admission, 2/1 (priority date). Applications are processed on a rolling basis. *Application fee:* $45.

Expenses: Tuition, state resident: full-time $3,564. Tuition, nonresident: full-time $10,230.

Financial aid: In 1999–00, 5 research assistantships, 7 teaching assistantships were awarded; Federal Work-Study, institutionally sponsored loans, traineeships, and tuition waivers (full) also available. Financial aid application deadline: 2/1; financial aid applicants required to submit FAFSA.

Faculty research: Neuropharmacology and toxicology, cardiovascular pharmacology, molecular pharmacology, biochemical toxicology, biochemical pharmacology. Robert E. Stityel, Chair, 304-293-5761.

Application contact: Charles R. Craig, Director of Graduate Studies, 304-293-5795, *E-mail:* ccraig@wvu.edu.

Find an in-depth description at www.petersons.com/graduate.

■ WEST VIRGINIA UNIVERSITY

School of Pharmacy, Program in Pharmaceutical Sciences, Morgantown, WV 26506

AWARDS Administrative pharmacy (PhD); behavioral pharmacy (MS, PhD); biopharmaceutics/phamacokinetics (PhD); biopharmaceutics/pharmacokinetics (MS); industrial pharmacy (MS); medicinal chemistry (MS, PhD); pharmaceutical chemistry (MS, PhD); pharmaceutics (MS, PhD); pharmacology and toxicology (MS); pharmacy (MS); pharmacy administration (MS).

Students: 25 full-time (16 women), 4 part-time (2 women), 16 international. Average age 28. In 1999, 4 degrees awarded. Terminal master's awarded for partial completion of doctoral program.

Degree requirements: For master's, thesis required, foreign language not required; for doctorate, one foreign language (computer language can substitute), dissertation, written and oral comprehensive exams required.

Entrance requirements: For master's and doctorate, GRE General Test, TOEFL, minimum GPA of 2.75. *Application deadline:* For fall admission, 3/1 (priority date). *Application fee:* $45.

Expenses: Tuition, state resident: full-time $3,932. Tuition, nonresident: full-time $11,200.

Financial aid: In 1999–00, 12 research assistantships, 8 teaching assistantships were awarded; career-related internships or fieldwork, Federal Work-Study, institutionally sponsored loans, and tuition waivers (full and partial) also available. Financial aid application deadline: 2/1; financial aid applicants required to submit FAFSA.

Faculty research: Pharmaceutics, medicinal chemistry, behavioral and administration pharmacy, medical informatics, biopharmaceutics/pharmacokinetics.

Application contact: Dr. Patrick S. Callery, Assistant Dean for Research and Graduate Programs, 304-293-1482, *Fax:* 304-293-5483, *E-mail:* pcallery@hsc.wvu.edu.

Physiology

CARDIOVASCULAR SCIENCES

■ ALBANY MEDICAL COLLEGE

Graduate Programs in the Biological Sciences, Program in Cardiovascular Sciences, Albany, NY 12208-3479

AWARDS MS, PhD. Part-time programs available.

Students: 16 full-time (8 women); includes 3 minority (all Asian Americans or Pacific Islanders), 1 international. Terminal master's awarded for partial completion of doctoral program.
Degree requirements: For master's, thesis required, foreign language not required; for doctorate, dissertation, oral qualifying exam, written preliminary exam required, foreign language not required.
Entrance requirements: For master's and doctorate, GRE General Test, TOEFL. *Application deadline:* Applications are processed on a rolling basis. *Application fee:* $0 ($60 for international students).
Expenses: Tuition: Full-time $13,367; part-time $446 per credit hour.
Financial aid: Fellowships, Federal Work-Study, grants, scholarships, and tuition waivers (full) available.
Faculty research: Cell and molecular approaches to endothelial cell biology, vascular permeability, pathogenesis of hypertension, extracellular matrix, cardiovascular disease.
Dr. Peter A. Vincent, Graduate Director, 518-262-6296, *Fax:* 518-262-5669, *E-mail:* vincenp@mail.amc.edu.

Find an in-depth description at www.petersons.com/graduate.

■ BAYLOR COLLEGE OF MEDICINE

Graduate School of Biomedical Sciences, Program in Cardiovascular Sciences, Houston, TX 77030-3498

AWARDS PhD, MD/PhD.

Faculty: 31 full-time (3 women).
Students: 23 full-time (7 women); includes 4 minority (1 African American, 3 Asian Americans or Pacific Islanders), 14 international. Average age 30. *26 applicants, 31% accepted.* In 1999, 5 degrees awarded (100% entered university research/teaching).
Degree requirements: For doctorate, dissertation, public defense, qualifying exam required, foreign language not required.

Average time to degree: Doctorate–6.21 years full-time.
Entrance requirements: For doctorate, GRE General Test, GRE Subject Test (strongly recommended), TOEFL, minimum GPA of 3.0, strong background in biology and biochemistry. *Application deadline:* For fall admission, 2/1 (priority date). Applications are processed on a rolling basis. *Application fee:* $30. Electronic applications accepted.
Expenses: Tuition: Full-time $8,200. Required fees: $175. Full-time tuition and fees vary according to student level.
Financial aid: In 1999–00, 23 students received aid, including 15 fellowships (averaging $16,000 per year), 8 research assistantships (averaging $16,000 per year); Federal Work-Study, institutionally sponsored loans, and tuition waivers (full) also available. Financial aid applicants required to submit FAFSA.
Faculty research: Cell biology of the vascular wall, cell biology of cardiac tissue, biology and models of specific cardiovascular diseases.
Dr. Julius C. Allen, Director, 713-798-4977.
Application contact: Donna Campbell, Graduate Program Administrator, 713-798-4977, *Fax:* 713-790-0681, *E-mail:* donnac@bcm.tmc.edu.

Find an in-depth description at www.petersons.com/graduate.

■ MCP HAHNEMANN UNIVERSITY

School of Medicine, Biomedical Graduate Programs, Department of Biochemistry, Program in Cardiovascular Biology, Philadelphia, PA 19102-1192

AWARDS MS, PhD, MD/PhD.

Degree requirements: For master's, thesis, comprehensive exam required, foreign language not required; for doctorate, dissertation, qualifying exam required, foreign language not required.
Entrance requirements: For master's, GRE General Test, TOEFL, minimum GPA of 2.75; for doctorate, GRE General Test, TOEFL, minimum GPA of 3.2.
Faculty research: Lipid and lipoprotein metabolism: cholesterol homeostasis in health and disease, membrane structure, lipid peroxidation and atherosclerosis, growth factors and the vessel wall, cell biology of cardiovascular tissue.

■ NORTHEASTERN UNIVERSITY

Bouvé College of Health Sciences Graduate School, Department of Cardiopulmonary Sciences, Program in Cardiopulmonary Science (Perfusion Technology), Boston, MA 02115-5096

AWARDS MS.

Students: 11 full-time (4 women). *13 applicants, 77% accepted.* In 1999, 6 degrees awarded.
Degree requirements: For master's, comprehensive exam required, thesis optional, foreign language not required.
Entrance requirements: For master's, GRE General Test or MAT. *Application deadline:* Applications are processed on a rolling basis. *Application fee:* $50.
Expenses: Tuition: Full-time $16,560; part-time $460 per quarter hour. Required fees: $150; $25 per year. Tuition and fees vary according to course load and program.
Financial aid: Career-related internships or fieldwork, Federal Work-Study, tuition waivers (partial), and unspecified assistantships available. Aid available to part-time students. Financial aid application deadline: 3/1; financial aid applicants required to submit FAFSA.
Faculty research: Compliment activation of extra corporeal circuits, application of neural networks, pathophysiology of cardio bypass.
Dr. Eric B. Pepin, Director, 617-373-4183, *Fax:* 617-373-2968, *E-mail:* e.pepin@nunet.neu.edu.
Application contact: Bill Purnell, Director of Graduate Admissions, 617-373-2708, *Fax:* 617-373-4701, *E-mail:* w.purnell@nunet.neu.edu.

■ UNIVERSITY OF MEDICINE AND DENTISTRY OF NEW JERSEY

School of Health Related Professions, Department of Interdisciplinary Studies, Program in Health Sciences, Newark, NJ 07107-3001

AWARDS Cardiopulmonary sciences (PhD); clinical laboratory sciences (PhD); health sciences (MSHS); interdisciplinary studies (PhD); nutrition science (PhD); physical therapy/movement science (PhD).

Degree requirements: For doctorate, dissertation required.
Entrance requirements: For doctorate, TOEFL (minimum score of 550 required; 213 for computer-based), interview, writing sample.

University of Medicine and Dentistry of New Jersey (continued)

Expenses: Tuition, state resident: part-time $270 per credit hour. Tuition, nonresident: part-time $407 per credit hour. Part-time tuition and fees vary according to campus/location and program.

Find an in-depth description at www.petersons.com/graduate.

■ UNIVERSITY OF SOUTH DAKOTA

School of Medicine and Graduate School, Biomedical Sciences Graduate Program, Cardiovascular Research Center, Vermillion, SD 57069-2390

AWARDS MA, PhD.

Faculty: 5 full-time (0 women), 1 part-time/adjunct (0 women).
Students: 5 full-time (2 women); includes 1 minority (Asian American or Pacific Islander), 2 international. Average age 25. *4 applicants, 75% accepted.*
Degree requirements: For master's, thesis required, foreign language not required; for doctorate, variable foreign language requirement, dissertation required.
Entrance requirements: For master's and doctorate, GRE General Test, TOEFL, minimum GPA of 3.0. *Application deadline:* For fall admission, 3/15 (priority date). Applications are processed on a rolling basis. *Application fee:* $15.
Expenses: Tuition, state resident: full-time $2,126; part-time $89 per credit. Tuition, nonresident: full-time $6,270; part-time $261 per credit. Required fees: $1,194; $50 per credit. Tuition and fees vary according to course load and reciprocity agreements.
Financial aid: In 1999–00, 5 students received aid, including 2 research assistantships with full tuition reimbursements available (averaging $13,500 per year); fellowships, teaching assistantships, tuition waivers (partial) also available.
Faculty research: Cardiovascular disease.
Dr. A. Martin Gerdes, Group Coordinator, 605-357-1314, *Fax:* 605-357-1539, *E-mail:* mgcerdes@usd.edu.
Application contact: Dr. Scott E. Campbell, Associate Professor, 605-357-1314, *Fax:* 605-357-1539, *E-mail:* scampbel@usd.edu.

PHYSIOLOGY

■ ARIZONA STATE UNIVERSITY

Graduate College, College of Liberal Arts and Sciences, Department of Biology, Program in Physiology, Tempe, AZ 85287

AWARDS MS, PhD. Terminal master's awarded for partial completion of doctoral program.

Degree requirements: For master's, thesis required, foreign language not required; for doctorate, dissertation, oral exam required, foreign language not required.
Entrance requirements: For master's and doctorate, GRE General Test, GRE Subject Test. *Application deadline:* For fall admission, 12/15. *Application fee:* $45.
Expenses: Tuition, state resident: part-time $115 per credit hour. Tuition, nonresident: part-time $389 per credit hour. Required fees: $18 per semester. Tuition and fees vary according to program.
Financial aid: Application deadline: 12/15.

■ AUBURN UNIVERSITY

College of Veterinary Medicine and Graduate School, Graduate Program in Veterinary Medicine, Department of Anatomy, Physiology and Pharmacology, Auburn, Auburn University, AL 36849-0002

AWARDS Anatomy and histology (MS); physiology and pharmacology (MS). Part-time programs available.

Students: *9 applicants, 44% accepted.*
Degree requirements: For master's, thesis required, foreign language not required.
Entrance requirements: For master's, GRE General Test. *Application deadline:* For fall admission, 7/1; for spring admission, 12/1. Applications are processed on a rolling basis. *Application fee:* $25 ($50 for international students).
Expenses: Tuition, state resident: full-time $2,895; part-time $80 per credit hour. Tuition, nonresident: full-time $8,685; part-time $240 per credit hour.
Financial aid: Research assistantships, teaching assistantships available. Aid available to part-time students. Financial aid application deadline: 3/15.
Faculty research: Chemosensory systems, embryo transfer, osteoarthritis, neurosenses, audiology, cardiovascular physiology, molecular endocrinology. *Total annual research expenditures:* $400,000.
Dr. Philip Posner, Head, 334-844-4427.
Application contact: Dr. John F. Pritchett, Dean of the Graduate School, 334-844-4700.

■ BALL STATE UNIVERSITY

Graduate School, College of Sciences and Humanities, Department of Physiology and Health Science, Program in Physiology, Muncie, IN 47306-1099

AWARDS MA, MS.

Students: 9 full-time (3 women), 8 part-time (5 women); includes 4 minority (all African Americans), 2 international. Average age 25. *13 applicants, 85% accepted.* In 1999, 5 degrees awarded.
Degree requirements: For master's, foreign language not required.
Application fee: $25 ($35 for international students).
Expenses: Tuition, state resident: full-time $3,024. Tuition, nonresident: full-time $7,482. Tuition and fees vary according to course load.
Financial aid: Teaching assistantships with full tuition reimbursements available. Financial aid application deadline: 3/1.
Jeffrey Clark, Head.

■ BAYLOR COLLEGE OF MEDICINE

Graduate School of Biomedical Sciences, Department of Molecular Physiology and Biophysics, Houston, TX 77030-3498

AWARDS PhD, MD/PhD.

Faculty: 21 full-time (2 women).
Students: 13 full-time (8 women); includes 1 minority (Hispanic American), 8 international. Average age 27. *26 applicants, 27% accepted.*
Degree requirements: For doctorate, dissertation, public defense, qualifying exam required, foreign language not required.
Entrance requirements: For doctorate, GRE General Test (average 80th percentile), GRE Subject Test (strongly recommended), TOEFL, minimum GPA of 3.0. *Application deadline:* For fall admission, 2/1 (priority date). *Application fee:* $30. Electronic applications accepted.
Expenses: Tuition: Full-time $8,200. Required fees: $175. Full-time tuition and fees vary according to student level.
Financial aid: In 1999–00, 13 students received aid, including 8 fellowships (averaging $16,000 per year), 5 research assistantships (averaging $16,000 per year); Federal Work-Study, institutionally sponsored loans, and tuition waivers (full) also available. Financial aid applicants required to submit FAFSA.
Faculty research: Membrane ion channels, ion transport, recombinant DNA.
Dr. Brian Knoll, Director, 713-798-5107, *Fax:* 713-798-3475.

Application contact: Becky Pullen, Graduate Program Administrator, 713-798-5107, *Fax:* 713-798-3475, *E-mail:* rpullen@bcm.tmc.edu.

Find an in-depth description at www.petersons.com/graduate.

■ BOSTON UNIVERSITY

Graduate School of Arts and Sciences, Department of Biology, Boston, MA 02215

AWARDS Botany (MA, PhD); cell and molecular biology (MA, PhD); cell biology (MA, PhD); ecology (PhD); ecology, behavior, and evolution (MA, PhD); ecology/physiology, endocrinology and reproduction (MA); marine biology (MA, PhD); molecular biology, cell biology and biochemistry (MA, PhD); neurobiology, neuroendocrinology and reproduction (MA, PhD); physiology, endocrinology, and neurobiology (MA, PhD); zoology (MA, PhD). Part-time programs available.

Faculty: 41 full-time (8 women).
Students: 131 full-time (74 women), 11 part-time (7 women); includes 10 minority (7 Asian Americans or Pacific Islanders, 3 Hispanic Americans), 33 international. Average age 27. *238 applicants, 39% accepted.* In 1999, 61 master's, 45 doctorates awarded. Terminal master's awarded for partial completion of doctoral program.
Degree requirements: For master's, one foreign language, thesis not required; for doctorate, one foreign language, dissertation, qualifying exam required. *Average time to degree:* Master's–1 year full-time, 3 years part-time; doctorate–5.75 years full-time.
Entrance requirements: For master's and doctorate, GRE General Test, GRE Subject Test, TOEFL. *Application deadline:* For fall admission, 1/1 (priority date); for spring admission, 11/1. *Application fee:* $50.
Expenses: Tuition: Full-time $23,770; part-time $743 per credit. Required fees: $220. Tuition and fees vary according to class time, course level, campus/location and program.
Financial aid: In 1999–00, 82 students received aid, including 1 fellowship with full tuition reimbursement available (averaging $12,000 per year), 28 research assistantships with full tuition reimbursements available (averaging $11,500 per year), 43 teaching assistantships with full tuition reimbursements available (averaging $11,500 per year); Federal Work-Study, grants, institutionally sponsored loans, scholarships, and traineeships also available. Financial aid application deadline: 1/15; financial aid applicants required to submit FAFSA.

Faculty research: Marine science, endocrinology, behavior. *Total annual research expenditures:* $5 million.
Geoffrey M. Cooper, Chairman, 617-353-2432, *Fax:* 617-353-6340, *E-mail:* gmcooper@bu.edu.
Application contact: Yolanta Kovalko, Senior Staff Assistant, 617-353-2432, *Fax:* 617-353-6340, *E-mail:* yolanta@bu.edu.

Find an in-depth description at www.petersons.com/graduate.

■ BOSTON UNIVERSITY

Sargent College of Health and Rehabilitation Sciences, Department of Health Sciences, Boston, MA 02215

AWARDS Applied anatomy and physiology (MS, D Sc); nutrition (MS). Part-time programs available.

Faculty: 8 full-time (4 women), 2 part-time/adjunct (1 woman).
Students: 49 full-time (41 women), 7 part-time (5 women); includes 4 minority (3 Asian Americans or Pacific Islanders, 1 Hispanic American), 6 international. Average age 26. *86 applicants, 66% accepted.* In 1999, 8 master's, 3 doctorates awarded.
Degree requirements: For master's, thesis or alternative required, foreign language not required; for doctorate, computer language, dissertation required, foreign language not required.
Entrance requirements: For master's, GRE General Test, minimum GPA of 3.0; for doctorate, GRE General Test, master's degree. *Application deadline:* For fall admission, 4/1 (priority date); for spring admission, 10/1. Applications are processed on a rolling basis. *Application fee:* $60.
Expenses: Tuition: Full-time $23,770; part-time $743 per credit. Required fees: $220. Tuition and fees vary according to class time, course level, campus/location and program.
Financial aid: In 1999–00, 20 fellowships, 7 research assistantships, 6 teaching assistantships were awarded; career-related internships or fieldwork, Federal Work-Study, institutionally sponsored loans, and scholarships also available. Aid available to part-time students. Financial aid application deadline: 4/15.
Faculty research: Muscle metabolism, body acid-base balance, human performance, physical conditioning, diabetes.
Dr. Gary Skrinar, Chairman, 617-353-2717.
Application contact: Judy Skeffington, Senior Admissions Coordinator, 617-353-2713, *Fax:* 617-353-7500, *E-mail:* jaskeff@bu.edu.

■ BOSTON UNIVERSITY

School of Medicine, Division of Graduate Medical Sciences, Department of Physiology and Biophysics, Boston, MA 02118

AWARDS MA, PhD, MD/PhD.

Faculty: 28.
Students: 27 full-time. Average age 28. In 2000, 1 master's, 1 doctorate awarded.
Degree requirements: For master's and doctorate, thesis/dissertation, qualifying exam required, foreign language not required.
Entrance requirements: For master's and doctorate, GRE General Test, GRE Subject Test in related fields (strongly recommended), TOEFL. *Application deadline:* For fall admission, 1/15 (priority date); for spring admission, 10/15 (priority date). *Application fee:* $50. Electronic applications accepted.
Expenses: Tuition: Full-time $24,700; part-time $772 per credit. Required fees: $220.
Financial aid: Fellowships with tuition reimbursements, research assistantships with tuition reimbursements, scholarships and traineeships available.
Faculty research: X-ray scattering, NMR spectroscopy, protein crystallography, structural electron microscopy, molecular modeling.
Dr. Donald M. Small, Chairman.
Application contact: Dr. Christopher Akey, Chair of Admissions and Student Affairs Committee, 617-638-4051, *Fax:* 617-638-4041, *E-mail:* cakey@bu.edu.

Find an in-depth description at www.petersons.com/graduate.

■ BROWN UNIVERSITY

Graduate School, Division of Biology and Medicine, Program in Molecular Pharmacology and Physiology, Providence, RI 02912

AWARDS MA, Sc M, PhD, MD/PhD.

Faculty: 16 full-time (6 women).
Students: 7 full-time (2 women); includes 2 minority (both Asian Americans or Pacific Islanders), 4 international. Average age 24. *21 applicants, 14% accepted.* In 1999, 1 degree awarded. Terminal master's awarded for partial completion of doctoral program.
Degree requirements: For master's, foreign language not required; for doctorate, dissertation, preliminary exam required.
Entrance requirements: For master's, GRE General Test, GRE Subject Test; for doctorate, GRE General Test, GRE Subject Test, TOEFL. *Application deadline:* For fall admission, 1/2 (priority date).

Brown University (continued)
Applications are processed on a rolling basis. *Application fee:* $60.
Financial aid: In 1999–00, 2 fellowships, 1 research assistantship, 5 teaching assistantships were awarded; traineeships also available. Financial aid application deadline: 1/2.
Faculty research: Structural biology, antiplatelet drugs, nicotinic receptor structure/function.
Dr. John Marshall, Director, 401-863-1596.
Application contact: Cheryl Pariseau, Administrative Assistant, 401-863-1596, *Fax:* 401-863-1595, *E-mail:* cheryl_pariseau@brown.edu.
Find an in-depth description at www.petersons.com/graduate.

■ CASE WESTERN RESERVE UNIVERSITY

School of Medicine and School of Graduate Studies, Graduate Programs in Medicine, Department of Physiology and Biophysics, Cleveland, OH 44106

AWARDS Biophysics and bioengineering (PhD); cell physiology (PhD); physiology and biophysics (PhD); systems physiology (PhD).

Degree requirements: For doctorate, dissertation required, foreign language not required.
Entrance requirements: For doctorate, GRE General Test, TOEFL. Electronic applications accepted.
Faculty research: Cardiovascular physiology, calcium metabolism, epithelial cell biology.

Find an in-depth description at www.petersons.com/graduate.

■ CLEMSON UNIVERSITY

Graduate School, College of Agriculture, Forestry and Life Sciences, School of Animal, Biomedical and Biological Sciences, Department of Animal, Dairy and Veterinary Sciences, Program In Animal Physiology, Clemson, SC 29634

AWARDS MS, PhD. Offered in cooperation with the Department of Poultry Science.

Students: 14 full-time (10 women), 2 part-time (both women); includes 1 minority (African American), 1 international. Average age 26. *9 applicants, 33% accepted.* In 1999, 5 master's, 1 doctorate awarded.
Degree requirements: For master's and doctorate, thesis/dissertation required, foreign language not required. *Average time to degree:* Master's–2.2 years full-time; doctorate–3.3 years full-time.
Entrance requirements: For master's and doctorate, GRE General Test, TOEFL.

Application deadline: For fall admission, 6/1 (priority date); for spring admission, 11/1 (priority date). Applications are processed on a rolling basis. *Application fee:* $40. Electronic applications accepted.
Expenses: Tuition, state resident: full-time $3,480; part-time $174 per credit hour. Tuition, nonresident: full-time $9,256; part-time $388 per credit hour. Required fees: $5 per term. Full-time tuition and fees vary according to course level, course load and campus/location.
Financial aid: Fellowships, research assistantships, teaching assistantships, career-related internships or fieldwork available. Financial aid application deadline: 6/1; financial aid applicants required to submit FAFSA.
Faculty research: Reproductive physiology, endocrinology, stress physiology, immunology.
Dr. John R. Diehl, Coordinator, 864-656-5166, *Fax:* 864-656-3131, *E-mail:* jdiehl@clust1.clemson.edu.

■ COLORADO STATE UNIVERSITY

College of Veterinary Medicine and Biomedical Sciences and Graduate School, Graduate Programs in Veterinary Medicine and Biomedical Sciences, Department of Physiology, Fort Collins, CO 80523-0015

AWARDS MS, PhD.

Faculty: 18 full-time (3 women).
Students: 29 full-time (15 women), 10 part-time (4 women); includes 4 minority (1 African American, 1 Asian American or Pacific Islander, 2 Hispanic Americans), 6 international. Average age 28. *38 applicants, 34% accepted.* In 1999, 8 master's, 3 doctorates awarded.
Degree requirements: For master's and doctorate, thesis/dissertation required, foreign language not required.
Entrance requirements: For master's and doctorate, GRE General Test, TOEFL. *Application deadline:* For fall admission, 2/1 (priority date). Applications are processed on a rolling basis. *Application fee:* $30. Electronic applications accepted.
Expenses: Tuition, state resident: full-time $2,694; part-time $150 per credit. Tuition, nonresident: full-time $10,460; part-time $581 per credit. Required fees: $32 per semester.
Financial aid: In 1999–00, 26 research assistantships, 2 teaching assistantships were awarded; fellowships, Federal Work-Study, institutionally sponsored loans, and traineeships also available.
Faculty research: Cardiopulmonary physiology, neurophysiology, renal physiology, reproductive physiology. *Total annual research expenditures:* $3.8 million.

Dr. Alan Tucker, Head, 970-491-6187, *Fax:* 970-491-7569, *E-mail:* atucker@cvmbs.colostate.edu.
Application contact: Dr. C. W. Miller, Graduate Coordinator, 970-491-6187, *E-mail:* cwmiller@cvmbs.colostate.edu.

■ COLUMBIA UNIVERSITY

College of Physicians and Surgeons and Graduate School of Arts and Sciences, Graduate School of Arts and Sciences at the College of Physicians and Surgeons, Department of Physiology and Cellular Biophysics, New York, NY 10032

AWARDS M Phil, MA, PhD, MD/PhD. Only candidates for the PhD are admitted. Terminal master's awarded for partial completion of doctoral program.

Degree requirements: For master's, foreign language and thesis not required; for doctorate, dissertation required, foreign language not required.
Entrance requirements: For master's and doctorate, GRE General Test, TOEFL.
Expenses: Tuition: Full-time $25,072.
Faculty research: Membrane physiology, cellular biology, cardiovascular physiology, neurophysiology.

■ CORNELL UNIVERSITY

Graduate School, Graduate Fields of Veterinary Medicine, Field of Physiology, Ithaca, NY 14853-0001

AWARDS Behavioral physiology (MS, PhD); cardiovascular and respiratory physiology (MS, PhD); endocrinology (MS, PhD); environmental and comparative physiology (MS, PhD); gastrointestinal and metabolic physiology (MS, PhD); membrane and epithelial physiology (MS, PhD); molecular and cellular physiology (MS, PhD); neural and sensory physiology (MS, PhD); reproductive physiology (MS, PhD).

Faculty: 44 full-time.
Students: 20 full-time (12 women); includes 3 minority (2 Hispanic Americans, 1 Native American), 12 international. *10 applicants, 80% accepted.* In 1999, 2 master's, 2 doctorates awarded.
Degree requirements: For master's, thesis required, foreign language not required; for doctorate, dissertation, 1 semester of teaching experience, seminar presentation required, foreign language not required.
Entrance requirements: For master's, GRE General Test, GRE Subject Test (biochemistry, cell and molecular biology, biology or chemistry), TOEFL; for doctorate, GRE General Test, GRE Subject Test (biochemistry, cell and molecular biology, biology, or chemistry), TOEFL. *Application deadline:* For fall

admission, 1/15. *Application fee:* $65. Electronic applications accepted.
Expenses: Tuition: Full-time $12,400.
Financial aid: In 1999–00, 19 students received aid, including 4 fellowships with full tuition reimbursements available, 7 research assistantships with full tuition reimbursements available, 8 teaching assistantships with full tuition reimbursements available; institutionally sponsored loans, scholarships, tuition waivers (full and partial), and unspecified assistantships also available. Financial aid applicants required to submit FAFSA.
Faculty research: Behavioral physiology, endocrinology and reproductive physiology, cardiovascular and respiratory physiology, gastrointestinal and metabolic physiology, membrane and epithelial physiology.
Application contact: Graduate Field Assistant, 607-253-3552, *Fax:* 607-253-3541, *E-mail:* physzoograd@cornell.edu.

■ DARTMOUTH COLLEGE

School of Arts and Sciences, Department of Physiology, Lebanon, NH 03756

AWARDS PhD, MD/PhD.

Faculty: 19 full-time (3 women), 4 part-time/adjunct (1 woman).
Students: 15 full-time (8 women); includes 2 minority (both Asian Americans or Pacific Islanders). Average age 25. *14 applicants, 57% accepted.* In 1999, 3 doctorates awarded (100% entered university research/teaching).
Degree requirements: For doctorate, dissertation required.
Entrance requirements: For doctorate, GRE General Test, GRE Subject Test. *Application deadline:* Applications are processed on a rolling basis. *Application fee:* $0.
Expenses: Tuition: Full-time $24,624. Required fees: $916. One-time fee: $15 full-time. Full-time tuition and fees vary according to program.
Financial aid: In 1999–00, 13 students received aid, including 2 fellowships with full tuition reimbursements available, 10 research assistantships with full tuition reimbursements available; Federal Work-Study, grants, and institutionally sponsored loans also available. Financial aid application deadline: 4/15.
Faculty research: Respiratory control, endocrinology of reproduction and immunology, regulation of receptors and channels, electrophysiology of membranes, renal function.
Dr. Donald Bartlett, Chairman, 603-650-7717, *Fax:* 603-650-6130, *E-mail:* donald.bartlet.jr@dartmouth.edu.
Application contact: Dr. Valerie Anne Galton, Coordinator of Graduate Studies, 603-650-7717, *Fax:* 603-650-6130, *E-mail:* valerie.a.galton@dartmouth.edu.

Find an in-depth description at www.petersons.com/graduate.

■ DUKE UNIVERSITY

Graduate School, Department of Cell Biology, Division of Physiology and Cellular Biophysics, Durham, NC 27708-0586

AWARDS PhD.

Degree requirements: For doctorate, dissertation required, foreign language not required.
Entrance requirements: For doctorate, GRE General Test, GRE Subject Test (recommended). *Application deadline:* For fall admission, 12/31. *Application fee:* $75.
Expenses: Tuition: Full-time $21,406; part-time $760 per unit. Required fees: $3,136; $3,136 per year. One-time fee: $30. Tuition and fees vary according to program.
Financial aid: Federal Work-Study available. Financial aid application deadline: 12/31.
G. Vann Bennett, Director of Graduate Studies, Department of Cell Biology, 919-684-3538, *Fax:* 919-684-3590, *E-mail:* b.sampson@cellbio.duke.edu.

■ EAST CAROLINA UNIVERSITY

School of Medicine, Department of Physiology, Greenville, NC 27858-4353

AWARDS PhD.

Faculty: 9 full-time (0 women).
Students: 3 full-time (1 woman), 8 part-time, 1 international. Average age 29. *8 applicants, 13% accepted.* In 1999, 1 degree awarded.
Degree requirements: For doctorate, computer language, dissertation required, foreign language not required.
Entrance requirements: For doctorate, GRE General Test, GRE Subject Test, TOEFL. *Application deadline:* For fall admission, 6/1 (priority date). Applications are processed on a rolling basis. *Application fee:* $40.
Expenses: Tuition, state resident: full-time $1,012. Tuition, nonresident: full-time $8,578. Required fees: $1,006. Full-time tuition and fees vary according to degree level. Part-time tuition and fees vary according to course load.
Financial aid: Fellowships available. Financial aid application deadline: 6/1.
Faculty research: Cell and nerve biophysics; neurophysiology; cardiovascular, renal, endocrine, and gastrointestinal physiology.
Dr. Robert Lust, Chairman, 252-816-2762, *Fax:* 252-816-3460, *E-mail:* lust@brody.med.ecu.edu.

Application contact: Dr. Richard Ray, Coordinator for Graduate Studies, 252-816-2776, *Fax:* 252-816-3460, *E-mail:* rray@brody.med.ecu.edu.

■ EAST TENNESSEE STATE UNIVERSITY

James H. Quillen College of Medicine and School of Graduate Studies, Biomedical Science Graduate Program, Johnson City, TN 37614

AWARDS Anatomy and cell biology (MS, PhD); biochemistry and molecular biology (MS, PhD); biophysics (MS, PhD); microbiology (MS, PhD); pharmacology (MS, PhD); physiology (MS, PhD). Part-time programs available.

Faculty: 40 full-time (9 women).
Students: 23 full-time (10 women), 2 part-time (1 woman); includes 1 minority (Asian American or Pacific Islander), 4 international. Average age 30. *83 applicants, 20% accepted.* In 1999, 5 master's, 4 doctorates awarded. Terminal master's awarded for partial completion of doctoral program.
Degree requirements: For master's, one foreign language (computer language can substitute), thesis, comprehensive qualifying exam required; for doctorate, 2 foreign languages (computer language can substitute for one), dissertation required.
Entrance requirements: For master's, GRE General Test, minimum GPA of 3.0, bachelor's degree in biological or related science; for doctorate, GRE General Test, GRE Subject Test. *Application deadline:* For fall admission, 3/15 (priority date); for spring admission, 3/1. *Application fee:* $25 ($35 for international students).
Expenses: Tuition, state resident: full-time $10,342. Tuition, nonresident: full-time $21,080. Required fees: $532.
Financial aid: In 1999–00, 16 research assistantships, 5 teaching assistantships were awarded; fellowships, career-related internships or fieldwork, Federal Work-Study, grants, institutionally sponsored loans, and tuition waivers (full) also available.
Dr. Mitchell Robinson, Assistant Dean for Graduate Studies, 423-439-4658, *E-mail:* robinson@etsu.edu.

■ FINCH UNIVERSITY OF HEALTH SCIENCES/THE CHICAGO MEDICAL SCHOOL

School of Graduate and Postdoctoral Studies, Department of Physiology and Biophysics, Program in Applied Physiology, North Chicago, IL 60064-3095

AWARDS MS.

Finch University of Health Sciences/The Chicago Medical School (continued)

Students: 67 full-time (20 women); includes 39 minority (38 Asian Americans or Pacific Islanders, 1 Hispanic American). Average age 24. In 1999, 45 degrees awarded.

Degree requirements: For master's, computer language required, foreign language and thesis not required.

Entrance requirements: For master's, GRE General Test, TOEFL, TWE. *Application deadline:* For fall admission, 6/1 (priority date). *Application fee:* $25.

Expenses: Tuition: Full-time $32,027.

Financial aid: Application deadline: 6/9. Dr. Timothy Hansen, Director, 847-578-3314.

Application contact: Dana Frederick, Admissions Officer, 847-578-3209.

Find an in-depth description at www.petersons.com/graduate.

■ FINCH UNIVERSITY OF HEALTH SCIENCES/THE CHICAGO MEDICAL SCHOOL

School of Graduate and Postdoctoral Studies, Department of Physiology and Biophysics, Program in Physiology, North Chicago, IL 60064-3095

AWARDS MS, PhD, MD/MS, MD/PhD.

Students: 11 full-time (3 women); includes 3 minority (all Asian Americans or Pacific Islanders), 2 international. Average age 23. In 1999, 1 master's, 1 doctorate awarded.

Degree requirements: For master's and doctorate, computer language, thesis/dissertation required, foreign language not required.

Entrance requirements: For master's and doctorate, GRE General Test, TOEFL, TWE. *Application deadline:* For fall admission, 6/1 (priority date). Applications are processed on a rolling basis. *Application fee:* $25.

Expenses: Tuition: Full-time $14,054; part-time $391 per credit hour. Tuition and fees vary according to program.

Financial aid: Fellowships available. Financial aid application deadline: 6/9; financial aid applicants required to submit FAFSA.

Application contact: Dana Frederick, Admissions Officer, 847-578-3209.

Find an in-depth description at www.petersons.com/graduate.

■ FLORIDA STATE UNIVERSITY

Graduate Studies, College of Arts and Sciences, Department of Biological Science, Program in Physiology, Tallahassee, FL 32306

AWARDS MS, PhD.

Faculty: 11 full-time (1 woman).

Students: 16 full-time (8 women); includes 5 minority (3 Asian Americans or Pacific Islanders, 2 Hispanic Americans), 1 international.

Degree requirements: For master's and doctorate, thesis/dissertation, teaching experience required.

Entrance requirements: For master's, GRE General Test, TOEFL; for doctorate, GRE General Test, GRE Subject Test, TOEFL. *Application deadline:* For fall admission, 1/15; for spring admission, 10/15. *Application fee:* $20.

Expenses: Tuition, state resident: full-time $3,504; part-time $146 per credit hour. Tuition, nonresident: full-time $12,162; part-time $507 per credit hour. Tuition and fees vary according to program.

Financial aid: In 1999–00, fellowships with full tuition reimbursements (averaging $13,740 per year), research assistantships with full tuition reimbursements (averaging $13,740 per year), teaching assistantships with full tuition reimbursements (averaging $13,740 per year) were awarded. Financial aid application deadline: 1/15; financial aid applicants required to submit FAFSA.

Faculty research: Comparative physiology, reproductive physiology, endocrinology, neurophysiology, sensory physiology. Dr. Thomas C. S. Keller, Associate Professor and Associate Chairman, 850-644-3023, *Fax:* 850-644-9829.

Application contact: Judy Bowers, Coordinator, Graduate Affairs, 850-644-3023, *Fax:* 850-644-9829, *E-mail:* bowers@bio.fsu.edu.

■ FLORIDA STATE UNIVERSITY

Graduate Studies, College of Human Sciences, Department of Nutrition, Food, and Exercise Sciences, Tallahassee, FL 32306

AWARDS Exercise science (MS, PhD), including exercise physiology, motor learning and control; nutrition and food science (PhD); nutrition and food sciences (MS), including clinical nutrition, food science, nutrition and sport, nutrition science, nutrition, education and health promotion.

Faculty: 13 full-time (9 women).

Students: 67 full-time (37 women), 16 part-time (10 women); includes 14 minority (6 African Americans, 2 Asian Americans or Pacific Islanders, 6 Hispanic Americans), 8 international. *93 applicants, 70% accepted.* In 1999, 21 master's awarded (100% found work related to degree); 4 doctorates awarded.

Degree requirements: For master's, thesis optional, foreign language not required; for doctorate, dissertation required, foreign language not required.

Entrance requirements: For master's and doctorate, GRE General Test, minimum

GPA of 3.0. *Application fee:* $20. Electronic applications accepted.

Expenses: Tuition, state resident: full-time $3,504; part-time $146 per credit hour. Tuition, nonresident: full-time $12,162; part-time $507 per credit hour. Tuition and fees vary according to program.

Financial aid: In 1999–00, 33 students received aid, including 2 fellowships (averaging $10,000 per year), 3 research assistantships with partial tuition reimbursements available (averaging $8,000 per year), 28 teaching assistantships with partial tuition reimbursements available (averaging $8,000 per year); career-related internships or fieldwork, Federal Work-Study, institutionally sponsored loans, scholarships, and unspecified assistantships also available. Financial aid applicants required to submit FAFSA.

Faculty research: Nutrition and exercise, vitamin A deficiency, protein biochemistry, cardiovascular responses to exercises, physiological effects of cigarette smoking related to health and wellness. *Total annual research expenditures:* $320,386.

Dr. Robert Moffatt, Chair, 850-644-1828, *Fax:* 850-644-0700, *E-mail:* rmoffatt@mailer.fsu.edu.

Application contact: Dr. Natholyn Harris, Graduate Coordinator, 850-644-4800, *Fax:* 850-644-0700, *E-mail:* nharris@mailer.fsu.edu.

■ GEORGETOWN UNIVERSITY

Graduate School of Arts and Sciences, Programs in Biomedical Sciences, Department of Physiology and Biophysics, Washington, DC 20057

AWARDS MS, PhD, MD/PhD.

Degree requirements: For master's, foreign language not required; for doctorate, dissertation required, foreign language not required.

Entrance requirements: For master's, GRE General Test, MCAT, TOEFL; for doctorate, GRE General Test, TOEFL.

■ GEORGIA STATE UNIVERSITY

College of Arts and Sciences, Department of Biology, Program in Cell Biology and Physiology, Atlanta, GA 30303-3083

AWARDS MS, PhD.

Degree requirements: For master's, one foreign language (computer language can substitute), thesis or alternative, exam required; for doctorate, dissertation required.

Entrance requirements: For master's and doctorate, GRE General Test, TOEFL, minimum GPA of 3.0. *Application deadline:*

For fall admission, 7/18; for spring admission, 2/13. Applications are processed on a rolling basis. *Application fee:* $25.
Expenses: Tuition, state resident: full-time $2,896; part-time $121 per credit hour. Tuition, nonresident: full-time $11,584; part-time $483 per credit hour. Required fees: $228. Full-time tuition and fees vary according to course load and program.
Financial aid: Application deadline: 2/6.
Application contact: Latesha Morrison, Graduate Administrative Coordinator, 404-651-2759, *Fax:* 404-651-2509, *E-mail:* biolxm@langate.gsu.edu.

Find an in-depth description at www.petersons.com/graduate.

■ HARVARD UNIVERSITY

School of Public Health, Department of Environmental Health, Boston, MA 02115-6096

AWARDS Environmental epidemiology (SM, DPH, SD); environmental health (SM); environmental science and engineering (SM, SD); occupational health (MOH, SM, DPH, SD); physiology (SD). Part-time programs available.

Faculty: 24 full-time (4 women), 30 part-time/adjunct (4 women).
Students: 67 full-time (36 women), 7 part-time (3 women); includes 5 minority (2 African Americans, 3 Asian Americans or Pacific Islanders), 33 international. Average age 32. *80 applicants, 54% accepted.* In 1999, 11 master's, 12 doctorates awarded.
Degree requirements: For master's, thesis not required; for doctorate, dissertation, qualifying exam required.
Entrance requirements: For master's and doctorate, GRE, TOEFL. *Application deadline:* For fall admission, 1/3. *Application fee:* $60.
Expenses: Tuition: Full-time $22,950; part-time $574 per credit.
Financial aid: Fellowships, research assistantships, teaching assistantships, career-related internships or fieldwork, Federal Work-Study, grants, scholarships, traineeships, tuition waivers (partial), and unspecified assistantships available. Aid available to part-time students. Financial aid application deadline: 2/12; financial aid applicants required to submit FAFSA.
Faculty research: Industrial hygiene and occupational safety, population genetics, indoor and outdoor air pollution, cell and molecular biology of the lungs, infectious diseases.
Dr. Joseph D. Brain, Chairman, 617-432-1272.
Application contact: Stanley Hudson, Assistant Dean for Enrollment Services,

617-432-1031, *Fax:* 617-432-2009, *E-mail:* admisofc@hsph.harvard.edu.

Find an in-depth description at www.petersons.com/graduate.

■ HOWARD UNIVERSITY

Graduate School of Arts and Sciences, Department of Physiology and Biophysics, Program in Physiology, Washington, DC 20059-0002

AWARDS PhD.

Degree requirements: For doctorate, dissertation, comprehensive exam required, foreign language not required.
Entrance requirements: For doctorate, GRE General Test, minimum B average in field.
Expenses: Tuition: Full-time $10,500; part-time $583 per credit hour. Required fees: $405; $203 per semester.

■ ILLINOIS STATE UNIVERSITY

Graduate School, College of Arts and Sciences, Department of Biological Sciences, Normal, IL 61790-2200

AWARDS Biological sciences (MS); biology (PhD); botany (PhD); ecology (PhD); genetics (PhD); microbiology (PhD); physiology (PhD); zoology (PhD). Part-time programs available.

Faculty: 24 full-time (4 women).
Students: 57 full-time (25 women), 27 part-time (10 women); includes 6 minority (1 African American, 3 Asian Americans or Pacific Islanders, 2 Hispanic Americans), 13 international. *58 applicants, 59% accepted.* In 1999, 12 master's, 3 doctorates awarded.
Degree requirements: For master's, thesis or alternative required; for doctorate, variable foreign language requirement (computer language can substitute for one), dissertation, 2 terms of residency required.
Entrance requirements: For master's, GRE General Test, minimum GPA of 2.6 in last 60 hours; for doctorate, GRE General Test. *Application deadline:* Applications are processed on a rolling basis. *Application fee:* $0.
Expenses: Tuition, state resident: full-time $2,526; part-time $105 per credit hour. Tuition, nonresident: full-time $7,578; part-time $316 per credit hour. Required fees: $1,082; $38 per credit hour. Tuition and fees vary according to course load and program.
Financial aid: In 1999–00, 8 research assistantships, 63 teaching assistantships were awarded; Federal Work-Study, tuition waivers (full), and unspecified assistantships also available. Financial aid application deadline: 4/1.

Faculty research: Phenotypic plasticity in reproduction: molecular mechanisms, physiological control, adaptive significance; molecular stress physiology of *Listeria monocytogenes*; enzymology of eggshell formation in *Schistosoma mansoni*, compensatory adaptation in the rat midbrain after neurodegeneration, analysis of staphylococcal virulence germs. *Total annual research expenditures:* $586,648.
Dr. Hou Cheung, Chairperson, 309-438-3669.
Application contact: Derek A. McCracken, Graduate Adviser, 309-438-3664.

Find an in-depth description at www.petersons.com/graduate.

■ INDIANA STATE UNIVERSITY

School of Graduate Studies, College of Arts and Sciences, Department of Life Sciences, Terre Haute, IN 47809-1401

AWARDS Clinical laboratory sciences (MS); ecology (MA, MS, PhD); microbiology (MA, MS, PhD); physiology (MA, MS, PhD).

Degree requirements: For doctorate, computer language, dissertation required.
Entrance requirements: For master's and doctorate, GRE General Test. Electronic applications accepted.
Expenses: Tuition, state resident: full-time $3,552; part-time $148 per hour. Tuition, nonresident: full-time $8,088; part-time $337 per hour.

Find an in-depth description at www.petersons.com/graduate.

■ INDIANA UNIVERSITY BLOOMINGTON

Medical Sciences Program, Bloomington, IN 47405

AWARDS Anatomy and cell biology (MA, PhD); pharmacology (MS, PhD); physiology (MA, PhD).

Faculty: 12 full-time (3 women).
Students: 10 full-time (6 women), 6 part-time (2 women); includes 1 minority (Asian American or Pacific Islander), 3 international. In 1999, 1 master's, 1 doctorate awarded.
Entrance requirements: For master's, GRE, TOEFL, minimum GPA of 3.0; for doctorate, GRE, TOEFL. *Application deadline:* For fall admission, 1/15. *Application fee:* $45.
Expenses: Tuition, state resident: full-time $3,853; part-time $161 per credit hour. Tuition, nonresident: full-time $11,226; part-time $468 per credit hour. Required fees: $360 per year. Tuition and fees vary according to course load and program.
Dr. Talmage Bosin, Assistant Dean, 812-855-8118, *E-mail:* bosin@indiana.edu.

Indiana University Bloomington (continued)

Application contact: Kimberly Bunch, Director of Graduate Admissions, 812-855-1119, *E-mail:* kbunch@indiana.edu.

■ **INDIANA UNIVERSITY–PURDUE UNIVERSITY INDIANAPOLIS**

School of Medicine, Graduate Programs in Medicine, Department of Physiology and Biophysics, Indianapolis, IN 46202-2896

AWARDS MS, PhD, MD/MS, MD/PhD.

Students: 39 full-time (17 women), 5 part-time (3 women). Average age 24. *22 applicants, 18% accepted.* In 1999, 2 doctorates awarded (100% entered university research/teaching). Terminal master's awarded for partial completion of doctoral program.
Degree requirements: For master's, foreign language and thesis not required; for doctorate, dissertation required, foreign language not required. *Average time to degree:* Master's–2 years full-time; doctorate–5.5 years full-time.
Entrance requirements: For master's, GRE General Test, GRE Subject Test, previous course work in biology, calculus, physical chemistry, and physics; for doctorate, GRE General Test, GRE Subject Test. *Application deadline:* For fall admission, 4/15 (priority date). Applications are processed on a rolling basis. *Application fee:* $35 ($55 for international students).
Expenses: Tuition, state resident: full-time $13,245; part-time $158 per credit hour. Tuition, nonresident: full-time $30,330; part-time $455 per credit hour. Required fees: $121 per year. Tuition and fees vary according to course load and degree level.
Financial aid: In 1999–00, 14 students received aid, including 4 fellowships with partial tuition reimbursements available (averaging $16,000 per year), 7 research assistantships with partial tuition reimbursements available (averaging $16,000 per year); Federal Work-Study and institutionally sponsored loans also available.
Faculty research: Cardiovascular physiology, cell growth and development, respiratory biology, cell signaling mechanisms, cytoskeleton function.
Dr. Rodney A. Rhoades, Chair, 317-274-7772, *Fax:* 317-274-3318.
Application contact: Dr. Fred M. Pavalko, Graduate Director, 317-274-3140, *Fax:* 317-274-3318, *E-mail:* fpavalko@iupui.edu.
Find an in-depth description at www.petersons.com/graduate.

■ **IOWA STATE UNIVERSITY OF SCIENCE AND TECHNOLOGY**

College of Veterinary Medicine and Graduate College, Graduate Programs in Veterinary Medicine, Department of Biomedical Sciences, Ames, IA 50011

AWARDS Veterinary anatomy (MS, PhD); veterinary physiology (MS, PhD).

Faculty: 11 full-time, 1 part-time/adjunct.
Students: 14 full-time (7 women), 4 part-time (3 women); includes 1 minority (Asian American or Pacific Islander), 12 international. *12 applicants, 33% accepted.* In 1999, 1 master's, 4 doctorates awarded.
Degree requirements: For master's, thesis or alternative required; for doctorate, dissertation required.
Entrance requirements: For master's and doctorate, GRE General Test, TOEFL. *Application deadline:* For fall admission, 6/1 (priority date); for spring admission, 11/1 (priority date). *Application fee:* $20 ($50 for international students). Electronic applications accepted.
Expenses: Tuition, state resident: full-time $3,308. Tuition, nonresident: full-time $9,744. Tuition and fees vary according to course load and program.
Financial aid: In 1999–00, 7 research assistantships with partial tuition reimbursements (averaging $10,355 per year), 6 teaching assistantships with partial tuition reimbursements (averaging $9,450 per year) were awarded; scholarships also available.
Dr. Richard J. Martin, Chair, 515-294-2440, *Fax:* 515-294-2315, *E-mail:* biomedsci@iastate.edu.

■ **JOAN AND SANFORD I. WEILL MEDICAL COLLEGE AND GRADUATE SCHOOL OF MEDICAL SCIENCES OF CORNELL UNIVERSITY**

Graduate School of Medical Sciences, Program in Physiology, Biophysics, and Molecular Medicine, New York, NY 10021

AWARDS PhD, MD/PhD.

Faculty: 21 full-time (4 women).
Students: 24 full-time (9 women); includes 3 minority (2 Asian Americans or Pacific Islanders, 1 Hispanic American), 10 international. *63 applicants, 10% accepted.* In 2000, 2 degrees awarded.
Degree requirements: For doctorate, dissertation, final exam required.
Entrance requirements: For doctorate, GRE General Test, GRE Subject Test, MCAT (MD/PhD), introductory courses in biology, inorganic and organic chemistry, physics, and mathematics.

Application deadline: For fall admission, 1/15. *Application fee:* $50.
Expenses: All students in good standing receive an annual stipend of $22,880.
Financial aid: Fellowships, stipends available.
Lawrence Palmer, Director, 212-746-6350.

■ **JOHNS HOPKINS UNIVERSITY**

School of Hygiene and Public Health, Department of Environmental Health Sciences, Division of Physiology, Baltimore, MD 21218-2699

AWARDS MHS, Sc M, PhD.

Degree requirements: For master's, thesis required (for some programs), foreign language not required; for doctorate, dissertation, 1 year full-time residency, oral and written exams required, foreign language not required.
Entrance requirements: For master's and doctorate, GRE General Test, TOEFL. *Application deadline:* For fall admission, 2/1 (priority date). Applications are processed on a rolling basis. *Application fee:* $60. Electronic applications accepted.
Expenses: Tuition: Full-time $23,660; part-time $493 per unit. Full-time tuition and fees vary according to degree level, campus/location and program.
Financial aid: Federal Work-Study, institutionally sponsored loans, scholarships, traineeships, and stipends available. Aid available to part-time students. Financial aid application deadline: 4/15.
Faculty research: Chemoreceptor mechanisms, pulmonary immunology, mechanics, medical ethics, cerebral circulation, pulmonary circulation.
Dr. Wayne Mitzner, Director, 410-955-3515, *E-mail:* wmitzner@jhsph.edu.

Find an in-depth description at www.petersons.com/graduate.

■ **JOHNS HOPKINS UNIVERSITY**

School of Medicine, Graduate Programs in Medicine, Department of Physiology, Baltimore, MD 21218-2699

AWARDS Cellular and molecular physiology (PhD); physiology (PhD).

Faculty: 29 full-time (7 women).
Students: 6 full-time (all women); includes 4 minority (all Asian Americans or Pacific Islanders). Average age 28. In 1999, 2 degrees awarded.
Degree requirements: For doctorate, dissertation, oral and qualifying exams required, foreign language not required.
Entrance requirements: For doctorate, GRE General Test, TOEFL, previous course work in biology, calculus, chemistry, and physics. *Application deadline:* For fall admission, 2/1. *Application fee:* $50.
Expenses: Tuition: Full-time $23,660.

Financial aid: In 1999–00, 6 students received aid. Institutionally sponsored loans and stipends available. Financial aid application deadline: 2/1.

Faculty research: Membrane biochemistry and biophysics; signal transduction; developmental genetics and physiology; physiology and biochemistry; transporters, carriers, and ion channels.

Dr. William S. Agnew, Chairman, 410-955-4581.

■ KANSAS STATE UNIVERSITY

College of Veterinary Medicine and Graduate School, Graduate Programs in Veterinary Medicine, Department of Anatomy and Physiology, Manhattan, KS 66506

AWARDS Anatomy (MS); physiology (MS, PhD).

Degree requirements: For master's, thesis required, foreign language not required; for doctorate, dissertation required.

Expenses: Tuition, state resident: part-time $103 per credit hour. Tuition, nonresident: part-time $338 per credit hour. Required fees: $17 per credit hour. One-time fee: $64 part-time.

■ KANSAS STATE UNIVERSITY

Graduate School, College of Arts and Sciences, Division of Biology, Manhattan, KS 66506

AWARDS Cell biology (MS, PhD); developmental biology and physiology (MS, PhD); microbiology and immunology (MS, PhD); molecular biology and genetics (MS, PhD); systematics and ecology (MS, PhD); virology and oncology (MS, PhD). Terminal master's awarded for partial completion of doctoral program.

Degree requirements: For master's and doctorate, thesis/dissertation required, foreign language not required.

Entrance requirements: For master's and doctorate, GRE General Test. Electronic applications accepted.

Expenses: Tuition, state resident: part-time $103 per credit hour. Tuition, nonresident: part-time $338 per credit hour. Required fees: $17 per credit hour. One-time fee: $64 part-time.

Faculty research: Immune cell function, prairie ecology.

■ KENT STATE UNIVERSITY

College of Arts and Sciences, Department of Biological Sciences, Kent, OH 44242-0001

AWARDS Botany (MA, MS, PhD); ecology (MS, PhD); physiology (MS, PhD); zoology (MA, PhD).

Faculty: 28 full-time.

Students: 31 full-time (17 women), 10 part-time (9 women); includes 1 minority (African American), 12 international. *30 applicants, 77% accepted.* In 1999, 6 master's, 5 doctorates awarded.

Degree requirements: For master's and doctorate, thesis/dissertation required, foreign language not required.

Entrance requirements: For master's, GRE General Test, minimum GPA of 2.75; for doctorate, GRE General Test, minimum GPA of 3.0. *Application deadline:* For fall admission, 7/12. Applications are processed on a rolling basis. *Application fee:* $30.

Expenses: Tuition, state resident: full-time $5,334; part-time $243 per hour. Tuition, nonresident: full-time $10,238; part-time $466 per hour.

Financial aid: Fellowships, research assistantships, teaching assistantships, Federal Work-Study, institutionally sponsored loans, and tuition waivers (full) available. Financial aid application deadline: 2/1.

Dr. Brent C. Bruot, Chairman, 330-672-3613, *Fax:* 330-672-3713.

Application contact: Dr. John R. D. Stalvey, Coordinator of Graduate Studies, 330-672-2819.

■ KENT STATE UNIVERSITY

School of Biomedical Sciences, Program in Physiology, Kent, OH 44242-0001

AWARDS MS, PhD. Offered in cooperation with Northeastern Ohio Universities College of Medicine. Terminal master's awarded for partial completion of doctoral program.

Degree requirements: For master's and doctorate, thesis/dissertation required, foreign language not required.

Entrance requirements: For master's and doctorate, GRE General Test.

Expenses: Tuition, state resident: full-time $5,334; part-time $243 per hour. Tuition, nonresident: full-time $10,238; part-time $466 per hour.

Faculty research: Reproductive neuroendocrinology, steroid synthesis, seasonal reproduction, exercise physiology, hormonal and neural vascular control, fluid balance.

Find an in-depth description at www.petersons.com/graduate.

■ LOMA LINDA UNIVERSITY

Graduate School, Graduate Programs in Medicine, Program in Physiology, Loma Linda, CA 92350

AWARDS MS, PhD. Part-time programs available.

Degree requirements: For master's, thesis or alternative required, foreign language not required; for doctorate, 2 foreign languages (computer language can substitute for one), dissertation required.

Entrance requirements: For master's and doctorate, GRE General Test. *Application deadline:* Applications are processed on a rolling basis. *Application fee:* $40.

Expenses: Tuition: Part-time $395 per unit.

Financial aid: Tuition waivers (full and partial) available. Aid available to part-time students.

Faculty research: Fetal physiology, neurophysiology, endocrinology, reproductive physiology, neuroendocrinology.

Dr. John Leonora, Coordinator, 909-824-4564.

■ LOUISIANA STATE UNIVERSITY HEALTH SCIENCES CENTER

School of Graduate Studies in New Orleans, Department of Physiology, New Orleans, LA 70112-2223

AWARDS MS, PhD, MD/PhD.

Faculty: 23 full-time (4 women).

Students: 7 full-time (1 woman), 1 international. Average age 26. *22 applicants, 23% accepted.* In 1999, 1 degree awarded (100% entered university research/teaching). Terminal master's awarded for partial completion of doctoral program.

Degree requirements: For master's and doctorate, thesis/dissertation required, foreign language not required. *Average time to degree:* Doctorate–5 years full-time.

Entrance requirements: For master's and doctorate, GRE General Test, TOEFL. *Application deadline:* For fall admission, 4/1 (priority date); for spring admission, 10/1. Applications are processed on a rolling basis. *Application fee:* $30.

Expenses: Tuition, state resident: full-time $2,878; part-time $126 per hour. Tuition, nonresident: full-time $6,003; part-time $265 per hour. Required fees: $2,272. Tuition and fees vary according to course load, degree level and program.

Financial aid: In 1999–00, 7 students received aid, including fellowships with full tuition reimbursements available (averaging $16,000 per year), teaching assistantships with full tuition reimbursements available (averaging $16,000 per year); tuition waivers (full) also available.

Faculty research: Host defense, lipoprotein metabolism, regulation of cardiopulmonary function, alcohol and drug abuse, cell to cell communication, cytokinesis, physiologic functions of nitric oxide.

Dr. John J. Spitzer, Head, 504-568-6172.

Louisiana State University Health Sciences Center (continued)

Application contact: Dr. Michele A. Meneray, Graduate Coordinator, 504-948-8585, *Fax:* 504-619-8617, *E-mail:* mmener@lsumc.edu.

Find an in-depth description at www.petersons.com/graduate.

■ LOUISIANA STATE UNIVERSITY HEALTH SCIENCES CENTER

School of Graduate Studies in Shreveport, Department of Physiology and Biophysics, Shreveport, LA 71130-3932

AWARDS Physiology (MS, PhD).

Degree requirements: For master's and doctorate, thesis/dissertation required, foreign language not required.
Entrance requirements: For master's and doctorate, GRE General Test, TOEFL.
Expenses: Tuition, state resident: full-time $2,878; part-time $126 per hour. Tuition, nonresident: full-time $6,003; part-time $265 per hour. Required fees: $2,272. Tuition and fees vary according to course load, degree level and program.
Faculty research: Cardiovascular, gastrointestinal, renal, and neutrophil function; cellular detoxication systems; hypoxia and mitochondria function.

■ LOYOLA UNIVERSITY CHICAGO

Graduate School, Program in Cell and Molecular Physiology, Maywood, IL 60153

AWARDS MS, PhD. MS offered only to students enrolled in a first professional degree program.

Faculty: 14 full-time (1 woman).
Students: 9 full-time (4 women); includes 5 minority (4 Asian Americans or Pacific Islanders, 1 Hispanic American), 3 international. Average age 30. *53 applicants, 11% accepted.* In 1999, 2 doctorates awarded (100% entered university research/teaching). Terminal master's awarded for partial completion of doctoral program.
Degree requirements: For master's, thesis required, foreign language not required; for doctorate, dissertation, comprehensive exam required, foreign language not required.
Entrance requirements: For master's and doctorate, GRE General Test. *Application deadline:* For fall admission, 5/1. *Application fee:* $35.
Expenses: Tuition: Part-time $500 per credit hour. Required fees: $42 per term.
Financial aid: In 1999–00, 8 students received aid, including 7 fellowships with

tuition reimbursements available (averaging $17,500 per year).
Faculty research: Cardiovascular system–emphasis in neural and metabolic control of circulation, ion channels, excitation contraction coupling.
Dr. Donald M. Bers, Chair, 708-216-6305, *Fax:* 708-216-6308.
Application contact: Dr. Stephen B. Jones, Graduate Program Director, 708-327-2470, *Fax:* 708-216-6308, *E-mail:* sjones@luc.edu.

Find an in-depth description at www.petersons.com/graduate.

■ MAHARISHI UNIVERSITY OF MANAGEMENT

Graduate Studies, Program in Physiology, Molecular, and Cell Biology, Fairfield, IA 52557

AWARDS MS, PhD. Program admits applicants every other year. Terminal master's awarded for partial completion of doctoral program.

Degree requirements: For master's, thesis or alternative required, foreign language not required; for doctorate, dissertation required, foreign language not required.
Entrance requirements: For master's, GMAT or GRE, minimum GPA of 3.0; for doctorate, GRE General Test, GRE Subject Test, bachelor's degree in biology, chemistry, or related quantitative science; minimum GPA of 3.0.
Faculty research: Developmental neurobiology, aging, neurochemistry.

■ MARQUETTE UNIVERSITY

Graduate School, College of Arts and Sciences, Department of Biology, Milwaukee, WI 53201-1881

AWARDS Cell biology (MS, PhD); developmental biology (MS, PhD); ecology (MS, PhD); endocrinology (MS, PhD); evolutionary biology (MS, PhD); genetics (MS, PhD); microbiology (MS, PhD); molecular biology (MS, PhD); muscle and exercise physiology (MS, PhD); neurobiology (MS, PhD); reproductive physiology (MS, PhD).

Faculty: 16 full-time (4 women), 2 part-time/adjunct (0 women).
Students: 34 full-time (20 women), 3 part-time; includes 3 minority (all Asian Americans or Pacific Islanders), 2 international. Average age 31. *42 applicants, 29% accepted.* In 1999, 1 master's, 4 doctorates awarded. Terminal master's awarded for partial completion of doctoral program.
Degree requirements: For master's, thesis, 1 year of teaching experience or equivalent, comprehensive exam required,

foreign language not required; for doctorate, dissertation, 1 year of teaching experience or equivalent, qualifying exam required, foreign language not required.
Entrance requirements: For master's and doctorate, GRE General Test, GRE Subject Test, TOEFL. *Application fee:* $40.
Expenses: Tuition: Part-time $510 per credit hour. Tuition and fees vary according to program.
Financial aid: In 1999–00, 4 fellowships, 22 teaching assistantships were awarded; research assistantships, Federal Work-Study, institutionally sponsored loans, scholarships, and tuition waivers (full and partial) also available. Aid available to part-time students. Financial aid application deadline: 2/15.
Faculty research: Microbial and invertebrate ecology, evolution of gene function, DNA methylation, DNA arrangement. *Total annual research expenditures:* $1.5 million.
Dr. Brian Unsworth, Chairman, 414-288-7355, *Fax:* 414-288-7357.
Application contact: Barbara DeNoyer, Graduate Studies Coordinator, 414-288-7355, *Fax:* 414-288-7357.

Find an in-depth description at www.petersons.com/graduate.

■ MCP HAHNEMANN UNIVERSITY

School of Medicine, Biomedical Graduate Programs, Department of Physiology, Philadelphia, PA 19102-1192

AWARDS PhD, MD/PhD.

Degree requirements: For doctorate, dissertation, qualifying exam required, foreign language not required.
Faculty research: Cardiovascular physiology, cell and receptor physiology, gastrointestinal physiology, neurophysiology.

■ MEDICAL COLLEGE OF GEORGIA

School of Graduate Studies, Department of Physiology and Endocrinology, Augusta, GA 30912-1500

AWARDS MS, PhD.

Faculty: 13 full-time (0 women).
Students: 8 full-time (3 women), 3 international. *7 applicants, 43% accepted.* In 1999, 1 degree awarded. Terminal master's awarded for partial completion of doctoral program.
Degree requirements: For master's and doctorate, thesis/dissertation required, foreign language not required.
Entrance requirements: For master's and doctorate, GRE General Test, TOEFL. *Application deadline:* For fall admission,

6/30 (priority date). Applications are processed on a rolling basis. *Application fee:* $25.

Expenses: Tuition, state resident: full-time $2,896; part-time $121 per hour. Tuition, nonresident: full-time $11,584; part-time $483 per hour. Required fees: $286; $143 per semester. Tuition and fees vary according to program.

Financial aid: In 1999–00, research assistantships with partial tuition reimbursements (averaging $15,500 per year); fellowships, grants, institutionally sponsored loans, and traineeships also available. Aid available to part-time students. Financial aid application deadline: 3/31; financial aid applicants required to submit FAFSA.

Faculty research: Ovarian follicular development, Leydig cells, microsomal androgen metabolism, ATP hydrolysis. Dr. Clinton Webb, Chair, 706-721-2781, *Fax:* 706-721-7299, *E-mail:* cwebb@ mail.mcg.edu.

Application contact: Dr. Chris Wingard, Director, 706-721-3741, *Fax:* 706-721-7299, *E-mail:* cwingard@mail.mcg.edu.

■ MEDICAL COLLEGE OF OHIO

Graduate School, Department of Physiology, Toledo, OH 43614-5805

AWARDS MS. Part-time programs available.

Faculty: 13 full-time (3 women).
Students: 6 full-time (3 women), 2 part-time (both women), 1 international. Average age 25. In 1999, 2 degrees awarded (100% found work related to degree).
Degree requirements: For master's, thesis, qualifying exam required, foreign language not required. *Average time to degree:* Master's–2 years full-time.
Entrance requirements: For master's, GRE General Test, minimum undergraduate GPA of 3.0. *Application fee:* $30.
Expenses: Tuition, state resident: part-time $193 per hour. Tuition, nonresident: part-time $445 per hour. Tuition and fees vary according to degree level.
Financial aid: In 1999–00, 2 fellowships were awarded; Federal Work-Study and institutionally sponsored loans also available. Financial aid applicants required to submit FAFSA.
Faculty research: Blood flow regulation, cell physiology, psychophysiology, contraceptive and reproductive physiology, hypertension. *Total annual research expenditures:* $1.4 million.
Dr. John P. Rapp, Chairman, 419-383-4117, *Fax:* 419-383-6140, *E-mail:* mcogradschool@mco.edu.
Application contact: Joann Braatz, Clerk, 419-383-4117, *Fax:* 419-383-6140, *E-mail:* mcogradschool@mco.edu.

■ MEDICAL COLLEGE OF WISCONSIN

Graduate School of Biomedical Sciences, Department of Physiology, Milwaukee, WI 53226-0509

AWARDS MS, PhD, MD/MS, MD/PhD.

Degree requirements: For master's and doctorate, thesis/dissertation required, foreign language not required.
Entrance requirements: For master's and doctorate, GRE General Test, TOEFL.
Expenses: Tuition, state resident: full-time $9,318. Tuition, nonresident: full-time $9,318. Required fees: $115.
Faculty research: Cardiovascular, respiratory, renal, and exercise physiology; mathematical modeling; molecular and cellular biology.

Find an in-depth description at www.petersons.com/graduate.

■ MEDICAL UNIVERSITY OF SOUTH CAROLINA

College of Graduate Studies, Department of Physiology, Charleston, SC 29425-0002

AWARDS MS, PhD, MD/PhD.

Faculty: 8 part-time/adjunct (0 women).
Students: 13 full-time (4 women); includes 4 minority (1 African American, 3 Asian Americans or Pacific Islanders). Average age 32. *15 applicants, 73% accepted.* In 1999, 2 doctorates awarded. Terminal master's awarded for partial completion of doctoral program.
Degree requirements: For master's, thesis, research seminar required; for doctorate, dissertation, teaching and research seminar, oral and written exams required.
Entrance requirements: For master's and doctorate, GRE General Test, TOEFL, interview. *Application deadline:* Applications are processed on a rolling basis. *Application fee:* $55. Electronic applications accepted.
Expenses: Tuition, state resident: full-time $3,470; part-time $160 per semester hour. Tuition, nonresident: full-time $4,426; part-time $213 per semester hour. Required fees: $408 per semester. One-time fee: $160. Tuition and fees vary according to program.
Financial aid: In 1999–00, 1 fellowship (averaging $16,000 per year), 2 research assistantships were awarded; teaching assistantships, Federal Work-Study and tuition waivers (partial) also available. Financial aid application deadline: 4/1; financial aid applicants required to submit FAFSA.
Faculty research: Molecular and cellular physiology, circulatory shock, neuroendocrinology, mammalian

myocardium, spinal cord injury. *Total annual research expenditures:* $905,507. Peter Kalivas, Chairman, 843-792-2005.
Application contact: Julie Johnston, Director of Admissions, 843-792-8710, *Fax:* 843-792-3764.

Find an in-depth description at www.petersons.com/graduate.

■ MEHARRY MEDICAL COLLEGE

School of Graduate Studies, Department of Anatomy and Physiology, Nashville, TN 37208-9989

AWARDS Physiology (PhD).

Faculty: 12 full-time (2 women).
Students: 15 full-time (11 women); includes 14 minority (all African Americans). Average age 29. *5 applicants, 0% accepted.* In 1999, 2 degrees awarded.
Degree requirements: For doctorate, dissertation, oral and written comprehensive exams required, foreign language not required. *Average time to degree:* Doctorate–6.4 years full-time.
Entrance requirements: For doctorate, GRE. *Application deadline:* For fall admission, 6/1. Applications are processed on a rolling basis. *Application fee:* $45.
Expenses: Tuition: Full-time $8,732. Required fees: $2,133.
Financial aid: Fellowships, institutionally sponsored loans available. Financial aid application deadline: 4/15; financial aid applicants required to submit FAFSA.
Faculty research: Neurochemistry, pain, smooth muscle tone, HP axis and peptides neural plasticity.
Dr. Hubert K. Rucker, Chairman, 615-327-6288, *Fax:* 615-327-6655, *E-mail:* rrucker@mmc.edu.

■ MICHIGAN STATE UNIVERSITY

College of Human Medicine and Graduate School, Graduate Programs in Human Medicine and College of Natural Science and Graduate Programs in Osteopathic Medicine, Department of Physiology, East Lansing, MI 48824

AWARDS MS, PhD.

Faculty: 18.
Students: 7 full-time (4 women), 3 part-time (2 women), 7 international. Average age 30. *21 applicants, 24% accepted.* In 1999, 4 degrees awarded. Terminal master's awarded for partial completion of doctoral program.
Degree requirements: For master's, thesis required, foreign language not required; for doctorate, dissertation, comprehensive exams required, foreign language not required.
Entrance requirements: For master's and doctorate, GRE General Test, GRE

Michigan State University (continued)
Subject Test, TOEFL, minimum GPA of 3.0. *Application deadline:* For fall admission, 1/15 (priority date). *Application fee:* $30 ($40 for international students). Electronic applications accepted.

Expenses: Tuition, state resident: full-time $10,868; part-time $229 per credit. Tuition, nonresident: full-time $23,168; part-time $464 per credit.

Financial aid: In 1999–00, research assistantships (averaging $12,720 per year); fellowships, institutionally sponsored loans also available. Aid available to part-time students. Financial aid application deadline: 4/1.

Faculty research: Retinal changes in glaucoma and with neuro protection, role of stroma in mammary gland proliferation, physiologic basis of muscle functional MRI, hepatic gene expression. *Total annual research expenditures:* $1.8 million.
Dr. William Spielman, Chairperson, 517-353-4539, *Fax:* 517-432-1967.
Application contact: Dr. Donald B. Jump, Director of Research and Graduate Studies, 517-355-6475 Ext. 1246, *Fax:* 517-355-5125, *E-mail:* shaft@psl.msu.edu.

Find an in-depth description at www.petersons.com/graduate.

■ **MICHIGAN STATE UNIVERSITY**

College of Osteopathic Medicine and Graduate School, Graduate Programs in Osteopathic Medicine, East Lansing, MI 48824

AWARDS Anatomy (MS, PhD); biochemistry (MS, PhD); microbiology (PhD); pathology (MS, PhD); pharmacology/toxicology (MS, PhD), including pharmacology; physiology (MS, PhD), including environmental toxicology (PhD), neuroscience (PhD). Part-time programs available.

Students: 17 full-time (9 women), 1 part-time; includes 5 minority (3 Asian Americans or Pacific Islanders, 1 Hispanic American, 1 Native American), 5 international. Average age 26. *33 applicants, 9% accepted.* In 1999, 2 doctorates awarded.

Degree requirements: For doctorate, dissertation required.

Entrance requirements: For master's and doctorate, GRE. *Application deadline:* For fall admission, 3/1 (priority date). Applications are processed on a rolling basis. *Application fee:* $30 ($40 for international students).

Expenses: Tuition, area resident: Full-time $15,879. Tuition, state resident: full-time $33,797.

Financial aid: Fellowships, research assistantships, teaching assistantships, career-related internships or fieldwork, Federal Work-Study, and institutionally

sponsored loans available. Financial aid application deadline: 4/2. *Total annual research expenditures:* $5 million.
Application contact: Dr. Veronica M. Maher, Associate Dean for Graduate Studies, 517-353-7785, *Fax:* 517-353-9004, *E-mail:* maher@com.msu.edu.

■ **MISSISSIPPI STATE UNIVERSITY**

College of Agriculture and Life Sciences, Program in Animal Physiology, Mississippi State, MS 39762

AWARDS MS, PhD. Part-time programs available.

Faculty: 21 full-time (2 women).
Students: 12 full-time (5 women), 2 part-time (1 woman); includes 1 minority (African American), 3 international. Average age 29. *8 applicants, 63% accepted.* In 1999, 2 master's, 3 doctorates awarded. Terminal master's awarded for partial completion of doctoral program.

Degree requirements: For master's and doctorate, thesis/dissertation, comprehensive oral or written exam required, foreign language not required.

Entrance requirements: For master's, GRE, TOEFL, minimum GPA of 2.75; for doctorate, GRE, TOEFL. *Application deadline:* For fall admission, 7/1; for spring admission, 11/1. Applications are processed on a rolling basis. *Application fee:* $25 for international students.

Expenses: Tuition, state resident: full-time $3,017; part-time $168 per credit. Tuition, nonresident: full-time $6,119; part-time $340 per credit. Part-time tuition and fees vary according to course load and program.

Financial aid: In 1999–00, 10 students received aid, including 5 research assistantships with full tuition reimbursements available (averaging $9,900 per year); Federal Work-Study, institutionally sponsored loans, and unspecified assistantships also available. Financial aid applicants required to submit FAFSA.

Faculty research: Ovarian physiology, embryo technology, growth physiology, digestive physiology, toxicology. *Total annual research expenditures:* $550,000.
Dr. Terry E. Kiser, Director of Admissions, 662-325-2802, *Fax:* 662-325-8873, *E-mail:* tkiser@ads.msstate.edu.

■ **MOUNT SINAI SCHOOL OF MEDICINE OF NEW YORK UNIVERSITY**

Graduate School of Biological Sciences, Department of Physiology and Biophysics, New York, NY 10029-6504

AWARDS PhD, MD/PhD.

Faculty: 19 full-time.
Students: 20 full-time (9 women); includes 1 minority (Hispanic American), 9 international. *25 applicants, 12% accepted.* In 1999, 1 degree awarded.

Degree requirements: For doctorate, dissertation required, foreign language not required.

Entrance requirements: For doctorate, GRE General Test, GRE Subject Test, TOEFL. *Application deadline:* For fall admission, 4/15. *Application fee:* $35.

Expenses: Tuition: Full-time $21,750; part-time $725 per credit. Required fees: $750; $25 per credit. Full-time tuition and fees vary according to student level.

Financial aid: Fellowships, grants available.
Dr. Harel Weinstein, Chairman, 212-241-7018.
Application contact: C. Gita Bosch, Administrative Manager and Assistant Dean, 212-241-6546, *Fax:* 212-241-0651, *E-mail:* grads@mssm.edu.

Find an in-depth description at www.petersons.com/graduate.

■ **NEW YORK MEDICAL COLLEGE**

Graduate School of Basic Medical Sciences, Program in Physiology, Valhalla, NY 10595-1691

AWARDS MS, PhD, MD/PhD. Part-time and evening/weekend programs available.

Faculty: 11 full-time (0 women).
Students: 12 full-time (5 women), 9 part-time (4 women); includes 4 minority (all Asian Americans or Pacific Islanders), 4 international. Average age 30. *21 applicants, 38% accepted.* In 1999, 5 master's awarded. Terminal master's awarded for partial completion of doctoral program.

Degree requirements: For master's and doctorate, computer language, thesis/dissertation required, foreign language not required.

Entrance requirements: For master's, GRE General Test, TOEFL; for doctorate, GRE General Test, GRE Subject Test, TOEFL. *Application deadline:* For fall admission, 7/1 (priority date); for spring admission, 12/1 (priority date). Applications are processed on a rolling basis. *Application fee:* $35 ($60 for international students).

Expenses: Tuition: Part-time $430 per credit. Required fees: $15 per semester. One-time fee: $100.
Financial aid: In 1999–00, 5 research assistantships with full tuition reimbursements were awarded; career-related internships or fieldwork, Federal Work-Study, grants, institutionally sponsored loans, and tuition waivers also available. Aid available to part-time students. Financial aid applicants required to submit FAFSA.
Faculty research: Cardiovascular physiology, renal physiology, endocrine physiology, neurophysiology, cell biology.
Dr. Carl I. Thompson, Director, 914-594-4106.

Find an in-depth description at www.petersons.com/graduate.

■ NEW YORK UNIVERSITY

Graduate School of Arts and Science, Department of Basic Medical Sciences, New York, NY 10012-1019
AWARDS Biochemistry (MS, PhD); cell biology (MS, PhD); microbiology (MS, PhD); parasitology (PhD); pathology (MS, PhD); pharmacology (PhD); physiology (MS, PhD). Part-time programs available.
Faculty: 23 full-time (2 women), 3 part-time/adjunct (1 woman).
Students: 182 full-time (73 women), 4 part-time (1 woman); includes 52 minority (11 African Americans, 34 Asian Americans or Pacific Islanders, 7 Hispanic Americans), 48 international. Average age 26. *671 applicants, 11% accepted.* In 1999, 15 master's, 32 doctorates awarded. Terminal master's awarded for partial completion of doctoral program.
Degree requirements: For master's, thesis or alternative, written comprehensive exam required, foreign language not required; for doctorate, one foreign language, dissertation, oral and written comprehensive exams required.
Entrance requirements: For master's and doctorate, GRE General Test, GRE Subject Test, TOEFL. *Application deadline:* For fall admission, 2/1 (priority date). *Application fee:* $60.
Expenses: Tuition: Full-time $17,880; part-time $745 per credit. Required fees: $1,140; $35 per credit. Tuition and fees vary according to course load and program.
Financial aid: Fellowships with tuition reimbursements, research assistantships with tuition reimbursements, teaching assistantships with tuition reimbursements, career-related internships or fieldwork, Federal Work-Study, institutionally sponsored loans, and tuition waivers (full and partial) available. Financial aid application deadline: 2/1; financial aid applicants required to submit FAFSA.

Dr. Joel D. Oppenheim, Director, 212-263-5648, *Fax:* 212-263-7600, *E-mail:* sackler-info@nyumed.med.nyu.edu.

■ NEW YORK UNIVERSITY

Graduate School of Arts and Science, Department of Biology, New York, NY 10012-1019
AWARDS Applied recombinant DNA technology (MS); biochemistry (PhD); biomedical journalism (MA); cell biology (PhD); computers in biological research (MS); environmental biology (PhD); general biology (MS); neural sciences and physiology (PhD); oral biology (MS); population and evolutionary biology (PhD). Part-time programs available.
Faculty: 22 full-time (5 women), 8 part-time/adjunct.
Students: 99 full-time (48 women), 61 part-time (31 women); includes 30 minority (4 African Americans, 22 Asian Americans or Pacific Islanders, 4 Hispanic Americans), 52 international. Average age 24. *371 applicants, 41% accepted.* In 1999, 54 master's, 4 doctorates awarded. Terminal master's awarded for partial completion of doctoral program.
Degree requirements: For master's, thesis or alternative, qualifying paper required, foreign language not required; for doctorate, dissertation, oral and written comprehensive exams required, foreign language not required.
Entrance requirements: For master's, GRE General Test, TOEFL; for doctorate, GRE General Test, GRE Subject Test, TOEFL. *Application deadline:* For fall admission, 1/4 (priority date). *Application fee:* $60.
Expenses: Tuition: Full-time $17,880; part-time $745 per credit. Required fees: $1,140; $35 per credit. Tuition and fees vary according to course load and program.
Financial aid: Fellowships with tuition reimbursements, research assistantships with tuition reimbursements, teaching assistantships with tuition reimbursements, career-related internships or fieldwork, Federal Work-Study, institutionally sponsored loans, and tuition waivers (full and partial) available. Financial aid application deadline: 1/4; financial aid applicants required to submit FAFSA.
Faculty research: Development and genetics, neurobiology, plant sciences, molecular and cell biology.
Philip Furmanski, Chairman, 212-998-8200.
Application contact: Gloria Coruzzi, Director of Graduate Studies, 212-998-8200, *Fax:* 212-995-4015, *E-mail:* biology@nyu.edu.

Find an in-depth description at www.petersons.com/graduate.

■ NEW YORK UNIVERSITY

School of Medicine and Graduate School of Arts and Science, Sackler Institute of Graduate Biomedical Sciences, Department of Neuroscience and Physiology, New York, NY 10012-1019
AWARDS Neuroscience (PhD); physiology (PhD).
Faculty: 22 full-time (4 women).
Students: 16 full-time (2 women); includes 3 minority (2 Asian Americans or Pacific Islanders, 1 Hispanic American), 4 international. Average age 25. In 1999, 2 degrees awarded.
Degree requirements: For doctorate, one foreign language, dissertation, qualifying exam required. *Average time to degree:* Doctorate–5.5 years full-time.
Entrance requirements: For doctorate, GRE General Test, GRE Subject Test, TOEFL. *Application deadline:* For fall admission, 2/1 (priority date). Applications are processed on a rolling basis. *Application fee:* $60.
Expenses: Tuition: Full-time $17,880; part-time $745 per credit. Required fees: $1,140; $35 per credit. Tuition and fees vary according to course load and program.
Financial aid: In 1999–00, 6 research assistantships were awarded; fellowships, teaching assistantships Financial aid application deadline: 1/15.
Faculty research: Synaptic transmission, retinal physiology, signal transduction, CNS intrinsic properties, cerebellar function. *Total annual research expenditures:* $2.3 million.
Dr. Rodolfo R. Llinás, Chairman, 212-263-5415.
Application contact: Dr. Stewart A. Bloomfield, Graduate Adviser, 212-263-5770, *Fax:* 212-263-8072, *E-mail:* blooms01@med.nyu.edu.

Find an in-depth description at www.petersons.com/graduate.

■ NORTH CAROLINA STATE UNIVERSITY

Graduate School, College of Agriculture and Life Sciences and College of Veterinary Medicine, Program in Physiology, Raleigh, NC 27695
AWARDS MLS, MS, PhD.
Students: 25 full-time (14 women), 14 part-time (6 women); includes 9 minority (7 African Americans, 2 Hispanic Americans), 5 international. Average age 31. *18 applicants, 33% accepted.*
Degree requirements: For master's and doctorate, thesis/dissertation required, foreign language not required.

North Carolina State University (continued)

Entrance requirements: For master's and doctorate, GRE General Test. *Application deadline:* For fall admission, 6/25; for spring admission, 11/25. Applications are processed on a rolling basis. *Application fee:* $45.

Expenses: Tuition, state resident: full-time $1,578. Tuition, nonresident: full-time $10,744. Required fees: $892. Full-time tuition and fees vary according to program.

Financial aid: Fellowships, research assistantships, teaching assistantships available.

Faculty research: Reproduction, neurophysiology, cell gastrointestinal physiology.

Dr. James N. Petitte, Director of Graduate Programs, 919-829-5389, *Fax:* 919-829-2625, *E-mail:* j_petitte@ncsu.edu.

■ NORTHWESTERN UNIVERSITY

The Graduate School, Judd A. and Marjorie Weinberg College of Arts and Sciences, Department of Neurobiology and Physiology, Evanston, IL 60208

AWARDS MS. Admissions and degrees offered through The Graduate School. Part-time programs available.

Faculty: 12 full-time (3 women), 11 part-time/adjunct (3 women).
Students: 7 full-time (4 women), 5 part-time (2 women). *31 applicants, 52% accepted.* In 1999, 8 degrees awarded.
Degree requirements: For master's, thesis required, foreign language not required.
Entrance requirements: For master's, GRE General Test and MCAT (strongly recommended), TOEFL. *Application deadline:* For fall admission, 8/1. Applications are processed on a rolling basis. *Application fee:* $50 ($55 for international students).
Expenses: Tuition: Full-time $23,301. Full-time tuition and fees vary according to program.
Financial aid: Career-related internships or fieldwork, Federal Work-Study, and institutionally sponsored loans available. Financial aid application deadline: 1/15; financial aid applicants required to submit FAFSA.
Faculty research: Sensory neurobiology and neuroendocrinology, reproductive biology, vision physiology and psychophysics, cell and developmental biology.
Lawrence Pinto, Chair, 847-491-5521.

Application contact: Michael Kennedy, Assistant Chair, 847-491-5521, *Fax:* 847-491-5211, *E-mail:* m-kennedy@northwestern.edu.

Find an in-depth description at www.petersons.com/graduate.

■ NORTHWESTERN UNIVERSITY

The Graduate School, Judd A. and Marjorie Weinberg College of Arts and Sciences, Interdepartmental Biological Sciences Program (IBiS), Concentration in Integrative Biology, Evanston, IL 60208

AWARDS PhD.

Faculty: 59 full-time (11 women).
Students: 79 full-time (46 women); includes 5 minority (2 African Americans, 3 Hispanic Americans), 14 international. *236 applicants, 19% accepted.*
Degree requirements: For doctorate, dissertation, 2 quarters of teaching experience required, foreign language not required.
Entrance requirements: For doctorate, GRE General Test, TOEFL, TSE. *Application deadline:* For fall admission, 1/15. Applications are processed on a rolling basis. *Application fee:* $50 ($55 for international students).
Expenses: Tuition: Full-time $23,301. Full-time tuition and fees vary according to program.
Financial aid: In 1999–00, 15 fellowships with full tuition reimbursements (averaging $12,078 per year), 64 research assistantships with tuition reimbursements (averaging $17,000 per year), 15 teaching assistantships with tuition reimbursements (averaging $16,620 per year) were awarded; Federal Work-Study, institutionally sponsored loans, and traineeships also available. Financial aid application deadline: 12/31; financial aid applicants required to submit FAFSA.
Application contact: Latonia Trimuel, Program Assistant, 800-546-1761, *E-mail:* ibis@northwestern.edu.

Find an in-depth description at www.petersons.com/graduate.

■ THE OHIO STATE UNIVERSITY

College of Medicine and Public Health and Graduate School, Graduate Programs in the Basic Medical Sciences, Department of Physiology and Cell Biology, Columbus, OH 43210

AWARDS PhD.

Faculty: 17 full-time (3 women), 7 part-time/adjunct (1 woman).
Students: 14 full-time (7 women); includes 1 minority (Hispanic American), 9 international. Average age 27. *27 applicants, 19% accepted.* In 1999, 3 doctorates awarded.

Degree requirements: For doctorate, dissertation required, foreign language not required.
Entrance requirements: For doctorate, GRE General Test. *Application deadline:* For fall admission, 3/15 (priority date). Applications are processed on a rolling basis. *Application fee:* $30 ($40 for international students).
Expenses: Tuition, state resident: full-time $5,400. Tuition, nonresident: full-time $14,535. Part-time tuition and fees vary according to course load.
Financial aid: In 1999–00, 13 students received aid, including fellowships with full tuition reimbursements available (averaging $12,224 per year), research assistantships with full tuition reimbursements available (averaging $15,504 per year); institutionally sponsored loans also available. Financial aid application deadline: 4/1.
Faculty research: Neurobiology of cell and muscle, intestinal mucosal control, autonomic neurophysiology, cardiovascular regulation, cell biology. *Total annual research expenditures:* $2 million.
Jack A. Rall, Chair, 614-292-5448, *Fax:* 614-292-4888, *E-mail:* rall.1@osu.edu.
Application contact: P. E. Ward, Graduate Committee Chair, 614-292-5448, *Fax:* 614-292-4888, *E-mail:* ward.10@osu.edu.

■ THE OHIO STATE UNIVERSITY

College of Veterinary Medicine and Graduate School, Graduate Programs in Veterinary Medicine, Department of Veterinary Biosciences, Columbus, OH 43210

AWARDS Anatomy and cellular biology (MS, PhD); pathobiology (MS, PhD); pharmacology (MS, PhD); toxicology (MS, PhD); veterinary physiology (MS, PhD).

Faculty: 28 full-time (8 women).
Students: 49 full-time (20 women); includes 2 minority (1 African American, 1 Asian American or Pacific Islander), 18 international.
Degree requirements: For master's and doctorate, thesis/dissertation, final exam required.
Entrance requirements: For master's, GRE General Test; for doctorate, GRE General Test, master's degree. *Application fee:* $25.
Expenses: Tuition, state resident: full-time $5,757. Tuition, nonresident: full-time $14,892.
Financial aid: Fellowships, research assistantships, teaching assistantships available.
Faculty research: Microvasculature, muscle biology, neonatal lung and bone development.
Charles C. Capen, Interim Chair, 614-292-4489.

Application contact: Graduate Admission Committee, 614-292-4489.

■ OREGON HEALTH SCIENCES UNIVERSITY

School of Medicine, Graduate Programs in Medicine, Department of Physiology and Pharmacology, Portland, OR 97201-3098

AWARDS Pharmacology (PhD); physiology (PhD).

Degree requirements: For doctorate, dissertation required, foreign language not required.

Entrance requirements: For doctorate, GRE General Test, MCAT, TOEFL. *Application deadline:* For fall admission, 1/15. Applications are processed on a rolling basis. *Application fee:* $60.

Expenses: Tuition, state resident: full-time $3,132; part-time $174 per credit hour. Tuition, nonresident: full-time $5,256; part-time $292 per credit hour. Required fees: $8.5 per credit hour. $146 per term. Part-time tuition and fees vary according to course load.

Financial aid: Fellowships, research assistantships, Federal Work-Study, institutionally sponsored loans, scholarships, and tuition waivers (full) available. Financial aid application deadline: 3/1; financial aid applicants required to submit FAFSA.

Dr. John A. Resko, Chairman, 503-494-8262, *Fax:* 503-494-4352.

Application contact: Dr. Charles Roselli, Graduate Program Director, 503-494-5837, *Fax:* 503-494-4352.

■ THE PENNSYLVANIA STATE UNIVERSITY MILTON S. HERSHEY MEDICAL CENTER

Graduate School, Intercollege Graduate Program in Physiology, Hershey, PA 17033-2360

AWARDS MS, PhD, MD/PhD.

Students: 16 full-time (7 women). Average age 27. In 1999, 2 master's, 4 doctorates awarded.

Entrance requirements: For master's and doctorate, GRE General Test, MCAT. *Application deadline:* For fall admission, 7/26. *Application fee:* $50.

Expenses: Tuition, state resident: full-time $6,886; part-time $291 per credit. Tuition, nonresident: full-time $14,118; part-time $588 per credit. Required fees: $43 per semester. Part-time tuition and fees vary according to course load.

Dr. Leonard S. Jefferson, Chair, 717-531-8566.

■ THE PENNSYLVANIA STATE UNIVERSITY UNIVERSITY PARK CAMPUS

Graduate School, Intercollege Graduate Programs, Intercollege Graduate Program in Physiology, State College, University Park, PA 16802-1503

AWARDS MS, PhD.

Students: 12 full-time (6 women), 7 part-time (5 women). In 2000, 5 master's, 6 doctorates awarded.

Entrance requirements: For master's and doctorate, GRE General Test. *Application fee:* $45.

Expenses: Tuition, state resident: full-time $6,886; part-time $291 per credit. Tuition, nonresident: full-time $14,118; part-time $588 per credit. Required fees: $46 per semester. Part-time tuition and fees vary according to course load and program. Dr. Daniel Deaver, Chairman, 814-863-3664.

Application contact: Dr. James S. Ultamn, Chair, 814-865-5557, *Fax:* 814-865-9451.

■ PURDUE UNIVERSITY

School of Veterinary Medicine and Graduate School, Graduate Programs in Veterinary Medicine, Department of Basic Medical Sciences, West Lafayette, IN 47907

AWARDS Anatomy (MS, PhD); pharmacology (MS, PhD); physiology (MS, PhD). Part-time programs available.

Faculty: 19 full-time (3 women).

Students: 26 full-time (13 women). Average age 27. *26 applicants, 27% accepted.* In 1999, 4 master's, 5 doctorates awarded. Terminal master's awarded for partial completion of doctoral program.

Degree requirements: For master's and doctorate, thesis/dissertation required, foreign language not required. *Average time to degree:* Master's–3 years full-time; doctorate–3 years full-time.

Entrance requirements: For master's and doctorate, GRE General Test, TOEFL. *Application deadline:* For fall admission, 7/1 (priority date); for spring admission, 12/1 (priority date). *Application fee:* $30. Electronic applications accepted.

Expenses: Tuition, state resident: full-time $3,732. Tuition, nonresident: full-time $8,732.

Financial aid: In 1999–00, 6 fellowships, 14 research assistantships, 3 teaching assistantships were awarded. Financial aid application deadline: 3/1; financial aid applicants required to submit FAFSA.

Faculty research: Development and regeneration, tissue injury and shock, biomedical engineering, ovarian function,

bone and cartilage biology, cell and molecular biology. *Total annual research expenditures:* $764,843.

Dr. Gordon L. Coppoc, Head, 765-494-8632, *Fax:* 765-494-0781, *E-mail:* coppoc@vet.purdue.edu.

Application contact: Dr. Ronald Hullinger, Chairman, Graduate Committee, 765-494-8580, *Fax:* 765-494-0781, *E-mail:* bmsgrad@vet.purdue.edu.

■ RUSH UNIVERSITY

Graduate College, Department of Molecular Biophysics and Physiology, Chicago, IL 60612-3832

AWARDS Physiology (PhD).

Faculty: 7 full-time (0 women), 5 part-time/adjunct (0 women).

Students: 2 full-time (0 women); both minorities (both Hispanic Americans). Average age 26. *25 applicants, 4% accepted.* In 1999, 1 degree awarded.

Degree requirements: For doctorate, dissertation required, foreign language not required. *Average time to degree:* Doctorate–4 years full-time.

Entrance requirements: For doctorate, GRE, TOEFL. *Application deadline:* For fall admission, 4/15 (priority date). Applications are processed on a rolling basis. *Application fee:* $25.

Expenses: Tuition: Full-time $13,020; part-time $390 per credit. Tuition and fees vary according to program.

Financial aid: In 1999–00, 2 fellowships with full tuition reimbursements (averaging $25,000 per year) were awarded; Federal Work-Study and institutionally sponsored loans also available. Aid available to part-time students. Financial aid application deadline: 4/15.

Faculty research: Membrane structure and function, synaptic physiology, ion channels, excitation-contraction coupling membrane fusion. *Total annual research expenditures:* $2.2 million.

Dr. Fredric Cohen, Director, 312-942-6753, *Fax:* 312-942-8711, *E-mail:* fcohen@rush.edu.

Application contact: Thyra Jackson, Coordinator of Admissions, 312-942-6247, *Fax:* 312-942-2100, *E-mail:* tjackson@rushu.rush.edu.

■ RUTGERS, THE STATE UNIVERSITY OF NEW JERSEY, NEW BRUNSWICK

Graduate School, Program in Physiology and Neurobiology, New Brunswick, NJ 08901-1281

AWARDS PhD.

Faculty: 62 full-time (10 women).

Students: 18 full-time (9 women), 18 part-time (14 women); includes 8 minority (3

Rutgers, The State University of New Jersey, New Brunswick (continued)
African Americans, 1 Asian American or Pacific Islander, 4 Hispanic Americans), 13 international. *44 applicants, 20% accepted.* In 1999, 1 degree awarded.

Degree requirements: For doctorate, dissertation, qualifying exam required, foreign language not required. *Average time to degree:* Doctorate–4.9 years full-time.

Entrance requirements: For doctorate, GRE General Test, GRE Subject Test (biology or chemistry), TOEFL, minimum undergraduate GPA of 3.0. *Application deadline:* For fall admission, 3/1 (priority date). Applications are processed on a rolling basis. *Application fee:* $50.

Expenses: Tuition, state resident: full-time $6,776; part-time $279 per credit. Tuition, nonresident: full-time $9,936; part-time $412 per credit. Required fees: $20 per credit. $89 per semester. Tuition and fees vary according to course load, campus/location and program.

Financial aid: In 1999–00, 7 students received aid, including 1 fellowship, 5 research assistantships with full tuition reimbursements available, 1 teaching assistantship with full tuition reimbursement available; tuition waivers (full) also available. Financial aid applicants required to submit FAFSA.

Faculty research: Neuronal growth factors, neuronal gene expression, neurogenetics, circulation controls, reproduction. *Total annual research expenditures:* $3.4 million.

Dr. Ira B. Black, Director, 732-235-5388, *Fax:* 732-235-5885.

Application contact: David Egger, Director of Admissions, 732-235-4522, *Fax:* 732-235-4029.

■ SAINT LOUIS UNIVERSITY

School of Medicine and Graduate School, Graduate Programs in Biomedical Sciences, Department of Pharmacological and Physiological Science, St. Louis, MO 63103-2097

AWARDS MS(R), PhD.

Faculty: 27 full-time (9 women), 8 part-time/adjunct (2 women).

Students: 9 full-time (6 women), 9 part-time (3 women); includes 2 minority (1 African American, 1 Hispanic American). Average age 28. In 1999, 5 doctorates awarded.

Degree requirements: For master's, thesis, comprehensive oral exam required; for doctorate, dissertation, departmental qualifying exams required.

Entrance requirements: For master's and doctorate, GRE General Test, minimum B average in undergraduate course work; previous course work in biology, chemistry, mathematics, and physics. *Application*

deadline: Applications are processed on a rolling basis. *Application fee:* $50. Electronic applications accepted.

Expenses: Tuition: Part-time $507 per credit hour. Required fees: $38 per term.

Financial aid: In 1999–00, 17 students received aid, including 8 fellowships, 1 teaching assistantship; research assistantships, traineeships and tuition waivers (partial) also available. Aid available to part-time students. Financial aid application deadline: 8/1; financial aid applicants required to submit FAFSA.

Faculty research: Molecular endocrinology, neuropharmacology, cardiovascular science, drug abuse, neurotransmitter and hormonal signaling mechanisms.

Dr. Thomas C. Westfall, Chairman, 314-577-8551, *Fax:* 314-577-8554.

Application contact: Director of Admissions, 314-577-8551, *Fax:* 314-577-8554, *E-mail:* inquiry@slu.edu.

Find an in-depth description at www.petersons.com/graduate.

■ SALISBURY STATE UNIVERSITY

Graduate Division, Program in Applied Health Physiology, Salisbury, MD 21801-6837

AWARDS MS. Part-time and evening/weekend programs available.

Faculty: 7 part-time/adjunct (3 women).

Degree requirements: For master's, fieldwork required.

Entrance requirements: For master's, minimum GPA of 2.75. *Application deadline:* For fall admission, 8/1; for spring admission, 1/1. Applications are processed on a rolling basis. *Application fee:* $30.

Expenses: Tuition, state resident: part-time $162 per credit. Tuition, nonresident: part-time $318 per credit. Required fees: $4 per credit.

Faculty research: Body image and self-concept, nutrition supplements.

Susan Muller, Graduate Director, 410-548-5555, *E-mail:* smmuller@ssu.edu.

■ SAN FRANCISCO STATE UNIVERSITY

Graduate Division, College of Science and Engineering, Department of Biology, Program in Physiology and Behavioral Biology, San Francisco, CA 94132-1722

AWARDS MA.

Entrance requirements: For master's, minimum GPA of 2.5 in last 60 units.

Expenses: Tuition, nonresident: full-time $5,904; part-time $246 per unit. Required fees: $1,904; $637 per semester. Tuition and fees vary according to course load.

■ SOUTHERN ILLINOIS UNIVERSITY CARBONDALE

School of Medicine and Graduate School, Graduate Program in Medicine, Department of Physiology, Carbondale, IL 62901-6806

AWARDS MS, PhD.

Faculty: 18 full-time (4 women).

Students: 18 full-time (8 women), 5 part-time (1 woman); includes 3 minority (1 African American, 1 Asian American or Pacific Islander, 1 Native American), 3 international. Average age 28. *3 applicants, 67% accepted.* In 1999, 3 master's, 2 doctorates awarded. Terminal master's awarded for partial completion of doctoral program.

Degree requirements: For master's, thesis required, foreign language not required; for doctorate, dissertation required. *Average time to degree:* Master's–2.5 years full-time; doctorate–5 years full-time.

Entrance requirements: For master's, GRE General Test, TOEFL, minimum GPA of 3.0; for doctorate, GRE General Test, TOEFL, minimum GPA of 3.25. *Application deadline:* For fall admission, 6/1 (priority date). Applications are processed on a rolling basis. *Application fee:* $0.

Expenses: Tuition, state resident: full-time $2,604. Tuition, nonresident: full-time $5,208. Required fees: $380 per semester.

Financial aid: In 1999–00, 18 students received aid, including 3 fellowships with full tuition reimbursements available, 1 research assistantship with full tuition reimbursement available, 10 teaching assistantships with full tuition reimbursements available; institutionally sponsored loans and tuition waivers (full) also available.

Faculty research: Hormones, neurotransmitters, cell biology, membrane protein, membranes transport.

Dr. Andrzej Bartke, Chair, 618-453-1512, *Fax:* 618-453-1517.

Application contact: Graduate Program Committee, 618-453-1544, *Fax:* 618-453-1517.

■ STANFORD UNIVERSITY

School of Medicine, Graduate Programs in Medicine, Department of Molecular and Cellular Physiology, Stanford, CA 94305-9991

AWARDS PhD.

Faculty: 8 full-time (0 women).

Students: 2 full-time (1 woman), 4 part-time (1 woman); includes 1 minority (African American). Average age 28. *24 applicants, 8% accepted.* In 1999, 3 doctorates awarded.

Degree requirements: For doctorate, dissertation, qualifying exam required, foreign language not required.
Entrance requirements: For doctorate, GRE General Test, GRE Subject Test (biology), TOEFL, bachelor's degree in science; 1 year of course work in chemistry, calculus, physics, biology. *Application deadline:* For fall admission, 12/15. *Application fee:* $65 ($80 for international students). Electronic applications accepted.
Expenses: Tuition: Full-time $23,058. Required fees: $152. Part-time tuition and fees vary according to course load.
Financial aid: Research assistantships available. Financial aid application deadline: 12/15.
Faculty research: Signal transduction, ion channels, intracellular calcium, synaptic transmission.
W. James Nelson, Chair, 650-725-7596, *Fax:* 650-725-8021, *E-mail:* wjnelson@leland.stanford.edu.
Application contact: Student Services Coordinator, 650-725-7554.

■ STATE UNIVERSITY OF NEW YORK AT BUFFALO

Graduate School, School of Medicine and Biomedical Sciences, Graduate Programs in Medicine and Biomedical Sciences, Department of Physiology and Biophysics, Buffalo, NY 14260

AWARDS Biophysical sciences (MS, PhD); physiology (MA, PhD).

Faculty: 26 full-time (3 women), 2 part-time/adjunct (both women).
Students: 6 full-time (2 women), 17 part-time (9 women), 8 international. Average age 26. *7 applicants, 57% accepted.* In 1999, 5 master's, 1 doctorate awarded. Terminal master's awarded for partial completion of doctoral program.
Degree requirements: For master's, thesis (for some programs), oral exam, project required, foreign language not required; for doctorate, dissertation, oral and written qualifying exam or 2 research proposals required, foreign language not required.
Entrance requirements: For master's and doctorate, GRE General Test, TOEFL. *Application deadline:* For fall admission, 2/1 (priority date). *Application fee:* $35.
Expenses: Tuition, state resident: full-time $5,100. Tuition, nonresident: full-time $8,416. Required fees: $935.
Financial aid: In 1999–00, 17 research assistantships with tuition reimbursements (averaging $16,000 per year), 1 teaching assistantship with tuition reimbursement were awarded; fellowships, Federal Work-Study, institutionally sponsored loans, and unspecified assistantships also available.

Financial aid application deadline: 2/1; financial aid applicants required to submit FAFSA.
Faculty research: Neurosciences, ion channels, cardiac physiology, renal/epithelial transport, cardiopulmonary exercise. *Total annual research expenditures:* $4.9 million.
Dr. Harold C. Strauss, Chair, 716-829-2738, *Fax:* 716-829-2344, *E-mail:* hstrauss@buffalo.edu.
Application contact: Dr. Malcolm M. Slaughter, Director of Graduate Studies, 716-829-3240, *Fax:* 716-829-2344, *E-mail:* pgy-bph@acsu.buffalo.edu.

■ STATE UNIVERSITY OF NEW YORK AT STONY BROOK

Health Sciences Center, School of Medicine and Graduate School, Graduate Programs in Medicine, Department of Molecular Physiology and Biophysics, Stony Brook, NY 11794

AWARDS Physiology and biophysics (PhD).

Faculty: 38.
Students: 12 full-time (1 woman), 18 part-time (8 women); includes 7 minority (2 African Americans, 4 Asian Americans or Pacific Islanders, 1 Hispanic American), 11 international. Average age 28. *33 applicants, 21% accepted.* In 1999, 2 degrees awarded.
Degree requirements: For doctorate, computer language, dissertation, comprehensive exams required, foreign language not required.
Entrance requirements: For doctorate, GRE General Test, GRE Subject Test, TOEFL, BS in related field, minimum GPA of 3.0. *Application deadline:* For fall admission, 1/15. *Application fee:* $50.
Expenses: Tuition, state resident: full-time $5,100. Tuition, nonresident: full-time $8,416. Required fees: $492.
Financial aid: In 1999–00, 3 fellowships, 12 research assistantships were awarded; teaching assistantships, Federal Work-Study also available. Financial aid application deadline: 3/15.
Faculty research: Cellular electrophysiology, membrane permeation and transport, metabolic endocrinology. *Total annual research expenditures:* $3.7 million.
Dr. Peter Brink, Chair, 631-444-2287, *Fax:* 631-444-3432.
Application contact: Dr. Leon C. Moore, Graduate Adviser, 631-444-2287, *Fax:* 631-444-3432, *E-mail:* moore@pofvax.pnb.sunysb.edu.

Find an in-depth description at www.petersons.com/graduate.

■ STATE UNIVERSITY OF NEW YORK HEALTH SCIENCE CENTER AT BROOKLYN

School of Graduate Studies, Program in Biophysics and Physiology, Brooklyn, NY 11203-2098

AWARDS PhD, MD/PhD.

Degree requirements: For doctorate, one foreign language, dissertation required.
Entrance requirements: For doctorate, GRE.
Expenses: Tuition, state resident: full-time $5,100; part-time $213 per credit. Tuition, nonresident: full-time $8,416; part-time $351 per credit. Required fees: $200. Full-time tuition and fees vary according to program and student level.
Faculty research: Cardiovascular physiology, neurophysiology, developmental physiology, membrane transport, molecular basis of muscle contraction.

■ STATE UNIVERSITY OF NEW YORK UPSTATE MEDICAL UNIVERSITY

College of Graduate Studies, Department of Neuroscience and Physiology, Syracuse, NY 13210-2334

AWARDS Physiology (MS, PhD).

Faculty: 34.
Students: 4 full-time (1 woman), 1 part-time; includes 2 minority (both Asian Americans or Pacific Islanders), 1 international. *19 applicants, 11% accepted.* In 1999, 1 doctorate awarded (100% continued full-time study). Terminal master's awarded for partial completion of doctoral program.
Degree requirements: For master's, thesis required, foreign language not required; for doctorate, dissertation, comprehensive exam required, foreign language not required.
Entrance requirements: For master's and doctorate, GRE General Test, GRE Subject Test, TSE. *Application deadline:* For fall admission, 4/1 (priority date). Applications are processed on a rolling basis. *Application fee:* $40.
Expenses: Tuition, state resident: full-time $5,100; part-time $213 per credit. Tuition, nonresident: full-time $8,416; part-time $351 per credit. Required fees: $410; $25 per credit. Part-time tuition and fees vary according to course load and program.
Financial aid: Fellowships, research assistantships, Federal Work-Study and institutionally sponsored loans available. Aid available to part-time students. Financial aid application deadline: 4/15.
Faculty research: Neurophysiological mechanisms involved in learning,

State University of New York Upstate Medical University (continued)
mechanisms basic to olfactory discrimination, cardiac muscle physiology.
Dr. Russell Durkovic, Interim Chairperson, 315-464-4413.
Application contact: Dr. Larry Stoner, Professor, 315-464-4413.

■ TEACHERS COLLEGE, COLUMBIA UNIVERSITY

Graduate Faculty of Education, Department of Biobehavioral Studies, Program in Applied Physiology, New York, NY 10027-6696

AWARDS Ed M, MA, MS, Ed D.

Faculty: 2 full-time (0 women), 1 part-time/adjunct.
Students: 10 full-time (4 women), 39 part-time (18 women); includes 8 minority (2 African Americans, 3 Asian Americans or Pacific Islanders, 3 Hispanic Americans), 2 international. Average age 32. *23 applicants, 83% accepted.* In 1999, 17 master's, 1 doctorate awarded.
Degree requirements: For master's, integrative paper required; for doctorate, dissertation required.
Entrance requirements: For doctorate, GRE General Test. *Application deadline:* For fall admission, 5/15; for spring admission, 12/1. *Application fee:* $50.
Expenses: Tuition: Part-time $670 per credit. Required fees: $161 per semester. Part-time tuition and fees vary according to program.
Financial aid: Teaching assistantships, career-related internships or fieldwork, Federal Work-Study, institutionally sponsored loans, and tuition waivers (full and partial) available. Aid available to part-time students. Financial aid application deadline: 2/1.
Faculty research: Exercise physiology, body composition, metabolism, vagal-cardiac activity, measurement of the dicrotic notch.
Application contact: Debbie Lesperance, Office of Admissions, 212-678-3710, *Fax:* 212-678-4171.

■ TEMPLE UNIVERSITY

Health Sciences Center, School of Medicine and Graduate School, Graduate Programs in Medicine, Department of Physiology, Philadelphia, PA 19122-6096

AWARDS MS, PhD, MD/PhD.

Faculty: 18 full-time (1 woman).
Students: 26 full-time (8 women), 1 part-time; includes 8 minority (1 African American, 7 Asian Americans or Pacific Islanders), 1 international. *120 applicants, 4% accepted.* In 1999, 5 master's, 1 doctorate awarded.

Degree requirements: For master's and doctorate, thesis/dissertation, research seminars required, foreign language not required. *Average time to degree:* Master's–3 years full-time, 4 years part-time; doctorate–5.5 years full-time, 6.5 years part-time.
Entrance requirements: For master's and doctorate, GRE General Test, GRE Subject Test, minimum GPA of 3.0. *Application deadline:* For fall admission, 7/1; for spring admission, 11/1. *Application fee:* $40. Electronic applications accepted.
Expenses: Tuition, state resident: full-time $6,030. Tuition, nonresident: full-time $8,298. Required fees: $230. One-time fee: $10 full-time.
Financial aid: Fellowships available.
Faculty research: Pulmonary, microvascular, and molecular physiology; cardiac electrophysiology.
Dr. Ronald Tuma, Chair, 215-707-2560, *Fax:* 215-707-4003, *E-mail:* tumarf@astro.ocis.temple.edu.
Application contact: Dr. John A. Drees, Admissions Coordinator, 215-707-3272, *Fax:* 215-707-4003, *E-mail:* jdrees@astro.ocis.temple.edu.

■ TEXAS A&M UNIVERSITY

College of Veterinary Medicine and Office of Graduate Studies, Graduate Programs in Veterinary Medicine, Department of Veterinary Physiology and Pharmacology, College Station, TX 77843

AWARDS Physiology (MS, PhD); toxicology (MS, PhD).

Faculty: 21 full-time (2 women), 3 part-time/adjunct (1 woman).
Students: 14 full-time (9 women), 6 part-time (2 women). Average age 32. *11 applicants, 27% accepted.* In 1999, 7 master's, 6 doctorates awarded.
Degree requirements: For master's and doctorate, foreign language and thesis not required.
Entrance requirements: For master's and doctorate, GRE General Test, TOEFL. *Application fee:* $50 ($75 for international students).
Expenses: Tuition, state resident: part-time $76 per semester hour. Tuition, nonresident: part-time $292 per semester hour. Required fees: $11 per semester hour.
Financial aid: Fellowships, research assistantships, teaching assistantships available. Financial aid application deadline: 4/1; financial aid applicants required to submit FAFSA.
Faculty research: Gamete and embryo physiology, endocrinology, equine laminitis.

Glen Laine, Head, 979-845-7261.
Find an in-depth description at www.petersons.com/graduate.

■ TEXAS A&M UNIVERSITY SYSTEM HEALTH SCIENCE CENTER

College of Medicine, Graduate School of Biomedical Sciences, Department of Medical Physiology, College Station, TX 77840-7896

AWARDS PhD.

Faculty: 10 full-time (1 woman).
Students: 11 full-time (3 women); includes 2 minority (both Asian Americans or Pacific Islanders), 2 international. Average age 28. *32 applicants, 13% accepted.*
Degree requirements: For doctorate, dissertation required, foreign language not required. *Average time to degree:* Doctorate–4 years full-time.
Entrance requirements: For doctorate, GRE General Test. *Application deadline:* For fall admission, 2/1 (priority date). *Application fee:* $50 ($75 for international students).
Expenses: Tuition, area resident: Full-time $1,368. Tuition, state resident: part-time $76 per credit. Tuition, nonresident: full-time $5,256; part-time $292 per credit. International tuition: $5,256 full-time. Required fees: $678; $38 per credit. Full-time tuition and fees vary according to course load and student level.
Financial aid: In 1999–00, 10 students received aid; fellowships, research assistantships, institutionally sponsored loans available. Financial aid application deadline: 4/1; financial aid applicants required to submit FAFSA.
Faculty research: Cardiovascular physiology, vascular cell and molecular biology. *Total annual research expenditures:* $1.5 million.
Dr. Harris J. Granger, Head, 409-845-7816, *Fax:* 409-847-8635.
Application contact: Dr. David C. Zawieja, Assistant Professor, 409-845-7816, *Fax:* 409-847-8635, *E-mail:* dcz@tamu.edu.

Find an in-depth description at www.petersons.com/graduate.

■ TEXAS TECH UNIVERSITY HEALTH SCIENCES CENTER

Graduate School of Biomedical Sciences, Department of Physiology, Lubbock, TX 79430

AWARDS MS, PhD, MD/PhD.

Faculty: 19 full-time (4 women).
Students: 8 full-time (1 woman); includes 4 minority (2 Asian Americans or Pacific Islanders, 2 Hispanic Americans), 3

international. Average age 29. *8 applicants, 50% accepted.* In 1999, 2 master's awarded (100% entered university research/teaching); 1 doctorate awarded (100% entered university research/teaching). Terminal master's awarded for partial completion of doctoral program.

Degree requirements: For master's and doctorate, thesis/dissertation required, foreign language not required.

Entrance requirements: For master's and doctorate, GRE General Test, TOEFL, minimum GPA of 3.0. *Application deadline:* For fall admission, 4/15 (priority date). Applications are processed on a rolling basis. *Application fee:* $30 ($55 for international students).

Expenses: Tuition, state resident: part-time $38 per credit hour. Tuition, nonresident: part-time $254 per credit hour. Part-time tuition and fees vary according to program.

Financial aid: In 1999–00, 1 fellowship (averaging $14,500 per year), 7 research assistantships (averaging $14,500 per year) were awarded.

Faculty research: Cardiovascular physiology, neurophysiology, renal physiology, respiratory physiology. *Total annual research expenditures:* $996,358.

Dr. John Orem, Chair, 806-743-2520, *Fax:* 806-743-1512, *E-mail:* phyjmo@ttuhsc.edu.

Application contact: Dr. Jean Strahlendorf, Graduate Adviser, 806-743-2520, *Fax:* 806-743-1512, *E-mail:* phyjcs@ttuhsc.edu.

■ THOMAS JEFFERSON UNIVERSITY

College of Graduate Studies, Program in Physiology, Philadelphia, PA 19107

AWARDS PhD.

Faculty: 12 full-time (2 women).
Students: 2 full-time (0 women). *20 applicants, 15% accepted.* In 1999, 2 degrees awarded.

Degree requirements: For doctorate, dissertation required, foreign language not required.

Entrance requirements: For doctorate, GRE General Test, TOEFL, minimum GPA of 3.2. *Application deadline:* For fall admission, 3/1 (priority date). Applications are processed on a rolling basis. *Application fee:* $40.

Expenses: Tuition: Full-time $12,670. Tuition and fees vary according to degree level and program.

Financial aid: Fellowships with full tuition reimbursements, research assistantships, Federal Work-Study, institutionally sponsored loans, traineeships, and training grants available. Aid available to part-time students. Financial aid application

deadline: 5/1; financial aid applicants required to submit FAFSA.

Faculty research: Cardiovascular physiology, smooth muscle physiology, pathophysiology of myocardial ischemia, endothelial cell physiology, molecular biology of ion channel physiology.

Dr. Marilyn J. Woolkalis, Chair, Graduate Committee, 215-503-6715, *Fax:* 215-503-2073, *E-mail:* woolkall@jeflin.tju.edu.

Application contact: Jessie F. Pervall, Director of Admissions, 215-503-4400, *Fax:* 215-503-3433, *E-mail:* cgs-info@mail.tju.edu.

■ TUFTS UNIVERSITY

Sackler School of Graduate Biomedical Sciences, Graduate Program in Cellular and Molecular Physiology, Medford, MA 02155

AWARDS PhD.

Faculty: 17 full-time (5 women).
Students: 18 full-time (10 women); includes 5 minority (all Asian Americans or Pacific Islanders), 5 international. Average age 27. *42 applicants, 10% accepted.* In 1999, 2 degrees awarded (100% entered university research/teaching).

Degree requirements: For doctorate, dissertation required, foreign language not required. *Average time to degree:* Doctorate–6 years full-time.

Entrance requirements: For doctorate, GRE General Test, GRE Subject Test, TOEFL. *Application deadline:* For fall admission, 1/15 (priority date). Applications are processed on a rolling basis. *Application fee:* $45.

Expenses: Tuition: Full-time $19,325.

Financial aid: In 1999–00, research assistantships with full tuition reimbursements (averaging $18,805 per year); scholarships and tuition waivers (full) also available. Financial aid application deadline: 2/15.

Dr. Irwin M. Arias, Chairman, 617-636-6739, *Fax:* 617-636-0445, *E-mail:* iarias@infonet.tufts.edu.

Application contact: Secretary, 617-636-6767, *Fax:* 617-636-0375, *E-mail:* sackler-school@tufts.edu.

Find an in-depth description at www.petersons.com/graduate.

■ TULANE UNIVERSITY

School of Medicine and Graduate School, Graduate Programs in Medicine, Department of Physiology, New Orleans, LA 70118-5669

AWARDS MS, PhD, MD/PhD. MS and PhD offered through the Graduate School.

Students: 9 full-time (5 women), 1 part-time; includes 1 minority (Asian American or Pacific Islander), 7 international. *23 applicants, 22% accepted.*

Degree requirements: For master's, one foreign language, thesis required; for doctorate, 2 foreign languages (computer language can substitute for one), dissertation required.

Entrance requirements: For master's, GRE General Test, TOEFL, or TSE, minimum B average in undergraduate course work; for doctorate, GRE General Test, TOEFL, or TSE. *Application deadline:* For fall admission, 2/1. *Application fee:* $45.

Expenses: Tuition: Full-time $23,030.

Financial aid: Fellowships, research assistantships available. Financial aid application deadline: 2/1.

Dr. L. Gabriel Navar, Chairman, 504-588-5251.

Find an in-depth description at www.petersons.com/graduate.

■ UNIFORMED SERVICES UNIVERSITY OF THE HEALTH SCIENCES

School of Medicine, Division of Basic Medical Sciences, Department of Physiology, Bethesda, MD 20814-4799

AWARDS PhD.

Faculty: 13 full-time (1 woman), 10 part-time/adjunct (2 women).
Students: 3 full-time (1 woman). Average age 28.

Degree requirements: For doctorate, one foreign language (computer language can substitute), dissertation, comprehensive and qualifying exams required. *Average time to degree:* Doctorate–5 years full-time.

Entrance requirements: For doctorate, GRE General Test, GRE Subject Test, minimum GPA of 3.0, U.S. citizenship. *Application deadline:* For fall admission, 1/15 (priority date). Applications are processed on a rolling basis. *Application fee:* $0.

Financial aid: In 1999–00, 2 fellowships with full tuition reimbursements (averaging $15,000 per year) were awarded; tuition waivers (full) also available.

Faculty research: Peripheral circulation, blood, radiation, hypertension, exercise physiology.

Dr. Harvey Pollard, Interim Chair, 301-295-3200, *E-mail:* hpollard@usuhs.mil.

Application contact: Janet M. Anastasi, Graduate Program Coordinator, 301-295-9474, *Fax:* 301-295-6772, *E-mail:* janastasi@usuhs.mil.

■ THE UNIVERSITY OF ALABAMA AT BIRMINGHAM

Graduate School and School of Medicine and School of Dentistry, Graduate Programs in Joint Health Sciences, Department of Physiology and Biophysics, Birmingham, AL 35294

AWARDS PhD.

Students: 26 full-time (13 women); includes 3 minority (all African Americans), 9 international. *59 applicants, 25% accepted.* In 1999, 6 degrees awarded.
Entrance requirements: For doctorate, GRE General Test, TOEFL, interview, minimum GPA of 3.0. *Application deadline:* For fall admission, 3/1. Applications are processed on a rolling basis. *Application fee:* $35 ($60 for international students). Electronic applications accepted.
Expenses: All doctoral students receive a full fellowship, stipend, and tuition.
Financial aid: Fellowships available.
Faculty research: Standard physiology (neurological, endocrine, cardiovascular, respiratory, and renal), cell and membrane biology.
Dr. Dale J. Benos, Interim Chair, 205-934-6220, *Fax:* 205-934-2377.
Application contact: Director of Graduate Studies, 205-934-3969, *Fax:* 205-975-9028.

Find an in-depth description at www.petersons.com/graduate.

■ THE UNIVERSITY OF ALABAMA AT BIRMINGHAM

Graduate School, School of Natural Sciences and Mathematics, Department of Biology, Birmingham, AL 35294

AWARDS Comparative and cellular biology (PhD); comparative and cellular physiology (MS); marine science (MS, PhD); microbial ecology and physiology (MS, PhD); reproduction and development (MS, PhD).

Students: 34 full-time (19 women), 6 international. *61 applicants, 38% accepted.* In 1999, 1 master's, 3 doctorates awarded. Terminal master's awarded for partial completion of doctoral program.
Degree requirements: For master's and doctorate, thesis/dissertation required.
Entrance requirements: For master's and doctorate, GRE General Test, TOEFL, previous course work in biology, calculus, organic chemistry, physics. *Application deadline:* Applications are processed on a rolling basis. *Application fee:* $35 ($60 for international students). Electronic applications accepted.
Expenses: Tuition, state resident: part-time $104 per semester hour. Tuition, nonresident: part-time $208 per semester

hour. Required fees: $17 per semester hour. $57 per quarter. Tuition and fees vary according to program.
Financial aid: In 1999–00, 22 students received aid, including 3 fellowships with full tuition reimbursements available (averaging $14,000 per year), 19 teaching assistantships with full tuition reimbursements available (averaging $14,000 per year); research assistantships, career-related internships or fieldwork, Federal Work-Study, institutionally sponsored loans, and tuition waivers (full) also available. Aid available to part-time students.
Faculty research: Invertebrate physiology, marine biology, environmental biology.
Dr. Daniel D. Jones, Chairman, 205-934-4290, *Fax:* 205-975-6097, *E-mail:* ddjones@uab.edu.

Find an in-depth description at www.petersons.com/graduate.

■ THE UNIVERSITY OF ARIZONA

Graduate College, Graduate Interdisciplinary Programs, Graduate Interdisciplinary Program in Physiological Sciences, Tucson, AZ 85721

AWARDS PhD.

Degree requirements: For doctorate, dissertation required.
Entrance requirements: For doctorate, GRE General Test, TOEFL.
Expenses: Tuition, nonresident: full-time $4,814; part-time $274 per unit. Required fees: $1,094; $115 per unit. Tuition and fees vary according to course load and program.
Faculty research: Cellular transport and signaling, receptor and messenger modulation, neural interaction and biomechanics, fluid network regulation, environmental adaptation.

Find an in-depth description at www.petersons.com/graduate.

■ UNIVERSITY OF ARKANSAS FOR MEDICAL SCIENCES

College of Medicine and Graduate School, Graduate Programs in Medicine, Department of Physiology and Biophysics, Little Rock, AR 72205-7199

AWARDS MS, PhD, MD/PhD.

Faculty: 24 full-time (3 women).
Students: 23 full-time (9 women), 6 part-time (2 women); includes 7 minority (4 African Americans, 3 Asian Americans or Pacific Islanders), 4 international. In 1999, 2 master's awarded.
Degree requirements: For master's and doctorate, thesis/dissertation required, foreign language not required.

Entrance requirements: For master's and doctorate, GRE General Test. *Application fee:* $0.
Expenses: Tuition: Full-time $8,928.
Financial aid: In 1999–00, 16 research assistantships were awarded. Aid available to part-time students.
Dr. Michael L. Jennings, Chairman, 501-686-5123.
Application contact: Dr. Parimal Chowdhury, Graduate Coordinator, 501-686-5123.

■ UNIVERSITY OF CALIFORNIA, BERKELEY

Graduate Division, College of Letters and Science, Department of Integrative Biology, Berkeley, CA 94720-1500

AWARDS Endocrinology (PhD); integrative biology (PhD).

Degree requirements: For doctorate, dissertation, oral qualifying exam required.
Entrance requirements: For doctorate, GRE General Test, GRE Subject Test (biology).
Expenses: Tuition, nonresident: full-time $9,804. Required fees: $4,268. Tuition and fees vary according to program.
Faculty research: Morphology, physiology, development of plants and animals, behavior, ecology.

■ UNIVERSITY OF CALIFORNIA, BERKELEY

Graduate Division, Group in Endocrinology, Berkeley, CA 94720-1500

AWARDS MA, PhD.

Degree requirements: For doctorate, dissertation, oral qualifying exam required, foreign language not required.
Entrance requirements: For master's, GRE General Test, GRE Subject Test (biology), minimum GPA of 3.0; for doctorate, GRE General Test, GRE Subject Test (biology), minimum GPA of 3.4.
Expenses: Tuition, nonresident: full-time $9,804. Required fees: $4,268. Tuition and fees vary according to program.

■ UNIVERSITY OF CALIFORNIA, DAVIS

Graduate Studies, Programs in the Biological Sciences, Program in Physiology, Davis, CA 95616

AWARDS MS, PhD.

Faculty: 66 full-time (11 women).
Students: 44 full-time (24 women), 1 part-time; includes 8 minority (1 African American, 5 Asian Americans or Pacific Islanders, 2 Hispanic Americans), 3

international. Average age 29. *53 applicants, 36% accepted.* In 1999, 4 master's, 10 doctorates awarded. Terminal master's awarded for partial completion of doctoral program.
Degree requirements: For master's, thesis optional; for doctorate, dissertation required.
Entrance requirements: For master's and doctorate, GRE General Test. *Application deadline:* For fall admission, 4/1. Applications are processed on a rolling basis. *Application fee:* $40. Electronic applications accepted.
Expenses: Tuition, nonresident: full-time $9,804. Tuition and fees vary according to program and student level.
Financial aid: In 1999–00, 43 students received aid, including 20 fellowships with full and partial tuition reimbursements available, 16 research assistantships with full and partial tuition reimbursements available, 9 teaching assistantships with partial tuition reimbursements available; Federal Work-Study, grants, institutionally sponsored loans, and scholarships also available. Financial aid application deadline: 1/15; financial aid applicants required to submit FAFSA.
Faculty research: Systemic physiology, cellular physiology, neurophysiology, cardiovascular physiology, endocrinology. *Total annual research expenditures:* $226,827. Jack Rutledge, Chair, 530-752-0717, *E-mail:* jcrutledge@ucdavis.edu.
Application contact: Tori Hollowell, Administrative Assistant, 530-752-9092, *Fax:* 530-752-8822, *E-mail:* trhollowell@ucdavis.edu.
Find an in-depth description at www.petersons.com/graduate.

■ **UNIVERSITY OF CALIFORNIA, IRVINE**
College of Medicine and Office of Research and Graduate Studies, Graduate Programs in Medicine and School of Biological Sciences, Department of Physiology and Biophysics, Irvine, CA 92697
AWARDS Biological sciences (PhD). Students apply through the Graduate Program in Molecular Biology, Genetics, and Biochemistry.
Faculty: 13 full-time (2 women), 4 part-time/adjunct (1 woman).
Students: 10 full-time (3 women); includes 1 minority (Hispanic American), 1 international. *292 applicants, 25% accepted.* In 1999, 1 doctorate awarded.
Degree requirements: For doctorate, dissertation required.
Entrance requirements: For doctorate, GRE General Test, GRE Subject Test. *Application deadline:* For fall admission, 2/1

(priority date). *Application fee:* $40. Electronic applications accepted.
Expenses: Tuition, nonresident: full-time $10,322; part-time $1,720 per quarter. Required fees: $5,354; $1,300 per quarter. Tuition and fees vary according to program.
Financial aid: Fellowships, research assistantships, institutionally sponsored loans and tuition waivers (full and partial) available. Financial aid application deadline: 3/2; financial aid applicants required to submit FAFSA.
Faculty research: Membrane physiology, exercise physiology, regulation of hormone biosynthesis and action, endocrinology, ion channels and signal transduction.
Dr. Janos K. Lanyi, PhD, Chair, 949-824-7788, *Fax:* 949-824-8540, *E-mail:* jklanyi@uci.edu.
Application contact: Kimberly McKinney, Administrator, 949-824-8145, *Fax:* 949-824-7407, *E-mail:* gp-mbgb@uci.edu.

■ **UNIVERSITY OF CALIFORNIA, LOS ANGELES**
Graduate Division, College of Letters and Science, Department of Physiological Science, Los Angeles, CA 90095
AWARDS MS, PhD.
Students: 78 full-time (32 women); includes 23 minority (18 Asian Americans or Pacific Islanders, 5 Hispanic Americans), 4 international. *87 applicants, 48% accepted.*
Degree requirements: For master's, comprehensive exam or thesis required; for doctorate, dissertation, oral and written qualifying exams required, foreign language not required.
Entrance requirements: For master's, GRE General Test, minimum GPA of 3.0; for doctorate, GRE General Test, minimum undergraduate GPA of 3.0. *Application fee:* $40. Electronic applications accepted.
Expenses: Tuition, nonresident: full-time $9,804. Required fees: $4,405. Full-time tuition and fees vary according to program and student level.
Financial aid: In 1999–00, 63 students received aid, including 29 fellowships, 41 research assistantships; teaching assistantships, Federal Work-Study, institutionally sponsored loans, scholarships, and tuition waivers (full and partial) also available.
Faculty research: Diet and exercise in the prevention and management of degenerative diseases, neuromuscular physiology and plasticity, neural control of movement and homeostatis.
Dr. Allen D. Grinnell, Chair, 310-825-3891.

Application contact: Departmental Office, 310-825-3891, *E-mail:* mcarr@physci.ucla.edu.

■ **UNIVERSITY OF CALIFORNIA, LOS ANGELES**
School of Medicine and Graduate Division, Graduate Programs in Medicine, Department of Physiology, Los Angeles, CA 90095
AWARDS MS, PhD.
Faculty: 14.
Students: 9 full-time (1 woman); includes 4 minority (2 African Americans, 2 Asian Americans or Pacific Islanders), 3 international. *3 applicants, 67% accepted.*
Degree requirements: For master's, foreign language not required; for doctorate, dissertation, oral and written qualifying exams required, foreign language not required.
Entrance requirements: For master's, GRE General Test; for doctorate, GRE General Test, GRE Subject Test. *Application fee:* $40.
Expenses: Tuition, nonresident: full-time $9,804. Required fees: $4,405.
Financial aid: Fellowships, research assistantships, teaching assistantships, scholarships available. Financial aid application deadline: 3/1.
Faculty research: Membrane physiology, cell physiology, muscle physiology, neurophysiology, cardiopulmonary physiology.
Dr. Ernest Wright, Chair, 310-825-6717.
Application contact: UCLA Access Coordinator, 800-284-8252, *Fax:* 310-206-5280, *E-mail:* uclaaccess@lbes.medsch.ucla.edu.

■ **UNIVERSITY OF CALIFORNIA, SAN DIEGO**
School of Medicine and Graduate Studies and Research, Graduate Studies in Biomedical Sciences, Physiology Program, La Jolla, CA 92093
AWARDS PhD.
Faculty: 75.
Degree requirements: For doctorate, dissertation, qualifying exam required, foreign language not required.
Entrance requirements: For doctorate, GRE General Test, TOEFL. *Application deadline:* For fall admission, 1/5. *Application fee:* $40.
Expenses: Program pays tuition, fees, health insurance, and stipend for all students in good standing.
Faculty research: Cell and organ physiology, eukaryotic regulatory and molecular biology, molecular and cellular pharmacology.

University of California, San Diego (continued)
Frank Powell, Director.
Application contact: Gina Butcher, Graduate Program Representative, 858-534-3982.

■ UNIVERSITY OF CALIFORNIA, SAN FRANCISCO

Graduate Division, Biomedical Sciences Graduate Group, Program in Endocrinology, San Francisco, CA 94143

AWARDS PhD.

Degree requirements: For doctorate, dissertation required.
Entrance requirements: For doctorate, GRE General Test. *Application deadline:* For fall admission, 2/15. *Application fee:* $40.
Financial aid: Application deadline: 1/10.
Application contact: Pamela Humphrey, Program Administrator, 415-476-8467.

■ UNIVERSITY OF CALIFORNIA, SAN FRANCISCO

Graduate Division, Biomedical Sciences Graduate Group, Program in Physiology, San Francisco, CA 94143

AWARDS PhD.

Faculty: 18 full-time (2 women).
Students: In 1999, 3 degrees awarded.
Degree requirements: For doctorate, dissertation required, foreign language not required.
Entrance requirements: For doctorate, GRE General Test, GRE Subject Test, TOEFL. *Application deadline:* For fall admission, 2/1. *Application fee:* $40.
Financial aid: Application deadline: 1/10.
Faculty research: Circulation and respiration, enzyme kinetics, fetal and neonatal physiology, nephrology.
Application contact: Pamela Humphrey, Program Administrator, 415-476-8467.

■ UNIVERSITY OF CHICAGO

Division of the Biological Sciences, Neurobiology, Pharmacology, and Cell Physiology, Department of Neurobiology, Pharmacology and Physiology, Chicago, IL 60637-1513

AWARDS Cell physiology (PhD); pharmacological and physiological sciences (PhD).

Faculty: 33 full-time (8 women).
Students: 11 full-time (6 women). In 1999, 2 degrees awarded.
Degree requirements: For doctorate, dissertation, preliminary exam required. *Average time to degree:* Doctorate–6 years full-time.

Entrance requirements: For doctorate, GRE General Test, TOEFL. *Application deadline:* For fall admission, 1/5 (priority date). Applications are processed on a rolling basis. *Application fee:* $55.
Expenses: Tuition: Full-time $24,804; part-time $3,422 per course. Required fees: $390. Tuition and fees vary according to program.
Financial aid: Fellowships, grants and institutionally sponsored loans available. Financial aid application deadline: 6/1.
Faculty research: Psychopharmacology, neuropharmacology.
Application contact: Dr. Sangiam S. Sisodia, Chairman, 773-834-2900.

■ UNIVERSITY OF CHICAGO

Division of the Biological Sciences, Neurobiology, Pharmacology, and Cell Physiology, Program in Cellular and Molecular Physiology, Chicago, IL 60637-1513

AWARDS PhD.

Faculty: 31 full-time (5 women).
Students: 12 full-time (5 women). *0 applicants, 0% accepted.* In 1999, 1 degree awarded.
Degree requirements: For doctorate, dissertation, preliminary exam required. *Average time to degree:* Doctorate–6 years full-time.
Entrance requirements: For doctorate, GRE General Test, TOEFL. *Application deadline:* For fall admission, 1/5 (priority date). *Application fee:* $55.
Expenses: Tuition: Full-time $24,804; part-time $3,422 per course. Required fees: $390. Tuition and fees vary according to program.
Financial aid: Fellowships, grants and institutionally sponsored loans available.
Faculty research: Molecular genetics, biochemical biological and physical approaches to cell physiology.
Application contact: Chair, 773-705-3849.

■ UNIVERSITY OF CINCINNATI

Division of Research and Advanced Studies, College of Medicine, Graduate Programs in Medicine, Department of Molecular and Cellular Physiology, Cincinnati, OH 45267

AWARDS Biophysics (MS, PhD); physiology (MS, PhD).

Faculty: 13 full-time.
Students: 19 full-time (6 women), 1 (woman) part-time; includes 4 minority (2 African Americans, 2 Asian Americans or Pacific Islanders), 3 international. *64 applicants, 16% accepted.* In 1999, 1 master's, 1 doctorate awarded.
Degree requirements: For master's, thesis required, foreign language not

required; for doctorate, dissertation, qualifying exam required, foreign language not required. *Average time to degree:* Master's–2.8 years full-time; doctorate–4.8 years full-time.
Entrance requirements: For master's and doctorate, GRE General Test, GRE Subject Test. *Application deadline:* For fall admission, 2/1 (priority date). Applications are processed on a rolling basis. *Application fee:* $30.
Expenses: Tuition, state resident: full-time $5,139; part-time $196 per credit hour. Tuition, nonresident: full-time $10,326; part-time $369 per credit hour. Required fees: $561; $187 per quarter.
Financial aid: Unspecified assistantships available. Financial aid application deadline: 5/1.
Faculty research: Neurobiology, electrophysiology, muscle physiology, cardiovascular physiology, endocrinology. *Total annual research expenditures:* $5.1 million.
Dr. David E. Millhorn, Head, 513-558-5636, *Fax:* 513-558-5738, *E-mail:* david.millhorn@uc.edu.
Application contact: John Dedman, Graduate Program Director, 513-558-4145, *Fax:* 513-558-5738, *E-mail:* john.dedman@uc.edu.

■ UNIVERSITY OF CINCINNATI

Division of Research and Advanced Studies, College of Medicine, Graduate Programs in Medicine, Program in Molecular and Cellular Pathophysiology, Cincinnati, OH 45267

AWARDS D Sc.

Faculty: 6 full-time.
Degree requirements: For doctorate, one foreign language, dissertation, qualifying exam required.
Entrance requirements: For doctorate, GRE General Test, surgical residency. *Application deadline:* For fall admission, 2/1 (priority date). Applications are processed on a rolling basis. *Application fee:* $30.
Expenses: Tuition, state resident: full-time $5,139; part-time $196 per credit hour. Tuition, nonresident: full-time $10,326; part-time $369 per credit hour. Required fees: $561; $187 per quarter.
Financial aid: Unspecified assistantships available. Financial aid application deadline: 5/1. *Total annual research expenditures:* $2.6 million.
Dr. Josef Fischer, Chairman, 513-558-4202, *Fax:* 513-872-6999, *E-mail:* josef.fischer@uc.edu.
Application contact: Dr. George Babcock, Director, 513-872-6231, *Fax:* 513-872-6999, *E-mail:* george.babcock@uc.edu.

UNIVERSITY OF COLORADO AT BOULDER

Graduate School, College of Arts and Sciences, Department of Environmental, Population, and Organic Biology, Boulder, CO 80309

AWARDS Animal behavior (MA, PhD); aquatic biology (MA, PhD); behavioral genetics (MA, PhD); ecology (MA, PhD); microbiology (MA, PhD); neurobiology (MA, PhD); plant and animal physiology (MA, PhD); plant and animal systematics (MA, PhD); population biology (MA, PhD); population genetics (MA, PhD).

Faculty: 39 full-time (9 women).
Students: 84 full-time (36 women), 14 part-time (7 women); includes 17 minority (6 Asian Americans or Pacific Islanders, 10 Hispanic Americans, 1 Native American). Average age 29. *147 applicants, 14% accepted.* In 1999, 7 master's, 13 doctorates awarded. Terminal master's awarded for partial completion of doctoral program.
Degree requirements: For master's, thesis or alternative, comprehensive exam required, foreign language not required; for doctorate, dissertation, comprehensive exam required, foreign language not required. *Average time to degree:* Master's–3 years full-time; doctorate–5 years full-time.
Entrance requirements: For master's, GRE General Test, GRE Subject Test, minimum undergraduate GPA of 3.0; for doctorate, GRE General Test, GRE Subject Test. *Application deadline:* For fall admission, 1/15 (priority date). *Application fee:* $40 ($60 for international students).
Expenses: Tuition, state resident: part-time $181 per credit hour. Tuition, nonresident: part-time $542 per credit hour. Required fees: $99 per term. Tuition and fees vary according to course load and program.
Financial aid: Fellowships, research assistantships, teaching assistantships, Federal Work-Study, institutionally sponsored loans, and tuition waivers (full) available. Financial aid application deadline: 3/1.
Faculty research: Evolution, developmental biology, behavior and neurobiology. *Total annual research expenditures:* $1.8 million.
Michael Breed, Chair, 303-492-8981, *Fax:* 303-492-8699, *E-mail:* michael.breed@colorado.edu.
Application contact: Jill Skarstadt, Graduate Secretary, 303-492-7654, *Fax:* 303-492-8699, *E-mail:* jill.skarstadt@colorado.edu.

UNIVERSITY OF COLORADO HEALTH SCIENCES CENTER

Graduate School, Programs in Biological and Medical Sciences, Department of Physiology and Biophysics, Denver, CO 80262

AWARDS Cellular and molecular physiology (PhD), including cellular, molecular, and developmental neuroscience.

Degree requirements: For doctorate, dissertation required, foreign language not required.
Entrance requirements: For doctorate, GRE, minimum GPA of 2.75.
Expenses: Tuition, state resident: full-time $1,512; part-time $56 per hour. Tuition, nonresident: full-time $7,209; part-time $267 per hour. Full-time tuition and fees vary according to course load and program.
Find an in-depth description at www.petersons.com/graduate.

UNIVERSITY OF CONNECTICUT

Graduate School, College of Liberal Arts and Sciences, Biological Sciences Group, Department of Physiology and Neurobiology, Storrs, CT 06269

AWARDS Neurobiology (MS, PhD); physiology (MS, PhD).

Degree requirements: For doctorate, dissertation required.
Entrance requirements: For master's and doctorate, GRE General Test, GRE Subject Test, TOEFL.
Expenses: Tuition, state resident: full-time $5,118. Tuition, nonresident: full-time $13,298. Required fees: $1,022.
Find an in-depth description at www.petersons.com/graduate.

UNIVERSITY OF DELAWARE

College of Arts and Science, Department of Biological Sciences, Newark, DE 19716

AWARDS Biotechnology (MS, PhD); cell and extracellular matrix biology (MS, PhD); cell and systems physiology (MS, PhD); ecology and evolution (MS, PhD); microbiology (MS, PhD); molecular biology and genetics (MS, PhD); plant biology (MS, PhD).

Faculty: 37 full-time (10 women).
Students: 22 full-time (11 women), 1 part-time; includes 2 minority (both African Americans), 9 international. Average age 25. *37 applicants, 27% accepted.* In 2000, 9 doctorates awarded.
Degree requirements: For master's and doctorate, thesis/dissertation required, foreign language not required. *Average time to degree:* Master's–2.5 years full-time; doctorate–6 years full-time.

Entrance requirements: For master's and doctorate, GRE General Test, GRE Subject Test (advanced biology). *Application deadline:* For fall admission, 6/15. Applications are processed on a rolling basis. *Application fee:* $50. Electronic applications accepted.
Expenses: Tuition, state resident: full-time $4,380; part-time $243 per credit. Tuition, nonresident: full-time $12,750; part-time $708 per credit. Required fees: $15 per term. Tuition and fees vary according to program.
Financial aid: In 2000–01, 18 students received aid, including 2 fellowships with full tuition reimbursements available (averaging $18,000 per year), 4 research assistantships with full tuition reimbursements available (averaging $18,000 per year), 11 teaching assistantships with full tuition reimbursements available (averaging $18,000 per year); tuition waivers (partial) also available. Financial aid application deadline: 6/15.
Faculty research: Cell interactions, molecular mechanisms, microorganisms, embryo implantation. *Total annual research expenditures:* $1.8 million.
Dr. Daniel D. Carson, Chair, 302-831-6977, *Fax:* 302-831-2281, *E-mail:* dcarson@udel.edu.
Application contact: Norman Karin, Graduate Coordinator, 302-831-1841, *Fax:* 302-831-2281, *E-mail:* ccoletta@udel.edu.
Find an in-depth description at www.petersons.com/graduate.

UNIVERSITY OF FLORIDA

College of Medicine and Graduate School, Interdisciplinary Program in Biomedical Sciences, Concentration in Physiology and Pharmacology, Gainesville, FL 32611
AWARDS PhD.

Degree requirements: For doctorate, dissertation required, foreign language not required.
Entrance requirements: For doctorate, GRE General Test, minimum GPA of 3.0. Electronic applications accepted.
Expenses: Tuition, state resident: part-time $144 per credit hour. Tuition, nonresident: part-time $505 per credit hour. Tuition and fees vary according to course level, course load and program.
Find an in-depth description at www.petersons.com/graduate.

UNIVERSITY OF FLORIDA

College of Medicine and Graduate School, Interdisciplinary Program in Biomedical Sciences, Department of Physiology, Gainesville, FL 32610
AWARDS PhD.

University of Florida (continued)

Degree requirements: For doctorate, dissertation required, foreign language not required.

Entrance requirements: For doctorate, GRE General Test, TOEFL, minimum GPA of 3.0. Electronic applications accepted.

Expenses: Tuition, state resident: part-time $144 per credit hour. Tuition, nonresident: part-time $505 per credit hour. Tuition and fees vary according to course level, course load and program.

Faculty research: Cell and general endocrinology, neuroendocrinology, neurophysiology, respiration, membrane transport and ion channels.

■ UNIVERSITY OF GEORGIA

College of Veterinary Medicine and Graduate School, Graduate Programs in Veterinary Medicine, Department of Physiology and Pharmacology, Athens, GA 30602

AWARDS Pharmacology (MS, PhD); physiology (MS, PhD).

Degree requirements: For master's, thesis required, foreign language not required; for doctorate, one foreign language (computer language can substitute), dissertation required.

Entrance requirements: For master's and doctorate, GRE General Test. Electronic applications accepted.

Expenses: Tuition, state resident: full-time $7,516; part-time $431 per credit hour. Tuition, nonresident: full-time $12,204; part-time $793 per credit hour.

■ UNIVERSITY OF HAWAII AT MANOA

John A. Burns School of Medicine and Graduate Division, Graduate Programs in Biomedical Sciences, Department of Physiology, Honolulu, HI 96822

AWARDS MS, PhD.

Faculty: 19 full-time (1 woman), 5 part-time/adjunct (1 woman).

Students: 8 full-time (4 women), 5 part-time (4 women). Average age 32. *14 applicants, 79% accepted.* In 1999, 2 master's, 2 doctorates awarded.

Degree requirements: For master's, computer language, thesis required (for some programs), foreign language not required; for doctorate, computer language, dissertation required, foreign language not required. *Average time to degree:* Master's–2 years full-time; doctorate–5.2 years full-time.

Entrance requirements: For master's and doctorate, GRE or MCAT. *Application deadline:* For fall admission, 2/1; for spring

admission, 9/1. *Application fee:* $25 ($50 for international students).

Expenses: Tuition, state resident: part-time $168 per credit. Tuition, nonresident: part-time $415 per credit. Required fees: $51 per semester. Part-time tuition and fees vary according to course load.

Financial aid: In 1999–00, 10 students received aid, including 4 research assistantships (averaging $16,065 per year); teaching assistantships, tuition waivers (full) also available.

Faculty research: Neurophysiology, endocrinology, cardiovascular, respiration. *Total annual research expenditures:* $65,000. Dr. David Lally, Chairperson, 808-956-8640, *Fax:* 808-956-9722, *E-mail:* lally@hawaii.edu.

■ UNIVERSITY OF ILLINOIS AT CHICAGO

College of Medicine and Graduate College, Graduate Programs in Medicine, Department of Physiology and Biophysics, Chicago, IL 60607-7128

AWARDS MS, PhD.

Faculty: 19 full-time (6 women).

Students: 23 full-time (12 women), 1 (woman) part-time; includes 4 minority (1 African American, 3 Asian Americans or Pacific Islanders), 4 international. Average age 29. *55 applicants, 22% accepted.* In 1999, 5 doctorates awarded. Terminal master's awarded for partial completion of doctoral program.

Degree requirements: For master's and doctorate, thesis/dissertation required, foreign language not required.

Entrance requirements: For master's and doctorate, GRE General Test, TOEFL. *Application deadline:* For fall admission, 6/1; for spring admission, 11/1. *Application fee:* $40 ($50 for international students).

Expenses: Tuition, state resident: full-time $3,750. Tuition, nonresident: full-time $10,588. Tuition and fees vary according to course load.

Financial aid: In 1999–00, 10 students received aid; fellowships, research assistantships, teaching assistantships, Federal Work-Study and traineeships available. Financial aid application deadline: 3/1; financial aid applicants required to submit FAFSA.

Faculty research: Neuroscience, endocrinology and reproduction, cell physiology, exercise physiology, NMR. R. John Solaro, Head, 312-996-7620.

Find an in-depth description at www.petersons.com/graduate.

■ UNIVERSITY OF ILLINOIS AT URBANA–CHAMPAIGN

Graduate College, College of Liberal Arts and Sciences, School of Life Sciences, Program in Molecular and Integrative Physiology, Urbana, IL 61801

AWARDS MS, PhD.

Faculty: 13 full-time (2 women), 9 part-time/adjunct (1 woman).

Students: 43 full-time (18 women); includes 9 minority (1 African American, 7 Asian Americans or Pacific Islanders, 1 Hispanic American), 13 international. *35 applicants, 23% accepted.* In 1999, 6 master's, 4 doctorates awarded.

Degree requirements: For master's, foreign language and thesis not required; for doctorate, dissertation required, foreign language not required.

Entrance requirements: For master's, minimum GPA of 3.0. *Application deadline:* Applications are processed on a rolling basis. *Application fee:* $40 ($50 for international students).

Expenses: Tuition, state resident: full-time $4,616. Tuition, nonresident: full-time $11,768. Full-time tuition and fees vary according to course load.

Financial aid: Fellowships, research assistantships, teaching assistantships available. Financial aid application deadline: 2/15.

Philip Best, Chair, 217-333-1734, *Fax:* 217-333-1133, *E-mail:* p-best@uiuc.edu.

Application contact: Denice Wells, Director of Graduate Studies, 217-333-1734, *Fax:* 217-333-1133, *E-mail:* d-wells2@uiuc.edu.

Find an in-depth description at www.petersons.com/graduate.

■ THE UNIVERSITY OF IOWA

College of Medicine and Graduate College, Graduate Programs in Medicine, Department of Physiology and Biophysics, Iowa City, IA 52242-1316

AWARDS Physiology and biophysics (PhD); physiology and biophysiology (MS).

Faculty: 21 full-time (2 women), 3 part-time/adjunct (1 woman).

Students: 16 full-time (4 women); includes 2 minority (1 African American, 1 Asian American or Pacific Islander), 5 international. Average age 25. *61 applicants, 7% accepted.* In 1999, 4 doctorates awarded (100% entered university research/teaching). Terminal master's awarded for partial completion of doctoral program.

Degree requirements: For master's and doctorate, thesis/dissertation, comprehensive exam, teaching experience required, foreign language not required. *Average time to degree:* Doctorate–5 years full-time.

Entrance requirements: For master's and doctorate, GRE General Test, minimum GPA of 3.0. *Application deadline:* For fall admission, 4/1. Applications are processed on a rolling basis. *Application fee:* $20 ($30 for international students).

Expenses: Tuition, state resident: full-time $3,308. Tuition, nonresident: full-time $10,662. Tuition and fees vary according to course load and program.

Financial aid: In 1999–00, 7 fellowships with full tuition reimbursements (averaging $16,787 per year), 9 research assistantships with full tuition reimbursements (averaging $16,787 per year) were awarded; traineeships also available. Financial aid application deadline: 4/1.

Faculty research: Cellular and molecular endocrinology, membrane structure and function, cardiac cell electrophysiology, regulation of gene expression, neurophysiology.

Robert E. Fellows, Head, 319-335-7802, *Fax:* 319-335-7330, *E-mail:* robert-fellows@uiowa.edu.

Application contact: Dr. Toshinori Hoshi, Chairman of Graduate Admissions, 319-335-7845, *Fax:* 319-335-7330, *E-mail:* toshinori-hoshi@uiowa.edu.

Find an in-depth description at www.petersons.com/graduate.

■ UNIVERSITY OF KANSAS

Graduate Studies Medical Center, Graduate Programs in Biomedical and Basic Sciences, Department of Molecular and Integrative Physiology, Kansas City, KS 66160

AWARDS MS, PhD, MD/PhD. Part-time programs available.

Faculty: 18 full-time (1 woman), 1 part-time/adjunct (0 women).

Students: 1 (woman) full-time, 10 part-time (5 women); includes 1 minority (Asian American or Pacific Islander), 2 international. Average age 28. *0 applicants, 0% accepted.* In 1999, 2 master's, 2 doctorates awarded. Terminal master's awarded for partial completion of doctoral program.

Degree requirements: For master's, thesis, comprehensive oral exam required, foreign language not required; for doctorate, dissertation, comprehensive oral exam required.

Entrance requirements: For master's and doctorate, GRE General Test, TOEFL, TSE. *Application deadline:* For fall admission, 1/31 (priority date). Applications are processed on a rolling basis. *Application fee:* $0.

Expenses: Tuition, state resident: full-time $2,482; part-time $103 per credit hour.

Tuition, nonresident: full-time $8,104; part-time $338 per credit hour. Required fees: $428; $31 per credit hour. Tuition and fees vary according to program.

Financial aid: Fellowships, research assistantships, teaching assistantships, Federal Work-Study, institutionally sponsored loans, and scholarships available. Aid available to part-time students. Financial aid application deadline: 3/31; financial aid applicants required to submit FAFSA.

Faculty research: Neurophysiology-motor control, developmental biology, molecular neurobiology, neuroplasticity, ovarian organization. *Total annual research expenditures:* $4.3 million.

Dr. James Voogt, Chairman, 913-588-7025, *Fax:* 913-588-7430, *E-mail:* jvoogt@kumc.edu.

Application contact: Dr. Thomas Imig, Director of Graduate Studies, 913-588-7407, *Fax:* 913-588-7430, *E-mail:* timig@kumc.edu.

■ UNIVERSITY OF KENTUCKY

Graduate School and College of Medicine, Graduate Programs in Medicine, Program in Physiology and Biophysics, Lexington, KY 40506-0032

AWARDS Molecular, cellular and integrative physiology (PhD).

Entrance requirements: For doctorate, GRE General Test.

Expenses: Tuition, state resident: full-time $3,596; part-time $188 per credit hour. Tuition, nonresident: full-time $10,116; part-time $550 per credit hour.

■ UNIVERSITY OF LOUISVILLE

School of Medicine and Graduate School, Integrated Programs in Biomedical Sciences, Department of Physiology and Biophysics, Louisville, KY 40292-0001

AWARDS MS, PhD.

Degree requirements: For master's and doctorate, thesis/dissertation required, foreign language not required.

Entrance requirements: For master's, GRE General Test, 36 hours of graduate course work; for doctorate, GRE General Test, minimum GPA of 3.0. Electronic applications accepted.

Expenses: Tuition, state resident: full-time $3,260; part-time $182 per hour. Tuition, nonresident: full-time $9,780; part-time $544 per hour. Required fees: $143; $28 per hour.

Faculty research: Control of small blood vessels, neuroendocrine physiology, renal blood flow, regulation of heart and lungs, neurophysiological control of breathing and circulation.

■ UNIVERSITY OF MARYLAND

Graduate School, Graduate Programs in Medicine, Department of Physiology (Medicine), Baltimore, MD 21201-1627

AWARDS Neuroscience (PhD); physiology (PhD); reproductive endocrinology (PhD).

Degree requirements: For doctorate, dissertation required, foreign language not required.

Entrance requirements: For doctorate, GRE General Test, GRE Subject Test, TOEFL, minimum GPA of 3.0.

Expenses: Tuition, state resident: part-time $261 per credit hour. Tuition, nonresident: part-time $468 per credit hour. Tuition and fees vary according to program.

Faculty research: Membrane physiology, biophysics and morphology, central nervous system physiology, EEG analysis, information theory.

■ UNIVERSITY OF MASSACHUSETTS WORCESTER

Graduate School of Biomedical Sciences, Department of Physiology, Worcester, MA 01655-0115

AWARDS Cellular and molecular physiology (PhD).

Faculty: 35 full-time (3 women).

Students: 8 full-time (4 women); includes 1 minority (Hispanic American), 2 international.

Degree requirements: For doctorate, dissertation required.

Entrance requirements: For doctorate, GRE General Test, GRE Subject Test, 1 year of calculus, physics, organic chemistry and biology. *Application deadline:* For fall admission, 1/15 (priority date). Applications are processed on a rolling basis. *Application fee:* $25 ($50 for international students).

Expenses: Tuition, state resident: full-time $2,640. Tuition, nonresident: full-time $9,756. Required fees: $825. Full-time tuition and fees vary according to program.

Financial aid: In 1999–00, research assistantships with full tuition reimbursements (averaging $17,500 per year); unspecified assistantships also available.

Faculty research: Endocrinology, regulation of cellular and tissue metabolism, electrophysiology, muscle physiology.

Dr. Maurice Goodman, Chair, 508-856-2101.

Application contact: Dr. James G. Dobson, Graduate Director, 508-856-2102.

Find an in-depth description at www.petersons.com/graduate.

■ UNIVERSITY OF MEDICINE AND DENTISTRY OF NEW JERSEY

Graduate School of Biomedical Sciences, Graduate Programs in Biomedical Sciences, Department of Pharmacology and Physiology, Newark, NJ 07107

AWARDS MS, PhD. Terminal master's awarded for partial completion of doctoral program.

Degree requirements: For master's, thesis required; for doctorate, dissertation, qualifying exam required, foreign language not required.

Entrance requirements: For master's and doctorate, GRE General Test, TOEFL. *Application deadline:* For fall admission, 2/1; for spring admission, 10/1. Applications are processed on a rolling basis. *Application fee:* $40.

Expenses: Tuition, state resident: part-time $270 per credit hour. Tuition, nonresident: part-time $407 per credit hour. Part-time tuition and fees vary according to campus/location and program.

Financial aid: Fellowships, research assistantships, Federal Work-Study and institutionally sponsored loans available. Financial aid application deadline: 5/1. Dr. Andrew Philip Thomas, Acting Chairman, 973-972-4660, *E-mail:* thomasap@umdnj.edu.

Application contact: Dr. Henry E. Brezenoff, Dean, Graduate School of Biomedical Sciences, 973-972-5333, *Fax:* 973-972-7148, *E-mail:* hbrezeno@umdnj.edu.

Find an in-depth description at www.petersons.com/graduate.

■ UNIVERSITY OF MIAMI

School of Medicine and Graduate School, Graduate Programs in Medicine, Department of Physiology and Biophysics, Coral Gables, FL 33124

AWARDS PhD, MD/PhD.

Faculty: 16 full-time (2 women). **Students:** 13 full-time (9 women); includes 5 minority (4 Asian Americans or Pacific Islanders, 1 Hispanic American). Average age 25. In 1999, 1 doctorate awarded (100% entered university research/teaching).

Degree requirements: For doctorate, dissertation, qualifying exam required, foreign language not required. *Average time to degree:* Doctorate–5 years full-time.

Entrance requirements: For doctorate, GRE General Test, TOEFL, minimum GPA of 3.0 in sciences. *Application deadline:* Applications are processed on a rolling basis. *Application fee:* $0.

Expenses: Tuition, area resident: Part-time $899 per credit.

Financial aid: Fellowships, research assistantships, career-related internships or fieldwork and institutionally sponsored loans available.

Faculty research: Cell and membrane physiology, cell-to-cell communication, molecular neurobiology, neuroimmunology, neural development. Dr. Karl Magleby, Chairman, 305-243-6821, *Fax:* 305-243-6898, *E-mail:* kmagleby@mednet.med.miami.edu.

Application contact: Dr. Nirupa Chaudhari, Adviser, 305-243-3187, *Fax:* 305-243-5931, *E-mail:* nchaudha@newssun.med.miami.edu.

Find an in-depth description at www.petersons.com/graduate.

■ UNIVERSITY OF MICHIGAN

Medical School and Horace H. Rackham School of Graduate Studies, Program in Biomedical Sciences (PIBS), Department of Physiology, Ann Arbor, MI 48109

AWARDS PhD.

Faculty: 28 full-time (6 women). **Students:** 19 full-time (10 women); includes 6 minority (2 African Americans, 3 Asian Americans or Pacific Islanders, 1 Hispanic American). 20 applicants, 40% accepted. In 1999, 6 doctorates awarded (50% entered university research/teaching, 33% continued full-time study).

Degree requirements: For doctorate, oral defense of dissertation, preliminary exam required. *Average time to degree:* Doctorate–5 years full-time.

Entrance requirements: For doctorate, GRE General Test. *Application deadline:* For fall admission, 1/10 (priority date). Applications are processed on a rolling basis. *Application fee:* $55.

Expenses: Tuition, state resident: full-time $10,316. Tuition, nonresident: full-time $20,922.

Financial aid: In 1999–00, 13 fellowships with full tuition reimbursements (averaging $17,000 per year), 5 research assistantships with full tuition reimbursements (averaging $17,000 per year) were awarded; grants, traineeships, tuition waivers (full), and departmental funding also available.

Faculty research: Ion transport, cardiovascular physiology, gene expression, hormone action, gastrointestinal physiology, endocrinology, muscle, signal transduction. *Total annual research expenditures:* $5.1 million. Dr. John A. Williams, Chair, 734-936-2355, *Fax:* 734-936-8813, *E-mail:* jawillms@umich.edu.

Application contact: Anne Many, Student Services Assistant, 734-936-2355, *Fax:* 734-936-8813, *E-mail:* ahm@umich.edu.

Find an in-depth description at www.petersons.com/graduate.

■ UNIVERSITY OF MINNESOTA, DULUTH

School of Medicine, Graduate Program in Physiology, Duluth, MN 55812-2496

AWARDS MS, PhD.

Faculty: 5 full-time (1 woman). **Students:** Average age 25. *5 applicants, 20% accepted.* In 1999, 1 doctorate awarded (100% entered university research/teaching). Terminal master's awarded for partial completion of doctoral program.

Degree requirements: For master's, thesis required, foreign language not required; for doctorate, computer language, dissertation required, foreign language not required. *Average time to degree:* Doctorate–4 years full-time.

Entrance requirements: For master's, GRE or MCAT, TOEFL; for doctorate, GRE or MCAT, TOEFL, 1 year of course work in calculus, physics, and biology; 2 years of course work in chemistry; minimum GPA of 3.0 in science. *Application deadline:* For fall admission, 4/15 (priority date). Applications are processed on a rolling basis. *Application fee:* $40.

Expenses: Tuition, state resident: full-time $5,040; part-time $420 per credit. Tuition, nonresident: full-time $9,900; part-time $825 per credit. Required fees: $509. Tuition and fees vary according to course load and program.

Financial aid: In 1999–00, research assistantships with partial tuition reimbursements (averaging $14,768 per year); Federal Work-Study also available. Financial aid application deadline: 4/15.

Faculty research: Neural control of posture and locomotion, transport and metabolic phenomena in biological systems, control of organ blood flow, intracellular means of communication. *Total annual research expenditures:* $19,000. Edwin W. Haller, Associate Director of Graduate Studies, 218-726-8551, *Fax:* 218-726-6356, *E-mail:* ehaller@d.umn.edu.

■ UNIVERSITY OF MINNESOTA, TWIN CITIES CAMPUS

Medical School and Graduate School, Graduate Programs in Medicine, Department of Physiology, Minneapolis, MN 55455-0213

AWARDS Cellular and integrative physiology (MS, PhD).

Entrance requirements: For doctorate, GRE General Test or MCAT.

Expenses: Tuition, state resident: full-time $11,984; part-time $1,498 per semester. Tuition, nonresident: full-time $22,264; part-time $2,783 per semester. Full-time tuition and fees vary according to program and student level. Part-time tuition and fees vary according to course load and program.
Faculty research: Cell physiology, neuroscience, molecular biology, cardiovascular physiology.

■ UNIVERSITY OF MISSISSIPPI MEDICAL CENTER

Graduate Programs in Biomedical Sciences, Department of Physiology and Biophysics, Jackson, MS 39216-4505

AWARDS MS, PhD, MD/PhD.

Faculty: 16 full-time (1 woman).
Students: 8 full-time (6 women). Average age 25. *6 applicants, 50% accepted.*
Degree requirements: For master's, thesis required, foreign language not required; for doctorate, dissertation, first authored publication required, foreign language not required. *Average time to degree:* Master's–4 years full-time; doctorate–5 years full-time.
Entrance requirements: For master's and doctorate, GRE General Test, minimum GPA of 3.0. *Application deadline:* For fall admission, 8/1. Applications are processed on a rolling basis. *Application fee:* $10.
Expenses: Tuition, state resident: full-time $2,378; part-time $132 per hour. Tuition, nonresident: full-time $4,697; part-time $261 per hour. Tuition and fees vary according to program.
Financial aid: In 1999–00, 5 students received aid, including 5 research assistantships (averaging $16,234 per year); Federal Work-Study also available. Financial aid application deadline: 4/1.
Faculty research: Cardiovascular, renal, endocrine, and cellular neurophysiology; molecular physiology. *Total annual research expenditures:* $1.4 million.
Dr. John E. Hall, Chairman, 601-984-1801, *Fax:* 601-984-1817.
Application contact: Dr. Michael W. Brands, Director of Graduate Programs, 601-984-1820, *Fax:* 601-984-1817, *E-mail:* mbrands@physiology.umsmed.edu.

■ UNIVERSITY OF MISSOURI–COLUMBIA

College of Veterinary Medicine and Graduate School, Graduate Programs in Veterinary Medicine, Department of Veterinary Biomedical Sciences, Columbia, MO 65211

AWARDS Physiology (PhD); veterinary biomedical sciences (MS).

Degree requirements: For master's, thesis required; for doctorate, dissertation required.
Entrance requirements: For master's and doctorate, GRE General Test, minimum GPA of 3.0.
Expenses: Tuition, state resident: full-time $3,020; part-time $168 per hour. Tuition, nonresident: full-time $6,066; part-time $505 per hour. Required fees: $445; $18 per hour. Tuition and fees vary according to course load and program.

■ UNIVERSITY OF MISSOURI–COLUMBIA

School of Medicine and Graduate School, Graduate Programs in Medicine, Department of Physiology, Columbia, MO 65211

AWARDS MS, PhD. Terminal master's awarded for partial completion of doctoral program.

Degree requirements: For master's and doctorate, thesis/dissertation required, foreign language not required.
Entrance requirements: For master's and doctorate, GRE General Test, TOEFL, minimum GPA of 3.0.
Expenses: Tuition, state resident: full-time $3,020; part-time $168 per hour. Tuition, nonresident: full-time $6,066; part-time $505 per hour. Required fees: $445; $18 per hour.
Faculty research: Hypertension, microcirculation, membrane biophysics, cardiac muscle dynamics, regulation of vascular control.

■ UNIVERSITY OF MISSOURI–ST. LOUIS

Graduate School, College of Arts and Sciences, Department of Biology, St. Louis, MO 63121-4499

AWARDS Biology (MS, PhD), including animal behavior (MS), biochemistry, biotechnology (MS), conservation biology (MS), development (MS), ecology (MS), environmental studies (PhD), evolution (MS), genetics (MS), molecular biology and biotechnology (PhD), molecular/cellular biology (MS), physiology (MS), plant systematics, population biology (MS), tropical biology (MS); biotechnology (Certificate); tropical biology and conservation (Certificate). Part-time programs available.

Faculty: 46.
Students: 21 full-time (11 women), 75 part-time (44 women); includes 13 minority (2 African Americans, 2 Asian Americans or Pacific Islanders, 8 Hispanic Americans, 1 Native American), 23 international. In 1999, 14 master's, 4 doctorates awarded.
Degree requirements: For master's, thesis or alternative required, foreign

language not required; for doctorate, one foreign language, dissertation, 1 semester of teaching experience required.
Entrance requirements: For doctorate, GRE General Test. *Application deadline:* For fall admission, 7/1 (priority date); for spring admission, 11/1 (priority date). Applications are processed on a rolling basis. *Application fee:* $25 ($40 for international students). Electronic applications accepted.
Expenses: Tuition, state resident: full-time $4,932; part-time $173 per credit hour. Tuition, nonresident: full-time $13,279; part-time $521 per credit hour. Required fees: $775; $33 per credit hour. Tuition and fees vary according to degree level and program.
Financial aid: In 1999–00, 8 research assistantships with partial tuition reimbursements (averaging $10,635 per year), 14 teaching assistantships with partial tuition reimbursements (averaging $11,488 per year) were awarded; career-related internships or fieldwork and Federal Work-Study also available. Aid available to part-time students. Financial aid application deadline: 2/1. *Total annual research expenditures:* $908,828.
Application contact: Graduate Admissions, 314-516-5458, *Fax:* 314-516-6759, *E-mail:* gradadm@umsl.edu.

■ UNIVERSITY OF NEBRASKA MEDICAL CENTER

Graduate College, Department of Physiology and Biophysics, Omaha, NE 68198

AWARDS Physiology (PhD).

Faculty: 12 full-time.
Students: 4 full-time (3 women), (all international). Average age 29. In 1999, 2 doctorates awarded.
Degree requirements: For doctorate, dissertation required, foreign language not required.
Entrance requirements: For doctorate, GRE General Test or MCAT, previous course work in biology, chemistry, mathematics, and physics. *Application fee:* $35.
Expenses: Tuition, state resident: part-time $116 per semester hour. Tuition, nonresident: part-time $270 per semester hour. Tuition and fees vary according to program.
Financial aid: Fellowships, research assistantships, teaching assistantships, institutionally sponsored loans available. Aid available to part-time students. Financial aid application deadline: 3/1.
Faculty research: Cardiovascular, renal, and neurophysiology/neuroendocrine cellular physiology.

University of Nebraska Medical Center (continued)

Dr. Steven C. Sansom, Chairman, Graduate Committee, 402-559-2919, *Fax:* 402-559-4438, *E-mail:* ssansom@unmc.edu.
Application contact: Jo Wagner, Associate Director of Admissions, 402-559-6468.

■ UNIVERSITY OF NEVADA, RENO

School of Medicine and Graduate School, Graduate Programs in Medicine, Interdisciplinary Program in Cellular and Molecular Pharmacology and Physiology, Reno, NV 89557

AWARDS MS, PhD.

Faculty: 23.
Students: 8 full-time (5 women), 2 part-time (1 woman); includes 2 minority (both Asian Americans or Pacific Islanders), 2 international. Average age 27. *9 applicants, 33% accepted.* In 1999, 3 doctorates awarded (100% continued full-time study). Terminal master's awarded for partial completion of doctoral program.
Degree requirements: For master's, thesis required, foreign language not required; for doctorate, one foreign language (computer language can substitute), dissertation required.
Entrance requirements: For master's, GRE General Test, TOEFL, minimum GPA of 2.75; for doctorate, GRE General Test, TOEFL, minimum GPA of 3.0. *Application deadline:* For fall admission, 3/1 (priority date); for spring admission, 11/1. Applications are processed on a rolling basis. *Application fee:* $40.
Expenses: Tuition, state resident: full-time $7,782. Tuition, nonresident: full-time $22,808. Required fees: $1,918.
Financial aid: Research assistantships, teaching assistantships available. Aid available to part-time students. Financial aid application deadline: 3/1.
Faculty research: Neuropharmacology, toxicology, cardiovascular pharmacology, neuromuscular pharmacology.
Dr. Burton Horowitz, Director, 775-784-1462.
Find an in-depth description at www.petersons.com/graduate.

■ UNIVERSITY OF NEW MEXICO

Graduate School, College of Arts and Sciences, Department of Biology, Albuquerque, NM 87131-2039

AWARDS Biology (MS, PhD), including air land ecology, behavioral ecology, botany, cellular and molecular biology, community ecology, comparative immunology, comparative physiology, conservation biology, ecology, ecosystem ecology, evolutionary biology, evolutionary genetics, microbiology, molecular genetics, parasitology, physiological ecology, physiology, population biology, vertebrate and invertebrate zoology. Part-time programs available.

Faculty: 35 full-time (5 women), 18 part-time/adjunct (11 women).
Students: 71 full-time (37 women), 28 part-time (11 women); includes 8 minority (2 Asian Americans or Pacific Islanders, 5 Hispanic Americans, 1 Native American), 11 international. Average age 33. *93 applicants, 30% accepted.* In 1999, 11 master's, 12 doctorates awarded. Terminal master's awarded for partial completion of doctoral program.
Degree requirements: For master's, one foreign language (computer language can substitute), thesis required (for some programs); for doctorate, 2 foreign languages (computer language can substitute for one), dissertation required.
Entrance requirements: For master's and doctorate, GRE General Test, GRE Subject Test, minimum GPA of 3.2. *Application deadline:* For fall admission, 1/15. *Application fee:* $25.
Expenses: Tuition, state resident: full-time $2,514; part-time $105 per credit hour. Tuition, nonresident: full-time $10,304; part-time $417 per credit hour. International tuition: $10,304 full-time. Required fees: $516; $22 per credit hour. Tuition and fees vary according to program.
Financial aid: In 1999–00, 58 students received aid, including 24 fellowships (averaging $1,645 per year), 26 research assistantships with tuition reimbursements available (averaging $8,921 per year), 40 teaching assistantships with tuition reimbursements available (averaging $11,066 per year); career-related internships or fieldwork, Federal Work-Study, institutionally sponsored loans, and tuition waivers (full and partial) also available. Aid available to part-time students. Financial aid applicants required to submit FAFSA.
Faculty research: Developmental biology, immunobiology. *Total annual research expenditures:* $4.5 million.
Dr. Kathryn Vogel, Chair, 505-277-3411, *Fax:* 505-277-0304, *E-mail:* kgvogel@unm.edu.
Application contact: Vivian Kent, Information Contact, 505-277-1712, *Fax:* 505-277-0304, *E-mail:* vkent@unm.edu.

■ THE UNIVERSITY OF NORTH CAROLINA AT CHAPEL HILL

School of Medicine and Graduate School, Graduate Programs in Medicine, Department of Cell and Molecular Physiology, Chapel Hill, NC 27599

AWARDS PhD.

Faculty: 21 full-time (5 women).
Students: 18 full-time (10 women); includes 2 minority (both Asian Americans or Pacific Islanders), 3 international. Average age 28. *20 applicants, 55% accepted.* In 1999, 3 doctorates awarded.
Degree requirements: For doctorate, dissertation required, foreign language not required. *Average time to degree:* Doctorate–4.8 years full-time.
Entrance requirements: For doctorate, GRE General Test. *Application deadline:* For fall admission, 1/1 (priority date). Applications are processed on a rolling basis. *Application fee:* $55. Electronic applications accepted.
Expenses: Tuition, state resident: full-time $1,966. Tuition, nonresident: full-time $11,026. Required fees: $8,940. One-time fee: $15 full-time. Part-time tuition and fees vary according to course load.
Financial aid: In 1999–00, 5 fellowships with full tuition reimbursements (averaging $16,000 per year), 12 research assistantships with full tuition reimbursements (averaging $16,000 per year) were awarded; Federal Work-Study and institutionally sponsored loans also available. Aid available to part-time students. Financial aid application deadline: 3/1; financial aid applicants required to submit FAFSA.
Faculty research: Signal transduction; growth factors; cardiovascular diseases; neurobiology; hormones, receptors, and ion channels. *Total annual research expenditures:* $5.4 million.
Dr. Stanley C. Froehner, Chairman, 919-966-1239, *Fax:* 919-966-6927, *E-mail:* froehner@med.unc.edu.
Application contact: Dr. Sharon Milgram, Director of Graduate Program, 919-966-3935, *Fax:* 919-966-6927, *E-mail:* physgrad@med.unc.edu.
Find an in-depth description at www.petersons.com/graduate.

■ UNIVERSITY OF NORTH DAKOTA

School of Medicine and Graduate School, Graduate Programs in Medicine, Department of Physiology, Grand Forks, ND 58202

AWARDS MS, PhD.

Faculty: 6 full-time (1 woman).
Students: 4 full-time (1 woman), 1 part-time. *2 applicants, 0% accepted.* In 1999, 1 master's awarded.
Degree requirements: For master's, thesis, final exam required, foreign language not required; for doctorate, one foreign language (computer language can substitute), dissertation, final exam, comprehensive exam required.

Entrance requirements: For master's, GRE General Test, TOEFL, minimum GPA of 3.0; for doctorate, GRE General Test, GRE Subject Test, TOEFL, minimum GPA of 3.5. *Application deadline:* For fall admission, 3/1 (priority date). Applications are processed on a rolling basis. *Application fee:* $25.

Expenses: Tuition, state resident: full-time $2,690; part-time $112 per credit. Tuition, nonresident: full-time $7,182; part-time $299 per credit. Required fees: $46 per semester.

Financial aid: In 1999–00, 3 students received aid, including 3 research assistantships with full tuition reimbursements available (averaging $10,586 per year); fellowships, teaching assistantships, Federal Work-Study, institutionally sponsored loans, scholarships, and tuition waivers (full and partial) also available. Aid available to part-time students. Financial aid application deadline: 3/15; financial aid applicants required to submit FAFSA.

Faculty research: Neuroendocrinology, renal pathophysiology, central nervous system, cytokines in immune function, vitamin C.

Dr. Holly Brown-Borg, Chairperson, 701-777-3975, *Fax:* 701-777-4490, *E-mail:* brownbrg@mail.med.und.nodak.edu.

■ UNIVERSITY OF NORTH TEXAS HEALTH SCIENCE CENTER AT FORT WORTH

Graduate School of Biomedical Sciences, Fort Worth, TX 76107-2699

AWARDS Anatomy and cell biology (MS, PhD); biochemistry and molecular biology (MS, PhD); biotechnology (MS); integrative physiology (MS, PhD); microbiology and immunology (MS, PhD); pharmacology (MS, PhD).

Faculty: 65 full-time (9 women), 11 part-time/adjunct (1 woman).

Students: 59 full-time (27 women), 44 part-time (20 women); includes 30 minority (10 African Americans, 9 Asian Americans or Pacific Islanders, 11 Hispanic Americans), 23 international. *70 applicants, 70% accepted.* In 1999, 5 master's awarded (40% found work related to degree, 60% continued full-time study); 5 doctorates awarded (80% entered university research/teaching, 20% found other work related to degree).

Degree requirements: For doctorate, computer language, dissertation required. *Average time to degree:* Master's–2.5 years full-time, 4 years part-time; doctorate–5 years full-time.

Entrance requirements: For master's and doctorate, GRE General Test, TOEFL. *Application deadline:* For fall admission, 5/1; for spring admission, 11/1. Applications

are processed on a rolling basis. *Application fee:* $25 ($50 for international students).

Expenses: Tuition, state resident: full-time $1,188; part-time $66 per credit. Tuition, nonresident: full-time $5,058; part-time $281 per credit. Required fees: $366; $183 per semester.

Financial aid: In 1999–00, 11 fellowships, 70 research assistantships (averaging $16,500 per year) were awarded; teaching assistantships, career-related internships or fieldwork, Federal Work-Study, grants, institutionally sponsored loans, and traineeships also available. Aid available to part-time students. Financial aid application deadline: 4/1; financial aid applicants required to submit FAFSA.

Faculty research: Alzheimer's disease, diabetes, eye diseases, cancer, cardiovascular physiology.

Dr. Thomas Yorio, Dean, 817-735-2560, *Fax:* 817-735-0243, *E-mail:* yoriot@hsc.unt.edu.

Application contact: Jan Sharp, Administrative Assistant, 817-735-0258, *Fax:* 817-735-0243, *E-mail:* gsbs@hsc.unt.edu.

Find an in-depth description at www.petersons.com/graduate.

■ UNIVERSITY OF NOTRE DAME

Graduate School, College of Science, Department of Biological Sciences, Program in Physiology, Notre Dame, IN 46556

AWARDS MS, PhD.

Degree requirements: For master's and doctorate, thesis/dissertation required, foreign language not required.

Entrance requirements: For master's and doctorate, GRE General Test, GRE Subject Test, TOEFL. *Application deadline:* For fall admission, 2/1 (priority date); for spring admission, 11/1. Applications are processed on a rolling basis. *Application fee:* $50.

Expenses: Tuition: Full-time $21,930; part-time $1,218 per credit. Required fees: $95. Tuition and fees vary according to program.

Financial aid: Fellowships with full tuition reimbursements, research assistantships with full tuition reimbursements, teaching assistantships with full tuition reimbursements, tuition waivers (full) available. Financial aid application deadline: 2/1.

Faculty research: Comparative endocrinology-vertebrate/invertebrate; cold tolerance in insects; amphibian, fish, and avian reproductive physiology; parasite biochemistry and physiology; insect neurobiology.

Application contact: Dr. Terrence J. Akai, Director of Graduate Admissions, 219-631-7706, *Fax:* 219-631-4183, *E-mail:* gradad@nd.edu.

■ UNIVERSITY OF OKLAHOMA HEALTH SCIENCES CENTER

College of Medicine and Graduate College, Graduate Programs in Medicine, Department of Physiology, Oklahoma City, OK 73190

AWARDS MS, PhD. Part-time programs available.

Faculty: 17 full-time (1 woman).

Students: 5 full-time (3 women), 1 (woman) part-time; includes 1 minority (Asian American or Pacific Islander). *13 applicants, 38% accepted.* In 1999, 2 master's, 2 doctorates awarded. Terminal master's awarded for partial completion of doctoral program.

Degree requirements: For master's, thesis required (for some programs), foreign language not required; for doctorate, dissertation required, foreign language not required.

Entrance requirements: For master's and doctorate, GRE General Test, TOEFL, minimum B average. *Application deadline:* For spring admission, 12/1. *Application fee:* $25 ($50 for international students).

Expenses: Tuition, state resident: part-time $90 per semester hour. Tuition, nonresident: part-time $264 per semester hour. Tuition and fees vary according to program.

Financial aid: Fellowships, research assistantships, teaching assistantships, career-related internships or fieldwork available.

Faculty research: Cardiopulmonary physiology, neurophysiology, exercise physiology, cell and molecular physiology.

Dr. Robert D. Foreman, Head, 405-271-2226.

Application contact: Dr. Sinya Benyajati, Chairman, Graduate Studies Committee, 405-271-2226, *Fax:* 405-271-3181, *E-mail:* gradphys@uokhsc.edu.

■ UNIVERSITY OF PENNSYLVANIA

School of Medicine, Biomedical Graduate Studies, Graduate Group in Cell and Molecular Biology, Program in Cell Biology and Physiology, Philadelphia, PA 19104

AWARDS PhD, MD/PhD, VMD/PhD.

Degree requirements: For doctorate, dissertation required, foreign language not required.

Entrance requirements: For doctorate, GRE General Test, TOEFL. *Application deadline:* For fall admission, 1/2 (priority

University of Pennsylvania (continued)
date). Applications are processed on a rolling basis. *Application fee:* $65.

Expenses: Tuition: Full-time $17,256; part-time $2,991 per course. Required fees: $2,588; $363 per course. $726 per term.

Financial aid: Fellowships, research assistantships, teaching assistantships, grants and institutionally sponsored loans available. Financial aid application deadline: 1/2.

Dr. Morris Birnbaum, Head, 215-898-5001.

Application contact: Mary Webster, Coordinator, 215-898-4360, *E-mail:* camb@mail.med.upenn.edu.

■ UNIVERSITY OF PITTSBURGH

School of Medicine, Graduate Programs in Medicine, Program in Cell Biology and Molecular Physiology, Pittsburgh, PA 15260

AWARDS MS, PhD.

Faculty: 26 full-time (6 women).
Students: 9 full-time (3 women); includes 3 minority (all Asian Americans or Pacific Islanders), 1 international. *325 applicants, 22% accepted.* In 1999, 1 master's, 1 doctorate awarded.
Degree requirements: For doctorate, dissertation required, foreign language not required. *Average time to degree:* Doctorate–5.3 years full-time.
Entrance requirements: For doctorate, GRE General Test, GRE Subject Test, TOEFL, minimum QPA of 3.0. *Application deadline:* For fall admission, 1/15 (priority date). Applications are processed on a rolling basis. *Application fee:* $30 ($40 for international students).
Expenses: Tuition, state resident: full-time $9,778; part-time $403 per credit. Tuition, nonresident: full-time $20,146; part-time $830 per credit. Required fees: $480; $90 per semester.
Financial aid: Research assistantships with tuition reimbursements, teaching assistantships with tuition reimbursements, Federal Work-Study, institutionally sponsored loans, scholarships, traineeships, and unspecified assistantships available.
Faculty research: Epithelial cell biology, developmental biology of muscle, reproductive physiology and neuroendocrinology. *Total annual research expenditures:* $76.3 million.
Application contact: Graduate Studies Administrator, 412-648-8957, *Fax:* 412-648-1236, *E-mail:* biomed_phd@fs1.dean-med.pitt.edu.

Find an in-depth description at www.petersons.com/graduate.

■ UNIVERSITY OF PUERTO RICO, MEDICAL SCIENCES CAMPUS

School of Medicine, Division of Graduate Studies, Department of Physiology, San Juan, PR 00936-5067

AWARDS MS, PhD.

Faculty: 13 full-time (2 women).
Students: 9 full-time (5 women); all minorities (all Hispanic Americans). Average age 28. *9 applicants, 33% accepted.* In 1999, 1 doctorate awarded (100% entered university research/teaching). Terminal master's awarded for partial completion of doctoral program.
Degree requirements: For master's and doctorate, one foreign language, thesis/dissertation required. *Average time to degree:* Doctorate–6 years full-time.
Entrance requirements: For master's and doctorate, GRE General Test, GRE Subject Test, interview. *Application deadline:* For fall admission, 2/15. *Application fee:* $15.
Expenses: Tuition, state resident: full-time $5,500. Tuition, nonresident: full-time $8,400. Required fees: $600. Tuition and fees vary according to class time, course load, degree level and program.
Financial aid: In 1999–00, 3 research assistantships were awarded; fellowships, teaching assistantships, career-related internships or fieldwork and institutionally sponsored loans also available. Financial aid application deadline: 4/30.
Faculty research: Transport, respiration, neuroendocrinology, cellular and molecular physiology.
Dr. Nelson Escobales, Director, 787-758-2525 Ext. 1606.
Application contact: Dr. Walter I. Silva, Coordinator, 787-758-2525 Ext. 1608, *Fax:* 787-753-0120, *E-mail:* w_silva@rcmaca.upr.clu.edu.

■ UNIVERSITY OF ROCHESTER

School of Medicine and Dentistry, Graduate Programs in Medicine and Dentistry, Department of Pharmacology and Physiology, Program in Physiology, Rochester, NY 14627-0250

AWARDS MS, PhD.

Students: 5 full-time (2 women); includes 1 minority (Asian American or Pacific Islander), 1 international. In 1999, 1 master's, 1 doctorate awarded. Terminal master's awarded for partial completion of doctoral program.
Degree requirements: For master's, thesis required, foreign language not required; for doctorate, dissertation, qualifying exam required, foreign language not required.

Entrance requirements: For master's and doctorate, GRE General Test. *Application deadline:* For fall admission, 2/1. *Application fee:* $25.
Expenses: Tuition: Part-time $697 per credit hour. Tuition and fees vary according to program.
Financial aid: Fellowships, research assistantships, teaching assistantships, tuition waivers (full and partial) available. Financial aid application deadline: 2/1.
Application contact: Linda Fullington, Graduate Program Secretary, 716-275-0447.

Find an in-depth description at www.petersons.com/graduate.

■ UNIVERSITY OF SOUTH ALABAMA

College of Medicine and Graduate School, Program in Basic Medical Sciences, Specialization in Physiology, Mobile, AL 36688-0002

AWARDS PhD.

Faculty: 8 full-time (1 woman).
Students: In 1999, 2 degrees awarded.
Degree requirements: For doctorate, dissertation required, foreign language not required.
Entrance requirements: For doctorate, GRE General Test or MCAT. *Application deadline:* For fall admission, 4/1. Applications are processed on a rolling basis. *Application fee:* $25.
Expenses: Tuition, state resident: part-time $116 per semester hour. Tuition, nonresident: part-time $230 per semester hour. Required fees: $121 per semester.
Financial aid: Fellowships, research assistantships, institutionally sponsored loans available. Financial aid application deadline: 4/1.
Faculty research: Cardiovascular physiology.
Dr. Aubrey E. Taylor, Chair, 334-460-7004.
Application contact: Lanette Flagge, Coordinator, 334-460-6153.

■ UNIVERSITY OF SOUTH DAKOTA

School of Medicine and Graduate School, Biomedical Sciences Graduate Program, Physiology and Pharmacology Group, Vermillion, SD 57069-2390

AWARDS MA, PhD.

Faculty: 9 full-time (3 women).
Students: 6 full-time (2 women), 3 international. Average age 25. *65 applicants, 5% accepted.* In 1999, 1 doctorate awarded (100% entered university research/teaching). Terminal master's awarded for partial completion of doctoral program.

Degree requirements: For master's and doctorate, thesis/dissertation required, foreign language not required. *Average time to degree:* Doctorate–4.5 years full-time.

Entrance requirements: For master's and doctorate, GRE General Test, TOEFL, minimum GPA of 3.0. *Application deadline:* For fall admission, 3/15 (priority date). Applications are processed on a rolling basis. *Application fee:* $15.

Expenses: Tuition, state resident: full-time $2,126; part-time $89 per credit. Tuition, nonresident: full-time $6,270; part-time $261 per credit. Required fees: $1,194; $50 per credit. Tuition and fees vary according to course load and reciprocity agreements.

Financial aid: Fellowships, research assistantships available.

Faculty research: Pulmonary physiology and pharmacology, drug abuse, reproduction, signal transduction, cardiovascular physiology and pharmacology.

Dr. Steven B. Waller, Head, 605-677-5157, *Fax:* 605-677-6381, *E-mail:* swaller@usd.edu.

■ UNIVERSITY OF SOUTHERN CALIFORNIA

Keck School of Medicine and Graduate School, Graduate Programs in Medicine, Department of Physiology and Biophysics, Los Angeles, CA 90089

AWARDS MS, PhD, MD/PhD.

Faculty: 12 full-time (4 women).

Students: 22 full-time (13 women); includes 5 minority (all Asian Americans or Pacific Islanders), 10 international. Average age 25. *26 applicants, 23% accepted.* In 1999, 4 degrees awarded (100% entered university research/teaching). Terminal master's awarded for partial completion of doctoral program.

Degree requirements: For master's, foreign language not required; for doctorate, dissertation required, foreign language not required. *Average time to degree:* Doctorate–5 years full-time.

Entrance requirements: For master's and doctorate, GRE General Test, minimum GPA of 3.0. *Application deadline:* For fall admission, 2/1 (priority date). Applications are processed on a rolling basis. *Application fee:* $55. Electronic applications accepted.

Expenses: Tuition: Full-time $22,198; part-time $748 per unit. Required fees: $406.

Financial aid: In 1999–00, 3 fellowships with partial tuition reimbursements (averaging $18,598 per year), 14 research assistantships with full tuition reimbursements (averaging $18,598 per year), 2 teaching assistantships with full tuition reimbursements (averaging $18,598 per

year) were awarded; grants, traineeships, and tuition waivers (full) also available. Financial aid application deadline: 2/1.

Faculty research: Endocrinology, metabolism, cell transport, molecular biology, mathematical modelling. *Total annual research expenditures:* $2.1 million.

Dr. Alicia McDonough, Director, Graduate Studies, 323-442-1238, *Fax:* 323-442-2283, *E-mail:* mcdonoug@hsc.usc.edu.

Application contact: Elena Camarena, Graduate Coordinator, 323-442-1039, *Fax:* 323-442-2283, *E-mail:* physiol@hsc.usc.edu.

Find an in-depth description at www.petersons.com/graduate.

■ UNIVERSITY OF SOUTH FLORIDA

College of Medicine and Graduate School, Graduate Programs in Medical Sciences, Department of Physiology and Biophysics, Tampa, FL 33620-9951

AWARDS PhD.

Degree requirements: For doctorate, 2 foreign languages (computer language can substitute for one), dissertation required.

Entrance requirements: For doctorate, GRE General Test, minimum GPA of 3.0.

Expenses: Tuition, state resident: part-time $148 per credit hour. Tuition, nonresident: part-time $509 per credit hour.

Faculty research: Cardiovascular, neurorespiratory, and endocrine physiology; Alzheimer's disease; cell membrane biophysics.

Find an in-depth description at www.petersons.com/graduate.

■ UNIVERSITY OF SOUTH FLORIDA

Graduate School, College of Arts and Sciences, Department of Biology, Tampa, FL 33620-9951

AWARDS Biology (PhD); botany (MS); ecology (PhD); marine biology (MS, PhD); microbiology (MS); physiology (PhD); zoology (MS). Part-time programs available.

Degree requirements: For master's, foreign language not required; for doctorate, 2 foreign languages (computer language can substitute for one), dissertation required.

Entrance requirements: For master's, GRE General Test, minimum GPA of 3.0 in last 60 hours; for doctorate, GRE General Test, GRE Subject Test in biology. Electronic applications accepted.

Expenses: Tuition, state resident: part-time $148 per credit hour. Tuition, nonresident: part-time $509 per credit hour.

■ THE UNIVERSITY OF TENNESSEE

Graduate School, College of Agricultural Sciences and Natural Resources, Department of Animal Science, Knoxville, TN 37996

AWARDS Animal anatomy (PhD); breeding (MS, PhD); management (MS, PhD); nutrition (MS, PhD); physiology (MS, PhD). Part-time programs available.

Faculty: 19 full-time (2 women).

Students: 24 full-time (15 women), 10 part-time (4 women); includes 3 minority (1 African American, 1 Asian American or Pacific Islander, 1 Hispanic American), 7 international. *27 applicants, 56% accepted.* In 1999, 5 master's, 4 doctorates awarded.

Degree requirements: For master's and doctorate, thesis/dissertation required, foreign language not required.

Entrance requirements: For master's and doctorate, GRE General Test, TOEFL, minimum GPA of 2.7. *Application deadline:* For fall admission, 2/1 (priority date). Applications are processed on a rolling basis. *Application fee:* $35. Electronic applications accepted.

Expenses: Tuition, state resident: full-time $3,806; part-time $184 per credit hour. Tuition, nonresident: full-time $9,874; part-time $522 per credit hour. Tuition and fees vary according to program.

Financial aid: In 1999–00, 19 research assistantships, 3 teaching assistantships were awarded; fellowships, career-related internships or fieldwork, Federal Work-Study, institutionally sponsored loans, and unspecified assistantships also available. Financial aid application deadline: 2/1; financial aid applicants required to submit FAFSA.

Dr. Kelly Robbins, Head, 865-974-7286, *Fax:* 865-974-7297, *E-mail:* krobbins@utk.edu.

Application contact: Dr. James Godkin, Graduate Representative, *E-mail:* jgodkin@utk.edu.

■ THE UNIVERSITY OF TENNESSEE HEALTH SCIENCE CENTER

College of Graduate Health Sciences, Department of Physiology, Memphis, TN 38163-0002

AWARDS MS, PhD.

Degree requirements: For master's, thesis, oral and written comprehensive exams required, foreign language not required; for doctorate, dissertation, oral and written preliminary and comprehensive exams required, foreign language not required.

Entrance requirements: For master's, GRE General Test, TOEFL, minimum

GPA of 3.0; for doctorate, GRE General Test, GRE Subject Test, TOEFL, minimum GPA of 3.0.

Find an in-depth description at www.petersons.com/graduate.

■ THE UNIVERSITY OF TEXAS HEALTH SCIENCE CENTER AT SAN ANTONIO

Graduate School of Biomedical Sciences, Department of Physiology, San Antonio, TX 78229-3900

AWARDS MS, PhD. Terminal master's awarded for partial completion of doctoral program.

Degree requirements: For master's and doctorate, computer language, thesis/dissertation required, foreign language not required.

Entrance requirements: For master's and doctorate, GRE General Test.

Expenses: Tuition, state resident: part-time $38 per credit hour. Tuition, nonresident: part-time $249 per credit hour.

Faculty research: Systemic and cellular aspects of physiology.

■ THE UNIVERSITY OF TEXAS–HOUSTON HEALTH SCIENCE CENTER

Graduate School of Biomedical Sciences, Program in Integrative Biology, Houston, TX 77225-0036

AWARDS Cell biology (MS, PhD); physiology (MS, PhD).

Faculty: 25 full-time (8 women).
Students: 4 full-time (1 woman); includes 1 minority (Hispanic American), 1 international. Average age 27. *12 applicants, 50% accepted.* In 1999, 3 master's awarded. Terminal master's awarded for partial completion of doctoral program.
Degree requirements: For master's and doctorate, thesis/dissertation required, foreign language not required.
Entrance requirements: For master's and doctorate, GRE General Test, TOEFL, TWE. *Application deadline:* For fall admission, 1/15 (priority date); for spring admission, 11/1. Applications are processed on a rolling basis. *Application fee:* $10. Electronic applications accepted.
Financial aid: Fellowships, research assistantships, teaching assistantships, institutionally sponsored loans available. Financial aid application deadline: 1/15.
Faculty research: Cell signaling, regulation of gene expression, cell growth and

adaptation, cell injury and repair/protection, membrane trafficking and ion channels.
Dr. Roger G. O'Neil, Director, 713-500-6316, *Fax:* 713-500-7444, *E-mail:* roneil@girch1.med.uth.tmc.edu.
Application contact: Anne Baronitis, Director of Admissions, 713-500-9860, *Fax:* 713-500-9877, *E-mail:* abaron@gsbs.gs.uth.tmc.edu.

Find an in-depth description at www.petersons.com/graduate.

■ THE UNIVERSITY OF TEXAS MEDICAL BRANCH AT GALVESTON

Graduate School of Biomedical Sciences, Program in Cellular Physiology and Molecular Biophysics, Galveston, TX 77555

AWARDS MS, PhD.

Faculty: 32 full-time (4 women).
Students: 14 full-time (5 women), 1 part-time; includes 1 minority (Asian American or Pacific Islander), 11 international. Average age 30. *15 applicants, 47% accepted.* In 1999, 1 master's, 3 doctorates awarded.
Degree requirements: For master's, thesis or alternative required, foreign language not required; for doctorate, dissertation required, foreign language not required.
Entrance requirements: For master's and doctorate, GRE General Test. *Application deadline:* For fall admission, 3/2 (priority date). Applications are processed on a rolling basis. *Application fee:* $25 ($50 for international students). Electronic applications accepted.
Expenses: Tuition, state resident: full-time $684; part-time $38 per credit hour. Tuition, nonresident: full-time $4,572; part-time $254 per credit hour. Required fees: $29; $7.5 per credit hour. One-time fee: $55. Tuition and fees vary according to program.
Financial aid: Fellowships, research assistantships, Federal Work-Study and institutionally sponsored loans available. Financial aid applicants required to submit FAFSA.
Faculty research: Molecular biology of surface proteins, membrane transportation, channel biophysics, cell signaling, structural biology.
Dr. Simon A. Lewis, Director, 409-772-3397, *Fax:* 409-772-3381, *E-mail:* slewis@utmb.edu.
Application contact: Lori Ann Stewart, Secretary, 409-772-5445, *Fax:* 409-772-3381, *E-mail:* lastewart@utmb.edu.

■ UNIVERSITY OF THE PACIFIC

School of Pharmacy and Graduate School, Graduate Programs in Pharmaceutical Sciences, Department of Physiology/Pharmacology, Stockton, CA 95211-0197

AWARDS Pharmacology (MS, PhD); physiology (MS, PhD).

Faculty: 5 full-time (1 woman).
Students: 2 full-time (1 woman), 6 part-time (all women), 4 international. Terminal master's awarded for partial completion of doctoral program.
Degree requirements: For master's, thesis required, foreign language not required; for doctorate, dissertation, qualifying exam required, dissertation, qualifying exam required.
Entrance requirements: For master's and doctorate, GRE General Test, TOEFL, BS in biology, chemistry, or pharmacy. *Application deadline:* For fall admission, 3/1; for spring admission, 10/15. *Application fee:* $50.
Expenses: Tuition: Full-time $19,570; part-time $612 per unit. Required fees: $260. Tuition and fees vary according to program.
Financial aid: Teaching assistantships, institutionally sponsored loans available. Financial aid application deadline: 3/1.
Faculty research: Polyamines, cardiovascular pharmacology, pharmacology of naturally-occurring drugs.
James Blankenship, Chairman, 209-946-3167.
Application contact: Dr. Tim Smith, Graduate Studies Coordinator, 209-946-2487, *Fax:* 209-946-2410, *E-mail:* tsmith@uop.edu.

■ UNIVERSITY OF UTAH

School of Medicine and Graduate School, Graduate Programs in Medicine, Department of Physiology, Salt Lake City, UT 84112-1107

AWARDS PhD.

Faculty: 8 full-time (1 woman), 19 part-time/adjunct (2 women).
Students: 10 full-time (1 woman); includes 1 minority (Asian American or Pacific Islander), 2 international. Average age 29. *10 applicants, 20% accepted.*
Degree requirements: For doctorate, computer language, dissertation, comprehensive qualifying exam, preliminary exam required, foreign language not required. *Average time to degree:* Doctorate–5.5 years full-time.
Entrance requirements: For doctorate, GRE General Test, GRE Subject Test, minimum GPA of 3.0. *Application deadline:* For fall admission, 6/1. Applications are processed on a rolling basis. *Application fee:* $50 ($70 for international students).

Expenses: Tuition, state resident: full-time $2,105. Tuition, nonresident: full-time $6,312.
Financial aid: In 1999–00, 10 research assistantships with tuition reimbursements (averaging $12,700 per year) were awarded; fellowships, institutionally sponsored loans and tuition waivers (full and partial) also available.
Faculty research: Cell neurobiology, chemosensory systems, cardiovascular and kidney physiology, endocrinology. *Total annual research expenditures:* $1.2 million.
Dr. Salvatore J. Fidone, Acting Chair, 801-581-6354, *Fax:* 801-581-3476.
Application contact: Dr. H. Mack Brown, Graduate Adviser, 801-581-6317, *Fax:* 801-581-3476.

Find an in-depth description at www.petersons.com/graduate.

■ UNIVERSITY OF VERMONT

College of Medicine and Graduate College, Graduate Programs in Medicine, Department of Molecular Physiology and Biophysics, Burlington, VT 05405
AWARDS MS, PhD, MD/MS, MD/PhD.

Degree requirements: For master's and doctorate, thesis/dissertation required.
Entrance requirements: For master's and doctorate, GRE General Test, TOEFL.
Expenses: Tuition, state resident: full-time $7,464; part-time $311 per credit. Tuition, nonresident: full-time $18,672; part-time $778 per credit. Full-time tuition and fees vary according to degree level and program.

Find an in-depth description at www.petersons.com/graduate.

■ UNIVERSITY OF VIRGINIA

College and Graduate School of Arts and Sciences, Department of Molecular Physiology and Biological Physics, Charlottesville, VA 22903
AWARDS PhD.

Faculty: 22 full-time (5 women).
Students: 20 full-time (9 women); includes 1 minority (Asian American or Pacific Islander), 12 international. Average age 28. *6 applicants, 50% accepted.* In 1999, 3 degrees awarded.
Degree requirements: For doctorate, dissertation required.
Entrance requirements: For doctorate, GRE General Test, GRE Subject Test. *Application deadline:* For fall admission, 7/15; for spring admission, 12/1. Applications are processed on a rolling basis. *Application fee:* $40. Electronic applications accepted.
Expenses: Tuition, state resident: full-time $3,832. Tuition, nonresident: full-time

$15,519. Required fees: $1,084. Tuition and fees vary according to course load and program.
Financial aid: Application deadline: 2/1. Dr. Andrew P. Somlyo, Chairman, 804-924-5108.

Find an in-depth description at www.petersons.com/graduate.

■ UNIVERSITY OF WASHINGTON

School of Medicine and Graduate School, Graduate Programs in Medicine, Department of Physiology and Biophysics, Seattle, WA 98195
AWARDS PhD.

Faculty: 37 full-time (6 women), 3 part-time/adjunct (1 woman).
Students: 21 full-time (11 women); includes 3 minority (1 African American, 1 Asian American or Pacific Islander, 1 Hispanic American), 2 international. Average age 28. *39 applicants, 18% accepted.* In 1999, 2 doctorates awarded.
Degree requirements: For doctorate, dissertation required, foreign language not required. *Average time to degree:* Doctorate–5.5 years full-time.
Entrance requirements: For doctorate, GRE General Test, TOEFL. *Application deadline:* For fall admission, 2/1.
Expenses: Tuition, state resident: full-time $9,210; part-time $236 per credit. Tuition, nonresident: full-time $23,256; part-time $596 per credit.
Financial aid: In 1999–00, 20 students received aid, including fellowships with tuition reimbursements available (averaging $17,304 per year), research assistantships with tuition reimbursements available (averaging $17,304 per year); Federal Work-Study, institutionally sponsored loans, and tuition waivers (full) also available. Financial aid applicants required to submit FAFSA.
Faculty research: Membrane and cell biophysics, neuroendocrinology, cardiovascular and respiratory physiology, systems neurophysiology and behavior, molecular physiology. *Total annual research expenditures:* $4.7 million.
Dr. Albert Berger, Acting Chair, 206-543-0954, *Fax:* 206-685-0619, *E-mail:* berger@u.washington.edu.
Application contact: Alyssa Kern, Graduate Coordinator, 206-685-0519, *Fax:* 206-685-0619, *E-mail:* pbio@u.washington.edu.

■ UNIVERSITY OF WISCONSIN–MADISON

Graduate School, College of Agricultural and Life Sciences, Department of Animal Health and Biomedical Sciences, Program in Comparative Biosciences, Madison, WI 53706-1380
AWARDS Anatomy (MS, PhD); biochemistry (MS, PhD); cellular and molecular biology (MS, PhD); environmental toxicology (MS, PhD); neurosciences (MS, PhD); pharmacology (MS, PhD); physiology (MS, PhD).

Degree requirements: For doctorate, dissertation required.
Expenses: Tuition, state resident: full-time $5,406; part-time $339 per credit. Tuition, nonresident: full-time $17,110; part-time $1,071 per credit. Full-time tuition and fees vary according to program and reciprocity agreements. Part-time tuition and fees vary according to course load and program.

■ UNIVERSITY OF WISCONSIN–MADISON

Graduate School, Endocrinology-Reproductive Physiology Program, Madison, WI 53706-1380
AWARDS MS, PhD.

Faculty: 25 full-time (6 women).
Students: 29 full-time (13 women); includes 1 minority (Hispanic American), 14 international. Average age 29. *33 applicants, 12% accepted.* In 1999, 1 master's, 7 doctorates awarded.
Degree requirements: For doctorate, dissertation required.
Entrance requirements: For master's and doctorate, GRE, TOEFL. *Application deadline:* For fall admission, 7/1 (priority date); for spring admission, 11/15 (priority date). Applications are processed on a rolling basis. *Application fee:* $45.
Expenses: Tuition, state resident: full-time $5,406; part-time $339 per credit. Tuition, nonresident: full-time $17,110; part-time $1,071 per credit. Full-time tuition and fees vary according to program and reciprocity agreements. Part-time tuition and fees vary according to course load and program.
Financial aid: Research assistantships with tuition reimbursements, teaching assistantships with tuition reimbursements available.
Faculty research: Ovarian physiology and endocrinology, fertilization and gamete biology, hormone action and cell signaling, placental function and pregnancy, embryo and fetal development.
Dr. Barry D. Bavister, Director, 608-262-3222, *Fax:* 608-262-7420, *E-mail:* erp@ahabs.wisc.edu.

University of Wisconsin–Madison (continued)

Application contact: Jeanette Rutschow, Coordinator, 608-262-3222, *Fax:* 608-262-7420, *E-mail:* erp@ahabs.wisc.edu.

■ UNIVERSITY OF WISCONSIN–MADISON

Medical School and Graduate School, Graduate Programs in Medicine, Department of Physiology, Madison, WI 53706-1380

AWARDS Neurophysiology (PhD); physiology (PhD).

Faculty: 24 full-time (6 women).
Students: 26 full-time (8 women); includes 1 minority (Hispanic American), 8 international.
Degree requirements: For doctorate, dissertation, written exams required, foreign language not required. *Average time to degree:* Doctorate–5 years full-time.
Entrance requirements: For doctorate, GRE, TOEFL. *Application deadline:* For fall admission, 1/15 (priority date). Applications are processed on a rolling basis. *Application fee:* $45. Electronic applications accepted.
Expenses: Tuition, state resident: full-time $5,406; part-time $339 per credit. Tuition, nonresident: full-time $17,110; part-time $1,071 per credit.
Financial aid: In 1999–00, fellowships with tuition reimbursements (averaging $16,400 per year), research assistantships with tuition reimbursements (averaging $16,400 per year), teaching assistantships with tuition reimbursements (averaging $16,400 per year) were awarded.
Faculty research: Studies in molecular cellular systems, cardiovascular, neuroscience.
Dr. Richard Moss, Chair, 608-262-1939, *Fax:* 609-265-5072, *E-mail:* rlmoss@facstaff.wisc.edu.
Application contact: Sue Krey, Program Assistant, 608-262-9114, *Fax:* 608-265-5512, *E-mail:* sskrey@facstaff.wisc.edu.

■ UNIVERSITY OF WYOMING

Graduate School, College of Arts and Sciences, Department of Zoology and Physiology, Laramie, WY 82071

AWARDS MS, PhD.

Faculty: 24.
Students: 42 full-time (19 women), 23 part-time (4 women); includes 2 minority (1 Asian American or Pacific Islander, 1 Hispanic American), 2 international. *64 applicants, 22% accepted.* In 1999, 14 master's, 4 doctorates awarded.
Degree requirements: For master's and doctorate, thesis/dissertation required, foreign language not required.

Entrance requirements: For master's and doctorate, GRE General Test, minimum GPA of 3.0. *Application deadline:* For fall admission, 2/15 (priority date); for spring admission, 9/15. Applications are processed on a rolling basis. *Application fee:* $40. Electronic applications accepted.
Expenses: Tuition, state resident: full-time $2,520; part-time $140 per credit hour. Tuition, nonresident: full-time $7,790; part-time $433 per credit hour. Required fees: $440; $7 per credit hour. Full-time tuition and fees vary according to course load and program.
Financial aid: In 1999–00, 23 research assistantships with full tuition reimbursements (averaging $8,667 per year), 23 teaching assistantships with full tuition reimbursements (averaging $8,667 per year) were awarded. Financial aid application deadline: 3/1.
Faculty research: Cell biology, ecology/wildlife, organismal physiology. *Total annual research expenditures:* $2.5 million.
Dr. Nancy L. Stanton, Head, 307-766-4207.

Find an in-depth description at www.petersons.com/graduate.

■ VANDERBILT UNIVERSITY

Graduate School and School of Medicine, Department of Molecular Physiology and Biophysics, Nashville, TN 37240-1001

AWARDS PhD, MD/PhD.

Faculty: 20 full-time (2 women).
Students: 23 full-time (11 women), 1 part-time; includes 3 minority (1 Asian American or Pacific Islander, 2 Hispanic Americans), 4 international. Average age 29. In 1999, 11 degrees awarded.
Degree requirements: For doctorate, dissertation, preliminary, qualifying, and final exams required, foreign language not required.
Entrance requirements: For doctorate, GRE General Test, GRE Subject Test (recommended). *Application deadline:* For fall admission, 1/15. *Application fee:* $40.
Expenses: Tuition: Full-time $17,244; part-time $958 per hour. Required fees: $242; $121 per semester. Tuition and fees vary according to program.
Financial aid: In 1999–00, fellowships with full tuition reimbursements (averaging $17,000 per year), research assistantships with full tuition reimbursements (averaging $17,000 per year) were awarded; Federal Work-Study, institutionally sponsored loans, traineeships, and tuition waivers (partial) also available. Financial aid application deadline: 1/15.
Faculty research: Molecular endocrinology, membrane transport biophysics, metabolic regulation, neurobiology.

Alan D. Cherrington, Chair, 615-322-7000, *Fax:* 615-322-7236, *E-mail:* alan.cherrington@mcmail.vanderbilt.edu.
Application contact: Roger J. Colbran, Director of Graduate Studies, 615-322-7000, *Fax:* 615-322-7236, *E-mail:* roger.j.colbran@vanderbilt.edu.

Find an in-depth description at www.petersons.com/graduate.

■ VIRGINIA COMMONWEALTH UNIVERSITY

School of Graduate Studies, School of Allied Health Professions, Department of Physical Therapy, Program in Physiology and Physical Therapy, Richmond, VA 23284-9005

AWARDS PhD.

Degree requirements: For doctorate, dissertation required, foreign language not required.
Entrance requirements: For doctorate, GRE General Test. *Application fee:* $30.
Expenses: Tuition, state resident: full-time $4,031; part-time $224 per credit hour. Tuition, nonresident: full-time $11,946; part-time $664 per credit hour. Required fees: $1,081; $40 per credit hour. Tuition and fees vary according to campus/location and program.
Application contact: Dr. Sheryl Finucane, Assistant Professor, 804-828-0234.

■ VIRGINIA COMMONWEALTH UNIVERSITY

School of Graduate Studies and School of Medicine, School of Medicine Graduate Programs, Department of Physiology, Richmond, VA 23284-9005

AWARDS Neurosciences (PhD); physiology (MS, PhD, CBHS).

Students: 2 full-time (0 women), 6 part-time; includes 3 Asian Americans or Pacific Islanders. In 1999, 4 master's, 4 doctorates, 20 other advanced degrees awarded. Terminal master's awarded for partial completion of doctoral program.
Degree requirements: For master's, thesis required, foreign language not required; for doctorate, dissertation, comprehensive oral and written exams required, foreign language not required.
Entrance requirements: For master's, DAT, GRE General Test, or MCAT; for doctorate, GRE General Test. *Application fee:* $30.
Expenses: Tuition, state resident: full-time $4,031; part-time $224 per credit hour. Tuition, nonresident: full-time $11,946; part-time $664 per credit hour. Required fees: $1,081; $40 per credit hour. Tuition and fees vary according to campus/location and program.

Financial aid: Fellowships, research assistantships, teaching assistantships, career-related internships or fieldwork and tuition waivers (full) available.
Dr. Margaret C. Biber, Chair, 804-828-9756, *Fax:* 804-828-7382.
Application contact: Dr. James L. Poland, Graduate Program Director, 804-828-9557, *Fax:* 804-828-7382, *E-mail:* jlpoland@vcu.edu.
Find an in-depth description at www.petersons.com/graduate.

■ WAKE FOREST UNIVERSITY

School of Medicine and Graduate School, Graduate Programs in Medicine, Department of Physiology and Pharmacology, Program in Physiology, Winston-Salem, NC 27109
AWARDS PhD.

Degree requirements: For doctorate, dissertation required.
Entrance requirements: For doctorate, GRE General Test, GRE Subject Test.
Expenses: Tuition: Full-time $18,300.
Faculty research: Cardiovascular-renal physiology, endocrine physiology, neurophysiology.

■ WAYNE STATE UNIVERSITY

School of Medicine and Graduate School, Graduate Programs in Medicine, Department of Physiology, Detroit, MI 48202
AWARDS MS, PhD, MD/PhD.

Degree requirements: For master's and doctorate, thesis/dissertation required, foreign language not required.
Entrance requirements: For master's, GRE General Test, GRE Subject Test, minimum GPA of 2.6; for doctorate, GRE General Test, GRE Subject Test, minimum GPA of 3.0.
Faculty research: Regulation of brain blood flow, mechanism of hormone action, regulation of pituitary hormone secretion, regulation of renin secretion, regulation of islet cell function.

■ WESLEYAN UNIVERSITY

Graduate Programs, Department of Biology, Middletown, CT 06459-0260
AWARDS Cell biology (PhD); comparative physiology (PhD); developmental biology (PhD); genetics (PhD); neurophysiology (PhD); population biology (PhD).

Faculty: 12 full-time (3 women).
Students: 24 full-time (12 women); includes 1 minority (African American), 11 international. Average age 28. *125 applicants, 4% accepted.* In 1999, 2 doctorates awarded.

Degree requirements: For doctorate, one foreign language (computer language can substitute), dissertation required.
Entrance requirements: For doctorate, GRE Subject Test. *Application deadline:* For fall admission, 1/15. Applications are processed on a rolling basis. *Application fee:* $0.
Expenses: Tuition: Full-time $24,876. Required fees: $650. Tuition and fees vary according to program.
Financial aid: Research assistantships, teaching assistantships, stipends available.
Faculty research: Microbial population genetics, genetic basis of evolutionary adaptation, genetic regulation of differentiation and pattern formation in *drosophila*.
Dr. Fred Cohan, Chairman, 860-685-3489.
Application contact: Marina J. Melendez, Director of Graduate Student Services, 860-685-2390, *Fax:* 860-685-2439, *E-mail:* mmelendez@wesleyan.edu.

Find an in-depth description at www.petersons.com/graduate.

■ WEST VIRGINIA UNIVERSITY

College of Agriculture, Forestry and Consumer Sciences, Division of Animal and Veterinary Sciences, Program in Animal and Food Sciences, Morgantown, WV 26506
AWARDS Agricultural biochemistry (PhD); animal nutrition (PhD); animal physiology (PhD); production management (PhD).

Students: 12 full-time (6 women), 5 part-time (2 women), 9 international. Average age 36. In 1999, 2 degrees awarded.
Degree requirements: For doctorate, dissertation, oral and written exams required, foreign language not required.
Entrance requirements: For doctorate, TOEFL. *Application fee:* $45.
Expenses: Tuition, state resident: full-time $2,910; part-time $154 per credit hour. Tuition, nonresident: full-time $8,368; part-time $457 per credit hour.
Financial aid: In 1999–00, 9 research assistantships, 1 teaching assistantship were awarded; Federal Work-Study, institutionally sponsored loans, and tuition waivers (full and partial) also available. Financial aid application deadline: 2/1; financial aid applicants required to submit FAFSA.
Faculty research: Ruminant nutrition, metabolism, forage utilization, physiology, reproduction.
Dr. Hillar Klandorf, Coordinator, 304-293-2631 Ext. 4436.

■ WEST VIRGINIA UNIVERSITY

College of Agriculture, Forestry and Consumer Sciences, Division of Animal and Veterinary Sciences, Program in Animal and Veterinary Sciences, Morgantown, WV 26506
AWARDS Animal sciences (MS); breeding (MS); food sciences (MS); nutrition (MS); physiology (MS); production (MS). Part-time programs available.

Students: 10 full-time (3 women), 1 (woman) part-time, 3 international. Average age 28. In 1999, 4 degrees awarded.
Degree requirements: For master's, thesis, oral and written exams required, foreign language not required.
Entrance requirements: For master's, TOEFL, minimum GPA of 2.5. *Application fee:* $45.
Expenses: Tuition, state resident: full-time $2,910; part-time $154 per credit hour. Tuition, nonresident: full-time $8,368; part-time $457 per credit hour.
Financial aid: In 1999–00, 9 research assistantships, 1 teaching assistantship were awarded; Federal Work-Study, institutionally sponsored loans, and tuition waivers (full and partial) also available. Financial aid application deadline: 2/1; financial aid applicants required to submit FAFSA.
Faculty research: Animal nutrition, reproductive physiology, food science.
Dr. Hillar Klandorf, Coordinator, 304-293-2631 Ext. 4436.

■ WEST VIRGINIA UNIVERSITY

College of Agriculture, Forestry and Consumer Sciences, Interdisciplinary Program in Reproductive Physiology, Morgantown, WV 26506
AWARDS MS, PhD. Part-time programs available.

Students: 6 full-time (3 women), 3 part-time (2 women), 3 international. Average age 29. *8 applicants, 25% accepted.* Terminal master's awarded for partial completion of doctoral program.
Degree requirements: For master's, thesis required, foreign language not required; for doctorate, dissertation, comprehensive exam required, foreign language not required.
Entrance requirements: For master's, TOEFL, minimum GPA of 2.75; for doctorate, TOEFL, minimum GPA of 3.0. *Application deadline:* Applications are processed on a rolling basis. *Application fee:* $45.
Expenses: Tuition, state resident: full-time $2,910; part-time $154 per credit hour. Tuition, nonresident: full-time $8,368; part-time $457 per credit hour.
Financial aid: In 1999–00, 5 research assistantships, 2 teaching assistantships were awarded; Federal Work-Study,

West Virginia University (continued)
institutionally sponsored loans, and tuition waivers (full and partial) also available. Financial aid application deadline: 2/1; financial aid applicants required to submit FAFSA.

Faculty research: Uterine prostaglandins, luteal function, neural control of luteinizing hormone and follicle-stimulating hormone, follicular development.

■ **WEST VIRGINIA UNIVERSITY**

School of Medicine, Graduate Programs in Health Sciences, Department of Microbiology and Immunology, Morgantown, WV 26506

AWARDS Genetics (MS, PhD); immunology (MS, PhD); mycology (PhD); parasitology (MS, PhD); pathogenic bacteriology (MS, PhD); physiology (PhD); psysiology (MS); virology (MS, PhD).

Students: 25 full-time (11 women), 1 (woman) part-time; includes 2 minority (both Asian Americans or Pacific Islanders), 4 international. Average age 27. *62 applicants, 8% accepted.* In 1999, 1 master's, 5 doctorates awarded. Terminal master's awarded for partial completion of doctoral program.

Degree requirements: For master's, thesis required, foreign language not required; for doctorate, dissertation, comprehensive exam required, foreign language not required. *Average time to degree:* Doctorate–4.5 years full-time.

Entrance requirements: For master's and doctorate, GRE General Test, GRE Subject Test, TOEFL, minimum GPA of 3.0. *Application deadline:* For fall admission, 3/1 (priority date). Applications are processed on a rolling basis. *Application fee:* $45. Electronic applications accepted.

Expenses: Tuition, state resident: full-time $3,564. Tuition, nonresident: full-time $10,230.

Financial aid: In 1999–00, 7 research assistantships, 14 teaching assistantships were awarded; fellowships, Federal Work-Study and institutionally sponsored loans also available. Financial aid application deadline: 3/1; financial aid applicants required to submit FAFSA.

Faculty research: Mechanisms of pathogenesis, microbial genetics, molecular virology, immunotoxicology, oncogenes. *Total annual research expenditures:* $2.6 million.

John Barnett, Chair, 304-293-2649.
Application contact: James Sheil, Graduate Coordinator, 304-293-3559, *Fax:* 304-293-7823, *E-mail:* jsheil@wvu.edu.

■ **WEST VIRGINIA UNIVERSITY**

School of Medicine, Graduate Programs in Health Sciences, Department of Physiology, Morgantown, WV 26506

AWARDS Cardiovascular physiology (MS, PhD); cell physiology (MS, PhD); endocrine physiology (MS, PhD); ernal physiology (PhD); muscle physiology (MS, PhD); neural physiology (MS, PhD); renal physiology (MS); respiratory physiology (MS).

Students: 11 full-time (3 women). Average age 26. In 1999, 2 doctorates awarded. Terminal master's awarded for partial completion of doctoral program.

Degree requirements: For master's and doctorate, thesis/dissertation required, foreign language not required.

Entrance requirements: For master's, GRE, TOEFL, minimum GPA of 2.5; for doctorate, GRE, TOEFL. *Application deadline:* For fall admission, 3/1 (priority date). Applications are processed on a rolling basis. *Application fee:* $45.

Expenses: Tuition, state resident: full-time $3,564. Tuition, nonresident: full-time $10,230.

Financial aid: In 1999–00, 1 research assistantship, 9 teaching assistantships were awarded; Federal Work-Study, institutionally sponsored loans, and tuition waivers (full) also available. Financial aid application deadline: 2/1; financial aid applicants required to submit FAFSA.

Faculty research: Mechanisms of hypertension and kidney disease, pulmonary cell physiology, spinal cord and auditory physiology, microvascular control. Acting Chair, 304-293-4991, *Fax:* 304-293-3850.

Application contact: Dr. Matthew Boegehold, Graduate Studies Committee Chair, 304-293-4992, *Fax:* 304-293-3850.

Find an in-depth description at www.petersons.com/graduate.

■ **WILLIAM PATERSON UNIVERSITY OF NEW JERSEY**

College of Science and Health, Department of Biology, General Biology Program, Wayne, NJ 07470-8420

AWARDS General biology (MA); limnology and terrestrial ecology (MA); molecular biology (MA); physiology (MA). Part-time and evening/weekend programs available.

Students: 1 (woman) full-time, 6 part-time (3 women); includes 1 minority (Hispanic American). Average age 25. *9 applicants, 44% accepted.* In 1999, 1 degree awarded.

Degree requirements: For master's, comprehensive exam, independent study or thesis required.

Entrance requirements: For master's, GRE General Test, minimum GPA of 2.75. *Application deadline:* For fall admission, 4/1; for spring admission, 10/15. Applications are processed on a rolling basis. *Application fee:* $35. Electronic applications accepted.

Expenses: Tuition, state resident: part-time $244 per credit. Tuition, nonresident: part-time $350 per credit.

Financial aid: In 1999–00, 2 teaching assistantships (averaging $9,000 per year) were awarded; career-related internships or fieldwork and unspecified assistantships also available. Financial aid application deadline: 4/1; financial aid applicants required to submit FAFSA.

Application contact: Office of Graduate Studies, 973-720-2237, *Fax:* 973-720-2035.

Find an in-depth description at www.petersons.com/graduate.

■ **WRIGHT STATE UNIVERSITY**

School of Graduate Studies, College of Science and Mathematics, Department of Physiology and Biophysics, Dayton, OH 45435

AWARDS MS.

Students: 3 full-time (2 women), 1 part-time; includes 1 minority (Asian American or Pacific Islander), 2 international. Average age 29. *5 applicants, 100% accepted.* In 1999, 3 degrees awarded.

Degree requirements: For master's, thesis required, foreign language not required.

Entrance requirements: For master's, TOEFL. *Application deadline:* For fall admission, 7/15. *Application fee:* $25.

Expenses: Tuition, state resident: full-time $5,568; part-time $175 per quarter hour. Tuition, nonresident: full-time $9,696; part-time $302 per quarter hour. Full-time tuition and fees vary according to course load, campus/location and program.

Financial aid: Fellowships, research assistantships, teaching assistantships, unspecified assistantships available. Aid available to part-time students. Financial aid applicants required to submit FAFSA.

Faculty research: Membrane transport, ion channels, pH regulation, lipid dynamics, neurophysiology.

Dr. Peter K. Lauf, Chair, 937-775-2360, *Fax:* 937-775-3769, *E-mail:* peter.lauf@wright.edu.

Application contact: Dr. Noel S. Nussbaum, Director, 937-775-2081, *Fax:* 937-775-3769, *E-mail:* noel.nussbaum@wright.edu.

■ **YALE UNIVERSITY**

Graduate School of Arts and Sciences, Department of Cellular and Molecular Physiology, New Haven, CT 06520

AWARDS PhD.

Faculty: 28 full-time (7 women), 1 (woman) part-time/adjunct.
Students: 10 full-time (6 women). In 1999, 2 degrees awarded.
Degree requirements: For doctorate, dissertation required. *Average time to degree:* Doctorate–5 years full-time.
Entrance requirements: For doctorate, GRE General Test, GRE Subject Test. *Application deadline:* For fall admission, 1/4. *Application fee:* $65.
Expenses: Tuition: Full-time $22,300. Full-time tuition and fees vary according to program.
Financial aid: Fellowships, research assistantships, teaching assistantships, Federal Work-Study, institutionally sponsored loans, and traineeships available. Aid available to part-time students.
Application contact: Admissions Information, 203-432-2770.

Find an in-depth description at www.petersons.com/graduate.

■ **YALE UNIVERSITY**

School of Medicine and Graduate School of Arts and Sciences, Combined Program in Biological and Biomedical Sciences (BBS), Cell Biology and Molecular Physiology Track, New Haven, CT 06520

AWARDS PhD, MD/PhD.

Degree requirements: For doctorate, dissertation required.
Entrance requirements: For doctorate, GRE General Test, TOEFL. *Application deadline:* For fall admission, 1/2. *Application fee:* $65. Electronic applications accepted.
Expenses: All students receive full tuition of $22,330 and an annual stipend of $17,600 .
Financial aid: Fellowships, research assistantships available.
Dr. Susan Ferro-Novick, Co-Director of Graduate Studies, 203-787-5207, *E-mail:* bbs.cbmp@yale.edu.

Find an in-depth description at www.petersons.com/graduate.

■ **YESHIVA UNIVERSITY**

Albert Einstein College of Medicine, Sue Golding Graduate Division of Medical Sciences, Department of Physiology and Biophysics, Bronx, NY 10461

AWARDS PhD, MD/PhD.

Faculty: 30 full-time.

Students: 8 full-time (3 women); includes 1 minority (Asian American or Pacific Islander), 4 international. In 1999, 2 degrees awarded.
Degree requirements: For doctorate, dissertation required, foreign language not required.
Entrance requirements: For doctorate, GRE General Test, TOEFL. *Application deadline:* For fall admission, 1/15. *Application fee:* $0.
Expenses: Tuition: Part-time $525 per credit. Tuition and fees vary according to degree level and program.
Financial aid: In 1999–00, 8 fellowships were awarded.
Faculty research: Biophysical and biochemical basis of body function at the subcellular, cellular, organ, and whole-body level.
Dr. Denis M. Rousseau, Chairperson, 718-430-4264.
Application contact: Sheila Cleeton, Assistant Director, 718-430-2128, *Fax:* 718-430-8655, *E-mail:* phd@aecom.yu.edu.

Find an in-depth description at www.petersons.com/graduate.

Zoology

ANIMAL BEHAVIOR

■ **ARIZONA STATE UNIVERSITY**

Graduate College, College of Liberal Arts and Sciences, Department of Biology, Program in Behavior, Tempe, AZ 85287

AWARDS MS, PhD. Terminal master's awarded for partial completion of doctoral program.

Degree requirements: For master's, thesis required, foreign language not required; for doctorate, dissertation, oral exam required, foreign language not required.
Entrance requirements: For master's and doctorate, GRE General Test, GRE Subject Test. *Application deadline:* For fall admission, 12/15. *Application fee:* $45.
Expenses: Tuition, state resident: part-time $115 per credit hour. Tuition, nonresident: part-time $389 per credit hour. Required fees: $18 per semester. Tuition and fees vary according to program.
Financial aid: Application deadline: 12/15.

■ **BUCKNELL UNIVERSITY**

Graduate Studies, College of Arts and Sciences, Department of Animal Behavior, Lewisburg, PA 17837

AWARDS MA, MS.

Students: 4 full-time (3 women).
Degree requirements: For master's, thesis required.

Entrance requirements: For master's, GRE General Test, GRE Subject Test, TOEFL, minimum GPA of 2.8. *Application deadline:* For fall admission, 6/1 (priority date); for spring admission, 12/1 (priority date). Applications are processed on a rolling basis. *Application fee:* $25.
Expenses: Tuition: Part-time $2,600 per course. Tuition and fees vary according to course load.
Financial aid: Unspecified assistantships available. Financial aid application deadline: 3/1.
Dr. Douglas Keith Candland, Head, 570-577-1200.

■ UNIVERSITY OF CALIFORNIA, DAVIS

Graduate Studies, Programs in the Biological Sciences, Program in Animal Behavior, Davis, CA 95616

AWARDS MS, PhD.

Faculty: 33 full-time (8 women).
Students: 38 full-time (24 women); includes 2 minority (both Asian Americans or Pacific Islanders). Average age 28. *70 applicants, 24% accepted.* In 1999, 7 master's, 5 doctorates awarded.
Degree requirements: For doctorate, dissertation required. *Average time to degree:* Doctorate–6 years full-time.
Entrance requirements: For doctorate, GRE General Test. *Application deadline:* For fall admission, 1/15. *Application fee:* $40. Electronic applications accepted.
Expenses: Tuition, nonresident: full-time $9,804. Tuition and fees vary according to program and student level.
Financial aid: In 1999–00, 35 students received aid, including 17 fellowships with full and partial tuition reimbursements available, 6 research assistantships with full and partial tuition reimbursements available, 15 teaching assistantships with partial tuition reimbursements available; Federal Work-Study, grants, institutionally sponsored loans, and scholarships also available. Financial aid application deadline: 1/15; financial aid applicants required to submit FAFSA.
Faculty research: Wildlife behavior, conservation biology, companion animal behavior, behavioral endocrinology, animal communication.
Nicola Clayton, Chair, 530-752-8786, *Fax:* 530-752-5582, *E-mail:* nsclayton@ucdavis.edu.
Application contact: Jeni Trevitt, Graduate Staff, 530-752-4863, *Fax:* 530-752-8391, *E-mail:* jmtrevitt@ucdavis.edu.

■ UNIVERSITY OF COLORADO AT BOULDER

Graduate School, College of Arts and Sciences, Department of Environmental, Population, and Organic Biology, Boulder, CO 80309

AWARDS Animal behavior (MA, PhD); aquatic biology (MA, PhD); behavioral genetics (MA, PhD); ecology (MA, PhD); microbiology (MA, PhD); neurobiology (MA, PhD); plant and animal physiology (MA, PhD); plant and animal systematics (MA, PhD); population biology (MA, PhD); population genetics (MA, PhD).

Faculty: 39 full-time (9 women).
Students: 84 full-time (36 women), 14 part-time (7 women); includes 17 minority (6 Asian Americans or Pacific Islanders, 10 Hispanic Americans, 1 Native American).

Average age 29. *147 applicants, 14% accepted.* In 1999, 7 master's, 13 doctorates awarded. Terminal master's awarded for partial completion of doctoral program.
Degree requirements: For master's, thesis or alternative, comprehensive exam required, foreign language not required; for doctorate, dissertation, comprehensive exam required, foreign language not required. *Average time to degree:* Master's–3 years full-time; doctorate–5 years full-time.
Entrance requirements: For master's, GRE General Test, GRE Subject Test, minimum undergraduate GPA of 3.0; for doctorate, GRE General Test, GRE Subject Test. *Application deadline:* For fall admission, 1/15 (priority date). *Application fee:* $40 ($60 for international students).
Expenses: Tuition, state resident: part-time $181 per credit hour. Tuition, nonresident: part-time $542 per credit hour. Required fees: $99 per term. Tuition and fees vary according to course load and program.
Financial aid: Fellowships, research assistantships, teaching assistantships, Federal Work-Study, institutionally sponsored loans, and tuition waivers (full) available. Financial aid application deadline: 3/1.
Faculty research: Evolution, developmental biology, behavior and neurobiology. *Total annual research expenditures:* $1.8 million.
Michael Breed, Chair, 303-492-8981, *Fax:* 303-492-8699, *E-mail:* michael.breed@colorado.edu.
Application contact: Jill Skarstadt, Graduate Secretary, 303-492-7654, *Fax:* 303-492-8699, *E-mail:* jill.skarstadt@colorado.edu.

■ UNIVERSITY OF MINNESOTA, TWIN CITIES CAMPUS

Graduate School, College of Biological Sciences, Department of Ecology, Evolution, and Behavior, Minneapolis, MN 55455-0213

AWARDS Ecology, animal behavior, and evolution (MS, PhD).

Faculty: 55 full-time (11 women).
Students: 45 full-time (24 women), 1 (woman) part-time; includes 2 minority (1 Asian American or Pacific Islander, 1 Hispanic American). Average age 26. *71 applicants, 17% accepted.* In 1999, 2 master's awarded (50% found work related to degree, 50% continued full-time study); 3 doctorates awarded (33% entered university research/teaching, 67% found other work related to degree). Terminal master's awarded for partial completion of doctoral program.
Degree requirements: For master's, thesis or projects, comprehensive exam

required; for doctorate, dissertation, comprehensive exam required, foreign language not required. *Average time to degree:* Master's–6.5 years full-time; doctorate–9 years full-time.
Entrance requirements: For master's and doctorate, GRE General Test, minimum GPA of 3.2. *Application deadline:* For fall admission, 1/4. *Application fee:* $50 ($55 for international students). Electronic applications accepted.
Expenses: Tuition, state resident: full-time $5,040; part-time $420 per credit. Tuition, nonresident: full-time $9,900; part-time $825 per credit. Full-time tuition and fees vary according to course load, program and reciprocity agreements.
Financial aid: In 1999–00, 44 students received aid, including 5 fellowships with tuition reimbursements available (averaging $12,000 per year), 12 research assistantships with tuition reimbursements available (averaging $10,920 per year), 14 teaching assistantships with tuition reimbursements available (averaging $10,920 per year); Federal Work-Study, institutionally sponsored loans, and tuition waivers (partial) also available. Financial aid application deadline: 1/4. *Total annual research expenditures:* $2.7 million.
Prof. Robert W. Sterner, Head, 612-625-5700, *Fax:* 612-624-6777, *E-mail:* stern007@tc.umn.edu.
Application contact: Prof. Elmer C. Birney, Director of Graduate Studies, 612-625-5713, *Fax:* 612-624-6777, *E-mail:* ecbirney@cbs.umn.edu.

■ UNIVERSITY OF MISSOURI–ST. LOUIS

Graduate School, College of Arts and Sciences, Department of Biology, St. Louis, MO 63121-4499

AWARDS Biology (MS, PhD), including animal behavior (MS), biochemistry, biotechnology (MS), conservation biology (MS), development (MS), ecology (MS), environmental studies (PhD), evolution (MS), genetics (MS), molecular biology and biotechnology (PhD), molecular/cellular biology (MS), physiology (MS), plant systematics, population biology (MS), tropical biology (MS); biotechnology (Certificate); tropical biology and conservation (Certificate). Part-time programs available.

Faculty: 46.
Students: 21 full-time (11 women), 75 part-time (44 women); includes 13 minority (2 African Americans, 2 Asian Americans or Pacific Islanders, 8 Hispanic Americans, 1 Native American), 23 international. In 1999, 14 master's, 4 doctorates awarded.
Degree requirements: For master's, thesis or alternative required, foreign language not required; for doctorate, one

foreign language, dissertation, 1 semester of teaching experience required.
Entrance requirements: For doctorate, GRE General Test. *Application deadline:* For fall admission, 7/1 (priority date); for spring admission, 11/1 (priority date). Applications are processed on a rolling basis. *Application fee:* $25 ($40 for international students). Electronic applications accepted.
Expenses: Tuition, state resident: full-time $4,932; part-time $173 per credit hour. Tuition, nonresident: full-time $13,279; part-time $521 per credit hour. Required fees: $775; $33 per credit hour. Tuition and fees vary according to degree level and program.
Financial aid: In 1999–00, 8 research assistantships with partial tuition reimbursements (averaging $10,635 per year), 14 teaching assistantships with partial tuition reimbursements (averaging $11,488 per year) were awarded; career-related internships or fieldwork and Federal Work-Study also available. Aid available to part-time students. Financial aid application deadline: 2/1. *Total annual research expenditures:* $908,828.
Application contact: Graduate Admissions, 314-516-5458, *Fax:* 314-516-6759, *E-mail:* gradadm@umsl.edu.

■ THE UNIVERSITY OF MONTANA–MISSOULA

Graduate School, College of Arts and Sciences, Department of Psychology, Programs in Experimental Psychology, Missoula, MT 59812-0002
AWARDS Animal behavior (PhD); developmental psychology (PhD).

Faculty: 7 full-time (3 women).
Students: 6 full-time (5 women), 1 (woman) part-time; includes 1 minority (Native American). Average age 30. *6 applicants, 0% accepted.*
Degree requirements: For doctorate, computer language, dissertation, oral exam required, foreign language not required.
Entrance requirements: For doctorate, GRE General Test, minimum GPA of 3.2. *Application deadline:* For fall admission, 1/15. *Application fee:* $45.
Expenses: Tuition, state resident: full-time $2,484; part-time $151 per credit. Tuition, nonresident: full-time $8,000; part-time $305 per credit. Required fees: $1,600. Full-time tuition and fees vary according to degree level and program.
Financial aid: In 1999–00, teaching assistantships with full tuition reimbursements (averaging $8,400 per year); fellowships, research assistantships, Federal Work-Study also available. Financial aid application deadline: 3/1.

Faculty research: Animal learning and behavior, neurological assessment, developmental psychology.
Application contact: Graduate Secretary, 406-243-4523, *Fax:* 406-243-6366, *E-mail:* psycgrad@selway.umt.edu.

■ THE UNIVERSITY OF TENNESSEE

Graduate School, College of Arts and Sciences, Department of Ecology and Evolutionary Biology, Knoxville, TN 37996
AWARDS Behavior (MS, PhD); ecology (MS, PhD); evolutionary biology (MS, PhD). Part-time programs available.

Faculty: 21 full-time (4 women), 1 part-time/adjunct (0 women).
Students: 30 full-time (18 women), 26 part-time (10 women); includes 1 minority (Asian American or Pacific Islander), 9 international. *54 applicants, 17% accepted.* In 1999, 6 master's, 9 doctorates awarded.
Degree requirements: For master's and doctorate, thesis/dissertation required, foreign language not required.
Entrance requirements: For master's and doctorate, GRE General Test, TOEFL, minimum GPA of 2.7. *Application deadline:* For fall admission, 2/1 (priority date). Applications are processed on a rolling basis. *Application fee:* $35. Electronic applications accepted.
Expenses: Tuition, state resident: full-time $3,806; part-time $184 per credit hour. Tuition, nonresident: full-time $9,874; part-time $522 per credit hour. Tuition and fees vary according to program.
Financial aid: In 1999–00, 4 fellowships, 9 research assistantships, 26 teaching assistantships were awarded; Federal Work-Study, institutionally sponsored loans, and unspecified assistantships also available. Financial aid application deadline: 2/1; financial aid applicants required to submit FAFSA.
Dr. Tom Hallam, Head, 865-974-3065, *Fax:* 865-974-3067, *E-mail:* thallam@utk.edu.
Application contact: Dr. C. R. Boake, Graduate Representative, *E-mail:* cboake@utk.edu.

■ THE UNIVERSITY OF TEXAS AT AUSTIN

Graduate School, College of Natural Sciences, School of Biological Sciences, Program in Ecology, Evolution and Behavior, Austin, TX 78712-1111
AWARDS MA, PhD.

Entrance requirements: For master's and doctorate, GRE General Test. *Application deadline:* Applications are processed on a

rolling basis. *Application fee:* $50 ($75 for international students). Electronic applications accepted.
Expenses: Tuition, state resident: part-time $114 per semester hour. Tuition, nonresident: part-time $330 per semester hour. Tuition and fees vary according to program.
Financial aid: Fellowships, research assistantships, teaching assistantships available. Financial aid application deadline: 1/5.
Dr. Klaus O. Kalthoff, Chairman, 512-471-1152.
Application contact: Michael J. Ryan, Graduate Adviser, 512-471-7131, *E-mail:* mryan@mail.utexas.edu.

ZOOLOGY

■ AUBURN UNIVERSITY

Graduate School, College of Sciences and Mathematics, Department of Biological Sciences, Auburn, Auburn University, AL 36849-0002
AWARDS Botany (MS, PhD); microbiology (MS, PhD); zoology (MS, PhD).

Faculty: 31 full-time (6 women).
Students: 38 full-time (23 women), 35 part-time (18 women); includes 6 minority (5 African Americans, 1 Hispanic American), 11 international. *66 applicants, 48% accepted.* In 1999, 12 master's, 1 doctorate awarded.
Entrance requirements: For master's and doctorate, GRE General Test, TOEFL. *Application deadline:* For fall admission, 7/7; for spring admission, 11/24. Electronic applications accepted.
Expenses: Tuition, state resident: full-time $2,895; part-time $80 per credit hour. Tuition, nonresident: full-time $8,685; part-time $240 per credit hour.
Financial aid: Research assistantships, teaching assistantships available.
Dr. Alfred E. Brown, Interim Chair, 334-844-4830, *Fax:* 334-844-1645.

Find an in-depth description at www.petersons.com/graduate.

■ BOSTON UNIVERSITY

Graduate School of Arts and Sciences, Department of Biology, Boston, MA 02215
AWARDS Botany (MA, PhD); cell and molecular biology (MA, PhD); cell biology (MA, PhD); ecology (PhD); ecology, behavior, and evolution (MA, PhD); ecology/physiology, endocrinology and reproduction (MA); marine biology (MA, PhD); molecular biology, cell biology and biochemistry (MA, PhD); neurobiology, neuroendocrinology and reproduction (MA, PhD); physiology,

Boston University (continued)
endocrinology, and neurobiology (MA, PhD); zoology (MA, PhD). Part-time programs available.

Faculty: 41 full-time (8 women).
Students: 131 full-time (74 women), 11 part-time (7 women); includes 10 minority (7 Asian Americans or Pacific Islanders, 3 Hispanic Americans), 33 international. Average age 27. *238 applicants, 39% accepted.* In 1999, 61 master's, 45 doctorates awarded. Terminal master's awarded for partial completion of doctoral program.
Degree requirements: For master's, one foreign language, thesis not required; for doctorate, one foreign language, dissertation, qualifying exam required. *Average time to degree:* Master's–1 year full-time, 3 years part-time; doctorate–5.75 years full-time.
Entrance requirements: For master's and doctorate, GRE General Test, GRE Subject Test, TOEFL. *Application deadline:* For fall admission, 1/1 (priority date); for spring admission, 11/1. *Application fee:* $50.
Expenses: Tuition: Full-time $23,770; part-time $743 per credit. Required fees: $220. Tuition and fees vary according to class time, course level, campus/location and program.
Financial aid: In 1999–00, 82 students received aid, including 1 fellowship with full tuition reimbursement available (averaging $12,000 per year), 28 research assistantships with full tuition reimbursements available (averaging $11,500 per year), 43 teaching assistantships with full tuition reimbursements available (averaging $11,500 per year); Federal Work-Study, grants, institutionally sponsored loans, scholarships, and traineeships also available. Financial aid application deadline: 1/15; financial aid applicants required to submit FAFSA.
Faculty research: Marine science, endocrinology, behavior. *Total annual research expenditures:* $5 million.
Geoffrey M. Cooper, Chairman, 617-353-2432, *Fax:* 617-353-6340, *E-mail:* gmcooper@bu.edu.
Application contact: Yolanta Kovalko, Senior Staff Assistant, 617-353-2432, *Fax:* 617-353-6340, *E-mail:* yolanta@bu.edu.

Find an in-depth description at www.petersons.com/graduate.

■ BRIGHAM YOUNG UNIVERSITY

Graduate Studies, College of Biological and Agricultural Sciences, Department of Zoology, Provo, UT 84602-1001

AWARDS Biological science education (MS); molecular biology (MS, PhD); wildlife and range resources (MS, PhD); zoology (MS, PhD). Part-time programs available.

Faculty: 32 full-time (2 women), 3 part-time/adjunct (0 women).
Students: 54 full-time (20 women), 8 part-time (3 women); includes 6 minority (2 Asian Americans or Pacific Islanders, 4 Hispanic Americans), 6 international. Average age 25. *32 applicants, 63% accepted.* In 1999, 17 master's awarded (23% continued full-time study); 5 doctorates awarded (80% entered university research/teaching, 20% found other work related to degree).
Degree requirements: For master's and doctorate, thesis/dissertation required, foreign language not required. *Average time to degree:* Master's–2.5 years full-time; doctorate–6 years full-time.
Entrance requirements: For master's, GRE General Test, minimum GPA of 3.0 during previous 2 years; for doctorate, GRE General Test, minimum GPA of 3.0 overall. *Application deadline:* For fall admission, 2/1 (priority date). *Application fee:* $30. Electronic applications accepted.
Expenses: Tuition: Full-time $3,330; part-time $185 per credit hour. Tuition and fees vary according to program and student's religious affiliation.
Financial aid: In 1999–00, 54 students received aid, including fellowships with partial tuition reimbursements available (averaging $11,000 per year), 12 research assistantships with full tuition reimbursements available (averaging $11,000 per year), 50 teaching assistantships with partial tuition reimbursements available (averaging $11,000 per year); career-related internships or fieldwork, institutionally sponsored loans, scholarships, tuition waivers (partial), unspecified assistantships, and tuition awards also available. Financial aid application deadline: 2/1.
Faculty research: Sex differentiation of brain, exercise physiology, toxicology, phylogenetic systematics, population biology.
Dr. John D. Bell, Chair, 801-378-2006, *Fax:* 801-378-7423, *E-mail:* jdb32@email.byu.edu.
Application contact: Dr. R. Ward Rhees, Graduate Coordinator, 801-378-2158, *Fax:* 801-378-7423, *E-mail:* ward_rhees@byu.edu.

■ CLEMSON UNIVERSITY

Graduate School, College of Agriculture, Forestry and Life Sciences, School of Animal, Biomedical and Biological Sciences, Department of Biological Sciences, Program in Zoology, Clemson, SC 29634

AWARDS MS, PhD.

Students: 26 full-time (15 women), 13 part-time (7 women), 3 international. *15 applicants, 33% accepted.* In 1999, 3 master's, 4 doctorates awarded.
Degree requirements: For master's, thesis optional, foreign language not required; for doctorate, dissertation, comprehensive exam required, foreign language not required.
Entrance requirements: For master's and doctorate, GRE General Test, TOEFL. *Application deadline:* For fall admission, 6/1. *Application fee:* $40.
Expenses: Tuition, state resident: full-time $3,480; part-time $174 per credit hour. Tuition, nonresident: full-time $9,256; part-time $388 per credit hour. Required fees: $5 per term. Full-time tuition and fees vary according to course level, course load and campus/location.
Financial aid: Research assistantships, teaching assistantships available. Financial aid application deadline: 3/15; financial aid applicants required to submit FAFSA. *Total annual research expenditures:* $600,000.
Dr. J. M. Colacino, Coordinator of Graduate Studies, 864-656-3581, *Fax:* 864-656-0435, *E-mail:* jmclc@clemson.edu.

■ COLORADO STATE UNIVERSITY

Graduate School, College of Natural Sciences, Department of Biology, Program in Zoology, Fort Collins, CO 80523-0015

AWARDS MS, PhD. Part-time programs available.

Faculty: 15 full-time (3 women).
Students: 17 full-time (10 women), 9 part-time (5 women); includes 3 minority (2 Asian Americans or Pacific Islanders, 1 Hispanic American). Average age 30. *60 applicants, 20% accepted.* In 1999, 13 master's, 3 doctorates awarded.
Degree requirements: For master's, one foreign language required, (computer language can substitute); for doctorate, 2 foreign languages (computer language can substitute for one), dissertation required.
Entrance requirements: For master's and doctorate, GRE General Test, GRE Subject Test, TOEFL, minimum GPA of 3.0. *Application deadline:* For fall admission, 2/1 (priority date); for spring admission, 11/1. Applications are processed on a rolling basis. *Application fee:* $30. Electronic applications accepted.
Expenses: Tuition, state resident: full-time $2,694; part-time $150 per credit. Tuition, nonresident: full-time $10,460; part-time $581 per credit. Required fees: $32 per semester. Tuition and fees vary according to program.
Financial aid: In 1999–00, 2 research assistantships, 15 teaching assistantships were awarded; fellowships, career-related

internships or fieldwork, Federal Work-Study, institutionally sponsored loans, and traineeships also available.

Faculty research: Physiological and behavioral ecology, animal behavior, cell physiology, aquatic and terrestrial ecology.
Application contact: Tina Sund, Graduate Coordinator, 970-491-1923, *Fax:* 970-491-0649, *E-mail:* tsund@lamar.colostate.edu.

■ CONNECTICUT COLLEGE

Graduate School, Department of Zoology, New London, CT 06320-4196

AWARDS MA, MAT. Part-time programs available.

Degree requirements: For master's, thesis required, foreign language not required.
Entrance requirements: For master's, GRE Subject Test or MAT.
Faculty research: Ultrastructure, comparative physiology, marine ecology, behavioral ecology, population genetics.

■ CORNELL UNIVERSITY

Graduate School, Graduate Fields of Agriculture and Life Sciences, Field of Zoology, Ithaca, NY 14853-0001

AWARDS Animal cytology (PhD); comparative and functional anatomy (PhD); developmental biology (PhD); ecology (PhD).

Faculty: 18 full-time.
Students: 5 full-time (all women), 1 international. 7 *applicants, 0% accepted.* In 1999, 1 doctorate awarded.
Degree requirements: For doctorate, dissertation, 2 semesters of teaching experience required, foreign language not required.
Entrance requirements: For doctorate, GRE General Test, GRE Subject Test (biology), TOEFL. *Application deadline:* For fall admission, 2/1. *Application fee:* $65. Electronic applications accepted.
Expenses: Tuition: Full-time $12,100.
Financial aid: In 1999–00, 3 students received aid, including 1 fellowship with full tuition reimbursement available, 1 research assistantship with full tuition reimbursement available, 1 teaching assistantship with full tuition reimbursement available; institutionally sponsored loans, scholarships, tuition waivers (full and partial), and unspecified assistantships also available. Financial aid applicants required to submit FAFSA.
Faculty research: Organismal biology, functional morphology, biomechanics, comparative vertebrate anatomy, comparative invertebrate anatomy, paleontology.
Application contact: Graduate Field Assistant, 607-253-3552, *E-mail:* physzoograd@cornell.edu.

■ EASTERN ILLINOIS UNIVERSITY

Graduate School, College of Sciences, Program in Biological Sciences, Charleston, IL 61920-3099

AWARDS Biological sciences (MS); botany (MS); environmental biology (MS); zoology (MS).

Degree requirements: For master's, exam required, foreign language and thesis not required.

■ EMPORIA STATE UNIVERSITY

School of Graduate Studies, College of Liberal Arts and Sciences, Division of Biological Sciences, Emporia, KS 66801-5087

AWARDS Botany (MS); environmental biology (MS); general biology (MS); microbial and cellular biology (MS); zoology (MS). Part-time programs available.

Faculty: 15 full-time (3 women), 4 part-time/adjunct (0 women).
Students: 20 full-time (10 women), 8 part-time (3 women), 2 international. *8 applicants, 75% accepted.* In 1999, 12 degrees awarded.
Degree requirements: For master's, comprehensive exam or thesis required.
Entrance requirements: For master's, TOEFL, written exam. *Application deadline:* For fall admission, 8/15 (priority date). Applications are processed on a rolling basis. *Application fee:* $30 ($75 for international students). Electronic applications accepted.
Expenses: Tuition, state resident: full-time $2,410; part-time $108 per credit hour. Tuition, nonresident: full-time $6,212; part-time $266 per credit hour.
Financial aid: In 1999–00, 2 fellowships (averaging $1,396 per year), 1 research assistantship (averaging $5,390 per year), 11 teaching assistantships with full tuition reimbursements (averaging $5,047 per year) were awarded; career-related internships or fieldwork, Federal Work-Study, and institutionally sponsored loans also available. Financial aid application deadline: 3/15; financial aid applicants required to submit FAFSA.
Faculty research: Fisheries, range, and wildlife management; aquatic, plant, grassland, vertebrate, and invertebrate ecology; mammalian and plant systematics, taxonomy, and evolution; immunology, virology, and molecular biology.
Dr. Marshall Sundberg, Chair, 316-341-5311, *Fax:* 316-341-5607, *E-mail:* sundberm@emporia.edu.

■ ILLINOIS STATE UNIVERSITY

Graduate School, College of Arts and Sciences, Department of Biological Sciences, Normal, IL 61790-2200

AWARDS Biological sciences (MS); biology (PhD); botany (PhD); ecology (PhD); genetics (PhD); microbiology (PhD); physiology (PhD); zoology (PhD). Part-time programs available.

Faculty: 24 full-time (4 women).
Students: 57 full-time (25 women), 27 part-time (10 women); includes 6 minority (1 African American, 3 Asian Americans or Pacific Islanders, 2 Hispanic Americans), 13 international. *58 applicants, 59% accepted.* In 1999, 12 master's, 3 doctorates awarded.
Degree requirements: For master's, thesis or alternative required; for doctorate, variable foreign language requirement (computer language can substitute for one), dissertation, 2 terms of residency required.
Entrance requirements: For master's, GRE General Test, minimum GPA of 2.6 in last 60 hours; for doctorate, GRE General Test. *Application deadline:* Applications are processed on a rolling basis. *Application fee:* $0.
Expenses: Tuition, state resident: full-time $2,526; part-time $105 per credit hour. Tuition, nonresident: full-time $7,578; part-time $316 per credit hour. Required fees: $1,082; $38 per credit hour. Tuition and fees vary according to course load and program.
Financial aid: In 1999–00, 8 research assistantships, 63 teaching assistantships were awarded; Federal Work-Study, tuition waivers (full), and unspecified assistantships also available. Financial aid application deadline: 4/1.
Faculty research: Phenotypic plasticity in reproduction: molecular mechanisms, physiological control, adaptive significance; molecular stress physiology of *Listeria monocytogenes*; enzymology of eggshell formation in *Schistosoma mansoni*, compensatory adaptation in the rat midbrain after neurodegeneration, analysis of staphylococcal virulence germs. *Total annual research expenditures:* $586,648.
Dr. Hou Cheung, Chairperson, 309-438-3669.
Application contact: Derek A. McCracken, Graduate Adviser, 309-438-3664.

Find an in-depth description at www.petersons.com/graduate.

■ INDIANA UNIVERSITY BLOOMINGTON

Graduate School, College of Arts and Sciences, Department of Biology, Program in Evolution, Ecology, and Behavior, Bloomington, IN 47405

AWARDS Ecology (MA, PhD); evolutionary biology (MA, PhD); zoology (MA, PhD). PhD offered through the University Graduate School. Part-time programs available.

Students: 66 full-time (35 women), 2 part-time (1 woman); includes 8 minority (2 African Americans, 2 Asian Americans or Pacific Islanders, 4 Hispanic Americans), 3 international. In 1999, 2 master's, 7 doctorates awarded. Terminal master's awarded for partial completion of doctoral program.
Degree requirements: For master's, thesis or alternative required; for doctorate, dissertation required.
Entrance requirements: For master's and doctorate, GRE General Test, TOEFL. *Application deadline:* For fall admission, 1/5 (priority date); for spring admission, 9/1. Applications are processed on a rolling basis. *Application fee:* $45. Electronic applications accepted.
Expenses: Tuition, state resident: full-time $3,853; part-time $161 per credit hour. Tuition, nonresident: full-time $11,226; part-time $468 per credit hour. Required fees: $360 per year. Tuition and fees vary according to course load and program.
Financial aid: Fellowships with tuition reimbursements, research assistantships with tuition reimbursements, teaching assistantships with tuition reimbursements, scholarships available. Financial aid application deadline: 1/15.
Faculty research: Ecosystem of community, plant and animal population biology, avian sociobiology, fish ethology, evolutionary genetics.
Dr. Gerald Gastony, Head, 812-855-3333, *Fax:* 812-855-6705, *E-mail:* gastony@indiana.edu.
Application contact: Gretchen Clearwater, Advisor for Graduate Affairs, 812-855-1861, *Fax:* 812-855-6705, *E-mail:* biograd@bio.indiana.edu.
Find an in-depth description at www.petersons.com/graduate.

■ IOWA STATE UNIVERSITY OF SCIENCE AND TECHNOLOGY

Graduate College, College of Liberal Arts and Sciences and College of Agriculture, Department of Zoology and Genetics, Ames, IA 50011

AWARDS MS, PhD.

Faculty: 31 full-time, 7 part-time/adjunct.
Students: 59 full-time (27 women), 6 part-time (2 women); includes 2 minority (1 African American, 1 Hispanic American), 30 international. *12 applicants, 50% accepted.* In 1999, 3 doctorates awarded.
Degree requirements: For master's and doctorate, thesis/dissertation required.
Entrance requirements: For master's and doctorate, GRE General Test, TOEFL. *Application deadline:* For fall admission, 2/1 (priority date). *Application fee:* $20 ($50 for international students). Electronic applications accepted.
Expenses: Tuition, state resident: full-time $3,308. Tuition, nonresident: full-time $9,744. Part-time tuition and fees vary according to course load, campus/location and program.
Financial aid: In 1999–00, 65 students received aid, including 1 fellowship with full tuition reimbursement available (averaging $20,000 per year), 34 research assistantships with partial tuition reimbursements available (averaging $15,000 per year), 18 teaching assistantships with partial tuition reimbursements available (averaging $13,000 per year); scholarships also available. Financial aid application deadline: 2/1.
Faculty research: Animal behavior, animal models of gene therapy, cell biology, comparative physiology, developmental biology.
Dr. M. Duane Enger, Chair, 515-294-3908, *Fax:* 515-294-8457, *E-mail:* zg@iastate.edu.
Application contact: Dr. Eileen Muff, Graduate Coordinator, 515-294-3909, *E-mail:* zg@iastate.edu.
Find an in-depth description at www.petersons.com/graduate.

■ KENT STATE UNIVERSITY

College of Arts and Sciences, Department of Biological Sciences, Kent, OH 44242-0001

AWARDS Botany (MA, MS, PhD); ecology (MS, PhD); physiology (MS, PhD); zoology (MA, PhD).

Faculty: 28 full-time.
Students: 31 full-time (17 women), 10 part-time (9 women); includes 1 minority (African American), 12 international. *30 applicants, 77% accepted.* In 1999, 6 master's, 5 doctorates awarded.
Degree requirements: For master's and doctorate, thesis/dissertation required, foreign language not required.
Entrance requirements: For master's, GRE General Test, minimum GPA of 2.75; for doctorate, GRE General Test, minimum GPA of 3.0. *Application deadline:* For fall admission, 7/12. Applications are processed on a rolling basis. *Application fee:* $30.
Expenses: Tuition, state resident: full-time $5,334; part-time $243 per hour. Tuition, nonresident: full-time $10,238; part-time $466 per hour.
Financial aid: Fellowships, research assistantships, teaching assistantships, Federal Work-Study, institutionally sponsored loans, and tuition waivers (full) available. Financial aid application deadline: 2/1.
Dr. Brent C. Bruot, Chairman, 330-672-3613, *Fax:* 330-672-3713.
Application contact: Dr. John R. D. Stalvey, Coordinator of Graduate Studies, 330-672-2819.

■ LOUISIANA STATE UNIVERSITY AND AGRICULTURAL AND MECHANICAL COLLEGE

Graduate School, College of Basic Sciences, Department of Biological Sciences, Baton Rouge, LA 70803

AWARDS Biochemistry (MS, PhD); microbiology (MS, PhD); plant biology (MS, PhD); zoology (MS, PhD). Part-time programs available.

Faculty: 79 full-time (7 women), 2 part-time/adjunct (0 women).
Students: 91 full-time (44 women), 22 part-time (10 women); includes 9 minority (3 African Americans, 1 Asian American or Pacific Islander, 5 Hispanic Americans), 29 international. Average age 28. *98 applicants, 28% accepted.* In 1999, 13 master's, 7 doctorates awarded. Terminal master's awarded for partial completion of doctoral program.
Degree requirements: For master's, foreign language not required; for doctorate, dissertation required.
Entrance requirements: For master's and doctorate, GRE General Test, minimum GPA of 3.0. *Application deadline:* Applications are processed on a rolling basis. *Application fee:* $25.
Expenses: Tuition, state resident: full-time $2,881. Tuition, nonresident: full-time $7,081. Part-time tuition and fees vary according to course load and program.
Financial aid: In 1999–00, 12 fellowships, 20 research assistantships with partial tuition reimbursements, 62 teaching assistantships with partial tuition reimbursements were awarded; Federal Work-Study, institutionally sponsored loans, and unspecified assistantships also available. Aid available to part-time students. *Total annual research expenditures:* $2.2 million.
Dr. Harold Silverman, Chairman, 225-388-2601, *Fax:* 225-388-2597, *E-mail:* cxsiv@unix1.sncc.lsu.edu.
Find an in-depth description at www.petersons.com/graduate.

■ MIAMI UNIVERSITY

Graduate School, College of Arts and Sciences, Department of Zoology, Oxford, OH 45056

AWARDS MA, MS, PhD. Part-time programs available.

Faculty: 27 full-time (9 women).
Students: 19 full-time (7 women), 35 part-time (20 women); includes 3 minority (all African Americans), 5 international. *25 applicants, 76% accepted.* In 1999, 11 master's, 4 doctorates awarded.
Degree requirements: For master's, thesis, final exam required; for doctorate, dissertation, comprehensive and final exams required.
Entrance requirements: For master's, GRE General Test, minimum undergraduate GPA of 3.0 during previous 2 years or 2.75 overall; for doctorate, GRE General Test, minimum GPA of 2.75 (undergraduate), 3.0 (graduate). *Application deadline:* For fall admission, 3/1 (priority date); for spring admission, 12/15. Applications are processed on a rolling basis. *Application fee:* $35.
Expenses: Tuition, state resident: part-time $260 per hour. Tuition, nonresident: full-time $3,125; part-time $538 per hour. International tuition: $6,452 full-time. Required fees: $18 per semester. Tuition and fees vary according to campus/location.
Financial aid: In 1999–00, 47 fellowships, 16 research assistantships, 12 teaching assistantships were awarded; Federal Work-Study and tuition waivers (full) also available. Financial aid application deadline: 3/1.
Dr. James T. Oris, Director of Graduate Studies, 513-529-3100, *Fax:* 513-529-6900, *E-mail:* zoology@muohio.edu.

■ MICHIGAN STATE UNIVERSITY

Graduate School, College of Natural Science, Department of Zoology, East Lansing, MI 48824

AWARDS Zoology (MS, PhD); zoology-environmental toxicology (PhD).

Faculty: 26.
Students: 54 full-time (24 women), 26 part-time (15 women); includes 7 minority (1 African American, 5 Asian Americans or Pacific Islanders, 1 Hispanic American), 9 international. Average age 27. *106 applicants, 18% accepted.* In 1999, 6 master's, 4 doctorates awarded.
Degree requirements: For master's, thesis required (for some programs), foreign language not required; for doctorate, dissertation required.
Entrance requirements: For master's and doctorate, GRE General Test, TOEFL. *Application deadline:* For fall admission, 1/1

(priority date). *Application fee:* $30 ($40 for international students).
Expenses: Tuition, state resident: part-time $229 per credit. Tuition, nonresident: part-time $464 per credit. Required fees: $241 per semester. Tuition and fees vary according to course load, degree level and program.
Financial aid: In 1999–00, 15 research assistantships with tuition reimbursements (averaging $11,473 per year), 35 teaching assistantships with tuition reimbursements (averaging $11,501 per year) were awarded; fellowships. Financial aid application deadline: 1/1; financial aid applicants required to submit FAFSA.
Faculty research: Aquatic ecosystem ecology, dioxin research, biotic integrity for great lakes costal wetlands. *Total annual research expenditures:* $851,020.
Dr. Thomas Burton, Chairperson, 517-355-4640, *Fax:* 517-432-2789, *E-mail:* nsczol@msu.edu.
Application contact: Dr. Richard W. Hill, Associate Chair, 517-355-4640.

Find an in-depth description at www.petersons.com/graduate.

■ NORTH CAROLINA STATE UNIVERSITY

Graduate School, College of Agriculture and Life Sciences, Department of Zoology, Raleigh, NC 27695

AWARDS MAWB, MLS, MS, PhD.

Faculty: 53 full-time (2 women), 20 part-time/adjunct (3 women).
Students: 55 full-time (22 women), 19 part-time (9 women); includes 4 minority (1 Asian American or Pacific Islander, 2 Hispanic Americans, 1 Native American), 4 international. Average age 29. *75 applicants, 19% accepted.* In 1999, 24 master's, 3 doctorates awarded. Terminal master's awarded for partial completion of doctoral program.
Degree requirements: For master's, thesis (for some programs), oral exam required, foreign language not required; for doctorate, dissertation, oral and written exams required, foreign language not required.
Entrance requirements: For master's and doctorate, GRE General Test, TOEFL, minimum GPA of 3.0. *Application deadline:* For fall admission, 6/25; for spring admission, 11/25. Applications are processed on a rolling basis. *Application fee:* $45.
Expenses: Tuition, state resident: full-time $1,578. Tuition, nonresident: full-time $10,744. Required fees: $892. Full-time tuition and fees vary according to program.
Financial aid: In 1999–00, 7 fellowships (averaging $4,836 per year), 29 research

assistantships (averaging $4,656 per year), 17 teaching assistantships (averaging $4,950 per year) were awarded; career-related internships or fieldwork, Federal Work-Study, and institutionally sponsored loans also available. Financial aid application deadline: 5/15.
Faculty research: Fish and aquatic ecology, herpetology, behavioral ecology, neurobiology, avian ecology, limnology, conservation biology. *Total annual research expenditures:* $6.3 million.
Dr. Thurman L. Grove, Interim Head, 919-515-2741, *Fax:* 919-515-5327, *E-mail:* thurman_grove@ncsu.edu.
Application contact: Dr. B. J. Copeland, Director of Graduate Programs, 919-515-4589, *Fax:* 919-515-5327, *E-mail:* bj_copeland@ncsu.edu.

■ NORTH DAKOTA STATE UNIVERSITY

Graduate Studies and Research, College of Science and Mathematics, Department of Zoology, Fargo, ND 58105

AWARDS Cellular and molecular biology (PhD); natural resources management (MS); zoology (MS, PhD).

Faculty: 8 full-time (1 woman).
Students: 14 full-time (4 women), 3 part-time (1 woman), 1 international. Average age 24. *14 applicants, 43% accepted.* In 1999, 6 master's awarded (50% found work related to degree, 33% continued full-time study); 3 doctorates awarded (33% entered university research/teaching, 67% found other work related to degree).
Degree requirements: For master's and doctorate, thesis/dissertation required, foreign language not required. *Average time to degree:* Master's–2.75 years full-time, 5 years part-time; doctorate–3.5 years full-time, 5 years part-time.
Entrance requirements: For master's and doctorate, GRE General Test, GRE Subject Test (biology), TOEFL. *Application deadline:* For fall admission, 3/15 (priority date); for spring admission, 3/31. Applications are processed on a rolling basis. *Application fee:* $25. Electronic applications accepted.
Expenses: Tuition, state resident: full-time $3,096; part-time $112 per credit hour. Tuition, nonresident: full-time $7,588; part-time $299 per credit hour. Tuition and fees vary according to course load, campus/location and reciprocity agreements.
Financial aid: In 1999–00, 4 fellowships with full tuition reimbursements (averaging $3,000 per year), 6 research assistantships with full tuition reimbursements (averaging $9,550 per year), 8 teaching assistantships with full tuition reimbursements (averaging

North Dakota State University (continued)
$9,550 per year) were awarded; career-related internships or fieldwork, Federal Work-Study, and institutionally sponsored loans also available. Aid available to part-time students. Financial aid application deadline: 4/15; financial aid applicants required to submit FAFSA.
Faculty research: Comparative endocrinology, physiology, behavioral ecology, vertebrate pest management, aquatic biology. *Total annual research expenditures:* $325,700.
Dr. W. J. Bleier, Head, 701-231-8421, *Fax:* 701-231-7149, *E-mail:* bleier@ plains.nodak.edu.

■ OHIO UNIVERSITY

Graduate Studies, College of Arts and Sciences, Department of Biological Sciences, Athens, OH 45701-2979
AWARDS Biological sciences (MS, PhD); microbiology (MS, PhD); zoology (MS, PhD).
Faculty: 24 full-time (7 women), 11 part-time/adjunct (6 women).
Students: 53 full-time (17 women), 5 part-time (2 women); includes 1 minority (African American), 18 international. Average age 24. *85 applicants, 21% accepted.* In 1999, 3 master's awarded (100% continued full-time study); 5 doctorates awarded (100% found work related to degree).
Degree requirements: For master's, thesis, 1 quarter of teaching experience required, foreign language not required; for doctorate, dissertation, 2 quarters of teaching experience required.
Entrance requirements: For master's and doctorate, GRE General Test. *Application deadline:* For fall admission, 1/15. *Application fee:* $30.
Expenses: Tuition, state resident: full-time $5,754; part-time $238 per credit hour. Tuition, nonresident: full-time $11,055; part-time $457 per credit hour. Tuition and fees vary according to course load, degree level and campus/location.
Financial aid: In 1999–00, 1 fellowship with full tuition reimbursement, 1 research assistantship with full tuition reimbursement (averaging $15,000 per year), 33 teaching assistantships with full tuition reimbursements (averaging $13,500 per year) were awarded; Federal Work-Study, institutionally sponsored loans, and tuition waivers (full) also available. Financial aid application deadline: 1/15.
Faculty research: Ecology and evolutionary biology, exercise physiology and muscle biology, endocrinology and metabolic physiology, neurobiology. *Total annual research expenditures:* $2.8 million.
Dr. Anne Loucks, Chair, 740-593-2290, *Fax:* 740-593-0300, *E-mail:* loucks@ ohiou.edu.

Application contact: Dr. William R. Holmes, Graduate Chair, 740-593-2334, *Fax:* 740-593-0300, *E-mail:* holmes@ ohiou.edu.
Find an in-depth description at www.petersons.com/graduate.

■ OKLAHOMA STATE UNIVERSITY

Graduate College, College of Arts and Sciences, Department of Zoology, Stillwater, OK 74078
AWARDS Wildlife and fisheries ecology (MS, PhD); zoology (MS, PhD).
Faculty: 16 full-time (2 women), 2 part-time/adjunct (1 woman).
Students: 21 full-time (11 women), 58 part-time (22 women); includes 5 minority (1 African American, 2 Hispanic Americans, 2 Native Americans), 4 international. Average age 29. In 1999, 4 master's, 2 doctorates awarded.
Degree requirements: For master's, thesis required, foreign language not required; for doctorate, dissertation required.
Entrance requirements: For master's and doctorate, GRE General Test, GRE Subject Test, TOEFL. *Application deadline:* For fall admission, 7/1 (priority date). *Application fee:* $25.
Expenses: Tuition, state resident: part-time $86 per credit hour. Tuition, nonresident: part-time $275 per credit hour. Required fees: $17 per credit hour. $14 per semester. One-time fee: $20 full-time. Tuition and fees vary according to course load.
Financial aid: In 1999–00, 53 students received aid, including 17 research assistantships (averaging $15,583 per year), 48 teaching assistantships (averaging $14,526 per year); career-related internships or fieldwork, Federal Work-Study, and tuition waivers (partial) also available. Aid available to part-time students. Financial aid application deadline: 3/1.
Dr. Jim Shaw, Head, 405-744-5555.

■ OREGON STATE UNIVERSITY

Graduate School, College of Science, Department of Zoology, Corvallis, OR 97331
AWARDS MA, MAIS, MS, PhD.
Faculty: 15 full-time (2 women).
Students: 34 full-time (20 women); includes 5 minority (1 African American, 2 Asian Americans or Pacific Islanders, 1 Hispanic American, 1 Native American), 2 international. Average age 28. In 1999, 2 master's, 3 doctorates awarded. Terminal master's awarded for partial completion of doctoral program.

Degree requirements: For master's, foreign language and thesis not required; for doctorate, dissertation required, foreign language not required.
Entrance requirements: For master's and doctorate, GRE General Test, GRE Subject Test, TOEFL, minimum GPA of 3.0 in last 90 hours. *Application deadline:* For fall admission, 1/15. *Application fee:* $50.
Expenses: Tuition, state resident: full-time $6,489. Tuition, nonresident: full-time $11,061. Tuition and fees vary according to program.
Financial aid: Fellowships, research assistantships, teaching assistantships, Federal Work-Study and institutionally sponsored loans available. Aid available to part-time students. Financial aid application deadline: 2/1.
Faculty research: Cell and developmental biology, population biology and marine community ecology, behavioral physiology, comparative immunology, plant-herbivore interaction.
Dr. Stevan J. Arnold, Chair, 541-737-5337, *Fax:* 541-737-0501, *E-mail:* arnolds@ bcc.orst.edu.
Application contact: Valarie Breda, Chair of Graduate Admissions, 541-737-3705, *Fax:* 541-737-0501, *E-mail:* bredav@ bcc.orst.edu.

■ SOUTHERN ILLINOIS UNIVERSITY CARBONDALE

Graduate School, College of Science, Department of Zoology, Carbondale, IL 62901-6806
AWARDS MS, PhD.
Faculty: 25 full-time (1 woman).
Students: 53 full-time (20 women), 17 part-time (7 women); includes 5 minority (2 Asian Americans or Pacific Islanders, 3 Hispanic Americans), 3 international. Average age 25. *50 applicants, 34% accepted.* In 1999, 14 master's awarded.
Degree requirements: For master's, thesis required, foreign language not required; for doctorate, dissertation required.
Entrance requirements: For master's, GRE, TOEFL, minimum GPA of 2.7; for doctorate, GRE, TOEFL, minimum GPA of 3.25. *Application deadline:* Applications are processed on a rolling basis. *Application fee:* $20.
Expenses: Tuition, state resident: full-time $2,902. Tuition, nonresident: full-time $5,810. Tuition and fees vary according to course load.
Financial aid: In 1999–00, 49 students received aid, including 5 fellowships with full tuition reimbursements available, 30 research assistantships with full tuition reimbursements available, 24 teaching

assistantships with full tuition reimbursements available; Federal Work-Study, institutionally sponsored loans, and tuition waivers (full) also available. Aid available to part-time students.

Faculty research: Ecology, fisheries and wildlife, systematics, behavior, vertebrate and invertebrate biology.

William Muhlach, Chairperson, 618-536-2314.

Application contact: Carey Krajewski, Director of Graduate Studies, 618-453-4132, *Fax:* 618-453-2806, *E-mail:* careyk@siu.edu.

■ TEXAS A&M UNIVERSITY

College of Science, Department of Biology, Program in Zoology, College Station, TX 77843

AWARDS MS, PhD.

Students: 19 full-time (7 women), 8 part-time (7 women). Average age 30. *22 applicants, 41% accepted.* In 1999, 5 master's, 4 doctorates awarded.

Degree requirements: For master's, foreign language not required; for doctorate, dissertation required, foreign language not required.

Entrance requirements: For master's and doctorate, GRE General Test, TOEFL. *Application fee:* $50 ($75 for international students).

Expenses: Tuition, state resident: part-time $76 per semester hour. Tuition, nonresident: part-time $292 per semester hour. Required fees: $11 per semester hour. Tuition and fees vary according to program.

Financial aid: Application deadline: 4/1.

Dr. Mark Zoran, Head, 979-845-7755.

■ TEXAS TECH UNIVERSITY

Graduate School, College of Arts and Sciences, Department of Biological Sciences, Lubbock, TX 79409

AWARDS Biology (MS, PhD); environmental toxicology (MS); microbiology (MS); zoology (MS, PhD). Part-time programs available.

Faculty: 36 full-time (4 women), 1 part-time/adjunct (0 women).

Students: 86 full-time (41 women), 26 part-time (10 women); includes 5 minority (1 Asian American or Pacific Islander, 4 Hispanic Americans), 28 international. Average age 30. *54 applicants, 46% accepted.* In 2000, 26 master's, 11 doctorates awarded.

Degree requirements: For master's, thesis required (for some programs), foreign language not required; for doctorate, dissertation required, foreign language not required.

Entrance requirements: For master's and doctorate, GRE General Test. *Application deadline:* For fall admission, 4/15 (priority

date); for spring admission, 11/1 (priority date). Applications are processed on a rolling basis. *Application fee:* $25 ($50 for international students). Electronic applications accepted.

Expenses: Tuition, state resident: full-time $2,376; part-time $99 per credit hour. Tuition, nonresident: full-time $7,560; part-time $315 per credit hour. Required fees: $464 per semester. Part-time tuition and fees vary according to course load, program and reciprocity agreements.

Financial aid: In 2000–01, 60 students received aid, including 40 research assistantships (averaging $10,405 per year), 53 teaching assistantships (averaging $11,067 per year); fellowships, career-related internships or fieldwork, Federal Work-Study, and institutionally sponsored loans also available. Aid available to part-time students. Financial aid application deadline: 5/15; financial aid applicants required to submit FAFSA.

Faculty research: Development of strains of transgenic plants, ecological studies of Arctic tundra and Puerto Rican rain forests, genome organization and evolution. *Total annual research expenditures:* $2.1 million.

Dr. Carleton Phillips, Chairman, 806-742-2715, *Fax:* 806-742-2963.

Application contact: Graduate Adviser, 806-742-2715, *Fax:* 806-742-2963.

■ UNIFORMED SERVICES UNIVERSITY OF THE HEALTH SCIENCES

School of Medicine, Division of Basic Medical Sciences, Bethesda, MD 20814-4799

AWARDS Anatomy and cell biology (PhD), including cell biology, developmental biology, and neurobiology; biochemistry (PhD), including emerging infectious diseases; emerging infectious diseases (PhD); medical and clinical psychology (PhD), including clinical psychology, medical psychology; medical history (MMH); microbiology and immunology (PhD); molecular and cell biology (PhD); neuroscience (PhD); pathology (PhD), including molecular pathobiology; pharmacology (PhD); physiology (PhD); preventive medicine/biometrics (MPH, MSPH, MTMH, Dr PH, PhD), including public health (MPH, MSPH, Dr PH), tropical medicine and hygiene (MTMH), zoology (PhD).

Faculty: 142 full-time (40 women), 335 part-time/adjunct (73 women).

Students: 119 full-time (62 women), 15 part-time (2 women); includes 20 minority (10 African Americans, 6 Asian Americans or Pacific Islanders, 3 Hispanic Americans, 1 Native American). Average age 26. *183 applicants, 28% accepted.* In 1999, 37 master's, 9 doctorates awarded. Terminal

master's awarded for partial completion of doctoral program.

Degree requirements: For master's, comprehensive exam required; for doctorate, dissertation, qualifying exam required. *Average time to degree:* Master's–1 year full-time.

Entrance requirements: For master's and doctorate, GRE General Test, U.S. citizenship. *Application deadline:* For fall admission, 1/15 (priority date). Applications are processed on a rolling basis. *Application fee:* $0.

Financial aid: In 1999–00, fellowships with full tuition reimbursements (averaging $15,000 per year), research assistantships with full tuition reimbursements (averaging $15,000 per year) were awarded; career-related internships or fieldwork and tuition waivers (full) also available.

Dr. Michael N. Sheridan, Associate Dean, 800-772-1747, *Fax:* 301-295-6772, *E-mail:* msheridan@usuhs.mil.

Application contact: Janet M. Anastasi, Graduate Program Coordinator, 301-295-9474, *Fax:* 301-295-6772, *E-mail:* janastasi@usuhs.mil.

Find an in-depth description at www.petersons.com/graduate.

■ UNIFORMED SERVICES UNIVERSITY OF THE HEALTH SCIENCES

School of Medicine, Division of Basic Medical Sciences, Department of Preventive Medicine/Biometrics, Program in Zoology, Bethesda, MD 20814-4799

AWARDS PhD.

Faculty: 45 full-time (8 women), 186 part-time/adjunct (35 women).

Students: 4 full-time (1 woman). Average age 29. *2 applicants, 0% accepted.* In 1999, 1 degree awarded (100% found work related to degree).

Degree requirements: For doctorate, dissertation, qualifying exam required, foreign language not required.

Entrance requirements: For doctorate, GRE General Test, GRE Subject Test, minimum GPA of 3.0, U.S. citizenship. *Application deadline:* For fall admission, 1/15 (priority date). Applications are processed on a rolling basis. *Application fee:* $0.

Financial aid: In 1999–00, 3 fellowships with full tuition reimbursements (averaging $15,000 per year) were awarded; tuition waivers (full) also available.

Faculty research: Epidemiology, biostatistics, tropical public health, parasitology, vector biology.

Col. Gary Gackstetter, Graduate Program Director, 301-295-3050, *Fax:* 301-295-1933, *E-mail:* ggackstetter@usuhs.mil.

Uniformed Services University of the Health Sciences (continued)
Application contact: Janet M. Anastasi, Graduate Program Coordinator, 301-295-9474, *Fax:* 301-295-6772, *E-mail:* janastasi@usuhs.mil.

■ UNIVERSITY OF ALASKA FAIRBANKS

Graduate School, College of Science, Engineering and Mathematics, Department of Biology and Wildlife, Program in Biological Sciences, Fairbanks, AK 99775

AWARDS Biology (MAT, MS, PhD); botany (MS, PhD); zoology (MS, PhD). Part-time programs available.

Faculty: 24 full-time (2 women), 2 part-time/adjunct (0 women).
Students: 39 full-time (19 women), 11 part-time (6 women); includes 1 minority (Native American), 6 international. Average age 31. *32 applicants, 38% accepted.* In 1999, 5 master's, 5 doctorates awarded.
Degree requirements: For master's, thesis, comprehensive exam required, foreign language not required; for doctorate, one foreign language (computer language can substitute), dissertation, comprehensive exam required.
Entrance requirements: For master's and doctorate, GRE General Test, GRE Subject Test, TOEFL. *Application deadline:* For fall admission, 8/1. Applications are processed on a rolling basis. *Application fee:* $35.
Expenses: Tuition, state resident: full-time $3,006; part-time $167 per credit. Tuition, nonresident: full-time $5,868; part-time $326 per credit. Required fees: $370; $10 per credit. $140 per semester.
Financial aid: Research assistantships, teaching assistantships, career-related internships or fieldwork available. Financial aid application deadline: 6/1.
Faculty research: Plant insect interactions, wildlife ecology, adaptations to winter/cold, cell/molecular biology, ecology.
Dr. Ed Murphy, Acting Dean, College of Science, Engineering and Mathematics, 907-474-7941.

■ UNIVERSITY OF CALIFORNIA, DAVIS

Graduate Studies, Program in Avian Sciences, Davis, CA 95616

AWARDS MS. Part-time programs available.

Faculty: 24 full-time (9 women).
Students: 7 full-time (4 women). Average age 29. *9 applicants, 44% accepted.* In 1999, 5 degrees awarded (100% continued full-time study). *Average time to degree:* Master's–2 years full-time.

Entrance requirements: For master's, GRE General Test, minimum GPA of 3.0. *Application deadline:* For fall admission, 3/1 (priority date). Applications are processed on a rolling basis. *Application fee:* $40. Electronic applications accepted.
Expenses: Tuition, nonresident: full-time $9,804. Tuition and fees vary according to program and student level.
Financial aid: In 1999–00, 6 students received aid, including 4 fellowships with full and partial tuition reimbursements available, 1 teaching assistantship with partial tuition reimbursement available; research assistantships with full and partial tuition reimbursements available, career-related internships or fieldwork, Federal Work-Study, institutionally sponsored loans, and scholarships also available. Financial aid application deadline: 1/15; financial aid applicants required to submit FAFSA.
Faculty research: Reproduction, nutrition, toxicology, food products, ecology of avian species.
Kirk C. Klasing, Graduate Chair, 530-752-1901, *E-mail:* kcklasing@ucdavis.edu.

■ UNIVERSITY OF CHICAGO

Division of the Biological Sciences, Darwinian Sciences: Ecological, Integrative and Evolutionary Biology, Department of Organismal Biology and Anatomy, Chicago, IL 60637-1513

AWARDS Functional and evolutionary biology (PhD); organismal biology and anatomy (PhD).

Faculty: 12 full-time (1 woman), 5 part-time/adjunct (3 women).
Students: 18 full-time (5 women); includes 2 minority (1 Hispanic American, 1 Native American), 4 international. Average age 23. *20 applicants, 20% accepted.* In 1999, 3 doctorates awarded (100% entered university research/teaching).
Degree requirements: For doctorate, dissertation required, foreign language not required. *Average time to degree:* Doctorate–6 years full-time.
Entrance requirements: For doctorate, GRE General Test, TOEFL. *Application deadline:* For fall admission, 1/5 (priority date). *Application fee:* $55.
Expenses: Tuition: Full-time $24,804; part-time $3,422 per course. Required fees: $390. Tuition and fees vary according to program.
Financial aid: In 1999–00, 18 students received aid, including 18 fellowships with tuition reimbursements available (averaging $17,050 per year). Financial aid application deadline: 6/1.
Faculty research: Ecological physiology, evolution of fossil reptiles, vertebrate paleontology.

Dr. Neil Shubin, Chairman, 773-702-9228, *Fax:* 773-702-0037.
Application contact: Carolyn Johnson, Graduate Administrative Director, 773-702-9474, *Fax:* 773-702-4699, *E-mail:* cs-johnson@uchicago.edu.

■ UNIVERSITY OF COLORADO AT BOULDER

Graduate School, College of Arts and Sciences, Department of Environmental, Population, and Organic Biology, Boulder, CO 80309

AWARDS Animal behavior (MA, PhD); aquatic biology (MA, PhD); behavioral genetics (MA, PhD); ecology (MA, PhD); microbiology (MA, PhD); neurobiology (MA, PhD); plant and animal physiology (MA, PhD); plant and animal systematics (MA, PhD); population biology (MA, PhD); population genetics (MA, PhD).

Faculty: 39 full-time (9 women).
Students: 84 full-time (36 women), 14 part-time (7 women); includes 17 minority (6 Asian Americans or Pacific Islanders, 10 Hispanic Americans, 1 Native American). Average age 29. *147 applicants, 14% accepted.* In 1999, 7 master's, 13 doctorates awarded. Terminal master's awarded for partial completion of doctoral program.
Degree requirements: For master's, thesis or alternative, comprehensive exam required, foreign language not required; for doctorate, dissertation, comprehensive exam required, foreign language not required. *Average time to degree:* Master's–3 years full-time; doctorate–5 years full-time.
Entrance requirements: For master's, GRE General Test, GRE Subject Test, minimum undergraduate GPA of 3.0; for doctorate, GRE General Test, GRE Subject Test. *Application deadline:* For fall admission, 1/15 (priority date). *Application fee:* $40 ($60 for international students).
Expenses: Tuition, state resident: part-time $181 per credit hour. Tuition, nonresident: part-time $542 per credit hour. Required fees: $99 per term. Tuition and fees vary according to course load and program.
Financial aid: Fellowships, research assistantships, teaching assistantships, Federal Work-Study, institutionally sponsored loans, and tuition waivers (full) available. Financial aid application deadline: 3/1.
Faculty research: Evolution, developmental biology, behavior and neurobiology. *Total annual research expenditures:* $1.8 million.
Michael Breed, Chair, 303-492-8981, *Fax:* 303-492-8699, *E-mail:* michael.breed@colorado.edu.
Application contact: Jill Skarstadt, Graduate Secretary, 303-492-7654, *Fax:*

303-492-8699, *E-mail:* jill.skarstadt@colorado.edu.

■ UNIVERSITY OF CONNECTICUT

Graduate School, College of Liberal Arts and Sciences, Biological Sciences Group, Storrs, CT 06269

AWARDS Ecology and evolutionary biology (MS, PhD), including botany, ecology, entomology, systematics, zoology; molecular and cell biology (MS, PhD), including biochemistry, biophysics, biotechnology (MS), cell and developmental biology, genetics, microbiology, plant molecular and cell biology; physiology and neurobiology (MS, PhD), including neurobiology, physiology.

Degree requirements: For doctorate, dissertation required.

Entrance requirements: For master's and doctorate, GRE General Test, GRE Subject Test, TOEFL.

Expenses: Tuition, state resident: full-time $5,118. Tuition, nonresident: full-time $13,298. Required fees: $1,022.

■ UNIVERSITY OF CONNECTICUT

Graduate School, College of Liberal Arts and Sciences, Biological Sciences Group, Department of Ecology and Evolutionary Biology, Field of Zoology, Storrs, CT 06269

AWARDS MS, PhD.

Degree requirements: For doctorate, dissertation required.

Entrance requirements: For master's and doctorate, GRE General Test, GRE Subject Test, TOEFL.

Expenses: Tuition, state resident: full-time $5,118. Tuition, nonresident: full-time $13,298. Required fees: $1,022.

Find an in-depth description at www.petersons.com/graduate.

■ UNIVERSITY OF FLORIDA

Graduate School, College of Liberal Arts and Sciences, Department of Zoology, Gainesville, FL 32611

AWARDS MS, MST, PhD.

Faculty: 43.
Students: 51 full-time (23 women), 3 part-time (2 women); includes 5 minority (1 Asian American or Pacific Islander, 4 Hispanic Americans), 8 international. *111 applicants, 16% accepted.* In 1999, 11 master's, 6 doctorates awarded.
Degree requirements: For master's, thesis or alternative required, foreign language not required; for doctorate, one foreign language, dissertation required.
Entrance requirements: For master's and doctorate, GRE General Test, minimum GPA of 3.0. *Application deadline:* For fall admission, 1/14. Applications are processed on a rolling basis. *Application fee:* $20. Electronic applications accepted.
Expenses: Tuition, state resident: part-time $144 per credit hour. Tuition, nonresident: part-time $505 per credit hour. Tuition and fees vary according to course level, course load and program.
Financial aid: In 1999–00, 48 students received aid, including 6 fellowships, 10 research assistantships, 31 teaching assistantships; unspecified assistantships also available.
Faculty research: Behavior, ecology, evolutionary biology, comparative physiology.
Dr. Jane Brockmann, Chair, 352-392-1297, *Fax:* 352-392-3704, *E-mail:* hjb@zoo.ufl.edu.
Application contact: Dr. Colin Chapman, Graduate Coordinator, 352-392-1196, *Fax:* 352-392-3704, *E-mail:* cachapman@zoo.ufl.edu.

■ UNIVERSITY OF HAWAII AT MANOA

Graduate Division, College of Arts and Sciences, College of Natural Sciences, Department of Zoology, Honolulu, HI 96822

AWARDS MS, PhD.

Faculty: 34 full-time (6 women), 6 part-time/adjunct (1 woman).
Students: 44 full-time (2 women), 6 part-time (2 women). Average age 30. *107 applicants, 16% accepted.* In 1999, 6 master's, 5 doctorates awarded.
Degree requirements: For master's, thesis not required; for doctorate, one foreign language, dissertation, seminar required. *Average time to degree:* Master's–3 years full-time; doctorate–7.3 years full-time.
Entrance requirements: For master's and doctorate, GRE General Test, GRE Subject Test. *Application deadline:* For fall admission, 2/1. Applications are processed on a rolling basis. *Application fee:* $25 ($50 for international students).
Expenses: Tuition, state resident: full-time $4,032; part-time $168 per credit. Tuition, nonresident: full-time $9,960; part-time $415 per credit. Required fees: $51 per semester. Part-time tuition and fees vary according to course load and program.
Financial aid: In 1999–00, 36 research assistantships (averaging $15,871 per year), 22 teaching assistantships (averaging $13,297 per year) were awarded; tuition waivers (full) also available. Financial aid application deadline: 2/1.
Faculty research: Molecular evolution, reproductive biology, animal behavior, conservation biology, avian biology.

Samuel Haley, Chairperson, 808-956-8617, *Fax:* 808-956-9812, *E-mail:* rhaley@zoogate.zoo.hawaii.edu.
Application contact: Dr. Andrew Taylor, Graduate Field Chairperson, 808-956-4706, *Fax:* 808-956-9812, *E-mail:* ataylor@lala.zoo.hawaii.edu.

■ UNIVERSITY OF IDAHO

College of Graduate Studies, College of Letters and Science, Department of Biological Sciences, Programs in Zoology, Moscow, ID 83844-4140

AWARDS MS, PhD.

Students: 18 full-time (11 women), 9 part-time (4 women); includes 2 minority (both Asian Americans or Pacific Islanders), 7 international. In 1999, 3 master's, 1 doctorate awarded.
Degree requirements: For master's, foreign language not required; for doctorate, dissertation required.
Entrance requirements: For master's, GRE, minimum GPA of 2.8; for doctorate, GRE, minimum undergraduate GPA of 2.8, 3.0 graduate. *Application deadline:* For fall admission, 8/1; for spring admission, 12/15. *Application fee:* $35 ($45 for international students).
Expenses: Tuition, nonresident: full-time $6,000; part-time $239 per credit hour. Required fees: $2,888; $144 per credit hour. Tuition and fees vary according to program.
Financial aid: Application deadline: 2/15.
Dr. Rolf L. Ingermann, Interim Chair, Department of Biological Sciences, 208-885-7764.

■ UNIVERSITY OF MAINE

Graduate School, College of Natural Sciences, Forestry, and Agriculture, Department of Biological Sciences, Program in Zoology, Orono, ME 04469

AWARDS MS, PhD. Terminal master's awarded for partial completion of doctoral program.

Degree requirements: For master's, thesis required; for doctorate, one foreign language, computer language, dissertation required.
Entrance requirements: For master's and doctorate, GRE General Test, TOEFL.
Expenses: Tuition, state resident: full-time $3,564. Tuition, nonresident: full-time $10,116. Required fees: $378. Tuition and fees vary according to course load.

■ UNIVERSITY OF NEW HAMPSHIRE

Graduate School, College of Life Sciences and Agriculture, Graduate Programs in the Biological Sciences and Natural Resources, Department of Zoology, Durham, NH 03824

AWARDS MS, PhD. Part-time programs available.

Faculty: 27 full-time.
Students: 16 full-time (8 women), 17 part-time (9 women); includes 3 minority (2 Asian Americans or Pacific Islanders, 1 Hispanic American), 3 international. Average age 28. *30 applicants, 33% accepted.* In 1999, 8 master's, 1 doctorate awarded. Terminal master's awarded for partial completion of doctoral program.
Degree requirements: For master's, thesis required, foreign language not required; for doctorate, dissertation required.
Entrance requirements: For master's and doctorate, GRE General Test, GRE Subject Test. *Application deadline:* For fall admission, 4/1 (priority date); for winter admission, 12/1. Applications are processed on a rolling basis. *Application fee:* $50.
Expenses: Tuition, area resident: Full-time $5,750; part-time $319 per credit. Tuition, state resident: full-time $8,625; part-time $478. Tuition, nonresident: full-time $14,640; part-time $598 per credit. Required fees: $224 per semester. Tuition and fees vary according to course load, degree level and program.
Financial aid: In 1999–00, 2 fellowships, 5 research assistantships, 17 teaching assistantships were awarded; career-related internships or fieldwork, Federal Work-Study, scholarships, and tuition waivers (full and partial) also available. Aid available to part-time students. Financial aid application deadline: 2/15.
Faculty research: Behavior development, ecology, endocrinology, fisheries, invertebrates.
Dr. Larry Harris, Chairperson, 603-862-3897, *E-mail:* lharris@cisunix.unh.edu.
Application contact: Dr. Jim Haney, Graduate Coordinator, 603-862-2105, *E-mail:* jfhaney@cisunix.unh.edu.

■ UNIVERSITY OF NEW MEXICO

Graduate School, College of Arts and Sciences, Department of Biology, Albuquerque, NM 87131-2039

AWARDS Biology (MS, PhD), including air land ecology, behavioral ecology, botany, cellular and molecular biology, community ecology, comparative immunology, comparative physiology, conservation biology, ecology, ecosystem ecology, evolutionary biology, evolutionary genetics, microbiology, molecular genetics, parasitology, physiological ecology, physiology, population biology, vertebrate and invertebrate zoology. Part-time programs available.

Faculty: 35 full-time (5 women), 18 part-time/adjunct (11 women).
Students: 71 full-time (37 women), 28 part-time (11 women); includes 8 minority (2 Asian Americans or Pacific Islanders, 5 Hispanic Americans, 1 Native American), 11 international. Average age 33. *93 applicants, 30% accepted.* In 1999, 11 master's, 12 doctorates awarded. Terminal master's awarded for partial completion of doctoral program.
Degree requirements: For master's, one foreign language (computer language can substitute), thesis required (for some programs); for doctorate, 2 foreign languages (computer language can substitute for one), dissertation required.
Entrance requirements: For master's and doctorate, GRE General Test, GRE Subject Test, minimum GPA of 3.2. *Application deadline:* For fall admission, 1/15. *Application fee:* $25.
Expenses: Tuition, state resident: full-time $2,514; part-time $105 per credit hour. Tuition, nonresident: full-time $10,304; part-time $417 per credit hour. International tuition: $10,304 full-time. Required fees: $516; $22 per credit hour. Tuition and fees vary according to program.
Financial aid: In 1999–00, 58 students received aid, including 24 fellowships (averaging $1,645 per year), 26 research assistantships with tuition reimbursements available (averaging $8,921 per year), 40 teaching assistantships with tuition reimbursements available (averaging $11,066 per year); career-related internships or fieldwork, Federal Work-Study, institutionally sponsored loans, and tuition waivers (full and partial) also available. Aid available to part-time students. Financial aid applicants required to submit FAFSA.
Faculty research: Developmental biology, immunobiology. *Total annual research expenditures:* $4.5 million.
Dr. Kathryn Vogel, Chair, 505-277-3411, *Fax:* 505-277-0304, *E-mail:* kgvogel@unm.edu.
Application contact: Vivian Kent, Information Contact, 505-277-1712, *Fax:* 505-277-0304, *E-mail:* vkent@unm.edu.

■ UNIVERSITY OF NORTH DAKOTA

Graduate School, College of Arts and Sciences, Department of Biology, Grand Forks, ND 58202

AWARDS Botany (MS, PhD); ecology (MS, PhD); entomology (MS, PhD); environmental biology (MS, PhD); fisheries/wildlife (MS, PhD); genetics (MS, PhD); zoology (MS, PhD).

Faculty: 18 full-time (3 women).
Students: 21 full-time (8 women). *13 applicants, 62% accepted.* In 1999, 3 master's awarded. Terminal master's awarded for partial completion of doctoral program.
Degree requirements: For master's, thesis, final exam required; for doctorate, dissertation, comprehensive exam, final exam required.
Entrance requirements: For master's, GRE General Test, GRE Subject Test, TOEFL, minimum GPA of 3.0; for doctorate, GRE General Test, GRE Subject Test, TOEFL, minimum GPA of 3.5. *Application deadline:* For fall admission, 3/1 (priority date). Applications are processed on a rolling basis. *Application fee:* $25.
Expenses: Tuition, state resident: full-time $3,166; part-time $158 per credit. Tuition, nonresident: full-time $7,658; part-time $345 per credit. International tuition: $7,658 full-time. Required fees: $46 per credit. Tuition and fees vary according to program and reciprocity agreements.
Financial aid: In 1999–00, 6 research assistantships with full tuition reimbursements (averaging $11,250 per year), 15 teaching assistantships with full tuition reimbursements (averaging $11,250 per year) were awarded; fellowships, Federal Work-Study, institutionally sponsored loans, scholarships, and tuition waivers (full and partial) also available. Aid available to part-time students. Financial aid application deadline: 3/15; financial aid applicants required to submit FAFSA.
Faculty research: Population biology, wildlife ecology, RNA processing, hormonal control of behavior.
Dr. Jeff Lang, Director, 701-777-2621, *Fax:* 701-777-2623, *E-mail:* jlang@badlands.nodak.edu.

■ UNIVERSITY OF OKLAHOMA

Graduate College, College of Arts and Sciences, Department of Zoology, Norman, OK 73019-0390

AWARDS M Nat Sci, MS, PhD.

Faculty: 30 full-time (9 women), 7 part-time/adjunct (0 women).
Students: 32 full-time (9 women), 11 part-time (5 women); includes 2 minority (both Native Americans), 3 international. *32 applicants, 41% accepted.* In 1999, 9 master's, 9 doctorates awarded.
Degree requirements: For master's, thesis defense required; for doctorate, dissertation defense, general exam required.
Entrance requirements: For master's and doctorate, GRE General Test, GRE Subject Test, TOEFL, TSE. *Application*

deadline: For fall admission, 1/15; for spring admission, 11/15. Applications are processed on a rolling basis. *Application fee:* $25.

Expenses: Tuition, state resident: full-time $2,064; part-time $86 per credit hour. Tuition, nonresident: full-time $6,588; part-time $275 per credit hour. Required fees: $468; $12 per credit hour. $94 per semester. Tuition and fees vary according to course level, course load and program.

Financial aid: In 1999–00, 10 research assistantships, 27 teaching assistantships were awarded; fellowships, career-related internships or fieldwork, Federal Work-Study, institutionally sponsored loans, and tuition waivers (partial) also available. Aid available to part-time students.

Faculty research: Animal behavior, cellular/molecular/developmental biology, ecology, and systematics, physiology. *Total annual research expenditures:* $549,832.

Dr. James N. Thompson, Chair, 405-325-4821, *Fax:* 405-325-6202, *E-mail:* jthompson@ou.edu.

Application contact: Dr. William Shelton, Chair, Graduate Selection Committee, 405-325-1058, *Fax:* 405-325-6202, *E-mail:* wshelton@ou.edu.

■ UNIVERSITY OF PUERTO RICO, MEDICAL SCIENCES CAMPUS

School of Medicine, Division of Graduate Studies, Department of Microbiology and Medical Zoology, San Juan, PR 00936-5067

AWARDS Medical zoology (MS, PhD); microbiology and medical zoology (MS, PhD).

Faculty: 16 full-time (8 women).

Students: 30 full-time (23 women); all minorities (1 Asian American or Pacific Islander, 29 Hispanic Americans). Average age 23. *15 applicants, 40% accepted.* In 1999, 4 master's awarded.

Degree requirements: For master's and doctorate, one foreign language, thesis/dissertation required.

Entrance requirements: For master's and doctorate, GRE General Test, GRE Subject Test, interview, minimum GPA or GPS of 3.0. *Application deadline:* For fall admission, 2/15; for spring admission, 9/15. *Application fee:* $15.

Expenses: Tuition, state resident: full-time $5,500. Tuition, nonresident: full-time $8,400. Required fees: $600. Tuition and fees vary according to class time, course load, degree level and program.

Financial aid: Fellowships, research assistantships, teaching assistantships, institutionally sponsored loans and tuition waivers (full) available. Financial aid application deadline: 4/30.

Faculty research: Molecular and general parasitology, immunology, development of viral vaccines and antiviral agents, fungal dimorphism, AIDS, pathogenesis.

Dr. Guillermo Vázquez, Director, 787-758-2525 Ext. 1309, *Fax:* 787-758-4808.

Application contact: Dr. Adelfa Serrano, Coordinator, 787-758-2525 Ext. 1313, *Fax:* 787-758-4808.

■ UNIVERSITY OF RHODE ISLAND

Graduate School, College of Arts and Sciences, Department of Zoology, Kingston, RI 02881

AWARDS MS, PhD.

Application deadline: For fall admission, 4/15 (priority date). Applications are processed on a rolling basis. *Application fee:* $35.

Expenses: Tuition, state resident: full-time $3,540; part-time $197 per credit. Tuition, nonresident: full-time $10,116; part-time $197 per credit. Required fees: $1,352; $37 per credit. $65 per term.

Dr. Harold Bibb, Chairperson, 401-874-2372.

■ UNIVERSITY OF SOUTH FLORIDA

Graduate School, College of Arts and Sciences, Department of Biology, Program in Zoology, Tampa, FL 33620-9951

AWARDS MS. Part-time programs available.

Degree requirements: For master's, thesis required, foreign language not required.

Entrance requirements: For master's, GRE General Test, minimum GPA of 3.0 in last 60 hours. Electronic applications accepted.

Expenses: Tuition, state resident: part-time $148 per credit hour. Tuition, nonresident: part-time $509 per credit hour.

■ UNIVERSITY OF WASHINGTON

Graduate School, College of Arts and Sciences, Department of Zoology, Seattle, WA 98195-1800

AWARDS PhD.

Faculty: 31 full-time (11 women), 18 part-time/adjunct (4 women).

Students: 80 full-time (35 women); includes 5 minority (3 Asian Americans or Pacific Islanders, 1 Hispanic American, 1 Native American), 11 international. Average age 30. *149 applicants, 7% accepted.* In 1999, 9 degrees awarded (99% entered university research/teaching).

Degree requirements: For doctorate, dissertation required, foreign language not required. *Average time to degree:* Doctorate–6 years full-time.

Entrance requirements: For doctorate, GRE General Test, TOEFL, minimum GPA of 3.0. *Application deadline:* For fall admission, 1/15. *Application fee:* $50. Electronic applications accepted.

Expenses: Tuition, state resident: full-time $5,196; part-time $495 per credit. Tuition, nonresident: full-time $13,485; part-time $1,285 per credit. Required fees: $387; $36 per credit. Tuition and fees vary according to course load and program.

Financial aid: In 1999–00, 3 fellowships with full tuition reimbursements (averaging $10,440 per year), 20 research assistantships with full tuition reimbursements (averaging $10,440 per year), 18 teaching assistantships with full tuition reimbursements (averaging $10,440 per year) were awarded; traineeships and tuition waivers (full) also available. Financial aid application deadline: 2/15.

Faculty research: Ecology and evolution, developmental and cellular biology, physiology and ethology, mathematical biology and biomechanics.

Dr. John C. Wingfield, Chair, 206-543-1620, *Fax:* 206-543-3041, *E-mail:* jwingfie@u.washington.edu.

Application contact: Judy Farrow, Graduate Program Assistant, 206-685-8240, *Fax:* 206-543-3041, *E-mail:* farrowj@u.washington.edu.

■ UNIVERSITY OF WISCONSIN–MADISON

Graduate School, College of Letters and Science, Department of Zoology, Madison, WI 53706-1380

AWARDS MA, MS, PhD. Part-time programs available.

Degree requirements: For master's, thesis required, foreign language not required; for doctorate, one foreign language, dissertation required.

Entrance requirements: For master's and doctorate, GRE General Test, TOEFL. Electronic applications accepted.

Expenses: Tuition, state resident: full-time $5,406; part-time $339 per credit. Tuition, nonresident: full-time $17,110; part-time $1,071 per credit. Full-time tuition and fees vary according to program and reciprocity agreements. Part-time tuition and fees vary according to course load and program.

Faculty research: Developmental biology, ecology, neurobiology, aquatic ecology, animal behavior.

Find an in-depth description at www.petersons.com/graduate.

■ UNIVERSITY OF WISCONSIN–OSHKOSH

Graduate School, College of Letters and Science, Department of Biology and Microbiology, Oshkosh, WI 54901

AWARDS Biology (MS), including botany, microbiology, zoology.

Degree requirements: For master's, thesis required, foreign language not required.

Entrance requirements: For master's, GRE General Test, minimum GPA of 3.0, BS in biology.

Expenses: Tuition, state resident: full-time $3,917; part-time $219 per credit. Tuition, nonresident: full-time $12,375; part-time $684 per credit. Part-time tuition and fees vary according to course load and program.

■ UNIVERSITY OF WYOMING

Graduate School, College of Arts and Sciences, Department of Zoology and Physiology, Laramie, WY 82071

AWARDS MS, PhD.

Faculty: 24.

Students: 42 full-time (19 women), 23 part-time (4 women); includes 2 minority (1 Asian American or Pacific Islander, 1 Hispanic American), 2 international. *64 applicants, 22% accepted.* In 1999, 14 master's, 4 doctorates awarded.

Degree requirements: For master's and doctorate, thesis/dissertation required, foreign language not required.

Entrance requirements: For master's and doctorate, GRE General Test, minimum GPA of 3.0. *Application deadline:* For fall admission, 2/15 (priority date); for spring admission, 9/15. Applications are processed on a rolling basis. *Application fee:* $40. Electronic applications accepted.

Expenses: Tuition, state resident: full-time $2,520; part-time $140 per credit hour. Tuition, nonresident: full-time $7,790; part-time $433 per credit hour. Required

fees: $440; $7 per credit hour. Full-time tuition and fees vary according to course load and program.

Financial aid: In 1999–00, 23 research assistantships with full tuition reimbursements (averaging $8,667 per year), 23 teaching assistantships with full tuition reimbursements (averaging $8,667 per year) were awarded. Financial aid application deadline: 3/1.

Faculty research: Cell biology, ecology/wildlife, organismal physiology. *Total annual research expenditures:* $2.5 million. Dr. Nancy L. Stanton, Head, 307-766-4207.

Find an in-depth description at www.petersons.com/graduate.

■ VIRGINIA POLYTECHNIC INSTITUTE AND STATE UNIVERSITY

Graduate School, College of Arts and Sciences, Department of Biology, Program in Zoology, Blacksburg, VA 24061

AWARDS MS, PhD.

Faculty: 22 full-time (2 women).

Degree requirements: For master's and doctorate, thesis/dissertation required, foreign language not required.

Entrance requirements: For master's and doctorate, GRE General Test, TOEFL. *Application deadline:* For fall admission, 12/1 (priority date). *Application fee:* $25.

Expenses: Tuition, state resident: full-time $4,122; part-time $229 per credit hour. Tuition, nonresident: full-time $6,930; part-time $385 per credit hour. Required fees: $828; $107 per semester. Part-time tuition and fees vary according to course load.

Financial aid: Fellowships, research assistantships, teaching assistantships, unspecified assistantships available. Financial aid application deadline: 4/1.

Dr. Joe R. Cowles, Chairman, Department of Biology, 540-231-8928, *E-mail:* cowlesjr@vt.edu.

■ WASHINGTON STATE UNIVERSITY

Graduate School, College of Sciences, School of Biological Sciences, Department of Zoology, Pullman, WA 99164

AWARDS MS, PhD.

Faculty: 17 full-time (2 women), 6 part-time/adjunct (0 women).

Students: 24 full-time (11 women), 2 part-time (both women); includes 1 minority (Asian American or Pacific Islander), 1 international. Average age 29. In 1999, 8 master's, 3 doctorates awarded.

Degree requirements: For master's, thesis or alternative, oral exam required, foreign language not required; for doctorate, dissertation, oral exam, written exam required, foreign language not required. *Average time to degree:* Master's–2 years full-time; doctorate–4 years full-time.

Entrance requirements: For master's and doctorate, GRE General Test, GRE Subject Test, minimum GPA of 3.0. *Application deadline:* For fall admission, 3/1 (priority date); for spring admission, 2/1. Applications are processed on a rolling basis. *Application fee:* $35.

Expenses: Tuition, state resident: full-time $5,654. Tuition, nonresident: full-time $13,850. International tuition: $13,850 full-time. Tuition and fees vary according to program.

Financial aid: In 1999–00, 3 research assistantships with full and partial tuition reimbursements, 23 teaching assistantships with full and partial tuition reimbursements were awarded; Federal Work-Study, institutionally sponsored loans, and tuition waivers (partial) also available. Financial aid application deadline: 4/1; financial aid applicants required to submit FAFSA. *Total annual research expenditures:* $638,295. Dr. Gary Thorgaard, Chairman, 509-335-3553, *Fax:* 509-335-3184, *E-mail:* zoology@wsu.edu.

School Index

School Index

Index of
Directories in This Book

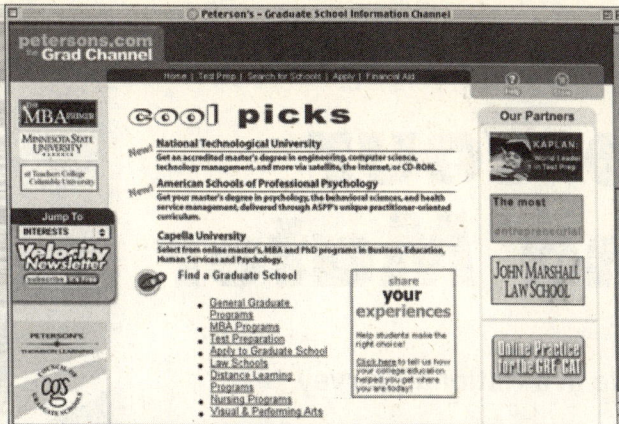